John E Stewar
18 Pasture Lane
Poughkeepsie, NY 12603

PIPING HANDBOOK

OTHER McGRAW-HILL HANDBOOKS OF INTEREST

PIPING HANDBOOK

RENO C. KING, B.M.E., M.M.E., D.Sc., P.E.

Professor of Mechanical Engineering and Assistant Dean,
School of Engineering and Science, New York University
Registered Professional Engineer

The first four editions of this Handbook were edited by

SABIN CROCKER, M.E.

Fellow, ASME; Registered Professional Engineer

Fifth Edition

M c G R A W - H I L L   B O O K   C O M P A N Y

New York   San Francisco   Toronto   London   Sydney

PREFACE

Since its First Edition in 1930, the purpose of the Piping Handbook has been to provide authoritative and accessible data for the engineer interested in piping design. This present edition, the Fifth, has been prepared with the same aim in mind: its scope, contents and arrangement reflect the tremendous advances which have been made in piping design in the twenty-odd years that have elapsed since the original publication of the Fourth Edition.

The first nine chapters deal with aspects of piping that are generally common to all piping systems. The remaining fourteen chapters treat the design considerations which are unique to specific applications. Much of the material is entirely new, some has been completely rewritten, and a good deal of the excellent material of former editions has been retained.

The twenty-three chapters have been prepared by eighteen authors, each of whom is an authority in his area of specialty. Many sources have been drawn upon and, throughout, credit is given to the original source of material. In some cases, material from an original source is condensed and a reference is given in the text or in the bibliography at the chapter end. In other cases, material may be quoted directly. If the quote is lengthy, the material is set in smaller type.

In this Fifth Edition of the Piping Handbook, the scope has been extended to cover relatively new fields such as those of nuclear and cryogenic piping systems. The flow of non-Newtonian fluids, the one-dimensional flow of compressible fluids and the design of sewerage systems are dealt with. The sections on manufacture, fabrication, specification and inspection of piping have been greatly enlarged and are well illustrated.

In recognition of the advantages in speed and accuracy afforded by the digital computer, an entirely new chapter on expansion and flexibility has been written. The fundamentals of matrix manipulation are introduced,

tables of compliances of piping system components are presented, examples are given of analytical solutions of expansion problems, and the economies in time and money of computer solutions are discussed. Simplified flexibility charts and graphs of the previous edition are partially retained in order that approximate, rapid solutions may be obtained.

Throughout the chapters which deal with design of specific piping systems, reference has been made to the latest edition of the governing code which was available at the time of writing. In instances in which no final code has been approved, test cases are often cited.

The names and affiliations of the several authors appear in the List of Contributors and, further, the name of each author appears at the heading of each chapter. To all of them, grateful acknowledgment is made for their having performed a most notable contribution to the art of piping design. Credit is due them for their achievement. For errors in form and typography, and for errors of omission, I accept full responsibility.

Reno C. King

CONTRIBUTORS

John E. Brock Professor of Mechanical Engineering, U.S. Naval Post-graduate School, Monterey, California

Arthur L. Brown Mechanical Engineer, 195 St. Paul St., Brookline, Massachusetts 02146

C. J. Danowitz Mechanical Engineer, Ebasco Services, Inc., Two Rector Street, New York, New York 10006

William E. Dobbins Professor of Civil Engineering, New York University, University Heights, New York 10453

Stanley W. Ehrlich Chemical Engineer, Hydrocarbon Research, Inc., 115 Broadway, New York, New York 10006

Matthew L. Hepburn Chemical Engineer, Ebasco Services, Inc., Two Rector Street, New York, New York 10006

Harry M. Howarth Dean Witter and Co., 50 West Adams St., Chicago, Illinois 60603

Robert L. Jones, Jr. Safety Code Coordinator, Carrier Air Conditioning Company, Carrier Parkway, Syracuse, New York 13201

Frank M. Kamarck Staff Engineer, Burns and Roe, Inc., 720 Kinderkamack Road, Oradell, New Jersey

William H. Kapfer Associate Professor of Chemical Engineering, New York University, University Heights, New York 10453

Reno C. King Professor of Mechanical Engineering, New York University, University Heights, New York 10453

Joseph Klapper Metallurgical Engineer, Ebasco Services, Inc., Two Rector Street, New York, New York 10006

Vincent T. Manas Consulting Engineer, 4513 Potomac Avenue, NW, Washington, D.C. 20007

Manufacturers Standardization Society of the Valve and Fittings Industry Subcommittee consisting of Walter A. Schlamp, Chairman (Fee and Mason Manufacturing Co., Inc.); Harold Erikson, Vice Chairman (Bergen-Paterson Pipesupport Corp.); George E. Crowell (The Grabler Manufacturing Co.); H. J. Marik (Automatic Sprinkler Corporation of America); George E. Paterson (Carpenter and Paterson, Inc.); P. C. Sherburne (Grinnell Co., Inc.); Robert V. Warrick (Executive Secretary, Manufacturers Standardization Society)

C. George Segeler Director of Technical Services, David Sage, Inc., New York, New York

James A. Sheppard, Jr. Assistant Division Engineer, Consolidated Edison Company, 4 Irving Place, New York, New York 10003

Helmut Thielsch Manager, Research and Development Division, Grinnell Company, Inc., Providence, Rhode Island 02901

Samuel Z. Weiner Supervising Mechanical Engineer, Gibbs and Hill, Inc., 393 Seventh Avenue, New York, New York 10001

Edwin J. Wesemann 35 Deerpath, Roslyn Heights, New York 11577

CONTENTS

* This chapter was written by a subcommittee of the Manufacturers Standardization Society of the Valve and Fittings Industry.   The membership of that committee appears in the List of Contributors.

PIPING HANDBOOK

1

DEFINITIONS, WEIGHTS, AND MEASURES

Reno C. King*

DEFINITIONS OF PIPING TERMS

Alloy Steel. A steel which owes its distinctive properties to elements other than carbon. Alloy-steel billets are rolled into stud-bolt stock and pipe and forged into large-headed bolts, pipe flanges, etc. Cast alloy steel is used in valve bodies and fittings where high tensile strength at elevated temperatures is desired or where resistance to corrosion is a factor.

Annealing. A process involving heating to and holding at a suitable temperature and then cooling at a suitable rate to reduce hardness, facilitate cold working, produce a desired microstructure, or obtain other properties.

Arc Welding. A group of welding processes wherein coalescence is produced by heating with an electric arc or arcs, with or without the addition of filler metal.

Automatic Welding. Welding with equipment which performs the entire welding operation without constant observation and adjustment of the controls by an operator. The equipment may or may not perform the loading and unloading of the work.

Backing Ring. A strip of metal used to prevent weld splatter from entering a pipe when making a butt-welded joint and to assure complete penetration of the weld to the inside of the pipe wall.

Back-pressure Valve. A valve similar to a low-pressure safety valve which is set to maintain a certain back pressure on feed heaters, oiling systems, or other devices requiring a constant operating pressure irrespective of pressure variations of the supply. The back-pressure valve is arranged to relieve any excess supply to

* Professor of Mechanical Engineering, New York University, New York, N.Y.

atmosphere or elsewhere, and it opens and closes automatically as required to produce this result.

Base Metal. (or parent metal). The metal to be welded or cut.

Bevel. A type of end or edge preparation.

Bevel Angle. The angle formed between the prepared edge of a member and a plane perpendicular to the surface of the member (see Fig. 1).

Blank Flange. A flange that is not drilled but is otherwise complete.

Bleeder. A small cock or valve to draw off water of condensation from a run of piping. A small connection to obtain circulation in warming up a line.

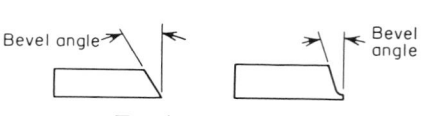

Fig. 1. Bevel angle.

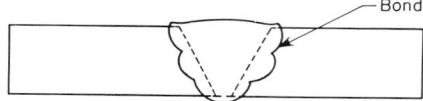

Fig. 2. Bond between base metal and weld metal.

Blind Flange. A flange used to close the end of a pipe. It produces a blind end which is also known as a dead end.

Bond. The junction of the weld metal and the base metal or the junction of the base metal parts when weld metal is not present (see Fig. 2).

Branch Tee. A header or manifold having many side branches. A common example is the branch tee used in making up pipe coil radiators for building heating systems.

Brazed. Connected by hard solder which usually is copper and zinc, half and half. Such solder requires a full red heat and is commonly used with borax flux. Pure copper is sometimes used for brazing.

Butt Joint. A joint between two members which lie in approximately the same plane. Joints may be square butts or beveled butts and may be parallel to or perpendicular to the axis of the pipe.

Butt Weld. Welded along a seam that is butted edge to edge. A term used to designate pipe made by this process. Also applied to circumferential pipe joints made by the fusion-welding process (see Fig. 3).

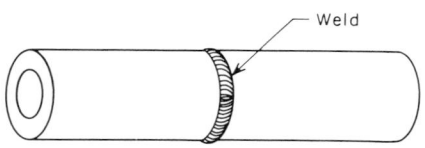

Fig. 3. A circumferential butt-welded joint.

Bypass. A small passage around a large valve for warming up a line. An emergency connection around a reducing valve, trap, etc., to use in case they are out of commission.

Carbon Steel. A steel which owes its distinctive properties chiefly to the carbon (as distinguished from the other elements) which it contains.

Chamfering. The preparation of a contour, other than for a square groove weld, on the edge of a member for welding.

Cold Bending. The bending of pipe to a predetermined radius at any temperature below some specified phase change or transformation temperature, but especially at or near room temperature. Frequently, pipe is bent to a radius of five times the nominal pipe diameter.

Cold Working. Deformation of a metal plastically. Although ordinarily done at room temperature, cold working may be done at a temperature and rate at which strain hardening occurs. Bending of steel piping at 1300 F would be considered a cold-working operation.

Companion Flange. A pipe flange suited to connect with another flange or with a flanged valve or fitting. A loose flange which is attached to a pipe by threading, Van Stoning, welding, or similar method as distinguished from a flange which is cast integrally with a fitting or pipe.

Consumable Insert Ring. A ring which is inserted in the root of a weld joint and which is melted (consumed) by the initial weld pass usually made by inert-gas tungsten-arc welding. (The composition of insert rings generally corresponds to that of the pipe base metal but may be of a composition corresponding to welding filler metals.) See Fig. 4.

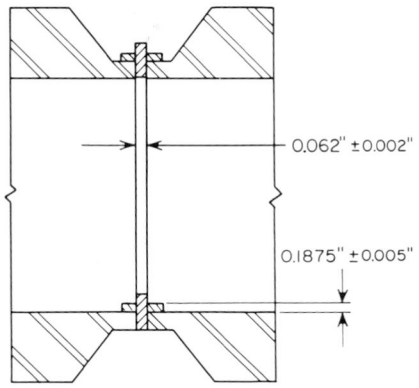

Controlled Cooling. A process of cooling from an elevated temperature in a predetermined manner to avoid hardening, cracking, or internal damage, or to produce a desired metallurgical microstructure. This cooling usually follows the final hot forming or post-heating operation.

Corner Joint. A joint between two members located approximately at right angles to each other in the form of an "L" (see Fig. 5).

Coupling. A threaded sleeve used to connect two pipes. Commercial couplings have internal threads to fit external threads on pipe.

FIG. 4. Consumable insert ring inserted in pipe joint eccentrically for welding in horizontal position.

Covered Electrode. A filler metal electrode, used in arc welding, consisting of a metal core wire with a relatively thick covering which provides protection for the molten metal from the atmosphere, improves the properties of the weld metal, and stabilizes the arc. Covered electrodes are extensively used in the shop fabrication and field erection of piping of carbon, alloy, and stainless steels.

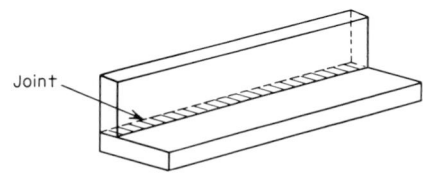

FIG. 5. Corner joint.

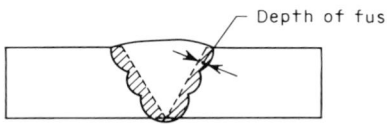

FIG. 6. Depth of fusion.

Creep or Plastic Flow of Metals. At sufficiently high temperatures all metals flow under stress. The higher the temperature and stress, the greater the tendency to plastic flow for any given metal.

Cutting Torch. A device used in oxygen, air, or powder cutting for controlling and directing the gases used for preheating and the oxygen or powder used for cutting the metal.

Deposited Metal. Filler metal that has been added during a welding operation.

Depth of Fusion. The distance that fusion extends into the base metal from the surface melted during welding (see Fig. 6).

Double Extra-strong refers to a schedule or wrought-pipe weights in common use (see Table 1, Chap. 7).

Edge Joint.   A joint between the edges of two or more parallel or nearly parallel members.

Edge Preparation.   The contour prepared on the edge of a member for welding (see Fig. 7).

Electrode.   See Covered Electrode.

End Preparation.   The contour prepared on the end of a pipe for welding. The particular preparation is prescribed by the governing Code and may be as shown in Fig. 1, or the contour may be curved.

Extra-heavy.   The term "extra-heavy" was formerly used to designate cast-iron flanges, fittings, valves, etc., suitable for a maximum working steam pressure of 250 lb gage. Owing to the general use of much higher pressures, the term "extra-heavy" in this sense has become inappropriate. The use of the term "extra-heavy" to denote "extra-strong" pipe is incorrect.

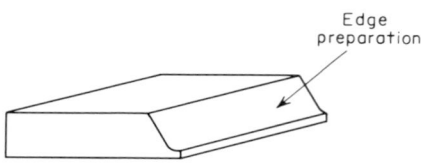

FIG. 7. Edge preparation.

Extra-strong.   A schedule of wrought-pipe weights in common use. In sizes 8 in. and smaller, extra-strong pipe is identical with Schedule 80 pipe (see Table 1, Chap. 7).

Extruded Nozzles.   Forming of nozzle (tee) outlets in pipe by pulling hemispherically or conically shaped dies through a circular hole from the inside of the pipe. Although some cold extruding is done, it is generally performed on steel after the area to be shaped has been heated to temperatures between 2000 and 1600 F.

Filler Metal.   Metal to be added in making a weld.

Fillet Weld.   A weld of approximately triangular cross section joining two surfaces at about right angles to each other in a lap joint, tee joint, or corner joint (see Fig. 8).

FIG. 8. Fillet weld.   FIG. 9. Flat land bevel.

Flat Land Bevel.   A square extended root face preparation extensively used in inert-gas, root-pass welding of piping (see Fig. 9).

Flat Position.   The position of welding wherein welding is performed from the upper side of the joint and the face of the weld is approximately horizontal (see Fig. 10).

Forge Weld.   A method of manufacture similar to hammer welding. The

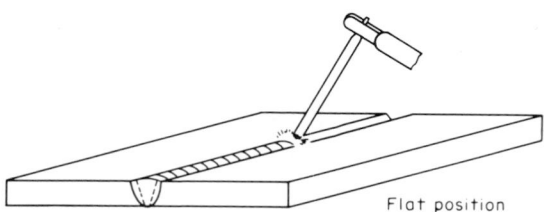

FIG. 10. Welding in the flat position.

term "forge welded" is applied more particularly to headers and large drums, while "hammer welded" usually refers to pipe.

Full Annealing. A heating and cooling process that results in maximum softening of the metal involved. On carbon and alloy steel, it defines a softening process in which the steel is heated to a temperature above the transformation range and, after being held for a sufficient time at this temperature, it is cooled slowly to a temperature below the transformation range. Ordinarily cooling is done in the furnace.

Full-furnace Annealing. Full annealing of a complete steel section, such as a fabricated piping assembly, in a furnace.

Furnace Weld. A term applied to the process of making butt-welded or lap-welded pipe in which the skelp is heated in a furnace preparatory to welding by passing through rolls.

Fusion. The melting together of filler metal and base metal or of base metal only which results in coalescence. See Depth of Fusion.

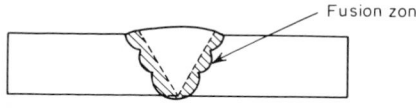

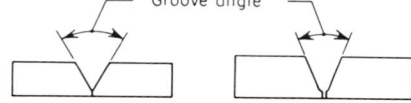

FIG. 11. Fusion zone, the section of the parent metal which melts during the welding process.

FIG. 12. The groove angle is twice the bevel angle.

Fusion Zone. The area of base metal melted as determined on the cross section of a weld.

Galvanizing. A process by which the surface of iron or steel is covered with a layer of zinc.

Gas Welding. A group of welding processes wherein coalescence is produced by heating with a gas flame or flames, with or without the use of filler metal.

Groove. The opening provided for a groove weld.

Groove Angle. The total included angle of the groove between parts to be joined by a groove weld (see Fig. 12).

Groove Face. That surface of a member included in the groove (see Fig. 13).

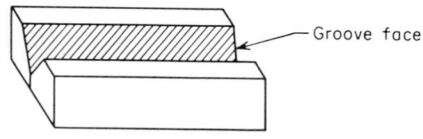

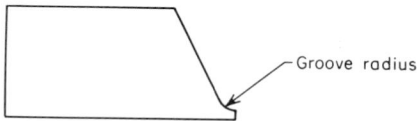

FIG. 13. A groove face.

FIG. 14. A groove radius.

Groove Radius. The radius of a J or U groove (see Fig. 14).

Groove Weld. A weld made in the groove between two members to be joined. The standard types of groove welds are square groove weld, single V-groove weld, single-bevel groove weld, single U-groove weld, single J-groove weld, double V-groove weld, double U-groove weld, double-bevel groove weld, double J-groove weld, flat-land single and double V-groove welds. See Fig. 15 for typical groove weld.

Hammer Weld. Method of manufacturing large pipe (usually 20 in. and and larger) by bending a plate into circular form, heating the overlapped edges to a

welding temperature, and welding the longitudinal seam with a power hammer applied to the outside of the weld while the inner side is supported on an overhung anvil.

Heat-affected Zone. That portion of the base metal which has not been

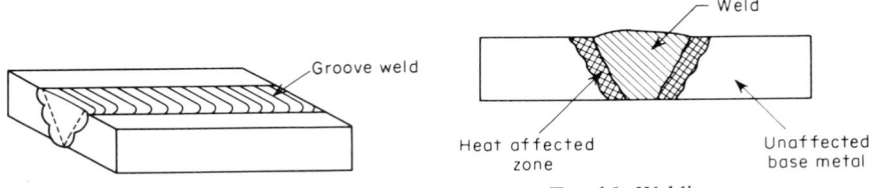

FIG. 15. Groove weld.　　　　　　FIG. 16. Welding zones.

melted but whose mechanical properties or microstructures have been altered by the heat of welding or cutting (see Fig. 16).

Horizontal Fixed Position. In pipe welding, the position of a pipe joint in which the axis of the pipe is approximately horizontal and the pipe is not rotated during the welding operation.

Horizontal-position Fillet Weld. The position of welding in which welding is performed on the upper side of an approximately horizontal surface and against an approximately vertical surface (see Fig. 18).

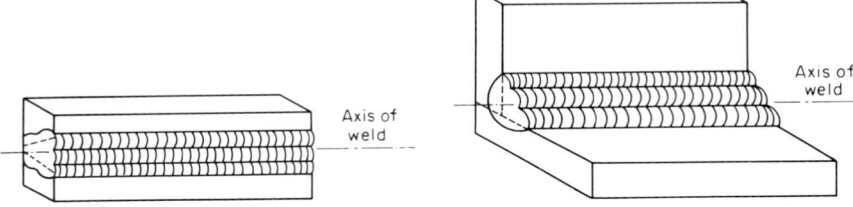

FIG. 17. Horizontal position groove weld.　　FIG. 18. Horizontal position fillet weld.

Horizontal-position Groove Weld. The position of welding in which the weld axis lies in an approximately horizontal plane and the face of the weld lies in an approximately vertical plane (see Fig. 17).

Horizontal Rolled Position. The position of a pipe joint in which welding is performed in the flat position by rotating the pipe (see Fig. 19).

Hot Bending. Bending of piping to a predetermined radius after heating to a suitable high temperature for hot working. On many pipe sizes, the pipe is firmly packed with sand to avoid wrinkling and excessive out-of-roundness.

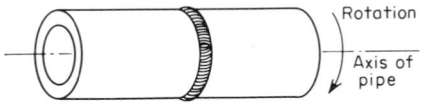

FIG. 19. Horizontal rolled position.

Hot Working. The plastic deformation of metal at such a temperature and rate that strain hardening does not occur. Extruding or swaging of chrome-moly piping at temperatures between 2000 and 1600 F would be considered hot-forming or hot-working operations.

Incomplete Fusion. Fusion which is less than complete and which does not result in melting completely through the thickness of the joint.

Induction Heating. Heat treatment of completed welds in piping by means of placing induction coils around the piping. This type of heating is usually performed during field erection in those cases where stress relief of carbon- and alloy-steel field welds is required by applicable ASA or ASME Codes.

Inert-gas Tungsten-arc Welding. An arc-welding process in which coalescence is produced by heating with an electric arc between a tungsten electrode and the work. Shielding is obtained by the use of helium or argon. Filler metal may or may not be used.

Insert Ring. See Consumable Insert Ring.

Interpass Temperature. In a multiple-pass weld, the minimum or maximum temperature of the deposited weld metal before the next pass is started.

Interrupted Welding. Interruption of welding and preheat by allowing the weld area to cool to room temperature as generally permitted on carbon-steel and on chrome-moly alloy-steel piping after sufficient weld passes equal to at least one third of the pipe wall thickness or two weld layers, whichever is greater, have been deposited.

Joint Geometry. The shape and dimensions of a joint in cross section prior to welding.

Joint Penetration. The minimum depth a groove weld extends from its face into a joint, exclusive of reinforcement (see Fig. 20).

Lapped Joint. A type of pipe joint made by using loose flanges on lengths of pipe whose ends are lapped over to give a bearing surface for a gasket or metal-to-metal joint.

Lap Weld. Welded along a longitudinal seam in which one part is overlapped by the other. A term used to designate pipe made by this process.

Local Preheating. Preheating a specific portion of a structure.

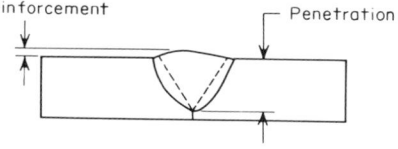

FIG. 20. Weld penetration.

Local Stress-relief Heat Treatment. Stress-relief heat treatment of a specific portion of a weldment. This is done extensively with induction coils, resistance coils, or propane torches in the field erection of steel piping.

Machine Welding. Welding with equipment which performs the welding operation under the observation and control of an operator. The equipment may or may not perform the loading and unloading of the work.

Malleable Iron. Cast iron which has been heat-treated in an oven to relieve its brittleness. The process somewhat improves the tensile strength and enables the material to stretch to a limited extent without breaking.

Mill Length. Also known as "random length." The usual run-of-mill pipe is 16 to 20 ft in length. Line pipe and pipe for power-plant use are sometimes made in double lengths of 30 to 35 ft.

Nipple. A piece of pipe less than 12 in. long and threaded on both ends. Pipe over 12 in. long is regarded as cut pipe. Common types of nipples are: close nipple, about twice the length of a standard pipe thread and without any shoulder; shoulder nipple, of any length and having a shoulder between the pipe threads; short nipple, a shoulder nipple slightly longer than a close nipple and of a definite length for each pipe size which conforms to manufacturer's standard; long nipple, a shoulder nipple longer than a short nipple which is cut to specific length.

Nonreturn Valve. A stop valve whose disk can move independently of the stem so that the valve can act as a check. Such valves are largely used between boilers and headers to prevent steam from the header entering the boiler in case of

tube failure or other trouble necessitating shutdown. The name "stop and check valve" is often applied to this type.

Normalizing. A process in which a carbon or alloy steel is heated to a suitable temperature above the transformation range and is subsequently cooled in still air to room temperature. Hot bent 5 Cr-$\frac{1}{2}$ Mo alloy-steel pipe is frequently normalized by heating to 1650 F and holding for 1 hr per inch of wall thickness followed by cooling in still air.

Nozzle. As applied to piping, this term refers usually to a flanged connection on a boiler, tank, or manifold consisting of a pipe flange, a short neck, and a welded attachment to the boiler or other vessel. A short length of pipe, one end of which is welded to the vessel with the other end chamfered for butt welding, is also referred to as a "welding" nozzle.

Overhead Position. The position of welding in which welding is performed from the underside of the joint.

Oxyacetylene Cutting. An oxygen-cutting process in which the severing of metals is effected by means of the chemical reaction of oxygen with the base metal at elevated temperatures. The necessary temperature is maintained by means of gas flames obtained from the combustion of acetylene with oxygen.

Oxyacetylene Welding. A gas-welding process in which coalescence is produced by heating with a gas flame or flames obtained from the combustion of acetylene with oxygen and with or without the addition of filler metal.

Pass. A single longitudinal progression of a welding operation along a joint or weld deposit. The result of a welding pass is a weld bead.

Peening. The mechanical working of metals by means of hammer blows.

Pickle. The chemical or electrochemical removal of surface oxides. Following welding operations, piping is frequently "pickled" in order to remove mill scale, oxides formed during storage, and the weld discolorations.

Pipe. The name "pipe" is applied to tubular products of dimensions and materials commonly used for pipe lines and connections, formerly designated as "iron pipe size" (IPS). The outside diameter of all weights and kinds of IPS pipe is of necessity the same for a given pipe size on account of threading.

Polarity. The direction of flow of current with respect to the welding electrode and the workpiece.

Porosity. Presence of gas pockets or voids in metal.

Positioned Weld. A weld made in a joint which has been so placed as to facilitate the making of the weld.

Postheating. The application of heat to a fabricated or welded section subsequent to a fabrication, welding, or cutting operation. Postheating may be done locally, as by induction heating, or the entire assembly may be postheated in a furnace.

Purging. The displacement during welding, by an inert or neutral gas, of the air inside the piping underneath the weld area in order to avoid oxidation or contamination of the underside of the weld. Gases most commonly used are argon, helium, and nitrogen (the last principally limited to austenitic stainless steel). Purging can be done within a complete pipe section or, by means of purging fixtures, of a small area underneath the pipe weld (see Fig. 21).

Reinforcement Weld. Weld metal on the face of a groove weld in excess of the metal necessary for the specified weld size.

Relief Valve. A valve arranged to provide an automatic relief in case of excess pressure. It may be either spring loaded or of the dead-weight type.

Resistance Weld. Method of manufacturing pipe by bending a plate into circular form and passing electric current through the material to obtain a welding temperature.

Reynolds Number. The Reynolds number is a dimensionless group used in many fluid mechanics calculations and is expressed as the product of density, velocity, and diameter divided by the viscosity of the fluid.

Root Edge. A root face of zero width.

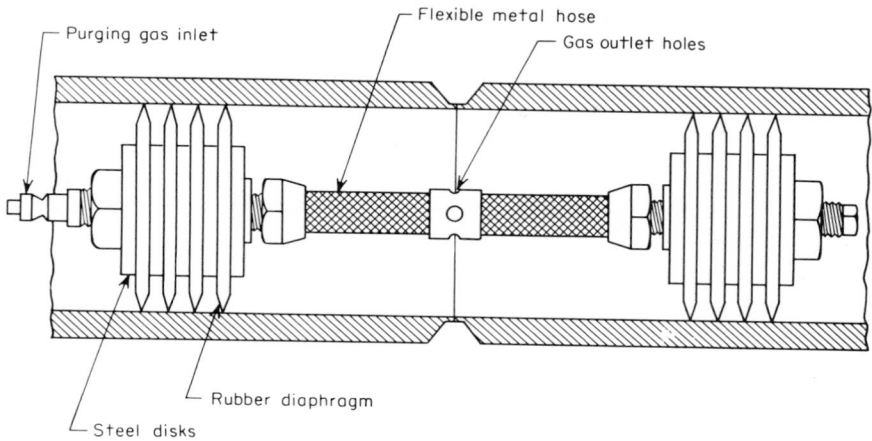

Purging gas inlet

Flexible metal hose

Gas outlet holes

Rubber diaphragm

Steel disks

FIG. 21. Purging.

Root Face. That portion of the groove face adjacent to the root of the joint. This portion is also referred to as the root land (see Fig. 22).

Root of Joint. That portion of a joint to be welded where the members approach nearest to each other. In cross section, the root of a joint may be a point, a line, or an area (see Fig. 22).

Root of a Weld. The points, as shown in cross section, at which the bottom of the weld intersects the base metal surfaces (see Fig. 22).

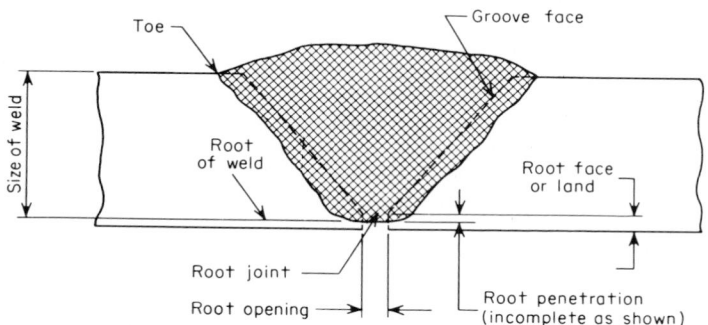

Toe

Groove face

Size of weld

Root of weld

Root face or land

Root joint

Root opening

Root penetration (incomplete as shown)

FIG. 22. Nomenclature at joint of groove weld.

Root Opening. The separation, between the members to be joined, at the root of the joint (see Fig. 22).

Root Penetration. The depth which a groove weld extends into the root of a joint as measured on the center line of the root cross section. Welds made under the ASME Boiler and Pressure Vessel Codes are considered unacceptable if they

show incomplete penetration. See Fig. 22 for a groove weld of incomplete penetration.

Run.   The portion of a fitting having its end in line, or nearly so, as distinguished from branch connections, side outlets, etc.

Saddle Flange.   Also known as "tank flange" or "boiler flange." A curved flange shaped to fit a boiler, tank, or other vessel and receive a threaded pipe. A saddle flange is usually riveted or welded to the vessel.

Safety Valve.   A relief valve for expansive fluids provided with a huddling ring and chamber to control the amount of blow-back before the valve reseats.

Sargol.   A special type of joint in which a lip is provided for welding to make the joint fluid tight, while mechanical strength is provided by bolted flanges. The Sargol joint is used with both Van Stone pipe and fittings.

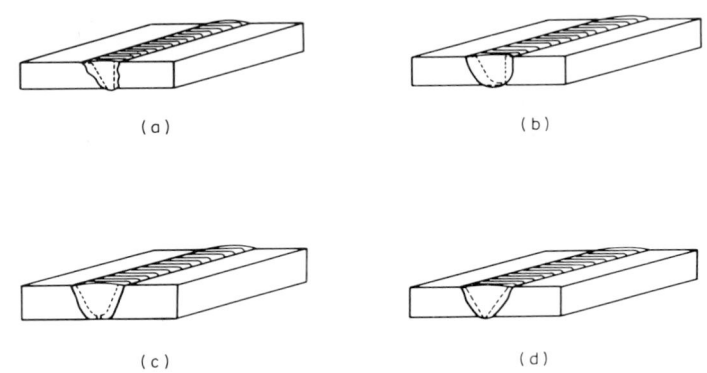

FIG. 23. Groove welds.   (*a*) Single bevel, (*b*) single J, (*c*) double U, (*d*) double V.

Sarlun.   An improved type of Sargol joint.

Schedule Numbers.   Schedule numbers indicate approximate values of the expression $1{,}000 \times P/S$ where P is the service pressure and S is the allowable stress, both expressed in pounds per square inch.

Seal Weld.   Any weld used primarily to obtain tightness, such as one used to seal the closure plugs on radiographic access holes.

Seamless.   Pipe formed by piercing and rolling a solid billet or by cupping from a plate is termed "seamless."

Semiautomatic Arc Welding.   Arc welding with equipment which controls only the filler metal feed. The advance of the welding is manually controlled.

Semisteel.   A high grade of cast iron made by the addition of steel scrap to pig iron in the cupola or electric furnace. More correctly described as "high-strength gray iron." It is used to some extent for valve bodies and fittings.

Service Fitting.   A street ell or street tee having a male thread at one end.

Shielded Metal-arc Welding.   An arc-welding process in which coalescence is produced by heating with an electric arc between a covered metal electrode and the work. Shielding is obtained from decomposition of the electrode covering. Filler metal is obtained from the electrode.

Shot Blasting.   Mechanical removal of surface oxides and scale on the pipe inner and outer surfaces by the abrasive impingement of small steel pellets.

Single Bevel-, Single J-, Single U-, Single V-groove Welds.   All are specific types of groove welds and are illustrated in Fig. 23.

Size of Weld.   For a groove weld, the joint penetration which is the depth of chamfering plus the root penetration (see Fig. 22). For fillet welds, the leg length

of the largest isosceles right triangle which can be inscribed within the fillet weld cross section (see Fig. 24).

Skelp. A piece of plate prepared by forming and bending, ready for welding into pipe. Flat plates when used for butt-welded pipe are called skelp.

Slag Inclusion. Nonmetallic solid material entrapped in weld metal or between weld metal.

Socket Weld. Fillet-type seal weld used in power-plant and other piping systems to join pipe to valves and fittings or to other sections of pipe. Generally used for piping whose nominal diameter is 2 in. or smaller.

Source Nipple. A short length of heavy-walled pipe between high-pressure mains and the first valve of bypass, drain, or instrument connections.

Spatter. In arc and gas welding, the metal particles expelled during welding and which do not form a part of the weld.

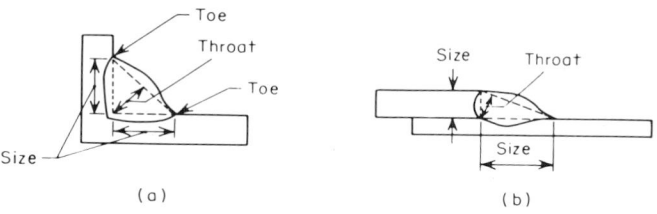

Fig. 24. Size of weld (a) in fillet weld of equal legs, (b) in fillet weld of unequal legs.

Spatter Loss. Difference in weight between the amount of electrode consumed and the amount of electrode deposited.

Spiral Riveted. A method of manufacturing pipe by coiling a plate into a helix and riveting together the overlapped edges.

Spiral Welded. A method of manufacturing pipe by coiling a plate into a helix and fusion-welding the overlapped or abutted edges.

Square-groove Weld. A groove weld in which the pipe ends are not chamfered. Square-groove welds are generally used on piping and tubing of wall thickness no greater than ⅛ in.

Stainless Steel. An alloy steel having unusual corrosion-resisting properties, usually imparted by nickel and chromium.

Standard. The term "standard" was formerly used to designate cast-iron flanges, fittings, valves, etc., suitable for a maximum working steam pressure of 125 lb gage. The multiplicity of standards which have come into use in connection with high pressures makes the old use of this word inappropriate.

Standard Weight. A schedule of wrought-pipe weights in common use. Standard-weight pipe in sizes 10 in. and smaller is identical with Schedule 40 pipe.

Stop and Check Valve. See Nonreturn Valve.

Stop Valve. A valve of the gate or globe type used to shut off a line.

Stress-relief Heat Treatment (Stress Relieving). The uniform heating of a structure or a portion thereof to a sufficiently high temperature below the critical range to relieve the major portion of the residual stresses. Stress relieving is employed to relieve the stresses induced by welding, cold or hot bending or forming, normalizing, and similar operations. Chrome-moly piping is usually stress relieved at temperatures between 1300 and 1400 F for 1 hr per inch of wall thickness, with 1 hr as a minimum.

String Bead. A type of weld bead made by moving the electrode in a direction essentially parallel to the axis of the bead. There is no appreciable transverse oscillation of the electrode. The deposition of a number of string beads is known

as string beading and is used extensively in the welding of austenitic stainless-steel materials.

Submerged-arc Welding. An arc-welding process in which coalescence is produced by heating with an electric arc or arcs between a bare metal electrode or electrodes and the work. The welding is shielded by a blanket of granular fusible material on the work. Filler metal is obtained from the electrode or from a supplementary welding rod.

Swaging. Reducing the ends of pipe and tube sections with rotating dies which are pressed intermittently against the pipe or tube end.

Tack Weld. A small weld made to hold parts of a weldment in proper alignment until the final welds are made.

Tee Joint. A welded joint between two members located approximately at right angles to each other in the form of a "T."

Tempering. A process of heating a normalized or quench-hardened steel to a temperature below the transformation range and, from there, cooling at any rate desired. This operation is also frequently called stress relieving.

Throat of a Weld. A term applied to fillet welds: it is the perpendicular distance from the beginning of the root of a joint to the hypotenuse of the largest right triangle that can be inscribed within the fillet-weld cross section (see Fig. 24).

Toe of Weld. The junction between the face of a weld and the base metal (see Fig. 24).

Trepanning. The removal by destructive means of a small section of piping (usually containing a weld) for an evaluation of weld and base-metal soundness. The operation is frequently performed with a hole saw.

Tungsten Electrode. A nonfiller-metal (nonconsumable) electrode used in inert-gas arc welding, consisting of a tungsten wire.

Turbinizing. Mechanical removal of scale from the pipe inside by means of air-driven centrifugal rotating cleaners. The operation is performed on steel pipe bends after hot bending to remove loose scale and sand.

Underbead Crack. A crack in the heat-affected zone or in previously deposited weld metal paralleling the underside contour of the deposited weld bead and usually not extending to the surface.

Undercut. A groove melted into the base metal adjacent to the toe of a weld and left unfilled by weld metal.

Van Stoning. Hot upsetting of lapping pipe ends to form integral lap flanges, the lap generally being of the same diameter as that of the raised face of standard ASA flanges.

Vertical Position. With respect to pipe welding, the position in which the axis of the pipe is vertical with the welding being performed in the horizontal position. The pipe may or may not be rotated.

Weave Bead. A type of weld bead made with oscillation of the electrode transverse to the axis of the weld. Contrast to string bead.

Weldability. The ability of a metal to be welded under the fabrication conditions imposed into a specific, suitably designed structure and to perform satisfactorily in the intended service.

Weld Bead. A weld deposit resulting from a pass.

Weld Metal. That portion of a weld which has been melted during welding. The portion may be the filler metal or base metal or both.

Weld Metal Area. The area of the weld metal as measured on the cross section of a weld.

Weld Penetration. See Joint Penetration and Root Penetration.

Welded Joint. A localized union of two or more members produced by the application of a welding process.

Welder.   One who is capable of performing a manual or semiautomatic welding operation.

Welder Qualification.   Acceptance test determining the ability of a welder to make satisfactory pipe welds in the horizontal rolled, horizontal fixed, and vertical pipe positions.   The test is required under several Codes, including the ASME Boiler and Pressure Vessel Codes and the ASA Code for Pressure Piping.

Welding Current.   The current which flows through the electrical welding circuit during the making of a weld.

Welding Fittings.   Wrought- or forged-steel elbows, tees, reducers, and similar pieces for connection by welding to each other or to pipe.   In small sizes, these fittings are available with counterbored ends for connection to pipe by fillet welding, and are known as socket-weld fittings.   In larger sizes, the fittings are supplied with ends chamfered for connection to pipe by means of butt welding, and are known as butt-welding fittings.

Welding Generator.   The electrical generator used for supplying welding current.

Welding Machine.   Equipment used to perform the welding operation.

Welding Operator.   One who operates a welding machine or automatic welding equipment.

Welding Procedure.   The detailed methods and practices, including joint welding procedures, or parameters, involved in the production of a weldment.

Welding Process.   A metal-joining process in which coalescence is produced by heating to suitable temperatures, with or without the use of filler metal.   See Gas Welding and Arc Welding.

Welding Rod.   Filler metal, in wire or rod form, used in gas-welding and brazing procedures and those arc-welding processes where the electrode does not furnish the filler metal.

Welding Sequence.   The order of making the welds in a weldment.

Welding-end Valves.   Valves which are to be welded into a piping assembly. As is the case with welding fittings, welding-end valves are furnished with either socket-weld or butt-weld ends.

Weldment.   An assembly whose component parts are to be joined by welding.

Weld-prober Sawing.   Removal of a boat-shaped sample from a pipe weld for examination of the weld and its adjacent base-metal area.   This operation is usually performed in graphitization studies.

Wrought Iron.   Iron refined in a plastic state in a puddling furnace.   It is characterized by the presence of about 3 per cent of slag irregularly mixed with pure iron and about 0.5 per cent carbon and other elements in solution.

Wrought Pipe.   The term "wrought pipe" refers to both wrought steel and wrought iron.   Wrought in this sense means "worked" as in the process of forming furnace-welded pipe from skelp, or seamless pipe from plates or billets.   The expression "wrought pipe" is thus used as a distinction from cast pipe.   Wrought pipe in this sense should not be confused with "wrought-iron pipe," which is only one variety of wrought pipe.   When "wrought-iron pipe" is referred to, it should be designated by its complete name.

CENTER OF GRAVITY

General.   The center of gravity of a body or of a system of bodies connected rigidly is that point about which there would be no tendency for the body to rotate if it were used as the point of suspension.   In bodies of homogeneous material the center of gravity and the center of magnitude are identical.   The center of gravity

of a regular figure is the same as its geometrical center, as in a circle, cylinder, circular ring, etc.

Center of Gravity of a Triangle. The intersection of the lines joining the vertices with the mid-points of the sides and at a perpendicular distance from any side equal to one-third of the corresponding altitude.

Of a Trapezoid. Draw a diagonal dividing it into two triangles. Draw a line joining their centers of gravity. Draw the other diagonal, making two other triangles and a line joining their centers. The intersection of these two lines is the center of gravity.

Of a Sector of a Circle. On the radius bisecting the arc, a distance $2cr \div 3l$ from the center, c being the chord, r the radius, and l the length of arc.

Of a Semicircle. On the middle radius $0.4244r$ from the center.

Of a Quadrant. On the middle radius $0.6002r$ from the center.

Of a Segment of a Circle. $c^3/12a$ from the center, c being the chord, a the area of the segment.

Of a Cone or Pyramid. In the axis one-fourth of its length from the base.

The common center of gravity for two bodies is that point which divides the distance between their respective centers of gravity in inverse proportion to their weights. For more than two bodies, find the common center of gravity of any two as above. Then find the common center of these two jointly with a third, and so on.

FORCES, MOMENTS, AND EQUILIBRIUM

Simple Forces. When two or more forces act upon a body at one point, they may be single or combined into a resultant force. Conversely, any force may be resolved into component forces. In Fig. 25, let the vectors F_1 and F_2 represent two forces acting on a point O. The resultant force F is represented in direction and magnitude by the diagonal of the parallelogram of which F_1 and F_2 are the sides.

Conversely, any force F may be resolved into component forces by a reverse of the above operation.

Moments. The moment of a force with respect to a given point is the tendency of that force to produce rotation around it. The magnitude of the moment is represented by the product of the force and the perpendicular

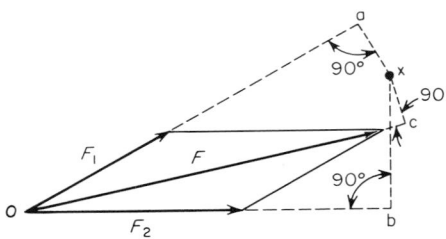

FIG. 25. Vectors and moments.

distance from its line of action to the point or center of moment. In the English system of weights and measures, moments are expressed as the product of the force in pounds and the length of the moment arm in feet or inches, the unit of the moment being termed the pound-foot or the pound-inch. Moments acting in a clockwise direction are designated as positive, and those acting in a counterclockwise direction are negative. They may be added and subtracted algebraically, as moments, regardless of the direction of the forces themselves.

With respect to Fig. 25, moments about an arbitrary point x are calculated as follows: Extend the line of action of F_1 until its extension intersects the perpendicular ax drawn from point x. Draw bx from x perpendicular to F_2. The sum of moments about point x due to the two forces is then

$$\Sigma \, M_x = F_1 \times ax - F_2 \times bx$$

Alternatively, since F_1 and F_2 have been shown to be the vector equivalent of the resultant F, the moments about x can be calculated as

$$\Sigma M_x = F \times cx$$

Couples. Two parallel forces of equal magnitude acting in opposite directions constitute a couple. The moment of the couple is the product of one of the forces and the perpendicular distance between the two. A couple has no single resultant and can be balanced only by another couple of equal moment of opposite sign.

Law of Equilibrium. When a body is at rest, the external forces acting upon it must be in equilibrium and there must be a zero net moment on the body. This means that (1) the algebraic sums of the components of all forces with reference to any three axes of reference at right angles with each other must each be zero and (2) the algebraic sum of all moments with reference to any three such axes must be zero. When the forces all lie in the same plane, the algebraic sums of their components with respect to any two axes must be equal to zero and the algebraic sum of all moments with respect to any point in the plane must be zero.

WORK, POWER, AND ENERGY

Work. When a body is moved against a resistance, work must be done upon it. The amount of work done is the product of the force and the distance through which it acts. The unit of work in the English system is the foot-pound, which is the amount of work done by a force of 1 lb acting through a distance of 1 ft. The following symbols are used in this section in defining the interrelation of work, power, and energy:

A = area, sq in. or sq ft as noted.
F = force, lbf
g = local acceleration of gravity, ft per sec^2
g_c = conversion constant, ft lbf/(lbm sec^2)
h = vertical distance, ft
H = enthalpy, Btu
hp = horsepower
kw = kilowatts
KE = kinetic energy, ft-lbf
PE = potential energy, ft-lbf
p = pressure, psi
l = distance, ft
T = time, sec
v = velocity, fps
V = volume, cu ft
w = weight, lb
W = work, ft-lb

According to the above definition of work, the following expressions may be written to represent work:

$$W = \int F \, dl = \int \frac{w}{g_c} g \, dh = \int pA \, dl = \int p \, dV$$

If the force is independent of distance, if the process takes place at sea level, if pressure and area are independent of distance, and if pressure is independent of volume, respectively, the above expressions reduce to

$$W = F(l_2 - l_1) = w(h_2 - h_1) = pA(l_2 - l_1) = p(V_2 - V_1)$$

where the subscripts 2 and 1 refer to final and initial states, respectively.   The above expressions contain no term involving time, since the measure of work is independent of the time interval during which it is performed.

Power.   Power is the time rate of performing work.   The English unit of power is the horsepower, which is defined as 33,000 ft-lb/min or 550 ft-lb/sec.   Electrical power is commonly expressed in watts or kilowatts, 1 kw being equivalent to 1.34 hp and 1 hp to 0.746 kw.   The expressions for horsepower corresponding to those given above for work are

$$hp = \frac{W}{550T} = \frac{Fl}{550T} \text{ etc.}$$

Electrical power is the product of volts × amperes, i.e.,

$$kw = \frac{\text{volts} \times \text{amperes}}{1,000}$$

The above expression for the determination of electrical power is strictly true for direct current and for alternating current with a zero power factor.   For the latter case, if the power factor is different from zero, the expression becomes

$$kw = \frac{\text{volts} \times \text{amperes}}{1,000} \times \text{power factor}$$

Energy.   Energy is the capacity for doing work possessed by a system through virtue of work having previously been done upon it.   Whenever work has been done upon a system in producing a *change* in either its *motion*, its *position*, or its *molecular condition*, the system has acquired the capacity for doing work.   Energy may be that due to motion, termed "kinetic energy;" that due to position, termed "potential energy;" or that due to molecular activity or configuration and is manifest as a change in its internal or stored energy.   These three forms of energy are mutually convertible.   In the English system, the units of energy are the foot-pound and the Btu, which are related by the fact that 1 Btu is equivalent to 778 ft-lb.   Some of the more common expressions for energy are as follows:

1. The potential energy of a body of weight w lb mass which has been raised h ft against gravity is $PE = (wg/g_c)h$.

2. The kinetic energy possessed by a body of weight w lb mass moving at a velocity v fps is $KE = wv^2/2g_c$.

3. If the body of 1, initially at rest, were to fall freely through the distance h, its potential energy would be converted to kinetic energy and it would acquire a velocity v determined as follows:

$$PE = wh = KE = wv^2/2g_c \quad \text{hence } wh = wv^2/2h \quad \text{and} \quad v = \sqrt{2g_c h}$$

4. The energy, resulting from its temperature, of a gas in motion is measured by its specific enthalpy h with units of Btu per pound mass.   This energy is available for conversion to kinetic energy, as given by

$$\Delta H = \Delta KE$$

$$w \, \Delta h = \frac{w}{778 \times 2g_c} (v_2^2 - v_1^2)$$

If the initial velocity v_1 is negligible, there is obtained

$$v_2 = 223.7 \sqrt{\Delta h}$$

5. Energy is measured in the English system in horsepower-hours, kilowatthours, Btu, and foot-pounds.   The relation among these units is as follows:

$$1 \text{ hp-hr} = 0.746 \text{ kwhr} = 2,546 \text{ Btu} = 1,980,788 \text{ ft-lb}$$
$$1 \text{ kwhr} = 1.34 \text{ hp-hr} = 3,412 \text{ Btu} = 2,654,536 \text{ ft-lb}$$

HEAT AND TEMPERATURE

Units of Heat. The unit of heat commonly used in the English system is the British thermal unit, or Btu, and is approximately equal to the quantity of heat that must be transferred to one pound of water in order that its temperature be raised one degree Fahrenheit. In laboratory work and throughout much of the world, the calorie is the common unit of heat. A gram calorie is the approximate quantity of heat that must be transferred to one gram of water in order to raise its temperature by one degree centigrade. The kilocalorie, sometimes called the kilogram calorie, is equal to 1,000 gram calories.

The definitions above are indicated as being approximate because, over the temperature range from freezing to boiling points of water, different quantities of heat are required to produce a unit temperature change. For this reason, the calorie and the Btu have been defined in international units as

$$1 \text{ IT calorie} = 1/860 \text{ international watthour}$$
$$1 \text{ Btu} = 251.996 \text{ IT calories}$$

In most engineering work, it is sufficiently accurate to use the relations that 1 kg-cal = 3.968 Btu and that 1 Btu = 0.252 kg-cal.

Units of Temperature. The relative "hotness" or "coldness" of a body is denoted by the term "temperature." The temperature of a substance is measured by noting its effect upon a thermometer or pyrometer whose thermal properties are known. The mercury thermometer is suitable for measuring temperatures from −39 to about 600 F. This limit may be extended to 1000 F if the capillary tube above the mercury is filled with nitrogen or carbon dioxide under pressure. High temperatures must be measured with thermocouples or optical pyrometers. The most commonly used thermometer scales are the Fahrenheit and the centigrade. Thermometer scales have as their bases the melting and boiling points of water, both measured at atmospheric pressure. The relation of the Fahrenheit and centigrade scales is as follows:

	Absolute zero	Freezing point of water	Boiling point of water
Degrees Fahrenheit.........	−459.6	32	212
Degress centigrade	−273	0	100

The relation between the two scales is

$$C = \tfrac{5}{9}(F - 32) \quad \text{and} \quad F = \tfrac{9}{5}C + 32$$

in which C is the reading on the centigrade scale and F is the reading on the Fahrenheit scale.

In certain calculations, it is necessary to express the temperature in "absolute" units. The absolute temperature associated with the Fahrenheit scale is called the Rankine temperature, and that associated with the centigrade scale is termed the Kelvin temperature. The relationships among these scales are as follows:

$$R = F + 459.6$$
$$K = C + 273$$
$$R = 1.8 K$$
$$K = \tfrac{5}{9} R$$

where R and K designate absolute temperatures on the Rankine and Kelvin scales, respectively.

Specific Heat. The specific heat of a substance is the quantity of heat required to produce a unit temperature change in a unit mass of that substance. Typical units are calories per gram per degree centigrade and Btu per pound per degree Fahrenheit. The numerical value of the specific heat is a function of the process by which the unit temperature change is effected; if a gas expands at constant pressure owing to the addition of heat, work is done by the walls of the containing vessel on the surrounding atmosphere and the heat addition must be greater than would have been required to cause the same temperature change at constant volume. The two most frequently used specific heats are those at constant volume and constant pressure, and they are represented symbolically as c_v and c_p, respectively.

The definition of specific heat given in the preceding paragraph is convenient for engineering applications. By thermodynamic analysis, it can be shown that the two specific heats referred to are given by

$$c_v = \left(\frac{\partial u}{\partial T} \right)_v$$

$$c_p = \left(\frac{\partial h}{\partial T} \right)_p$$

where u and h represent internal energy and enthalpy, respectively, and v and p indicate that volume or pressure remains constant during the measurement of the corresponding specific heat.

The specific heats of most substances vary with temperature. For a general functional relationship, the mean value of specific heat over a temperature range from T_1 to T_2 is given by

$$c_{\text{mean}} = \frac{\int_{T_1}^{T_2} c(T)\, dT}{T_2 - T_1}$$

If the algebraic relationship between specific heat and temperature is not known but the relation is available in form of a graph or table, it is usually sufficiently accurate to evaluate the average or mean specific heat at the average of temperature over the temperature range in question.

LENGTHS, AREAS, SURFACES, AND VOLUMES

List of Symbols

A = angle, deg[1]
C = length of chord
d = diameter of circle or sphere = $2r$
h = height of segment, altitude of cone, etc., as explained in context
π = ratio of circumference to diameter of circle = 3.1416
θ = angle in radian measure[1]
S = length of arc, slant height, etc., as explained in context
r = radius of circle or sphere = $d/2$
R = mean radius of curvature for pipe bends

[1] Degrees can be converted to radian measure by multiplying by 0.0175, since 2π radians = 360 deg. Hence, $\theta = 0.0175A$.

Areas are expressed in square units and volumes in cubical units of the same system in which lengths are measured.

Triangle.   Area = one-half base × altitude.

Circle.   (Fig. 26):

Circumference = $\pi d = 2\pi r$.   Area = $\pi r^2 = \pi d^2/4$.

Length of arc, $S = \theta r = 0.0175Ar$.

Length of chord, $C = 2r \sin \theta/2 = 2r \sin A/2$.

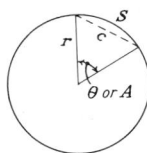

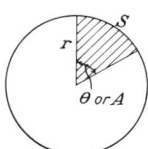

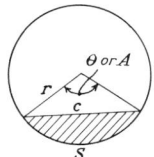

Fig. 26. Length of arc and chord.	Fig. 27. Area of sector.	Fig. 28. Area of segment, method I.

Area of Sector.   (Fig. 27):

$$\text{Area} = \tfrac{1}{2}rS = \tfrac{1}{2}r^2\theta = \pi r^2 A/360 = 0.008727r^2 A$$

Area of Segment (Method I, Fig. 28).   Find area of sector having same arc, and area of triangle formed by chord and radii of sector.   Area of segment equals sum of these two areas if segment is greater than a semicircle, and it equals their difference if segment is less than a semicircle.

$$\text{Area} = \tfrac{1}{2}r^2(\theta \pm \sin \theta)$$
$$= \tfrac{1}{2}r^2(0.0175A + \sin A)$$

Area of Segment [Method II (approximate)]:[1]

When $h = 0$ to $\tfrac{1}{4}d$, area = $h\sqrt{1.766dh - h^2}$.

When $h = \tfrac{1}{4}d$ to $\tfrac{1}{2}d$, area = $h\sqrt{0.017d^2 + 1.7dh - h^2}$.

When $h = \tfrac{1}{2}d$ to d, subtract area of empty sector from area of entire circle.

Offset Bends (Fig. 29).   The relation of D, R, H, and L is determined by geometry for the general case shown in Fig. 29a and b as follows: Consider the diagonal line joining the centers of curvature of the two arcs in either figure as forming the hypotenuse of three right-angled triangles and write an equation between the squares of the two other sides, thus,

$$2\sqrt{(D/2)^2 + R^2} = \sqrt{(2R - H)^2 + L^2}$$

squaring both sides and solving for each term in turn,

$$D = \sqrt{H^2 - 4HR + L^2}$$
$$L = \sqrt{D^2 + 4HR - H^2}$$
$$H = 2R - \sqrt{4R^2 - L^2 + D^2}$$
$$R = \frac{L^2 + H^2 - D^2}{4H}$$

$\theta = 2 \tan^{-1}\left(\dfrac{H}{L + D}\right)$ (from similarity of triangles, see Fig. 29a).

[1] For a sketch and table of volumes in partly full horizontal tanks, see Table 4.   The greatest error possible by this method is 0.23 per cent.

When $D = 0$ (Fig. 29c),

$$L = \sqrt{(4R - H)H}$$

$$R = \frac{L^2 + H^2}{4H}$$

$$C = \sqrt{RH}$$

Length of pipe in offset:

$$L = 2R\theta + D$$
$$= 0.035RA + D$$
$$= 4R \tan^{-1}\left(\frac{H}{L + D}\right) + D$$

(where the angle is expressed in radians).

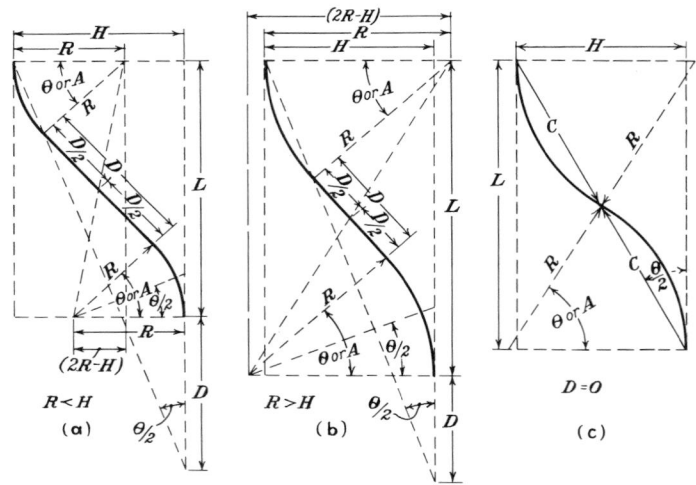

FIG. 29. Offset bends.

Cylinder:

$$\text{Area} = 2\pi rh + 2\pi r^2 = 2\pi r(h + r)$$

where r = radius of base
h = height

$$\text{Volume} = \pi r^2 h$$

Pyramid.   Right pyramid (i.e., vertex directly above center of base):

Lateral area = one-half slant height × perimeter of base
Volume = one-third altitude × area of base

Cone:

Volume = one-third area of base × perpendicular distance from vertex to plane of base.

Right circular cone:

$$\text{Lateral area} = \pi rs$$
$$\text{Volume} = \frac{1}{3}\pi r^2 h$$

where s = slant height
r = radius of base
h = perpendicular distance from vertex to plane of base

Frustum of right circular cone:

$$\text{Lateral area} = \pi s(r + r')s = \sqrt{(r - r')^2 + h^2}$$

where r = radius of lower base
$\quad r'$ = radius of upper base
$\quad h$ = height of frustum
$\quad s$ = slant height of frustum
Volume = $\frac{1}{3}\pi h(r^2 + rr' + r'^2)$

Sphere:

$$\text{Area} = 4\pi r^2 = \pi d^2$$
$$\text{Volume} = \frac{4}{3}\pi r^3 = \frac{1}{6}\pi d^3$$

UNITS OF WEIGHT AND MEASURE

From Carl Hering's "Conversion Tables."

The values of units used in the United States are given unless otherwise indicated. ap = apothecary; av = avoirdupois; Br = British; cm = centimeter; g = gram; kg = kilogram; km = kilometer; m = meter; ml = milliliter; mm = millimeter; U. S. = United States.

Acre = 208.710 ft sq (length of each side of square) = 43,560 sq ft = 4,840 sq yd = 0.00156250 sq mile = 0.404687 hectare = 4,046.887 sq m.

Acre-foot (irrigation) = 43,560 cu ft = 325,851 gal (U. S.) = 1,233.49 cu meter.

Barrel (liquid, U. S.) = No legal value. The following value is customary to some extent: 42 gal (U. S.—Standard Oil Co.).

Butt or pipe, see pipe.

Centigram = 0.01 g = 0.154324 grain.

Centimeter = 0.01 m = 0.0328083 ft = 0.393700 in. = 393.700 mils.

Centimeter, circular = 100 cir mm = 78.5398 sq mm = 0.785398 sq cm = 0.155000 cir in. = 0.121736 sq in. = 155,000 cir mils.

Centimeter^2 (square centimeter) = 127.324 cir mm = 100 sq mm = 1.27324 cir cm = 0.01 sq decimeter = 197,352 cir mils = 155,000 sq mils = 0.197352 cir in. = 0.155000 sq in.

Centimeter^3 (cubic centimeter) = 0.999973 ml = 0.000999973 liter = 0.001 cu decimeter = 0.0610234 cu in.

Circular mil, circular inch, circular centimeter, etc., see mil, inch, centimeter, etc.

Cubic inch, cubic centimeter, etc., see inch, centimeter, etc.

Decigram = 0.1 g = 1.54324 grains

Deciliter = 0.1 liter = 0.1000027 cu decimeter = 6.10250 cu in.

Decimeter^2 (square decimeter) = 0.01 sq meter = 15.5000 sq in. = 0.1076387 sq ft = 0.0119599 sq yd.

Decimeter^3 (cubic decimeter) = 0.999973 liter.

Dekameter = 10 meter = 10.9361 yd.

Dram (av) = $\frac{1}{16}$ oz (av) = 27.34375 grains = 0.455729 dram (ap) = 1.77185 g.

Dram (ʒ) (ap) = $\frac{1}{8}$ oz (troy or ap) = 3 scruples = 60 grains = 2.19429 drams (av) = 3.8879351 g.

Foot (Br) = 0.9999971 ft (U. S.) = 30.4800 cm.

Foot (U. S.) = 12,000 mils = 12 in. = $\frac{1}{3}$ yd = 1/5,280 or 0.000189394 statute mile (U. S.) = 30.4801 cm = 1.0000029 ft (Br).

Foot, circular = 144 cir in. = 113.097 sq in. = 0.785398 sq ft = 929.034 cir cm = 729.662 sq cm.

Foot^2 (square foot) (Br) = 0.9999942 sq ft (U. S.) = 0.0929029 sq m.

Foot^2 (square foot) (U. S.) = 144 sq in. = $\frac{1}{9}$ or 0.111111 sq yd = 183.346 cir in. = 1.27324 cir ft = 929.034 sq cm = 1.0000057 sq ft (Br).

Foot³ (cubic foot) = 1,728 cu in. = 0.0370370 cu yd = 28.316265 liter = 28.3170 cu decimeters = 7.48052 gal (U. S.) = 0.803564 bu (U. S.).

Gallon, Imperial (Br) = 32 gills (Br) = 8 pt (Br) = 4 qt (Br) = 0.5 pk (Br) = $\frac{1}{8}$ or 0.125 bu (Br) = 4.5458269 liter = 4.5459631 cu decimeters = 1.20091 gal (U. S.) = 277.410 cu in. = 4,545.9631 cu cm.

Gallon (liquid; U. S.) = 231 cu in. = 0.133681 cu ft = 3.78533 liter = 3.78543 cu decimeters = 3,785.43 cu cm = 32 gills (U. S.) = 8 pt (liquid; U. S.) = 4 qt (liquid; U. S.) = 0.8327024 gal (Br.)

Gill (Br) = $\frac{1}{4}$ pt (Br) = $\frac{1}{32}$ or 0.03125 gal (Br).

Gill (liquid; U. S.) = $\frac{1}{4}$ pt (liquid; U. S.) = $\frac{1}{32}$ gal (U. S.).

Grain (same in av, troy, and ap weights) = 1/7,000 lb (av) = 0.00228571 oz (av) = 1/5,760 lb (troy or ap) = 0.0647989 g.

Gram = 0.001 kg = 15.43235639 grains = 0.564383 dram (av) = 0.0352740 oz (av) = 0.00220462 lb (av) = 0.771618 scruple = 0.257206 dram (ap) = 0.0321507 oz (troy or ap).

Gram-molecule of any gas at 0 C and 760 mm of mercury pressure has a volume of 22,380 cu cm.

Hectare = 100 ares = 2.47104 acres = 10,000 sq m = 11,959.9 sq yd.

Hectoliter = 100 liter = 3.53145 cu ft.

Hectometer = 100 m = 109.361 yd.

Hogshead (liquid; U. S.) = 63 gal (U. S.) = 2 bbl of 31$\frac{1}{2}$ gal (U. S.) = 238.4759 liter.

Hundredweight, short = 100 lb (av) = $\frac{1}{20}$ or 0.05 short or net ton = 45.35924 kg.

Hundredweight, long = 112 lb (av) = $\frac{1}{20}$ or 0.05 long or gross ton = 50.8024 kg.

Inch = 1,000 mils = $\frac{1}{12}$ ft = $\frac{1}{36}$ yd = 2.540005 cm.

Inch, circular = 1,000,000 cir mils = $\frac{1}{144}$ or 0.00694444 cir ft = 785,398 sq mils = 0.785398 sq in. = 0.00545415 sq ft = 645.163 cir mm = 6.45163 cir cm = 506.709 sq mm = 5.06709 sq cm.

Inch² (square inch) = $\frac{1}{144}$ or 0.00694444 sq ft = 1,000,000 sq mils = 1,273,240 cir mils = 1.27324 cir in. = 6.45163 sq cm = 8.21447 cir cm.

Inch³ (cubic inch) = $\frac{1}{1728}$ or 0.000578704 cu ft = 16.38716 cu cm = 16.3867 ml = 0.0163867 liter = 0.01638716 cu decimeter.

Kilogram or kilo = 1,000 g = 0.001 metric ton = 15,432.4 grains = 35.2740 oz (av) = 2.20462 lb (av) = 0.0220462 cwt (short) = 0.0196841 cwt (long) = 0.00110-231 short or net ton = 0.000984206 long or gross ton = 32.1507 oz (troy or ap).

Kilometer = 1,000 meters = 3,280.83 ft = 1,093.61 yd = 0.621370 statute mile (U. S.) = 0.539593 nautical mile (U. S.).

Kilometer² (square kilometer) = 100 hectares = 247.104 acres = 1,195,985 sq yd = 0.386101 sq mile.

Liter = 1.000027 cu decimeters = 10 deciliters = 1,000.027 cu cm = 0.01 hecto-liter = 0.001000027 cu meter = 61.0250 cu in. = 0.0353134 cu ft = 2.11342 pt (liquid; U. S.) = 1.05671 qt (liquid; U. S.) = 0.264178 gal (U. S.) = 1.81620 pt (dry; U. S.) = 0.908102 qt (dry; U. S.) = 0.113513 pk (U. S.) = 0.028378 bu (U. S.) = 1.7598475 pt (Br) = 0.8798475 qt (Br) = 0.2199809 gal (Br) = 0.10999097 pk (Br) = 0.02749764 bu (Br).

Meter (international) = 0.001 km = 0.01 hectometer = 0.1 dekameter = 10 deci-meters = 100 cm = 1,000 mm = 1,000,000 micrometers = 39.370113 in. (Br) = 39.37 in. exact legal value (U. S.) = 3.28083 ft (U. S.) = 1.09361 yd (U. S.) = 0.000621370 statute mile (U. S.).

Meter² (square meter) = 10,000 sq cm = 100 sq decimeters = 0.01 are = 1,550 sq in. = 10.76387 sq ft (U. S.) = 1.19599 sq yd (U. S.) = 10.76393 sq ft (Br) = 1.19599 sq yd (Br) = 0.000247104 acre.

Meter³ (cubic meter) = 999.973 liter = 1.30794 cu yd.

Micro-meter, micron, or microne = 0.001 mm.

Mil = 0.001 in. = 0.02540005 mm.

Mil, circular = 0.000001 cir in. = 0.785398 sq mil = 0.000000785398 sq in. = 0.000645163 cir mm = 0.000506709 sq mm.

Mil² (square mil) = 0.000001 sq in. = 0.00000127324 cir in. = 1.27324 cir mils = 0.000821447 cir mm = 0.000645163 sq mm.

Mile (Br) = 5,280 ft = 1,760 yd = practically the same as statute mile (U. S.).

Mile, international geographical = 24,350.3 ft = 8,116.77 yd = 7.422 km = 7,422 m = 4.6118 statute miles (U. S.) = $\frac{1}{15}$ of 1 deg at the equator.

Mile, international nautical = 6,076.10 ft = 2,025.37 yd = 1.852 km = 1,852 meter = $\frac{1}{60}$ of 1 deg of meridian at the equator.

Mile, nautical (Br) = 6,080 ft = 2,026.67 yd = 0.999966 nautical mile (U. S.) = 1.15152 statute miles (U. S.) = 1.85319 km.

Mile, nautical (U. S.) same as sea mile and geographical mile (all U. S.) = 6,080.20 ft = 1.85325 km = 1.15155 statute miles (U. S.) = 1.000034 nautical miles (Br) = 1 min of earth's circumference.

Mile, statute or land (U. S.) = 5,280 ft = 1.60935 km.

Mile² (square mile) = 640 acres = 3,097,600 sq yd = 2.59000 sq km.

Milligram = 0.001 g = 0.0154324 grain.

Millimeter = 0.001 meter = 39.370 mils = 0.039370 in.

Millimeter, circular = 0.01 cir cm = 0.785398 sq mm = 1,550 cir mils = 1,217.36 sq mils = 0.00121736 sq in.

Millimeter² (square millimeter) = 0.01 sq cm = 1.27324 cir mm = 1.973.52 cir mils = 1,550 sq mils = 0.001550 sq in.

Ounce (℥) same as troy ounce = 480 grains = 24 scruples = 8 drams (ap) = $\frac{1}{12}$ or 0.0833333 lb (troy or ap) = 3.1035 g.

Ounce (av) = 16 drams (av) = $\frac{1}{16}$ or 0.062500 lb (av) = 437.500 grains = 28.3495 g = 0.911458 oz (troy or ap).

Ounce (troy, gold and silver) same as ap ounce = 480 grains = 20 dwt = $\frac{1}{12}$ or 0.0833333 lb (troy or ap) = 0.0684714 lb (av) = 31.1035 g = 1.09714 oz (av).

Ounce, fluid (ap U. S.) = 480 minims (ap U. S.) = 8 fluid drams (ap U. S.) = $\frac{1}{16}$ pt (ap U. S.) = 1.80469 cu in. = 0.0295729 liter.

Pennyweight (troy) = 24 grains = 1.55517 g = $\frac{1}{20}$ oz (troy or ap).

Pint (ap U. S.) same as ordinary U. S. liquid pint = 128 fluid drams (ap U. S.) = 16 fl oz (ap U. S.) = 0.473167 liter.

Pint (Br) = 0.125 gal (Br) = 0.015625 bu (Br) = 0.568230047 liter.

Pint (dry; U. S.) = 0.5 qt (dry; U. S.) = 0.550599 liter.

Pint (liquid; U. S.) = 0.125 gal (U. S.) = 0.473167 liter.

Pipe or **butt** (liquid; U. S.) = 126 gal (U. S.) = 2 hogsheads (U. S.) = 476.9471 liters.

Pound (av) = 7,000 grains = 16 oz (av) = 256 drams (av) = 14,5833 oz (troy or ap) = 453.5924277 g = 0.4535924277 kg = 7,000/5,760 or 1.21528 lb (troy or ap).

Pound (troy or ap) = 5,760 grains = 12 oz (troy or ap) = 0.373242 kg = 5,760/7,000 or 0.822857 lb (av).

Quart (Br) = 0.25 gal (Br) = 1.03202 qt (dry; U. S.) 1.20091 qt (liquid; U. S.) = 1.1364593 liter.

Quart (dry; U. S.) = 2 pt (dry; U. S.) = $\frac{1}{8}$ or 0.125 pk (U. S.) = $\frac{1}{32}$ or 0.031250 bu (U. S.) = 67.200625 cu in. = 0.0388893 cu ft = 1.101198 liter = 1,101.23 cu cm = 11.01198 deciliters = 1.16365 qt (liquid; U. S.) = 0.968972 qt (Br) = 0.242243 gal (Br).

Quart (liquid; U. S.) = 0.25 gal (U. S.) = 0.946333 liter.

Scruple (℈) (ap) = 20 grains = $\frac{1}{3}$ dr (ap) = $\frac{1}{24}$ oz (troy or ap) = 1.295978 gr.

Section (of land) = 1 mile sq = 640 acres.

Square (building) = 100 sq ft.

Square inch, square centimeter, square mil, etc., see inch, mil, centimeter, etc.

Stone (Br) = 14 lb (av) = 6.35029 kg.

Ton (gross) **displacement of water** = 35.8813 cu ft = 1.01605 cu meters.

Ton, register (shipping, for whole vessels) = 100 cu ft = 2.8317 cu meters.

Ton, long or gross = 2,240 lb (av) = 1.12 short or net tons = 1016.05 kg = 1.01605 metric tons.

Ton, short or net = 2,000 lb (av) = 20 cwt (short) = 907.185 kg = 0.907185 metric ton = 17.8571 cwt (long) = 0.892857 long or gross ton.

Ton, metric, tonne, tonneau, millier, or bar = 2,204.62 lb (av) = 1.10231 short or net tons = 0.984206 long or gross ton = 1,000 kg.

Yard (Br) = 0.9999971 yd (U. S.) = 0.9143992 meter.

Yard (U. S.) = 36 in. = 3 ft = 1.0000029 yd (Br) = 0.914402 meter.

Yard^2 (square yard) (Br) = 0.9999943 sq yd (U. S.) = 0.836126 sq meter.

Yard^2 (square yard) (U. S.) = 1,296 sq in. = 9 sq ft = 1/4,840 or 0.000206612 acre = 0.836131 sq meter = 1.0000057 sq yd (Br).

Yard^3 (cubic yard) = 27 cu ft = 46,656 cu in. = 0.764559 cu meter.

Table I.   Decimal Equivalents of Eighths, Sixteenths, Thirty-seconds, and Sixty-fourths of an Inch

Eighths	$9/32 = 0.28125$	$19/64 = 0.296875$
$1/8 = 0.125$	$11/32 = 0.34375$	$21/64 = 0.328125$
$1/4 = 0.250$	$13/32 = 0.40625$	$23/64 = 0.359375$
$3/8 = 0.375$	$15/32 = 0.46875$	$25/64 = 0.390625$
$1/2 = 0.500$	$17/32 = 0.53125$	$27/64 = 0.421875$
$5/8 = 0.625$	$19/32 = 0.59375$	$29/64 = 0.453125$
$3/4 = 0.750$	$21/32 = 0.65625$	$31/64 = 0.484375$
$7/8 = 0.875$	$23/32 = 0.71875$	$33/64 = 0.515625$
Sixteenths	$25/32 = 0.78125$	$35/64 = 0.546875$
$1/16 = 0.0625$	$27/32 = 0.84375$	$37/64 = 0.578125$
$3/16 = 0.1875$	$29/32 = 0.90625$	$39/64 = 0.609375$
$5/16 = 0.3125$	$31/32 = 0.96875$	$41/64 = 0.640625$
$7/16 = 0.4375$	Sixty-fourths	$43/64 = 0.671875$
$9/16 = 0.5625$	$1/64 = 0.015625$	$45/64 = 0.703125$
$11/16 = 0.6875$	$3/64 = 0.046875$	$47/64 = 0.734375$
$13/16 = 0.8125$	$5/64 = 0.078125$	$49/64 = 0.765625$
$15/16 = 0.9375$	$7/64 = 0.109375$	$51/64 = 0.796875$
Thirty-seconds	$9/64 = 0.140625$	$53/64 = 0.828125$
$1/32 = 0.03125$	$11/64 = 0.171875$	$55/64 = 0.859375$
$3/32 = 0.09375$	$13/64 = 0.203125$	$57/64 = 0.890625$
$5/32 = 0.15625$	$15/64 = 0.234375$	$59/64 = 0.921875$
$7/32 = 0.21875$	$17/64 = 0.265625$	$61/64 = 0.953125$
		$63/64 = 0.984375$

Table 2. Wire and Sheet-metal Gages
(Diameters and thicknesses in decimal parts of an inch)

Gauge No.	American wire gauge, or Brown and Sharpe (for copper wire)	Steel wire gauge, or Washburn and Moen or Roebling (for steel wire)	Birmingham wire gauge (B.W.G.) or Stubs' iron wire (for steel wire or sheets)	Stubs steel wire gauge	British Imperial standard wire gauge (S.W.G.)	U.S. standard gauge for sheet metal (iron and steel) 480 lb per cu ft	AISI inch equivalent for U.S. steel sheet thickness	British standard for iron and steel, sheets and hoops 1914 (B.G.)
0000000	......	0.4900	.....	.....	0.500	0.500	......	0.6666
000000	......	0.4615	.....	.....	0.464	0.469	......	0.625
00000	......	0.4305	.....	.....	0.432	0.438	......	0.5883
0000	0.460	0.3938	0.454	.....	0.400	0.406	......	0.5416
000	0.410	0.3625	0.425	.....	0.372	0.375	......	0.5000
00	0.365	0.3310	0.380	.....	0.348	0.344	......	0.4452
0	0.325	0.3065	0.340	.....	0.324	0.312	......	0.3964
1	0.289	0.2830	0.300	0.227	0.300	0.281	......	0.3532
2	0.258	0.2625	0.284	0.219	0.276	0.266	......	0.3147
3	0.229	0.2437	0.259	0.212	0.252	0.250	0.2391	0.2804
4	0.204	0.2253	0.238	0.207	0.232	0.234	0.2242	0.2500
5	0.182	0.2070	0.220	0.204	0.212	0.219	0.2092	0.2225
6	0.162	0.1920	0.203	0.201	0.192	0.203	0.1943	0.1981
7	0.144	0.1770	0.180	0.199	0.176	0.188	0.1793	0.1764
8	0.128	0.1620	0.165	0.197	0.160	0.172	0.1644	0.1570
9	0.114	0.1483	0.148	0.194	0.144	0.156	0.1495	0.1398
10	0.102	0.1350	0.134	0.191	0.128	0.141	0.1345	0.1250
11	0.091	0.1205	0.120	0.188	0.116	0.125	0.1196	0.1113
12	0.081	0.1055	0.109	0.185	0.104	0.109	0.1046	0.0991
13	0.072	0.0915	0.095	0.182	0.092	0.094	0.0897	0.0882
14	0.064	0.0800	0.083	0.180	0.080	0.078	0.0747	0.0785
15	0.057	0.0720	0.072	0.178	0.072	0.070	0.0673	0.0699
16	0.051	0.0625	0.065	0.175	0.064	0.062	0.0598	0.0625
17	0.045	0.0540	0.058	0.172	0.056	0.056	0.0538	0.0556
18	0.040	0.0475	0.049	0.168	0.048	0.050	0.0478	0.0495
19	0.036	0.0410	0.042	0.164	0.040	0.0438	0.0418	0.0440
20	0.032	0.0348	0.035	0.161	0.036	0.0375	0.0359	0.0392
21	0.0285	0.0317	0.032	0.157	0.032	0.0344	0.0329	0.0349
22	0.0253	0.0286	0.028	0.155	0.028	0.0312	0.0299	0.0313
23	0.0226	0.0258	0.025	0.153	0.024	0.0281	0.0269	0.0278
24	0.0201	0.0230	0.022	0.151	0.022	0.0250	0.0239	0.0248
25	0.0179	0.0204	0.020	0.148	0.020	0.0219	0.0209	0.0220
26	0.0159	0.0181	0.018	0.146	0.018	0.0188	0.0179	0.0196
27	0.0142	0.0173	0.016	0.143	0.0164	0.0172	0.1064	0.0175
28	0.0126	0.0162	0.014	0.139	0.0148	0.0156	0.0149	0.0156
29	0.0113	0.0150	0.013	0.134	0.0136	0.0141	0.0135	0.0139
30	0.0100	0.0140	0.012	0.127	0.0124	0.0125	0.0120	0.0123
31	0.0089	0.0132	0.010	0.120	0.0116	0.0109	0.0105	0.0110
32	0.0080	0.0128	0.009	0.115	0.0108	0.0102	0.0097	0.0098
33	0.0071	0.0118	0.008	0.112	0.0100	0.0094	0.0090	0.0087
34	0.0063	0.0104	0.007	0.110	0.0092	0.0086	0.0082	0.0077
35	0.0056	0.0095	0.005	0.108	0.0084	0.0078	0.0075	0.0069
36	0.0050	0.0090	0.004	0.106	0.0076	0.0070	0.0067	0.0061
37	0.0045	0.0085	.....	0.103	0.0068	0.0066	0.0064	0.0054
38	0.0040	0.0080	.....	0.101	0.0060	0.0062	0.0060	0.0048
39	0.0035	0.0075	.....	0.099	0.0052	......	......	0.0043
40	0.0031	0.0070	.....	0.097	0.0048	......	......	0.0039
41	......	0.0066	.....	0.095	0.0044	......	......	0.0034
42	......	0.0062	.....	0.092	0.0040	......	......	0.0031
43	......	0.0060	.....	0.088	0.0036	......	......	0.0027
44	......	0.0058	.....	0.085	0.0032	......	......	0.0024
45	......	0.0055	.....	0.081	0.0028	......	......	0.0022
46	......	0.0052	.....	0.079	0.0024	......	......	0.0019
47	......	0.0050	.....	0.077	0.0020	......	......	0.0017
48	......	0.0048	.....	0.075	0.0016	......	......	0.0015
49	......	0.0046	.....	0.072	0.0012	......	......	0.0014
50	......	0.0044	.....	0.069	0.0010	......	......	0.0012

Table 3.   Contents, in Cubic Feet and U. S. Gallons, of Pipes and Cylindrical Tanks of Various Diameters and I Ft in Length, When Completely Filled[1]

(1 gal = 231 cu in.   1 cu ft = 7.4805 gal)

Diameter, in inches	For 1 ft. in length		Length, in inches, of cylinder of 1 cu. ft. capacity	Diameter, in inches	For 1 ft. in length		Length, in inches, of cylinder of 1 cu. ft. capacity
	Cubic feet; also, area in square feet	U. S. gal., 231 cu. in.			Cubic feet; also, area in square feet	U. S. gal., 231 cu. in.	
12½	0.8522	6.375	14.080	21¼	2.463	18.42	4.872
12⅝	0.8693	6.503	13.800	21½	2.521	18.86	4.760
12¾	0.8866	6.632	13.530	21¾	2.580	19.30	4.651
12⅞	0.9041	6.763	13.270	22	2.640	19.75	4.545
13	0.9218	6.895	13.020	22¼	2.700	20.20	4.445
13⅛	0.9395	7.028	12.780	22½	2.761	20.66	4.347
13¼	0.9575	7.163	12.530	22¾	2.823	21.12	4.251
13⅜	0.9757	7.299	12.300	23	2.885	21.58	4.160
13½	0.994	7.436	12.070	23¼	2.948	22.05	4.070
13⅝	1.013	7.578	11.850	23½	3.012	22.53	3.990
13¾	1.031	7.712	11.640	23¾	3.076	23.01	3.901
13⅞	1.051	7.855	11.420	24	3.142	23.50	3.819
14	1.069	7.997	11.230	25	3.409	25.50	3.520
14⅛	1.088	8.139	11.030	26	3.678	27.58	3.263
14¼	1.107	8.281	10.840	27	3.976	29.74	3.018
14⅜	1.127	8.431	10.650	28	4.276	31.99	2.806
14½	1.147	8.578	10.460	29	4.587	34.31	2.616
14⅝	1.167	8.730	10.280	30	4.909	36.72	2.444
14¾	1.187	8.879	10.110	31	5.241	39.21	2.290
14⅞	1.207	9.029	9.940	32	5.585	41.78	2.149
15	1.227	9.180	9.780	33	5.940	44.43	2.020
15⅛	1.248	9.336	9.620	34	6.305	47.16	1.903
15¼	1.268	9.485	9.460	35	6.681	49.98	1.796
15⅜	1.289	9.642	9.310	36	7.069	52.88	1.698
15½	1.310	9.801	9.160	37	7.467	55.86	1.607
15⅝	1.332	9.964	9.010	38	7.876	58.92	1.527
15¾	1.353	10.121	8.870	39	8.296	62.06	1.446
15⅞	1.374	10.278	8.730	40	8.727	65.28	1.375
16	1.396	10.440	8.600	41	9.168	68.58	1.309
16¼	1.440	10.772	8.330	42	9.621	71.91	1.247
16½	1.485	11.11	8.081	43	10.085	75.44	1.190
16¾	1.530	11.45	7.843	44	10.559	78.99	1.136
17	1.576	11.79	7.511	45	11.045	82.62	1.087
17¼	1.623	12.14	7.394	46	11.541	86.33	1.040
17½	1.670	12.49	7.186	47	12.048	90.13	0.996
17¾	1.718	12.85	6.985	48	12.566	94.00	0.955
18	1.768	13.22	6.787	49	13.095	97.96	0.916
18¼	1.817	13.59	6.604	50	13.635	102.00	0.880
18½	1.867	13.96	6.427	51	14.186	106.12	0.846
18¾	1.917	14.34	6.259	52	14.748	110.32	0.814
19	1.969	14.73	6.094	53	15.320	114.60	0.783
19¼	2.021	15.12	5.938	54	15.904	118.97	0.755
19½	2.074	15.51	5.786	55	16.499	122.82	0.727
19¾	2.128	15.92	5.639	56	17.104	127.95	0.702
20	2.182	16.32	5.500	57	17.720	132.55	0.677
20¼	2.237	16.73	5.365	58	18.347	137.24	0.654
20½	2.292	17.15	5.236	59	18.985	142.02	0.632
20¾	2.348	17.56	5.110	60	19.637	146.89	0.611
21	2.405	17.99	4.989				

[1] To find the capacity of pipes greater than the largest given in the table, look in the table for a pipe one-half the given size and multiply its capacity by 4, or one of one-third its size, and multiply its capacity by 9, etc.

Table 4. Contents of Pipes and Cylindrical Tanks—Axis Horizontal—Flat Ends—Contents per Foot of Length for any Depth of Liquid

| h = Depth of liquid inches | \multicolumn{10}{c}{d = diameter of tank, inches} | | | | | | | | |
| | 12 | | 18 | | 24 | | 30 | | 36 | |
	Gal.	Cu. ft.	Gal.	Cu. ft.	Gal.	Cu. ft.	Gal.	Cu. ft.	Gal.	Cu. ft.
2	0.64	0.0860	0.80	0.1072	0.93	0.1244	1.05	0.1400	1.15	0.154
4	1.73	0.2317	2.18	0.2920	2.57	0.3440	2.90	0.3878	3.21	0.429
6	2.94	0.3927	3.85	0.5149	4.59	0.6140	5.23	0.6988	5.80	0.775
8	4.14	0.5537	5.67	0.7578	6.85	0.9152	7.85	1.049	8.75	1.17
10	5.23	0.6994	7.55	1.009	9.26	1.238	10.72	1.432	12.0	1.60
12	5.87	0.7854	9.38	1.252	11.75	1.571	13.72	1.833	15.4	2.03
14	.....	.....	11.04	1.476	14.24	1.903	16.82	2.248	19.0	2.54
16	.....	.....	12.43	1.659	16.65	2.226	19.90	2.660	22.6	3.02
18	.....	.....	13.22	1.767	18.91	2.527	23.00	3.075	26.4	3.53
20	.....	.....	.....	.....	20.93	2.797	26.00	3.476	29.6	3.95
22	.....	.....	.....	.....	22.57	3.017	28.85	3.859	33.4	4.46
24	.....	.....	.....	.....	23.50	3.1416	31.49	4.209	37.4	5.00
26	.....	.....	.....	.....	.....	.....	33.82	4.521	40.4	5.40
28	.....	.....	.....	.....	.....	.....	35.67	4.768	43.7	5.84
30	.....	.....	.....	.....	.....	.....	36.72	4.908	46.6	6.23
32	.....	.....	.....	.....	.....	.....	.....	.....	49.1	6.55
34	.....	.....	.....	.....	.....	.....	.....	.....	51.2	6.85
36	.....	.....	.....	.....	.....	.....	.....	.....	52.9	7.07
38										
40										
42										
44										
46										
48										
50										
52										
54										
56										
58										
60										
64										
68										
72										
76										
80										
84										

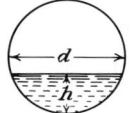

Formulas for determination of approximate capacity of horizontal cylindrical tanks for any depth. Given:—diameter of tank d and height of segment h.

To find area of segment

when $h = 0$ to $\frac{1}{4}d$; area $= h\sqrt{1.766dh - h^2}$

when $h = \frac{1}{4}d$ to $\frac{1}{2}d$; area $= h\sqrt{0.017d^2 + 1.7dh - h^2}$

1 cu ft = 7.4805 U. S. gal

Table 4. (Continued)

h = depth of liquid inches	\multicolumn{8}{c}{d = diameter of tank, inches}
	\multicolumn{2}{c}{42}	\multicolumn{2}{c}{48}	\multicolumn{2}{c}{54}	\multicolumn{2}{c}{60}
	Gal.	Cu. ft.	Gal.	Cu. ft.	Gal.	Cu. ft.	Gal.	Cu. ft.
2	1.25	0.167	1.36	0.182	1.43	0.191	1.47	0.197
4	3.49	0.465	3.72	0.496	3.98	0.531	4.19	0.560
6	6.31	0.843	6.90	0.921	7.25	0.967	7.48	1.00
8	9.57	1.28	10.3	1.37	11.0	1.47	11.6	1.55
10	13.3	1.77	14.2	1.89	15.2	2.02	16.2	2.16
12	16.9	2.26	18.6	2.48	19.7	2.63	21.0	2.81
14	21.0	2.80	22.8	3.04	24.2	3.23	26.3	3.52
16	25.2	3.36	27.4	3.66	29.4	3.92	31.4	4.19
18	29.4	3.92	32.3	4.31	34.8	4.64	36.9	4.93
20	33.8	4.51	37.0	4.94	40.1	5.35	42.8	5.72
22	38.2	5.10	42.0	5.61	45.6	6.08	48.8	6.53
24	42.5	5.67	47.0	6.27	51.0	6.80	54.7	7.30
26	46.8	6.25	52.0	6.94	56.7	7.56	61.0	8.15
28	50.9	6.80	57.0	7.61	62.3	8.33	66.9	8.94
30	55.0	7.34	61.7	8.23	67.8	9.05	73.4	9.81
32	58.8	7.86	66.6	8.89	73.4	9.79	79.7	10.7
34	62.4	8.19	71.3	9.52	78.8	10.5	85.9	11.5
36	65.6	8.75	75.7	10.1	84.2	11.2	92.6	12.4
38	68.4	9.13	79.9	10.7	89.4	11.9	98.0	13.1
40	70.7	9.44	83.7	11.2	94.4	12.6	105	14.0
42	72.0	9.61	87.4	11.7	99.5	13.3	110	14.7
44	.....	.....	90.3	12.1	104	13.9	115	15.4
46	.....	.....	92.7	12.4	108	14.4	121	16.2
48	.....	.....	94.0	12.6	112	14.9	126	16.8
50	.....	.....	.....	.....	115	15.4	131	17.5
52	.....	.....	.....	.....	117	15.6	135	18.0
54	.....	.....	.....	.....	119	15.9	139	18.6
56	.....	.....	.....	.....	.....	.....	143	19.1
58	.....	.....	.....	.....	.....	.....	145	19.4
60	.....	.....	.....	.....	.....	.....	147	19.6
64								
68								
72								
76								
80								
84								

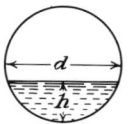

Formulas for determination of approximate capacity of horizontal cylindrical tanks for any depth.   Given:—diameter of tank d and height of segment h.

To find area of segment

when $h = 0$ to $\tfrac{1}{4}d$; area $= h\sqrt{1.766dh - h^2}$

when $h = \tfrac{1}{4}d$ to $\tfrac{1}{2}d$; area $= h\sqrt{0.017d^2 + 1.7dh - h^2}$

1 cu ft $= 7.4805$ U. S. gal

Table 4.   (*Concluded*)

h = Depth of liquid inches	\(d = \) diameter of tank, inches							
	66		72		73		84	
	Gal.	Cu. ft.	Gal.	Cu. ft.	Gal.	Cu. ft.	Gal.	Cu. ft.
2	1.57	0.210	1.65	0.220	1.73	0.229	1.77	0.236
4	4.42	0.580	4.64	0.618	4.81	0.641	4.95	0.661
6	8.04	1.07	8.10	1.16	8.78	1.17	9.13	1.22
8	12.2	1.63	12.8	1.71	13.4	1.78	13.9	1.86
10	17.0	2.27	17.7	2.36	18.6	2.48	19.7	2.63
12	22.1	2.96	23.6	3.14	24.2	3.22	25.7	3.43
14	27.6	3.68	28.9	3.85	30.2	4.03	31.5	4.21
16	33.3	4.45	35.0	4.66	36.1	4.81	38.2	5.10
18	39.3	5.25	41.7	5.55	43.3	5.78	45.8	6.12
20	45.5	6.08	48.0	6.40	50.3	6.72	52.5	7.01
22	51.8	6.93	54.7	7.29	57.6	7.69	60.0	8.02
24	58.3	7.79	61.9	8.25	64.8	8.65	68.8	9.19
26	65.0	8.68	68.7	9.15	72.1	9.63	75.7	10.1
28	71.8	9.59	76.0	10.2	80.0	10.7	84.1	11.2
30	78.6	10.5	83.5	11.1	88.0	11.8	91.6	12.3
32	85.4	11.4	90.7	12.1	96.0	12.8	101	13.5
34	92.3	12.3	98.2	13.1	104	13.9	109	14.5
36	99.1	13.2	106	14.1	112	14.9	117	15.6
38	106	14.2	113	15.1	120	16.0	126	16.8
40	113	15.1	121	16.1	128	17.1	135	18.0
42	119	15.9	128	17.1	136	18.2	144	19.2
44	126	16.8	136	18.2	144	19.2	153	20.4
46	132	17.6	143	19.1	152	20.3	162	21.6
48	138	18.4	150	20.0	160	21.4	171	22.8
50	144	19.2	157	21.0	168	22.4	179	23.9
52	150	20.0	164	21.9	176	23.5	187	25.0
54	156	20 8	170	22.7	183	24.4	196	26.2
56	161	21.5	176	23.5	191	25.5	204	27.2
58	165	22.0	182	24.3	198	26.4	212	28.3
60	169	22.6	188	25.1	205	27.4	219	29.2
64	176	23.5	198	26.4	218	29.1	235	31.4
68	.....	.....	207	27.6	230	30.7	250	33.4
72	.....	.....	211	28.2	239	31.9	262	35.0
76	.....	.....	.....	.....	246	32.8	274	36.6
80	.....	.....	.....	.....	.....	.....	283	37.8
84	.....	.....	.....	.....	.....	.....	288	38.5

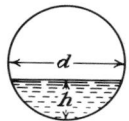

Formulas for determination of approximate capacity of horizontal cylindrical tanks for any depth.   Given:—diameter of tank d and height of segment h.

To find area of segment

when $h = 0$ to $\tfrac{1}{4}d$; area $= h\sqrt{1.766dh - h^2}$

when $h = \tfrac{1}{4}d$ to $\tfrac{1}{2}d$; area $= h\sqrt{0.017d^2 + 1.7dh - h^2}$

1 cu ft = 7.4805 U. S. gal

2

PHYSICAL AND METALLURGICAL PROPERTIES OF PIPING MATERIALS

Helmut Thielsch*

The application of pipe in a commercial steam power plant, nuclear power plant, refinery, chemical plant, paper mill, or gas-transmission piping system is based generally on identical or very similar design considerations. Materials of the same mechanical and physical properties, chemical compositions and metallurgical structures are used. Fabrication operations such as bending, shaping, welding, and heat treatment involve identical procedures which depend on the final quality desired and not on the application. Finally, inspection by radiographic, ultrasonic, or other techniques does not differentiate, for example, between nuclear and commercial steam power plant piping applications.

The designation "piping" normally includes pipe, tubing, fittings (elbows, tees, flanges, reducers, etc.), valves, headers, expansion joints, flow nozzles, and other components in the piping system.

SELECTION OF PIPING MATERIALS

The selection of the piping materials for any given application should follow the recommendations of the respective applicable codes, dimensional standards and established material specifications. However, the design engineer must take into consideration also the service requirements and consider the effects of conditions such as corrosion, scaling, thermal or mechanical fatigue, creep, notch toughness, and metallurgical instability at elevated temperatures.

* Manager, Research and Development Division, Grinnell Corp., Providence, R.I.

Basing the requirements in specifications for piping materials on the results of laboratory tests made on small sections of material may lead to unrealistic interpretations.   For example, the results of extensive small-section creep tests conducted at elevated temperatures on Type 347 stainless-steel specimens led initially to the assumption that this grade was one of the best suitable for service at temperatures of from 1050 to 1200 F.   The application of this specific material in steam power piping, however, has revealed a tendency towards serious cracking of the heat-affected zone in the pipe base metal adjacent to the weld.

Before laboratory test data alone are used for the specification of fabrication or heat-treating procedures, the effects of normal pipe-manufacturing, shop-fabrication and field-erection conditions should be evaluated.   Creep tests conducted on small laboratory specimens, for example, have shown that chromium-molybdenum steel normalized at 1650 F and tempered subsequently at 1250 F exhibits greater resistance to creep than hot-worked steel that has been tempered only.

The uniformity of heat treatment in a laboratory furnace is not attained under shop conditions where a number of pipe sections are heat-treated at the same time. Temperature gradients are likely to exist in the large furnaces necessary to handle several bars of pipe.   Moreover, when the movable car is pulled out of the furnace after the normalizing heat treatment at 1650 F and the pipe is permitted to air cool, the pipe side at the top will naturally cool most rapidly.   The pipe side nearest the heat-retaining brick lining at the bottom of the furnace car will cool more slowly. As a result, the metallurgical structure of the pipe at the top may differ considerably from that of the bottom side.

Large pipe sections removed from a furnace for air cooling after the normalizing heat treatment at 1650 F may be distorted somewhat because of one side being cooled more rapidly than the other.   Welding, hot fabrication, and heat treatment of welds may also cause local distortions.   Such sections are then straightened in the mill or fabricating plant by heating the pipe locally with gas burners to orange heat (1600 to 1900 F) and pulling the sections back into shape (see section on "Flame Straightening" in Chap. 7).   This procedure, if it is not controlled properly, will remove from the heated areas any effects the normalizing heat treatment might have had.   This practice, however, has been performed extensively on carbon and low-alloy steel piping and, when carefully controlled, deleterious results have not been caused.

It must be recognized also that perfect piping materials are not made commercially.   Defects may range from atom-size dislocations in the metal which cannot even be observed under the highest power microscopes, to major metal discontinuities visible to the eye.   Some visible defects may not reduce the service life of the piping component.   On the other hand, many visible surface defects and invisible subsurface defects can be harmful and lead to service failures.

Good engineering practice must be based on realism and must require a level of quality of materials, fabrication and welding which meets the demand of the particular service.   Many engineers who are not familiar with materials, fabrication, welding and inspection write requirements into specifications which involve extremely costly materials and procedures, but which do not improve the reliability or the life expectancy of the product.   By concerning themselves with unimportant (and frequently costly) requirements, important considerations involving pipe manufacture, forming, welding, heat treating, or inspection procedures are often confused or neglected.

Satisfactory piping materials are readily available for the ordinary service conditions encountered in equipment and operations involving temperatures ranging from subfreezing to high-temperature environments and pressures ranging from a vacuum to high pressures.   The mechanical and physical properties and metallurgical

characteristics of such materials are well known, and their limitations are generally recognized.

To insure that the piping component gives satisfactory service, it is generally necessary to select the proper material and to specify certain well-established physical or chemical properties, usually in accordance with applicable materials specification such as are issued by the American Society for Testing and Materials, the American Petroleum Institute, or others (see Chap. 8).

The acceptability of a piping material for any given service hinges on whether the operating conditions exceed the safe working temperatures and pressures of that material. Although there are exceptions, it is generally considered acceptable practice to assume that the physical properties of the material at the working temperature normally have a definite relation to the same properties at atmospheric temperatures. Thus, with few exceptions, the physical properties at atmospheric temperatures are used as a basis of design and acceptance testing.

The chemical composition and final heat treatments performed on the piping materials frequently require consideration as they can affect significantly the physical properties and endurance of the material. Frequently, considerable variation in chemical composition and heat treatment will nevertheless result in equally satisfactory service performance.

Where it is definitely known that certain elements are objectionable in the piping material, it is advisable to specify limits on the percentage content of these elements which may be tolerated.

For example, on steel castings, forgings and bolts maximum allowable sulfur and phosphorus contents may be specified differing from the limits applicable to pipe. Limits may include even trace elements not normally analyzed for in the piping material. For example, piping materials for some nuclear applications may have written into the specification limits on cobalt, which is considered to become radioactive. Where weldability of the material is a consideration, special limits on other elements such as boron may be established. Some elements, which steel foundries sometimes add to improve casting properties, may make the steel difficult to weld without cracking, or may reduce the hot working characteristics.

Specification in exact terms of the chemical requirements, physical properties and final heat treatments should be based only on sound metallurgical experience and definite knowledge that those requirements are necessary to provide the properties desired.

The properties that are used to determine the suitability of a certain material for a particular service application may be either physical or metallurgical in nature. It is difficult, if not impossible, to delineate precisely between physical and metallurgical properties inasmuch as an alteration in metallurgy is often accompanied by a change in physical properties. Generally speaking, physical properties as referred to in this chapter will denote those characteristics which may be observed visually on a macroscopic level: properties such as grain size and precipitation at boundaries, which usually can be observed on a microscopic level only, will be referred to as metallurgical properties. As will be noted subsequently, not only does a change in metallurgy result in a variation in physical properties but, also, a physical or mechanical working of the material will result frequently in a change in the metallurgical characteristics.

MECHANICAL PROPERTIES OF PIPING MATERIALS

The physical properties of piping materials include specific well-established mechanical properties, such as tensile strength, yield strength, and elongation which are generally covered in the applicable material specifications. However, many

other physical properties may be of interest, as they may have an important bearing on the selection of materials, workability of the piping material, weldability, and service behavior. These include toughness, hardness, modulus of elasticity, creep strength, short-time high-temperature strength, coefficient of expansion, hot shortness and others.

The mechanical properties[1] of piping materials are those properties that are associated with the elastic and inelastic reaction when force is applied, or that involve the relationship between stress and strain.

Stress-Strain Relations

A tensile stress is one which resists a force tending to pull a body apart; a compressive stress is one which resists a force tending to crush or squeeze a body; a shearing stress is one which resists a force tending to make one layer of a body slide across another layer; a torsional stress is a form of shearing stress which resists a moment tending to twist a body. Tensile and compressive stress may be developed in a material either by a direct-acting force or by a bending moment. These various types of stress frequently occur in combination.

The following major terms involving stress-strain relations are extensively used in piping terminology[2]:

Stress. The intensity at a point in a body of the internal forces or components of force that act on a given plane through the point. Stress is expressed in force per unit of area (pounds per square inch, kilograms per square millimeter, etc.)

NOTE 1. As used in tension, compression, or shear tests prescribed in product specifications, stress is calculated on the basis of the original dimensions of the cross section of the specimen. The word "original" refers to dimensions, area, etc., of specimens before the beginning of testing.

NOTE 2. The stress or component of stress acting perpendicular to a given plane is called the normal stress. A normal stress may be either a tensile stress or a compressive stress, depending upon the direction of the force. The stress or component of stress acting tangential to the plane is called the shear stress.

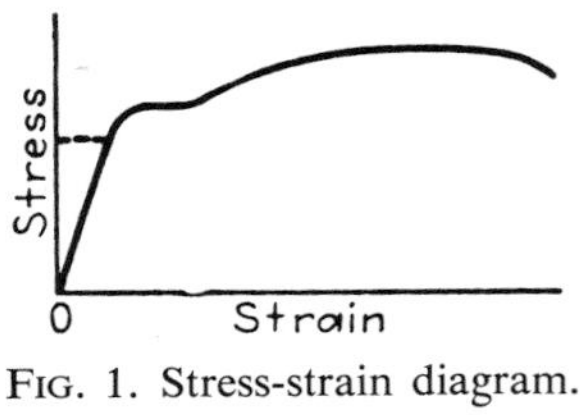

NOTE 3. "True Stress" in a tension or compression test differs from stress as defined previously in that it is calculated from the area at the time a given stress exists rather than the original area.

Strain. The change, due to force, in the size or shape of a body referred to its original size or shape. Strain is a nondimensional quantity, but it is frequently expressed in inches per inch, etc.

FIG. 1. Stress-strain diagram.

NOTE 1. Strains may be either linear (tensile strain or compressive strain), or angular (shear strain). In the ordinary tension, compression, or torsion test, it is usual to measure only one component of strain and to refer to this as "the strain." Linear strain is defined as the change per unit of length in an original linear dimension. Shear strain is defined as the tangent of the angular change between two lines originally perpendicular to each other.

NOTE 2. Linear thermal expansion per unit length, sometimes called "thermal strain," is not to be considered strain in mechanical testing.

Stress-Strain Diagram. A diagram in which corresponding values of stress and strain are plotted against each other. Values of stress are usually plotted as ordinates (vertically) and values of strain as abscissas (horizontally), see Fig. 1.

Elastic Limit. The greatest stress which a material is capable of sustaining without any permanent strain remaining upon complete release of the stress.

---

[1] These properties have often been referred to as "physical properties," but the term "mechanical properties" is much to be preferred.

[2] From ASTM Standard E6-63, "Standard Definitions of Terms Relating to Testing."

NOTE. Due to practical considerations in determining the elastic limit, measurements of strain, using a small load rather than zero load, are usually taken as the initial and final reference.

Proportional Limit. The greatest stress which a material is capable of sustaining without any deviation from proportionality of stress to strain (Hooke's law).

NOTE. Many experiments have shown that values observed for the proportional limit vary greatly with the sensitivity and accuracy of the testing equipment, eccentricity of loading, the scale to which the stress-strain diagram is plotted, and other factors. When determination of proportional limit is required, the procedure and the sensitivity of the test equipment should be specified.

Yield Strength. The stress at which a material exhibits a specified limiting deviation from the proportionality of stress to strain. The deviation is expressed in terms of strain.

NOTE. It is customary to determine yield strength by:

(a) Offset method (usually a strain of 0.2 per cent is specified),

(b) Total-extension-under-load method (usually a strain of 0.5 per cent is specified although other values of strain may be used). A formula for calculating strain is:

$$X = \left(\frac{Y}{E} + \text{specified offset/strain}\right) \times 100$$

where X = limiting strain, per cent
Y = specified yield strength, psi
E = nominal modulus of elasticity of the material, psi
(This strain, X, is multiplied by the gage length to determine the total extension under load.)

(c) "Drop of the Beam" Method—In this method the load is applied to the specimen at a steady rate of increase and the operator keeps the beam in balance by running out the poise at approximately a steady rate. When the yield strength of the material is reached, the increase of load stops, but the operator runs the poise a trifle beyond the balance position, and the beam of the machine drops for a brief but appreciable interval of time. In a machine fitted with a self-indicating load-measuring device, there is a sudden halt of the load-indicating pointer corresponding to the drop of the beam. The load at the "halt" or the "drop" is recorded, and the corresponding stress is taken as the yield strength.

For nearly all materials, if at any point on the stress-strain diagram such as at r in Fig. 2, the load is released, the diagram for decreasing load will follow a line, rm, approximately parallel to the initial portion, OA, of the diagram for increasing load. The offset Om will then give the approximate value of the permanent set after release of stress OR. The value of this set is reported in percentage of the original gage length. Thus, to determine the yield strength by the "offset method," it is necessary to secure data (autographic or numerical) from which a stress-strain diagram may be drawn. Then with the stress-strain diagram Om is laid off equal to the specified value of the set and mn is drawn parallel to OA, thus locating r, the intersection of mn with the stress-strain diagram. By drawing Rr parallel to the X axis, the value of the yield strength is given as OR.

In reporting values of yield strength obtained by this method, the specified value of "offset" used should be stated in parentheses after the term "yield strength."

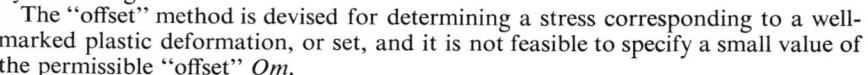

FIG. 2. Offset method of determining yield strength.

The "offset" method is devised for determining a stress corresponding to a well-marked plastic deformation, or set, and it is not feasible to specify a small value of the permissible "offset" Om.

In the specification for seamless and welded austenitic stainless-steel pipe (ASTM Designation A312), a yield strength by the offset method of 0.2 per cent is specified for materials which do not exhibit a definite yield point.

Yield Point. The first stress in a material, less than the maximum attainable stress, at which an increase in strain occurs without an increase in stress.

NOTE. It should be noted that only materials that exhibit the unique phenomenon of yielding have a "yield point."

Tensile Strength. The maximum tensile stress which a material is capable of sustaining. Tensile strength is calculated from the maximum load during a tension test carried to rupture and the original cross-sectional area of the specimen.

Compressive Strength. The maximum compressive stress which material is capable of sustaining. Compressive strength is calculated from the maximum load during a compression test and the original cross-sectional area of the specimen.

NOTE. In the case of a material which fails in compression by a shattering fracture, the compressive strength has a very definite value. In the case of materials which do not fail in compression by a shattering fracture, the value obtained for compressive strength is an arbitrary value depending upon the degree of distortion which is regarded as indicating complete failure of the material.

Shear Strength. The maximum shear stress which a material is capable of sustaining. Shear strength is calculated from the maximum load during a shear or torsion test and is based on the original dimensions of the cross-section of the specimen.

Breaking Load. The load at which fracture occurs.

NOTE. When used in connection with tension tests of thin materials or materials of small diameter for which it is often difficult to distinguish between the breaking load and the maximum load developed, the latter is considered to be the breaking load.

Proof Strength. The load required to produce a permanent set, or an elongation of 0.0005″ in overall length, under axial stress, as in a tensile testing machine (widely applied to bolting).

Ductility. The property of elongation above the plastic limit, under tensile stress. The measure of ductility is the percentage of elongation of the fractured test bar over an initial length (usually 2″ or 8″).

Elongation. The increase in gage length (Note 1) of a tension test specimen, usually expressed as percentages of the original gage length.

NOTE 1. The increase in gage length may be determined either at or after fracture, as specified for the material under test.

NOTE 2. The term elongation, when applied to metals, generally means measurement after fracture; when applied to plastics and elastomers, measurement at fracture. Such interpretation is usually applicable to values of elongation reported in the literature when no further qualification is given.

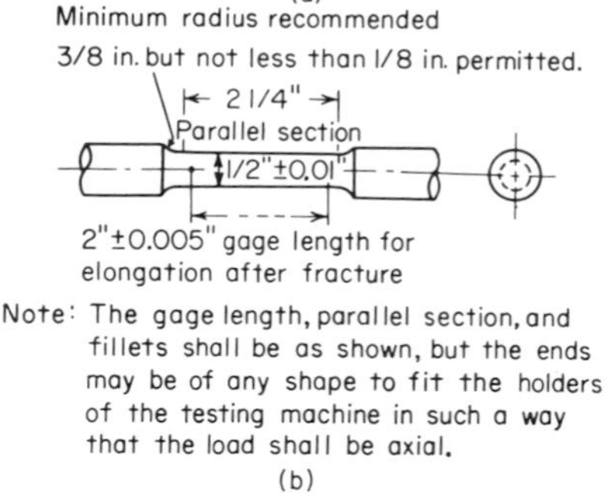

The measurement of elongation after fracture of test specimens can be made with sufficient accuracy by means of a pair of dividers and a scale. The elongation should not be reported for any tension-test specimen which breaks outside the middle third of the gage length (see Fig. 3). It has been found approximately true that the same results will be obtained from two test pieces if they are of similar form

FIG. 3. Tension-test specimens. (*a*) Strip specimen showing measurements which are taken to determine elongation. (*b*) Standard round specimen with 2-in. gage length.

when it is not possible to make them of identical dimensions. This is known as Barba's law, or the law of proportionality or similarity, and is expressed by the formula $P = l/\sqrt{a}$,

where P = proportionality

l = gaged length

a = cross-sectional area of diameter d

For the standard ASTM test specimen,

where l = 2 in.

d = 0.505 in.

P = 4.5

Hence, $l = 4.5\sqrt{a} = 4d$ (approximately), and a proportional test specimen of other dimensions should have a length approximately four times its diameter.

Reduction of Area. The difference between original cross-sectional area of a tension test specimen and the area of its smallest cross section.

Modulus of Elasticity. The ratio of stress to corresponding strain below the proportional limit.

NOTE 1. The stress-strain relations of many materials do not conform to Hooke's law throughout the elastic range, but deviate therefrom even at stresses well below the elastic limit. For such materials, the slope of either the tangent to the stress-strain curve at the origin or at a low stress, the secant drawn from the origin to any specified point on the stress-strain curve, or the chord connecting any two specified points on the stress-strain curve is usually taken to be the "modulus of elasticity." In these cases, the modulus should be designated as the "tangent modulus," the "secant modulus," or the "chord modulus," and the stress or stresses stated. Thus, for materials where the stress-strain relationship is curvilinear rather than linear one of the four following terms may be used:

a. Initial Tangent Modulus. The slope of the stress-strain curve at the origin.

b. Tangent Modulus. The slope of the stress-strain curve at any specified stress.

c. Secant Modulus. The slope of the secant drawn from the origin to any specified point on the stress-strain curve.

d. Chord Modulus. The slope of the chord drawn between any two specified points on the stress-strain curve.

NOTE 2. Modulus of elasticity, like stress, is expressed in force per unit of area (pounds per square inch, etc.). As for stress, there is a modulus of elasticity in tension, in compression, and in shear.

The modulus of elasticity is a measure of the stiffness of any material and determines the slopes of its stress-strain line; the higher the modulus of elasticity, the stiffer is the material and the steeper is the line. Elasticity is the property that enables deformed bodies to resume their original shape after removal of load. A point that is frequently overlooked is that the modulus of elasticity changes with temperature. In general, as the temperature increases, there is a tendency for the modulus to decrease. This change is of importance in considering the elastic deformation under the load and is of special significance in connection with the study of stresses and reactions resulting from thermal expansion.

Number of Tests. A tension test from which the above mechanical properties can be determined is usually made on one or more specimens from each melt or pour or heat-treatment charge of the material. A test of one specimen is sufficient to determine these properties or to plot a stress-strain diagram; one or more additional specimens are frequently tested as a check. Attention is called to the simplicity of such a test at atmospheric temperature as compared with the extensive work required to determine these values over an extended temperature range, as discussed in later paragraphs.

Effects of Change in Temperature. The test results described above are such as can be obtained from pulling a single tension-test specimen at atmospheric

temperature. This test is the kind called for in ordinary specifications where a report is required on tensile strength, yield point, reduction in area, and elongation. Usually two or more tests are required from each lot for check, but the tests are all made on supposedly similar samples under approximately identical conditions. When information is desired on the changes in mechanical properties with temperature, a series of identical tensile-test specimens is normally pulled at various temperatures under very carefully controlled conditions. Such tests are normally

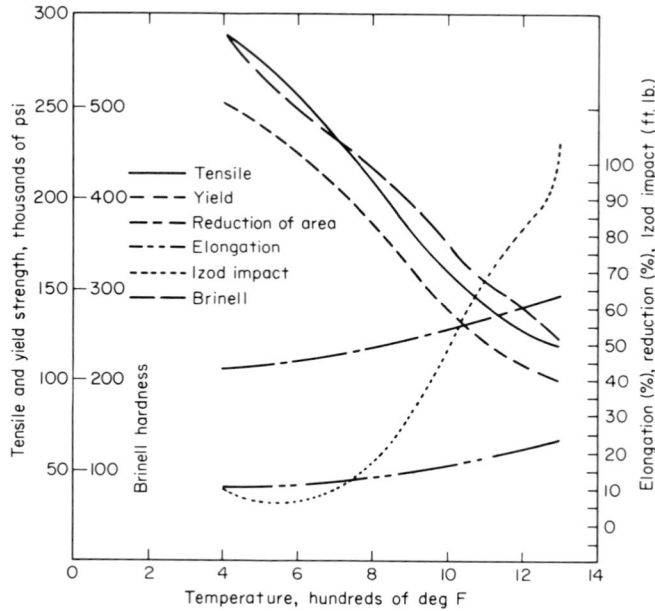

FIG. 4. A typical illustration of results obtained from short-time tensile tests run on alloy steel AISI 4140 bolting material which has been normalized at 1600 F and quenched at 1550 F in agitated oil.

considered as research rather than as a required plant routine which would be quite expensive for ordinary inspection purposes; such tests are necessary, however, on some piping materials to determine their suitability for hot forming. Tests of this kind are also useful in studying the mechanical properties of a metal over its operating temperature range. The curves of Fig. 4, which are plotted from the results of such a test on an alloy-steel bolting material, are included here as a typical illustration of this type of test. A number of identical specimens were made up from the same lot of bolting material coming from the same heat and having undergone the same heat treatment. Each specimen was then pulled at a different temperature and all the mechanical characteristics shown in the curves were recorded. By repeating these tests at 50 or 100 F steps throughout the temperature range, sufficient data were obtained to plot the curves.

Hardness and Toughness

Hardness is the property of metals which enables them to resist indentation, scratching, or abrasion. In valve seats and disks and similar parts, hardness

coupled with toughness is of great importance in resisting abrasion and erosion. There are several instruments available for measuring hardness, among which are the Brinell and the Rockwell hardness testers, and the Shore scleroscope. For the determination of microstructural components, Knoop hardness values are measured usually with the aid of a microscope.

In each test, impressions are generally made in the metal surface, resulting in a so-called "indentation." From this, numbers or indentation hardness values are obtained which are related to the area or to the depth of the indentation made by the "indenter" of fixed geometry under a known static load.

The *Brinell hardness number* (BHN) relates to the area of the permanent impression made by a ball indenter of specified size, usually 10 mm in diameter, pressed into the surface of the material under a specified load. The surface area of the impression is determined from the average measured diameter of the rim of the impression and from the ball diameter. The Brinell hardness number is calculated from the load on the ball divided by this area.

$$\text{BHN} = \frac{P}{\pi(D/2)[D - (D^2 - d^2)^{1/2}]} \tag{1}$$

where P = applied load
D = diameter of ball
d = diameter of impression

In reporting the Brinell hardness number, the test load is stated.

The *Rockwell hardness number* (RHN) is derived from the net increase in depth of impression as the load on an indenter is increased from a fixed minimum load to a higher load and then returned to the minimum load. Indenters for the Rockwell test include steel balls of several specified diameters and a diamond cone penetrator having an included angle of 120 deg with a spherical tip having a radius of 0.2 mm. Rockwell hardness numbers are always quoted with a scale symbol representing the indenter, load, and dial used.

The *diamond pyramid hardness number* (DPH) is obtained by dividing the load in kilograms applied to a square-based pyramidal diamond indenter having included face angles of 136 deg by the surface area of the impression in square millimeters calculated from the measured diagonals of the impression. The diamond pyramid hardness number is calculated from the following formula:

$$\text{DPH} = \frac{1.8544P}{d^2} \tag{2}$$

where P = applied load, kg
d = diagonal of impression, mm

In reporting diamond pyramid hardness, the test load is stated.

Another hardness determination also occasionally used on piping materials is the Knoop hardness number (KHN). It is primarily used to study the hardness of microstructural constituents in the metallurgical structure of the material.

While the nature of the tests made with these instruments varies considerably, there exists some relation between the results. The Brinell test is better known, is more widely used than the others, and is frequently given among the requirements in ASTM and other specifications. The conversions among the more common test results applicable to steel are given in Table 1.

Hardness—Tensile Strength Relation. The fact that there is a rather definite relation between the hardness and the tensile strength of metals is of importance in that the tensile strength can be approximately estimated from a very simple and inexpensive hardness test. This is of great value in making individual tests on large numbers of small objects, such as bolts, etc. While tensile strengths

Table I. Conversion Table for Various Hardness Scales

Approximate Equivalent Hardness Numbers for Diamond Pyramid Hardness (Vickers, DPH) for Steel[1]

Diamond pyramid hardness No.	Brinell hardness No. 10-mm ball, 3,000-kg load			Rockwell hardness No.				Rockwell superficial hardness No., superficial brale penetrator			Shore scleroscope hardness No.	Tensile strength (approx.) 1,000 psi
	Standard ball	Hultgren ball	Tungsten carbide ball	A-scale, 60-kg load, brale penetrator	B-scale, 100-kg load, 1/16-in.-diam ball	C-scale, 150-kg load, brale penetrator	D-scale, 100-kg load, brale penetrator	15-N scale, 15-kg load	30-N scale, 30-kg load	45-N scale, 45-kg load		
940	...	...	...	85.6	...	68.0	76.9	93.2	84.4	75.4	97	...
920	...	...	...	85.3	...	67.5	76.5	93.0	84.0	74.8	96	...
900	...	...	767	85.0	...	67.0	76.1	92.9	83.6	74.2	95	...
880	...	...	757	84.7	...	66.4	75.7	92.7	83.1	73.6	93	...
860	...	...	...	84.4	...	65.9	75.3	92.5	82.7	73.1	92	...
840	...	...	745	84.1	...	65.3	74.8	92.3	82.2	72.2	91	...
820	...	...	733	83.8	...	64.7	74.3	92.1	81.7	71.8	90	...
800	...	...	722	83.4	...	64.0	73.8	91.8	81.1	71.0	88	...
780	...	...	710	83.0	...	63.3	73.3	91.5	80.4	70.2	87	...
760	...	...	698	82.6	...	62.5	72.6	91.2	79.7	69.4	86	...
740	...	...	684	82.2	...	61.8	72.1	91.0	79.1	68.6	84	...
720	...	...	670	81.8	...	61.0	71.5	90.7	78.4	67.7	83	...
700	...	615	656	81.3	...	60.1	70.8	90.3	77.6	66.7	81	...
690	...	610	647	81.1	...	59.7	70.5	90.1	77.2	66.2	...	...
680	...	603	638	80.8	...	59.2	70.1	89.8	76.8	65.7	80	329
670	...	597	630	80.6	...	58.8	69.8	89.7	76.4	65.3	...	324
660	...	590	620	80.3	...	58.3	69.4	89.5	75.9	64.7	79	319
650	...	585	611	80.0	...	57.8	69.0	89.2	75.5	64.1	...	314
640	...	578	601	79.8	...	57.3	68.7	89.0	75.1	63.5	77	309
630	...	571	591	79.5	...	56.8	68.3	88.8	74.6	63.0	...	304
620	...	564	582	79.2	...	56.3	67.9	88.5	74.2	62.4	75	299
610	...	557	573	78.9	...	55.7	67.5	88.2	73.6	61.7	...	294
600	...	550	564	78.6	...	55.2	67.0	88.0	73.2	61.2	74	289
590	...	542	554	78.4	...	54.7	66.7	87.8	72.7	60.5	...	284
580	...	535	545	78.0	...	54.1	66.2	87.5	72.1	59.9	72	279
570	...	527	535	77.8	...	53.6	65.8	87.2	71.7	59.3	...	274
560	...	519	525	77.4	...	53.0	65.4	86.9	71.2	58.6	71	269
550	505	512	517	77.0	...	52.3	64.8	86.6	70.5	57.8	...	264
540	496	503	507	76.7	...	51.7	64.4	86.3	70.0	57.0	69	260
530	488	495	477	76.4	...	51.1	63.9	86.0	69.5	56.2	...	254
520	480	487	488	76.1	...	50.5	63.5	85.7	69.0	55.6	67	250
510	473	479	479	75.7	...	49.8	62.9	85.4	68.3	54.7	...	244
500	465	471	471	75.3	...	49.1	62.2	85.0	67.7	53.9	66	240
490	456	460	460	74.9	...	48.4	61.6	84.7	67.1	53.1	...	234
480	448	452	452	74.5	...	47.7	61.3	84.3	66.4	52.2	64	230

C1	C2	C3	C4	C5	C6	C7	C8	C9	C10	C11	C12	C13
470	441	**442**	442	74.1		46.9	60.7	83.9	65.7	51.3		224
460	433	**433**	433	73.6		46.1	60.1	83.6	64.9	50.4	62	219
450	425	**425**	425	73.3		45.3	59.4	83.2	64.3	49.4		214
440	415	**415**	415	72.8		44.5	58.8	82.8	63.5	48.4	59	210
430	405	**405**	405	72.3		43.6	58.2	82.3	62.7	47.4		204
420	397	**397**	397	71.8		42.7	57.5	81.8	61.9	46.4	57	200
410	388	**388**	388	71.4		41.8	56.8	81.4	61.1	45.3		195
400	379	**379**	379	70.8		40.8	56.0	81.0	60.2	44.1	55	190
390	369	**369**	369	70.3		39.8	55.2	80.3	59.3	42.9		185
380	360	**360**	360	69.8	(110.0)	38.8	54.4	79.8	58.4	41.7	52	180
370	350	**350**	350	69.2		37.7	53.6	79.2	57.4	40.4		175
360	341	**341**	341	68.7	(109.0)	36.6	52.8	78.6	56.4	39.1	50	170
350	331	**331**	331	68.1		35.5	51.9	78.0	55.4	37.8		166
340	322	**322**	322	67.6	(108.0)	34.4	51.1	77.4	54.4	36.5	47	161
330	313	**313**	313	67.0		33.3	50.2	76.8	53.6	35.2		156
320	303	**303**	303	66.4	(107.0)	32.2	49.4	76.2	52.3	33.9	45	151
310	294	**294**	294	65.8		31.0	48.4	75.6	51.3	32.5		146
300	284	**284**	284	65.2	(105.5)	29.8	47.5	74.9	50.2	31.1	42	141
295	280	**280**	280	64.8		29.2	47.1	74.6	49.7	30.4		139
290	275	**275**	275	64.5	(104.5)	28.5	46.5	74.2	49.0	29.5	41	136
285	270	**270**	270	64.2		27.8	46.0	73.8	48.4	28.7		134
280	265	**265**	265	63.8	(103.5)	27.1	45.3	73.4	47.8	27.9	40	131
275	261	**261**	261	63.5		26.4	44.9	73.0	47.2	27.1		129
270	256	**256**	256	63.1	(102.0)	25.6	44.3	72.6	46.4	26.2	38	126
265	252	**252**	252	62.7		24.8	43.7	72.1	45.7	25.2		124
260	247	**247**	247	62.4	(101.0)	24.0	43.1	71.6	45.0	24.3	37	121
255	243	**243**	243	62.0		23.1	42.2	71.1	44.2	23.2		119
250	238	**238**	238	61.6	99.5	22.2	41.7	70.6	43.4	22.2	36	116
245	233	**233**	233	61.2		21.3	41.1	70.1	42.5	21.1		114
240	228	**228**	228	60.7	98.1	20.3	40.3	69.6	41.7	19.9	34	111
230	219	**219**	219		96.7	(18.0)					33	106
220	209	**209**	209		95.0	(15.7)					32	101
210	200	**200**	200		93.4	(13.4)					30	97
200	190	**190**	190		91.5	(11.0)					29	92
190	181	**181**	181		89.5	(8.5)					28	88
180	171	**171**	171		87.1	(66.0)					26	84
170	162	**162**	162		85.0	(3.0)					25	79
160	152	**152**	152		81.7	(0.0)					24	75
150	143	**143**	143		78.7						22	71
140	133	**133**	133		75.0						21	66
130	124	**124**	124		71.2						20	62
120	114	**114**	114		66.7							57
110	105	**105**	105		62.3							
100	95	**95**	95		56.2							
95	90	**90**	90		52.0							
90	86	**86**	86		48.0							
85	81	**81**	81		41.0							

1 The values in **bold-face type** correspond to the values in the joint SAE-ASM-ASTM hardness conversions as printed in ASTM, E140 Table 1. Values in parentheses are beyond normal range and are given for information only.

Table I. (Continued)

Approximate Equivalent Hardness Numbers for Brinell Hardness Numbers for Steel[1]

Brinell indentation diam, mm	Brinell hardness No.[2] 10-mm ball, 3,000-kg load			Diamond pyramid hardness No.	Rockwell hardness No.[3]				Rockwell Superficial hardness No., superficial ball penetrator			Shore scleroscope hardness No.	Tensile strength (approx) 1,000 psi
	Standard ball	Hultgren ball	Tungsten carbide ball		A-scale, 60-kg load, brale penetrator	B-scale, 100-kg load, 1/16-in.-diam ball	C-scale, 150-kg load, brale penetrator	D-scale, 100-kg load, brale penetrator	15-N scale, 15-kg load	30-N scale, 30-kg load	45-N scale, 45-kg load		
...	...	...	...	940	85.6	...	68.0	76.9	93.2	84.4	75.4	97	...
...	...	...	...	920	85.3	...	67.5	76.5	93.0	84.0	74.8	96	...
...	...	...	...	900	85.0	...	67.0	76.1	92.9	83.6	74.2	95	...
...	...	...	767	880	84.7	...	66.4	75.7	92.7	83.1	73.6	93	...
...	...	...	757	860	84.4	...	65.9	75.3	92.5	82.7	73.1	92	...
2.25	...	...	745	840	84.1	...	65.3	74.8	92.3	82.2	72.2	91	...
...	...	...	733	820	83.8	...	64.7	74.3	92.1	81.7	71.8	90	...
...	...	...	722	800	83.4	...	64.0	73.8	91.8	81.1	71.0	88	...
2.30	...	...	712	780	83.0	...	63.3	73.3	91.5	80.4	70.2	87	...
...	...	...	710	760	82.6	...	62.5	72.6	91.2	79.7	69.4	86	...
2.35	...	...	698	740	82.2	...	61.8	72.1	91.0	79.1	68.6	84	...
...	...	...	684	737	82.2	...	61.7	72.0	91.0	79.0	68.5	83	...
...	...	...	682	720	81.8	...	61.0	71.5	90.7	78.4	67.7	...	...
...	...	...	670	700	81.3	...	60.1	70.8	90.3	77.6	66.7	81	...
2.40	...	...	653	697	**81.2**	...	**60.0**	**70.7**	**90.2**	**77.5**	**66.5**	81	...
...	...	...	647	690	81.1	...	59.7	70.5	90.1	77.2	66.2	80	329
...	...	...	638	680	80.8	...	59.2	70.1	89.8	76.8	65.7	...	324
...	...	...	630	670	80.6	...	58.8	69.8	89.7	76.4	65.3	79	323
2.45	...	...	627	667	**80.5**	...	**58.7**	**69.7**	**89.6**	**76.3**	**65.1**	...	328
2.50	...	601	601	657	80.7	...	59.1	70.0	89.8	76.8	65.7	77	309
...	...	...	...	640	79.8	...	57.3	68.7	89.0	75.1	63.5	...	309
2.55	...	578	578	627	79.8	...	57.3	68.7	89.0	75.1	63.5	75	297
...	...	...	...	615	**79.1**	...	**56.0**	**67.7**	**88.4**	**73.9**	**62.1**	...	293
2.60	...	555	555	607	78.8	...	55.6	67.4	88.1	73.5	61.6	73	285
...	...	...	...	591	78.4	...	54.7	66.7	87.8	72.7	60.6	...	279
2.65	...	534	534	579	78.0	...	54.0	66.1	87.5	72.0	59.8	71	274
...	...	...	...	569	**77.8**	...	**53.5**	**65.8**	**87.2**	**71.6**	**59.2**	...	266
2.70	...	514	514	553	77.1	...	52.5	65.0	86.7	70.7	58.0	70	263
...	...	...	...	547	76.9	...	52.1	64.7	86.5	70.3	57.6	...	259
2.75	495	495	495	539	76.7	...	51.6	64.3	86.3	69.9	56.9	...	254
...	...	...	...	530	**76.3**	...	**51.0**	**63.8**	**85.9**	**69.4**	**56.1**	68	253
2.80	477	477	...	516	75.9	...	50.3	63.2	85.6	68.7	55.2	66	247
...	...	...	...	508	75.6	...	49.6	62.7	85.3	68.2	54.5	...	243
2.85	461	461	...	495	75.1	...	48.8	61.9	84.9	67.4	53.5	65	237
...	...	...	...	491	**74.9**	...	**48.5**	**61.7**	**84.7**	**67.2**	**53.2**	...	235

Hardness conversion table (Brinell indentation diameter basis).

Dia.[2] (mm)	Brinell HB	Vickers	Rockwell A	Rockwell B	Rockwell C	Rockwell D	15-N	30-N	45-N	Shore	Tensile strength (1000 psi)
2.90	(444)	474	74.3	⋮	47.2	61.0	84.1	66.0	51.7	⋮	226
2.90	444	472	74.2	⋮	47.1	60.8	84.0	65.8	51.5	63	225
2.95	429	455	73.8	⋮	45.7	59.7	83.4	64.6	49.9	61	217
3.00	415	440	72.8	⋮	44.5	58.8	82.8	63.5	48.4	59	210
3.05	401	425	72.0	⋮	43.1	57.8	82.0	62.3	46.9	58	202
3.10	388	410	71.4	⋮	41.8	56.8	81.4	61.1	45.3	56	195
3.15	375	396	70.6	⋮	40.4	55.7	80.6	59.9	43.6	54	188
3.20	363	383	70.0	⋮	39.1	54.6	80.0	58.7	42.0	52	182
3.25	352	372	69.3	(110.0)	37.9	53.8	79.3	57.6	40.5	51	176
3.30	341	360	68.7	(109.0)	36.6	52.8	78.6	56.4	39.1	50	170
3.35	331	350	68.1	(108.5)	35.5	51.9	78.0	55.4	37.8	48	166
3.40	321	339	67.5	(108.0)	34.3	51.0	77.3	54.3	36.4	47	160
3.45	311	328	66.9	(107.5)	33.1	50.0	76.7	53.3	34.4	46	155
3.50	302	319	66.3	(107.0)	32.1	49.3	76.1	52.2	33.8	45	150
3.55	293	309	65.7	(106.0)	30.9	48.3	75.5	51.2	32.4	43	145
3.60	285	301	65.3	(105.5)	29.9	47.6	75.0	50.3	31.2	⋮	141
3.65	277	292	64.6	(104.5)	28.8	46.7	74.4	49.3	29.9	41	137
3.70	269	284	64.1	(104.0)	27.6	45.9	73.7	48.3	28.5	40	133
3.75	262	276	63.6	(103.0)	26.6	45.0	73.1	47.3	27.3	39	129
3.80	255	269	63.0	(102.0)	25.4	44.2	72.5	46.2	26.0	38	126
3.85	248	261	62.5	(101.0)	24.2	43.2	71.7	45.1	24.5	37	122
3.90	241	253	61.8	100.0	22.8	42.0	70.9	43.9	22.8	36	118
3.95	235	247	61.4	99.0	21.7	41.4	70.3	42.9	21.5	35	115
4.00	229	241	60.8	98.2	20.5	40.5	69.7	41.9	20.1	34	111
4.05	223	234	⋮	97.3	(18.8)	⋮	⋮	⋮	⋮	⋮	⋮
4.10	217	228	⋮	96.4	(17.5)	⋮	⋮	⋮	⋮	33	105
4.20	207	218	⋮	94.6	(15.2)	⋮	⋮	⋮	⋮	32	100
4.30	197	207	⋮	92.8	(12.7)	⋮	⋮	⋮	⋮	30	95
4.40	187	196	⋮	90.7	(10.0)	⋮	⋮	⋮	⋮	⋮	90
4.50	179	188	⋮	89.0	(8.0)	⋮	⋮	⋮	⋮	27	87
4.60	170	178	⋮	86.8	(5.4)	⋮	⋮	⋮	⋮	26	83
4.70	163	171	⋮	85.0	(3.3)	⋮	⋮	⋮	⋮	25	79
4.80	156	163	⋮	82.9	(0.9)	⋮	⋮	⋮	⋮	⋮	76
4.90	149	156	⋮	80.8	⋮	⋮	⋮	⋮	⋮	23	73
5.00	143	150	⋮	78.7	⋮	⋮	⋮	⋮	⋮	22	71
5.10	137	143	⋮	76.4	⋮	⋮	⋮	⋮	⋮	21	67
5.20	131	137	⋮	74.0	⋮	⋮	⋮	⋮	⋮	⋮	65
5.30	126	132	⋮	72.0	⋮	⋮	⋮	⋮	⋮	20	63
5.40	121	127	⋮	69.8	⋮	⋮	⋮	⋮	⋮	19	60
5.50	116	122	⋮	67.6	⋮	⋮	⋮	⋮	⋮	18	58
5.60	111	117	⋮	65.7	⋮	⋮	⋮	⋮	⋮	15	56

[1] The values in bold-face type correspond to the values in the joint SAE-ASM-ASTM hardness conversions as printed in ASTM E140, Table 3.

[2] Brinell numbers are based on the diameter of impressed indentation. If the ball distorts (flattens) during test, Brinell numbers will vary in accordance with the degree of such distortion when related to hardnesses determined with a Vickers diamond pyramid, Rockwell brale, or other penetrator which does not sensibly distort. At high hardnesses, therefore, the relationship between Brinell and Vickers or Rockwell scales is affected by the type of ball used. Steel balls (standard or Hultgren) tend to flatten slightly more than carbide balls, resulting in larger indentation and lower Brinell number than shown by a carbide ball. Thus, on a specimen of 640 Vickers, a Hultgren ball will leave a 2.55-mm impression (578 Bhn), and the carbide ball a 2.50-mm impression (601 Bhn). Conversely, identical impression diameters for both types of ball will correspond to different Vickers or Rockwell values. Thus, if both impressions are 2.55 mm (578 Bhn), material tested with a Hultgren ball has a Vickers hardness of 640, while material tested with a carbide ball has a Vickers hardness of 615.

[3] Values in parentheses are beyond normal range

Table I. (Continued)

Approximate Equivalent Hardness Numbers for Rockwell C Hardness Numbers for Steel[1]

| Rockwell C-scale hardness No. | Diamond pyramid hardness No. | Brinell hardness No., 10-mm ball, 3,000-kg load | | | Rockwell hardness No. | | | Rockwell superficial hardness No., superficial brale penetrator | | | Shore scleroscope hardness No. | Tensile strength (approx) 1,000 psi |
		Standard ball	Hultgren ball	Tungsten carbide ball	A-scale, 60-kg load, brale penetrator	B-scale, 100-kg load, 1/16-in-diam ball	D-scale, 100-kg load, brale penetrator	15-N scale, 15-kg load	30-N scale, 30-kg load	45-N scale, 45-kg load		
68	940	...	...	...	85.6	...	76.9	93.2	84.4	75.4	97	...
67	900	...	...	...	85.0	...	76.1	92.9	83.6	74.2	95	...
66	865	...	...	...	84.5	...	75.4	92.5	82.8	73.3	92	...
65	832	...	...	739	83.9	...	74.5	92.2	81.9	72.0	91	...
64	800	...	...	722	83.4	...	73.8	91.8	81.1	71.0	88	...
63	772	...	...	705	82.8	...	73.0	91.4	80.1	69.9	87	...
62	746	...	...	688	82.3	...	72.2	91.1	79.3	68.8	85	...
61	720	...	...	670	81.8	...	71.5	90.7	78.4	67.7	83	...
60	697	...	613	654	81.2	...	70.7	90.2	77.5	66.6	81	326
59	674	...	599	634	80.7	...	69.9	89.8	76.6	65.5	80	315
58	653	...	587	615	80.1	...	69.2	89.3	75.7	64.3	78	305
57	633	...	575	595	79.6	...	68.5	88.9	74.8	63.2	76	295
56	613	...	561	577	79.0	...	67.7	88.3	73.9	62.0	75	287
55	595	...	546	560	78.5	...	66.9	87.9	73.0	60.9	74	278
54	577	...	534	543	78.0	...	66.1	87.4	72.0	59.8	72	269
53	560	...	519	525	77.4	...	65.4	86.9	71.2	58.6	71	262
52	544	500	508	512	76.8	...	64.6	86.4	70.2	57.4	69	253
51	528	487	494	496	76.3	...	63.8	85.9	69.4	56.1	68	...
50	513	475	481	481	75.9	...	63.1	85.5	68.5	55.0	67	245
49	498	464	469	469	75.2	...	62.1	85.0	67.6	53.8	66	239
48	484	451	455	455	74.7	...	61.4	84.5	66.7	52.5	64	232
47	471	442	443	443	74.1	...	60.8	83.9	65.8	51.4	63	225
46	458	432	432	432	73.6	...	60.0	83.5	64.8	50.3	62	219
45	446	421	421	421	73.1	...	59.2	83.0	64.0	49.0	60	212
44	434	409	409	409	72.5	...	58.5	82.5	63.1	47.8	58	206
43	423	400	400	400	72.0	...	57.7	82.0	62.2	46.7	57	201
42	412	390	390	390	71.5	...	56.9	81.5	61.3	45.5	56	196
41	402	381	381	381	70.9	...	56.2	80.9	60.4	44.3	55	191

40	392	**371**	**371**	371	70.4	…	55.4	80.4	59.5	43.1	54	186
39	382	**362**	**362**	362	69.9	…	54.6	79.9	58.6	41.9	52	181
38	372	**353**	**353**	353	69.4	…	53.8	79.4	57.7	40.8	51	176
37	363	**344**	**344**	344	68.9	…	53.1	78.8	56.8	39.6	50	172
36	354	**336**	**336**	336	68.4	(109.0)	52.3	78.3	55.9	38.4	49	168
35	345	**327**	**327**	327	67.9	(108.5)	51.5	77.7	55.0	37.2	48	163
34	336	**319**	**319**	319	67.4	(108.0)	50.8	77.2	54.2	36.1	47	159
33	327	**311**	**311**	311	66.8	(107.5)	50.0	76.6	53.3	34.9	46	154
32	318	**301**	**301**	301	66.3	(107.0)	49.2	76.1	52.1	33.7	44	150
31	310	**294**	**294**	294	65.8	(106.0)	48.4	75.6	51.3	32.5	43	146
30	302	**286**	**286**	286	65.3	(105.5)	47.7	75.0	50.4	31.3	42	142
29	294	**279**	**279**	279	64.7	(104.5)	47.0	74.5	49.5	30.1	41	138
28	286	**271**	**271**	271	64.3	(104.0)	46.1	73.9	48.6	28.9	41	134
27	279	**264**	**264**	264	63.8	(103.0)	45.2	73.3	47.7	27.8	40	131
26	272	**258**	**258**	258	63.3	(102.5)	44.6	72.8	46.8	26.7	38	127
25	266	**253**	**253**	253	62.8	(101.5)	43.8	72.2	45.9	25.5	38	124
24	260	**247**	**247**	247	62.4	(101.0)	43.1	71.6	45.0	24.3	37	121
23	254	**243**	**243**	243	62.0	100.0	42.1	71.0	44.0	23.1	36	118
22	248	**237**	**237**	237	61.5	99.0	41.6	70.5	43.2	22.0	35	115
21	243	**231**	**231**	231	61.0	98.5	40.9	69.9	42.3	20.7	35	113
20	238	**226**	**226**	226	60.5	97.8	40.1	69.4	41.5	19.6	34	110
(18)	230	**219**	**219**	219	…	96.7	…	…	…	…	33	106
(16)	222	**212**	**212**	212	…	95.5	…	…	…	…	32	102
(14)	213	**203**	**203**	203	…	93.9	…	…	…	…	31	98
(12)	204	**194**	**194**	194	…	92.3	…	…	…	…	29	94
(10)	196	**187**	**187**	187	…	90.7	…	…	…	…	28	90
(8)	188	**179**	**179**	179	…	89.5	…	…	…	…	27	87
(6)	180	**171**	**171**	171	…	87.1	…	…	…	…	26	84
(4)	173	**165**	**165**	165	…	85.5	…	…	…	…	25	80
(2)	166	**158**	**158**	158	…	83.5	…	…	…	…	24	77
(0)	160	**152**	**152**	152	…	81.7	…	…	…	…	24	75

[1] The values in bold-face type correspond to the values in the joint SAE-ASM-ASTM hardness conversions as printed in ASTM E140, Table 2. Values in parentheses are beyond normal range and are given for information only.

Table I. (Continued)

Brinell Hardness Numbers, (10-mm Ball Diameter)

Indentation diam, mm	Load, kg					
	500	1,000	1,500	2,000	2,500	3,000
2.00	158	316	473	632	788	945
2.05	150	300	450	600	750	899
2.10	143	286	428	572	714	856
2.15	136	272	409	544	681	817
2.20	130	260	390	520	650	780
2.25	124	248	373	496	621	745
2.30	119	238	356	476	593	712
2.35	114	228	341	456	568	682
2.40	109	218	327	436	545	653
2.45	104	208	314	416	522	627
2.50	100	200	301	400	500	601
2.55	96.3	193	289	385	482	578
2.60	92.6	185	278	370	462	555
2.65	89.0	178	267	356	445	534
2.70	85.7	171	257	343	429	514
2.75	82.6	165	248	330	413	495
2.80	79.6	159	239	318	398	477
2.85	76.8	154	231	307	384	461
2.90	74.1	148	222	296	371	444
2.95	71.5	143	215	286	358	429
3.00	69.1	138	208	276	346	415
3.05	66.8	134	201	267	334	401
3.10	64.6	129	194	258	324	388
3.15	62.5	125	188	250	313	375
3.20	60.5	121	182	242	303	363
3.25	58.6	117	176	234	293	352
3.30	56.8	114	171	227	284	341
3.35	55.1	110	166	220	276	331
3.40	53.4	107	161	214	267	321
3.45	51.8	104	156	207	259	311
3.50	50.3	101	151	201	252	302
3.55	48.9	97.8	147	196	244	293
3.60	47.5	95.0	143	190	238	285
3.65	46.1	92.2	139	184	231	277
3.70	44.9	89.8	135	180	225	269
3.75	43.6	87.2	131	174	218	262
3.80	42.4	84.8	128	170	212	255
3.85	41.3	82.6	124	165	207	248
3.90	40.2	80.4	121	161	201	241
3.95	39.1	78.2	118	156	196	235
4.00	38.1	76.2	115	152	191	229
4.05	37.1	74.2	112	148	186	223
4.10	36.2	72.4	109	145	181	217
4.15	35.3	70.6	106	141	177	212
4.20	34.4	68.8	104	138	172	207
4.25	33.6	67.2	101	134	167	201
4.30	32.8	65.6	98.5	131	164	197
4.35	32.0	64.0	96.0	128	160	192
4.40	31.2	62.4	93.5	125	156	187
4.45	30.5	61.0	91.5	122	153	183
4.50	29.8	59.6	89.5	119	149	179
4.55	29.1	58.2	87.0	116	145	174
4.60	28.4	56.8	85.0	114	142	170
4.65	27.8	55.6	83.5	111	139	167
4.70	27.1	54.2	81.5	108	136	163
4.75	26.5	53.0	79.5	106	133	159
4.80	25.9	51.8	78.0	104	130	156
4.85	25.4	50.8	76.0	102	127	152
4.90	24.8	49.6	74.5	99.2	124	149
4.95	24.3	48.6	73.0	97.2	122	146
5.00	23.8	47.6	71.5	95.2	119	143
5.05	23.3	46.6	70.0	93.2	117	140
5.10	22.8	45.6	68.5	91.2	114	137
5.15	22.3	44.6	67.0	89.2	112	134
5.20	21.8	43.6	65.5	87.2	109	131
5.25	21.4	42.8	64.0	85.6	107	128
5.30	20.9	41.8	63.0	83.6	105	126
5.35	20.5	41.0	61.5	82.0	103	123
5.40	20.1	40.2	60.5	80.4	101	121
5.45	19.7	39.4	59.0	78.8	98.5	118
5.50	19.3	38.6	58.0	77.2	96.5	116
5.55	18.9	37.8	57.0	75.6	95.0	114
5.60	18.6	37.2	55.5	74.4	92.5	111
5.65	18.2	36.4	54.5	72.8	90.8	109
5.70	17.8	35.6	53.5	71.2	89.2	107
5.75	17.5	35.0	52.5	70.0	87.5	105
5.80	17.2	34.4	51.5	68.8	85.8	103
5.85	16.8	33.6	50.5	67.2	84.2	101
5.90	16.5	33.0	49.6	66.0	82.5	99.2
5.95	16.2	32.4	48.7	64.8	81.2	97.3
6.00	15.9	31.8	47.8	63.6	79.5	95.5
6.05	15.6	31.2	46.9	62.4	78.0	93.7
6.10	15.3	30.6	46.0	61.2	76.7	92.0
6.15	15.1	30.2	45.2	60.4	75.3	90.3
6.20	14.8	29.6	44.4	59.2	73.8	88.7
6.25	14.5	29.0	43.6	58.0	72.6	87.1
6.30	14.2	28.4	42.8	56.8	71.3	85.5
6.35	14.0	28.0	42.0	56.0	70.0	84.0
6.40	13.7	27.4	41.3	54.8	68.8	82.5
6.45	13.5	27.0	40.5	54.0	67.5	81.0

estimated in this way are not as reliable as those determined by pulling a specimen in a tension-testing machine, they are extensively used as check tests, especially on large lots of small objects, such as alloy-steel bolts.   If the results of such Brinell tests seem to show that the metal does not have the desired tensile strength, it is the usual practice to pull one or more tension-test specimens to determine the actual tensile strength.   When Brinell hardness and tensile strength are both used as requirements in a specification, both tests, of course, must be conducted.

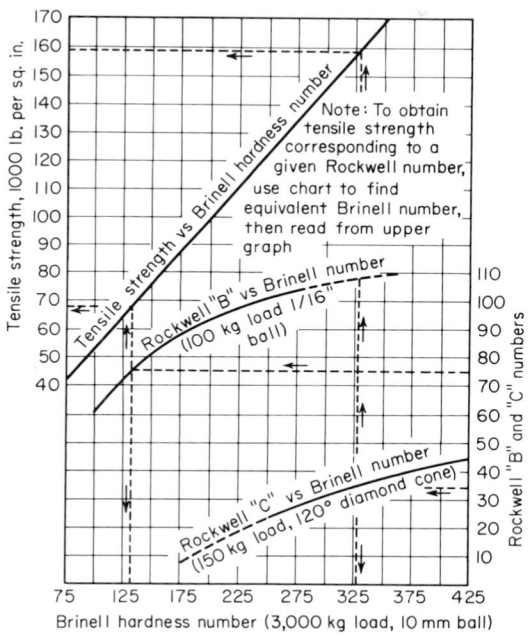

Note: To obtain tensile strength corresponding to a given Rockwell number, use chart to find equivalent Brinell number, then read from upper graph

Fig. 5. Conversion chart for Brinell and Rockwell hardness numbers, giving corresponding tensile strength for steel.   Based on hardness conversion table.   (*SAE Handbook*, 1964.)

Brinell tests of this kind are applicable to a considerable variety of metals, but they are most frequently used in estimating the tensile properties of steel (see Fig. 5).

Toughness.   Toughness is the ability of a material to resist impact, or to withstand repeated reversals of stress, or to absorb energy when stressed beyond the elastic limit.   In piping work, the resistance to shock is a desirable attribute of valve bodies and fittings which are subject to water hammer and similar forms of surge.   An approximate indication of the toughness of metals can be obtained by means of a pendulum impact machine.   A small notched bar of the material is prepared and fixed in position in the path of a relatively heavy pendulum.   In the ·Charpy type of impact test, the specimen is supported as a simple beam, whereas in the Izod type, the specimen is clamped at one end to form a cantilever beam. The general requirements as to rigidity of machine, dimensions of specimens, methods of notching, and necessary precautions in testing are covered in Methods of Impact Testing of Metallic Materials, ASTM Specification E23.

PHYSICAL PROPERTIES

Physical properties are normally understood to represent those properties, other than mechanical properties, that pertain to the physics of a material. Such properties include density, thermal expansion (or contraction), electrical conductivity, and thermal conductivity.

The more important physical properties applicable to common piping material are summarized in Tables 2 and 3.

Table 2 shows density, magnetic permeability, specific heat, melting range and modulus of elasticity in tension and in torsion. On some alloys, particularly the stainless-steel types, the magnetic permeability can vary substantially with the amount of cold work performed. On many stainless-steel grades, the magnetic permeability increases with cold work.

The coefficient of thermal expansion data are of interest in fabricating, welding and service. Distortion is usually greater with alloys of higher coefficients of expansion. Dissimilar metal welds may fail in service involving thermal fatigue when the different materials differ significantly in coefficients of expansion. Where dissimilar piping materials have to be joined and involve thermal-fatigue service environments, some welding filler metal materials may have advantages if the coefficients of expansion lie between those of the base materials to be joined.

METALLURGICAL PROPERTIES

An understanding of the metallurgical properties of piping materials is just as important as knowing their mechanical and physical properties.

Metallurgical considerations cover the whole range from the extraction of the metals from their ores to their ultimate utilization. This includes the melting of the metals, the casting, forging or drawing of the piping materials, their working, forming and joining characteristics, their response to heat treatment and the effects of the service environment.

Even after the piping component has been installed and is in service, metallurgical considerations may enter involving chemical cleaning, high-temperature creep, notch brittleness at low temperature, etc. When failures have occurred, metallurgical techniques are among the major tools employed to evaluate the causes of failure and to determine solutions and remedies.

Thus, a knowledge of metallurgy is necessary to the proper economic utilization of metals in the establishment of proper fabrication, welding, heat-treating and cleaning procedures. It is also helpful to the inspectors of piping to understand the metallurgical characteristics of the materials which are being inspected and to recognize their susceptibility toward embrittlement, cracking, crack propagation and corrosion.

Structure of metals

Basic to the metallurgy of materials is an understanding of the structure of metals. In the solid state, this involves the orderly arrangements of atoms of which all metals are composed. The atoms are arranged in so-called space lattices. Of the fourteen possible types of space lattices, only three are of primary importance. These represent the metals and alloys normally involved in commercial piping materials. These are BCC—*body-centered cubic*, FCC—*face-centered cubic*, and HCP—*hexagonal close-packed* (Fig. 6).

A few metals such as iron, titanium, cobalt, and tin may occur in two or more lattice structures. These metals will undergo changes from one type of structure

to another when heated (or cooled) above (or below) specific temperatures.   This change in lattice structures is a principal reason for the importance of the iron-base alloys.   A great variety of properties can be obtained by causing changes in the lattice structure.   Heat treatment is the primary technique used in accomplishing such changes.

The addition of a "foreign" atom to a pure metal may have several effects.   The atom may locate between existing atoms in the lattice and thus occupy an interstitial position.   Small atoms such as hydrogen and carbon often enter in this

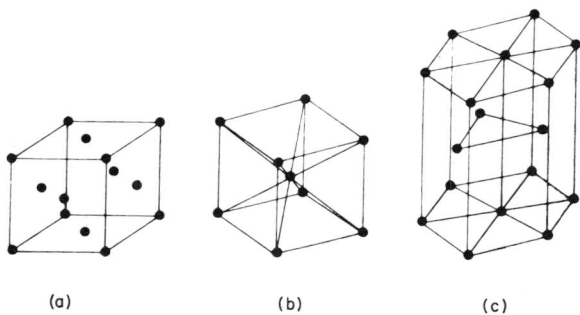

(a)   (b)   (c)

Fig. 6. The three most common crystal structures in metals and alloys in metals and alloys.   (*a*) Face-centered cubic (FCC).   (*b*) Body-centered cubic (BCC).   (*c*) Hexagonal close-packed (HCP).

manner.   A foreign atom may also replace one of the atoms in the lattice structure of the "pure" metal.   In either case, the foreign atom is considered "dissolved" or "in solution" and forms an interstitive or a substitutional solution, respectively. The third form involves direct combinations of the added atom with the atoms in the space lattice to form a distinctly different crystal structure.

The foreign atoms may be present unintentionally as residual or trace elements. They may have been added purposely for specific reasons as alloying elements, for deoxidation, for better fluidity during casting, for improved machinability or weldability, or to produce still other properties or effects desired.   Sometimes, minute fractions of a percentage of one element, when added to a metal, may produce major changes in the mechanical, physical or metallurgical properties of the metal.   In other alloys, major additions are necessary.

Commercial metals generally consist of many individual grains.   These form originally upon solidification from the molten state.   In the "as-cast" or "as-welded" condition, these grains frequently appear internally as dendrites (Fig. 7).   Depending on factors such as the purity of the metal, the melting practice, the rate of solidification, subsequent hot or cold working, or heat treatment, these grains may be

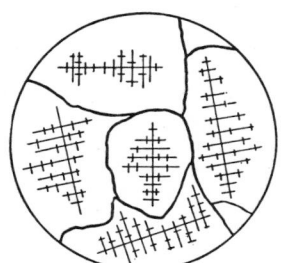

Fig. 7. Sketch illustrating individual grain growth from nuclei and dendrites.

small and can be seen only at magnification of 50, 100, or 500 diameters, or they may be very large.   Examples of both are illustrated in Fig. 8.

The boundaries between grains, in essence, represent imperfections in the orderly lattice arrangement.

Table 2. Some Physical Properties of Piping Materials

Material	Density, lb/in.^{3}	Specific heat, mean (temperature, F)	Melting temperature, F	Modulus of elasticity (room temperature) Tension, psi $\times 10^{-6}$	Torsion, psi $\times 10^{-6}$	Magnetic permeability
Pure iron	0.2845	0.112 (122–212) 0.170 (1562–1652)	2781–2799	29.8	11.77	250 to 7,000
Gray cast iron	0.251–0.265	. . . .	2150–2360	9.6–14 (Grade 20) 20.4–23.5 (Grade 60)	3.9–5.6 (Grade 20) 7.8–8.5 (Grade 60)	600
Malleable cast iron ferritic	0.260–0.265	0.11 (at 70) 0.165 (at 800)	2750	25	12.5	Ferromagnetic
Malleable cast iron pearlitic	0.264	0.11 (at 70) 0.165 (at 800)	2750	25.5	9.4–10.2	Ferromagnetic
Nodular iron	0.257	. . . .	2050–3150	22	. . . .	Ferromagnetic
Wrought iron.	0.28	0.115 (122–212) 0.264 (1292–1382)	2750	26–29	. . . .	Ferromagnetic
0.06 C, 0.38 Mn.	0.2844	0.116 (122–212) 0.342 (1292–1382)	2600	30	11	Ferromagnetic
0.23 C, 0.35 Mn.	0.2839	0.116 (122–212)	2600	30	11	Ferromagnetic
0.43 C, 0.69 Mn.	0.2834	0.116 (122–212) 0.227 (1292–1472)	2600	. . . . .	. . . .	Ferromagnetic
1.22 C, 0.35 Mn.	0.2839	0.116 (122–212) 0.499 (1292–1382)	2600	. . . . .	. . . .	Ferromagnetic

					Annealed	10% cold work	Severe cold work
Carbon–½ Mo	0.28	....	2600–2800	30.8	....		Ferromagnetic
1¼ Cr–½ Mo	0.283	0.114 (122–212)	2600–2800	30.4	....		Ferromagnetic
2¼ Cr–1 Mo	0.283	0.11	2600–2800	31	....		Ferromagnetic
5 Cr–½ Mo	0.28	0.11	2700–2800	30	....		Ferromagnetic
9 Cr–1 Mo	0.28	0.11	2700–2800	30	....		Ferromagnetic
3½% Ni steel	0.28	0.115 (212)	....	30	....		Ferromagnetic
Type 304 wrought	0.29	0.12 (32–212)	2550–2650	28	12.5	1.02	1.10	7.0
CF-8 cast	0.28	0.12	2600	28	....	1.00 to 1.3 after heat treatment thin section ¾ in.	1.005 for 0% ferrite 2.8 for 16.2% ferrite	
Type 316 wrought	0.29	0.12 (32–212)	2500–2550	28	....	1.003	...	10
CF-8M cast	0.28	0.12	2550	....	....	As a function of composition, can be 1.0, 1.5 to 2.5		
Type 321 wrought	0.29	0.12 (32–212)	2550–2600	28	....	1.02		
Type 347 wrought	0.29	0.12 (32–212)	2550–2660	28	....	1.02		
CF-8C cast	0.28	0.12	2550–2600	28	....	1.2–1.8		
405 wrought	0.28	0.11	2700–2790	29	....			Ferromagnetic

Table 2. (Continued)

Material	Density, lb/in.³	Specific heat (mean) temperature, F	Melting temperature, F	Modulus of elasticity (room temperature)		Magnetic permeability
				Tension, psi $\times$ 10⁻⁶	Torsion, psi $\times$ 10⁻⁶	
CA15 cast..............	0.275	0.11	2750	29	....	Ferromagnetic
410 wrought..........	0.28	0.11	2700–2790	29	....	Ferromagnetic
446 wrought..........	0.273	0.144	2550–2750	29	....	Ferromagnetic
CC50 cast.............	0.272	0.12	2725	29	....	Ferromagnetic
Aluminum 1100........	0.098	0.23 (212)	1190–1215	10.0	3.8	Dimagnetic
Aluminum 6061........	0.098	0.23 (212)	1080–1200	....	....	Dimagnetic
Aluminum 4043........	0.097	0.23 (212)	1065–1170	10.3	....	Dimagnetic
Aluminum 356.........	0.097	0.23 (212)	1035–1135	10.5	....	Dimagnetic
Copper (DHP).........	0.323	0.092	1981	17	....	Dimagnetic
Red brass (wrought)....	0.316	0.09	1810–1880	17	....	Dimagnetic
Yellow brass (wrought)..	0.306	0.09	1660–1710	15	....	Dimagnetic
Admiralty brass (wrought)	0.308	0.09	1650–1720	15	....	Dimagnetic
Manganese bronze (wrought)	0.302	0.09	1590–1630	15	....	Dimagnetic
Cupronickel (70–30) (wrought)	0.323	0.09	2140–2260	22	....	Dimagnetic

Aluminum bronze (3) (wrought)	0.281	0.09	1910–1940	18	....	Dimagnetic
Beryllium copper (wrought)	0.297	0.10 (86–212)	1587–1750	17 (annealed) 19 (heat treated)	6.5 (annealed) 7.3 (heat treated)	Dimagnetic
Chemical lead	0.4097	0.0309	618	2	....	Dimagnetic
50/50 Sn Pb solder	0.321	0.046	361–421	....	....	Dimagnetic
Nickel (A) (wrought)	0.321	0.13	2615–2635	30	11	110 initially 600 max.
Monel (70 Ni–30 Cu) (wrought)	0.319	0.127	2370–2460	26	9.5	2,000–10,000 cgs
Inconel (wrought)	0.307	0.11	2540–2600	31	11	1.005 cgs max.
Incoloy (wrought)	0.290	0.12	2540–2600	29.5	9.5	1.0137 (hot rolled) 1.0092 (annealed)
Hastelloy B (wrought)	0.334	0.091	2410–2460	26.4	....	Paramagnetic over 1.000 and less than 1.001
Hastelloy C (wrought)	0.323	0.092	2320–2380	29.8	....	Paramagnetic over 1.000 and less than 1.001
Tin	0.26	0.0534	449.4	6 (cast)	....	Dimagnetic
Titanium (99.0%)	0.163	0.125	3002–3038	16	....	1.00005 at 20 oersteds
Tantalum	0.600	0.034	5425	27	....	

Table 3. Thermal Conductivity and Expansion of Piping Material

Material	Thermal conductivity		Linear thermal expansion average	
	$\text{Btu}/\left(\text{hr F} \dfrac{\text{ft}^2}{\text{ft}}\right)$	Temperature, F	Microinch/(in. F)	Temperature range, F
Pure iron	43 28	70 752	6.83 8.97	68–212 932–1112
Gray cast iron	27 23.7	70 752	5.83	32–212
Malleable cast iron ferritic	....	......	6.6	70–750
Malleable cast iron pearlitic	....	......	6.6	70–750
Nodular iron............	18 (Pearlitic) 20 (Ferritic)	212 212	6.46 6.97 7.49 7.69	68–212 68–600 68–1000 68–1400
Wrought iron	34 26	212 752		
Wrought carbon steel: 0.06 C, 0.38 Mn	34	32	7.83 7.95 8.02 8.21 8.36	70–800 70–900 70–1000 70–1100 70–1200
0.23 C, 0.635 Mn	30 25 17	70 752 2192	6.50	68–212
0.435 C, 0.69 Mn	....	......	6.44 8.39	68–212 68–1292
1.22 C, 0.35 Mn	26 22 16	70 752 2192	5.89 9.33	122–212 932–1832
Carbon–½ Mo	25.8	212	7.70 7.85 7.95 8.07	68–800 68–1000 68–1100 68–1200
1¼ Cr–½ Mo	17.9	212	7.32 7.44 7.56 7.63 7.74 7.82	70–800 70–900 70–1000 70–1100 70–1200 70–1300
2¼ Cr–1 Mo	16.3	212	7.49 7.65 7.72 7.78 7.84 7.88	70–800 70–900 70–1000 70–1100 70–1200 70–1300

Table 3. (*Continued*)

Material	Thermal conductivity		Linear thermal expansion average	
	$Btu \Big/ \Big(hr\ F\ \dfrac{ft^2}{ft} \Big)$	Temperature, F	Microinch/(in. F)	Temperature range, F
5 Cr–½ Mo ....,......	21.2	212	6.44	0–212
	20.8	392	6.91	70–800
	20.4	572	7.02	70–900
	19.8	752	7.10	70–1000
	19.5	932	7.19	70–1100
			7.31	70–1200
			7.35	70–1300
9 Cr–1 Mo..............	....	......	6.28	70–300
			6.60	70–800
			6.75	70–900
			6.81	70–1000
			6.95	70–1100
			7.07	70–1200
			7.13	70–1300
3½% Ni steel	21	212		
	14	1472		
Type 304 wrought	9.4	212	9.6	32–212
	10.3	392	9.9	32–600
	11.0	572	10.2	32–1000
	11.8	752	10.4	32–1200
	12.5	932	11.2	32–1800
CF-8 cast..............	9.2	212		
	12.1	1000		
Type 316 wrought	9.0	212	8.9	32–212
	12.1	932	9.0	32–600
			9.7	32–1000
			10.3	32–1200
			11.1	32–1500
CF-8M cast.............	9.4	212	8.9	68–212
	12.3	1000	9.7	68–1000
Type 321 wrought	9.3	212	9.3	32–212
	10.2	392	9.5	32–600
	11.1	752	10.3	32–1000
	11.9	932	10.7	32–1200
	12.8		11.2	32–1500
Type 347 wrought	9.3	212	9.3	32–212
	10.2	392	9.5	32–600
	11.1	572	10.3	32–1000
	11.9	752	10.6	32–1200
	12.8	932	11.1	32–1500
CF-8C cast	9.3	212	9.3	68–212
	12.8	1000	10.3	68–1000
405 wrought	....	......	6.0	32–212
			6.4	32–600
			6.7	32–1000
			7.5	32–1200

Table 3. (*Continued*)

Material	Thermal conductivity		Linear thermal expansion average	
	$Btu \big/ \left(hr\ F\ \frac{ft^2}{ft} \right)$	Temperature, F	Microinch/(in. F)	Temperature range, F
CA15 cast	14.5	212	5.5	68–212
	16.7	1000	6.4	68–1000
			6.7	68–1300
410 wrought	14.4	212	6.1	32–212
	16	752	7.2	32–1000
			7.6	32–1832
446 wrought	12.1	212	5.9	32–212
	14.1	932	6.3	32–1000
			7.6	32–1832
CC50 cast	12.6	212		
	17.9	1000	5.9	68–212
	20.3	1500	6.4	68–1000
	24.2	2000		
Aluminum 1100	128	70	12.2	−58 to +68
			13.1	68–212
			13.7	68–392
			14.2	68–572
Aluminum 6061	99 (0 temper)	70	12.1	−76 to +68
	90 (T4 temper)	70	13.0	68–212
	90 (T6 temper)	70	13.5	68–392
			14.1	68–572
Aluminum 43	82 (as cast)	70	12.2	68–212
	94 (annealed)	70	12.8	68–392
			13.3	68–572
Aluminum 356	97 sand cast T51	70	11.9	68–212
	88 sand cast T6	70	12.8	68–392
			13.0	68–572
Copper (DHP)	196	68	9.8	68–572
Red brass	92	68	10.4 cold rolled	68–572
Yellow brass	67	68	11.3	68–572
Admiralty brass	64	68	11.2	68–572
Manganese bronze	61	68	11.8	68–572
Cupronickel (70–30)	17	68	9.0	68–572
Aluminum bronze (3)	44	68	9.0	68–572
Beryllium copper	33–41 cold worked	68	9.3	68–212
	48–68 precipitation hardened		9.4	68–392
			9.9	68–572

Table 3. *(Continued)*

Material	Thermal conductivity		Linear thermal expansion average	
	$Btu / \left(hr\ F\ \frac{ft^2}{ft} \right)$	Temperature, F	Microinch/(in. F)	Temperature range, F
Chemical lead	. . . .	. . . . . .	16.3 14.7	65–212 −130 to +66
50/40 SnPb solder	27	129	13.0	60–230
Nickel (A) wrought	35	32–212	7.4	77–212
Monel (70 Ni–30 Cu) (wrought).	15	32–212	7.8	32–212
Inconel.	8.4	70–212	6.4 8.3	68–212 70–1000
Incoloy.	6.8	32–212	8.0	32–212
Hastelloy B	6	. . . . . .	5.3 7.8	70–200 70–1600
Hastelloy C	5	70	6.6 8.2	70–200 70–1600
Tin	36	32	12.8	32–212
Titanium (99.0 %).	9.0–11.5 12.4	68 1500	4.8 5.6 5.7	68–200 68–1200 68–1600
Tantalum.	31	68	3.6	

(a)   (b)

FIG. 8. (*a*) Typical large grain structure from slowly cooled cast stainless steel. (Magnification 1 ×, size reduced in printing.) (*b*) Typical grain structure of rapidly cooled stainless steel. (Magnification 250 ×, size reduced in printing.)

Some elements, present either as alloying elements or as impurities, tend to segregate along the grain boundaries.   Under some conditions, these segregations along grain boundaries may reduce significantly the strength or ductility of the metal or may lower its resistance to corrosion or oxidation.

Under some conditions of hot forming or welding, cracking may start from the grain boundaries (Fig. 9).

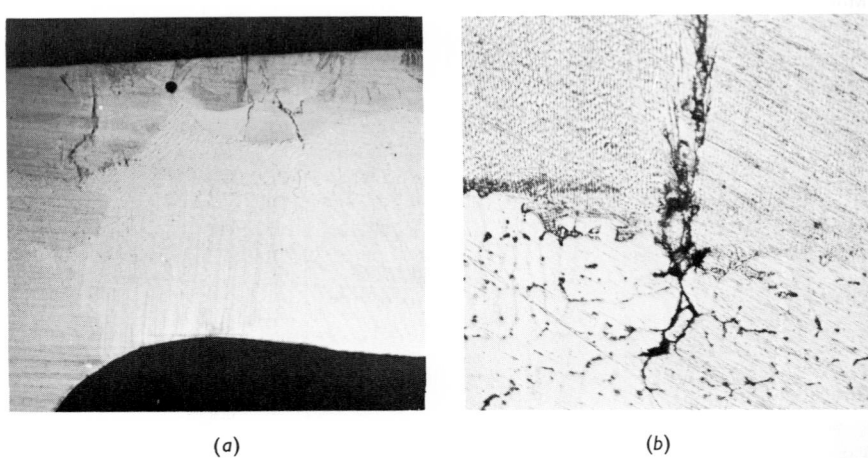

(*a*)	(*b*)

FIG. 9. Cracking starting at grain boundaries.   [Magnification (*a*) 5× and (*b*) 100×, size reduced in printing.]

Cold or hot working tends to break up the original grains and produce refinement, particularly at elevated temperatures.   Excessive temperatures may effect grain growth causing coarsening.   With metals and alloys subject to lattice-structure changes, grain refinement can be also accomplished by heat treatment.

Equilibrium Diagrams

Equilibrium or phase diagrams are utilized extensively by metallurgists to show in graphical form the phase relation which alloys exhibit at different temperatures.

Numerous equilibrium diagrams have been determined for many commercial and experimental metal combinations.   The best known and most widely used is the iron-carbon diagram, shown in Fig. 10.   At room temperature, the iron atoms are arranged in the body-centered cubic lattice shown in Fig. 6.   This arrangement is called alpha iron.   It is magnetic.   At a temperature of 1670 F, pure iron transforms from a body-centered to a face-centered lattice, which is referred to as gamma iron.   It is nonmagnetic.   At 2535 F, the face-centered lattice structure reverts back to the body-centered arrangement and is called delta ferrite.   Such changes are called *allotropic modifications*.

The temperature at which alpha-iron changes to gamma-iron is referred to as the A_3 *transformation* or critical point.   The addition of carbon lowers the A_3 transformation temperature until the carbon content reaches 0.85 per cent.   When alpha-iron transforms to austenite, the iron carbides Fe_3C go into solution.   This is referred to as the A_1 transformation.   This transformation is reversible.   On cooling, the iron carbides tend to form again.   This iron-carbide constituent is generally described as *pearlite*.

Under equilibrium conditions, the transformation temperatures should occur at the same temperature whether the steel is being heated or cooled. However, a lag generally occurs in the attainment of equilibrium conditions. On heating, the A_1 and A_3 points tend to be higher, whereas on cooling, they tend to be lower.

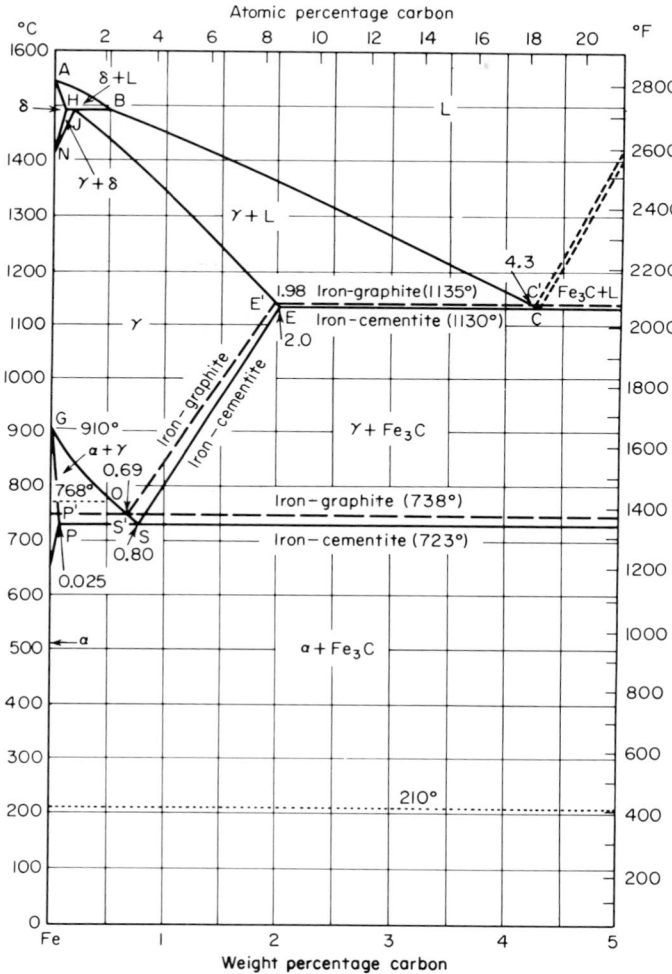

FIG. 10. Iron-carbon equilibrium diagram.

Then the critical points A_1 and A_3 are further modified by the letters[1] r and c to indicate cooling or heating, respectively. On slow heating, the transformation starts at Ac_1 and is completed at the Ac_3 point. When cooling, the transformation starts at the Ar_3 critical point and is completed at the Ar_1 point.

[1] After the French words *refroidissement* and *chauffrage*.

If the rate of cooling is fast enough, the austenite does not undergo transformation to ferrite and pearlite on passing through the $Ar_3 - Ar_1$ range. Under such conditions, the austenite is retained until temperatures of 600 F or below are reached. At these temperatures, the austenite begins to transform to the hard acicular constituent described as *martensite*. The transformation of austenite to martensite on cooling starts at the M_s point and is completed at the M_f point. The temperature range of martensite formation is characteristic for a given steel. It is not affected by the cooling rate.

Alloying elements can alter significantly the A_1 and A_3 and M_s points, and the equilibrium relations.

Recognition of the critical points is important in pipe fabrication. It is generally not recommended that hot forming be done at temperatures between the A_1 and A_3 points. Normalizing must be done above the Ac_3 temperatures, whereas stress relieving or tempering should be done below the Ac_1 temperatures.

Effects of Alloying Materials

Two of the most important alloying elements added to piping steel materials are chromium and nickel. Phase diagrams illustrating the effects of these elements in the ranges normally added to steel are shown in Figs. 11, 12, and 13.

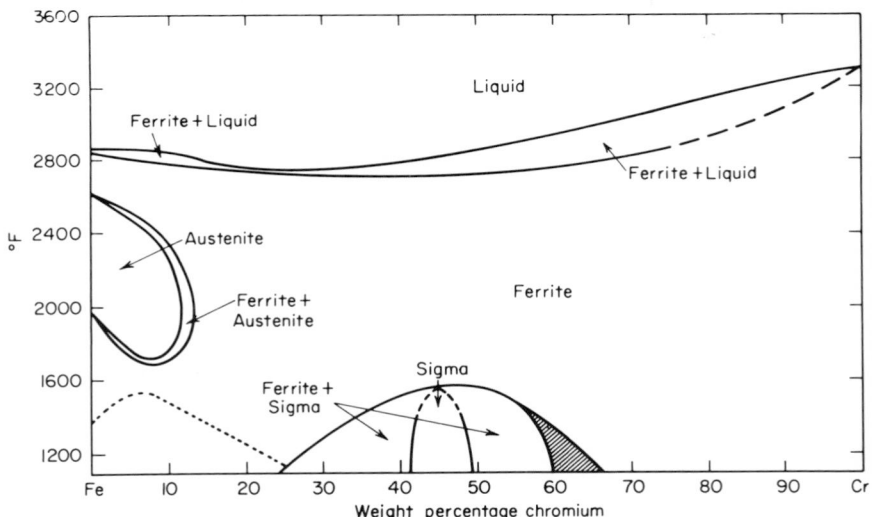

FIG. 11. Phase diagram for iron-chromium system.

On cooling, alloys of iron and chromium form a continuous series of solid solutions (Fig. 11). Whereas a small amount of chromium lowers the A_3 point, larger percentages raise this point. Chromium lowers the austenite to delta-ferrite transformation temperature (referred to as A_4 points). As a result of this, the two transformation points A_3 and A_4 merge at 12 per cent chromium to form the so-called gamma loop.

On heating an alloy containing 11 to 12 per cent chromium, the ferrite begins to transform to austenite at about 1500 F (Fig. 12). This continues with increasing temperature until the alloy is fully austenitic. Chromium carbides which are

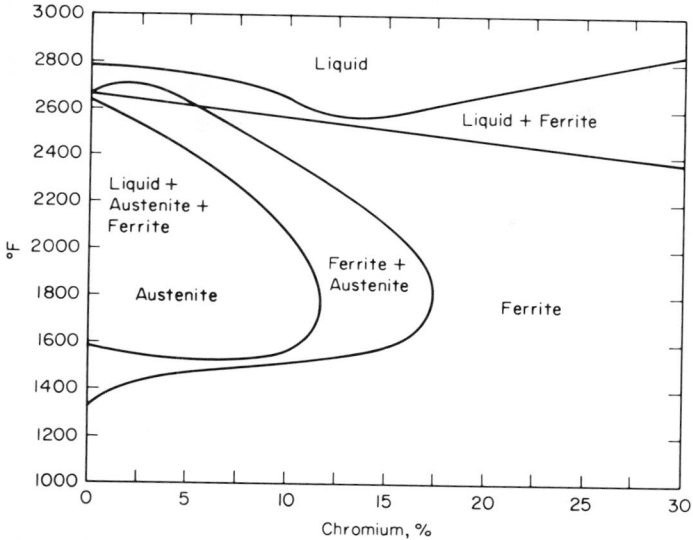

FIG. 12. Phase diagram for iron-chromium system containing 0.1 per cent carbon.

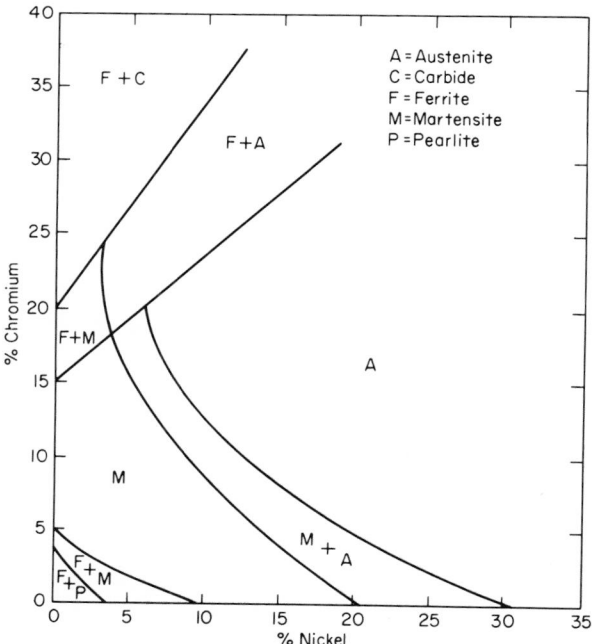

FIG. 13. Phase diagram for iron-nickel-chromium, showing change in phases with respect to nickel and chromium content.

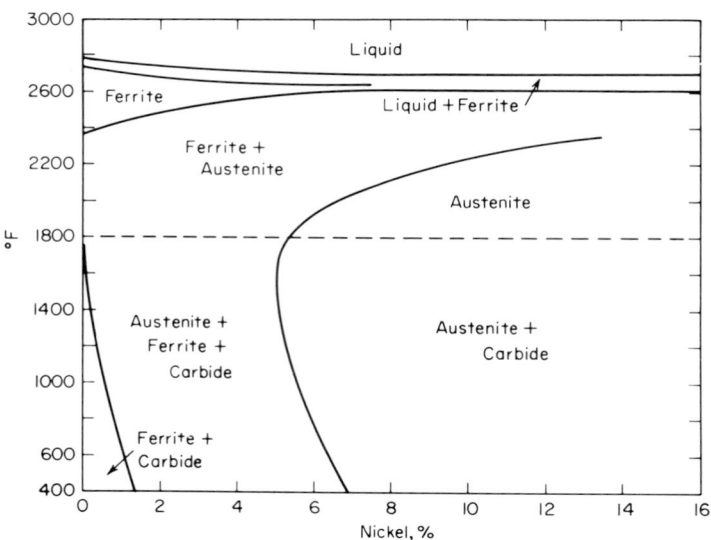

FIG. 14. The effect of nickel additions on the iron-carbon equilibrium phase diagram for steel containing 0.1 per cent carbon and 18 per cent chromium.

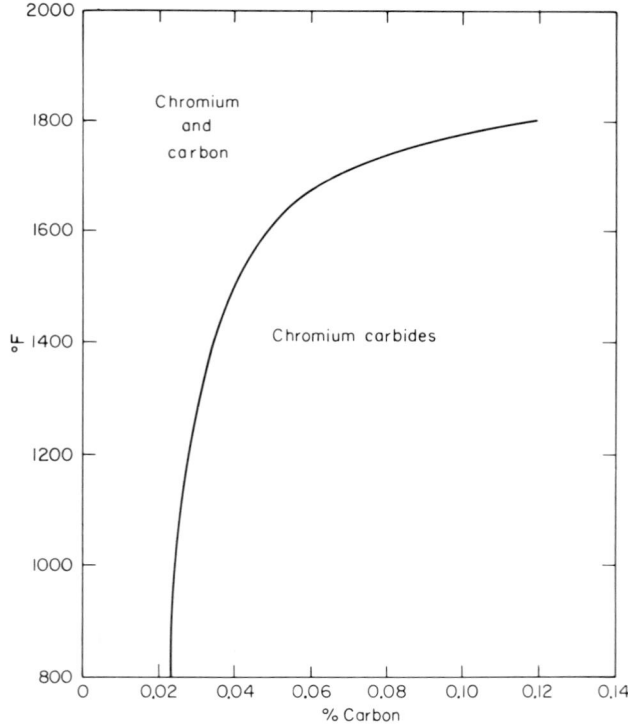

FIG. 15. Effects of temperature on solubility of chromium carbides.

present dissolve in the austenite, depending on the time the 11 to 12 per cent Cr alloy is in the austenite range.   When rapidly cooling this alloy from approximately 1800 F, martensite is formed.   On the other hand, an 18 per cent chromium alloy is not subject to these phase transformations.   Thus, this alloy retains its ferrite structure.   The 18 per cent chromium alloy, therefore, cannot be hardened when quenched from elevated temperatures.

The addition of carbon extends the gamma loop to the right.

These characteristics of iron-chromium alloys have resulted in two important groups of stainless steels referred to as the *martensitic* types and the *ferritic* types.

The effects of nickel addition to steel (0.1 per cent carbon alloyed with 18 per cent chromium) are illustrated in Fig. 14.   Progressive additions of nickel extend the austenite range until the alloy becomes completely austenitic, even at room temperatures.   This has resulted in the development of the very important group of austenitic stainless steels.   The chromium-carbide phase (usually in the form of small particles) is present under equilibrium conditions to approximately 1800 F, as is indicated by the carbide solubility line in Fig. 15.

Molybdenum and silicon act as *ferrite stabilizers*.   They tend to shift to the right the curve along which the chromium-nickel alloy steel becomes fully austenitic.   Carbon and manganese tend to shift this curve to the left.

FIG. 16. Hot-extruded outlet from centrifugally cast stainless-steel pipe. Ferrite composition exceeded 10 per cent.   Cracks resulted from excessive hot working.

These phase relations can be very important in the working and forming of piping materials.   For the extrusion and severe forming of piping, a fully austenitic structure is generally desirable in stainless steels.   For that reason, commercial hot-forged or wrought pipe is generally fully austenitic (or almost so).

On the other hand, the retention of some ferrite in the microstructure will increase significantly the tensile and yield strength.   Sand-cast stainless-steel pipe fittings and valves, and centrifugally cast pipe have been produced to take advantage of these higher strength properties.   However, cracking may occur when these materials with over 10 per cent ferrite are severely hot worked, as involved in the hot extrusion of outlets (Fig. 16), in lapping or in swaging operations.

Welding of most stainless steels is also improved when the weld metal is of a composition which results in 4 to 7 per cent ferrite in structure.   Where fully austenitic, the structure is more susceptible to cracking.   This is particularly true in the columbium-stabilized stainless grades such as Types 347 and 348.

The composition in weld deposits is generally estimated from the Schaeffler diagram, shown in Fig. 17.   Some specifications for welding filler metals indicate that weld deposits with a small percentage of ferrite (often 4 to 7 per cent) should be produced.

Most commercial nonferrous piping materials do not undergo allotropic transformations.   Most alloying elements form solid solutions with the respective metal. However, special phases are also formed which may produce additional strengthening by precipitation of these phases from solid solutions.

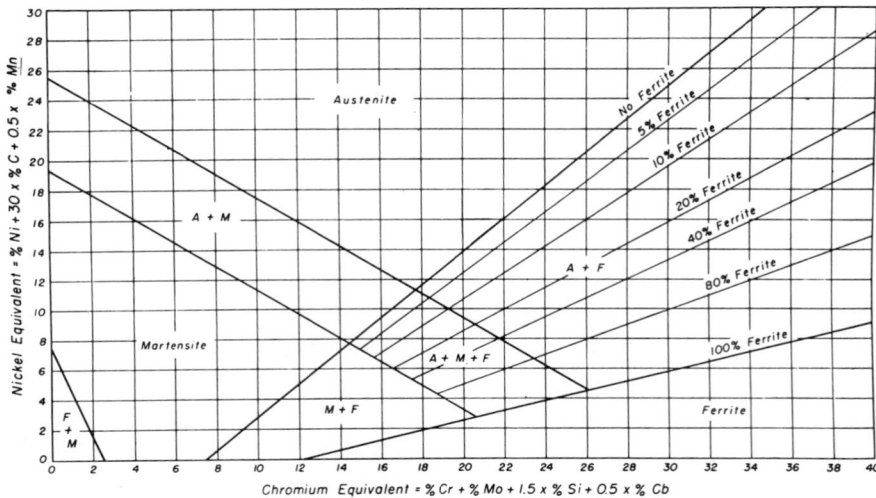

FIG. 17. Schaeffler diagram showing the structure of various stainless steels.

Metallurgical Characteristics of Commercial Piping Materials

The large number of ferrous and nonferrous metals and alloys available in the form of piping and tubing preclude any detailed discussion of the metallurgical characteristics and mechanical and physical properties of all materials. Changes in the metallurgical characteristics generally result in changes in the mechanical properties. To obtain desired mechanical or physical properties, control over the metallurgical structure is often important. Such control may be accomplished by operations such as heat treatment, hot working, cold reduction or expansion.

Significant differences in metallurgical, mechanical and physical properties may even exist in commercial materials produced to the same specifications. This is

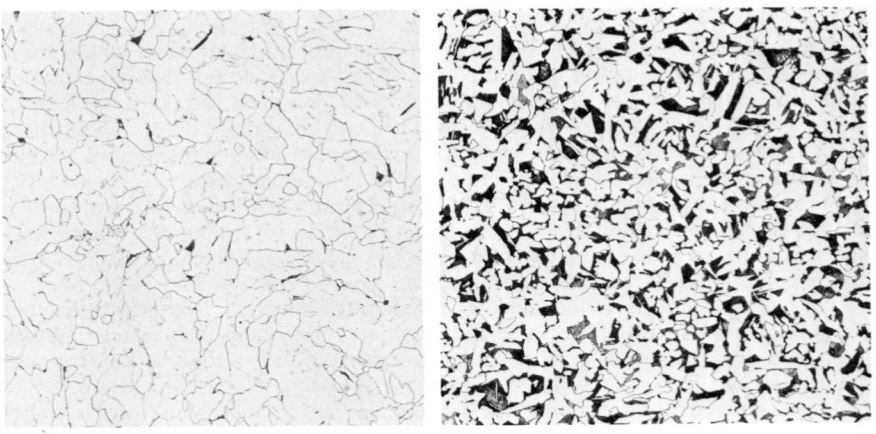

FIG. 18. Structural differences in A-53 carbon-steel pipe. (Magnification 100×, size reduced in printing.)

sometimes critical and can affect adversely the fabricating characteristics or service behavior.

Carbon Steels. Even carbon-steel piping materials may exhibit significant differences. For example, piping to ASTM Specification A-53 is not subject to specific limits on carbon content and does not require deoxidation during the melting practice. Considerable variations in carbon content may occur (between 0.05 and 0.23 per cent). This will produce significant differences in the microstructures (Fig. 18) and in the response to heat treatment.

Carbon-steel piping to ASTM Specification A-106 is designed for the more critical applications. Deoxidation requirements and closer control over carbon content produce a more identical structure in commercial heats (Fig. 19).

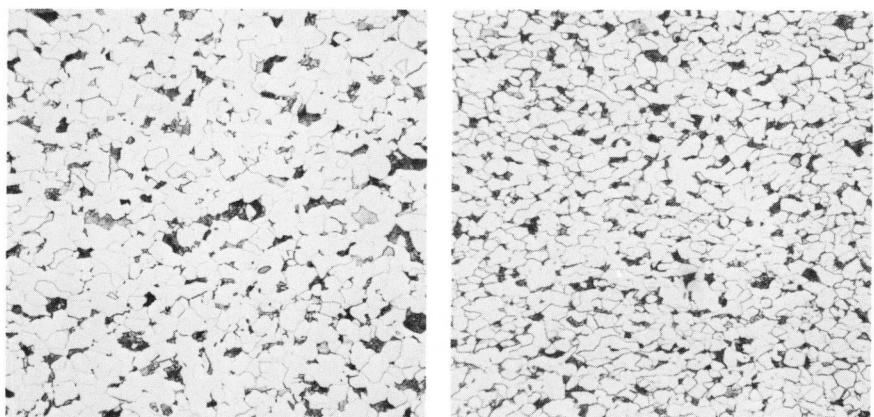

FIG. 19. More identical structure in ASTM A106 carbon piping is obtained from closer control over carbon content and deoxidation requirements. (Magnification 100×, size reduced in printing.)

When tube and pipe materials are cold worked, their strength and hardness are increased while their ductility is decreased. Cold expansion of pipe and tube materials is done intentionally to obtain the higher strength values as in some API grades. In other service environments, cold-worked materials may not be desirable.

The effects of cold work can be removed by heat treatment. The effects of heat treatment on the hardness and microstructures of cold-worked carbon steels are illustrated in Fig. 20. Up to 800 F, the decrease in hardness is very slight. Sometimes an almost imperceptible increase occurs in the hardness. Above 800 F, the hardness begins to decrease. This is known as the *recovery stage*. Nevertheless, the steel continues to exhibit a cold-worked microstructure until temperatures of about 1050 F are reached. At temperatures above 1000 F, the hardness decreases rapidly. At 1100 F, small equiaxial grains form in the microstructure. At this stage, recovery is considered complete and recrystallization begins to take place. At 1150 F, the evidence of cold work disappears completely. The microstructure is then considered to be completely recrystallized.

Alloy Steels. As the alloy content increases, the effects of heating operations and heat treatment play an increasing effect on the microstructure and mechanical properties.

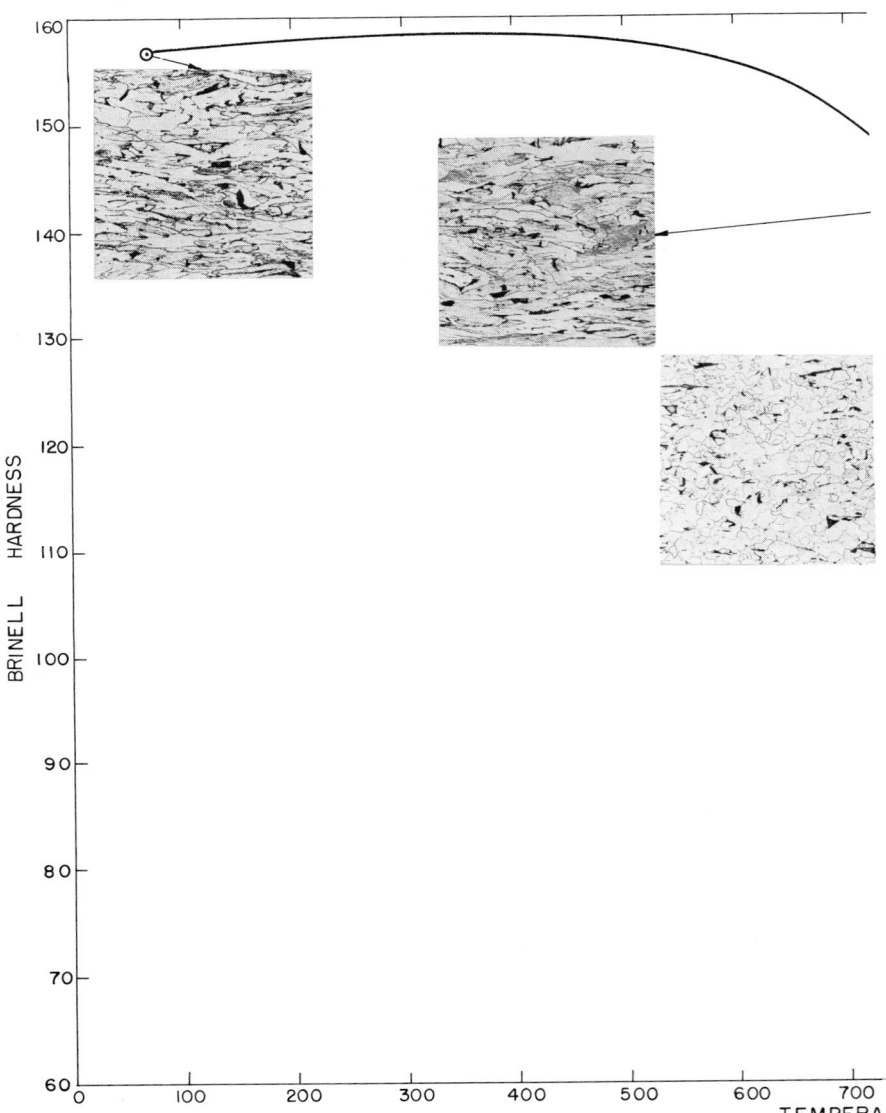

FIG. 20. Effects of heat treatment on the hardness

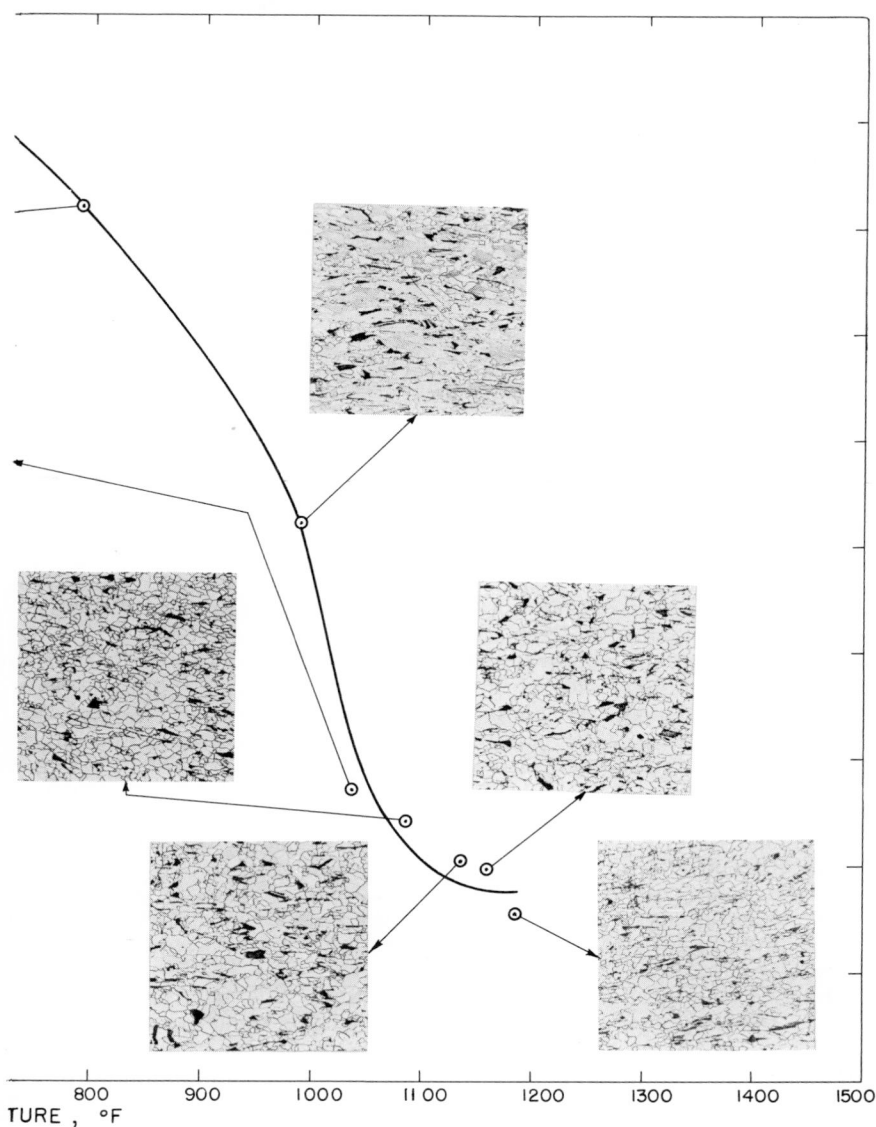

and microstructures of cold-worked carbon steels.

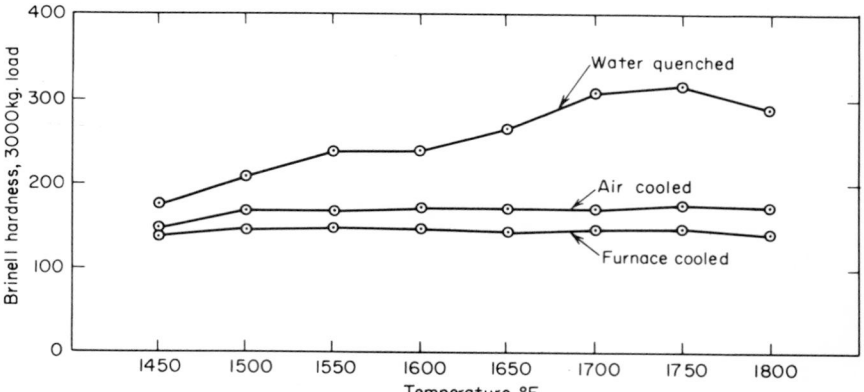

FIG. 21. Effects of cooling rate from different elevated temperatures on hardness values of 1½ Cr–½ Mo alloy-steel pipe.

Figure 21 illustrates the effects of cooling rate from different elevated temperatures on hardness values of a 1¼ Cr–½Mo alloy-steel pipe. Corresponding microstructures at 1450, 1550, 1700 and 1800 F are shown in Fig. 22. Subsequent stress relieving for 1 hr at 1325 F will reduce the hardness values (Fig. 23). Nevertheless, significant differences remain in the microstructures (Fig. 24).

Structural Changes at Elevated Temperatures. When steels are exposed to temperatures above 800 F, some changes may take place in the microstructure. The higher the temperatures, the more rapid are the rates at which the changes occur.

In low-carbon steels, which have been normalized or annealed, the iron carbide phase (Fe_3C) appears normally as the lamellar pearlite shown in Fig. 25. Upon prolonged heating above 900 F, the carbide gradually changes to the spherical form (Fig. 26); this phenomenon is termed *spheroidization*.

Chromium tends to stabilize the carbides and delay spheroidization until higher temperatures are reached.

Carbon steels, when exposed for long periods to temperatures over 800 F, and carbon-molybdenum steels when exposed to temperatures over 825 F, have been subject to a condition of carbide instability known as *graphitization*.[1] Primarily susceptible have been fine-grained aluminum-killed steels.

Graphitization involves the breakup of the iron carbide into iron and carbon, and the combination of carbon into graphite. When it occurs in steels, graphite formation has been most pronounced in the heat-affected zone parallel to the weld. Whereas small, well-dispersed particles of graphite have not been critical, severe concentrations of graphite in a plane parallel to the weld have caused very severe embrittlement.

Examples of "mild," "moderate" and "severe" degrees of graphitization are illustrated in Figs. 27 to 29. The "chain" or "clustered" graphite severely reduces the ductility, as measured by bend tests. Figure 30 illustrates how bend specimens taken across zones of "severe" graphitization fail upon bending by a few degrees. The relation of bend angle to degree of graphitization is shown in Table 4.

[1] For a more detailed discussion of graphitization, see "Considerations in the Evaluation of Graphitization in Piping Systems" by H. Thielsch, E. M. Phillips and E. R. Jerome, Jr., *The Welding J.*, vol. 35, Research Suppl., pp. 286s–294s, June, 1955.

Not all carbon and carbon-molybdenum steels are susceptible to graphite formation when involved in long-time service at temperatures over 800 F. In fact, improved current steelmaking practices virtually have eliminated this problem. However, there have been several extremely brittle failures in pipe installed in the 1940s and early 1950s.

Where graphite formation is mild or moderate it can be removed successfully by a 2-hr solution heat treatment at 1750 F. Subsequent new graphite formation can generally be suppressed by a so-called metallurgical stabilization heat treatment (of the iron-carbides) involving a 4-hr holding period at 1325 to 1350 F.

Another form of metallurgical instability involving carbon occurs in some types of chromium-nickel austenitic stainless steels. The result of chromium carbide

Table 4. Relationship between Bend Angle and Degree of Graphitization

Metallographic evaluation	Bend angle, deg
None	180
Very mild	180
Mild	90–180
Moderate	50–90
Heavy	30–50
Severe	15–30
Extremely severe	Below 15

precipitation along the grain boundaries, this instability is referred to as *intergranular carbide precipitation.*

When the common austenitic stainless steels are exposed, during fabrication or in service, in the temperature range between 900 and 1500 F the carbon tends to diffuse to the grain boundaries and combine with chromium to form chromium carbide particles. The process of intergranular carbide precipitation is illustrated in Fig. 31. A stainless steel in which it has occurred usually is described as "sensitized" by the exposure in the critical 900 to 1500 F sensitizing temperature range. Precipitation of these chromium carbide particles at the grain boundaries reduces the resistance of the stainless steel to certain corrosive solutions. Since the corrosive attack occurs preferentially at the grain boundaries, it is generally known as *intergranular corrosion* (Fig. 32).

The severity of intergranular carbide precipitation primarily depends upon the carbon content of the steel, the temperature, and the length of time it is held in the critical sensitizing temperature range. Rate of precipitation is greatest at about 1200 F.

Welding can cause the precipitation of the chromium carbides in the grain boundaries adjacent to that part of the heat-affected zone heated to temperatures between 900 and 1500 F. This is often referred to as "weld decay."

Types 301 and 302 stainless steels which contain 0.15 per cent maximum carbon are more susceptible to intergranular carbide precipitation than is Type 304 which contains less than 0.08 per cent carbon. Rate of intergranular carbide precipitation is slightly retarded by a higher chromium and/or molybdenum content as in Type 310, 309 or 316. Intergranular corrosion may be prevented by the addition of columbium or columbium-tantalum (Types 347, 348, 318) or titanium (Type 321). This is described as "stabilization." Intergranular corrosion may also be prevented by a suitable annealing heat treatment between about 1850 and 2050 F and quenching in water or a water spray following the final fabricating operation. If carbon content is limited to 0.03 per cent maximum, time for carbide precipitation can be lengthened and the amount diminished so that all normal welding operations

1450°F.   1550°F.

Water
Quenched

Air
Cooled

Furnace
Cooled

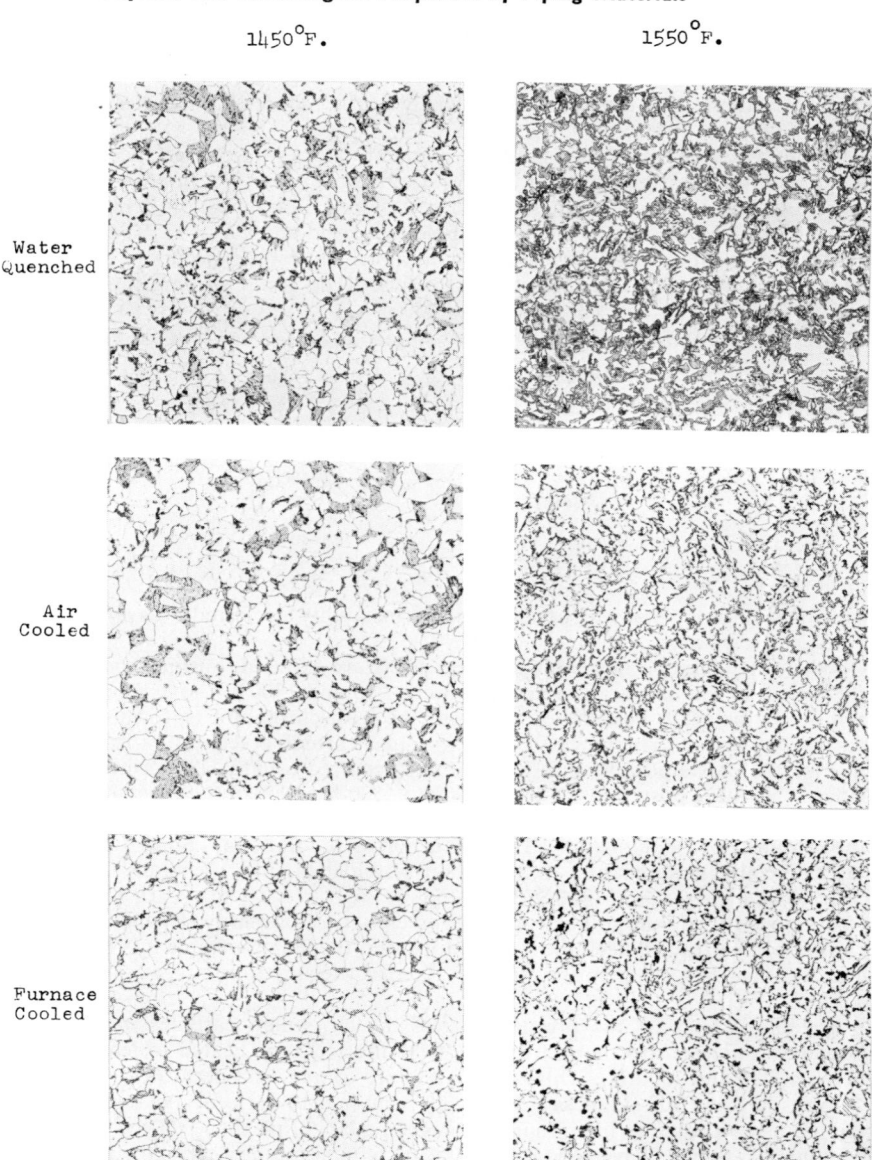

FIG. 22. Effects of cooling rates from different tempera-

1700°F.   1800°F.

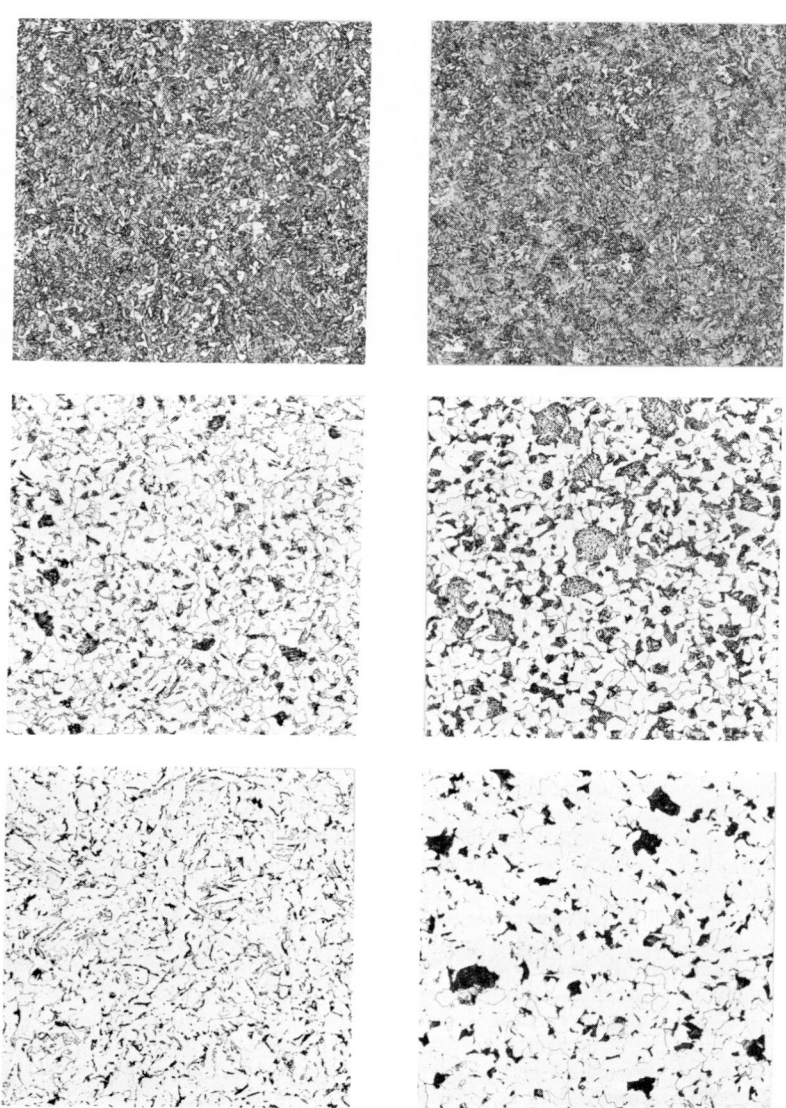

tures on the microstructures of 1¼ Cr–½ Mo alloy-steel pipe.

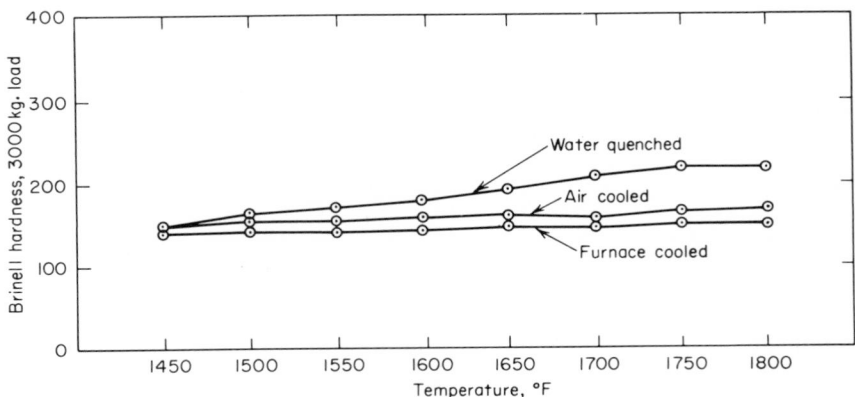

Fig. 23. Effects of stress-relief heat treatment for 1 hr at 1325 F on specimens previously cooled at different cooling rates from different elevated temperatures on hardness values of 1¼ Cr–½ Mo alloy-steel pipe.

may be carried out without deleterious effects.   Stainless steels with 0.03 per cent maximum carbon content are referred to as extra-low-carbon grades.

Intergranular carbide precipitation resulting from welding or heating operations generally will not impair the mechanical properties of the steel.

Since welding exposes the stainless steel for only a relatively brief period in the sensitizing temperature range, the amount of intergranular carbide precipitation after welding is less than it would be after an extended service exposure at these elevated temperatures.

Unless the stainless steel is subsequently exposed to active corrosive solutions, the presence of a slight amount of precipitated carbides usually does not affect the life of the equipment.   Thus, in mildly corrosive environments, as in dairy machinery, kitchen and cafeteria equipment, architectural decorative work and heat-resisting parts, welded structures made of stainless Types 301 and 302 generally will give entirely satisfactory service.   Exceptions to this rule are occasionally found in heavily industrialized areas containing chemical plants where some corrosive attack adjacent to the weld deposit has sometimes been noted in these stainless-steel grades.

In many other mildly corrosive applications, the 0.08 per cent maximum carbon Types 304, 308, 309S and 310S, 316 and 317, may be completely satisfactory in the as-welded condition.   In these grades, intergranular carbide precipitation in the heat-affected zone generally becomes appreciable only when the cooling time through the 1500 to 900 F temperature range exceeds approximately 1 to 1½ min.

When these nonstabilized grades are being welded, and when the service environment requires the absence of intergranular chromium carbide precipitation in the steel, special precautions should be taken to allow the weldment to cool as rapidly as possible.   This is generally accomplished by: (1) use of small-diameter electrodes, (2) low welding current, (3) deposition of the filler metal with a stringer-bead technique (no weaving), (4) use of chill bars in the fixtures and/or (5) immediate application of an air blast or a water quench or spray following the welding operation.

When the necessary rapid cooling rates cannot be provided, or when the corrosive application is extremely severe, extra-low-carbon or stabilized stainless-steel grades and electrodes must be used.   With less than 0.03 per cent carbon available in the

extra-low-carbon grades, intergranular carbide precipitation is usually prevented. Although extra-low-carbon stainless-steel electrodes are available commercially, the extra-low-carbon grades are often welded with Type 347 (18-8 Cb) electrodes. The extra-low-carbon grades generally are not recommended for parts which are exposed during service at temperatures above 800 F.

Equipment which is operated continuously or intermittently at temperatures above 800 F, and then exposed to a corrosive environment, is generally made of the stabilized stainless grades, Types 321, 347 or 348.

In the fabrication of *dissimilar metal combinations*, the significant metallurgical effects consist of (1) dilution of the weld metal during welding and (2) diffusion across the dissimilar joint as a result of heat treatment or of high-temperature service at temperatures exceeding approximately 800 F.

Dilution describes the mixing of the molten welding filler metal being deposited with that portion of the base metal which is melted ("fused") by the welding operation. When the weld is made without the addition of filler metal, the "dilution" would consist of the melting and mixing of portions of the two dissimilar base metals being joined. The amount of dilution varies with the different welding processes and welding conditions. Dilution of the weld by the base metal may be as low as 10 per cent or as high as 60 per cent.

In dissimilar-metal joints between austenitic stainless steels and ferritic carbon or low-alloy steels, a large amount of dilution may produce a rather hard and brittle zone adjacent to the base metal which is usually associated with martensite formation. The undesirable effects of this may be minimized by careful electrode selection and preheat and postheat treatments.

Diffusion in dissimilar-metal joints describes the movement or migration of atoms across the bond. In steels of dissimilar composition the diffusion of carbon atoms across the bond tends to have the most pronounced effect. *Carbon migration*, as it is usually called, depends on time and temperature.

At temperatures below 800 F, carbon migration is not considered sufficiently significant to have a harmful effect upon the service properties of the dissimilar-metal joint. With increasing temperatures, the diffusion rates increase. Thus, in the average carbon-steel or carbon-molybdenum-steel joint at 850 F, the embrittlement may become significant after a period of about 5 to 10 years. But at 950 F, only about 1 year may be involved to effect the same degree of embrittlement. At 1200 F, the critical time factor may be reduced to the order of days. From a practical point of view, this is not too important, since carbon and carbon-molybdenum steels are rarely used at temperatures exceeding 1000 F. In fact, in high-temperature steam plants carbon steels are now limited to service below 750 F and carbon-molybdenum steels to service below 800 F. For chemical plants and refinery high-temperature applications, these limits are usually set somewhat higher.

The rate of carbon migration depends also upon the "degree of dissimilarity." It is more rapid from a carbon steel to a $2\frac{1}{4}$ Cr–1 Mo steel than to a $\frac{1}{2}$ Cr–$\frac{1}{2}$ Mo or to a carbon-molybdenum steel. The direction of carbon migration usually is from the lower alloy to the higher alloy steel. More precisely, the carbon atoms migrate towards the steel containing the stronger carbide-forming elements or the greater quantity of them.

ELEVATED-TEMPERATURE PROPERTIES

The behavior of piping materials at elevated temperatures has a bearing both on their suitability for fabrication and erection and on their service life.

The effects on piping materials of high temperatures as a result of hot forging,

1450°F.   1550°F.

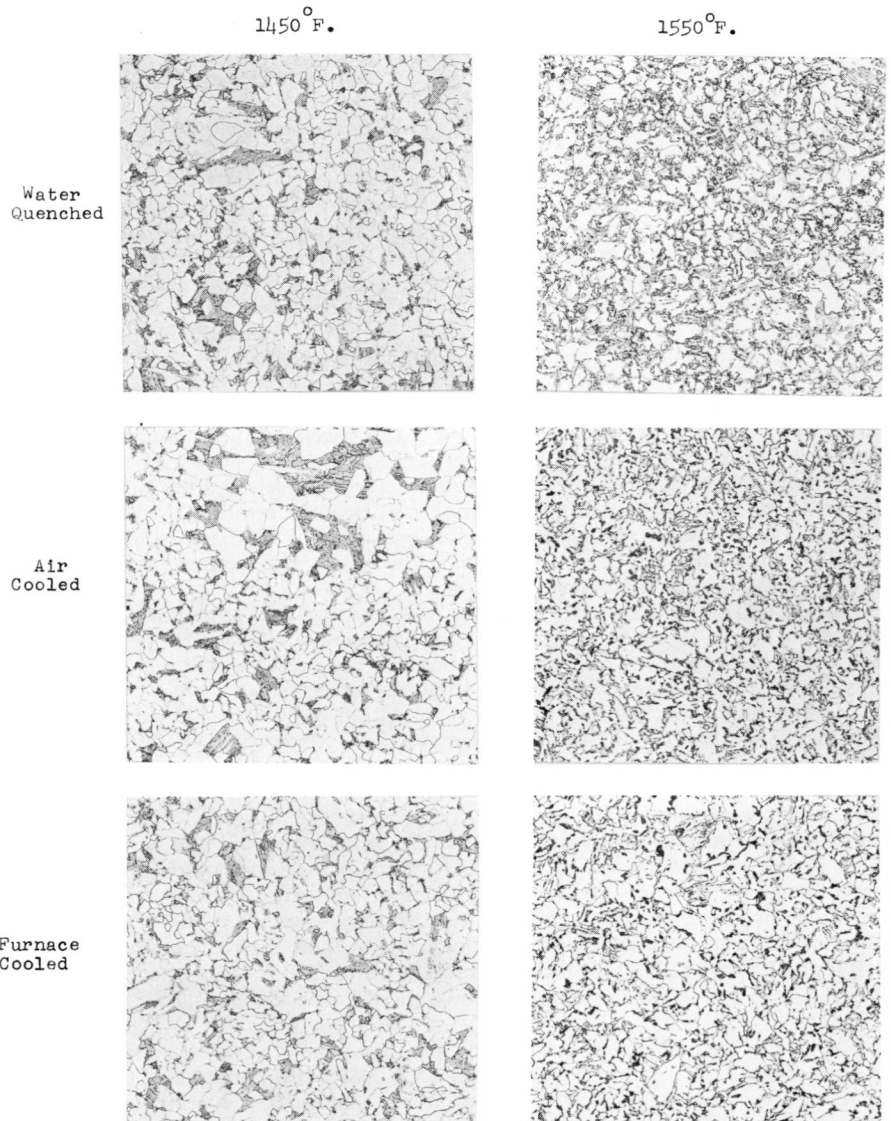

Water
Quenched

Air
Cooled

Furnace
Cooled

FIG. 24. Microstructures of furnace-cooled, air-cooled, and water-quenched 1¼ Cr–
at 1325 F for one hour.   (Magnification 100×, size reduced in printing.)

1700°F.   1800°F.

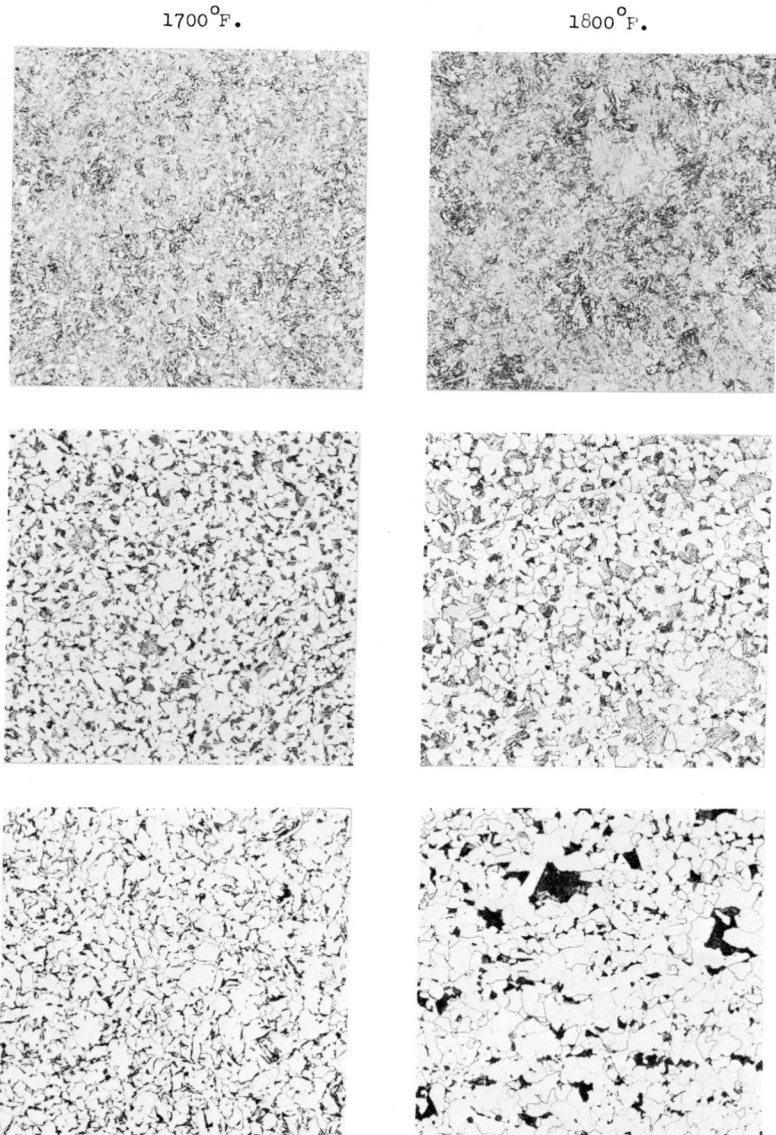

½ Mo alloy-steel pipe sections shown in Fig. 22 after being subsequently stress relieved

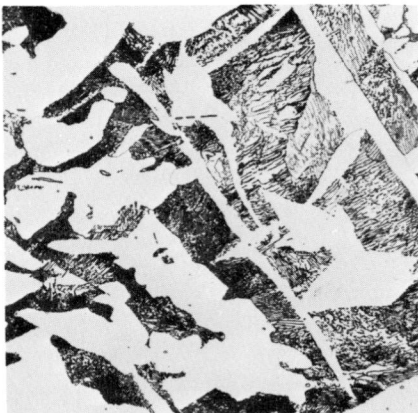

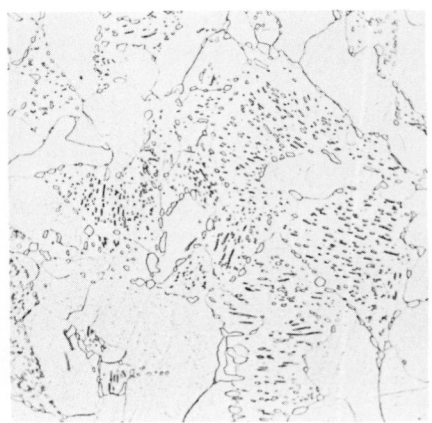

Fig. 25. Iron carbide (Fe₃C) appears normally as lamellar pearlite in annealed carbon steels. (Magnification 500×, size reduced in printing.)

FIG. 26. Spheroidization: Upon prolonged heating above 900 F, the iron carbide (Fe₃C) gradually changes to spherical form. (Magnification 1,000×, size reduced in printing.)

hot bending, welding and heat treatment are of great importance. If not understood, failures may result during fabrication, or defects may be introduced into the piping materials. Among the properties of importance are high tensile strength and ductility properties at hot-working or hot-bending temperatures, resistance to scaling, and resistance to thermal shock such as is caused by rapid quenching.

In selecting piping materials for elevated-temperature service, consideration must be given to the principal mechanical and physical properties of the materials. This should include the entire temperature range to which the piping may be exposed during service on a continuous, cyclic or intermittent basis. Mechanical properties of importance are tensile strength, proportional limit, thermal- fatigue or

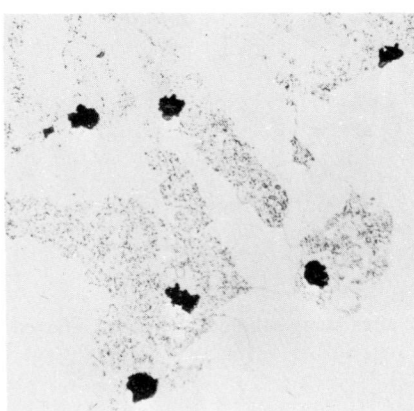

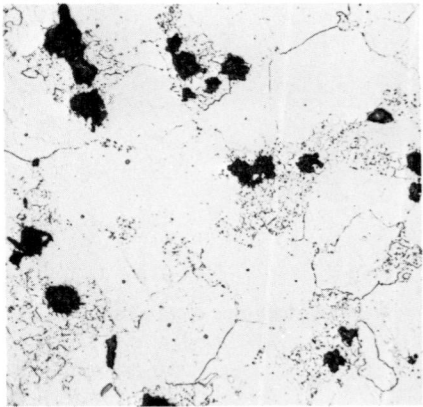

FIG. 27. An example of mild graphitization. (Magnification 500×, size reduced in printing.)

FIG. 28. An example of moderate graphitization. (Magnification 500×, size reduced in printing.)

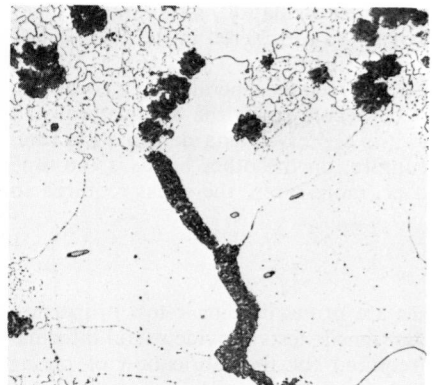
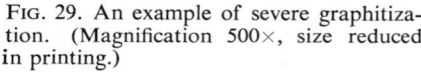

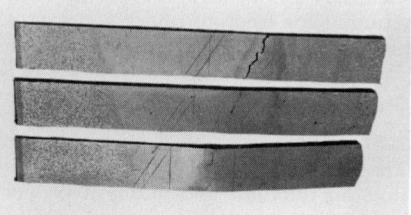

FIG. 29. An example of severe graphitization. (Magnification 500×, size reduced
in printing.)

FIG. 30. Failure of bend specimens taken
across zones of severe graphitization.

shock resistance, mechanical-fatigue resistance at elevated temperatures, torsional
elastic limit, toughness or wearing qualities and resistance to erosion.

Of primary importance is the ability of the material to withstand continued load
over periods of long duration without distortion or undue plastic flow. This
involves consideration of the creep and stress-rupture properties of the material.
Excellent mechanical properties at elevated temperatures are not the only criteria
to be used in selecting materials, as scaling and oxidation are deciding factors. In
high-temperature service, it is essential that the necessary mechanical and physical
properties be combined with sufficient chemical stability so as to obtain satisfactory
performance during the useful life of the piping system or component.

In the design of piping and tube assemblies for service applications at elevated
temperatures, various factors may impose limitations on the service life. Such

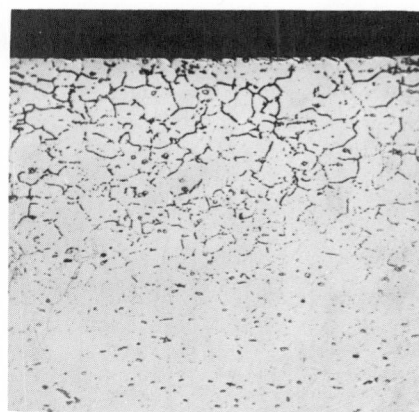

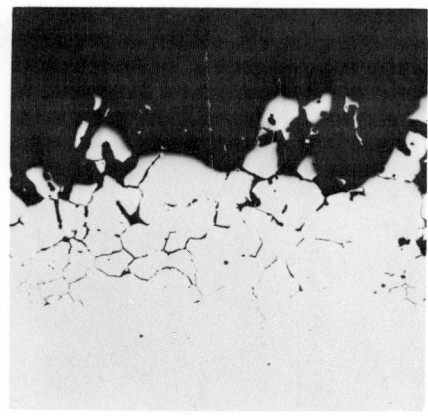

FIG. 31. Intergranular carbide precipitation
in a Type 304 stainless steel. (Magnification 100×, size reduced in printing.)

FIG. 32. Intergranular corrosion as a result
of intergranular carbide precipitation.
(Magnification 100×, size reduced in printing.)

factors as creep and stress rupture, fatigue, surface oxidation, structural changes within the material, corrosion and others have to be considered in the selection of materials and in the design of piping. Among these, creep and stress rupture represent two of the more important factors. Creep is generally defined as the time-dependent part of the deformation which accompanies the application of a constant load to a solid material. Usually, it is expressed as a deformation rate, such as 1 per cent in 100,000 hr. Stress rupture, on the other hand, is the time required for fracture under a constant load or, more often, the stress required to produce fracture in a specified time.

Short-time Properties

Short-time elevated-temperature properties are primarily tensile-test properties. The results obtained by means of so-called hot-tensile tests provide useful information on ductility and strength properties required for the fabrication of piping

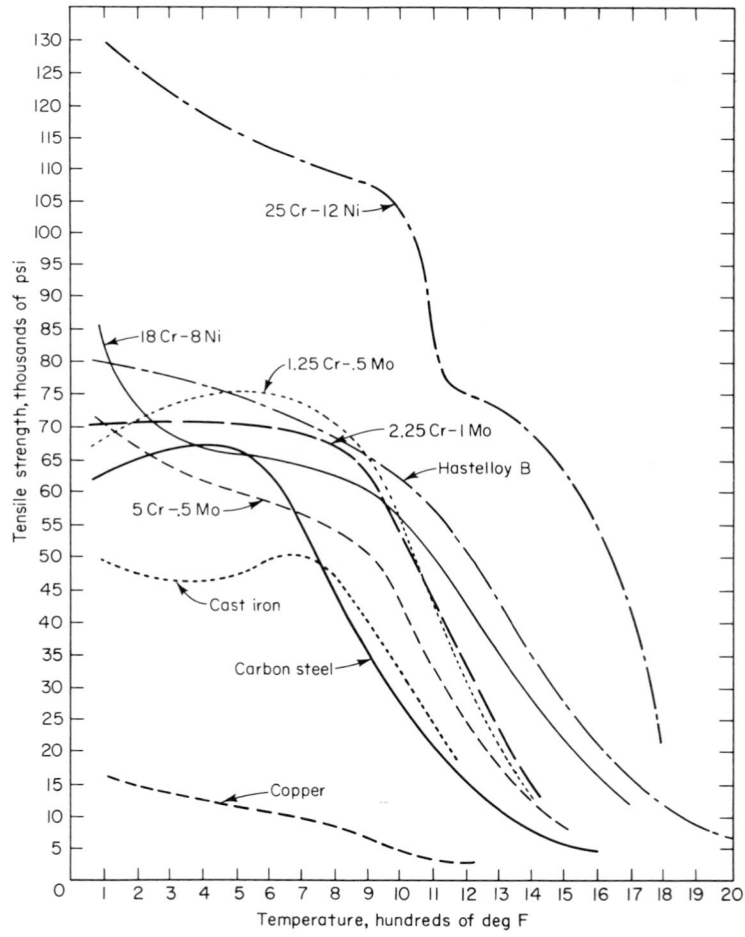

FIG. 33. Tensile strengths of some common piping materials.

components.   Insufficient ductility may lead to rupture during hot-forming, forging or bending operations.

Hot-tensile-test data are also useful in determining the effects of composition on elevated-temperature properties.   This test is recognized as a check on quality control in several ASTM Materials Specifications.

The tensile strengths of some of the common piping materials tend to increase with respect to the corresponding values at ambient temperatures for a few hundred degrees.   With further rise in temperature they tend to fall off rapidly (see Fig. 33). Other materials show a continuous decrease in strength with increase in temperature.

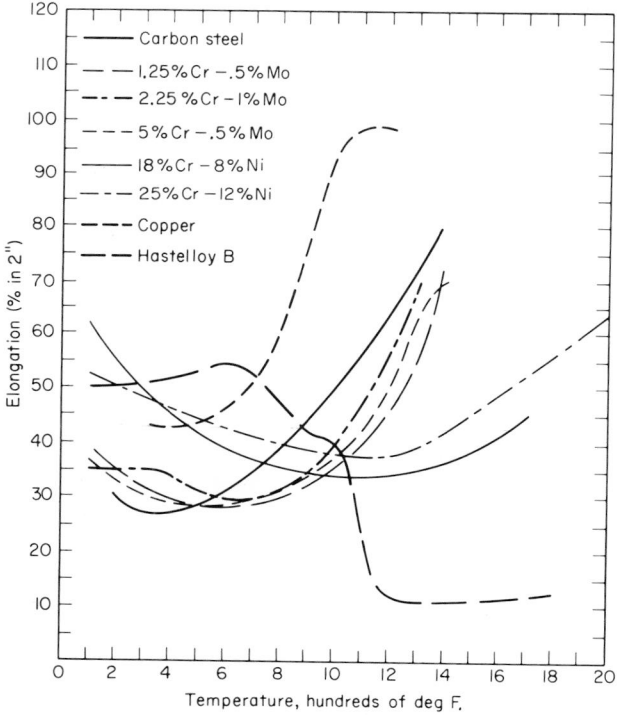

FIG. 34. Ductility of various alloys with respect to temperature.

Sometimes, when creep and stress-rupture data are not available, short-time hot-tensile tests are also used as a basis for design.   This may be adequate, particularly when a relatively short-time service life is anticipated, as in some missile components.   However, such short-time tensile-test data may bear little or no relation to the resistance to creep flow under long maintained load.   Some materials, especially when heat-treated by quenching and tempering, may exhibit relatively high short-time tensile-strength characteristics, whereas their creep resistance is relatively poor.

Ductility properties, which are also very important, are illustrated in Fig. 34. For hot bending to a 5-diameter radius, the piping materials should exhibit a ductility of at least 20 per cent over the temperature range at which hot bending is done.   For the extrusion of outlets or the swaging of reducing ends, a ductility of 25 to 30 per cent is desirable at the forming temperatures.

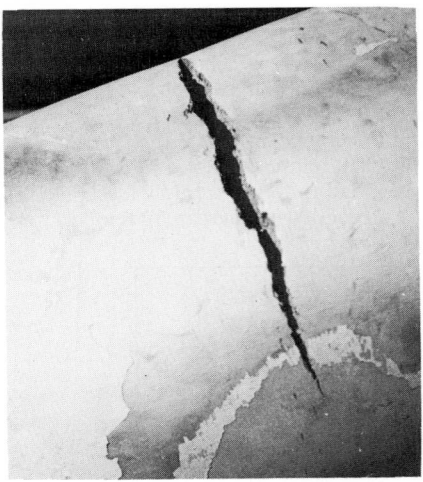

FIG. 35. Hot shortness resulted in severe cracking in 30-in.-OD 1⅛-in. wall 1¼ Cr–½ Mo pipe during hot bending.

Insufficient hot ductility, which may be the result of hot-shortness, may lead to failure during hot forming. Figure 35 illustrates severe cracking of a 30-in.-OD 1⅛-in.-thick pipe. The rupture occurred during hot bending through an angle of approximately 15 deg on a 20 ft radius at 1900 F. The pipe was produced from wrought 1¼ Cr–½ Mo plate steel material that had been seam welded. Subsequent hot-tensile tests shown in Fig. 36 confirmed that the steel was hot short and did not possess sufficient or normal ductility at the temperatures at which hot bending or hot forging of steel is generally done.

Creep and Stress-rupture Properties

When a load is applied to a metal at an elevated temperature over a prolonged period of time, the metal may undergo continuous plastic deformation, that is, it may experience a progressive change in its dimensions. The amount of gradual deformation or "flow" depends on the composition, the processing and heat

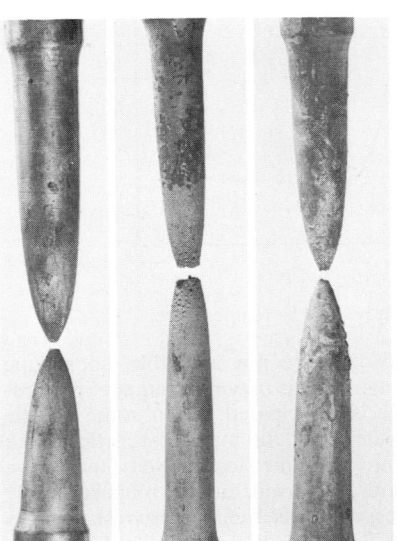
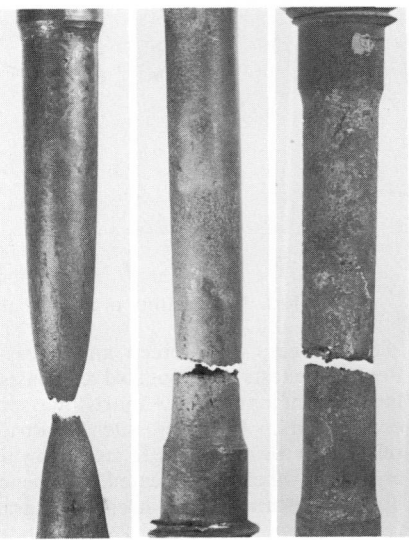

FIG. 36. Hot tensile tests of normal steel (left three specimens) are compared with those of hot short steel (right three specimens). Testing temperatures for both normal and hot short steel were, from left to right: 1400 F, 1600 F, and 1850 F. Elongation in 2 in. for normal steel was, from left to right; 56 per cent, 65 per cent, and 75 per cent. Elongation in 2 in. for hot short steel was, from left to right; 43 per cent, 20 per cent, and 5 per cent.

treatment of the material, the temperature, the shape of the section, the initially applied stress and the time at temperature.   At certain temperature levels creep, which is the term used to describe this progressive dimensional change, may occur in metals even at stresses below the short-time yield strength or proportional limit. Thus, the yield strength and proportional limit, which are determined by short-time tensile tests at room or at elevated temperatures, do not represent satisfactory criteria for the design of piping systems over the entire temperature range.   Even for those cases where the total deformation is considered relatively unimportant, data obtained from short-time elevated-temperature tensile tests must still be augmented by long-time sustained load tests since creep at elevated temperatures may terminate in fracture even at loads considerably below the short-time tensile strength.   Such high-temperature fractures are commonly referred to as stress-rupture failures.   Long-time tests, generally under constant load, carried out to fracture are called stress-to-rupture tests.

Short-time elevated-temperature tensile tests generally do not provide data which are suitable to the design engineer for high-temperature applications.   Thus a large number of long-time tensile tests have been and are being made on various materials at elevated temperatures to determine the rate of deformation or *creep rate*, or the fracture time, as a function of applied stress and temperature.   It may seem most desirable to be able to determine so-called "creep limits," which for each material would represent the stress at a particular temperature below which plastic deformation or creep either would not occur or would occur at too slow a rate to be significant.   Such a limit, however, is merely a practical assumption, not a true value, since creep may be found to occur at increasingly lower temperatures and stresses as the accuracy and sensitivity of the testing equipment improve.

Since a definite creep limit does not exist, engineers must take into consideration some degree of plastic yielding during service and design in accordance with such maximum applied stresses as will result in no more than certain specific limiting deformations during the contemplated service life of the piping material.   The amount of permissible deformation and the contemplated service life may vary for different applications.   In most high-temperature piping systems as, for example, those built to ASME Boiler and Pressure Vessel Code rules, the stress to produce a creep rate of 1.0 per cent elongation in 100,000 hr determined from tests lasting only a few thousand hours, has been considered a limiting value.   This, however, is generally considered a practical limit based on experience.   It is not necessarily the maximum deformation the piping could satisfactorily sustain.   In the design of mechanical equipment, such as turbines and valves, required clearances might necessitate the use of lower stresses to limit the creep to less than 1.0 per cent in the desired lifetime.   Where large deformations can be tolerated, as for example in still tubes, the results of stress-to-rupture tests may have a greater significance than those of creep tests.

Thus, in the design of piping, it is necessary to know the effects of the three independent variables—stress, time and temperature—on the plastic properties and fracture strength of the materials from which the piping is to be constructed.   Such information is obtained from creep tests.

Creep rates, as determined by creep tests, will not indicate the order of elongation with which a material will fail.   Then it becomes desirable to know the stresses which will result in rupture as a function of time and temperature.   This information is obtained from stress-to-rupture tests and has been used together with creep and short-time tensile-test data by the ASME Boiler and Pressure Vessel Code Committee, the American Petroleum Institute and by similar organizations to establish design stresses for the materials which are used in piping intended for service at elevated temperatures.

Creep and Stress-rupture Tests

One of the major problems in the design of piping systems intended for long service (such as 10, 15 or 25 years) at elevated temperatures is that design values must be based upon extrapolation of relatively short-time laboratory test data.

In addition to the determination of the relation between deformation and time (the creep rate), it is also important to know the total amount of deformation at the elevated temperature which the material can experience before it ruptures. Otherwise, the piping, which is designed on the basis of extrapolated creep data, may fail because the particular material ruptured before attaining the contemplated (permissible design) strain. Thus, the engineer should consider stress-to-rupture data as well as creep data.

The creep and stress-to-rupture tests are very similar. In each test, a tensile specimen is loaded in uniform axial tension at a constant temperature. In the creep test, the changes in length (elongation) with respect to time are recorded; in the stress-to-rupture test the time required to cause failure is of primary interest. In order to cause rupture in practical testing periods (up to 10,000 hr), higher applied stresses are usually used in the stress-to-rupture test than in the creep test. In current testing equipment both test criteria are usually measured. Determination of changes in length (strain rates) are made during the stress-to-rupture test.

In stress-rupture tests, the stresses causing rupture and elongation at rupture may be plotted on log-log paper, against time to rupture, as illustrated in Fig. 37. The short-time tensile-strength value represents the initial point on the curve.

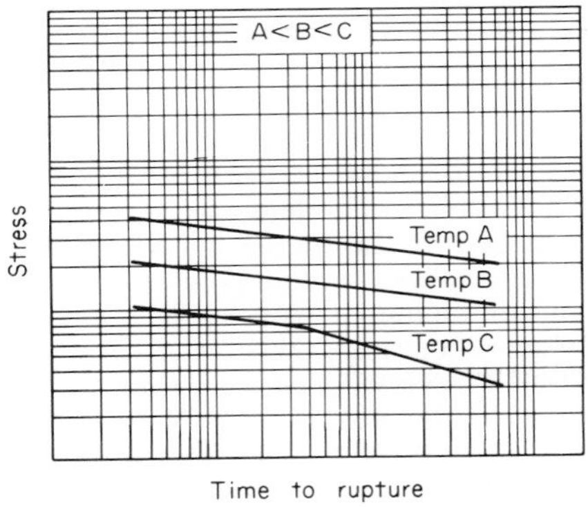

Creep-test Data. The conventional creep test represents a precise measurement of the deformation of a tensile specimen exposed under a constant load at a particular elevated temperature. The tests are performed with very close temperature control and they are usually conducted for periods of from 1,000 to 10,000 or 20,000 hr. The elongation is read at more or less regular time intervals.

Fig. 37. Typical stress-to-rupture curves.

The results of typical creep tests for several different loads (or, instead, for different temperatures) are shown in Fig. 38. The curves may be divided into four sections designated by *OA*, *AB*, *BC* and *CD*.

1. *OA* represents the elongation which occurs as soon as the load is applied. It consists partly of an elastic strain (recoverable) and partly of a plastic strain (not recoverable). Generally, this stage (*OA*) is referred to as a *rapid straining*.

2. *AB*, also called *first-stage creep*, is a period characterized by a decreasing creep rate.

3. *BC*, or *second-stage creep*, is characterized by a relatively constant creep rate which also represents the minimum rate.

4. *CD*, or *third-stage creep*, corresponds to an increasing creep rate which ends in the rupture of the specimen.

The logarithm of the deformation rate, in the second stage of creep (*BC*), frequently expressed in per cent elongation per hour, usually is either a natural or

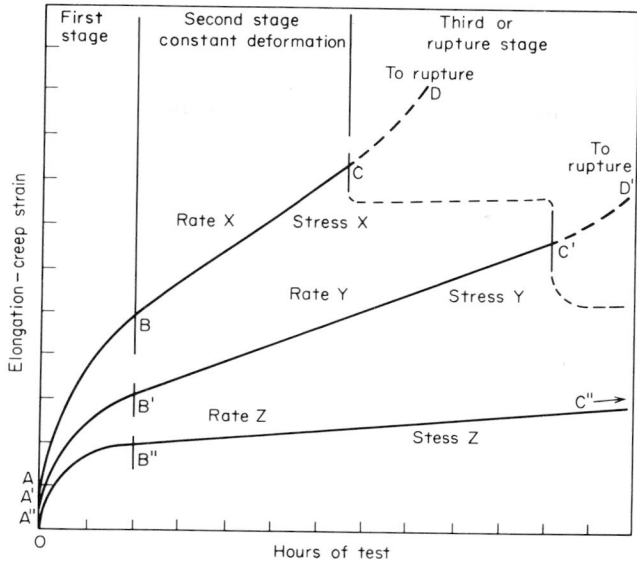

Fɪɢ. 38. Creep time vs. elongation curves at a given temperature.

logarithmic function of the applied stress at a given temperature. The latter relationship is illustrated in Fig. 39. In plotting this curve, it is well to have data for at least three loads (stress levels) at a given temperature, so that it will be possible to interpolate or extrapolate with some degree of assurance. A plot such as this is used to determine the normally accepted design criteria as, for example, the stress Y to produce a second-stage (constant) creep rate of 1 per cent in 10,000 hr (1⅛ year), corresponding to a creep rate of 0.0001 per cent per hour or the stress Z to produce a creep of 1 per cent in 100,000 hr (11¼ years), equivalent to a creep rate of 0.00001 per cent per hour. These respective points are indicated in Fig. 39. The usual custom of considering a creep rate of 0.0001 per cent per hour as equivalent to one of 1 per cent in 10,000 hr or of 0.00001 per cent per hour as equivalent to one of 1 per cent in 100,000 hr is an assumption which may or may not be entirely valid.

The third stage of creep (CD) is characterized by an increase in the rate of elongation. This stage of creep is caused by the reduction in cross-sectional area, by a damaging structural change or by intergranular oxidation. Once the material has entered this stage it is not possible to restore its original properties by further mechanical or thermal treatment. Consequently, creep rates of this stage are never used in the design of piping. Thus, it is important for the designer to know that, at the design stress, the material will not enter the third stage of creep before the expected service life of the piping system has been completed.

By plotting, on logarithmic paper, as in Fig. 40, the applied stress versus the time at which the specimen enters the third stage of creep, the maximum stress at which the specimen will still exhibit a constant rate of creep is obtained for given testing times. Since the change from second- to third-stage creep generally is not abrupt, this curve represents an empirical interpretation of the maximum second-stage creep stress. Data of this type often are not as useful to the design engineer as the information obtained from stress-to-rupture tests.

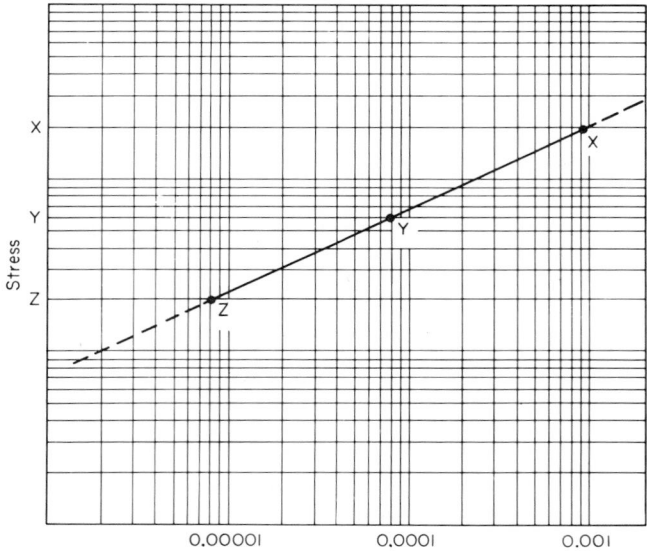

Fig. 39. Stress vs. deformation-rate curve.

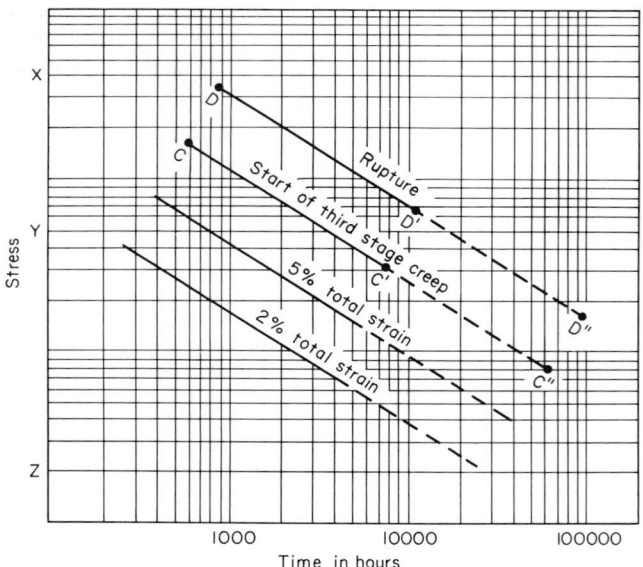

Fig. 40. Stress vs. initial third-stage creep curve.

Stress-to-rupture[1] Test Data.   Stress-to-rupture data are obtained from high-temperature, long-time tensile tests in which the specimen is tested to rupture. Higher stresses (or loads) usually are applied to stress-to-rupture tests than in creep tests in order to obtain rupture in reasonable times of the order of 100 to 5,000 hr. However, many tests have been conducted over much longer time periods and the same time range could be used as in the creep test.

In order to obtain a straight-line relationship, the stress is generally plotted against the time to rupture at a constant temperature on log-log coordinates as illustrated in Fig. 37.   For many materials, a discontinuity or sudden change in slope (curve *C*) often is observed in the stress-to-rupture curves.   This change is sometimes associated with a change in the type of fracture from transgranular at relatively short fracture times to intergranular at longer times.   However, steels

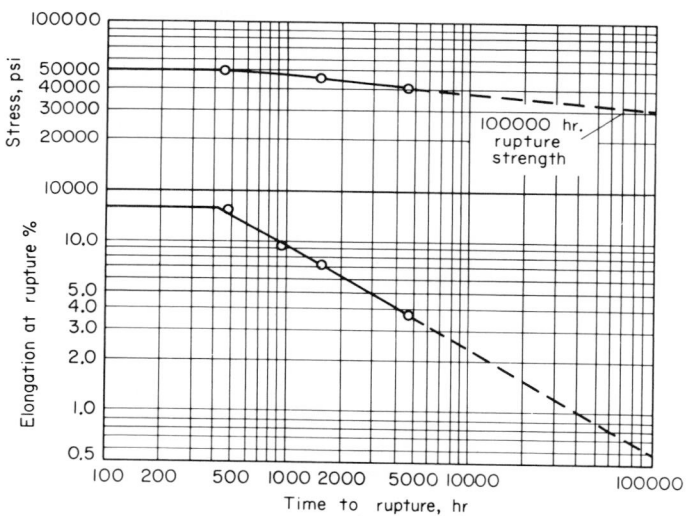

FIG. 41. Rupture tests at 900 F, normalized carbon-molybdenum pipe.   Size: 6 in.; composition: C 0.14, Mn 0.44, Mo 0.46, Si 0.16.   (*After E. L. Robinson.*)

that do not show a discontinuity may also fail intergranularly in the more prolonged tests.   The transition from transgranular to intergranular fracture is not abrupt, and mixed fractures occur at intermediate rupture times.   Since intergranular fractures generally start at the surface of the specimen, they are considerably affected by the stability of the material with respect to the atmosphere which surrounds the specimen.   Other types of phenomena such as sigma phase formation in certain stainless steels, some types of embrittlement, rates of strain hardening, recrystallization, and spheroidization and tempering in quenched or normalized steels may also cause a break in the rupture curve.

Along with stress-to-rupture values, the elongation at rupture may be plotted against time to rupture, as illustrated in Fig. 41.   The short-time tensile-strength value represents the initial point on the curve.

Creep and stress-to-rupture data, expressed as stress to produce a specific creep rate or rupture time, are sometimes replotted on rectangular coordinate or semi-logarithmic coordinate paper as stress, or log stress, for a constant strain rate versus

[1] The term creep-to-rupture, although used less, is technically more accurate.   Some authors use this term to denote a continuation of the creep test to rupture.

test temperature, as shown in Fig. 42. This procedure allows extrapolation or interpolation between the test data for various temperatures. Generally, however, the log-log curves, which are usually straight lines, for constant test temperatures (Fig. 37) are preferred.

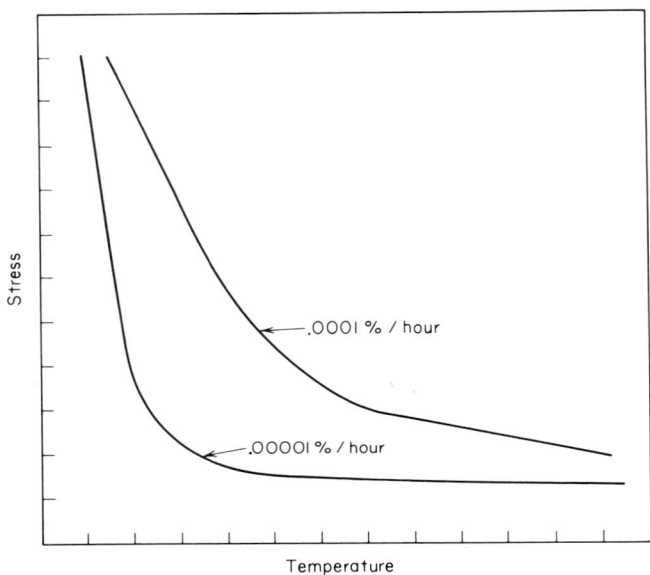

Fig. 42. Creep stress vs. temperature curves.

Factors Which Affect Creep Strength

The creep and stress-to-rupture properties are considerably influenced by the metallurgical characteristics of the material and the testing and service environments to which it is exposed.

The most important metallurgical characteristics of the material are *chemical composition*, *structure* and *grain size* which are primarily controlled by the prior processing and heat treatment, and *structural stability* at testing or service temperature. Close control of many of the metallurgical factors is not always possible, since some of them are dependent upon the chemical composition, the melting practice, the prior heat treatment and fabrication process.

Effects of Composition. Composition is the most obvious and important variable which affects the high-temperature strength of piping materials. Improvement in creep and stress-to-rupture properties by alloying additions generally may be related either to the amount and size of fine particles which are distributed within the structure of the metal or to a general strengthening effect of the over-all (matrix) structure without the formation of such particles. However, the mere addition of alloying elements does not insure higher creep and stress-to-rupture properties. Some elements may reduce these properties. Other elements which may act in a highly beneficial manner in certain quantities may be detrimental in other quantities. The correct usage of alloying elements by the alloy producer is highly important.

Typical creep and stress-to-rupture curves for a carbon steel, two low-alloy steels and an 18-8 (Cr-Ni) stainless steel are shown in Figs. 43 and 44, respectively. The

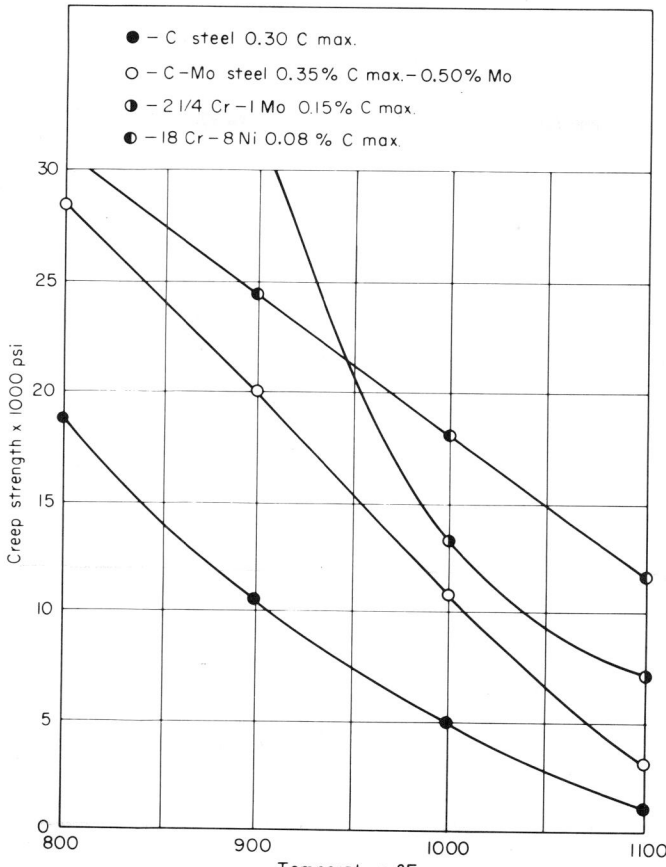

FIG. 43. Effect of alloy additions on creep strength (stress for creep rate of 1 per cent per 10,000 hr).

curves show that the elevated-temperature strength of steel may be effectively increased by addition of molybdenum or by highly alloying as with the stainless alloy.

Effects of Structure and Grain Size.   Definite values of creep and stress-to-rupture properties cannot be predicted accurately from composition alone.   This is because these properties are also structure sensitive.

At temperatures above the so-called equicohesive temperature[1] which varies for

---

[1] The equicohesive temperature is the temperature at which the grain strength (strength of the material within the grains) is equal to the strength of the material at the grain boundaries.   Above the equicohesive temperature, the grain strength is higher and below the equicohesive temperature, the grain strength is lower than the grain boundary strength. Thus, above the equicohesive temperature, failure will be intergranular and below the equicohesive temperature, failure will be transgranular.   The equicohesive temperature is affected differently by various factors.   For example, increases in the rate of straining will raise the equicohesive temperature.   Thus, this term, although being of considerable convenience, describes only a relative temperature depending upon the specimen material and testing conditions.

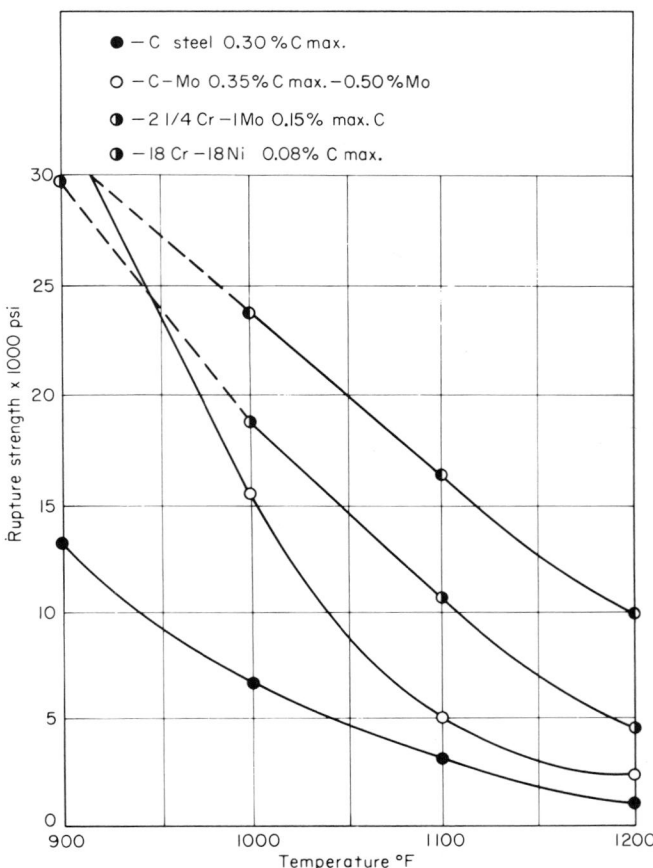

FIG. 44. Effect of alloy additions on the rupture strength (rupture time of 10,000 hr).

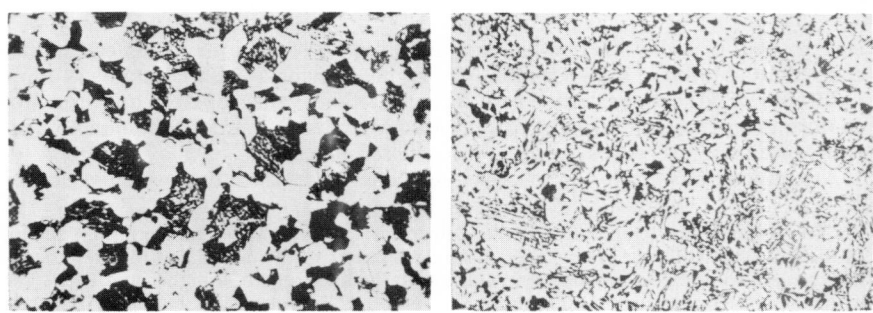

FIG. 45. The microstructures represent two pipe sections from the same original steel ingot meeting the requirements of ASTM A335 Gr. P-11 (1¼ Cr–½ Mo).

different steels (usually between 800 and 1000 F for low-alloy steels), coarse-grained steels generally exhibit better creep and stress-to-rupture properties than do fine-grained steels. Below the equicohesive temperature, the trend reverses and steels with a fine grain size usually exhibit superior creep and stress-to-rupture properties.

Effects of Heat Treatment. Heat treatments are important primarily because of their effects on the structure of the metal. For example, heat-treating temperatures and techniques which produce a larger grain size generally will improve the creep and stress-to-rupture properties of steels at the more elevated temperatures. Cooling rates from the temperatures of the heat treatment are of similar importance. Other factors, such as distribution of the structural constituents, as carbides, also are important as they are affected by heat-treating temperatures and cooling rates. Thus, a normalized and tempered steel is often superior to the same steel in the fully annealed condition. When heat-treated steels are employed at elevated temperatures, it is customary to use a tempering temperature at least 150 F above the expected service temperature.

Some commercial heat-treating furnaces do not provide uniform temperature control and metallurgical structures of considerable difference may result. The microstructures shown in Fig. 45 represent two pipe sections from the same original steel ingot. They were normalized and tempered in the same heat-treating range. The differences in microstructure are considerable, and will result in different resistance to creep. Even with such differences, failures in heavy piping materials have been rare because of the relatively conservative safety factors applied on design. However, engineers are becoming increasingly concerned with the effects of heat-treating temperatures and the control over uniform heat treatments.

Effects of Testing Variables. Among the testing variables, it is most important that the testing temperature be kept constant. Short durations at higher temperatures may considerably reduce the creep and stress-to-rupture properties.

Stress-to-rupture and creep tests lasting 5,000 or 10,000 hr or longer are more significant than shorter tests because the longer tests minimize the errors introduced by the extrapolations necessary for practical use of the data. Thus, creep or stress-to-rupture data can be considerably in error when extrapolated to 100,000 hr from relatively short-time tests, for example, tests of less than 5,000 hr.

The results of tests obtained in different laboratories are affected by the mechanical operation of the test equipment and the design of the test specimen. The precision of temperature control, the precision of stress application and the precision of the strain-measuring equipment have an important effect on the final test results. A less understood, but probably important factor, also may be the variation in the dimensions of the creep and rupture specimens used in different laboratories. Reference should be made to ASTM Specification E139[1] for recommended testing procedures.

After the engineer and metallurgist are assured that the data on hand meet the qualifications necessary for the job—that is, the right material at the right temperature for a long enough duration of time to give a certain degree of assurance to extrapolation—a survey should then be made of the intended application for possible sources of trouble. Some of the important variables in service are temperature cycling, stress cycling, corrosive environment and thermal stresses over and above the major applied stress.

It is well to consider that most creep and stress-to-rupture tests are performed at a constant temperature, and that thermal cycling temperatures may materially

---

[1] ASTM Specification E139, "Recommended Practice for Conducting Creep and Time-for-Rupture Tension Tests of Materials" (Tentative).

reduce the strength of the metal from fatigue when in service. Because of these additional thermal stresses introduced, the life of the piping system may be shortened. Possibly, some test data may, in the future, serve as a guide in establishing safety factors for applications in which materials are exposed to specific operating conditions, although thermal stresses of the same order of magnitude and distribution as are encountered in service are difficult, if not impossible, to reproduce in the laboratory.

Application of Creep Data to Piping Design

For most piping applications, the allowable stress value to use for a given material at a given temperature is given by the ASME Boiler and Pressure Vessel Code under which the piping is to be built. In the United States, the values for design stresses for steels are established by the Subcommittee on Stress Allowances for Ferrous Materials of the ASME Boiler and Pressure Vessel Code Committee. This subcommittee collects all available data and establishes tables of maximum allowable design stress values.

In the determination of stress values, the ASME Subcommittee on Stress Allowances for Ferrous Materials is guided by successful experience in service with the particular steels in question insofar as evidence of satisfactory performance is available, such evidence being preferred to all forms of test data. However, in the evaluation of new materials, it is always necessary to be guided, to a certain extent, by the comparison of test information with similar data on established materials. In carrying out this process on examination and evaluation, the ASME Subcommittee found useful the considerations as given in the following paragraphs.

At temperatures below the creep range, allowable stress values were established at the lowest value of stress obtained from using 25 per cent of the specified minimum ultimate strength at room temperature or 25 per cent of the minimum expected ultimate strength at temperature, or $62\frac{1}{2}$ per cent of the minimum expected yield strength for 0.2 per cent offset, at temperature. For bolting material, the stress values were based on 20 per cent of the minimum tensile strength, or 25 per cent of the yield strength for 0.2 per cent offset, whichever is lower. (It is recognized that bolts are always expected to function at stresses above the design value as distinguished from other parts.)

No credit was allowed for any improvement in tensile properties by special heat treatment.

At higher temperatures, where creep governs, the stress values were based on 100 per cent of the stress to produce a creep rate of 0.01 per cent for 1,000 hr, the values so chosen being based on a conservative average of many reported tests as evaluated by the Subcommittee, greater weight being given to longer-time tests in evaluating data. In addition to the above-stated creep-strength requirement, stress values were also limited to 100 per cent[1] of the stress to produce rupture at the end of 100,000 hr, the values so chosen being based on a conservative average as evaluated by the Subcommittee. However, in most cases, the creep strength is far below the rupture strength. Also, in a few cases, the Subcommittee has provided stress values without any rupture test data on the specific composition, such approval being based on tests of materials of similar composition.

In the transition range of temperatures, the stress allowances were limited to

[1] This 100 per cent value pertains to the Unfired Pressure Vessel Code (Table UCS-23 and UHA-23). In the Power Boiler Code (Table PG 23.1) this stress limitation is 60 per cent of the average or 80 per cent of the minimum stress to produce rupture in 100,000 hr as reported by test data.

values obtained from a smooth curve joining the values for the low- and high-temperature ranges, the curve lying on or below the curve of 62½ per cent of the minimum expected yield strength at temperature.

In the choice of stress values in the range where a percentage of the tensile strength or yield strength governs, the limitations indicated above have been waived in certain cases, identified by footnote, as it was felt that higher stress values might be justified when deformation was not in itself objectionable, provided all other requirements were met.

In the design of equipment not covered by codes, the design stress values may be decided upon by the manufacturer and purchaser of the piping, and should be based on the best available data, plus a knowledge of the expected life of the equipment as well as the operating conditions and the possible hazard to personnel. Rules generally followed are:

1. Up to 750 or 850 F, 25 per cent of the short-time tensile strength and not exceeding 62½ per cent yield strength
2. Above 900 F, 100 per cent of the stress to produce a second-stage creep rate of 0.01 per cent in 1,000 hr, or 80 per cent of the stress to produce rupture in 100,000 hr, whichever is lower.

Carbon, Molybdenum and Chromium Steels. Over the past 20 years, extensive high-temperature test data involving short-time tensile tests and creep and stress-rupture tests have become available.

Among the most detailed sources are the data collected by the ASTM-ASME Joint Committee on Effect of Temperature on the Properties of Metals. Important reports issued by this Committee and published by the American Society for Testing and Materials as Special Technical Bulletins on Elevated Temperature Properties are:

No. 180   Carbon Steels (1955)
No. 199   Wrought Medium-carbon Alloy Steels (1957)
No. 151   Chromium-Molybdenum Steels (1953)
No. 226   Weld-deposited Metal and Weldments (1958)

The available high-temperature test data are summarized in graphical form, with the original data sheets included in appendixes. The graphical presentation includes curves for short-time tensile and yield strengths; elongation and reduction of area; creep rates of 0.0001 and 0.00001 per cent per hour and stress-to-rupture curves in 100, 1,000, 10,000 and 100,000 hr.

Since the method of processing of the steel and its chemistry are both important, data are included to show where applicable that the material was produced from open-hearth, electric-furnace, or other melting practice, and that the final shape was a casting, forging, or wrought pipe, or that the material had been normalized, annealed, etc. Some of these variables are interrelated so that their individual effects have been difficult to determine.

Whenever possible, the tensile-strength data and the 100- and 1,000-hr rupture-strength data are used to plot Larson-Miller[1] master curves. This method of plotting rupture data, sometimes generally referred to as the Larson-Miller parameter method, has gained wide acceptance in recent years. Time and temperature are related by the following equation:

$$P = T(20 + \log t) \tag{3}$$

[1] F. R. Larson and James Miller, "A Time-Temperature Relationship for Rupture and Creep Stresses," *Trans. ASME*, vol. 74, pp. 765–775, July, 1952.

where T = absolute temperature, deg Rankine
  t = time to rupture, hr
  P = Larson-Miller parameter
and the log is to base 10.

Thus, if rupture stress is plotted as a function of the parameter $T(20 + \log t)$, all of the points should fall on a single curve regardless of the test temperature. This method should not be used to extrapolate data to temperatures higher than those at which tests were actually performed, except to a limited extent in a temperature range where structural changes and consequent changes in properties are

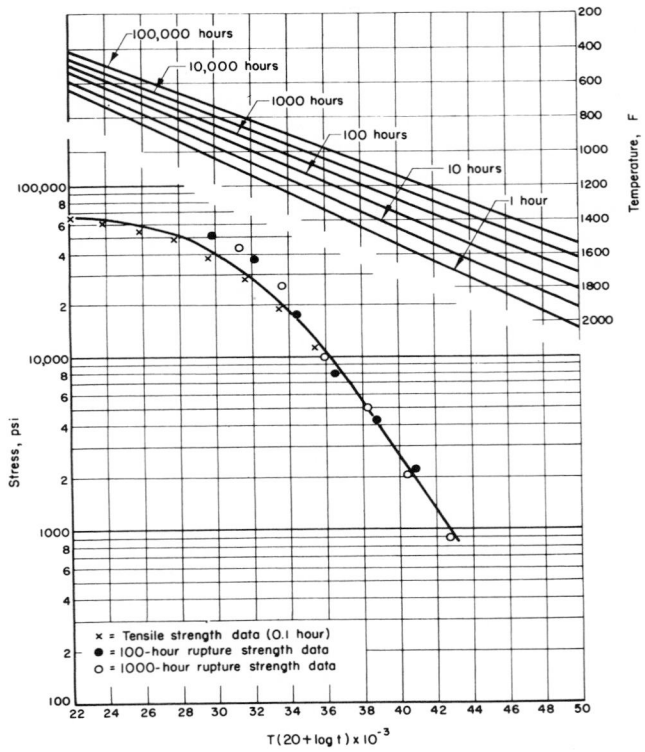

Fig. 46. Larson-Miller graph for annealed ½ Mo steel.

at a minimum. For example, in annealed ½ per cent Mo steel (Fig. 46) the Larson-Miller graph rupture could be expected at a stress of 14,000 psi at 1000 F in 10,000 hr and at 1130 F in 100 hr.

High-temperature tensile strengths, and Larson-Miller parameters are shown for some of the common piping materials in Figs. 47 to 51.

Carbon steels generally are not used at temperatures over 775 F.

The ½ per cent Mo steels (Fig. 46) with approximately 0.15 per cent C are no longer widely used in piping, except on shipboard, where they are used at temperatures as high as 875 F. However, as tubing, the ½ per cent Mo steel grades are used in superheaters to about 850 F.

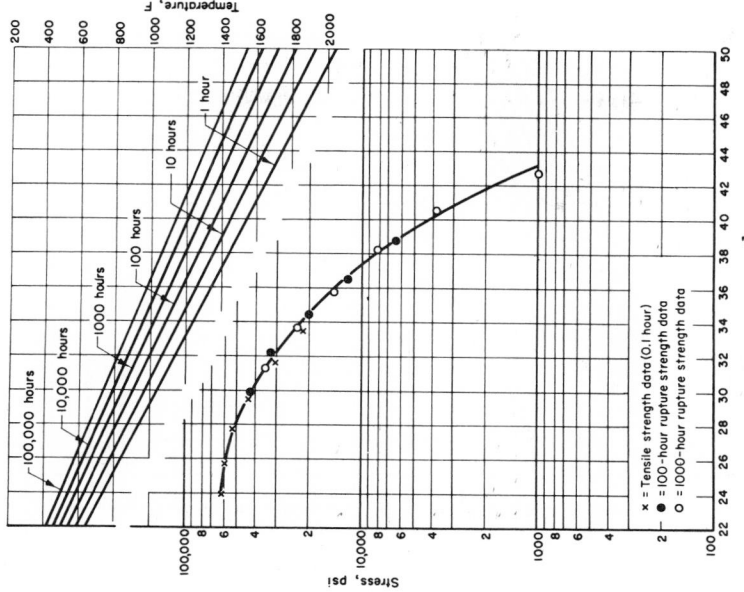

FIG. 48. Larson-Miller graph for annealed 2¼ Cr–1 Mo steel.

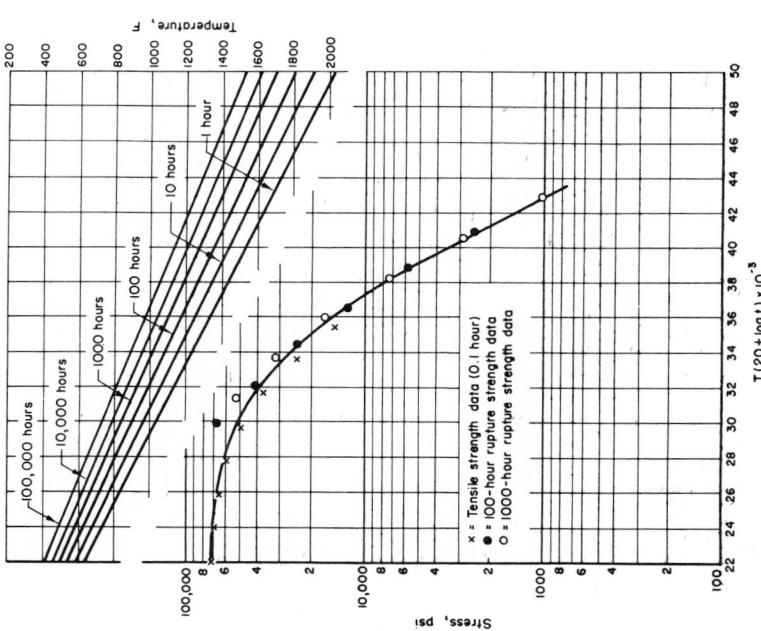

FIG. 47. Larson-Miller graph for annealed 1¼ Cr–½ Mo steel.

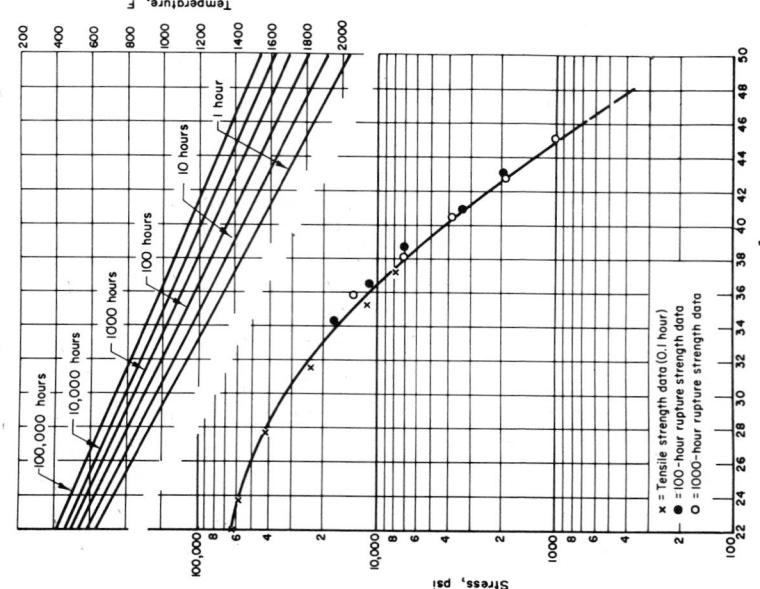

FIG. 50. Larson-Miller graph for annealed 7 Cr-½ Mo steel.

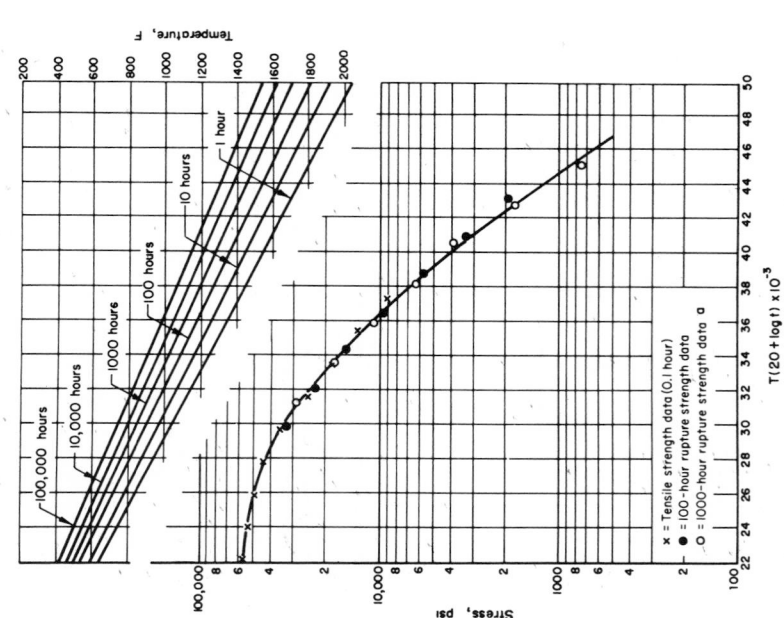

FIG. 49. Larson-Miller graph for annealed 5 Cr-½ Mo steel.

The 1¼ Cr–½ Mo steels (Fig. 47) are used in steam plant piping, boiler tubes and cracking-still tubes for service up to 950 or 1000 F.

The 2¼ Cr–1 Mo grades (Fig. 48) are used extensively in steam power plants to temperatures of about 1060 F.   They exhibit slightly better oxidation resistance than the 1¼ Cr–½ Mo grades.

The 5 Cr–½ Mo (Fig. 49), 7 Cr–½ Mo (Fig. 50), and 9 Cr–1 Mo grades (Fig. 51), are used primarily in oil refineries and chemical plants because of their

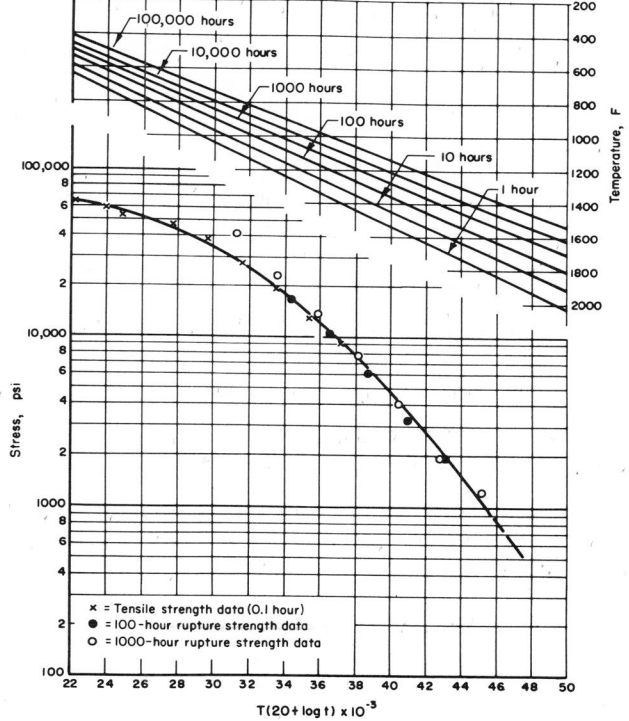

Fig. 51. Larson-Miller graph for annealed 9 Cr–1 Mo steel.

better oxidation and corrosion resistance at temperatures up to 1500 F.   The elevated-temperature strength properties of the 5 Cr–½ Mo and 9 Cr–1 Mo grades are somewhat lower than those of the low-chromium-alloy steels.   Applications include piping, heat exchangers and superheater tubing.

Some use is made, also, of Cr–Mo–V alloy steels.   Although used occasionally as pipe and tubing, these steels are more widely used for bolting, turbine blades and similar applications.   In Europe they have been accepted more widely as piping material.   These steels, where normalized and tempered or quenched and tempered, exhibit substantially stronger yield and creep strength properties than do the chromium-molybdenum grades, as is shown in Fig. 52.

Chromium Stainless Steels.   The high-temperature properties of the martensitic and ferritic chromium stainless steels have been summarized in *Special Technical Publication* 228 published in 1958 by the American Society for Testing and Materials.

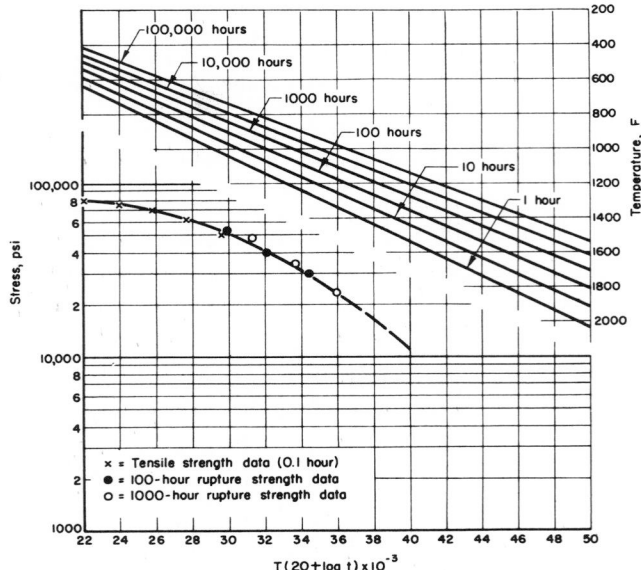

FIG. 52. Larson-Miller graph for normalized 1 Cr–1 Mo + V steel.

Although not used as widely in pipe and tube grades as the Cr–Mo alloy and austenitic stainless-steel materials, some use is made of these materials in special applications.

Among the martensitic stainless steels, the Type 410 alloy is most readily available. It is used at intermediate temperatures up to about 950 F in applications requiring strength. High-temperature properties are given in Fig. 53.

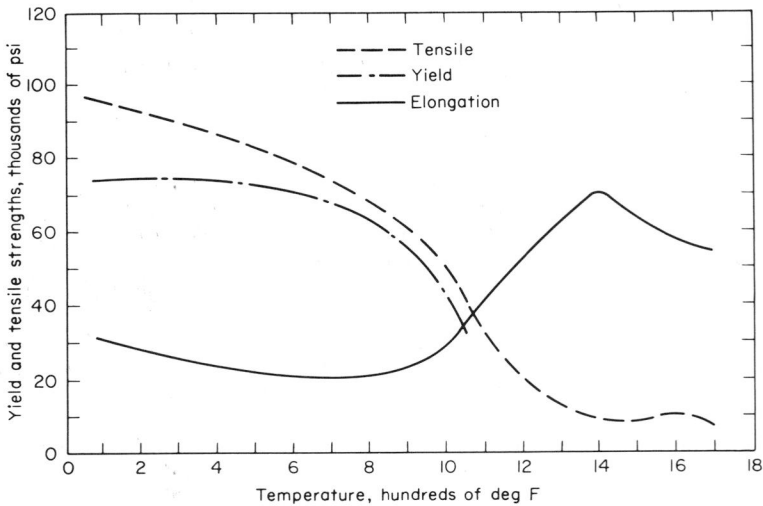

FIG. 53. High-temperature tensile, yield, and elongation properties of 410 stainless steel.

Among the ferritic stainless steels, the Type 430 grade is used in heat exchangers, condensers and special chemical applications. High-temperature properties are summarized in Fig. 54.

Chromium-Nickel Stainless Steels. The high-temperature properties of the austenitic stainless steels have been summarized in *Special Technical Publication* 100 published by the American Society for Testing and Materials in 1950.

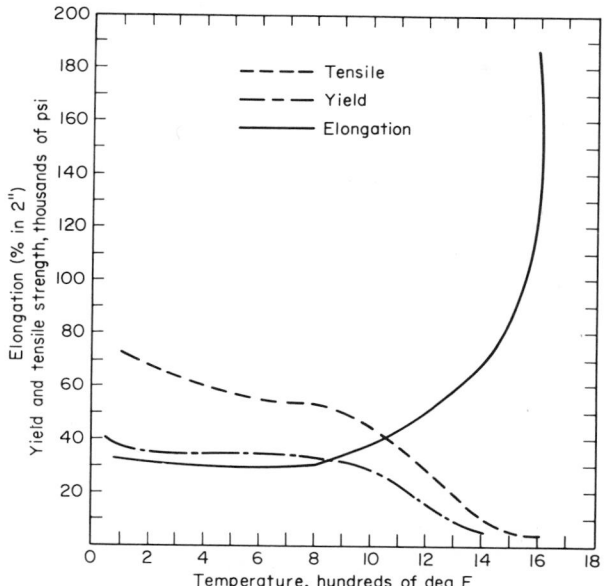

FIG. 54. High-temperature tensile, yield, and elongation properties of 430 stainless steel.

The short-time tensile properties at elevated temperatures are summarized in Fig. 55 for Types 304, 316, 347, 310, 309 and Kromarc-58.[1] At 1200 F, the short-time tensile strength is highest for Kromarc-58 and intermediate for Types 310 and 316. Also shown are elongation at elevated temperatures in Fig. 56. The stress-rupture properties of Kromarc-58 are given in Fig. 57.

Creep strength data for Types 304, 309, 310, 316 and 347 stainless steel are given in Fig. 58. Type 316 is generally superior to other commercial stainless-steel grades. It has been used for piping in steam power plants at temperatures between 1050 and 1200 F.

However, the other stainless-steel grades have also been used at elevated temperatures as either cost considerations or other high-temperature properties may indicate advantages for specific grades. Type 304 itself exhibits good resistance

[1] Typical composition:

Carbon, max.	0.04%	Chromium	14.00 –16.00%
Manganese	8.00–11.00%	Molybdenum	1.75 – 2.50%
Phosphorus, max.	0.020%	Nitrogen	0.19 – 0.27%
Sulfur, max.	0.020%	Vanadium	0.15 – 0.35%
Silicon, max.	0.25%	Zirconium	0.003– 0.030%
Nickel	21.00–24.00% with iron the remainder	Boron	0.003– 0.020%

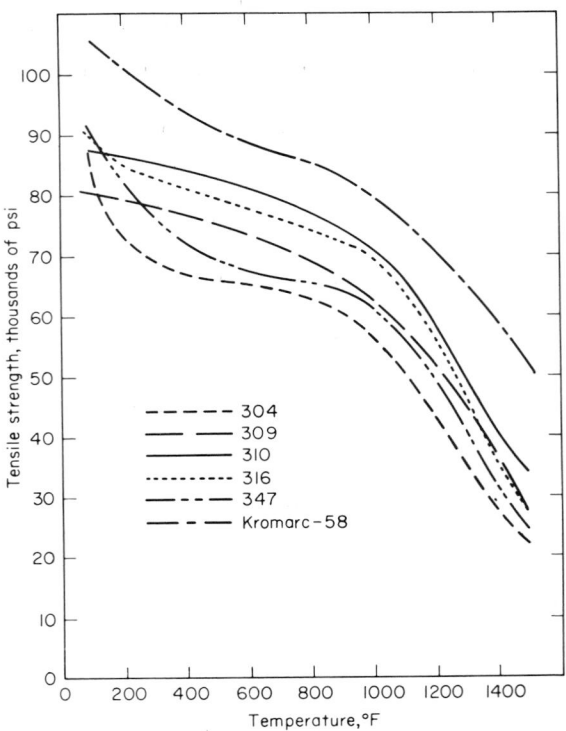

FIG. 55. High-temperature tensile strength of various stainless steels.

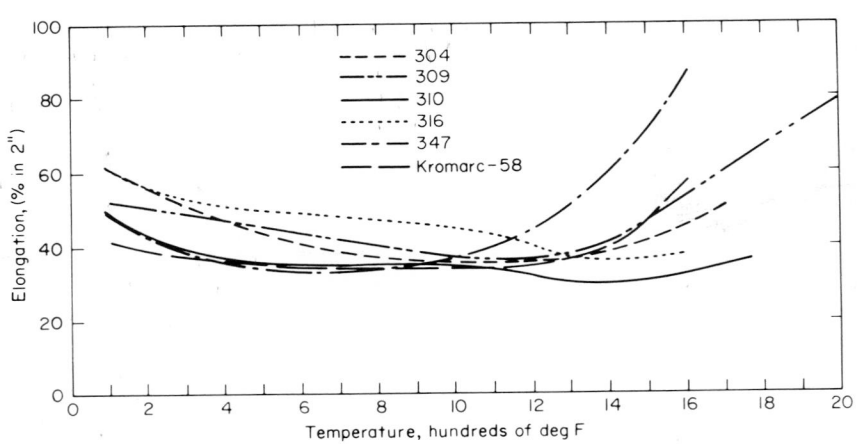

FIG. 56. High-temperature elongation properties of various stainless steels.

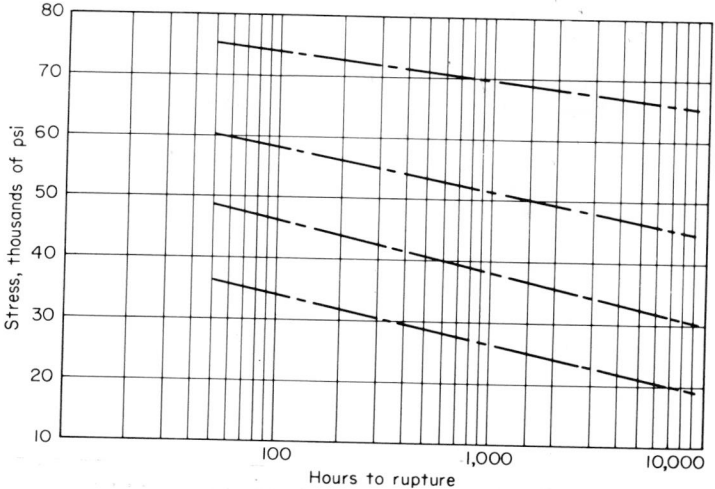

FIG. 57. Stress-rupture diagram of Kromarc-58.

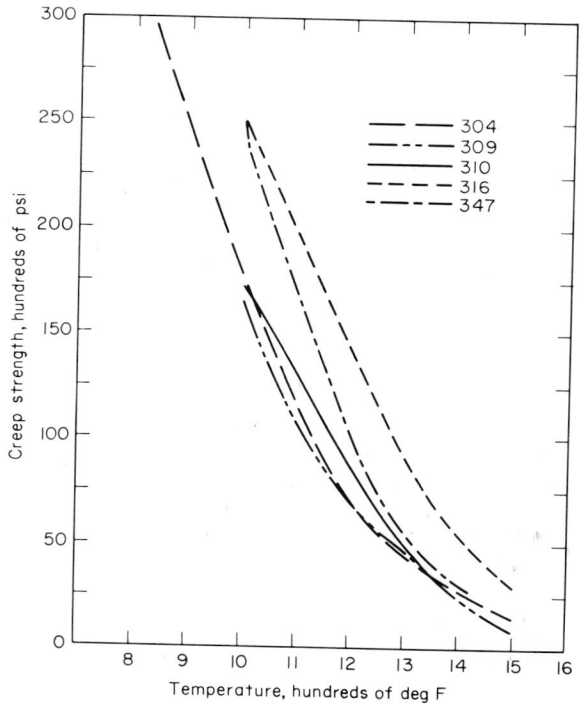

FIG. 58. Creep strength of various stainless steels.

to atmospheric corrosion and oxidation. Types 309 and 310 exhibit still greater resistance to oxidation because of their higher chromium and nickel contents. Type 310 is particularly preferred where intermittent heating and cooling are encountered since it forms a more adherent scale than Type 309.

For high-temperature service, tube materials frequently receive special solution heat treatments to provide a coarse grain size resulting in improved creep strength properties. Stainless steels so heat-treated are designated as H grades, such as TP 321H.[1]

If cold worked, the minimum solution treating temperature for grades TP321H, TP347H, and TP348H is 2000 F. If the H grade is hot rolled, the minimum solution treating temperature for these grades is 1925 F. For grades TP304H and TP316H, the minimum temperature is 1800 regardless of whether the tube is cold or hot rolled.

A number of special high-alloy austenitic stainless-steel compositions have also been developed for high-temperature service. They generally contain higher nickel compositions than those of the ordinary austenitic stainless steels. Some of them have become available as piping materials. These include Incoloy, Incoloy T, and Kromarc-58.

Up to 800 F, the short-time tensile properties of these alloys are similar to those of the standard stainless steels. Over 1000 to 1200 F, the short-time tensile, and long-time creep strength properties become higher.

Precipitation Hardening Stainless Steels. Some of the precipitation hardening stainless steels are also used at intermediate and high temperatures. In these steels, the nickel content is normally lower than in the standard Type 300 series, which contain at least 8 per cent nickel. Hardening is usually accomplished by means of alloying with aluminum, copper, molybdenum or columbium. To obtain maximum ductility for cold forming, bending, machining or other operations, these steels are normally solution annealed above 1900 F.

After fabrication and welding, some of the materials require another solution anneal. Subsequently, on some alloys, an intermediate austenitizing heat treatment between 1400 and 1750 F may be required, followed by rapid cooling, sometimes to subfreezing temperatures. Precipitation hardening is accomplished subsequently at temperatures of 850 to 1050 F.

Care must be exercised with large piping assemblies or rapid and nonuniform cooling from the solution annealing and austenitizing temperatures may result in severe distortion.

Short-time tensile and creep-rupture properties for 17-4PH are illustrated in Fig. 59.

Superalloys. Numerous so-called superalloys have been developed for high-temperature application. However, so far, only a few of them have been produced as pipe in tube shapes, and then only in relatively small quantities. Their applications rarely involve actual piping systems. Instead, in tubular form, these materials are used as exhaust tubing in jet engines, fuel tubing in high-temperature furnaces and combustion chambers, missiles, and in similar applications.

Cast Gray Iron. The short-time tensile strength of gray iron shows no significant decrease until approximately 800 F (Fig. 60).

Ordinary cast irons are limited for elevated temperature applications due to breakdown of carbides and growth of the component. Growth refers to the permanent increase in volume which occurs under certain conditions of heating

---

[1] ASTM Specification A213-65, "Tentative Specification for Seamless Ferritic and Austenitic Alloy Steel Boiler, Superheater and Heat-Exchanger Tubes."
   ASTM Specification A249-65, "Tentative Specification for Welded Austenitic Steel Boiler, Superheater, Heat-Exchanger and Condenser Tubes."

and cooling, and is considered to be independent of stress. It is affected to a significant degree by the presence of superheated steam and certain corrosive fluids. In severe cases of growth, the volume may increase as much as 50 per cent with an attendant loss of strength and development of brittleness which make it worthless as an engineering material under the service conditions involved.

Below a temperature of 700 to 800 F, at which graphitization is of importance, corrosion appears to be the important factor. In the case of coarse-grained cast iron, severe growth and disintegration have been observed with superheated steam at temperatures below 600 F. Apparently, the graphite flakes in combination with porosity allow penetration of oxidizing gases. The resulting products of oxidation

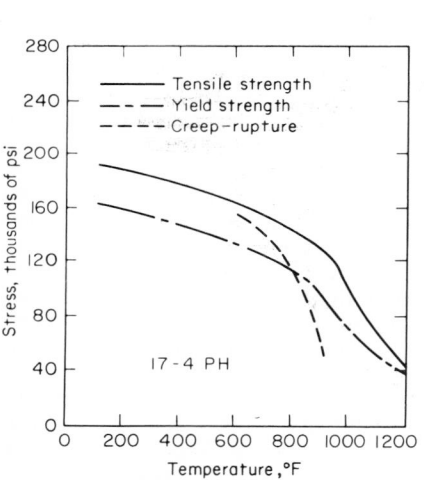

Fig. 59. Short-time tensile and creep-rupture properties for 17-4PH.

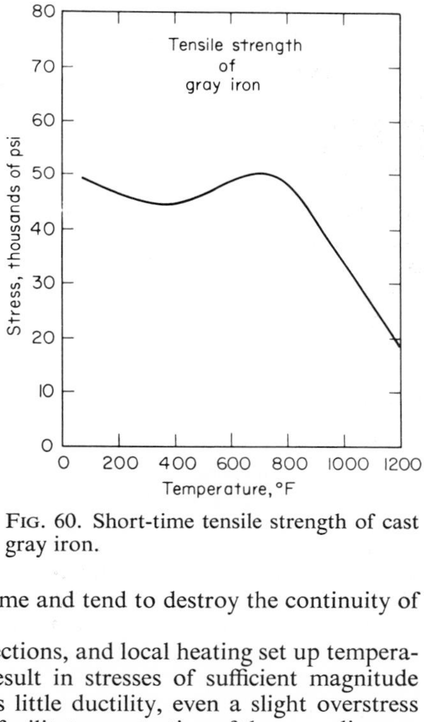

Fig. 60. Short-time tensile strength of cast gray iron.

of the silicon and iron occupy greater volume and tend to destroy the continuity of the structure.

Rapid heating or cooling, nonuniform sections, and local heating set up temperature gradients or differences which may result in stresses of sufficient magnitude to cause cracks. Since gray cast iron has little ductility, even a slight overstress may result in cracks. These cracks further facilitate penetration of the corroding gas or liquid. Pressure from occluded gases may be a further contributing factor in the formation of such cracks.

Growth due to graphitization in the temperature range 900 to 1300 F and growth due to allotropic changes (change of the pearlite-ferrite mixture to austenite), which occur at about 1340 F, have little practical significance in piping practice. The ASME Boiler Code limits the use of cast iron made to ASTM A-278 for pressure-containing parts up to 250 psi to temperatures below 450 F. The ASA Code for Pressure Piping likewise limits the use of cast-iron pipe, valves, and fittings to temperatures not in excess of 450 F. A further limitation that cast-iron pipe shall not be used for oil at a temperature above 300 F is made in the latter code in both the "Power Piping" and "Oil Piping" sections.

Specific limitations for cast-iron pipe and fittings as to pressure, pipe sizes, and allowance for water hammer are contained in the various sections of the Code

for Pressure Piping. These restrictions in conjunction with the temperature limitations are designed to prevent the use of cast iron under conditions which might constitute a hazard to life or property by reason of its low ductility and tendency to growth.

The resistance of gray cast iron to growth is improved by care in design to avoid nonuniform sections, by securing a dense grain, by reducing the total carbon content so that thinner graphite flakes will result, and by the addition of small amounts of nickel, chromium and molybdenum, or other alloys. It is possible to double the room-temperature tensile strength by these modifications, and even to secure a certain degree of ductility. However, cast iron is not normally considered as a high-temperature engineering material. High-alloy cast irons containing 14 per cent nickel with additions of copper, chromium, or silicon have been developed which resist growth and oxidation at temperatures up to 1500 F. These alloy cast irons are termed "austenitic" cast irons since they retain a portion of the carbon in solid solution. The thermal expansion of these austenitic cast irons is approximately 50 per cent greater than that of ordinary cast irons.

One high-temperature application of gray iron in piping involves its use as valve seat inserts at temperatures as high as 1300 F.

Malleable Iron. Malleable cast iron is produced by prolonged heating at temperatures up to 1750 F and subsequent slow cooling of cast iron, which in the as-cast condition, has a white fracture and is extremely hard and brittle. The heat treatment converts the combined carbon which exists as iron carbide into nodules of graphite surrounded by ferrite. Because the graphite in fully malleablized cast iron exists as rounded nodules of free carbon rather than heavy flakes as in gray cast iron, it does not afford paths for the penetration of oxidizing gases. The lower silicon content of malleable cast iron reduces the effect of oxidation of that element while its greater ductility minimizes the effect of temperature gradients.

By different heat treatments, it is possible to produce malleable iron castings with widely different mechanical properties. These different properties are, in general, obtained by arresting the graphitizing process at different stages. The matrix surrounding the globules of graphite or temper carbon will then retain a greater or less amount of carbon in the combined state. These matrices may be ferritic, as in fully graphitized iron, pearlite, or sorbitic in character, or they may be in the spheroidized condition. In this latter condition, tensile strengths in excess of 100,000 psi with an elongation of some 7 per cent in 2 in. have been obtained.

The short-time tensile properties of ferritic and pearlite malleable iron show no significant change up to 700 F (Fig. 61). Stress-rupture characteristics are also presented graphically in Fig. 62.

The Code for Pressure Piping permits the use of malleable cast iron at temperatures up to 500 F, provided the pressure is limited to 300 psi. The ASME Boiler and Pressure Vessel Code, Section I, paragraph P12(c) permits the use of malleable cast iron made to ASTM A-47 for boiler and superheater connections such as pipe fittings, water columns, and valves, and their bonnets, for pressures not to exceed 350 psi, provided the steam temperature does not exceed 450 F.

Malleable cast iron, if properly malleablized and if not embrittled by improper galvanizing, possesses good resistance to shock. Its ductility as indicated by an elongation in 2 in. of 10 to 15 per cent, and a reduction of area of 15 to 20 per cent makes it a useful engineering material for small-pipe fittings and other light sectioned parts.

High-nickel Alloys. The temperature limits for the nickel and the ordinary nickel alloys are approximately 800 to 1000 F. Annealed nickel at 800 F at a stress of 2,000 psi would creep at a rate of approximately 3 per cent in 100,000 hr. No appreciable creep occurs at a stress of 10,000 psi at 550 F.

Hot-rolled Monel supports a stress of 26,000 psi at 600 F, and 19,000 psi at 800 F, with creep at the rate of 1 per cent in 100,000 hr. However, at 1000 F, the stress producing the same rate of creep is reduced to only 1,650 psi.

Nickel, Monel (70 per cent nickel, 30 per cent copper), and various modifications of these materials are used extensively in turbine blading, valve trim, and miscellaneous power-plant accessories handling steam.

Monel metal gaskets have been used successfully in the flanged joints of a line supplying steam to a 10,000-kw 1000 F turbine.

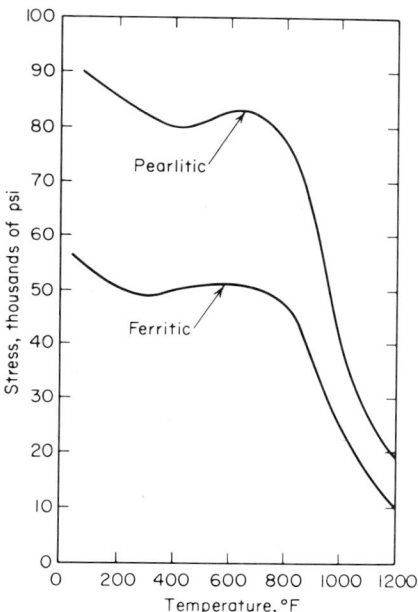 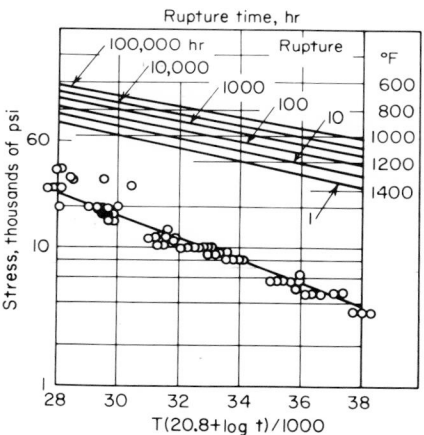

FIG. 61. Short-time tensile properties of ferritic and pearlite malleable iron.

FIG. 62. Larson-Miller characteristics of malleable iron.

The presence of even small quantities of sulfur in a reducing or neutral furnace atmosphere will result in surface embrittlement of nickel and Monel at temperatures of 700 to 1200 F.

Numerous copper-nickel alloys (60 to 70 per cent copper, 30 to 40 per cent nickel) with varying amounts of iron, zinc, tin or aluminum additions have been developed for valve trim, turbine nozzle blocks, tubing, and the like. Some of these alloys provide good performance up to 850 F on superheated-steam service. However, some have failed by intergranular cracking and pitting at less than 750 F.

By the addition of Cr, Co, Mo, Ti, Al or Cb, the high-temperature strength and creep resistance of the nickel-base materials has been increased substantially. The short-time tensile properties of Inconel X, Inconel 700, Hastelloy X, and Hastelloy B are summarized in Fig. 63. However, low-ductility values at elevated temperatures require special care in forming these materials (Fig. 64). Stress-rupture values of Hastelloy X, Hastelloy B, and Inconel X are shown in Fig. 65.

Copper and Copper Alloys. A detailed report on the "Elevated-temperature Properties of Coppers and Copper Base Alloys" has been issued by the American Society for Testing and Materials as *Special Technical Publication* 181 (1956).

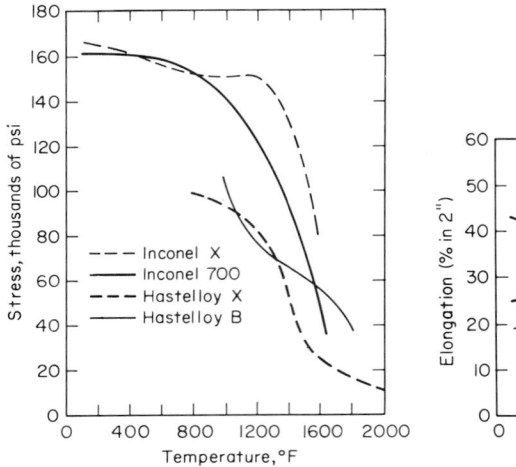

FIG. 63. Short-time tensile properties of four nickel alloys.

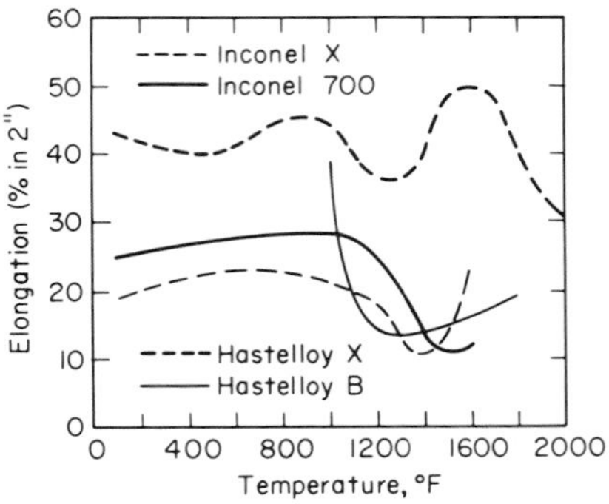

FIG. 64. High-temperature ductility properties of four nickel alloys.

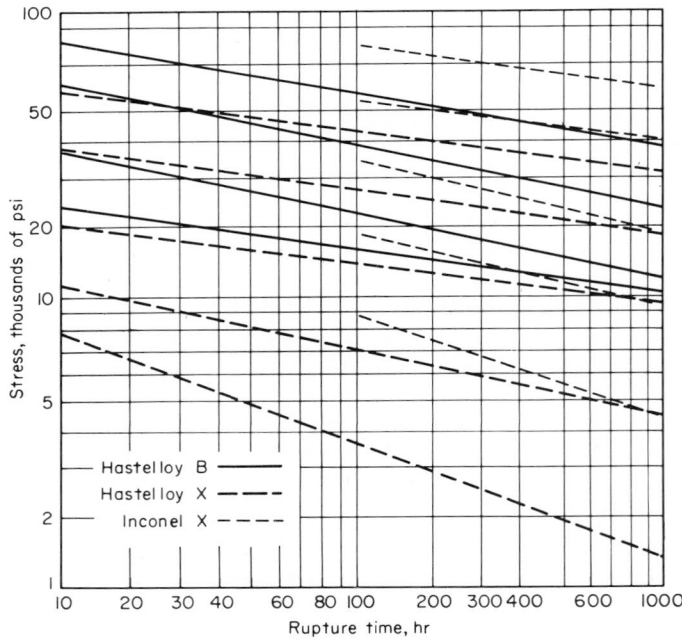

FIG. 65. Stress-rupture values of three nickel alloys.

Short-time tensile-strength and creep-strength properties are illustrated in Figs. 66 and 67.   All alloys shown are readily available as tube materials, except the 0.5 Te-Ni-P alloy.   90-10 and 70-30 cupronickel, brass and copper are also readily available in pipe sizes.

The use of copper and copper alloys for elevated temperatures is limited to temperatures below the lower recrystallization temperature for the particular alloy. This is the temperature at which cold-worked specimens begin to soften.   This recrystallization is usually accompanied by a marked reduction in tensile strength

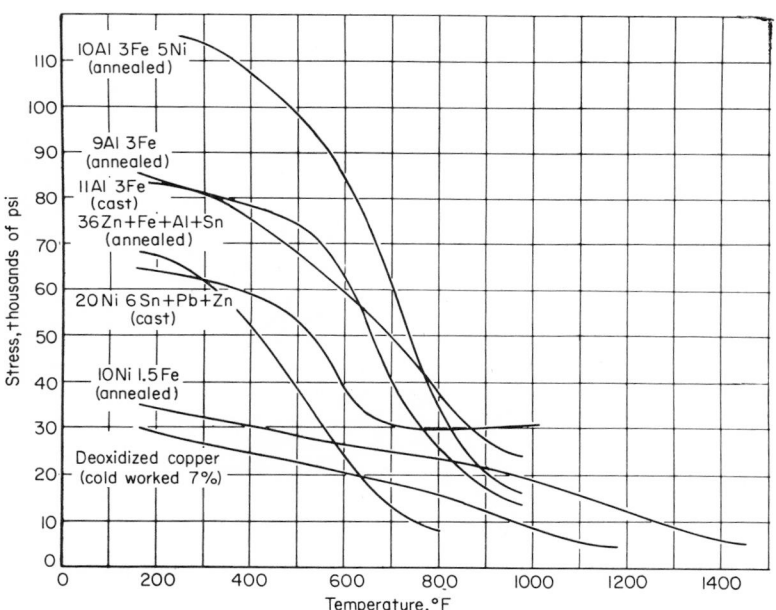

Fig. 66. Short-time tensile strength of selected coppers and copper-base alloys.

and a corresponding increase in ductility.   Grain size and degree of cold working have an important effect on creep resistance.   Exceptionally coarse-grained non-ferrous material, either annealed or cold worked, may become embrittled and fail suddenly, even though exhibiting a high creep resistance.   For annealed 70-30 brass, creep resistance increases with grain size at 300 to 500 F for stresses within the range normally used.   Finer grain metal may be superior at temperatures below 300 F.   At least it has a greater load-carrying capacity without appreciable deformation because of a higher yield strength.

Brasses containing 70 per cent or more of copper may be used successfully at temperatures up to 400 F, while those containing only 60 per cent of copper should not be used at temperatures over 300 F.

The ASME Boiler and Pressure Vessel Code limits the use of brass and copper pipe and tubing intended for piping service, as distinguished from heater tubes, to temperatures not exceeding 406 F.   The ASA Code for Pressure Piping limits the use of brass and copper pipe and tubing to temperatures not over 406 F for steam, gas, and air piping.

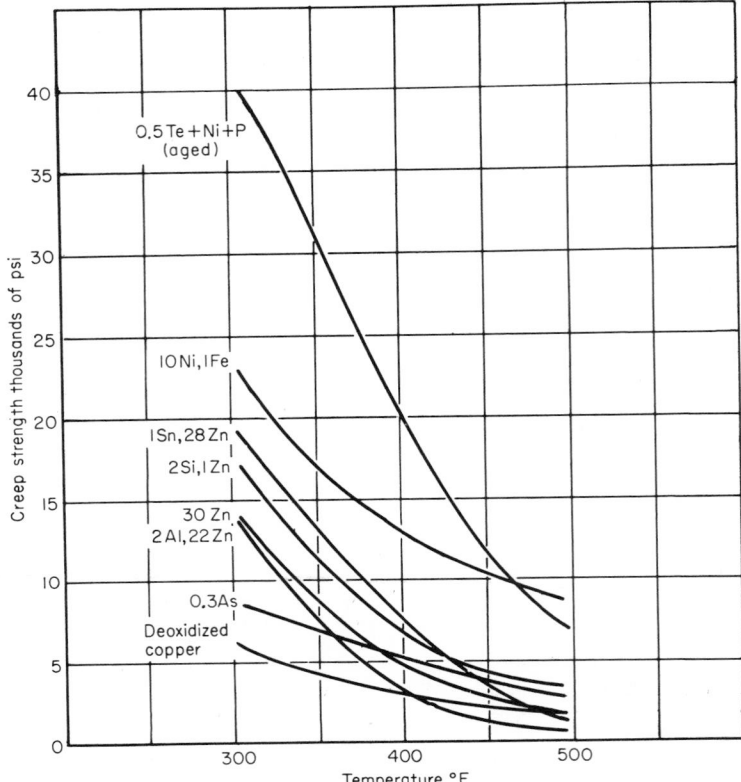

Fig. 67. Creep strength of coppers and copper base alloys at 0.00001 per cent per hour.

LOW-TEMPERATURE PROPERTIES

Introduction. Unexpected and sudden failures in piping, pressure vessels, bridges and other structures and failures in welded steel ships have made engineers and metallurgists aware that steels which behave ordinarily in a ductile manner may, under certain conditions, exhibit highly brittle characteristics. Although relatively few failures of this type have occurred in piping, the consequences of such sudden failures are extremely disturbing and can be tragic.

It is not unusual for engineers to ascribe the brittle fractures associated with these failures to such metallurgical causes as poor steel quality or an impairment of quality from welding operations. However, actual examination of most sudden failures of steel fabrication has revealed that superimposed mechanical factors were intimately associated with these brittle failures. In general, brittle failures have been found to be associated with inferior design, faulty workmanship including careless or improper welding practice, use of a steel which is notch-sensitive under the particular operating conditions, or a combination of these influences. Particularly in piping which is welded from the OD only, severe notches may result from lack of penetration or cracks in the root of the weld even though radiographic inspection may not reveal lack of penetration or cracking in the root of the weld along the pipe inside surface.

Brittle Behavior of Metals

Steel is generally considered to be a ductile material. When overloaded, it usually gives warning by flowing plastically; i.e., by bulging, stretching, bending or necking before rupturing. Contrary to expectation, however, steels sometimes rupture without prior evidence of distress. Such brittle failures are accompanied by but little plastic deformation, and the energy required to propagate the fracture appears to be quite low. Under certain conditions, steel may shatter like glass, but in piping, this extreme behavior generally occurs only at low temperatures.

The conditions which control this tendency for steel to behave in a brittle fashion include (1) high stress concentrations; i.e., notches, nicks, scratches, internal flaws or sharp changes in geometry; (2) a high rate of straining and (3) a low environmental temperature. These three factors are so interrelated that a determination of the effect of any one of them gives indication of how the steel will react to intensification of either or both of the others. The effect of lowering the testing temperature is the condition most convenient to measure quantitatively. Consequently, the transition from ductile to brittle behavior of a steel is generally expressed in terms of temperature.

The *transition temperature* for any steel is the temperature above which the steel behaves in a predominantly ductile manner and below which it behaves in a predominantly brittle manner. Steel with a high transition temperature is more likely to behave in a brittle manner during fabrication or in service. It follows that a steel with a low transition temperature is more likely to behave in a ductile manner, and therefore, steels with low transition temperatures are generally preferred for service involving severe stress concentrations, impact loading, low temperatures or combinations of the three.

Metallurgical factors, such as deoxidation practice, chemical composition, rolling, forging or extruding practice and subsequent heat treatment influence the transition temperature of steel. In carbon-steel piping materials, under the worst conditions, the transition temperature may be above 200 F or, under the best conditions, below minus 200 F.

Steels treated in accordance with most favorable deoxidation practice are those which are fully killed. In pipe steels deoxidation is generally accomplished with sufficient silicon to provide about 0.10 to 0.20 per cent of silicon in the steel. Aluminum has been used also as a deoxidizer, although less in recent years than several decades ago. The amount of aluminum used caused the retention of only a few hundredths of a per cent residual aluminum in the steel. Such steels are sometimes referred to as being made in accordance with fine-grain melting practice.

Carbon influences unfavorably the transition temperature of as-rolled or normalized steels. Its upper limit in plain carbon steels is generally accepted as about 0.25 per cent, and in low-alloy steels as about 0.20 per cent or even lower. High ratios of manganese to carbon may be beneficial. Most elements other than those used for deoxidation raise the transition temperature with the notable exception of nickel, which lowers appreciably the transition temperature of carbon steel. Austenitic chromium-nickel stainless steel and some high-nickel steels show no transition at temperatures lower than minus 325 F.

Steels which have been fully annealed are in the poorest condition to resist embrittlement. Normalizing offers improvement. Frequently further benefit is derived from tempering or stress relieving after welding. Optimum properties are obtained by fully quenching and tempering to moderate strength levels. Such treatment is seldom applied to pipe steels and has not received recognition in the ASA Code for Pressure Piping and the ASME Boiler and Pressure Vessel Code.

The presence of notches or other stress-concentration factors is of very considerable importance. Sharp notches in welded joints may encourage failure at loads well below those permitted by design. Any aggravated notch, just like a severe geometric shape change, can result in a crack at relatively low loads.

Considerable piping of rimmed or semikilled steels (for example, ASTM A53) has operated satisfactorily in service for long periods although it has a higher transition temperature than fully killed steels (for example, ASTM A106). These steels offer cost advantages and are usually in shock-free service where failure is unlikely. However, even in the absence of shock, piping operating under high stress should be made of steels which are tough at the lowest service temperature, such as those commercially available under ASTM Specification A106. This is

Table 5. Low-temperature Limitations for Various Piping Materials

Low-temp limit, F	Material and suitable ASTM designation	Comments
Zero	Mild steel (A53, A120, A135)	No requirements other than suitable pressure rating
−20	Mild steel (A53, A135)	Reduce pressure rating 1 % for each F deg below zero, or Charpy impact test, 15 ft-lb at design temperature
−50	Killed or limited carbon steel (A333, GR-1)	Charpy impact test, 15 ft-lb at design temperature
−150	3½ % Ni-steel (A333, GR-3)	Charpy impact test, 15 ft-lb at design temperature
−325	Austenitic stainless steel (Types 304, 316, etc.)	Limited carbon content
No limit	Nonferrous copper, brass, aluminum	Aluminum, copper, brass

particularly important for piping located where failure would endanger life and property.

Brittle failures rarely occur in piping. Nevertheless, their prevention is of paramount importance because economic and fabrication limitations rarely favor construction of a perfect piping system. Primary emphasis should be placed upon the selection of steels of suitable quality for their intended use, and followed by design and fabrication practices that will hold stress raisers to an acceptable minimum.

Low-temperature Service. Table 5 indicates the low-temperature limitations for various piping materials. Low-alloy steels may be used at temperatures below 0 F when they have Charpy keyhole impact values of at least 15 ft-lb at the lowest design temperature. Austenitic stainless steels with a limited carbon content, copper and copper alloys, and aluminum do not experience transitions in impact strength from ductile to brittle fracture and, therefore, may be used for low temperatures without pressure rating penalties.

Low-temperature piping is generally covered with thermal insulation which helps provide protection from external impact blows or shock. This, however, is not sufficient insurance against the type of damage that could result if a pipe should fracture.

Nature of Brittle Behavior. Brittle behavior in steel may result from a number of factors including (1) rapid rate of straining, (2) the presence of multi-directional

stresses, or notches causing restraint such as surface defects, discontinuities, incomplete welds, underbead cracks, microcracks and sharp reentrant corners, (3) low operating temperature, or (4) a combination of these factors.

The brittle behavior of a steel is generally evidenced by sudden failure and sometimes by shattering. The initial fracture usually will propagate rapidly and, under certain conditions, the rate of propagation is practically infinite. At failure, the surface of the fracture appears bright, granular and crystalline with the cross section showing little or no evidence of necking or plastic deformation. A typical example is the notched specimen shown in Fig. 68. Such fractures are generally called cleavage fractures.

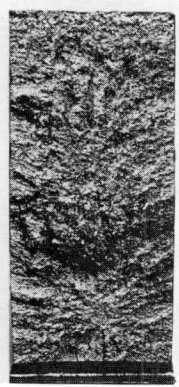

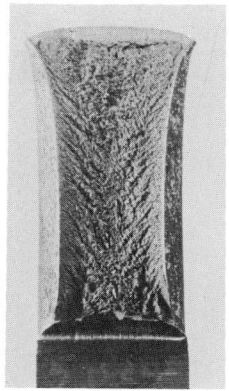

FIG. 68. Brittle fracture in mild steel: bright, granular, and crystalline plastic deformation is absent.

FIG. 69. Ductile fracture in mild steel: dull and fibrous plastic deformation is present.

In contrast, a steel which behaves in a ductile manner generally will fail gradually, i.e., at a much slower rate. Moreover, ductile failures in pipe or tubes usually are preceded by local bulging, i.e. by plastic deformation, or flow in the material because of shearing forces. Such shear fractures, as in the notched specimen shown in Fig. 69, will appear relatively dull and fibrous and show a measurable and often considerable contraction of area across the fracture section.

Evaluation of Brittle Behavior of Steels. The conventional static tension and bend tests do not differentiate adequately between steels of varying susceptibilities to brittle behavior. This has created the desire for a simple test that could be used to select steels that would be considered "safe" for use under any predictable conditions. Such a test should permit selection of a particular steel for a particular type of element in the structure for use under normal and, possibly, for an occasionally unusual service condition or environment which the structure may encounter.

Unfortunately, there is no single test that fulfills these requirements. This is because the common tests measure the mechanical properties of the steel under the particular conditions imposed by the test method, and not its behavior in an actual structure as influenced by such factors as design, workmanship, surface notches, welding quality, restraints and stress distribution. Nevertheless, there are several accepted tests that are extremely useful because they permit comparison among the various types and grades of steels. Moreover, these tests, which usually require

notched-bar specimens, provide some knowledge of brittle behavior and the approximate relation of such factors as thickness of the steel, chemical composition, deoxidation practice, sharpness of notches and heat treatment, to fracture characteristics.

The determination of the temperature at which a steel may become susceptible to brittle failure under certain conditions is based on one of three testing categories: (1) impact energy, (2) notch ductility and (3) fracture appearance.

The transition temperature of any one grade or type of steel as determined by one type of test generally should not be compared to that of another grade of steel unless the same criterion of brittleness has been used.   Under the same type of test, the transition temperatures obtained are of only qualitative value in assuring that the

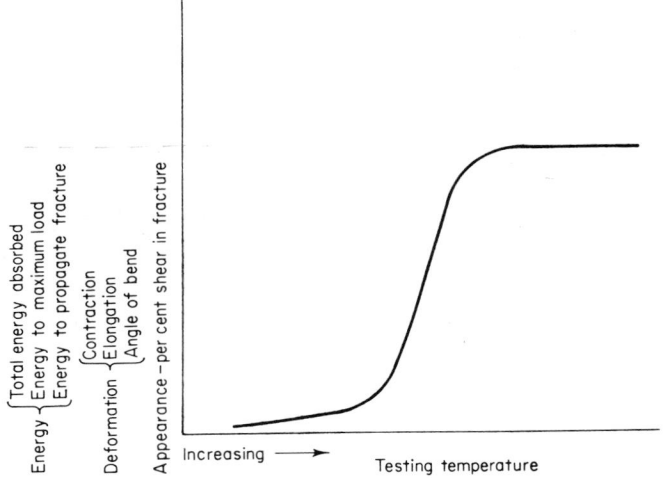

FIG. 70. Schematic illustration of transition-temperature range.

steel which exhibits the lower transition temperature is less likely to fail in a brittle manner than the steel showing a higher transition temperature.   For any particular service, however, other factors being equal, the steel with the lower transition temperature should be preferred because it offers additional safety, even though the limiting low-service temperature of the structure may not be precisely known.

Effects of Temperature.   The transition from ductile to brittle behavior of carbon and low-alloy steels is not a phenomenon only occasionally noted; it is an inherent property which can be demonstrated by a number of accepted test methods. For example, Fig. 70 illustrates schematically this transition as brought about by a decrease in the testing temperature.   It should be noted that since several criteria may be employed to evaluate the transition temperature, its position with respect to the temperature scale may be expected to vary with the criterion selected.

Transition Temperature.   Various tests have been developed to determine whether a steel under certain test conditions will fail definitely in a ductile or brittle manner.   Each of these conditions actually shows a scatter region or a transition-temperature range (Fig. 71) within which a steel under the same set of test conditions will exhibit either ductile or brittle behavior, or both.

Usually, some empirically established point within the transition-temperature range is selected and is thereafter designated as the transition temperature for the steel tested.

For example, when the appearance of the fracture is the criterion, the transition temperature may be arbitrarily designated as the temperature at which half of the fracture surface shows a cleavage failure while the other half exhibits a shear failure. When impact energy absorption is the criterion which represents the most widely accepted method, some arbitrary energy value is generally used. For example, in the Charpy test, the temperature corresponding to the 15 ft-lb value is often designated arbitrarily as the transition temperature. It must be noted that at this value, the fracture may be predominately of the cleavage type. This last method is illustrated in Fig. 71.

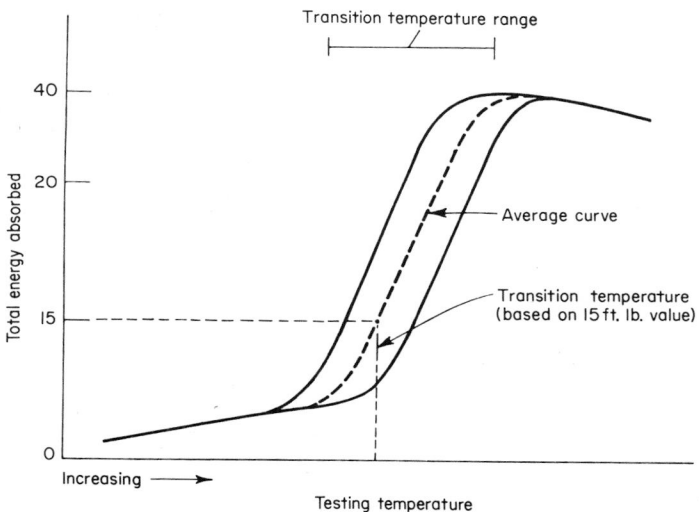

Fig. 71. Transition temperature range and transition temperature in Charpy impact test.

Other criteria are the temperature at which the initial appearance of cleavage (brittle) fracture is observed and the temperature at which the energy absorbed is half of the maximum energy value obtained over the testing range.

It should be recognized that there exists no one "transition temperature" for a given steel except for a particular set of conditions and one criterion of brittleness. Each particular forged or extruded pipe or plate thickness rolled from a given steel will exhibit a range of temperatures in which the fracture shifts from tough to brittle. This range may be varied considerably by changing the metallurgical and mechanical properties of a steel.

Effects of Notches: Notch Sensitivity. The sudden brittle failures which occur without measurable deformation are generally ascribed to notch sensitivity of steel at the operating temperatures to which it was exposed. Although not always obvious, actual notches or notch effects are present in all structures. They may consist of minute surface or subsurface cracks, surface laps or scabs, visible scratches, abrupt shape changes such as sharp corners, tool marks, grooves from drawing dies, edges, etc., or fabrication defects, such as from arc strikes or similar causes.

Notch sensitivity is usually associated with an inability of the steel to deform in a plastic manner (that is to flow) underneath the notch. This resistance to flow is increased by the triaxial state of stress induced under a notch by a tensile stress.

The condition is accentuated as the thickness of the steel increases.   Notches also are stress raisers.   The greater the sharpness of the notch, the greater will be the degree of restraint, the more severe will be the stresses as to both triaxiality and magnitude, and the higher will be the transition temperature.   The effects of notches of varying severity on the brittle behavior of Charpy test specimens are illustrated in Fig. 72.   This is the basis of the general concept that the transition-temperature range or the arbitrarily selected transition temperature will increase with the severity of the notch.   In other words, a steel which contains extremely severe notches will fail in a brittle manner at higher ambient temperatures than if less severe notches were present.

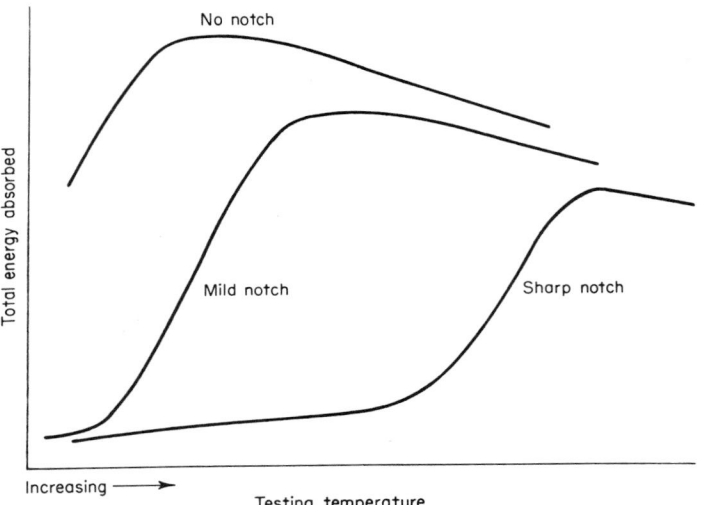

FIG. 72. Effects of notch severity on the transition temperature.

It should also be noted that in instances where service conditions impose a triaxial state of stress, as is true of internally loaded piping, brittle behavior may be encountered even in the absence of notches.

Testing Methods

Various "small-scale" tests have been developed to evaluate the notch sensitivity of steel.   These tests may be divided roughly into three categories consisting of single-blow notched-bar impact tests, slow-bend notch tests and notched tensile tests.   The latter include the symmetrically loaded type of specimen such as the edge-notched plate test and the asymmetrically loaded type of specimen such as the Navy Tear Test.

Small-scale notch toughness tests are by far the most widely used to evaluate steel piping materials.   They are generally made for one of three primary purposes: (1) to examine and correlate the metallurgical characteristics of various types of steels; (2) to inspect the steel (approve or reject) in accordance with specification and code requirements and (3) to indicate in so far as possible the service behavior of a particular steel in a finished structure.

The examination and correlation of metallurgical characteristics is of interest primarily to the metallurgists who, by means of these tests, can evaluate the effects

of steelmaking and rolling practices, composition, heat treatment, etc., on the mechanical properties of steel.

The "acceptance" tests made in accordance with requirements of design codes and materials specifications, and the tests made to obtain an indication of service behavior are of major interest to the engineer. They are of course also useful to the metallurgist, as they guide him in his efforts to obtain steel of the desired properties.

Determination of the effects of metallurgical variables on the test properties of steels is relatively simple but to interpret the effect of these variables in terms of probable service behavior is extremely difficult. One reason for this difficulty is that, in many respects, the steel in a relatively rigid piping system does not behave in the same manner as it does in small-scale laboratory test specimens. Moreover, in piping systems, certain mechanical shape and assembly factors and adverse effects of fabrication procedures and workmanship may come into play. Consequently,

Table 6. Minimum Impact-testing Temperatures for Various Low-temperature Steels

Material	Grade	Temperature, min, F
Carbon steel	1	−50
3½ Ni steel	3	−150
Cr-Cu-Ni steel	4	−150
4¼ Ni steel	5	−150

the results obtained with the conventional notched-bar testing procedures on small specimens are not necessarily the same that would be obtained from tests of full-scale structures of the same steel. Nevertheless, some large-scale tests, such as direct explosion tests have been suggested for evaluating the effects of welding and other fabricating processes. The initial results of direct explosion tests indicate significant differences in the performance of steels which have been welded by different procedures. Ultimately, these tests may prove to be useful for determining the proper welding procedure to obtain low-transition temperatures in pipe weldments.

Code-test Requirements. For carbon and low-alloy steels for use at temperatures below minus 20 F, the 1962 edition of the Unfired Pressure Vessel Code and Section VIII of the ASME Boiler and Pressure Vessel Code require that three Charpy keyhole specimens exhibit an average impact value of 15 ft-lb at the lowest operating temperature, with only one of the specimens permitted to show a minimum of 10 ft-lb. The V-notch Charpy and the Izod specimens are not recognized by the 1962 Unfired Pressure Vessel Code. Similarly ASTM Specifications A333-63T, A334-63T, A350-61T and A420-63T for carbon and alloy-steel pipe materials for service at low temperatures require impact testing with Charpy specimens having either a Keyhole or equivalent slot notch. Tests must be made at the minimum temperatures shown in Table 6. These tests also recognize the effects of specimen size, as shown in Table 7, and the average of three impact test specimens must meet the values given in this table.

Significance of Test Data. Mechanical-property requirements of standard specifications for steel piping supply only limited information to the designer, although such requirements are useful in purchasing and in classifying materials on the basis of tensile and flattening properties. One reason tensile-test data are of limited application in design is that the tension test is uniaxial, whereas the hazard

Table 7. Impact Requirements in ASTM Specifications A333, A334, A350, A420

Size of specimen, mm	Minimum average notched bar impact value of each set of three specimens, ft-lb	Minimum notched bar impact value of one specimen only of a set, ft-lb
10 by 10	15	10
10 by 7.5	12.5	8.5
10 by 5	10	7.0
10 by 2.5	5	3.5

of brittle failure is due to the existence of multiaxial stresses and stress concentrations. The static tensile strength, yield strength and ductility of the unnotched tensile bars at ambient temperatures and the flattening characteristics represent a relative quality factor rather than a value that applies to design. This is substantiated by the fact that certain special alloy steels, which in the conventional room-temperature tensile test have elongation values of slightly under 5 per cent, have given satisfactory service in many applications. However, these special steels had been processed to develop transition temperatures considerably lower than that of mild steel having conventional tensile ductility values above 25 per cent. In the applications involved, steels of higher transition temperatures would not have performed satisfactorily no matter how high their tensile-test ductility at room temperature.

There is no intent here to discount the need for ductility in steel for satisfactory performance in piping systems, but ductility at room temperature as determined in the static tensile test does not necessarily indicate that the steel will behave in a ductile manner under different conditions of stressing. For instance, large cross-sectional areas and undesirable design features may cause mechanical restraint. As mechanical restraint opposes plastic deformation, the steel will fail in a brittle manner if restraint is sufficiently severe and if the temperature is sufficiently low.

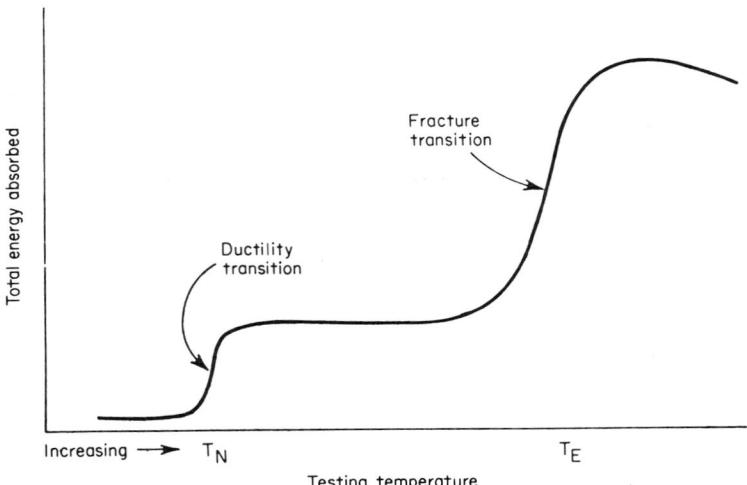

FIG. 73. Ductility and fracture transitions.

The major difficulty is that there is no accepted method of correlating ductility and toughness with other mechanical properties obtained from standard tensile or notched-bar tests that can be applied to assure the most efficient use of a steel in a piping system.

Ductility and Fracture Transition. Test results obtained with many types of specimens usually will show two transition temperatures instead of one. This is shown schematically by the solid curve in Fig. 73. The two transition temperatures are identified as fracture transition and ductility transition. The drop in energy at T_e (the fracture transition) is generally associated with the change in fracture

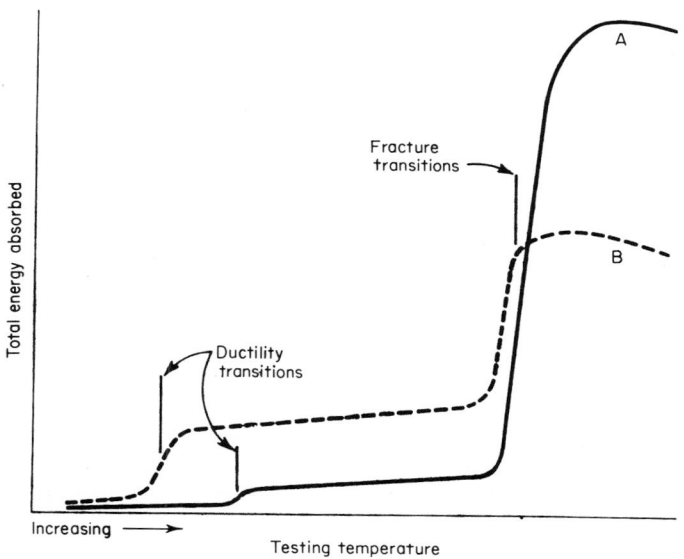

FIG. 74. Transition temperature of large, centrally notched, wide-plate specimens (curve A) and of small notched specimens (curve B).

appearance over the fractured surface which occurs in this temperature region. The drop at T_n (the ductility transition) is accompanied often by a change in fracture appearance of a small zone at the base of the notch.

The temperature at which the ductility transition occurs is very much dependent upon notch geometry and the influence of plastic straining prior to failure. As the notch is made sharper and deeper, the strain is more localized and the degree of triaxiality of the state of stress is greater. This leads to higher ductility-transition temperatures.

The fracture-transition-temperature range, on the other hand, is less sensitive to notch geometry as specimens with different size notches may show widely different ductility transitions without exhibiting similar differences in the fracture-transition temperatures. As an example, in Fig. 74 are shown schematically the transition curves for (A) large, centrally notched, wide-plate specimens and for (B) small notched specimens.

Both transition-temperature ranges have a bearing on the behavior of the steel. Whereas the ductility transition temperature, T_n, is associated with the initiation of a crack, the fracture transition temperature, T_e, is associated with its propagation.

FACTORS DETERMINING TRANSITION TEMPERATURES

The factors which influence the transition-temperature range of a steel may be classified as being either metallurgical or mechanical in nature.   The metallurgical factors are characteristic of the steel itself, whereas the mechanical factors depend upon the service or testing environments as well as on many effects produced by fabricating procedures.   An analysis and understanding of the various factors is often highly important, particularly when it is desirable to make the most efficient use of a particular steel.

Metallurgical Factors: Effects of Composition.   Carbon and nitrogen are considered to be among the most important elements which raise the transition temperature of steel in the as-rolled or normalized condition.   Oxygen and phosphorus in quantities greater than normally tolerated, and silicon in percentages greater than required for deoxidation, also may raise the transition temperature. In fact, most additions usually made to steel raise the transition temperature.   The additions that will lower the transition temperature are nickel and, under certain conditions, manganese.   Nickel is frequently added to low-carbon steel when a transition temperature below that obtainable with carbon steel is desired.   Aluminum, in the range of quantities used for deoxidation, is very potent in lowering the transition temperature and is used in steels that are made specifically for low-temperature service.

Commercial mild steels, even those within the same grade, may exhibit considerable differences in their notch-sensitive characteristics.   These differences, which are not predictable from the conventional chemical analyses, may be attributed to various factors such as steelmaking practice, deoxidation, grain size, finishing temperature, thickness of section, heat treatment and others.

In the fully quenched and tempered condition, composition, other than high-carbon content, does not appear to be much of a factor provided the steels harden fully on quenching.   This does not appear to hold true in the normalized or annealed conditions more commonly applied to piping materials.   Steels that are susceptible to temper embrittlement are also exceptions.

Effect of composition on transition temperature of steel is still the subject of considerable study.   There is some indication that nitrogen present as aluminum nitride may be beneficial, but detrimental if present as other nitrides.

Homogeneity.   It is generally recognized that the concentration of carbon, alloying elements and impurities varies throughout the ingot and, consequently, throughout the finished steel pipe.   These variations in composition affect the conventional mechanical properties as well as the notch characteristics and the transition temperature.   For example, the ductility in the tension test may increase, whereas the strength may decrease from the top to the bottom and from the center to the edge of the ingot from which pipe is subsequently extruded.   In a set of Charpy keyhole and V-notch impact tests, the toughness increased and the transition temperature decreased from the top to the bottom of an ingot.   These effects all were consistent with chemical variations.

Some extruded pipe end plate steels show marked directional properties such as higher values for ductility and notched-bar impact when tests are made on specimens taken parallel to the direction of extruding or rolling, as compared to specimens taken transverse to this direction.   On plate steels, (as used for making seam-welded pipe), directional properties may be minimized by cross rolling and by certain heating procedures prior to rolling.   This is not possible on extruded pipe.   An example of the effect of directionality in an A106 pipe is shown in Fig. 75 with Charpy V-notch impact-test specimens removed parallel and perpendicular to the axis of the pipe.

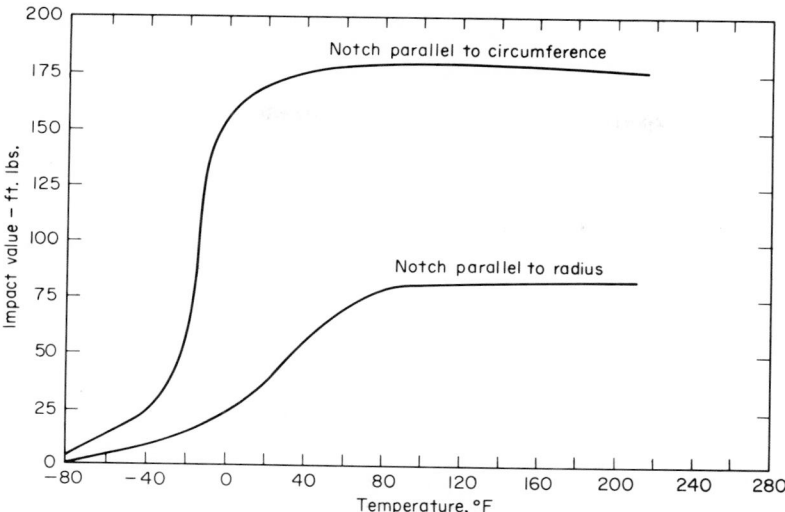

Fig. 75. Effect of directionality on impact bar results.

Effects of Grain Size. The smaller the ferrite grain size, the lower generally will be the transition temperature. Thus, the lower the final rolling temperature of the steel and the higher the cooling rate, the smaller will be the grain size and the lower will be the transition temperature unless the final rolling temperature is so low that the steel is, to some degree, cold rolled.

Aluminum and silicon additions during final deoxidation are useful in providing fine grain size in normalized material. A normalizing treatment is particularly effective if the pipe or plate was finished at a high rolling temperature.

Effects of Straining. Cold deformation or straining generally raises the transition temperature of a steel (Fig. 76), especially the ductility transition. The fracture transition is not affected until the ductility transition has been raised to

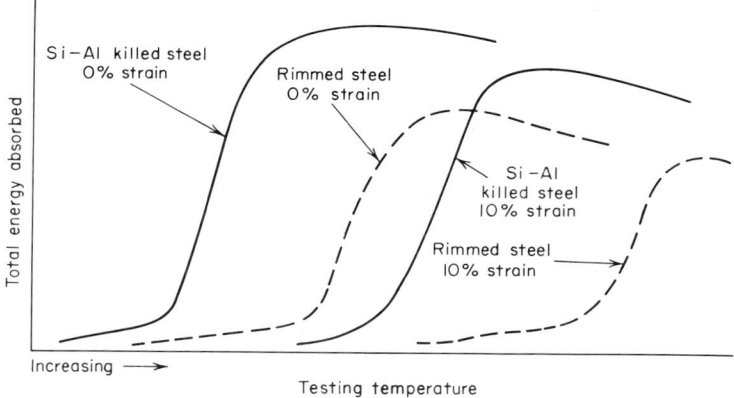

Fig. 76. Effects of straining on the transition temperatures of Si-Al killed and rimmed steels.

the level of the fracture transition. This may have adverse effects on pipe which has been cold expanded or cold bent.

Effects of Straining and Aging. The transition temperature of a steel which is susceptible to strain aging is raised to a higher temperature after straining and aging than after straining alone. This is illustrated in Fig. 77 which shows results which may be expected in a rimmed steel.

Effects of Heat Treatment. Some carbon-steel pipe is furnished in the hot-finished condition. Hot finishing is generally performed between 2200 and 1600 F and is followed by air cooling. Under these conditions, these steels can be compared to normalized steels, although it should be recognized that the temperature of finishing is an important factor.

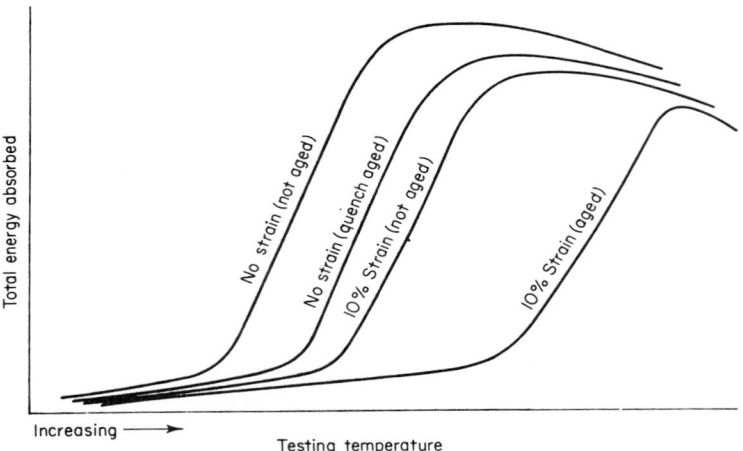

FIG. 77. Effects of straining and aging on the transition temperature of a rimmed steel.

Because large sections are usually finished at higher temperatures and cool more slowly than light sections, they exhibit higher transition temperatures. This is at least in part due to the coarser grain size, and to the coarseness and distribution of pearlite.

The form and distribution of the carbide have a profound influence on the transition temperature of steel. In general, lowest transition temperatures are obtained with liquid quenching and tempering heat treatments. Normalizing or normalizing and tempering is the second most effective treatment. Conventional annealing cycles consisting of heating to above the critical range followed by slow cooling to ambient temperatures result in high transition temperatures, especially if the annealing temperature is so high as to cause grain coarsening. Such treatments are not recommended for pipe for low-temperature service. Normalizing at about 1650 F and tempering or a stress-relief heat treatment at 1100 F or above, but not higher than 1250 to 1350 F, are the preferred treatments. The range 1350 to 1450 F sometimes raises transition temperature and should be avoided; likewise, low-temperature stress-relief heat treatments (below 900 F) are to be avoided in some instances, as they raise the transition temperatures.

Various *mechanical factors* may also exert considerable influence on the transition temperature of a steel. Most important are the stress system, section size, design, workmanship, welding practice and others.

Effect of the Stress System.   Multiaxial tensile stresses raise the transition temperature.   This is particularly true at the base of a notch or crack where multi-axial tensile stresses of considerable magnitude may develop.

Effects of Section Size.   If the section size is increased without other changes in geometry, the transition temperature will also be increased.   As is illustrated in Fig. 78, the amount of energy absorbed rises as the section size is first increased and the fracture remains ductile, but beyond a certain critical size, there is a sharp drop in energy accompanied by a change to brittle fracture.   Further increase in size results in a very slight increase in energy absorption, but the fracture remains brittle.   These results are partially due to the greater restraint of the wider specimens which increases the severity of the triaxial stresses.   The standard size Charpy keyhole or V-notch impact specimen requires that the pipe to be tested have a wall

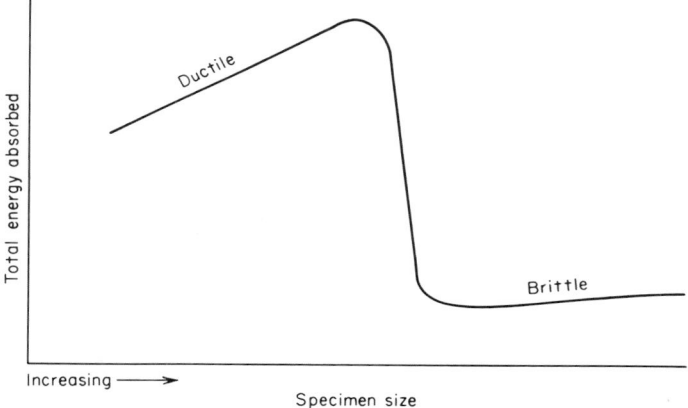

Fig. 78. Relationship between size of specimen and type of fracture.

thickness of at least 2.5 mm.   For thinner pipe walls, reduced section tensile tests are frequently made.   Typical test results on A106 pipe are illustrated in Fig. 79 involving full, three-quarter- and half-size Charpy V-notch specimens.

Effect of Design.   Design based upon conventional tensile-test data gives no assurance that piping will not fail in a brittle manner.   Nor can such assurance be obtained by simply increasing the section size with the intent of increasing the "factor of safety."   In the presence of notches, increase in section size will most likely increase restraint and may even lead to failure at lower applied loads.

The designer must take into account the conditions which promote notch sensitivity and must attempt to control these conditions so that the ductility transition temperature of the piping, under the particular conditions of design and service, will be below the temperature at which it operates.   This may be accomplished by selecting the proper steel, by using generous radii, by limiting the extent of surface defects and by controlling welding and other fabrication procedures. When it is necessary to incorporate undesirable design features, or when economic considerations restrict selection of the steel, the adverse factors may be at least partly offset by using a low design stress and, where possible, by avoiding shock conditions in service.   It also can be assumed that if piping fails under these conditions, it will fail in a brittle manner.   The hazards associated with sudden failure of piping will dictate whether it should be used.   If these hazards are great

enough, the fracture transition temperature should be below the minimum service temperature.

Effect of Workmanship and Welding. Workmanship represents an intangible but highly important factor, the effect of which is difficult to evaluate in the completed piping component or system. Inspection utilizing radiographic or ultrasonic methods or magnetic particle or liquid penetrant techniques at critical locations, as discussed in Chap. 8 have generally been helpful, in judging and controlling the quality of workmanship. These inspection procedures, when judiciously applied, also have the psychological effect of influencing the welder to

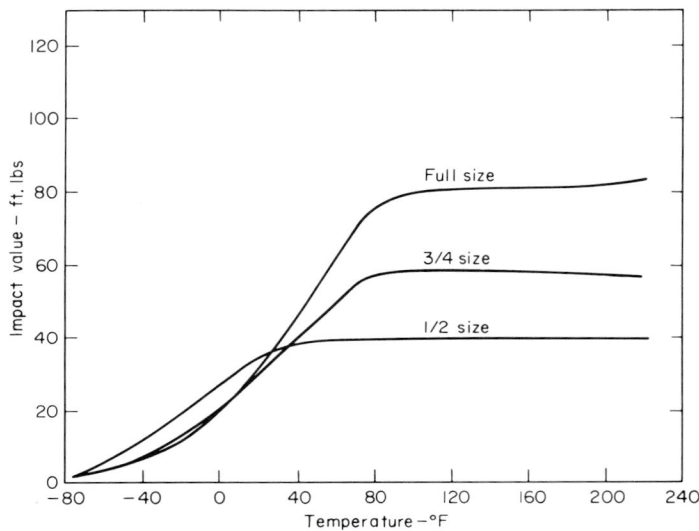

Fig. 79. Effect of specimen size on Charpy V-notch test results.

improve his workmanship and thus reduce the probability of incorporating notches into the structure.

It is even more difficult to evaluate the brittle behavior of weld deposits than to evaluate the properties of wrought seamless piping. This is so primarily because the conventional testing procedures have not been correlated with the service behavior of the material after it has been welded. For example, one electrode may be preferred over another because the weld metal produced from it develops higher tensile strength. However, it might well be that an incipient surface crack will propagate readily in the weld deposit made from the one electrode but not in the deposit from the other, because the latter has the ability to adjust itself plastically at the service temperature.

In welded piping involving service where notch-brittle behavior is a consideration, it is advisable to keep the width of the heat-affected zone as narrow as possible by means of rapid rates of electrode travel. When the layer thickness of the weld metal is held constant, faster rates of electrode travel reduce the width of the heat-affected zone. Apparently, the effect of the heat from the retreating arc is minimized, thus reducing the width of the heat-affected zone. This is desirable because transition temperatures tend to rise as the heat-affected zone becomes wider. The effects of width of the heat-affected zone on the transition temperature are illustrated in Fig. 80.

Welded structures in the as-welded state always contain residual welding stresses, which in certain sections may be as high as the yield strength of the steels. Under certain conditions, these residual stresses may not be particularly harmful. This is confirmed by the fact that many structures, piping and vessels have given highly satisfactory service without having received stress-relieving treatments subsequent to welding operations. Moreover, when failures have occurred, it was generally found that they were brought on by causes other than residual stresses alone.

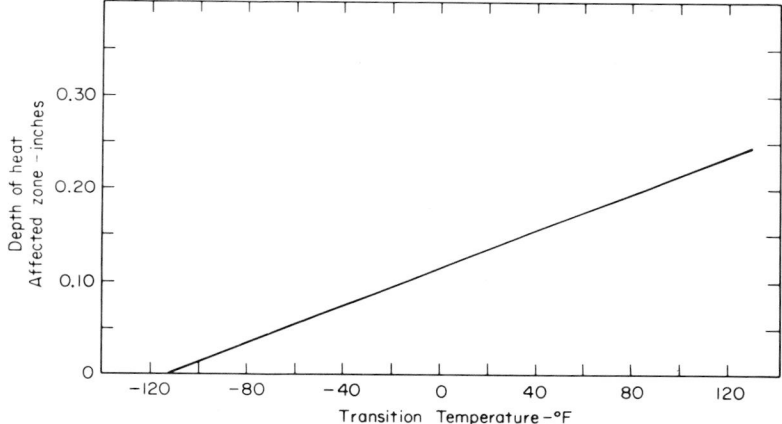

Fig. 80. Effect of depth of heat-affected zone on the transition temperature of a welded rimmed steel (composition: 0.16 per cent C, 0.45 per cent Mn, 0.01 per cent Si).

When residual stresses are believed to be a factor in the service life of piping, a stress-relief heat treatment should be applied. In some instances, however, stress relieving may raise the transition temperature.

It has been demonstrated by certain restraint tests, such as the circular-groove tests, that for specific preheat and interpass temperatures, the degree of restraint during the cooling cycle affects the transition temperature of weld deposits. When controlling preheat and interpass temperatures at 70 to 100 F, the transition temperature of low-alloy steel weld-metal deposits has been raised from −90 to +10 F by increase in restraint. Where the degree of restraint is held constant, increasing the preheat and interpass temperatures effectively lowers the transition temperature of the weld deposit.

Welding at subfreezing temperatures is not permitted by most codes. Such practice will produce weld deposits having reduced ductility and impact toughness. The drop in ductility becomes pronounced at welding temperatures below −50 F. Results of impact tests show that specimens welded at −50 F exhibit a toughness 15 or 20 per cent lower than specimens welded at ordinary atmospheric temperatures (75 F).

Extensive investigations of the last few years indicate that, due to a combination of hydrogen and straining, ordinary electrodes of the E6010 type may produce microcracks as the weld cools from 500 F to ambient temperature, particularly when welding heavy sections of pipe, or when welding under severe restraint. Preheating, or the use of E7015, E6015, E6018, E7016 or E7018 type electrodes circumvents this difficulty.

OXIDATION AND SCALING

When piping and tubing are exposed at elevated temperature, oxidation of the metal surface tends to occur. As oxidation continues, a scale will form from the initial oxide film on the metal surface.

Oxidation and scaling represent a corrosive mechanism in which the oxygen diffuses into the metal. The result is a gradual decrease of the wall thickness of the piping or tube material. Although oxidation occurs over the whole exposed metal surface and attacks the metal surface uniformly (Fig. 81), localized attack may also occur, often preferentially along the grain boundaries of the metal.

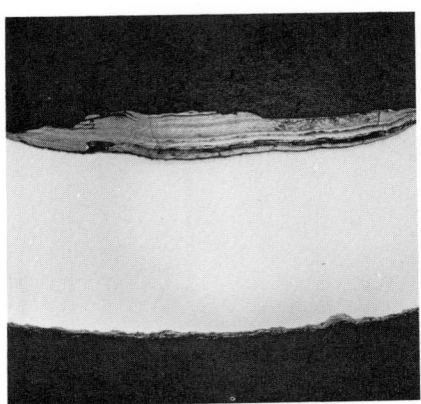

Sometimes both oxidation and localized corrosion may occur. An example of localized pitting attack underneath scale in boiler tubing is shown in Fig. 82.

Oxidation and scaling initially involves the chemical combination of oxygen with the metal to form the oxide or oxides. As the scale forms and grows in thickness, the rate of the oxide-metal reactions is gradually retarded and, in time, may stop. However, if part of the scale breaks or flakes off, or otherwise comes off by chemical or mechanical treatment or by erosion, then fresh metal surface will be exposed and may become subject to oxidation.

FIG. 81. Cross section of boiler tube, illustrating oxidation scaling on fire side. (Magnification 5×, size reduced in printing.)

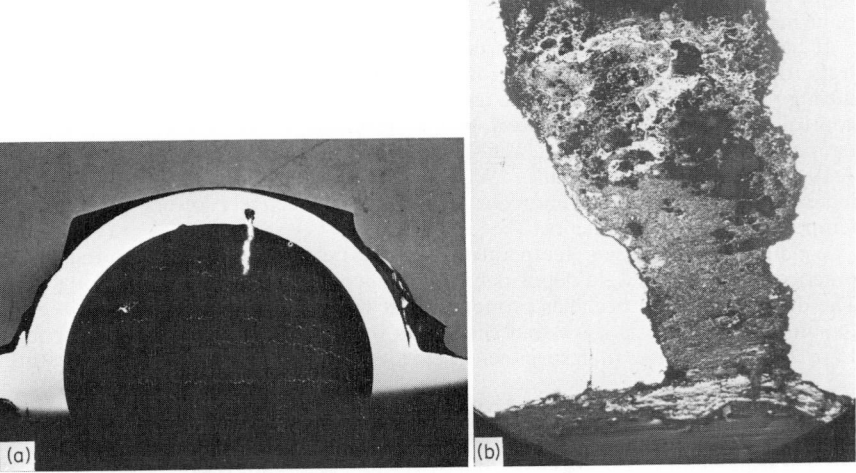

FIG. 82. Cross section of boiler tube, illustrating localized corrosion pitting and scaling. (*a*) Magnification 1×. (*b*) Magnification 30×, enlargement of localized pitted area of (*a*). (Size reduced in printing.)

Effects of Composition.   The resistance of metals to oxidation can be significantly improved by alloying.   The oxidation resistance of steel in air is substantially improved by chromium, silicon and aluminum which form stable oxide films on the metal surface.   These elements oxidize preferentially and form a tightly adherent surface film which tends to retard inward diffusion of oxygen, thus inhibiting further oxidation.

The effect of chromium on the oxidation resistance of steel is illustrated in Fig. 83.   Increased service temperatures require alloy steels of increasing chromium

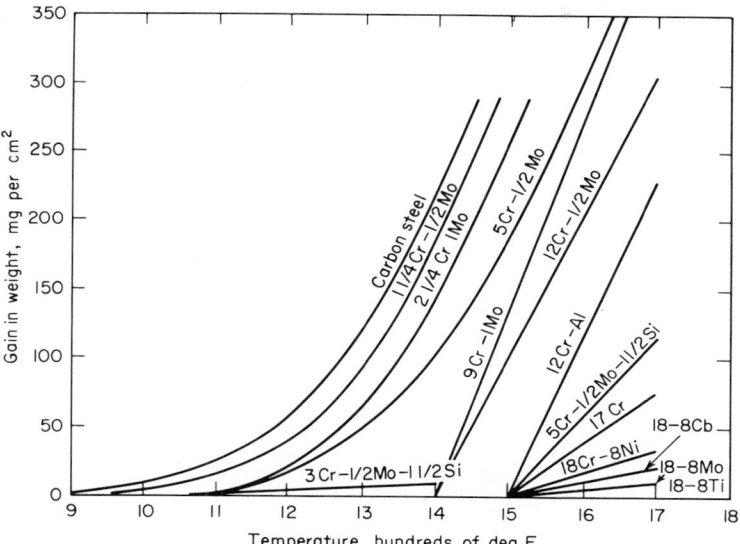

FIG. 83. The effect of chromium on the oxidation resistance of steel.

content to insure adequate oxidation resistance.   Figure 83 indicates that oxidation deposits on 9 per cent and higher chromium composition pipe do not occur until temperatures of at least 1400 F are encountered.

The resistance to oxidation in air of the chromium-molybdenum steels may be increased further by the addition of 1 to 2 per cent silicon, as illustrated in Fig. 83. In steam atmospheres, the 1 to 2 per cent silicon addition is not as effective.   In some applications, silicon additions may be undesirable, as silicon tends to decrease resistance to secondary creep.

Aluminum also increases oxidation resistance.   However, since it affects high-temperature strength adversely, it is not generally added to commercial pipe and tube steel materials used in elevated-temperature applications.

Of course, various other alloying elements are also added to metals to improve or provide specific properties.   Some of these may not improve, or may even reduce, oxidation resistance.   Thus, the behavior of alloys in oxidizing atmospheres may become complex.

Effects of Atmosphere.   Although oxygen is most generally responsible for oxidation, superheated steam, combustion atmospheres, hot gases and similar environments may cause or contribute to oxidation and scaling.

The composition of the high-temperature atmosphere is also very important. It is generally difficult to predict the resistance of many materials to specific

atmospheres, as various contaminants can affect very significantly the rate of oxidation and the type of scale formed.   Sulfur or vanadium in the fuel oil being burned in boilers can increase substantially oxidation rates.   Figure 84[1] illustrates the effect of temperature on the rate of oxidation in the presence of sulfur.   Vanadium as vanadium pentoxide[2] in the oil ash attacks all metals at 1000 F.   At higher temperatures, the ash becomes primarily liquid, and causes heavy attack which may increase to the "catastrophic" level and may amount to several inches per year.

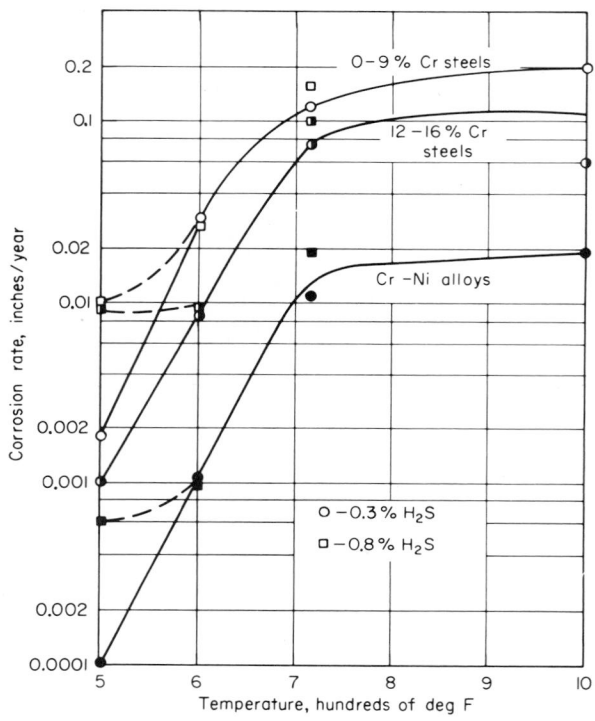

Fig. 84. Effect of temperature and concentration on hydrogen sulfide corrosion at 465 psig hydrogen pressure.   (*Backensto, Drew, and Stapleford*).

Nickel-free alloys are less attacked in this environment than high-nickel alloys. Whereas in nickel-free or low-nickel alloys the attack generally is uniformly over the surface, in high-nickel alloys intergranular penetration also occurs frequently.
Effects of Surface Contaminations.   Sometimes, copper or other metals are embedded within the scale.   This may have been the result of metal particles or metal ions having been flowed through the piping system.   For example, in heat exchangers, copper tubing gradually gives up some copper to the feed system and this travels through the boiler tubing.   Iron and other metals may similarly be flowed through the piping system.   Sometimes, these elements may even plate out.

[1] E. B. Backensto, R. D. Drew, C. C. Stapleford, Corrosion, 12, 6T, 1956.
[2] C. T. Evans, Jr., "Oil Ash Corrosion of Materials at Elevated Temperatures," ASTM Special Technical Publication No. 108, *Symposium on Corrosion of Materials at Elevated Temperatures*, June, 1950.

Effects of Fatigue. Mechanical or thermal fatigue may cause breaks in the scale or cause it to flake off. Particularly in thermal fatigue, the scale may flake off as a result of the different coefficients of expansion of the scale and base metal which results in stresses at the scale-metal interface. The freshly exposed metal surface is then likely to become subject to further oxidation.

Service Shutdowns. In some chemical and refinery high-temperature applications, sulfur dioxide and sublimed sulfur may be contained in the scale. Several stainless steels and Inconel materials are suitable for continued exposure involving these conditions. During shutdown periods, however, condensate forming over the scale may produce acidity and cause corrosion.

During shutdown periods, the exposure to moist atmospheric conditions may cause pitting, general corrosion, intergranular corrosion or even stress corrosion cracking. The susceptibility toward such deterioration may be further enhanced by pickling or other cleaning operations, particularly if pickling is not followed by a thorough neutralizing rinse.

It is very important, therefore, to recognize the complete service requirements. In some process operations, the materials selected must be suitable for resistance to high-temperature oxidation as well as to possible corrosive conditions prevailing during shutdown periods. Sometimes, it may be desirable to maintain in the piping system during shutdown periods a purge involving neutral or safe gases. Occasionally, noncorrosive liquids are also utilized.

SERVICE FAILURES[1]

Although numerous conditions may cause or lead to service failures, the responsibility for a failure can generally be assigned to one of five classifications:

1. Design (structural flexibilities, design notches, joint location of weld joint configuration)
2. Materials (selection and handling of base and welding materials)
3. Base-metal defects (introduced during manufacture and shaping of cast, forged or wrought piping components such as pipe, valves and fittings)
4. Fabrication (fabrication, welding heat treatment or cleaning of piping during shop fabrication or field erection)
5. Service (excessively severe service conditions)

In some instances, the responsibility can be related to a combination of several of these classifications. For example, a pipe weld containing root defects, such as lack of penetration, may not fail during service until thermal or mechanical fatigue of unexpected severity causes cracking and crack propagation of the existing defect across the wall thickness. If complete 100 per cent weld-root penetration was originally specified, then the failure would be classified as follows:

> *Primary Classification:* Fabrication
> *Condition:* Improper welding, lack of penetration in root of weld
> *Secondary Classification:* Service
> *Condition:* Thermal fatigue

In the above example if, on the other hand, the weld quality was not specified at the time the weld was made, and some incomplete weld-root penetration was considered acceptable since thermal-fatigue service conditions were not anticipated,

[1] Thielsch, H., "*Defects and Failures in Pressure Vessels and Piping*," Reinhold Publishing Corporation, New York, 1965.

then the failure would be classified as follows:

Primary Classification:	Service—excessively severe
Condition:	Thermal fatigue
Secondary Classification:	Fabrication
Condition:	Lack of penetration in root of weld, considered acceptable under applicable specifications.

A failure occurs by cracking or corrosion or, sometimes, by a combination of these two mechanisms. By far the majority of failures occur gradually. Upon traversing the wall thickness of the piping component, a surface opening develops through which may leak some of the steam, liquid or gas carried in the system. This, in effect, produces a warning of the potentially hazardous condition. Very rare, indeed, are failures where partial cracking across the wall occurs, followed by a sudden pipe rupture, or where a sudden rupture occurs which is not preceded by detectable prior cracking. These sudden-rupture-type failures are viewed with extreme concern, as they may result in injury of personnel or loss of life, and may become extremely costly to the plant.

The *initiation* of a failure may relate to three conditions which must be recognized in quality control and inspection. They are:

1. Structural defects and nonhomogeneities present in the original components
2. Metallurgical defects and nonhomogeneities present in the original components
3. Defects or embrittlement introduced during service in structurally and metallurgically sound materials.

Generally, only the first condition is searched for by means of nondestructive inspection techniques.

In any discussion of defects, it must be recognized that perfect piping materials and welds are not made commercially.

Structural defects may range from atom-size dislocations in the metal, which cannot be observed even under the highest power microscopes, to major metal discontinuities which are visible to the eye. Many gross defects, such as small levels of weld porosity discernible on radiographic films, may not reduce the service life of the piping component. However, many other visible surface and invisible subsurface defects can result in service failures.

Metallurgical defects represent differences in properties of sufficient magnitude to become a potential cause of failure. Such conditions can occur within the same metal, for example, where localized heat treatment produces an area of high hardness adjacent to an area of lower hardness. Metallurgical notches also can represent a higher hardness level in the heat-affected zone adjacent to base metal or weld metal, or may be due to differences in properties between a wrought base metal and a weld deposit, which can be considered to represent essentially a cast material. Localized cold work, or residual stresses also are included. More apparent metallurgical notches occur between dissimilar-metal combinations, such as clad materials, weld overlays and dissimilar weld joints.

Defects introduced during service involve cracks, embrittlement, or corrosion pits and grooves which may occur at locations where the material itself was structurally and metallurgically sound in the originally erected condition.

The ability of a defect to be propagated into a failure will depend, of course, upon the service conditions and is difficult to predict. Some obviously serious defects have not propagated at all in over 10 years of service at critical high-temperature, high-pressure steam power plant operating conditions. Other very minor or inconsequential-appearing defects have failed very rapidly under the same or very similar operating conditions.

In failure-analysis studies and attempted correlations with mechanical or laboratory simulation tests, it is also extremely important to recognize that each specific test will evaluate a particular material or a designed component in comparison with another material or component design tested under the same conditions. The interpretations of test results and their correlation with actual service characteristics are, at best, only approximate. Quite frequently, such test results have little or no relation to the actual service performance of the piping component, system, material or weld joint.

Relation of Design to Service Failures

Failures associated primarily with design may be the result of improper structural design, often involving insufficient flexibility in the piping system. Involved also are failures which start from design notches, such as sharp corners, reinforcements, or attachments to the piping which cause severe localized stresses or restraints.

Finally, these types of failures may start in welds where the weld-joint design or machined weld end preparations make difficult the deposition of a sound weld.

Piping-system Design. A failure by cracking in a steam line, associated with insufficient flexibility, is shown in Fig. 85. During periodic shutdowns of the boiler, which caused cooling of the piping system, thermal contraction occurred in the piping. Start-up of the boiler, of course, caused thermal expansion again. Insufficient flexibility in the system to absorb the stresses resulted in the initiation of cracks that gradually propagated across the pipe wall. Such cracks occur most frequently near anchor points, major branch connections, or other restraints which restrict the free movement of the piping system or section.

FIG. 85. Failure in steam piping associated with insufficient flexibility in the piping system.

Design Notches. When a piping system or its components are subjected to thermal or mechanical fatigue, cracking is likely to start at the location of notches at the outside or inside surfaces. Sharp corners and sudden changes in section wall thickness should be especially avoided where thermal or mechanical fatigue is involved. Failures have occurred in welds between heavy-wall valves and pipe of light-wall thicknesses, at socket welds (Fig. 86) and in headers with reinforcing saddles or rings where the weld does not blend gradually into the pipe wall. Similarly, attachments for pipe hangers or supports with heavy welds along the ends have failed, as they act as excessive restraints interfering with normal expansion during service involving thermal fatigue. Figure 87 shows a failure of this type.

Weld-joint Design. Unfavorable weld-joint designs have been responsible, also, for service failures. Designers who specify weld-joint preparations often do not recognize that the primary concern in weld-joint design should be joint weldability, making it as easy as possible for the welder to produce sound welds. Some weld-joint preparations are designed so that they are either difficult to produce

Fig. 86. Thermal-fatigue failure in socket weld on thermometer well is the result of design notch.

under shop or field erection environments, or are susceptible to defect conditions. This is the case not only in butt welds where good joint design should be obvious, but also in many other types of joints.   Particularly poor joint designs are encountered in valves, flow nozzles, expansion joints and similar piping components.

Some joint designs for welding involve configurations which are unsuitable for subsequent nondestructive examination of the weld deposits.   Defects or notch conditions present in the original weld joint, which were not detected, have subsequently resulted in service failures.

A number of failures can be associated with oversimplified joint designs selected primarily for the low cost of the joint preparation involved.   The 37½-deg V bevel used in butt welds, which is the cheapest to prepare and with which standard fittings

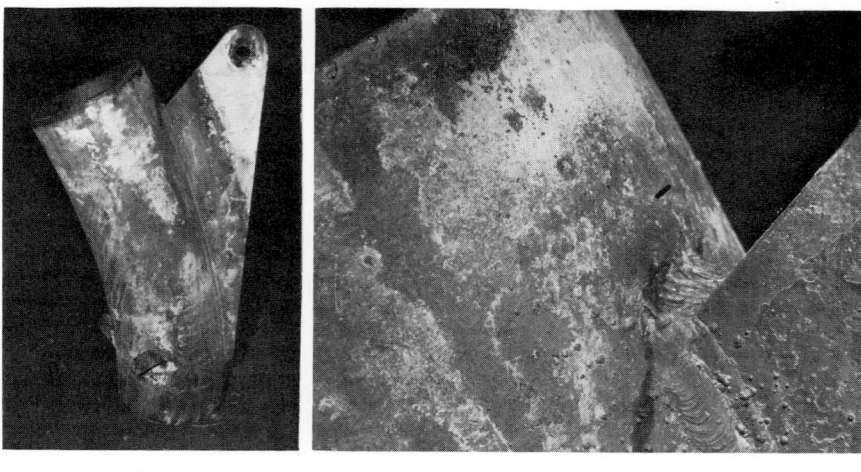

| (a) | (b) |

Fig. 87. Failure by thermal fatigue at welded hanger attachment because of sharp change in weldment design. (*a*) Reduced to approximately ⅟₂₀ size. (*b*) Reduced to approximately ⅓ size.

are usually furnished, has not been, in general, the best for critical piping systems. This has been particularly true where the first weld pass (root pass) is made by the inert-gas tungsten-arc welding process normally used in the fabrication of the piping systems of nuclear power plants. Since cracking in the first (root) weld pass is not always detected by radiographic examination of the completed weld (Fig. 88) it is very important that weld-joint designs be specified which provide the greatest assurance of weld soundness. In certain critical-use areas, an unproved joint design should be evaluated by tests that simulate as best as possible the service conditions.

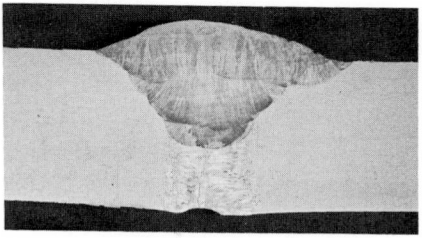

FIG. 88. Crack in root pass of stainless-steel pipe weld which was not detected by radiographic examination.

Relation of Materials to Service Failure

Base Materials. The selection of piping materials on the basis of laboratory creep, stress-rupture, fatigue or corrosion tests has been considered the primary cause of a number of failures. Piping of heavy wall thicknesses may fail in some service environments, even though in accelerated laboratory tests on small-scale specimens, failures did not occur. Examples of such failures have been the well-publicized failures of carbon-molybdenum steel piping[1] at temperatures of over 800 F because of graphitization adjacent to weld deposits in the heat-affected zone (Fig. 89). Similarly, unexpected failures have occurred in Type 347 stainless-steel piping in service at temperatures over 1,050 F. Cracking generally propagated in the heat-affected zone parallel to the weld deposited.

A substantial number of failures have occurred also in weld joints between austenitic stainless-steel thermometer wells and ferritic steel pipe, particularly where the service involves thermal fatigue at temperatures above 500 F. At temperatures above 800 F, decarburization of the ferritic steel base metal adjacent to the stainless-steel weld has further tended to accelerate crack propagation.

Weld Metal. Improper selection of the welding filler metals can be similarly critical. This may become particularly important in the joining of dissimilar-metal combinations, especially where the

FIG. 89. Crack in heat-affected zone in carbon–½ per cent molybdenum steel piping caused by graphitization after service at temperatures above 800 F.

[1] Thielsch, H., E. M. Phillips, and E. R. Jerome, Jr., "Considerations in the Evaluation of Graphitization in Piping Systems," *The Welding J.* vol. 34, no. 6, Research Suppl., pp. 286s–294s, June, 1955.

service temperatures exceed 800 F.   In most dissimilar-metal combinations, changes may take place across the interface zone, during welding and during service.

That these changes can be extremely critical, even when the chemical compositions of the materials involved are very similar, is illustrated in Fig. 90.   This shows a near failure by cracking in a repair weld which was made in the shop on a carbon–molybdenum steel valve (0.16 per cent C, 0.47 per cent Mo).   A plain carbon-steel electrode was used.   The valve was then field welded with carbon–½ per cent

FIG. 90. Near failure of carbon-molybdenum steel valve repair weld resulted from the use of carbon-steel electrodes in making the repair weld.

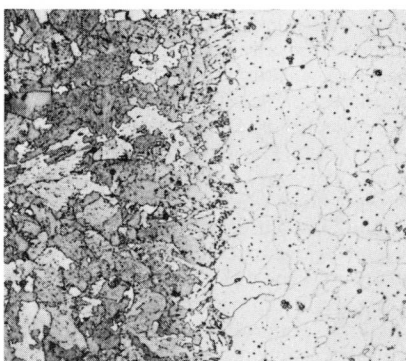

FIG. 91. Microstructure of bond area between decarburized carbon-steel weld metal and carbon–½ molybdenum steel base metal. (Magnification 500×, size reduced in printing.)

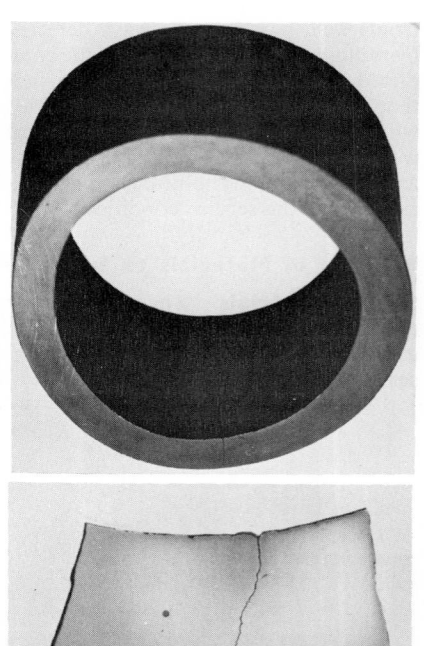

FIG. 92. Failure of tube was caused by notch formed along the inside by tube-drawing die. Cross section of tube is shown in top photo; cross section of tube wall containing crack is shown in bottom photo at a magnification of approximately 5×.

molybdenum steel electrodes into a 900 F 900-psi main-steam piping system. Although the ½ per cent molybdenum dissimilarity in the repair weld might normally be considered inconsequential, the elevated service temperature caused carbon diffusion (migration) from the carbon-steel repair weld into the adjacent carbon-molybdenum steel, as is shown in Fig. 91.   Subsequent crack initiation and propagation occurred in the decarburized zone in the steel repair weld.

Similar failures have occurred when chromium-molybdenum-alloy steel piping has been welded with carbon-molybdenum steel or with stainless-steel welding filler metals.   Although strength welds would be more critically affected, failures have occurred also where localized weld repairs have been made with dissimilar electrodes.

Relation of Base-metal Defects to Service Failure

Base-metal defects are introduced by the manufacturer or processor of the metal. Such defects may have been introduced during the casting or ingot stage, or they may have developed during the subsequent rolling, forming or forging of the ingot or billet. They may also be caused during the extrusion of pipe in a pipe mill or forging shop, or during the production of elbows, tees, or flanges by a fitting producer.

The defect conditions may be related to the following:

1. Mechanical notches
2. Metallurgical notches
3. Material identification

Mechanical notches involve actual discontinuities or separations in the material, either along the surface or on the inside of the material.

Metallurgical notches pertain to differences in properties within the metal that are of sufficient magnitude to become potential causes of failure.

Material identification should be considered an important phase of the manufacture and processing of piping materials. A substantial number of failures have occurred in piping components that were either identified incorrectly or on which no identification was made.

Mechanical Notches. Laminations, laps, scabs, tears and similar imperfections represent very common types of metal separations in steel piping. Intergranular surface defects along the inside of steel tubing, often referred to as "bark," are also common. In most cases, these conditions are not critical. They are not likely to affect the service characteristics of the material, particularly when they occur parallel to the pipe surface. In fact, seam-welded plate pipe with laminations parallel to the surface can be compared to the layered vessel construction employed by some pressure vessel manufacturers. Laminations that are perpendicular or diagonal to the pipe

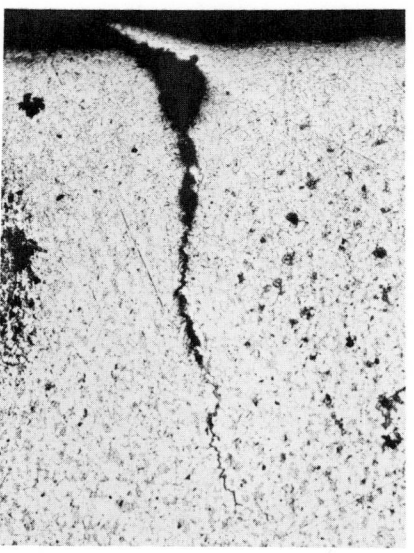

FIG. 93. Stress corrosion initiated cracking in draw-die groove in boiler tube. (Magnification 50×, size reduced in printing.)

surface should be considered more critical notches.   Heavy laminations, regardless of directionality, have led to weld metal cracking that originates at the point where the weld deposit is fused into the laminations.

Mechanical notches caused by pipe or tube drawing or extruding, by seam welding of pipe or fittings, or by machining and other production operations can have a detrimental effect on the service life of the fabricated material.   Figure 92 illustrates a tube failure that originated at a draw-die groove formed on the inside of the tube by the drawing operation. During service, mechanical fatigue initiated and propagated cracking at the notch formed by the die groove.   Environments involving stress corrosive conditions are even more likely to initiate cracking in die grooves.   An example is shown in Fig. 93.

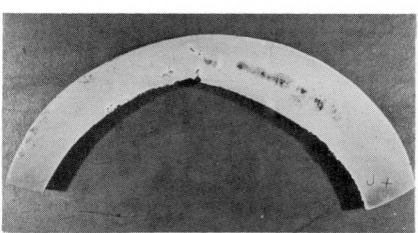

Fig. 94. Cross section of 5 Cr–½ Mo casting furnished to ASTM A217 for high-temperature refinery service, containing cracking, shrinkage defects, and slag inclusions.

Fig. 95. Cracking through ferritic steel casting delineated by dye penetrant examination and resulting from notches along inside surface.

Metal separations are particularly common in static castings. In fact, a more liberal interpretation is generally applied to radiographs of castings than of weld deposits, even though the cast components and welds may be located in the same piping system. For example, larger and more severe groupings of gas pockets are acceptable in cast valve bodies than would be permissible in weld deposits. Most internal defects rarely, if ever, result in service failures. However, on occasion, extremely defective castings have been furnished commercially (Fig. 94) and have failed subsequently in service (Fig. 95).

Metallurgical Notches. The hardness of a steel varies primarily with its chemical composition and the heat treatments given. A higher hardness tends to increase the yield and tensile strengths and reduce ductility. Failures have occurred when the difference in hardness exceeded approximately 70 to 100 points Brinell, and thermal and/or mechanical fatigue were involved in the service. Figure 96 illustrates cracking along the weld-to-base metal interface in a carbon–½ per cent molybdenum main-steam piping system. The base metal was a valve body with a

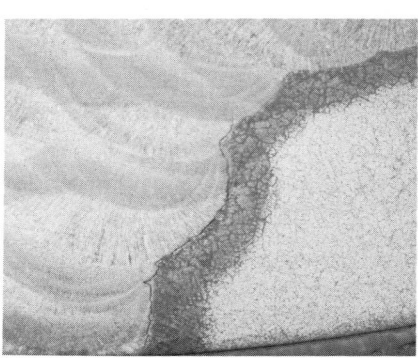

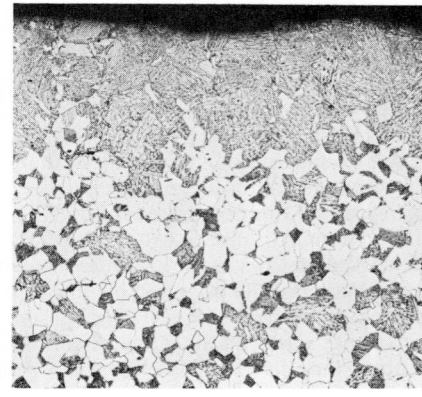

Fig. 96. Failure at the weld-to-base metal interface was caused by metallurgical notch resulting from hardness differences.

Fig. 97. Surface carburization in seamless elbow of 1¼ Cr–½ Mo steel (magnification 100×, size reduced in printing).

somewhat higher than normal carbon content (0.26 to 0.31 per cent). As is typical with castings, variations in chemical composition may intensify the susceptibility to localized hardening. As was normal procedure at the time of welding 10 years ago, the weld area was stress relieved at approximately 1150 F. The Brinell hardness values were found to be as follows:

Area	Brinell-hardness values
Valve base metal	165–189
Valve heat-affected zone	231–235
Valve heat-affected zone adjacent to weld	268–290
Weld deposit	177–181

A lower carbon content in the valve base metal or a higher heat-treating temperature, as might now be recommended, would have resulted in a reduced hardness differential along the weld metal to valve base-metal interface.

Surface metallurgy is rarely considered in evaluating plate, pipe, or similar materials. Nevertheless, carburization or decarburization may affect critically the welding and service characteristics of materials.

During hot forming into plate, pipe and fittings, care should be exercised to insure that excessive carburization or decarburization of the steel surface does not occur. For example, in the forming of steel elbows by conventional techniques, tubular sections are heated by means of gas burners to temperatures on the order of 1500 to 1850 F. The sections are subsequently "pushed" over special mandrels to form the elbows. If the gas used to heat the tubular sections is not "neutral" but instead, has a composition that tends to be carburizing, the steel surface may become severely carburized. This may occur, also, in the gas heating of flat steel shapes for hot forming into half elbows and tee sections subsequently welded together and marketed as welded fittings. Gas heating for pipe bending and straightening and many other hot-forming and shaping operations may produce similar results. Figure 97 illustrates surface carburization in a seamless elbow of a 1¼ per cent Cr–½ per cent Mo low-alloy steel composition, such as is widely used in critical high-pressure, high-temperature piping systems. Butt welds were subsequently made on this material; they included proper preheating to 500 F and stress relieving at 1325 F. These welds failed upon bending by as little as 45 deg. This is shown in Fig. 98. If a lug or guide plate for hanger attachments had been welded to this fitting or if a nozzle connection had been made, failure might have resulted in service

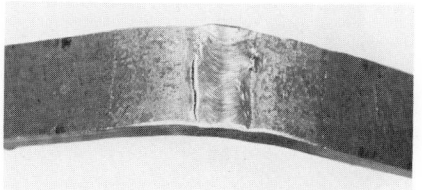

Fig. 98. Cracked bend-test specimen across butt weld in surface-carburized 1¼ Cr–½ Mo elbow.

from severe stresses caused by pipe movement. If such a failure were to occur in service, it is likely that the fabricator who made the attachment weld, rather than the fitting manufacturer, would have been held responsible. Even if a chemical-analysis certificate had been furnished, the surface carburization would not have been discovered. Only weldability tests and, possibly, photomicrographs of the surface area would have revealed this condition. To require such tests on all materials furnished would have been prohibitive in cost.

Other metallurgical conditions which have caused or contributed to service failures have involved *hot shortness*. When piping materials, heated to hot-working

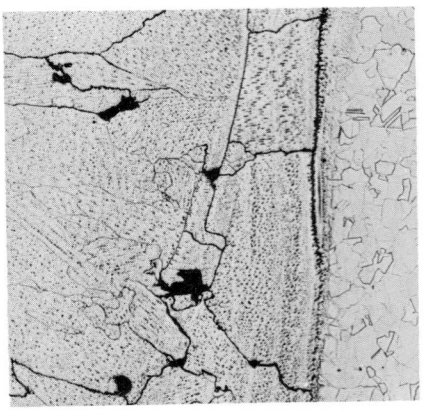

FIG. 99. Failure of two 3,000-lb couplings identified as Type 304 stainless steel occurred after being welded to Type 304 header in nuclear installation. They were actually Monel.

FIG. 100. Photomicrograph of weld metal (*left*) to base metal interface zone between Type 308 weld deposit and 70 per cent Ni, 30 per cent Cu base metal illustrates hot-short cracking caused by copper penteration. (Magnification 100×, size reduced in printing.)

FIG. 101. Failure in carbon-steel boiler tube which had been identified as 1¼ Cr–½ Mo steel. The service involved temperatures of 1000 F for which a 1¼ per cent Cr–½ Mo steel was specified.

FIG. 102. Failure by corrosion of carbon-steel pipe in boiler feedwater piping system as result of localized heating during fabrication.

or hot-forming temperatures, come in contact with nonferrous metals such as copper, lead, zinc or tin, the surface of the steel may become embrittled because of penetration of these metals into the steel. Improper control of the chemistry of steelmaking, or residual elements present has also resulted in steel that is inherently brittle, hot short, and susceptible to cracking.

Incorrect Identification. Incorrect identification of alloys has been a far more frequent cause of failure than might be expected. The danger of mixup is particularly great when markings have not been applied directly to the surface of the alloy by die stamping. However, even on die-stamped products, a significant number of errors have occurred.

Figure 99 illustrates two 3,000-lb couplings that failed after being welded to a Type 304 stainless-steel header in a nuclear installation. The couplings had been purchased as Type 304 stainless steel. To avoid possible mixup in the fabricator's shop, the manufacturer had applied the Type 304 markings directly to the surface of the couplings. Certified mill test reports also attested that the chemical composition conformed to ASTM requirements. The couplings failed in a service test. Subsequent examination revealed that the couplings actually were Monel. Welding of Monel with stainless-steel electrodes results in brittle joints, as shown in Fig. 100, because of the copper content in Monel. It is fortunate that the pressure test revealed the defective joint. It might have held sufficiently during the brief test, and then failed later in service. In a radioactive piping system, such a failure might render a plant inoperative for a long time.

A number of service failures have occurred in steam power plants when carbon-steel sections, identified incorrectly as chromium-molybdenum-alloy steel materials, have been welded into critical high-temperature high-pressure pipelines or into boiler tube systems requiring alloy steels. An example of a service failure in an incorrectly identified boiler tube is shown in Fig. 101.

Relation of Fabrication to Service Failure

Fabrication of piping involves (1) forming by hot or cold bending, extruding, or swaging; (2) cutting; (3) welding; (4) heat-treating; and (5) mechanical or chemical cleaning. When performed improperly, any of these operations may result in defects that, in high-temperature high-pressure applications, can be and have been responsible for service failures.

Not enough can be said about the importance of know-how, experience, and responsibility on the part of the fabricator. Many failures are the direct result of the lack of one or all of these qualifications. Fabricators who are involved in the fabrication of piping for critical applications should not lack the personnel experienced in welding, metallurgy, quality control, and inspection. It is also essential that the fabricator understands fully the applicable codes and specifications and is familiar with the service requirements.

Fabricators, consulting engineers, and purchasers share alike in the blame for any defective workmanship that occurs; *fabricators* for overeagerness to expand and excessive price cutting; *consulting engineers* for carelessly prepared specifications; and *purchasers* for failure to check specifications, lack of appreciation of proper quality, and misplaced cost consciousness.

Cutting. The preparation of pipe ends for welding may involve machining operations or flame or arc cutting. Frequently, a pipe is sectioned initially by flame cutting, and the pipe ends are then machined in post mills.

When machining is done, care must be taken that deep notches or grooves are not produced on the inside of the pipe. Moreover, if pipe ends are machined internally for better fit-up, the change of contour should be gradual along the inside surface.

Particular care should be exercised when flame- or arc-cutting procedures are employed. Excessive localized heating can be harmful to the material involved. Special subsequent heat treatments may then be necessary to remove the harmful heating effects. A failure by corrosion of a carbon-steel pipe used in a boiler feedwater piping system is shown in Fig. 102. Etching of the cross section evidences a change in the metallurgical structure of the steel, indicating that the pipe had been heated for the purpose of making a cut (Fig. 103). Examination of the metallurgical structure provided further evidence that the localized area had been heated to approximately 2200 F.

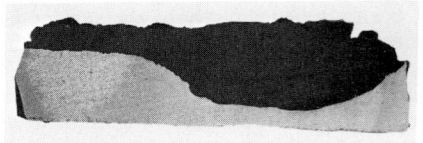

FIG. 103. Etching of cross section sliding across edge of failure in Fig. 102 reveals change in metallurgical structure caused by local heating to about 2200 F.

Welding. Pipe butt welds, of course, are usually welded from the outside only. The first weld pass in the root of the pipe joint is thus particularly critical. It is generally more difficult to make sound welds in pipe than in plates or pressure vessels which normally can be welded from both the inside and outside. Thus, in pipe welds, it is very important to avoid defects in the first or root pass.

In high-temperature high-pressure piping systems, the first pass is normally made by shielded-metal-arc welding against backing rings, or by inert-gas tungsten-arc welding. When the latter welding process is used, backing rings generally are not employed. The same or other welding processes may then be used for the subsequent welding passes.

Regardless of the welding process, the use of good joint fit-up and proper end preparation is very important.

Figure 104 illustrates a failure in a weld in a main steam piping system carrying steam at a temperature of 700 F. Improper end preparations of the 1/2-in.-thick

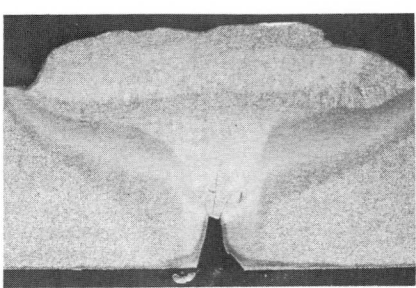

FIG. 104. Lack of penetration in root of butt weld led to failure of this carbon-steel main steam pipe joint. (Magnification 4×, size reduced in printing.)

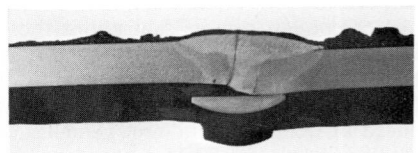

FIG. 105. Poor fit-up produced notches that resulted in failure of carbon steel pipe in process steam application, even though a backing ring was employed.

pipe, lack of a backing ring, and careless fit-up and welding resulted in considerable lack of penetration in the root of the butt weld.

Even though a backing ring was used, poor fit-up resulted in the failure shown in Fig. 105. The service involved steam in a process piping system. Notch conditions formed by the poor fit-up led to stress-corrosion cracking.

Where the root pass is made without backing rings by the inert-gas tungsten-arc welding process, cracking has occurred more rapidly than in backing-ring welds

made by shielded metal-arc welding. In the latter method, the initial weld pass provides more metal of a favorable composition and greater thickness, both of which contribute to a lesser susceptibility to cracking.

It is often not recognized, even though a weld joint has been inspected by various techniques, that such inspection is not a guarantee of soundness. Small cracks, inherent brittleness, or other defects may have been introduced by the particular welding process or filler material used.

The difficulty in determining root-pass cracks by X-ray radiography is illustrated in Fig. 106. The root pass in the 6-in. $1\frac{1}{4}$ per cent chromium–$\frac{1}{2}$ per cent molybdenum schedule 80 (0.432-in. wall) pipe was welded by the inert-gas tungsten-arc

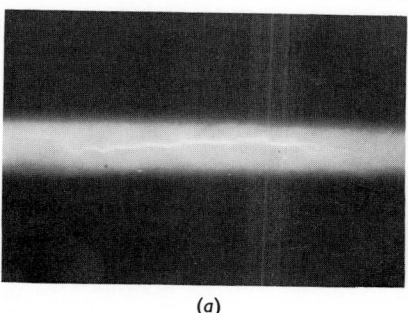 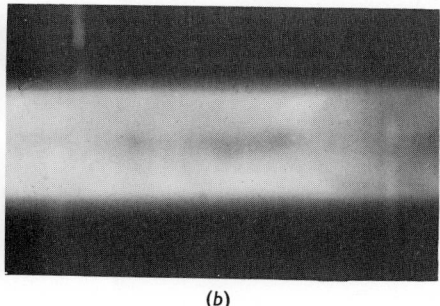

| (a) | (b) |

FIG. 106. Crack in root pass of pipe butt weld made by inert-gas tungsten-arc welding. (a) Radiograph of root pass containing a $2\frac{1}{2}$-in.-long centerline crack. (b) Crack no longer visible in the radiograph after second metal-arc pass with covered electrode.

process and contained the 2.5-in.-long center-line crack shown in Fig. 106a. As is normally practiced, the subsequent weld passes were made by shielded-metal-arc welding with $1\frac{1}{4}$ per cent chromium–$\frac{1}{2}$ per cent molybdenum-covered welding electrodes. After the second cover, or fill, pass with the covered electrodes, the root-pass crack was no longer visible by X-ray radiography (Fig. 106b). Obviously, the completed weld also appeared sound and without cracks on the radiographic films. Subsequent sectioning of the completed weld, which involved approximately 10 shielded-metal-arc welding passes revealed, nevertheless, that the root-pass crack was still present (Fig. 107).

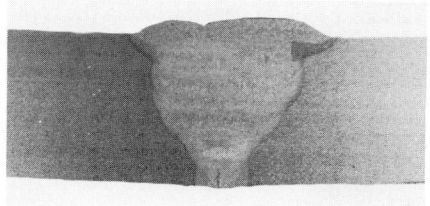

FIG. 107. Cross section of weld illustrating crack in root pass not visible by radiographic examinations after subsequent weld passes.

The "disappearance" of the crack was caused by the shrinkage of the first covered-electrode pass deposited over the inert-gas tungsten-arc root pass. The shrinkage drew the crack in the root pass tightly together. Certainly, the potential danger of such a crack is not lessened by its being tightened mechanically.

Other weld defects in the first pass, such as lack of penetration, slag, or other conditions that cause notch effects on the inner surfaces of the pipe joints, have also resulted in weld failures. Even arc strikes on pipe surfaces have provided conditions for cracking during service, particularly where the service involves thermal or mechanical fatigue and the arc strike produces a highly hardened zone. Heat

treatments that reduce the base-metal hardness adjacent to the arc strike may reduce and eliminate the tendency for this type of cracking.

Actually, there is no pipe-welding process that guarantees weld soundness. For each application, the welding techniques should be evaluated carefully under conditions identical with, or even more critical than, those that will apply to the production weld. A weld that is sound when made on a relatively short test section may contain cracks or other defects when made under conditions of restraint.

FIG. 108. Lack of penetration in root of pipe weld made by consumable method arc welding process under 75 per cent argon–25 per cent CO_2 gas protection.

Consumable metal-arc welding under the protection of CO_2, argon, or helium is becoming increasingly employed in piping applications. Although this process produces welds adequate for many noncritical services, it does not produce consistently high-quality butt welds. Such welds are susceptible to weld craters, cracks, or lack of penetration in the root pass. With the equipment and techniques presently employed, it is particularly difficult to avoid craters in the weld root adjacent to the location of the tack welds, or to produce good root fusion into the

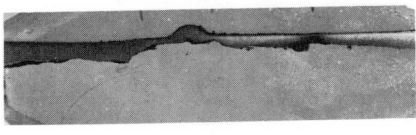

FIG. 109. Corrosion failure of boiler tube butt welds resulted from weld flash remaining on the inside of tube.

pipe base metal. The resulting lack of penetration, as shown in Fig. 108, frequently cannot be detected by means of radiographic inspection. A number of pipe failures have occurred in welds made by this process. The origin of the crack has generally involved defects in the root of the weld along the pipe inside surface.

Flash and resistance butt welding of boiler tubing can also result in problems. For example, insufficient atmosphere protection has caused improper fusion of the tube ends.

Weld flash remaining on the inside of joints can permit entrapment of corrosives or introduce other conditions enhancing localized corrosion, particularly on the fireside of the boiler tubing. A failure is illustrated in Fig. 109.

Heat Treatment. Proper heat treatment is as essential to a soundly fabricated piping system as are good design, forming and welding procedures. Many service failures can be ascribed to careless or improper heat treatments, or to the elimination of heat treatment where it should have been called for.

Among the failures ascribed to improper heat treatments are excessive hardness differences, variations in metallurgical structures, localized stresses, and surface defects.

On carbon and alloy steels that have been heated to high temperatures, rapid cooling rates may significantly increase the strength and hardness in the heated area. This can be the result of actual heat treatments, or it may be caused by cutting, welding, hot-forming, or flame-straightening operations. Differences in chemical compositions or carbon content of the steel or weldment may also produce variations in the response to heat treatments (see Fig. 96).

Conditions for failure may occur in steels when significant hardness differences are not reduced by subsequent heat treatment.

FIG. 110. Stress corrosion crack across ¼-in. wall of 2½-in.-diameter boiler tube occurred in cold bent area, starting on the surface inside directly under hanger lugs that had been welded on the outside. No stress relieving was employed after welding. (Magnification 8×, size reduced in printing.)

There have been, also, a number of cracking failures in chromium-molybdenum steel piping when repair welding was done on butt welds that had been previously stress relieved. In these cases, the repair welding was not followed by another final stress-relieving operation.

The fact that codes permit the elimination of postheat treatments does not imply that they should be eliminated in every instance. For example, stress relieving is generally not required by the major codes for 1¼ per cent chromium–½ per cent molybdenum, and 2¼ per cent chromium–1 per cent molybdenum alloy steel piping that is 4 in. in diameter or smaller, and has a wall thickness less than ½ in.[1]

A number of failures have occurred in boiler tubing which could have been avoided by a postheat treatment. In one such instance, the tubing had been cold bent, and lugs were subsequently welded onto the outside. The tubing was then cleaned by pickling and finally erected into the boiler. The cause of failure was from stress-corrosion cracking in the cold-bent area. The cracking started at the inside of the tubing, underneath the lug welds, as illustrated in Fig. 110. Stress relieving of the tubing after the cold bending and welding would have avoided these failures.

Severity of Service Application

Failures have occurred when the service became more severe than the conditions for which the piping was originally designed. Thermal or mechanical fatigue are usually the most common causes of failures in high-temperature piping systems. Severe localized mechanical stresses have caused or contributed to failures. For example, when the free movement of the piping system is restrained, binding,

[1] ASME Boiler and Pressure Vessel Code, Section I, Paragraph P-88, Notes 11 & 14; Section VIII, Table USC-56, Notes 11 and 13; ASA Code for Pressure Piping, Case 41. PFI Standard ES-8.

seizing or bending may result. Failures have been caused also by overheating, hydrogen embrittlement, creep, and still other service conditions considered excessively severe for the design application or for the materials involved. At times, several of these conditions have combined to produce the failure. For example, oxidation inside cracks will resist the normal thermal contraction when the metal cools and may contribute to crack propagation.

Thermal Fatigue. For best metallurgical conditions, the temperature of the high-temperature piping systems should be maintained continuously and uniformly. Operation economics often make it advisable, however, to shut down certain plant operations during specific periods. For example, steam power plants often have lowered demand requirements at night and, particularly, on week ends. The reduced load requirements necessitate the shutdown of some boilers at night, and of a larger number of boilers on week ends. For example, during a week-end shutdown, the piping carrying the steam from a boiler to the turbine may cool from the 900 to 1100 F normal operating temperatures to about 500 to 600 F. This causes contraction of the piping system, the boiler tubing and welded attachments in accordance with the coefficients of expansion of the materials involved. On Sunday night, when the power-plant unit is returned to normal operations, the boiler tubing, steam piping, and turbine casings are heated up again by the steam produced in the boiler. Thermal expansion occurs in all metal components and the piping systems expand to their normal hot position.

FIG. 111. Cracking in main steam pipe section caused by thermal fatigue.

Over a period of months, the resulting thermal fatigue may be more harmful to the materials involved, particularly at mechanical or metallurgical notches, than continuous operation would be at a particular temperature over a period of many years.

Cracking in a main steam line caused by thermal fatigue is shown in Fig. 111. Sometimes single cracks occur; however, more often, a whole family of parallel cracks occur which usually start from the inside of the pipe. The inner surface is subject to more severe thermal

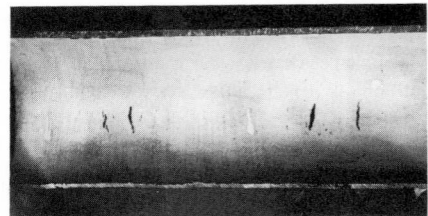

FIG. 112. Parallel cracks on inner surface of pipe section subject to thermal and mechanical fatigue.

shock than is the outer surface. These parallel cracks (Fig. 112) gradually propagate until they reach the outer surface.

Many failures have occurred in temperature-control sections in which water is sprayed on the pipe inside. The quenching effect results in temperature cycling over the range 300 to 500 F and causes failures across the pipe wall, as in Fig. 113.

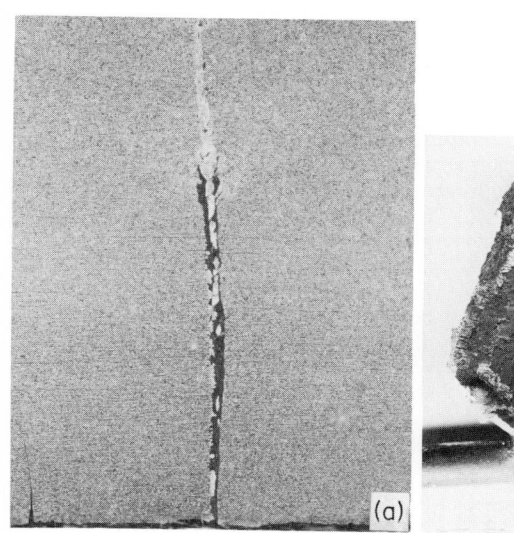

FIG. 113. Cracks caused by thermal fatigue in pressure-reducing section. (*a*) Photograph of inner pipe surface. (*b*) Cross section magnified approximately 5×.

The insertion of stainless-steel liner plates inside the pipe generally has not provided an improvement. These liners have failed, also (Fig. 114), usually just as readily as the pipe has failed prior to the use of the liners.

The vast majority of thermal-fatigue failures occurs gradually. In time, a short crack becomes visible on the other side of the pipe wall. Steam or water will then usually leak through and provide a warning of the hazardous condition. Though rarely, sudden ruptures, however, do occasionally occur. Figure 115 illustrates a sudden rupture of a stainless-steel pipe on the outside of a vessel resulting in a serious injury. The stainless-steel pipe carried steam at approximately 550 F.

Mechanical Fatigue. Even under continuous high-temperature operation, some failures have occurred. Usually, these can be related to some degree of

FIG. 114. Cracks in stainless-steel liner plate welded on inside of pressure-reducing pipe section.

FIG. 115. Rupture of stainless-steel pipe at nozzle weld preceded by severe thermal-fatigue cracking on inside of vessel.

mechanical fatigue resulting from pipe movement, vibrations, restraints preventing free movement, or other conditions. Such failures often show a number of parallel cracks.

Overheating. Failures caused by overheating of boiler tubing have been already extensively discussed in the literature. Such failures are usually confirmed by considerable differences in the microstructure.

Hydrogen Embrittlement. Hydrogen-embrittlement failures in boiler tubing have occurred in a number of power plants. An example of such a brittle failure is shown in Fig. 116. These failures involve service at temperatures between 600 and 1000 F, and are usually preceded by considerable localized scaling. Moreover, the microstructure shows decarburization.

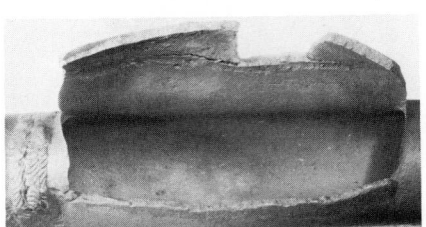

FIG. 116. Failure in carbon-steel boiler tube as result of hydrogen embrittlement. Reduced to approximately ¼ size.

The failures illustrated by no means exhaust the conditions leading to service failures in piping and tube materials. In addition, a substantial number of failures have been caused by corrosion.[1] Frequently, it is quite difficult to diagnose the basic cause involved in a failure. Often, service requirements necessitate an immediate repair, which makes it impossible to conduct an objective and conclusive analysis, even though such an analysis may be desirable to determine the causes involved in the failure and, by so doing, to prevent a recurrence.

[1] Thielsch, H., "Combating Corrosion in Service Environments for the Prevention of Pipe Failures," *Heating, Piping and Air Conditioning*, vol. 36, pp. 139–147, January, 1964.

3

FLUID MECHANICS

Reno C. King*

PROPERTIES OF FLUIDS

A *fluid* is a substance which cannot sustain a shear stress in a condition of static equilibrium. Fluids offer no resistance to distortion of form; they yield continuously to tangential forces, no matter how small. Ordinarily, fluids are classified as being liquids or gases. Some classifications also include the vapor form among the group of fluids.

A *liquid* is a fluid which occupies a definite volume independent of the shape or size of the vessel in which it is contained. Liquids are characterized by their high resistance to compression; they change volume very slightly with considerable variation in pressure, and when the pressure is removed, they do not dilate significantly.

A *gas* is a fluid which tends to expand to fill completely any vessel in which it is contained. A change in pressure is accompanied by a change in volume.

A *perfect gas* satisfies two conditions: (1) it obeys the gas law $PV = mRT$, and (2) its specific heats are constant. In (1) above, P, V, and T are pressure, volume, and temperature, respectively, and m and R represent the quantity of gas and the gas constant, all in proper units. From (2) it follows also that the specific-heat ratio, $k = c_p/c_v$, is also constant.

A real gas, at a pressure low with respect to sea-level barometric pressure and at a temperature high with respect to its critical temperature, obeys with reasonable accuracy the perfect-gas law. If higher degrees of accuracy are required, textbooks in thermodynamics list corrections (as, for example, van der Waals' equation) which take into account deviations from ideal conformance.

* Professor of Mechanical Engineering, New York University, New York, N.Y.

All liquids under certain conditions of pressure and temperature will become vapor. Depending on the amount of energy associated with the change in state, varying amounts of vapor will be formed. The addition of precisely the correct quantity of energy will vaporize all the liquid, and the vapor formed is referred to as *saturated vapor*. If still more energy is added, the vapor will increase in temperature, even though its pressure may remain constant, and such vapor which exists at a temperature above that of its saturation point is termed *superheated vapor*.

Vapor which is in contact with liquid at the same temperature, saturated vapor, and slightly superheated vapor all deviate markedly from the perfect-gas law. A highly superheated vapor at a low pressure very often obeys the perfect-gas law with sufficient accuracy for many engineering calculations. Properties of vapors are given generally in graphical or in tabular form.

General Fluid Properties. Fluids are identified by certain properties such as are listed and defined below.

The *bulk modulus* of a fluid is a direct measure of its compressibility and is defined as the ratio of the pressure change to the unit change in volume resulting from the change in pressure. Mathematically,

$$B = -\frac{dp}{dv/v} = -v\frac{dp}{dv}$$

where B represents the bulk modulus. The negative sign appears in order that the bulk modulus be positive, since a positive pressure change produces a negative change in volume.

For incompressible fluids such as oil or water, the above definition suffices. It is convenient, in the case of gases, to refer to the conditions which existed at the time the pressure and volume measurements were made. If the temperature was maintained constant by means of heat transfer to or from the fluid, the corresponding bulk modulus is termed the *isothermal* bulk modulus B_T. If no heat was transferred to or from the fluid and the measuring process was carried out slowly, the corresponding ratio is termed the *adiabatic* (or isentropic) bulk modulus B_S. Thus

$$B_T = -v\left(\frac{\partial p}{\partial v}\right)_T \quad \text{and} \quad B_S = -v\left(\frac{\partial p}{\partial v}\right)_S$$

The *capillarity* of a liquid relates to a discontinuity in its free surface at or near a bounding surface. If the liquid wets the solid (water and glass), *adhesion* of the liquid to the solid particles is greater than is *cohesion* between the liquid particles. Conversely, if the liquid does not wet the solid (mercury and glass), cohesive forces between the liquid particles are greater than are the adhesive forces between the liquid and solid particles. The effects of capillarity are most noticeable when small-diameter glass tubes are used as manometer gages. These effects will cause water to stand higher and mercury to stand lower than would be the case if larger diameter tubes were used. In both examples, the curved upper surface of liquid within the tube is referred to as the *meniscus*, and all manometer readings should be taken at the level of the middle of the meniscus. The capillary effect can be reduced greatly if tubes of at least ⅜ in. diameter are used as manometers.

The *density* of a fluid is defined as the mass of that fluid per unit volume. Ordinarily, this definition is sufficiently precise, but when one is working with gases in rarefied atmospheres, it is emphasized that the volume must be large enough to contain a representative number of particles. If ΔV is the volume of a fluid mass Δm and $\Delta V'$ is the smallest volume for which the fluid can be considered a continuum, then the density ρ is

$$\rho = \lim_{\Delta V \to \Delta V'} \Delta m/\Delta V$$

Density may be expressed in a variety of units. In fluid mechanics and in English units, density is often stated in slugs per cubic foot where a slug is defined as pounds mass divided by g_c, the gravitational constant equal to 32.17 ft lb/(lbf sec²). Table 1 lists density conversion factors.

The reciprocal of density is termed the *specific volume*. Thus specific volume v is given by

$$v = \lim_{\Delta V \to \Delta V'} \Delta V / \Delta m = 1/\rho$$

Specific weight is a term often used instead of density. It is defined as the weight (as contrasted to mass) per unit of volume and in English units is expressed as pound force per cubic foot. It is noted that, if the weight at sea level is used in the determination of specific weight, the density and specific weight will be identical numerically.

Table I. Density Equivalents

Grams/cu cm	Lb/cu in.	Lb/cu ft	Slugs/cu ft
1	0.03613	62.43	1.94
27.68	1	1,728	53.71
0.01602	0.0005787	1	0.031
0.5148	0.0186	32.17	1

The *free surface* of a liquid at rest is a surface of equal pressure normal to the line of action of the force of gravity and, in ordinary engineering work, may be considered as horizontal.

The *specific heat* of a fluid, for ordinary engineering purposes, can be considered as the quantity of heat required to change the temperature of a unit mass of the fluid by one degree. In English units, the specific heat is usually expressed in Btu per pound mass per degree Fahrenheit. It can be shown easily that a specific heat value in these units is numerically equal to the specific heat in gram calories per gram per degree centigrade [cal/(g C)].

The numerical value of specific heat is dependent upon the thermodynamic process employed during its measurement. Two values are of principal engineering importance, and these are the constant-pressure and constant-volume specific heats. They are defined as

$$c_p = \left(\frac{\Delta Q}{\Delta T}\right)_p$$

and

$$c_v = \left(\frac{\Delta Q}{\Delta T}\right)_v$$

where c_p and c_v are the constant-pressure and constant-volume specific heats. ΔQ represents the quantity of heat transferred to a unit mass of fluid while its temperature changes ΔT deg. The subscripts p and v represent that the pressure or the volume remains constant during the interval of heat addition.

Values of specific heat for water are given in Fig. 24a and b, for liquid petroleum products in Table 35, and for gases, in Table 33.

The *viscosity* of a fluid is one measure of its deviation from the concept of an ideal fluid. It is of great significance in fluid mechanics, since it indicates the ease with which the fluid flows through a pipe or conduit. A high value of viscosity, all other factors being constant, results in a high friction drop.

The force F required to move a fluid layer of surface area A located a distance D above a stationary surface with a velocity V is related as follows (see Fig. 1):

$$F\alpha(V/D)A$$

or
$$F = \mu(V/D)A \tag{1}$$

The coefficient μ is termed the *absolute viscosity* or *dynamic viscosity*. Its dimensions, in the English system, are usually given either as

$$|\mu| = \text{lbf sec/sq ft} \qquad \text{or} \qquad |\mu| = \text{lbm/(ft sec)}$$

The *kinematic viscosity* ν is obtained by dividing the absolute viscosity by the fluid density ρ. Thus, if density is expressed in pound mass per cubic foot, the kinematic viscosity can be expressed as

$$|\nu| = \text{sq ft/sec}$$

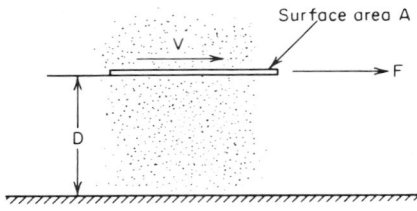

Surface area A

FIG. 1. Concept of absolute viscosity.

In the cgs system of units, the absolute viscosity is expressed in *poises* or in *centipoises*. If in Eq. (1), the velocity is in centimeters per second, the distance is in centimeters, the area is in square centimeters, and the force is in dynes, then the dimensions of absolute or dynamic viscosity will be in poises or in grams per centimeter per second. A more frequently used expression is the *centipoise*, which is equal to 1/100 poise. To convert centipoises into other units, multiply the viscosity in centipoises by constants as given below:

0.01 for viscosity in poises
0.000672 for viscosity in lbm/(ft/sec)
3.60 for viscosity in kg mass (hr meter)
2.068×10^{-5} for viscosity in (lbf sec)/sq ft

In the cgs system of units, the kinematic viscosity is expressed in *stokes* or *centistokes* in accordance with

$$\nu = \mu/\rho$$

For μ in poises (grams per centimeter per second) and density in grams per cubic centimeter, the dimensions of ν will be square centimeter per second or in stokes. To obtain centistokes, multiply stokes by 100. Table 2 lists values for conversion between centistokes and the corresponding indications of various viscometers.

FLUID PRESSURE

Pressure at a Point. Pressure, here assumed to be a continuous function of time and space within a fluid, is defined as the force per unit area between adjacent fluid surfaces or between a fluid and the face of a solid. If a total force ΔF acts on an elementary area ΔA, then the pressure at a point is defined as

$$P = \lim_{\Delta A \to 0} \Delta F/\Delta A$$

and if ΔA is considered to be very small, as at a point in the fluid, the above expression becomes

$$dF = P\,dA$$

or
$$F = \int_A P\,dA$$

Table 2. Kinematic Viscosity Conversion Table

(Centistokes to Engler, Saybolt, and Redwood units)

Centi-stokes	Engler degrees	Saybolt seconds at 130 F	Redwood seconds at 140 F	Centi-stokes	Engler degrees	Saybolt seconds at 130 F	Redwood seconds at 140 F	Centi-stokes	Engler degrees	Saybolt seconds at 130 F	Redwood seconds at 140 F
2.0	1.140	32.66	30.95	18.0	2.644	89.37	78.45	41.0	5.465	190.6	168.3
2.5	1.182	34.46	32.20	19.0	2.755	93.48	82.10	42.0	5.590	195.1	172.3
3.0	1.224	36.07	33.45	20.0	2.870	97.69	85.75	43.0	5.720	199.6	176.4
3.5	1.266	37.67	34.70	21.0	2.984	101.9	89.50	44.0	5.845	204.2	180.4
4.0	1.308	39.17	35.95	22.0	3.100	106.2	93.25	45.0	5.975	208.8	184.5
4.5	1.350	40.78	37.20	23.0	3.215	110.5	97.05	46.0	6.105	213.4	188.5
5.0	1.400	42.38	38.45	24.0	3.335	114.8	100.9	47.0	6.235	218.0	192.6
5.5	1.441	43.98	39.80	25.0	3.455	119.1	104.7	48.0	6.365	222.6	196.6
6.0	1.481	45.59	41.05	26.0	3.575	123.5	108.6	49.0	6.495	227.2	200.7
6.5	1.521	47.19	42.40	27.0	3.695	127.9	112.5	50.0	6.630	231.8	204.7
7.0	1.563	48.79	43.70	28.0	3.820	132.4	116.5	52.0	6.890	241.1	212.8
7.5	1.605	50.44	45.00	29.0	3.945	136.8	120.4	54.0	7.106	250.3	221.0
8.0	1.653	52.10	46.35	30.0	4.070	141.2	124.4	56.0	7.370	259.5	229.1
8.5	1.700	53.80	47.75	31.0	4.195	145.6	128.3	58.0	7.633	268.7	237.2
9.0	1.746	55.51	49.10	32.0	4.320	150.0	132.3	60.0	7.896	277.9	245.3
9.5	1.791	57.21	50.55	33.0	4.445	154.5	136.3	62.0	8.159	287.2	253.5
10.0	1.837	58.91	52.00	34.0	4.570	159.0	140.2	64.0	8.422	296.4	261.6
11.0	1.928	62.42	55.00	35.0	4.695	163.5	144.2	66.0	8.686	305.6	269.8
12.0	2.020	66.03	58.10	36.0	4.825	168.0	148.2	68.0	8.949	314.8	277.9
13.0	2.120	69.73	61.30	37.0	4.955	172.5	152.2	70.0	9.212	324.0	286.0
14.0	2.219	73.54	64.55	38.0	5.080	177.0	156.2	72.0	9.475	333.3	294.1
15.0	2.323	77.35	67.95	39.0	5.205	181.5	160.3	74.0	9.738	342.5	302.2
16.0	2.434	81.25	71.40	40.0	5.335	186·0	164.3	75.0	9.870	347.2	306.3
17.0	2.540	85.26	74.85								

Supplementary Kinematic Viscosity Conversion Table

Centistokes..............	2	6	10	20	30	40	50	60	70
Saybolt at 100 F	32.60	45.50	58.80	97.50	40.9	185.7	231.4	277.4	323.4
Saybolt at 210 F	32.83	45.82	59.21	98.18	141.9	187.0	233.0	279.3	325.7
Redwood at 70 F	30.20	40.50	51.70	85.40	123.7	163.2	203.3	243.5	283.9
Redwood at 200 F	31.20	41.50	52.55	86.90	126.0	166.7	208.3	250.0	291.7

Hydrostatic Pressure. The pressure existing at a point within a fluid body due to the weight of the fluid above that point is known as the hydrostatic pressure. In the case of gaseous fluids, the weight of the fluid column is relatively small unless great vertical heights are involved. With denser fluids such as liquids, the increase in pressure due to depth within the liquid can be of great significance.

If P_0 is the pressure at the free surface of a liquid whose specific weight is w, then at a depth h within the fluid the pressure is

$$P = P_0 + \int_0^h w \, dh \qquad (2)$$

which, for a single homogeneous fluid reduces to

$$P = P_0 + wh$$

Equation (2), properly interpreted, is applicable to determination of hydrostatic pressure in any circumstance. When applied to the two-fluid mixture in a container as shown in Fig. 2, the expression for hydrostatic pressure at a horizontal plane located a distance h below the free surface is

$$P = P_0 + w_1 h_1 + w_2 h_2$$

Dimensions of Pressure. Pressure is often expressed in terms of an equivalent height of fluid column. In Fig. 2, for example, if the vessel were open to atmosphere, P_0 would be atmospheric pressure; the pressure at a horizontal plane located a distance h_1 beneath the free surface could then be said to be h_1 units of fluid of specific weight w_1.

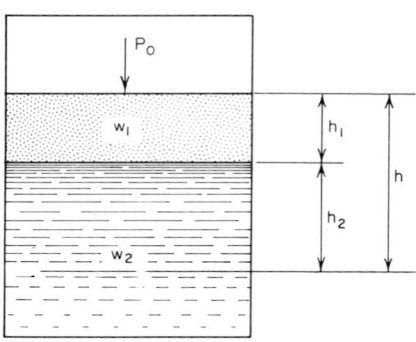

In the English system of units, pressure is generally assigned the dimensions of pounds per square inch or per square foot, of so many feet or inches of water, or of inches of mercury or any other specific fluid that is used as the measuring medium in a manometer. Table 3 lists in a convenient form factors by which pressures in different units can be converted.

FIG. 2. Hydrostatic pressure in two-fluid mixture.

Gage and Absolute Pressure. Most pressure and vacuum gages read the difference between the *absolute* pressure of the fluid and the pressure of the atmosphere, and this difference is referred to as the *gage* pressure. Thus, to determine the absolute pressure of a fluid from the reading of a pressure gage, it is necessary to add to the gage pressure the pressure of the atmosphere at the location of the pressure gage.

Standard atmospheric pressure is defined as that pressure produced by a column of mercury of 760 mm length at a mercury density of 13.5951 gm/cu cm and at an acceleration due to gravity of 32.1740 ft/sec². On this basis, then,

$$1 \text{ standard atmosphere} = 14.6959 \text{ psi}$$
$$= 1.01324 \times 10^6 \text{ dynes/sq cm}$$

For many engineering calculations, it is sufficiently accurate to use 14.7 psi as being equivalent to one standard atmosphere.

In high-vacuum work (very low absolute pressures), pressures are often measured in microns of mercury. A micron is one-millionth of a meter and

$$1 \text{ micron mercury} = 10^{-6} \text{ meter} = 1.933 \times 10^{-5} \text{ psi}$$

Definition of Heads. The term "head" has various special meanings in connection with fluids. According to accepted usage, these conceptions of head may be defined as follows:

The *velocity head* of a fluid moving with a given velocity is the equivalent height through which a body must fall to acquire the same velocity. It may be expressed in symbols[1] as

---

[1] Alternatively, velocity head can be expressed as $h = V^2/2g_c$. A dimensional check will disclose that the units of h will then be ft lbf/lbm; that is, a velocity head can be interpreted as having the dimension of energy.

Table 3. Conversion Factors for Units of Pressure and Head

To convert from	Pounds per square inch	Ounces per square inch	Pounds per square foot	Kilograms per square meter	Kilograms per square centimeter	Atmospheres	Feet of fresh water	Feet of sea water	Inches of fresh water	Meters of fresh water	Inches of mercury	Millimeters of mercury
						Multiply by						
Pounds per square inch	1.000	16.000	144.00	703.067	0.07031	0.06804	2.3067	2.2504	27.68	0.70307	2.036	51.712
Ounces per square inch	0.0625	1.000	9.000	43.942	0.004394	0.004253	0.14417	0.14065	1.7300	0.04394	0.1272	3.2320
Pounds per square foot	0.006945	0.1111	1.000	4.8824	0.0004882	0.0004725	0.01602	0.01563	0.1922	0.004882	0.01414	0.3591
Kilograms per square meter	0.001422	0.02275	0.2048	1.000	0.0001	0.00009676	0.00328	0.003201	0.03937	0.0010	0.002896	0.07355
Kilograms per square centimeter	14.223	227.57	2,048.2	10.000	1.000	0.9676	32.81	32.01	393.72	10.000	28.96	735.51
Atmospheres	14.696	235.136	2,116.35	10,332.9	1.0333	1.000	33.90	33.07	406.80	10.329	29.921	760.00
Feet in height of fresh water	0.4335	6.9360	62.428	304.801	0.03048	0.02949	1.000	0.9756	12.000	0.3048	0.8826	22.419
Feet in height of sea water	0.4443	7.1094	63.9887	312.420	0.03124	0.03024	1.0250	1.000	12.300	0.3124	0.9047	22.979
Inches in height of fresh water	0.03613	0.5780	5.2023	25.400	0.002540	0.002458	0.08333	0.08130	1.000	0.0254	0.07355	1.8682
Meters in height of fresh water	1.422	22.756	204.82	1,000	0.1000	0.09678	3.2808	3.2008	39.3696	1.000	2.8957	73.5514
Inches in height of mercury	0.4912	7.859	70.731	345.34	0.03453	0.03342	1.1329	1.1053	13.596	0.3453	1.000	25.400
Millimeters in height of mercury	0.01934	0.3094	2.7847	13.596	0.001359	0.001316	0.04461	0.04352	0.5353	0.01359	0.03937	1.000

Example.—To convert from atmospheric pressure in pounds per square inch (14.696) to inches of mercury or feet of water: 14.696 × 2.036 = 29.921 in. Hg and 14.696 × 2.3067 = 33.90 ft water.

Note.—The following values were used in calculating this table: Specific gravity of mercury = 13.596; specific gravity of sea water = 1.025 (average); density of fresh water = 62.428 lb per cu ft.

$$h = V^2/2g \tag{3}$$

in which V represents velocity and g the acceleration due to gravity, both expressed in the same units of space and time. In common engineering practice, h is the vertical feet of head of the particular fluid; V is its velocity in feet per second, g is the acceleration due to gravity of 32.2 ft per sec². Velocity head is that head which represents kinetic energy due to directional motion of the fluid through a pipe, flume, or similar conduit.

The *pressure head* or *static head* of a fluid is that head which is due to the application of external force, such as is produced by a pump or a compressible fluid (gas or vapor) with which it is in contact, or by the weight of a column of fluid which it supports.

The *elevation head* is that head representing vertical distance above the elevation taken as datum.

The *friction head* represents that loss of head in a moving fluid expended in overcoming frictional resistance to flow.

Heads or *pressures* are measured from some reference point or datum fixed by conditions of the problem. Intensity of pressure, unless otherwise stated, is measured from atmospheric pressure as zero and is called "gage pressure." *Absolute pressure* is gage pressure plus atmospheric pressure. The use of the terms "head" and "pressure" is interchangeable.

STRESS DUE TO FLUID PRESSURE

Stress Reactions Produced in Cylinders by Fluid Pressure. In analyzing the stress conditions set up in the walls of a vessel by internal or external fluid pressure, consideration must be given to the way the walls of the vessel react to support the pressure. The most common form of vessel supporting fluid pressure

(a)   (b)

FIG. 3. Stress conditions in (a) cylinder with open ends, (b) cylinder with closed ends.

has, in general, the properties of a right circular hollow cylinder. Pipes, tubes, and other circular conduits come under this classification, and, for convenience, the term "cylinder" will be here considered as embracing all such conduits. It should be clearly understood at the start, in order to avoid needless repetition, that the stress formulas here derived apply to sections through the cylinder walls so far removed from closed ends, end flanges, or other stiffening elements as not to receive any reinforcement from them. Ordinarily, this condition is met at a distance of four to six cylinder diameters from the reinforcement.

Internal or external fluid pressure may give rise (1) to a tangential or hoop stress within the wall of the cylinder, (2) to both a hoop and a longitudinal stress acting jointly. In either case the wall is also under a direct radial compressive stress whose magnitude is usually so small with regard to the other two that it may be neglected. Case 1 is illustrated in Fig. 3a, and Case 2, which represents the usual condition, is illustrated in Fig. 3b.

The following symbols are used throughout this section to designate the corresponding items in the same system of measure:

d = internal diameter of cylinder (or pipe), linear units
D = external diameter of cylinder, linear units
t = thickness of cylinder wall, same linear units
p = internal fluid pressure, weight units per area unit
P = external fluid pressure, weight units per area unit
S = fiber stress, same weight unit per area unit
λ = Poisson's ratio (coefficient of lateral contraction) for values see Table 4
F = force, same weight units, produced by a pressure p acting on a given area

Table 4.   Poisson's Ratio and Average Values of the Modulus of Elasticity for Common Piping Materials at Room Temperature

	Poisson's ratio λ	Modulus of elasticity in tension E
Brass or bronze	0.333	14,000,000
Copper	0.333	16,000,000
Cast iron	0.270	12,000,000
Cement asbestos	0.10 to 0.20	3,400,000
Concrete	0.10 to 0.20	3,000,000
Lead	0.450	2,500,000
Steel	0.303	30,000,000
Wrought iron	0.278	28,000,000
Wood	....	1,500,000

Where additional symbols or subscripts are used, their significance is explained in the immediate context.

Figure 3a illustrates a cylinder with frictionless plungers fitted into its ends, the plungers being kept in place by external forces F_2 which exactly balance the internal fluid pressure tending to force them outward.   In this case the tube wall is subjected to only the internal forces shown as acting at right angles to its inner surface.   It is obvious that these forces can give rise to radial and hoop stresses only in the tube wall.

Figure 3b illustrates the ordinary case of a cylinder with both ends closed.   In this case the tube wall is subjected not only to the hoop and radial stresses mentioned above but, at the same time, to a longitudinal stress at right angles to the tangential or hoop stress.

Derivation of Formula for Hoop Stress.   The derivation of the formula for hoop or tangential stress is as follows: Let Fig. 4 represent one-half portion of a cylinder of length l.   Consider the half cylinder as a free body and resolve all the elementary radial forces due to internal pressure into their components perpendicular to the cutting plane.   The component of these

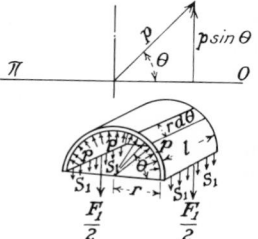

Fig. 4. Derivation of loop stress.

elementary forces perpendicular to the cutting plane is represented by $plr\,d\theta \sin \theta$. The total force normal to this plane is the summation of these elementary forces over one-half circumference, or from 0 to π in radian measure.   This summation may be written as follows:

$$F_1 = \int_0^\pi plr\, d\theta \sin \theta = plr \int_0^\pi \sin \theta\, d\theta = 2plr = pld$$

This normal force is opposed by an equal force on the other half of the cylinder, acting in the opposite direction 180 deg from the first force. These opposing forces F_1 produce a hoop stress S_1 in the cylinder walls equal to the force F_1 divided by the area over which the stress is distributed. This area is $2tl$, and $S_1 = F_1/2tl$ or $F_1 = 2tlS_1$. Equating the two values obtained for F_1, $pld = 2tlS_1$, canceling l, and transposing,

$$S_1 = pd/2t \tag{4}$$

These relations are independent of the length of the cylinder, provided the section is far enough removed from the flanges or other reinforcements to receive no support from them. Equation (4) is generally known as the hoop-stress formula and is frequently used in calculations of bursting strength for thin-walled cylinders without taking into account the effect of lateral contraction; it is assumed that the hoop stress is uniformly distributed across the cylinder wall. This condition does not exist, except in the case of cylinders having walls of infinitesimal thickness.

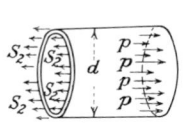

FIG. 5. Derivation of longitudinal stress.

Empirical formulas such as that of Barlow and its modifications used in various codes differ from the hoop-stress formula in the substitution of the outside diameter for the inside diameter.

Derivation of Formula for Longitudinal Stress. The derivation of the formula for longitudinal stress in a cylinder with closed ends is as follows: Let Fig. 5 represent a section of a cylinder with closed ends. The total force F_2 acting against each end of the cylinder is $F_2 = p\pi d^2/4$. This force is distributed over an area of metal in the cylinder wall equal to $\pi(d + t)t$ and the unit stress is $S_2 = F_2 \div \pi(d + t)t$. Substituting the value above,

$$S_2 = p\pi d^2/4 \div \pi(d + t)t$$

In the case of thin-walled cylinders the value of $(d + t)$ approaches d, and an approximate equation may be written

$$S_2 = p\pi d^2/4 \div \pi dt = pd/4t \tag{5}$$

which is equal to one-half the hoop stress of Eq. (4).

Radial Stress. There is a direct compression on the inside surface of a pipe equal to the internal pressure. As this stress is small in relation to the hoop and longitudinal stresses, it is usually neglected although it has some bearing on the theories of bursting strength applicable to pipe. It is designated as S_3 for convenience in combining with the hoop and longitudinal stresses in Eq. (6).

Combination of Hoop, Longitudinal, and Radial Stresses. When a material is subjected to mutually perpendicular stresses, as in the case of pipe under internal pressure. it is necessary to employ some theory of strength to relate failure of material under the action of such stresses to failure of the same material under simple tensile stress in a testing machine. Four principal theories have been evolved based on different concepts of the manner in which material fails under a combination of stresses. This failure may be due to exceeding a given value of: (1) the maximum tensile stress; (2) the maximum strain; (3) the maximum shear stress; or (4) a function of the differences of the principal stresses.

In the first, or maximum-stress theory, only hoop tension is considered to be related to failure of the pipe, the longitudinal and radial stresses having no effect.

In the second, or maximum-strain theory, the longitudinal stress is considered to strengthen the material, the amount of the strengthening depending upon the value of Poisson's ratio for the particular material. Clavarino's formula [see Eq. (9)] is a direct application of the maximum-strain theory. The third theory, that of

maximum shear, holds that failure is dependent only on the greatest and least stresses acting. Since the longitudinal tension is intermediate in value between the hoop tension and the radial compression, it does not, according to this theory, affect the stress at which failure will occur.

The fourth theory, known as the Mises-Hencky maximum-energy-of-distortion theory, holds that yielding failure takes place in accordance with a relation given by the differences of the principal stresses. This relation is expressed by the following equation:

$$(S_1 - S_2)^2 + (S_2 - S_3)^2 + (S_3 - S_1)^2 = 2S_0^2 \tag{6}$$

in which S_1, S_2, and S_3 are the principal stresses, hoop, longitudinal, and radial, respectively, and S_0 is the yield strength of the material determined from simple tensile tests.

Collapsing Pressure. If a thin-walled ring or short section of pipe is subjected to external fluid pressure, a compressive stress is set up in the walls. By an analysis similar to that used in the case of internal pressure, it can be shown that here again $S = pD/2t$, S being a compressive stress and D the outside diameter.

Taking the effect of lateral contraction into account, the collapsing pressure of a long circular tube having a perfectly circular form is given by the following equation:

$$P = \frac{2E}{1 - \lambda^2}\left(\frac{t}{D}\right)^3 \tag{7}$$

where P is the critical value of external pressure, E is the modulus of elasticity, λ is Poisson's ratio, and t and D are wall thickness and mean diameter of tube, respectively, all dimensions being expressed in inches. For usual pipe-wall thicknesses, D may be taken as the outside diameter. This formula can be used for calculating the critical value of the collapsing pressure provided that the corresponding compressive stress does not exceed the yield strength of the material. The limiting value of the ratio t/D may be found from the relation

$$S = \frac{E}{1 - \lambda^2}\left(\frac{t}{D}\right)^2 \tag{8}$$

The above equations apply to ideal tubes of perfectly circular form. Commercial pipe may deviate as much as 1 per cent over or under the specified outside diameter,[1] and the failure of pipe under external pressure is profoundly affected by initial eccentricity. Maximum values of external pressure on the basis of the initial eccentricity may be calculated by elastic theory.[2]

Values of the critical stress and the *maximum* external pressures which steel pipe having an initial eccentricity of 1 per cent can support without collapsing are given in Table 5. Depending on the likelihood of shock and other considerations, a factor of safety of from 2 to 4 should be allowed in setting *safe* external pressures. For example, if the *maximum* external pressure difference given in Table 5 is 20 psi, the *safe* external pressure difference for design purposes would be 5 to 10 psi. The figures in this table are for steel having a modulus of elasticity, $E = 30 \times 10^6$ psi. The safe external pressure for brass, copper, and other materials with a lower modulus of elasticity and yield strength may be obtained for values of D/t greater than 50 by multiplying by the ratio of their modulus of elasticity to that of steel. For values of D/t less than 50 the external pressure for the lower strength material should be adjusted in the ratio of the yield strengths of the materials.

---

[1] For tolerances on pipe outside-diameter variations see ASTM A120, A139, and A106.

[2] "Theory of Elastic Stability," by S. Timoshenko, McGraw-Hill Book Company, New York.

Table 5. Maximum Uniform External Pressure for Steel Pipe, Psi

$\dfrac{D}{t}$	Critical[1] pressure, round pipe	Maximum pressure, pipe with 1% initial eccentricity[2] Yield stress, psi			$\dfrac{D}{t}$	Critical[1] pressure, round pipe	Maximum pressure, pipe with 1% initial eccentricity[2] Yield stress, psi		
		30,000	40,000	60,000			30,000	40,000	60,000
10	8,000	4,600	6,000	6,980	160	16.0	13.2	13.7	15.3
20	4,000	1,800	2,200	2,850	170	13.6	11.4	11.7	12.9
30	2,130	910	1,060	1,280	180	11.5	9.7	10.0	11.0
40	1,000	500	550	630	190	9.75	8.32	8.55	9.28
50	528	284	320	380	200	8.25	7.12	7.32	7.88
60	322	190	200	240	210	7.02	6.16	6.30	6.70
70	200	130	143	160	220	6.10	5.40	5.55	5.90
80	129	90	100	110	230	5.36	4.80	4.90	5.20
90	91	65	66	80	240	4.76	4.30	4.40	4.62
100	66	48	53	56	250	4.22	3.80	3.90	4.15
110	50	38	41	43	260	3.80	3.40	3.50	3.80
120	38	30	33	35	270	3.40	3.04	3.22	3.40
130	31	24	26	28	280	3.05	2.70	2.80	2.95
140	25	20	21	23	290	2.72	2.40	2.50	2.60
150	20	15	16	19	300	2.44	2.18	2.20	2.30

[1] Computed by equation (7) for values of D/t above the limits given by equation (8). For values of D/t below 20 the critical stress is taken as equal to the yield stress. Intermediate values are based on straight-line interpolation between the yield stress and the critical stress corresponding to the limiting D/t ratio given by equation (8).

[2] Computed by the quadratic equation

$$P_e{}^2 - \left[2S\left(\frac{t}{D}\right) + \left(1 + 0.03\,\frac{D}{t}\right)P\right]P_e + 2S\left(\frac{t}{D}\right)P = 0$$

where $P_e =$ maximum external pressure for pipe initial eccentricity of 1 per cent psi. For other eccentricities adjust the factor 0.03 in the ratio of the percentages.

$S =$ yield stress, psi.

$P =$ critical value of collapsing pressure for perfectly circular pipe from equation (7), psi.

From "Theory of Elastic Stability," by S. Timoshenko.

For comparison with the above values of maximum working pressure for an assumed initial eccentricity of 1 per cent, the collapsing pressures shown in Table 6 have been computed from empirical formulas.

The collapsing pressures deduced from tests are higher than the safe working external pressures computed for an initial eccentricity of 1 per cent for the thicker pipes and slightly lower for pipe thinner than those having a D/t ratio greater than 70. The values of safe working pressure based on the formulas developed by Dr. S. Timoshenko represent a more rational solution of the problem.

Table 6. Collapsing Pressure of Seamless-steel Tubes Deduced from Collapsing Tests

D/t	10	20	30	40	50	60	70	80	90	100
Carmen[1]	7,462	2,690	1,100	785	402	232	146	78	69	50
Stewart[2]	7,281	2,954	1,504	782	402	232	146	78	69	50

[1] "Tests at University of Illinois Experiment Station," *Bull.* 17, June, 1906.

[2] "Tests on Lap-welded Bessemer Steel Tubes," by Reid T. Stewart, *Trans. ASME*, vol. 27.

Thick-walled Cylinders—Clavarino's Formula. In the case of a thick-walled pipe or cylinder subjected to internal or external fluid pressure, it can no longer be assumed that the stress due to bursting or collapsing pressure is uniformly distributed across the wall. In either case the stress will be greatest at the inner surface and decrease to a minimum at the outer surface. As with thin-walled cylinders, the stress is tensile with internal pressure, and compressive with external. According to Clavarino,[1] for either thin- or thick-walled cylinders, if

p = internal unit pressure, psi
P = external unit pressure, psi
r = the radius to any point in the wall, in.
d = internal diameter of cylinder, in.
D = external diameter of cylinder, in.
S = unit stress at any radius r, psi
S_m = unit stress at the inner surface (point of maximum stress)
λ = Poisson's ratio (for values see Table 4)

For simultaneous internal and external pressure

$$S = \frac{[(1 - 2\lambda)(d^2p - D^2P)] + [(1 + \lambda)(d^2D^2/4r^2)(p - P)]}{D^2 - d^2} \tag{9}$$

In the case of internal pressure alone where $P = 0$ and $r = d/2$, the maximum stress may be written

$$S_m = p\,\frac{d^2(1 - 2\lambda) + D^2(1 + \lambda)}{D^2 - d^2} \tag{10}$$

or

$$p = S_m\,\frac{D^2 - d^2}{d^2(1 - 2\lambda) + D^2(1 + \lambda)}$$

For convenience in the solution of the above formula, the expression

$$\frac{D^2 - d^2}{d^2(1 - 2\lambda) + D^2(1 + \lambda)}$$

may be termed k and the formula written $p = kS_m$ or $S_m = p/k$. In the case of steel pipe or cylinders where λ is approximately 0.3, the expression for k becomes $k = \dfrac{10(D^2 - d^2)}{4d^2 + 13D^2}$. Since the value of k depends on the ratio of t/D, a schedule of approximate values for k based on this relation may be worked out, as has been done for steel cylinders in Table 7. Through the use of this table, the solution of Clavarino's formula for the case of internal pressure in steel pipes or cylinders with closed ends is greatly simplified.

Example 1. Required the safe working internal fluid pressure p when D is 5.0 in., wall thickness t is 0.625 in., maximum working fiber stress of steel S_m is 10,000 psi.
Solution. $t/D = 0.625/5 = 0.125$; referring to Table 7, the value of k corresponding to $t/D = 0.125$ is 0.2869. Substituting this value in $p = kS_m$,

$$p = 0.2869 \times 10,000 = 2,869 \text{ psi}$$

Example 2. Required the fiber stress in a steel cylinder wall when D is 10.0 in., t is 0.75 in. and p is 350 psi.

[1] For a derivation of Clavarino's formula see "Mechanics of Materials," by M. and T. Merriman, John Wiley & Sons, Inc., New York.

Table 7. Internal Fluid Pressure Factors k for Conditions Shown in Fig. 3b[1]

[Calculated by Clavarino's formula, assuming for steel a "coefficient of lateral contraction" (Poisson's ratio) equal 0.3]

Rule. Divide thickness of tube or pipe by its outside diameter, both being expressed in inches, then multiply the tabular value corresponding to this quotient by the working-fiber stress in pounds per square inch. The result will be the safe internal pressure in pounds per square inch.

For further use of table, see examples which follow.

t/D	0.000	0.001	0.002	0.003	0.004	0.005	0.006	0.007	0.008	0.009
0.01	0.0235	0.0259	0.0282	0.0306	0.0329	0.0352	0.0376	0.0399	0.0423	0.0446
0.02	0.0470	0.0493	0.0517	0.0540	0.0564	0.0587	0.0610	0.0634	0.0657	0.0681
0.03	0.0704	0.0727	0.0751	0.0774	0.0797	0.0321	0.0344	0.0867	0.0391	0.0914
0.04	0.0937	0.0961	0.0984	0.1007	0.1031	0.1054	0.1077	0.1100	0.1123	0.1147
0.05	0.1170	0.1193	0.1216	0.1239	0.1263	0.1286	0.1309	0.1332	0.1355	0.1378
0.06	0.1401	0.1424	0.1448	0.1471	0.1494	0.1517	0.1540	0.1563	0.1586	0.1609
0.07	0.1632	0.1655	0.1678	0.1700	0.1723	0.1746	0.1769	0.1792	0.1815	0.1838
0.08	0.1861	0.1883	0.1906	0.1929	0.1952	0.1974	0.1997	0.2020	0.2043	0.2065
0.09	0.2088	0.2111	0.2133	0.2156	0.2178	0.2201	0.2223	0.2246	0.2269	0.2291
0.10	0.2314	0.2336	0.2358	0.2381	0.2403	0.2425	0.2448	0.2470	0.2493	0.2515
0.11	0.2537	0.2559	0.2582	0.2604	0.2626	0.2648	0.2670	0.2692	0.2715	0.2737
0.12	0.2759	0.2781	0.2803	0.2825	0.2847	0.2869	0.2890	0.2912	0.2934	0.2956
0.13	0.2978	0.3000	0.3022	0.3043	0.3065	0.3087	0.3108	0.3130	0.3152	0.3173
0.14	0.3195	0.3216	0.3238	0.3259	0.3231	0.3302	0.3323	0.3345	0.3366	0.3388
0.15	0.3409	0.3430	0.3451	0.3472	0.3494	0.3515	0.3536	0.3557	0.3578	0.3599
0.16	0.3620	0.3641	0.3662	0.3683	0.3704	0.3724	0.3745	0.3766	0.3787	0.3803
0.17	0.3828	0.3849	0.3869	0.3890	0.3910	0.3931	0.3951	0.3972	0.3992	0.4013
0.18	0.4033	0.4053	0.4073	0.4094	0.4114	0.4134	0.4154	0.4174	0.4194	0.4214
0.19	0.4234	0.4254	0.4274	0.4294	0.4314	0.4333	0.4353	0.4373	0.4393	0.4412
0.20	0.4432	0.4452	0.4471	0.4490	0.4510	0.4529	0.4548	0.4568	0.4587	0.4606
0.21	0.4626	0.4645	0.4664	0.4683	0.4702	0.4721	0.4740	0.4758	0.4777	0.4796
0.22	0.4815	0.4834	0.4852	0.4871	0.4889	0.4908	0.4926	0.4945	0.4964	0.4982
0.23	0.5001	0.5019	0.5037	0.5055	0.5073	0.5091	0.5109	0.5127	0.5145	0.5163
0.24	0.5181	0.5199	0.5216	0.5234	0.5252	0.5269	0.5287	0.5304	0.5322	0.5340
0.25	0.5357	0.5374	0.5391	0.5408	0.5426	0.5443	0.5460	0.5477	0.5494	0.5511
0.26	0.5528	0.5545	0.5561	0.5578	0.5594	0.5611	0.5628	0.5644	0.5661	0.5677
0.27	0.5694	0.5710	0.5726	0.5742	0.5758	0.5774	0.5790	0.5806	0.5822	0.5838
0.28	0.5854	0.5870	0.5885	0.5901	0.5916	0.5932	0.5947	0.5963	0.5978	0.5994
0.29	0.6009	0.6024	0.6039	0.6054	0.6069	0.6084	0.6099	0.6114	0.6129	0.6143
0.30	0.6158	0.6173	0.6187	0.6201	0.6216	0.6230	0.6244	0.6259	0.6273	0.6287

NOTE. Curves for ready solution of Clavarino's formula for internal pressures of 1,000 lb and above were given in an article on "Design of Thick Walled Tubes and Cylinders," by F. E. Wertheim, *Heating, Piping and Air Conditioning*, October, 1931.

[1] Reproduced from "National Pipe Standards," by permission of The National Tube Co.

Solution. $t/D = 0.75/10.0 = 0.075$; referring to Table 7, the value of k corresponding to $t/D = 0.075$ is 0.1746. Substituting this value in $S_m = p/k$

$$S_m = \frac{350}{0.1746} = 2,000 \text{ psi}$$

Example 3. Required the thickness of the pipe wall t, when D is 6.0 in., S_m is 9,000 psi, and the working fluid pressure p is 540 psi.

Solution. The factor $k = p/S_m = 540/9,000 = 0.060$; referring to Table 7, the ratio of t/D corresponding to 0.060 is 0.0256. Substituting: $t = D \times 0.0256 = 6.0 \times 0.0256 = 0.1536$ in.

In the case of *external pressure* alone where $p = 0$

$$S_m = -P \frac{D^2(2 - \lambda)}{D^2 - d^2} \tag{11}$$

The minus sign before the right-hand member of this equation denotes compressive stress. While this equation offers a rational solution for cylinders under external pressure, Eq. (7) should be used in preference, to be on the side of safety.

Division between Thick- and Thin-walled Pipe.   The dividing line between the thick-walled and thin-walled pipe depends upon the ratio of thickness to outside diameter t/D. For pipes subjected to internal pressure, formulas for thin-walled pipe are applicable to nearly all commercial steel pipe as this ratio seldom exceeds 0.10. Formulas developed by Miller[1] based on the maximum-energy-of-distortion theory show that the correction for thickness is of the order of 2 per cent for a pipe with a t/D ratio of 0.10. These formulas yield relatively simple expressions for the case in which the pipe wall carries the load due to internal pressure on the closed ends.

$$p = 2.31KS(1 - K) \tag{12}$$

where p = internal pressure
   S = stress in inner surface of pipe
   K = wall thickness/outside diameter = t/D

For a thin-walled pipe, the factor $(1 - K)$ differs by a negligible amount from unity so the expression for stress in a closed-end pipe reduces to

$$S = pD/2.31t \tag{13}$$

While based on different concepts of failure or yielding, these formulas show much the same reduction in hoop stress due to the presence of an axial tension as is given by Clavarino's formula, which is based on the maximum-strain theory. By either theory the computed stress in the inner skin of a thin-walled pipe, with closed ends free to move axially, is approximately 15 per cent less than if there is no longitudinal tension acting. If there is an axial compression acting equal to the longitudinal tension due to internal pressure, then a condition equivalent to an open-ended pipe obtains and the stress at the inner surface is correspondingly increased. The condition of an open-ended pipe is covered by Barlow's formula.

Barlow's Formula.   An empirical relation for internal fluid pressure which gives results on the side of safety for all practical thickness ratios is that known as *Barlow's formula*. This formula resembles the hoop-stress formula except that the outside diameter of the cylinder is used instead of the inside. Barlow's formula is

$$S = pD/2t \tag{14}$$

While Barlow's formula is widely used because of its convenience of solution, it was not generally considered to have any theoretical justification until formulas based on the maximum-energy-of-distortion theory showed that for *thin-walled* pipe with no axial tension Barlow's formula actually is theoretically correct. Since most commercially important pipe has a ratio of wall thickness to outside diameter less than 0.10, Barlow's formula for *thin-walled* tubes is of great significance. Comprehensive bursting tests[2] on commercial steel pipe have demonstrated that Barlow's formula predicts the pressure at which the pipe will rupture with an accuracy well within the limits of uniformity of commercial pipe thickness. In general, failure occurred at a pressure about 3 per cent higher than predicted from Barlow's formula.

Barlow's formula with modifications has been employed in such well-recognized standards as the ASME Boiler Construction Code, the ASA Standard for Steel

---

[1] See discussion by Benjamin Miller of paper "Theories of Strength," by A. Nádai, *Trans. ASME*, vol. 1, no. 3, APM 55–15, p. 124, July–September, 1933.
[2] Tests by The National Tube Co.

Flanges and Flanged Fittings, the ASA Standard for Wrought Iron and Wrought Steel Pipe and the ASA Code for Pressure Piping.

Distinction between Cylinders with Open and Closed Ends. According to the foregoing review of current theory thin-walled pipe, which is defined as having a t/D ratio of less than 0.10 and thus includes nearly all commercial steel pipe, can be designed reliably for bursting strength according to Barlow's formula since this formula is on the safe side for either open-end or closed-end conditions. While the assumption of closed ends, where justified, would indicate the possibility of using 15 per cent lighter wall thicknesses in some cases for thin-walled pipes, such an assumption seldom is warranted for pipelines although it might be for cylinders with closed ends such as tanks or receivers. The uncertain conditions imposed on thin-walled pipe lines through the presence of valves, expansion joints, bends, anchors, supports, and attachment to equipment vitiate any consistent attempt to distinguish between open- and closed-end conditions and point out the need for designing all for the more severe assumption, viz., as having open ends. In the case of thick-walled pipes the difference between open-end and closed-end conditions is not so great and either Clavarino's formula or the following expression known as "Lame's formula" for maximum hoop stress[1] at the inner surface can be used

$$S = p \, \frac{D^2 + d^2}{D^2 - d^2}$$

Code Formulas for Pipe Wall Thickness. Early in 1951 a task force of ASA Sectional Committee B31 on Code for Pressure Piping undertook to resolve the different ideas on how to calculate wall thickness for power piping. This effort resulted in the adoption of a single formula[2] which serves for heavy-wall pipe as well as for thin-wall and which at the same time, through the use of suitable parameters, adjusts for elastic or plastic properties of the respective materials over the expected range of operating temperatures. This formula as adopted for the Power and District Heating Sections of the B31 Piping Code and for the Power Section of the ASME Boiler Construction Code is given later. For other piping systems, the Code has somewhat different rules, which are abstracted herein in their respective chapters.

Thickness of Steam Piping. (a) For inspection purposes, the minimum thickness of pipe wall to be used for piping at different pressures and for temperatures not exceeding those given for the various materials in Tables 8 and 9 shall be determined by the formula given below:

$$t_m = \frac{PD}{2S + 2yP} + C \tag{16}$$

$$P = \frac{S}{\frac{1}{2}[D/(t_m - C)] - y} \tag{17}$$

where t_m = minimum[3] pipe wall thickness, in. The minimum thickness used in the formula shall in no case be more than the minimum thickness resulting from

[1] See "Strength of Materials," by S. Timoshenko, D. Van Nostrand Company, Inc.

[2] See "A Wall Thickness Formula for High-pressure High-temperature Piping," by W. R. Burrows, R. Michel, and A. W. Rankin, *ASME Paper* 52-A-151, *Trans. ASME,* vol. 76, no. 3, p. 427, April, 1954.

[3] If pipe is ordered by its nominal wall thickness, as is customary in trade practice, the manufacturing tolerance on wall thickness must be taken into account. After the minimum pipe wall thickness t_m is determined by the formula, this minimum thickness shall be increased by an amount sufficient to provide the manufacturing tolerance allowed in the applicable pipe specification. The next heavier commercial wall thickness may then be selected from standard thickness schedules as contained in ASA B36.10. The manufacturing tolerances are given in the several pipe specifications listed in Tables 8 and 9.

Table 8.   Allowable Stresses (S Values) Psi for Carbon-steel and Wrought-iron Pipe in Power and District-heating Systems[1]

(Abstracted from Code for Pressure Piping, ASA B31.1, and ASME Power Boiler Code)

Material	ASTM specification	Grade or symbol	Minimum tensile strength, psi	S values, psi, for metal temperatures not to exceed[2]			
				−20 to 650 F	700 F	750 F	800 F
Electric-fusion-welded steel[3,4]	A134	A245A	48,000	8,800			
		A245B	52,000	9,600			
		A245C	55,000	10,100			
		A283A	45,000	8,300			
		A283B	50,000	9,200			
		A283C	55,000	10,100			
		A283D	60,000	10,100			
Electric-fusion-welded steel[3,4]	A139	A	48,000	9,600	9,250	8,300	
		B	60,000	12,000	11,350	9,950	
Electric-fusion-welded steel.[3]	A155, Class 2	C45	45,000	10,100	9,800	8,700	7,500
Note.—The values tabulated are for Class 2 pipe. For Class 1 pipe which is heat treated and radiographed, these stresses may be increased by the ratio 0.95/0.90.		C50	50,000	11,250	10,900	9,900	8,450
		C55	55,000	12,400	11,900	10,850	9,200
		KC55	55,000	12,400	11,900	10,850	9,200
		KC60	60,000	13,500	12,900	11,650	9,700
		KC65	65,000	14,600	13,950	12,450	10,250
		KC70	70,000	15,750	14,950	13,250	10,800
Electric-resistance-welded steel[3]	A53 or A135	A	48,000	10,200	9,900	9,100	7,650[5]
		B	60,000	12,750	12,200	11,000	9,200[5]
Lap-welded steel.	A53	. . . .	45,000	9,000			
Lap-welded wrought iron.	A72	. . . .	40,000	8,000			
Butt-welded steel.	A53	. . . .	45,000	6,750			
Butt-welded wrought iron. . . .	A72	. . . .	40,000	6,000			
Seamless steel.	A53 or A106	A	48,000	12,000	11,650	10,700	9,000[6]
		B	60,000	15,000	14,350	12,950	10,800[6]
	A83	Type A	47,000	11,750	11,500	10,700	9,000[6]
	A179[4]	Low carbon	. . . . .	11,750	11,450	10,550	9,000[6]
	A192	. . . .	47,000	11,750	11,500	10,700	9,000[6]
	A210	. . . .	60,000	15,000	14,350	12,950	10,800[6]

[1] For other materials, higher temperatures, or possible changes in S values, see latest revisions of Codes.
[2] Where welded construction is used, consideration should be given to the possibility of graphite formation with carbon steels at temperatures above 775 F.
[3] S values for welded-seam pipe include joint efficiencies.
[4] Not included in Boiler Code.
[5] Not included in Piping Code above 750 F.
[6] Not included in District Heating Section of Piping Code above 750 F.

the application of the tolerances given in the specification for the material to be used, including tube material which is to be used for piping.

P = maximum internal service pressure, psig, at the operating metal temperature for which the value of S is taken from Tables 8 and 9.[1]   The value of P in the formula shall not be taken at less than 100 for any condition of service or material.

D = outside diameter of pipe, in.

S = allowable stress due to internal pressure at the operating temperature in material, psi (see Tables 8 and 9 and note[2] below).

C = allowance for threading, mechanical strength, and/or corrosion, in.

y = a coefficient having values as given in Table 10.

[1] When computing the allowable pressure for a pipe of a definite minimum wall thickness, the value obtained by the formulas may be rounded out to the next higher unit of 10.

[2] The factors of safety and allowance for joint efficiencies of welded pipe have been taken into account in the S values given in Tables 8 and 9.   Where pipe is subjected to more than one "pressure-temperature" operating condition, the value of t_m shall be determined from the "pressure-temperature" operating condition which results in the maximum t_m, except as provided in (f).

Table 9. Allowable Stresses (S Values) Psi for Alloy-steel Pipe in Power Piping Systems[1]

(Abstracted from ASME Power Boiler Code and Code for Pressure Piping, ASA B31.1)

Material	ASTM specification	Grade or symbol	Minimum tensile strength, psi	S values, psi, for metal temperatures not to exceed[2]									
				−20 to 650 F	700 F	750 F	800 F	850 F	900 F	950 F	1000 F	1050 F	1100 F
Electric-fusion-welded:[3]													
Carbon-molybdenum	A155	A204A	65,000	14,600	14,600	14,600	14,100	12,950	11,250				
Carbon-molybdenum	A155	A204B	70,000	15,750	15,750	15,750	15,200	13,500	11,450				
Carbon-molybdenum	A155	A204C	75,000	16,850	16,850	16,850	16,200	14,300	11,700				
½% Cr, ½% Mo	A155	A301A	65,000	14,600	14,600	14,600	14,100	12,950	11,250	9,000	5,600		
1% Cr, ½% Mo	A155	A301B	60,000	13,500	13,500	13,500	13,250	12,750	11,800	9,900	6,750	4,500	2,500
1¼% Cr, ½% Mo	A155	P11	60,000	13,500	13,500	13,500	13,500	12,950	11,800	9,900	7,000	4,950	3,600
2¼% Cr, 1% Mo	A155	P22	60,000	13,500	13,500	13,500	13,500	12,950	11,800	9,900	7,000	5,200	3,750
Seamless ferritic steels:													
Carbon-molybdenum	A335	P1	55,000	13,750	13,750	13,750	13,450	13,150	12,500				
0.65 Cr, 0.55 Mo	A335	P2	55,000	13,750	13,750	13,750	13,450	13,150	12,500				
1.00 Cr, 0.55 Mo	A335	P12	60,000	15,000	15,000	15,000	14,750	14,200	13,100	10,000	6,250	5,000	2,800
1.25 Cr, 0.55 Mo	A335	P11	60,000	15,000	15,000	15,000	15,000	14,400	13,100	11,000	7,800	5,500	4,000
2.25 Cr, 1.00 Mo	A335	P22	60,000	15,000	15,000	15,000	15,000	14,400	13,100	11,000	7,800	5,800	4,200
3.00 Cr, 0.90 Mo	A335	P21	60,000	15,000	14,800	14,500	13,900	13,200	12,000	11,000	7,000	5,500	4,000
5% Cr, ½% Mo	A335	P5a	60,000	See Code	13,400	13,100	12,800	12,400	11,500	9,000	7,000	5,200	3,300
5% Cr, ½% Mo + Si	A335	P5b	60,000	See Code	13,400	13,100	12,800	12,400	10,900	9,000	5,500	3,500	2,500
Seamless austenitic steels:													
18% Cr, 10% Ni + Ti	A158	P8b(321)	75,000	See Code	14,800	14,700	14,550	14,300	14,100	13,850	13,500	13,100	10,300
18% Cr, 10% Ni + Cb	A158	P8d(347)	75,000	See Code	14,800	14,700	14,550	14,300	14,100	13,850	13,500	13,100	10,300

[1] For other materials, higher temperatures, or possible changes in S values, see latest revisions of Codes.

[2] Where welded construction is used, consideration should be given to the possibility of graphite formation in carbon-molybdenum steel above 875 F or in chromium-molybdenum steel containing less than 0.60% chromium above 975 F.

[3] The values given are for Class 2 A155 pipe. For Class 1 A155 pipe which is radiographed, these stresses may be increased by the ratio 0.95/0.90.

Table 10. Values of Coefficient y

	Temperature, deg F					
	900 and below	950	1000	1050	1100	1150 and above
Ferritic steels..........	0.4	0.5	0.7	0.7	0.7	0.7
Austenitic steels........	0.4	0.4	0.4	0.4	0.5	0.7

(b) The value of C used in Eqs. (16) and (17) shall be not less than that given in Table 11.

(c) The value of S used in Eqs. (16) and (17) shall not exceed that given in Tables 8 and 9 for the respective material and temperature contemplated, but when the temperature differs from that given in the table, S may be determined by interpolation.

(d) Pipe lighter than standard or Schedule 40 of American Standard ASA B36.10 shall not be threaded.

(e) While the thickness given by the formula is theoretically ample to take care of both bursting pressure and material removed in threading, when steel pipe is threaded and used for steam pressures of 250 psi and over or for water pressures above 100 psi gage at temperatures of 220 F and over, it shall be seamless of a quality at least equal to ASTM Specifications A53 or A83, and of a weight at least equal to Schedule 60 or 80 (extra strong) in order to furnish added mechanical strength.

Table 11. Values of C[1]

Type of pipe	Value of C, in.
Threaded pipe:	
⅜ in. and smaller.............................	0.05
½ in. and larger	Depth of thread
Plain-end pipe or tubes for 1 in. size and smaller......	0.05
Plain-end pipe or tubes for sizes above 1 in...........	0.065

[1] The values of C stipulated above are such that the actual stress due to internal pressure in the wall of the pipe is less than the values of S given in Tables 8 and 9 as applicable in the formulas.

(f) *Variations in Pressure and Temperature.* Piping systems shall be considered safe for operation if the maximum sustained pressure and temperature on any part do not exceed the pressure and temperature allowed by the Code for all component parts of the system. It is recognized that occasional departures from the nominal operating pressure and temperature inevitably occur and therefore the piping system shall also be considered safe for occasional operation for short periods at higher pressures and temperatures.

Either pressure or temperature, or both, may exceed the nominal values if the corresponding stress in the pipe wall calculated by the formulas does not exceed the S value allowable for the actual expected temperature by more than the following allowances for the periods of duration indicated:

(1) Up to 15 per cent increase above the S value during 10 per cent of the operating period;

(2) Up to 20 per cent increase of the S value during 1 per cent of the operating period.

Allowable Stresses.[1] Allowable stresses (S values) adopted for the Power Section of the ASME Boiler Construction Code are about 25 per cent higher than those previously obtaining and were set with regard to tensile strengths of the respective materials at temperatures within their elastic ranges and with regard to their creep rates and rupture strengths at temperatures within their plastic ranges. The new allowable stresses have been adopted by ASA Sectional Committee B31

[1] For allowable stresses (S values) for use in other sections of the ASA B31.1 Code, see the chapter which deals with the specific section of interest.

for use in the Power and District Heating Sections of the American Standard Code for Pressure Piping.

Cast-iron Pipe. The Power and District Heating Sections of the ASA B31 Piping Code provide that the thickness of cast-iron pipe conveying liquids may be determined by selection from ASTM Specification A44 (ASA A21.2), or from Federal Specification WW-P-421. using the class of pipe for the pressure next higher than the maximum internal service pressure in pounds per square inch. These thicknesses include allowances for foundry tolerances and water hammer. Where the thickness for liquid service is calculated, the methods of ASA A21.1 shall be followed.

Where cast-iron pipe is used for steam service under these code sections, the thickness shall be calculated from Eqs. (16) and (17), using allowable working stresses stated in the Code.

Design of Special Fittings. Although intended primarily for computing pipe wall thickness required for resisting bursting pressure, Eqs. (16) and (17) can be extended to the design of cast or forged fittings and manifolds. Since an exact stress analysis for irregular shapes is impractical,[1] the usual rule for such design is to increase the computed wall thickness in the affected section of cast fittings by 50 per cent to compensate for the weakening effect of branch outlets or irregular shapes. Rules for the reinforcement of welded branch connections will be found in the codes.

In designing *special steel castings* in accordance with the rules of the ASME Boiler Construction Code, the following quality factors shall be applied to the values of allowable stress for castings as given in Table 9.

1. A factor of 80 per cent for castings inspected only in accordance with the minimum requirements of the specification for the material.

2. A factor of 90 per cent for minimum inspection plus the following additional requirements:

(*a*) Thorough surface inspection, plus at least three pilot castings of the first lot of five shall be sectioned at all critical sections, or completely radiographed, without revealing injurious defects.

(*b*) One additional casting from every lot of five shall be sectioned or radiographed without showing defects.

(*c*) All other castings shall have critical sections examined using magnetic powder, fluorescent or colored penetrating oil, or grinding and etching.

(*d*) In the case of a single casting, or any casting of a lot that has been completely radiographed without revealing defects, the 90 per cent factor may be applied.

3. Where all vital areas are exposed for inspection of the full wall thickness by machining, the inspection afforded may be taken in lieu of destructive or radiographic examination required in (2) and a casting factor of 90 per cent may be applied.

In applying the foregoing casting quality factors to allowable stresses for cast-steel welding-end valves and fittings, the cylindrical welding ends may be proportioned with a quality factor of 100 per cent provided these areas are finish-machined both inside and outside and carefully inspected. In no case, however, shall the thickness of the ends be less, or more than 15 per cent greater, than that of the adjoining pipe. The machined ends are required to conform to a maximum slope line given in ASA B16.5. These casting quality factors shall not be applied to the regular line of castings made according to the American Standard for Steel Pipe

---

[1] See "Some Aspects of the Design of Large High-pressure Steel Valves," by F. D. Cotterman and R. E. Falls, *J. Am. Soc. Naval Engrs.*, vol. 52, no. 3, p. 23, August, 1940.

Flanges and Flanged Fittings, ASA B16.5. Both regular and special forgings are generally assumed to have a 100 per cent quality factor.

Hollow Sphere. The force F due to internal or external pressure which tends to force apart or push together the two halves of a sphere is $F = p\pi d^2/4$. The stress in the wall of a thin-walled sphere is $S = pd/4t$, while $S = p(D^3 + 2d^3)/2(D^3 - d^3)$ is Lame's formula for any sphere.

FLOW OF FLUIDS

In this section, the general energy equation which governs all fluid flows is given. This equation is then specialized for the case of steady flow and, also, for the case of steady flow in the absence of heat transfer and of work. The following nomenclature is used:

$\dot{Q}$ = heat-transfer rate (for example, Btu/hr) and is to be taken as positive if heat is added to the flow system and as negative if heat is transferred out

$\dot{W}$ = time rate at which work is done (for example, Btu/hr or any similar consistent unit) and is taken as positive if work is done by the system and as negative if work is done on the system

$\dot{m}$ = mass-flow rate (for example, lb/hr) entering or leaving the system

m = mass of fluid within system at any instant of time

h = enthalpy of fluid (for example, Btu/lb)

u = internal energy of fluid (for example, Btu/lb)

V = average velocity of the fluid (for example, fps)

Z = height of fluid element above some datum (for example, ft)

t = time and d/dt represents the derivative with respect to time

g_c = 32.17 ft lbm/(lbf sec^2), the gravitational constant

J = 778 ft lbf/Btu, the conversion from mechanical to thermal dimensions

g = local acceleration due to gravity, ft/sec^2

Subscript 1 denotes fluid properties at inlet to system. Subscript 2 denotes fluid properties at exit from system. Subscript σ denotes the system under consideration.

The general energy equation, using the above nomenclature, is

$$\dot{Q} + \dot{m}_1\left(h_1 + \frac{V_1^2}{2g_cJ} + \frac{g}{g_cJ}Z_1\right) = W + \dot{m}_2\left(h_2 + \frac{V_2^2}{2g_cJ} + \frac{g}{g_cJ}Z_2\right)$$
$$+ \frac{d}{dt}\left(mu + \frac{mV^2}{2g_cJ} + \frac{g}{g_cJ}Z\right)_\sigma$$

For *steady-flow systems,* the time-dependent term vanishes and the equation becomes

$$\dot{Q} + \dot{m}\left(h_1 + \frac{V_1^2}{2g_cJ} + \frac{g}{g_cJ}Z_1\right) = W + \dot{m}\left(h_2 + \frac{V_2^2}{2g_cJ} + \frac{g}{g_cJ}Z_2\right)$$

in which $\dot{m}_1$ and $\dot{m}_2$ are equal by definition of steady flow, and hence no subscripts in the mass-rate terms appear above.

For steady-flow systems in the *absence of heat-transfer and work,* the equation reduces to

$$h_1 + \frac{V_1^2}{2g_cJ} + \frac{g}{g_cJ}Z_1 = h_2 + \frac{V_2^2}{2g_cJ} + \frac{g}{g_cJ}Z_2 \tag{18}$$

For the steady-flow of a fluid in the absence of heat transfer and work and, further, in which the fluid is incompressible and frictionless, there is obtained the

well-known Bernoulli equation[1]

$$\frac{P_1}{\rho} + \frac{V_1^2}{2g_c} + \frac{g}{g_c} Z_1 = \frac{P_2}{\rho} + \frac{V_2^2}{2g_c} + \frac{g}{g_c} Z_2 = \text{constant} \qquad (19)$$

The pressure and density terms appear from the definition of enthalpy ($h = u + p/\rho$), and since there is no heat transfer, the internal energy terms cancel in frictionless flow.

If the above equation is multiplied by the ratio g_c/g, there is obtained

$$\frac{P_1}{w} + \frac{V_1^2}{2g} + Z_1 = \frac{P_2}{w} + \frac{V_2^2}{2g} + Z_2 \qquad (20)$$

in which the density ρ has been replaced by the specific weight w.

Classification of Flows. Flows of fluids, in the very general case, may be functions of both time and space. Both density and velocity, most generally, may be dependent on time or on location within the fluid body. This dependence, using velocity as an example, may be expressed implicitly as

$$V = V(x,y,z,t)$$

where x, y, and z are space coordinates and t represents time. When the above expression is differentiated with respect to time, there is obtained

$$\frac{dV}{dt} = u\frac{\partial V}{\partial x} + v\frac{\partial V}{\partial y} + w\frac{\partial V}{\partial z} + \frac{\partial V}{\partial t}$$

A *steady flow* is one in which $\partial v/\partial t$ and $\partial \rho/\partial t$ are both zero. That is, in the steady flow of a fluid, velocity and density are entirely independent of time, although each may vary from point to point within the fluid.

An *unsteady flow* is one in which either or both velocity and density are variable with time. Unsteady flows occur in ordinary pieces of engineering apparatus such as a steam boiler during the start-up period when steam and water flows are not balanced. In certain flow phenomena, as, for example, in water hammer, the flow is unsteady. In some cases, as in a Comprex compressor, the phenomenon of unsteady flow is utilized to advantage in order to gain compression.

In piping systems, flow is frequently steady or at least steady for sufficient engineering accuracy. Throughout this chapter, particular emphasis will be placed on steady flow.

A *compressible flow* is one in which the density of the fluid varies with respect to location in the flow channel. All fluids are compressible to a certain extent; in fact, the compressibility of liquids is a property that renders them useful in "liquid spring" applications. However, use of the term "compressible" flow implies a significant change in density. Gases and vapors which flow at relatively high velocities are compressible fluids. Of great significance in compressible flow is the fact that the term $\int 1/\rho \, dP$ in the Bernoulli equation cannot be evaluated unless the functional relationship is known between P and ρ.

An *incompressible* fluid is one in which there is no appreciable change in density throughout the flow channel under consideration. Practically all liquids fit into this category; certain gases, such as air at Mach numbers of 0.2 and less, also may be considered incompressible with sufficient engineering accuracy. In the Bernoulli equation, since there is no significant density change, the term $\int 1/\rho \, dP$ evaluates directly into $P_2/\rho - P_1/\rho$.

A *one-dimensional flow* is one in which all fluid properties, such as pressure and velocity, vary with respect to one length dimension only. In Fig. 6, the velocity

[1] This equation is strictly applicable to a streamline in the flow path.

is a function of x but not of y; in Fig. 7, the velocity is a function of y but not of x. In a *two-dimensional flow*, any or all of the fluid properties may vary with respect to two dimensions, and in *three-dimensional flow*, the fluid properties vary in each coordinate direction.

The flow of an actual fluid is most likely to be affected by at least two length dimensions. Viscous effects at the boundary walls of the conduit retard the fluid, and in fact, the velocity of the fluid is zero immediately adjacent to the walls. The distance into the fluid stream in which the decelerating effects of wall friction are evident is known as the *boundary layer*. As the length of the flow path increases, the thickness of the boundary layer also increases, and with sufficient length of the flow path, the boundary layer will fill the conduit completely.

For many engineering calculations, particularly those which involve low-velocity flows of incompressible fluids, it is sufficiently accurate to assume one-dimensional

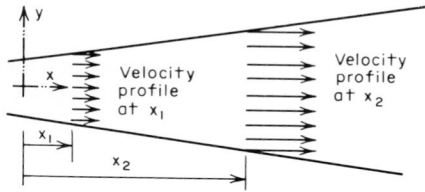

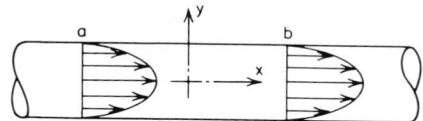

FIG. 6. One-dimensional flow: at any cross section, velocity is uniform but velocity varies with location along flow channel.

FIG. 7. One-dimensional flow: flow pattern at section a is identical with that at b, but at each section, velocity varies with distance from pipe center.

flow in which the average velocity (volume flow rate per cross-sectional area) is used in the calculations.

An important distinction between flows is whether they are *laminar* or *turbulent*. In laminar flow, the fluid flows in layers which slide one upon another. In turbulent flow, the fluid has velocity components normal as well as parallel to the axis of the conduit in which it flows. Turbulent flow has been likened in appearance to that of a large railway station during the commuter rush hour; people skirt about in all different directions, but the general flow is toward the gate which leads to the train.

Laminar flow occurs for fluid Reynolds numbers of about 2,000 and less. Flows with Reynolds numbers of greater than about 10,000 are generally turbulent. In the range of Reynolds numbers from 2,000 to 10,000, the flow may be laminar or turbulent, and in design, it is well to avoid this range. Pressure drops due to friction differ markedly in the two flow regimes.

Continuity Equation. The equation of continuity is one of the most important relationships between properties of a flowing fluid. Under the assumptions that the fluid is a continuum and that no nuclear reactions occur, the most general form of the continuity equation is

$$\frac{d}{dt} \int_{\text{Vol.}} \rho \, dx \, dy \, dz + \int_{\text{Area}} \rho V \cos \theta \, dA = 0$$

in which ρ = fluid density
x, y, z = space coordinates
t = time
A = area
V = velocity
$\cos \theta$ = cosine of angle between fluid velocity vector and the outward normal to the areas at which fluid crosses control surfaces

For steady flow, the time-dependent first term vanishes and the continuity equation becomes

$$\int_A \rho V \cos \theta \, dA = 0$$

If, in addition, the fluid is incompressible, the density term will vanish, and hence for steady, incompressible flow, the continuity equation is

$$\int_A V \cos \theta \, dA = 0$$

If the control surfaces are selected so that the fluid velocity is normal to them, and if the flow is one-dimensional, the continuity equation assumes a very convenient form:

$$\Sigma(\rho A V)_{\text{inlets}} = \Sigma(\rho A V)_{\text{outlets}} \tag{21}$$

FLOW OF LIQUIDS

In the form in which it appears in Eq. (20), the energy equation is identical with the Bernoulli equation and is of great usefulness in dealing with the flow of liquids. It is reproduced here in slightly altered form.

$$H = V^2/2g + P/w + z = \text{constant} \tag{22}$$

where H is referred to as the total "head" of the fluid and is made up of three components. First, the term $V^2/2g$ has the dimensions of feet and is called the

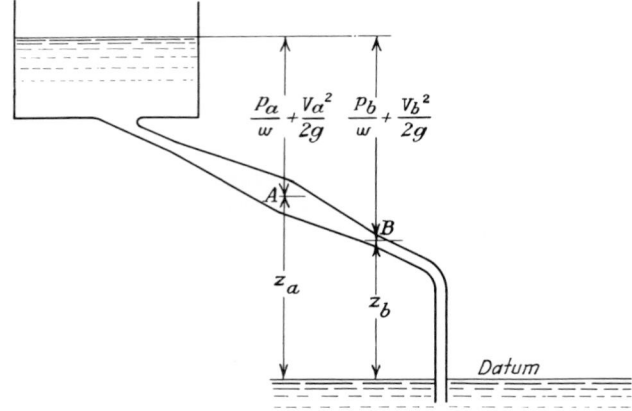

FIG. 8. Illustration of Bernoulli's theorem.

"velocity head." The second term, P/w, has the dimension of feet if p is expressed in pound force per square foot and w in pound force per cubic foot; the ratio is referred to as the "pressure head." The third term, z, represents the "static head" of the fluid, and its dimension is also feet.

Expressed in words, Eq. (22) states that in the steady, incompressible flow of a frictionless fluid, the sum of velocity head, pressure head, and static head is constant. This is the well-known Bernoulli theorem, and in Fig. 8, it is illustrated schematically. Liquid flows by gravity from an elevated reservoir through a conduit to a lower elevation which is taken as the reference plane, or datum. The subscripts a and b

denote conditions at sections A and B. For this flow condition, the Bernoulli equation is

$$Va^2/2g + P_a/w + z_a = V_b^2/2g + P_b/w + z_b = \text{constant}$$

In actual practice, friction does occur as the fluid flows from section A to section B. The effect of friction is taken into account by the introduction of a "friction head" term h_f. For the flow condition of Fig. 8, if friction were to be taken into account, the modified Bernoulli equation would be

$$V_a^2/2g + P_a/w + z_a$$

$$= V_b^2/2g + P_b/w + z_b + h_f \quad (23)$$

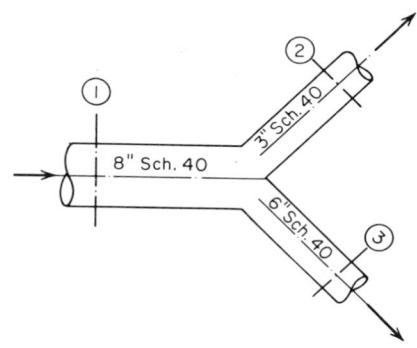

FIG. 9. Example illustrating continuity of flow.

As an application of the Bernoulli and continuity equations, refer to the flow situation depicted in Fig. 9. An incompressible fluid of density 60 lbm/cu ft enters through a horizontal 8-in. Schedule 40 inlet at the rate of 3,000 gpm and at a pressure of 100 psig. The fluid leaves through the horizontal 3-in. Schedule 40 outlet at a pressure of 95 psig and through the horizontal 6-in. Schedule 40 outlet at an unknown velocity and pressure. It is required to find the velocity and pressure at section 3 (the 6-in. Schedule 40 connection) so that the flow situation is possible.

Solution:

$$V_1 = \frac{3{,}000 \text{ gpm}}{7.48 \text{ gal/cu ft} \times 60 \text{ sec/min} \times 0.3474 \text{ sq ft}}$$

$$= 19.25 \text{ fps}$$

Applying the Bernoulli equation between sections 1 and 3 (in so doing, it is assumed that flow between these sections is frictionless),

$$P_1/\rho + V_1^2/2g_c = P_2/\rho + V_2^2/2g_c$$

$$100(144)/60 + 19.25^2/2(32.17) = 95(144)/60 + V_2^2/2(32.17)$$

and

$$V_2 = 34.7 \text{ fps}$$

If one-dimensional flow of this incompressible fluid is assumed, the continuity equation (21) takes the form

$$A_1 V_1 = A_2 V_2 + A_3 V_3$$

and

$$V_3 = 24.5 \text{ fps}$$

The Bernoulli equation between sections 1 and 3 is

$$P_1/\rho + V_1^2/2g_c = P_3/\rho + V_3^2/2g_c$$

whence $P_3 = 98.5$ psia

That is, *if* the pressures at sections 1, 2, and 3 are maintained, respectively, at 100, 95, and 98.5 psia, the velocity at section 2 will be 34.7 fps and the velocity at section 3 will be 24.5 fps as long as fluid continues to be supplied at the rate of 3,000 gpm.

Entrance and Exit Losses for Liquids. At the free surface in the reservoir of Fig. 8, the liquid has a very low velocity, and in many cases, the velocity may be

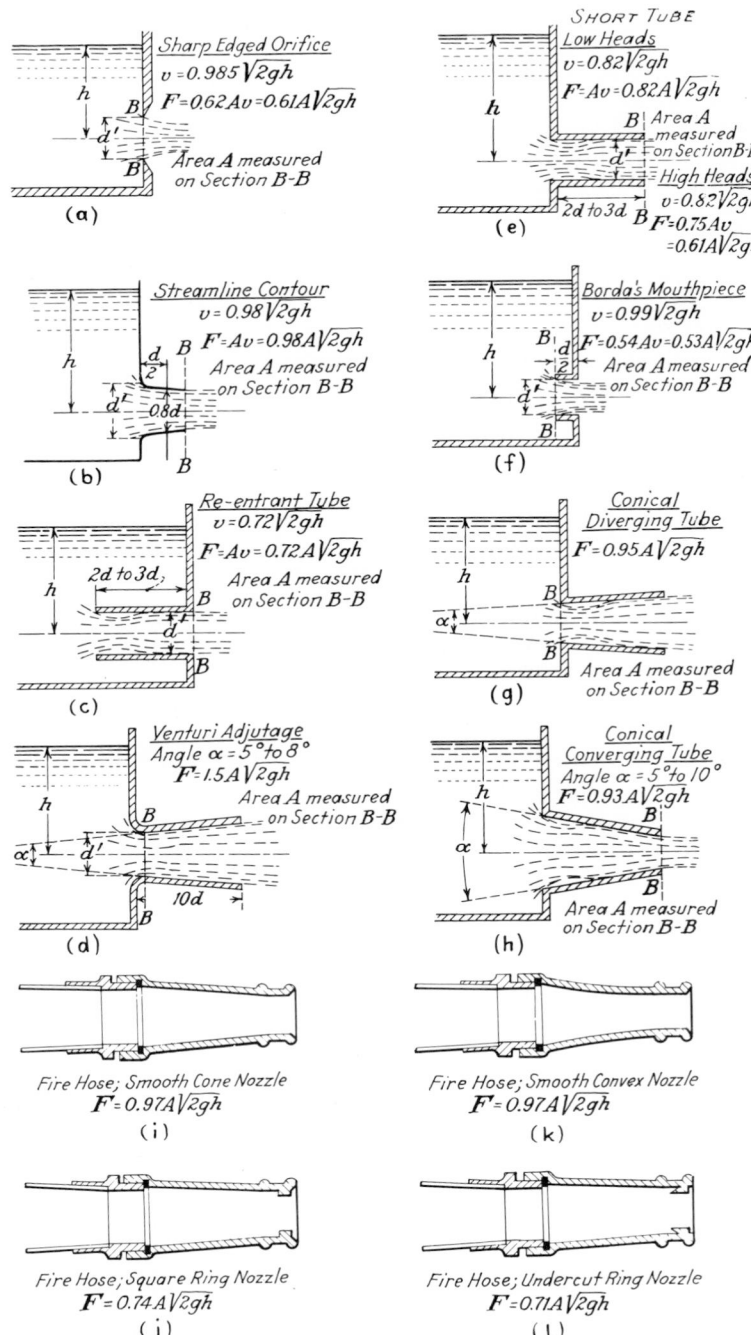

(a) *Sharp Edged Orifice*
$v = 0.985\sqrt{2gh}$
$F = 0.62Av = 0.61A\sqrt{2gh}$
Area A measured on Section B-B

(b) *Streamline Contour*
$v = 0.98\sqrt{2gh}$
$F = Av = 0.98A\sqrt{2gh}$
Area A measured on Section B-B

(c) *Re-entrant Tube*
$v = 0.72\sqrt{2gh}$
$F = Av = 0.72A\sqrt{2gh}$
Area A measured on Section B-B

(d) *Venturi Adjutage*
Angle $\alpha = 5°$ to $8°$
$F = 1.5A\sqrt{2gh}$
Area A measured on Section B-B

(e) SHORT TUBE
Low Heads
$v = 0.82\sqrt{2gh}$
$F = Av = 0.82A\sqrt{2gh}$
Area A measured on Section B-B
High Heads
$v = 0.82\sqrt{2gh}$
$F = 0.75Av = 0.61A\sqrt{2gh}$

(f) *Borda's Mouthpiece*
$v = 0.99\sqrt{2gh}$
$F = 0.54Av = 0.53A\sqrt{2gh}$
Area A measured on Section B-B

(g) *Conical Diverging Tube*
$F = 0.95A\sqrt{2gh}$
Area A measured on Section B-B

(h) *Conical Converging Tube*
Angle $\alpha = 5°$ to $10°$
$F = 0.93A\sqrt{2gh}$
Area A measured on Section B-B

(i) *Fire Hose; Smooth Cone Nozzle*
$F = 0.97A\sqrt{2gh}$

(k) *Fire Hose; Smooth Convex Nozzle*
$F = 0.97A\sqrt{2gh}$

(j) *Fire Hose; Square Ring Nozzle*
$F = 0.74A\sqrt{2gh}$

(l) *Fire Hose; Undercut Ring Nozzle*
$F = 0.71A\sqrt{2gh}$

FIG. 10. Types of orifices and nozzles.

3–26

assumed equal to zero.   At the pipe inlet, the velocity is not zero, and its magnitude can be calculated easily from the continuity equation.   The energy required to accelerate the fluid from an initial zero velocity to that which exists at the pipe inlet is supplied by the static head (or by the static and pressure head if the reservoir is subjected to a pressure above atmospheric).

The velocity head of the fluid flowing through the pipe is lost at exit unless a suitable expanding nozzle is provided there to reconvert the velocity head back to static head.   This reconversion is very effectively accomplished in the draft tube of water turbines, etc., where in many cases from 80 to 95 per cent of the velocity energy is reclaimed.   Where a pipe discharges into the atmosphere or a reservoir without the equivalent of a draft tube, the entire velocity head may be considered as lost. Examples of this kind are the water boxes on condensers, feedwater heaters, etc., where the velocity head in the supply pipe and that in the tubes of each pass are successively lost in the water boxes of the equipment.

Table 12.   Entrance Losses for Liquids

Name	Letter designation in Fig. 10	k value	$n =$ equivalent resistance in number of diameters of straight pipe[1]
Sharp-edged orifice............	a	1.00	33
Streamline contour...........	b	0.04	1.3
Reentrant tube...............	c	0.93	31
Venturi adjutage	d	0.04	1.3
Square entrance	e	0.49	16
Borda mouthpiece............	f	1.00	33
Conical diverging tube........	g	0.18	6
Conical converging tube.......	h	0.18	6

  [1] Computed for a friction factor $f = 0.0075$.

Considerable investigation has been made of entrance losses associated with various designs of inlet and outlet configurations.   Results of many such investigations are summarized in Fig. 10 in which several different types of orifices and nozzles are shown.   For each orifice or nozzle, the velocity obtained by test as a function of reservoir level h is indicated.   Also, the volume-flow rate F as a function of area at nozzle outlet is given.   The velocity coefficient includes the effect of both acceleration and of friction due to turbulence.   For example, in Fig. 10c, the test velocity is $0.72 \sqrt{2gh}$.   Solving this equation for V, there is obtained $h = 1.93(V^2/2g)$.   At outlet, since there is free discharge to atmosphere, none of the velocity head is recovered and the velocity head loss is $h_v = 1.00(V^2/2g)$.   By difference, the head loss due to friction is $h_f = 0.93(V^2/2g)$.

The frictional head loss is of the form $h_f = k(V^2/2g)$.   When similar calculations are carried out for each of the nozzles or orifices of Fig. 10, values of the coefficient k are determined and are listed in Table 12.   Listed also are the approximate equivalent resistances in number of straight-pipe diameters.

Table 13 lists the values of h_v which correspond to various velocities.   It must be emphasized that these heads are expressed in feet head of the flowing fluid.   For example, if water of 62.4 lbm/cu ft density were flowing in a conduit at 5 fps velocity, the velocity head would be 0.389 ft of water and the corresponding pressure equivalent would be 24.2 lbf/sq ft.   However, if air of density 0.075 lbm/cu ft were the flow medium, the pressure equivalent would be only 0.029 lbf/sq ft, although the velocity head would be 0.389 ft of air.

The ASME Special Research Committee on Fluid Meters has proposed the following distinction between orifices and nozzles.   A *nozzle* has a converging

Table 13.   Velocity Heads Corresponding to Various Velocities

Velocity, feet per second, V	Velocity head, feet, $h = V^2/2g$	Velocity, feet per second, V	Velocity head, feet, $h = V^2/2g$
0.5	0.004	5.5	0.466
1.0	0.016	6.0	0.560
1.5	0.034	6.5	0.650
2.0	0.062	7.0	0.762
2.5	0.097	7.5	0.876
3.0	0.140	8.0	0.995
3.5	0.190	8.5	1.120
4.0	0.248	9.0	1.26
4.5	0.314	9.5	1.40
5.0	0.389	10.0	1.55

approach of sufficient length and curvature to suppress substantially all contraction. An *orifice* has so little, if any, approach curvature that contraction is fully developed or only partially suppressed.   The orifice, if circular, may be further limited to an opening having a width as measured parallel to its axis and including any approach curvature that is less than one-fifth its diameter.

Coefficients of Velocity, Contraction, and Discharge. The jet of any fluid, whether liquid, gas, or vapor, passing through an orifice usually contracts to an area smaller than the opening itself, depending on the nature of the orifice.   This section of minimum area is called the "vena contracta."   With liquids it is the critical section where the actual velocity approaches the theoretical.   Figure 10 illustrates the more usual types of nozzles and orifices and shows in which types the vena contracta is most pronounced.   The coefficient of contraction C_c is the ratio of the area of the jet at the vena contracta to the area of the orifice.   The numerical values of C_c range from about 0.54 to unity, depending on the nature of the orifice and the fluid.

The coefficient of velocity C_v is the ratio of the actual mean velocity of the jet at the vena contracta to the theoretical velocity due to the whole head on the orifice. Its value varies from about 0.95 to 0.995.   The vena contracta is the critical section at which the actual velocity most nearly approaches the theoretical corresponding to $\sqrt{2gh}$.

The coefficient of discharge C is the ratio of the actual discharge to the theoretical discharge for the entire orifice area, or

$$C = F/A \sqrt{2gh} \quad \text{or} \quad F = CA \sqrt{2gh} \tag{24}$$

C is also equal to the product of the coefficient of velocity C_v and the coefficient of contraction C_c;

$$C = C_v C_c$$

Approximate values of C_c, C_v, and C for other common types of orifices and nozzles are shown in connection with the conventional illustrations of Fig. 10.   It will be noted that, for convenience, these orifices are shown in the side of a reservoir with the head on the orifice appearing as h, the distance from the center of the orifice to the surface of the liquid.   Substantially the same values for C apply, however, if the orifices are placed between pipe flanges or on the end of a pipe discharging into the atmosphere or for submerged discharge, provided the velocity of approach is taken into account.   The head on the orifice is, in each case, the differential head or pressure between the inlet and outlet sides measured on the center line of the orifice.

Effect of Velocity of Approach. The actual quantity discharged is still further modified by the velocity of approach if this is appreciable with respect to the velocity

through the orifice or nozzle. In dealing with the effect of velocity of approach, it is necessary to distinguish between compressible and incompressible fluids. In the case of compressible fluids the specific volume changes with change in pressure through the nozzle, and the correction factor for velocity of approach is treated somewhat differently.

Referring again to the types of orifices and nozzles illustrated in Fig. 10 and assuming them to be on the end of a pipe or between a pair of flanges, the effect of velocity of approach for liquids may be determined as follows: For convenience in comparing this formula with those later derived for the flow of gases under similar conditions, the subscript a will be used to denote conditions in the pipe approaching the orifice, and subscripts t and b to denote conditions in the throat at Sec. B and at vena contracta, respectively.

From the continuity equation

$$F = A_a V_a = A_t V_t = A_b V_b$$

In the case of liquids, the velocity V_b attained in the vena contracta approximates that corresponding to the entire effective head H on the orifice, viz., $H = (h + V_a^2)/2g$, where h is the difference in static head across the orifice and $V_a^2/2g$ is the head due to the velocity of the liquid approaching the orifice. Hence, if the efficiency of the orifice in converting the effective head into velocity is expressed by C_v,

$$V_b = C_v \sqrt{2gH} = C_v \sqrt{2gh + V_a^2}$$

and $$A_b = C_c A_t$$

Substituting in the continuity equation,

$$F = A_a V_a = C_c A_t C_v \sqrt{2gh + V_a^2}$$

Squaring and substituting for V_a its equivalent $V_a = F/A_a$,

$$F^2 = C_c^2 C_v^2 A_t^2 (2gh + F^2/A_a^2)$$

Solving for F,

$$F = C_c C_v A_t \sqrt{2gh} \times \frac{1}{\sqrt{1 - (C_c C_v A_t/A_a)^2}} \qquad (25)$$

It will be noted that $C_c C_v = C$ and that Eq. (25) differs from Eq. (24) only by the factor

$$K = \frac{1}{\sqrt{1 - (CA_t/A_a)^2}}$$

which is designated the *correction factor for velocity of approach K*. Hence,

$$F = CA_t K \sqrt{2gh} \qquad (26)$$

Numerical values for K corresponding to different ratios of CA_t/A_a are given in Table 14.

Table 14. Correction Factor K for Velocity of Approach

$\dfrac{CA_t}{A_a}$	K	$K = \dfrac{1}{\sqrt{1 - (CA_t/A_a)^2}}$
0.04	1.0008	
0.16	1.0130	$C = C_v C_c =$ product of velocity coefficient and the coefficient of
0.36	1.0719	contraction (see Fig. 10).
0.64	1.3014	$A_t/A_a =$ ratio of area of orifice to area of pipe.

Sharp-edged Orifice. In its simplest and most familiar form the *sharp-edged orifice*, otherwise known as the *orifice in a thin plate*, is merely a circular hole in a thin flat diaphragm that is clamped between the flanges of a pipe joint so that its plane is perpendicular to the axis of the pipe and the hole is concentric with the pipe. A thicker plate is sometimes used, in which case it is chamfered around the hole on the outlet side so as to leave only a thin edge, but the inlet face of the plate remains flat with a sharp 90-deg corner at the edge of the hole.

In the case of average conditions for water, the velocity at the vena contracta is 0.98 to 0.99 of $\sqrt{V_0^2 + 2gh}$, and the coefficient of contraction around 0.62. Hence the quantity of water discharged through this orifice is

$$F = C_v C_c A_t K \sqrt{2gh} = C A_t K \sqrt{2gh}$$

or substituting approximate numerical values for C_v, C_c, and C,

$$F = 0.985 \times 0.62 A_t K \sqrt{2gh} = 0.61 A_t K \sqrt{2gh}$$

Where the diameter of a *thin*-plate orifice is less than two-tenths of the pipe diameter, the effect of velocity of approach is negligible. But, as the orifice diameter approaches eight-tenths of the pipe diameter, the quantity discharged is increased by as much as 10 per cent, and the results become discordant when the ratio exceeds 0.8.

Head Lost over Orifice. Since the acceleration and deceleration of the fluid stream in passing through an orifice are accompanied by considerable turbulence and consequent dissipation of energy, it is to be expected that the overall pressure

Table 15. Factor m for Determining Overall Friction Loss across Orifice

d_t/d_a	A_t/A_a	m	
0.2	0.04	0.96	$m = 1 - \dfrac{A_t}{A_a}$
0.4	0.16	0.84	
0.6	0.36	0.64	
0.8	0.64	0.36	

loss across an orifice will be much greater than where the stream is guided as in a venturi. The portion of the velocity head not recovered has been found by experiment to agree very nearly with the relation $1 - A_t/A_a$, where A_t and A_a represent areas of throat and approaching pipe, respectively. Values determined from this relation are given in Table 15 in connection with loss of head over an orifice used for measurement of flow.

Venturi Meters. The venturi meter is a direct application of Bernoulli's theorem to the measurement of liquids flowing in pipes under pressure. It is frequently applied to the measurement of water flow in cases where the intensity of pressure or the volume of flow renders impracticable the use of the common disk type of water meter or similar devices. Venturi meters are also applied to the measurement of flow of gases as explained in a later paragraph. A description of the venturi meter and its measurement of liquid flow can be given by reference to the conventional diagram of Fig. 11 and the following symbols which are used in conjunction with those previously defined. The upstream section at A is called the "inlet," the section at B the "throat," and that at C the "outlet." The section at C does not necessarily have to be of the same diameter as that at A, but in order to show graphically the friction head loss h_λ, it was convenient to assume it such in this instance. Static pressures in the pipe at points A, B, and C are designated as p_a, p_b, and p_c, respectively; the internal diameter d' is measured in feet.

From the continuity equation,

$$F = A_a V_a = A_b V_b = A_c V_c$$

and
$$A_a = \pi d_a'^2/4 \qquad A_b = \pi d_b'^2/4 \qquad A_c = \pi d_c'^2/4$$

Then,
$$F = \pi d'^2/4 \times V_a = \pi d_b'^2/4 \times V_b = (\pi d_c'^2/4)\, V_c$$

and
$$V_a = (d_b'^2/d_a'^2) V_b \qquad \text{or} \qquad V_a^2 = (d_b'^4/d_a'^4) V_b^2$$

From Bernoulli's theorem,

$$V_a^2/2g + p_a/w + h_a = V_b^2/2g + p_b/w + h_b + h_{\lambda}' = V_c^2/2g + p_c/w + h_c + h_{\lambda}$$

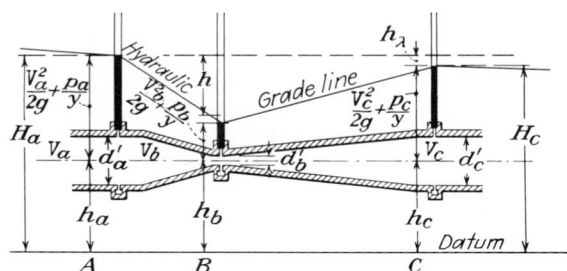

FIG. 11. Conventional diagram of venturi meter.

If the axis of the meter is set level, the head due to elevation is constant and the terms h_a, h_b, and h_c can be dropped.

Therefore, $\quad V_a^2/2g + p_a/w = V_b^2/2g + p_b/w + h_{\lambda}' = V_c^2/2g + p_c/w + h_{\lambda}$

Hence,
$$\frac{V_b^2 - V_a^2}{2g} = \frac{p_a - p_b}{w} - h_{\lambda}'$$

Dropping h_{λ}', which is insignificant, and substituting p/w for $(p_a - p_b)/w$,

$$\frac{V_b^2 - V_a^2}{2g} = \frac{p}{w} = h$$

Substituting the value of $V_a^2 = (d_b'^4/d_a'^4)V_b^2$,

$$V_b = \frac{d_a'^2}{\sqrt{d_a'^4 - d_b'^4}} \sqrt{2gh}$$

$$F = \frac{\pi d_b^2}{4} \times V_b = \frac{\pi d_a'^2 d_b'^2}{4\sqrt{d_a'^4 - d_b'^4}} \sqrt{2gh}$$

It is customary to make allowance for friction by applying a coefficient of discharge C. The value of C varies with velocity at the throat, ratio of inlet and outlet diameters, and departures from exact dimensions. Under ordinary conditions the value of C lies between 0.97 and unity. Therefore, inserting C, the value of F becomes

$$F = C \frac{\pi d_a'^2 d_b'^2}{4\sqrt{d_a'^4 - d_b'^4}} \sqrt{2gh}$$

h in this case represents the difference in the static pressure head at points A and B. It frequently is convenient to measure this differential head with a mercury U tube or manometer and convert the readings into equivalent head of the liquid flowing in the pipe.

The use of a venturi meter in measuring quantity of fluid flow involves very little friction loss if the slopes of the converging and diverging nozzles are properly selected. The diverging nozzle on the outlet side serves to convert the increased velocity head back again into static pressure with a very small pressure drop over the entire meter. In this respect it is superior to the flowmeters described in the next paragraph. Its considerable length is a drawback in some instances, however, and the flowmeter is frequently employed in preference for this reason. Venturi meters are ordinarily equipped with suitable indicating, totalizing, and recording or integrating devices to register the amount of flow.

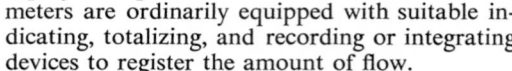
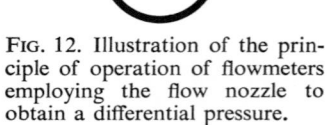
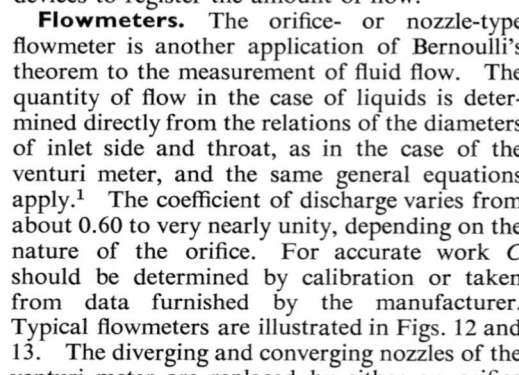

FIG. 12. Illustration of the principle of operation of flowmeters employing the flow nozzle to obtain a differential pressure.

Flowmeters. The orifice- or nozzle-type flowmeter is another application of Bernoulli's theorem to the measurement of fluid flow. The quantity of flow in the case of liquids is determined directly from the relations of the diameters of inlet side and throat, as in the case of the venturi meter, and the same general equations apply.[1] The coefficient of discharge varies from about 0.60 to very nearly unity, depending on the nature of the orifice. For accurate work C should be determined by calibration or taken from data furnished by the manufacturer. Typical flowmeters are illustrated in Figs. 12 and 13. The diverging and converging nozzles of the venturi meter are replaced by either an orifice plate or short converging nozzle, as shown. No special device is employed to regain the additional velocity head acquired in passing through the restricted opening, and consequently, much of this head is permanently lost.

The loss of head over a thin-plate or sharp-edged orifice, such as shown in Fig. 13,

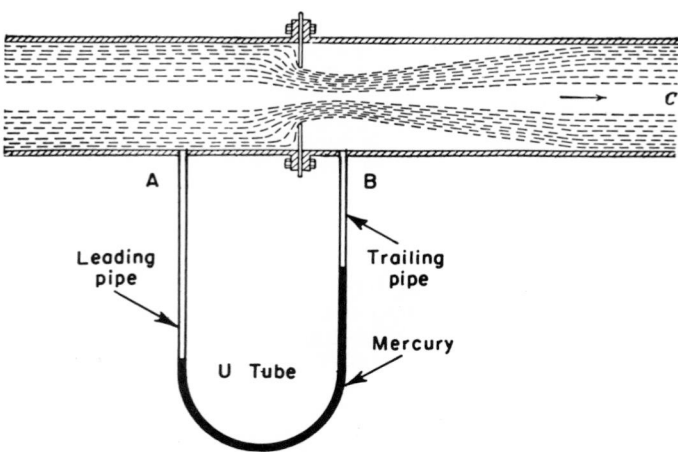

FIG. 13. Illustrating the principle of operation of flowmeters employing the orifice plate to obtain a differential pressure.

[1] See "Flow Measurement by Means of Standardized Nozzles and Orifice Plates," "Flow Measurement," ASME Power Test Codes.

is found from the following relation. All heads are expressed in feet of liquid and velocities in feet per second.

$$h = h_a - h_c = m(h_a - h_b) = m(V_b^2 - V_a^2)/2g$$

Values of the coefficient m are given in Table 15.

Figures 12 and 13 indicate that a mercury-filled U tube is utilized to measure the pressure differential across the flow nozzle or flow orifice. For recording purposes, secondary elements are available commercially; these devices by electrical, pneumatic, or mechanical means automatically give a chart indication of flow over some prescribed time period and, in addition, may be provided with a feature which integrates total flow.

Pitot Tubes. The pitot tube is a device for determining the velocity head of any fluid flowing through any conduit. In the case of the pitot tube, the same general laws apply to the flow of both liquids and gases and vapors. Its use is more common in the measurement of water and air flow. From the velocity head, the corresponding velocity is readily determined from the relation $V = \sqrt{2gh}$. Figure 14 illustrates in a conventional way the principles used in determining the velocity

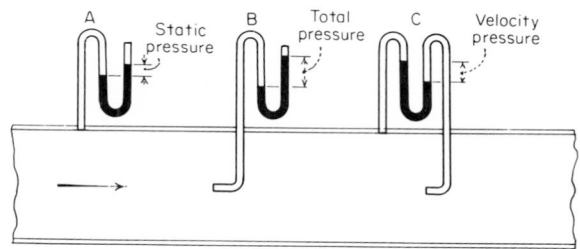

FIG. 14. Principle of the pitot tube.

head with a pitot tube. Manometer A is connected so as to indicate the static or internal fluid head or pressure in the conduit; manometer B is connected to indicate the total head, which consists of static pressure plus velocity head; manometer C is connected to indicate directly the velocity head, which is the differential head between total head and static head. Figure 15 illustrates a compound pitot tube equivalent to the arrangement used with manometer C. The inclined manometer shown in Fig. 15 is used in measuring the flow of air or gases where the heads are too small to read accurately with an ordinary U tube. Colored kerosene, water, mercury, or other liquid can be used in such a manometer, depending on its calibration and the heads to be measured.

In Fig. 15, the tube is inserted into the conduit in such a way that its part A-B is parallel to the flow of fluid, with the end A toward the flow. The part A-B consists of an inner tube which transmits the total pressure to the tube D and an outer jacket through which the static pressure is transmitted to the tube C. This outer jacket contains several small holes through which the static pressure acts. The two pressures are transmitted to the ends of the differential inclined manometer E, which is a U tube arranged with one leg at an angle so that the linear deflection of the liquid per unit of head is increased.

The velocity of flow is not constant at all points in the cross section of the conduit. Near the walls, flow is retarded by friction, and it reaches a maximum at the center. It is, therefore, necessary to measure the velocity head at several points in the cross section and calculate the velocity at each by the formula $V = \sqrt{2gh_2}$ in order to

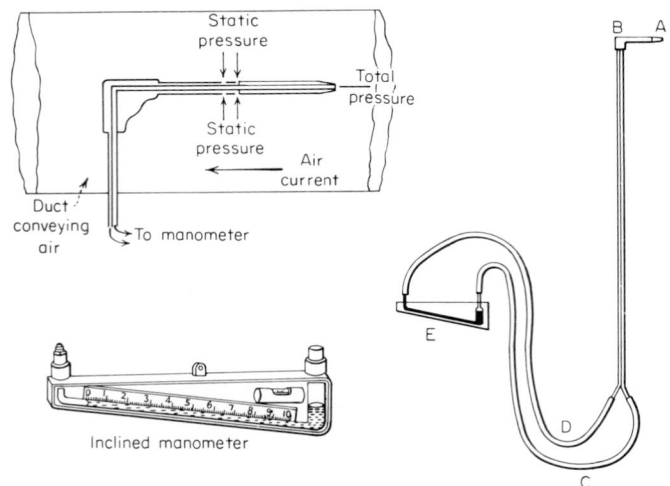

FIG. 15. Pitot tube.

obtain an average figure. The numerical average of these velocities may then be used as the average velocity for the entire cross section. The quantity of fluid flowing can be computed readily from the average velocity and the cross-sectional area of the conduit. Since the velocity head is usually read in linear units of head h_1 of some liquid such as water or mercury, it is necessary to convert these readings into linear units of head, h_2, of the fluid in the conduit. This is accomplished by multiplying h_1 by a conversion factor z which is defined as

$$z = \frac{\text{density of manometer fluid}}{\text{density of flowing fluid}}$$

The relation between h_1 and h_2 then is $h_2 = zh_1$, and an equation may be written

$$V = 2gzh_1 \tag{27}$$

For a round conduit, the cross-sectional area should be divided into a number of annular zones of *equal area* and a traverse of the conduit made in both a horizontal and a vertical direction as shown in Fig. 16. For each foot of pipe diameter, the cross section should be divided into at least three of these zones. Table 16 gives the distance from the center of the pipe at which each reading should be taken in percentage of pipe diameter. It is important that the velocities be computed

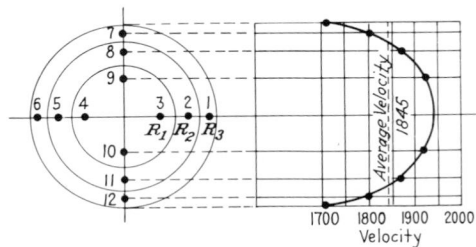

FIG. 16. Division of round pipe into annular zones for pitot tube traversal.

Table 16. Pipe Traverse for Pitot Tube Readings

(Distance from center of pipe to point of reading
in percentage of pipe diameter)

Number of equal areas in traverse	Number of readings	1st R_1	2d R_2	3d R_3	4th R_4	5th R_5	6th R_6	7th R_7	8th R_8
3	12	20.4	35.3	45.5					
4	16	17.7	30.5	39.4	46.6				
5	20	15.5	27.2	35.3	41.7	47.4			
6	24	14.5	25.0	32.3	38.2	43.3	47.9		
7	28	13.4	23.1	29.9	35.3	40.1	44.3	48.2	
8	32	12.5	21.6	28.0	33.2	37.6	41.5	45.1	48.4

separately and averaged, for the velocity varies as the square root of the pressure, and accurate results cannot be obtained by averaging the pressure readings. The method outlined above is essentially the same as recommended by the ASME Special Research Committee on Fluid Meters.[1]

Similar methods are used by hydraulic engineers although even with large pipes it is usually considered sufficiently accurate to make the traverse in the center circle and three or four rings only.

The following sample problem will aid in making clear the use of this formula and method. Problem: Air at atmospheric pressure and 70 F is flowing through a 10-in. ID conduit. The velocity head at point 2 was observed as 0.25 in. of water at 70 F. The corresponding velocity in feet per second is determined as follows:

$$z = \frac{\text{density of water in lb per cu ft}}{(\text{density of air in lb per cu ft}) \times (\text{number of in. in 1 ft})}$$

$$z = \frac{62.4}{0.07493 \times 12} = 69.5$$

$$V = \sqrt{2gzh_1} = \sqrt{2 \times 32.2 \times 69.5 \times 0.25} = \sqrt{1{,}120}$$
$$= 33.46 \text{ fps}$$

or
$$V = 33.46 \times 60 = 2{,}008 \text{ fpm}$$

For the purpose of this illustration assume that the average velocity for all the points 1, 2, 3, etc., has been calculated and found to be 2,500 fpm. The quantity Q_a of air flowing per minute then is

$$Q_a = 2{,}500 \frac{\pi 10^2}{4 \times 144} = 1{,}363 \text{ cfm} \tag{28}$$

For convenience in repeated calculations, Eq. (28) can be written in simplified form to suit the particular conditions. Suppose, for instance, that the velocity of any fluid is desired in feet per minute or per second when the velocity head readings are given in inches of water. By a solution similar to that given above,

$$V = 18.275 \sqrt{h_1/w} = \text{velocity, fps}$$

or
$$V = 1{,}096.5 \sqrt{h_1/w} = \text{velocity, fpm}$$

where w is the density of the fluid in pounds per cubic foot under the conditions at which h_1 was measured.

Where extreme accuracy is not required in measuring the flow of fluids with a pitot tube, an average velocity may be determined by applying a coefficient to the

[1] "Fluid Meters, Their Theory and Application," published by the American Society of Mechanical Engineers, 345 East 47th St., New York, 10017.

velocity measured at the center of the conduit.   Various authorities give a coefficient of from 0.81 to 0.82 for air or gas flow in circular conduits, by which the *velocity head* readings taken at the center of the pipe should be multiplied to obtain the corrected average head.   Consequently, the velocity based on the observed pressure readings may be multiplied by the coefficient 0.91 to obtain the corrected average velocity for air or gas, since the square root of 0.815 is 0.91.   Similarly, in the case of water flow, the mean velocity may be determined, approximately, by multiplying the center velocity by 0.83 or 0.84.   This ratio of average velocity to maximum velocity should be accurate within about 3 per cent plus or minus in the case of water.

The principles and construction of pitot tubes are discussed at some length in "Fluid Meters, Their Theory and Application," Part I, *Report* of ASME Special Research Committee on Fluid Meters, 4th ed., 1937,[1] from which the following is taken:

> The Pitot tube is suitable for use only where the conditions of flow, as indicated by the readings, are reasonably steady.   Steady readings are indicative of steady flow.   The question of symmetrical distribution of flow in the pipe is not as important as steady flow. Well designed bends of long radius in the pipe will affect the distribution of velocity across the diameter of the pipe for a considerable distance beyond the bend, but this is not necessarily objectionable.   It is desirable to have a run of straight pipe, at least 15 pipe diameters long, following short bends, valves, and poorly designed intakes, and immediately preceding the Pitot tube to minimize the disturbances to flow which may result from such features. The greater the length of straight pipe ahead of the Pitot tube the greater is the probability that parallel stream flow will exist, which will increase the accuracy of the velocity measurement.   The uniformity of the readings obtained at the various points on repeated trials is the most reliable criterion by which to judge the character of the indications.

Measurement by Volume or Weight.   The methods described above for measurement of flow in most cases involve the use of coefficients directly or indirectly derived by comparison with volumetric or weight measurements.   In the case of measuring water, for instance, the first and last word in checking the accuracy of the measuring device is to take its discharge for a given time interval into a tank mounted on scales or into a tank of known volume.   Such volumetric or weight measurements are considered the most authentic methods of testing boilers, steam turbines, water pumps, and other power-plant equipment.   Where weigh tanks are used in power-plant testing, it is customary to install them in pairs, so that the measurement can be made continuous by emptying one tank while the other is being filled.

FLOW OF GASES

In this section, we deal with the one-dimensional steady flow of a perfect gas. A perfect gas is one which:

1. Obeys the equation of state $Pv = (\bar{R}/M_w)T$ (29)
2. Its constant-pressure and constant-volume specific heats are constant.

Nomenclature, as follows, will be used:

P = pressure, lbf/sq ft
v = specific volume, cu ft/lbm
$\bar{R}$ = universal gas constant, 1,546 ft lbf/(lb mole R)
R = gas constant for a particular gas, ft lbf/(lbm R)
M_w = molecular weight of the gas, lbm/lb mole

---

[1] This report also contains much valuable information regarding venturi meters, flowmeters, and other types of meters, with an extensive bibliography.

c_p = constant-pressure specific heat, Btu/(lbm R)
c_v = constant-volume specific heat, Btu/(lbm R)
k = specific-heat ratio, c_p/c_v
ρ = density of the gas, lbm/cu ft
c = acoustic velocity, fps
c^* = acoustic velocity, at the point where the Mach number is unity, fps
h = static enthalpy, Btu/lbm
T = absolute static temperature, deg Rankine
V = fluid velocity, fps
M = Mach number of the flowing fluid = V/c
$M^* = V/c^*$
A = area, sq ft
A^* = area at the point in flow path where $M = 1$
$\dot{m}$ = mass flow rate, lbm/sec
G = mass velocity = $\dot{m}/A$, lbm/(sec sq ft)
t = time, sec
g = local acceleration due to gravity, ft/sec²
g_c = gravitational constant, ft lbm/(lbf sec²)
J = conversion constant (mechanical equivalent of heat) and is taken as equal to 778 ft lbf/Btu
s = entropy, Btu/(lbm R)

Subscript o represents the value of the indicated property under stagnation conditions. Thus, P_o represents stagnation pressure, T_o represents stagnation temperature, etc.

Superscript * represents the value of the indicated property at the point where the Mach number is unity. Thus T^* represents the static temperature at the point in the flow channel where the Mach number is unity.

Isentropic Flow

An isentropic flow is one in which the entropy of the flowing medium remains constant. If no heat is transferred to or from the fluid during the flow process, the absence of entropy change requires that the flow be mechanically reversible, that is, that the flow be frictionless.

In most piping systems the flow velocity is sufficiently high that the amount of heat transfer between a unit mass of fluid and the surroundings is so low as to be considered negligible. That is to say, the flow in many cases can be assumed without significant error to be adiabatic.

Generally, the lengths of flow channels involved in isentropic-flow calculations are short. Thus, in many cases, the assumption of frictionless (mechanically reversible) flow is justified.

Adiabatic Flow: Stagnation Properties. In dealing with the flow of gases in which the density is low as compared with liquids, the potential-energy changes are usually very small and are here considered negligible. For an adiabatic steady flow in the absence of shaftwork and potential-energy changes, the energy equation written between sections 1 and 2 in the flow channel is

$$h_1 + V_1^2/2g_cJ = h_2 + V_2^2/2g_cJ = \text{constant} \tag{30}$$

Since h and the kinetic energy terms are in the same dimensions, they are additive directly, and their sum is denoted as the stagnation enthalpy h_o. Thus,

$$h_o = h + V^2/2g_cJ \tag{31}$$

and the stagnation enthalpy is constant in any adiabatic flow, regardless of any friction effects.

For a perfect gas, $\Delta h = c_p \Delta T$ and $c_p - c_v = R/J$. Combining these relationships with the definition of stagnation enthalpy, there is obtained

$$V = \sqrt{[2g_c k/(k-1)]R(T_o - T)}$$

in which T_o is the stagnation temperature and represents the temperature reached whenever a flowing fluid at a static temperature T is brought to rest adiabatically.

When the above expression is combined with the definition of the Mach number in which c is the acoustic velocity at the point where velocity is V and temperature is T, that is,

$$M = V/c = V/\sqrt{g_c kRT}$$

there is obtained

$$T_o/T = 1 + [(k-1)/2]M^2 \tag{32}$$

Measurement of Temperature in Gas Streams. Thermocouples are frequently inserted into high-velocity gas streams in order to measure temperature. The flowing gas comes to rest adiabatically against the thermocouple, and if heat transfer did not occur, the thermocouple would read the stagnation temperature as indicated by Eq. (32). However, as soon as the thermocouple temperature increases ever so slightly above that of the flowing fluid, heat transfer by radiation takes place to the walls of the flow channel. Finally an equilibrium temperature T_c between T and T_o will be reached. To take this apparent discrepancy into account a recovery factor α is used:

$$\alpha = (T_c - T)/(T_o - T)$$

and by inserting the ratio into Eq. (32), there is obtained

$$T_c/T = 1 + \alpha[(k-1)/2]M^2$$

The recovery factor generally ranges from 0.75 for unshielded couples to about 0.95 for thermocouples which are installed with shields to eliminate almost entirely the radiation effects.

Example. A perfect gas of molecular weight 40 lbm/lb mole, a constant-pressure specific heat of 0.28 Btu/(lbm R), a static pressure of 30 psia, and a static temperature of 340 F passes in steady flow through a 3- by 2-ft duct at the rate of 800 lbm/sec. (a) What is the stagnation temperature of the gas? (b) If an unshielded thermocouple installed in the flowing stream indicates a temperature of 390 F, what is the thermocouple recovery factor?
Solution.

$$R = \bar{R}/M_w = 1,546/40 = 38.6 \text{ ft lbf/(lbm R)}$$

$$c_v = c_p - R/J = 0.28 - 38.6/778 = 0.2304 \text{ Btu/(lbm R)}$$

$$k = c_p/c_v = 0.28/0.2304 = 1.215$$

$$\rho = P/RT = 30 \times 144/38.6 (340 + 460) = 0.14 \text{ lbm/cu ft}$$

$$V = \dot{m}/\rho A = 800/(0.14 \times 3 \times 2) = 952 \text{ fps}$$

$$c = \sqrt{kg_c RT} = \sqrt{1.215 \times 32.17 \times 38.6 \times 800} = 1,100 \text{ fps}$$

$$M = V/c = 952/1,100 = 0.864$$

(a) $\displaystyle T_o = T\left(1 + \frac{k-1}{2}M^2\right) = 800\left(1 + \frac{1.215 - 1}{2} \times 0.864^2\right) = 864R$

(b) $\displaystyle \alpha = \frac{T_c - T}{T_o - T} = \frac{850 - 800}{864 - 800} = 0.78$

Subsonic and Supersonic Flows. When the continuity equation, the Euler equation, and the equation for acoustic velocity are combined, there is obtained in differential form

$$dA/A = (dP/\rho V^2)g_c(1 - M^2) \tag{33}$$

This equation relates pressure variations to changes in the dimensions of the flow channel for the general case of isentropic flow. Clearly, if there is no change in area of flow path ($dA = 0$) in isentropic flow, there will be no change in pressure.

The Euler equation for steady flow of a frictionless fluid in the absence of potential-energy changes is

$$dP = -\rho(V\,dV/g_c) \tag{34}$$

It is seen that a positive pressure change requires a negative change in velocity. Conversely, a velocity increase must be accompanied by a decrease in pressure. This is in conformity also with the Bernoulli equation; in fact, the Bernoulli equation results directly from integration of Eq. (34) for the incompressible flow case.

A device in which the fluid velocity increases is referred to here as a *nozzle*, and a flow device in which the fluid velocity decreases will be termed a *diffuser*. In

Table 17.   Subsonic and Supersonic Flows

(Property changes as affected by change in flow area)

	Subsonic flow (M < 1)	Supersonic flow (M > 1)
Diffuser	Area increases Pressure increases Velocity decreases	Area decreases Pressure increases Velocity decreases
Nozzle	Area decreases Pressure decreases Velocity increases	Area increases Pressure decreases Velocity increases

both devices, an area change is accompanied by changes in pressure and velocity in accordance with Eqs. (33) and (34). It is noted that the parenthetical terms $(1 - M^2)$ in Eq. (33) will be positive for subsonic flows ($M < 1$) and will be negative for supersonic flows ($M > 1$). Thus an area increase can result in either pressure increases or decreases, as dictated by the Mach number of the flow. The effects of area increase for nozzles and diffusers are shown in Table 17.

Property Relations in Isentropic Flow. When Eq. (32) is combined with the isentropic relation

$$pv^k = \text{constant}$$

Table 18. One-dimensional Isentropic Compressible Flow Functions for a Perfect Gas with Constant Specific Heat

$$(k = 1.4)$$

M	M_*	$\dfrac{A}{A_*}$	$\dfrac{p}{p_0}$	$\dfrac{\rho}{\rho_0}$	$\dfrac{T}{T_0}$	$\dfrac{F}{F_*}$	$\left(\dfrac{A}{A_*}\right)\left(\dfrac{p}{p_0}\right)$
0.05	0.05476	11.592	0.99825	0.99875	0.99950	9.1584	11.571
0.10	0.10943	5.8218	0.99303	0.99502	0.99800	4.6236	5.7812
0.15	0.16395	3.9103	0.98441	0.98884	0.99552	3.1317	3.8493
0.20	0.21922	2.9635	0.97250	0.98027	0.99206	2.4004	2.8820
0.25	0.27216	2.4027	0.95745	0.96942	0.98765	1.9732	2.3005
0.30	0.32572	2.0351	0.93947	0.95638	0.98232	1.6979	1.9119
0.35	0.37879	1.7780	0.91877	0.94128	0.97608	1.5094	1.6336
0.40	0.43133	1.5901	0.89562	0.92428	0.96899	1.3749	1.4241
0.45	0.49326	1.4487	0.87027	0.90552	0.96108	1.2763	1.2607
0.50	0.53452	1.3398	0.84302	0.88517	0.95238	1.2027	1.12951
0.55	0.58506	1.2550	0.81416	0.86342	0.94295	1.1472	1.02174
0.60	0.63480	1.1882	0.78400	0.84045	0.93284	1.10504	0.93155
0.65	0.68374	1.1356	0.75283	0.81644	0.92208	1.07314	0.85493
0.70	0.73179	1.09437	0.72092	0.79158	0.91075	1.04915	0.78896
0.75	0.77893	1.06242	0.68857	0.76603	0.89888	1.03137	0.73155
0.80	0.82514	1.03823	0.65602	0.74000	0.88652	1.01853	0.69110
0.85	0.87037	1.02067	0.62351	0.71361	0.87374	1.00966	0.63640
0.90	0.91460	1.00886	0.59126	0.68704	0.86058	1.00399	0.59650
0.95	0.95781	1.00214	0.55946	0.66044	0.84710	1.00093	0.56066
1.0	1	1	0.52828	0.63394	0.83333	1	0.52828
1.1	1.08124	1.00793	0.46835	0.58169	0.80515	1.00305	0.47206
1.2	1.1583	1.03044	0.41238	0.53114	0.77640	1.01082	0.42493
1.3	1.2311	1.06631	0.36092	0.48291	0.74738	1.02170	0.38484
1.4	1.2999	1.1149	0.31424	0.43742	0.71839	1.03458	0.35036
1.5	1.3646	1.1762	0.27240	0.39498	0.68965	1.04870	0.32039
1.6	1.4254	1.2502	0.23527	0.35573	0.66138	1.06348	0.29414
1.7	1.4825	1.3376	0.20259	0.31969	0.63372	1.07851	0.27099
1.8	1.5360	1.4390	0.17404	0.28682	0.60680	1.09352	0.25044
1.9	1.5861	1.5552	0.14924	0.25699	0.58072	1.1083	0.23211
2.0	1.6330	1.6875	0.12780	0.23005	0.55556	1.1227	0.21567
2.2	1.7179	2.0050	0.09352	0.18405	0.50813	1.1500	0.18751
2.4	1.7922	2.4031	0.06840	0.14720	0.46468	1.1751	0.16437
2.6	1.8572	2.8960	0.05012	0.11787	0.42517	1.1978	0.14513
2.8	1.9140	3.5001	0.03685	0.09462	0.38941	1.2182	0.12897
3.0	1.9640	4.2346	0.02722	0.07623	0.35714	1.2366	0.11528
3.2	2.0079	5.1210	0.02023	0.06165	0.32808	1.2530	0.10359
3.4	2.0466	6.1837	0.01512	0.05009	0.30193	1.2676	0.09353
3.6	2.0808	7.4501	0.01138	0.04089	0.27840	1.2807	0.08482
3.8	2.111	8.9506	0.00863	0.03355	0.25720	1.2924	0.07723
4.0	2.1381	10.719	0.00658	0.02766	0.23810	1.3029	0.07059
4.2	2.1622	12.792	0.00506	0.02292	0.22085	1.3123	0.06475
4.4	2.1837	15.210	0.00392	0.01909	0.20525	1.3208	0.05959
4.6	2.2030	18.018	0.00305	0.01597	0.19113	1.3284	0.05500
4.8	2.2204	21.264	0.00240	0.01343	0.17832	1.3354	0.05091
∞	2.4495	∞	0	0	0	1.4289	0

there are obtained many useful relationships between fluid properties. For easy reference they are listed below:

$$\frac{P_o}{P} = \left(1 + \frac{k-1}{2} M^2\right)^{k/k-1} \tag{35}$$

$$\frac{\rho_o}{\rho} = \left(1 + \frac{k-1}{2} M^2\right)^{1/k-1} \tag{36}$$

$$\frac{T_o}{T^*} = \frac{k+1}{2} \tag{37}$$

$$\frac{P_o}{P^*} = \left(\frac{k+1}{2}\right)^{k/k-1} \tag{38}$$

$$\frac{\dot{m}}{A} = \sqrt{\frac{kg_c}{R}} \frac{P_o}{\sqrt{T_o}} \frac{M}{\left(1 + \dfrac{k-1}{2} M^2\right)^{(k+1)/2(k-1)}} \tag{39}$$

In particular, for a perfect gas with a specific-heat ratio k equal to 1.4,

$$T^*/T_o = 0.833 \tag{40}$$
$$P^*/P_o = 0.528 \tag{41}$$
$$(\dot{m}\sqrt{T_o})/(A^*P_o) = 0.532 \tag{42}$$

The relations of Eqs. (35) to (39) are seen to be functions of k and of Mach number only. Thus, for selected values of k and M, numerical values can be readily calculated. Tables 30 to 35 of "Gas Tables" by Keenan and Kaye (John Wiley & Sons, Inc.) list values of the above property ratios for values of k ranging from 1 to 1.67 and for Mach numbers ranging from zero to infinity. Table 18 below lists selected values of the ratios for k equal to 1.4.

To illustrate the use of Table 18, the following example is presented.

Example. Air, to be considered a perfect gas with a specific ratio of 1.4, is to be expanded from an initial Mach number of 0.1 to a final Mach number of 0.8 in a frictionless adiabatic nozzle. At nozzle inlet, static pressure and static temperature are 100 psia and 800 R, respectively. The flow rate is to be 10 lbm/sec.

It is required to determine the inlet and outlet areas and the static pressure, static temperature, and velocity at nozzle outlet.

Solution. At the nozzle inlet (subscript 1)

$\rho_1 = P_1/RT_1 = 100 \times 144/53.3 \times 800 = 0.337$ lbm/ cu ft

$c_1 = \sqrt{g_c kRT} = 49.1\sqrt{T_1}$ for air with $k = 1.4$

$c_1 = 1,390$ fps

$V_1 = M_1 c_1 = 0.1(1,390) = 139$ fps

$A_1 = \dot{m}/\rho_1 V_1 = 10/0.337 \times 139 = 0.214$ sq ft $= 30.8$ sq in.

Then, by Table 18, using subscript 2 to denote properties at the nozzle outlet,

$$A_2 = \frac{A_2/A^*}{A_1/A^*} A_1 = \frac{1.03823}{5.8218} \times 30.8 = 5.48 \text{ sq in.}$$

$$P_2 = \frac{P_2/P_o}{P_1/P_o} P_1 = \frac{0.65602}{0.99303} \times 100 = 66 \text{ psia}$$

$$T_2 = \frac{T_2/T_o}{T_1/T_o} T_1 = \frac{0.88652}{0.99800} \times 800 = 710 \text{ R}$$

$c_2 = 49.1\sqrt{T_2} = 49.1\sqrt{710} = 1,308$ fps

$V_2 = c_2 M_2 = 1,308 \times 0.8 = 1,045$ fps

As a check on the above calculations, the density can be calculated at the exit conditions and then combined with area and velocity at exit in the continuity equation to obtain a mass-flow rate for comparison with the given rate of 10 lbm/sec.

Thus, $$\dot{m} = \rho_2 A_2 V_2 = \frac{P_2}{RT_2} A_2 V_2 = \frac{66 \times 5.48 \times 1{,}045}{53.3 \times 710} = 10 \text{ lbm/sec}$$

Friction Flow

In this section there will be treated the friction flow of a perfect gas in a channel of constant cross-sectional area in the absence of heat transfer to or from the flowing

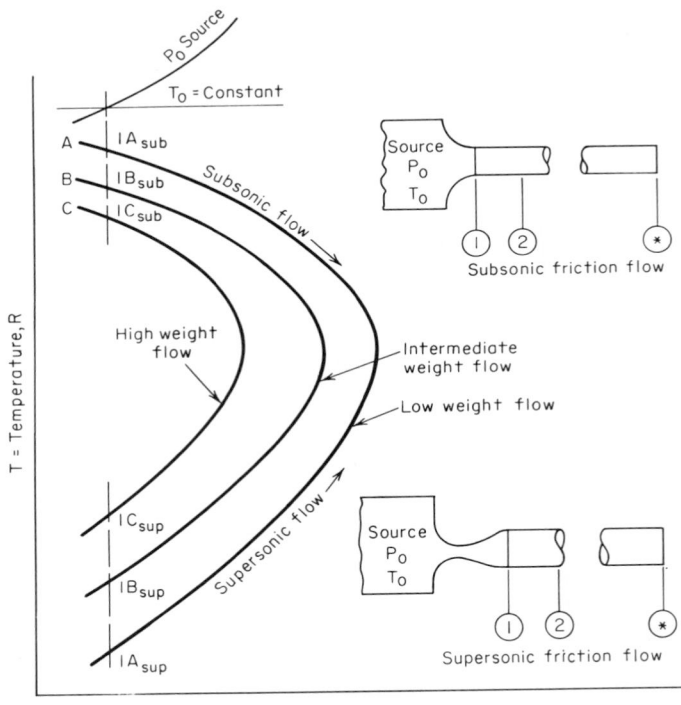

FIG. 17. Subsonic and supersonic friction flow.

fluid. For the more general and more complex case of friction flow with heat transfer in a channel of variable area, the reader is referred to "Dynamics and Thermodynamics of Compressible Fluid Flow" by A. H. Shapiro (Ronald Press).

The locus of state points on a temperature-entropy plot for the one-dimensional, adiabatic, friction flow of a perfect gas is called a *Fanno* line. For a duct of given cross-sectional area, each different weight-flow rate corresponds to a different Fanno line, as shown in Fig. 17. With reference to that figure, three Fanno lines have been shown and would correspond to three different rates of flow.

Fanno line *A* depicts the simultaneous changes in temperature and entropy for a low weight rate of flow; the upper, or subsonic, portion of the curve would be reached by a fluid which entered the duct from a reservoir maintained at stagnation

conditions P_o and T_o; the lower, or supersonic, portion would correspond to expansion through a convergent-divergent nozzle prior to duct entry. These curves are sketched for a duct of fixed cross-sectional area. The vertical distance between point o (the source) and the point which corresponds to duct inlet ($1A$, $1B$, $1C$) is directly proportional to the change in static temperature and, therefore, static enthalpy. Thus the velocity at $1A_{sub}$ is less than that at $1B_{sub}$; this accounts for the fact that curve A corresponds to a lower flow rate than does curve B or C.

Flow can occur only in the direction as indicated by the arrows; that is, a flow which is initially subsonic can, in the limit, become sonic but can never become supersonic because, to do so, the entropy must decrease and such an adiabatic drop in entropy would be a violation of the second law of thermodynamics.

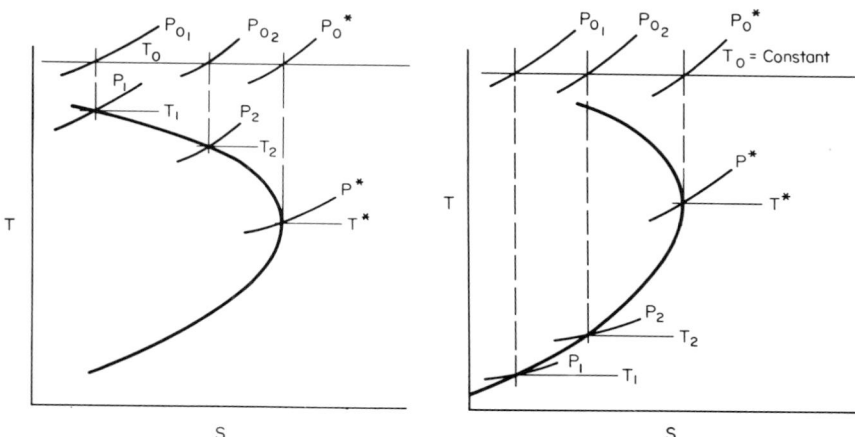

FIG. 18. Subsonic, one-dimensional, adiabatic flow process of a perfect gas.

FIG. 19. Supersonic, one-dimensional, adiabatic flow of a perfect gas.

Similarly, a flow which is initially supersonic will decrease in Mach number and increase in temperature during friction flow; if the duct is long enough, in the limit, the duct exit velocity can be sonic but can never become subsonic.

Figure 18 shows, on T-S coordinates, a Fanno line for an arbitrary but constant weight flow per unit area. Throughout, by definition, the stagnation temperature remains constant. At duct inlet, static pressure and static temperature are P_1 and T_1 respectively. At some arbitrary point, say 2, in the duct the static pressure and static temperature have been reduced by friction to P_2 and T_2. If point 2 corresponds to the end of the duct, then the exit state is defined by P_2 and T_2. It is noted that the stagnation pressure at point 2, P_{o2}, is lower than that at the duct inlet, P_{o1}. If the duct length is equal to a critical length, L_{max}, then at the duct exit the conditions will be sonic; that is the fluid velocity will be the sonic velocity corresponding to T^*.

For subsonic, one-dimensional, adiabatic friction flow of a perfect gas it is important to recognize that (1) static pressure decreases, (2) static temperature decreases, (3) velocity increases, (4) Mach number decreases, and (5) stagnation pressure decreases all in the direction of flow.

Figure 19 shows, on T-S coordinates, flow on the supersonic branch of a Fanno line. Fluid enters the duct with static pressure P_1 and static temperature T_1; such a state could be reached by expansion through a convergent-divergent nozzle from

source conditions P_{o1} and T_o. As flow progresses in the duct, the corresponding state points will lie on the Fanno line. At some arbitrary location along the duct length, the pressure and temperature are P_2 and T_2. If the duct length is equal to a critical length L_{max}, the state at the exit end will be sonic with the static temperature being T^*, the static pressure being P^*, the velocity being V^* and the Mach number being unity. Throughout the entire flow process, by definition of adiabatic flow, the stagnation temperature remains constant; however, as will be noted by reference to the figure, the stagnation pressure decreases in direction of flow.

For supersonic, one-dimensional, adiabatic friction flow of a perfect gas it is important to recognize that (1) static pressure increases, (2) static temperature increases, (3) stagnation pressure decreases, and (4) Mach number decreases all in the direction of flow.

For both subsonic and supersonic flow there exists a critical length of duct referred to as L_{max}; if that length is exceeded in subsonic flow, there will be a reduction in mass-flow rate; if L_{max} is exceeded in supersonic flow, there will be a readjustment in flow through a process known as *shock* that will result in a discontinuous change in type of flow from supersonic to subsonic.

Compressible Friction-flow Relationships. By simultaneous solution of

1. The continuity equation: $\dot{m} = \rho A V = $ constant
2. The energy equation: $h_o = $ constant
3. The equation of state: $Pv = mRT$

it can be shown that the following dimensionless ratios apply to one-dimensional, adiabatic, friction flow of a perfect gas:

$$\frac{T}{T^*} = \frac{k+1}{2\left(1 + \dfrac{k-1}{2} M^2\right)}$$

$$\frac{P}{P^*} = \frac{1}{M} \sqrt{\frac{k+1}{2\left(1 + \dfrac{k-1}{2} M^2\right)}}$$

$$\frac{V}{V^*} = M \sqrt{\frac{k+1}{2\left(1 + \dfrac{k-1}{2} M^2\right)}}$$

$$\frac{P_o}{P_o^*} = \frac{1}{M} \sqrt{\left[\frac{2\left(1 + \dfrac{k-1}{2} M^2\right)}{k+1}\right]^{(k+1)/(k-1)}}$$

$$4f\frac{L_{max}}{D} = \frac{1-M^2}{kM^2} + \frac{k+1}{2k} \ln \frac{(k+1)M^2}{2\left(1 + \dfrac{k-1}{2} M^2\right)}$$

In all the foregoing, it is noted that each dimensionless ratio is a function of only the specific-heat ratio and the Mach number of the flowing fluid. For specified values of specific-heat ratio ranging from 1.0 to 1.67, using the Mach number as the argument, values of the ratios have been calculated and are listed in "Gas Tables"

by Keenan and Kaye (John Wiley & Sons, Inc.). For easy reference an abridgment of those tables is presented, for $k = 1.4$, in Table 19. All terms with the exception of f and D have been defined previously. The term f is the average friction factor for the entire flow process. Since the friction factor f is a function of Reynolds number, it will vary along the entire length of duct. For accurate work, it will be necessary to iterate the calculations in order to obtain a solution. However, since compressible flow at relatively high velocity is being dealt with here, it is quite

Table 19. Fanno Line

$(k = 1.4,$ subsonic branch only)

M	$\dfrac{T}{T_*}$	$\dfrac{p}{p_*}$	$\dfrac{p_0}{p_{0*}}$	$\dfrac{V}{V_*}$	$\dfrac{F}{F_*}$	$4\dfrac{fL_{max}}{D}$
0	1.2000	∞	∞	0	∞	∞
0.05	1.1994	21.903	11.5914	0.05476	9.1584	280.02
0.10	1.1976	10.9435	5.8218	0.10943	4.6236	66.922
0.15	1.1946	7.2866	3.9103	0.16395	3.1317	27.932
0.20	1.1905	5.4555	2.9635	0.21822	2.4004	14.533
0.25	1.1852	4.3546	2.4027	0.27217	1.9732	8.4834
0.30	1.1788	3.6190	2.0351	0.32572	1.6979	5.2992
0.35	1.1713	3.0922	1.7780	0.37880	1.5094	3.4525
0.40	1.1628	2.6958	1.5901	0.43133	1.3749	2.3085
0.45	1.1533	2.3865	1.4486	0.48326	1.2763	1.5663
0.50	1.1429	2.1381	1.3399	0.53453	1.2027	1.06908
0.55	1.1315	1.9341	1.2549	0.58506	1.1472	0.72805
0.60	1.1194	1.7634	1.1882	0.63481	1.10504	0.49081
0.65	1.10650	1.6183	1.1356	0.68374	1.07314	0.32460
0.70	1.09290	1.4934	1.09436	0.73179	1.04915	0.20814
0.75	1.07865	1.3848	1.06242	0.77893	1.03137	0.12728
0.80	1.06383	1.2892	1.03823	0.82514	1.01853	0.07229
0.85	1.04849	1.2047	1.02067	0.87037	1.00966	0.03632
0.90	1.03270	1.12913	1.00887	0.91459	1.00399	0.014513
0.95	1.01652	1.06129	1.00215	0.95782	1.00093	0.003280
1.00	1	1	1	1	1	0

probable that the flow will be entirely within the completely turbulent region and, thus, there will be usually very little change in the friction factor.

It is emphasized here that the average friction factor f above is one-fourth the Fanning or Darcy friction factor. For flow in smooth ducts, f has been found to be of the order of 0.003.

The term D in Table 19 is the hydraulic diameter which is here defined as

$$D = \frac{4 \times \text{cross-sectional area}}{\text{wetted perimeter}}$$

It is evident that, for round ducts, the hydraulic diameter is exactly equal to the geometrical duct diameter.

Example. Air, to be considered a perfect gas with a specific-heat ratio of 1.4, passes in steady, one-dimensional adiabatic flow through a duct of cross-sectional dimensions 1 ft 6 in. by 2 ft 4 in. At a particular point in the flow path, the static pressure is 100 psia and the static temperature is 340 F. At a point 560 ft downstream of the first point, the static pressure is 61.8 psia. If the flow rate is 410 lb/sec, determine the value of $\bar{f}$.

Solution:

$$\rho_1 = P_1/RT_1 = 100 \times 144/53.3 \times 800 = 0.337 \text{ lbm/cu ft}$$

$$A = 1.5 \times 2.33 = 3.5 \text{ sq ft}$$

$$V_1 = \dot{m}/\rho_1 A = 410/0.337 \times 3.5 = 348 \text{ fps}$$

$$c_1 = 49.1\sqrt{T_1} = 49.1\sqrt{800} = 1,390 \text{ fps}$$

$$M_1 = V_1/c_1 = 348/1,390 = 0.25$$

By reference to Table 19, it is seen that the value of $4\bar{f}L_{\max}/D$ is 8.4834 at a Mach number of 0.25

$$P_2/P^* = (P_2/P_1)(P_1/P^*) = 61.8/100 \times 4.3546 = 2.7$$

Table 19 lists a value of P/P^* of 2.6958 at a Mach number of 0.40. If this be taken as approximately equal to 2.7, then the Mach number at section 2 in the duct will be 0.40 and the corresponding value of $4\bar{f}L_{\max}/D$ is 2.3805. By reference to Fig. 20, it is seen that the actual length of duct between section 1 and 2 is given by

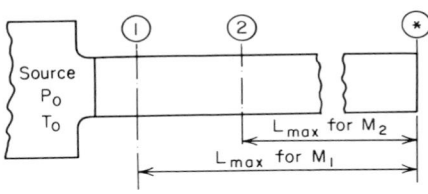

$$L_{\text{actual}} = L_{\max(M_1)} - L_{\max(M_2)}$$

and if the same average friction factor $\bar{f}$, is assumed to be applicable to flows originating at either section 1 or 2, then

$$\frac{4\bar{f}L_{\text{act}}}{D} = \left(\frac{4\bar{f}L_{\max}}{D}\right)_{M_1} - \left(\frac{4\bar{f}L_{\max}}{D}\right)_{M_2}$$

Fig. 20. Concept of $L_{\max}$.

or $\dfrac{4\bar{f}L_{\text{act}}}{D} = 8.4834 - 2.3085 = 6.1749$

In the above relation, the hydraulic diameter must be used. Recalling that hydraulic diameter is defined as four times the cross-sectional area divided by the wetted perimeter,

$$D = \frac{4 \times 3.5}{2(1.5 + 2.33)} = 1.825 \text{ ft}$$

then $\bar{f} = \dfrac{6.1749 \times D}{4 \times L_{\text{act}}} = \dfrac{6.1749 \times 1.825}{4 \times 560} = 0.00503$

The corresponding Fanning friction factor is four times that amount, or 0.0201.

Flow with Heating or Cooling

In this section there will be treated the frictionless flow of a perfect gas in a channel of constant cross-sectional area; the fluid is heated or cooled during passage through the duct. For the more general and more complex case of fluid flow in a conduit of variable area with friction and heat transfer, the reader is referred to "Dynamics and Thermodynamics of Compressible Fluid Flow" by A. H. Shapiro (Ronald Press).

The locus of state points on a temperature-entropy plot for the one-dimensional frictionless flow of a perfect gas which is being heated or cooled in a duct of constant area is called a Rayleigh line. Since flow without friction in a constant-area duct

satisfies identically the momentum equation, a Rayleigh line represents the locus of all points of equal momentum.

Figure 21 shows a Rayleigh line. An infinite number of such lines exists, and for each line

$$F/A = P + \rho V^2/g_c$$

assumes an arbitrary constant value dependent on the simultaneous values of P, ρ, and V. The upper portion of the curve corresponds to subsonic flow, and the lower portion to supersonic flow. The Mach number corresponding to the condition at the right-hand nose of the curve is unity; that is, at the point where the subsonic and supersonic branches meet, the fluid is sonic and the pressure, temperature, and velocity are denoted by P^*, T^*, and V^*, respectively.

A heating process is sketched on the subsonic portion of the Rayleigh line in Fig. 21. It is noted that static temperature increases with heating up to a certain point; continued heat addition subsequent to that corresponding to a maximum point will result in a decrease in static temperature. This phenomenon is due to the fact that the velocity increases very rapidly beyond the point of maximum static temperature; that is, the heat addition increases kinetic energy at the expense of internal energy and static enthalpy. However, the stagnation temperature and enthalpy increase with heat addition.

The converse is true in cooling. If a flow just less than sonic is cooled, there will first be an increase in static temperature and a decrease in kinetic energy. After a temperature corresponding to the maximum point on the curve has been attained, continued cooling results in a decrease in static temperature. In all cases of cooling, the stagnation temperature and enthalpy decrease.

For subsonic, one-dimensional, steady, and frictionless flow of a perfect gas in a conduit of constant cross-sectional area, it is important to recognize that when heat is added to the fluid (1) stagnation temperature always increases, (2) static temperature increases to some maximum value and then decreases as the fluid Mach number approaches unity, (3) the static pressure decreases, (4) the stagnation pressure decreases, and (5) the velocity increases. For subsonic cooling under the same limitations, the converse property changes occur.

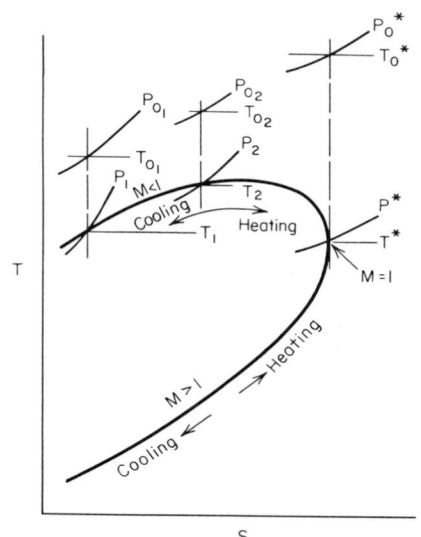

FIG. 21. Rayleigh line.

Compressible Heating and Cooling Flow Relationships. By simultaneous solution of the

1. Continuity equation: $\dot{m} = \rho A V = $ constant
2. Momentum equation: $F/A = P + \rho V^2/g_c = $ constant
3. Equation of state: $P v = m R T$

it can be shown that the following dimensionless ratios apply to the one-dimensional, steady, frictionless flow of a perfect gas in a conduit of constant area with heat

transfer to or from the fluid:

$$\frac{T_o}{T_o{}^*} = \frac{2(k+1)M^2\left(1 + \dfrac{k-1}{2}M^2\right)}{(1 + kM^2)^2}$$

$$\frac{T}{T^*} = \frac{M^2(1+k)^2}{(1 + kM^2)^2}$$

$$\frac{P}{P^*} = \frac{k+1}{1 + kM^2}$$

$$\frac{P_o}{P_o{}^*} = \frac{k+1}{1 + kM^2}\left[\frac{2\left(1 + \dfrac{k-1}{2}M^2\right)}{k+1}\right]^{k/k-1}$$

$$\frac{V}{V^*} = \frac{(k+1)M^2}{1 + kM^2}$$

Other relations of great utility follow from the first and second laws of thermodynamics:

$$\dot{Q} = \dot{m}(h_{o_2} - h_{o_1}) = \dot{m}c_p(T_{o_2} - T_{o_1}) \tag{43}$$

$$\frac{S - S^*}{c_p} = \ln M^2\left(\frac{k+1}{1 + kM^2}\right)^{(k+1)/k} \tag{44}$$

Table 20. Rayleigh Line

$(k = 1.4, \text{ subsonic branch only})$

M	$\dfrac{T_o}{T_o{}^*}$	$\dfrac{T}{T^*}$	$\dfrac{P}{P^*}$	$\dfrac{P_o}{P_o{}^*}$	$\dfrac{V}{V^*}$
0	0	0	2.4000	1.2679	0
0.05	0.01192	0.01430	2.3916	1.2657	0.00598
0.10	0.04678	0.05602	2.3669	1.2591	0.02367
0.15	0.10196	0.12181	2.3267	1.2486	0.05235
0.20	0.17355	0.20661	2.2727	1.2346	0.09091
0.25	0.25684	0.30440	2.2069	1.2177	0.13793
0.30	0.34686	0.40887	2.1314	1.1985	0.19183
0.35	0.43894	0.51413	2.0487	1.1779	0.25096
0.40	0.52903	0.61515	1.9608	1.1566	0.31372
0.45	0.61393	0.70803	1.8699	1.1351	0.37865
0.50	0.69136	0.79012	1.7778	1.1140	0.44445
0.55	0.75991	0.85987	1.6860	1.09397	0.51001
0.60	0.81892	0.91670	1.5957	1.07525	0.57447
0.65	0.86833	0.96081	1.5080	1.05820	0.63713
0.70	0.90850	0.99289	1.4235	1.04310	0.69751
0.75	0.94009	1.01403	1.3427	1.03010	0.75525
0.80	0.96394	1.02548	1.2658	1.01934	0.81012
0.85	0.98097	1.02854	1.1931	1.01091	0.86204
0.90	0.99207	1.02451	1.1246	1.00485	0.91097
0.95	0.99814	1.01463	1.0603	1.00121	0.95692
1.00	1	1	1	1	1

It will be observed that the five dimensionless ratios presented are all functions of the specific-heat ratio k and of Mach number M. Thus, for assumed values of M at an arbitrarily chosen value of k, it is a matter of simple calculation to determine numerical values for each of the ratios. Such calculations are carried out (for values of k ranging from 1.00 to 1.67 and for values of M ranging from zero to infinity) in "Gas Tables" by Keenan and Kaye published by John Wiley & Sons, Inc. For convenience, there is presented in Table 20 an abridged version of one such tabulation prepared for $k = 1.4$.

Maximum Heat Transfer in Rayleigh-line Flow. The Rayleigh line of Fig. 21 shows that there is a maximum possible entropy, and therefore, from any initial point a maximum possible entropy increase. Thus, for any given Rayleigh line, there is a maximum possible heat addition. The heat addition between any two points, say 1 and 2, is given by Eq. (43). The maximum possible heat addition is thus

$$\dot{Q}_{max} = \dot{m} c_p (T_o^* - T_{o_1})$$

For example, if 100 lb/sec of air $[c_p = 0.24 \text{ Btu/(lbm R)}]$ flows in a duct and heat is added to the fluid which, before heat addition, is at a Mach number of 0.5, a static pressure of 50 psia, and a static temperature of 500 R, Q_{max} can be obtained as follows:

From Table 18 at $M_1 = 0.5$, $T_1/T_o = 0.95238$.
Hence, $T_{o_1} = 500/0.95238 = 525$ R.
From Table 20 at $M_1 = 0.5$, $T_{o_1}/T_o^* = 0.69136$ and $T_o^* = 525/0.69136 = 758$ R.
Then $\dot{Q}_{max} = 100 \times 0.24 \times (758 - 525) = 5590$ Btu/sec.

If heat is added to the fluid at a rate less than 5590 Btu/sec, then after the heat addition, the fluid stagnation temperature will be less than 758 R. If heat is added precisely at the rate of 5590 Btu/sec, the fluid after heating will be at a Mach number of unity and at a stagnation temperature of 758 R. If heat is added at a rate greater than 5590 Btu/sec, there will be an adjustment in flow rate.

Suppose that heat were added to the fluid at the rate of 4000 Btu/sec. Then, from Eq. (43),

$$T_{o_2} - T_{o_1} = \dot{Q}/\dot{m} \, C_p = 4{,}000/(100 \times 0.24) = 166.7 \text{ R}$$

and

$$T_{o_2} = 166.7 + T_{o_1} = 166.7 + 525 = 691.7 \text{ R}$$

Then

$$T_{o_2}/T_o^* = (T_{o_2}/T_{o_1})(T_{o_1}/T_o^*) = 691.7/525 \times 0.69136 = 0.909 \text{ approximately}$$

From Table 20, it is seen that at a Mach number of 0.7, T_o/T_o^* is read as 0.90850. Thus a heat addition of 4000 Btu/sec will accelerate 100 lb/sec of air from a Mach number of 0.5 to a Mach number of 0.7.

Now, suppose that heat were added to the fluid at the rate of 8000 Btu/sec, that is, at a rate greater than $\dot{Q}_{max}$. Such a heat addition would correspond to $\dot{Q}_{max}$ for another flow rate, and this new flow rate can be determined easily by a trial-and-error process, as follows:

For an assumed M_1 of 0.20, $T_o^* = 525/0.17355 = 3020$ R.
Then $\dot{m} = 8{,}000/0.24 \, (3{,}020-525) = 13.3$ lbm/sec.
At inlet to the heating chamber, from the initial condition as stated in the problem,

$$P_{o_1} = P_1 \times P_{o_1}/P_1 = 50 \times 1/0.84302 = 59.3 \text{ psia}$$

Then, at M_1 of 0.20,

$$P_1 = 59.3 \times 0.9725 = 57.5 \text{ psia}$$

$$T_1 = 525 \times 0.99206 = 521 \text{ R}$$

$$\rho_1 = 57.5 \times 144/53.3 \times 521$$

$$= 0.297 \text{ lb/cu ft}$$

$$V_1 = M_1 C_1 = 0.2 \times 49.1 \sqrt{521} = 224 \text{ fps}$$

It is necessary to return to the initially stated problem to determine the value of the cross-sectional area A. To avoid confusion, the subscript i will be used.

$$\rho_i = P_i/RT_i = 50 \times 144/53.3 \times 500 = 0.269 \text{ lb/cu ft}$$

$$V_i = 0.5 \times 49.1 \sqrt{500} = 550 \text{ fps}$$

$$A = \dot{m}_i/\rho_i V_i = 100/0.269 \times 550 = 0.675 \text{ sq ft}$$

Returning now to the problem under consideration,

$$\dot{m}_1 = \rho_i A V_i = 0.297 \times 0.675 \times 224 = 44.8 \text{ lb/sec}$$

Obviously, this flow rate does not agree with the 13.3 lb/sec which corresponds to the assumed Mach number of 0.20.

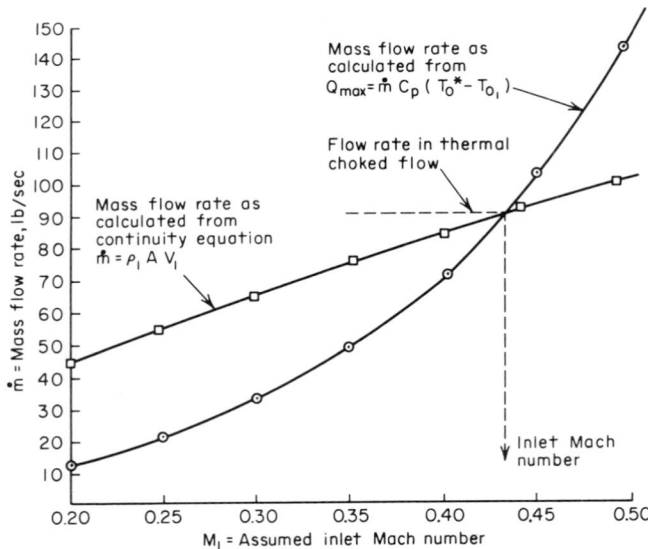

Fig. 22. Graphical solution to problem in subsonic choked Rayleigh-line flow.

Calculations as above are carried out for Mach numbers of 0.25, 0.30, 0.35, 0.40, 0.45, and 0.50, and the results plotted in Fig. 22. End results of the calculations are listed at the top of the next page.

Assumed Mach No.	Mass-flow rate calculated from energy equation, lbm/sec	Mass-flow rate calculated from continuity equation, lbm/sec
0.20	13.3	45.0
0.25	22.0	55.8
0.30	33.6	65.6
0.35	49.4	75.4
0.40	71.4	84.0
0.45	104.0	92.2
0.50	143.0	100.0

The reduction in mass-flow rate which results from a heat addition in excess of $\dot{Q}_{max}$ is referred to as "thermal choking." Calculations carried out above involved subsonic flow; the phenomenon of thermal choking also takes place in supersonic flow. In that case, through a process known as *shock*, adjustment takes place by a change in some portion of the convergent-divergent nozzle which precedes the heating chamber; the initially supersonic fluid becomes subsonic after shock, and $\dot{Q}_{max}$ is greater for the subsonic stream.

One-dimensional Normal Shock

In this section there will be treated very briefly the phenomenon of one-dimensional normal shock. By normal it is implied that the shock wave is normal, or at right angles, to the one-dimensional flow. That is, the problem of curved shocks, bow waves, and detached waves encountered in external flow over submerged bodies will not be discussed.

Shock occurs in a very short length of flow path; measurements across a shock wave have indicated that the standing wave is thousandths or tens of thousandths of an inch in thickness. This leads to the conclusion that, in such a short distance, heat transfer and friction effects can be disregarded. If analysis is restricted to the steady flow of a perfect gas, the following relationships would be satisfied across a shock:

Energy equation:	$h_o = $ constant
Continuity equation:	$\dot{m} = \rho A V$
Equation of state:	$Pv = mRT$
Momentum equation:	$F/A = P + \rho V^2/g_c$

By reference to the discussion on friction flow, it will be seen that the first three of the above four equations govern the analysis of one-dimensional, steady, and adiabatic friction flow. The last three of the above equations are the governing relations in the analysis of frictionless, one-dimensional, steady flow with heating or cooling. Therefore, in shock flow, the requirements of adiabatic friction flow and non-adiabatic frictionless flow are met simultaneously. Graphically, then, shock flow must take place at the points of intersection of a given Fanno and Rayleigh line.

Figure 23 shows a Fanno line and a Rayleigh line each for the same flow rate per unit area. It will be noted that these two curves intersect at two points one of which (Y) corresponds to subsonic flow, with the other (X) corresponding to supersonic flow. Since the flow process by definition is adiabatic, the second law of thermodynamics requires that the irreversible discontinuity be at an increase in entropy. Therefore, the process must be in the direction of from X to Y. Clearly,

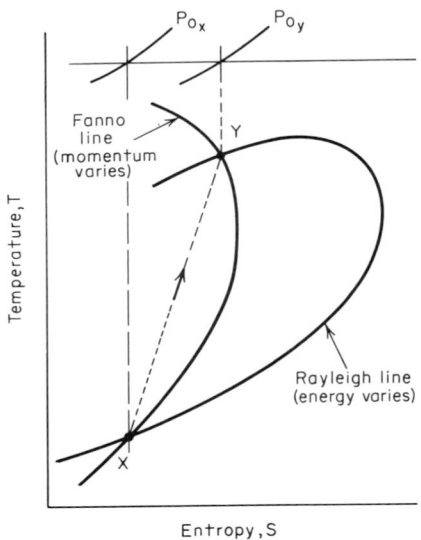

P_{ox} P_{oy}

Fanno line (momentum varies)

Temperature, T

Rayleigh line (energy varies)

Entropy, S

FIG. 23. Shock process in flow of a perfect gas.

also, it can be concluded that shock occurs only in flows which are initially supersonic and that, after shock, the flow must be subsonic.

Shock flow is of importance in each of the three types of compressible flow which have been discussed previously. In isentropic flow, surface irregularities in the nozzle contour or an unfavorable back-pressure condition can result in shock. In friction flow, if the fluid is initially supersonic and the duct length is greater than L_{max} for the entering Mach number, or if the duct exit back pressure is unfavorable, readjustment in flow through shock will occur. In Rayleigh-line flow, if the fluid is initially supersonic and the heat addition exceeds $\dot{Q}_{max}$ for the entering Mach number, or if the heating-chamber exit back pressure is unfavorable, a readjustment by means of shock will occur.

The Compressible-flow Relations across a Shock. By simultaneous solution of the four equations which govern shock flow, it can be shown that the following dimensionless ratios will be obtained:

$$\frac{M_x}{M_y}\sqrt{\frac{2 + (k-1)M_x^2}{2 + (k-1)M_y^2}} = \frac{1 + kM_x^2}{1 + kM_y^2}$$

$$\frac{P_y}{P_x} = \frac{2kM_x^2 - (k-1)}{k+1}$$

$$\frac{\rho_y}{\rho_x} = \frac{V_x}{V_y} = \frac{(k+1)M_x^2}{2 + (k-1)M_x^2}$$

$$\frac{T_y}{T_x} = \frac{2 + (k-1)M_x^2}{2 + (k-1)M_y^2}$$

$$\frac{P_{oy}}{P_{ox}} = \left[\frac{(k+1)M_x^2/2}{1 + (k-1)M_x^2/2}\right]^{k/k-1} \bigg/ \left[\frac{2kM_x^2 - (k-1)}{k+1}\right]^{1/k-1}$$

$$\frac{P_{oy}}{P_x} = \left(\frac{k+1}{2}M_x^2\right)^{k/k-1} \bigg/ \left[\frac{2kM_x^2 - (k-1)}{k+1}\right]^{1/k-1}$$

In the above, the subscript x refers to the fluid properties just prior to shock and the subscript y refers to fluid properties immediately following the shock process. Thus M_x is the Mach number before and M_y the Mach number after shock.

Each dimensionless ratio is a function of the specific-heat ratio k and the Mach number M_x before shock. Thus, for a given specific-heat ratio, all the dimensionless ratios can be easily calculated for an initial Mach number M_x. Table 21, which lists selected values for $k = 1.4$, is abridged from "Gas Tables" by Keenan and Kaye (John Wiley & Sons, Inc.). The following example will illustrate the use of the shock tables.

Table 21. One-dimensional Normal Shock Functions

$$(k = 1.4)$$

M_x	M_y	$\dfrac{p_y}{p_x}$	$\dfrac{\rho_y}{\rho_x}$	$\dfrac{T_y}{T_x}$	$\dfrac{p_{o_y}}{p_{o_x}}$	$\dfrac{p_{o_y}}{p_x}$
1.0	1.00000	1.00000	1.00000	1.00000	1.00000	1.8929
1.1	0.9117	1.2450	1.1691	1.06494	0.99892	2.1328
1.2	0.84217	1.5133	1.3416	1.1280	0.99280	2.4075
1.3	0.78596	1.8050	1.5157	1.1909	0.97935	2.7135
1.4	0.73971	2.1200	1.6896	1.2547	0.95819	3.0493
1.5	0.70109	2.4583	1.8621	1.3202	0.92978	3.4133
1.6	0.66844	2.8201	2.0317	1.3880	0.89520	3.8049
1.7	0.64055	3.2050	2.1977	1.4583	0.85573	4.2238
1.8	0.61650	3.6133	2.3592	1.5316	0.81268	4.6695
1.9	0.59562	4.0450	2.5157	1.6079	0.76735	5.1417
2.0	0.57735	4.5000	2.6666	1.6875	0.72088	5.6405
2.1	0.56128	4.9784	2.8119	1.7704	0.67422	6.1655
2.2	0.54706	5.4800	2.9512	1.8569	0.62812	6.7163
2.3	0.53441	6.0050	3.0846	1.9468	0.58331	7.2937
2.4	0.52312	6.5533	3.2119	2.0403	0.54015	7.8969
2.5	0.51299	7.1250	3.3333	2.1375	0.49902	8.5262
2.6	0.50387	7.7200	3.4489	2.2383	0.46012	9.1813
2.7	0.49563	8.3383	3.5590	2.3429	0.42359	9.8625
2.8	0.48817	8.9800	3.6635	2.4512	0.38946	10.569
2.9	0.48138	9.6450	3.7629	2.5632	0.35773	11.302
3.0	0.47519	10.333	3.8571	2.6790	0.32834	12.061
4.0	0.43496	18.500	4.5714	4.0469	0.13876	21.068
5.0	0.41523	29.000	5.0000	5.8000	0.06172	32.654
∞	0.37796	∞	6.0000	∞	0	∞

Example. A convergent-divergent nozzle has a throat area of 6 sq in. It passes air which is to be considered a perfect gas with a specific-heat ratio k of 1.4. Flow through the nozzle is frictionless and adiabatic. The air has a constant stagnation temperature of 1000 R, and except for shock, the stagnation pressure is 500 psia. A shock occurs at a Mach number of 2.0. Find: (a) the Mach number after shock, (b) the static and stagnation pressure after shock, (c) the mass-flow rate, and (d) the required exit area if flow after shock is isentropic and the back pressure is 15 psia.

Solution. (a) From Table 21, the Mach number after shock is $M_y = 0.57735$.

(b) $P_y = (P_y/P_x)(P_x/P_{o_x})P_{o_x} = (4.5)(0.12780)(500) = 289.5$ psia

$\quad P_{o_y} = (P_{o_y}/P_{o_x})P_{o_x} = (0.72088)(500) = 360.5$ psia

(c) $T^* = (T^*/T_o)T_o = 0.83333(1000) = 833.33$ R

$\quad P^* = (P^*/P_{o_1})P_{o_1} = 0.52828(500) = 264.14$ psia

$\quad \rho^* = P^*/RT^* = 264.14 \times 144/53.3 \times 833.33 = 0.855$ lb cu ft

$\quad V^* = c^* = 49.1\sqrt{T^*} = 49.1\sqrt{833.33} = 1,415$ fps

$\quad m = \rho^* A V^* = 0.855 \times 6/144 \times 1,415 = 50.4$ lb/sec

(c) Alternate: the solution can also be based on properties at shock.

$$T_y = (T_y/T_x)(T_x/T_o)T_o = (1.6875)(0.55556)(1000) = 937 \text{ R}$$
$$\rho_y = P_y/RT_y = 289.5 \times 144/53.3 \times 937 = 0.833 \text{ lb/cu ft}$$
$$c_y = 49.1\sqrt{T_y} = 49.1\sqrt{937} = 1,500 \text{ fps}$$
$$V_y = M_y c_y = 0.57735 \times 1,500 = 865 \text{ fps}$$
$$A_y = A_x = (A_x/A^*)A^* = 1.6875(6/144) = 0.0703 \text{ sq ft}$$
$$\dot{m} = \rho_y A_y V_y = (0.0703)(0.833)(865) = 50.4 \text{ lb/sec}$$

(c) Alternate: It can be shown that, for critical flow through an isentropic nozzle,

$$\frac{\dot{m}}{A^*} = \sqrt{\frac{kg_c}{R}\left(\frac{2}{k+1}\right)^{(k+1)/(k-1)}} \frac{P_o}{\sqrt{T_o}}$$

and, for a value of k equal to 1.4,

$$\frac{\dot{m}}{A^*}\frac{\sqrt{T_o}}{P_o} = 0.532$$

Using the latter relation,

$$\dot{m} = \frac{0.532A^*P_o}{\sqrt{T_o}} = \frac{0.532(6)(500)}{\sqrt{1,000}} = 50.4 \text{ lb/sec}$$

(d) At $M_y = 0.57735$, $P_y = 289.5$ psia, $P_{o_y} = 360.5$ psia $T_{o_y} = 1000$ R, $T_y = 937$ R, $A_y = 0.0703$ sq ft.

$$P_2/P_{o_2} = (P_2/P_y)(P_y/P_{o_y}) = (15/289.5)(0.79766) = 0.0413$$

This pressure ratio corresponds to a Mach number of about 2.7.
$A_2/A^* = $ about 3.2 by interpolation in Table 18. Then $A_2 = (A_2/A^*)(A^*/A_y)A_y = (3.2)(1.22)(0.0703) = 0.274$ sq ft.

FLOW OF VAPORS

A vapor is a fluid which, as does a gas, fills completely any vessel which contains it. A perfect gas lends itself to easy mathematical treatment because of the constancy of its properties; a vapor, on the other hand, is a gaseous-like substance whose physical properties change with pressure, temperature, and composition. For this reason, the flow of a vapor is more complex than that of a gas, and exact solution of flow problems may sometimes require iterative procedures.

Basic to the treatment of any vapor-flow problem are the Bernoulli equation (23), the continuity equation (21), and the energy equation (18); in many instances Eq. (30), in which potential-energy terms are neglected, may be a more useful form of the energy equation.

Fortunately, the flow process involved with many vapors is of such nature that the specific-heat ratio may be considered constant over the associated range of property changes. For example, the specific-heat ratio of steam does not differ appreciably from $k = 1.3$ over a considerable range of temperature and pressure. In such a case, an approximate solution can be readily obtained by reference to the "Gas Tables" by Keenan and Kaye. It will be recalled when those tables are used that the gas constant for a vapor may be approximated by

$$R = \bar{R}/M_w$$

For carbon dioxide, for example, of molecular weight 44, the approximate gas constant is

$$R = 1546/44 = 35.1 \text{ ft lbf/(lbm R)}$$

and for steam,

$$R = 1546/18 = 85.9 \text{ ft lbf/(lbm R)}$$

Flow of a Vapor through a Nozzle. Here the term nozzle includes also the venturi and the thick- or thin-plate orifice. The energy equation need not include potential-energy terms, and if the nozzle is well insulated, heat-transfer effects are also negligible. Thus, for nozzle flow,

$$h_1 + V_1^2/2g_cJ = h_2 + V_2^2/2g_cJ = h_o = \text{constant} \tag{45}$$

If subscripts 1 and 2 refer to the nozzle inlet and outlet, respectively, the discharge velocity can easily be shown to be

$$V_2 = \sqrt{V_1^2 + 2g_cJ(h_1 - h_2)} \tag{46}$$

In many instances, the square of the inlet velocity V_1 is small as compared with $2g_cJ(h_1 - h_2)$, and in such cases of negligible inlet kinetic energy,

$$V_2 = \sqrt{2g_cJ(h_1 - h_2)} = 223.7\sqrt{h_1 - h_2} \tag{47}$$

The flow rate is calculated from the continuity equation

$$\dot{m} = \rho AV = \rho_1 A_1 V_1 = \rho_2 A_2 V_2 = \rho^* A^* V^* \tag{48}$$

in which the superscript * refers to conditions at the throat of a convergent-divergent (venturi) nozzle. Obviously, in a convergent nozzle which has critical flow at the exit, states 2 and * in Eq. (48) are identical. In a convergent nozzle or in a thin-plate orifice, owing to contraction of the jet at exit, the area of the jet may be less than that of the orifice, or nozzle, and it is then necessary to apply a coefficient of discharge C to the calculation. Thus

$$\dot{m} = CA_2 V_2$$

Coefficients of discharge range from approximately 0.60 for sharp-edged orifices to nearly unity for well-rounded nozzles.

In Eqs. (46) and (47), state 2 refers to the actual state of the fluid. Frictional effects are usually accounted for by the introduction of a nozzle efficiency e_n, which is defined as

$$e_n = (h_1 - h_2)/(h_1 - h_2') \tag{49}$$

where h_2' would represent the ideal static enthalpy of the compressible fluid at exit from the nozzle, that is, the static enthalpy that would ensue from an isentropic drop from the initial to the final pressure. The exit velocity, taking frictional effects into account, would then be

$$V_2 = \sqrt{V_1^2 + 2g_cJe_n(h_1 - h_2')} \tag{50}$$

and if the inlet velocity were negligible,

$$v_2 = 223.7\sqrt{e_n(h_1 - h_2')} \tag{51}$$

The effect of nozzle friction is to increase the specific volume of the flowing fluid and to decrease its velocity. Structurally, this requires that the sectional area of an actual nozzle be greater than that of one in which there occurs frictionless flow. Thermodynamically, the internal-energy increases resulting from friction will raise the quality of a wet vapor or will elevate the temperature of a superheated vapor. The phenomenon is referred to as "nozzle reheat," and its magnitude is defined in terms of the "reheat factor." In the case of multistage expansion, as in a steam turbine, the effect of reheat results in a cumulative efficiency for the entire series of expansions which is higher than the efficiencies of the individual stages.

Reheat Factor. The reheat factor R for a series of nozzles is the ratio of the sum of the individual isentropic enthalpy drops for the series, divided by the overall

isentropic enthalpy drop. The reheat factor is always greater than unity but approaches unity as the individual nozzle efficiencies approach 100 per cent, or in other words as the actual expansion line approaches the vertical line representing perfect isentropic expansion on an enthalpy-entropy diagram. These relations are expressed in symbols as follows:

e_s = average of individual efficiencies of the nozzles in series
e_c = cumulative efficiency for the entire series based on the overall isentropic drop in enthalpy
H_s = isentropic drop in any individual nozzle, Btu/lb
ΣH_s = summation of isentropic drops in all nozzles taken individually, Btu/lb
H_s' = overall isentropic enthalpy drop, Btu/lb
R = reheat factor, which is defined as

$$R = e_c/e_s = \Sigma H_s/H_s'$$

Entrance and Exit Losses for Compressible Fluids. In the case of compressible fluids flowing from a reservoir into a pipe and out again, the entrance and exit losses are represented by the energy corresponding to the velocity head. In this case velocity is created at the expense of the internal energy of the fluid. From Eq. (47) the isentropic enthalpy drop required to produce a velocity V in the pipe is $\Delta h = V^2/50,070$. The corresponding pressure drop can be read from an enthalpy entropy diagram. In case an expanding nozzle is provided at the outlet of the pipe, a good share of this velocity energy can be reconverted into pressure. In case no expanding nozzle is provided at the outlet, the combined entrance and exit losses are equal to the energy required to produce the velocity in the pipe as explained above.

Contour of Nozzles for Compressible Fluids. The function of a nozzle is to generate velocity, and it is desirable, therefore, to proportion the contour of the nozzle so as to obtain the most efficient conversion from thermal head to velocity. The shape of nozzles for compressible fluids should be of the convergent-divergent form provided the back pressure does not exceed the critical pressure at the throat. The reason for this contour is the changing relation of velocity and specific volume during expansion. The proper contour can be determined from the continuity equation. The ratio of V/v determines the proper area A at any cross section. If the values of V/v are plotted for successive steps in expansion, it will be found that the ratio first increases to a maximum corresponding to conditions at the throat of the nozzle and then decreases from this section on. Since the value of A varies inversely as V/v it follows that the nozzle contour should converge to a throat and then diverge to the outlet. These relations are readily worked out for steam with reference to a steam table or Mollier diagram or by the perfect-gas laws in the cases of gases.

The rates of convergence and divergence of a nozzle depend on its length as well as on the theoretical area required at any section. The convergent section may be reduced until it is little more than a well-rounded entrance without affecting appreciably the efficiency of the nozzle. The divergent section is usually a truncated cone or pyramid with its mouth tangent at the throat to the rounded entrance portion. The angle between the sides of the divergent portion should not exceed 12 deg for high efficiency. The cross sections of nozzles are usually circular, square, or rectangular as best suits the particular application. In the case of nozzles where the back pressure exceeds the critical pressure at the throat, the divergent section is not required.

Practice shows that the cross section of a nozzle, whether circular, elliptical, square, or rectangular (providing the two last have rounded corners), has very

little effect on the efficiency, provided the inner surfaces are smooth and the ratio of the area at the throat to that of the mouth is correctly proportioned. The *velocity* efficiency of a properly proportional nozzle with straight-line divergence varies from 95 to 97 per cent, corresponding to an *energy* efficiency of 92 to 94 per cent, so that usually it is not considered worthwhile to attempt to follow the more difficult but theoretically exact divergence curve.

Critical Pressure. The section of smallest area in a nozzle is called the "throat." In the case of an orifice, the vena contracta is equivalent to the throat of a nozzle proper. For each compressible fluid, the throat pressure bears a certain definite and constant relation to the initial pressure, provided the back (or discharge) pressure does not exceed the throat pressure. The throat pressure which satisfies these conditions is termed the *critical* pressure and is expressed as a certain decimal fraction of the initial pressure. For saturated steam the critical pressure is approximately $0.578P_o$, for superheated steam 0.55 to $0.57P_o$ depending on the amount of superheat, and for a perfect gas of $k = 1.4$ is $0.528P_o$.

The critical pressure referred to here is the P^* which was encountered in the section "Flow of Gases." It is the pressure of the fluid in the flow path at which the Mach number is unity, and it will be recalled that, at unity Mach number, the fluid velocity is equal to the acoustic velocity. With respect to the flow of vapors, the critical pressure can be approximated by use of a representative average value of k for the process between nozzle inlet and the throat. With this understanding, then

$$P^* = P_o[2/(k + 1)]^{k/(k-1)}$$

Values of specific-heat ratio k along with other properties are listed for various vapors (which may be treated as gases at least as a first approximation) in Table 22. For saturated and superheated steam the value of k is approximately 1.3, but for accurate work, the average value as obtained from Fig. 8 on page 82 of "Thermodynamic Properties of Steam" by Keenan and Keyes should be used.

If the nozzle back pressure is equal to or greater than the critical pressure, the nozzle contour must be divergent only. For supersonic flow throughout the entire diverging portion of a convergent-divergent nozzle, the back pressure must be equal to or less than P_y corresponding to a one-dimensional normal shock in the exit plane; if the back pressure exceeds the aforementioned P_y, the shock will proceed upstream in the flow path and, downstream of the shock, flow will be subsonic.

As has been pointed out in connection with Fanno- and Rayleigh-line flow of a perfect gas, critical pressure may be associated with friction flow and flow with heat addition. The same phenomenon can occur with the flow of a vapor.

Supersaturation. If steam is superheated throughout the entire expansion path in a nozzle, equilibrium conditions can be assumed to exist at all points. Likewise, if the steam is initially saturated with expansion into the wet region, equilibrium can be assumed. However, if steam is initially superheated at nozzle inlet and expansion occurs past the saturation line into the wet region, a phenomenon known as supersaturation will occur. This effect, attributed to a delay in the formation of liquid droplets associated with surface tension, results in a flow which may be slightly greater than that which would be obtained in an isentropic flow without supersaturation.

Supersaturation occurs only in the diverging portions of steam nozzles, and this is one of the reasons that steam turbine nozzles are usually convergent only. In a nozzle in which supersaturation occurs, liquid droplets should begin to form just as the expansion process crosses the saturation curve. However, there is a delay in the formation of liquid, and expansion takes place to a point at which the vapor

Table 22. Properties of Miscellaneous Gases and Vapors[1]

Gas	Chemical symbol	Number of atoms N	Molecular weight Approximate	Molecular weight Exact, $O_2 = 32$	Weight in pounds of 1 cu ft at atmospheric pressure At 62 F	At 32 F	Specific gravity (air = 1)	Gas constant, R, $\frac{\text{ft lbf}}{\text{lbm R}}$	Specific heat, Btu per lb deg F c_p	c_v	Specific heat, Btu per cu ft per deg F at atmos. pres. and 62 F c_p	c_v	$k = \frac{c_p}{c_v}$
Helium.	He	1	4.0	4.0	0.0105	0.0112	0.137	386.0	1.25	0.75	0.0131	0.0079	1.66
Argon.	A	1	40.0	39.9	0.1048	0.1112	1.378	38.70	0.124	0.075	0.0131	0.0079	1.66
Air.	. . .	. . .	29.0	28.95	0.0761	0.0807	1	53.34	0.241	0.171	0.0183	0.0130	1.40
Oxygen.	O_2	2	32.0	32	0.0840	0.0892	1.105	48.25	0.217	0.155	0.0182	0.0130	1.40
Nitrogen.	N_2	2	28.0	28.02	0.0737	0.0783	0.970	54.99	0.247	0.176	0.0182	0.0130	1.40
Hydrogen	H_2	2	2.0	2.016	0.00529	0.00562	0.0696	765.86	3.42	2.44	0.0181	0.0129	1.40
Nitric oxide.	NO	2	30.0	30.04	0.0789	0.0838	1.038	51.40	0.231	0.165	0.0183	0.0130	1.40
Carbon monoxide. .	CO	2	28.0	28.00	0.0734	0.0780	0.968	55.14	0.243	0.172	0.0180	0.0126	1.41
Hydrochloric acid. .	HCl	2	36.5	36.45	0.0958	0.1017	1.260	42.35	0.191	0.136	0.0183	0.0130	1.40
Carbon dioxide. . .	CO_2	3	44.0	44.00	0.1156	0.1227	1.520	35.09	0.210	0.160	0.0243	0.0185	1.31
Nitrous oxide	N_2O	3	44.0	44.03	0.1157	0.1229	1.522	35.03	0.221	0.171	0.0256	0.0198	1.26
Sulfur dioxide	SO_2	3	64.0	64.06	0.1684	0.1786	2.213	24.10	0.154	0.123	0.0260	0.0207	1.25
Ammonia.	NH_3	4	17.0	17.06	0.04483	0.0476	0.590	90.50	0.523	0.399	0.0234	0.0178	1.31
Acetylene.	C_2H_2	4	26.0	26.02	0.0684	0.0725	0.899	59.34	0.350	0.270	0.024	0.0185	1.28
Methyl chloride . . .	CH_3Cl	5	50.5	50.47	0.1326	0.1407	1.744	30.59	0.24	0.20	0.032	0.0265	1.20
Methane.	CH_4	5	16.0	16.03	0.0421	0.0447	0.554	96.31	0.593	0.450	0.025	0.019	1.32
Ethylene.	C_2H_4	6	28.0	28.03	0.0738	0.0780	0.969	55.08	0.40	0.33	0.029	0.024	1.20

[1] Reproduced, by permission, from Marks's "Mechanical Engineers' Handbook."

temperature is less than the saturation temperature for the given pressure. The fluid in this metastable state then suddenly experiences a condensation shock, and liquid droplets are formed.

Fortunately, the phenomenon of supersaturation occurs only where expansion is across the saturation curve. Hence, it may be of consequence in only one stage of a multistage steam turbine. The following empirical equation for the calculation of weight flow under supersaturation conditions is attributed to Goudie:

$$\dot{m} = 0.0173A^*P_1^{0.969}$$

In this relation, the weight flow is in pounds per second if the throat area is in square inches and the initial static pressure is in pounds per square inch absolute.

Weight of Superheated Steam Discharged through a Nozzle. In the cases where supersaturation does not occur, the flow rate of superheated steam may be approximated by (1) assuming that the steam acts as a perfect gas with $k = 1.3$ and (2) neglecting the effects of inlet or approach velocity. In such a case, the stagnation pressure P_o at the inlet may be replaced by the static pressure P_1. For a nozzle in which the sonic velocity is reached, substitution into Eq. (39) yields

$$\frac{\dot{m}}{A^*} = \sqrt{\frac{1.3 \times 32.17}{85.9}} \; \frac{P_o}{\sqrt{P_o \, v_o/85.9}} \; \frac{1}{\left(1 + \dfrac{1.3 - 1}{2}\right)^{3.833}}$$

and

$$\dot{m} = 3.786A^* \sqrt{P_1/v_1} \tag{52}$$

in which $\dot{m}$ = flow rate of superheated steam, lb/sec
$\quad A^*$ = throat area, sq ft
$\quad P_1$ = inlet pressure, lb/sq ft abs
$\quad v_1$ = inlet specific volume, cu ft/lbm

If the throat area is expressed in square inches and the pressure in pounds per square inch absolute, Eq. (52) becomes

$$\dot{m} = 0.3155A^* \sqrt{P_1/v_1}$$

For the case of a convergent nozzle in which the acoustic velocity is not reached at exit, similar substitution into Eq. (39) will yield

$$\dot{m} = 1.395(A_2/v_2) \sqrt{P_1v_1 \, [1 - (P_2/P_1)^{0.231}]} \tag{53}$$

It is emphasized that Eqs. (52) and (53) are approximate inasmuch as they are predicated on a unique value of specific heat and, also, because the effects of initial velocity are ignored.

If the steam at the nozzle inlet is not superheated, then since the weight of saturated steam discharged through a nozzle depends on the initial quality of the steam, it is expedient to resort to empirical equations of which a number are available. For Case 1, *where the back pressure does not exceed the critical pressure,* the following formula is sometimes used. x_1 represents the initial quality of the steam expressed as a decimal fraction (for instance, for steam of 98 per cent quality, $x_1 = 0.98$)

$$w = a_t p_1^{0.97}/(60 \sqrt{x_1}) \tag{54}$$

This is practically equivalent to the formula generally known as "Grashof's."

The most commonly used formula for calculating the quantity of steam discharged through a nozzle is attributed to Napier. Its simple form makes it convenient to apply under any condition of Case 1. For steam initially dry and saturated, Napier's formula is

$$w = a_t p_1/70 \tag{55}$$

For other conditions, either superheated or wet, it is necessary to apply a correction factor. For superheated steam, the following correction factor is commonly used in which D_1 is the number of Fahrenheit degrees superheat under initial conditions:

$$\text{Correction factor} = 1/(1 + 0.00065 D_1)$$

For steam initially wet, Emswiler in "Thermodynamics" gives

$$\text{Correction factor} = 1/(1 - 0.012 m_1)$$

where m_1 is the per cent moisture under initial conditions. The correction factor $1/\sqrt{x_1}$ used in Eq. (54) can also be applied with slightly different results. Napier's equation for the flow of wet steam then becomes

$$w = a_t p_1/(70 \sqrt{x_1}) \tag{56}$$

Grashof derived the following formula for dry saturated steam which is very nearly equivalent to Eq. (54):

$$w = 0.01654 a_t p_1^{0.9696} \tag{57}$$

Weight of Saturated Steam Discharged through a Nozzle. Flow rates of saturated steam may be calculated by use of analytical expressions such as those represented by Eq. (5) and the continuity equation. Frequently, however, it is convenient to resort to the use of empirical relations. For critical flows, that is, for discharge to regions in which the back pressure is less than the critical pressure at the nozzle throat, the flow rate is dictated by the throat condition; for this type of flow with the steam being initially dry and saturated, Napier's equation gives

$$\dot{m} = A^* P_1/70$$

in which the flow rate is expressed in pounds per second and the throat area and the initial pressure are in consistent units.

If the steam is initially superheated D F, the flow rate is given by

$$\dot{m} = 1/(1 + 0.00065 D) (A^* P_1/70)$$

If the steam is initially wet with a quality x_1, Napier's equation may be expressed as

$$\dot{m} = A^* P_1/(70 \sqrt{x_1})$$

For critical flow of wet steam, Mayer gives the relationship

$$\dot{m} = A^* P_1^{0.97}/(60 \sqrt{x_1})$$

The following relation for the critical flow of dry saturated steam is due to Grashof:

$$\dot{m} = 0.01654 A^* P_1^{0.9696}$$

For cases in which the back pressure is greater than the critical pressure, Napier gives the following relation for steam which is initially dry and saturated:

$$\dot{m} = A^* P_2/42 \sqrt{3(P_1 - P_2)/2P_2}$$

in which P_2 represents the pressure in the exit plane of the nozzle. The dimensions selected for area and pressure must be consistent with each other; if so selected, the flow rate is in pounds per second.

The Napier and Grashof equations for critical flow of dry, saturated steam can be extended to the subcritical case by multiplying the critical-flow relations by a factor

K which is defined as

$$K = 2.182 \sqrt{r(1 - 1.19r)}$$

in which $r = 1 - P_2/P_1$.

For convenience, values of K as a function of P_2/P_1 are tabulated below:

P_2/P_1	0.98	0.96	0.94	0.92	0.90	0.88	0.86	0.84	0.82	0.80
K	0.321	0.428	0.512	0.585	0.646	0.698	0.744	0.784	0.818	0.850
P_2/P_1	0.78	0.76	0.74	0.72	0.70	0.68	0.66	0.64	0.62	0.60
K	0.877	0.901	0.922	0.940	0.956	0.970	0.981	0.988	0.995	0.998

The Napier and Grashof relations for subcritical flow of dry and saturated steam then become, respectively:

$$\dot{m} = K(A^*P_1/70)$$

and

$$\dot{m} = 0.01654KA^*P_1{}^{0.9696}$$

For subcritical flow of steam which is initially wet, the Napier equation becomes

$$\dot{m} = K(A^*P_1/70 \sqrt{x_1})$$

Examples in which calculations of flow of steam through nozzles are made will be found in the section "Properties of Steam."

PROPERTIES OF SPECIFIC FLUIDS

Properties of Water

Composition. Pure water is a liquid which is composed of one part by weight of hydrogen combined with eight parts of oxygen. In the formation of water 2 atoms of hydrogen are chemically combined with 1 atom of oxygen as follows:

	Molecular weight	Percentage by weight
Hydrogen, 2 atoms.......	2.00	11.11
Oxygen, 1 atom..........	16.00	88.89
Molecular weight........	18.00	100.00

Water is never found pure in nature owing to the readiness with which it absorbs impurities from the air and soil. The usual impurities absorbed by water are organic matter, salts, and soluble gases. The presence of dissolved salts causes water to become "hard." A distinction is made between two kinds of hardness, i.e., temporary and permanent. Temporary hardness is due to the presence of the bicarbonates of lime and magnesium which may be precipitated by boiling at 212 F, and water containing no other scale-forming elements becomes "soft" under such treatment. Permanent hardness is due mainly to the presence of sulfate of lime, which is precipitated only at temperatures above 300 F but may be removed by chemical treatment.

The compositions of salt water, which vary in the different oceans, the Dead Sea, and Great Salt Lake, are given in Van Nostrand's "Scientific Encyclopedia."[1] These compositions may be used as bases for computing the density or specific gravity of the respective sea water. In general, the total dissolved solids in sea water are of the order of 2½ to 5 per cent, and the specific gravity usually is given as 1.025 to 1.030. In the Great Salt Lake and the Dead Sea, however, the total dissolved solids are in the range of 15 to 25 per cent by weight.

Dissolved gases can be removed by boiling, which is frequently done under

[1] Published by D. Van Nostrand Company, Inc., Princeton, N.J.

Table 23. Boiling Point of Water at Various Altitudes

Boiling point, degrees Fahrenheit	Altitude above sea level, feet	Atmospheric pressure, pounds per square inch	Barometer reduced to 32°F., inches	Boiling point, degrees Fahrenheit	Altitude above sea level, feet	Atmospheric pressure, pounds per square inch	Barometer reduced to 32°F., inches
184	15,221	8.20	16.70	199	6,843	11.29	22.90
185	14,649	8.38	17.96	200	6,304	11.52	23.47
186	14,075	8.57	17.45	201	5,764	11.76	23.95
187	13,498	8.76	17.83	202	5,225	12.01	24.45
188	12,934	8.95	18.22	203	4,697	12.26	24.96
189	12,367	9.14	18.61	204	4,169	12.51	25.48
190	11,799	9.34	19.02	205	3,642	12.77	26.00
191	11,243	9.54	19.43	206	3,115	13.03	26.53
192	10,685	9.74	19.85	207	2,589	13.30	27.08
193	10,127	9.95	20.27	208	2,063	13.57	27.63
194	9,579	10.17	20.71	209	1,539	13.85	28.19
195	9,031	10.39	21.15	210	1,025	14.13	28.76
196	8,481	10.61	21.60	211	512	14.41	29.33
197	7,932	10.83	22.05	212	Sea level	14.70	29.92
198	7,381	11.06	22.52				

vacuum, as in deaerators, etc. Organic matter can be removed by filtration, coagulation, etc.   An evaporator and condenser are frequently employed to obtain pure water through the elimination of any or all of the above mentioned impurities. Where dissolved gases are among the impurities to be removed by evaporation, some method of deaeration is essential in connection with the condensing apparatus.

Boiling Temperature of Water at Various Altitudes.   The boiling point of water varies with the barometric pressure, which, in turn, varies with the altitude above sea level.   Assuming standard atmospheric conditions at sea level, the boiling points at various altitudes are as given in Table 23.

Table 24. Relative Volume and Weight per Cubic Foot and per Gallon of Water at Various Temperatures and Corresponding Saturation Pressures

Temperature degrees Fahrenheit	Relative volume	Weight in pounds per cubic foot	Weight in pounds per gallon	Temperature degrees Fahrenheit	Relative volume	Weight in pounds per cubic foot	Weight in pounds per gallon	Temperature degrees Fahrenheit	Relative volume	Weight in pounds per cubic foot	Weight in pounds per gallon
32	1.00013	62.42	8.34	250	1.061	58.83	7.86	480	1.256	49.7	6.64
39.1	1.00000	62.428	8.345	260	1.066	58.55	7.83	490	1.269	49.2	6.58
40	1.00001	62.42	8.34	270	1.072	58.26	7.79	500	1.283	48.7	6.51
50	1.00027	62.42	8.34	280	1.077	57.96	7.75	510	1.297	48.1	6.43
60	1.00096	62.37	8.34	290	1.083	57.65	7.71	520	1.312	47.6	6.36
70	1.00201	62.30	8.33	300	1.089	57.33	7.66	530	1.329	47.0	6.28
80	1.00338	62.22	8.32	310	1.095	57.00	7.62	540	1.35	46.3	6.19
90	1.00504	62.11	8.30	320	1.102	56.66	7.57	550	1.37	45.6	6.10
100	1.00698	62.00	8.29	330	1.109	56.30	7.53	560	1.39	44.9	6.00
110	1.0092	61.86	8.27	340	1.116	55.94	7.48	570	1.42	44.1	5.90
120	1.0116	61.71	8.25	350	1.124	55.57	7.43	580	1.44	43.3	5.79
130	1.0142	61.55	8.23	360	1.131	55.18	7.38	599	1.46	42.6	5.69
140	1.0171	61.38	8.21	370	1.140	54.78	7.32	600	1.49	41.8	5.59
150	1.020	61.20	8.18	380	1.148	54.36	7.27	610	1.52	41.0	5.48
160	1.023	61.00	8.15	390	1.157	53.94	7.21	620	1.55	40.2	5.37
170	1.027	60.80	8.13	400	1.167	53.5	7.15	630	1.59	39.4	5.27
180	1.030	60.58	8.10	410	1.177	53.0	7.09	640	1.63	38.5	5.15
190	1.034	60.36	8.07	420	1.187	52.6	7.03	650	1.67	37.5	5.01
200	1.038	60.12	8.04	430	1.197	52.2	6.98	660	1.72	36.4	4.87
210	1.043	59.88	8.00	440	1.208	51.7	6.91	670	1.78	35.2	4.71
212	1.044	59.83	8.00	450	1.220	51.2	6.84	680	1.86	33.8	4.52
220	1.047	59.63	7.97	460	1.232	50.7	6.78	690	1.95	32.1	4.29
230	1.052	59.37	7.94	470	1.244	50.2	6.71	706.1	3.11	20.1	2.69
240	1.056	59.11	7.90								

Boiling Temperature of Water at Various Pressures. The boiling temperature of water at various pressures is given in the steam tables. Thus water at a nominal atmospheric pressure of 14.7 psia boils at 212 F.

Density. Water has its greatest density at 39.1 F (4 C) when in the pure state it weighs 62.428 lb cu ft. Above this temperature the density decreases as the temperature increases. For ordinary engineering calculations with fresh water at atmospheric temperature, the density is taken as 62.4 lb cu ft. Table 24 gives the densities, relative volumes, and pounds per gallon of water from 32 to 706 F. The relative volumes serve to indicate the volumetric expansion with change in temperature.

Specific Heat. The specific heat of water varies at different temperatures and pressures. Figures 24a and 24b gives the instantaneous specific heats from

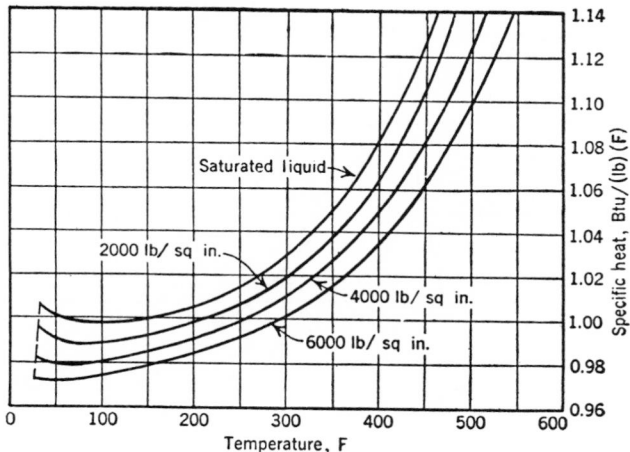

FIG. 24a. Instantaneous specific heat of water at low temperatures.

32 F to about 500 F. The average specific heat for the range between 32 and 212 F is unity. For rough calculations at temperatures below 250 F it is customary to consider the specific heat as one.

Compressibility. Water at its maximum density (39.1 F) has a coefficient of compressibility, according to the Smithsonian Physical Tables, of about 0.00005 for each atmosphere of added pressure and a modulus of elasticity in compression of about 300,000. The coefficient of compressibility decreases and the modulus increases with a rise in temperature; at 212 F they are about 0.00004 and 360,000, respectively. Thus 1 cu ft of water, which weighs 62.425 lb under 1 atm, under a pressure of 11 atm weighs $62.425 \times (1 + 0.00005 \times 10)$ or 62.456 lb, a very slight increase. Except where very accurate work is required or high pressures are involved, water may be considered as incompressible.

Freezing Point. Water freezes at 32 F under ordinary atmospheric pressure, and ice melts at the same temperature. The *latent heat of fusion of ice* at atmospheric pressure is 144 Btu/lb. The latent heat of fusion of a solid may be defined as the heat required in Btu to convert 1 lb of the substance from the solid to the liquid state or vice versa, without change in temperature. The *specific heat* of ice varies with temperature and pressure; for many ordinary engineering calculations, a value of 0.5 Btu/(lb F) may be used with sufficient accuracy.

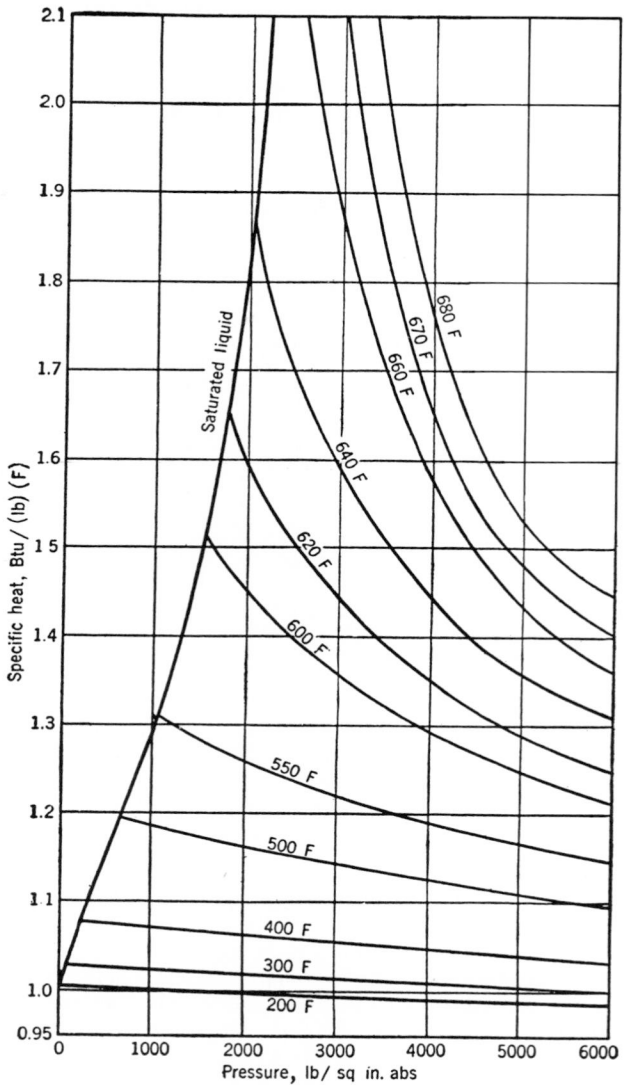

FIG. 24b. Instantaneous specific heat of water at various pressures and temperatures.

In freezing solid in a confined space water expands about 10 per cent in volume with an irresistible pressure which no pipe material can support without being permanently stretched or burst. The effect of increased pressure is to lower the freezing point to the extent of 0.0126 F/atm. According to Bridgman,[1] ice in the temperature range 0 to +32 F has a coefficient of compressibility of about 0.000023 for each atmosphere of added pressure. This value is about one-half that given for water. The modulus of elasticity for ice in compression is approximately 400,000

[1] P. W. Bridgman, *Proc. Am. Acad. Arts. Sci.*, vol. 47, pp. 439–558, 1912.

Table 25.   Solubility in Water of Various Substances

(Volume absorbed per unit volume of water)

	Temperature at atmospheric pressure, F		
	32	68	212
Air.................	0.032	0.020	0.012
Acetylene.............	1.89	1.12	
Ammonia.............	1250	700	
Carbon dioxide........	1.87	0.96	0.26
Carbon monoxide......	0.039	0.025	
Chlorine.............	5.0	2.5	0.00
Hydrogen.............	0.023	0.020	0.018
Hydrogen sulfide.......	5.0	2.8	0.87
Hydrochloric acid......	560	480	
Nitrogen.............	0.026	0.017	0.0105
Oxygen..............	0.053	0.034	0.0185
Sulfuric acid..........	87	43	

psi as determined by static methods.   Values three times as great have been found by dynamic methods and are regarded by physicists as being the true values.[1] However, from engineering considerations the static values seem more useful.

Absorption of Gases.   Many gases are readily absorbed by water.   Other liquids also possess this power to a greater or less degree.   Water will, for example, absorb its own volume of carbonic acid gas, 430 times its volume of ammonia, $2\frac{1}{3}$ times its volume of chlorine, and only about one-twentieth its volume of oxygen.

The weight of gas that is absorbed by a given volume of liquid is proportional to the pressure.   But as the volume of a mass of gas is less as the pressure is greater, the volume which a given amount of liquid can absorb at a certain temperature will be constant, whatever the pressure.   Water, for example, can absorb its own volume of carbonic acid gas at atmospheric pressure; it will also dissolve its own volume if the pressure is twice as great, and consequently twice the weight of gas is dissolved.

The solubility of substances in water varies with both pressure and temperature. Table 25 lists the fractional volume for a number of solutes that can be absorbed by a unit volume of water at atmospheric pressure but at three different temperatures.

Table 26.   Viscosity of Water

Temp, F	Viscosity, lbm/(ft sec)	Temp, F	Viscosity, lbm/(ft sec)
32	1.2×10^{-3}	200	0.205×10^{-3}
40	1.04	250	0.158
50	0.88	300	0.126
60	0.76	350	0.105
70	0.658	400	0.091
80	0.578	450	0.080
90	0.514	500	0.071
100	0.458	550	0.064
150	0.292	600	0.058

Viscosity of Water.   Water is less viscous at high than at low temperatures. Apparently, as the temperature increases, the cohesive forces between the tightly packed molecules decrease.   Table 26 shows the viscosity of saturated water as a function of its temperature.

[1] See "Properties of Ordinary Water Substance," by N. E. Dorsey, Reinhold Publishing Corporation, New York, N.Y.

Thermal Conductivity of Water. The thermal conductivity of water frequently is involved in heat-transfer problems in piping systems. It has a maximum value at about 250 F, conductivities at higher and lower temperatures being smaller. Table 27 lists values of thermal conductivity over the range 40 to 600 F.

Table 27. Thermal Conductivity of Air-free Water

Temp, F	Thermal conductivity, Btu/(hr ft F)	Temp, F	Thermal conductivity, Btu/(hr ft F)
40	0.325	200	0.394
50	0.332	250	0.396
60	0.340	300	0.395
70	0.347	350	0.391
80	0.353	400	0.381
90	0.359	450	0.367
100	0.364	500	0.349
150	0.384	600	0.292

Properties of Steam

As a function of the pressure and temperature to which it is subjected, the substance H_2O, consisting of two parts by weight of hydrogen and eight parts of oxygen, can exist in the solid, liquid, or vapor form. This section treats the properties of the substance in vapor form, which is commonly referred to as steam.

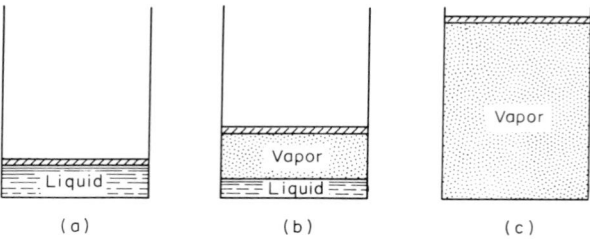

FIG. 25. Formation of steam at constant pressure.

The Formation of Steam. Figure 25a shows liquid water contained within a vessel and at a certain temperature (say 60 F) and a certain pressure (say 14.7 psia). If heat is transferred to the water, its temperature will rise and its specific volume will increase slightly while the pressure remains constant. Until the temperature reaches 212 F, which is the boiling point at 14.7 psia, no change other than an in-crease in temperature and specific volume will occur. Once, however, the boiling-point temperature is reached, additional heat transfer to the water will result in the boiling or vaporization of some of the water.

Figure 25b shows the vessel and its contents. Sufficient heat has been added to vaporize some, but not all, of the liquid. The vapor and the liquid water are intimately in contact with each other at the same pressure and the same temperature. In this state, the vapor is referred to as *saturated* vapor and the liquid, as *saturated* liquid. The temperature and the pressure are referred to as the *saturation temperature* and *saturation pressure*.

With reference still to Fig. 25b, when most but not all of the liquid has vaporized, the liquid may be dispersed more or less uniformly throughout the steam. The entire mixture, vapor plus entrained liquid, would then be referred to as *wet* steam

or as *wet, saturated* steam. The composition of the mixture is designated by its quality *x*, defined as

$$x = \frac{\text{mass of saturated vapor}}{\text{mass of total mixture (liquid + vapor)}}$$

In Fig. 25c, additional heat has been transferred to the vessel contents and all the liquid, even that entrained in the steam, has evaporated. The vessel contains nothing but steam at 212 F and 14.7 psia. The steam is 100 per cent saturated and is referred to as saturated steam or, to emphasize the absence of liquid, as *dry, saturated* steam.

If, after the steam has become dry and saturated, additional heat is transferred to it, its temperature will increase while the pressure remains constant. Whenever the temperature is at a value greater than the saturation temperature (in this case, 212 F), the steam is said to be *superheated*. For example, if the steam temperature had been raised to 300 F, the vessel contents would be referred to as superheated steam at 300 F and 14.7 psia with 88 F of *superheat*.

The following is to be emphasized:

1. The state of a *saturated vapor* or of a *saturated liquid* is defined by specification only of either the temperature or the pressure.

2. The state of a *wet-steam* mixture is defined by specification of either the temperature or the pressure *plus* the specification of one other independent property such as the quality or specific volume.

3. The state of a *superheated vapor* is defined by the specification of two independent properties. Most often, pressure and temperature are the two defining properties. However, temperature and specific volume or pressure and enthalpy would define the state just as precisely.

Critical Pressure. The discussion relative to the formation of steam was based on an assumed mixture pressure of 14.7 psia. Actually, the pressure could have been at, say, 200 psia and the same end results would have been reached except that the saturation temperature would not have been 212 F. For each different pressure, there is a corresponding saturation temperature, as shown in Fig. 26. At higher pressures, it is found that lesser heat additions are necessary to effect a change in state from saturated liquid to saturated vapor. For example, at 14.7 psia the necessary heat addition is 970.3 Btu/lb while at 200 psia, only 843.0 Btu are required to change 1 lb of saturated liquid to saturated vapor. As pressures are increased, the heat-addition requirement becomes successively smaller until, finally, at 3,206.2 psia, the required

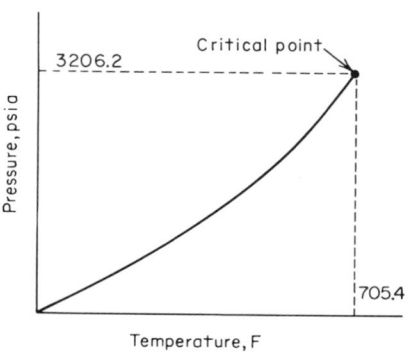

Fig. 26. Vapor-pressure curve for water.

heat addition is zero. This value of pressure is referred to as the *critical pressure*, and the corresponding temperature, 705.4 F, is the *critical temperature*.

The vapor-pressure curve of Fig. 26 ends discontinuously at the critical pressure and temperature. At this unique point, specific volume, internal energy, and enthalpy of the liquid are identical with those of the vapor, and the liquid and vapor are indistinguishable from each other.

Steam Tables. If the pressure of steam is low with respect to its critical pressure and its temperature is high with respect to its critical temperature, it has been observed that this low-pressure vapor obeys with reasonable accuracy the perfect-gas laws. However, except for applications in such fields as air conditioning, this phenomenon is not encountered. Thus it is necessary to resort to empirical data, supplemented where advantageous by analytical treatment, in order to describe the properties of steam.

Various tables of thermodynamic steam data are available. Probably the most widely used and accepted set of data is that contained in the "Thermodynamic Properties of Steam" by Keenan and Keyes and published by John Wiley & Sons, Inc. Tables 28 to 30 are abridged from the Keenan and Keyes tables.

Table 28 gives values of the properties pressure, specific volume, enthalpy, and entropy of saturated liquid and saturated vapor at saturation temperatures from 32 F to the critical temperature 705.4 F. Changes in specific volume, enthalpy, and entropy during the vaporization process are also listed.

Table 29 lists values of temperature, specific volume, enthalpy, entropy, and internal energy of saturated liquid and saturated vapor at saturation pressures from 1 in. mercury absolute (0.491 psia) to the critical pressure 3,206.2 psia. Changes which occur in enthalpy and entropy during the vaporization process are also indicated.

Table 30 lists values of specific volume, enthalpy, and entropy for pressures ranging from 1 to 3,500 psia at temperatures from 200 to 1600 F.

In all three of the above tables, nomenclature is used as follows:

$$t = \text{temperature, deg F}$$
$$p = \text{pressure, psia}$$
$$v = \text{specific volume, cu ft/lbm}$$
$$h = \text{enthalpy, Btu/lbm}$$
$$u = \text{internal energy, Btu/lbm}$$
$$s = \text{entropy, Btu/(lbm R)}$$

Subscript f refers to the saturated liquid. Subscript g refers to the saturated vapor. Subscript fg refers to the change in the indicated property as the fluid changes phase from a saturated liquid to a saturated vapor.

The following will be helpful in connection with use of the Steam Tables:

$$h = u + 144pv/J \tag{58}$$

$$h = h_f + xh_{fg} \tag{59}$$

$$u = u_f + xu_{fg} \tag{60}$$

$$s = s_f + xs_{fg} \tag{61}$$

$$v = v_f + xv_{fg} \tag{62}$$

In the above, J is the conversion factor from thermal to mechanical dimensions and has the value of 778 ft lbf/Btu. x is the quality of steam expressed as a fraction or decimal.

Graphical Representation of Steam Properties. Processes which involve the flow of steam are very conveniently displayed in graphical form on an enthalpy-entropy plot. Such a plot is referred to as a Mollier diagram, and an abbreviated version is presented as Fig. 27. Detailed diagrams are included in "Thermodynamic Properties of Steam" by Keenan and Keyes and in the "Steam Tables" of Combustion-Engineering-Superheater, Inc., 200 Madison Avenue, New York.

With reference to Fig. 27, it will be seen that the superheat and wet regions are separated by the saturation line. Specifications of temperature and pressure identify states in the superheat region, and in the wet region, the specification of pressure and quality determines the state. Once a state is located on the diagram, simultaneous value of enthalpy and entropy may be read.

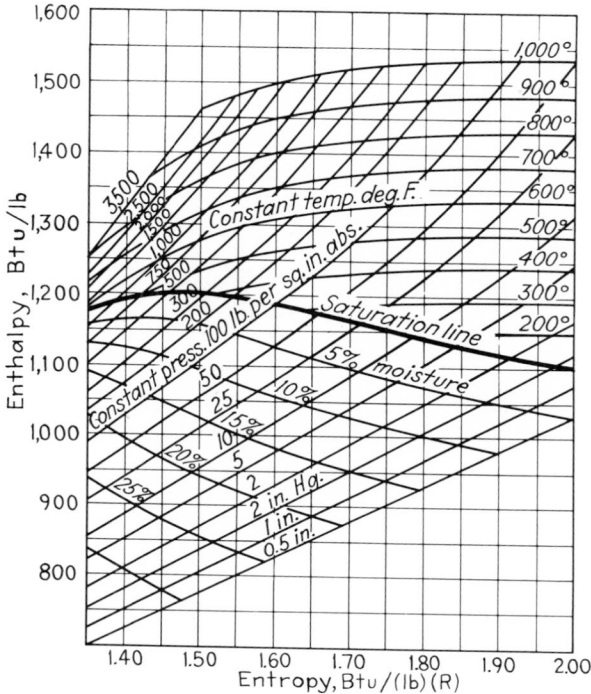

FIG. 27. Abbreviated Mollier diagram.

Examples in the Use of Steam Tables and Mollier Diagram

Example 1. Steam is to be expanded in a convergent-divergent nozzle from initial conditions of 200 psia, 800 F, to a final pressure of 100 psia at the rate of 30,000 lb/hr. If the nozzle efficiency is 95 per cent, determine the required throat and exit areas. Also, what must be the inlet area if approach velocity is to be limited to 100 fps?

Solution. From Table 30, at 300 psia and 800 F

$$v_1 = 2.442 \text{ cu ft/lbm}$$
$$h_1 = 1420.6 \text{ Btu/lbm}$$
$$s_1 = 1.7184 \text{ Btu/(lbm R)}$$

Similar values are read, although with not the same degree of precision, from Fig. 27   Once the state point 1 is located on Fig. 27, the state following isentropic expansion is found by dropping vertically downward to 100 psia and h_2' is read as 1286 Btu/lbm.

Then, using Eq. (50)

$$V_2 = \sqrt{V_1^2 + 2g_c J e_n (h_1 - h_2')}$$
$$= \sqrt{100^2 + 2(32.17)(778)(0.95)(1420.6 - 1286)}$$
$$= 2{,}530 \text{ fps}$$

Table 28. Steam Table: Saturation Temperatures[1]

Temp, F t	Abs press., psi p	Specific volume, cu ft/lb			Enthalpy Btu/lb			Entropy Btu/(lb F)		
		Sat. liquid v_f	Evap. v_{fg}	Sat. vapor v_g	Sat. liquid h_f	Evap. h_{fg}	Sat. vapor h_g	Sat. liquid s_f	Evap. s_{fg}	Sat. vapor s_g
32	0.08854	0.01602	3306	3306	0.00	1075.8	1075.8	0.0000	2.1877	2.1877
35	0.09995	0.01602	2947	2947	3.02	1074.1	1077.1	0.0061	2.1709	2.1770
40	0.12170	0.01602	2444	2444	8.05	1071.3	1079.3	0.0162	2.1435	2.1597
45	0.14752	0.01602	2036.4	2036.4	13.06	1068.4	1081.5	0.0262	2.1167	2.1429
50	0.17811	0.01603	1703.2	1703.2	18.07	1065.6	1083.7	0.0361	2.0903	2.1264
60	0.2563	0.01604	1206.6	1206.7	28.06	1059.9	1088.0	0.0555	2.0393	2.0948
70	0.3631	0.0606	867.8	867.9	38.04	1054.3	1092.3	0.0745	1.9902	2.0647
80	0.5069	0.01608	633.1	633.1	48.02	1048.6	1096.6	0.0932	1.9428	2.0360
90	0.6982	0.01610	468.0	468.0	57.99	1042.9	1100.9	0.1115	1.8972	2.0087
100	0.9492	0.01613	350.3	350.4	67.97	1037.2	1105.2	0.1295	1.8531	1.9826
110	1.2748	0.01617	265.3	265.4	77.94	1031.6	1109.5	0.1471	1.8106	1.9577
120	1.6924	0.01620	203.25	203.27	87.92	1025.8	1113.7	0.1645	1.7694	1.9339
130	2.2225	0.01625	157.32	157.34	97.90	1020.0	1117.9	0.1816	1.7296	1.9112
140	2.8886	0.01629	122.99	123.01	107.89	1014.1	1122.0	0.1984	1.6910	1.8894
150	3.718	0.01634	97.06	97.07	117.89	1008.2	1126.1	0.2149	1.6537	1.8685
160	4.741	0.01639	77.27	77.29	127.89	1002.3	1130.2	0.2311	1.6174	1.8485
170	5.992	0.01645	62.04	62.06	137.90	996.3	1134.2	0.2472	1.5822	1.8293
180	7.510	0.01651	50.21	50.23	147.92	990.2	1138.1	0.2630	1.5480	1.8109
190	9.339	0.01657	40.94	40.96	157.95	984.1	1142.0	0.2785	1.5147	1.7932
200	11.526	0.01663	33.62	33.64	167.99	977.9	1145.9	0.2938	1.4824	1.7762
210	14.123	0.01670	27.80	27.82	178.05	971.6	1149.7	0.3090	1.4508	1.7598
212	14.696	0.01672	26.78	26.80	180.07	970.3	1150.4	0.3120	1.4446	1.7566
220	17.186	0.01677	23.13	23.15	188.13	965.2	1153.4	0.3239	1.4201	1.7440
230	20.780	0.01684	19.365	19.382	198.23	958.8	1157.0	0.3387	1.3901	1.7288
240	24.969	0.01692	16.306	16.323	208.34	952.2	1160.5	0.3531	1.3609	1.7140
250	29.825	0.01700	13.804	13.821	216.48	945.5	1164.0	0.3675	1.3223	1.6998
260	35.429	0.01709	11.746	11.763	228.64	938.7	1167.3	0.3817	1.3043	1.6860
270	41.858	0.01717	10.044	10.061	238.84	931.8	1170.6	0.3958	1.2769	1.6727
280	49.203	0.01726	8.628	8.645	249.06	924.7	1173.8	0.4096	1.2501	1.6597
290	57.556	0.01735	7.444	7.461	259.31	917.5	1176.8	0.4234	1.2238	1.6472

300	67.013	0.01745	6.449	6.466	269.59	910.1	1179.7	0.4369	1.1980	1.6350
310	77.68	0.01755	5.609	5.626	279.92	902.6	1182.5	0.4504	1.1727	1.6231
320	89.66	0.01765	4.896	4.914	290.28	894.9	1185.2	0.4637	1.1478	1.6115
330	103.06	0.01776	4.289	4.307	300.68	887.0	1187.7	0.4769	1.1233	1.6002
340	118.01	0.01787	3.770	3.788	311.13	879.0	1190.1	0.4900	1.0992	1.5891
350	134.63	0.01799	3.324	3.342	321.63	870.7	1192.3	0.5029	1.0754	1.5783
360	153.04	0.01811	2.939	2.957	332.18	862.2	1194.4	0.5158	1.0519	1.5677
370	173.37	0.01823	2.606	2.625	342.79	853.5	1196.3	0.5286	1.0287	1.5573
380	195.77	0.01836	2.317	2.335	353.45	844.6	1198.1	0.5413	1.0059	1.5471
390	220.37	0.01850	2.0651	2.0836	364.17	835.4	1199.6	0.5539	0.9832	1.5371
400	247.31	0.01864	1.8447	1.8633	374.97	826.0	1201.0	0.5664	0.9608	1.5272
410	276.75	0.01878	1.6512	1.6700	385.83	816.3	1202.1	0.5788	0.9386	1.5174
420	308.83	0.01894	1.4811	1.5000	396.77	806.3	1203.1	0.5912	0.9166	1.5078
430	343.72	0.01910	1.3308	1.3499	407.79	796.0	1203.8	0.6035	0.8947	1.4982
440	381.59	0.01926	1.1979	1.2171	418.90	785.4	1204.3	0.6158	0.8730	1.4887
450	422.6	0.0194	1.0799	1.0993	430.1	774.5	1204.6	0.6280	0.8513	1.4793
460	466.9	0.0196	0.9748	0.9944	441.4	763.2	1204.6	0.6402	0.8298	1.4700
470	514.7	0.0198	0.8811	0.9009	452.8	751.5	1204.3	0.6523	0.8083	1.4606
480	566.1	0.0200	0.7972	0.8172	464.4	739.4	1203.7	0.6645	0.7868	1.4513
490	621.4	0.0202	0.7221	0.7423	476.0	726.8	1202.8	0.6766	0.7653	1.4419
500	680.8	0.0204	0.6545	0.6749	487.8	713.9	1201.7	0.6887	0.7438	1.4325
520	812.4	0.0209	0.5385	0.5594	511.9	686.4	1198.2	0.7130	0.7006	1.4136
540	962.5	0.0215	0.4434	0.4649	536.6	656.6	1193.2	0.7374	0.6568	1.3942
560	1133.1	0.0221	0.3647	0.3868	562.2	624.2	1186.4	0.7621	0.6121	1.3742
580	1325.8	0.0228	0.2989	0.3217	588.9	588.4	1177.3	0.7872	0.5659	1.3532
600	1542.9	0.0236	0.2432	0.2668	617.0	548.5	1165.5	0.8131	0.5176	1.3307
620	1786.6	0.0247	0.1955	0.2201	646.7	503.6	1150.3	0.8398	0.4664	1.3062
640	2059.7	0.0260	0.1538	0.1798	678.6	452.0	1130.5	0.8679	0.4110	1.2789
660	2365.4	0.0278	0.1165	0.1442	714.2	390.2	1104.4	0.8987	0.3485	1.2472
680	2708.1	0.0305	0.0810	0.1115	757.3	309.9	1067.2	0.9351	0.2719	1.2071
700	3093.7	0.0369	0.0392	0.0761	823.3	172.1	995.4	0.9905	1.1484	1.1389
705.4	3206.2	0.0503	0	0.0503	902.7	0	902.7	1.0580	0	1.0580

[1] Abridged from "Thermodynamic Properties of Steam" by Joseph H. Keenan and Frederick G. Keyes. Copyright, 1937, by Joseph H. Keenan and Frederick G. Keyes. Published by John Wiley & Sons, Inc., New York

Table 29. Steam Table: Saturation Pressures[1]

Abs press., psi p	Temp., F t	Specific volume cu ft/lb		Enthalpy, Btu/lb			Entropy, Btu/(lb F)			Internal energy, Btu/lb	
		Sat. liquid v_f	Sat. vapor v_g	Sat. liquid h_f	Evap. h_{fg}	Sat. vapor h_g	Sat. liquid s_f	Evap. s_{fg}	Sat. vapor s_g	Sat. liquid u_g	Sat. vapor u_g
0.491	79.03	0.01608	652.3	47.05	1049.2	1096.3	0.0914	1.9473	2.0387	47.05	1037.0
0.736	91.72	0.01611	444.9	59.71	1042.0	1101.7	0.1147	1.8894	2.0041	59.71	1041.1
0.982	101.14	0.01614	339.2	69.10	1036.6	1105.7	0.1316	1.8481	1.9797	69.10	1044.0
1.227	108.71	0.01616	274.9	76.65	1032.3	1108.9	0.1449	1.8160	1.9609	76.65	1046.4
1.473	115.06	0.01618	231.6	82.99	1028.6	1111.6	0.1560	1.7896	1.9456	82.99	1048.5
1.964	125.43	0.01622	176.7	93.34	1022.7	1116.0	0.1738	1.7476	1.9214	93.33	1051.8
2.455	133.76	0.01626	143.25	101.66	1017.7	1119.4	0.1879	1.7150	1.9028	101.65	1054.3
5	162.24	0.01640	73.52	130.13	1001.0	1131.1	0.2347	1.6094	1.8441	130.12	1063.1
10	193.21	0.01659	38.42	161.17	982.1	1143.3	0.2835	1.5041	1.7876	161.14	1072.2
14.696	212.0	0.01672	26.80	180.07	970.3	1150.4	0.3120	1.4446	1.7566	180.02	1077.5
15	213.03	0.01672	26.29	181.11	969.7	1150.8	0.3135	1.4415	1.7549	181.06	1077.8
16	216.32	0.01674	24.75	184.42	967.6	1152.0	0.3184	1.4313	1.7497	184.37	1078.7
18	222.41	0.01679	22.17	190.56	963.6	1154.2	0.3275	1.4128	1.7403	190.50	1080.4
20	227.96	0.01683	20.089	196.16	960.1	1156.3	0.3356	1.3962	1.7319	196.10	1081.9
25	240.07	0.01692	16.303	208.42	952.1	1160.6	0.3533	1.3606	1.7139	208.34	1085.1
30	250.33	0.01701	13.746	218.82	945.3	1164.1	0.3680	1.3313	1.6993	218.73	1087.8
35	259.28	0.1708	11.898	227.91	939.2	1167.1	0.3807	1.3063	1.6870	227.80	1090.1
40	267.25	0.01715	10.498	236.03	933.7	1169.7	0.3919	1.2844	1.6763	235.90	1092.0
45	274.44	0.01721	9.401	243.36	928.6	1172.0	0.4019	1.2650	1.6669	243.22	1093.7
50	281.01	0.01727	8.515	250.09	924.0	1174.1	0.4110	1.2474	1.6585	294.93	1095.3
55	287.07	0.01732	7.787	256.30	919.6	1175.9	0.4193	1.2316	1.6509	256.12	1095.7
60	292.71	0.01738	7.175	262.09	915.5	1177.6	0.4270	1.2168	1.6438	261.90	1097.9
65	297.97	0.01743	6.655	267.50	911.6	1179.1	0.4342	1.2032	1.6374	267.29	1099.1
70	302.92	0.01748	6.206	272.61	907.9	1180.6	0.4409	1.1906	1.6315	272.38	1100.2
75	307.60	0.01753	5.816	277.43	904.5	1181.9	0.4472	1.1787	1.6259	277.19	1101.2
80	312.03	0.01757	5.472	282.02	901.1	1183.1	0.4531	1.1676	1.6207	281.76	1102.1
85	316.25	0.01761	5.168	286.39	897.8	1184.2	0.4587	1.1571	1.6158	286.11	1102.9
90	320.27	0.01766	4.896	290.56	894.7	1185.3	0.4641	1.1471	1.6112	290.27	1103.7
100	327.81	0.01774	4.432	298.40	888.8	1187.2	0.4740	1.1286	1.6026	298.08	1105.2
110	334.77	0.01782	4.049	305.66	883.2	1188.9	0.4832	1.1117	1.5948	305.30	1106.5

Abs. press., lbf/in²	Temp, °F	v_f	v_g	h_f	h_{fg}	h_g	s_f	s_{fg}	s_g	u_f	u_g
120	341.25	0.01789	3.728	312.44	877.9	1190.4	0.4916	1.0962	1.5878	312.05	1107.6
130	347.32	0.01796	3.455	318.81	872.9	1191.7	0.4995	1.0817	1.5812	318.38	1108.6
140	353.02	0.01802	3.220	324.82	868.2	1193.0	0.5069	1.0682	1.5751	324.35	1109.6
150	358.42	0.01809	3.015	330.51	863.6	1194.1	0.5138	1.0556	1.5694	330.01	1110.5
160	363.53	0.01815	2.834	335.95	859.2	1195.1	0.5204	1.0436	1.5640	335.39	1111.2
170	368.41	0.01822	2.675	341.09	854.9	1196.0	0.5266	1.0324	1.5590	340.52	1111.9
180	373.06	0.01827	2.532	346.03	850.8	1196.9	0.5325	1.0217	1.5542	345.42	1112.5
190	377.51	0.01833	2.404	350.79	846.8	1197.6	0.5381	1.0116	1.5497	350.15	1113.1
200	381.79	0.01839	2.288	355.36	843.0	1198.4	0.5435	1.0018	1.5453	354.68	1113.7
250	400.95	0.01865	1.8438	376.00	825.1	1201.1	0.5675	0.9588	1.5263	375.14	1115.8
300	417.33	0.01890	1.5433	393.84	809.0	1202.8	0.5879	0.9225	1.5104	392.79	1117.1
350	431.72	0.01913	1.3260	409.69	794.2	1203.9	0.6056	0.8910	1.4966	408.45	1118.0
400	444.59	0.0193	1.1613	424.0	780.5	1204.5	0.6214	0.8630	1.4844	422.6	1118.5
450	456.28	0.0195	1.0320	437.2	767.4	1204.6	0.6356	0.8378	1.4734	435.5	1118.7
500	467.01	0.0197	0.9278	449.4	755.0	1204.4	0.6487	0.8147	1.4634	447.6	1118.6
550	476.94	0.0199	0.8424	460.8	743.1	1203.9	0.6608	0.7934	1.4542	458.8	1118.2
600	486.21	0.0201	0.7698	471.6	731.6	1203.2	0.6720	0.7734	1.4454	469.4	1117.7
650	494.90	0.0203	0.7083	481.8	720.5	1202.3	0.6826	0.7548	1.4374	479.4	1117.1
700	503.10	0.0205	0.6554	491.5	709.7	1201.2	0.6925	0.7371	1.4296	488.8	1116.3
750	510.86	0.0207	0.6092	500.8	699.2	1200.0	0.7019	0.7204	1.4223	598.0	1115.4
800	518.23	0.0209	0.5687	509.7	688.9	1198.6	0.7108	0.7045	1.4153	506.6	1114.4
850	525.26	0.0210	0.5327	518.3	678.8	1197.1	0.7194	0.6891	1.4085	515.0	1113.3
900	531.98	0.0212	0.5006	526.6	668.8	1195.4	0.7275	0.6744	1.4020	523.1	1112.1
950	538.43	0.0214	0.4717	534.6	659.1	1193.7	0.7355	0.6602	1.3957	530.9	1110.8
1,000	544.61	0.0216	0.4456	542.4	649.4	1191.8	0.7430	0.6467	1.3897	538.4	1109.4
1,100	556.31	0.0220	0.4001	557.4	630.4	1187.8	0.7575	0.6205	1.3780	552.9	1106.4
1,200	567.22	0.0223	0.3619	571.7	611.7	1183.4	0.7711	0.5956	1.3667	566.7	1103.0
1,300	577.46	0.0227	0.3293	585.4	593.2	1178.6	0.7840	0.5719	1.3559	580.0	1099.4
1,400	587.10	0.0231	0.3012	598.7	574.7	1173.4	0.7963	0.5491	1.3454	592.7	1095.4
1,500	596.23	0.0235	0.2765	611.6	556.3	1167.9	0.8082	0.5269	1.3351	605.1	1091.2
2,000	635.82	0.0257	0.1878	671.7	463.4	1135.1	0.8619	0.4230	1.2849	662.2	1065.6
2,500	668.13	0.0287	0.1307	730.6	360.5	1091.1	0.9126	0.3197	1.2322	717.3	1030.6
3,000	695.36	0.0346	0.0858	802.5	217.8	1020.3	0.9731	0.1885	1.1615	783.4	972.7
3,206.2	705.40	0.0503	0.0503	902.7	0	902.7	1.0580	0	1.0580	872.9	872.9

[1] Abridged from "Thermodynamic Properties of Steam" by Joseph H. Keenan and Frederick G. Keyes. Copyright, 1937, by Joseph H. Keenan and Frederick G. Keyes. Published by John Wiley & Sons, Inc., New York.

Table 30. Steam Table: Properties of Superheated Steam[1]

Abs Press., psi (sat. temp.)		200	300	400	500	600	700	800	900	1000	1100	1200	1400	1600
1 (101.74)	v	392.6	452.3	512.0	571.6	631.2	690.8	750.4	809.9	869.5	869.5	988.7	1107.8	1227.0
	h	1150.4	1195.8	1241.7	1288.3	1335.7	1383.8	1432.8	1482.7	1533.5	1585.8	1637.7	1745.7	1857.5
	s	2.0512	2.1153	2.1720	2.2233	2.2702	2.3137	2.3542	2.3923	2.4283	2.4625	2.4952	2.5566	2.6137
5 (162.24)	v	78.16	90.25	102.26	114.22	126.16	138.10	150.03	161.95	173.87	185.79	197.71	221.6	245.4
	h	1148.8	1195.0	1241.0	1288.0	1335.4	1383.6	1432.7	1482.6	1533.4	1585.1	1637.7	1745.7	1857.4
	s	1.8718	1.9370	1.9942	2.0456	2.0927	2.1361	2.1767	2.2148	2.2509	2.2851	2.3178	2.3792	2.4363
10 (193.21)	v	38.85	45.00	51.04	57.05	63.03	69.01	74.98	80.95	86.92	92.88	98.84	110.77	122.69
	h	1146.6	1193.9	1240.6	1287.5	1335.1	1383.4	1432.5	1482.4	1533.2	1585.0	1637.6	1745.6	1857.3
	s	1.7927	1.8595	1.9172	1.9689	2.0160	2.0596	2.1003	2.1383	2.1744	2.2086	2.2413	2.3028	2.3598
14.696 (212.00)	v	......	30.53	34.68	38.78	42.86	46.94	51.00	55.07	59.13	63.19	67.25	75.37	83.48
	h	......	1192.8	1239.9	1287.1	1334.8	1383.2	1432.3	1482.3	1533.1	1584.8	1637.5	1745.5	1857.3
	s	......	1.8160	1.8743	1.9261	1.9734	2.0170	2.0576	2.0958	2.1319	2.1662	2.1989	2.2603	2.3174
20 (227.96)	v	......	22.36	25.43	28.46	31.47	34.47	37.46	40.45	43.44	46.42	49.41	55.37	61.34
	h	......	1191.6	1239.2	1286.6	1334.4	1382.9	1432.1	1482.1	1533.0	1584.7	1637.4	1745.4	1857.2
	s	......	1.7808	1.8396	1.8918	1.9392	1.9829	2.0235	2.0618	2.0978	2.1321	2.1648	2.2263	2.2834
40 (267.25)	v	......	11.040	12.628	14.168	15.688	17.198	18.702	20.20	21.70	23.20	24.69	27.68	30.66
	h	......	1186.8	1236.5	1284.8	1333.1	1381.9	1431.3	1481.4	1532.4	1584.3	1637.0	1745.1	1857.2
	s	......	1.6994	1.7608	1.8140	1.8619	1.9058	1.9467	1.9850	2.0212	2.0555	2.0883	2.1498	2.2069
60 (292.71)	v	......	7.259	8.357	9.403	10.427	11.441	12.449	13.452	14.454	15.453	16.451	18.446	20.44
	h	......	1181.6	1233.6	1283.0	1331.8	1380.9	1430.5	1480.8	1531.9	1583.8	1636.6	1744.8	1856.7
	s	......	1.6492	1.7135	1.7678	1.8162	1.8605	1.9015	1.9400	1.9762	2.0106	2.0434	2.1049	2.1621
80 (312.03)	v	......	......	6.220	7.020	7.797	8.562	9.322	10.077	10.830	11.582	12.332	13.830	15.523
	h	......	......	1230.7	1281.1	1330.5	1379.9	1429.7	1480.1	1531.3	1583.4	1636.2	1744.5	1856.5
	s	......	......	1.6791	1.7346	1.7836	1.8281	1.8694	1.9079	1.9442	1.9787	2.0115	2.0731	2.1303
100 (327.81)	v	......	......	4.937	5.589	6.218	6.835	7.446	8.052	8.656	9.259	9.860	11.060	12.258
	h	......	......	1227.6	1279.1	1329.1	1378.9	1428.9	1479.5	1530.8	1582.9	1635.7	1744.2	1856.2
	s	......	......	1.6518	1.7085	1.7581	1.8029	1.8443	1.8829	1.9193	1.9538	1.9867	2.0484	2.1056
120 (341.25)	v	......	......	4.081	4.636	5.165	5.683	6.195	6.207	7.207	7.710	8.212	9.214	10.213
	h	......	......	1224.4	1277.2	1327.7	1377.8	1428.1	1478.8	1530.2	1582.4	1635.3	1743.9	1856.0
	s	......	......	1.6287	1.6869	1.7370	1.7822	1.8237	1.8625	1.8990	1.9335	1.9664	2.0281	2.0854

Temperature

psia (sat. temp.)		1	2	3	4	5	6	7	8	9	10	11
140 (353.02)	r	8.752	7.895	7.035	6.604	6.172	5.738	5.301	4.861	4.413	3.954	3.468
	h	1855.7	1743.5	1634.9	1581.9	1529.7	1478.2	1427.3	1376.8	1326.4	1275.2	1221.1
	s	2.0683	2.0110	1.9493	19.163	1.8817	1.8451	1.8063	1.7645	1.7190	1.6683	1.6087
160 (363.53)	r	7.656	6.906	6.152	5.775	5.396	5.015	4.631	4.244	3.849	3.443	3.008
	h	1855.5	1743.2	1634.5	1581.4	1529.1	1477.5	1426.4	1375.7	1325.0	1273.1	1217.6
	s	2.0535	1.9962	1.9344	1.9014	1.8667	1.8301	1.7911	1.7491	1.7033	1.6519	1.5908
180 (373.06)	r	6.804	6.136	5.466	5.129	4.792	4.452	4.110	3.764	3.411	3.044	2.649
	h	1855.2	1742.9	1634.1	1581.0	1528.6	1476.8	1425.6	1374.7	1323.5	1271.0	1214.0
	s	2.0404	1.9831	1.9212	1.8882	1.8534	1.8167	1.7776	1.7355	1.6894	1.6373	1.5745
200 (381.79)	r	6.123	5.521	4.917	4.613	4.309	4.002	3.693	3.380	3.060	2.726	2.361
	h	1855.0	1742.6	1633.7	1580.5	1528.0	1476.2	1424.8	1373.6	1322.1	1268.9	1210.3
	s	2.0287	1.9713	1.9094	1.8763	1.8415	1.8048	1.7655	1.7232	1.6767	1.6240	1.5594
220 (389.86)	r	5.565	5.017	4.467	4.191	3.913	3.634	3.352	3.066	2.772	2.465	2.125
	h	1854.7	1742.3	1633.3	1580.0	1527.5	1475.5	1424.0	1372.6	1320.7	1266.7	1206.5
	s	2.0181	1.9607	1.8987	1.8656	1.8308	1.7939	1.7545	1.7120	1.6652	1.6117	1.5453
240 (397.37)	r	5.100	4.597	4.093	3.839	3.584	3.327	3.068	2.804	2.533	2.247	1.9276
	h	1854.5	1742.0	1632.9	1579.6	1526.9	1474.8	1423.2	1371.5	1319.2	1264.5	1202.5
	s	2.0084	1.9510	1.8889	1.8558	1.8209	1.7839	1.7444	1.7017	1.6546	1.6003	1.5319
260 (404.42)	r	4.707	4.242	3.776	3.541	3.305	3.067	2.827	2.582	2.330	2.063	……
	h	1854.2	1741.7	1632.5	1579.1	1526.3	1474.2	1422.4	1370.4	1317.7	1262.3	……
	s	1.9995	1.9420	1.8799	1.8467	1.8118	1.7748	1.7352	1.6922	1.6447	1.5897	……
280 (411.05)	r	4.370	3.938	3.504	3.286	3.066	2.845	2.621	2.392	2.156	1.9047	……
	h	1854.0	1741.4	1632.1	1578.6	1525.8	1473.5	1421.5	1369.4	1316.2	1260.0	……
	s	1.9912	1.9337	1.8716	1.8383	1.8033	1.7662	1.7265	1.6834	1.6354	1.5796	……
300 (417.33)	r	4.078	3.674	3.269	3.065	2.859	2.652	2.442	2.227	2.005	1.7675	……
	h	1853.7	1741.0	1631.7	1578.1	1525.2	1472.8	1420.6	1368.3	1314.7	1257.6	……
	s	1.9835	1.9260	1.8638	1.8305	1.7954	1.7582	1.7184	1.6751	1.6268	1.5701	……
350 (431.72)	r	3.493	3.147	2.798	2.622	2.445	2.266	2.084	1.8980	1.7036	1.4923	……
	h	1853.1	1740.3	1630.7	1577.0	1523.8	1471.1	1418.5	1365.6	1310.9	1251.5	……
	s	1.9663	1.9086	1.8463	1.8130	1.7777	1.7403	1.7002	1.6563	1.6070	1.5481	……
400 (444.59)	r	3.055	2.751	2.445	2.290	2.134	1.9767	1.8161	1.6508	1.4770	1.2851	……
	h	1852.5	1739.5	1629.6	1575.8	1522.4	1469.4	1416.4	1362.7	1306.9	1245.1	……
	s	1.9513	1.8936	1.8311	1.7977	1.7623	1.7247	1.6842	1.6398	1.5894	1.5281	……

Table 30. (Continued)

Abs Press., psi (sat. temp.)		Temperature													
		500	550	600	620	640	660	680	700	800	900	1000	1200	1400	1600
450 (456.28)	v	1.1231	1.2155	1.3005	1.3332	1.3652	1.3967	1.4278	1.4584	1.6074	1.7516	1.8928	2.170	2.443	2.714
	h	1238.4	1272.0	1302.8	1314.6	1326.2	1337.5	1348.8	1359.9	1414.3	1467.7	1521.0	1628.6	1738.7	1851.9
	s	1.5095	1.5437	1.5735	1.5845	1.5951	1.6054	1.6153	1.6250	1.6699	1.7108	1.7486	1.8177	1.8803	1.9381
500 (467.01)	v	0.9927	1.0800	1.1591	1.1893	1.2188	1.2478	1.2763	1.3044	1.4405	1.5715	1.6996	1.9504	2.197	2.442
	h	1231.3	1266.8	1298.6	1310.7	1322.6	1334.2	1345.7	1357.0	1412.1	1466.0	1519.6	1627.6	1737.9	1851.3
	s	1.4919	1.5280	1.5588	1.5701	1.5810	1.5915	1.6016	1.6115	1.6571	1.6982	1.7363	1.8056	1.8683	1.9262
550 (476.94)	v	0.8852	0.9686	1.0431	1.0714	1.0989	1.1259	1.1523	1.783	1.3038	1.4241	1.5414	1.7706	1.9957	2.219
	h	1223.7	1261.2	294.3	1306.8	1318.9	1330.8	1342.5	1354.0	1409.9	1464.3	1518.2	1626.6	1737.1	1850.6
	s	1.4751	1.5131	1.5451	1.5568	1.5680	1.5787	1.5890	1.5991	1.6452	1.6868	1.7250	1.7946	1.8570	1.9155
600 (486.21)	v	0.7947	0.8753	0.9463	0.9729	0.9988	1.0241	1.0489	1.0732	1.1899	1.3013	1.4196	1.6208	1.8279	2.033
	h	1215.7	1255.5	1289.9	1302.7	1315.2	1327.4	1339.3	1351.1	1407.7	1462.5	1516.7	1625.5	1736.3	1850.0
	s	1.4586	1.4990	1.5323	1.5443	1.5558	1.5667	1.5773	1.5875	1.6343	1.6762	1.7147	1.7846	1.8476	1.9056
700 (503.10)	v	......	0.7277	0.7934	0.8177	0.8411	0.8639	0.8860	0.9077	1.0108	1.1082	1.2024	1.3853	1.5641	1.7405
	h	......	1243.2	1280.6	1294.3	1307.5	1320.3	1332.8	1345.0	1403.2	1459.0	1513.9	1623.5	1734.8	1848.8
	s	......	1.4722	1.5084	1.5212	1.5333	1.5449	1.5559	1.5665	1.6147	1.6573	1.6963	1.7666	1.8299	1.8881
800 (518.23)	v	......	0.6154	0.6779	0.7006	0.7223	0.7433	0.7635	0.7833	0.8763	0.9633	1.0470	1.2088	1.3662	1.5214
	h	......	1229.8	1270.7	1285.4	1299.4	1312.9	1325.9	1338.6	1398.6	1455.4	1511.4	1621.4	1733.2	1847.5
	s	......	1.4467	1.4863	1.5000	1.5129	1.5250	1.5366	1.5476	1.5972	1.6407	1.6801	1.7510	1.8146	1.8729
900 (531.98)	v	......	0.5364	0.5873	0.6089	0.6294	0.6491	0.6680	0.6863	0.7716	0.8506	0.9262	1.0714	1.2124	1.3509
	h	......	1215.0	1260.1	1275.9	1290.9	1305.1	1318.8	1332.1	1393.9	1451.8	1508.1	1619.3	1731.6	1846.3
	s	......	1.4216	1.4653	1.4800	1.4938	1.5066	1.5187	1.5303	1.5814	1.6257	1.6656	1.7371	1.8009	1.8595
1000 (544.61)	v	......	0.4533	0.5140	0.5350	0.5546	0.5733	0.5912	0.6084	0.6878	0.7604	0.8294	0.9615	1.0893	1.2146
	h	......	1198.3	1248.8	1265.9	1281.9	1297.5	1311.4	1325.3	1389.2	1448.2	1505.1	1617.3	1730.0	1845.0
	s	......	1.3961	1.4450	1.4610	1.4757	1.4893	1.5021	1.5141	1.5670	1.6121	1.6525	1.7245	1.7885	1.8474
1100 (556.31)	v	......	......	0.4532	0.4738	0.4929	0.5100	0.5281	0.5445	0.6191	0.6866	0.7503	0.8716	0.9885	1.1031
	h	......	......	1236.7	1255.3	1272.4	1288.5	1303.7	1318.3	1384.4	1444.5	1502.2	1615.2	1728.4	1843.8
	s	......	......	1.4251	1.4425	1.4583	1.4728	1.4862	1.4989	1.5535	1.5995	1.6405	1.7130	1.7775	1.8363
1200 (567.22)	v	......	......	0.4016	0.4222	0.4410	0.4586	0.4752	0.4909	0.5617	0.6250	0.6843	0.7967	0.9046	1.0101
	h	......	......	1223.5	1243.9	1262.4	1279.6	1295.7	1311.0	1379.3	1440.7	1499.3	1613.1	1726.9	1842.5
	s	......	......	1.4052	1.4243	1.4413	1.4568	1.4710	1.4843	1.5409	1.5879	1.6293	1.7025	1.7672	1.8263

Abs. Press. Lb/Sq In. (Sat. Temp)															
1400 (587.10)	v	....	....	0.3174	0.3390	0.3580	0.3753	0.3912	0.4062	0.4714	0.5281	0.5805	0.6789	0.7727	0.8640
	h	....	....	1193.0	1218.4	1240.4	1260.3	1278.5	1295.5	1369.1	1433.1	1493.2	1608.9	1723.7	1840.0
	s	....	....	1.3639	1.3877	1.4079	1.4258	1.4419	1.4567	1.5177	1.5666	1.6093	1.6836	1.7489	1.8083
1600 (604.90)	v	....	....	....	0.2733	0.2936	0.3112	0.3271	0.3417	0.4034	0.4553	0.5027	0.5904	0.6738	0.7545
	h	....	....	....	1187.8	1215.2	1238.7	1259.6	1278.7	1358.4	1425.3	1487.0	1604.6	1720.5	1837.5
	s	....	....	....	1.3489	1.3741	1.3952	1.4137	1.4303	1.4964	1.5476	1.5914	1.6669	1.7328	1.7926
1800 (621.03)	v	....	....	....	....	0.2407	0.2597	0.2760	0.2907	0.3502	0.3986	0.4421	0.5218	0.5968	0.6693
	h	....	....	....	....	1185.1	1214.0	1238.5	1260.3	1347.2	1417.4	1480.8	1600.4	1717.3	1835.0
	s	....	....	....	....	1.3377	1.3638	1.3855	1.4044	1.4765	1.5301	1.5752	1.6520	1.7185	1.7786
2000 (635.82)	v	....	....	....	....	0.1936	0.2161	0.2337	0.2489	0.3074	0.3532	0.3935	0.4668	0.5352	0.6011
	h	....	....	....	....	1145.6	1184.9	1214.8	1240.0	1335.5	1409.2	1474.5	1596.1	1714.1	1832.5
	s	....	....	....	....	1.2945	1.3300	1.3564	1.3783	1.4576	1.5139	1.5603	1.6384	1.7055	1.7660
2500 (668.13)	v	....	....	....	....	....	....	0.1484	0.1686	0.2294	0.2710	0.3061	0.3678	0.4244	0.4784
	h	....	....	....	....	....	....	1132.3	1176.6	1303.6	1387.8	1458.4	1585.3	1706.1	1826.2
	s	....	....	....	....	....	....	1.2687	1.3073	1.4127	1.4772	1.5273	1.6088	1.6775	1.7389
3000 (695.36)	v	....	....	....	....	....	....	....	0.0984	0.1760	0.2159	0.2476	0.3018	0.3505	0.3966
	h	....	....	....	....	....	....	....	1060.1	1267.2	1365.0	1441.8	1574.3	1698.0	1819.9
	s	....	....	....	....	....	....	....	1.1966	1.3690	1.4439	1.4984	1.5837	1.6540	1.7163
3206.2 (705.40)	v	....	....	....	....	....	....	....	....	0.1583	0.1981	0.2288	0.2806	0.3267	0.3703
	h	....	....	....	....	....	....	....	....	1250.5	1355.2	1434.7	1569.8	1694.6	1817.2
	s	....	....	....	....	....	....	....	....	1.3508	1.4309	1.4874	1.5742	1.6452	1.7080
3500	v	....	....	....	....	....	....	....	0.0306	0.1364	0.1762	0.2058	0.2546	0.2977	0.3381
	h	....	....	....	....	....	....	....	780.5	1224.9	1340.7	1424.5	1563.3	1689.8	1813.6
	s	....	....	....	....	....	....	....	0.9515	1.3241	1.4127	1.4723	1.5615	1.6336	1.6968
4000	v	....	....	....	....	....	....	....	0.0287	0.1052	0.1462	0.1743	0.2192	0.2581	0.2943
	h	....	....	....	....	....	....	....	763.8	1174.8	1314.4	1406.8	1552.1	1681.7	1807.2
	s	....	....	....	....	....	....	....	0.9347	1.2757	1.3827	1.4482	1.5417	1.6154	1.6795
4500	v	....	....	....	....	....	....	....	0.0276	0.0798	0.1226	0.1500	0.1917	0.2273	0.2602
	h	....	....	....	....	....	....	....	753.5	1113.9	1286.5	1388.4	1540.8	1673.5	1800.9
	s	....	....	....	....	....	....	....	0.9235	1.2204	1.3529	1.4253	1.5235	1.5990	1.6640
5000	v	....	....	....	....	....	....	....	0.0268	0.0593	0.1036	0.1303	0.1696	0.2027	0.2329
	h	....	....	....	....	....	....	....	746.4	1047.1	1256.5	1369.5	1529.5	1665.3	1794.5
	s	....	....	....	....	....	....	....	0.9152	1.1622	1.3231	1.4034	1.5066	1.5839	1.6499
5500	v	....	....	....	....	....	....	....	0.0262	0.0463	0.0880	0.1143	0.1516	0.1825	0.2106
	h	....	....	....	....	....	....	....	741.3	985.0	1224.1	1349.2	1518.2	1657.0	1788.1
	s	....	....	....	....	....	....	....	0.9090	1.1093	1.2930	1.3821	1.4908	1.5699	1.6369

[1] Abridged from "Thermodynamic Properties of Steam," by Joseph H. Keenan and Frederick G. Keyes. Copyright, 1937, by Joseph H. Keenan and Frederick G. Keyes. Published by John Wiley & Sons, Inc., New York. Temperature in deg F; specific volume v in cu ft/lb; enthalpy h in Btu/lb; entropy s in Btu/(lb F).

The actual enthalpy h_2 is determined by the use of definition of nozzle efficiency, Eq. (49):

$$h_2 = h_1 - e_n(h_1 - h_2') = 1420.6 - 0.95(1420.6 - 1286)$$
$$h_2 = 1292.8 \text{ Btu/lbm}$$

By linear interpolation in Table 30, at an enthalpy of 1292.8 Btu/lbm and a pressure of 100 psia, the specific volume is found to be

$$v_2 = 5.73 \text{ cu ft/lbm}$$

The exit area is determined by use of the continuity equation (21):

$$A_2 = \dot{m}/\rho_2 V_2 = \dot{m}v_2/V_2$$
$$A_2 = 30,000/3,600 \times 5.73/2,530 = 0.0188 \text{ sq ft} = 2.72 \text{ sq in.}$$

In order to determine the required throat area, the critical pressure must be obtained. Since the expansion is entirely within the superheat region, the critical pressure ratio will be about 0.55. Then

$$P^* = 0.55P_1 = 0.55(300) = 165 \text{ psia}$$

By Fig. 27, for expansion to $P^* = 165$ psia, there is obtained

$$h^{*'} = 1345 \text{ Btu/lbm}$$

If the nozzle efficiency of 0.95 applies to the section between inlet and throat as well as to the entire nozzle, the throat enthalpy can be calculated as

$$h^* = 1420.6 - 0.95(1420.6 - 1345) = 1348.9 \text{ Btu/lbm}$$

The specific volume at $P^* = 165$ psia, $h^* = 1348.9$ Btu/lbm is obtained by interpolation in Table 30 as

$$v^* = 3.96 \text{ cu ft/lbm}$$
$$V^* = 223.7\sqrt{h_1 - h^*} = 1,895 \text{ fps}$$

Then $$A^* = \dot{m}v^*/V^* = \frac{30,000}{3,600} \times \frac{3.96}{1,895} \times 144 = 2.51 \text{ sq in.}$$

Alternatively, A^* can be calculated from Eq. (52):

$$A^* = \frac{\dot{m}}{0.3155}\sqrt{\frac{v_1}{P_1}} = \frac{30,000}{3,600 \times 0.3155}\sqrt{\frac{2.442}{300}} = 2.38 \text{ sq in.}$$

In view of inaccuracies associated with reading values from the abbreviated Mollier diagram and with interpolation in the superheat tables, this is a close check.

The inlet area is found easily, as follows:

$$A_1 = \frac{\dot{m}}{\rho_1 V_1} = \frac{\dot{m}v_1}{V_1} = \frac{30,000 \times 2.442}{3,600 \times 100} \times 144 = 29.3 \text{ sq in.}$$

Obviously, this area is excessive and the final nozzle design should be altered. The most practical solution would involve raising the initial velocity to, say, 300 fps. If this were done, the throat and exit areas would be different from those calculated above.

Example 2. Steam with an initial pressure and quality of 200 psia and 95 per cent, respectively, is to be expanded to a pressure of 150 psia at the rate of 30,000 lb/hr. If the nozzle efficiency is 92 per cent, determine the required exit area.

Solution. From Table 29, at 200 psia

$$h_f = 355.36 \text{ Btu/lb}$$
$$h_{fg} = 843.0 \text{ Btu/lb}$$
$$s_f = 0.5435 \text{ Btu/lb}$$
$$s_{fg} = 1.0018 \text{ Btu/lb}$$

Then $h_1 = h_f + x_1 h_{fg} = 355.36 + 0.95(843) = 1156 \text{ Btu/lb}$
$s_1 = s_f + x_1 + x_1 s_{fg} = 0.5435 + 0.95(1.0018) = 1.4937 \text{ Btu/lb F}$

If expansion to 150 psia were isentropic, then $s_2 = s_1$

$$s_2 = 1.4937 = s_{f2} + x_2's_{fg2} = 0.5138 + 1.0556x_2'$$

and
$$x_2' = 0.9799/1.0556 = 0.928$$

Then
$$h_2' = h_{f2} + x_2'h_{fg} = 330.51 + 0.928(863.6) = 1131 \text{ Btu/lb}$$

and from Eq. (51),
$$v_2 = 223.7\sqrt{e_n(h_1 - h_2')} = 1{,}070 \text{ fps}$$

From Eq. (49),
$$h_2 = h_1 - e_n(h_1 - h_2') = 1133 \text{ Btu/lb}$$

Then, since $h_2 = h_{f2} + x_2h_{fg2}$, there is obtained

$$x_2 = (h_2 - h_{f2})/h_{fg2} = (1133 - 330.51)/863.6 = 0.929$$

From Eq. (62) and Table 29
$$v_2 = v_f + x_2v_{fg} = 2.795 \text{ cu ft/lb}$$

Then, by the equation of continuity,

$$A_2 = \dot{m}/\rho_2V_2 = \dot{m}v_2/V_2 = \frac{30{,}000 \times 2.795}{3{,}600 \times 1{,}070} = 0.0217 \text{ ft}^2$$

or
$$A_2 = 3.13 \text{ sq in.}$$

If the unmodified Napier equation (55) has been used, the exit area would be

$$a_2 = \frac{70w}{p_1} = \frac{70 \times 30{,}000}{3{,}600 \times 200} = 2.91 \text{ sq in.}$$

If the modified Napier equation were used, the pressure ratio P_2/P_1 would be $150/200$ or 0.75. By interpolation, the correction factor K is found to be 0.912. There

$$a_2 = 70\dot{m}/KP_1 = 2.91/0.912 = 3.19 \text{ sq in.}$$

If the modified Grashof equation (54) were used, the exit area A_2 would be

$$A_2 = \frac{\dot{m}}{0.01654KP_1^{0.9696}} = \frac{30{,}000}{0.01654 \times 3{,}600 \times 0.912 \times 200^{0.9696}}$$

or
$$A_2 = 3.25 \text{ sq in.}$$

Viscosity of Steam. The viscosity of steam increases with temperature and pressure. Table 31, reproduced from the Keenan and Keyes "Thermodynamic Properties of Steam," lists the viscosity of saturated and superheated vapor at certain discrete points. It is noted that the dimensions of viscosity in that table

Table 31. Viscosity of Steam

(Tabular values multiplied by 10^{-7} yield viscosity in lbf sec/sq ft)

Pressure, psia	Saturated vapor	Temperature, F					
		200	400	600	800	1000	1200
0	....	2.59	3.49	4.35	5.19	5.99	6.76
500	5.90	...	...	6.47	7.30	8.10	8.87
1,000	8.17	...	...	8.41	9.24	10.0	10.8
1,500	10.2	...	...	10.2	11.0	11.8	12.6
2,000	11.9	...	...	....	12.6	13.4	14.2
2,500	13.5	...	...	....	14.0	14.8	15.6
3,000	14.8	...	...	....	15.2	16.0	16.8
3,500	....	...	...	....	16.3	17.1	17.9

are in lbf sec/sq ft.   To express viscosity in other units, the following is useful:

Viscosity in lbf sec/sq ft $\times$ 32.17 ft lbm/(lbf sec^2) = viscosity in lbm/(ft sec)
Viscosity in centipoises $\times$ 0.672 $\times$ 10^{-3} = viscosity in lbm/(ft sec)
Viscosity in centipoises $\times$ 20.88 $\times$ 10^{-6} = viscosity in lbf sec/sq ft

Properties of Gases

As used in this section, the term *gases* refers to the multitude of compressible fluids other than steam.   In many cases, particularly those which involve processes at low pressures and relatively high temperatures, the compressible fluid may be treated as a *perfect gas* or as a *semi perfect gas*.

A *perfect* gas is one which satisfies two criteria.   First, it obeys the equation of state

$$Pv = (\bar{R}/M_w)T$$

in which P = pressure
$\quad v$ = volume
$\quad \bar{R}$ = universal gas constant
$\quad M_w$ = molecular weight
$\quad T$ = temperature

Second, a perfect gas has invariant constant-pressure and constant-volume specific heats.   It follows, then, that the specific-heat ratio $k = c_p/c_v$ is constant in a perfect gas.

A *semiperfect* gas obeys the equation of state, but the specific heats are functions of temperature only.   That is, the specific heat of a semiperfect gas may be presented in the general form as

$$c = A + BT^n + CT^m + \cdots$$

where A, B, C, n, and m are constants and T is the temperature.

The universal gas constant $\bar{R}$ is taken as being equal to 1.986 Btu/(lb mol R) which is equivalent to 1545.3 ft lb/(lb mole R).   Table 32 compares values for common gases of the gas constant R as obtained by dividing the molecular weight into the universal gas constant with those obtained by measurements of simultaneous values of pressure, volume, and temperature.   It will be noted that deviation of

Table 32.   Gas Constants and Specific Heats

Gas	Mol. wt, lb/mole, M_w	Sp ht, Btu/(lbm R) at 1 atm, 70 F			Gas constant, R, ft lbf/(lbm R)	
		c_p	c_v	k	$\dfrac{\bar{R}}{M_w} = R$ at zero press.	$Pv/T = R$ 1 atm 32 F
Air	28.97	0.240	0.171	1.40	53.35	53.34
Argon (A)	39.94	0.123	0.074	1.67	38.68	38.65
Helium (He)	4.003	1.25	0.75	1.66	386.2	386.3
Carbon monoxide (CO) .	28.01	0.249	0.178	1.40	55.18	55.13
Carbon dioxide (CO$_2$) ...	44.01	0.202	0.156	1.30	35.12	34.88
Hydrogen (H$_2$)	2.016	3.42	2.43	1.41	766.6	767.0
Nitrogen (N$_2$)	28.02	0.248	0.177	0.40	55.16	55.13
Oxygen (O$_2$)	32.00	0.219	0.156	1.40	48.29	48.24
Sulfur dioxide (SO$_2$).....	64.07	0.154	0.122	1.26	24.12	23.55
Acetylene (C$_2$H$_2$)	26.04	0.383	0.303	1.26	59.35	58.77
Methane (CH$_4$).........	16.04	0.532	0.403	1.32	96.35	96.07
Ethane (C$_2$H$_6$)..........	30.07	0.409	0.342	1.22	51.40	50.82
Iso-butane	58.12	0.398	0.358	1.11	26.59	25.79

experimental from analytical values is small and within the range of ordinary engineering accuracy.

The specific heats listed in Table 32 are measured at 1 atm and at room temperature. The effect of temperature on constant-pressure specific heat as measured at a pressure of 1 atm is given in Table 33. Note that values listed are in Btu/(lb mole R);

Table 33. Specific Heats of Gases at 1 Atm

Gas	Symbol	Equation for C_p, Btu/mole R	Temp range, R	Source
Oxygen.............	O_2	$11.515 - \left(\dfrac{172}{\sqrt{T}}\right) + \left(\dfrac{1530}{T}\right)$	540–4000	a
		$11.515 - \left(\dfrac{172}{\sqrt{T}}\right) + \left(\dfrac{1530}{T}\right)$ $+ \left(\dfrac{0.05(T - 4000)}{1000}\right)$	4000–9000	a
Nitrogen	N_2	$9.47 - \left(\dfrac{3.47 \times 10^3}{T}\right)$ $+ \left(\dfrac{1.16 \times 10^6}{T^2}\right)$	540–5000	a
Carbon monoxide......	CO	$9.46 - \left(\dfrac{3.96 \times 10^3}{T}\right)$ $+ \left(\dfrac{1.07 \times 10^6}{T^2}\right)$	540–5000	a
Hydrogen	H_2	$5.76 + \left(\dfrac{0.578T}{1000}\right) + \left(\dfrac{20}{\sqrt{T}}\right)$	540–4000	a
		$5.76 + \left(\dfrac{0.578T}{1000}\right) + \left(\dfrac{20}{\sqrt{T}}\right)$ $- \left(\dfrac{0.33(T - 4000)}{1000}\right)$	4000–9000	a
Water.............	H_2O	$19.86 - \left(\dfrac{597}{\sqrt{T}}\right) + \left(\dfrac{7500}{T}\right)$	540–5000	a
Carbon dioxide........	CO_2	$16.2 - \left(\dfrac{6.53 \times 10^3}{T}\right)$ $+ \left(\dfrac{1.41 \times 10^6}{T^2}\right)$	540–6300	a
Methane.............	CH_4	$4.22 + 8.211 \times 10^{-3}T$	492–1800	b
		$27.0 - \dfrac{14.400}{T}$	1800–5940	b
Ethylene.............	C_2H_4	$6.0 + 8.33 \times 10^{-3}T$	720–1440	c
Ethane	C_2H_6	$6.6 + 13.33 \times 10^{-3}T$	720–1440	c
Ethyl alcohol.........	C_2H_6O	$4.5 + 21.1 \times 10^{-3}T$	680–1120	c
Methyl alcohol	CH_4O	$2.0 + 16.67 \times 10^{-3}T$	680–1100	c
Benzene	C_6H_6	$6.5 + 28.9 \times 10^{-3}T$	520–1120	c
Octane	C_8H_{18}	$14.4 + 53.3 \times 10^{-3}T$	720–1440	c
Dodecane	$C_{12}H_{26}$	$19.6 + 80.0 \times 10^{-3}T$	720–1440	c

[a] Sweigert and Beardsley, "Empirical Specific Heat Equations Based upon Spectroscopic Data," *Ga. School Tech., State Eng. Expt. Sta. Bull.* 2, 1938.
[b] Schwarz, "Die Spezifischen Wärmen der Gase als Hilfswerte zur Beruchung von Gleichgewichten," *Arch. Eisenhüttenw.*, vol. 9, p. 389, 1936.
[c] Parks and Huffman, *Am. Chem. Soc., Mon.* 60. 1932.

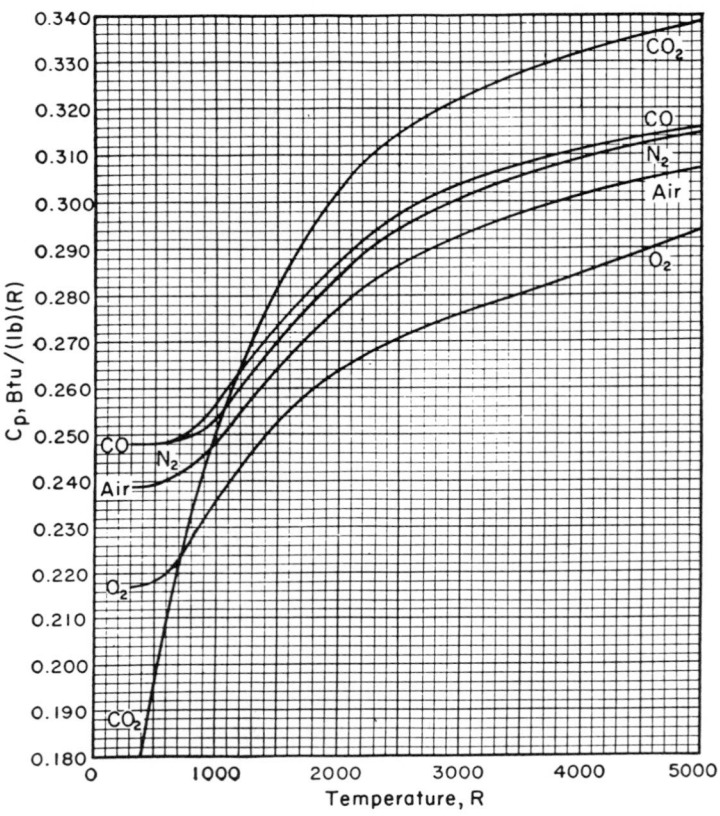

FIG. 28. Constant-pressure specific heats of certain gases at low pressure. ("*Thermodynamic Properties of Steam*" *by Keenan and Keyes.*)

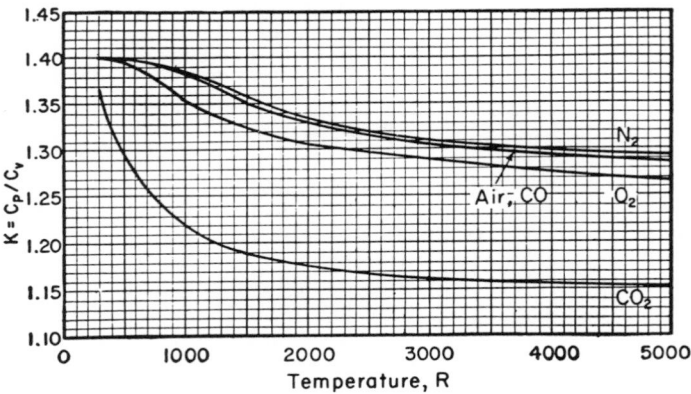

FIG. 29. Specific-heat ratios for certain gases at low pressure. ("*Thermodynamic Properties of Steam*" *by Keenan and Keyes.*)

to convert to Btu/(lbm R) it is necessary first to compute the specific heat on a molar basis and then to divide by the molecular weight. For example, let it be required to determine the zero-pressure specific heat at constant pressure for oxygen at a temperature of 400 F. The following is the procedure:

$$T = 400 + 460 = 860 \text{ R}$$

From Table 33,

$$C_{po} = 11.515 - 172/\sqrt{T} + 1530/T$$

or $\qquad C_{po} = 11.515 - 172/\sqrt{860} + 1{,}530/860 = 7.44 \text{ Btu/(mole R)}$

If this value is compared with the room-temperature constant-pressure specific heat listed in Table 32, it will be seen that specific heat apparently increases with temperature. This trend actually does occur as shown in the curves of Fig. 28.

The values of specific heat in Table 33 and in Fig. 28 are instantaneous values. In order to determine the average or mean value of specific heat over a range in temperature, it is necessary to integrate mathematically in Table 33 or graphically in Fig. 28.

Example. Determine the heat transfer necessary to change the temperature of 10 lb of oxygen from 100 to 400 F.

Solution:

$$Q = n(h_2 - h_1) = nc_{po}(T_2 - T_1)$$

$$Q = n\int_{T_1}^{T_2} c_{po}\, dT = {}^{10}\!\!/_{32} \int (11.515 - 172^{-1/2} + 1{,}530 T^{-1})\, dT$$

$$Q = {}^{10}\!\!/_{32}\,[11.515 T - 172(2)T^{1/2} + 1{,}530 \ln T]_{T_1}^{T_2}$$

$$Q = {}^{10}\!\!/_{32}\,[11.515(860 - 560) - 344(\sqrt{860} - \sqrt{560}) + 1{,}530 \ln (860/530)$$

$$= 721 \text{ Btu}$$

If, in the above, oxygen had been regarded as a perfect gas with a specific heat of 0.219 Btu/(lb R) as listed in Table 32, there would have been obtained

$$Q = 10 \times 0.219 \times 300 = 657 \text{ Btu}$$

and the error would have been about 9 per cent.

Example. What is the average or mean value of specific heat over the temperature range of the above example?

Solution:

$$\overline{C_p} = \int_{T_1}^{T_2} C_p\, dT \Big/ (T_2 - T_1) = 2{,}310/300 = 7.7 \text{ Btu/(mole R)}$$

or $\qquad \overline{C_p} = 7.7/32 = 0.24 \text{ Btu/(lb R)}$

Figure 29 shows the variation in specific-heat ratio k with temperature. The curves shown are based on test data conducted at a pressure of 1 atm. It will be noted that, in general, specific-heat ratio decreases with temperature. This is due to the fact that the constant-volume specific heat C_v increases with temperature at a rate greater than that of C_p.

Viscosities for some of the common gases are listed in Table 34. The dimensions given are lbm/(ft sec); to convert to centipoises, divide the tabular values by 0.672×10^{-3}.

Common Fuel and Illuminating Gases. The name "illuminating gas," used at one time to designate the common commercial product, is descriptive of the use for which gas was originally intended. For a great many years after the first commercial production of illuminating gas, it was used for no other purpose than that which its name implies. From an economy standpoint it was out of the

question to consider it as a fuel at that time. Manufacturing methods were crude and expensive, necessitating a high price for gas, while wood, the prevailing fuel, was plentiful and cheap. The flat-flame burner, which was the only gas lamp known at that time, produced light through combustion of the illuminating constituents of the gas. Then came the incandescent gas lamp of the mantle type, of which the Welsbach burner is an example. The incandescent lamp consumes gas in the same manner as a range or water-heater burner, depending on the energy released in a bunsen-type burner to heat a refractory mantle to incandescence. In a bunsen (or atmospheric burner) a certain proportion of air is drawn in and mixed with the gas before combustion to produce a blue, nonluminous flame. Such a

Table 34. Viscosities of Common Gases

[Tabular values $\times$ 10^{-5} yield viscosity in lbm/(ft sec)]

Temp, F	Oxygen	Nitrogen	Hydrogen	Helium	Carbon monoxide	Carbon dioxide
0	1.215	1.055	0.540	1.140	1.065	0.88
100	0.420	1.222	0.620	....	....	1.05
200	1.610	1.380	0.692	1.480	1.390	1.22
400	1.955	1.660	....	1.780	1.670	
600	2.260	1.915	....	2.020	1.910	
800	2.530	2.145	....	2.285	2.134	
1000	2.780	2.355	1.160	2.520	2.336	2.30
1500	3.320	2.800	1.415	3.160	2.783	2.86
2000	....	....	1.640	....	....	3.30
3000	....	....	1.720	....	....	3.92

flame contains none of the incandescent carbon particles which furnish the greater part of the light produced by a luminous flame. In a bunsen flame the hydrocarbons are burned before they can decompose, owing to the intermingling of air and gas before combustion sets in, and as a consequence, no luminous carbon particles are present. Such a burner will not smoke or deposit soot upon the mantle of an incandescent lamp or a cooking vessel, as would happen if a luminous flame were used, and combustion is more likely to be complete. The amount of heat produced by the consumption of equal amounts of gas is practically the same whether the gas be burned in a luminous or bunsen burner, provided complete combustion is obtained in both cases.

The *illuminating power* of a gas comes from the hydrocarbons, called "illuminants," contained in the gas. The luminosity of a flame depends on the presence of solid particles, the density of the burning gas, and its temperature. The solid particles are of very finely divided carbon and are heated to incandescence before they meet sufficient air for their combustion. These carbon particles are produced by the decomposition of hydrocarbons in the gas while in the center of the flame and out of contact with air, but subject to the temperature of combustion on the surface of the flame. In general the more dense the gases undergoing combustion, the more luminous they are when heated to high temperatures, and anything which tends to raise the temperature of the flame increases the luminosity, other things being equal. Smoke and soot are produced by the escape of unburned particles of carbon. The standard of illuminating value in general use is the sperm candle of a size running six to the pound and burning at the rate of 120 grains an hour. The illuminating power of a gas is expressed by stating the number of times a flame burning at the rate of 5 cu ft of gas per hour is greater than the standard candle.

The *heating value* of a gas is expressed in Btu per cubic foot. In the case of hydrogen and substances containing hydrogen, a distinction must be drawn between

gross or total heating value and net or available heating value. The complete combustion of such substances produces water which must be evaporated into steam at the expense of a part of the heat produced in combustion. It is customary in all ordinary work to express quantities of heat in terms of gross value, unless otherwise stated. The higher heating value is determined by cooling the products of combustion to the original temperature of air and gas. Since the amount of heat rejected by the water vapor depends on the temperature to which the cooling is carried, it is possible that various interpretations of the lower heating value may be made. Lucke and Flather state ("Text Book of Engineering Thermodynamics") that it is sufficiently close for engineering work to accept, as the difference between the high and low heat values, 970.4 times the weight of water vapor formed per fuel unit burned.[1]

NOTE: The latent heat of water vapor at atmospheric pressure is 970.4 Btu/lb, and each pound of hydrogen burned will form 9 lb of water vapor; hence

High heat value − low heat value = 970.4 × (weight of water vapor formed)

= 9 × 970.4 × (weight of hydrogen per fuel unit)

See also the method for computing moisture loss given in connection with Table 35.

Standard Conditions. The standard cubic foot of artificial gas is measured at a temperature of 60 F when under an absolute pressure of 30 in. of mercury column at 32 F which is equivalent to 30.028 in., or 14.73 psia, at 60 F. The gas is considered as saturated with water vapor at 60 F.

The *density* of a gas is usually expressed as the specific gravity relative to air under the same conditions of temperature and pressure. The specific gravities relative to air of the more common individual gases are given in Table 22, while those for the commercial heating and illuminating gases are given in Table 35. Specific gravities relative to water at 39 F of some of the common individual gases are shown in Fig. 30 over a considerable temperature range.

Coal gas is produced by the destructive distillation of bituminous coal, at high temperature, in a retort or oven externally heated. Coal gas is usually the most profitably manufactured of the artificial gases, owing to the credit obtained from the sale of by-products and residual coke. Nine to ten thousand cubic feet of gas are obtained per ton of coal distilled. Owing to municipal or state regulation of heating value and candlepower, it is generally necessary to employ some method of enrichment whereby these qualities are increased. The most common form of enrichment is by the addition of a portion of carbureted water gas which brings the mixed gas to the proper quality to meet these requirements. Two typical percentage compositions of coal gas by volume are:

Carbon dioxide (CO_2)	1.60	2.71[1]
Oxygen (O_2)	0.39	0.78
Benzine vapor (C_6H_6)	0.50	
Heavy hydrocarbons (C_2H_4)	4.25	4.51 illuminants
Carbon monoxide (CO)	8.04	8.38
Hydrogen (H_2)	47.04	47.81
Methane (CH_4)	36.02	32.00
Nitrogen	2.16	3.81
Total	100.00	100.00

[1] From *Bur. Mines Bull.* 6.

The average heating value runs from 500 to 600 Btu/cu ft, and the illuminating value about 15 candles. Owing to its large percentage of hydrogen, coal gas has a low specific gravity of approximately 0.45 (air = 1).

[1] For more accurate work, Table 10 of "Gas Tables" by Keenan and Keyes (John Wiley) contains a listing of heating values for various hydrocarbon fuels.

Table 35. Typical Compositions and Properties of Fuel and Illuminating Gases

	Percentage composition by volume										Density,[2] lb per cu ft	Specific gravity (air = 1)	Specific heat at constant pressure, Btu per lb per F	Specific heat at constant pressure, Btu per cu ft per F[2]	High-heat value,[3] Btu per lb	High-heat value,[3] Btu per cu ft[3]
	H_2	CH_4	C_2H_6	Illuminants,[1] C_nH_{2n}	C_3H_8	C_4H_{10}	CO	CO_2	N_2	O_2						
Specific heat in Btu per lb per degree F at constant pressure	3.42	0.593	0.400	0.400	0.390	0.390	0.243	0.210	0.247	0.217						
Oil gas	32.0	48.0		16.5					3.0	0.5	0.03666	0.48	0.63	0.023	23,077	846
Refinery gas: Dubbs	5.0	33.0	15.0	33.0	14.0						0.07332	0.96	0.44	0.040	20,458	1,500
Refinery gas: Houdrie		28.6	17.5	0.6	33.4	19.9					0.11533	1.51	0.41	0.047	17,342	2,000
Natural gas		87.0	9.6					0.4	3.0		0.04659	0.61	0.54	0.025	23,932	1,115
Natural gas		74.0	14.5					0.1	11.2	0.2	0.05230	0.685	0.38	0.020	19,311	1,010
Natural gas		79.4			20.0				0.6		0.08604	1.13	0.40	0.034	21,513	1,851
Pintsch gas	12.6	45.4		35.7			0.6	0.7	3.0	2.0	0.06416	0.84	0.47	0.031	23,379	1,500
Blast-furnace gas	2.0	2.0					26.0	12.0	58.0		0.07638	1.00	0.25	0.019	1,309	100
Producer gas, coal	12.0	2.6		0.4			29.0	4.0	52.0		0.06645	0.87	0.28	0.019	2,453	163
Coke-oven gas	50.0	36.0		4.0			6.0	1.5	2.0	0.5	0.02902	0.38	0.73	0.021	20,779	603
Coal-retort gas	52.5	31.4		2.2			8.6	1.5	3.5	0.3	0.03208	0.42	0.70	0.021	17,924	575
Blue water gas	51.3						43.4	3.5	1.3	0.5	0.04048	0.53	0.46	0.018	7,411	300
Carbureted water gas	35.6	16.5		14.6			28.0	3.8	1.0	0.5	0.04888	0.64	0.46	0.023	11,457	560

[1] Includes ethylene C_2H_4, propylene C_3H_6, etc.

[2] For gas measured under standard conditions, see p.3-85 "Piping Handbook."

[3] Owing to the varying hydrogen content of different types of gas the low-heat value will vary considerably, depending also on the flue gas temperature. Correction for loss due to water vapor can be made by the following formula (see ASME Power Test Code for Stationary Steam Generating Units).

Heat loss in Btu per fuel unit $= 9H_2(1{,}066 + 0.5t_g - t_a)$, when t_g is more than 550 F

$$= 9H_2(1{,}089 + 0.46t_g - t_a), \text{ when } t_g \text{ is less than 550 F}$$

where H_2 = weight of hydrogen in the fuel unit (lb H_2 per lb of gas, or lb H_2 per cu ft of gas as the case may be).

t_g = temperature of the flue gas, degrees F.

t_a = temperature of the atmosphere, degrees F.

For low heat values of various gases at different stack temperatures see Table 38.

Water gas is produced by the decomposition of steam by contact and union with carbon at a high temperature. *Carbureted water gas* is made from water gas by adding to and mixing with a certain proportion of oil gases, the richness of the carbureted gas depending on the proportions in which the water and oil gases are mixed. In this way a wide range of luminosity and heating values may be obtained, which make it a simple and flexible means of meeting quality requirements. Water gas is usually made from gas coke, although anthracite also can be used for this

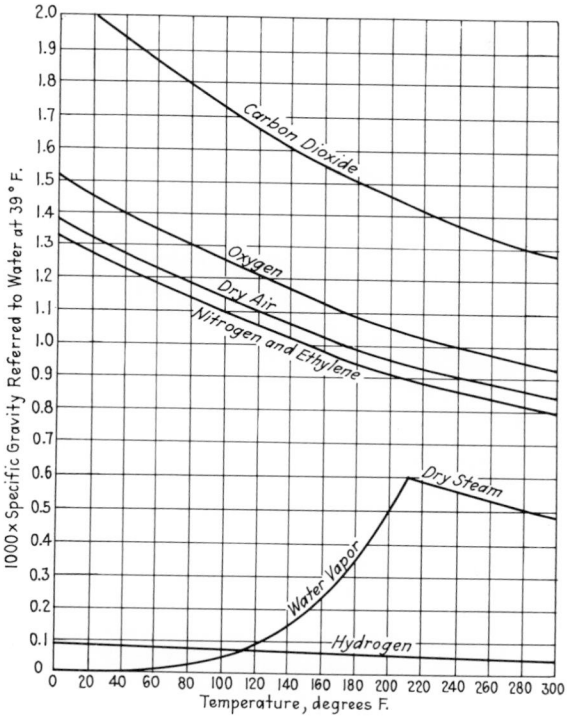

FIG. 30

purpose. Good operating results should show a fuel consumption of 30 to 35 lb of coke per 1,000 cu ft of gas made. Uncarbureted water gas has a calorific value of only about 300 Btu/cu ft and no illuminating value. It may be used for lighting, however, by causing it to heat to incandescence some solid substance, as a Welsbach or other incandescent mantle. An illuminating gas having a higher calorific value is made from water gas by adding to it hydrocarbon gases or vapors which are obtained from crude petroleum or its distillate known as "gas oil." The gas so made should be called "carbureted water gas," but it is ordinarily known in the United States simply as "water gas." Gas oil is most commonly used for this purpose. The results obtained from an oil depend primarily upon its specific gravity and composition, but an average production of 1,000 cu ft from 3 gal of oil is considered good practice. The illuminating value of carbureted water gas may run as high as 20 candles, and the heating value 640 Btu cu ft, although in actual practice these values are reduced to meet the local requirements. It is common

Table 36. Properties of Hydrocarbons Found in Natural Gas and Casinghead Gas

	Methane	Ethane	Propane	Butane	Pentane	Hexane	Heptane	Octane
Formula	CH_4	C_2H_6	C_3H_8	C_4H_{10}	C_5H_{12}	C_6H_{14}	C_7H_{16}	C_8H_{18}
Molecular weight	16.03	30.05	44.07	58.08	72.10	86.12	100.13	114.15
Specific gravity of liquid		0.432 = 194° Bé	0.515 = 142° Bé	0.585 = 109° Bé	0.630 = 92.2° Bé	0.670 = 78.9° Bé	0.697 = 70.9° Bé	0.718 = 65.0° Bé
Specific gravity of gas	0.555	1.049	1.526	2.008	2.496	2.982	3.467	3.952
Boiling point at atmospheric pressure	−165 C / −265 F	−93 C / −135 F	−45 C / −49 F	+1 C / 34 F	36.3 C / 97 F	69 C / 156 F	98.4 C / 200 F	125.5°C / 258 F
Pressure to liquefy at 60 F, psi	100+	475	105	35	6.5	1.8	0.5	0.15
Vapor pressure 70 F in percentage of atmosphere		100+	100+	100+	55	10	2.7	0.7
Gallons per 1,000 cu ft at B.P. reduced to 60 F		22.13	27.01	31.28	36.13	40.56	45.34	50.11
Weight 1,000 cu ft vapor at B.P. reduced to 60 F, lb.	42	79.7	116	152.6	189.7	226.6	263.5	300
Shrinkage in volume by 1 gal liquid removed per 1,000 cu ft.					2.8 per cent	2.5 per cent	2.2 per cent	2.0 per cent
Maximum possible removable gal. per 1,000 cu ft at 70 F gal.					19.87	4.06	1.22	0.35
High heat value, Btu per cu ft	1,065	1,861	2,685	3,447	4,250	5,012	5,780	6,542
Btu per lb.	25,360	23,350	23,150	22,590	22,400	22,120	21,935	21,807
Cubic feet air to burn 1 cu ft gas.	9.57	16.72	23.92	31.10	38.28	46.46	53.6	60.8
Carbon percentage	75.0	80.0	81.8	82.8	83.3	83.7	84.0	84.2
Explosive mixture percentage in air, Maximum	14.5	5.0	3.5	3.0	2.5	2.2	1.9	1.6
Minimum	5.6	3.0	2.1	1.6	1.3			
Specific heats Btu per lb per °F at 760 mm and 60 F. At constant pressure, C_p	0.530	0.415	0.376	0.357	0.347	0.339	0.335	0.330
At constant volume, C_v	0.405	0.346	0.324	0.312	0.309	0.305	0.302	0.300
Ratio $C_p/C_v = k$ (adiabatic exponent)	1.309	1.198	1.161	1.144	1.121	1.113	1.109	1.100

[1] For computation of low heat values, see footnote 3, Table 35.

practice in many plants to make both coal and water gas and mix them in the proper proportions to meet the local requirements. A typical composition of carbureted water gas by volume is given in Table 35.

Natural Gas. Pure natural gas is odorless and colorless, burns with a luminous flame, and is highly explosive when mixed with air. Its chief constituent is marsh gas, or methane, a member of the paraffin series. Natural gas is classified as either "wet" or "dry," according to its content of gasoline. Wet gas contains not only ethane, propane, butane, and pentane, the lighter members of the methane series which predominate in the dry gas, but also some heavier hydrocarbons. Dry gas contains chiefly methane, the lightest known hydrocarbon, which has a specific gravity of 0.559. Natural gas usually occurs in connection with petroleum, but it is also found in places far removed from oil fields. A considerable amount of casinghead gasoline is obtained from natural gas by compression, refrigeration, or absorption methods. The composition of natural gas varies considerably in different fields. Typical compositions by volumes of the natural gas used in eight cities of the United States are as follows:

City	Methane, percentage CH$_4$	Ethane, percentage C$_2$H$_6$	Nitrogen, percentage N$_2$
Pittsburgh, Pa.	79.2	19.6	1.2
Louisville, Ky.	77.8	20.4	1.8
Buffalo, N. Y.	79.9	15.2	4.9
Cincinnati, Ohio.	79.8	19.5	0.7
Cleveland, Ohio.	80.5	18.2	1.3
Springfield, Ohio.	80.3	14.7	5.0
Columbus, Ohio.	80.4	18.1	1.5
Chelsa, Okla.	75.4	17.7	6.6
Detroit, Mich.	74.0	14.5	11.2

These analyses were made by the ordinary combustion method, and, hence, show only the two predominating paraffin hydrocarbons.

The usual high heat value is from 800 to 1000 Btu/cu ft, (see Table 35), and the specific gravity (relative to air) 0.60 to 0.70. In exceptional cases the specific gravity may run as high as 0.8 or 1.0 and the heating value up to 1500 Btu cu ft. Such gases are usually rich in casinghead gasoline. Properties of the constituents of natural gas are given in Table 36. No generally accepted set of standard conditions has been adopted in the natural-gas industry, although temperature of 60 F and pressure 1/4 lb or 4 oz above an assumed mean atmospheric pressure of 14.4 psi is used to a considerable extent. This is equivalent to 520 F abs and 14.65 psia.

References for the properties, production, and distribution of natural gas will be found in "Measurement, Compression, and Transmission of Natural Gas," by Lester Clyde Lichty, John Wiley & Sons, Inc., New York; "Problems in Natural Gas Engineering," by Thomas R. Weymouth, *Trans. ASME*, Vol. 34, 1912; "Gas Engineers' Handbook" of the Pacific Coast Gas Association, McGraw-Hill Book Company, New York; "American Gas Practice, Vol. 2, Distribution and Utilization of City Gas," by Jerome J. Morgan, published by Jerome J. Morgan, Maplewood, N.J. Considerable interesting data regarding the amount of natural gas required to accomplish different tasks are given in Table 37 which is reproduced, by permission, from *Bull.* 25 of the Kansas City Testing Laboratory.

Oil Gas. This is the least used of the artificial gases under consideration. Strictly speaking, it is the gas resulting from the destructive distillation of oil (liquid hydrocarbons). Pintsch gas, which is used in coach lighting, and Blau gas (liquid), which is used for domestic purposes, are the principal kinds of pure-oil

Table 37.   About Natural Gas and Its Usefulness

(An average sample of natural gas has 950 Btu/cu ft)

1 lb mill coal will evaporate 9 lb water
1 gal oil will evaporate 100 lb water
1 cu ft gas will evaporate 0.85 lb water
1 ton coal used under boilers = 18,500 cu ft of gas
1 bbl oil (42 gal) under boilers = 5,000 cu ft of gas
40 to 50 cu ft of gas per hr = 1 boiler hp

Gas engines:
   Highest grade gas engines develop 1 brake hp on 8500 Btu/hr
   Average large engine develops 1 hp on 10500 Btu/hr
   Oil-well engine develops 1 hp on 20000 Btu/hr
   In a small steam turbine plant 20 cu ft gas per kwhr is a fair average.
   It requires 40,000 cu ft of gas to pump 1,000,000 gal of water against 200-ft head

Brick plants—gas used per thousand brick made:
   1,800 cu ft for powder
   1,800 cu ft for drying
   15,000 cu ft for kilns

Ice plants:
   2,000 ft gas per ton of refrigeration

Zinc plants:
   15,000 cu ft for roasting per ton of metal produced
   65,000 cu ft for smelting per ton of metal produced
   20,000 cu ft for power and miscellaneous uses per ton of metal produced

Cement plants:
   60 to   100 cu ft/bbl for power
   80 to   100 cu ft/bbl for roasters
   1,800 to 2,000 cu ft/bbl for kilns

Salt plants:
   Direct-fire pans, 9,000 cu ft/ton
   Steam pans, 10,000 cu ft/ton
   Single-effect vacuum pan, 15,000 cu ft/ton
   Double-effect vacuum pan, 10,000 cu ft/ton
   Triple-effect vacuum pan, 6,000 cu ft/ton

Flour mills:
   200 to 400 cu ft/bbl

Gas compressors:
   Horsepower required to compress 1,000 cu ft/min of gas:

To 15 lb	50 hp
To 30 lb	85 hp
To 45 lb	111 hp
To 60 lb	134 hp
To 80 lb	117 hp (2 stages)
To 100 lb	151 hp (2 stages)
To 200 lb	212 hp (2 stages)

   Horsepower required to compress 1,000 cu ft/hr of gas:

To 15 lb	1 hp
To 30 lb	1.75 hp
To 45 lb	2.25 hp
To 60 lb	2.75 hp

gas, but under the heading of oil gas there must be included also that which is manufactured and used in California and other west-coast states. Pacific coast oil gas is made by a process similar to the manufacture of carbureted water gas, except that oil instead of solid fuel is used to heat the apparatus.   In Pintsch and Blau gases the high-heat value ranges from 1350 to 1500 Btu and the illuminating value from 50 to 60 candles.   California oil gas has an average heating value of 680 Btu and an illuminating value of 20 candles.   It is made from crude oil obtained in the fields close to the point of manufacture, so that a low price prevails.   Gas coal is not found in the Pacific states, and oil gas is, of necessity, the prevailing kind.   A large quantity of lampblack is produced in the manufacture of California oil gas, which is made into briquets, constituting a valuable byproduct.

Adiabatic Compression of Natural Gas.[1] The following table gives the rise in temperature due to the adiabatic compression of natural gas. p_1 is the absolute initial and p_2 the absolute final pressure, p_2/p_1 being, therefore, the ratio of compression. The initial temperature of the gas is assumed to be 60 F.

p_2/p_1	Rise in temperature, degrees Fahrenheit	p_2/p_1	Rise in temperature, degrees Fahrenheit	p_2/p_1	Rise in temperature, degrees Fahrenheit
1.0	0	6.0	238	14	386
1.5	47	6.5	251	16	412
2.0	82	7.0	263	18	435
2.5	110	7.5	274	20	456
3.0	135	8.0	285	25	503
3.5	157	8.5	296	30	543
4.0	177	9.0	305	35	578
4.5	194	10.0	324	40	609
5.0	210	11.0	341	45	638
5.5	224	12.0	357	50	664

Effect of Altitude on Gas Pressure. The effect of altitude on low-pressure gas distribution becomes significant in hilly country or where there are tall office buildings or waterless piston-type gas holders. On account of the altitude or stack effect, pressure on the mains is somewhat greater when such holders are nearly empty than when they are almost full. Likewise the available pressure differential is greater on the upper floors of tall buildings than it is at street level. In many cases the increase of pressure will more than cover the friction drop in a given riser. The difference is greater in the case of a gas of low specific gravity than with a heavier gas. Contrary to water distribution practice where head tanks are located on relatively high ground, waterless piston-type gas holders to be most effective should be located at points of low elevation.

The change in water column corresponding to a given difference in elevation can be computed conveniently as follows:

1 ft of air at 60 F $= 0.07638/62.37 = 0.0012246$ ft of water column

100 ft of air at 60 F $= 0.0012246 \times 100 \times 12 = 1.47$ in. of water column

100 ft of gas at 60 F $= 1.47s$ in. of water column where s is the specific gravity

The altitude effect of a 100-ft change in elevation is, therefore, $1.47 - (1.47s) = 1.47(1 - s)$ in. of water column, where gas, air, and water columns all are at 60 F. Should any of these be at some other temperature, this can be taken into account readily in the foregoing solution through substituting corresponding values for the density of air or water. The altitude effects at 60 F for gas of different specific gravities are given in the following table. The figures listed represent the depression in inches of water column for each 100-ft difference in elevation.

Specific gravity......	0.1	0.2	0.3	0.4	0.5	0.6	0.7	0.8	0.9
Inches H_2O.........	1.32	1.18	1.03	0.88	0.74	0.59	0.44	0.29	0.15

Composition, Heating Value, and Physical Constants. These properties for common fuel and illuminating gases will be found in Table 35. Because of the varying hydrogen content of different types of gas, the realizable heating value may

---

[1] Reproduction from "National Pipe Standards," by permission of The National Tube Co. For data on the work requirement and temperature rise in adiabatic compression, see "A Series of Enthalpy-entropy Charts for Natural Gases," by George Granger Brown, *AIMME Tech. Publ.* 1747.

differ considerably. The available energy, Btu per cubic foot, for given flue-gas temperatures when burning different types of gas under theoretically perfect conditions is given in Table 38.

Table 38. Heating Value, Btu per Cubic Foot, When Burning Different Gases with Complete Combustion and the Theoretical Amount of Air

	Natural gas	Natural gas	Coke-oven gas	Coke-oven gas	Carbureted water gas	Blue water gas
Higher heat value...............	1232	967	600	490	534	310
Lower heating value:						
Flue temperature, 300 F......	1050	825	510	420	420	270
Flue temperature, 400 F......	1025	800	500	405	440	260
Flue temperature, 600 F......	970	760	475	380	420	245

The properties of *other miscellaneous gases* commonly encountered will be found in Table 22.

Properties of Air

Air is a physical combination of several gaseous constituents. Its composition may vary slightly from point to point on the earth's surface and also with altitude. The volumetric composition of dry air at sea level is listed in the "International Critical Tables" as

N_2—78.03 per cent   Ne—0.00123 per cent

O_2—20.99 per cent   He—0.0004 per cent

A— 0.94 per cent   Kr—0.00005 per cent

CO_2— 0.03 per cent   Xe—0.0000006 per cent

For ordinary engineering calculations, the composition of air is taken as 79 per cent nitrogen and 21 per cent oxygen by volume; on a weight basis, the composition is usually taken as 76.8 per cent nitrogen and 23.2 per cent oxygen.

The molecular weight of air is 28.970 and is frequently taken as 29.0 in many engineering calculations. Based on the more precise molecular weight, the gas constant R for air is 53.35 ft lbf/(lbm R). The critical pressure and temperature of air are 546 psia and -221 F, respectively. Thus atmospheric air, being at a low pressure with respect to the critical pressure and at a high temperature with respect to the critical temperature, would be expected to obey the perfect-gas law with a fair degree of precision. Reference to Fig. 28 shows that the constant-pressure specific heat of air is about 0.239 Btu/(lbm R) at temperatures of 400 R and lower; at room temperature, say 530 R, the value of c_p is about 0.240 Btu/(lbm R). Figure 29 indicates that the specific-heat ratio k is constant at $k = 1.4$ for temperatures below about 500 R but that the value of k decreases at higher temperatures. The deviation from constant values of c_p and k of air at higher temperatures thus classifies it as a semiperfect gas; for computational accuracy it is thus necessary to take into account the variation in specific heats.

Table 39 lists the variation in c_p and k for temperatures ranging from 100 to 6000 R. The acoustic velocity corresponding to each temperature is also tabulated. This table gives instantaneous values of the various properties; in calculations it is necessary to use a mean or average value over the temperature range in question.

Table 39.   Air at Low Pressures

T, R	t,F	c_p, Btu/(lb R)	k	Acoustic velocity, fps
100	−360	0.2392	1.402	490.5
200	−260	0.2392	1.402	693.6
300	−160	0.2392	1.402	849.4
400	− 60	0.2393	1.402	980.9
500	40	0.2396	1.401	1,096.4
600	140	0.2403	1.399	1,200.3
700	240	0.2416	1.396	1,295.1
800	340	0.2434	1.392	1,382.5
900	440	0.2458	1.387	1,463.6
1000	540	0.2486	1.381	1,539.4
1200	740	0.2547	1.368	1,678.6
1400	940	0.2611	1.356	1,805.0
1600	1140	0.2671	1.345	1,922.0
1800	1340	0.2725	1.336	2,032
2000	1540	0.2773	1.328	2,135
2200	1740	0.2813	1.322	2,234
2400	1940	0.2848	1.317	2,329
2600	2140	0.2878	1.313	2,420
2800	2340	0.2905	1.309	2,508
3000	2540	0.2929	1.306	2,593
3200	2740	0.2950	1.303	2,675
3400	2940	0.2969	1.300	2,755
3600	3140	0.2986	1.298	2,832
3800	3340	0.3001	1.296	2,907
4000	3540	0.3015	1.294	2,981
5000	4540	0.3072	1.287	3,323
6000	5540	0.3114	1.282	3,634

Air Tables.   Table 40, abridged from "Gas Tables" by Keenan and Kaye published by John Wiley & Sons, Inc., is of great utility in connection with expansion and compression processes of air.   The tables are based on reference conditions of 1 psia and 492 R.   Values of pertinent properties are listed; meaning of these properties is explained below.

The *relative pressure* P_r is the pressure which would be reached by isentropic compression from 1 psia and 492 R to the indicated temperature.   For example, at 700 R, a value of 3.446 is listed for P_r.   This means that if air is compressed isentropically from 1 psia, 492 to 700 R, its pressure will be 3.446 psia.   If the air were *incorrectly* assumed to be a perfect gas with a constant k of 1.4, the corresponding pressure would be

$$P_2 = P_{ref}(T_2/T_{ref})^{k/k-1} = 1(700/492)^{3.5} = 3.41 \text{ psia}$$

The discrepancy between the above incorrect value and the correct value of 3.446 psia is due to the deviation of the properties of actual air from those of ideal air.

The *relative volume* v_r is the specific volume that would exist following isentropic compression from 1 psia, 492 R, to the indicated temperature.   At the reference temperature and pressure, the relative specific volume would be

$$v_{rel} = RT_{ref}/P_{ref} = (53.35 \times 492)/(1 \times 144) = 182 \text{ cu ft/lbm}$$

Table 40. Thermodynamic Properties of Air

T, R	h	P_r	u	v_r	ϕ
400	95.53	0.4858	68.11	305.0	0.52890
410	97.93	0.5295	69.82	286.8	0.53481
420	100.32	0.5760	71.52	270.1	0.54058
430	102.71	0.6253	73.23	254.7	0.54621
440	105.11	0.6776	74.93	240.6	0.55172
450	107.5	0.7329	76.65	227.45	0.55710
460	109.90	0.7913	78.36	215.33	0.56235
470	112.30	0.8531	80.07	204.08	0.56751
480	114.69	0.9182	81.77	193.65	0.57255
490	117.08	0.9868	83.49	183.94	0.57749
500	119.48	1.0590	85.20	174.90	0.58233
510	121.87	1.1349	86.92	166.46	0.58707
520	124.27	1.2147	88.62	158.58	0.59173
530	126.66	1.2983	90.34	151.22	0.59630
540	129.06	1.3860	92.04	144.32	0.60078
550	131.46	1.4779	93.76	137.85	0.60518
560	133.86	1.5742	95.47	131.78	0.60950
570	136.26	1.6748	97.19	126.08	0.61376
580	138.66	1.7800	98.90	120.70	0.61793
590	141.06	1.8899	100.62	115.65	0.62204
600	143.47	2.005	102.34	110.88	0.62607
610	145.88	2.124	104.06	106.38	0.63005
620	148.28	2.249	105.78	102.12	0.63395
630	150.68	2.379	107.50	98.11	0.63781
640	153.09	2.514	109.21	94.30	0.64159
650	155.50	2.655	110.94	90.69	0.64533
660	157.92	2.801	112.67	87.27	0.64902
670	160.33	2.953	114.40	84.03	0.65263
680	162.73	3.111	116.12	80.96	0.65621
690	165.15	3.276	117.85	78.03	0.65973
700	167.56	3.446	119.58	75.25	0.66321
710	169.98	3.623	121.32	72.60	0.66664
720	172.39	3.806	123.04	70.07	0.67002
730	174.82	3.996	124.78	67.67	0.67335
740	177.23	4.193	126.51	65.38	0.67665
750	179.66	4.396	128.25	63.20	0.67991
760	182.08	4.607	129.99	61.10	0.68312
770	184.51	4.826	131.73	59.11	0.68629
780	186.94	5.051	133.47	57.20	0.68942
790	189.38	5.285	135.22	55.38	0.69251

Table 40. Thermodynamic Properties of Air (*Continued*)

T, R	h	P_r	u	v_r	ϕ
800	191.81	5.526	136.97	53.63	0.69558
810	194.25	5.775	138.72	51.96	0.69860
820	196.69	6.033	140.47	50.35	0.70160
830	199.12	6.299	142.22	48.81	0.70455
840	201.56	6.573	143.98	47.34	0.70747
850	204.01	6.856	145.74	45.92	0.71037
860	206.46	7.149	147.50	44.57	0.71323
870	208.90	7.450	149.27	43.26	0.71606
880	211.35	7.761	151.02	42.01	0.71886
890	213.80	8.081	152.80	40.80	0.72163
900	216.26	8.411	154.57	39.64	0.72438
910	218.72	8.752	156.34	38.52	0.72710
920	221.18	9.102	158.12	37.44	0.72979
930	223.64	9.463	159.89	36.41	0.73245
940	226.11	9.834	161.68	35.41	0.73509
950	228.58	10.216	163.46	34.45	0.73771
960	231.06	10.610	165.26	33.52	0.74030
970	233.53	11.014	167.05	32.63	0.74287
980	236.02	11.430	168.83	31.76	0.74540
990	238.50	11.858	170.63	30.92	0.74792
1000	240.98	12.298	172.43	30.12	0.75042
1010	243.48	12.751	174.24	29.34	0.75290
1020	245.97	13.215	176.04	28.59	0.75536
1030	248.45	13.692	177.84	27.87	0.75778
1040	250.95	14.182	179.66	27.17	0.76019
1050	253.45	14.686	181.47	26.48	0.76259
1060	255.96	15.203	183.29	25.82	0.76496
1070	258.47	15.734	185.10	25.19	0.76732
1080	260.97	16.278	186.93	24.58	0.76964
1090	263.48	16.838	188.75	23.98	0.77196
1100	265.99	17.413	190.58	23.40	0.77426
1110	268.52	18.000	192.41	22.84	0.77654
1120	271.03	18.604	194.25	22.30	0.77880
1130	273.56	19.223	196.09	21.78	0.78104
1140	276.08	19.858	197.94	21.27	0.78326
1150	278.61	20.51	199.78	20.771	0.78548
1160	281.14	21.18	201.63	20.293	0.78767
1170	283.68	21.86	203.49	19.828	0.78985
1180	286.21	22.56	205.33	19.377	0.79201
1190	288.76	23.28	207.19	18.940	0.79415

Table 40. Thermodynamic Properties of Air (*Continued*)

T, R	h	P_r	u	v_r	ϕ
1200	291.30	24.01	209.05	18.514	0.79628
1210	293.86	24.76	210.92	18.102	0.79840
1220	296.41	25.53	212.78	17.700	0.80050
1230	298.96	26.32	214.65	17.311	0.80258
1240	301.52	27.13	216.53	16.932	0.80466
1250	304.08	27.96	218.40	16.563	0.80672
1260	306.65	28.80	220.28	16.205	0.80876
1270	309.22	29.67	222.16	15.857	0.81079
1280	311.79	30.55	224.05	15.518	0.81280
1290	314.36	31.46	225.93	15.189	0.81481
1300	316.94	32.39	227.83	14.868	0.81680
1310	319.53	33.34	229.73	14.557	0.81878
1320	322.11	34.31	231.63	14.253	0.82075
1330	324.69	35.30	233.52	13.958	0.82270
1340	327.29	36.31	235.43	13.670	0.82464
1350	329.88	37.35	237.34	13.391	0.82658
1360	332.48	38.41	239.25	13.118	0.82848
1370	335.09	39.49	241.17	12.851	0.83039
1380	337.68	40.59	243.08	12.593	0.83229
1390	340.29	41.73	245.00	12.340	0.83417
1400	342.90	42.88	246.93	12.095	0.83604
1410	345.52	44.06	248.86	11.855	0.83790
1420	348.14	45.26	250.79	11.622	0.83975
1430	350.75	46.49	252.72	11.394	0.84158
1440	353.37	47.75	254.66	11.172	0.84341
1450	356.00	49.03	256.60	10.954	0.84523
1460	358.63	50.34	258.54	10.743	0.84704
1470	361.27	51.68	260.49	10.537	0.84884
1480	363.89	53.04	262.44	10.336	0.85062
1490	366.53	54.43	264.38	10.140	0.85239
1500	369.17	55.86	266.34	9.948	0.85416
1510	371.82	57.30	268.30	9.761	0.85592
1520	374.47	58.78	270.26	9.578	0.85767
1530	377.11	60.29	272.23	9.400	0.85940
1540	379.77	61.83	274.20	9.226	0.86113
1550	382.42	63.40	276.17	9.056	0.86285
1560	385.08	65.00	278.13	8.890	0.86456
1570	387.74	66.63	280.11	8.728	0.86626
1580	390.40	68.30	282.09	8.569	0.86794
1590	393.07	70.00	284.08	8.414	0.86962

Table 40. Thermodynamic Properties of Air (*Continued*)

T, R	h	P_r	u	v_r	ϕ
1600	395.74	71.73	286.06	8.263	0.87130
1610	398.42	73.49	288.05	8.115	0.87297
1620	401.09	75.29	290.04	7.971	0.87462
1630	403.77	77.12	292.03	7.829	0.87627
1640	406.45	78.99	294.03	7.691	0.87791
1650	409.13	80.89	296.03	7.556	0.87954
1660	411.82	82.83	298.02	7.424	0.88116
1670	414.51	84.80	300.03	7.295	0.88278
1680	417.20	86.82	302.04	7.168	0.88439
1690	419.89	88.87	304.04	7.045	0.88599
1700	422.59	90.95	306.06	6.924	0.88758
1710	425.29	93.08	308.07	6.805	0.88916
1720	428.00	95.24	310.09	6.690	0.89074
1730	430.69	97.45	312.10	6.576	0.89230
1740	433.41	99.69	314.13	6.465	0.89387
1750	436.12	101.98	316.16	6.357	0.89542
1760	438.83	104.30	318.18	6.251	0.89697
1770	441.55	106.67	320.22	6.147	0.89850
1780	444.26	109.08	322.24	6.045	0.90003
1790	446.99	111.54	324.29	5.945	0.90155
1800	449.71	114.03	326.32	5.847	0.90308
1810	452.44	116.57	328.37	5.752	0.90458
1820	455.17	119.16	330.40	5.658	0.90609
1830	457.90	121.79	332.45	5.566	0.90759
1840	460.63	124.47	334.50	5.476	0.90908
1850	463.37	127.18	336.55	5.388	0.91056
1860	466.12	129.95	338.61	5.302	0.91203
1870	468.86	132.77	340.66	5.217	0.91350
1880	471.60	135.64	342.73	5.134	0.91497
1890	474.35	138.55	344.78	5.053	0.91643
1900	477.09	141.51	346.85	4.974	0.91788
1910	479.85	144.53	348.91	4.896	0.91932
1920	482.60	147.59	350.98	4.819	0.92076
1930	485.36	150.70	353.05	4.744	0.92220
1940	488.12	153.87	355.12	4.670	0.92362
1950	490.88	157.10	357.20	4.598	0.92504
1960	493.64	160.37	359.28	4.527	0.92645
1970	496.40	163.69	361.36	4.458	0.92786
1980	499.17	167.07	363.43	4.390	0.92926
1990	501.94	170.50	365.53	4.323	0.93066

Table 40. Thermodynamic Properties of Air (Continued)

T, R	h	P_r	u	v_r	ϕ
2000	504.71	174.00	367.61	4.258	0.93205
2010	507.49	177.55	369.71	4.194	0.93343
2020	510.26	181.16	371.79	4.130	0.93481
2030	513.04	184.81	373.88	4.069	0.93618
2040	515.82	188.54	375.98	4.008	0.93756
2050	518.61	192.31	378.08	3.949	0.93891
2060	521.39	196.16	380.18	3.890	0.94026
2070	524.18	200.06	382.28	3.833	0.94161
2080	526.97	204.02	384.39	3.777	0.94296
2090	529.75	208.06	386.48	3.721	0.94430
2100	532.55	212.1	388.60	3.667	0.94564
2110	535.35	216.3	390.71	3.614	0.94696
2120	538.15	220.5	392.83	3.561	0.94829
2130	540.94	224.8	394.93	3.510	0.94960
2140	543.74	229.1	397.05	3.460	0.95092
2150	546.54	233.5	399.17	3.410	0.95222
2160	549.35	238.0	401.29	3.362	0.95352
2170	552.16	242.6	403.41	3.314	0.95482
2180	554.97	247.2	405.53	3.267	0.95611
2190	557.78	251.9	407.66	3.221	0.95740
2200	560.59	256.6	409.78	3.176	0.95868
2210	563.41	261.4	411.92	3.131	0.95996
2220	566.23	266.3	414.05	3.088	0.96123
2230	569.04	271.3	416.18	3.045	0.96250
2240	571.86	276.3	418.31	3.003	0.96376
2250	574.69	281.4	420.46	2.961	0.96501
2260	577.51	286.6	422.59	2.921	0.96626
2270	580.34	291.9	424.74	2.881	0.96751
2280	583.16	297.2	426.87	2.841	0.96876
2290	585.99	302.7	429.01	2.803	0.96999
2300	588.82	308.1	431.16	2.765	0.97123
2310	591.66	313.7	433.31	2.728	0.97246
2320	594.49	319.4	435.46	2.691	0.97369
2330	597.32	325.1	437.60	2.655	0.97489
2340	600.16	330.9	439.76	2.619	0.97611
2350	603.00	336.8	441.91	2.585	0.97732
2360	605.84	342.8	444.07	2.550	0.97853
2370	608.68	348.9	446.22	2.517	0.97973
2380	611.53	355.0	448.38	2.483	0.98092
2390	614.37	361.3	450.54	2.451	0.98212
2400	617.22	367.6	452.70	2.419	0.98331

In Table 40, for example, at 700 R a value of 75.25 is listed for v_r.   This means that, if air is compressed isentropically from 1 psia, 492 to 700 R, its volume will be 75.25 cu ft/lbm.   If the air were *incorrectly* assumed to be a perfect gas with a constant value of k equal to 1.4, the specific volume following isentropic compression would be

$$v_2 = v_{ref}(T_{ref}/T_2)^{1/k-1} = 182(492/700)^{2.5} = 75.1 \text{ cu ft/lbm}$$

Once again, the discrepancy between the above incorrect value and the correct value of 75.25 cu ft/lbm is due to the fact that air is not a perfect gas.

For specification of values for *enthalpy h* and *internal energy u*, in order to avoid the existence of negative numbers in the tables, a base or reference temperature of 0 R was chosen and a value of zero assigned to the enthalpy and internal energy at that point.   For example, at 700 R, Table 40 lists a value of 167.56 Btu/lbm for enthalpy.   If the air were *incorrectly* assumed to be a perfect gas with a constant-pressure specific heat of 0.24 Btu/(lb R), the enthalpy rise referred to 0 R would be

$$h_2 = h_{ref} + c_p(T_2 - T_{ref}) = 0 + 0.24(700-0) = 168.0 \text{ Btu/lbm}$$

This calculation indicates how nearly equal to 0.24 Btu/lbm is the constant-pressure specific heat over the range 0 to 700 R.

It can be shown easily that the entropy change during a reversible process is

$$ds = c_p(dT/T) - (R/J)(dP/P)$$

or, between any two state points, 1 and 2,

$$s_2 - s_1 = \int_{T_1}^{T_2} c_p(dT/T) - (R/J)\ln(P_2/P_1) \tag{63}$$

The integral term of Eq. (63) can be evaluated if the relationship between c_p and T is known.   For a semiperfect gas at low pressure, c_p is a function of T only; if T corresponds to the reference temperature of 492 R, and if c_p is known at successive values of temperature between 492 R and T_2, then the integral can be evaluated easily. This is done in Table 40, and its value is assigned the designation ϕ, or

$$\phi = \int_{T_{ref}}^{T_2} c_p \, dT$$

where T_2 is entered in the first column of Table 40.   Then, between two state points

$$s_2 - s_1 = \phi_2 - \phi_1 - (R/J)\ln(P_2/P_1) \tag{64}$$

Examples in Use of Air Tables

Example 1. Air is expanded in a frictionless and adiabatic nozzle from an initial static pressure of 100 psia and an initial static temperature of 1040 F to a final pressure of 70 psia. The nozzle inlet area is sufficiently large that inlet velocity is negligible.   It is required to find the temperature at nozzle outlet and the required outlet area for a discharge coefficient of 0.98 if the flow rate is 5 lb/sec.

Solution. Table 40 is entered at a value of T_1 equal to $(1040 + 460)$ or 1500 R.   There are obtained

$$h_1 = 291.30 \text{ Btu/lb} \qquad P_{r1} = 24.01 \qquad v_{r1} = 18.514$$

Then, since $P_{r1}, P_{r2}, P_1,$ and P_2 all lie on the same isentropic,

$$P_{r2} = P_{r1}(P_2/P_1) = 24.01 \times {}^{70}\!/_{100} = 16.807$$

Reference to Table 40 indicates that at 1080 R, P_r is 16.278 and that at 1090 R, P_r is 16.838.   Interpolation between these values will yield $T_2 = 1089.5$ R, the nozzle outlet temperature.

Similarly, it is found that

$$h_2 = 263.35 \text{ Btu/lbm} \qquad v_{r_2} = 23.95 \text{ cu ft/lbm}$$

At the initial state, v_1 is calculated from the perfect-gas equation of state

$$v_1 = RT_1/P_1 = 53.35 \times 1,500/100 \times 144 = 5.55 \text{ cu ft/lbm}$$

Then $v_2 = v_1(v_{r_2}/v_{r_1}) = 5.55 \times 23.95/18.514 = 7.18$ cu ft/lbm
From Eq. (30), with negligible inlet velocity,

$$v_2 = \sqrt{2g_cJ(h_1 - h_2)} = 223.7\sqrt{291.30 - 263.65}$$
$$= 1,178 \text{ fps}$$

By use of the continuity equation

$$A_2 = \dot{m}v_2/V_2C = 5 \times 7.18/1,178 \times 0.98 = 0.0311 \text{ sq ft}$$

or the exit area is 4.48 sq in.

If the air had been assumed incorrectly to behave as a perfect gas with $k = 1.4$, a solution could be obtained by use of Table 18, as follows:

$$P_2/P_o = 70/100 = 0.7 \qquad \text{and} \qquad M_2 = 0.73$$

From Eq. (42), $A^* = \dot{m}\sqrt{T_o}/0.532P_o$

or $\qquad\qquad A^* = 5\sqrt{1,500}/0.532 \times 100 = 3.64$ sq in.

at $\qquad\qquad M_2 = 0.73, A_2/A^* = 1.07$ approximately

and $\qquad\qquad A_2 = 1.07(3.64) = 3.9$ sq in. for $C = 1$

For $\qquad\qquad C = 0.98, A_2 = 3.9/0.98 = 3.96$ sq in.

Thus, use of Table 18 with $k = 1.4$ involves an error of about 10 per cent. This is due to the fact that, as shown by Fig. 29, the value of k for the average temperature of this flow process is about 1.37 and not 1.4 as assumed.

Example 2. Air is compressed from 15 psia, 540 R, to 150 psia and 600 F at the rate of 10 lb/sec. During the compression process, 5 gpm of cooling water is circulated through the compressor jackets and intercooler, and this water increases in temperature 30 F.
(a) Determine the work in horsepower done by the compressor on the fluid.
(b) What will be the pipe size needed at compressor outlet to limit the velocity to 50 fps?
(c) What is the change in entropy of the air during the compression process?
Solution. (a) Heat transfer $Q = 5 \times 8.33 \times 30/60 = 20.83$ Btu/sec. By substitution into the steady-flow energy equation preceding Eq. (18) and neglecting changes in potential and kinetic energy, there is obtained

$$\dot{Q} + \dot{m}(h_1 - h_2) = \dot{W} \tag{65}$$

From Table 40, $h_1 = 129.06$ Btu/lb and $h_2 = 255.96$ Btu/lb

Then $\qquad\qquad \dot{W} = 20.83 + 10(129.06 - 255.96) = -1248$ Btu/sec

or $\qquad\qquad \dot{W} = -1248$ Btu/sec $\times 3,600$ sec/hr $\times 1/2,545$ hp-hr/Btu
$$= -1,765 \text{ hp}$$

It is noted that this value represents the work done by the compressor on the air; this is the significance of the negative sign. Numerically, this would represent also the power input to a compressor of 100 per cent efficiency. If the compressor isentropic (adiabatic) efficiency were, say 70 per cent, then the required compressor input would be 1,765/0.7 or 2,520 bhp.

(*b*) At the outlet, the density is

$$\rho_2 = P_2/RT_2 = 150 \times 144/53.35 \times 1,060 = 0.382 \text{ lb/ft}^3$$

$$A_2 = \dot{m}/\rho_2 V_2 = 10/0.382 \times 50 = 0.522 \text{ ft}^2 = 75.4 \text{ in.}^2$$

$$d_2 = \sqrt{\frac{A_2}{\pi/4}} = \sqrt{\frac{75.4}{0.7854}} = 9.6$$

or use 10-in. Schedule 40 pipe.

(*c*) By substitution into Eq. (64)

$$s_2 - s_1 = \phi_2 - \phi_1 - (R/J) \ln (P_2/P_1)$$

$$= 0.76496 - 0.60078 - \frac{53.35}{778} \ln \frac{150}{15}$$

$$= 0.00668 \text{ Btu/(lb R)}$$

Table 41. Viscosity of Dry Low-pressure Air

[Tabular values $\times 10^{-6}$ yield viscosity in lbm/(ft sec)]

Temp, R	Viscosity, millionths of lbm/(ft sec)	Temp, R	Viscosity, millionths of lbm/(ft sec)
400	10.0	1300	23.0
500	11.8	1400	24.2
600	13.5	1500	25.3
700	15.1	1600	26.4
800	16.6	1700	27.4
900	17.9	1800	28.4
1000	19.2	1900	29.3
1100	20.5	2000	30.2
1200	21.8	2100	31.1

Viscosity of Air. The viscosity of air increases with temperature. Instantaneous values are listed in Table 41; it is emphasized that, if there is a considerable temperature change during a flow process, the mean or average value of viscosity should be used.

If a flow process occurs in which air temperature changes from, say, 800 to 1000 R, it would probably be sufficiently accurate to use 17.9×10^{-6} lbm/(ft sec) as the average viscosity during the process. As shown in Fig. 31, the variation in viscosity is not linear. Thus, if a flow process takes place over a wide range in temperature, accuracy in calculation would require graphical integration of the area beneath the viscosity curve between the two temperature end points and then a division of that area by the temperature variation.

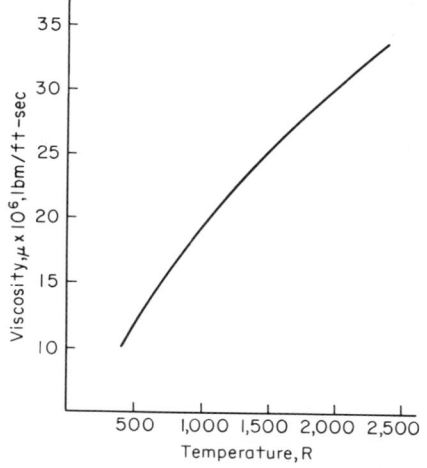

FIG. 31. Viscosity of dry air.

Properties of Air-vapor Mixtures

In the previous section, the properties of dry air were discussed. Many

processes, however, involve the flow of air mixed with a vapor. The most common example of such a process occurs in the air-conditioning field, in which water vapor, in reality low-pressure steam, is mixed physically with dry air. If it is assumed that the two constituents of such a mixture, both singly and jointly, obey the perfect-gas laws, the treatment of a flow process is readily carried out.

Thermodynamic Relations. Dalton's law of partial pressure applies. If P_a is the partial pressure of dry air and P_v is the partial pressure of water vapor, then the pressure P of a mixture containing these two constituents is

$$P = P_a + P_v \tag{66}$$

If the mixture contains m_a lb of dry air and m_v lb of vapor, the mass fraction of vapor is

$$mf_v = m_v/(m_v + m_a) \tag{67}$$

Similarly, if the mixture contains n_a moles of dry air and n_v moles of vapor, the volume or mole fraction of the vapor is

$$x_c = n_v/(n_v + n_a)$$

It can be shown that

$$x_v = P_v/P \quad \text{and} \quad x_a = P_a/P \tag{68}$$

Properties can be determined easily as weighted averages. Thus

$$c_{p\text{(mixture)}} = mf_v c_{pv} + mf_a c_{pa} \tag{69}$$

$$h_{\text{(mixture)}} = mf_v h_v + mf_a h_a \tag{70}$$

Relative Humidity. The relative humidity ϕ is defined as the ratio of the mole fraction of the vapor in a mixture to the mole fraction of vapor in a saturated mixture at the same temperature and total pressure. If P_s represents the saturation pressure of water vapor (low-pressure steam) at the mixture temperature, then

$$\phi = \frac{P_v/P}{P_s/P} = \frac{P_v}{P_s} \tag{71}$$

Example. One-half pound of water vapor and 9.5 lb of dry air are in direct contact with each other at a mixture temperature of 120 F and at atmospheric pressure. What is the relative humidity of the mixture?

Solution:

$$n_v = 0.5/18 = 0.0278 \text{ mole vapor}$$
$$n_a = 9.5/28.97 = 0.328 \text{ mole air}$$
$$n = n_v + n_a = 0.3558 \text{ mole mixture}$$

Then
$$x_v = n_v/n = 0.0278/0.3558 = 0.0782$$

and by Eq. (68),

$$P_v = x_v P = 0.0782(14.7) = 1.145 \text{ psia}$$

at 120 F, the pressure of saturated steam is (see Table 28) 1.6924 psia. The relative humidity is

$$\phi = P_v/P_s = 1.145/1.6924 = 67.7\%$$

Humidity Ratio. The humidity ratio, sometimes referred to as the specific humidity, is defined as the ratio of the mass of water vapor to the mass of dry air, or

$$\omega = mv/ma$$

where ω is the humidity ratio. In the previous example,

$$\omega = 0.5/9.5 = 0.0526 \text{ lb vapor/lb dry air}$$

It can be shown also that

$$\omega = 0.622 P_v/P_a$$

This can be checked by using the data of the previous example.

$$\omega = 0.622 \times 1.145/(14.7 - 1.145) = 0.0526 \text{ lb vapor/lb dry air}$$

Wet-bulb Temperature. The temperature indicated by an ordinary thermometer may be referred to as the dry-bulb temperature. The wet-bulb temperature refers to the temperature indicated by a thermometer whose bulb is located in a water-vapor-saturated environment. In practice, a piece of gauze containing a few drops of water is placed around the thermometer bulb and the entire instrument is rotated vigorously through the room air. This causes evaporation of some of the water, and that requires a transfer of heat from the thermometer bulb. Owing to the heat transfer, the thermometer bulb reaches a temperature below that of the room air, and the thermometer reading at equilibrium is known as the wet-bulb temperature.

Psychrometric Chart. The properties of mixtures of air and water vapor are most conveniently shown in graphical form. Figure 32 is a psychrometric chart which indicates pertinent properties of air-vapor mixture. The following example will clarify the method of usage.

Example. Air, initially at 95 dry-bulb temperature and with a humidity ratio of 160 grains per lb of dry air, is cooled and dehumidified to 75 F dry bulb and a relative humidity of 50 per cent. Flow is at the rate of 50,000 lb/hr of dry air, and the duct is 2 by 3 ft in cross section. (*a*) What is the velocity before and after cooling? (*b*) What is the weight of water, in pounds per hour, removed by the cooling coil? (*c*) What is the heat-transfer rate in Btu per hour in the cooling coil?

Solution. Initially, at state 1, there is read from the chart of Fig. 32,

$$h_1 = 48.05 \text{ Btu/lb air} \qquad \text{Wet-bulb temperature} = 84 \text{ F}$$
$$v_1 = 14.49 \text{ cu ft/lb dry air} \qquad \omega_1 = 160 \text{ grains/lb dry air}$$
$$\phi_1 = 64\% \qquad P_{v1} = 0.521 \text{ psia}$$

Finally, at state 2, there is read from Fig. 32,

$$h_2 = 28.1 \text{ Btu/lb dry air} \qquad \text{Wet bulb temperature} = 62.5 \text{ F}$$
$$v_2 = 13.67 \text{ cu ft/lb dry air} \qquad \omega_2 = 64.1 \text{ grains/lb dry air}$$
$$\phi_2 = 50\% \qquad P_{v2} = 0.21 \text{ psia}$$

(*a*) $V_1 = \dot{m}v_1/A = 50,000 \times 14.49/3,600 \times 2 \times 3 = 33.5 \text{ fps}$

$V_2 = \dot{m}v_2/A = 50,000 \times 13.67/3,600 \times 6 = 31.6 \text{ fps}$

(*b*) Water removed $= \omega_1 - \omega_1 = 160 - 64.1 = 95.9 \text{ grain/lb dry air}$

$= 95.9 \text{ (grains)/(lb dry) air} \times (1/7,000) \text{ lb/grains} \times 50,000 \text{ lb dry air/hr}$

$= 685 \text{ lb water/hr}$

(*c*) From the energy equation, in the absence of significant kinetic- and potential-energy changes and assuming that heat losses from the duct to atmosphere are negligible,

$$\dot{Q} = \dot{m}(h_1 - h_2)$$
$$= 50,000 (48.05 - 28.1)$$
$$= 898,000 \text{ Btu/hr}$$
$$= 263 \text{ kw}$$

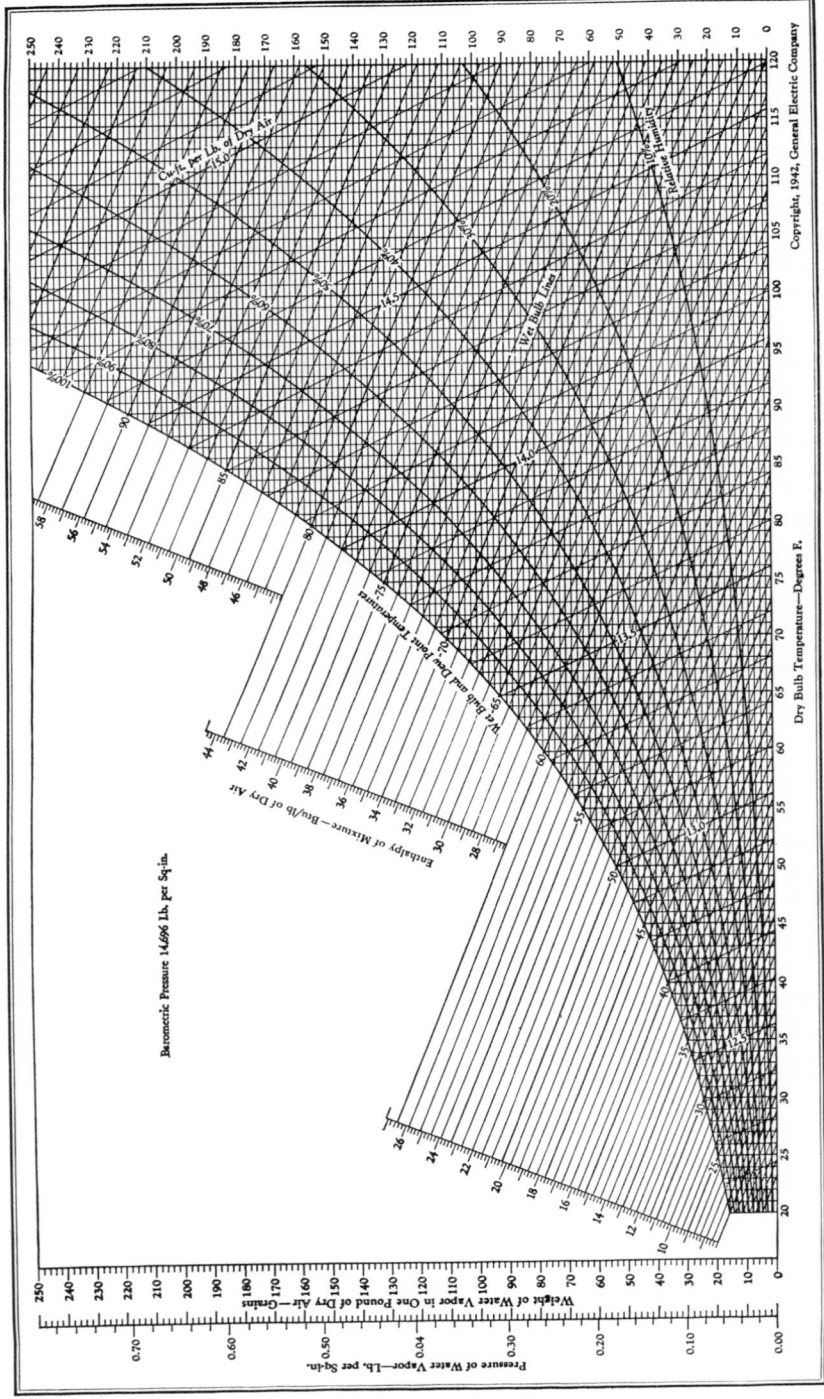

FIG. 32. Psychrometric chart.

Properties of Oils[1]

The lubricating oils and liquid fuels commonly encountered in engineering work are refined from "crude" oil or petroleum. The word petroleum is derived from two Latin terms, *petra* meaning rock and *oleum* meaning oil. Rock oil, which was an early name given it in America, is accounted for by the fact that certain shales and coals contain oil as a constituent. Petroleum is one of the family of bitumens which in their natural state assume many forms and are of world-wide distribution. It usually occurs in connection with pockets of natural gas. The occurrence of petroleum has been observed and recorded from the earliest times. The first extensive commercial exploitation of petroleum started about 1850.

Gasoline is now the most valuable product of petroleum. Gasoline was originally used for lighting purposes and for domestic stoves, but since the invention of the internal-combustion engine, production has increased tremendously to supply the fuel requirements of automobiles, motor boats, and airplanes. Prior to 1920 most of the gasoline produced was obtained by the fractional distillation of crude oil which yielded only 5 to 6 gal from a 42-gal barrel of crude oil.

The following explanation of the fractional distillation of crude oil with typical percentages of products obtained is quoted from *U.S. Department of Agriculture Circular* 405:

Crude oils obtained from different fields vary considerably. They are divided into two main groups, paraffin base oils and asphaltic base oils. The paraffin oils vary in color from a dark green to a light amber and are found principally in the Appalachian regions and mid-continent fields. The asphaltic types are heavier oils and are found in large quantities in California and the Gulf-coast regions. They are darker in color than the paraffin oils and vary from a red brown to black. A great many products are derived from crude oil by slowly heating it and collecting the products given off. (This is known as fractional distillation.) At the lower temperatures gasoline and kerosene are obtained, while at higher temperatures fuel oil, lubricating oils, lubricating grease, and other products are given off, each within a fixed range of temperature. The number of products obtained depends upon the market demand. Figure 33 shows a typical distillation of a paraffin oil and an asphaltic oil in which fuel oil was obtained. Either distillation

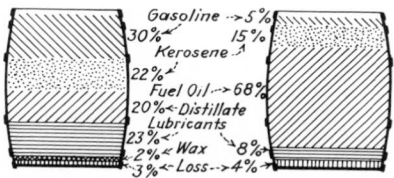

Gasoline →5%
30% — 15%
Kerosene
22%
Fuel Oil →68%
20% ← Distillate
Lubricants
23% ←
2% ← Wax 8%
3% ← Loss →4%

Paraffin Base Oil   Asphaltum Base Oil

FIG. 33. Typical examples of products resulting from ordinary refining of paraffin and asphaltum base oils (cracking process not used).

[1] The following references are listed as furnishing a more complete discussion of the properties of various oils than is possible here:

(a) "Handbook of the Petroleum Industry," vol. I, David T. Day, Editor in Chief, John Wiley & Sons, Inc., New York.

(b) "Industrial Oil Engineering," by John Rome Battle, J. B. Lippincott Company, Philadelphia.

(c) "A Handbook of Petroleum, Asphalt, and Natural Gas," by Roy Cross, *Bull.* 25, Kansas City Testing Laboratory, 1928.

(d) "National Standard Petroleum Oil Tables," *Natl. Bur. Std. Circ.* 154.

(e) "ASTM Standards on Petroleum Products and Lubricants" (methods of testing, specifications, definitions, charts, and tables), prepared by ASTM Committee D2. Issued annually by the American Society for Testing Materials, 260 S. Broad St., Philadelphia, Pa. See also ASA and API standards.

(f) "The Chemical Technology of Petroleum," 3d ed., by W. A. Gruse and D. R. Stevens, McGraw-Hill Book Company, New York, 1960.

might have been varied to give a different range of products. In the case of asphaltic oil, for example, the residue after the gasoline and kerosene had been driven off might have been used for road building purposes or further distilled for a number of grades of lubricants without producing any fuel oil.

By 1920 cracking processes had been introduced and developed to a point where the yield was over 10 gal per 42-gal barrel. In the cracking processes the heavier fractions like kerosene, gas oil, and fuel oil which were produced along with gasoline by fractional distillation are rerun in specially built stills under high temperature and pressure, thus cracking the molecules and getting a further proportion of gasoline as a result. The recovery of casinghead gasoline from natural gas has been another important source of high-volatile fuel.

The polymerization process which is the reverse of cracking came into use in the early 1930's. It gathers petroleum refinery gases, formerly largely wasted and blown away, and converts them into liquid fuel. Hydrogenation, or the hydroforming process, originated in Europe and was perfected technically in the United States. Hydroforming was used during World War II to produce synthetic toluene for TNT at a rate equivalent to about twice that produced by the whole coal-tar industry. When the proper naphtha fraction is used, hydroforming will produce high-octane aviation gasoline.

By 1938 gasoline yield by all methods employed in the United States had risen to 18½ gal per 42-gal barrel of crude. In 1938 announcement was made of the Houdry catalytic cracking process which gives 33 to 34 gal of gasoline per barrel of crude. The development of isopropyl and alkylate for blending with leaded gasoline has contributed to the development of 100-octane fuel on a commercial scale.

Advances in the art of refining ultimately promise almost complete flexibility of output, thereby permitting production of any desired percentage of any one of the most important products and thus relieving storage problems created by variation in seasonal demands.

Liquid fuels and lubricants are generally distinguished one from another by their density or heaviness, commonly referred to as the "gravity" and by their viscosity.

Density of Oil. It is standard American practice to indicate specific gravities of petroleum products at 60 F. The variation of density with temperature is important with petroleum products, since most of them are sold by volume and the specific gravity is determined at the prevailing temperature rather than at 60 F. The Bureau of Standards has issued *Circular* 410 containing tables which enable (1) reducing observed degrees API to degrees API at 60 F, (2) obtaining volume at 60 F occupied by unit volume at indicated temperature, and (3) reducing observed specific gravities to specific gravities at 60/60 F. Hydrometers used for determining the density of oil are graduated with either the Baumé scale (Bureau of Standards) or the API scale (American Petroleum Institute). Both of these scales are arbitrary and differ but little from each other. They are identical at the 10-deg mark, which corresponds to a specific gravity of 1.00, or the density of water. In order to overcome the confusion that has existed in the petroleum-oil industry by reason of the use of two so-called "Baumé scales" for light liquids, the American Petroleum Institute, the Bureau of Mines, and the Bureau of Standards in December, 1921, agreed to recommend that in the future only the scale based on the modulus of 141.5 be used in the petroleum industry and that it be known as the "API scale."

Table 43 gives the specific gravities, for liquids both lighter than and heavier than water, corresponding to each degree Baumé of either scale. Table 44 gives temperature corrections for Baumé hydrometers. Specific gravity and pounds per gallon for each API degree, with the corresponding Bureau of Standards Baumé

Table 42.   Degrees Baumé, Specific Gravity, and Pounds per Gallon Corresponding to Each Degree API

Degrees A.P.I. modulus 141.5	Degrees Baumé modulus 140	Specific gravity at 60/60 F.	Pounds per gallon at 60 F	Degrees A.P.I. modulus 141.5	Degrees Baumé modulus 140	Specific gravity at 60/60 F	Pounds per gallon at 60 F	Degrees A.P.I. modulus 141.5	Degrees Baumé modulus 140	Specific gravity at 60/60 F	Pounds per gallon at 60 F
10	10.00	1.0000	8.328	40	39.68	0.8251	6.870	70	69.36	0.7022	5.845
11	10.99	0.9930	8.270	41	40.67	0.8203	6.830	71	70.35	0.6988	5.817
12	11.98	0.9861	8.212	42	41.66	0.8155	6.790	72	71.34	0.6953	5.788
13	12.97	0.9792	8.155	43	42.65	0.8109	6.752	73	72.33	0.6919	5.759
14	13.96	0.9725	8.099	44	43.64	0.8063	6.713	74	73.32	0.6886	5.731
15	14.95	0.9659	8.044	45	44.63	0.8017	6.675	75	74.31	0.6852	5.703
16	15.94	0.9593	7.989	46	45.62	0.7972	6.637	76	75.30	0.6819	5.676
17	16.93	0.9529	7.935	47	46.61	0.7927	6.600	77	76.29	0.6787	5.649
18	17.92	0.9465	7.882	48	47.60	0.7883	6.563	78	77.28	0.6754	5.622
19	18.90	0.9402	7.830	49	48.59	0.7839	6.526	79	78.27	0.6722	5.595
20	19.89	0.9340	7.778	50	49.58	0.7796	6.490	80	79.26	0.6690	5.568
21	20.88	0.9279	7.727	51	50.57	0.7753	6.455	81	80.25	0.6659	5.542
22	21.87	0.9218	7.676	52	51.56	0.7711	6.420	82	81.24	0.6628	5.516
23	22.86	0.9159	7.627	53	52.54	0.7669	6.385	83	82.23	0.6597	5.491
24	23.85	0.9100	7.578	54	53.53	0.7628	6.350	84	83.22	0.6566	5.465
25	24.84	0.9042	7.529	55	54.52	0.7587	6.316	85	84.20	0.6536	5.440
26	25.83	0.8984	7.481	56	55.51	0.7547	6.283	86	85.19	0.6506	5.415
27	26.82	0.8927	7.434	57	56.50	0.7507	6.249	87	86.18	0.6476	5.390
28	27.81	0.8871	7.387	58	57.49	0.7467	6.216	88	87.17	0.6446	5.365
29	28.80	0.8816	7.341	59	58.48	0.7428	6.184	89	88.16	0.6417	5.341
30	29.79	0.8762	7.296	60	59.47	0.7389	6.151	90	89.15	0.6388	5.316
31	30.78	0.8708	7.251	61	60.46	0.7351	6.119	91	90.14	0.6360	5.293
32	31.77	0.8654	7.206	62	61.45	0.7313	6.087	92	91.13	0.6331	5.269
33	32.76	0.8602	7.163	63	62.44	0.7275	6.056	93	92.12	0.6303	5.246
34	33.75	0.8550	7.119	64	63.43	0.7238	6.025	94	93.11	0.6275	5.222
35	34.74	0.8498	7.076	65	64.42	0.7201	5.994	95	94.10	0.6247	5.199
36	35.72	0.8448	7.034	66	65.41	0.7165	5.964	96	95.09	0.6220	5.176
37	36.71	0.8398	6.993	67	66.40	0.7128	5.934	97	96.0.	0.6193	5.154
38	37.70	0.8348	6.951	68	67.39	0.7093	5.904	98	97.07	0.6166	5.131
39	38.69	0.8299	6.910	69	68.38	0.7057	5.874	99	98.06	0.6139	5.109
								100	99.05	0.6112	5.086

For more complete data see Table 5 of the "National Standard Petroleum Oil Tables," *Bur. Standards Circ.* 410. Also ASTM Standard D287.

reading, are given in Table 42.   Specific gravity vs. temperature curves are given in Fig. 38 for petroleum oils.   For characteristics of typical oils, see Table 45.

To convert degrees Baumé Bureau of Standards to specific gravity at 60/60 F use the formulas

$$S = \frac{140}{130 + °\text{Bé}} \text{ for liquids lighter than water}$$

and

$$S = \frac{145}{145 - °\text{Bé}} \text{ for liquids heavier than water}$$

To convert degrees Baumé Bureau of Standards to weight per cubic foot at 60/60 F use the formulas

$$\rho = \frac{62.37 \times 140}{130 + °\text{Bé}} \text{ for liquids lighter than water}$$

and

$$\rho = \frac{62.37 \times 145}{145 - °\text{Bé}} \text{ for liquids heavier than water}$$

The above conversion modulus is that known as the "U.S. Bureau of Standards." A modification commonly used in the petroleum industry and elsewhere for liquids

Table 43. Relation of Baume Scales to Specific Gravities at 60/60 F

(Bureau of Standards and API Scales)

Hydrometer readings, degrees	Sp. gr. at 60°/60° F. Liquids heavier than water — Bu. of Stds.	Sp. gr. at 60°/60° F. Liquids lighter than water — Bu. of Stds.	A.P.I.	Hydrometer readings, degrees	Sp. gr. at 60°/60° F. Liquids heavier than water — Bu. of Stds.	Sp. gr. at 60°/60° F. Liquids lighter than water — Bu. of Stds.	A.P.I.	Hydrometer readings, degrees	Sp. gr. at 60°/60° F. Liquids heavier than water — Bu. of Stds.	Sp. gr. at 60°/60° F. Liquids lighter than water — Bu. of Stds.	A.P.I.
0	1.0000	......	......	34	1.3063	0.8537	0.8550	68	1.8831	0.7071	0.7093
1	1.0069	......	......	35	1.3182	0.8485	0.8498	69	1.9079	0.7035	0.7057
2	1.0140	......	......	36	1.3303	0.8434	0.8448	70	1.9333	0.7000	0.7022
3	1.0211	......	......	37	1.3426	0.8383	0.8398	71	1.9595	0.6965	0.6988
4	1.0284	......	......	38	1.3551	0.8333	0.8348	72	1.9863	0.6931	0.6952
5	1.0357	......	......	39	1.3679	0.8284	0.8299	73	2.0139	0.6897	0.6919
6	1.0432	......	......	40	1.3810	0.8235	0.8251	74	2.0423	0.6863	0.6886
7	1.0507	......	......	41	1.3942	0.8187	0.8203	75	2.0714	0.6829	0.6852
8	1.0584	......	......	42	1.4078	0.8140	0.8155	76	2.1015	0.6796	0.6819
9	1.0662			43	1.4216	0.8093	0.8109	77	2.1324	0.6763	0.6787
10	1.0741	1.0000	1.0000	44	1.4356	0.8046	0.8063	78	2.1642	0.6731	0.6754
11	1.0821	0.9929	0.9930	45	1.4500	0.8000	0.8017	79	2.1970	0.6699	0.6722
12	1.0902	0.9859	0.9861	46	1.4647	0.7955	0.7972	80	......	0.6667	0.6690
13	1.0985	0.9790	0.9792	47	1.4796	0.7910	0.7927	81	......	0.6635	0.6659
14	1.1069	0.9722	0.9725	48	1.4949	0.7865	0.7883	82	......	0.6604	0.6628
15	1.1154	0.9655	0.9659	49	1.5104	0.7821	0.7839	83	......	0.6573	0.6597
16	1.1240	0.9589	0.9593	50	1.5263	0.7778	0.7796	84	......	0.6542	0.6566
17	1.1328	0.9524	0.9529	51	1.5426	0.7735	0.7753	85	......	0.6512	0.6536
18	1.1417	0.9460	0.9465	52	1.5591	0.7692	0.7711	86	......	0.6482	0.6506
19	1.1508	0.9396	0.9402	53	1.5761	0.7650	0.7669	87	......	0.6452	0.6476
20	1.1600	0.9333	0.9340	54	1.5934	0.7609	0.7628	88	......	0.6422	0.6446
21	1.1694	0.9272	0.9279	55	1.6111	0.7568	0.7587	89	......	0.6393	0.6417
22	1.1789	0.9211	0.9218	56	1.6292	0.7527	0.7547	90	......	0.6364	0.6388
23	1.1885	0.9150	0.9159	57	1.6477	0.7487	0.7507	91	......	0.6335	0.6360
24	1.1984	0.9091	0.9100	58	1.6667	0.7447	0.7467	92	......	0.6306	0.6331
25	1.2083	0.9032	0.9042	59	1.6861	0.7407	0.7428	93	......	0.6278	0.6303
26	1.2185	0.8974	0.8984	60	1.7059	0.7368	0.7389	94	......	0.6250	0.6275
27	1.2288	0.8917	0.8927	61	1.7262	0.7330	0.7351	95	......	0.6222	0.6247
28	1.2393	0.8861	0.8871	62	1.7470	0.7292	0.7313	96	......	0.6195	0.6220
29	1.2500	0.8805	0.8816	63	1.7683	0.7254	0.7275	97	......	0.6167	0.6193
30	1.2609	0.8750	0.8762	64	1.7901	0.7217	0.7238	98	......	0.6140	0.6166
31	1.2719	0.8696	0.8708	65	1.8125	0.7180	0.7201	99	......	0.6114	0.6139
32	1.2832	0.8642	0.8654	66	1.8354	0.7143	0.7165	100	......	0.6087	0.6112
33	1.2946	0.8589	0.8602	67	1.8590	0.7107	0.7128				

NOTE. For more complete data see Table 5, *Circ.* 410, Bureau of Standards, "National Standard Petroleum Oil Tables," also ASTM Standard D287.

lighter than water employs 141.5/(131.5 + °Bé) instead of 140/(130 + °Bé). This modification, formerly known as the "Baume Tagliabue scale," is now designated as the API scale. It has been adopted as standard by the American Petroleum Institute and is now generally followed. Owing to the previous use of both scales it is always advisable to specify which is used in stating Baumé readings. The general practice with liquids heavier than water is to state specific gravity rather than degrees Baumé, or API. The relation among specific gravity, specific heat, and temperature for petroleum oils is given in Fig. 34. The specific gravities of oil corresponding to different degrees Baumé are given in Tables 45 and 46.

Specific Heat. The following discussion of the specific heat, etc., of the petroleum oils, has been briefed from a paper on "Specific Heat—Specific Gravity—Temperature Relations of Petroleum Oils," by William Rankine Eckart, published in *Mechanical Engineering* for July, 1925, pages 535–540:

The value of the specific heat commonly used in practice for liquid hydrocarbons is 0.50 Btu per pound. As values as low as 0.40 and as high as 0.60 have been determined, an error may be involved within these limits as great as ±20 per cent. Any relationship, empirical or otherwise, which may be established so as to make it possible to confine this error within narrower limits should be of estimable value.

Table 44. Temperature Corrections to Readings of Baumé Hydrometers in American Petroleum Oils at Various Temperatures[1]

(Standard at 60 F, modulus 140)

Observed temperature, degrees Fahrenheit	Observed degrees Baumé							
	20.0	30.0	40.0	50.0	60.0	70.0	80.0	90.0
	Add to observed degrees Baumé							
30	1.7	2.0	2.4	3.0	3.7	4.3	5.0	5.7
32	1.6	1.9	2.3	2.8	3.4	4.0	4.7	5.3
34	1.5	1.8	2.1	2.6	3.1	3.7	4.3	4.9
36	1.4	1.6	2.0	2.4	2.9	3.4	4.0	4.6
38	1.3	1.5	1.8	2.2	2.6	3.1	3.6	4.2
40	1.2	1.4	1.6	2.0	2.4	2.8	3.2	3.8
42	1.1	1.2	1.5	1.8	2.2	2.5	2.9	3.4
44	0.9	1.1	1.3	1.6	2.0	2.2	2.6	3.0
46	0.8	0.9	1.1	1.4	1.7	1.9	2.3	2.7
48	0.7	0.8	0.9	1.2	1.4	1.6	2.0	2.3
50	0.6	0.7	0.8	1.0	1.2	1.4	1.6	1.9
52	0.5	0.6	0.7	0.8	1.0	1.1	1.3	1.5
54	0.3	0.4	0.5	0.6	0.8	0.9	1.0	1.1
56	0.2	0.3	0.3	0.4	0.5	0.6	0.6	0.7
58	0.1	0.1	0.1	0.2	0.3	0.3	0.3	0.4
	Subtract from observed degrees Baumé							
60	0.0	0.0	0.0	0.0	0.0	0.0	0.0	0 0
62	0.1	0.1	0.1	0.2	0.2	0.3	0.3	0.4
64	0.2	0.3	0.3	0.4	0.4	0.6	0.6	0.7
66	0.3	0.4	0.5	0.6	0.7	0.8	0.9	1.0
68	0.5	0.6	0.6	0.7	0.9	1.1	1.3	1.4
70	0.6	0.7	0.8	0.9	1.1	1.4	1.6	1.7
72	0.7	0.8	0.9	1.1	1.3	1.6	1.9	2.1
74	0.8	0.9	1.1	1.3	1.6	1.8	2.2	2.5
76	0.9	1.1	1.3	1.5	1.8	2.1	2.5	2.8
78	1.0	1.2	1.4	1.7	2.0	2.4	2.8	3.1
80	1.1	1.3	1.5	1.8	2.2	2.6	3.1	3.5
82	1.2	1.4	1.7	2.0	2.5	2.9	3.4	3.9
84	1.3	1.5	1.8	2.2	2.7	3.2	3.7	4.3
86	1.4	1.7	2.0	2.4	2.9	3.4	4.0	4.6
88	1.6	1.8	2.1	2.6	3.1	3.7	4.2	4.9
90	1.7	2.0	2.3	2.7	3.3	3.9	4.5	5.2
92	1.8	2.1	2.4	2.9	3.5	4.2	4.8	5.6
94	1.9	2.2	2.6	3.1	3.8	4.4	5.1	5.9
96	2.0	2.3	2.7	3.3	4.0	4.4	5.4	6.3
98	2.1	2.4	2.9	3.4	4.2	4.9	5.7	6.6
100	2.2	2.6	3.0	3.6	4.4	5.1	6.0	6.9
102	2.3	2.7	3.2	3.8	4.6	5.4	6.3	7.2
104	2.4	2.9	3.3	4.0	4.8	5.7	6.6	7.5
106	2.5	3.0	3.5	4.2	5.0	5.9	6.9	7.9
108	2.7	3.1	3.6	4.3	5.2	6.2	7.2	8.2
110	2.8	3.2	3.7	4.4	5.4	6.4	7.5	8.5
112	2.9	3.3	3.9	4.6	5.6	6.7	7.7	8.8
114	3.0	3.4	4.0	4.7	5.8	6.9	7.9	9.1
116	3.1	3.6	4.1	4.9	6.0	7.1	8.2	9.4
118	3.2	3.7	4.3	5.1	6.2	7.3	8.5	9.8
120	3.3	3.8	4.4	5.3	6.4	7.5	8.8	10.1

This table is calculated from the same data as Table II, *Circ.* 57, Bureau Standards.

[1] Reprinted by permission from "Handbook of Petroleum Industry," Vol. II, by the late David T. Day, published by John Wiley & Sons, Inc., New York, 1922.

Table 45.   Characteristics of Tested Oils[1]

| Oils tested | Gravity | | Pounds per gallon |
	Degrees Baumé	Specific gravity	
California, Bakersfield............................	16.0	0.9595	7.994
Louisiana, Jennings.................................	24.0	0.9105	7.585
Ohio, Lima..	37.0	0.8395	6.994
Oklahoma residuum................................	28.0	0.8870	7.390
Oklahoma residuum................................	24.0	0.9105	7.585
Oklahoma crude, (G. P.)...........................	32.2	0.8631	7.198
Oklahoma crude....................................	36.2	0.8423	7.023
Oklahoma crude....................................	35.4	0.8464	7.055
Pennsylvania......................................	43.7	0.8059	6.722
Pennsylvania......................................	38.2	0.8323	6.944
Russian, Baku.....................................	29.0	0.8815	7.344
Texas, Beaumont..................................	22.0	0.9220	7.681
Texas, Sour Lake..................................	20.0	0.9340	7.781
Texas, residuum...................................	18.0	0.9465	7.885
Texas, gas oil.....................................	27.9	0.8866	7.394
West Virginia.....................................	40.0	0.8250	6.873

[1] Reproduced from the "Union Engineering Handbook," 5th ed., by permission of the Union Steam Pump Co.

The material used for the preparation of the charts presented in this paper has been gathered from all available sources. It is of a truly representative character, and is not limited in any sense to the work of any single investigator, to any particular method, or to samples from any one field.

No positive relationship is claimed herein between the specific heat and specific gravity of petroleum oils, nor is inaccuracy imputed to data which do not agree with the average curves presented. From a practical and engineering standpoint, if not from that of the physicist, it is believed that results warrant the use of these mean data for purposes of design, where exact experimental data are not available for the particular petroleum oil being considered.

Method of Approach.—The study of the problem may be conveniently divided into five steps:

1. The variation of specific gravity with temperature.

2. The variation of specific heat with temperature.

3. The reduction of all data pertaining to specific gravity and specific heat to a standard temperature of 60 F and the determination of a mean relationship between these quantities at this temperature.

4. The preparation of a chart showing the mean relationship between specific gravity and specific heat for the range of the available data.

5. The extrapolation of these data to temperatures as high as 800 F, and the preparation of a chart showing the variation of both specific gravity and specific heat with temperature for this range, or as far as the critical temperature where this is lower than 800 F.

The above-mentioned steps are developed in detail in Eckart's paper until he presents the data in the final form, reproduced here as Fig. 34, regarding which he says:

Below 200 F the specific gravity curves agree with the Bureau of Standards Tables (*Circular* 154) .... Within this range of temperature (0 to 200 F) the variation of specific gravity with temperature is to all intents and purposes a straight-line function. Wilson and Bahlke have shown that this linear relation does not extend to high temperatures and that, with the exception of the heavier oils, it is not safe to use the linear relation beyond 200 F.

To meet the necessity of data for the design of high-temperature apparatus Fig. 34 [Fig. 3 in Professor Eckart's paper] is presented in which the specific heat and specific gravity have been plotted against temperature up to 880 F. From this chart reasonably accurate values may be obtained of both the specific heat and specific gravity of any oil up to a temperature of 800 F (or up to the critical temperature if this is less than 800 F) when the specific gravity at 60 F is known.

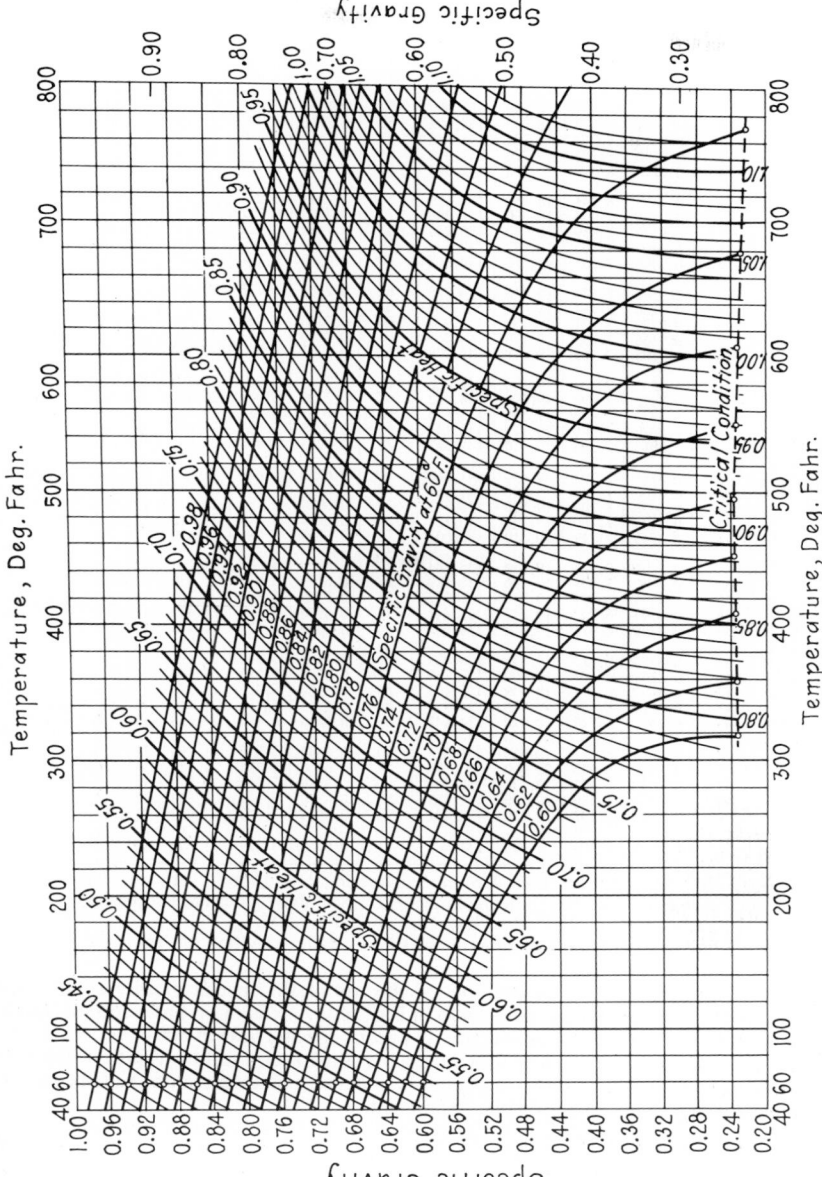

FIG. 34. Relation between specific gravity, specific heat, and temperature for petroleum oils.

Table 46.   Approximate Heating Value vs. Density of Average Fuel Oils[1]

Degrees Baumé[2]	Specific gravity at 60/60 F	Pounds per gallon at 60 F	B.t.u. per pound	B.t.u. per gallon at 60 F	Degrees Baumé[2]	Specific gravity at 60/60 F	Pounds per gallon at 60 F	B.t.u. per pound	B.t.u. per gallon at 60 F
10	ONE	8.331	18,650	158,090	51	0.7750	6.457	20,290	131,013
11	0.9930	8.273	18,690	154,622	52	0.7710	6.423	20,330	130,580
12	0.9860	8.214	18,730	153,848	53	0.7670	6.390	20,370	130,164
13	0.9790	8.156	18,770	153,088	54	0.7630	6.357	20,410	129,746
14	0.9720	8.098	18,810	152,323	55	0.7585	6.319	20,450	129,224
15	0.9655	8.044	18,850	151,629	56	0.7545	6.286	20,490	128,804
16	0.9595	7.994	18,890	151,006	77	0.7505	6.252	20,530	128,354
17	0.9530	7.939	18,930	150,285	58	0.7470	6.223	20,570	128,007
18	0.9465	7.885	18,970	149,578	59	0.7430	6.190	20,610	127,576
19	0.9400	7.831	19,010	148,867	60	0.7390	6.157	20,650	127,142
20	0.9340	7.781	19,050	148,228	61	0.7355	6.127	20,690	126,768
21	0.9230	7.731	19,090	147,585	62	0.7315	6.094	20,730	126,229
22	0.9220	7.681	19,130	146,928	63	0.7280	6.065	20,770	125,970
23	0.9165	7.635	19,170	146,363	64	0.7240	6.032	20,810	125,526
24	0.9105	7.585	19,210	145,708	65	0.7205	6.002	20,850	125,142
25	0.9045	7.536	19,250	145,068	66	0.7165	5.969	20,890	124,692
26	0.8990	7.490	19,290	144,482	67	0.7130	5.940	20,930	124,342
27	0.8930	7.440	19,330	143,815	68	0.7095	5.911	20,970	123,954
28	0.8870	7.390	19,370	143,144	69	0.7060	5.882	21,010	123,580
29	0.8815	7.344	19,410	142,547	70	0.7025	5.853	21,050	123,206
30	0.8755	7.294	19,450	141,868	71	0.6990	5.823	21,090	122,807
31	0.8700	7.248	19,490	141,263	72	0.6955	5.794	21,130	122,327
32	0.8650	7.206	19,530	140,733	73	0.6920	5.765	21,170	122,045
33	0.8595	7.160	19,570	140,121	74	0.6890	5.740	21,210	121,745
34	0.8545	7.119	19,610	139,604	75	0.6855	5.711	21,250	121,369
35	0.8490	7.073	19,650	138,984	76	0.6820	5.682	21,290	120,970
36	0.8440	7.031	19,690	138,440	77	0.6790	5.657	21,330	120,664
37	0.8395	6.994	19,730	137,992	78	0.6755	5.628	21,370	120,270
38	0.8345	6.952	19,770	137,441	79	0.6720	5.598	21,410	119,843
39	0.8295	6.911	19,810	136,907	80	0.6690	5.573	21,450	119,541
40	0.8250	6.873	19,850	136,429	81	0.6655	5.544	21,490	119,141
41	0.8205	6.836	19,890	135,960	82	0.6620	5.515	21,530	118,739
42	0.8155	6.794	19,930	135,405	83	0.6585	5.486	21,570	118,333
43	0.8110	6.756	19,970	134,917	84	0.6545	5.453	21,610	117,839
44	0.8065	6.719	20,010	134,447	85	0.6510	5.423	21,650	117,480
45	0.8015	6.677	20,050	133,874	86	0.6480	5.398	21,690	117,083
46	0.7970	6.640	20,090	133,398	87	0.6450	5.373	21,730	116,755
47	0.7925	6.602	20,130	132,398	88	0.6420	5.349	21,770	116.448
48	0.7885	6.569	20,170	132,497	89	0.6390	5.324	21,810	116,116
49	0.7840	6.632	20,210	132,012	90	0.6365	5.303	21,850	115,871
50	0.7795	6.494	20,250	131,504	91	0.6330	5.274	21,890	115,448

[1] Reproduced from the "Union Engineering Handbook," 5th ed., by permission of the Union Steam Pump Co.   See also "Thermal Properties of Petroleum Products," *Bur. Standards, Misc. Pub.* 97.

[2] For relation of API and Baumé scales to specific gravities, see Table 42.

Example. The following is an example of the use of the specific-gravity–specific-heat temperature chart for petroleum oils shown in Fig. 34.   A certain oil has a specific gravity of 0.64 at 60 F.   What are its specific gravity and its specific heat at 300 F?

Solution. Find the curve for 0.64 specific gravity at 60 F and follow it until it intersects the vertical ruling for 300 F.   The specific gravity at 300 F is then seen to be 0.48 and the specific heat 0.75 Btu per lb per degree Fahrenheit.

It should be noted that the specific heats read from the chart are *instantaneous* rather than *mean* specific heats.   If it is desired to calculate the quantity of heat required to raise the temperature of 1 lb of oil from temperature *A* to temperature *B*, it is necessary to determine by graphical integration the average or mean value of the specific heat between the limits *A* and *B*.   In the above example, for instance, if it

is desired to compute the number of Btu required to raise the temperature of 1 lb of that particular oil from 60 to 300 F, proceed as follows: The temperature increase is $300 - 60 = 240$ F. Divide this increase into any convenient number of steps, say six in this case, and average the specific heats read at the mid-points of the successive steps. Instantaneous specific heats will be read then at 80, 120, 160, 200, 240, and 280 F from Fig. 34 and averaged as follows:

At 80 F the instantaneous specific heat is 0.534 Btu/(lb F)
At 120 F the instantaneous specific heat is 0.573 Btu/(lb F)
At 160 F the instantaneous specific heat is 0.613 Btu/(lb F)
At 200 F the instantaneous specific heat is 0.652 Btu/(lb F)
At 240 F the instantaneous specific heat is 0.691 Btu/(lb F)
At 280 F the instantaneous specific heat is 0.730 Btu/(lb F)
Average specific heat 60 to 300 F = $\overline{3.793} \div 6 = 0.632$ Btu/(lb F)

NOTE. It so happens that the example chosen falls in the portion of the curve where the specific-heat–specific-gravity relation is very nearly a straight-line function. Consequently, in this instance the average of the specific heats at 60 and 300 F happens to given an identical result with that obtained by the more lengthy solution above:

At 60 F the instantaneous specific heat is 0.514 Btu/(lb F)
At 300 F the instantaneous specific heat is 0.750 Btu/(lb F)
Average specific heat 60 to 300 F = $\overline{1.264} \div 2 = 0.632$ Btu/(lb F)

The heat transfer required to raise the temperature of 1 lb of this oil from 60 to 300 F then is $240 \times 0.632 = 151.68$ Btu.

Viscosity of Oils. *Viscosity* is the internal friction within a fluid which tends to resist the sliding of one particle or layer over another. Low viscosity means that the particles slide easily, as in water, gasoline, and thin, light oils. High viscosity means that the particles do not slide easily, as in tar and asphalt-base crude oils. Viscosity is a property which is also manifested by gases and vapors, where, of course, it is relatively very low. The viscosity of liquids decreases as the temperature increases. Since viscosity has a very marked effect on the pressure drop occurring when oil flows through a pipe, it is sometimes desirable to heat the more viscous oils to obtain easier flow. Similarly, where oil-transmission pipelines extend through regions having cold winters, it is advisable to lay the pipe below the frost line.

For most commercial purposes it is desirable to make an independent determination of the viscosity of the liquid in question. This can be done in a practical way by employing a standard instrument, such as a Saybolt Universal viscometer or Saybolt Furol, Engler, or other commonly used viscometer.[1] A convenient chart for converting such viscometer readings into kinematic viscosity is given in Fig. 35.

The kinematic viscosity of a fluid is equal to its absolute viscosity divided by its density at the temperature at which the viscosity is measured. These relations expressed in metric and English units are, respectively, as follows:

$$\text{Kinematic viscosity in sq cm/sec} = \frac{\text{absolute viscosity in poises}}{\text{density in g/cu cm}}$$

$$= \frac{\text{absolute viscosity in centipoises}}{100 \text{ density in g/cu cm}} = \frac{z}{100y}$$

$$= \frac{\text{absolute viscosity in centipoises}}{\text{specific gravity}} = \frac{z}{100S}$$

[1] See ASTM D88 for "Standard Method of Test for Viscosity by Means of the Saybolt Viscosimeter," or ASTM D445 for "Method of Test for Kinematic Viscosity."

where poise = 1 g mass/(cm sec) = cgs unit of absolute viscosity
  centipoise = 0.01 poise
  stoke = 1 sq cm/sec = cgs unit of kinematic viscosity
  centistoke = 0.01 stoke
  centipoises = centistokes × density (at designated temperature)
  y = density, g/cu cm

$$\text{Kinematic viscosity in sq ft/sec} = \frac{\text{absolute viscosity in lbm/(ft sec)}}{\text{density in lb/cu ft}}$$

No name for the English unit of kinematic viscosity is generally recognized. Conversion from one set of units of kinematic viscosity to the other may be

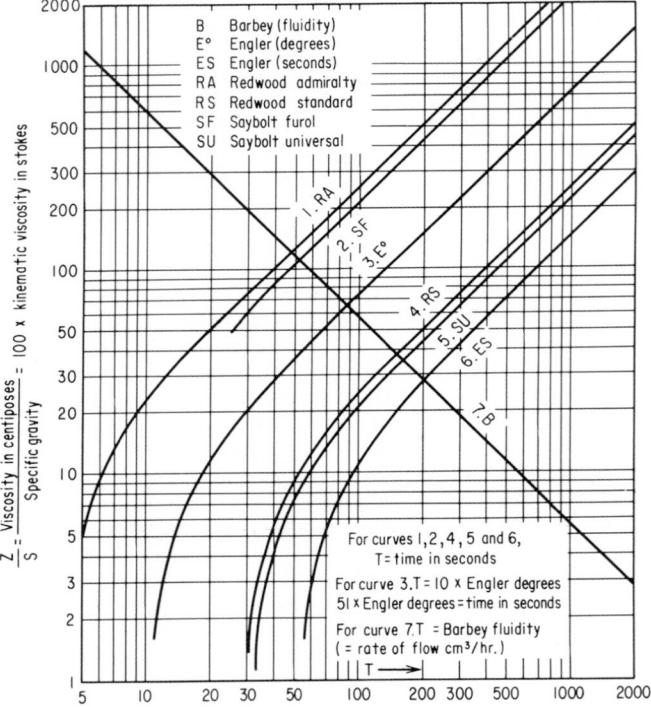

FIG. 35. Viscometer time of efflux vs. z/S. NOTE: The value of z/S when time of efflux T exceeds 2,000 is 10 times the value of z/S corresponding to $T/10$. (*Reproduced by permission from International Critical Tables.*)

accomplished by the following relation: a kinematic viscosity of 1 sq ft/sec equals 929 times the kinematic viscosity in square centimeters per second.

The following formula may be used to determine kinematic viscosity from Saybolt Universal viscometer readings:

$$\text{Kinematic viscosity in sq cm/sec} = 0.0022T - 1.8/T$$

where T is the Saybolt time in seconds.

The corresponding formula for Saybolt Furol is

$$\text{Kinematic viscosity in sq cm/sec} = 0.022T - 2.03/T$$

(See ASTM D446 for Standard Method for Conversion of Kinematic Viscosity to Saybolt Universal Viscosity.)

To convert viscometer readings into English units, the following will be helpful. In the equations t represents time of efflux in seconds and v represents the kinematic viscosity in square feet per second.

Saybolt Universal:	$v = 0.00000237t - 0.00194/t$	(72)
Saybolt Furol:	$v = $ approximately $\frac{1}{10}$ of Saybolt Universal	
Engler:	$v = 0.00000158t - 0.00403/t$	(73)
Redwood:	$v = 0.00000280t - 0.00185/t$	(74)

Temperature-viscosity Relations of Petroleum Oils. Temperature-viscosity charts on semilogarithmic paper are a convenient means for ascertaining the viscosity of a petroleum oil at any temperature within the liquid range, provided viscosities at two temperatures are known. Conversely, such charts may be used to ascertain the temperature at which a desired viscosity is attained. On semilogarithmic charts the viscosity-temperature points for any given petroleum oil lie on a straight line. The procedure is to plot the two known temperature-viscosity points on the chart and to draw a sharply defined straight line carefully through them. A point on this line, if within the liquid range, shows the viscosity at the corresponding desired temperature and vice versa. Viscosity-temperature points may be used on the extrapolated portion of the line, but for satisfactory extrapolation the two known points should be far apart.

Blank charts for this purpose on semilogarithmic paper may be purchased from the American Society for Testing Materials, 1916 Race St., Philadelphia 3, Pa., in connection with ASTM Standard Designation D341, which also has been approved as American Standard ASA Z11.39 and American Petroleum Institute Standard API 533. These standards are all known as "Standard Viscosity-temperature Charts for Liquid Petroleum Products." Four styles of chart blanks are available in different ranges, two being for temperature vs. Saybolt Universal viscosity, and two for temperature vs. kinematic viscosity. For the purposes of this handbook, however, the combined chart shown in Fig. 36 is considered more suitable and can be made by superposing additional scales on the standard charts. The chart of Fig. 36 was developed by Neil McCoull of the Texas Company and with the following examples is reproduced here by permission from *Bull.* 25 of the Kansas City Testing Laboratory.

Example. Oil "5" has a viscosity of 500 sec Saybolt Universal at 100 F and 55 sec at 210 F. Drawing the line shows the viscosity at 130 F to be 200. Or if a viscosity of 100 is desired it will be obtained by heating this oil to 161 F. Given the viscosity of an oil at only one temperature and knowing the source of the oil, it is possible to draw the viscosity-temperature line with a fair degree of accuracy.

Example. Oils 3, 4, 5, and 6 are all 500 viscosity at 110 F; 3 is a Pennsylvania blend; 4 is a mid-continent paraffin-base blend; 5 is a Gulf-coastal viscous neutral, and 6 is a California oil; 3 has the highest Baumé gravity and 6 the lowest of the the group. The gravity will often help to identify the source of the oil and assist in eatablishing the slope.

The straight-line feature does not hold good below the temperature of wax separation.

Example. 7 is a paraffin-base spindle oil having a cloud test of 45 F and a pour test of 35 F. The viscosity-temperature line begins to bend upward at 45 and is vertical before reaching 30. Heavy residual stocks will have lines similarly bent upward but not so abruptly; 2 is a heavy residue and begins to bend at above 100 F becoming vertical at about 40 F. The long gradual bend is characteristic of slightly cracked stocks resulting from inadequate steaming.

Oils containing no paraffin do not bend within the limits of the chart but may be considered to become vertical at the pour test which for these oils corresponds to a viscosity of 50,000 Saybolt Furol.

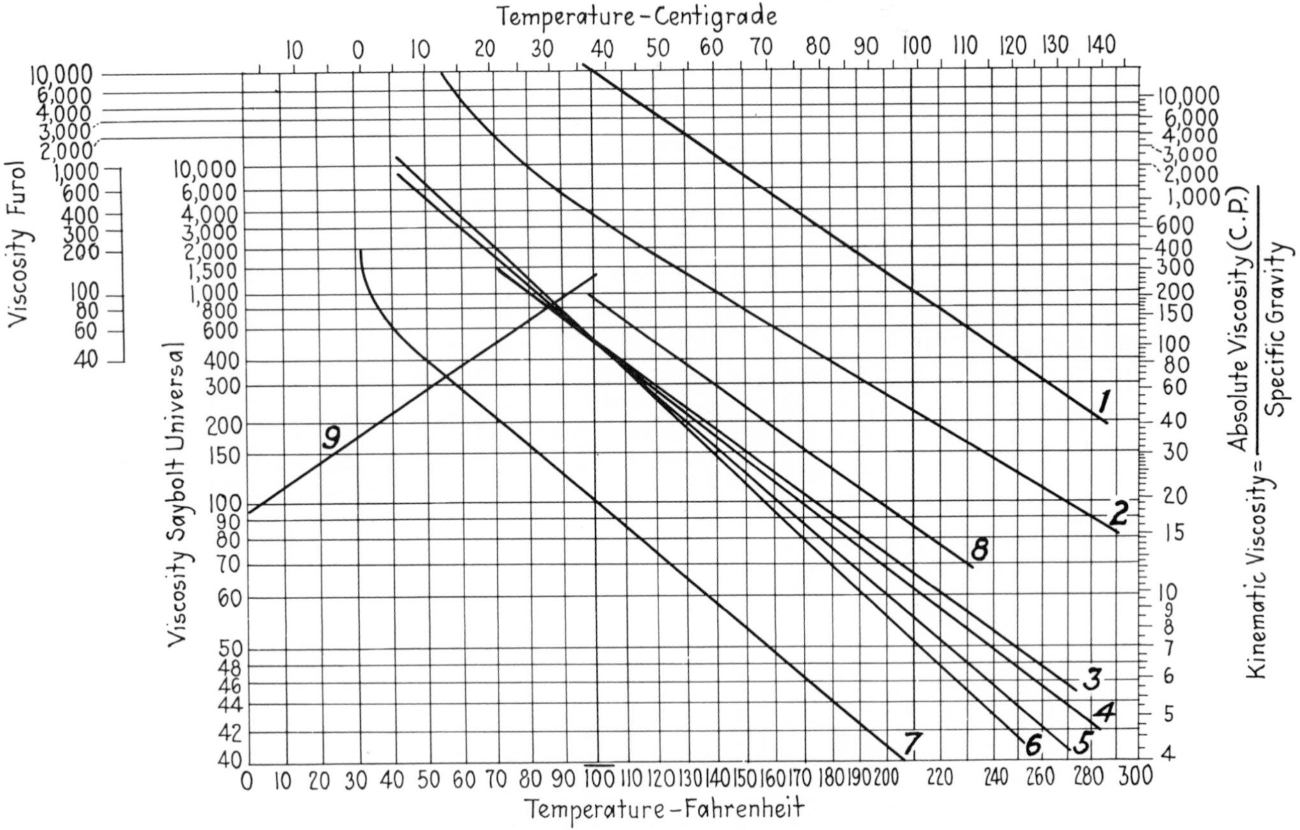

FIG. 36. Temperature-viscosity relations for petroleum oils.

$$\text{Kinematic Viscosity} = \frac{\text{Absolute Viscosity (C.P.)}}{\text{Specific Gravity}}$$

Temperature – Centigrade

Temperature – Fahrenheit

Viscosity Saybolt Universal

Viscosity Furol

Blends may be determined on the basis of this chart. Given the viscosity-temperature lines of two oils, their blends will be inversely proportional to the relative distances to the known lines.

Example. 8 is a blend of two parts of 4 with one part of 2, the distance from 2 being twice as far as from 4.

To determine the proportions to be used to get a viscosity of 135 at 130 F using stocks having viscosities of 95 and 1,300 at 130.

Example. Line 9 connects 95 viscosity with 1,300 viscosity crossing all intervening viscosities. Using the scale at the bottom as a percentage scale (0–100) the desired viscosity is shown as calling for 19 per cent of the heavy oil and 81 per cent of the lighter. Actual viscosities (not stock viscosities) must be had for accurate percentage results when using the chart for this purpose.

Conversely, the makeup of an oil may be known if its viscosity and its constituent oils are known. In fact, due to the straight-line characteristics of the chart, many problems which have previously depended on formulas more or less approximate can be settled directly by translation to the chart.

The use of absolute viscosities is becoming noticeable in the commercial as well as the engineering world. The right margin of the chart provides the values from which the absolute viscosity in terms of centipoises may be obtained by multiplying the value given by the specific gravity of the subject oil at the given temperature. To convert to poises divide the centipoises by 100.

Example. Determine the absolute viscosity in dynes of oil represented by line 5 at 120 F. Chart shows the kinematic viscosity of this oil at 120 F to be 56. Specific gravity at 120 F is 0.906. Therefore the absolute viscosity is $56 \times 0.906 = 49.7$ centipoises or 0.497 poise or dyne sec/sq cm.

The variation of viscosity as a function of temperature is shown for several different liquids in Fig. 37. The variation of specific gravity with temperature is shown for a number of liquids, among them some light oils, in Fig. 38.

FRICTION LOSS IN PIPE AND DUCTS

For the steady flow of any fluid which obeys Newton's second law,[1] the Euler equation can be written in the form

$$h_f = (z_1 - z_2) + (V_1^2 - V_2^2)/2g - \int_1^2 dP/\rho \qquad (75)$$

where h_f = head loss due to friction, ft head of flowing fluid
z = elevation above datum plane, ft
V = velocity, fps
P = static pressure of fluid, lb/sq ft abs
ρ = density of fluid, lb/cu ft

Subscripts 1 and 2 refer to sections along the pipeline.

If the relationship between P and ρ is known, the integral expression in Eq. (75) can be evaluated; such integration was carried out in connection with final results presented for the compressible isentropic flow of a perfect gas and for the compressible flow of a perfect gas with friction and with heat transfer.

If the flowing fluid is incompressible, the density term may be removed from beneath the integral sign; the integration can then be performed directly with results as follow:

$$h_f = (z_1 - z_2) + (V_1^2 - V_2^2)/2g + (P_1 - P_2)/\rho \qquad (76)$$

[1] This, of course, excludes all non-Newtonian fluids such as sludges and slurries. Treatment of the flow problem for such substances is more complex and is discussed in Chap. 19, "Flow of Sludges and Slurries."

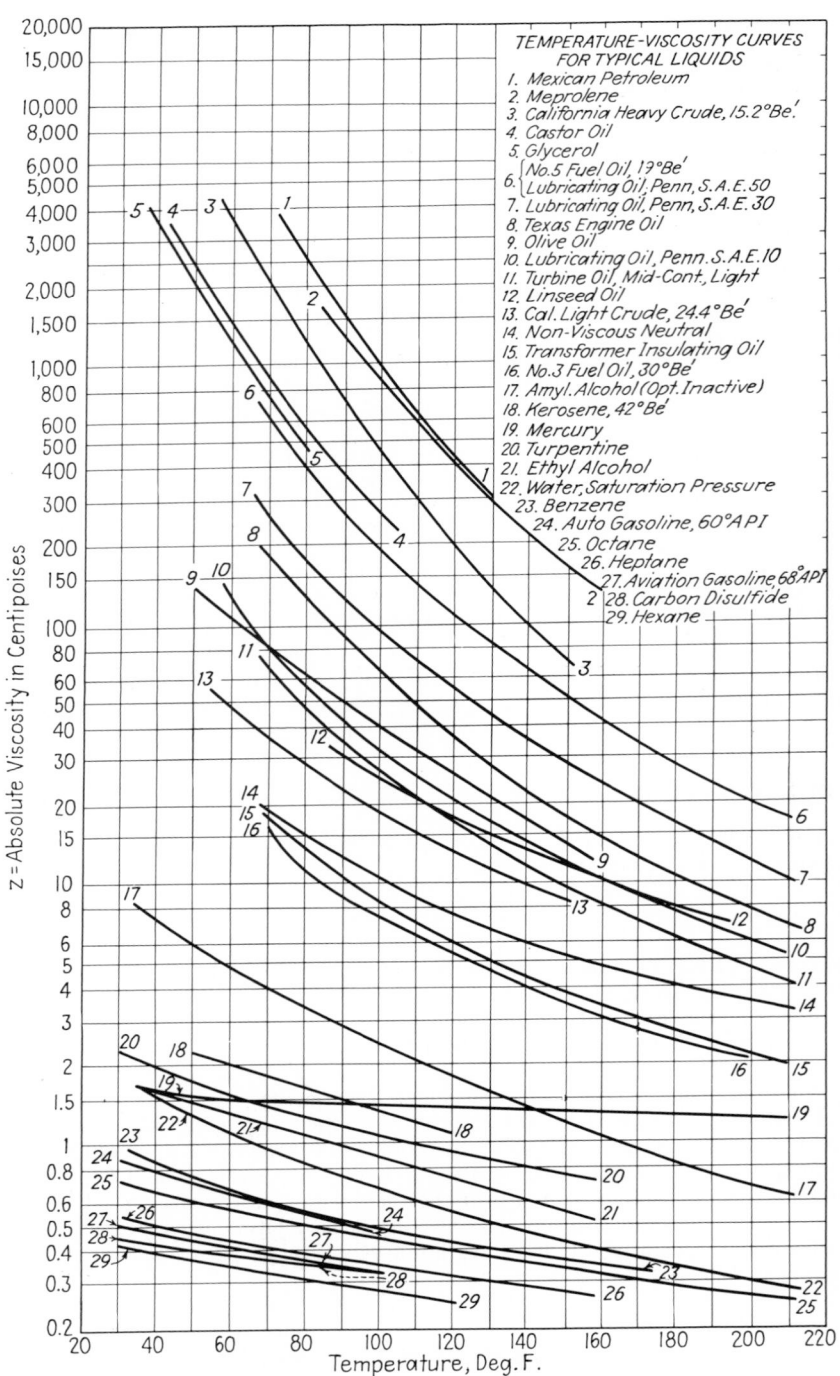

FIG. 37. Temperature vs. viscosity curves for liquids.

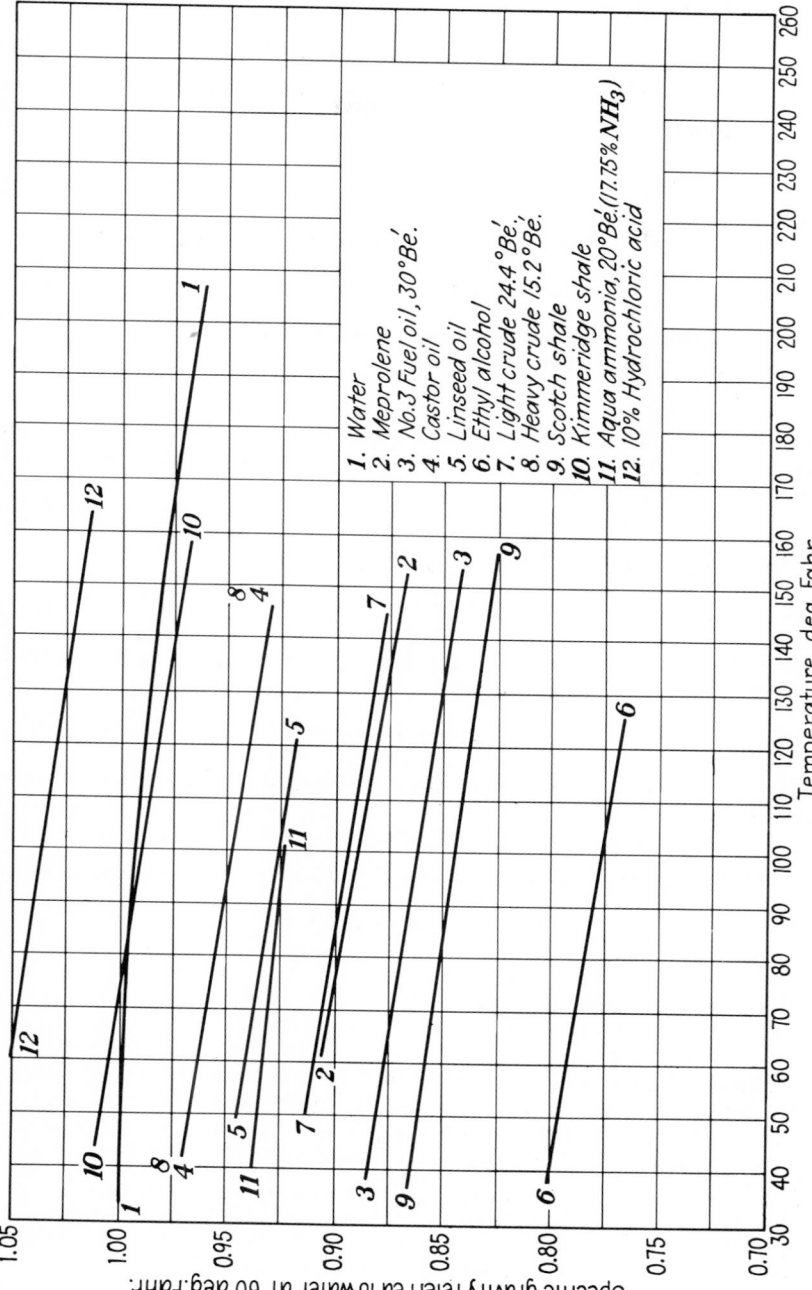

FIG. 38. Temperature–specific-gravity curves for typical liquids.

1. Water
2. Meprolene
3. No.3 Fuel oil, 30°Bé.
4. Castor oil
5. Linseed oil
6. Ethyl alcohol
7. Light crude 24.4°Bé.
8. Heavy crude 15.2°Bé.
9. Scotch shale
10. Kimmeridge shale
11. Aqua ammonia, 20°Bé.(17.75% NH_3)
12. 10% Hydrochloric acid

Temperature, deg. Fahr.

Specific gravity referred to water at 60 deg. Fahr.

Thus, for an incompressible frictionless fluid in steady flow, it is seen that the sums of

$$-(P_1 - P_2)/\rho, \text{ the net vapor pressure head}$$

$$-(V_1^2 - V_2^2)/2g, \text{ the net kinetic-energy head}$$

$$-(z_1 - z_2), \text{ the net potential energy head}$$

and h_f, the friction head

are equal to zero.

It can be shown that the friction-head term may be expressed also as

$$h_f = f(L/D)(V_2/2g) \tag{77}$$

where L = length of conduit associated with flow process
       D = diameter (hydraulic) of conduit
       V = velocity of flow, fps
       f = friction factor

It is often convenient to have available expressions, as follow, for equivalent friction loss in other dimensions and expressed as functions of mass flow, rather than velocity.   For example,

$$\Delta P = f\rho(L/D)(V^2/2g) \tag{77a}$$

where ΔP = pressure drop due to friction, lb/sq ft
       ρ = density, lb/cu ft

$$\Delta P = 0.0252 f v L \dot{m}^2 / D^5 \tag{77b}$$

where P is in pounds per square foot, v is in cubic feet per pound, $\dot{m}$ is in pounds per second, L is in feet, and D is in feet.

$$\Delta P = 0.000175 f v L \dot{m}^2 / d^5 \tag{77c}$$

where ΔP is in pounds per square inch and all other terms are in the same dimensions as used in Eq. (77b).

$$\Delta P = 43.4 f v L \dot{m}^2 / d^5 \tag{77d}$$

where ΔP is in pounds per square inch, d is pipe diameter in inches, and all other terms are unchanged.

$$\Delta P = f v L M^2 / 298,000 d^5 \tag{77e}$$

where ΔP = pressure drop due to friction, psi
       v = specific volume, cu ft/lb
       M = mass flow rate, lb/hr
       d = pipe diameter, in.

The solution of any problem involving the steady friction flow of any incompressible Newtonian fluid requires the determination of the friction factor, f. This will be the subject of succeeding paragraphs. However, to assure understanding of the development thus far, some illustrative examples will be given now.

Example 1. A boiler-feed pump takes its supply from a deaerating heater (see Fig. 39) which is maintained at a pressure of 100 psia.   The interconnecting piping is 8-in. Schedule 40, and the flow rate is 400,000 lb/hr of saturated water.   Water level in the heater is maintained at elevation 75 ft 6 in., and the pump center line is installed at elevation 3 ft 6 in. If the friction drop in the suction piping is estimated to be 8 psia, what would be the reading of (a) a Bourdon-tube pressure gage and (b) a pitot-tube stagnation pressure gage each of which is located right at the pump inlet?

Solution. Taking section 1 as the maintained level in the heater and section 2 as the pump inlet, Eq. (76) may be written as

$$(P_1 - P_2)/\rho = h_f - (z_1 - z_2)$$
$$- (V_1^2 - V_2^2)/2g$$

The tank cross-sectional area is large as compared with that of the piping. Thus V_1 may be considered to be negligible. The pipe velocity V_2 is calculated as follows:

$$V_2 = \dot{m}v_2/A_2 = (400{,}000/3{,}600)$$
$$\times 0.01774/0.3474 = 5.68 \text{ fps}$$

in which v_2 is v_f read from Table 29 and A_2 is the cross-sectional area of an 8-in. Schedule 40 pipe as read from Table 47.

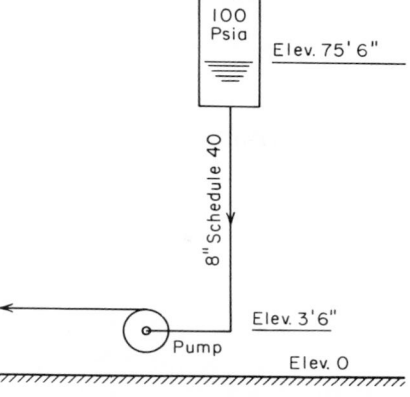

FIG. 39. Sketch for illustrative example.

The potential-energy term $z_1 - z_2$ is obviously the difference in elevation between the water level in the heater and the pump center line. Then

$$P_1 - P_2 = \rho h_f - \rho(z_1 - z_2) + \frac{\rho(V_1^2 - V_2^2)}{2g}$$

Table 47. Inside-diameter Functions of Steel Pipe[1] Conforming to ASA B36.10

Nominal pipe size	Outside diameter D	Wall thickness t	Inside diameter d	Inside diameter squared d^2	Inside diameter cubed d^3	Inside diameter, fourth power d^4	Inside diameter, fifth power d^5	Internal cross-sectional area Square inches	Internal cross-sectional area Square feet^2
Schedule 10:									
14 O.D.	14.0	0.250	13.50	182.25	2,460	33,215	448,403	143.14	0.994
16 O.D.	16.0	0.250	15.50	240.25	3,724	57,720	894,660	188.69	1.310
18 O.D.	18.0	0.250	17.50	306.25	5,359	93,789	1,641,307	240.53	1.670
20 O.D.	20.0	0.250	19.50	380.25	7,415	144,590	2,819,505	298.65	2.074
24 O.D.	24.0	0.250	23.50	552.25	12,978	304,980	7,167,030	433.74	3.012
30 O.D.	30.0	0.312	29.376	862.95	25,350	744,680	21,875,768	677.76	4.706
Schedule 20:									
8	8.625	0.250	8.125	66.02	536.4	4,358	35,409	51.85	0.3601
10	10.75	0.250	10.25	105.06	1,077	11,038	113,141	82.52	0.5731
12	12.75	0.250	12.25	150.06	1,838	22,519	275,855	117.86	0.8185
14 O.D.	14.0	0.312	13.376	178.92	2,393	32,011	428,185	140.52	0.9758
16 O.D.	16.0	0.312	15.376	236.42	3,635	55,895	859,442	185.69	1.290
18 O.D.	18.0	0.312	17.376	301.92	5,246	91,159	1,583,978	237.13	1.647
20 O.D.	20.0	0.375	19.25	370.56	7,133	137,317	2,643,344	290.04	2.014
24 O.D.	24.0	0.375	23.25	540.56	12,568	292,208	6,793,832	424.56	2.948
30 O.D.	30.0	0.500	29.00	841.00	24,389	707,281	20,511,149	660.52	4.587
Schedule 30:									
8	8.625	**0.277**	8.071	65.14	525.7	4,243	34,248	51.16	0.3553
10	10.75	**0.307**	10.136	102.74	1,041	10,555	106,987	80.69	0.5603
12	12.75	**0.330**	12.09	146.17	1,767	21,365	258,304	114.80	0.7972
14 O.D.	14.0	0.375	13.25	175.56	2,326	30,822	408,394	137.88	0.9575
16 O.D.	16.0	0.375	15.25	232.56	3,546	54,085	824,801	182.65	1.268
18 O.D.	18.0	0.437	17.126	293.30	5,023	86,025	1,473,261	230.36	1.600
20 O.D.	20.0	0.500	19.0	361.00	6,859	130,321	2,476,099	283.53	1.969
24 O.D.	24.0	0.562	22.876	523.31	11,971	273,855	6,264,703	411.00	2.854
30 O.D.	30.0	0.625	28.75	826.56	23,764	683,206	19,642,160	649.18	4.508

Table 47. Inside-diameter Functions of Steel Pipe[1]
Conforming to ASA B36.10 (Continued)

Nominal pipe size	Outside diameter D	Wall thickness t	Inside diameter d	Inside diameter squared d^2	Inside diameter cubed d^3	Inside diameter fourth power d^4	Inside diameter fifth power d^5	Internal cross-sectional area Square inches	Internal cross-sectional area Square feet^2
Schedule 40:									
⅛	0.405	0.068	0.269	0.0724	0.0195	0.00524	0.00141	0.0569	0.00040
¼	0.540	0.088	0.364	0.1325	0.0482	0.01755	0.00639	0.1041	0.00072
⅜	0.675	0.091	0.493	0.2430	0.1198	0.05907	0.02912	0.1909	0.00133
½	0.840	0.109	0.622	0.3869	0.2406	0.1497	0.09310	0.3039	0.00211
¾	1.050	0.113	0.824	0.679	0.5595	0.4610	0.3799	0.5333	0.00371
1	1.315	0.133	1.049	1.100	1.154	1.211	1.270	0.8639	0.00600
1¼	1.660	0.140	1.380	1.904	2.628	3.627	5.005	1.495	0.01040
1½	1.900	0.145	1.610	2.592	4.173	6.719	10.82	2.036	0.01414
2	2.375	0.154	2.067	4.272	8.831	18.25	37.72	3.356	0.02330
2½	2.875	0.203	2.469	6.096	15.05	37.16	91.75	4.788	0.03322
3	3.5	0.216	3.068	9.413	28.88	88.60	271.8	7.393	0.05130
3½	4.0	0.226	3.548	12.59	44.66	158.5	562.2	9.888	0.06870
4	4.5	0.237	4.026	16.21	65.26	262.7	1,058	12.73	0.08840
5	5.563	0.258	5.047	25.47	128.6	648.8	3,275	20.01	0.1390
6	6.625	0.280	6.065	36.78	223.1	1,353	8,206	28.89	0.2006
8	8.625	0.322	7.981	63.70	508.4	4,057	32,380	50.03	0.3474
10	10.75	0.365	10.02	100.4	1,006	10,080	101,000	78.85	0.5475
12	12.75	0.375	12.00	144.0	1,728	20,736	248,832	113.1	0.7854
12	12.75	0.406	11.938	142.5	1,701	20,311	242,470	111.93	0.7773
14 O.D.	14.0	0.437	13.126	172.3	2,262	29,684	389,638	135.32	0.9397
16 O.D.	16.0	0.500	15.000	225.0	3,375	50,625	759,375	176.72	1.2272
18 O.D.	18.0	0.562	16.876	284.8	4,806	81,110	1,368,820	223.68	1.5533
20 O.D.	20.0	0.593	18.814	354.0	6,650	125,292	2,357,244	278.00	1.9305
24 O.D.	24.0	0.687	22.626	511.9	11,583	262,078	5,929,784	402.07	2.7921
Schedule 60:									
8	8.625	0.406	7.813	61.04	476.9	3,726	29,113	47.94	0.3329
10	10.75	0.500	9.750	95.06	926.9	9,037	88,110	74.66	0.5185
12	12.75	0.500	11.75	138.1	1,622	19,061	223,966	108.4	0.7527
12	12.75	0.562	11.626	135.16	1,571	18,269	212,399	106.16	0.7372
14 O.D.	14.0	0.593	12.814	164.20	2,104	26,961	345,480	128.96	0.8956
16 O.D.	16.0	0.656	14.688	215.74	3,169	46,543	683,618	169.44	1.1766
18 O.D.	18.0	0.750	16.500	272.25	4,492	74,120	1,222,947	213.82	1.4848
20 O.D.	20.0	0.812	18.376	337.68	6,205	114,026	2,095,342	265.21	1.8417
24 O.D.	24.0	0.968	22.064	486.82	10,741	236,989	5,228,934	382.34	2.6550

Rearranging, $$P_2 = P_1 - \rho h_f + \rho(z_1 - z_2) + \frac{\rho(V_1^2 - V_2^2)}{2g}$$

or $$P_2 = P - \frac{h_f}{v} + \frac{z_1 - z_2}{v} + \frac{V_1^2 - V_2^2}{2gv}$$

$$P_2 = 100(144) - 8(144) + \frac{75.5 - 3.5}{0.01774} + \frac{0 - 5.68^2}{(0.01774)(2)(32.17)}$$

(a) $P_2 = 17,270$ lb/sq ft $= 120$ psia, or a static pressure gage located at the pump inlet would read about $(120 - 14.7)$ or 105 psi. It is noted here that, despite friction and acceleration losses, pressure at the pump inlet is greater than the vapor pressure at the water temperature.   The excess, in this case 20 psi, is referred to as the available net positive suction head (NPSH).   It is essential when pumping hot liquids to make available to the pump a value of NPSH as specified by the manufacturer.   As is evident in this problem, this specification can be met by installation of the suction storage vessel a minimum distance above the pump.

(b) The stagnation pressure can be easily determined by application of the Bernoulli theorem at the pump inlet, thus

$$(P_{02} - P_2)/\rho = V_2^2$$

or $$P_{02} = P_2 + \rho(V_2^2/2g) = 17,270 + 5.68^2/(0.01774)(2)(32.17)$$

Table 47. Inside-diameter Functions of Steel Pipe[1] Conforming to ASA B36.10 (Continued)

Nominal pipe size	Outside diameter D	Wall thickness t	Inside diameter d	Inside diameter squared d^2	Inside diameter cubed d^3	Inside diameter, fourth power d^4	Inside diameter, fifth power d^5	Internal cross-sectional area Square inches	Internal cross-sectional area Square feet^2
Schedule 80:									
⅛	0.405	0.095	0.215	0.0462	0.0099	0.00214	0.000459	0.0363	0.00025
¼	0.540	0.119	0.302	0.0912	0.0275	0.00832	0.002513	0.0716	0.00050
⅜	0.675	0.126	0.423	0.1789	0.0757	0.03201	0.01354	0.1405	0.00098
½	0.840	0.147	0.546	0.2981	0.1628	0.08887	0.04852	0.2341	0.00163
¾	1.050	0.154	0.742	0.5506	0.4085	0.3031	0.2249	0.4324	0.00300
1	1.315	0.179	0.957	0.9158	0.8765	0.8388	0.8027	0.7193	0.00499
1¼	1.660	0.191	1.278	1.633	2.087	2.667	3.409	1.283	0.00891
1½	1.900	0.200	1.500	2.250	3.375	5.062	7.594	1.767	0.01225
2	2.375	0.218	1.939	3.760	7.290	14.14	27.41	2.953	0.02050
2½	2.875	0.276	2.323	5.396	12.54	29.12	67.64	4.238	0.02942
3	3.5	0.300	2.900	8.410	24.39	70.73	205.1	6.605	0.04587
3½	4.0	0.318	3.364	11.32	38.07	128.1	430.8	8.891	0.06170
4	4.5	0.337	3.826	14.64	56.00	214.3	819.8	11.50	0.07986
5	5.563	0.375	4.813	23.16	111.5	536.6	2,583	18.19	0.1263
6	6.625	0.432	5.761	33.19	191.2	1,102	6,346	26.07	0.1810
8	8.625	0.500	7.625	58.14	443.3	3,380	25,775	45.66	0.3171
10	10.75	0.593	9.564	91.47	874.8	8,367	80,020	71.84	0.4989
12	12.75	0.687	11.376	129.41	1,472	16,784	190,523	101.64	0.7058
14 O.D.	14.0	0.750	12.500	156.25	1,953	24,414	305,176	122.72	0.8522
16 O.D.	16.0	0.843	14.314	204.89	2,933	41,980	600,904	160.92	1.1175
18 O.D.	18.0	0.937	16.126	260.05	4,194	67,625	1,090,518	204.24	1.4183
20 O.D.	20.0	1.031	17.938	321.77	5,772	103,537	1,857,248	252.72	1.7550
24 O.D.	24.0	1.218	21.564	465.01	10,027	216,230	4,662,798	365.22	2.5362
Schedule 100:									
8	8.625	0.593	7.439	55.34	411.7	3,062	22,781	43.46	0.3018
10	10.75	0.718	9.314	86.75	799.5	7,526	69,357	68.13	0.4732
12	12.75	0.843	11.064	122.41	1,354	14,985	165,791	96.14	0.6677
14 O.D.	14.0	0.937	12.126	147.04	1,783	21,621	262,173	115.49	0.8020
16 O.D.	16.0	1.031	13.938	194.27	2,708	37,740	526,020	152.58	1.0596
18 O.D.	18.0	1.156	15.688	246.11	3,861	60,572	950,250	193.30	1.3423
20 O.D.	20.0	1.281	17.438	304.08	5,303	92,474	1,612,536	238.82	1.6584
24 O.D.	24.0	1.531	20.938	438.40	9,179	192,190	4,024,074	344.32	2.3910

and
$$P_{02} = 17,306 \text{ lb/sq ft} = 120.2 \text{ psia}$$

That is, the stagnation pressure is only slightly more than the static pressure.

Example 2. Two measuring stations, 1 and 2, separated by a distance of 500 ft are used for determining the friction drop and friction factor in a 12-in. Schedule 80 steam line which conveys 750,000 lb/hr of steam. At station 1, the static pressure and static temperature are 1,000 psia and 1000 F. At station 2, the static pressure is 950 psia and the static temperature is 1000 F. It is required to determine the measured friction factor for this condition.

Solution. By use of Eq. (77),
$$f = h_f(D/L)(2g/V^2)$$
where h_f must be in feet.

Clearly, the velocity at station 1 is different from that at station 2 because of the difference in static pressure at the two locations. Also, because of the difference in density, the velocity of the fluid will vary along the pipe length. If it is essential to know precisely the friction factor, the method outlined previously for Fanno-line flow could be used. For most ordinary engineering purposes, however, sufficient accuracy is obtained by basing calculation on specific volume and velocity at the arithmetic average of pressure and temperature. Proceeding on this basis,

$$P_{avg} = (P_1 + P_2)/2 = (1,000 + 950)/2 = 975 \text{ psia}$$
$$T_{avg} = 1000 \text{ F throughout}$$

Table 47.   Inside-diameter Functions of Steel Pipe[1]
Conforming to ASA B36.10 (*Continued*)

Nominal pipe size	Outside diameter D	Wall thickness t	Inside diameter d	Inside diameter squared d^2	Inside diameter cubed d^3	Inside diameter, fourth power d^4	Inside diameter, fifth power d^5	Internal cross-sectional area Square inches	Internal cross-sectional area Square feet[2]
Schedule 120:									
4	4.5	0.437	3.626	13.15	47.67	172.9	626.8	10.33	0.0717
5	5.563	0.500	4.563	20.82	95.00	433.5	1,978	16.35	0.1136
6	6.625	0.562	5.501	30.26	166.5	915.7	5,037	23.77	0.1650
8	8.625	0.718	7.189	51.68	371.6	2,671	19,202	40.59	0.2819
10	10.75	0.843	9.064	82.16	744.7	6,750	61,179	64.53	0.4481
12	12.75	1.000	10.750	115.56	1,242	13,355	143,563	90.76	0.6303
14 O.D.	14.0	1.093	11.814	139.57	1,649	19,480	230,137	109.62	0.7612
16 O.D.	16.0	1.218	13.564	183.98	2,496	33,849	459,133	144.50	1.0035
18 O.D.	18.0	1.375	15.250	232.56	3,547	54,092	824,890	182.65	1.2683
20 O.D.	20.0	1.500	17.000	289.00	4,913	83,521	1,419,857	226.98	1.5762
24 O.D.	24.0	1.812	20.376	415.18	8,460	172,381	3,512,338	326.08	2.2643
Schedule 140:									
8	8.625	0.812	7.001	49.01	343.1	2,402	16,819	38.50	0.2673
10	10.75	1.000	8.750	76.56	669.9	5,862	51,291	60.13	0.4176
12	12.75	1.125	10.500	110.25	1,158	12,155	127,628	86.59	0.6013
14 O.D.	14.0	1.250	11.500	132.25	1,521	17,490	201,136	103.87	0.7213
16 O.D.	16.0	1.437	13.126	172.29	2,262	29,686	389,670	135.32	0.9397
18 O.D.	18.0	1.562	14.876	221.30	3,292	48,972	728,502	173.80	1.2070
20 O.D.	20.0	1.750	16.500	272.25	4,492	74,120	1,222,981	213.82	1.4849
24 O.D.	24.0	2.062	19.876	395.06	7,852	156,069	3,102,022	310.28	2.1547
Schedule 160:									
½	0.840	0.187	0.466	0.2172	0.1012	0.04716	0.02197	0.1706	0.00118
¾	1.050	0.218	0.614	0.3370	0.2315	0.1421	0.08726	0.2961	0.00206
1	1.315	0.250	0.815	0.6642	0.5413	0.4412	0.3596	0.5217	0.00362
1¼	1.660	0.250	1.160	1.346	1.561	1.811	2.110	1.057	0.00734
1½	1.900	0.281	1.338	1.790	2.395	3.205	4.288	1.406	0.00976
2	2.375	0.343	1.689	2.853	4.818	8.138	13.74	2.241	0.01556
2½	2.875	0.375	2.125	4.516	9.596	20.39	43.33	3.546	0.02463
3	3.5	0.437	2.626	6.896	18.11	47.55	124.9	5.416	0.03761
4	4.5	0.531	3.438	11.82	40.64	139.7	480.3	9.283	0.06447
5	5.563	0.625	4.313	18.60	80.23	346.0	1,492	14.61	0.1015
6	6.625	0.718	5.189	26.93	139.7	725.0	3,762	21.15	0.1469
8	8.625	0.906	6.813	46.42	316.2	2,155	14,679	36.46	0.2532
10	10.75	1.125	8.500	72.25	614.1	5,220	44,371	56.75	0.3941
12	12.75	1.312	10.126	102.54	1,038	10,514	106,461	80.53	0.5592
14 O.D.	14.0	1.406	11.188	125.17	1,400	15,668	175,292	98.31	0.6827
16 O.D.	16.0	1.593	12.814	164.20	2,104	26,961	345,474	128.96	0.8955
18 O.D.	18.0	1.781	14.438	208.46	3,009	43,444	627,256	163.72	1.1369
20 O.D.	20.0	1.968	16.064	258.05	4,145	66,585	1,069,617	202.67	1.4073
24 O.D.	24.0	2.343	19.314	373.03	7,205	139,157	2,687,681	292.98	2.0345

[1] All dimensions are in inches except where otherwise noted.
[2] This column also represents contents in cubic feet per foot of length.

From Table 30, by interpolation, or by using an unabridged set of Steam Tables,

$$v_{avg} = 0.8518 \text{ cu ft/lb}$$

$$V_{avg} = \dot{m}v_{avg}/A = (750{,}000/3{,}600) \times 0.8518/0.7508 = 252 \text{ fps}$$

$$h_f = 50 \times 144 \text{ lb/sq ft} \times 0.8518 \text{ cu ft/lb} = 6{,}140 \text{ ft}$$

Then

$$f = 6{,}140 \times \frac{11.376/12}{500} \times \frac{2 \times 32.17}{252^2} = 0.017$$

Friction Factor f. Equation (77), generally referred to as the Darcy equation, was proposed by Darcy and others in the nineteenth century. The friction factor f, usually termed the Darcy friction factor, has been subsequently shown to be a function of the Reynolds number and the relative roughness of the pipe. The Reynolds number $DV\rho/\mu$ is based on the pipe diameter D and the fluid properties, velocity, density, and viscosity. The relative roughness ϵ/D is a function of the

pipe alone; ϵ is the root mean square of the height of protruberances of the pipe-wall irregularities, and D is the mean pipe diameter in the same units as ϵ, so that the ratio is dimensionless.

Stanton and Nikuradse are credited with original experimental work which culminated in a logarithmic plot of friction factor f vs. the Reynolds number at various parametric values of ϵ/D. Figure 40 is a reproduction of the graphical presentation of Nikuradse. In the laminar flow range, Reynolds numbers varying

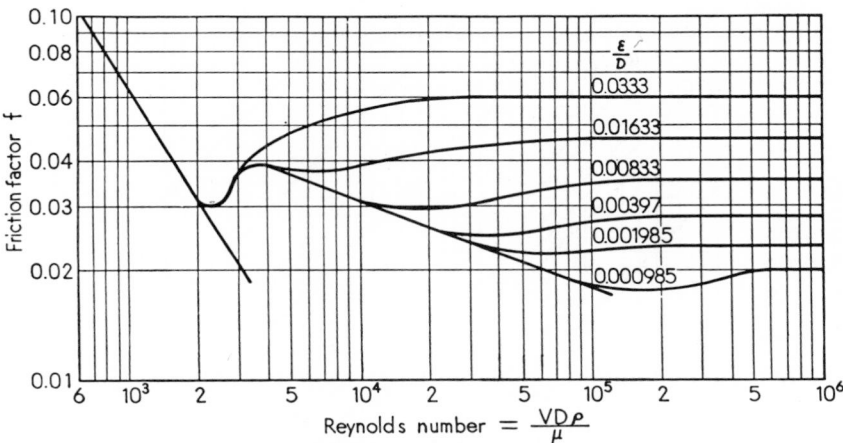

Fig. 40. Relation of Reynolds number, friction factor, and relative roughness for similar pipes. (*J. Nikuradse, "Strommung Gesetze in rauhen Rohren," VDI Forschungsh.,* 1933, p. 361.)

from zero to 2,000, it is seen that the friction factor varies linearly with Reynolds number. In fact, it can be shown analytically that, for laminar flow, the friction factor is given by

$$f = 64/N_r$$

where N_r represents the Reynolds number for the laminar flow. If this definition of friction factor is combined with Eq. (77), there is obtained the Hagen-Poiseuille equations for friction drop in laminar flow

$$\Delta P = 32L\mu V/D^2 = 128\mu LQ/\pi D^4 \tag{78}$$

in which Q represents the volume-flow rate and dimensions of other variables are taken so as to afford consistency.

In current practice, selection of friction factors is based on work done by L. F. Moody and as presented by him in 1944. Figure 41 is a chart from which can be obtained values of the relative roughness ϵ/D. As is seen from Table 48, there is a

Table 48. Absolute Roughness Values for Commercial Pipe

Pipe	Absolute roughness ϵ, ft
Commercial steel or wrought iron	0.00015
Asphalted cast iron.	0.0004
Galvanized iron.	0.0005
Cast iron.	0.00085
Wood stave	0.0006–0.003
Concrete.	0.001–0.01
Riveted steel.	0.003–0.03

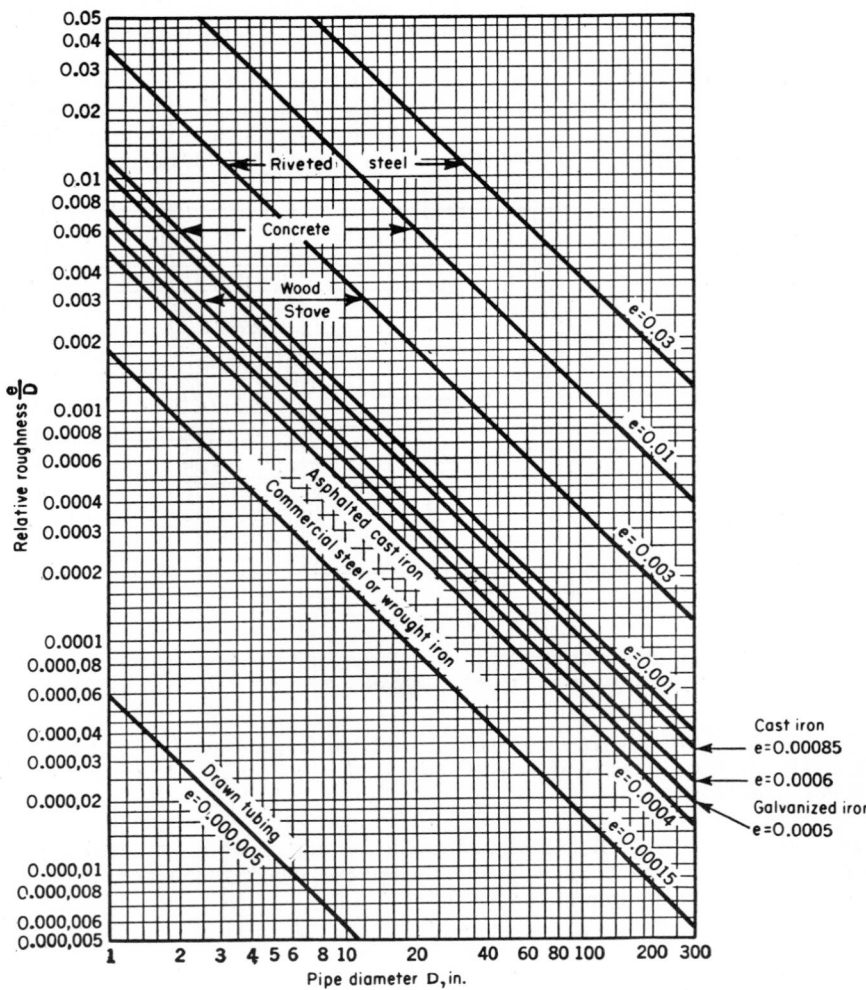

FIG. 41. Chart for determining relative roughness of pipes. (L. F. Moody, *Trans. ASME,* *November*, 1944.)

considerable range in absolute roughness for many commercial materials. Thus, in Fig. 41, some judgment must be used in selection of values.

Figure 42, taken from Moody, shows variation in friction factor over the range of Reynolds numbers from the laminar to completely turbulent regions. It is noted that, at values of Reynolds number to the right of the dotted line, an increase in Reynolds number has no effect on the friction factor; flow in this region is referred to as completely turbulent flow.

Reynolds Number Calculations. The Reynolds number N_r is defined as

$$N_r = DV\rho/\mu = DV/\nu \tag{79}$$

where D = some significant length which, in pipe flow, is taken as the hydraulic diameter

V = velocity of fluid flow

ρ = density of fluid

μ = absolute or dynamic viscosity

ν = kinematic viscosity

The hydraulic diameter D_h is defined as

$$D_h = \frac{4 \times \text{cross-sectional area}}{\text{wetted perimeter}}$$

For a cylindrical pipe, $D_h = 4 \times (\pi/4)D^2/\pi D$, or the geometrical and hydraulic diameters are identical in the case of the usual round piping.

In Eq. (79), all the terms must be selected in such a way as to make the Reynolds number dimensionless. For D in feet, V in feet per second, ρ in pound mass per cubic foot, μ must be in pound mass per foot per sec. That this selection results in a dimensionless ratio may be verified as follows:

$$N_r = \text{ft} \times \text{fps(lbm/cu ft)/(lbm/ft sec)} = \text{dimensionless}$$

Density equivalents, useful for converting from some other dimension to pounds per cubic foot, are given in Table 1. Absolute viscosity in centipoises multiplied by 0.000672 will yield absolute viscosity in pound mass per foot second.

The kinematic viscosity ν for use in Reynolds number calculations in the English system of units has the dimensions of

$$|\nu| = \text{(lbm/ft sec)/(lbm/cu ft)} = \text{sq ft/sec}$$

Kinematic viscosity is often expressed in *stokes* (square centimeters per second). A more common dimension is the *centistoke* ($= \frac{1}{100}$ stoke). To convert from the metric to the English system, multiply stokes by 1.075×10^{-3} to get kinematic viscosity in square foot per second. Table 2 lists factors by which the readings of various viscometers can be converted into centistokes.

Example 1. A certain liquid has a viscosity of 0.70 centipoise and a specific gravity of 0.72. This liquid flows in a 6-in. Schedule 40 pipeline with a velocity of 12 fps. It is required to determine the Reynolds number and the friction factor.

Solution. From Table 47,

$$D = 6.065/12 = 0.505 \text{ ft}$$
$$V = 12 \text{ fps}$$
$$\rho = 0.72 \times 62.4 = 44.95 \text{ lbm/cu ft}$$
$$\mu = 0.70 \text{ cp} \times 0.000672 \text{ lbm/(ft sec cp)} = 0.00047 \text{ lbm/(ft sec)}$$

Then
$$N_r = DV\rho/\mu = 0.505 \times 12 \times 44.95/47 \times 10^{-5} = 5.8 \times 10^5$$

If the pipeline is "commercial steel," reference to Fig. 41 indicates a relative roughness ϵ/D of about 0.0003. Entering the Moody graph of Fig. 42 at $N_r = 5.8 \times 10^5$ and proceeding vertically upward until the intersection with the relative roughness of 0.0003, the friction factor f is read as approximately 0.016.

The frictional drop per, say, 1,000 ft of piping could then be calculated easily from Eq. (77).

$$h_f = 0.016 \times (1,000 \times 12/6.065) \times 12^2/(2 \times 32.17) = 70.8 \text{ ft/1,000 ft}$$

The corresponding pressure drop is

$$\Delta P = \rho h_f/144 = 44.95 \times 70.8/144 = 22.1 \text{ psi/1,000 ft}$$

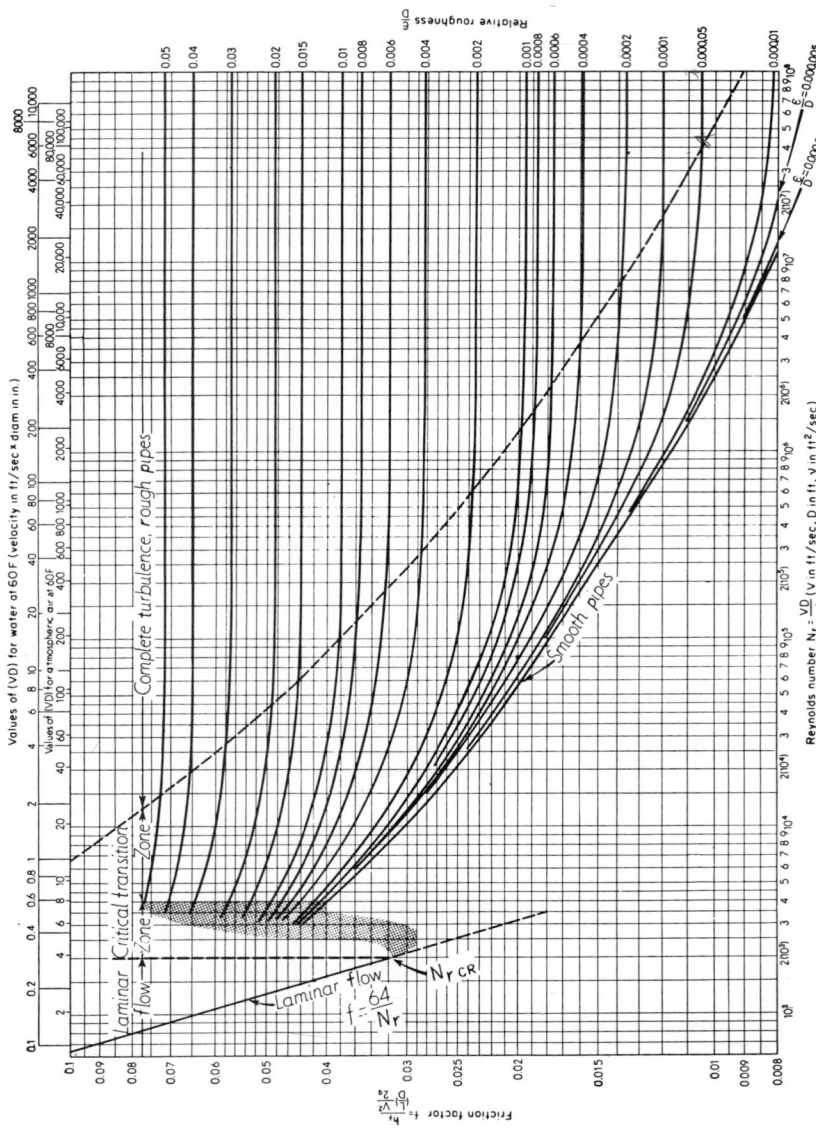

Fig. 42. Friction factor vs. Reynolds number with roughness as parameter. (*L. F. Moody, Trans. ASME, November, 1944.*)

Example 2. The kinematic viscosity of a certain oil is 56 Saybolt seconds at 130 F. Determine the Reynolds number of this oil when flowing in a 10-in. Schedule 80 pipeline at a velocity of 8 fps, and its pressure drop in feet per 1,000 ft.

From Table 47, $D = 9.564/12 = 0.797$ ft.

In Table 2, at 55.51 SS, viscosity is 9.0 centistokes and at 57.21 SS, viscosity is 9.5 centistokes. By interpolation, it is found that a viscosity of 56 SS corresponds to 9.14 centistokes or 0.0914 stokes. Then

$$\nu = 1.075 \times 10^{-3} \times 0.0914 = 98.3 \times 10^{-6} \text{ sq ft/sec}$$
$$N_r = 0.797 \times 8/(98.3 \times 10^{-6}) = 64,800$$

By Fig. 41, $\epsilon/D = 0.00016$, and from Fig. 42, $f =$ approximately 0.02. Then

$$h_f = 0.02 \times 1,000/0.797 \times 8^2/(2 \times 32.17) = 25 \text{ ft/1,000 ft}$$

Example 3. Steam at an average pressure and temperature of 1,000 psia and 950 F flows in a 12-in. Schedule 120 steel pipeline at the rate of 350,000 lb/hr. It is required to determine the pressure drop in pound per square inch per 100 ft.

Solution. From Table 30, interpolation yields $v = 0.7949$ cu ft/lbm. By interpolation in Table 31, the viscosity is found to be 9.81×10^{-7} lbf sec/sq ft. Table 47 gives the diameter and area of 12-in Schedule 120 pipe as 10.75 in. and 0.6303 sq ft, respectively. Then

$$D = 10.75/12 = 0.895 \text{ ft}$$
$$V = 350,000/3,600 \times 0.7949/0.6303 = 122 \text{ fps}$$
$$\rho = 1/0.7949 = 1.26 \text{ lbm/cu ft}$$
$$\mu = 9.81 \times 10^{-7} \text{ lbf sec/sq ft} \times 32.17 \text{ ft lbm/(lbf sec)}$$
$$= 31.6 \times 10^{-6} \text{ lbm/(ft sec)}$$

Then $\qquad N_r = 0.895 \times 122 \times 1.26/(31.6 \times 10^{-6}) = 4.35 \times 10^6$

From Fig. 41, $\epsilon/D = 0.00016$ and by use of Fig. 42, there is obtained $f = 0.013$.

Then $\qquad \Delta P = \rho f(L/D)(V^2/2g) = 1.26 \times 0.013 \times 100/0.895 \times 122^2/(2 \times 32.17)$
or $\qquad \Delta P = 422 \text{ psf} = 2.94 \text{ psi}$

All the above examples indicate that a straightforward calculation results in pressure-drop determination. However, it is necessary to perform conversion between different systems of units and from different dimensions of properties. Certain dimensions are peculiar to particular industries; for example, oil flows in barrels per day are commonly specified in the oil industry and flows in millions of cubic feet per day are used in the gas industry. For such reasons, specialized equations suitable to specific industries have been developed over the years. Such equations and associated tables and charts will be presented in the sections which follow later in this chapter.

Equivalent Resistance of Fittings and Valves. An actual piping installation consists of straight pipe, bends, elbows, tees, valves, and various other obstructions to flow. Thus, it is necessary to take into account the frictional resistance of the fittings involved. The usual approach is to express the loss through a fitting as being the equivalent of the loss through a certain number of linear feet of straight pipe. Alternatively, the loss through a fitting may be expressed as being a fraction (not necessarily less than unity) of the velocity head of the flowing fluid. Table 89 lists values of k for various fittings for use in the equations

$$h_f = k(V^2/2g) \qquad \text{for pressure drop, ft of fluid}$$
or $\qquad \Delta P = k\rho(V^2/2g) \qquad \text{for pressure drop, psf}$

Figure 43, taken from the Crane Co. "Flow of Fluids through Valves, Fittings and Pipe," is a convenient means of approximating the equivalent resistance of the indicated fittings. The discrepancies involved probably are of a lesser order than

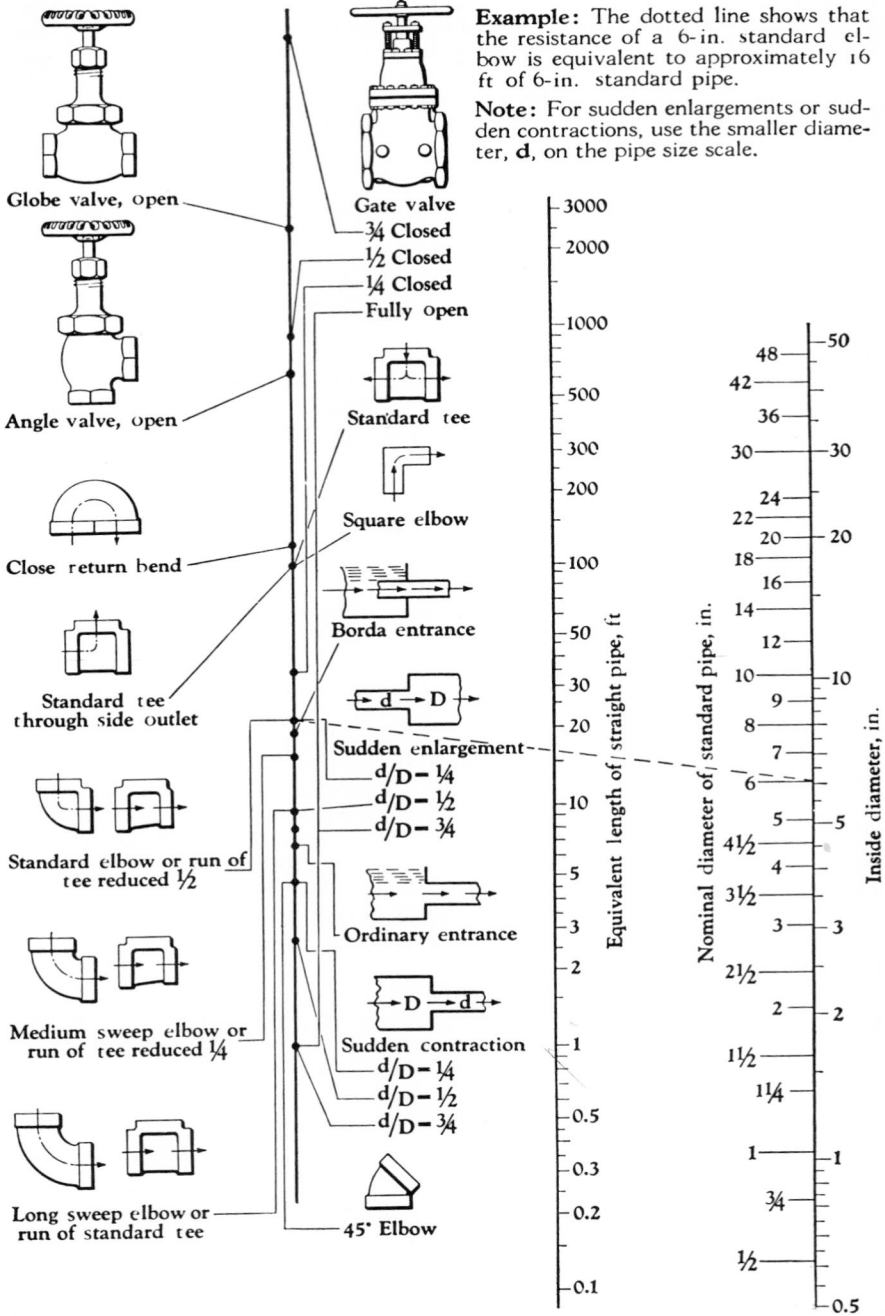

Example: The dotted line shows that the resistance of a 6-in. standard elbow is equivalent to approximately 16 ft of 6-in. standard pipe.

Note: For sudden enlargements or sudden contractions, use the smaller diameter, **d**, on the pipe size scale.

Globe valve, open

Angle valve, open

Close return bend

Standard tee through side outlet

Standard elbow or run of tee reduced ½

Medium sweep elbow or run of tee reduced ¼

Long sweep elbow or run of standard tee

Gate valve
¾ Closed
½ Closed
¼ Closed
Fully open

Standard tee

Square elbow

Borda entrance

Sudden enlargement
d/D – ¼
d/D – ½
d/D – ¾

Ordinary entrance

Sudden contraction
d/D – ¼
d/D – ½
d/D – ¾

45° Elbow

Equivalent length of straight pipe, ft

3000
2000
1000
500
300
200
100
50
30
20
10
5
3
2
1
0.5
0.3
0.2
0.1

Nominal diameter of standard pipe, in.

48
42
36
30
24
22
20
18
16
14
12
10
9
8
7
6
5
4½
4
3½
3
2½
2
1½
1¼
1
¾
½

Inside diameter, in.

50
30
20
10
5
3
2
1
0.5

FIG. 43. Resistance of valves and fittings. (*Crane Co.*)

variations due to indeterminate factors such as degree of roughness of the pipe interior or exact geometrical similitude of shape.   In using Fig. 43, it is noted that the right-hand scale has two different graduations, one corresponding to actual inside diameter of the pipe and the other corresponding to the inside diameter of Schedule 40 pipe.   Thus, for example, the drop through a 10-in. Schedule 160 wide-open gate valve (by Table 47, inside diameter is 8.5 in.) is found to be equivalent to that of about 4.8 ft of 10-in. Schedule 160 straight pipe.

Alternatively, equivalent lengths may be expressed in terms of standard (Schedule 40) 90-deg elbows, as shown in Fig. 44.   The equivalent length of an 8.5-in. elbow

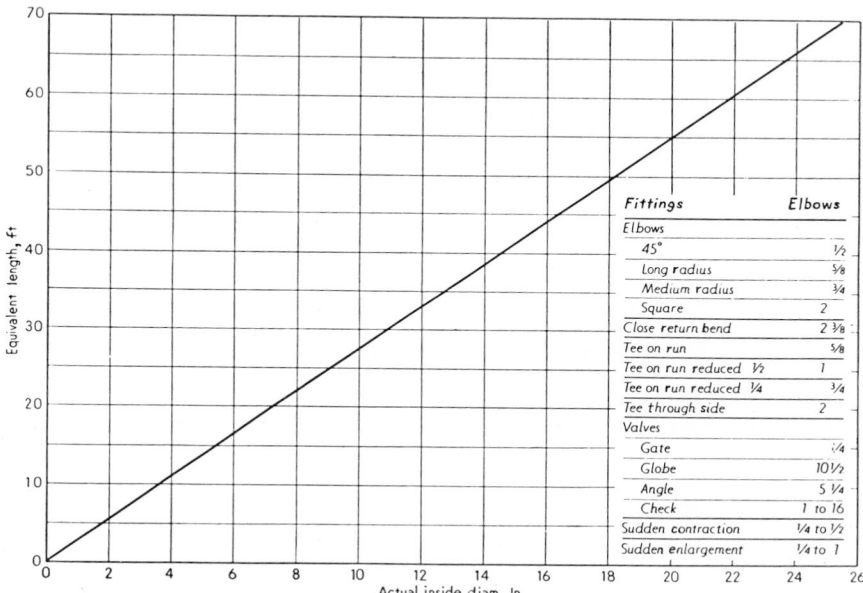

Fig. 44. Equivalent lengths of standard elbows.   (*Adapted from data by Crane Co. and others.*)

is indicated to be about 23 ft of straight 10-in. Schedule 160 pipe.   The inset table shows that a gate valve (fully open by inference) has a resistance of one quarter that of an elbow.   Thus, the resistance to flow of a wide-open 10-in. gate valve in a Schedule 160 pipeline would be equivalent to

$$\tfrac{1}{4} \times 23 = 5.8 \text{ ft of 10-in. Schedule 160 pipe}$$

The discrepancy between the above value and that of 4.8 ft obtained by use of Fig. 43 indicates that the determination of equivalent resistance is not an exact art.

Effect of Curvature on Resistance of Bends.   A misconception has been more or less prevalent to the effect that all short-radius bends and fittings necessarily cause greater pressure drop than do long-radius bends.   Published results show that the least overall resistance is produced by bends having a radius of two to four pipe diameters and that elbows or bends having a radius of $1\tfrac{1}{2}$ to $2d$ do not compare unfavorably with those having a radius of 4 to $6d$ or greater.   The agreement of nearly all observers that a curvature between 2 and $4d$ gives a minimum overall pressure drop is explained by the supposition that too short turns on the one

extreme give excessive pressure drops, whereas on the other extreme of easy curvature the disturbance persists over a greater length of travel.

Losses Due to Sudden Enlargement or Contraction.   As with the equivalent resistance of bends, fittings, and valves, it is convenient to express the losses due to sudden enlargement or contraction as a fractional part of a velocity head or as the equivalent of so many diameters of straight pipe.   The relations involved are similar to those for entrance and exit losses.   In the case of a sudden enlargement the loss cannot exceed the head required to produce the given change in velocity, while with a sudden contraction the loss usually is somewhat less owing to a lesser amount of turbulence.   This is in keeping with the well-known fact that the energy losses accompanying a decrease in velocity generally are greater than those associated with an increase.

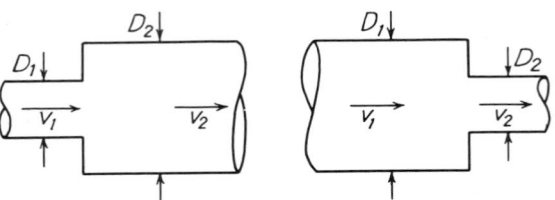

(a) Sudden enlargement   (b) Sudden contraction

FIG. 45. Loss due to sudden enlargement or contraction.

Referring to Fig. 45, the loss produced by a given change in velocity $(V_1 - V_2)$ caused by a sudden *enlargement* from diameter D_1 to diameter D_2 may be expressed as

$$h_f = (V_1 - V_2)^2/2g$$

which also may be written

$$h_f = \left(1 - \frac{D_1^2}{D_2^2}\right)\frac{V_1^2}{2g} = k\frac{V_1^2}{2g}$$

from which the losses due to a *sudden enlargement* are

D_1/D_2	0.1	0.2	0.3	0.4	0.5	0.6	0.7	0.8	0.9
k factor.........	0.98	0.92	0.83	0.71	0.56	0.41	0.28	0.13	0.04

Considerable difference of opinion exists as to the magnitude of the losses occasioned by *sudden contractions*.   Nearly every hydraulics text gives a different set of values.   Brightmore[1] concluded that the loss due to a sudden contraction is 0.7 of the loss from a sudden enlargement as given in the preceding paragraph. This is conceded by some to give satisfactory results where velocities are low, say below what would correspond to a loss of 1 ft of head in hydraulics.   For higher velocities a more complicated expression seems to be required for k.   The following expression has been used by Merriman and others for the loss of head in *sudden contractions:*

$$h_f = \left(\frac{1}{C_c} - 1\right)^2 \frac{V_2^2}{2g} = k\frac{V_2^2}{2g}$$

where

$$C_c = 0.582 + \frac{0.0418}{1.1 - D_2/D_1}$$

[1] A. W. Brightmore, *Proc. Inst. Civil Engrs.*, vol. 169, p. 323.

The following values for k and n for *sudden contractions* seem a satisfactory compromise:

D_2/D_1	0.1	0.2	0.3	0.4	0.5	0.6	0.7	0.8	0.9
k factor......	0.46	0.45	0.42	0.40	0.36	0.28	0.19	0.10	0.04

Relative Carrying Capacity of Pipes. Frequently it is desirable, particularly in calculations which involve a complex configuration of piping, to express the flow-carrying capability of different-size pipes. That is to say, for equal pressure drop, equal lengths of pipes, and equal flow rates, it is of interest to know the number N of small pipes of diameter D_1 that will carry the same amount of fluid as one large pipe of diameter D_2.

If a precise determination is to be made, a trial-and-error solution in which Reynolds number effects are taken into account would be necessary. However, if it can be assumed that flow in both cases is completely turbulent and that differences in relative roughness ϵ/D are negligible, an approximate relationship can be easily derived as follows, in which the subscript 1 refers to the smaller and the subscript 2 to the larger pipe.

From Eq. (77),

$$h_{f1} = f_1(L_1/D_1)(V_1^2/2g)N = f_2(L_1/D_2)(V_2^2/2g) = h_{f2}$$

By hypothesis, the friction drops, friction factors and lengths are all identical. For equal volume-flow rates Q of the fluid, which is here considered to be incompressible,

$$Q_1 = (\pi/4)D_1^2V_1N = (\pi/4)D_2^2V_2 = Q_2$$

By substitution, there is obtained $N = (D_2/D_1)^{5/2}$. Experiment indicates that this relation predicts quite accurately for smaller diameter pipes the flow-carrying capabilities. However, for larger pipes the results are more accurate if the relation is modified as follows:

$$N = \frac{D_1^3\sqrt{D_2 + 3.6}}{D_2^3\sqrt{D_1 + 3.6}}$$

where the D's must be in inches.

Based on the above modified relationship, the number of small pipes equivalent to one large pipe have been calculated and are listed in Table 49. It must be emphasized that the values listed there are for approximate use only; an accurate calculation must take into account the variation in f with Reynolds number and with pipe diameter.

Special Considerations. Until the work of Stanton, Nikuradse, and Moody, friction flow in piping was treated empirically and investigation was confined primarily to the characteristics of individual fluids. As a result many formulas of an empirical form were developed and adjusted approximately to fit the characteristics of some one particular fluid under certain limiting conditions. Examples of such empirical formulas are those attributed to Babcock, Carpenter, Fritzsche, and Unwin, or others for steam; Hazen and Williams, Saph and Schoder, Chézy, Darcy, and Fanning, or others for water; Unwin and Harris, or others for air; Spitzglass or Weymouth for gas, etc. These formulas have a certain general similarity and usually can be reduced to forms in which the friction factor f is experimentally determined. In some of the formulas mentioned above for steam, air, or gas, the friction factor f involves a term containing the pipe diameter, such as

Table 49. Relative Carrying Capacity of Standard-weight (Schedule 40)[1] Wrought Pipe and Actual Inside Diameters

(For air, steam, and gas)

Diameter	½	¾	1	1½	2	2½	3	4	5	6	7	8	9	10	11	12	13	14	15	16	17	Diameter
½	....	2.27	4.88	15.8	31.7	52.9	96.9	205	377	620	918	1,292	1,767	2,488	3,014	3,786	4,904	5,927	7,321	8,535	9,717	½
¾	2.60	......	2.05	6.97	14.0	23.3	42.5	90.4	166	273	405	569	779	1,096	1,328	1,668	2,161	2,615	3,226	3,761	4,282	¾
1	7.55	2.90	......	3.45	6.82	11.4	20.9	44.1	81.1	133	198	278	380	536	649	815	1,070	1,263	1,576	1,837	2,092	1
1½	24.2	9.30	3.20	......	2.00	3.34	6.13	13.0	23.8	39.2	58.1	81.7	112	157	190	239	310	375	463	539	614	1½
2	54.8	21.0	7.25	2.26	.....	1.67	3.06	6.47	11.9	19.6	29.0	40.8	55.8	78.5	95.1	119	155	187	231	269	307	2
2½	102	39.4	13.6	4.23	1.87	.....	1.83	3.87	7.12	11.7	17.4	24.4	33.4	47.0	56.9	71.5	92.6	112	138	161	184	2½
3	170	65.4	22.6	7.03	3.11	1.66	.....	2.12	3.89	6.39	9.48	13.3	20.9	23.7	31.2	39.1	50.6	61.1	75.5	88.0	100	3
4	376	144	49.8	15.5	6.87	3.67	2.21	....	1.84	3.02	4.48	6.30	8.61	12.1	14.7	18.5	23.9	28.9	35.7	41.6	47.4	4
5	686	263	90.9	28.3	12.5	6.70	4.03	1.83	....	1.65	2.44	3.43	4.69	6.60	8.00	10.0	13.0	15.7	19.4	22.6	25.8	5
6	1,116	429	148	46.0	20.4	10.9	6.56	2.97	1.63	....	1.48	2.09	2.85	4.02	4.86	6.11	7.91	9.56	11.8	13.8	15.6	6
7	1,707	656	226	70.5	31.2	16.6	10.0	4.54	2.49	1.51	....	1.41	1.93	2.71	3.28	4.12	5.34	6.45	7.97	9.31	10.6	7
8	2,435	936	322	101	44.5	23.8	14.3	6.48	3.54	2.18	1.43	.....	1.35	1.93	2.33	2.92	3.79	4.57	5.67	6.60	7.52	8
9	3,335	1,281	440	137	60.8	32.5	19.5	8.85	4.85	2.98	1.95	1.37	.....	1.41	1.71	2.14	2.77	3.35	4.14	4.83	5.50	9
10	4,393	1,688	582	181	80.4	42.9	25.8	11.7	6.40	3.93	2.57	1.80	1.32	.....	1.21	1.52	1.97	2.38	2.94	3.43	3.91	10
11	5,642	2,168	747	233	103	55.1	33.1	15.0	8.22	5.05	3.31	2.32	1.70	1.28	.....	1.26	1.63	1.88	2.43	2.83	3.22	11
12	7,087	2,723	938	293	129	69.2	41.6	18.8	10.3	6.34	4.15	2.91	2.13	1.61	1.26	.....	1.30	1.57	1.93	2.26	2.58	12
13	8,657	3,326	1,146	358	158	84.5	50.7	23.0	12.6	7.75	5.07	3.56	2.60	1.98	1.53	1.22	.....	1.21	1.49	1.74	1.98	13
14	10,600	4,070	1,403	438	193	103	62.2	28.2	15.4	9.48	6.21	4.35	3.18	2.41	1.88	1.50	1.22	.....	1.24	1.44	1.64	14
15	12,824	4,927	1,698	530	234	125	75.3	34.1	18.7	11.5	7.52	5.27	3.85	2.92	2.27	1.81	1.48	1.21	.....	1.17	1.35	15
16	14,978	5,758	1,984	619	274	146	88.0	39.9	21.8	13.4	8.78	6.15	4.51	3.41	2.66	2.12	1.73	1.42	1.18	.....	1.14	16
17	17,537	6,738	2,322	724	320	171	103	46.6	25.6	15.7	10.3	7.20	5.27	3.99	3.11	2.47	2.03	1.66	1.37	1.17	.....	
18	20,327	7,810	2,691	840	371	198	119	54.1	29.6	18.2	11.9	8.35	6.11	4.63	3.60	2.87	2.35	1.92	1.59	1.36	1.16	
20	26,676	10,249	3,532	1,102	487	260	157	70.9	38.9	23.9	15.6	10.9	8.02	6.07	4.73	3.76	3.08	2.52	2.08	1.78	1.52	
24	42,624	16,376	5,644	1,761	778	416	250	113	62.1	38.2	25.0	17.5	12.8	9.70	7.55	6.01	4.92	4.02	3.32	2.84	2.43	
30	75,453	28,990	9,990	3,117	1,378	736	443	201	110	67.6	44.2	31.0	22.7	17.2	13.4	10.7	8.72	7.14	5.88	5.03	4.30	
36	120,100	46,143	15,902	4,961	2,193	1,172	705	319	175	108	70.4	49.3	36.1	27.3	21.3	16.9	13.9	11.3	9.37	8.01	6.85	
42	177,724	68,282	23,531	7,341	3,245	1,734	1,044	473	259	159	104	73.0	53.4	40.5	31.5	25.1	20.5	16.8	13.9	11.9	10.1	
48	249,351	95,818	33,020	10,301	4,554	2,434	1,465	663	363	223	146	102	75.0	56.8	44.2	35.2	28.8	23.5	19.4	16.6	14.2	
Diameter	½	¾	1	1½	2	2½	3	4	5	6	7	8	9	10	11	12	13	14	15	16	17	

The upper portion above the diagonal lines of blanks pertains to nominal pipe sizes while the lower portion is for pipe of the actual internal diameter given. Equivalent number of pipes = $(d_1^3/d_2^3)\sqrt{(d_2 + 3.6)/(d_1 + 3.6)}$. *Ex.* 1.—Carrying capacity of ¾-in. standard-weight pipe is equivalent to 2.27 half-inch standard-weight pipes. *Ex.* 2.—Carrying capacity of pipe with inside diameter of 13 in. is equivalent to 3.56 pipes with inside diameter of 8 in.

[1] Standard-weight pipe 10 in. and smaller is identical with Schedule 40 pipe.

The data in this table are reproduced by permission from "Steam Power Plant Engineering," by G. F. Gebhardt, 4th ed., John Wiley & Sons, Inc., New York.

[1 + (3.6/d)]. In certain other formulas for friction in pipes, the investigators have tried to obtain agreement with test results by using fractional exponents for *v* and *d*, with fair success in a limited range of application. Examples of formulas involving fractional exponents are that of Fritzsche for steam and that of Saph and Schoder for water.

The utility of such specialized equations is due to at least two factors. First, they are generally expressed in terms of the variables which are peculiar to a particular industry; for example, in the flow of low-pressure air, it is convenient to work with flow rates expressed as standard cubic feet per minute. Second, frequently coefficients are incorporated in the equations which eliminate iteration in calculations.

Those specialized equations may be criticized on the basis that their use does not result in optimum accuracy. However, errors introduced in application of the analytical (Moody) relation, such as out-of-roundness of pipe and knowledge of the internal pipe roughness, may frequently be as great as those encountered by use of the less exact relations. In the sections which follow, specialized equations and methods for various of the principal fluids will be discussed.

FRICTION FLOW OF STEAM

Pressure drop due to friction in steam pipes can be determined by the use of Eqs. (77) together with Figs. 41 and 42. In certain instances it may be advantageous to use the empirical equations and the charts based thereon, which are presented below. No claim for accuracy is made in their presentation; conversely, accuracy requires the use of analytical relations.

Unwin's Equation. The drop in pressure ΔP in pounds per square inch is given by the following:

$$P = KvLW^2/d^5 \tag{80}$$

in which v = specific volume, cu ft/lb
L = equivalent length of pipe, ft
W = steam flow, lb/hr
d = internal pipe diameter, in.
$K = 3{,}625(1 + 3.6/d)/10^{11}$

A graphical solution of Eq. (79) is given in Fig. 46. As an application of the use of the equation the sample problem, solved graphically in Fig. 46, is here solved by equation.

From Table 47, $d = 12.00$ in. Then

$$K = 3625/(1 + 3.6/12.00)10^{11} = 4.71 \times 10^{-8}$$

$$L = 100 \text{ ft}$$

From Table 30, at 225 psia and 692 F (200 F superheat), specific volume $v = 2.965$ cu ft/lb. Then

$$\Delta P = 4.71 \times 10^{-8} \times 2.965 \times 100 \times (60 \times 2{,}000)^2/12^5$$

From Table 47, 12^5 may be read as 248,832

$$\Delta P = 0.81 \text{ psi}$$

as compared to 0.82 psi by solution using Fig. 46.

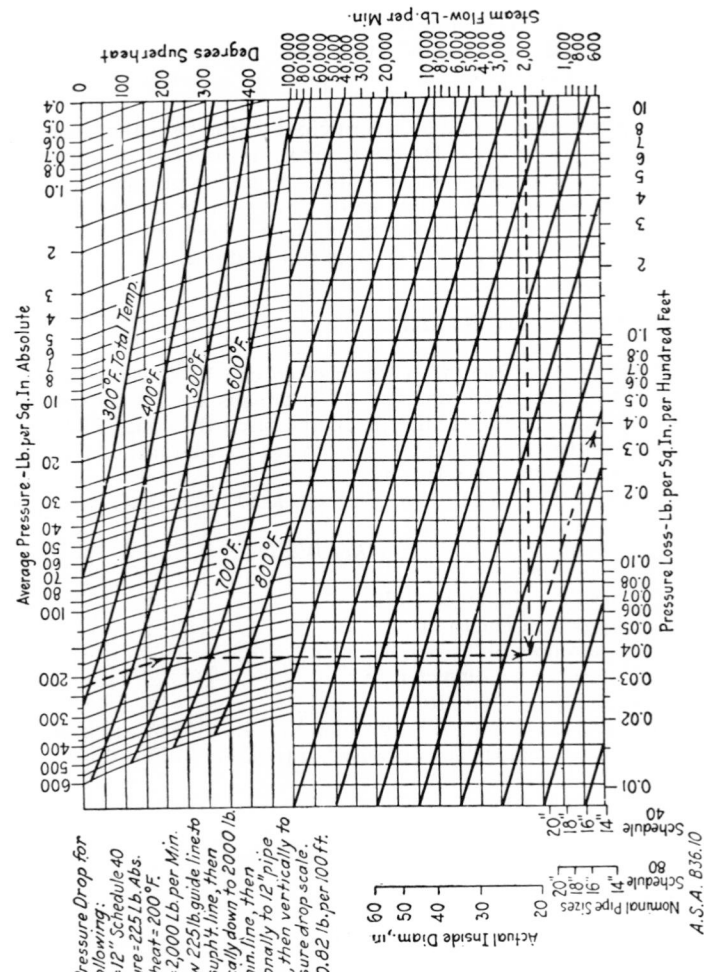

Find Pressure Drop for
the following:
Pipe=12" Schedule 40
Pressure=225 Lb. Abs.
Superheat=200°F.
Flow=2,000 Lb. per Min.
Follow 225 lb. guide line to
200°superht. line, then
vertically down to 2000 lb.
per min. line, then
diagonally to 12"pipe
diam., then vertically to
pressure drop scale.
Ans. 0.82 lb. per 100 ft.

A.S.A. B36.10

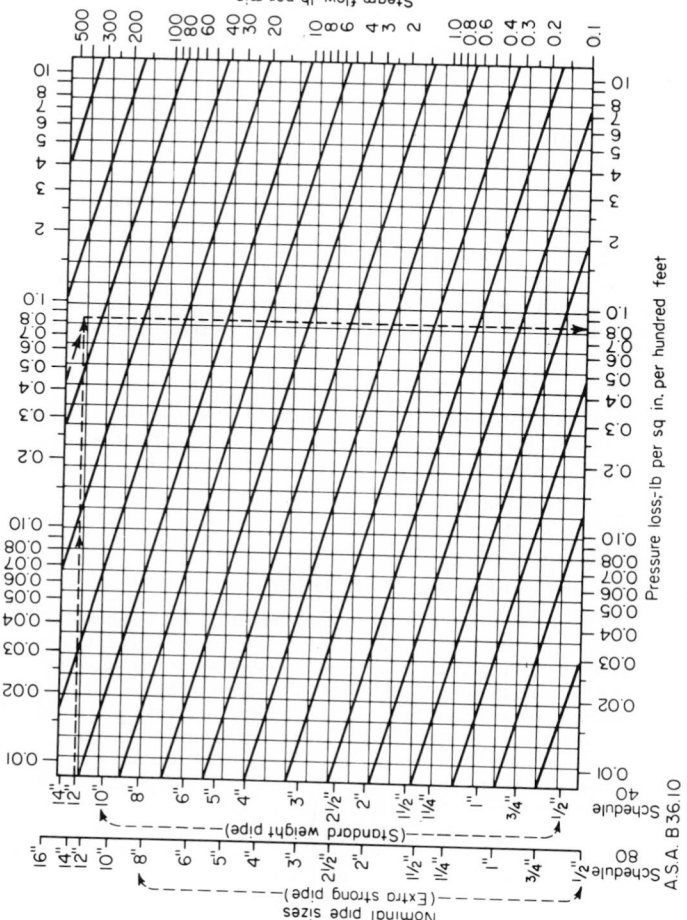

Fig. 46. Graphical solution of Unwin's formula for flow of steam in pipes. All formulas for flow of steam in pipes are of the form $\Delta p = vKLW^2 \div d^5$. Values of K ascribed by Unwin, Babcock, Carpenter, and Martin give practically the same result.

It may be of interest to solve this problem, also, by use of the analytical expression and the Moody chart.   Proceeding,

$$V = 2,000/60 \times 2.965 \times 1/0.7854 = 125.5 \text{ fps}$$
$$D = 12/12 = 1 \text{ ft}$$
$$\rho = 1/2.965 = 0.337 \text{ lb/cu ft}$$
$$\mu = 5.72 \times 10^{-7} \text{ lb sec/sq ft by interpolation in Table 31}$$
$$= 18.4 \times 10^{-6} \text{ lbm/(ft sec)}$$
$$N_r = 1 \times 125.5 \times 0.337/(18.4 \times 10^{-6}) = 2.29 \times 10^6$$
$$\epsilon/D = 0.00014 \text{ by Fig. 41, using commercial steel pipe}$$
$$f = 0.0135 \text{ by Fig. 42}$$

Then $\Delta P = f\rho(L/D)(V^2/2g) = 0.0135 \times 0.337 \times (100/1) \times 125.5^2/(2 \times 32.17)$
$$\Delta P = 112 \text{ lb/sq ft}$$
or $\Delta P = 0.78 \text{ psi}$

The discrepancy is noted as compared with solutions obtained by use of the Unwin equation.   Some of this discrepancy may be attributed to the loss in accuracy associated with interpolation for viscosity in Table 31.   However, as will be noted by reference to Fig. 42, the flow involved is so turbulent that the friction factor is almost independent of Reynolds number.

Fritzche's Equation.   The pressure drop ΔP in pounds per square inch is given by

$$\Delta P = 2.1082vLW^{1.85} \times d^{4.97}/10^7 \tag{81}$$

in which the symbols have the same meaning as those in Eq. (80).   Figure 47 is a graphical solution of Eq. (81).   The example below represents an explanation of the solution for the problem which is presented in the upper right corner of Fig. 47.

Example. Steam pressure = 700 psia
Enthalpy = 1275 Btu/lb
Weight of steam flowing = 100,000 lb/hr
Pipe diameter = 10-in. ASA Schedule 100

Solution. Entering the chart at the correct point of pressure (700 lb) shown by the scale at the top, proceed along a constant pressure line to the point where it intersects the enthalpy, total temperature, or quality line given for the particular problem at hand. (1275 Btu enthalpy.)   From there proceed vertically downward to the flow desired (100,000 lb/hr) noting the velocity coefficient at the point of crossing the velocity co-efficient scale (0.90).   Thence, proceed to the right or left (right in this case) along the diagonal guide lines which slope from upper left to lower right to the pipe-diameter line (10-in. ASA Schedule 100).   At that point read pressure drop in pounds per square inch per 100 ft of pipe (0.45) and the equivalent velocity in feet per minute (3,100 ft).   On applying the velocity coefficient the actual velocity in the pipe is found to be 3,100 × 0.9 = 2,790 fpm.   With a little practice in the use of the chart, it may be worked in any direction with ease and entered at the points for which data are available.

It may be observed that the representation of steam conditions is nothing but an enthalpy vs. volume chart for steam and that the only scale involved which essentially belongs to the pressure-drop chart proper is the specific-volume scale.   In this connection, attention is called to the fact that any steam condition not covered by the chart may still be worked through the chart by obtaining the specific volume from a steam table and starting from that point on the specific-volume scale.

It should be noted that the equivalent-velocity scale is not laid out for a given pipe size, as is the case with a different type of steam-pressure drop chart in common use.   The equiv-alent velocity here given is the velocity which would exist at a given pressure drop per 100 ft if the specific volume of the steam were unity.   The velocity coefficient is then a correction for specific volume of the steam flowing in the pipe and is independent of the pipe diameter.

The same problem can be solved by direct substitution into Eq. (81).   From Table 30, the specific volume at 700 psia and 1275 Btu/lb is 0.7835 cu ft/lb.   From Table 47, the inside diameter of 10-in. Schedule 100 pipe is 9.314 in.   Then

$$\Delta P = 2.1082 \times 0.7835 \times 100 \times 100,000^{1.85} \times 9.314^{4.97}/10^7$$
$$= 0.47 \text{ psi}$$

If the same problem is solved by use of Eq. (77) and Figs. 41 and 42, there is obtained $\Delta P = 0.32$ psi/100 ft.   Once again there is noted a discrepancy between results obtained by use of analytical and empirical equations.

Spitzglass Chart.   Figure 48 is a chart of rather limited use.   It is prepared for the determination of pressure drops in Schedule 40 piping involving flow of saturated steam in pressures up to 200 psig.

Alternate Forms.   It has been seen that an expression for friction drop in pipelines is

$$P = \rho f(L/D)(V^2/2g)$$

Alternatively, this can be expressed as

$$\Delta P = 0.001295 f \rho L V^2/d \qquad (82a)$$

or as

$$\Delta P = 0.01214 f v L W^2/d^5 \qquad (82b)$$

in which ΔP = pressure drop, psi
ρ = density, lb/cu ft
V = velocity, fps
v = specific volume, cu ft/lb
W = mass flow rate, lb/min
d = pipe diameter, in.
L = pipe length, ft

Equation (82b) is of the form $P = KvLW^2/d^5$, in which the constant K includes the numerical constant 0.01214.   Different investigators, in past times, have published values of K as listed below.

Investigator	K
Babcock.................	$0.0001321\ (1 + 3.6/d)$
Babcock and Wilcox	$0.0001310\ (1 + 3.6/d)$
Carpenter-Unwin.........	$0.0001305\ (1 + 3.6/d)$
Gutermuth	0.0003557
Hawksley...............	0.000337

The tabulated values of K include, except in those due to Gutermuth and Hawksley, a friction factor f which is a function of diameter only.   Those of Gutermuth and Hawksley assume a constant friction factor for all diameters. Thus no equation based on any of the above values of K can be expected to give accurate results over the entire range of diameters and Reynolds numbers.   Such equations are, however, of value as the first approximation to many complex problems in fluid flow.

Another approximation which is reasonably accurate for flows of superheated steam involves the assumption that the steam acts as a perfect gas and that flow is isothermal.   Both these assumptions are approximately valid.   It can be shown that, with these assumptions,

$$P_2 = \sqrt{P_1^2 - CW^2L} \qquad (83)$$

where P_2 and P_1 = final and initial pressures, psia
W = flow rate, lb/hr
L = total equivalent pipe length, ft
$C = 6.7444 \times 10^{-6} f P_1 v_1/d^5$

PRESSURE DROP IN STEAM PIPES – BY FRITZSCHE'S FORMULA

AVERAGE LINE PRESSURE, POUNDS PER SQ IN. ABSOLUTE

$$P = \frac{0.8\,L F^{1.85}}{Y d^{4.97}} = \frac{2.1082 \times SP.VOL. \times L \times W^{1.85}}{10^{7} \times d^{4.97}}$$

F = STEAM FLOW, LB PER SECOND   W = FLOW, LB PER HOUR
P = PRESSURE DROP, LB PER SQ INCH
L = EQUIVALENT LENGTH OF STRAIGHT PIPE, FEET
Y = AVERAGE DENSITY OF STEAM, LB PER CU FT
d = ACTUAL INSIDE DIAMETER OF PIPE, INCHES

EXAMPLE

WITH STEAM AT 700 LB ABS 1275 B T U PER LB AND 100,000 LB
PER HOUR FLOWING IN A 10 IN SCHEDULE 100 PIPE, FIND PRESSURE
DROP PER 100 FT AND THE VELOCITY IN FT PER MIN
PRESSURE DROP IS READ DIRECTLY FROM THE CHART =
0.45 LB PER 100 FT EQUIVALENT VELOCITY IS READ
DIRECTLY FROM THE CHART = 3,100 FT PER MIN
ACTUAL VELOCITY = 3100 x 0.9 = 2790 FT PER MIN

NOTE: IF ANY THREE OF THE FIVE QUANTITIES F, d, P, Y, AND EQUIVALENT
VELOCITY, ARE KNOWN OR CAN BE FOUND, A COMPLETE SOLUTION IS POSSIBLE

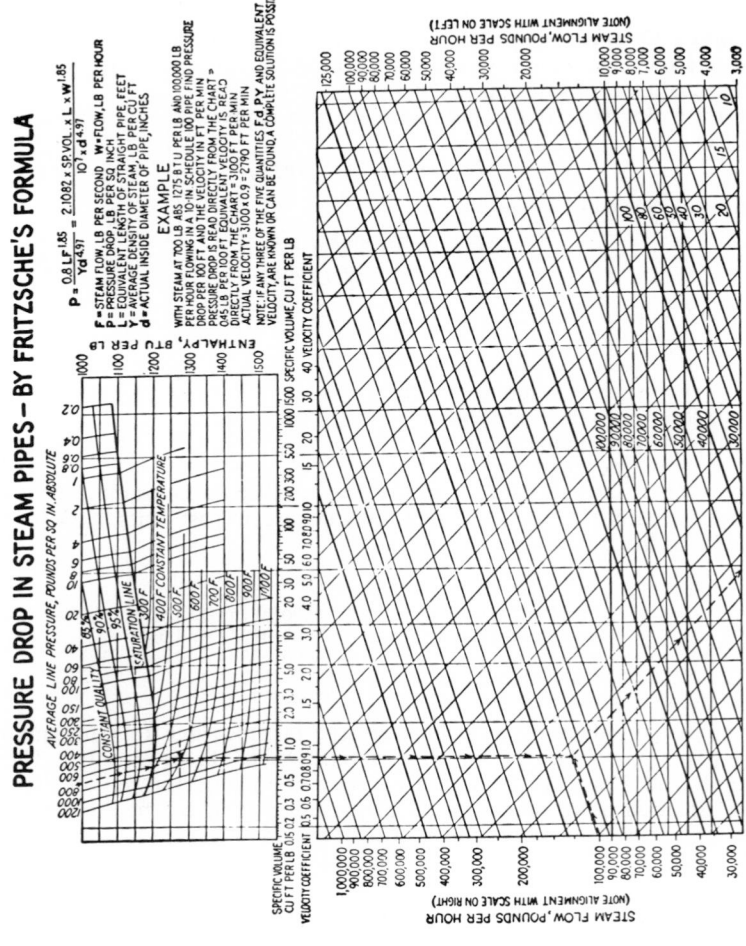

STEAM FLOW, POUNDS PER HOUR
(NOTE ALIGNMENT WITH SCALE ON LEFT)

STEAM FLOW, POUNDS PER HOUR
(NOTE ALIGNMENT WITH SCALE ON RIGHT)

ENTHALPY, BTU PER LB

SPECIFIC VOLUME, CU FT PER LB
VELOCITY COEFFICIENT

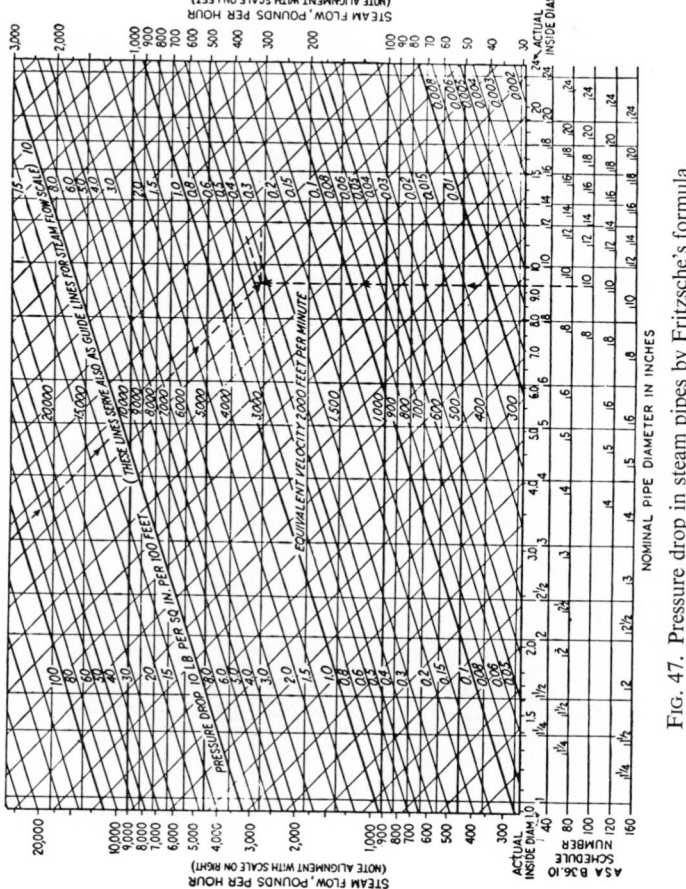

FIG. 47. Pressure drop in steam pipes by Fritzsche's formula.

3–141

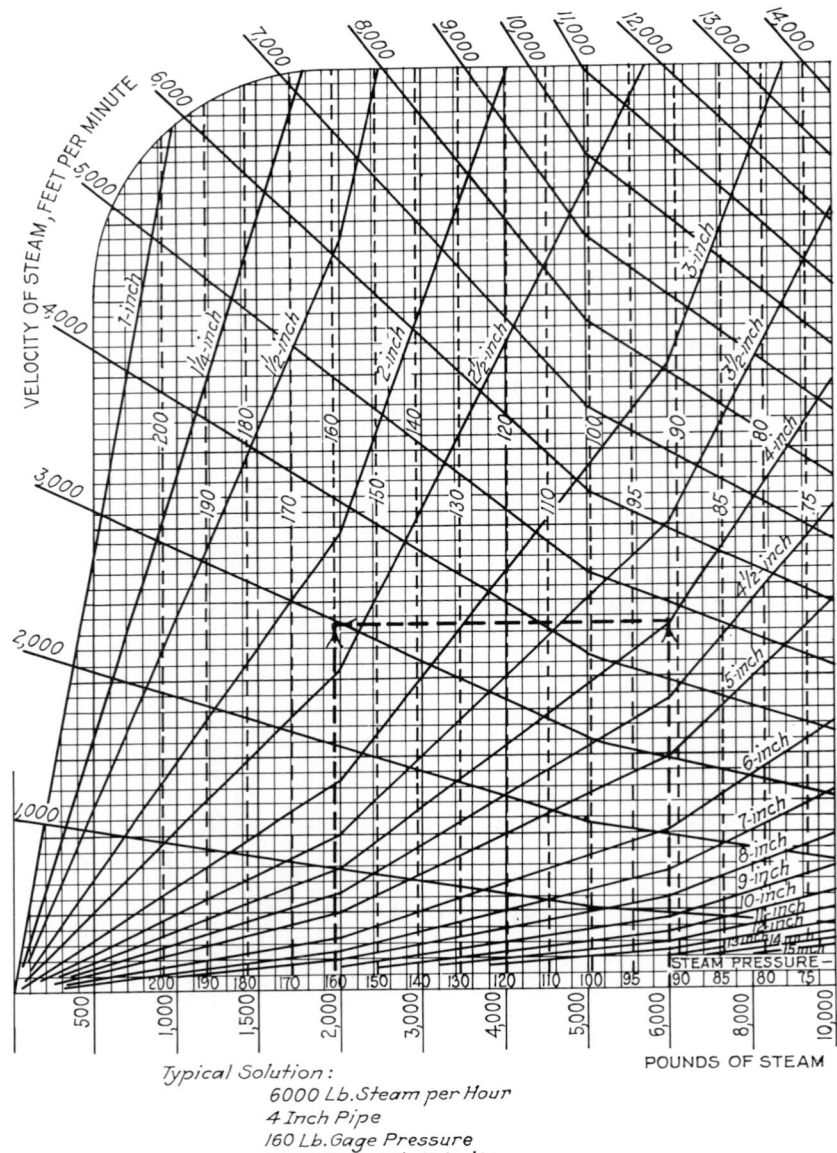

Typical Solution:
6000 Lb. Steam per Hour
4 Inch Pipe
160 Lb. Gage Pressure
3000 Ft. per Min. Velocity

When any three are known, the fourth can be determined

FIG. 48. Saturated-steam flow chart for standard weight (Schedule 40) pipe. (*Drawn*

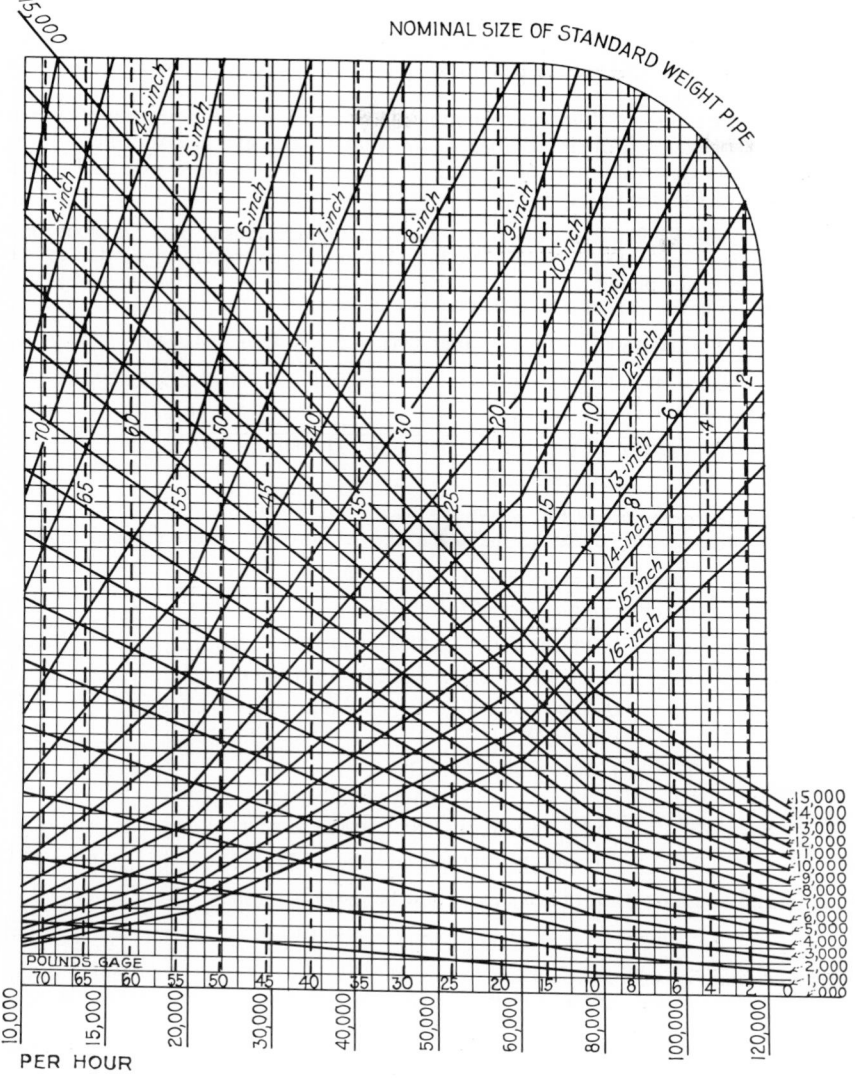

Use Steam Quantity Lines to intersect Pipe Sizes
Use Steam Pressure Lines to intersect Velocities

by J. M. Spitzglass for "Steam Power Plant Engineering," by Prof. G. F. Gebhardt.)

In the determination of C, the Babcock and Wilcox expression for friction factor, $f = 0.01080(1 + 3.6/d)$, is used frequently; also, in the expression for C, P_1 is in pounds per square inch absolute, v_1 is in cubic feet per pound, and d is in inches of pipe diameter.

The obvious utility of this approximate equation rests on the fact that a knowledge of the pressure drop is not necessary; that is, the average specific volume required in the determination of average Reynolds number need not be known. An example of the use of Eq. (83) will now be given.

Example. A main steam line is of 12-in. Schedule 160 pipe and is to pass 1,500,000 lb/hr of steam initially at 2,000 psia and 1,000 F. Determine the pressure drop due to 300 equivalent feet of the piping.

Solution. Using Eq. (83) as a first approximation,

$$f = 0.00180(1 + 3.6/d) = 0.01080(1 + 3.6/10.126) = 0.0146$$

Then $C = 6.7444 \times 10^{-6} fP_1v_1/d^5 = 6.7444 \times 10^{-6} \times 0.0146 \times 2,000 \times 0.3935/10,6461$

or $C = 730 \times 10^{-12}$

$$P_2 = \sqrt{P_1^2 - CW^2L} = \sqrt{(2,000)^2 - 730 \times 10^{-12}(1.5)^2 10^{12} \times 300}$$
$$= 1,875 \text{ psia}$$

or the pressure drop by use of Eq. (83) is 125 psia.

Now this value should be used as the first trial in a more accurate solution as follows:

$$P_{avg} = (P_1 + P_2)/2 = (2,000 + 1,875)/2 = 1,938 \text{ psi}$$

By interpolation in Table 30, $v_{avg} = 0.4078$ cu ft/lb.

$$V_{avg} = 1,500,000/3,600 \times 0.4078 \times 1/0.5592 = 304 \text{ fps}$$
$$v_{avg} = 13.2 \times 10^{-7} \times 32.17 = 4.24 \times 10^{-6} \text{ lbm/(ft sec)}$$
$$N_r = DV\rho/\mu = 10.126 \times 304 \times 10^6/(12 \times 0.4078 \times 4.24)$$
$$= 1.48 \times 10^8$$
$$\epsilon/D = 0.00016 \text{ and } f = 0.013$$

Then $$\Delta P = \rho f \frac{L}{D} \frac{V^2}{2g} \times \frac{1}{144} = \frac{0.013 \times 300 \times 12 \times 304^2}{0.4708 \times 10.126 \times 64.34 \times 144}$$

or $$\Delta P = 113 \text{ psi as against the trial value of 125 psi.}$$

For accurate results, the problem should now be reworked, using

$$P_{avg} = 2,000 - 113/2 = 1,943.5 \text{ psi}$$

Reynolds number effects would not change the friction factor as is evident from an inspection of the Moody graph in the pertinent region. The specific volume and velocity are altered very slightly, and the correct pressure drop is found to be 115 psi.

Table 50 has been prepared for convenience in the solution of Eq. (83). The table lists values of the constant C for different pipe sizes and schedule numbers all for a value of P_1v_1 equal to 600. For all other values of P_1v_1, multiply the tabular value of C by $P_1v_1/600$. For example, in the problem just worked out, P_1v_1 was $2,000 \times 0.3935$, or 787. Then

$$C = 556.0 \times 10^{-12} \times 787/600 = 730 \times 10^{-12}$$

Pressure Drop with Very High Steam Velocity. In some applications, the flow of steam at or near acoustic velocity is encountered. Among such applications are the flow through safety-valve vents, flow through starting-up lines to the condenser, and flow through temporary blowout lines to the atmosphere. In any of the above instances, the change in steam density is so large as to render invalid the assumption of incompressible flow. Such problems can be treated analytically by assuming that the steam follows a Fanno-line expansion as a perfect

gas with a specific heat ratio k of 1.3. This subject is dealt with in the section on "Friction Flow" (of a perfect gas) starting on page 42.

If it is known that the flow is critical or "choked," as in the case of a main-steam blowout line, the problem is relatively simple, as shown by the following example. The same sort of calculation could be made for a safety-valve discharge line.

Table 50. Values of C in Eq. (83) for $P_1v_1 = 600$

Nominal pipe size, in.	Schedule number		
	40	80	160
½	3.184×10^{-3}	6.838×10^{-3}	17.34×10^{-3}
1	152.4×10^{-6}	259.0×10^{-6}	657.8×10^{-6}
1½	13.07×10^{-6}	19.55×10^{-6}	37.58×10^{-6}
2	3.173×10^{-6}	4.552×10^{-6}	9.948×10^{-6}
2½	1.170×10^{-6}	1.646×10^{-6}	2.716×10^{-6}
3	350.2×10^{-9}	477.2×10^{-6}	829.1×10^{-6}
4	78.20×10^{-9}	103.4×10^{-9}	186.1×10^{-9}
6	8.479×10^{-9}	11.18×10^{-9}	19.66×10^{-9}
8	1.957×10^{-9}	2.494×10^{-9}	4.547×10^{-9}
10	577.6×10^{-12}	751.0×10^{-12}	1.400×10^{-9}
12	234.4×10^{-12}	301.7×10^{-12}	556.0×10^{-12}
14 OD	142.8×10^{-12}	184.3×10^{-12}	329.3×10^{-12}
16 OD	71.30×10^{-12}	90.95×10^{-12}	157.9×10^{-12}
18 OD	38.71×10^{-12}	48.98×10^{-12}	85.04×10^{-12}
20 OD	22.07×10^{-12}	28.24×10^{-12}	48.98×10^{-12}
24 OD	8.536×10^{-12}	10.93×10^{-12}	18.96×10^{-12}

Example. A main-steam blowout line is of 10-in. Schedule 40 commercial steel and has an equivalent length of 300 ft. Estimate the mass-flow rate and exit velocity if the line is blown at a constant boiler pressure of 1,200 psia, dry and saturated.

Solution. Let it be assumed that the Darcy friction factor is 0.02, corresponding to fully developed, completely turbulent flow. The validity of that assumption can be checked later. Recalling that the Darcy friction factor is four times the gas-dynamic friction factor $\bar{f}$, then the parameter $4\bar{f}(L_{max}/D)$ is

$$4\bar{f}(L_{max}/D) = 0.02 \times (300 \times 12)/10.02 = 7.2$$

In the above, L has been taken to be L_{max}, since flow through the line will be "choked" and, at the pipe exit, the pressure will be P^* and the velocity will be $V^* = C^*$.

Using unabridged Fanno-line tables (Table 46 in Keenan and Keyes "Gas Tables") for $k = 1.3$, at $4\bar{f}(L_{max}/D)$ of 7.2, there is obtained $M_1 = 0.279$, $P_1/P^* = 3.85$, and $V_1/V^* = 0.1975$, all by interpolation.

From Table 34 in the above reference, $P_1/P_{o1} = 0.9505$ and $T_1/T_o = 0.9884$.

At $P_{o1} = 1,200$ psia, $T_o = 460 + 567.22$ or, say, 1,027 R. Then $T_1 = 0.9884 \times T_o = 1,010R$ and $V_1 = M_1\sqrt{g_c k R T_1}$

and $V_1 = 0.279\sqrt{32.17 \times 1.3 \times 85.9 \times 1,010}$

or $V_1 = 532$ fps   and $V^* = (V^*/V_1)V_1 = 532/0.2975 = 1,790$ fps

At $P_{o1} = 1,200$ psia, $P_1 = 0.9505(1,200) = 1,140$ psia.

Then $P^* = (P^*/P_1)P_1 = 1,140/3.85 = 296$ psia.

The mass-flow rate may be calculated from conditions which exist either at the inlet or at the exit. Choosing the inlet,

$$\dot{m} = \rho_1 A_1 V_1 = A V_1/v_1 = 0.5475 \times 532/0.3619 = 805 \text{ lb/sec}$$

or mass-flow rate $= 805 \times 3,600 = 2,890,000$ lb hr. From the above calculations, it is seen that exit pressure and velocity are 296 psia and 1,790 fps, respectively.

When the inlet conditions P_1 and T_1 and the outlet conditions P^* and T^* are known, the average specific volume and average velocity can be calculated and an average Reynolds number obtained.   By use of Figs. 41 and 42, an average friction factor can be determined as a check on that assumed.

In connection with safety-valve vent stacks, the problem is a bit more complex. The total pressure drop, from boiler pressure to stack discharge pressure, is taken in two steps.   First, there is a drop across the valve itself to a pressure corresponding to the vent stack inlet.   Second, there is a pressure drop through the vent pipe.   In a small-diameter vent pipe or in one of great length, a large portion of the total pressure drop might occur in the piping.   If a very large vent pipe were installed, or if its length were very short, most of the pressure drop would occur across the valve.

The solution of such a problem is quite tedious if done by manual calculations, involving as it does iterative procedures to make valve discharge pressure identical with that at pipe inlet.   However, the problem can be arranged quite easily for digital computation.   In such calculations, the safety valve is often assumed to have a nozzle efficiency of 30 per cent due to the large degree of turbulence associated with flow through the valve seat and around the disk.   Based on an assumed nozzle efficiency of 30 per cent, the data of Table 51 have been included to assist

Table 51.   Minimum Values of W/d^2 at Safety-valve Outlets Required for Satisfactory Operation with Umbrella-fitted Vents[1]

Initial steam tempera- ture, F	Valve outlet pressure, psi gauge	Initial steam pressure, psi gauge								
		20	50	100	200	400	600	800	1,200	1,500
Satura- tion	0	710	960	1,140	1,340	1,450	1,550	1,600	1,830	1,780
	5	...	1,140	1,340	1,570	1,810	1,880	2,040	2,360	2,310
	10	...	.....	1,610	1,850	2,160	2,320	2,440	2,830	2,790
	20	...	.....	.....	2,630	2,870	3,120	3,260	3,740	3,680
	30	...	.....	.....	.....	**3,500**	3,730	3,930	4,650	4,550
	40	...	.....	.....	.....	**5,350**	**4,480**	4,700	5,460	5,350
	50	...	.....	.....	.....	.....	**5,970**	**5,460**	6,330	6,210
600	0	...	.....	890	1,000	1,160				
	5	...	.....	1,120	1,240	1,450				
	10	...	.....	1,280	1,490	1,790				
	20	...	.....	.....	2,530	2,360				
	30	...	.....	.....	**4,580**	3,260				
	40	...	.....	.....	.....	4,780				
700	0	...	.....	850	1,020	1,140	1,240			
	5	...	.....	1,060	1,260	1,410	1,530			
	10	...	.....	1,260	1,530	1,730	1,870			
	20	...	.....	.....	2,380	2,320	2,520			
	30	...	.....	.....	.....	**3,160**	3,020			
	40	...	.....	.....	.....	**4,650**	**4,100**			
800	0	...	.....	.....	1,010	1,080	1,200	1,240	1,300	
	5	...	.....	.....	1,280	1,410	1,540	1,590	1,680	
	10	...	.....	.....	1,490	1,690	1,830	1,930	2,000	
	20	...	.....	.....	2,280	2,230	2,420	2,540	2,700	
	30	...	.....	.....	.....	**3,030**	2,990	3,120	3,320	
	40	...	.....	.....	.....	**4,450**	**3,930**	3,720	3,870	
	50	...	.....	.....	.....	.....	.....	**5,000**	4,520	
900	0	...	.....	.....	.....	1,080	1,120	1,170	1,230	1,230
	5	...	.....	.....	.....	1,380	1,440	1,520	1,630	1,650
	10	...	.....	.....	.....	1,610	1,710	1,790	1,930	1,970
	20	...	.....	.....	.....	2,140	2,300	2,420	2,560	2,620
	30	...	.....	.....	.....	**2,910**	2,830	2,960	3,140	3,200
	40	...	.....	.....	.....	**4,320**	**3,740**	3,560	3,720	3,850
	50	...	.....	.....	.....	.....	**5,050**	**4,760**	4,380	4,380

[1] Values of W/d^2 in plain type indicate conditions in which the limiting factor is the valve outlet pressure.   Boldface type indicates conditions in which the limiting factor is the assumed maximum nozzle efficiency of 30 per cent.

in the selection of safety-valve vent stacks.   In this table, W represents steam flow in pounds per hour and d represents the diameter in inches of the vent stack.

Friction Flow of Gases and Air

In those cases in which compressibility effects can be neglected, pressure drop due to friction in pipes which convey gas or air can be determined by the use of Eq. (77) together with Figs. 41 and 42.   For flow of gases or air at high velocity, in which compressibility effects assume significance, use of the compressible flow relations developed previously for Fanno-line flow is recommended.   In certain instances it may be expeditious to use the empirical and specialized equations and the charts and tables based thereon, which are presented below.   No special claim for accuracy is made in their presentation; conversely, accuracy requires the adoption of a more analytical approach.

Standard Cubic Feet.   The flow of gas and air is often expressed in terms of standard cubic feet per minute (sfcm) or as standard cubic feet per hour.   Standard conditions are here taken as being at a pressure of 14.7 psia and a temperature of 60 F.

Density at standard conditions may be calculated for air and for most gases from the perfect-gas equation of state, as follows:

$$\rho_0 = P_0/RT_0 = P_0 \left/ \frac{R}{M_w} \right. T_0 \qquad (84)$$

where the subscript o refers to standard density, pressure, and temperature; R is the universal gas constant equal to 1,545 ft lbf/(1bm R); and M_w is the molecular weight.   For air, standard density is

$$\rho_0 = \frac{14.7 \times 144}{1{,}545/28.97 \times 520} = 0.07638 \text{ lb/cu ft}$$

If Q_c represents the volume flow at pressure P_c and temperature T_c, then the corresponding volume flow Q_0 at standard conditions is obtained from the perfect-gas relation

$$Q_0 = Q_c (T_0/T_c)(P_c/P_0) = 35.4 (P_c/T_c) Q_c \qquad (85)$$

The mass flow M in pounds per minute of a gas of specific gravity s can be determined from the relation

$$M = Q_0 \rho_0 s = Q_c \rho_c s \qquad (86)$$

Thus a flow of 500 scfm of air would correspond to a mass flow of 500 (0.07638) or 38.19 lb/min.

Specialized Equations.   The basic equations for pressure drop due to friction are

$$h_f = f(L/D)(V^2/2g)$$

and

$$\Delta P = \rho f(L/D)(V^2/2g) \qquad (87)$$

In the second of these, if the substitution $V = Q_c/60A$ is made, there will result

$$\Delta P = f\rho_c L Q_c^2/82.8d^5 = f\rho_a s L Q_c^2/82.8d^5 \qquad (88)$$

where ΔP = pressure drop, psi
  ρ_a = density of air at flow conditions, lb/cu ft
  ρ_c = density of actual fluid at flow conditions, lb/cu ft
  s = specific gravity relative to air, of flowing fluid
  d = pipe diameter, in.
  Q_c = actual (not standard) flow rate, cfm.

When Eq. (85) is combined with Eq. (88), there is obtained

$$\Delta P = (P_0/T_0)^2 \times fsT_cQ_o^2/(30.7P_cd^5) \tag{89}$$

Recognizing that $P_c = (P_1 + P_2)/2$ and that $\Delta P = P_1 - P_2$, where P_1 and P_2 are the initial and final flow states, and substituting into Eq. (89), there is obtained

$$Q_0 = 3.92 \, T_0/P_0 \, \sqrt{(P_1^2 - P_2^2)d^5/fsT_cL} \tag{90}$$

and

$$d = \left[\frac{fsT_cL}{15.4(P_1^2 - P_2^2)} \times \left(\frac{P_0Q_0}{T_0} \right)^2 \right]^{1/5} \tag{91}$$

For standard conditions of 14.7 psia and 520 R, the two above equations become

$$Q_0 = 6.08 \, \sqrt{(P_1^2 - P_2^2)d^5/fsL} \tag{92}$$

and

$$d = [fsLQ_o^2/37(P_1^2 - P_2^2)]^{1/5} \tag{93}$$

For flow rate expressed in pounds per minute, M, there is obtained

$$M = 0.464 \, \sqrt{(P_1^2 - P_2^2)sd^5/fL} \tag{94}$$

Flow at Near-atmospheric Pressure. When gas or air flows through a conduit at a pressure only slightly above atmospheric pressure, as is the case in low-pressure gas-distribution systems and in ventilating work, it is convenient to express pressure drops in inches of water rather than in pounds per square inch or in feet. Upon substituting the conversion factor 0.03613 psi per inch of water column and recognizing that the fluid pressure is approximately atmospheric, substitution into Eq. (89) yields

$$h = (13.25/T_c^2) \, fsT_cLQ_o^2/d^5 \tag{95}$$

or

$$Q_0 = T_0/3.64 \, \sqrt{hd^5/fsT_cL} \tag{96}$$

where h represents the pressure drop expressed in inches of water column and all the other symbols have the same meaning as formerly.

In *ventilating* work where standard conditions for air are 14.7 psia at 70 F instead of at 60 F, the absolute temperatures of 530 F corresponding to 70 F can be substituted for T_0 and s dropped for *air*. This gives

$$Q_a = 530/3.64 \, \sqrt{hd^5/fT_cL} = 145.6 \, \sqrt{hd^5/fT_cL} \tag{97}$$

If standard conditions, as in the *gas* industry, are approximately 14.7 psia at 60 F and the flow temperature also is assumed to be 60 F, the equation can be simplified still further by substituting $T_0 = T_c = 520$, which gives for gas *or* air,

$$h = (13.25/520) \times fsLQ_a^2/d^5 = fsLQ_a^2/39.2d^5 \tag{98}$$

$$Q_a = 6.26 \, \sqrt{hd^5/fsL} \tag{99}$$

and since the flow in *pounds* per minute $M = 0.07638Q_a$,

$$h = (fsL/39.2d^5)(M/0.07638)^2 = 4.38 \, fsLM^2/d^5 \tag{100}$$

$$M = 0.478 \, \sqrt{hd^5/fsL} \tag{101}$$

NOTE: In *ventilating* work where *standard conditions* for air are 14.7 psia at 70 F instead of at 60 F, a density of 0.07493 lb/cu ft should be substituted for 0.07638 in the above formulas.

Likewise in the *gas* industry where flow is desired in *cubic feet* per *hour* instead of per *minute*, the corresponding substitutions can be made in Eq. (64) for *any* standard conditions:

$$h = (p_0/T_0)^2 fs T_c L Q_{60}^2 / 4{,}000 p_c d^5 \qquad (102)$$

$$Q_{60} = 63.2 \, T_0/p_0 \, \sqrt{h p_c d^5 / fs T_c L} \qquad (103)$$

If standard conditions are taken as approximately 14.7 psia at 60 F and the flow temperature also is assumed to be 60 F, these equations can be simplified by substituting $T_0 = T_c = 520$, and $p_0 = p_c = 14.7$ as follows:

$$h = (14.7/520)^2 \, fs520LQ^2/4{,}000 \times 14.7d^5 = fsLQ^2/141{,}600d^5 \qquad (104)$$

$$Q = 376 \, \sqrt{hd^5/fsL} \qquad (105)$$

And where weight of flow is desired in pounds per hour, $W = 0.07638Q = 60M$ can be substituted in Eq. (100) or (101) from which

$$h = fsLW^2/820d^5 \qquad (106)$$

$$W = 28.6 \, \sqrt{hd^5/fsL} \qquad (107)$$

In low-pressure work the *velocity head* or change in static pressure required to produce a given velocity may be determined from the fundamental relation $H = V^2/2g$ as follows: If H represents the head in feet of gas or air and h the corre-sponding inches of water column, then $H = h/12 \times 62.4/\rho$ where ρ is the weight of gas or air in pounds per cubic foot at the given static pressure, temperature, and humidity. Substituting $H = V^2/2g = h/12 \times 62.4/\rho$ and transposing and solving for h,

$$h = 0.1925\rho \, (V^2/2g) = 0.00299 \rho V^2, \qquad (108)$$

and

$$V = 18.27 \, \sqrt{h/\rho} \qquad (109)$$

where V is the velocity in feet per *second*.

In *ventilating* work the formula may be reduced to standard air conditions of 70 F and 14.7 psia by substituting for ρ its numerical value of 0.07493 which gives

$$h = 62.4 \times 10^{-9} V_m^2 \qquad (110)$$

and

$$V_m = 4005 \, \sqrt{h} \qquad (111)$$

where V_m represents air velocity in feet per minute.

Kinetic Energy Changes.   In the flow of gas or air where there is a considerable energy conversion from static pressure to velocity head, particularly with high-velocity flow at low pressure, the effect on the computed result may be enough to warrant taking the compressibility effects into account.

Adjusting Flow Factor to Test Results.   The principal point of variance between the formulas proposed by different investigators of the flow of air, gas, and steam through pipes lies in their method of fitting test results into the basic formula structure.   Early tests were limited in scope, and the initial tendency was to assign for use in the basic flow formula the same coefficients of friction and flow factors for all internal diameters and roughness of pipe.   Later this was found to give discordant results, and various methods of adjustment were tried in fitting the formula to other conditions.   The three principal means of adjustment are listed

below in connection with some of the well-known formulas in which each is employed:

1. *Flow Factor a Function of Pipe Diameter.* In the Unwin and Spitzglass formulas the flow factor is made a function of pipe diameter. Another way of accomplishing an equivalent result is to assign arbitrarily a different flow factor to each pipe size.

2. *Fractional Exponents.* In the Williams-Hazen and the Saph-Schoder formulas for the flow of water; the Fritzsche formulas for air, gas, or steam; the Weymouth formula for gas; and the Harris formula for air, fractional exponents are assigned to d, V, or Q as the case may be.

SPECIFIC FLOW FORMULAS—GAS AND AIR

In the order of their importance, the empirical formulas principally used at the present time for computing the flow of gas are those attributed to Weymouth, Spitzglass, Fritzsche, and Unwin. These formulas are applicable also to compressed air and in some cases to ventilating work. The Harris formula, which is widely used for compressed air, closely resembles the Weymouth formula and gives much the same results within its own field. All these well-known formulas are presented in this section in a succession of rearrangements intended to facilitate solving for pressure drop, for flow in different terms, or for pipe diameter as the respective independent variable.

Principal Formulas in Common Use

Unwin Formulas. For what he termed "exceptionally smooth" pipes Unwin[1] determined a friction factor for *air* from tests on the Paris compressed-air mains which gives somewhat lower pressure drops than he found to exist in gas mains (see below). Unwin's friction factor for air as expressed for use in the general formula is $f = 0.0025 (1 + 3.6/d)$. Substitution of this value in Eqs. (92) and (94) and noting that $s = 1$ and $\rho_0 = 0.07638$ for air gives the following results for air:

Unwin Formulas for Compressed-air Flow—Standard Conditions

(For definition of symbols, see text)

$$\Delta P = \frac{(1 + 3.6/d)LQ_a^2}{7{,}400p_c d^5} \qquad \Delta P = \frac{(1 + 3.6/d)LM^2}{43p_c d^5}$$

$$Q_a = 86\sqrt{\frac{p_c d^5 \times \Delta P}{(1 + 3.6/d)L}} \qquad M = 6.56\sqrt{\frac{p_c d^5 \times \Delta P}{(1 + 3.6/d)L}}$$

$$= 43\sqrt{\frac{(p_1{}^2 - p_2{}^2)d^5}{(1 + 3.6/d)L}} \qquad = 4.63\sqrt{\frac{(p_1{}^2 - p^2{}_2)d^5}{(1 + 3.6/d)L}}$$

If standard conditions are other than 14.7 psia at 60 F or the flow temperature is not 60 F, the corresponding substitutions can be made.

On an assumption of a rougher pipe interior, Unwin assigned a somewhat higher friction factor for *gas* than for air or steam. Expressed in terms of f for substitution in the general flow formulas of this handbook, this can be reduced

[1] "Flow of Gas in Mains and Distribution at High Pressure," by W. C. Unwin, *Proc. Inst. Gas Engrs.*, published in *J. Gas Lighting, Water Supply*, etc., June 21, 1904, pp. 852–867.

to $f = 0.0176 (1 + 1.714/d)$. Substitution of this expression for f in the appropriate equations gives the following results for gas:

Unwin Formulas for Gas Flow—Standard Conditions

(For definition of symbols, see text)

$$\Delta P = \frac{(1 + 1.714/d)sLQ_a^2}{4,200p_cd^5}$$

$$\Delta P = \frac{(1 + 1.714/d)sLQ^2}{151 \times 10^5 p_cd^5}$$

$$Q_a = 64.8 \sqrt{\frac{p_cd^5 \times \Delta P}{(1 + 1.714/d)sL}}$$

$$Q = 3,885 \sqrt{\frac{p_cd^5 \times \Delta P}{(1 + 1.714/d)sL}}$$

$$= 45.8 \sqrt{\frac{(p_1^2 - p_2^2)d^5}{(1 + 1.714/d)sL}}$$

$$= 2,745 \sqrt{\frac{(p_1^2 - p_2^2)d^5}{(1 + 1.714/d)sL}}$$

$$\Delta P = \frac{(1 + 1.714/d)LM^2}{24.45sp_cd^5}$$

$$\Delta P = \frac{(1 + 1.714/d)LW^2}{88,000sp_cd^5}$$

$$M = 4.95 \sqrt{\frac{sp_cd^5 \times \Delta P)}{(1 + 1.714/d)L}}$$

$$W = 296.5 \sqrt{\frac{sp_cd^5 \times \Delta P}{(1 + 1.714/d)L}}$$

$$= 3.49 \sqrt{\frac{(p_1^2 - p_2^2)sd^5}{(1 + 1.714/d)L}}$$

$$= 209.5 \sqrt{\frac{(p_1^2 - p_2^2)sd^5}{(1 + 1.714/d)L}}$$

Fritzsche Formulas. Modified slightly and expressed in English units for substituting in the general formulas of this handbook, Fritzsche's[1] friction factor can be written for *any* standard conditions designated by the subscript 0.

Fritzsche Formulas for Gas or Air Flow in Cubic Feet per Hour for Any Standard Conditions at Any Flow Temperature

(For definition of symbols, see text)

$$f = 0.0192 \left(\frac{3,600 \times 53.33T_0}{144sp_0Q_{60}}\right)^{1/7} = 0.05372 \left(\frac{T_0}{sp_0Q_{60}}\right)^{1/7}$$

NOTE: Examination of Fritzsche's expression for f shows it to involve empirically established relationships which represent the partial equivalent of a Reynolds criterion. Hence the Fritzsche formulas within their own field may be regarded as approaching the adaptability of the rational formula solution, but without the ability to take into account varying degrees of roughness or the pipe interior or to be extended to embrace fluids having viscosities other than those for which the empirical constants were determined.

Substituting the above value for f in the Fritzsche equation gives

$$\Delta P = \frac{T_cLs^{0.857}}{204 \times 10^4 p_cd^5} \times \left(\frac{p_0Q_{60}}{T_0}\right)^{1.857}$$

$$Q_{60} = \frac{1,750}{s^{0.462}} \times \frac{T_0}{p_0} \left[\frac{(p_1^2 - p_2^2)d^5}{T_cL}\right]^{0.538}$$

$$d = \left[\frac{T_cLs^{0.857}}{102 \times 10^4(p_1^2 - p_2^2)} \times \left(\frac{p_0Q_{60}}{T_0}\right)^{1.857}\right]^{1/5}$$

NOTE: $\frac{1}{7} = 0.143$; $\frac{7}{13} = 0.538$; $\frac{13}{7} = 1.857$.

[1] Fritzsche's formulas were published in German in 1908. For an account in English, see "Principles of Thermodynamics," by G. A. Goodenough, Henry Holt and Company, Inc., New York, 3d ed., 1925, p. 159.

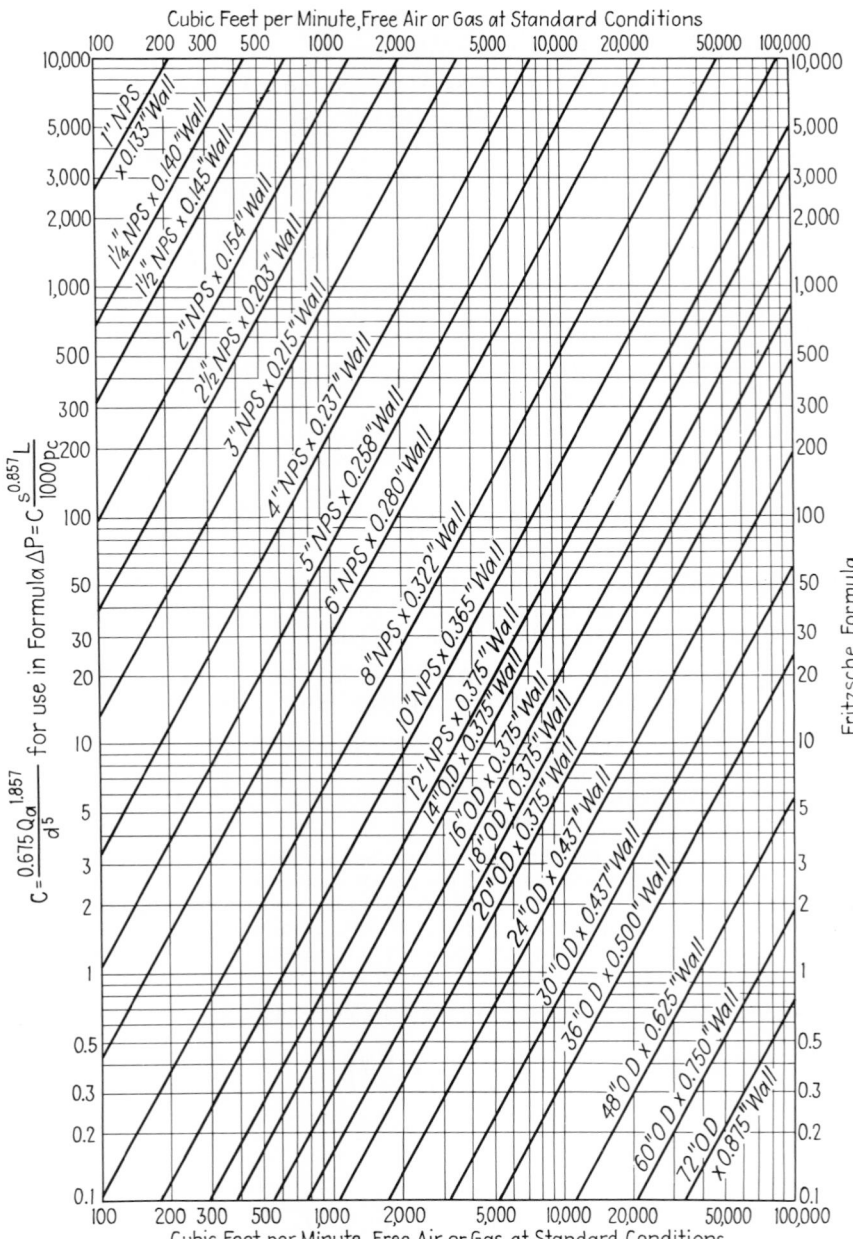

FIG. 49. Chart for determining flow coefficient C used in simplifying the Fritzsche formula for air and gas where Δp is given in terms of s, L, Q_a, and d.

If desired, similar derivations can be made from Eq. (89) for flow in cubic feet per minute, or the flow can be converted to an hourly quantity and substituted in the above formulas.

Where *standard conditions* are those usual for compressed air or gas which approximate 14.7 psia at 60 F and the flow temperature also is assumed to be 60 F, a considerable simplification can be made through substituting the following numerical values in the foregoing formulas: $p_0 = p_a = 14.7$ psia; $T_0 = T_a = T_c = 460 \times 60 = 520$ F. Also under these conditions the Fritzsche friction factor simplifies to $f = 0.0896/(sQ)^{1/7}$; $f = 0.05/(sQ_a)^{1/7}$; $f = 0.0345/M^{1/7}$; and $f = 0.0624/W^{1/7}$. These values for f can be substituted readily in the general flow formulas involving Q, Q_a, M, and W, respectively, to get

Fritzsche Formulas for Air and Gas Flow—Standard Conditions of 14.7 Psi at 60 F

(For definition of symbols, see text)

$$\Delta P = \frac{s^{0.857}LQ_a^{1.857}}{1{,}480p_cd^5}$$

$$\Delta P = \frac{s^{0.857}LQ^{1.857}}{296 \times 10^4 p_cd^5}$$

$$Q_a = \frac{51}{s^{0.462}}\left(\frac{p_c\Delta Pd^5}{L}\right)^{0.538}$$

$$Q = \frac{3{,}050}{s^{0.462}}\left(\frac{p_c\Delta Pd^5}{L}\right)^{0.538}$$

$$= \frac{s^{0.462}}{35}\left[\frac{(p_1^2 - p_2^2)d^5}{L}\right]^{0.538}$$

$$= \frac{2{,}100}{s^{0.462}}\left[\frac{(p_1^2 - p_2^2)d^5}{L}\right]^{0.538}$$

$$d = 0.2333s^{0.1714}\left(\frac{LQ_a^{1.857}}{\Delta Pp_c}\right)^{0.2}$$

$$d = 0.0507s^{0.1714}\left(\frac{LQ^{1.857}}{p_c\Delta P}\right)^{0.2}$$

$$= 0.267s^{0.1714}\left(\frac{LQ_a^{1.857}}{p_1^2 - p_2^2}\right)^{0.2}$$

$$= 0.0538s^{0.1714}\left(\frac{LQ^{1.857}}{p_1^2 - p_2^2}\right)^{0.2}$$

$$\Delta P = \frac{LM^{1.857}}{12.5sp_cd^5}$$

$$\Delta P = \frac{LW^{1.857}}{24{,}900sp_cd^5}$$

$$M = 3.89\left(\frac{sp_c\Delta Pd^5}{L}\right)^{0.538}$$

$$W = 233\left(\frac{sp_c\Delta Pd^5}{L}\right)^{0.538}$$

$$= 2.68\left[\frac{(p_1^2 - p_2^2)sd^5}{L}\right]^{0.538}$$

$$= 161\left[\frac{(p_1^2 - p_2^2)sd^5}{L}\right]^{0.538}$$

$$d = 0.603\left(\frac{LM^{1.857}}{sp_c\Delta P}\right)^{0.2}$$

$$d = 0.132\left[\frac{LW^{1.857}}{sp_c\Delta P}\right]^{0.2}$$

$$= 0.692\left[\frac{LM^{1.857}}{s(p_1^2 - p_2^2)}\right]^{0.2}$$

$$= 0.152\left[\frac{sLW^{1.857}}{s(p_1^2 - p_2^2)}\right]^{0.2}$$

Like other fractional exponent formulas, the Fritzsche equations have to be solved with logarithms or a log-log slide rule. Figure 49 has been included to get around this difficulty, insofar as the equation for ΔP in terms of Q_a under standard conditions is concerned, by providing a chart from which a flow coefficient incorporating the fractional power of Q_a can be read and substituted in the following simplified expressions:

$$\Delta P = C \times T_c s^{0.857}L/1.924p_c$$

where

$$C = 0.675Q_a^{1.857}/d^5$$

or $\qquad \Delta P = C \times s^{0.857}L/1{,}000p_c$, if the flow temperature is 60 F (520 F abs).

The flow coefficient C is read from the chart opposite the corresponding Q_a and pipe size intersection. As in other applications of this type of formula where flow is associated with appreciable pressure drop, p_c should be taken as the average pressure in the line, viz., $p_c = (p_1 + p_2)/2$, and T_c, if assumed other than 60 F, should be the average temperature in the line. In the case of air the specific gravity is unity and $s^{0.857}$ drops out of the formula. For gas where the specific gravity has to be taken into account, the tabular values of $s^{0.857}$ given in Table 52 will be an aid to solution.

Table 52.   Values of $s^{0.857}$ for Use in Fritzsche Formula for Gas

s	$s^{0.857}$	s	$s^{0.857}$	s	$s^{0.857}$
0.10	0.139	0.40	0.456	0.70	0.737
0.15	0.198	0.45	0.504	0.75	0.782
0.20	0.252	0.50	0.553	0.80	0.826
0.25	0.305	0.55	0.559	0.85	0.870
0.30	0.356	0.60	0.646	0.90	0.914
0.35	0.408	0.65	0.692	0.95	0.957

Fritzsche Formulas for Low-pressure Air and Gas Flow

(For definition of symbols, see text)

The Fritzsche formulas can be converted for ready use in computing the flow of air or gas in round ducts or pipes under low pressure, as in ventilating work or gas distribution. Pressures only slightly above atmosphere are measured in inches of water column. Converting pressure drops in pounds per square inch to corresponding drops in inches of water column in the respective Fritzsche formulas and considering $p_c = 14.7$ psia

$$h = \frac{s^{0.857}LQ_a^{1.857}}{785d^5} \qquad Q_a = \left(\frac{786hd^5}{s^{0.857}L}\right)^{0.538} \qquad d = \left(\frac{s^{0.857}LQ_a^{1.857}}{786h}\right)^{1/5}$$

$$h = \frac{s^{0.857}LQ^{1.857}}{1,572,000d^5} \qquad Q = \left(\frac{1,572,000hd^5}{s^{0.857}L}\right)^{0.538} \qquad d = \left(\frac{s^{0.857}LQ^{1.857}}{1,572,000h}\right)^{1/5}$$

$$h = \frac{LM^{1.857}}{6.62sd^5} \qquad M = \left(\frac{6.62sd^5}{L}\right)^{0.538} \qquad d = \left(\frac{LM^{1.857}}{6.62sh}\right)^{1/5}$$

$$h = \frac{LW^{1.857}}{13,240sd^5} \qquad W = \left(\frac{13,240hsd^5}{L}\right)^{0.538} \qquad d = \left(\frac{LW^{1.857}}{13,240sh}\right)^{1/5}$$

The values given in some fan and blower catalogues are from 25 to 50 per cent higher than those obtained by the above formulas. In ventilating work the conditions for *standard air* differ from those used in compressed air and gas work in that the air temperature is taken at 70 F rather than 60 F, the pressure being 14.7 psia (29.921 in. Hg barometer) in both cases. The above formulas were derived on the basis of 60 F air at 14.7 psia where the density is 0.07638 lb/cu ft. For 70 F air the density would be 0.07493 lb/cu ft which can be substituted if desired. The difference involved is small, however, in comparison with discrepancies between formulas and with test results and can well be neglected for most purposes.

The relation between *velocity head* pressure and in air or gas flow is discussed in connection with Eq. (108).

Harris Formulas. In the Harris[1] formula the friction factor f for *compressed air*, converted for substitution in the general flow formulas of this handbook, becomes $f = 0.0308/d^{0.31}$. This expression can be substituted for f and the specific gravity term s dropped in the case of air.

Since air flow usually is expresssed in cubic feet or pounds per minute and this formula is commonly used for standard conditions of 14.7 psia at 60 F with the flow temperature likewise assumed to be 60 F, the following will suffice:

Harris Formulas for Compressed-air Flow—Standard Conditions

(For definition of symbols, see text)

$$\Delta P = \frac{LQ_a^2}{2{,}390 p_c d^{5.31}} \qquad\qquad \Delta P = \frac{LM^2}{13.95 p_c d^{5.31}}$$

$$Q_a = 48.9 \sqrt{\frac{p_c d^{5.31} \Delta P}{L}} \qquad\qquad M = 3.74 \sqrt{\frac{p_c d^{5.31} \Delta P}{L}}$$

$$= 34.5 \sqrt{\frac{(p_1^2 - p_2^2) d^{5.31}}{L}} \qquad\qquad = 2.635 \sqrt{\frac{(p_1^2 - p_2^2) d^{5.31}}{L}}$$

$$d = 0.232 \left(\frac{LQ_a^2}{p_c \Delta P} \right)^{0.188} \qquad\qquad d = 0.61 \left(\frac{LM^2}{p_c \Delta P} \right)^{0.188}$$

$$= 0.255 \left(\frac{LQ_a^2}{p_1^2 - p_2^2} \right)^{0.188} \qquad\qquad = 0.694 \left(\frac{LM^2}{p_1^2 - p_2^2} \right)^{0.188}$$

Another convenient approximation of the Harris formulas is furnished in Table 53 reproduced by permission of the Ingersoll Rand Company from "Compressed Air Data," by F. W. O'Neil (handbook published by *Compressed Air Magazine*, New York, N.Y.). The following variations of the same problem illustrate possibilities in using this table:

Example 1. Given a flow Q_a of 3,000 cfm of free air, an initial pressure of 120 psia, and an assumed pipe diameter of 6 in., what is the pressure drop per 1,000 ft of pipe?
Solution. The ratio of compression is $r = 120/14.7 = 8.15$. Read opposite 3,000 cfm and below 6 in. in diameter a tabular value $N = 17.7$ The pressure drop per 1,000 ft of pipe is $N/r = 17.7/8.15 = 2.17$ psi.

Example 2. Given a flow of 3,000 cfm, an initial pressure of 120 psia, and a desired pressure drop of about 2 psi per 1,000 ft of pipe, what diameter pipe will be required?
Solution. Here again the ratio of compression $r = 8.15$. The approximate tabular number to look for will be $N = 2 \times 8.15 = 16.3$. In the table opposite 3,000 cfm it appears that a 6-in. pipe has an N of 17.7 which comes closest to the 16.3 desired. The answer is, use of a 6-in. pipe.

Example 3. Given a required flow of 3,000 cfm, a pipe diameter of 6 in., and a desired pressure drop of about 2.17 psi per 1,000 ft of pipe, what initial pressure will be required?

[1] See *Univ. of Missouri Bull.* 4, vol. 1, 1912, or "Compressed Air," by Elmo G. Harris, McGraw-Hill Book Company, New York, out of print.

Table 53. Flow of Compressed Air through Schedule 40 (Standard-weight) Steel Pipe—Harris Formula[1]

Free air, cfm	Nominal diameter in inches									
	2	2½	3	3½	4	4½	5	6	8	10
320	61.1	23.8	7.5	3.5						
340	69.0	26.8	8.4	3.9						
360	77.3	30.1	9.5	4.4	2.0					
380	86.1	33.5	10.5	4.9	2.2					
400	94.7	37.1	11.7	5.4	2.5					
420	105.2	40.9	12.9	6.0	3.1					
440	115.5	44.9	14.1	6.6	3.4					
460	125.6	48.8	15.4	7.1	3.7	2.0				
480	137.6	53.4	16.8	7.8	4.0	2.2				
500	150.0	58.0	18.3	8.5	4.3	2.4				
525	165.0	64.2	20.2	9.4	4.8	2.6				
550	181.5	70.2	22.1	10.2	5.2	2.9				
575	197	76.7	24.2	11.2	5.7	3.1				
600	215	83.5	26.3	12.2	6.2	3.4				
625	233	92.7	28.5	13.2	6.8	3.7				
650	253	98.0	30.9	14.3	7.3	4.0	2.2			
675	272	105.7	33.3	15.4	7.9	4.3	2.4			
700	294	113.7	35.8	16.6	8.5	4.6	2.6			
750	337	130.5	41.1	19.0	9.7	5.3	2.9			
800	382	148.4	46.7	21.7	11.1	6.1	3.3			
850	433	168	52.8	24.4	12.5	6.8	3.8			
900	468	188	59.1	27.4	14.0	7.7	4.2			
950	541	209.4	65.9	30.5	15.7	8.6	4.7			
1,000	600	232.0	73.0	33.8	17.3	9.5	5.2	1.9		
1,050	658	256	80.5	37.8	19.1	10.4	5.8	2.1		
1,100	723	280.6	88.4	40.9	21.0	11.5	6.3	2.4		
1,150	790	306.8	96.6	44.7	22.9	12.5	6.9	2.6		
1,200	850	344.0	105.2	48.8	25.0	13.7	7.5	3.3		
1,300	.....	392.0	123.4	57.2	29.3	16.0	8.8	3.8		
1,400	.....	.....	.....	66.3	33.9	18.6	10.2	3.8		
1,500	.....	.....	.....	76.1	39.0	21.3	11.8	4.4		
1,600	.....	.....	.....	86.6	44.3	24.2	13.4	5.1		
1,700	.....	.....	.....	97.8	50.1	27.4	15.1	5.7		
1,800	.....	.....	.....	110.0	56.1	30.7	16.9	6.4		
1,900	.....	.....	.....	122	62.7	34.2	18.9	7.1	1.6	
2,000	.....	.....	.....	135	69.3	37.9	20.9	7.8	1.8	
2,100	.....	.....	.....	149	76.4	40.8	23.0	8.7	2.0	
2,200	.....	.....	.....	166	83.0	45.8	25.3	9.5	2.2	
2,300	.....	.....	.....	179	91.6	50.1	27.6	10.4	2.4	
2,400	.....	.....	.....	195	99.8	54.6	30.1	11.3	2.6	
2,500	.....	.....	.....	212	108.3	59.2	32.6	12.3	2.9	
2,600	.....	.....	.....	229	117.2	64.9	35.3	13.3	3.1	
2,700	.....	.....	.....	247	126	69.1	38.1	14.3	3.3	
2,800	.....	.....	.....	265	136	74.3	41.0	15.4	3.6	
2,900	.....	.....	.....	285	146	79.8	43.9	16.5	3.9	
3,000	.....	.....	.....	305	156	85.2	47.0	17.7	4.1	
3,200	.....	.....	.....	347	177	97.1	53.5	20.1	4.7	
3,400	.....	.....	.....	391	200	109.5	60.4	22.7	5.3	
3,600	.....	.....	.....	438	224	122.8	67.6	25.4	5.6	1.8
3,800	.....	.....	.....	488	250	137	75.5	28.4	6.6	2.0
4,000	.....	.....	.....	542	277	151	83.6	31.4	7.3	2.2
4,200	.....	.....	.....	.....	305	168	92.1	34.6	8.1	2.4
4,400	.....	.....	.....	.....	335	183	101.2	38.1	8.9	2.7
4,600	.....	.....	.....	.....	366	200	110.5	41.5	9.7	2.9
4,800	.....	.....	.....	.....	399	218	120.4	45.2	10.5	3.2
5,000	.....	.....	.....	.....	433	236	131	49.1	11.5	3.4
5,250	.....	.....	.....	.....	477	260	144	54.1	12.6	3.8

[1] Based on the formulas of Elmo G. Harris, *op. cit.* This table is reproduced by permission of the Ingersoll Rand Co. from "Compressed Air Data." The tabular values N divided by the ratio of compression r represent the pressure drop in psi per 1,000 ft of pipe.

Solution. In the table opposite 3,000 cfm for a 6-in. pipe, the value of N is read to be 17.7. The compression ratio corresponding to a pressure drop of 2 psi per 1,000 ft of pipe is $r = N/2.17 = 8.15$, from which $p_1 = 14.7 \times 8.15 = 120$ psia.

Example 4. Given a 6-in.-diameter pipe, an initial pressure of 120 psia, and a pressure drop of about 2.17 psi per 1,000 ft of pipe, what will be the flow in cubic feet per minute of free air?

Solution. The tabular number can be determined as follows: $r = 120/14.7 = 8.15$, and $N = r1,000\Delta P/L = 8.15 \times 2.17 = 17.7$. Looking down the column for 6-in.-diameter pipe, an N value of 17.7 is found opposite 3,000 cfm. Hence the answer is 3,000 cfm.

Comparison of Example 1 of the foregoing tabular solution with results of various formulas:[1]

NOTE: The specific gravity term s has been dropped from these solutions owing to its being unity for air.

Harris formula:

$$\Delta P = \frac{LQ_a^2}{2,390 p_c d^{5.31}} = \frac{1,000 \times 3,000^2}{2,390 \times 119 \times 14,300} = 2.21 \text{ psi}$$

Unwin formula:

$$\Delta P = \frac{(1 + 3.6/d)LQ_a^2}{7,400 p_c d^5} = \frac{1.593 \times 1,000 \times 3,000^2}{7,400 \times 119 \times 8,206} = 1.98 \text{ psi}$$

Fritzsche Formula:

$$\Delta P = \frac{LQ_a^{1.857}}{1,480 p_c d^5} = \frac{1,000 \times 3,000^{1.857}}{1,480 \times 119 \times 8,206} = 1.94 \text{ psi}$$

NOTE: The same value be can be obtained by the chart solution of Fig. 49.

Spitzglass Formula:

$$\Delta P = \frac{3,600 L Q_a^2}{2,333 \times 10^4 p_c K^2} = \frac{3,600 \times 1,000 \times 3,000^2}{2,333 \times 10^4 \times 119 \times 67.97^2} = 2.53 \text{ psi}$$

Weymouth Formula:

$$\Delta P = \frac{lQ^2}{1,572 p_c d^{5.33}} = \frac{1,000}{5,280} \times \frac{(3,000 \times 60)^2}{1,572 \times 119 \times 14,970} = 2.19 \text{ psi}$$

Equation (87):

$$\Delta P = \frac{fLQ_a^2}{74 p_c d^5} = \frac{0.0172 \times 1,000 \times 3,000^2}{74 \times 119 \times 8,206} = 2.14 \text{ psi}$$

Spitzglass Formulas. In 1912 Spitzglass[2] published the results of his tests for the Peoples Gas Light and Coke Company of Chicago and suggested a new formula for the flow of gas or air through pipes. Spitzglass concluded that the effect of internal friction on the viscosity of the fluid must cause variation of the friction factor with pipe diameter. He reasoned that the resistance to flow caused by internal friction bore some relation to the cross-sectional area, whereas skin friction varied with the wetted perimeter. This pointed to the possibility that when the pipe diameter increased over a certain limit, the internal friction would become so preponderant that the friction factor, instead of diminishing, would increase with the diameter of the pipe. This consideration led Spitzglass to the adoption of a friction factor having the form $f = a(1 + b/d + cd)$ "which takes care of variation in both directions" as he put it. As expressed for use in the general flow formulas of this handbook, the Spitzglass friction factor for gas and air flow

[1] For a comparison of the volume of gas flow computed by different formulas see "Flow of Natural Gas through High-pressure Transmission Lines," a joint report by T. W. Johnson and W. B. Berwald, *Monograph* 6, Bureau of Mines, 1935. Reprints may be obtained through the American Gas Association, 420 Lexington Ave., New York, N.Y.

[2] "Flow of Gas Formulae, Derived, Analyzed and Checked by Experimental Data, with Diagrams for Figuring the Flow of Gas in Street Mains and Services," by J. M. Spitzglass, *Am. Gas Light J.*, vol. 96, pp. 269–271, 274–276, 290–296, 312–315, 1912.

becomes $f = 0.0112(1 + 3.6/d + 0.03d)$. This expression gives a rather rapid decrease in f with increase in diameter up to about 6-in. pipe size, a slow decrease in f from 6 in. to 10 or 12 in., and a slow rise in f from 12 in. up.

This form of expression does not lend itself readily to a direct solution for pipe diameter, but it can be solved conveniently for pressure drop or volume of flow. Solution is facilitated by grouping all the diameter terms into one expression, customarily designated as $K = \sqrt{d^5/(1 + 3.6/d + 0.03d)}$, for which numerical values for different pipe diameters can be read from tables. The Spitzglass formulas set up for use with K are as follows:

For pressures over 1 psig:

$$\Delta P = \frac{sLQ^2}{2,333 \times 10^4 p_c K^2}$$

$$Q = 4,830K \sqrt{\frac{p_c \Delta P}{sL}}$$

$$= 3,410K \sqrt{\frac{p_1^2 - p_2^2}{sL}}$$

For pressures not over 1 psig:

$$h = \frac{sLQ^2}{126 \times 10^5 K^2}$$

$$Q = 3,550K \sqrt{\frac{h}{sL}}$$

According to Johnson and Berwald in *Bureau of Mines Monograph 6*,[1] the Spitzglass formula gives a somewhat lower gas flow than would be computed by the Weymouth formula.[2] The difference varies with pipe diameter, ranging from 10 per cent below Weymouth flow quantities for a 6-in. pipe to 30 per cent below for a 24-in. pipe. Whereas the Weymouth formula is regarded by Berwald and Johnson as the most accurate of the empirical formulas, especially in the larger pipe sizes and with clean steel pipe, the deviation of the Spitzglass formula is on the conservative side. Hence some gas engineers prefer to use the Spitzglass formula where much rust, scale, or condensates are present in the line, especially in the diameters of cast-iron pipe 12 in. or smaller. The present trend, however, is toward the rational form of solution by use of equations of the type of (77) or (87) and Figs. 41 and 42.

Weymouth Formulas. In 1912 Thomas R. Weymouth presented in an ASME paper his formulas which are generally considered to be the best suited of the empirical formulas for computing the flow of gas through intermediate and high-pressure transmission lines. Johnson and Berwald concluded from the results of flow tests made on 29 natural gas lines totaling 757 miles of pipe of 6 to 22 in. in diameter and operating at pressures of 30 to 600 psi that, "The curve based on the Weymouth formula . . . agreed closely with the average of the experimental results and satisfied the tests data as well as any curve that could be drawn through the plotted points."

The Weymouth formulas employ fractional exponents and have to be solved with logarithms or a log-log slide rule except where numerical values for the fractional powers of d can be read from a table. As expressed for use in the general flow formulas of this handbook, the Weymouth friction factor becomes

[1] See *Monograph 6*, "Flow of Natural Gas through High-pressure Transmission Lines," Bureau of Mines, a joint report by T. W. Johnson and W. B. Berwald, 1935. Copies may be obtained through the American Gas Association, 420 Lexington Ave., New York 17, N. Y.

[2] "Problems in Natural-gas Engineering," by Thomas R. Weymouth, *Trans. ASME*, vol. 34, pp. 185–206, 1912.

$f = 0.032/d^{1/3}$. The Weymouth formula for generalized conditions is expressed in the symbols of this handbook as follows (see text for definition of symbols and note that l is used here to denote the length of the line in *miles* instead of L in feet and that the subscript 0 is used to denote *any* standard condition):

Weymouth Formulas for Gas or Air Flow in Cubic Feet per Hour for Any Standard Conditions at Any Flow Temperature

(See text for definition of symbols, l is in miles)

$$\Delta P = \frac{sT_c l}{652 p_c d^{16/3}} \left(\frac{p_0 Q_{60}}{T_0} \right)^2$$

$$Q_{60} = 18.062 \frac{T_0}{p_0} \sqrt{\frac{(p_1^2 - p_2^2)d^{16/3}}{sT_c l}}$$

$$d = \left[\frac{sT_c l}{326(p_1^2 - p_2^2)} \times \left(\frac{p_0 Q_{60}}{T_0} \right)^2 \right]^{3/16}$$

NOTE: $\frac{8}{3} = 2.667$; $\frac{16}{3} = 5\frac{1}{3} = 5.333$; $\frac{3}{16} = 0.1875$.

Where *standard conditions* are those usual for gas or compressed air which approximate 14.7 psia at 60 F and the flow temperature also is assumed to be 60 F, a considerable simplification can be made through substituting the following numerical values in the foregoing formulas:

$$p_0 = p_a = 14.7 \text{ psia} \qquad T_0 = T_a = T_c = 460 + 60 = 520 \text{ F}$$

Weymouth Formulas for Gas or Air Flow in Cubic Feet per Hour for Standard Conditions and 60 F Flow Temperature

(See text for definition of symbols, l is in miles)

$$\Delta P = \frac{slQ^2}{1,572 p_c d^{16/3}}$$

$$Q = 28.05 \sqrt{\frac{(p_1^2 - p_2^2)d^{16/3}}{sl}}$$

$$d = \left[\frac{slQ^2}{786(p_1^2 - p_2^2)} \right]^{3/16}$$

Sample Problems in Gas Flow. The following examples have been chosen as a means of demonstrating the principal gas flow formulas and of showing how to use the aids to solution provided in this handbook, at the same time affording a spot comparison of the results.[1] Owing to the frequency with which the expression $\sqrt{p_1^2 - p_2^2}$ appears in gas flow formulas, a schedule of values for the expression has been provided in Table 54.

Example 1. What is the carrying capacity in cubic feet of natural gas per hour measured at standard conditions of a 60-mile line of 8-in. Schedule 30 pipe if the initial pressure is 300 psig, the final pressure is 30 psig, the flow temperature is 60 F, and the specific gravity is 0.7?

[1] For a thoroughgoing comparison of gas flow formulas see Johnson and Berwald, *op. cit.*

Table 54. Computed Values of the Expression $\sqrt{p_1^2 - p_2^2}$ for Use in Gas and Air-flow Formulas[1]

(NOTE.—For convenience in use, the column and line designations are given in psi *gauge*, whereas the tabular values were computed from the corresponding *absolute* pressures)

Tabular values $= \sqrt{p_1^2 - p_2^2}$, where p_1 and p_2 are absolute pressures in psi, *viz.*, gauge pressure $+ 14.7$

Inlet pressure, psi, gauge	Discharge pressure, psi, gauge																						
	1	2	4	6	8	10	20	30	40	50	60	70	80	90	100	150	200	250	300	350	400	450	500
2	5.69																						
4	10.2	8.31																					
6	13.5	12.2	8.87																				
8	16.4	15.4	12.9	9.31																			
10	19.1	18.4	16.1	13.5	9.74																		
12	21.6	20.8	19.1	16.9	14.1	10.1																	
14	24.0	23.3	21.8	19.9	17.6	14.6																	
16	26.4	25.8	24.4	22.7	20.7	18.2																	
18	28.7	28.1	26.8	25.3	23.5	21.4																	
20	30.9	30.4	29.2	27.8	26.2	24.4																	
30	41.9	41.5	40.6	39.6	38.5	37.3	28.2																
40	52.4	52.1	51.4	50.6	49.8	48.8	42.3	31.5															
50	62.8	62.5	61.9	61.3	60.6	59.8	54.6	46.8	34.6														

60	73.0	72.8	72.3	71.8	71.2	70.5	66.2	59.9	50.9	36.3													
70	83.2	83.0	82.6	82.1	81.6	81.0	77.3	71.9	64.7	54.7	39.9												
80	93.4	93.2	92.8	92.4	91.9	91.4	88.1	83.5	77.3	69.2	58.2	42.4											
90	104	103	103	103	102	102	98.8	94.7	89.3	82.3	73.4	61.6	44.7										
100	114	114	113	113	112	112	109	106	101	94.7	87.0	77.3	64.7	46.8									
150	164	164	164	163	163	163	161	159	155	152	147	141	135	127	118								
200	214	214	214	214	214	213	212	210	208	205	201	197	193	187	182	138							
250	264	264	264	264	264	264	262	261	259	257	254	251	247	243	239	207	155						
300	314	314	314	314	314	314	313	312	310	308	306	303	300	297	293	268	230	170					
350	364	364	364	364	364	364	363	362	361	359	357	355	352	349	346	325	295	251	184				
400	414	414	414	414	414	414	413	412	410	409	408	406	404	401	399	381	355	319	270	197			
450	464	464	464	464	464	464	463	463	462	460	459	457	455	453	450	435	412	382	342	288	210		
500	515	514	514	514	514	514	514	513	512	511	509	508	506	504	502	488	468	441	407	363	305	221	
600	615	615	615	614	614	614	614	613	612	611	610	609	607	606	604	592	576	555	528	494	454	402	336
700	715	715	715	714	714	714	714	713	713	712	711	710	708	707	705	695	678	664	642	615	572	533	496
800	815	815	815	814	814	814	814	814	813	812	811	810	809	808	807	798	786	771	752	729	701	669	622
900	915	915	915	915	914	914	914	914	913	912	912	911	910	909	908	900	889	876	859	839	815	788	756
1,000	1,015	1,015	1,015	1,015	1,014	1,014	1,014	1,014	1,013	1,013	1,012	1,011	1,010	1,009	1,008	1,001	992	980	965	947	926	902	875

[1] Irrespective of the available pressure difference, the maximum capacity of a pipeline is limited by the acoustic velocity of the fluid at the downstream end of the line.

Solution Aids. The following aids to solution were used in connection with the formulas listed below:

$$\sqrt{p_1{}^2 - p_2{}^2} = \sqrt{(300 + 14.7)^2 + (30 + 14.7)^2} = 312 \text{ (from Table 54)}$$

$$p_1{}^2 - p_2{}^2 = 312^2 = 97,344$$

$$l = 60 \text{ miles}, \ L = 60 \times 5,280 = 316,800 \text{ ft}$$

$$d = 8.071; \ \sqrt{d^5} = 185.1; \ \sqrt{d^{16/3}} = 262.1; \ K = 142.5$$

$$d^3 = 525.7; \ d^4 = 4,243; \ d^5 = 34,248 \text{ (from Table 47)}.$$

Unwin Solution:

$$Q = 2,745 \sqrt{\frac{(p_1{}^2 - p_2{}^2)d^5}{(1 + 1.714/d) sL}} = \frac{2,745 \times 312 \times 185.1}{\sqrt{(1 + 1.714/8.071) \, 0.7 \times 316.800}}$$

$$= 306,000 \text{ cu ft/hr}$$

Fritzsche Solution:

$$Q = \frac{2,100}{s^{0.462}} \left[\frac{(p_1{}^2 - p_2{}^2)d^5}{L} \right]^{0.538} = \frac{2,100}{0.7^{0.462}} \left(\frac{97,344 \times 34,248}{316,800} \right)^{0.538}$$

$$= \frac{2,100 \times 146}{0.849} = 361,000 \text{ cu ft/hr}$$

Spitzglass Solution:

$$Q = 3,410K \sqrt{\frac{p_1{}^2 - p_2{}^2}{sL}} = \frac{3,410 \times 142.5 \times 312}{\sqrt{0.7 \times 316,800}} = 321,000 \text{ cu ft/hr}$$

Weymouth Solution:

$$Q = 28.05 \sqrt{\frac{(p_1{}^2 - p_2{}^2)d^{16/3}}{sl}} = \frac{28.05 \times 312 \times 261.1}{0.7 \times 60} = 352,200 \text{ cu ft/hr}$$

Solution [using Eq. (92), and Figs. 41 and 42]:

$$Q = 364.8 \sqrt{\frac{(p_1{}^2 - p_2{}^2)d^5}{fsL}}$$

$$= \frac{364.8 \times 312 \times 185.1}{\sqrt{0.0162 \times 0.7 \times 316,800}} = 350,000 \text{ cu ft/hr}$$

NOTE: Since Q is unknown, it is necessary to assume a value for Q in order to determine the friction factor f. It is most reasonable to assume, as a trial value of Q, any of the results of the above empirical equations. In this case, the Fritzsche solution was used and the final value of f determined by iteration.

Example 2. Required, the diameter of a pipe 10,000 ft in length for a flow of 3,000 cfm free air; initial pressure 120 psia with a pressure drop of 20 psi; flow temperature 60 F.
Harris Solution:

$$d = 0.255 \left(\frac{LQ_a{}^2}{p_1{}^2 - p_2{}^2} \right)^{0.188} = 0.255 \left(\frac{10,000 \times 3,000^2}{120^2 - 100^2} \right)^{0.188} = 6.03 \text{ in.}$$

Fritzsche Solution:

$$d = 0.267 s^{0.1714} \left(\frac{LQ^{1.857}}{p_1{}^2 - p_2{}^2} \right)^{0.2} = 0.267 \times 1 \left(\frac{10,000 \times 3,000^{1.857}}{120^2 - 100^2} \right)^{0.2} = 6.14 \text{ in.}$$

Weymouth Solution:

$$d = \left[\frac{slQ^2}{786(p_1{}^2 - p_2{}^2)} \right]^{3/16} = \left[\frac{1 \times 10,000 \times (60 \times 3,000)^2}{786 \times 5,280(120^2 - 100^2)} \right]^{3/16} = 6.25 \text{ in.}$$

Solution by Analytical Expressions: This types of problem, in which the desired pressure drop is given, does not lend itself to analytical solution. From the empirical equation solutions, it appears that a 6-inch Schedule 40 pipe is satisfactory. In order to check this size, the analytical expressions will now be used.

$$P_{avg} = P_1 - \tfrac{1}{2}\Delta P = 120 - 20/2 = 100 \text{ psia}$$

$$T_{avg} = 60 \text{ F} = 520 \text{ R}$$

$$\rho_{avg} = P_{avg}/RT_{avg} = 110 \times 144/53.3 \times 520 = 0.572 \text{ lbm/cu ft}$$

$$D = 6.065/12 = 0.505 \text{ ft (by Table 47)}$$

$$\mu = 12.14 \times 10^{-6} \text{ lbm/(ft sec) (Table 41)}$$

$$V = \dot{m}/\rho A = 3,000 \times 0.07638/60 \times 0.572 \times 0.2006$$

$$V = 33.3 \text{ fps}$$

$$N_r = DV\rho/\mu = 0.505 \times 33.3 \times 0.572 \times 10^6/12.14 = 774,000$$

$$\epsilon/D = 0.0003 \text{ and } f = 0.016$$

$$\Delta P = f\rho(L/D)(V^2/2g) = 0.016 \times 0.572 \times \frac{10,000}{0.505} \times \frac{33.3^2}{2 \times 32.17} \times \frac{1}{144}$$

$$\Delta p = 21.6 \text{ psi}$$

If greater accuracy is desired, the problem can be reworked using $P_{avg} = 120 - 21.6/2$. It does appear that a 6-in. pipe will satisfy the condition within about 5 per cent. If the pressure drop of 20 psi is considered as the absolute maximum at the stated flow rate, an 8-in. pipe should be used.

Complex Pipelines. The formulas given above for flow in simple pipelines can be extended to apply to divided circuits, series circuits consisting of different size pipes, combinations of divided and series circuits, and distribution networks. In order to simplify the calculations, it often is convenient to neglect the fact that the friction factor f is a function of Reynolds number and the pipe diameter and assume that f is the same for all pipe diameters. This is not strictly correct, but it avoids involving the diameter in a complex term and greatly facilitates the solution. The general method here indicated can be employed readily with any of the fractional exponent type of empirical formulas commonly applied to various fluids. The assumption that carrying capacity varies as the fifth power of the diameter used in the present solution is not strictly accurate in some instances. In the case of water at ordinary atmospheric temperatures, the fifth power relation gives quite accurate results, as shown by Saph and Schoder's[1] tests which indicated a fractional exponent for d of 4.97. In the case of natural gas, Weymouth[2] arrives at a value of $5\tfrac{1}{3}$ as the exponent of d. The majority of empirical equations for all fluids use exponents for d very closely approximating 5. Fritzsche in his formula for the flow of steam gives 4.97.

Cross[3] has analyzed the flow problem in distribution networks for water-supply systems and has indicated how to extend his methods to computing the flow of air

[1] "An Experimental Study of the Flow of Water in Pipes," *Trans. ASCE Paper 964*, vol. 51, p. 253, 1903.

[2] "Problems in Natural Gas Engineering," *Trans. ASME*, vol. 34, p. 185, 1912; see pp. 187, 197, "Piping Handbook." For another solution of complex pipelines based on the Weymouth formula see pp. 45–52 of "Flow of Natural Gas through High-pressure Transmission Lines," by T. W. Johnson and W. B. Berwald, *Monograph 6*, Bureau of Mines. Reprints may be obtained from the American Gas Association, 420 Lexington Ave., New York, N.Y.

[3] "Analysis of Flow in Networks of Conduits or Conductors," by Prof. Hardy Cross, *Univ. Illinois, Bull.* 286, November, 1936, Engineering Experiment Station, Urbana, Ill.

gas, or steam in networks and to the design of electrical circuits. Simplifications and short cuts for this method have been evolved by others.[1]

Complex-series Pipeline. A complex-series pipeline consists of two or more runs of different size pipe in series. The problem in this case is to determine the length of some one size of pipe having a frictional resistance equivalent to the series. The first step is to select the size pipe in which the equivalent length is to be expressed, and the next to convert each run in turn into the corresponding length of the size pipe selected. This is readily done from the general friction-loss formula, since the pressure drop, quantity of discharge, and density are the same by definition for both actual and equivalent pipes. The subscripts 1, 2, 3, etc., are used to designate the different sections of pipe in series, with subscript 1 referring to the size pipe selected to which the others are to be converted. Small letters l and d are used to denote actual lengths and diameters, while capital letters L and D are used for equivalent lengths and diameters. By definition $l_1 = L_1$, and $d_1 = D_1 = D_2 = D_3$, etc. For each successive section considered, the following relations hold:

From Eq. (77)
$$\Delta P_{act.} = f_2 l_2 v M^2 / 298,000 d_2{}^5$$
and
$$\Delta P_{equiv.} = f_1 L_2 v M^2 / 298,000 D_2{}^5$$

Dividing one equation by the other, $1 = f_2 l_2 D_2{}^5 / f_1 L_2 d_2{}^5$,

or
$$L_2 = l_2 (f_2/f_1)(D_2/d_2)^5$$
but
$$D_2 = d_1$$
hence,
$$L_2 = l_2 (f_2/f_1)(d_1/d_2)^5$$
and, similarly,
$$L_3 = l_3 (f_3/f_1)(d_1/d_3)^5$$

The total equivalent length L of the series of connected pipes is $L = L_1 + L_2 + L_3 \cdots + L_n$ where L_1, L_2, L_3, etc., are the equivalent lengths reduced to diameter d_1 for the respective sections that make up the series.

Example. A series pipeline shown in Fig. 50a consists of three runs of 6-, 8-, and 10-in. standard-weight (Schedule 40) wrought pipe of 300-, 400-, and 500-ft lengths, respectively. What is the equivalent length of 6-in. pipe assuming the same friction factor for each, i.e., $f_1 = f_2 = f_3$?

From Table 47, the fifth powers of the inside diameters of 6-, 8-, and 10-in. standard-weight pipe are 8,206, 32,380, and 101,000, respectively.

The equivalent length in 6-in. pipe of section (1) is, of course,   300 ft

The equivalent length of section (2) is $L_2 = 400(8,206/32,380)$ = 101 ft

The equivalent length of section (3) is $L_3 = 500(8,206/101,000)$ =   41 ft

Total equivalent length of 6-in. pipe   = 442 ft

The answer is that 442 ft of 6-in. standard-weight pipe have a frictional resistance equal to that of the complex-series pipe line shown in Fig. 50a.

[1] (a) "The Hardy Cross Method, Its Practical Application in Determining Flow and Heads in Pipe Systems," by D. R. Taylor, *Water Works and Sewerage* vol. 90, no. 3, March and April, 1943. Contains numerous aids to solution and a bibliography of references to the Hardy Cross method.

(b) *Eng. News-Record*, Oct. 1, 1936, and Mar. 3, 1938, and *Civil Civil Eng.*, May, 1938.

(c) "Computation of Flows in Distribution Systems," by Weston Gavett, *Tour. AWWA*, vol. 35, no. 3, pp. 267–287, March, 1943.

Divided Circuit or Loop. In the divided circuit shown in Fig. 50b, the solution is made in two steps: (1) each branch is reduced to an equivalent diameter D_n corresponding to a common length which is taken to agree with the length of that branch whose diameter and friction factor are made the basis for expressing equivalent lengths; (2) substituting the equivalent diameters and their corresponding common length in the flow formula, the flow through each branch can be computed and the total taken for all branches.

Step (2) above can be simplified from the following considerations. The condition existing in a loop system is that the initial and final pressures are the same

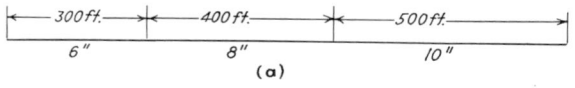

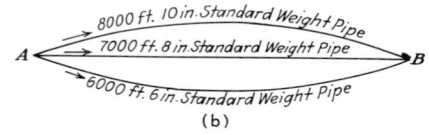

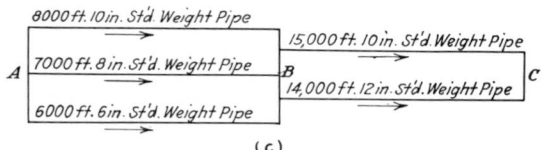

Fig. 50

for each pipe, and the only difference between the computations for flow is the $\sqrt{D_n{}^5}$. A summation of all the quantities $\sqrt{D_n{}^5}$ can be made, therefore, and the summation used in the flow formula. This makes it possible to obtain the answer with a single solution of the flow formula.

A solution for step (1) is derived as follows from the general flow formula. For any one branch of the loop system, M, ΔP and ρ are the same for both actual and equivalent conditions. The subscripts 1, 2, 3, etc., are used to designate the different branch pipes of the loop with subscript 1 referring to the size pipe selected to which the others are to be converted. Small letters l and d are used to denote actual lengths and diameters, while capital letters L and D are used for equivalent lengths and diameters. By definition $d_1 = D_1$ and $l_1 = L_1 = L_2 = L_3$, etc. Writing the flow equations for one of the branches in terms of both actual and equivalent dimensions and dividing one equation by the other:

$$\Delta P = f_2 l_2 W_2{}^2 v / 298{,}000 d_2{}^5 \quad \text{and} \quad \Delta P = f_1 L_2 W_2{}^2 v / 298{,}000 D_2{}^5$$

the quotient is
$$1 = f_2 l_2 D_2{}^5 / f_1 L_2 d_2{}^5$$
or
$$D_2 = d_2 \, (f_1 L_2 / f_2 l_2)^{1/5}$$
But
$$L_2 = l_1,$$

hence,
$$D_2 = d_2\,(f_1l_1/f_2l_2)^{1/5}$$
and similarly
$$D_3 = d_3\,(f_1l_1/f_3l_3)^{1/5}$$

This completes the derivation required for step (1).

The two methods of accomplishing step (2) mentioned above are executed as follows: The pressure drop, density, and friction factor for each branch of the loop system are identical. Assuming a common length l_1 for all branches, their corresponding equivalent diameters have just been computed. It is now possible to write the following series of equations (based on the general flow equation) for flow in the different branches:

$$M_1 = \sqrt{298{,}000(\Delta P)d_1{}^5/f_1l_1v} \qquad M_2 = \sqrt{298{,}000(\Delta P)D_2{}^5/f_1l_1v}$$

$$M_3 = \sqrt{298{,}000(\Delta P)D_3{}^5/f_1l_1v} \text{ etc.}$$

The total flow M through the loop is the summation of the flow through the separate branches, or $M = M_1 + M_2 + M_3 \cdots + M_n$. An examination of the above equations shows, however, that the right-hand members are identical with the exception of the terms $d_1^{5/2}$, $D_2^{5/2}$, $D_3^{5/2}$, etc. The summation may then be written

$$M = [298{,}000(\Delta P)/f_1l_1v]^{1/2}(d_1^{5/2} + D_2^{5/2} + D_3^{5/2} \cdots + D_n^{5/2})$$

from which it is apparent that the equivalent diameter for the entire loop reduced to an equivalent length l_1 is

$$d = (d_1^{5/2} + D_2^{5/2} + D_3^{5/2} \cdots + D_n^{5/2})^{2/5}$$

Example. One million cubic feet of natural gas per hour, measured at standard pressure and temperature, are to be transmitted through a loop system consisting of three pipelines of about equal smoothness (see Fig. 50*b*). The pressure of the gas at the delivery end (*B*) of the system is to be 50 psia. The specific gravity of the gas is 0.65 and the temperature in the pipe is 60 F. Determine the required initial pressure of the gas at *A*. Although the effect of friction factor is of only minor importance in this case, it is carried through for the purpose of illustration.

The actual inside diameters of 6-, 8-, and 10-in. standard-weight steel pipe are 6.065, 7.981, and 10.02 in., respectively (see Table 47). The friction factors used below were computed by assuming an approximate inlet pressure of 70 psia. Using 8,000 ft as the equivalent length for each pipe,

$$D_1 = d_1 = 10.02$$
$$D_2 = 7.981(0.00393/0.00395 \times 8{,}000/7{,}000)^{1/5} = 8.19$$
$$D_3 = 6.065(0.00393/0.00410 \times 8{,}000/6{,}000)^{1/5} = 6.37$$
$$D_1{}^{5/2} = 10.02^{5/2} = 317.9$$
$$D_2{}^{5/2} = 8.19^{5/2} = 191.9$$
$$D_3{}^{5/2} = 6.37^{5/2} = 102.4$$
$$\Sigma D^{5/2} = \overline{612.2}$$
$$D = 13.02 \text{ in.}$$

Or, in other words, the resistance of the entire loop is equivalent to that of a single pipe having an inside diameter of 13.02 in. and a length of 8,000 ft.

The balance of the answer can be obtained readily through substitution in Eq. (92) where $Q = 364.4\sqrt{(p_1{}^2 - p_2{}^2)d^5/fsl}$ and solving as follows: The friction factor f is determined to be 0.016. Substituting and solving for p_1

$$1{,}000{,}000 = 364.4\sqrt{\frac{(p_1{}^2 - 50^2)13.02^5}{0.016 \times 0.65 \times 8{,}000}}$$

$$p_1 = 64.62$$

Answer. The required initial pressure at *A* is 64.62 or in round numbers 65 psia.

Series-loop System.   A piping system made up of two sections or more of loop systems in series, as shown in Fig. 50c, is known as a "series-loop system." A large variety of such systems exists.   The general method of solution is as follows: (1) first determine a single equivalent line for each loop as explained above under "Divided Circuits;" (2) next determine a single equivalent line for the series of single equivalent lines, following the method explained under "Complex-series Pipelines."   The explanation under the above headings, supplemented by the illustrations given there, is sufficient for handling problems of this sort.   An example illustrating an actual solution for a series-loop system is given later in connection with natural gas.   In the example, the carrying capacity is assumed to vary as the $\frac{8}{3}$ power of the diameter instead of the $\frac{5}{2}$ power.   As was previously mentioned, the use of the $\frac{8}{3}$ power of d in connection with natural gas was proposed by Thomas R. Weymouth to take into account the variation in friction factor with diameter.

Pipeline Storage Capacity.   The storage capacity of a pipeline is the excess gas, over and above the average rate of supply, that can be packed in the line during the period when the demand is less than the supply.   The supply rate is the rate of average daily delivery, and since the volume of gas contained in the line is a function of the pressure, it is neces-sary, in order to determine the storage capacity, to develop an expression for the total line contents which will take into account the variable pressure conditions at all points along the line.   Having such an expression, the total contents under the "packed" and "unpacked" conditions may be computed and the storage ascer-tained by taking their difference.

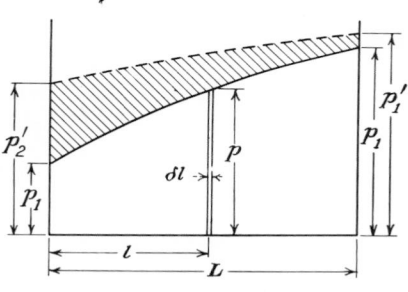

Fig. 51.

The unpacked condition of a pipeline obtains when the gas is flowing at the average daily rate, for when the consump-tion is below this point, all excess gas is being stored, and this point, therefore, is the lower limit, or base of the peaks in the daily delivery curve for the day.

The upper limit, or packed-line condition, is not so readily determined, but if it is considered to be such that the intake pressure is at its maximum point, as fixed by considerations of safety, station-pump pressure limits, etc., and the flow is a mean between the minimum and average rates for the day, the result will be very nearly actual conditions, and whatever error may thereby be involved will be on the side of safety in estimating the storage capacity.

Figure 51 shows the variation of intake pressure p_1 for varying values of L with the discharge pressure p_2 constant.   Two curves are plotted: The solid line repre-sents the conditions with an assumed average rate of flow and the discharge pressure p_2 fixed by the minimum value allowable by the requirements of the distributing system.   The dotted line represents the pressure conditions in the same line with the intake pressure p_1' at the maximum allowable value and the flow at an assumed mean of the minimum and average rates for the day.   This latter curve represents the packed line, and the shaded area the quantity of gas stored in the line available for peak loads.

Let p be the absolute pressure in pounds per square inch at any point, Q the discharge in cubic feet per hour, l or L the length of the line in miles, and d the internal diameter in inches; then for any given line and flow condition, from

the Weymouth equation and assuming $s = 0.573$ in the line:

$$p^2 - p_2^2 = l(Q^2/37^2 d^{5\frac{1}{3}}) = Kl$$

wherein

$$K = Q^2/37^2 d^{5\frac{1}{3}}$$

Hence,

$$p = \sqrt{Kl + p_2^2}$$

The gas contained in a length δl based on pressure $p_0 = 14.65$ will be in cubic feet measured at standard conditions

$$\delta V = 5{,}280 A \times \delta l \times p/p_0$$

where A is the cross-sectional area of the pipe in square feet. The total quantity of gas in the line under the given conditions will then be

$$V = \int \delta V = 5{,}280 \frac{A}{p_0} \int_0^L \sqrt{Kl + p_2^2}\, \delta l$$

$$= \frac{3{,}520}{K} \frac{A}{p_0} [(KL + p_2^2)^{3/2} - p_2^3]$$

But

$$KL = p_1^2 - p_2^2 \text{ and } K = \frac{p_1^2 - p_2^2}{L}$$

Hence,

$$V = \frac{3{,}520 AL}{p_0(p_1^2 - p_2^2)} (p_1^3 - p_2^3)$$

$$= 3{,}520 \frac{AL}{p_0} \left(p_1 + p_2 - \frac{p_1 p_2}{p_1 + p_2} \right)$$

$$= 19.20 \frac{d^2 L}{p_0} \left(p_1 + p_2 - \frac{p_1 p_2}{p_1 + p_2} \right)$$

In the case of a complex system, it is necessary, first, to ascertain the common pressures at all junction points where the lines are tied together and then to compute separately the capacity of each section. To do this it is necessary to use the actual diameter, or area, and length of each of the pipes in the loop. The gas content of a looped section thus becomes

$$V = \frac{3{,}520}{p_0} \left(p_1 + p_2 - \frac{p_1 p_2}{p_1 + p_2} \right) \Sigma AL$$

or

$$V = \frac{19.20}{p_0} \left(p_1 + p_2 - \frac{p_1 p_2}{p_1 + p_2} \right) \Sigma d^2 L$$

The gas content of the entire system will thus be the sum of the contents obtained for the several sections of line. If the quantity thus obtained for average flow conditions is deducted from that similarly computed for minimum flow with maximum intake pressure p_1', the result will be the total available storage capacity of the line.

If the letters used in the two preceding equations be taken to represent the pressure conditions of the unpacked lines and p_1' and p_2' those of the packed line, the available storage capacity of the system will be

$$S = \frac{3{,}520}{p_0} \left(p_1' - p_1 + p_2' - p_2 - \frac{p_1' p_2'}{p_1' + p_2'} + \frac{p_1 p_2}{p_1 + p_2} \right) \Sigma AL$$

or

$$S = \frac{19.20}{p_0} \left(p_1' - p_1 + p_2' - p_2 - \frac{p_1' p_2'}{p_1' + p_2'} + \frac{p_1 p_2}{p_1 + p_2} \right) \Sigma d^2 L \qquad (112)$$

The computation of pipeline storage capacity often involves the principles of flow through complex pipelines and divided circuits or loops. The following example, taken from "Problems in Natural Gas Engineering," by Thomas R. Weymouth, serves to illustrate one of the more complicated situations involving the storage capacity of multiple loops.

Example. Assume a complex line made up as shown in Fig. 52, and let it be required to find the equivalent line of single pipe and the intake, or initial, pressure p_1 necessary to force 1,000,000 cu ft/hr measured under standard conditions, through the system, with a terminal pressure $p_2 = 50$ psia, a flow temperature of 40 F, and a specific gravity of 0.60.

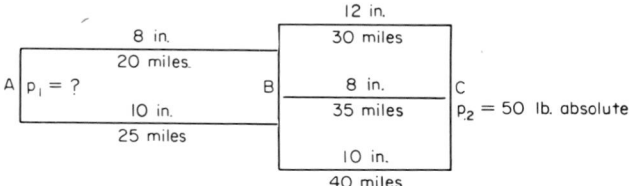

FIG. 52. Diagram illustrating applications of pipe-line formulas.

Taking first the section from *A* to *B*, we have 20 miles of 8-in. (net) pipe and 25 miles of 10-in. By the divided-circuit solution the diameter of a pipe 25 miles long that will be equivalent to the 20 miles of 8-in. pipe is, in inches, using the Weymouth exponents,

$$d_{eq.} = 8(25/20)^{3/16} = 8.342$$

Section *AB* is then equivalent to two parallel or looped lines each 25 miles long and having respective inside diameters of 8.342 in. and 10.

For $d = 8.342$ in., $d^{8/3} = 286.2$ in.

For $d = 10.0$, $d^{8/3} = 464.2$

$\Sigma d^{8/3} = 750.4$ in.

and $d_0 = (\Sigma d^{8/3})^{3/8} = (750.4)^{3/8} = 11.97$

or, section *AB* is equivalent to one pipe 11.97 in. in diameter and 25 miles long.

By the divided-circuit solution, the 8-in. pipe, 35 miles long, in section *BC* is equivalent to a 7.772-in. pipe 30 miles long, since

$$d_{eq.} = 8(30/35)^{3/16} = 7.772$$

In the same manner, the 10-in. line 40 miles long is equivalent to a 9.475-in. line 30 miles long. Then summing up

for $d = 12.0$ $d^{8/3} = 754.8$

for $d = 7.772$, $d^{8/3} = 237.0$

for $d = 9.475$, $d^{8/3} = 402.0$

$\Sigma d^{8/3} = 1,393.8$

$$d_{eq.} = (\Sigma d^{8/3})^{3/8} = (1,393.8)^{3/8} = 15.10 \text{ in.}$$

Hence, section *BC* is equivalent to one pipe 15.10 in. in diameter, 30 miles long. The equivalent length of section *BC* in terms of the 11.97-in. pipe to which section *AB* was reduced is obtained by the complex-series method.

$$L_{eq.} = 30(11.97/15.10)^{5\frac{1}{3}} = 8.68 \text{ miles}$$

Thus, section *BC* is equivalent to a single line 11.97 in. in diameter and 8.68 miles long, and the whole system is equivalent to a single line 11.97 in. in diameter and 33.68 miles long.

For ascertaining the required initial pressure for a flow of 1,000,000 cu ft/hr, the known line constants are, therefore, $p_2 = 50$, $d = 11.97$, $L = 33.68$, $Q = 1,000,000$. By the Weymouth equation

$$1,000,000 = 37 \sqrt{\frac{p_1{}^2 - 50^2}{33.68}} \; 11.975^{1/3} \quad p_1 = 215.1 \text{ psia}$$

The relative quantities of gas passed by the several pipes of the loops are to each other as the $\frac{8}{3}d$ of their equivalent diameters for equal length. The pressure at junction *B* can be ascertained by means of the Weymouth equation using for *L* and *d* the equivalent values derived above for section *AB*. This pressure is required to be known in order to determine the storage capacity of the line. After having ascertained the above quantities, the pressure drops being known, the computation can be checked by computing the flow through the several lines as they actually exist.

To compute the storage capacity of the system shown in Fig. 52, first consider section *AB*. Assume the maximum allowable pressure at *A* to be $p_1' = 225$ psia and the flow $Q = 800,000$ cu ft/hr. This section was shown to be equivalent to a single line 11.97 in. in diameter and 25 miles long. Substituting these known values in the Weymouth equation, the pressure at *B* is found to be $p_2' = 172.7$ lb.

When the line is unpacked, i.e., under average-flow conditions of 1,000,000 cu ft/hr, p_1 was shown to be 215.1 lb and by formula, p_2 is found to be 117.4 lb.

Substituting these values in Eq. (112) together with the actual dimensions of the pipe in the section of line under consideration, the available storage capacity of this section *AB* in cubic feet is

$$S = \frac{19.20}{14.65}\Big(225 - 215.1 + 17.27 - 117.4 - \frac{225 \times 172.7}{225 + 172.7}$$

$$+ \frac{215.1 \times 117.4}{215.1 + 117.4}\Big)[(8^2 \times 20) + (10^2 \times 25)] = 215,000$$

In like manner the storage capacity of *BC* is computed, the sum of the two results being the total storage capacity of the whole system available for peak demands.

FRICTION FLOW OF WATER

General Principles of Flow. The flow of fluids in general has been discussed at some length and is directly applicable to water. In practically all cases water may be considered as incompressible, and for calculations at approximately atmospheric temperature, it is customary to assume that it has a uniform density of 62.4 lb/cu ft which holds very nearly constant throughout the temperature range from 32 to 60 F. In calculations for hot-water heating systems, boiler-feed pump discharge heads, etc., it is necessary to take into account the change in density and viscosity with temperature. Application of the common empirical equations for water flow is limited to water at usual atmospheric temperatures of, say, 32 to 100 F. At higher temperatures the changes in density and viscosity have a considerable bearing on flow relations, and where close results are desired, the use of Eq. (77) together with Figs. 41 and 42 is recommended in preference to any of the commonly used empirical formulas.

For convenience, Eqs. (77) have been rearranged below to be most useful in connection with the solution of problems involving flow of water at ordinary temperatures. Density has been taken as equal to 62.4 lb/cu ft and the term $2g$ in the denominator of Eq. (77) has been incorporated in the numerical constants

of Eqs. (113) to (115).

$$\Delta P = fLV^2/12.4d \qquad\qquad (113)$$

$$\Delta P = 2{,}715 fLQ^2/d^5 \qquad\qquad (114)$$

$$\Delta P = 0.01345 fLG^2/d^5 \qquad\qquad (115)$$

in which ΔP = pressure drop due to friction, psi
$\quad f$ = friction factor from Fig. 42
$\quad V$ = velocity, fps
$\quad Q$ = volume flow rate, cfs
$\quad G$ = volume flow rate, gpm
$\quad d$ = pipe diameter, in.
$\quad L$ = equivalent length, ft

Before an analytical method of solution was developed, numerous investigators undertook modifications of the basic hydraulic formulas with a view to securing better agreement with test results throughout the usual range of operating conditions. Several of the empirical formulas commonly used in hydraulic work are given in succeeding paragraphs.

Saph and Schoder Formula. One of the most carefully worked out empirical formulas for water flow is that of Saph and Schoder.[1] The experimental work on which they based their formulas was carried out with unusual care and frequently has been quoted by other investigators and writers. Saph and Schoder gave the following values based on an analysis of their experimental data for velocities above the critical:

$$h_\lambda = \frac{0.296 \text{ to } 0.469}{d'^{1.25}} \, v^{1.74 \text{ to } 2.00}$$

where h_λ = head lost, ft of water per 1,000 ft of pipe
$\quad v$ = velocity, fps
$\quad d'$ = inside diameter of pipe, ft

The low values of the constant and exponent given above apply to smooth pipes and the high values to rough pipes. Their criterion for critical velocity was $v = 0.37/d^{0.85}$, d in this case being the inside diameter of the pipe in inches.

In a later article[2] Schoder gave the following formulas as representing average conditions for clean cast-iron and wrought-iron or wrought-steel pipe:

$$h_\lambda = 0.38v^{1.86}/d'^{1.25}$$

This formula is the basis for the equations and charts shown in Fig. 53a and b from which, in turn, the flow data given in Table 55 were tabulated. These charts and tables, which correspond closely to the Williams and Hazen values for "smooth new iron pipe" and for $C = 130$ to 140, are recommended for designing aboveground service-water piping in power and industrial plants.

Where the interior surface is *decidedly rough* and for *spiral riveted* pipe, Schoder proposed $h_\lambda = 0.5v^{1.95}/d'^{1.25}$. In the case of ordinary spiral-riveted pipe, the friction loss is much greater than for cast-iron or steel pipes of the same size and diameter, because the overlapping plates and projecting rivet heads usually offer more resistance than the joints in cast-iron or steel pipe. With new spiral-*welded* pipe or new spiral-*riveted* pipe having taper joints with beveled edges, a friction loss comparable with "smooth new iron pipe" is to be expected (see values given under Scobey formulas).

Change of Interior Condition with Age. The extent to which the interior wall surface of a pipe changes with age has been the subject of much discussion.

[1] "An Experimental Study of the Flow of Water in Pipes," *Trans. ASCE, Paper* 964, vol. 51, presented at meeting of Sept. 2, 1903, by Augustus V. Saph and Ernest H. Schoder.
[2] *Eng. Record*, Sept. 3, 1904.

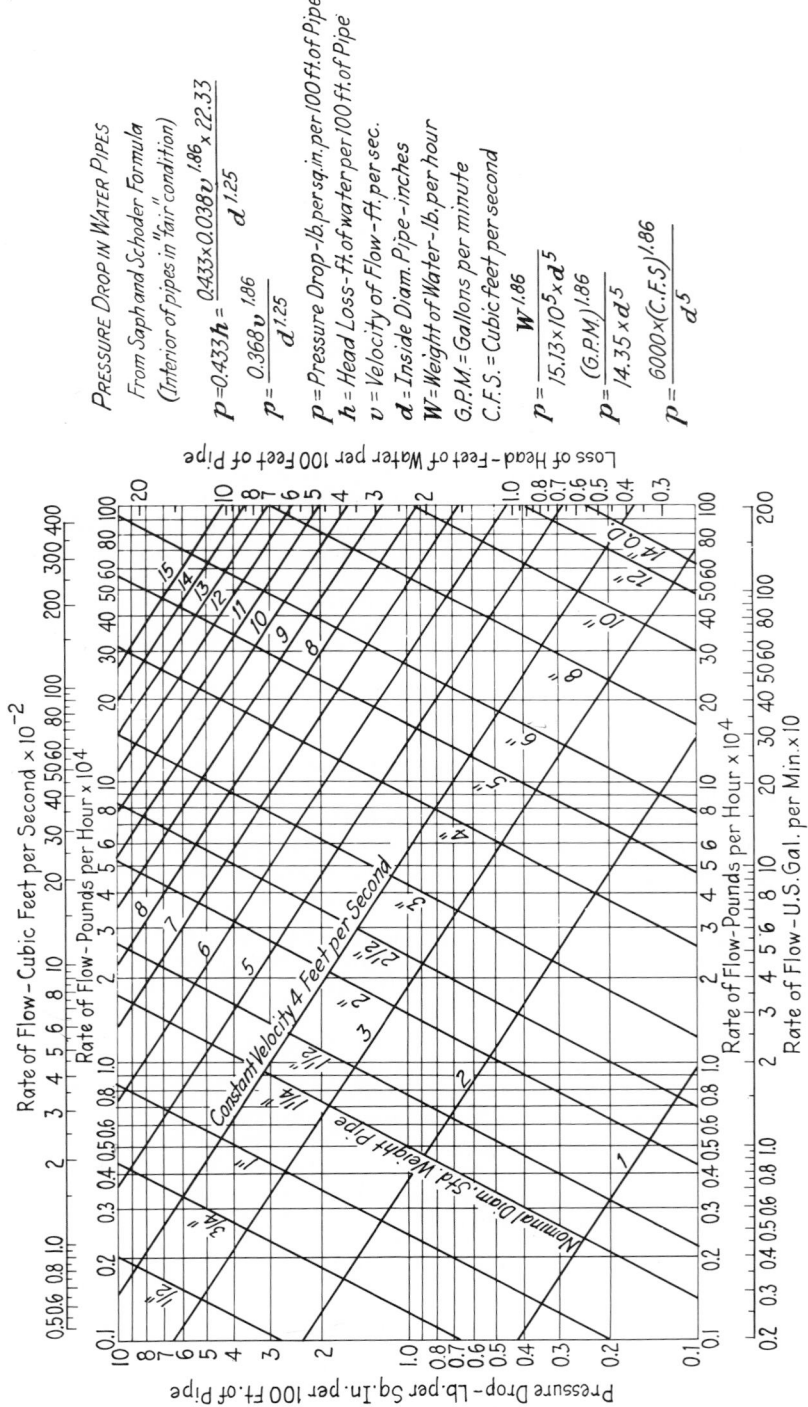

FIG. 53a. Flow of water in pipes ½ to 14 in. in diameter.

PRESSURE DROP IN WATER PIPES

From Saph and Schoder Formula
(Interior of pipes in "fair condition")

$$P = 0.433h = \frac{0.433 \times 0.038v^{1.86} \times 22.33}{d^{1.25}}$$

$$P = \frac{0.368v^{1.86}}{d^{1.25}}$$

P = Pressure Drop-lb. per sq. in. per 100 ft. of Pipe
h = Head Loss-ft. of water per 100 ft. of Pipe
v = Velocity of Flow-ft. per sec.
d = Inside Diam. Pipe-inches
W = Weight of Water-lb. per hour
$G.P.M.$ = Gallons per minute
$C.F.S.$ = Cubic feet per second

$$P = \frac{W^{1.86}}{15.13 \times 10^5 \times d^5}$$

$$P = \frac{(G.P.M.)^{1.86}}{14.35 \times d^5}$$

$$P = \frac{6000 \times (C.F.S.)^{1.86}}{d^5}$$

Rate of Flow—Cubic Feet per Second ×10⁻²

Rate of Flow—Pounds per Hour ×10⁴

Loss of Head—Feet of Water per 100 Feet of Pipe

Rate of Flow—Pounds per Hour ×10⁴

Rate of Flow—U.S. Gal. per Min. ×10

Pressure Drop—Lb. per Sq. In. per 100 Ft. of Pipe

Constant Velocity 4 Feet per Second

Nominal Diam.—Std. Weight Pipe

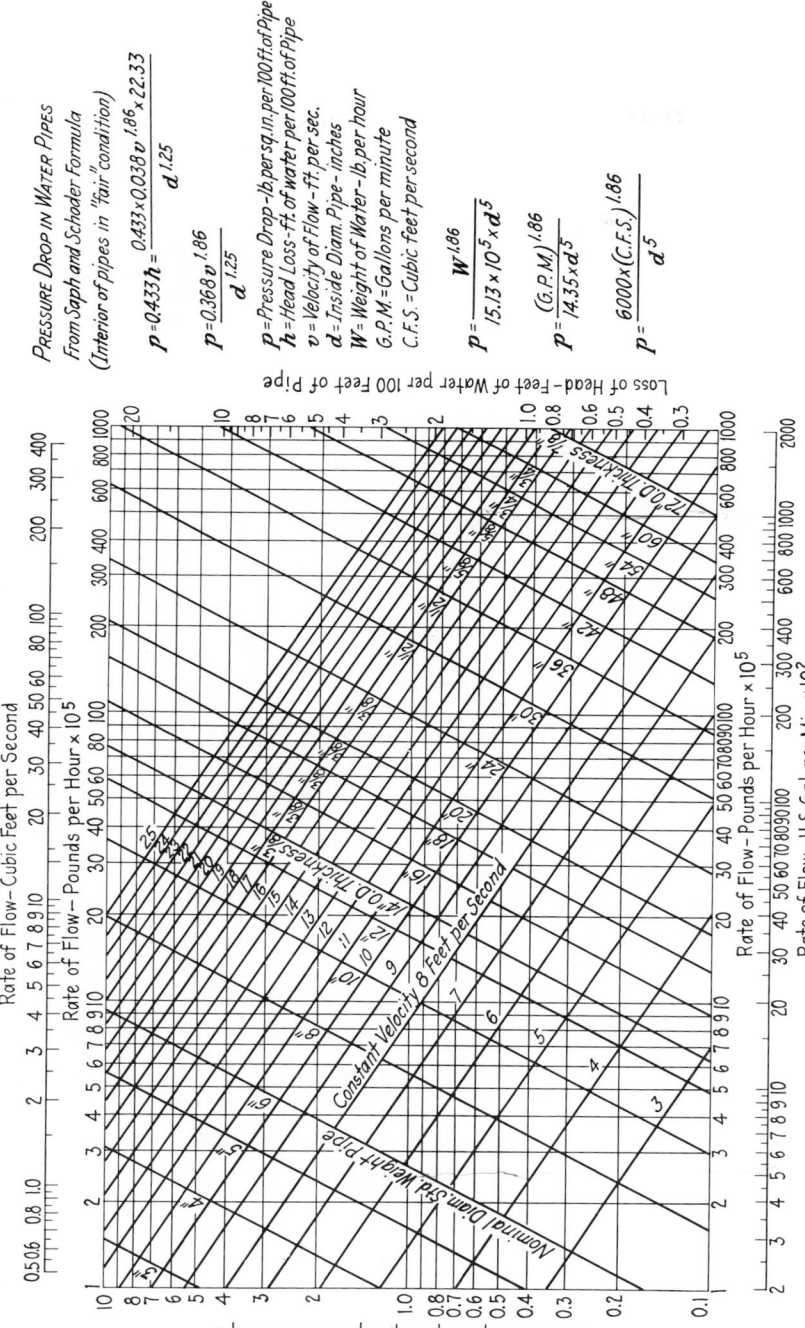

PRESSURE DROP IN WATER PIPES
From Saph and Schoder Formula
(Interior of pipes in "fair" condition)

$$p = 0.433 \times h = \frac{0.0038 \, v^{1.86} \times 22.33}{d^{1.25}}$$

$$p = \frac{0.368 \, v^{1.86}}{d^{1.25}}$$

p = Pressure Drop – lb.per sq.in.per 100 ft.of Pipe
h = Head Loss – ft.of water per 100 ft.of Pipe
v = Velocity of Flow – ft.per sec.
d = Inside Diam. Pipe – inches
W = Weight of Water – lb.per hour
G.P.M. = Gallons per minute
C.F.S. = Cubic feet per second

$$p = \frac{W^{1.86}}{15.13 \times 10^5 \times d^5}$$

$$p = \frac{(G.P.M.)^{1.86}}{14.35 \times d^5}$$

$$p = \frac{6000 \times (C.F.S.)^{1.86}}{d^5}$$

FIG. 53b. Flow of water in pipes 3 to 72 in. in diameter.

3–173

Table 55. Carrying Capacity and Friction Loss for Standard-weight Wrought-iron and Steel Water Pipes, also Approximate for Type K Copper Water Tubes of Same Nominal Size

(Independent variables: gallons per minute and pipe size. Dependent variables: velocity, friction head, and pressure drop. Friction head and pressure drop per 100 ft of pipe, interior "fair" condition)

In each pipe-size group the three sub-columns are: Velocity, feet per second; Friction head, feet; Friction loss, pounds per square inch. Where two nominal sizes share a column (e.g. "1½ in. / 8 in."), the small-size data appear in the lower gpm rows and the large-size data in the higher gpm rows.

Gallons per minute	3 in. Vel.	3 in. Head	3 in. Loss	2½ in. Vel.	2½ in. Head	2½ in. Loss	2 in. Vel.	2 in. Head	2 in. Loss	1½ in. / 8 in. Vel.	1½ / 8 in. Head	1½ / 8 in. Loss	1¼ in. / 6 in. Vel.	1¼ / 6 in. Head	1¼ / 6 in. Loss	1 in. / 5 in. Vel.	1 / 5 in. Head	1 / 5 in. Loss	¾ in. / 4 in. Vel.	¾ / 4 in. Head	¾ / 4 in. Loss
5				0.33	0.04	0.02	0.48	0.09	0.04	0.79	0.30	0.13	1.07	0.65	0.28	1.86	2.54	1.10	3.00	8.30	3.60
10	0.43	0.04	0.02	0.67	0.12	0.05	0.96	0.30	0.13	1.57	1.06	0.46	2.14	2.31	1.00	3.71	9.00	3.90	6.00	23.7	10.3
15	0.65	0.09	0.04	1.00	0.27	0.12	1.43	0.65	0.28	2.36	2.24	0.97	3.20	4.90	2.12	5.56	18.9	8.20	9.00	62.3	27.0
20	0.87	0.16	0.07	1.34	0.46	0.20	1.91	1.08	0.47	3.14	3.81	1.65	4.27	8.44	3.65	7.42	33.2	14.4	12.0	113.0	49.0
25	1.08	0.23	0.10	1.67	0.69	0.30	2.39	1.66	0.72	3.93	5.77	2.50	5.35	12.7	5.50	9.27	49.8	21.6	15.0	166.0	72.0
30	1.30	0.32	0.14	2.01	0.97	0.42	2.87	2.31	1.00	4.72	8.08	3.50	6.42	18.0	7.80	11.1	71.5	31.0			
35	1.51	0.44	0.19	2.34	1.29	0.56	3.34	3.05	1.32	5.50	10.85	4.70	7.48	24.9	10.8	13.0	95.0	41.2			
40	1.73	0.55	0.24	2.68	1.66	0.72	3.82	3.92	1.70	6.30	13.85	6.00	8.56	31.7	13.7	14.9	121.0	52.6	1.01	0.14	0.06
45	1.95	0.69	0.30	3.01	2.08	0.90	4.29	4.88	2.12	7.07	17.10	7.40	9.62	39.0	16.9				1.13	0.18	0.08
50	2.16	0.85	0.37	3.35	2.49	1.08	4.77	5.95	2.58	7.86	21.0	9.10	10.7	47.4	20.5				1.26	0.21	0.09
60	2.59	1.20	0.52	4.02	3.51	1.52	5.73	8.32	3.60	9.44	30.0	13.0	12.8	66.6	28.8	0.96	0.09	0.04	1.51	0.30	0.13
70	3.02	1.57	0.68	4.68	4.66	2.02	6.68	11.1	4.80	11.0	39.7	17.2	14.9	88.4	38.2	1.12	0.14	0.06	1.76	0.42	0.18
75	3.24	1.80	0.78	5.02	5.31	2.30	7.16	12.6	5.45	11.8	44.8	19.4	16.0	100.0	43.2	1.20	0.16	0.07	1.89	0.46	0.20
80	3.47	2.03	0.88	5.37	6.00	2.60	7.64	14.1	6.10	12.6	51.8	22.4				1.28	0.18	0.08	2.01	0.53	0.23
90	3.89	2.54	1.10	6.03	7.38	3.20	8.59	17.5	7.60	14.1	63.2	27.4				1.44	0.21	0.09	2.27	0.65	0.28
100	4.34	3.05	1.32	6.71	9.00	3.90	9.55	21.5	9.30	15.7	76.2	33.1	1.11	0.09	0.04	1.60	0.26	0.11	2.52	0.78	0.34
125	5.42	4.62	2.00	8.37	13.8	6.00	11.9	33.2	14.4	19.6	116.2	50.4	1.39	0.16	0.07	2.00	0.39	0.17	3.14	1.20	0.52
150	6.50	6.47	2.80	10.1	19.4	8.40	14.3	46.4	20.1				1.67	0.21	0.09	2.40	0.55	0.24	3.77	1.68	0.73
175	7.58	8.76	3.80	11.7	25.6	11.1	16.7	61.5	26.7				1.94	0.30	0.13	2.80	0.74	0.32	4.42	2.26	0.98
200	8.67	11.2	4.85	13.4	33.2	14.4	19.1	78.3	33.9	1.28	0.09	0.04	2.22	0.37	0.16	3.20	0.92	0.40	5.04	2.88	1.25
250	10.8	17.1	7.40	16.7	49.8	21.6	23.9	119.5	51.8	1.60	0.14	0.06	2.78	0.58	0.25	4.00	1.41	0.61	6.30	4.38	1.89
300	13.0	23.5	10.2	20.1	69.7	30.2	28.6	166.0	72.0	1.92	0.18	0.08	3.33	0.81	0.35	4.80	2.01	0.87	7.55	6.07	2.63

Friction Head and Pressure Drop in Pipe

The table below gives, for each pipe size, the velocity (ft/sec), friction head (ft per 100 ft) and pressure drop (psi per 100 ft) as a function of flow (gal per min).

10 in. (velocity, friction head ft/100 ft, pressure drop psi/100 ft)

G.P.M.	Velocity	Head	Psi
450	1.83	0.14	0.06
500	2.03	0.16	0.07
750	3.05	0.35	0.15
1,000	4.07	0.60	0.26
1,250	5.07	0.90	0.39
1,500	6.10	1.25	0.54
2,000	8.13	2.17	0.94
2,500	10.2	3.23	1.40
3,000	12.2	4.62	2.00
3,500	14.2	6.24	2.70
4,000	16.3	7.99	3.46
4,500	18.3	9.93	4.30

12 in. (velocity, friction head ft/100 ft, pressure drop psi/100 ft)

G.P.M.	Velocity	Head	Psi
500	1.42	0.07	0.03
750	2.12	0.14	0.06
1,000	2.83	0.25	0.11
1,250	3.53	0.37	0.16
1,500	4.25	0.51	0.22
2,000	5.70	0.90	0.39
2,500	7.09	1.34	0.58
3,000	8.50	1.89	0.82
3,500	9.93	2.52	1.09
4,000	11.3	3.23	1.40
4,500	12.8	4.04	1.75
5,000	14.2	4.92	2.13
5,500	15.6	5.87	2.54
6,000	17.0	6.88	2.98

14 in. O.D., 3/8 in. thick — friction head (ft/100 ft): 0.23, 0.32, 0.55, 0.89, 1.15, 1.55, 1.99, 2.47, 3.00, 3.58, 4.20, 4.85, 5.59, 6.35, 7.16; pressure drop (psi/100 ft): 0.10, 0.14, 0.24, 0.36, 0.50, 0.67, 0.86, 1.07, 1.30, 1.55, 1.82, 2.10, 2.42, 2.75, 3.10

16 in. O.D., 3/8 in. thick — velocity (ft/sec): 2.63, 3.51, 4.39, 5.27, 6.15, 7.03, 7.91, 8.78, 9.66, 10.5, 11.4, 12.3, 13.1, 15.8

18 in. O.D., 3/8 in. thick — friction head (ft/100 ft): 0.14, 0.23, 0.30, 0.42, 0.53, 0.65, 0.81, 0.95, 1.13, 1.29, 1.50, 1.71, 1.92, 2.38, 2.91, 3.47, 4.07

20 in. O.D., 3/8 in. thick — friction head (ft/100 ft): 0.23, 0.30, 0.37, 0.46, 0.55, 0.65, 0.74, 0.85, 0.99, 1.11, 1.39, 1.69, 2.01, 2.36, 2.73, 3.14, 3.56; pressure drop (psi/100 ft): 0.10, 0.13, 0.16, 0.20, 0.24, 0.28, 0.32, 0.37, 0.43, 0.48, 0.60, 0.73, 0.87, 1.02, 1.18, 1.36, 1.54

24 in. O.D., 3/8 in. thick — friction head (ft/100 ft): 0.18, 0.21, 0.25, 0.30, 0.35, 0.39, 0.44, 0.53, 0.65, 0.79, 0.92, 1.06, 1.22, 1.39, 2.38; pressure drop (psi/100 ft): 0.08, 0.09, 0.11, 0.13, 0.15, 0.17, 0.19, 0.23, 0.28, 0.34, 0.40, 0.46, 0.53, 0.60, 1.03

30 in. O.D., 3/8 in. thick — velocity (ft/sec): 3.16, 3.40, 3.64, 3.89, 4.37, 4.86, 5.34, 5.83, 6.32, 6.80, 7.29, 9.71, 12.1, 14.6, 17.0; friction head (ft/100 ft): 0.09, 0.11, 0.12, 0.13, 0.15, 0.17, 0.25, 0.30, 0.35, 0.39, 0.46, 0.79, 1.20, 1.66, 2.22; pressure drop (psi/100 ft): 0.04, 0.05, 0.05, 0.06, 0.08, 0.09, 0.11, 0.13, 0.15, 0.17, 0.20, 0.34, 0.52, 0.72, 0.96

36 in. O.D., 3/8 in. thick — velocity (ft/sec): 2.75, 3.09, 3.43, 3.78, 4.12, 4.46, 4.80, 5.15, 6.86, 8.58, 10.3, 13.7; friction head (ft/100 ft): 0.07, 0.07, 0.09, 0.09, 0.12, 0.14, 0.16, 0.18, 0.32, 0.51, 0.69, 0.92, 1.20; pressure drop (psi/100 ft): 0.03, 0.03, 0.04, 0.05, 0.06, 0.07, 0.08, 0.14, 0.22, 0.30, 0.40, 0.52

42 in. O.D., 3/8 in. thick — velocity (ft/sec): 2.99, 3.24, 3.49, 3.74, 4.98, 6.23, 7.47, 8.72, 9.96; friction head (ft/100 ft): 0.05, 0.07, 0.07, 0.09, 0.14, 0.23, 0.32, 0.42, 0.53; pressure drop (psi/100 ft): 0.02, 0.03, 0.03, 0.04, 0.06, 0.10, 0.14, 0.18, 0.23

Friction head and pressure drop from curves based on Saph and Schoder formula: $p_\lambda = \dfrac{(\text{G.P.M.})^{1.86}}{14.35 \times d^5}$ in which p_λ = pressure drop, psi per 100-ft pipe

and d = inside diameter of pipe, inches. Applies to pipe with interior in "fair" condition. Based on 62.428 lb per cu ft. To convert gallons per minute to cubic feet per second, divide by 448.8; to convert gallons per minute to pounds per hour, multiply by 500.7.; to convert gallons per minute to millions of gallons per day, divide by 694.4.

The following statement is quoted from the "Handbook of Cast Iron Pipe," published by the Cast Iron Pipe Research Association:

In attempting to compute the capacity of a pipe or to figure the probable loss in head after the pipe has reached a certain age, it is absolutely essential to know something about the water to be conveyed. In most cases the quality of the water is such that the carrying capacity is affected very little by the age of the pipe. In other cases, the water may be so soft as to cause tuberculation and consequent loss in carrying capacity, or so turbid as to cause deposits of sand or mud with the same effect. Waters that cause tuberculation are the rare exception, and outside of a few raw water conduits, muddy water is also unusual.

In spite of this fact, many of the books and articles on hydraulics and water supply make the bold statement that a definite correction factor must be applied to flow formulas as the pipe increases in age. It is evident that this is incorrect, since, first of all, a large number of experiments have been made that show quite definitely that in many places there is no change whatever in carrying capacity with age. Secondly, assume that a layer of tubercles 2 in. thick is produced as a result of many year's use of a pipe, it is evident that the carrying capacity of a 12-in. pipe would be considerably more reduced than would a 48-in. pipe with the same thickness of tubercles, a fact that is not taken into account in the formulas in common use.

The formulas, charts, and tables shown in Fig. 53a and b and Table 55 give conservative values for clean iron or steel pipe having a smooth interior un-obstructed with rivet heads or the like. These charts furnish safe design data for such pipe under ordinary conditions of service, irrespective of the age of the pipe. Where unusual conditions are expected, other constants and exponents may be substituted in the Saph and Schoder formula as previously indicated, or one of the hydraulic formulas discussed in succeeding paragraphs may be used instead. Approximately equivalent values of the roughness coefficients used in these formulas will be found in Table 59.

Williams and Hazen Formula.   Among water-supply and hydraulic engineers the empirical formula developed by Williams and Hazen[1] in the early 1900's is extensively used for pipes 4 in. in diameter and larger. This formula is written:

$$v = 1.318CR^{0.63}s^{0.54} \quad \text{and} \quad s = 3.0226v^{1.852}/C^{1.852}d'^{1.167}$$

from which

$$h_\lambda = 3022.6v^{1.852}/C^{1.852}d'^{1.167} = Kv^{1.852}/d'^{1.167}$$

Values of K corresponding to different values of C are as follows:

C	60	65	70	75	80	85	90	95	100
K	1.540	1.329	1.154	1.021	0.905	0.828	0.727	0.657	0.598

C	105	110	115	120	125	130	135	140
K	0.546	0.501	0.461	0.426	0.395	0.367	0.343	0.321

where v = mean velocity of flow, fps
C = a coefficient representing the roughness of the interior surface of the pipe
$K = 3022.6/C^{1.852}$
d' = inside diameter of pipe, ft
R = mean hydraulic radius in feet, which for a circular pipe is $\frac{1}{4}$ the inside diameter d' expressed in feet
s = hydraulic grade or slope, ft per ft of length of a pipe of uniform size
h_λ = loss of head, ft per 1,000 ft of pipe

[1] See "Hydraulic Tables," by G. S. Williams and Allen Hazen, 3d ed., John Wiley & Sons, Inc., New York, 1920. For the sake of uniformity with the Chézy formula, the expression for velocity frequently is written in the following equivalent form:

$$v = CR^{0.63}s^{0.54}0.001^{-0.04}$$

Values of C recommended for use in the Williams-Hazen formula will be found in Table 56. The direct solution of formulas having fractional exponents requires the use of logarithms or a log-log slide rule. Various aids to solution are available for use with the respective formulas which are of material assistance where repeated

Table 56A. Values of C Recommended for Use in the Williams and Hazen Formula

$C = 140$ for "extremely smooth and straight pipes" with "continuous interior" and welded or coupled joints, such as
New brass, copper, lead, tin
New cast iron
New welded or seamless steel
Smooth concrete (see Scobey concrete formula for full details on various degrees of roughness)
Smooth cement-lined cast-iron or steel pipe
Cement asbestos
$C = 130$ for "very smooth" pipes, such as
Welded or seamless steel with "continuous interior" in "fair" condition
New welded-steel pipe with riveted girth joints
New cast iron, usual value
Old brass, copper, lead, tin
$C = 120$ for "smooth" pipes, such as
Smooth wooden pipes or wood-stave pipes
Ordinary concrete
$C = 110$-130 for "new full-riveted" steel or wrought-iron pipe, depending on thickness of plate and extent to which rivets are countersunk (see also Scobey formula)
$C = 110$ for old cement-lined pipe, or vitrified-crock sewers in good condition
$C = 100$ for old cast-iron or "old continuous interior" steel pipes where the carrying capacity over a long period of years is somewhat impaired through tuberculation or sedimentation. For sizes below 6 in., somewhat lower values should be used. Velocities in feet per second and loss of head in feet per 1,000 ft of pipe for $C = 100$ are given in Table 57
$C = 95$ for "old full-riveted" steel under the same conditions
$C = 90$ for brick sewers
$C = 60$ for "corrugated"[1] pipe or "badly tuberculated" iron or steel pipes

[1] For flow through corrugated pipe see "Flow of Water through Culverts," by D. L. Yarnell, F. A. Nagler, and S. M. Woodward, University of Iowa, Studies in Engineering, 1926, which gives results of experiments on concrete, vitrified clay, and corrugated-metal pipe culverts 12, 18, 24, and 30 in. in diameter, and on box culverts.

solutions are required. In the case of the Williams-Hazen formula their own hydraulic tables can be used, or numerous charts and tables are available in texts on hydraulics and in pipe manufacturers' catalogues.[1]

The carrying capacities for different pipe sizes given in Table 57 were compiled for a C coefficient of 100 which Williams and Hazen considered a fair value to use

[1] See, for instance,
(*a*) "Hydro-Electric Handbook," by W. P. Creager and J. D. Justin, John Wiley & Sons, Inc., New York, 1927.
(*b*) "American Sewerage Practice," Vol I, "Design of Sewers," by L. Metcalf and H. P. Eddy, McGraw-Hill Book Company, Inc., New York, 1928.
(*c*) Pipe Section of "General Catalog," Taylor Forge and Pipe Works, P.O. Box 485, Chicago, Ill.

Table 56B.   Approximate Values of C Recommended by National Board of Fire Underwriters for Use in the Williams and Hazen Formula[1]

Kind of pipe or hose	Value of C
Cast-iron or other iron or steel pipe with smooth interior surface:	
New pipe	120
10 years old[2]	110
15 years old	100
20 years old	90
30 years old	80
50 years old	70
75 years old	60
Riveted steel pipe, new	110
Smooth brass, copper, or lead pipe	140
Cement-lined pipe	140
Cement-asbestos pipe	140
Good quality cotton rubber-lined hose	140
Unlined linen hose	90

[1] See "Crosby-Fiske-Forster Handbook of Fire Protection," published by National Fire Protection Association, Boston, Mass.

[2] The corrosive effect of various kinds of water may have much greater or less effect than indicated by these factors; the choice of factor should be guided by experience with the particular water.

for general computation with cast-iron and steel pipes after some years in service. The greater degree of obstruction with pipes smaller than 6 in. in diameter or with all sizes of pipes where the water is particularly conducive to corrosion sometimes calls for using C coefficients considerably less than 100. Under these circumstances or where the use of a higher C coefficient is warranted, adjustment of the values of Table 57 up or down to the desired basis is readily made through applying the factors given below the table.

Scobey Formulas. Another formula widely used for computing flow in large *iron and steel* water conduits is that of Fred C. Scobey[1] which is written in the following forms:

$$h_\lambda = K_s(v^{1.9}/d'^{1.1}) \qquad v = h_\lambda^{0.53}d'^{0.58}K_s^{0.53} \qquad F = 0.78h_\lambda^{0.53}d'^{2.53}/K_s^{0.53}$$

where h_λ = feet of head lost per 1,000 ft of pipe

K_s = a coefficient representing the roughness of the interior surface of the pipe

v = mean velocity of flow, fps

d' = inside diameter of pipe, ft

F = flow, cu ft per sec

In order to define K_s for this formula, Scobey divided iron and steel pipe into three classes in accordance with the smoothness of its interior surface as follows:

Class 1.   Full-riveted pipe, having both longitudinal and girth seams held by one or more lines of rivets with projecting heads. From a capacity standpoint, pipe with countersunk rivet heads on the interior belongs in Class 3.

[1] See "The Flow of Water in Riveted Steel and Analogous Pipes," by Fred C. Scobey, *Tech. Bull.* 150, U.S. Department of Agriculture. For sale by the Superintendent of Documents, Washington, D.C. Contains a bibliography of 184 references on the flow of water in pipes. At the time of these investigations Mr. Scobey was principal irrigation engineer, U.S. Department of Agriculture.

Class 2. Girth-riveted pipe, having no retarding rivet heads in the longitudinal seams but having the same girth seams as full-riveted pipe.

Class 3. Continuous-interior pipe, having the interior unmarred by plate offsets or by projecting rivet heads in either longitudinal or girth seams. Not necessarily described as "smooth." Pipe having longitudinal-welded seams, girth-welded seams, or spiral-welded seams belongs in Class 3.

For the three foregoing classes of pipe, Scobey gave the following coefficients:

Class 1a. $K_s = 0.38$ for new sheet-metal full-riveted pipes up to $\frac{3}{16}$ in. thickness.

Class 1b. $K_s = 0.44$ for new plate-metal full-riveted pipes from $\frac{3}{16}$ to $\frac{7}{16}$ in. thickness, with either taper or cylinder joints.

Class 1c. $K_s = 0.48$ for new plate-metal full-riveted pipes from $\frac{1}{2}$ in. thickness up with either taper or cylinder joints, and for pipes from $\frac{1}{4}$ to $\frac{7}{16}$ in. thickness when butt jointed.

Class 1d. $K_s = 0.52$ for new butt-strap full-riveted plate-metal pipes from $\frac{1}{2}$ in. thickness up.

Class 2. $K_s = 0.34$ for new girth-riveted pipes.

Class 3. $K_s = 0.32$ for new continuous-interior pipes. Welded steel pipes with welded field joints or connected with bolted couplers of the Dresser type belong in this class.

Convenient tables and charts for aid in solving the Scobey formula for iron and steel pipe will be found in *Bull.* 150 and in the "Handbook of Welded Steel Pipe."[1]

Scobey's formula for *wood-stave pipe*[2] written in three different forms is

$$h_\lambda = 0.419(v^{1.8}/d'^{1.17}) \qquad v = 1.62d'^{0.65}h_\lambda^{0.555} \qquad F = 1.272d'^{2.65}h_\lambda^{0.555}$$

Scobey's formula for *concrete pipe*[3] written in three different forms with the inside diameter *d* given in inches is

$$h_\lambda = (v^2/C_s^2 d^{1.25}) \qquad v = C_s h_\lambda^{0.5} d^{0.625} \qquad F = 0.00546 C_s d^{2.625} h_\lambda^{0.5}$$

The following values are suggested for the coefficient C_s for concrete pipe:

Class 1. $C_s = 0.267$. For concrete pipes laid with a generous supply of mortar without removal of mortar squeeze. This coefficient is recommended also for pipes of Class 2 when used to convey sewage.

Class 2. $C_s = 0.310$. For "dry-mix" concrete pipes and monolithic concrete pipes or tunnel linings made over rough wood forms. Also for surfaces as left by cement-gun process.

Class 3. $C_s = 0.345$. For small "wet-mix" concrete pipes in short units; for "dry-mix" pipes in long units; for average monolithic pipes made on steel forms.

Class 4. $C_s = 0.370$. For glazed-interior concrete pipelines; for large cement-lined iron pipes; for monolithic pipelines where joint scars and all interior surface irregularities are removed. Particularly applicable to jointed lines of units made from wet, well-spaded concrete, deposited against oiled-steel forms.

Kutter and Manning Formulas. These formulas, published in 1869 and 1890, respectively, are used extensively by civil engineers for gravity-flow design in both open channels and buried conduits carrying either water or sewage, particularly where the conduit runs only partly full.[4] The numerical constant $1.486/n$ in

[1] Published by the Welded Pipe Division of the California Corrugated Culvert Co., Berkeley, Calif. See also "Handbook of Water Control," published by the same company.

[2] See "The Flow of Water in Wood Stave Pipe," by Fred C. Scobey, *Bull.* 376, U.S. Department of Agriculture, 1916, rev., 1926.

[3] See "The Flow of Water in Concrete Pipe," by Fred C. Scobey, *Bull.* 852, U.S. Department of Agriculture, 1920. See also "Concrete Pipe Lines," by M. W. Loving, published 1942 by the American Concrete Pipe Association, Chicago.

[4] See "Determination of Kutter's *n* for Sewers Partly Filled," by C. Frank Johnson, *Proc. ASCE* vol. 69, February, 1943, with discussion April, 1943.

Table 57. Carrying Capacity and Friction Loss per 1,000 Ft of Pipe Having Actual Inside Diameters Shown, Williams and Hazen Formula for C = 100[1]

(See Table 60 for converting gallons per minute to millions of gallons per day, cubic feet per second and thousands of pounds per hour[2])

Gpm	3 in. / 16 in. Vel ft/sec	Head ft	Loss psi	4 in. / 18 in. Vel ft/sec	Head ft	Loss psi	6 in. / 20 in. Vel ft/sec	Head ft	Loss psi	8 in. / 24 in. Vel ft/sec	Head ft	Loss psi	10 in. / 30 in. Vel ft/sec	Head ft	Loss psi	2 in. / 12 in. Vel ft/sec	Head ft	Loss psi	2½ in. / 14 in. Vel ft/sec	Head ft	Loss psi
100	4.54	50	22	2.56	12	5.20	1.13	1.65	0.72	0.64	0.42	0.18	0.41	0.14	0.06	10.2	358	155	6.55	120	52
120	5.45	70	30	3.07	17	7.36	1.36	2.30	1.00	0.77	0.58	0.25	0.49	0.20	0.09	12.2	500	217	7.86	168	73
140	6.35	92	40	3.58	23	9.97	1.58	3.12	1.35	0.89	0.78	0.34	0.57	0.26	0.11	14.3	670	290	9.17	223	97
160	7.26	118	51	4.10	29	12.6	1.81	3.95	1.71	1.02	0.98	0.43	0.65	0.33	0.14	16.3	860	373	10.5	290	126
180	8.16	148	64	4.60	36	15.6	2.03	4.96	2.15	1.15	1.20	0.52	0.74	0.42	0.18	18.4	1070	464	11.7	357	155
200	9.08	178	78	5.12	44	19.1	2.26	6.00	2.60	1.28	1.50	0.65	0.82	0.51	0.22	20.4	1290	560	13.1	431	187
250	11.3	271	120	6.40	67	29	2.82	9.15	3.97	1.60	2.20	0.95	1.04	0.77	0.33	25.5	1965	840	16.4	655	284
300	13.6	380	160	7.68	93	40	3.39	12.8	5.55	1.92	3.15	1.37	1.22	1.09	0.47	0.85	0.59	0.26	19.6	920	399
350	15.9	505	220	8.96	124	54	3.96	17.0	7.36	2.24	4.42	1.92	1.43	1.45	0.63				22.9	1220	530
400	18.2	650	280	10.2	160	69	4.52	22	9.53	2.56	5.40	2.34	1.63	1.94	0.84	1.14	0.75	0.33			
450	20.4	805	350	11.5	198	86	5.08	28	12.1	2.88	6.70	2.90	1.84	2.30	1.00	1.28	0.93	0.40	0.94	0.45	0.20
500				12.8	240	104	5.65	36	15.6	3.20	8.20	3.56	2.04	2.80	1.21	1.42	1.13	0.49	1.05	0.57	0.25
600	0.96	0.39	0.17	15.4	337	146	6.78	47	22	3.84	11.8	5.11	2.45	3.95	1.71	1.70	1.59	0.69	1.25	0.75	0.33
700	1.12	0.52	0.23	17.9	449	194	7.91	62	27	4.48	15.4	6.68	2.86	5.30	2.30	1.99	2.10	0.91	1.46	1.00	0.43
800	1.28	0.68	0.29				9.04	80	35	5.12	20	8.67	3.27	6.80	2.95	2.27	2.70	1.17	1.67	1.27	0.55
900	1.44	0.84	0.36	1.13	0.48	0.21	10.2	100	43	5.75	25	10.8	3.67	8.40	3.64	2.56	3.45	1.50	1.88	1.57	0.68
1,000	1.60	1.02	0.44	1.26	0.55	0.24	11.3	118	51	6.40	30	13.0	4.08	10.2	4.42	2.84	4.05	1.76	2.09	1.90	0.82
1,200	1.92	1.45	0.63	1.51	0.79	0.34	13.5	169	73	7.68	42	18.2	4.90	14.5	6.28	3.41	5.85	2.54	2.51	2.73	1.18
1,400	2.24	1.90	0.82	1.76	1.05	0.46	15.8	222	96	8.95	55	24	5.70	19.0	8.23	3.98	7.80	3.38	2.92	3.60	1.56
1,600	2.56	2.35	1.04	2.02	1.33	0.58				10.2	72	31	6.50	24	10.4	4.55	9.80	4.25	3.34	4.60	2.00
1,800	2.88	2.98	1.29	2.27	1.67	0.72	1.84	1.01	0.44	11.5	90	39	7.40	31	13.4	5.11	12.4	5.37	3.76	5.80	2.51
2,000	3.20	3.60	1.56	2.52	2.03	0.88	2.04	1.22	0.53	12.8	110	48	8.20	37	16.0	5.68	15.0	6.50	4.18	7.00	3.03
2,500	4.00	5.48	2.37	3.15	3.10	1.34	2.55	1.83	0.79				10.4	57	25	7.10	23	10.0	5.23	10.4	4.51
3,000	4.80	7.70	3.34	3.78	4.30	1.86	3.06	2.57	1.11	2.13	1.07	0.46	12.3	77	33	8.52	32	14.0	6.27	15.0	6.50
3,500	5.60	10.4	4.50	4.41	5.75	2.49	3.57	3.45	1.50	2.48	1.45	0.63			—	9.94	42	18.0	7.32	19.5	8.45

Flow																			
4,000	6.40	13.3	5.76	5.04	7.40	3.21	4.08	4.40	1.91	2.84	1.83	0.79	1.82	0.62	0.27	11.4	53	8.37	26 · 11.3
4,500	7.20	16.5	7.15	5.67	9.20	3.99	4.59	5.50	2.38	3.19	2.30	1.00	2.04	0.78	0.34	1.58	0.38	9.40	31 · 13.4
5,000	8.00	20	8.66	6.30	11.3	4.90	5.10	6.70	2.91	3.55	2.76	1.20	2.27	0.93	0.40	1.90	0.53	10.5	38 · 16.5
6,000	9.60	28	12.1	7.56	15.5	6.71	6.12	9.40	4.07	4.26	3.90	1.69	2.72	1.30	0.56	2.22	0.72	12.5	54 · 23.4
7,000	11.2	37	16.0	8.82	22	9.53	7.14	11.2	4.85	4.97	5.20	2.25	3.18	1.75	0.76				42 in.

8,000	1.60	48 in.	1.68	10.1	27	11.7	8.16	16.0	6.93	5.68	6.70	2.90	3.63	2.23	0.97	2.54	0.92	1.86	0.43	0.19
9,000	1.78		1.96	11.3	33	14.3	9.18	19.5	8.45	6.38	8.30	3.60	4.08	2.80	1.21	2.85	1.13	2.09	0.54	0.23
10,000	2.14		0.47	54 in.	0.12	0.27	10.2	24	10.4	7.10	10.0	4.33	4.54	3.50	1.52	3.17	1.38	2.32	0.64	0.28
12,000	2.14		1.21	0.68	0.36	0.42	12.2	33	14.3	8.52	14.5	10.4	5.45	4.75	2.06	3.80	2.00	2.78	0.90	0.39
14,000	2.49		1.96	1.05	0.27	0.63		60 in.	14.3	9.94	19.0	8.23	6.35	6.25	2.71	4.44	2.60	3.25	1.20	0.52

16,100	2.85	48 in.	1.68	1.90	0.82	0.20	1.82	1.20	0.52	0.66	0.76	0.33	7.26	8.00	3.47	5.07	3.35	3.71	1.55	0.67
18,000	3.20		0.99	2.52	1.13	0.24	2.05	1.53	0.66	0.79	0.99	0.43	8.17	9.90	4.30	5.70	4.20	4.17	1.90	0.82
20,000	3.56		1.21	2.80	0.68	0.30	2.28	1.83	0.79	1.00	0.40	0.52	9.08	12.0	5.20	6.34	5.00	4.64	2.30	1.00
25,000	4.45		1.82	3.20	1.05	0.46	2.85	2.30	1.00	1.43	1.46	0.63		72 in.		7.92	7.70	5.80	3.60	1.56
30,000	5.34		1.17	3.50	1.45	0.63	3.42	3.30	1.43	0.95	0.90	0.56	2.36	0.36	0.16	9.51	10.7	6.96	5.00	2.17

35,000	6.23		3.40	1.90	0.82	66 in.	3.99	4.30	1.86	0.52	0.76	0.33	2.76	0.48	0.21	4.06	0.80	8.12	6.70	2.91
40,000	7.12		4.50	2.60	1.13	1.95	4.56	5.50	132 in.	0.66	0.99	0.43	3.15	0.59	0.26	4.64	1.04	9.28	8.40	3.64
45,000	8.01		5.50	3.20	1.39	2.38	5.13	2.30		0.79	1.46	0.52	3.55	0.76	0.33	5.22	1.30		96 in.	
50,000	8.90		6.70	4.10	1.78	2.91	5.70	2.30	1.00	1.00	1.46	0.63	3.94	0.94	0.41	5.80	1.60	2.22	0.23	0.10
60,000	10.7		9.60	5.40	2.34	4.16	6.84	3.30	1.43	1.43	2.18	0.95	4.73	1.29	0.56	3.48	2.25	2.66	0.32	0.14

70,000	2.80	108 in.	4.80	7.10	3.08	7.98	4.30	1.86	1.20	2.77	0.76	5.51	1.79	0.78	4.06	0.80	3.11	0.42	0.18
80,000	3.15		0.30	120 in.	5.60	9.11	5.50	2.38	1.52	3.50	0.99	6.30	2.30	1.00	4.64	1.04	3.55	0.54	0.23
90,000	3.50		0.37	3.20	132 in.	2.34	0.18	0.08	4.42	4.44	1.46	7.09	2.80	1.21	5.22	1.30	3.99	0.68	0.29
100,000	4.20		0.46	3.41	2.81	0.24	0.10	144 in.	4.73	5.63	1.52	7.88	3.40	1.47	5.80	1.60	4.44	0.83	0.36
120,000	4.20		0.64	5.40	2.34	0.16		156 in.	2.37		0.07				6.96	2.25	5.32	1.16	0.50

140,000	4.80		0.84	3.98	168 in.	3.28	0.23	0.14	2.76	0.22	0.10	2.35	0.15	0.07	4.06	0.13	6.21	1.53	0.66
160,000	5.60		1.10	4.55	0.66	3.75	0.28	0.18	3.15	0.27	0.12	2.69	0.19	0.08	2.32	0.16	2.28	180 in.	0.05
180,000	6.30		1.39	5.11	0.83	4.22	0.36	0.23	3.55	0.35	0.15	3.02	0.23	0.10	2.61	0.20	2.53	0.12	0.06
200,000	7.00		1.68	5.68	1.00	4.68	0.43	0.27	3.94	0.42	0.18	3.36	0.28	0.12	2.90	0.30	3.16	0.14	0.09
250,000	8.75		2.60	7.10	1.55	5.86	0.67	0.42	4.83	0.63	0.27	4.20	0.42	0.18	3.62			0.21	

300,000	10.5		3.65	8.52	2.20	7.03	0.95	1.37	5.91	0.89	0.39	5.04	0.60	0.26	4.35	0.42	3.79	0.30	0.13
350,000	12.3		4.80	9.95	2.82	8.19	1.22	1.79	6.90	1.20	0.52	5.88	0.80	0.35	5.07	0.55	4.43	0.39	0.17
400,000	14.0		6.15	11.4	3.75	9.36	1.63	2.28	7.88	1.52	0.66	6.72	1.05	0.46	5.80	0.73	5.06	0.50	0.22

[1] For other values of C, multiply the friction heads or pressure drops by the following factors:

C value	60	65	70	75	80	85	90	95	100	105	110	115	120	125	130	135	140	145
Factor	2.58	2.22	1.93	1.71	1.51	1.38	1.22	1.10	1.00	0.913	0.838	0.771	0.712	0.661	0.615	0.573	0.536	0.502

[2] To convert gallons per minute to millions of gallons per day, divide by 694.4; to cubic feet per second, divide by 448.8; to thousands of pounds per hour, multiply by 500.4.

Table 58.   Horton's Values of *n* for Use with Kutter and Manning Formulas

Surface	Best	Good	Fair	Bad
Uncoated cast-iron pipe...........................	0.012	0.013	0.014	0.015
Coated cast-iron pipe.............................	0.011	0.012[1]	0.013[1]	
Commercial wrought pipe, black...................	0.012	0.013	0.014	0.015
Commercial wrought pipe, galvanized..............	0.013	0.014	0.015	0.017
Smooth brass and glass pipe......................	0.009	0.010	0.011	0.013
Smooth lockbar and welded "OD" pipe..............	0.010	0.011[1]	0.013[1]	
Spiral riveted steel pipe...........................	0.013	0.015[1]	0.017[1]	
Corrugated steel pipe[2]............................	......	0.021[1]		
Vitrified sewer pipe...............................	0.011	0.013[1]	0.015	0.017
Common clay drainage tile........................	0.011	0.012[1]	0.014[1]	0.017
Glazed brickwork.................................	0.011	0.012	0.013[1]	0.015
Brick in cement mortar; brick sewers...............	0.012	0.013	0.015[1]	0.017
Neat cement surfaces.............................	0.010	0.011	0.012	0.013
Cement mortar surfaces...........................	0.011	0.012	0.013[1]	0.015
Concrete pipe....................................	0.012	0.013	0.015[1]	0.016
Wood-stave pipe.................................	0.010	0.011	0.012	0.013
Plank flumes:				
Planed.......................................	0.010	0.012[1]	0.013	0.014
Unplaned.....................................	0.011	0.013[1]	0.014	0.015
With battens..................................	0.012	0.015[1]	0.016	
Concrete-lined channels..........................	0.012	0.014[1]	0.016[1]	0.018
Cement-rubble surface............................	0.017	0.020	0.025	0.030
Dry-rubble surface...............................	0.025	0.030	0.033	0.035
Dressed-ashlar surface............................	0.013	0.014	0.015	0.017
Semicircular metal flumes, smooth.................	0.011	0.012	0.013	0.015
Semicircular metal flumes, corrugated.............	0.0225	0.025	0.0275	0.030
Canals and ditches:				
Earth, straight and uniform....................	0.017	0.020	0.0225[1]	0.025
Rock cuts, smooth and uniform.................	0.025	0.030	0.033[1]	0.035
Rock cuts, jagged and irregular................	0.035	0.040	0.045	
Winding sluggish canals.......................	0.0225	0.025[1]	0.0275	0.030
Dredged earth channels.......................	0.025	0.0275[1]	0.030	0.033
Canals with rough stony beds, weeds on earth banks..	0.025	0.030	0.035[1]	0.040
Earth bottom, rubble sides.....................	0.028	0.030[1]	0.033[1]	0.035
Natural stream channels:				
(1) Clean, straight bank, full stage, no rifts or deep pools	0.025	0.0275	0.030	0.033
(2) Same as (1), but some weeds and stones.........	0.030	0.033	0.035	0.040
(3) Winding, some pools and shoals, clean...........	0.033	0.035	0.040	0.045
(4) Same as (3), lower stages, more ineffective slope				
and sections..................................	0.040	0.045	0.050	0.055
(5) Same as (3), some weeds and stones.............	0.035	0.040	0.045	0.050
(6) Same as (4), stony sections....................	0.045	0.050	0.055	0.060
(7) Sluggish river reaches, rather weedy or with very				
deep pools..................................	0.050	0.060	0.070	0.080
(8) Very weedy reaches..........................	0.075	0.100	0.125	0.150

[1] Values commonly used in designing.
[2] Added from other sources.

Table 59.   Approximately Equivalent Values of Roughness Coefficients Used in Various Hydraulic Flow Formulas[1]

Williams and Hazen, C	Scobey		Manning and Kutter, n
	Steel, K_s	Concrete, C_s	
140	0.30	0.40	0.010
130	0.33	0.38	0.011
120	0.38	0.35	0.012
110	0.44	0.33	0.013
100	0.52	0.30	0.014
90	0.64	0.28	0.015
80	0.80	0.25	0.016
70	0.96	0.20	0.017

[1] For average conditions of about 5 ft per sec velocity and pipe diameters of 12 to 48 in.   For equivalent values for individual pipe size and velocities see Table 13, pp. 98–99 of Scobey *Bull.* 150.

the Manning formulas is designed to suit the corresponding values of Kutter's n. Owing to the involved nature of Kutter's formula and the fact that equivalent results can be obtained from Manning's formula,[1] the latter only is reproduced here. It is suggested that those having occasion to use Kutter's formula refer to the logarithmic flow charts prepared by Prof. John H. Gregory.[2]

Using the same terms as in the preceding formulas, the Manning formula can be written:

$$v = (1.486/n)R^{2/3}h_\lambda^{1/2} \quad \text{and} \quad h_\lambda = n^2v^2/2.208R^{4/3}$$

where   hydraulic radius $R = \dfrac{\text{cross-sectional area in sq ft}}{\text{wetted perimeter in ft}}$

And for round pipes running full $R = d'/4$, from which

$$v = 0.589d'^{2/3}h_\lambda^{1/2}/n \quad \text{and} \quad h_\lambda = n^2v^2/0.347d'^{4/3}$$

Horton's[3] values of n shown in Table 58 are generally used with the Kutter and Manning formulas. According to Metcalf and Eddy, however, the value of $n = 0.013$ for vitrified crock sewers should be increased to $n = 0.015$ to provide a margin for losses in manholes, branches, and changes in direction and for possible retardation by air currents in small sewers only partly filled.

Comparison of Hydraulic Coefficients. Approximately equivalent values of the coefficients used in the foregoing hydraulic flow formulas are given in Table 59. This comparison will be found useful in converting the coefficients for interior roughness stated for any one of these formulas into equivalent values usable in the others. The average conversions probably are as accurate as are warranted where the roughness of the interior surface is as indeterminate over a period of years as usually is the case, especially with iron or steel pipes.

Relative Carrying Capacity of Water Pipes. A derivation of the formula for determining the relative carrying capacity of different diameter pipes having the same length was made in connection with the preparation of Table 49. The generally accepted relation for water pipes is that the carrying capacity varies as the square root of the fifth power of the inside diameter. The relative carrying capacities of standard-weight wrought water pipes as given by the above-mentioned formula are shown in Table 61. The decimal fractions in the upper right-hand portion of the table are the reciprocals of the corresponding values in the lower left-hand portion. Through the use of this table, rather involved problems in carrying capacity can be worked out readily, as shown by the following examples:

Example 1. What size water main has a carrying capacity equal to 80 1-in. standard-weight pipes? Referring to Table 61, go down the 1-in. pipe column to 80, and opposite 80 read the corresponding pipe size as 6 in.

Example 2. An 8-in. main supplies four 2-in. pipes and one 3-in. pipe. How many 4-in. pipes can be added without exceeding the relative carrying capacity of the 8-in. main? The fractional carrying capacities of the small pipes in terms of an 8-in. main, as taken from

---

[1] See "Handbook of Hydraulics," by H. W. King, *op. cit.* This text contains extensive Manning formula tables which are supplemented by others published in "Hydraulic Tables," of U.S. War Department, Corps of Engineers (copies of the latter can be obtained from Superintendent of Documents, Washington, D.C.). King shows that the maximum carrying capacity of a circular conduit exists when the depth of water in the conduit is 0.938 of its inside diameter.

[2] Reproduced in "American Sewage Practice," Vol. I, "Design of Sewers," *op. cit.*

[3] "Some Better Kutter's Flow Coefficients," by R. E. Horton, *Eng. News*, February 24, 1916; May 4, 1916.

Table 60.   Conversions from Gallons per Minute to Millions of Gallons per Day, Cubic Feet per Second, and Thousands of Pounds per Hour at a Density of 8.34 Lb per Gal (7.48 Gal per Cu Ft)

Gal per min	Millions of gal per day	Cu ft per sec	Thousands of lb per hr	Gal per min	Millions of gal per day	Cu ft per sec	Millions of lb per hr
100	0.144	0.223	50.04	10,000	14.40	22.28	5.004
120	0.173	0.267	60.05	12,000	17.28	26.74	6.005
140	0.202	0.312	70.06	14,000	20.16	31.19	7.006
160	0.230	0.356	80.06	16,000	23.04	35.65	8.006
180	0.259	0.401	90.07	18,000	25.92	40.10	9.007
200	0.288	0.446	100.1	20,000	28.80	44.56	10.01
250	0.360	0.557	125.1	25,000	36.00	55.70	12.51
300	0.432	0.668	150.1	30,000	43.20	66.84	15.01
350	0.504	0.780	175.1	35,000	50.40	77.98	17.51
400	0.576	0.891	200.2	40,000	57.60	89.12	20.02
450	0.648	1.003	225.2	45,000	64.80	100.3	22.52
500	0.720	1.114	250.2	50,000	72.00	111.4	25.02
600	0.864	1.337	300.2	60,000	86.40	133.7	30.02
700	1.008	1.560	350.3	70,000	100.8	156.0	35.03
800	1.152	1.762	400.3	80,000	115.2	178.2	40.03
900	1.296	2.005	450.4	90,000	129.6	200.5	45.04
1,000	1.440	2.228	500.4	100,000	144.0	222.8	50.04
1,200	1.728	2.674	600.5	120,000	172.8	267.4	60.05
1,400	2.016	3.119	700.6	140,000	201.6	311.9	70.06
1,600	2.304	3.564	800.6	160,000	230.4	356.4	80.06
1,800	2.592	4.010	900.7	180,000	259.2	401.0	90.07
2,000	2.880	4.456	1,001	200,000	288.0	445.6	100.1
2,500	3.600	5.570	1,251	250,000	360.0	557.0	125.1
3,000	4.320	6.684	1,501	300,000	432.0	668.4	150.1
3,500	5.040	7.798	1,751	350,000	504.0	779.8	175.1
4,000	5.760	8.912	2,002	400,000	576.0	891.2	200.2
4,500	6.480	10.03	2,252	450,000	648.0	100.3	225.2
5,000	7.200	11.14	2,502	500,000	720.0	1,114	250.2
6,000	8.640	13.37	3,002	600,000	864.0	1,337	300.2
7,000	10.08	15.60	3,503	700,000	1,008	1,560	350.3
8,000	11.52	17.82	4,003	800,000	1,152	1,782	400.3
9,000	12.96	20.05	4,504	900,000	1,296	2,005	450.4

Table 61, add up as follows:

$$\begin{aligned} \text{Four 2-in. pipes} &= 4 \times 0.0345 = 0.1380 \\ \text{One 3-in. pipe} &= 1 \times 0.0917 = \underline{0.0917} \\ & \hspace{3.2cm} 0.2297 \end{aligned}$$

Subtracting 0.2297 from unity gives 0.7703 of the total capacity of the 8-in. main still available to supply 4-in. pipes. From Table 61 one 4-in. pipe has 0.1818 of the carrying capacity of an 8-in. pipe. Therefore, 0.7703 ÷ 0.1818 = 4.25 4-in. pipes, which could be added.

The answer, therefore, is four.

Example 3. A building is supplied by three water service pipes: one 2-in., one 1½-in., and one 1¼-in. It is desired to add additional equipment requiring the capacity of a 1-in. pipe and at the same time replace the various water-supply pipes with a single service. What size pipe will have the same relative carrying capacity as the four mentioned above? Express the carrying capacity of each pipe in terms of the smallest by reference to Table 61.

$$\begin{aligned} \text{One 2-in. pipe} &= 5.5 \text{ 1-in. pipes} \\ \text{One 1½-in. pipe} &= 2.9 \text{ 1-in. pipes} \\ \text{One 1¼-in. pipe} &= 2.0 \text{ 1-in. pipes} \\ \text{One 1-in. pipe} &= \underline{1.0} \text{ 1-in. pipe} \\ \text{Total capacity} &= \overline{11.4} \text{ 1-in. pipes} \end{aligned}$$

From Table 61 a 2½-in. pipe has the carrying capacity of eight 1-in. pipes, while a 3-in has the capacity of fifteen 1-in. pipes. It will be necessary, therefore, to use a 3-in. pipe.

Water Horsepower. By definition a horsepower is the ability to do work at the rate of 33,000 ft-lb/min. The total horsepower developed by water in falling from a given height is the product of the weight of flow W in pounds per minute times the height h in feet divided by 33,000, or water horsepower $= Wh/33,000$. The actual horsepower available for doing useful work is less than the total by the amount consumed in friction and lost in velocity head. If the water is used to run a *water turbine or other engine*, the shaft horsepower is further diminished by the mechanical losses in the turbine. If h represents the total elevation head in feet, h_λ the head lost in friction and velocity head at exit, and E the mechanical efficiency of the water turbine, the shaft horsepower developed is

$$\text{Shaft horsepower} = EW(h - h_\lambda)/33,000$$

The water flow to turbines is commonly measured in cubic feet per second. At 50 F water weighs 62.4 lb/cu ft. If Q represents water flow in cubic feet per second, $W = 60Q \times 62.4$ and the above formulas may be written: water horsepower $= (60Q \times 62.4h) \div 33,000 = (Qh) \div 8.81$; and shaft horsepower $= EQ(h - h_\lambda) \div 8.81$.

Horsepower Required to Pump Water. In a similar way the water horsepower required to pump water against a total head h is water horsepower $= Wh/33,000$. If E is the mechanical efficiency of the pump, the shaft horsepower $= Wh/(E \times 33,000)$, which is also the horsepower output of the motor or other engine driving the pump. If E' represents the efficiency of the driving engine, the input of the latter is line horsepower $= Wh/(EE' \times 33,000)$. If the input of a motor is desired in kilowatts, it can be obtained readily from the relation 1 hp $= 0.746$ kw. The electrical input of the motor then is

$$\text{Line kw} = \frac{0.746 Wh}{EE' \times 33,000} = \frac{Wh}{EE' \times 44,230}$$

In case it is desired to solve pumping problems in terms of gallons per minute rather than weight, the above formulas can be adapted readily from the relation that 1 gal of water weighs 8.34 lb at 50 F, hence $W = 8.34 \times$ gpm, and the above formulas can be rewritten: water horsepower $= (8.34 \times \text{gpm} \times h) \div 33,000 = (\text{gpm} \times h) \div 3,957 = 0.000252 \text{ gpm} \times h$; shaft horsepower $= (8.34 \times \text{gpm} \times h) \div (E \times 33,000) = (\text{gpm} \times h) \div (E \times 3,957)$. Due account must be taken of the pump-and-motor or the engine efficiency in converting work in foot-pounds into horsepower-hours or kilowatthours.

Time Required to Empty Tanks. An open tank whose cross-sectional area is uniform throughout its depth (i.e., a vertical cylindrical or prismatic tank) will empty itself through an orifice or a pipe (the discharge point being at the same level as the bottom of the tank), if there is no inflow, in just twice the time required to discharge the same amount of liquid under the initial head. For other shapes of reservoirs, the time may be computed by dividing the volume into several layers and computing the time for each layer to discharge under the average head for that layer. In case the outlet of the pipe is located some distance below the bottom of the tank, the time required will be that corresponding to the average head at the outlet during the process of emptying.

Water Hammer[1]

Occurrence of Water Hammer. If the velocity of water or other liquid flowing in a pipe is suddenly diminished, the energy given up by the liquid will be

[1] See also the Water Hammer section in Chap. 11, Water Supply Piping.

Table 61. Relative Carrying Capacities of Standard-weight[2] and Outside Diameter Wrought Pipe for Water[1]

| Actual inside diameter, inches | 0.269 | 0.364 | 0.493 | 0.622 | 0.824 | 1.049 | 1.380 | 1.610 | 2.067 | 2.469 | 3.068 | 3.548 |
Nominal size	⅛	¼	⅜	½	¾	1	1¼	1½	2	2½	3	3½
⅛	1	0.475	0.222	0.122	0.0625	0.0333	0.0167	0.0114	0.0061	0.00392	0.00228	0.00158
¼	2.1	1	0.475	0.263	0.130	0.0714	0.0357	0.0244	0.0130	0.00833	0.00485	0.00336
⅜	4.5	2.1	1	0.555	0.278	0.151	0.077	0.0526	0.0278	0.0178	0.0103	0.0072
½	8.2	3.8	1.8	1	0.500	0.270	0.139	0.091	0.050	0.022	0.0185	0.0128
¾	16	7.7	3.6	2	1	0.555	0.278	0.189	0.100	0.0645	0.0370	0.0263
1	30	14	6.6	3.7	1.8	1	0.500	0.344	0.182	0.1180	0.0666	0.0476
1¼	60	28	13	7.2	3.6	2	1	0.666	0.357	0.232	0.143	0.0910
1½	88	41	19	11	5.3	2.9	1.5	1	0.526	0.345	0.200	0.139
2	164	77	36	20	10	5.5	2.8	1.9	1	0.625	0.370	0.256
2½	255	120	56	31	15.5	8.5	4.3	2.9	1.6	1	0.588	0.400
3	439	206	97	54	27	15	7	5	2.7	1.7	1	0.715
3½	632	297	139	78	38	21	11	7.2	3.9	2.5	1.4	1
4	867	407	191	107	53	29	15	9.9	5.3	3.4	2	1.4
5	1,525	716	335	188	93	51	26	17	9.3	6	3.5	2.4
6	2,414	1,133	531	297	147	80	41	28	15	9.5	5.5	3.8
8	4,795	2,251	1,054	590	292	160	80	54	29	19	10.9	7.6
10	8,468	3,976	1,862	1,042	516	282	142	97	52	33	19	13.4
12	13,292	6,240	2,923	1,635	809	443	223	152	81	52	30	21
14 in. O.D.	17,028	7,994	3,745	2,094	1,037	567	286	194	104	67	39	27
16 in. O.D.	24,199	11,361	5,322	2,976	1,474	806	406	276	148	95	55	38
18 in. O.D.	31,750	14,906	6,982	3,905	1,933	1,057	533	362	194	124	72	50
20 in. O.D.	41,928	19,685	9,221	5,157	2,553	1,396	703	478	256	164	95	66
24 in. O.D.	67,599	31,737	14,866	8,315	4,116	2,251	1,134	771	413	265	154	107

Table 61. *(Continued)*

Actual inside diameter inches	4.026	5.047	6.065	7.981	10.02	12.00	13.25	15.25	17.00	19.00	23.00
Nominal size	4	5	6	8	10	12	14 in. O.D.	16 in. O.D.	18 in. O.D.	20 in. O.D.	24 in. O.D.
⅛	0.00115	0.000655	0.000415	0.000209	0.000118	0.000075	0.000059	0.000041	0.000032	0.000024	0.000015
¼	0.00246	0.00139	0.000882	0.000445	0.000252	0.000160	0.000125	0.000088	0.000067	0.000051	0.0000315
⅜	0.00523	0.00298	0.00188	0.000948	0.000536	0.000342	0.000267	0.000188	0.000143	0.000108	0.0000672
½	0.00934	0.00532	0.00336	0.00169	0.000959	0.000611	0.00047	0.000336	0.000256	0.000193	0.000120
¾	0.0189	0.01075	0.00680	0.00342	0.00194	0.00124	0.000965	0.000678	0.000517	0.000392	0.000243
1	0.0345	0.0196	0.0125	0.00625	0.00355	0.00226	0.00176	0.00124	0.000945	0.000716	0.000444
1¼	0.0666	0.0384	0.0244	0.0125	0.00705	0.00449	0.00350	0.00246	0.00188	0.00142	0.000881
1½	0.1010	0.0588	0.0357	0.0185	0.0103	0.00657	0.00515	0.00362	0.00276	0.00209	0.001300
2	0.1887	0.1075	0.0566	0.0345	0.0192	0.0123	0.00961	0.00675	0.00515	0.00390	0.00242
2½	0.294	0.1665	0.1050	0.0526	0.0301	0.0192	0.0149	0.0105	0.00805	0.00610	0.00377
3	0.500	0.296	0.182	0.0917	0.0526	0.0333	0.0256	0.0182	0.0139	0.01050	0.00650
3½	0.715	0.417	0.263	0.1315	0.0746	0.0476	0.0370	0.0263	0.0200	0.01515	0.00935
4	1	0.555	0.357	0.1820	0.1020	0.0666	0.0500	0.0357	0.0270	0.0208	0.0128
5	1.8	1	0.625	0.322	0.1785	0.1150	0.0909	0.0625	0.0476	0.0370	0.0227
6	2.8	1.6	1	0.500	0.286	0.1820	0.1430	0.1000	0.0769	0.0555	0.0357
8	5.5	3.1	2	1	0.555	0.3570	0.278	0.2000	0.1515	0.1150	0.0714
10	9.8	5.6	3.5	1.8	1	0.6250	0.500	0.3450	0.2630	0.2000	0.1250
12	15	8.7	5.5	2.8	1.6	1	0.769	0.555	0.416	0.322	0.196
14 in. O.D.	20	11	7	3.6	2	1.3	1	0.714	0.526	0.400	0.250
16 in. O.D.	28	16	10	5.0	2.9	1.8	1.4	1	0.769	0.588	0.357
18 in. O.D.	37	21	13	6.6	3.8	2.4	1.9	1.3	1	0.769	0.476
20 in. O.D.	48	27	18	8.7	5	3.1	2.5	1.7	1.3	1	0.625
24 in. O.D.	78	44	28	14	8	5.1	4	2.8	2.1	1.6	1

[1] The carrying capacity for water varies as the square root of the fifth power of the actual inside diameter. The relative carrying capacity for air, steam, and gas is given in Table 49. The relative carrying capacities listed for standard-weight pipe also are approximate for Type *K* copper water tubes of same nominal size.
Example. A 2-in. pipe has a water-carrying capacity equal to ten ¾-in. pipes, or 0.1887 that of a 4-in. pipe.

[2] Schedule 40 pipe 10 in. and smaller is identical with standard-weight pipe.

divided among compressing the liquid itself, stretching the pipe walls, and frictional resistance to wave propagation. Water hammer is manifest as a series of shocks, sounding like hammer blows, which may have sufficient magnitude to rupture the pipe or damage connected equipment. It may be caused by the nearly instantaneous or too rapid closing of a valve in the line or by an equivalent stoppage of flow such as would take place with the sudden failure of electricity supply to a motor-driven pump. The shock pressure is not concentrated at the valve, and if rupture occurs, it may take place near the valve simply because it acts there first. The pressure wave due to water hammer travels back upsteam to the inlet end of the pipe, where it reverses and surges back and forth through the pipe, getting weaker on each successive reversal. The velocity of the wave is that of an acoustic wave in an elastic medium, the elasticity of the medium in this case being a compromise between that of the liquid and the pipe. The excess pressure due to water hammer is additive to the normal hydrostatic pressure in the pipe and depends on the elastic properties of the liquid and pipe and on the magnitude and rapidity of change in velocity. Complete stoppage of flow is not necessary to produce water hammer, as any sudden change in velocity will create it to a greater or lesser degree, depending on the conditions mentioned above.

The phenomena of water hammer have been analyzed by a number of investigators. In stating the formulas for velocity of wave travel, pressure rise, etc., the following symbols and definitions of units are used:

E = modulus of elasticity in tension for pipe material, psi
g = acceleration due to gravity = 32.2 ft per sec^2
K = bulk modulus of elasticity of liquid, psi (approximately 300,000)
K' = virtual modulus of elasticity for liquid and pipe combination, psi
l = length of pipe from valve to inlet end, ft
λ = Poisson's ratio for pipe material (see Table 4)
p = normal or initial pressure of liquid in pipe, psi
p_0 = excess pressure due to water hammer, psi
r = inside radius of pipe, in.
S = velocity of wave travel in rigid pipe, fps
S' = velocity of wave travel in elastic pipe, fps
t = thickness of pipe wall, in.
T = time, sec
v = initial velocity in pipe, fps
v_0 = reduction in velocity causing water hammer, fps
y = density of liquid, lb/cu ft

Wave Propagation in a Rigid Pipe. If a column of liquid flowing at a steady velocity v through a *rigid* pipe of uniform diameter and of length l ft has its motion instantaneously retarded by the partial or complete closure of a rigid valve (or equivalent stoppage), the phenomena experienced are due to the elasticity of the column alone and are analogous to those obtaining in the case of the longitudinal impact of an elastic bar against a rigid wall.

At the instant of closure, the velocity of the layer of water in contact with the valve is retarded by an amount designated as v_0, and the kinetic energy corresponding to $v_0^2/2g$ is converted into resilience or energy of strain, with a consequent sudden rise in pressure. This checks the adjacent layer, with the result that a state of reduced velocity and increased pressure (this at any point being designated as p_0 above the pressure p obtaining at that point with steady flow) is propagated as a wave along the pipe with velocity S. Assuming instantaneous retardation, a *rigid* pipe, and *complete reflection* of the wave at the valve, which represent worst conditions of water hammer, the velocity of wave travel is $S = 12\sqrt{gK/y}$. (This is

the well-known formula of physics for the velocity of an acoustic wave in an elastic medium.)   This wave reaches the inlet end of the pipe after T sec, where $T = l \div S$. At this instant the column of liquid has been retarded to its new uniform velocity of $v - v_0$ and is in a state of maximum compression of $p + p_0$ where $p_0 = (v_0/12)\sqrt{yK/g}) = ySv_0/144g$. This is derived from the relation that if a cube of unit side be subjected to a pressure increasing from p to $p + p_0$, the change in volume will be $p_0 \div K$, and since the mean increase in pressure during compression is $p_0 \div 2$, the work done in inch-pounds is $p_0^2 \div 2K$.   Hence, equating the change in energy $yv_0^2/(2g \times 144) = p_0^2/2K$ and $p_0 = (v_0/12)\sqrt{yK/g}$.   The second expression for p_0 is obtained directly by substituting the value of K obtained from the preceding equation.

This is not a state of equilibrium, since the pressure $p + p_0$ existing inside the pipe is greater than the applied pressure p from the external medium.   In consequence, the strain energy due to compression is reconverted into kinetic energy, the internal pressure falls to that of the external source, and the liquid rebounds with a velocity almost equal to $v - 2v_0$ in a direction opposite to that of initial flow.   At this instant the whole of the column is at normal pressure p and is moving with a velocity of approximately $v - 2v_0$ toward the inlet end of the pipe. The end of the column tends to leave the valve but cannot do so without causing a reduction in pressure at that point which in turn acts to retard the reversed flow. Motion is consequently checked, and the kinetic energy of reversed flow goes to reduce the strain energy to a value below that corresponding to normal pressure. The pressure drops suddenly by an amount equal to that through which it originally rose, and a wave of zero velocity and of pressure p_0 below normal is transmitted along the pipe, to be reflected from the inlet end as a wave of normal pressure and velocity toward the valve.   When this wave reaches the valve $4l/S$ sec after the instant of closure, the conditions are the same as at the beginning of the cycle, and the whole is repeated.   This cycle would continue indefinitely were it not for viscosity of the liquid and friction against the pipe walls.   The completeness of wave reflection at the valve and at the inlet end of the pipe will vary in different cases—the above explanation applies to worst conditions, consisting of complete reflection at both ends, a rigid pipe, and instantaneous reduction in velocity at the valve.   The condition of instantaneous closure applies, provided the time of closure (whether partial or complete) does not exceed that required for the wave to travel from the valve to the inlet end of the pipe and back again, i.e., provided the time of closure does not exceed $(2l/S)$.   It is apparent, therefore, that if severe water hammer is to be avoided, the time of closure should exceed the time corresponding to $(2l/S)$, and the longer the time taken in closure, the less will be the shock. The excess pressure resulting from water hammer can be kept within any desired limit by computing the change in velocity which will produce that excess pressure and then arranging to have the rate of valve closure such that the time T exceeds $2l/S$.

Wave Propagation in an Elastic Pipe.   Owing to elasticity of the pipe walls, part of the kinetic energy of the moving column is expended in stretching these, with a resulting increase in the complexity of the phenomena, a reduction in the maximum pressure attained, and an increase in the rate at which the pressure waves die out.   The elasticity of the pipe may modify the results in two ways:

1. If the pipe is free to move in a longitudinal direction, the impact of the moving column of water against the closed or partially closed valve will tend to drive the pipe ahead in the direction of flow.   As the compression wave in the liquid reflects and surges in the opposite direction, there is a tendency for the pipe to rebound in the opposite direction.   This pulsation will continue until the disturbance inside the pipe dies down.   Hence, the necessity to secure boiler-feed lines and similar

piping with anchors or shock absorbers, especially where reciprocating pumps are used.   The longitudinal stretching of a pipe free to move in this direction super-imposes the effect of a wave in the pipe material on the wave in the liquid, although the net effect on the liquid wave is slight and usually may be neglected.

2. The second effect of the elasticity of the pipe line is due to the fact that, since the walls extend both longitudinally and circumferentially under pressure, the apparent diminution of volume of the fluid under a given increment of pressure is greater than in a rigid pipe.   The effect of this is to reduce the value of K obtaining with a rigid pipe to a virtual value of K' for elastic pipe.   The value of K' is influenced by the lateral contraction of the pipe material under longitudinal strain, which is taken into account by introducing Poisson's ratio in the equation.   The equation for determining K' is[1]

$$K' = \frac{1}{(1/K) + (r/2tE)(5 - 4\lambda)}$$

If the pipe is so anchored that all longitudinal extension is prevented but that circumferential extension is free, this becomes

$$K' = \frac{1}{1/K + 2r/tE}$$

The velocity of wave travel through a liquid in an elastic pipe then is, as explained in connection with rigid pipes,

$$S' = 12\sqrt{gK'/y}$$

and the excess pressure due to water hammer is

$$p_0 = v_0/12\sqrt{yK'/g} = yS'v_0/144g$$

Under identical conditions, the velocity of wave travel and the excess pressure due to water hammer are always less in an elastic pipe than in a rigid pipe.

Numerical Values of Coefficients.   The bulk modulus of elasticity K of water and other liquids is approximately 300,000 psi, within a limit of about 10 per cent plus or minus, varying somewhat with the pressure and temperature, which is sufficiently accurate in view of the uncertainty of other factors entering the problem.   Poisson's ratio and average values of the modulus of elasticity at room temperature for common piping materials are given in Table 4.

Numerical Values for Wave Velocity and Excess Pressure Due to Water Hammer.   The numerical values for steel, wrought-iron, and cast-iron pipe shown in Table 62 were computed by the above formulas.   The values of p_0 are for instantaneous reduction in velocity of 1 fps and complete reflection of the wave. Corresponding values of p_0 for any other reduction in velocity can be obtained by multiplying these values by that reduction in velocity.   The values of K', S', and p_0 given opposite different ratios of r/t are for pipe which is free to extend both longitudinally and circumferentially.

Numerical values for maximum excess pressure due to water hammer p_0 can be computed directly from the table without recourse to the formula as follows: divide the radius of the pipe by its thickness to obtain the ratio r/t; opposite this value of r/t read the excess pressure p_0 due to a change in velocity of 1 fps; multiply this value by the change in velocity v_0 produced by valve closure to obtain the excess pressure p_0 caused by v_0 change in velocity.   In case of complete stoppage of flow in a time less than $2l/S'$, v_0 should be taken as the initial velocity v.

[1] The derivation of this equation is too long to include here.

Example. What will be the excess pressure due to water hammer in a 12-in. standard-weight steel pipe for an instantaneous reduction in velocity of 5 fps?

Solution. In this case $r = 6$ in., $t = 0.375$ in., and $r/t = 16$. Interpolating in the table to find the value of p_0 for $r/t = 16$ gives an excess pressure of 55.7 psi for a change in velocity of 1 fps. For an instantaneous change in velocity of 5 fps the excess pressure will be $55.7 \times 5 = 279$ psi.

For cement-asbestos and concrete pipe the maximum shock pressure can be computed from the formulas and coefficients given, or for the usual r/t ratios of about 5 obtaining with these products, the excess pressure can be assumed to be of

**Table 62. Numerical Values of Terms Used in Formulas
for Water Hammer**

(Pipe free to extend both longitudinally and circumferentially)

r/t	Cast iron $E = 15,000,000$ $\lambda = 0.270$			Wrought iron $E = 28,000,000$ $\lambda = 0.278$			Steel $E = 30,000,000$ $\lambda = 0.303$		
	K'	S'	$p_0{}^1$	K'	S'	$p_0{}^1$	K'	S'	$p_0{}^1$
Rigid pipe	300,000	4,720	63.5	300,000	4,720	63.5	300,000	4,720	63.5
5	251,000	4,320	58.1	271,500	4,490	60.5	274,000	4,510	60.7
10	215,700	4,000	53.8	248,300	4,293	57.7	252,000	4,325	58.2
15	189,100	3,750	50.5	228,500	4,119	55.5	233,900	4,165	56.1
20	168,000	3,535	47.5	212,000	3,964	53.4	217,900	4,025	54.2
25	151,500	3,355	45.2	197,100	3,827	51.6	203,500	3,890	52.4
30	137,500	3,195	43.0	184,700	3,702	49.9	191,300	3,775	50.8
35	126,500	3,063	41.2	173,700	3,588	48.4	180,400	3,665	49.3
40	116,900	2,945	39.6	163,600	3,485	47.0	170,700	3,565	48.0
45	108,500	2,840	38.2	155,000	3,390	45.6	161,800	3,463	46.6
50	101,300	2,745	36.9	146,900	3,302	44.5	154,000	3,385	45.6

[1] Values of p_0 per unit change in velocity ($v_0 = 1$) where time of valve closure T is less than $2l \div S$ and with complete reflection of the wave.

the order of 40 to 45 psi for each 1-fps change in velocity within the critical period. In the case of reinforced concrete pipe the values of E, λ, K', and S' are difficult to determine accurately owing to its composite structure, but here again the r/t values are about 5 and it seems reasonable to assume a possible maximum shock pressure of 40 to 45 psi for each 1-fps change in velocity within the critical period.

End Restraint. The effect of end restraint is apparent from a comparison of the first and second formulas for K'. For $\lambda = 0.250$ these formulas become identical and it follows that the excess pressure due to water hammer would be the same for either free or restrained ends. For materials having λ greater than 0.250 the excess pressure would be slightly higher for the assumption of pipe free to extend longitudinally. By the same token, the excess pressure with materials having λ less than 0.250 would be slightly less for the free end condition. Actually the differences are too small to be significant in the face of the possible variations in λ, E, or the rate of valve closure. Hence for practical purposes it usually does not matter whether or not the pipe is assumed to be free to extend longitudinally.

Water-hammer Suppressors. The rise in pressure caused by water hammer may be minimized by the use of relief valves, surge tanks, or air chambers. To obtain greatest effectiveness the relief valve or other form of suppressor should be located as close as possible to the source of the disturbance. The necessity for replenishing the air in air chambers should be recognized in considering their use as water-hammer suppressors. In some cases, restricting the passages between the pipeline and the air chambers increases the effectiveness of a given size of air chamber. Suppressors, as a general rule, do not eliminate shock entirely but will

reduce it by 10 to 40 per cent, which often is sufficient to remove the clanking sound.

The amount of kinetic energy set up by water hammer that has to be at least partially dissipated by a suppressor may be computed from the relation that the kinetic energy of the liquid which has been transformed into increased pressure has a value of $p_0{}^2/2K$ in.-lb/cu in. of liquid involved between the point of arrested motion and the point of normal pressure. In the case of a branch pipe, for instance, this may involve the entire column of liquid flowing in the branch. The suppressor must be designed to absorb a good part of this energy and return it to the liquid in the proper part of the cycle or, in the case of a relief valve, to spill a certain amount of liquid from the pipe.

Design Values for Water Hammer. In the case of pipe made of steel or other relatively elastic materials it is sometimes permissible to assume that occasional shock pressures can be looked after satisfactorily within the designed factor of safety allowed for working pressure. This is the usual practice with boiler-feed lines and other steel piping used to convey water in power and industrial plants. The adequacy of this practice depends on the ratio of shock pressure to working pressure, and while it might do well enough to have a shock pressure of 200 psi in a boiler-feed line designed for 1,000-psi working pressure, this might not do at all in a large city-water line designed for 50-psi working pressure.

With relatively brittle materials such as cast iron and cement or concrete, it is customary to design for an internal pressure which represents the sum of the nominal working pressure plus a reasonable allowance for shock pressure. This may be desirable also with *steel* pipe used for underground water service, owing to the likelihood of wastage of the pipe wall through corrosion. What shock pressure should be designed for may be a matter of engineering judgment, or it may be determined by empirical rules. Devices are often employed, particularly in large-diameter underground water-supply systems, which act to prevent building up the maximum shock pressures theoretically possible if all the change in velocity took place within the critical time period. As a first step, an effort should be made to lengthen the time interval for valve closure or other stoppage of flow; as a second step, where needed, means may be provided for cushioning or relieving shock as described under "Water-hammer Suppressors."

Hence in waterworks practice it is customary to design for shock pressures considerably less than would be expected for the instantaneous loss in velocity of 5 fps corresponding to usual maximum flow conditions in such lines. Many years ago the following schedule of shock pressures based on experience and varying with pipe diameter was suggested by Dexter Brackett,[1] when engineer of the Boston water works.

Allowances Proposed by Dexter Brackett for Shock Pressure Due to Water Hammer, Psi

Nominal pipe diameter, in....	3 to 10	12 and 14	16 and 18	20	24	30	36	42 to 60
Allowance for water hammer, psi........................	120	110	100	90	85	80	75	70

These allowances have found wide acceptance in the waterworks industry and are the basis for rules set up for the design of cast-iron pipe in the American Standard Code for Pressure Piping, ASA B31.1, and in the American Recommended Practice

[1] See "A Proposed New Method for Determining Barrel Thickness of Cast Iron Pipe," by T. H. Wiggin, M. L. Enger, and W. J. Schlick, *J. AWWA*, vol. 31, p. 843, May, 1939. Brackett is said to have recommended these allowances prior to 1895.

Manual for the Computation of Strength and Thickness of Cast Iron Pipe, ASA A21.1.

In long-distance gravity-flow water-supply conduits and other places where economic considerations forbid designing the line for water-hammer allowances as generous as those suggested by Brackett, it becomes necessary to hold shock pressures[1] to satisfactory limits through the use of cushioned check valves and overflow pipes, surge tanks, or relief valves of either the direct-acting or relay-operated types.

Entirely different conditions are encountered with high-pressure piping for operating hydraulic presses and similar equipment where water velocities of 20 to 40 fps are customary with almost instantaneous stoppage of the press plunger at the end of each stroke. Under these circumstances shock pressures of 2,000 psi or more may be experienced unless adequate suppressors are provided.

FRICTION FLOW OF OILS

The calculation of oil flow through pipes is much more complicated than in the case of water. Water is a fluid of well-defined and almost constant physical characteristics within ordinary temperature limits; oil is the opposite; no two oils are exactly alike, and even any one oil is subject to important physical changes under variable temperatures. While the flow of water through pipes offers in itself a sufficiently difficult problem, still the laws governing it have been determined empirically with an exactness sufficient to meet all ordinary practical requirements. The principal difference between oil and water from a pipe-flow point of view lies in the variable viscosity. Ordinary crude oil is not a simple homogeneous liquid such as water—it is a very complex substance composed of compounds of carbon and hydrogen which exist in petroleum in an almost bewildering number.

For the reasons stated above, it is impossible to obtain any simple empirical equation which is universally applicable to the flow of oil in pipes. Satisfactory results can be obtained, however, by the use of pertinent temperature-viscosity relations and the friction factor vs. Reynolds number correlation of Fig. 42. This method gives remarkably consistent results, and is at the same time universally applicable to the flow of all fluids, provided their physical properties are sufficiently known. Its chief disadvantage is the number of factors which must be taken into account in the solution. In the present text sufficient tables, charts, etc. have been provided which vastly simplify the steps leading up to the ultimate answer. After the various factors are determined step by step from these data, it is comparatively easy to substitute the values in the governing equation and obtain an answer. The final solution is not complicated, and if reasonable care is taken in looking up the factors in the tables, charts, etc., there is small probability of error. In this connection attention is called to the following data included in this text:

1. Temperature corrections for Baumé hydrometer readings, Table 44
2. Relation of Bureau of Standards and API Baumé scales to specific gravities at 60/60 F (Table 43)
3. Degrees Baumé, specific gravities at 60/60 F and pounds per gallon corresponding to each degree API (Table 42)
4. Friction-factor charts of Figs. 41 and 42
5. Temperature vs. viscosity curves for various liquids (Fig. 38*a*)
6. Temperature-viscosity relations for petroleum oils (Fig. 37) and for typical oils used in hydraulic systems (Fig. 20 of chap. 14)

[1] See also "Water Hammer Correctives," by Richard Bennett, *Water Works & Sewerage*, vol. 89, pp. 92–97, 1942. This article describes a number of relieving and suppressing devices with illustrations.

7. Viscometer time of efflux vs. z/S (Fig. 36)

8. Temperature vs. specific-gravity curves for various liquids (Fig. 38*b*)

9. Relation between specific gravity, specific heat, and temperature for petroleum oils (Fig. 35)

10. Inside-diameter functions of steel pipe (Table 47)

11. Pipe-friction charts for flow of oil (Fig. 54*a* through *g*)

The data on pressure drop given in Table 63 were calculated for oil of 20 Bé gravity and the viscosities indicated.   This table is reproduced from "Lubrication," by permission of The Texas Company.   See also Table 2.

Table 63. Flow of Oils through Commercial Pipes

Size of pipe and average inside diameter, inches	Capacity, gallons per minute	Pressure drop in pounds per square inch per 100 ft. of pipe based on oils of 20° Bé. gravity					
		Viscosity in Saybolt Universal seconds					
		100	200	300	400	500	600
½ 0.622	2	6.59	14.3	21.7	29.1	36.9	43.6
	5	21.1	34.8	53.4	71.5	89.5	107.0
	7	37.7	49.2	74.6	101.0	129.0	152.0
	10	69.8	85.8	107.0	144.0	177.0	217.0
	15	143.0	174.0	194.0	217.0	267.0	325.0
¾ 0.824	2	2.16	4.74	7.04	9.48	11.6	14.2
	5	5.6	11.5	17.1	23.1	28.9	35.2
	7	9.96	16.1	24.5	32.8	40.5	48.9
	10	18.4	22.3	34.9	46.0	58.5	70.9
	15	37.9	45.7	51.5	70.4	88.3	106.0
1 1.05	5	2.06	4.32	6.68	8.85	11.0	13.4
	10	5.9	8.74	13.4	17.9	22.0	26.6
	15	11.9	14.5	19.7	26.6	33.2	40.3
	20	19.7	23.8	26.8	35.4	44.1	53.5
	25	29.0	35.1	39.3	44.2	54.5	66.8
1½ 1.61	10	0.783	1.57	2.40	3.22	3.98	4.86
	20	2.61	3.17	4.85	6.52	8.02	9.70
	30	5.30	6.43	7.15	9.68	12.1	14.7
	40	8.72	10.6	11.9	12.8	16.0	19.4
	50	12.7	15.7	17.4	18.8	19.8	24.2
2 2.067	10	0.266	0.578	0.875	1.17	1.47	1.79
	20	0.79	1.15	1.75	2.34	2.92	3.52
	30	1.60	1.93	2.62	3.50	4.40	5.30
	40	2.63	3.18	3.56	4.66	5.85	7.10
	50	3.86	4.68	5.28	5.88	7.26	8.85
4 4.026	50	0.174	0.198	0.307	0.412	0.494	0.612
	100	0.550	0.668	0.744	0.810	1.01	1.22
	150	1.09	1.36	1.50	1.62	1.74	1.84
	200	1.81	2.22	2.49	2.68	2.85	2.98
	250	2.67	3.26	3.67	3.97	4.20	4.41
6 6.065	100	0.0788	0.0956	0.115	0.158	0.194	0.236
	200	0.259	0.315	0.359	0.388	0.408	0.480
	500	1.27	1.55	1.74	1.88	2.01	2.20
	700	2.29	2.80	3.15	3.36	3.60	3.78
	1000	4.24	5.21	5.82	6.32	6.68	7.05
8 8.03	200	0.0696	0.0834	0.0947	0.102	0.128	0.155
	500	0.340	0.418	0.459	0.500	0.530	0.556
	1000	1.13	1.37	1.55	1.67	1.78	1.87
	1500	2.27	2.74	3.07	3.30	3.59	3.75
	2000	3.88	4.55	5.15	5.56	5.92	6.22
12 12.05	1000	0.165	0.203	0.228	0.242	0.258	0.272
	2000	0.552	0.670	0.750	0.810	0.866	0.905
	3000	1.15	1.34	1.50	1.63	1.73	1.83
	4000	1.91	2.22	2.50	2.69	2.84	3.00
	5000	2.90	3.34	3.63	3.97	4.22	4.41

To change capacities from gallons per minute to barrels (42 gal) per hour, multiply pressure drops by 1.43.   To change from pressure drops per square inch per 100 ft to pressure drops per square inch per mile, multiply by 52.8.

The charts shown in Fig. 54a through *g* furnish a simplified method of solving Eq. (77) for *oil* or *water* for pipe sizes to which they apply. These charts are reproduced from "Standards of The Hydraulic Society," 4th ed., by permission of The Hydraulic Society and the Worthington Pump and Machinery Corporation. Straight lines on the charts cover viscous or stream-line flow, and curved lines cover turbulent flow. It should be noted that the upper lines on each chart have their humps smoothed out and appear as unbroken curves. As a result, the portions of such curves lying to the left of where the humps should be are somewhat inaccurate.

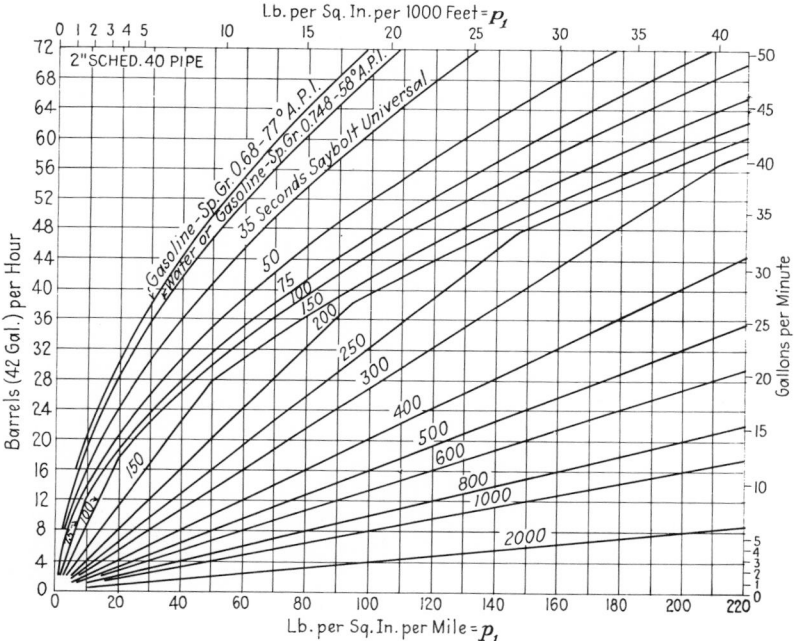

FIG. 54a. Pipe friction based on average Saybolt Universal viscosity and specific gravity of fluid in pipeline. Pressure drop $\Delta p = p_1 \times$ sp gr.

When close results are required in this region, direct substitution into Eq. (77) should be performed. Vertical section lines on the charts of Fig. 54 give friction loss p_1 in pounds per square inch per 1,000 ft and per mile of pipe for an oil having an assumed specific gravity of unity. This value p_1 when multiplied by the actual specific gravity of the oil under the conditions of flow, gives the actual pressure drop per 1,000 ft or per mile of pipe. Horizontal section lines represent the quantity of oil to be pumped in barrels per hour and in gallons per minute. The Saybolt Universal seconds and specific gravities used on the charts are those obtaining at the average temperatures actually existing during flow. The following example serves to illustrate the use of these charts:

Example. Six hundred barrels per hour of oil are to be pumped through an 8-in. pipeline 10 miles long. At 60 F the viscosity is 2,500 SSU and the gravity is 20 deg API. What is the pressure drop with the flow temperature at 60 F and what will be the drop if the oil is heated so as to maintain an average temperature of 100 F?

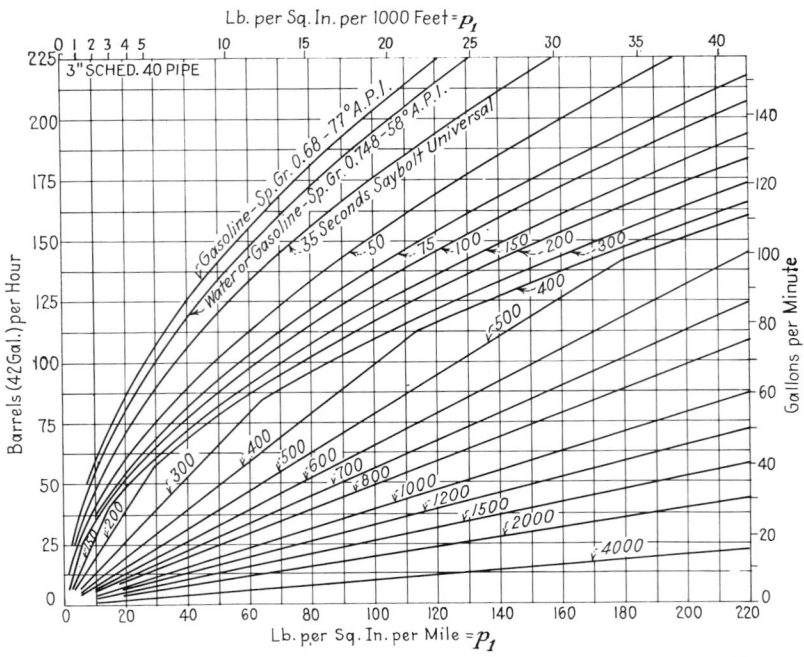

Fig. 54b. Pipe friction based on average Saybolt Universal viscosity and specific gravity of fluid in pipeline. Pressure drop $\Delta p = p_1 \times$ sp gr.

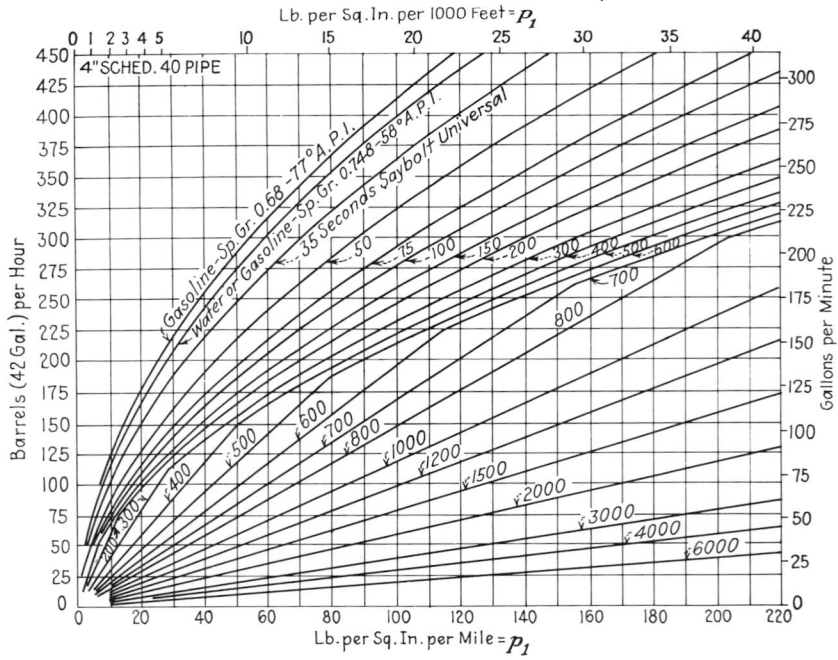

Fig. 54c. Pipe friction based on average Saybolt Universal viscosity and specific gravity of fluid in pipeline. Pressure drop $\Delta p = p_1 \times$ sp gr.

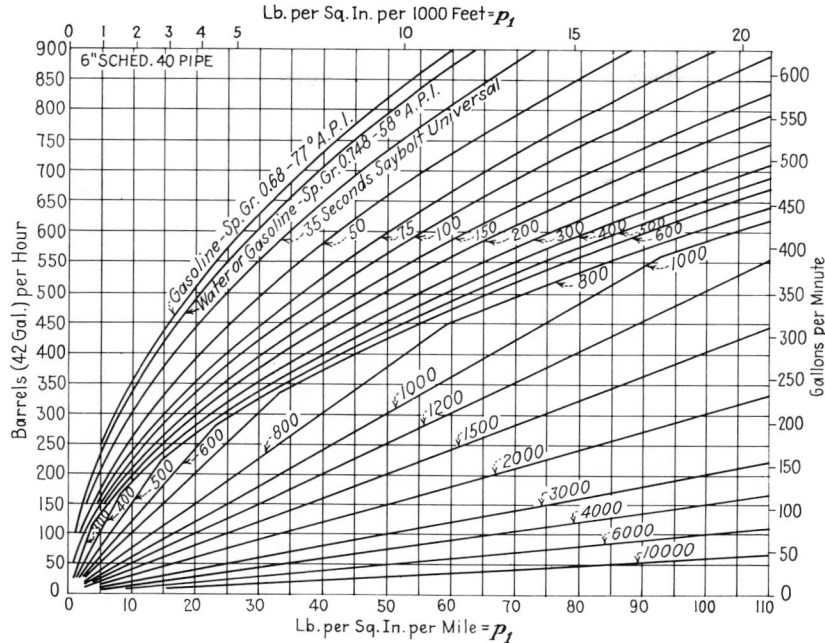

FIG. 54d. Pipe friction based on average Saybolt Universal viscosity and specific gravity of fluid in pipeline. Pressure drop $\Delta p = p_1 \times$ sp gr.

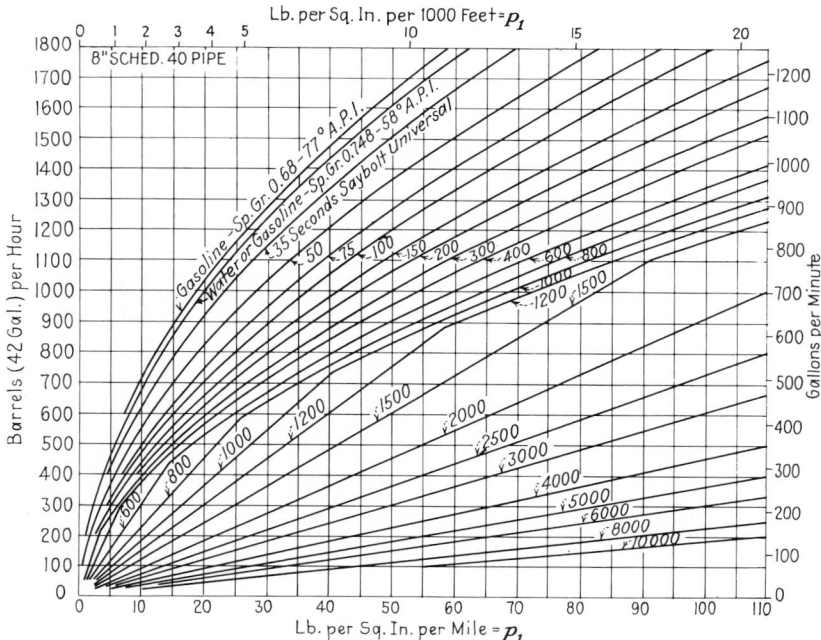

FIG. 54e. Pipe friction based on average Saybolt Universal viscosity and specific gravity of fluid in pipeline. Pressure drop $\Delta p = p_1 \times$ sp gr.

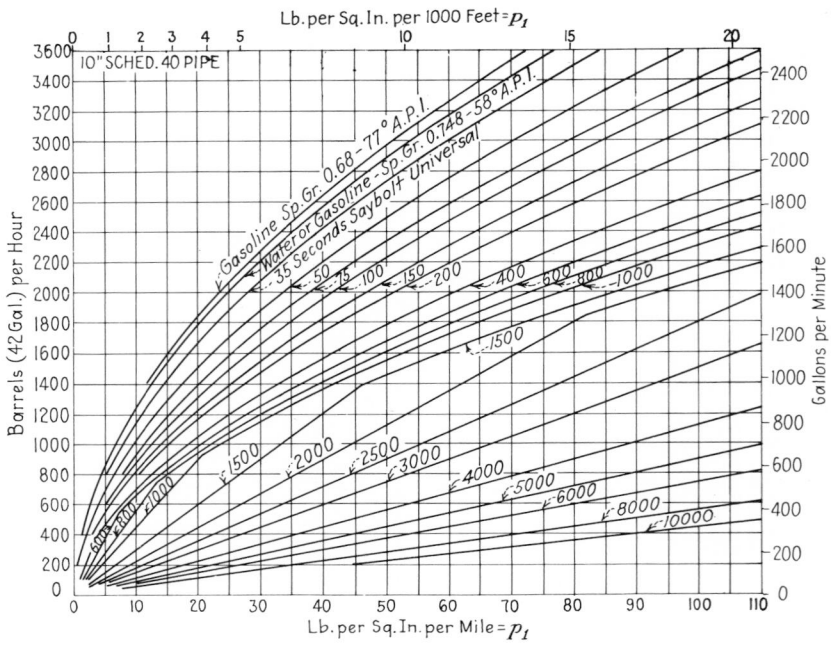

FIG. 54*f*. Pipe friction based on average Saybolt Universal viscosity and specific gravity of fluid in pipeline.   Pressure drop $\Delta p = p_1 \times$ sp gr.

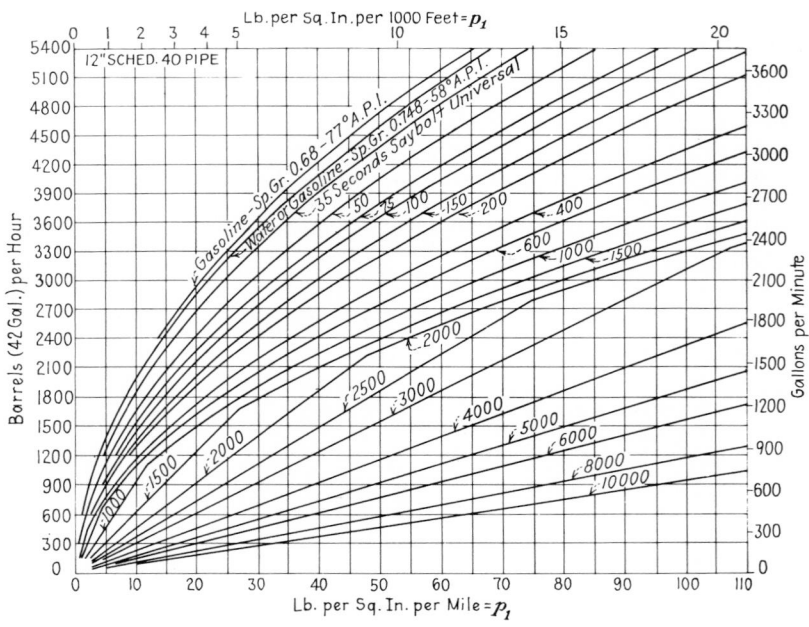

FIG. 54*g*. Pipe friction based on average Saybolt Universal viscosity and specific gravity of fluid in pipeline.   Pressure drop $\Delta p = p_1 \times$ sp gr.

Solution for Flow Temperature of 60 F. From Table 43 the specific gravity corresponding to 20 deg API at 60/60 F is 0.9340. Referring to Fig. 54e, the value of p_1 for 600 bbl/hr and 2,500 SSU is 82.5 psi per mile. The pressure drop is $\Delta P = p_1 \times$ specific gravity $\times$ length in miles $= 82.5 \times 0.9340 \times 10 = 771$ psi.

Solution for Flow Temperature of 100 F. From Fig. 35, the specific gravity of this oil at 100 F is 0.920. From Fig. 37, the viscosity at 100 F is approximately 500 SSU. Referring to Fig. 54e, the value of p_1 for 600 bbl/hr and 500 SSU is 23.75 psi per mile. The pressure drop is $\Delta P = p_1 \times$ specific gravity $\times$ length in miles $= 23.75 \times 0.920 \times 10 = 218$ psi.

NOTE: The solution for a flow temperature of 100 F comes in the smoothed-over region of the chart as explained above, and for this reason is not particularly accurate. A pressure drop of 150 psi was obtained by solution of Eq. (77e).

An *empirical formula* for the flow of oil through pipes may be employed for preliminary calculations where accurate results are not essential. If accurate results are desired, it is recommended that the final calculations be made by use of one of the Eq. (77) together with Fig. 42. The following empirical formula is taken from an early edition of Marks' "Mechanical Engineers' Handbook":

The resistance to the flow of oils lighter than about 30 deg Bé (specific gravity 0.875) is not much different from that of water. A convenient formula used in the oil fields in designing pipe lines for crude oil (42 to 43 deg Bé, specific gravity 0.814 to 0.819) is given below. The quantity discharged increases about 1 per cent for each 3 deg Bé, i.e., lighter oil flows more easily. $B = 1.125 \, d^{2.5} \sqrt{p_\lambda / l}$ where B is expressed in barrels (of 42 U.S. gal) per hour; d is actual inside diameter in inches; p_λ is the total friction loss in pounds per square inch; and l is the length of the line in miles. Friction increases with cold oils in winter and decreases in summer. Deposits of paraffin in the pipes conveying crude oil reduce the effective diameter and increase the friction. Scrapers are, therefore, driven through the pipes periodically to clean away the paraffin. Exceedingly viscous asphalt-base oils having a specific gravity of about 0.97 (14 deg Bé) may be pumped through a long line by using a "rifled" pipe and mixing 10 per cent of water with the oil.

The foregoing empirical formula and that which follows in the succeeding paragraph are approximate in that they do not take into account the variation in friction factor with pipe diameter and velocity and that viscosity is considered only indirectly in connection with density. As has been pointed out, all of these have a bearing on the numerical value of the friction factor.

The following empirical equation for oil flow in pipes is taken from the "Union Engineering Handbook" of the Union Steam Pump Company. It is similar to that quoted above from Marks but is accompanied by a more extensive schedule of constants and by Table 64 for the flow of 38 deg Bé oil.

The friction loss in oil pipelines may be found by the following formula:

$$\Delta P = (C \times B^2)/(10 \times d^5)$$

where ΔP = friction, psi per mile (5,280 ft).

C = a constant from Table 64 which depends on the character of the oil.

B = barrels of oil per hour (42 gal per barrel).

d = inside diameter of pipe, in.

Using 9.00 as a *constant* for 38 deg Bé oil, subtract 0.06 for each degree above 38 deg, and add 0.06 for each degree below 38 deg.

For every 10 deg above 60 F, subtract 1 deg from the Baumé reading, and for every 10 deg below 60 F add 1 deg to the Baumé reading.

Example. Raise 38 deg Bé oil from 60 F to 80 F; then $80 - 60 = 20$ and $20 \div 10 = 2$: and $38 - 2 = 36$ deg Bé, the new Baumé reading for 38 deg Bé oil heated from 60 to 80 F.

The *economics* of oil flow through pipes is of prime importance in cross-country transmission lines where the capital costs of different size lines must be balanced against their respective pumping costs. Another interesting problem arises in

Table 64.   Friction of Oil[1]—38 Deg Bé

(Pounds per square inch in pipes 1 mile long)

Barrels per hour	Diameter of pipe						
	2 in.	3 in.	4 in.	5 in.	6 in.	8 in.	10 in.
10	2.8	0.37	0.088	0.0288	0.0116	0.00274	0.00091
15	6.2	0.83	0.198	0.063	0.026	0.0064	0.00203
20	11.2	1.48	0.352	0.115	0.0462	0.011	0.0036
25	17.6	2.3	0.55	0.18	0.0725	0.0172	0.0056
30	25.3	3.32	0.792	0.26	0.104	0.0247	0.0081
35	34.5	4.52	1.08	0.353	0.142	0.0337	0.0112
40	45.0	5.9	1.43	0.46	0.185	0.044	0.0144
45	57.0	7.49	1.78	0.58	0.235	0.0556	0.0181
50	70.0	9.24	2.2	0.72	0.29	0.0686	0.0225
55	85.0	11.2	2.26	0.87	0.35	0.084	0.0273
60	101.0	13.3	3.16	1.04	0.416	0.099	0.0325
65	118.4	15.6	3.72	1.22	0.49	0.116	0.0381
70	138.0	18.1	4.3	1.41	0.568	0.136	0.0442
75	158.0	20.8	4.95	1.62	0.651	0.155	0.0508
80	180.0	23.6	5.62	1.84	0.74	0.176	0.0578
85	203.0	26.7	6.35	2.08	0.836	0.198	0.0652
90	227.0	30.0	7.11	2.33	0.939	0.224	0.073
95	253.0	33.4	7.92	2.6	1.045	0.248	0.0814
100	280.0	37.0	8.8	2.88	1.16	0.275	0.0902
125	440.0	57.8	13.7	4.5	1.81	0.43	0.142
150	630.0	83.0	19.8	6.5	2.6	0.618	0.203
175	.....	113.0	27.0	8.8	3.54	0.84	0.275
200	.....	148.0	35.2	11.5	4.63	1.1	0.36
225	.....	187.0	42.5	14.6	5.85	1.39	0.455
250	.....	230.0	55.0	18.0	7.22	1.72	0.56
275	.....	280.0	66.5	21.7	8.75	2.08	0.68
300	.....	332.0	79.2	26.0	10.4	2.47	0.81
325	.....	390.0	93.0	30.4	12.2	2.9	0.95
350	.....	453.0	10.2	35.3	14.2	3.37	1.12
375	.....	.....	124.0	40.5	16.3	3.85	1.26
400	.....	.....	141.0	46.0	18.5	4.4	1.44
425	.....	.....	159.0	52.0	20.9	4.96	1.63
450	.....	.....	178.0	58.2	23.5	5.56	1.83
475	.....	.....	198.0	65.0	26.1	6.25	2.04
500	.....	.....	220.0	72.0	28.9	6.88	2.25
525	.....	.....	242.0	79.5	31.9	7.58	2.48
550	.....	.....	266.0	87.0	35.0	8.3	2.72

[1] Based on empirical formula of Union Steam Pump Co. described above.   When ΔP is given and B computed, add 1 per cent to B for every 3 deg above and subtract 1 per cent from B for every 3 deg below 38 deg Bé.   When B is given and ΔP computed subtract 2 per cent from ΔP for every 3 deg above and add 2 per cent for every 3 deg below—or interpolate for 9.00.

designing loading systems for tank ships, cars, or trucks where a three-way balance should be struck between pipe cost, pumping cost, and cost of delay to the transport equipment.   Owing to the relatively high cost of valves, it sometimes is advantageous to select them one or two sizes smaller than the pipeline.[1]

Flow of Gasoline through Pipes.   Gasoline has a relatively low viscosity compared to other liquids (see Fig. 38a).   Relative to water which by definition has a viscosity of 1 centipoise at 68 F, gasoline has a viscosity of 0.45 to 0.85 centipoise depending upon its density.   This fact, together with the decreasing effect which viscosity has on fluid motion in the turbulent range, makes it possible

[1] For a discussion of all these phases of the economic problem, reference may be made to "Selection of the Most Economical Pipe and Valve Size and Rate of Flow in Piping Systems," by S. P. Johnson and F. L. Maker, *Proc. API*, Division of Refining, Mid-year Meeting, May 29, 1940.   Also published in *Refiner and Natural Gasoline Manufacturer*, June, 1940, vol. 19, no 6, pp. 169–180.   *Author's Note:* Where the factor (3.6^3) appears in the formulas of this reference it seems to have been transposed between numerator and denominator and should be reversed, with corresponding corrections in subsequent expressions.   As far as has been determined, these errata do not affect the diagrams which, though ostensibly plotted from these formulas, are understood to have been actually plotted from earlier and somewhat different expressions.

Table 64. *(Continued)*

Barrels per hour	Diameter of pipe				Barrels per hour	Diameter of pipe	
	5 in.	6 in.	8 in.	10 in.		8 in.	10 in.
550	87.0	35.0	8.3	2.72	3,600	356	117
575	95.0	38.4	9.1	2.98	3,800	397	130
600	104.0	41.7	9.9	3.24	4,000	440	144
625	112.0	45.4	10.7	3.5	4,200	...	158
650	122.0	49.0	11.6	3.8	4,400	...	174
675	131.0	53.0	12.5	4.1	4,600	...	192
700	141.0	56.9	13.5	4.42	4,800	...	207
725	152.0	61.0	14.4	4.73	5,000	...	225
750	162.0	65.3	15.5	5.06	5,200	...	243
775	173.0	69.8	16.5	5.41	5,400	...	252
600	184.0	74.3	17.6	5.76	5,600	...	282
825	196.0	79.0	18.7	6.12	5,800	...	303
850	208.0	83.9	19.8	6.5	6,000	...	324
875	221.0	88.8	21.1	6.89	6,200	...	346
900	233.0	94.0	22.3	7.3	6,400	...	368
925	247.0	99.2	23.5	7.7	6,600	...	392
950	260.0	104.5	24.8	8.12	6,800	...	416
975	274.0	111.0	26.2	8.55	7,000	...	441
1,000	288.0	115.8	27.5	9.0			
1,100	348.0	140.0	33.2	10.9			
1,200	415.0	166.5	39.5	13.0			
1,300	486.0	195.5	46.4	15.2			
1,400	.....	227.0	53.8	17.6			
1,500	.....	261.0	61.8	20.3			
1,600	.....	296.0	70.4	23.1			
1,700	.....	335.0	79.3	26.1			
1,800	.....	375.0	89.0	29.2			
1,900	.....	418.0	99.0	32.5			
2,000	.....	463.0	110.0	36.1			
2,200	.....	.....	133.0	43.6			
2,400	.....	.....	158.0	52.0			
2,600	.....	.....	186.0	61.0			
2,800	.....	.....	216.0	70.6			
3,000	.....	.....	247.5	81.1			
3,200	.....	.....	284.0	92.2			
3,400	.....	.....	318.0	104.0			
3,600	.....	.....	356.0	117.0			

Constant C for Different Oils

Degrees, Baumé API	Constants	Degrees, Baumé API	Constants	Degrees, Baumé API	Constants	Degrees, Baumé API	Constants
65	7.38	53	8.10	41	8.82	29	9.54
62	7.56	50	8.28	38	9.00	26	9.72
59	7.74	47	8.46	35	9.18	23	9.90
56	7.92	44	8.64	32	9.36	20	10.08

to solve gasoline flow problems with reasonable accuracy by empirical formulas. A modification of the Williams-Hazen hydraulic formula has been employed extensively for this purpose.[1]   Modified to include the specific gravity of the gasoline and expressing the friction loss ΔP as pounds per square inch per 1,000 ft of line B in barrels (42 gal) per hour, the Williams-Hazen formula may be written:

$$\Delta P = \frac{2,324B^{1.85}S}{C^{1.85}d^{4.87}}$$

[1] See (a) "A Discussion of Flow Formulas Used for Design of Gasoline Pipe Lines," by W. E. Rea, *Oil Gas J.*, Jan. 23, 1941, pp. 38–47.

(b) "Application of Pipe-line Efficiency Concept to Gasoline Pipe-line Calculations," by Benjamin Miller, *Oil Gas J.*, vol. 41 (35), pp. 122–123, Jan. 7, 1943.

(c) "Use of Scrapers Maintains High C-Factors on Sohio-operated Lines," by R. L. Harris and E. F. Morrill, *Oil Gas J.*, vol. 42 (27), pp. 244–248, Nov. 11, 1943.

In this form the Williams-Hazen equation contains all the pertinent factors except the viscosity of the gasoline which is assumed to be constant with minor variations neglected.

Numerical values of the Williams-Hazen coefficient C usually obtained in clean gasoline lines range from 125 to 150. Good average values for 4, 6, and 8 in. nominal diameter lines are said to be 140, 135, and 130, respectively, which is contrary to most published data which show an increase in C with pipe diameter.

Equations (77) are used extensively in designing gasoline lines, but their successful application depends on the selection of a suitable friction factor f. This is accomplished readily by knowledge of viscosity and density of the gasoline.

Flow of Mixtures of Oil and Gas. Two-phase flow involving the simultaneous passage of a gas or vapor and liquid in a pipe is encountered in tube stills, refrigeration systems, condensate return lines, etc. A method of computing the pressure drop in a pipe in which known quantities of vapor and oil or other liquid are flowing simultaneously has been presented by Martinelli[1] et al. In an example in which 0.50 lb of air and 0.30 lb of oil per second were flowing in a 2-in. pipe, it was shown that the resulting pressure drop for the two-phase flow was about fifteen times the pressure drop of either the air or oil flowing alone. In this example the air flow was turbulent, while the oil flow was viscous. Similar results were found for an air and water combination.

Useful Information Regarding Oil. The following useful information regarding oil is taken from the "Union Engineering Handbook":

1 bbl = 42 U.S. gal.
1 bbl/hr = 0.7 U.S. gpm.
Barrels/hr × 0.7 = gpm.
Gallons/min divided by 0.7 = bbl/hr.
Barrels/hr × 24 = bbl/day.
1 bbl/day = 0.0292 gpm.
Barrels/day × 0.0292 = gpm.
Gallons/min divided by 0.0292 = bbl/day.
Number of bbl in pipe 1 mile long equals inside diameter of pipe in inches squared × 5⅛.
Velocity in fps = 0.0119 × bbl/day divided by diameter of pipe in inches squared; or velocity equals 0.2856 × bbl/hr divided by the diameter of pipe in inches squared, or velocity equals 0.408 × gpm divided by the diameter of pipe in inches squared.
Net horsepower = the theoretical horsepower necessary to do the work.
Net horsepower = barrels/day × pressure × 0.000017.
Net horsepower = barrels/hour × pressure × 0.000408.
Net horsepower = gallons/minute × pressure × 0.000583.

The characteristics of tested oils and the characteristics of averaged oils given in Tables 45 and 46 were also taken from the same source.

[1] "Isothermal Pressure Drop for Two-phase Two-component Flow in a Horizontal Pipe," by R. C. Martinelli, L. M. K. Boelter, T. H. M. Taylor, E. G. Thomsen, and E. H. Morrin, *Trans. ASME*, vol. 66, no. 2, 1944, pp. 139–151. Contains 4 references and a bibliography of 15 related articles.

4

EXPANSION AND FLEXIBILITY

John E. Brock*

Engineering materials experience changes in dimensions as a result of application of stress, change in temperature, passage of time, change in internal configuration, changes in dissolved moisture content, and possibly a few other causes. Of these the first two are by far the most important as far as piping is concerned. Piping should be designed so that time-dependent inelastic effects such as creep and relaxation are of negligible importance (although relaxation of bolting may be important in high-temperature flanged joints). Creep will be discussed briefly later in this chapter. Materials selected for piping service should be stabilized and should be used at temperatures for which there is no significant change in internal structure. Moisture content is obviously not of significance in metallic piping. Thus, essentially, dimensional changes in piping material depend upon changes in temperature and stress.

The complex problem of determining the stress distribution in a solid body subjected to a change in temperature is treated in books by Timoshenko and Goodier,[245]† Gatewood,[116] and Boley and Wiener.[45] Generally speaking, these methods and results are applicable to stress determination in individual piping components, such as valve bodies, straight cylindrical runs of pipe, etc., rather than to an entire piping system. Advances in nuclear engineering have produced an

* Professor of Mechanical Engineering, U.S. Naval Post-graduate School, Monterey, Calif.

This chapter was contributed in June, 1963, and, except for certain changes made by the editor, represents the professional position of the author at that time.

† Superscript numbers refer to bibliographical references at the end of this chapter.

acceleration in technically applicable studies of this type. Nuclear installations require more than customary assurance against failure or malfunction, so that it is necessary to perform analyses which would not be called for in other technologies. Also, the fact that certain fluids employed in nuclear technology, particularly the liquid metals, involve high coefficients of heat transfer accounts for the emphasis which has been given in recent years to transient thermal-stress problems and thermal shock. Methods and techniques of most direct applicability are given in Ref. 255.

The peculiar geometry of piping systems as a whole, i.e., their elongated, filamental character, implies that special methods may be employed successfully for their analysis. Indeed, the similarity between a piping system and a rigid framed steel structure is evident, and the methods which have been developed for the analysis of framed structures can be and have been applied to the analysis of piping systems.

The history of these and similar applications is discussed in Refs. 150 and 182. Since a knowledge of the history of various methods of piping-flexibility analysis is by no means necessary to the understanding of their modern successors, no further remarks will be made concerning history.

MECHANICAL PROPERTIES OF PIPING MATERIALS

Although the processes of piercing, extrusion, cupping and drawing, rolling, etc., that are used to produce pipe might be expected to lead to anisotropy, there is no evidence in the literature that the consequences have been studied or regarded as of particular significance. Thus, confining the theory to small strains and presuming that the material is essentially isotropic, the principal elastic constants of interest are the Young's modulus E and the Poisson's ratio μ. Of equal importance is the coefficient of thermal expansion, usually denoted by the symbol α.

Although the usual developments of the theory of thermal stresses as given in Boley and Wiener,[45] Gatewood,[116] Timoshenko and Goodier,[245] and others, regard these parameters as independent of temperature, they are in fact dependent upon the temperature in a way which must be determined by experiment. Since the literature contains inconsistent data on the variation of these parameters with change in temperature, a Task Subgroup was appointed in 1951 to evaluate existing data. The report of the Task Group was made by Michel,[192] and the recommended data were incorporated into the 1955 edition of the Code for Pressure Piping, Ref. 1, as Tables 22, 23, and 24. These data included information about Young's modulus E and about the modulus of rigidity G, while the body of the code asserted that Poisson's ratio could be taken as 0.3 for all materials at all temperatures. There was some slight inconsistency between these data and the customary theoretical relationship $G = E/2(1 + \mu)$, and in view of the difficulty of obtaining precise consistent experimental results and the relatively slight influence upon final results that arise from variations in Poisson's ratio, the present policy of the Mechanical Design Committee, which has Code cognizance over expansion and flexibility, is to omit values of G from the Code and give $\mu = 0.3$ uniformly, with permission to use better values if they are available.

These data, extended also into regions of low temperature, together with values of linear thermal expansion, are presented in Tables 1 and 2, which are reproduced from ASA B31.3-1962, Petroleum Refinery Piping; cf. Ref. 1. See also Refs. 187, 192, and 114. The values listed for α are based on tests made at stresses very near zero.

For other materials or for temperatures outside the ranges indicated in these tabulations, information must be obtained by communication with the manufacturer of the material or by individual tests. However, in many cases valid

Table I. Moduli of Elasticity† (million psi)‡

Material	\-325	\-200	\-100	70	100	200	300	400	500	600	700	800	900	1000	1100	1200	1300	1400
Carbon steels with carbon content 0.30% or less	30.0	29.5	29.0	27.9	…	27.7	27.4	27.0	26.4	25.7	24.8	23.4	18.5	15.4	13.0			
Carbon steels with carbon content above 0.30%	31.0	30.6	30.4	29.9	…	29.5	29.0	28.3	27.4	26.7	25.4	23.8	21.5	18.8	15.0	11.2		
Carbon-moly steels, low cromoly steels through 3% Cr	31.0	30.6	30.4	29.9	…	29.5	29.0	28.6	28.0	27.4	26.6	25.7	24.5	23.0	20.4	15.6		
Intermediate Cr-moly steels (5%–9% Cr), austenitic stainless steel	29.4	28.5	28.1	27.4	…	27.1	26.8	26.4	26.0	25.4	24.9	24.2	23.5	22.8	21.9	20.8	19.5	18.1
Straight chromium stainless steel (12 Cr, 17 Cr, 27 Cr)	30.8	30.3	29.8	29.2	…	28.7	28.3	27.7	27.0	26.0	24.8	23.1	21.1	18.6	15.6	12.2		
Wrought iron	30.6	30.2	30.0	29.5	…	28.6	28.2	27.7	27.0	26.5	25.8	23.0						
Gray cast iron	…	…	…	13.4		13.2	12.9	12.6	12.2	11.7	11.0	10.2						
Monel (67 Ni–30 Cu, 66 Ni–29 Cu, Al)	26.8	26.6	26.4	26.0	26.0	26.0	25.8	25.6	25.4	24.7	23.1	21.0	18.6	16.0	14.3	13.0		
Copper-nickel (80–20, 70–30)	20.5	20.0	19.5	18.9	18.8	18.4	18.0	17.6	17.2	16.7	16.2	15.3						
Aluminum	11.3	11.1	10.9	10.6	10.6	10.4	10.2	9.5	8.5									
Copper (99.98% Cu)	17.0	16.7	16.5	16.0	15.8	15.6	15.4	15.1	14.7	14.2	13.7							
Commercial brass (66 Cu, 34 Zn)	15.0	14.7	14.5	14.0	13.9	13.7	13.5	13.0	12.7	12.2	11.8							
Leaded tin bronze (88 Cu, 6 Sn, 1.5 Pb, 4.5 Zn)	14.2	13.8	13.5	13.0	12.9	12.7	12.4	12.0	11.7	11.3	10.9							

Temperature, F

† These data are for information, and it is not to be implied that materials are suitable for all the temperature ranges shown. Poisson's ratio may be taken as 0.3 at all temperatures for all metals; however, more accurate and authoritative data may be used if available.

‡ Extracted from Petroleum Refinery Piping Code (ASA B31.3-1962), a Section of the American Standard Code for Pressure Piping, with the permission of the publisher, The American Society of Mechanical Engineers, 345 East 47th St., New York 17, N.Y.

Table 2. Unit Linear Thermal Expansion† (in./100 ft)‡

Material	−325	−150	−50	70	Temperature range 70 F to												
					200	300	400	500	600	700	800	900	1000	1100	1200	1300	1400
Carbon steel; carbon-moly steel; low-chrome steels (through 3% Cr)	−2.37	−1.45	−0.84	0	0.99	1.82	2.70	3.62	4.60	5.63	6.70	7.81	8.89	10.04	11.10	12.22	13.34
Intermediate alloy steels (5 Cr Mo through 9 Cr Mo)	−2.22	−1.37	−0.79	0	0.94	1.71	2.50	3.35	4.24	5.14	6.10	7.07	8.06	9.05	10.00	11.06	12.05
Austenitic stainless steels	−3.85	−2.27	...	0	1.46	2.61	3.80	5.01	6.24	7.50	8.80	10.12	11.48	12.84	14.20	15.56	16.92
Straight chromium stainless steels; (12 Cr, 17 Cr, and 27 Cr)	−2.04	−1.24	−0.72	0	0.86	1.56	2.30	3.08	3.90	4.73	5.60	6.49	7.40	8.31	9.20	10.11	11.01
25 Cr–20 Ni	−3.00	−1.81	−0.98	0	1.21	2.18	3.20	4.24	5.33	6.44	7.60	8.78	9.95	11.12	12.31	13.46	14.65
Monel (67 Ni–30 Cu)	−2.62	−1.79	...	0	1.22	2.21	3.25	4.33	5.46	6.64	7.85	9.12	10.42	11.77	13.15	14.58	16.02
Monel (66 Ni–29 Cu Al)	−2.53	−1.70	−0.98	0	1.17	2.12	3.13	4.17	5.28	6.43	7.62	8.86	10.16	11.50	13.00	14.32	15.78
Aluminum	−4.68	−2.88	−1.67	0	2.00	3.66	5.39	7.17	9.03								
Gray cast iron	...	...	...	0	0.90	1.64	2.42	3.24	4.11	5.03	5.98	6.97	8.02				
Bronze	−3.98	−2.31	−1.32	0	1.56	2.79	4.05	5.33	6.64	7.95	9.30	10.68	12.05	13.47	14.92		
Brass	−3.88	−2.24	−1.29	0	1.52	2.76	4.05	5.40	6.80	8.26	9.78	11.35	12.98	14.65	16.39		
Wrought iron	−2.70	−1.67	−0.96	0	1.14	2.06	3.01	3.99	5.01	6.06	7.12	8.26	9.36				
Copper-nickel (70–30)	−3.15	−1.95	−1.13	0	1.33	2.40	3.52										

† These data are for information, and it is not to be implied that materials are suitable for all the temperature ranges shown.
‡ Extracted from Petroleum Refinery Piping Code (ASA B31.3-1962), a Section of the American Standard Code for Pressure Piping, with the permission of the publisher, The American Society of Mechanical Engineers, 345 East 47th St., New York 17, N.Y.

conclusions may be reached by use of data which are only approximately accurate.

Physical theory does not explain satisfactorily the variations with temperature of the physical properties E, G, μ, σ, and α. Truesdell[248] and Brilbuin[56] discuss this matter. However, essentially, only experimental studies offer information which is immediately useful in piping-flexibility studies.

The dependence of strain on both temperature and stress creates only slight complication. For the simple case of uniaxial stress, by superposition methods, it can be shown that as long as elastic action takes place,

$$e = -\sigma_0/E_0 + \int_{T_0}^{T_F} \sigma(0,T)\,dT + \sigma_F/E_F$$

in which e = axial strain

σ_0 = axial stress at initial temperature T_0
σ_F = axial stress at final temperature T_F
E_0 = Young's modulus at initial temperature
E_F = Young's modulus at final temperature
$\sigma(0,T)\,dT$ = the strain, at zero value of stress, in a differential temperature increment, dT

The values of linear expansion of Table 2 are, in reality, the values of strain at the zero stress condition. Neglect of the variation in linear expansion with stress results in an error of the order of 2.

INTRODUCTORY EXAMPLE AND DISCUSSION OF METHODS

A structure which is subjected to a change in temperature will change its physical dimensions if it is free to do so and, if it is not, will be placed in a condition of stress and will exert reactive forces and moments on the equipment at its ends. The basic problem of piping-flexibility analysis is to determine the magnitude of these stresses internal to the piping system itself and the reactions it exerts upon the terminal equipment and decide whether or not they are tolerable.

Before undertaking a discussion of the problem in the general case, consider the simplest nontrivial case, that of an L-shaped configuration in which the change of direction is by means of a miter reinforced in such a way that it represents no locally flexible component. This assumption will greatly simplify and facilitate the analysis immediately to follow; the properties of pipe bends and welding elbows are more complicated and will be treated later.

Figure 1a shows such a configuration, the lengths of the legs being a and b. The final configuration, shown in Fig. 1d, may be arrived at through the following sequence. The system is cut apart at point B, and the temperature is increased from the initial temperature to the final temperature, the legs increasing in length by amounts e_a and e_b, where e, the unit thermal strain, may be obtained from Table 2. For example, if the material is carbon steel and the temperature increases from 70 to 500 F, the value of e is 3.62 in. per 100 ft, so that if length a is 40 ft 9 in., say, its increase in length is 1.475 in. Next, in Fig. 1c is shown the effect of movements of the termini of the piping system, such as would result from the expansion of the terminal equipment itself. Suppose that these movements are δx_A, δy_A, δx_C, and δy_C, all positive as shown.

Finally, forces F_x and F_y and a moment M_B are applied to the free ends of these cantilever beams in such a way as to restore the continuity which was lost when the system was cut apart at point B. It may be seen that this system of forces must be such as to cause tip rotations θ as shown and tip deflections $\Delta x = ea + \delta x_A - \delta x_C$ and $\Delta y = eb + \delta y_A - \delta y_C$ (see Fig. 1e and f).

Using elementary beam theory, the following four equations are obtained:

$$EI\,\Delta y = F_y a^3/3 - M_B a^2/2$$
$$EI\,\Delta x = F_x b^3/3 - M_B b^2/2$$
$$EI\,\theta = M_B a - F_y a^2/2$$
$$EI\,\theta = F_x b^2/2 - M_B b$$

These equations imply that flexural rigidity EI is constant, that axial deformations and cross shearing deformations are small compared with bending deformations in

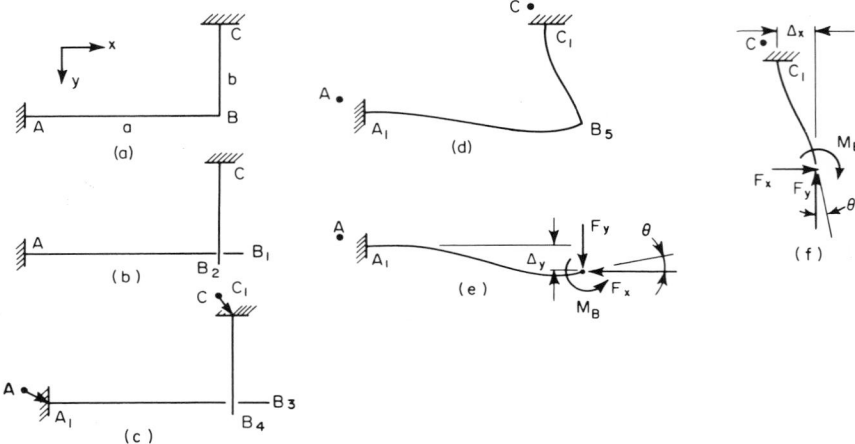

FIG. 1. Introductory example, showing sequence of operations leading to problem in structural theory.

long, slender structures, and that the deformations are themselves small compared with the lengths of the legs. They may be solved for the unknown quantities F_x, F_y, and M_B as follows:

$$F_x = \frac{3EI}{ab(a+b)}\frac{(4ab + a^2)\,\Delta x + 3b^2\,\Delta y}{b^2}$$

$$F_y = \frac{3EI}{ab(a+b)}\frac{3a^2\,\Delta x + (4ab + b^2)\,\Delta y}{a^2}$$

$$M_B = \frac{6EI}{ab(a+b)}(a\,\Delta x + b\,\Delta y)$$

and knowing these, the terminal moments may be found by the expressions

$$M_A = aF_y - M_B$$
$$M_c = bF_x - M_B$$

Several things should be noted from this simple example. First, it is evident that the results are rather complicated, even though this is the simplest possible case. If a bend were used rather than the right-angle miter postulated, the results would have been very much more complicated, and it should be clear that any attempt to find formulas, in literal terms as above, for more complicated cases is out of the

question; if such formulas could be obtained by unremitting labor, they would be far too complicated to employ successfully.

Second, it may be observed that the first part of the analysis was devoted to obtaining a system of equations representing the relations between deflections some of which (Δx and Δy) were known and one of which (θ) was not known and the unknown forces and moments F_x, F_y, and M_B. The second part of the analysis involved solving these equations simultaneously for the unknowns; in this case θ was not determined, since it is not of particular interest even though it is involved in the theory. Finally, by use of the equations of statics, resisting moments were found at other points of the structure. This sequence of operations is involved in every method of flexibility analysis that attempts to make reasonably precise determinations of piping flexibility, except for some chart or graph solutions in which, essentially, answers may be picked by comparison or interpolation from a body of presolved problems.

Third, it may be observed that the solution presented some difficulties even though certain idealizations had been made. By neglecting the product of axial force by lateral displacement (i.e., neglecting the "beam-column" effect), by neglecting axial and cross shearing deflections, by regarding all deflections as "small," and, in particular, by neglecting the difference between arc and chord of the bent pipe, the equations were simplified. In ordinary structural analysis these assumptions are almost always made, but there are some cases, of practical importance in piping-stress analysis, where it may be necessary to look more closely at these assumptions and possibly to make allowances for the inaccuracy they introduce. In the general theoretical development which will follow shortly, axial and cross shearing deformations (as well as torsional deformation which did not enter into this two-dimensional problem) will be taken into account. In the "Compliances of Piping Elements" section of this chapter, it will be indicated how the "beam-column" effect may be taken into account, at least approximately, and some remarks will also be made about the effect of slopes large enough to result in significant difference between arc and chord.

The solution given above to a quite elementary problem in structural analysis was based upon the properties of cantilever beams and is probably the simplest way to approach the problem for a person who has not made a special study of the methods of rigid frame analysis. However, there is a large number of other methods of doing essentially the same thing. A partial list would include these procedures:

1. Area-moment method
2. Elastic center method
3. Conjugate beam (or conjugate frame) method
4. Slope-deflection method
5. Moment-distribution method
6. Least work method (and various other strain energy methods)
7. Column analogy method

and possibly others. The methods named above have proved to be very useful in the analysis of engineering structures, and there are advantages and disadvantages to each. Most of these methods were employed originally for the analysis of plane structures and require some modification or augmenting in order to be useful for three-dimensional structures in which torsion is of importance. The various energy methods are inherently capable of taking torsional, axial, and cross shearing deformations into account without any modifications. However, the other methods have also been used successfully, and each is capable of equal accuracy and each deserves the name of "exact analysis" in the sense that most careful structural

analyses are exact.   Those who wish to learn more about these methods may refer to the bibliography at the end of the present chapter or consult the excellent bibliographies to be found in Refs. 131, 150, and 182.   Reference 150 also contains an interesting history of piping-flexibility analysis.   It would be well to cite the principal sources for three of the most widely used methods.   First, the area-moment method as augmented and improved for use in piping-flexibility analyses by Todd, Gould, Crocker, and McCutchan and retitled the "graphoanalytical method" is most fully described in earlier editions of this "Piping Handbook."   Second, the elastic center method as developed by Spielvogel became the basis for his book[241] and is generally referred to as Spielvogel's method.   Finally, the method of least work is the point of departure for the method devised largely by Wallstrom and described in the M. W. Kellogg book[150] in which it is called the "general analytical method;" as this method is presented, its power depends to a large extent on a very systematic use of preintegrated forms, cyclic permutation of coordinates, and other modern notational devices.

The most influential developments in piping-flexibility analysis in the past ten years have been the simultaneous advent of matrix methods of structural analysis and the high-speed digital computer.   The former provide analytical methods for describing and dealing with even the most complicated structural problems and provide a compact notation and strong computational motivation, that is, a built-in way of outlining the entire sequence of a complicated solution.   However, complicated problems require prodigious computational effort, no matter how they are formulated, and if it were not for the computational power of the modern computer, the potential benefits of matrix methods would continue to lie out of reach.

Matrix methods in structural analysis have been developed through three main avenues: analysis of civil engineering type of structures, aircraft structural analysis, and piping-flexibility analysis.

A formidable amount of work has been done in these areas by many competent authorities.   At the risk of having overlooked some contribution of significant importance, a partial list of bibliography references follows: Refs. 10, 17, 20, 21, 23, 27, 41, 44, 46, 50 to 52, 57, 58, 61, 63, 66, 67, 70, 71, 75, 81 to 83, 89, 96, 103, 111, 118, 119, 122, 129 to 131, 133, 135, 151, 155 to 157, 161 to 165, 167, 170, 178, 180, 186, 189, 195, 197, 199 to 201, 207, 210, 212 to 214, 224 to 226, 238 to 240, 242, 254, 260, 261.

The application of matrix methods to piping-flexibility analysis is made possible by the labors of many of the contributors referred to above.   Explanation of these methods and their adaptation to flexibility analysis follows in the succeeding section.

ELEMENTARY PROPERTIES OF MATRICES

Before undertaking an exposition of matrix applications to the analysis of piping flexibility, it will be useful to review very briefly the applicable properties of matrices which will be employed.   This is not intended to be more than a review and reminder.   The reader who does not already have some familiarity with matrices and matrix operations should refer to a more extended treatment.   Most modern books on elementary college mathematics have a chapter on matrices.   One standard treatment that can be recommended highly is that of Frazer, Duncan, and Collar.[111]

The examples which have been chosen to illustrate the definitions and properties given below have purposely been made quite simple so as not to obscure the essential ideas by a mass of computation.

A *matrix* is a rectangular array of numbers, e.g.,

$$A = \begin{bmatrix} A_{11} & A_{12} & A_{13} & A_{14} \\ A_{21} & A_{22} & A_{23} & A_{24} \\ A_{31} & A_{32} & A_{33} & A_{34} \end{bmatrix}$$

This matrix A has three *rows* and four *columns* and is said to be of *order* 3 by 4 (written 3×4) or is said to be a 3 by 4 matrix.   More generally, a matrix is an m by n matrix if it has m rows and n columns.   An *element* A_{ij} is distinguished by subscripts which indicate the row and column, respectively, in which it is found. Elements of a matrix will be enclosed in brackets [ ] in this treatment; other notations which are frequently used are parentheses ( ) and double vertical lines ‖   ‖.   However, since an n by 1 matrix or *column matrix* such as

$$A = \begin{bmatrix} A_{11} \\ A_{21} \\ A_{31} \\ A_{41} \\ A_{51} \end{bmatrix}$$

occupies so much space on a page when written as above, a special notation will be used, viz:

$$A = \{A_1, A_2, A_3, A_4, A_5\}$$

the braces { } indicating a column matrix.   Note also that the superfluous second subscript has been omitted.   A column matrix is frequently called a *vector*.
   A *diagonal matrix*, such as

$$A = \begin{bmatrix} \alpha & 0 & 0 & 0 \\ 0 & \beta & 0 & 0 \\ 0 & 0 & \gamma & 0 \\ 0 & 0 & 0 & \delta \end{bmatrix}$$

in which all elements not situated on the *principal diagonal* are zeros, will be written as

$$A = \text{diag } \{\alpha, \beta, \gamma, \delta\}$$

Addition and subtraction are defined for two matrices of the same order.   If matrices A and B, of the same order, are given, their sum S and difference D are written

$$S = A + B$$
$$D = A - B$$

where the ij element of S is the sum of the ij elements of A and B, viz., $S_{ij} = A_{ij} + B_{ij}$, and similarly for the difference.

Example:

$$A = \begin{bmatrix} 3 & 1 & 5 \\ 6 & 2 & -4 \end{bmatrix} \qquad B = \begin{bmatrix} 2 & -1 & -3 \\ 3 & 1 & 3 \end{bmatrix}$$

$$S = \begin{bmatrix} 5 & 0 & 2 \\ 9 & 3 & -1 \end{bmatrix} \qquad D = \begin{bmatrix} 1 & 2 & 8 \\ 3 & 1 & -7 \end{bmatrix}$$

Scalar multiplication of a matrix is defined as follows: If A is a matrix and α is a scalar, i.e., an ordinary number rather than a matrix, the product αA is a matrix of the same order as A each of whose elements is α times the corresponding element of A.

Example:

$$\alpha = 1.7 \qquad A = \begin{bmatrix} 3 & 1 & 5 \\ 6 & 2 & -4 \end{bmatrix} \qquad \alpha A = \begin{bmatrix} 5.1 & 1.7 & 8.5 \\ 10.2 & 3.4 & -6.8 \end{bmatrix}$$

Note that this includes the case where α is an integer; thus, $A + A + A + A = 4A$.

Matrix multiplication, i.e., multiplication of two matrices, is a fairly involved process a full understanding of which is essential to an understanding of the structural applications which will follow. If A is an $m \times n$ matrix and B is an $n \times r$ matrix, they are said to be *conformable* in the order AB and a matrix product $C = AB$ is defined such that the ij element of C is obtained as a sum of products according to the formula

$$C_{ij} = \sum_{k=1}^{n} A_{ik} B_{kj}$$

The product C is a matrix of order $m \times r$. Note that multiplication is indicated by writing the factors in juxtaposition just as in scalar algebra.

Example:

$$A = \begin{bmatrix} 2 & -1 & 0 \\ 4 & 3 & -2 \end{bmatrix} \qquad B = \begin{bmatrix} 6 & -3 & -4 & -3 \\ 2 & 1 & 0 & 4 \\ -5 & -2 & 3 & 6 \end{bmatrix}$$

$$C = AB = \begin{bmatrix} 2 & -1 & 0 \\ 4 & 3 & -2 \end{bmatrix} \begin{bmatrix} 6 & -3 & -4 & -3 \\ 2 & 1 & 0 & 4 \\ -5 & -2 & 3 & 6 \end{bmatrix} = \begin{bmatrix} 10 & -7 & -8 & -10 \\ 40 & -5 & -22 & -12 \end{bmatrix}$$

Note, for example, that $C_{23} = -22$ is obtained as

$$A_{21}B_{13} + A_{22}B_{23} + A_{23}B_{33} = (4)(-4) + (3)(0) + (-2)(3) = -22$$

In this example, A is of order 2×3, B is of order 3×4, and $C = AB$ is of order 2×4.

In the product AB, A is the *prefactor* and is said to *premultiply* B while B is the *postfactor* and *postmultiplies* A. Note that, in this example, matrices A and B are not conformable in the order BA so that no matrix product BA is defined. Even though two matrices may be conformable in both orders, the two products AB and BA, may not be the same.

Example: If

$$A = \begin{bmatrix} 2 & 1 \\ 0 & -3 \\ 6 & 2 \end{bmatrix} \qquad \text{and} \qquad B = \begin{bmatrix} 5 & 2 & -3 \\ 3 & 4 & 1 \end{bmatrix}$$

then

$$AB = \begin{bmatrix} 13 & 8 & -5 \\ -9 & -12 & -3 \\ 36 & 20 & -16 \end{bmatrix} \qquad \text{and} \qquad BA = \begin{bmatrix} -8 & -7 \\ 12 & -7 \end{bmatrix}$$

Thus, generally, matrix products are *not commutative*, and it is necessary to be careful not to alter the order of matrix factors inadvertently in matrix algebraic operations. In the example immediately above, changing the order of the factor gives two different products which have different orders, 3×3 and 2×2, but even

in multiplying two square matrices of the same order $m \times m$ which evidently are conformable in both orders and yield products of order $m \times m$, the two products may be and usually are different from each other.

Even though the commutative law of ordinary algebra does not hold for matrices (i.e., for matrices, $AB \neq BA$, generally), the other fundamental laws do hold, viz.:

$$A(BC) = (AB)C$$
$$A + (B + C) = (A + B) + C$$
$$A(B + C) = AB + AC$$

Division of a matrix by a scalar is included in the discussion of multiplication by a scalar, for if α is a scalar, the quotient A/α is the same as the product $(1/\alpha)A$. Division of one matrix by another is simply not a defined operation. However, an analogous operation, multiplication by the reciprocal or inverse matrix, will be discussed later.

Certain special matrices are frequently encountered, and it is useful to discuss their properties beforehand. Consider products of the *diagonal* matrix $A = \text{diag } \{\alpha,\beta,\gamma\}$ with the square matrix

$$B = \begin{bmatrix} a & b & c \\ d & e & f \\ g & h & i \end{bmatrix}$$

The two possible products are

$$AB = \begin{bmatrix} \alpha a & \alpha b & \alpha c \\ \beta d & \beta e & \beta f \\ \gamma g & \gamma h & \gamma i \end{bmatrix} \quad \text{and} \quad BA = \begin{bmatrix} \alpha a & \beta b & \gamma c \\ \alpha d & \beta e & \gamma f \\ \alpha g & \beta h & \gamma i \end{bmatrix}$$

Note how the rows of B are multiplied, respectively, by the elements of the diagonal matrix A when A is the prefactor. The columns are similarly multiplied when A is the postfactor. If all the elements of a diagonal matrix are equal, the matrix is called a *scalar* matrix. Thus, in the present example if $A = \text{diag } \{\alpha,\alpha,\alpha\}$, then $AB = BA = \alpha B$.

In particular, if the nonzero elements of the diagonal matrix A are all unity, then A is called a *unit matrix*, written I, viz.:

$$I = \begin{bmatrix} 1 & 0 & 0 \\ 0 & 1 & 0 \\ 0 & 0 & 1 \end{bmatrix}$$

Sometimes a subscript is used to indicate the order of a unit matrix; thus, the unit matrix exhibited above might be denoted as I_3. Usually, the order may be inferred from the context in which the symbol appears. For any matrix A, it is true that

$$AI = IA = A$$

it being understood that the I's in this equation are of suitable order to be conformable with A. Thus, for example, if A is of order 3×4, then the first I is I_3 and the second I is I_4.

Any matrix all of whose elements are zero is called a *null matrix*. It is designated by the symbol 0, and its order, if not apparent from context, may be indicated by subscripts; thus

$$0_{23} = \begin{bmatrix} 0 & 0 & 0 \\ 0 & 0 & 0 \end{bmatrix}$$

A null matrix may result from the multiplication of two non-null matrices; e.g.,

$$\begin{bmatrix} 1 & 0 & 3 \\ 2 & 1 & 12 \end{bmatrix} \begin{bmatrix} 6 & -3 & 9 \\ 12 & -6 & 18 \\ -2 & 1 & -3 \end{bmatrix} = \begin{bmatrix} 0 & 0 & 0 \\ 0 & 0 & 0 \end{bmatrix}$$

whereas, in ordinary arithmetic, if a product is zero, at least one of the factors must be zero.

Transposition of a matrix is an important operation. If A is a matrix, its *transposed matrix* (most frequently abbreviated to its *transposed*) is formed by interchanging rows and columns of A and is designated, in this exposition, by a superior tilde, thus, $\tilde{A}$. Another widely used notation, but one which will not be used here, is to indicate the transposed by a superscript T, thus, A^T, or by a subscript T, thus, A_T. Note that $\tilde{A}_{ij} = A_{ji}$.

Example:

$$A = \begin{bmatrix} 1 & 2 & 3 \\ 4 & 5 & 6 \end{bmatrix} \qquad \tilde{A} = \begin{bmatrix} 1 & 4 \\ 2 & 5 \\ 3 & 6 \end{bmatrix} \qquad \tilde{A}_{21} = A_{12} = 2$$

If matrices A and B are conformable in order AB, then $\tilde{A}$ and $\tilde{B}$ are conformable in order $\tilde{B}\tilde{A}$, and indeed, if $AB = C$, then $\tilde{B}\tilde{A} = \tilde{C}$. The proof of this and other similar elementary properties is omitted here but may be found in any more extensive treatment of matrices. For a string of factors a similar relationship is true, the transposed of a product being the product of the transposed of the individual factors, taken in reverse order: if $X = ABC \ldots G$, then $\tilde{X} = \tilde{G}\tilde{F}\tilde{E} \ldots \tilde{A}$.

The terms *symmetric* and *skew* are also frequently encountered. The matrix A is symmetric if $\tilde{A} = A$. The matrix A is skew if $\tilde{A} = -A$. Obviously these terms can be applied only to square matrices.

If A is a square matrix, its *determinant*, written $|A|$, is the determinant formed by the elements of A. It should be recalled that a determinant may be evaluated, according to rules which are given in most books on elementary algebra, and thus may be expressed as a single scalar number.

Example:

$$\text{if } A = \begin{bmatrix} 9 & 6 & 2 \\ 7 & 1 & 3 \\ 4 & 5 & 8 \end{bmatrix} \text{ then } |A| = \begin{vmatrix} 9 & 6 & 2 \\ 7 & 1 & 3 \\ 4 & 5 & 8 \end{vmatrix} = -265$$

If the determinant of A is zero, A is said to be singular.

Corresponding to the element A_{ij} of $|A|$ is its *minor*, obtained by striking out of $|A|$ the ith row and the jth column. This minor is itself a determinant of one lower order than that of $|A|$ and has a certain numerical value, denoted here by n_{ij}. The number $(-1)^{i+j}n_{ij}$ is called the *cofactor* of A_{ij} in $|A|$. The matrix B whose element B_{ij} is $(-1)^{i+j}n_{ji}$ (note the order of the subscripts) is called the adjoint matrix of A (or more briefly, the adjoint of A) and is written Adj A. Thus for the matrix A immediately above, the adjoint is

$$\text{Adj } A = \begin{bmatrix} -7 & -38 & 16 \\ -44 & 64 & -15 \\ 31 & -21 & -33 \end{bmatrix}$$

Multiplying A by its adjoint in either order gives an interesting product, viz.:

$$A (\text{Adj } A) = (\text{Adj } A)A = \begin{bmatrix} -265 & 0 & 0 \\ 0 & -265 & 0 \\ 0 & 0 & -265 \end{bmatrix} = |A|I$$

This is generally true as a consequence of the laws governing the expansion of a determinant in terms of its cofactors, but the proof will not be given here.

If A is nonsingular (i.e., if $|A| \neq 0$), the *inverse* (or *reciprocal*) of A, written A^{-1}, is defined as

$$A^{-1} = (\text{Adj } A)/|A|$$

and has the property that

$$AA^{-1} = A^{-1}A = I$$

If $X = ABC \ldots EFG$ and none of the factors is singular, then

$$X^{-1} = G^{-1}F^{-1}E^{-1} \ldots C^{-1}B^{-1}A^{-1},$$

a result which is similar to one given above for the operation of transposition.

In matrix arithmetic, the operation corresponding to ordinary division consists of premultiplying or postmultiplying (as the situation may require) by an inverse matrix. The calculation of inverse matrices is one of the most difficult arithmetic tasks involved in dealing with matrices. Obviously, there are a great many elementary arithmetic operations to be performed in calculating the inverse of a large matrix. Many methods have been devised for doing this. Dwyer [96] discusses many of these in a very lucid fashion. The present author has had very satisfactory experience with the "square-root" method described by Dwyer. The subject of inversion of matrices is one which is of great basic importance to the success of a computer facility, and there is an extensive and constantly growing literature on the subject. Questions of round-off, of random error, of cyclic time, etc., become very important and very difficult for high-order matrices, and the procedure which is actually used in computer applications should be selected keeping in mind the peculiar properties of the individual computer in question. Further discussion on this topic is outside the scope of this chapter. It is certainly obvious, however, that those cases which would give an electronic computer difficulty would also be difficult for a human operator.

So far, the elements A_{ij} of a matrix A have been considered to be ordinary numbers. However, for some purposes it is convenient to consider matrices whose elements are themselves matrices. Thus, the matrices

$$A = \begin{bmatrix} 3 & 7 & 2 & 1 & 4 \\ 6 & 5 & -2 & 3 & 6 \\ 5 & 1 & 2 & -3 & 3 \\ 2 & -1 & 4 & 7 & 1 \end{bmatrix} \quad \text{and} \quad B = \begin{bmatrix} 1 & 3 & 4 \\ -2 & 1 & 3 \\ 3 & 2 & -2 \\ 4 & -5 & 6 \\ 5 & 4 & 2 \end{bmatrix}$$

may be written as

$$A = \begin{bmatrix} \alpha_{11} & \alpha_{12} \\ \alpha_{21} & \alpha_{22} \end{bmatrix} \quad B = \begin{bmatrix} \beta_{11} & \beta_{12} \\ \beta_{21} & \beta_{22} \end{bmatrix}$$

where

$$\alpha_{11} = \begin{bmatrix} 3 & 7 & 2 \\ 6 & 5 & -2 \\ 5 & 1 & 2 \end{bmatrix} \qquad \alpha_{12} = \begin{bmatrix} 1 & 4 \\ 3 & 6 \\ -3 & 3 \end{bmatrix}$$

$$\alpha_{21} = \begin{bmatrix} 2 & -1 & 4 \end{bmatrix} \qquad \alpha_{22} = \begin{bmatrix} 7 & 1 \end{bmatrix}$$

$$\beta_{11} = \begin{bmatrix} 1 \\ -2 \\ 3 \end{bmatrix} \qquad \beta_{12} = \begin{bmatrix} 3 & 4 \\ 1 & 3 \\ 2 & -2 \end{bmatrix} \qquad \beta_{21} = \begin{bmatrix} 4 \\ 5 \end{bmatrix} \qquad \beta_{22} = \begin{bmatrix} -5 & 6 \\ 4 & 2 \end{bmatrix}$$

Then

$$AB = \left[\begin{array}{c:c} \alpha_{11}\beta_{11} + \alpha_{12}\beta_{21} & \alpha_{11}\beta_{12} + \alpha_{12}\beta_{22} \\ \hdashline \alpha_{21}\beta_{11} + \alpha_{22}\beta_{21} & \alpha_{22}\beta_{12} + \alpha_{22}\beta_{22} \end{array} \right]$$

$$= \left[\begin{array}{c:c} \begin{bmatrix} -5 \\ -10 \\ 9 \end{bmatrix} + \begin{bmatrix} 24 \\ 42 \\ 3 \end{bmatrix} & \begin{bmatrix} 20 & 29 \\ 19 & 43 \\ 20 & 19 \end{bmatrix} + \begin{bmatrix} 11 & 14 \\ 9 & 30 \\ 27 & -12 \end{bmatrix} \\ \hdashline 16 + 33 & \begin{bmatrix} 13 & -5 \end{bmatrix} + \begin{bmatrix} -31 & 44 \end{bmatrix} \end{array} \right]$$

$$= \begin{bmatrix} 19 & 31 & 43 \\ 32 & 28 & 73 \\ 12 & 47 & 7 \\ 49 & -18 & 39 \end{bmatrix}$$

which is the same result as is obtained by direct multiplication. (In this example, there was no advantage involved employing the alternate procedure.) When the matrices A and B are divided as indicated into *submatrices* α_{11}, α_{12}, α_{21}, α_{22}, β_{11}, etc. they are said to be *partitioned,* and their elements, when they are written in such a manner, are matrices rather than ordinary scalar numbers. When using partitioned matrices, it is important not inadvertently to change the sequence of the factors which are involved; thus, in this example, the product $\alpha_{11}\beta_{11}$ is well determined but the product $\beta_{11}\alpha_{11}$ is not even defined. Finally, the *integral* of a matrix A is a matrix whose elements are the integrals of the corresponding elements of A, thus

$$\int \begin{bmatrix} a_{11} & a_{12} & a_{13} \\ a_{21} & a_{22} & a_{23} \end{bmatrix} dx = \begin{bmatrix} \int a_{11}\,dx & \int a_{12}\,dx & \int a_{13}\,dx \\ \int a_{21}\,dx & \int a_{22}\,dx & \int a_{23}\,dx \end{bmatrix}$$

A similar definition applies for differentiation of a matrix; however, we shall have no occasion to employ it.

MATRICES APPLIED TO PIPING-FLEXIBILITY ANALYSIS

Forces Expressed as Column Matrices. For the purposes of the following developments, let a right-handed, rectangular coordinate system be specified.

Consider any deformable structure S which is so constrained that it cannot freely move as a rigid body under the application of the force system to be described. At a point P of S, there can be applied a force of magnitude F_x in the (positive) x direction, another of magnitude F_y in the y direction, and another of magnitude F_z in the z direction.   At the same time, torques (or moments) may be applied to S at the point P: one of magnitude M_x about an axis through P and parallel to the x axis and, similarly, moments of magnitudes M_y and M_z about the other two axes through P.   In dealing with moments (and later with rotations) we employ the "right-hand rule" for positive signs; that is, a moment (rotation) is considered positive if it twists (turns) in the direction in which the fingers of the right hand are curled when the axis in question is grasped by the right hand with the right thumb pointing in the positive direction of the axis.   All these force and moment components are to be thought of as being applied to S at the point P, and they will produce a certain distortion of S.   If they were applied at some other point of S, they would produce a different distortion.

This combination of three force components and three moment components can be described in a number of different ways, and one of the branches of theoretical mechanics deals with these different modes of description.

The force components may be thought of as comprising a force *vector*, which frequently (because of the principle of transmissibility in rigid-body mechanics) is called a *sliding vector* or an *axial vector*.   Similarly the moment components may be thought of as comprising a single moment vector which, in rigid-body mechanics, can be displaced in any fashion parallel to itself without altering its effect and is, for this reason, frequently called a *polar vector*.   Thus, the force system under discussion may be thought of as a combination of a sliding vector and a polar vector.   If some transformations are made on this system, it may be reduced uniquely (in general) to a *wrench*, which consists of a force vector (axial) and a colinear moment vector.   The system under discussion may also be thought of as a *motor*, and an elaborate and quite elegant system of motor algebra has been developed by Mises[193,194] and Brand[49] to deal with such systems.   The motor concept is particularly illuminating and permits compact specification of conditions which require separate statements in terms of vectors.   Thus, for example, when vectors are used, the requirements for equilibrium of a rigid body are usually given in two vector equations: $\Sigma\,\mathbf{F}\,=0$ and $\Sigma\,\mathbf{M}_A\,=\,0$, where A is a conveniently chosen point.   When the motor concept is used, these two separate statements can be combined into a single statement: The sum of the motors (normalized in a convenient way) acting on the body must be zero.

However, for present purposes, another mode of description will be used, namely that of representing the force and moment components by the six components of a column matrix F, viz.:

$$F = \{F_x, F_y, F_z, M_x, M_y, M_z\}$$

and further to simplify notation, this will be rewritten as a column matrix:

$$F = \{F_1, F_2, F_3, F_4, F_5, F_6\}$$

where $F_1 = F_x$, $F_2 = F_y$, $F_3 = F_z$, $F_4 = M_x$, $F_5 = M_y$, and $F_6 = M_z$.   It should not be thought that the combination of forces and moments expressed in this way is different or distinct from the other descriptions mentioned above.   It is simply that for present purposes it is most convenient to express or exhibit the combination in the form of a column matrix.

Transformations of the Force Matrix.   Suppose that two different right-handed, rectangular coordinate systems xyz and $x'y'z'$ are specified, both having

the same origin.   The forces and moments that can be applied to structure S at point P will have different components as expressed in coordinates xyz (say $F = \{F_1, F_2, F_3, F_4, F_5, F_6\}$) and in coordinates $x'y'z'$ (say $F' = \{F_1', F_2', F_3', F_4', F_5', F_6'\}$), and it is of interest to be able to transform from one representation to the other. If K denotes the 3×3 matrix,

$$K = \begin{bmatrix} \cos(x,x') & \cos(y,x') & \cos(z,x') \\ \cos(x,y') & \cos(y,y') & \cos(z,y') \\ \cos(x,z') & \cos(y,z') & \cos(z,z') \end{bmatrix}$$

where the symbol $\cos(x,x')$ indicates the cosine of the angle between the positive x and x' axes, etc., then the matrix L

$$L = \begin{bmatrix} K & 0 \\ 0 & K \end{bmatrix}$$

which is of order 6×6 (the null matrices 0 are of order 3×3 as may be determined from the context), is such that

$$F' = LF$$

and this result is presented without proof.

The matrix L has the peculiar property that its transposed and its inverse are the same.   It is obvious that the formula $F = \tilde{L}F'$ is true, for it evidently may be obtained in the same manner as the formula $F' = LF$.   But it is also true (as may easily be verified by performing the indicated operations) that $F = L^{-1}LF = L^{-1}F'$. Comparing these results leads to the conjecture that $L\tilde{L} = \tilde{L}L = I$, so that $\tilde{L} = L^{-1}$, and that this is indeed the case may be verified immediately by direct multiplication, recalling the properties of the direction cosines of a line in space.

Before treating the next important transformation, that of translation of a force, it is convenient to introduce a 3×3 square matrix which is formed from the elements of a 3×1 column matrix.   If the column matrix is

$$V = \{V_1, V_2, V_3\}$$

then the square matrix in question, called the "tensor" of the "vector" V, is given by the expression

$$T = \begin{bmatrix} 0 & V_3 & -V_2 \\ -V_3 & 0 & V_1 \\ V_2 & -V_1 & 0 \end{bmatrix}$$

If, as is usually the case, it is necessary to note specifically the dependence of T upon V, the notation $T(V)$ is used.   This concept permits a matrix operation analogous to vector or cross multiplication of vectors, for, if $U = \{U_1, U_2, U_3\}$ is another column matrix, then the product

$$W = T(V)U = \begin{bmatrix} U_2 V_3 & -U_3 V_2 \\ U_3 V_1 & -U_1 V_3 \\ U_1 V_2 & -U_2 V_1 \end{bmatrix} = \{W_1, W_2, W_3\}$$

is essentially the same thing as the cross product $\mathbf{w} = \mathbf{u} \times \mathbf{v}$, where $\mathbf{u} = u_1\mathbf{e}_1 + u_2\mathbf{e}_2 + u_3\mathbf{e}_3$, $\mathbf{v} = v_1\mathbf{e}_1 + v_2\mathbf{e}_2 + v_3\mathbf{e}_3$, $\mathbf{w} = w_1\mathbf{e}_1 + w_2\mathbf{e}_2 + w_3\mathbf{e}_3$.   Just as in vector algebra, changing the order of the two factors of a vector product changes the algebraic sign of the product; in this context, the relation

$$T(V)U = -T(U)V$$

is true.

Now consider the situation shown in Fig. 2a. Force F_P (in the generalized sense of three force components and three moment components) is applied at point P to sub-structure I which is rigidly joined to substructure II at point Q. It is desired to determine the force F_Q applied by substructure I to substructure II at point Q. Figure 2a also shows an origin and a coordinate system in terms of which the forces F_P and F_Q will be described. In Fig. 2b is shown a free-body diagram of substructure I, the force and moment components applied at point P being those of F_P while the force and moment components applied (to substructure I) at point Q are those of $-F_Q$, since II applies to I a system of forces and moments equal and opposite to those applied by I to II. (Note that all components illustrated in the figure act in the positive sense of the co-ordinate system; this is done so as to avoid confusion with regard to sign.) Writing the equations of equilibrium, of which the following two are typical

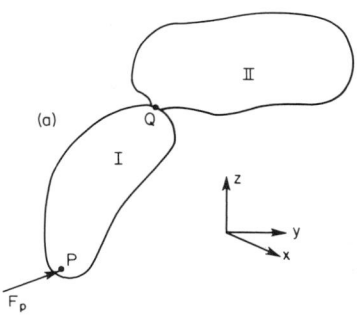

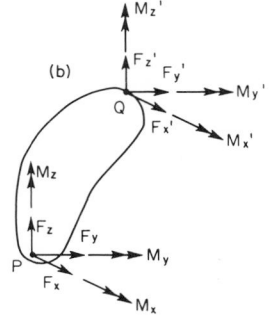

$$F_x + F_{x'} = 0$$

$$M_x + M_{x'} + F_z(y_Q - y_p) - F_y(z_Q - z_p) = 0$$

and interpreting the values F_x, F_y, etc., as F_{P1}, F_{P2}, etc., and the values $F_{x'}$, $F_{y'}$, etc., as $-F_{Q1}$, $-F_{Q2}$, etc., the following equations are obtained:

$$F_{P1} = F_{P1}$$

$$F_{Q2} = F_{P2}$$

$$F_{Q3} = F_{P3}$$

$$F_{Q4} = F_{P4} - F_{P3}(y_Q - y_p) + F_{P2}(z_Q - z_p)$$

$$F_{Q5} = F_{P5} - F_{P1}(z_Q - z_p) + F_{P3}(x_Q - x_p)$$

$$F_{Q6} = F_{P6} - F_{P2}(x_Q - x_p) + F_{P1}(y_Q - y_p)$$

FIG. 2. Transformation of a force from one point to another: (a) Force F_p applied at point P to substructure I is transmitted to substructure II at point Q; (b) free-body diagram of substructure I.

These somewhat complicated relations may be conveniently expressed in matrix form as

$$F_Q = B_{QP}F_P$$

where

$$B_{QP} = \begin{bmatrix} I_3 & 0_3 \\ T(Q - P) & I_3 \end{bmatrix}$$

and vectors (column matrices) P and Q are matrices of the coordinates of points P and Q, viz.:

$$P = \{x_P, y_P, z_P\}$$

$$Q = \{x_Q, y_Q, z_Q\}$$

It is useful at this point to establish some properties of the matrix B_{QP} which effects the translation of a force (system) from one point to another. In verifying

these properties, it is convenient to see the matrix written *in extensio*, viz.:

$$B_{QP} = \begin{bmatrix} 1 & 0 & 0 & 0 & 0 & 0 \\ 0 & 1 & 0 & 0 & 0 & 0 \\ 0 & 0 & 1 & 0 & 0 & 0 \\ 0 & (z_Q-z_P) & -(y_Q-y_P) & 1 & 0 & 0 \\ -(z_Q-z_P) & 0 & (x_Q-x_P) & 0 & 1 & 0 \\ (y_Q-y_P) & -(x_Q-x_P) & 0 & 0 & 0 & 1 \end{bmatrix}$$

The properties are

$$B_{QP}^{-1} = B_{PQ} \qquad \text{(commute subscripts to invert)}$$
$$B_{QP}B_{PR} = B_{QR} \qquad \text{(transitivity)}$$
$$\tilde{B}_{QP} = \begin{bmatrix} I_3 & T(P-Q) \\ 0_3 & I_3 \end{bmatrix} \qquad [\text{since } T(P-Q) = \tilde{T}(Q-P)]$$

Displacements Expressed as Column Matrices. An elementary volume of material of a flexible structure which is acted upon by a force system may experience displacements, due to the distortion of the structure, which may consist of three components of translation D_x, D_y, D_z and certain rotations as well. If the rotations are finite, it requires a rather exotic mode of description to deal with them: Euler angles, Cayley-Klein parameters, Rodrigues multipliers, or quaternions are generally used for finite rotations, since finite rotations do not obey the fundamental law of vector addition. If a book, say, is rotated 90 deg about the x axis, then 90 deg about the y axis, and finally 90 deg about the z axis, it ends up in a completely different orientation from these rotations if applied in a different order. However, infinitesimal rotations do, indeed, satisfy the law of vector addition. The right-hand rule is used to represent such rotations by vectors. If the thumb of the right hand is pointed along the representative vector, the fingers curl in the sense of the rotation. In terms of a right-handed rectangular coordinate system xyz, the rotation about the x axis will be denoted by D_4, that about the y axis by D_5, and that about the z axis by D_6.

It is presumed that the deformations of the structure are small and in particular that the rotations are infinitesimal.

Thus, a generalized displacement is expressed by the column matrix

$$D = \{D_1, D_2, D_3, D_4, D_5, D_6\}$$

As was the case with generalized forces, there are two important transformations applicable to generalized displacements. First, one and the same physical displacement may have different expressions in differently oriented coordinate systems with common origin, as

$$D = \{D_1, D_2, D_3, D_4, D_5, D_6\}$$

in system xyz and

$$D' = \{D_1', D_2', D_3', D_4', D_5', D_6'\}$$

in system $x'y'z'$. It may easily be verified that these are related by the formula

$$D' = LD$$

where $L = \text{diag}\,\{K, K\}$ has been described in the section "Transformations of the Force Matrix."

Second, referring to Fig. 3, if structure PQ is considered to be rigid (for the moment) and the element at P experiences a displacement (in the present generalized sense) described by the column matrix D_P, then the element at Q will experience a displacement D_Q which differs from D_P because, although the rotation components will be the same, the translation components will differ by quantities involving the products of coordinate differences and rotation components. For example,

$$D_{Q2} = D_{P2} - D_{P4}(z_Q - z_P) + D_{P6}(x_Q - x_P)$$

Using a notation for B_{QP} introduced earlier, all such relations may be combined into the matrix equation

$$D_Q = \tilde{B}_{PQ}D_P$$

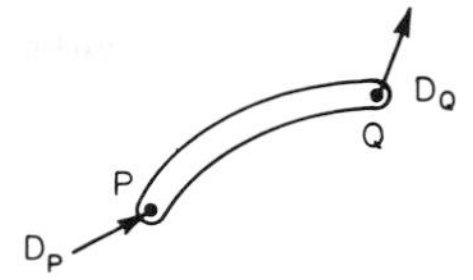

FIG. 3. Transformation of a displacement by a rigid structure.

Summary of Transformations. The four transformation equations obtained so far are sufficiently important to be exhibited again as follows:

Rotation of Coordinates. F and D expressed in terms of xyz, F' and D' expressed in terms of $x'y'z'$ (xyz and $x'y'z'$ having common origin), $L = \text{diag }\{K,K\}$

$$K = \begin{bmatrix} \cos(x',x) & \cos(x',y) & \cos(x',z) \\ \cos(y',x) & \cos(y',y) & \cos(y',z) \\ \cos(z',x) & \cos(z',y) & \cos(z',z) \end{bmatrix}$$

then

$$F' = LF$$
$$D' = LD$$

Translation. Using the notations $P = \{x_P, y_P, z_P\}$, $Q = \{x_Q, y_Q, z_Q\}$

$$T(Q - P) = \begin{bmatrix} 0 & (z_Q - z_P) & -(y_Q - y_P) \\ -(z_Q - z_P) & 0 & (x_Q - x_P) \\ (y_Q - y_P) & -(x_Q - x_P) & 0 \end{bmatrix}, \text{ etc.}$$

$$B_{QP} = \begin{bmatrix} I_3 & 0_3 \\ T(Q - P) & I_3 \end{bmatrix}, \text{ etc.}$$

then

$$F_Q = B_{QP}F_P$$
$$D_Q = \tilde{B}_{PQ}D_P$$

Relation between Force and Deflection. Figure $4a$ shows an elastic element to which a force F_P is applied at point P causing deflections D_P also at point P. (Investigation of deflections at some other point such as Q or R will be discussed later.) It will be presumed that a generalized Hooke's law holds, so that the components of deflection are related linearly to those of applied force. That is, there are constants C_{ij} such that six equations of the form

$$D_i = C_{1i}F_1 + C_{2i}F_2 + C_{3i}F_3 + C_{4i}F_4 + C_{5i}F_5 + C_{6i}F_6 \qquad i = 1, \ldots, 6$$

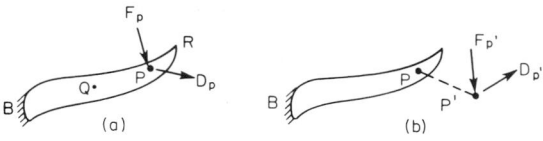

FIG. 4. (a) Force applied to point P of structure causing deflection measured at point P; (b) shift of point to arbitrary "base point" P'. Force and deflection are in generalized sense.

may be established. These relations may most compactly be exhibited in matrix form, thus:

$$D = CF$$

in which C is a square matrix consisting of 36 elements C_{ij};

$$C = \begin{bmatrix} C_{11} & C_{12} & C_{13} & C_{14} & C_{15} & C_{16} \\ C_{21} & C_{22} & C_{23} & C_{24} & C_{25} & C_{26} \\ C_{31} & C_{32} & C_{33} & C_{34} & C_{35} & C_{36} \\ C_{41} & C_{42} & C_{43} & C_{44} & C_{45} & C_{46} \\ C_{51} & C_{52} & C_{53} & C_{54} & C_{55} & C_{56} \\ C_{61} & C_{62} & C_{63} & C_{64} & C_{65} & C_{66} \end{bmatrix}$$

The task of determining these elements may be more or less difficult depending upon the nature of the elastic element. In the case of a slender steel bar of length a and solid circular cross section of radius r fixed at one end and loaded at the other, as in Fig. 5, the matrix C takes the form

$$C = \begin{bmatrix} a^3/3 & 0 & 0 & 0 & 0 & -a^2/2 \\ 0 & ar^2/4 & 0 & 0 & 0 & 0 \\ 0 & 0 & a^3/3 & a^2/2 & 0 & 0 \\ 0 & 0 & a^2/2 & a & 0 & 0 \\ 0 & 0 & 0 & 0 & 1.3a & 0 \\ -a^2/2 & 0 & 0 & 0 & 0 & a \end{bmatrix} \div EI$$

the factor 1.3 being the ratio EI/GI_P for a steel solid circular section. In the section "Compliances of Piping Elements," there will be established formulas for the matrix C for elements (such as straight lengths and bends or elbows) important in piping flexibility. For the present, it may be presumed that such expressions are available for simple structural elements, and the immediately following analysis will be concerned with combining such expressions so as to describe more complicated structures.

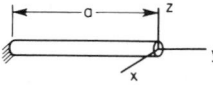

FIG. 5. Illustrative case of solid cylindrical steel cantilever bar.

Compliance Matrix. The matrix C is called the *compliance* matrix for the element to which it refers, the term arising from analogy to electric circuit theory and expressing the fact that large values of the elements in matrix C produce rather large deflections under the action of small forces, a structure behaving in such a way being said to be compliant.

All such compliance matrices C are symmetrical (cf. the matrix above for the cantilever bar) as a consequence of the reciprocal relations of Maxwell and Betti. Thus, of the 36 elements they contain, 30 are equal in pairs so that there are only 21 (i.e., 6 plus 15) independent elements. The dimensions of these elements (the units in terms of which they are expressed) depend on their position within the matrix and are such as to accord with the units in terms of which F and D are

expressed. Exhibiting units only, for the partitioned matrices, one has (symbolically)

$$
\begin{bmatrix} \text{in.} \\ \hline - \end{bmatrix} = \begin{bmatrix} \text{in. lb}^{-1} & \vdots & \text{lb}^{-1} \\ \hline \text{lb}^{-1} & \vdots & \text{in.}^{-1}\,\text{lb}^{-1} \end{bmatrix} \begin{bmatrix} \text{lb} \\ \hline \text{in. lb} \end{bmatrix}
$$

for the matrix equation $D = CF$.

Stiffness Matrix. If both sides of the equation $D = CF$ are premultiplied by the reciprocal of C, there is obtained

$$ F = C^{-1}D = SD $$

where the reciprocal matrix C^{-1} is denoted by the symbol S and is called the stiffness matrix.

For the steel cantilever whose compliance matrix was exhibited above, calculation of the inverse gives the stiffness matrix as

$$
S = \begin{bmatrix}
12/a^3 & 0 & 0 & 0 & 0 & 6a/^2 \\
0 & 4/ar^2 & 0 & 0 & 0 & 0 \\
0 & 0 & 12/a^3 & -6a/^2 & 0 & 0 \\
0 & 0 & -6/a^2 & 4/a & 0 & 0 \\
0 & 0 & 0 & 0 & 1/1.3a & 0 \\
6/a^2 & 0 & 0 & 0 & 0 & 4/a
\end{bmatrix} EI
$$

As will shortly become apparent when formulas for the compliance of more complicated elements are exhibited, the difficulties of determining and exhibiting the reciprocal, or stiffness, matrices in literal form become overwhelming, so that, in effect, it simply is not possible to obtain useful literal formulas for the forces and moments which must be applied to a configuration in order to produce prescribed deflections. Instead, each individual problem must be treated from a numerical point of view.

Just as the inverse or reciprocal of a compliance matrix C is a stiffness matrix S, the reciprocal of a stiffness matrix S is a compliance matrix C.

Base Point for Specification of Compliance or Stiffness. In dealing with piping elements it is well to observe that they are generally slender, so that it will be clear which portions are under stress (and thus distort) and which portions are unstressed (and thus maintain their shape). In Fig. 4a, only the portion between the point P and the root B is loaded and only this portion distorts. The compliance matrix C which relates D_P and F_P describes the elastic properties of this portion only and does not tell about the portion from P to R. Neither does it separate the properties from P to B into two portions, say from P to Q and from Q to B, so that from a knowledge of this C alone, one cannot infer anything about the displacements observed at point Q.

The elastic properties from P to B, however, do fully determine the relation between displacement $D_{P'}$ and applied force $F_{P'}$ at some other point P' connected to point P by means of an actual or conceptual perfectly rigid level PP', shown in Fig. 4b by the heavy dashed line. Using the transformation formulas given earlier, the force applied by the rigid lever at point P becomes $F_P = B_{PP'}F_{P'}$ and the deflection $D_P = CF_P$ observed at point P becomes $D_{P'} = \tilde{B}_{PP'}D_P$ at the end P' of the lever. Combining these relations and setting off the significant

combination by parentheses, it may be seen that

$$D_{P'} = (\tilde{B}_{PP'}CB_{PP'})F_{P'}$$
$$= C'F_{P'}$$

The matrix C' gives essentially the same information as is given by the matrix C, but it may be simpler or more convenient to use in calculation. In any case, when the compliance of a certain element is given as a matrix C, it is necessary to specify also what *base point* is used for reference, that is, at what point the force F is applied and the deflection D is measured, in the equation $D = CF$. A subscript may be used to specify the base point, and the formula for transferring from one base point to another is

$$C_{P'} = \tilde{B}_{PP'}C_{P}B_{PP'}$$

If a certain piping element, such as AC in Fig. 6, is given and a definite base point (point B in Fig. 6) is specified, the elastic properties are given fully by the matrix C relating deflection and force at B, regardless of *which* end of the element is regarded as fixed. That is, the same matrix C describes both situations shown in Fig. 6. To see this, suppose that the matrices were C and C^* and conceive of

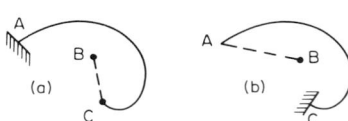

FIG. 6. Compliance of element is independent of which end is regarded as anchored provided the base point is unchanged.

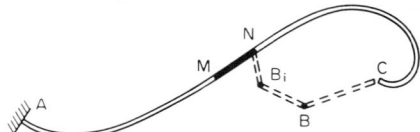

FIG. 7. Series assembly of elements. Compliance of the entire structure is obtained by considering the sum of contributions of the flexibilities of subelements such as MN.

AC as free in space with rigid levers AB and CB which meet but are not joined together at point B. At point B apply force F to AB and force $-F$ to CB. Then, as measured in a coordinate system fixed with respect to lever CB, the B end of AB will experience the deflections $D = CF$, and as measured in a coordinate system fixed with respect to lever AB, the B end of CB will experience deflections $-D = C^*(-F)$, and this would be true for *any* value of F. From this it may be inferred that $C^* = C$.

Series Assembly of Elements. Presuming that it is possible thus to describe the elastic properties of an individual piping element by specifying its base point B_i and giving its compliance matrix C_i, the next stage in the development of the theory is to construct from such information the compliance matrix C, with respect to a chosen base point B, of a series assembly of such individual elements.

Figure 7 shows such a series assembly AC composed of individual elements, the ith one of which is MN for which the base point is B_i and the compliance matrix is C_i. It is desired to find the compliance matrix C of AC with respect to base point B. To obtain this result, first think of AC being entirely rigid except only for element MN. The force F is applied at point B and is transmitted to AC via the rigid lever BC. There is stress in AC, but only portion MN distorts, portions AM and NC of AC acting (for the present) as rigid levers. Thus, the force applied at B_i may be thought of as being transmitted by the rigid path $BCNB_i$ or, equivalently, by the rigid path BB_i, and it has the value $F_{B_i} = B_{B_i B}F_B$ and results in displacement $D_{B_i} = C_i F_{B_i}$, which, transmitted back to point B via the rigid lever system, becomes $D_B = \tilde{B}_{B_i B}D_{B_i} = \tilde{B}_{B_i B}C_i B_{B_i B}F_B$. In this process the distortion of MN causes NC to be rotated and translated slightly out of its

original position so that it does not present the same orientation to applied forces as it did originally. However, it is inherent in the small deflection theory which is used here to neglect such deviations and to regard NC as being unchanged. Thus, the total deflection at B due to the distortion of *all* the elements of which AC is composed is simply the summation of the individual components of deflection due to the distortion of the individual elements of AC, viz.:

$$D_B = \sum_i \tilde{B}_{B_iB}C_iB_{B_iB}F_B = CF_B$$

where

$$C = \sum_i \tilde{B}_{B_iB}C_iB_{B_iB}$$

Normalization. The preceding operation is considerably simplified if the same base point is chosen for all elements, for if all points B_i coincide with point B, then all the transfer matrices B_{B_iB} reduce to unit matrices and the formula for C becomes simply $C = \sum_i C_i$. The process of initially transferring all forces, compliance matrices, and deflections to a common reference point or base point is called *normalization*, and its systematic use was introduced into the literature by Chen.[70] In what follows, it will be presumed that this has been done, the common reference point being denoted as point O. All forces and all deflections will be "at" point O in the sense described above, that is, as applied to or observed at the O end of a rigid lever from point O to the point in question in the piping system. These levers, of course, are not physically present and do not interfere with each other or with the piping system itself; they are merely a convenient device intended to add physical significance to a mathematical simplification. The actual force acting in a pipe element at a point P is given by the formula

$$F_P = B_{PO}F_O$$

and the actual deflection experienced at this point in the piping system is given by the formula

$$D_P = \tilde{B}_{OP}D_O$$

where F_O and D_O are the force and deflection given by the analysis which follows and in which henceforth the subscript O is no longer written.

This process of normalization in effect separates the geometry from the algebra of the problem. Geometry enters into finding the compliances C with respect to convenient local base points and also into normalizing such compliances with respect to the common base point (also commonly called the origin of "global" coordinates). Geometry also enters into the process of normalizing whatever forces and/or displacements may be given in the problem. Accomplishing this much poses a purely algebraic problem for solution. After this problem is solved, a final geometrical step is required in reducing the normalized solutions back to results which have intended physical significance—a process of "denormalization," so to speak. In general, in the remainder of the present section, it will be presumed that normalization has been accomplished and this all coordinates are global coordinates. In the section "Compliances of Piping Elements" which deals with the calculation of individual compliance matrices, the coordinates are local coordinates, located as indicated in the individual subsections. The brief "Solution of Flexibility Equations" deals with the algebraic problem; by the time the problem reaches this stage, all quantities will have been normalized and the results of the computations will be in normalized form. The section headed "Calculations of Stresses and Deflections" deals with the calculation of stresses and deflections from the algebraic solutions, and its first subsection points out how to "denormalize" so as to obtain results with the desired physical significance.

Branched Systems without Loops. Consider a branched system without loops, such as that shown in Fig. 8, which is rigidly constrained at one terminus (such as point A), and observe the deflection D_i at a point P_i resulting from the application of force F_j at point P_j. (All forces, deflections, and compliances are presumed to be normalized to a common point O as described in the preceding section, so that, for example, the force actually applied to the system at point P_j is really the product $B_{P_jO}F_j$, but for simplicity and convenience we shall deal only with F_j.) Owing to the application of force F_j, all portions of the structure along the (only) path from point P_j to the constrained point A experience distortions. However, only the distortion of portion AQ will have any effect on the displacement which takes place at point P_i. Noting that there is no stress and thus no distortion in portion QP_i, it follows that

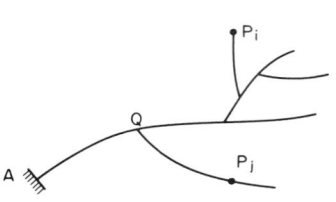

$$D_i = C_{ij}F_j$$

where C_{ij} indicates the summation of the compliance matrices for all portions of the structure which are common to the two paths AP_i and AP_j. Note that $C_{ij} = C_{ji}$.

FIG. 8. Branched system without loops. Illustrates significance of quantities C_{ij}, compliances having two subscripts.

If there are n points $P_1 \ldots P_n$ in such a system at which forces may be applied and deflections observed (in the normalized sense), the deflection at any particular point P_i is the sum of the deflections at P_i due to individual forces F_j, viz.:

$$D_i = \sum_{j=i}^{n} C_{ij}F_j \qquad i = 1, \ldots, n$$

and the totality of these relations may conveniently be expressed in the form of a matrix equation $D = CF$ where the elements of these matrices are themselves matrices, viz.:

$$
\begin{bmatrix} D_1 \\ D_2 \\ \cdot \\ \cdot \\ \cdot \\ D_n \end{bmatrix}
=
\begin{bmatrix}
C_{11} & C_{12} & \cdots & C_{1n} \\
C_{21} & C_{22} & \cdots & C_{2n} \\
\cdot & \cdot & \cdots & \cdot \\
\cdot & \cdot & \cdots & \cdot \\
\cdot & \cdot & \cdots & \cdot \\
C_{n1} & C_{n2} & \cdots & C_{nn}
\end{bmatrix}
\begin{bmatrix} F_1 \\ F_2 \\ \cdot \\ \cdot \\ \cdot \\ F_n \end{bmatrix}
$$

Here D and C are $6n \times 1$ matrices while C is a $6n \times 6n$ square and symmetrical matrix. The points P_i may lie at termini of the system, they may be interior to portions of the system, or they may even lie at junctions where several portions of the system are joined together. In the basic problem of piping flexibility, all the column matrices D_i may be easily calculated by methods which will be given under "Calculation of the Deflections D_i," and the problem is to find the column matrices F_i. Formally this is easily done by finding the inverse of the matrix C, and there results the solution

$$F = C^{-1}D$$

Constraints. Some of the points P_i at which forces and deflections are of interest may be points of constraint at which there is a hanger, a guide, a partial anchor, etc. If this constraining element may be described by a compliance matrix, nothing new is involved. Thus, presume that the constraining element

P_rP_s indicated in Fig. 9, although it is not actually composed of pipe, has flexibility properties describable by a compliance matrix C (normalized to the common reference or base point O as are all the other compliance matrices). Then any of the C_{ij}'s in which either of the subscripts is the letter s (corresponding to point P_s) can be determined as previously. Also D_s, describing the prescribed motion of point P_s, can be determined in the same way (yet to be described) as any of the other D's.

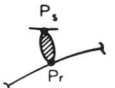

However, more often than not, the constraining element is describable by a singular (rather than a nonsingular) stiffness matrix. Such a singular stiffness matrix does not possess an inverse or reciprocal. Thus there is no compliance matrix capable of describing such a constraining element.

Fig. 9. Element P_rP_s, applying a constraint to the pipe at point P_r; usually such elements have a singular stiffness matrix.

For example, an ordinary spring hanger oriented in the vertical z direction imposes a z force equal to the product of spring constant times z deflection but is not capable of imposing any x or y forces or any x, y, or z moments, even though corresponding deflection components may be experienced. Such a constraint has a singular stiffness matrix, viz.:

$$S_B = \begin{bmatrix} 0 & 0 & 0 & 0 & 0 & 0 \\ 0 & 0 & 0 & 0 & 0 & 0 \\ 0 & 0 & k & 0 & 0 & 0 \\ 0 & 0 & 0 & 0 & 0 & 0 \\ 0 & 0 & 0 & 0 & 0 & 0 \\ 0 & 0 & 0 & 0 & 0 & 0 \end{bmatrix}$$

in which k denotes the spring constant. The determinant of this matrix is zero, and its inverse may not be calculated. As written above, this matrix has *not* been normalized with respect to the common base point O, the subscript B referring to its individual base point located at any convenient place on the axis of the hanger. The normalized form is $S = \tilde{B}_{BO}S_B B_{BO}$, and in this analysis it will be presumed that the normalized form is meant unless, as in the present case, a contrary statement is made. However, it is clear that, if S_B is singular, then S is singular and indeed with the same degree of degeneracy.

The General Case of Singular Constraints. The following section, dealing with the general case of a singular constraint stiffness matrix S, may be omitted without loss in continuity. The operations are indicated without giving the physical motivation leading to them, and certain notation, used temporarily for this development only, is used without full and explicit definitions. The special cases which occur most frequently in practice may be treated by the considerably simpler argument given in the following section.

Presume that the normalized symmetrical matrix S is q-fold degenerate. This implies that there exists an orthogonal matrix N such that $S = \tilde{N}S'N$, where S' is of the form

$$S' = \begin{bmatrix} 0 & 0 \\ 0 & \sigma \end{bmatrix}$$

and σ is a nonsingular matrix of order $p = 6 - q$ and with inverse c. The statement that N is orthogonal means simply that $\tilde{N} = N^{-1}$. We are interested particularly in finding the forces applied by the constraining element P_rP_s to the

piping system at point P_r. Suppose that the deflections at these points are D_r and D_s and that F_r is the force applied by the constraining element to the piping system at point P_r. By the methods given earlier the relation $D = CF$ may be established where

$$D = \{D_1, D_2, \ldots, D_r, \ldots, D_n\}$$

$$F = \{F_1, F_2, \ldots, F_r, \ldots, F_n\}$$

$$C = \begin{bmatrix} C_{11} & C_{12} & \cdots & C_{1r} & \cdots & C_{1n} \\ C_{21} & C_{22} & \cdots & C_{2r} & \cdots & C_{2n} \\ \cdot & \cdot & & \cdot & & \cdot \\ \cdot & \cdot & & \cdot & & \cdot \\ \cdot & \cdot & & \cdot & & \cdot \\ C_{r1} & C_{r2} & \cdots & C_{rr} & \cdots & C_{rn} \\ \cdot & \cdot & & \cdot & & \cdot \\ \cdot & \cdot & & \cdot & & \cdot \\ \cdot & \cdot & & \cdot & & \cdot \\ C_{n1} & C_{n2} & \cdots & C_{nr} & \cdots & C_{nn} \end{bmatrix}$$

Let $J = \mathrm{diag}\,\{I_6, I_6, \ldots, N, \ldots, I_6\}$ and observe that $\tilde{J} = J^{-1}$. Premultiplying the relation $D = CF$ by J gives $JD = JC\tilde{J}JF$, which may be written out as

$$\begin{bmatrix} D_1 \\ D_2 \\ \cdot \\ \cdot \\ \cdot \\ ND_r \\ \cdot \\ \cdot \\ \cdot \\ D_n \end{bmatrix} = \begin{bmatrix} C_{11} & C_{12} & \cdots & C_{1r}\tilde{N} & \cdots & C_{1n} \\ C_{21} & C_{22} & \cdots & C_{2r}\tilde{N} & \cdots & C_{2n} \\ \cdot & \cdot & & \cdot & & \cdot \\ \cdot & \cdot & & \cdot & & \cdot \\ \cdot & \cdot & & \cdot & & \cdot \\ NC_{r1} & NC_{r2} & \cdots & NC_{rr}\tilde{N} & \cdots & NC_{rn} \\ \cdot & \cdot & & \cdot & & \cdot \\ \cdot & \cdot & & \cdot & & \cdot \\ \cdot & \cdot & & \cdot & & \cdot \\ C_{n1} & C_{n2} & \cdots & C_{nr}\tilde{N} & \cdots & C_{nn} \end{bmatrix} \begin{bmatrix} F_1 \\ F_2 \\ \cdot \\ \cdot \\ \cdot \\ NF_r \\ \cdot \\ \cdot \\ \cdot \\ F_n \end{bmatrix}$$

Now let X be the matrix $\begin{bmatrix} 0 & 0 \\ 0 & c \end{bmatrix}$ and to the above add the identity

$$\begin{bmatrix} 0 \\ 0 \\ \cdot \\ \cdot \\ \cdot \\ XNF_r \\ \cdot \\ \cdot \\ \cdot \\ 0 \end{bmatrix} = \begin{bmatrix} 0 & 0 & \cdots & 0 & \cdots & 0 \\ 0 & 0 & \cdots & 0 & \cdots & 0 \\ \cdot & & & \cdot & & \cdot \\ \cdot & & & \cdot & & \cdot \\ \cdot & & & \cdot & & \cdot \\ 0 & 0 & \cdots & X & \cdots & 0 \\ \cdot & \cdot & & \cdot & & \cdot \\ \cdot & \cdot & & \cdot & & \cdot \\ \cdot & \cdot & & \cdot & & \cdot \\ 0 & 0 & \cdots & 0 & \cdots & 0 \end{bmatrix} \begin{bmatrix} F_1 \\ F_2 \\ \cdot \\ \cdot \\ \cdot \\ NF_r \\ \cdot \\ \cdot \\ \cdot \\ F_n \end{bmatrix}$$

The sum of these is $D^* = C^*F^*$, where

$$F^* = \{F_1, F_2, \ldots, NF_r, \ldots, F_n\}$$

$$D^* = \{D_1, D_2, \ldots, ND_r + XNF_r, \ldots, D_n\}$$

and C^* is the same as the large square matrix exhibited *in extensio* except that the rr term is now $NC_{rr}\tilde{N} + X$ rather than $NC_{rr}\tilde{N}$. The relation $C^* = C^*F^*$ is between $6n \times 1$ column matrices and a $6n \times 6n$ square matrix. However, the order may be reduced by q as follows. Since the pipe applies force $-F_r$ to the constraining element at point P_r, the relation $F_r = S(D_s - D_r)$ is established. Premultiply by N and introduce appropriately the relations $S = \tilde{N}S'N$ and $N\tilde{N} = I$ to obtain the formula

$$NF_r = S'(ND_s - ND_r)$$

From the nature of S' it is clear that the first q elements of NF_r must be zeros. Thus the coefficients (from the matrix C^*) which multiply these zeros in the product C^*F^* may be omitted. Also, note that the rth term in D^* is $ND_r + XNF_r$, which may be simplified as follows:

$$ND_r + XNF_r = ND_r + \begin{bmatrix} 0 & 0 \\ 0 & c \end{bmatrix}\begin{bmatrix} 0 & 0 \\ 0 & \sigma \end{bmatrix}(ND_s - ND_r)$$

$$= ND_r + \begin{bmatrix} 0 & 0 \\ 0 & I_p \end{bmatrix}(ND_s - ND_r)$$

$$= \begin{bmatrix} I_q & 0 \\ 0 & 0 \end{bmatrix}(ND_r) + \begin{bmatrix} 0 & 0 \\ 0 & I_p \end{bmatrix}(ND_s)$$

The elements of D_r are all unknown. In the displacement ND_s only the last p elements may be specified; the first q elements are unknowns. Thus the system is reduced to the $6n - q$ system

$$
\begin{bmatrix} D_1 \\ D_2 \\ \cdot \\ \cdot \\ \cdot \\ \overline{ND_s} \\ \cdot \\ \cdot \\ \cdot \\ D_n \end{bmatrix} =
\begin{bmatrix} C_{11} & C_{12} & \cdots & | \, C_{1r}\tilde{N} & \cdots & C_{1n} \\ C_{21} & C_{22} & \cdots & | \, C_{2r}\tilde{N} & \cdots & C_{2n} \\ \cdot & \cdot & & \cdot & & \cdot \\ \cdot & \cdot & & \cdot & & \cdot \\ \cdot & \cdot & & \cdot & & \cdot \\ \overline{NC_{r1}} & \overline{NC_{r2}} & \cdots & | \, \overline{(NC_{rr}\tilde{N} + X)} & & \overline{NC_{rn}} \\ \cdot & \cdot & & \cdot & & \cdot \\ \cdot & \cdot & & \cdot & & \cdot \\ \cdot & \cdot & & \cdot & & \cdot \\ C_{n1} & C_{n2} & \cdots & | \, C_{nr}\tilde{N} & & C_{nn} \end{bmatrix}
\begin{bmatrix} F_1 \\ F_2 \\ \cdot \\ \cdot \\ \cdot \\ NF_r \\ \cdot \\ \cdot \\ \cdot \\ F_n \end{bmatrix}
$$

in which a bar written over a matrix indicates that the first q rows are to be omitted and a bar written at the left indicates that the first q columns are to be omitted. In the solution one obtains $F_1, F_2, \ldots, \overline{NF_r}, \ldots, F_n$. The matrix NF_r may be recovered from $\overline{NF_r}$ by inserting q initial zeros after which $F_r = \tilde{N}(NF_r)$.

Special Cases of Singular Constraints. In the most important practical cases, a constraint at a point P_r of a piping system may be considered to be composed

of from one to five individual constraints of the type to be described in the following. If there are more than one, each is assumed to act at a distinct point $(P_{r1}, P_{r2}, \ldots)$ of the system all of which, however, geometrically coincide with the point P_r. Thus, we may consider only a single such constraint in the following analysis. It may be described in terms of a scalar "spring constant" k and a "unit vector" V, which is a column matrix the sum of the squares of whose elements is unity. The last property may also be expressed by saying that V is such that $\tilde{V}V = 1$. (Example: in the case of the spring hanger of constant k oriented in the z direction, $V = \{0, 0, 1, 0, 0, 0\}$; however, it is not required that any of the elements vanish in order to satisfy the present requirements on V.)

It is presumed that the constraint is of the nature that the only force it applies to the piping system is of the form $f_r = gV$ where g is a scalar constant and that, in this case, under the action of the force $-f_r$ applied to the constraint at point P_r by the piping, the point P_r experiences a displacement $d_r = -f_r/k + B_{sr}e_s$. The lower-case letters d_r, f_r, and e_s are used to signify that these are *local* rather than normalized quantities. The column matrix e_s denotes the specified displacements at the P_s end of the constraining element. Replacing the local quantities f_r and d_r in terms of normalized quantities E_r and D_r, one obtains the relations

$$F_r = B_{0r}Vg$$

$$D_r = \tilde{B}_{r0}\tilde{B}_{sr}e_s - \tilde{B}_{r0}Vg/k$$

and we shall henceforth use the abbreviation $E = \tilde{B}_{r0}\tilde{B}_{sr}e_s$ and $H = B_{0r}V$.

Now consider the rth matrix equation of the set implied by $D = CF$. It is

$$D_r = C_{r1}F_1 + C_{r2}F_2 + \cdots + C_{rr}F_r + \cdots + C_{rn}F_n$$

and make the substitutions $F_r = Hg$, $D_r = E - (\tilde{B}_{r0}B_{r0}/k)Hg$; transpose the term $\tilde{B}_{r0}B_{r0}Hg/k$; and premultiply by $\tilde{H}$. There is obtained

$$\tilde{H}E = \tilde{H}C_{r1}F_1 + \tilde{H}C_{r2}F_2 + \cdots + \tilde{H}C_{rr}{}^*Hg + \cdots + \tilde{H}C_{rn}F_n$$

wherein $C_{rr}{}^* = C_{rr} + \tilde{B}_{r0}B_{r0}/k$. Any other typical equation of the set, say the kth, upon the substitution $F_r = Hg$, becomes

$$D_k = C_{k1}F_1 + C_{k2}F_2 + \cdots + C_{kr}Hg + \cdots + C_{kn}F_n$$

and all these equations may be combined into a single system of order $6n$-5, viz.:

$$
\begin{bmatrix}
D_1 \\
D_2 \\
\cdot \\
\cdot \\
\cdot \\
\tilde{H}E \\
\cdot \\
\cdot \\
D_n
\end{bmatrix}
=
\begin{bmatrix}
C_{11} & C_{12} & \cdots & C_{1r}H & \cdots & C_{1n} \\
C_{21} & C_{22} & \cdots & C_{2r}H & \cdots & C_{2n} \\
\cdot & \cdot & & \cdot & & \cdot \\
\cdot & \cdot & & \cdot & & \cdot \\
\cdot & \cdot & & \cdot & & \cdot \\
\tilde{H}C_{r1} & \tilde{H}C_{r2} & \cdots & \tilde{H}C_{rr}{}^*H & \cdots & \tilde{H}C_{rn} \\
\cdot & \cdot & & \cdot & & \cdot \\
\cdot & \cdot & & \cdot & & \cdot \\
C_{n1} & C_{n2} & \cdots & C_{nr}H & \cdots & C_{nn}
\end{bmatrix}
\begin{bmatrix}
F_1 \\
F_2 \\
\cdot \\
\cdot \\
\cdot \\
g \\
\cdot \\
\cdot \\
F_n
\end{bmatrix}
$$

If there are two such constraints, say corresponding to the rth and the tth positions,

the system is of order $6n$-10, viz.:

$$
\begin{bmatrix}
D_1 \\
D_2 \\
\cdot \\
\cdot \\
\cdot \\
H_r E_r \\
\cdot \\
\cdot \\
\cdot \\
H_t E_t \\
\cdot \\
\cdot \\
\cdot \\
D_n
\end{bmatrix}
=
\begin{bmatrix}
C_{11} & C_{12} & \cdots & C_{1r}H_r & \cdots & C_{1t}H_t & \cdots & C_{1n} \\
C_{21} & C_{22} & \cdots & C_{2r}H_r & \cdots & C_{2t}H_t & \cdots & C_{2n} \\
\cdot & \cdot & & \cdot & & \cdot & & \cdot \\
\cdot & \cdot & & \cdot & & \cdot & & \cdot \\
\cdot & \cdot & & \cdot & & \cdot & & \cdot \\
H_rC_{r1} & H_rC_{r2} & \cdots & H_rC_{rr}{}^*H_r & \cdots & H_rC_{rt}H_t & \cdots & H_rC_{rn} \\
\cdot & \cdot & & \cdot & & \cdot & & \cdot \\
\cdot & \cdot & & \cdot & & \cdot & & \cdot \\
\cdot & \cdot & & \cdot & & \cdot & & \cdot \\
H_tC_{t1} & H_tC_{t2} & \cdots & H_rC_{tr}H_r & \cdots & H_tC_{tt}{}^*H_t & \cdots & H_tC_{tn} \\
\cdot & \cdot & & \cdot & & \cdot & & \cdot \\
\cdot & \cdot & & \cdot & & \cdot & & \cdot \\
\cdot & \cdot & & \cdot & & \cdot & & \cdot \\
C_{n1} & C_{n2} & \cdots & C_{nr}H_r & \cdots & C_{nt}H_t & \cdots & C_{nn}
\end{bmatrix}
\begin{bmatrix}
F_1 \\
F_2 \\
\cdot \\
\cdot \\
\cdot \\
g_r \\
\cdot \\
\cdot \\
\cdot \\
g_t \\
\cdot \\
\cdot \\
\cdot \\
F_n
\end{bmatrix}
$$

in which the notation should by now be self-explanatory.

Relation between Preceding Analyses: Example. In considering the general case of singular constraints, it was stated that the transformed matrix S' was of the form $S' = \text{diag } \{O_q, \sigma_p\}$ where σ_p is nonsingular. Actually, more than this can be asserted: The orthogonal matrix N may be chosen so that σ_p is *itself* diagonal so that, if $\sigma_p = \text{diag } \{k_1, k_2, \ldots, k_p\}$, then the inverse c may easily be calculated as $c = \text{diag } \{k_1^{-1}, k_2^{-1}, \ldots, k_p^{-1}\}$. This choice of N effectively "uncouples" the stiffness matrix so that it may be regarded as p separate constraints of the type considered in the preceding section on "Special Cases." Thus, in reality there is no essential loss of generality in confining attention to the special case. The essential difficulty of uncoupling the separate constraints appears in different guises in the two presentations. In the general case, one has the problem of determining the orthogonal matrix N. In the special case, it was presumed that each constraint considered is of the special form specified; however, to take a general singular constraint and exhibit it as p special constraints would require, in effect, determining the transformation matrix N. Actually, the general analysis is, in general, somewhat simpler than the special case, since the matrix N is required only to permit writing S' as $\text{diag } \{O_q, \sigma_p\}$ and we do not require that σ_p be diagonal, only that its inverse may be found.

All physically possible constraints have stiffness matrix elements which are finite or zero. However, in idealizing a physical problem it may be assumed that a constraint is, in effect, perfectly rigid. This implies an unboundedly large element (or elements) in the stiffness matrix. To treat this situation, one may simply regard the corresponding spring constant as being infinitely large and its reciprocal as being zero. To do this neatly within the framework of what has been presented above, it is best either to regard the constraint to be of the special type with $k^{-1} = 0$ or to presume that the orthogonal matrix N indeed does reduce S' to diagonal form, one of the diagonal terms, k_i, say, being such that $k_i^{-1} = 0$.

To illustrate these notions, consider the example indicated in Fig. 10 wherein there is a constraint applied to the piping at point P_r which has the effect of a spring hanger with constant K in the z direction together with a spring guide, with constant k, in the y direction. For simplicity normalizing to the point P_r, the stiffness

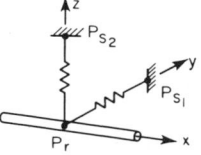

FIG. 10. Example illustrating treatment of constraints.

matrix is $S = \text{diag}\{O, k, K, O, O, O\}$. This is already disjoint, so that there is no real advantage in rearranging the values to give initial zeros; however, this can be done by using the orthogonal matrix

$$N = \begin{bmatrix} 1 & 0 & 0 & 0 & 0 & 0 \\ 0 & 0 & 0 & 1 & 0 & 0 \\ 0 & 0 & 0 & 0 & 1 & 0 \\ 0 & 0 & 0 & 0 & 0 & 1 \\ 0 & 1 & 0 & 0 & 0 & 0 \\ 0 & 0 & 1 & 0 & 0 & 0 \end{bmatrix}$$

to get $S' = \text{diag}\{O, O, O, O, k, K\}$. The matrix X becomes

$$X = \text{diag}\{O, O, O, O, k^{-1}, K^{-1}\}$$

The two-element column matrices ND_s and NF_r become $\{D_2, D_3\}$ and $\{F_2, F_3\}$, where D_2 is the y deflection of point P_{s1}, D_3 is the z deflection of point P_{s2}, F_2 is the y force applied to the pipe, and F_3 is the z force applied to the pipe; D_2 and D_3 are presumed to be given quantities, and F_2 and F_3 are the unknown constraint forces which are among the unknown quantities which must be determined. In the large square matrix of coefficients, terms such as $\overline{NC}$ are of the form

$$\overline{NC} = \begin{bmatrix} C_{21} & C_{22} & C_{23} & C_{24} & C_{25} & C_{26} \\ C_{31} & C_{32} & C_{33} & C_{34} & C_{35} & C_{36} \end{bmatrix}$$

while terms such as $\mid C\tilde{N}$ are the transposed of such matrices. Finally, the term $\overline{\mid(NC_{rr}\tilde{N} + X)}$ is simply the 2×2 square matrix

$$\begin{bmatrix} C_{22} + k^{-1} & C_{23} \\ C_{32} & C_{33} + K^{-1} \end{bmatrix}$$

If the vertical (z direction) constraint is presumed to be infinitely rigid, the only change is to set $K^{-1} = 0$.

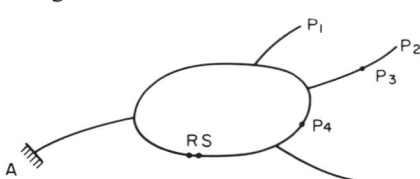

Looped Systems. In actual piping systems the requirement of bypassing heat exchangers, pressure-regulating valves, or other equipment frequently introduces loops, so that there is no single and unique path from a specified point of the configuration back to the constrained terminus A. Thus, for example, in Fig. 11, a path from point P_1 to A could include either the upper or the lower part of the loop illustrated; compare this configuration with that shown in Fig. 8. Accordingly, the matrices C_{ij} defined earlier cannot be formed uniquely without a procedure equivalent to the following.

FIG. 11. Looped system characterized by having more than one path back to basic anchor; loop is removed by cutting the loop at a convenient point $R = S$ and introducing self-equilibrating forces to the cut ends.

Conceive of the system being cut apart at some point at which no external forces are applied so as to reduce it to a no-loop system. (If the system is multiply looped, it must be conceptually cut apart at several such cut points.) Let the two termini formed at the cut be designated as R and S. Also, since there is no net force exerted externally on the system at the cut point, it is true that $F_R + F_S = 0$.

However, regarding points R and S as termini of an unlooped system, all matrices C_{ij} are determined uniquely, so that the large matrix equation $D = CF$ may be formed as previously.   Without loss of mathematical generality and for convenience of notation, presume that there are n ordinary termini, that force and deflection corresponding to point R occupy the $n +$ first position, and that force and deflection corresponding to point S occupy the $n +$ second position and consider the large matrices to be partitioned, thus:

$$
\begin{bmatrix} D_1 \\ \cdot \\ \cdot \\ \cdot \\ D_n \\ \hline D_R \\ D_S \end{bmatrix}
=
\begin{bmatrix}
C_{11} & \cdots & C_{1n} & C_{1R} & C_{1S} \\
\cdot & \cdots & \cdot & \cdot & \cdot \\
\cdot & \cdots & \cdot & \cdot & \cdot \\
\cdot & \cdots & \cdot & \cdot & \cdot \\
C_{n1} & \cdots & C_{nn} & C_{nR} & C_{nS} \\
\hline
C_{R1} & \cdots & C_{Rn} & C_{RR} & C_{RS} \\
C_{S1} & \cdots & C_{Sn} & C_{SR} & C_{SS}
\end{bmatrix}
\begin{bmatrix} F_1 \\ \cdot \\ \cdot \\ \cdot \\ F_n \\ \hline F_R \\ F_S \end{bmatrix}
$$

With this scheme of partitioning, let

$$
A = \begin{bmatrix}
I_{nn} & O_{n2} \\
\hline
& 1 \quad -1 \\
O_{2n} & \\
& 1 \quad\ \ 0
\end{bmatrix}
$$

$$
B = \begin{bmatrix}
I_{nn} & O_{n2} \\
\hline
& 1 \quad 0 \\
O_{2n} & \\
& 1 \quad 1
\end{bmatrix}
$$

and note that

$$
B^{-1} = \begin{bmatrix}
I_{nn} & O_{n2} \\
\hline
& 1 \quad\ \ 0 \\
O_{2n} & \\
& -1 \quad 1
\end{bmatrix}
$$

Then, performing the operations indicated by

$$AD = ACB^{-1}BF$$

one obtains

$$
\begin{bmatrix} D_1 \\ \cdot \\ \cdot \\ D_n \\ \hline D_R - D_S \\ D_R \end{bmatrix}
$$

$$
=
\begin{bmatrix}
C_{11} & \cdots & C_{1n} & (C_{1R} - C_{1S}) & C_{1S} \\
\cdot & \cdots & \cdot & \cdot & \cdot \\
\cdot & \cdots & \cdot & \cdot & \cdot \\
\cdot & \cdots & \cdot & \cdot & \cdot \\
C_{n1} & \cdots & C_{nn} & (C_{nR} - C_{nS}) & C_{nS} \\
\hline
(C_{R1} - C_{S1}) & \cdots & (C_{Rn} - C_{Sn}) & (C_{RR} + C_{SS} - 2C_{RS}) & (C_{RS} - C_{SS}) \\
C_{R1} & \cdots & C_{Rn} & (C_{RR} - C_{RS}) & C_{RS}
\end{bmatrix}
\begin{bmatrix} F_1 \\ \cdot \\ \cdot \\ F_n \\ \hline F_R \\ F_R + F_S \end{bmatrix}
$$

The $n + 1$st diagonal term originally is $C_{RR} + C_{SS} - C_{RS} - C_{SR}$, but because $C_{SR} = C_{RS}$, it may be written as above. As will be shown in the next section, the term $D_R - D_S$ may be evaluated, even though D_R itself is unknown, and since $F_R + F_S = 0$, the last element in the column matrix at the right is zero. Thus, by eliminating the last row of the column matrices and the last row and column of the square matrix, one has a system of $6n + 6$ equations which can be solved. After the solution has been obtained, the value of D_R may be obtained from the $n +$ second relation above.

The foregoing matrix procedure is simply a formal and quite generalized way of doing what is quite reasonable anyhow, namely, equating D_S to D_R and F_S to $-F_R$ and simplifying the system of equations. If the piping system is multiply looped, one should make sufficient cuts of the sort described above to reduce it to an unlooped system as considered before. Corresponding to each such cut there are two matrix relations of the sort just considered, one relating the deflections on either side of the cut and the other equilibrating the forces on either side of the cut.

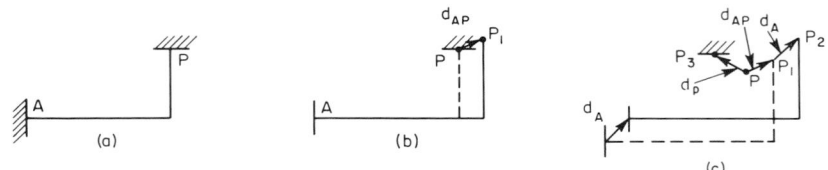

FIG. 12. Nomenclature and calculation of the deflections D_i.

Calculation of the Deflections D_i. The usual case of determining deflections D_i is illustrated by the configuration shown in Fig. 12. Point A is regarded as the basic anchor. Initially, the pipe is considered to be cut free from other anchors such as P and permitted to expand, which causes the free end to move from P to P_1 as shown; this motion d_{AP} is illustrated as a simple translation, which it usually is, but it may have rotatory parts under some circumstances. At any rate d_{AP} is a column matrix which depends on the temperature distribution in the pipe and the physical properties of the pipe material. Because of the heating of terminal equipment (such as boilers, etc.) the anchors themselves do not remain absolutely fixed in position but rather undergo certain deflections themselves, such as is illustrated by the vectors d_A and d_P in Fig. 12c. Now the forces actually applied at the terminus P_2 must be such as to force this terminus to conform to the modified anchor position P_3; that is, the forces are such as to cause the deflection

$$D_P = d_P - d_A - d_{AP}$$

and it is values such as these which should appear in the column matrix of column matrices which has been previously designated as D.

The values (column matrices) d_A and d_P are frequently called the "free expansions" of the terminal equipment. They may be supplied by the manufacturers of the equipment or may be calculated from a knowledge of the temperature distribution in the equipment and of the location of its anchorages to the building structure. (It is presumed that the building structure is itself unyielding; otherwise the structural problem is vastly more complicated if deflections of the building must be reckoned with.)

The free expansion of the pipe itself is usually calculated by assuming that the pipe is at a uniform temperature, determining the unit thermal strain e from Table 2, and obtaining

$$d_{AP} = e\{x_P - x_A, y_P - y_A, z_P - z_A, O, O, O\}$$

If the temperature of the pipe varies from point to point along its length but each cross section may be considered to be at uniform temperature, the matrix d_{AP} may be calculated as

$$d_{AP} = \int_A^P e\{\alpha, \beta, \gamma, 0, 0, 0\}\, ds$$

where α, β, and γ are the direction cosines of an elemental length for which the unit thermal strain is e, and the integration is along the axis of the pipe from A to P. In the case of a looped system (Fig. 10), we have

$$D_R = d_R - d_A - d_{AR}$$
$$D_S = d_S - d_A - d_{AS}$$

and although the quantities d_{AR} and d_{AS} may be determined as above (note that they are computed along different paths from point A to the common point R, S), the deflection $d_R = d_S$ is not known. However, in the formulation of the preceding subsection, it is the difference $D_R - D_S$ which is needed, and this is

$$D_R - D_S = d_{AS} - d_{AR}$$

If the piping along the route AR is at the same temperature as the piping along route AS, and if the coefficient of thermal expansion is the same for the two portions, then $d_{AS} - d_{AR}$ is zero; otherwise a nonzero value applies. This may be thought of as the relative motion of the two points when cut apart and the pipe is subjected to temperature change only without mechanical loading. In any case, the contributions of common portions of the two paths cancel out.

Bowing Due to Nonuniform Temperature. If each individual cross section is not itself under uniform temperature, "bowing" may take place and complicate the calculation of d_{AP}. Flieder, Loria, and Smith[108] have discussed bowing of cryogenic pipelines, where the phenomenon is of particular importance. The following treatment is parallel to theirs, more inclusive in that it contemplates curved as well as straight elements and discusses the overall effect on d_{AB} but less inclusive in not treating particular temperature distributions as does their paper. For low-temperature materials properties, see Refs. 114 and 187.

Figure 13a shows a length of (curved) pipe under uniform reference temperature conditions, and Fig. 13b shows it after the temperature has been changed. The temperature change is presumed to be a function of position in the cross section, but the temperature distribution at all cross sections of the illustrated element is presumed to be the same. Later, variations along the length will be permitted as an integration is performed in which the integrand may be a function of arc length. It is presumed that both conditions (a) and (b) are free from external load except that in condition (b) there has been applied a self-equilibrating system of axial and shearing stresses which cause the plane bounding cross sections AB and CD to remain plane after the temperature change. It is this self-equilibrating system which gives rise to local thermal stress, and such a system or an approximation to it must exist except locally near free ends. Considering the longitudinal strain of element EF,

$$\epsilon = \frac{(\rho + y)\delta - (R + y)\beta + \gamma x}{(R + y)\beta} = \frac{\sigma}{E} + \alpha T$$

where σ denotes longitudinal tensile stress. It is presumed that radial and circumferential stresses due to this free thermal expansion are nonexistent or small. This is a reasonable assumption for thin-wall pipe, but for thick-wall pipe under

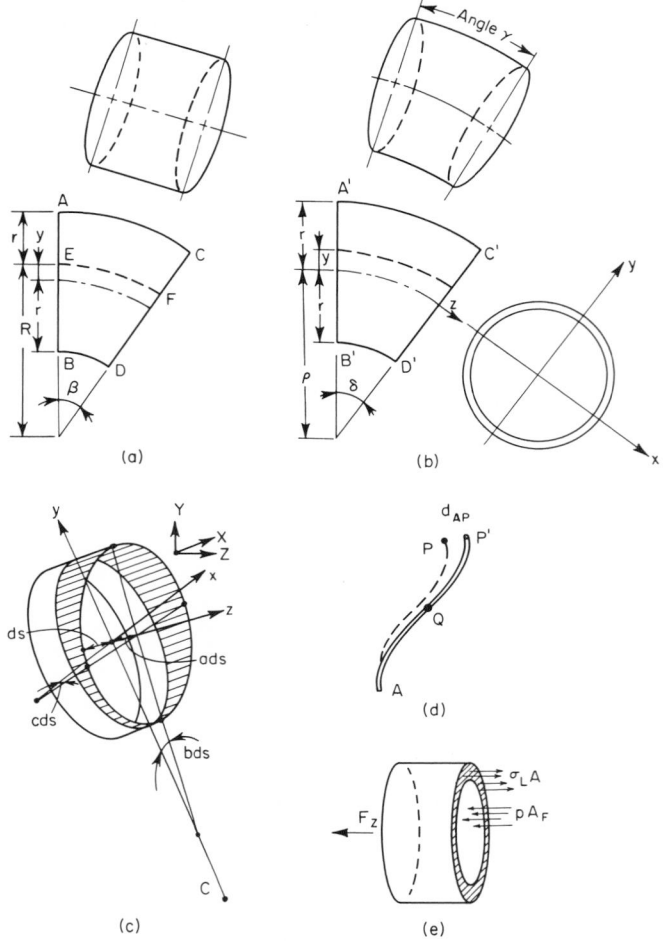

FIG. 13. Effects of nonuniform heating and of internal pressure. Unstrained element (*a*) is nonuniformly heated and subjected to self-equilibrating stress system causing plane bounding sections to remain plane as in (*b*). These local distortions expressed in local *xyz* coordinate system (*c*) must be transformed into global *XYZ* coordinates and integrated along the pipe to obtain terminal thermal displacement d_{AP}. (*e*) shows axial equilibrium of a short straight element under the action of structural load F_z, longitudinal stress o_L and internal pressure *p*.

certain temperature distributions it may be in considerable error. Using the notations $a = \rho\delta/\beta R - 1$, $b = (\delta - \beta)/R\beta$, $c = \gamma/R\beta$, and $L = R\beta$, the last being the original center-line length of the element,

$$\epsilon = \frac{a + by + cx}{1 + y/R}$$

so that $\sigma = -E\alpha T + E(a + by + cx)/(1 + y/R)$

For self-equilibration it is required that this system have no resultant axial force or moment about either the x or the y axis; that is, $\int \sigma \, dA = 0, \int x\sigma \, dA = 0, \int y\sigma \, dA = 0$.

Thus
$$\int E\alpha T \, dA = a\int E \, dA^* + b\int yE \, dA^* + c\int xE \, dA^*$$
$$\int yE\alpha T \, dA = a\int yE \, dA^* + b\int y^2 E \, dA^* + c\int xyE \, dA^*$$
$$\int xE\alpha T \, dA = a\int xE \, dA^* + b\int xyE \, dA^* + c\int x^2 E \, dA^*$$

where
$$dA^* = \frac{dA}{1 + y/R}$$

From these three equations, the constants a, b, and c may be determined and thus the self-equilibrating stress component σ introduced here may be evaluated. This should later be added to the stress system determined (as in the section on "Calculation of Stresses and Deflections") from an analysis of the structural problem, i.e., the fundamental flexibility problem.

While the preceding equations have the advantage of generality and the integrals may be evaluated by numerical or other approximate means if analytic evaluations should prove too burdensome, in practice it is usually possible to perform a significant simplification by assuming that E is constant over the cross section. The most important cases where nonuniform heating is significant are in cryogenic systems, where there may be substantial temperature differences over the cross section but where E is relatively insensitive to temperature change, and in high-temperature systems in which the variation itself is rather small compared with the total temperature change. Thus if E is assumed to be constant in the preceding integrals, if symmetry with respect to the y axis is assumed, then for the case of uniform wall thickness, there is obtained

$$\int \alpha T \, dA = aJ_1 + bJ_2$$
$$\int y\alpha T \, dA = (a - bR)J_2$$
$$\int x\alpha T \, dA = cJ_3$$

where
$$J_1 = 2\pi R(\sqrt{R^2 - r_i^2} - \sqrt{R^2 - r_o^2})$$
$$J_2 = R(A - J_1)$$
$$J_3 = R^2 A + 2\pi R[(R^2 - r_o^2)^{3/2} - (R^2 - r_i^2)^{3/2}]/3$$
$$A = \pi(r_o^2 - r_i^2)$$

and where r_o and r_i denote outside and inside radii, respectively, of the pipe cross section. If the wall thickness is not uniform as may be the case in bends and elbows, the integrals do not permit the ready evaluation given here. However, if the integrals can be evaluated, it is a simple matter to solve for a, b, and c.

If the beam is straight, or if R is much larger than r_o, and if the wall thickness is uniform, an explicit solution is as follows:

$$a = \left(\int \alpha T \, dA\right) / A$$
$$b = \left(4\int y\alpha T \, dA\right) / A(r_o^2 + r_i^2)$$
$$c = \left(4\int x\alpha T \, dA\right) / A(r_o^2 + r_i^2)$$

From their definitions it is easy to see that the constants a, b, and c are, respectively, unit center-line elongation and curvatures in the principal normal and the binormal planes. This statement may be made clearer by a reference to Fig. 13c, which shows to an enlarged scale a short element of curved pipe, center of curvature at C, having length ds. Because of the distortions being considered here, the end nearest the reader has moved, relative to its former position, as indicated, experiencing a translation $a\ ds$ in the z direction and changes in slope, positive as indicated, of amounts $b\ ds$ and $c\ ds$. In the case of a straight length of pipe, the x and y directions may be chosen conveniently but at right angles to each other and to the z axis, which is, locally, along the axis of the pipe.

Thus the element which has length ds has experienced distortions which we denote by $G\ ds$, where $G = \{0, 0, a, b, -c, 0\}$ in the local x, y, z coordinate system. Expressed in the universal X, Y, Z coordinate system illustrated, this becomes $LG\ ds$, where $L = \text{diag}\ \{K, K\}$ and

$$K = \begin{bmatrix} \cos(X, x) & \cos(X, y) & \cos(X, z) \\ \cos(Y, x) & \cos(Y, y) & \cos(Y, z) \\ \cos(Z, x) & \cos(Z, y) & \cos(Z, z) \end{bmatrix}$$

This local distortion at point Q (cf. Fig. 13d) gives rise to the deflection $\tilde{B}_{QP}LG\ ds$ at P, and the total deflection at the free end P is, thus,

$$d_{AP} = \int_A^P \tilde{B}_{QP}LG\ ds$$

where each component of the integrand may vary along the length. In the case that b and c are zero, this reduces to

$$d_{AP} = \int_A^P a\{\alpha, \beta, \gamma, 0, 0, 0\}\ ds$$

where $\alpha = \cos(X, z)$, $\beta = \cos(Y, z)$, and $\gamma = \cos(Z, z)$, and this may be compared with the simpler expression given earlier in this subsection, noting that a is, in effect, a weighted average value of the unit thermal elongation e.

There are two sources of error in the preceding analysis, only one of which can be compensated easily. This analysis has presumed that the originally circular cross section remains circular. However, the ability of pipe cross sections to distort implies that pipe bends and pipe elbows are effectively more flexible than ordinary theory would predict and, in reality warping will take place in such a way as to render the bend more flexible than ordinary theory would predict. As will be developed in "Compliances of Piping Elements," a flexibility factor n should apply in the present case. The formula for this is

$$n = \frac{1.65}{tR/r^2} \qquad \text{(but not less than 1.0)}$$

where t is wall thickness, R is bend radius, and r is mean pipe radius. This flexibility factor applies to bending but not to longitudinal extension. Accordingly, the formula for the column matrix G which appears above should be modified to read as follows:

$$G = \{0, 0, a, nb, -nc, 0\}$$

The other source of error is more difficult to treat. It results from writing the longitudinal strain as $\sigma/E + \alpha T$ rather than as $\sigma/E + \alpha T - (\nu/E)(\sigma_r + \sigma_t)$, where

σ_r and σ_t are radial and tangential stress components.   Although it may be reasonable to suppose that σ_r remains small, it is by no means clear that σ_t will be small. Neglecting this stress component leads to a simple analysis, but the degree of error is difficult to assess.   In the case of pipe bends and elbows, the effect of σ_r and σ_t is neglected.   However, in the case of straight pipe, it may be possible to investigate the effect of the simplification.   Flieder, Loria, and Smith have used this simplification in their work,[108] and it would be of interest to compare their simplified results with exact analyses which could be based upon equations given by Boley and Wiener[45] based upon work by Gatewood, Goodier, Biot, and others.   As far as the author is aware, such comparisons have not been made.

Effects of Pressure on Free Expansions.   One more practical factor may enter into the calculations of the deflections D_i.   If internal pressure is acting, particularly in thin-wall pipes, an additional longitudinal strain, here called pressure strain and analogous to thermal strain in its effect on the analysis, may become of significance.   Figure 13e shows a short segment of (straight) pipe together with indications of longitudinal force systems acting on it.   At the left is shown the "structural" reaction F_z (the z axis being taken as that of the pipe) which has entered into the theory given earlier in this chapter.   At the other end, equilibrating F_z, is shown the metallic tensile force $\sigma_L A$, where as before A stands for the area of the metallic cross section $\pi(r_o^2 - r_i^2)$.   There is also applied a fluid compressive force pA_F where p is internal pressure and A_F is the internal or flow area πr_i^2. The equation of longitudinal equilibrium leads to the evaluation

$$\sigma_L = F_Z/A + pA_F/A$$

In this formula, the first term represents the longitudinal stress due to direct structural axial load (the total structural longitudinal stress includes the effect of structural bending loads as well), and its effect is included in the structural compliance matrices to be given later.   The second term is usually called the longitudinal pressure stress and is provided for in design both by the formulas for pipe-wall thickness and by the formula which establishes the value of the allowable expansion stress (range) and the circumstances under which this allowable value may be increased.   That is, in effect, the Code separates the consideration of the stresses arising from internal pressure from those resulting from thermal expansion, providing separate design policies for assuring integrity of the piping system.   There are cogent reasons for performing this separation, but there may be circumstances under which the designer may actually wish to evaluate the combined stress picture resulting from both internal pressure and thermal expansion at a certain epoch.

Whether or not the stress evaluation is performed separately, the longitudinal pressure stress gives rise to an axial extension of the pipe pA_F/AE.   Furthermore (in straight pipe) the circumferential and radial stress components resulting from internal pressure cause an (Poisson) axial contraction of amount $-2\nu pA_F/AE$. This result follows from the Lamé analysis of thick-wall cylinders and is equally valid for thick- or for thin-wall pipe.   Thus the total axial extensional strain resulting from internal pressure is $(1 - 2\nu)pA_F/AE$, and this should be added to the thermal strain αT; that is, one should use

$$e = \alpha T + (1 - 2\nu)pA_F/AE$$

rather than simply $e = \alpha T$, in determining the quantities d_{AP} which are of fundamental importance in the deflections D_P.

The immediately foregoing discussion of the effect of internal pressure is applicable strictly only to straight pipe elements.   For elbows and bends the situation is

considerably more complicated.   Even for thin-wall bends there is no fully satisfactory stress analysis for the effects of internal pressure.   A customary approximation to the actual stress distribution may be obtained by what is called membrane
analysis.   This neglects discontinuity stresses which must obtain, in addition, at
and near junctions of bends and straight segments.   However, even though these
discontinuity stresses may be of significance as far as the overall stress picture is
concerned, pressure design formulas assure integrity of the pipe and the discontinuity
stresses probably are of quite secondary importance compared with the membrane
components in influencing gross distortion.   The analysis which follows is confined
to thin-wall bends of uniform wall thickness.   Membrane analysis leads to circumferential (hoop) pressure stress given by

$$\sigma_c = (pr/t)(2R - r \sin \theta)/(2r \sin \theta)$$

this stress component varying with θ, an angle measured perpendicular from the
plane of the bend, so that $\theta = 90$ deg at the intrados and $\theta = -90$ deg at the
extrados.   The longitudinal stress does not depend on θ but has the constant value
$pr/2t$ which it has for straight pipe.   Noting that $r/2t$ is the same as A_F/A in the case
of thin-wall pipe, it is reasonable to take the pressure strain as being

$$\left(1 - 2v \frac{R - \frac{1}{2}r \sin \theta}{R - r \sin \theta}\right) \frac{pA_F}{AE}$$

which reduces to the previous formula if R is much greater than r.   If we consider
an element of bend radius R and included angle β, so that the center-line length is
βR, and calculate the lengthening of the arc portion which is at the distance $y = r \sin \theta$ from the pipe center line (in toward the center of the bend), we find the value

$$\left(1 - 2v \frac{R - y/2}{R - y}\right)(pA_F/AE)\beta(R - y).$$

Dividing by the original center-line length βR, we get $[(1 - 2v) - y(1 - v)/R]$
(pA_F/AE).   The first term is the unit center-line elongation, and the second, when
divided by y, is the angle change per unit arc length, i.e., the curvature resulting from
deformation.   Thus, the corresponding contribution to d_{AP} is

$$d_{AP} = \int_A^P \tilde{B}_{QP} LG^* \, ds$$

where $G^* = \{0, 0, 1 - 2v, (1 - v)/R, 0, 0\} (pA_F/AE)$

If the wall thickness is not uniform, as assumed here, similar but more complicated
calculations may be performed.

   It may be remarked parenthetically that, while the pressure effects discussed here
could actually be taken into account in a flexibility analysis and in some cases
might have a significant effect upon the final results, the author knows of no case in
which this actually has been done.   The fomula appearing in paragraph 419.6.4 of
the ASA Oil Transportation Code (ASA B31.4-1959), Ref. 1, for determining what
is there called the net longitudinal compressive stress due to the combined effects of
temperature rise and fluid pressure, does indeed, in effect, deduct the Poisson contraction due to hoop tension but does not consider the other influences discussed in
this section.

   Important Variations in Viewpoint.   The foregoing theoretical development
is one which is capable of treating the most complex piping system and is rather
easily programmed for high-speed automatic digital computation.   However,

there are alternate formulations which have been or could be employed, and these are now treated briefly.

The formulation presented in this chapter involves calculating the matrix C, which may be called (using a nomenclature of F. H. Branin) the "mesh-flexibility matrix." This matrix is relatively easy to calculate on the basis of physical and mechanical principles. Solution of the flexibility problem involves, next, the algebraic chore of inverting the matrix C to obtain $F = C^{-1}D$ (or roughly equivalent tasks). The matrix C^{-1} may be called the mesh-stiffness matrix, and its determination is by no means a trivial task from an algebraic viewpoint. An almost inverse point of view may be adopted, however. In this alternate procedure, the joint displacements are regarded as the fundamental unknowns, a "joint-stiffness matrix" is readily obtained, and the algebraic difficulties are concentrated in inverting this matrix so as to obtain the joint-flexibility matrix in terms of which the final results are readily obtained. This viewpoint appears to have been applied to piping structures for the first time almost simultaneously by Chen[70] and Kjeldgaard.[155]

To explain this method it will be assumed that all quantities appearing below have been normalized to a common base point. The following notation will be used:

$$F_{PQ} = \text{force exerted by member } PQ \text{ on joint } Q$$
$$f_Q = \text{force exerted on joint } Q \text{ by agencies external to the system}$$
$$d_P = \text{actual deflection of point } P \text{ due to all causes}$$
$$d_{PQ} = \text{free thermal motion of point } Q \text{ considering point } P \text{ as fixed}$$

Then the equations of member equilibrium, joint equilibrium, and member distortion are

$$F_{PQ} + F_{QP} = 0$$
$$\sum_Q F_{PQ} + f_Q = 0$$
$$d_P - d_Q + d_{PQ} = C_{PQ}F_{PQ}$$

and when the last equation is premultiplied by the stiffness $S_{PQ} = C_{PQ}^{-1}$, there results

$$S_{PQ}(d_P - d_Q) + f_{PQ} = F_{PQ}$$

where $f_{PQ} = S_{PQ}d_{PQ}$ is a known quantity. Making proper substitutions, there is obtained

$$SD + f = 0$$

where $D = \{d_A, d_B, \ldots\}$, $f = \left\{ f_A + \sum_I f_{IA}, f_B + \sum_I f_{IB}, \ldots \right\}$, the summations extending only over those subscripts corresponding to members framing into the given joint, and where S is a square symmetrical matrix whose off-diagonal terms are the stiffnesses S_{PQ} and whose diagonal terms are the sums of the negatives of the other elements in the same row. S is thus a singular matrix, but in the usual case it is only simply singular, so that all deflections d_A may be expressed in terms of a chosen deflection, say d_N. That is, the equations above deal with the distortions of the elastic structure; upon this a rigid-body motion may be superimposed. To hold the body in place, so to speak, the deflection of the "ground" joint may be specified. It should be noted that all anchor points are topologically equivalent to one and the same ground joint.

The joint-stiffness matrix S or the contraction of it resulting from the specification of the displacement of the ground joint is generally rather sparse; that is, the 6×6 null matrix occurs in every position corresponding to two joints which are not connected by a flexible member. There are techniques for inverting and otherwise dealing

with sparse matrices, but it should be noted that, although S may be sparse, its inverse, the joint-flexibility matrix, is generally dense.

While some configurations may be quite complex when treated by the mesh-flexibility method and comparitively simple when treated by this alternate joint-stiffness method, the converse is true for other configurations. Generally, the topology of free-floating piping systems is such that one method involves about as much difficulty as does the other. For systems having constraints, Chen[70] has concluded that the mesh-flexibility method is better.

The methodology of Kron leads to procedures which may be thought of as lying between the extremes represented by the mesh-flexibility and the joint-stiffness methods. Branin (51) has examined and discussed these procedures (using electrical engineering nomenclature) and attributes to A. S. Householder the observation of the relation between Kron's methods and the purely algebraic identity

$$(F + GHK)^{-1} = F^{-1} - F^{-1}G(KF^{-1}G + H^{-1})^{-1}KF^{-1}$$

Here F and its inverse are presumed known and F is modified by the addition of the triple product GHK, the inverse of this sum being desired. The inverse of H must be obtainable, and the matrices G, H, and K must have conformable dimensions but are otherwise arbitrary. When H is selected of small dimension, the task of obtaining the modified inverse may be quite trivial. A rather widely known procedure, first published by Peck et al.[207] and called the 6×6 method, may be explained in terms of the preceding relation and may also be described in more fundamental terms as follows, referring to

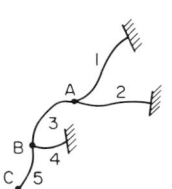

FIG. 14. Example used to illustrate "6 × 6" viewpoint.

Fig. 14 for an example configuration. The compliance matrices at point A for the two branches 1 and 2 may be obtained as described in the subsection "Series Assembly of Elements." Their inverses are the stiffnesses of the two branches. Adding these stiffnesses together gives the combined stiffness (at A) of the two branches, and the inverse of this sum is the compliance (also at A) of the combination of branches 1 and 2. Using the appropriate transformation matrices, this compliance can be added to that of portion 3 to get overall compliance at B of the subconfiguration composed of branches 1, 2, and 3. The inverse of this is the stiffness at B of 1, 2, and 3. To this may be added the inverse of the compliance (at B) of branch 4 so as to get the total stiffness at B of portions 1, 2, 3, and 4. When the inverse of this is combined with the compliance of portion 5, the total compliance at C of the entire system is obtained. In this whole sequence of steps only 6×6 compliance and stiffness matrices are involved. The thermal-expansion matrices and the anchor-motion matrices can also be formalized. In brief, a complete treatment using only 6×6 and 6×1 matrices has been developed and successfully programmed for automatical digital computation and has been used to solve a number of piping problems of considerable complexity.

Other recent work, not necessarily in matrix form, which the reader may find useful are found in Refs. 40, 115, 120, 121, and 143.

COMPLIANCES OF PIPING ELEMENTS

Compliance Matrix of a General Elastic Piping Element. For an infinitesimally short element with axis oriented along the x axis as shown in Fig. 15, the (differential) compliance matrix is

$$C' = ds \times \text{diag}\,\{1/AE,\ \zeta_y/AG,\ \zeta_z/AG,\ \zeta_\theta/GI,\ 1/EI,\ 1/EI\} = Y\,ds$$

In this expression the ζ's are coefficients which account for the actual nonuniform stress distribution.   The first term is the axial extension, and the next two are the cross shearing deformations.   If the shearing stress were uniformly distributed over the cross section, the values of ζ_y and ζ_z would be unity, but under actual conditions their value is nearly 2.0 as will be developed in a later section of this analysis.   The fourth term is the torsional deformation I_p denoting the polar moment of inertia of the cross section and ζ_θ accounting for the torsional shearing-stress distribution; for circular pipes $\zeta_\theta = 1$.   The last two terms give the angular deformation due to bending and derive directly from the customary Euler-Bernoulli relation between moment and curvature, namely, that the quotient M/EI equals curvature, defined as rate of change of slope with respect to arc length.   Force components in the y and z directions do not produce contributions to rotation terms since, in customary bending theory, these involve squares of the length of the element, and these are

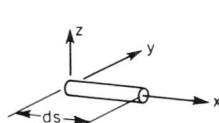

FIG. 15.   Infinitesimal element used in calculating compliance matrices.

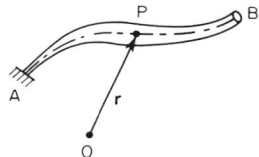

FIG. 16. Center line of structural element is generated by vector r, which is a function of a parameter t.

infinitesimals of higher order.   A similar explanation may be made for the absence of other terms one might expect to see.   As will become apparent in what follows, these contributions will be correctly obtained by means of the integration process which will be employed.

Suppose that the center line of the piping element lies along a curve, the general point P of which is given by a vector (from a convenient origin) of the form

$$\mathbf{r}(t) = \mathbf{i}x(t) + \mathbf{j}y(t) + \mathbf{k}z(t)$$

where $\mathbf{i}$, $\mathbf{j}$, and $\mathbf{k}$ are the customary unit vectors along the x, y, and z directions, respectively, and the parenthetical t indicates that $\mathbf{r}$ and its components are variable functions of some parameter t, so that the center-line curve from A to B (see Fig. 16) is traversed by the variable point P as t varies between some initial value t_1 and some final value t_2.   It is presumed that the functions $x(t)$, $y(t)$, and $z(t)$ are sufficiently differentiable for what follows.

The methods of differential geometry permit the establishment of a convenient *local* coordinate system at the general point P; this may be described by the unit tangent vector $\mathbf{T}$, the unit normal vector $\mathbf{N}$, and the unit binormal vector $\mathbf{B}$.   For simplicity, some notation from dynamics may be borrowed; although the parameter t does not signify time, let a superior dot denote differentiation with respect to t, write $\dot{\mathbf{r}} = \mathbf{v}$, $\dot{\mathbf{v}} = \mathbf{a}$, and let the symbol v denote the magnitude of the vector $\mathbf{v}$. Also note that the general infinitesimal arc length ds may be written as

$$ds = (\dot{x}^2 + \dot{y}^2 + \dot{z}^2)\,dt$$

From differential geometry, the following expressions may be obtained for the local coordinate vectors

$$\mathbf{T} = \mathbf{v}/v \qquad \mathbf{B} = \mathbf{v} \times \mathbf{a}/|\mathbf{v} \times \mathbf{a}| \qquad \mathbf{N} = \mathbf{B} \times \mathbf{T}$$

Now consider the element of length ds at point P and presume that it has been convenient to employ a local coordinate system given by unit vectors **I**, **J**, and **K**, the first of these being along the center line of the element. The vector **T** defined by the geometry of the curve lies along **I** (and without loss of generality, it may be assumed to have the same rather than the opposite sense), but the vectors **J** and **K** may be rotated with respect to **N** and **B**, the angle of rotation, denoted by the symbol θ in Fig. 17, possibly varying in a known way along the length of the curve. The differential compliance matrix given at the beginning of this section relates force F' and D' by the formula $D' = C'F'$, the primes denoting that these quantities are measured in the **I**, **J**, **K** system. To obtain the relation between F and K, measured in the (fixed) **i**, **j**, **k** system it is noted that $F' = LF$, $D' = LD$, $L^{-1} = \tilde{L}$, where $L = \text{diag}\{K, K\}$ and K is the matrix of direction cosines which may be expressed in terms of scalar (dot) products as

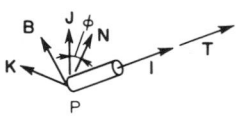

FIG. 17. Relationship between local unit vectors I, J, K, and T, N, B. I and T are tangent to the center line; J and K may be along the principal axes of the cross section; N and B are unit normal and unit binormal to the curve.

$$K = \begin{bmatrix} \mathbf{i}\cdot\mathbf{I} & \mathbf{j}\cdot\mathbf{I} & \mathbf{k}\cdot\mathbf{I} \\ \mathbf{i}\cdot\mathbf{J} & \mathbf{j}\cdot\mathbf{J} & \mathbf{k}\cdot\mathbf{J} \\ \mathbf{i}\cdot\mathbf{K} & \mathbf{j}\cdot\mathbf{K} & \mathbf{k}\cdot\mathbf{K} \end{bmatrix}$$

Noting that $\mathbf{I} = \mathbf{T}$, $\mathbf{J} = \mathbf{N}\cos\theta + \mathbf{B}\sin\theta$, and $\mathbf{K} = -\mathbf{N}\sin\theta + \mathbf{B}\cos\theta$, the transformation matrix L is completely given as a function of the parameter t. Then

$$D = \tilde{L}LD = \tilde{L}D' = \tilde{L}C'F' = \tilde{L}C'LF$$

so that the desired compliance matrix in the fixed coordinate system is $\tilde{L}C'L$. Finally, it is noted that the compliance matrix given at the start of this section and transformed as immediately above has a base point located somewhere in or near the infinitesimal element itself. Accordingly it is convenient to transfer to a common base point so that the contributions of the individual infinitesimal elements may easily be summed. If the transfer is to point O, a B transformation is called for giving the final result $\tilde{B}_{PO}\tilde{L}CLB_{PO}$ for the differential compliance matrix; here B_{PO} is the matrix

$$\left[\begin{array}{ccc|c} & I_3 & & O_3 \\ \hline 0 & z & -y & \\ -z & 0 & x & I_3 \\ y & -x & 0 & \end{array}\right] = \begin{bmatrix} I_3 & O_3 \\ T & I_3 \end{bmatrix}$$

in which x, y, and z are the coordinates of the variable point P. Presuming that the dimensions and properties of the cross section may also vary from point to point along the curve as given functions of the parameter t, the total compliance of the element may be expressed as

$$C = \int_{t_1}^{t_2} \tilde{B}_{PO}\tilde{L}\,YLB_{PO}(\dot{x}^2 + \dot{y}^2 + \dot{z}^2)\,dt$$

where Y is the diagonal matrix exhibited at the beginning of this section.

It should be noted here and elsewhere in this section on compliance that actual dimensions of piping elements may differ from the ideal dimensions cited in specifications or given in appropriate engineering drawings. For example, there is a 12½ per cent mill (under) tolerance for pipe-wall thickness under most piping

specifications, with no specific overtolerance being given. Since flexibility calculations must usually be performed long before the piping material is available for inspection, it is a long-established custom, sanctioned by the ASA B31 Code for Pressure Piping, to use nominal dimensions in calculating piping flexibility and expansion stresses; this applies not only to pipe itself but also to fittings and other piping elements.

Simplification of the Diagonal Matrix Y and of the Integrand. For a straight circular pipe, $\zeta_y = \zeta_z$ because of symmetry, and we presume that this is also (approximately) true for curved pipe and denote the common value by ζ; more will be said later about the value of ζ. Also, for a circular section $\zeta_\theta = 1$ and $I_p = 2I$. Let ρ denote the radius of gyration of the cross section so that $I = A\rho^2$. Using the notation $m = \zeta E/G$, the diagonal matrix Y of the preceding section may be written as

$$Y = \text{diag}\,\{M, N\}$$

where
$$M = (\rho^2/EI)\,\text{diag}\,\{1, m, m\}$$

and
$$N = (1/EI)\,\text{diag}\,\{1 + \mu, 1, 1\}$$

where μ is Poisson's ratio and the customary relationship $E = 2(1 + \mu)G$ has been used.

Because of the zero matrices 0_3 which appear when partitioning is employed, the matrix integrand $\tilde{B}_{PO}\tilde{L}\,YLB_{PO}\,ds$ may be written as

$$\begin{bmatrix} I & \tilde{T} \\ 0 & I \end{bmatrix}\begin{bmatrix} \tilde{K} & 0 \\ 0 & \tilde{K} \end{bmatrix}\begin{bmatrix} M & 0 \\ 0 & N \end{bmatrix}\begin{bmatrix} K & 0 \\ 0 & K \end{bmatrix}\begin{bmatrix} I & 0 \\ T & I \end{bmatrix}ds$$

and simplified to

$$\begin{bmatrix} \tilde{K}MK + \tilde{T}\tilde{K}NK & \tilde{T}\tilde{K}NK \\ \tilde{K}NKT & \tilde{K}NK \end{bmatrix}ds$$

Compliance of a Straight Piece of Pipe. Let the direction cosines of the element be α, β, and γ with respect to the fixed coordinate axes x, y, z. The matrix K may be written as

$$K = \begin{bmatrix} \alpha & \beta & \gamma \\ \alpha' & \beta' & \gamma' \\ \alpha'' & \beta'' & \gamma'' \end{bmatrix}$$

where the primed and double-primed cosines are not uniquely determined. However, because of the orthogonality relations, there are available such relations as

$$\alpha'^2 + \alpha''^2 = 1 - \alpha^2$$
$$\alpha'\beta' + \alpha''\beta'' = -\alpha\beta$$

and so forth. Thus it is possible to evaluate

$$\tilde{K}MK = \frac{\rho^2}{EI}\begin{bmatrix} \alpha^2 + m(1 - \alpha^2) & (1 - m)\alpha\beta & (1 - m)\alpha\gamma \\ (1 - m)\alpha\beta & \beta^2 + m(1 - \beta^2) & (1 - m)\beta\gamma \\ (1 - m)\alpha\gamma & (1 - m)\beta\gamma & \gamma^2 + m(1 - \gamma^2) \end{bmatrix}$$

and, introducing the notation

$$\sigma = [\alpha \quad \beta \quad \gamma]$$

this may be written more compactly as $\tilde{K}MK = [(1 - m)\tilde{\sigma}\sigma + mI]\,\rho^2/EI$ and in a similar fashion there is obtained $\tilde{K}NK = (I + \mu\tilde{\sigma}\sigma)/EI$.

Taking the base point as the mid-point of the straight pipe itself, the coordinates are $x = \alpha s$, $y = \beta s$, and $z = \gamma s$, so that by introducing the notation

$$\eta = \begin{bmatrix} 0 & \gamma & -\beta \\ -\gamma & 0 & \alpha \\ \beta & -\alpha & 0 \end{bmatrix}$$

we have $T = \eta s$.   Thus

$$\tilde{K}NKT = (\eta + \tilde{\mu}\sigma\sigma\eta)(s/EI) = \eta s/EI$$

since, as may readily be shown, $\sigma\eta = 0$.   Finally,

$$\tilde{T}\tilde{K}NKT = \tilde{\eta}\eta s^2/EI = -\eta\eta s^2/EI = (I - \tilde{\sigma}\sigma)s^2/EI$$

the last equality being easy to show by direct evaluation.   Thus the desired compliance is

$$C = \int_{-l/2}^{l/2} \left[\begin{array}{c|c} (I - \tilde{\sigma}\sigma)s^2 + [(1 - m)\tilde{\sigma}\sigma + mI]\rho^2 & -\eta s \\ \hline \eta s & I + \mu\tilde{\sigma}\sigma \end{array} \right] \frac{ds}{EI}$$

$$C = \operatorname{diag} \{(I - \tilde{\sigma}\sigma)l^3/12 + [(1 - m)\tilde{\sigma}\sigma + mI]l\rho^2,\ (I + \mu\tilde{\sigma}\sigma)l\}/EI$$

It may be well to review the notation employed in this important formula.   This C is the compliance of a straight piece of pipe of length l oriented with direction cosines α, β, γ with respect to the fixed x, y, z coordinate system (see Fig. 18).

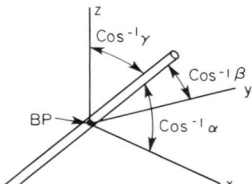

FIG. 18. Straight pipe element having general orientation and with base point at its geometric center.

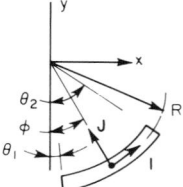

FIG. 19. Circular bend having bend radius R and central arc $\theta_2 - \theta_1$; note the particular orientation.

The symbol I represents both the unit matrix I_3 and the moment of the cross section. E is Young's modulus, and μ is Poisson's ratio.   The row matrix σ is defined as $[\alpha\ \beta\ \gamma]$, and $m = \zeta E/G = 2\zeta(1 + \mu)$, where ζ is the shear distribution factor which will be discussed later.   The base point for this compliance matrix is at the mid-point of the element.

In the case where the straight element is oriented along the x axis, the preceding formula for C may be written as

$$C = \operatorname{diag} \{l\rho^2,\ l^3/12 + ml\rho^2,\ l^3/12 + ml\rho^2,\ (1 + \mu)l,\ l,\ l\}/EI$$

Although outside the scope of the linear theory, the work described in Refs. 22, 104, and 218 is also of interest.

Compliance of a Circular Bend or Elbow.   Figure 19 shows a circular bend having bend radius R and central arc $\theta_2 - \theta_1$.   In order to keep the formulas from becoming too complicated, the compliance matrix will be obtained with respect to the coordinate system shown.   The base point is at the geometric center

of the bend.   Other choices of base point would simplify the final result to a certain extent, but the advantages of such a simplification would be negated by the fact that the base points to accomplish this are inconveniently located and a separate calculation is called for to determine their positions.

Such a piping element does *not* obey the Euler-Bernoulli-Navier theory of bending. Instead, the cross section is able to warp from its originally circular shape in such a way that the relationship between moment and curvature is: curvature $= nM/EI$, where n is a factor greater than unity.   It is a remarkable fact that in the first approximation at least the value of n, called the bend-flexibility factor, is the same for bending in the plane of the pipe (so as to increase or decrease the curvature of the bend) as it is for bending out of the plane of the pipe bend.   In a subsequent section of this analysis a simple derivation will be given for the bend-flexibility factor and a discussion will be offered of more elaborate theoretical investigations and of their experimental verification.   Briefly, a synthesis of all information available at this time suggests the simple formula specified by the ASA B31 Code for Pressure Piping;[1] viz.: $n = 1.65r^2/tR$, where r is the mean radius of cross section t is the wall thickness, and R is the center-line radius of the bend, with the further understanding that no value of n less than unity is to be used.

Thus the diagonal matrix N introduced previously takes the form

$$N = (1/EI) \, \text{diag} \, \{1 + \mu, \, nn\}$$

where n's have replaced unities in the last two elements of the diagonal.   Using fixed unit vectors $\mathbf{i}$, $\mathbf{j}$, $\mathbf{k}$, and using the abbreviations $S = \sin \phi$, $C = \cos \theta$, it is seen that

$$\mathbf{r} = R(S\mathbf{i} + C\mathbf{j})$$
$$\mathbf{T} = \mathbf{I} = C\mathbf{i} + S\mathbf{j}$$
$$\mathbf{N} = \mathbf{J} = -S\mathbf{i} + C\mathbf{j}$$
$$\mathbf{B} = \mathbf{K} = \mathbf{k}$$

so that the matrix K takes the form

$$K = \begin{bmatrix} C & S & 0 \\ -S & C & 0 \\ 0 & 0 & 1 \end{bmatrix}$$

It is convenient also to introduce temporarily a matrix

$$A = \begin{bmatrix} 0 & 0 & 0 \\ 0 & 0 & 0 \\ C & S & 0 \end{bmatrix}$$

Then it is not difficult to calculate

$$\tilde{K}MK = (\rho^2/EI)[mI - (m - 1)\tilde{A}A]$$
$$\tilde{K}NK = (1/EI)[nI + (1 + \mu - n)\tilde{A}A]$$

Also, the matrix T may be written as follows

$$T = R \begin{bmatrix} 0 & 0 & C \\ 0 & 0 & S \\ -C & -S & 0 \end{bmatrix} = R(\tilde{A} - A)$$

and note that $\tilde{T} = -T$. The next several steps are simplified by introducing temporarily a matrix $Q = \text{diag}\{0, 0, 1\}$ and noting the following identities:

$$A(\tilde{A} - A) = Q \qquad\qquad \tilde{A}Q = \tilde{A} \qquad\qquad (A - \tilde{A})\tilde{A} = Q = Q$$

$$(A - \tilde{A})A = AA - \tilde{A}A = -\tilde{A}A$$

Thus, it may easily be shown that

$$\tilde{K}NKT = (R/EI)[(1 + \mu)\tilde{A} - nA]$$
$$\tilde{T}\tilde{K}NKT = (R^2/EI)[(1 + \mu)Q + n\tilde{A}A]$$

Finally, noting that $ds = R\,d\phi$, the compliance matrix integral may be written as

$$C = \frac{1}{EI}\int_{\theta_1}^{\theta_2}$$

$$\left[\begin{array}{c|c} R^3[(1 + \mu)Q + n\tilde{A}A] + R\rho^2[mI - (m - 1)\tilde{A}A] & R^2[(1 + \mu)A - n\tilde{A}] \\ \hline R^2[(1 + \mu)\tilde{A} - nA] & R[nI + (1 + \mu - n)\tilde{A}A] \end{array}\right] d\theta$$

Using the abbreviations $S' = \sin\theta_1$, $S'' = \sin\theta_2$, $C' = \cos\theta_1$, and $C'' = \cos\theta_2$, the several integrals involved may be evaluated as follows:

$$\int_{\theta_1}^{\theta_2} A\,d\phi = \begin{bmatrix} 0 & 0 & 0 \\ 0 & 0 & 0 \\ S'' - S' & C' - C'' & 0 \end{bmatrix}$$

$$\int_{\theta_1}^{\theta_2} \tilde{A}\,d\phi = \begin{bmatrix} 0 & 0 & S'' - S' \\ 0 & 0 & C' - C'' \\ 0 & 0 & 0 \end{bmatrix}$$

$$\int_{\theta_1}^{\theta_2} \tilde{A}A\,d\phi = \begin{bmatrix} \theta_2 - \theta_1 + S''C'' - S'C' & C'^2 - C''^2 & 0 \\ C'^2 - C''^2 & \theta_2 - \theta_1 - S''C'' + S'C' & 0 \\ 0 & 0 & 0 \end{bmatrix}$$

$$\int_{\theta_1}^{\theta_2} I\,d\phi = (\theta_2 - \theta_1)I$$

$$\int_{\theta_1}^{\theta_2} Q\,d\phi = (\theta_2 - \theta_1)Q$$

Finally, performing all the indicated operations and introducing double angles for simplification, the formulas given in Table 3a are obtained for the elements of the compliance matrix. The letters A through L indicate a nomenclature of convenient brevity which will be used shortly in considering orientations other than that shown in Fig. 19.

The cases most frequently encountered are with $\theta_1 = 0$ deg and $\theta_2 = 45$, 90, or 180 deg. If the radius of gyration of the cross section is presumed to be small compared with the radius of the bend, so that $(\rho/R)^2$ is a very small number, the approximations given in Table 3b are useful, particularly for manual computation.

While it might at first appear that the bend-flexibility factor n should somehow enter into the formula for $F = C_{33}$, a little reflection will show that with the base point located at the center of the bend, each element is placed in pure torsion under the action of an F_3 loading, so that neither in-plane nor out-of-plane bending is involved.

Table 3a.   Formulas for Compliance of Pipe Bends and Elbows

$A = C_{11} = (R/4EI)\{(2\theta_2 - 2\theta_1 + \sin 2\theta_2 - \sin 2\theta_1)[nR^2 - (m - 1)\rho^2] + 4\rho^2 m(\theta_2 - \theta_1)\}$

$B = C_{12} = C_{21} = (R/4EI)(\cos 2\theta_1 - \cos 2\theta_2)[nR^2 - (m - 1)\rho^2]$

$C = C_{16} = C_{61} = (nR^2/EI)(\sin \theta_1 = \sin \theta_2)$

$D = C_{22} = (R/4EI)\{(2\theta_2 - 2\theta_1 - \sin 2\theta_2 + \sin 2\theta_1)[nR^2 - (m - 1)\rho^2] + 4\rho^2 m(\theta_2 - \theta_1)\}$

$E = C_{26} = C_{62} = (nR^2/EI)(\cos \theta_2 - \cos \theta_1)$

$F = C_{33} = (R/EI)[(1 + \mu)R^2 + m\rho^2](\theta_2 - \theta_1)$

$G = C_{34} = C_{43} = (R^2/EI)(1 + \mu)(\sin \theta_2 - \sin \theta_1)$

$H = C_{35} = C_{53} = (R^2/EI)(1 + \mu)(\cos \theta_1 = \cos \theta_2)$

$I = C_{44} = (R/4EI)[4n(\theta_2 - \theta_1) + (1 + \mu - n)(2\theta_2 - 2\theta_1 + \sin 2\theta_2 - \sin 2\theta_1)]$

$J = C_{45} = C_{54} = (R/4EI)(1 + \mu - n)(\cos 2\theta_1 - \cos 2\theta_2)$

$K = C_{55} = (R/4EI)[4n(\theta_2 - \theta_1) + (1 + \mu - n)(2\theta_2 - 2\theta_1 - \sin 2_{31} + \sin 2\theta_1)]$

$L = C_{66} = (nR/EI)(\theta_2 - \theta_1)$

All other elements C_{ij} are zero.

If the pipe bend actually has a general orientation with respect to the coordinate system chosen for describing the piping system as a whole, an L transformation may be made from the coordinates (of Fig. 19) used to describe the bend to those used to describe the piping system as a whole, i.e., the global coordinates.   However, since most bends in practice actually lie in planes parallel to the coordinate planes (for the system), a systematic changing of signs and interchanging the elements in the matrix gives the desired compliance matrix, or, to put it in another way, the elements of the L-transformation matrix are all zeros, plus ones, or minus ones, so that the L transformation itself amounts to a shifting and shuffling of elements.

In order to employ this simple method of performing an L transformation, an "orientation symbol" will be used to describe the orientation of an actual bend. This symbol consists of two of the letters x, y, z, each preceded by a plus or minus sign, the two letters indicating the coordinate plane to which the bend is parallel. The method of determining the symbol is most easily given by the use of examples. In Fig. 20, one may say that the bend "starts" out in the $+x$ direction (at angle θ_1) and bends toward the $+y$ direction (finishing at angle θ_2); the symbol is $(+x, +y)$.

Table 3b.   Numerical Approximations for Compliances of Pipe Bends and Elbows

	$\theta_1 = 0°, \theta_2 = 45°$	$\theta_1 = 0°, \theta_2 = 90°$	$\theta_1 = 0°, \theta_2 = 180°$	Common factor
A	0.64270	0.78540	1.57080	nR^3/EI
B	0.25000	0.50000	0.00000	nR^3/EI
C	−0.70711	−1.00000	0.00000	nR^2/EI
D	0.14270	0.78540	1.57080	nR^3/EI
E	−0.29289	−1.00000	−2.00000	nR^2/EI
F	1.02102	2.04204	4.08407	R^3/EI
G	0.91924	1.30000	0.00000	R^2/EI
H	0.38076	1.30000	2.60000	R^2/EI
I	$\begin{pmatrix}0.83551 +\\ 0.14270n\end{pmatrix}$	$\begin{pmatrix}1.02102 +\\ 0.78540n\end{pmatrix}$	$\begin{pmatrix}2.04204 +\\ 1.57080n\end{pmatrix}$	R/EI
J	0.325–0.25n	0.65–0.5n	0.00000	R/EI
K	$\begin{pmatrix}0.18551 +\\ 0.64270n\end{pmatrix}$	$\begin{pmatrix}1.02102 +\\ 0.78540n\end{pmatrix}$	$\begin{pmatrix}2.04204 +\\ 1.57080n\end{pmatrix}$	R/EI
L	0.78540	1.57080	3.14159	nR/EI

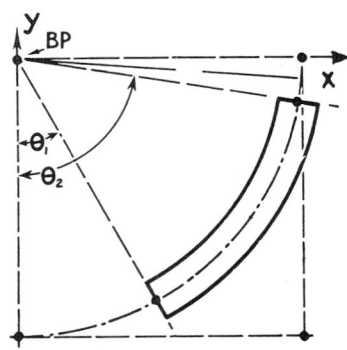

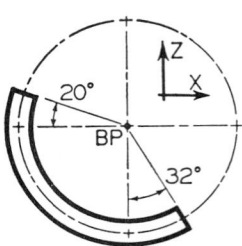

Fig. 20. Notation used in defining orientation symbol for pipe bend in planes parallel to global coordinate planes but not oriented as in Fig. 19.

Fig. 21. Example showing that many different descriptions of the same bend are possible.

This case corresponds to the orientation shown in Fig. 19, and no changes of sign or interchanges of the matrix elements are necessary.   Alternately, this same bend can be considered as starting out in the $-y$ direction (at angle $90 - \theta_2$ deg) and bending toward the $-x$ direction (ending with angle $90 - \theta_1$ deg), and thought of in this way, the bend is described by the orientation symbol $(-y, -x)$, with appropriate interpretation of the angles involved.   The case shown in Fig. 21 may have any of the following descriptions:

Symbol	θ_1, deg	θ_2, deg
$+x, +z$	-110	32
$+x, -z$	148	290
$-x, +z$	-32	110
$-x, -z$	58	200
$-z, +x$	-122	20
$+z, -x$	160	302
$-z, +x$	-20	122
$-z, -x$	58	200

Of these the third and the seventh are most obvious, but any one is satisfactory. Corresponding to any one of these choices, using the orientation symbol shown (and noting the values of θ_1 and θ_2) one refers next to Table 4 to see how the elements of the desired compliance matrix are obtained from those of the standard case $(+x, +y)$ for which formulas have already been derived.

The Shear-distribution Factor ζ.   One of the important areas of investigation within the framework of the mathematical theory of elasticity is the matter of the behavior of beams of arbitrary cross section.   Only a few solutions are known, among them the solution of a straight tip-loaded cantilever beam, with the load distributed on the tip cross section in a certain manner determined by the geometry of the cross section.   For this case, presuming that the lateral forces of restraint at the constrained end are also distributed in the same manner, it is possible to obtain a complete solution which shows the effect of both bending and cross shear.   Actual beams are not loaded in this ideal fashion, and the principle of St. Venant must be employed to explain that the actual stress distributions and deformations are not "much" different from the local stress distributions and deformations.

Table 4. Interchange of Matrix Elements for 24 Basic Orientations of Bends and Elbows‡‡

Case No.	Orient. symbol	11	12	13	15	16	22	23	24	26	33	34	35	44	45	46	55	56	66
1	+x, +y	A	B	.	.	C	D	.	.	E	F	G	H	I	J	.	K	.	L
2	+x, −y	A	−B	.	.	−C	D	.	.	E	F	−G	H	I	−J	.	K	.	L
3	−x, +y	A	−B	.	.	−C	D	.	.	−E	F	−G	−H	I	−J	.	K	.	L
4	−x, −y	A	B	.	.	C	D	.	.	−E	F	G	−H	I	J	.	K	.	L
5	+y, +z	F	.	.	G	H	A	B	C	.	D	E	.	L	.	.	I	J	K
6	+y, −z	F	.	.	−G	H	A	−B	−C	.	D	E	.	L	.	.	I	−J	K
7	−y, +z	F	.	.	−G	−H	A	−B	−C	.	D	−E	.	L	.	.	I	−J	K
8	−y, −z	F	.	.	G	−H	A	B	C	.	D	−E	.	L	.	.	I	J	K
9	+z, +x	D	.	B	E	.	F	.	H	G	A	.	C	K	.	J	L	.	I
10	+z, −x	D	.	−B	E	.	F	.	H	−G	A	.	−C	K	.	−J	L	.	I
11	−z, +x	D	.	−B	−E	.	F	.	−H	−G	A	.	−C	K	.	−J	L	.	I
12	−z, −x	D	.	B	−E	.	F	.	−H	G	A	.	C	K	.	J	L	.	I
13	+x, +z	A	.	B	C	.	F	.	G	H	D	E	.	I	.	J	L	.	K
14	+x, −z	A	.	−B	−C	.	F	.	G	H	D	−E	.	I	.	−J	L	.	K
15	−x, +z	A	.	−B	−C	.	F	.	−G	−H	D	−E	.	I	.	−J	L	.	K
16	−x, −z	A	.	B	C	.	F	.	−G	−H	D	E	.	I	.	J	L	.	K
17	+y, +x	D	B	.	.	G	A	.	.	C	F	H	E	K	J	.	I	.	L
18	+y, −x	D	−B	.	.	−G	A	.	.	C	F	−H	E	K	−J	.	I	.	L
19	−y, +x	D	−B	.	.	−G	A	.	.	−C	F	−H	−E	K	−J	.	I	.	L
20	−y, −x	D	B	.	.	G	A	.	.	−C	F	H	−E	K	J	.	I	.	L
21	+z, +y	F	.	.	.	.	D	B	E	C	A	H	G	L	.	.	K	J	I
22	+z, −y	F	.	.	.	.	D	−B	−E	C	A	H	−G	L	.	.	K	−J	I
23	−z, +y	F	.	.	.	.	D	−B	−E	−C	A	−H	−G	L	.	.	K	−J	I
24	−z, −y	F	.	.	.	.	D	B	E	−C	A	−H	G	L	.	.	K	J	I

† Opposite desired case number and under symbol for desired element find letter together with algebraic sign giving desired entry. The dots denote zeros.

‡ The numbers at the top of Table (11, 12, 13, 15, etc.) are subscripts for elements in the compliance matrix C_{ij} as listed in Table 3a.

Making use of the solution which is available for the case of a straight tip-loaded cantilever of hollow cylindrical cross section and presuming ideal loading in the sense just described, there is still some ambiguity regarding the influence of cross shear. One method of analysis makes use of the average value of the shearing stress over the neutral surface, and a second method makes use of the shearing strain energy in the cross section. In this discussion, the shear-distribution factor as determined by these methods will bear the subscripts 1 and 2, respectively. Customary formulas for evaluation are

$$\zeta_1 = \frac{\text{average } (\tau_{xz})_{x=0}}{P/A}$$

where A is the cross-sectional area, P is the total load, and τ_{xz} is the component of shearing stress acting on a normal section and in the direction of the loads, and

$$\zeta_2 = (A/P^2)\int\int(\tau_{xz}^2 + \tau_{yz}^2)dx\,dy$$

in which τ_{yx} is the component of shearing stress on a normal section in a direction perpendicular to that of the loads. The contribution of the τ_{yz}^2 term is generally much smaller than that of the τ_{xz}^2 term, and, moreover, while it generally is rather easy to get reasonably good values for τ_{xy}, it is quite difficult to get good values for τ_{yz}. Accordingly, ζ_2 is frequently approximated by taking $\tau_{yz} = 0$. Customarily reported values for several interesting cross sections are as follows:

Cross section	ζ_1	ζ_2
Solid rectangle......	$\frac{3}{2}$	$\frac{6}{5}$†
Solid circle.........	$\frac{4}{3}$	$\frac{10}{9}$†
Thin annulus.......	2	2

† Indicates neglect of τ_{yz}.

The fact that $\zeta_1 = \zeta_2$ for a thin-walled tube is accounted for by the somewhat remarkable but rather easily verified lack of any warping of the cross section. The differences between ζ_1 and ζ_2 in the other cases is attributable to the fact that each method of evaluation makes certain implicit assumptions regarding the constraints which are applied, and if there is warping, the difference in constraint gives rise to a difference in deflection.

An exact evaluation is possible for the case of a solid or hollow cylinder, and making use of bending solutions available in the literature on elasticity [cf. Sokolnikoff[236]], the following formulas have been obtained by the author.[64]

$$\zeta_1 = (4/3)[1 + q/(1 + q^2)]$$

$$\zeta_2 = \frac{(7 + 14\mu + 8\mu^2) + [2q^2/(1 + q^2)^2](10 + 20\mu + 8\mu^2)}{6(1 + \mu^2)}$$

In these formulas μ denotes Poisson's ratio and q denotes the ratio of inside diameter to outside diameter. Figure 22 shows a plot of ζ_1 and ζ_2 as functions of q, ζ_2 being evaluated using $\mu = 0.3$. It is quite evident that even for rather thick-walled tubes, the values of ζ_1 and ζ_2 do not differ much from 2.0, and when it is realized that the shearing contribution to deflection is quite small anyhow, the customary use of $\zeta = 2$ is fully justified. However, it is not at all difficult to incorporate either of the above formulas in a computer program for piping flexibility. Although the two evaluations do not differ much in the most important range of values of q, the second formula is probably to be preferred.

In summarizing this section it may be said that theoretical analysis of the bending of a straight cantilever beam of hollow cylindrical cross section which is subjected to ideally disposed lateral loads at its two ends leads to two different values of ζ which differ because of subtle differences in the constraints implicitly incorporated in the two evaluations. Then one or the other (or a compromise) value is universally applied in piping analysis to curved sections as well as straight sections and to cases where it is known that the loads are not applied in the ideal manner initially presumed. Even so, the results are probably very accurate.

The Bend-flexibility Factor n. Customarily, methods of strength of materials are used to analyze the stress distribution in and deformation of pipe bends. For compact cross sections, such as rectangles and solid circles, the customary Winkler-Bach analysis for curved beams leads to a hyperbolic stress distribution and the corresponding deformations may be evaluated. For noncompact cross sections such as thin flanged sections or hollow ducts and pipes, a much more important reason for deviation from the usual Euler-Bernoulli-Navier analysis is that the warping of the cross section from its initial shape permits some of the fibers to escape the axial deformation which would be required under either the linear strain of the usual theory or the hyperbolic strain of the Winkler-Bach theory.

This results in a flexibility greater than that which would be calculated under the usual theory, and this increased ratio of

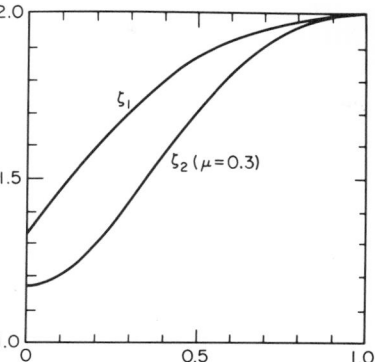

FIG. 22. Values of the shear distribution factor ζ. Abcissas are value of the ratio $q =$ (inside diam)/(outside diam) while ordinates are values of the factor. ζ_1 is based on average shear strain on neutral axis, and ζ_2, which is to be preferred, is based on strain energy considerations.

flexibility is called the bend-flexibility factor and is denoted here by the symbol n. (In other works, it may be denoted by K or k.) At the same time, since the cross section is being bent out of its original shape, there are introduced important transverse bending stresses, so that the "effective" stress acting in the bend is greater, by a factor i called the bend stress-intensification factor, than would be calculated using the usual theory.

Much theoretical work has been done by several investigators in the determination of expressions for n. A full account is provided in Refs. 150 and 252.

A satisfactory approximation, introduced by Beskin[26] and simplified by Markl[184] and incorporated in the Code for Pressure Piping is

$$n = 1.65/h \qquad n \text{ not less than } 1.0$$

where $\qquad h = tR/r^2 \qquad h$ is dimensionless

The above expression for n neglects the effects of internal pressure P. If these are included the revised bend-flexibility factor becomes

$$n_p = n/[1 + 6(P/E)(r/t)^{7/3}(R/r)^{1/3}]$$

For 90-deg bends and elbows which are flanged at one end, the Code for Pressure Piping gives

$$n' = 1.65/h^{5/6}$$

and if the 90-deg bend or elbow is flanged at both ends

$$n'' = 1.65/h^{\frac{2}{3}}$$

Those who are concerned with the theoretical development of the first-approximation von Karman equation

$$n = (12h^2 + 10)(12h^2 + 1)$$

are referred to the derivation of den Hartog.[87]

Other Sources of Compliance.   Although the two principal components of piping systems are straight lengths and bends (or elbows), other elements are also frequently involved.   These include valves, strainers, reducers, tees and nozzles, etc.   Furthermore, there is frequently appreciable flexibility in the terminal equipment such as turbine shells, heat-exchanger shells, headers and tube bundles, etc.   Presently there are very few data available concerning the compliance of such elements, and the piping designer must employ considerable ingenuity in estimating the effect of these sources of flexibility.

Usually the length of such elements is rather small compared with the total length of the entire piping system, so that a reasonable estimate of flexibility will not compromise the validity of the flexibility analysis.   For example, the flexibility of a valve may be approximated as that of a length of straight pipe having the same length as the valve but having greater wall thickness and diameter so as to approximate the actual geometry of the valve.   The same procedure may be used with reducers, strainers, and other such elements.

In the usual procedure for flexibility analysis, the geometry of the system is idealized to a line.   Pipe branches are considered to intersect at the point of intersection of their center lines.   In effect, this gives a somewhat greater length of outlet than is actually present (by one run radius); however, this additional flexibility tends to account for the fact that there are local distortions at the intersection which are otherwise not accounted for.   In the case of a large, thin pipe or of a large cylindrical or spherical vessel, the local deformations may be significant.   In an important series of papers (Refs. 28 to 38), Bijlaard has studied the deformations and stresses at the intersection of piping and cylindrical and spherical shells, and his results may be used to estimate the additional compliance resulting from local deformations.   Stevens, Groth, and Bell[243] have shown how these results may be employed in piping-flexibility analysis, using sample problems and referring to the several papers of Bijlaard for the necessary coefficients.

There is no good reason why the compliance matrix for a definite piece of terminal equipment, such as a pump, could not be determined experimentally and made use of in a flexibility analysis of the piping system in which the pump is terminal equipment.   However, this does not yet appear to have been done.

Another important type of terminus is connection to a header which in turn is connected to an anchor only via a bundle of relatively flexible tubes.   The overall flexibility of such an assembly may be appropriate and could contribute substantially to the overall compliance of the system.   It may be taken into account simply by regarding the header and tubes to be a part of the piping system and employing the procedures previously described.   However, the degree of redundance would, in the usual case, be greatly increased. and some precalculating or prearranging may be useful.   Figure 23 shows a header and tube bundle, each tube itself having one or more constraints.   If C denotes the compliance matrix (including the effect of constraints) for a single tube at its convenient base point P, and if the tubes are all alike and the base point of each is in the same geometrical location relative to the tube, and if, furthermore, the header itself is relatively stiff compared with the tubes,

the compliance of the entire assembly at some convenient base point Q may be obtained as

$$C_{\text{total}} = [\Sigma \, (\tilde{B}_{P_iQ} C B_{P_iQ})^{-1}]^{-1}$$

In many cases, the computational labor may be considerably simplified, since the geometry of the bundle may be such that several of the generally nonvanishing terms in B_{P_iQ} may be made zero by a convenient choice of base points. It should be observed that gas seals, etc., frequently involve unknown restraints, and these restraints should not be underestimated in estimating the compliance of individual tubes.

Corrugated piping has been employed to provide increased flexibility as compared with straight cylindrical piping. While piping fabricators are prepared to produce corrugated piping, corrugated bends, or what are called creased bends, most of them advise that there has been almost no call for such elements within the last ten years or so. Based on what can hardly be considered as adequate or convincing test data, a bend-flexibility factor $n = 5.0$ for both in-plane and out-of-plane bending has been recommended in the ASA B31 Code for Pressure Piping for corrugated straight pipe and corrugated or creased bends. Since the geometry of the corrugations is not specified it is clear that this value is some sort of overall average for the geometries represented in the few experiments which were conducted. Some studies of corrugated piping are reported in Refs. 7, 91, and 202.

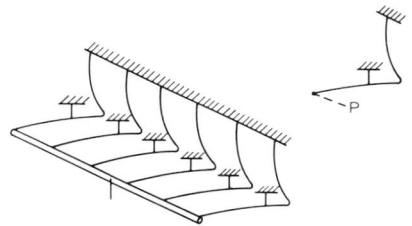

FIG. 23. Header and tube bundle. Small figure shows one of many identical tubes. If header is stiff, it is easy to obtain a single compliance matrix for the entire assembly.

Miter bends are also used occasionally in piping, and these are subjected to the same influences tending to cause ovalization as are involved with circular bends and elbows, although the analysis is considerably more complicated and less certain. Formulas for the flexibility factors (and other properties) of miter bends are given in Fig. 29 of this chapter.

There are two other sources of additional compliance which may be mentioned. First, it has been postulated so far that the material of which the piping system is composed obeys a generalized Hooke's law so that, briefly, stress is proportional to strain. However, for most engineering materials, this is true (leaving geometrical effects aside for the moment) only until the stress has reached the proportional limit, after which strain increases at an accelerating rate compared with stress. Thus, if the thermal expansion to be accommodated is sufficiently large, to take into account only the compliances mentioned so far may grossly overestimate the reactions and stresses in the system. While generally it is undesirable for the pipe to yield plastically (except perhaps on initial operation), there may be some comfort in the knowledge that in the usual case wherein the piping is much weaker than the terminal equipment, the pipe will yield before it applies to the terminal equipment the exceedingly large reactions which might be calculated. However, the designer must be alert to the consequences of plastic behavior. The allowable stress ranges which are incorporated in the Code for Pressure Piping have been selected to assure that even though plastic behavior may take place (under certain circumstances) during initial operation, shortly thereafter all such plastic action will cease and the system will behave elastically in subsequent thermal cycling.

The last source of additional compliance that will be described here results from the fact that the analysis so far has oversimplified the actual geometry of the

system in order to obtain linear relationships which are more susceptible to satisfactory mathematical analysis.   There are at least two additional compliance effects of this sort which may be of importance in special cases even though their neglect is quite customary.   These effects are such as to reduce actual reactions and stresses below the values that are calculated by the customary analyses.   Specifically they are the neglect of the bending-moment contributions that result from axial forces acting through the lateral displacements and the neglect of the difference between arc and chord as a structure is distorted.

To illustrate the first of these effects, namely, the contribution of axial force to bending moment, consider the plane system illustrated in Fig. 1a. The horizontal member may be considered to be loaded as in Fig. 24. The customary Euler-Bernoulli relationship is

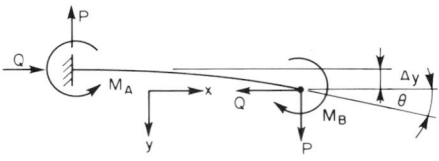

FIG. 24. Horizontal member of an L configuration, subjected to moment, lateral, and axial loading at the tip.   The axial load has a nonlinear intensifying effect.

$$EIy'' = M_A - Qy - Px$$

subject to the conditions that $y = y' = 0$ at $x = 0$.   Letting

$$\alpha^2 = Q/EI$$

it may be determined that

$$y = (P/\alpha Q)(\sin \alpha x - \alpha x) + (M_A/Q)(1 - \cos \alpha x)$$

satisfies these conditions.   Evaluating at $x = a$,

$$\Delta y = (P/\alpha Q)(\sin u - u) + (M_A/Q)(1 - \cos u)$$
$$\theta = -(P/Q)(1 - \cos u) + (\alpha M_A/Q) \sin u$$

where $u = \alpha a$.   However, $M_A = M_B + Q\Delta y + Pa$, so that the following relationships may be obtained:

$$\Delta y = (P/\alpha Q)(\tan u - u) + (M_B/Q)(\sec u - 1)$$
$$\theta = (P/Q)(\sec u - 1) + (\alpha M_B/Q) \tan u$$

and writing $Q = EI\alpha^2 = EIu^2/a^2$, these may be written

$$\begin{bmatrix} \Delta y \\ \theta \end{bmatrix} = \frac{1}{EI} \begin{bmatrix} \tfrac{1}{3}a^3 R_1 & \tfrac{1}{2}a^2 R_2 \\ \tfrac{1}{2}a^2 R_2 & a R_3 \end{bmatrix} \begin{bmatrix} P \\ M_B \end{bmatrix} = \begin{bmatrix} a_{11} & a_{12} \\ a_{21} & a_{22} \end{bmatrix} \begin{bmatrix} P \\ M_B \end{bmatrix}$$

where

$$R_1 = 3(\tan u - u)/u^3$$
$$R_2 = 2(\sec u - 1)/u^2$$
$$R_3 = \tan u/u$$

Similarly, by considering the vertical leg, one may obtain

$$\begin{bmatrix} \Delta x \\ -\theta \end{bmatrix} = \frac{1}{EI} \begin{bmatrix} \tfrac{1}{3}b^3 S_1 & \tfrac{1}{2}b^2 S_2 \\ \tfrac{1}{2}b^2 S_2 & b S_3 \end{bmatrix} \begin{bmatrix} Q \\ M_B \end{bmatrix} = \begin{bmatrix} b_{11} & b_{12} \\ b_{21} & b_{22} \end{bmatrix} \begin{bmatrix} Q \\ M_B \end{bmatrix}$$

where

$$S_1 = 3(\tan v - v)/v^3$$
$$S_2 = 2(\sec v - 1)/v^2$$
$$S_3 = \tan v/v$$
$$v = \beta b$$
$$\beta^2 = P/EI$$

Combining these results so as to eliminate θ and M_B, one obtains

$$\begin{bmatrix} \Delta_t \\ \Delta x \end{bmatrix} = \begin{bmatrix} C_{11} & C_{12} \\ C_{21} & C_{22} \end{bmatrix} \begin{bmatrix} P \\ Q \end{bmatrix}$$

where
$$C_{11} = a_{11} - a_{12}^2/(a_{22} + b_{22})$$
$$C_{12} = C_{21} = -a_{12}b_{12}/(a_{22} + b_{22})$$
$$C_{22} = b_{11} - b_{12}^2/(a_{22} + b_{22})$$

it being observed that $a_{ij} = a_{ji}$ and $b_{ij} = b_{ji}$. As in "Introductory Example and Discussion Methods"

$$\Delta y = eb + \delta y_A - \delta y_C$$
$$\Delta x = ea + \delta x_A - \delta x_C$$

may readily be determined and the solution for P and Q could easily be obtained by inverting the C_{ij} matrix except only that its elements are functions of u and v and thus of the unknowns P and Q themselves.

Livesley,[174] among the first to discuss problems of this type, employs an iterative procedure in which the effect of axial load is neglected in an initial evaluation so that the first solution corresponds to that obtained previously. By considering limiting values, it may be verified that all R's and S's become unity for zero axial load; accordingly, the matrix solution presently under discussion will provide these original results. Next, the initial values of P and Q are used to determine values of u and v which, in turn, are used to evaluate the R's and S's, from which a second evaluation is made of P and Q. This iterative procedure is continued until convergence is obtained.

In Livesley's formulation, the stiffness rather than the compliance matrix is used, and the functions corresponding to the present R's and S's are better behaved, so that the question of convergence seems to give less trouble than with the present formulation in terms of compliance. The nature of the difficulty is easy to see upon considering the nature of the functions R_i. If Q becomes negative, no real difficulty is involved. The trigonometric functions are replaced in an appropriate way by hyperbolic functions. However, for $u = \pi/2$, i.e., for $Q = \pi^2 EI/4a^2$, the R's become infinite. This is to be expected, since this value of the axial load corresponds to the Euler buckling load on a prismatic beam or column fixed at one end and freely hinged at the other. (Livesley uses the corresponding behavior of his functions to determine the critical buckling loads on an entire structure.) In the case of a piping structure, it is the deformations that are determined by the thermal expansion, and the loads must adjust themselves to these deformations. Accordingly, since the deformations remain finite, the loads under no circumstances may exceed the Euler loads. The iterative procedure indicated above may accordingly fail to converge to the correct value. If the expansions are substantial, the values of P or Q calculated initially neglecting the effect under discussion may exceed the Euler loads. This would throw the determinations of the R's or S's onto another and physically meaningless branch beyond the point where these functions become infinite.

The author has been successful in some experiments with solutions of this sort by modifying the functions in a physically unimportant way for the present purpose (although this modification would deny the possibility of determining buckling loads for the structure). For values of u larger than, say, $u = 1.5$, the actual curves are to be replaced by their tangents drawn at $u = 1.5$. With this modification, the iterative procedure has converged in several test problems. No proof is given for this convergence, and it is conceivable that some concavity should be

introduced in the replacement curve rather than the straight-line tangent mentioned above.

All three of these functions are very closely approximated by the function $R_4 = (1 - Q/Q_e)^{-1}$, where Q is the Euler load $\pi^2 EI/4a^2$. A graph of this substitute function is shown in Fig. 25, since the individual functions R_1, R_2, and R_3 would lie too close together over most of their range to permit distinguishing them in a plot of this scale. However, the correct functions may be calculated and employed easily in digital computer programs. In the preceding, direct axial and shearing deformations were neglected. To take them into account, Δy and Δx should be decreased, respectively, by the quantities $Pb/AE + 2Pa/AG$ and $Qa/AE + 2Qb/AG$. Thus, the set of equations to be dealt with becomes

$$\begin{bmatrix} \Delta y \\ \Delta x \end{bmatrix} = \begin{bmatrix} C_{11} + b/AE + 2a/AG & -C_{12} \\ -C_{21} & C_{22} + a/AE + 2b/AG \end{bmatrix} \begin{bmatrix} P \\ Q \end{bmatrix}$$

Even so, these equations do not include the effect of the difference between the curved lengths of the members and their straight-line projections on the coordinate axes. Since

$$ds - dx = (\sqrt{1 + (y')^2} - 1)\, dx \approx \tfrac{1}{2} y'^2\, dx$$

this difference may be approximated by obtaining as simple an expression for y' as possible and integrating. From the initial solution, it may be shown that

$$y' = \frac{2 \sin z}{Q \cos u} [P \sin (u - z) + \alpha M_B \cos z]$$

where $z = \tfrac{1}{2}\alpha x$. The desired difference in length λ may then be calculated as

$$\lambda = \int_0^a \tfrac{1}{2} y'^2\, dx = \frac{1}{\alpha} \int_0^{u/2} y'^2\, dz$$

$$= P^2 T_1(u) + P M_B T_2(u) + M_B^2 T_3(u)$$

where

$$T_1(u) = \frac{u + 2u \cos^2 u - \sin u \cos u\, (3 - 4 \cos u + 4 \cos^2 u)}{4u^5 \cos^2 u E^2 I^2/a^5}$$

$$T_2(u) = \frac{u \sin u - 2 \cos u + 2 \sin u \cos^2 u}{2u^4 \cos^2 u E^2 I^2/a^4}$$

$$T_3(u) = \frac{u - \sin u \cos u}{4u^3 E^2 I^2/a^3}$$

and noting that $u = a\sqrt{Q/EI}$, these can be evaluated in terms of Q. Thus the value of $\Delta y = eb + \delta y_A - \delta y_C$ should be decreased by this value of λ, and the value of $\Delta x = ea + \delta x_A - \delta x_C$ should be decreased by a correspondingly calculated amount. In these calculations, M_B should be replaced by its value in terms of P and Q; namely,

$$M_B = -(a_{21} P + b_{21} Q)/(a_{22} + b_{22})$$

Now both the Δ's and the coefficients are functions of the unknowns P and Q, but for cases where the first approximation to λ is a good one, a convergent iterative solution can be obtained easily with a modern digital computer. For tensile loads, the formulas for the T's should be manipulated in an appropriate fashion to obtain hyperbolic rather than trigonometric functions.

It does not appear that much work has been done with computational schemes such as that described above. Certainly these nonlinear effects are of minor

importance in most configurations, that is, in those in which one dimension does not greatly exceed the other two.   To see this, consider the equal-leg case of the foregoing; this can be handled very easily by invoking symmetry and noting that $a = b$, $P = Q$, and $\theta = 0$, from which the simpler formula

$$\Delta y/a = (\csc u - \cot u)/\tfrac{1}{2}u - 1$$

may be obtained.

Taking the case where the terminal deflections δ_A and δ_C are zero, writing $\Delta y = ea - PA/AE - 2Pa/AG$, noting that $I/A = D^2/8$ for a thin-wall pipe taking $\mu = 0.3$, and performing some simplifications lead to the equation

$$12e = u^2(1 + 9.3D^2/a^2)$$

terms of order u^4 and higher being neglected.   First, it may be noted that it would be a very stubby pipe indeed for which the second term in parentheses would be of importance.   Neglecting this term, which arises from direct axial and cross shearing deformations, there is obtained the equation $(e + 1)u/2 = \csc u - \cot u$.   Expanding the right side in series leads to $e = u^2/12 + u^4/120 + \cdots$.   Noting that $u^2 = Pa^2/EI$ and solving, there results to a high degree of accuracy

$$P = \frac{12eEI/a^2}{1 + 1.2e}$$

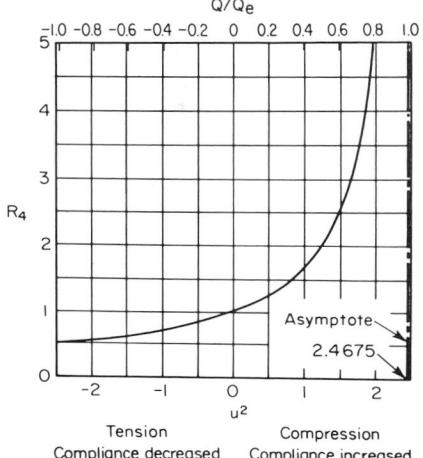

FIG. 25. Graph of the intensifying functions R_1, R_2, and R_3 used in discussing the influence of axial loading.   Actually only the substitute function $R_4 = (1 - Q/Q_e)^{-1}$ has been plotted; R_1, R_2, and R_3 lie within plotting error of R_4.   Scale of abcissas at the top is in terms of the ratio Q/Q_e; that at the bottom is in terms of $u^2 = Qa^2/EI$.

the numerator of which is the result obtained by the usual methods of structural analysis and the denominator of which indicates the effect of the increased compliance due to axial compression.   In practical cases, for even the highest temperature ranges, the value of e does not exceed about 0.01, so that the increased compliance reduces the axial force by only about 1 per cent or less.   Similar reductions may be expected in the bending moments and the stresses.

The preceding paragraph described a case in which the two major dimensions of the piping configuration were equal.   However, if one dimension greatly exceeds the other two, the nonlinear effects of additional compliance due to axial force and the difference between arc and chord may be of some importance in admitting configurations which would appear too stiff by usual calculation methods.   Some recent work in this area will be reported in Ref. 65.   Another methodology is given in Ref. 113.

On the other hand, consideration of such effects may also reveal that stresses and deflections are considerably *greater* than ordinary computations would indicate. As an example, consider a straight 100-ft length of 8-in. Schedule 40 ASTM A-106 Grade B pipe, installed with no stress at 70 F and rigidly constrained at its ends. To simplify the analysis, constant values of $\alpha = 6.4 \times 10^{-6}/\deg$ F and $E = 28.0 \times 10^6$ psi will be used.   Pertinent pipe data are $I = 72.5$ in.^{4}, $A = 8.40$ in.^{2}, $c = D/2 = 4.31$ in., and $k^2 = 72.5/8.40 = 8.63$ in.^{2}.   (k is the radius of gyration.)

The usual formula from strength of materials presumes axial loading and strain; it does not indicate any lateral deflection at all and gives a direct compressive stress of

$$S = E\alpha(T - 70) = 179.2(T - 70)$$

This relation is correct only up to a temperature of 107 F, since at this temperature the calculated stress is 6,625 psi which corresponds to the Euler critical stress $S_e = 4\pi^2 Ek^2/L^2$ for a fixed-end column. As the temperature increases above 107 F, the net axial force remains constant (equal to AS_e) and the pipe begins to buckle to the

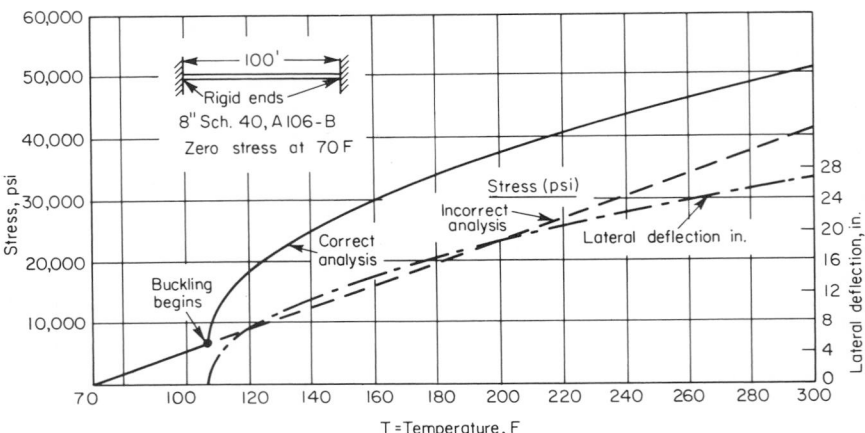

FIG. 26. Stress and deflection in 8-in. pipe example.

side. The shape of this buckle is

$$y = B[1 - \cos(2\pi x/L)]$$

where B is half the mid-length lateral deflection. From this,

$$y' = (2\pi B/L)\sin(2\pi x/L) \quad \text{and} \quad \lambda = \pi^2 B^2/L.$$

Also, the maximum curvature is (approximately) $4\pi^2 B/L^2$, which gives a bending stress of $4\pi^2 EcB/L^2 = S_e cB/k^2$. From this, the total stress is $S_T = S_e(1 + cB/k^2)$. Also, the expansion is accommodated by both the net axial deformation $S_e L/E$ and the difference λ between arc and chord, so that $L\alpha(T - 70) = \lambda + S_e L/E$. From this it may be established that $T = 70 + (\pi^2/\alpha L^2)(B^2 + 4k^2)$. Inserting numerical values, $S_T = 6,625 + 3,310B$ and $T = 107 + 1.071B^2$. Eliminating B, $S_T = 6,625 + 3,200\sqrt{T - 107}$. The last formula should replace $S = 179.2(T - 70)$ for $T > 107$ F. These relationships are shown in Fig. 26. If the allowable stress range were, say, 22,500 psi, the superficial analysis would indicate that the temperature could rise to 195 F and there would be no lateral deflection. Actually, at 195 F the stress would be 36,500 psi, right at the ordinary yield stress, and the lateral deflection would be 18 in.! Not to belabor this rather impractical example much more, one may note the effect of placing a rigid restraint right at the 50-ft position. This has the effect of breaking the previous case up into two cases in which the length is half its previous value. It may be calculated that the linear relation $S = 179.2(T - 70)$ now holds up to $T = 218$ F, corresponding to an Euler critical stress of 26,500 psi.

However, any increase in temperature above 218 F sends both total stress and lateral deflection (at the 25- and 75-ft positions) up at an intolerably rapid rate. The value of this example is not solely pedagogical if it be noted that this line of reasoning permits deciding the maximum spacing of lateral restraints in low-temperature piping systems.   The basic idea is to assure that there is sufficient restraint to prevent buckling at the maximum temperature, since buckling generally causes inadmissable total stress and lateral deflection.   If buckling does not take place, the stress may be calculated from the formula $S = E\alpha \, \Delta T$.   The anchors at the ends must be able to sustain a compressive load of amount SA.

In all this discussion of the effect of axial loads and the effect of the difference between arc and chord, attention was confined to initially straight elements, rigidly joined so as to form a structure.   There do not seem to be any similar studies available which specifically refer to curved elements such as pipe bends or are directly applicable to structures which are composed of both straight and curved portions.   Evidently there is much work to be done in this area that might profitably occupy the attention of able theoretical structural analysts.

SOLUTION OF THE FLEXIBILITY EQUATIONS

After having determined the various compliances, transforming them, and combining them, and after having determined the deformation constants D_i, the flexibility analyst is confronted with the problem of solving a large set of simultaneous equations.   If the number of anchors is N and the number of individual constraints is C, the number of equations is $6(N - 1) + C$, and this can be a formidable task.   (In alternate procedures, the corresponding operation may consist of inverting a large number of 6×6 matrices; this task is also formidable.) In a very large number of applications in science and technology this problem of solving large linear systems is encountered, and there is a considerable literature on the subject.

The difficulties are generally of two types: those concerned with the astronomically large number of arithmetic operations required and those concerned with round-off and other errors which, with so many operations involved, can lead to very serious errors in the final results.   It is quite beyond the scope of this chapter to go into detail on this subject; it is a job for a competent and specially trained mathematician to deal with these problems and to select and modify computational procedures appropriate to the nature of the problem and to the facilities available for computation.   The nature of the problem and the difficulties surrounding its solution are entertainingly discussed by Forsythe.[109]   Dwyer's excellent book[96] considers entire hierarchies of methods of solving linear equations and inverting matrices.   The author has most frequently employed the "square-root method" described by Dwyer, although probably most piping-flexibility analysts have used Crout's method,[84] which is also described by Dwyer.   Other references are 44, 101, 111, and 136.

For systems having any appreciable degree of complexity, manual solutions become by far too uneconomically time consuming and recourse must be had to automatic computation using high-speed electronic digital computers.   While the use of computers in piping-flexibility analysis will be discussed later in this chapter, no discussion will be given there of the task which confronts the computer programmer in choosing the optimum computation procedure for solving the simultaneous equations, for assuring adequate accuracy, and for minimizing required computer time.   If one is faced with the problem, it is well to consult an able mathematician or to avail oneself of the experience and accumulated procedures of the manufacturer of the computer to be employed.

CALCULATION OF STRESSES AND DEFLECTIONS

Determination of Internal Loading at a Point.   The solution of the system of linear equations provides terminal, constraint, and loop-cut forces, all normalized to the common reference or base point.   For each point internal to the configuration at which it is desired to evaluate significant stresses, it is necessary to determine the loading transmitted by one portion of the system to the other portion via the point under consideration.   Typically, such a point is point J of Fig. 27.   The solution

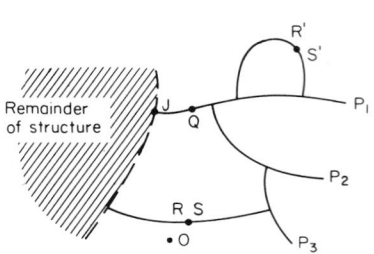

FIG. 27. Determination of force system transmitted from one portion of the system to another via a particular point (e.g., point J) of the system.

gives anchor forces for anchors P_1, P_2, and P_3; the restraint force magnitude (and direction) at Q; and the loop-cut forces at the point R, S and at the point R', S', all normalized to point O.   Consider for the moment only the anchor force corresponding to anchor P_1.   The solution gives its value normalized to the point O.   Denote this value by the symbol F_O which may be thought of as the force which would have to be applied to the O end of a rigid lever connected to pipe terminus P_1 in order to exert on this terminus the same force as is actually exerted by the anchor itself.   Thus, the actual anchor force, *at the anchor*, denoted by F_{P_1}, is $F_{P_1} = B_{P_1O}F_O$.   Next,

acting through the pipe itself, this exerts a force $F_J = B_{JP_1}F_{P_1} = B_{JP_1}B_{P_1O}F_O = B_{JO}F_O$ at point J; this is the force exerted on the piping to the left of point J by the piping to the right of point J due to the anchor reaction at P_1.   (Note that the symbol F_O used in the preceding discussion is the normalized force which was denoted by the symbol F_{P_1} in the sections on singular constraints.)   Similarly, the normalized anchor reactions for anchors P_2 and P_3 may be transferred to J.

Likewise, the constraint forces may be similarly transferred.   For the loop which is cut at $R'S'$ the loop-cut forces equilibrate each other and need not be counted.   However, point J is a point on the loop which is cut at point RS and the normalized loop-cut force at S should be transferred in the same way.   The sum of all these transferred forces is the force actually transferred via point J by the piping to the right of J to the piping to the left of J.

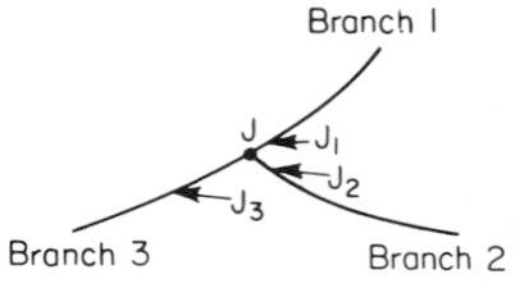

It usually is particularly important to evaluate the internal force system at a branch, bifurcation, or nozzle, as in Fig. 28.   If the branch point is J, consider three immediately adjacent points J_1, J_2, and J_3 located an infinitesimal distance from J along the three branches respectively.   If F_1 and F_2 are the forces actually

Branch I

Branch 3   Branch 2

FIG. 28. Evaluation of the internal force system at a branching point. J_1, J_2, and J_3 are points on branches 1, 2, and 3, respectively, and are immediately adjacent to the branching point J.

transferred via points J_1 and J_2 (these may be determined as above) into the branch point J, then the force transmitted from branch point J into branch number 3 is the sum of F_1 and F_2.   The force transferred from branch 3 via point J_3 into branch point J is, of course, the negative of this sum.

Evaluation of the internal force system at a point must precede a calculation of the stresses at the point.   Such a calculation must be made for every point in the system at which the stresses are likely to be of interest.   Specifically, each point

at which the stresses may be of governing or limiting magnitude must be investigated.

Remarks Concerning the Expansion Stress (Range) S_E. At each such potentially limiting position it is necessary to evaluate what is known as the expansion stress S_E. (S_E is actually what is known as the stress *range*, which will be discussed in detail later; for the time being, it may be thought of as a stress rather than a stress range or range of stresses.) It is actually true, of course, that simultaneously there are acting a number of force systems, and, presuming reasonably linear behavior, the actual stress system resulting is the superposition of the stress systems corresponding to each of the systems of loading. These loading systems include internal pressure, local thermal-stress systems, dead weight of piping including contents, and covering (only imperfectly equilibrated by the system of hangers and supports), thrusts due to accelerations of moving contents, thrusts due to action of pressure-relief devices, externally applied mechanical loads (painters' scaffolds, for example), and last, but not least, the thermally induced anchor and constraint reactions which have been the subject of discussion in this chapter so far. These several loading systems are not necessarily applied simultaneously; at the least, their relative importance varies with time. Furthermore, there are other essential differences between these loadings. For example, the circumferential hoop stress resulting from internal pressure causes strains which slightly enlarge the diameter of the pipe and slightly decrease its thickness; these effects do not at all decrease or mitigate the (circumferential) stress which caused them. On the other hand, the thermal-expansion stresses with which this chapter is principally concerned are such that the very process of straining tends to relieve the reaction forces inducing the strains. For these and other reasons, not the least of which is the practical impossibility of combining all the effective stress systems in all possible ways, it has become established practice and a part of the requirements of the ASA B31 Code for Pressure Piping that the flexibility analysis should evaluate the stresses resulting from gross thermal expansion *alone* and compare a certain calculated figure of merit, the expansion stress S_E, with an allowable value determined by the properties of the piping material, the extreme temperatures through which it operates, the type of service (power piping, petroleum refinery piping, etc.), the number of full thermal cycles of operation during the expected service life of the system, and the value of the sum of longitudinal stresses due to internal pressure, weight, and other sustained external loadings.

Thus, at this point attention can be confined to evaluating the stress system resulting from gross thermal expansion, that is, resulting from the internal force systems discussed in the preceding section.

At any point along the length of a straight piece of pipe there are moments and stresses which can be resolved into the following components: one axial stress causing axial tension or compression and two cross shearing stresses which can be combined into a single lateral force causing shear in the cross section and one axial or torsional moment and two lateral moments which can be combined into a single lateral moment causing bending of the pipe.

Using subscripts b for bending, s for cross shear, t for torsion, and a for axial, the following stresses may be determined:

$$S_a = F_a/A$$
$$S_s = \zeta F_s/A$$
$$S_t = M_t r_0/I_p$$
$$S_b = M_b r_0/I$$

In these formulas A is the area of cross section, r_0 is the outside radius of the pipe, I is the ordinary moment of inertia of the cross section, and $I_p = 2I$ is the polar moment of inertia of the cross section. The shearing stress S_s varies throughout the cross section in a quite complicated manner; to a close approximation, the maximum value of the shearing stress is $2F_s/A$. In the case of strictly two-dimensional configurations, it is known that the shearing stress S_s as calculated above is a maximum at the neutral axis and that the bending stress is a maximum at the extreme fibers so that the two do not have to be combined. In the general case of piping stresses, however, under particular circumstances, the orientations of the F_s force vector and the M_b moment vector may be such that the maxima occur at the same place. Even so, however, for simplicity and based on many years of experience as reflected in a report of an ASA Task Force of Flexibility (see the excellent paper by Markl,[182] which gives and augments the report of this Task Force), it is recommended in the ASA B31 Code for Pressure Piping that the expansion stress be based only on the combination of the torsional stress S_t and the bending stress S_b given above, viz.:

$$S_E = \sqrt{S_b{}^2 + 4S_t{}^2}$$

This very simple formula is based on the maximum shearing-stress failure theory associated with the names of Coulomb, Tresca, and Guest. Although an alternate theory, that of maximum distortion energy or maximum octahedral shearing stress and frequently called the Hencky-Mises theory in honor of the two persons who independently derived it on quite different bases, is generally conceded to give slightly better correlation with multiaxial tests on most ductile materials than does the maximum shearing-stress theory, the differences are slight and the latter may be considered not only to be perfectly satisfactory but also to have the sanction of successful application in piping-flexibility problems.

It should be noted that these theories of failure describe failure only in the rather limited sense of incipient yielding or incipient inelastic action rather than in the sense of tearing or separating of the material. Thus they apply to ductile materials and not to those which fail in a brittle fashion. Although some materials which are ordinarily thought of as being quite ductile may under certain circumstances fail in a catastrophic, brittle manner, proper selection and fabrication of piping materials should assure that this does not take place. The entire topic of brittle failure is summarized admirably in the book by Parker.[203] See also Refs. 76, 77, 78, 153, 171, 191, and 220. Failure in the sense of these failure theories does not mean denial of function for a piping system. The allowable values for piping expansion stress and the philosophy of application contemplates the possibility of some plastic deformation taking place in normal service. If such inelastic deformations are actually experienced, a stress analysis carried out by the methods given in this chapter evaluates subsequent elastic behavior.

Although the simple formula given above is intended by the B31 Code to be used in obtaining values of S_E which then must be compared with allowable values also to be calculated by Code formulas, in any particular case where there is reason to feel especial concern about stress distribution or otherwise to depart from the evaluation procedure implied by and incorporated in the Code, the usual methods of combining stresses as developed in the theory of elasticity may be employed. The following simplified procedure includes the effect of bending and torsional moments and internal pressure, the latter producing both circumferential hoop tension and, in some cases, what is called a longitudinal pressure stress, which is equivalent to the longitudinal tensile stress which would be required to hold the head or cap onto a pressurized cylinder. Other than this, axial and cross shearing forces are not considered. The formulas are based on the maximum shearing-stress theory and

include evaluation of "worst" combinations, without, however, being too conservative.   To evaluate the expansion stress (including the effect of internal pressure P as described above) at the outside surface, one evaluates in sequence the quantities $k =$ (wall thickness)/(outside diameter), $f = (1 - 2k^2)/2k(1 - k)$, $g = (\pi/4) \times (1 - 2k + 2k^2)(1 - 2k)^2$, $A = M_b/PD^3g$, $B = M_t/PD^3g$, $F = 2(A + \frac{1}{2})^2 + B^2$, and $S_E = PfF$.   Usually the stress at the outer surface is greatest, but if the calculated value is close to an allowable or limiting value, the stress at the inner surface should be calculated.   The only difference in procedure is that for the inner surface $f = 2 + (1 - 2k^2)/2k(1 - k)$ and $g = (\pi/4)(1 - 2k - 2k^2)/(1 - 2k)$. Another possible variation in the use of these formulas is to employ the maximum distortion energy theory (maximum octahedral shearing stress theory) rather than the maximum shearing-stress theory; in this case the formula for F becomes $F = \sqrt{4A^2 + 3B^2 + 2A + 1 + q}$, where $q = 0$ for the outside surface and $q = k^2(1 - k)^2 + k(1 - k)(\frac{1}{2} - A)$ for the inside surface.   Under no practical circumstances does q become larger than 0.1, so that its neglect is quite justified.

Stress Intensification in Pipe Bends and Elbows.   The preceding section provides procedures for evaluating the expansion stress in straight runs of pipe.   In the case of bends and elbows the more complicated stress distribution produces stress combinations which are larger than those found in adjacent straight pipe.

Carefully instrumented experiments of pipe bends under static load conditions correlate well with theoretically determined stress-intensification factors.   However, under most service conditions, failure or distress of a piping system seems to result from repeated cycling of load (or temperature), and careful fatigue tests[183–185] of full-size piping components comparing bends, etc., with straight commercial pipe show failure of bends and elbows to correlate with stress-intensification factors half as large as the theoretical values.   That is, the dependence of the stress-intensification factors upon the bend characteristic h is indicated by the more complete theoretical analyses to be in proportion to $h^{-2/3}$, but the constant of proportionality is just half what theory would indicate.   A reasonable explanation of the difference postulates that ordinary straight pipe as commercially manufactured, fabricated, supported, and actually constrained in service has an inherent fatigue stress-intensification factor of about 2.0, which neither is nor need be explicitly recognized in assigning allowable stress values or computing equivalent stresses, since the philosophy involved in the assignment of allowable stress values is based on successful experience with piping systems in cyclic service.   This matter was not taken into account until recently, since earlier theoretical studies had erred in giving attention only to the longitudinal bending stress σ_L rather than the actually more severe circumferential bending stress σ_T and had thus obtained results about half as great as they should have been.   This matter is discussed in Ref. 150; see also the series of papers by Markl[182–184] and by Markl and George[185] and the discussions thereof.

On the basis of the aforementioned fatigue tests and other considerations, Markl proposed some formulas for effective stress intensification in bends and elbows which have since received substantial theoretical confirmation via quite different analytical approaches.   Markl's original recommendations are

$$i_i = 0.90h^{-2/3} \qquad i_o = 0.75h^{-2/3}$$

(but in no case less than 1.0) for in-plane and out-of-plane bending, respectively. However, for the purpose of simplifying the Code, the same formula ($i = 0.90h^{-2/3}$) was used for both in-plane and out-of-plane bending in 1962 editions of the Code. The Mechanical Design Committee of the ASA B31 Pressure Piping Code Committee is now considering listing both formulas as originally proposed by Markl (see

Appendix). The separate listing has been shown in Figs. 29 and 30 of this chapter.

If internal pressure is acting in thin-wall pipe, the ovalization of bends and elbows is inhibited with the result, previously described, of reducing the bend-flexibility factor n. Internal pressure also reduces the stress-intensification factors i_i and i_o, the work of Rodabaugh and George[221] indicating that these factors should more accurately be given as

$$i_{i,o;p} = i_{i,o} \div [1 + 3.25(P/E)(r/t)^{5/2}(R/r)^{2/3}]$$

(but in no case less than 1.0) where the subscript i, o indicates that this reduction is equally applicable to in-plane and out-of-plane bending and the subscript p indicates that internal pressure is being taken into account.

These stress-intensification factors, which do not so much measure a static stress intensification as they do an effective intensification tending to increase susceptibility to fatigue failure, are intended to be used in the following manner in calculating the expansion stress S_E. The computed bending moment is to be broken up into vector components, one, M_{bi}, causing in-plane bending, and the other M_{bo}, causing out-of-plane bending. Then, with the use of the formulas

$$S_{bi} = M_{bi}i_i/Z \qquad S_{bo} = M_{bo}i_o/Z \qquad S_t = M_t/2Z$$

where Z is the pipe section modulus, the basic formula $S_E = \sqrt{S_b^2 + 4S_t^2}$ may be written as

$$S_E = \sqrt{(M_{bi}i_i)^2 + (M_{bo}i_o)^2 + (M_t)^2}/Z$$

If the simplification is used that both i_i and i_o are the same, $i = 0.90h^{-2/3}$, this formula is further simplified to

$$S_E = \sqrt{(M_b i)^2 + (M_t)^2}/Z$$

since $M_{bi}^2 + M_{bo}^2 = M_b^2$. Finally, for quick calculating purposes, since i cannot be less than unity, one may note that

$$S_E \leq iM_R/Z$$

where M_R is the resultant bending moment, calculated from the relations $M_R^2 = M_b^2 + M_t^2 = M_{bi}^2 + M_{bo}^2 + M_t^2$. The values i_i and i_o given in this subsection are also shown under item 1 of Fig. 29.

Stress Intensification in Tees, Nozzles, and Other Piping Components. The fatigue tests of full-scale piping components, previously cited, also provided some information on flanged bends and elbows, on miter bends, on tees and nozzles, and on various types of pipe joints. Reasonable and convenient formulas for stress intensification (and flexibility factor) for a number of important piping components are presented in Fig. 29† which represents latest (1962) recommendations of the subgroup within the Mechanical Design Committee having cognizance over such matters (see Appendix). Notations in Fig. 29 differ in some respects from those used in ASA publications but are consistent with the notation employed in this chapter.

Tabulation of these factors is not intended to countenance use of miter bends in jurisdictions which prohibit or limit their use; values given here merely describe the behavior of miter bends used where they are permitted. The factors C_f for bends and miters having one or both ends flanged are intended to reflect the fact

† The bulk of evidence presented in Fig. 29 is due to A. R. C. Markl.

that such flanges inhibit ovalization and thus reduce n and i; accordingly, values of C_f larger than unity should not be used.

The tabulation is intended to represent effective stress intensification in customary piping components manufactured according to usual practice and fabricated or joined by good techniques. Sudden changes of section, porous or imperfectly penetrating welds, etc., cause additional stress intensification which is not represented in this tabulation.

Tees and nozzles are a very important class of piping component. All these are collectively called tees in Fig. 29. Reflecting standard dimensions, controlled factory manufacture, and successful experience, tees conforming to ASA Standard B16.9 are assigned more favorable values than fabricated tees. Values pertaining to the latter are predicted upon the use of qualified materials and good workmanship. S_E may be calculated for all cases by use of the formula

$$S_E = \sqrt{(M_{bi}i_i)^2 + (M_{bo}i_o)^2 + (M_t i_t)^2}/Z$$

where M_{bi} and M_{bo} are in-plane and out-of-plane components of bending moment, respectively, and M_t is the torsional moment; i_i, i_o, and i_t are in-plane bending, out-of-plane bending, and torsional stress-intensification factors, respectively; and Z is effective or equivalent section modulus as defined below.

For items 1 to 11, $i_t = 1.0$.   For items 7 to 12 of Fig. 29, the bending moment cannot be divided into in-plane and out-of-plane components, and one has

$$S_E = \sqrt{(M_b i_b)^2 + (M_t i_t)^2}/Z$$

For item 12, $i_b = 5$ and $i_t = 0.9$; for items 7 to 11, i_b is the tabulated value.

For tees, values of S_E should be computed for each of the three branches, that is, the outlet branch and the two run branches, and, of course, the largest value computed in this way is the one of interest.

In all these formulas, the section modulus Z is defined and calculated as follows: Z is the section modulus of the significant tubular cross section of the component or branch of the component involved; where the cross section is not uniform throughout the body, as in the case of manufactured welding tees, the section modulus of the matching pipe shall be used; where the component includes superposed reinforcement such as pads, saddles, or collars, the section modulus of the header (run) pipe or branch pipe shall be used without considering such nonintegral reinforcement.

The above formula may be too conservative in the case of many reducing outlet tees.   In such cases, the value of S_E for the branch or outlet may be evaluated by using the formula

$$S_E = \sqrt{(M_b i_i)^2 + (M_{bo} i_o)^2 + (M_t i)^2}/Z i$$

where i is the smaller of the two quantities i_i and t_H/t_B, t_H and t_B denoting significant thickness of header and branch, respectively.

Sometimes in order to reinforce against internal pressure a fabricated tee or nozzle makes use of a run or barrel portion of significantly greater thickness than that of the main run pipe.   In this case, the value of t (for the header or run) may be taken as the actual barrel thickness rather than that of the main run pipe if the thickened portion extends a significant distance, say at least one branch radius, outside the OD of the branch on either side and is smoothly faired into the main run pipe.

The reader should be constantly aware of the fact that the values given in Fig. 29 represent the best available information applicable to a large class of piping components and that special geometries and unusual designs are not intended to be covered

Item	Description	Flexibility factor n	Stress int. factor i	
			In-plane i	Out-of-plane i_o
1	Welding elbow[1,2,3,5,8] or pipe bend	$\dfrac{1.65}{h}$	$\dfrac{0.9}{h^{2/3}}$	$\dfrac{0.75}{h^{2/3}}$
2	Closely spaced miter bend[1,2,3,8] $s < r(1 + \tan\theta)$	$\dfrac{1.52}{h^{5/6}}$	$\dfrac{0.9}{h^{2/3}}$	$\dfrac{0.75}{h^{2/3}}$
3	Widely spaced miter bend,[1,2,4,8] $s \geq r(1 + \tan\theta)$	$\dfrac{1.52}{h^{5/6}}$	$\dfrac{0.9}{h^{2/3}}$	$\dfrac{0.75}{h^{2/3}}$
4	Welding toe[1,2,4] per ASA B16.9	1	$\tfrac{3}{4}i_o + \tfrac{1}{4}$	$\dfrac{0.9}{h^{2/3}}$
5	Reinforced fabricated tee,[1,2,6] with pad or saddle	1	$\tfrac{3}{4}i_o + \tfrac{1}{4}$	$\dfrac{0.9}{h^{2/3}}$
6	Unreinforced fabricated tee[1,2,6]	1	$\tfrac{3}{4}i_o + \tfrac{1}{4}$	$\dfrac{0.9}{h^{2/3}}$
7	Butt-welded joint, reducer, or welding neck flange	1	1.0	
8	Double-welded slip-on or socket weld flange	1	1.2	
9	Fillet-welded joint, or single-welded socket weld flange	1	1.3	
10	Lap-joint flange (with ASA B16.9 lap-joint stub)	1	1.6	
11	Screwed pipe joint, or screwed flange	1	2.3	
12	Corrugated straight pipe, or corrugated or creased bend[7]	5	2.5	

FIG. 29. Flexibility factor n and stress-intensification factors i_i and i_o. This figure has been Design Committee of the American Standards Association Committee B31 on Code for are similar except that some of the wordings of definitions are somewhat older and no out-of-plane loadings.

Flexibility characteristic h	Sketch
$\dfrac{tR}{r^2}$	R=Bend radius
$\dfrac{\cot\theta \; ts}{2 \cdot r^2}$	$R = \dfrac{s\,\cot\theta}{2}$
$\dfrac{1+\cot\theta}{2}\dfrac{t}{r}$	$R = \dfrac{r(1+\cot\theta)}{2}$
$4.4\dfrac{t}{r}$	
$\dfrac{(t+\frac{1}{2}T)^{5/2}}{t^{5/2}r}$	Pad Saddle
$\dfrac{t}{r}$	

Meaning of symbols:

r = mean radius of matching pipe

t = (for elbows, bends, and miter bends nominal wall thickness of elbow, bend, or miter bend; see Note 5

t = (for tees and nozzles), the nominal wall thickness of the matching pipe

R = bend radius of pipe bend or elbow

θ = one-half angle between adjacent miter axes

s = miter spacing at center line

T = pad or saddle thickness, in.

NOTES:

1. The flexibility factors n and the stress-intensification factors i in the table apply to bending and in no case shall be taken less than unity. Factors for torsion equal unity. Both factors apply over the effective arc length (shown by heavy center lines in the sketches) for curved and miter elbows and to the intersection point for tees.

2. Those flexibility and stress intensification factors which are proportional to a power of the characteristic h may be read from the graph of Fig. 30.

3. Where flanges are attached at one or both ends, the values of n and i in the table shall be corrected by the multiplicative factors C_f given by the formulas:

$$C_f = h^{1/6} \text{ one end flanged)}$$
$$C_f = h^{1/3} \text{ (both ends flanged)}$$

4. Also includes single-miter joint.

5. Cast butt-welding elbows may have considerably heavier walls than that of the pipe with which they are used. Large errors may be introduced unless the effect of such greater thickness is considered.

6. $h = 4.05t/r$ for $T > 1\frac{1}{2}t$

7. Factors shown apply to bending; flexibility factor for torsion = 0.9.

8. In large-diameter thin-wall elbows and bends internal pressure can significantly decrease both flexibility and stress intensification factors. To correct values obtained from above tabulation so as to account for internal pressure, divide the flexibility factor by the quantity

$$[1 + 6(P/E)(r/t)^{7/3}(R/r)^{1/3}]$$

and divide the stress intensification factor by the quantity

$$[1 + 3.25(P/E)(r/t)^{5/2}(R/r)^{2/3}]$$

where P = internal gage pressure and E = Young's modulus.

adapted from the report of the Subgroup on Expansion and Flexibility of the Mechanical Pressure Piping (see Appendix to this chapter). Current requirements in the Code itself separate values of i_i are given, those listed above for i_o being used for both in-plane and

by this collection of formulas. Generally, the manufacturer of valves, strainers, and similar equipment is the person who should be best aware of the effective stress concentrations which apply to his product and of the possible effects of overstressing. In cases of doubt, a consultation with the manufacturer is surely called for.   In other cases, it may be necessary to conduct destructive tests or carefully instrumented nondestructive tests in order to obtain the necessary design information.   Finally, in many cases of doubt, it may prove most economical to overdesign so that even though it may not be possible to make accurate estimates of what the stresses actually are, it may yet be possible to assert that they are quite small and definitely

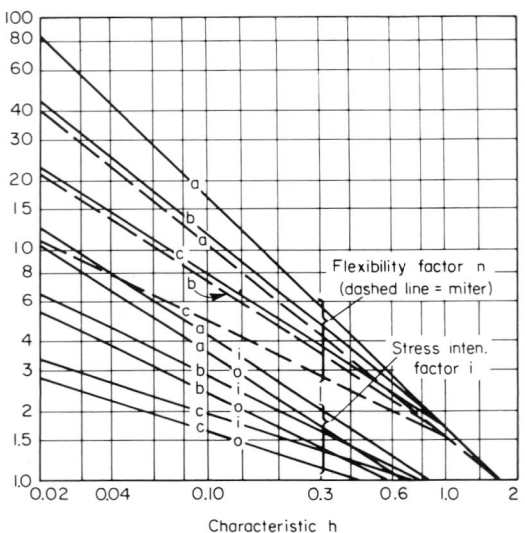

Characteristic h

FIG. 30. Flexibility and stress-intensification factors for those cases where result is a power of the characteristic h, defined in Fig. 29.   Key: a = no flanges; b = one flange; c = two flanges; i = in plane; o = out of plane.

not of limiting magnitude.   The reader should note items 7 to 11 of Fig. 29 and should not overlook the presence of these stress-intensifying joints and elements. In particular, the termini of a system are usually subjected to a stress intensification owing to the discontinuous geometry of the transition from pipe to terminal equipment whatever the latter may be.   The items given in Fig. 29 may serve as a guide for estimating an appropriate terminal stress-intensification factor.   In most cases, if a better estimate cannot be made on the basis of a particular geometry at the termini, it is reasonable to take a value $i = 1.3$ for bending (in-plane and out-of-plane) and $i = 1.0$ for torsion.

Although the formulas given in Fig. 29 may easily be evaluated by use of a log-log slide rule, it may also be convenient to make use of Fig. 30, which presents graphically those formulas which are proportional to a power of the characteristic h. The captions on the curves describe their applicability; however, to assure that there is no misunderstanding, it may be noted that these curves, reading down along the left edge of the graph, respectively represent the formula $1.65h^{-1}$, $1.65h^{-5/6}$, $1.52h^{-5/6}$, $1.65h^{-2/3}$, $1.52h^{-2/3}$, $0.9h^{-2/3}$, $1.52h^{-1/2}$, $0.75h^{-2/3}$, $0.9h^{-1/2}$, $0.75h^{-1/2}$, $0.9h^{-1/3}$ and $0.75h^{-1/3}$.

It should also be recalled that the basic formula for calculating S_E which is given in this section takes account only of bending and torsional moments and does not

include the effects of direct axial and shearing loads due to expansion. Likewise, this formula does not immediately include the effect of other loadings such as internal pressure (although such other loadings are considered in determining the allowable values). Experience with a large variety of piping configurations, composed of a number of different materials and operating through a large gamut of thermal cycles and conditions, indicates that this formula evaluates satisfactorily overall system life and performance. However, in particular cases, there may be reason to attempt to include other considerations or to combine all loadings simultaneously, etc. Aside from pointing out the difficulty of such calculations, all that may be said is that the general principles of the theories of elasticity and plasticity may be applied. References 86, 158, and 190 may also be cited.

The "Range" Concept, Reaction Ranges and Stress Ranges, Allowable Values. The results of investigations[223] into the significance of the stress systems which result from internal pressure, thermal expansion, and other causes have been incorporated in the 1955 and subsequent issues of the Code for Pressure Piping. Reference 182 summarizes, in this regard, the work of the Task Force on Flexibility, which was established in 1951.

The formal piping-flexibility problem, discussed so far in this chapter, is to be solved formally as if the system indeed behaved linearly and elastically. The value of Young's modulus E in the cold or installed condition is to be used in this calculation, and no account is to be made of any cold spring which may have been put into the system during its erection. The calculated terminal reaction and constraint components represent a *range* of values through which the actual, physical values vary as the temperature changes. If, in actuality, the system does behave linearly and elastically and is completely without stress at the low-temperature condition of the thermal cycle, then a range value does indeed represent the actual value of (the appropriate) reaction or constraint component at the high-temperature condition of the thermal cycle. However, if the system was installed with cold spring (that is, initial stress or loading designed to counteract that induced by the thermal expansion), or if there has been some plastic behavior, the calculated values may not represent actual reaction or constraint components at any given specific time in the thermal cycle. Nevertheless, they are of significance in predicting the long-term performance of the system.

Corresponding to these reaction *ranges*, *stress ranges* may be calculated by use of the methods given immediately above. That is, the quantity S_E, which hitherto in this chapter has been called the expansion stress, in reality is the expansion *stress range*. If the worst or largest value of this expansion stress range is sufficiently small, one may be assured that the piping system will behave elastically. If this worst value is somewhat larger than this and there has been no cold springing, then it is possible that actual stresses may at one or more points become large enough that plastic behavior may take place. This plastic behavior may be simple yielding by application of a stress system which causes plastic flow, or it may be gradual through the action of creep or relaxation. In either case, the inelastic behavior causes deformations which are not accounted for by usual methods of structural analysis (including the matrix methods given in this chapter), so that such analysis is not capable of evaluating the actual reaction loads and stresses. However, unless the reaction range is too large, the amount of plastic deformation will be limited. Plastic yielding will take place only the first one or two cycles the system is in operation, and after such initial plastic self-springing, the system will thereafter behave elastically. Likewise, under these conditions, creep will take place so slowly or be relieved so adequately that denial of function through creep action need not be of concern. In such a case, even though it is not possible to be sure what the actual reaction components are at any given time or what the expansion

stress system really is at any particular point on the full thermal cycle, it is possible to assert that the system will give satisfactory service.

However, if the expansion stress range S_E as calculated in the above-described manner exceeds a certain limiting or allowable value at some point in the configuration, then it must be concluded that the likelihood of excessive creep or of repeated plastic yielding, cycle after cycle, is so great that the system is unlikely to perform in a satisfactory manner.

The selection of appropriate limiting stress-range values involves many considerations, not the least of which is an ingrained conservatism which makes many piping engineers reluctant to employ what they regard as unreasonably high values for stress in a piping system.   However, there is a growing realization that stress, per se, is a poor criterion of probable performance, that stress range and stress are two different things, and that the growing successful experience with critical systems designed using this viewpoint must be given proper weight.   Other considerations are the criteria which are employed by the various committees (under several code jurisdictions) to establish basic allowable stress values (principally for hoop stress calculations), the difference in nature between those stress systems which are mitigated via deformation and those which are not, the effective stress-intensification factors which result from normal imperfections in commercial pipe and components, etc.

Based on criteria normally used by stress allowance committees for establishing the basic allowable stresses or allowable "S values," for example, cf. Ref. 18, the Task Force Report makes plausible an allowance of $1.6(S_c + S_h)$, where S_c and S_h are tabulated values of allowable stress in appropriate ASA Piping Code documents. However, so as to account for unpredictable behavior and so as to incorporate experience gained with systems designed under earlier criteria, the Code Committee adopted the following somewhat more conservative formula for allowable stress range S_A:

$$S_A = f(1.25S_c + 0.25S_h)$$

where S_c and S_h are tabulated allowable stress values (under proper jurisdiction) at the minimum and maximum (respectively) normal operating temperatures. Values for seamless material should be used even though the actual material has a seam which, for hoop tension purposes, is assigned a joint factor less than unity. The word "normal" in the definition of S_c and S_h implies the extreme temperatures expected under controlled operating conditions and does not include unusual or unanticipated temperature excursions.   As in all cases of trying to anticipate emergency or runaway conditions, different designers will take such conditions into account in different manners and all that can be done here is to call attention to the necessity of giving thought to the possibilities.

The factor f in the preceding formula is the stress-range reduction factor for cyclic conditions and may be read from the graph of Fig. 31.   The number of cycles N is the number of full temperature cycles for which the expansion stress S_E has been calculated, that is, from minimum to maximum temperature and back. If there are also lesser temperature variations, or if thermal conditions vary from cycle to cycle, an equivalent value of N for use with Fig. 31 may be obtained as follows:

$$N = N_E + (S_1/S_E)^5 N_1 + (S_2/S_E)^5 N_2 + (S_3/S_E)^5 N_3 + \cdots$$

where S_E and N_E, respectively, are the maximum computed expansion stress range and the corresponding actual number of cycles and S_1 and N_1, S_2 and N_2, etc., are the expansion stress ranges and cycles of successively lesser amplitude.   The fifth-power form of correction implied by this formula results from fatigue tests of Markl, cited previously, which indicate a relation of the form $iSN^{0.2} = C$, for a wide

variety of pipe and components, where i is the stress-intensification factor, S the nominal endurance strength, N the number of stress reversals to failure, and C a materials constant. Those concerned with fatigue phenomena in general are still seeking a universally applicable and generally acceptable law of cumulative fatigue damage, and there are good reasons to suspect that a law of the sort implied by the formula given above fails to represent some important features of fatigue behavior (such, for example, as increasing fatigue resistance by "exercising" at low stress levels). However, such laws have been widely used and are probably the simplest way to attempt to evaluate the sum of fatigue damage resulting from working through stress cycles of varying magnitude.

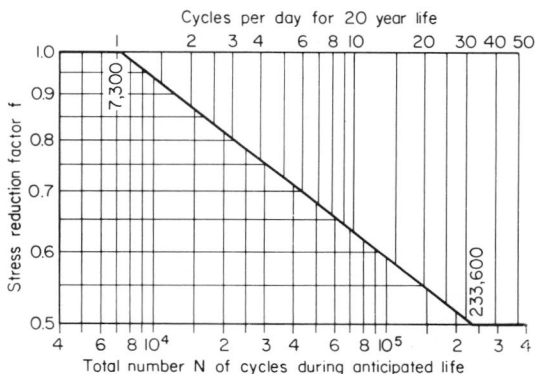

FIG. 31. Stress-range reduction factors. Extracted from Petroleum Refinery Piping Code (ASA B31.3–1959), a section of the American Standard Code for Pressure Piping, with the permission of the publisher, The American Society of Mechanical Engineers, 345 East 47th Street, New York 17, N.Y.

Where the variable conditions apply equally to the entire piping system, the ratios S_1/S_E, etc., may be simply computed to be the ratios $\Delta T_1/\Delta T_E$ of lesser temperature variation to the maximum temperature variation ΔT_E for which the flexibility calculations have been made. Where temperature cycles for different parts of the system differ either in frequency or in amplitude and the number of equivalent full cycles is expected to exceed 7,300, separate calculations should be made for each condition and the preceding equations should be used for each point at which the stress range becomes a maximum in any of the calculations.

Said in another way, it is necessary to assure that S_E does not exceed S_A at any point of the system, and both of these quantities may vary from point to point. S_E varies from point to point both as a consequence of the static equilibrium of the individual parts of the system for any particular temperature cycle and also because there are different temperature cycles which place different parts of the system under different states of stress at different times. S_A varies from point to point in the system because the value S_h (and sometimes S_c) may be different in different parts of the system, because the equivalent N may be different in different parts of the system, and also for the reason given in the next paragraph. For complicated piping systems which go through a variety of different thermal cycles, it requires considerable calculation and ingenuity to determine the worst, and therefore the limiting, circumstances.

Piping codes require that the wall thickness of the pipe be sufficient to sustain the internal pressure without the hoop tensile stress exceeding the tabulated stress

value S_h at the operating temperature.   Most of them also require that the longitudinal tensile stress resulting from pressure, weight, and other sustained external loadings shall also not exceed S_h, and the ASA Code for Pressure Piping stipulates that where the sum of these longitudinal stresses is less than the stress value S_h, the difference may be added to the term $0.25S_h$ in the formula for S_A, giving

$$S_A = f[1.25S_c + 0.25S_h + (S_h - \Sigma\, S_L)]$$
$$= f[1.25(S_c + S_h) - \Sigma\, S_L]$$

where $\Sigma\, S_L$ denotes the sum of such longitudinal stresses.   (It is clearly implied in the Codes that this sum does not include longitudinal stress components resulting from the thermal expansion with which this chapter is concerned.)   Generally speaking, pipe-wall thickness is selected so that the hoop tensile stress is not appreciably less than S_h and the longitudinal pressure stress is very nearly exactly half of this.   Essentially, this leaves an allowance of about $\frac{1}{2}S_h$ for "weight and other sustained loadings."   Since even a liberally engineered hanger and suspension system does not perfectly equilibrate dead weight, it is only in rather unusual circumstances that the term $S_h - \Sigma\, S_L$ permits a significant increase in the value of S_A, and the formula for S_A given earlier, namely, $S_A = f(1.25S_c + 0.25S_h)$, is the one normally employed.

The concept of reaction range (range of reactions due to thermal expansion) and stress range (range of stresses due to thermal expansion), although introduced to piping engineers over twenty years ago (Ref. 223), is subtle and has gained acceptance only gradually.   Although the Task Force Report mentioned above explicitly made use of the concept, the resulting material in the 1955 edition of the ASA B31 Code for Pressure Piping avoided the use of the word "range" in most of the places where current Code documents employ it.   The range concept is a significant change in design philosophy from earlier viewpoints and takes cognizance of the fact that most piping materials have sufficient ductility to accommodate slight yielding, if necessary, without denial of function.   It may be fairly said that many early piping engineers possessed sufficient sound engineering sense to design systems which made full use of the virtues of the piping materials with which they were dealing but were previously hard put to justify their designs to critics armed with careful stress analyses based on purely elastic theory.   In one sense, then, the stress-range concept may be regarded as a theoretical justification for successful practice which violated earlier (and evidently inadequate) theoretical criteria.

It is the intention of the ASA B31 Code that the design criterion afforded by assuring that S_E nowhere exceeds S_A (both calculated according to the formulas given above) also specifically requires that the calculations leading to S_E be based upon the cold (Young's) modulus of elasticity E_c and upon the full temperature range, regardless of whether or not cold spring has been employed in erecting the system.

Reactions on Terminal Equipment, Cold Spring.   The criterion that $S_E \leq S_A$, both formally calculated as in the preceding subsection, is intended to assure the integrity of the pipe itself and its internal components (valves, etc.) over the desired operating life span.   However, the designer must also assure that the reactions applied by the piping to the terminal equipment do not deny the function of the latter.   Most pumps and other terminal equipment themselves expand as they are heated and may not be constrained by massive anchors at their inlet and outlet connections.   Accordingly, it is a matter of importance to assure not only that the stress range in the pipe itself is within acceptable limits but also that the forces and moments exerted between pipe and terminal equipment do not

cause damage, curtailment of service life, misalignment, malfunction, or other denial of terminal equipment function.

Many manufacturers of such equipment state their limitations on reactions directly in terms of reaction range components. Most terminal equipment is manufactured from materials which are closely related to the adjoining pipe, and stress criteria appropriate for the pipe are, in many cases, also appropriate for the equipment.

In other cases, equipment manufacturers may state their limitations in terms of reactions rather than reaction ranges. In certain simple cases it requires only a formal calculation to provide rather good estimates of what the extreme values of reaction components actually are, but in cases where there has been plastic yielding or creep, it may be considerably more difficult to obtain reliable estimates.

A part of the computational difficulty is that many systems are cold-sprung; that is, the pipe is fabricated to slightly shorter dimensions than is required to extend between its terminal points and, during erection, is pulled into place. This sets up a system of internal stresses which are opposite to those which will later be induced by the thermal expansion of the piping material as it heats from installation temperature up to its operating temperature. Thus, the expansion tends (initially at least) to relieve internal piping stresses and terminal reactions. If the system is 100 per cent cold-sprung, the cut-shorts, that is, the amounts by which the pipe has been fabricated short of projection dimensions, are made equal to the computed expansion of piping (plus expansion of terminal equipment; that is, the cut-shorts are made equal to the D_i of the section "Calculations of the Deflections D_i") so that, theoretically, at the operating temperature, the system is without stress and the terminal and constraint reactions all vanish. Smaller percentages of cold spring imply that the zero stress condition takes place at some temperature above the installation temperature and below the operating temperature. Finally, zero cold spring implies that stresses and reactions are zero at the installation temperature and a maximum at the operating temperature. The amount of cold spring is usually expressed as the single constant C, called the cold-spring factor, which is equal to the ratio between actual cut-shorts and the cut-shorts required for 100 per cent cold spring; that is, $C = 0$ for zero cold spring and $C = 1$ for 100 per cent cold spring.

There are no obvious good reasons to employ cold spring greater than $C = 1$. The use of values of C greater than unity in order to evade some of the consequences of the formulas for terminal reactions to be given shortly is denied specifically by the Code. Some designers employ different cold-spring factors C in different coordinate directions.

Different designers assume different attitudes with regard to cold spring. In the petroleum-refining industry very few high-temperature systems are cold-sprung, while in the electric utility industry almost all high-temperature lines have from 50 to 100 per cent cold spring. Those who recommend cold springing to 100 or nearly 100 per cent feel that it is of importance that the line have as little as possible additional stress due to thermal expansion over and above the stresses due to internal pressure and dead weight which must be sustained by the pipe during its high-temperature operation. Cogent arguments for large amounts of cold spring are cited by Robinson,[220] who argues that most piping systems under thermal-expansion loading contain large amounts of elastic strain energy which is available to cause continued creep in locally highly stressed regions. This phenomenon, which Robinson calls "elastic follow-up," can cause appreciable creep strains to take place locally, and even though these may be so localized as not to result in any gross permanent distortion of the system, they may result in increased susceptibility to brittle fracture. Some materials which ordinarily behave in a very ductile fashion and

show quite satisfactory total elongations in ordinary static tension tests are able to sustain a much smaller creep elongation before they fracture catastrophically. Even though Robinson's paper may perhaps overemphasize this danger, it is clear that certain configurations may exhibit undesirable elastic follow-up, and although experienced designers may be able to avoid such configurations, it does not appear simple to establish objective criteria which identify them. Thus, a designer, in routing his pipe so as to conform to a building structure, might possibly stumble into trouble.

Cold springing to less than 100 per cent, however, may effectively mitigate the undesirable effects of elastic follow-up. Multiaxial creep is a very complicated phenomenon, but there are some results[106,107] available which show that a very pronounced reduction in creep rate accompanies a relatively slight reduction in stress level. Robinson's simplified analysis[220] was based on uniaxial creep data which showed that creep rate was dependent upon the fourth power of stress. With elastic follow-up, the stress does not relax appreciably and the creep strain experienced in a given time (say, a hoped-for 20-year operating life) thus depends on the fourth power of actual stress. Using 50 per cent cold spring ($C = 0.5$) effectively reduces actual high-temperature stress to half its value with zero cold spring and thus reduces creep (resulting from elastic follow-up) to one-sixteenth the value it would have with no cold spring. While such calculations must of necessity be highly approximate, based as they are upon a uniaxial representation of a complicated multiaxial phenomenon which is itself only imperfectly understood, there is no denying their overall general validity. A small amount of cold spring significantly reduces creep; 50 per cent cold spring reduces it to less than 10 per cent of what it would be with no cold spring; 100 per cent cold spring (if it could actually be achieved) would completely eliminate creep. This brief analysis postulates that there is no plastic yielding of any consequence; there are two reasons for this, first, if yielding should take place, it is equivalent to a certain degree of cold spring as far as reducing creep rate is concerned, and second, with 50 per cent cold spring and satisfying the requirement that $S_E \leq S_A$, it is highly unlikely that appreciable yielding could take place even locally at points where there might be a high stress intensification.

Another reason for employing cold spring in the mid-range of about 50 per cent, is that this effectively splits the reaction range on terminal equipment, so that hot loadings are approximately equal to cold loadings and in the opposite sense. In cases where allowable static loads on terminal equipment may prove to be the limiting consideration, it may be very desirable to do this. More will be said presently about methods of estimating actual static values of terminal reactions for hot and cold conditions when they have their extreme values, but it may be noted that the power engineer more frequently is involved with terminal equipment which is completely designed by an outside group and the strength of which is more or less beyond his control. Manufacturers of boiler and turbine equipment are in a position to stipulate to the piping engineer just what loading they will permit being placed on their product. On the other hand, the refinery engineer more frequently has control over the strength of the vessels to which his high-temperature piping connects and may not have to resort to cold spring in order to limit loadings thereon.

Finally, it is perfectly clear that it is easier to erect a piping system without cold spring than it is to perform this additional and usually rather difficult step. In systems which have adequate flexibility, there is no point in going to the trouble and expense of performing cold spring. Specifically, if the stress range is such that no plastic yielding will take place the first time the system is brought up to operating temperature, and if this stress range (which thus is also the actual operating stress)

is substantially below the stress values for which creep takes place at an undesirable rate at operating temperature, then there is no reason to cold spring.

The ultimate effect of inelastic action is quite similar to that achieved by cold springing and the term self-springing[182] is used to describe this behavior. It may be assumed that many lines which are installed without cold spring do become self-sprung during operation. Indeed, it is a not uncommon occurrence upon breaking a flanged joint in a line which is out of service to find that the mating flanges spring apart as the last bolts are removed, and this may be dangerous to personnel if large lines are involved. If the inelastic behavior is not all confined to one quite localized portion of the piping system, self-springing, in effect, automatically achieves some of the desirable results of cold springing. Actually, because of the superposition of dead-weight stresses, pressure stresses, thermal-expansion stresses, local thermal stresses, and yet other loadings, high-temperature systems frequently do alter their shapes in ways which are somewhat different from what might be predicted on the basis of the analyses given in this chapter, and upon breaking a joint of a line which has been in service for some time, it is seldom found that the two ends occupy the same position as they did before the system went into operation.

It is of the greatest order of difficulty to achieve during field erection of a piping system precisely the cold spring which may be called for by the designing engineer. The reasons for this are:

1. The mathematical analysis assumes that the piping system is weightless.

2. The piping system, even if weight were to be considered in the analysis, is supported by a finite number of hangers.

3. After piping erection, the installation of insulation and jacketing, and the subsequent addition of fluid contents, alter the loading and its distribution.

4. It is almost impossible to specify exactly the magnitudes and directions of field-applied forces which are required to result in the desired cold spring.

Although cold springing is far from an exact process in the field, the author is convinced that in cases where cold springing is theoretically desirable, it should be done and reasonable controls should be exerted to accomplish a job in the field that approximates the theoretically ideal job.

One successful procedure may be described briefly as follows: The erection specification for critical piping requires that each additional piece of pipe be suspended by its permanent hangers in blocked position and by additional supports as required and brought into welding position with as little mechanical strain as possible. Transit measurements are taken and recorded during this and subsequent operations both to serve as a control and to provide useful records. Erection takes place from one anchor to terminal point sequentially through the line to other anchors or termini. At points where there is to be a cold-spring gap, a fabricated spacer or template is temporarily installed which enforces the desired gaps and orientations. Such templates are removed just prior to the actual cold-spring operation, and if the erection has been ideally successful, there will be no relative motion of the two pipe ends when the template is removed and before the cold-spring efforts are applied. It is clear that such a procedure is extravagant of personnel, and field circumstances may require that less elaborate precautions be observed.

In the selection of an optimal point (or points) in the system to close the cold-spring gap (or gaps) accessibility is of greatest importance. Also, it is generally considered desirable that bending and torsional moments be relatively small at the cold-spring gap. If the forces and moments required to close the gap for welding (or other joint assembly) are to be applied only locally, that is, near the gap, it is evident why this should be so, but it is frequently the case that application of effort at

points remote from the gap may be more effective. The discussion in Ref. 150 indicates how this may be evaluated. In closing a cold-spring gap, there should be only parallel translation of the matching ends, no relative rotations. The template which was referred to in the preceding paragraph may contain guides and gages to assist in assuring that no rotations take place. Since a relative rotation about the axis of the pipe itself is most difficult to detect (and easiest for field personnel to shrug off, since it does not hinder the welding process) and also most difficult to do anything about, the author has usually selected the point of minimum torsional moment as the point for the cold-spring gap.

The restraints which are used to close up the cold-spring gap should remain in position until after any postheat treatment has been completed. Many field situations will call for specially devised procedures. For example, it is not uncommon to include special bypasses or closures which are used only during preliminary testing or cleanout procedures. For scheduling purposes, it may be necessary to have completed the cold-spring operation before these temporary facilities are removed. In such a case, rigid anchors may be installed temporarily so that the cold-springing operation may be completed, and these anchors may be removed after the bypasses, etc., have served their purposes and been removed and the final configuration has been erected.

Calculation of Terminal Reactions. The terminal and constraint forces which result from the solution of the flexibility equations are, as has been pointed out in the last several subsections of this chapter, actually *ranges* which tell by how much the appropriate reaction components change as the piping system goes from its cold to its hot condition. In these calculations, the cold modulus of elasticity E_c has been used. Since E_c is generally greater than E_h, this may be regarded simply as incorporation of an additional safety factor in the calculation for S_E, the expansion stress range which is of primary interest. Also, when the flexibility equations are being set up and solved, no account is made of any cold spring which may have been incorporated in the system. A formal solution to the flexibility problem thus provides what may be regarded as a slightly conservative estimate of reaction ranges from which expansion stress ranges may be computed and compared with the allowable stress range.

In those cases where it is necessary or of interest to obtain estimates of the actual extreme values of reaction components at termini and constraints, no essential difficulty is involved *if* there is no inelastic action either by plastic yielding or by creep. The following few paragraphs will discuss the theory of the calculations, and the remainder of this subsection will discuss the practical aspects, particularly in cases where inelastic action may take place.

The symbol ϑ will stand for a temperature *distribution*, possibly varying from point to point along the configuration. ϑ_0 will denote the initial distribution, ϑ_F will denote the final distribution, and ϑ_M will denote an "intermediate" distribution, which will be discussed in what follows. Suppose that there exists a temperature distribution ϑ_M for which the piping system is continuous with itself and with its constraints (that is, no gaps or other discontinuities) and is completely unstressed. Zero cold spring implies that ϑ_M is the same as ϑ_0, while 100 per cent cold spring implies that ϑ_M is the same as ϑ_F. Making use of the notation previously developed

$$d_{AP}' = \int_A^P \left[\int_{\vartheta_M}^{\vartheta_F} \alpha(T, 0) \, dT \right] \{\alpha, \beta, \gamma, 0, 0, 0\} \, ds$$

and if this column matrix is used also with terminal expansions d_P' and d_A' corresponding to the thermal displacements of the terminal equipment (or constraint points) in going from ϑ_M to ϑ_F in the formula $D_P' = d_P' - d_A' - d_{AP}'$, the hot modulus E_h is used in the compliances, and the corresponding flexibility equations

are solved, the solutions will represent actual hot reactions (not reaction ranges). Similarly, using

$$d_{AP}{''} = -\int_A^P \left[\int_{\vartheta_0}^{\vartheta_M} \alpha(T, 0)\, dT \right] \{\alpha, \beta, \gamma, 0, 0, 0\}\, ds$$

and $d_P{''}$ and $d_A{''}$ (with proper algebraic signs) corresponding to the thermal displacement of the terminal equipment in going between the temperature distributions ϑ_0 and ϑ_M and incorporating the cold modulus (i.e., at distribution ϑ_0) in the compliance, the solution of the flexibility equations yields actual cold reactions. Note that these two problems are generally nontrivially different, since the ratios E_h/E_c may differ from point to point and the alphas may differ in such a way that the deflections D_i do not bear a constant ratio to each other in the two problems.

However, practically, the variations in alpha as a result of variations in stress are of rather minor importance. Roughly, $\partial\alpha/\partial\sigma$ is approximately 7×10^{-12} in²/ (lb F) and α itself is about 7×10^{-6} per F, so that, say, a stress level of 20,000 psi results in only a 2 per cent variation in alpha. Thus, while theoretically the calculations should always have a zero-stress condition (at temperature distribution ϑ_M) as a point of departure, the zero stress values of alpha may be used without much error even at considerable stress levels. Hot reactions should be based on compliances involving E_h, and cold reactions should be based on compliances involving E_c.

If the symbol R stands for any reaction component as calculated directly from the flexibility equations using the full expansion from low extreme temperature to high extreme and using E_c in the compliances, that is, if R stands for any of the reaction ranges determined in the original flexibility calculation, and if there is purely elastic action, the corresponding actual cold reaction component has the value $-CR$, where C is the cold-spring factor. That is, for zero cold spring, the cold reaction is itself zero, while for 100 per cent cold spring, the cold reaction corresponds to full expansion using the cold modulus. Usually, in published formulas the minus sign is suppressed, it being understood that the cold reaction is opposite in sense to the reaction range resulting from going from a cold to a hot condition. By a similar token, the hot reactions would be $(1 - C)(E_h/E_c)R$. However, perfect cold springing is practically impossible to achieve, and the experience of the industry has led to the use of formulas which, in effect, give conservative (that is, safe, or upper) estimates of hot reactions by the expedient of allowing credit for only two-thirds the cold spring one tries to achieve. Thus, the usual formula is given as

$$R_c = CR$$
$$R_h = (1 - 2C/3)(E_h/E_c)R$$

the subscripts c and h designating cold and hot conditions, respectively.

If no inelastic action takes place, these are good formulas to use; they permit estimating actual extreme reactions with the same degree of accuracy as is implicit in the entire computational process. There is, of course, some theoretical ambiguity involved in systems using more than one material, since, for example, in such a system there would be no unique value for the ratio E_h/E_c. Furthermore, there is additional ambiguity in systems in which the temperatures are not uniform throughout. Basically, the following discussion is properly limited to one-material uniform-temperature systems. In particular cases where these conditions are not satisfied, it may be possible to make an unequivocal determination of the actual extreme values of the reactions, but it should be noted that the difficulties of understanding and interpretation are considerable. As a practical thing, it must be remembered that the purpose of calculating actual reactions is to assist in determining whether or not terminal equipment will be statically overloaded and in completing the design

of such equipment. The theoretical ambiguities considered in this paragraph usually involve small practical variations compared with the results of the other uncertainties which enter into such an evaluation.

One way of assuring that little or no inelastic action takes place, and thus assuring that the formulas given above are applicable, is to verify that the maximum expansion stress range S_E satisfies the two conditions

$$S_E \leq \frac{E_c}{E_h} \frac{S_h}{1 - 2C/3}$$

$$S_E \leq 1.6 S_c / C$$

The rationale of these conditions is as follows. If the first is satisfied, the actual extreme hot stress, $(1 - 2C/3)(E_h/E_c)(S_E)$, does not exceed the Code-approved S value S_h, which has been selected so that neither gross yielding nor undesirable creep can take place at the higher temperature. Even though, as has been previously discussed, the actual state of stress is a complicated one whereas the creep data upon which S_h is based come from uniaxial tests, there is sufficient built-in safety factor that inelastic or creep action will not take place at the high temperature. The second condition assures no yielding at the cold temperature, for the Code-approved S_c does not exceed five-eighths the room-temperature yield stress; thus the second condition assures that the actual maximum cold stress CS_E does not exceed the yield stress.

If inelastic action does take place, the problem of estimating actual extreme reactions on terminal equipment may prove to be very difficult. At the very worst, the cold extreme cannot exceed R and the hot extreme cannot exceed $E_h R/E_c$ in magnitude, but these may be grossly conservative estimates.

There is, however, one case where a reasonable estimate can be made even though inelastic action (self-springing) does take place. This is the case of two anchor systems without intermediate constraints (also satisfying the conditions, given above, of consisting of only one material and being subject to uniform temperature variations throughout). In such a case, if self-springing does occur, it has much the same effect as additional cold spring. However, in order to obtain a safe estimate for design purposes, no credit for additional cold spring should be taken in the formula $R_h = (1 - 2C/3)(E_h/E_c)R$, which therefore may be employed even though the physical behavior is not elastic. On the other hand, self-springing tends to increase cold reactions. For the simple systems described in this paragraph, the ASA Code for Pressure Piping suggests the following estimate for cold reactions:

$$R_c = CR \text{ or } C_1 R$$

(whichever is greater) where C is the cold-spring factor, as before, and C_1 is an estimated self-spring factor. Reference 182 gives a rationale leading to the formula

$$C_1 = 1 - (S_h/S_E)(E_c/E_h)$$

(with the further understanding that, of course, no negative value of C_1 should be used; that is, small values of S_E do not imply negative self-spring).

The formulas given in this subsection for estimating extreme values of actual terminal reactions are intended as a guide to the designer and in the absence of alternate agreements may serve as a common basis of understanding between piping designer and equipment manufacturer.

In the absence of a statement of reactions permitted at equipment termini, the following general criterion may be of some use. Calculate the expansion stress range S_E at the terminus in question, including an appropriate stress-intensification factor (1.3, if no better estimate is available). Multiply by the factors R_h/R and

R_c/R, as determined by the formulas given above or other appropriate procedure as discussed above; these give what might be regarded as extreme estimates of hot and cold (respectively) actual stresses in the pipe in the neighborhood of the terminus. If values are "reasonable," the design should be satisfactory. In deciding what is reasonable, one should make some sort of mental evaluation of the relative ruggedness of the equipment as compared with the piping. For example, a heavy forging or casting should be able to withstand greater values than a thin-shell structure. If the terminal equipment contains moving parts (for example, pumps, turbines, valves), so that elastic distortion is a problem, or if the terminal equipment is inherently weak (for example, floating headers for furnace-heated tube bundles), a more generous allowance may be made.

If, as is frequently the case, the extreme loadings which a manufacturer permits would result in a stress in the adjacent pipe of but a small fraction of the S_h Code-allowable stress value, it may be that the limits are not realistic or that the equipment has not been properly designed to discharge its function under reasonable service conditions. In certain cases involving important and special equipment, structural changes in the equipment at small additional cost would eliminate the need for a costly and circuitous pipe configuration. In other cases, discussion with the manufacturer may reveal that the limitation is based upon a condition which, in the given installation, cannot actually occur in practice. In these and other cases where a generally reputable and progressive manufacturer imposes limitations which on their face appear to be unreasonable, it is obviously desirable to try to achieve a mutual understanding by discussion of the problems involved on both sides. Some of the larger manufacturers have published brochures which are intended to assist the designer in understanding the manufacturer's limitations and their rationale.[117,264] Even in cases where the limitations may imply costly rerouting of pipe, understanding of the manufacturer's problems will assist discussions leading to the most economical solution.

AUTOMATIC DIGITAL COMPUTATION OF PIPING FLEXIBILITY

Historical Development. If the great volume of computation implied by the developments in preceding sections of this chapter is considered, it will be immediately apparent either that gross simplifications must be made, the implications of which would be increasingly difficult to estimate as the systems analyzed grew in complexity, or that personnel and time requirements for the staggeringly great number of arithmetic operations would soon get completely out of hand. Some highly compact manual computation schemes have been devised, and some analysts developed an uncanny insight into ways of breaking complicated problems up into a number of simplified problems and recombining the solutions of these so as to make valid conclusions about the original system. (The author wishes to acknowledge here his amazement at and respect for the highly successful analyses which were made by L. Labriola using such procedures.)

However, just as modern linear algebraic methods were in the air in the early 1950's, this time also marked an awakening of engineers to the possibilities of the high-speed automatic digital computing machines which were becoming available. Several independent efforts at programming the piping-flexibility problem for solution by use of automatic calculation came into being in the early years of the past decade. Because of publication delays and the growing but nonuniform practice of "publication" by means of internal company reports, which nevertheless receive considerable outside circulation, it is not intended to assign any priorities in remarking that among the earliest of these efforts were those of Johnson,[145–147]

Jackson and Johnson,[142] and Peiser and Katz.[208]   However, it is clear that many other efforts were in progress at that time or shortly after, even though they may not have been reported in the literature until later.   Among these is the 1955 computer program of the M. W. Kellogg Company, which afforded a continuation of the piping-flexibility analysis service which had previously been offered to the piping industry by manual calculation.   The author also devised a computer program for the matrix method in 1955, but this came after a similar effort of Luke and Lubkin at Midwest Research Institute in 1954.[178]   A program jointly sponsored by Blaw-Knox and Arthur D. Little Companies began commercial operation in 1956.[9]   In 1956 a service bureau division of IBM (now the separate Service Bureau Corporation) offered a program by Pickerell, Rogers, and Woo.[209]   Besides the work by Johnson, mentioned above, the General Electric Company also devised a program for use at KAPL and another program by Lela Keough,[152] which is available to SHARE members.   C. F. Braun Company offered a computer service for piping flexibility beginning in 1959.[48]   Electric Boat Division of General Dynamics Corporation has offered an extensive service since about the same time; this program by now has expanded to include analysis of other types of structures and dynamic analysis as well.   The Service Bureau Corporation (an IBM subsidiary) devised a program which became available at about this time.   Pittsburgh Piping Company was the original sponsor of this work.   Somewhat earlier, Coldham had devised a program which made use of an English computer DEUCE.[79]   Bechtel Corporation has developed a program under the direction of N. Blair, with the cooperation of C. Johnson and Virginia Palm (Nadler) then of Remington-Rand Univac.   There have been reports of work at Oak Ridge National Laboratory on a piping-flexibility program employing the computer ORACLE.   Recently a very general structural program has been described by Friedrich at the Westinghouse BETTIS laboratory.[112]   Similarly, Eisemann, Woo, and Namyet[97] describe a very powerful program.   The steam division of Westinghouse Electric Corporation uses an IBM 7090 program of the Kellogg "general analytical method" for analysis of piping connected to pumps, turbines, and similar equipment of its manufacture. An excellent and very powerful program has been employed by K. Gordis of Kent Scientific Services, New York.   Work has also been done by Bridge;[55] Ellis, Schmidt, and Watkins;[98] Olson and Cramer,[199] and many others to whom apologies are extended for thus failing specifically to acknowledge their contributions.   Absence of a literature citation above implies that the information has come from private correspondence or from company bulletins and brochures; in the latter case, such brochures should be available upon request to the company involved.   Other literature citations include Refs. 62, 141, 159, 175, 176, 177, and 262.

The Nature of Presently Available Computer Programs and Services for Piping-flexibility Analysis.   The importance of obtaining reliable analyses, even though approximate, of the flexibility of important piping systems in the past demanded that these problems be taken under attack by personnel sufficiently competent to deal with the quite complex structural and mechanical aspects involved.   This usually involved the services of at least one of the best engineers available plus the time of junior engineers, clerks, and desk-calculator operators. For complicated systems or projects involving a number of piping systems literally weeks of effort were required of such a group, usually precisely at the stage of development of the project that these personnel were most vitally needed for other responsibilities.

The present availability of advanced structural methods (described in the foregoing sections of this chapter) in combination with that of modern high-speed digital computers implies that this volume of computation can be handled more quickly, much more surely, and very much more economically by permitting the

machines to make the calculations and freeing the personnel of all responsibilities except that of (1) deciding what systems require formal analysis, (2) preparing input data for the computer program, and (3) deciding, on the basis of the computer output, whether the system is satisfactory or, if not, what changes are called for. When one observes that a modern computer can store tens of hundreds of thousands of numbers and partial answers and can perform thousands of arithmetic operations with these numbers in every second, it is easy to see why it is usually better to employ the machine at, say, \$1,000 per hour than a man at, say, \$10 per hour (both figures including all overhead items) to perform these voluminous routing operations.

There are two philosophies and two procedures for doing just this. Many companies have installed their own computers which are used for all sorts of in-house problems, including billing, inventory control, payroll, and other business-type operations as well as heat-transfer evaluations, reduction of experimental data, analysis of duct losses, and other engineering and scientific operations.   Other companies prefer to avoid the investment in such a computer setup and its highly specialized personnel and, instead, to farm their problems out to specialists who have both computer facilities and a knowledge of how to employ them for business, engineering, scientific, and other problems.   There are many reasons which can be cited both for and against either practice, and the individual company alone is in a position to decide which is the more advantageous procedure.

The author, however, has a very strong and long-standing conviction that the piping-flexibility problem can best be handled by use of such outside specialists. While it is true that many companies should for various reasons acquire computer facilities of their own, the present or future possession of such facilities should not be too strongly adduced in favor of internal handling of piping-flexibility problems. To devise a modern "problem-oriented" program requires extensive and expensive effort by highly sophisticated and hard-to-get programmers, who, experience indicates, frequently fall behind promised deadlines. A problem which is fully disposed of in a matter of 10 sec on a large-scale computer may require many hours on a small computer on which both the accuracy and the completeness of the output may have to be compromised.   Even though a piping program may already be available, evolution of flexibility programs plus economic competition between those offering a piping service implies that it may soon become outdated in the sense that newer programs will offer more information or better guides to design.

Probably the oldest piping-flexibility service is that of the M. W. Kellogg Company, which for many years has solicited flexibility problems for analytical solution, originally employing manually operated desk-type calculating machines and model tests but changing procedures as new technology became available, first using the old card-programmed computers and more recently the modern high-speed electronic digital computers.   The Kellogg program makes use of the so-called general analytical method of analysis, described in detail in Ref. 150 but by now cast into matrix algebraic form for convenience in internal computer operations.[151]   This program has been adapted to the Datatron and more recently to the IBM 7070 computers.   The Service Bureau Corporation (SBC), an IBM subsidiary, and the Pittsburgh Piping Company wrote the first program for a really high-speed computer, devising a very powerful and general program for IBM 704 which has since been accommodated to the 7090 system.   Several man-years of programming work were then required to flesh out the specifications and procedures which resulted from the initial planning.   It is creditable that this program is highly "customer oriented," that is, aimed at the user's convenience by such devices as many alternates of dimensional input (for example, one may write 34-5-3/16, 34-5.1875, 34.4323, 0-413.1875, or 0-413-3/16, as one chooses, for a length of 34 ft $5\frac{3}{16}$ in. and the computer will interpret correctly).

In a similar fashion, the Power Piping Division of Blaw-Knox Company obtained the cooperation of Arthur D. Little Company engineers and programmers to devise, originally for UNIVAC but more recently for several other high-speed computers including the IBM 700 series, a program using matrix procedures involving only 6×1 and 6×6 matrices. This service appears to have been the first commercially available service which employed what is now called a high-speed electronic digital computer. It is more fully described in Ref. 233.

A fourth important service is available from the Electric Boat Division of General Dynamics Corporation, where a quite general flexibility program has been made available for commercial service. This program is also for the IBM 700 series and with a certain amount of special programming has been also used to determine dynamic behavior of complicated piping systems, especially for submarine-propulsion system piping. Although these four services are probably the most important ones, there are a number of others available, and a potential user of flexibility-analysis service may obtain information by inquiring locally from consulting engineering firms having access to a computer.

The rates that these service organizations charge are quite low and competetive with one another. In round numbers the average is about \$5 per item of pipe in 1963, although this may vary depending on the projective complexity of the piping configuration, the degree of engineering control and surveillance to be exercised by the service, the volume of business transacted, and other similar considerations. The term item of pipe which appears above refers to a straight length or arc of pipe. A constraint point, a tee or nozzle, or a point where external forces are applied, in other words, a point at which anything "unusual" happens to the geometry or the loading of the pipe, separates the pipe at that point into two or more "items." Thus, what a piping designer frequently calls a "spool" may actually consist of more than one item. Even so, the processing cost charged by one of these service organizations is small compared with a realistic evaluation of actual in-house costs. In the section "Simplified Procedures for Desk Computation," a very simple example will be worked out using a simplified procedure (see Fig. 40). It consists of nine items and would cost in the neighborhood of \$45 to solve by use of a piping-flexibility service. Such a solution would give deflections and rotations as well as force and moment values at every significant point in the configuration and possibly other significant information as well. Manual calculation using the simplified but reasonably accurate method to be described would require about 2 man-hours. For reasonable assurance that no significant errors were involved, a second independent solution should be made (errors in algebraic sign are difficult to detect unless two quite independent solutions are made and are compared). This solution so obtained manually does not, however, afford information about displacements, so that separately calculated estimates are required in order to guide the selection and sizing of supports. These are referred to as estimates, since the simplified method itself does not provide for their calculation and the partial data in the simplified solution cannot easily be employed. Thus, the overall cost of in-house solution is comparable even though the accuracy, reliability, and completeness of the in-house solution compares very unfavorably with the service organization solution. In cases of greater complexity, the disparity immediately becomes much greater. If a rigid constraint were called for at one point of this system, in order to protect terminal equipment, the service organization charge would be increased by about \$5 while the in-house cost would be at least tripled.

The following description of the input data required, the steps in processing, and the nature of the output is intended to be a composite of the available services taken from company brochures and personal knowledge of the programs. Individual programs may differ in detail from this description, but these differences are not

major. The input data for a piping problem involve the following types of information:

1. *System geometry*, including specification of coordinate system, dimensions and/or coordinate specification of all pipe items, location of and nature of constraints, etc.

2. *Pipe and material data*, i.e., pipe size and schedule number or wall thickness, pipe material and/or physical constants, etc.

3. *Special element data*, i.e., dimensions and rigidity of valves, spring constants for restraints, compliances of terminal equipment, etc.

4. *Free expansions* of termini and constraint points

5. *Temperature conditions* corresponding to various operating and emergency conditions for which analysis is desired

These data must be reduced eventually to a standard format from which key-punch operators or personnel using similar equipment prepare the computer input data.   Reduction to standard format may be done either by the customer or by the service, and there are some obvious advantages and disadvantages to either procedure.   Reduction by the service organization's personnel involves extra costs and the possibility of communication errors (misinterpreting drafting-room conventions, etc.) as well as possible delays in processing but has the advantages of requiring no special knowledge by the customer and of affording critical scrutiny by personnel accustomed to a wide variety of piping configurations and their problems. Reduction by customer personnel decreases service billings and may save time in processing; it also helps assure that drafting errors are caught.   This procedure does, however, require that customer personnel become acquainted with the input format, and this requirement varies from program to program and with the degree of modern professional background represented by customer personnel.   Most services have prepared instruction brochures which guide the personnel in input preparation, and it would require of the order of 2 hr for a reader of such a brochure to become adequately adept in the discipline; an occasional special question can be cleared up by telephone.

Whether the data sheets are prepared in-house or by the service is a matter of individual user policy.

Following the preparation of the data format sheets, clerical personnel (key-punch operators, for example) prepare the primary input medium, usually punched cards or punched paper tape.   At this stage of the operation it is customary to verify the correctness of this clerical step by duplicate punching or some similar discipline. Following this, the primary medium is most frequently loaded by the use of a peripheral special-purpose device into high-speed magnetic tape; this is done so that the main computing device will not be burdened with such a trivial task.   This magnetic tape now has the data stored on it in a very compact form.   On the same tape will be stored a number of problems, some of them piping-flexibility problems and some of them problems of entirely different nature, each containing an appropriately coded instruction concerning its own nature.   The tape unit containing the tape in question is then switched from the peripheral equipment to the main computer control, and when the computer is finished with the task presently occupying it or when the operator instructs it to do so, the problems are transferred one at a time and each in a fraction of a second to the main computer.   The main computer ascertains the nature of the problem and calls in from an appropriate waiting tape unit the previously prepared program required to solve it.   The solution which is obtained in a matter of a few seconds is then recorded on another magnetic-tape unit, and the main computer then calls in another problem.   After a sequence of results has been stored on an output magnetic tape, it is switched from main computer control to output control and a peripheral control unit causes it to print its information in desired output format at the rate of several hundred to a thousand

full lines per minute. This printed output is assembled according to customer or job code into booklets or problem brochures and is ready for overall analysis and interpretation by the service organization's engineers or by the customer's engineers as the contractual agreement may call for.

In general, the output contains the following items:

1. Input data for record and verification purposes
2. Verification of geometric consistency of input data
3. Physical and/or dimensional properties obtained by program via computation or table look-up rather than as a part of direct input
4. Terminal and constraint reaction ranges
5. Force and moment ranges at desired points throughout the configuration
6. Stress ranges at these desired points
7. Deflection and rotation components at these desired points

If the agreement calls for engineering control, these output data are further inspected by service organization personnel and comments and suggestions are made. By prior arrangement, overall findings may be communicated by telephone and appropriate new problems accepted for solution so as optimally and rapidly to proceed, in the case of a difficult routing problem or a tight configuration, through a limited sequence of trial configurations in order to arrive at an acceptable and economical final solution.

It has already been indicated in previous sections of this chapter that different mathematical treatments of the problem may yield identical correct results. Accordingly, the inner workings of the several programs may be quite different. Furthermore, even if the same basic theory is used, different programs may use different programming devices to handle some of the difficulties which arise. One of the principal such difficulties follows from the fact that human analysts have an innate sense for the topology of a piping configuration and can instantly make appropriate decisions based on this topology whereas the computer has difficulty with this sort of decision, which accordingly must be written in terms of the greater-than, equal-to, or less-than decisions which the computer can make quickly.

In Fig. 32 is shown the "flow chart" of a typical large-scale piping-flexibility computer program. Arrows indicate the *logic flow* between the various *boxes* which are labeled with the function they perform. The program is divided into five *phases* which operate in sequence. Phase I involves initial internal data conversion and storage. Phase II involves computation of compliances matrices for each branch of the system. Phase III involves inversion of compliances to obtain stiffnesses and vice versa, and their addition and combination. Phase IV involves the calculation of force and deflection matrices and the expansion stress S_E at each significant point of the structure. Phase V involves the final internal editing of results and the output of answers. The reader will notice from the captions printed on the individual boxes that most of them involve highly sophisticated and complex operations. A box marked NEW BRANCH?—YES—NO may involve only one or at most a few fundamental machine operations, but the box labeled INVERSION, COMBINATION, AND REINVERSION OF MATRICES probably involves several hundred machine language commands most of which are cycled through a very great number of times during the solution of a problem. Continual evolution of programs probably implies that this particular program has already been modified and improved to an extent that may render Fig. 32 obsolete.

In the formulation of a flexibility problem that is presented for computer analysis and solution, there are certain features which must be taken into account, among which are:

1. The user of the computer must understand the bases of stress, reaction, and moment calculations.

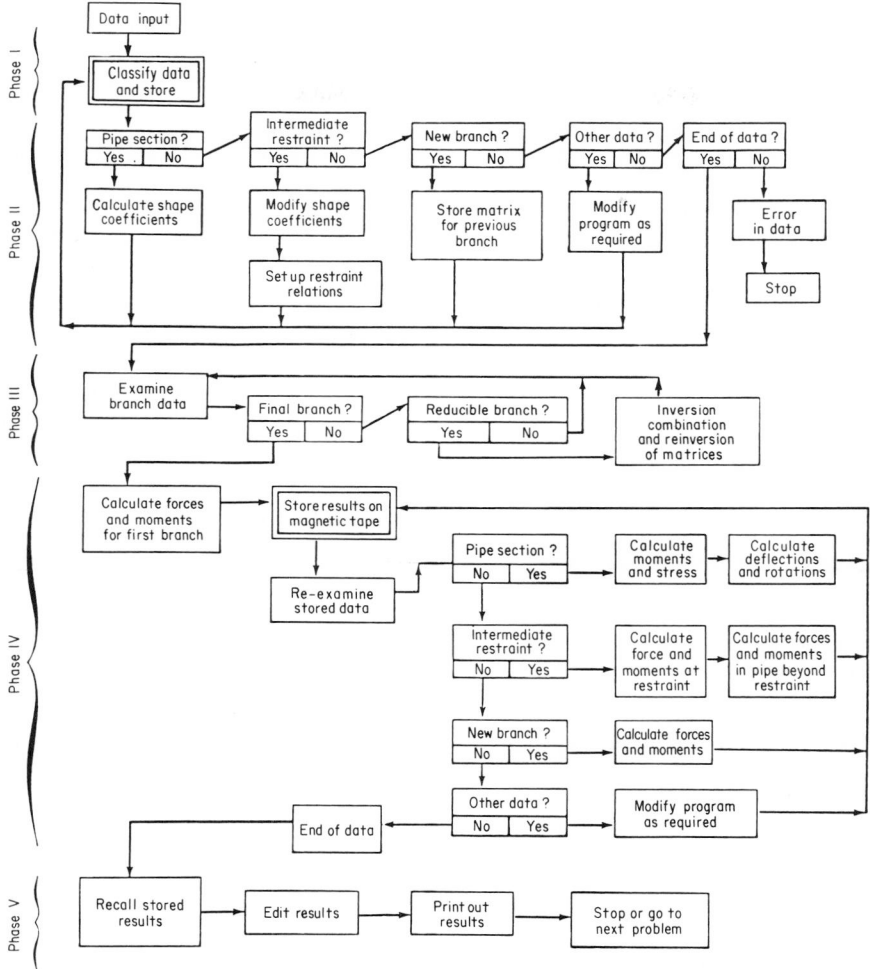

FIG. 32. Logic flow chart for typical modern large-scale computer program for piping flexibility analysis. (*Courtesy of The M. W. Kellogg Company.*)

2. The problem given to the computer must be clear of ambiguity, and its solution must be possible.

3. The proper problem or question must be stated to the computer. The computer will supply the correct answer to the problem which it is given, and it is incumbent upon the user to assure that the proper question is asked.

4. All computers utilize right-handed coordinate systems. Thus a problem posed in a left-handed system will confuse the computer, and meaningless jargon will result.

5. All relevant data must be supplied. Such data have to include properties of the piping system and of the terminal equipment. In order to calculate S_E, for

example, details of tees and nozzles, stress-intensification factors, and bend-flexibility factors must all be fed into the computer logic and memory system.

Example of Computer Input for Flexibility Problem. Figure 33 shows a configuration the purpose of which is to illustrate the type of projective and constraint complications which can be very easily and quickly dealt with by use of a modern digital computer program. The reader will see from Note 3 that part of the piping is composed of a special alloy steel having the properties listed. Motions of termini are indicated. There are spring hangers at points 18 and 43 having spring scales of 2,000 and 1,500 lb/in., respectively. At point 10, y deflections and z rotations are prevented, while at point 37, there is a limit stop which permits a downward motion of 0.12 in. At point 32 there are applied by some external agency a force and a moment. All valves are considered to be sufficiently massive that their flexibility is negligible compared with that of a similar piece of pipe.

The input to the key punch operators is shown in Table 5. It is not feasible in the space that can reasonably be assigned to this description to attempt to explain this input in full detail; however, a few explanatory remarks are in order. The "type" designations are mnemonic, thus: HED = heading, COM = comments, MAP = material properties (for an exotic material whose properties are not built into the program), BAN = basic anchor (giving its coordinates), BRA indicates a branching, TIN indicates the intersection point of the tangents to a bend or elbow, CAD = constraint against deflection, CAR = constraint against rotation, ANC = anchor, CAP = check absolute position (to guard against dimensional error in the input), MAN = movement of anchors, REM = remarks, TER = terminus, FOR = forces applied by external agency, MOM = moments applied by an external agency, NOP = not ordinary point (signaling that something like a CAD is coming next), and END = end. There are some very simple and nearly obvious rules about numbering of points and about the positions in which special information must be filled. Exotic "types" may call for referring to a short set of instructions, but most of the input data are quite obvious. For example, in going from point 2 to point 3, one proceeds 12 ft 6½ in. in the negative z direction, from point 3 to the tangent intersection 4 one goes 9 in. in the negative z direction, in going from point 4 to point 5 one goes 9 in. in the positive x direction, and so on. At point 18 it is indicated that there is a constraint with spring scale 2,000 lb/in. with direction cosines 0, 1, 0. Incidentally, the writer would have included a REM (remark) following the CAD at point 37 to remind himself to check the output and verify that the limit stop was actually engaged as postulated; the output shows that it is.

Closing loops is done more or less obviously in proceeding through the system; for example, following point 25 is point 2, after which one goes back and picks up again at some earlier branch. The sequence of entries is not unique. Quite different sequences could describe the same problem and generate the same output information. In spite of its complexity, this configuration has only 35 "pieces" which are counted for billing purposes and would cost about $200 to run. This system has five anchors and five additional constraints (two of them at point 10) and therefore is 29-fold redundant. The output format, which runs to nine pages of double-spaced type-script and is clearly too voluminous to reproduce here, contains a print of input, a listing of anchor and constraint reaction ranges; listings of force, moment, and stress; ranges at every numbered point; and a listing of displacement and rotation components at every listed point. The input and output described above are for the Service Bureau Corporation program, cf. Ref. 233; the other modern and complete services mentioned previously differ in detail as regards input and output but are quite similar in their ability to provide all these results.

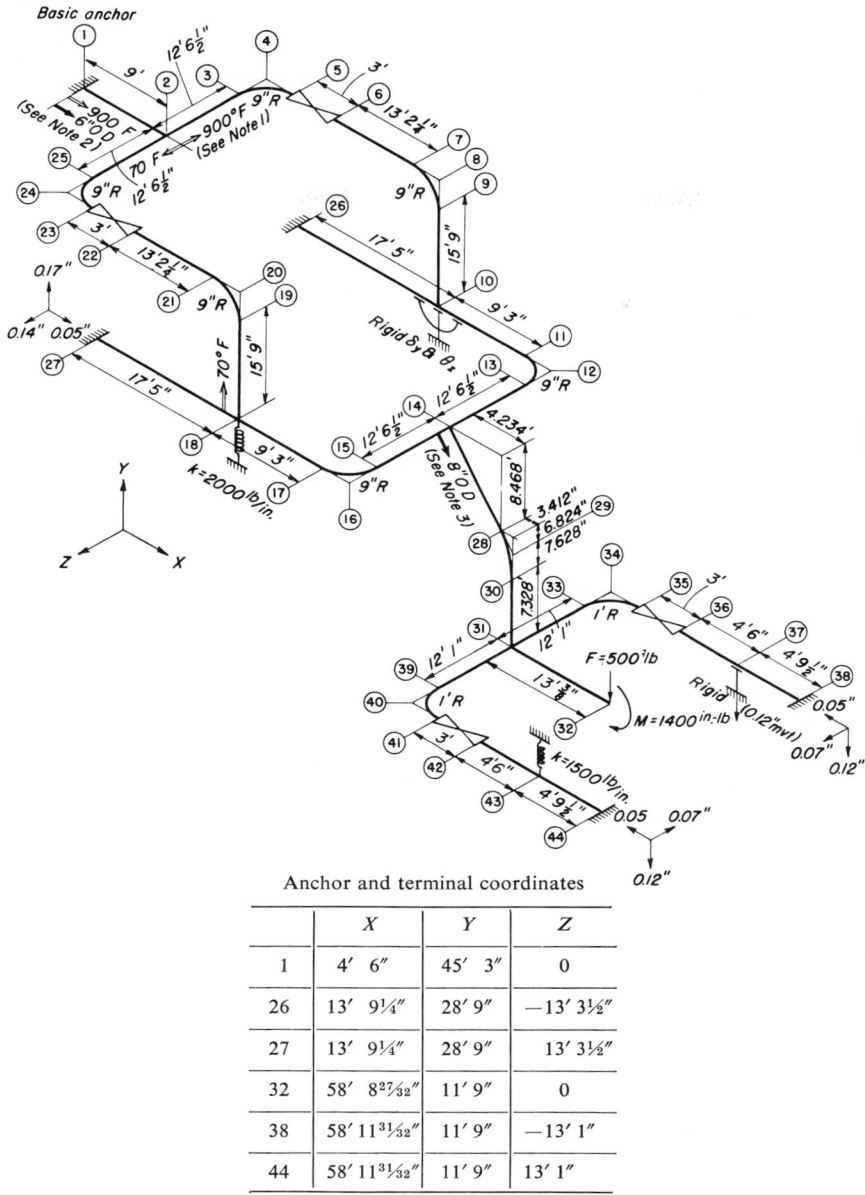

Fig. 33. Configuration illustrating digital computer input. (*Courtesy of The Service Bureau Corporation.*)

Anchor and terminal coordinates

	X	Y	Z
1	4′ 6″	45′ 3″	0
26	13′ 9¼″	28′ 9″	−13′ 3½″
27	13′ 9¼″	28′ 9″	13′ 3½″
32	58′ 8²⁷⁄₃₂″	11′ 9″	0
38	58′ 11³¹⁄₃₂″	11′ 9″	−13′ 1″
44	58′ 11³¹⁄₃₂″	11′ 9″	13′ 1″

NOTES:

1. All pipe is at 900 F except as shown.

2. All pipe between points ① and ㉗ is low-carbon steel, 6 in., schedule 40 (OD = 6.625 in., wall thickness = 0.280 in.).

3. All pipe in the lower part of the system starting at point ⑭ is special alloy steel, 8 in., sch. 40 (OD = 8.625 in., wall thickness = 0.322 in.). $E_c = 32.7 \times 10^6$ psi; $E_h = 29.1 \times 10^6$ psi; $\alpha = 7.145$ in./100 ft; $\nu = 0.31$.

Table 5. Sample Input Data† for Digital Computer Calculations‡

S B C The Service Bureau Corporation, a Subsidiary of IBM

PIPING FLEXIBILITY ANALYSIS

TYPE	FROM	TO	FROM → TO ΔX	ΔY	ΔZ	OUTSIDE DIAMETER	WALL THICKNESS	METAL CODE	TEMP. ERATURE
HED			THE SERVICE BUREAU CORPORATION, 3/21/60						
COM			ILLUSTRATIVE PIPING SYSTEM NO. 3						
COM			NOTE THAT COORDINATE SYSTEM ORIGIN IS NOT AT BASIC ANCHOR						
MAP			32.7E6	29.1E6	7.145	0.31		XXX	900
BAN		1	4 – 6	45 – 3					
BRA		2	9 – 0			6.625	0.237	LCS	900
		3			– 12 – 6 – ½				
TIN		4		– 0 – 9					
		5	0 – 9						
		6	3 – 0			RIGID			
		7	13 – 2 – ¼						
TIN		8	0 – 9						
		9		– 0 – 9					
BRA		10		– 15 – 9					
CAD	1	10	0	1	0	RIGID			
CAR	2	10	0	0	1	RIGID			
		11	9 – 3						
TIN		12	0 – 9						
		13			0 – 9				
BRA		14			12 – 6 – ½				
		15			12 – 6 – ½				
TIN		16			0 – 9				
		17	– 0 – 9						
BRA		18	– 9 – 3						
CAD		18	0	1	0	2000			
		19		15 – 9					70
TIN		20		0 – 9					
		21	– 0 – 9						
		22	– 13 – 2 – ¼						
		23	– 3 – 0			RIGID			
TIN		24	– 0 – 9						
		25		– 0 – 9					
BRA		2		– 12 – 6 – ½					
ANC	10	26	– 17 – 5						900
CAP		26	13 – 9 – ¼	28 – 9	– 13 – 3 – ½				
ANC	18	27	– 17 – 5						
CAP		27	13 – 9 – ¼	28 – 9	13 – 3 – ½				
MAN		27	0.05	0.17	0.14				
	14	28	4.234	– 8.468		8.625	0.322	XXX	
TIN		29	0 – 3.412	– 0 – 6.824					
		30		– 0 – 7.628					
BRA		31		– 7.328					
REM			CHECK VERT. DISPL. AT POINT 31 – NOT TO EXCEED 0.75 IN. DOWN						
TER		32	13 – 0 – ⅜						
CAP		32	58 – 8 – ²⁷⁄₃₂	11 – 9					
FOR	1	32	0	1	0	– 500			
MOM	2	32	0	0	1	– 1400			
	31	33			– 12 – 1				
TIN		34			– 1 – 0				
		35	1 – 0						
		36	3 – 0			RIGID			
NOP		37	4 – 6						
CAD		37	0	1	0	RIGID	– 0.12		
ANC		38	4 – 9 – ½						
CAP		38	58 – 11 – ³¹⁄₃₂	11 – 9	– 13 – 1				
MAN		38	– 0.05	– 0.12	0.07				
	31	39			12 – 1				
TIN		40			1 – 0				
		41	1 – 0						
		42	3 – 0			RIGID			
NOP		43	4 – 6						
CAD		43	0	1	0	1500			
ANC		44	4 – 9 – ½						
CAP		44	58 – 11 – ³¹⁄₃₂	11 – 9	13 – 1				
MAN		44	– 0.05	– 0.12	– 0.07				
END									

† To be used in conjunction with Fig. 33.
‡ Courtesy of The Service Bureau Corporation.

SIMPLIFIED PROCEDURES FOR DESK COMPUTATIONS

General Discussion.   Prior to the application of high-speed electronic digital computers to the problem of piping-flexibility analysis, one of the major virtues claimed for any new method was its speed, accuracy, and convenience for manual computation by slide-rule or desk calculating machine.   Since, as is evident from what has gone before, any exact solution (no matter how the word *exact* may be used in such a context) is out of the question for any except the simplest of geometries, all such methods are "simplified methods" in some sense of the word.   They may be divided into three general classes as follows⌡   The first class contains overall criteria which are intended only as a rapid way of assuring that a piping system either is definitely flexible enough or is definitely not flexible enough; these criteria are not capable of deciding about a wide latitude of practical cases, for which, accordingly, somewhat more powerful methods must be employed, but they afford a very rapid and economical way of treating many cases which do arise in practice. The second class of simplified methods makes use of accurate solutions to a group of problems the results of which are usually presented in graphical form; the user of such a method in effect compares the geometry and circumstances of his problem with those of the precalculated problem and attempts to determine significant answers by comparing and interpolating the precalculated answers.   The third class of methods employs various devices to simplify the actual mathematical analysis implied by the theory given earlier in this chapter.

None of the simplified methods is capable of dealing with systems having more than two termini, and those discussed in this section are applicable only to *two-anchor systems*.   However, since the great majority of actual piping systems are essentially two-anchor systems, these simplified methods still have great practical utility.

Conservative Overall Criteria.   Two examples will be given of the overall criteria which form the first class of simplified method.   The first criterion was proposed by the M. W. Kellogg Company representation on the Flexibility Task Force on the basis of design and operating experience acquired by the company, and it appeared in the 1955 and subsequent editions of the Code for Pressure Piping.   Although it has met with considerable objection on the two bases, first, that its genesis remains unknown so that its range of validity may not conveniently be examined by theoretical means and, second, that it may fail to give the correct result for some rather unusual configurations (against which a designer's insight should presumably warn him anyway), nevertheless the consensus of piping engineers charged with ASA Code responsibilities is that the generally successful experience which has been accumulated with use of the formula together with its great simplicity warrants its continued citation in the Code.   Briefly, according to this criterion, a piping configuration may be considered to be adequately flexible to prevent damage to the pipe itself if the inequality

$$DY/(L - U)^2 \leq 0.03$$

is satisfied, whereas if the expression on the left exceeds 0.03, analysis by a more accurate method is called for.   In this formula the symbols have the following meanings:

L = total length of pipe centerline, ft
U = straight-line distance between anchors, ft
D = nominal pipe diameter, in.
Y = (geometrical) resultant of expansion and terminal movements, in.

This criterion does not directly yield information that can be used in estimating reaction ranges or reactions on terminal equipment but, instead, deals only with the integrity of the pipe itself.

A rather similar criterion has been proposed by W. Enders,[99] who finds that, within a wide range of geometrical ratios, a fairly inclusive variety of two-anchor configurations may be conservatively evaluated by use of the criterion

$$L/U = 1 + 6\sqrt{(L^*/U)}$$

where, in addition to the symbols given above, the quantity L^* is a fictitious length defined by the formula

$$L^* = (E/S)(Y/U)(D/12)$$

E and S, respectively, denote Young's modulus and allowable stress, which, within the framework of the American ASA Code, should be interpreted as E_c and allowable stress range S_A, respectively. That is, Enders' criterion asserts that, as long as the actual total length is at least as great as $U[1 + 6\sqrt{(L^*/U)}]$, the piping is not overstressed. (Actually, the inequality in Ender's paper is of the opposite sense, indicating that, in the cases he has considered, the limiting condition occurs for somewhat shorter total lengths than the formula gives.)

Both these criteria generally give quite conservative results. Accordingly, it is a common experience that more exact analyses of systems which fail to satisfy one or both may still reveal that the system is indeed quite satisfactory. It is interesting to note also that a limiting relation, somewhat similar to the two given above, was evolved by Markl[148] from a comparison of a variety of square-corner plane shapes with an equal-leg L-shape configuration; Markl, however, did not propose his findings in the form of a criterion for actual use. Both Markl[182] and the M. W. Kellogg engineers[150] offer considerable discussion about the circumstances under which these criteria could conceivably lead to nonconservative conclusions.

Graphical and Chart Solutions. A great number of actual solutions have been made for those two-anchor systems which are most frequently used in practice. It would not be possible to list either here or in the bibliography all the graphs and charts which have been published in one place or another on the basis of such solutions, but mention will be made of some of the more important and useful collections. In view of the fact that the most significant evolution in piping-flexibility analysis in recent years has been with automatic computation using quite exact methods, there has been little significant evolution of graph and chart solutions.

Probably the most significant group of graph and chart solutions is based on the work of Wert, Smith, and Cope, which has since been augmented and modified by one or more of the original authors and others and published with various titles and under various company imprimatures.[42,211,263] Useful graphical results are also given by the M. W. Kellogg Company,[150] Juergensonn,[148] and the Tube Turns Division of Chemetron Company[181] and are also incorporated in any of a number of company catalogues and bulletins. In general, these methods may be considered to be satisfactory for the evaluation of relatively noncritical two-anchor systems which conform reasonably to the geometry and other limitations of the system of charts employed.

It is also possible to employ such graphical methods for the approximate evaluation of systems which differ from the ideal systems used in the preparation of the graphs by introducing modifications in actual dimensions so as to produce an "equivalent" system having what appears to be substantially the same behavior as that of the actual systems and conforming in geometry to the limitations of the graph. It is evident that considerable insight and ingenuity are required to be

successful in this approach, and the user of such "approximate" techniques must be aware that the results obtained may be of doubtful value.

Another approach, useful also for multianchor and constrained systems, is to divide a piping system into two or more subsystems each of which is regarded as a two-anchor system and analyzed by the use of the appropriate graph. Since the junction of the subsystems does not possess the degree of fixity postulated in the solution, the complete system is overall more flexible than any of its parts. It is even possible to permit some displacement at such a fictitious anchor, the same displacement in each of the two solutions in which the anchor plays a role, so as better to match the reactions. In any such conjuring, insight and ingenuity must be heavily relied upon and no general rules can be given for extending the usefulness of chart and graph solutions beyond the range for which they were originally devised.

Results may be given in tabular rather than graphical form,[126] and many industrial magazines frequently contain a graphical or tabular presentation appropriate to some useful configuration. The calculations involved in using solutions of this type have even been reduced to the form of a slide chart which incorporates in one breast-pocket-size device a series of coefficient charts for various configurations and a special slide-rule with a number of scales. Such a device was copyrighted in 1956 by the Blaw-Knox Company. It is fair to state that many analyses which formerly would have been attempted by graphical, chart, or tabular methods are now done by precise structural analysis using automatic computation.

The remainder of this subsection will present material prepared by S. Crocker and A. McCutchan for earlier editions of this "Piping Handbook." The earlier material did not make use of the stress-*range* concept, so that the value of 12,000 psi, now listed as the assumed allowable stress range, is somewhat lower and therefore more conservative than would usually be permissible. However, the text and example show how the solutions obtained may be modified for other values of allowable stress range. Strictly speaking, since these tables and figures were originally determined using the hot modulus of elasticity, one should employ a corrected allowable stress range S_A' obtained by use of the formula $S_A' = (E_h/E_c)S_A$. Also, one should note that the *reciprocal* bend-flexibility factor, denoted in the figures by the symbol K, is used rather than the bend-flexibility factor n which has been employed heretofore in this chapter; $Kn = 1.0$. Although the quantities q and m which are used in the figures are defined thereon, for convenience, the definitions are also given here: $q = (l_1 + l_2)/2H$; $m = W/H$.

A convenient solution for conventional expansion bends used with straight runs of pipe is furnished in Tables 6 to 9, which give the overall lengths of pipe cared for by various radii of bends and the corresponding reacting forces at different operating temperatures. Bends should be located midway between anchor points or at least within the middle third of the span to agree with the assumptions made in preparing these tables. Chart valves have been adjusted for change of E with temperature.

The overall lengths given in Tables 6 and 7 for expansion U bends and double-offset expansion U bends, respectively, are strictly accurate for Schedule 80 (extra-strong) pipe. But these tables may be used with reasonable accuracy for all pipe-thickness schedules, since bending stress is independent of the wall thickness in straight pipe and is affected only to a minor degree in the case of long-radius bends by variations in the factors that correct for flattening of the circular cross section of curved pipe during flexure. For wall thicknesses less than Schedule 80, the lengths given in Tables 6 and 8 are conservative, but not unduly so, because of the relatively long-radius bends considered in these tables. For thickness greater than Schedule 80, the lengths tabulated are slightly but not significantly greater than found by a strict analysis.

Table 6. Expansion U Bends, all Pipe-thickness Schedules

(Over-all length of pipe in feet permissible with expansion U bends
of radii and for operating temperatures shown)

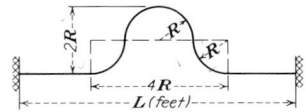

Nominal pipe size, inches d_n	$\dfrac{R}{d_n}$	Mean radius of bend inches, R	L = over-all length of pipe in feet permissible with expansion U bends of radius indicated (Based on an allowable stress range of 12,000 psi)						
			200 F	300 F	400 F	500 F	600 F	700 F	800 F
2	6	12	28.6	15.9	10.8	7.9			
	8	16	45.8	25.7	17.6	13.1			
	10	20	65.0	36.5	24.9	20.3			
2½	6	15	37.0	20.6	13.9	10.4			
	8	20	59.5	33.3	22.9	17.3			
	10	25	83.5	47.5	31.7	24.5			
3	6	18	46.7	26.1	17.5	13.2	10.7		
	8	24	72.0	40.3	28.1	20.8	16.7		
	10	30	102.0	58.5	40.0	31.0	25.0		
4	6	24	67.0	37.5	25.6	19.0	15.3		
	8	32	100.0	57.2	39.6	29.6	23.7		
	10	40	141.0	80.0	55.0	42.5	34.6		
6	6	36	100.7	60.0	41.2	30.4	24.1		
	8	48	158.3	90.5	61.7	46.7	37.1		
	10	60	207.0	118.0	81.0	63.0	51.0		
8	6	48	150.0	84.2	58.0	42.9	34.6		
	8	64	221.0	124.2	86.7	66.2	53.3	44.2	
	10	80	301.0	170.0	116.0	88.0	71.0	58.0	
10	6	60	200.0	112.1	72.1	57.9	45.8	38.7	
	8	80	297.0	167.2	115.7	86.7	69.2	59.2	51.7
	10	100	400.0	225.0	155.0	120.0	98.0	83.0	72.0
12	6	72	243.0	136.8	93.3	70.0	55.8	47.5	40.8
	8	96	357.0	201.0	139.2	104.2	84.2	71.7	63.0
	10	120	485.0	282.0	190.0	143.0	118.0	100.0	85.0
14 O.D.	6	84	290.0	168.2	115.2	86.3	69.2	57.9	51.7
	8	112	416.0	243.0	171.6	130.0	104.2	89.2	77.5
	10	140	615.0	345.0	234.0	178.0	144.0	127.0	105.0
16 O.D.	5	80	252.2	147.1	100.3	74.6	59.6	50.4	43.8
	6	96	327.0	191.0	132.0	98.8	79.7	67.5	58.7
	7	112	392.0	225.0	157.0	118.0	95.0	80.0	70.0
18 O.D.	5	90	289.0	168.4	115.5	85.8	68.4	57.6	50.4
	6	108	378.0	220.0	152.2	114.2	91.7	84.8	74.7
	7	126	460.0	258.0	175.0	133.0	106.0	90.0	79.0
20 O.D.	5	100	318.0	185.5	127.5	94.2	75.8	63.3	55.0
	6	120	414.0	241.5	169.2	126.3	101.6	95.8	75.4
	7	140	488.0	280.0	195.0	148.0	118.0	100.0	88.0
24 O.D.	5	120	375.0	218.5	157.5	115.5	92.5	78.0	68.0
	6	144	517.0	302.0	213.0	156.6	126.3	107.2	94.2
	7	168	610.0	357.0	252.0	195.0	126.0	136.2	120.0

Table is based on Fig. 35. Pipe is assumed to be installed at 60 F. See text for method of adjusting lengths for allowable stress ranges above or below 12,000 psi. Lengths are based on values of q and i for Schedule 80 pipe but are sufficiently accurate for all pipe-thickness schedules.

Example.—What length of 16-in. O.D. pipe can be used under the following conditions with an expansion U bend having a radius of 6-pipe diameters if the line is installed at 60 F and the operating temperature is 400 F?
(a) For the basic conditions of Table 6?
(b) If the allowable stress range is 16,000 psi instead of 12,000 psi?
Solution.—(a) The length of line L read from Table 6 is 132 ft.
(b) For an allowable stress range of 16,000 psi the length would be $132 \times {}^{16}\!/_{12} = 176$ ft.

Table 7. Expansion U Bend, Schedule 80 Pipe

(Reacting force ranges in pounds for expansion U bends of radii and for operating temperatures shown when absorbing expansion of over-all lengths of pipe given in Table 6)

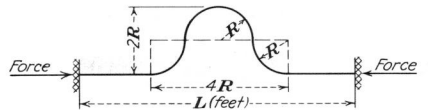

Nominal pipe size, inches d_n	$\dfrac{R}{d_n}$	Mean radius of bend, inches R	Reacting force ranges in pounds for Schedule 80 pipe expansion U bends of radius indicated when absorbing expansion of over-all lengths of pipe given in Table 6 which are based on an allowable stress range of 12,000 psi						
			200 F	300 F	400 F	500 F	600 F	700 F	800 F
$I = 0.8679$ 2	6	12	458	504	550	615			
	8	16	332	342	367	396			
	10	20	246	263	282	310			
$I = 1.924$ 2½	6	15	658	720	790	852			
	8	20	467	500	535	576			
	10	25	361	381	407	435			
$I = 3.894$ 3	6	18	912	986	1090	1168	1260		
	8	24	652	700	747	804	863		
	10	30	502	532	570	597	632		
$I = 9.61$ 4	6	24	1335	1435	1560	1700	1850		
	8	32	950	1012	1078	1156	1232		
	10	40	722	767	822	860	900		
$I = 40.49$ 6	6	36	2550	2740	2930	3180	3420		
	8	48	1800	1920	2050	2170	2330		
	10	60	1325	1410	1516	1580	1670		
$I = 105.7$ 8	6	48	3810	4080	4400	4770	5060		
	8	64	2750	2920	3100	3290	3500	3680	
	10	80	2100	2210	2350	2480	2580	2780	
$I = 244.8$ 10	6	60	5650	6070	6620	6980	7500	7830	
	8	80	4120	4360	4620	4910	5210	5330	5680
	10	100	3180	3340	3540	3720	3890	4080	4270
$I = 475.1$ 12	6	72	7680	8220	8860	9470	10100	10660	11080
	8	96	5590	5920	6320	6680	7050	7410	7680
	10	120	4300	4570	4820	5020	5330	5460	5700
$I = 687.3$ 14 O.D.	6	84	9100	9670	10350	11060	11700	12000	12900
	8	112	6280	6640	7000	7450	7860	8130	8480
	10	140	4870	5080	5470	5640	6120	6130	6340
$I = 1,156$ 16 O.D.	5	80	13600	14800	16000	17100	18250	19300	20100
	6	96	11150	11950	12750	13650	14400	15200	15830
	7	112	9520	10050	10750	11450	12050	12650	13200
$I = 1,833$ 18 O.D.	5	90	16960	18400	19860	21400	22700	23700	25000
	6	108	14000	14950	16000	17000	18100	18500	19250
	7	126	11830	12600	13500	14350	15150	15830	16500
$I = 2,772$ 20 O.D.	5	100	20900	22600	24400	26200	27900	29300	30600
	6	120	17150	18280	19500	20800	22400	23100	24000
	7	140	14560	15400	16400	17400	18400	19500	20300
$I = 5,672$ 24 O.D.	5	120	29700	32100	34300	36900	39300	41200	43100
	6	144	24300	26000	27700	29600	31400	33000	34300
	7	168	20600	21900	23300	24500	25600	26900	27900

Table is based on Fig. 35. Reacting forces correspond to Schedule 80 pipe of lengths given in Table 6. For reacting forces for Schedule 40 or other pipe thickness schedules, multiply tabulated forces by the ratio of their respective moments of inertias. See text for method of determining forces for lengths adjusted for allowable stress ranges above or below 12,000 psi.

Example.—What will be the reacting force under the following conditions for a 16-in. O.D. Schedule 80 pipe line containing an expansion U bend having a radius of six pipe diameters if the line is installed at 60 F and the operating temperature is 400 F?
(a) For the basic conditions of Table 6?
(b) If the allowable stress range is 16,000 psi instead of 12,000 psi?
(c) If Schedule 40 pipe having a moment of inertia of $I = 732$ is used instead of Schedule 80 pipe having a moment of inertia of $I = 1,156$?
Solution.—(a) The reacting force for the basic conditions of Table 6 and a length of line of 132 ft is read from Table 7 as 12,750 lb (see also example under Table 6).
(b) For an allowable stress range of 16,000 psi and a corresponding length of 176 ft, the reacting force would be $12,750 \times {}^{16}\!/_{12} = 17,000$ lb.
(c) For Schedule 40 16-in. O.D. pipe the reacting forces would be reduced in proportion to the ratio of the moments of inertia, *viz.*, to $732/1,156 = 0.634$ of the forces existing with Schedule 80 pipe.

Table 8. Double-offset (540 Deg) Expansion U Bend, All Pipe-thickness Schedules

(Over-all lengths of pipe in feet permissible with double-offset expansion U bends of radii and for operating temperatures shown)

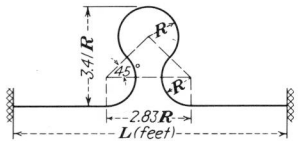

Nominal pipe size, inches d_n	$\dfrac{R}{d_n}$	Mean radius of bend, inches R	L = over-all length of pipe in feet permissible with double-offset expansion U bend of radius indicated (Based on an allowable stress range of 12,000 psi)							
			200 F	300 F	400 F	500 F	600 F	700 F	800 F	850 F
2	5	10	55.8	31.2	22.5	16.6	12.9	11.2	10.0	9.2
	6	12	72.0	40.8	29.1	21.7	17.5	15.0	13.3	12.5
	7	14	91.0	51.7	36.6	27.5	22.1	19.6	17.5	16.2
2½	5	12.5	72.5	40.8	28.3	20.8	17.1	15.0	13.3	12.1
	6	15	93.8	52.5	37.1	27.9	29.9	19.6	17.5	15.8
	7	17.5	117.5	66.6	46.7	35.0	29.1	25.0	22.1	20.8
3	5	15	91.3	51.3	35.9	27.5	22.1	18.3	16.7	15.4
	6	18	117.5	66.3	46.3	36.8	28.3	24.2	21.7	20.4
	7	21	142.5	82.5	57.5	44.2	35.9	30.8	27.5	25.8
4	5	20	131.2	73.3	50.8	38.3	31.3	26.7	23.3	22.5
	6	24	168.2	95.0	66.3	50.5	40.8	35.0	31.3	29.6
	7	28	203.0	118.4	82.5	62.5	50.8	44.2	39.2	37.5
6	5	30	208.5	117.8	81.2	62.5	50.4	42.8	38.3	36.2
	6	36	245.0	151.5	105.7	80.0	65.4	55.8	50.0	47.5
	7	42	320	186.6	132.5	99.2	81.3	69.7	62.5	59.2
8	5	40	285	166.6	114.2	86.7	70.8	60.4	53.3	50.8
	6	48	362	211.5	149.2	113.2	92.4	79.2	70.3	67.5
	7	56	443	262	185.0	141.5	115.8	99.2	88.3	84.2
10	5	50	372	216.5	152.5	115.8	85.0	80.8	70.8	67.5
	6	60	462	270	190	149.0	122.5	105.0	93.8	89.3
	7	70	590	343	242	187.0	152.0	131.2	116.2	110.8
12	5	60	438	256	181	139.5	114.2	97.5	86.7	82.5
	6	72	560	330	232	181	149.2	128.6	114.6	109.2
	7	84	693	400	280	220	181.5	160.0	142.5	136.0
14 O.D.	5	70	542	316	172.6	159.2	142	121.2	107.5	101.6
	6	84	687	400	283	219	179	159.2	141.6	133.2
	7	98	850	495	350	270	226.5	195.8	175.0	165.0
16 O.D.	5	80	624	365	258	196	161.6	138.2	122.5	116.2
	6	96	782	456	322	249	205	179.2	160.8	153.3
	7	112	990	577	407	315	257	216	187	175
18 O.D.	5	90	697	407	287	222	180	156.8	140	133.5
	6	108	940	548	387	299	243	204	183.3	175
	7	126	1,110	647	457	353	287	240	206	193
20 O.D.	5	100	777	453	319	247	202	174	157	148
	6	120	1010	588	405	305	248	209	181	170
	7	140	1230	717	507	392	320	267	228	214
24 O.D.	5	120	930	542	383	293	237	203	186	179
	6	144	1270	742	523	390	311	262	227	213
	7	168	1580	922	642	495	403	340	291	273

Table is based on Fig. 36. Pipe is assumed to be installed at 60° F. See text for method of adjusting lengths for allowable stress ranges above or below 12,000 psi. Lengths are based on values of q and i for Schedule 80 pipe but are sufficiently accurate for all pipe-thickness schedules.

Example.—What length of 16-in. O.D. pipe can be used under the following conditions with a double-offset expansion U bend having a radius of six pipe diameters if the line is installed at 60 F and the operating temperature is 400 F?

(*a*) For the basic conditions of Table 8.

(*b*) If the allowable stress range is 16,000 psi instead of 12,000 psi?

Solution.—(*a*) The length of line L read from Table 8 is 322 ft.

(*b*) For an allowable stress range of 16,000 psi the length would be $322 \times {}^{16}\!/_{12} = 429$ ft.

4-94

Table 9. Double-offset (540 Deg) Expansion U Bend, Schedule 80 Pipe

(Reacting force ranges in pounds for double-offset expansion U bends of
radii and for operating temperatures shown when absorbing
expansion of over-all lengths of pipe given in Table 8)

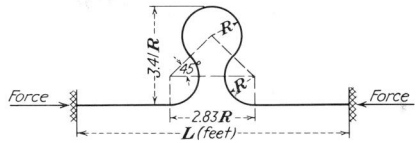

Nominal pipe size, inches d_n	$\dfrac{R}{d_n}$	Mean radius of bend, inches R	Reacting force ranges in pounds for Schedule 80 pipe double-offset expansion U bends of radius indicated when absorbing expansion of over-all lengths of pipe given in Table 8 which are based on an allowable stress range of 12,000 psi							
			200 F	300 F	400 F	500 F	600 F	700 F	800 F	850 F
$I = 0.8679$	5	10	299	316	331	347	365	375	384	393
2	6	12	251	264	275	287	298	309	316	320
	7	14	211	220	229	239	247	253	259	263
$I = 1.924$	5	12.5	455	493	503	530	550	565	578	592
$2\frac{1}{2}$	6	15	367	385	402	418	432	448	457	467
	7	17.5	305	319	332	345	357	366	376	379
$I = 3.894$	5	15	631	663	695	727	757	784	892	817
3	6	18	508	532	556	577	602	621	635	642
	7	21	427	447	464	482	498	512	522	529
$I = 9.61$	5	20	903	952	1000	1048	1086	1122	1152	1162
4	6	24	745	778	815	873	873	900	922	932
	7	28	628	652	678	707	728	747	762	770
$I = 40.49$	5	30	1720	1810	1910	1990	2070	2110	2160	2190
6	6	36	1420	1480	1540	1600	1660	1710	1740	1760
	7	42	1200	1240	1280	1340	1380	1410	1440	1450
$I = 105.7$	5	40	2590	1720	2850	2970	3090	3180	3260	3290
8	6	48	2130	2230	2310	2410	1490	2560	2600	2630
	7	56	1820	1890	1960	2030	2120	2150	2190	2200
$I = 244.8$	5	50	3850	4050	4230	4420	4640	4720	4840	4870
10	6	60	3190	3330	3460	3590	3710	3810	3870	3901
	7	70	2700	2810	2910	3000	3100	3170	3240	3270
$I = 475.1$	5	60	5260	5520	5770	6000	6210	6400	6500	6600
12	6	72	4330	4530	4720	4870	5020	5170	5270	5320
	7	84	3690	3830	3970	4080	4210	4300	4390	4430
$I = 687.3$	5	70	5920	6220	6720	6820	6950	7170	7300	7380
14 O.D.	6	84	4870	5070	5270	5450	5620	5720	5860	5920
	7	98	4060	4200	4350	4480	4590	4720	4790	4850
$I = 1,156$	5	80	7630	8000	8350	8670	8950	9220	9440	9570
16 O.D.	6	96	6300	6520	6820	7050	7240	7420	7540	7630
	7	112	5330	5530	5720	5880	6060	6230	6390	6470
$I = 1,883$	5	90	9520	10000	10420	10820	11230	11480	11740	11840
18 O.D.	6	108	7920	8200	8500	8820	9090	9370	9540	9620
	7	126	6690	6960	7180	7420	7620	7830	8070	8170
$= 2,772$	5	100	11700	12040	12780	13320	13750	14120	14440	14650
20 O.D.	6	120	9600	10000	10400	10800	11180	11520	11850	12000
	7	140	8180	8480	8780	9040	9320	9580	9840	9940
$I = 5,672$	5	120	16600	17420	18140	18960	19600	20200	20500	20650
24 O.D.	6	144	13660	14250	14780	15350	15920	16380	16810	17050
	7	168	11620	12040	12480	12880	13260	13640	14000	14160

Table is based on Fig. 36. Reacting forces correspond to Schedule 80 pipe of lengths given in Table 8. For reacting forces for Schedule 40 or other pipe-thickness schedules, multiply tabulated forces by the ratio of their respective moments of inertias. See text for method of determining forces for lengths adjusted for allowable stress ranges above or below 12,000 psi.

Example.—What will be the reacting force under the following conditions for a 16-in. O.D. Schedule 80 pipe-line containing a double-offset expansion U bend having a radius of six pipe diameters if the line is installed at 60 F and the operating temperature is 400 F?

(*a*) For the basic conditions of Table 8?

(*b*) If the allowable stress range is 16,000 psi instead of 12,000 psi?

(*c*) If Schedule 40 pipe having a moment of inertia of $I = 732$ is used instead of Schedule 80 pipe having a moment of inertia of $I = 1,156$?

Solution.—(*a*) The reacting force for the basic conditions of Table 8 and a length of line of 322 ft is read from Table 9 as 6820 lb (see also example under Table 8).

(*b*) For a bending stress of 16,000 psi and a corresponding length of 429 ft, the reacting force would be 6,820 × $^{16}/_{12}$ = 9,093 lb.

(*c*) For Schedule 40 16-in. O.D. pipe the reacting forces would be reduced in proportion to the ratio of the moments of inertia, *viz.*, to 732/1,156 = 0.634 of the forces existing with Schedule 80 pipe.

The lengths given in Tables 6 and 8 are based on an assumed allowable stress range of 12,000 psi. If, for example, it is desired to permit a stress range of 16,000 psi, it is only necessary to multiply the lengths given in the tables by $^{16}/_{12}$. When adjusting for stress ranges above 12,000 psi, the resulting lengths are somewhat conservative, while for stresses below 12,000 psi, the lengths thus found are slightly longer than given by a strict analysis. In neither case are the inaccuracies significant within the usual range of stress variation encountered.

The reacting force ranges on the anchors for Schedule 80 pipe bends corresponding to the overall lengths determined from the tables described in the foregoing are given in Tables 7 and 9 for expansion U bends and double-offset expansion U bends, respectively. For pipe-wall thicknesses greater or less than Schedule 80, the tabulated forces should be multiplied by the ratio of their respective moments of inertias. The moments of inertia for Schedule 80 pipe used in calculating the forces given in Tables 7 and 9 are reproduced on these tables.

The tabulated forces are based on the lengths of pipe determined for an allowable stress range of 12,000 psi. The forces for the lengths obtained by adjustment for allowable stress ranges other than 12,000 psi are found by applying the ratio of the stresses in exactly the same manner as the adjustment for lengths. The slightly greater flexibility of the longer lengths of pipe between bend and anchor points when lengths are adjusted to take advantage of cold spring results in only a minor reduction in forces, which may be neglected. Further examples of the use of these tables are given in Ref. 188.

The charts of Figs. 34 to 36 together with the formulas given therein enable the piping designer to determine the flexibility of any length of straight pipe in combination with a square bend made up of 90-deg bends or elbows and straight tangents, an expansion U bend, or a double-offset expansion U bend inserted in its length. From these charts and formulas the reacting forces, bending moments, and longitudinal bending stresses can be determined for the respective types of bends for any given amount of expansion and for the assumed overall line dimensions.

The chart for the square bend made up of 90-deg bends and straight lengths (Fig. 34) is particularly useful in estimating the flexibility of proposed piping layouts. The square-corner approximation gives results which closely agree with exact solutions if the reciprocal bend-flexibility factor K for the curved portions is approximately 0.6 and if the bend is located midway between the anchor points. If the longer run between the bend and anchor is not more than four times the shorter run, the results are within the manufacturing tolerances permitted with pipe.

The relations between deflection and reacting forces and restraining moments at the ends of the section of piping are found from the curves of Figs. 34 to 36 by forming ratios between the lengths of straight pipe between the ends of the bend and the anchors and the principal dimensions of the bends.

The charts of Fig. 37 are for a 90-deg bend with tangents fixed at the ends. The forces, moments, and bending stresses are obtained directly from the charts. The results are strictly correct for an arc where the center-line radius is five times the nominal diameter of the pipe and where the reciprocal bend-flexibility factor K is 0.5.

The charts of Figs. 34 to 37, inclusive, were originally developed in cooperation with the Walworth Company for use in its Catalogue 89.

Reference to Table 10 will serve to indicate where it is necessary to correct for variations in the value of K above and below 0.5. For the usual proportions encountered in piping-design problems, the correction will not exceed 10 per cent.

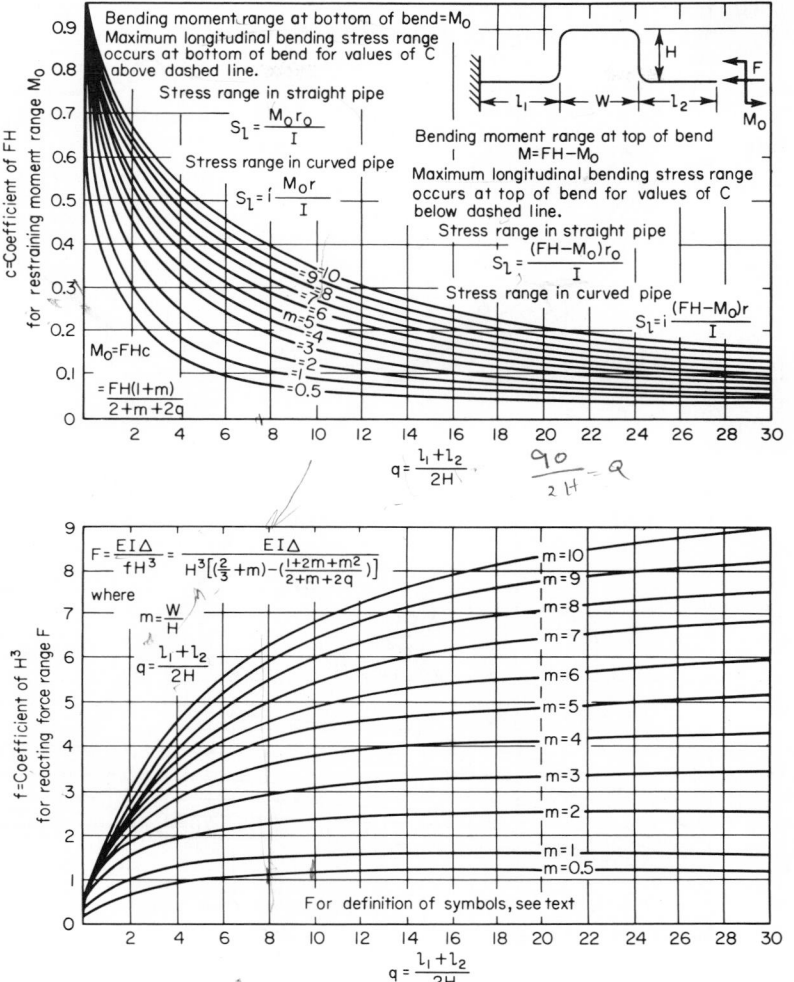

FIG. 34. Chart for determining relation between deflection and reacting force range for an *square bend* made up of 90-deg arcs and straight pipe inserted in a run of straight pipe having fixed anchors at the ends. Relations are accurately represented if the reciprocal flexibility factor K for the curved pipe is 0.59 and if $l_1 = l_2$. This chart may be used with reasonable accuracy, however, for the pipe bend ratios usually employed if l_1 is not greater than four times l_2.

The chart of Fig. 37, through the insertion of hypothetical anchor points as indicated in Fig. 38, makes possible an approximate solution of many problems encountered in piping design. While the forces and stresses determined will be approximate, it can be shown that the line will possess adequate flexibility if the sections are themselves sufficiently flexible. In the three-dimensional pipeline shown, the true forces will not be found by this means, although adequate flexibility can be assured by such an approximation.

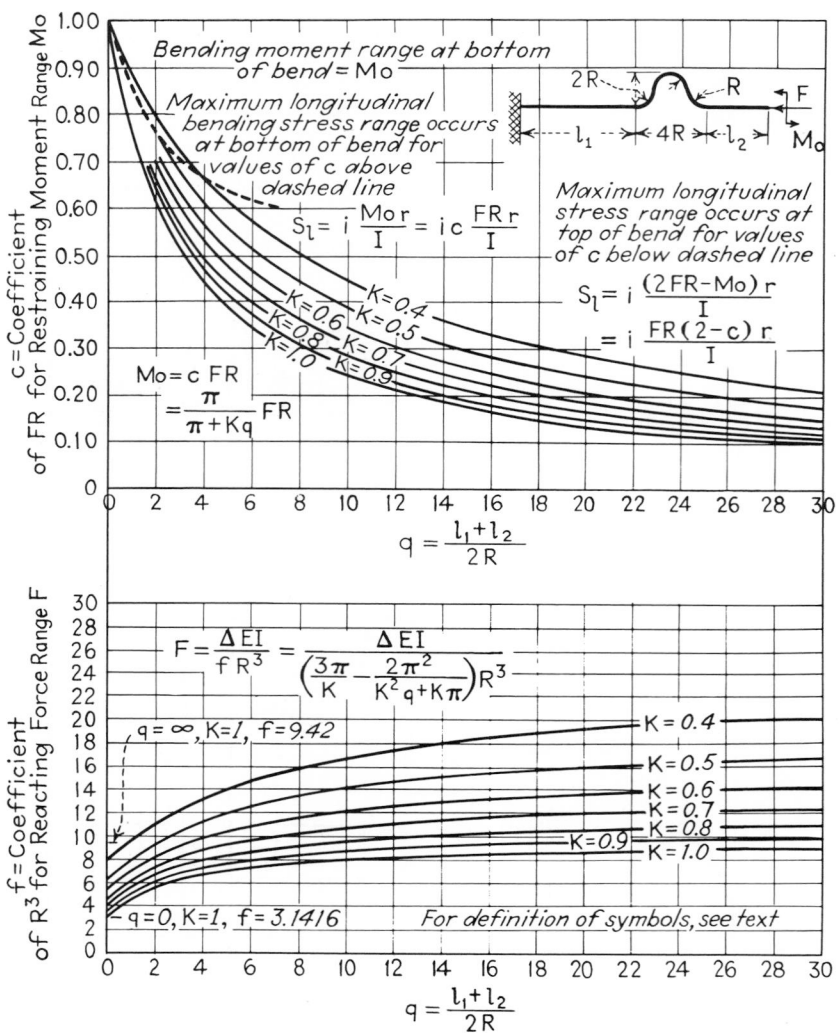

FIG. 35. Chart for determining relation between deflection and reacting force range for an *expansion U bend* inserted in a run of straight pipe having fixed anchors at the ends. Relations are strictly correct where $l_1 = l_2$. This chart may be used with reasonable accuracy, however, if l_1 is not greater than four times l_2.

Simplified Computational Schemes. There are also many methods which involve the actual calculation of a flexibility problem appropriate to the problem at hand but with short cuts and simplifications made a part of the calculating procedure in such a manner as to reduce the computational labor involved without, hopefully, effecting an inadmissible reduction in the reliability of the answers. Again, these methods are essentially limited to two-anchor problems.

In a two-dimensional piping system, there is a well-defined point, usually called the "elastic center," through which the reaction force acts; i.e., the reaction moment

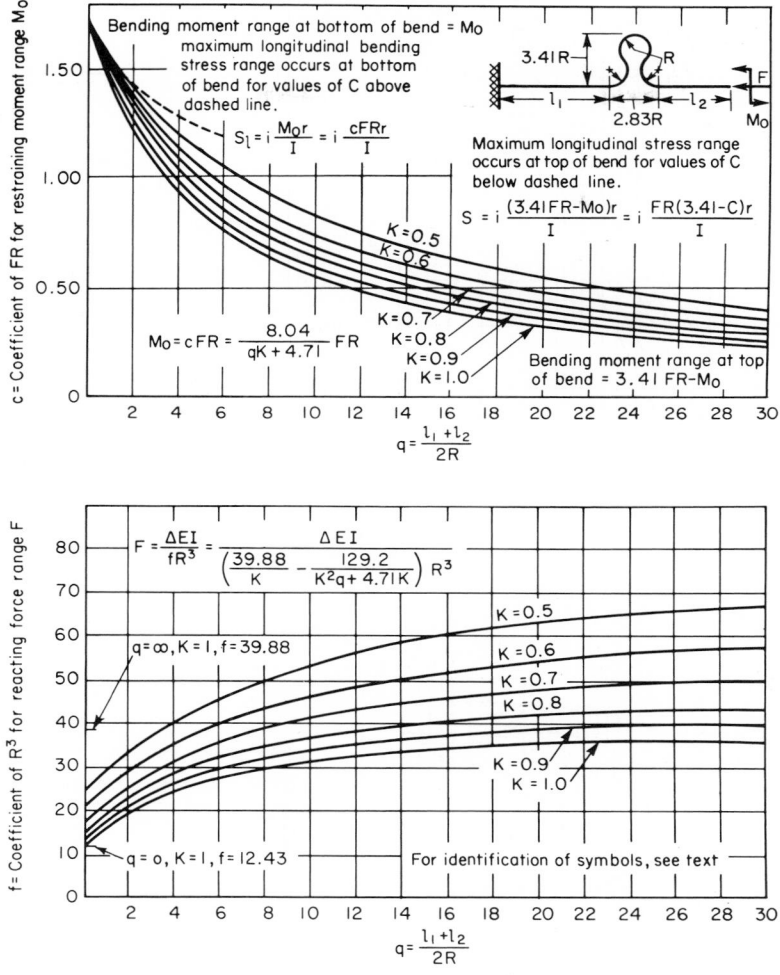

FIG. 36. Chart for determining relation between deflection and reacting force range for a *double-offset expansion U bend* inserted in a run of straight pipe having fixed anchors at the ends. Relations are strictly correct where $l_1 = l_2$. This chart may be used with reasonable accuracy, however, if l_1 is not greater than four times l_2.

about this point is zero. Den Hartog[87] gives a very compact proof of this fact which he calls Spielvogel's theorem after the late S. W. Spielvogel, a well-known piping designer who made substantial contributions to piping-flexibility analysis. If the direction of the reaction force can be determined, the line passing through the elastic center and having this direction, known as the neutral axis of the system, locates points having zero bending moment. If the magnitude of the reaction can be determined, the bending moment at any point is determined by multiplying reaction magnitude by distance of point from the neutral axis. Essentially, the problem becomes one of finding magnitude and direction of the reaction force.

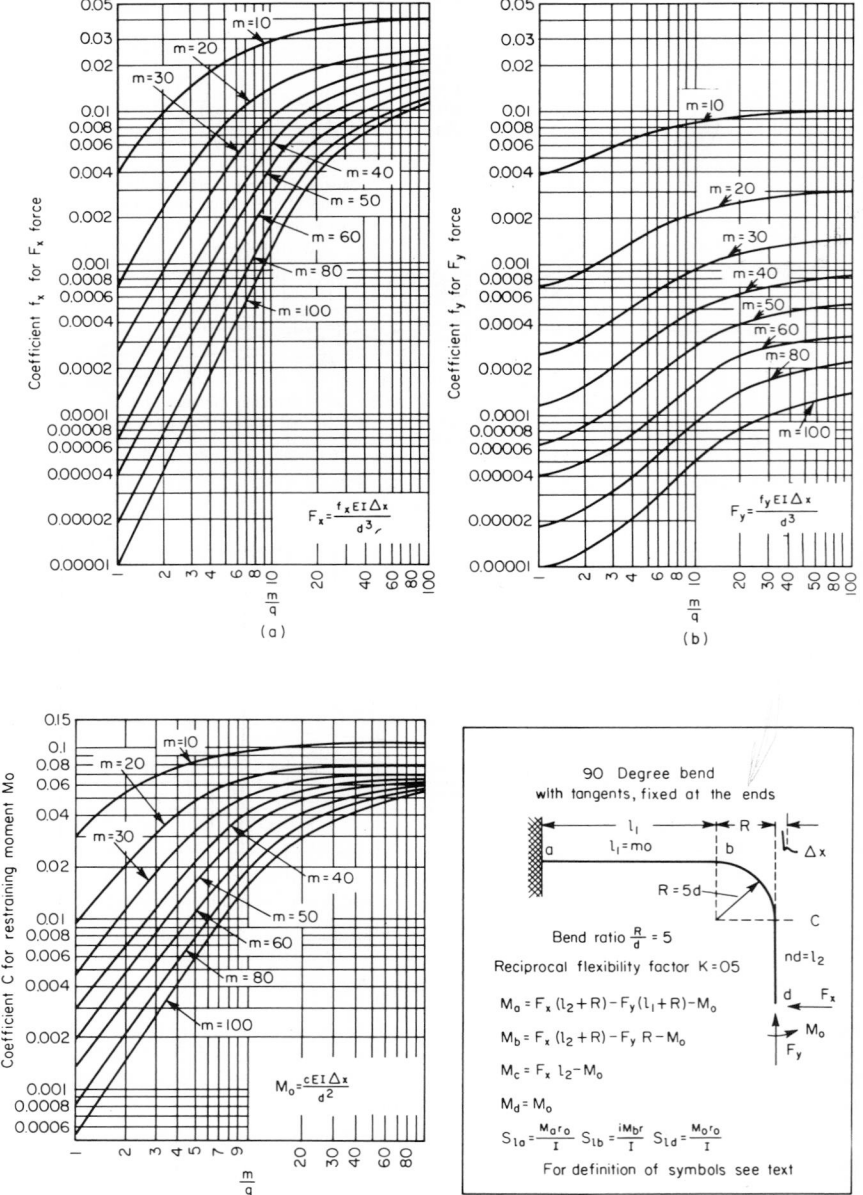

Fig. 37. Chart for determining reacting force bending moment, and stress ranges in a *90-deg bend with tangents fixed at the ends*. Relations are strictly correct for bends formed to a center-line radius five times the nominal pipe diameter and where the reciprocal flexibility factor for curved pipe $K = 0.5$.

Reference to Table 10 will serve to indicate where it is necessary to correct for variations in the value of K above and below 0.5. For the usual proportions encountered in piping-design problems, the correction will not exceed 10 per cent.

Table 10. Correction Factors for Results Obtained From Fig. 37 for 90-deg Bend if K is not 0.5

DIRECTIONS: If K is not 0.5, multiply the results obtained from Fig. 37 for F_x, F_y, or M_o as the case may be, by the correction factor shown below for the desired value of K, interpolating where necessary.

		$K = 0.4$			$K = 0.6$			$K = 0.8$		
		m/n 1	m/n 10	m/n 100	m/n 1	m/n 10	m/n 100	m/n 1	m/n 10	m/n 100
	F_x	0.915	0.873	0.820	1.070	1.114	1.141	1.200	1.300	1.440
$m = 10$	F_y	0.915	0.897	0.873	1.070	1.096	1.117	1.200	1.260	1.333
	M_o	0.955	0.873	0.801	1.030	1.111	1.148	1.100	1.290	1.485
	F_x	0.938	0.963	0.840	1.048	1.023	1.147	1.119	1.064	1.433
$m = 50$	F_y	0.938	0.952	0.917	1.048	1.036	1.061	1.119	1.094	1.151
	M_o	0.963	0.968	0.816	1.028	1.018	1.173	1.072	1.052	1.519
	F_x	0.962	0.983	0.880	1.033	1.007	1.107	1.075	1.021	1.311
$m = 100$	F_y	0.962	0.960	0.971	1.033	1.022	1.019	1.075	1.061	1.059
	M_o	0.977	0.989	0.866	1.022	1.004	1.121	1.048	1.012	1.359

In a widely used simplification, Mitchell[196] assumed that the direction of the neutral axis is parallel to the line joining the anchors, a condition which is obviously true for symmetrical systems. The method of Mitchell has been widely used, but its use may lead to unconservative evaluation of systems which depart greatly from a condition of symmetry. Refinements have been introduced in the procedure by Bridge[53,54] and Lee[173] for the purpose of quickly estimating the departure of the neutral axis orientation from that of the line between anchors, and with this improvement the method is considerably more reliable. The references cited should enable the reader to become acquainted with the use of this method.

Since, however, the Mitchell method is essentially applicable only to two-dimensional cases and since even with the modifications mentioned it may give results of doubtful accuracy for extended and nonsymmetrical systems, it is useful to have available a simplified method which is capable of obtaining reliable results for even rather complicated two-anchor problems. In a three-dimensional piping system the concept of the elastic center becomes quite ambiguous. Spielvogel, however, managed to make use of the notion by defining what may be thought of as a "moving elastic center." What he did, in effect, is to work with three different elastic centers, each of which is appropriate to the corresponding planar projection of the piping system. Development of this idea together with certain other simplifications provides the basis for his simplified elastic center method.[241]

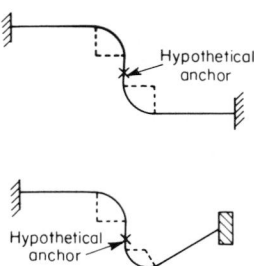

FIG. 38. Use of hypothetical anchors to reduce complicated cases to that of Fig. 37.

Hao Hsiao has made additional simplifications to this procedure and recently published a form of an approximate analysis which appears to be the best available combination of reliability and simplicity in a manual computational procedure. His contribution is sufficient to warrant calling the procedure Hao Hsiao's method.[137–140]

The method will be elucidated by means of a simple example, and remarks will be added suggesting how the method may also be applied to more complicated cases. Figure 39 shows a simple piping configuration consisting of an 80-ft run and 10-ft

offset, relieved by a square loop at the beginning of the run, 10 ft high and 10.5 ft long.   The pipe is 12 in. (nominal) diameter, Schedule 80 thickness, conforming to ASTM specification A106 Grade B.   Installation is at 70 F without cold spring, and the line works between this temperature and a maximum operating temperature of 450 F.   Directional changes are accomplished by ASA B16.9 butt-welding elbows, which have a center-to-end distance of 18 in.   There is no movement of the anchors due either to temperature change or to distortion of the terminal equipment.

The calculations are performed on previously prepared data sheets as shown in Tables 11 and 12.   The heavy lettering and numbering indicate the preliminary preparation, and the light numbers indicate the calculations that are performed

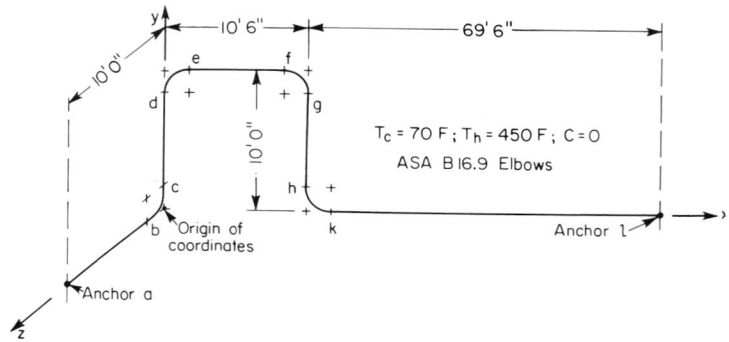

FIG. 39. Problem illustrating Hao Hsiao's method of approximate flexibility analysis. Pipe 12 in. Schedule 80, ASTM A106 Grade B.

during the solution of the problem.   First, the box labeled **LINE DATA** is filled in.   Outside diameter D and wall thickness t come from ASA Standard B36.10 or from other compilations which also give cross-sectional moment of inertia I and section modulus S^* (to avoid confusion with reaction range component Z).   The maximum operating temperature 450 F is entered, and the unit expansion e is obtained by interpolating in Table 2 to get 3.16 in. per 100 ft for a temperature variation of 70 to 450 F.   Values of E_h and E_c are obtained from Table 1.   The thermal expansions to be accommodated are determined by considering anchor a to be held fixed, releasing anchor l, and increasing the temperature.   The quantities Δx, Δy, and Δz are the distances that end l would move, the algebraic signs being in accordance with the right-handed coordinate system shown in Fig. 39.   The quantities should be in the units shown so that conversion factors implicit in later computations will give correct final results.   Next, the box labeled **FACTORS** is filled in. h is the bend characteristic Rt/r^2, n is the bend-flexibility factor, and i_i and i_o are the stress-intensification factors.   In calculating h, units should be consistent, say R, r, and t, all in inches.   However, in the next step, R is shown in *feet*, and all other dimensions entered in later steps are also in *feet*.   The quantities λ' and τ are calculated according to the formulas

$$\lambda' = 5\pi R(n - 1.3)/6$$
$$\tau = \pi R(n + 1.3)/4$$

Next, the main body of the form is filled out.   Note that there is a major system of divisions, indicated by captions "XY PLANE," etc., at the left.   In each of

Table II

	DESIG-NATION	L³	L	Lx	Lx²	Ly	Ly²	Lz	Lz²	Lxy	Lxz	Lyz	x	y	z
BRAN. ∥Z	ab	614.1	8.5	0	0	0	0	48.88	281.0	0	0	0	0	0	5.75

XY PLANE

		L³	L	Lx	Lx²	Ly	Ly²	Lz	Lz²	Lxy	Lxz	Lyz		
(1)	SUM $I_1=614.1$		8.5	0	0	0	0	48.88	281.0	0	0	0	LINE DATA	
ELL ∥XY	de		L=1	.54	.29	9.46	89.49	0	0	5.11	0	0		
	fg			9.96	99.20	9.46	89.49	0	0	94.22	0	0	$D=12.75''$ $E_h=26.7$ ×10⁶psi	
	hk			11.04	121.88	.54	.29	0	0	5.96	0	0	$t=.687''$ $E_c=27.9$ ×10⁶psi	
(2)	SUM		No. of Ells 3	21.54	221.37	19.46	179.27	0	0	105.29	0	0	$I=475in^4$	
(3)	x'×(2)		41.85	300.5	3088	271.5	2501			1469			$S^*=74.5in^3$ $\Delta x=2.528''$	
(4)	Σ=(1)+(3)+(25)		473.65	1122.5	491280	1238.1	10507			6548			T 450F $\Delta y=0$	
(5)	(4)÷ΣL		x̄=24.538			ȳ=2.614							$e=3.16‰'$ $\Delta z=-.316''$	
(6)	(5)×Σ			x̄ΣLx=285193		ȳΣLy=3236				ȳΣLx=30382				
(7)	(4)-(6)			I'ᵧ=206090		I'ₓ=7271				Iₓᵧ=23834				

XZ PLANE

		L³	L	Lx	Lx²	Ly	Ly²	Lz	Lz²	Lxy	Lxz	Lyz	x	y	z
BRAN. ∥Y	cd	343	7	0	0	35	175	0	0	0	0	0	0	5	0
	gh	343	7	73.5	771.8	35	175	0	0	367.5	0	0	10.5	5	0
(8)	SUM $I_2=686$		14	73.5	771.8	70	350	0	0	367.5	0	0	FACTORS		
ELL ∥XZ			L=1										h=.34		
													n=4.85		
(9)	SUM		No. of Ells 0	0	0	0	0	0	0	0	0	0	R=1.5		
(10)	x'×(9)		0	0	0	0	0	0	0						
(11)	Σ=(8)+(10)+(25)		437.3	11396	488963			175.9	943.6		0	0	x'=13.95		
(12)	(11)÷ΣL		x̄=26.060					z̄=.402					τ=7.25		
(13)	(12)×Σ			x̄ΣLx=296980				z̄ΣLz=70.7		z̄ΣLx=4581					
(14)	(11)-(13)			I'ₓ=191980				I'ₓ=87.9		Iₓz=-4581			i₁=1.85 i₀=1.54		

YZ PLANE

		L³	L	Lx	Lx²	Ly	Ly²	Lz	Lz²	Lxy	Lxz	Lyz	x	y	z
BRAN. ∥X	ef	422	7.5	39.4	207	75	750	0	0	394	0	0	5.25	10	0
	bl	314432	68	3128	143888	0	0	0	0	0	0	0	46	0	0
(15)	SUM $I_3=314854$		75.5	3167	144095	75	750	0	0	394	0	0	TOTAL MOMENTS OF INERTIA		
ELL ∥YZ	bc		L=1	0	0	.54	.29	.54	.29	0	0	.29	I'ₓ=7271 I'ᵧ=206090		
													I'ₓ=872 I'ᵧ=882		
													I₁/3.6=171 I₁/3.6=171		
(16)	SUM		No. of Ells 1	0	0	.54	.29	.54	.29	0	0	.29			
(17)	x'×(16)		13.95			7.5	4.0	7.5	4.0			4.0	I₂/3.6=191 I₃/3.6=87459		
(18)	Σ=(15)+(17)+(25)		512.75			1049.1	8760	183.4	947.6			11.00	Iₓ=8505 Iᵧ=294602		
(19)	(18)÷ΣL					ȳ=2.046		z̄=-.358					I'z=191980		
(20)	(19)×Σ					ȳΣLy=2146		z̄ΣLz=65.6		z̄ΣLy=376			I'z=6614		
(21)	(18)-(20)					I'z=6614		I'ᵧ=882		Iᵧz=365			I₂/3.6=191		

ALL PLANES

		L³	L	Lx	Lx²	Ly	Ly²	Lz	Lz²	Lxy	Lxz	Lyz		
(22)	(2)+(9)+(16)		4	21.54	221.37	20.0	179.56	.54	.29	105.29	0	.29	I₃/3.6=87459	
(23)	τ×(22)		29.0	156.17	1604.9	145.0	1301.8	3.91	2.10	763.4	0	2.10	Iz=286244	
(24)	(1)+(8)+(15)+(23)		127.0	3397	146472	290	2402	52.79	283.1	1524	0	2.10		
(25)	(24)×3.333		423.3	11322	488191	966.6	8006	175.9	943.6	5079	0	7.00	COEFFICIENTS	

REACTION FORCE RANGES

(1)		Ⓐ 2.802	Ⓑ .5386	Ⓒ 7597	$E_c I÷518.4=2.556×10^7$
(2)		Ⓓ 12.3606	Ⓔ -.0153	Ⓕ 0	
(3)		Ⓖ -.080	Ⓗ -62.485	Ⓚ 1763	$\dfrac{I_{xy}}{I'_x}=2.802$ Ⓐ $-\dfrac{I'_{xz}}{I'_x}=.5386$ Ⓑ
(4)	(1)+(2)	-9.559	Ⓛ .5233	7597	
(5)	(1)+(3)	2.722	Ⓜ -61.946	9360	$\dfrac{I'_y}{I_{xy}}=12.3606$ Ⓓ $-\dfrac{I'_{yz}}{I_{xy}}=-.0153$ Ⓔ
(6)	(5)+(8)	Ⓝ -.0230	Ⓟ -79.07		
(7)	(4)-(6)	Ⓠ -9.536	Ⓡ 7676		$\dfrac{I'_{yz}}{I_{xz}}=-.080$ Ⓖ $\dfrac{I'_z}{I_{xz}}=-62.485$ Ⓗ
(8)	Ⓛ÷Ⓜ = -.008448				$\dfrac{\Delta_x E_c I}{I_x(518.4)}=7597$ Ⓒ
	Y = Ⓡ÷Ⓠ = -804.9				$\dfrac{\Delta_y E_c I}{I_{xy}(518.4)}=0$ Ⓕ
	Z = (Ⓟ-ⓃY)÷Ⓛ = -186.5				$\dfrac{\Delta_z E_c I}{I_{xz}(518.4)}=1763$ Ⓚ
	X = Ⓒ-ⒶY-ⒷZ = 9953				

these divisions fill in the lengths (in feet) of straight pipes parallel to the indicated coordinate direction. Thus, section ab is of 8.5 ft length and is parallel to the Z direction. At the far right of the line in which this length is entered, the co-ordinates (also in feet) of its geometric center should be entered. For the elbows, initially their length is taken as unity, and after certain computations have been performed, multiplication by λ', the "virtual length of an elbow," takes care of their actual dimensions and extra flexibility. With the elbow length initially 1.0, the entries Lx, Ly, and Lz are actually the coordinates of the centroid of the elbow located as shown in Fig. 40, which indicates the position of the centroid of a general bend or elbow as well as that of a 90-deg bend or elbow.

The lines of the form are filled in in the following order: first those lines leading to lines (1), (8), and (15); next lines (1) through (3), lines (8) through (10), and lines (15) through (17); next, lines (22) through (25); next lines (4) through (7), (11) through (14), and (18) through (21). In this process, the locations of the three elastic centers (one in each projective plane) are determined and certain quantities designated by upper case I's with various subscripts and primes are calculated.

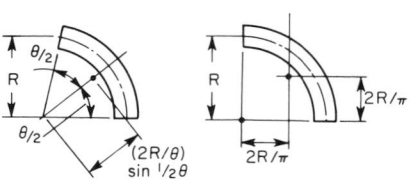

FIG. 40. Location of centroid of bend or elbow, for use in Hao Hsiao's method.

The next step is to go to the box marked TOTAL MOMENTS OF INERTIA and make the indicated calculations. Next the coefficients at the lower right are calculated. The combination $E_c I/518.4$ occurs three times and is conveniently calculated separately as 2.556×10^7. Next, the calculations at the lower left are accomplished; the format indicates the necessary arithmetic operations. Finally, the quantities Y, Z, and X, determined in that order, are the reactions force ranges (pounds) exerted by the pipe on anchor l.

Final evaluations are made on the form shown in Table 12. These calculations should be completely self-explanatory except for the stress-intensification factors i_z, i_y, and i_x and subsequent calculations. These intensification factors are essentially for bends or elbows and for termini. At termini a and l the torsional moment component is not subjected to any effective intensifications, whereas both bending-moment components are subjected to an intensification appropriate to the connection between pipe and equipment. Lacking better information, a value of 1.3 is used. For the elbows, the torsional moment is not subjected to an intensification, the in-plane bending moment is subjected to the intensification $i_i = 1.85$, and the out-of-plane bending moment is subjected to the intensification $i_o = 1.54$. This is consistent with what is shown in Fig. 29 and anticipates that the ASA Code for Pressure Piping will distinguish between i_i and i_o values. At present writing (August, 1962), a strict interpretation of the Code would require using 1.85 for both i_i and i_o, which would not affect the final answer to the present problem. Table 12 contains the formula

$$S_E = \frac{\sqrt{M_z^2 i_z^2 + M_y^2 i_y^2 + M_x^2 i_x^2}}{(S^*/12)}$$

from which the stress range is calculated, the factor 12 accounting for the fact that the moments in Table 12 are in units of pound-feet. Calculations were made for points a, e, f, g, k, and l, but calculations could be made for other points if it appeared as if the maximum stress range would occur at some other point. Inspection of the M_i values indicates that the worst condition is at point a, so that there is no need to calculate S_E for any of the other points. Finally, this value of 20,440 psi must be

Table 12

POINT			a	e	f	g	k	l		
X= 9953	(1)	x	0	1.5	9	10.5	12	80		
Y= -805	(2)	y	0	10	10	8.5	0	0		
X= -186	(3)	z	10	0	0	0	0	0		
XY PLANE	(4)	x-x̄	-24.54	-23.04	-15.54	-14.04	-12.54	55.46		
x̄ = 24.54	(5)	y-ȳ	-2.61	7.39	7.39	5.89	-2.61	-2.61		
ȳ = 2.61	(6)	X·(5)	-25980	73550	73550	58620	-25980	-25980		
	(7)	Y·(4)	19750	18550	12510	11300	10090	-44650		
	M_z	(6)-(7)	-45730	55000	61040	47320	-36070	18670		
XZ PLANE	(8)	z-z̄	9.60	-.40	-.40	-.40	-.40	-.40		
x̄ = 26.06	(9)	x-x̄	-26.06	-24.56	-17.06	-15.56	-14.06	53.94		
z̄ = .40	(10)	Z·(9)	48 50	45 70	3170	2890	2620	-10030		
	(11)	X·(8)	95550	-3980	-3980	-3980	-3980	-3980		
	M_y	(10)-(11)	-90700	8550	7150	6870	6600	-6050		
YZ PLANE	(12)	y-ȳ	-2.05	7.95	7.95	6.45	-2.05	-2.05		
ȳ = 2.05	(13)	z-z̄	9.64	-.36	-.36	-.36	-.36	-.36		
z̄ = .36	(14)	Y·(13)	-7760	290	290	290	290	290		
	(15)	Z·(12)	380	-1480	-1480	-1200	380	380		
	M_x	(14)-(15)	-8140	1770	1770	1490	-90	-90		
S*= 74.5		i_z	1	1.85	1.85	1.85	1.85	1.3		
S*/12 = 6.21		i_y	1.3	1.54	1.54	1	1.54	1.3		
S_E =		i_x	1.3	1	1	1.54	1	1		
		$M_z i_z$	-45730	101750	112920	87540	-66730	24270		
$\dfrac{\sqrt{M_z^2 i_z^2 + M_y^2 i_y^2 + M_x^2 i_x^2}}{(S^*/12)}$		$M_y i_y$	-117910	13170	11010	6870	10160	-7860		
		$M_x i_x$	-10580	1770	1770	2290	-90	-90		
		S_E	20440	16530	18280	14160	10870	4120		

compared with the allowable stress range $(1.0)[(1.25)(150,000) + (0.25)(15,000)] = 22,500$ psi, and the conclusion is that this piping system satisfies the ASA Code requirements for stress range in the pipe.

In his own exposition of the method, Hsiao proceeds somewhat differently, calculating moments and stress ranges using *relative* values which eliminate a few arithmetic operations but conceal the significance of the quantities involved. Also, he explains a way of selecting the possible points of maximum stress range without doing quite as much calculating as is indicated in Table 12; this alternate viewpoint is not really a part of the calculating method itself. A comparison of the results obtained above with those obtained by use of a modern precise digital computer analysis is shown in Table 13. This comparison is for point a only, but similar excellent comparisons may also be demonstrated for other points of the configuration. Hsiao's recent paper also contains a different example which may assist the reader in applying the method to his own problem.

Table 13. Comparison of Approximate and Precise Reaction and Stress Ranges at Point *a* of Fig. 39

	Approx. analysis	Precise analysis
F_x	−9,953 lb	−9,921 lb
F_y	805 lb	808 lb
F_z	186 lb	225 lb
M_x	8,140 lb-ft	8,356 lb-ft
M_y	90,700 lb-ft	88,059 lb-ft
M_z	45,730 lb-ft	45,839 lb-ft
S_E	20,440 psi	19,942 psi

The formats of Tables 11 and 12 are essentially for cases in which all straight lengths are parallel to one or another of the coordinate directions and in which directional changes are accomplished by elbows or by bends having radii relatively small compared with the overall dimensions of the configuration. However, the method may be applied to more general situations by the following procedure and still achieve good accuracy. For an inclined straight portion having direction cosines α, β, γ, divide into several short sections, each of which is to be regarded as a separate piece of pipe. For the piece having length L enter a reduced length $\gamma^2 L$ in the position marked "BRAN. ‖ Z," a reduced length $\beta^2 L$ in the portion marked "BRAN. ‖ Y," and a reduced length $\alpha^2 L$ in the portion marked "BRAN. ‖ X," and proceed as before. For a long, sweeping bend, a similar procedure should be used. The bend should be divided up into several shorter pieces. Values of α, β, and γ will be different for different pieces. Each such piece should be treated as a straight section and entered as described above. The errors in this extension of Hsiao's method should be tolerable. The principal systematic error would seem to be the neglect of the additional flexibility of the bend, but for bends having gentle enough radius that their dimensions are not small compared with the overall size of the configuration, the value of the bend-flexibility factor should be unity or very nearly unity. Although somewhat more theoretically correct procedures may be devised to extend Hao Hsiao's method to include inclined tangents and sweeping bends, that described above seems most in keeping with the simplified spirit of the basic method.

It may be mentioned that the methods of Blick[43] and Markl[249] are also quick and convenient. The latter has been arranged so as always to give reliably conservative results; that is, the reaction and stress ranges are always overestimated.

Mention will also be made of one more method, called by its users the "6EI Delta" method. It is not simple enough to be classified among the overall criteria discussed previously, but it is not comparable as far as reliability and accuracy are concerned to the other simplified computational methods discussed in this present section, depending, as it does, upon considerable judgment on the part of its users. This method uses the formula

$$F_x = 6EI\Delta_x \div \Sigma L_{ix}{}^2$$

where Δ_x is the x component of the total expansion which must be accommodated and the sum in the denominator is of the squares of the lengths of whatever legs and offsets can absorb an x-direction expansion. Similar formulas hold for F_y and F_z, and the same length squared may appear in two such denominators, since, for example, a length oriented in the z direction can absorb both x and y expansions. The coefficient 6 was selected as being a compromise between the values 3 for a free

cantilever and 12 for a guided cantilever. If the reactions (actually reaction ranges) estimated in this way are sufficiently small, and if their products by representative distances in the configuration are also sufficiently small, then the line may be considered to be satisfactory. It is clear that this is a very rough procedure, and reliable results can be obtained only with the exercise of considerable judgment; for these reasons, the users of this method also refer to it as the "ouija board."

Flexibility Analysis by Use of Model Tests. Before the advent of the high-speed digital computer, one way of avoiding coming to grips manually with the voluminous calculations required for a fairly complicated flexibility analysis was by constructing and testing a physical model of the system. In this country, the work of Andrews[3-5] is best known. However, in recent years, the convenience and reliability of automatic digital computation have resulted in decreasing use of model testing. This section, accordingly, will deal briefly only with aspects of model testing which are of sufficiently recent vintage that they are not well known in the piping literature.

The Belgian IRSIA (Institut pour l'Encouragement de la Recherche Scientifique dans l'Industrie et l'Agriculture) has maintained an interest in piping design and analysis (see papers by Blanjean, Dony, and Goethals in the bibliography), and, as a part of the work of its Research Commission on Steam Piping, reported[8] on a comparative study of different model analysis procedures which had been used by others, discussing the equipment and techniques involved and suggesting some rather elaborate counterweighing embellishments to increase the accuracy that can be obtained. This comparative study seems to be the only one of its kind.

The only more recent significant work of which the author is aware is that of Rauch,[26] who obtained excellent accuracy with a novel experimental procedure which should have applicability to more general classes of structural analysis. In most model tests much attention is devoted to the design of the "measuring head" which serves as an anchor and also measures the reaction force vector components applied by the piping to the anchor. In order to preserve the rigidity of the restraint, the compliance of this device must be very small, but small compliance makes for rather difficult measuring techniques. The more successful procedures have employed strain-gage rosettes and circuitry capable of distinguishing among the several reaction components which contribute to the strains experienced and measured in the device. Rauch's method does not use such a measuring head but rather makes measurement of displacements and rotations observed at a convenient point near the end of the first straight length from the anchor. Five displacement components are observed (the axial deformation is too small to measure), and by inversion of the 5×5 matrix of corresponding compliances, it is possible to calculate the values of the five force and moment components (axial force is not determined) which cause them. By reversing the model, end for end, it is possible to obtain another five reaction components, and in most cases this affords an evaluation of all six unknown components plus four additional relations which can be used for verifying and checking the accuracy of the determination.

While theoretically it should prove easy to take into account the peculiar properties of pipe bends and elbows that result in additional flexibility, in fact this is almost impossible to do. It would superficially appear that all one need do is to choose tubing diameter and wall thickness to simulate the pipe and then to choose the model scale so that the bend radii would give the same bend characteristic h. However, the range of commercially available tubing is somewhat limited, and it proves quite difficult to fabricate bends in such small tubing which have physical characteristics similar to full-size bends fabricated in pipe. The difficulty of simulating elbows is even greater.

Although it can hardly be regarded as "model testing," there has been some work done on the analysis of piping flexibility by the use of electrical analogue procedures. As the reader is aware, analogue methods depend on the fact that two different physical phenomena may be described by the same sets of equations (algebraic equations, differential equations, etc.). Some of Kron's earlier papers[66,67,161,162] were devoted to this analogy between electric circuit theory and structural theory, and more recently Goethals[123] has done some piping-flexibility work with such analogues. However, it may be fairly said that digital computation is so much better a way of proceeding that analogue methods (or model methods) are not presently of much importance.

EXPANSION JOINTS

Introduction. All the other sections of this chapter are aimed at the problem of accommodating the expansion of a piping system and the movement of its terminal points by the flexible action of the piping itself. This section will deal with the employment of special devices, usually called expansion joints, in which, so to speak, flexibility is concentrated. If service conditions are not too severe, such devices may afford economical and attractive solutions to problems of thermal expansion and terminal motions. In some cases the thermal expansion may be so large or the space available in which to route the piping configuration may be so limited that it is not possible to accommodate the strains involved by the inherent flexibility of the pipe itself without imposing excessive reaction loading on the terminal equipment or upon the valves and other internal appurtenances or without inducing excessive stress range in the piping. In certain cases the economic penalty associated with excessive pressure drop may indicate a very direct routing for the configuration, one which is not sufficient to provide requisite inherent flexibility. In other cases, the cost of providing, erecting, and supporting additional piping required for an adequately devious routing may exceed the cost of installing and maintaining an expansion joint. In the case of dockside piping to vessels which rise and fall considerably because of tidal action and changes in lading or in the case of steam and condensate lines to mixers and calender rolls or other similar cases, the motion of termini may be so great that accommodation by inherent flexibility is completely out of the question. In these and a variety of other situations the use of expansion joints may be the best solution to a difficult problem.

The theory and literature concerning expansion joints may not be considered to be at the same stage of advancement as that dealing with the analysis of "inherent" flexibility. There are many reasons for this, chief among which are:

1. The geometrical details of pipe and elbows are uniformly simple, whereas there are literally hundreds of different geometries for expansion-joint details.

2. The need for analysis has been greatest in those cases where the severity of service conditions has more or less immediately ruled out the use of expansion joints; the chief example which comes to attention is the case of main steam piping in electric utility stations.

3. Expansion joints are manufactured by a spectrum of companies ranging from those with a minimum of machines, equipment, and technical competence to those which are capable of conducting extensive development and testing programs and of maintaining uniform high quality in a product.

4. The incentive for the production of most existing literature on expansion joints has been to promote sales of individual designs and manufacturers, and for this reason it sometimes appears lacking in objectivity.

Manufacturers' literature provides the fullest source of available information on expansion joints, commercial sizes, service conditions which may be accommodated, installation and maintenance instructions, case studies of prior successful experience, etc.   Occasionally the technical literature includes an article by an engineer representing a responsible manufacturer of expansion joints which makes a reasonably objective appraisal of the advantages and disadvantages of various types and designs; some of these articles are cited in the bibliography.[6,19,219,229,232]   The Expansion Joint Manufacturers Association, founded in 1954 by a group of manufacturers of bellows expansion joints, publishes a set of standards[100] which contains much useful information.   This document also repeatedly points out that differences in detail design and manufacturing methods among the various member companies imply that certain important parameters cannot be standardized even for bellows designs and refers the user to the manufacturer's literature describing the particular item he is planning to use.   A useful design manual[250] for bellows joints is available. Generally, each reputable manufacturer of expansion joints, whether of bellows or other type, provides useful literature of which the designer is well advised to avail himself.   On the other hand, some manufacturers who are unable to maintain adequate standards of design, manufacture, and inspection also offer highly attractive literature.   Thus, in the purchase of expansion joints it is quite necessary for the buyer to beware and to assure himself by every reasonable method that he is dealing with a reputable manufacturer or jobber, that there is a full understanding of what the applications problem really is, and that the purchase contract or specification clearly sets forth all pertinent requirements and obligations.

Other sources of information on expansion joints include an excellent series of articles by A. Kovacs,[160] an extensive and highly useful chapter in the M. W. Kellogg manual,[150] and earlier editions of this Handbook.

Types of Expansion Joint—Types Involving Sliding.   Expansion joints may be divided into two basic categories: first, *sliding* joints, in which there is relative motion of adjacent parts, and second, *flexible* joints, in which there is no such relative motion but rather a distributed distortion of the device.   Another pair of terms sometimes used for this distinction is *packed* and *packless*, since the former type, involving relative motion, must have some sort of packing to prevent leakage. However, this second pair of words involves an invidious distinction and is not descriptive of the joint itself but rather of an adjunct.

Sliding joints may further be subdivided into slip joints, swivel joints, ball joints, special couplings, and screwed joints.   Flexible joints may be further divided into those depending on the flexibility of rubber or other highly pliable material, bellows joints, metal hose, and corrugated pipe.

Of the sliding joints, slip, swivel, and ball types are especially made to accommodate thermal expansion and movement of the piping.   However, there is inherent in certain patented coupling devices for quick or convenient assembly of pipe in the field a degree of flexibility which will accommodate a limited amount of expansion or terminal movement.   Also, it has long been a practice in small-size screwed piping to provide for thermal expansion by use of offsets composed of nipples and elbows.   This practice proves useful in plumbing systems, but it is not considered as a reliable expedient in the case of larger or more highly pressurized piping.

All the sliding joints involve relative motion of adjacent parts, and this implies the necessity of packing to contain internal pressure without leakage and of lubrication to minimize the probability of seizing and to reduce wear on mating surfaces. This requirement for packing and lubrication tends to limit the upper ranges of temperature and pressure of the fluids which may be transported by pipelines containing such joints.   However, some joints are available in which the packing material is inorganic (for example, graphite and asbestos) and provides its own

lubrication.   Other types involve direct metal-to-metal contact of metals and alloys chosen in pairs so as to reduce the chance of seizing and galling.   Also, the rather recent development of the polyfluorethylene compounds has given rise to a group of new packing materials which are capable of withstanding relatively high temperatures, of providing self-lubrication, and of resisting various types of chemical attack.

Depending on the amount and frequency of relative motion; the adequacy of guides so as to relieve unwanted loads upon the expansion joint; the temperature, pressure, and corrosive properties of the transported fluid; the ambient atmosphere and its dust or impurities content; and similar considerations, the necessity of lubrication and of repacking may vary widely.   Most devices which have provision for lubrication permit this being done when the line is in service, usually by the use of grease pressure fittings.   In most cases, replacement of packing requires complete disassembly of the joint, but in some cases, where frequent joint maintenance is called for by the severity of operating conditions or the corrosive nature of the fluid, provision is made for repacking with a minimum of service interruption. Some slip joints permit injection during operation of a plastic packing compound which backs up and applies pressure to the packing rings which actually contain the pressure, thus permitting the latter to function for extended periods without replacement.   Many other devices have take-up sleeves or cap screws by means of which the packing may be tightened to compensate for wear.   The problems with packing of expansion joints are similar to those with valves, except that in expansion joints the diameters which are involved are substantially larger and the guidance (so as to avoid nonsymmetrical loading) is likely to be considerably poorer than in the case of valves.

It is obvious that expansion joints which require periodic inspection and maintenance must be installed in accessible locations, and this may involve the expense of ladders, platforms, manholes, etc.   It also involves the stocking of appropriate maintenance supplies and the availability of maintenance personnel.   However, some users of joints of this nature assert that under proper conditions, total maintenance costs assignable to fairly large slip joints can be of the order of a dollar per year.

The most widely used joint which permits relative motion is the slip joint, typical details of which are shown in Fig. 41.   As is the case for most other types of expansion joint, this type also is susceptible to lateral buckling, or squirming as it is sometimes called, due to the action of internal pressure, and suitable guides must be provided to assure that such buckling does not take place and that the male and female members are and remain concentric at all times and at all degrees of traverse.

Anchors must be capable of withstanding all the various forces imposed upon them; calculation of these forces will be discussed later in this section.   Guides also must be proportioned adequately, since not only must they serve to direct the motion axially and prevent buckling but they must also resist accidental moments due to such causes as nonuniform friction around the periphery of the packing. Slip joints have the great advantage of being capable of absorbing relatively large amounts of expansion in a single device and doing so in the most direct way possible. They are available in sizes through 30-in. diameter, capable of absorbing up to 24-in. expansion and applicable for services with pressures at 400 psi and higher and for temperatures up to 650 F in well-ventilated installations.   As has been suggested immediately above, the slip joint accommodates expansion in the most direct possible way—by direct axial motion.   In this, it is unlike ball-and-swivel joints and like certain types of bellows-joint devices.   However, it does enjoy one advantage with respect to the latter which may be of considerable importance in

certain applications. Like the swivel-and-ball joints, the slip joint easily accommodates torsional motion, whereas the bellows type of joint must usually be protected against any but quite nominal torsional action. On the other hand, the slip joint is not capable of absorbing any except direct axial and torsional motion components, and the greatest of care must be observed to assure that it is not called upon to absorb any bending or offset type of motions which could cause binding, galling, and other forms of imperfect behavior which would impose intolerable

Fig. 41. Typical slip-type expansion joints. Note lubrication fittings in both figures. Both types permit adjustment of packing gland; in addition, the lower design has provision for packing by pressure gun insertion of plastic material. The lower design shows a double-ended slip joint. (*Courtesy of Yarnall-Waring Company.*)

loadings upon the slip joint or other components of the system. Adequate guides, however, will assure that only the tolerable components are imposed upon the slip joint and will provide that slight misalignments and lateral or offset motions are absorbed by the inherent flexibility of the piping.

Typical ball-and-swivel joints are illustrated in Figs. 42 and 43. The design of most such devices is such as to assure that the direct pressure loading is transmitted directly through the device and not forced back upon anchors. In this respect, ball-and-swivel joints make for simpler installation than other types of expansion joint except possibly for some special types involving bellows elements. However, it is clear that ball-and-swivel joints do not directly absorb axial motions but instead must rely on lateral motions to accommodate the imposed motions. Thus they are not suitable for single straight lines unless offsets are added. However, in other than simple straight lines, an offset of some sort is present in the geometry

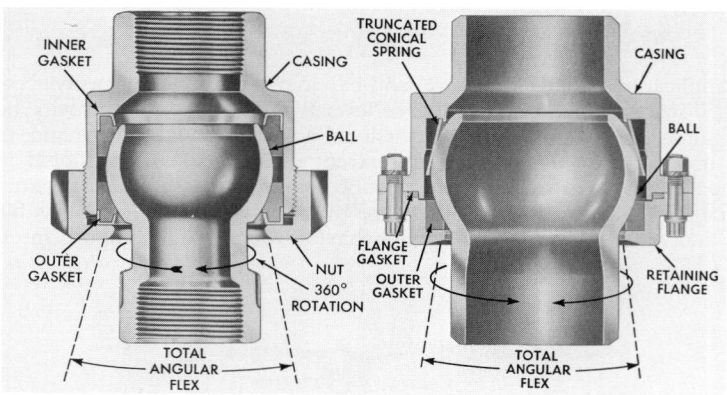

Fig. 42. Ball joints. (*a*) Standard unit with inner and outer composition gaskets; other end connections are available, and designs with one or two right-angle direction changes may be obtained. (*b*) Unit for high-temperature service; the outer gasket is metallic and is lapped to the ball; the conical spring is used to guide the ball, hold it in position, allow for differential expansion, and adjust for wear. (*Courtesy of Aeroquip Corporation, Barco Division.*)

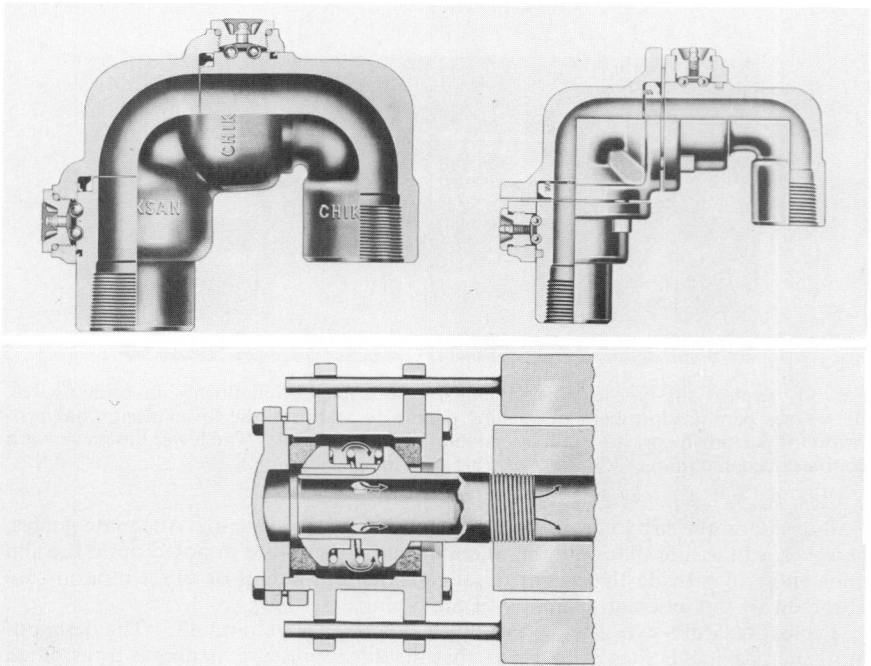

Fig. 43. Typical swivel and rotary joints. The upper left figure shows a swivel joint for 6,000 psi and 225 F (400 F with special packing). Similar joints are available for pressures up to 15,000 psi. The upper right figure shows a design which may be repacked without disassembling the joint. These figures were supplied through the courtesy of Chiksan Division, FMC Corporation. The lower figure shows what is usually called a rotary joint, using carbon-graphite seal rings and primarily intended for continuous rotary service. The inlet is in the side of the body, not shown; body lugs and support rods are turned 90 deg to show construction. Designs capable of handling two fluids simultaneously are available. Lower figure courtesy of the Johnson Corporation.

of the line, and these devices afford a way of handling the thermal expansion and terminal motion without the necessity, in the usual case, of providing special anchors and guides. Ball-and-swivel joints provide a rather simple solution to any flexibility problem where the motion of termini is large or continuous. In many designs packing seals are engineered so that the direct pressure loading is applied through the packing material and serves to seal the joint more tightly against leakage. Some swivel devices are able to handle two fluids at once by use of concentric internal elements; such a design is useful in cases where a steam supply and condensate drain are involved. Lubrication is frequently provided by the nature of the sealing or packing materials; in other cases grease fittings may be provided. Some swivels are provided with single or multiple rows of ball bearings to reduce friction, provide guidance, and transmit all force and moment components (except torsional) without the necessity of providing guides or anchors to sustain them.

Ball joints are available in sizes through 16 in. diameter. High-temperature designs are capable of handling service through 1000 F. Pressure ratings depend on size and temperature requirements, ranging from 4,500 psi in small sizes and low temperatures to such conditions as, for example, 550 psi at 500 F for a 14-in. size. One manufacturer supplies ball joints capable of accommodating any amount of torsional action plus a total angular flex of $7\frac{1}{2}$ deg each side of straight; for some of his models, the angular flex may be as large as $19\frac{1}{2}$ deg each side of straight. They are available either for straight-through flow or with a 90-deg elbow intrinsic with the ball joint. The latter type may most easily be used to provide an offset configuration which is capable of absorbing a total axial traverse of amount $2L \sin \theta$, where L is the offset distance and θ is the permissible angular flex each side of straight. This motion is accompanied by a lateral offset motion of amount $L(1 - \cos \theta)$, which is easily accommodated by the inherent flexibility of the piping in most applications. For example, with the $7\frac{1}{2}$-deg model, a 2-ft offset (between ball centers) provides the capacity to absorb up to 6.26 in. expansion, the lateral deflection being only 0.21 in.

Types of Expansion Joint—Types Involving Flexible Elements. The second and fundamentally different type of expansion joint does not involve relative motion of adjacent parts and for this reason may be designed so that there is a perfect seal against leakage of contents, the need for packing, lubrication, and maintenance thus being eliminated. There is a wide variety of particular designs and methods of manufacture. In some, a number of formed annular plates are circumferentially welded so as to form the bellows. In others, the bellows is formed from a tube, either seamless or longitudinally welded, the corrugations being either distinct from one another, which is the more common construction, or spirally advancing, which is a type of construction found frequently in metal hoses. The basic bellows itself, which is merely a series of formed or fabricated corrugations, must have end connections which may be flanged, screwed, welding end, or of other design. In addition there may be other adjuncts such as the following:

1. Outside covers, usually telescoping metal cylinders, which serve to protect the corrugations against corrosion and mechanical damage and to prevent the lodging of extraneous material and debris in the corrugations in such a way as to damage them or impede their motion.

2. Liners or internal sleeves, which are intended to minimize the contact between the interior surface of the bellows and the fluid which is being handled.

3. Limit stops, which are intended to limit the expansion which may be applied to the expansion joint. These may be of various types including ties or rods to sustain axial pressure thrust, to limit the rotation of a hinged expansion joint, to

distribute the movement among several parts of an expansion joint, to transmit the weight of adjacent piping, etc.

4. Reinforcing rods, which are closed rings of solid or tubular construction which fit snugly on the outside of the bellows in the small-diameter portion of the corrugations. They are intended to provide additional reinforcement against the gross hoop stresses due to internal pressure, and they obviously serve this function. Additionally, they serve to equalize the motion of adjacent corrugations (so that one does not close up preferentially in comparison with another) and to guide the local distortion of the small diameter of the corrugation as it flexes.

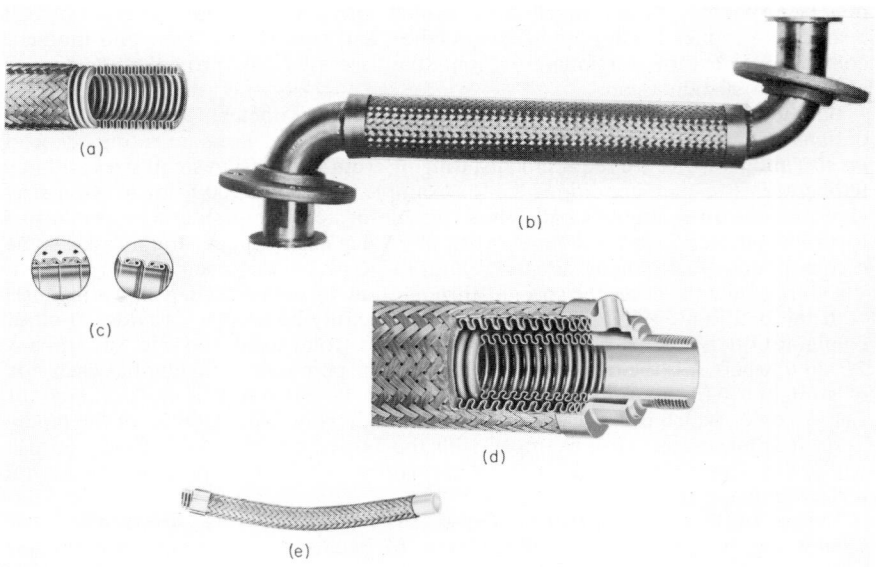

Fig. 44. Typical details of flexible metal hoses. (a) Typical detail showing braid cover and (magnified) detail of coil construction. These coils are helically advancing; other constructions have independent corrugations as in most bellows joints. (b) Typical assembly. This has 90-deg flanged ends; many other types and configurations are available. (c) Another type of construction. This requires a sealant which in this specific case is by means of asbestos cord packing; direction of flow is indicated. (d) Jacketed hose, an example of the variety of products which are available. (e) Wire braided teflon (E.I. du Pont T.M.) hose showing a combination of materials, the teflon carrying corrosive fluid with minimal pressure loss and the braid providing additional strength. (*Courtesy of Allied Metal Hose Company, Inc.*)

As is the case with most of the other types of expansion joint, the axial-pressure thrust is usually capable of destroying a bellows joint unless it is properly provided for. In the case of metal hose, an exterior braid, usually of bronze or stainless-steel wire, is applied in such a way as to provide minimal interference with the flexibility of the inner bellows but so as to carry and sustain the axial pressure thrust if necessary. There is a wide variety of metal hose designs available commercially; some typical details are shown in Fig. 44. Additionally, there may be inner linings of flexible noncorroding material, exterior coverings designed to resist abrasion and mechanical forces, etc. The exterior coverings may contain heavy wire reinforcement which serves to sustain high internal pressure and to protect against external loadings. Metal hose is commercially available in sizes

up through 24 in. diameter and by, special order, in sizes larger than this. Of course, the small diameters are capable of withstanding more severe service conditions than are the large diameters. General remarks about metal hoses and precautions for the selection, installation, and usage are given in a useful recent article.[6]

It may be remarked that some metal hoses are composed of formed spiral strips with pressure sealing accomplished by nonmetallic inserts; however, whether they are made in this way or are some form of bellows construction, the principles of application are the same.

For other than low-pressure service metal hoses are essentially intended to absorb only those strains which are perpendicular to their own axis. For this reason, generally they should be installed at right angles to the movement involved, and if there is movement in more than one direction, it may be necessary to install two sections of metal hose at right angles to each other. Metal hose is also very good for accommodating angular rotations. Like other metal bellows expansion joints, however, metal hoses should be protected from torsional moments.

Another special form of expansion absorbing design which can be classified as a bellows joint is corrugated piping and/or creased bends.[7,91,202] The analysis of piping systems containing corrugated piping and creased bends may be accomplished by the analytical procedures given elsewhere in this chapter provided that reasonably accurate compliance components can be obtained. Such elements, if they are made from other than very thin-wall pipe, usually have sufficient inherent stiffness to sustain and resist the axial-pressure thrust, and literature accounts of the analysis of systems containing such elements do not appear to be concerned with the possibility of instability due to the pressure thrust. However, as has been indicated, even though major fabricators are prepared to furnish corrugated piping and creased bends, there has been very little call for such items in recent years.

Returning to bellows-type expansion joints proper, we may remark that, although there is a rather extensive literature concerning these joints,[47,73,85,95,102,144,172,179, 228,229,251,253] it may be fairly said that most of this work is on a more highly idealized level than is of practical interest to most engineers concerned with employing these joints. In the first place, almost without exception, theoretical analyses have been confined to cases wherein axial symmetry has been preserved. Thus the important cases of lateral movement, torsion, and flexing remain relatively unexplored, and only rather approximate methods are available to handle them. Likewise, the geometry of the corrugations is usually idealized to an extent which renders it suspect. Moreover, although some manufacturers are inclined to dispute this point, it seems to be the consensus that, unless bellows joints are overdesigned to the point where they are no longer competitive, some plastic action takes place during each cycle of operation. This may be observed either by obtaining extremely high calculated elastic stresses using theoretical analysis or by actually testing bellows under well-instrumented simulated service conditions. A recent paper by Marcal and Turner[179] uses an elastic solution as an approximation to the plastic solution, using the upper and lower bound theorems of the theory of limit analysis to obtain a range into which the theoretically correct but practically unobtainable solution should fall. These range values were compared with experiment, and good comparisons were obtained.

In remarking that the literature concerning the theory of bellows is not of particular immediate value to one who must select and employ expansion joints, it is not intended to suggest that this work is of other than the highest value in adding progressively to the ability of theoreticians to deal constructively with this most difficult aspect of shell theory and in providing a sound theoretical background to assist in correlating experimental observations and in guiding the analysis of extreme designs for which there is no service experience.

Generally, most manufacturers have relied upon relatively elementary theoretical considerations backed up by and correlated with experimental and test observations and with successful experience under field conditions. Any user of an expansion bellows for a critical service should assure himself by close contact with the manufacturer that the test and service experience which the manufacturer adduces in support of his product is indeed sufficient to provide the necessary assurance of acceptable future performance.

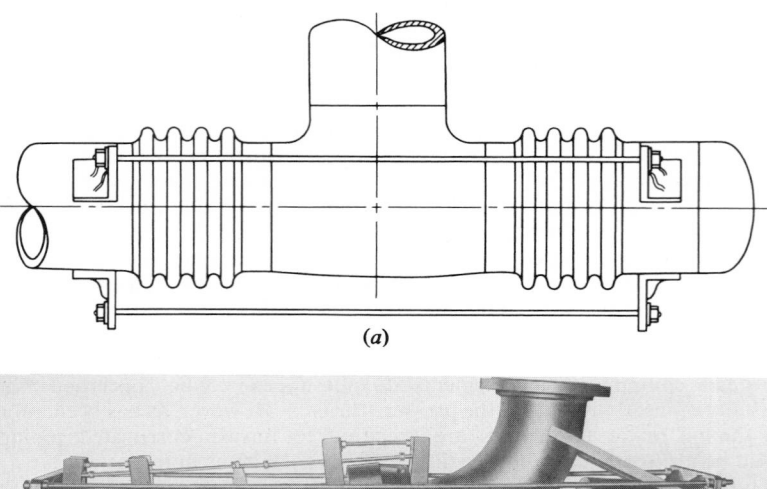

(a)

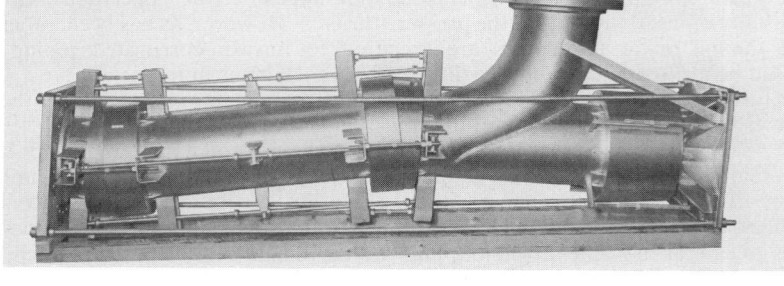

(b)

Fig. 45. Pressure-compensated expansion joint for use where a directional change is available. (*a*) Schematic drawing of basic concept. This design uses a welding tee and two bellows elements; the tee and the riser can "float" between the horizontal portions which are tied together to resist the longitudinal pressure force. (*b*) Composite joint which provides for offset displacement as well as pressure-compensated axial motion. Three bellows elements (in protective external casings) are used along with a fabricated swept elbow; the device is shown in the cold-sprung configuration. (*Courtesy of The Badger Company, Inc.*)

A number of somewhat special expansion joints have been designed for special applications or for service where customary anchoring and guidance cannot be applied. If there is a change in the direction of the piping in the vicinity of the desired location of the expansion joint, a device called a pressure-balanced expansion joint may be installed. As shown in Fig. 45 this consists of two bellows elements, one of which is located in a closed-off dummy leg. The pressure thrust is completely balanced, and the only forces involved are those required to flex the bellows convolutions or, in the case of a slip joint, to overcome friction. A number of other special devices are described in manufacturers' catalogues. Still other types,

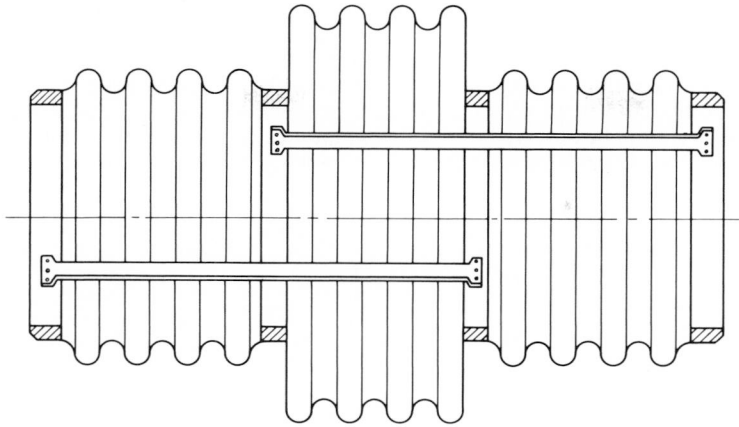

Fɪɢ 46. J. L. Donahue's pressure-compensated expansion joint.   Schematic view showing internal ties.   The pressure area of the central portion is twice that of the end portions. If an axial shortening x is imposed, each end portion shortens by amount x and the central portion extends by amount x, the total volume remaining unchanged so that pressure does not work.

including a very ingenious straight-line pressure-balanced type are described in a paper by J. L. Donahue.[90]   This last mentioned device is also illustrated schematically in Fig. 46.

Equivalent Movements of Expansion Joints.   Table 14 gives a rough overall indication of the types of motion which the various basic expansion joints are capable of handling.   Special design features may permit a certain design to accommodate motions to a better or to a poorer degree than this table indicates.   In any case, in selecting a joint for a particular service, consideration of all the factors

Table 14.   Types of Motion That Can Be Accommodated by Various Expansion Joints†

Type of joint	Displacements			Rotations		
	X	Y	Z	X	Y	Z
Rubber, etc.	G	S	S	S	S	S
Bellows	G	S*	S*	0	S*	S*
Metal hose	0**	G	G	0	G	G
Slip joints	G	0	0	I	0	0
Swivel joints	0	0	0	I	0	0
Ball joints	0	0	0	I	G	G

† It is assumed that the axis of the joint element is in the x direction.   The symbol I denotes the capacity to handle infinite torsional motion.   The symbol G indicates a "good" capacity to handle the motion in question.   The symbol S indicates "some" capacity.   The symbol 0 indicates (essentially) no capacity at all. One asterisk indicates that formulas for the equivalent axial motion may be found in the text.   Two asterisks indicate that under some circumstances (in particular, under low pressure) metal hoses can handle some axial motion.   All these listings are relative and approximate and do not take into account other important factors such as service conditions, cost, maintenance, etc.

involved (service conditions, corrosive action, leakage, etc.) will probably demand that all types be considered at least initially, and literature describing competing types should be compared.   It should be noted in particular that ball-and-swivel joints can accommodate no relative *displacement* of the mating parts; however, used in combinations, they can, of course, accommodate quite general motions. Similarly, it should be observed that bellows (and to a smaller extent, metal hose) are fundamentally incapable of accommodating torsional motions.   Also, it should be

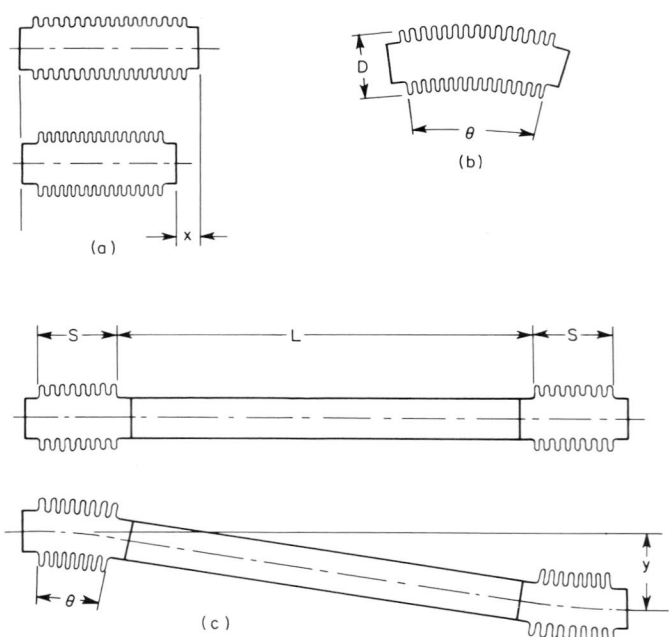

FIG. 47. Types of deformation which may be imposed on bellows-type expansion joints (*a*) Axial deformation.   (*b*) Angular deformation.   (*c*) Offset deformation.

noted that slip joints are quite incapable of handling other than the two motion components indicated in the table and that to impose other types of motion upon them is to assure that difficulties will be encountered.

Because combinations of motions may be involved or because the same joint may be called upon to handle different types of motion, it is necessary to be able to relate the various types of motion to one standard in terms of which the joint may be rated.   Manufacturers' literature usually contains instructions for doing this.   In the case of bellows joints, the basic capability of the joint is frequently given in terms of a rated axial extension for each corrugation, and it is required that the total extreme displacement per corrugation not exceed the rated value.   The types of motion considered are illustrated in Fig. 47 and consist of straight axial extension of total amount x, relative angular displacement of the two ends by an amount θ (but with no change in center-line arc length), and parallel offset between end axes of amount y.   If there are a total of N corrugations involved, the axial displacement x contributes an amount

$$e_x = x/N$$

to the effective axial motion per corrugation.   The angular distortion contributes an amount

$$e_\theta = \theta D/2N$$

where D is an "effective" diameter for the purposes of this calculation.   This formula is that of the Expansion Joint Manufacturers Association and implies that the angle θ is expressed in radians.   If θ is expressed in degrees the formula becomes

$$e_\theta = 0.0087\theta D/N$$

The offset motion is more complicated. Figure 47c shows the configuration before and after the offset deflection has taken place; however, it is presumed that any net axial deflection has already taken place so that the dimension S is the operational length of the bellows element.   Presuming that the configuration is symmetrical and that the two end portions are indeed constrained to be parallel to each other, there is a point of counterflexure at the geometric center of the assembly so that the reaction exerted at this point by one portion on the other

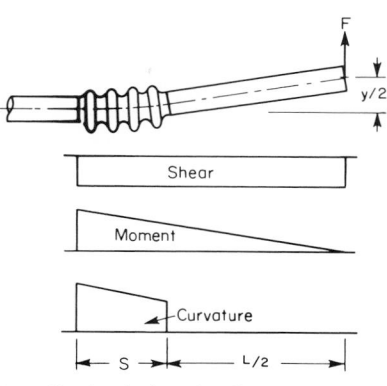

FIG. 48. Analysis of offset motion of bellows-type expansion joint.

is simply a lateral force F.   This is shown acting on one-half the assembly in Fig. 48.   Shear, bending moment, and curvature diagrams are also shown, the flexibility of the pipe itself being neglected in comparison to that of the bellows.   Measuring x positive to the right of the left end of the assembly, as shown, there is obtained

$$\text{Shear} = -F$$
$$\text{Moment} = (F/2)(2S + L - 2x)$$
$$\text{Curvature} = (F/2EI)(2S + L - 2x) \text{ in bellows region only}$$
$$\text{Slope} = (F/2EI)(2Sx + Lx - x^2) \text{ in bellows region only}$$
$$\text{Deflection} = (F/12EI)(6Sx^2 + 3Lx^2 - 2x^3) \text{ in bellows region only}$$

From these it follows that

$$y/2 = \text{deflection}(S) + \tfrac{1}{2}L \cdot \text{slope}(S)$$

the evaluations being made, as indicated, for $x = S$.   Thus

$$y = (FS/6EI)(4S^2 + 6SL + 3L^2)$$

Now the greatest rate of extension per convolution is at the ends where the curvature is greatest and

$$e_y = (DS/N) \times \text{maximum curvature}$$
$$= (DS/N)(F/2EI)(2S + L)$$

Eliminating F/EI between these relations, there is obtained

$$e_y = \frac{3Dy(2S + L)}{N(4S^2 + 6SL + 3L^2)}$$

The Expansion Joint Manufacturers Association writes this formula in the form

$$e_y = KDy/N(L + S)$$

where $K = (6S^2 + 9SL + 3L^2)/(4S^2 + 6SL + 3L^2)$, except that the original rather than the operation length S is used in determining K. Reference 250 uses the formula as derived here. Of course, the notation and actual form of the result are different, in both cases, from that used here, and the reader should be sure to use consistent notation when he refers to these other sources of information.

The Expansion Joint Manufacturers Association uses the largest diameter of the bellows assembly for D, while Ref. 250 uses a somewhat smaller effective diameter. Also, Ref. 250 modifies the expression for e_θ to a slight degree by adding another safety factor, so that the formula reads $e_\theta = \theta D/90N$ for θ in degrees.

There are many reasons for these and still other differences in the various procedures that might be recommended for establishing equivalent axial traverse per convolution. Essentially, the problem is one of the deformation of a very complicated shell structure, and any procedure which attempts to treat such a structure as an equivalent beam cannot be expected to give highly accurate and self-consistent results. The formulas given above, however, have proved to be satisfactory in actual use.

Cases where there is total angular deflection, as well as an offset handled by a combination of two bellows elements separated by a spacer, are more complicated yet. If the total angular deflection can be divided in some definite proportion between the two bellows elements, then a simple analysis leads to unequivocal results. There appears to be sufficient inherent factor of safety in commercial bellows that the combining formula recommended by most manufacturers, namely,

$$|e_x| + |e_\theta| + |e_y| \leq e_{\text{rated}}$$

will serve except under quite unusual circumstances; in the latter case, direct consultation with the manufacturer is indicated. (In the preceding formula, the vertical bars indicate that magnitudes are to be added.)

Different sources of information employ different notations or use the same symbol to mean different things, and thus care must be exercised in comparing different literature references.

Cyclic Operation of Expansion Joints. Although expansion joints are frequently used in situations where they are called upon to act only quite infrequently, for example, start-up, shutdown, bypass, or emergency situations, more often than not they are expected to function repeatedly, cycle after cycle, for the service life of the plant. An operation involving an hourly cycle rate would impose 2,000 cycles per year, or 40,000 cycles in a 20-year life. A slip or ball joint would have to be repacked and maintained many times during this service and might have to be replaced. A bellows joint might fail by fatigue action. The costs of failure and replacement must be weighed against the costs of a device assuredly capable of providing uninterrupted service for the desired service life of the plant. The manufacturer, either in his literature or by direct contact, should be expected to advise concerning the ability of his product to withstand highly repetitive service. Unfortunately, there is little standardization with regard to the assertions which are made, and different manufacturers may make divergent claims for devices which in practice would have essentially the same service life, or conversely, identical claims may be made for devices which have widely differing service life. It may be expected that in the future the user of expansion joints will have a more quantitative basis for decision than now exists.

Customarily, service-life expectancy seems to be based on average performance. Some manufacturers, however, use other measures; for example, one uses a standard

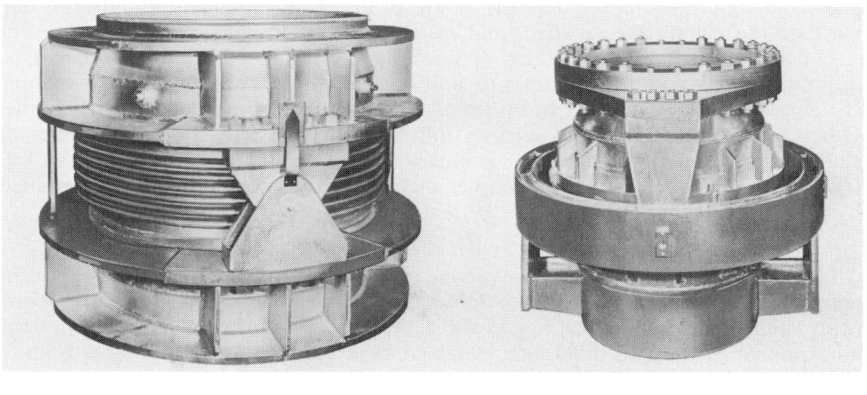

(a)   (b)

(c)

FIG. 49. Typical expansion joints employing bellows elements. (a) Hinged joint for angular rotation in one plane only; note arrow and scale to indicate extent of motion. (b) Gimbal joint for angular motion in two planes. (c) "Universal" type for offset or lateral motion (cf. Fig. 47c). (*Courtesy Zallea Brothers, Inc.*)

of 99 per cent survival; that is, he asserts that 99 per cent of his joints survive the stipulated number of service cycles, rather than only 50 per cent.   Under this type of service rating, the average item will survive many more cycles than the stipulated number.   Naturally, this increased survival probability and increased capability of the average item are not achieved without cost.   Thus, the user is faced with the

problem of interpreting the differing claims of different manufacturers, of deciding whether the claims are justified, and of attaching economic value to the result.

The user also has the problem of deciding the number of cycles that will be imposed even before he attempts to find a product that will meet the demands. If there is a definite number of cycles of identical traverse, the problem is simple. Otherwise, he must convert cycles of differing amplitude to equivalent cycles of standard amplitude. Usually, the standard should be taken as 100 per cent of full traverse. Then the equivalent number of full traverse cycles may be determined by a formula of the type

$$N_{eq} = \sum_r N_r r^a$$

where N_r is the number of cycles involving fractional traverse r and a is an exponent which depends upon design and material. A similar formula was presented previously, and the exponent was 5. In the case of metallic bellows, Ref. 250 employs an exponent of 4. This difference, obtained by experiment, is attributable both to

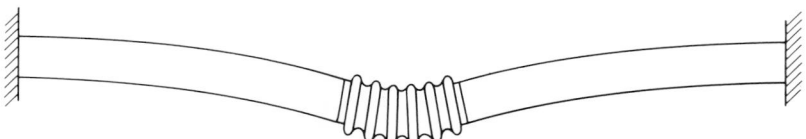

FIG. 50. Lateral buckling or "squirming" of a pipeline containing an expansion joint and without adequate guides.

material and geometry differences and also to the fact that bellows usually do not behave in a purely elastic manner for full or nearly full traverse. Thus, as an example, if a bellows expansion joint rated at 5-in. full traverse is subjected to 1,200 cycles of 4.5-in. traverse, 8,000 cycles of 2.5 in., and 20,000 cycles of 1.0 in., the equivalent number of 5-in. cycles may be determined to be $1,200(0.9)^4 + 8,000(0.5)^4 + 20,000(0.2)^4 = 787.3 + 500 + 32 = 1,320$ cycles.

It should be emphasized, however, that there is no generally agreed upon procedure for determining an effective number of cycles, and in the selection of a device to handle a certain combination of service cycles, the manufacturer should be consulted if the situation is at all unusual. It is becoming increasingly common to require, in a purchase specification for expansion joints for critical or demanding service, that prototype joints be cycled in order to verify their life expectancy. Because of the statistical nature of failure of such items, the precise details of the testing program and the criteria of judgment should be carefully agreed upon in advance.

Lateral Buckling and Squirming of Lines Containing Expansion Joints. The fact that an expansion joint yields readily to axial loads combined with the fact that internal pressure imposes what may be a very large axial force may lead to a lateral instability destructive of the line, the joint, and other appurtenances. The first part of this subsection will consider the case of a length of pipe which contains a bellows joint at its center and which is unrestrained except at its ends, which are assumed to be fully anchored. For simplicity in analysis, assume the expansion joint to be at the center of the span. Clearly there is a possibility that the configuration may deform as indicated in Fig. 50. An analysis by energy methods may be accomplished as follows. The total length of the deflected system is greater than the original length by an amount λ, which may be computed as

$$\lambda = \frac{1}{2} \int_0^l (y')^2 \, dx$$

and this increase in length takes place in the bellows portion where the effective pressure area is somewhat greater than that of the pipe itself, corresponding to, say, the mean diameter of the convolutions. Thus, during the deformation, pressure performs an amount of work $U_p = pV = pA_p\lambda$. If there is to be equilibrium in the deflected position, this work must be absorbed by the structure in the form of strain energy—in the pipe itself to the extent of $U_1 = \frac{1}{2}EI\int_0^l (y'')^2\, dx$ and in the bellows both because of the axial deformation of the bellows, $U_2 = \frac{1}{2}f_x\lambda^2$, where f_x is axial spring constant, and because of flexure of the bellows $U_3 = \frac{1}{2}m_a\theta^2$, where m_a is a flexure spring constant and θ is the angular distortion of the bellows. The equation $U_p = U_1 + U_2 + U_3$ determines the pressure at which such an equilibrium could obtain. A lower pressure would be insufficient to cause the lateral distortion, and a higher pressure would be capable not only of causing the distortion but also of getting up some speed in the process.

The method outlined above tacitly assumes that the shape of the deflected center line is known so that y' and y'' may be obtained. Of course, in general this is not the case. However, a reasonably good approximation to the theoretical correct curve, i.e., an approximation that satisfies continuity, end restraints, etc., will give a satisfactory evaluation of the buckling or squirming pressure. Analysis of a spectrum of practical cases leads to the conclusion that U_2 and U_3 are small compared with U_1. In other words, the restraining effect of the flexible bellows is small compared with that of the pipe. If one equates $U_p = U_1$, the result turns out to be that the critical buckling pressure is such that pressure times pressure area equals Euler buckling load for a pipe column fixed at one end and free at the other and having a length equal to the total span indicated in Fig. 50; namely,

$$pA_p = \pi^2 EI/l^2$$

This calculation presumes that the ends are fully restrained against rotation. If the guidance at the ends is not perfectly rigid, buckling will take place under smaller pressures than those calculated above. Also, this analysis has presumed that the expansion joint is located at the center of the span. If it is located elsewhere, the shearing forces which must be transmitted through the expansion joint complicate the analysis.

In practice not only the axial pressure force, discussed above, but also whatever force is required to flex a bellows joint or to overcome friction in a slip joint together, act so as to cause buckling. Accordingly, the guide spacing should be determined from a formula of the type

$$l = (\pi/n)\sqrt{EI/F}$$

where n is an appropriate factor of safety and F is the maximum possible combined axial compressive load. As a practical matter, it is essential to provide guidance so as to prevent such lateral buckling if the fluid pressure force is any appreciable fraction of the calculated Euler buckling load, and generally, it is best to provide guidance under all circumstances. The efficacy of guidance is augmented if the joint is located near one of the anchored ends, and the Expansion Joint Manufacturers Association recommends[100] that this be done, that the first guide be located a maximum distance of 4 pipe diameters beyond the expansion joint, that the second guide be located a maximum distance of 14 pipe diameters beyond this, and that subsequent guides, if required, be located at a spacing not to exceed values obtained from Fig. 51. These spacings which correspond to about one-fifth of a constrained Euler length are quite conservative. The spacing is sufficiently far apart so as not to impose significant economic penalty in providing more guides

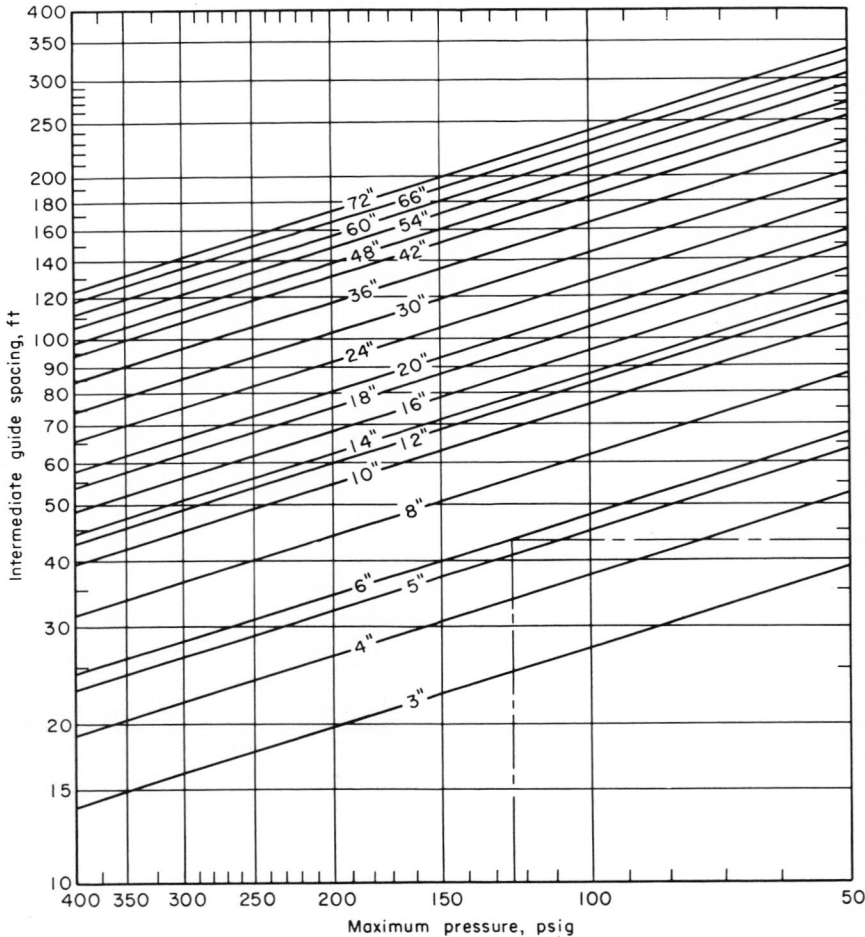

FIG. 51. Recommended maximum expansion-joint guide spacing for standard weight carbon steel pipe. NOTE: First pipe guide must be located within a distance of 4 pipe diameters from the expansion joint, and the second guide must be located within a distance of 14 pipe diameters from the first guide. (*Courtesy of Expansion Joint Manufacturers Association.*)

than might be considered absolutely necessary but sufficiently close that there is minimum danger of binding at one or more of the guides. If there should be binding, the expansion would not be absorbed in the expansion joint provided for it but would rather result in lateral buckling and probable severe overstressing of a section of pipe at some distance from the section containing the expansion joint. Reference 250 recommends guide spacing very slightly less than that obtainable from Fig. 51.

The preceding discussion applies strictly only to bellows-type expansion joints. It is clear that this type of instability cannot take place in ball or swivel joints. In the case of axial slip joints, the problem is in some ways similar to the case of bellows

joints and in some ways different.   Pressure times area, acting through a distance, does work, and this work can provide the strain energy for a laterally deflected configuration.   However, instead of providing negligible restraint against free motion of the pipe ends to which it is connected, a slip joint differs from a bellows joint in affording considerable restraint to all except very small distortions.   However, in offering this restraint it may impose upon itself lateral forces sufficient to bind it completely against performing the service for which it was intended or to destroy it.   Thus, in the case of slip joints it is even more important than it is for bellows joints to assure that there is no lateral deflection; guides for this purpose must be manufactured and installed to rather careful tolerances.

Guide spacing recommendations made by manufacturers of slip joints are more liberal than those recommended by bellows joint manufacturers.   In the case of guides adjacent to the expansion joint itself, this may possibly be attributed to the degree of flexural constraint provided by a slip-type joint, but for guides twice or more times removed from the expansion joint itself, it is difficult to see why there should be any difference in spacing requirements.

It should be noted that bellows joints that are completely restrained against axial extension (by tie rods or cradle rods or similar devices) may, if they are sufficiently long, limber, and pressurized, squirm without involving any distortion of the mating pipe.   In other words, using the energy approach introduced earlier in this subsection, the equation $U_p = U_2 + U_3$ gives a value of pressure for what might be called self-squirming.   Haringx[132] has studied this phenomenon by using direct solution of the differential equation rather than an energy method.

Installation of Expansion Joints; Preset, Guides, Anchors.   In all cases an expansion joint may be considered to be a structural "weak link" in the sense that the distortion which must be accommodated is concentrated in the expansion joint rather than being spread out along the length of the pipe as is the case for elastic deformation of the pipe itself. Although the manufacturers of these expansion joints provide adequate strength in their products to withstand the normal loads imposed in service and although instructions regarding proper installation, operation, and maintenance of the devices are made readily available, these devices cannot help but tend to be susceptible to damage if they are improperly used. There are three overridingly important design considerations which apply to the selection, installation, and operation of all these devices.   These are:

1. Under no circumstances must the motion which an expansion joint is required to accommodate exceed, in either sense, the capability of the joint to absorb this motion.   This may imply that a certain degree of preset be accomplished during installation so that both expansion and contraction may take place.

2. Suitable anchors must be provided to sustain the direct pressure thrust (as well as any other mechanical loading which may be imposed upon them).

3. Appropriate guides must be provided to assure that the joint is protected from all loads except those which it is designed to carry and accommodate.

The specific design considerations which these three fundamental rules may call for depend, naturally, upon the circumstances: the type of expansion joint or joints being considered, the layout of the piping, the amount of expansion which must be absorbed, etc.   For purposes of illustration, however, consider the case of an 8-in. standard carbon-steel line to carry saturated steam at a maximum pressure of 250 psig.   The line must run straight for a distance of 134 ft 6 in. in a pipe tunnel which does not permit any devious routing, and the expansion attributable to this length must be absorbed within the length.   Also, the minimum out-of-service temperature will be 20 F during the winter months.   The installation will take place when the ambient temperature is 80 F.   The expansions tabulated in Table 2 are

for 100-ft lengths, so that tabular entries must be multiplied, in this problem, by 1.354. The following values may be calculated:

Expansion from 70 to 406 F $= (1.345)(2.70 + 0.06 \times 0.92) = 3.706$ in.

Expansion from 70 to 80 F $= (1.345)(10/130)(0.99) = 0.102$ in.

Expansion from 70 to 20 F $= (1.345)(50/120)(-0.84) = -0.471$.in

The total expansion to be accommodated is $3.706 - (-0.471) = 4.177$ in. This is the barest of minima, and any error in predicting operating extremes or in installation or any slightest degree of overpressure or superheat could result in damage of the expansion joint. Accordingly, a joint capable of absorbing 5.00 in. total traverse will be selected. This available traverse should be divided as follows:

Installation to cold condition $= (5.000/4.177)(0.102 + 0.471) = 0.686$ in.

Installation to hot condition $= (5.000/4.177)(3.706 - 0.102) = 4.314$ in.

and during its installation in the line the joint should be held at an extension of 0.686 in. from its fully extended position until the installation is complete.

An approximate division of the total 5-in. available traverse into the two parts as above may be accomplished as follows:

$$\text{Installation to cold condition} = (5) \; \frac{80 - 20}{406 - 20} = 0.777 \text{ in.}$$

$$\text{Installation to hot condition} = (5) \; \frac{406 - 80}{406 - 20} = 4.223 \text{ in.}$$

The error in the approximate evaluation is due to the fact that the coefficient of linear expansion of the pipe is not a constant but rather itself depends upon the temperature. However, if the traverse available in the selected device is reasonably in excess of the theoretical minimum required, the error is of no consequence.

One should guard against the possibility of damaging conditions being imposed which are not accounted for in the selection of the device. For example, in one case the water used for the hydrostatic test of the line was taken from a deep well and was considerably colder than the lowest temperature for which the expansion joint was designed; this resulted in a failure of the joint, which was, however, adequately sized for the expansions it was supposed to absorb during actual service.

Next, returning to the steam line which is being considered as an example, suppose that the pipe is 12 in. standard. The internal flow area is 113.1 sq in., so that the axial pressure loading is $113.1 \times 250 = 28,275$ lb. However, the system must be capable of withstanding the greater pressure applied during hydrostatic testing, and for purposes of this example, this is taken to be $(1.5)(250) = 375$ psi. Also, the effective pressure area of the expansion joint may be substantially greater than that of the pipe itself. For example, in one popular type of bellows joint, the pressure thrust area is 189 sq in. for the 12-in. nominal diameter joint. This product is $(189)(1.5)(250) = 70,875$ lb. In addition, the anchor should be designed to carry also whatever force is necessary to compress the expansion joint to its fully compressed state. This may be as high as 6,000 or 7,000 lb, depending on the manufacture and type of expansion joint, resulting from packing friction in the case of slip joints and from the force required to deform the convolutions of bellows joints. Also, the anchors must also be capable of accommodating any frictional force which may be imposed by the guides. If there is a directional change at the anchors, as is frequently the case, the anchor must be able to sustain the thrusts coming from both directions and, in addition, the fluid impulse thrust

$(2 \sin \theta/2)(\dot{M}v)$, where the total angle change is θ, the flow velocity is v, and the mass rate is $\dot{M}$. All these anchor forces should be added vectorially to arrive at the total anchor loading, and in the design of the anchor it should be assured that not only are the structural details of attachment to the pipe adequate to sustain this force but also that the structure to which it is attached is itself sufficiently strong and immobile. Probably most of the failures of expansion joints have been the result of

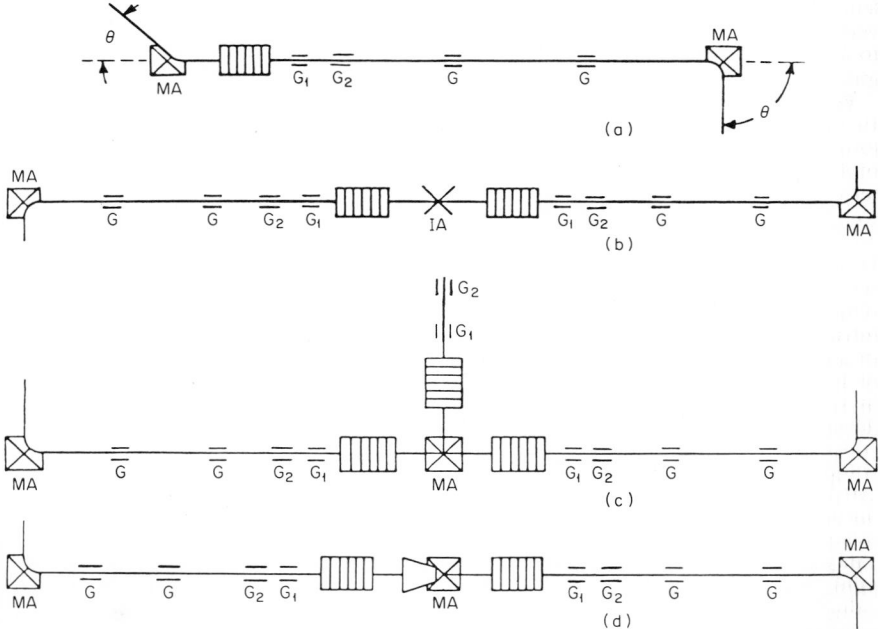

FIG. 52. Illustrations of good practice in the use of expansion joints. Note the use of one expansion joint between each pair of main anchors (MA), the nearness of an expansion joint to an anchor, the closeness of the first alignment guide (G_1), the spacing between the first alignment guide and the second alignment guide (G_2), and the spacing of additional alignment guides (G); see also Fig. 51. In b, note the use of an intermediate anchor (IA) in conjunction with the two main anchors to divide the pipeline into individual expanding sections with one expansion joint in each section, In d, note that a main anchor is necessary when the line contains a reducer, in order to withstand the differential in the thrust of the expansion joints on each side of the reducer. (*Illustrations courtesy of the Expansion Joint Manufacturers Association.*)

faulty installation in which anchor loads were grossly underestimated and inadequate guidance and anchorage provided. Forces of the magnitude described above require that substantial anchorage be provided.

The question of guides was discussed at some length in the preceding subsection, and procedures for determining the appropriate spacing of guides were outlined; see also Ref. 125. Also, some typical recommended installation suggestions are shown in Fig. 52. In general, the error in practical cases, if any, is of inadequate or insufficient guidance. However, there are some cases where difficulties result from overstringent guidance. An example is given by the case of a long horizontal line the expansion of which is taken up by a pair of bellows (as in Fig. 47c) in a

dog-leg riser, the axial pressure force being constrained by tie rods or cradle rods which permit offset motion in the riser but which do not permit any change in the length of the riser.   As the expansion takes place, the riser cants, thus altering the vertical intercept of its end points.   If guides in the horizontal line have been spaced to close to the top and bottom of the riser, excessive forces may be placed on these guides and on the tierods.

Combination of Expansion Joints with Inherent Flexibility of Piping. In installations where ball or swivel joints or appropriately constrained bellows joints are employed, a sufficient number of them are usually specified so that all motion components may be absorbed and the only forces imposing reactions on the terminal equipment and inducing stress in the line itself are those of the gravity weight of the piping and components and the friction in the ball or swivel joints and any guides which may be used and/or the force required to flex the bellows, if used. However, it is clear that a smaller number of such joints may alter significantly the reactions calculated on the basis of inherent flexibility alone without completely removing them.   The theoretical analysis given in other portions of this chapter is intrinsically capable of treating such articulated systems by use of calculating processes in which certain flexibility or compliance components become very large or infinitely large.   Infinity is difficult to deal with computationally, however, even in the most expensive modern digital computer, so that additional theoretical developments over and above those described in this chapter are necessary for the construction of a programmed procedure for dealing with such cases.   A computational trick which has proved successful in practical cases is simply to use astronomically large compliance components to account for the freedoms introduced by ball-and-swivel joints.   Thus, for example, in the case of a swivel joint, the freedom from torsional moment can be accomplished by using a very large fictitious value of μ in the final formulas for C for a straight element adjacent to the swivel joint.   The value of m in these formulas should be calculated using the correct value of Poisson's ratio μ, but the value of μ which appears in the expression $(I + \mu\bar\sigma\sigma)$ or in $(1 + \mu)$ should be made quite large, say 10,000,000.   The formal calculations will then be accomplished with a very slight torsional stiffness implicit at this point, and the final results will involve a very small torsional moment which may be neglected.

In an important series of papers,[160] Kovacs has studied systems which contain such articulations and shows how significant reductions in terminal reactions and maximum stress (range) may be accomplished by introducing an insufficient number of such freedoms to render the configuration into a mechanism.   He has studied ball joints, swivels, and universal joints of the Hooke type.   The last type of articulation approximates the case of a bellows which is permitted to flex in either lateral plane but is constrained against displacement or torsional motion.   Although analysis of particular cases involving such articulations may be accomplished within the framework of the theory developed in this chapter or of what would appear to be only minor extensions of it, a systematic investigation of the design possibilities for economies which may be accomplished by the optimal introduction of articulations appears to be a very promising avenue for research on the part of piping engineers and designers.   See also Ref. 234.

Guide for Specifying Expansion Joints.   This discussion of expansion joints will be brought to a close with a listing of specific points or aspects on which the user may wish to satisfy himself or with respect to which he may wish to come to agreement with the expansion-joint manufacturer.   It is not intended to suggest that each such item actually become the subject matter for a clause in a purchase specification; for example, it is usually better to permit the manufacturer of the device to employ his own initiative and ingenuity in determining the best methods of

fabricating an expansion joint which will meet certain service requirements. However, in some cases, it may be appropriate to come to agreement concerning details of fabrication and manufacture. The following also lists items of information which should be communicated to the manufacturer in order adequately to delineate the applications problem. This listing is intended to be suggestive rather than exhaustive, and the full details of each item in the list are left to the reader to complete according to his interests and needs.

1. Material being handled, its extreme temperature and pressure conditions, its corrosiveness, its content of erosive particles. Does it freeze or change phase? Will it cavitate? What is flow-rate maximum? Will a vacuum ever be imposed?

2. Motion or motions that must be accommodated. How many cycles of each type will be imposed during plant lifetime? Whose responsibility is it to convert this to equivalent full traverse cycles? What degree of assurance is necessary that the expansion joint(s) supplied will survive stipulated number of cycles? How can this assurance be obtained? Should cyclic tests of prototype models be specified? Can one group of tests qualify a hierarchy of joints? Are there cycles of pressure and/or thermal stress superimposed upon the cycles of traverse operation?

3. Who has responsibility of designing anchors and guides? Is it necessary to know forces and moments required to operate the joint? Can this come from the manufacturer's literature or files, or must it be established by test?

4. Details of ends, welding end preparation, flanges and bolting; etc. Note (as indicated in Ref. 100) that ambiguity in specifying flanges should be avoided. Note also that some tied or gimbaled joints have the tie, trunion, or gimbal structure connecting to the flanged ends which thus must withstand the corresponding mechanical loads as well as bolt and gasket loadings without distress or leakage.

5. Are pressure-drop limitations involved? If so, a way of determining them should be agreed upon and the manufacturer should design interior sleeve to meet the limitations.

6. Support of dead weight during operation; shall user or manufacturer provide for supporting the weight of the joint? The manufacturer must know of all loading, from whatever source, which will ever be imposed on the joint.

7. What details, accessories, appurtenances, etc., are required, and what service conditions must they meet? This includes internal sleeves, external covers, insulation lugs, ties, trunions, gimbals, cradles, drains, bleeds, jacketing, steam or electric tracing lines, etc.

8. What testing is required at the manufacturer's plant? If a prototype joint is cycled to failure, should it be sectioned, inspected for thinning of walls during the manufacturing operation, etc? Should a vacuum test be called for? What testing is called for after installation? Hydrotest pressure, hydrotest fluid temperature, etc.

9. Fabrication requirements. Qualification of welding operators and procedures. Heat treatment. Restrictions on method of forming such as hot-formed, cold-formed, hydraulically formed, press-formed, explosively formed, etc.

10. Nondestructive tests required such as radiography of welded joints, dye penetrant tests, ultrasonic inspection of plate and tube materials for flaws and laminations, metallographic tests, other laboratory tests. Tests establishing physical properties of material coupons.

11. Specification and certification of materials to be used. Mill test reports. Test laboratory reports. Other qualifying test reports. Supporting calculations. Other supporting documentation.

12. Maintenance requirements. Lubrication and packing fittings and procedures. Should maintenance be possible without taking the joint out of service?

13. Should the joint be equipped with indicators showing current configuration of the joint with marks showing extreme traverse positions?

14. Should the joint have stamped or scribed match marks to assure proper orientation for installation?

15. Is a special name plate required, and if so, what data should it contain? Should there be a line number or item number tag to prevent mixups?

16. Requirements for packaging, boxing, protection against weather, palletizing, sling hooks, etc.

17. Spacer bars and similar devices to protect during shipment and installation and to assure that desired preset (if established at factory) is maintained until installation is complete. Should such parts to be removed prior to operation be painted a distinguishing color?

18. Procedure for inspection after installation and prior to operation to assess damage by mechanical action, by welding in the vicinity, and by other causes. Assure that shipping stops are removed. Assure that foreign matter is not lodged in such a way as to impede free traverse of the joint.

19. Access for inspection and maintenance during operating life. Maintenance supplies, instructions, schedules, etc.

20. Last, but by no means least, what basic type of expansion joint will do the job best—a slip joint, swivel joint, ball joint, rubber or composition material, bellows joint, metal hose, or combination of these?

A listing such as the preceding cannot, however, do more than alert the piping designer to some of the various aspects of the selection and specification process which deserve thought and attention. In general, the total responsibility of assuring that the proper expansion joint is properly installed and operated is one which must be shared between piping designer and joint manufacturer, and if essential information is not communicated, malfunction is likely to result.

Acknowledgments

The author wishes here to express his appreciation to the persons and institutions, too numerous to name individually, who have assisted him in the preparation of this chapter. In particular, he wishes to offer thanks for the high quality of secretarial assistance offered by Mrs. B. B. Allen and for the great help offered to him by all members of the staff of the libraries at the U.S. Naval Post-graduate School. He wishes also to acknowledge assistance by Mr. A. R. C. Markl, with whom it has been his good fortune to work on Piping Code Committees for a number of years, and of Drs. J. L. Lubkin and L. H. Chen who provided useful reference materials.

Both simple truth and convention require, however, that the author himself shoulder the blame for errors and omissions. He would consider it a great kindness if these would be called to his attention by interested readers.

REFERENCES

This last section of the present chapter on "Expansion and Flexibility" consists of a number of references to the literature, cited by number in the text of the chapter. At the time the assignment of writing this chapter was undertaken, it was intended to present in this place a complete bibliography on the subject. However, it soon became apparent that this was utterly out of the question. The number of papers that should be included in such a complete listing is, indeed, astronomical, and the number of out-of-the way journals in which they appear is bewildering. Moreover, as the writing progressed, it became more and more difficult to draw a fine line separating those papers which pertained to piping flexibility from those which did not. Accordingly, what is presented here consists simply of a large

number of papers, most of them fairly recent, which relate to the subject and from which, it is hoped, only a small number of significant efforts may be missing. Almost all of these are specifically cited in the text. No effort has been made to include papers merely because of their historical significance unless they have, for one reason or another, been cited in the text. As has been noted in the first section of this chapter, historical discussions are available elsewhere.

BIBLIOGRAPHY

1. American Standards Association, Code for Pressure Piping (B31.1-1955) (augmented by "Interpretations" issued as "Cases" at irregular intervals). Also, separate industry documents including Petroleum Refinery Piping (B31.3-1962); Oil Transportation Piping, with addenda (B31.4-1959, B31.4A-1961); Refrigeration Piping (B31.5-1962); Gas Transmission and Distribution Piping Systems, with addenda (B31.8-1958, B31.8a-1961), published by American Standards Association, Inc., 10 East 40th Street, New York 10, N.Y., and jointly by other engineering societies and groups.
2. American Standards Association, Steel Butt-welding Fittings (B16.9-1958), American Standards Association, New York, 1959.
3. Andrews, L. C., "Analyzing Complex Piping Systems," *Consulting Engr.*, vol. 10, no. 3, pp. 123–128, 130, March, 1958.
4. Andrews, L. C., "Analyzing Piping Stresses by Tests of Models," *Heating, Piping, Air Conditioning*, vol. 17, no. 8, pp. 425–429, August, 1945.
5. Andrews, L. C., "Piping Flexibility Analysis by Model Test," *Trans. ASME*, vol. 74, no. 1, 123–133, January, 1952.
6. Anon., "Flexible Metal Hose, Properly Used, Can Take Strain out of Piping Problems," *Heating, Piping, Air Conditioning*, vol. 33, no. 11, pp. 120–125, November, 1961.
7. Anon., "Is Corrugated Piping Still Suitable?" *Power*, vol. 98, no. 11, pp. 136–137, November, 1957.
8. Anon., "Theoretical and Experimental Study of Piping Subjected to Thermal Expansion under Constraint," *Compt. Rend. Rech. IRSIA*, No. 16, November, 1955 (Belgium, in French).
9. Anon., "Briefcase Job for a Computer," *Business Week*, July 28, 1956, pp. 66–68, 73.
10. Argyris, J. H., "Energy Theorems and Structural Analysis," *Aircraft Engr.*, misc. pages, 1954–1955.
11. Argyris, J. H., "The Matrix Analysis of Structures with Cut-outs and Modifications," *Proc. Ninth Intern Congr. Appl. Mech.*, Brussels, vol. VI, pp. 131–142, 1957.
12. Argyris, J. H., "Matrix Theory of Statics," *Ing.-Arch.*, vol. 25, no. 3, pp, 174–192, 1957 (in German).
13. Argyris, J. H., "On the Analysis of Complex Elastic Structures," *Appl. Mech. Rev.*, vol. 11, no. 7, pp. 331–338, July, 1958.
14. Argyris, J. H., and S. Kelsey, "The Matrix Force Method of Structural Analysis and Some New Applications," *Aeron. Res. Council London, Res. Mem.* 3034, 33 pp., 1957.
15. Argyris, J. H., and S. Kelsey, "Note on the Theory of Aircraft Structural Analysis," *Z. Flugwiss.*, vol., 7, no. 3, pp. 73–77, March, 1959.
16. Argyris, J. H., and S. Kelsey, "On the Matrix Theory of Structures," *Z. Flugwiss.*, vol. 8, no. 6, pp. 169–172, June, 1960.
17. Argyris, J. H., and S. Kelsey, "Initial Strains in the Matrix Force Method of Structural Analysis," *J. Roy. Aeron. Soc.*, vol. 64, no. 596, pp. 493–495 (tech. notes), August, 1960.
18. ASME Boiler Code Committee, "Unfired Pressure Vessels," Section VIII of the ASME Boiler and Pressure Vessel Code, Appendix P, Basis for Establishing Stress Values for Ferrous Materials, American Society of Mechanical Engineers, New York, 1962.
19. Bagnard, G. M., "Use Swivel Joints for Flexibility," *Chem. Eng.*, vol. 67, no. 19, pp. 175–180, September 19, 1960.

20. Baron, F., "Matrix Analysis of Structures Curved in Space," *Proc. ASCE*, vol. 87, no. ST.3 (*J. Struct. Div.*), pp. 17–38, March, 1961.

21. Bazant, Z., "Analysis of Framed Structures," *Appl. Mech. Rev.*, vol. 8, no. 3, pp. 85–86, March 1955.

22. Belluzzi, O., "A Case of Instability by Ovalization of a Tube Subjected to Flexure," *Ric. Ing.*, anno I, no. 3, 1933-XI (in Italian).

23. Benscoter, S. V., "The Partitioning of Matrices in Structural Analysis," *J. Appl. Mech.*, vol. 15, no. 4, pp. 303–307, December, 1948.

24. Bergman, D. J., "Some Cases of Stress Due to Temperature Gradient," *Trans. ASME*, vol. 78, no. 5, pp. 1011–1019, July, 1956.

25. Berman, F. R., "Some Basic Concepts in Matrix Structural Analysis," *Proc. ASCE*, vol. 86, no ST8 (*J. Struct. Div.*), pp. 59–85, August, 1960.

26. Beskin, L., "Bending of Curved Thin Tubes," *J. Appl. Mech.*, vol. 12, no. 1, pp. 1–7, March, 1945.

27. Besseling, J. F., "Application of Matrix Calculus in Adjusting Stiffness and Vibration Properties of Redundant Structures," *Ninth Intern. Congr. Appl. Mech.*, Brussels, vol. VI, pp. 168–175, 1957.

28. Bijlaard, P. P., "Stresses from Local Loads in Cylindrical Pressure Vessels," *Welding J.*, vol. 33, no. 12, pp. 615S–623S, December, 1954.

29. Bijlaard, P. P., "Stresses from Local Loadings in Cylindrical Pressure Vessels," *Trans. ASME*, vol. 77, no. 6, pp. 805–816, August, 1955.

30. Bijlaard, P. P., "Stresses from Radial Loads and External Moments in Cylindrical Pressure Vessels," *Welding J.*, vol. 34, no. 12, pp. 608S–617S, December, 1955.

31. Bijlaard, P. P., "Computation of the Stresses from Local Loads in Spherical Pressure Vessels or Pressure Vessel Heads," *Welding Res. Council Bull.* 34, March, 1957.

32. Bijlaard, P. P., "Local Stresses in Spherical Shells from Radial or Moment Loading," *Welding J.*, vol. 36, no 5, pp. 240S–43S, May, 1957.

33. Bijlaard, P. P., "Stresses in Spherical Vessel from Radial Loads Acting on a Pipe," *Welding Res. Council Bull.* 49, pp. 1–30, April, 1959.

34. Bijlaard, P. P., "Stresses in a Spherical Vessel from External Moments Acting on a Pipe," *Welding Res. Council Bull.* 49, pp. 31–62, April, 1959.

35. Bijlaard, P. P., "Influence of a Reinforcing Pad on the Stresses in a Spherical Vessel Under Local Loading," *Welding Res. Council Bull.* 49, pp. 63–73, April, 1959.

36. Bijlaard, P. P., "Stresses in Spherical Vessels from Local Loads Transferred by a Pipe," *Welding Res. Council Bull.* 50, pp. 1–9, May, 1959.

37. Bijlaard, P. P., "Additional Data on Stresses in Cylindrical Shells under Local Loading," *Welding Res. Council Bull.* 50, pp. 10–50, May, 1959.

38. Bijlaard, P. P., and E. T. Cranch, "Interpretive Commentary on the Application of Theory to Experimental Results for Stresses and Deflections Due to Local Loads on Cylindrical Shells," *Welding Res. Council Bull.* 60, pp. 1–2, May, 1960.

39. Blanjean, L., "Study of the Behavior of a General Pipe Line (Planar or in Space) under Constrained Thermal Expansion," Maison Desoer, Liége, Belgium, 1955 (in French).

40. Blanjean, L., "Summary of the Studies of the Research Committee on Steam Piping in the Field of Analytical Calculation," *Rev. Tijdscher. Mecan. Werk.* vol. 4, no. f, pp. 245–249, 1958 (in French).

41. Blanjean, L., and E. Dony, "Matrix Method for the Analysis of Piping Subject to Restrained Thermal Expansion," *Acier-Stahl-Steel*, vol. 22, no. 9, pp. 365–376, September, 1957; no. 10, pp. 413–419, October, 1957; no. 11, pp. 463–472, November, 1957, no. 12, pp. 515–522, December, 1957.

42. Blaw-Knox Company, "Design of Piping for Flexibility with Flex-Anal Charts," Blaw-Knox Company, Power Piping Division, 5th ed., revised, Pittsburgh, 1947.

43. Blick, R. G., "Pipe Flexibility Calculations," *Petrol. Refiner*, vol. 33, no. 2, 122–130, February, 1954.

44. Bodewig, E., "Matrix Calculus," 2d ed., Interscience Publishers, Inc., New York, 1959.

45. Boley, B. A., and J. H. Wiener, "Theory of Thermal Stresses," John Wiley & Sons, Inc., New York, 1960.

46. Borg, S. F., "Matrix-tensor Methods in Continuum Mechanics," D. Van Nostrand Company, Inc., Princeton, N.J., 1963.
47. Bowden, A. T., and B. E. Drumm, "Design and Testing of Large Gas Ducts," *Proc. Inst. Mech. Engrs.*, vol. 174, pp. 119–157 (including discussion), 1960.
48. Bowen, J. B., E. H. Fredrickson, and J. E. Soehrens, "Flexibility Analyses of Power Plant Piping," Braun Computer Programs, C. F. Braun and Co., Alhambra, Calif., July, 1959.
49. Brand, L., "Vector and Tensor Analysis," John Wiley & Sons, Inc., New York, 1947.
50. Branin, Jr., F. H., "Kron's Method of Tearing and Its Applications," *IBM Tech. Note* TN 009,000.260, Mar. 11, 1958.
51. Branin, F. H., "Machine Analysis of Networks and Its Applications," *IBM Tech. Publ.* TR 00.855, Mar. 30, 1962 (consisting of four papers).
52. Branin, F. H., "Computer Analysis of Networks and Its Applications," to be published by Prentice-Hall, Inc., Englewood Cliffs, N.J.
53. Bridge, T. E., "How to Design Piping with Desired Flexibility," *Heating, Piping, Air Conditioning*, vol. 22, no. 10, p. 94, October, 1950; no. 11, p. 94, November, 1950; no. 12; p. 92, December 1950; vol. 23, no. 1, p. 136, January, 1951; no. 2, p. 107, February, 1951.
54. Bridge, T. E., "New Formulas Aid Computer Solution of Piping Design Stress Problems," *Heating, Piping, Air Conditioning*, vol. 32, no. 1, pp. 196–198, January, 1960; correction, no. 4, p. 106, April, 1960.
55. Bridge, T. E., "How to Write, Use a Computer Program for Calculating Pipe Stresses," *Heating, Piping, Air Conditioning*, vol. 32, no. 3, pp. 125–128, March, 1960; discussion and closure, no. 6, pp. 110–114, June, 1960.
56. Brillouin, L., "On Thermal Dependence of Elasticity in Solids," *Phys. Rev.*, vol. 54, no. 11, p. 916, Dec. 1, 1938.
57. Brock, J. E., "Matrix Analysis of Flexible Filaments," *Proc. First U.S. Natl. Congr. Appl. Mech.*, ASME, 1952, pp. 285–290.
58. Brock, J. E., "A Matrix Method for Flexibility Analysis of Piping Systems," *J. Appl. Mech.*, vol. 19, no. 4, pp. 501–516, December, 1952.
59. Brock, J. E., "Flexibility of Piping Systems Supported by Equally Spaced Rigid Hangers," *J. Appl.. Mech.*, vol. 21, no. 1, pp. 11–18, March, 1954.
60. Brock, J. E., "Try These Rules for Pipe Bend Design," *Heating, Piping, Air Conditioning*, vol. 27, no. 4, p. 99, 1955.
61. Brock, J. E., "Matrix Analysis of Piping Flexibility," *J. Appl. Mech.*, vol. 22, no. 3, pp. 361–362, September, 1955.
62. Brock, J. E., "How Giant Brains Aid Piping Design," *Heating, Piping, Air Conditioning*, vol. 29, no. 5, pp. 106–108, May, 1957; no. 6, pp. 106–109, June, 1957; no. 7, pp. 116–119, July, 1957.
63. Brock, J. E., "Some Formulas for Piping Design," *J. Am. Soc. Naval Engrs.*, vol. 73, no. 2, 395–397, May, 1961.
64. Brock, J. E., "Shear Distribution in Piping," *Heating, Piping, Air Conditioning*, vol. 35, no. 1, pp. 141–143, January, 1963.
65. Brock, J. E., "Some Secondary Effects in a Simple Piping Structure under Heating," to be published in *J. Appl. Mech.*
66. Carter, G. K., "Numerical and Network Analyzer Solution of the Equivalent Circuits for the Elastic Field," *J. Appl. Mech.*, vol. 11, no. 3, pp. 162–167, September, 1944.
67. Carter, G. K., and G. Kron, "Network Analyzer Solution of the Equivalent Circuits for Elastic Structures," *J. Franklin Inst.*, vol. 238, no. 6, pp. 443–452, December, 1944.
68. Change, C. C., and W. H. Chu, "Stresses in a Metal Tube under Both High Radial Temperature Variation and Internal Pressure," *J. Appl. Mech.*, vol. 21, no. 2, 101–108, June, 1954.
69. Chapman, W. P., "Are Thermal Stresses a Problem in Snow Melting Systems?" *Heating, Piping, Air Conditioning*, vol. 27, no. 6, pp. 92–95, June, 1955.
70. Chen, L. H., "Piping Flexibility Analysis by Stiffness Matrix," *J. Appl. Mech.*, vol. 26, no. 4, pp. 608–612, December, 1959.
71. Chen, L. H., and T. F. Myles, "Generalized Loop Forces and Displacement in Redundant Structures," *Proc. Fourth U.S. Natl. Congr. Appl. Mech.*, to be published by ASME.

72. Chernika, V. S., ";Stress Analysis of an Expansion Bend under a Flexural Load," *Inzh. Sbornik, Akad. Nauk USSR*, vol. 22, pp. 133–149, 1955 (in Russian).
73. Clark, R. A., "On the Theory of Thin Elastic Toroidal Shells," *J. Math. Phys.*, vol. XXIX, no. 3, pp. 146–178, October, 1950.
74. Clark, R. A., and E. Reissner, "Bending of Curved Tubes," *Advan. Appl. Mech.*, vol. II, Academic Press, Inc., New York, 1951.
75. Clough, R. W., "Matrix Analysis of Beams," *Proc. ASCE*, vol. 84, EM1 (*J. Eng. Mech.*), 24 pp., January, 1958.
76. Coffin, Jr., L. F., "A Study of the Effects of Cyclic Thermal Stresses on Ductile Metal," *Trans. ASME*, vol. 76, no. 6, pp. 931–950, August, 1954.
77. Coffin, Jr., L. F., "Design Aspects of High-temperature Fatigue with Particular Reference to Thermal Stresses," *Trans. ASME*, vol. 78, no. 3, pp. 527–532, April, 1956.
78. Coffin, Jr., L. F., "An Investigation of Thermal Stress Fatigue as Related to High Temperature—Piping Flexibility," *Trans. ASME*, vol. 79, no. 7, pp. 1637–1651, October, 1957.
79. Coldham, V., "Programming for Electric Computation of Stresses in Piping Systems," *J. Mech. Eng. Sci.*, vol. 1, no. 2, pp. 93–102, September, 1959.
80. Crandall, S. H., and N. C. Dahl, "The Influence of Pressure on the Bending of Curved Tubes," *Proc. Ninth Intern. Congr. Appl. Mech.*, Brussels, vol. VI, pp. 101–111, 1957.
81. Crawford, L., "Piping under Dynamic Loading," *J. Am. Soc. Naval Engrs.*, vol. 68, no. 2, pp. 345–370, May, 1956.
82. Crawford, L., "Pipe Stress Analysis for Static and Dynamic Loading," *Trans. Soc. Naval Arch. Marine Engrs.*, vol. 65, 1957.
83. Critchlow, W. J., and G. W. Haggenmacher, "The Analysis of Redundant Structures by the Use of High Speed Digital Computers," *J. Aerospace Sci.*, vol. 27, no. 8, pp. 595–606, 614, August, 1960.
84. Crout, P. D., "A Short Method of Evaluating Determinants and Solving Systems of Linear Equations with Real or Complex Coefficients," *Trans. AIEE*, vol. 60, pp. 1235–1240, 1941.
85. Dahl, N. C., "Toroidal-shell Expansion Joints," *J. Appl. Mech.*, vol. 20, no. 4, pp. 497–503, December, 1953.
86. Deagle, L., "Pressure and Thermal Stresses in a Pipe Attachment to a Sphere," KAPL-2109, General Electric Co., Sept. 21, 1959.
87. Den Hartog, J. P., "Advanced Strength of Materials," McGraw-Hill Book Company, New York, 1952.
88. Denke, P. H., "A Matric Method of Structural Analysis," *Proc. Second U.S. Natl. Congr. Appl. Mech.*, June, 1954, pp. 445–451.
89. Denke, P. H., "The Matric Solution of Certain Nonlinear Problems in Structural Analysis," *J. Aeron. Sci.*, vol. 23, no. 3, pp. 231–236, March, 1956.
90. Donahue, J. E., "Analysis of Pipe Systems with Special Expansion Features," *ASME Paper* 54-SA-70, not published, available, at Engineering Society Library.
91. Donnell, L. H., "The Flexibility of Corrugated Pipes under Longitudinal Forces and Bending," *Trans. ASME* (*J. Appl. Mech.*), vol. 54, no. 11, pp. 69–75, June 15, 1932.
92. Dopp, "Bill," "How *Not* to Anchor U Pipe bends," *Heating, Piping, Air Conditioning*, vol. 27, no. 3, pp. 136–139, March, 1955.
93. Dopp, "Bill," "Fixed Pipe Hanger Shifts Load," *Heating, Piping, Air Conditioning*, vol. 27, no. 6, pp. 114–117, June, 1955.
94. Dopp, "Bill," "Providing Flexibility in Piping," *Heating, Piping, Air Conditioning*, vol. 27, no. 11, pp. 128–131, November, 1955.
95. Dubinskii, S. L., E. I. Rusanova, and B. F. Stoll, "Calculation of Toroidal Expansion Joints for Low Pressure Piping," *Sudostroenie*, vol. 5, pp. 14–16, 1956 (in Russian).
96. Dwyer, P. S., "Linear Computations," John Wiley & Sons, Inc., New York, 1951.
97. Eisemann, K., L. Woo, and S. Namyet, "Space Frame Analysis by Matrices and Computer," *ASCE Preprint* 3365 ST6, December, 1962. To be published in *Proc. ASCE*.
98. Ellis, H. B., N. Schmidt, and E. R. Watkins, "Computer Program KO's Stress Problems," *Heating, Piping, Air Conditioning*, vol. 31, no. 12, pp. 87–91, December, 1959.

99. Endres, W., "Thermal Stresses in Pipelines," *Forsch. Gebiete Ingenieurw.*, vol. 23, no. 1/2, pp. 33–37, 1957 (in German).
100. Expansion Joint Manufacturers Association, "Standards of the EJMA," EJMA, 53 Park Place, New York 7, 1958.
101. Faddeeva, V. N., "Computational Methods of Linear Algebra," translated by C. D. Benster, Dover Publications, Inc., New York, 1959.
102. Feely, Jr., F. J., and W. M. Goryl, "Stress Studies on Piping Expansion Bellows," *J. Appl. Mech.*, vol. 17, no. 2, pp. 135–141, June, 1950.
103. Fenves, S. J., and F. H. Branin, "A Network Topological Formulation of Structural Analysis," to be published by ASCE.
104. Ferrarese, G., "Finite Deformation of a Tubular Solid with Curved Generator," *Accad. Nazl. Lincei (Rome), Rend., Cl. Sc. fis. Mat. nat.*, vol. 27, no. 6, pp. 347–355, December, 1959 (in Italian).
105. Filho, F. V., "Matrix Analysis of Plane Rigid Frames," *Proc. ASCE*, vol. 86, no. ST 7 (*J. Struct. Div.*), pp. 95–108, July, 1960.
106. Finnie, I., "Stress Analysis in the Presence of Creep," *Appl. Mech. Rev.*, vol. 13, no. 10, pp. 705–712, October, 1960.
107. Finnie, I., and W. R. Heller, "Creep of Engineering Materials," McGraw-Hill Book Company, New York, 1959.
108. Flieder, W. G., J. C. Loria, and W. J. Smith, "Bowing of Cryogenic Pipelines," *Trans. ASME*, vol. 83 E (*J. Appl. Mech.*) no. 3, pp. 409–416, September, 1961.
109. Forsythe, G. E., "Solving Linear Algebraic Equations Can be Interesting," *Bull. Am. Math. Soc.*, vol. 59, no. 4, pp. 299–329, July, 1953.
110. Frame, J. S., "A Continuous Fraction Solution of the Karman Tube Bending Problem," *Öst. Ing.-Arch.*, vol. 12, no. 1/2, pp. 95–101, November, 1958 (in German).
111. Frazer, R. A., W. J. Duncan, and A. R. Collar, "Elementary Matrices," Cambridge University Press and The Macmillan Company, New York, 1947.
112. Friedrich, C. M., "KOAD-1-A Digital Computer Program to Calculate Stresses and Deflections in Linear Elastic Structures under Thermal Distortion, Pressure, and Applied Loads," Bettis WAPD-TM-243 (Contract AT-11-GEN-14), Westinghouse Electric Corp., March, 1961.
113. Frisch-Fay, R., "Flexible Bars," Butterworths, Washington, 1962.
114. Furman, D. E., "Thermal Expansion Characteristics of Stainless Steels between −300 F and 1000 F," *J. Metals*, vol. 188, no. 4, pp. 688–691, April, 1950.
115. Gascoyne, J., "Analysis of Pipe Structures for Flexibility," John Wiley & Sons, Inc. New York, 1959.
116. Gatewood, B. E., "Thermal Stresses," McGraw-Hill Book Company, New York, 1957.
117. General Electric Company, "Steam Piping Systems Connected to Turbines," GET-1911C, Schenactady, N.Y., April, 1954.
118. Gerstle, K. H., "Flow Graphs in Structural Analysis," *Proc. ASCE*, vol. 86, no. ST.10 (*J. Struct Div.*), pp. 125–137, October, 1960.
119. Gheorghin, A., "Conjugate Systems and Their Utilization in the Preparation of Influence Lines," *Ann. Ponts Chaus.*, vol. 128, no. 6, pp. 807–861, November–December, 1958 (in French).
120. Ghosh Hazra, S. K., "Flexibility Analysis of Piping," *J. Inst. Engrs. India*, vol. 40, no. 1, p. 3, November, 1959.
121. Ghosh Hazra, S. K., "Design of Piping Systems for High Pressure and High Temperature," *J. Inst. Engrs. India*, vol. 40, no. 2, 6, pp. 219–230, February, 1960.
122. Glodkowski, R., "Determination of Piping Flexibility by Matrix Calculus," *Chaleur Ind.*, vol. 42, no. 430, pp. 135–152, May, 1961 (in French).
123. Goethals, J., "Analogical Determination of Thermal Expansion in Piping," *Rev. Soc. Roy. Belge Ingrs. Industriels*, no. 3, Brussels, 1958.
124. Granville, W. A., P. F. Smith, and W. R. Longley, "Elements of Calculus," Ginn and Company, Boston, 1946.
125. Gregory, W. P., and W. C. White, "The Installation and Use of Pipe Guides," *Heating, Piping, Air Conditioning*, vol. 33, no. 9, pp. 133–136, September, 1961.
126. Grinnell Company, Inc., "Piping Design and Engineering," Grinnell Company, Inc., Providence, R. I., 1951.

127. Gross, N., "Experiments on Short-radius Pipe-bends," *Proc. Inst. Mech. Engrs. (B)*, vol. 1B, no. 10, pp. 465–479, 1952–1953.
128. Gross, N., and H. Ford, "The Flexibility of Short-radius Pipe Bends," *Proc. Inst. Mech. Engrs. (B)*, vol. 1B, no. 10, pp. 480–491, 1952–1953.
129. Grzedzielski, A. Z. M., "Note on Some Applications of the Matric Force Method of Structural Analysis," *J. Roy Aeron. Soc.*, vol. 64, no. 594, pp. 354–357 (tech. notes), June, 1960.
130. Hall, A. S., and R. W. Woodhead, "Frame Analysis," John Wiley & Sons, Inc., New York, 1961.
131. Hanson, K. L., and W. E. Jahsman, "An Evaluation of Piping Analysis Methods," KAPL-1384, AEC R AND D Report Engineering, TID-4500, 10th ed., Office of Technical Services, U.S. Dept. of Commerce, Washington 25, D.C.
132. Haringx, J. A., "Stability of Bellows Subjected to Internal Pressure," *Phillips Res. Rept.*, vol. 7, no. 3, pp. 189–196, June, 1952.
133. Higgins, T. J., "A Classified Selected Bibliography of the Tensor Theory of Electrical Networks, Machines and Systems (and of Associated Generalized Techniques)," *Matrix Tensor Quart.*, vol. 12, no. 2, pp. 50–60, December, 1961.
134. Hilton, H. H., "Thermal Stresses in Bodies Exhibiting Temperature-dependent Elastic Properties," *J. Appl. Mech.*, vol. 19, no. 3, pp. 350–354, September, 1952.
135. Hoath, P. T., "Steam Pipework Design in Ships," *Trans. Inst. Marine Engrs.*, vol. 59, p. 10, October, 1947.
136. Householder, A. S., "Principles of Numerical Analysis," McGraw-Hill Book Company, New York, 1953.
137. Hsiao, K. H., "Time Saving Procedure for Piping Stress Analysis," *Air Conditioning, Heating, Ventilating*, vol. 56, no. 6, pp. 81–96, June, 1959; no. 9, pp. 94–97, September, 1959.
138. Hsiao, K. H., "Calculating Pipe Stress Efficiently," *Heating, Piping, Air Conditioning*, vol. 29, no. 9, pp. 98–101, September, 1957; no. 12, pp. 114–117, December, 1957; vol. 30, no. 1, pp. 172–175, January, 1958; no. 9, pp. 106–109, September, 1959.
139. Hsiao, K. H., "Calculating Contraction Stresses in Missile Piping," *Heating, Piping, Air Conditioning*, vol. 31, no. 10, p. 122, October, 1959.
140. Hsiao, K. H., "Calculating Pipe Stresses Efficiently—a Further Simplification," *Heating, Piping, Air Conditioning*, vol. 34, no. 1, pp. 194–200, January, 1962.
141. Hunt, P. M., "The Electronic Digital Computer in Aircraft Structural Analysis," *Aircraft Eng.*, vol. 28, no. 325, pp. 70–76, March, 1956; no. 326, pp. 111–118, April, 1956; no. 327, pp. 155–164, May, 1956.
142. Jackson, R. L., and L. H. Johnson, "Designing Steam Piping System for Large Central Station Applications," *Heating, Piping, Air Conditioning*, vol. 26, no. 11, pp. 112–115, November, 1954.
143. Jacquemin, R., "Method of Calculating Stresses and Displacements in Pipelines under Thermal Expansion," *Rev. Tijdschr. Mecan. Werk.*, vol. 4, no. 4, pp. 250–258, 1958 (in French).
144. Jennings, F. B., "Theories on Bourdon Tubes," *Trans. ASME*, vol. 78, no. 1, pp. 55–64, January, 1956.
145. Johnson, L. H., "The Solution of Pipe Expansion Problems by Punched-card Machines," *ASME Paper* 53-F-23, not published, available at Engineering Society Library.
146. Johnson, L. H., Memoranda dated Feb. 25, 1953, Feb. 25, 1953, and Aug. 7, 1953, and several undated memoranda distributed to personnel in General Electric Company Large Steam Turbine Engineering Department, on several aspects of punched-card solution of piping flexibility problems.
147. Johnson, L. H., "How to Solve Pipe Expansion Problems by Punched Card Machines," *Heating, Piping, Air Conditioning*, vol. 28, no. 11, pp. 122–124, November, 1956.
148. Juergensonn, H. V., "Elastizitaet u. Festigkeit im Rohrleitungsbau" (Elasticity and strength in piping construction), 2d ed., Springer-Verlag, OHG, Berlin, 1953 (in German).
149. Kafka, P. G., and M. B. Dunn, "Stiffness of Curved Circular Tubes with Internal Pressure," 55-A-32, 8 pp., *J. Appl. Mech.*, vol. 23, no. 1, pp. 247–254, June, 1956.

150. Kellogg, M. W., Company, "Design of Piping Systems," 2d ed., John Wiley & Sons, Inc., New York, 1956.
151. Kellogg, M. W., Company, "A Matrix Method of Piping Analysis and the Use of Digital Computers," M. W. Kellogg Company, New York, 1961.
152. Keough, L. M., "Piping Flexibility Analysis LO3812" (PA 713), SHARE program by General Electric Co., June, 1959.
153. Kerkof, W. P., "Stresses for Pressure Vessels and Boilers up to 650 F," *Welding J.*, vol. 33, no. 5, pp. 239S–251S, May, 1954.
154. Ketchum, E. R., "Calculating Forces Due to a "Snake" of Piping," *Heating, Piping, Air Conditioning*, vol. 27, no. 2, pp. 108–110, February, 1955.
155. Kjeldgaard, P. D., "Matrix Analysis of Elastic Structures," M.S. Thesis, U.S. Naval Postgraduate School, Monterey, Calif. June, 1959.
156. Klein, B., "Application of the Witte Rearranging Method to a Typical Structural Matrix," *J. Aeron. Sci.*, vol. 25, no. 5, pp. 342–343, May, 1958.
157. Klein, B., "A simple Method of Matrix Structural Analysis," Parts I-V, *J. Aeron. Sci.*, vol. 24, no. 1, pp. 39–46, January, 1957; no. 11, pp. 813–820, November, 1957; vol. 25, no. 6, pp. 385–394, June, 1958; vol. 26, no. 6, pp. 351–359, June, 1958; vol. 27, no. 11, pp. 859–866, November, 1960.
158. Kloth, W., "Atlas der Spannungsfelder in technischen Bauteilen" (Catalog of stress distributions in elements used in technical construction), Verlag Stahleisen mbH, Düsseldorf, 1961 (in German).
159. Kolflat, T., and K. E. Knapp, "Using Digital Computers to Calculate Piping Flexibility," *Heating, Piping, Air Conditioning*, vol. 29, no. 8, pp. 106–108, August, 1957; no. 11, pp. 124–127, November, 1957.
160. Kovacs, A., "Mechanics and Analysis of Articulated Pipe Lines," Parts 1–3, *Chaleur Ind.* vol. 40, no. 411, pp. 301–316, October, 1959; no. 412, pp. 347–361, November, 1959; no, 413, pp. 381–400, December, 1959.
161. Kron, G., "Equivalent Circuits of the Elastic Field," *J. Appl. Mech.*, vol. 11, no. 3, pp. 149–161, September, 1944.
162. Kron, G., "Tensorial Analysis and Equivalent Circuits of Elastic Structures," *J. Franklin Inst.*, vol. 238, no. 6, pp. 399–442, December, 1944.
163. Kron, G., "Solving Highly Complex Elastic Structures in Easy Stages," *J. Appl. Mech.*, vol. 22, no. 2, pp. 235–244, June, 1955.
164. Kron, G., "Tearing and Interconnecting as a Form of Transformation," *Quart. Appl. Math.*, vol. 13, no. 2, pp. 147–159, July, 1955.
165. Kron, G., "Solution of Complex Nonlinear Plastic Structures by the Method of Tearing," *J. Aeron. Sci.*, vol. 23, no. 6, pp. 557–562, June, 1956.
166. Kunin, N. F., and V. N. Kunin, "Influence of Stresses on the Thermal Expansion of Deformed Metal," *Fiz. Metal. i Metalloved.*, vol. 5, pp. 173–174, 1957.
167. Langefors, B., "Exact Reduction and Solution by Parts of Equation for Elastic Structures," *Tech. Note* 24, SAAB Airc. Co., Linköping, Sweden, 9 pp., 1953.
168. Langefors, B., "Algebraic Methods for the Numerical Analysis of Built-up Structures," *Tech. Note* **38**, SAAB Airc. Co., Linköping, Sweden, 55 pp., 1957.
169. Langefors, B., "Practical Aspects of Structural Analysis," *Z. Flugwiss.*, vol. 8, no. 6, pp. 161–168, June, 1960.
170. Langefors, B., "Theory of Aircraft Structural Analysis," *Z. Flugwiss,*. vol. 6, no. 10, pp. 281–291, October, 1958.
171. Langer, B. F., L. F. Coffin, Jr., R. U. Blaser, and W. E. Cooper, Proceedings of a panel discussion on thermal stresses, *Proc. Soc. Exptl. Stress Anal.*, vol. 15, no. 2, pp. 115–147, 1958.
172. Laupa, A., and N. A. Weil, "Analysis of U-shaped Expansion Joints," *J. Appl. Mech.*, vol. 29, no. 1, pp. 115–123, March, 1962; discussion and closure, vol 30, no. 1, pp. 153–155, March, 1963.
173. Lee, C. A., "Calculate Expansion Stress in Pipe Bends for Single Plane Structures," *Heating, Piping, Air Conditioning*, vol, 33, no. 7, pp. 109–112, July, 1961.
174. Livesley, R. K., "The Application of an Electronic Digital Computer to Some Problems of Structural Analysis," *Structural Engr.*, vol. 34, no. 1, pp. 1–12, January, 1956.
175. Livesley, R. K., "Analysis of Rigid Frames by an Electronic Digital Computer," *Engineering*, vol. 176, no. 4,569, pp. 230–233; no. 4,570, pp. 277–278, August, 1953.

176. Livesley, R. K., and T. M. Charlton, "Analysis of Rigid-jointed Plane Frameworks by Use of an Electronic Digital Computer," *Engineering*, vol. 177, no. 4,595, pp. 239–241, February, 1954.
177. Livesley, R. K., and T. M. Charlton, "The Use of a Digital Computer with Particular Reference to the Analysis of Structures," *N. E. Cat. Inst. Engrs. Shipb. Trans.*, vol. 71, part 2, pp. 67–89, December, 1954.
178. Lubkin, J. L., and Y. L. Luke, "Analysis of Piping Systems," *Midwest Res. Inst. Proj. 302-P-374 Final Rept.*, revised Aug. 1, 1956, Kansas City, Mo., 1956.
179. Marcal, P. V., and C. E. Turner, "Elastic Solution in the Limit Analysis of Shells of Revolution with Special Reference to Expansion Bellows," *J. Mech. Eng. Sci.*, vol. 3, no. 3, pp. 252–257, 1961.
180. Marcus, M., "Basic Theorems in Matrix Theory," *Natl. Bur. Std. Appl. Math. Ser.*, No. 57, Government Printing Office, Washington, D.C., 1960.
181. Markl, A. R. C., "Z-, L-, U-, and Expansion U-bends," *Piping Engr. paper* 4.02, Tube Turns Division of Chemetron Corp., December, 1950 (cf. Ref. 249).
182. Markl, A. R. C., "Piping-flexibility Analysis," *Trans. ASME*, vol. 77, no. 2, pp. 127–143 (discussion pp. 143–149), February, 1955.
183. Markl, A. R. C., "Fatigue Tests of Welding Elbows and Comparable Double-mitre Bends," *Trans. ASME*, vol. 69, no. 8, pp. 869–879, November, 1947.
184. Markl, A. R. C., "Fatigue Tests of Piping Components," *Trans. ASME*, vol. 74, no. 3, pp. 287–299 (discussion pp. 299–303), April, 1952.
185. Markl, A. R. C., and H. H. George, "Fatigue Tests on Flanged Assemblies," *Trans. ASME*, vol. 72, no. 1, pp. 77–87, January, 1950.
186. Matheson, J. A. L., et al., "Hyperstatic Structures," 2 vols, Academic Press, Inc., New York, 1959–1960.
187. McClintock, R. M., and H. P. Gibbons, "Mechanical Properties of Structural Materials at Low Temperatures" (a compilation from the literature), *Nat. Bur. Std. Monograph* 13, 180 pp., Washington, D.C., June, 1960.
188. McCutchan, A., "Design of Steam Transmission Piping," *Heating, Piping, Air Conditioning*, vol. 15, no. 8, pp. 401–404, August, 1943; no. 9, pp. 450–455, September, 1943.
189. McMinn, S. J., "Matrices for Structural Analysis," John Wiley & Sons, Inc., 1962.
190. Mehringer, F. J., and W. E. Cooper, "Experimental Determination of Stresses in the Vicinity of Pipe Appendages to a Cylindrical Shell," *Proc. Soc. Exptl. Stress Anal.*, vol. 14, no. 2, pp. 159–174, 1957.
191. Merckx, K. R., "Cyclic Operation of Pressure Piping with Gamma Heating," *Trans. ASME*, vol. 82D (*J. Basic Eng.*,) no. 2, pp. 447–452, June, 1960.
192. Michel, R., "Elastic Constants and Coefficients of Thermal Expansion of Piping Materials Proposed for 1954 Code for Pressure Piping," *Trans. ASME*, vol. 77, no. 2: 151–157 (discussion pp. 157–159), February, 1955.
193. von Mises, R. "Motor Calculus: a New Tool for Physicists," *Z. Angew. Math. Mech.*, vol. 4, no. 2, pp. 155–181, April, 1924 (in German).
194. von Mises, R. "Applications of Motor Calculus," *Z. Angew. Math. Mech.*, vol. 4, no. 3, pp. 193–213, June, 1924 (in German).
195. von Mises, R. "Remarks on Mr. Schleicher's Paper on Elastic Deformation of Curved Bars," *Z. Angew. Math. Mech.*, vol. 4, no. 6, pp. 486–487, December, 1924 (in German).
196. Mitchell, C. T., "Graphic Method for Determining Expansion Stresses in Pipe Lines," *Trans. ASME*, vol. 52, no. 12, pp. 167–176, May–August, 1930.
197. Morice, P. B., "Linear Structural Analysis," The Ronald Press Company, New York, 1959.
198. Nowinski, J., "Thermoelastic Problem for an Isotropic Sphere with Temperature Dependent Properties," *Z. Angew. Math. Phys.*, vol. 10, no. 6, pp. 565–567, November, 1959.
199. Olson, J. and Cramer, R., "Pipe Flexibility Analysis Using IBM 705 Computer," *MEC 21 Program Rept.* 277–59, revision 1, Design Division Engineering Computer Application Section, Mare Island Naval Shipyard, June, 1959.
200. Owens, R. H., "Flexibility Analysis of Piping Systems Formulated for Digital Computer Solution," *Proc. third U.S. Natl. Congr. Appl. Mech.*, June, 1958, pp. 419–430.

201. Owens, R. H., "An Elementary Development of Piping Flexibility Analysis with Illustrated Examples," *J. Am. Soc. Naval. Engrs.*, vol. 72, no. 1, pp. 159–170, February, 1960.

202. Palmer, P. J., "An Approximate Analysis Giving Design Data for Corrugated Pipes," *Proc. Inst. Mech. Engrs.*, vol. 174, pp. 635–641, 1960.

203. Parker, E. R., "Brittle Behavior of Engineering Structures," John Wiley & Sons, Inc., New York, 1957.

204. Parkes, E. W., "Stresses in Boiler Tubes, Routine Procedures for Desk Calculating Machine," *Engineering*, vol. 181, no. 4,695, pp. 84–87, January, 1956.

205. Partch, L. E., "Calculating Flexibility for Straight Line Piping Elements," *Heating, Piping, Air Conditioning*, vol. 25, no. 12, pp. 74–77, December, 1953.

206. Partch, L. E., "New Method Speeds Design of Pipeline Expansion Loops," *Heating, Piping, Air Conditioning*, vol. 31, no. 7, pp. 123–129, July, 1959.

207. Peck, L. G., R. F. Meyer, P. F. Strong, and H. Kalson, "The Automatic Calculation of Forces and Deflections in Piping Systems," *Trans. ASME*, vol. 80, no. 1, pp. 235–244, January, 1958.

208. Peiser, A. M., and S. Katz, "Pipe Stress Calculations," *Petrol. Refiner*, vol. 33, no. 2, pp. 153–155, February, 1954.

209. Pickrell, W. S., J. H. Rogers, and L. S. Woo, "Pipe Stress Program Specifications," *IBM Appl. Sci.* Library Routine Doc., 650, Los Angeles, Aug. 1, 1956.

210. Pipes, L. A., "Matrix Methods for Engineering," Prentice-Hall, Inc., Englewood Cliffs, N.J., 1963.

211. Pittsburgh Piping and Equipment Company, "Pittsburgh Piping Design Manual," Pittsburgh Piping and Equipment Company, Pittsburgh, 1949.

212. Plunkett, R., and M. E. Gurtin, "Use of Excess Constraints in Structural Analysis," *J. Mech. Eng. Sci.*, vol. 2, no. 2, pp. 101–104, June, 1960.

213. Poyner, J. F., "Dead-weight Stresses in Piping Systems and Their Inclusion in Thermal Stress Analysis," *Proc. Instn. Mech. Engrs.*, vol. 172, pp. 513–530, 1958.

214. Rabinovich, I. M., "Structural Mechanics in the USSR 1917–1957," (English translation edited by G. Herrmann), Pergamon Press, New York, 1960.

215. Randolph, L. F., "End Reactions and Stresses in Pipelines," *Engineering*, vol. 182, no. 4,728, pp. 492–496, October, 1956, no. 4,730, pp. 555–558, November, 1956.

216. Rauch, Jr., C. F., "A Method of Piping Flexibility Analysis by Deflection Measurements in Scale Models," M.S. Thesis, U.S. Naval Postgraduate School, June, 1957.

217. Reece, R. H., "Piping Flexibility Invades 'ivy' Walls," *Heating, Piping, Air Conditioning*, vol. 28, no. 7, pp. 102–104, July, 1956.

218. Reissner, E., and H. J. Weinischke, "Finite Pure Bending of Circular Cylindrical Tubes," *Quart. Appl. Math.*, vol. 20, no. 4, pp. 305–319, January, 1963.

219. Roberts, K. S., "Design Procedures for Slip-type Expansion Joints," *Air Conditioning, Heating, Ventilating*, vol. 56, no. 4, pp. 65–68, April, 1959.

220. Robinson, E. L., "Steam-piping Design to Minimize Creep Concentrations," *Trans. ASME*, vol. 77, no. 7, pp. 1,147–1,158 (discussion 1,158–1,162), October, 1955.

221. Rodabaugh, E. C., and H. H. George, "Effect of Internal Pressure on the Flexibility and Stress Intensification Factors of Curved Pipe or Welding Elbows," *Trans. ASME*, vol. 79, no. 4, pp. 939–948, May, 1957.

222. Rosenfeld, A. R., and B. L. Averbach, "Effect of Stress on the Expansion Coefficient," *J. Appl. Phys.*, vol. 27, no. 2, pp. 154–156, February, 1956.

223. Rossheim, D. B., and A. R. C. Markl, "The Significance of, and Suggested Limits for, the Stress in Pipe Lines Due to the Combined Effects of Pressure and Expansion," *Trans. ASME*, vol. 62, no. 5, pp. 443–464, July, 1940.

224. Roth, J. P., "An Application of Algebraic Topology: Kron's Method of Tearing," *Quart. Appl. Math.*, vol. 17, no. 2, pp. 1–24, April, 1959.

225. Roth, J. P., "The Validity of Kron's Method of Tearing," *Proc. Natl. Acad., Sci.*, vol. 41, no. 8, pp. 599–600, August, 1955.

226. Roussopsoulos, A., "Theory of Elastic Complexes and Elastic Connections," *Tech. Chronika*, vol. 35, no 403/4, pp. 1–28, January–February, 1958; no. 405/6, pp. 121–159, March–April, 1958; no. 407/8, pp. 249–265, May–June, 1958 (in Greek).

227. Rubey, R. J., "How to Determine Expansion in Carbon and Alloy Piping," *Heating, Piping, Air Conditioning*, vol. 27, no. 3, pp. 145–146, March, 1955.

228. Samans, W., and L. Blumberg, "Endurance Testing of Expansion Joints," *ASME Paper* 54-A-103, not published, available at Engineering Society Library.
229. Samoiloff, A., "Evaluation of Expansion-joint Behavior," *Power*, vol. 105, no. 1, pp. 57–59, January, 1961.
230. Schoene, O., and E. Schwenk, "Rohrleitungen in Neuzeitlichen Waermekraftanlagen" (Piping systems in modern power plants), Springer-Verlag OHG, Berlin, 1961 (in German).
231. Selig, F., "The Potential Concept in the Motor Method and Its Application to Thin Rods," *Z. Angew. Math. Mech.*, vol. 34, no. 8/9, pp. 327–328, August–September, 1954 (in German).
232. Sellers, F. T., "Designing Flexible Teflon Bellows," *Design News*, vol. 17, no. 3, pp. 4–9, Feb. 7, 1962.
233. Service Bureau Corporation, "Piping Flexibility Users Manual," Service Bureau Corporation, New York, 1960.
234. Sestini, G., "Thermomechanical Calculations of Piping Systems Articulated with Expansion Joints," *Termotecnica (Milan)*, vol. 9, no. 9, pp. 399–406, September, 1953 (in Italian).
235. Silbering, L., "Obliczanie Wytizymalosciowe Rurociagow" (Calculation of stresses in pipelines), Panstwowe Wydawnictwa Techniczne, 133 pp., Warszawa, 1959 (in Polish).
236. Sokolnikoff, I. S., "Mathematical Theory of Elasticity," 2d ed., McGraw-Hill Book Company, New York, 1956.
237. Sokolowski, M., "One-dimensional Thermoelastic Problems for Elastic Bodies with Material Constants Dependent on Temperature," *Bull. Acad. Polon. Sci.*, vol. 8, no. 4, pp. 153–160, 1960.
238. Soule, J. W., "The Solution of Multiple-branch Piping Flexibility Problems by Tensor Analysis," *J. Appl. Mech.*, vol. 23, no. 2, pp. 176–180, June, 1956.
239. Soule, J. W., "Tensor-flexibility Analysis of Pipe Supporting Systems," *J. Appl. Mech.*, vol. 23, no. 2, pp. 181–184, June, 1956.
240. Soule, J. W., "Tensor Flexibility Analysis of Closed-loop Piping Systems," *J. Appl. Mech.*, vol. 25, no. 1, pp. 11–16, March, 1958.
241. Spielvogel, S. W. "Piping Stress Calculations Simplified," published by author, Lake Success, N.Y., 1951, (Previous editions published by McGraw-Hill Book Company).
242. Spijkers, A. A., and B. H. Boonstra, "Piping Flexibility Analysis," *Ingenieur*, vol. 72, no. 22, pp. 93–97, June, 1961.
243. Stevens, P. G., V. J. Groth, and R. B. Bell, "Vessel Nozzles and Piping Flexibility Analysis," *Trans. ASME*, vol. 84B (*J. Eng. Ind.*), no. 2, pp. 225–236, May, 1962.
244. Swanson, S. A. V., and H. Ford, "Stresses in Thick Wall Plane Pipe Bends," *J. Mech. Eng. Sci.*, vol. 1, no. 2, pp. 102–112, September, 1959.
245. Timoshenko, S., and J. N. Goodier, "Theory of Elasticity," 2d ed., McGraw-Hill Book Company, New York, 1951.
246. Trostel, R., "Thermal Stresses in Tubes with Temperature-dependent Material Properties," *Ing.-Arch.*, vol. 26, no. 2, pp. 134–142, 1958 (in German).
247. Trostel, R., "Steady State Thermal Stresses with Temperature Dependent Material Properties," *Ing.-Arch.*, vol. 26, no. 6, pp. 416–434, 1958 (in German).
248. Truesdell, C., "The Mechanical Foundations of Elasticity and Fluid Dynamics," *J. Rational Mech. Analysis*, vol. 1, no. 1, pp. 125–171; no. 2, pp. 172–300, 1952.
249. Tube Turns, "Piping Engineering," a series of monographs prepared by the Tube Turns research staff and issued at irregular intervals, Tube Turns Division of Chemetron Corp., Louisville, Ky.
250. Tube Turns, "Bellows Expansion Joint Design Manual," Tube Turns Division of Chemetron Corp., Louisville, Ky., 1962.
251. Turner, C. E., "Stress and Deflection Studies of Flat Plate and Toroidal Expansion Bellows," *J. Mech. Eng. Sci.*, vol. 1, no. 2, pp. 130–143, September, 1959.
252. Turner, C. E., and H. Ford, "Examination of the Theories for Calculating the Stresses in Pipe Bends Subjected to In-plane Bending," *Proc. Inst. Mech. Engrs.*, vol. 171, no. 15, pp. 513–525, 1957.

253. Turner, C. E., and H. Ford, "Stress and Deflection Studies of Pipeline Expansion Bellows," *Proc. Inst. Mech. Engrs.*, vol. 171, no. 15, pp. 526–552, 1957.

254. Turner, M. J., E. H. Dill, H. C. Martin, and R. J. Melosh, "Large Deflections of Structures Subjected to Heating and External Loads," *J. Aerospace Sci.*, vol. 27, no. 2, pp. 97–106, 127, February, 1960.

255. United States Department of Commerce, Office of Technical Services, "Tentative Structural Design Basis for Reactor Pressure Vessels and Directly Associated Components" (pressurized, water cooled systems), PB151987, Washington, D.C., Dec. 1, 1958 (revised).

256. Utecht, E. A., "Stresses in Curved Circular, Thin-wall Tubes," *J. Appl. Mech.*, vol. 30, no. 1, p. 134, March, 1963.

257. Vigness, I., "Elastic Properties of Curved Tubes," *Trans. ASME*, vol. 65, no. 2, pp. 105–120, February, 1943.

258. Vinieratos, S. D., and D. R. Zeno, "Piping Flexibility and Stresses," Cornell Maritime Press, Cambridge, Md., 1941.

259. Vissat, P. L., and A. J. Del Buono, "In-plane Bending Properties of Welding Elbows," *Trans. ASME*, vol. 77, no. 2, pp. 161–175, February, 1955.

260. Wang, C. K., "Matric Formulation of the Slope-deflection," *Proc. ASCE*, vol. 84, no. ST6 (*J. Struct. Div.*), 19 pp., October, 1958.

261. Weinzweig, A. J., "The Kron Method of Tearing and the Dual Method of Identification," *Quart. Appl. Math.*, vol. 18, no., 2, pp. 183–190, July, 1960.

262. Wert, E. A., "Advanced Approach Simplifies Piping Flexibility Analysis," *Heating, Piping, Air Conditioning*, vol. 30, no. 9, pp. 130–133, September, 1958; no. 10, pp. 86–88, October, 1958.

263. Wert, E. A., S. Smith, and E. T. Cope, "A Manual for the Design of Piping for Flexibility by Use of Graphs," The Detroit Edison Company, Detroit, 1934.

264. Westinghouse Electric Corp., "Installation Instruction Leaflet for Type E Turbines," IL-110-75, Westinghouse Electric Corp., South Philadelphia, Pa., December, 1950.

265. Wolosewick, F. E., "Equipment Stresses Imposed by Piping," *Petrol. Refiner*, vol. 29, no. 8, pp. 89–91, August, 1950.

APPENDIX

Proposed Text of "Clauses Pertinent to Piping Design for Expansion and Flexibility" for use by the Mechanical Design Committee of the ASA B31 Committee on Code for Pressure Piping.

[The Mechanical Design Committee of ASA B31 Code for Pressure Piping has nominal cognizance over mechanical design aspects of the various ASA Code for Pressure Piping documents, this cognizance being exercised via recommendations which are transmitted through the Executive Committee to the various Sectional Committees which have direct cognizance over and responsibility for the Code documents themselves. The Mechanical Design Committee usually presents its recommendations in the form of what is known as the MDC Document, which consists of prepared clauses recommended for adoption by the various Sectional Committees (to the extent they may deem appropriate). The MDC Document is periodically revised and approved by the Mechanical Design Committee on the basis of efforts by the various subgroups of which it is composed. The "Clauses Pertinent to Piping Design for Expansion and Flexibility" which appear below were prepared by the MDC subgroup on Expansion and Flexibility, consisting of A. R. C. Markl, chairman, and N. Blair and J. E. Brock, members, the draft being put into final form by Markl and submitted under date of July 17, 1962, to the Mechanical Design Committee. The original form of these clauses is due to the Task Force Groups referred to in this chapter and comprised Chap. 3 of Sect. 6 of the 1955 edition of the ASA Code for Pressure Piping [(ASA B31.1-1955).]

The tables and figures referred to in these "Clauses" are not reproduced here, since they are identical with or almost identical with tables or figures appearing in the text of this chapter. These correspondences are as follows: Chart 002.3.2 is the same as Fig. 31 of this chapter; Table 019.3.1 is the same as Table 2 of this chapter; Table 019.3.2 is the same as Table 1 of this chapter; Table 019.3.6 includes all of Fig. 29 of this chapter and the same

information as is given in slightly different format, in Fig. 30 of this chapter.   There are also some inconsequential differences in notation.

The proposed text follows.

002   DESIGN CRITERIA

002.3.2   (*c*) Allowable Stress Range for Expansion Stresses: The expansion stress range S_E (see par. 019.4.5) shall not exceed the allowable stress range S_A given by:

$$S_A = f(1.25S_c + 0.25S_h) \qquad (1)$$

where S_c = basic[1] material allowable stress at minimum (cold) normal[2] operating temperature or at installation or shut-down temperature, whichever is lower (from Table 002.3.1)

S_h = basic[1] material allowable stress at maximum (hot) normal[2] operating temperature (from Table 002.3.1)

f = stress-range reduction factor for cyclic conditions[3] for total number N of full temperature cycles over total number of years during which system is expected to be in active operation (read or interpolate from Chart 002.3.2). By full temperature cycles is meant the number of cycles of temperature change from minimum to maximum temperature expected to be encountered.[4]

(*d*) Additive Stresses: The sum of the longitudinal stresses due to pressure, weight and other sustained external loadings shall not exceed S_h.

Where the sum of these stresses is less than S_h, the difference between S_h and this sum may be added to the term in parentheses in Equation (1).[5]

019   EXPANSION AND FLEXIBILITY

019.1   *General:* The following clauses define the objectives of piping flexibility analysis and alternate ways in which these can be realized.

[1] Joint efficiencies need not be applied.

[2] Normal maximum temperature means maximum temperature expected under controlled operating conditions and does not include unusual or unanticipated departures therefrom.   Similarly for normal minimum temperature.

[3] Applies to essentially non-corrosive services.   Corrosion can sharply decrease cyclic life.   Corrosion resistant materials should be used where a large number of major stress cycles is anticipated.

[4] If the range of temperature changes varies, i.e., if cycles of lesser stress amplitude are superimposed on the major cycles (usually those incident to start-up and shut down of the system), an equivalent number of major cycles may be computed from the equation

$$N = N_E + (S_1/S_E)^5 N_1 + (S_2/S_E)^5 N_2 + \cdots$$

where S_E and N_E, respectively, are the maximum computed expansion stress range and the corresponding cycles, and S_1, and N_1, S_2, and N_2, etc., are corresponding expansion stress ranges and cycles of lesser amplitude.

Where the variable conditions apply equally to the entire system, the ratios of lesser temperature variation ΔT_1, ΔT_2, etc., to the temperature range ΔT_E for which the flexibility calculations have been made may be substituted for the ratios of the lesser stress ranges S_1, S_2, etc., to the maximum stress range S_E.

Where temperature cycles for different parts of the system differ in either frequency or amplitude, and the number of equivalent cycles is expected to exceed 7,300 (the maximum for which $f = 1$ in Chart 002.3.2), separate calculations shall be made for each condition, and the above equation shall be applied to all points for which the stress range becomes a maximum in any of the calculations.

[5] The sum of the stresses may thus initially exceed the yield strength, but this will be relieved by plastic deformation with the result that the piping will not be repeatedly strained beyond the yield point strain.

019.1.1 OBJECTIVES: Piping systems shall be designed to have sufficient flexibility to prevent thermal expansion from causing

(1) failure of piping or anchors from overstress or overstrain,

(2) leakage at joints, or

(3) detrimental distortion of connected equipment (such as pumps, turbines, or valves) resulting from excessive thrusts and moments.

019.1.2 EXPANSION STRAINS may be taken up in two ways, either primarily by bending or torsion in which case only the extreme fibers at the critical location are stressed to the limit, or by compression and tension in which case the entire cross-sectional area over the entire length is substantially equally stressed.

(1) Bending or torsional flexibility may be provided by the use of bends, loops, or offsets; or by swivel-joints, corrugated pipe or expansion joints of the bellows type permitting rotational movement, suitably anchored, tied or otherwise connected to resist end forces from fluid pressure, frictional or other resistance to joint movement and other causes.

(2) Axial flexibility may be provided by expansion joints of the slip-joint or bellows types, suitably anchored, tied or otherwise connected to resist end forces from fluid pressure, frictional or other resistance to joint movement and other causes.

019.2 *Concepts* peculiar to piping flexibility analysis and requiring special consideration are explained in the following clauses.[1]

019.2.1 STRESS RANGE: As contrasted with stresses from sustained loads (such as internal pressure or weight), stresses caused by thermal expansion are permitted to attain sufficient initial magnitude to cause local yielding or creep. The attendant relaxation or reduction of stress in the hot condition leads to the creation of a stress of reverse sign when the component returns to the cold condition. This phenomenon is designated as self-springing of the line and is similar in effect to cold-springing. The amount of self-springing depends on the initial magnitude of the expansion stress, the material, the temperature, and the elapsed time. While the expansion stress in the hot condition tends to diminish with time, the arithmetic sum of the expansion stresses in the hot and cold conditions during any one cycle remains substantially constant. This sum, referred to as the stress range, is the determining factor in the thermal design of piping.

019.2.2 EXPANSION RANGE: In computing the stress range, the full thermal expansion range from the minimum to maximum metal temperature normally expected during operation and shut-down (see Footnote 2 to par. 002.3.2); shall be used whether the piping is coldsprung or not. Linear or angular movements of the equipment to which the piping is attached shall be included. For values of the unit thermal expansion range, refer to par. 019.3.1. Where substantial terminal (or anchor) movement may result from causes other than thermal expansion (e.g., tidal displacements of dockside piping or wind sway of piping attached to slender towers), such terminal movements may be considered analogous to those caused by thermal expansion (see also par. 019.5.7).

019.2.3 COLD SPRINGING is recognized as beneficial[2] in that it serves to balance hot and cold stresses without drawing on the ductility of the material, for which reason it is recommended in particular for materials of relatively low ductility, as also for systems where creep concentrations are considered likely to occur (see par. 019.2.4 and 019.5.2); it can also be used to limit departure from as-erected hanger settings and meet cold and hot load limits on connected equipment; it is not generally required because ductile systems of relatively uniform flexibility tend to adjust themselves without adverse effects to a more balanced stress level in service as a result of yielding or creep. Because of this, and because the life of a system under cyclic conditions depends primarily on the stress range rather than the stress level at any one time, no credit for cold spring is allowed for stress

[1] For an account of the development of the present Code rules and an explanation of the basic concepts and theoretical and experimental basis, see "Piping Flexibility Analysis" by A. R. C. Markl, *Trans. ASME*, vol. 77, no. 2, pp. 127–149, February, 1955.

[2] The intended benefits of cold spring may be vitiated by inadequate specifications and supervision of field erection procedures.

range calculations. In calculating end thrusts and moments where actual reactions at any one time rather than their range are considered significant, cold
spring is taken into account (see par. 019.4.6).

019.2.4 LOCAL OVERSTRAIN: All the commonly used methods of piping flexibility analysis
assume elastic behavior of the entire piping system. This assumption is sufficiently
accurate for systems where plastic straining occurs at many points or over relatively
wide regions, but fails to reflect the actual strain distribution in unbalanced systems
where only a small portion of the piping undergoes plastic strain, or where, in
piping operating in the creep range, the strain distribution is very uneven. In
these cases, the weaker or higher stressed portions will be subjected to strain
concentrations due to elastic follow-up of the stiffer or lower stressed portions.[1]
Unbalance can be produced:

(1) By use of small pipe runs in series with larger or stiffer pipe, with the small
    lines relatively highly stressed.

(2) By local reduction in size or cross-section, or local use of a weaker material.

(3) In a system of uniform size, by use of a line configuration for which most of
    the piping lies near the straight line drawn between the anchors or terminals,
    with only a very small portion projecting away from this line absorbing most
    of the expansion strain. Conditions of this type should preferably be avoided,
    particularly where materials of relatively low ductility are used: if unavoid-
    able, they should be mitigated by the judicious application of cold spring.

019.3 *Properties:* The following clauses deal with materials and geometric properties
of pipe and piping components and the manner in which they are to be used
in piping flexibility analysis.

019.3.1 UNIT THERMAL EXPANSION RANGE: The unit thermal expansion range (in./100 ft)
used in calculating the expansion range (par. 019.2.2) shall be determined from
Table 019.3.1 as the algebraic difference between the unit expansion values shown
for the maximum and minimum metal temperatures normally expected during
operation and shutdown. For materials not included in this table, reference
shall be made to authoritative source data, such as publications of the National
Bureau of Standards.

019.3.2 MODULI OF ELASTICITY: The cold and hot moduli of elasticity, E_c and E_h (psi),
respectively, shall be taken as the values shown for the minimum and maximum
normal-operating metal temperatures in Table 019.3.2 For materials not included in these tables, reference shall be made to authoritative source data, such
as publications of the National Bureau of Standards.

019.3.3 POISSON'S RATIO may be taken as 0.3 at all temperatures for all metals. However,
more accurate and authoritative data may be used if available.

019.3.4 ALLOWABLE STRESS RANGE: The allowable basic expansion stress range S_A and
permissible additive stresses shall be as specified in pars. 002.3.2(c) and (d) for
systems primarily stressed in bending or torsion.

019.3.5 DIMENSIONS: Nominal dimensions of pipe and fittings, and cross-sectional areas,
moments-of-inertia, and section-moduli based thereon, shall be used in flexibility
calculations.

019.3.6 FLEXIBILITY AND STRESS INTENSIFICATION FACTORS: Calculations shall take into
account stress intensification factors found to exist in components other than
plain straight pipe. Credit may be taken for the extra flexibility of such components. In the absence of more directly applicable data, the flexibility and
stress intensification factors shown in Table 019.3.6 may be used. For piping
components or attachments (such as valves, strainers, anchor rings or bands)
not covered in the table, suitable stress intensification factors may be assumed by
comparison of their significant geometry with that of the components shown.

019.4 *Analysis for Bending Flexibility:* The following clauses establish under what
circumstances and in what manner piping flexibility analyses are to be made
where the system primarily derives its flexibility from bending or torsional strains.

---

[1] For more information on this and related embrittling effects (from causes other than
low temperature or corrosion), refer to "Steam-piping Design to Minimize Creep Concentrations" by Ernest L. Robinson, *Trans. ASME*, pp. 1,147–1,158, October, 1955, and
"Stresses in Welded Pressure Vessels" by W. P. Kerkhof, *Welding J.* vol. 35, January, 1956.

019.4.1 FORMAL CALCULATIONS or model tests shall be required only where reasonable doubt exists as to the adequate flexibility of a system.

019.4.2 ADEQUATE FLEXIBILITY may generally be assumed to be available in systems which

    (1) are duplicates of successfully operating installations or replacements of systems with a satisfactory service record,

    (2) can be readily adjudged adequate by comparison with previously analyzed systems,

    (3) are of uniform size, have no more than two points of fixation and no intermediate restraints, are designed for essentially non-cyclic service (less than 7,300 total cycles), and satisfy the following approximate criterion:[1]

$$\frac{DY}{(L-U)^2} \leqslant 0.03$$

where D = nominal pipe size, in.

    Y = resultant of movements to be absorbed by pipe line, in.

    U = anchor distance (length of straight line joining anchors) ft

    L = developed length of line axis, ft

However, compliance with this formula does not assure that terminal reactions will be satisfactory.

019.4.3 METHODS OF ANALYSIS: Systems which do not meet the requirements of par. 019.4.2 shall be analyzed by a method appropriate to the hazard entailed by failure of the line, the importance of maintaining continuous service, the complexity of the layout, and the strain sensitivity of the pipe material. Simplified or approximate methods may be applied without correction only if they are used for the range of configurations for which their adequate accuracy has been demonstrated. Accompanying any flexibility calculation, there shall be an adequate statement of the method and any simplifying assumptions used.

019.4.4 STANDARD ASSUMPTIONS, specified in par. 019.3, shall be followed in all cases.[2] In calculating the flexibility of a piping system between anchor points, the system shall be treated as a whole. The significance of all parts of the line and of all restraints such as solid hangers or guides, including intermediate restraints introduced for the purpose of reducing moments and forces on equipment or small branch lines, as also the restraint introduced by support friction, shall be recognized. Not only the expansion of the line itself, but also linear and angular movements of the equipment to which it is attached shall be considered.[3]

019.4.5 STRESSES: Stresses from expansion shall be computed using the cold modulus of elasticity E_c and combined in accordance with Equation (24):

$$S_E = \sqrt{S_b{}^2 + 4S_t{}^2} \tag{24}$$

where S_E = computed expansion stress range, psi

$$S_b = \frac{\sqrt{(i_i M_i)^2 + (i_o M_o)^2}}{Z} = \text{equivalent bending stress}$$

$$S_t = M_t/2Z = \text{torsional stress}$$

---

[1] The user is cautioned that this criterion, while satisfactory for the bulk of applications, can err on the unsafe side for abnormal configurations such as near-straight saw-tooth configurations or U-bends with unequal legs where the spacing between the two legs is small in relation to their height.

[2] Where line size, thickness, material, or temperature vary within the system, flexibility calculations shall be based on the cold modulus E_c and moment of inertia I of the major (usually stiffest) part of the system, except that a flexibility-change factor $K' = E_h I / E_h' I'$ shall be applied to all members having a flexibility $1/E_h' I'$ (where E_h = hot elasticity modulus) differing from that of the major part in the same manner as flexibility factors K are applied to curved members (accordingly, the product of K and K' will apply to curved members).

[3] For a study of the effect of supports on line stresses, refer to "Flexibility of Piping Systems Supported by Equally Spaced Rigid Hangers" by J. E. Brock, *J. Appl. Mech., Trans. ASME*, vol. 76, pp. 11–18, March, 1954. For a method of including the flexibility of terminal equipment, refer to "Vessel Nozzles and Piping Flexibility Analysis" by P. G. Stevens, V. J. Groth and R. B. Bell, *Trans. ASME*, vol. 84, ser. B, no. 2, May, 1962.

M_i = bending moment in plane of member (for members having significant orientation, such as elbows or tees; for the latter the moments in the run and branch portions are to be considered separately), in.-lb

M_o = bending moment out of, or transverse to, plane of member, in.-lb

M_t = torsional moment, in.-lb

i_i = stress intensification factor under bending in plane of member (from Table 019.2.6)

i_o = stress intensification factor under bending out of, or transverse to, plane of member (from Table 019.2.6)

Z = section modulus, in.^{3}, of the significant tubular cross-section of the component or branch of the component involved; where the cross-section is non-uniform throughout the body, as in the case of manufactured welding tees, the section modulus of the matching pipe shall be used; where the component includes superimposed reinforcement, such as pads, saddles, or collars, the section modulus of header or branch pipe, respectively, shall be used without considering non-integral reinforcement; except that the effective section modulus used in calculating the bending stress S_b in the branch of any type of reducing-outlet branch intersection may be increased by a factor i_i or t_H/t_B, whichever is the lesser, where t_H and t_B, designate the significant thicknesses of header and branch pipe, respectively.

019.4.6 REACTIONS: In the design of anchors and restraints and in the evaluation of some mechanical effects of expansion on terminal equipment (such as pumps, turbines, boilers, heat exchangers etc.) either reaction ranges R (see definition below) or instantaneous values of reaction forces and moments in the hot and cold condition may be of significance. Determination of the latter may be complicated by plastic yielding, by creep, by the difficulty of performing the desired coldspring, and by other factors. Thus their determination may imply an elaborate engineering calculation the basis of which should be clearly set forth. In the absence of a better procedure, in the case of one-material uniform-temperature two-anchor systems without intermediate constraints,[1] the reaction ranges may be estimated from the formulas:[2]

$$R_h = (1 - \tfrac{2}{3}C)(E_h/E_c)R \tag{25}$$

$$R_0 = CR \text{ or } C_1R, \text{ whichever is greater} \tag{26}$$

where C = coldspring factor (assumed uniform in all directions) varying from zero for no coldspring to one for 100 per cent coldspring.

C_1 = $1 - (S_h/S_E)(E_c/E_h)$, estimated self-spring or relaxation factor; use zero if calculated value is negative.

E_h and E_c = modulus of elasticity of the piping material in the hot and cold conditions, respectively, psi (from Table 019.3.2).

R = reaction range, i.e., range of reaction forces or moments corresponding to the full expansion range (par. 019.2.2) based on E_c and as derived from the flexibility analysis, lb or in.-lb.

R_h and R_c = maximum estimated reaction forces or moments in the hot and cold conditions, respectively, lb or in.-lb.

S_h and S_c = tabulated stress values (S-values) for the piping material in the hot and cold conditions, respectively, psi.

S_E = maximum computed expansion stress range.

[1] Formulas (25) and (26) may be applied to more complicated configurations for one-material uniform-temperature systems if the calculated stress range S_E does not exceed the smaller of the two quantities
$$\frac{E_c}{E_h} - \frac{S_h}{1 - \tfrac{2}{3}C} \quad \text{and} \quad \frac{1.6S_c}{C}$$

[2] The factor of $\tfrac{2}{3}$ appearing in Equation (25) accounts for the observation that specified coldspring cannot be fully assured, even with elaborate precautions.

5

HANGERS AND SUPPORTS*

DESIGN ENGINEERING OF SUPPORT SYSTEMS

The correct and economical selection of the hangers for any piping system usually presents difficulties of varying degree, some relatively minor and others of a more critical nature. Their proper selection is the concern of all those connected with the design and installation of such systems and should be considered during all phases of construction.

Many pipe-support problems may be minimized or avoided if serious consideration is given to the means of support during the layout phases of design. The piping designer's familiarity with support problems, accepted practices, and commercially available pipe-support components and their application can be extremely important at this point. For example, other considerations being equal, pipelines should be routed to use the surrounding structure to advantage to provide logical points of support, anchorage, guidance, or restraint, with room available at such points for use of the proper component. Vertical banks of horizontal piping which operate at different temperatures should be avoided. Parallel lines, both vertical and horizontal, should be spaced sufficiently apart to allow room for independent pipe attachments for each line.

* Prepared by a Committee of the Manufacturers Standardization Society of the Valve and Fittings Industry. The Committee consisted of Walter A. Schlamp, Chairman (Fee and Mason Manufacturing Co., Inc.); Harold Erikson, Vice Chairman (Bergen-Paterson Pipesupport Corp.); George E. Crowell (The Grabler Manufacturing Co.); H. J. Marik (Automatic Sprinkler Corp. of America); Walter V. Bowen. Jr. (Carpenter and Paterson, Inc.); P. C. Sherburne (Grinnell Co., Inc.); Robert V. Warrick, Executive Secretary Manufacturers Standardization Society.

Pipe-hanger specifications for individual jobs must be written in such a way as to assure proper support and to provide for pitch, expansion, anchorage, and covering protection.   Familiarity with standard practices, customs of the trade, types and functions of commercial hangers (and an understanding of their individual advantages and limitations), together with references to existing specifications such as MSS SP-58, pertaining to pipe-hanger components, and SP-69, pertaining to their application, can be of great help in achieving the desired results.

Unless complete design details are provided by the consulting engineer or piping-system layout designer, the final responsibility for selection of pipe hangers, capable of completely satisfying the system requirements and job specifications, rests with the piping erector.   Obviously any piping system is inoperable until the pipe hangers and supports have been selected and installed.   Experience shows that a high percentage of pipe-support problems cannot be recognized during a cursory examination of the piping drawings but await a detailed analysis by those responsible for preparing specific material lists and details.   In the interest of properly coordinating hanger installations with the piping erection schedule, early priority should be given to the selection and procurement of pipe hangers.

The dollar value of the support system is generally outweighed many times by the value of pipe, valves, and fittings which are to be supported.   Failure to allow sufficient time for the procurement, design, and fabrication of the hangers can lead to costly erection delays and the use of temporary hangers.

Pipe hangers are generally identified by a specification type number (as in MSS SP-58) or a manufacturer's figure number, a more or less descriptive name, and size number.   The last depends upon the type of component; for example, pipe attachments, such as rolls, are prescribed by pipe size, upper attachments are identified by rod size, and spring hangers are sized by the calculated load to be supported.   In addition to the above, pipe-hanger components vary in their load-carrying capacity.   The load capacity of different types of hangers is given in most manufacturers' catalogues.   Various components and their functions will be discussed at more length in subsequent paragraphs.

General Design Considerations

Hangers and supports must be designed to meet all static and operational conditions to which the piping and equipment may be subjected.   The supporting systems must provide for and control, subject to the requirements of the piping configuration, the free or intended movement due to the thermal expansion and construction of the piping and connected equipment.   Proper design necessitates a thorough knowledge of the complete cyclic behavior of each section of the piping to be supported and a diligent awareness of the proximity of the line with respect to building structure, other piping systems, and equipment in the immediate vicinity.

A substantial reduction in the complexity of hanger design can be effected when piping-support requirements are coordinated properly with the plant and piping design phases.   Initial plant design should take cognizance that the hanger designer requires access to the piping, sufficient space in which to install the supporting equipment, and adequate structure to support from or frame into.   The hanger designer should be afforded the opportunity to offer his comments on the design from the piping-support viewpoint beginning with initial design phases.

Starting with any given set of piping and structural drawings, it is necessary for the hanger designer first to become familiar with the overall design concept and any special requirements that may be called for in the specifications.   When dealing with a number of piping systems in any area, it is advisable to superimpose the

piping (using a single-line representation) on the structural drawings.   A preliminary study very often reveals that consideration has been given to the supporting phase and provision has been made in the form of a pattern of steelwork, anchor bolts or inserts in concrete, or runs of piping purposely coordinated with suitable supporting structure.   Naturally, full advantage should be taken of such conditions. The piping should also be studied from the standpoint of possibly coordinating hangers for one system or pipe run with those of another, thereby attaining an orderly supporting arrangement.   On the other hand, it may also be necessary because of loading conditions purposely to stagger hangers in order to distribute the weight better on the supporting structure.   All this should be determined prior to the start of detailed hanger design for any one system.

A tentative overall pattern of hangers for any particular area having been established, it is advisable to begin specific hanger design with the most critical or largest piping systems, thereby reserving the best possible supporting conditions for the most important lines.

Of prime consideration in hanger design is the determination of hanger location. Although supports are located ideally to suit the requirements of the piping configuration, some degree of compromise may be required to take the fullest advantage of the available supporting structure.   The piping system or run should first be investigated as a whole.   With the use of the allowable hanger spacing as dictated by code, practice, or special calculation, the support points are located tentatively, taking into consideration division of straight runs, concentrated loads, elimination of excessive overhung sections or bends, and load reaction on terminal connections.

These locations should then be compared with the available supporting structure, modified as required and recorded on the superimposed piping which had been sketched previously on the structural drawings.   The recording of each hanger location, along with an indication of any supplementary steel required for the hanger, serves as a valuable aid in checking clearances and coordinating hangers during the course of design, especially when a number of designers are working on the same project.

A satisfactory pattern of support locations having been established, the next step is to determine the loading and movement conditions existing at each hanger point.   Here coverage will be confined to the general considerations only regarding loads and movements.   Recommended methods for obtaining or calculating specific loads and movements are covered later under "Calculation of Loads and Movements."

Loading Considerations

Regardless of the specific method used in determining the loads at each hanger point, the calculations must take into account all the following: the weight of the piping, valves, fittings, insulating materials, suspended hanger components, and all appurtenances along with the weight of normal operating contents.   This basic hanger load is used in determining the normal operating supporting effect required at each support point.

The actual design load used in determining required strength of the various components often exceeds the normal operating load because of one or a possible combination of the following conditions: hydrostatic test, special cleaning procedures, vertical seismic forces, external shock loadings, reaction forces, or possible specific overdesign as in the case of safety hangers.

Hydrostatic-test Loading.   The added weight of water used in the hydrostatic-testing procedure must be considered for lines which do not carry water during normal cyclic operation.   Hanger loadings for lines which normally carry water

remain unaffected by the hydrostatic test. For hangers which incorporate spring units, the springs would be calibrated to carry the normal operating load and therefore, means must be provided to accommodate the added weight of water. Spring supports are available with hydrostatic test stops which, in effect, transform the units into rigid supports. If desired, the added weight can also be accommodated by the addition of a number of strategically placed blocks or temporary supports. Some sections of the ASA Code for Pressure Piping make special provision for increased working stresses during hydrostatic test.

The added weight of a cleaning medium whose density might be greater than that of the normal contents of a line should also be taken into consideration.

Seismic Shock Loading. In the case of rigid supports, the design of hanger components must include the seismic overloading. Sections of piping carried on spring supports should be evaluated to determine if overloading can be tolerated, and, if not, then additional means such as limit stops or hydraulic snubbers may be required.

Overloading from Various Causes. Any other possible overloading condition must be anticipated and considered as part of the design loading. In addition to those treated separately above, some of the more common conditions which will serve as examples of others to be aware of are reaction forces from the operation of safety or relief valves, hydraulic shock as a result of water hammer, intentional restraint against normal thermal growth of a section of piping, constraint of expansion joints, or, as is quite often the case, overdesign of the rigid support on a long riser to serve as a safety support capable of carrying the entire weight of the riser.

Some sections of the ASA Code for Pressure Piping allow a short-time overloading of 20 per cent which can be applied against any form of short-time overloading. However, it is well to point out that, where there is a possibility of two or more causes to occur simultaneously, any portion of the combined overloading in excess of 20 per cent must be added to the design load.

Determination of Hanger Locations

Support locations are dependent on pipe size, piping configuration, the location of heavy valves and fittings, and the structure that is available for the support of the piping.

Table I. Recommended Maximum Pipe Support Spacing

(1,500-psi stress, $\frac{1}{10}$-in. deflection, water-filled pipe)

Nominal pipe size	1	1½	2	2½	3	3½	4	5	6	8	10	12	14	16	18	20	24
Span	7	9	10	11	12	13	14	16	17	19	22	23	25	27	28	30	32

No firm rules or limits exist which will positively fix the location of each support on a piping system. Instead, the engineer must exercise his own judgment in each case to determine the appropriate hanger location.

The suggested maximum spans between hangers listed in Table 1 reflect the practical considerations involved in determining support spacings on straight runs of standard wall pipe. They are normally used for the support spacings of critical systems.

The spans are based on a combined bending and shear stress of 1,500 psi when the pipe is filled with water and $\frac{1}{10}$-in. deflection is allowed between supports. They do not apply where concentrated weights such as valves or heavy fittings, or where changes in direction of the piping system, occur between hangers.

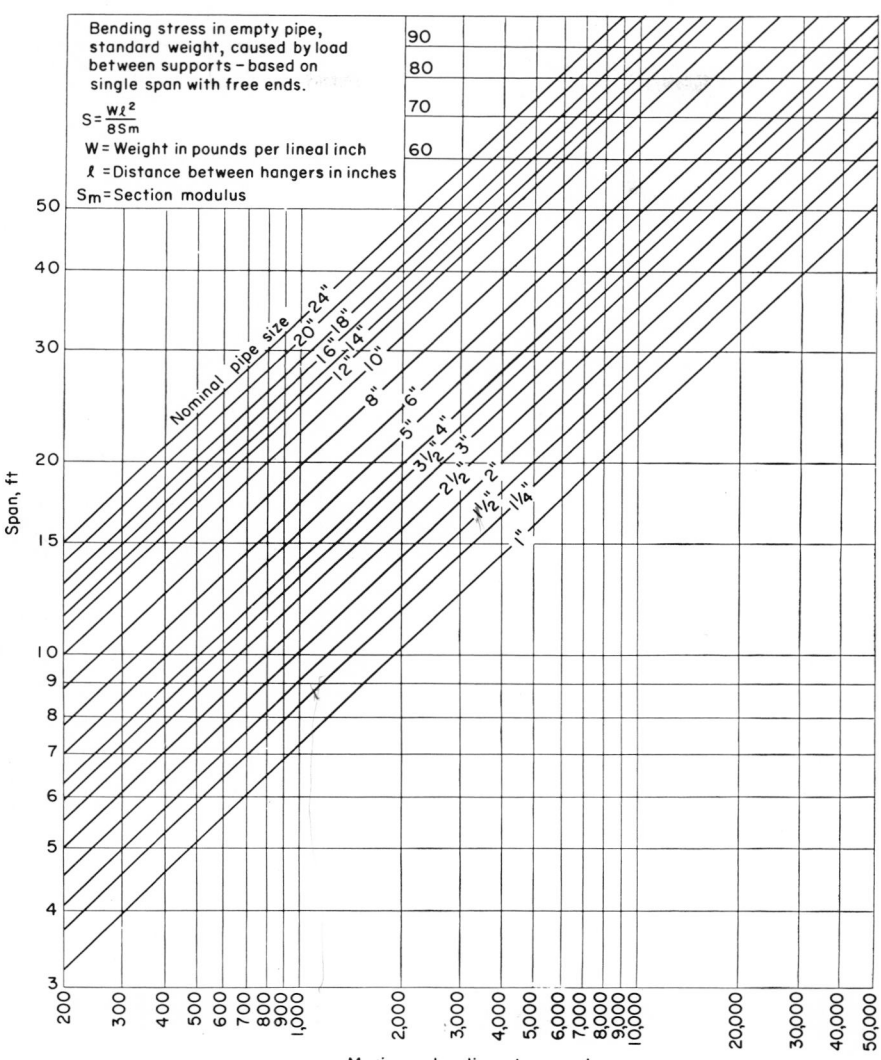

FIG. 1. Bending stress in empty pipe.

Supports should be placed as close as possible to concentrated loads in order to keep bending stresses to a minimum.   Where changes in direction of the piping of any critical system occur between hangers, it is good practice to keep the total length of pipe between the supports less than three-fourths the full spans listed in Table 1. When practical, a hanger should be located immediately adjacent to any change in direction of the piping.

For economy in the support of low-pressure, low-temperature systems and long outdoor transmission lines, hanger spans may be based on the allowable total stresses of the pipe and the amount of allowable deflection between supports.

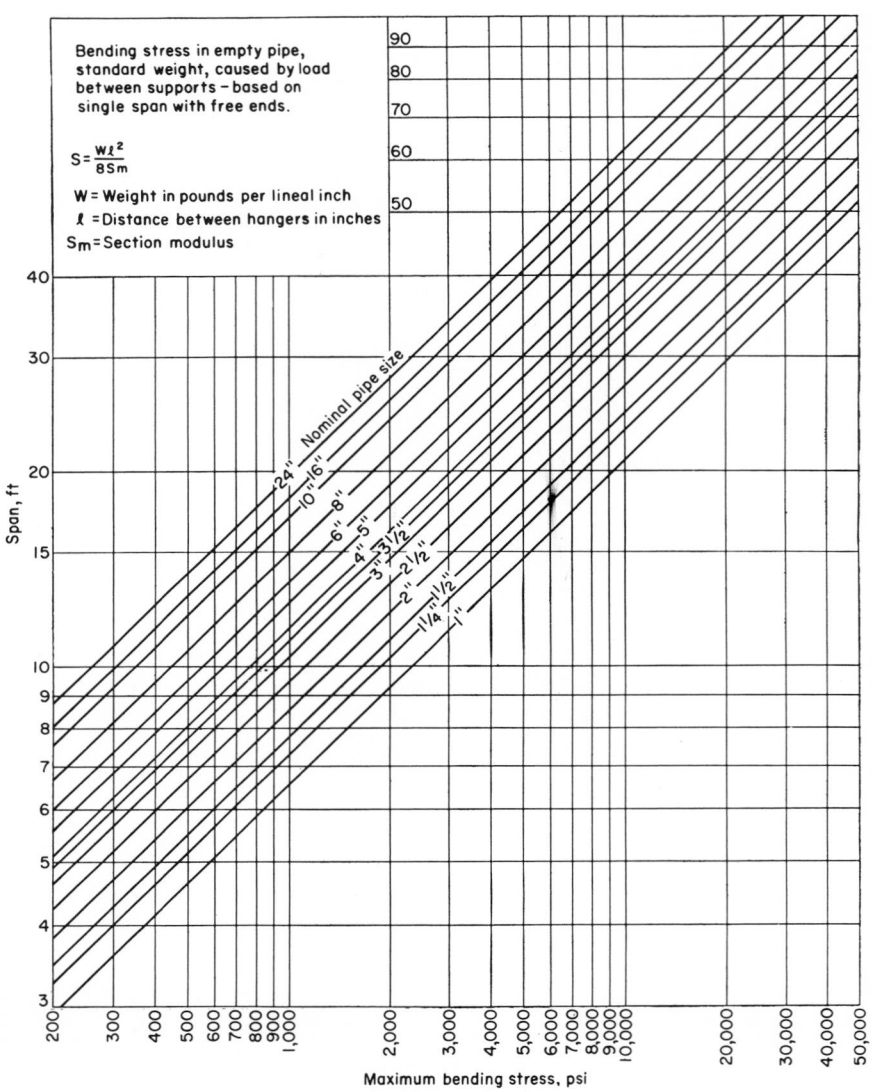

Fig. 2. Bending stress in water-filled pipe.

In steam lines with long spans, the deflection caused by the weight of the pipe may be large enough to cause an accumulation of condensate at the low points of the line. Water lines, unless properly drained, can be damaged by freezing. These conditions can be avoided by erecting the line with a downward pitch in such a manner that succeeding supports are lower than the points of maximum deflection in preceding spans.

The stresses indicated in Figs. 1 and 2 are bending stresses resulting from the weight of the pipe between supports. It should be realized that this stress must be

considered with all other stresses in the piping, such as those due to the pressure of the fluid within the pipe and the bending and torsional stresses resulting from thermal expansion, in order to determine the *total* stress in the system.

The stresses indicated in Figs. 1 and 2 and the deflections shown in Fig. 3 are based on a single span of pipe with free ends and make no allowances for concentrated loads of valves, flanges, etc., between hangers.

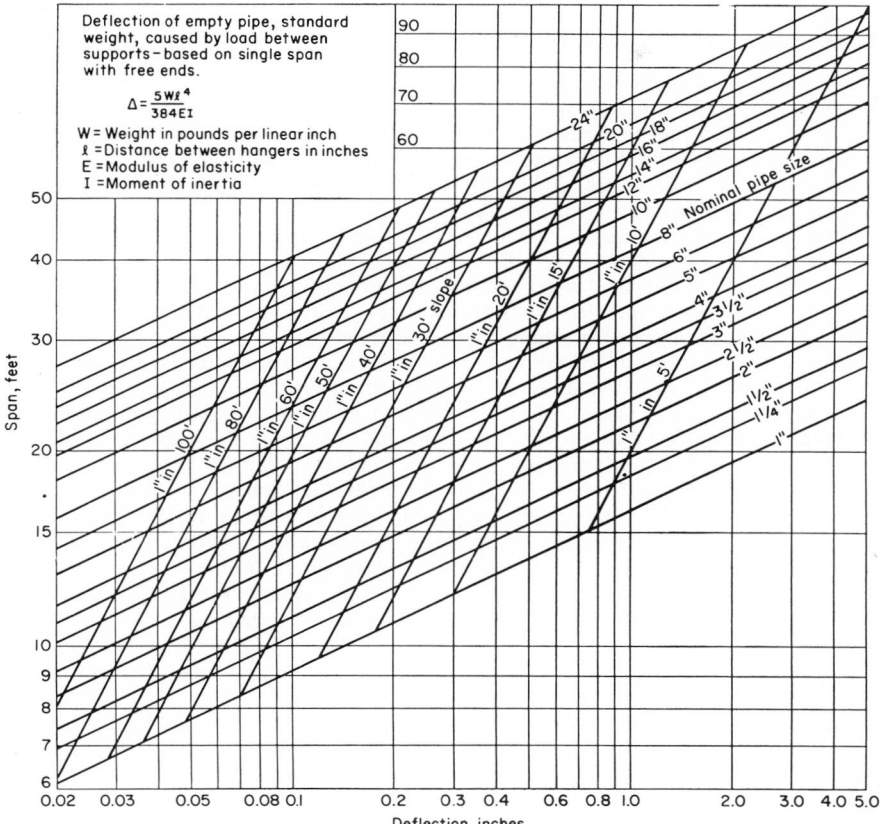

Deflection of empty pipe, standard weight, caused by load between supports – based on single span with free ends.

$$\Delta = \frac{5Wl^4}{384EI}$$

W = Weight in pounds per linear inch
l = Distance between hangers in inches
E = Modulus of elasticity
I = Moment of inertia

FIG. 3. Deflection of empty pipe. Values are plotted for the pipes empty, since this more nearly approaches the condition that exists for pocketing of condensation. Although the weight of fluid carried by the pipe will cause an increase in the deflection of the pipe between supports, this increased sag disappears during drainage. Therefore, the deflection produced by the weight of empty pipe should be considered in determining slope for drainage.

The stress and deflection values are based on a free-end beam formula and reflect a conservative analysis of the piping. Actually, the pipeline is a continuous structure partially restrained by the pipe supports, and the true stress and deflection values lie between those calculated for the free-end beam and a fully restrained structure.

The deflections and bending stress values indicated represent safe values for any schedule pipe from Schedule 10 to XS pipe.

For fluids other than water, the bending stress can be found by first finding the added stress caused by water from Figs. 1 and 2 and multiplying by the specific gravity of the fluid. This additional stress is added to the stress value of the empty pipe.

For lines which are thickly insulated, the deflection or bending stress resulting from the weight of pipe bare is determined and multiplied by a ratio of the weight of pipe plus insulation per foot to the weight of bare pipe per foot.

Determination of Loads and Movements

The movement at each support point dictates the basic type of hanger required. Each type of support selected must be capable of accommodating movements established by one of the methods outlined later in this section. It is a good logical practice to select first the most simple or basic rigid type of hanger and then add to the complexity only as conditions warrant. No advantage will be gained in upgrading a hanger when a simpler, more economical type can be shown to satisfy all the design requirements. Both vertical and horizontal movement must be considered and evaluated.

Zero or negligible vertical and horizontal movement permits the use of simple rod hangers for piping suspended from above or base, bracket, stanchion, structural cross member, or sleeper for piping supported from below.

With zero vertical and definite horizontal movement, the simple rod hanger will still suffice provided the overall length is sufficient to keep the angular swing of the rod within reasonable limits, normally accepted as being 4 deg from the vertical. Advantage should be taken of the possibility of advancing the upper connection some portion of the total movement as a means for cutting down the angularity. For piping supported from below, some form of slide must be incorporated to provide for the horizontal movement or, in the case of assured longitudinal movement, a pipe roll may be used. Suspended hangers with considerable horizontal movement and low headroom will require either single- or double-direction trolleys or rollers. Where both longitudinal and lateral movements are large, consideration may be given to the use of a single-direction trolley oriented on the resultant movement vector.

Piping which is subject to vertical movement requires the incorporation of spring cushions, variable springs or constant support units.

Spring cushions are reserved for use on lines where formal load and movement calculations are normally not made but on which observation shows that small movements do exist. The cushioning effect permits the piping to normalize itself over the entire system or section and thereby to eliminate the possibility of localized overstressing or imposition of excessive loading on terminal equipment connections. It is usual practice to leave the load setting of cushion-type hangers to the installer's discretion and experience.

Variable supports will normally suffice for all other required spring hangers except for certain critical systems in central power plants, refineries, and some specialized lines in process and chemical plants which may possibly have large vertical movements requiring that constant supports be used. As the name "variable support" implies, the supporting effect varies in a relation to the vertical movement of the suspended piping and the resultant compression or extension of the spring coil. The change in supporting force is in proportion to the spring stiffness or rate and to the amount of movement. When variable supports are used, the loss or gain in supporting forces should be considered as to its effect on the stress added to the piping system and the total accumulated loading that may possibly be thrown onto terminal connections.

The loading on a vessel nozzle connection naturally would not cause so much concern as an equal loading on a pump or turbine connection where the possibility of distortion or misalignment is present.   A true and complete evaluation of the loading would require formal stress-analysis calculations which would include effects of the change in supporting force.   This may be practical and warranted in critical cases involving extreme temperature and stresses; for normal installations, certain industries have established conventions concerning the use of variable supports and constant supports which may be regarded as good practice.   For example, the following would be conventional power field usage:

At points of support subject to vertical movement, springs of suitable design to prevent excessive variation in supporting effect shall be used to provide for the movement.   The amount of variation that can be tolerated shall be based on such considerations as bending effect, control of piping elevation, allowable terminal connection loadings, etc.   In general the deviation in supporting effect shall be limited to plus or minus 6% for critical systems such as main steam, hot and cold reheat, extraction lines over 750°F and the portions of the boiler feed discharge in the vicinity of the pumps and boiler terminal connection. On non-critical systems, the variations in supporting effect shall be limited to 25%.

For all systems a greater allowance in per cent variation is permissible at points of support where the variation in supporting effect is transferred directly to a rigid support or terminal connection specifically designed for the resulting loading condition.

When the load and movement at each selected hanger location are determined, detailed design of the individual hangers can be undertaken.   The movement determines basic type (such as spring or rigid, slide or roller); the piping, its temperature, and ambient temperature determine the pipe attachment and remaining hanger material, respectively; the proximity to supporting structure and available clearances determines the hanger configuration such as single- or double-rod suspended, base, stanchion, or bracket-type support; and the load determines the required strength of each component.

The design should take full advantage of commercially available load-rated and -tested hanger components to the greatest extent possible.   An effort should be made to maintain uniformity and simplicity in design.   The hanger should be functional, provide means for piping elevation adjustment, and be readily installable using normal field labor and equipment.

For economic reasons, the use of commercial catalogue hanger products available from piping-support manufacturers eliminates the need for detail design of the numerous types and varying sizes of components required to complete a typical piping-support installation.   Substantial reductions in total hanger cost are effected through volume runs, standardized fabricating procedures, and available existing tooling, forms, and dies.

Thermal Movement Calculations.   The calculation of piping movements to a high degree of accuracy necessitates a very complicated study of the piping system.   Fortunately, in the case of high-temperature critical systems subjected to formal machine stress analysis, movements at the various reference points are given along with the stress results.   For systems not subjected to machine analysis, thermal movements at the hanger points are normally determined by approximate methods.   The simplified method shown below is one which gives satisfactory approximations of the piping movements.   Whenever differences occur between the approximations and actual movements, the approximation of the movement will always be the greater amount.

Step 1.   The piping system of Fig. 4 is drawn and on it are shown all known vertical movements of the piping from its cold to hot, or operating, position. These movements will include those supplied by the equipment manufacturers for the terminal point connections.   For the illustrated problem, the following vertical

movements are known:

   Point *A*—2 in. up, cold to hot
   Point *B*—$\frac{1}{16}$ in. up, cold to hot
   Point *C*—$\frac{1}{8}$ in. down, cold to hot
   H-4—0 in, cold to hot

The operating temperature of the system is given as 1050 F, and the coefficient of expansion for low-chrome steel at 1050 F is 0.0946 in./ft.

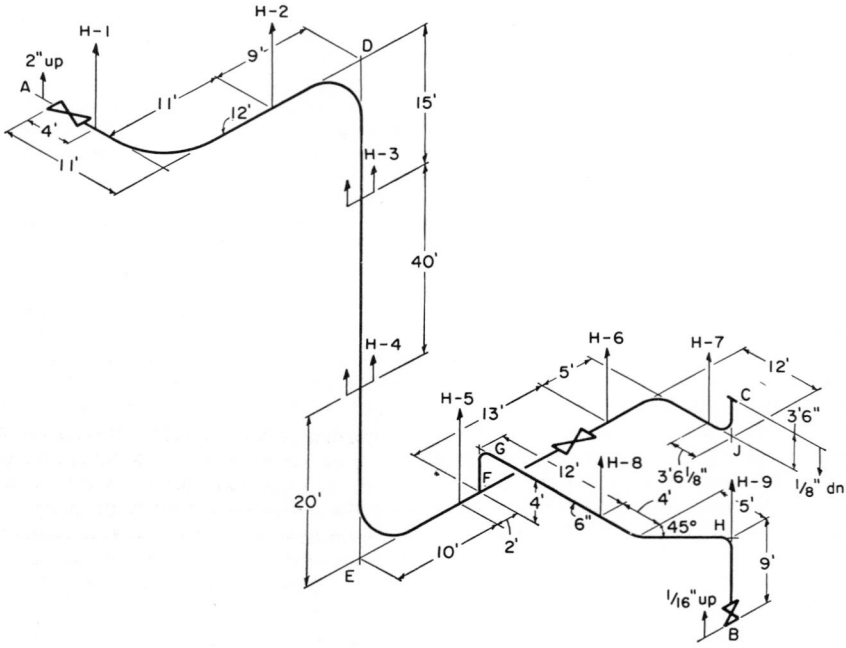

FIG. 4. One-line piping diagram for calculation of hanger movements.   Points *A*, *B*, and *C* are equipment connections.   *H*-1, *H*-2, etc., represent hanger locations.

The movements at points *D* and *E* are calculated by multiplying the coefficient of expansion by the vertical distance of each point from the position of zero movement on the riser *DE*:

$$55 \text{ ft} \times 0.0946 \text{ in./ft} = 5.20 \text{ in. } up \text{ at } D$$

$$20 \text{ ft} \times 0.0946 \text{ in./ft} = 1.89 \text{ in. } down \text{ at } E$$

Step 2.   A simple drawing is made of the piping between two adjacent points of known movement, extending the piping into a single plane as shown for the portion of the system between *A* and *D*.

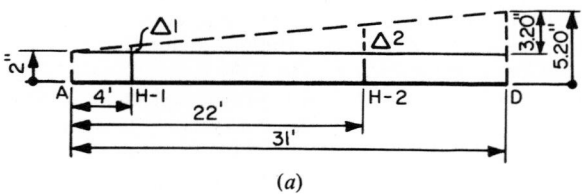

(*a*)

The vertical movement at any hanger location will be proportional to its distance from the end points:

$$\Delta_1 = \tfrac{4}{31} \times 3.20$$
$$\Delta_1 = 0.41 \text{ in.}$$

The vertical movement at H-1 $= 0.41$ in. $+ 2$ in.

$$\Delta H\text{-}1 = 2.41 \text{ in. up}$$
$$\Delta_2 = \tfrac{22}{31} \times 3.20$$
$$\Delta_2 = 2.27 \text{ in.}$$

The vertical movement at H-2 $= 2.27$ in. $+ 2$ in.

$$\Delta H\text{-}2 = 4.27 \text{ in. up}$$

Step 3. To calculate the vertical movement at H-3, multiply its distance from H-4 by the coefficient of expansion.

$$40 \text{ ft} \times 0.0946 \text{ in./ft} = 3.78 \text{ in.}$$
$$\Delta H\text{-}3 = 3.78 \text{ in. up}$$

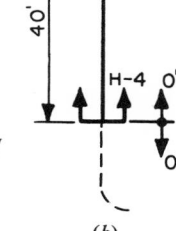

(b)

Step 4. The next section of pipe on which there are two points of known movement is the length E-J. The movement at E was calculated as 1.89 in. down.

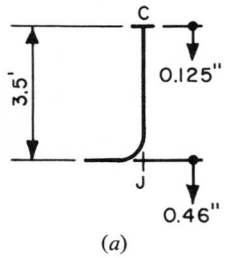

(a)

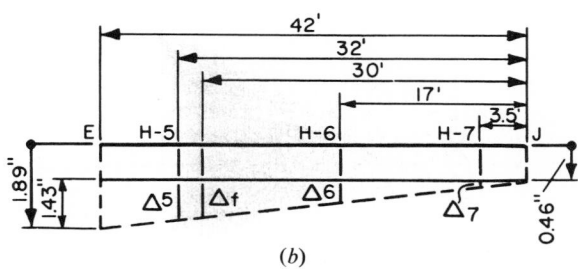

(b)

The movement at J is equal to the movement at the terminal point C ($\tfrac{1}{8}$ in. down) plus the amount of expansion of the leg C-J:

$$\Delta J = 0.125 \text{ in.} + 3.5 \text{ ft} \times 0.0946 \text{ in./ft}$$
$$\Delta J = 0.46 \text{ in. down}$$
$$\Delta_7 = 3.5/42 \times 1.43 = 0.12 \text{ in.}$$
$$\Delta H\text{-}7 = 0.12 \text{ in.} + 0.46 \text{ in.}$$
$$\Delta H\text{-}7 = 0.58 \text{ in. down}$$
$$\Delta_6 = \tfrac{17}{42} \times 1.43 = 0.58 \text{ in.}$$
$$\Delta H\text{-}6 = 0.58 + 0.46 \text{ in.}$$
$$\Delta H\text{-}6 = 1.04 \text{ in. down}$$
$$\Delta_f = \tfrac{30}{42} \times 1.43 = 1.02 \text{ in.}$$
$$\Delta F = 1.02 + 0.46$$
$$\Delta F = 1.48 \text{ in. down}$$
$$\Delta_5 = \tfrac{32}{42} \times 1.43 = 1.09 \text{ in.}$$
$$\Delta H\text{-}5 = 1.09 + 0.46$$
$$\Delta H\text{-}5 = 1.55 \text{ in. down}$$

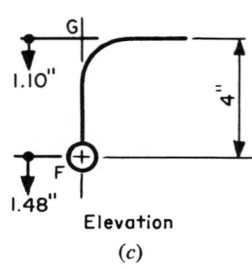

Elevation

(c)

Step 5. In the section *G-H*, the movement at *G* is equal to the movement at *F minus* the expansion of the leg *GF*:

$$\Delta G = 1.48 \text{ in. down} - 4 \text{ ft} \times 0.0946 \text{ in./ft}$$
$$\Delta G = 1.10 \text{ in. down}$$

The movement at *H* is equal to the movement of the terminal point $B(\frac{1}{16}$ in. up$)$ *plus* the expansion of the leg *B-H*:

$$\Delta H = 0.0625 \text{ in. up} + 9 \text{ ft} \times 0.0946 \text{ in./ft}$$
$$\Delta H = 0.91 \text{ in. up}$$

Since *H-9* is located at point *H*,

$$\Delta H\text{-}9 = \Delta H = 0.91 \text{ in. up}$$
$$\Delta_y = 12/23.1 \times 2.01 = 1.04 \text{ in.}$$
$$\Delta H\text{-}8 = 1.10 - 1.04$$
$$\Delta H\text{-}8 = 0.60 \text{ in. down}$$

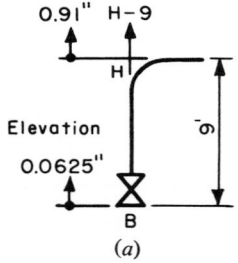

(a)

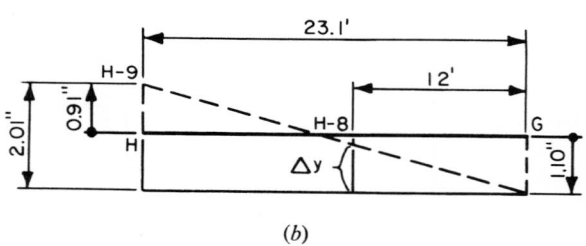

(b)

After calculating the movement at each hanger location it is often helpful, for easy reference when selecting the appropriate type hanger, to make a simple table of hanger movements.

Hanger number	Movement, in.
H-1	2.41 up
H-2	4.27 up
H-3	3.78 up
H-4	0
H-5	1.55 down
H-6	1.04 down
H-7	0.58 down
H-8	0.06 down
H-9	0.91 up

Reaction Calculations. A formal solution for movement and loads would approach that of pipe-stress calculations in complexity. In general, these costly, time-consuming calculations can be replaced by simple, practical solutions that give results well within the pipe tolerances that can be expected in a normal installation. It is usual practice to calculate loads by use of the simple beam method in which each segment of pipe, fitting, and valve is distributed between two adjacent hangers by taking moments and solving for the reactions at each hanger point. Calculation of moments can be simplified by representing the piping in plan and scaling off the distances to the center of gravity of the various elements. Hanger load calculations

must account for all the elements that make up the total weight of the piping system. These elements include pipe, valves, fittings, insulation, and operating contents. Weights of the more common piping elements are shown in Table 4, Chap. 7. Weights of other or special items should be obtained from the manufacturer.   In calculating loads on critical systems such as those for main steam and hot reheat, which may be mainly spring supported, it is recommended that actual piping weights be used when available.   The following typical examples serve to illustrate both the method and systematic presentation which prove to be a valuable aid to the user in eliminating errors and of assistance to others in checking.

Load Calculation by Weight Balance.   The following example is used to illustrate a method by which hanger loadings may be determined.   The method consists of locating the center of gravity of the specific piping configuration and then, by equating moments, to determine the resultant loads at particular hangers.

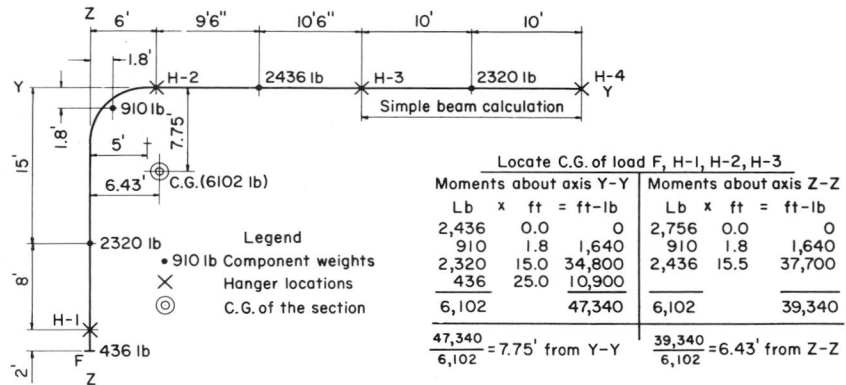

FIG. 5. One-line piping diagram for illustration of load calculation by weight balance.

A single-plane bend is shown in Fig. 5.   Hangers are indicated as *H*-1, *H*-2, *H*-3, and *H*-4.   The effects of uniform and concentrated loads are indicated at the points at which these loads act; it is noted that the weight of the 90-deg bend acts at the centroid of a quarter circle which, in this example, is located 1.8 ft distant from the center line of the pipe run.   The straight pipe length between hangers *H*-3 and *H*-4 is not included in this calculation because it can be analyzed by simple straight-beam theory.

For the piping section which lies between equipment flange *F* and hanger *H*-3, moments are taken about the *Y-Y* and *Z-Z* axes.   As an example, let the center of gravity of this configuration be located *Y* ft from the *Y-Y* axis.   Then, from equilibrium considerations, the following equation may be written:

$$2{,}436(0) + 910(1.8) + 2{,}320(15) + 436(25) = 6{,}102\,Y$$

A solution to this equation results in $Y = 7.75$ ft.

Similarly, the distance from the *Z-Z* axis to the center of gravity is found to be 6.43 ft.

For convenience, the calculations are made frequently in a tabular fashion as shown on Fig. 5.

Let it be now required to determine hanger loadings for the piping configuration of Fig. 5 with the stipulation that no load due to weight be imposed on the

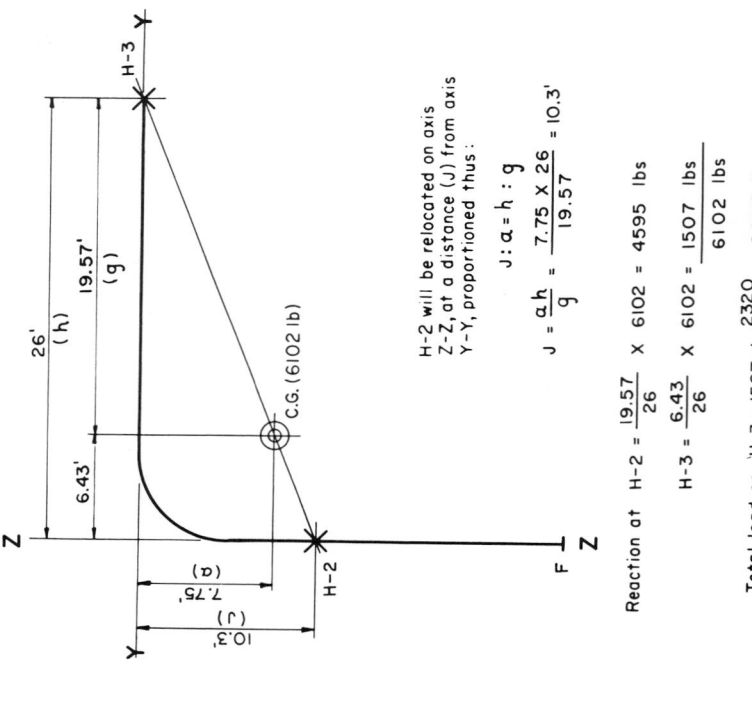

FIG. 7. Hanger load calculations for system of Fig. 5 except that one hanger has been eliminated.

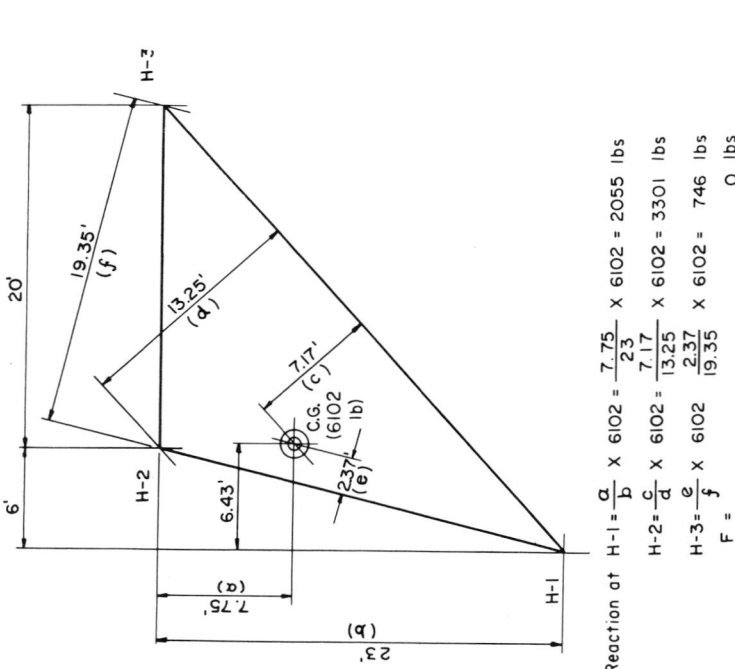

FIG. 6. Hanger load calculations for system of Fig. 5. Three hangers with zero reaction at flange F.

equipment flange *F*. This is accomplished easily by use of simple geometrical relationships, and the solution is as indicated in Fig. 6.

If it were desired to support the piping with two, rather than three, hangers, it would be convenient to eliminate *H*-1 and to relocate *H*-2 to a position at which it would be colinear with the center of gravity and hanger *H*-3. The construction for this arrangement and the associated hanger-load calculations are shown in Fig. 7.

In each of the two above cases, one-half of the 2,320-lb load between *H*-3 and *H*-4 has been included in the calculations for hanger loading on *H*-3. Thus *H*-4 would be required to support 1,160 lb plus, of course, any additional piping load to the right of *H*-4 in Fig. 5.

Calculation of Hanger Loads: Complete System. A 6-in. medium-temperature steam piping system is shown in Fig. 8. Terminal movements at

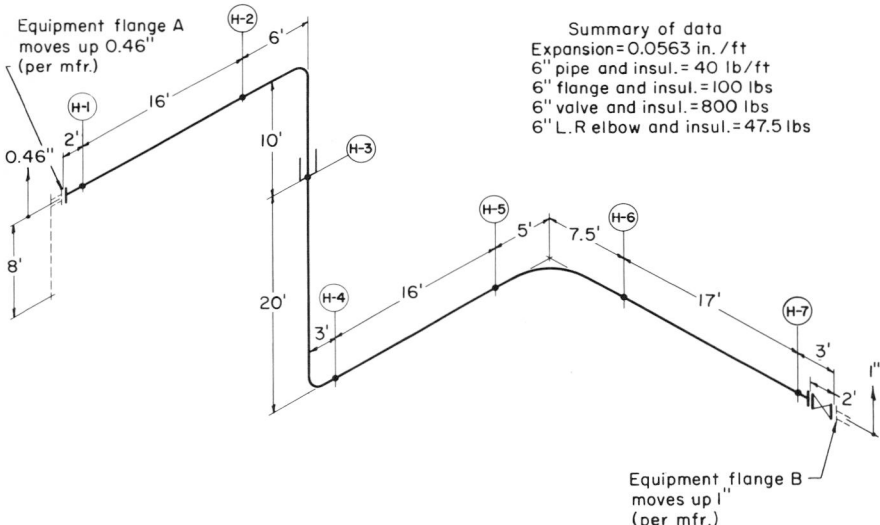

FIG. 8. One-line piping diagram for calculation of hanger loadings and deflections.

equipment flanges *A* and *B* are indicated; dimensions of system components and physical data are also given. It is required to determine hanger loadings and also to determine movements at each of the hangers *H*-1 through *H*-7.

It is noted that hanger *H*-3 on the vertical leg has been located 20 ft above the lower horizontal pipe run. Calculations would indicate that the center of gravity of the vertical leg is 16.16 ft above the lower horizontal run. It would not be desirable to place the hanger at the center of gravity because the hanger would then act as a pivot point and would not resist sway. If the hanger *H*-3 were placed below the center of gravity, an unstable turnover condition would result. The most desirable location is above the center of gravity; hanger *H*-3 has thus been placed arbitrarily a distance of 20 ft above the lower horizontal piping run.

Starting with equipment flange *A*, the system is broken up into component parts between hangers and hanger reactions are calculated. The procedure is indicated in Figs. 9*a* to *g*, and the results are listed in Table 2. Hanger deflections, or movements, are determined as shown in Figs. 10*a* and *b*.

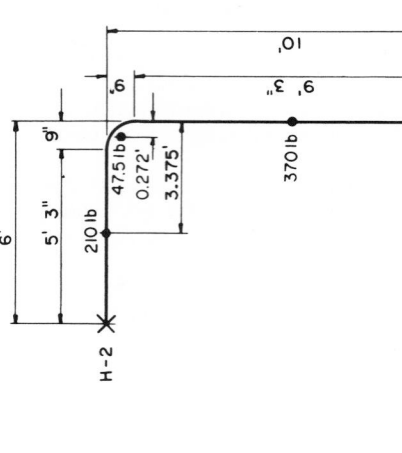

Elevation

$$\text{Reaction @ H-2} = \frac{721.7}{6.0} = 120.3 \text{ lbs}$$
$$\text{Reaction @ H-3} = 627.5 - 120.3 = 507.2 \text{ lbs}$$

Taking moments about H-3

Ft.	x	lb	=	ft-lb
0.0		370.0	=	0
0.272		47.5	=	12.9
3.375		210.0	=	708.8
		627.5		721.7

FIG. 9c. Distribution of weight between H-2 and H-3.

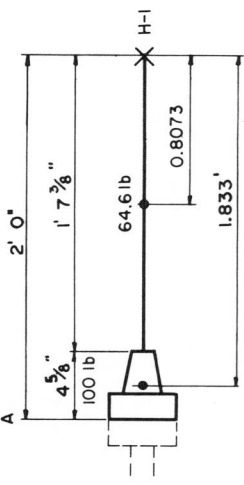

Taking moments about H-1

Ft.	x	lb	=	ft-lb
0.8073		64.6	=	52.2
1.833		100.0	=	183.3
		164.6		235.5

$$\text{Reaction @ flange A} = \frac{235.5}{2.0} = 117.8 \text{ lbs}$$
$$\text{Reaction @ H-1} = 164.6 - 117.8 = 46.8 \text{ lbs}$$

FIG. 9a. Distribution of weight between equipment flange A and H-1.

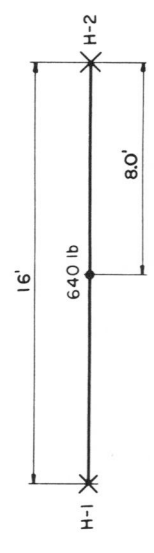

$$\text{Reactions H-1 and H-2} = \frac{640}{2} = 320 \text{ lbs}$$

FIG. 9b. Distribution of weight between H-1 and H-2.

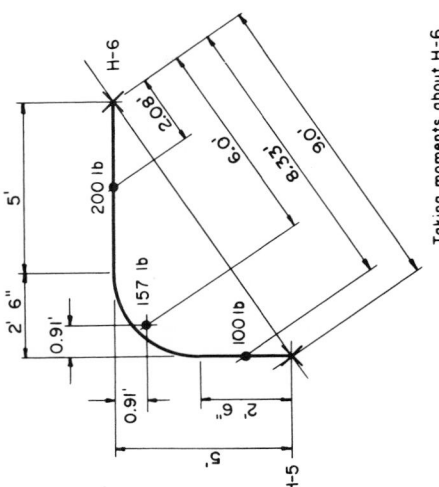

Taking moments about H-6

Ft.	×	lb	=	ft-lb
2.08		200	=	416.0
6.00		157	=	942.0
8.33		100	=	833.0
		457		2191.0

Reaction @ H-5 = $\frac{2191}{90}$ = 243.4 lbs

Reaction @ H-6 = 457 – 243.4 = 213.6 lbs

FIG. 9f. Distribution of weight between H-5 and H-6.

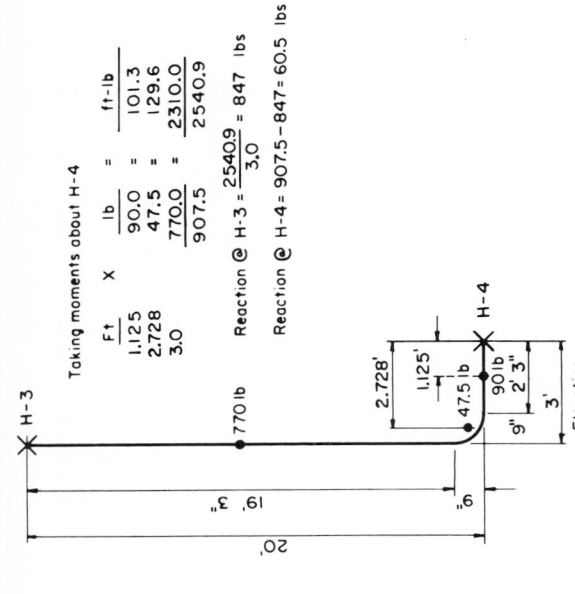

Taking moments about H-4

Ft	×	lb	=	ft-lb
1,125		90.0	=	101.3
2,728		47.5	=	129.6
3.0		770.0	=	2310.0
		907.5		2540.9

Reaction @ H-3 = $\frac{2540.9}{3.0}$ = 847 lbs

Reaction @ H-4 = 907.5 – 847 = 60.5 lbs

Elevation

FIG. 9d. Distribution of weight between H-3 and H-4.

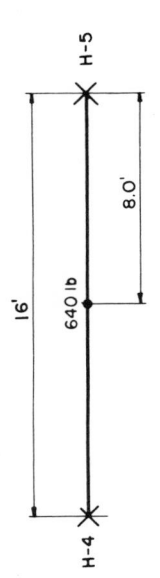

Reaction @ H-4 and H-5 = $\frac{640}{2}$ = 320 lbs

FIG. 9e. Distribution of weight between H-4 and H-5.

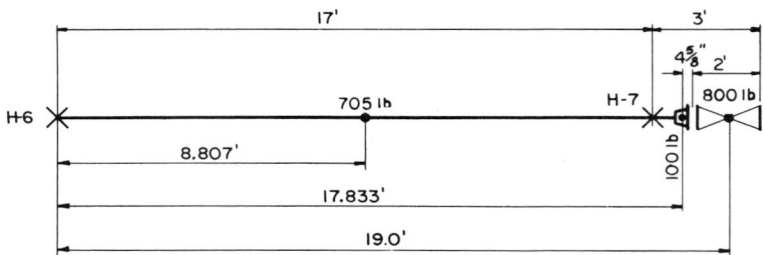

Taking moments about H-6

Ft	X	lb	=	ft-lb
8.807		705	=	6210
17.833		100	=	1783
19.0		800	=	15200
		1605		23193

$$\text{Reaction @ H-7} = \frac{23193}{17.0} = 1364 \text{ lbs}$$

$$\text{Reaction @ H-6} = 1605 - 1364 = 241 \text{ lbs}$$

FIG. 9g. Distribution of weight between *H*-6 and *H*-7 to maintain zero reaction on flange *B*.

Table 2.   Summary of Hanger Loadings

Hanger Mark	Reactions, lb							Hanger load, lb
	A to H-1	H-1 to H-2	H-2 to H-3	H-3 to H-4	H-4 to H-5	H-5 to H-6	H-6 to H-7	
Flange *A*	117.8	....	....	....	....	....	....	117.8
H-1	46.8	320.0	....	....	....	....	....	366.8
H-2	....	320.0	120.3	....	....	....	.....	440.3
H-3	....	....	507.2	847.0	....	....	.....	1,354.2
H-4	....	....	....	60.5	320.0	....	.....	380.5
H-5	....	....	....	....	320.0	243.4	.....	563.4
H-6	....	....	....	....	....	213.6	241.0	454.6
H-7	....	....	....	....	....	....	1364.0	1,364.0
Flange *B*	....	....	....	....	....	....	0.0	0.0

Flatten out pipe shape into plane and establish movement at
top and bottom of vertical leg.
Use method for one vertical leg.

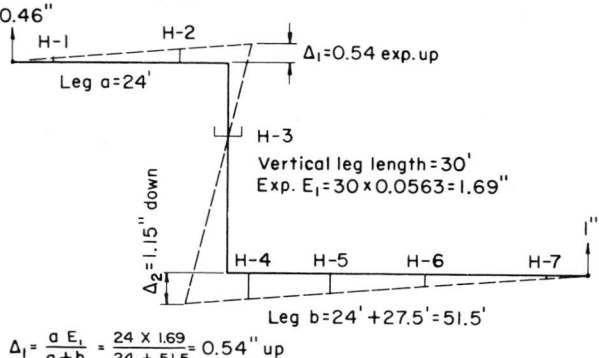

$$\Delta_1 = \frac{a\,E_1}{a+b} = \frac{24 \times 1.69}{24 + 51.5} = 0.54'' \text{ up}$$

$$\Delta_2 = E_1 - \Delta_1 = 1.69 - 0.54 = 1.15'' \text{ down}$$

FIG. 10*a*. Deflections of vertical leg of Fig. 8.

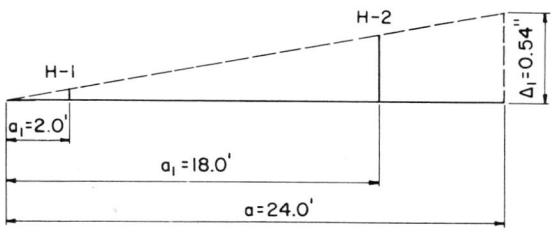

$$\Delta_x = \frac{a_1 \Delta_1}{a}$$

$$\Delta_x @ H\text{-}1 = \frac{2.0 \times 0.54}{24.0} = 0.045''$$

$$\Delta_x @ H\text{-}2 = \frac{18.0 \times 0.54}{24.0} = 0.405'''$$

FIG. 10*b*. Determination of deflections at *H*-1 and *H*-2 of Fig. 8.

SELECTION OF PIPE-SUPPORTING DEVICES

Piping Systems: Temperature Classification. Piping systems, for the purposes of this chapter, are divided into the following temperature categories in order to provide a basis for the selection of hangers, anchors, or supports.

1. Hot systems
   a. The temperature range is from 120 to 450 F.   Typical examples are low-pressure steam, hot water, and certain process piping.
   b. The temperature range is from 450 to 650 F.   Typical examples are boiler plant and industrial steam and hot-water piping systems.
   c. The temperature ranges from 750 F and upward.   A typical example is a high-pressure steam power-plant piping system.
   d. In the temperature range 650 F and higher, there is the possibility of metallurgical change if unalloyed carbon steel is used.   It is suggested that

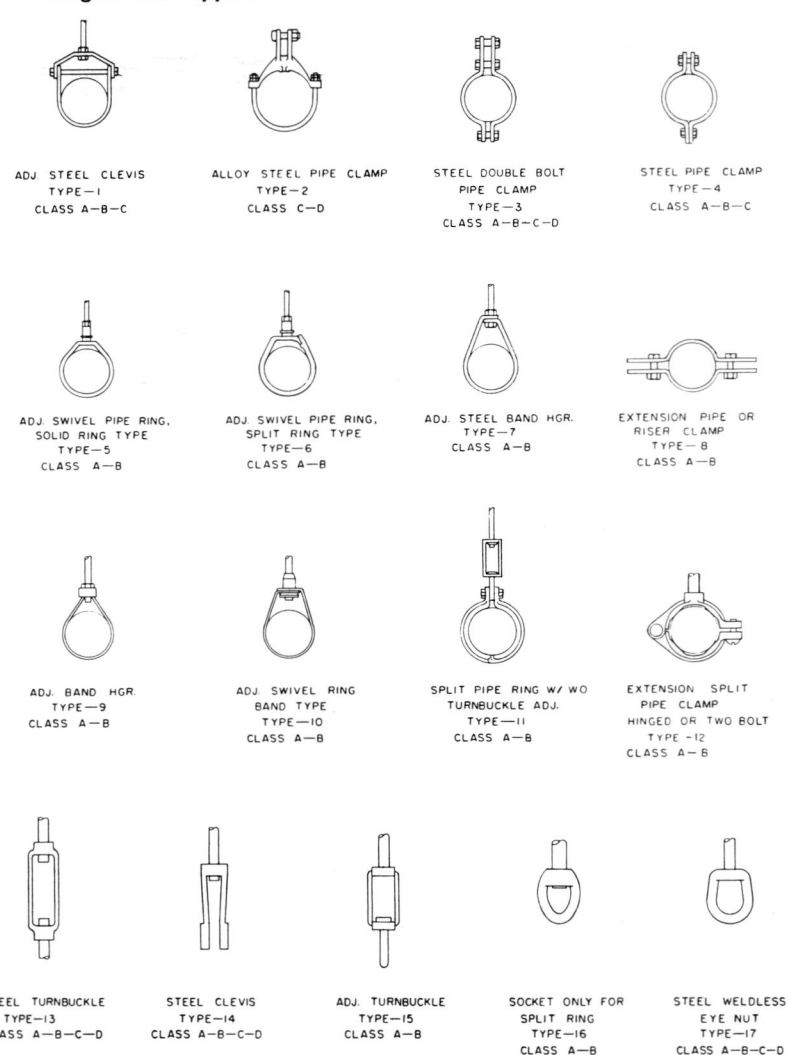

FIG. 11. Pipe hanger types.

hangers, anchors, and supports for piping which operates at above 650 F be of materials at least equal to those of the piping system itself.

2. Ambient systems in which the contents of the line are not heated or cooled by mechanical means. Temperatures would range up to 120 F. Plant air and service water would be typical systems.

3. Cold systems
   a. The temperatures range upward from 32 F. A typical example would be chilled water piping.
   b. The temperature ranges downward from 32 to minus 20 F, as in brine systems.
   c. Below minus 20 F, as in cryogenic systems.

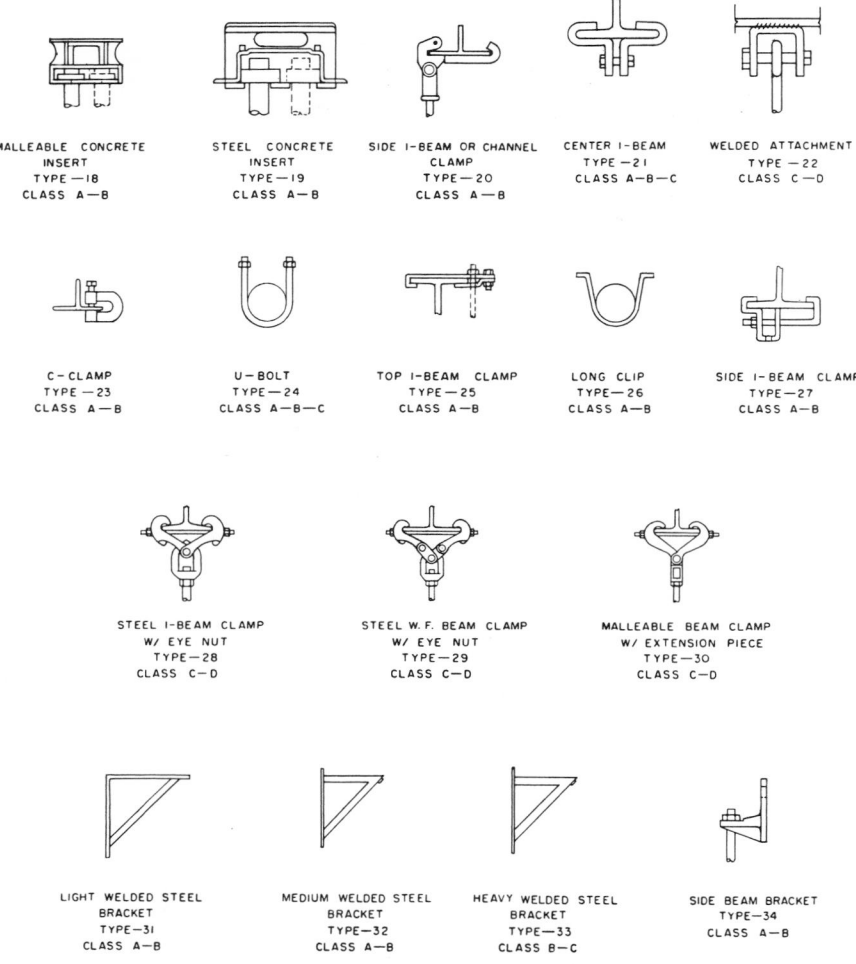

FIG. 11. Pipe hanger types (*continued*).

Pipe Attachments.   Hangers for the various systems described above may be selected from Fig. 11 in accordance with the following recommendations:

For Type 1*a* systems, hangers Types 1 and 3 through 12 are used.   Rollers should be Types 41 through 47 with appropriate saddles of Type 39, items 1 and 2. Supports should be of Types 35 through 38.

For Type 1*b* systems, hangers Types 1, 3, 4, and 8 are used.   Rollers should be of Types 41 through 47 with appropriate saddles of Type 39, items 1 and 2.

For Type 1*c* systems, alloy hangers are used as required by the line temperature. Hangers should be of Types 2, 3, or 8 with saddles of Type 39, items 1 or 2, and rollers of Types 41 through 47.

For Type 2 systems, hangers can be of Types 1 and 3 through 12 with supports of Types 24, 26, and 35 through 38.

For Type 3 systems, the hanger or support must be outside the insulation and the

FIG. 11. Pipe hanger types (*continued*).

vapor barrier must be left undisturbed (see "Pipe-covering Protection Saddles"). A Type 40 insulation protection shield must be used to distribute the loading on the insulation. Hangers sized for the outside diameter of the insulation can be of Type 1, 4, 6, 7, 9, 10, or 11. For Type 3c systems, special consideration must be given to the type and nature of the piping and its layout.

Consideration may be given to the use of welded lug attachments. Where used on Types 1c and 3c, the welded attachment must be of an alloy material which is compatible with the material of the piping system itself.

Pipe-covering Protection Saddles. Where insulation is used in a piping system, it is frequently necessary to make some modification at the point of attachment of the hanger. Since different methods and practices exist for hot lines as

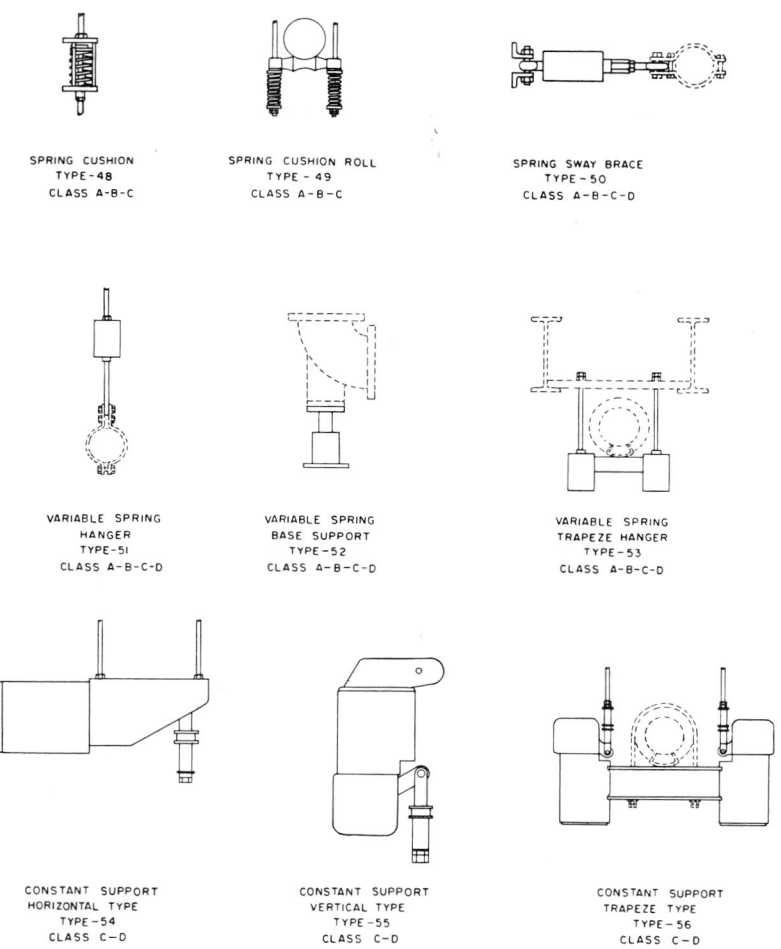

SPRING CUSHION
TYPE-48
CLASS A-B-C

SPRING CUSHION ROLL
TYPE - 49
CLASS A-B-C

SPRING SWAY BRACE
TYPE-50
CLASS A-B-C-D

VARIABLE SPRING
HANGER
TYPE-51
CLASS A-B-C-D

VARIABLE SPRING
BASE SUPPORT
TYPE-52
CLASS A-B-C-D

VARIABLE SPRING
TRAPEZE HANGER
TYPE-53
CLASS A-B-C-D

CONSTANT SUPPORT
HORIZONTAL TYPE
TYPE-54
CLASS C-D

CONSTANT SUPPORT
VERTICAL TYPE
TYPE-55
CLASS C-D

CONSTANT SUPPORT
TRAPEZE TYPE
TYPE-56
CLASS C-D

FIG. 11. Pipe hanger types (*continued*).

contrasted to low-temperature lines (chilled water and brine), they should be treated separately.

Insulation is available in thicknesses as shown in Table 3. Caution must be observed in the supporting of insulated piping to assure that a firm attachment is secured. Particularly, with respect to high-temperature lines, provision in the form of pipe-protection saddles is made to assure that the pipe supports do not become over-heated. Protection saddles, Type 39, items 1 and 2, for use with high-temperature insulation should be 12 in. long and of approximately 60 deg of arc. Metal should be of $1/_8$-in. thickness for pipe sizes up to 5 in. and of $3/_{16}$-in. thickness for larger pipe sizes. These saddles and matching rollers are available commercially for most sizes and thicknesses listed in Table 3.

For support from above, either a conventional type of pipe clamp may be used directly on the pipe and the arms of the clamp (Type 3) extended outside the

Table 3. Simplified Thicknesses for Pipe Insulation

Nominal pipe size, in.	Nominal thickness, in.														
	1		1 1/2		2		2 1/2		3		3 1/2		4	4 1/2	5
	Approximate thickness and outer layer pipe size, in.[1]														
	Thickness	Pipe size	Thickness	Pipe size	Thickness	Pipe size	Thickness	Pipe size	Thickness	Pipe size	Thickness	Pipe size	Thickness	Thickness	Thickness
1/2	1	2 1/2	1 9/16	3 1/2	2 1/16	4 1/2	2 7/8	6	3 11/32	7	3 7/8	8	4 3/8	4 29/32	5 13/32
3/4	29/32	2 1/2	1 15/32	3 1/2	1 31/32	4 1/2	2 3/4	6	3 1/4	7	3 3/4	8	4 9/22	4 13/16	5 5/16
1	1 3/32	3	1 19/32	4	2 1/8	5	2 21/32	6	3 5/32	7	3 21/32	8	4 5/32	4 11/16	5 3/16
1 1/4	29/32	3	1 21/32	4 1/2	1 15/16	5	2 15/32	6	2 31/32	7	3 15/32	8	3 31/32	4 1/2	5
1 1/2	1 1/32	3 1/2	1 17/32	4 1/2	2 5/16	6	2 27/32	7	3 11/32	8	3 27/32	9	4 3/8	4 7/8	5 3/8
2	1 1/16	4	1 19/32	5	2 1/8	6	2 5/8	7	3 1/8	8	3 5/8	9	4 5/32	4 21/32	5 5/32
2 1/2	1 1/16	4 1/2	1 7/8	6	2 11/32	7	2 7/8	8	3 3/8	9	3 29/32	10	4 13/32	4 29/32	5 9/16
3	1 1/32	5	1 9/16	6	2 1/16	7	2 9/16	8	3 1/16	9	3 19/32	10	4 3/32	4 19/32	5 1/4
3 1/2	1 9/32	6	1 13/16	7	2 9/32	8	2 25/32	9	3 11/32	10	3 27/32	11	4 5/16	....	4 31/32
4	1 1/16	6	1 9/16	7	2 1/16	8	2 9/16	9	3 3/32	10	3 19/32	11	4 3/32	4 3/4	5 3/16
4 1/2	1 9/32	7	1 13/16	8	2 5/16	9	2 13/16	10	3 11/32	11	3 27/32	12	....	4 1/2	4 15/16
5	1 1/32	7	1 17/32	8	2 1/32	9	2 9/16	10	3 1/16	11	3 9/16	12	4 7/32	4 21/32	5 5/32
6	1	8	1 1/2	9	2 1/32	10	2 17/32	11	3 1/32	12	3 11/16	14	4 1/8	4 5/8	5 3/16
7	....	....	1 17/32	10	2 1/32	11	2 17/32	12	3 1/8	14	3 5/8	15	4 1/8	4 5/8	5 1/8
8	....	....	1 17/32	11	2 1/32	12	2 11/16	14	3 1/8	15	3 5/8	16	4 3/16	4 11/16	5 3/16
9	....	....	1 17/32	12	2 5/32	14	2 21/32	15	3 1/8	16	3 21/32	17	4 5/32	4 21/32	5 5/32
10	....	....	1 19/32	14	2 3/32	15	2 19/32	16	3 3/32	17	3 19/32	18	4 3/32	4 19/32	5 3/32
11	....	....	1 19/32	15	2 3/32	16	2 5/8	17	3 3/32	18	3 19/32	19	4 3/32	4 19/32	5 3/32
12	....	....	1 19/32	16	2 3/32	17	2 5/8	18	3 3/32	19	3 19/32	20	4 1/8	4 5/8	5 1/8
14	....	....	1 1/2	17	2	18	2 1/2	19	3	20	3 1/2	21	4	4 1/2	5
16	....	....	1 1/2	19	2	20	2 1/2	21	3	22	3 1/2	23	4	4 1/2	5
18	....	....	1 1/2	21	2	22	2 1/2	23	3	24	3 1/2	25	4	4 1/2	5

[1] Subject to manufacturing tolerances.

insulation or a larger clamp (Type 1 or 4) may be used and lined with saddles or noncrushable insulation material.

Lugs may be welded to the pipe and extended outside the insulation to an attachment (Type 14) or, if supported from below, to a sliding shoe or guided roller. The important point is to have the actual attachment or supporting member outside the insulation so that movement of the line will not result in insulation damage.

For low-temperature service, in addition to heat loss or gain, the problem of atmospheric condensation must be considered, and such lines are usually insulated with a material that has an outer covering or seal called a vapor barrier. This barrier prevents the insulation from absorbing moisture. For this reason it is not permissible to penetrate the insulation with load-carrying members such as the legs of a conventional high-temperature saddle or a pipe clamp. Since most low-temperature insulations have low compressive strength, it is necessary to provide

Table 4. Minimum Dimensions: Shields for Insulation Protection

Nominal pipe of tubing size, in.	Length, in.	Gage thickness
½–3½	12	18
4	12	16
5	15	16
6	18	16
8	24	14

shields to line the hangers and to spread out the bearing area sufficiently to prevent crushing of the insulation. Such shields should fit the outer diameter of the insulation and cover 180 deg of arc. Metal weights and lengths are given in Table 4 taken from MSS SP-58.

Attachments to Buildings or Other Structures. When pipe is to be hung from steel, Types 20 through 23, 25, and 27 through 30 beam clamps should be used.

When pipe is to be hung from concrete, malleable iron or steel inserts of Types 18 and 19 or a continuous strip insert may be used. Embedded anchor bolts may also be used under specific conditions. For wall supports on either concrete or steel brackets, Types 31 through 33 may be used.

In many cases, it is necessary to provide additional structure as a means of upper attachment. Such structure must be designed for the particular load and can be of structural angle, beam or channel.

Multiple Pipe Runs. It is advantageous to support in groups those multiple runs of piping where the bottoms of the various lines are approximately at the same elevation. The support takes the form of a trapeze hanger fabricated from either steel or preformed channel framing. Multiple pipe runs are also supported on fabricated bents or frames. This is quite common on oil-refinery or tunnel work.

Where multiple runs of pipe are relatively near grade, sleepers in the form of wood, steel, or concrete are used.

On all multiple pipe runs, provisions should be made to keep the lines in their relative positions to each other by the use of either clamps or clips. Lines subject to thermal expansion must be free to slide or to roll.

Spring Supports. When there is an appreciable temperature difference between the operating and nonoperating condition of a piping system, there is a resultant expansion or contraction of the pipe due to the thermal change. When the system consists entirely of horizontal piping, the differential expansion can be taken care of

entirely by means of rollers or by swinging rods of sufficient length. When there are vertical portions of the piping system, the thermal change in length causes elevation changes such as are illustrated by Figs. 12 and 13. Movement of terminal points or of the structure to which the hangers are attached will also result in elevation changes. These elevation changes must be accommodated by some sort of resilient support.

Helical-coiled springs are commonly used for such resilient supports. The degree of resilience to be provided by the spring support depends on the conditions of

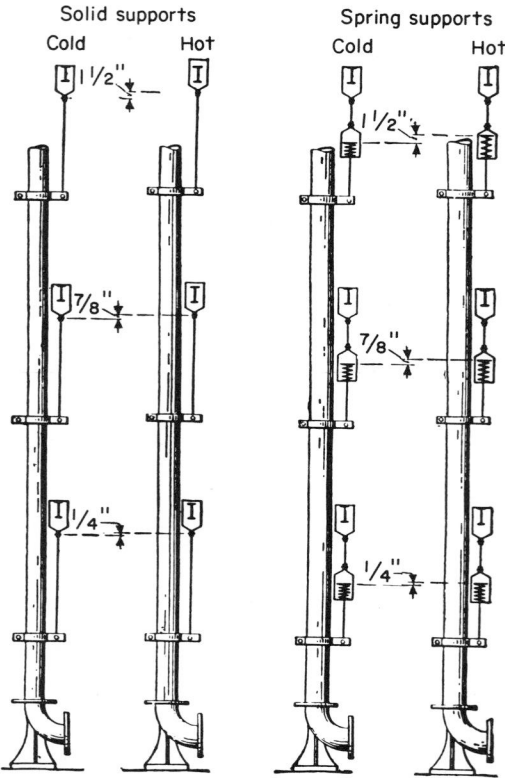

FIG. 12. Action of vertical pipe with and without spring supports.

the piping system. For systems which operate at temperatures below 750 F, a good rule is that the variation in supporting force be limited to 25 per cent of the load. This means that the vertical movement multiplied by the spring modulus and divided by the load supported at that point shall be 25 per cent or less. When the suggestions are followed for stress limits set forth in MSS SP-58, paragraph 11, and in ASTM Specification A125, springs to suit specific conditions may be designed. However, the price of a specially designed spring includes engineering and setup charges, and unless a large quantity of a particular size of spring is to be used, it is not economical to design individual springs. A more prudent approach is to select spring devices which are available commercially.

Commercial spring supports are made in three general types. Spring cushion

hangers, Types 48 and 49, are made for light to medium loads and for $\frac{1}{4}$-in. maximum vertical movement.   Generally, this type is used in systems where the temperature does not exceed 450 F.

Variable spring supports, Types 51, 52, and 53, are available for a wide range of loads, from around 50 to 30,000 lb.   This type of hanger is used on piping systems where the resulting variation in supporting force can be tolerated.   The commercial varieties available provide a selection of variability.

Constant load supports, Types 54, 55, and 56, are spring devices in which the varying force of the spring is compensated so that the support variability is within the range of plus or minus 6 per cent.   This type of support is used on systems where there are large thermal movements or critical stress conditions or a combination of both.   Such conditions exist on high-pressure, high-temperature steam lines in electric generating stations. These supports are made for loads of approximately 50 to 50,000 lb and for vertical movements up to 16 in.   The variability of these supports is not dependent on vertical movement, and the support is adjusted at the factory for specified load and travel.   Therefore, extreme care should be taken in determining loads and travel for the selection of hanger size.

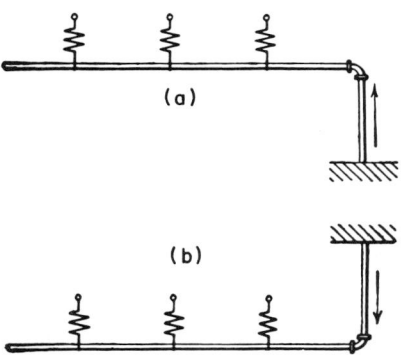

Fig. 13. Conditions where spring supports are required.

Vibration arising from pump pulse and similar conditions is a problem in piping systems.   Where such vibration exists and is in resonance with a spring-supported system, the results can be serious.   Such conditions can usually be avoided by judicious use of commercially available spring supports.   Systems that respond to exciting vibrations can be controlled satisfactorily by the use of dampening devices. There are two general types to consider: the coiled spring and the hydraulic vibration dampener.

There are two types of coiled-spring vibration dampeners: the opposed-spring type and the double-acting spring type (Type 50).   These devices should be arranged so that the springs are in the neutral position during normal operating conditions of the system.

The hydraulic vibration control is a unit which operates by means of a controlled flow of fluid through an orifice.   Resistance to movement increases with the speed of displacement.   One distinct advantage of the hydraulic device is that there is a minimum of resistance to thermal movement of the piping.

Both spring and hydraulic cylinder devices may be used to control sway and absorb shocks.   Seismic shock, in the areas where this must be considered, causes side thrust that must be provided for.   This side thrust is from one fifth to one tenth of the gravitational load.   Public buildings should have the protection afforded by use in design of the side thrust being taken as one fifth of the gravitational load.

These same devices may be used to resist reactions from safety-valve discharges and similar applications.   Both the spring and hydraulic types are available commercially in several sizes.   Manufacturers' literature gives comprehensive data for selection and performance characteristics.

Hanger Rod.   Rod used for pipe-support purposes is usually hot-rolled steel with cut threads conforming to *National Bureau of Standards Handbook H-28,*

Table 5.   Load Rating of Threaded Hot-rolled Steel Conforming to ASTM A107

Nominal rod diam. in.	Root area of thread, sq in.	Maximum safe load at rod temp of 650 F, lb
1/4	0.027	240
5/16	0.046	410
3/8	0.068	610
1/2	0.126	1,130
5/8	0.202	1,810
3/4	0.302	2,710
7/8	0.419	3,770
1	0.552	4,960
1 1/8	0.693	6,230
1 1/4	0.889	8,000
1 3/8	1.053	9,470
1 1/2	1.293	11,630
1 5/8	1.515	13,630
1 3/4	1.744	15,690
1 7/8	2.048	18,430
2	2.292	20,690
2 1/4	3.021	27,200
2 1/2	3.716	33,500
2 3/4	4.619	41,600
3	5.621	50,600
3 1/4	6.720	60,500
3 1/2	7.918	71,260

Class 2A, for the coarse thread series. Rolled threads to the same standard may be used. It must be pointed out that the length of a rolled thread cannot be increased by running a die over it, since the basic diameter of the rod is less than the size of the threaded portion.

Safe load capacities of rods are based on the area at the root of the thread. A generally accepted standard for such capacities is given by Table 5, taken from MSS SP-58.

For convenience in ordering and assembling various items, most catalogued pipe-hanger products have a definite relationship between rod size and pipe size as shown in Table 6.

Table 5 conforms to Underwriters Laboratories and Factory Mutual requirements in pipe sizes up to 12 in. The Code for Pressure Piping, ASA B31, does not allow the use of rod whose diameter is less than 3/8 in. in diameter.

On piping 1 in. and smaller in size, which is not subject to jurisdiction of the above regulatory bodies, 1/4 and 5/16 in. threaded rod may be used. However, these sizes are easily damaged during installation, and care should be taken in their use.

Table 6.   Relation between Pipe and Rod Sizes

Pipe size, in.	Rod size, in.
2 and smaller	3/8
2 1/2–3 1/2	1/2
4 and 5	5/8
6	3/4
8–12	7/8
14 and 16	1
18	1 1/8
20	1 1/4
24	1 1/2

Anchors, Guides, and Restraints. In addition to supporting gravitational loads, the designer must also be concerned with the provision of a suitable system of anchors, guides, restraints, stops, and braces to control intended movement, maintain piping position, and protect equipment from possible excessive loading or shock forces. The piping designer will normally call for and locate all anchor and restraint points that may be required to hold stresses and reactions within allowable limits, thereby assuring that the system will be installed in accordance with the overall piping-design concept. These requirements must naturally be fulfilled, and any intention for the relocation, elimination, or addition of anchors or restraints should be brought to the attention of the piping designer for his consideration and approval.

Supplementing the dictates of the piping designer, the layout of each system or section of piping should be reviewed, taking note of such factors as configuration, branches, expansion joints and loops, pipe sizes, terminal connections, relative stiffness of each leg in all planes, and system operating conditions. The digestion of all these factors, coupled with a visualization of the normal thermal movement of the system under consideration, enables an evaluation of the specific requirements necessary to assure positive control during all phases of operation.

Anchors and restraints may be required to establish definite movement patterns, counteract thrust forces, or, as in the case of vibration-imposing equipment used, prevent transmittal and possible build-up of the vibration throughout the entire system. Specific examples are the need for properly located anchors in a steam-distribution system to prevent overloading of the smaller branches, anchors and guides to actuate and align expansion joints and loops properly, and restraints or fixed points in the vicinity of compressor equipment or quick-closing control valves. Long, straight runs or sections of piping that are obviously weak in some plane may require additional guiding or bracing to provide lateral structural stability.

Piping installations which are located in seismic areas require special consideration regarding anchoring and guiding. During an earthquake disturbance the building structure and all mounted equipment move with the earth. Owing to inertia, the piping and all suspended equipment tends to remain fixed in space, causing a relative displacement between piping and structure. This displacement would be resolved into forces and moments at all terminal connections which, depending on the severity of the disturbance, could prove to exceed allowable design loadings. The terminal connection loading can be kept within allowable limits by providing a system of anchors, guides, and braces that will cause the piping either to move with the building structure or limit the relative displacement.

As is the case in all applications of anchors and guides, the overall installation must provide sufficient flexibility to accommodate thermal growth. For sections where the movement does not permit the use of rigid struts, guides with sufficient clearance to accommodate the normal movement may suffice by limiting the displacement. Positive strut action can be obtained at points subject to movement through the use of special devices such as hydraulic snubbers. Cylinders fitted with either fixed orifices or spring-loaded poppet valves are available which will accommodate the normal rate of thermal growth without restriction but will act as rigid members against a suddenly imposed loading.

Anchors, guides, and restraints must be designed for the imposed loading as determined by formal, approved calculations. Loading imposed by free bellows and slip-type expansion joints may be calculated as the product of the line pressure acting on the maximum unbalanced cross-sectional area plus a compression or friction factor as advised by the manufacturer. Systems operating under a vacuum require restraints to keep the joints from collapsing. The factor for

earthquake design is expressed as a percentage of gravity and specific values are available for each seismic area throughout the world. A typical value would be 0.2g horizontal and 0.1g vertical, in which case bracing capable of resisting two-tenths of the entire suspended weight in all directions must be provided. The one-tenth vertical force can normally be accommodated by accounting for the added load in the design of rigid supports.

Anchors may be connected to pipe or fittings by means of bolted clamps, welded attachments, or cast pads in the case of cast fittings. Where welded attachments are used, attention is called to the factors discussed under "Pipe Attachments" and "Welded Fabrication."

Riser Supports. A riser or vertical section of a piping system often creates special support problems. The support of service water lines and fire-protection systems is governed in the first case by the local building codes and in the latter case by National Fire Codes. These codes dictate the type and frequency of riser

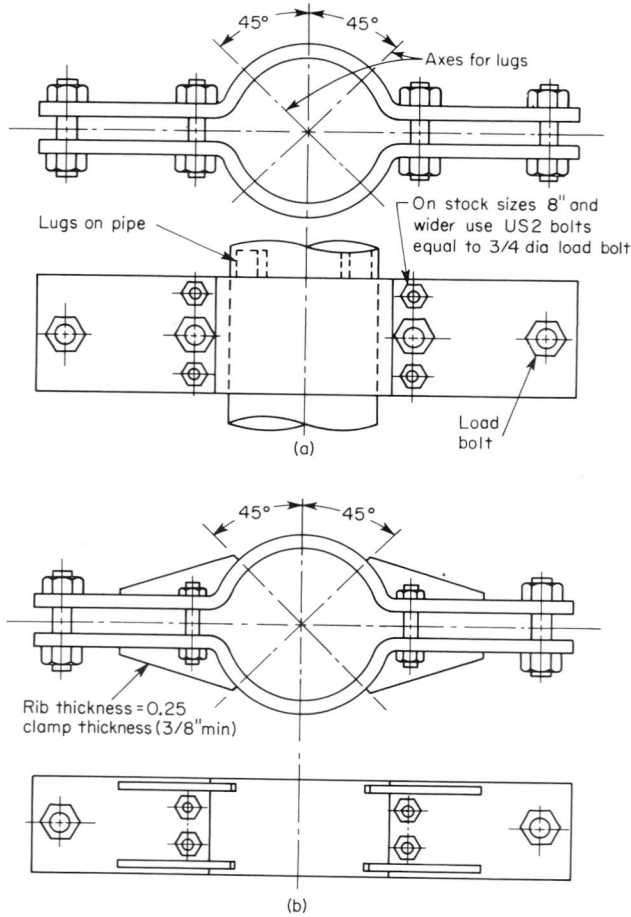

FIG. 14. Pipe clamps for vertical risers: (*a*) Unribbed clamp. (*b*) Ribbed clamp.

supports used on those systems; however, piping systems that are subject to thermal movement present more complicated support problems, and each case must be considered separately.

Risers are equivalent to concentrated loads; however, in the support of this load, several important points must be considered. These are:

1. Is the support to take the entire riser weight, or is this weight to be distributed among several supports?

2. Are hydrostatic-test conditions more severe than service conditions; that is, will the cold-water-filled condition impose stresses on the support higher than allowable (in cold condition) as compared with the hot operating condition and the imposed stresses? When this decision is made, the system erection sequence should be considered and a determination made whether other supports are effective or ineffective during hydrostatic testing.

3. Is the support to be located at a point of zero vertical movement and hence to be considered a rigid support? If this is the case, then the horizontal and flexural movements must be analyzed. Pure horizontal movement can be provided for by long support rods which are allowed to swing. However, if flexural movement exists, it may cause tipping, and then it must be assumed that the entire load can transfer to one support rod. In this case, the riser support must be designed for double the calculated load.

The support arms must extend out from the pipe far enough to clear any insulation. If springs are a part of the support assembly, then the arms must be long enough to provide clearance for the springs used.

The final consideration in the design of a riser support is its attachment to the pipe. Since a thermal change occurs between the operating and nonoperating condition of the pipe, the frictional grip of the support on the pipe cannot be relied upon and there must be some positive means of engagement between the pipe and the clamp. Sometimes a complete ring is welded to the pipe. The more normal procedure is to use four lugs welded to the pipe, which are called "hold-down" lugs and are disposed on a 45-deg axis as illustrated in Fig. 14. These lugs are to be made of the same material as the pipe, are to be the same thickness as the clamp support, and are to be welded on three sides.

Support Requirements for Specific Piping Materials

Cast-iron Pressure Pipe. Hanger sizes depend upon the outer diameter of pipe which is to be supported. The diameter of cast-iron pressure pipe exceeds that of nominal sized steel pipe and varies with its class or intended service.

Hanger spacing should provide at least one support for each length of pipe, with the hanger preferably located adjacent to the joint, and the hanger spacing should be no more than 12 ft. Also, each change of direction or branch connection should be supported.

Buried lines are usually supported by backfill tamped with wood blocks, except that lines buried under building slabs are sometimes supported from the slab with hairpin straps or rods suitable to the soil characteristics.

Special consideration is required in cases where movement occurs either in the terminal points or in the structure to which hangers are attached. Spring hangers may be required to allow for deflection.

Copper Tubing. Since hangers made for copper tubing are designed to fit the tubing diameter, it is not recommended that hangers which have been formed for piping be improvised as tubing supports. Most hangers for copper tubing are

electroplated with copper for easy identification.   Types 1, 8, 9, 10, 11 and 12[1] are available in tubing sizes.

Where there are indications that electrolysis may occur due to the fact that the hanger and tubing are made of dissimilar metals, the hanger should be lined to prevent such action.

Cast-iron Soil Pipe.   Hanger sizes correspond to sizes for steel pipe.   Types 1, 7, 8, and 11[1] are normally used.

Hanger spacing should provide at least one support for each length of pipe, with the hanger located adjacent to the joint, and a maximum spacing of 10 ft. Also, each change of direction or branch connection should be supported.

Buried lines are usually supported by backfill tamped with wood blocks, except that lines buried under building slabs are sometimes supported from the slab with hairpin straps or rods suitable to the soil characteristics.

Special consideration is required in cases where movement occurs either in the terminal points or in the structure to which hangers are attached.   Spring hangers may be required to allow for deflection.

Fire-protection Systems.   Sprinkler hangers will usually be subject to the consideration and approval of the same insurance agencies that have jurisdiction over the sprinkler-system piping and layout.   The supports will be required to conform closely to the standards specified by the Fixed Extinguishing Equipment sections of the National Fire Codes published by the National Fire Protection Association or in pamphlet form by the National Board of Fire Underwriters. These pamphlets are *NBFU* 11 *for Foam, NBFU* 12 *for Carbon Dioxide, NBFU* 13 *for Sprinkler Systems, NBFU* 14 *for Dry Chemical Systems* and *NBFU* 15 *for Water Spray Systems.*   Hangers are covered in *Pamphlet* 13.   However, if the system is other than a standard building water-sprinkler system, the particular standard should also be consulted; for example, in the design or selection of hangers for a foam fire extinguishing system, the standard *NBFU* 11 should be consulted.

The rules are explicit with respect to hanger spacing and fasteners and method of fastening and also regulate the material used in the construction of supports. Underwriters' Laboratories, Inc., Factory Mutual Engineering Division, and Underwriters' Laboratories, Inc., of Canada have tested and listed or approved all types of hangers necessary to meet the various conditions of construction.

The placing of hangers along the pipe is an important consideration to assure adequate support for the piping.   In addition to the consideration of pipe support, consideration must also be given to sway bracing (especially in deluge systems), riser supports, soundness of the attachment, vibration, and pipe slope for drainage. The stability of the support during a fire, corrosion, and aging is also of prime importance.   A sprinkler system may be installed for many years before it is operated, and it must always operate in the manner for which it was designed.

Plastic Pipe.   There are many types of plastic pipe, both rigid and semirigid. Under normal conditions, rigid plastic pipe can be supported using conventional hangers with the spacing half that used with steel pipe.   The support of plastic pipe or tubing should be continuous if, owing to the nature of the plastic, it will become flexible from elevated temperatures or from line contents.   The continuous support can be in the form of a light angle or channel into which the plastic pipe is laid.   Special hangers are required to support the combination.   It is suggested that recommendations of the manufacturer of the specific plastic pipe be followed.

Glass Pipe.   Glass pipe is used for laboratory service, food processing, and many industrial applications which require the durability or chemical resistance of

---

[1] The type numbers refer to Fig. 11.

glass pipe. Because of the nature of glass, special consideration of hangers and attachments is necessary.

Glass pipe conforms closely to IPS standards, and standard iron-pipe size hangers can be used. However, in all cases the hanger, even if it is painted or electroplated, must have a padding or cushion to avoid scratching the pipe. The hanger should fit loosely around the pipe yet contact the pipe in a manner to distribute the load over the largest possible area. Point loading must be avoided. The system of hangers must be designed with the least number of rigid anchor points.

Glass pipe will generally require two hangers per each 10-ft section. One extra hanger will be required if there are three or more couplings in a 10-ft section. For maximum protection and accessibility, hangers should be placed about 1 ft from each joint or coupling.

Glass pipe hangers and layouts should be designed in accordance with general fundamentals applicable to other piping materials. However, extreme care must be taken to minimize strain and scratching of the glass. The pipe manufacturer and reputable hanger manufacturers should be consulted in the design of hanger systems for glass pipe.

Asbestos-cement Pressure Pipe. Hanger sizes depend upon the class of pipe to be supported. The outer diameter of this pipe exceeds that of steel pipe and varies with its class or service.

Hanger spacing should provide at least one support for each length of pipe, with the support being located adjacent to the joint and with a maximum spacing of 12 ft. Each change of direction, each branch connection, and both sides of each expansion joint should be supported.

Buried lines are usually supported by backfill tamped with wood blocks, except that lines buried under building slabs are sometimes supported from the slab with hairpin straps or rods suitable to the soil characteristics.

Special consideration is required in cases where movement occurs either in the terminal points or in the structure to which hangers are attached. Spring hangers may be required to allow for deflection.

DESIGN DETAIL CONSIDERATIONS

ASA Code for Pressure Piping. Specific design requirements for the support piping are included in the ASA Code for Pressure Piping in the sections covering Power Piping (B31.1), Petroleum Refinery Piping (B31.3), and Refrigeration Piping (B31.5). These requirements should be adhered to on piping installations when piping must conform to these Codes. In most cases, hangers conforming to MSS SP-58 are acceptable for these installations.

Government Specifications. The Federal Specification WWH-171, of the latest issue and entitled Hangers and Supports, Pipe, is the governing specification for all Federal agencies. This specification illustrates types of pipe hangers and supports and lists the requirements of certain applications.

Design Temperature. Design temperatures for parts of hangers and supports in direct contact with pipe shall be the temperature of the contained fluid. Parts of hangers and supports not in direct contact with pipe and exterior to any insulation may be designed for one-third fluid temperature or ambient temperature, whichever is greater.

Allowable stresses for materials commonly used in the design of pipe hangers and supports are listed in Table 7.

Welded Fabrication. Welded fabrication shall be designed and proportioned in accordance with good engineering practice as prescribed by the American Welding Society or other recognized authorities. Attachments welded directly to the pipe

Table 7. Maximum Allowable Stress Values in Tension in Pounds per Square Inch for Metal Temperatures Not Exceeding Deg. F[1]

Material and spec. number	Grade	Spec. min. tensile	−20 to 450	650	700	750	800	850	900	950	1,000	1,050	1,100	1,150	1,200	1,250	1,300	1,350	1,400	1,450	1,500
Structural steel:																					
ASTM A7		60,000	15,000	15,000																	
ASTM A36		60,000	15,000	15,000																	
Rods and bars:																					
ASTM A107	1,015	50,000	11,500	11,500																	
ASTM A107	1,020	55,000	12,650	12,650																	
ASTM A107	1,025	58,000	12,650	12,650																	
ASTM A322	4,140	75,000	15,000	15,000	15,000	15,000	14,850	14,200	13,100	11,000	7,500	5,000	2,800	1,550	1,000						
ASTM A276	304	75,000	11,200	11,200	10,800	10,400	10,000	9,700	9,400	9,100	8,800	8,500	7,500	5,750	4,500	3,250	2,450	1,800	1,400	1,000	750
ASTM A276	321	75,000	14,850	14,850	14,800	14,700	14,550	14,350	14,100	13,850	13,500	13,100	10,300	7,600	5,000	3,600	2,700	2,000	1,550	1,200	1,000
ASTM A276	347	75,000	14,850	14,850	14,800	14,700	14,550	14,300	14,100	13,850	13,500	13,100	10,300	7,600	5,000	3,600	2,700	2,000	1,550	1,200	1,000
Bolting:																					
ASTM A307	A	55,000	13,750	13,750																	
ASTM A193	B-7	125,000	20,000	20,000	20,000	20,000	20,000	16,250	12,500	8,500	4,500										
ASTM A193	B-8	75,000	10,450	8,950	8,650	8,300	8,000	7,750	7,500	7,250	7,050	6,800	6,300	5,750	4,500	3,250	2,450	1,800	1,400	1,000	750
ASTM A193	B-8C	75,000	12,450	11,850	11,800	11,750	11,650	11,450	11,300	11,100	10,800	10,500	10,000	8,000	5,000	3,600	2,700	2,000	1,550	1,200	1,000
ASTM A193	B-8T	75,000	12,450	11,850	11,800	11,750	11,650	11,450	11,300	11,100	10,800	10,500	10,000	8,000	5,000	3,600	2,700	2,000	1,550	1,200	1,000
Plates and sheet:																					
ASTM A283	C	55,000	12,650	12,650																	
ASTM A283	D	60,000	12,650	12,650																	
ASTM A285	B	50,000	12,500	12,500	12,100	11,150	9,600	8,050	6,500												
ASTM A285	C	55,000	13,750	13,750	13,250	12,050	10,200	8,350	6,500												
ASTM A387	C	60,000	15,000	15,000	15,000	15,000	15,000	14,400	13,100	11,000	7,800	5,500	4,000	2,500	1,200						
ASTM A387	D	60,000	15,000	15,000	15,000	15,000	15,000	14,400	13,100	11,000	7,800	5,800	4,200	3,000	2,000						
ASTM A240	304	75,000	11,200	11,200	10,800	10,400	10,000	9,700	9,400	9,100	8,800	8,500	7,500	5,750	4,500	3,250	2,450	1,800	1,400	1,000	750
ASTM A240	321	75,000	14,850	14,850	14,800	14,700	14,550	14,300	14,100	13,850	13,500	13,100	10,300	7,600	5,000	3,600	2,700	2,000	1,550	1,200	1,000
ASTM A240	347	75,000	14,850	14,850	14,800	14,700	14,550	14,300	14,100	13,850	13,500	13,100	10,300	7,600	5,000	3,600	2,700	2,000	1,550	1,200	1,000
Pipe and tubing:																					
ASTM A53	A	48,000	12,000	12,000	11,650	10,700	9,000	7,100	5,000												
ASTM A53	B	60,000	15,000	15,000	14,350	12,950	10,800	7,800	5,000												
ASTM A335	P-11	60,000	15,000	15,000	15,000	15,000	15,000	14,000	13,100	11,000	7,800	5,500	4,000	2,500	1,200						
ASTM A335	P-22	60,000	15,000	15,000	15,000	15,000	15,000	14,000	13,100	11,000	7,800	5,800	4,200	3,000	2,000						
ASTM A312	TP-304	75,000	11,200	11,200	10,800	10,400	10,000	9,700	9,400	9,100	8,800	8,500	7,500	5,750	4,500	3,250	2,450	1,800	1,400	1,000	750
ASTM A312	TP 321	75,000	14,850	14,850	14,800	14,700	14,550	14,300	14,100	13,850	13,500	13,100	10,300	7,600	5,000	3,600	2,700	2,000	1,550	1,200	1,000
ASTM A312	TP 347	75,000	14,850	14,850	14,800	14,700	14,550	14,300	14,100	13,850	13,500	13,100	10,300	7,600	5,000	3,600	2,700	2,000	1,550	1,200	1,000

Allowable stress values, psi, for the materials and temperatures shown (temperature columns increase left‑to‑right from the minimum‑tensile column).

Material	Min. tensile																		
Castings:²																			
ASTM A48 20	20,000	2,000																	
ASTM A48 25	25,000	2,500																	
ASTM A48 30	30,000	3,000																	
ASTM A47 32,510	50,000	10,000																	
ASTM A47 35,018	53,000	10,600																	
ASTM A197	40,000	8,000																	
ASTM A216 WCA	60,000	15,000	14,350	12,950	10,800	8,650	6,500	4,500	2,500										
ASTM A216 WCB	70,000	17,500	16,600	14,750	12,000	9,250	6,500	4,500	2,500										
ASTM A217 WC6	70,000	17,500	17,500	17,500	17,000	15,800	14,000	11,000	7,800	5,500	4,000	3,000	2,000						
ASTM A217 WC9	70,000	17,500	17,500	17,500	17,000	15,800	14,000	11,000	7,800	5,800	4,200	3,000	2,000						
ASTM A351 CF8	70,000	11,500	11,300	11,100	10,900	10,650	10,400	10,100	9,850	9,600	7,500	5,750	4,500	3,250	2,450	1,800	1,400	1,000	750
ASTM A351 CF8C	70,000	15,800	13,700	13,300	12,900	12,600	12,300	11,900	11,600	11,200	10,800	8,000	5,000	3,600	2,700	2,000	1,550	1,200	1,000
ASTM A351 CF8M	70,000	16,100	14,700	14,350	14,000	13,500	13,000	12,350	11,700	10,600	9,400	8,000	6,800	5,300	4,000	3,000	2,350	1,850	1,500
ASTM A395	60,000	10,800																	
Forgings:																			
ASTM A105 I	60,000	15,000	14,350	12,950	10,800	8,650	6,500	4,500	2,500										
ASTM A105 II	70,000	17,500	16,600	14,750	12,000	9,250	6,500	4,500	2,500										
ASTM A182 F-11	70,000	16,800	16,150	15,500	15,000	14,400	13,100	11,000	7,800	5,500	4,000	3,000	2,000						
ASTM A182 F-22	70,000	17,500	17,500	17,500	17,500	16,000	14,000	11,000	7,800	5,800	4,200	3,000	2,000						
ASTM A182 F-304	75,000	11,200	10,800	10,400	10,000	9,700	9,400	9,100	8,800	8,500	7,600	5,750	5,000	3,250	2,450	1,800	1,400	1,000	750
ASTM A182 F-321	75,000	14,850	14,800	14,700	14,550	14,300	14,100	13,850	13,500	13,100	10,300	7,600	5,000	3,300	2,200	2,000	1,550	1,200	1,000
ASTM A182 F-347	75,000	14,850	14,800	14,700	14,550	14,300	14,100	13,850	13,500	13,100	10,300	7,600	5,000	3,300	2,200	2,000	1,550	1,200	1,000
ASTM A235 A	47,000	11,750	10,400																
ASTM A235 C	60,000	15,000	13,300																
Rivets and nuts:																			
ASTM A141	52,000	13,000																	
ASTM A195	68,000	17,000																	
Springs																				

See Figure II under Paragraph 11 for Materials and Stresses.

NOTE: Values for stresses taken from material proposed for inclusion in American Standard Code for Pressure Piping, ASA B31.1, Power Piping, except those underlined which are taken from the ASME Boiler and Pressure Vessel Code.

[1] Reproduced by permission from Table II of MSS-SP58.

[2] Except for cast iron (ASTM A48) and malleable iron (ASTM A47) a casting quality factor of 0.8 shall be applied to stress values given, or the instructions in paragraph UG-24 of Section VIII of the ASME Boiler Code shall be followed.

must be of appropriate (compatible) chemical composition, and the process of attachment must conform to the requirements for fabrication of the pipe as regards preheating, welding, and stress relieving.

Cold Spring. The cold-springing procedure for piping systems involves, basically, the cutting short of each segment of piping in an amount equal to some specific percentage of the normal thermal growth of the segment with proper allowance to compensate for possible terminal connection movements. The cumulative effect results in an offset gap between the piping ends at the point of final field closure. The drawing together and alignment of the piping ends by the application of required forces and moments are called cold pull and result in a slight relocation of all points along the entire piping system. This relocation is the only effect cold springing has on hangers. The piping is then considered as being in the cold-sprung position, which is equivalent to the cold operating position. The movement at all hanger points from the erected to the cold-sprung position must be calculated, and provision made for this movement in the form of hanger rod adjustment.

Adjustment. It is necessary to provide vertical adjustment to attain the desired elevation of the piping system. On piping supported from above, it may suffice to adjust a Type 1, 5, 6, 7, 9, or 10[1] hanger through the yoke portion by raising or lowering the nut on the rod. For larger ranges of adjustments, it is necessary to provide a turnbuckle, Type 13 or 15[1], in the hanger rod. It is also practical to select a top attachment whereby adjustment can be made at the top of the hanger rod.

On piping which is supported from below, provision for adjustment can be made with screw thread stanchions, Type 35 or 38[1], or by shims or grout.

Protective Coatings. Protective coatings are normally applied to pipe hangers for corrosion resistance. Metallic coatings may be applied by either the electroplating or hot-dip process. Nonmetallic coatings, if selected for specific purposes, should be applied as recommended by the manufacturer of such coating. Consideration must be given to coatings used on threaded parts that are to be assembled after coating.

Piping Weights. A step in the design of piping supports is the calculation of the weight of the piping to be supported. This will necessarily include weight of pipe, water or other fluid being transported, fittings, flanges, valves, insulation, and any other related items the weights of which are also to be supported as part of the piping system.

Pipe material weights are shown in bold-face type in Table 4, Chap. 7. The weights are subject to tolerances of applicable manufacturing specifications.

Weights of insulation are for conventional 85 per cent magnesia and diatomaceous silica and asbestos compositions and include approximate weights of canvas, cement, paint, wire, and bands but not weatherproofing or other special protection. Weights of other compositions of pipe covering will vary and should be obtained from the insulation manufacturer. Weights of weatherproof protection, if specified, must be added. Weights shown for insulation are related to conventional thickness recommendations by insulation manufacturers and do not necessarily agree with insulation specifications for a particular job. Insulation specifications should be reviewed prior to development of final weights of piping.

To determine weights of insulation to be added to weights of flanges, valves, and fittings, multiply the weight per foot of pipe covering by the appropriate factor shown in light-face subscript.

Flange, flanged valve, and flanged fitting weights include approximate proportional weights of bolts and nuts.

---

[1] Type numbers refer to Fig. 11.

Weights shown for butt-welding reducers are for one pipe size reduction.

Weights of 125- and 250-lb cast-iron valves are for valves with standard flange ends.   Weights of steel valves are for welding-end type.   To calculate weights of flanged-end steel valves, add the tabulated weight of the valve to the weight of two corresponding slip-on flanges.

Valve weights vary among particular manufacturer's designs.   Weights shown in Table 4, Chap. 7 are approximate only and do not include weights of electric-motor operators or other devices which may be specified for particular valves.   It is suggested that, wherever possible, valve weights should be obtained from the manufacturer of the particular valves which are to be installed in the piping.

6

THERMAL INSULATION

E. J. Wesemann*

Electric power generation and steam production for diverse industrial applications rely on the conversion of the potential energy in a fuel into useful work. Steam-generating power stations using complex equipment and heat cycles are successful in converting only approximately 40 per cent of the total potential energy in the fuel into useful work. Heat-insulating materials make a major contribution toward the capture of this percentage of the total potential energy.

The importance of insulation as a means of reducing heat transfer is emphasized by recognizing the magnitude of losses from bare heated surfaces as compared with the relatively small losses from such surfaces when properly insulated. This is illustrated in Fig. 1 in which heat losses per degree temperature difference from bare surfaces are shown by the upper curve and from insulated surfaces by the lower curve. The area between the two curves represents the saving by insulation.

In addition to the saving of fuel, insulation performs other valuable functions. Notable among these are the minimizing of undesirable temperature drop in superheated steam lines, hot-air ducts, etc.; the maintaining of desired temperatures in equipment; the prevention of leaky flange joints; and the assurance of more comfortable working conditions in the vicinity of heated surfaces.

Heat Losses from Bare Surfaces. The rates of heat losses from bare surfaces at temperature differences up to 1000 F are shown in Fig. 2. These are average values for still-air conditions, and although there is some variation for different pipe sizes and for different absolute temperatures of surroundings, these variations are small as compared with those caused by comparatively low air velocities; therefore, these average values are sufficiently accurate for engineering purposes.

* Gibbs and Hill, Inc., New York, N.Y.

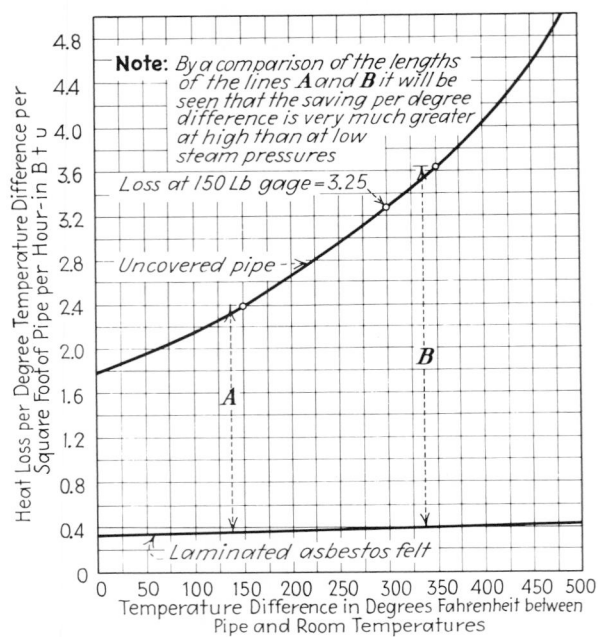

FIG. 1. Heat loss from insulated and bare pipe surfaces. (*Univ. Illinois Eng. Expt. Sta. Circ.* 7.)

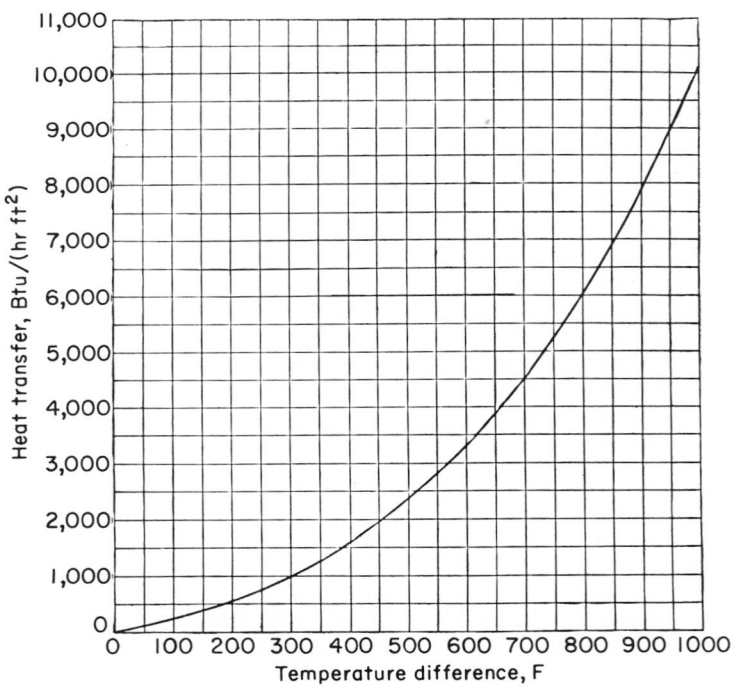

FIG. 2. Heat transfer in still air from bare surfaces at various temperatures.

Heat losses expressed only in Btu are not usually so significant as when expressed in the equivalent actual costs in dollars.   In Fig. 3 the equivalent losses in dollars per square foot per year (8,760 hr) have been shown for various temperatures per 1,000,000 Btu.   (1,000,000 Btu is approximately equivalent to 1,000 lb of steam.)

The value of energy per 1,000,000 Btu is either known or may be computed readily for a given fuel of known cost and a given efficiency; thus, this procedure renders Fig. 3 applicable to a wide range of fuels and conditions.   Taking as a specific example a square foot of surface on an equipment operating 7,200 hr per year at a temperature of 300 F above that of the surrounding air, the loss per year,

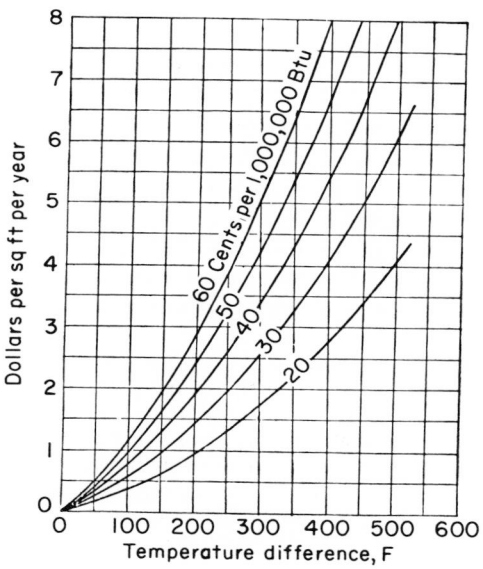

Fig. 3.   Cost of heat losses per year at various values of energy.

at an energy value of 30 cents per 1,000,000 Btu, is $2.60 × 7,200/8,760 = $2.14. Suitable insulation saving upward of 90 per cent of this loss may be applied at a cost considerably less than one year's saving.   This illustrates the desirability of insulating such surfaces as boiler heads, flanges, and fittings which are frequently left uninsulated, even though adjacent piping is provided with effective insulation.

Effect of Air Velocity on Losses from Bare Surfaces.   The rate of heat loss from a surface maintained at constant temperature is greatly increased by air circulation over the surface.   This is illustrated in Fig. 4.

In the case of well-insulated surfaces the increases in losses caused by air circulation are very small compared with increases shown above for bare surfaces.   The maximum increase in heat loss due to air velocity ranges from about 30 per cent in the case of 1-in.-thick insulation to about 10 per cent in the case of 3-in.-thick insulations, provided that the insulation is thoroughly sealed so that air can flow only over the surface and does not penetrate the insulating material.

Radiating Surface of Pipes.   In order to determine heat losses per linear foot of pipe from known losses per square foot, it is necessary to know the number of square feet area per linear foot of pipe.   Tables 1a and 1b give these areas for various standard pipe and tubing sizes.

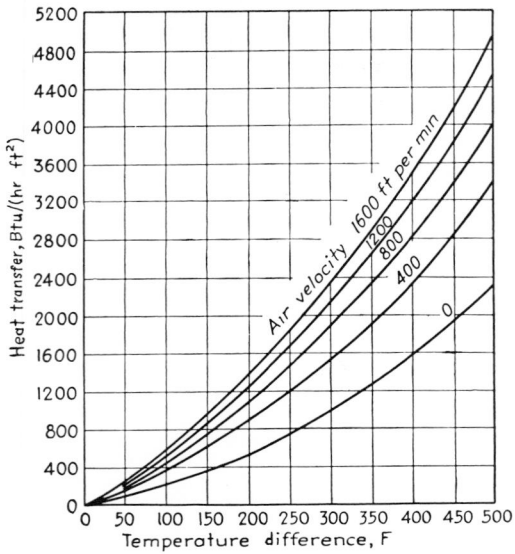

Fɪɢ. 4. Effect of air velocity on rates of heat transfer.

Heat Losses from Bare Fittings.   Very often, even where pipes are thoroughly insulated, flanges and fittings are left bare because of the belief that the losses from these parts are not large.   However, the fact that a pair of 10-in. standard flanges having an area of 3.43 sq ft would lose at 100-lb steam pressure an amount of heat equivalent to more than 1 ton of coal per year shows the necessity for insulating such surfaces.   Table 2 shows the areas of both 125- and 250-lb (or 150- and 300-lb steel) flanged fittings including the accompanying flanges bolted to the fitting.

Table Ia.   Radiating Surface per Linear Foot of Steel Pipe

Nominal pipe size, inches	External surface, square feet	Nominal pipe size, inches	External surface, square feet	Nominal pipe size, inches	External surface, square feet
½	0.22	2	0.622	5	1.456
¾	0.275	2½	0.753	6	1.734
1	0.344	3	0.917	8	2.257
1¼	0.435	3½	1.047	10	2.817
1½	0.498	4	1.178	12	3.338

Table Ib.   Radiating Surface per Linear Foot of Copper Tubing

(Outside diameter ⅛ in. greater than nominal size)

Tubing size, inches	External surface, square feet	Tubing size, inches	External surface, square feet	Tubing size, inches	External surface, square feet
½	0.164	2	0.556	5	1.342
¾	0.229	2½	0.687	6	1.604
1	0.295	3	0.818	8	2.128
1¼	0.360	3½	0.949		
1½	0.426	4	1.080		

Table 2.   Areas of Flanged Fittings,[1] Square Feet

Nominal pipe size, inches	Flanged joint		90-deg ell		Long radius ell		Tee		Cross	
Pressure	125	250	125	250	125	250	125	250	125	250
1	0.320	0.438	0.795	1.015	0.892	1.083	1.235	1.575	1.622	2.07
1¼	0.383	0.510	0.957	1.098	1.084	1.340	1.481	1.925	1.943	2.53
1½	0.477	0.727	1.174	1.332	1.337	1.874	1.815	2.68	2.38	3.54
2	0.672	0.848	1.65	2.01	1.84	2.16	2.54	3.09	3.32	4.06
2½	0.841	1.107	2.09	2.57	2.32	2.76	3.21	4.05	4.19	5.17
3	0.945	1.484	2.38	3.49	2.68	3.74	3.66	5.33	4.77	6.95
3½	1.122	1.644	2.98	3.96	3.28	4.28	4.48	6.04	5.83	7.89
4	1.344	1.914	3.53	4.64	3.96	4.99	5.41	7.07	7.03	9.24
5	1.622	2.18	4.44	5.47	5.00	6.02	6.81	8.52	8.82	10.97
6	1.82	2.78	5.13	6.99	5.99	7.76	7.84	10.64	10.08	13.75
8	2.41	3.77	6.98	9.76	8.56	11.09	10.55	14.74	13.44	18.97
10	3.43	5.20	10.18	13.58	12.35	15.60	15.41	20.41	19.58	26.26
12	4.41	6.71	13.08	17.73	16.35	18.76	19.67	26.65	24.87	34.11
14 O.D.	5.39	8.30	16.38	22.31	20.17	25.70	24.81	33.63	31.48	43.15
16 O.D.	6.69	10.05	20.17	27.18	25.41	31.73	30.32	40.94	38.34	52.35

[1] Includes areas of adjoining companion flanges bolted on each outlet.   Areas for 150- and 300-lb steel fittings are same as 125- and 250-lb cast iron, respectively.

Heat Transfer through Insulation.   The rate of heat transfer through insulation is dependent upon the internal resistance offered by the insulation, the surface resistance, and the temperature of the hot surface and of the surrounding air.

The internal resistance to heat flow is dependent upon the thickness x and the conductivity k.   In the case of flat surfaces it varies directly as the thickness and inversely as the conductivity and is equal to x/k.   Heat transfer through a material having flat surfaces is represented by Eq. (1), in which U is the overall rate of heat-transfer coefficient in Btu per square foot per hour per degree difference in temperature, x is the insulation thickness in feet, k is the thermal conductivity of the insulating material in Btu per hour per foot per degree Fahrenheit and f is the convective conductance in Btu per hour per square foot per degree Fahrenheit between the outer surface and the surrounding air.

$$U = \frac{1}{x/k + 1/f} \tag{1}$$

The total heat transmitted is $H = UA(t_1 - t_2)$ in which

H = total heat transmitted, Btu/hr
A = area of surface, sq ft
t_1 = temperature of hot surface, F
t_2 = temperature of surrounding air, F

Conductivity.   Thermal conductivity is defined as rate of heat transfer in one direction (perpendicular to an area) per unit area, per unit temperature differential per unit thickness, per unit time.   In the English system of units, typical dimensions are:

$$|k| = \text{Btu/(hr F sq ft/ft)} = \text{Btu/(hr ft F)}$$

or

$$|k| = \text{Btu/(hr F sq ft/in.)}$$

Conductivity is a specific property of a material.   It is not dependent on the area, thickness, or shape of the material.   It is a rate, not a quantity.   The total quantity

Table 3. Thermal Conductivities, Densities, and Recommended Use-temperature Limits for Various Types of Heat-insulating Materials

(Coefficient k, Btu/(hr F sq ft/in.).)

Type of material (For descriptions, see pages 710–714)	Mean temperature of insulation, F									Approximate density, lb per cu ft	Use-temperature limit, F
	100 F	200 F	300 F	400 F	500 F	600 F	700 F	800 F	900 F		
1. High-temperature pipe covering and blocks (diatomaceous earth and asbestos)	0.624	0.659	0.694	0.729	0.764	0.799	0.834	0.869	.....	24 to 26	1600 to 1900
2. Laminated asbestos pipe covering and sheets (approximately 30 to 40 laminations per inch)	0.360	0.415	0.470	0.525	0.585	.....	.....	.....	.....	30 to 40	500
3. Laminated indented asbestos pipe covering and sheets (approximately 20 laminations per inch)	0.366	0.450	0.534	0.618	0.702	.....	.....	.....	.....	17 to 21	500
4. 85 per cent magnesia pipe covering and blocks	0.398	0.437	0.476	0.516	0.555	0.599	.....	.....	.....	15 to 17	600
5. Corrugated asbestos pipe covering and sheets (4 plies per 1 in. thickness)	0.496	0.621	0.746	0.871	.....	.....	.....	.....	.....	11 to 15	400
6. Corrugated asbestos pipe covering and sheets (8 plies per 1 in. thickness)	0.506	0.601	0.696	0.791	.....	.....	.....	.....	.....	21	400
7. Mineral wool (rock wool, slag wool, glass wool):											
(a) Loose	0.261[1]	0.353	0.445	0.537	0.629	0.721	0.813	0.905	.....	6 to 10	1200
(b) Blankets	0.300	0.368	0.436	0.504	0.575	0.640	0.708	0.776	.....	8 to 15	800
(c) Block	0.347	0.393	0.439	0.485	0.531	0.577	0.623	0.669	.....	15 to 23	1200 to 1800
8. Amosite asbestos pipe covering	0.402	0.450	0.500	0.550	0.596	0.652	0.690	0.738	.....	22	1200
9. Expanded vermiculite (mica) pipe covering and blocks	0.350	0.405	0.460	0.515	0.570	0.625	0.680	.....	.....	13	700
10. Aluminum foil (3 crumpled layers per inch)	0.880	0.930	0.980	1.030	1.080	1.130	1.180	1.230	.....	20	1200
11. Cork pipe covering and blocks	0.358	0.423	0.488	0.553	0.619	0.684	0.749	0.815	.....	0.5	1200
12. Hair-felt sheet	0.30 to 0.39	.....	.....	.....	.....	.....	.....	.....	.....	7 to 14	200
13. Tar-lined wool-felt pipe covering	0.300	.....	.....	.....	.....	.....	.....	.....	.....	6 to 11	150
14. Insidine blocks	0.300	.....	.....	0.630	0.720	0.800	0.880	0.965	1.050	18	150

NOTES.—When a range of values is given for the Approximate Density or Use-temperature Limit, it indicates that these properties vary among the products of different manufacturers. The thermal conductivity values given were selected as a result of a comprehensive survey of published and unpublished data from numerous sources. The majority of the values given are based upon tests by R. H. Heilman, Mellon Institute (private communication).

[1] May vary from 0.240 to 0.350 at 100 F, depending upon the quality and density of the product.

of heat transmitted is dependent upon the area, shape, and length of path (thickness of material), but conductivity is not. Conductivity is dependent upon temperature, but this is also true of other specific properties of material, density, for example. Standard test methods for determining conductivity are specified in ASTM Specification C177, Test for Thermal Conductivity of Materials by Guarded Hot Plate.

In Table 3 conservative conductivities are shown. In this table conductivities are shown as functions of the mean temperatures or the mean between the inner and outer surface temperatures of the insulation. This method of expressing conductivities permits their use in the calculation of heat transfer through materials whether used singly or in combination with other materials.

Surface Resistance. The term $1/f$ in Eq. (1) represents the surface resistance. When heat flows through a solid material and then out into air (or any other fluid), a resistance to heat flow is encountered at the surface separating the solid from the

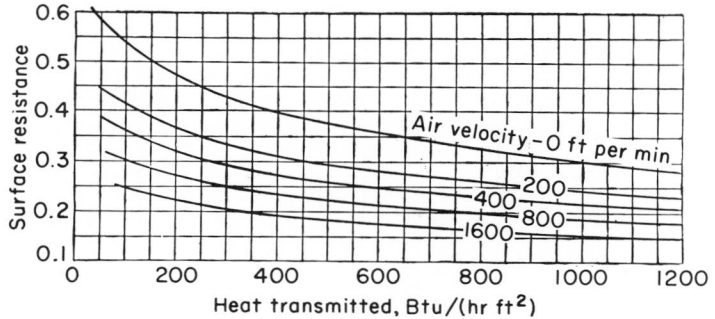

Fig. 5. Surface resistance $1/f$ at various air velocities.

air. Less heat will flow from the surface, therefore, than if no resistance were offered at this point.

In the case of good conductors of heat, surface resistance is the greater part of the total resistance to heat flow. In connection with efficient insulating materials, however, surface resistance is small compared with the internal resistances of the materials.

Numerically, surface resistance is the reciprocal of the rate of heat transmission from surface to air. That is, if the rate of heat transmission from surface to air is 2.0 Btu/sq ft per degree temperature difference per hour, the surface resistance is 0.5 hr sq ft/Btu. A higher rate of heat transmission from surface indicates a lower surface resistance and vice versa.

Surface resistances are very materially reduced by air velocity, and Fig. 5 shows values of surface resistances sufficiently accurate for use in insulation calculations where the surface resistance is usually less than 25 per cent and frequently less than 10 per cent of the total resistance.

Effect of Thickness of Insulation on Rate of Heat Transfer. The manner in which the rate of heat transfer varies as the thickness of the insulation is increased is shown in Fig. 6, which is based on the conditions of Eq. (1). It will be noted that the curve appears relatively flat beyond 2 in. thickness. This is by no means an indication, however, that the additional savings by thicker insulations will not be large where temperatures and value of heat demand the use of such greater thickness. The curves show losses per degree temperature difference, and the total losses are determined by multiplying the values from the curve by the temperature difference. For example, the saving by the additional inch of thickness from 3 to

4 in. looks small as compared with that from 1 to 2 in., but the greater thicknesses are used at the higher temperatures; therefore, actually at 500 F temperature difference, the saving, by increasing the thickness from 3 to 4 in., is much greater than that which would result from increasing it from 1 to 2 in. where the temperature difference was 100 F.

The heat-transfer coefficient through a combination of any number of materials on a flat surface may be determined through the use of the following equation, in which

$$U = \frac{1}{1/f_1 + x_1/k_1 + x_2/k_2 + x_3/k_3 + \cdots + 1/f_2} \tag{2}$$

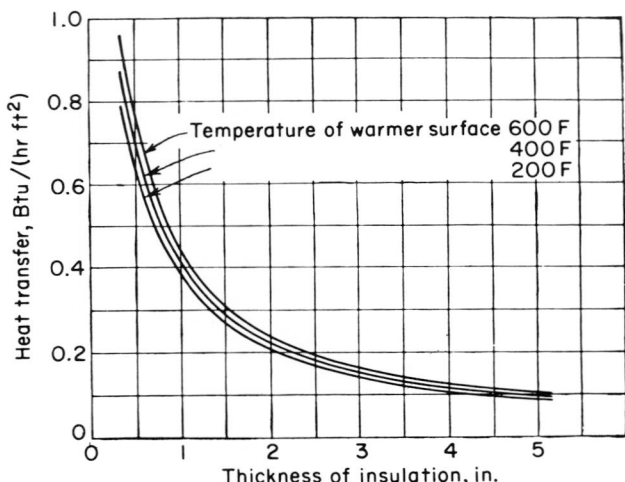

FIG. 6. Effect of thickness on heat transfer through flat surfaces.

x_1, x_2, x_3, etc., are the thicknesses and k_1, k_2, k_3, etc., are the conductivities of the respective materials; f_1 is the convective conductance at the warmer surface; and f_2 is the convective conductance at the cooler surface.

The inside-surface resistance $1/f_1$ is not used when the temperature of the warmer surface is known. Also, its magnitude is often negligible where effective insulation is placed directly against a heated surface, the temperature of which is known. It is included in these general equations, however, in order that it be not neglected in cases where it should be taken into account.

Example. Suppose it is required to calculate the heat transmission through a composite wall consisting of 4 in. of a material (brick, for example) with a conductivity of 5.0, and 2 in. of material (insulation) with a conductivity of 0.5, and that the other conditions are: inside-surface temperature, 200 F; outside-air temperature, 70 F; and rate of heat transfer from surface to air, 2.0 Btu/sq ft per degree temperature difference per hour. Substituting in Eq. (2) gives

$$U = \frac{1}{\frac{4}{5} + 2/0.5 + \frac{1}{2}} = \frac{1}{5.3} = 0.189$$

Heat transmission, $H = AU(t_1 - t_2)$.
For 1 sq ft, $H = 0.189(200 - 70) = 24.5$ Btu

The percentages of the total insulating value contributed by the various items are as follows:

4-in. material (brick):
  0.8/5.3 = 15.1 per cent of total
2-in. material (insulation):
  4.0/5.3 = 75.5 per cent of total
Surface resistance:
  0.5/5.3 = 9.4 per cent of total

It is apparent, therefore, that the 2-in. material of low conductivity contributed five times as much insulating value as did the 4-in. material with relatively high conductivity and the surface resistance is to be credited with less than 10 per cent of the total insulating value. This illustrates the relative importance of the various items.

Cylindrical Surfaces. Except on flat surfaces, the internal resistance of a material does not vary directly as the thickness. In the case of cylindrical surfaces,

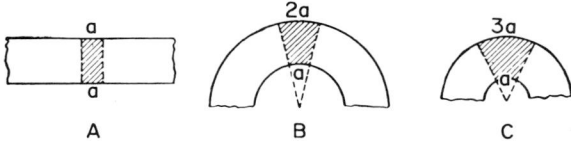

FIG. 7. Areas of paths for heat transfer through insulations on flat and cylindrical surfaces.

increasing the thickness supplies additional resistance through which the heat must flow but at the same time increases the area of the path through which the heat may flow. This is illustrated in Fig. 7, which shows the areas of paths for flat and cylindrical surfaces. It is clearly apparent from this diagram that the heat transfer per unit of area of inner surface will be greater for insulation on a curved than on a flat surface and that the smaller the radius of curvature, the greater will be the rate of heat transfer per unit of inner-surface area.

The heat loss per hour per degree difference in temperature per square foot of *outer surface* of the insulation on a pipe or other cylindrical surface is given by Eq. (3) in which r_1 is the external radius of the pipe or cylinder, r_2 is the radius of the outer surface of the insulation, and the other terms are as defined previously.

$$U_2 = \cfrac{1}{\cfrac{r_2 \log_e r_2/r_1}{k} + \cfrac{1}{f}} \tag{3}$$

The loss per square foot of *pipe surface* per degree difference in temperature is given by Eq. (3a) in which U_1 and U_2 represent the rates of heat transfer per hour per degree difference in temperature per square foot of pipe surface and outer surface of insulation, respectively.

$$U_1 = r_2/r_1 \times U_2 \tag{3a}$$

The loss per linear foot of pipe per degree is found by multiplying the rate of heat transfer per hour per square foot of pipe surface U_1 by the square feet of external surface per foot of pipe given in Table 1.

The effect of pipe size on the rates of heat transfer through insulation per square foot of pipe surface under the conditions covered by Eqs. (3) and (3a) is illustrated in Fig. 8. It will be noted that the rate of heat transfer through 2-in.-thick insulation on 1-in. pipe is more than twice as great as that through the same thickness of insulation on 12-in. pipe.

The heat-transfer coefficient through combinations of two or more insulations on a pipe or other cylindrical surface may be calculated from Eq. (4) and (4a)

$$U_s = \cfrac{1}{\cfrac{r_s}{r_1} \times \cfrac{1}{f_1} + \cfrac{r_s \log_e r_2/r_1}{k_1} + \cfrac{r_s \log_e r_3/r_2}{k_2} + \cfrac{r_s \log_e r_4/r_3}{k_3} + \cdots + \cfrac{1}{f_2}} \tag{4}$$

$$U_1 = r_s/r_1 \times U_s, \tag{4a}$$

in which r_s is the radius of the outer surface of insulation and U_s is rate of heat transfer per hour per degree difference in temperature per square foot of this surface.   Other terms have the same significance as in previous equations.

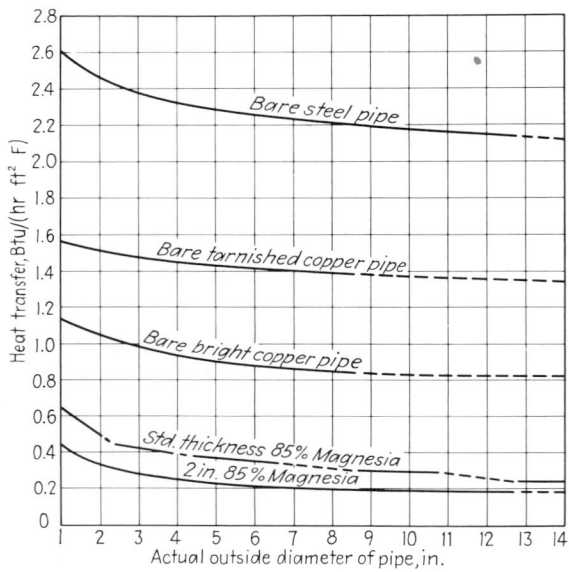

Fig. 8. Variation with pipe size of rate of heat loss from bare-steel and copper pipe compared with various thicknesses of 85 per cent magnesia insulation.   Pipe temperature 210 F, surrounding air at 70 F.   Dash lines indicate interpolated or extrapolated data.

Effect of Air Velocity on Heat Losses from Insulated Surfaces.   In the case of well-insulated surfaces the increases in heat losses due to air velocity are very small as compared with the increases previously shown for bare surfaces.   This is due to the fact that air flowing over the surface of the insulation can increase only the rate of heat transfer from surface to air and cannot change the internal resistance to heat flow inherent in the insulation itself.   The effect of the air circulation, therefore, is to cool the surface of the insulation to a temperature lower than it would have under still-air conditions, thereby increasing the temperature drop through the insulation.

In the case of surfaces located out of doors, the combined effect of wind and rain may bring the surface temperature of the insulation practically down to the air temperature, yet even in this extreme case the increase in heat loss through the insulation is not so great as might be expected.   This is illustrated by the following example, based on a flat surface insulated with 2-in.-thick material, having a conductivity of 0.5 Btu/sq ft per degree temperature difference per 1 in. thickness

per hour and a rate of heat transfer from its surface to air under still-air conditions of 1.8 Btu/sq ft per degree temperature difference per hour.

$$\text{Internal resistance of insulation} = 2/0.5 = 4.0$$
$$\text{Surface resistance} = 1/1.8 = 0.556$$
$$\text{Total resistance} = \overline{4.556}$$
$$\text{Rate of heat transfer} = 1/4.556 = 0.22 \ \text{Btu/(sq ft per degree temperature}$$
$$\text{difference per hour)}$$

If the surface resistance were completely eliminated, owing to the cooling action of wind and rain, the internal resistance of 4.0 would still remain, and the rate of heat transfer would be 1/4.0 = 0.25 Btu/(sq ft per degree temperature difference per hour). Therefore, the increase in loss due to wind and rain, eliminating surface resistance, would be

$$(0.25 - 0.22)/0.22 = 13.6 \ \text{per cent}$$

In like manner, it may be shown that the increase for 1 in. thickness of the same material under the same conditions is 27.9 per cent, and in the case of 3 in. thickness, 9.2 per cent. It is therefore, apparent that the thicker or the more efficient an insulation is, the less its heat transfer will be affected by air circulation.

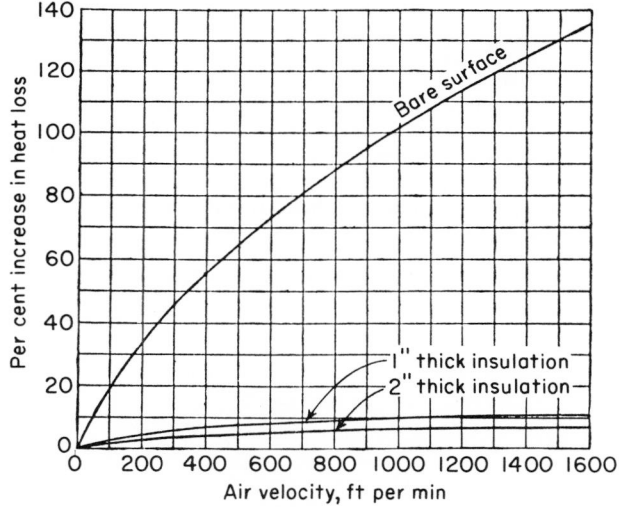

FIG. 9. Increase in heat losses due to air circulation.

Figure 9 shows graphically the relative increases in rates of heat transfer due to air circulation in the case of a bare surface maintained at 400 F and the same surface insulated with 1- and 2-in. thickness of an insulation with a conductivity of 0.48 Btu/sq ft per degree temperature difference per 1-in. thickness per hour.

All of the above discussion as to effect of air circulation on losses through insulation is based on flow of air over the surface of the insulation and applies to cases where the insulation is tightly sealed. If the condition of the insulation is such that the air may circulate through cracks and crevices in the insulation, the increases may be far greater than those given above. It is essential, therefore, that all insulation be sealed as tightly as possible, and this is most particularly true of insulation located out of doors. On the latter, effective weatherproofing should be provided which should be so designed and applied as to seal the insulation against infiltration of air, as well as protecting it from the weather.

Cold Pipes. Insulation on cold surfaces also should be sealed thoroughly against the penetration of moist air. The temperature within the insulation frequently will be, and in the case of insulated cold water, brine, and ammonia piping usually will be, below the dew point of the surrounding air. If air leaks into the insulation, therefore, the air will be chilled below its dew point and moisture will be deposited in the insulation. Aside from damage which may result from the consequent drippage of water on equipment, goods, etc., such condensation is undesirable since the wetted pipe surface promotes increased heat transfer from the cooling fluid, and the cold effect of the latter is reduced through absorption of the latent heat liberated by the condensing moisture, thus reducing the useful cooling effect of the refrigerant. Also, water soaking will seriously impair the effectiveness of the insulation, not only because of the increased heat transfer through the insulation, but also because of actual mechanical damage to the material itself. Where the surfaces insulated are below 32 F, moisture that enters the insulation freezes, and frequently the expansive forces due to the formation of frost are sufficient to disintegrate the insulation. In connection with insulation on such cold surfaces, therefore, the provision and maintenance of a moistureproof seal are two of the more essential requirements.

HEAT-INSULATING MATERIALS

Piping insulations are furnished molded or cut into cylinders, half cylinders, or curved segments. They are furnished in nominal thickness of 1- and ½-in. increments through a 4-in. total thickness.

Equipment insulations are furnished in flat or curved blocks, flexible blankets, semirigid felts, and loose or granular form.

In Table 4 are shown the standard industrial insulations indexed to ASTM and government specifications.

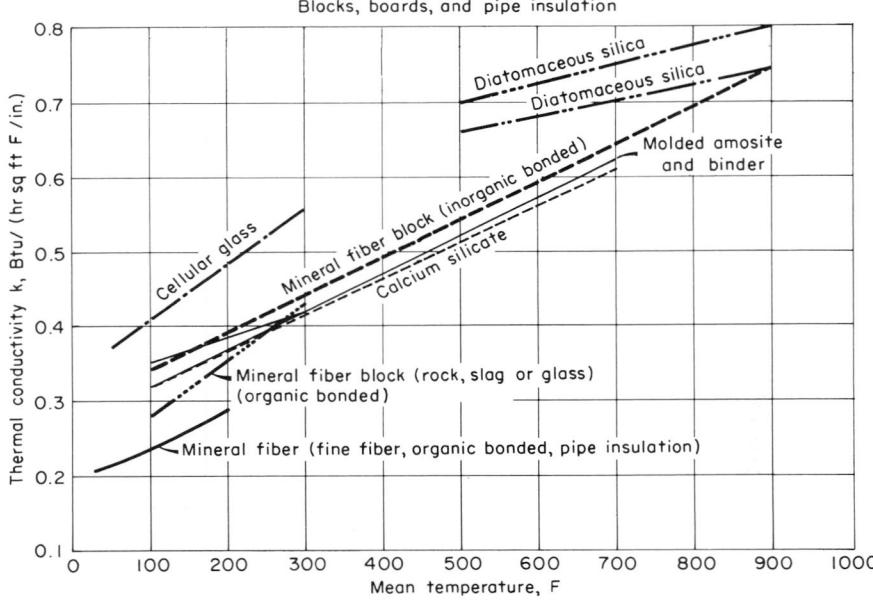

Fig. 10. Conductivity k design values. Blocks, boards, and pipe insulation.

Figures 10 to 12 indicate average conductivity design values for insulations and cements.

These data are as compiled by the National Insulation Manufacturers Association (NIMA). For the k values of specific insulations and for temperature limits in specific applications, manufacturers should be consulted.

Mineral fiber (rock, slag, or glass)

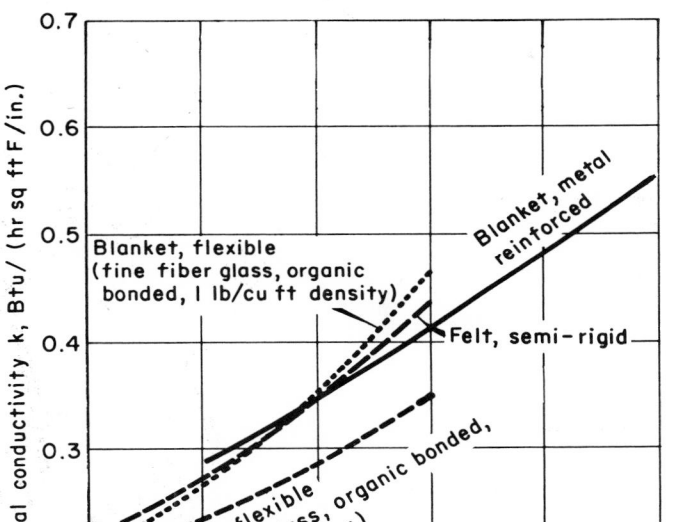

Fig. 11. Conductivity k design values. Mineral fiber (rock, slag, or glass).

In addition to the listed heat-insulating material data, other types for possible consideration to specific problems where installed cost is secondary to heat conservation are:

1. *Heat-reflector Type.* An all-metal product consisting of an outer case for weather resistance and many closely spaced reflective inner sheets supported to form isolated air chambers. This type of insulation in refined form has particular application in the field of cryogenics.

2. *Foam Plastic Type.* Plastic manufactured with minute air cells so designed that in service the cells do not allow escape of gas to impair the thermal conductivity of the insulation.

3. *Bonded-silica Type.* Developed for outer-space travel, this insulation is an extremely dense structure of bonded silica with fine pores less than four-millionths of an inch in diameter with a thermal conductivity of 0.14 Btu/(hr sq ft F/in.) at 400 F.

Table 4. Index to ASTM and Government Specifications for Industrial Insulations

Product form	Reference name and general description	Accepted max temp.[1] F	ASTM specifications[2]	Military (MIL) and Maritime (MA) specifications[2]	Federal specifications[2]
Pipe insulation....	Asbestos (molded amosite and binder)	1200		MIL-I-2781 Grade II, Class C Grade III, Class F	HH-I-561, Type IV
	Calcium silicate (calcium silicate and asbestos)	1200	C345	MIL-I-2781 Grade I, Class B Grade II, Class D Grade III, Class E, Type I	HH-I-523, Class 2
	Cellular glass (foamed, fabricated from block)	800	C381		HH-I-551
	Cellular silica (foamed, fabricated from block)	1600 (cyclic) 2200 (cont.)			
	Diatomaceous silica (diatomaceous silica and asbestos)	1600 1900	C334 C334	MIL-I-2781 Grade III, Class E, Type II	
	85% Magnesia (Basic magnesium carbonate and asbestos)	600	C320	MIL-I-2781 Grade I, Class B	HH-I-554, Type II
	Mineral fiber (rock, slag or glass) low-temp (organic binder)	250	C300		HH-I-562, Type II HH-I-552, Type I
	Low- and medium-temp (fine fiber, organic binder)	370	C281 C300	MIL-I-22344	HH-I-562, Type II HH-I-552, Type I
	High-temp (blanket-type metal reinforced)	1200	C280		HH-I-552, Type II
Blocks and boards........	Calcium silicate (calcium silicate and asbestos)	1200	C344	MIL-I-2819, Classes A and B MIL-I-002819, Classes 1 and 2	HH-I-523, Class 1
	Cellular glass (foamed)	800	C343		HH-I-551
	Cellular silica (foamed)	1600 (cyclic) 2200 (cont.)			
	Diatomaceous silica (diatomaceous silica and asbestos)	1600 1900	C333 C333	MIL-I-2819, Classes B and C MIL-I-002819, Class 3 MIL-I-002819, Class 4	

Category	Material	Max temp[1]	ASTM	Federal / Military spec	HH-I spec
	85% magnesia (basic magnesium carbonate and asbestos)	600	C319	MIL-I-2819, Class A / MIL-I-002819, Class 1	HH-I-554, Type I
	Mineral fiber (rock, slag or glass) low-temp (organic binder)	250	C378	32-MC-2[3] MIL-I-742[4]	HH-I-546, Class A / HH-I-526, (for Roofs) / HH-I-562, Type I
	Low-temp (fine fiber, organic binder)	400	C378, C392, Class 1		HH-I-562, Type I
	High-temp (inorganic binder)	1800	C392, Class 2		HH-I-564, Class B, C, D
Cements	Calcium silicate (calcium silicate, asbestos, and binders)	1200			HH-I-523, Class 3
	Diatomaceous silica (diatomaceous silica and asbestos)	1900	C197	MIL-C-2861, Type A	HH-I-00500, Type V
	Magnesia (basic magnesium carbonate, asbestos, and binder)	600	C193	MIL-P-2886	HH-I-00500, Type I
	Mineral fiber (rock, slag, or glass) insulating (colloidal clay and binder)	1800	C195	MIL-C-2861, Type B	HH-I-00500, Type III / HH-C-168
	Finishing (hydraulic-setting cement and binder)	1200	C449	MIL-C-2908, Type II	
Blankets and felts	Mineral fiber (rock, slag, or glass) metal-reinforced blanket	1200	C263	MIL-I-2818	HH-I-563, Type I
	Flexible (organic bonded)	400	C264 / C382	MIL-I-942 MIL-I-22023[6] / MIL-B-5924 } [5] MIL-I-16688 / MIL-I-7171 } 32-MA-3[3] / MIL-I-15475	HH-I-542, Type I
	Industrial batt (no binder)	1200	C262	MIL-W-15427	
	Felt (semi rigid, organic bonded)	400	C264 / C382	MIL-I-16688 } [7] MIL-I-942 / MIL-I-15475 } 32-MC-1[3]	HH-I-521, Type I / HH-I-542, Type II / HH-I-563, Type II
Loose and granulated	Mineral fiber (rock, slag or glass)	1200			HH-I-521, Type I—Loose / Type II—Gran

1 These temperatures are generally accepted as maximum. For specific applications consult the manufacturer. In columns where no specification number appears, no specifications have yet been adopted.
2 See latest revisions of these specification numbers.
3 Specifications 32-MC-1 and -2 and 32-MA-3 apply to Maritime Administration use only.
4 Specification MIL-I-742 applies to special mineral fiber board used by the Navy for hull and duct insulation.
5 Specifications MIL-B-5924 and MIL-I-7171 apply to glass fiber batting and blanket for thermal and acoustical insulation of aircraft compartments.
6 Supersedes MIL-I-15365 and MIL-I-16022.
7 Specifications MIL-I-15475 and MIL-I-16688 apply to Naval cold storage and refrigeration spaces, etc.

Table 5. Recommended Thicknesses for Calcium Silicate

Nominal pipe size, in.	100–199	200–299	300–399	400–499	500–599	600–699	700–799	800–899	900–999	1000–1099	1100–1200
	Temperature of pipe, F										
	Nominal thickness										
Utility—steam generation											
1½ and less	1	1	1½	2[1]	2[1]	2½	2½	2½	3	3	3
2	1	1	1½	2[1]	2[1]	2½	2½	3	3	3	3½
2½	1	1	1½	2[1]	2[1]	2½	2½	3	3	3½	3½
3	1	1	1½	2[1]	2[1]	2½	3	3	3	3½	3½
3½	1	1	1½	2[1]	2½[1]	2½	3	3	3½	3½	3½
4	1	1	1½	2[1]	2½[1]	2½	3	3	3½	3½	4
4½	1	1	1½	2	2½[1]	2½	3	3	3½	3½	4
5	1	1½	1½	2	2½	2½	3	3½	3½	4	4
6	1	1½	2	2	2½	3	3	3½	3½	4	4
7	1½	1½	2	2	2½	3	3	3½	3½	4	4
8	1½	1½	2	2	2½	3	3½	3½	4	4	4½
9	1½	1½	2	2½	2½	3	3½	3½	4	4	4½
10	1½	1½	2	2½	2½	3	3½	3½	4	4	4½
11	1½	1½	2	2½	2½	3	3½	3½	4	4½	4½
12	1½	1½	2	2½	3[1]	3	3½	4	4	4½	4½
14 and up	1½	1½	2	2½	3[1]	3	3½	4	4	4½	4½
Process											
1½ and less	1	1	1	1	1	1	1½	1½	1½	2	2
2	1	1	1	1	1	1½	1½	1½	2	2	2
2½	1	1	1	1	1	1½	1½	1½	2	2	2
3	1	1	1	1	1	1½	1½	2	2	2	2½
3½	1	1	1	1	1	1½	1½	2	2	2	2½
4	1	1	1	1	1	1½	1½	2	2	2	2½
4½	1	1	1	1	1½	1½	1½	2	2	2½	2½
5	1	1	1	1	1½	1½	1½	2	2	2½	2½
6	1	1	1	1	1½	1½	2	2	2	2½	2½
7	1½	1½	1½	1½	1½	1½	2	2	2	2½	2½
8	1½	1½	1½	1½	1½	1½	2	2	2½	2½	2½
9	1½	1½	1½	1½	1½	1½	2	2	2½	2½	2½
10	1½	1½	1½	1½	1½	1½	2	2	2½	2½	2½
11	1½	1½	1½	1½	1½	1½	2	2	2½	2½	3
12	1½	1½	1½	1½	1½	1½	2	2	2½	2½	3
14 and up	1½	1½	1½	1½	1½	2	2	2	2½	2½	3

Nominal pipe size, in.	100–199	200–299	300–399	400–499	500–599	600–699	700–799	800–899	900–999	1000–1099	1100–1200
	Temperature of pipe, F										
	Nominal thickness, in.										
Commercial—full year operation											
1½ and less	1	1½	2	2½[1]	3[1]						
2	1	1½	2	2½[1]	3[1]						
2½	1	2	2½	3[1]	3[1]						
3	1	2	2½	3[1]	3½[1]						
3½	1	2	2½	3[1]	3½[1]						
4	1	2	2½	3[1]	3½[1]						
4½	1	2	2½	3[1]	3½[1]						
5	1	2	2½	3[1]	3½[1]						
6	1	2	3	3½[1]	4[1]						
7	1½	2	2	3½[1]	4[1]						
8	1½	2	3	3½[1]	4[1]						
9	1½	2½	3	3½[1]	4[1]						
10	1½	2½	3	3½[1]	4[1]						
11	1½	2½	3	3½[1]	4[1]						
12	1½	2½	3	3½[1]	4						
14 and up	1½	2½	3½	4[1]	4½						
Commercial—seasonal operation											
1½ and less	1	1	1	1½	1½						
2	1	1	1	1½	1½						
2½	1	1	1	1½	2[1]						
3	1	1	1½	1½	2[1]						
3½	1	1	1½	1½	2[1]						
4	1	1	1½	1½	2[1]						
4½	1	1	1½	1½	2						
5	1	1	1½	1½	2						
6	1	1	1½	2	2						
7	1½	1½	1½	2	2						
8	1½	1½	1½	2	2						
9	1½	1½	1½	2	2						
10	1½	1½	1½	2	2½						
11	1½	1½	1½	2	2½						
12	1½	1½	1½	2	2½						
14 and up	1½	1½	1½	2	2½						

[1] Available in single or double layer.

For insulation of piping which operates at temperatures above 600 F, pipe expansion is a significant factor. Single-layer insulation joints may deteriorate under such conditions, and the pipe covering might char or burn. Therefore for such service, double-layer construction is recommended.

Insulation of austenitic stainless-steel piping requires consideration of possible chloride stress corrosion of the piping. Tests have indicated that insulations with a

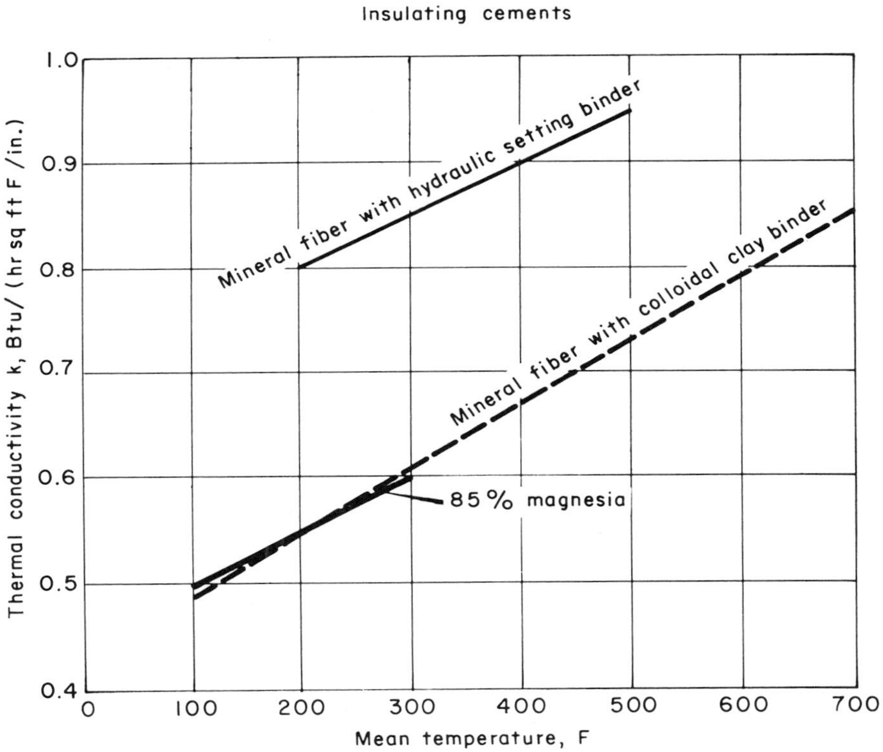

Insulating cements

FIG 12. Conductivity k design values. Insulating cements.

chloride content of as little as 0.05 per cent by weight, because of leaching out of chlorides in wet insulation, can result in a chloride concentration of 2 to 5 per cent at the point of stress corrosion. Laboratory tests have indicated that sodium silicate tends to inhibit the chloride effect. It is recommended that the insulation binder be sodium silicate for such insulation installations. Furthermore, consideration should be given to coating the piping with sodium silicate prior to installation of insulation.

Piping in trenches or underground conduits should be insulated with materials that will not disintegrate if submerged in water for periods of time. Calcium silicate insulations meet this requirement.

Selections of Insulation Thickness. Typical selection of nominal pipe insulation thicknesses recommended by the insulation manufacturers for utility, process, and commercial services are shown in Tables 5 and 6. Tabulated thicknesses are optimum thicknesses calculated on an economic basis for heat conservation

under average operating conditions and assure adequate temperature control. Unusual conditions may warrant use of other thicknesses.

Table 7 gives typical selections for the various services encountered in underground steam-distribution piping. The insulation of low-pressure steam and hot-water

Table 6. ‧ Recommended Thicknesses for 85 Per Cent Magnesia

	Temperature of pipe, F						Temperature of pipe, F				
Nominal pipe size, in.	100–199	200–299	300–399	400–499	500–600	Nominal pipe, size in.	100–199	200–299	300–399	400–499	500–600
	Nominal thickness, in.[1]										
Utility—steam generation						**Process**					
1½ and less	1	1	1½	2	2	1½ and less	1	1	1	1	1
2	1	1	1½	2	2	2	1	1	1	1	1
2½	1	1	1½	2	2	2½	1	1	1	1	1
3	1	1	1½	2	2	3	1	1	1	1	1
3½	1	1	1½	2	2½	3½	1	1	1	1	1
4	1	1	1½	2	2½	4	1	1	1	1	1
4½	1	1	1½	2	2½	4½	1	1	1	1	1½
5	1	1½	1½	2	2½	5	1	1	1	1	1½
6	1	1½	2	2	2½	6	1	1	1	1	1½
7	1½	1½	2	2	2½	7	1½	1½	1½	1½	1½
8	1½	1½	2	2	2½	8	1½	1½	1½	1½	1½
9	1½	1½	2	2½	2½	9	1½	1½	1½	1½	1½
10	1½	1½	2	2½	2½	10	1½	1½	1½	1½	1½
11	1½	1½	2	2½	2½	11	1½	1½	1½	1½	1½
12	1½	1½	2	2½	3	12	1½	1½	1½	1½	1½
14 and up	1½	1½	2	2½	3	14 and up	1½	1½	1½	1½	1½
Commercial—full year operation						**Commercial—seasonal operation**					
1½ and less	1	1½	2	2½	3	1½ and less	1	1	1	1½	1½
2	1	1½	2	2½	3	2	1	1	1	1½	1½
2½	1	2	2½	3	3	2½	1	1	1	1½	2
3	1	2	2½	3	3½	3	1	1	1½	1½	2
3½	1	2	2½	3	3½	3½	1	1	1½	1½	2
4	1	2	2½	3	3½	4	1	1	1½	1½	2
4½	1	2	2½	3	3½	4½	1	1	1½	1½	2
5	1	2	2½	3	3½	5	1	1	1½	1½	2
6	1	2	3	3½	4	6	1	1	1½	2	2
7	1½	2	3	3½	4	7	1½	1½	1½	2	2
8	1½	2	3	3½	4	8	1½	1½	1½	2	2
9	1½	2½	3	3½	4	9	1½	1½	1½	2	2
10	1½	2½	3	3½	4	10	1½	1½	1½	2	2½
11	1½	2½	3	3½	4	11	1½	1½	1½	2	2½
12	1½	2½	3	3½	4	12	1½	1½	1½	2	2½
14 and up	1½	2½	3½	4	4½	14 and up	1½	1½	1½	2	2½

[1] Tabulated thicknesses are optimum thicknesses calculated on an economic basis for heat conservation under average operating conditions and assure adequate temperature control. Unusual conditons may warrant the use of other thicknesses.

piping is given in Table 9. Typical selections for insulation of low-temperature refrigerating and air-conditioning piping are given in Table 8.

Recently, using a computer to solve the complicated formulas of heat transfer through insulation, it was generally found that steam-generating stations had more than the economically required insulation thickness. Conversely many industrial installations were under insulated.

Table 7. Typical Selection of Insulation for Underground Steam-distribution Piping

Nominal pipe size, inches	Thickness of insulation, inches					
	Surface mains and feeders[1]	Lines in tunnels and feeders	Service connection, buried construction[1]	Service connection, tunnels and tunnel shafts	Surface and tunnel return lines in basements and areaways	Drip connections in tunnels, manholes, basements, and areaways[7]
¾	....	....	....	....	....	1
1	....	....	....	....	....	1
1¼	....	....	....	....	....	1
1½	....	....	....	....	....	1
2	....	....	....	1½[2]	1[5]	1
2½	....	....	1	1½[2]	1[5]	
3	....	....	1	1½[2]	1[5]	
4	1	1½[2]	1	1½[2]	1[5]	
6	1	1½[2]	1	1½[2]	1[5]	
8	1	1½[2]	1	1½[2]	1[5]	
10	1	3[3]	1	3[3]	1[5]	
12	1½	3[3]	1½	3[4]	1½[1,6]	
14 O.D.	1½	3[3]	1½	3[4]	1½[1,6]	
16 O.D.	1½	3[3]	1½	3[4]	1½[1,6]	

[1] Laminated asbestos with waterproof jacket.
[2] Eighty-five per cent magnesia pipe covering.
[3] Eighty-five per cent magnesia, 1½-in. inner layer and 1½ × 3 × 36 in. flat-block outer layer asbestos cement, and 9-oz duck.
[4] Double-layer flat block, 1½ × 3 × 36 in. with asbestos cement and 9-oz duck.
[5] Standard thickness, 85 per cent magnesia, except for damp conditions where laminated asbestos is used. See Table IV for standard thicknesses.
[6] Single-layer flat block, 1½ × 3 × 36 in.
[7] Laminated asbestos with roofing cement coating.
NOTE.—Amosite asbestos sectional pipe covering, with or without waterproof jacket, may be used as an alternate for any of the services and in the same thicknesses as designated in this table.

An economic study to determine the justified thickness should consider the following factors.

1. Hours operation per year
2. Fuel cost including labor and maintenance, Btu value, and efficiency of combustion
3. Pipe diameter and operating temperature and ambient still-air temperature
4. Estimated cost of installed insulation at actual plant site
5. Amortization period
6. Heat loss (gain in cold system) per lineal foot as obtained from manufacturer's tables

Step 1.   Determine the cost of heat loss per 100 lin ft of pipe for a range of insulation thicknesses above and below the recommended thickness by use of the following equation:

Hr. operation × fuel cost × 100 ft × heat loss

÷ fuel heating value × eff. of combustion

Step 2.   Divide installation cost per 100 ft by the amortization period.

Step 3.   For each insulation thickness, total the dollar value obtained from steps 1 and 2.   One insulation thickness total cost will be lower than lesser or greater thicknesses.

The labor cost to install insulation, in a survey, was found to be from 1½ to three times the cost of the insulation material.

Table 8. Typical Selection of Insulation for Low-temperature Refrigerating and Air-conditioning Piping

Service	Temperature range, deg. F	Insulation	Number of layers	Total thickness (nominal), inches	Waterproofed
Cold water (antisweat).	50 to 75	Hair or wool felt	1	5 in. and less— 3/4 6 in. and over— 1	Inside and outside
Ice water.............	32 to 50	Hair or wool felt	1 2	3 in. and less— 3/4 4 in. and over— 1½	Inside and outside
		Cork	1	Ice water thick[1]	Seams and edges
Brine and ammonia....	25 to 35	Cork Hair felt	1 2	Ice water thick[1] 2	Seams and edges Inside and outside
Brine and ammonia....	0 to 25	Cork Hair felt	1 3	Brine thickness[1] 3	Seams and edges Inside and outside
Brine and ammonia....	−25 to 0	Cork Hair felt	1 4	Heavy brine th.[1] 4	Seams and edges Inside and outside

[1].—"Ice water thickness" cork is approximately 1½ in. thick.

"Standard brine thickness" cork is approximately 2 in. thick on pipes ½ to 1 in., 2½ in. thick on pipes 1¼ to 3 in., and 3 in. thick on larger sizes.

"Heavy brine thickness" cork is approximately 3 in. thick on pipes ½ to 3 in., 3½ in. thick on pipes 3½ to 4 in., and 4 in. thick on larger sizes.

NOTE 1.—Application:

Correct application of insulation on cold piping is absolutely essential to the maintenance of lasting efficiency. The insulation must be thoroughly sealed against admission of moisture from the air. The lower the pipe temperature, the more imperative the necessity for perfect sealing, because at pipe temperatures below freezing the slightest opening at the joints will allow moisture to enter, condense, and freeze, and this formation of frost tends to damage or destroy the insulation.

NOTE 2. Where water pipes are exposed to air temperatures below 32 F, they are usually insulated to prevent freezing. A common specification consists of three layers of standard hair felt (total nominal thickness 3 in.) protected by means of a weatherproof roofing jacket lapped and sealed at all joints. This specification is suitable where water circulation is maintained continuously or where circulation is interrupted for only brief periods. Where conditions of exposure are more severe or where periods of no flow are more prolonged, the line is usually protected by means of a small steam line alongside the water line, and both pipes are then insulated together by wrapping with two plies of ¼-in. asbestos air-cell paper and two layers of standard hair felt finally protected with a weatherproof roofing jacket sealed at all joints.

For additional data on insulation and velocities of water to prevent freezing, see Chap. 32 on Pipe Insulation, "1959 Guide of the American Society of Heating and Ventilating Engineers."

NOTE 3. Glass-wool sectional pipe covering may be used as an alternate for low-temperature refrigerating and air-conditioning piping. Recommended thicknesses for each type of service should be obtained from the manufacturer.

Table 9. Typical Selection of Insulation for Low-pressure Steam and Hot-water Piping

Pipe sizes	Insulation, hot water	Steam	
		25 psi, maximum	100 psi, maximum
½ to 30 in.	Air-cell type, ¾-in.	Air-cell type, ¾ in.	85% magnesia, standard thickness

Heat-insulation Coverings. The durability and the possibility of reuse of an insulation covering are matters of extreme importance in many installations.

Tar-paper and canvas coverings in general will have a service life of approximately 5 to 10 years. However, they cannot be reused after a piping alteration, are not fire resistant, deteriorate rapidly in corrosive atmospheres, and require frequent maintenance.

Aluminum jacket coverings, while approximately seven times more expensive than tar-paper coverings, have an infinite life when furnished with a suitable moisture barrier, are easier to install, can be reused, and require little or no maintenance.

A possible new covering of recent origin is a polyvinyl fluoride film. Test data indicate a service life of approximately 15 years and resistance to fire, weather, and corrosive atmospheres.

Steam or Electrically Traced Piping. For steam or electrically traced piping, insulation manufacturers now offer an out-of-round nominal-pipe-size insulation at the same price as standard round insulation. This obround insulation eliminates the need for oversized, more expensive standard-type insulation to serve the same purpose.

Electrical tracing of piping, after a promising start a few years ago, has not been used to any great extent of late. Future design movements could restore this type of tracing to popularity.

7

MANUFACTURE, FABRICATION, AND JOINING OF COMMERCIAL PIPING

Helmut Thielsch*

PIPE MANUFACTURE

Many different processes are used in the manufacture of pipe. They are grouped by definition into several classifications such as wrought seamless pipe, forged pipe, welded pipe, and cast pipe. Several processes may be used in the manufacture. For example, centrifugally cast pipe may be cold-worked by hydraulic expansion which alters the cast metallurgical structure. Seam-welded pipe is also cold-expanded or cold-reduced by some manufacturers to produce more uniform tolerances. Cold expansion also increases the transverse yield strength. Within each classification, a number of specific processes are employed.

WROUGHT SEAMLESS PIPE

Ferrous Pipe Materials

Steel pipe is usually made from steel produced in the open-hearth, basic oxygen furnace, Bessemer converter, or electric arc furnace. Where relatively small melts are required, as in the casting of steel pipe with special compositions, electric induction furnaces are often used.

Although the manufacture of pipe from steel melted in Bessemer converters lessened substantially in the years following World War II, the introduction of

* Manager, Research and Development Division, Grinnell Corporation, Providence, R.I.

oxygen and air-oxygen into the converter is now substantially increasing the use of this equipment, particularly on carbon steels.

In the United States, wrought seamless carbon and alloy pipe is made in sizes up to 26 in. OD. It is normally manufactured by one of four different methods.

Hot Rotary Piercing. The most common method of manufacture (Fig. 1) involves piercing a heated round in either one or two piercing mills, which consist of a pair of cylindrical rolls rotating in the same direction with their axes inclined to each other.

The steel billet, at a forging temperature from 2200 to 2400 F, is pushed into the piercing mill, where it is gripped by two rolls which rotate and advance it over the piercer point to form a hole throughout its length. For large pipe, a similar second operation reduces the wall thickness and increases the diameter and length of the pierced billet.

The pierced billet represents a rough tube larger in diameter and of a wall thickness in excess of that required in the finished pipe. A further reduction in diameter

Piercing mill   Plug rolling mill   Reeling mill   Sizing mill

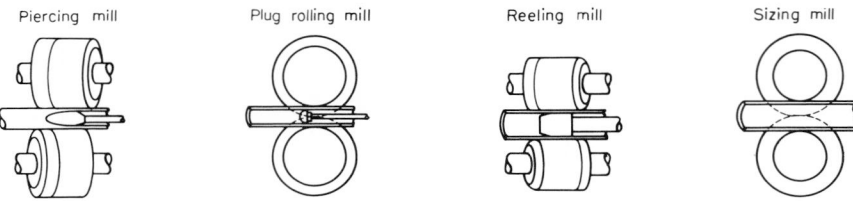

Fig. 1. Hot rotary piercing process (National Tube Co.).

and wall thickness and an increase in length are effected by rolling the pierced billet over a mandrel in a plug-rolling mill.

The function of the reeling mill is to burnish the inside and outside surfaces and to true up the tube, which is slightly oval in shape as it leaves the plug-rolling mill. The reeling mill rolls are inclined at an angle with one another, and, as in the piercing mill, the rolls grip the pipe and advance it over the mandrel.

The pipe, reheated if necessary, is finally passed through two or more sets of sizing rolls. The grooves in the sizing rolls have a diameter smaller than that of the pipe which comes from the reeling mill. The reduction in diameter induces uniform size and roundness throughout the length of pipe.

Pilger-mill Process. In the Pilger-mill process, a round cast steel ingot, rolled-steel billet, or thick-walled hollow bloom of either rimmed or deoxidized steel is heated and pierced on a heavy type of Mannesman roll piercer (Fig. 2). The hollow bloom may have been produced in an oblique rolling mill.

A mandrel about 10 ft long and of a diameter approximating the inside diameter of the finished pipe is then forced through the pierced ingot or billet by a hydraulic ram. The mandrel encased in the ingot is placed between the rolls of the Pilger mill. These rolls have a cam-shaped contour and revolve counter to the direction in which the ingot is forced by a hydraulic ram and air-cylinder mechanism. The rolls first engage the red-hot billet. As they rotate farther, a small length of the steel is nipped, forming a shaft. Simultaneously, the pressure exerted by the rolls causes the billet to be forced backwards. The tube section resulting from the metal that is squeezed out is smoothed in the adjacent part of the roll groove.

Rotation of the rolls produces the equivalent effect of forging-hammer blows, which reduce the ingot wall by forging down against the mandrel and drive the ingot and mandrel back against the ram. For this reason, the process is also called the "rotary-forged process." As the relieved portion of the rolls comes into position

with further rotation of the rolls, the plunger of the air-cylinder and hydraulic ram again forces the ingot into position between the rolls.

The tube is subsequently reheated and passed to the reeler to improve the uniformity of the wall thickness. The tube may then be passed through a sizing mill to provide a more uniform diameter.

Push-bench (Cupping) Process. In the push-bench process, plates 2 to 7 ft square and $\frac{3}{8}$ to 4 in. thick are cut into circular disks. They are heated to a forging temperature of about 2300 F, placed concentric with a bottom die, and

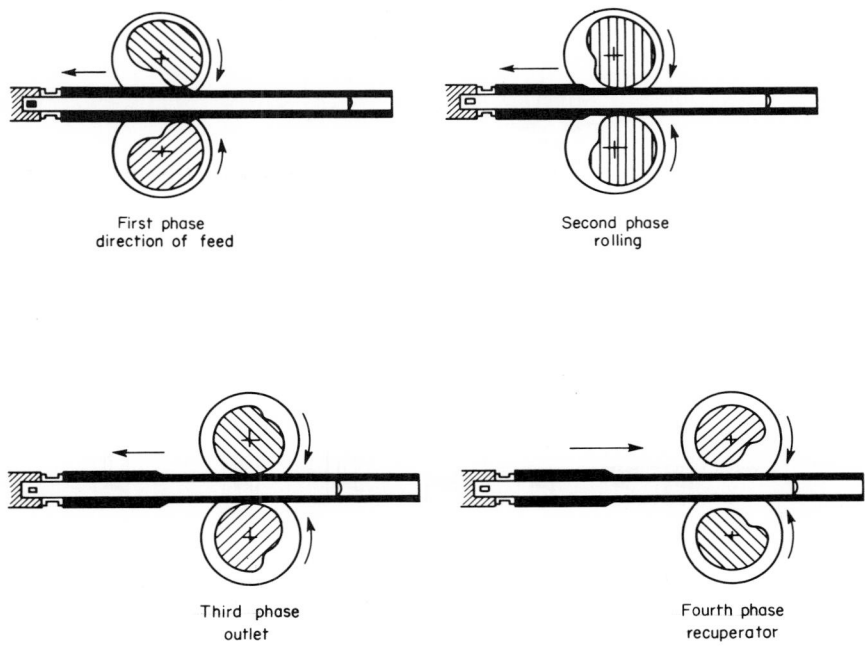

First phase
direction of feed

Second phase
rolling

Third phase
outlet

Fourth phase
recuperator

Fɪɢ. 2. Pilger-mill process.

pushed through with a round-nose plunger. The cup thus formed is reheated to the forging temperature and is forced through a smaller die, producing an essentially cylindrical cup.

In some pipe mills, a square steel ingot, instead of the plate, is heated to a temperature of 2300 F. It is then placed in a circular container and punched so that a closed end (cup) is left (Fig. 3). The punch is so dimensioned that the material forced out follows the contours of the container, filling the hollow spaces between the container wall and the square ingot.

The closed-end cylinder is reheated and pushed, with the closed end foremost, through a series of 3 to 12 dies, of successively decreasing diameters, mounted on a horizontal bench. Reheating between these drawing operations may be necessary. A mandrel-extracting machine then loosens the mandrel and pulls it out of the tube. The closed end (cup) is cut off in a circular saw. Final working then involves cold-rolling or straightening of the tubing. This process is particularly suitable for small-diameter (to 4 in.) seamless tubing of relatively light wall thicknesses (below $\frac{1}{2}$ in.).

Extrusion Process. Seamless steel tubes of small diameter and pipes of larger diameters are also produced with mechanical extrusion presses. Commercial equipment involves vertical presses (as shown in Fig. 4) or horizontal presses (Fig. 5). A descaled steel billet, heated to approximately 2300 F, is placed in the container of the press, in the bottom of which the extrusion die is located. A ram is applied on the billet without, however, pushing the billet downward. A piercing mandrel, which moves in the bore of the ram, then punches the billet, which rises slightly. This produces a cylinder, from which the punched-out piece is ejected

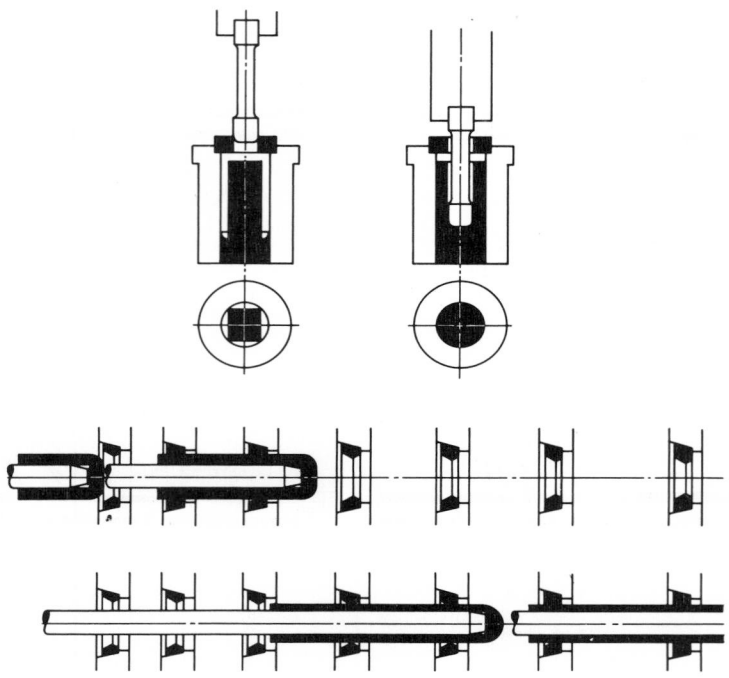

Fig. 3. Push-bench (cupping) process.

through the extrusion die. When pressure is applied with the ram, the billet is extruded through the annular gap which the extrusion die forms with the piercing mandrel, and a tube results.

In horizontal presses (Fig. 5) the piercing may be done first as a separate step, or a hollow is used with mandrel and die. Tungsten-chromium-carbon and chromium-tungsten-molybdenum-alloy steels with hardnesses of approximately 46 Rockwell C are used for mandrels, dies, and other tools. Glass is the most effective lubricant. The hot, forged, and machined billet is rolled over glass which partially fuses to it, or, preferably, it is coated by a layer of glass powder spread over the asbestos blanket of the chute which delivers the billet from the furnace to the press.

On tubing where the extruding operation is completed in a few seconds, the tubes are generally transferred to a reducing mill while still at the hot forging temperature. In this mill, the tube diameter is further reduced to the proper size.

Carbon-steel, alloy-steel, and stainless-steel tubings are produced commercially by this method in diameters from $\frac{3}{8}$ to 4 in. and from 30 to 60 ft in length. Pipe is produced in sizes from 8 to 24 in. diameter and in wall thicknesses from $\frac{1}{2}$ to 3 in.

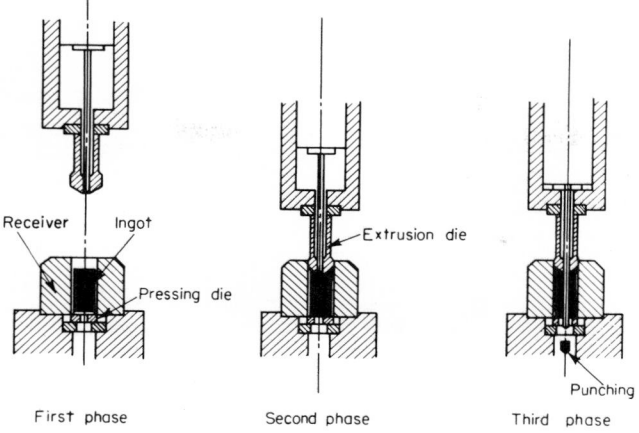

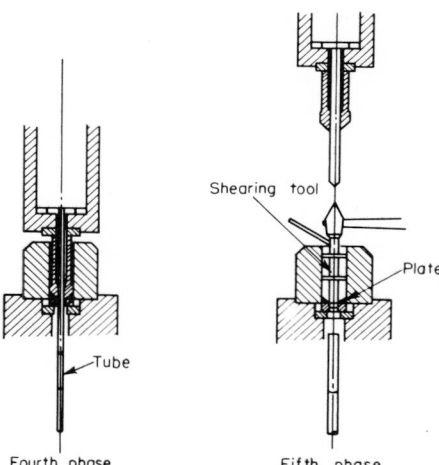

Fig. 4. Extrusion process.

Aluminum and Aluminum Alloys

Seamless aluminum-alloy tubing or pipe for pressure service is made by the die-and-mandrel extrusion process in sizes 1 in. and larger. Hollow round ingots are used either cast hollow or made by boring a solid round ingot. After the ingot is preheated to an appropriate temperature (depending on the alloy), it is loaded in the extrusion press cylinder. A mandrel extends through the ingot and through the extrusion die and thus provides an annular space through which the aluminum flows when pressure is applied to the ingot.

When extrusion is complete, the tube is sheared off at the die, leaving a butt end of ingot which is removed before the next ingot is positioned.

Pipe for pressure service in sizes smaller than 1 in. is usually produced by drawing larger stock extruded by the die-and-mandrel process. Outside diameters as large as 20 in. can be extruded on existing presses.

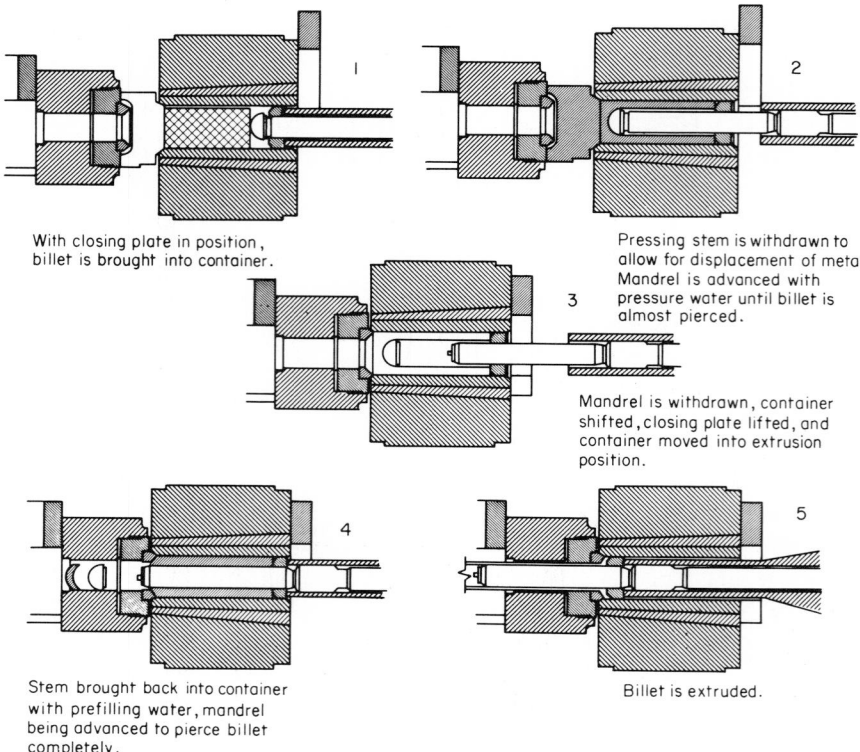

With closing plate in position,
billet is brought into container.

Pressing stem is withdrawn to
allow for displacement of metal.
Mandrel is advanced with
pressure water until billet is
almost pierced.

Mandrel is withdrawn, container
shifted, closing plate lifted, and
container moved into extrusion
position.

Stem brought back into container
with prefilling water, mandrel
being advanced to pierce billet
completely.

Billet is extruded.

FIG. 5. Horizontal extrusion by Ugine-Sejournet process.

Some alloys require heat treatment to achieve the desired mechanical properties.
This may be accomplished by quenching with air or water as the pipe emerges from
the extrusion press, or a separate heat-treating furnace may be employed.

A wide variety of wall thicknesses can be made by extrusion or by a combination
of extrusion and drawing.   Generally speaking, the thinner walls are available only
in the smaller sizes.   In general, for pipe, the wall thicknesses range from those for
Schedule 5 and higher.   Lengths are limited by ingot size and heat-treating equipment.   Forty-foot lengths of pipe are usual, and lengths of 70 to 80 ft or more can be
furnished in some alloys and tempers.

Copper and Copper Alloys

In the production of seamless copper and copper-alloy pipe, copper is by far the
greater tonnage item.   Copper and alloy pipe are produced by the same processes
and on the same equipment.   They are hot- and cold-worked and, therefore, are
wrought.   Very little cast or welded pipe is produced in copper or copper alloys.

Copper is refined and remelted in reverberatory furnaces and cast into solid
billets of various diameters.   Copper alloys are melted in electric furnaces and cast
into solid billets in the mills of the brass industry.

Seamless pipe is produced by one of five principal processes.

Hot Piercing Process. Most copper pipe and some copper-alloy pipe is produced on Mannesmann machines. A 3- to 12-in.-diameter solid copper or alloy billet is heated to 800 or 900 C and is passed between two rolls whose axes are set at an angle. The rolls rotate in the same direction and give the billet a forward spiral action. The pressure on the outside diameter of the hot billet produces an internal break along the centerline which is the start of a longitudinal cavity. The billet, as it rotates between the rolls, is forced over a pointed mandrel that has an anvil effect in which the rolls work and smooth out the inside diameter of the pierced billet.

The set between the rolls and the mandrel determines the wall thickness of the pierced shell.

The pierced shell is drawn on drawbenches between a die which determines the outer diameter and a plug which controls the inner diameter of the pipe. There may be several draws and intermediate anneals with subsequent cleaning and pickling operations before the pipe is finished. The pipe is sawed to length, tested, inspected, and packed.

Extrusion Process. Copper and copper-alloy pipe are also made from extruded shells. The shells are produced on hydraulic presses which are capable of exerting 3,000 tons per square inch pressure on heated billets. The billets are heated from 700 to 900 C, depending on the alloy, and placed in a container or massive steel cylinder which confines the hot billet while high pressures are exerted to force the metal through the dies. Pressure is applied, and the billet is slightly upset and held in the container while a mandrel within the ram punches out the center of the billet; the mandrel is left protruding through the die. High pressure is applied by the ram to the billet, and the metal is extruded through the die and over the mandrel. The die determines the outer diameter, and the mandrel the inner diameter, of the extruded shell.

Several draws, intermediate anneals, and pickles produce the finished pipe.

Cup-and-draw Process. The cup-and-draw process is a practice by which large-diameter pipe is produced. Both copper and alloy pipe 12 in. and over in outer diameter are made in this manner.

Large circles are cut from wrought copper or alloy plates. The circles are placed on a die, and a plug pushes through the die to draw a cup. The cup is further drawn through several successive small dies and plugs with a resulting reduction in wall thickness and an increase in length. The bottom of the cup or shell is cut off, and the remainder is drawn further to finished pipe.

Drawing Process. Bench drawing to smaller sizes is the principal cold-working process that is performed on copper or copper-alloy tubes that are made by extrusion or piercing methods. The drawbench is a machine for pulling tubes through a die to reduce the diameter or wall area. Drawbenches may be single, or they may be multiple and able to handle a number of tubes simultaneously. In most cold-drawing operations, an outside die and an inside mandrel reduce the outside and inside diameters and, at the same time, reduce the wall thickness and circular area. The inside diameter, generally, is reduced slightly less than the outside diameter, and the operation results in a smooth inside and outside surface finish.

During the drawing operation, the metal becomes work-hardened so that at intermediate stages the tubes are annealed. Copper is so ductile that it can be reduced severely without intermediate annealing, but to restore ductility the tube is usually annealed after two or three drawing operations. Brasses, however, work-harden more rapidly and are annealed more frequently.

Tube-rolling Process. Another process used for reducing copper tubing is tube-rolling, sometimes referred to as "rocking." This is a cold-working process

in which semicircular, tapered, and grooved dies are rocked rapidly back and forth over the tube on a mandrel that controls the inside diameter. The groove at the larger end of the die tapers to the opposite end, and the mandrel has a corresponding taper so that in the rocking motion, the tube is pinched between the die and mandrel, reducing the diameter and wall and increasing the tube length. At the end of each rocking stroke, the tube is turned so as to distribute the work uniformly over the entire circumference. Tubing produced by this method is very concentric and uniform in dimension.

Nickel and Nickel Alloys

Nickel and nickel-alloy wrought seamless pipe and tubing generally are produced by extrusion and cold-drawing. Products of the extrusion process are extruded tubing or shells that are converted into pipe or tubing by cold-drawing and additional processing.

Extrusion Process. The extrusion process most commonly used is the Ugine-Sejournet method shown in Fig. 5. The press is a four-column horizontal type of 4,000-ton capacity distributed as 2,500 tons on the main ram and 1,500 tons on the piercer ram. The hydraulic medium is water from an air-loaded accumulator system. The pressure water at 4,250 psi is supplied by 2 three-throw single-stage pumps driven by 400-hp motors. A unique feature of the press is the cupping plate attachment, whereby a solid die block is dropped between the front end of the container and the die, thus sealing the container. For stiffer alloys, the billet may be initially drilled to facilitate extrusion. Following extrusion, the tubes are water-quenched, deglassed in hydrofluoric acid, and trimmed.

Extruded tubing is produced in outside diameters of $2\frac{1}{2}$ to $9\frac{1}{4}$ in.; wall thicknesses vary from $\frac{1}{4}$ to 1 in.; maximum lengths range from 3 to 30 ft, depending on the other parameters. In addition to cylindrical tubes, it is practical to extrude tubular shapes of relatively simple geometry that are symmetrical about their longitudinal axes.

Cold-drawing. In the conversion of extruded shells to cold-drawn tubes, the first step is centerless grinding to remove any rough surface and chromium oxide that may exist; this step is not required on all alloys. The shells are then pickled, and any defects are removed by grinding with manually controlled electric swing grinders with 40-grit grinding wheels. The pickling, inspection, and grinding operations are repeated at least once to insure that all defects have been removed.

Extruded tubes over 4 in. in outside diameter are pointed on a hydraulic swaging machine. Smaller ones are usually pointed on a mechanical hammer.

Tubes less than 3 in. in outside diameter are reduced nearly to size on a tube reducer, with the final passes being made on draw benches. All cold-reduction of larger tubes is performed on drawbenches of a variety of sizes.

After cold-reduction, the tubes are annealed under time and temperature conditions dictated by their composition and size and are again pickled. In the manufacture of hard tubing, annealing and pickling are usually omitted after the final reduction.

The final condition of pipe and tubing varies according to composition of the alloy, its end use, and the fabrication it will receive. If little or no subsequent machining or bending will be required, pipe and tubing are furnished in the annealed condition, with additional age-hardening if required. Where moderate bending or coiling will be performed, the tubing is stress-relieved. Tubing to be redrawn or otherwise severely formed is fully annealed; a pickling or grinding operation may follow annealing.

After the final thermal treatment, tubes are straightened on rotary machines and cut to length with abrasive wheels or on lathes.

Cold-drawn seamless tubing is furnished to average wall requirements ranging in thicknesses from 0.025 to 0.500 in., depending on the alloy composition and the other dimensions. Outside diameters range from 0.500 to 8.250 in. and lengths up to 30 ft. Cold-drawn seamless pipe is furnished in sizes and schedules that conform to the requirements of the American Standards Association. Nominal pipe sizes range from $\frac{1}{8}$ to 8 in. in lengths up to 30 ft.

Titanium and Titanium Alloys

The extrusion process is the most commonly used technique for producing titanium pipe. Although other methods have been successfully used with titanium, the small volume of titanium pipe produced in one size at one time does not justify setting up these high-volume mills. The extrusion process is very adaptable to the numerous die changes required for titanium.

Extrusion presses used for titanium are predominantly hydraulic, with the ram operating in a horizontal plane. Titanium billets are pierced or drilled prior to extrusions. The billets are heated to approximately 1800 F and extruded through a die over a mandrel. After extrusion, the pipe may or may not be further reduced to finished size by various cold-reducing operations.

Hot titanium is very reactive with die materials, and several techniques are employed to protect the dies such as copper-cladding the billets or coating them with molten glass. The coating is chemically removed from the pipe after extruding.

FORGED PIPE

Forged pipe is made primarily in pipe sizes of larger diameters and heavier wall thicknesses, where other seamless grades are not readily available because of costs or equipment limitation.

Two types are recognized commercially in the specifications of the American Society for Testing and Materials. They are *forged and bored* and *hollow forged*.

Forged-and-bored Pipe. A steel billet or ingot is first heated to 2300 F and then is elongated by forging in heavy presses or under forging hammers to a diameter approximately 1 in. greater than the diameter desired for the finished pipe. The billet is then turned in a lathe, which removes the excess steel and produces the actual outside diameter required. By means of a special boring or trepanning tool, the inside is then bored out from one or both ends to the specified inside diameter.

By this process, pipe in diameters between 10 and 30 in. and in wall thicknesses from $1\frac{1}{2}$ to 4 in. is produced commercially. The machining permits the average pipe wall thickness to be held closer to the minimum wall requirements specified by the designers of the piping system. This may result in weight savings. The price advantage of forged-and-bored pipe over pipe made by other manufacturing methods increases with the diameter and the wall thickness. Pipe sections as long as 50 ft and weighing as much as 100 tons have been produced.

Hollow Forged Pipe. Hollow forged pipe is produced directly from steel ingots melted in an electric arc furnace. The ingots are hot pierced at about 2000 to 2200 F in a vertical press. They are then transferred to a horizontal draw bench where the pierced bloom is placed over a mandrel and is worked through a series of ring dies to produce the desired size.

Pipe in diameters from 10 to 30 in. and in wall thickness from about $1\frac{1}{2}$ to 4 in. is normally produced commercially. Special sections can be produced. The product is machined on the outer and inner diameters.

Nonferrous pipe, in commercial quantities, has not generally been produced by forging methods.

WELDED PIPE

Welding of plates, skelp, or coils of strip steel into pipe is done either by furnace heating and forge welding for butt-weld pipe or by fusion welding employing electric resistance, flash, submerged-arc welding, inert-gas tungsten-arc welding, or gas-shielded consumable metal-arc welding. The welded seam may be either a longitudinal seam parallel to the axis of the pipe or a spiral weld.

Ferrous Pipe Materials: Furnace-welded Pipe

Furnace-welded pipe, also known as continuous-welded or butt-welded pipe, is available only in carbon steel grades. The pipe generally is made from open-hearth and basic oxygen bessemer steel. Rephosphorized open-hearth and basic oxygen steels also are widely used because of improved machinability.

Furnace-welded pipe normally is considered the lowest-cost steel pipe. It is used in relatively noncritical service such as in low-pressure gas piping, water piping, air lines, low-pressure steam systems, and similar service environments. Furnace-welded pipe is generally limited to pipe sizes of 4 in. diameter and smaller.

In the butt-welding process, a strip or plate referred to as "skelp" is passed through a continuous furnace where it is heated to a "welding" temperature of approximately 2450 F. The strip is then passed immediately through forming and welding rolls, simultaneously forming the tube and welding the edges of the strip. In the continuous butt-weld process, the end of one strip is resistance-welded to the next coil.

Ferrous Pipe Materials: Fusion-welded Pipe

Fusion welding of pipe is done by resistance welding, by induction welding, or by arc-welding methods.

Resistance Welding. Four methods of resistance welding are extensively used. These are flash welding, low-frequency resistance welding, high-frequency induction welding, and high-frequency resistance welding.

In every process, the strip or plate is initially formed by rolling or ring forming into a circular shape.

In the **flash-welding** process, the edges of preformed pipe are held together by copper contact shoes. Just enough pressure is applied to produce very light contact between the butting edges of the pipe seam. Application of the electric current produces short circuits between the edges, causing metal flashing. The short circuits occur at a speed indicating visually a continuous flashing operation for the entire seam length. Because of the extremely high rate of heating, a very steep temperature gradient is produced. When the temperature at a certain distance is proper for forming, the pipe seam edges are pressed together. This forces out the molten metal and upsets a slight amount of the heated base metal. The resulting weld represents a pressure fusion weld in which the fusion line is produced between the pipe ends at a temperature below the melting point of the steel.

The "flashed-out" and upset metal is subsequently removed by machining with a hot scarfing tool to leave a very slight metal excess along the inside and outside seam-weld surfaces.

The pipe-manufacturing process is applied particularly to high-strength carbon steel pipe of large diameters between 4 and 36 in.

In the manufacture of seam-welded pipe by the **low-frequency resistance** method, the electric current and pressure are applied simultaneously as the pipe seam passes through the welding station. The heat developed brings the abutting edges to a welding temperature and causes slight melting along the steel edges. The pressure applied during the heating squeezes the plastic metal out of the joint and upsets the heated edges to produce a joint similar in appearance to a flash-welded seam. The metal excess is trimmed off by machining immediately following welding or is removed subsequently.

When necessary, a postheat treatment may be applied after welding to stress-relieve, temper, or recrystallize the heat-affected metal.

Low-frequency resistance heating is applied to pipe of outside diameters up to 22 in. Welding and flash trimming generally are done in a continuous process. The pipe is cut to length after welding.

This process is also used in some plants for welding single lengths of plate skelp, particularly where large pipe diameters are produced.

In **high-frequency induction** welding, the abutting plate edges are brought together under a high pressure. Welding is effected by means of an induction coil which raises the temperature of the seam edges to a welding heat. The applied pressure forces the abutting edges together and causes fusion and some upsetting. The very high frequency currents normally used produce an almost instantaneous heat, permitting a high rate of production. The resulting close control over heating and fusion keeps metal upsetting to a minimum.

This process is used primarily for the production of small sizes of pipe.

High-frequency resistance welding, in principle, is similar to the low-frequency resistance-welding process described earlier, except that the welding heat is produced by a high-frequency alternating current of about 450,000 cps, which follows the low-inductance path, instead of the path of low resistance. This longer and shallower inductance path produces a high temperature in a very shallow band adjacent to the abutting edges that minimizes metal upsetting when the contact edges are brought together by the applied external force.

This process is applied to the manufacture of pipe of various wall thicknesses up to 42 in.

Arc-welding Processes. Arc welding is done commercially by the submerged-arc-welding process, the inert-gas tungsten-arc-welding process, and the gas-shielded consumable metal-arc-welding process. Submerged-arc welding is applied to carbon, alloy steel, stainless steel, and high-nickel-alloy pipe, usually of diameters of 8 in. and over. Inert-gas tungsten-arc welding is done primarily on light-wall stainless steel pipe in sizes from 1 to 24 in. The gas-shielded consumable metal-arc-welding processes are used on stainless steel and nonferrous pipe.

In the **submerged-arc**-welding process, bare consumable electrodes in wire form are used which add metal to the weld being made. The arc and molten metal are produced under a blanket of granulated flux. The heat of the arc melts a portion of the flux, creating a protective atmosphere and a slag covering the weld metal. The molten flux solidifies at a temperature below that of the weld metal. This provides a liquid protective shield until the metal has solidified.

In many plants, the seam welds are made with twin submerged arcs, where the second arc follows about 1 in. behind the first. One twin arc bead is laid inside the pipe and one bead on the outside. In heavy wall pipe, several beads may be required.

In pipe-manufacturing plants, the welding heads may be stationary and the pipe moved at a controlled rate, or vice versa. Completely automatic equipment controls the rate of pipe or welding-head movement, the rate of wire feed, the welding current, and the flux feed.

Submerged-arc seam-welded pipe may be used in critical high-temperature or high-pressure applications. Many standard specifications for such pipe now require nondestructive examination of the weld seam by radiographic methods which employ either film or fluorescent screen techniques.

Welded carbon-steel and considerable stainless-steel pipe, particularly in lighter wall thickness, is made by the **inert-gas tungsten-arc**-welding process. On light wall thicknesses, welding is generally done without filler metal. However, some processes employ filler wire which is fed automatically into the arc zone protected under a blanket of argon, helium, or mixtures of these. A number of variations of this process exist. Welding may be done from the outside or top of the pipe, from the outside at the bottom of the pipe, or from the inside. Moreover, either the welding equipment is stationary and the pipe is advanced at a uniform speed through the arc zone or the pipe is held stationary and the welding head is moved forward.

In most mills, the pipe is formed prior to welding by molding or press braking. However, a number of modern mills use strip which is rolled into pipe and welded in a continuous operation.

Spiral-welded Pipe. Spiral-welded pipe may be made by many of the electric-welding processes. Spiral welding generally is applied to lightweight pipe for transmission lines, dredging piping, irrigation systems, and other services. Recently, double submerged-arc-welded pipe has been produced for line pipe service.

Spiral-welded pipe usually is made in 40-ft lengths from strip, corresponding to U.S. Standard gage numbers. The use of 20-ft lengths of spiral-welded pipe, in conjunction with Victaulic, Dresser, or similar couplings, has made possible the economic erection of temporary piping systems, or systems in which considerable maintenance and inspection are required.

Spiral-welded pipe is manufactured by winding coils of rather narrow strip steel into cylinders with the abutting or overlapping edges of the strip forming a helix. In some processes, the edges of the strip are matched with a square or lap-end preparation. They are then fused by the electric arc, or one of the resistance welding processes, into a butt or lap weld. Another process involves overlapping of the strip ends. The outside edge is then fillet-welded to the strip underneath by the electric-arc process. The inner edge is frequently scarfed and the outer edge flared, so that the helix fits together flush on the inside and provides a projecting ring on the outside, facilitating deposition of the fillet weld.

Nonferrous Pipe Materials

Fusion-welding processes, similar to those applied to ferrous materials, are also extensively used for the production of welded nonferrous pipe.

Most extensively used are the arc-welding processes. On light wall pipe, the inert-gas tungsten-arc process is most widely used, and on heavier wall pipe (primarily the nickel-alloy materials), the submerged-arc process is used.

Aluminum and Aluminum Alloys. Piping of these materials is manufactured by one of the resistance-welding or arc-welding processes. *Resistance welding* may be of either the high-frequency-induction or high-frequency-resistance types. The same procedures as described for ferrous materials apply also to aluminum and its alloys, except that high-frequency induction welding is not employed for piping of less than $3/4$ in. diameter.

Arc welding is done by either the inert-gas tungsten-arc or the inert-gas metal-arc processes. The first of these methods is employed on aluminum-alloy pipe and tubing of thicknesses 0.030 to $1/4$ in., while the second is used on aluminum alloys of wall thicknesses greater than $1/4$ in.

Spiral-welded aluminum-alloy pipe is manufactured by methods previously described for ferrous materials. The butt welds are made by either of the two above-described arc-welding processes.

Copper and Copper Alloys. Welding is not widely used in the manufacture of copper and copper-alloy pipe and tubing, except that cupronickel alloys are produced as seam-welded pipe by the inert-gas tungsten-arc-welding process.

Nickel Alloys. The arc-welding methods described above for ferrous materials are also used extensively in the manufacture of nickel and nickel-alloy piping.

Titanium and Titanium Alloys. Welded titanium pipe is produced by the inert-gas arc-welding process. Both tungsten-arc and plasma-arc techniques are employed. Some pipe is made by first forming the metal on a press brake and then welding either by moving the welding head over the stationary pipe or by moving the pipe under a stationary welding arc. The modern operation whereby strip from coils is rolled to shape and welded continuously is being used with increasing frequency.

WROUGHT-IRON PIPE

Wrought-iron pipe is a two-component ferrous metal consisting of high-purity iron and a glasslike iron silicate slag amounting to about 3 per cent by weight of the material. The iron silicate is distributed evenly throughout the base metal in the form of elongated fibers as many as 250,000 per cross-sectional square inch. The threadlike inclusions are independent of the grain structure. The uniformly distributed networks of iron silicate are considered primarily responsible for the corrosion resistance of wrought iron.

Wrought-iron pipe is manufactured in diameters from $\frac{1}{8}$ to 24 in. In diameters from $\frac{1}{8}$ through 12 in., the outside diameters normally follow IPS sizes.

CAST STEEL PIPE

Cast pipe is made either by static casting or by centrifugal casting. Static pipe castings generally are limited to relatively short lengths. Valves, fittings, and other components are produced extensively by sand casting.

Centrifugally Cast Pipe. Centrifugally cast steel pipe is produced by introducing molten steel made in electric arc or induction furnaces into a spinning mold and allowing the metal to solidify under the pressure of the centrifugal force. The mold is usually rotated about a horizontal axis and is turned at speeds that provide a centrifugal force of 50 to 200 times that of gravity. In current practice, a density equivalent to that of wrought steel is generally obtained.

Molds containing rammed sand with binders, molds with ceramic surfaces, or permanent metal molds are used. Pipe made from these three types of molds is performing satisfactorily in high-temperature service applications. Some authorities claim that sand molds are preferable where the molten steel must flow a considerable distance over the mold surface and where chilling of the cast metal is undesirable. Others feel that with permanent metal molds, there is no danger of sand inclusions in the steel, and shrinkage voids are minimized because of the higher cooling rates provided by metal molds. With either production method, to assure a perfectly sound wall section, the feed metal at the inner diameter of the pipe which corresponds to the riser of a static casting must be removed by machining.

Centrifugally cast pipe has been produced in outside diameters from about 4 to 54 in. and in lengths of up to about 30 ft. Its economical advantage increases with pipe of larger diameter and heavier wall thickness. Initially, applications of centrifugally cast pipe involved paper mill rolls, gun barrels, etc. However, this pipe is now also being used in high-temperature high-pressure service applications,

particularly in refineries involving temperatures as high as 1050 F and pressures up to 800 psi.

Some nonferrous materials are also produced as pipe by static and centrifugal casting methods. Commercially, this involves primarily nickel and high-nickel-alloy materials.

Cold-wrought Pipe. Centrifugally cast stainless-steel pipe, in one process known as hydroforging, is subsequently cold-expanded by means of hydraulic pressure applied internally. In austenitic stainless steels, this technique eliminates the large dendritic grains of the centrifugal casting by producing recrystallization and grain refinement and improves tensile-strength properties.

Although, in effect, such pipe might be considered wrought pipe, code committees so far are classifying the material in a separate product group.

CAST-IRON PIPE

Cast-iron pipe has a relatively long life because of its heavy wall and inherently good resistance to internal and external corrosion. This pipe is used extensively for water and gas distribution systems and sewage lines in cities and towns, particularly under paved streets where it is important to use a material having a long life in order to avoid tearing up and repairing pavements to fix leaks or renew pipe.

Cast-iron pipe is commonly specified under Federal Specification WW-P-421b, Pipe, Cast-iron, Pressure (for Water and Other Liquids).

Bell-and-spigot pipe and fittings with the AWWA water bell or the AGA gas bell have been used extensively in most underground cast-iron pipe applications. Cast-iron pipe with mechanical bell and slip-on joints is finding increasing use, especially in distribution systems for natural gas and dry manufactured gas.

Flanged cast-iron pipe and mechanical bell-joint pipe of various types are used to a considerable extent aboveground in gas works and oil refineries and in process industries for carrying certain fluids for which cast iron will withstand corrosion better than steel. Cast-iron pipe is made, also, to the outside diameter of wrought pipe for making up with American Standard taper pipe threads and screwed couplings, and it can be obtained with plain ends for compression couplings of the Dayton or Dresser types or with grooved ends for Victaulic couplings.

Cast-iron pipe is manufactured by four distinct processes: (1) pit-cast vertically in dry-sand molds, (2) horizontally cast in green-sand molds, (3) centrifugally cast in sand-lined molds, and (4) centrifugally cast in permanent metal molds. At the present time, over 75 per cent of all the pipe manufactured is produced by either the centrifugally or horizontally cast process. The pipe is available with bell-and-spigot joints ordinarily used for underground water-supply installations, with flanged joints for construction work aboveground, and with mechanical (gland-type) joints for underground gas construction as well as for water-works and industrial installations.

Vertical Pit Process. The pit process consists essentially in forming a mold by ramming sand around a pattern, drying this mold in an oven, inserting a previously made core, and pouring the iron in the space between the core and the mold. After the iron is cooled and the core bar is removed, the pipe is shaken out of the flask, cleaned, dipped in a tar bath, and subjected to an internal test pressure; it is then ready for shipping. Pit-cast pipe for water-supply systems may be ordered to American Standard Specifications for Cast Iron Pit Cast Pipe for Water or Other Liquids (ASA A21.2) or to ASTM Specification A377. In the gas industry, pit-cast pipe is ordered to AGA Standard Specifications for Cast Iron Pipe and Special Castings. Pit-cast culvert pipe can be bought to ASTM Designation A142. Under Federal Specification WWP-421b, pit-cast pipe shall not be furnished for any specified nominal diameter less than 30 in.

Horizontal Process. In the horizontal method of making cast-iron pipe, molds are placed on ramming machines and then filled with properly prepared sand and mechanically rammed. Cores are made by building up tempered sand around the core bar which is perforated to allow for the escape of steam and gas. The cores are set, gauged, and located in the proper position in the mold and clamped into position to prevent movement. The top of the mold is then set in place, and metal is poured from a multiple-lipped ladle that is designed to draw the iron from the bottom. In this manner, the introduction of impurities into the mold is eliminated, and the molten iron travels a minimum distance into the mold. After the pipes are cooled, they are cleaned, dipped, and tested.

Centrifugal Casting in Sand Molds. In making pipe by the centrifugal process in sand-lined molds, a mold is made by ramming sand around the proper sized pattern. This mold is then placed horizontally in a centrifugal casting machine and, while the mold is revolving, the exact quantity of metal required to make the pipe is introduced. The speed of rotation sets up sufficient centrifugal force to distribute the molten metal on the wall of the mold, producing a completely formed pipe within a few seconds. The spinning liquid metal solidifies gradually in from 2 to 10 min. Within 30 to 60 min, the pipe is removed from the flask, the spinning and cooling time being determined by the diameter and thickness of the pipe.

Manufacture of this pipe is covered by American Standard Specification for Cast Iron Pipe Centrifugally Cast in Sand-lined Molds, for Water or Other Liquids (ASA A21.8).

Under Federal Specification WW-P-421b, pipe with a specified nominal diameter of more than 48 in. shall not be centrifugally cast in sand-lined molds.

Centrifugal Casting in Metal Molds. In making centrifugally cast pipe in permanent metal molds, the molten metal is poured into a rapidly rotating water-cooled metal mold to the surface of which a thin refractory coating is applied to control the cooling of the iron. A sand core is used for forming the inside contour of the bell. The mold is mounted on rollers in a water jacket so that it can be rotated at high speeds, and the water jacket is on wheels so that the entire assembly can be moved by a hydraulic cylinder in the direction of the axis of the mold. The molten iron is fed into the mold through a trough that has a spout curved toward a side wall. The tube used to spray the refractory material is built into the trough with the nozzle set just behind the pouring spout, and iron is poured from a tilting ladle at a uniform rate. As the assembly moves over the spout, the iron is distributed from one end of the mold to the other and is held against the inner wall by centrifugal force. After completion of the pour, the mold is kept rotating until the pipe has cooled below the critical temperature. It is then removed and placed in a thermostatically controlled heat-treating furnace where the pipe is reheated; after this the temperature is gradually lowered. After leaving the furnace, the pipe is cleaned, dipped, and hydraulically tested.

Pipe spun in metal molds is covered by American Standard Specification for Cast Iron Pipe Centrifugally Cast in Metal Molds, for Water or other Liquids (ASA A21.6).

Under Federal Specification WW-P-421b, pipe with a specified nominal diameter of more than 24 in. shall not be centrifugally cast in metal molds. It is required, also, that this pipe be heat-treated after withdrawal from the molds.

COMMERCIAL PIPE SIZES

Table 1 lists the pipe sizes and wall thicknesses currently established as standard, or specifically:

1. The traditional standard weight, extra strong, and double extra strong pipe

Table I. Standard Commercial Pipe Sizes Commonly Used[1]

Nominal pipe size	Outside diameter	Sch 5S[2]	Sch 10S[2]	Sch 10	Sch 20	Sch 30	Standard[3]	Sch 40	Sch 60	Extra-strong[4]	Sch 80	Sch 100	Sch 120	Sch 140	Sch 160	XX strong
1/8	0.405	....	0.049	....	....	....	0.068	0.068	....	0.095	0.095	....	....	....	....	
1/4	0.540	....	0.065	....	....	....	0.088	0.088	....	0.119	0.119	....	....	....	....	
3/8	0.675	....	0.065	....	....	....	0.091	0.091	....	0.126	0.126	....	....	....	....	
1/2	0.840	0.065	0.083	....	....	....	0.109	0.109	....	0.147	0.147	....	....	....	0.188	0.294
3/4	1.050	0.065	0.083	....	....	....	0.113	0.113	....	0.154	0.154	....	....	....	0.219	0.308
1	1.315	0.065	0.109	....	....	....	0.133	0.133	....	0.179	0.179	....	....	....	0.250	0.358
1 1/4	1.660	0.065	0.109	....	....	....	0.140	0.140	....	0.191	0.191	....	....	....	0.250	0.382
1 1/2	1.900	0.065	0.109	....	....	....	0.145	0.145	....	0.200	0.200	....	....	....	0.281	0.400
2	2.375	0.065	0.109	....	....	....	0.154	0.154	....	0.218	0.218	....	....	....	0.344	0.436
2 1/2	2.875	0.083	0.120	....	....	....	0.203	0.203	....	0.276	0.276	....	....	....	0.375	0.552
3	3.5	0.083	0.120	....	....	....	0.216	0.216	....	0.300	0.300	....	....	....	0.438	0.600
3 1/2	4.0	0.083	0.120	....	....	....	0.226	0.226	....	0.318	0.318	....	....	....		
4	4.5	0.083	0.120	....	....	....	0.237	0.237	....	0.337	0.337	....	0.438	....	0.531	0.674
5	5.563	0.109	0.134	....	....	....	0.258	0.258	....	0.375	0.375	....	0.500	....	0.625	0.750
6	6.625	0.109	0.134	....	....	....	0.280	0.280	....	0.432	0.432	....	0.562	....	0.719	0.864
8	8.625	0.109	0.148	....	0.250	0.277	0.322	0.322	0.406	0.500	0.500	0.594	0.719	0.812	0.906	0.875
10	10.75	0.134	0.165	....	0.250	0.307	0.365	0.365	0.500	0.500	0.594	0.719	0.844	1.000	1.125	1.000
12	12.75	0.156	0.180	....	0.250	0.330	0.375	0.406	0.562	0.500	0.688	0.844	1.000	1.125	1.312	1.000

Nominal wall thickness

Size (OD)														
14.0	0.156	0.188	0.250	0.312	0.375	0.375	0.438	0.594	0.500	0.750	0.938	1.094	1.250	1.406
16.0	0.165	0.188	0.250	0.312	0.375	0.375	0.500	0.656	0.500	0.844	1.031	1.219	1.438	1.594
18.0	0.165	0.188	0.250	0.312	0.438	0.375	0.562	0.750	0.500	0.938	1.156	1.375	1.562	1.781
20.0	0.188	0.218	0.250	0.375	0.500	0.375	0.594	0.812	0.500	1.031	1.281	1.500	1.750	1.969
22.0	0.188	0.218	0.250	0.375	0.500	0.375	.....	0.875	0.500	1.125	1.375	1.625	1.875	2.125
24.0	0.218	0.250	0.250	0.375	0.562	0.375	0.688	0.969	0.500	1.218	1.531	1.812	2.062	2.344
26.0	.....	.....	0.312	0.500	.....	0.375	.....	.....	0.500					
28.0	.....	.....	0.312	0.500	0.625	0.375	.....	.....	0.500					
30.0	0.250	0.312	0.312	0.500	0.625	0.375	.....	.....	0.500					
32.0	.....	.....	0.312	0.500	0.625	0.375	0.688	.....	0.500					
34.0	.....	.....	0.312	0.500	0.625	0.375	0.688	.....	0.500					
36.0	.....	.....	0.312	0.500	0.625	0.375	0.750	.....	0.500					
42.0	.....	.....	.....	.....	.....	0.375	.....	.....	0.500					

All dimensions are given in inches.

The decimal thicknesses listed for the respective pipe sizes represent their nominal or average wall dimensions. The actual thicknesses may be as much as 12.5% under the nominal thickness because of mill tolerance. Thicknesses shown in light face for Schedule 60 and heavier pipe are not currently supplied by the mills, unless a certain minimum tonnage is ordered.

[1] Additional nominal wall thicknesses established as standard sizes are listed in ASA Standard B36.10-1959. Also, Schedule 10S is available in carbon steel in sizes 12 in. and smaller.

[2] Schedules 5S and 10S are available in corrosion-resistant materials and Schedule 10S is also available in carbon steel.

[3] Thicknesses shown in italics are available also in stainless steel, under the designation Schedule 40S.

[4] Thicknesses shown in italics are available also in stainless steel, under the designation Schedule 80S.

2. The pipe wall thickness schedules listed in American Standard B36.10, which are applicable to carbon steel and alloys other than stainless steels

3. The pipe wall thickness schedules listed in American Standard B36.19 are applicable only to corrosion-resistant materials.

PIPE AND TUBE PRODUCTS

Commercial pipe and tube products are grouped into various classifications generally based on the application or use and not on the manufacturing method. Most tubular products fall into one of three very broad classifications: (1) pipe, (2) pressure tubes, and (3) mechanical tubes. Each classification falls into various subgroupings, which may be defined and standardized differently by the different trade or user groups. Moreover, the same standard materials specifications may apply to several of the (user) classifications. For example, ASTM A120 or A53 pipe may be used for applications representing refrigeration, pressure, and nipple service. Cost considerations enter also into the selection of specific piping materials. In some sizes, prices of pipe made to different materials specifications may vary, whereas in other sizes, they may be identical.

Within the broad use classifications listed above, the method classifications described in the preceding section are often referred to. These primarily involve (1) seamless wrought, (2) seamless cast, and (3) seam-welded tubular products. The large variety of single and combination of pipe- or tube-forming methods can produce different characteristics and properties. In addition, the final finishing can involve *hot-finished* and *cold-finished products*. Cold-finishing may have been done by *reducing* or by *expanding*.

PIPING

On the basis of *user* classification, the more commonly used types of pipe are tabulated in Table 2. This listing ignores method of manufacture, size range, wall thickness, and finish, for which the different user groups may have developed different standard requirements.

Standard Pipe. *Mechanical service pipe* is produced in three classes of wall thickness—standard weight, extra strong, and double extra strong. It is available as welded or seamless pipe of ordinary finish and dimensional tolerances, produced in sizes up to 12 in. nominal OD inclusive. This pipe is used for structural and mechanical purposes. Certain applications have other requirements for size, surface finish, or straightness.

Refrigeration pipe (also known as ice-machine pipe or ammonia pipe) may be butt-welded, lap-welded, electric-resistance-welded, or seamless and is intended for use as a conveyor of refrigerants. This pipe is suitable for coiling, bending, and welding. The sizes commonly used are ¾ to 2 in. inclusive, produced in random and double random lengths in standard line pipe sizes and weights. Double random lengths are used as ice-rink pipe. It can be produced with plain ends, with threaded ends only, or with threaded ends and line pipe couplings, as desired.

Dry-kiln pipe is butt-welded, electric-resistance-welded, or seamless pipe, for use in the lumber industry. It is produced in standard-weight pipe sizes of ¾, 1, and 1¼ in. Joints are designed to permit subsequent "makeup" after expansion has occurred. Dry-kiln pipe is commonly produced with threaded ends and couplings and in random lengths.

Pressure Pipe. Pressure pipe is used for conveying fluids or gases at normal, subzero, or elevated temperatures or pressures. It generally is not subjected to

Table 2. Major Pipe Classifications and Examples of Applications

Type of pipe	Uses
Standard.....................	Mechanical (structural) service pipe, low-pressure service pipe, refrigeration (ice machine) pipe, ice-rink pipe, dry-kiln pipe
Pressure	Liquid, gas or vapor service pipe, service for elevated temperature or pressure, or both
Line........................	Threaded or plain ends, gas, oil, and steam pipe
Water well	Reamed and drifted, water-well casing, drive pipe, driven well pipe, pump pipe, turbine pump pipe
Oil country tubular goods......	Casing, well tubing, drill pipe
Other pipe	Conduit, piles, nipple pipe, sprinkler pipe, bedstead tubing

external heat application.   The range of sizes is $\frac{1}{8}$ in. nominal size to 36 in. actual outside diameter in various wall thicknesses.   Pressure piping is furnished in random lengths, with threaded or plain ends, as required.   Jointers are not customarily produced.   Pressure pipe generally receives a hydrostatic test by the mill.

Line Pipe.   Line pipe is welded or seamless pipe produced in sizes from $\frac{1}{8}$ in. nominal OD to 36 in. actual OD, inclusive.   It is used principally for conveying gas, oil, or water.   Line pipe is produced with ends plain, threaded, beveled, grooved, flanged, or expanded, as required for various types of mechanical couplers or for welded joints.   When threaded ends and couplings are required, recessed couplings are normally supplied.

Water-well Pipe.   Water-well pipe is welded or seamless steel pipe used for conveying water for municipal and industrial applications.   Pipelines for such purposes involve flow mains, transmission mains, force mains, water mains, or distribution mains.   The mains are generally laid underground.   Sizes range from $\frac{1}{8}$ to 96 in. in a variety of wall thicknesses.   Pipe is produced with ends suitably prepared for mechanical couplers, with plain ends beveled for welding, with ends fitted with butt straps for field welding, or with bell-and-spigot joints with rubber gaskets for field joining.   Pipe is produced in double random lengths of about 40 ft, single random lengths of about 20 ft, or in definite cut lengths, as specified.   Wall thicknesses vary from 0.068 in. for $\frac{1}{8}$ in. nominal outer diameter to 1.00 in. for 96 in. actual outer diameter.

When required, water-well pipe is produced with a specified coating or lining or both.   For example, cement-mortar coatings are extensively used.

Oil Country Goods.   *Casing* is used as a structural retainer for the walls of oil or gas wells and is also used to exclude undesirable fluids, and to confine and conduct oil or gas from productive subsurface strata to the ground level.   Casing is produced in sizes $4\frac{1}{2}$ to 20 in. outside diameter inclusive.   Size designations refer to actual outside diameter and weight per foot.   Ends are commonly threaded and furnished with couplings.   When required, the ends are prepared to accommodate other types of joints.

Table 3. Principal Properties of Commercial Piping Materials

Nominal pipe size, outside diam., in.	Schedule number*			Wall thickness, in.	Inside diam., in.	Inside area, sq in.	Metal area, sq in.	Sq ft outside surface, per ft	Sq ft inside surface, per ft	Weight per ft, lb†	Weight of water per ft, lb	Moment of inertia, in.⁴	Section modulus, in.³	Radius gyration, in.
	a	b	c											
1 *1.315*	..	..	5S	0.065	1.185	1.103	0.2553	0.344	0.310	0.868	0.478	0.0500	0.0760	0.443
	..	..	10S	0.109	1.097	0.945	0.413	0.344	0.2872	1.404	0.409	0.0757	0.1151	0.428
	40	Std	40S	0.133	1.049	0.864	0.494	0.344	0.2746	1.679	0.374	0.0874	0.1329	0.421
	80	XS	80S	0.179	0.957	0.719	0.639	0.344	0.2520	2.172	0.311	0.1056	0.1606	0.407
	160	..	..	0.250	0.815	0.522	0.836	0.344	0.2134	2.844	0.2261	0.1252	0.1903	0.387
	..	XXS	..	0.358	0.599	0.2818	1.076	0.344	0.1570	3.659	0.1221	0.1405	0.2137	0.361
1½ *1.900*	..	..	5S	0.065	1.770	2.461	0.375	0.497	0.463	1.274	1.067	0.1580	0.1663	0.649
	..	..	10S	0.109	1.682	2.222	0.613	0.497	0.440	2.085	0.962	0.2469	0.2599	0.634
	40	Std	40S	0.145	1.610	2.036	0.799	0.497	0.421	2.718	0.882	0.310	0.326	0.623
	80	XS	80S	0.200	1.500	1.767	1.068	0.497	0.393	3.631	0.765	0.391	0.412	0.605
	160	..	..	0.281	1.338	1.406	1.429	0.497	0.350	4.859	0.608	0.483	0.508	0.581
	..	XXS	..	0.400	1.100	0.950	1.885	0.497	0.288	6.408	0.412	0.568	0.598	0.549
2 *2.375*	..	..	5S	0.065	2.245	3.96	0.472	0.622	0.588	1.604	1.716	0.315	0.2652	0.817
	..	..	10S	0.109	2.157	3.65	0.776	0.622	0.565	2.638	1.582	0.499	0.420	0.802
	40	Std	40S	0.154	2.067	3.36	1.075	0.622	0.541	3.653	1.455	0.666	0.561	0.787
	80	XS	80S	0.218	1.939	2.953	1.477	0.622	0.508	5.022	1.280	0.868	0.731	0.766
	160	..	..	0.343	1.689	2.240	2.190	0.622	0.442	7.444	0.971	1.163	0.979	0.729
	..	XXS	..	0.436	1.503	1.774	2.656	0.622	0.393	9.029	0.769	1.312	1.104	0.703
3 *3.500*	..	..	5S	0.083	3.334	8.73	0.891	0.916	0.873	3.03	3.78	1.301	0.744	1.208
	..	..	10S	0.120	3.260	8.35	1.274	0.916	0.853	4.33	3.61	1.822	1.041	1.196
	40	Std	40S	0.216	3.068	7.39	2.228	0.916	0.803	7.58	3.20	3.02	1.724	1.164
	80	XS	80S	0.300	2.900	6.61	3.02	0.916	0.759	10.25	2.864	3.90	2.226	1.136
	160	..	..	0.437	2.626	5.42	4.21	0.916	0.687	14.32	2.348	5.03	2.876	1.094
	..	XXS	..	0.600	2.300	4.15	5.47	0.916	0.602	18.58	1.801	5.99	3.43	1.047

Nom.	O.D.	Sched.	Desig.	Sched. S	Wall	I.D.									
4	4.500			5S	0.083	4.334	14.75	1.152	1.178	1.135	3.92	6.40	2.811	1.249	1.562
				10S	0.120	4.260	14.25	1.651	1.178	1.115	5.61	6.17	3.96	1.762	1.549
		40	Std	40S	0.237	4.026	12.73	3.17	1.178	1.054	10.79	5.51	7.23	3.21	1.510
		80	XS	80S	0.337	3.826	11.50	4.41	1.178	1.002	14.98	4.98	9.61	4.27	1.477
		120			0.437	3.626	10.33	5.58	1.178	0.949	18.96	4.48	11.65	5.18	1.445
		160			0.531	3.438	9.28	6.62	1.178	0.900	22.51	4.02	13.27	5.90	1.416
			XXS		0.674	3.152	7.80	8.10	1.178	0.825	27.54	3.38	15.29	6.79	1.374
6	6.625			5S	0.109	6.407	32.2	2.231	1.734	1.677	5.37	13.98	11.85	3.58	2.304
				10S	0.134	6.357	31.7	2.733	1.734	1.664	9.29	13.74	14.40	4.35	2.295
		40	Std	40S	0.280	6.065	28.89	5.58	1.734	1.588	18.97	12.51	28.14	8.50	2.245
		80	XS	80S	0.432	5.761	26.07	8.40	1.734	1.508	28.57	11.29	40.5	12.23	2.195
		120			0.562	5.501	23.77	10.70	1.734	1.440	36.39	10.30	49.6	14.98	2.153
		160			0.718	5.189	21.15	13.33	1.734	1.358	45.30	9.16	59.0	17.81	2.104
			XXS		0.864	4.897	18.83	15.64	1.734	1.282	53.16	8.17	66.3	20.03	2.060
8	8.625			5S	0.109	8.407	55.5	2.916	2.258	2.201	9.91	24.07	26.45	6.13	3.01
				10S	0.148	8.329	54.5	3.94	2.258	2.180	13.40	23.59	35.4	8.21	3.00
		20			0.250	8.125	51.8	6.58	2.258	2.127	22.36	22.48	57.7	13.39	2.962
		30			0.277	8.071	51.2	7.26	2.258	2.113	24.70	22.18	63.4	14.69	2.953
		40	Std	40S	0.322	7.981	50.0	8.40	2.258	2.089	28.55	21.69	72.5	16.81	2.938
		60			0.406	7.813	47.9	10.48	2.258	2.045	35.64	20.79	88.8	20.58	2.909
		80	XS	80S	0.500	7.625	45.7	12.76	2.258	1.996	43.39	19.80	105.7	24.52	2.878
		100			0.593	7.439	43.5	14.96	2.258	1.948	50.87	18.84	121.4	28.14	2.847
		120			0.718	7.189	40.6	17.84	2.258	1.882	60.63	17.60	140.6	32.6	2.807
		140			0.812	7.001	38.5	19.93	2.258	1.833	67.76	16.69	153.8	35.7	2.777
			XXS		0.875	6.875	37.1	21.30	2.258	1.800	72.42	16.09	162.0	37.6	2.757
		160			0.906	6.813	36.5	21.97	2.258	1.784	74.69	15.80	165.9	38.5	2.748
10	10.750			5S	0.134	10.482	86.3	4.52	2.815	2.744	15.15	37.4	63.7	11.85	3.75
				10S	0.165	10.420	85.3	5.49	2.815	2.728	18.70	36.9	76.9	14.30	3.74
		20			0.250	10.250	82.5	8.26	2.815	2.683	28.04	35.8	113.7	21.16	3.71
		30			0.279	10.192	81.6	9.18	2.815	2.668	31.20	35.3	125.9	23.42	3.70
		40	Std	40S	0.307	10.136	80.7	10.07	2.815	2.654	34.24	35.0	137.5	25.57	3.69
		60	XS	80S	0.365	10.020	78.9	11.91	2.815	2.623	40.48	34.1	160.8	29.90	3.67
		80			0.500	9.750	74.7	16.10	2.815	2.553	54.74	32.3	212.0	39.4	3.63
		100			0.593	9.564	71.8	18.92	2.815	2.504	64.33	31.1	244.9	45.6	3.60
		120			0.718	9.314	68.1	22.63	2.815	2.438	76.93	29.5	286.2	53.2	3.56
		140			0.843	9.064	64.5	26.24	2.815	2.373	89.20	28.0	324	60.3	3.52
		160			1.000	8.750	60.1	30.6	2.815	2.291	104.13	26.1	368	68.4	3.47
					1.125	8.500	56.7	34.0	2.815	2.225	115.65	24.6	399	74.3	3.43

Table 3. *(Continued)*

Nominal pipe size, *outside diam.,* in.	Schedule number* a	b	c	Wall thick-ness, in.	Inside diam., in.	Inside area, sq in.	Metal area, sq in.	Sq ft outside surface, per ft	Sq ft inside surface, per ft	Weight per ft, lb†	Weight of water per ft, lb	Moment of inertia, in.⁴	Section modulus, in.³	Radius gyration, in.
	...	...	5S	0.156	12.438	121.4	6.17	3.34	3.26	20.99	52.7	122.2	19.20	4.45
	...	...	10S	0.180	12.390	120.6	7.11	3.34	3.24	24.20	52.2	140.5	22.03	4.44
	20	...	...	0.250	12.250	117.9	9.84	3.34	3.21	33.38	51.1	191.9	30.1	4.42
	30	...	...	0.330	12.090	114.8	12.88	3.34	3.17	43.77	49.7	248.5	39.0	4.39
	...	Std	40S	0.375	12.000	113.1	14.58	3.34	3.14	49.56	49.0	279.3	43.8	4.38
12 *12.750*	40	...	...	0.406	11.938	111.9	15.74	3.34	3.13	53.53	48.5	300	47.1	4.37
	...	XS	80S	0.500	11.750	108.4	19.24	3.34	3.08	65.42	47.0	362	56.7	4.33
	60	...	...	0.562	11.626	106.2	21.52	3.34	3.04	73.16	46.0	401	62.8	4.31
	80	...	...	0.687	11.376	101.6	26.04	3.34	2.978	88.51	44.0	475	74.5	4.27
	100	...	...	0.843	11.064	96.1	31.5	3.34	2.897	107.20	41.6	562	88.1	4.22
	120	...	...	1.000	10.750	90.8	36.9	3.34	2.814	125.49	39.3	642	100.7	4.17
	140	...	...	1.125	10.500	86.6	41.1	3.34	2.749	139.68	37.5	701	109.9	4.13
	160	...	...	1.312	10.126	80.5	47.1	3.34	2.651	160.27	34.9	781	122.6	4.07
	10	...	...	0.250	13.500	143.1	10.80	3.67	3.53	36.71	62.1	255.4	36.5	4.86
	20	...	...	0.312	13.376	140.5	13.42	3.67	3.50	45.68	60.9	314	44.9	4.84
	30	Std	...	0.375	13.250	137.9	16.05	3.67	3.47	54.57	59.7	373	53.3	4.82
	40	...	...	0.437	13.126	135.3	18.62	3.67	3.44	63.37	58.7	429	61.2	4.80
	...	XS	...	0.500	13.000	132.7	21.21	3.67	3.40	72.09	57.5	484	69.1	4.78
	...	...	...	0.562	12.876	130.2	23.73	3.67	3.37	80.66	56.5	537	76.7	4.76
14 *14.000*	60	...	...	0.593	12.814	129.0	24.98	3.67	3.35	84.91	55.9	562	80.3	4.74
	...	...	...	0.625	12.750	127.7	26.26	3.67	3.34	89.28	55.3	589	84.1	4.73
	80	...	...	0.687	12.626	125.2	28.73	3.67	3.31	97.68	54.3	638	91.2	4.71
	...	...	...	0.750	12.500	122.7	31.2	3.67	3.27	106.13	53.2	687	98.2	4.69
	...	...	...	0.875	12.250	117.9	36.1	4.67	3.21	122.66	51.1	781	111.5	4.65
	100	...	...	0.937	12.126	115.5	38.5	3.67	3.17	130.73	50.0	825	117.8	4.63
	120	...	...	1.093	11.814	109.6	44.3	3.67	3.09	150.67	47.5	930	132.8	4.58
	140	...	...	1.250	11.500	103.9	50.1	3.67	3.01	170.22	45.0	1127	146.8	4.53
	160	...	...	1.406	11.188	98.3	55.6	3.67	2.929	189.12	42.6	1017	159.6	4.48

Nom. Size	Sched. No.	Desig.	Wall	ID									
16 16.000	10		0.250	15.500	188.7	12.37	4.19	4.06	42.05	81.8	384	48.0	5.57
	20		0.312	15.376	185.7	15.38	4.19	4.03	52.36	80.5	473	59.2	5.55
	30	Std	0.375	15.250	182.6	18.41	4.19	3.99	62.58	79.1	562	70.3	5.53
			0.437	15.126	179.7	21.37	4.19	3.96	72.64	77.9	648	80.9	5.50
	40	XS	0.500	15.000	176.7	24.35	4.19	3.93	82.77	76.5	732	91.5	5.48
			0.562	14.876	173.8	27.26	4.19	3.89	92.66	75.4	813	106.6	5.46
			0.625	14.750	170.9	30.2	4.19	3.86	102.63	74.1	894	112.2	5.44
	60		0.656	14.688	169.4	31.6	4.19	3.85	107.50	73.4	933	116.6	5.43
			0.687	14.626	168.0	33.0	4.19	3.83	112.36	72.7	971	121.4	5.42
			0.750	14.500	165.1	35.9	4.19	3.80	122.15	71.5	1047	130.9	5.40
	80		0.843	14.314	160.9	40.1	4.19	3.75	136.46	69.7	1157	144.6	5.37
			0.875	14.250	159.5	41.6	4.19	3.73	141.35	69.1	1193	154.1	5.36
	100		1.031	13.938	152.5	48.5	4.19	3.65	164.83	66.1	1365	170.6	5.30
	120		1.218	13.564	144.5	56.6	4.19	3.55	192.29	62.6	1556	194.5	5.24
	140		1.437	13.126	135.3	65.7	4.19	3.44	223.64	58.6	1760	220.0	5.17
	160		1.593	12.814	129.0	72.1	4.19	3.35	245.11	55.9	1894	236.7	5.12
18 18.000	10		0.250	17.500	240.5	13.94	4.71	4.58	47.39	104.3	549	61.0	6.28
	20		0.312	17.376	237.1	17.34	4.71	4.55	59.03	102.8	678	75.5	6.25
		Std	0.375	17.250	233.7	20.76	4.71	4.52	70.59	101.2	807	89.6	6.23
	30		0.437	17.126	230.4	24.11	4.71	4.48	82.06	99.9	931	103.4	6.21
		XS	0.500	17.000	227.0	27.49	4.71	4.45	93.45	98.4	1053	117.0	6.19
	40		0.562	16.876	223.7	30.8	4.71	4.42	104.75	97.0	1172	130.2	6.17
			0.625	16.750	220.5	34.1	4.71	4.39	115.98	95.5	1289	143.3	6.15
			0.687	16.626	217.1	37.4	4.71	4.35	127.03	94.1	1403	156.3	6.13
	60		0.750	16.500	213.8	40.6	4.71	4.32	138.17	92.7	1515	168.3	6.10
			0.875	16.250	207.4	47.1	4.71	4.25	160.04	89.9	1731	192.8	6.06
	80		0.937	16.126	204.2	50.2	4.71	4.22	170.75	88.5	1834	203.8	6.04
	100		1.156	15.688	193.3	61.2	4.71	4.11	207.96	83.7	2180	242.2	5.97
	120		1.375	15.250	182.6	71.8	4.71	3.99	244.14	79.2	2499	277.6	5.90
	140		1.562	14.876	173.8	80.7	4.71	3.89	274.23	75.3	2750	306	5.84
	160		1.781	14.438	163.7	90.7	4.71	3.78	308.51	71.0	3020	336	5.77

Table 3. (Continued)

Nominal pipe size, outside diam., in.	Schedule number*			Wall thickness, in.	Inside diam., in.	Inside area, sq in.	Metal area, sq in.	Sq ft outside surface, per ft	Sq ft inside surface, per ft	Weight per ft, lb†	Weight of water per ft, lb	Moment of inertia, $in.^4$	Section modulus, $in.^3$	Radius gyration, in.
	a	b	c											
20 20.000	10			0.250	19.500	298.6	15.51	5.24	5.11	52.73	129.5	757	75.7	6.98
				0.312	19.376	294.9	19.30	5.24	5.07	65.40	128.1	935	93.5	6.96
	20	Std		0.375	19.250	291.0	23.12	5.24	5.04	78.60	126.0	1114	111.4	6.94
				0.437	19.126	287.3	26.86	5.24	5.01	91.31	124.6	1286	128.6	6.92
	30	XS		0.500	19.000	283.5	30.6	5.24	4.97	104.13	122.8	1457	145.7	6.90
				0.562	18.876	279.8	34.3	5.24	4.94	116.67	121.3	1624	162.4	6.88
	40			0.593	18.814	278.0	36.2	5.24	4.93	122.91	120.4	1704	170.4	6.86
				0.625	18.750	276.1	38.0	5.24	4.91	129.33	119.7	1787	178.7	6.85
				0.687	18.626	272.5	41.7	5.24	4.88	141.71	118.1	1946	194.6	6.83
				0.750	18.500	268.8	45.4	5.24	4.84	154.20	116.5	2105	210.5	6.81
	60			0.812	18.376	265.2	48.9	5.24	4.81	166.40	115.0	2257	225.7	6.79
				0.875	18.250	261.6	52.6	5.24	4.78	178.73	113.4	2409	240.9	6.77
	80			1.031	17.938	252.7	61.4	5.24	4.70	208.87	109.4	2772	277.2	6.72
	100			1.281	17.438	238.8	75.3	5.24	4.57	256.10	103.4	3320	332	6.63
	120			1.500	17.000	227.0	87.2	5.24	4.45	296.37	98.3	3760	376	6.56
	140			1.750	16.500	213.8	100.3	5.24	4.32	341.10	92.6	4220	422	6.48
	160			1.968	16.064	202.7	111.5	5.24	4.21	379.01	87.9	4590	459	6.41
24 24.000	10			0.250	23.500	434	18.65	6.28	6.15	63.41	188.0	1316	109.6	8.40
				0.312	23.376	430	23.20	6.28	6.12	78.93	186.1	1629	135.8	8.38
	20	Std		0.375	23.250	425	27.83	6.28	6.09	94.62	183.8	1943	161.9	8.35
				0.437	23.126	420	32.4	6.28	6.05	109.97	182.1	2246	187.4	8.33
		XS		0.500	23.000	415	36.9	6.28	6.02	125.49	180.1	2550	212.5	8.31
	30			0.562	22.876	411	41.4	6.28	5.99	140.80	178.1	2840	237.0	8.29
				0.625	22.750	406	45.9	6.28	5.96	156.03	176.2	3140	261.4	8.27
	40			0.687	22.626	402	50.3	6.28	5.92	171.17	174.3	3420	285.2	8.25
				0.750	22.500	398	54.8	6.28	5.89	186.24	172.4	3710	309	8.22
	60			0.968	22.064	382	70.0	6.28	5.78	238.11	165.8	4650	388	8.15
	80			1.218	21.564	365	87.2	6.28	5.65	296.36	158.3	5670	473	8.07
	100			1.531	20.938	344	108.1	6.28	5.48	367.40	149.3	6850	571	7.96
	120			1.812	20.376	326	126.3	6.28	5.33	429.39	141.4	7830	652	7.87
	140			2.062	19.876	310	142.1	6.28	5.20	483.13	134.5	8630	719	7.79
	160			2.343	19.314	293	159.4	6.28	5.06	541.94	127.0	9460	788	7.70
30 30.000	10			0.312	29.376	678	29.1	7.85	7.69	98.93	293.8	3210	214	10.50
	20			0.500	29.000	661	46.3	7.85	7.59	157.53	286.3	5040	336	10.43
	30			0.625	28.750	649	57.6	7.85	7.53	196.08	281.5	6220	415	10.39

The following formulas were used in the computation of the values shown in the table:

Weight† of pipe per foot (pounds) $= 10.6802t(D - t)$
Weight of water per foot (pounds) $= 0.3405d^2$
Square feet outside surface per foot $= 0.2618D$
Square feet inside surface per foot $= 0.2618d$
Inside area (square inches) $= 0.785d^2$
Area of metal (square inches) $= 0.785(D^2 - d^2)$
Moment of inertia (inches⁴) $= 0.0491(D^4 - d^4)$

Section modulus (inches³) $= \dfrac{A_M R_g^2}{\dfrac{D}{D}}$

$= \dfrac{0.0982(D^4 - d^4)}{D}$

Radius of gyration (inches) $= 0.25\sqrt{D^2 + d^2}$

A_M = area of metal (square inches)
d = inside diameter (inches)
D = outside diameter (inches)
R_g = radius of gyration (inches)
t = pipe wall thickness (inches)

Standard weight pipe and Schedule 40 are the same in all sizes through 10 in.; from 12 in. through 24 in., standard weight pipe has a wall thickness of ³⁄₈ in.
Extra-strong weight pipe and Schedule 80 are the same in all sizes through 8 in.; from 8 in. through 24 in., extra-strong weight pipe has a wall thickness of ½ in.
Double extra-strong weight pipe has no corresponding schedule number.
a: ASA B36.10 Steel Pipe Schedule Numbers.
b: ASA B36.10 Steel Pipe Nominal Wall Thickness Designations.
c: ASA B36.19 Stainless Steel Pipe Schedule Numbers.
† The ferritic stainless steels may be about 5% less and the austenitic stainless steels about 2% greater than the values shown in this table which are based on weights for carbon steel.

Drill pipe is used to transmit power by rotary motion from ground level to a rotary drilling tool below the surface and also to convey flushing mediums to the cutting face of the tool.   Drill pipe is produced in sizes $2\frac{3}{8}$ to $6\frac{5}{8}$ in. outside diameter, inclusive.   Size designations refer to actual outside diameter and weight per foot.   Drill pipe is generally upset, either internally or externally, or both, and is furnished with threaded ends and couplings, threaded only, or prepared to accommodate other types of joints.

Tubing is used within the casing of oil wells to conduct oil to ground level.   It is produced in sizes 1.050 to 4.500 in. OD inclusive, in several weights per foot. Ends are threaded and fitted with couplings and may or may not be upset externally.

Other Classifications.   *Rigid conduit pipe* is welded or seamless pipe intended especially for the protection of electrical wiring systems.   Conduit pipe is not subjected to hydrostatic tests unless so specified.   It is furnished in standard-weight pipe sizes from $\frac{1}{4}$ to 6 in. in 10-ft lengths,[1] with plain ends or with threaded ends and couplings, as specified.

Piling pipe is welded or seamless pipe for use as piles, where the cylinder section acts as a permanent load-carrying member or where it acts as a shell to form cast-in-place concrete piles.   Specifications provide for the choice of three grades by minimum tensile strength in which the sizes listed are $8\frac{5}{8}$ to 24 in. outside diameters in a variety of wall thicknesses and in two length ranges.   Ends are plain or beveled for welding.

Nipple pipe is standard-weight, extra-strong, or double-extra-strong welded or seamless pipe produced for the manufacture of pipe nipples.   Standard-weight pipe with threaded ends is also used in sprinkler systems.   Nipple pipe is commonly produced in random lengths with plain ends in nominal sizes $\frac{1}{8}$ to 12 in.   Close OD tolerances, sound welds, good threading properties, and surface cleanliness are essential in this product.   It is commonly coated with oil or zinc and well protected in shipment.   When reference is made to ASTM Specifications, Specification A120 is generally specified for diameters to 5 in. OD and A53 for diameters of 5 in. and over.

Standard Sizes.   Standard pressure, line, and other pipe with plain ends for welding or with threaded ends are standardized in accordance with two scales. Diameters of 12 in. and less have a *nominal* size which represents approximately that of the inside diameter of standard-weight pipe.   The nominal outside diameter is standard, regardless of weight.   The increase in wall thickness results in a decrease of the inside diameter.

The standardization of pipe sizes over 12 in. is based on the actual outside diameter, the wall thickness, and the weight per foot.

The principal dimensions of commercial piping materials are summarized in Table 3.

Standard Weights.   The weight of standard piping materials is tabulated in Table 3.   The weights of butt-welding tees and laterals are given in Tables 19 through 24.   The weights of reducing fittings are approximately the same as for full-size fittings.

The weights of welding reducers are for one size reduction and are thus only approximately correct for other reductions.

For special materials, the equations listed below for weights of tubes and weights of contents of tubes may be helpful.

---

[1] Although some specifications of rigid conduit pipe list lengths to 20 ft, the National Electric Code, 1965, limits lengths to 10 ft.

$$\text{Weight of tube, lb/ft} = F \times 10.68 \times T \times D - T$$

where T = wall thickness, in.
$\quad D$ = outside diameter, in.
$\quad F$ = relative weight factor

The weight of tube calculation is based on low-carbon steel weighing 0.2833 lb/cu in. and is extended to other materials through the factor F.

Relative Weight Factor F

Aluminum 0.35
Brass 1.12
Cast iron.................... 0.91
Copper 1.14
Ferritic stainless steel 0.95
Austenitic stainless steel........ 1.02
Steel........................ 1.00
Wrought iron................. 0.98

$$\text{Weight of contents of a tube, lb/ft} = G \times 0.3405 \times (D - 2T)^2$$

where G = specific gravity of contents
$\quad T$ = tube wall thickness, in.
$\quad D$ = tube outside diameter, in.

The weight per foot of steel pipe is subject to the tolerances listed in Table 4.

Table 4. Weight Tolerances of Steel Piping

Specification	Size	Tolerance
ASTM A53	Std wt	$+5\%,\quad -5\%$
	XS wt	$+5\%,\quad -5\%$
ASTM A120	XXS wt	$+10\%,\quad -10\%$
ASTM A106	Sch 10-120	$+6.5\%,\quad -3.5\%$
	Sch 140-160	$+10\%,\quad -3.5\%$
ASTM A335	12″ and under	$+6.5\%,\quad -3.5\%$
	Over 12″	$+10\%,\quad -5\%$
ASTM A312 ASTM A376	12″ and under	$+6.5\%,\quad -3.5\%$
API 5L	Std wt Reg wt XS wt XXS wt	$+10\%,\quad -3.5\%$
	Spec. plain end pipe	$+10\%,\quad -5\%$

PRESSURE TUBING

Pressure-tube applications commonly involve external heat applications, as in boilers or superheaters.

Table 5. Representative Weights of Tubing, lb per foot length

OD, in.	Wall thickness, in.															
	0.050	0.095	0.150	0.200	0.250	0.300	0.360	0.400	0.460	0.500	0.625	0.750	0.875	1.000	1.125	1.250
0.500	0.2403	0.4109	0.5607													
0.750	0.3738	0.6646	0.9612	1.175	1.335											
1.000	0.5073	0.9182	1.362	1.709	2.003	2.243	2.461									
1.250	0.6408	1.172	1.762	2.243	2.670	3.044	3.422	3.631								
1.500	0.7743	1.426	2.163	2.777	3.338	3.845	4.383	4.699	5.109	5.340						
1.750	0.9078	1.679	2.563	3.311	4.005	4.646	5.344	5.767	6.338	6.675	7.509					
2.000	1.041	1.933	2.964	3.845	4.673	5.447	6.305	6.835	7.566	8.010	9.178	10.01				
2.250	1.175	2.186	3.364	4.379	5.340	6.248	7.267	7.903	8.794	9.345	10.85	12.02				
2.500	1.308	2.440	3.765	4.913	6.008	7.049	8.228	8.971	10.02	10.68	12.52	14.02	15.19			
2.750	1.442	2.694	4.165	5.447	6.675	7.850	9.189	10.04	11.25	12.02	14.18	16.02	17.52	18.69		
3.000	1.575	2.947	4.566	5.981	7.343	8.651	10.15	11.11	12.48	13.35	15.85	18.02	19.86	21.36		
3.250	1.709	3.201	4.966	6.515	8.010	9.452	11.11	12.18	13.71	14.69	17.52	20.03	22.19	24.03		
3.500	1.842	3.455	5.367	7.049	8.678	10.25	12.07	13.24	14.93	16.02	19.19	22.03	24.53	26.70		
3.750	1.976	3.708	5.767	7.583	9.345	11.05	13.03	14.31	16.16	17.36	20.86	24.03	26.87	29.37	31.54	
4.000	2.109	3.962	6.168	8.117	10.01	11.85	14.00	15.38	17.39	18.69	22.53	26.03	29.20	32.04	34.54	
4.250	2.243	4.216	6.568	8.651	10.68	12.66	14.96	16.45	18.62	20.03	24.20	28.04	31.54	34.71	37.55	
4.500	2.376	4.469	6.969	9.185	11.35	13.46	15.92	17.52	19.85	21.36	25.87	30.04	33.88	37.38	40.55	
4.750	2.510	4.723	7.369	9.719	12.02	14.26	16.88	18.58	21.08	22.70	27.53	32.04	36.21	40.05	43.55	46.73

5.000	2.643	4.977	7.770	10.25	12.68	15.06	17.84	19.65	22.30	24.03	29.20	34.04	38.55	42.72	46.56	50.06
5.250	....	5.230	8.170	10.79	13.35	15.86	18.80	20.72	23.53	25.37	30.87	36.05	40.88	45.39	49.56	53.40
5.500	....	5.484	8.571	11.32	14.02	16.66	19.76	21.79	24.76	26.70	32.54	38.05	43.22	48.06	52.57	56.74
5.750	....	5.738	8.971	11.85	14.69	17.46	20.72	22.86	25.99	28.04	34.21	40.05	45.56	50.73	55.57	60.08
6.000	....	5.991	9.372	12.39	15.35	18.26	21.68	23.92	27.22	29.37	35.88	42.05	47.89	53.40	58.57	63.41
6.250	....	6.245	9.772	12.92	16.02	19.06	22.65	24.99	28.45	30.71	37.55	44.06	50.23	56.07	61.58	66.75
6.500	....	....	10.17	13.46	16.69	19.86	23.61	26.06	29.67	32.04	39.22	46.06	52.57	58.74	64.58	70.09
6.750	....	....	10.57	13.99	17.36	20.67	24.57	27.13	30.90	33.38	40.88	48.06	54.90	61.41	67.58	73.43
7.000	....	....	10.97	14.52	18.02	21.47	25.53	28.20	32.13	34.71	42.55	50.06	57.24	64.08	70.59	76.76
7.250	....	....	11.37	15.06	18.69	22.27	26.49	29.26	33.36	36.05	44.22	52.07	59.57	66.75	73.59	80.10
7.500	....	....	....	15.59	19.36	23.07	27.45	30.33	34.59	37.38	45.89	54.07	61.91	69.42	76.60	83.44
7.750	....	....	....	16.13	20.03	23.87	28.41	31.40	35.81	38.72	47.56	56.07	64.25	72.09	79.60	86.78
8.000	....	....	....	16.66	20.69	24.67	29.37	32.47	37.04	40.05	49.23	58.07	66.58	74.76	82.60	90.11
8.250	....	....	....	17.19	21.36	25.47	30.34	33.54	38.27	41.39	50.90	60.08	68.92	77.43	85.61	93.45
8.500	....	....	....	17.73	22.03	26.27	31.30	34.60	39.50	42.72	52.57	62.08	71.26	80.10	88.61	96.79
8.750	....	....	....	....	22.70	27.07	32.26	35.67	40.73	44.06	54.23	64.08	73.59	82.77	91.61	100.1
9.000	....	....	....	....	23.36	27.87	33.22	36.74	41.96	45.39	55.90	66.08	75.93	85.44	94.62	103.5
9.250	....	....	....	....	24.03	28.68	34.18	37.81	43.18	46.73	57.57	68.09	78.26	88.11	97.62	106.8
9.500	....	....	....	....	24.70	29.48	35.14	38.88	44.41	48.06	59.24	70.09	80.60	90.78	100.6	110.1
9.750	....	....	....	....	25.37	30.28	36.10	39.94	45.64	49.40	60.91	72.09	82.94	93.45	103.6	113.5
10.000	....	....	....	....	26.03	31.08	37.06	41.01	46.87	50.73	62.58	74.09	85.27	96.12	106.6	116.8

Pressure tubing is produced to the actual outside diameter and minimum wall or average wall thickness specified by the purchaser. Pressure tubing may be hot- or cold-finished.

The wall thickness is normally given in decimal parts of an inch rather than as a fraction or gage number. When gage numbers are given without reference to a gage system, Birmingham Wire Gage (BWG) is implied. Weights of commercial tubing are given in Table 5.

Pressure tubing is usually made from steel produced by the open-hearth, basic oxygen, or electric-furnace processes.

Seamless pressure tubing may be either hot-finished or cold-drawn. Cold-drawn steel tubing is frequently process-annealed at temperatures above 1200 F. To assure quality, maximum hardness values are frequently specified. For example, in ASTM Specification A192, Tentative Specification for Seamless Carbon Steel Boiler Tubes for High-pressure Service, the following maximum hardness values are given.

Boiler tubes	Brinell hardness No. (tubes 0.200 in. and over in wall thickness)	Rockwell hardness No. (tubes less than 0.200 in. in wall thickness)
Hot-finished tubes	137	B77
Cold-drawn (normalized) tubes	125	B72

Hot-finished or cold-drawn seamless low-alloy steel tubes generally are process-annealed at temperatures between 1200 and 1350 F.

Austenitic stainless-steel tubes are usually annealed at temperatures between 1800 and 2100 F, with specific temperatures varying somewhat with each grade. This is generally followed by pickling, unless bright-annealing was done.

MECHANICAL TUBING

Unlike pipe and pressure tubes, mechanical tubing is generally classified by the method of manufacture and the degree of finish. Examples of classifications are "seamless hot-finished," "cold-drawn welded," or "flash-in-grade," etc.

Seamless Tubes. Seamless tubes are available as either *hot-* or *cold*-finished. They are normally made in sizes from 0.187 in. OD to 10.750 in. OD.

Dimensions for hot-finished mechanical tubes are given in Table 6. Dimensions for cold-finished tubes are listed in Table 7.

Welded Tubes. Welded tubes generally are produced by electric resistance methods. Where required, the welding flash is removed with a cutting tool. Industry practice normally recognizes a number of finish conditions which are summarized in Table 8.

Flash-in-type tubing is generally limited to applications where nothing is inserted in the tube.

Flash-controlled tubing is used where moderate control of the inside diameter is required. Generally the outside and inside diameters are specified.

The designation *sink-draw tubes* is specified where close control over the outer diameter is required with normal tolerance applying to the wall thickness. Smoothness of the inside surface is not controlled, except that the flash is generally controlled to a height of 0.005 or 0.010 in. maximum.

Mandrel-drawn tubes usually are normalized after welding by passing the tubes through a continuous atmosphere-controlled furnace. After descaling, the tubes are cold-drawn through a die with a mandrel on the inside of the tube. These

Table 6. Diameter and Wall-thickness Tolerances for Seamless Hot-finished Mechanical Tubing of Carbon and Alloy Steel (AISI)[1]

Specified size, OD, in.	Ratio of wall thickness to OD	OD tolerance		Wall thickness tolerance, %							
				0.109 in. and under		Over 0.109 to 0.172 in., incl.		Over 0.172 to 0.203 in., incl.		Over 0.203 in.	
		Over	Under	Over	Under	Over	Under	Over	Under	Over	Under
Under 3	All wall thicknesses	0.023	0.023	16.5	16.5	15	15	14	14	12.5	12.5
3–5½, excl.	All wall thicknesses	0.031	0.031	16.5	16.5	15	15	14	14	12.5	12.5
5½–8, excl.	All wall thicknesses	0.047	0.047	...	...	...	...	14	14	12.5	12.5
8–10¾, incl.	5% and over	0.047	0.047	...	...	...	...	...	...	12.5	12.5
8–10¾, incl.	Under 5%	0.063	0.063	...	...	...	...	...	...	12.5	12.5

[1] The common range of sizes of hot-finished tubes is 1½ to and including 10¾ in. outside diameter with wall thickness not less than 0.095 in. (No. 13 BWG) or 3 per cent or more of the outside diameter. For sizes under 1½ or over 10¾ in. outside diameter, the tolerances are commonly negotiated between the purchaser and producer. (From AISI Steel Products Manual.)

Table 7. Diameter and Wall-thickness Tolerances for Seamless Cold-worked Mechanical Tubing of Carbon and Alloy Steel (AISI)[1]

Size, OD, in.	Unannealed or finish-annealed				Soft annealed or normalized				Quenched and tempered				Wall thickness all conditions, %	
	OD, in.		ID, in.		OD, in.		ID, in.		OD, in.		ID, in.			
	Over	Under	Over	Under	Over	Under	Over	Under	Over	Under	Over	Under	Over	Under
3/16–1/2, excl.[2,3]	0.004	0	....	....	0.006	0.002	....	....	0.010	0.010	....	....	15	15
1/2–1 1/2, excl.[2,3,4,5]	0.005	0	0	0.005	0.008	0.002	0.002	0.008	0.015	0.015	0.015	0.015	10	10
1 1/2–3 1/2, excl.[2,3,4,5]	0.010	0	0	0.010	0.015	0.005	0.005	0.015	0.030	0.030	0.030	0.030	10	10
3 1/2–5 1/2, excl.[4,5]	0.015	0	0.005	0.015	0.023	0.007	0.015	0.025	0.045	0.045	0.045	0.045	10	10
5 1/2–8, excl.[5] wall less than 5% OD	0.030	0.030	0.035	0.035	0.060	0.060	0.070	0.070	....	....	....	....	10	10
5 1/2–8, excl. wall from 5 to 7.5% OD	0.020	0.020	0.025	0.025	0.040	0.040	0.050	0.050	....	....	....	....	10	10
5 1/2–8, excl.[4] wall over 7.5% OD	0.030	0	0.015	0.030	0.045	0.015	0.037	0.053	....	....	....	....	10	10
8–10 3/4, incl.[5] wall less than 5% OD	0.045	0.045	0.050	0.050	....	....	....	....	....	....	....	....	10	10
8–10 3/4, incl. wall from 5 to 75% OD	0.035	0.035	0.040	0.040	....	....	....	....	....	....	....	....	10	10
8–10 3/4, incl.[4] wall over 7.5% OD	0.045	0	0.015	0.040	....	....	....	....	....	....	....	....	10	10

[1] For tolerances closer than those indicated, availability, and applicable tolerances for tubing less than 3/16 in. OD or larger than 10 3/4-in. OD, the producer should be consulted.

[2] For those tubes with inside diameter less than 1/2 in. (or less than 5/8 in. when the wall thickness is more than 20% of the outside diameter), which are not commonly drawn over a mandrel. Note [4] is not applicable. Unless otherwise agreed upon by the purchaser and producer, the wall thickness may vary 15% over and under that specified, and the inside diameter is governed by the outside diameter and wall-thickness tolerances shown.

[3] For tubes with inside diameter less than 1/2 in. (or less than 5/8 in. when the wall thickness is more than 20% of the outside diameter), which can be produced by the rod or bar mandrel process, the tolerances are as shown in the above table except that the wall-thickness tolerances are 10% over and under the specified wall thickness.

[4] Many tubes with inside diameter less than 50% of outside diameter, or with wall thickness more than 25% of outside diameter, or with wall thickness over 1 1/4 in., or weighing more than 90 lb per ft, are difficult to draw over a mandrel. Unless otherwise agreed upon by the purchaser and producer the inside diameter may vary over or under by an amount equal to 10% of the wall thickness and the wall thickness may vary 12 1/2% over and under and that specified.

[5] Tubing having a wall thickness less than 3% of the outside diameter cannot be straightened properly without a certain amount of distortion. Consequently, such tubes, while having an average outside diameter and inside diameter within the tolerances shown in the above table, require an ovality tolerance of 0.5% over and under nominal outside diameter, this being in addition to the tolerances indicated in the above table.   From AISI Steel Products Manual.

Table 8. Finish Classifications Normally Recognized as Welded Mechanical Tubing

	Type	Condition
A	Flash-in, hot-rolled	Made by longitudinally forming and welding hot-rolled strip; the interior welding flash remains in place
B	Flash-in, cold-rolled	Made by longitudinally forming and welding cold-rolled strip; the interior flash remains in place
C	Flash controlled—0.010 max, hot-rolled	Same as (A), except that the interior flash is partially removed so that the height remaining does not exceed 0.010 in.
D	Flash controlled—0.010 max, cold-rolled	Same as (C), except that cold-rolled strip has been used
E	Flash controlled—0.005 max, hot-rolled	Same as (C), except that flash height on tube inside does not exceed 0.005"
F	Flash controlled—0.005 max, cold-rolled	Same as (E), except that cold-rolled strip has been used
G	Sink-drawn, hot-rolled	Tubing with flash controlled to 0.010" max, which has been descaled and cold-drawn through a die to cold-finish the exterior surface
H	Sink-drawn, cold-rolled	Tubing with flash controlled to 0.010" max, which has been cold-drawn through a die to cold-finish the exterior surface
I	Mandrel-drawn	Cold-rolled tubing with the interior flash removed and drawn through a die and over a mandrel to cold-finish the exterior and interior surfaces

tubes provide maximum control over surface finish, outside or inside diameters, and wall thickness.   The normalizing heat treatment removes the effects of welding and provides a uniform microstructure around the tube circumference.

The different finish classifications may result in substantial differences in the mechanical properties of the steel material.

Typical examples for low-carbon steel material are given in Table 9.   Differences in carbon content and other chemistry, heat treatment, etc., may significantly change these typical values.

Other Tubing Types. Among other tube classifications are sanitary tubing usually made of 18 Cr–8 Mo stainless steel and available as seamless or welded. This tubing is used extensively in the dairy, beverage, and food industries.   Sanitary tubing is generally available in sizes from 1 to 4 in. OD.   It may be furnished either hot- or cold-finished.   It is normally heat-treated at temperatures above 1900 F.

Some welded tube is also produced by fusion-welding methods of either the

Table 9.   Typical Mechanical Properties of Resistance-welded Mechanical Carbon-steel Tubing

Type	Yield strength, psi	Tensile strength, psi	Elongation, %	Hardness, Rockwell B
Flash-in or flash controlled, hot-rolled.	48,000	58,000	29	68
Flash-in or flash controlled, cold-rolled.	68,000	76,000	17	84
Normalized.	34,000	52,000	39	61
Sink-drawn	73,000	76,000	20	84
Mandrel-drawn.	80,000	83,000	15	86

inert-gas tungsten-arc-welding or gas shielded consumable metal-arc-welding types. This tubing is generally more expensive than the resistance-welded types.

Single-strip type   Double-strip type

Class I brazed tubing, double wall, 360-deg brazed construction

Butt-brazed type   Bevel-brazed type

Class II brazed tubing, single wall construction

FIG. 6. Construction of copper-brazed steel tubing, as specified in ASTM A254.

The butt-welded cold-finished tubes are made from hot-rolled or cold-rolled strip and fusion welded. This tubing is usually furnished as sink-drawn or mandrel-drawn.

Butt-welded tubing is made in heavier wall thicknesses than the resistance-welded tube.

Several tubing materials used in the automobile industry are covered by specifications of the Society of Automotive Engineers, SAE Handbook.

Brazed Steel Tubing. In the automotive, refrigeration, and stove industries, extensive use is made of copper-brazed steel tubes. Applications include fuel and brake lines, oil lines, heating and cooling lines, and similar services.

The tubing is produced from copper-coated steel strip. Single or double strip is formed into tubing which is heated in a reducing atmosphere to effect a brazed joint between the mating surfaces.

ASTM Specification A254, Specification for Copper Brazed Steel Tubing, recognizes two classifications, each consisting of the two types shown in Fig. 6.

This tubing is available in sizes from ³⁄₁₆-in. OD to ⅝-in. OD. Wall thicknesses range from 0.025 in. to 0.035 in. To improve surface finish and tolerances, this tubing is sometimes sink-drawn.

PIPE FITTINGS

The major piping materials are produced also in the form of standard fittings. Among the more widely used are cast-iron fittings, malleable-iron fittings, brass and copper fittings, cast-steel fittings, and wrought-steel fittings. Other major non-ferrous piping materials are also produced in the form of cast and wrought fittings.

Cast-iron fittings are made by conventional founding methods for a variety of joints including bell-and-spigot, flanged, and mechanical (gland-type), or other proprietary designs.

Cast-iron fittings are covered by a number of ASA and Federal Standards:

ASA B16b1   Cast Iron Pipe Flanges and Flanged Fittings for 800 Psig Hydraulic Pressure

ASA B16b2   Cast Iron Pipe Flanges and Flanged Fittings, 25 lb

ASA B16.1   Cast Iron Flanges and Flanged Fittings, Class 125 (Standard includes also bolt, nut and gasket data)

ASA B16.2   Cast Iron Pipe Flanges and Flanged Fittings, Class 250 (bolt, nut and gasket data are also included)

ASA B16.4   Cast Iron Screwed Fittings, 125 and 250 lb

ASA B16.12   Cast Iron Screwed Drainage Fittings

WW-P-491   Pipe Fittings, Cast Iron, Drainage

WW-P-501   Pipe Fittings, Cast Iron, Screwed 125 and 250 pounds

Cast-iron Screwed Fittings. Cast-iron screwed fittings are covered by ASA Standard B16.4.

Tables 10 and 11 give the American Standard dimensions for cast-iron screwed fittings for maximum working saturated steam pressures of 125 and 250 psi gage,

Table 10.   Dimensions of 125-lb Cast-iron Screwed 90- and 45-deg Elbows, Tees, and Crosses (Straight Sizes) (ASA B16.4-1963)

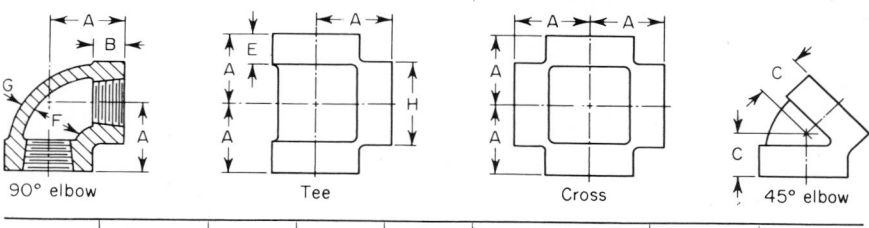

90° elbow   Tee   Cross   45° elbow

Nominal pipe size	Center to end, elbows, tees, and crosses A	Center to end, 45-deg elbows C	Length of thread, min B	Width of band, min E	Inside diameter of fitting F		Metal thickness G	Outside diameter of band, min H
					Max	Min		
1/4	0.81	0.73	0.32	0.38	0.584	0.540	0.110	0.93
3/8	0.95	0.80	0.36	0.44	0.719	0.675	0.120	1.12
1/2	1.12	0.88	0.43	0.50	0.897	0.840	0.130	1.34
3/4	1.31	0.98	0.50	0.56	1.107	1.050	0.155	1.63
1	1.50	1.12	0.58	0.62	1.385	1.315	0.170	1.95
1 1/4	1.75	1.29	0.67	0.69	1.730	1.660	0.185	2.39
1 1/2	1.94	1.43	0.70	0.75	1.970	1.900	0.200	2.68
2	2.25	1.68	0.75	0.84	2.445	2.375	0.220	3.28
2 1/2	2.70	1.95	0.92	0.94	2.975	2.875	0.240	3.86
3	3.08	2.17	0.98	1.00	3.600	3.500	0.260	4.62
3 1/2	3.42	2.39	1.03	1.06	4.100	4.000	0.280	5.20
4	3.79	2.61	1.08	1.12	4.600	4.500	0.310	5 79
5	4.50	3.05	1.18	1.18	5.663	5.563	0.380	7.05
6	5.13	3.46	1.28	1.28	6.725	6.625	0.430	8.28
8	6.56	4.28	1.47	1.47	8.725	8.625	0.550	10.63
10	8.08*	5.16	1.68	1.68	10.850	10.750	0.690	13.12
12	9.50*	5.97	1.88	1.88	12.850	12.750	0.800	15.47

All dimensions given in inches.
* This applies to elbows and tees only.

Table 11.   Dimensions of 250-lb Cast-iron Screwed 90- and 45-deg Elbows, Tees, and Crosses (Straight Sizes) (ASA B16.4-1963)

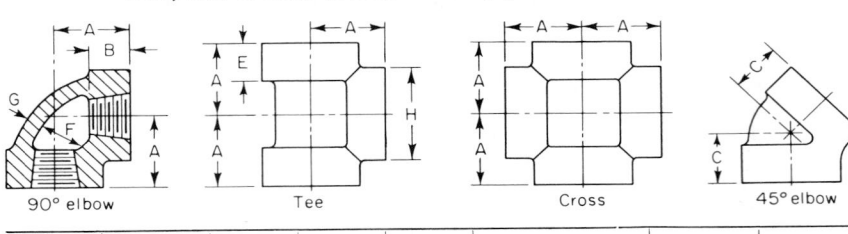

Nominal pipe size	Center to end, elbows, tees, and crosses A	Center to end, 45-deg elbows C	Length of thread, min B	Width of band, min E	Inside diameter of fitting F		Metal thickness G	Outside diameter of band, min H
					Max	Min		
¼	0.94	0.81	0.43	0.49	0.584	0.540	0.18	1.17
⅜	1.06	0.88	0.47	0.55	0.719	0.675	0.18	1.36
½	1.25	1.00	0.57	0.60	0.897	0.840	0.20	1.59
¾	1.44	1.13	0.64	0.68	1.107	1.050	0.23	1.88
1	1.63	1.31	0.75	0.76	1.385	1.315	0.28	2.24
1¼	1.94	1.50	0.84	0.88	1.730	1.660	0.33	2.73
1½	2.13	1.69	0.87	0.97	1.970	1.900	0.35	3.07
2	2.50	2.00	1.00	1.12	2.445	2.375	0.39	3.74
2½	2.94	2.25	1.17	1.30	2.975	2.875	0.43	4.60
3	3.38	2.50	1.23	1.40	3.600	3.500	0.48	5.36
3½	3.75	2.63	1.28	1.49	4.100	4.000	0.52	5.98
4	4.13	2.81	1.33	1.57	4.600	4.500	0.56	6.61
5	4.88	3.19	1.43	1.74	5.663	5.563	0.66	7.92
6	5.63	3.50	1.53	1.91	6.725	6.625	0.74	9.24
8	7.00	4.31	1.72	2.24	8.725	8.625	0.90	11.73
10	8.63	5.19	1.93	2.58	10.850	10.750	1.08	14.37
12	10.00	6.00	2.13	2.91	12.850	12.750	1.24	16.84

All dimensions given in inches.

The 250-lb standard for screwed fittings covers only the straight sizes of 90- and 45-deg elbows, tees, and crosses.

ASA B16.4-1963.   The maximum hydraulic service pressure ratings, including shock, given in ASA B16.4 are 175 and 400 psi gage, respectively, in the range from −20 to +150 F.   Minimum material must conform to Class A of ASTM A126. Body thicknesses at no point are to be less than 90 per cent of the specified minimum metal thickness.   Cast-iron screwed fittings customarily are furnished in black finish only.

Malleable-iron Screwed Fittings.   Malleable-iron fittings are also extensively produced.   They are generally made with screwed joints.

The following applicable standards have been issued by the American Standards Association:

ASA B16.3   Malleable-iron Screwed Fittings, 150 lb and 300 lb

Malleable-iron screwed fittings for 150- and 300-lb gage are standardized in ASA B16.3.   Malleable-iron fittings are furnished either black or galvanized. Standard dimensions of elbows, tees, and crosses are shown in Table 12.   The galvanized screwed fittings commonly used in water piping for homes are 150-lb

Table 12. Dimensions of 150-lb Malleable-iron Screwed 90- and 45-deg Elbows, Tees, and Crosses (Straight Sizes) (ASA B16.3-1963)

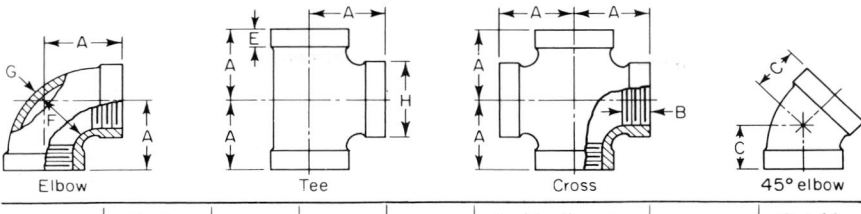

| Elbow | Tee | Cross | 45° elbow |

Nominal pipe size	Center to end, elbows, tees, and crosses A	Center to end, 45-deg elbows C	Length of thread, min B	Width of band, min E	Inside diameter of fitting F		Metal thickness, min G	Outside diameter of band, min H
					Min	Max		
⅛	0.69	. . .	0.25	0.200	0.405	0.435	0.090	0.693
¼	0.81	0.73	0.32	0.215	0.540	0.584	0.095	0.844
⅜	0.95	0.80	0.36	0.230	0.675	0.719	0.100	1.015
½	1.12	0.88	0.43	0.249	0.840	0.897	0.105	1.197
¾	1.31	0.98	0.50	0.273	1.050	1.107	0.120	1.458
1	1.50	1.12	0.58	0.302	1.315	1.385	0.134	1.771
1¼	1.75	1.29	0.67	0.341	1.660	1.730	0.145	2.153
1½	1.94	1.43	0.70	0.368	1.900	1.970	0.155	2.427
2	2.25	1.68	0.75	0.422	2.375	2.445	0.173	2.963
2½	2.70	1.95	0.92	0.478	2.875	2.975	0.210	3.589
3	3.08	2.17	0.98	0.548	3.500	3.600	0.231	4.285
3½	3.42	2.39	1.03	0.604	4.000	4.100	0.248	4.843
4	3.79	2.61	1.08	0.661	4.500	4.600	0.265	5.401
5	4.50	3.05	1.18	0.780	5.563	5.663	0.300	6.583
6	5.13	3.46	1.28	0.900	6.625	6.725	0.336	7.767

All dimensions in inches.

malleable iron. Minimum properties of malleable iron are required to meet those of ASTM A197, "Cupola Malleable Iron."

Cast-brass Screwed Fittings. Standard-weight and extra-heavy brass screwed fittings are made in accordance with American Standard ASA B16.15 and Federal Specification WW-P-460. These cover fittings with working steam pressures of 150 and 250 psig.

The standard dimensions of cast brass or bronze fittings are given in Table 13.

Soldered-joint Fittings. Soldered-joint wrought-metal and cast-brass or bronze fittings for use with copper water tubes are covered by ASTM B88 and ASA H23.1. They are made in accordance with American Standard ASA B16.22 and ASA B16.18, respectively. Standard cast-brass or bronze fittings are illustrated in Fig. 7. The laying length of cast-brass fittings (center-to-shoulder distance) is included in the specification, but because of the variety of forming methods, the laying length of wrought-metal fittings has not been standardized.

Fittings with this type of joint made with 50-50 tin-lead solder, 95-tin 5-antimony, and solder melting above 1100 F have the pressure-temperature ratings shown in Table 14.

Wrought fittings normally have a minimum copper content of 83 per cent. Cast-brass fittings conform to ASTM B62 and have the following nominal composition: 85 per cent copper, 5 per cent tin, 5 per cent lead, and 5 per cent zinc.

Table 13. Dimensions of 125- and 250-lb Cast Bronze or Brass Screwed Elbows, Tees, and Crosses (Straight Sizes) (ASA B16.15-1964)[2,3]

125-lb fittings

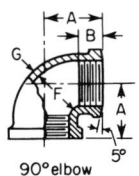

90° elbow

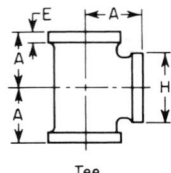

Tee

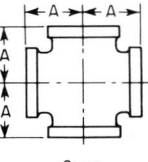

Cross

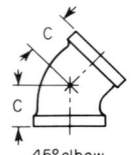

45° elbow

Nominal pipe size	Center to end elbows, tees, crosses	Length of thread, min	Center to end 45 deg elbows	Band length, min	Inside diameter of cast fitting		Metal thickness[1]	Band diameter, min
					Min	Max		
	A	B	C	E	F		G	H
1/8	0.54	0.25	0.42	0.14	0.41	0.44	0.08	0.67
1/4	0.71	0.32	0.56	0.16	0.54	0.58	0.08	0.81
3/8	0.82	0.36	0.63	0.17	0.68	0.72	0.09	1.00
1/2	1.01	0.43	0.78	0.19	0.84	0.90	0.09	1.17
3/4	1.18	0.50	0.89	0.23	1.05	1.11	0.10	1.42
1	1.43	0.58	1.06	0.27	1.32	1.39	0.11	1.72
1 1/4	1.69	0.67	1.22	0.31	1.66	1.73	0.12	2.10
1 1/2	1.84	0.70	1.30	0.34	1.90	1.97	0.13	2.38
2	2.12	0.75	1.45	0.41	2.38	2.45	0.15	2.92
2 1/2	2.70	0.92	1.95	0.48	2.88	2.98	0.17	3.49
3	3.08	0.98	2.17	0.55	3.50	3.60	0.19	4.20
4	3.79	1.08	2.61	0.66	4.50	4.60	0.22	5.31

250-lb fittings

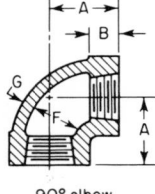

90° elbow

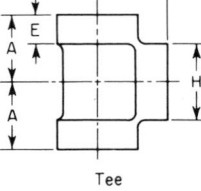

Tee

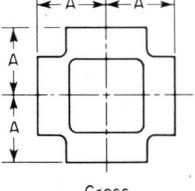

Cross

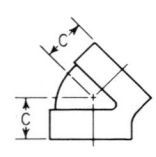

45° elbow

Nominal pipe size	Center to end, elbows, tees, and crosses[1]	Length of thread, min	Center to end 45-deg elbows[1]	Width of band, min	Inside diameter of fitting		Metal thickness[1]	Outside diameter of band, min
					Min	Max		
	A	B	C	E	F		G	H
1/4	0.81	0.32	0.73	0.38	0.54	0.58	0.11	0.93
3/8	0.95	0.36	0.80	0.44	0.68	0.72	0.12	1.12
1/2	1.12	0.43	0.88	0.50	0.84	0.90	0.13	1.34
3/4	1.31	0.50	0.98	0.56	1.05	1.11	0.16	1.63
1	1.50	0.58	1.12	0.62	1.32	1.38	0.17	1.95
1 1/4	1.75	0.67	1.29	0.69	1.66	1.73	0.19	2.39
1 1/2	1.94	0.70	1.43	0.75	1.90	1.97	0.20	2.68
2	2.25	0.75	1.68	0.84	2.38	2.45	0.22	3.28
2 1/2	2.70	0.92	1.95	0.94	2.88	2.98	0.24	3.86
3	3.08	0.98	2.17	1.00	3.50	3.60	0.26	4.62
4	3.79	1.08	2.61	1.12	4.50	4.60	0.31	5.79

[1] The actual metal thickness must be at least 90% of the indicated values.
[2] All dimensions are in inches.
[3] Several other fittings produced to the requirements of ASA B16.15 are available from various manufacturers. These types include couplings, reducers, return bends, 4-branches, caps and reducing elbows and tees. Manufacturers should be consulted for availability and dimensions.

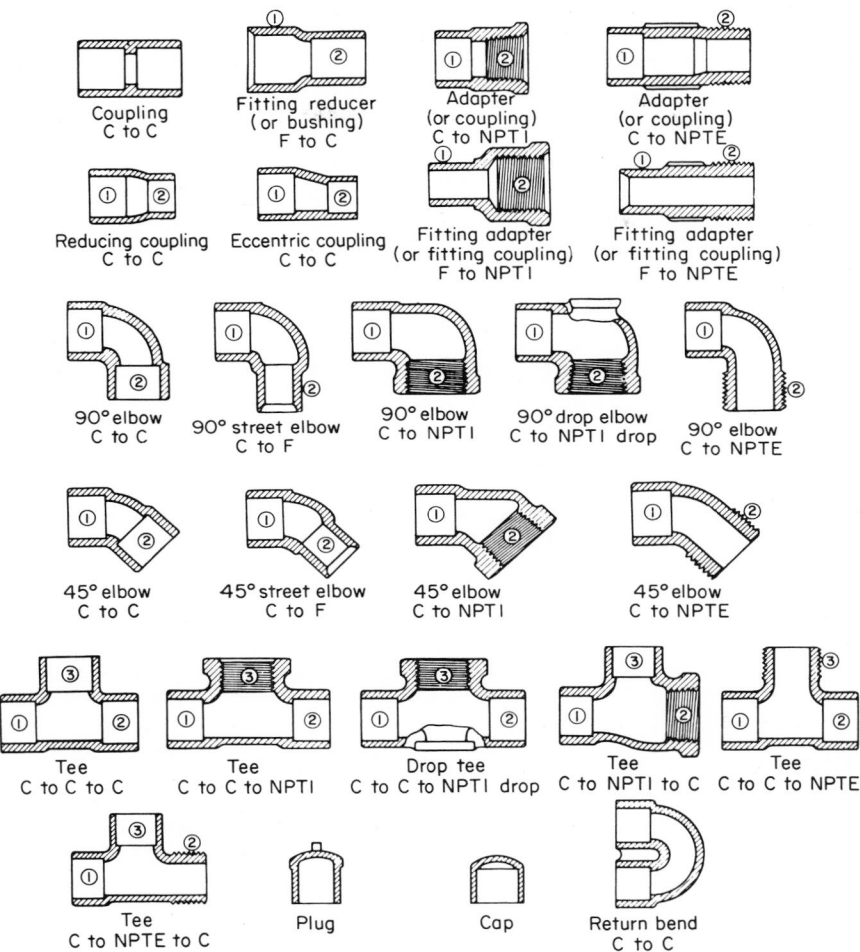

Note : Reducing fittings are designated by size in the order ① x ② x ③

Fig. 7. Standard cast-bronze solder-joint pressure fittings, as covered by American Standard ASA B16.18-1963. C, Solder-joint fitting end (female) made to receive copper tube diameter. F, Solder-joint fitting end (male) made to copper tube diameter. NPTI, Internal American Standard taper pipe thread. NPTE, External American Standard taper pipe thread.

The minimum requirements for 50-50 tin-lead solder are covered in ASTM B32 alloy grade 50A.   Metal thickness tolerances and general dimensions of fittings are given in Table 15.   For laying length dimensions of reducing elbows, tees, crosses, and couplings in both straight and reducing sizes, references should be made to ASA B16.18.   Elbows and coupling adapters with male and female pipe thread ends also are available.

An estimate of the amount of solder required for 100 joints can be made on the basis of 1.25 to 1.50 lb of solder per inch of water tube diameter for sizes up to and

Table 14. Pressure Ratings of Solder Joints (ASA B16.18-1963)

Maximum working pressure, psi

Solder used in joints	Working temperatures, F	⅛–1 in., incl.[1]	1¼–2 in., incl.[1]	2½–4 in., incl.[1]	5–8 in., incl.[1]
50-50, tin-lead[2]	100	200	175	150	135
	150	150	125	100	90
	200	100	90	75	70
	250	85	75	50	45
95-5, tin-antimony	100	500	400	300	270
	150	400	350	275	250
	200	300	250	200	180
	250	200	175	150	135
Solders melting at or above 1100 F	350	270	190	155	140

[1] Standard water tube sizes.
[2] ASTM B32-60aT Alloy Grade 50A.

including $3\frac{1}{2}$ in., and 1.75 to 3 lb/in. for 4 to 6 in. inclusive. An allowance of 2.0 oz of flux per pound of solder is suggested by one manufacturer of soldered-joint fittings.

Cast-steel Flanged Fittings. Cast-steel flanged fittings comply with the specifications of ASA B16.5. Dimensions are given in Table 16.

Center-to-face and face-to-face dimensions include the height of raised faces of flanges.

Reducing fittings with raised-face flanges have the same center-to-face dimensions as those of straight-size fittings of the largest opening; face-to-face dimensions for reducers are as listed for the larger opening.

Center-to-face dimensions shown for fittings with ring-joint flanges apply to straight sizes only. For reducing fittings and reducers the dimensions shown for raised face flanges of the largest opening should be used. On 400-lb and higher classes, the $\frac{1}{4}$-in. raised face for each flange should be subtracted, but do not subtract the $\frac{1}{16}$-in. raised face in 150- and 300-lb classes. The height of the ring-joint raised face L applying to each flange should be added.

For calculating the "laying length" of fittings with ring joints to the center-to-face dimensions in these tables must be added the approximate distance D between flange faces when the ring is compressed.

Forged-steel Screwed Fittings. Forged-steel screwed fittings are made in accordance with ASA Standard B16.11. Material for carbon steel fittings conforms to ASTM Specification A181, Grade I (except that minimum yield strength is 35,000 psi). Alloy steel fittings conform to ASTM Specification A182. Dimensions are listed in Table 17.

Forged-steel Socket-welding Fittings. Forged-steel socket-welding fittings are made in accordance with ASA Standard B16.11. Material for carbon steel fittings conforms to ASTM Specification A181, Grade I (except for a minimum yield strength of 35,000 psi). Material for alloy and stainless steel fittings conforms to ASTM Specification A182.

Dimensions of standard fittings are given in Table 18.

Wrought-steel Butt-welding Fittings. Wrought-steel welding fittings include elbows, tees, crosses, reducers, laterals, lap-joint stub ends, caps, and saddles.

Wrought-steel fittings are made to the dimensional requirements of ASA Standards B16.9 and B16.28 in sizes 24 in. and smaller, and to the requirements of

Table 15. Dimensions of Soldered-joint Fittings (ASA B16.22-1963), (ASA B16.18-1963)

All dimensions in inches.

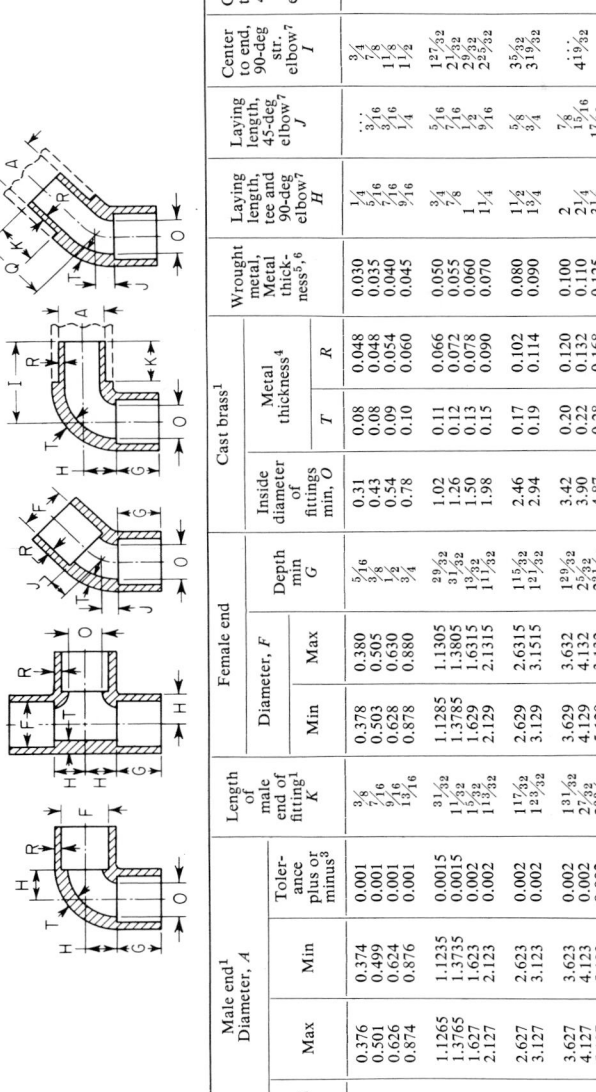

Nominal size²	Male end¹ Diameter, A				Length of male end of fitting¹ K	Female end Diameter, F		Female end Depth min G	Cast brass¹ Inside diameter of fittings min, O	Cast brass Metal thickness⁴		Wrought metal Metal thickness⁵,⁶	Laying length, tee and 90-deg elbow⁷ H	Laying length, 45-deg elbow⁷ J	Center to end, 90-deg str. elbow⁷ I	Center to end, 45-deg str. elbow⁷ Q	Inspection tolerances, applied to H, I, J, and Q
	Mean⁴	Max	Min	Tolerance plus or minus³		Min	Max			T	R						
1/4	0.375	0.376	0.374	0.001	3/8	0.378	0.380	5/16	0.31	0.08	0.048	0.030	1/4	1/8	3/4	.....	±3/64
3/8	0.500	0.501	0.499	0.001	7/16	0.503	0.505	3/8	0.43	0.08	0.048	0.035	5/16	3/16	7/8	.....	±1/16
1/2	0.625	0.626	0.624	0.001	9/16	0.628	0.630	1/2	0.54	0.09	0.054	0.040	5/16	3/16	1 1/8	3/4	±1/16
3/4	0.875	0.876	0.874	0.001	13/16	0.878	0.880	3/4	0.78	0.10	0.060	0.045	9/16	1/4	1 1/2	7/8	±1/16
1	1.125	1.1265	1.1235	0.0015	31/32	1.1285	1.1305	29/32	1.02	0.11	0.066	0.050	3/4	5/16	1 27/32	1 3/16	±5/64
1 1/4	1.375	1.3765	1.3735	0.0015	1 3/32	1.3785	1.3805	1 1/32	1.26	0.12	0.072	0.055	7/8	1/2	2 1/32	1 5/16	±5/64
1 1/2	1.625	1.627	1.623	0.002	1 9/32	1.629	1.6315	1 7/32	1.50	0.13	0.078	0.060	1	1/2	2 3/32	1 9/16	±5/64
2	2.125	2.127	2.123	0.002	1 13/32	2.129	2.1315	1 11/32	1.98	0.15	0.090	0.070	1 1/4	9/16	2 23/32	2 1/4	±5/64
2 1/2	2.625	2.627	2.623	0.002	1 17/32	2.629	2.6315	1 15/32	2.46	0.17	0.102	0.080	1 1/2	5/8	3 5/32	2 1/8	±7/64
3	3.125	3.127	3.123	0.002	1 23/32	3.129	3.1515	1 21/32	2.94	0.19	0.114	0.090	1 3/4	3/4	3 19/32	.....	±7/64
3 1/2	3.625	3.627	3.623	0.002	1 31/32	3.629	3.632	1 29/32	3.42	0.20	0.120	0.100	2	7/8	4 1/32	.....	±7/64
4	4.125	4.127	4.123	0.002	2 7/32	4.129	4.132	2 5/32	3.90	0.22	0.132	0.110	2 1/4	1 7/16	.....	.....	±1/8
5	5.125	5.127	5.123	0.002	2 23/32	5.129	5.132	2 21/32	4.87	0.28	0.168	0.125	3 1/8	1 7/16	.....	.....	±1/8
6	6.125	6.127	6.123	0.002	3 7/32	6.129	6.132	3 5/32	5.84	0.34	0.204	0.140	3 5/8	1 5/8	.....	.....	±5/32

1 These dimensions are used for wrought-metal fittings as well as for cast-brass fittings.
2 This size is the nominal bore of the tube.
3 From American Standard Specifications for Copper Water Tube, ASA H23.1-1963 (ASTM B88).
4 Patterns shall be designed to produce body thicknesses given in the table. Metal thickness at no point shall be less than 90 per cent of the thicknesses given in the table.
5 This dimension has the same thickness as Class L tubing.
6 These dimensions are nominal, but in every case the thickness of wrought fittings should be at least as heavy as the tubing with which it is to be used. The dimensions R and T are equal for wrought fittings.
7 Apply to cast fittings only. Laying lengths have not been standardized for wrought fittings.

NOTE: Wrought fittings, as well as cast fittings, must be provided with a shoulder or stop at the bottom end of the socket.

Table 16. Dimensions of Typical Commercial Cast-steel Flanged Fittings

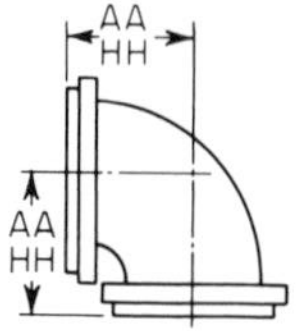

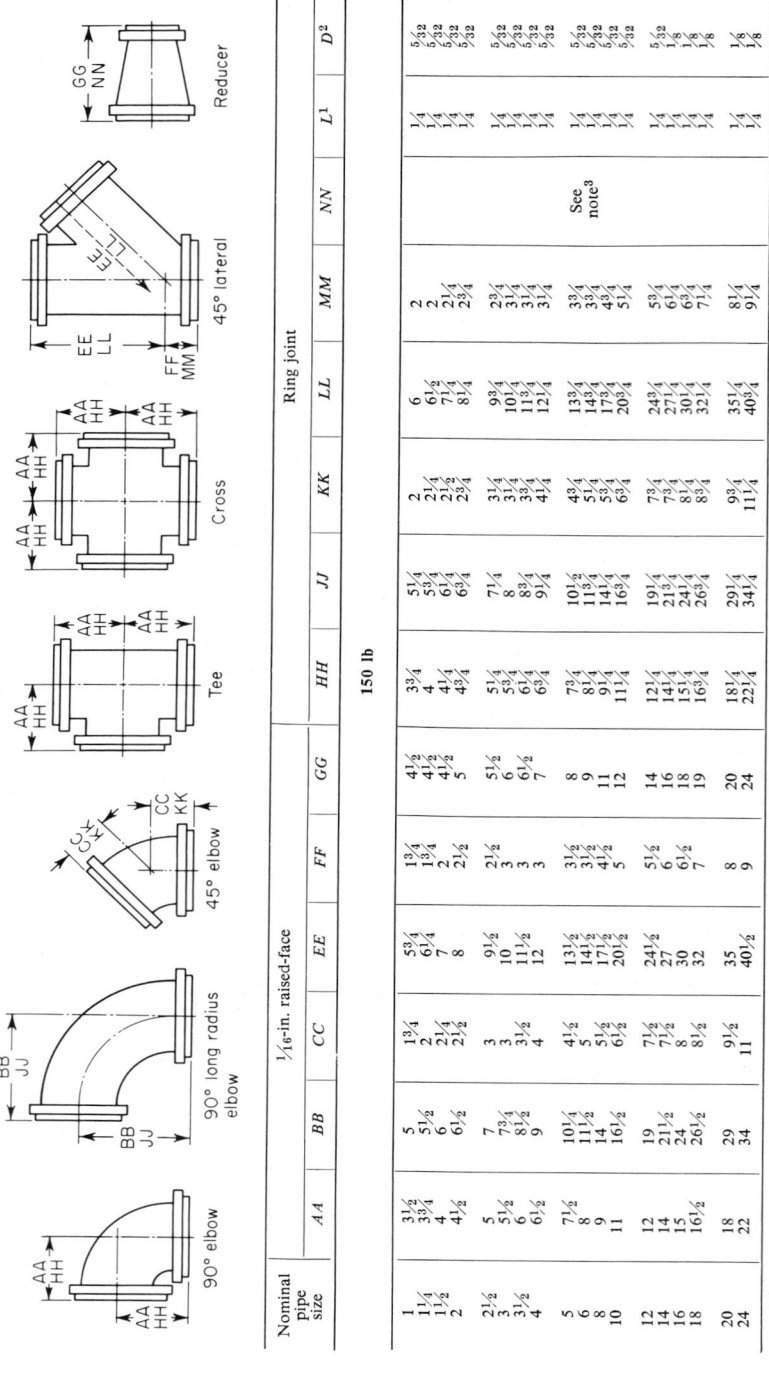

90° elbow

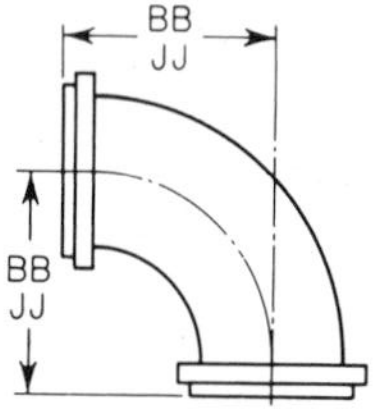

90° long radius elbow

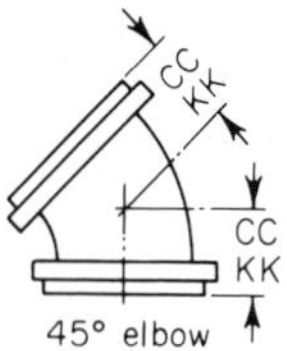

45° elbow

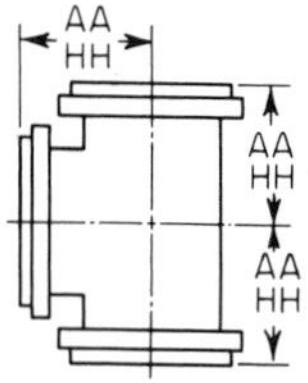

Tee

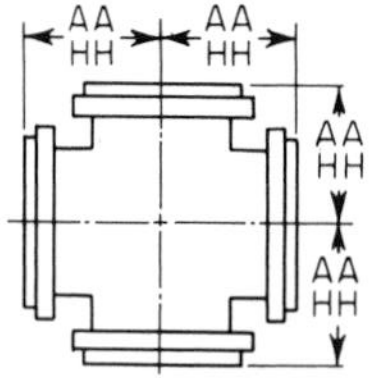

Cross

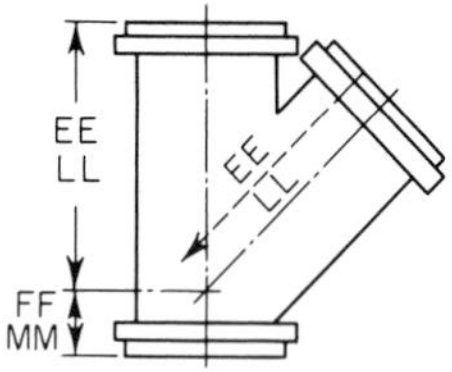

45° lateral

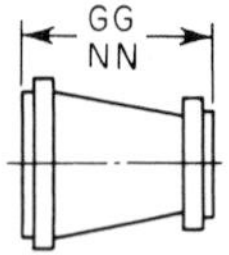

Reducer

Nominal pipe size	1/16-in. raised-face						Ring joint							
	AA	*BB*	*CC*	*EE*	*FF*	*GG*	*HH*	*JJ*	*KK*	*LL*	*MM*	*NN*	L^1	D^2
150 lb														
1	3½	5	1¾	5¾	1¾	4½	3¾	5¼	2	6	2		1/4	5/32
1¼	3¾	5½	2	6¼	1¾	4½	4	5¾	2¼	6½	2		1/4	5/32
1½	4	6	2¼	7	2	4½	4¼	6¼	2½	7¼	2¼		1/4	5/32
2	4½	6½	2½	8	2½	5	4¾	6¾	2¾	8¼	2¾		1/4	5/32
2½	5	7	3	9½	2½	5½	5¼	7¼	3¼	9¾	2¾		1/4	5/32
3	5½	7¾	3	10	3	6	5¾	8	3¼	10¼	3¼		1/4	5/32
3½	6	8½	3½	11½	3	6½	6¼	8¾	3¾	11¾	3¼		1/4	5/32
4	6½	9	4	12	3	7	6¾	9¼	4¼	12¼	3¼		1/4	5/32
5	7½	10¼	4½	13½	3½	8	7¾	10½	4¾	13¾	3¾		1/4	5/32
6	8	11½	5	14½	3½	9	8¼	11¾	5¼	14¾	3¾	See note[3]	1/4	5/32
8	9	14	5½	17½	4½	11	9¼	14¼	5¾	17¾	4¾		1/4	5/32
10	11	16½	6½	20½	5	12	11¼	16¾	6¾	20¾	5¼		1/4	5/32
12	12	19	7½	24½	5½	14	12¼	19¼	7¾	24¾	5¾		1/4	5/32
14	14	21½	7½	27	6	16	14¼	21¾	7¾	27¼	6¼		1/4	1/8
16	15	24	8	30	6½	18	15¼	24¼	8¼	30¼	6¾		1/4	1/8
18	16½	26½	8½	32	7	19	16¾	26¾	8¾	32¼	7¼		1/4	1/8
20	18	29	9½	35	8	20	18¼	29¼	9¾	35¼	8¼		1/4	1/8
24	22	34	11	40½	9	24	22¼	34¼	11¼	40¾	9¼		1/4	1/8

Nominal pipe size	1/16-in. raised-face						Ring joint							
	AA	*BB*	*CC*	*EE*	*FF*	*GG*	*HH*	*JJ*	*KK*	*LL*	*MM*	*NN*	*L*[1]	*D*[2]
300 lb														
1	4	5	2 1/4	6 1/2	2	4 1/2	4 1/4	5 1/4	2 1/2	6 3/4	2 1/4		1/4	5/32
1 1/4	4 1/4	5 1/2	2 1/2	7 1/4	2 1/4	4 1/2	4 1/2	5 3/4	2 3/4	7 1/2	2 1/2		1/4	5/32
1 1/2	4 1/2	6	2 3/4	8 1/2	2 1/2	4 1/2	4 3/4	6 1/4	3	8 3/4	2 3/4		1/4	5/32
2	5	6 1/2	3	9	2 1/2	5	5 5/16	6 13/16	3 5/16	9 5/16	2 13/16		5/16	7/32
2 1/2	5 1/2	7	3 1/2	10 1/2	2 1/2	5 1/2	5 13/16	7 5/16	3 13/16	10 13/16	2 13/16		5/16	7/32
3	6	7 3/4	3 1/2	11	3	6	6 5/16	8 1/16	3 13/16	11 5/16	3 5/16		5/16	7/32
3 1/2	6 1/2	8 1/2	4	12 1/2	3	6 1/2	6 13/16	8 13/16	4 5/16	12 13/16	3 5/16		5/16	7/32
4	7	9	4 1/2	13 1/2	3	7	7 5/16	9 5/16	4 13/16	13 13/16	3 5/16		5/16	7/32
5	8	10 1/4	5	15	3 1/2	8	8 5/16	10 9/16	5 5/16	15 5/16	3 13/16		5/16	7/32
6	8 1/2	11 1/2	5 1/2	17 1/2	4	9	8 13/16	11 13/16	5 13/16	17 13/16	4 5/16		5/16	7/32
8	10	14	6	20 1/2	5	11	10 5/16	14 5/16	6 5/16	20 13/16	5 5/16	See note[3]	5/16	7/32
10	11 1/2	16 1/2	7	24	5 1/2	12	11 13/16	16 13/16	7 5/16	24 5/16	5 13/16		5/16	7/32
12	13	19	8	27 1/2	6	14	13 5/16	19 5/16	8 5/16	27 13/16	6 5/16		5/16	7/32
14	15	21 1/2	8 1/2	31	6 1/2	16	15 5/16	21 13/16	8 13/16	31 5/16	6 13/16		5/16	7/32
16	16 1/2	24	9 1/2	34 1/2	7 1/2	18	16 13/16	24 5/16	9 13/16	34 13/16	7 13/16		5/16	7/32
18	18	26 1/2	10	37 1/2	8	19	18 5/16	26 13/16	10 5/16	37 13/16	8 5/16		5/16	7/32
20	19 1/2	29	10 1/2	40 1/2	8 1/2	20	19 7/8	29 3/8	10 7/8	40 7/8	8 7/8		3/8	7/32
24	22 1/2	34	12	47 1/2	10	24	22 15/16	34 7/16	12 7/16	47 15/16	10 7/16		7/16	1/4

Table 16. *(Continued)*

Nominal pipe size	1/4-in. raised-face					Ring joint						
	AA	*CC*	*EE*	*FF*	*GG*	*HH*	*KK*	*LL*	*MM*	*NN*	*L*[1]	*D*[2]
400 lb (for sizes smaller than 4 in., use 600 lb)												
4	8	5 1/2	16	4 1/2	8 1/4	8 1/16	5 9/16	16 1/16	4 9/16		5/16	7/32
5	9	6	16 3/4	5	9 1/4	9 1/16	6 1/16	16 13/16	5 1/16		5/16	7/32
6	9 3/4	6 1/4	18 3/4	5 1/4	10	9 13/16	6 5/16	18 13/16	5 5/16		5/16	7/32
8	11 3/4	6 3/4	22 1/4	5 3/4	12	11 13/16	6 13/16	22 5/16	5 13/16		5/16	7/32
10	13 1/4	7 3/4	25 3/4	6 1/4	13 1/2	13 5/16	7 13/16	25 13/16	6 5/16		5/16	7/32
12	15	8 3/4	29 3/4	6 1/2	15 1/4	15 1/16	8 13/16	29 13/16	6 9/16	See note[3]	5/16	7/32
14	16 1/4	9 1/4	32 3/4	7	16 1/2	16 5/16	9 5/16	32 13/16	7 1/16		5/16	7/32
16	17 3/4	10 1/4	36 1/4	8	18 1/2	17 13/16	10 5/16	36 5/16	8 1/16		5/16	7/32
18	19 1/4	10 3/4	39 1/4	8 1/2	19 1/2	19 5/16	10 13/16	39 5/16	8 9/16		5/16	7/32
20	20 3/4	11 1/4	42 3/4	9	21	20 7/8	11 3/8	42 7/8	9 1/8		3/8	7/32
24	24 1/4	12 3/4	50 1/4	10 1/2	24 1/2	24 7/16	12 15/16	50 7/16	10 11/16		7/16	1/4
600 lb												
1/2	3 1/4	2	5 3/4	1 3/4	5	3 7/32	1 31/32	5 23/32	1 23/32		7/32	1/8
3/4	3 3/4	2 1/2	6 3/4	2	5	3 3/4	2 1/2	6 3/4	2		1/4	5/32
1	4 1/4	2 1/2	7 1/4	2 1/4	5	4 1/4	2 1/2	7 1/4	2 1/4		1/4	5/32
1 1/4	4 1/2	2 3/4	8	2 1/2	5	4 1/2	2 3/4	8	2 1/2		1/4	5/32
1 1/2	4 3/4	3	9	2 3/4	5	4 3/4	3	9	2 3/4		1/4	5/32
2	5 3/4	4 1/4	10 1/4	3 1/2	6	5 13/16	4 5/16	10 5/16	3 9/16		5/16	3/16
2 1/2	6 1/2	4 1/2	11 1/2	3 1/2	6 3/4	6 9/16	4 9/16	11 9/16	3 9/16		5/16	3/16
3	7	5	12 3/4	4	7 1/4	7 1/16	5 1/16	12 13/16	4 1/16		5/16	3/16
3 1/2	7 1/2	5 1/2	14	4 1/2	7 3/4	7 9/16	5 9/16	14 1/16	4 9/16		5/16	3/16
4	8 1/2	6	16 1/2	4 1/2	8 3/4	8 9/16	6 1/16	16 9/16	4 9/16	See note[3]	5/16	3/16
5	10	7	19 1/2	6	10 1/4	10 1/16	7 1/16	19 9/16	6 1/16		5/16	3/16
6	11	7 1/2	21	6 1/2	11 1/4	11 1/16	7 9/16	21 1/16	6 9/16		5/16	3/16
8	13	8 1/2	24 1/2	7	13 1/4	13 1/16	8 9/16	24 9/16	7 1/16		5/16	3/16
10	15 1/2	9 1/2	29 1/2	8	15 3/4	15 9/16	9 9/16	29 9/16	8 1/16		5/16	3/16
12	16 1/2	10	31 1/2	8 1/2	16 3/4	16 9/16	10 1/16	31 9/16	8 9/16		5/16	3/16
14	17 1/2	10 3/4	34 1/4	9	17 3/4	17 9/16	10 13/16	34 5/16	9 1/16		5/16	3/16
16	19 1/2	11 3/4	38 1/2	10	19 3/4	19 9/16	11 13/16	38 9/16	10 1/16		5/16	3/16
18	21 1/2	12 1/4	42	10 1/2	21 3/4	21 9/16	12 5/16	42 1/16	10 9/16		5/16	3/16
20	23 1/2	13	45 1/2	11	23 3/4	23 5/8	13 1/8	45 5/8	11 1/8		3/8	3/16
24	27 1/2	14 3/4	53	13	27 3/4	27 11/16	14 15/16	53 3/16	13 3/16		7/16	7/32

Nominal pipe size	1/4-in. raised-face					Ring joint						
	AA	*CC*	*EE*	*FF*	*GG*	*HH*	*KK*	*LL*	*MM*	*NN*	*L*[1]	*D*[2]
900 lb (for sizes smaller than 3 in., use 1,500 lb)												
3	7 1/2	5 1/2	14 1/2	4 1/2	7 3/4	7 9/16	5 9/16	14 9/16	4 9/16		5/16	5/32
4	9	6 1/2	17 1/2	5 1/2	9 1/4	9 1/16	6 9/16	17 9/16	5 9/16		5/16	5/32
5	11	7 1/2	21	6 1/2	11 1/4	11 1/16	7 9/16	21 1/16	6 9/16		5/16	5/32
6	12	8	22 1/2	6 1/2	12 1/4	12 1/16	8 1/16	22 9/16	6 9/16		5/16	5/32
8	14 1/2	9	27 1/2	7 1/2	14 3/4	14 9/16	9 1/16	27 9/16	7 9/16		5/16	5/32
10	16 1/2	10	31 1/2	8 1/2	16 3/4	16 9/16	10 1/16	31 9/16	8 9/16	See note[3]	5/16	5/32
12	19	11	34 1/2	9	17 3/4	19 1/16	11 1/16	34 9/16	9 1/16		5/16	5/32
14	20 1/4	11 1/2	36 1/2	9 1/2	19	20 7/16	11 11/16	36 11/16	9 11/16		7/16	5/32
16	22 1/4	12 1/2	40 3/4	10 1/2	21	22 7/16	12 11/16	40 15/16	10 11/16		7/16	5/32
18	24	13 1/4	45 1/2	12	24 1/2	24 1/4	13 1/2	45 3/4	12 1/4		1/2	3/16
20	26	14 1/2	50 1/4	13	26 1/2	26 1/4	14 3/4	50 1/2	13 1/4		1/2	3/16
24	30 1/2	18	60	15 1/2	30 1/2	30 7/8	18 3/8	60 3/8	15 7/8		5/8	7/32
1,500 lb												
1/2	4 1/4	3	....	....	....	4 1/4	3	.....	.....		1/4	5/32
3/4	4 1/2	3 1/4	....	....	....	4 1/2	3 1/4	.....	.....		1/4	5/32
1	5	3 1/2	9	2 1/2	5	5	3 1/2	9	2 1/2		1/4	5/32
1 1/4	5 1/2	4	10	3	5 3/4	5 1/2	4	10	3		1/4	5/32
1 1/2	6	4 1/4	11	3 1/2	6 1/4	6	4 1/4	11	3 1/2		1/4	5/32
2	7 1/4	4 3/4	13 1/4	4	7 1/4	7 5/16	4 13/16	13 5/16	4 1/16		5/16	1/8
2 1/2	8 1/4	5 1/4	15 1/4	4 1/2	8 1/4	8 5/16	5 5/16	15 5/16	4 9/16		5/16	1/8
3	9 1/4	5 3/4	17 1/4	5	9 1/4	9 5/16	5 13/16	17 5/16	5 1/16		5/16	1/8
4	10 3/4	7 1/4	19 1/4	6	10 3/4	10 13/16	7 5/16	19 5/16	6 1/16		5/16	1/8
5	13 1/4	8 3/4	23 1/4	7 1/2	13 3/4	13 5/16	8 13/16	23 5/16	7 9/16	See note[3]	5/16	1/8
6	13 7/8	9 3/8	24 7/8	8 1/8	14 1/2	14	9 1/2	25	8 1/4		3/8	1/8
8	16 3/8	10 7/8	29 7/8	9 1/8	17	16 9/16	11 1/16	30 1/16	9 5/16		7/16	5/32
10	19 1/2	12	36	10 1/4	20 1/4	19 11/16	12 3/16	36 3/16	10 7/16		7/16	5/32
12	22 1/4	13 1/4	40 3/4	12	23	22 9/16	13 9/16	41 1/16	12 5/16		9/16	3/16
14	24 3/4	14 1/4	44	12 1/2	25 3/4	25 1/8	14 5/8	44 3/8	12 7/8		5/8	7/32
16	27 1/4	16 1/4	48 1/4	14 3/4	28 1/4	27 11/16	16 11/16	48 11/16	15 3/16		11/16	5/16
18	30 1/4	17 3/4	53 1/4	16 1/2	31 1/2	30 11/16	18 3/16	53 11/16	16 15/16		11/16	5/16
20	32 3/4	18 3/4	57 3/4	17 3/4	34	33 3/16	19 3/16	58 3/16	18 3/16		11/16	3/8
24	38 1/4	20 3/4	67 1/4	20 1/2	39 3/4	38 13/16	21 5/16	67 13/16	21 1/16		13/16	7/16

[1] *L* = height of raised face of ring joint flanges.
[2] *D* = approximate distance between flange faces when ring is compressed.
[3] Center-to-face dimensions shown for fittings with ring-joint flanges apply to straight sizes only. For reducing fittings and reducers, use dimensions shown for raised-face flanges of largest opening; 400 lb and higher classes, subtract the 1/4-in. raised face for each flange (do not subtract the 1/16-in. raised face in 150- and 300-lb classes); add height of ring-joint raised face (*L*) applying to each flange.

For calculating the "laying length" of fittings with ring joints, add the approximate distance (*D*) between flange faces when ring is compressed to the center-to-face dimensions in these tables.

Table 17. Dimensions of Typical Commercial Forged-steel Screwed Fittings (ASA B16.11-1966)[1]

Fitting illustrations (left to right): 90° elbow, Tee, 45° elbow, Cross, Coupling, Reducer, Half coupling, Pipe cap

Dimensions, in.

2,000 lb

	1/8	1/4	3/8	1/2	3/4	1	1 1/4	1 1/2	2	2 1/2	3	4
A	13/16	31/32	1 1/8	1 1/8	1 5/16	1 1/2	1 3/4	2	2 3/8	3	3 3/8	4 3/16
B	7/8	1	1 5/16	1 5/16	1 1/2	1 13/16	2 7/16	2 7/16	2 31/32	3 5/8	4 5/16	5 3/4
C	11/16	3/4	7/8	7/8	1	1 1/8	1 5/16	1 3/8	1 11/16	2 1/16	2 1/2	3 1/8
D	7/8	1	1 5/16	1 1/2	1 1/2	1 13/16	2 3/16	2 7/16	2 31/32	3 5/8	4 5/16	5 3/4
E	13/16	31/32	1 1/8	1 1/8	1 5/16	1 1/2	1 3/4	2	2 3/8	3	3 3/8	4 3/16
F	7/8	1	1 5/16	1 5/16	1 1/2	1 13/16	2 3/16	2 7/16	2 31/32	3 5/8	4 5/16	5 3/4

3,000 lb

	1/8	1/4	3/8	1/2	3/4	1	1 1/4	1 1/2	2	2 1/2	3	4
A	13/16	31/32	1 1/8	1 1/8	1 1/2	1 3/4	2	2 3/8	2 1/2	3 1/4	3 3/4	4 1/2
B	7/8	1	1 5/16	1 7/8	1 13/16	2 3/16	2 7/16	2 31/32	3 5/16	4	4 3/4	6
C	11/16	3/4	7/8	15/16	1 1/8	1 5/16	1 3/8	1 11/16	1 23/32	2 1/16	2 1/2	3 1/8
D	7/8	1	1 5/16	1 1/2	1 13/16	2 3/16	2 7/16	2 31/32	3 5/16	4	4 3/4	6
E	13/16	31/32	1 1/8	1 1/8	1 1/2	1 3/4	2	2 3/8	2 1/2	3 1/4	3 3/8	4 1/2
F	7/8	1	1 5/16	1 7/8	1 13/16	2 3/16	2 7/16	2 31/32	3 5/16	4	4 3/4	6
N	5/8	3/4	7/8	1 1/8	1 3/8	1 3/4	2 1/4	2 1/2	3	3 5/8	4 1/4	5 1/2
P	1 1/4	1 3/8	1 1/2	1 7/8	2	2 3/8	2 5/8	3 1/8	3 3/8	3 5/8	4 1/4	4 3/4
Q	5/8	11/16	3/4	15/16	1	1 3/16	1 5/16	1 9/16	1 11/16	1 13/16	2 1/8	2 3/8
R	3/4	1	1	1 1/4	1 7/16	1 5/8	1 3/4	1 3/4	1 7/8	2 3/8	2 9/16	2 11/16

A	4½	4 3/16	3¾	3¼	2½	2⅜	2	1¾	1½	1 5/16	1⅛	31/32
B	6	5¾	4¾	4	3 5/16	2 31/32	2 7/16	2 3/16	1 13/16	1½	1 5/16	1
C	3⅛	3⅛	2½	2 1/16	1 23/32	1 11/16	1⅜	1 5/16	1⅛	1	⅞	¾
D	6	5¾	4⅝	4	3 5/16	2 31/32	2 7/16	2 3/16	1 13/16	1½	1 5/16	1
E	4½	4 3/16	3¾	3¼	2½	2⅜	2	1¾	1½	1 5/16	1⅛	31/32
F	6	5¾	4⅝	4	3 5/16	2 31/32	2 7/16	2 3/16	1 13/16	1½	1 5/16	1
N	6¼	5	4¼	3⅝	3	2½	2¼	1¾	1½	1¼	1	⅞
P	4¾	4¼	3⅝	3⅜	3⅛	2⅝	2⅜	2	1⅞	1½	1⅜	1¼
Q	2⅜	2⅛	1 13/16	1 11/16	1 9/16	1 5/16	1 3/16	1	15/16	¾	11/16	⅝
R	2 15/16	2 11/16	2½	2	1⅞	1 13/16	1 11/16	1½	1 5/16	1 1/16	1 1/16	.....

[1] Manufacturers catalogs should be consulted for dimensions of street elbows and of laterals. These two types of fittings are no longer covered by ASA Standards.

Table 18. Dimensions of Typical Commercial Forged-steel Socket-welding Fittings (ASA B16.11-1966)[1]

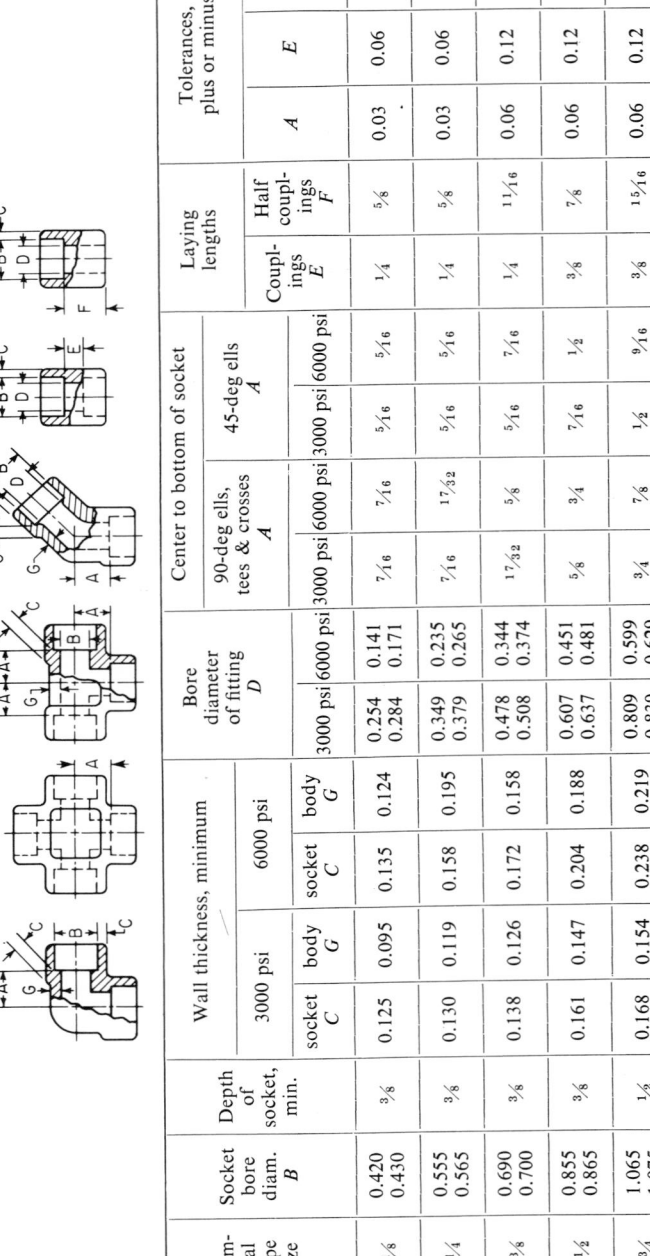

Nominal pipe size	Socket bore diam. B	Depth of socket, min.	Wall thickness, minimum 3000 psi socket C	3000 psi body G	6000 psi socket C	6000 psi body G	Bore diameter of fitting D 3000 psi	Bore diameter D 6000 psi	Center to bottom of socket 90-deg ells, tees & crosses A 3000 psi	90-deg A 6000 psi	45-deg ells A 3000 psi	45-deg A 6000 psi	Laying lengths Couplings E	Half couplings F	Tolerances, plus or minus A	E	F
1/8	0.420 0.430	3/8	0.125	0.095	0.135	0.124	0.254 0.284	0.141 0.171	7/16	7/16	5/16	5/16	1/4	5/8	0.03	0.06	0.03
1/4	0.555 0.565	3/8	0.130	0.119	0.158	0.195	0.349 0.379	0.235 0.265	7/16	17/32	5/16	5/16	1/4	5/8	0.03	0.06	0.03
3/8	0.690 0.700	3/8	0.138	0.126	0.172	0.158	0.478 0.508	0.344 0.374	17/32	5/8	5/16	7/16	1/4	11/16	0.06	0.12	0.06
1/2	0.855 0.865	3/8	0.161	0.147	0.204	0.188	0.607 0.637	0.451 0.481	5/8	3/4	7/16	7/16	3/8	7/8	0.06	0.12	0.06
3/4	1.065 1.075	1/2	0.168	0.154	0.238	0.219	0.809 0.839	0.599 0.629	3/4	7/8	1/2	9/16	3/8	15/16	0.06	0.12	0.06

1	1.330 1.340	½	0.196	0.179	0.273	0.250	1.034 1.064	0.800 0.830	⅞	1 1/16	9/16	11/16	½	1⅛	0.08	0.16	0.08
1¼	1.675 1.685	½	0.208	0.191	0.273	0.250	1.365 1.395	1.145 1.175	1 1/16	1¼	11/16	13/16	½	1 3/16	0.08	0.16	0.08
1½	1.915 1.925	½	0.218	0.200	0.307	0.281	1.595 1.625	1.323 1.353	1¼	1½	13/16	1	½	1¼	0.08	0.16	0.08
2	2,406 2.416	⅝	0.238	0.218	0.374	0.344	2.052 2.082	1.674 1.704	1½	1⅝	1	1⅛	¾	1⅝	0.08	0.16	0.08
2½	2.906 2.921	⅝	0.301	0.276	. . .	0.375	2.439 2.499	. . .	1⅝	. . .	1⅛	. . .	¾	1 11/16	0.10	0.20	0.10
3	3.535 3.550	⅝	0.327	0.300	. . .	0.438	3.038 3.098	. . .	2¼	. . .	1¼	. . .	¾	1¾	0.10	0.20	0.10
4	4.545 4.560	¾	0.368	0.337	. . .	0.531	3.996 4.056	. . .	2⅝	. . .	1⅝	. . .	¾	1⅞	0.10	0.20	0.10

[1] Dimensions for caps and reducers are not standardized. For such information, reference should be made to manufacturers' catalogs.

MSS SP-48 in sizes 26 to 36 in., inclusive.   Pressure-temperature ratings are identical with those of seamless pipe of the same size, thickness or schedule, and material grade.

Dimensions of long-radius and short-radius 90-deg elbows are given in Tables 19 and 20, respectively.   Elbows of uniform 1-diameter (short) radius and 1½-diameter (long) radius offer a substantially smaller resistance to flow than mitered elbows.   Figure 8 illustrates friction losses in 4-in. Schedule 40 pipe elbows, miters, bends, and straight pipe.

Dimensions of straight tees and reducers are given in Tables 21 and 22.

Table 19. Dimensions of Typical Commercial 90-deg Long-radius Butt-welding Elbows (ASA B16.9–1964 and MSS SP 48–1956)

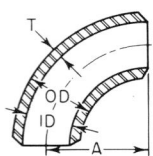

Nominal pipe size	Outside diameter, OD	Inside diameter, ID	Wall thickness, T	Center to face, A	Pipe schedule number[1]	Weight (approx), lb
			Standard			
½	0.840	0.622	0.109	1½	40	0.2
¾	1.050	0.824	0.113	1⅛	40	0.2
1	1.315	1.049	0.133	1½	40	0.4
1¼	1.660	1.380	0.140	1⅞	40	0.6
1½	1.900	1.610	0.145	2¼	40	0.9
2	2.375	2.067	0.154	3	40	1.4
2½	2.875	2.469	0.203	3¾	40	2.9
3	3.500	3.068	0.216	4½	40	4.5
3½	4.000	3.548	0.226	5¼	40	6.4
4	4.500	4.026	0.237	6	40	8.7
5	5.563	5.047	0.258	7½	40	14.7
6	6.625	6.065	0.280	9	40	22.9
8	8.625	7.981	0.322	12	40	46.0
10	10.750	10.020	0.365	15	40	81
12	12.750	12.000	0.375	18	▲²	119
14	14.000	13.250	0.375	21	30	154
16	16.000	15.250	0.375	24	30	201
18	18.000	17.250	0.375	27	▲²	256
20	20.000	19.250	0.375	30	20	317
22	22.000	21.250	0.375	33	20	385
24	24.000	23.250	0.375	36	20	458
26	26.000	25.250	0.375	39	▲²	539
28	28.000	27.250	0.375	42	▲²	626
30	30.000	29.250	0.375	45	▲²	720
32	32.000	31.250	0.375	48	▲²	818
34	34.000	33.250	0.375	51	▲²	926
36	36.000	35.250	0.375	54	▲²	1,040
42	42.000	41.250	0.375	63	▲²	1,420

Table 19. (*Continued*)

Nominal pipe size	Outside diameter, OD	Inside diameter, ID	Wall thickness, T	Center to face, A	Pipe schedule number[1]	Weight (approx), lb
			Extra strong			
½	0.840	0.546	0.147	1½	80	0.3
¾	1.050	0.742	0.154	1⅛	80	0.3
1	1.315	0.957	0.179	1½	80	0.5
1¼	1.660	1.278	0.191	1⅞	80	0.8
1½	1.900	1.500	0.200	2¼	80	1.0
2	2.375	1.939	0.218	3	80	2.0
2½	2.875	2.323	0.276	3¾	80	3.8
3	3.500	2.900	0.300	4½	80	6.1
3½	4.000	3.364	0.318	5¼	80	8.7
4	4.500	3.826	0.337	6	80	11.9
5	5.563	4.813	0.375	7½	80	20.6
6	6.625	5.761	0.432	9	80	34.1
8	8.625	7.625	0.500	12	80	69
10	10.750	9.750	0.500	15	60	109
12	12.750	11.750	0.500	18	▲[2]	157
14	14.000	13.000	0.500	21	▲[2]	202
16	16.000	15.000	0.500	24	40	265
18	18.000	17.000	0.500	27	▲[2]	338
20	20.000	19.000	0.500	30	30	419
22	22.000	21.000	0.500	33	30	508
24	24.000	23.000	0.500	36	▲[2]	606
26	26.000	25.000	0.500	39	20	713
28	28.000	27.000	0.500	42	20	829
30	30.000	29.000	0.500	45	20	953
32	32.000	31.000	0.500	48	20	1,090
34	34.000	33.000	0.500	51	20	1,230
36	36.000	35.000	0.500	54	20	1,380
42	42.000	41.000	0.500	63	▲[2]	1,880
			Schedule 160[1]			
1	1.315	0.815	0.250	1½	160	0.6
1¼	1.660	1.160	0.250	1⅞	160	1.0
1½	1.900	1.338	0.281	2¼	160	1.4
2	2.375	1.689	0.343	3	160	2.9
2½	2.875	2.125	0.375	3¾	160	4.9
3	3.500	2.624	0.438	4½	160	8.3
4	4.500	3.438	0.531	6	160	17.6
5	5.563	4.313	0.625	7½	160	32.2
6	6.625	5.189	0.718	9	160	53
8	8.625	6.813	0.906	12	160	117
10	10.750	8.500	1.125	15	160	226
12	12.750	10.126	1.312	18	160	375

Table 19. (*Continued*)

Nominal pipe size	Outside diameter, OD	Inside diameter, ID	Wall thickness, T	Center to face, A	Pipe schedule number[1]	Weight (approx), lb
Double extra strong						
¾	1.050	0.434	0.308	1⅛	▲[2]	0.4
1	1.315	0.599	0.358	1½	▲[2]	0.7
1¼	1.660	0.896	0.382	1⅞	▲[2]	1.2
1½	1.900	1.100	0.400	2¼	▲[2]	1.8
2	2.375	1.503	0.436	3	▲[2]	3.4
2½	2.875	1.771	0.552	3¾	▲[2]	6.5
3	3.500	2.300	0.600	4½	▲[2]	10.7
3½	4.000	2.728	0.636	5¼	▲[2]	15.4
4	4.500	3.152	0.674	6	▲[2]	21.2
5	5.563	4.063	0.750	7½	▲[2]	37.2
6	6.625	4.897	0.864	9	▲[2]	61
8	8.625	6.875	0.875	12	▲[2]	114

[1] Pipe schedule numbers in accordance with ASA B36.10.
[2] This size and thickness does not correspond with any schedule number.

Table 20. Dimensions of Typical Commercial 90-deg Short-radius Elbows (ASA B16.28–1964)

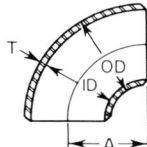

Nominal pipe size	Outside diameter, OD	Inside diameter, ID	Wall thickness, T	Center to face, A	Pipe schedule number[1]	Weight (approx), lb
Standard						
1	1.315	1.049	0.133	1	40	0.3
1¼	1.660	1.380	0.140	1¼	40	0.4
1½	1.900	1.610	0.145	1½	40	0.6
2	2.375	2.067	0.154	2	40	1.0
2½	2.875	2.469	0.203	2½	40	1.9
3	3.500	3.068	0.216	3	40	3.0
3½	4.000	3.548	0.226	3½	40	4.2
4	4.500	4.026	0.237	4	40	5.7
5	5.563	5.047	0.258	5	40	9.7
6	6.625	6.065	0.280	6	40	15.2
8	8.625	7.981	0.322	8	40	30.5
10	10.750	10.020	0.365	10	40	54
12	12.750	12.000	0.375	12	▲[2]	79
14	14.000	13.250	0.375	14	30	102
16	16.000	15.250	0.375	16	30	135
18	18.000	17.250	0.375	18	▲[2]	171
20	20.000	19.250	0.375	20	20	212
22	22.000	21.250	0.375	22	▲[2]	256
24	24.000	23.250	0.375	24	20	305
26[3]	26.000	25.250	0.375	26	▲[2]	359
28	28.000	27.250	0.375	28	▲[2]	415
30	30.000	29.250	0.375	30	▲[2]	480
32	32.000	31.250	0.375	32	▲[2]	546
34	34.000	33.250	0.375	34	▲[2]	617
36	36.000	35.250	0.375	36	▲[2]	692
42	42.000	41.250	0.375	48	▲[2]	1,079

Table 20. *(Continued)*

Nominal pipe size	Outside diameter, OD	Inside diameter, ID	Wall thickness, T	Center to face, A	Pipe schedule number[1]	Weight (approx), lb
			Extra strong			
1½	1.900	1.500	0.200	1½	80	0.7
2	2.375	1.939	0.218	2	80	1.3
2½	2.875	2.323	0.276	2½	80	2.5
3	3.500	2.900	0.300	3	80	4.0
3½	4.000	3.364	0.318	3½	80	5.7
4	4.500	3.826	0.337	4	80	7.8
5	5.563	4.813	0.375	5	80	13.7
6	6.625	5.761	0.432	6	80	22.6
8	8.625	7.625	0.500	8	80	45.6
10	10.750	9.750	0.500	10	60	72
12	12.750	11.750	0.500	12	▲[2]	104
14	14.000	13.000	0.500	14	▲[2]	135
16	16.000	15.000	0.500	16	40	177
18	18.000	17.000	0.500	18	▲[2]	225
20	20.000	19.000	0.500	20	30	278
22	22.000	21.000	0.500	22	30	338
24	24.000	23.000	0.500	24	▲[2]	404
26[3]	26.000	25.000	0.500	26	20	474
28	28.000	27.000	0.500	28	20	581
30	30.000	29.000	0.500	30	20	634
32	32.000	31.000	0.500	32	20	722
34	34.000	33.000	0.500	34	20	817
36	36.000	35.000	0.500	36	20	913
42	42.000	41.000	0.500	48	▲[2]	1,430

[1] Pipe schedule numbers in accordance with ASA B36.10.
[2] This size and thickness has no corresponding schedule number.
[3] Dimensional data for pipe sizes 26 in. and larger are not included in ASA B16.28.

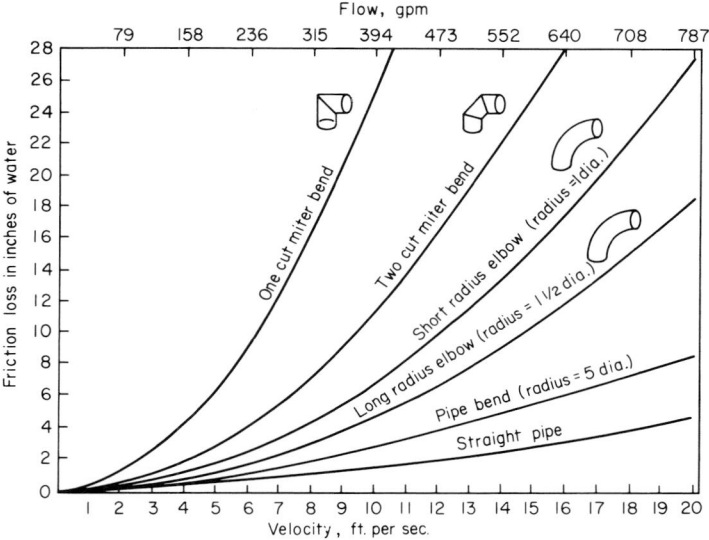

FIG. 8. Comparative friction losses in 4-in. Schedule 40 pipe and fittings (after G. T. Caspary).

Table 21.   Dimensions of Typical Commercial Straight Butt-welding Tees (ASA B16.9–1964 and MSS SP 48–1956)

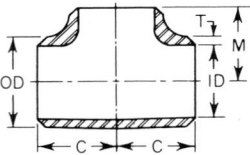

Nominal pipe size	Outside diameter, OD	Inside diameter, ID	Wall thickness, T	Center to end, C	Center to end, M	Pipe schedule number[1]	Weight, (approx) lb
			Standard				
1/2	0.840	0.622	0.109	1	1	40	0.3
3/4	1.050	0.824	0.113	1 1/8	1 1/8	40	0.4
1	1.315	1.049	0.133	1 1/2	1 1/2	40	0.8
1 1/4	1.660	1.380	0.140	1 7/8	1 7/8	40	1.3
1 1/2	1.900	1.610	0.145	2 1/4	2 1/4	40	2.0
2	2.375	2.067	0.154	2 1/2	2 1/2	40	2.9
2 1/2	2.875	2.469	0.203	3	3	40	5.2
3	3.500	3.068	0.216	3 3/8	3 3/8	40	7.4
3 1/2	4.000	3.548	0.226	3 3/4	3 3/4	40	9.8
4	4.500	4.026	0.237	4 1/8	4 1/8	40	12.6
5	5.563	5.047	0.258	4 7/8	4 7/8	40	19.8
6	6.625	6.065	0.280	5 5/8	5 5/8	40	29.3
8	8.625	7.981	0.322	7	7	40	53
10	10.750	10.020	0.365	8 1/2	8 1/2	40	91
12	12.750	12.000	0.375	10	10	▲[2]	132
14	14.000	13.250	0.375	11	11	30	172
16	16.000	15.250	0.375	12	12	30	219
18	18.000	17.250	0.375	13 1/2	13 1/2	▲[2]	282
20	20.000	19.250	0.375	15	15	20	354
22	22.000	21.250	0.375	16 1/2	16 1/2	20	437
24	24.000	23.250	0.375	17	17	20	493
26	26.000	25.250	0.375	19 1/2	19 1/2	▲[2]	634
28[3]	28.000	27.250	0.375	20 1/2	20 1/2	▲[2]	729
30	30.000	29.250	0.375	22	22	▲[2]	855
32[3]	32.000	31.250	0.375	23 1/2	23 1/2	▲[2]	991
34	34.000	33.250	0.375	25	25	▲[2]	1,136
36	36.000	35.250	0.375	26 1/2	26 1/2	▲[2]	1,294

Table 21. *(Continued)*

Nominal pipe size	Outside diameter, OD	Inside diameter, ID	Wall thickness, T	Center to end, C	Center to end, M	Pipe schedule number[1]	Weight (approx), lb
			Extra strong				
½	0.840	0.546	0.147	1	1	80	0.3
¾	1.050	0.742	0.154	1⅛	1⅛	80	0.5
1	1.315	0.957	0.179	1½	1½	80	0.9
1¼	1.660	1.278	0.191	1⅞	1⅞	80	1.6
1½	1.900	1.500	0.200	2¼	2¼	80	2.4
2	2.375	1.939	0.218	2½	2½	80	3.7
2½	2.875	2.323	0.276	3	3	80	6.4
3	3.500	2.900	0.300	3⅜	3⅜	80	9.4
3½	4.000	3.364	0.318	3¾	3¾	80	12.6
4	4.500	3.826	0.337	4⅛	4⅛	80	16.4
5	5.563	4.813	0.375	4⅞	4⅞	80	26.4
6	6.625	5.761	0.432	5⅝	5⅝	80	42.0
8	8.625	7.625	0.500	7	7	80	76
10	10.750	9.750	0.500	8½	8½	60	118
12	12.750	11.750	0.500	10	10	▲[2]	167
14	14.000	13.000	0.500	11	11	▲[2]	203
16	16.000	15.000	0.500	12	12	40	271
18	18.000	17.000	0.500	13½	13½	▲[2]	351
20	20.000	19.000	0.500	15	15	30	442
22	22.000	21.000	0.500	16½	16½	30	548
24	24.000	23.000	0.500	17	17	20	607
26	26.000	25.000	0.500	19½	19½	20	794
28[3]	28.000	27.000	0.500	20½	20½	20	910
30	30.000	29.000	0.500	22	22	20	1,065
32[3]	32.000	31.000	0.500	23½	23½	20	1,230
34	34.000	33.000	0.500	25	25	20	1,420
36	36.000	35.000	0.500	26½	26½	20	1,610
			Schedule 160[1]				
½	0.840	0.466	0.187	1	1	160	0.4
¾	1.050	0.614	0.218	1⅛	1⅛	160	0.6
1	1.315	0.815	0.250	1½	1½	160	1.1
1¼	1.660	1.160	0.250	1⅞	1⅞	160	1.9
1½	1.900	1.338	0.281	2¼	2¼	160	3.0
2	2.375	1.689	0.343	2½	2½	160	4.9
2½	2.875	2.125	0.375	3	3	160	7.8
3	3.500	2.626	0.438	3⅜	3⅜	160	12.2
4	4.500	3.438	0.531	4⅛	4⅛	160	22.8
5	5.563	4.313	0.625	4⅞	4⅞	160	38.5
6	6.625	5.189	0.718	5⅝	5⅝	160	59
8	8.625	6.813	0.906	7	7	160	120
10	10.750	8.500	1.125	8½	8½	160	222
12	12.750	10.126	1.312	10	10	160	360

Nominal pipe size	Outside diameter, OD	Inside diameter, ID	Wall thickness, T	Center to end, C	Center to end, M	Pipe schedule number[1]	Weight (approx), lb
			Double extra strong				
½	0.840	0.252	0.294	1	1	▲[2]	0.4
¾	1.050	0.434	0.308	1⅛	1⅛	▲[2]	0.6
1	1.315	0.599	0.358	1½	1½	▲[2]	1.3
1¼	1.660	0.896	0.382	1⅞	1⅞	▲[2]	2.4
1½	1.900	1.100	0.400	2¼	2¼	▲[2]	3.7
2	2.375	1.503	0.436	2½	2½	▲[2]	5.7
2½	2.875	1.771	0.552	3	3	▲[2]	9.8
3	3.500	2.300	0.600	3⅜	3⅜	▲[2]	14.8
3½	4.000	2.728	0.636	3¾	3¾	▲[2]	20.2
4	4.500	3.152	0.674	4⅛	4⅛	▲[2]	26.6
5	5.563	4.063	0.750	4⅞	4⅞	▲[2]	43.4
6	6.625	4.897	0.864	5⅝	5⅝	▲[2]	68
8	8.625	6.875	0.875	7	7	▲[2]	118

[1] Pipe schedule numbers in accordance with ASA B36.10.   Other thicknesses available.
[2] This size and thickness does not correspond with any schedule number.
[3] Dimensional data for pipe sizes 28 and 32 in. are not included in MSS SP 48.

Table 22. Dimensions of Typical Commercial Concentric and Eccentric Butt-welding Reducers (ASA B16.9-1964 and MSS SP 48-1956)

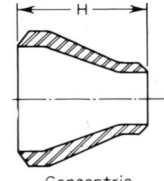
Concentric

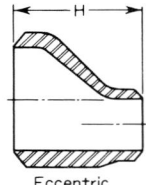
Eccentric

Nominal pipe size	Length, H	Weight (approx), lb (concentric or eccentric)			
		Standard	Extra strong	Schedule 160	Double extra strong
¾ × ⅜	1½	0.2	0.3	0.3	...
¾ × ½	1½	0.2	0.3	0.3	0.4
1 × ⅜	2	0.3	0.4	0.4	0.4
1 × ½	2	0.3	0.4	0.5	0.5
1 × ¾	2	0.3	0.4	0.5	0.5
1¼ × ½	2	0.5	0.5	0.6	0.7
1¼ × ¾	2	0.5	0.5	0.6	0.7
1¼ × 1	2	0.5	0.6	0.7	0.8
1½ × ½	2½	0.5	0.6	0.8	1.0
1½ × ¾	2½	0.5	0.6	0.9	1.0
1½ × 1	2½	0.6	0.7	0.9	1.0
1½ × 1¼	2½	0.6	0.8	1.0	1.2
2 × ¾	3	0.8	1.0	1.4	1.7
2 × 1	3	0.9	1.0	1.4	1.6
2 × 1¼	3	0.9	1.1	1.4	1.8
2 × 1½	3	0.9	1.2	1.6	1.9
2½ × 1	3½	1.3	1.7	2.3	3.0
2½ × 1¼	3½	1.4	1.7	2.2	3.1
2½ × 1½	3½	1.5	1.8	2.2	3.0
2½ × 2	3½	1.6	2.0	2.7	3.3

Table 22. **(Continued)**

Nominal pipe size	Length, H	Weight (approx), lb (concentric or eccentric)			
		Standard	Extra strong	Schedule 160	Double extra strong
3 × 1¼	3½	1.7	2.2	3.1	4.1
3 × 1½	3½	1.8	2.1	3.1	4.0
3 × 2	3½	2.0	2.6	3.4	4.0
3 × 2½	3½	2.1	2.8	3.7	4.6
3½ × 1¼	4	2.3	3.2	...	5.8
3½ × 1½	4	2.5	3.1	...	5.8
3½ × 2	4	2.7	3.5	...	5.7
3½ × 2½	4	2.9	3.8	...	5.9
3½ × 3	4	3.0	4.0	...	6.8
4 × 1½	4	2.7	3.8	5.6	6.6
4 × 2	4	3.1	3.9	5.6	6.6
4 × 2½	4	3.3	4.4	5.5	6.3
4 × 3	4	3.5	4.7	6.5	7.7
4 × 3½	4	3.6	4.8	...	8.2
5 × 2	5	5.0	6.6	10.6	12.2
5 × 2½	5	5.5	7.2	10.2	11.7
5 × 3	5	5.7	7.8	10.2	11.1
5 × 3½	5	5.8	8.0	...	13.3
5 × 4	5	5.9	8.3	12.4	14.2
6 × 2½	5½	7.6	9.9	15.8	18.8
6 × 3	5½	8.0	11.1	15.1	18.5
6 × 3½	5½	8.1	11.6	...	17.3
6 × 4	5½	8.1	12.0	17.2	19.1
6 × 5	5½	8.6	12.6	18.8	21.4
8 × 3½	6	12.8	16.1	...	27.9
8 × 4	6	13.1	18.6	26.9	25.7
8 × 5	6	13.4	19.5	29.6	29.2
8 × 6	6	13.9	20.4	32.1	32.7
10 × 4	7	21.1	25.3	50	
10 × 5	7	21.8	28.7	48	
10 × 6	7	22.3	29.8	50	
10 × 8	7	23.2	31.4	58	
12 × 5	8	30.5	39.1	78	
12 × 6	8	31.1	40.6	75	
12 × 8	8	32.1	37.4	86	
12 × 10	8	33.4	43.6	94	
14 × 6	13	55	74		
14 × 8	13	57	76		
14 × 10	13	60	79		
14 × 12	13	63	83		
16 × 8	14	70	93		
16 × 10	14	72	96		
16 × 12	14	75	99		
16 × 14	14	77	102		
18 × 10	15	86	114		
18 × 12	15	89	118		
18 × 14	15	90	120		
18 × 16	15	94	123		
20 × 12	20	134	176		
20 × 14	20	135	179		
20 × 16	20	138	182		
20 × 18	20	142	186		
22 × 14	20	148	195		
22 × 16	20	151	198		
22 × 18	20	154	202		
22 × 20	20	157	207		

Table 22. *(Continued)*

Nominal pipe size	Length, H	Weight (approx), lb (concentric or eccentric)			
		Standard	Extra strong[1]	Schedule 160	Double extra strong
24 × 14	20	158	209		
16	20	160	211		
18	20	163	215		
20	20	167	220		
26 × 18	24	200	272		
20	24	200	272		
22	24	200	272		
24	24	200	272		
28[2] × 18	24	210	290		
20	24	210	290		
24	24	210	290		
26	24	210	290		
30 × 20	24	220	315		
24	24	220	315		
26	24	220	315		
28	24	220	315		
32[2] × 24	24	255 / *221*	335 / *293*		
26	24	255 / *229*	335 / *304*		
28	24	255 / *237*	335 / *315*		
30	24	255 / *245*	335 / *325*		
34 × 24	24	270 / *229*	355 / *304*		
26	24	270 / *237*	355 / *315*		
30	24	270 / *253*	355 / *336*		
32	24	270 / *261*	355 / *347*		
36 × 24	24	340 / *237*	360 / *315*		
26	24	340 / *245*	360 / *325*		
30	24	340 / *261*	360 / *347*		
32	24	340 / *269*	360 / *357*		
34	24	340 / *277*	360 / *368*		
42[2] × 24	24	260	350		
26	24	270	360		
30	24	285	380		
32	24	295	390		
34	24	300	400		
36	24	310	410		

[1] Weights in italics are for eccentric reducers in sizes 32, 34, and 36 in. For all other sizes, weights shown are for either concentric or eccentric reducers.

[2] Data for 28-, 32- and 42-in. reducers are not included in MSS SP 48.

Table 23.   Dimensions of Typical Commercial Butt-welding Laterals

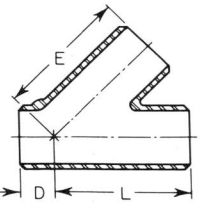

Nominal pipe size	Standard		Weight (approx), lb	Extra strong		Weight (approx), lb
	L and E	D		L and E	D	
Straight						
1	5¾	1¾	1.7	6½	2	2.5
1¼	6¼	1¾	2.4	7¼	1¼	3.8
1½	7	2	3.2	8½	2½	5.4
2	8	2½	5.0	9	2½	7.7
2½	9½	2½	9.2	10½	2½	13.5
3	10	3	12.6	11	3	18.8
3½	11½	3	17.2	12½	3	25.6
4	12	3	20.8	13½	3	32.8
5	13½	3½	31.4	15	3½	49.8
6	14½	3½	42.4	17½	4	79
8	17½	4½	76	20½	5	140
10	20½	5	124	24	5½	202
12	24½	5½	180	27½	6	273
14	27	6	218	31	6½	340
16	30	6½	275	34½	7½	433
18	32	7	326	37½	8	526
20	35	8	396	40½	8½	628
24	40½	9	544	47½	10	882

Dimensions of common laterals are given in Table 23.   Working pressures are rated at 40 per cent of the allowable working pressure established for pipe from which laterals are made.   Where full allowable pipe pressures must be met, the laterals are generally made from heavier pipe with ends machined to match standard pipe dimensions.   Dimensional tolerances of laterals vary not more than $\pm\frac{1}{16}$ in. for sizes up to and including 8 in. and $\pm\frac{1}{32}$ in. for sizes 10 through 24 in.

Standard dimensions of caps are given in Table 24.   Pressure-temperature ratings are identical with those of seamless pipe for the same size, thickness or schedule, and material grades.   Caps conform to ASME Boiler and Pressure Vessel Code requirements.

Welding caps are formed from steel plate and are stress-relieved after forming. They are ellipsoidal in shape, the minor axis being equal to half the major axis.

Radii R and r closely approximate the actual semielliptical shape.

Saddles are not designed to contain internal pressure but are to be used as reinforcements which are slipped over the welded joint or branch pipe and header and then fillet-welded all around.   A vent hole prevents build-up of pressure under the saddle.

Table 24.   Dimensions of Typical Commercial Butt-welding Standard Caps

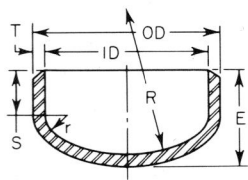

Nominal pipe size	Outside diameter, OD	Inside diameter, ID	Wall thickness, T	Length, E	Tangent, S	Dish radius, R	Knuckle radius, r	Pipe schedule number[1]	Weight (approx), lb
½	0.840	0.622	0.109	1	0.74	0.54	0.10	40	0.1
¾	1.050	0.824	0.113	1¼	0.93	0.72	0.14	40	0.2
1	1.315	1.049	0.133	1½	1.10	0.92	0.17	40	0.3
1¼	1.660	1.380	0.140	1½	1.02	1.35	0.23	40	0.4
1½	1.900	1.610	0.145	1½	0.95	1.41	0.27	40	0.4
2	2.375	2.067	0.154	1½	0.83	1.81	0.34	40	0.6
2½	2.875	2.469	0.203	1½	0.68	2.15	0.41	40	0.9
3	3.500	3.068	0.216	2	1.02	2.69	0.51	40	1.4
3½	4.000	3.548	0.226	2½	1.39	3.11	0.59	40	2.1
4	4.500	4.026	0.237	2½	1.26	3.52	0.67	40	2.5
5	5.563	5.047	0.258	3	1.48	4.42	0.84	40	4.2
6	6.625	6.065	0.280	3½	1.70	5.31	1.01	40	6.4
8	8.625	7.981	0.322	4	1.68	6.98	1.33	40	11.3
10	10.750	10.020	0.365	5	2.13	8.77	1.67	40	20.0
12	12.750	12.000	0.375	6	2.62	10.50	2.00	▲[2]	29.5
14	14.000	13.250	0.375	6½	2.81	11.60	2.21	30	35.3
16	16.000	15.250	0.375	7	2.81	13.34	2.54	30	44.3
18	18.000	17.250	0.375	8	3.31	15.08	2.88	▲[2]	57
20	20.000	19.250	0.375	9	3.81	16.84	3.21	20	71
22	22.000	21.250	0.375	10	4.31	18.60	3.54	20	86
24	24.000	23.250	0.375	10½	4.31	20.35	3.88	20	102
26	26.000	25.250	0.375	10½	3.81	22.10	4.21	▲[2]	110
28	28.000	27.250	0.375	10½	3.31	23.85	4.54	▲[2]	120
30	30.000	29.250	0.375	10½	2.81	25.60	4.88	▲[2]	125
32	32.000	31.250	0.375	10½	2.31	27.35	5.21	▲[2]	145
34	34.000	33.250	0.375	10½	1.81	29.10	5.54	▲[2]	160
36	36.000	35.250	0.375	10½	1.31	30.85	5.88	▲[2]	175
42	42.000	41.250	0.375	12	1.31	36.10	6.88	▲[2]	230

[1] Pipe schedule numbers in accordance with ASA B36.10.
[2] This size and thickness does not correspond with any schedule number.

Forged Branch Fittings.   Weldolets, Sweepolets, and Latrolets are illustrated in Fig. 9.

The Weldolet is a 90-deg fitting with the reinforcement required to meet ASME and ASA codes included in the design of the forged fitting.

The Sweepolet is also a 90-deg fitting with integral reinforcement but with a sweeping portion tying in horizontally (in longitudinal cross section) with the run pipe with an insert butt weld.   Thus, the edge of the Sweepolet is the same thickness as that of the pipe into which it is welded.

The Latrolet is a 45-deg forged outlet connection attached in much the same manner as the Weldolet.   These fittings are available in outlet sizes up through 12 in.   They provide good flow conditions.   The Latrolet construction allows inexpensive 45-deg branch fabrication involving applications where 45-deg laterals might have been used but were considered unsuitable because of the high cost of a lateral fitting or the poor quality of field fabricated laterals.

Alloy-steel and Nonferrous Fittings.   The more common types of fittings, such as elbows, tees, reducers, and caps, are also available in most alloy steels and nonferrous piping materials.

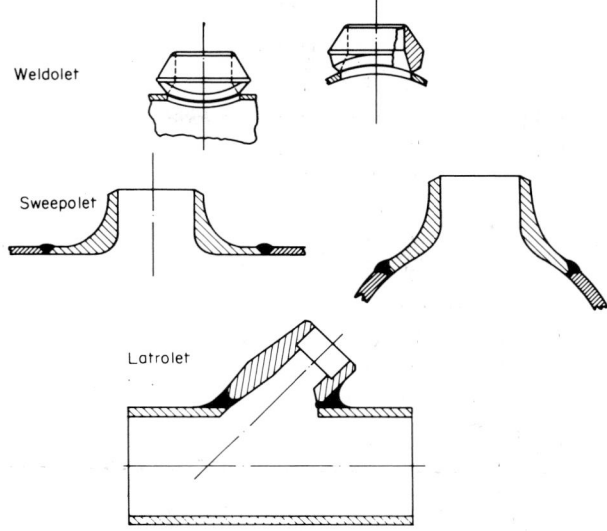

FIG. 9. Weldolet; Sweepolet; Latrolet.

Table 25 provides a listing of the sizes of fittings usually available. Other materials, sizes, and wall thicknesses can also be furnished on special order.

VALVES

Valves are extensively used in piping systems and on equipment to which piping is connected. Some valves are used continuously, others intermittently, and some, like safety values, are utilized only in rare instances.

Most basic valve designs either embody a bonnet to join the moving part to the body or utilize plugs or balls without a bonnet.

VALVE CLASSIFICATIONS AND TYPES

Start or Stop Flow. The interruption or start of flow is the most widely performed function of valves. A basic requirement in the design of these valves is that they offer minimum flow restriction and pressure loss when open. In many applications, tight shut-off at closure is also an essential feature. In some applications, this may not be critical.

Gate, plug, ball, or butterfly valves are most widely used for the interruption or start of flow. Diaphragm valves may be preferred in corrosive applications or in service where contamination of the fluid is not permissible, such as in the drug and beverage industry.

Regulation of Flow. Many applications require that the flow of the fluid or gas be regulated (or throttled) in various steps between closed and open limits. This is generally done by introducing resistance to flow either by a change in direction, or by causing a restriction, or by combinations of these. Commonly used valve designs are of the globe, angle, needle, and butterfly types. Diaphragm

Table 25. Alloy-steel and Nonferrous Piping Materials Normally Available as Fittings and Flanges

Material	Fittings—Nominal pipe sizes				Flanges—Nominal pipe sizes						
	Sch 5s	Sch 10s	Sch 40s (st)	Sch 80s (xs)	150 lb	300 lb	400 lb	600 lb	900 lb	1,500 lb	2,500 lb
Stainless steel:											
304	½–24	½–24	½–24	½–24	½–24	½–24	½–24	½–24	½–24	½–24	½–12
304L	½–24	½–24	½–24	½–24	½–24	½–24	½–24	½–24	½–24	½–24	½–12
316	½–24	½–24	½–24	½–24	½–24	½–24	½–24	½–24	½–24	½–24	½–12
316L	½–24	½–24	½–24	½–24	½–24	½–24	½–24	½–24	½–24	½–24	½–12
347	½–24	½–24	½–24	½–24	½–24	½–24	½–24	½–24	½–24	½–24	½–12
Aluminum:											
3003	.....	.....	½–24	½–24	½–24	½–24	½–24	½–24	½–24	½–24	½–12
5052	.....	.....	½–24	½–24							
6061	.....	.....	½–24	½–24	½–24	½–24	½–24	½–24	½–24	½–24	½–12
Low and intermediate alloys:											
Carbon Moly	.....	.....	½–24	½–24	½–24	½–24	½–24	½–24	½–24	½–24	½–12
1 Cr–½ Mo	.....	.....	½–24	½–24	½–24	½–24	½–24	½–24	½–24	½–24	½–12
1¼ Cr–½ Mo	.....	.....	½–24	½–24	½–24	½–24	½–24	½–24	½–24	½–24	½–12
2¼ Cr–1 Mo	.....	.....	½–24	½–24	½–24	½–24	½–24	½–24	½–24	½–24	½–12
5 Cr–½ Mo	.....	.....	½–24	½–24	½–24	½–24	½–24	½–24	½–24	½–24	½–12
7 Cr–½ Mo	.....	.....	½–24	½–24	½–24	½–24	½–24	½–24	½–24	½–24	½–12
3½ Ni	.....	.....	½–24	½–24	½–24	½–24	½–24	½–24	½–24	½–24	½–12
Cr–Cu–Ni	.....	.....	½–24	½–24	½–24	½–24	½–24	½–24	½–24	½–24	½–12
Nickel and nickel alloys:											
Nickel	.....	½–24	½–24	½–24	½–24	½–24	½–24	½–24	½–24	½–24	½–12
Monel	.....	½–24	½–24	½–24	½–24	½–24	½–24	½–24	½–24	½–24	½–12
Inconel	.....	½–24	½–24	½–24	½–24	½–24	½–24	½–24	½–24	½–24	½–12
Hastelloy B	.....	½–12	½–12	½–12	½–24	½–24	½–24	½–24	½–24	½–24	½–12
Hastelloy C	.....	½–12	½–12	½–12	½–24	½–24	½–24	½–24	½–24	½–24	½–12
Copper and copper alloys:											
Deoxidized copper	.....	.....	½–12	½–12							
Copper nickel, 70/30	.....	.....	½–24	½–24	½–24	½–24	½–24	½–24	½–24	½–24	½–12
Silicon bronze	.....	.....	½–12	½–12	½–24	½–24	½–24	½–24	½–24	½–24	½–12

valves are also suitable. Closest control over flow is most readily obtained with globe- or angle-type valves with characterized plugs or with needle valves.

Back-flow Prevention. In some applications, it is essential to prevent back flow. For this purpose, check valves are generally used. These valves are kept open by fluid flow. They close either by gravity or by a reversal of flow. The two major types in use are generally designated as *lift checks* and *swing checks*.

Pressure Regulation. In some applications, the incoming or line pressure must be reduced. The required service pressure will then have to be achieved and maintained uniformly, even when slight variations occur in the line pressure. Valves used for this purpose are referred to as pressure-reducing valves or regulators.

Pressure Relief. Relief valves are used in applications where excessive pressure in the system can cause damage or failure. Some failures have been extremely disastrous. The safety, or pop, valves generally are spring-loaded. They open when the pressure exceeds the limits set for the valve. A special type of safety relief device involves rupture disks which rupture when the disk's pressure limit is exceeded. Rupture disk relief devices have the advantages of being leak-tight up to the rupture pressure and being able to open and relieve quickly large cross-sectional areas.

Special Valves. Many other types of special valves have also been developed for special applications in the missile, nuclear, and numerous other industries.

Gate Valves

Application. Gate valves are primarily designed to start or stop flow. Thus, they provide "free-flow" service. In service, these valves generally are either fully open or completely closed. When fully open, the fluid or gas flows through the valve in a straight line with very little resistance to flow. As a result, the pressure loss between the valve ends is very small.

Gate valves should not be used in the regulation or throttling of flow because accurate control is not possible. High-flow velocity in partially opened valves may cause erosion of the disks and seating surfaces and of the downstream side of the valve seats. Vibration may also result in shattering of the partially opened valve disk.

However, there are some applications where special gate valves are used for throttling. An example involves guillotine gate valves for pulp stock.

Design Features. Gate valves consist of three major components: body, bonnet, and trim. The body is generally connected to the piping by means of flanged, screwed, or welded connections. The bonnet, containing the moving parts, is joined to the body—generally with bolts to permit cleaning and maintenance. The valve trim includes the stem, the "gate," the wedge, or disk, and the seat rings.

Two basic types of gate valves are the *wedge type* and the *double disk type*.

Wedge Type. In the wedge types or disk-wedge types (Figs. 10 to 12) either a tapered solid or tapered split wedge is used. When the valve is closed, the gate disk is wedged on both sides against the seat. The solid wedge is most widely used and is favored where high flow rates or turbulent flow is involved, as in steam service. In split-wedge gate valves (Fig. 12) the two-piece wedge disk is seated between matching tapered seats in the body. This type is preferred where the body seats might be distorted from undue pipeline strains.

In the rising stem valves (Fig. 10) the operating threads are out of direct contact with the fluid or gas. The non-rising-stem type, Fig. 11, is preferred where space is limited and where the fluid passing through the valve will not corrode or erode the threads or leave deposits on the threads.

In the rising-stem type of valve, the upper part of the stem is threaded and a nut is fastened solidly to the handwheel and held in the yoke by thrust collars.   As the handwheel is turned, the stem moves up or down.   In the non-rising-stem valve, the lower end of the stem is threaded and screws into the disk, vertical motion of the stem being restrained by a thrust collar.   The rising-stem valve requires a greater amount of space when opened.   It is generally to be preferred, however, because the position of the stem indicates at once whether the valve is open or closed. Non-rising-stem valves are sometimes provided with an indicator for this purpose.

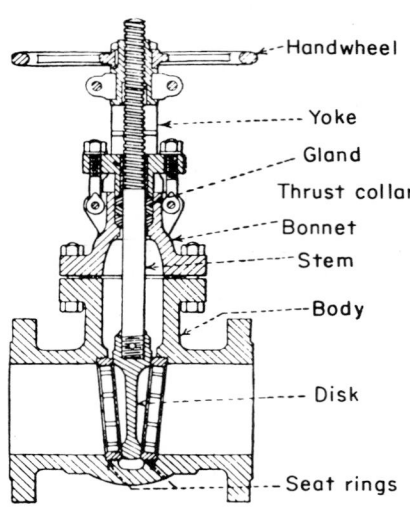

FIG. 10. Rising-stem solid-wedge gate valve for 250-psig steam service.

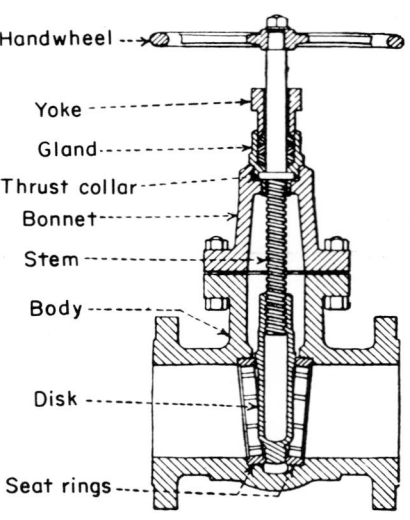

FIG. 11. Non-rising-stem solid-wedge gate valve for 250-psig steam service.

Up to a steam working pressure of 250 psig, gate valves are usually made with an oval bonnet flange to reduce the face-to-face dimension.   In the valves designed for higher pressures, a bonnet flange of circular shape is preferred to provide a recessed gasket joint.   A typical valve design for 600-lb steam pressure is illustrated in Fig. 12.

Pressure-sealed bonnet joints are recommended for high-pressure, high-tempera-ture steam and feed water service.   They are normally available in the 600-, 900-, 1500-, and 2,500-lb classes.

Double-disk Type.   In the double-disk parallel-seat gate valves (Figs. 13 and 14), the disks are forced against the valve seats by a wedging mechanism as the stem is tightened.

Some double-disk parallel-seat valves employ a design (Fig. 15) which depends for its tightness mainly upon the fluid pressure exerted against one side or the other of the disk.   The major advantage of this type is that the disk cannot be jammed into the body, which otherwise might make it difficult to open the valve subse-quently.   This is particularly important where motors are used for opening and closing.   However, the tightness of the solid-wedge gate valve may also result to a significant degree from the action of fluid pressure against one side or the other. Thus, the two types are essentially similar in this respect.

Each disk in the parallel-seat types slides against its seat while the valve is being opened or closed.   Consequently, these components must be made of metals which

do not tend to gall or tear when in sliding contact with each other. By means of guides, the wedge gate does not come into contact with the seat until the closing point is almost reached. The double-disk parallel-seat gate valve is often favored for high-temperature steam service because it is less likely to stick in the closed position as a result of change in temperature. Gate valves are used where a straight-through flow is desired with a minimum amount of pressure loss.

The valve shown in Fig. 15 has been assembled by welding. This valve, with welding ends, a welded bonnet joint, and welded-in seat rings, is particularly suited for high-temperature service where trouble might be experienced in keeping flanged

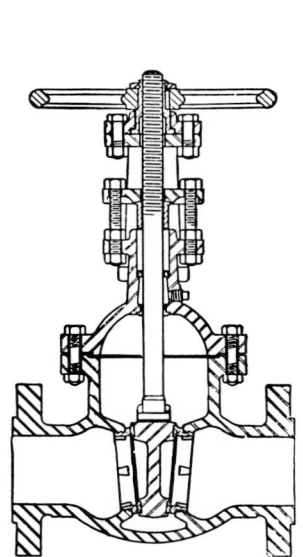

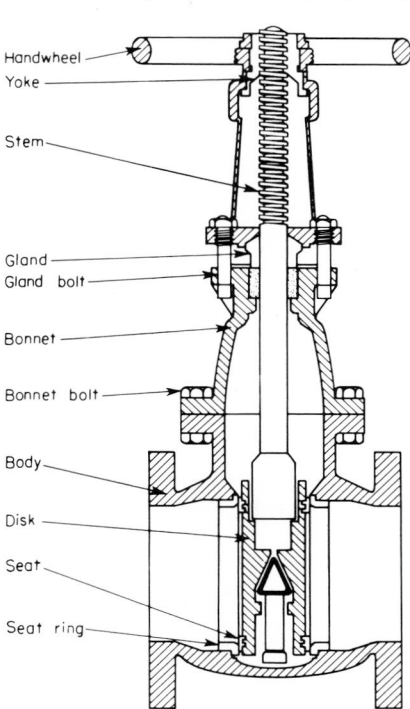

Fig. 12. Rising-stem solid-wedge gage valve with circular bonnet flange; for 600-psig service.

Fig. 13. Double-disk rising-stem gate flanged-end gate valve for 150-psig service.

joints tight. The use of welded-in seat rings with stellite facing is deemed to insure sufficient durability to warrant welding the bonnet joint as well as the line joints.

Lubricant-seal Valves. Some gate valves are furnished with lubricant-seal systems to fill damaged valve seats which have become scored or pitted. In these valves, a sealing material, which is insoluble in the liquid or gas flowing through the valve, is pressed through a small port in the valve body into a groove under the seat ring (Fig. 16). From this groove, the sealing material passes through small holes in the seat ring into an annular groove in the face of the ring. The holes in the seat ring are made small in order to prevent the sealant from being washed out.

The lubricant also reduces sliding friction between the seat and wedge.

Lubricant-seal systems are available in cast-iron and cast-steel bodies.

Special Features. Another type of gate valve employs a solid plug disk of conical shape. It has the advantage of the solid wedge gate and may be utilized for regulation of flow. Its application is primarily limited to corrosive service involving stainless steel or nonferrous materials.

Standard Pressure Ratings. Bronze gate valves are available for 125-, 150-, 200-, 300-, and 350-lb steam-pressure ratings.

Cast-iron bodies are extensively used for gate valves in sizes 1½ in. and up for working pressures of 25-, 125-, 150-, 250-, and 300-lb steam. Nodular iron gate

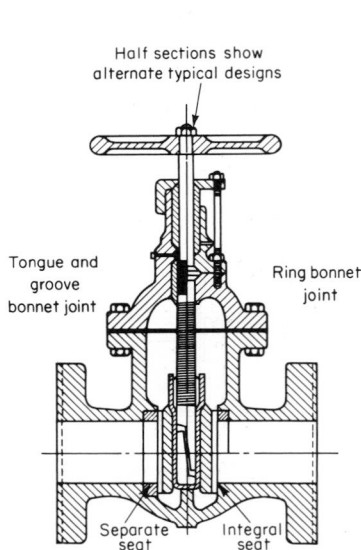

FIG. 14. Double-disk non-rising-stem gate valve.

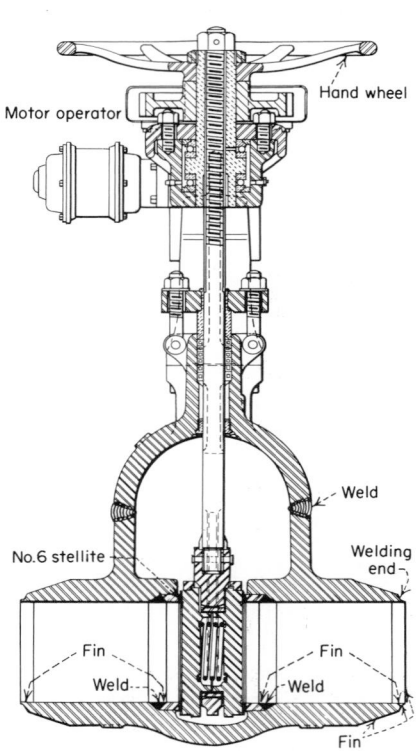

FIG. 15. Parallel seat gate valve showing welded construction for high-temperature service with welded-in seat ring.

valves for refinery service in the 150- and 300-lb classifications are covered by API Standard 604, Flanged Nodular Iron Gate and Plug Valves for Refinery Use. Steel valves normally are available for the ASA working-steam-pressure ratings of 150, 300, 400, 600, 900, 1,500, and 2,500 lb. The higher pressure ratings are also used with water. Steel gate valves are covered by the following API Standards:

Standard 600   Flanged and Butt-Welding End Steel Gate and Plug Valves for Refinery Use

Standard 6D   Specification for Steel Gate, Plug, Ball, and Check Valves for Pipeline Service

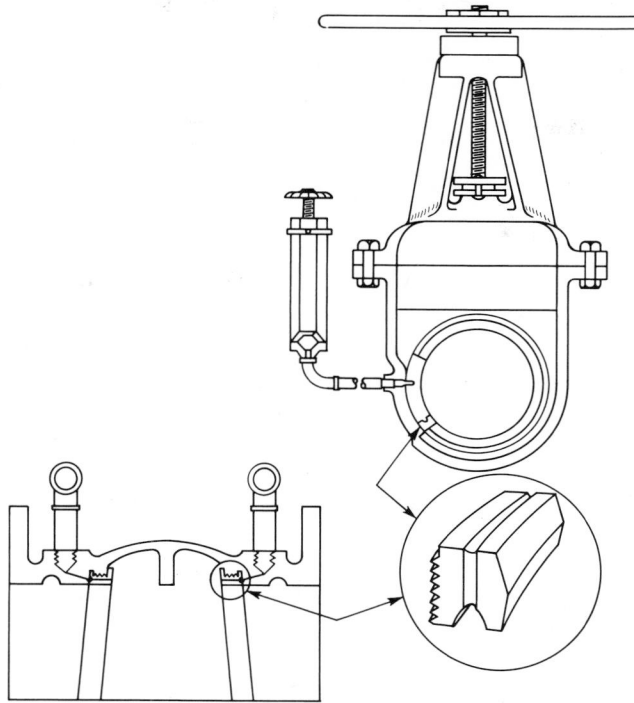

Fig. 16. Gate valve with lubricant-seal system.

Dimensions. Dimensions of welding- and flanged-end valves are given in Table 26.

American Standard ASA B16.10 establishes face-to-face and center-to-end dimensions for ferrous flanged- and welding-end valves. Applicable dimensions conform to those established by American Petroleum Institute Standards 6C, 6D, and 600 for the corresponding types of valves. Dimensions for control valves conform with those established in Instrument Society of America Tentative Standard RP 4.1. Applicable dimensions for corrosion-resistant valves conform with those established in Manufacturers' Standardization Society Standard Practice MSS-SP 42. Reference should be made to the above API standards for dimensional data for valves whose pressure ratings differ from those shown in Table 246.

Welding-end valves [8 in. and smaller (150-lb rating and all sizes 300-, 400-, 600-, 900-, 1,500, and 2,500-lb rating)] have the same face-to-face dimensions as raised-face flanged valves.

Plug Valves

Applications. Plug valves, also called cocks, generally are used for the same full-flow service as gate valves, where quick shut-off is required. They are used for steam, water, oil, gas, and chemical liquid service.

Plug valves are not generally designed for the regulation of flow. Nevertheless, in some applications, these valves are used for this purpose, particularly for gas-flow throttling. For low-flow regulation, special plugs are available.

Table 26. Face-to-face and End-to-end Dimensions[1] of Ferrous Valves (ASA B16.10-1957)

Class 125 Cast Iron and 150-lb Steel Valves

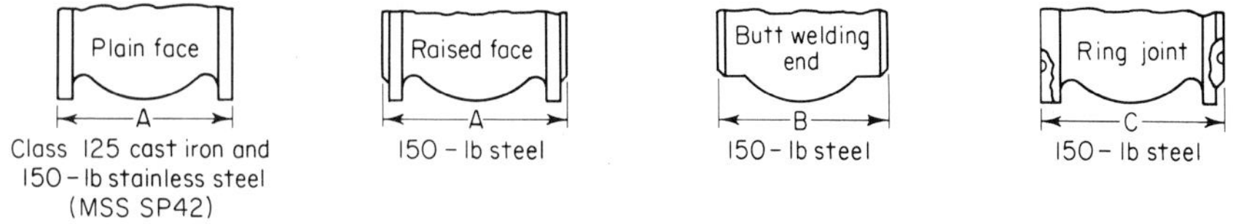
Plain face
A
Class 125 cast iron and
150-lb stainless steel
(MSS SP42)
Raised face
A
150-lb steel
Butt welding end
B
150-lb steel
Ring joint
C
150-lb steel

Valves with Class 125 or 150-lb end flanges or with welding ends[1]

1	2	3	4	5	6	7	8	9	10	11	12	13	14	15	16	17	18
	Class 125 cast iron									150-lb steel							
	Flanged-end—plain face									Flanged-end[4] (1/16-in. raised-face) and welding-end							
Nominal valve size	Gate		Plug			Globe and lift check	Angle and lift check	Swing check[2]	Control[3]	Gate[5]				Plug[5]			
	Solid wedge A	Double disk A	Short pattern A	Regular A	Venturi A	A	D	A	A	Solid wedge A	Double disk A	Solid wedge B	Double disk B	Short pattern A	Regular A	Venturi A	Round port full bore A
1/4	....	....	....	....	....	....	....	....	....	4	4	4	4	....	....	....	....
3/8	....	....	....	....	....	....	....	....	....	4	4	4	4	....	....	....	....
1/2	....	....	....	....	....	....	....	....	....	4 1/4	4 1/4	4 1/4	4 1/4	....	....	....	....
3/4	....	....	....	....	....	....	....	....	....	4 5/8	4 5/8	4 5/8	4 5/8	....	....	....	....
1	....	....	5 1/2	5 1/2	....	....	....	....	7 1/4	5	5	5	5	5 1/2	....	....	7
1 1/4	....	....	....	6 1/2	....	....	....	....	....	5 1/2	5 1/2	5 1/2	5 1/2	....	....	....	....
1 1/2	....	....	6 1/2	6 1/2	....	....	....	....	8 3/4	6 1/2	6 1/2	6 1/2	6 1/2	6 1/2	....	....	8 3/4
2	7	7	7	7 1/2	....	8	4	8	10	7	7	8 1/2	8 1/2	7	....	....	10 1/2
2 1/2	7 1/2	7 1/2	7 1/2	8 1/4	....	8 1/2	4 1/4	8 1/2	10 7/8	7 1/2	7 1/2	9 1/2	9 1/2	7 1/2	....	....	11 3/4
3	8	8	8	9	....	9 1/2	4 3/4	9 1/2	11 3/4	8	8	11 1/8	11 1/8	8	....	....	13 1/2
3 1/2	8 1/2	8 1/2	....	....	....	....	....	....	....	....	....	....	....	....	....	....	....
4	9	9	9	9	....	11 1/2	5 3/4	11 1/2	13 7/8	9	9	12	12	9	....	....	17
5	10	10	10	14	....	13	6 1/2	13	....	10	10	15	15				
6	10 1/2	10 1/2	10 1/2	15 1/2	15 1/2	14	7	14	17 3/4	10 1/2	10 1/2	15 7/8	15 7/8	10 1/2	15 1/2	....	21
8	11 1/2	11 1/2	11 1/2	18	18	19 1/2	9 3/4	19 1/2	21 3/8	11 1/2	11 1/2	16 1/2	16 1/2	11 1/2	18	....	25
10	13	13	13	21	21	24 1/2	12 1/4	24 1/2	....	13	13	18	18	13	21	21	31
12	14	14	14	24	24	27 1/2	13 3/4	27 1/2	....	14	14	19 3/4	19 3/4	14	24	24	36
14	15	....	....	27	27	31	15 1/2	31	....	15	15	22 1/2	22 1/2	....	....	27	....
16	16	....	....	30	30	36	18	....	....	16	16	24	24	....	....	30	....
18	17	....	....	34	34	....	....	....	....	17	17	26	26	....	....	34	....
20	18	....	....	36	36	....	....	....	....	18	18	28	28	...	...	36	....
24	20	....	....	....	42	....	....	....	....	20	20	32	32	....	....	42	....
30	....	....	....	....	51	....	....	....	....	....	....	....	....	....	....	....	....

[1] API Standards 6D and 600 conform to the dimensions shown for corresponding sizes, valve type, and flange class or welding-end.

[2] These dimensions are not intended to cover the type of check valve having the seat angle at approximately 45 deg to the run of the valve or the "Underwriter Pattern" or other patterns where large clearances are required.

[3] Instrument Society of America Tentative Standard ISA RP4.1 face-to-face conforms to dimensions shown in column 10 and column 23.

[4] 150-lb Corrosion-resistant valves covered by MSS SP-42 have plain faces.

[5] 150-lb steel conduit-type gate valves are not standardized.

[6] These valves are flanged end only.

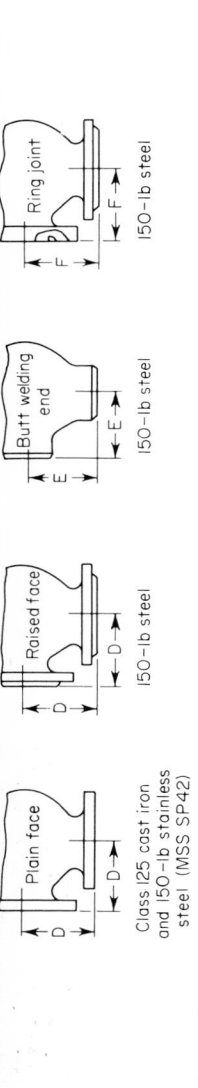

Plain face — 150-lb steel
Raised face — 150-lb steel
Butt welding end — 150-lb steel
Ring joint — 150-lb steel

Class 125 cast iron and 150-lb stainless steel (MSS SP42)

Valves with Class 125 or 150-lb end flanges or with welding ends[1]

Columns 19–23: **Flanged end[4] (1/16-in. raised-face) and welding-end** — 150-lb steel
Columns 24–32: **Flanged-end (ring joint)** — 150-lb steel (Gate[5] = cols 24–26; Plug = cols 27–29)

1	19	20	21	22	23	24	25	26	27	28	29	30	31	32
Nominal valve size	Globe and lift check	Angle and lift check	Swing check	"Y" pattern globe	Control[3,6]	Gate[5] Solid wedge	Gate[5] Double disk	Short pattern	Plug Regular	Plug Venturi	Round port full bore	Globe and lift check	Angle and lift check	Swing check
	A & B	D & E	A & B	A & B	A	C	C	C	C	C	C	C	F	C
1/4	4	2	4	....	....	....	....	....	....	....	....	....	....	....
3/8	4	2	4	....	....	....	....	....	....	....	....	....	....	....
1/2	4 1/4	2 1/4	4 1/4	5 1/2	....	4 11/16	4 11/16	....	....	....	....	4 11/16	2 15/32	4 11/16
3/4	4 5/8	2 1/2	4 5/8	6	....	5 1/8	5 1/8	....	....	....	....	5 1/8	2 3/4	5 1/8
1	5	2 3/4	5	6 1/2	7 1/4	5 1/2	5 1/2	6	....	....	7 1/2	5 1/2	3	5 1/2
1 1/4	5 1/2	3	5 1/2	7 1/4	....	6	6	....	....	....	....	6	3 1/4	6
1 1/2	6 1/2	3 1/4	6 1/2	8	8 3/4	7	7	7	....	....	9 1/4	7	3 1/2	7
2	8	4	8 1/2	9	10	7 1/2	7 1/2	7 1/2	....	....	11	8 1/2	4 1/4	8 1/2
2 1/2	8 1/2	4 1/4	8 1/2	11	10 7/8	8	8	8	....	....	12 1/4	9	4 1/2	9
3	9 1/2	4 3/4	9 1/2	12 1/2	11 3/4	8 1/2	8 1/2	8 1/2	....	....	14	....	5	10
3 1/2	11 1/2	5 3/4	11 1/2	....	....	9 1/2	9 1/2	....	....	....	....	12	6	12
4	14	7	13	14 1/2	13 7/8	10 1/2	10 1/2	9 1/2	....	....	17 1/2	14 1/2	7 1/4	13 1/2
5	16	8	14	18 1/2	....	11	11	....	....	....	....	16 1/2	8 1/4	14 1/2
6	19 1/2	9 3/4	19 1/2	23 1/2	17 3/4	12	12	11	16	....	21 1/2	20	10	20
8	24 1/2	12 1/4	24 1/2	26 1/2	21 3/8	13 1/2	13 1/2	12	18 1/2	....	25 1/2	25	12 1/4	25
10	27 1/2	13 3/4	27 1/2	30 1/2	....	14 1/2	14 1/2	13 1/2	21 1/2	21 1/2	31 1/2	28	14	28
12	31	15 1/2	31	....	....	15 1/2	15 1/2	14 1/2	24 1/2	24 1/2	36 1/2	31 1/2	15 3/4	31 1/2
14	36	18	....	....	....	16 1/2	16 1/2	....	....	27 1/2	....	36 1/2	18 1/4	....
16	....	....	....	....	....	17 1/2	17 1/2	....	....	30 1/2	....	....	....	....
18	....	....	....	....	....	....	....	....	....	34 1/2	....	....	....	....
20	....	....	....	....	....	18 1/2	18 1/2	....	....	36 1/2	....	....	....	....
24	....	....	....	....	....	20 1/2	20 1/2	....	....	40 1/2	....	....	....	....
30	....	....	....	....	....	....	....	....	....	....	....	....	....	....

Table 26. (Continued)

Class 250 Cast Iron and 300-lb Steel Valves

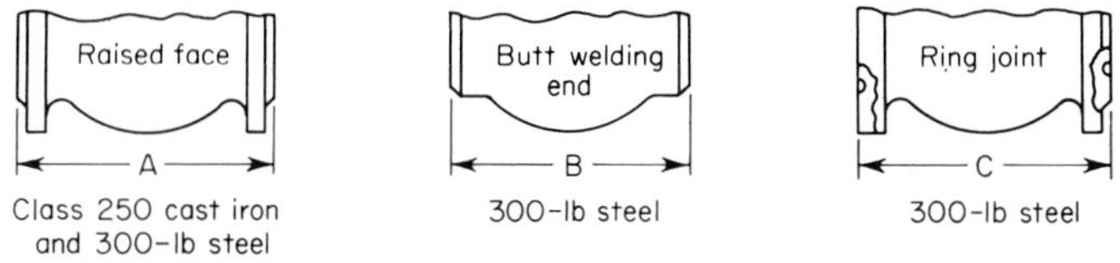

Valves with Class 250 or 300-lb end flanges or with welding ends[1]

1	2	3	4	5	6	7	8	9	10	11	12	13	14	15	16
	Class 250 cast iron									300-lb steel					
	Flanged-end (1/16-in. raised-face)									Flanged-end (1/16-in. raised-face) and welding-end					
Nominal valve size	Gate		Plug			Globe and lift check	Angle and lift check	Swing check	Control[2]	Gate[3]		Plug			
	Solid wedge	Double disk	Short pattern	Regular	Venturi					Solid wedge	Double disk	Short pattern		Regular	Venturi
	A	*A*	*A*	*A*	*A*	*A*	*D*	*A*	*A*	*A* and *B*	*A* and *B*	*A*	*B*	*A*	*A*
½	....	....	....	....	....	....	....	....	7½	....	....	....	....	....	....
¾	....	....	....	....	....	....	....	....	7⅝	....	....	....	....	....	....
1	....	....	....	6¼	....	....	....	....	7¾	....	....	6¼	....	....	....
1¼	....	....	....	....	....	....	....	....	....	....	....	....	....	....	....
1½	....	....	....	7½	....	....	....	....	9¼	7½	7½	7½	....	....	....
2	8½	8½	7¼	8½	....	10½	5¼	10½	10½	8½	8½	8½	10½	....	....
2½	9½	9½	8	9½	....	11½	5¾	11½	11½	9½	9½	9½	12	....	....
3	11⅛	11⅛	9¼	11⅛	....	12½	6¼	12½	12½	11⅛	11⅛	11⅛	13	....	....
3½	....	....	....	....	....	....	....	....	....	....	....	....	....	....	....
4	12	12	10½	12	....	14	7	14	14½	12	12	12	14	....	....
5	15	15	....	....	....	15¾	7⅞	15¾	....	15	15	....	....	....	....
6	15⅞	15⅞	....	16¾	15⅞	17½	8¾	17½	18⅝	15⅞	15⅞	15⅞	18	15⅞	15⅞
8	16½	16½	....	19¾	16½	21	10½	21	22⅜	16½	16½	16½	20½	19¾	16½
10	18	18	....	23½	18	24½	12¼	24½	....	18	18	18	22	22⅜	18
12	19¾	19¾	....	28	19¾	28	14	28	....	19¾	19¾	19¾	25	....	19¾
14	22½	22½	....	....	30	....	....	....	....	30	30	....	....	....	30
16	24	24	....	....	33	....	....	....	....	33	33	....	....	....	33
18	26	26	....	....	36	....	....	....	....	36	36	....	....	....	36
20	28	28	....	....	39	....	....	....	....	39	39	....	....	....	39
22	....	....	....	....	44	....	....	....	....	43	43	....	....	....	43
24	31	31	....	....	45	....	....	....	....	45	45	....	....	....	45
26	....	....	....	....	....	....	....	....	....	49	49	....	....	....	....
28	....	....	....	....	....	....	....	....	....	53	53	....	....	....	....
30	....	....	....	....	....	....	....	....	....	55	55	....	....	....	....

[1] API Standards 6C, 6D, and 600 conform to the dimensions shown for corresponding sizes, valve type, and flange class or welding-end.
[2] Instrument Society of America Tentative Standard ISA RP4.1 face to face conforms to dimensions shown in column 10 and column 22.
[3] 300-lb steel conduit-type gate valves are not standardized.

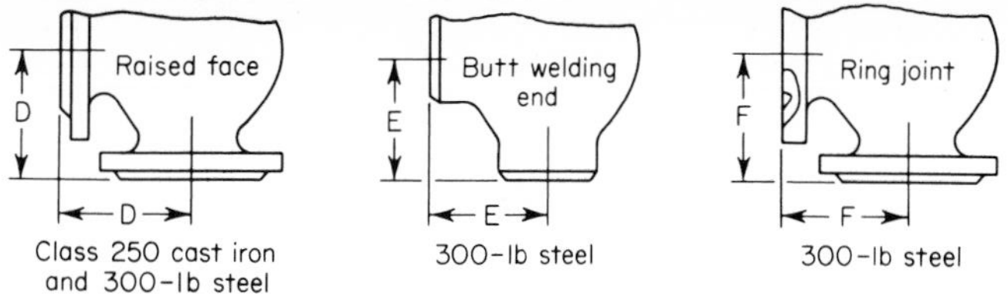

Valves with Class 250 or 300-lb end flanges or with welding ends[1]

1	17	18	19	20	21	22	23	24	25	26	27	28	29	30	31
	300-lb steel														
	Flanged-end (1/16-in. raised face) and welding-end						Flanged-end (ring joint)								
Nominal valve size	Plug		Globe and lift check	Angle and lift check	Swing check	Control[2]	Gate[3]		Plug				Globe and lift check	Angle and lift check	Swing check
	Venturi	Round port full bore					Solid wedge	Double disk	Short pattern	Regular	Venturi	Round port full bore			
	B	*A* and *B*	*A* and *B*	*D* and *E*	*A* and *B*	*A*	*C*	*C*	*C*	*C*	*C*	*C*	*C*	*F*	*C*
½	....	....	6	3	....	7½	....	....	....	....	....	....	6 7/16	3 7/32	....
¾	....	....	7	3½	....	7⅝	....	....	....	....	....	....	7½	3¾	....
1	....	7½	8	4	8½	7¾	....	....	6¾	....	....	8	8½	4¼	9
1¼	....	....	8½	4¼	9	....	....	....	....	....	....	....	9	4½	9½
1½	....	9½	9	4½	9½	9¼	8	8	8	....	....	10	9½	4¾	10
2	....	11⅛	10½	5¼	10½	10½	9⅛	9⅛	9⅛	....	....	11¾	11⅛	5 9/16	11⅛
2½	....	13	11½	5¾	11½	11½	10⅛	10⅛	10⅛	....	....	13⅝	12⅛	6 1/16	12⅛
3	....	15¼	12½	6¼	12½	12½	11¾	11¾	11¾	....	....	15⅞	13⅛	6 9/16	13⅛
3½	....	....	....	....	....	....	....	....	....	....	....	....	....	....	....
4	....	18	14	7	14	14½	12⅝	12⅝	12⅝	....	....	18⅝	14⅝	7 5/16	14⅝
5	....	....	14¾	7⅞	15¾	....	15⅝	15⅝	....	....	....	....	16⅜	8 3/16	16⅜
6	18	22	17½	8¾	17½	18⅝	16½	16½	16½	16½	16½	22⅝	18⅛	9 1/16	18⅛
8	20½	27	22	11	21	22⅜	17⅛	17⅛	17⅛	20⅜	17⅛	27⅝	22⅝	11 5/16	21⅝
10	22	32½	24½	12¼	24½	....	18⅝	18⅝	18⅝	23	18⅝	33⅛	25⅛	12 9/16	25⅛
12	25	38	28	14	28	....	20⅜	20⅜	20⅜	....	20⅜	38⅝	28⅝	14 5/16	28⅝
14	30	....	....	....	....	....	30⅝	30⅝	....	....	30⅝	....	....	....	....
16	33	....	....	....	....	....	33⅝	33⅝	....	....	33⅝	....	....	....	....
18	36	....	....	....	....	....	36⅝	36⅝	....	....	36⅝	....	....	....	....
20	39	....	....	....	....	....	39¾	39¾	....	....	39¾	....	....	....	....
22	43	....	....	....	....	....	43⅞	43⅞	....	....	43⅞	....	....	....	....
24	45	....	....	....	....	....	45⅞	45⅞	....	....	45⅞	....	....	....	....
26	....	....	....	....	....	....	50	50	....	....	....	....	....	....	....
28	....	....	....	....	....	....	54	54	....	....	....	....	....	....	....
30	....	....	....	....	....	....	56	56	....	....	....	....	....	....	....

Table 26. (Continued)

400-lb Steel Valves

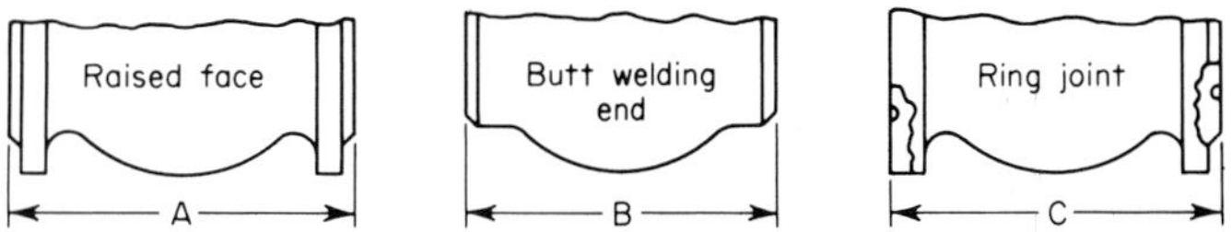
Raised face
A
Butt welding end
B
Ring joint
C

Valves with 400-lb end flanges or with welding ends*

1	2	3	4	5	6	7	8	9	10	11
	400-lb steel									
	Flanged-end (¼-in. raised face) and welding-end									
Nominal valve size	Gate		Plug					Globe and lift check	Angle and lift check	Swing check
	Solid wedge	Double disk	Regular		Venturi	Round port full bore				
	A and *B*	*A* and *B*	*A*	*B*	*A* and *B*	*A*	*B*	*A* and *B*	*D* and *E*	*A* and *B*
½†	6½	....	....	....	....	....	....	6½	3¼	6½
¾†	7½	....	....	....	....	....	....	7½	3¾	7½
1†	8½	8½	8½	....	....	10	....	8½	4¼	8½
1¼†	9	9	9	....	....	....	....	9	4½	9
1½†	9½	9½	9½	9½	....	12½	....	9½	4¾	9½
2†	11½	11½	11½	11½	....	13	....	11½	5¾	11½
2½†	13	13	13	13	....	15	....	13	6½	13
3†	14	14	14	14	....	17½	....	14	7	14
3½†	....	....	....	....	....	....	....	....	....	....
4	16	16	16	16	....	19	22	16	8	16
5	18	18	....	....	....	....	....	18	9	18
6	19½	19½	19½	19½	19½	24	28	19½	9¾	19½
6 × 7	21½	21½	....	....	....	....	....	....	....	....
8	23½	23½	23½	....	23½	29	33¼	23½	11¾	23½
8 × 9	....	....	....	....	....	....	....	....	....	....
10	26½	26½	26½	....	26½	35	....	26½	13¼	26½
10 × 11	....	....	....	....	....	....	....	....	....	....
12	30	30	30	....	30	40	....	30	15	30
14	32½	32½	....	....	32½	....	....	....	....	....
16	35½	35½	....	....	35½	....	....	....	....	....
18	38½	38½	....	....	38½	....	....	....	....	....
20	41½	41½	....	....	41½	....	....	....	....	....
22	45	45	....	....	45	....	....	....	....	....
24	48½	48½	....	....	48½	....	....	....	....	....

* API Standards 6C, 6D, and 600 conform to the dimensions shown for corresponding sizes, valve type, and flange class or welding-end.
† The face-to-face dimensions and connecting end flanges for 400-lb valve are ½″ and smaller are identical with those for 600-lb valves.

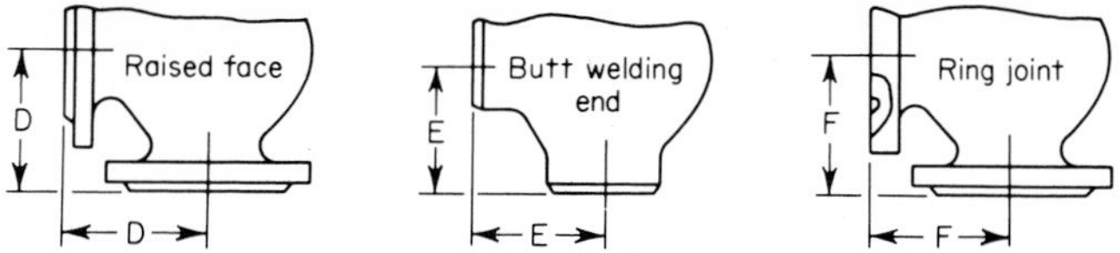

Raised face
D
D
Butt welding end
E
E
Ring joint
F
F

Valves with 400-lb end flanges or with welding ends*

1	12	13	14	15	16	17	18	19	20	21
	400-lb steel									
	Flanged-end (ring joint)									
Nominal valve size	Gate				Plug			Globe and lift check	Angle and lift check	Swing check
	Solid wedge	Double disk	Drilling through full bore: Solid wedge	Drilling through full bore: Double disk	Regular	Venturi	Round port full bore			
	C	C	C	C	C	C	C	C	F	C
1/2†	7 7/16	....	....	....	....	....	....	6 7/16	3 7/32	6 7/16
3/4†	7 1/2	....	....	....	....	....	....	7 1/2	3 3/4	7 1/2
1†	8 1/2	8 1/2	....	....	8 1/2	....	10	8 1/2	4 1/4	8 1/2
1 1/4†	9	9	....	....	9	....	....	9	4 1/2	9
1 1/2†	9 1/2	9 1/2	....	....	9 1/2	....	12 1/2	9 1/2	4 3/4	9 1/2
2†	11 5/8	11 5/8	....	....	11 5/8	....	13 1/8	11 5/8	5 13/16	11 5/8
2 1/2†	13 1/8	13 1/8	....	....	13 1/8	....	15 1/8	13 1/8	6 9/16	13 1/8
3†	14 1/8	14 1/8	....	....	14 1/8	....	17 5/8	14 1/8	7 1/16	14 1/8
3 1/2†	....	....	....	....	....	....	....	....	....	....
4	16 1/8	16 1/8	....	....	16 1/8	....	19 1/8	16 1/8	8 1/16	16 1/8
5	18 1/8	18 1/8	20 1/8	20 1/8	....	....	....	18 1/8	9 1/16	18 1/8
6	19 5/8	19 5/8	21 5/8	21 5/8	19 5/8	19 5/8	24 1/8	19 5/8	9 13/16	19 5/8
6 × 7	21 5/8	21 5/8	23 5/8	23 5/8	....	....	....	....	....	....
8	23 5/8	23 5/8	25 5/8	25 5/8	23 5/8	23 5/8	29 1/8	23 5/8	11 13/16	23 5/8
8 × 9	....	....	....	....	....	....	....	....	....	....
10	26 5/8	26 5/8	28 5/8	28 5/8	26 5/8	26 5/8	35 1/8	26 5/8	13 5/16	26 5/8
10 × 11	....	....	....	....	....	....	....	....	....	....
12	30 1/8	30 1/8	32 1/8	32 1/8	30 1/8	30 1/8	40 1/8	30 1/8	15 1/16	30 1/8
14	32 5/8	32 5/8	....	....	....	32 5/8	....	....	....	....
16	35 5/8	35 5/8	37 5/8	37 5/8	....	35 5/8	....	....	....	....
18	38 5/8	38 5/8	....	....	....	38 5/8	....	....	....	....
20	41 3/4	41 3/4	....	....	....	41 3/4	....	....	....	....
22	45 3/8	45 3/8	....	....	....	45 3/8	....	....	....	....
24	48 3/8	48 7/8	....	....	....	48 7/8	....	....	....	....

Table 26. (*Continued*)

800-lb Hydraulic Cast Iron and 600-lb Steel Valves

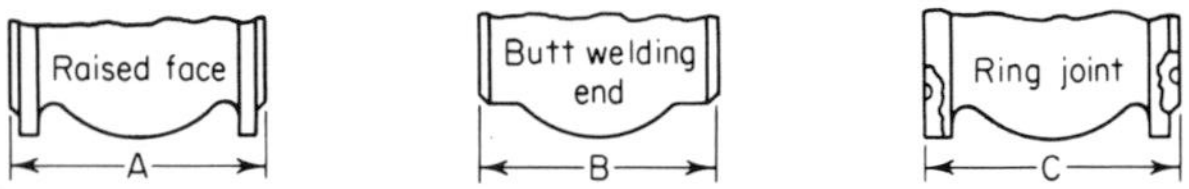
Raised face
A
Butt welding end
B
Ring joint
C

800-lb hydraulic cast iron and 600-lb steel — 600-lb steel — 600-lb steel

Valves with 800-lb hydraulic or 600-lb end flanges or with welding ends*

1	2	3	4	5	6	7	8	9	10	11	12	13	14	15
Nominal valve size	800-lb hyd. cast iron				600-lb steel									
	Flanged-end (¼-in. raised face)				Flanged-end (¼-in. raised face) and welding-end									
	Gate		Plug	Swing check	Gate			Plug					Globe and lift check	
	Solid wedge	Double disk			Solid wedge	Double disk	Short† pattern	Regular		Venturi	Round port full bore		Regular	Short† pattern
	A	*A*	*A*	*A*	*A* and *B*	*A* and *B*	*B*	*A*	*B*	*A* and *B*	*A*	*B*	*A* and *B*	*B*
½	....	....	....	....	6½	....	....	....	....	....	....	....	6½	....
¾	....	....	....	....	7½	....	....	....	....	....	....	....	7½	....
1	....	....	....	....	8½	8½	5¼	8½	....	....	10	....	8½	5¼
1¼	....	....	....	....	9	9	5¾	9	....	....	....	....	9	5¾
1½	....	....	....	....	9½	9½	6	9½	9½	....	12½	....	9½	6
2	11½	11½	11½	11½	11½	11½	7	11½	11½	....	13	....	11½	7
2½	13	13	13	13	13	13	8½	13	13	....	15	....	13	8½
3	14	14	14	14	14	14	10	14	14	....	17½	....	14	10
3½	....	....	....	....	....	....	....	....	....	....	....	....	....	....
4	17	17	17	17	17	17	12	17	17	....	20	22	17	12
5	....	....	....	....	20	20	15	....	....	....	....	....	20	15
6	22	22	22	22	22	22	18	22	22	22	26	28	22	18
6 × 7	....	....	....	....	24	24	....	....	....	....	....	....	....	....
8	26	26	26	26	26	26	23	26	26	26	31¼	33¼	26	23
8 × 9	....	....	....	....	28	28	....	....	....	....	....	....	....	....
10	31	31	31	31	31	31	28	31	31	31	37	40	31	28
10 × 11	....	....	....	....	....	....	....	....	....	....	....	....	....	....
12	33	33	33	33	33	33	32	....	....	33	42	....	33	32
12 × 13	....	....	....	....	....	....	....	....	....	....	....	....	....	....
14	....	....	....	....	35	35	35	....	....	35	....	....	....	....
16	....	....	....	....	39	39	39	....	....	39	....	....	....	....
18	....	....	....	....	43	43	43	....	....	43	....	....	....	....
20	....	....	....	....	47	47	47	....	....	47	....	....	....	....
22	....	....	....	....	51	51	....	....	....	51§	....	....	....	....
24	....	....	....	....	55	55	55	....	....	55	....	....	....	....
26	....	....	....	....	57	57	....	....	....	57	....	....	....	....
28	....	....	....	....	61	61	....	....	....	....	....	....	....	....
30	....	....	....	....	65	65	....	....	....	65	....	....	....	....

* API Standards 6C, 6D, and 600 conform to the dimensions shown for corresponding sizes, valve type, and flange class or welding-end.
† These dimensions apply to pressure-seal or flangeless bonnet valves only.
‡ Instrument Society of America Tentative Standard ISA RP4.1 face-to-face conforms to dimensions shown in column 19.
§ [illegible]

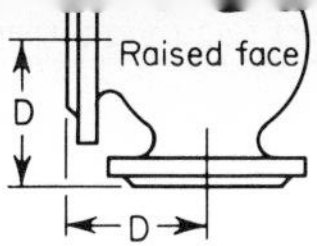

800-lb hydraulic cast iron and 600-lb steel

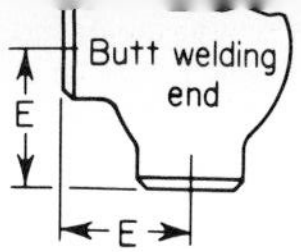

600-lb steel

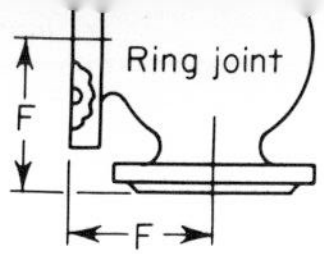

600-lb steel

Valves with 800-lb hydraulic or 600-lb end flanges or with welding ends*

1	16	17	18	19	20	21	22	23	24	25	26	27	28	29	30
	600-lb steel														
	Flanged-end (1/4-in. raised face) and welding-end				Flanged-end (ring joint)										
Nominal valve size	Angle and lift check		Swing check	Control‡	Gate				Plug				Globe and lift check	Angle and lift check	Swing check
	Regular	Short† pattern			Solid wedge	Double disk	Drilling through full bore: Solid wedge	Drilling through full bore: Double disk	Regular	Venturi	Round port full bore	Drilling through full bore			
	D and *E*	*E*	*A* and *B*	*A*	*C*	*C*	*C*	*C*	*C*	*C*	*C*	*C*	*C*	*F*	*C*
1/2	3 1/4	....	6 1/2	8	6 7/16	....	....	....	....	....	....	....	6 7/16	3 7/32	6 7/16
3/4	3 3/4	....	7 1/2	8 1/8	7 1/2	....	....	....	....	....	....	....	7 1/2	3 3/4	7 1/2
1	4 1/4	....	8 1/2	8 1/4	8 1/2	8 1/2	....	....	8 1/2	....	10	....	8 1/2	4 1/4	8 1/2
1 1/4	4 1/2	....	9	....	9	9	....	....	9	....	....	....	9	4 1/2	9
1 1/2	4 3/4	....	9 1/2	9 7/8	9 1/2	9 1/2	....	....	9 1/2	....	12 1/2	....	9 1/2	4 3/4	9 1/2
2	5 3/4	4 1/4	11 1/2	11 1/4	11 5/8	11 5/8	....	....	11 5/8	....	13 1/8	....	11 5/8	5 13/16	11 5/8
2 1/2	6 1/2	5	13	12 1/4	13 1/8	13 1/8	....	....	13 1/8	....	15 1/8	....	13 1/8	6 9/16	13 1/8
3	7	6	14	13 1/4	14 1/8	14 1/8	....	....	14 1/8	....	17 5/8	....	14 1/8	7 1/16	14 1/8
3 1/2	....	....	....	....	....	....	....	....	....	....	....	....	....	....	....
4	8 1/2	7	17	15 1/2	17 1/8	17 1/8	....	....	17 1/8	....	20 1/8	....	17 1/8	8 9/16	17 1/8
5	10	8 1/2	20	....	20 1/8	20 1/8	22 1/8	22 1/8	....	....	....	....	20 1/8	10 1/16	20 1/8
6	11	10	22	20	22 1/8	22 1/8	24 1/8	24 1/8	22 1/8	22 1/8	26 1/8	28 5/8	22 1/8	11 1/16	22 1/8
6 × 7	....	....	....	....	24 1/8	24 1/8	26 1/8	26 1/8	....	....	....	29 1/8	....	....	....
8	13	....	26	24	26 1/8	26 1/8	28 1/8	28 1/8	26 1/8	26 1/8	31 3/8	34 5/8	26 1/8	13 1/16	26 1/8
8 × 9	....	....	....	....	28 1/8	28 1/8	30 1/8	30 1/8	....	....	....	36 1/8	....	....	....
10	15 1/2	....	31	....	31 1/8	31 1/8	33 1/8	33 1/8	31 1/8	31 1/8	37 1/8	40 5/8	31 1/8	15 9/16	31 1/8
10 × 11	....	....	....	....	....	....	35 1/8	35 1/8	....	....	....	42 5/8	....	....	....
12	16 1/2	....	33	....	33 1/8	33 1/8	35 1/8	35 1/8	....	33 1/8	42 1/8	....	33 1/8	16 9/16	33 1/8
12 × 13	....	....	....	....	....	....	37 1/8	37 1/8	....	....	....	....	....	....	....
14	....	....	....	....	35 1/8	35 1/8	....	....	....	35 1/8	....	....	....	....	....
16	....	....	....	....	39 1/8	39 1/8	....	....	....	39 1/8	....	....	....	....	....
18	....	....	....	....	43 1/8	43 1/8	....	....	....	43 1/8	....	....	....	....	....
20	....	....	....	....	47 1/4	47 1/4	....	....	....	47 1/4	....	....	....	....	....
22	....	....	....	....	51 3/8	51 3/8	....	....	....	51 3/8	....	....	....	....	....
24	....	....	....	....	55 3/8	55 3/8	....	....	....	55 3/8	....	....	....	....	....
26	....	....	....	....	57 1/2	57 1/2	....	....	....	57 1/2	....	....	....	....	....
28	....	....	....	....	61 1/2	61 1/2	....	....	....	....	....	....	....	....	....
30	....	....	....	....	65 1/2	65 1/2	....	....	....	65 1/2	....	....	....	....	....

Table 26. (Continued)

900-lb Steel Valves

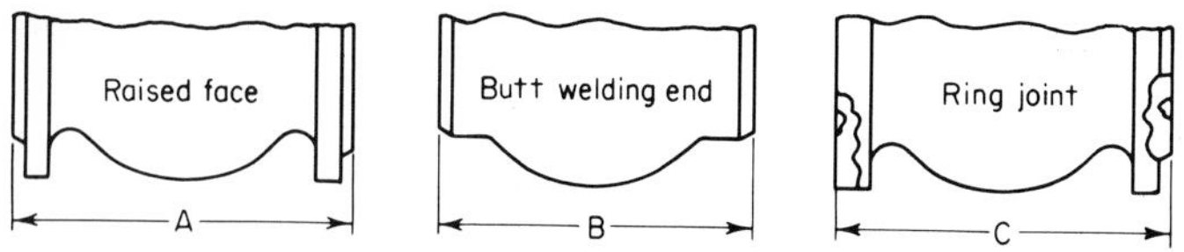
Raised face
A
Butt welding end
B
Ring joint
C

Valves with 900-lb end flanges or with welding ends*

900-lb steel — Flanged-end (¼-in. raised face) and welding-end

1	2	3	4	5	6	7	8	9	10	11	12
Nominal valve size	Gate			Plug				Globe and lift check		Angle and lift check	
	Solid wedge	Double disk	Short† pattern	Regular		Venturi	Round port full bore	Regular	Short† pattern	Regular	Short† pattern
	A and B	A and B	B	A	B	A and B	A	A and B	B	D and E	E
¾‡	....	....	....	....	....	....	....	9	....	4½	....
1‡	10	....	5½	10	....	....	....	10	....	5	....
1¼‡	11	....	6½	11	....	....	....	11	....	5½	....
1½‡	12	....	7	12	....	....	14	12	....	6	....
2‡	14½	14½	8½	14½	....	....	15	14½	....	7¼	....
2½‡	16½	16½	10	16½	....	....	17	16½	....	8¼	....
3	15	15	12	15	15	....	18½	15	12	7½	6
3½	....	....	....	....	....	....	....	....	....	....	....
4	18	18	14	18	18	....	22	18	14	9	7
5	22	22	17	....	....	....	....	22	17	11	8½
6	24	24	20	24	24	24	29	24	20	12	10
6 × 7	26	26	....	....	....	....	....	....	....	....	....
8	29	29	26	29	29	29	32	29	26	14½	13
8 × 9	31	31	....	....	....	....	....	....	....	....	....
10	33	33	31	33	33	33	38	33	31	16½	15½
10 × 11	....	....	....	....	....	....	....	....	....	....	....
12	38	38	36	....	....	38	44	38	36	19	18
12 × 13	....	....	....	....	....	....	....	....	....	....	....
14	40½	40½	39	....	....	....	....	40½	39	20¼	19½
16	44½	44½	43	....	....	44½	....	....	....	....	....
18	48	48	....	....	....	....	....	....	....	....	....
20	52	52	....	....	....	....	....	....	....	....	....
22	....	....	....	....	....	....	....	....	....	....	....
24	61	61	....	....	....	....	....	....	....	....	....

* API Standards 6C, 6D, and 600 conform to the dimensions shown for corresponding sizes, valve type, and flange class or welding-end, except the 3- and 4-in. full-bore flow-line valves of 6C which are 17⅛ and 20⅛ in., respectively.

† These dimensions apply to pressure-seal or flangeless bonnet valves only.

‡ The connecting end flanges for 900-lb valves 2½ in. and smaller are identical with those of 1,500-lb valves. The face-to-face dimensions for all 900-lb valves 2½ and smaller, except round-port full-bore plug valves (columns 8 and 20) are identical with those of 1,500-lb valves.

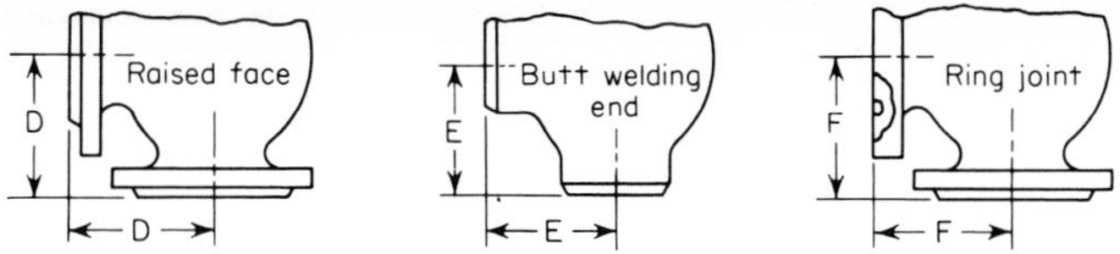
Raised face
D
D
Butt welding end
E
E
Ring joint
F
F

Valves with 900-lb end flanges or with welding ends*

1	13	14	15	16	17	18	19	20	21	22	23	24
	900-lb steel											
	Flanged-end (ring joint)											
		Gate				Plug						
				Drilling through full bore								
Nominal valve size	Swing check	Solid wedge	Double disk	Solid wedge	Double disk	Regular	Venturi	Round port full bore	Drilling through full bore	Globe and lift check	Angle and lift check	Swing check
	A and *B*	*C*	*C*	*C*	*C*	*C*	*C*	*C*	*C*	*C*	*F*	*C*
$\frac{3}{4}$‡	9	....	....	....	....	....	....	....	....	9	$4\frac{1}{2}$	9
1‡	10	10	....	....	....	10	....	....	....	10	5	10
$1\frac{1}{4}$‡	11	11	....	....	....	11	....	....	....	11	$5\frac{1}{2}$	11
$1\frac{1}{2}$‡	12	12	....	....	....	12	....	14	....	12	6	12
2‡	$14\frac{1}{2}$	$14\frac{5}{8}$	$14\frac{5}{8}$	....	....	$14\frac{5}{8}$	....	$15\frac{1}{8}$	....	$14\frac{5}{8}$	$7\frac{5}{16}$	$14\frac{5}{8}$
$2\frac{1}{2}$‡	$16\frac{1}{2}$	$16\frac{5}{8}$	$16\frac{5}{8}$	....	....	$16\frac{5}{8}$	....	$17\frac{1}{8}$	....	$16\frac{5}{8}$	$8\frac{5}{16}$	$16\frac{5}{8}$
3	15	$15\frac{1}{8}$*	$15\frac{1}{8}$*	....	....	$15\frac{1}{8}$	....	$18\frac{5}{8}$	....	$15\frac{1}{8}$	$7\frac{9}{16}$	$15\frac{1}{8}$
$3\frac{1}{2}$	....	....	....	....	....	....	....	....	....	....	....	....
4	18	$18\frac{1}{8}$*	$18\frac{1}{8}$*	....	....	$18\frac{1}{8}$	....	$22\frac{1}{8}$	....	$18\frac{1}{8}$	$9\frac{1}{16}$	$18\frac{1}{8}$
5	22	$22\frac{1}{8}$	$22\frac{1}{8}$	$24\frac{1}{8}$	$24\frac{1}{8}$	....	....	....	....	$22\frac{1}{8}$	$11\frac{1}{16}$	$22\frac{1}{8}$
6	24	$24\frac{1}{8}$	$24\frac{1}{8}$	$26\frac{1}{8}$	$26\frac{1}{8}$	$24\frac{1}{8}$	$24\frac{1}{8}$	$29\frac{1}{8}$	$30\frac{1}{8}$	$24\frac{1}{8}$	$12\frac{1}{16}$	$24\frac{1}{8}$
6 × 7	....	$26\frac{1}{8}$	$26\frac{1}{8}$	$28\frac{1}{8}$	$28\frac{1}{8}$	....	....	....	$31\frac{5}{8}$	....	....	....
8	29	$29\frac{1}{8}$	$29\frac{1}{8}$	$31\frac{1}{8}$	$31\frac{1}{8}$	$29\frac{1}{8}$	$29\frac{1}{8}$	$32\frac{1}{8}$	$36\frac{5}{8}$	$29\frac{1}{8}$	$14\frac{9}{16}$	$29\frac{1}{8}$
8 × 9	....	$31\frac{1}{8}$	$31\frac{1}{8}$	$33\frac{1}{8}$	$33\frac{1}{8}$	....	....	....	$38\frac{1}{8}$	....	....	....
10	33	$33\frac{1}{8}$	$33\frac{1}{8}$	$35\frac{1}{8}$	$35\frac{1}{8}$	$33\frac{1}{8}$	$33\frac{1}{8}$	$38\frac{1}{8}$	$42\frac{5}{8}$	$33\frac{1}{8}$	$16\frac{9}{16}$	$33\frac{1}{8}$
10 × 11	....	....	....	$37\frac{1}{8}$	$37\frac{1}{8}$	....	....	....	$44\frac{1}{8}$	....	....	....
12	38	$38\frac{1}{8}$	$38\frac{1}{8}$	$40\frac{1}{8}$	$40\frac{1}{8}$	....	$38\frac{1}{8}$	$44\frac{1}{8}$	....	$38\frac{1}{8}$	$19\frac{1}{16}$	$38\frac{1}{8}$
12 × 13	....	....	....	$42\frac{1}{8}$	$42\frac{1}{8}$	....	....	....	....	....	....	....
14	$40\frac{1}{2}$	$40\frac{7}{8}$	$40\frac{7}{8}$	....	....	....	....	....	....	$40\frac{7}{8}$	$20\frac{7}{16}$	$40\frac{7}{8}$
16	....	$44\frac{7}{8}$	$44\frac{7}{8}$	....	....	....	$44\frac{7}{8}$	....	....	....	....	....
18	....	$48\frac{1}{2}$	$48\frac{1}{2}$	....	....	....	....	....	....	....	....	....
20	....	$52\frac{1}{2}$	$52\frac{1}{2}$	....	....	....	....	....	....	....	....	....
22	....	....	....	....	....	....	....	....	....	....	....	....
24	....	$61\frac{3}{4}$	$61\frac{3}{4}$	....	....	....	....	....	....	....	....	....

Table 26. (*Continued*)

1500-lb Steel Valves

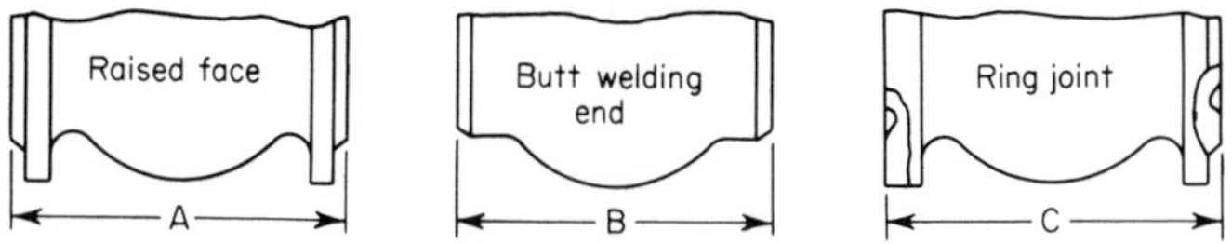
Raised face
A
Butt welding end
B
Ring joint
C

Valves with 1,500-lb end flanges or with welding ends[1]

1	2	3	4	5	6	7	8	9	10	11
	1,500-lb steel									
	Flanged-end (¼-in. raised face) and welding-end									
Nominal valve size	Gate			Plug				Globe and lift check	Angle and lift check	Swing check
				Regular		Venturi	Round port full bore			
	Solid wedge	Double disk	Short[2] pattern							
	A and *B*	*A* and *B*	*B*	*A*	*B*	*A* and *B*	*A*	*A* and *B*	*D* and *E*	*A* and *B*
¾	....	....	....	....	....	....	....	9	4½	9
1	10	....	5½	10	....	....	....	10	5	10
1¼	11	....	6½	11	....	....	....	11	5½	11
1½	12	....	7	12	....	....	....	12	6	12
2	14½	14½	8½	14½	....	....	15⅜	14½	7¼	14½
2½	16½	16½	10	16½	....	....	17⅞	16½	8¼	16½
3	18½	18½	12	18½	18½	....	20⅝	18½	9¼	18½
3½	....	....	....	....	....	....	....	....	....	....
4	21½	21½	16	21½	21½	....	24⅝	21½	10¾	21½
5	26½	26½	19	....	....	....	....	26½	13¼	26½
6	27¾	27¾	22	27¾	27¾	27¾	31	27¾	13⅞	27¾
6 × 7	....	....	....	....	....	....	....	....	....	....
8	32¾	32¾	28	32¾	32¾	32¾	35	32¾	16⅜	32¾
8 × 9	....	....	....	....	....	....	....	....	....	....
10	39	39	34	39	39	39	42	39	19½	39
10 × 11	....	....	....	....	....	....	....	....	....	....
12	44½	44½	39	44½	44½	44½	48	44½	22¼	44½
14	49½	49½	42	....	....	....	....	49½	24¾	49½
14 × 13	....	....	....	....	....	....	....	....	....	....
16	54½	54½	47	....	....	....	....	....	....	....
18	60½	60½	53	....	....	....	....	....	....	....
20	65½	65½	58	....	....	....	....	....	....	....
22	....	....	....	....	....	....	....	....	....	....
24	76½	76½	....	....	....	....	....	....	....	....

[1] API Standards 6C, 6D, and 600 conform to the dimensions shown for corresponding sizes, valve type, and flange class or welding-end.
[2] These dimensions apply to pressure-seal or flangeless bonnet valves only.

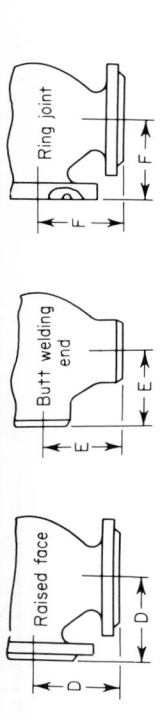

Raised face

Butt welding end

Ring joint

1,500-lb steel

Flanged-end (ring joint)

1	12	13	14	15	16	17	18	19	20	21	22
	Gate				**Plug**						
	Solid wedge	Double disk	Drilling through full bore		Regular	Venturi	Round port full bore	Drilling through full bore	Globe and lift check	Angle and lift check	Swing check
			Solid wedge	Double disk							
Nominal valve size	C	C	C	C	C	C	C	C	C	F	C
3/4									9	4 1/2	9
1	10				10				10	5	10
1 1/4	11				11				11	5 1/2	11
1 1/2	12				12				12	6	12
2	14 5/8	14 5/8			14 5/8		15 1/2		14 5/8	7 5/16	14 5/8
2 1/2	16 5/8	16 5/8			16 5/8		18		16 5/8	8 5/8	16 5/8
3	18 5/8	18 5/8			18 5/8		20 3/4		18 5/8	9 5/16	18 5/8
3 1/2											
4	21 5/8	21 5/8			21 5/8		24 3/4		21 5/8	10 13/16	21 5/8
5	26 5/8	26 5/8	28 5/8	28 5/8					26 5/8	13 5/16	26 5/8
6	28	28	30	30	28	28	31 1/4	36 1/8	28	14	28
6 × 7	33 1/8	33 1/8	32	32	33 1/8	33 1/8	35 5/8	38 1/8	33 1/8	16 9/16	33 1/8
8			35 1/8	35 1/8				43 1/8			
8 × 9	39 3/8	39 3/8	37 1/8	37 1/8	39 3/8	39 3/8	42 5/8	45 1/8	39 3/8	19 11/16	39 3/8
10											
10 × 11	45 1/8	45 1/8			45 1/8	45 1/8	48 5/8		45 1/8	22 9/16	45 1/8
12	50 1/4	50 1/4							50 1/4	25 5/8	50 1/4
14 × 13	55 3/8	55 3/8									
16											
18	61 3/8	61 3/8									
20	66 3/8	66 3/8									
22											
24	77 5/8	77 5/8									

Table 26. (Continued)

2500-lb Steel Valves

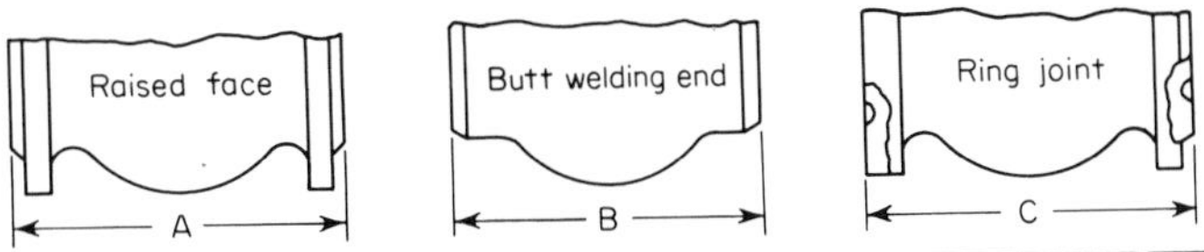

Valves with 2,500-lb end flanges or with welding ends[1]

2,500-lb steel — Flanged-end (1/4-in. raised face) and welding-end

1	2	3	4	5	6	7	8
Nominal valve size	Gate			Plug	Globe and lift check	Angle and lift check	Swing check
	Solid wedge *A* and *B*	Double disk *A* and *B*	Short[2] pattern *B*	Regular *A*	*A* and *B*	*D* and *E*	*A* and *B*
1/2	10 3/8	....	....	....	10 3/8	5 3/16	10 3/8
3/4	10 3/4	....	....	....	10 3/4	5 3/8	10 3/4
1	12 1/8	....	7 5/16	....	12 1/8	6 1/16	12 1/8
1 1/4	13 3/4	....	9 1/8	....	13 3/4	6 7/8	13 3/4
1 1/2	15 1/8	....	9 1/8	....	15 1/8	7 9/16	15 1/8
2	17 3/4	17 3/4	11	17 3/4	17 3/4	8 7/8	17 3/4
2 1/2	20	20	13	20	20	10	20
3	22 3/4	22 3/4	14 1/2	22 3/4	22 3/4	11 3/8	22 3/4
3 1/2	....	....	....	....	....	....	....
4	26 1/2	26 1/2	18	26 1/2	26 1/2	13 1/4	26 1/2
5	31 1/4	31 1/4	21	31 1/4	31 1/4	15 5/8	31 1/4
6	36	36	24	36	36	18	36
8	40 1/4	40 1/4	30	40 1/4	40 1/4	20 1/8	40 1/4
10	50	50	36	50	50	25	50
12	56	56	41	56	56	28	56
14	....	....	44	....	....	....	....
16	....	....	49	....	....	....	....
18	....	....	55	....	....	....	....

1 API Standard 600 conforms to the dimensions shown for corresponding sizes, valve type, and flange class or welding end.
2 These dimensions apply to pressure-seal or flangeless bonnet valves only.

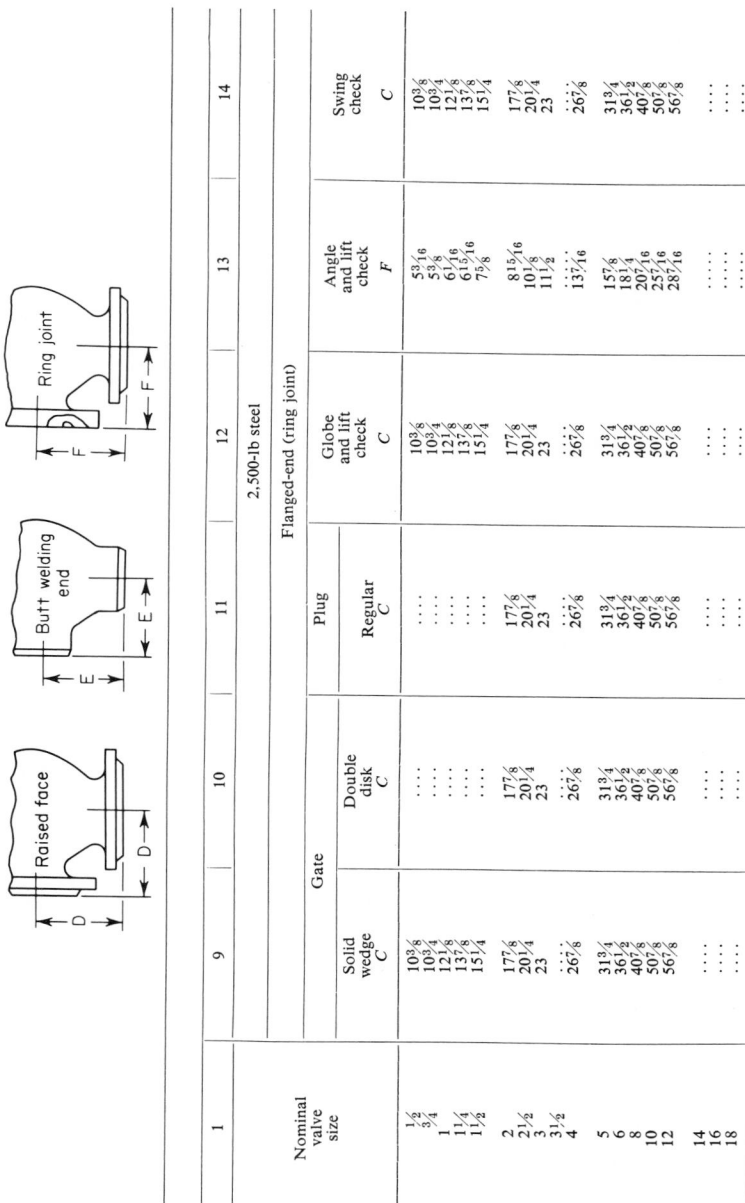

2,500-lb steel

Flanged-end (ring joint)

1	9	10	11	12	13	14
	Gate		Plug	Globe and lift check	Angle and lift check	Swing check
Nominal valve size	Solid wedge C	Double disk C	Regular C	C	F	C
$\frac{1}{2}$	$10\frac{3}{8}$	$\ldots$	$\ldots$	$10\frac{3}{8}$	$5\frac{3}{16}$	$10\frac{3}{8}$
$\frac{3}{4}$	$10\frac{3}{4}$	$\ldots$	$\ldots$	$10\frac{3}{4}$	$5\frac{3}{8}$	$10\frac{3}{4}$
1	$12\frac{1}{8}$	$\ldots$	$\ldots$	$12\frac{1}{8}$	$6\frac{1}{8}$	$12\frac{1}{8}$
$1\frac{1}{4}$	$13\frac{7}{8}$	$\ldots$	$\ldots$	$13\frac{7}{8}$	$6\frac{15}{16}$	$13\frac{7}{8}$
$1\frac{1}{2}$	$15\frac{1}{4}$	$\ldots$	$\ldots$	$15\frac{1}{4}$	$7\frac{5}{8}$	$15\frac{1}{4}$
2	$17\frac{7}{8}$	$17\frac{7}{8}$	$17\frac{7}{8}$	$17\frac{7}{8}$	$8\frac{15}{16}$	$17\frac{7}{8}$
$2\frac{1}{2}$	$20\frac{1}{4}$	$20\frac{1}{4}$	$20\frac{1}{4}$	$20\frac{1}{4}$	$10\frac{1}{8}$	$20\frac{1}{4}$
3	23	23	23	23	$11\frac{1}{2}$	23
$3\frac{1}{2}$						
4	$26\frac{7}{8}$	$26\frac{7}{8}$	$26\frac{7}{8}$	$26\frac{7}{8}$	$13\frac{7}{16}$	$26\frac{7}{8}$
5	$31\frac{1}{4}$	$31\frac{1}{4}$	$31\frac{1}{4}$	$31\frac{1}{4}$	$15\frac{7}{8}$	$31\frac{1}{4}$
6	$36\frac{1}{2}$	$36\frac{1}{2}$	$36\frac{1}{2}$	$36\frac{1}{2}$	$18\frac{1}{4}$	$36\frac{1}{2}$
8	$40\frac{7}{8}$	$40\frac{7}{8}$	$40\frac{7}{8}$	$40\frac{7}{8}$	$20\frac{7}{16}$	$40\frac{7}{8}$
10	$50\frac{7}{8}$	$50\frac{7}{8}$	$50\frac{7}{8}$	$50\frac{7}{8}$	$25\frac{7}{16}$	$50\frac{7}{8}$
12	$56\frac{7}{8}$	$56\frac{7}{8}$	$56\frac{7}{8}$	$56\frac{7}{8}$	$28\frac{7}{16}$	$56\frac{7}{8}$
14	$\ldots$	$\ldots$	$\ldots$	$\ldots$	$\ldots$	$\ldots$
16	$\ldots$	$\ldots$	$\ldots$	$\ldots$	$\ldots$	$\ldots$
18						

Raised face — Butt welding end — Ring joint

Design Features.   The basic design of plug valves (Fig. 17) is similar to the old-fashioned wooden spigot.   Full flow is obtained when the opening in the tapered plug faces in the direction of flow.   When the plug is rotated a quarter of a turn, flow is stopped.

The body and tapered plug represent the essential features in plug valves.   Careful design of the internal contours of the valve produces maximum flow efficiency.

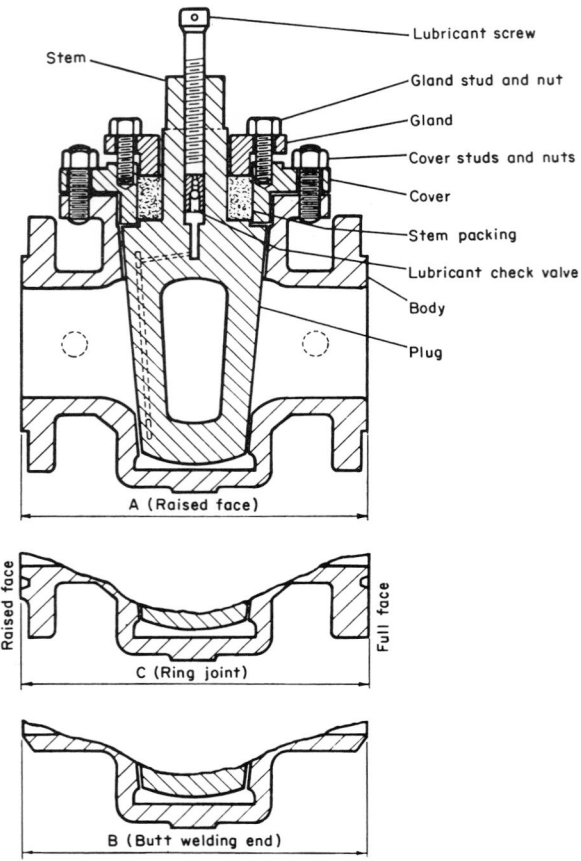

FIG. 17. Plug valve with lubricant system, as specified in API Standard 600.

The hole, or port, in the tapered plug is generally rectangular.   However, valves are also available with round port design.

Major valve patterns or types are identified as *regular, venturi, short, roundport,* and *multiport.*

The regular pattern employs the tapered form of port opening, the area of which is from 70 to 100 per cent of the internal pipe area.   In some cases, the face-to-face lengths are greater than those of standard gate valves.

The venturi pattern, available in flanged sizes 6 in. and larger, provides stream-lined flow and thus permits reduction in the port size.   The port opening area is approximately 35 per cent of the internal pipe area.

The short pattern has face-to-face lengths that conform with 150- and 300-lb steel gate valves.

The roundport full-bore pattern has a circular port through the plug and body equal to or greater than the inside diameter of the pipe or fitting. Operating efficiency is equal to or greater than that of gate valves of the same size.

Use of multiport valves (Fig. 18) is advantageous in many installations because it provides simplification of piping and convenience in operation. One 3-way or 4-way multiport valve may be used in place of two, three, or four straightway valves.

Major types of plug valves involve *lubricated* and *nonlubricated* designs.

Lubricant-seal plug valves are less subject to seizing or wear and may exhibit somewhat greater resistance to corrosion in some service environments.

Nonlubricated plug valves are used where maintenance must be kept to a minimum. Both types of valves provide a bubbletight closure and are of compact size.

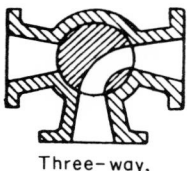 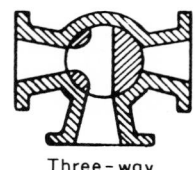 

Three-way,   Three-way,   Four-way,
Two-port   Three-port   Four-port

Fig. 18. Multiport valves.

These valves generally can be readily repaired or cleaned without necessitating removal of the body from the piping system. They are available for pressure service from vacuum to 10,000 psi and temperatures from −50 to 1500 F.

Standard Pressure Ratings. Nodular iron-plug valves in the 150- and 300-lb classifications are covered by API Standard 604, Flanged Nodular Iron Gate and Plug Valves for Refinery Use. High-strength gray iron (also called semisteel) valves are available for 175-, 200-, 400-, 500-, and 800-psi service. Steel valves are made for 150-, 300-, 600-, 900-, 1,500- and 2,500-lb service. Steel plug valves are covered by the following API Standards:

Standard 600   Flanged- and Butt-welding-end Steel Gate and Plug Valves for Refinery Use

Standard 6D   Specification for Steel Gate, Plug, Ball, and Check Valves for Pipeline Service

Lubricant-seal Valves. In lubricant-seal valves, channels for the admission of the lubricant surround the ports to insure positive sealing against internal or external leakage. The lubricant pressure developed by a turn of the lubricant screw or injection of lubricant with a pressure gun exerts a powerful hydraulic jacking action on the plug, momentarily lifting it from the seat and making it easy to turn. Since the lubricant pressure is greater than the line pressure, it is virtually impossible for solids to lodge between the valve body and plug.

The functions of pressure lubrication in plug valves are (1) hydraulic action, keeping the plug in free working condition, (2) maintenance of positive seal against internal and external leakage, (3) free turning even of the largest sizes and against heavy differential pressure, and (4) protection of working surfaces from wear and corrosion. This principle of pressure lubrication makes it possible to take full advantage of the inherent simplicity, compactness, and positive rotary action of the tapered plug valve.

The stem or shank used to rotate the plug is sealed by screwed or bolted packing glands.

Lubricants. The word "lubricant" does not define precisely the part this material plays in the efficient functioning of lubricated plug valves. More properly, such valves might be called "plastic sealed valves" and the lubricant could better be designated "plastic sealant." The use of an effective lubricant is most important since, in operation, the valve structure and plastic sealing film are an integral unit, and each component is dependent on the other for ultimate performance.

The lubricant in effect becomes a structural part of the valve, as it provides a flexible and renewable seat. This eliminates the necessity of "force fits" and metal-to-metal "distortable-seat" contacts to effect a seal. For this purpose, the lubricant must exhibit proper plasticity as well as resistance to solvents and chemicals to avoid the destructive action of the line fluid and to form an impervious seal around each body port even under pressure. The film of lubricant also protects the metal surfaces between the plug and body from corrosion. The seal formed by the lubricant transmitted in a system of lubricant grooves circuiting each part aids in maintaining the essential film on the metal closure surfaces.

This lubricant film also eases valve operation. By preventing metal-to-metal contact on internal valve parts, galling and seizing are minimized. The lubricant must provide a high degree of lubrication to the bearing surfaces of the valve over a wide temperature range.

Table 27. Dimensions of 150-lb Flanged-end Plug Valves (API 600-1961)

(Letter dimensions refer to Fig. 17.)

| Nominal size | Body thickness, min[1] | Face-to-face dimension A, end-to-end dimension C | | | | | | |
| | | Short pattern | | Regular | | Venturi | | Round-port full-bore | |
		A[2] raised-face[3]	C[2] ring joint[3]	A[2] raised-face[3]	C[2] ring joint[3]	A[2] raised-face[3]	C[2] ring joint[3]	A[2] raised-face[3]	C[2] ring joint[3]
1	....	5½	6	....	....	..	....	7	7½
1¼									
1½	....	6½	7	....	....	..	....	8¾	9¼
2	0.344	7	7½	....	....	..	....	10½	11
2½	0.375	7½	8	....	....	..	....	11¾	12¼
3	0.406	8	8½	....	....	..	....	13½	14
4	0.438	9	9½	....	....	..	....	17	17½
6	0.469	10½	11	15½	16	..	....	21	21½
8	0.500	11½	12	18	18½	..	....	25	25½
10	0.563	13	13½	21	21½	21	21½	31	31½
12	0.625	14	14½	24	24½	24	24½	36	36½
14	0.656	....	....	....	....	27	27½		
16	0.687	....	....	....	....	30	30½		
18	0.718	....	....	....	....	34	34½		
20	0.750	....	....	....	....	36	36½		
24	0.812	....	....	....	....	42	42½		

All dimensions in inches.

[1] Minimum wall thickness in throat between end flanges and body.

[2] Permissible tolerance $\pm \frac{1}{16}$ in. on sizes 10 in. or smaller, and $\pm \frac{1}{8}$ in. on sizes 12 in. or larger. Dimensions A and C shown are the same as given in Table 26.

[3] Flange and facing proportions and drilling templates shall conform to ASA B16.5.

Among the newer lubricants used in some applications are molybdenum disulfide and fluorinated plastics.

Dimensions.   Dimensions of 150-lb plug valves are given in Table 27.

Diaphragm Valves

Application.   Diaphragm valves offer advantages not possible with other types of valves.   They provide smooth, streamlined fluid passages without pockets, as well as flow control and leaktight closure, even with suspended solids in the pipeline.   In certain positions, these valves are selfdraining.

Absolute isolation of the working parts from the fluid stream prevents product contamination and corrosion of the operating mechanism.   Maintenance is extremely simple.   Flexibility of assembly and a wide choice of materials for bodies,

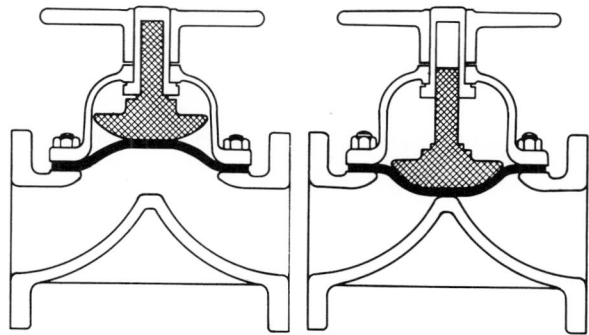

Fig. 19.   Sketch of weir-type diaphragm valve in open and closed positions.

body linings, and diaphragms make these valves extraordinarily adaptable to diverse applications.

Diaphragm valves are commonly used in a wide variety of industries to overcome problems of corrosion, abrasion, contamination, clogging, leakage, and valve maintenance.   The valves are particularly suitable for lines handling corrosive fluids, viscous materials, fibrous slurries, sludges, solids in suspension, beverages, semifluid foods, water, gases, and compressed air.

Design Features.   In diaphragm valves (Fig. 19) the resilient, flexible diaphragm is connected to the "compressor" plug by a stud molded into the diaphragm.   The compressor plug is moved up and down by the valve stem.   Thus, the diaphragm is lifted high when the compressor is raised and is pressed tight against the body weir when the compressor is lowered.   The benefits inherent in this type of operation are important.

The operating mechanism is not subjected to the corrosive action of chemicals or other pipeline material since the diaphragm completely isolates the working parts from the fluid stream.   There is no possibility of stem leakage.   Conversely, lubricants cannot contaminate the fluid in the line.

Only the diaphragm is normally subject to wear.   Depending on the type of service, it may last for very long periods of time.   However, it can be replaced without removing the valve body from the pipeline.   Packing glands, disk holders, and seats involving metal-to-metal contact are eliminated in these types of valves.

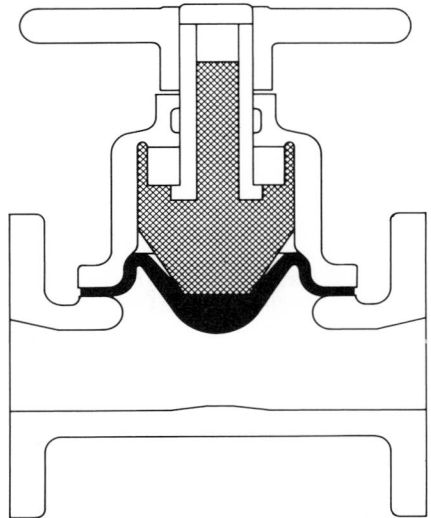

FIG. 20. Straightway-type diaphragm valve.

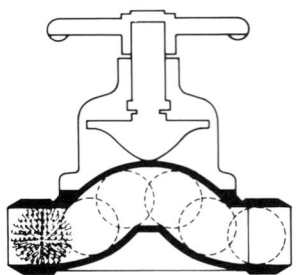

FIG. 21. Full-bore-type dia-phragm valve illustrating pas-sage of ball-brush cleaner through valve.

Table 28. Typical Materials Used for Diaphragms

Valve type	Service	Material	Temp, F Min	Temp, F Max
Conventional weir	Abrasive	Soft natural rubber	−30	180
	Water	Natural rubber	−30	180
	Food and beverage	White natural rubber	0	160
	Weak chemical, air, oil	Neoprene	−30	200
	Weak chemical, high vacuum	Reinforced Neoprene	−30	200
	Other chemicals, gases	Black chlorinated butyl	−20	250
	Food and beverage	White chlorinated butyl	−10	225
	Special for hydrogen peroxide	Clear Tygon	0	150
	Oils and gasoline	Hycar (gen. purpose)	10	180
	Oxidizing services	Hypalon	0	225
	Brewery services	Pure gum rubber	−30	160
	Special service on temperature	Silicone	50	350
	Radioactive conditions	G.R.S.	−10	225
	Severe chemicals, solvents	Teflon[1]	−30	325
	Severe chemicals	Kel-F[1]	60	250
	Specific acids	Polyethylene	10	135
Full flow	Cold beer	White rubber	−30	160
	Hot wort and cold beer	White chlorinated butyl	−10	225
	Cold beer	Pure gum rubber	−30	160
Straightway	Water	Natural rubber	−30	180
	Chemical, air, oil	Neoprene	0	180
	Oils and gasoline	Hycar (gen. purpose)	10	180
	Fatty acids	Black chlorinated butyl	0	225
	Oxidizing services	Hypalon	0	200
	Food and beverage	White chlorinated butyl	−10	200

[1] With backing cushion.

Commercial valve types include the *conventional weir* type (Fig. 19), the *straight-way* type (Fig. 20), and the *full-bore* type (Fig. 21).

When the *straightway* valve is open, its diaphragm lifts high for full streamlined flow in either direction. When the valve is closed, the diaphragm seals tight for positive closure even with gritty or fibrous materials in the line.

The *full-bore* type of valve is most extensively used in the beverage industry. It permits ball-brush cleaning with either steam or caustic soda, without opening or removal of the valve from the line.

For severe corrosive applications, diaphragm valves are made of stainless steel or PVC plastics, or they are lined with glass, rubber, lead, plastics, titanium, or still other materials. Some of the common materials used for diaphragms are listed in Table 28. Diaphragm life, however, depends not only upon the nature of the material handled but also upon the temperature, pressure, and frequency of operation. End-to-end dimensions of diaphragm valves are listed in Table 29.

Ball Valves

In the past, the use of ball valves (Fig. 22) has been rather limited as bubbletight service was not possible because of problems in the sealing ability of metal-to-metal seats.

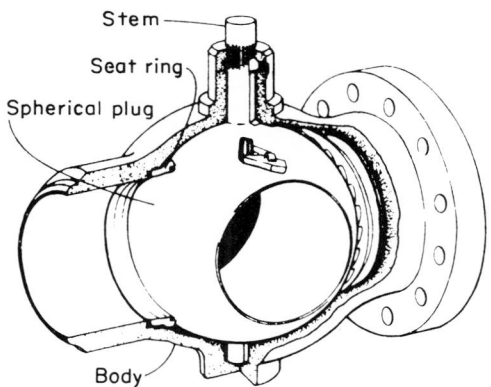

Stem
Seat ring
Spherical plug
Body

Fig. 22. Ball valve in closed position.

In recent years, the use of plastics, such as nylon, delrin, synthetic rubbers, and fluorinated polymers for seating, has substantially increased the use of ball valves.

With fluorinated polymer seats, ball valves are used for service temperatures ranging from −450 to 500 F. With graphite seats, temperatures as high as 1000 F are possible. Ball valves, similar to plug valves, are quick opening, needing only a quarter turn from full open to full close. Ball valves are nonsticking, and they provide tight closure. They also exhibit a negligible pressure drop because of their smooth, full-opening port. These valves are easy to repair, and maintenance costs are low.

Major components of the ball valve are the body, spherical plug, and seats. Ball valves are made in three general patterns—*venturi port*, *full port*, and *reduced port*. The full-port valve has an inside diameter equal to the inside diameter of the pipe. The reduced port generally involves one pipe size smaller than the line size. Stem sealing is by bolted packing glands and O-ring seals. A lubricant-seal system in a ball valve is illustrated in Fig. 23.

Table 29. End-to-end Dimensions of Weir-type Diaphragm Valves[1]

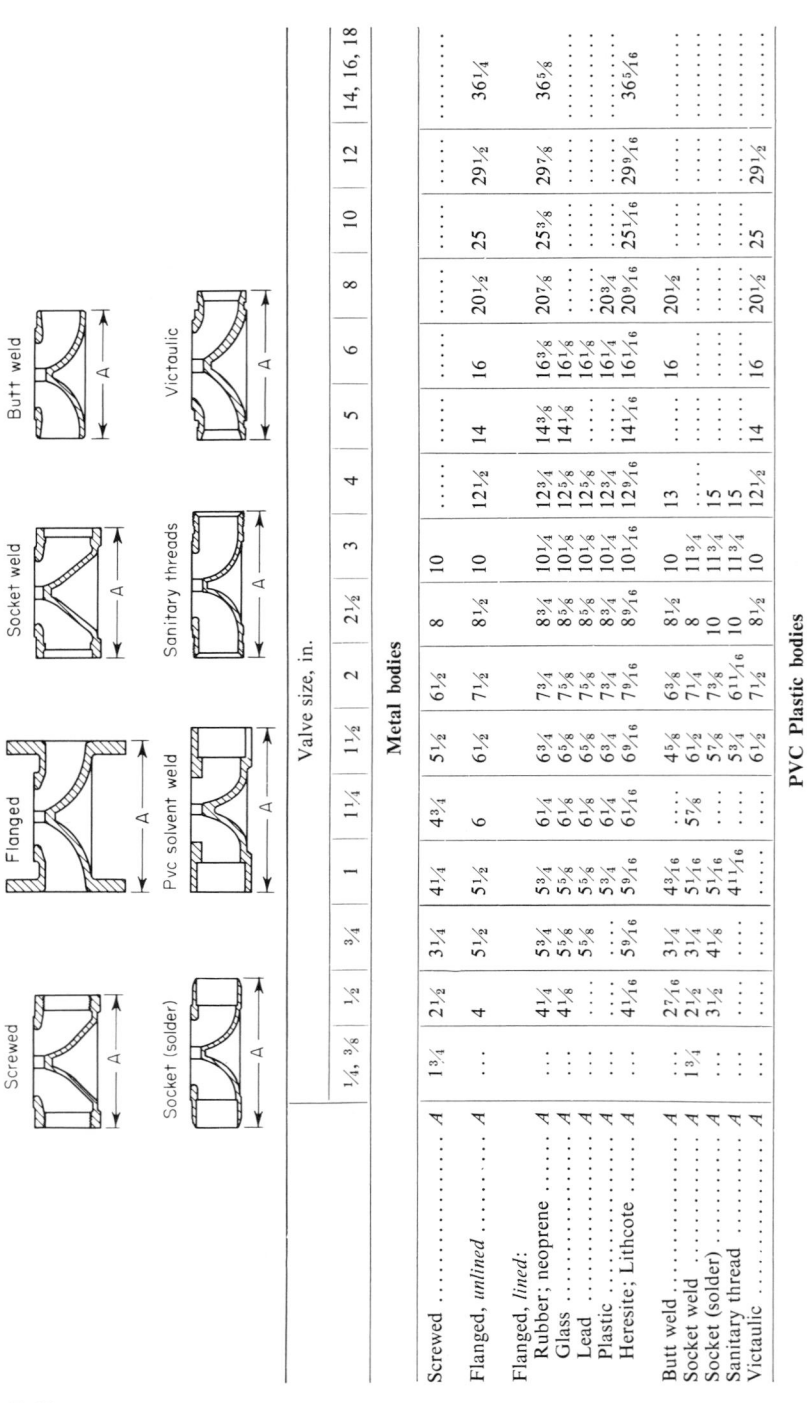

Valve cross-section types illustrated: Screwed, Flanged, Socket weld, Butt weld, Socket (solder), Pvc solvent weld, Sanitary threads, Victaulic (dimension A = end-to-end).

Metal bodies

		1/4, 3/8	1/2	3/4	1	1 1/4	1 1/2	2	2 1/2	3	4	5	6	8	10	12	14, 16, 18
Screwed	A	1 3/4			4 1/4	4 3/4	5 1/2	6 1/2	8	10							
Flanged, *unlined*	A		4	5 1/2	5 1/2	6	6 1/2	7 1/2	8 1/2	10	12 1/2	14	16	20 1/2	25	29 1/2	36 1/4
Flanged, *lined:*																	
Rubber; neoprene	A		4 1/4	5 3/4	5 3/4	6 1/4	6 3/4	7 3/4	8 3/4	10 1/4	12 3/4	14 3/8	16 3/8	20 7/8	25 3/8	29 7/8	36 5/8
Glass	A		4 1/8	5 5/8	5 5/8	6 1/8	6 5/8	7 5/8	8 5/8	10 1/8	12 5/8	14 1/8	16 1/8				
Lead	A			5 5/8	5 5/8	6 1/8	6 5/8	7 5/8	8 5/8	10 1/8	12 5/8		16 1/8	20 3/4			
Plastic	A				5 3/4	6 3/4	6 3/4	7 3/4	8 3/4	10 1/4	12 3/4		16 1/4	20 3/4			
Heresite; Lithcote	A		4 1/16	5 9/16	5 9/16	6 9/16	6 9/16	7 9/16	8 9/16	10 1/16	12 9/16	14 1/16	16 1/16	20 9/16	25 1/16	29 9/16	36 5/16
Butt weld	A	2 7/16		3 1/4	4 3/16		4 5/8	6 3/8	8 1/2	10	13		16	20 1/2			
Socket weld	A	1 3/4	2 1/2	3 1/4	5 1/16		6 1/2	7 1/4	8	11 3/4	15		16				
Socket (solder)	A		3 1/2	4 1/8	5 1/16		5 7/8	7 3/8	10	11 3/4	15						
Sanitary thread	A						5 3/4	6 11/16	10	11 3/4	15						
Victaulic	A				4 11/16			7 1/2	8 1/2	10	12 1/2	14	16	20 1/2	25	29 1/2	

PVC Plastic bodies

		1/4, 3/8	1/2	3/4	1	1 1/4	1 1/2	2	2 1/2	3	4	5	6	8	10	12	14, 16, 18
Screwed	A		3 3/4	4 5/16	4 3/4	6	6	7 1/2									
Socket (solvent weld)	A		4 1/4	5	5 5/8	7	7	8 1/4									

[1] Diaphragm valves are not standardized by any code or regulatory agency. The manufacturer should be consulted for specific information.

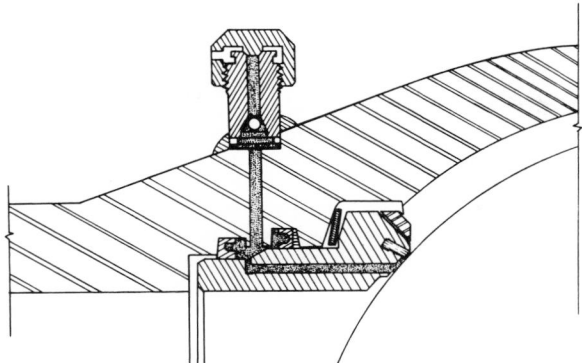

FIG. 23. Lubricant-seal system in a ball valve.

Globe Valves

Applications.   Although available in sizes up to 12 in., globe valves are most frequently used in smaller sizes to about 3 in.   They are extensively employed for the control of flow and where positive shut-off is required.   Close control over flow is readily accomplished.

The change in direction of the fluid as it flows through the valve results in increased resistance to the flow.   Complete drainage of the piping system is also not readily accomplished with globe valves.

General Design Features.   A typical example of a large globe valve with flanged ends is shown in Fig. 24.   A small screwed-end valve is illustrated in Fig. 25.

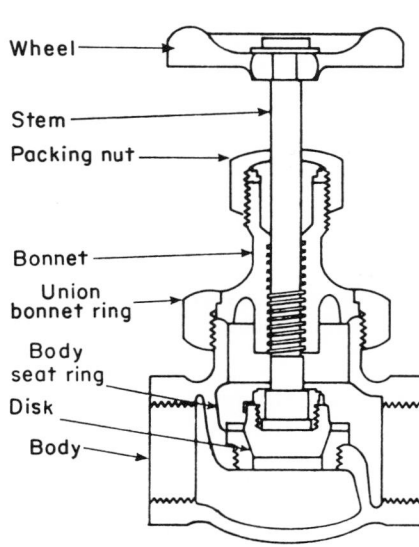

Wheel

Stem

Packing nut

Bonnet

Union bonnet ring

Body seat ring

Disk

Body

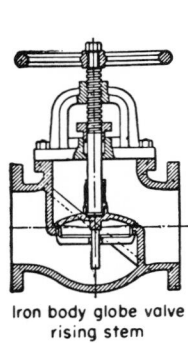

Iron body globe valve
rising stem

FIG. 24. A typical large globe valve with flanged ends.

FIG. 25. A typical small screwed-end globe valve.

Globe valves usually have rising stems, and the larger sizes are of the outside screw-and-yoke construction.

Components of the globe valve are similar to those of the gate valve. This type of valve has seats in a plane parallel to the line of flow. Seat erosion is minimized because the direct contact between seat and disk ends as soon as flow begins.

The disks and seats are normally readily replaced. This makes globe valves particularly suitable for services which require frequent valve maintenance. Where valves are operated manually, the shorter disk travel offers advantages in saving operator time, especially if the valves are adjusted frequently.

The principal variations in globe-valve design are in the types of disks employed. Plug-type disks have a long, tapered configuration with a wide bearing surface. They provide maximum resistance to the erosive action of the fluid stream.

In the composition disk, the disk has a flat face that is pressed against the seat opening like a cap. Foreign material that might prevent tight closure or might score seating surfaces of metal-to-metal seated valves does not ordinarily affect the tightness of valves with composition disks. The unit consists of a metal disk holder, a composition disk, and a retaining nut. A number of different compositions are available to meet a wide variety of services. Asbestos is extensively used as disk material suitable for saturated steam at 150 psi. Valves can be changed from one kind of service to another by simply changing the type of disk.

The conventional disk, in contrast to the plug-type, provides a thin contact between the taper of the conventional seat and the face of the disk. This narrow contact area tends to break down hard deposits that may form on the seats and thus facilitates pressure-tight closure operation.

In cast-iron globe valves, disk and seat rings are regularly furnished of bronze.

In steel-globe valves for temperatures up to 750 F, the trim is regularly furnished of stainless steel and so provides resistance to seizing and galling. The mating faces are normally heat-treated to obtain differential hardness values. Other trim materials are also used.

The seating surface is ground to insure full bearing surface contact when the valve is closed. For lower pressure classes, alignment is maintained by a long disk lock-nut. For higher pressures, disk guides are cast into the valve body. The disk turns freely on the stem to prevent galling of the disk face and seat ring. The stem bears against a hardened thrust plate, eliminating galling of the stem and disk at the point of contact.

Seat rings of all sizes of steel-globe valves in the 150-, 300-, 400-, and 600-lb classes are screwed and tack-welded and are renewable. In the 900-, 1,500-, and 2,500-lb classes, they are screwed in and seal welded. All steel valves have full-size port openings. A variation of the globe valve is the *angle valve*, illustrated in Fig. 26.

Standard Pressure Ratings. On cast-steel globe and angle valves, the bonnets generally are round. The bonnet joint in the 150-, 300-, 400-, and 600-lb classes is usually male and female. Tongue-and-groove-type bonnet joints are generally utilized in the 900- and 1,500-lb classes. Pressure-sealed bonnet joints are recommended for high-pressure, high-temperature steam and feed water service. They are available in the 600-, 900-, 1,500, and 2,500-lb classes.

Needle Valves

Needle valves generally are used for instrument, gage, and meter line service. Very accurate throttling is possible with needle valves, and they are also extensively used in applications involving high pressures and/or high temperatures. In needle

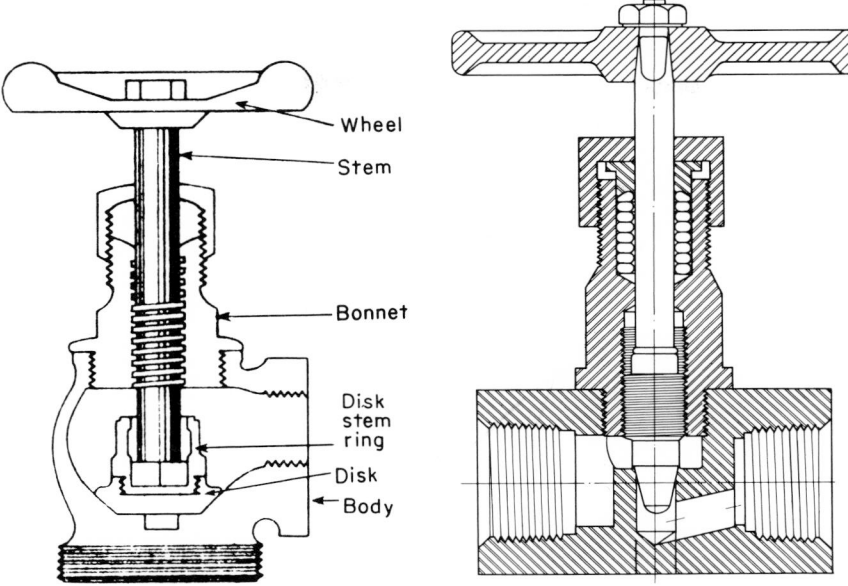

FIG. 26. Angle globe valve with screwed ends.

FIG. 27. Needle valve for accurate throttling of flow.

valves (Fig. 27) the end of the stem is needle pointed. The needle fits accurately into its seat and thus assures tight closure with the least effort.

Butterfly Valves

Butterfly valves are low-pressure valves of extremely simple design, which are used to control and regulate flow. They are characterized by fast operation and low differential pressure drop. They require only a quarter turn from closed to full-open position. Butterfly valves are not intended for pressure-tight service. A typical flanged butterfly valve is illustrated in Fig. 28.

Check Valves

Check valves are designed to prevent backflow in lines. The three principal types of check valves used are the lift check, the swing check, and the ball check, illustrated in Figs. 29 to 31. The swing check is the more commonly used.

Lift, *poppet*, or *piston check* valves are frequently desirable in vertical pipelines. The force of gravity plays an obvious part in the functioning of a check valve, and the position of the valve must always be given consideration. Lift- and ball-check valves must always be placed so that the direction of lift is exactly vertical. Swing checks must be located to insure that the flapper will always be closed freely and positively by gravity.

Lift check valves are particularly adapted for high-pressure service where velocity of flow is high. Pressure-sealed bonnet joints can be obtained in the 600-, 900-, 1,500-, and 2,500-lb classes.

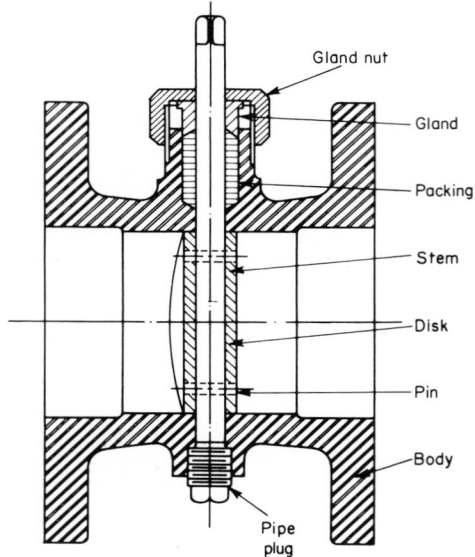

FIG. 28. Typical flanged-end butterfly valve.

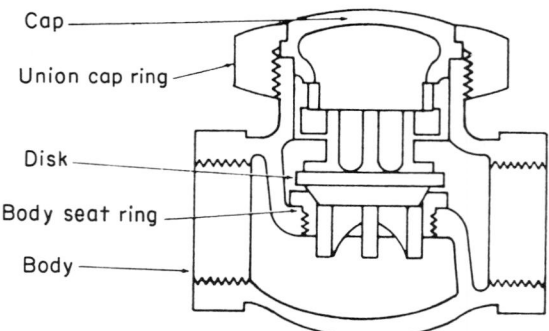

FIG. 29. Lift check valve.

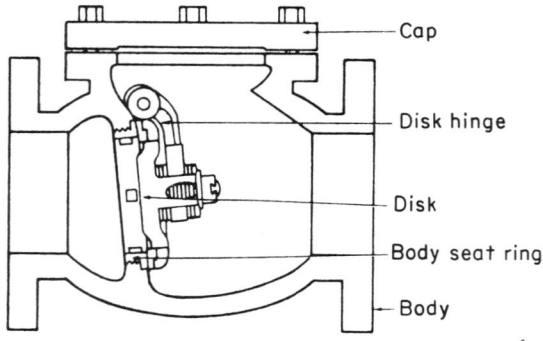

FIG. 30. Swing check valve.

In lift check valves, the piston disk is accurately guided by long contact and a close sliding fit with the perfectly centered dash pot. The walls of the piston and dash pot are of approximately equal thickness. Large steam jackets are located outside of the dash pot and inside the piston to eliminate sticking due to differential expansion. The seat ring is of a barrel-type design of heavy uniform cross section. It is normally screwed in and seal-welded. The flow opening is full port size.

In swing check valves, the disks on smaller sizes are of the same material as the trim. On larger valve sizes, the disks are faced with the trim material. The disk is swung from lugs cast integrally with the body. Arm-and-disk assembly may, if necessary, be repaired or renewed by removing the bolted valve cover.

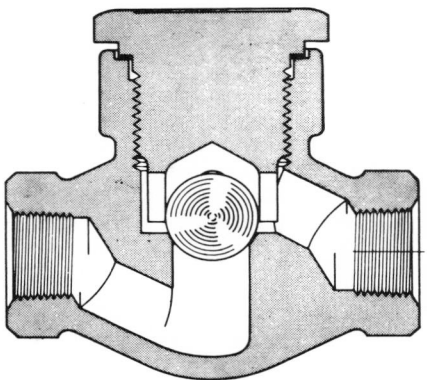

FIG. 31. Ball check valve.

When swing check valves are used in a system in which frequent flow reversals may be encountered, the valves may have a tendency to chatter. This can be avoided by using a swing check valve equipped with an outside weight and lever.

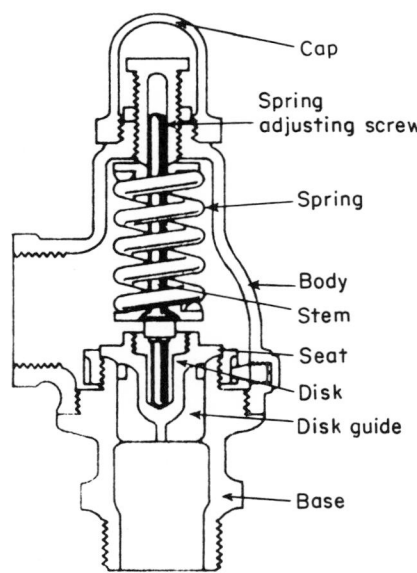

Cap

Spring
adjusting screw

Spring

Body
Stem

Seat

Disk

Disk guide

Base

FIG. 32. Relief valve opens when line pressure exceeds preset loading on the spring.

Pressure-relief Devices

Application. Safety valves are used to prevent excessive overpressure in process and piping systems and thus protect piping and equipment from failure. The two major types of safety valves are the *relief valve* and the *pop valve*. Both valve types should open quickly.

Relief valves are primarily used to relieve excessive pressure in liquid service. Pop valves are used in high-pressure applications (balanced type). Neither of these types of safety valves should be used in service which (1) is corrosive to the valve components, (2) involves back pressure, (3) requires piping of discharge to remote locations, or (4) involves pressure control or by-pass valves.

General Design Features. Relief and pop valves are both spring-loaded. In the *relief valve* (shown in Fig. 32) the valve is activated by static pressure under the seat. When the line pressure exceeds the preset pressure of the spring, the seat or feather is opened.

The essential element in *relief valves* is a disk which is held against a seat by a heavy spring and a so-called "huddling chamber." The huddling chamber is so

constructed that when the valve is slightly opened a static pressure is built up in the chamber which immediately forces the valve open. The pressure below the valve must drop a few pounds below the opening pressure before the valve will close. This is known as the *blowdown* and is from 2 to 4 per cent of the set pressure but not less than 2 lb in any case. The amount of blowdown can be changed by means of the adjusting ring of the huddling chamber. Since reducing the blowdown affects the capacity and operation of the valve, it is mandatory that the blowdown adjustment be made and sealed by the manufacturer.

The *pop valve* is also an automatic spring-loaded pressure-relieving valve actuated by static pressure under the seat. At a set pressure, it is opened fully. Depending on back-pressure action, pop valves are also designated as *balanced* or *conventional* types. The former are used for high-pressure relief.

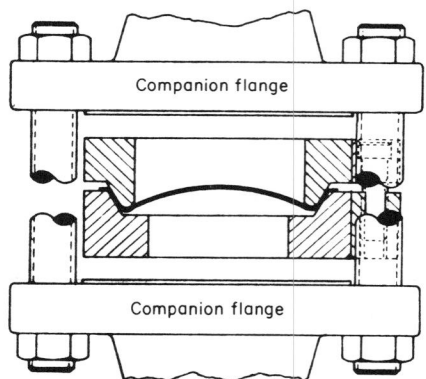

Fig. 33. The rupture disk, a type of relief device.

Relieving Capacity. The approximate relieving capacities of high-lift pop valves for saturated steam service are given in Table 30. For compressed air, the relieving capacities are given in Table 31.

Because of inability to predict exactly the relieving capacity of valves by formulas, the ASME Boiler and Pressure Vessel Code requires that the maximum rated capacity of a safety valve shall be 90 per cent of the flow determined by actual steamflow tests. The lift in inches, the popping and blowdown pressures, and the relieving capacity at a pressure 3 per cent over the set pressure must be plainly stamped on each valve. The relieving capacity must be determined by tests with steam on valves of corresponding design and construction.

Pressure and temperature limits of Flanged Steel Safety Relief Valves for Use in Petroleum Refineries are given in American Petroleum Standard 526, and recommendations for the installation of pressure-relief valves in refineries are covered in American Petroleum Institute Standard RP520.

Rupture Disks. A special type of pressure-relief device involves rupture disks (Fig. 33). The design generally involves flanges with special machined hold-down seats to prevent slippage of the disks. The disks are designed to rupture automatically at a predetermined pressure. These devices have a particular advantage when large volumes of gas or liquid must be relieved quickly.

Rupture-disk devices are also used in conjunction with the regular spring-loaded safety valve. By employing rupture disks for relief at a pressure set approximately 5 to 10 per cent above the relief pressure of the safety valve, the rupture disk will fail if the regular relief valve does not operate properly.

Gate or plug valves may be installed, also, ahead of rupture disks. With the rupture disk in place, these valves are normally left open. Closing of these valves may be necessary when the disk is replaced after rupture or for maintenance.

Aluminum, copper, and stainless steels are most widely used as disk materials. Metal disks coated with neoprene or other plastics are also utilized. In highly corrosive service, the precious metals silver, gold, and platinum are also employed.

Table 30. Approximate Relieving Capacities of High-lift Pop Valves for Saturated Steam[1] Service, lb/hr

Gage pressures, psi

Valve size, in.	5	10	15	20	25	30	40	50	60	75	100	125	150	175	200
1½	750	1,000	1,200	1,300	1,400	1,600	2,000	2,400	2,800	3,500	4,500	5,000	6,000	7,000	8,000
2	1,200	1,500	1,750	2,000	2,000	2,200	2,700	3,200	3,700	4,100	5,000	6,000	7,000	8,000	9,000
2½	1,650	2,100	2,550	2,850	3,000	3,300	3,900	4,500	5,100	6,000	7,500	8,500	9,500	10,500	12,000
3	2,300	2,850	3,400	3,950	4,000	4,400	5,200	6,000	6,800	8,000	10,000	11,500	13,000	14,500	16,000
3½	2,850	3,500	4,200	4,850	5,000	5,500	6,500	7,500	8,500	10,000	12,000	14,000	16,000	18,000	20,000
4	3,500	4,400	5,300	6,050	6,500	7,200	8,600	10,000	11,400	13,000	16,000	18,500	21,000	23,500	26,000
4½	4,300	5,300	6,250	7,250	8,000	8,800	10,400	12,000	13,600	15,500	20,000	23,000	26,000	29,000	32,000
6			9,800	11,750	22,000	25,000	31,000	37,000	44,000	51,000	65,000	79,000	94,000	108,000	122,000

Valve size, in.	250	300	350	400	450	500	550	600	650	700	750	800	850	900
1½	9,000	10,000	11,000	12,000	13,000	14,000	15,000	16,000	17,000	18,000	19,000	20,000	22,000	24,000
2	10,500	12,000	14,500	17,000	19,000	21,000	23,000	25,000	27,000	29,000	31,000	33,000	35,000	37,000
2½	14,000	16,000	22,000	28,000	31,000	34,000	37,000	40,000	44,000	48,000	52,000	56,000	71,000	87,000
3	20,000	24,000	31,000	38,000	43,000	48,000	53,000	58,000	63,000	68,000	73,000	78,000	106,000	135,000
3½	25,000	30,000	40,000	50,000	60,000	70,000	80,000	90,000	97,500	105,000	112,500	120,000	162,000	204,000
4	32,000	38,000	51,000	64,000	85,000	106,000	127,000	148,000	178,000	208,000	238,000	268,000	284,000	300,000
4½	39,000	50,000	63,000	80,000	110,000	140,000	170,000	200,000						
6	150,000	179,000	207,000	236,000										

[1] The maximum relieving capacities given are typical of the results guaranteed by several manufacturers for their high-capacity external-spring-type valves at 3 % overpressure and where the back pressure does not exceed 60 % of the initial pressure.   These values represent about the maximum discharge obtainable with the best design of valves made at the present time.   In old-style valves the relieving capacities may not exceed one-third to one-half of those given above. Where accurate values are desired, reference should be made to the manufacturer's catalogue.

Table 31. Approximate Relieving Capacities, cfm of Free Air, for High-lift Pop Valves for Compressed-air Service

Diameter D, in.	Gage pressure, psi															
	50	100	150	200	250	300	350	400	500	600	800	1,000	1,200	1,600	2,000	2,400
¼	12	20	27	33	38	43	48	53	61	70	84	97	109	128	147	160
⅜	17	27	36	44	51	58	65	72	83	95	115	133	149	176	197	215
½	20	32	42	51	59	67	74	111	129	147	177	205	230	270	304	330
¾	37	59	78	96	112	127	141	176	224	232	242	346	386	423	474	518
1	58	94	124	152	178	202	224	248	286	324	390	450	500	586		
1 ¼	84	135	180	221	259	293	325	352	400	443	509					
1 ½	114	186	248	302	354	400	444	478	528	568	634					
2	189	306	410	501	592	668	741									
2 ½	282	457	613	750	880	998	1,114									
3	393	638	856	1,050	1,230	1,398	1,557									

The foregoing table is based on the following formulas,

$$Q = 28PDl \text{ for 45-deg bevel-seat valves}$$
$$Q = 40PDl \text{ for flat-seat valves}$$

where Q = discharge measured at 14.7 psia and 60 F, cfm of free air
P = absolute pressure at which the safety valve opens (gage pressure plus 14.7 psi at sea level)
D = diam of the inside edge of the bearing surface between the disk and seat, in.
l = vertical lift of the safety-valve disk from its seat, representing the lift for min discharge capacity for satisfactory operation of the valve, in.

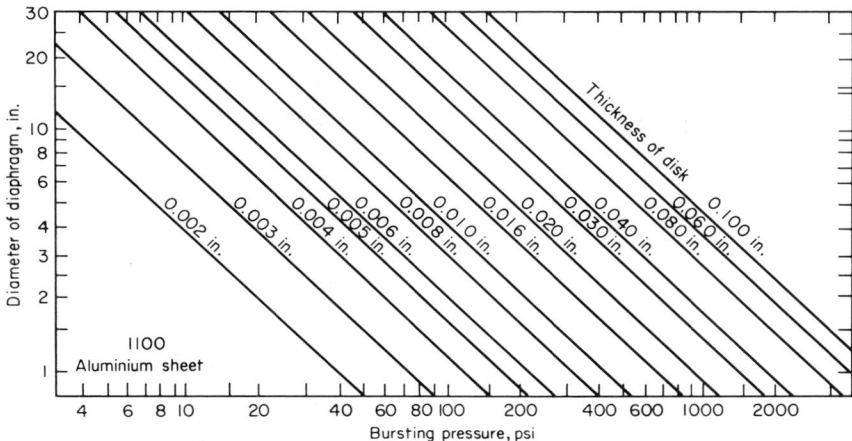

FIG. 34. Bursting pressure of Type 1100 aluminum sheet.

The pressure at which the particular disk material ruptures is a direct function of the inside flange diameter and the thickness of the metal. For Type 1100 aluminum, this is illustrated in Fig. 34.

Pressure-reducing Valves

The primary purpose of pressure-reducing valves is to maintain a constant and reduced pressure in a piping system which is supplied from a higher-pressure source. In the simple rubber-diaphragm type (Fig. 35) the amount of opening of the valve is controlled by the pressure on the reduced-pressure side, which is conducted through a pilot line to a rubber diaphragm attached to the valve stem. The pressure exerted on the diaphragm is opposed by a weighted lever (or sometimes a spring). The pressure to be maintained is adjusted by shifting the weight on the lever, by adding more weight, or by adjusting the spring force.

The pilot type of reducing valve (shown in Fig. 36) is used where the reduced pressure is more than 10 lb or when the service is severe. In this type of valve, the reduced pressure brought back through a pilot line to the diaphragm *B* is opposed by the spring *C*. The small pilot valve *E* is thus controlled to admit just enough high-pressure fluid above the piston *F* to maintain the necessary opening of the main valve. Many variations of this design exist. All are based on the same principle involving a pilot valve which is controlled by the reduced pressure and which operates the main valve.

The pilot type of reducing valve is more complicated than the rubber-diaphragm type. It is more difficult to keep in good working order, particularly when the steam contains some solid matter. Pilot-type valves should not be used when the rubber-diaphragm valves are adequate.

Reducing valves tend to wear at the disk and seat because of continual throttling action. They cannot be depended upon to close tightly after being in service. This cutting is often made worse because of a tendency to select a valve which is too large for the service.

Size of Reducing Valves. The proper size of reducing valve may be determined on the basis of the pipe size which will give a reasonable velocity for the amount of flow. A velocity of 6,000 fpm on the inlet side is good practice. This

method, though rather arbitrary, is a simple method recommended by some manufacturers. In any case, the pressure and temperature conditions and the desired fluid flow rate should be specified.

Installation. Reducing valves should be installed with a gate valve on either side and a globe valve bypass which permits steam to be supplied while the valve is out of service. Improper operation of the reducing valve may result in the building up of high pressure on the low-pressure side. Since this may be dangerous, a relief valve should be installed. The ASA Code for Pressure Piping requires that, except

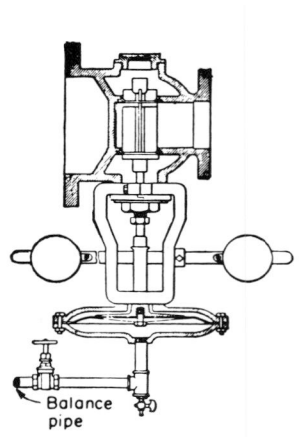

FIG. 35. Simple rubber diaphragm pressure-reducing valve.

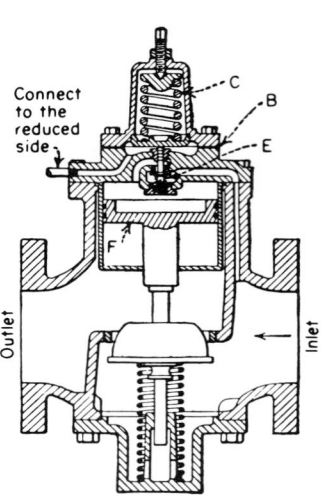

FIG. 36. Pilot-type reducing valve for severe service.

in a steam plant used for district heating, one or more relief valves are to be provided in case the piping of equipment on the low-pressure side does not meet the requirements for the full initial pressure. It is mandatory that a pressure gage be installed on the low-pressure side of a reducing valve.

Traps

The function of a steam trap is to discharge condensate from steam piping without permitting steam to escape. Three principal types are used.

The float trap, (Fig. 37) has a hollow float which rises as water enters. Through a system of levers, the valve is opened and the water is discharged.

In the bucket trap (Fig. 38) water spills over the top of the bucket, causing it to sink and to open the valve. The float trap gives a more or less continuous discharge if the flow of water to it is steady, whereas the bucket trap always discharges intermittently. An inverted-bucket-type trap is illustrated in Fig. 39. The size of a trap should be chosen on the basis of its effective valve area or its actual discharge capacity, rather than on the size of the inlet and outlet connections.

Another type of trap known as the impulse trap is illustrated in Fig. 40. In this type, flashing of the hot condensate tends to force a small piston into the discharge opening when the temperature of the condensate approaches within about 30 F

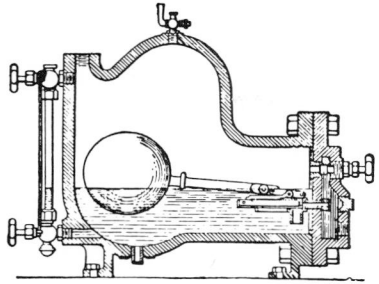

Fig. 37. Float-type trap.

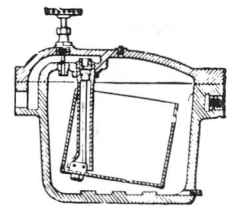

Fig. 38. Bucket-type trap.

of the saturation temperature.   As soon as the condensate collected in the drain system cools sufficiently below the flash temperature, the trap opens and discharges the accumulated water until the temperature of the condensate once more approaches the saturation temperature and flashes, thereby closing the trap and again repeating the cycle.   A small orifice permits a continuous discharge of steam or flashed vapor when the trap is closed.

Single orifices are sometimes used to remove condensate from high-pressure, high-temperature steam lines.   Where the drains are required only in bringing the line up to temperature, the use of orifices is particularly desirable.

Miscellaneous Valve Types

Numerous variations exist of the basic valve types discussed in the previous sections.   There are also many special types of valves for special applications in the nuclear weapons and missile industry.

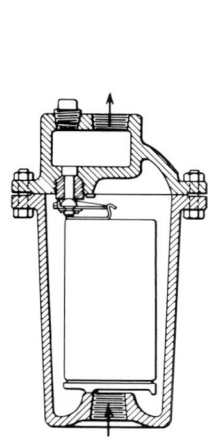

Fig. 39. Inverted-bucket trap.

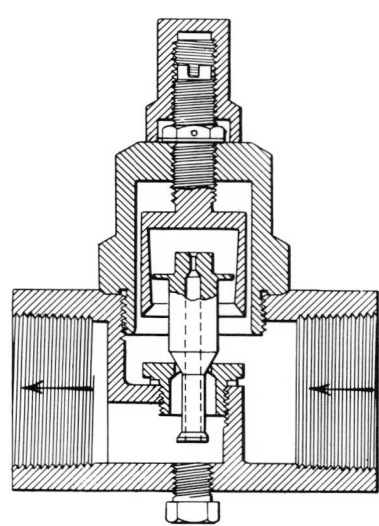

Fig. 40. Impulse trap.

Special valves and variations of basic valves include foot valves, stop check valves, slide valves, sampling valves, line blind valves, pinch-clamp valves, solenoid valves, and quick-opening or self-closing gate and globe valves. Where these specialized valves are to be used, reference should be made to specific descriptions in the manufacturers' literature.

VALVE BODY MATERIALS

Valve bodies are produced in as many different materials as are available in pipe. Moreover, within a valve, a number of different materials may be used. Table 32 lists typical materials in a cast-iron gate valve, and Table 33 illustrates the materials used in a stainless-steel globe valve.

Standard Body Materials. Valve bodies are made of brass or bronze mainly in the smaller sizes and for moderate pressures and temperatures. Cast iron is used in all sizes up to working steam pressures of 250 lb, temperatures of 450 F, and hydraulic pressures of 800 lb. Cast steels are used for more severe service. For high-temperature service, valve bodies of the chromium-molybdenum alloy steels are available. Forged steel is used in small valve bodies which are machined and drilled out. This is not a practicable method in the larger sizes, although valves up to 8 in. have been made from solid stock.

For the more corrosive services, valves made of AISI Type 304 and Type 316 stainless steels are available as standard sizes.

Special Body Materials. For special service requirements, valves have been made of all of the major metals and alloys available in the form of piping. These include special stainless-steel grades, such as Type 347, No. 20, aluminum alloys, Monel, Inconel, cupronickel, titanium, etc.

In smaller sizes, the bodies are generally made as castings or are machined from

Table 32. Typical Materials Used in a Cast-iron Gate Valve

Part	Material
Handwheel	Cast iron
Yoke nut	Bronze
Stem	Bronze rod
Hex nuts	Bronze
Gland	Bronze
Eyehead bolts	Rust-proofed steel
Bonnet	Cast iron
Test plug (not shown)	Cast iron
Throat flange nuts and bolts	Rust-proofed steel
Body	Cast iron
Disks	Cast iron
Disk rings	Bronze
Seat rings	Bronze

Table 33. Typical Materials Used in a Stainless-steel Globe Valve

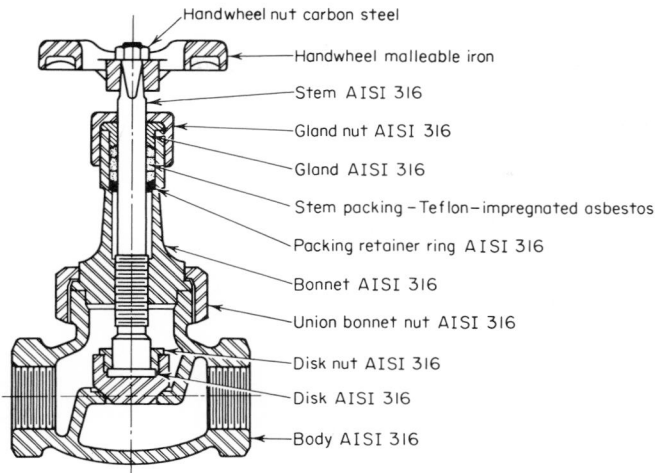

Handwheel nut carbon steel

Handwheel malleable iron

Stem AISI 316

Gland nut AISI 316

Gland AISI 316

Stem packing – Teflon–impregnated asbestos

Packing retainer ring AISI 316

Bonnet AISI 316

Union bonnet nut AISI 316

Disk nut AISI 316

Disk AISI 316

Body AISI 316

forgings or from bar stock.   In larger sizes, they may be assembled by welding cast, forged, or wrought components.

VALVE TRIM MATERIALS

Valve trim materials include the seat ring, disk or facing, and stem.   Commonly used trim materials in several major valve types are summarized in Table 34. Aluminum-silicon bronze and naval bronze are the favored bronze grades used as stem material.   Stems are also made of stainless-steel grade No. 20.   Special disk materials include nitrided steel, 17 per cent chromium, Hastelloy B and C, and stainless-steel grade No. 20.

Among the principal factors which influence the performance of trim metals are (1) tensile properties, chemical stability, and corrosion resistance at the operating temperature; (2) hardness and toughness; (3) a coefficient of expansion which corresponds closely to that of the valve body; and (4) a difference of sufficient magnitude in the properties of seat and disk facings to prevent seizing of their surfaces when one slides over the other.

It is essential that a seat metal retain adequate compressive properties[1] at the operating temperature and that it be immune from "growth" or other chemical change due to temperature.   If it is to retain the smooth surface which is essential to tightness, a seat material must resist oxidation and corrosion under the service conditions involved.

Brass or bronze are entirely satisfactory as seat and disk materials for water service.   Medium-hard rubber is extensively used for disk rings in bronze valves. Saturated steam may cause erosion in standard brass or bronze seats and disks.

For temperatures in the range 500 to 750 F, Monel (70% Ni–30% Cu) is extensively used.   Also suitable but less used is cupronickel (70% Cu–30% Ni).

[1] For ductile materials, compressive properties are essentially the same as tensile properties. Thus, tensile-strength data are normally utilized.

Table 34. Commonly Used Valve-trim Materials

Trim metal	Gate valves			Globe and angle valves		
	Disk or facing	Seat ring or facing	Stem	Disk or facing	Seat ring or facing	Stem
Monel	Monel	Monel	Monel	Monel cupronickel	Monel cupronickel	Monel
Bronze	Bronze	Bronze	Bronze	Bronze	Bronze	Bronze
Stainless steel, 12–14% Cr.	12–14% Cr	12–14% Cr	12–14% Cr	12–14% Cr	12–14% Cr	12–14% Cr
Stellite	Stellite	Stellite	12–14% Cr	Stellite	Stellite	12–14% Cr
Stainless steel, 18–8 (Type 304)	18–8	18–8	18–8	18–8	18–8	18–8
Stainless steel, 18–8 Mo (Type 316)	18–8 Mo	18–8 Mo	18–8 Mo	18–8 Mo	18–8 Mo	18–8 Mo
Stainless steel (12–14% chromium), To Stellite	12–14% Cr	Stellite	12–14% Cr	12–14% Cr	Stellite	12–14% Cr
Stainless steel (12–14% chromium), To nickel-copper-tin alloy	12–14% Cr	Ni–Cu–Sn	12–14% Cr	12–14% Cr	Ni–Cu–Sn	12–14% Cr

Modifications of these alloys with additions of tin, iron, zinc, chromium, silicon, or aluminum have been used up to 850 F with reasonably good success.

Straight chromium stainless steels (12 to 14% Cr hardened to min 250 Brinell) have found wide acceptance as valve-trim materials. Austenitic stainless steels Type 304 (18% Cr–8% Ni) and Type 316 (18% Cr–8% Ni-Mo) have been used where corrosive conditions are severe or for applications to which they seem particularly suited, e.g., steam safety-valve nozzles. For some corrosive environments, seat rings are made of stainless-steel grade No. 20 or 20% Ni–8% Cr compositions. Nitralloy seats and disks have been used on valves for superheated steam and hot oil with good success. Some pitting has occurred where Nitralloy has been used on intermittent saturated-steam service. For the more severe service conditions, welded-on overlays of Stellite or Colmonoy hard-surfacing alloys have given excellent results. Seats rings of Hastelloy B and C are also available.

Hardness and toughness are properties which tend to prevent wear caused by sliding one surface over the other and erosion due to cutting action of the fluid. The proper balance between hardness and toughness to secure good resistance to cutting and other abrading influences is dependent on a number of factors which differ with each material. Steam-impingement tests conducted on a large number of materials tend to show that greater hardness does not necessarily indicate greater resistance to erosion. A forged 12 to 14 per cent chromium stainless steel, heat-treated to give a Brinell hardness of 302, showed only about one-sixth the depth of erosion measured for the same material with a Brinell hardness of 444. Nitrided-steel seats having an equivalent Brinell of 1,150 did not resist erosion as well as the moderately hard stainless material. The cobalt-chromium-tungsten (Stellite) deposited material likewise showed superior resistance in these steam-impingement tests. Corrosion may also influence these results.

Similar coefficients of expansion for seat metals and valve bodies are desirable in order to prevent loosening of screwed-in rings after repeated heating and cooling. Leakage around seat rings is one of the more common difficulties with valves in high-temperature service. Despite the fact that the combination of brass and bronze used for valve trim on cast-iron valves expands approximately 50 per cent more than cast iron, it is used quite successfully at temperatures up to 450 F. Under conditions of frequent heating and cooling, however, such rings occasionally loosen. The expansion of Monel and modified grades of cupronickel used for valve trim is about 20 per cent greater than that of carbon cast steel, although the straight chromium stainless steels expand about 15 per cent less than the carbon- and low-alloy steels used for valve bodies. The austenitic stainless steels (18% Cr–8% Ni) expand about one-third more than carbon- and low-alloy steels.

It is apparent from consideration of the coefficients of expansion of the foregoing seating materials that it is improbable that a screwed seat ring made of these materials will act as a unit with the valve body. Nevertheless, screwed seat rings made of Monel and cupronickel have performed reasonably satisfactorily in 850 F service because the rings deform elastically and plastically to compensate for these differences in expansion. To avoid loosening in high-temperature service, screwed-in seat rings are often seal welded in the valve body.

In respect to similarity of expansion, screwed seat rings of nitrided steel or of carbon or alloy steel with welded-on overlays should be preferred. The greater permanency of the welded-on-overlay seats makes the need for replacement less necessary.

Galling or seizing occurs when certain metals are in sliding contact. The use of dissimilar materials for seats and disks, such as seats of a ferrous alloy and disks of a nickel-copper alloy, has been a common method of avoiding galling. The use of the same materials heat-treated to give different hardness values also has been

used successfully with the straight chromium–stainless-steel valve-trim materials. For example, the valve seat, wedge or facing, and stem made of 12 per cent chromium stainless steel may have been heat-treated to exhibit the properties shown in Table 35.

The development of free-machining stainless steels with sulfur and selenium additions has also made it possible to use seats and disks of the same hardness, with no tendency to seize in some applications. Nitrided seats and disks do not show a tendency to gall or seize at temperatures below about 900 F, although this trouble has been experienced at 1100 F. Because of its extremely low coefficient of friction

Table 35. Typical Properties of Valve-trim Materials of 12% Chromium Stainless Steel

Mechanical property	Seat ring	Wedge or facing	Stem
Tensile strength, psi	130,000	185,000	110,000
Yield strength, psi.	115,000	160,000	85,000
Elongation, % in 2″	20	16	23
Reduction in area, %	68	60	65
Brinell hardness.	277	388	220
Rockwell hardness	C-28	C-41	B-93

and other properties peculiar to its composition, evidence of galling has not been found in the seats and disks faced with welded-on overlays of Stellite. The following composition of Stellite has been used extensively for welded-on overlays: 65% Co, 30% Cr, 4% W. The physical properties of the more common seat metals are summarized in Table 36, and the chemical analyses are listed in Table 37.

In ball valves, Teflon or Buna "N", synthetic rubber impregnated with molybdenum disulfide, are used as seat materials.

BYPASSES, TAPS, AND DRAINS

Some valves are furnished with bypasses used in steam service to warm up a line before the main valve is opened and, in steam and other services, to facilitate the operation of the main valve by balancing the pressure on both sides of the wedge in gate valves or on both sides of the disk in globe valves.

In gate valves (Fig. 41) the bypass is normally socket welded into the side of the main body with both pipe sections horizontal. In globe and angle valves (Fig. 42) the socket welded bypass pipe connections generally are made on the right-hand side of the valve with the inlet and outlet pipes vertical. The nominal diameter of the bypass piping used is given in Table 38.

In some applications, it is desirable that valves be furnished with drain connections. Such connections are standardized in Manufacturers' Standardization Society Standard MSS SP 45, By-Pass and Drain Connection Standard.

Standard types of drains are illustrated in Fig. 43, examples of tap locations in typical gate, globe, angle, and check valves are shown in Fig. 44, and standard sizes of drain connections are given in Table 39.

Table 36. Physical Properties of the More Common Seat Materials

Trade name	Tensile strength, psi	Yield strength, psi	Elongation in 2 in., %	Reduction of area, %	Brinell hardness number range, 10-mm ball	Rockwell hardness number range
Aluminum, cast.............	12,000	....	40	...	15–20	
Colmonoy No. 6, cast	55,000	....	...	...	600	C55–C60
Davis metal, cast..........	67,000	40,000	15	...	120	
Everbrite, cast	75,000	45,000	14	...	160–180	
Everdur 1,000, cast........	50,000	20,000	20	25	70–90	
Hastelloy B, cast...........	85,000	50,000	10	10	....	B93
Hastelloy C, cast...........	72,000	45,000	10	11	175–215	
Inconel, wrought, annealed	80,000	25,000	35	60	110–170	
Lead	1,900	700	50	...	4.5–6	
Monel, cast	65,000	33,000	25	...	125–150	
Monel, silicon, cast........	90,000	70,000	...	...	280–325	
Nickel, cast	55,000	20,000	15	...	100–130	
Ni-resist cast iron	25,000	....	2	...	120–170	
Nitralloy, N125.............	130,000	115,000	18	54	Surface 1,000, core 250	
Platnam metal, cast	72,000	60,000	...	...	225	
Silicon cast iron (Duriron, etc.)	17,000	....	...	...	....	C52
Stellite....................	105,000	....	1	...	....	C30–C45
Stainless steel: AISI type 410, Annealed .... Hardened	65,000 100,000	35,000 60,000	25 10	60 25	135–165 290–400	B75–B85 C31–C43
AISI type 420, Annealed .... Hardened	85,000 225,000	40,000 185,000	25 3	40 5	170–200 450–500	C47–C51
AISI type 440, Annealed .... Hardened	90,000 250,000	40,000 190,000	20 1.5	40 2	170–210 500–600	C51–C59
AISI type 304	75,000	36,000	50	45	137	B75
AISI type 316	82,000	42,000	50	45	170	B86
Ni–Cu–Si	70,000	50,000	10	15	192	B91
Tin......................	2,000	200	...	...	4–6	
Zinc, annealed	18,000	....	30	...	....	

Table 37. Chemical Analysis of the More Common Seat Materials

Trade name	Carbon	Manganese	Silicon	Phosphorus	Sulfur	Iron
Aluminum, cast............	....	....	....	....	....	....
Colmonoy No. 6, cast......	0.75*	....	4.25*	....	....	4.75
Davis metal, cast..........	0.5*	1.5*	....	....	....	2*
Everbrite, cast.............	....	0.5*	....	....	....	7*
Everdur 1,000, cast.........	....	1.1*	4*	....	....	....
Hastelloy B, cast...........	0.05	1.00	1.00	0.04	0.03	4.0–7.0
Hastelloy C, cast...........	0.12	1.00	1.00	0.04	0.03	4.0–7.0
Inconel, wrought, annealed................	0.15	1.00	0.5	....	0.015	6.0–10.0
Lead.....................	....	....	....	....	....	....
Monel, cast...............	0.35	0.5–1.5	1–2.24	....	....	2.5
Monel, silicon, cast........	0.25	0.5–1.5	3.5–5.0	....	....	3.5
Nickel, cast...............	1.00	1.5	2.0	....	....	1.25
Ni-resist cast iron..........	2.60–2.80	0.8–0.9	1.4–1.6	0.09–0.13	0.3–0.5	Bal.
Nitralloy, N125.............	0.20–0.29	....	....	....	....	....
Platnam metal, cast........	....	....	....	....	....	0.5*
Silicon cast iron (Duriron, etc.)...........	0.85*	0.65*	14.5	....	....	....
Stellite....................	....	....	....	....	....	....
Stainless steel: AISI type 410, annealed, hardened.......	0.15	1.00	1.00	0.04	0.03	Bal.
AISI type 420, annealed, hardened.......	0.15	1.00	1.00	0.04	0.03	Bal.
AISI type 440, annealed, hardened.......	0.60–1.20	1.00	1.00	0.04	0.03	Bal.
AISI type 304..........	0.08	2.00	1.00	0.04	0.03	Bal.
AISI type 316...........	0.10	2.00	1.00	0.04	0.03	Bal.
Ni–Cu–Si................	....	....	....	....	....	....
Tin.....................	....	....	....	....	....	....
Zinc, annealed............	....	....	....	....	....	....

Maximum amounts unless a range is specified.
* Typical analysis.

Nickel	Cobalt	Molybdenum	Chromium	Copper	Other	Trade name
....	....	....	....	....	Alum. 99+	...........Aluminum, cast
....	....	....	13.5*	....	B–3*	...... Colmonoy No. 6, cast
29*	....	....	....	67*	....	...........Davis metal, cast
30.5*	....	....	....	62*	....	.... Everbrite, cast
....	....	....	....	Bal.	....	.... Everdur 1,000, cast
Bal.	2.50	26.0–30.0	1.0	....	V 0.20–0.60	.... Hastelloy B, cast
Bal.	2.5	16.0–18.0	15.5–17.5	....	W 3.75–5.25 V 0.2–0.4	.... Hastelloy C, cast
72.0 min	....	....	14.0–17.0	0.5	....	Inconel, wrought, ... annealed
....	....	....	....	....	Pb 99+	.... Lead
62–68	....	....	....	Bal.	....	...............Monel, cast
62–68	....	....	....	Bal.	....	.... Monel, silicon, cast
Bal.	....	....	....	....	....	................Nickel, cast
20	....	0.50	5	6.0	Ti 0.25	.... Ni-resist cast iron
....	....	0.15–0.25	0.9–1.4	....	Al 0.85–1.2	.... Nitralloy, N125
54*	....	....	....	33*	Sn 12–Al 0.3	........Platnam metal, cast
....	....	....	....	....	....	Silicon cast iron .... (Duriron, etc.)
....	65*	....	30	....	W 4*	...................Stellite
....	....	....	11.5–13.5	....	....	Stainless steel: AISI type 410, ... annealed, hardened
....	....	....	12–14	....	....	AISI type 420, ...annealed, hardened
....	....	0.75	16.0–18	....	....	AISI type 440, ...annealed, hardened
8–11	....	....	18–20	....	....	...........AISI type 304
10–14	....	2.0–3.0	16–18	....	....	...........AISI type 316
....	....	....	....	....	....	.... Ni–Cu–Si
....	....	....	....	....	Sn 99+	.......................Tin
....	....	....	....	....	Zinc 99+	.... Zinc, annealed

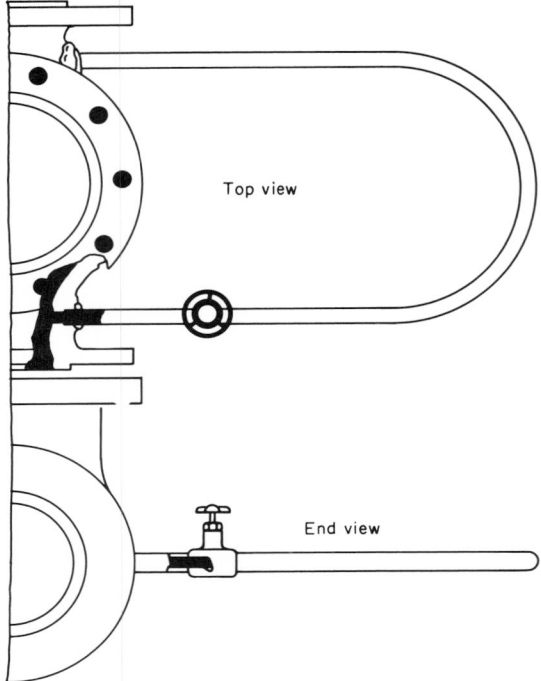

Fig. 41. Gate-valve bypass.

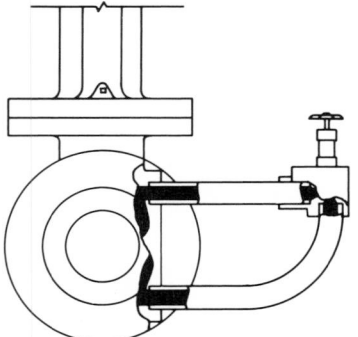

Fig. 42. Bypass for globe and angle
valves.

Table 38. Nominal Diameter of Bypass Piping as Standardized in MSS SP 45-1953

Main valve size	Bypass size	
	Series A[1]	Series B[2]
4	$\frac{1}{2}$	1
5	$\frac{3}{4}$	$1\frac{1}{4}$
6	$\frac{3}{4}$	$1\frac{1}{4}$
8	$\frac{3}{4}$	$1\frac{1}{2}$
10	1	$1\frac{1}{2}$
12	1	2
14	1	2
16	1	3
18	1	3
20	1	3
24	1	4

All dimensions given in inches.
[1] Series A comprehends steam service for warming up before the main line is opened and for balancing pressures where the lines are of limited volume.   They are in common use on ferrous globe and angle valves and on steel gate valves.
[2] Series B comprehends lines conveying gases or liquids, where bypassing may facilitate the operation of the main valve through balancing the pressure on both sides of the disk or disks thereof. These, in the larger sizes, may be of the bolted-on type.

Table 39. Standard Size of Drain Connections (MSS SP 45-1953)

Standard drain sizes

Valve size, in.	Size of drain tapping, in.
2–4	$\frac{1}{2}$
5–8	$\frac{3}{4}$
10–24	1

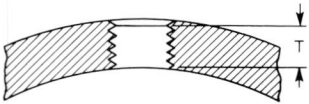

Thread length for drain tapping

1	2	3	4	5	6	7	8
Size of drain	⅜	½	¾	1	1¼	1½	2
Length of thread T'	0.41	0.51	0.55	0.68	0.71	0.72	0.76

All dimensions given in inches.

In no case shall the effective length of thread T be less than that shown in table above. These lengths are equal to the effective thread length of American Standard External Pipe Threads (ASA B2.1).

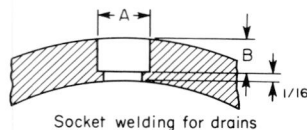

Socket welding for drains

1	2	3	4	5	6	7	8
Size of drain	⅜	½	¾	1	1¼	1½	2
Min diameter of socket A	0.690	0.855	1.065	1.330	1.675	1.915	2.406
Min depth of socket B	³⁄₁₆	³⁄₁₆	¼	¼	¼	¼	⁵⁄₁₆

All dimensions given in inches.

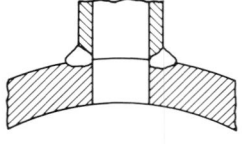

Butt welding for drains

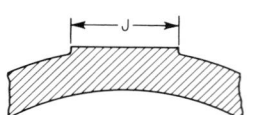

Bosses for drains

1	2	3	4	5	6	7	8
Size of drain	⅜	½	¾	1	1¼	1½	2
Diameter of boss	1¼	1½	1¾	2⅛	2½	2¾	3⅜

Fig. 43. Standard types of drains as standardized in Manufacturers Standardization Society Standard MSS SP 45-1953.

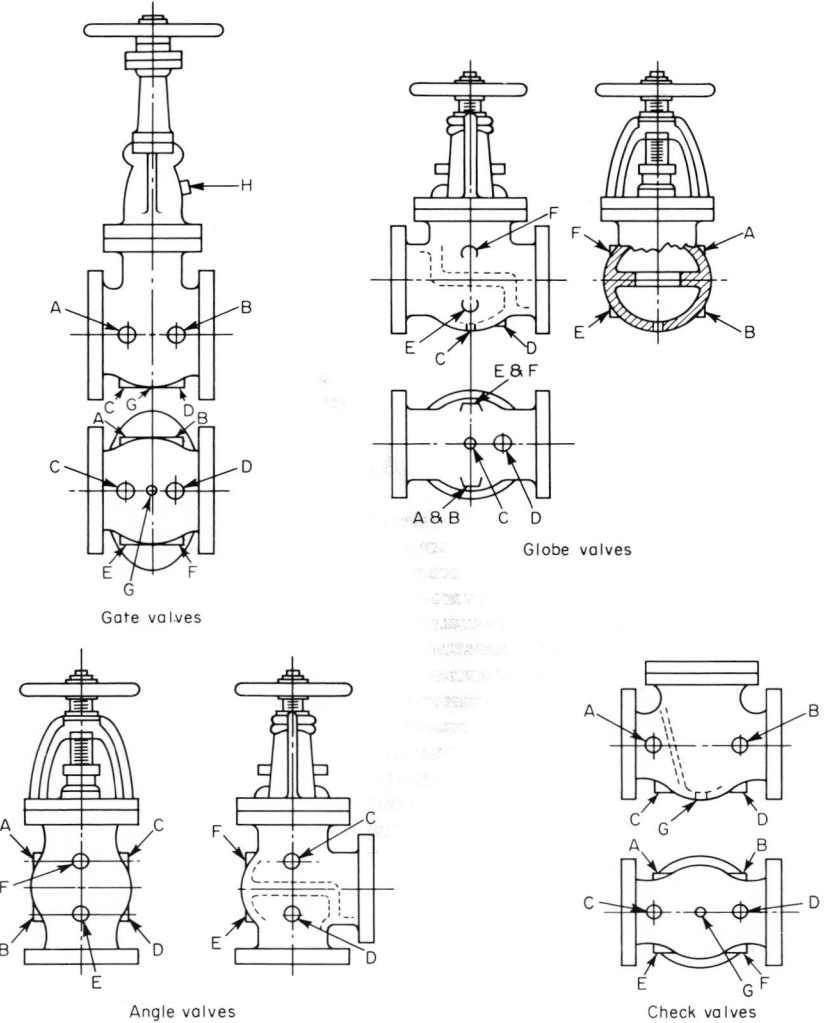

Gate valves

Globe valves

Angle valves

Check valves

FIG. 44. Standard locations of openings in valve bodies for drains and bypasses as set forth in MSS SP 45-1953.

On most cast-steel valves, bosses are normally furnished as part of the standard body.

VALVE INSPECTION

Valves should be dismantled periodically for inspection of the internal parts. This is particularly important in pressure-relief devices; for example, in pressure-relief valves, seating surfaces must be inspected to insure tight contact, as leakage

might cause the valve to function improperly.   Similarly, spring failures due to cracking or corrosion might make the valves inoperative.

Gate, plug, and ball valves which have been used for throttling may have suffered significant corrosion or erosion.   In gate valves, the seat area is particularly prone to erosion due to steam or liquid turbulence.

Reassembled valves should be tested hydrostatically or pneumatically.   Recommendations for valve inspection in refinery service have been published by the American Petroleum Institute Guide for Inspection of Refinery Equipment, Chapter XI, "Pipe Valves, and Fittings," and Chapter XVI, "Pressure-relieving Devices," and in Standard RP 525, "Testing Procedure for Pressure-relieving Devices Discharging Against Variable Back Pressure."

PIPE FABRICATION

Pipe fabrication involves the various forming, shaping, machining, welding, cleaning, and heat-treatment operations necessary to convert initially straight pipe sections, valves, and fittings into a finished piping system or into components which may become integral parts of piping systems.

Fig. 45. Fabricated piping components for nuclear reactor application.

Considerable prefabrication is generally done in fabricating plants where specialized equipment is available for the production of piping components under carefully controlled supervision.   The fabrication of piping may follow any of several patterns, including assembly and joining of all pieces at the erection site location. As a general rule, prefabrication of as much piping as practicable in suitable shops

is preferable, followed by on-site assembly and joining of the prefabricated components. This is because shop assembly, where specialized equipment is available and working conditions are favorable, is conducive to easier, more consistent, and more dependable fabrication than field assembly. The advantages of shop fabrication become particularly pronounced with increasing diameter and wall thickness of the piping components. Assemblies which require precision fit-up such as headers, nuclear reactor components, and other complex configurations also are best prefabricated in properly equipped pipe-fabricating plants (see Fig. 45). Many shaping and forming operations such as bending, extruding, swaging, automatic welding, and full-furnace heat treatment are normally done only in fabricating plants.

Since standardization is relatively recent in the pipe-fabrication industry, many different terms have been coined to describe identical operations. The numerous specific trade names used further add to the confusion. To assist in the standardization of terminology, the terms most commonly used are defined in Chap. 1.

CODES AND STANDARDS

Various codes and standards applicable to fabricated and welded piping systems have been prepared by committees of leading engineering societies and standardization groups. These are generally written to establish minimum requirements of quality and safety. Many applications, however, may require more conservative design and welding requirements than prescribed in these standards. Recognition of service requirements in the evaluation of the applicable standards is essential to a properly engineered and constructed piping job. Service failures frequently result when this is not done. Many standards also specify inspection requirements in order to establish proper levels of quality. Code-writing committees are beginning to recognize that the level of quality attained in a fabricated and erected piping system, and the cost of producing this level, must be related to specific inspection requirements. For example, where full-penetration weld deposits are required, 100 per cent radiographic examination must be specified. Spot radiographic examination of 10 per cent of the weld joints in a piping system should permit a lower level of weld quality, allowing, for example, some lack of penetration. This is recognized in Section 31.3 of the American Standards Association Code for Pressure Piping Applicable to Petroleum Refinery Piping, which states, for incomplete penetration:

100% radiographic examination—not acceptable.

Spot radiographic examination—the total joint penetration shall not be less than the thinner of the two components being joined, except to the extent that incomplete root penetration is permitted. The depth of incomplete root penetration or lack of fusion at the weld root shall not exceed $\frac{1}{32}''$ or half the thickness of the weld reinforcement, whichever is smaller. The total length of such incomplete root penetration or lack of fusion at the root shall not exceed $1\frac{1}{2}$ in. in any 6 inches of weld length.

Standards and codes applicable to fabricated and erected industrial piping systems and their respective levels of quality in these systems have been prepared by a number of engineering and trade associations.

American Society of Mechanical Engineers Boiler and Pressure Vessel Code.[1] This code covers piping connected to boilers (Section I) or pressure vessels (Section VIII). Insurance companies and most state and municipal authorities recognize this code and make it a prerequisite for acceptance and installation of such power equipment within their jurisdiction.

[1] American Society of Mechanical Engineers, United Engineering Center, 345 East 47th Street, New York, New York 10017.

This code involves nine sections of which the following are concerned with piping:

Section I   *Power Boilers:* Formulates minimum specifications for the construction of piping directly connected to steam power boilers within prescribed limits.   Where specific rules are not given in other sections, the rules in Section I may be used.

Section III   *Nuclear Vessels:* Covers minimum construction requirements for the materials, design, fabrication, inspection and testing, and certification of vessels in nuclear power plants.   This section recommends that the increase in the brittle fracture transition temperature of steels due to neutron irradiation be checked periodically.

Section VIII   *Unfired Pressure Vessels:* Contains rules, fabrication, and inspection requirements for the attachment of piping to pressure vessels.

Procedures for the qualification of welding procedures, welders, and welding operators are established in

Section IX   *Qualification Standard* for welding and brazing procedures, welders, brazers, and welding and brazing operators.

Special requirements not incorporated into the respective sections of the Boiler and Pressure Vessel Code are covered by Code Cases.   These are issued and published by the Boiler and Pressure Vessel Code Committee, usually as the result of questions received from industry, requesting clarification of an operation or application not yet covered by the Code or of a controversial issue.   Periodically, these cases are published in *Mechanical Engineering* and in booklet form by the American Society of Mechanical Engineers.

To substantiate code compliance, the ASME Boiler and Pressure Vessel Code requires that partial-data reports be furnished by the fabricator to the purchaser or his designated inspection agency.

American Petroleum Institute.[1]   The American Petroleum Institute generally recognizes the Specifications by the ASA Code for Pressure Piping, Section 3, Refinery Piping and, when applicable, the ASME Boiler and Pressure Vessel Code.

One additional specification has been issued, also, to cover welding and inspection practices.

1104   *Standard for Welding Pipe Lines and Related Facilities.*   This Standard includes weld quality acceptability limits, radiographic inspection requirements, and welder and welding procedure test requirements.

American Standards Association.[2]   The American Standards Association is preparing or has issued the following sections of the Code for Pressure Piping:

Section 1   Power Piping
Section 2   Industrial Gas & Air Piping
Section 3   Petroleum Refinery Piping (B31.3-1966)
Section 4   Oil Transportation Piping (B31.4-1959)
Section 5   Refrigeration Piping (B31.5-1962)
Section 6   Chemical Industry Process Piping
Section 8   Gas Transmission and Distribution Piping Systems (B31.8-1963)
Section —   Nuclear Piping

These sections deal primarily with minimum safety requirements for the selection and designation of dimensional standards, materials, design, fabrication, erection,

[1] American Petroleum Institute, 1271 Avenue of the Americas, New York, New York 10020.

[2] Published by the American Society of Mechanical Engineers, United Engineering Center, 345 East 47th Street, New York, New York 10017.

**Table 40a.   Summary of Major Requirements of the ASA Code for Pressure
Piping (Power Piping, ASA B 31.1-1955)**

Determining minimum pipe-wall thickness ..	Wall-thickness requirements in Para. 104
Allowable stress in materials	Same as ASME Boiler Code Sec. I-1962, except for ASME Case 1319
Pipe-wall allowance for thinning in bending .	Allowance required in Table 102.4.5. On five-diameter bends, allowance is $1.08t_m$
Limitations of pipe flattening in bending....	Limitation is $\pm 8\%$ (tentative)
Permissible ID and OD misalignment of abutting ends for butt welding...............	$\frac{1}{16}''$ max internal misalignment of abutting walls. External misalignment is not defined (Para. 127.3)
Limitations on use of slip-on and socket-weld flanges...............................	No limitations, provided pressure-temperature ratings do not exceed those in ASA Standards approved by Sec. I
Preheating for welding of carbon-steel materials..............................	50 F min. When over 0.30% C, minimum is 175 F (Table 131)
Preheating (above hand-hot) for welding of carbon moly, low- and medium-Cr-Mo alloy steels...........................	Required by Table 131
Stress relieving of welds, carbon steel.......	Required when wall thickness is $\frac{3}{4}''$ and over (Para. 131.3)
Stress-relieving of welds, carbon moly, low- and medium-Cr-Mo alloy steels..........	Required (Para. 131.3)
Postheat treatment of hot-formed bends and shapes in low- and medium-Cr-Mo steels and in austenitic stainless steels..........	No requirement by Sec. I but required by industry practice on Cr-Mo steels
Hydrostatic or pressure testing after erection of piping systems	Required wherever practical at 1.5 times the design pressure (Para. 137.4)
Hardness limitations on welds and heat-affected zones	Nothing defined
Radiographic and other inspection, when required	Required on welds in pipe when steam temperatures exceed 925 F or for pipe sizes listed in Table 40g for Sec. I, ASME Boiler Code
Welding procedure and welder qualification .	Same as Sec. IX, ASME Boiler Code
Special quality considerations..............	....

testing, and inspection of piping systems.   The principal purpose of the code is to
serve as a guide to state and municipal authorities in drawing up their regulations
and to serve architects, engineers, equipment manufacturers, fabricators, and
erectors as a standard of reference for fabricated and erected piping systems.   As
the above sections are prepared by different committees who represent somewhat
differing interests, the various code sections may vary somewhat in specific requirements.   These are summarized in Tables 40a through h.   For comparison,
requirements in Sections I and III of the ASME Boiler and Pressure Vessel Code are
also included in this summary.

Special requirements not yet incorporated into the Standard Specifications are
covered by Code Cases.

American Water Works Association.[1]   The American Water Works
Association has issued jointly with the American Welding Society the following
Standard:

C206-62 (AWS D7.0-62)   AWWA Standard for Field Welding of Steel Water
Pipe Joints.

[1] American Water Works Association, 2 Park Avenue, New York, New York 10016.

Table 40b. Summary of Major Requirements of the ASA Code for Pressure Piping (Petroleum Piping, ASA B31.3-1966)

Determining minimum pipe-wall thickness ..	Includes allowances for corrosion, threading, and weld-joint factors (Para. 302.4)
Allowable stress in materials	Some values higher than Sec. I
Pipe-wall allowance for thinning in bending .	Allowance required (Para. 304.2.1)
Limitations of pipe flattening in bending	Allowance required (Paras. 329.1 and 304.2.1)
Permissible ID and OD misalignment of abutting ends for butt welding.	$\frac{1}{16}''$ max internal misalignment of abutting walls preferred but not mandatory. External misalignment not defined. (Para. 327.3.1C)
Limitations on use of slip-on and socket-weld flanges.	Same as Sec. I, except approval of standards by Sec. III is required
Preheating for welding of carbon-steel materials.	To hand-warm condition
Preheating (above hand-hot) for welding of carbon moly, low- and medium-Cr-Mo alloy steels	Required for materials with P numbers 3, 4, 5, 6, 9, and 10 (Para. 331.2)
Stress relieving of welds, carbon steel	Same as Sec. I (Para. 331.3)
Stress-relieving of welds, carbon moly, low- and medium Cr-Mo alloy steels.	Required for materials with P numbers 1, 3, 4, 5, 6, 9, and 10 in thickness of ¾ in. or more (Para. 331.3)
Postheat treatment of hot-formed bends and shapes in low- and medium-Cr-Mo steels and in austenitic stainless steels	Required as necessary to restore desired properties (Paras. 330.1 and 331.3.2)
Hydrostatic or pressure testing after erection of piping systems	Pressure testing required on all installed piping (Para. 337.1)
Hardness limitations on welds and heat-affected zones	Hardness limits apply on some materials when postheat treatment differs from Table 331.2.1 in Sec. III (Para. 331.3)
Radiographic and other inspection, when required	Limitations defined for spot and for 100% radiographic examination (Para. 327.4)
Welding procedure and welder qualification .	ASME Boiler Code, Sec. IX, except that performance qualification papers are transferable
Special quality considerations.	. . . .

American Welding Society.[1]   The American Welding Society has published monographs and several recommended welding practices applicable to pipe welding. The following are concerned primarily with pipe welding:

D10.2-54   Recommended Practices for Repair Welding of Cast Iron Pipe, Valves and Fittings.

D10.3-55   Recommended Practices for Interruption of Heat Treatment Cycles for Low Chromium-Molybdenum Steel Piping Materials.

D10.4-55   The Welding of Austenitic Chromium-Nickel Steel Piping and Tubing.

D10.5.59   Welding Ferrous Materials for Nuclear Power Piping.

D10.6-59   Gas Tungsten-arc Welding of Titanium Piping and Tubing.

D10.7-60   Recommended Practices for Gas Shielded-arc Welding of Aluminum and Aluminum Alloy Pipe.

D10.8-61   Welding of Chromium-Molybdenum Steel Piping.

[1] American Welding Society, United Engineering Center, 345 East 47th Street, New York, New York 10017.

Table 40c. Summary of Major Requirements of the ASA Code for Pressure Piping (Oil Transportation Piping, ASA B31.4-1959)

Determining minimum pipe-wall thickness ..	Differs from other ASA sections (Para. 404.1.2)
Allowable stress in materials	Not included. Yield is used instead of stress values
Pipe-wall allowance for thinning in bending .	No allowance is required
Limitations of pipe flattening in bending....	None
Permissible ID and OD misalignment of abutting ends for butt welding...............	Nothing included on misalignment
Limitations on use of slip-on and socket-weld flanges...............................	Same as Sec. I, except approval of standards by Sec. IV is required
Preheating for welding of carbon-steel materials............................	Not required
Preheating (above hand-hot) for welding of carbon moly, low- and medium-Cr-Mo alloy steels...........................	No requirement (Sec. IV is limited to carbon steels)
Stress relieving of welds, carbon steel.......	No requirement
Stress-relieving of welds, carbon moly, low- and medium-Cr-Mo alloy steels..........	No requirement (section is limited to carbon steels)
Postheat treatment of hot-formed bends and shapes in low and medium-Cr-Mo steels and in austenitic stainless steels..........	No requirement
Hydrostatic or pressure testing after erection of piping systems	Hydrostatic testing required on portions of piping systems (Para. 437.4.1)
Hardness limitations on welds and heat-affected zones	Nothing defined
Radiographic and other inspection, when required	None required by Sec. IV, but specifying such tests is optional with the designer
Welding procedure and welder qualification .	....
Special quality considerations.............	....

Pipe Fabrication Institute.[1] Standards have been prepared by the Pipe Fabrication Institute covering various fabrication, design, cleaning, and inspection operations. The following have been issued to date:

ES1   End Preparation and Machined Backing Rings for Butt Welds
ES2   Method of Dimensioning Welded Assemblies
ES3   Linear Tolerances, Bending Radii, Minimum Tangents
ES4   Shop Hydrostatic Testing of Fabricated Piping
ES5   Cleaning Fabricated Piping
ES7   Minimum Length and Spacing for Welded Nozzles
ES8   Preheat and Postheat Welding Practices for Low Chromium–Molybdenum Steel Pipe
ES9   Shielded Metal Arc-welding Dissimilar Ferritic Steels
ES10   Stress Relieving Welded Attachments
ES11   Permanent Identification of Piping Materials
ES12   Preheat and Postheat Welding Practices for Medium Chromium–Molybdenum Steel Pipe
ES13   Classification of Shop Testing, Inspection and Cleaning
ES14   Recommended Practice for Magnetic Particle Inspection
ES15   Recommended Radiographic Interpretation of Tungsten Inert Gas Welds
ES16   Access Holes and Plugs for Radiographic Inspection of Pipe Welds
ES17   Liquid Penetrant Inspection
ES18   Ultrasonic Inspection of Seamless Piping

[1] Pipe Fabrication Institute, 992 Perry Highway, Pittsburgh, Pa. 15237.

Table 40d. Summary of Major Requirements of the ASA Code for Pressure Piping (Refrigeration Piping, ASA B31.5-1962)

Determining minimum pipe-wall thickness ..	Differs from other ASA sections (Para. 504.1.2)
Allowable stress in materials	Differs from other ASA sections (Para. 502.3.1 and Table 502.3.1)
Pipe-wall allowance for thinning in bending .	Allowance required (Para. 504.2.1)
Limitations of pipe flattening in bending	Limitations do apply ±10% from nominal diameter (External pressure service ±3% from nominal diameter) (Para. 504.2.1)
Permissible ID and OD misalignment of abutting ends for butt welding	Either 1/16″ or one-fourth nominal internal misalignment of abutting walls whichever is smaller. External misalignment not defined (Para. 527.3.1C)
Limitations on use of slip-on and socket-weld flanges...............................	Same as Sec. I, except approval of standards by Sec. V is required
Preheating for welding of carbon-steel materials...........................	Same as Sec. III
Preheating (above hand-hot) for welding of carbon moly, low- and medium-Cr-Mo alloy steels..........................	Required for materials with P numbers 3, 4, 5, 6, 9, and 10 (Para. 531.2)
Stress relieving of welds, carbon steel.......	Same as Sec. I (Para. 531.3)
Stress-relieving of welds, carbon moly, low- and medium-Cr-Mo alloy steels..........	Required for materials with P numbers 3, 4, 5, 6, 9, and 10 (Para. 531.3)
Postheat treatment of hot-formed bends and shapes in low- and medium-Cr-Mo steels and in austenitic stainless steels..........	Required as necessary to restore desired properties (Paras. 530.1 and 531.3.2)
Hydrostatic or pressure testing after erection of piping systems	Pressure testing required on all piping erected on the premises (Para. 537.2)
Hardness limitations on welds and heat-affected zones	Hardness limits mandatory on some materials regardless of postheat treatment used (Para. 531.3.3) (same as Sec. III)
Radiographic and other inspection, when required	None required by Sec. V, but specifying such tests is optional with the designer
Welding procedure and welder qualification .	ASME Boiler Code, Sec. IX, except that qualification papers are transferable
Special quality considerations.............	....

Other Commercial Codes. Specific requirements of various regulatory state and municipal codes have been summarized by the National Bureau of Casualty Underwriters, Boiler and Machinery Division, in its "Synopsis of Boiler and Pressure Vessel Laws, Rules, and Regulations."

Governmental Agencies. Many standards and specifications applicable to pipe fabrication, welding, and inspection have been issued by governmental agencies.

Of the many specifications concerned with piping, the following represent the more widely used standards:

General Specifications for Ships for the United States Navy. This specification covers general requirements and is divided into various sections concerned with design, testing, and fabrication.

538-1　Steam Heating Systems
546-1　Condensers
548-1　General Piping Systems

Table 40e. Summary of Major Requirements of the ASA Code for Pressure Piping (Gas Transmission Piping, ASA B31.8-1963)

Determining minimum pipe-wall thickness .. — Differs from other ASA sections. Includes allowances for construction type, longitudinal joint, temperature, and corrosion

Allowable stress in materials .... — Not included. Yield is used instead of stress values

Pipe-wall allowance for thinning in bending — No allowance required

Limitations of pipe flattening in bending .... — Limitations do apply ±2.5% from nominal diameter

Permissible ID and OD misalignment of abutting ends for butt welding.............. — Nothing included on misalignment

Limitations on use of slip-on and socket-weld flanges................................ — Same as Sec. I, except approval of standards by Sec. VIII is required

Preheating for welding of carbon-steel materials............................. — Required when carbon exceeds 0.32% or carbon equivalent exceeds 0.65% (Para. 826)

Preheating (above hand-hot) for welding of carbon moly, low- and medium-Cr-Mo alloy steels........................... — Alloy steels not included in Sec. VIII. Refers to B31.1, Sec. VI, Chap. IV.

Stress relieving of welds, carbon steel....... — Required when wall thickness exceeds 1¼″ and when the carbon exceeds 0.32% or carbon equivalent exceeds 0.65% (Para. 827)

Stress-relieving of welds, carbon moly, low- and medium-Cr-Mo alloy steels.......... — Alloy steels not included in Sec. VIII. Refers to B31.1, Sec. VI, Chap. IV

Postheat treatment of hot-formed bends and shapes in low- and medium-Cr-Mo steels and in austenitic stainless steels.......... — No requirement

Hydrostatic or pressure testing after erection of piping systems .... — Pressure testing required on some piping systems (Para. 841.4), except nonstandard assemblies may have to be leak-tested at the operating pressure of the erected line (Para. 831.34)

Hardness limitations on welds and heat-affected zones .... — Nothing defined

Radiographic and other inspection, when required .... — Radiographic examination API Standard 1104. Nondestructive tests or removal of completed welds required for some operations to check weld quality (Para. 828.2)

Welding procedure and welder qualification . — API Standard 1104. Bend tests shall apply to welders on compressor stations ASA B31.1, Sec. VI, Chap. IV. ASME Boiler and Pressure Vessel Code, Sec. IX

Special quality considerations.............. — Limitations on surface gouges and grooves, dents, and arc burns (Para. 841.24)

551-1 Fire Tube Boilers

551-2 Oil Fired Water Tube Boilers and Superheaters

S9-1 Requirements for Fabrication by Welding and Allied Processes of Ships and Their Components

Different levels of quality are recognized for piping welds, pressure-vessel welds, and machinery welds. Requirements are given on welding processes, procedures, procedure approval, preheat-postheat requirements, and inspection acceptance standards for radiographic, magnetic particle, and liquid penetrant testing.

Table 40f. Summary of Major Requirements of the ASA Code for Pressure Piping (Until Publication of Sec. B31, Nuclear Piping, ASA B31 Nuclear Code Cases Apply)

Determining minimum pipe-wall thickness	....
Allowable stress in materials	Allowable stress values for austenitic stainless steels given in Cases N-7 and N-8
Pipe-wall allowance for thinning in bending .	....
Limitations of pipe flattening in bending	....
Permissible ID and OD misalignment of abutting ends for butt welding	....
Limitations on use of slip-on and socket-weld flanges	....
Preheating for welding of carbon-steel materials	....
Preheating (above hand-hot) for welding of carbon moly, low- and medium-Cr-Mo alloy steels......................	....
Stress relieving of welds, carbon steel	....
Stress-relieving of welds, carbon moly, low- and medium-Cr-Mo alloy steels	....
Postheat treatment of hot-formed bends and shapes in low- and medium-Cr-Mo steels and in austenitic stainless steels..........	In sodium-reactor piping, ASA B31 Case N-8 requires postheat treatment at 1950 F on hot- and cold-formed pipe
Hydrostatic or pressure testing after erection of piping systems	....
Hardness limitations on welds and heat-affected zones	....
Radiographic and other inspection, when required	Cases N-7 and N-8 require radiography of all longitudinal and circumferential welds in aqueous-type and sodium-type reactor piping. Liquid penetrant inspection also required
Welding procedure and welder qualification .	....
Special quality considerations	....

MIL-STD-248   *Qualification Tests for Welders (other than aircraft weldments)* covers the requirements and tests necessary to qualify welders for production of military weldments other than aircraft weldments.

MIL-STD-271   *Nondestructive Testing for Metals* covers nondestructive testing techniques and qualification requirements for radiographic, liquid penetrant, magnetic particle, ultrasonic, and leak testing. This standard is applicable to piping systems.

MIL-STD-22   *Welded-joint Designs* covers welded-joint designs for manual and semiautomatic arc and gas welding. This standard does not apply to aeronautical equipment.

MIL-P-18178   *Fabrication of Pipe Bends for Shore Use* covers requirements for fabricated piping.

MIL-STD-278   *Welding and Inspection of Machinery, Piping and Pressure Vessels for Ships of the United States Navy*. Requirements are similar to those of Section S9-1 on welding processes, procedures, procedure approval, and preheat and postheat requirements and inspection. This standard of S9-1 is generally followed for welding piping for conventional surface naval vessels.

MIL-STD-418   *Mechanical Tests for Welded Joints* covers mechanical tests used for evaluation of welded joints.

MIL-STD-437   *X-ray Standards for Bare Aluminum Alloy Electrode Welds* states acceptability standards for semiautomatic welded aluminum joints.

Table 40g. Summary of Major Requirements of the ASME Power Boiler Code (Sec. I-1965)

Determining minimum pipe-wall thickness ..	Detailed requirements in Paras. PG 27 and PG 59
Allowable stress in materials	Stress values included in Sec. I. Modified for $1\frac{1}{4}\% \text{ Cr}-\frac{1}{2}\% \text{ Mo}$ in Code Case 1319
Pipe-wall allowance for thinning in bending .	No allowance required
Limitations of pipe flattening in bending....	None
Permissible ID and OD misalignment of abutting ends for butt welding..............	$1\frac{1}{4}$ in. maximum on wall, both internal and external, for butt welds. $\frac{1}{8}''$ max for longitudinal welds (Para. PW 33)
Limitations on use of slip-on and socket-weld flanges.............................	Slip-on flanges limited to 4″ max size. Socket-weld flanges limited to 3″ max size for 600 psi max and $2\frac{1}{2}''$ max size for 900 and 1,500 psi (Para. PW 41)
Preheating for welding of carbon-steel materials.............................	50 F min. When over 0.30% C, 175 F min
Preheating (above hand-hot) for welding of carbon moly, low- and medium-Cr-Mo alloy steels..........................	Min preheating requirements in Para. PW 38
Stress relieving of welds, carbon steel.......	Required above $\frac{3}{4}''$ thickness (Para. PW 39)
Stress-relieving of welds, carbon moly, low- and medium-Cr-Mo alloy steels..........	Required (Para. PW 39)
Postheat treatment of hot-formed bends and shapes in low- and medium-Cr-Mo steels and in austenitic stainless steels..........	No code requirement but required on Cr-Mo alloy steels by industry practice
Hydrostatic or pressure testing after erection of piping systems	Hydrostatic testing required after erection (Para. PG 99)
Hardness limitations on welds and heat-affected zones	Nothing defined
Radiographic and other inspection, when required	Required when part contains water and exceeds 10″ nominal pipe size or $1\frac{1}{8}''$ wall thickness or contains steam and exceeds 16″ nominal pipe size or $1\frac{5}{8}''$ wall thickness. Other testing at option of designer
Welding procedure and welder qualification .	ASME Boiler Code Sec. IX
Special quality considerations	....

NAVSHIPS 250-634-7 *Standard Terminology and Definitions for Weld Conditions and Defects* establishes a standard set of terminology and definitions for various weld-joint conditions and defects.

NAVSHIPS-250-582 *Fabrication and Control of Steel Pipe and Tubing for Welded Piping Systems* governs the material fabrication, welding, and heat treatment of seamless and welded carbon steel, carbon-molybdenum, chromium-molybdenum, and austenitic corrosion-resisting steel pipe and tubing for welding piping systems on naval vessels.

NAVSHIPS 250-692-2 *X-ray Standards for Production and Repair Welds* covers basic acceptance standards for radiographic inspection.

NAVSHIPS-250-1500-1 *Welding of Reactor Coolant and Associated Systems and Components for Naval Nuclear Power Plants.* The use of this standard is restricted to contractors involved in the fabrication of piping under Navy contracts. It is one of the most detailed specifications concerned with piping and contains extensive requirements on weld-joint design, qualification welding, quality control and inspection and acceptance standards.

Table 40h. Summary of Major Requirements of the ASME Nuclear Vessel Code (Sec. III-1965)

Determining minimum pipe-wall thickness ..	Detailed requirements include consideration of corrosion allowance and cyclic fatigue conditions
Allowable stress in materials	Same as ASME Boiler Code, Sec. I-1962
Pipe-wall allowance for thinning in bending .	No allowance required
Limitations of pipe flattening in bending	Calculated for shells and pipe where the difference between minor and major axes shall not exceed the smaller of $(D + 50)/200$ and $D/100$, where D is the nominal ID
Permissible ID and OD misalignment of abutting ends for butt welding..............	Allowable offset tolerances of up to $\frac{1}{8}''$ on longitudinal joints and $\frac{1}{4}''$ on circumferential joints in Para. N-525
Limitations on use of slip-on and socket-weld flanges	....
Preheating for welding of carbon-steel materials...........................	Not required
Preheating (above hand-hot) for welding of carbon moly, low- and medium-Cr-Mo alloy steels..........................	Specific requirements not included; recommends Sec. IX of ASME Boiler Code
Stress relieving of welds, carbon steel.......	Required above $\frac{3}{4}''$ thickness (Para. N-532)
Stress-relieving of welds, carbon moly, low- and medium-Cr-Mo alloy steels..........	Required (Para. N-532)
Postheat treatment of hot-formed bends and shapes in low and medium Cr-Mo steels and in austenitic stainless steels..........	No Code requirement but required on Cr-Mo alloy steels by industry practice
Hydrostatic or pressure testing after erection of piping systems	Required at not less than 1.25 times the design pressure (Para. N-714)
Hardness limitations on welds and heat-affected zones	Nothing defined
Radiographic and other inspection, when required	All castings and welds must be radiographed. Magnetic-particle and liquid-penetrant examination also required on castings. Similar requirements on pipe for reactor coolant systems (Paras. N-324 and N-323)
Welding procedure and welder qualification .	ASME Boiler Code, Sec. IX, with additional requirements in Para. N-540 for tube-to-tube sheet welds
Special quality considerations..............	Defect limitations defined in Para. N-324

FORMING

Forming of piping involves bending, expanding, extruding, swaging, and lapping. These operations generally are performed in pipe-fabricating plants where the specialized equipment necessary for these operations is available.

Pipe and tube bends generally are recognized to be the most economical means of changing directions while providing flexibility and end reactions of piping and tubing systems within the allowable limits. Standard pipe and tube bends are illustrated in Fig. 46.

By employing bends, the cost of materials, piping-system fabrication, and erection can usually be reduced significantly, compared with the use of welding elbows with attendant welds and greater footages of pipe.

Where fitting-to-valve or fitting-to-flange joints may require special spool pieces and two butt welds, an average of one spool piece and one butt weld may be eliminated by using bends with tangents of lengths that utilize total bar footages.

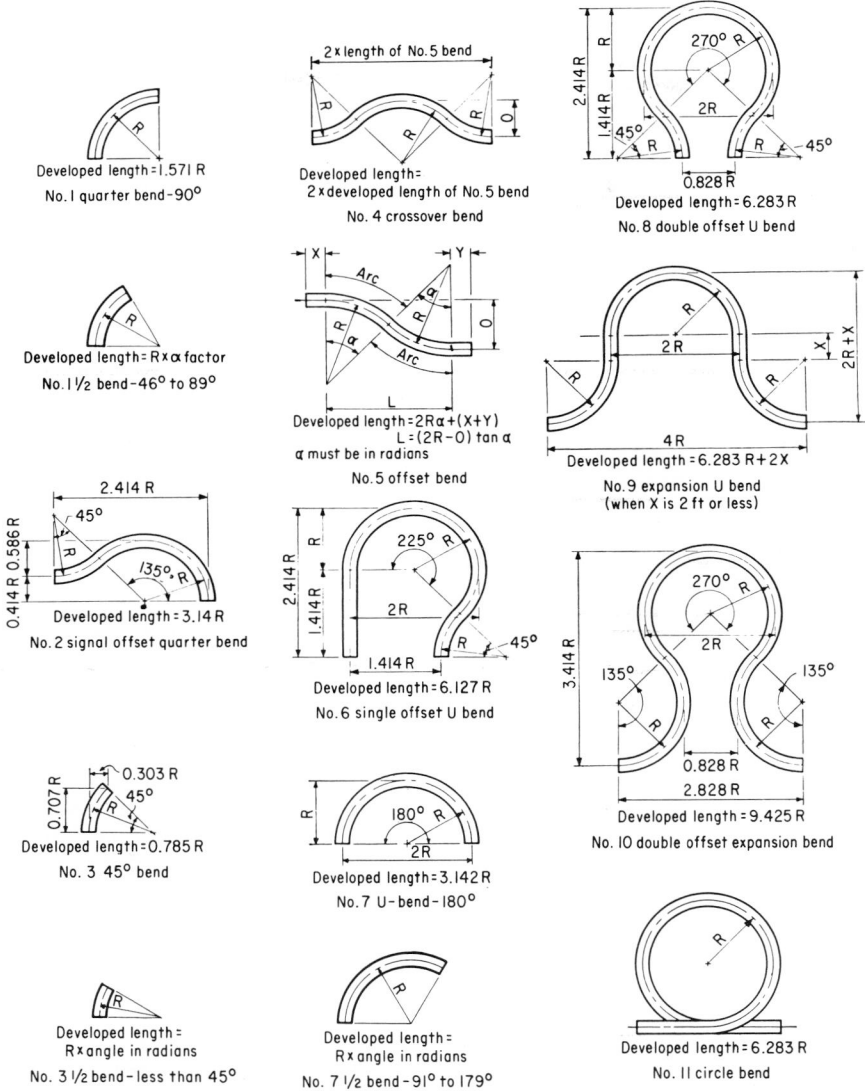

FIG. 46. Examples of various standard pipe bends.

Another advantage of bends over elbows is the reduction in resistance to flow, especially at high velocities, due to the larger radius (long sweep) of the bends.

The limitations on the bendability of pipe and tube materials depend on (1) the ductility of the metal or alloy, (2) the tendency of the outside of the bend, which is being stretched, to flatten or collapse, and (3) the tendency of the inside of the bend, which is being compressed, to buckle or wrinkle.

Ideally the metal should combine high ductility (elongation) with a low ratio of yield strength to ultimate strength. The metal should also exhibit a minimum rate of work hardening.

Bending of Ferrous Pipe and Tube Materials

Pipe and tube of small diameters (under $2\frac{1}{2}$ in. OD) of standard and relatively light wall thicknesses is usually bent cold. On the other hand, larger diameter and heavier wall pipe is generally bent hot. For pipe of diameters between 3 and 10 in. OD, the preference of hot bending over cold bending is a matter of cost consideration. Where the quantity of identical bends is substantial and the heavy bending equipment is available, cold bending may be more economical than hot bending. The bending radius, the number of identical bends required, and the chemical composition and metallurgical properties of the pipe material are the deciding factors.

On piping, industry experience has demonstrated the practicability of using a bending radius of five pipe diameters to keep expensive friction losses, erosion, and turbulence to a minimum. On tube, many different bending radii are produced in accordance with the desires of the designer.

FIG. 47. Portable hydraulic ram-type bender.

Essentially the same considerations apply to the bending of high-nickel alloys as are discussed here for the bending of ferrous pipe materials.

Cold Bending. Although cold bending is done extensively with pipe in nominal diameters up to $2\frac{1}{2}$ in. OD and with tube in diameters up to 4 in. OD, little use is made of cold-bent piping in larger diameters. In Europe, however, cold-bent piping in sizes up to 12 in. has been widely accepted and is often preferred to hot-bent piping because of savings involved for multiple bends of identical dimensions.

Cold-expanded pipe grades are covered by American Petroleum Institute specifications. This piping is used quite extensively in gas-transmission lines. Cold roto-rolled 4135 alloy-steel and Type 410 stainless-steel tubing having minimum specified yield strengths of 90,000 to 100,000 psi have been used successfully in the chemical industry's polyethylene plants. In the annealed condition, the 4135 and Type 410 steel grades exhibit yield strengths of approximately 60,000 and 30,000 psi, respectively.

Equipment for Cold Bending. Many years ago, cold bending was generally done in ram-type bending machines. They were particularly suited for pipe of heavier wall thickness (over $\frac{1}{2}$ in.) where some flattening in the bend area was not considered objectionable. Ram-type bending machines were not suitable for high production rates. Moreover, considerable skill was required to produce smooth, wrinkle-free bends.

In a ram-type bender (Fig. 47) two pressure dies are mounted in a fixed position on the frame of the machine. Their mounting pins, however, are free to rotate. The bending form is attached directly to the piston rod of the hydraulic cylinder. Although normally the pipe is bent to the radius of the bending form, bending to larger radii is possible by limiting the advancement of the ram.

Rotary-type bending machines of the type illustrated in Fig. 48 are now being increasingly used.

Fig. 48. Rotary-type bending machine for pipe with diameters to 10 in. OD.

As the cold working of the bending operation becomes more severe, the inside of the pipe must be supported with mandrels. Typical mandrels used in commercial pipe bending are illustrated in Fig. 49. An empirical graph illustrating the relation of mandrel support to bending radius, outside diameter, and wall thickness of the pipe to be bent is plotted in Fig. 50. This graph is widely used as a guide for cold bending of most ferrous and nonferrous piping materials.

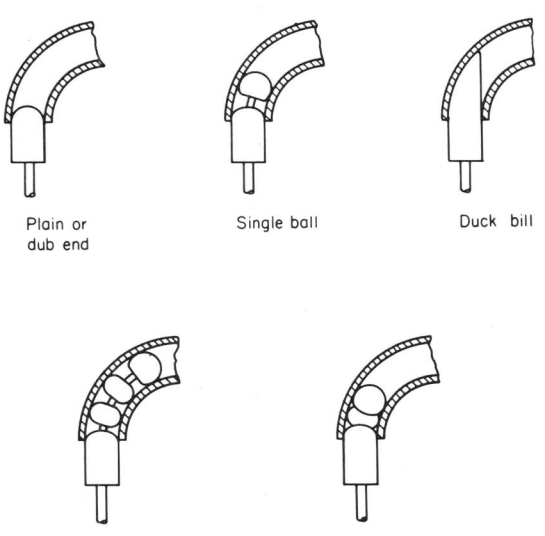

Plain or dub end   Single ball   Duck bill

Multi-ball   Shell or Nested ball

Fig. 49. Typical mandrels used in commerical cold-bending practice.

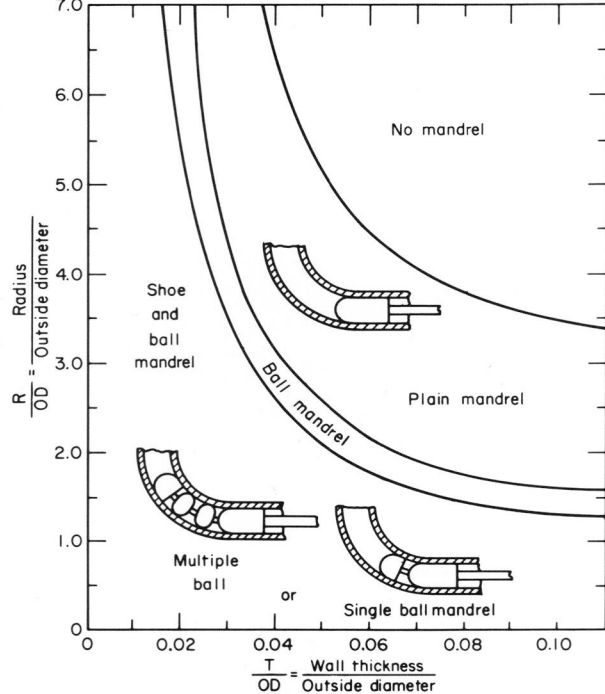

FIG. 50. Mandrel and shoe requirements for cold-bending of pipe.

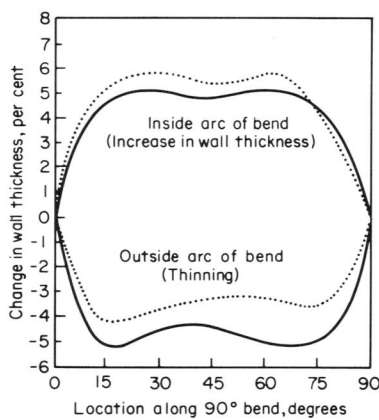

FIG. 51. Changes in wall thickness of two 8-in. OD Schedule 160 $2\frac{1}{4}$ Cr–1 Mo pipe cold bent to a five-diameter radius.

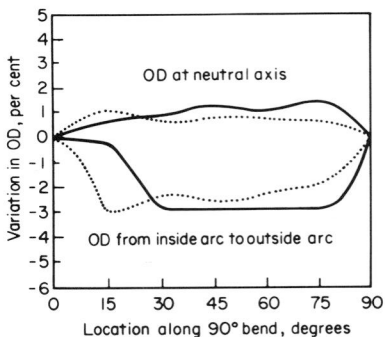

FIG. 52. Flattening in two 8-in. OD Schedule 160 $2\frac{1}{4}$ Cr–1 Mo pipe cold bent to a five-diameter radius.

Figure 51 illustrates the change in wall thickness for two sections of 8 in. OD Schedule 160 (0.906 in. wall) $2\frac{1}{4}\%$ Cr–1 % Mo alloy steel pipe cold bent to a five-diameter radius. For the same pipe, flattening is shown in Fig. 52.

Hot Bending. Hot bending is extensively used for making individual bends in piping of sizes $2\frac{1}{2}$ in. and larger. Prior to hot bending, the pipe is generally sand-filled. This facilitates more uniform bending, and minimizes excessive thinning and ovality. The sand also helps to maintain the pipe longer at the hot-bending temperatures and thus provides longer bending cycles within a limiting temperature range.

The pipe is normally heated in specially designed furnaces. The furnace (illustrated in Fig. 53) directs the gas flames in a circular path along the furnace wall.

FIG. 53. Gas-fired furnace for heating steel pipe at a controlled rate. Precision thermo-couples and pyrometers are used to monitor heating of pipe within specified temperature ranges through predetermined heating cycles. This precise method retains the special properties of modern steels.

This avoids direct impingement of the flames on the pipe surface, which minimizes hot spots, excessive oxidation, and scaling. Temperatures for heating prior to hot bending of ferrous piping materials normally range from 1900 to 2050 F. When it has reached the bending temperature, the pipe is placed on bending tables (Fig. 54), where it is bent to the specified radius. On ferrous materials, hot bending is generally not done below 1600 F. Sometimes, several heating and bending cycles may be required.

After the pipe is bent and cooled, the sand is removed from the inside of the piping. In carbon and low-alloy steel, the excessive scale on the pipe inside is removed by turbining, blasting, or other cleaning methods.

For particularly heavy wall thicknesses, the sand filling may be omitted. Although the absence of sand will increase ovality, the resulting ovality may still be within a range acceptable to the designing engineer. Flattening of ± 8 per cent is normally considered as the acceptable limit.

Fig. 54. Hot bending of steel pipe on bending table.

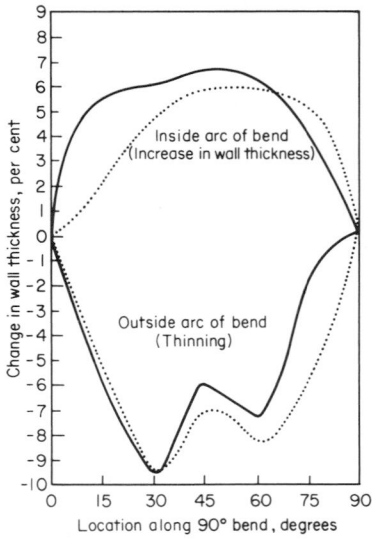

Fig. 55. Changes in wall thickness of two 8-in. OD Schedule 160 2¼ Cr–1 Mo pipes hot bent to a three-diameter radius.

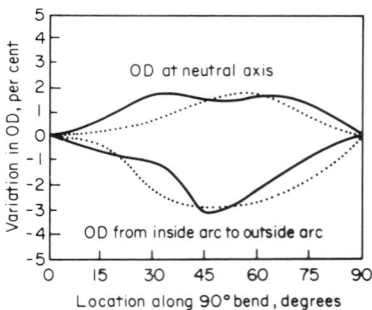

Fig. 56. Flattening in two 8-in. OD Schedule 160 2½ Cr–1 Mo pipes hot bent to a three-diameter radius.

Where sand filling is not done, it may be advantageous, in the case of ferrous and some nonferrous materials, to fill the pipe inside with argon or nitrogen gas. This will minimize scaling and eliminate subsequent cleaning operations.

The change in wall thickness of hot bends usually is slightly greater and more nonuniform than in cold pipe bends. The change in wall thickness for a specific 8-in. Schedule 160 (0.906 in. nominal wall) bend in $2\frac{1}{4}\%$ chromium-1% molybdenum-alloy steel piping bent to a three-diameter radius is illustrated in Fig. 55. However, no one bend is truly indicative of the thinning that results in all bends of all walls for the same bending radius.

To allow for the thinning in the outer arc of hot pipe bends, design engineers will usually increase the minimum wall thickness requirements t_m by the percentages listed in Table 41 for sand-filled hot bends with a radius of three to six diameters as noted.

Table 41. Thinning Allowance for Bending Steel-pipe Materials

Radius of bend	Min thickness required prior to bending
6 pipe diameters or greater	$1.06t_m$
5 pipe diameters or greater	$1.08t_m$
4 pipe diameters or greater	$1.14t_m$
3 pipe diameters or greater	$1.25t_m$

Where t_m is minimum wall thickness required as determined by Code formula.

Typical values of ovality in sand-filled 8-in. Schedule 160 pipe bends hot-bent to three-diameter radii are illustrated in Fig. 56.

For hot-bent piping, a five-diameter radius is most widely used. However, hot bends can be made to smaller radii. Recommended minimum radii and tangents are listed in Table 42 for various pipe sizes.

Creased bends are also produced in some pipe-fabrication plants. They are usually formed by utilizing excess pipe metal on the inner bending radii to form folds or creases extending outwardly along the pipe wall. Creased pipe bends permit radii to be as low as two pipe diameters.

Bending of Aluminum and Aluminum Alloys

In general, the same bending methods discussed for ferrous materials apply to aluminum pipe and tube materials. However, there is a greater difference in response to bending among the various aluminum alloys than among the steel materials normally available for pipe. Thus, for successful bending, it is important that the mechanical properties of the particular alloy be understood.

The stretch property (elongation) is a major factor for consideration in forming aluminum tubing. Tensile properties are also important. An alloy having good stretch properties and a reasonably high tensile strength is desirable for forming. Those with high elongation and low tensile strength are the easiest to bend but require careful tooling to avoid fracturing, flattening, and collapse of the inner wall. Such collapse will occur if the ability of the alloy to absorb compressive stresses is exceeded.

The more ductile alloys have better shrinkage properties but cause higher frictional loads and are more likely to flatten than the harder alloys. Therefore, one advantage is felt to offset the other, making it possible to establish a standard constant.

It is seldom necessary to bend aluminum tube hot. However, when this is necessary, the temperature of the aluminum should not exceed 375 F. Of course, hot bending will allow smaller radii than when bending cold.

Table 42.   Linear Tolerances, Bending Radii, and Minimum Tangents for Hot Bending of Ferrous Pipe (PFI Standard ES3-Nov. 1965)

Nominal pipe size, in.	Tolerances, in. All types of fabrication, E/E, E/F, F/F, C/E, C/F, C/C Column 1	Bending radii, in. Standard wall pipe Column 2	Bending radii, in. Extra-strong wall pipe and heavier Column 3	Min tangents required for pulling and holding ends in bending, in. Plain and beveled ends Column 4 One end	Min tangents required for pulling and holding ends in bending, in. Plain and beveled ends Column 4 Other end
1	±⅛	5	5	6	6
1¼	±⅛	6¼	6¼	6	6
1½	±⅛	7½	7½	6	6
2	±⅛	10	10	6	6
2½	±⅛	12½	12½	8	8
3	±⅛	15	15	8	8
3½	±⅛	17½	17½	8	8
4	±⅛	20	20	10	10
5	±⅛	25	25	10	10
6	±⅛	30	30	10	12
8	±⅛	40	40	12	16
10	±⅛	50	50	15	20
12	±³⁄₁₆	60	60	18	24
14 OD	±³⁄₁₆	70	70	21	28
16 OD	±³⁄₁₆	96	80	24	32
18 OD	±³⁄₁₆	108	90	27	36
20 OD	±³⁄₁₆	120	100	30	40
24 OD	±³⁄₁₆	144	120	36	48

Notes:

Column 1—The tolerances given in this column are not accumulative.   In the case of 180-deg welding fittings, ASA B16.9 tolerances shall apply.

Column 2—Standard wall pipe conforms with American Standard Wrought Steel and Wrought Iron Pipe, ASA B36.10.   Sizes 1 to 10 in. inclusive are identical with Schedule 40.   Sizes 12 to 24 in. inclusive have ⅜ in. wall thickness.

Column 3—Extra-strong wall pipe conforms with American Standard Wrought Steel and Wrought Iron Pipe, ASA B36.10.   Sizes 1 to 8 in. inclusive are identical with Schedule 80.   Sizes 10 to 24 in. inclusive have ½ in. wall thickness.

Column 4—The lengths of tangents given in this table are the minimum that are considered practical; longer tangents should be used whenever possible.   In special cases, bends with ends plain or beveled can be furnished without tangents.

With bending machines and proper dies and mandrels, the minimum bending radii normally applicable to aluminum pipe and tubing are calculated from the following formula:

$$R_{\min} = \left[\frac{4D - 4D \times 2S}{1.5708} - D\right] \times \left[\frac{W}{T} + (1.5708D \times C)\right]$$

where D = outside diameter of tube, in.

   S = stretch value (see Table 43)

   W = base wall factor (see Table 44)

   T = wall thickness of tube, in.

   C = a difficulty constant which is taken as equal to 0.15

The values for S and W are obtained from Tables 43 and 44, respectively.

Table 43.   Stretch Values for Bending and Forming Aluminum Tubing (Reynolds Metals Co.)

Alloy and temper[1]	Stretch value = S
1100-0, 3003-0, 5052-0	0.20
3004-0, 6063-0, 5154, 5454	0.18
6061-0, 6062-0	0.16
2014-0, 2024-0, and 6063-T42	0.12
1100-H12, 3003-H12, 5052-H32, 2024-T3, 2024-T4, and 7075-0	0.10
6061-T6, 6062-T6, 6063-T6	0.08
1100-H14, 3003-H14, 5052-H34, 2014-T6, and 7075-T6	0.07
1100-H16, 3003-H16, 5052-H36, 6063-T83, 6063-T831, and 6063-T832	0.05
1100-H18, 3003-H18, and 5052-H38	0.03

[1] For 2014-T4, 6061-T4, 6062-T4, 6063-T4, and 6063-T5, use the -T6 values for the respective alloy.

Table 44.   W Values (Reynolds Metals Co.)

Base wall factor $W = D/15$

Tube size, OD, in.	W value
1/4	0.017
3/8	0.025
1/2	0.033
5/8	0.042
3/4	0.050
7/8	0.058
1	0.067
1 1/4	0.083
1 1/2	0.100
1 3/4	0.117
2	0.133
2 1/4	0.150
2 1/2	0.167
2 3/4	0.183
3	0.200

Bending of Copper and Copper Alloys

Copper pipe and tubing and most copper alloys can be readily bent to relatively small radii—usually even smaller than possible on steel pipe.

The relative bending characteristics of copper pipe and tubing materials are summarized in Table 45.

Table 45.   Bending Characteristics of Copper Alloys (Copper and Brass Development Association)

Material	Cold	Hot	Suggested temperatures for hot bending, F
Copper	Excellent	Excellent	1400–1600
Silicon bronze	Excellent	Excellent	1300–1600
Red brass, 85%	Excellent	Good	1450–1650
Yellow brass..........	Excellent	Poor	
Muntz metal..........	Fair	Excellent	1150–1450
Cupronickel, 30%	Good	Good	1700–1900*

* Hot bending difficult and not recommended, but, if used, temperature should be as indicated.

Cold Bending. In commercial practice, the vast majority of bending applications on copper and copper-alloy pipe and tubing is done cold.

As with steel and aluminum, the cross section of the tube in bending should be supported either by insertion of a rounded-off mandrel, by the use of filler materials, or, where the bending radius approaches the lower limit, by the use of a rather snug-fitting forming block and shoe which together practically surround the tube at the point of bending and thereby minimize distortion and aid in preventing collapse.

Table 46 lists the minimum radii to which annealed copper and brass tubes may be bent cold for sizes ranging from $\frac{1}{8}$ to 4 in. outside diameter.

The data under conditions A and B are based on the use of proper bending equipment consisting of grooved bending blocks, bending shoes, and mandrels, as shown in Fig. 49. For the minimum radii under condition A, fit and adjustment are made with extreme care. The bending blocks and shoes must fit the contour of the tube accurately, and the mandrel must be placed at the exact spot where the bend is to take place. To make the minimum radii bends under condition B, fit and adjustments must be good, but they do not require the extreme care needed by those of condition A. For condition B, either a suitable supporting mandrel or filler material on the inside of the tube will suffice. The minimum radii bends under condition C can be made without mandrel or filler material or practically any other mechanical support. If, however, a filler material is used under conditions

Table 46. Minimum Radii for Cold Bending of Annealed Copper and Brass Tubes (Copper and Brass Development Association)

Actual outside diameter D, in.	Conditions		
	A	B	C
	Min bending radius, in.		
	Conforming bending block and mandrel		Cylindrical bending block
	For D/t not over 15*	For D/t not over 50*	For D/t not over 30*
	Optimum conditions	Commercial work	Without mandrel
$\frac{1}{8}$	$\frac{1}{16}$	$\frac{1}{4}$	$\frac{1}{2}$
$\frac{1}{4}$	$\frac{1}{8}$	$\frac{5}{16}$	1
$\frac{3}{8}$	$\frac{3}{16}$	$\frac{3}{8}$	2
$\frac{1}{2}$	$\frac{1}{4}$	$\frac{7}{16}$	3
$\frac{5}{8}$	$\frac{5}{16}$	$\frac{9}{16}$	4
$\frac{3}{4}$	$\frac{7}{16}$	$\frac{11}{16}$	6
$\frac{7}{8}$	$\frac{1}{2}$	$\frac{3}{4}$	8
1	$\frac{9}{16}$	$\frac{7}{8}$	10
$1\frac{1}{4}$	$1\frac{1}{16}$	$1\frac{3}{16}$	15
$1\frac{1}{2}$	$1\frac{3}{16}$	$1\frac{1}{8}$	20
$1\frac{3}{4}$	$1\frac{5}{16}$	$1\frac{1}{4}$	27
2	$1\frac{11}{16}$	$1\frac{3}{8}$	35
$2\frac{1}{2}$	$1\frac{3}{8}$	$1\frac{5}{8}$	
3	$1\frac{5}{8}$	$1\frac{7}{8}$	
$3\frac{1}{2}$	$1\frac{7}{8}$	$2\frac{1}{8}$	
4	$2\frac{1}{8}$	$2\frac{3}{8}$	

* D/t expresses the ratio of tube diameter to wall thickness.

generally similar to those of condition C, bends much sharper than those indicated can be made. This will depend considerably on the skill and experience of the operator.

Hot Bending. Almost any bend to any desired short radius can be made hot, provided the inside of the bend is sufficiently peened to prevent buckling and the bending is done slowly enough. The most suitable radii for commercial hot bending are those listed under condition C in Table 46. The larger-diameter tubes, principally those with heavy walls, are commonly bent hot. The use of bending blocks and shoes, shaped to the general contour of the tube, is recommended. In hot bending, it is customary to use either a sand filler or a steel mandrel.

EXTRUSIONS

In commercial practice, extrusions are generally produced only in ferritic steels, stainless steels, and nickel-alloy materials. However, the softer aluminum and copper alloys are also capable of being extruded—usually by cold methods.

Fig. 57. Extruding nozzle outlet in 8-in. OD Schedule 160 centrifugally cast 2½ Cr–1 Mo pipe.

Pipe extrusions represent outlets in pipe produced by pushing or pulling conical or hemispherical balls from the inside of the pipe to the outside (Fig. 57). Advantages of extruded outlets are that they provide a smoother and more gradual change in inside surface contour than intersection-welded nozzles. Since extruded nozzles (Fig. 58) employ butt welds, the fit-up and welding to branch connections is more readily accomplished and the branch may be of better quality than one in which welded nozzles are employed. Extruded nozzles can vary in diameter from a few inches to the size of the header itself.

Although some plants extrude nozzle outlets cold, by far the majority of plants perform this operation by heating the pipe section to be extruded to the hot-working temperatures of the respective alloy. A typical cross section of an extruded outlet is shown in Fig. 59. The thinning in the extruded outlet can be compensated for by starting initially with pipe of heavier wall thickness, usually by an amount up to 33⅓ per cent greater than the wall thickness required for the outlet of the full run size.

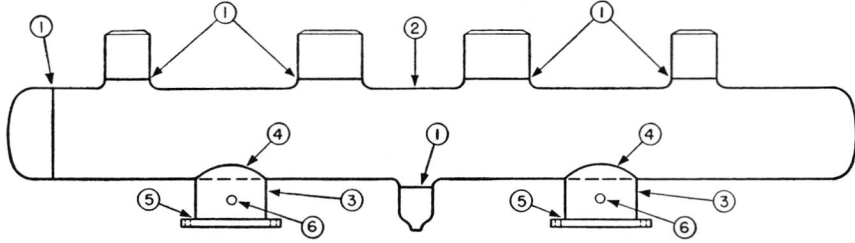

FIG. 58. Examples of extruded header design with extruded nozzles which are ideally suited for economical shop fabrication and welding: (1) All pressure welds are circumferential butt welds with backing rings. (2) Header may be seamless tube with caps welded to each end or turned and bored forging with integral cap on one end. (3) Anchor-and-guide base weldments from pipe of the same material as the header. (4) Strength nozzle weld only, not cut into the header. (5) Fillet welds to carbon-steel base plates with bolt holes for mechanical attachment to the building structure. (6) ¼-in. vent holes in pipe base.

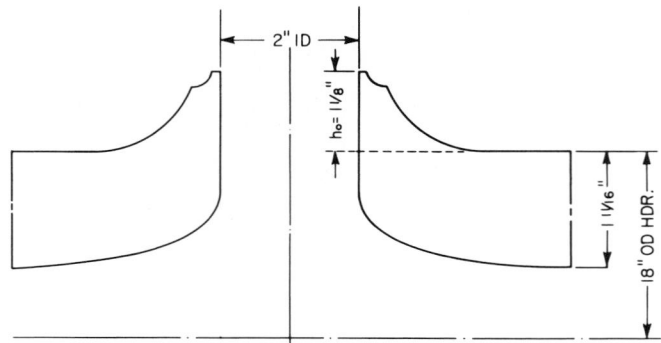

FIG. 59. Typical cross section of 2-in. ID nozzle in 18-in. OD header. Nozzle end machined for butt welding.

SWAGING, LAPPING, AND MISCELLANEOUS SHOP OPERATIONS

Ferrous Pipe and Tube. Swaging consists of reducing the ends of pipe with one or multiple rotating roller dies (Fig. 60) which are pressed intermittently against the pipe ends. The pipe material is normally heated into its hot-working range. For steels, swaging is done between 2000 and 1600 F.

In steels, depending on pipe wall thickness and type of carbon or alloy steel, each swaging operation can reduce the diameter by 5 to 15 per cent before the temperature falls below the normal hot-working range. After each reduction, the pipe end must be reheated to the upper end of its hot working temperature range. Swaged ends can be made either concentric or eccentric.

Although reductions of the diameter of steel pipe are normally limited to one to two pipe sizes, careful control over the technique makes possible reductions to less than one-half the original diameter on many steel alloys. These include carbon steels, the low chromium-molybdenum alloy steels, and the austenitic stainless steels.

Nonferrous Pipe and Tube. Aluminum, copper, and nickel alloys also can be readily swaged. Hot swaging is preferred on nickel alloys, and cold swaging is generally preferred on aluminum or copper, except that intermediate annealing may be desirable when severe reductions are involved.

Caution is necessary when swaging alloys that work-harden rapidly. Special procedures may be advisable for swaging alloys that are age-hardening or that are of a temper which restricts the workability of the alloy.

Lapping. In the Van Stone, or loose-flange joint, the flange is bored slightly larger than the outside diameter of the pipe. The flange is slipped over the end of the pipe, which is then heated to a forging temperature, upset and flared at right

FIG. 60. Swaging end of 8-ft OD Schedule 160 centrifugally cast 2¼ Cr–1 Mo pipe to 3-in. nominal OD.

angles to the pipe itself. To insure a tight joint, the flared ends of the different sections of the pipe are faced, so that when the pipe is in position, the adjacent ends will fit exactly against the gasket. Proper forging temperature is a prime requisite. Improper temperature control can result in cracks, laminations, and excessive internal stresses. Good lapping should result in flange thickness that will equal or exceed original pipe wall thickness. Temperature and pressure requirements in addition to joint or gasket material can be used to determine type of machine finish and flange width required.

Many other machines are needed in the properly equipped fabricating plant. These include heavy-duty lathes, horizontal and vertical boring mills, radial drills, post mills, flame cutting equipment, etc.

The machining operations on piping materials include cutting, grooving, turning, beveling, threading, facing, lining, drilling, and many other operations. Some of these are covered in subsequent sections.

HEAT TREATMENT

Almost all carbon- and alloy-steel piping and tubing materials are purchased under the piping specifications of the American Society for Testing Materials or corresponding governmental specifications.

Commercial carbon-steel pipe is normally supplied in the hot finished condition. Some carbon-steel pipe and the majority of the low- and medium-alloy-steel piping materials are furnished in the heat-treated condition, i.e., either in the normalized and tempered condition,[1] or in the annealed condition.[2]

Stainless-steel piping materials are generally furnished in the solution-heat-treated condition. Nonferrous piping also may have been heat-treated by the mill to obtain specific metallurgical properties.

Shop fabrication may then involve the hot or cold forming of bends, the extrusion of outlets, or the swaging down of ends to match pipe or fittings of a smaller diameter.

Proper heat treatment is as much an essential operation to produce a soundly fabricated piping system as are the various forming or welding operations.

In shop fabrication heat treatment is frequently done in furnaces because of the number of pipe sections which can be heat-treated at once, but localized heating is also done. In field erection, of course, all heating is done locally.

HEATING CYCLES AND METHODS

The heating temperatures and heating cycles normally recommended for carbon-, low-alloy, and high-alloy-steel materials are summarized in Table 47. On non-ferrous materials, heat treatments after forming or welding generally are not required. Exceptions may involve special age-hardening alloys, or service environments, which adversely affect the piping materials containing stresses.

Various methods are used in the heat treatment of piping. These include (1) furnace-heat treatment, (2) induction heating, (3) torch heating, involving various gases and gas mixtures such as propane, oxypropane, or butane, (4) electric-resistance heating, and (5) exothermal (chemical) heating. Each method has its own advantages and limitations.

Furnace-heat Treatment. Furnace-heat treatment is limited to shop fabrication. Large furnaces for this purpose are available in commercial pipe-fabrication plants.

Furnace-heat treatment generally is considered the most satisfactory method of postheat treatment of fabricated and welded pipe sections. Heating and cooling rates and holding temperatures are maintained uniformly by means of automatic furnace controls. Heating of such furnaces usually is done by either natural gas or oil of low vanadium and sulfur content.

Induction Heating. Induction-heat-treating methods are extensively used for the heat treatment of ferritic-steel pipe welds, particularly at field erection sites. This method is applied extensively on the heavier pipe wall thicknesses from about $\frac{3}{4}$ to 4 in. and over and in diameters of 6 in. and over. Relatively low frequencies of 25, 60, or 400 cycles generally are used. The 60-cycle equipment is most widely available. In torch and electric-resistance heating, the heat is slowly conducted to the inside wall of the pipe from the point of application on the outside surface. Induction heating generates heat essentially within the pipe wall. This has the advantage on pipe of the heavier wall thicknesses of providing more uniform temperature throughout the wall and a smaller temperature difference between the outside and inside surfaces. The lower the frequency, the deeper into the pipe wall the heating effect penetrates. Whereas the differences between 60- and 400-cycle heating are insignificant for wall thickness up to about $1\frac{1}{2}$ to 2 in., on wall

---

[1] Heating for 1 hr per in. of wall thickness at 1650 F, then cooling in air to below 600 F, followed by tempering at 1350 F for 1 hr per in. of wall thickness.

[2] Heating for 1 hr per in. of wall thickness at 1700 F, followed by slow furnace cooling.

Table 47. Recommended Preheat and Postheat Treatments for Piping Materials

P. No.	Material group	Preheat[a] temperature, F	Postheat treatment after welding[b] — Metal temperature, F	Time at temperature — Hrs/in. of wall thickness	Time at temperature — Hr (min)	Hot forming temperatures, F	Heat treatment after hot forming — Temperature, F	Heat treatment after hot forming — Holding time in hr per in. of wall thickness
1	Carbon steel	Not required[c]	1,100–1,250[d]	1	1	1,400–2,000	....	May be necessary on heavy wall thickness and where heating for hot forming was not done evenly
2	Wrought iron	Not required[c]	....	...	...	1,400–2,000	....	
3	Alloy steels, to 3/4% Cr max: total alloy 2% max	300–600	1,275–1,350[e,f]	1	1	1,500–2,000	1,275–1,350	1
4	Alloy steels, 3/4–2% Cr; total alloy 2–3/4% max	400–600	1,300–1,400[f]	1	1	1,600–2,000	1,300–1,400	1
	Alloy steels, 2–3% Cr	400–600	1,300–1,450[f]	1	1	1,600–2,000	1,300–1,400	1
5	Over 3% Cr–10% Cr total alloy 10% max	500–600	1,300–1,425	2	2	1,600–2,000	(A) 1,550–1,600[g] (N) 1,650–1,700[g] (T) 1,300–1,425[g]	1 hr with furnace cooling at 50 F/hr to 1,100 F 1 hr with air cooling
6	Alloy steel, martensitic, stainless	500–700	1,400–1,500	1	2	Varies with type of alloy and application	1,400–1,500	1
7	Alloy steel, ferritic, stainless	Not required	Not required[h]	...	...	1,500–1,900	Not required	
8	Ni-alloy steel, austenitic, stainless	Not required	Not required[i]	...	...	1,500–2,100	Not required[i]	
9	Ni, alloy steels	300–500	1,200–1,300	1	1	1,600–2,000	1,200–1,300	1

[a] Use lowest temperature for sections up to 1/2 in. thick and highest values for sections over 2 in. thick, with intermediate temperatures for intermediate thicknesses.

[b] Heating rate should be such as not to produce a temperature difference across the pipe wall of more than 150 F. Above 600 F, the heating rate is optional for light wall thicknesses. However, for pipe-wall thickness over 1 in. of the P1-P3 groups, over 3/4 in. of the P4-P5 groups (to 3 per cent Cr), and over 1/2 in. of the P5 and P6 groups, the following heating and cooling rates above 600 F apply: For induction heating, to 1 1/2 in. 60 and 400 cycle–1,000 F/hr max; 1 1/2 in. and over, 60 cycles, 500 F/hr max and for 400 cycles, 400 F/hr max. For furnace, gas, electric resistance, and other surface heating methods: 400 F/hr max or 400 F/hr divided by one-half the thickness, whichever is smaller.

[c] Where the ambient temperature is less than 32 F, local preheating to a condition warm to the hand is required.

[d] Postheat treatment is not required for carbon steels where the thickness is less than 3/4 in.

[e] Postheat treatment is not required for carbon–1/2% molybdenum steels where the thickness is less than 1/2 in.

[f] Postheat treatment is not mandatory on socket welds and butt welds of joints 4 in. OD and less, with wall thickness of less than 1/2 in. for all P3 and P4 groups or on the 2 1/4 and 3 per cent chromium materials of the P5 series.

[g] Heat treatment after hot forming involves either annealing (A) or normalizing (N) and tempering (T) at the temperatures shown.

[h] Postheat treatment after welding may be desirable on some grades to improve ductility.

[i] Annealing heat treatments may be necessary on some grades to improve resistance to corrosion in specific environments.

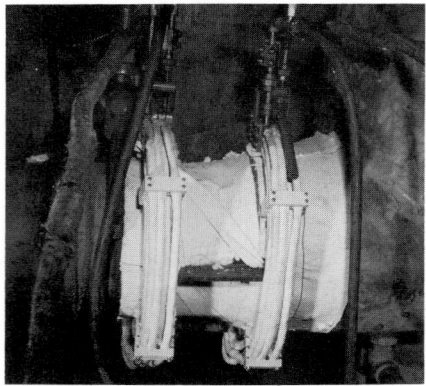

Fig. 61. Use of water-cooled coils to stress relieve heavy-wall piping by induction heating.

thicknesses over 2 in. more uniform and rapid heating is obtained with the 60-cycle or 25-cycle equipment.

The electrical field is normally obtained by wrapping insulated copper conductors around the weld to be heated. Special fixtures (Fig. 61) are also available to permit more rapid attachment or removal of the induction-heating coils.

Torch Heating. Torch heating is done either with single burners or with ring burners (Fig. 62). Heating with single burners normally is limited to piping of smaller diameters (below 3 in.) and lighter wall thicknesses. Methods employing "softer" flames, such as propane or butane torches, are normally preferred for pipe heat treatment, as contrasted to heating with oxyacetylene torches. Ring burners are also available for various gas mixtures including propane, oxypropane, and oxyacetylene.

Fig. 62. Postheat treatment of pipe weld with ring burner.

Sufficiently uniform preheating to temperatures of 500 to 600 F is readily done with single burner torches, but postweld heat treatment at stress-relieving temperatures in excess of 1000 F on circumferential joints should be done with ring burners. Uniform heating with single burner torches is difficult and may result in metal upsetting and even cracking.

Resistance Heating. Electric-resistance heating is usually done by wrapping the pipe with suitable lengths of nichrome wire and passing sufficient current through the wires to develop the necessary temperature. The current used should be direct current, and it may be taken from special power-supply units or from standard welding machines. Shorting or grounding of the resistance wires is prevented by means of insulating beads strung along the wires. The preheat and interpass temperatures can be controlled by means of thermocouple-actuated controls or magnetically attached surface-temperature-sensitive "on-off" electric contactors (Fig. 63).

This method of heating is normally limited to piping of wall thickness of less than 1 in.

Exothermal (Chemical) Heating. Exothermal heating of piping is limited to field erection and is done primarily on pipe of relatively light wall thicknesses, such as that used in refineries. The exothermal materials usually contain aluminum powder, metal oxides, refractory compounds, and binders, which are preformed into cylindrical shapes and placed around the pipe weld to be postheated. When ignited with an oxyacetylene or propane torch, the exothermic material burns with an exothermic reaction until the active constituents have been consumed.

The exothermic reaction occurs rapidly, bringing the temperature along the pipe outside from 1600 to 1700 F in 5 to 15 min. In fact, the initial surface temperatures may become substantially higher than 1700 F and may melt the thermocouple wires attached to the pipe surface.

Fig. 63. Automatic control of preheat and weld interpass temperature with magnetically attached "on-off" dial thermometer.

Because of the extremely rapid heating reaction along the pipe surface, the inside temperature of the pipe lags considerably. The differences between the outside and inside temperatures of the pipe increase with the pipe wall thickness. This can result in upsetting and distortion of the steel along the pipe surface.

One condition which arises from exothermal heating is that it austenitizes the weld metal. This lowers materially both yield strength and tensile strength. E7015, E7016, and E7018 deposited metal drop to about 55,000-psi tensile strength if austenitized (from about 75,000 psi when deposited or stress-relieved at subcritical temperatures). This loss can never be regained in either carbon-steel or low-alloy-steel weld metals. For this reason, particularly on heavier wall materials, low-frequency induction-heating methods are generally preferred by properly equipped field erectors.

HEAT TREATMENT OF WELDS

Heat-treating considerations include (1) preheating prior to welding, (2) interpass-temperature-control heating during welding, (3) postweld heat treatment upon completion of the weld, and (4) straightening of distorted pipe sections by heating. Satisfactory service performance of welded piping frequently depends upon the performance of proper heat treatments.

Proper heat treatment is affected by the heating and, even more critically, by the cooling rates. Specification of the heating and cooling rates should include consideration of the weld-joint configuration, the wall thickness involved, and the

heating methods employed. More rapid heating and cooling rates may be permissible on circumferential pipe butt welds than is possible on angular nozzle welds and other attachment welds between sections of dissimilar wall thicknesses. More rapid heating and cooling rates also may be permissible with induction heating in which the heating is accomplished within the pipe wall than is permissible with the surface heating methods which depend upon thermal conduction for heating throughout the thickness of the pipe wall.

Rapid rates of heating and cooling or uneven heating within the wall of heavy-wall pipe assemblies, particularly reinforced branch welds, may result in weld cracking. Rapid cooling of complicated pipe assemblies from high temperatures may also result in substantial distortion. This becomes of particular concern with fabrication involving austenitic stainless steels where water quenching from temperatures over 1900 F is required.

Occasionally, special consideration in specifying heat treatments of piping must be applied to valves, flow nozzle sections, etc., where the heat treatment may damage sensitive materials, such as valve seats and flow nozzle chambers.

Specifications and Codes. Because of the importance of heat treatments, they should be delineated clearly in the respective pipe-fabrication specifications. Many codes applicable to piping, such as the ASME Boiler and Pressure Vessel Code, section IX, the ASA Code for Pressure Piping and others, include in their procedure qualification and welder qualification requirements consideration of the preheat and postheat temperatures. Changes in the preheat and postheat temperatures by 100 F may require requalification of the applicable welding procedure.

Temperature Control. The control of temperature in both preheating and postweld heating operations is especially important and frequently influences selection of the equipment.

For *full-furnace and induction heating*, thermocouples are usually attached to the material to be heated. The thermocouple wires are then connected to control equipment, which may control automatically the time-temperature cycle and even program the heating and cooling rates.

For *torch heating*, temperature-indicating crayons are widely used. Preheating temperatures up to 700 F are also checked with direct-reading magnetically attached surface thermometers.

For *electric-resistance heating*, surface thermometers or electrically operated pyrometers are used to control automatically the current flow to the heating units. Thermocouples are also used.

When welds are being preheated or postheated in any of the alloys that have air-hardening tendencies, such as the chromium-molybdenum and nickel-alloy steels, special precautions should be taken to exclude as much as possible the circulation of air around the joint during the heating and cooling periods. This can be done by avoiding drafts and by blocking the ends of the parts being welded to prevent air circulation through the interior of the piping during the heating period.

During the cooling cycle, the joint should be covered with a suitable insulating jacket material to retard the cooling rate. Several layers of asbestos sheeting are often used for this purpose.

Metallurgical Heat Treatments. It is sometimes necessary to heat-treat complete welded piping assemblies not only to relieve stress, but also to restore the original structure of the metal or to attain other desirable properties. This may involve annealing, normalizing, or quenching treatments, depending on the alloy welded. The heat-treating conditions dictate the type of furnaces required. For example, the medium- and high-chromium-alloy steels often are normalized at 1650 F, air-cooled, and tempered subsequently at about 1300 F; or they are annealed at 1700 F, followed by furnace cooling. On the other hand, certain austenitic

stainless steels may require annealing at 1950 F, followed by rapid cooling to obtain maximum corrosion resistance.

Pipe Support during Heat Treatment.   The temperatures involved in stress relief and other heat-treatment operations are often in a range where the ability of the metals to resist deformation and distortion is lowered considerably.   It is, therefore, usually necessary to support welded pipe sections during the heat-treating operation in a manner that will reduce to a minimum the possibility of deflection in any direction.   This is accomplished in the shop by placing adjustable supports under the parts at assembly ends as near the joints as possible and allowing sufficient space for the placement of heating apparatus over joints, if necessary, for local heat application.

In field work, where the welds are made in position, chain falls or other suitable rigging secured to the building or other supporting structures are generally used to accomplish the same purpose.

Where stress relief or other heat treatments of a number of assemblies are done in a heat-treating furnace, special care should be exercised in placing the assemblies to insure that they are adequately and uniformly supported in all directions. Temporary structural bracing is sometimes employed to accomplish proper support of fabricated pipe sections.   The bracing may hold piping in alignment by means of collars around the piping or by its being tack-welded to the piping.

STRAIGHTENING OF WELDED ASSEMBLIES

Because of the normal expansion and contraction of the metals, the welding of piping assemblies tends to distort the original set-up and alignment of parts.   The amount and degree of distortion depends upon the metal or alloy welded; its size, shape, and thickness; the tacking and alignment; the welding process, procedure and sequence; and the care of the welder.   Rewelding of localized weld areas where defects had been removed by grinding, flame-gouging, or arc-gouging can also contribute to distortion.   Much of the distortion can be controlled during welding; some of it may be corrected during a postheat treatment.   Frequently, it is necessary to correct the effect of distortion independently.

The method most widely employed to correct misalignment in fabricated carbon- and alloy-steel pipe assemblies involves the controlled alternate heating and cooling of areas adjacent to the weld.   As a rule, this method of correcting misalignment is accomplished by rapidly heating relatively small localized areas on the side of the pipe toward which deflection is desired, followed by either slow or rapid cooling of the heated region.   This results in localized expansion and upsetting of the heated areas which, on cooling, contract and gradually draw the parts into proper alignment.   A somewhat similar method which is often used is to heat a complete band around a pipe and rapidly cool one side with a stream of water.   This causes rapid contraction of the water-cooled side and deflection of the free ends of the pipe toward the side of the pipe thus cooled.   The application of these local heat-straightening methods is limited to those materials and service applications where the effect of the alternate heating and cooling is not unduly detrimental to the properties or structure of the metal.

On carbon-steel piping, the application of water is considered permissible, except in services which involve severely stress-corrosive environments.   Water quenching of alloy-steel piping is rarely permitted because it can result in localized excessive hardness which may crack in environments involving thermal or mechanical fatigue.   Air cooling, therefore, must suffice.

Occasionally it is possible to correct piping assemblies that have been furnace-annealed or normalized by setting up the assembly in a facing head with the normal

axis of the piece in line with the center lines of the machine.   The flanges, or pipe ends, are then machined to produce flange faces or pipe ends which are square with the normal axis of the pipe and with each other and which meet the overall dimensional requirements of the piece section.

The foregoing methods of square-up and alignment are, of course, applicable in cases where the assemblies, after welding, are distorted to a considerable extent. Another method of correcting misalignment is by the gradual application of force, either by means of the cold method or by means of a section of pipe at the stress-relieving temperature.   The cold method has limited application, however, owing to the possibility of setting up excessive stress concentrations in an assembly and should be employed only in cases where the effect of such stress concentrations would be negligible.[1]

The final step in the straightening operation is the checking of all parts and dimensions of the assembly with the requirements of the drawing from which the assembly was constructed.   While this procedure is essentially a function of inspection, much additional handling can be avoided if the work is done before the assembly is removed from the straightening table.

CLEANING

Proper cleaning of fabricated, welded, or heat-treated piping frequently is very important.   The extent of cleaning and the methods used depend upon the service involved.   Cleaning requirements may range from merely blowing out piping with compressed air or steam to abrasive cleaning by shot or sand blasting to chemical methods involving pickling or degassing.

Standard shop practice[2] normally involves the removal of all "loose" foreign material such as scale, sand, weld-spatter particles, and cutting chips from the inside of the piping assembly by any suitable means, such as a mechanically driven rotary cleaning tool (turbining) or a wire brush.   Subsequently, the piping should be blown with compressed air or steam.

Turbine cleaners are mechanically driven rotary scraper tools which are pushed through the inside of the pipe in much the same manner as boiler-tube cleaners. Carbon- and low-alloy steel pipe sections which have been hot-bent are normally cleaned on the inside by means of turbining.

Service conditions which require piping to be free of scale on the inside may be cleaned by shot or sand blasting.   For shot blasting, special tools are available which, on passage through the pipe inside, spray shot particles in a rotary progression against the wall.   This will remove heat-treating scale and welding slag from the pipe inside surface.   Sometimes sand blasting of the outside of piping assemblies is required.

With all abrasive cleaning methods, care must be exercised that prolonged abrasion is not directed against a particular surface area since this may produce excessive localized thinning.

Chemical Cleaning.   Piping for some chemical plants, nuclear plants, cryogenic services, and other applications may necessitate chemical cleaning by pickling.   Degreasing with liquid or vapor solvents may also be required, particularly in some cryogenic service applications.

Pickling of completed piping and tubing assemblies is best done after final heat treatments by immersion in pickling tanks.   Sometimes swabbing of the piping with

[1] In the erection of some piping systems, cold pull is utilized to counteract expansion effects when the systems become heated in service.

[2] Pipe Fabrication Institute Standard ES-5, Cleaning Fabricated Piping.

pickling solutions may be adequate. Pickling of erected piping systems may be accomplished by pumping the pickling solution through the piping system. Rinsing after pickling with neutral or slightly alkaline solutions is frequently necessary to remove all acid traces. Some piping, such as austenitic stainless steels and nonferrous materials, is also passivated in oxidizing acids which may increase the corrosion resistance of the material in certain corrosive environments.

The solutions used for pickling or passivating vary with the metals used, and the methods employed are dependent on available facilities and practices of the manufacturer. Even on steel, various methods are used, and only general recommendations are made in applicable specifications. On steel surfaces, the Steel Structures Painting Council recommends the following methods:[1]

"1. Pickling in hot or cold solutions of sulfuric, hydrochloric (muriatic), or phosphoric acid, to which sufficient inhibitor has been added to minimize attack on the base metal, followed by adequate rinsing in hot water above 140 F.

2. Pickling in 5 to 10 per cent (by weight) sulfuric acid, containing an inhibitor, at a minimum of 140 F until all rust and scale are removed; then thorough rinsing in clean water and immersion for at least 3 to 5 min in about 2 per cent (by weight) phosphoric acid containing about 0.3 to 0.5 per cent iron phosphate, at a temperature of about 180 F.

3. Pickling in 5 per cent (by volume) sulfuric acid at 170 to 180 F with sufficient inhibitor added to minimize attack on the base metal, until all rust and scale are removed, followed by a 2-min rinse in hot water at 170 to 180 F. The pickled and rinsed steel shall then be immersed for at least 2 min in a hot, inhibitive solution maintained above 190 F and containing about 0.75 per cent sodium dichromate and about 0.5 per cent orthophosphoric acid.

4. Electrolytic pickling in an acid or an alkaline bath using alternating or direct current. When using direct current, the work is made the cathode and hydrogen embrittlement must be prevented or minimized by adequate treatment. If carried out in an alkaline bath, the electrolytic pickling must be followed by a thorough rinsing in hot water; it is then followed by a dip in a dilute solution of phosphoric acid, chromic acid, or solution of dichromate until no trace of alkali remains on the surface.

5. "Hydride" descaling, pickling in baths of acid salts, pickling in baths of molten salts, or pickling in any other manner than outlined in the preceding sections shall be permitted only when specified, since their details are beyond the scope of this specification."

During the pickling process, precautions as listed below must be observed.

"1. The dissolved iron content shall not be permitted to exceed 6 per cent in sulfuric acid baths or 10 per cent in hydrochloric (muriatic) acid baths.

2. Only fresh water or steam condensate shall be permitted for solutions and rinses. Rinse tanks shall be supplied continuously with new water, and the total amount of acid or dissolved salts due to carry-over shall not exceed 2 g/liter (0.2 per cent by weight).

3. To minimize carry-over, all steel shall be suspended briefly over the acid tank from which it has been withdrawn and the major portion of the acid solution shall be permitted to drain.

4. The pickled surfaces shall be examined for smut, metal deposits, and improperly cleaned areas, and any such contaminated areas shall be further cleaned."

[1] Standard SSPC-SP 8-527, "Surface Preparation Specifications No. 8 Pickling," Steel Structures Painting Council, 4400 Fifth Avenue, Pittsburgh, Pa. 15213.

Special Cleaning Requirements. Special cleaning requirements may apply to nuclear plants, cryogenic piping systems, piping for conducting beverages, pharmaceutical products, and others.

Cleaning methods or cleaning materials may not be strictly for the removal of particular types of contaminants that are injurious to mechanical seals and other mechanical parts or for the passivating of the pressure-bearing material. In nuclear plants, for example, cleaning may also be necessary for the prevention of a reactive condition or some other influence on the nuclear characteristics of the reactor or the primary coolant system.

MARKING AND IDENTIFICATION OF PIPING MATERIALS

To avoid the mixing of piping materials, it is necessary to apply permanent markings to the metal surface. In fact, many commercial piping components have markings applied as a normal part of the production methods of the mill. Wrought-steel fittings, for example, usually have the code, material, and manufacturer's identification symbols die-stamped into the steel prior to hot forming into the final fitting shapes. Pipe, on the other hand, rarely has stamped or impressed identification markings applied, but it is often marked with ink or paint-type materials as to code and specification number. For the identification of castings, markings are usually cast directly into the metal surface.

On fabricated piping assemblies, identification should be made as follows.[1]

Carbon and Chromium-Molybdenum-alloy Steels. Low-stress steel stamps (those which have a round or "U"-shaped cross section) should be used to identify the following materials:

Mild Carbon Steel or Wrought Iron—ASME Groups P1 and P2
Alloy Steel, up to 2% Cr—ASME Groups P3 and P4
Alloy Steels, $2\frac{1}{4}$% to 9% Cr—ASME Group P5

High-alloy and Nickel-alloy Steel Materials. Etching with an electric pencil[2] or by electrochemical means should be done to identify the following materials:

High Alloy Martensitic Steel—ASME Group P6
High Alloy Ferritic Steel—ASME Group P7
High Alloy Austenitic Steel—ASME Group P8
Nickel Alloy Steel—ASME Groups P9 and P10

Nonferrous Materials. Etching with an electric pencil or by electro-chemical means should be done to identify the following materials:

Aluminum and Aluminum Alloys—ASME Groups P21, P22 and P23
Copper and Copper Alloys—ASME Groups P31, P32, P33, P34 and P35
Nickel and Nickel Alloys—ASME Groups P41, P42 and P43

General Practice. All of the above methods are for general applications where the materials are used in compliance with the ASME or ASA Codes.

In some instances where extensive permanent markings are required, as in the case of ASME Code piping, it may be more economical, as well as metallurgically preferable, to stamp the required data on light gage plates. These plates may then

---

[1] Pipe Fabrication Standard ES-11, "Permanent Identification of Piping Materials."
[2] Electric-circuit-type marking tools that produce heating of the base metal or a spark should not be used. They may cause cracking of the base metal and high residual stresses in the region of the cracks.

be permanently affixed to the fabrication by strapping or by other approved methods. This method of marking is acceptable for all of the above materials.

Weld Identification. Some codes require that welds be die-stamped with the respective welder's identification symbol or number and any other identifying symbol stamp that may be required. Where carbon- or alloy-steel piping assemblies are subsequently stress-relieved, this stamping should be done prior to stress relieving.

SCREWED AND BOLTED PIPE JOINTS

SCREWED JOINTS

Thread Cutting. For first-class pipe fittings, it is necessary to have smooth, clean threads; otherwise it is difficult to produce joints that will go together easily and remain pressure-tight. Thread cutting should be regarded as a rather precise machining operation.

Threading Dies. A proper form for a threading die with four chasers is shown in Fig. 64. One of the most important features is the lip angle, which is the angle between the front face of the chaser and the actual cutting edge. This is also sometimes termed the *rake* or *hook*. For threading wrought iron, the lip angle should be not over 16 deg; for ordinary bessemer steel pipe, it should be from 15 to 20 deg; for open-hearth and basic steel, it should be at least 25 deg; and for brass, it should be very small. An improper lip angle produces a rough thread and tends to cause tearing of the metal. Proper

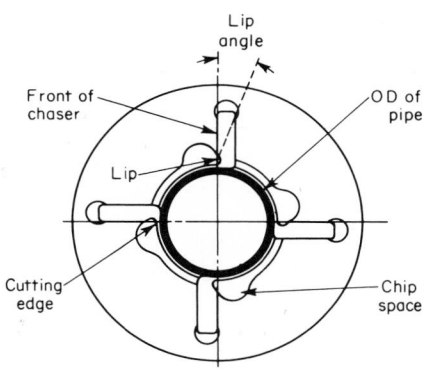

FIG. 64. Proper form of threading die.

clearance between the end or heel of the chaser and the work is also important. Excessive clearance results in chattering or in a wavy thread. Too little clearance is indicated by the entire heel of the chaser showing signs of wear instead of showing a decreasing amount of wear back of the cutting edge. Too small a clearance causes work-hardening in the die. The clearance is determined by the original machining of the chaser, and it cannot be changed in the field.

The first three or more threads on the chasers are cut away to enable the pipe to enter the die; the threads cut away are referred to as the *lead* or *throat*.

Another requirement in a well-constructed die is ample chip space. Insufficient space causes the chips to pack in and the die to bind.

Pipe Joint Compounds. Pipe threads are necessarily somewhat imperfect. Some form of compound is used to insure tightness as well as to lubricate threads while the joint is being screwed up. If the threads are very well made, a light oil such as linseed oil is sometimes a sufficient lubricant. In plumbing work, white lead or red lead is usually used. For steam piping, a prepared paste containing oil and a filling material is generally used. A compound containing powdered zinc has also been used with much success in making tight joints. For screwed oil piping, the joint compound frequently used is made up of molydisulfide and mineral oil. A

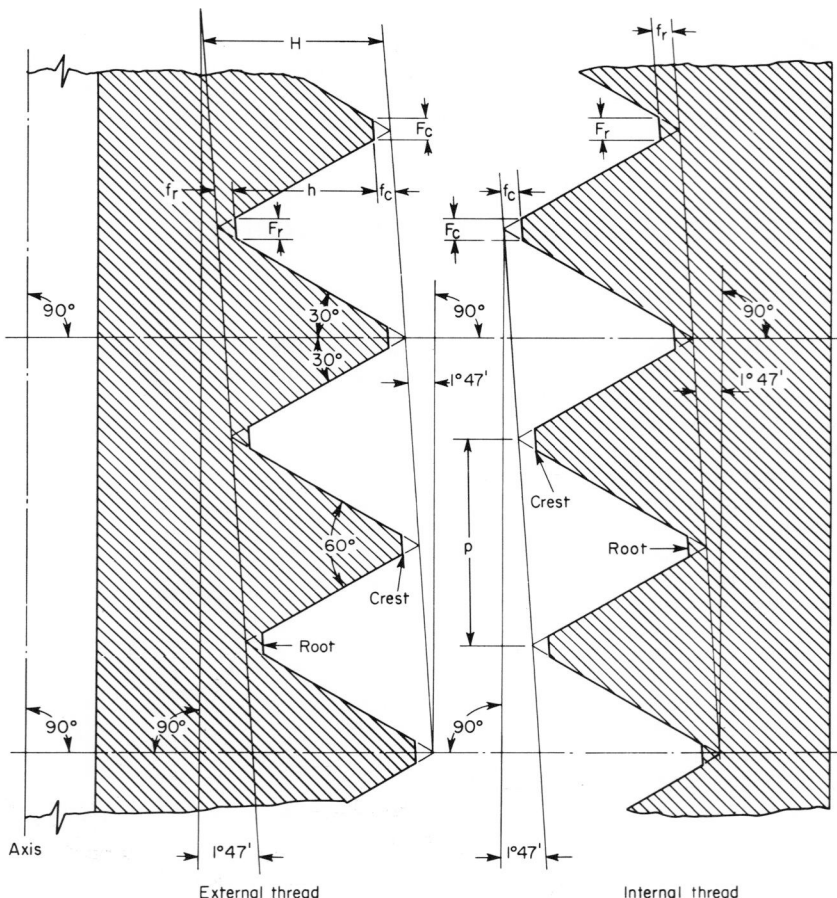

FIG. 65. Basic form of American Standard taper pipe threads as shown in ASA B2.1-1960. Notation: $H = 0.866025p$ = height of 60-deg sharp V thread; $h = 0.800000p$ = height of thread on product; $p = 1/n$ = pitch (measured parallel to axis); n = number of threads per inch; f_c = depth of truncation at crest; f_r = depth of truncation at root; F_c = width of flat at crest; F_r = width of flat at root.

NOTE: For a symmetrical straight screw thread, $H = \cot \alpha/2n$. For a symmetrical taper screw thread, $H = (\cot \alpha - \tan^2 \beta \tan \alpha)/2n$, so that the exact value for an American Standard taper pipe thread is $H = 0.865743p$ as against $H = 0.866025p$, the value given above. For an eight-pitch thread, which is the coarsest-standard taper-pipe-thread pitch, the corresponding values of H are 0.108218 and 0.108253 in., respectively, the difference being 0.000035 in. This difference being too small to be significant, the value of $H = 0.866025p$ continues in use for threads of three-fourths inch, or less, taper per foot on the diameter.

mixture of litharge and glycerine or glyptol is satisfactory for making joints tight against fluids, such as oil, which are difficult to hold. These materials are often preferred for screwed joints involving stainless-steel materials. However, it is difficult to separate joints which have been made up with these sealants.

Dimensional Standards. Dimensional standards are established in ASA Standard B2.1. This standard specifies dimensions, tolerances, and gaging for taper and straight pipe threads, including certain special applications. The normal type of pipe joint employs a tapered external and tapered internal thread, but straight pipe threads are used to advantage for certain types of pipe couplings, grease cup, fuel and oil fittings, mechanical joints for fixtures, and conduit and hose couplings.

The angle between the sides of the thread is 60 deg when measured in an axial plane, and the line bisecting this angle is perpendicular to the axis for both taper and straight threads (see Fig. 65). The crest and root are truncated a minimum of $0.033p$ and a maximum ranging from $0.096p$ for 27 threads per in. to $0.062p$ for eight threads. The basic maximum depth of the truncated thread, h, is based on factors entering into the manufacture of cutting tools and making tight joints. The *pitch* is the distance from a point on a screw thread to a corresponding point on the next thread measured parallel to the axis.

$$\text{Pitch in inches } (p) = \frac{1}{\text{number of threads per inch } (n)}$$

The *taper* of the thread is 1 in 16 or 0.75 in. per ft measured on the diameter and along the axis. The basic *pitch diameters* of the taper thread are determined by the following formulas based on the outside diameter of the pipe and the pitch of the thread:

$$E_0 = D - (0.050D + 1.1)\frac{1}{n} = D - (0.050D + 1.1)p$$

$$E_1 = E_0 + 0.0625L_1$$

where D = outside diameter of pipe
E_0 = pitch diameter of thread at end of pipe
E_1 = pitch diameter of thread at the gaging notch or large end of internal thread
L_1 = normal engagement by hand between external and internal threads
n = number of threads per inch
p = pitch of thread, in.

The increase in diameter per thread is $0.0625/n$.

The basic *length* L_2 of the effective external taper thread is determined by the following formula based on the outside of the pipe and the pitch of the thread:

$$L_2 = (0.80D + 6.8)\frac{1}{n} = (0.80D + 6.8)p$$

This formula determines directly the length of effective thread which includes approximately two usable threads that are slightly imperfect at the crest.

The normal *length of engagement* between external and internal taper threads when screwed together by hand is shown in column 6 in Table 48. This length is controlled by the construction and use of the gages. It is recognized that in special applications, such as flanges for high-pressure service, a longer thread engagement is used. In this case, the pitch diameter (dimension E_1 in Table 48) is maintained, and the pitch diameter at the end of pipe E_0 is proportionately smaller. All dimensions are given in inches, unless otherwise specified. The *maximum allowable*

Table 48. Basic Dimensions,ᵃ American Standard B2.I, Pipe Threads

1	2	3	4	5	Handtight engagement			Effective thread, external		11
					6	7	8	9	10	
Nominal pipe size, in.	Outside diameter of pipe, D	Threads per in. n	Pitch of thread, p	Pitch diameter at beginning of external thread, E_0	Length,ᵇ L_1 In.	Length,ᵇ L_1 Threads	Diameter,ᶜ in. E_1	Length,ᵈ L_2 In.	Length,ᵈ L_2 Threads	Diameter E_2
$\frac{1}{16}$	0.3125	27	0.03704	0.27118	0.160	4.32	0.28118	0.2611	7.05	0.28750
$\frac{1}{8}$	0.405	27	0.03704	0.36351	0.1615	4.36	0.37360	0.2639	7.12	0.38000
$\frac{1}{4}$	0.540	18	0.05556	0.47739	0.2278	4.10	0.49163	0.4018	7.23	0.50250
$\frac{3}{8}$	0.675	18	0.05556	0.61201	0.240	4.32	0.62701	0.4078	7.34	0.63750
$\frac{1}{2}$	0.840	14	0.07143	0.75843	0.320	4.48	0.77843	0.5337	7.47	0.79179
$\frac{3}{4}$	1.050	14	0.07143	0.96768	0.339	4.75	0.98887	0.5457	7.64	1.00179
1	1.315	11½	0.08696	1.21363	0.400	4.60	1.23863	0.6828	7.85	1.25630
1¼	1.660	11½	0.08696	1.55713	0.420	4.83	1.58338	0.7068	8.13	1.60130
1½	1.900	11½	0.08696	1.79609	0.420	4.83	1.82234	0.7235	8.32	1.84130
2	2.375	11½	0.08696	2.26902	0.436	5.01	2.29627	0.7565	8.70	2.31630
2½	2.875	8	0.12500	2.71953	0.682	5.46	2.76216	1.1375	9.10	2.79062
3	3.500	8	0.12500	3.34062	0.766	6.13	3.38850	1.2000	9.60	3.41562
3½	4.000	8	0.12500	3.83750	0.821	6.57	3.88881	1.2500	10.00	3.91562
4	4.500	8	0.12500	4.33438	0.844	6.75	4.38712	1.3000	10.40	4.41562
5	5.563	8	0.12500	5.39073	0.937	7.50	5.44929	1.4063	11.25	5.47862
6	6.625	8	0.12500	6.44609	0.958	7.66	6.50597	1.5125	12.10	6.54062
8	8.625	8	0.12500	8.43359	1.063	8.50	8.50003	1.7125	13.70	8.54062
10	10.750	8	0.12500	10.54531	1.210	9.68	10.62094	1.9250	15.40	10.66562
12	12.750	8	0.12500	12.53281	1.360	10.88	12.61781	2.1250	17.00	12.66562
14 OD	14.000	8	0.12500	13.77500	1.562	12.50	13.87262	2.2500	18.90	13.91562
16 OD	16.000	8	0.12500	15.76250	1.812	14.50	15.87575	2.4500	19.60	15.91562
18 OD	18.000	8	0.12500	17.75000	2.000	16.00	17.87500	2.6500	21.20	17.91562

1	12	13	14	15	16	17	18	19	20	21	22
Nominal pipe size, in.	Wrench makeup length for internal thread		Diameter, E_3	Vanish thread, V		Overall length external thread, L_4	Nominal perfect external threads[e]		Height of thread, h	Increase in diameter per thread, $0.0625/n$	Basic[f] minor diameter at small end of pipe, K_0
	Length, L_3						Length, L_5	Diameter, E_5			
	In.	Threads		In.	Threads						
$\frac{1}{16}$	0.1111	3	0.26424	0.1285	3.47	0.3896	0.1870	0.28287	0.02963	0.00231	0.2416
$\frac{1}{8}$	0.1111	3	0.35656	0.1285	3.47	0.3924	0.1898	0.37537	0.02963	0.00231	0.3339
$\frac{1}{4}$	0.1667	3	0.46697	0.1928	3.47	0.5946	0.2907	0.49556	0.04444	0.00347	0.4329
$\frac{3}{8}$	0.1667	3	0.60160	0.1928	3.47	0.6006	0.2967	0.63056	0.04444	0.00347	0.5676
$\frac{1}{2}$	0.2143	3	0.74504	0.2478	3.47	0.7815	0.3909	0.78286	0.05714	0.00446	0.7013
$\frac{3}{4}$	0.2143	3	0.95429	0.2478	3.47	0.7935	0.4029	0.99286	0.05714	0.00446	0.9105
1	0.2609	3	1.19733	0.3017	3.47	0.9845	0.5089	1.24543	0.06957	0.00543	1.1441
$1\frac{1}{4}$	0.2609	3	1.54083	0.3017	3.47	1.0085	0.5329	1.59043	0.06957	0.00543	1.4876
$1\frac{1}{2}$	0.2609	3	1.77978	0.3017	3.47	1.0252	0.5496	1.83043	0.06957	0.00543	1.7265
2	0.2609	3	2.25272	0.3017	3.47	1.0582	0.5826	2.30543	0.06957	0.00543	2.1995
$2\frac{1}{2}$	0.2500[g]	2	2.70391	0.4337	3.47	1.5712	0.8875	2.77500	0.100000	0.00781	2.6195
3	0.2500[g]	2	3.32500	0.4337	3.47	1.6137	0.9500	3.40000	0.100000	0.00781	3.2406
$3\frac{1}{2}$	0.2500	2	3.82188	0.4337	3.47	1.6837	1.0000	3.90000	0.100000	0.00781	3.7375
4	0.2500	2	4.31875	0.4337	3.47	1.7337	1.0500	4.40000	0.100000	0.00781	4.2344
5	0.2500	2	5.37511	0.4337	3.47	1.8400	1.1563	5.46300	0.100000	0.00781	5.2907
6	0.2500	2	6.43047	0.4337	3.47	1.9462	1.2625	6.52500	0.100000	0.00781	6.3461
8	0.2500	2	8.41797	0.4337	3.47	2.1462	1.4625	8.52500	0.100000	0.00781	8.3336
10	0.2500	2	10.52969	0.4337	3.47	2.3587	1.6750	10.65000	0.100000	0.00781	10.4453
12	0.2500	2	12.51719	0.4337	3.47	2.5587	1.8750	12.65000	0.100000	0.00781	12.4328
14 OD	0.2500	2	13.75938	0.4337	3.47	2.6837	2.0000	13.90000	0.100000	0.00781	13.6750
16 OD	0.2500	2	15.74688	0.4337	3.47	2.8837	2.2000	15.90000	0.100000	0.00781	15.6625
18 OD	0.2500	2	17.73438	0.4337	3.47	3.0837	2.4000	17.90000	0.100000	0.00781	17.6500
20 OD	0.2500	2	19.72188	0.4337	3.47	3.2837	2.6000	19.90000	0.100000	0.00781	19.6375
24 OD	0.2500	2	23.69688	0.4337	3.47	3.6837	3.0000	23.90000	0.100000	0.00781	23.6125

[a] The basic dimensions of the American Standard Taper Pipe Thread are given in inches to four or five decimal places. While this implies a greater degree of precision than is ordinarily attained, these dimensions are the basis of gage dimensions and are so expressed for the purpose of eliminating errors in computations.

[b] Also length of thin-ring gage and length from gaging notch to small end of plug gage.

[c] Also pitch diameter at gaging notch (handtight plane).

[d] Also length of plug gage.

[e] The length L_5 from the end of the pipe determines the plane beyond which the thread form is imperfect at the crest. The next two threads are perfect at the root. At this plane the cone formed by the crests of the thread intersects the cylinder forming the external surface of the pipe. $L_5 = L_2 - 2p$.

[f] Given as information for use in selecting tap drills. The E_3 dimensions are as follows: size $2\frac{1}{2}$ in., 2.69609; and size 3 in., 3.31719.

[g] Military Specification MIL—P—7105 gives the wrench make-up as three threads for 3 in. and smaller.

variation in the commercial product is one turn large or small from the basic dimensions.

On pipe fittings and valves (not steel) for steam pressure 300 lb and below, it is intended that plug-and-ring gage practice as set up in ASA B2.1 Standard provides for a satisfactory check on accumulated variations in such products due to taper, lead, and angle. Thus, tolerances on thread elements have not been established for this class. For service conditions where a more exacting check is required, a procedure as developed by industry and found practical, other than regulation plug-and-ring gage, may be used.

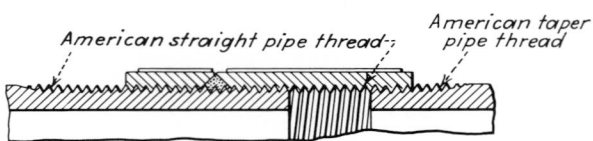

American straight pipe thread

American taper pipe thread

FIG. 66. Long screw joint between straight-threaded coupling and straight- and taper-threaded pipes.

For steel products and all pipe made of steel, wrought iron, or brass, the variation in thread elements should not exceed the following limits:

1. *Taper*—Sizes $\frac{1}{16}$ to $2\frac{1}{2}$ in. and larger, inclusive; maximum taper, $\frac{7}{8}$ in. per ft; minimum taper, $\frac{11}{16}$ in. per ft

2. *Lead*—*a.* 27 and 18 threads per inch, ±0.003 in. in length of effective thread
b. 14, $11\frac{1}{2}$, and 8 threads per inch, ±0.003 in. per in. ±0.006 in. cumulative

3. 60-*deg Angle*—*a.* 27 threads per inch, $\pm2\frac{1}{2}$ deg inclusive
b. 18 and 14 threads per inch, ±2 deg inclusive
c. $11\frac{1}{2}$ and 8 threads per inch, $\pm1\frac{1}{2}$ deg inclusive

4. *Height*—Dependent on minimum and maximum truncation of crest and root (Table 1 of ASA B2.1 not reproduced here)

5. *Pitch Diameter*—One turn large or small from gaging notch on plug or gaging face or ring when using working gages; also measured with wire gages

Pressure-tight Joints. Pressure-tight joints for low-pressure service are sometimes made with straight internal threads and the American Standard taper external threads. The ductility of the coupling enables the straight thread to conform to the taper of the pipe thread. In commercial practice straight-tapped couplings are furnished for standard-weight (Schedule 40) pipe 2 in. and smaller. If taper-tapped couplings are required for standard-weight pipe sizes 2 in. and smaller, line pipe in accordance with API 5L should be ordered. The thread lengths should be in accordance with the American Standard for Pipe Threads, ASA B2.1. Taper-tapped couplings are furnished on extra-strong (Schedule 80) pipe in all sizes and on standard-weight $2\frac{1}{2}$ in. and larger.

Couplings with straight pipe threads are sometimes used with straight external threads, in which case it is necessary to employ lock nuts and packing when it is desired to seal the joint. This type of joint is illustrated in Fig. 66.

Dryseal pipe threads are also employed for pressure-tight joints, particularly where the presence of a lubricant or sealer would contaminate the flow medium. They are similar to the pipe threads covered by Standard ASA B2.1: the essential difference is that, in dryseal pipe threads, the truncation of the crest and root is controlled to assure metal-to-metal contact coincident with or prior to flank contact, thus eliminating spiral leakage paths. Dryseal pipe threads are used in refrigerant systems and for fuel and hydraulic control lines in aircraft, automotive

Table 49. Basic Dimensions, American Standard B33.1 (Hose Couplings Screw Threads)

Service and nominal size, in.	Number of threads, in.	Outside diameter of nipple thread, max
Garden and similar hose:		
½, ⅝, ¾	11½	1.0625
Chemical, engine, and booster hose:		
¾, 1	8	1.3750
Fire protection hose:		
1½	9	1.9900
Steam, water, air, oil, and all other hose connections:		
½	14	0.8248
¾	14	1.0353
1	11½	1.2951
1¼	11½	1.6399
1½	11½	1.8788
2	11½	2.3528

and marine service. Thread sizes up to 3 in. are covered by American Standard ASA B2.2 and by Society of Automotive Engineers Standard SAE J-476.

Hose Nipples and Couplings. Hose coupling joints are ordinarily used with a gasket and made with straight internal and external loose-fitting threads.   There are several standards of hose threads having various diameters and pitches, one of which is based on the American Standard Pipe Thread. With this thread series, it is possible to join small hose sizes ½ to 2 in. inclusive to ends of standard pipe having American Standard External Taper Pipe Threads, using a gasket to seal the joint.

American Standard ASA B33.1 applies to the threaded parts of hose couplings, valves, nozzles, and all other fittings used in direct connection with hose intended for fire protection or for domestic and industrial general services in sizes ½, ⅝, ¾, 1, 1¼, 1½, and 2 in. Basic thread dimensions are listed in Table 49.

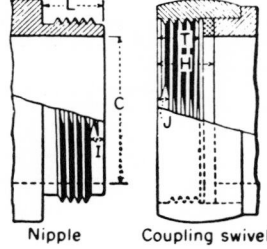

Nipple   Coupling swivel

FIG. 67. American Standard fire-hose coupling, as specified in ASA B26.

American Standard ASA B26 covers the threaded parts of fire-hose couplings, hydrant outlets, standpipe connections, and all other fittings on fire lines where fittings 2½ through 6 in. nominal diameter are used (Fig. 67). The threads conform to the American Standard (National) form having an included angle of 60 deg and are truncated top and bottom.

BOLTED JOINTS

Flanged joints are required in general where pipe lines, piping components, or equipment must be disassembled for maintenance work. Connections to cast-iron valves and fittings and to steel flanged-end valves, where such construction is considered more desirable, must be flanged, and valve-bonnet joints frequently are flanged.

Table 50a. **Dimensions of Typical Commercial Cast-iron Companion Flanges Manufactured in Accordance with ASA B16.1-1960 and ASA B16.2-1960, Which Also Conform to Federal Specification WW-F-406a**

Companion flange, 125-lb class

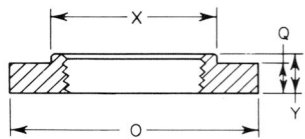

Size, in.	Diameter of flange O, in.	Thickness of flange[1] (min) Q, in.	Diameter of hub (min) X, in.	Length through hub[1] (min) Y, in.	Weight (approx) each, lb	
					Cast iron	Malleable[2]
1	4¼	7/16	1 15/16	11/16	1.75	
1¼	4⅝	½	2 5/16	13/16	2.00	
1½	5	9/16	2 9/16	⅞	2.25	2.25
2	6	⅝	3 1/16	1	4.00	4.00
2½	7	11/16	3 9/16	1⅛	6.00	6.00
3	7½	¾	4¼	1 3/16	7.63	7.63
3½	8½	13/16	4 13/16	1¼	9.00	9.00
4	9	15/16	5 5/16	1 5/16	11.75	11.75
5	10	15/16	6 7/16	1 7/16	14.00	14.00
6	11	1	7 9/16	1 9/16	16.50	16.50
8	13½	1⅛	9 11/16	1¾	26.00	26.00
10	16	1 3/16	11 15/16	1 15/16	37.75	
12	19	1¼	14 1/16	2 3/16	50.50	
14 OD	21	1⅜	15⅜	2¼	80.00	
16 OD	23½	1 7/16	17½	2½	100.00	
18 OD	25	1 9/16	19⅝	2 11/16	106.00	
20 OD	27½	1 11/16	21¾	2⅞	128.00	
24 OD	32	1⅞	26	3¼	202.00	

Companion flange, 250-lb class

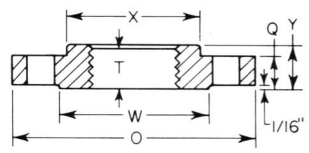

Size, in.	Diameter of flange O, in.	Thickness of flange (min) Q, in.	Diameter of hub (min) X, in.	Length through hub[3] (min) Y, in.	Length of threads (min) T, in.	Diameter of raised face W, in.	Weight (approx) each, lb	
							Cast iron	Malleable[2]
1½	6⅛	13/16	2¾	1⅛	0.87	3 9/16	5.75	
2	6½	⅞	3 5/16	1¼	1.00	4 3/16	6.50	6.50
2½	7½	1	3 15/16	1 7/16	1.14	4 15/16	9.50	9.50
3	8¼	1⅛	4⅝	1 9/16	1.20	5 11/16	12.33	12.33
3½	9	1 3/16	5¼	1⅝	1.25	6 5/16	16.00	
4	10	1¼	5¾	1¾	1.30	6 15/16	20.00	20.00
5	11	1⅜	7	1⅞	1.41	8 5/16	24.00	24.00
6	12½	1 7/16	8⅛	1 15/16	1.51	9 11/16	32.00	32.00
8	15	1⅝	10¼	2 3/16	1.71	11 15/16	51.00	51.00
10	17½	1⅞	12⅝	2⅜	1.92	14 1/16	77.00	
12	20½	2	14¾	2 9/16	2.12	16 7/16	103.00	

[1] All 125-lb cast-iron standard flanges have a plain face.

[2] Dimensional standards have not been established for malleable iron companion flanges; they are generally produced to the same dimensions as cast-iron flanges of the same class.

[3] Minimum thickness of 250-lb flanges includes 1/16-in. raised face.

Table 50b. Bolting Dimension for Cast-iron Flanges

125-lb class

Size, in.	Diameter of bolt circle	Number of bolts	Diameter of bolts	Diameter of bolt holes	Length of bolts
1	3⅛	4	½	⅝	1¾
1¼	3½	4	½	⅝	2
1½	3⅞	4	½	⅝	2
2	4¾	4	⅝	¾	2¼
2½	5½	4	⅝	¾	2½
3	6	4	⅝	¾	2½
3½	7	8	⅝	¾	2¾
4	7½	8	⅝	¾	3
5	8½	8	¾	⅞	3
6	9½	8	¾	⅞	3¼
8	11¾	8	¾	⅞	3½
10	14¼	12	⅞	1	3¾
12	17	12	⅞	1	3¾
14	18¾	12	1	1⅛	4¼
16	21¼	16	1	1⅛	4½
18	22¾	16	1⅛	1¼	4¾
20	25	20	1⅛	1¼	5
24	29½	20	1¼	1⅝	5½

250-lb class

Size, in.	Diameter of bolt circle	Diameter of bolt holes	Number of bolts	Size of bolts	Length of bolts
1½	4½	⅞	4	¾	2¾
2	5	¾	8	⅝	2¾
2½	5⅞	⅞	8	¾	3¼
3	6⅝	⅞	8	¾	3½
3½	7¼	⅞	8	¾	3½
4	7⅞	⅞	8	¾	3¾
5	9¼	⅞	8	¾	4
6	10⅝	⅞	12	¾	4
8	13	1	12	⅞	4½
10	15¼	1⅛	16	1	5¼
12	17¾	1¼	16	1⅛	5½

Cast-iron Flanges. Cast-iron flanges are produced to the following standards:

ASA B16b1   Cast Iron Pipe Flanges and Flanged Fittings for 800 psig Hydraulic Pressure

ASA B16b2   Cast Iron Pipe Flanges and Flanged Fittings 25 lbs.

ASA B16.1   Cast Iron Pipe Flanges and Flanged Fittings, Class 125

ASA B16.2   Cast Iron Pipe Flanges and Flanged Fittings, Class 250

ASTM A126   Class B Gray Iron Castings For Valves, Flanges and Pipe Fittings

Federal Specification WW-F-406 Flange Dimensions, Standard (Classes 125 and 250 cast-iron flanges; classes 150, 250, 300 bronze flanges; for land use).

Dimensions of standard and extra-heavy companion flanges are given in Table 50a. Bolting dimensions are given in Table 50b.

Table 51. Dimensions of Typical Commercial Forged-steel Welding-neck Flanges, Manufactured in Accordance with ASA B16.5-1961, Where Applicable

150 lb

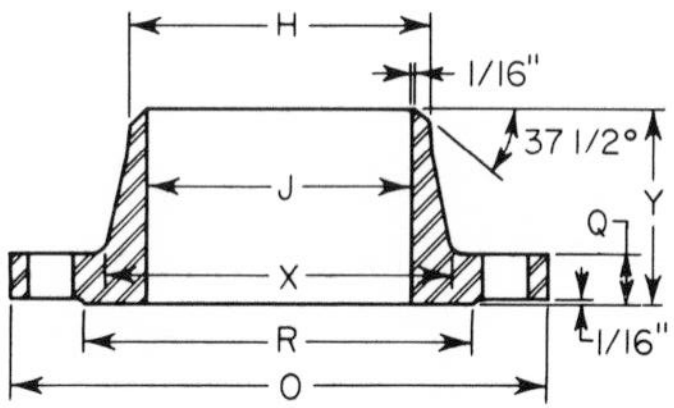

1/16" raised face included in dimensions Q and Y

Nominal pipe size	Outside diameter of flange O	Thickness of flange Q (min)	Diameter of raised-face R	Diameter of bore J[a]	Length through hub Y	Diameter of hub at base X	Diameter of hub at point of welding, H	Number of bolt holes	Diameter of bolts[c]	Diameter of bolt circle	Length of bolts: Stud[b] 1/16-in. raised-face	Length of bolts: Stud[b] Ring joint	Length of bolts: Machine 1/16-in. raised-face	Weight (approx), lb
150 lb														
1/2	3 1/2	7/16	1 3/8	0.62	1 7/8	1 3/16	0.84	4	1/2	2 3/8	2 1/4	...	1 3/4	2
3/4	3 7/8	1/2	1 11/16	0.82	2 1/16	1 1/2	1.05	4	1/2	2 3/4	2 1/4	...	2	2
1	4 1/4	9/16	2	1.05	2 3/16	1 15/16	1.32	4	1/2	3 1/8	2 1/2	3	2	2 1/2
1 1/4	4 5/8	5/8	2 1/2	1.38	2 1/4	2 5/16	1.66	4	1/2	3 1/2	2 1/2	3	2 1/4	2 1/2
1 1/2	5	11/16	2 7/8	1.61	2 7/16	2 9/16	1.90	4	1/2	3 7/8	2 3/4	3 1/4	2 1/4	4

2	6	$\frac{3}{4}$	$3\frac{5}{8}$	2.07	$2\frac{1}{2}$	$3\frac{1}{16}$	2.38	4	$\frac{5}{8}$	$4\frac{3}{4}$	3	$3\frac{1}{2}$	$2\frac{3}{4}$	6
$2\frac{1}{2}$	7	$\frac{7}{8}$	$4\frac{1}{8}$	2.47	$2\frac{3}{4}$	$3\frac{9}{16}$	2.88	4	$\frac{5}{8}$	$5\frac{1}{2}$	$3\frac{1}{4}$	$3\frac{3}{4}$	3	10
3	$7\frac{1}{2}$	$\frac{15}{16}$	5	3.07	$2\frac{3}{4}$	$4\frac{1}{4}$	3.50	4	$\frac{5}{8}$	6	$3\frac{1}{2}$	4	3	$11\frac{1}{2}$
$3\frac{1}{2}$	$8\frac{1}{2}$	$\frac{15}{16}$	$5\frac{1}{2}$	3.55	$2\frac{13}{16}$	$4\frac{13}{16}$	4.00	8	$\frac{5}{8}$	7	$3\frac{1}{2}$	4	3	12
4	9	$\frac{15}{16}$	$6\frac{3}{16}$	4.03	3	$5\frac{5}{16}$	4.50	8	$\frac{5}{8}$	$7\frac{1}{2}$	$3\frac{1}{2}$	4	3	15
5	10	$\frac{15}{16}$	$7\frac{5}{16}$	5.05	$3\frac{1}{2}$	$6\frac{7}{16}$	5.56	8	$\frac{3}{4}$	$8\frac{1}{2}$	$3\frac{3}{4}$	$4\frac{1}{4}$	$3\frac{1}{4}$	19
6	11	1	$8\frac{1}{2}$	6.07	$3\frac{1}{2}$	$7\frac{9}{16}$	6.63	8	$\frac{3}{4}$	$9\frac{1}{2}$	$3\frac{3}{4}$	$4\frac{1}{4}$	$3\frac{1}{4}$	24
8	$13\frac{1}{2}$	$1\frac{1}{8}$	$10\frac{5}{8}$	7.98	4	$9\frac{11}{16}$	8.63	8	$\frac{3}{4}$	$11\frac{3}{4}$	4	$4\frac{1}{2}$	$3\frac{1}{2}$	39
10	16	$1\frac{3}{16}$	$12\frac{3}{4}$	10.02	4	12	10.75	12	$\frac{7}{8}$	$14\frac{1}{4}$	$4\frac{1}{2}$	5	$3\frac{3}{4}$	52
12	19	$1\frac{1}{4}$	15	12.00	$4\frac{1}{2}$	$14\frac{3}{8}$	12.75	12	$\frac{7}{8}$	17	$4\frac{1}{2}$	5	4	80
14	21	$1\frac{3}{8}$	$16\frac{1}{4}$	13.25	5	$15\frac{3}{4}$	14.00	12	1	$18\frac{3}{4}$	5	$5\frac{1}{2}$	$4\frac{1}{4}$	102
16	$23\frac{1}{2}$	$1\frac{7}{16}$	$18\frac{1}{2}$	15.25	5	18	16.00	16	1	$21\frac{1}{4}$	$5\frac{1}{4}$	$5\frac{3}{4}$	$4\frac{1}{2}$	127
18	25	$1\frac{9}{16}$	21	17.25	$5\frac{1}{2}$	$19\frac{7}{8}$	18.00	16	$1\frac{1}{8}$	$22\frac{3}{4}$	$5\frac{3}{4}$	$6\frac{1}{4}$	$4\frac{3}{4}$	140
20^h	$27\frac{1}{2}$	$1\frac{11}{16}$	23	19.25	$5\frac{11}{16}$	22	20.00	20	$1\frac{1}{8}$	25	6	7	$5\frac{1}{4}$	170
$22\frac{1}{2}$	$29\frac{1}{2}$	$1\frac{13}{16}$	$25\frac{1}{4}$	21.25	$5\frac{7}{8}$	24	22.00	20	$1\frac{1}{4}$	$27\frac{1}{4}$	$6\frac{1}{2}$	7	$5\frac{1}{2}$	224
24	32	$1\frac{7}{8}$	$27\frac{1}{4}$	23.25	6	$26\frac{1}{8}$	24.00	20	$1\frac{1}{4}$	$29\frac{1}{2}$	$6\frac{3}{4}$	$7\frac{1}{4}$	$5\frac{3}{4}$	260

Table 51. (Continued)

300 lb

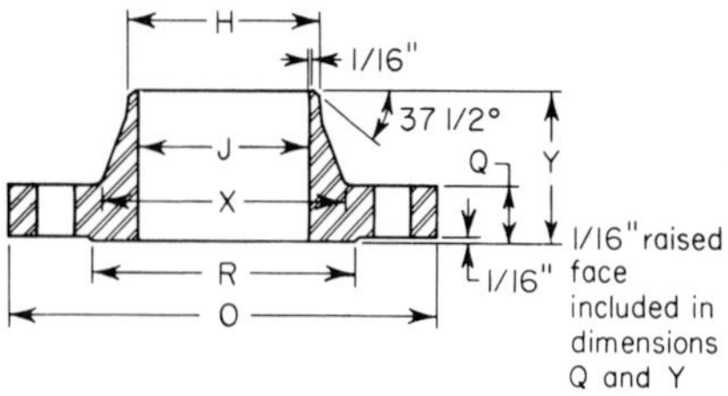
H
1/16"
37 1/2°
J
Q
Y
X
R
O
1/16"
1/16" raised face included in dimensions Q and Y

400 lb

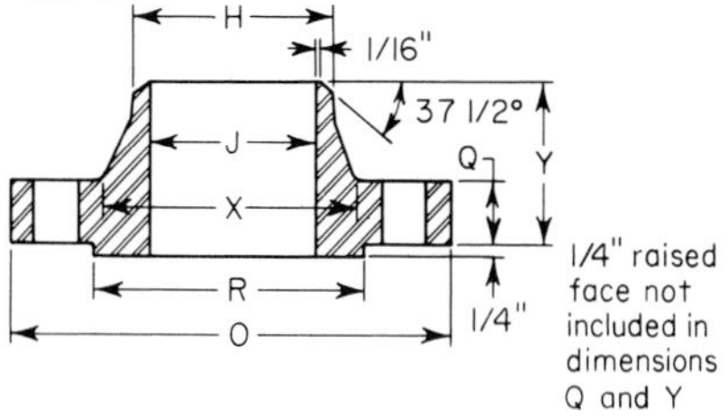
H
1/16"
37 1/2°
J
Q
Y
X
R
O
1/4"
1/4" raised face not included in dimensions Q and Y

Nominal pipe size	Outside diameter of flange O	Thickness of flange Q (min)	Diameter of raised-face R	Diameter of bore J[a]	Length through hub Y	Diameter of hub at base X	Diameter of hub at point of welding, H	Number of bolt holes	Diameter of bolts[c]	Diameter of bolt circle	Length of bolts: Stud[b] Raised-face[d]	Length of bolts: Stud[b] Ring joint	Length of bolts: Machine Raised face[d]	Weight (approx), lb
300 lb														
1/2	3 3/4	9/16	1 3/8	0.62	2 1/16	1 1/2	0.84	4	1/2	2 5/8	2 1/2	3	2	2
3/4	4 5/8	5/8	1 11/16	0.82	2 1/4	1 7/8	1.05	4	5/8	3 1/4	2 3/4	3 1/4	2 1/2	3
1	4 7/8	11/16	2	1.05	2 7/16	2 1/8	1.32	4	5/8	3 1/2	3	3 1/2	2 1/2	4
1 1/4	5 1/4	3/4	2 1/2	1.38	2 9/16	2 1/2	1.66	4	5/8	3 7/8	3	3 1/2	2 3/4	6
1 1/2	6 1/8	13/16	2 7/8	1.61	2 11/16	2 3/4	1.90	4	3/4	4 1/2	3 1/2	4	3	8
2	6 1/2	7/8	3 5/8	2.07	2 3/4	3 5/16	2.38	8	5/8	5	3 1/4	4	3	9
2 1/2	7 1/2	1	4 1/8	2.47	3	3 15/16	2.88	8	3/4	5 7/8	3 3/4	4 1/2	3 1/4	12
3	8 1/4	1 1/8	5	3.07	3 1/8	4 5/8	3.50	8	3/4	6 5/8	4	4 3/4	3 1/2	15
3 1/2	9	1 3/16	5 1/2	3.55	3 3/16	5 1/4	4.00	8	3/4	7 1/4	4 1/4	5	3 3/4	18
4	10	1 1/4	6 3/16	4.03	3 3/8	5 3/4	4.50	8	3/4	7 7/8	4 1/4	5	3 3/4	25
5	11	1 3/8	7 5/16	5.05	3 7/8	7	5.56	8	3/4	9 1/4	4 1/2	5 1/4	4	32
6	12 1/2	1 7/16	8 1/2	6.07	3 7/8	8 1/8	6.63	12	3/4	10 5/8	4 3/4	5 1/2	4 1/4	42
8	15	1 5/8	10 5/8	7.98	4 3/8	10 1/4	8.63	12	7/8	13	5 1/4	6	4 3/4	67
10	17 1/2	1 7/8	12 3/4	10.02	4 5/8	12 5/8	10.75	16	1	15 1/4	6	6 3/4	5 1/4	91
12	20 1/2	2	15	12.00	5 1/8	14 3/4	12.75	16	1 1/8	17 3/4	6 1/2	7 1/4	5 3/4	138
14	23	2 1/8	16 1/4	13.25	5 5/8	16 3/4	14.00	20	1 1/8	20 1/4	6 3/4	7 1/2	6	186
16	25 1/2	2 1/4	18 1/2	15.25	5 3/4	19	16.00	20	1 1/4	22 1/2	7 1/4	8	6 1/2	246
18	28	2 3/8	21	17.25	6 1/4	21	18.00	24	1 1/4	24 3/4	7 1/2	8 1/4	6 3/4	305
20	30 1/2	2 1/2	23	19.25	6 3/8	23 1/8	20.00	24	1 1/4	27	8	8 3/4	7	378
22	33	2 5/8	25 1/4	21.25	6 1/2	24 1/4	22.00	24	1 1/2	29 1/4	8 3/4	9 3/4	7 1/2	429
24	36	2 3/4	27 1/4	23.25	6 5/8	27 5/8	24.00	24	1 1/2	32	9	10	7 3/4	545

							400 lb						Stud[e,b]	
1/2[f]	3 3/4	9/16	1 3/8	0.55	2 1/16	1 1/2	0.84	4	1/2	2 5/8	3	3	2 3/4	3
3/4[f]	4 5/8	5/8	1 11/16	0.74	2 1/4	1 7/8	1.05	4	5/8	3 1/4	3 1/4	3 1/4	3	4
1[f]	4 7/8	11/16	2	0.96	2 7/16	2 1/8	1.32	4	5/8	3 1/2	3 1/2	3 1/2	3 1/4	5
1 1/4[f]	5 1/4	13/16	2 1/2	1.28	2 5/8	2 1/2	1.66	4	5/8	3 7/8	3 3/4	3 3/4	3 1/2	7
1 1/2[f]	6 1/8	7/8	2 7/8	1.50	2 3/4	2 3/4	1.90	4	3/4	4 1/2	4	4	3 3/4	10
2[f]	6 1/2	1	3 5/8	1.94	2 7/8	3 5/16	2.38	8	5/8	5	4	4 1/4	3 3/4	12
2 1/2[f]	7 1/2	1 1/8	4 1/8	2.32	3 1/8	3 15/16	2.88	8	3/4	5 7/8	4 1/2	4 3/4	4 1/4	18
3[f]	8 1/4	1 1/4	5	2.90	3 1/4	4 5/8	3.50	8	3/4	6 5/8	4 3/4	5	4 1/2	23
3 1/2[f]	9	1 3/8	5 1/2	3.36	3 3/8	5 1/4	4.00	8	7/8	7 1/4	5 1/4	5 1/2	5	26
4	10	1 3/8	6 3/16	3.83	3 1/2	5 3/4	4.50	8	7/8	7 7/8	5 1/4	5 1/2	5	35
5	11	1 1/2	7 5/16	4.81	4	7	5.56	8	7/8	9 1/4	5 1/2	5 3/4	5 1/4	43
6	12 1/2	1 5/8	8 1/2	5.76	4 1/16	8 1/8	6.63	12	7/8	10 5/8	5 3/4	6	5 1/2	57
8	15	1 7/8	10 5/8	7.63	4 5/8	10 1/4	8.63	12	1	13	6 1/2	6 3/4	6 1/4	89
10	17 1/2	2 1/8	12 3/4	9.75	4 7/8	12 5/8	10.75	16	1 1/8	15 1/4	7 1/4	7 1/2	7	126
12	20 1/2	2 1/4	15	11.75	5 3/8	14 3/4	12.75	16	1 1/4	17 3/4	7 3/4	8	7 1/2	177
14	23	2 3/8	16 1/4	13.00	5 7/8	16 3/4	14.00	20	1 1/4	20 1/4	8	8 1/4	7 3/4	233
16	25 1/2	2 1/2	18 1/2	15.00	6	19	16.00	20	1 3/8	22 1/2	8 1/2	8 3/4	8 1/4	294
18	28	2 5/8	21	17.00	6 1/2	21	18.00	24	1 3/8	24 3/4	8 3/4	9	8 1/2	360
20	30 1/2	2 3/4	23	19.00	6 5/8	23 1/8	20.00	24	1 1/2	27	9 1/2	9 3/4	9 1/4	445
22[h]	33	2 7/8	25 1/4	21.00	6 3/4	25 1/4	22.00	24	1 5/8	29 1/4	10	10 1/2	9 3/4	465
24	36	3	27 1/4	23.00	6 7/8	27 5/8	24.00	24	1 3/4	32	10 1/2	11	10 1/4	640

Table 51. (Continued)

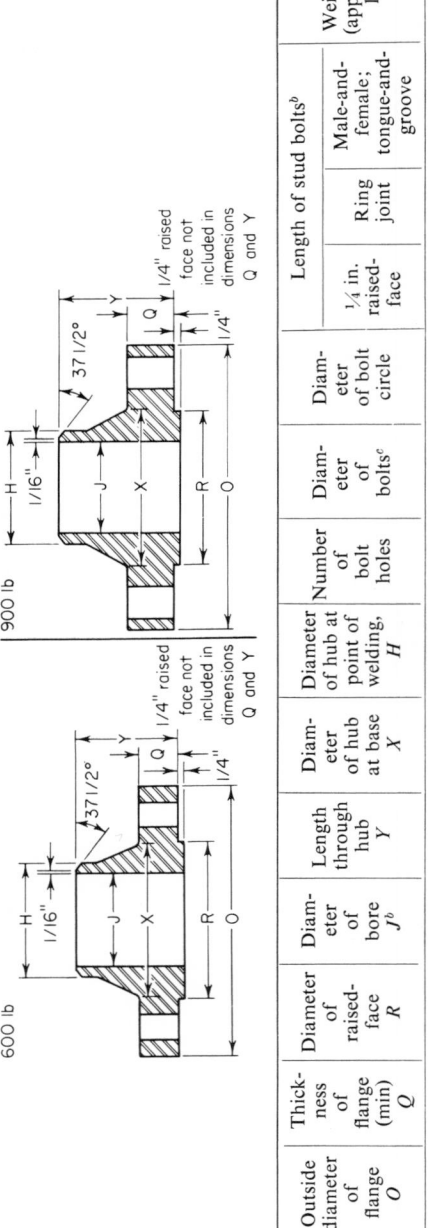

900 lb

600 lb

1/4" raised face not included in dimensions Q and Y

600 lb

Nominal pipe size	Outside diameter of flange O	Thickness of flange (min) Q	Diameter of raised-face R	Diameter of bore J[b]	Length through hub Y	Diameter of hub at base X	Diameter of hub at point of welding, H	Number of bolt holes	Diameter of bolts[c]	Diameter of bolt circle	Length of stud bolts[b]			Weight (approx), lb
											1/4 in. raised-face	Ring joint	Male-and-female; tongue-and-groove	
1/2	3 3/4	9/16	1 3/8	0.55	2 1/16	1 1/2	0.84	4	1/2	2 5/8	3	3	2 3/4	3
3/4	4 5/8	5/8	1 11/16	0.74	2 1/4	1 7/8	1.05	4	5/8	3 1/4	3 1/4	3 1/4	3	4
1	4 7/8	11/16	2	0.96	2 7/16	2 1/8	1.32	4	5/8	3 1/2	3 1/2	3 1/2	3 1/4	5
1 1/4	5 1/4	13/16	2 1/2	1.28	2 5/8	2 1/2	1.66	4	5/8	3 7/8	3 3/4	3 3/4	3 1/2	7
1 1/2	6 1/8	7/8	2 7/8	1.50	2 3/4	2 3/4	1.90	4	3/4	4 1/2	4	4	3 3/4	10
2	6 1/2	1	3 5/8	1.94	2 7/8	3 5/16	2.38	8	5/8	5	4	4 1/4	3 3/4	12
2 1/2	7 1/2	1 1/8	4 1/8	2.32	3 1/8	3 15/16	2.88	8	3/4	5 7/8	4 1/2	4 3/4	4 1/4	18
3	8 1/4	1 1/4	5	2.90	3 1/4	4 5/8	3.50	8	3/4	6 5/8	4 3/4	5	4 1/2	23
3 1/2	9	1 3/8	5 1/2	3.36	3 3/8	5 1/4	4.00	8	7/8	7 1/4	5 1/4	5 1/2	5	26
4	10 3/4	1 1/2	6 3/16	3.83	4	6	4.50	8	7/8	8 1/2	5 1/2	5 3/4	5 1/4	42
5	13	1 3/4	7 5/16	4.81	4 1/2	7 7/16	5.56	8	1	10 1/2	6 1/4	6 1/2	6	68
6	14	1 7/8	8 1/2	5.76	4 5/8	8 3/4	6.63	12	1	11 1/2	6 1/2	6 3/4	6 1/4	81
8	16 1/2	2 3/16	10 5/8	7.63	5 1/4	10 3/4	8.63	12	1 1/8	13 3/4	7 1/2	7 3/4	7 1/4	117
10	20	2 1/2	12 3/4	9.75	6	13 1/2	10.75	16	1 1/4	17	8 1/4	8 1/2	8	189
12	22	2 5/8	15	11.75	6 1/8	15 3/4	12.75	20	1 1/4	19 1/4	8 1/2	8 3/4	8 1/4	226
14	23 3/4	2 3/4	16 1/4	As specified by purchaser	6 1/2	17	14.00	20	1 3/8	20 3/4	9	9 1/4	8 3/4	347
16	27	3	18 1/8		7	19 1/2	16.00	20	1 1/2	23 3/4	9 3/4	10	9 1/2	481
18	29 1/4	3 1/4	21		7 1/4	21 1/2	18.00	20	1 5/8	25 3/4	10 3/4	10 3/4	10 1/4	555
20	32	3 1/2	23		7 1/2	24	20.00	24	1 5/8	28 1/2	11 1/4	11 1/4	11	690
22[h]	34 1/4	3 3/4	25 1/4		7 3/4	26 1/4	22.00	24	1 3/4	30 5/8	12	12 1/2	11 3/4	820
24	37	4	27 1/4		8	28 1/4	24.00	24	1 7/8	33	12 3/4	13 1/4	12 1/2	977

900 lb														
1/2[g]	4 3/4	7/8	1 3/8		2 3/8	1 1/2	0.84	4	3/4	3 1/4	4	4	3 3/4	7
3/4[g]	5 1/8	1	1 11/16		2 3/4	1 3/4	1.05	4	3/4	3 1/2	4 1/4	4 1/4	4	7
1[g]	5 7/8	1 1/8	2		2 7/8	2 1/16	1.32	4	7/8	4	4 3/4	4 3/4	4 1/2	9
1 1/4[g]	6 1/4	1 1/8	2 1/2		2 7/8	2 1/2	1.66	4	7/8	4 3/8	4 3/4	4 3/4	4 1/2	10
1 1/2[g]	7	1 1/4	2 7/8		3 1/4	2 3/4	1.90	4	1	4 7/8	5 1/4	5 1/4	5	14
2[g]	8 1/2	1 1/2	3 5/8		4	4 1/8	2.38	8	7/8	6 1/2	5 1/2	5 3/4	5 1/4	25
2 1/2[g]	9 5/8	1 5/8	4 1/8		4 1/8	4 7/8	2.88	8	1	7 1/2	6	6 1/4	5 3/4	36
3	9 1/2	1 1/2	5	As specified by purchaser	4	5	3.50	8	7/8	7 1/2	5 1/2	5 3/4	5 1/4	32
4	11 1/2	1 3/4	6 3/16		4 1/2	6 1/4	4.50	8	1 1/8	9 1/4	6 1/2	6 3/4	6 1/4	51
5	13 3/4	2	7 5/16		5	7 1/2	5.56	8	1 1/4	11	7 1/4	7 1/2	7	86
6	15	2 3/16	8 1/2		5 1/2	9 1/4	6.63	12	1 1/8	12 1/2	7 1/2	7 1/2	7 1/4	110
8	18 1/2	2 1/2	10 5/8		6 3/8	11 3/4	8.63	12	1 3/8	15 1/2	8 1/2	8 3/4	8 1/4	187
10	21 1/2	2 3/4	12 3/4		7 1/4	14 1/2	10.75	16	1 3/8	18 1/2	9	9 1/4	8 3/4	268
12	24	3 1/8	15		7 7/8	16 1/2	12.75	20	1 3/8	21	9 3/4	10	9 1/2	372
14	25 1/4	3 3/8	16 1/4		8 3/8	17 3/4	14.00	20	1 1/2	22	10 1/2	11	10 1/4	562
16	27 3/4	3 1/2	18 1/2		8 1/2	20	16.00	20	1 5/8	24 1/4	11	11 1/2	10 3/4	685
18	31	4	21		9	22 1/4	18.00	20	1 7/8	27	12 3/4	13 1/4	12 1/2	924
20	33 3/4	4 1/4	23		9 3/4	24 1/2	20.00	20	2	29 1/2	13 1/2	14	13 1/4	1,164
24	41	5 1/2	27 1/4		11 1/2	29 1/2	24.00	20	2 1/2	35 1/2	17	17 3/4	16 3/4	2,107

Table 51. (Continued)

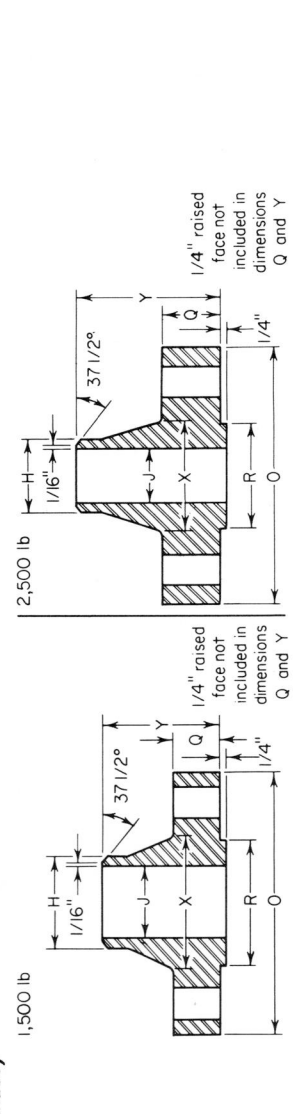

1,500 lb

2,500 lb

1/16" raised face not included in dimensions Q and Y

1/4" raised face not included in dimensions Q and Y

37 1/2°

1,500 lb

Nominal pipe size	Outside diameter of flange O	Thickness of flange Q (min)	Diameter of raised-face R	Diameter of bore J[a]	Length through hub Y	Diameter of hub at base X	Diameter of hub at point of welding H	Number of bolt holes	Diameter of bolts[c]	Diameter bolt of circle	Length of stud bolts[b]			Weight (approx), lb
											1/4-in. raised-face	Ring joint	Male-and-female; tongue-and-groove	
1/2	4 3/4	7/8	1 3/8		2 3/8	1 1/2	0.84	4	3/4	3 1/4	4	4	3 3/4	7
3/4	5 1/8	1	1 11/16		2 3/4	1 3/4	1.05	4	3/4	3 1/2	4 1/4	4 1/4	4	7
1	5 7/8	1 1/8	2		2 7/8	2 1/16	1.32	4	7/8	4	4 3/4	4 3/4	4 1/2	9
1 1/4	6 1/4	1 1/8	2 1/2	As speci-	2 7/8	2 1/2	1.66	4	7/8	4 3/8	4 3/4	4 3/4	4 1/2	10
1 1/2	7	1 1/4	2 7/8	fied by	3 1/4	2 3/4	1.90	4	1	4 7/8	5 1/4	5 1/4	5	14
2	8 1/2	1 1/2	3 5/8	pur-	4	4 1/8	2.38	8	7/8	6 1/2	5 1/2	5 3/4	5 1/4	25
2 1/2	9 5/8	1 5/8	4 1/8	chaser	4 1/8	4 7/8	2.88	8	1	7 1/2	6	6 1/4	5 3/4	36
3	10 1/2	1 7/8	5		4 5/8	5 1/4	3.50	8	1 1/8	8	6 3/4	7	6 1/2	48
4	12 1/4	2 1/8	6 3/16		4 7/8	5 3/8	4.50	8	1 1/4	9 1/2	7 1/2	7 3/4	7 1/4	73
5	14 3/4	2 7/8	7 5/16		6 1/8	7 3/4	5.56	8	1 1/2	11 1/2	9 1/2	9 3/4	9 1/4	132
6	15 1/2	3 1/4	8 1/2		6 3/4	9	6.63	12	1 3/8	12 1/2	10	10 1/4	9 3/4	164
8	19	3 5/8	10 5/8		8 3/8	11 1/2	8.63	12	1 5/8	15 1/2	11 1/4	11 3/4	11	273
10	23	4 1/4	12 3/4		10	14 1/2	10.75	12	1 7/8	19	13 1/4	13 1/2	13	454
12	26 1/2	4 7/8	15		11 1/8	17 3/4	12.75	16	2	22 1/2	14 3/4	15 1/4	14 1/2	690
14	29 1/2	5 1/4	16 1/4		11 3/4	19 1/2	14.00	16	2 1/4	25	16	16 3/4	15 3/4	940
16	32 1/2	5 3/4	18 1/2		12 1/4	21 3/4	16.00	16	2 1/2	27 3/4	17 1/4	18 1/2	17 1/4	1,250
18	36	6 3/8	21		12 7/8	23 1/2	18.00	16	2 3/4	30 1/2	19 1/4	20 1/4	19	1,625
20	38 3/4	7	23		14	25 1/4	20.00	16	3	32 3/4	21	22 1/4	20 3/4	2,050
24	46	8	27 1/4		16	30	24.00	16	3 1/2	39	24	25 1/4	23 3/4	3,325

2,500 lb														
1/2	5 1/4	1 3/16	1 3/8		2 7/8	1 11/16	0.84	4	3/4	3 1/2	4 3/4	4 3/4	4 1/2	8
3/4	5 1/2	1 1/4	1 11/16		3 1/8	2	1.05	4	3/4	3 3/4	4 3/4	4 3/4	4 1/2	9
1	6 1/4	1 3/8	2		3 1/2	2 1/4	1.32	4	7/8	4 1/4	5 1/4	5 1/4	5	13
1 1/4	7 1/4	1 1/2	2 1/2		3 3/4	2 7/8	1.66	4	1	5 1/8	5 3/4	6	5 1/2	20
1 1/2	8	1 3/4	2 7/8		4 3/8	3 1/8	1.90	4	1 1/8	5 3/4	6 1/2	6 3/4	6 1/4	28
2	9 1/4	2	3 5/8	As specified by purchaser	5	3 3/4	2.38	8	1	6 3/4	6 3/4	7	6 1/2	42
2 1/2	10 1/2	2 1/4	4 1/8		5 5/8	4 1/2	2.88	8	1 1/8	7 3/4	7 1/2	7 3/4	7 1/4	52
3	12	2 5/8	5		6 5/8	5 1/4	3.50	8	1 1/4	9	8 1/2	8 3/4	8 1/4	94
4	14	3	6 3/16		7 1/2	6 1/2	4.50	8	1 1/2	10 3/4	9 3/4	10 1/4	9 1/2	146
5	16 1/2	3 5/8	7 5/16		9	8	5.56	8	1 3/4	12 3/4	11 1/2	12 1/4	11 1/4	244
6	19	4 1/4	8 1/2		10 3/4	9 1/4	6.63	8	2	14 1/2	13 1/2	14	13 1/4	378
8	21 3/4	5	10 5/8		12 1/2	12	8.63	12	2	17 1/4	15	15 1/2	14 3/4	576
10	26 1/2	6 1/2	12 3/4		16 1/2	14 3/4	10.75	12	2 1/2	21 1/4	19	20	18 3/4	1,068
12	30	7 1/4	15		18 1/4	17 3/8	12.75	12	2 3/4	24 3/8	21	22	20 3/4	1,608

[a] Flanges bored to dimensions shown, unless otherwise specified. Dimensions shown correspond to ASA B36.10 inside diameter of standard wall pipe for 150- and 300-lb flanges; of extra-strong pipe for 400- and 600-lb flanges.

[b] Lengths of stud bolts include the thickness of two nuts but do not include height of points.

[c] Bolt holes are 1/8 in. larger than bolt diameter.

[d] 1/16-in. raised-face for 300-lb flange; 1/4 in. for 400-lb flange.

[e] Stud bolt lengths for male-and-female or tongue-and-groove facings.

[f] These dimensions same as those for 600-lb flanges.

[g] These dimensions same as those for 1,500-lb flange.

[h] Dimensions for 22 in. size are not covered by ASA B16.5.

Table 52. Dimensions of Typical Commercial Forged-steel Slip-on Flanges Manufactured in Accordance with ASA B16.5-1961, Where Applicable

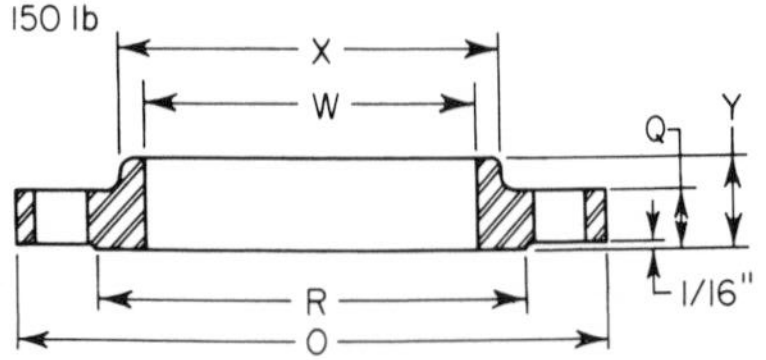

1/16" raised face included in dimensions Q and Y

Nominal pipe size	Outside diameter of flange O	Thickness of flange Q (min)	Diameter of raised-face R	Diameter of bore W	Length through hub Y	Diameter of hub at base X	Number of bolt holes	Diameter of bolts[a]	Diameter of bolt circle	Length of bolts: Stud[b], 1/16-in. raised-face	Length of bolts: Stud[b], Ring joint	Length of bolts: Machine, 1/16-in. raised-face	Weight (approx), lb
150 lb													
1/2	3 1/2	7/16	1 3/8	0.88	5/8	1 3/16	4	1/2	2 3/8	2 1/4	. . .	1 3/4	2
3/4	3 7/8	1/2	1 11/16	1.09	5/8	1 1/2	4	1/2	2 3/4	2 1/4	. . .	2	2
1	4 1/4	9/16	2	1.36	11/16	1 15/16	4	1/2	3 1/8	2 1/2	3	2	2
1 1/4	4 5/8	5/8	2 1/2	1.70	13/16	2 5/16	4	1/2	3 1/2	2 1/2	3	2 1/4	3
1 1/2	5	11/16	2 7/8	1.95	7/8	2 9/16	4	1/2	3 7/8	2 3/4	3 1/4	2 1/4	3
2	6	3/4	3 5/8	2.44	1	3 1/16	4	5/8	4 3/4	3	3 1/2	2 3/4	5
2 1/2	7	7/8	4 1/8	2.94	1 1/8	3 9/16	4	5/8	5 1/2	3 1/4	3 3/4	3	7
3	7 1/2	15/16	5	3.57	1 3/16	4 1/4	4	5/8	6	3 1/2	4	3	8
3 1/2	8 1/2	15/16	5 1/2	4.07	1 1/4	4 13/16	8	5/8	7	3 1/2	4	3	11
4	9	15/16	6 3/16	4.57	1 5/16	5 5/16	8	5/8	7 1/2	3 1/2	4	3	13

5	10	$\frac{15}{16}$	$7\frac{5}{16}$	5.66	$1\frac{7}{16}$	$6\frac{7}{16}$	8	$\frac{3}{4}$	$8\frac{1}{2}$	$3\frac{3}{4}$	$4\frac{1}{4}$	$3\frac{1}{4}$	15
6	11	1	$8\frac{1}{2}$	6.72	$1\frac{9}{16}$	$7\frac{9}{16}$	8	$\frac{3}{4}$	$9\frac{1}{2}$	$3\frac{3}{4}$	$4\frac{1}{4}$	$3\frac{1}{4}$	19
8	$13\frac{1}{2}$	$1\frac{1}{8}$	$10\frac{5}{8}$	8.72	$1\frac{3}{4}$	$9\frac{11}{16}$	8	$\frac{3}{4}$	$11\frac{3}{4}$	4	$4\frac{1}{2}$	$3\frac{1}{2}$	30
10	16	$1\frac{3}{16}$	$12\frac{3}{4}$	10.88	$1\frac{15}{16}$	12	12	$\frac{7}{8}$	$14\frac{1}{4}$	$4\frac{1}{2}$	5	$3\frac{3}{4}$	43
12	19	$1\frac{1}{4}$	15	12.88	$2\frac{3}{16}$	$14\frac{3}{8}$	12	$\frac{7}{8}$	17	$4\frac{1}{2}$	5	4	64
14	21	$1\frac{3}{8}$	$16\frac{1}{4}$	14.14	$2\frac{1}{4}$	$15\frac{3}{4}$	12	1	$18\frac{3}{4}$	5	$5\frac{1}{2}$	$4\frac{1}{4}$	85
16	$23\frac{1}{2}$	$1\frac{7}{16}$	$18\frac{1}{2}$	16.16	$2\frac{1}{2}$	18	16	1	$21\frac{1}{4}$	$5\frac{1}{4}$	$5\frac{3}{4}$	$4\frac{1}{4}$	93
18	25	$1\frac{9}{16}$	21	18.18	$2\frac{11}{16}$	$19\frac{7}{8}$	16	$1\frac{1}{8}$	$22\frac{3}{4}$	$5\frac{3}{4}$	$6\frac{1}{4}$	$4\frac{3}{4}$	120
20	$27\frac{1}{2}$	$1\frac{11}{16}$	23	20.20	$2\frac{7}{8}$	22	20	$1\frac{1}{8}$	25	6	$6\frac{1}{2}$	$5\frac{1}{4}$	155
22^{a}	$29\frac{1}{2}$	$1\frac{13}{16}$	$25\frac{1}{4}$	22.22	$3\frac{1}{8}$	24	20	$1\frac{1}{4}$	$27\frac{1}{4}$	$6\frac{1}{2}$	7	$5\frac{1}{2}$	159
24	32	$1\frac{7}{8}$	$27\frac{1}{4}$	24.25	$3\frac{1}{4}$	$26\frac{1}{8}$	20	$1\frac{1}{4}$	$29\frac{1}{2}$	$6\frac{3}{4}$	$7\frac{1}{4}$	$5\frac{3}{4}$	210

Table 52. (Continued)

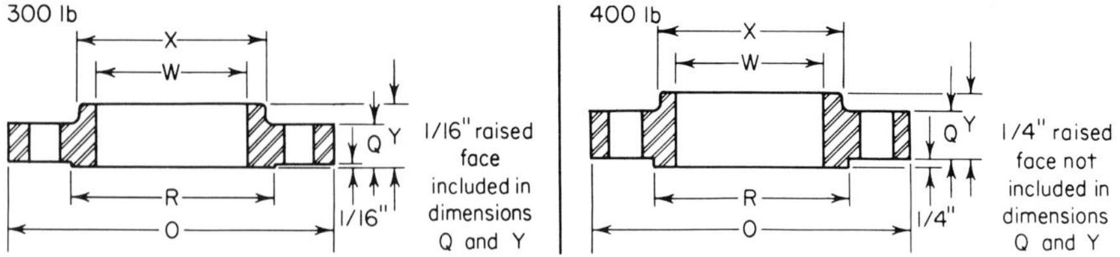

300 lb — 1/16" raised face included in dimensions Q and Y

400 lb — 1/4" raised face not included in dimensions Q and Y

300 lb

Nominal pipe size	Outside diameter of flange O	Thickness of flange Q (min)	Diameter of raised-face R	Diameter of bore W	Length through hub Y	Diameter of hub at base X	Number of bolt holes	Diameter of bolts[a]	Diameter of bolt circle	Stud[b] Raised-face[c]	Stud[b] Ring joint	Machine Raised-face[c]	Weight (approx), lb
½	3¾	9/16	1⅜	0.88	⅞	1½	4	½	2⅝	2½	3	2	3
¾	4⅝	⅝	1 11/16	1.09	1	1⅞	4	⅝	3¼	2¾	3¼	2½	3
1	4⅞	11/16	2	1.36	1 1/16	2⅛	4	⅝	3½	3	3½	2½	3
1¼	5¼	¾	2½	1.70	1 1/16	2½	4	⅝	3⅞	3	3½	2¾	4
1½	6⅛	13/16	2⅞	1.95	1 3/16	2¾	4	¾	4½	3½	4	3	6
2	6½	⅞	3⅝	2.44	1 5/16	3 5/16	8	⅝	5	3¼	4	3	7
2½	7½	1	4⅛	2.94	1½	3 15/16	8	¾	5⅞	3¾	4½	3¼	10
3	8¼	1⅛	5	3.57	1 11/16	4⅝	8	¾	6⅝	4	4¾	3½	13
3½	9	1 3/16	5½	4.07	1¾	5¼	8	¾	7¼	4¼	5	3¾	17
4	10	1¼	6 3/16	4.57	1⅞	5¾	8	¾	7⅞	4¼	5	3¾	22
5	11	1⅜	7 5/16	5.66	2	7	8	¾	9¼	4½	5¼	4	28
6	12½	1 7/16	8½	6.72	2 1/16	8⅛	12	¾	10⅝	4¾	5½	4¼	39
8	15	1⅝	10⅝	8.72	2 7/16	10¼	12	⅞	13	5¼	6	4¾	58
10	17½	1⅞	12¾	10.88	2⅝	12⅝	16	1	15¼	6	6¾	5¼	81
12	20½	2	15	12.88	2⅞	14¾	16	1⅛	17¾	6½	7¼	5¾	115
14	23	2⅛	16¼	14.14	3	16¾	20	1⅛	20¼	6¾	7½	6	164
16	25½	2¼	18½	16.16	3¼	19	20	1¼	22½	7¼	8	6½	220
18	28	2⅜	21	18.18	3½	21	24	1¼	24¾	7½	8¼	6¾	280
20	30½	2½	23	20.20	3¾	23⅛	24	1¼	27	8	8¾	7	325
22[p]	33	2⅝	25¼	22.22	4	25¼	24	1½	29¼	8¾	9¾	7½	433
24	36	2¾	27¼	24.25	4 3/16	27⅝	24	1½	32	9	10	7¾	490

Size					400 lb					Stud[d,b]			
1/2[e]	3 3/4	9/16	1 3/8	0.88	7/8	1 1/2	4	1/2	2 5/8	3	3	2 3/4	3
3/4[e]	4 5/8	5/8	1 11/16	1.09	1	1 7/8	4	5/8	3 1/4	3 1/4	3 1/4	3	3
1[e]	4 7/8	11/16	2	1.36	1 1/16	2 1/8	4	5/8	3 1/2	3 1/2	3 1/2	3 1/4	4
1 1/4[e]	5 1/4	13/16	2 1/2	1.70	1 1/8	2 1/2	4	5/8	3 7/8	3 3/4	3 3/4	3 1/2	6
1 1/2[e]	6 1/8	7/8	2 7/8	1.95	1 1/4	2 3/4	4	3/4	4 1/2	4	4	3 3/4	7
2[e]	6 1/2	1	3 5/8	2.44	1 7/16	3 5/16	8	5/8	5	4 1/4	4	3 3/4	9
2 1/2[e]	7 1/2	1 1/8	4 1/8	2.94	1 5/8	3 15/16	8	3/4	5 7/8	4 3/4	4 1/2	4 1/4	13
3[e]	8 1/4	1 1/4	5	3.57	1 13/16	4 5/8	8	3/4	6 5/8	5	4 3/4	4 1/2	16
3 1/2[e]	9	1 3/8	5 1/2	4.07	1 15/16	5 1/4	8	7/8	7 1/4	5 1/2	5 1/4	5	21
4	10	1 3/8	6 3/16	4.57	2	5 3/4	8	7/8	7 7/8	5 1/2	5 1/4	5	26
5	11	1 1/2	7 5/16	5.66	2 1/8	7	8	7/8	9 1/4	5 3/4	5 1/2	5 1/4	31
6	12 1/2	1 5/8	8 1/2	6.72	2 1/4	8 1/8	12	7/8	10 5/8	6	5 3/4	5 1/2	44
8	15	1 7/8	10 5/8	8.72	2 11/16	10 1/4	12	1	13	6 3/4	6 1/2	6 1/4	67
10	17 1/2	2 1/8	12 3/4	10.88	2 7/8	12 5/8	16	1 1/8	15 1/4	7 1/2	7 1/4	7	91
12	20 1/2	2 1/4	15	12.88	3 1/8	14 3/4	16	1 1/4	17 3/4	8	7 3/4	7 1/2	129
14	23	2 3/8	16 1/4	14.14	3 5/16	16 3/4	20	1 1/4	20 1/4	8 1/4	8	7 3/4	191
16	25 1/2	2 1/2	18 1/2	16.16	3 11/16	19	20	1 3/8	22 1/2	8 3/4	8 1/2	8 1/4	253
18	28	2 5/8	21	18.18	3 7/8	21	24	1 3/8	24 3/4	9	8 3/4	8 1/4	310
20	30 1/2	2 3/4	23	20.20	4	23 1/8	24	1 1/2	27	9 3/4	9 1/2	9 1/4	378
22[g]	33	2 7/8	25 1/4	22.22	4 1/4	25 1/4	24	1 5/8	29 1/4	10 1/2	10	9 3/4	464
24	36	3	27 1/4	24.25	4 1/2	27 5/8	24	1 3/4	32	11	10 1/2	10 1/4	539

Table 52. (Continued)

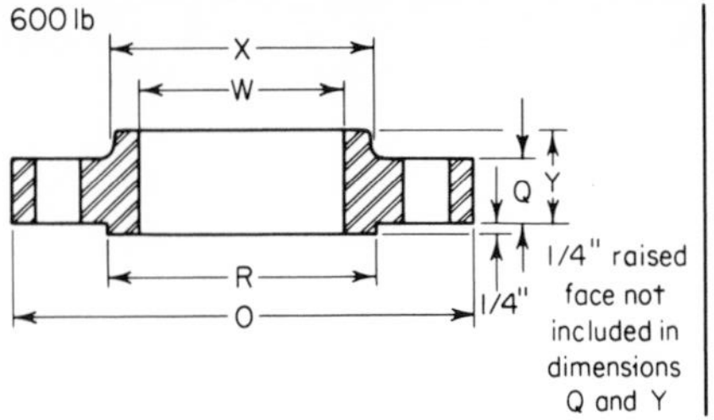
600 lb
X
W
Q
Y
R
O
1/4"
1/4" raised face not included in dimensions Q and Y

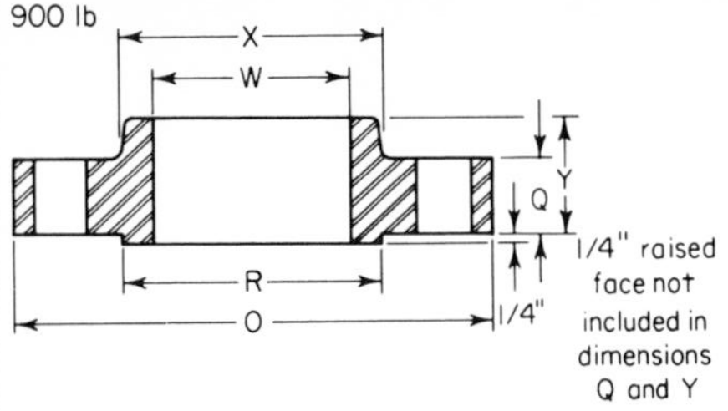
900 lb
X
W
Q
Y
R
O
1/4"
1/4" raised face not included in dimensions Q and Y

Nominal pipe size	Outside diameter of flange O	Thickness of flange Q (min)	Diameter of raised-face R	Diameter of bore W	Length through hub Y	Diameter of hub at base X	Number of bolt holes	Diameter of bolts[a]	Diameter of bolt circle	Length of stud bolts[b] ¼-in. raised-face	Length of stud bolts[b] Ring joint	Length of stud bolts[b] Male-and-female; tongue-and-groove	Weight (approx), lb
600 lb													
½	3¾	9⁄16	1⅜	0.88	⅞	1½	4	½	2⅝	3	3	2¾	3
¾	4⅝	⅝	1 11⁄16	1.09	1	1⅞	4	⅝	3¼	3¼	3¼	3	3
1	4⅞	11⁄16	2	1.36	1 1⁄16	2⅛	4	⅝	3½	3½	3½	3¼	4
1¼	5¼	13⁄16	2½	1.70	1⅛	2½	4	⅝	3⅞	3¾	3¾	3½	6
1½	6⅛	⅞	2⅞	1.95	1¼	2¾	4	¾	4½	4	4	3¾	7
2	6½	1	3⅝	2.44	1 7⁄16	3 5⁄16	8	⅝	5	4	4¼	3¾	9
2½	7½	1⅛	4⅛	2.94	1⅝	3 15⁄16	8	¾	5⅞	4½	4¾	4¼	13
3	8¼	1¼	5	3.57	1 13⁄16	4⅝	8	¾	6⅝	4¾	5	4½	16
3½	9	1⅜	5½	4.07	1 15⁄16	5¼	8	⅞	7¼	5¼	5½	5	21
4	10¾	1½	6 3⁄16	4.57	2⅛	6	8	⅞	8½	5½	5¾	5¼	37
5	13	1¾	7 5⁄16	5.66	2⅜	7 7⁄16	8	1	10½	6¼	6½	6	63
6	14	1⅞	8½	6.72	2⅝	8¾	12	1	11½	6½	6¾	6¼	80
8	16½	2 3⁄16	10⅝	8.72	3	10¾	12	1⅛	13¾	7½	7¾	7¼	115
10	20	2½	12¾	10.88	3⅜	13½	16	1¼	17	8¼	8½	8	177
12	22	2⅝	15	12.88	3⅝	15¾	20	1¼	19¼	8½	8¾	8¼	215
14	23¾	2¾	16¼	14.14	3 11⁄16	17	20	1⅜	20¾	9	9¼	8¾	259
16	27	3	18½	16.16	4 3⁄16	19½	20	1½	23¾	9¾	10	9½	366
18	29¼	3¼	21	18.18	4⅝	21½	20	1⅝	25¾	10½	10¾	10¼	476
20	32	3½	23	20.20	5	24	24	1⅝	28½	11¼	11½	11	612
22	34¼	3¾	25¼	22.22	5¼	26¼	24	1¾	30⅝	12	12½	11¾	643
24	37	4	27¼	24.25	5½	28¼	24	1⅞	33	12¾	13¼	12½	876

900 lb													
$\frac{1}{2}^{f}$	4¾	⅞	1⅜	0.88	1¼	1½	4	¾	3¼	4	4	3¾	9
$\frac{3}{4}^{f}$	5⅛	1	1 11/16	1.09	1⅜	1¾	4	¾	3½	4¼	4¼	4	9
1^{f}	5⅞	1⅛	2	1.36	1⅝	2 1/16	4	⅞	4	4¾	4¾	4½	9
$1\frac{1}{4}^{f}$	6¼	1⅛	2½	1.70	1⅝	2½	4	⅞	4⅜	4¾	4¾	4½	10
$1\frac{1}{2}^{f}$	7	1¼	2⅞	1.95	1¾	2¾	4	1	4⅞	5¼	5¼	5	14
2^{f}	8½	1½	3⅝	2.44	2¼	4⅛	8	⅞	6½	5½	5¾	5¼	25
$2\frac{1}{2}^{f}$	9⅝	1⅝	4⅛	2.94	2½	4⅞	8	1	7½	6	6¼	5¾	36
3	9½	1½	5	3.57	2⅛	5	8	⅞	7½	5½	5¾	5¼	41
4	11½	1¾	6 3/16	4.57	2¾	6¼	8	1⅛	9¼	6½	6¾	6¼	53
5	13¾	2	7 5/16	5.66	3⅛	7½	8	1¼	11	7¼	7½	7	83
6	15	2 3/16	8½	6.72	3⅜	9¼	12	1⅛	12½	7½	7½	7¼	108
8	18½	2½	10⅝	8.72	4	11¾	12	1⅜	15½	8½	8¾	8¼	172
10	21½	2¾	12¾	10.88	4¼	14½	16	1⅜	18½	9	9¼	8¾	245
12	24	3⅛	15	12.88	4⅝	16½	20	1⅜	21	9¾	10	9½	326
14	25¼	3⅜	16¼	14.14	5⅛	17¾	20	1½	22	10½	11	10¼	380
16	27¾	3½	18½	16.16	5¼	20	20	1⅝	24¼	11	11½	10¾	459
18	31	4	21	18.18	6	22¼	20	1⅞	27	12¾	13¼	12½	647
20	33¾	4¼	23	20.20	6¼	24½	20	2	29½	13½	14	13¼	792
24	41	5½	27¼	24.25	8	29½	20	2½	35½	17	17¾	16¾	1,480

Table 52. (*Continued*)

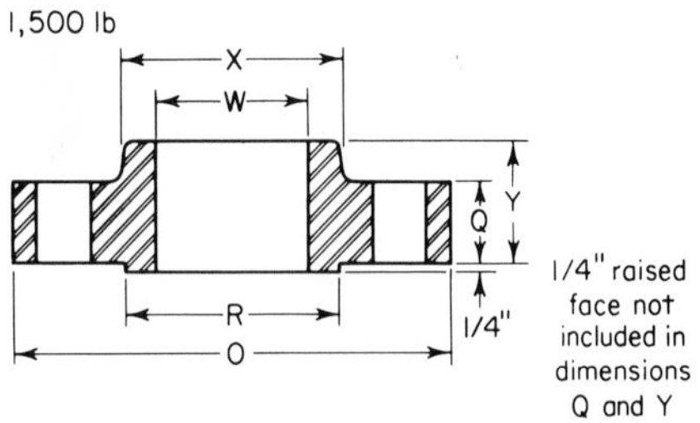
1,500 lb
X
W
Q
Y
R
O
1/4"
1/4" raised face not included in dimensions Q and Y

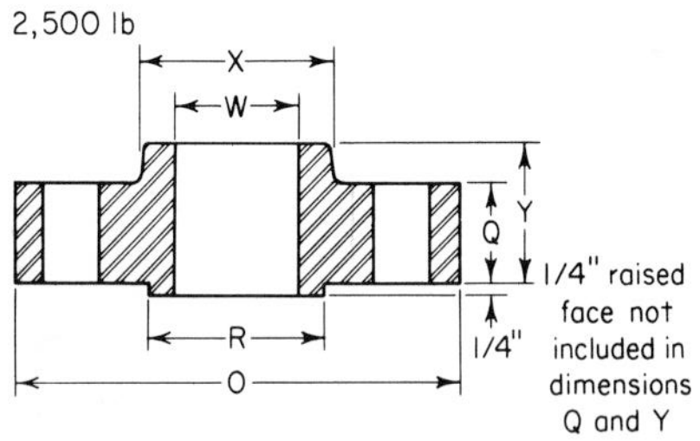
2,500 lb
X
W
Q
Y
R
O
1/4"
1/4" raised face not included in dimensions Q and Y

Nominal pipe size	Outside diameter of flange O	Thickness of flange Q (min)	Diameter of raised-face R	Diameter of bore W^h	Length through hub Y	Diameter of hub at base X	Number of bolt holes	Diameter of bolts[a]	Diameter of bolt circle	Length of stud bolts[b]: ¼-in. raised face	Length of stud bolts[b]: Ring joint	Length of stud bolts[b]: Male-and-female; tongue-and-groove	Weight (approx), lb
1,500 lb													
½	4¾	⅞	1⅜	0.88	1¼	1½	4	¾	3¼	4	4	3¾	9
¾	5⅛	1	1 11/16	1.09	1⅜	1¾	4	¾	3½	4¼	4¼	4	9
1	5⅞	1⅛	2	1.36	1⅝	2 1/16	4	⅞	4	4¾	4¾	4½	9
1¼	6¼	1⅛	2½	1.70	1⅝	2½	4	⅞	4⅜	4¾	4¾	4½	10
1½	7	1¼	2⅞	1.95	1¾	2¾	4	1	4⅞	5¼	5¼	5	14
2	8½	1½	3⅝	2.44	2¼	4⅛	8	⅞	6½	5½	5¾	5¼	25
2½	9⅝	1⅝	4⅛	2.94	2½	4⅞	8	1	7½	6	6¼	5¾	36
3	10½	1⅞	5	3.57	2⅞	5¼	8	1⅛	8	6¾	7	6½	48
4	12¼	2⅛	6 3/16	4.57	3 9/16	6⅜	8	1¼	9½	7½	7¾	7¼	73
5	14¾	2⅞	7 5/16	5.66	4⅛	7¾	8	1½	11½	9½	9¾	9¼	132
6	15½	3¼	8½	6.72	4 11/16	9	12	1⅜	12½	10	10¼	9¾	164
8	19	3⅝	10⅝	8.72	5⅝	11½	12	1⅝	15½	11¼	11¾	11	258
10	23	4¼	12¾	10.88	6¼	14½	12	1⅞	19	13¼	13½	13	436
12	26½	4⅞	15	12.88	7⅛	17¾	16	2	22½	14¾	15¼	14½	667
14	29½	5¼	16¼	14.14	. . .	19½	16	2¼	25	16	16¾	15¾	
16	32½	5¾	18½	16.16	. . .	21¾	16	2½	27¾	17½	18½	17¼	
18	36	6⅜	21	18.18	. . .	23½	16	2¾	30½	19¼	20¼	19	
20	38¾	7	23	20.20	. . .	25¼	16	3	32¾	21	22¼	20¾	
24	46	8	27¼	24.25	. . .	30	16	3½	39	24	25½	23¾	

2,500 lb

1/2	5 1/4	1 3/16	1 3/8	0.88	1 9/16	1 11/16	4	3/4	3 1/2	4 3/4	4 3/4	4 1/2	7
3/4	5 1/2	1 1/4	1 11/16	1.09	1 11/16	2	4	3/4	3 3/4	4 3/4	4 3/4	4 1/2	9
1	6 1/4	1 3/8	2	1.36	1 7/8	2 1/4	4	7/8	4 1/4	5 1/4	5 1/4	5	12
1 1/4	7 1/4	1 1/2	2 1/2	1.70	2 1/16	2 7/8	4	1	5 1/8	5 3/4	6	5 1/2	18
1 1/2	8	1 3/4	2 7/8	1.95	2 3/8	3 1/8	4	1 1/8	5 3/4	6 1/2	6 3/4	6 1/4	25
2	9 1/4	2	3 5/8	2.44	2 3/4	3 3/4	8	1	6 3/4	6 3/4	7	6 1/2	38
2 1/2	10 1/2	2 1/4	4 1/8	2.94	3 1/8	4 1/2	8	1 1/8	7 3/4	7 1/2	7 3/4	7 1/4	55
3	12	2 5/8	5	3.57	3 5/8	5 1/4	8	1 1/4	9	8 1/2	8 3/4	8 1/4	83
4	14	3	6 3/16	4.57	4 1/4	6 1/2	8	1 1/2	10 3/4	9 3/4	10 1/4	9 1/2	127
5	16 1/2	3 5/8	7 5/16	5.66	5 1/8	8	8	1 3/4	12 3/4	11 1/2	12 1/4	11 1/4	210
6	19	4 1/4	8 1/2	6.72	6	9 1/4	8	2	14 1/2	13 1/2	14	13 1/4	323
8	21 3/4	5	10 5/8	8.72	7	12	12	2	17 1/4	15	15 1/2	14 3/4	485
10	26 1/2	6 1/2	12 3/4	10.88	9	14 3/4	12	2 1/2	21 1/4	19	20	18 3/4	925
12	30	7 1/4	15	12.88	10	17 7/8	12	2 3/4	24 3/8	21	22	20 3/4	3,100

[a] Bolt holes are 1/8 in. larger than bolt diameter.

[b] Lengths of stud bolts include the thickness of two nuts but do not include height of points.

[c] 1/16-in. raised-face for 300-lb flange; 1/4 in. for 400-lb flange.

[d] Stud bolt lengths for male-and-female or tongue-and-groove facings.

[e] These dimensions same as those for 600-lb flanges.

[f] These dimensions same as those for 1,500-lb flanges.

[g] Dimensions for 22-in. size are not included in ASA B16.5.

[h] Diameter of bore for slip-on flanges in sizes 3 in. and larger has not been standardized. Typical dimensions are shown; in specific cases, manufacturer's catalog should be consulted.

Table 53. Dimensions of Typical Commercial Forged-steel Threaded Flanges Manufactured in Accordance with ASA B16.5-1965, Where Applicable

150 lb

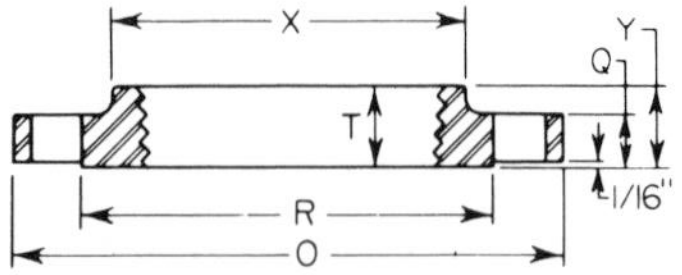

1/16″ raised face included in dimensions Q and Y

150 lb

Nominal pipe size	Outside diameter of flange O	Thickness of flange Q (min)	Diameter of raised-face R	Diameter of counter bore	Length through hub Y	Min thread length T^a	Diameter of hub at base X	Number of bolt holes	Diameter of bolts^b	Diameter of bolt circle	Stud^c 1/16-in. raised-face	Stud^c Ring joint	Machine 1/16-in. raised-face	Weight (approx), lb
½	3½	7/16	1⅜		5/8	5/8	1 3/16	4	½	2⅜	2¼	…	1¾	2
¾	3⅞	½	1 11/16		5/8	5/8	1½	4	½	2¾	2¼	…	2	2
1	4¼	9/16	2		11/16	11/16	1 15/16	4	½	3⅛	2½	3	2	2
1¼	4⅝	5/8	2½	No counter bore required	13/16	13/16	2 5/16	4	½	3½	2½	3	2¼	3
1½	5	11/16	2⅞		7/8	7/8	2 9/16	4	½	3⅞	2¾	3¼	2¼	3
2	6	¾	3⅝		1	1	3 1/16	4	⅝	4¾	3	3½	2¾	5
2½	7	⅞	4⅛		1⅛	1⅛	3 9/16	4	⅝	5½	3¼	3¾	3	7
3	7½	15/16	5		1 3/16	1 3/16	4¼	8	⅝	6	3½	4	3	8
3½	8½	15/16	5½		1¼	1¼	4 13/16	8	⅝	7	3½	4	3	11
4	9	15/16	6 3/16		1 5/16	1 5/16	5 5/16	8	⅝	7½	3½	4	3	13

				on 150-lb threaded flanges										
5	10	$^{15}/_{16}$	$7^5/_{16}$		$1^7/_{16}$	$1^7/_{16}$	$6^7/_{16}$	8	$^3/_4$	$8^1/_2$	$3^3/_4$	$4^1/_4$	$3^1/_4$	15
6	11	1	$8^1/_2$		$1^9/_{16}$	$1^9/_{16}$	$7^9/_{16}$	8	$^3/_4$	$9^1/_2$	$3^3/_4$	$4^1/_4$	$3^1/_4$	19
8	$13^1/_2$	$1^1/_8$	$10^5/_8$		$1^3/_4$	$1^3/_4$	$9^{11}/_{16}$	8	$^3/_4$	$11^3/_4$	4	$4^1/_2$	$3^1/_2$	30
10	16	$1^3/_{16}$	$12^3/_4$		$1^{15}/_{16}$	$1^{15}/_{16}$	12	12	$^7/_8$	$14^1/_4$	$4^1/_2$	5	$3^3/_4$	43
12	19	$1^1/_4$	15		$2^3/_{16}$	$2^3/_{16}$	$14^3/_8$	12	$^7/_8$	17	$4^1/_2$	5	4	64
14	21	$1^3/_8$	$16^1/_4$		$2^1/_4$	$2^1/_4$	$15^3/_4$	12	1	$18^3/_4$	5	$5^1/_2$	$4^1/_4$	85
16	$23^1/_2$	$1^7/_{16}$	$18^1/_2$		$2^1/_2$	$2^1/_2$	18	16	1	$21^1/_4$	$5^1/_4$	$5^3/_4$	$4^1/_4$	93
18	25	$1^9/_{16}$	21		$2^{11}/_{16}$	$2^{11}/_{16}$	$19^7/_8$	16	$1^1/_8$	$22^3/_4$	$5^3/_4$	$6^1/_4$	$4^3/_4$	120
20	$27^1/_2$	$1^{11}/_{16}$	23		$2^7/_8$	$2^7/_8$	22	20	$1^1/_8$	25	6	$6^1/_2$	$5^1/_4$	155
24	32	$1^7/_8$	$27^1/_4$		$3^1/_4$	$3^1/_4$	$26^1/_8$	20	$1^1/_4$	$29^1/_2$	$6^3/_4$	$7^1/_4$	$5^3/_4$	210

Table 53. (*Continued*)

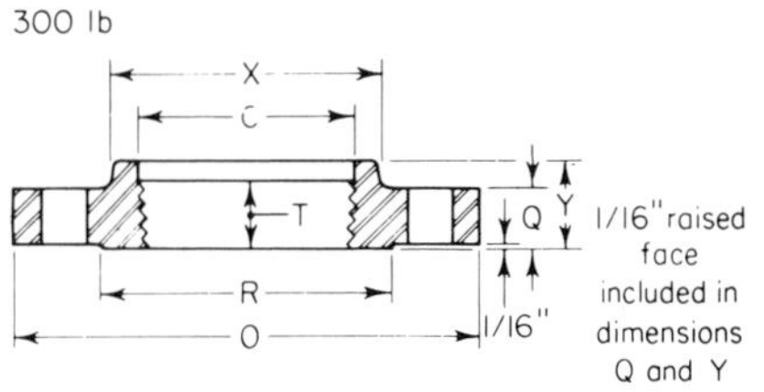
300 lb
X
C
T
R
O
Q
Y
1/16"
1/16" raised face included in dimensions Q and Y

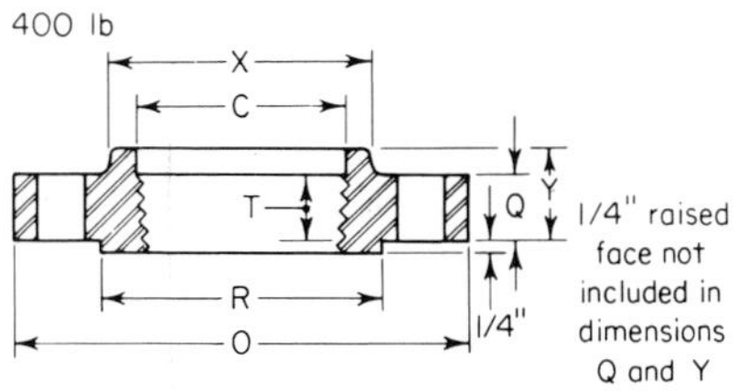
400 lb
X
C
T
R
O
Q
Y
1/4"
1/4" raised face not included in dimensions Q and Y

Nominal pipe size	Outside diameter of flange *O*	Thick-ness of flange *Q* (min)	Diameter of raised-face *R*	Diameter of counter bore *C*	Length through hub *Y*	Min thread length *T*[a]	Diameter of hub at base *X*	Number of bolt holes	Diameter of bolts[b]	Diameter of bolt circle	Length of bolts: Stud[c] Raised-face[d]	Length of bolts: Stud[c] Ring joint	Length of bolts: Machine Raised-face[d]	Weight (approx), lb
300 lb														
1/2	3 3/4	9/16	1 3/8	0.93	7/8	5/8	1 1/2	4	1/2	2 5/8	2 1/2	3	2	3
3/4	4 5/8	5/8	1 11/16	1.14	1	5/8	1 7/8	4	5/8	3 1/4	2 3/4	3 1/4	2 1/2	3
1	4 7/8	11/16	2	1.41	1 1/16	11/16	2 1/8	4	5/8	3 1/2	3	3 1/2	2 1/2	3
1 1/4	5 1/4	3/4	2 1/2	1.75	1 1/16	13/16	2 1/2	4	5/8	3 7/8	3	3 1/2	2 3/4	4
1 1/2	6 1/8	13/16	2 7/8	1.99	1 3/16	7/8	2 3/4	4	3/4	4 1/2	3 1/2	4	3	6
2	6 1/2	7/8	3 5/8	2.50	1 5/16	1 1/8	3 5/16	8	5/8	5	3 1/4	4	3	7
2 1/2	7 1/2	1	4 1/8	3.00	1 1/2	1 1/4	3 15/16	8	3/4	5 7/8	3 3/4	4 1/2	3 1/4	10
3	8 1/4	1 1/8	5	3.63	1 11/16	1 1/4	4 5/8	8	3/4	6 5/8	4	4 3/4	3 1/2	13
3 1/2	9	1 3/16	5 1/2	4.13	1 3/4	1 7/16	5 1/4	8	3/4	7 1/4	4 1/4	5	3 3/4	17
4	10	1 1/4	6 3/16	4.63	1 7/8	1 7/16	5 3/4	8	3/4	7 7/8	4 1/4	5	3 3/4	22
5	11	1 3/8	7 5/16	5.69	2	1 11/16	7	8	3/4	9 1/4	4 1/2	5 1/4	4	28
6	12 1/2	1 7/16	8 1/2	6.75	2 1/16	1 13/16	8 1/8	12	3/4	10 5/8	4 3/4	5 1/2	4 1/4	39
8	15	1 5/8	10 5/8	8.75	2 7/16	2	10 1/4	12	7/8	13	5 1/4	6	4 3/4	58
10	17 1/2	1 7/8	12 3/4	10.88	2 5/8	2 3/16	12 5/8	16	1	15 1/4	6	6 3/4	5 1/4	81
12	20 1/2	2	15	12.94	2 7/8	2 3/8	14 3/4	16	1 1/8	17 3/4	6 1/2	7 1/4	5 3/4	115
14	23	2 1/8	16 1/4	14.19	3	2 1/2	16 3/4	20	1 1/8	20 1/4	6 3/4	7 1/2	6	164
16	25 1/2	2 1/4	18 1/2	16.19	3 1/4	2 11/16	19	20	1 1/4	22 1/2	7 1/4	8	6 1/2	220
18	28	2 3/8	21	18.19	3 1/2	2 3/4	21	24	1 1/4	24 3/4	7 1/2	8 1/4	6 3/4	280
20	30 1/2	2 1/2	23	20.19	3 3/4	2 7/8	23 1/8	24	1 1/4	27	8	8 3/4	7	325
24	36	2 3/4	27 1/4	24.19	4 3/16	3 1/4	27 5/8	24	1 1/2	32	9	10	7 3/4	490

400 lb

														Stud[c,e]	
$\frac{1}{2}$ f	$3\frac{3}{4}$	$\frac{9}{16}$	$1\frac{3}{8}$	0.93	$\frac{7}{8}$	$\frac{5}{8}$	$1\frac{1}{2}$	4	$\frac{1}{2}$	$2\frac{5}{8}$	3	3	$2\frac{3}{4}$	3	
$\frac{3}{4}$ f	$4\frac{5}{8}$	$\frac{5}{8}$	$1\frac{11}{16}$	1.14	1	$\frac{5}{8}$	$1\frac{7}{8}$	4	$\frac{5}{8}$	$3\frac{1}{4}$	$3\frac{1}{4}$	$3\frac{1}{4}$	3	3	
1 f	$4\frac{7}{8}$	$\frac{11}{16}$	2	1.41	$1\frac{1}{16}$	$\frac{11}{16}$	$2\frac{1}{8}$	4	$\frac{5}{8}$	$3\frac{1}{2}$	$3\frac{1}{2}$	$3\frac{1}{2}$	$3\frac{1}{4}$	4	
$1\frac{1}{4}$ f	$5\frac{1}{4}$	$\frac{13}{16}$	$2\frac{1}{2}$	1.75	$1\frac{1}{8}$	$\frac{13}{16}$	$2\frac{1}{8}$	4	$\frac{5}{8}$	$3\frac{7}{8}$	$3\frac{3}{4}$	$3\frac{3}{4}$	$3\frac{1}{2}$	6	
$1\frac{1}{2}$ f	$6\frac{1}{8}$	$\frac{7}{8}$	$2\frac{7}{8}$	1.99	$1\frac{1}{4}$	$\frac{7}{8}$	$2\frac{3}{4}$	4	$\frac{3}{4}$	$4\frac{1}{2}$	4	4	$3\frac{3}{4}$	7	
2 f	$6\frac{1}{2}$	1	$3\frac{5}{8}$	2.50	$1\frac{7}{16}$	$1\frac{1}{8}$	$3\frac{5}{16}$	8	$\frac{5}{8}$	5	4	$4\frac{1}{4}$	$3\frac{3}{4}$	9	
$2\frac{1}{2}$ f	$7\frac{1}{2}$	$1\frac{1}{8}$	$4\frac{1}{8}$	3.00	$1\frac{5}{8}$	$1\frac{1}{4}$	$3\frac{15}{16}$	8	$\frac{3}{4}$	$5\frac{7}{8}$	$4\frac{1}{2}$	$4\frac{3}{4}$	$4\frac{1}{4}$	13	
3 f	$8\frac{1}{4}$	$1\frac{1}{4}$	5	3.63	$1\frac{13}{16}$	$1\frac{3}{8}$	$4\frac{5}{8}$	8	$\frac{3}{4}$	$6\frac{5}{8}$	$4\frac{3}{4}$	5	$4\frac{1}{2}$	16	
$3\frac{1}{2}$ f	9	$1\frac{3}{8}$	$5\frac{1}{2}$	4.13	$1\frac{15}{16}$	$1\frac{9}{16}$	$5\frac{1}{4}$	8	$\frac{7}{8}$	$7\frac{1}{4}$	$5\frac{1}{4}$	$5\frac{1}{2}$	5	21	
4	10	$1\frac{3}{8}$	$6\frac{3}{16}$	4.63	2	$1\frac{7}{16}$	$5\frac{3}{4}$	8	$\frac{7}{8}$	$7\frac{7}{8}$	$5\frac{1}{4}$	$5\frac{1}{2}$	5	26	
5	11	$1\frac{1}{2}$	$7\frac{5}{16}$	5.69	$2\frac{1}{8}$	$1\frac{11}{16}$	7	8	$\frac{7}{8}$	$9\frac{1}{4}$	$5\frac{1}{2}$	$5\frac{3}{4}$	$5\frac{1}{4}$	31	
6	$12\frac{1}{2}$	$1\frac{5}{8}$	$8\frac{1}{2}$	6.75	$2\frac{1}{4}$	$1\frac{13}{16}$	$8\frac{1}{8}$	12	$\frac{7}{8}$	$10\frac{5}{8}$	$5\frac{3}{4}$	6	$5\frac{1}{2}$	44	
8	15	$1\frac{7}{8}$	$10\frac{5}{8}$	8.75	$2\frac{11}{16}$	2	$10\frac{1}{4}$	12	1	13	$6\frac{1}{2}$	$6\frac{3}{4}$	$6\frac{1}{4}$	67	
10	$17\frac{1}{2}$	$2\frac{1}{8}$	$12\frac{3}{4}$	10.88	$2\frac{7}{8}$	$2\frac{3}{16}$	$12\frac{5}{8}$	16	$1\frac{1}{8}$	$15\frac{1}{4}$	$7\frac{1}{4}$	$7\frac{1}{2}$	7	91	
12	$20\frac{1}{2}$	$2\frac{1}{4}$	15	12.94	$3\frac{1}{8}$	$2\frac{3}{8}$	$14\frac{3}{4}$	16	$1\frac{1}{4}$	$17\frac{3}{4}$	$7\frac{3}{4}$	8	$7\frac{1}{2}$	129	
14	23	$2\frac{3}{8}$	$16\frac{1}{4}$	14.19	$3\frac{5}{16}$	$2\frac{1}{2}$	$16\frac{3}{4}$	20	$1\frac{1}{4}$	$20\frac{1}{4}$	8	$8\frac{1}{4}$	$7\frac{3}{4}$	191	
16	$25\frac{1}{2}$	$2\frac{1}{2}$	$18\frac{1}{2}$	16.19	$3\frac{11}{16}$	$2\frac{11}{16}$	19	20	$1\frac{3}{8}$	$22\frac{1}{2}$	$8\frac{1}{2}$	$8\frac{3}{4}$	$8\frac{1}{4}$	253	
18	28	$2\frac{5}{8}$	21	18.19	$3\frac{7}{8}$	$2\frac{3}{4}$	21	24	$1\frac{3}{8}$	$24\frac{3}{4}$	$8\frac{3}{4}$	9	$8\frac{1}{2}$	310	
20	$30\frac{1}{2}$	$2\frac{3}{4}$	23	20.19	4	$2\frac{7}{8}$	$23\frac{1}{8}$	24	$1\frac{1}{2}$	27	$9\frac{1}{2}$	$9\frac{3}{4}$	$9\frac{1}{4}$	378	
24	36	3	$27\frac{1}{4}$	24.19	$4\frac{1}{4}$	$3\frac{1}{4}$	$27\frac{5}{8}$	24	$1\frac{3}{4}$	32	$10\frac{1}{2}$	11	$10\frac{1}{4}$	539	

a Thread lengths for 150-lb flanges are American Standard for Pipe Threads ASA B2.1. For 300- and 400-lb flanges, thread lengths are American Standard for Pipe Threads ASA B2.1 (identical with API line pipe thread). For 600-, 900-, 1,500-, 2,500-lb flanges, thread lengths are proportionately longer than required for API or American Standard for Pipe Threads and satisfy the general requirement for longer threads for the higher pressures. Add depth or height of facing to thread length.

b Bolt holes are $\frac{1}{8}$ inch larger than bolt diameter.

c Lengths of alloy steel stud bolts include the thickness of two nuts, but do not include height of points.

d $\frac{1}{16}$-in. raised-face for 300-lb flange; $\frac{1}{4}$ in. for 400-lb flange.

e Stud bolt lengths for male-and-female or tongue-and-groove facings.

f These dimensions same as those for 600-lb flanges.

Table 54. Dimensions of Typical Commercial Forged-steel Socket-welding Flanges Manufactured in Accordance with ASA B16.5-1961, Where Applicable[3]

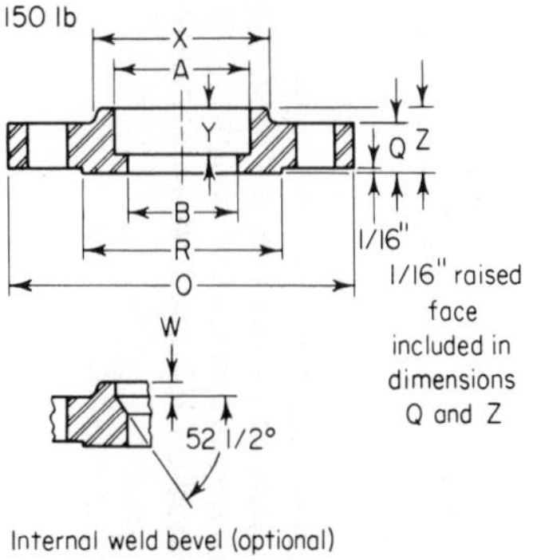
150 lb
X
A
Y
Q
Z
B
R
O
1/16"
1/16" raised face included in dimensions Q and Z
W
52 1/2°
Internal weld bevel (optional)

Nominal pipe size	Outside diameter of flange O	Thickness of flange Q (min)	Diameter of raised-face R	Diameter of bore B[1]	Length through hub Z	Diameter of hub at base X	Diameter of socket A	Depth of socket Y	Depth of socket W	Number of bolt holes	Diameter of bolts[2]	Diameter of bolt circle	Weight (approx), lb
150 lb													
1/4	3 1/2	7/16	1 3/8	0.36	5/8	1 3/16	0.58	3/8	3/16	4	1/2	2 3/8	1
3/8	3 1/2	7/16	1 3/8	0.49	5/8	1 3/16	0.72	3/8	3/16	4	1/2	2 3/8	1
1/2	3 1/2	7/16	1 3/8	0.62	5/8	1 3/16	0.88	3/8	3/16	4	1/2	2 3/8	2
3/4	3 7/8	1/2	1 11/16	0.82	5/8	1 1/2	1.09	7/16	1/8	4	1/2	2 3/4	2
1	4 1/4	9/16	2	1.05	11/16	1 15/16	1.36	1/2	1/8	4	1/2	3 1/8	2
1 1/4	4 5/8	5/8	2 1/2	1.38	13/16	2 5/16	1.70	9/16	3/16	4	1/2	3 1/2	3
1 1/2	5	11/16	2 7/8	1.61	7/8	2 9/16	1.95	5/8	3/16	4	1/2	3 7/8	3
2	6	3/4	3 5/8	2.07	1	3 1/16	2.44	11/16	1/4	4	5/8	4 3/4	5
2 1/2	7	7/8	4 1/8	2.47	1 1/8	3 9/16	2.94	3/4	1/4	4	5/8	5 1/2	7
3	7 1/2	15/16	5	3.07	1 3/16	4 1/4	3.57	13/16	1/4	4	5/8	6	8

3½	8½	15/16	5½	3.55	1¼	4 13/16	4.07	⅞	5/16	8	⅝	7	11
4	9	15/16	6 3/16	4.03	1 5/16	5 5/16	4.57	15/16	⅜	8	⅝	7½	13
5	10	15/16	7 5/16	5.05	1 7/16	6 7/16	5.66	15/16	½	8	¾	8½	15
6	11	1	8½	6.07	1 9/16	7 9/16	6.72	1 1/16	9/16	8	¾	9½	19
8	13½	1⅛	10⅝	7.98	1¾	9 11/16	8.72	1¼	⅝	8	¾	11¾	30
10	16	1 3/16	12¾	10.02	1 15/16	12	10.88	1 5/16	¾	12	⅞	14¼	43
12	19	1¼	15	12.00	2 3/16	14⅜	12.88	1 9/16	15/16	12	⅞	17	64
14	21	1⅜	16¼	13.25	2¼	15¾	14.14	1⅝	⅞	12	1	18¾	85
16	23½	1 7/16	18½	15.25	2½	18	16.16	1¾	1 1/16	16	1	21¼	93
18	25	1 9/16	21	17.25	2 11/16	19⅞	18.18	1 15/16	1⅛	16	1⅛	22¾	120
20	27½	1 11/16	23	19.25	2⅞	22	20.20	2⅛	1 3/16	20	1⅛	25	155
24	32	1⅞	27¼	23.25	3¼	26⅛	24.25	2½	1⅜	20	1¼	29½	210

Table 54. (Continued)

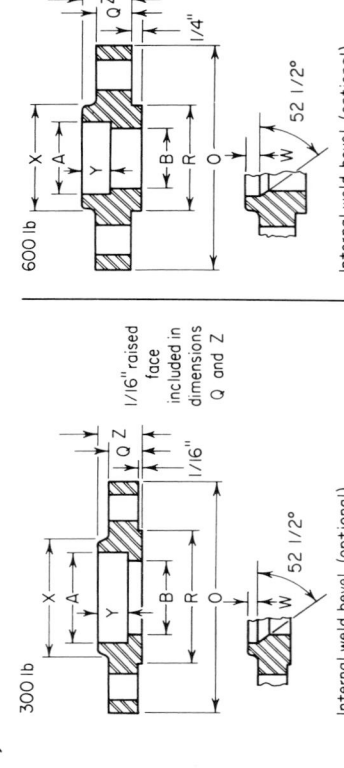

300 lb — 1/16" raised face included in dimensions Q and Z — Internal weld bevel (optional) — 52 1/2°

600 lb — 1/4" raised face not included in dimensions Q and Z — Internal weld bevel (optional) — 52 1/2°

300 lb

Nominal pipe size	Outside diameter of flange O	Thickness of flange Q (min)	Diameter of raised-face R	Diameter of bore B^1	Length through hub Z	Diameter of hub at base X	Diameter of socket A	Depth of socket Y	Depth of socket W	Number of bolt holes	Diameter of bolts^2	Diameter of bolt circle	Weight (approx), lb
1/4	3 3/4	9/16	1 3/8	0.36	7/8	1 1/2	0.58	3/8	5/16	4	1/2	2 5/8	1
3/8	3 3/4	9/16	1 3/8	0.49	7/8	1 1/2	0.72	3/8	5/16	4	1/2	2 5/8	1
1/2	3 3/4	9/16	1 3/8	0.62	7/8	1 1/2	0.88	3/8	5/16	4	1/2	2 5/8	3
3/4	4 5/8	5/8	1 11/16	0.82	1	1 7/8	1.09	7/16	3/8	4	5/8	3 1/4	3
1	4 7/8	11/16	2	1.05	1 1/16	2 1/8	1.36	1/2	3/8	4	5/8	3 1/2	3
1 1/4	5 1/4	3/4	2 1/2	1.38	1 1/16	2 1/2	1.70	9/16	5/16	4	5/8	3 7/8	4
1 1/2	6 1/8	13/16	2 7/8	1.61	1 3/16	2 3/4	1.95	5/8	3/8	4	3/4	4 1/2	6
2	6 1/2	7/8	3 5/8	2.07	1 5/16	3 5/16	2.44	11/16	7/16	8	5/8	5	7
2 1/2	7 1/2	1	4 1/8	2.47	1 1/2	3 15/16	2.94	3/4	1/2	8	3/4	5 7/8	10
3	8 1/4	1 1/8	5	3.07	1 11/16	4 5/8	3.57	13/16	9/16	8	3/4	6 5/8	13
3 1/2	9	1 3/16	5 1/2	3.55	1 3/4	5 1/4	4.07	7/8	9/16	8	7/8	7 1/4	17
4	10	1 1/4	6 3/16	4.03	1 7/8	5 3/4	4.57	15/16	5/8	8	7/8	7 7/8	22

600 lb													
1/4	3 3/4	9/16	1 3/8		7/8	1 1/2	0.58	3/8	5/16	4	1/2	2 5/8	1
3/8	3 3/4	9/16	1 3/8		7/8	1 1/2	0.72	3/8	5/16	4	1/2	2 5/8	1
1/2	3 3/4	9/16	1 3/8		7/8	1 1/2	0.88	3/8	5/16	4	1/2	2 5/8	3
3/4	4 5/8	5/8	1 11/16		1	1 7/8	1.09	7/16	3/8	4	5/8	3 1/4	3
1	4 7/8	11/16	2	As specified by purchaser	1 1/16	2 1/8	1.36	1/2	3/8	4	5/8	3 1/2	4
1 1/4	5 1/4	13/16	2 1/2		1 1/8	2 1/2	1.70	9/16	5/16	4	5/8	3 7/8	6
1 1/2	6 1/8	7/8	2 7/8		1 1/4	2 3/4	1.95	5/8	3/8	4	3/4	4 1/2	7
2	6 1/2	1	3 5/8		1 7/16	3 5/16	2.44	11/16	7/16	8	5/8	5	9
2 1/2	7 1/2	1 1/8	4 1/8		1 5/8	3 15/16	2.94	3/4	1/2	8	3/4	5 7/8	13
3	8 1/4	1 1/4	5		1 13/16	4 5/8	3.57	13/16	9/16	8	3/4	6 5/8	16
3 1/2	9	1 3/8	5 1/2		1 15/16	5 1/4	4.07	7/8	9/16	8	7/8	7 1/4	21

[1] Flanges bored to dimensions shown, unless otherwise specified. Lengths of stud bolts are identical to other types of flanges (except lap joint).

[2] Bolt holes are 1/8 in. larger than bolt diameter.

[3] Dimensions in sizes 1/2 through 3 in. are included in ASA Standard B16.5. Typical dimensions for other sizes are listed; supplier should be consulted for specific information.

Forged-steel Flanges. Forged-steel flanges are made in the welding-neck, slip-on, lap-joint, threaded, blind, socket, and reducing types.

Flanges are usually furnished to the following ASTM Specification, unless otherwise ordered:

150,	300 lb;	A181 grade I
400,	600 lb;	A105 grade I
900 to 2,500 lb;		A105 grade II

Dimensions of standard forged-steel flanges are covered by ASA Standard B16.5-1961. Dimensions of *welding-neck flanges* are given in Table 51. Dimensions of *slip-on flanges* are given in Table 52. Dimensions of *threaded flanges* are given in Table 53. Dimensions of *socket welding flanges* are given in Table 54.

Flange Types

Flanges differ in method of attachment to the pipe, that is, whether they are screwed, welded, or lapped. Contact surface facings may be plain, serrated, grooved for ring joints, seal-welded, or ground and lapped for metal-to-metal contact. The more important methods of attaching flanges to piping and some common types of joints and facings are shown in Fig. 68 and Table 55.

In Section VIII, Unfired Pressure Vessels, of the ASME Boiler and Pressure Vessel Code, three types of circular flanges are defined, and these are designated as loose type, integral-type, and optional-type flanges. Under the code, the welds and other details of construction shall satisfy the dimensional requirements stated therein.

Loose-type Flanges. This (slip-on) type covers those designs in which the flange has no direct connection to the nozzle neck or the vessel or pipe wall and designs where the method of attachment is not considered to give mechanical strength equivalent of integral attachment.

Integral-type Flanges. This type covers designs in which the flange is cast or forged integrally with the nozzle neck or the vessel or pipe wall, butt-welded thereto, or attached by other forms of arc or gas welding of such a nature that the flange and nozzle neck or vessel or pipe wall is considered to be the equivalent of an integral structure. In welded construction, the nozzle neck or the vessel or pipe wall is considered to act as a hub.

Optional-type Flanges. This type covers designs where the attachment of the flange to the nozzle neck or the vessel or pipe wall is such that the assembly is considered to act as a unit, which shall be calculated as an integral flange, except that for simplicity the designer may calculate the construction as a loose-type flange, provided that stipulated load values are not exceeded.

It is important in flange design to select materials and to proportion dimensions of bolts, flanges, and gaskets to insure that the necessary compression will be maintained on the joint faces over the expected life of the equipment.

Several distinct phases of the problem are involved: (1) type of flange facing, (2) finish of contact surfaces, (3) gasket type and proportions, (4) bolt load required to secure and maintain a tight joint, and (5) proportions of flange needed to support the bolt load.

Types of Flange Facing. There are numerous types of contact facings for flanges, the most simple of which is the plain face provided with a "smooth tool finish." Class 125 cast-iron flanged fittings are provided with this type of facing. For steel flanges and fittings, the typical facings shown in Table 55 are taken from the American Standard for Steel Pipe Flanges and Flanged Fittings, ASA B16.5. The raised face, the lapped, and the large male-and-female facings have the same

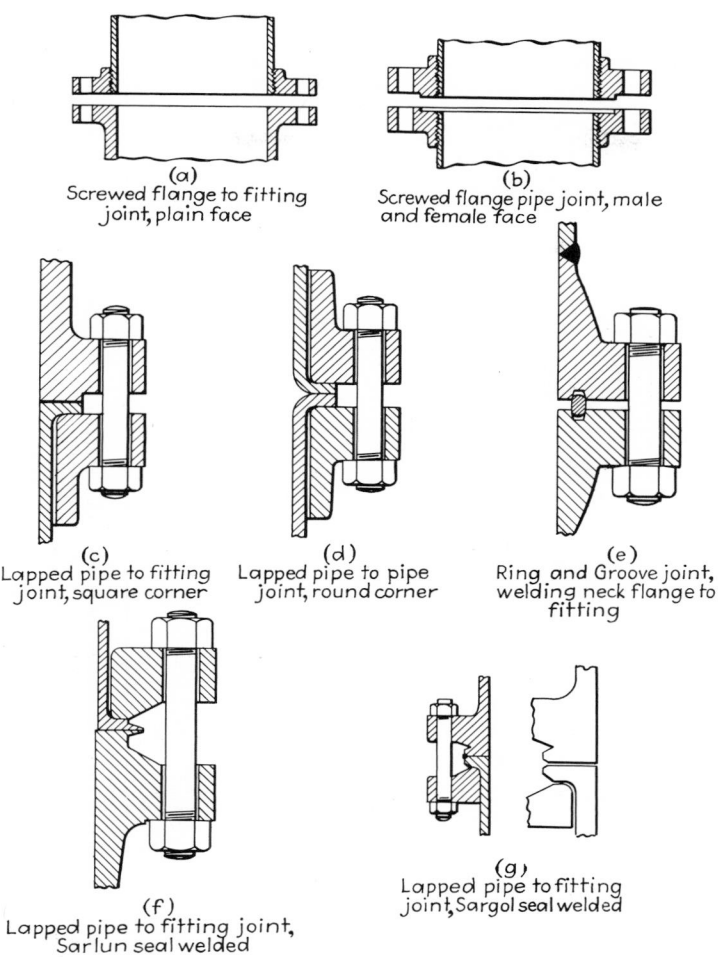

(a)
Screwed flange to fitting
joint, plain face

(b)
Screwed flange pipe joint, male
and female face

(c)
Lapped pipe to fitting
joint, square corner

(d)
Lapped pipe to pipe
joint, round corner

(e)
Ring and Groove joint,
welding neck flange to
fitting

(f)
Lapped pipe to fitting joint,
Sarlun seal welded

(g)
Lapped pipe to fitting
joint, Sargol seal welded

Fig. 68. Commonly used flanged joints.

dimensions, which provide a relatively large contact area. Where metal gaskets are used with these facings, the gasket area should be reduced to increase the gasket compression.

Commonly used flange-facing types are illustrated in Fig. 69. They range in size and contact area in the following order: large tongue-and-groove, small tongue-and-groove, small male-and-female, and ring joint. Because of the small gasket contact area, a tight joint may be secured with the ring-type facing using low bolting loads, thereby resulting in lowered flange stresses (ASA B16.5). The Sargol and Sarlun facings, which have lips for seal welding, are used frequently for severe service conditions. Seal welding is not always performed since, if properly made, a tight joint often can be maintained without the welded seal, thus facilitating disassembly. Typical facing dimensions for Sarlun and Sargol joints are shown in Fig. 70. Special types of facing of individual design intended for

Table 55. Dimensions of Facings for Flanges and Flanged Fittings (Other than Ring Joints and Not Lapped) as Specified in ASA B16.5-1961

Nominal pipe size	Outside diameter — Raised face; large male and large tongue R	Outside diameter — Small male[1] S	Outside diameter — Small tongue T	Inside diameter of large and small tongue U	Inside diameter of small male	Outside diameter — Large female and large groove W	Outside diameter — Small female X	Outside diameter — Small groove Y	Inside diameter of large and small groove Z	Height — Raised face	Height — Large and small male and tongue	Height — Depth of groove or female	Min diameter of raised portion — Small tongue and groove K^3	Min diameter of raised portion — Large tongue and groove L^3	Nominal pipe size
1/2	1 3/8	23/32	13/16	1	See Note 1	1 7/16	25/32	1 7/16	15/16	See Notes 4 and 5	See Notes 5 and 6	See Notes 2 and 5	1 3/4	1 13/16	1/2
3/4	1 11/16	15/16	1 1/16	1 5/16		1 3/4	1	1 3/4	1 11/16				2 1/16	2 1/16	3/4
1	2	1 3/16	1 5/16	1 1/2		2 1/16	1 1/4	1 15/16	2				2 5/16	2 7/16	1
1 1/4	2 1/2	1 1/2	2 1/4	1 7/8		2 9/16	1 9/16	2 5/16	2 1/4				2 5/8	2 15/16	1 1/4
1 1/2	2 7/8	1 3/4	2 1/2	2 1/8		2 15/16	1 13/16	2 9/16	2 9/16				2 7/8	3 3/16	1 1/2
2	3 5/8	2 1/4	3 1/4	2 7/8		3 11/16	2 5/16	3 5/16	2 13/16				3 5/8	4 1/16	2
2 1/2	4 1/8	2 11/16	3 3/4	3 3/8		4 3/16	2 3/4	3 13/16	3 5/16				4 1/8	4 9/16	2 1/2
3	5	3 5/16	4 5/8	4 1/8		5 1/16	3 3/8	4 11/16	4 3/16				5	5 7/16	3
3 1/2	5 1/2	3 13/16	5 1/8	4 3/4		5 9/16	3 7/8	5 3/16	4 11/16				5 1/2	5 15/16	3 1/2
4	6 3/16	4 5/16	5 11/16	5 3/16		6 1/4	4 3/8	5 3/4	5 3/16				6 7/16	6 5/8	4
5	7 5/16	5 3/8	6 13/16	6 5/16		7 3/8	5 7/16	6 7/8	6 1/4				7 5/16	7 3/4	5
6	8 1/2	6 3/8	8	7 1/2		8 9/16	6 7/16	8 1/16	7 7/16				8 1/2	8 15/16	6
8	10 5/8	8 3/8	10	9 3/8		10 11/16	8 7/16	10 1/16	9 5/16				10 5/8	11 1/16	8
10	12 3/4	10 1/2	12	11 1/4		12 13/16	10 9/16	12 1/16	11 3/16				12 3/4	13 3/16	10
12	15	12 1/2	14 1/4	13 1/2		15 1/16	12 9/16	14 5/16	13 7/16				15	15 7/16	12
14	16 1/4	13 3/4	15 1/2	14 3/4		16 5/16	13 13/16	15 9/16	14 11/16				16 1/4	16 11/16	14
16	18 1/2	15 3/4	17 5/8	16 3/4		18 9/16	15 13/16	17 1/16	16 11/16				18 1/2	18 15/16	16
18	21	17 3/4	20 1/8	19 1/4		21 1/16	17 13/16	20 3/16	19 3/16				21	21 7/16	18
20	23	19 3/4	22	21		23 1/16	19 13/16	22 1/16	20 15/16				23	23 7/16	20
24	27 1/4	23 3/4	26 1/4	25 1/4		27 5/16	23 13/16	26 5/16	25 3/16				27 1/4	27 11/16	24

All dimensions are given in inches.

1 For small male-and-female joints care should be taken in the use of these dimensions to insure that the inside diameter of fitting or pipe is small enough to permit sufficient bearing surface to prevent the crushing of the gasket. This applies particularly on lines where the joint is made on the end of the pipe. Inside diameter of fitting should match inside diameter of pipe as specified by purchaser. Screwed companion flanges for small male-and-female joints are furnished with plain face and are threaded with American Standard Lock-nut Thread (NPSL).

2 Depth of groove or female is 3/16 in.

3 Raised portion or full face may be furnished unless otherwise specified on order.

4 150- and 300-lb valves, fittings, and companion flanges are regularly furnished with 1/16-in. raised face which is included in the minimum flange thickness "C." 400-, 600-, 900-, 1,500- and 2,500-lb valves, fitting, and companion flanges are regularly furnished with 1/4-in. raised face which is additional to the minimum flange thickness "C."

5 For dimension of lapped joints consult ASA B16.5.

6 Height of large and small male and tongue is 1/4 in.

Table 55. (*Continued*)

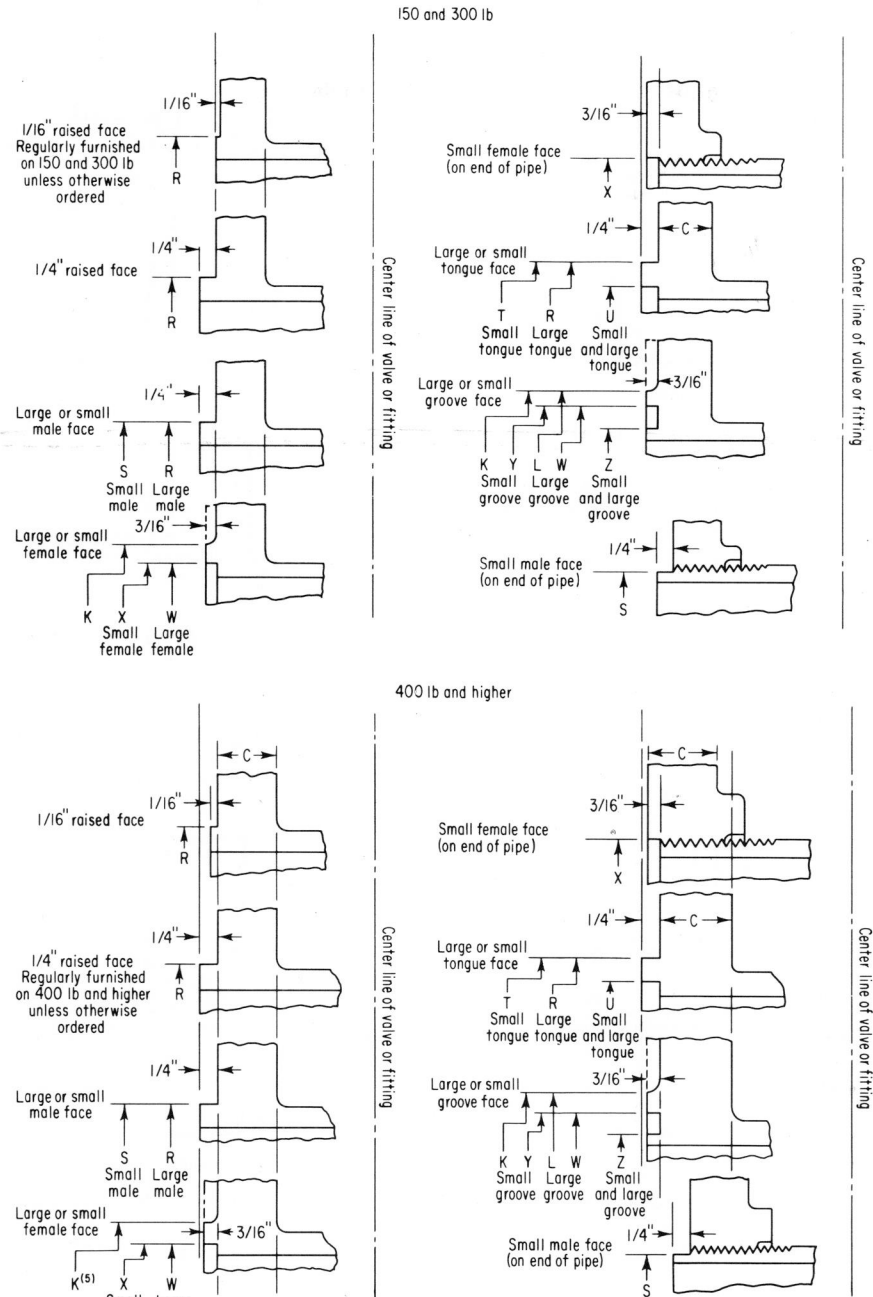

150 and 300 lb

400 lb and higher

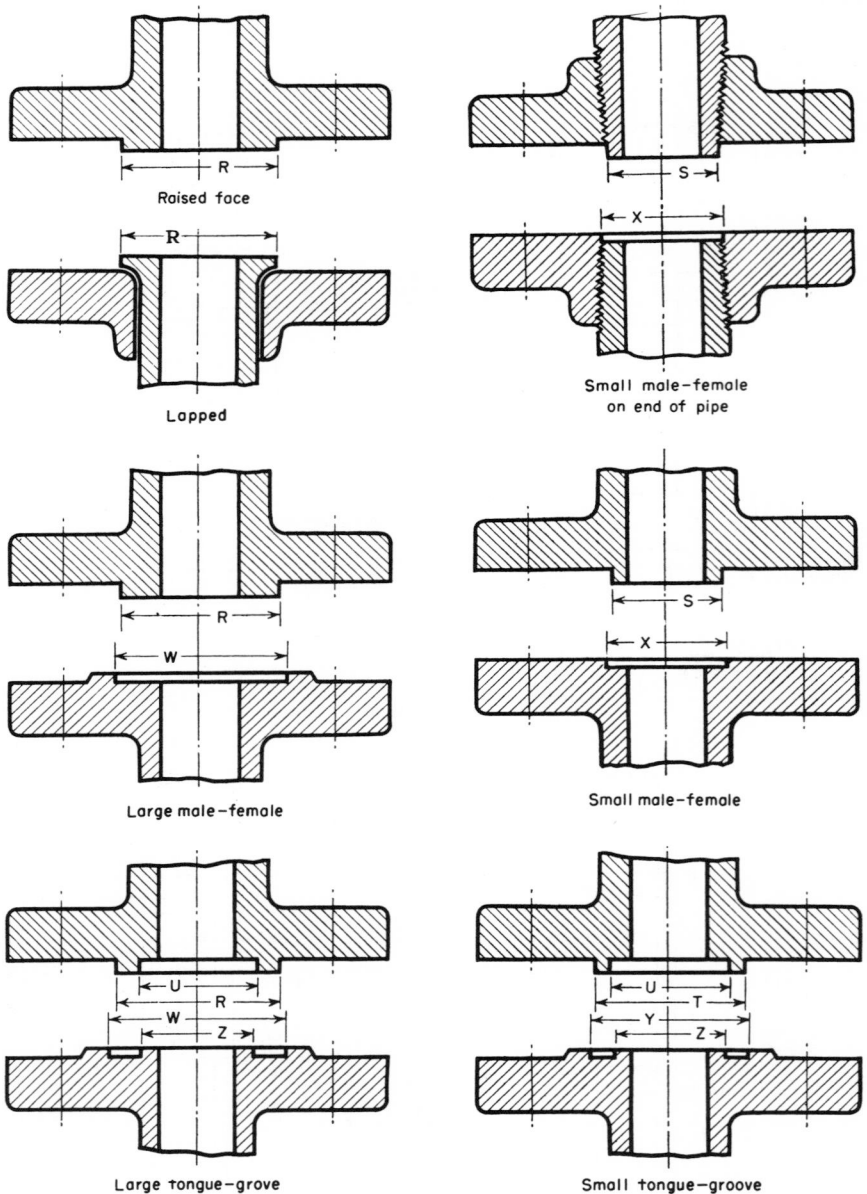

FIG. 69. Typical flange facings (for dimensions, see Table 55) (Section I, ASME Boiler and Pressure Vessel Code).

a specific service are numerous. Economic considerations generally make it desirable to use a standard facing wherever possible.

Selection of the type of facing depends to a considerable extent on the nature of the service. However, it is not possible to determine exactly which facing should be used. Prior experience usually is relied on as a guide. Plain-face joints with red-rubber gaskets have been found satisfactory for temperatures up to 220 F, whereas serrated raised-face joints with asbestos-composition gaskets are commonly used for temperatures up to 750 F. For high temperatures and pressures, faces giving a high contact pressure for a given bolt load are customary, such as the

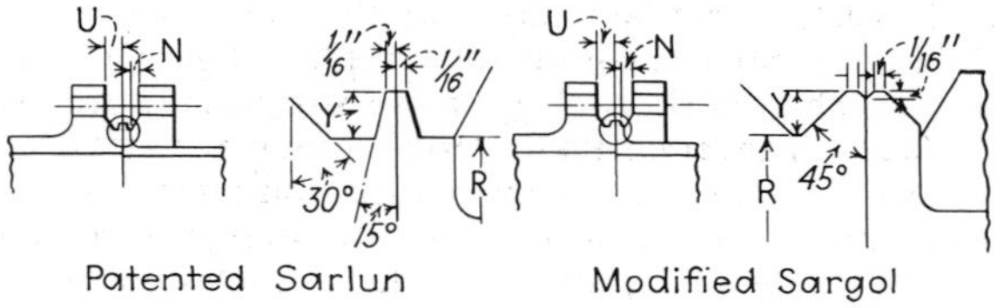

Nominal pipe size	Basic raised-face, outside diameter *R*	Height of face		Height of front hub		Height of welding projections	
		Sargol[1] *U*	Sarlun[2] *U*	Sargol[1] *N*	Sarlun[2] *N*	Sargol[1] *Y*	Sarlun[2] *Y*
2 1/2	4 1/8	1/2	11/16	1/4	5/16	1/8	5/16
3	5	5/8	11/16	5/16	5/16	3/16	5/16
4	6 3/16	3/4	3/4	7/16	13/32	5/16	5/16
5	7 5/16	7/8	7/8	1/2	1/2	3/8	3/8
6	8 1/2	1	1	9/16	9/16	3/8	3/8
8	10 5/8	1 1/8	1 1/8	5/8	5/8	3/8	3/8
10	12 3/4	1 1/4	1 1/4	5/8	3/4	3/8	3/8
12	15	1 5/16	1 5/16	5/8	13/16	3/8	3/8
14	16 1/2	1 3/8	1 3/8	5/8	7/8	3/8	3/8
16	18 1/2	1 3/8	1 3/8	5/8	7/8	3/8	3/8

All dimensions in inches.

[1] Dimensions of modified Sargol joint.

[2] Dimensions of Sarlun facings recommended by Sargent and Lundy, Inc.

FIG. 70. Typical facing dimensions for Sargol[1] and Sarlun[2] joints, 150- to 2,500-lb flanges.

tongue and groove and ring joints. An equally successful joint for most types of service can be made by using a profile-serrated metal gasket contacting the flange facing, which may be the plain male-to-male raised-face type.

Contact Surface Finish. The surface finish is an important factor in determining the extent to which a gasket must flow to secure an impervious seal. Bolting that results in adequate gasket flow to form a satisfactory seal with a smooth contact surface may be inadequate to secure a tight joint with a rough surface. The finish may vary from that produced by rough casting surfaces to that produced by grinding and lapping. Less gasket flow will be necessary for the latter, of course, than for the former. The finish most frequently provided on cast-iron and steel pipe flanges is the smooth tool finish. A serrated finish frequently is provided for steel flanges, particularly when using an asbestos-composition gasket with a wide contact area such as is furnished on raised, lapped, or large tongue-and-groove facings. The serrated finish consists of spiral or concentric grooves, usually about 1/64 in. deep with 32 serrations per inch.

Where metal gaskets are used, a smooth surface produced by grinding or lapping usually is provided.   The Sargol and Sarlun facings mate metal to metal without a gasket, in which case a mirrorlike finish is necessary.   This is usually produced by grinding and lapping.   It is evident that the surface finish varies with the type of contact face and gasket used and, therefore, should be specified accordingly.

Gaskets

Since it is expensive to grind and lap joint faces to obtain fluid-tight joints, a gasket of some softer material usually is inserted between contact faces.   Tightening the bolts causes the gasket material to flow into the minor machining imperfections, resulting in a fluid-tight seal.   A considerable variety of gasket types is in common use.   Soft gaskets, such as cork, rubber, or asbestos usually are plain with a relatively smooth surface.   The semimetallic design combines metal and a soft material, the metal to withstand the pressure, temperature, and attack of the confined fluid, and the soft material to impart resilience.   Various designs involving corrugations, strip-on-edge, metal jackets, etc., are available.   In addition to the plain, solid, and flat-surface metal gaskets, various modified designs and cross-sectional shapes of the profile, corrugated, serrated, and other types are used.   The object in general has been to retain the advantage of the metal gasket but to reduce the contact area to secure a seal without excessive bolting load.   Effective gasket widths are given in Section VIII, Unfired Pressure Vessels, of the ASME Boiler and Pressure Vessel Code.

Gasket Materials.   Gaskets are made of materials which are not chemically affected by the fluid in the pipe and which are resistant to deterioration by temperature.   Gasket materials may either be metallic or nonmetallic.   Metallic ring-joint gasket materials are covered by ASA Standard B16.20, Ring-joint Gaskets and Grooves for Steel Pipe Flanges.   Nonmetallic gaskets are covered in ASA Standard B16.21, Nonmetallic Gaskets for Pipe Flanges.   Typical selections of gasket materials for different services are shown in Table 56.

Gasket Compression.   In the usual type of high-pressure flange joint, a narrow gasket face or contact surface is provided to obtain higher unit compression on the gasket than is obtainable on full-face gaskets used with low-pressure joints. The compression on this surface and on the gasket if the gasket is used, before internal pressure is applied, depends on the bolt loading used.   In the case of standard raised-face joints of the ASA steel-flange standards, these gasket compressions range from 28 to 43 times the rated working pressure in the 150- to 400-lb standards, and from 11 to 28 times in the 600- to 2,500-lb standards for an assumed bolt stress of 60,000 psi.   For the lower-pressure standards, using composition gaskets, a bolt stress of 30,000 psi usually is adequate.   The effect of applying the internal pressure is to decrease the compression on the contact surfaces since part of the bolt tension is used to support the pressure load.

The initial compression required to force the gasket material into intimate contact with the joint faces depends upon the gasket material and the character of the joint facing.   For soft-rubber gaskets, a unit compression stress of 4,000 to 6,000 psi usually is adequate.   Laminated asbestos gaskets in serrated faced joints perform satisfactorily if compressed initially at 12,000 to 18,000 psi.   Metal gaskets such as copper, Monel, and soft iron should be given initial compressions considerably in excess of their yield strengths.   Unit pressures of 30,000 to 60,000 psi have been used successfully with metal gaskets.   Various forms of corrugated and serrated metal gaskets are available which enable high unit compression to be obtained without excessive bolt loads.   These are designed to provide a contact area that will flow under initial compression of the bolts so as to make an initially pressuretight

Table 56.   Selections of Gasket Materials for Different Services

Fluid	Application	Gasket material
Steam (high pressure)......	Temp up to 1000 F	Spiral-wound comp. asbestos
	Temp up to 1000 F	Steel, corrugated or plain
	Temp up to 1000 F	Monel, corrugated or plain
	Temp up to 1000 F	Hydrogen-annealed furniture iron
	Temp up to 1000 F	Stainless steel 12 to 14% chromium, corrugated
	Temp up to 1000 F	Ingot iron, special ring-type joint
	Temp up to 750 F	Comp. asbestos, spiral-wound
	Temp up to 600 F	Woven asbestos, metal asbestos
	Temp up to 600 F	Copper, corrugated or plain
Steam (low pressure)......	Temp up to 220 F	Red rubber, wire inserted
Water.................	Hot, medium, and high pressures	Black rubber, red rubber, wire inserted
	Hot, low pressures	Brown rubber, cloth inserted
	Hot	Comp. asbestos
Water.................	Cold	Red rubber, wire inserted
	Cold	Black rubber
	Cold	Soft rubber
	Cold	Asbestos
	Cold	Brown rubber, cloth inserted
Oils (hot)...............	Temp up to 750 F	Comp. asbestos
	Temp up to 1000 F	Ingot iron, special ring-type joint
Oils (cold).............	Temp up to 212 F	Cork-fiber
	Temp up to 300 F	Neoprene comp. asbestos
Air....................	Temp up to 750 F	Comp. asbestos
	Temp up to 220 F	Red rubber
	Temp up to 1000 F	Spiral-wound comp. asbestos
Gas...................	Temp up to 1000 F	Asbestos, metallic
	Temp up to 750 F	Comp. asbestos
	Temp up to 600 F	Woven asbestos
	Temp up to 220 F	Red rubber
Acids..................	(Varies, see section on corrosion)	Sheet lead or alloy steel
	Hot or cold mineral acids	Comp. blue asbestos
		Woven blue asbestos
Ammonia...............	Temp up to 1000 F	Asbestos, metallic
	Temp up to 700 F	Comp. asbestos
	Weak solutions	Red rubber
	Hot	Thin asbestos
	Cold	Sheet lead

joint, but at the same time the compressive stresses in the body of the gasket are sufficiently low as to be comparable to the long-time load-carrying ability of the bolting and flange material at high temperatures.

The residual compression on the gasket necessary to prevent leakage depends on how effective the initial compression has been in forming intimate contact with the flange joint faces.   Tests show that a residual compression on the gasket of only one to two times the internal pressure, with the pressure acting, may be sufficient to prevent leakage where the joint is not subjected to bending or to large and rapid temperature changes.   Since joints in piping customarily must withstand both these

disturbing influences, minimum residual gasket compressions of four to six times the working pressure should be provided for in the design of pipe joint.

Relation of Gaskets to Bolting. There is a tendency as indicated in the ASME Rules for Bolted Flanged Connections to assign lower residual contact-pressure ratios ranging from about 1 for soft-rubber gaskets to 6 or 7 for solid-metal gaskets. Whereas these are said to have proved satisfactory for heat-exchanger and pressure-vessel flanges, the more severe service encountered by pipe flanges due to bending moments and large temperature changes is considered by many to warrant designing on the basis of the larger residual gasket-compression ratios recommended in the previous paragraph. The lack of understanding of the mechanics of gasket action, the variety of gasket materials, shapes, widths, and thicknesses, the variety of facings used, the variation in flange stiffness, and the uncertainties in bolt pull-up are among the factors that render difficult a precise solution to the problem of gasket design.

The following rules have been established in Section VIII, Unfired Pressure Vessels, of the ASME Boiler and Pressure Vessel Code:

(a) Required Bolt Loads.
(1) The required bolt load for the operating conditions, W_{m1}, shall be sufficient to resist the hydrostatic end force, H, exerted by the maximum allowable working pressure on the area bounded by the diameter of gasket reaction and, in addition, to maintain on the gasket or joint-contact-surface a compression load, H_p, which experience has shown to be sufficient to assure a tight joint. (This compression load is expressed as a multiple m of the internal pressure. Its value is a function of the gasket material and construction.)

The required bolt load for the operating conditions, W_{m1}, is determined in accordance with Formula (1).

$$W_{m1} = H + H_p = 0.785G^2P + (2b \times 3.14GmP) \qquad (1)$$

(2) Before a tight joint can be obtained, it is necessary to seat the gasket or joint-contact surface properly by applying a minimum initial load (under atmospheric temperature conditions without the presence of internal pressure), which is a function of the gasket material and the effective gasket area to be seated. The minimum initial bolt load required for this purpose, W_{m2}, shall be determined in accordance with Formula (2).

$$W_{m2} = 3.14bGy \qquad (2)$$

The need for providing sufficient bolt load to seat the gasket or joint-contact surfaces in accordance with Formula (2) will prevail on many low-pressure designs and with facings and materials that require a high seating load, and where the bolt load computed by Formula (1) for the operating conditions is insufficient to seat the joint.

Accordingly, it is necessary to furnish bolting and to pre-tighten the bolts to provide a bolt load sufficient to satisfy both of these requirements, each one being individually investigated. When Formula (2) governs, flange proportions will be a function of the bolting instead of internal pressure.

(b) Total Required and Actual Bolt Areas, A_m and A_b. The total cross-sectional area of bolts, A_m, required for both the operating conditions and gasket seating is the greater of the values for A_{m1} and A_{m2} where $A_{m1} = W_{m1}/S_b$ and $A_{m2} = W_{m2}/S_a$. A selection of bolts to be used shall be made such that the actual total cross-sectional area of bolts, A_b, will not be less than A_m.

(c) Flange Design Bolt Load, W. The bolt loads used in the design of the flange shall be the values obtained from Formulas (3) and (4). For Operating Conditions

$$W = W_{m1} \qquad (3)$$

For Gasket Seating

$$W = \frac{(A_m + A_b)S_a}{2} \tag{4}$$

In addition to the minimum requirements for safety, Formula (4) provides a margin against abuse of the flange from overbolting. Since margin against such abuse is needed primarily for the initial bolting-up operating which is done at atmospheric temperature and before application of internal pressure, the flange design is required to satisfy this loading only under such conditions.** Bolting design should also provide for overloading during transient temperature conditions when the original stress may double.

Flange Design

Considerable attention has been given to the design of flanges for high-pressure work, in the effort to produce flanges that are amply safe and economically proportioned. The stresses which are set up by the tightening of the bolts and by other causes are rather complicated and require careful analysis.

Support from Hubs. There are two general types of flanges—the type that is attached to fittings and the loose type such as is used with a Van Stone joint or threaded to the pipe. The latter could be merely a flat ring, but the addition of a cylindrical hub increases its strength materially; all the higher-pressure flanges are so designed. The flange attached to a fitting is considerably reinforced by the wall of the fitting and is the more favorable case from a standpoint of strength. The tapered hubs of welding-neck flanges approach conditions existing with fitting flanges, although less reinforcement is offered by the attached pipe than is found with the heavier wall customary with cast fittings.

For loose flanges, such as lapped companion flanges, it is usually necessary to provide hubs on the flanges which serve to strengthen them in the same way that stiffening angles support flat surfaces on rectangular tanks. In the case of flanged fittings the body of the fitting furnishes the support which is derived from hubs on lapped flanges. It is possible, of course, in the case of loose flanges to increase the thickness of a flat flange sufficiently to obtain the required strength, but as in the case of many other flat surfaces, this is neither the most economical nor the most desirable method to use.

Standard Dimensions. Satisfactory proportions for cast-iron flanges have been evolved through long experience backed by computations and tests for 25-, 125-, 250-, and 800-lb service at temperatures not exceeding about 450 F (see listing of applicable ASA Standards on page 153).

The bolting of a steel lapped flange against a 125-lb cast-iron flange or flanged fitting is not recommended because there is danger of cracking the cast-iron flange. Bolting a steel lapped flange against a 250-lb cast-iron flange or flanged fitting is not recommended but is permitted by the Code for Pressure Piping, provided that the lap is extended to the inner edge of the bolt holes and the proper grade of carbon-steel bolts is used.[1] For satisfactory results in those cases where a cast-iron flange is bolted to a steel flange, it is desirable to use a flat-face steel flange and a full-gasket.

The proportions of steel flanges, bolting, and gaskets for service at pressures of 100, 300, 400, 600, 900, 1,500, and 2,500 psi at 750 F, when using carbon-steel

** Where additional safety against abuse is desired, or where it is necessary that the flange be suitable to withstand the full available bolt load $A_b \times S_a$, the flange may be designed on the basis of this latter quantity.

[1] The material should be in accordance with ASTM Specification A307 Grade B, Low-carbon Steel Externally and Internally Threaded Standard Fasteners.

flanges and raised-face joints, have been standardized in ASA Standard B16.5. Modifications in these ratings to take into account the use of alloy flange materials, the effect of temperatures above and below 750 F, and the reduction in bolt loading possible with the ring-type joint are given in the steel-flange standards.

Flange Materials and Design.   Flanges may be made of rolled or forged steel, cast steel, and in some cases, plate steel.   Materials in compositions corresponding to those of pipe materials are summarized in Chap. 8.

Numerous methods and formulas have been developed for the design of flanges. Some of these, although presenting a more rigorous and theoretically correct solution, are difficult to apply because of their complexity.

Most generally accepted is the method outlined in Section VIII of the ASME Boiler and Pressure Vessel Code, and that publication should be referred to by those interested in flange-design details.

Bolting

Materials.   For service temperatures up to approximately 700 F, carbon and low-alloy steel bolts are used.   Although there is considerable variation among bolting materials produced, common types include carbon-steel grades 1018, 1037, 1040, and 1045.   Alloy-steel bolts are extensively made of the 8640 grade, followed by the 4140 and 4340 grades.   Other grades used in bolting are 4135, 4142, 4145, 4150, 50B44, 8635, 8642, and 8735.   Relatively little use is made of the 1300 series, steel grades 1335 and 1340.   The 4037 grade used in the past has become practically obsolete.

Low-carbon-steel bolts are covered by ASTM Specification A307.[1]   Medium-carbon-steel bolts are covered by ASTM Specification A449.[2]   Rather than listing specific chemical compositions for low-alloy-steel bolting material, except for limits on phosphorus (0.05 per cent max) and sulfur (0.05 per cent max), ASTM Specification A354[3] bases its acceptability on mechanical properties.   The determination of the proof load is preferred over yield strength and tensile strength, though tensile and yield strength determinations may nevertheless be made as an alternate method.   Hexagonal head bolts are tested by the wedge-tension method. Machined specimens are also used to determine acceptability where tensile and yield strength are of importance.

Bolting materials for low-temperature service usually involve specific chemical compositions and must meet impact test requirements.[4]   Commonly used low-temperature bolting materials are 4140, 4142, 4145, 2317, 4340, $3\frac{1}{2}$ per cent Ni steel, and stainless steel Types 304, 347, 321, and 303.

In the SAE Handbook,[5] eight bolt grades (Table 57) are assigned to carbon- and low-alloy-steel bolting materials on the basis of proof load and tensile strength (Table 58).

Grades 0 and 1: Bolts usually represent hot-rolled carbon-steel materials.

Grade 2: In diameters to $\frac{3}{4}$ in. usually are cold-headed low-carbon steels.   Sizes over $\frac{3}{4}$ in. diameter generally involve hot-rolled low-carbon steel; however, they may also be made of cold-finished material.

[1] ASTM Specification A307, Specification for Low-carbon Steel Externally and Internally Threaded Standard Fasteners.

[2] ASTM Specification A449, Specification for Quenched and Tempered Steel Bolts and Studs.

[3] ASTM Specification A354, Specification for Quenched and Tempered Alloy-steel Bolts and Studs With Suitable Nuts.

[4] ASTM Specification A320-63, Specifications for Alloy-steel Bolting Materials for Low-temperature Service.

[5] "Society of Automotive Engineers," SAE Handbook, 1966 ed.

Table 57. Proof-load, Tensile-strength, Hardness, and Other Requirements of Bolts (SAE Handbook, 1966)

SAE grade	Bolt size diameter, in.	Proof load, psi	Tensile strength, psi min	Hardness Bhn	Hardness Rockwell	Steel and heat treatment	Identification (head marking)
0	All sizes	....	....	....	....	No requirements	None
1	All sizes	....	55,000	207 max	B95 max	Commercial steel	None
2	1/4–1/2	55,000	69,000	241 max	B100 max	Cold-headed product made from low-carbon steel of 0.28 C max, 0.04 P max, and 0.05 S max*	None
	1/2–3/4	52,000	64,000	241 max	B100 max		
	3/4–1 1/2	28,000	55,000	207 max	B95 max		
3	1/4–1/2	85,000	110,000	207–269	B95–104	Produced by cold heading, up to and including 6 in. in length, from medium-carbon steel of 0.28 to 0.55 C, 0.04 P max, and 0.05 S max	—
	1/2–5/8	80,000	100,000	207–269	B95–104		
5	1/4–3/4	85,000	120,000	241–302	C23–32	Medium-carbon steel of 0.28 to 0.55 C, 0.04 P max, and 0.05 S max; quenched and tempered at 800 F min	/
	3/4–1	78,000	115,000	235–302	C22–32		
	1–1 1/2	74,000	105,000	223–285	C19–30		
5.1	Up to 3/8 incl	85,000	120,000	....	C23–40	Low- or medium-carbon steel; quenched and tempered with assembled lock washer	—
7	1/4–1 1/2	105,000	133,000	269–321	C28–34	Medium-carbon fine-grained alloy steel †, ‡, §	✕
8	1/4–1 1/2	120,000	150,000	302–352	C32–38	Medium-carbon fine-grained alloy steel †, ‡	✕

* Lengths of over 6 in. may be hot-headed from medium-carbon steel of 0.55% C max. Deviation from composition by agreement between producer and consumer.
† Carbon steel may be used by agreement between producer and consumer. Carbon range is for check analysis of product.
‡ Alloy-steel contains 0.28 to 0.55% C, 0.04% P max, and 0.05% S max, providing sufficient hardenability to have a minimum oil-quenched hardness of Rockwell C 47 at the center of the threaded section, one diameter from the end of the bolt; oil-quenched and tempered at 800 F min.
§ Roll-threaded after heat treatment.

Table 58. Proof-load and Tensile-strength Requirements of Bolts (SAE Handbook, 1966)

Grade	Nominal bolt size, in.	Coarse thread		Fine thread	
		Proof load, lb	Tensile strength min, lb	Proof load, lb	Tensile strength min, lb
1	1/4	....	1,750	....	2,000
	5/16	....	2,900	....	3,200
	3/8	....	4,250	....	4,850
	7/16	....	5,850	......	6,550
	1/2	....	7,800	....	8,800
	9/16	....	10,000	....	11,150
	5/8	....	12,450	....	14,100
	3/4	....	18,350	....	20,500
	7/8	....	25,400	....	28,000
	1	....	33,350	....	36,450
	1 1/8	....	41,950	....	47,100
	1 1/4	....	53,300	....	59,000
	1 3/8	....	63,550	....	72,350
	1 1/2	....	77,300	....	86,950
2	1/4	1,750	2,200	2,000	2,500
	5/16	2,900	3,600	3,200	4,000
	3/8	4,250	5,350	4,850	6,050
	7/16	5,850	7,350	6,550	8,200
	1/2	7,800	9,800	8,800	11,050
	9/16	9,450	11,650	10,550	13,000
	5/8	11,750	14,450	13,300	16,400
	3/4	17,350	21,400	19,400	23,850
	7/8	12,900	25,400	14,250	28,000
	1	16,950	33,350	18,550	36,450
	1 1/8	21,350	41,950	23,950	47,100
	1 1/4	27,100	53,300	29,950	59,000
	1 3/8	32,300	63,550	36,800	72,350
	1 1/2	39,300	77,300	44,000	86,950
3	1/4	2,700	3,500	3,100	4,000
	5/16	4,450	5,750	4,950	6,400
	3/8	6,600	8,550	7,450	9,650
	7/16	9,050	11,700	10,100	13,050
	1/2	12,050	15,600	13,550	17,550
	9/16	14,550	18,200	16,250	20,300
	5/8	18,100	22,600	20,500	25,600
5	1/4	2,700	3,800	3,100	4,350
	5/16	4,450	6,300	4,950	6,950
	3/8	6,600	9,300	7,450	10,550
	7/16	9,050	12,750	10,100	14,250
	1/2	12,050	17,050	13,600	19,200
	9/16	15,450	21,850	17,250	24,350
	5/8	19,200	27,100	21,750	30,700
	3/4	28,400	40,100	31,700	44,750
	7/8	36,050	53,150	39,700	58,550
	1	47,250	69,700	51,700	76,250

Table 58. *(Continued)*

Grade	Nominal bolt size, in.	Coarse thread		Fine thread	
		Proof load, lb	Tensile strength min, lb	Proof load, lb	Tensile strength min, lb
5	1⅛	56,450	80,100	63,350	89,900
	1¼	71,700	101,750	79,400	112,650
	1⅜	85,450	121,300	97,300	138,100
	1½	103,950	147,550	117,000	166,000
5.1*	6	770	1,090	860	1,220
	8	1,190	1,680	1,250	1,770
	10	1,490	2,100	1,700	2,400
	12	2,060	2,900	2,190	3,100
7	¼	3,400	4,250	3,800	4,850
	5⁄16	5,500	6,950	6,100	7,700
	⅜	8,150	10,300	9,200	11,700
	7⁄16	11,150	14,150	12,450	15,800
	½	14,900	18,850	16,800	21,250
	9⁄16	19,100	24,200	21,300	27,000
	⅝	23,750	30,050	26,900	34,050
	¾	35,050	44,400	39,150	49,600
	⅞	48,500	61,450	53,450	67,700
	1	63,650	80,600	69,600	88,200
	1⅛	80,100	101,500	89,900	113,850
	1¼	101,750	128,900	112,650	142,700
	1⅜	121,300	153,600	138,100	174,900
	1½	147,550	186,850	166,000	210,250
8	¼	3,800	4,750	4,350	5,450
	5⁄16	6,300	7,850	6,950	8,700
	⅜	9,300	11,650	10,550	13,150
	7⁄16	12,750	15,950	14,250	17,800
	½	17,050	21,300	19,200	24,000
	9⁄16	21,850	27,300	24,350	30,450
	⅝	27,100	33,900	30,700	38,400
	¾	40,100	50,100	44,750	55,950
	⅞	55,450	69,300	61,100	76,350
	1	72,700	90,900	79,550	99,450
	1⅛	91,550	114,450	102,700	128,400
	1¼	116,300	145,350	128,750	160,950
	1⅜	138,600	173,250	157,800	197,250
	1½	168,600	210,750	189,700	237,150

* The values for ¼ and up are the same as for grade 5.   The nominal-size designations reflect screw sizes.

Grade 3: Bolts are normally made from cold-drawn medium-carbon steels, such as 1037.   They are most commonly made by cold heading and stress relieving.

Grade 4: Materials involve studs only.   They are produced by cold finishing medium-carbon steel, specially processed for higher than normal strength.

Grade 5: Normally are made from medium-carbon steels, such as 1037 or 1040. They are subsequently quenched and tempered.

Grade 5.1: Are made from low-carbon steels (carburizing grades).

Grades 7 and 8: Require alloy fine-grain steels of sufficient hardenability to produce a hardness of R_c 47 after oil quenching with over 80 per cent martensite at the

Table 59. Steels Recommended for Bolts to be Used at Temperatures between 400 and 700 F *

Bolt diameter, in.	Proof strength,† psi		
	75,000	100,000	125,000
¼–¾	1038	4037	4037
¾–1¼	1038	4140	4140
1¼–2	4140	4140	4145

* All selections are based on a minimum tempering temperature of 850 F.

† At room temperature.

center. These steels also exhibit normally good notched-bar impact-strength properties.

At temperatures over 400 F, the effects of relaxation should be recognized in bolt selection. Relaxation represents the effect of creep in which some of the elastic strain in the bolt changes to plastic strain during service with a corresponding reduction in stress. When bolts relax sufficiently, their total force no longer maintains joint tightness. Leakage may then occur.

For temperatures between 400 and 700 F, the commercial bolt steels which adequately resist relaxation are listed in Table 59. When the medium-carbon steels used are of the 1038 to 1045 types, they should be quenched and tempered.

The selection of nuts is not as critical as the selection of bolts. Frequently, a lower-grade material is suitable for the nuts than is required for the bolts. On carbon-steel nut materials, limitations on phosphorus and sulfur content are much more liberal, thereby improving machinability. Heat treatment of nuts is also generally not necessary. The nut dimensions usually provide a shear area of the threads more than double the tensile stress area of the bolts.

Threading. Bolts and nuts normally are threaded in accordance with ASA Standard for Unified Screw Threads B1.1. In diameters 1 in. and smaller, Class 2A fits on the bolt or stud and Class 2B on the nut applies with the coarse thread series. In diameters 1⅛ in. and larger, the eight-pitch thread series applies with the same fit.

Grade-7 bolts (Table 57) are threaded by roll threading after heat treatment. Roll threading cold works the surface uniformly. The resulting compressive stresses provide substantially increased fatigue strength at the thread root, which is usually the weakest point. The thread root is the weakest point because it is the smallest cross-sectional area in the bolt. The stressed area A_s of a bolt is computed from

$$A_s = 0.7854 \left(D - \frac{0.9743}{N} \right)^2$$

where D is the nominal bolt diameter and N represents the threads per inch.

Bolts with fine threads will exhibit a slightly higher proof strength (of about 10 per cent) than bolts with coarse threads (as illustrated in Fig. 71), provided that the length of engagement with the mating

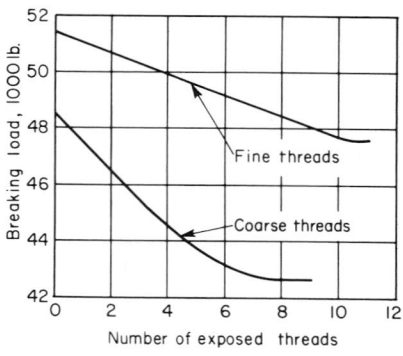

Fig. 71. Comparison of proof strength on fine and coarse threads, SAE Grade 5, ¾-in. bolts.

internal thread is sufficient to guarantee a tensile failure through the bolt rather than failure by thread stripping.

In practical bolt assemblies fine threads are considered weaker because of reduced thread height.   Fine threads have limited application for threaded assemblies. They should be used for adjustment rather than as a clamping force.

Dimensions.   The dimensions applicable to bolting materials are given in Table 60.   The information has been extracted from ASA Standard B16.5, American Standard Pipe Flanges and Flanged Fittings.

Securing and Tightening.   For the average low- and medium-pressure installations bolts are made in staggered sequence with open-end wrenches which will usually result in adequately tight joints.   For the high pressure and temperature joints, it becomes increasingly more important to make up each stud to a definite tension. Torque wrenches are sometimes used for this purpose.

In exceptional cases where a more positive method is desired, the studs may be tightened until a definite elongation has been attained.   For this condition, an initial cold tension of 30,000 to 35,000 psi in each stud is recommended.   Since the modulus of elasticity of stud material is about 30×10^6 psi, a tension of 30,000 psi would result in a unit elongation of $30,000/(30 \times 10^6)$ or 0.001 in./in. of effective length.   The effective length is the distance between nut faces plus one nut thickness. Special studs with ground ends are required to make micrometer measurements for this purpose.   After the joint has been in service, periodic checks of the actual cold lengths as compared with the tabulated lengths will detect any permanent elongation of the studs.   Permanent elongation will indicate overstressing, relaxation, and creep.   When these conditions become severe, new studs may be required to maintain the joint properly.

Special thread lubricants are available both for temperatures below 500 F and from 500 to 1000 F.   Such lubricants not only facilitate initial tightening but also permit easier disassembly after service.

Table 61 illustrates the turning effort required for tightening well-lubricated threads and bearing surfaces.   Tests with no lubricant on threads and bearing surfaces may increase torque requirements by 75 to 100 per cent to secure a given bolt stress.

WELDED AND BRAZED JOINTS

Welding of piping may be done in fabricating plants or at field erection locations. It encompasses weld-joint preparation, joint set-up, fixturing and tacking, pre-heating, welding, postheat treatment, dimensional checking, straightening, and cleaning.

To insure proper welding, welding procedures which spell out in detail all of the operations involved in the making of a sound weld are normally prepared for specific alloy groupings.

QUALIFICATION OF PROCEDURES AND WELDERS

To obtain a satisfactory welded-piping installation, it is considered necessary to establish and qualify a specific welding procedure and, subsequently, to qualify the necessary welders and welding operators to carry out that procedure.   The welding procedure should cover base metal specifications; filler metals; joint preparation; pipe position; welding process, techniques, and electrical details (current setting, electrode manipulation, welding sequence, etc.); preheat and inter-pass temperatures; and postheat treatments.   On very critical work, between-weld

Table 60. Dimensions of Bolting Materials for Flanged Fittings (American Standard Steel Pipe Flanges and Flanged Fittings, ASA B16.5-1961)

Nominal pipe size	150-lb flanges				300-lb flanges			
	Diameter of bolts[4]	Number of bolts	Length[1] of stud bolt		Diameter of bolts[4]	Number of bolts	Length[1] of stud bolt	
			Ring joint[2]	1/16-in.-raised-face[3] and flat face			Ring joint[2]	1/16-in.-raised-face[3] and flat face
½	½	4	...	2¼	½	4	3	2½
¾	½	4	...	2¼	⅝	4	3¼	2¾
1	½	4	3	2½	⅝	4	3½	3
1¼	½	4	3	2½	⅝	4	3½	3
1½	½	4	3¼	2¾	¾	4	4	3½
2	⅝	4	3½	3	⅝	8	4	3¼
2½	⅝	4	3¾	3¼	¾	8	4½	3¾
3	⅝	4	4	3½	¾	8	4¾	4
3½	⅝	8	4	3½	¾	8	5	4¼
4	⅝	8	4	3½	¾	8	5	4¼
5	¾	8	4¼	3¾	¾	8	5¼	4½
6	¾	8	4¼	3¾	¾	12	5½	4¾
8	¾	8	4½	4	⅞	12	6	5¼
10	⅞	12	5	4½	1	16	6¾	6
12	⅞	12	5	4½	1⅛	16	7¼	6½
14 OD	1	12	5½	5	1⅛	20	7½	6¾
16 OD	1	16	5¾	5¼	1¼	20	8	7¼
18 OD	1⅛	16	6¼	5¾	1¼	24	8¼	7½
20 OD	1⅛	20	6½	6	1¼	24	8¾	8
24 OD	1¼	20	7¼	6¾	1½	24	10	9

Nominal pipe size	400-lb flanges[1]: Diameter of bolts[4]	Number of bolts	Length[1] of stud bolt: Ring joint[2]	1/4-in.-raised-face	Flat face at flange edge[3]	Male-and-female; also tongue-and-groove	600-lb flanges[1]: Diameter of bolts[4]	Number of bolts	Length[1] of stud bolt: Ring joint[2]	1/4-in.-raised-face	Flat face at flange edge[3]	Male-and-female; also tongue-and-groove
1/2	1/2	4	3	3	2 1/2	2 3/4	1/2	4	3	3	2 1/2	2 3/4
3/4	5/8	4	3 1/4	3 1/4	2 3/4	3	5/8	4	3 1/4	3 1/4	2 3/4	3
1	5/8	4	3 1/2	3 1/2	3	3 1/4	5/8	4	3 1/2	3 1/2	3	3 1/4
1 1/4	5/8	4	3 3/4	3 3/4	3 1/4	3 1/2	5/8	4	3 3/4	3 3/4	3 1/4	3 1/2
1 1/2	3/4	4	4	4	3 1/2	3 3/4	3/4	4	4	4	3 1/2	3 3/4
2	5/8	8	4 1/4	4	3 1/2	3 3/4	5/8	8	4 1/4	4	3 1/2	3 3/4
2 1/2	3/4	8	4 3/4	4 1/2	4	4 1/4	3/4	8	4 3/4	4 1/2	4	4 1/4
3	3/4	8	5	4 3/4	4 1/4	4 1/2	3/4	8	5	4 3/4	4 1/4	4 1/2
3 1/2	7/8	8	5 1/2	5 1/4	4 3/4	5	7/8	8	5 1/2	5 1/4	4 3/4	5
4	7/8	8	5 1/2	5 1/4	4 3/4	5	7/8	8	5 3/4	5 1/2	5	5 1/4
5	7/8	8	5 3/4	5 1/2	5	5 1/4	1	8	6 1/2	6 1/4	5 3/4	6
6	7/8	12	6	5 3/4	5 1/4	5 1/2	1	12	6 3/4	6 1/2	6	6 1/4
8	1	12	6 3/4	6 1/2	6	6 1/4	1 1/8	12	7 3/4	7 1/2	7	7 1/4
10	1 1/8	16	7 1/2	7 1/4	6 3/4	7	1 1/4	16	8 1/2	8 1/4	7 3/4	8
12	1 1/4	16	8	7 3/4	7 1/4	7 1/2	1 1/4	20	8 3/4	8 1/2	8	8 1/4
14 OD	1 1/4	20	8 1/4	8	7 1/2	7 3/4	1 3/8	20	9 1/4	9	8 1/2	8 3/4
16 OD	1 3/8	20	8 3/4	8 1/2	8	8 1/4	1 1/2	20	10	9 3/4	9 1/4	9 1/2
18 OD	1 3/8	24	9	8 3/4	8 1/4	8 1/2	1 5/8	20	10 3/4	10 1/2	10	10 1/4
20 OD	1 1/2	24	9 3/4	9 1/2	9	9 1/4	1 5/8	24	11 1/2	11 1/4	10 3/4	11
24 OD	1 3/4	24	11	10 1/2	10	10 1/4	1 7/8	24	13 1/4	12 3/4	12 1/4	12 1/2

Table 60. (*Continued*)

Nominal pipe size	Diameter of bolts[4]	Number of bolts	Length[1] of stud bolt				Diameter of bolts[4]	Number of bolts	Length[1] of stud bolt			
			Ring joint[2]	1/4-in.-raised-face	Flat face at flange edge[3]	Male-and-female; also tongue-and-groove			Ring joint[2]	1/4-in.-raised-face	Flat face at flange edge[3]	Male-and-female; also tongue-and-groove
	900-lb flanges[1]						1,500-lb flanges[1]					
1/2	3/4	4	4	4	3 1/2	3 3/4	3/4	4	4	4	3 1/2	3 3/4
3/4	3/4	4	4 1/4	4 1/4	3 3/4	4	3/4	4	4 1/4	4 1/4	3 3/4	4
1	7/8	4	4 3/4	4 3/4	4 1/4	4 1/2	7/8	4	4 3/4	4 3/4	4 1/4	4 1/2
1 1/4	7/8	4	4 3/4	4 3/4	4 1/4	4 1/2	7/8	4	4 3/4	4 3/4	4 1/4	4 1/2
1 1/2	1	4	5 1/4	5 1/4	4 3/4	5	1	4	5 1/4	5 1/4	4 3/4	5
2	7/8	8	5 3/4	5 1/2	5	5 1/4	7/8	8	5 3/4	5 1/2	5	5 1/4
2 1/2	1	8	6 1/4	6	5 1/2	5 3/4	1	8	6 1/4	6	5 1/2	5 3/4
3	7/8	8	5 3/4	5 1/2	5	5 1/4	1 1/8	8	7	6 3/4	6 1/4	6 1/2
4	1 1/8	8	6 3/4	6 1/2	6	6 1/4	1 1/4	8	7 3/4	7 1/2	7	7 1/4
5	1 1/4	8	7 1/2	7 1/4	6 3/4	7	1 1/2	8	9 3/4	9 1/2	9	9 1/4
6	1 1/8	12	7 1/2	7 1/2	7	7 1/4	1 3/8	12	10 1/4	10	9 1/2	9 3/4
8	1 3/8	12	8 3/4	8 1/2	8	8 1/4	1 5/8	12	11 3/4	11 1/4	10 3/4	11
10	1 3/8	16	9 1/4	9	8 1/2	8 3/4	1 7/8	12	13 1/2	13 1/4	12 3/4	13
12	1 3/8	20	10	9 3/4	9 1/4	9 1/2	2	16	15 1/4	14 3/4	14 1/4	14 1/2
14 OD	1 1/2	20	11	10 1/2	10	10 1/4	2 1/4	16	16 3/4	16	15 1/2	15 3/4
16 OD	1 5/8	20	11 1/2	11	10 1/2	10 3/4	2 1/2	16	18 1/2	17 1/2	17	17 1/4
18 OD	1 7/8	20	13 1/4	12 3/4	12 1/4	12 1/2	2 3/4	16	20 1/4	19 1/4	18 3/4	19
20 OD	2	20	14	13 1/2	13	13 1/4	3	16	22 1/4	21	20 1/2	20 3/4
24 OD	2 1/2	20	17 3/4	17	16 1/2	16 3/4	3 1/2	16	25 1/2	24	23 1/2	23 3/4

Nominal pipe size	Diameter of bolts[4]	Number of bolts	Length[1] of stud bolt			
			Ring joint[3]	¼-in.-raised-face	Flat face at flange edge[2]	Male-and-female; also tongue-and-groove
			2,500-lb flange[1]			
½	¾	4	4¾	4¾	4¼	4½
¾	¾	4	4¾	4¾	4¼	4½
1	⅞	4	5¼	5¼	4¾	5
1¼	1	4	6	5¾	5¼	5½
1½	1⅛	4	6¾	6½	6	6¼
2	1	8	7	6¾	6¼	6½
2½	1⅛	8	7¾	7½	7	7¼
3	1¼	8	8¾	8½	8	8¼
4	1½	8	10¼	9¾	9¼	9½
5	1¾	8	12¼	11½	11	11¼
6	2	8	14	13½	13	13¼
8	2	12	15½	15	14½	14¾
10	2½	12	20	19	18½	18¾
12	2¾	12	22	21	20½	20¾

All dimensions given in inches.

[1] These lengths do not include the height of the points. A point is that part of a stud bolt beyond the thread, and it may be chamfered, rounded, or sheared.

[2] Bolt lengths for lapped joints may be determined as follows: for lapped to lapped, add thickness of both laps; for lapped to ¹⁄₁₆-in. raised-face, add one thickness of lap; for lapped to ¼-in. male face on flange, add thickness of lap and ¼ in.; for lapped to female face on flange, add thickness of one lap only, the minimum thickness of which must be ¼ in. to serve as a male face: for male-and-female lapped joint made in the laps, add two thicknesses of pipe; add thickness of pipe for each lap.

[3] When groove is made in the lap, add thickness of pipe, but the lap that serves as the male face must not be less than ¼ in.

[4] Diameters of bolt holes are ⅛ in. greater than diameter of bolt.

Table 61. Turning Efforts to Tighten Eight-pitch-thread Bolts

Nominal diameter of bolt, in.	Number of threads, per in.	Tensile stress area, A_s	Stress[1] 30,000 lb/sq in. Torque, ft-lb	Stress[1] 30,000 lb/sq in. Force per bolt, lb	Stress[1] 60,000 lb/sq in. Torque, ft-lb	Stress[1] 60,000 lb/sq in. Force per bolt, lb
½	13	0.1419	30	4,257	60	8,514
⁹⁄₁₆	12	0.182	45	4,560	90	10,920
⅝	11	0.226	60	6,780	120	13,560
¾	10	0.334	100	10,020	200	20,040
⅞	9	0.606	160	18,180	320	36,360
1	8	0.462	245	13,860	490	27,720
1⅛	8	0.790	355	23,700	710	47,400
1¼	8	1.000	500	30,000	1,000	60,000
1⅜	8	1.233	680	36,990	1,360	73,980
1½	8	1.492	800	44,760	1,600	89,520
1⅝	8	1.78	1,100	53,400	2,200	106,800
1¾	8	2.08	1,500	62,400	3,000	124,800
1⅞	8	2.41	2,000	72,300	4,000	144,600
2	8	2.77	2,200	83,100	4,400	166,200
2¼	8	3.56	3,180	106,800	6,360	213,600
2½	8	4.44	4,400	133,200	8,800	266,400
2¾	8	5.43	5,920	162,900	11,840	325,800
3	8	6.51	7,720	195,300	15,440	390,600
3¼	8	7.69	.....	230,700	......	461,400
3½	8	8.96	.....	268,800	......	537,600
3¾	8	10.34	.....	310,300	......	620,400
4	8	18.11	.....	354,300	......	708,600

[1] Stress has been calculated on basis of stressed area, A_s, where $A_s = 0.7854 (D - 0.9743/N)^2$ in which D is the nominal bolt diameter and N is threads per inch.

pass inspection requirements are essential and must be included in the proper welding procedure.

Regardless of the fact that a welder has been properly qualified, the maintenance of an acceptable quality level of welding still requires effective supervision to insure that all provisions of the applicable welding procedures are satisfied.   Several of the major code-writing and governmental organizations have established standard procedures for welder (manual and semi-automatic welding) and welding-machine operator (automatic welding) qualifications.

ASME Boiler and Pressure Vessel Code.   The requirements of the ASME Boiler and Pressure Vessel Code, Section IX, are most widely recognized by industry and by state and municipal regulating bodies.   Under this Code, the qualification of a welding procedure involving a specific welding process requires that tensile and bend tests be made on suitable sections removed from the test weld.

For the qualification of a welder involving a manual welding process, only bend testing is required.   For automatic welding processes, radiographic examination or sectioning of the test weld is required instead of bend testing.   The results of satisfactory tests must then be recorded on forms such as the one shown in Fig. 72. Under the ASME Boiler and Pressure Vessel Code, each contractor is responsible for testing and qualifying the welders whom he plans to employ on Code work.

DAUGHERTY COMPANY, INC.
3914 Oak Street
CINCINNATI, OHIO, 45227

RECORD OF WELDING QUALIFICATION TEST
FOR BUTT WELDS IN PIPE

Form of Test .... Welding Procedure .... Qualification No. ....

Welding Process .... Position ....

Material .... to .... Material Specifications .... to ....

P–No. .... to P–No. .... Pipe Size .... Thickness ....

Filler Metal .... Classification .... F–No. .... A–No. ....

Flux or Shielding Gas .... Back-up Ring or Purging Gas ....

Preheat .... Postheat ....

Welder .... Max. Thickness This Test Qualifies .... Identification No. ....

Test Conducted by .... Date ....

REDUCED–SECTION TENSILE TEST

Specimen No.	Dimensions		Area	Ultimate total load, lb.	Ultimate unit stress, lb. per sq. in.	Character of failure and location
	Width	Thickness				

GUIDED-BEND TESTS

Specimen No.	Type of Bend	Results—describe location, nature and size of any crack or tearing of specimen.

This is to certify that the statements made in this record are correct and that the test welds were prepared, welded, and tested in full accordance with the current requirements of Section IX of the ASME Boiler and Pressure Vessel Code and with the ASA Code for Pressure Piping.

Date ....   **Daugherty Company, Inc.**

By: ....
Helmut Thielsch
Metallurgical Engineer

DCO 4173-REVISED-2   PRINTED IN U.S.A.   8-64

FIG. 72. Example of welder test certificate.

ASA Code for Pressure Piping. The ASA Code for Pressure Piping has issued a welder- and procedure-qualification standard of essentially the same requirements as Section IX of the ASME Boiler and Pressure Vessel Code. However, this Code permits employers who use identical welding procedures to interchange qualified welders. This, however, may not be generally considered advisable because of abuses which may be practiced.

American Society for Testing and Materials. ASTM Specification A488 covers the Qualification Procedures and Personnel for the Welding of Steel Castings.

This specification is patterned after Section IX of the ASME Boiler and Pressure Vessel Code.

American Petroleum Institute.   Standard 1104 includes requirements for the qualification of welding procedures and welders applicable to the field welding of pipelines, usually low-pressure, noncritical piping.

Mechanical Contractors Association of America.   Mechanical Contractors Association of America has formulated standard welding procedures for manual shielded metal-arc and oxyacetylene welding of pipe.   The National Certified Pipe Welding Bureau is one of several proprietary organizations which supervises and certifies welder-qualification tests in a uniform manner in accordance with the Association's standard procedures.

The organization also supervises the interchange of qualified welders between contractors who have adopted the standard procedures without the necessity of retesting them.   This, however, is not permissible for welding under the ASME Boiler and Pressure Vessel Code.   Thus, for this type of work, a contractor is required to perform his own welder-qualification certification.

Governmental Requirements.   On most military contract work, the welder-qualification tests are normally made to the requirements of Military Specification MIL-STD-248.   These are similar in scope and detail to the requirements of the ASME Boiler and Pressure Vessel Code.

Very stringent qualification-test requirements have been written into the new Bureau of Ships Code NAVSHIPS 250-1500-1, covering pipe fabricated to nuclear standards for the Department of the Navy on nuclear contract work.

National Board of Boiler and Pressure Vessel Inspectors.   The National Board of Boiler and Pressure Vessel Inspectors was organized to administer uniformly the rules of the ASME Boiler and Pressure Vessel Code.   It is composed of the chief inspectors of states, provinces (Canada), and municipalities which have adopted the ASME Code.   A pamphlet issued by the National Board entitled, "Recommended Rules for Repairs by Fusion Welding to Power Boilers and Unfired Pressure Vessels" (over 15 psi), makes reference to seal welding and repairs to boiler tubes and pipe of low-carbon steels of known weldable quality and having a carbon content of less than 0.35 per cent.   For welding alloy steel, reference is made to the ASME Boiler and Pressure Vessel Code.   Appendices are included which relate to welder-qualification tests.   No repairs by welding may be made without the approval of the inspector.   Repairs must be recorded on a form referenced in the National Board Inspection Code Book.

Insurance Requirements.   Insurance companies and inspection agencies often require proof from the contractor that the pipe welds were made by properly experienced and qualified welders, unless the contractor himself can certify that all his welders were qualified and tested under his supervision or by a reliable testing laboratory for the specific job involved.

Repair of Base Materials by Welding.   The American Society for Testing and Materials in most of their piping materials specifications has a paragraph covering repair of defects by chipping or grinding and repair welding.   Reference is also made to the ASME Boiler and Pressure Vessel Code by stating that the applicable welding procedures and welders shall have been qualified in accordance with Section IX of the above code.

PIPE-WELD JOINT PREPARATION AND DESIGN

Butt Welds

The most common type of joint employed in the fabrication of welded-pipe systems is the circumferential butt joint.   It is the most satisfactory from the

standpoint of stress distribution. Its general field of application is pipe to pipe, pipe to flange, pipe to valve, and pipe to fitting joints. Butt joints may be used for all sizes, but fillet-welded joints can often be used to advantage for pipe 2 in. and smaller in diameter.

Welding fittings are usually furnished with end preparations prepared by machining in accordance with the American Standards Association Code for Pressure Piping (as shown in Fig. 73). These end preparations also appear in various other

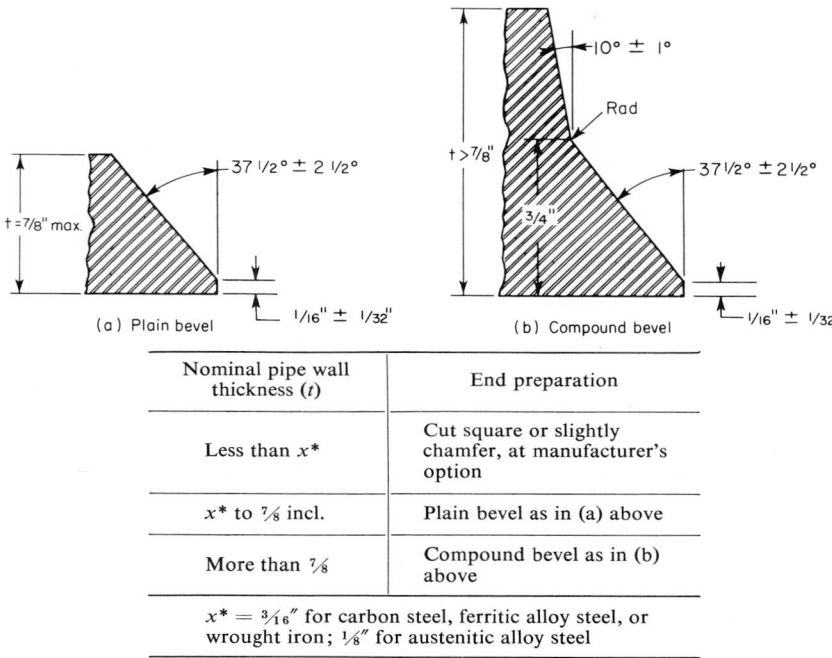

Nominal pipe wall thickness (t)	End preparation
Less than x^*	Cut square or slightly chamfer, at manufacturer's option
x^* to $\frac{7}{8}$ incl.	Plain bevel as in (a) above
More than $\frac{7}{8}$	Compound bevel as in (b) above

$x^* = \frac{3}{16}''$ for carbon steel, ferritic alloy steel, or wrought iron; $\frac{1}{8}''$ for austenitic alloy steel

FIG. 73. Basic welding bevel for all components (without backing ring, or with split ring).

ASA Standards as the American Standard for Steel Pipe Flanges and Flanged Fittings, ASA B16.5, and the American Standard for Butt Welding Fittings, ASA B16.9.

On piping, the end preparation is normally done by machining or grinding. Particularly on pipe of heavier wall thicknesses, machining is done generally in post mills. On carbon- and low-alloy steels, oxygen cutting and beveling is also used, particularly on pipe of wall thicknesses below $\frac{1}{2}$ in.

Because of fairly broad permissible eccentricity and size tolerances, considerable mismatch may be encountered on the inside of the piping. Limitations on fit-up tolerances are included in several piping codes (see Table 40). For critical service applications, internal machining may be required. Where this encroaches on the minimum wall thickness, weld metal build-up on the inside of the piping may be necessary prior to the internal machining operation. Such welding should be done in accordance with the procedure to be followed in welding the joint after fit-up.

Joints Between Unequal Wall Thicknesses. When piping components of unequal wall thickness are to be welded, care should be taken to provide a smooth

taper toward the edge of the thicker member. The length of the taper desirable is normally four times the offset between the pipe or vessel thicknesses, as outlined in Paras. PW-9 and UW-9 of the ASME Boiler and Pressure Vessel Code, Section I: Power Boilers and Section VIII: Unfired Pressure Vessels, respectively. The two methods of alignment which are recommended are shown in Fig. 74.

The wall thickness of cast-steel fittings and valve bodies is normally greater than that of the pipe to which they are joined. In order to provide equal thicknesses

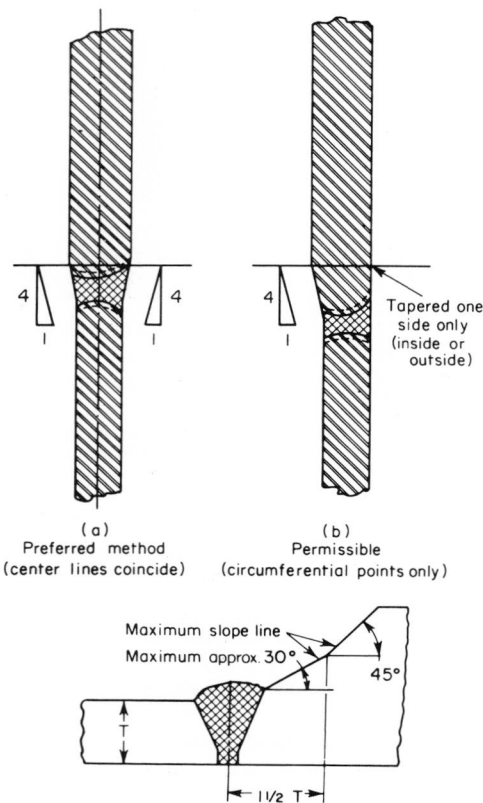

FIG. 74. Recommended welding-end sections for pipe, valves, and fittings of unequal thickness.

for welding, the ASME Boiler and Pressure Vessel Code and the ASA Code for Pressure Piping permit the machining of the cylindrical ends of cast-steel fittings and valve bodies to the nominal wall thickness of the adjoining pipe, provided that these areas are finish-machined both inside and outside and are carefully inspected. The machined ends may be extended back in any manner, provided that the longitudinal section comes within the maximum slope line indicated in Fig. 74. The transition from the pipe to the fitting or valve end at the joint must be such as to avoid sharp reentrant angles and abrupt changes in slope.

End Preparation for Inert-gas Tungsten-arc Root-pass Welding. The pipe end bevel preparations shown in Fig. 73 are considered adequate for shielded

metal-arc welding, but they pose a problem in inert-gas tungsten-arc welding. When this process is used, extended "U" or *flat-land* bevel preparations are considered more suitable to minimize excessive sink.   Specific standards have not been developed, but end preparations that are used extensively are illustrated in Fig. 75.

The end preparations apply to inert-gas tungsten-arc welding of carbon- and low-alloy steel piping, stainless steel piping, and most nonferrous piping materials. On aluminum piping, the flat-land bevel preparations are preferred by some fabricators.

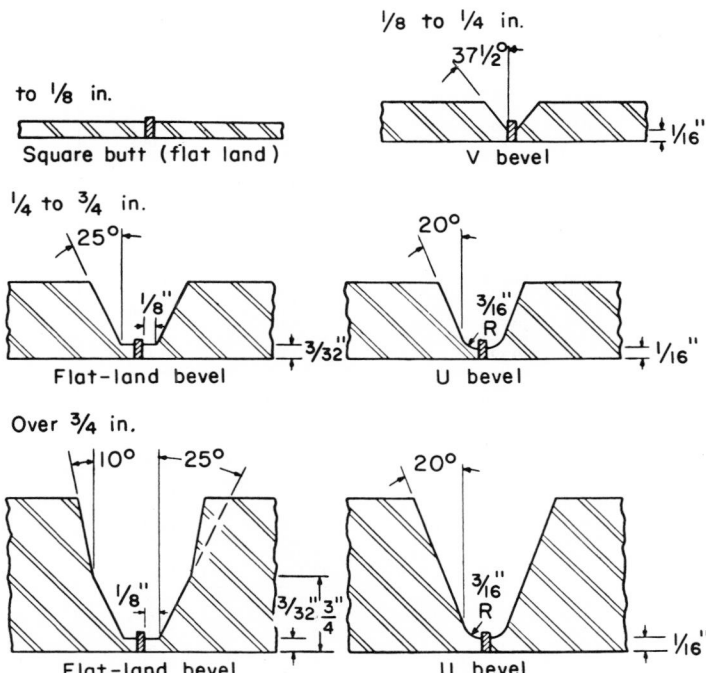

FIG. 75. Typical end preparations for pipe which is to be welded by the inert-gas tungsten-arc welding process.

Backing Rings.   Backing rings are employed in some piping systems, particularly where pipe joints are welded primarily by the shielded metal-arc welding process with covered electrodes.   For example, a significant number of pipe welds for steam power plants and several other applications are still made with the use of backing rings.   On the other hand, in many applications backing rings are not used, as they may restrict flow, provide crevices for the entrapment of corrosive substances, enhance susceptibility to stress corrosion cracking, or introduce still other objectionable features.   Thus, there is little, if any, use made of backing rings in most refinery piping systems or chemical process and similar plants.

The use of backing rings is primarily confined to carbon- and low-alloy steel and aluminum piping.   Carbon-steel backing rings are generally made of a mild carbon steel with a maximum carbon content of 0.20 per cent and a maximum sulfur

Table 62. Dimensions for Internal Machining and Backing Rings for Heavy-wall Pipe in Critical Applications

Nominal pipe size	Schedule No. or wall	Nominal OD "A"	Nominal ID "B"	Nominal wall thickness t	Machined ID of pipe "C" tolerance $+0.010$ -0.000	OD of backing ring	
						Tapered ring DT tolerance $+0.010$ -0.000	Straight ring DS tolerance $+0.000$ -0.010
3	XXS	3.500	2.300	0.600	2.409	2.419	2.409
4	XXS	4.500	3.152	0.674	3.279	3.289	3.279
5	160	4.500	4.313	0.625	4.428	4.438	4.428
	XXS	5.563	4.063	0.750	4.209	4.219	4.209
6	120	6.625	5.501	0.562	5.600	5.610	5.600
	160	6.625	5.187	0.719	5.326	5.336	5.326
	XXS	6.625	4.897	0.864	5.072	5.082	5.072
8	100	8.625	7.437	0.594	7.546	7.554	7.544
	120	8.625	7.187	0.719	7.326	7.336	7.326
	140	8.625	7.001	0.812	7.163	7.173	7.163
	XXS	8.625	6.875	0.875	7.053	7.063	7.053
	160	8.625	6.813	0.906	6.998	7.008	6.998
10	80	10.750	9.562	0.594	9.671	9.679	9.669
	100	10.750	9.312	0.719	9.451	9.461	9.451
	120	10.750	9.062	0.844	9.234	9.242	9.232
	140	10.750	8.750	1.000	8.959	8.969	8.959
	160	10.750	8.500	1.125	8.740	8.750	8.740
12	60	12.750	11.626	0.562	11.725	11.735	11.725
	80	12.750	11.374	0.688	11.507	11.515	11.505
	100	12.750	11.062	0.844	11.234	11.242	11.232
	120	12.750	10.750	1.000	10.959	10.969	10.959
	140	12.750	10.500	1.125	10.740	10.750	10.740
	160	12.750	10.126	1.312	10.413	10.423	10.413
14 OD	60	14.000	12.812	0.594	12.921	12.929	12.919
	80	14.000	12.500	0.750	12.646	12.656	12.646
	100	14.000	12.124	0.938	12.319	12.327	12.317
	120	14.000	11.812	1.094	12.046	12.054	12.044
	140	14.000	11.500	1.250	11.771	11.781	11.771
	160	14.000	11.188	1.406	11.498	11.508	11.498
16 OD	60	16.000	14.688	0.656	14.811	14.821	14.811
	80	16.000	14.312	0.844	14.484	14.492	14.482
	100	16.000	13.938	1.031	14.155	14.165	14.155
	120	16.000	13.562	1.219	13.826	13.836	13.826
	140	16.000	13.124	1.438	13.442	13.452	13.442
	160	16.000	12.812	1.594	13.171	13.179	13.169
18 OD	40	18.000	16.876	0.562	16.975	16.985	16.975
	60	18.000	16.500	0.750	16.646	16.656	16.646
	80	18.000	16.124	0.938	16.319	16.312	16.317
	100	18.000	15.688	1.156	15.936	15.946	15.936
	120	18.000	15.250	1.375	15.553	15.563	15.553
	140	18.000	14.876	1.562	15.225	15.235	15.225
	160	18.000	14.438	1.781	14.842	14.852	14.842
20 OD	40	20.000	18.812	0.594	18.921	18.929	18.919
	60	20.000	18.376	0.812	18.538	18.548	18.538
	80	20.000	17.938	1.031	18.155	18.165	18.155
	100	20.000	17.438	1.281	17.717	17.727	17.717
	120	20.000	17.000	1.500	17.334	17.344	17.334
	140	20.000	16.500	1.750	16.896	16.906	16.896
	160	20.000	16.062	1.969	16.515	16.523	16.513
22 OD	...	22.000	20.750	0.625	20.865	20.875	20.865
	60	22.000	20.250	0.875	20.428	20.438	20.428
	80	22.000	19.750	1.125	19.990	20.000	19.990
	100	22.000	19.250	1.375	19.553	19.563	19.553
	120	22.000	18.750	1.625	19.115	19.125	19.115
	140	22.000	18.250	1.875	18.678	18.688	18.678
	160	22.000	17.750	2.125	18.240	18.250	18.240
24 OD	30	24.000	22.876	0.562	22.975	22.985	22.975
	40	24.000	22.624	0.688	22.757	22.765	22.755
	60	24.000	22.062	0.969	22.265	22.273	22.263
	80	24.000	21.562	1.219	21.826	21.836	21.826
	100	24.000	20.938	1.531	21.280	21.290	21.280
	120	24.000	20.376	1.812	20.788	20.798	20.788
	140	24.000	19.876	2.062	20.350	20.360	20.350
	160	24.000	19.312	2.344	19.859	19.867	19.857

content of 0.05 per cent. The latter requirement is especially important since high sulfur in deposited weld metal (which could be created by an excessive sulfur content in such rings) may cause weld cracks. Split backing rings are satisfactory for non-critical piping systems.

For the more critical service applications involving carbon- and low-alloy steel piping, solid flat or taper-machined backing rings are preferred in accordance with the recommendations shown in Pipe Fabrication Institute Standard ES1 and illustrated in Fig. 76 and Table 62.

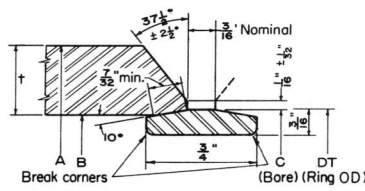

For wall thickness (t) 9/16 to 1 in. inclusive and tapered internal machining.

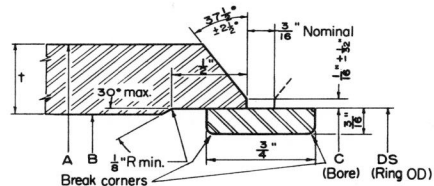

For wall thickness (t) 9/16 to 1 in. inclusive and straight internal machining.

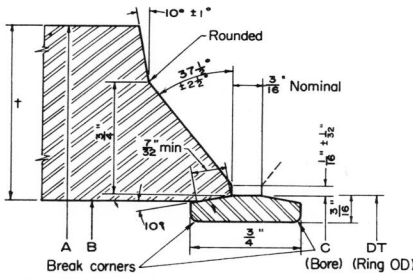

For wall thickness (t) greater than 1 in. and tapered internal machining.

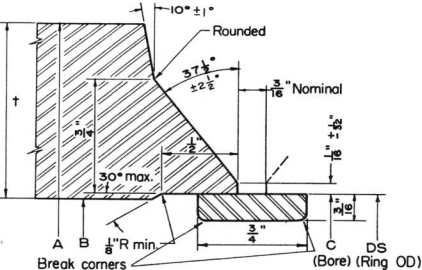

For wall thickness (t) greater than 1 in. and straight internal machining.

FIG. 76. End-preparation and backing-ring requirements for critical service applications employing flat or taper machined solid backing rings. See Table 62 for dimensional data.

When a machined backing ring is desired, it is a general recommendation that welding ends be machined on the inside diameter in accordance with the Pipe Fabrication Institute standard for the most critical services and then only when pierced seamless pipe is used that complies with the applicable specifications of the American Society for Testing and Materials. Such critical services include high-pressure steam lines between boiler and turbines and high-pressure boiler-feed discharge lines, as encountered in modern steam power plants. It is also recommended that the material of the backing ring be compatible with the chemical composition of the pipe, valve, fitting, or flange with which it is to be used. Where materials of dissimilar composition are being joined, the composition of the backing ring may be that of the lower alloy.

On turned-and-bored and fusion-welded pipe, the design of the backing ring and internal machining, if any, should be a matter of agreement between the customer and the fabricator. Regardless of the type of backing ring used, it is recommended that the general contour of the welding bevel shown in Fig. 76 be maintained.

When machining piping for backing rings, the resulting wall thickness should be not less than that required for the service pressure. Wherever internal machining

for machined backing rings is required on pipe and welding fittings in smaller sizes and lower schedule numbers than those listed in Table 62, weld metal may have to be deposited on the inside of the pipe in the area to be machined. This is to provide satisfactory contact between the machined surface on the pipe inside and the machined backing ring. For such cases, the machining dimension should be a matter of agreement between the fabricator and the purchaser.

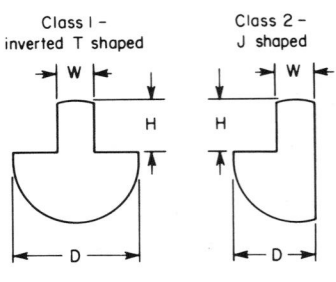

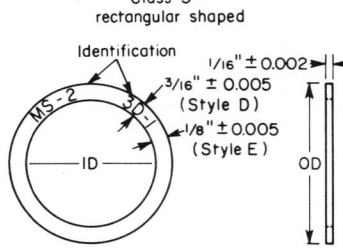

Fig. 77. Commercial consumable insert rings used in pipe welding (MIL-I-23413). Style D: for diameters 2 in. and larger. On Schedule 5 for diameters 5 in. and larger; Style E: for diameters less than 2 in. On Schedule 5 for diameters less than 5 in.

Whenever pipe and welding fittings in the sizes and schedule numbers listed in Table 62 have plus tolerance on the outside diameter, it also may be necessary to deposit weld metal on the inside of the pipe or welding fitting in the area to be machined. In such cases, sufficient weld metal should be deposited to result in an ID not greater than the nominal ID given in Table 62 for the particular pipe size and wall thickness involved.

Experience indicates that machining to dimension "C" for the pipe size and schedule number listed in Table 62 generally will result in a satisfactory seat contact of $7/_{32}$ in. minimum (approximately 75 per cent minimum length of contact) between pipe and the 10-degree backing ring. Occasionally, however, it will be necessary to deposit weld metal on the inside diameter of the pipe or welding fitting in order to provide sufficient material for machining a satisfactory seat.

In welding butt joints with backing rings, care should be exercised to insure good fusion of the first weld pass into the backing ring in order to avoid lack of weld penetration or other types of stress-raising notches.

Consumable Insert Rings. The chemical composition of a piping base metal is established primarily to provide it with certain mechanical, physical, or corrosion-resisting properties. Weldability characteristics, if considered at all, are of secondary concern. On the other hand, the chemical composition of most welding filler metals is determined with primary emphasis on producing a sound, high-quality weld.

The steel-making process employed in the manufacture of welding filler metals permits closer control of the composition range, which is usually considerably narrower than would be practical for the piping base metal where much larger tonnages of steel are involved.

On some base metals, the welding together by fusion of only the base-metal compositions may lead to such welding difficulties as cracking or porosity. The addition of filler metal tends to improve weld quality. However, in inert-gas tungsten-arc welding, the addition of welding filler metal from a separate wire which the welder feeds with one hand while manipulating the tungsten-arc torch with the other is a cumbersome process and interferes with welding ease. The welder may leave areas with lack of penetration, which generally are considered

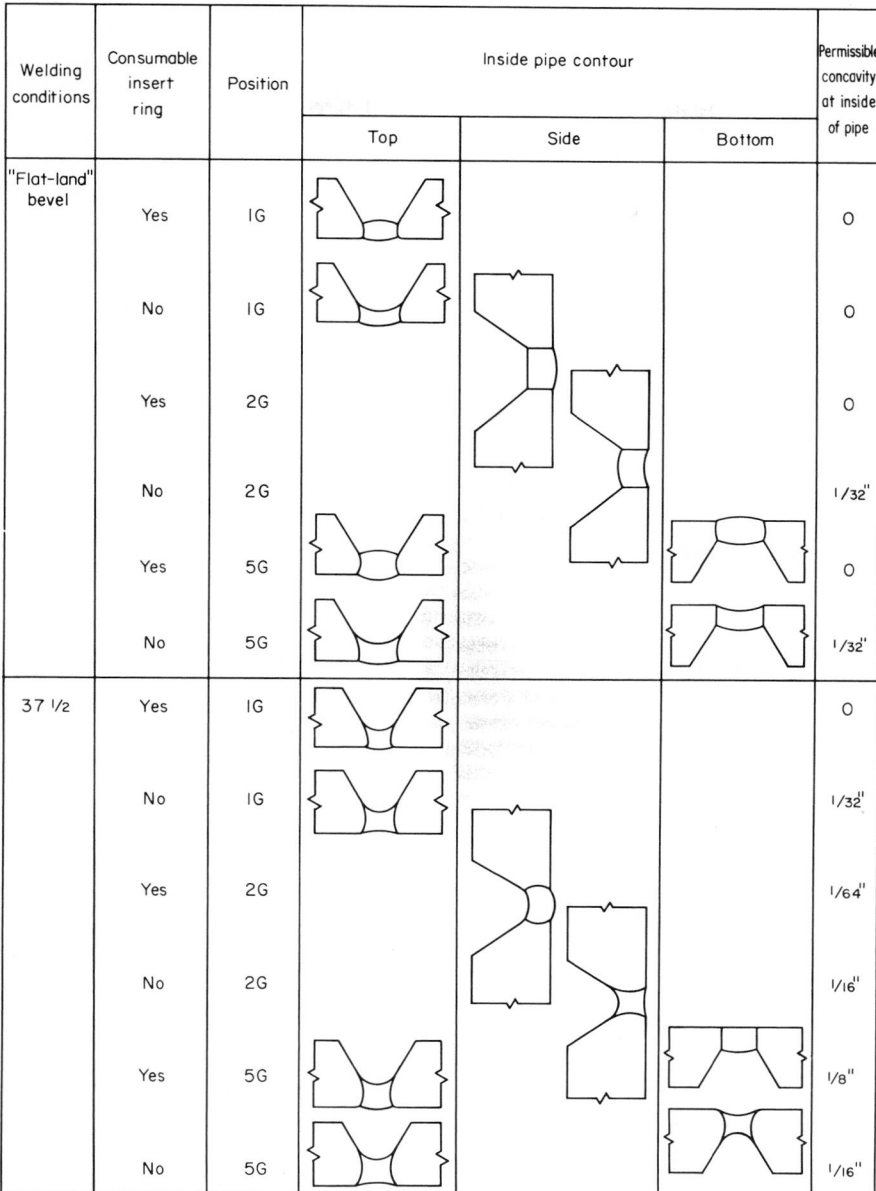

Welding conditions	Consumable insert ring	Position	Inside pipe contour			Permissible concavity at inside of pipe
			Top	Side	Bottom	
"Flat-land" bevel	Yes	IG				0
	No	IG				0
	Yes	2G				0
	No	2G				1/32"
	Yes	5G				0
	No	5G				1/32"
37 1/2	Yes	IG				0
	No	IG				1/32"
	Yes	2G				1/64"
	No	2G				1/16"
	Yes	5G				1/8"
	No	5G				1/16"

Fig. 78. Root-contour conditions which can be expected as the result of normal pipe welding with the gas tungsten-arc-welding process. In 5G (horizontal-fixed) position welding the insert ring is positioned eccentric to the centerline of the pipe, as illustrated in Fig. 79.

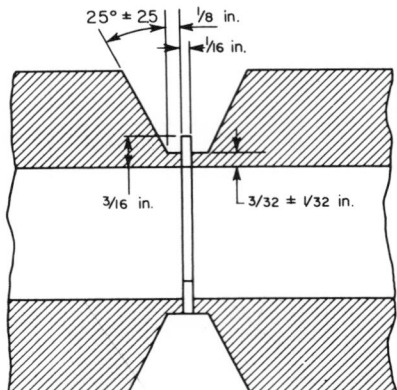

FIG. 79. Eccentric insertion of consumable insert ring in pipe welded in the fixed horizontal pipe position.

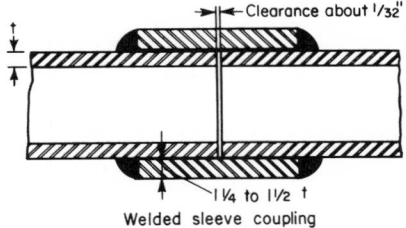

Welded sleeve coupling

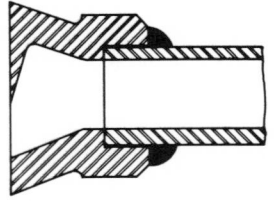

Socket detail for small welding and valve

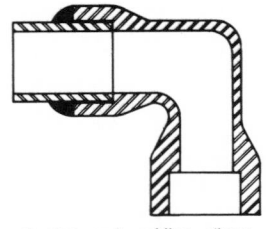

Socket end welding elbow

FIG. 80. Examples of typical fillet-welded joints.

unacceptable as can be seen, for example, in the rules of the ASME Boiler and Pressure Vessel Code. Since some types of serious weld defects are detected only with difficulty during inspection (if they are detected at all), it is extremely important to provide the easiest welding conditions for the welder to produce quality welds.

One technique to produce high-quality welds is to employ consumable insert rings of proper composition and dimensions. Consumable insert rings which are available commercially are shown in Fig. 77. The three primary functions of consumable insert rings are to: (1) provide the easiest welding conditions and thereby minimize the effects of undesirable welding variables caused by the "human" element, (2) give the most favorable weld contour to resist cracking resulting from weld-metal shrinkage and hot shortness, or brittleness, in hot metal, and (3) produce metallurgically the soundest possible weld-metal composition of desirable strength, ductility, and toughness properties.

The best welding conditions are obtained where the flat-land and extended "U"-bevel preparations are used. These joint preparations are particularly helpful where welding is done in the horizontal fixed pipe position (5G), since they insure a flat or slightly convex root contour and provide by far the greatest resistance to weld cracking in those alloys particularly susceptible to microfissuring.

The weld-root contour conditions to be expected from different bevel preparations and consumable insert rings are illustrated in Fig. 78. Where sink is not acceptable, it is considered obligatory to use consumable insert rings with the special flat-land or extended "U" bevel preparation. In horizontal-rolled (1G) and vertical-position (2G) welding, the insert ring should be placed concentrically into the beveled pipe.

In horizontal fixed-position (5G) welding, the insert ring should be placed eccentric to the centerline of the pipe (as shown in Fig. 79). In this position, the insert ring compensates for the downward

sag of the molten weld metal and aids in obtaining smooth, uniform root contour along the inner diameter of the joint.

Fillet Welds

Circumferential fillet-welded joints are generally used for joining pipe to pipe, pipe to flanges, pipe to valves, and pipe to socket joints in sizes 2 in. and smaller. Since backing rings appreciably reduce the internal area of small-diameter pipe, they are rarely used in small-diameter piping. Figure 80 illustrates three typical fillet-welded joints. These types of welds are subjected to shearing and bending stresses, and adequate penetration of the pieces being joined is essential. This is particularly important with the socket joint, since the danger of washing down the end of the hub may obscure, by reason of fair appearance, the lack of a full and sound fillet weld. This condition is one which cannot be detected in the finished weld by the usual visual inspection.

There are service applications where socket welds are not acceptable. Piping systems involving nuclear or radioactive service or corrosive service with solutions which promote stress corrosion cracking or concentration cell action generally require butt welds in all pipe sizes with complete weld penetration to the inside of the piping.

Intersection-type Joints

Intersection joints, such as medium- and large-size welded tees, laterals, wyes, and openings in vessels are usually the most difficult to weld soundly. Wrought carbon steel, carbon-molybdenum steel, and chromium-molybdenum steel fittings, all conforming to ASTM Specification A234, Factory-made Wrought Carbon-steel and Ferritic Alloy Steel Welding Fittings, are preferred for accomplishing the assembly of piping systems since they provide bursting strengths equivalent to that of pipe of the same weight or schedule and their installation involves butt welds only. Machines for oxygen cutting and beveling of header openings and branch ends are available but are not in general use. Cutting and beveling is usually done manually with an oxyacetylene cutting torch. Figure 81 illustrates two forms of preparation for 90-deg intersection joints. In one, the header opening is sufficient to permit insertion of the branch, and beveling of the header opening only is required. In the other, the header opening is equal to the inside diameter of the branch, and the branch only is beveled. The latter permits the use of a specially shaped backing ring, if desired. Unreinforced branches (as illustrated in Fig. 81) are adequate only where the pipe is to be used at pressures considerably below its full capability, usually between 50 and 75 per cent of the pressure which the run of the pipe can withstand safely. Reinforcing is required to make 90-deg intersections equivalent in strength to the pipe from which they are fabricated. The design of such reinforcing should be in accordance with the requirements of the ASME Boiler and Pressure Vessel Code or the American Standards Association Code for Pressure Piping, ASA B31.

Recommendations[1] for minimum lengths, distance from open end on run, and spacing dimensions of adjacent unreinforced and reinforced welded nozzles are given in Tables 63 and 64. Where attachments, such as flanges, fittings, valves, and pipe insulation are involved, the minimums shown may have to be revised to allow for the required clearances.

[1] Pipe Fabrication Institute Standard ES-7, Welded Nozzles—Recommended Minimum Lengths and Spacing.

Table 63. Dimensions Recommended for Nozzles without Saddles or Rings

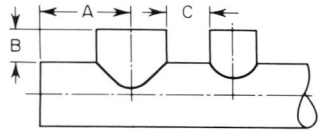

Nominal pipe size of nozzle	Min recommended dimensions		
	Center of nozzle to end of run A	OD of run to end of nozzle B	OD to OD of nozzle $C*$
4 and smaller	6	4	4
5	7	4½	4½
6	8	5	5
8	10	6	6
10	12	7	7
12	14	8	8
14	15	8½	8½
16	17	9	9
18	19	10	10
20	21	11	11
24	24	12	12

* Minimum dimension C should be taken as that which is tabulated for the larger of the two adjacent nozzles.

Table 64. Dimensions Recommended for Nozzles with Saddles or Reinforcements

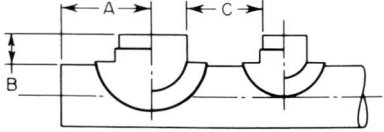

Nominal pipe size of nozzle	Min recommended dimensions		
	Center of nozzle to end of run A	OD of run to end of nozzle B	OD to OD of nozzle $C*$
4 and smaller	8	5½	8
5	9½	6	9½
6	11	6½	11
8	14	8	14
10	17	9½	17
12	20	11	20
14	22	12	22
16	25	13	25
18	28	14	28
20	31	15	31
24	36	16	36

* Minimum dimension C should be taken as that which is tabulated for the larger of the two adjacent nozzles.

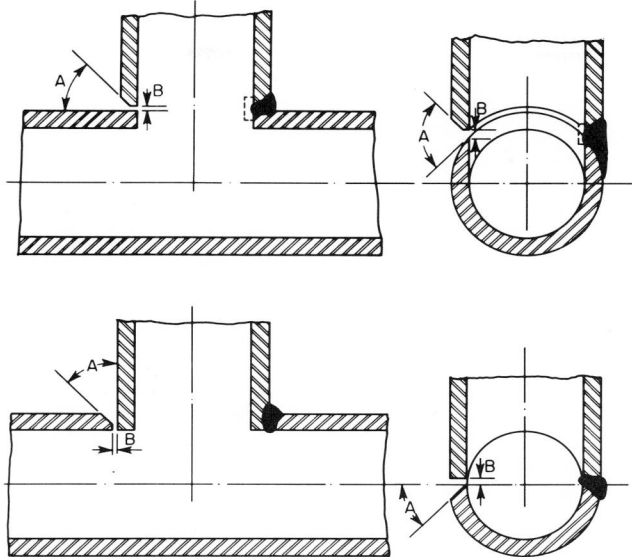

FIG. 81. Recognized types of preparation for 90-deg intersection welds. Angle *A* to be not less than 45 deg for any wall thickness of pipe. Dimension *B* to be not less than $\frac{1}{16}$ in. nor more than $\frac{1}{4}$ in.

The intersection of two pipes at an angle of 45 deg or less makes it difficult to secure the desired degree of penetration and soundness of the weld deposit. Complete weld penetration is essential for severe service conditions. Such intersection welds should be avoided for service where the pressure exceeds approximately 125 psi. Piping should be designed, whenever possible, so that all intersection joints, regardless of angle, can be made under shop conditions, where work can be positioned for welding in the flat position.

Manufacturers today have simplified the problems of the fabricator by providing factory-made welding-type nozzles, welding necks, flanges, and welding tees. Extruded outlets providing butt joints also eliminate a great deal of cutting, tacking, and fitting required by welded branch connections. Care must be taken in selecting some fittings, particularly laterals, if the full design pressure is to be maintained since some are not provided with the required reinforcement. Typical nozzle, neck, and welding outlets which can be used for almost all ranges of sizes are shown in Fig. 82.

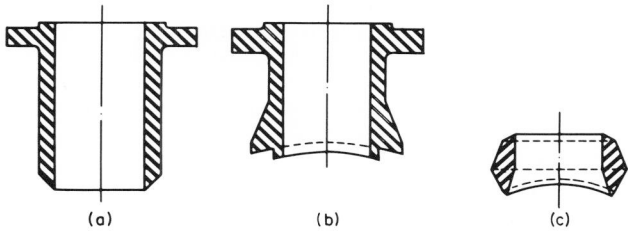

FIG. 82. Typical welded outlets. (*a*) Neck outlet; (*b*) nozzle outlet; (*c*) welding outlet.

Brazed Joints

Lap or shear-type joints generally are necessary to provide capillary attraction for brazing of connecting pipe. Square-groove butt joints may be brazed, but the results are unreliable unless the ends of the pipe or tube are accurately prepared, plane and square, and the joint is aligned carefully, as in a jig. High strengths may be obtained with butt joints if properly prepared and brazed. However, owing to the brittleness of the brazing alloy, they are not normally applicable.

The alloys generally used in brazing exhibit their greatest strength when the thickness of the alloy in the lap area is minimal. Thin alloy sections also develop the highest ductility. For brazing ferrous and nonferrous piping with silver- and copper-base brazing alloys, the thickness of the brazing alloy in the joint generally should not be over 0.006 in. and preferably not over 0.004 in. Thicknesses less than 0.003 in. may make assembly difficult, while those greater than 0.006 in. tend to produce joints having lowered strength. The brazing of certain aluminum alloys is similar in most respects to the brazing of other materials. However, joint clearances should be greater because of a somewhat more sluggish flow of the brazing alloys. For aluminum, a clearance of 0.005 to 0.010 in. will be found satisfactory. Care must be exercised in fitting dissimilar metals, as the joint clearance at brazing temperature is the controlling factor. In these cases, consideration must be given to the relative expansion rates of the materials that are being joined.

The length of lap in a joint, the shear strength of the brazing alloy, and the average percentage of the brazing surface area that normally bonds are the principal factors determining the strength of brazed joints. The shear strength may be calculated by multiplying the width by the length of lap by the percentages of bond area and by taking into consideration the shear strength of the alloy used. An empirical method of determining the lap distance is to take it as twice the thickness of the thinner or weaker member joined. Normally this will provide adequate strength, but in cases of doubt, the fundamental calculations should be employed.

Such detailed determinations are generally unnecessary for brazed piping, since commercial brazing fittings are available in which the length of lap is predetermined at a safe value. For brass and copper pipe, cast- or wrought-bronze and wrought-copper fittings are available. A bore of correct depth to accept the pipe is provided, and midway down this bore may be a groove into which, at the time of manufacture, a ring of brazing alloy is inserted. Since the alloy is preplaced in fittings with such a groove, separate feeding of brazing alloy by hand is generally unnecessary.

WELDING PROCESSES

Although increasing use is made of semiautomatic and automatic welding processes, the process still most extensively used in pipe shop welding, and particularly in field welding, is the manual shielded metal-arc process. Other processes (among which are automatic and semiautomatic submerged arc, inert-gas tungsten-arc, gas-shielded consumable metal arc, and oxyacetylene welding) are used to varying extents. These processes are extensively discussed in the trade literature and in manufacturers' catalogues. This section will be concerned only with the application of the various welding processes to the fabrication and erection of the more common piping materials.

Manual Shielded Metal-arc Welding

This process is utilized on nearly all ferrous and nonferrous metals used in piping systems. The equipment required for its application is comparatively simple and

compact, readily portable, safe to use and, with ordinary care, requires little maintenance. Electric power for the operation of either d-c or a-c machines is readily provided in fabricating plants and is available for field work. If electric power is not available, the use of gasoline or diesel-engine-driven welding machines gives satisfactory service.

The number of weld passes required for welding joints in ferrous piping varies with the wall thickness of the pipe, the welding position of the pipe, the diameter of electrode used, the presence of metal powders in the covering, and the welding currents employed. In welding low-carbon and low-alloy steel pipe in the rolled or fixed horizontal positions, under generally accepted procedures, the number of weld layers is approximately one per $\frac{1}{8}$ in. of pipe thickness.

In welding medium-carbon and higher-alloy steels in the rolled or horizontal position, the number of weld layers is sometimes increased 25 to 30 per cent by use of smaller electrodes to lessen the heat concentration and to insure grain refinement of the previously deposited weld metal.

Electrode manipulation also varies with the type of covering on the electrode. For example, electrodes with EXX18 coverings generally require a wider weave than the EXX10 or EXX16 types.

With the pipe in a vertical-fixed position, deposition as described above is not possible. Instead, the metal is deposited in the form of a series of overlapping string beads, the number of which when using $\frac{5}{32}$ in.-maximum-diameter electrodes may be approximated roughly by allowing 30 to 40 beads per square inch of weld area.

The number of passes for welding joints in nonferrous piping will vary to a considerable extent because of the wide differences in fusion temperatures and thermal conductivities of the metals, factors which have considerable bearing on the welding technique employed.

It is a common practice to do as much welding as possible in the flat, or downhand, position, using suitable power-driven rotators for continuously rotating the work at a speed consistent with the rate at which the filler metal can be deposited properly. This assures uniformity both in depositing the filler metal and in distributing the heat, thus lessening distortion. The extent to which flat-position welding can be applied is limited by the dimensions, shape, and weight of the component parts to be welded, the facilities for rotating the work, the amount of handling involved, and, sometimes, the heat-treating requirements.

Fixed-position welding generally applies to final field or in-plant erection. It requires special care in depositing successive beads, or layers, uniformly around the joints as called for by the welding procedure, in order to avoid excessive stress concentrations or distortion due to uneven temperature distribution. Wherever possible, the assembly or piping system should be arranged so that the joints are readily accessible to the welder from all essential points. When preheating or stress relieving is to be done, ample space should be provided for applying and removing the heating equipment.

In shielded metal-arc welding with covered electrodes, the preferred direction of welding is to start at the bottom and progress upward. Some welding is done in the opposite direction. Downward welding is often essential on thin-walled material or against thin backing rings in order to avoid burning through the underlying material. The danger in this procedure is that slag is more likely to be entrapped in the downward progression of welding. Ordinarily, more metal per layer is deposited when welding upward. Welding downward is considered to require a higher degree of manual skill to secure adequate fusion with the side walls and to avoid entrapment of slag.

Each layer of deposited weld metal must be thoroughly cleaned prior to the deposition of the following weld-metal layer. Wire brushing, especially when done

Table 65. Estimated Electrode Requirements for Pipe Butt Welds, lb

Nominal pipe size, in.	Schedule 40		Schedule 80		Schedule 100		Schedule 120		Schedule 160	
	$\frac{1}{8}$ in.	$\frac{5}{32}$ in.	$\frac{1}{8}$ in.	$\frac{5}{32}$ in.	$\frac{1}{8}$ in.	$\frac{5}{32}$ in.	$\frac{1}{8}$ in.	$\frac{5}{32}$ in.	$\frac{1}{8}$ in.	$\frac{5}{32}$ in.
$2\frac{1}{2}$	0.40	.....	0.30	0.40	....	....	....	....	0.30	0.60
3	0.50	.....	0.40	0.55	....	....	....	....	0.40	1.0
$3\frac{1}{2}$	0.65	.....	0.45	0.75	....	....	....	....	0.45	1.3
4	0.75	.....	0.50	0.90	....	....	0.40	1.3	0.40	1.8
5	0.50	0.50	0.50	1.20	....	....	0.50	2.1	0.50	3.2
6	0.60	0.70	0.60	1.80	....	....	0.60	2.9	0.60	5.0
8	1.80	1.20	0.80	3.20	0.80	4.3	0.80	6.5	0.80	8.8
10	1.00	2.00	1.00	5.6	1.00	8.1	1.00	9.0	1.00	16.0
12	1.20	2.80	1.20	8.2	1.20	11.3	1.20	15.3	1.20	22.8
14	1.30	3.6	1.20	10.0	1.20	15.3	1.20	18.6	1.20	28.6
16	1.50	6.0	1.50	14.0	1.50	20.0	1.50	26.0	1.50	39.0
18	1.70	8.2	1.70	20.0	1.70	27.0	1.70	33.5	1.70	51.0
20	1.90	10.0	1.90	26.0	1.90	35.0	1.90	43.0	1.90	66.0
24	2.30	17.0	2.30	38.0	2.30	53.0	2.30	68.0	2.30	105.0

with a power-driven brush, is effective in removing the slag deposited by most covered electrodes. Surface defects which would otherwise affect the soundness of the weld should be chipped or ground out.

The surface of the deposited metal and the faces of the base metal should be prepared for the following layer by removing bumps and sharp corners or grooves which might otherwise be difficult to fill without risk of slag inclusions. The grinding required, if any, will vary with the type of electrode used and the skill of the welder. Slag comes off readily on weld deposits properly made with EXX18 electrode types, and the weld deposit is smoother than with EXX10 electrode types. In welding for severe service conditions, between-pass cleaning may also be done with power-operated chipping tools.

Pipe welded by the shielded metal-arc-welding process is generally welded with $\frac{1}{8}$- and $\frac{5}{32}$-in.-diameter electrodes. Occasionally, $\frac{3}{32}$-in.-diameter electrodes are preferred for the first pass. Electrodes with a $\frac{3}{16}$ in. diameter are not readily handled by most welders, except where the pipe is rotated. However, $\frac{3}{16}$-in. or even $\frac{1}{4}$-in.-diameter electrodes are often used where heavy fillet welds must be made on the pipe material or where piping is large and rotated for flat-position welding. Estimated electrode requirements for butt welds in pipe of various diameters and sizes are given in Table 65.

Electrode requirements vary somewhat with the various commercial types of electrodes. For example, some electrodes exhibit high deposit efficiencies of about 70 per cent, whereas others may have a considerably lower efficiency characteristic of about 55 per cent. The electrode requirements also depend upon the welder who may or may not be careful in the amount of stub-end waste.

The totals shown in Table 65 are requirements of coated electrodes as furnished in boxes or cans by the various electrode manufacturers. They represent average conditions and include the normal efficiency loss due to spatter, grinding, and stub ends. The calculations are based on the end preparations shown in Fig. 73.

Inert-gas Tungsten-arc Welding

This process is used extensively on almost all of the ferrous and nonferrous piping materials. Even substantial quantities of carbon-steel piping are welded in

the root of butt joints with the inert-gas tungsten-arc-welding process. The thinnest section that can be hand-welded is approximately $\frac{1}{32}$ in. The maximum thickness that may be welded is limited by available equipment and will vary for different metals. For pipe wall thicknesses from $\frac{1}{4}$ to $\frac{3}{8}$ in. (depending on the pipe material), it is generally more economical to complete the pipe butt weld, after the inert-gas tungsten-arc root pass, with other welding processes, such as shielded metal-arc welding or submerged-arc welding.

Accurate preparation and good fit-up are particularly important considerations. The cleaning operations preparatory to welding should be carefully performed, and, in some cases, they may require the use of chemical treatments, especially on aluminum and titanium, because of their tendency to oxidize readily. Some types of gas or metal backing are beneficial, particularly on the higher-alloy steels and on most nonferrous alloys. Where the backing is done with a gas such as argon, helium, or nitrogen, this process is referred to as "purging." Care should be exercised in the selection of suitable backing materials or gases to avoid contamination or other undesirable effects on the weld.

In general, welding procedures follow the same general principles common to other processes, and equivalent welder skill is essential for satisfactory results. However, it is important that adequate attention be given to the conservation of the costly gas or gases used for shielding and purging.

Gas-shielded Consumable Metal-arc Welding

Gas-shielded consumable metal-arc welding has been done extensively on aluminum piping, and argon is usually used as the shielding gas. In recent years, the gas-shielded consumable metal-arc-welding processes have been adapted to the welding of steel piping. Because of differences in equipment, the welding characteristics and weld qualities obtained may differ substantially. With some equipment, carbon-steel filler wires cannot be used on pipe, and carbon-molybdenum steel wires are substituted. In the welding of steel gas shielding is normally done with pure argon, mixtures of argon and carbon dioxide, and mixtures of argon and oxygen. The type of gas that is best for a given application is dependent on the type of material to be welded, the thickness of the material, the position in which the welding is to be done, the penetration desired, and other factors.

Welding is done into open pipe joints with an open root spacing of approximately $\frac{1}{32}$ to $\frac{1}{16}$ in. In welding in the horizontal fixed pipe position, the first pass is normally started at the top and continued downward to the bottom. Subsequent weld passes can be started either at the top, then proceeding downward or at the bottom and proceeding upward.

In the case of tack welds, even with grinding of the edges, it is difficult with present equipment to obtain an X-ray quality joint, since at the edge of the tack welds, lack of penetration, weld craters, or burn-through irregularities often result. Accordingly, at the present time, the application of most of the commercial gas-shielded consumable metal-arc-welding equipment to pipe welding is limited to essentially noncritical applications. These involve weld quality requirements permitting defects, which would be unacceptable under the inspection requirements of the ASME Boiler and Pressure Vessel Code and other applicable codes involving critical applications, such as nuclear piping and some chemical-process piping.

Submerged-arc Welding

Automatic and semiautomatic submerged-arc welding are extensively used in shop welding of carbon-steel, chromium-molybdenum-alloy steel, and stainless-steel

piping.   This process is generally applicable to longitudinal seam welding of rolled plate piping, circumferential welds on piping 8 in. in diameter and larger, and straight fillet welds of fabricated plate attachments.   In addition, semiautomatic submerged-arc welding is also being used to some extent in welding intersection joints.   Shop facilities include 400- to 1,200-amp dc generators or ac transformers, work positioners, and, for the fully automatic method, equipment for supporting and positioning the welding head.   In some applications, backing rings are used for groove welds.   Where fit-up is poor, the first weld pass on the backing ring should be made by manual shielded metal-arc welding or by the inert-gas tungsten-arc-welding process.   Sometimes it may be necessary to deposit two or three weld passes by shielded metal-arc welding over the inert-gas tungsten-arc root pass before the submerged-arc welding is commenced in order to avoid excessive penetration by the submerged arc.   This is particularly true when the land at the root is less than $\frac{1}{16}$ in. in thickness.   Accurate preparation and good fit-up of the joints are essential since the process allows little, if any, flexibility to compensate for these types of irregularities.

On heavy-wall pipe butt welds, the interpass temperature tends to increase.   On alloy piping, this may necessitate air or water cooling or intermittent welding to avoid too high an interpass temperature.

In the welding of alloy piping, the alloying elements may be in the filler wire or in the flux.   For the greatest uniformity, they should be in the filler wire.   Where alloyed submerged-arc welding fluxes are used, increases in either the interpass temperature or the arc voltage may cause excessive alloy pick-up in the weld.

Oxyacetylene Welding

Pipe-welding applications which involve manual oxyacetylene welding are decreasing, although some 25 years ago, this process was extensively used for welding most of the ferrous and nonferrous metals then involved in piping systems.   Its use now is generally confined to small-sized steel piping.   Because of this limited usage, it is frequently difficult to obtain qualified pipe welders, particularly for field erection. The equipment required for oxyacetylene pipe welding is less expensive and more portable than that required for the shielded metal-arc-welding process.   Consequently, it provides an inexpensive means for welding piping systems in instances where the lack of electric power facilities or the cost of electric power-welding equipment precludes the use of arc welding.

As in the case of manual shielded metal-arc welding of piping, oxyacetylene-welding techniques may differ in some details.   However, the techniques which relate to joint design, number of passes, direction of welding, and cleaning are similar.   The number of passes required for welding joints in ferrous piping varies with the wall thickness of the pipe, the position of the pipe when welded, and the size of the welding rod used.

While the methods of depositing weld metal for the various pipe positions are similar to those employed in manual shielded metal-arc welding, the thickness of the deposited weld layers is usually greater.   For piping that can be rolled or when the work is in a horizontal position (using the backhand technique), one pass is usually used for wall thicknesses up to $\frac{3}{8}$ in., two for wall thicknesses $\frac{3}{8}$ to $\frac{5}{8}$ in., three for wall thicknesses $\frac{5}{8}$ to $\frac{7}{8}$ in., and four for wall thicknesses $\frac{7}{8}$ to $1\frac{1}{8}$ in.   With the forehand technique, the weld usually is deposited in a single pass.   With the pipe in the vertical fixed position, the number of passes varies widely.   However, for most applications, determination of the number of passes by the rule of one pass for each $\frac{1}{8}$ to $\frac{3}{16}$ in. thickness of pipe wall is considered good practice.

With regard to direction of welding, the backhand technique in oxyacetylene

welding usually involves starting at the top of the pipe and finishing at the bottom, whereas the forehand technique involves starting at the bottom of the pipe and working upward to the top. To obtain adequate ductility, as measured by the guided bend test, it is usually necessary to "torch anneal," or normalize, oxyacetylene welds.

Brazing, Braze Welding, and Soft Soldering

Brazing fittings are widely used. They are usually provided with a simple bore of correct diameter and depth. Brazing generally requires feeding of the alloy from a hand-held wire or strip when the filler alloy is not preplaced in a groove machined into the end of the fitting. With brass and bronze fittings, the copper-silver-phosphorus or the copper-phosphorus types of brazing alloy are generally used. For malleable fittings with ferrous piping, the brazing alloy should be of the silver-copper-cadmium-zinc type. Brass filler metal (60% Cu–40% Zn) of the type commonly employed in braze welding may also be used with ferrous fittings and pipe.

Cleanliness of the brazing area is an important consideration in brazing pipe joints. Thorough cleaning with steel wool, sandpaper, emery cloth, or a power-wire brush of the surfaces to be brazed is essential. Subsequently, the brazing flux suitable for the brazing alloy to be employed must be applied to both the fitting and the pipe in the braze area. After assembling the cleaned and fluxed pipe and fitting, heating may be accomplished with an oxyacetylene flame or other heat source that will provide the necessary heat and base-metal temperature for effecting the melting and flowing of the brazing alloy.

LAYOUT, ASSEMBLY, AND FIT-UP

Proper layout, assembly, and fit-up are essential for pipe welding of a quality which will assure acceptance under any specifications.

Joint Preparation. The end preparation of parts to be welded should conform to the joint design set up in the applicable welding-procedure specification. Cutting should be done with some form of machine tool. Where standard "V"-bevel preparations are applicable in wall thicknesses below $\frac{3}{4}$ in., oxyacetylene cutting may be employed. Mechanically guided torches traveling around the pipe or stationary torches with the pipe being rotated uniformly, are recommended. Oxyacetylene beveling is limited to metals which can be so cut and are not adversely affected by the temperature involved. On pipe of light wall thicknesses, cutting is also done with high-speed air or electric motor-driven saws with cut-off disks mounted in place of the saw blades. Stainless-steel piping, particularly in light wall thicknesses, is also cut with special inert-gas tungsten-arc torches.

All welding faces and adjoining surfaces for a distance of at least $\frac{1}{4}$ in. from the edge of the welding groove or from the toe of the fillet in the case of socket-welded or fillet-welded joints should be thoroughly cleaned of rust, scale, paint, oil, or grease. In addition, any contaminating or surface-coating material, which may reach the weld area as a result of weld heat, should be removed. Oxygen-cut surfaces should be ground reasonably smooth to remove all traces of scale and any irregularities incident to cutting. Depending on the metal being welded, chemical treatment may be necessary to clean the parts thoroughly and to avoid contamination during the welding. This is generally desirable on aluminum and titanium piping materials.

Layout and Assembly. Layout and assembly involve the fitting together of the various component parts comprising a subassembly or a complete piping

system in preparation for welding. Dimensional allowances may have to be made for shrinkage during welding, particularly on such materials as austenitic stainless steels which have high coefficients of expansion. The parts to be joined should be carefully spaced, aligned, and tack-welded together so that the final welded assembly will conform to the required dimensions within reasonable tolerances. Normally, in shop fabrication, an end-to-end tolerance of $\pm\frac{1}{8}$ in. is considered the maximum that is acceptable. However, more rigid tolerances may sometimes apply to specific piping components.

In making up subassemblies, the usual procedure is to set up the largest component, either on adjustable support "horses" or on a level-top "layout" table, with its longitudinal axis in a horizontal plane. The longitudinal axis and one end of the member are then used as base lines to which the locating dimensions and setting of the smaller parts can be referred, using a rule, steel tape, hand level, squares, straightedge, or bevel protractor as required.

In the layout and assembly of complete piping systems in place, the same general procedure applies. The largest horizontal components are erected, and then these members are used as reference points for fitting and assembling the smaller components. In such systems, the components, such as pipe, valves, fittings, or prefabricated assemblies, must be carefully aligned with respect to other structures, other equipment, and each other. Where thermal expansion, contraction, or other movement is not involved, the aligning operations are relatively simple. Where such movements are a consideration, care must be exercised in both the placement and alignment to make sure that the piping system, after it is welded and subjected to service conditions, is located consistent with the design and erection drawings.

In assembling butt-type joints, the use of accurately fitting backing rings facilitates spacing and alignment. Where backing rings are not used, the ends of the components to be welded should be carefully aligned and spaced preparatory to tack welding which, in open joints, may be facilitated with a spacing gage. Poor fit-up is a major cause of weld defects and service failures.

Intersecting joints should be carefully laid out with templates, standard layout curves, or jigs. After the parts of the joint have been accurately cut and before they are beveled, it is advisable to make a trial fitting of the joint. This will permit a visual gaging of the corrections that may have to be made for root spacing and alignment and will reveal any irregularities in the cutting of the parts that may require attention.

If the parts to be welded can be handled manually, the alignment, spacing, and, in many cases, assembly can be done without the use of external holding devices. In setting up heavy sections, the use of external clamps or other means of holding the parts in correct relationship to each other is essential for maintaining alignment and facilitating handling.

In production welding, where a number of identical assemblies are involved, as when welding flanges to pipe, the use of special set-up or alignment fixtures is of considerable advantage (Fig. 83). The design of such fixtures will vary, but the essential requirement is to provide a means for quickly setting component parts in proper relation to each other, ready for tack welding, without requiring individual clamps for each joint.

The dimensional allowance in the layout and assembly of piping systems to compensate for shrinkage during welding is established principally by experience. Many variables affect shrinkage, and it cannot be determined by calculation alone. Size of electrode, type and size of bevel opening, welding process and procedure, interpass temperature, and between-pass grinding affect shrinkage. Repair welding in local areas may intensify shrinkage and distortion.

In the usual types of butt-welded joints in carbon-steel or low-alloy steel piping of Schedule 40 and Schedule 80 thicknesses, experience indicates that the longitudinal shrinkage in each joint will equal about one-half the root spacing after tack welding.   Diametral shrinkage in carbon steel is negligible and may be disregarded. Austenitic stainless steels experience about twice the shrinkage and may also suffer from substantial diametral shrinkage.   In socket-welded or fillet-welded joints, both transverse and longitudinal shrinkage are negligible and generally are not considered in the layout and assembly operations.

Shrinkage in the welding of intersecting joints results principally in angular distortion of the branch pipe and bowing of the main pipe.   In these cases, it is virtually impossible to establish even approximate shrinkage values that would hold for all cases.   The steps taken to control shrinkage are quite as important as the making of dimensional allowances for such shrinkage in the layout and assembly operations.

In the training of welders, it is quite important to emphasize the factors affecting shrinkage.   As they become experienced in applying various welding procedures, they will recognize these factors and approximate their effects in terms of dimensional allowances during layout and assembly operations.   Not the least among these factors is the initial fit-up of the joint.   The necessity for careful and accurate preparation of the parts also cannot be overstressed in any training program.

Fig. 83. Set-up of slip-on flange with pipe-flange aligner.

Tack Welding. After the joint is properly lined up, short tack welds are made in the joint prior to the actual welding.   These welds should be sufficient in number and of suitable proportions to hold the part in place during ordinary handling. They need not be any larger than one-half the thickness of the pipe wall nor longer than twice the thickness.   Usually, on piping up to and including 4-in. nominal pipe size, two or three $\frac{1}{4}$- to $\frac{1}{2}$-in.-long tack welds should be made at locations that are about equally spaced around the pipe.   On larger pipe sizes, tack welds should be made at 4- to 6-in. intervals.   They should be of a quality equal to that specified by the welding procedure and should be made by a qualified pipe-welding operator.

For shielded metal-arc welding with covered electrodes, the tack welds should be fused through the full thickness of the land.   Where tack welding is done by the inert-gas tungsten-arc process, "skin" tacking is usually preferred without fusing through the land to the pipe inside.   Tack welds made by shielded metal-arc welding should be ground prior to welding to remove slag and possible crater cracks. Grinding may not be necessary on tack welds made by inert-gas tungsten-arc welding.

Where consumable insert rings are used, spot tack welding of the insert rings to the pipe with the smallest inside diameter in two or more equidistant places is sometimes desirable before assembly of the joint.   One method is to strike the arc on a small copper plate placed inside the pipe near the bevel and to "touch" the arc to the edge formed between the ring and the inside of the pipe.   The two pipe ends

are then brought together and tack-welded.   The tack welds should not be fused through the full thickness of the insert ring and land.   Before breaking the arc, the tungsten electrode should be manipulated slowly toward and up the side of the bevel.   This will help to avoid weld craters and cracks.   The use of a superimposed high-frequency current for arc starts and a foot-operated rheostat for variation of welding current also improves control.

Frequently, in setting up joints in steel pipe of heavy wall thickness, the need for tack welds in the joints is eliminated by the use of heavy steel bridge "C" clamps.

WELDING OF FERROUS PIPING MATERIALS

The term *weldability* is used to indicate the relative ease or difficulty with which a material can be welded.   Where specific welding conditions or other factors affect weldability adversely, greater precautions must be exercised.

In the welding of ferrous piping, the following factors should also be considered:

1. The tendency, if any, of the base or weld metals to crack during the welding operation or while cooling after welding

2. A change in mechanical properties caused by effects of the welding operation on base or weld metal

3. A marked change in the metallurgical structure of the base metal as a result of the welding operation

4. Changes in chemical composition by volatilization of alloying components, by reactions with air-oxidizing alloying elements, or by reactions with shielding or purging gases, electrode coverings, or slags

5. The quality of the deposited weld metal as affected by dilution, enrichment or flaws, or nonmetallic inclusion in the base metal

6. The formation of refractory coatings which may require special fluxes or shielding gases

Carbon-steel materials in the carbon ranges of less than 0.30% C and with wall thicknesses of less than $\frac{3}{4}$ in. generally do not crack during or after welding nor do significant changes take place in their physical properties or metallurgical structure. When the component parts have a thickness of $\frac{3}{4}$ in. or greater, additional precautions are required to produce welds free from cracks because of the mass of metal and the greater rigidity of the parts.

To reduce the adverse effects of the factors which contribute to cracking, preheating is considered advisable, particularly for steels whose carbon content exceeds 0.30 per cent.   Preheating is also considered advisable on most hardenable alloy steels.

Stress relieving is deemed necessary for most alloy steels and for carbon-steel piping of wall thicknesses of $\frac{3}{4}$ in. and over.   The principal beneficial effects of the stress-relief postheat treatment are reduced stresses in the weld area and a more ductile and uniform metallurgical structure.

Recommendations which represent commercial welding practice for preheat and stress-relieving cycles are given in the subsequent paragraphs.   Occasionally, specific job conditions may require deviations from them, and qualified welding engineers should be consulted where such changes are deemed necessary.

The effects of the preheat and postweld heat treatment may also vary with the heating methods and equipment employed.   The different heating methods used with pipe fabrication and erection are discussed in other sections of this chapter.

Carbon-steel-welding Processes

Although a number of semiautomatic and automatic welding processes are being used successfully on carbon-steel piping, the largest tonnage of commercial piping is still being welded by shielded metal-arc welding with covered electrodes.   EXX10

electrodes are widely used, although in recent years increasing use is being made of the EXX15, EXX16, and EXX18 low-hydrogen and low-hydrogen iron-powder types of electrodes.   The higher-strength carbon-steel piping grades of the 70,000-psi minimum tensile classifications generally are welded with E7010, E7015, E7016, or E7018 electrodes.

Welds in critical piping which requires full penetration are made with split or solid backing rings or, to an increasing degree, are made by welding the first pass by the inert-gas tungsten-arc root-pass welding process.   Since cracking, porosity, and bubbling tend to occur when carbon-steel piping is welded by the inert-gas tungsten-arc-welding process, consumable insert rings alloyed with deoxidizers are preferred to insure soundness of the root welds, to improve the inside contour of the underside of the weld, and to minimize the tendency toward cracking.

The gas-shielded consumable metal-arc-welding processes are also used on carbon-steel piping, particularly where full root penetration is not essential.   Some problems in proper root penetration are frequently encountered in commercial shop welding, particularly at the tack welds.   Shielding is normally done with gas mixtures of carbon dioxide and argon.

The use of oxyacetylene in the welding of piping continues in some phases of the small-process piping field and in specialized service applications.   Submerged-arc welding is used extensively in shop fabrication whenever it is practical to rotate piping in the larger diameters and heavier wall thicknesses—usually beginning with the 8-in. nominal pipe sizes.   Where backing rings are used and fit-up is good, the first pass may be made by submerged-arc welding.   Otherwise, the first pass should be made by shielded metal-arc welding with covered electrodes, by inert-gas tungsten-arc root-pass welding, or by shielded-gas consumable metal-arc welding.   Where carbon-steel grades to the 70,000-psi minimum tensile requirements are used, filler wires and/or fluxes should be used which produce weld deposits in the stress-relieved condition that match the tensile properties of the pipe base metal.

Heat Treatment.   Preheating of carbon-steel piping materials with carbon contents of less than 0.30% C is generally not considered necessary.   However, it is considered advisable by the ASA Code for Pressure Piping that carbon-steel materials having a tensile strength of over 70,000 psi be preheated to 250 to 400 F prior to welding.

Stress relieving generally is not considered necessary unless the wall thickness is $\frac{3}{4}$ in. or over.   This is required in the ASA Code for Pressure Piping and in the ASME Boiler and Pressure Vessel Code, Section I, Power Boilers.   Section VIII, Unfired Pressure Vessels, of the ASME Boiler and Pressure Vessel Code requires stress relieving when:

a. The wall thickness exceeds $1\frac{3}{4}$ in.

b. The material has been preheated to above 200 F, and the wall thickness exceeds $1\frac{1}{2}$ in.

c. Postweld heat treatment is not required for fillet welds attaching nonpressure parts to pressure parts that have a throat thickness of $\frac{1}{2}$ in. or less.

In the ASA Code for Pressure Piping, ASA B31.8, Gas Transmission and Distribution Systems, stress relieving is required when the wall thickness exceeds $1\frac{1}{2}$ in.   The stress-relieving cycle should consist of heating to 1100 to 1250 F and holding for 1 hr/in. of wall thickness with a minimum holding period of one-half hour.   Cooling in still air is normally considered adequate.

Wrought-iron-welding Processes

The carbon content of wrought iron is generally below 0.12 per cent, and its welding characteristics are similar to those of mild steel.   As little as possible of the

wrought iron should be melted by the welding operation in order to minimize pickup of the silicate slag by the weld deposit. Lower welding speeds and currents are used for proper welding to avoid porosity in the weld deposits. Preheating and postheating are not usually necessary.

All of the welding processes applicable to carbon steel are suitable for welding wrought-iron piping. Most extensively used is the shielded metal-arc-welding process. However, the oxyacetylene-welding process is still used in some field welding. Weld surfaces made by oxyacetylene welding may have a "greasy" appearance, caused by melting of the silicate slag.

Those welding processes which tend to melt considerable base metal, such as submerged-arc welding and shielded-gas consumable metal-arc welding with carbon dioxide shielding gas should employ at least two weld layers to insure sound pipe welds. The initial weld pass against the wrought iron may contain porosity produced by the silicate slag from the base metal. Upon the re-fusing of part of the somewhat porous initial weld pass, the soundness of the pipe weld is improved.

Welding Processes: Carbon-molybdenum Steels

In steam power plants, relatively little use is now being made in piping systems of carbon-molybdenum steels. Unfavorable experience with these steels in applications involving service temperatures exceeding 800 F have been attributed to graphitization of the steel. The graphite, which has formed in the shape of nodules or flakes in fine-grained steels, reduces very substantially the toughness of the steel, particularly in the heat-affected zone.

In merchant shipping, substantial use is still made of carbon-molybdenum steel piping for main steam-piping systems up to 875 F. Coarse-grained steels are used, and after welding, piping assemblies are normalized by heating for 4 hr at 1300 F. There has been no evidence of failures due to graphite formation in these piping systems.

The same welding processes discussed for carbon steels are applicable to the welding of carbon-molybdenum steels. Shielded metal-arc welding should be done with E7010-A1, E7016-A1, or E7018-A1 electrodes. Submerged-arc welding is done with either carbon-molybdenum steel filler wires and neutral fluxes or carbon-steel filler wires and fluxes containing molybdenum. Either method produces satisfactory weld deposits.

Heat Treatment. Preheating of carbon-molybdenum steels is considered advisable when the wall thickness is $\frac{1}{2}$ in. and greater. Temperatures between 300 and 500 F are normally employed.

Piping welded to the requirements of the ASA Code for Pressure Piping and the ASME Boiler and Pressure Vessel Code, Section I, must be stress-relieved if the wall thickness is $\frac{1}{2}$ in. or over or if the carbon content is over 0.25 per cent. Under Section VIII of the ASME Boiler and Pressure Vessel Code (Unfired Pressure Vessels), stress relieving is required:

A. On welded pipe, tubing, forged and cast flanges, fittings, and valves with wall thickness $1\frac{1}{4}$ in. or over.

B. On welded plate with wall thickness 0.58 in. and over.

C. On fillet welds when the weld leg dimension is 0.58 in. and over.

The stress-relieving cycle consists of heating to 1100 to 1200 F and holding for 1 hr/in. of wall thickness with a minimum of one-half hour followed by cooling in still air. Where the service temperature exceeds 800 F, the weld joints are stress relieved for 4 hr at 1300 to 1325 F. This heat treatment is often referred to as "metallurgical stabilization" and tends to suppress potential graphite formation.

Welding Processes: Chromium-molybdenum Alloy Steels

The chromium-molybdenum alloy steels represent an extensively used group of piping materials. These grades are almost exclusively used for service conditions in the 800 to 1060 F temperature range. The low-alloy group consists of the grades from $\frac{1}{2}$% Cr–$\frac{1}{2}$% Mo to 3% Cr–1% Mo, and the medium-alloy group is composed of the 5% Cr–$\frac{1}{2}$% Mo to the 9% Cr–1% Mo grades. Welding recommendations for chromium-molybdenum pipe materials are given in Table 66.

Critical pipe butt welds requiring full penetration weld deposits generally are made either by using backing rings or by welding the first pass with the inert-gas tungsten-arc-welding process. The weld is then completed, usually by shielded metal-arc welding or by submerged-arc welding. Shielded-gas consumable metal-arc welding is not widely used on the chromium-molybdenum piping steels.

Shielded metal-arc welding should be done with low-hydrogen type electrodes of the EXX15, EXX16, or EXX18 classification. In the low-alloy steel electrodes, the alloying elements generally are added to the covering which, in most commercial electrodes, does not affect adversely the quality or properties of the resulting weld deposit.

Submerged-arc welding is best done with chromium-molybdenum-alloy steel filler wires of compositions corresponding to the base metals to be welded, using essentially neutral fluxes. Although acceptable weld deposits may be obtained by submerged-arc welding with carbon-steel filler wires and with fluxes containing chromium and molybdenum, this practice may result in variable alloy contents in the finished welds since a build-up in the interpass temperature or an increase in the arc voltage tends to upset the alloy balance of the flux, and the chromium and molybdenum content of the weld deposit may be increased to more than twice that of the piping base material.

Heat Treatment. Preheating to the temperatures shown in Table 66 is considered essential. The higher temperatures in the preheating temperature range should be used for the higher alloy compositions, heavier wall thicknesses, and more complex structures.

Welding on the $\frac{1}{2}$% Cr–$\frac{1}{2}$% Mo to 3% Cr–1% Mo piping materials may be interrupted at any time, provided that the following conditions are maintained:[1]

1. A minimum of at least $\frac{3}{8}$ in. thickness of weld metal has been deposited or 25 per cent of the welding groove is filled (whichever is the greater), and
2. The weld is allowed to cool slowly from welding temperature to room temperature. A suggested manner of retarding cooling is to wrap the weld with asbestos and to allow the joint to cool in still air.

Welding on the 5% Cr–$\frac{1}{2}$% Mo to 9% Cr–1% Mo piping material may be interrupted, provided that:[2]

1. The recommended preheat temperature is maintained until welding is resumed, or
2. The partially completed weld is immediately stress relieved at a minimum temperature of 1200 F (but below 1425 F) for at least 30 min.

In every instance, the weld area must be at preheat temperature before welding is resumed, and stress relieving, in accordance with Table 66, generally is considered essential.

On small-diameter light-wall piping in the $\frac{1}{2}$% Cr–$\frac{1}{2}$% Mo to $2\frac{1}{4}$% Cr–1% Mo grades, stress relieving may be omitted, in accordance with the ASME Boiler and

[1] Pipe Fabrication Institute ES-8.
[2] Pipe Fabrication Institute ES-12.

Table 66. Welding Recommendations for Carbon-molybdenum and Chromium-molybdenum-alloy Steels

Steel	ASTM specification[a]		Welding recommendations		
	No.	Grade	Electrodes[b]	Preheat and interpass temperature,[c] F	Postheat treatment,[d] F
½ Mo	A155 A335	Several P1	E70XX-A1; E7016-A1; etc.[e]	300–400[f] (over ½" thick)	1,150–1,250 (over ½" thick)
½ Cr–½ Mo	A155 A387 A335	½ Cr A P2	E70XX-B1; E80XX-B1; 0	300–500[f]	1,275–1,350
1 Cr–½ Mo	A155 A213 A387 A335	1 Cr T12 B P12	E8015-B2; E9015-B2	300–500	1,300–1,350
1¼ Cr–½ Mo	A155 A199 A200 A213 A335	1¼ Cr T11 T11 T11 P11	E8015-B2; E9015-B2	300–500	1,300–1,350
1¾ Cr–¾ Mo	A335	P3	E8015-B3; E9015-B3	300–500	1,300–1,375
2 Cr–½ Mo	A199 A200 A213	T3b T3b T3b	E8015-B3; E9015-B3	400–600	1,300–1,400
2¼ Cr–1 Mo	A155 A199 A200 A213 A335	2¼ Cr T22 T22 T22 P22	E8015-B3; E9015-B3; etc[e]	400–600	1,325–1,400[g]
2½ Cr–½ Mo	A199 A200	T4 T4	E8015-B3; E9015-B3; etc.[e]	400–600	1,325–1,400[g]
3 Cr–1 Mo	A199 A200 A213 A335	T21 T21 T21 P21	5 Cr–½ Mo (E502)	400–600	1,350–1,400[g]
5 Cr–½ Mo	A155 A199 A200 A213 A335 A357	5 Cr T5 T5 T5 P5	5 Cr–½ Mo (E502)	400–700	1,350–1,425[c]
7 Cr–½ Mo	A199 A200 A213 A335	T7 T7 T7 P7	7 Cr–½ Mo; 9 Cr–1 Mo	400–700	1,350–1,425[g]
9 Cr–1 Mo	A199 A200 A213 A335	T9 T9 T9 P9	9 Cr–1 Mo; 12 Cr (E410)	400–700	1,350–1,425[g]

[a] This listing includes only the more important plate and pipe specifications. Several additional specifications also cover steels of these same compositions.

[b] Suffixes -B1, -B2, and -B3 refer to AWS-ASTM subclassifications specifying electrodes containing ½ Cr–½ Mo, 1¼ Cr–½ Mo, and 2¼ Cr–1 Mo, respectively.

[c] Use lowest temperature for sections up to ½ in. thick and highest values for sections over 2 in. thick, with intermediate temperatures for intermediate thicknesses.

[d] For steels to 3 per cent Cr, 1 hr/in. of wall thickness, 1 hr minimum; for steels over 3 per cent Cr, 2 hr/in. of wall thickness, 2 hr minimum. Longer holding periods may be necessary when low-hardness values are desired.

[e] Also the corresponding EXX16 and EXX18 low-hydrogen electrodes.

[f] May be necessary only to prevent cracks in root passes.

[g] After welding, the weld should be allowed to cool to below 600 F before the recommended postheat treatment is applied.

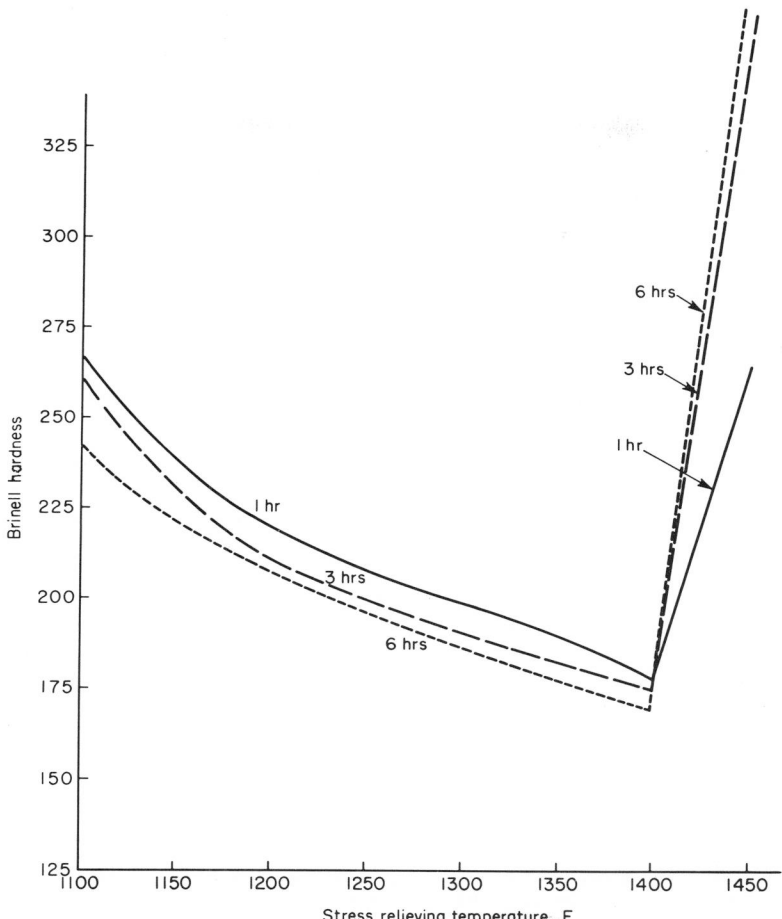

FIG. 84. Brinell hardness of 5 Cr–½ Mo base metal air-cooled from 1750 F and stress-relieved at temperatures between 1100 and 1450 F for 1, 3, and 6 hr.

Pressure Vessel Code, the ASA Code for Pressure Piping, and PFI Standard ES-8, provided that the outside pipe diameter is not over 4 in. and that the pipe wall thickness is less than ½ in.   Procedure qualifications are required to demonstrate acceptable welds without stress relieving.   Postheat treatment, however, may be advisable on piping materials when these may become susceptible to stress corrosion cracking as a result of subsequent acid cleaning or service environments.

The medium chromium-molybdenum alloy steels should always be stress relieved, and it is desirable to specify a maximum Brinell hardness of 240 to insure that the postheat treatment was properly made.   Typical average hardness values in commercial 5 % Cr–½ % Mo pipe base-metal grades and weld deposits are shown in Figs. 84 and 85.   Improperly heat-treated joints, resulting in excessive weld-joint hardness or differences in hardness between weld and base metal, may fail as a result of thermal or mechanical fatigue.

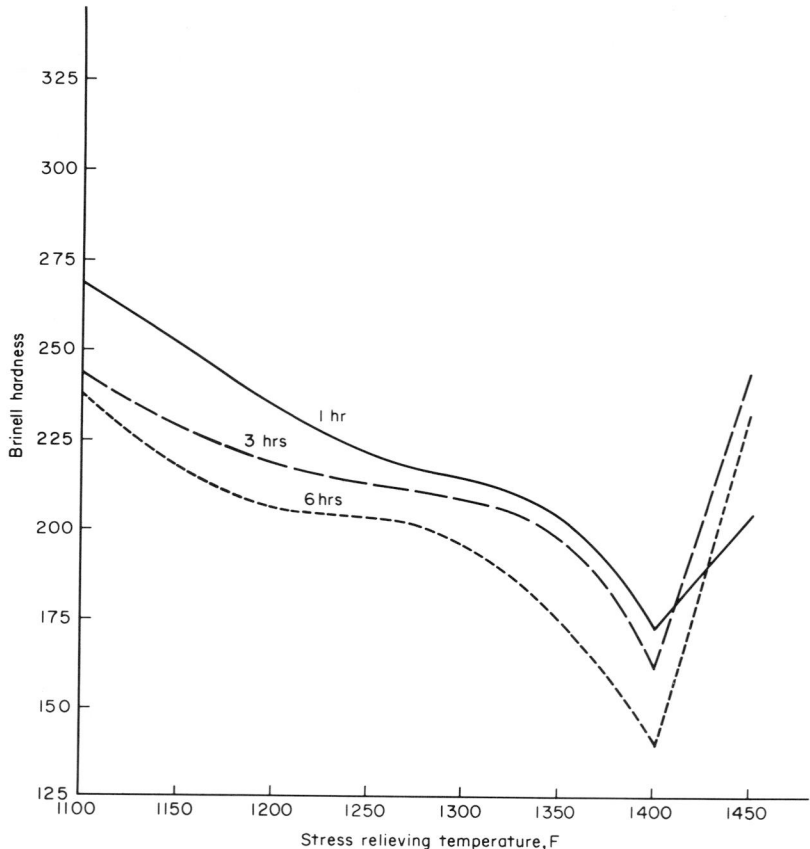

FIG. 85. Brinell hardness of 5 Cr–½ Mo weld-metal deposit preheated to 400 F and stress-relieved at temperatures between 1100 and 1450 F for 1, 3, and 6 hr.

Some codes specify permissible maximum hardness values. As an example, the requirements of the ASA Code for Pressure Piping ASA B31.3, Petroleum Refinery Piping, are listed in Table 67.   Other codes, and even other sections of the ASA Code for Pressure Piping, do not necessarily agree with Table 67.

When thermal or mechanical fatigue is severe, it is desirable to limit the hardness to 200 Brinell.   A weld hardness of 240 Brinell after stress relieving may correspond to a tensile strength of 115,000 psi.   Where the adjacent base metal has a hardness of only 180 to 200 Brinell, corresponding to a tensile strength of about 80,000 psi, a metallurgical notch of sufficient magnitude may be created which may propagate into a failure under severe fatigue.

Sometimes commercial heats of the 5 % Cr–½ % Mo to 9 % Cr–1 % Mo alloy steels do not respond to the 1350 to 1400 F stress-relieving heat treatment, and they exhibit hardness values over 240 Brinell.   On these, a double heat treatment involving normalizing and tempering, or annealing, must be employed.   This is particularly the case when a high preheat temperature prevents the austenite-to-marstensite transformation during welding.

Table 67. Postheat Treatments and Brinell-hardness Requirements for Petroleum Piping (ASA Code for Pressure Piping, Section B31.3-1962, Petroleum Refinery Piping)

P no.[3]	Min preheat temperature, deg F	Postheat treatment			Brinell hardness, max[1]
		Metal temperature, deg F	Soaking period[2]		
			Hr/in. of wall thickness, min	Time at temperature, hr, min	
1	Ambient[4]	1100–1200	1	$3/4$	...
2	Ambient[4]	None	...	...	...
3	175	1100–1350	1	1	215
4	300	1325–1375	1	2	215
5	350	1325–1375	1	2	241
6	400	1400–1450	1	2	241
7	Ambient[4]	None	...	...	...
8	Ambient[4]	None	...	...	...
9	300	1100–1200	$1/2$	1	...
10	300	1050–1100	$1/2$	1	...

[1] These hardness limits apply also to the heat affected zone of the welds. Any hardness test of the heat affected zone shall be made at a point as near as practicable to the edge of the weld.

[2] The piping being heat treated shall be maintained within the temperature limits shown for a time not less than the maximum of the two periods shown.

[3] See Appendix D of ASA 31.3 for definition of P numbers.

[4] See Para. 331.2.1 of ASA 31.3 for preheat required if ambient temperature is less than 32 F.

Welding Processes: Low-temperature Steels

For low-temperature-service piping, nickel-alloy steel, chromium-copper-nickel alloy steel, and several grades of fully deoxidized carbon steels and austenitic stainless steels are used. The lowest temperatures considered safe are:

Temperature, min, F	Type of steel
−50	Fully deoxidized steel
−75	$2\frac{1}{4}\%$ Ni steel
−150	$3\frac{1}{2}\%$ Ni steel, Cr–Cu–Ni steel
−320	9% Ni steel, austenitic stainless steel
−425	Austenitic stainless steel

Welding recommendations for these steels are given in Table 68 for the carbon- and nickel-bearing steels and in Table 69 for the austenitic stainless steels.

The $2\frac{1}{2}$ and $3\frac{1}{2}$ per cent nickel steels are the most widely used grades for service temperatures down to −150 F. Although more economical, the Cr-Cu-Ni steels are somewhat more difficult to weld than the $2\frac{1}{2}$ and $3\frac{1}{2}$ per cent nickel steels, and they may become sufficiently notch sensitive in the heat-affected zone so that the steel in this area would not meet the prescribed Charpy keyhole notch-impact test requirements between −100 and −150 F.

Nickel increases the air-hardening characteristics of steel significantly, so that preheating is necessary in the $2\frac{1}{2}$ and $3\frac{1}{2}$ per cent nickel steels when the carbon content exceeds 0.15 per cent.

The 9 per cent nickel steels are usually welded with the Inconel filler metals. They should be preheated to between 400 and 600 F. Stress relieving between 1050 and 1150 F for 1 hr/in. of wall thickness is essential. In automatic welding, it is essential to avoid excessive heat input.

Table 68.   Welding Recommendations for Mild-pipe Steels, Nickel Steels, and Chromium-copper-nickel Steels for Low-temperature Service

Steel	ASTM Spec.*		Electrode†	Preheat and interpass temperature,‡ F	Postheat treatment,§ F
	No.	Grade			
Deoxidized mild steel	A201 A212 A106	....	E6015;¶ E7015¶	70 min	1,175–1,250
2½ Ni	A203	A, B	E8015-C1¶	300–500	1,175–1,250 (over ½″ wall)
3½ Ni	A203 A333 A334	D, E 3 3	E8015-C2¶	400–500	1,175–1,250 (over ½″ wall)
9 Ni	A353	.....	E310 or Inconel 182	400–500	1,175–1,250
Cr–Cu–Ni	A410	.....	E8015-C1¶	300–500	1,000–1,100

* This listing includes only the more important plate and pipe specifications.   Several additional specifications also cover steels of these same classifications.

† Suffixes -C1 and -C2 refer to ASTM-AWS subclassifications specifying electrodes containing 2½% Ni and 3½% Ni, respectively.

‡ Use lowest temperature for sections up to ½ in. thick and highest values for sections over 2 in. thick, with intermediate thicknesses prorated.

§ 1 hr/in. of wall thickness, ½ hr minimum.

¶ Also, the corresponding EXX16 and EXX18 low-hydrogen electrodes.

Welding Processes: Austenitic Stainless Steels

Almost all of the major commercial stainless-steel alloys and grades and a number of special alloys are available as piping.   These include Types 304, 304L, 316, 316L, 347, and 310.   Welding recommendations for the common austenitic stainless-steel grades are given in Table 69.

Numerous problems can arise when welding austenitic stainless-steel materials. Cracking occurs frequently in fully austenitic weld deposits.   Because of this, careful control of the composition of the welding filler metals is essential.   Some stainless-steel alloys are susceptible to cracking in the heat-affected zone.   Where high-temperature service is involved, alloy compositions should be selected which are not susceptible to gradual embrittlement by a metallurgical structure referred to as "sigma" phase.   Where continual service-application trouble or problems arise in welding, a competent metallurgical or welding engineer should be consulted.

On stainless-steel piping materials, the first (root) pass is generally made by the inert-gas tungsten-arc-welding process to insure good root penetration.   For critical service applications, the inside of the pipe should be effectively purged with argon, helium, or nitrogen to prevent oxidation of the underside of the weld during root-pass welding.   Thus, stainless-steel pipe for nuclear and power plant service, most chemical-plant applications, etc., is normally purged.   However, in paper mills where underweld cleanliness is frequently not considered a requirement, purging is more often eliminated unless the piping system is considered critical and subject to severe corrosive or cleanliness requirements.   Consumable insert rings are also considered essential where critical service applications are involved—particularly where excessive concavity "sink" at the underside of the weld is undesirable.

Table 69. Welding Recommendations for Wrought Austenitic Stainless-steel Pipe

AISI designation	Popular designation	Chemical composition,[1] %				Recommended electrode or welding rod[2]
		C (max)	Cr	Ni	Others	
304	19-9	0.08	18-20	8-12	....	E, ER308
304 L	19-9 (extra-low carbon)	0.03	18-20	8-12	....	E, ER308 L or E, ER347
309	25-12	0.20	22-24	12-15	....	E, ER309
309 Cb	25-12 Cb	0.20	22-24	12-15	Cb = 10 × C min	E309 Cb
310	25-20	0.25	24-26	19-22	....	E, ER310
310 Cb	25-20 Cb	0.25	24-26	19-22	Cb = 10 × C min	E, ER310 Cb
310 Mo	25-20 Mo	0.25	24-26	19-22	2.0-3.0 Mo	E, ER310 Mo
316	18-12 Mo	0.08	16-18	10-14	2.0-3.0 Mo	E, ER316
316 L	18-12 Mo (extra-low carbon)	0.03	16-18	10-14	2.0-3.0 Mo	E, ER316 L or E 318
317	19-13 Mo	0.08	18-20	11-15	3.0-4.0 Mo	E, ER317
318	18-12 Mo Cb	0.10	18-20	10-14	2.0-3.0 Mo; Cb = 10 × C min	E318
321	18-8 Ti	0.08	17-19	9-12	Ti = 5 × C min	E, ER347
347	18-8 Cb	0.08	17-19	9-13	Cb-Ta = 10 × C min	E, ER347
348	18-8 Cb	0.08	17-19	9-13	Cb-Ta = 10 × C min 0.10 max Ta	E, ER347

[1] Unless otherwise specified, Mn is 2.00 max; Si, 1.00 max; P, 0.040 max; and S, 0.030 max.
[2] AWS and ASTM Specifications. E means grade recognized by AWS-ASTM as covered electrode and ER as bare electrode and welding rod.

On pipe of wall thickness up to $\frac{3}{16}$ in., the weld, once started by inert-gas tungsten-arc welding, is normally completed by this same process.   On pipe of heavier wall thicknesses, the balance of the weld, after the root pass, is generally completed by shielded metal-arc welding with covered electrodes, by the inert-gas shielded consumable electrode, or by the submerged-arc-welding process.

When submerged-arc welding is done, careful selection of the flux and wire is very important.   Some combinations tend to result in excessive ferrite.   This is not desirable for service temperatures above 800 F.   Where this cannot be avoided, submerged-arc welding should not be done.

Heat Treatment.   Postheat treatments of welds are not normally done on austenitic stainless-steel piping.   Where service environments tend to be stress corrosive, a stress-relieving heat treatment at a temperature between 1550 and 1600 F for approximately 1 hr/in. of thickness is desirable.   Weld deposits from some Type 316L electrodes, however, may suffer a severe loss of impact toughness when heated in this temperature range.   In such instances, a partial stress relief at 1200 to 1250 F may provide sufficient protection against stress-corrosion failures, unless pretested heats of 316L electrodes were used which do not embrittle as a result of the 1550 to 1600 F heat treatment.

In service environments where intergranular corrosion of certain stainless-steel grades is a factor, an annealing heat treatment at 1950 to 2050 F may restore the grade to its proper level of corrosive resistance, provided that the heat treatment is followed by rapid cooling, as in a water quench.   This may, however, result in severe distortion.   In such instances, it may be more advisable to select the stabilized stainless-steel grades or the extra-low-carbon grades, which exhibit little susceptibility, if any, to intergranular corrosion.

Martensitic and Ferritic Chromium Stainless Steels

The chromium stainless steels (also called the "straight chromium stainless steels") are generally separated into two further subclassifications: (1) the martensitic stainless steels, and (2) the ferritic stainless steels.

Martensitic or hardenable stainless steels usually contain up to about 15 per cent chromium.   Although these grades may be made ferritic by a final heat treatment between 1200 and 1400 F, they are primarily martensitic when cooled in air or when quenched into a liquid medium from temperatures above 1500 F.   Maximum hardening in most of these grades is achieved by cooling from temperatures above 1750 F to room temperature.   These steels, therefore, might be compared in their response to hardening and tempering to the ordinary hardenable carbon and alloy steels with slight temperature modifications.

Ferritic Stainless Steels.   Ferritic stainless steels usually contain between 15 and 30 per cent chromium.   Heating and cooling produce no significant structural changes in these steels, so that they remain essentially ferritic at all temperatures.   These steels, therefore, can be hardened only by cold working, except for slight age-hardening characteristics shown by certain specially alloyed types.

Between 14 and 18 per cent chromium, the separation between the martensitic and ferritic stainless-steel classifications is approximate and depends upon the composition of the particular stainless grade, primarily the carbon content.   Thus, a high-carbon content results in a hard martensitic structure.   With a lower-carbon content, these steels would be primarily ferritic, as in Type 430.   The addition of aluminum also tends to make the steel ferritic even at a chromium content of 11.5 to 14.5 per cent, as in Type 405 which contains 0.10 to 0.30 per cent aluminum. On the other hand, 1 to 3 per cent nickel makes these steels essentially martensitic, as in Type 431, which contains 15.0 to 17.0 per cent chromium and up to 2.5 per cent nickel.

Table 70. Welding Recommendations for Martensitic Stainless Pipe Steels

ASTM designation	Popular designation	Chemical composition, %		Welding recommendations		
		C (max)	Cr	Recommended electrode or welding rod[1]	Preheat interpass temperature,[3] F	Postheat treatment[2,3]
410	12 Cr	0.15	11.5–13.5	E, ER410 E, ER310 or E, ER309	600–700 400–600	Recommended Recommended
410 mod	12 Cr	0.08	11.5–13.5	E, ER410 E, ER310 or E, ER309	300–500 300–500	Highly recommended Recommended
420	13 Cr	Over 0.15	12.0–14.0	E, ER410 or E, ER430 E, ER310 or E, ER309	600–700 400–600	Highly recommended Recommended

[1] AWS and ASTM specifications. E means grade recognized by AWS-ASTM as covered electrode and ER as bare electrode and welding rod.

[2] 1,300 to 1,450 F for 1 hr/in. of thickness.

[3] When preheat treatments are not employed, use small-diameter electrodes.

Welding of Martensitic Stainless Steels. Welding recommendations for the commercial martensitic stainless pipe steels are given in Table 70. Without preheat treatments, the highly hardenable, fully martensitic stainless-steel grades generally are susceptible to cracking in the martensitic weld deposit and in the heat-affected zone, particularly when heavy sections are being welded. Light sections of the lower-carbon grades may be an exception. For example, Type 410 in wall thickness up to $\frac{1}{8}$ in. ordinarily exhibits good welding characteristics, so that preheating provides little improvement and may be omitted.

Cracking may be minimized or avoided by preheating these steels to temperatures between 400 and 700 F, depending upon the hardening characteristics of the base metal and weld metal and the intended service requirements. The preheat temperatures should be maintained during the welding operation.

After welding and cooling of the welds to near room temperature, a postheat treatment between 1300 and 1450 F for 1 hr/in. of thickness should follow (minimum treatment of 1 hr). The postheat-treated sections should be cooled in air. Under these conditions, very ductile weldments may be obtained. In some cases, normalizing and tempering postheat treatments have been given with equally good success. Occasionally, the postheat treatment may be omitted, and a satisfactory weld will still result. Such procedures, however, should be specified only after consulting with a properly qualified and experienced welding engineer.

The presence of some ferrite in the otherwise martensitic structure decreases the hardness developed in the steel as, for example, in the modified Type 410 grade (0.10% maximum C). This ferrite reduces cracking susceptibility. Nevertheless, cooling rates and interpass temperatures should be controlled. Preheat treatment between 300 and 500 F is generally advisable, and it should be followed by postheat treatment at 1300 to 1450 F. Only when welding thin pipe material below about $\frac{3}{16}$ in. thickness may these preheat and postheat treatments be omitted.

When commercial Type 410 electrodes or welding rods are used to weld the Type 410 stainless steels, a preheat temperature between 500 and 700 F is advisable, unless the joint thickness is less than $\frac{1}{8}$ in. When low-carbon (modified) Type 410 grades

Table 71. Welding Recommendations for Ferritic Stainless Pipe Steels

ASTM designation	Popular designation	Chemical composition, %			Welding recommendations		
		C (max)	Cr	Other[1]	Recommended electrode or welding rod[2]	Preheat and interpass temperature, F	Postheat treatment[3]
				Wrought stainless steels (AISI)			
405	12 Cr, Al	0.08	11.5–14.5	0.10–0.30 Al	E, ER430 E, ER310 or E, ER309	Not necessary Not necessary	Highly recommended Recommended
430	16 Cr	0.12	14.0–18.0	....	E, ER430 E, ER310, or E, ER309	Not necessary Not necessary	Highly recommended Recommended
446	27Cr	0.20	23.7–27.0	0.25 max N	446 E, ER310 or E, ER309	300–400 Not necessary	Essential Recommended

[1] Unless otherwise specified, Mn and Si are 1.00 maximum, P is 0.040 maximum and S is 0.030 maximum.
[2] AWS and ASTM Specifications. E means grade recognized by AWS-ASTM as covered electrode and ER as bare electrode and welding rod.
[3] 1,300–1,450 F for 1 hr/in. of thickness.

(0.08 % maximum C) are welded with Type 410 electrodes, this preheat temperature may be reduced.

Welding with either Type 310 or Type 309 austenitic stainless-steel filler metals is preferred if a postheat treatment is not possible. Nevertheless, the use of Type 410 filler metals is far better metallurgically. Regardless of the filler metal used, the hardness in the heat-affected zone in the Type 410 base metal will be quite high, depending on the preheat temperature and cooling rates.

Welding of Ferritic Stainless Steels. Welding recommendations for the common commercial ferritic stainless pipe steels are summarized in Table 71.

Since the ferritic stainless steels are not subject to air hardening, they are less susceptible to cracking in the welded section than are the martensitic stainless steels. However, because these steels may become embrittled, their welding characteristics should be understood.

The chromium stainless steels which become fully ferritic at temperatures above 2100 F are generally susceptible to an embrittlement which is associated with solution of the carbide particles. This embrittlement is accompanied by severe grain growth. The embrittlement can be removed by annealing the steel for 1 hr between about 1300 and 1450 F and following this by quenching or air cooling, even though the grain size remains coarse. Such a postheat treatment is particularly important in single-pass welding where, without this postheat treatment, the ferritic weld metal, as well as part of the heat-affected zone, would be extremely brittle and would crack readily on subsequent deformation or bending operations at room temperature.

When postheat treatment is not possible, multipass welding with small-diameter electrodes, low current, and stringer beads should be used to minimize this embrittlement. In these deposits, the subsequent weld beads produce annealing effects in the earlier beads and reduce brittleness in the weld and heat-affected zone. Since steels of Types 405 and 430 tend to contain an average of about 50 to 70 per cent ferrite (the balance being martensite), suitable preheat and postheat treatments are usually required to prevent hardening.

In the welding of chromium stainless steels containing up to 23 per cent chromium, satisfactory results are generally obtained with electrodes and welding rods whose composition is similar to that of the base metal. Preheat treatments are preferred. A postheat treatment at 1300 to 1450 F is essential if ductility is important, unless service at temperatures above 1000 F produces a similar effect. Chromium stainless steels containing more than 23 per cent chromium are welded with Types 310 or 309 electrodes.

WELDING OF NONFERROUS PIPING MATERIALS

The weldability characteristics of ferrous piping materials generally is related primarily to their tendency to crack during the welding operation or during the subsequent cooling period and, to a lesser but highly important degree, to their state of internal residual stress as a result of welding. These characteristics are usually of minor importance with nonferrous materials. In general, welded nonferrous metals, unlike some steels, do not tend to harden on cooling in air nor do they have suppressed transformations, accompanied by appreciable shrinkage, at temperatures below the plastic range. Thus, even though the ductility of the deposited metal is lower than that of the base metal, there is little danger of cracking in most alloys. The residual stresses usually are not of sufficient magnitude to warrant concern over possible stress-corrosion effects, although some nonferrous alloys are recognized as being susceptible to stress corrosion cracking. The nonferrous piping materials do not experience the intergranular carbide precipitation which may be caused by

high welding temperatures in unstabilized austenitic stainless steels. Stress relief or other heat treatment after welding usually is not necessary, except as may be required by the ASME Boiler and Pressure Vessel Code or by other applicable codes. Certain precipitation-hardening alloys are an exception and may require specific heat treatments after welding to obtain specified mechanical properties.

Aluminum and Aluminum Alloys

Relatively few of the many recognized aluminum alloys are available commercially as piping materials. Aluminum and aluminum alloys from the standpoint of weldability are characterized chiefly by their low melting points (1075 to 1210 F), high thermal conductivity, high coefficients of expansion, and high fluidity in the molten state. Training and experience are necessary before production welding should be attempted.

Table 72. Welding Recommendations for Aluminum Piping Materials

Nominal pipe size, in.	Position	Preheat temperature, F	Current, amp	Electrode size, in.	Argon flow, cu ft/hr	Filler rod diameter, in.	Number of passes
1	H (rolled)	None	130	$3/32$	25	$1/8$	1
	H (fixed)	None	120	$3/32$	25	$1/8$	1
	V	None	125	$3/32$	25	$1/8$	1
$1\frac{1}{2}$	H (rolled)	None	140	$3/32$	25	$1/8$	1
	H (fixed)	None	125	$3/32$	25	$1/8$	1
	V	None	130	$3/32$	25	$1/8$	1
2	H (rolled)	None	150	$1/8$	25	$1/8$	2
	H (fixed)	None	130	$1/8$	25	$1/8$	1
	V	None	140	$1/8$	25	$1/8$	2
$2\frac{1}{2}$	H (rolled)	Up to 400	170	$1/8$	25	$1/8$	2
	H (fixed)	350–400	140	$1/8$	25	$1/8$	1
	V	Up to 400	150	$1/8$	25	$1/8$	2
3	H (rolled)	Up to 400	190	$5/32$	25	$5/32$	2
	H (fixed)	350–450	150	$5/32$	30	$5/32$	1
	V	Up to 400	170	$5/32$	25	$5/32$	2
4	H (rolled)	Up to 400	225	$5/32$	30	$3/16$	2
	H (fixed)	400–500	175	$5/32$	35	$5/32$	1
	V	Up to 400	200	$5/32$	30	$5/32$	3
6	H (rolled)	Up to 500	250	$3/16$	30	$3/16$	2
	H (fixed)	400–600	190	$3/16$	40	$3/16$	1–2
	V	Up to 500	220	$3/16$	35	$3/16$	3
8	H (rolled)	Up to 500	275	$3/16$	35	$3/16$	3
	H (fixed)	400–600	225	$3/16$	45	$3/16$	2
	V	400–600	250	$3/16$	40	$3/16$	4
10	H (rolled)	Up to 500	300	$1/4$	40	$3/16$	3
	H (fixed)	400–600	225	$3/16$	50	$3/16$	2–3
	V	400–600	275	$3/16$	45	$3/16$	5
12	H (rolled)	Up to 500	325	$1/4$	40	$3/16$	3
	H (fixed)	400–600	250	$3/16$	50	$3/16$	3
	V	400–600	300	$3/16$	45	$3/16$	5–6

Some aluminum-alloy piping and tubing materials are supplied in either a work-hardened or precipitation-hardened condition (obtained by cold-working or heat treatment, depending upon the alloy) which gives the material greater strength. In these alloys, the welding operation softens the metal for a distance of one to five times the pipe thickness on either side of the weld and reduces materially the tensile and yield strengths in this zone. It is impractical to restore the original strength of fabricated assemblies by subsequent cold work or heat treatment. In spite of this, it is often desirable to use cold-worked or heat-treated pipe to obtain increased resistance to denting or bending in handling and erecting. The bursting strength of pipe joined by circumferential butt welds is not reduced from that of unwelded pipe to the extent that might be expected based on the actual local strength of the joint after welding.

Welding Processes. While aluminum and aluminum alloys can be joined by almost all of the commercial welding and brazing processes, those most commonly employed are the inert-gas tungsten-arc and the inert-gas consumable metal-arc processes. Welding recommendations for aluminum pipe materials welded by the inert-gas tungsten-arc process are given in Table 72.

Filler Metals. Most widely used as filler metals are commercially pure aluminum and an aluminum alloy containing 5 per cent silicon. More highly alloyed filler metals, however, are required for many aluminum alloys to provide a better match of mechanical and/or physical properties. The majority of commercial aluminum filler metals are covered by AWS-ASTM Specification, Aluminum Alloy Welding Rods and Bare Electrodes (AWS Specifications A5, A10, and ASTM Specification B285).

The filler metals normally recommended for some of the more common commercial pipe materials are listed in Table 73. The high fluidity of molten aluminum promotes sink, which tends to be most severe in horizontally-fixed-positioned piping. Aluminum backing rings are occasionally used. Collapsible stainless-steel back-up fixtures are also used, provided that they can be removed through open pipe ends. Consumable insert rings are also used when welding with the inert-gas tungsten-arc process, particularly when good root penetration inside the pipe is desired.

Heat Treatment. Preheating is desirable on aluminum pipe of diameters exceeding $2\frac{1}{2}$ in. Postheat treatment generally is not done.

Table 73. Welding Filler Metals for Aluminum Piping Materials

Piping base metal	Bare welding rod	Covered electrode
1100	1100 (R-990A)	1100 (A1-2)
3003	1100 (R-900A)	1100 (A1-2)
3004	4043 (R-S5B)	4043 (A1-43)
	5154 (R-GR40A)	
5050	4043 (R-S5B)	4043 (A1-43)
5052	4043 (R-S5B)	4043 (A1-43)
	5154 (R-GR40A)	
	5356	
5154	5154 (R-GR40A)	4043 (A1-43)
6061*	4043 (R-S5B)	4043 (A1-43)
	5356	
6062*	4043 (R-S5B)	4043 (A1-43)
6063*	4043 (R-S5B)	4043 (A1-43)
	5356	

* It is often very difficult to pass the guided-bend test in these materials unless 5356 filler metal is used.

Nickel and Nickel Alloys

This group of alloys is used primarily because of their strength properties, along with their good corrosion resistance to many acid environments. Nickel- and high-nickel-alloy pipe and tubing may be readily joined by welding the same materials to themselves, to other dissimilar alloys, or to headers, flanges, or fittings.

Nickel- and high-nickel-alloy castings of various compositions are obtainable. Often welding is limited to regular foundry-grade castings, which have to be repaired or salvaged by the shielded metal-arc- or inert-gas tungsten-arc-welding process. Castings made under other than controlled conditions may result in welds of questionable soundness; in particular, hairline cracking immediately adjacent to the weld may occur. Where the amount of welding is considerable or when castings require joining by structural or field welding to cast or wrought forms, such as fittings to pipe, casting chemistry must be modified to permit welding. In such cases, the order for castings should be clearly marked "Weldable-grade Castings Are Required" or some similar statement. Weldability tests may be desirable on Inconel forgings and Monel castings.

Welding Processes: Nickel. The welding processes used to join the nickel and high-nickel alloys are the same as those used for joining steel. However, some processes are not applicable to certain alloys. Since the nickel and high-nickel alloys are usually employed in severely corrosive or high-temperature service, the choice of welding process should be evaluated carefully. In some environments, it may be necessary to employ two different welding processes on the same joint, particularly on single-welded joints.

Nickel and high-nickel alloys are susceptible to embrittlement by lead, sulfur, phosphorus, and some low-melting metals and alloys. These contaminants, singly or in combination, are liable to be found around piping installations in the form of grease, oil, machining lubricants, lead-bearing thread lubricants on screwed fittings to be seal welded, paint, marking crayons, marking inks, some temperature-indicating crayons, and shop dirt. In the welding of nickel and high-nickel alloys, all foreign materials are considered as contaminants unless they have been proven harmless.

Before welding material which has been hot-worked in processing or fabrication, it is necessary to remove the thin, dark-colored oxide film from the immediate vicinity of the area to be welded either mechanically by machining, sand-blasting, grinding, or rubbing with an emery cloth or chemically by pickling. Difficulty in the form of an unstable arc may be encountered, if the oxide is not removed, due to the repelling effect created by the high-melting-point oxide.

Joint Design. In general, the nickel and high-nickel alloys possess limited "wetting" properties. The molten weld metal is more viscous, requiring joints to be opened up to allow for filler metal manipulation. Except for a wider included joint angle (80 deg minimum), joint designs approximate those used for steel welding. With some alloys, porosity and microfissuring are a problem, and they necessitate careful qualification of the welding filler metals.

Conventional back-up rings are very rarely used in nickel- and high-nickel-alloy piping. Crevices cannot be avoided, and these may promote corrosion in some environments. In some alloys, the notch effect can also cause root cracking. Consumable insert rings are used if quality pipe welds are required.

Filler Metals. The filler metals used to join the high-nickel alloys must exhibit equivalent, or nearly equivalent, properties if the weld is to perform satisfactorily in the same environment as the base metal. The nominal composition of the filler metal is dictated by the anticipated service performance, and it should match as closely as possible that of the base metal. In most alloys, it is necessary

to depart a little from the nominal composition in order to satisfy particular weldability requirements.   Additions are frequently made, for instance, to control porosity and hot-cracking tendencies.   To this extent, filler rod, wire, and electrode deposits may not be identical in composition to that of the base metal.   The nickel- and high-nickel-alloy filler metals contain some alloying elements whose specific function it is to control weld quality.

Nickel and high-nickel alloys may be welded in all positions with only slight modifications of the welding procedures used in welding steel.   As with any welding job, it is best to position for downhand welding.   Vertical welding in the horizontal fixed pipe position should progress in an upward direction.

Where traces of residual welding flux or slag might contaminate the product, the inert-gas tungsten-arc process is preferred for the root pass.   Consumable inserts have been used extensively in making root-pass welds in nickel and the high-nickel alloys.

Purging of the inside of piping underneath the weld is very important in root-pass welding of nickel and high-nickel alloys.   Helium, argon, hydrogen, and mixtures of these gases have been employed as purging gases.   Helium as a purging gas has merit over argon in that helium has the ability to increase "wetting."

The shielding gas is usually argon, helium, or a mixture of the two.   Some fabricators have used small amounts of hydrogen.   Hydrogen is thought to increase the temperature of the weld pool so that any gas in the molten metal is more readily evolved.   This is particularly advantageous in welding root passes as the chilling effect from the mass of base metal on either side of the joint can otherwise cause fusion-line porosity.

The procedure used for the manual form of inert-gas tungsten-arc welding duplicates, in many respects, the procedures used for oxyacetylene welding in all positions.   Filler metal is added in the same manner, and the precautions for avoiding excessive agitation of the weld puddle, keeping the hot end of the filler metal in the protective gas atmosphere, and maintaining the relation of wire diameter to the material being welded, are comparable.   Prior cleaning of the filler wire is essential to good welding.   Minimum porosity is obtained in manually welded joints when the joint consists of, at least, 50 per cent filler metal.

The inert-gas consumable electrode process may be used for welding the nickel and high-nickel alloys in heavy sections.   In common with other processes, procedure variables must be controlled if predictable results are to be obtained. Submerged-arc welding is rarely employed on nickel and high-nickel piping. Suitable wires and fluxes are not available.   The oxyacetylene process is occasionally used for welding nickel- and high-nickel-alloy pipe and tubing, especially when the wall thickness is too thin for shielded metal-arc welding.   Because of cleanliness requirements, the inert-gas tungsten-arc process is far more extensively used on thin piping materials.

The welding temperature does not produce deleterious effects in the base metal. The only possible effect is annealing and slight grain growth in a narrow band at the edges of the weld which is referred to as the heat-affected zone.   The Hastelloy alloys are an exception to this, and they must receive a solution heat treatment after welding.

Nickel and high-nickel alloys may be brazed as conveniently as are other common materials.   Before selecting a brazing alloy, it is well to ascertain whether the service requirements of the joint (strength or corrosion resistance) can be obtained with the use of a given brazing alloy.   Brazing alloys containing phosphorus should never be used on nickel and high-nickel alloys.   Silver alloys are by far the most frequently used, and the lower-melting-point alloys are preferred. Copper may also be used as a brazing material.

Fluxes satisfactory for other base materials are satisfactory for many of the nickel and high-nickel alloys. Special fluxes are required for the aluminum-containing-nickel and high-nickel alloys. If a suitable dry reducing sulfur-free furnace atmosphere is available, no flux is required.

The high-nickel alloys are subject to stress cracking in the presence of molten low-melting-point brazing alloys, such as those containing cadmium and zinc. Severely cold-worked parts should be annealed before brazing.

Heat Treatment. Preheat and postheat treatments generally are not required in welding nickel- and nickel-alloy pipe materials. Often, interpass temperature control of 300 F maximum is desirable.

Copper and Copper Alloys

Deoxidized or oxygen-free copper pipe and tubing have been standardized. Oxygen-bearing, electrolytic, or tough-pitch copper seamless tubes are also available. The chemical composition, high thermal conductivity, and high fluidity of copper and copper alloys have considerable influence on their weldability. Oxygen-bearing, electrolytic, or tough-pitch copper is susceptible to oxide embrittlement when welded or heated above 1292 F in an atmosphere which contains hydrogen.

Arc- and gas-welding processes are not usually recommended for copper. It can, however, be readily soft-soldered and can be brazed. Piping of deoxidized copper or oxygen-free copper, on the other hand, can be satisfactorily arc- or gas-welded, soldered, or brazed in any position.

Because of the high fluidity of molten deoxidized copper, arc and gas welds should generally be made with suitable backing rings, although the inert-gas tungsten-arc-welding process is being used to an increasing extent with good success. All-position welding of piping by oxyacetylene is possible but quite difficult. Soldering or brazing are also commonly used, particularly in beverage-industry applications. The high thermal conductivity of copper usually necessitates preheating for the welding of large-diameter or unusually heavy copper piping.

Oxyacetylene welding of red brass and yellow brass can be satisfactorily accomplished in all positions with suitable backing rings. Inert-gas tungsten-arc welding also produces good quality welds on red brass. Shielded metal-arc, carbon-arc, and inert-gas metal-arc welding are not usually employed, as the high arc temperature causes vaporization of the zinc in the metal and sweating out and oxidation of any lead that may be present. All-position welding of brass is possible with the oxyacetylene process. However, soldering or brazing are more generally employed. Large-diameter and heavy-wall pipe should generally be preheated to 300 to 500 F. Because of dezincification of yellow brass, the red-brass grades are generally used. Yellow-brass pipe and tubing are available only on special order.

Copper-silicon alloys have a thermal conductivity approaching that of steel. Their good welding characteristics are aided by a glasslike slag which forms on the surface of the molten weld pool. They generally do not contain chemical constituents which vaporize easily, and they thus exhibit better weldability characteristics than other copper alloys used in piping. The copper-silicon alloys are welded extensively by the inert-gas tungsten-arc- and oxyacetylene-welding processes. Shielded metal-arc welding is used occasionally when suitable covered electrodes are available. Welds can be made readily in the flat and vertical positions. The overhead position is more difficult. Backing rings are not required although they are sometimes used.

The principal copper-nickel alloy used for pipe and tubing is that known as "70-30 cupronickel" in which the numbers indicate the proportions of copper and nickel, respectively. Other cupronickel alloys with lower nickel proportions are

available. The 70-30 cupronickel alloy offers superior resistance to the corrosive action and velocity erosion of sea water. It is extensively used for water pipe and condenser tubing on ships. The most suitable process for welding is shielded metal-arc welding. Inert-gas tungsten-arc welding, and oxyacetylene welding with a special silicon-bearing welding rod, are also being used successfully.

While 70-30 cupronickel is one of the most readily weldable copper-base alloys, cracking due to hot shortness may, at times, be encountered. For brazing, BCuP-2 is frequently used. It has a brazing temperature range from 1350 to 1400 F. Silver alloys are also utilized as brazing filler metals. Care must be exercised to select a brazing alloy with a low flow point to prevent intergranular penetration which may occur at high temperatures, particularly if the base metal contains residual stresses set up during forming, shaping, or cold bending.

Titanium and Titanium Alloys

Titanium can be welded with weld efficiencies of 100 per cent by arc welding techniques. However, on titanium alloys (as on other refractory metals), it is extensively important that the molten and hot titanium metal be shielded from the atmosphere and from foreign materials. The molten weld metal and heat-affected zones must be shielded by a protective blanket of inert gases during welding. This can be done either by welding inside of special chambers or by means of a so-called trailing shield (as shown in Fig. 86). The inert-gas tungsten-arc- and inert-gas consumable metal-arc-welding processes are used most commonly. To provide rapid cooling, it is desirable to employ copper chill clamps.

Light-wall pipe and tubing are welded normally in a single pass without filler metals by the inert-gas tungsten-arc process. The use of filler wire may increase

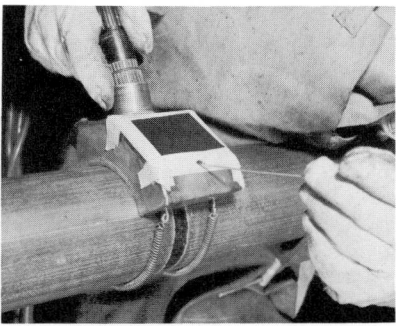

FIG. 86. Welding titanium pipe, using special tracking shield for effective gas shielding.

the possibility of contamination if the filler wire has not been thoroughly cleaned or if the hot end of the filler wire is withdrawn from the protective gas and exposed to the atmosphere during intermittent deposition of the filler metal. Since contamination of the weld would result in embrittlement, it is particularly important that pipe and tubing butt welds have complete penetration at the root of the weld.

Lead and Lead Alloys

The welding difficulties are related to the low-melting temperatures and high fluidity of lead and its alloys. The oxyhydrogen process is most widely used. The oxyacetylene process is not generally used for wall thicknesses below $\frac{1}{2}$ in. as this mixture of gases develops too high a temperature for the welding of thin pipe sections. Air-hydrogen, air-acetylene, oxy-city gas, or oxypropane gas may also be used. To avoid the toxic effects of lead fumes, provision for adequate ventilation should be made where lead is to be welded in closely confined quarters.

WELDING OF DISSIMILAR MATERIALS

The joining of dissimilar metals by any of the welding processes is dependent upon the relative melting points of the metals to be joined. If the melting points are

close (some authorities state within 100 F), welding techniques are generally used. For many dissimilar metal combinations where the respective melting points differ by over 100 F, brazing techniques may be preferable.

Qualification of the welding or brazing procedures with the same materials involved in the production joints is particularly important in dissimilar metal welding. It is a requirement of most codes concerned with welding-procedure qualification, such as the ASME Boiler and Pressure Vessel Code and the ASA Code for Pressure Piping.

Steels of Dissimilar Compositions

The practice normally followed for welding dissimilar material combinations between carbon- and low-alloy steel materials is summarized in Tables 74 to 76. The filler metals used for welding dissimilar metal joints may be chosen to produce deposited metal of the same nominal composition as either of the base materials or of an intermediate alloy content. Selection of weld-metal analysis should receive careful consideration if the dissimilar metal joint is to be exposed to heat treatment or service temperatures where significant carbon diffusion occurs. In general, low-hydrogen lime-type electrodes of the AWS-ASTM Classifications EXX15, EXX16, and EXX18 are preferred for welding dissimilar chromium-bearing ferritic alloy steels.

Austenitic stainless steel and other high-alloy compositions are sometimes employed in welding joints containing dissimilar ferritic steel. The possible stresses caused by the differences in the coefficients of expansion of ferritic and austenitic materials should be given careful consideration. The occurrence of diffusion must be recognized where service temperatures exceed 800 F, particularly if heavy sections or cycling temperatures are involved.

The selection of the welding filler metal should include consideration of the service temperatures. For nominal service, welds between the austenitic stainless-steel piping and ferritic alloy steel piping materials are usually made with stainless-steel electrodes. For service temperatures over 800 F, these dissimilar joints are frequently welded with Inconel-welding filler metals.

Nonferrous Materials of Dissimilar Compositions

For dissimilar welds involving nonferrous alloys, the selection of a welding process, filler metal, and welding procedure must be made with care. The choice of filler material should be based on an understanding of the metallurgical aspects of the proposed joint, and the welding procedure and joint design should be based upon the known welding characteristics of each of the dissimilar materials. The selection of welding process, filler metal, joint design, and welding procedure should be evaluated by adequate procedure-qualification tests to establish reproducibility.

Preparation of the welding procedure requires knowledge of the relative compatibility of the dissimilar base materials. This includes consideration of the dilution which can be tolerated without resulting in a weld which contains defects, brittleness, or other undesirable characteristics. For example, most nickel and high-nickel alloys can be joined to other metals with a tolerance for considerable dilution. The amount of dilution should be controlled to obtain consistent and satisfactory results.

If a combination is joined where dilution from one member will be detrimental, the arc should be directed toward the base metal which has the least adverse effect in contaminating or embrittling the weld deposit. For example, if stainless steel

Table 74. Filler Metals for Welding Dissimilar Steel Compositions

Service temperature of 800 F and above

Service temperature below 800 F	C-steel	C-Mo	½ Cr-½ Mo	1 Cr-½ Mo	1¼ Cr-½ Mo	2 Cr-½ Mo	2¼ Cr-1 Mo	5 Cr-½ Mo	7 Cr-½ Mo	9 Cr-1 Mo	18 Cr-8 Ni	18 Cr-12 Mo	18 Cr-8 Cb
C-steel		2	3	3	3	3	3	3	3	3	8	8	8
C-Mo	1		3	3	3	3	3	3	3	3	8	8	8
½ Cr-½ Mo	1	2		3	3	3	3	3	3	3	8	8	8
1 Cr-½ Mo	1	2	3		3	3	3	3	3	3	8	8	8
1¼ Cr-½ Mo	1	2	3	3		3	4	4	4	4	8	8	8
2 Cr-½ Mo	1	2	3	3	3		4	4	4	4	8	8	8
2¼ Cr-1 Mo	1	2	3	3	3	3		4	4	4	8	8	8
5 Cr-½ Mo	1	2	3	3	3	4	4		5	5	8	8	8
7 Cr-½ Mo	1	2	3	3	3	3	4	5		6	8	8	8
9 Cr-1 Mo	1	2	3	3	3	4	4	5	6		8	8	8
18 Cr-8 Ni (304)	7	7	7	7	7	7	7	7	7	7		8	8
18 Cr-12 Mo (316)	7	7	7	7	7	7	7	7	7	7	7		8
18 Cr-8 Cb (347)	7	7	7	7	7	7	7	7	7	7	7	7	

Example: When welding 2¼Cr-1 Mo to 1¼Cr-½ Mo at a service temperature of over 800 F, read 4 *above* diagonal line. For the same materials at service temperature less than 800 F, read 3 *below* diagonal line.

Welding filler metals:
1. carbon steel.
2. C-Mo.
3. 1¼ Cr-½ Mo.
4. 2¼ Cr-1 Mo.
5. 5 Cr-½ Mo.
6. 9 Cr-1 Mo.
7. 29 Cr-9 Ni (preferred); 25 Cr-12 Ni (alternate).
8. Inconel.

Table 75. Preheat- and Interpass-temperature Requirements for Welding Dissimilar Steel Compositions

	C-steel	C-Mo	½ Cr-½ Mo	1 Cr-½ Mo	1¼ Cr-½ Mo	2 Cr-½ Mo	2¼ Cr-1 Mo	5 Cr-½ Mo	7 Cr-½ Mo	9 Cr-1 Mo	18 Cr 8 Ni	18 Cr 12 Mo	18 Cr 8 Cb
C-steel	...	2	3	3	3	3	3	3	4	4	1	1	1
C-Mo	2	...	3	3	3	3	3	3	4	4	1	1	1
½ Cr-½ Mo	3	3	...	3	3	3	3	3	4	4	5	5	5
1 Cr-½ Mo	3	3	3	...	3	3	3	3	4	4	5	5	5
1¼ Cr-½ Mo	3	3	3	3	...	3	3	3	4	4	5	5	5
2 Cr-½ Mo	3	3	3	3	3	...	3	3	4	4	5	5	5
2¼ Cr-1 Mo	3	3	3	3	3	3	...	3	4	4	5	5	5
5 Cr-½ Mo	3	3	3	3	3	3	3	...	4	4	5		
7 Cr-½ Mo	4	4	4	4	4	4	4	4	...	4			
9 Cr-1 Mo	4	4	4	4	4	4	4	4	4	...			

Key:
1. No preheat (where atmospheric temperatures are below 70 F, the weld joint should be preheated to 100 F).
2. On butt welds, no preheat required for sections up to ½ in. nominal wall thickness. Sections with over ½ in. nominal wall thickness should be preheated to 400 to 500 F. On fillet welds no preheating required for any (throat) thickness. Where atmospheric temperatures are below 70 F, the weld joint should be preheated to 100 F.
3. 400–600 F.
4. 500–600 F.
5. 300–500 F for chrome-moly material only.

Table 76. Postheat Requirements for Welding Dissimilar Steel Compositions

	C-steel	C-Mo	½ Cr-½ Mo	1 Cr-½ Mo	1¼ Cr-½ Mo	2 Cr-½ Mo	2¼ Cr-1 Mo	5 Cr-½ Mo	7 Cr-½ Mo	9 Cr-1 Mo	18 Cr 8 Ni	18 Cr 12 Mo	18 Cr 8 Cb
C-steel	...	2	3	3	4	4	5	5	5	5	1	1	1
C-Mo	2	...	3	3	4	4	5	5	5	5	1	1	1
½ Cr-½ Mo	3	3	...	3	4	4	5	5	5	5	1	1	1
1 Cr-½ Mo	3	3	3	...	4	4	5	5	5	5	1	1	1
1¼ C-½ Mo	4	4	4	4	...	4	5	5	5	5	1	1	1
2 Cr-½ Mo	4	4	4	4	4	...	5	5	5	5	1	1	1
2¼ Cr-1 Mo	5	5	5	5	5	5	...	5	5	5	1		
5 Cr-½ Mo	5	5	5	5	5	5	5	...	5	5			
7 Cr-½ Mo	5	5	5	5	5	5	5	5	...	5			
9 Cr-1 Mo	5	5	5	5	5	5	5	5	5	...			

Key:
1. Do not postheat.
2. On butt welds, not required for sections up to ½ in. wall thickness. Sections with over ½ in. wall thickness shall be stress-relieved at 1,150 to 1,200 F for 1 hr/in. of wall thickness (1 hr min).
3. 1,200–1,300 F for 1 hr/in. of thickness (1 hr min).
4. 1,275–1,350 F for 1 hr/in. of thickness (1 hr min).
5. 1,300–1,375 F for 1 hr/in. of thickness (1 hr min).

is to be welded to Monel with Inconel electrodes or filler metal, the arc must be directed toward the Monel to insure melting of the least possible amount of stainless steel. Sometimes transition sections are used to ease welding problems and to obtain better service life, particularly where severe thermal fatigue is involved. For some metal combinations, "buttering" is done where one of the joint bevels is overlaid with a weld-metal alloy that is more compatible to the weld-metal compositions involved in the weld joint.

Brazing and Soldering Process: Dissimilar Materials. Where a wide difference (usually over 100 F) in melting point exists, it is frequently desirable or necessary to employ brazing, braze welding, or soldering techniques. Each of the metals involved in the joint must individually be capable of being welded, brazed, or soldered.

Brazing with silver-copper-cadmium-zinc alloys offers the best approach to joining dissimilar metals where the presence of these alloys is compatible with service requirements. Copper-phosphorus and copper-silver-phosphorus brazing filler metals are suitable only for copper and copper-base alloys. They cannot be used on ferrous metals, nickel, or nickel alloys. The standard joints employed in brazing similar metals apply equally well to dissimilar metals.

Soft soldering provides a means of connecting metals of widely differing melting points, particularly where one of the metals melts below the temperature necessary for brazing. Soldering is also helpful in those situations where it is necessary to avoid the high temperatures of welding and brazing. Joint designs for soldering dissimilar metals follow the principles and standards established for soft soldering similar metals.

CLAD PIPING

Clad piping is produced commercially by seam welding of clad plate, by the Pilger-mill process, or by centrifugally casting two metal layers. The last type is also referred to as "dual metal."

In most applications, clad steel piping is selected for its corrosion resistance in environments which would attack mild or low-alloy steel. There are also some applications where the clad pipe steels are selected for their thermal conductivity, oxidation resistance, and other characteristics. Although numerous metal combinations are possible, the majority of applications involve stainless clad steels because clad steels are more economical than solid stainless-steel pipe.

Joint Design. On piping where the weld must be made from one side only, special joint designs are required. Where backing strips can be used, suitable joint designs are illustrated in Fig. 87. If a smooth surface contour on the blind side is desired, as in most piping systems, the joint is prepared as shown in Fig. 88.

Welding Processes. Welding is done extensively by the shielded metal-arc process with covered electrodes. The root is usually welded by the inert-gas tungsten-arc process. On heavy pipe sizes, the balance of the weld is normally completed by shielded metal-arc welding. Where the clad side is blind, the cladding should be welded with electrodes or welding rods of the same type of composition as the cladding metal unless dilution must be considered. Special problems, however, may be involved in completing the weld.

On most materials of wall thicknesses up to $\frac{1}{2}$ in., the best practice is to complete the entire weld with the same electrodes or welding rods as were used to weld the cladding. This is not applicable to some nonferrous clad steels. On aluminum-clad steels, for example, the strength of the back-up steel must be maintained. On such materials, special types of joints with overlap-type root preparations may be desirable.

On wall thicknesses over ½ in., it is frequently desirable to complete the weld with mild or low-alloy steel electrodes after the clad root of the joint has been welded with the proper high-alloy steel or nonferrous electrodes. Extreme care must be exercised in selecting the electrodes or welding rods and the welding procedure. The electrodes or rods used should not produce a brittle transition zone in the finished weld. It is frequently desirable to deposit a transition weld having properties somewhat between those of the cladding and backing steel. This is called buttering. It is considered undesirable, for example, to deposit mild or low-alloy steel-weld metal against the austenitic stainless steels. The transition zone would be rather hard and brittle and very susceptible to cracking.

Heat Treatment. The proper heat-treating temperatures and procedures must be based on the characteristics of both the backing steel and the cladding

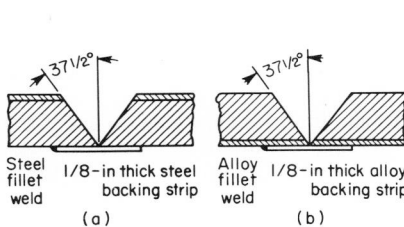

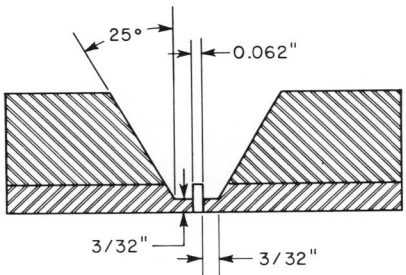

FIG. 87. Joint-design suitable for welding clad pipe from one side only, using a backing strip.

FIG. 88. Weld joint end preparation for clad pipe for applications in which a smooth contour is required on the blind side.

material. Care should be taken to specify a heat treatment which produces beneficial effects in one material and yet will not harden the other. Some heat treatments, although they may reduce stresses, may introduce adverse effects, such as decarburization, or the formation of brittle metallurgical structures. Cooling rates may have to be specified in cases where rapid cooling may adversely affect one of the materials. It must be recognized that the coefficients of expansion of the respective materials may differ considerably and may result in serious residual stresses.

Careful qualification of the welding procedure with the same materials, welding conditions, and preheat and postheat temperatures and cycles should be considered essential to insure that the production welds are likely to be sound.

LINED PIPING

Applied liners consist of sheets or relatively thin plates bonded intermittently by welding to backing material, usually carbon- or low-alloy steel. Applied liner construction thus combines a low-cost, high-strength carbon or low-alloy steel of relatively heavy wall thickness with a high-cost thin-walled stainless steel or nonferrous alloy.

In most applications, liners are selected for their corrosion resistance to environments which would attack mild- or low-alloy steel. There are also many applications where applied liners are selected because of their wear resistance, oxidation resistance, thermal conductivity, or other characteristics.

Although numerous liner combinations are possible, the greatest tonnage produced involves stainless steel liner materials. In piping, applied liners may become more economical than solid alloy materials or clad materials in wall thickness exceeding ¾ in.

On pipe where lining methods are somewhat more costly than on large-diameter tanks or pressure vessels, it is generally more economical to use solid stainless or other alloy materials. However, there are applications where rather costly materials are required, such as the Hastelloy-type alloys. When this is the case, the lining of carbon-steel pipe with preformed liner plates (as shown in Fig. 89) may be more economical than the use of solid Hastelloy piping. Elbows or other fittings of Hastelloy, Inconel, or other high-alloy materials frequently are not readily available. Special fabrication of such shapes would involve considerable time and cost. For such conditions, lining may be accomplished without undue delay and often at a significantly lower cost.

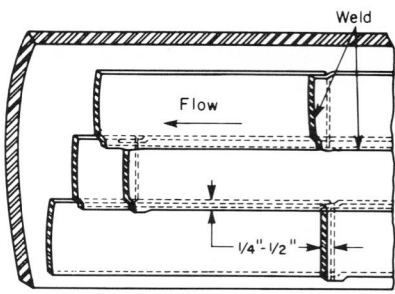

FIG. 89. Lining of carbon-steel pipe with Hastelloy preformed liner plates.

On pipe, lining is generally done by the strip-welding technique. Length of the strips usually varies from 2 to 10 ft. Width of the strips may vary between 2 and 6 in., depending upon the service requirements. The values given in Table 77 are generally followed as a guide by design engineers.

Narrower strips should be used when the service involves severe thermal cycling or steep temperature gradients. On the other hand, strips as wide as 24 in. have been used for essentially noncritical service applications at room temperature.

On pipe, the practical width of the liner strips will also depend on the inside curvature. For example, on 14-in. pipe, 4-in.-wide liner strips would generally be considered practical.

Figure 89 illustrates the shingle-strip lining technique recommended for 10- to 20-in.-diameter pipe, employing 4- to 5-in.-wide liner strips. The ends of the shingles overlapping previously welded shingles should be on the downstream side.

At certain inaccessible locations in lined pipe or fittings, it may not be practical to apply weld-metal overlays. The same applies to the ends of flanges where the gap between the liner strips and flange face plate may be overlayed or filled with weld metal. At pipe inlets, manhole openings, or wherever else necessary, the

Table 77. Strip Widths for Strip-welded Liners

Max operating temperature, F	Maximum strip widths of stainless steels	
	Type 400 series, in.	Type 300 series, in.
Up to 450	Up to 6	Up to 6
450–750	4½–5	3–4
750–850	4	2½–3
850–950	3¼	2
Over 950	3	2

exposed backing steel surfaces can be protected by weld-metal overlays with welding-rod alloys of proper composition.   The effects of dilution must be considered in the selection of the proper filler metal.

METAL-COATED PIPING

A number of different metal-coated pipe materials have been produced commercially.   Among the more common combinations are zinc-dipped (galvanized) steel pipe, aluminized (dipped or sprayed) steel pipe, and nickel-plated steel pipe.

Galvanized Steel Pipe.   Where prefabrication is possible, it is normally desirable to prefabricate the pipe sections prior to galvanizing, as the welding of zinc-coated pipe ends releases toxic vapors.   Where this is not possible, the zinc should be removed from the pipe ends and welding areas for a distance of $\frac{1}{4}$ in. Good ventilation should also be provided.

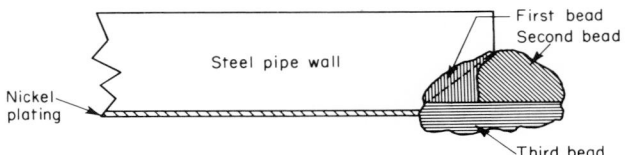

Fig. 90. Recommended buttering of pipe end for nickel-plated pipe, where weld dilution on pipe ID must be minimized.

The welding of galvanized steel pipe is done with the same procedures as carbon-steel pipe.   Regalvanizing of the weld areas is generally not possible, and this must be recognized when specifying the welding of galvanized piping.   The exposed steel generally, however, receives galvanic protection from the large surrounding anodic zinc areas.   Sometimes the exposed steel surface is coated by means of metal spraying or protective zinc-silicate coatings.

Aluminized Steel Pipe.   Aluminized steel piping is produced either by immersion in a bath of molten aluminum or by metal spraying.   The same welding processes apply to aluminized piping as to the base metals.

Postheat treatment, when necessary, may require special considerations.   Sprayed aluminum coating may blister or be loosened when stress relieving is done at temperatures over 1250 F.   Uniform stress relieving at 1100 to 1250 F may be necessary.

Nickel-plated Steel Pipe.   Nickel-plated steel pipe consists of a layer of pure nickel 0.008 to 0.015 in. thick which has been nickel-plated to the inside of the steel. Commercially, it is applied to mild steel pipe in diameters from $1\frac{1}{2}$ to 24 in. OD. The service environments in which the nickel-plated pipe is used generally require that the nickel coating be continuous so that no steel is exposed.

Special welding conditions, therefore, are generally necessary.   One method consists of buttering the pipe ends with nickel weld metal (as illustrated in Fig. 90). The pipe ends are then machined to one of the standard end preparations (as illustrated in Fig. 91) for shielded metal-arc welding or inert-gas tungsten-arc root-pass welding.   The balance of the weld is then usually completed with Inconel electrodes.

Normally, no preheat or stress-relief heat treatments are performed when welding nickel-plated pipe.   However, if specific code considerations or service requirements necessitate postheat treatments, they should be carried out at temperatures between 1150 and 1200 F for an hour per inch of wall thickness.   Overheating should be avoided in order to prevent excessive oxidation of the nickel coating.

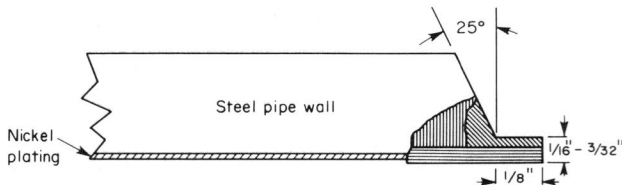

Fig. 91. Recommended preparation of pipe end for nickel-plated pipe, where weld dilution on pipe ID must be minimized.

REPAIR OF WELDS

When a pipe weld is to be repaired, removal of the defect is normally done by grinding, chipping, machining, oxygen or arc gouging, or inert-gas gouging, provided that these methods are suitable for the respective piping and welding materials involved. In heavy-wall steel pipe, weld defects near the weld root are normally removed by oxygen gouging, which generally permits closer control.

After removal of the defect, the surface upon which the repair weld is to be deposited should be ground and cleaned to insure that a sound weld repair may be obtained. Where the quality of the completed pipe weld is determined by radiographic examination, it may be desirable to make a radiographic exposure of the grooved-out cavity to insure that all of the defective material has been removed.

A substantial percentage of defects occurs in the root pass, particularly under field welding conditions. Their removal necessitates gouging or grinding completely through the wall. Such repairs are often difficult to make, as the welder tends to bridge an excessively wide gap with feather edges instead of the normal root gap and land. By carefully depositing a small amount of weld metal on each lip to reduce the gap and provide a land, a satisfactory repair can be made more readily on the first attempt. All repairs should be made using all of the requirements of the original procedure, including preheat and postheat treatments. Many codes include special sections covering the repair of weld defects, generally in conformance with the above paragraphs.

WELD SHRINKAGE AND DISTORTION

Precise calculations of shrinkage in welded pipe joints are difficult and are generally not made. As a general guide, some designers allow, on carbon-steel pipe butt welds, the following shrinkage: pipe under 2 in. thick = ⅛ in. per weld; pipe 2 in. thick and over = ¼ in. per weld.

Somewhat higher values would apply to austenitic stainless steels which have higher coefficients of expansion than carbon steels. Grinding between passes, weld repairs, preheat, etc., may substantially change the total shrinkage.

As a general rule, shrinkage values vary directly with the coefficients of expansion of the base and the weld metals involved, the thickness of the heaviest section being welded, and the heat input, and inversely with the degree of restraint imposed by fixtures. Another problem of shrinkage of pipe welds, particularly in materials having high coefficients of expansion, involves the upsetting of the pipe in the weld area and the gradual decrease in the inside diameter as the number of weld passes increases.

The angular distortion in circumferential and intersecting pipe welds can be controlled and reduced to a minimum by (1) properly applying the welding procedure,

particularly with respect to uniformity of heating (maintaining the minimum interpass temperature consistent with the material requirements), (2) planning a welding sequence which corrects angular distortion, (3) uniformly preheating the parts prior to and during the welding operation, and (4) using fixtures for holding the parts in fixed relation to each other during welding.   The first three of these methods permit free movement of the parts and generally result in lower strain during welding.

Residual stresses are normally of yield-point magnitude and generally do not differ with restraint or variations in the welding procedure used.   Restraint and the welding procedure do, however, influence greatly the shrinkage strain.   During welding, the tendency for cracks to develop depends on the shrinkage strain and the amount of distortion produced.

Residual stresses can be relieved by thermal treatment.   Peening is not considered an effective method for the relief of stresses, except when it is done on the last weld layer and base metal adjacent to the weld edges.   Distortion can be lessened by peening and can be corrected by the straightening operations described in another section of this chapter.

In fabricating headers or pipe assemblies containing various types of welded joints in close proximity to one another, the control of distortion due to shrinkage is quite complicated.   Major factors involved in distortion are the layout and fitting of assemblies, sequence of welding, positioning of the welds, and avoidance of heat concentrations.   The four steps employed to control such distortion have already been mentioned; however, their effectiveness is difficult to estimate and depends largely on the experience of the pipe-welding fabricator.

JOINING CAST-IRON PIPE

Bell-and-spigot Joint.   This joint for underground cast-iron pipe was developed as far back as 1785.   Standard dimensions as designated in Federal Specification WW-P-421 and ASA Standard A21.6 are shown in Table 78.   ASA Standard A21.8 also contains bell-and-spigot joints.

The joint may be made up with lead and oakum, sulfur compounds, or cement. Lead and oakum constitute the prevailing joint sealer for water-supply systems. Oakum with lead or cement backing is used commonly for low-pressure distribution systems for handling wet manufactured gas.   For joint tightness, it is necessary to keep the oakum wet either with moisture from the fluid conveyed or, in the case of dry-gas systems, by introducing an impregnating compound into the lines.   Whereas bell-and-spigot joints have proved satisfactory in water-supply systems operating at pressures of 200 psi or more, they are not regarded with favor for gas pressures in excess of 10 psi.   Regular cast-iron and soil pipe, where used in sewer systems, usually has bell-and-spigot joints.

The AWWA and AGA each have their own designs of bell-and-spigot joints which differ principally in the grooving of the bells.   When desired, bell-and-spigot pipe and fittings for fire lines and other high-pressure water service are furnished with bolting lugs.

Mechanical (Gland-type) Joint.   This modification of the bell-and-spigot joint, as designated in Federal Specification WW-P-421 and ASA Standard A21.11, is illustrated in Table 79.   This joint is commonly used for low- and intermediate-pressure gas-distribution systems, particularly those conveying natural gas or dry manufactured gas.   Mechanical joints are used also for water lines where excessive vibration may be expected and for aboveground water, gas, sewage, and process

Table 78. Standard Dimensions of Bell-and-spigot Joints for Pipe Centrifugally Cast in Metal Molds (Abstracted from ASA Standard A21.6-1962, except as noted)

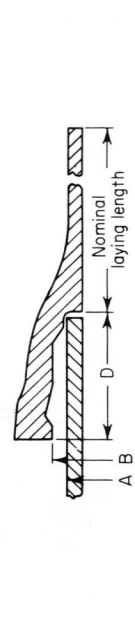

Nominal pipe size	Class	Thickness designation	Thickness of pipe	OD of pipe A	Diameter of socket B	Depth of socket D	Weight (approx), lb			Joint compound, lb per 2½ in. depth†	Jute, lb per joint†	Lead lb per 2½ in. depth†
							Barrel per ft	Bell	18 ft laying length*			
3	Through 350	22	0.32	3.96	4.76	3.30	11.4	11	215			
4	Through 350	22	0.35	4.80	5.60	3.30	15.3	14	290	2.00	0.21	8.00
6	Through 350	22	0.38	6.90	7.70	3.88	24.3	25	460	3.00	0.31	11.25
8	Through 350	22	0.41	9.05	9.85	4.38	34.7	41	665	4.00	0.44	14.50
10	Through 250 300 350	22 23 24	0.44 0.48 0.52	11.10 11.10 11.10	11.90 11.90 11.90	4.38 4.38 4.38	46.0 50.0 53.9	54 54 54	880 955 1,025	5.00	0.53	17.50
12	Through 200 250, 300 350	22 23 24	0.48 0.52 0.56	13.20 13.20 13.20	14.00 14.00 14.00	4.38 4.38 4.38	59.8 64.6 69.4	66 66 66	1,140 1,230 1,315	6.00	0.61	20.50

Size												
14	50	21	0.48	15.30	16.10	4.50	69.7	78	1,335			
	100	22	0.51	15.30	16.10	4.50	73.9	78	1,410			
	150	22	0.51	15.30	16.10	4.50	73.9	78	1,410	7.00	0.81	24.00
	200	23	0.55	15.30	16.10	4.50	79.5	78	1,510			
	250, 300	24	0.59	15.30	16.10	4.50	85.1	78	1,610			
	350	25	0.64	15.30	16.10	4.50	92.0	78	1,735			
16	50, 100	22	0.54	17.40	18.40	4.50	89.2	96	1,700			
	150	22	0.54	17.40	18.40	4.50	89.2	96	1,700			
	200	23	0.58	17.40	18.40	4.50	95.6	96	1,815	8.25	0.94	33.00
	250	24	0.63	17.40	18.40	4.50	103.6	96	1,960			
	300, 350	25	0.68	17.40	18.40	4.50	111.4	96	2,100			
18	50	21	0.54	19.50	20.50	4.50	100.4	114	1,920			
	100	22	0.58	19.50	20.50	4.50	107.6	114	2,050			
	150	22	0.58	19.50	20.50	4.50	107.6	114	2,050			
	200	23	0.63	19.50	20.50	4.50	116.5	114	2,210	9.25	1.00	36.90
	250	24	0.68	19.50	20.50	4.50	125.4	114	2,370			
	300	25	0.73	19.50	20.50	4.50	134.3	114	2,530			
	350	26	0.79	19.50	20.50	4.50	144.9	114	2,720			
20	50	21	0.57	21.60	22.60	4.50	117.5	133	2,250			
	100	22	0.62	21.60	22.60	4.50	127.5	133	2,430			
	150	22	0.62	21.60	22.60	4.50	127.5	133	2,430			
	200	23	0.67	21.60	22.60	4.50	137.5	133	2,610	10.50	1.25	40.50
	250	24	0.72	21.60	22.60	4.50	147.4	133	2,785			
	300	25	0.78	21.60	22.60	4.50	159.2	133	3,000			
	350	26	0.84	21.60	22.60	4.50	170.9	133	3,210			
24	50	21	0.63	25.80	26.80	4.50	155.4	179	2,975			
	100	22	0.68	25.80	26.80	4.50	167.4	179	3,190			
	150	23	0.73	25.80	26.80	4.50	179.4	179	3,410			
	200, 250	24	0.79	25.80	26.80	4.50	193.7	179	3,665	13.00	1.50	52.50
	300	25	0.85	25.80	26.80	4.50	207.9	179	3,920			
	350	26	0.92	25.80	26.80	4.50	224.4	179	4,220			

* Includes weight of bell.
† Not included in ASA B21.6-1962.

Table 79. Standard Dimensions of Mechanical (Gland-type) Joints (ASA Standard A21.11-1964) I-5

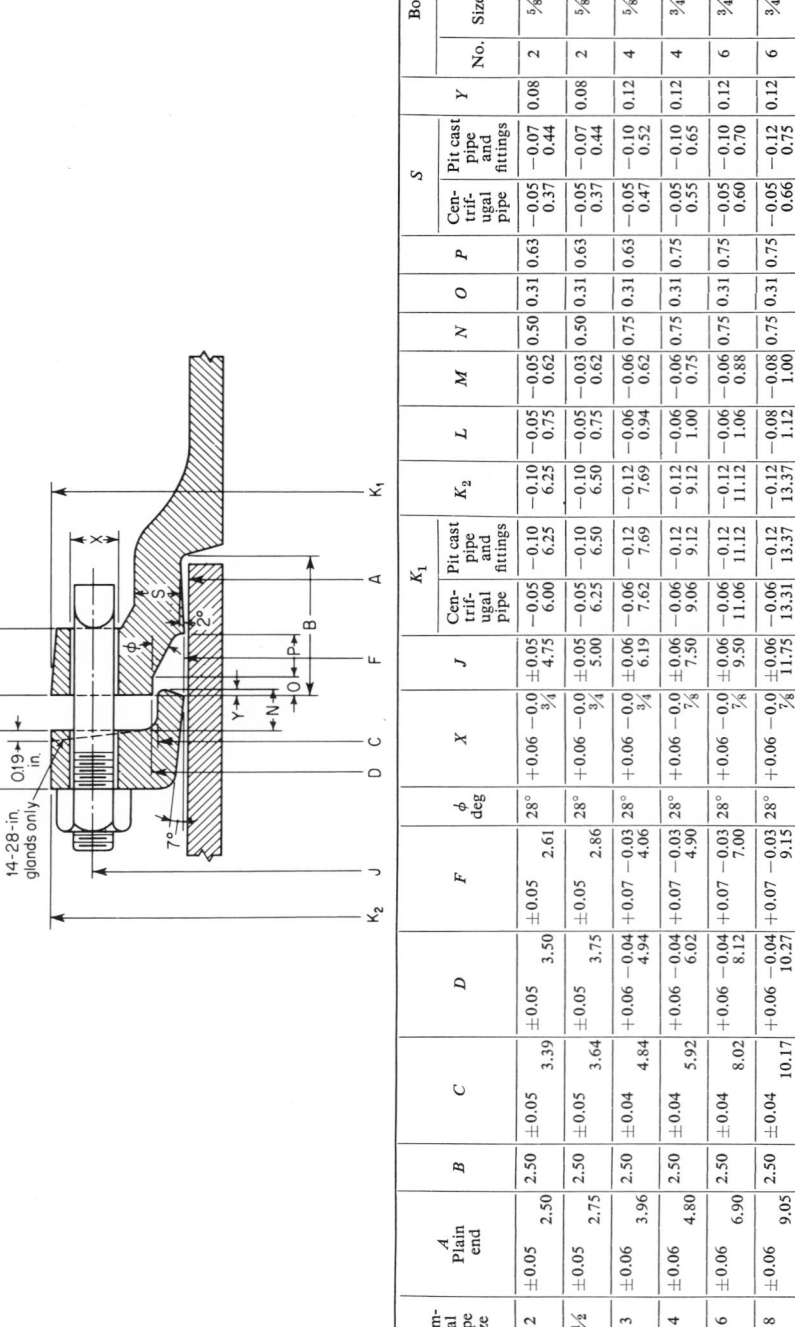

Nominal pipe size	A Plain end	B	C	D	F	φ deg	X	J	K₁ Centrifugal pipe	K₁ Pit cast pipe and fittings	K₂	L	M	N	O	P	S Centrifugal pipe	S Pit cast pipe and fittings	Y	Bolts No.	Bolts Size	Bolts Length
2	±0.05 2.50	2.50	±0.05 3.39	±0.05 3.50	±0.05 2.61	28°	+0.06 −0.0 ¾	±0.05 4.75	−0.05 6.00	−0.10 6.25	−0.10 6.25	−0.05 0.75	−0.05 0.62	0.50	0.31	0.63	−0.05 0.37	−0.07 0.44	0.08	2	⅝	2½
2½	±0.05 2.75	2.50	±0.05 3.64	±0.05 3.75	±0.05 2.86	28°	+0.06 −0.0 ¾	±0.05 5.00	−0.05 6.25	−0.10 6.50	−0.10 6.50	−0.05 0.75	−0.03 0.62	0.50	0.31	0.63	−0.05 0.37	−0.07 0.44	0.08	2	⅝	2½
3	±0.06 3.96	2.50	±0.04 4.84	+0.06 −0.04 4.94	+0.07 −0.03 4.06	28°	+0.06 −0.0 ¾	+0.06 −0.06 6.19	−0.06 7.62	−0.12 7.69	−0.12 7.69	−0.06 0.94	−0.06 0.62	0.75	0.31	0.63	−0.05 0.47	−0.10 0.52	0.12	4	⅝	3
4	±0.06 4.80	2.50	±0.04 5.92	+0.06 −0.04 6.02	+0.07 −0.03 4.90	28°	+0.06 −0.0 ⅞	+0.06 −0.06 7.50	−0.06 9.06	−0.12 9.12	−0.12 9.12	−0.06 1.00	−0.06 0.62	0.75	0.31	0.75	−0.05 0.55	−0.10 0.65	0.12	4	¾	3½
6	±0.06 6.90	2.50	±0.04 8.02	+0.06 −0.04 8.12	+0.07 −0.03 7.00	28°	+0.06 −0.0 ⅞	+0.06 −0.06 9.50	−0.06 11.06	−0.12 11.12	−0.12 11.12	−0.06 1.06	−0.06 0.88	0.75	0.31	0.75	−0.05 0.60	−0.10 0.70	0.12	6	¾	3½
8	±0.06 9.05	2.50	±0.04 10.17	+0.06 −0.04 10.27	+0.07 −0.03 9.15	28°	+0.06 −0.0 ⅞	+0.06 −0.06 11.75	−0.06 13.31	−0.12 13.37	−0.12 13.37	−0.08 1.12	−0.08 1.00	0.75	0.31	0.75	−0.05 0.66	−0.12 0.75	0.12	6	¾	4

Size																							
10	±0.06 / 11.10	2.50	+0.06 −0.04 / 12.22	+0.06 −0.04 / 12.34	+0.07 −0.03 / 11.20	28°	+0.06 −0.0 / ⅞	±0.06 / 14.00	−0.06 / 15.62	−0.12 / 15.69	−0.12 / 15.62	−0.08 / 1.19	−0.08 / 1.00	0.75	0.75	0.31	0.75	−0.06 / 0.72	−0.12 / 0.80	0.12	8	¾	4
12	±0.06 / 13.20	2.50	+0.06 −0.04 / 14.32	+0.06 −0.04 / 14.44	+0.07 −0.03 / 13.30	28°	+0.06 −0.0 / ⅞	±0.06 / 16.25	−0.06 / 17.88	−0.12 / 17.94	−0.12 / 17.88	−0.08 / 1.25	−0.08 / 1.00	0.75	0.75	0.31	0.75	−0.06 / 0.79	−0.12 / 0.85	0.12	8	¾	4
14	+0.05 −0.08 / 15.30	3.50	+0.07 −0.05 / 16.40	+0.06 −0.05 / 16.54	+0.06 −0.07 / 15.44	28°	+0.06 −0.0 / ⅞	±0.06 / 18.75	−0.08 / 20.25	−0.12 / 20.31	−0.12 / 20.25	−0.12 / 1.31	−0.12 / 1.25	0.75	0.75	0.31	0.75	−0.08 / 0.85	−0.12 / 0.89	0.12	10	¾	4
16	+0.05 −0.08 / 17.40	3.50	+0.07 −0.05 / 18.50	+0.06 −0.05 / 18.64	+0.06 −0.07 / 17.54	28°	+0.06 −0.0 / ⅞	±0.06 / 21.00	−0.08 / 22.50	−0.12 / 22.56	−0.12 / 22.50	−0.12 / 1.38	−0.12 / 1.31	0.75	0.75	0.31	0.75	−0.08 / 0.91	−0.12 / 0.97	0.12	12	¾	4½
18	+0.05 −0.08 / 19.50	3.50	+0.07 −0.05 / 20.60	+0.07 −0.05 / 20.74	+0.06 −0.07 / 19.64	28°	+0.06 −0.0 / ⅞	±0.06 / 23.25	−0.08 / 24.75	−0.15 / 24.83	−0.15 / 24.75	−0.12 / 1.44	−0.12 / 1.38	0.75	0.75	0.31	0.75	−0.08 / 0.97	−0.15 / 1.05	0.12	12	¾	4½
20	+0.05 −0.08 / 21.60	3.50	+0.07 −0.05 / 22.70	+0.07 −0.05 / 22.84	+0.06 −0.07 / 21.74	28°	+0.06 −0.0 / ⅞	±0.06 / 25.50	−0.08 / 27.00	−0.15 / 27.08	−0.15 / 27.00	−0.12 / 1.50	−0.12 / 1.44	0.75	0.75	0.31	0.75	−0.08 / 1.03	−0.15 / 1.12	0.12	14	¾	4½
24	+0.05 −0.08 / 25.80	3.50	+0.07 −0.05 / 26.90	+0.07 −0.05 / 27.04	+0.06 −0.07 / 25.94	28°	+0.06 −0.0 / ⅞	±0.06 / 30.00	−0.08 / 31.50	−0.15 / 31.58	−0.15 / 31.50	−0.12 / 1.62	−0.12 / 1.56	0.75	0.75	0.31	0.75	−0.08 / 1.08	−0.15 / 1.22	0.12	16	¾	5
30	+0.08 −0.06 / 32.00	4.00	+0.08 −0.06 / 33.29	+0.08 −0.06 / 33.46	+0.08 −0.06 / 32.17	20°	+0.06 −0.0 / 1⅛	±0.06 / 36.88	−0.12 / 39.12	−0.18 / 39.12	−0.18 / 39.12	−0.12 / 1.81	−0.12 / 2.00	0.75	0.75	0.38	1.00	−0.10 / 1.20	−0.15 / 1.50	0.12	20	1	6
36	+0.08 −0.06 / 38.30	4.00	+0.08 −0.06 / 39.59	+0.08 −0.06 / 39.76	+0.08 −0.06 / 38.47	20°	+0.06 −0.0 / 1⅛	±0.06 / 43.75	−0.12 / 46.00	−0.18 / 46.00	−0.18 / 46.00	−0.12 / 2.00	−0.12 / 2.00	0.75	0.75	0.38	1.00	−0.10 / 1.35	−0.15 / 1.80	0.12	24	1	6
42	+0.08 −0.06 / 44.50	4.00	+0.08 −0.06 / 45.79	+0.08 −0.06 / 45.96	+0.08 −0.06 / 44.67	20°	+0.06 −0.0 / 1⅜	±0.06 / 50.62	−0.12 / 53.12	−0.18 / 53.12	−0.18 / 53.12	−0.12 / 2.00	−0.12 / 2.00	0.75	0.75	0.38	1.00	−0.10 / 1.48	−0.15 / 1.95	0.12	28	1¼	6
48	+0.08 −0.06 / 50.80	4.00	+0.08 −0.06 / 52.09	+0.08 −0.06 / 52.26	+0.08 −0.06 / 50.97	20°	+0.06 −0.0 / 1⅜	±0.06 / 57.50	−0.12 / 60.00	−0.18 / 60.00	−0.18 / 60.00	−0.12 / 2.00	−0.12 / 2.00	0.75	0.75	0.38	1.00	−0.10 / 1.61	−0.15 / 2.20	0.12	32	1¼	6

1 The thickness of the bell, S, shall in all instances be equal to, and generally exceed by at least 10 per cent, the nominal wall thickness of the pipe or fitting of which it is a part.

2 Cored holes may be tapered an additional 0.06 in. in diameter.

3 In the event of ovalness of the plain end outside diameter, the mean diameter measured by a circumferential tape shall not be less than the minimum diameter shown in the table. The minor axis shall not be less than the above minimum diameter plus an additional minus tolerance of 0.04 in. for sizes 8–12 in., 0.07 in. for sizes 14–24 in. and 0.10 in. for sizes 30–48 in.

4 K_1 and K_2 are the dimensions across the bolt holes. For sizes 2 and 2¼ in., both flange and gland may be oval shaped. For sizes 3–48 in., the gland may be polygon shaped.

5 Mechanical joints require the use of specially designed bolts. See ASA 21.11.

piping. In the mechanical (gland-type) joint shown in Fig. 92 the lead and oakum of the conventional bell-and-spigot joint are supplanted by a stuffing box in which a rubber or composition packing ring, with or without a metal or canvas tip or canvas backing, is compressed by a cast-iron follower ring drawn up with bolts. In addition to making an inherently tight joint even under considerable pressure, this arrangement possesses the further advantage of permitting relatively large lateral deflections ($3\frac{1}{2}$ to 7 deg), as well as longitudinal expansion or contraction.

Tyton Joint. The Tyton joint is designed to contain an elongated grooved gasket. The inside contour of the socket bell provides a seat for the circular rubber in a modified bulb-shaped gasket. An internal ridge in the socket fits into the groove of the gasket. A slight taper on the plain end of the pipe facilitates assembly. Standard dimensions are given in Table 80. The maximum joint deflection angle is 5 deg for sizes through 12 in., 4 deg for 14 and 16 in., and 3 deg for 18, 20, and 24 in. Either all-bell U.S. standardized mechanical joint fittings or bell-and-spigot all-bell fittings with poured or cement joints can be used with Tyton-joint pipe.

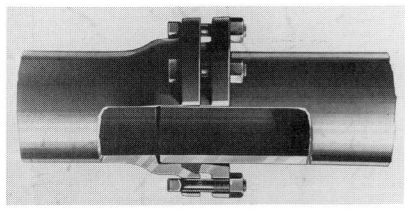

Fig. 92. Mechanical (gland-type) joint for cast-iron pipe.

Mechanical Lock-type Joint. For installations where the joints may tend to come apart owing to sag or lateral thrust in the pipeline, a mechanical joint having a self-locking feature is used to resist end pull. This joint is similar to the gland-type mechanical joint except that in the locked joint the spigot end of the pipe is grooved or has a recess to grip the gasket. Although only slight expansion or contraction can be accommodated in this type of joint, it does allow the usual $3\frac{1}{2}$- to 7-deg angular deflection. The lock-type joint finds application aboveground in the process industries and in river crossings on bridges or trestles, as well as in submarine crossings or in unusually loose or marshy soils. Where the locking feature is on the spigot rather than on the bell, this type of pipe can be used with the regular line of mechanical-joint fittings.

Mechanical Roll-on-type Joint. Where a low-cost mechanical joint is desired, the roll-on type can be used. In this joint, a round rubber gasket is placed over the spigot end which is pulled into the bell by mechanical means, thus pulling the ring into place in the bottom of the bell. Outside the rubber gasket, braided jute is wedged behind a projecting ridge in the bell. This serves to confine the gasket under pressure in the joint. A bituminous compound is used to seal the mouth of the bell and to aid in retaining the hemp and the rubber gasket. In addition, when pipe is laid in cold climates where electrical thawing of mains and services is sometimes necessary, a cold lead strip about $\frac{1}{4}$ in. square can be calked in between the hemp and bitumastic to provide an electrical circuit through the joint. Either bell-and-spigot or mechanical (gland-type) fittings are used with this line of pipe.

Mechanical Screw-gland-type Joint. This type of mechanical joint for cast-iron pipe makes use of a coarse-threaded screw gland drawn up by means of a spanner wrench to compress a standard rubber or composition packing gasket. The joint allows from 2 to 7 deg angular deflection, as well as expansion or contraction without danger of leaks. A lead ring, inserted in the bell ahead of the gasket, seals off the contents of the line from the gasket. The ring also provides an electrical circuit through the joint for thawing out frozen underground mains and service lines by the electrical method. The screw-gland joint is used in piping

Table 80. Standard Dimensions of Tyton Joints

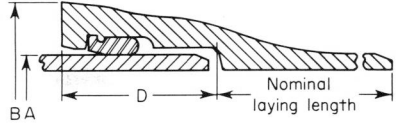

Nominal pipe size	Class	Thickness designation	Thickness of pipe	OD of pipe *A*	OD of bell *B*	Depth of socket *D*	Weight (approx), lb		
							Barrel per ft	Bell	18 ft length*
3	Through 350	22	0.32	3.96	6.08	3.00	11.4	11	215
4	Through 350	22	0.35	4.80	7.22	3.15	15.3	14	290
6	Through 350	22	0.38	6.90	9.47	3.38	24.3	25	460
8	Through 350	22	0.41	9.05	12.00	3.69	34.7	41	665
10	Through 250	22	0.44	11.10	14.20	3.75	46.0	54	880
	300	23	0.48				50.0		955
	350	24	0.52				53.9		1,025
12	Through 200	22	0.48	13.20	16.35	3.75	59.8	66	1,140
	250, 300	23	0.52				64.6		1,230
	350	24	0.56				69.4		1,315
14	50	21	0.48	15.30	19.15	5.00	69.7	78	1,335
	100, 150	22	0.51				73.9		1,410
	200	23	0.55				79.5		1,510
	250, 300	24	0.59				85.1		1,610
	350	25	0.64				92.0		1,735
16	Through 150	22	0.54	17.40	21.36	5.00	89.2	96	1,700
	200	23	0.58				95.6		1,815
	250	24	0.63				103.6		1,960
	300, 350	25	0.68				111.4		2,100
18	50	21	0.54	19.50	23.56	5.00	100.4	114	1,920
	100, 150	22	0.58				107.6		2,050
	200	23	0.63				116.5		2,210
	250	24	0.68				125.4		2,370
	300	25	0.73				134.3		2,530
	350	26	0.79				144.9		2,720
20	50	21	0.57	21.60	25.80	5.00	117.5	133	2,250
	100, 150	22	0.62				127.5		2,430
	200	23	0.67				137.5		2,610
	250	24	0.72				147.4		2,785
	300	25	0.78				159.2		3,000
	350	26	0.84				170.9		3,210
24	50	21	0.63	25.80	30.32	5.00	155.4	179	2,975
	100	22	0.68				167.4		3,190
	150	23	0.73				179.4		3,410
	200, 250	24	0.79				193.7		3,665
	300	25	0.85				207.9		3,920
	350	26	0.92				224.4		4,220

* Includes weight of bell.

which conveys water, gas, oil, and other fluids at considerable pressure. The gaskets and lead rings are interchangeable with those used in equivalent lines of mechanical joints of the bolted-gland type. A full line of fittings is available for use with screw-gland pipe.

Ball-and-socket Joints. For river crossings, submarine lines, or other places where great flexibility is necessary, cast-iron pipe can be obtained with ball-and-socket joints of either the lead-and-oakum or mechanical-gland types. The latter is shown in Fig. 93. Provision is made for longitudinal expansion and contraction in at least one mechanical-joint type, and a positive stop against disengagement of the joint is a feature of all designs. As much as 15-deg angular deflection can be accommodated without leakage. This pipe is heavy enough to remain underwater where laid without requiring river clamps or anchorage devices. The pipe may be

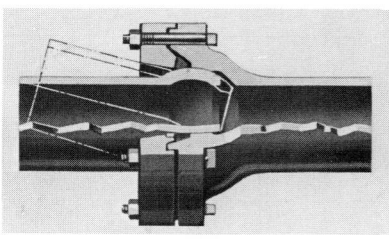

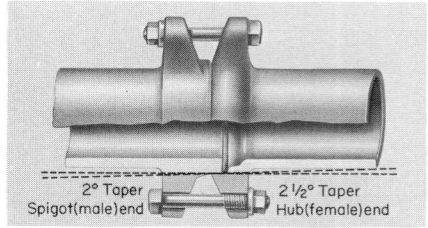

Fig. 93. Ball-and-socket mechanical joint for cast-iron pipe.

Fig. 94. Universal cast-iron pipe joint.

pulled across streams with a cable, as the joints are positively locked against separating, or it may be laid direct from a barge, bridge, or pontoons, without the services of a diver. The mechanical ball-and-socket joint is suitable for use with water, sewage, air, gas, oil, and other fluids at considerable pressure.

Either bell-and-spigot or mechanical (gland-type) fittings can be used with this line of pipe, although the integral ball present on the spigot end of some designs has to be cut off before the pipe can be inserted in a regular bell.

Universal Pipe Joints. This type of cast-iron pipe joint (shown in Fig. 94) has a machined taper seat which obviates the need for calking or for a compression gasket. The joint is pulled up snugly with two bolts, after which the nuts are backed off slightly, thus enabling the lock washers to give enough to avoid overstressing the socket or lugs. Pipe is made in 12- to 20-ft lengths to the usual pressure classes and can be bought as Type III under Federal Specification WW-P-421. Universal-joint fittings are available for use with the pipe. This type of joint is used to some extent in pipe diameters of 4 to 24 in. for underground water-supply systems, but it is not considered suitable for gas service and it does not permit much angular displacement or expansive movement.

Compression-sleeve Coupling. The type of joint shown in Fig. 95 is used with plain-end pipe of either cast iron or steel. It is widely known under the trade names of Dresser coupling and Dayton coupling. Compression-sleeve couplings are used extensively for air, gas, oil, water, and other services above- or underground. With a joint of this type, it is necessary to anchor or brace solidly at dead ends or turns to prevent the line from pulling apart. Compression couplings and fittings with screwed packing glands are available for use with small-size cast-iron or steel pipe. In welded transmission lines for oil or gas where any significant change in temperature is expected, a certain percentage of the joints may be made up with compression couplings instead of welding in order to allow for expansion.

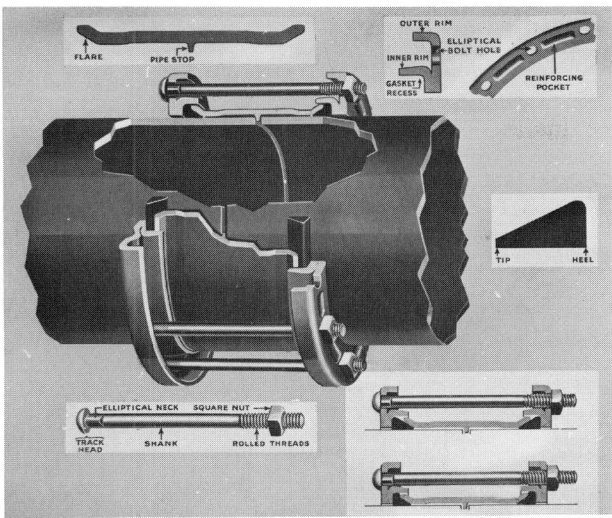

FIG. 95. Compression sleeve (Dresser) coupling for plain-end cast-iron or steel pipe.

Victaulic Coupling. The type of split coupling shown in Fig. 96 is used with either cast-iron or steel pipe having grooves near the ends which enable the coupling to grip the pipe in order to prevent disengagement of the joint. Victaulic couplings come in two segments in sizes from $3/4$ to 14 in. and in four or six segments in larger sizes. Grooved-end fittings are available to go with the couplings. With proper choice of gasket material, the Victaulic joint is suitable for use above- or underground with nearly any fluid or gas. Among its advantages are provision for considerable angular displacement and some expansive travel, pressure action in a direction that tends to seat the gasket, a positive stop against disengagement of the joint, and a design suited to rapid erection or tearing down of a line, as might be desired for temporary construction purposes. One of the disadvantages is the necessity for grooving the pipe ends nearly as deep as taper threads. Where this is objectionable as in the case of lightweight pipe, the ends can be upset or banded or can have heavier grooved nipples attached by welding.

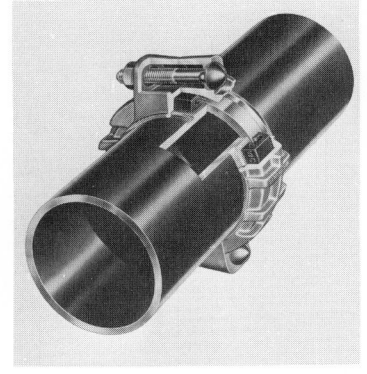

FIG. 96. Victaulic coupling for grooved-end cast-iron or steel pipe.

Flanged Joints. Flanged cast-iron pipe is used aboveground for low and intermediate pressures in water-pumping stations, gas works, power and industrial plants, oil refineries, booster stations for water, and gas and oil transmission lines. Flanges usually are faced and drilled according to the ASA Standards for Cast-iron Pipe Flanges and Flanged Fittings, ASA B16b2 (25 lb), B16.1 (125 lb), and B16.2 (250 lb). These are now being combined with the 800-lb (B16b1) class of cast iron into one standard.

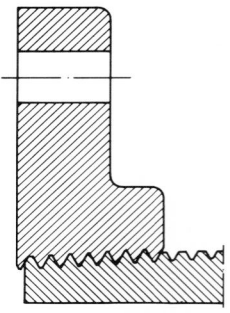

FIG. 97. Screwed-on
cast-iron flange.

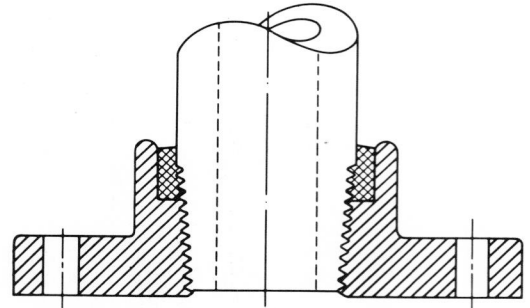

FIG. 98. High-hub cast-iron flanges with bitumastic
to protect the exposed threads.

Cast-iron pipe is made both with integrally cast flanges and with threaded companion flanges for screwing onto the pipe (as shown in Fig. 97).   In the latter case,
the outside diameter of the pipe conforms to "iron-pipe-size" (IPS) dimensions to
allow for the threads provided.   It is available in sizes 3 through 24 in. and in length
to 18 ft.   For lengths less than 3 ft in sizes 3 through 12 in. the flanges may be cast
integrally with the pipe, rather than screwed on the pipe, at the manufacturer's
option.

Standard dimensions of flanged joints for silver brazing are shown in Table 81.

**Table 81.   Standard Dimensions of 125-lb Flanged Joints for Silver Brazing with
Centrifugally Cast Pipe**

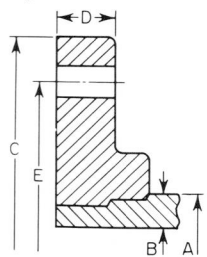

Nominal pipe size	Pipe	Flanges				Bolts	
	Outside diameter A*	Outside diameter C	Thickness D†	Bolt circle E	Weight, lb each	Number	Diameter
2	2.50	6	¾	4¾	4	4	⅝
3	3.96	7½	1	6	7	4	⅝
4	4.80	9	1⅛	7½	13	8	⅝
6	6.90	11	1¼	9½	17	8	¾
8	9.05	13½	1⅜	11¾	27	8	¾
10	11.10	16	1½	14¼	38	12	⅞
12	13.20	19	1½	17	58	12	⅞

   * For centrifugally cast pipe in metal molds in sizes 3 in. and larger, see ASA Standard A21.6
and Federal Specification WW-P-421.
   † Thickness D is slightly heavier than for standard cast-iron flanges in ASA B16.1-1960.

The end of the centrifugally cast pipe is machined to accept the flange which is press-fitted and silver-brazed to the pipe. The face of the flange and the pipe are machined simultaneously. The completed joint has a strength equal to or greater than that of the pipe itself.

A line of high-hub cast-iron flanges (see Fig. 98) is also manufactured for use with centrifugally cast iron pipe. This flange has the same bolt circle and size of bolts as the 125-lb cast-iron companion flange. It is threaded with the same form of thread as National Pipe Thread and the same taper. However, the pitch diameter is different. An American standard does not exist for these threads.

The purpose of the high hub is to provide a cup to fill with calking compound or bitumastic after the flange is made onto the pipe. This protects the exposed threads. The face is machined the same as a standard-weight cast-iron pipe flange and it uses a standard service gasket.

Screwed Joints. Small-size cast-iron pipe also is made to iron-pipe-size (IPS) dimensions for use with screwed couplings or fittings similar to those used with threaded steel pipe. Pressure pipe for water, gas, steam, and other services is supplied in standard, strong, and extra-strong weights and made to meet the requirements of Federal Specifications WW-P-421. A lightweight pipe for drain lines inside buildings and elsewhere is available which conforms to American Standard Cast-iron Pipe for Drainage, Vent and Waste Services, ASA A40.5. It is used with American Standard Cast-iron Screwed Drainage Fittings, ASA B16.12, or with other screwed fittings suiting particular applications.

JOINING METHODS

Lead Casting. Table 82 gives the weights of lead and hemp required for cast-iron soil pipe and fittings. For bell-and-spigot joints, these requirements are given in Table 83.

The requirements for minimum depth of jointing materials shown in Table 84 and the following specifications for jointing materials and making joints are taken from the Tentative AWWA Standard for Installation of Cast Iron Water Mains, AWWA C600-64T. Lead for calking purposes shall contain not less than 99.73 per cent pure lead. Impurities should not exceed the limits listed in Table 85.

Pipe joints should be sealed with cast lead, with cast sulfur compound, or with portland cement as required, following the preparation of the joint base with yarning

Table 82. Approximate Weights of Cast Lead and Hemp Needed per Joint for Cast-iron Soil Pipe and Fittings

Nominal pipe size, in.	Weight per joint, lb	
	Lead	Hemp
2	1½	1/7
3	2¼	⅙
4	3	⅕
5	3¾	¼
6	4½	⅓
8	6	3/7
10	8	½
12	9½	4/7
15	13	⅞

Table 83. Approximate Weights * of Cast Lead and Hemp Needed for Cast-iron Bell-and-spigot Water Pipe and Fittings (AWWA-C 102, ASA A21.2, ASTM A377 and Federal WW-P-421)

Nominal pipe diameter, in.	Weight of lead per joint, lb				Weight of hemp per joint, lb‡
	Depth of lead in bell, in.†				
	2	2¼	2½	3	
3	6.0	6.6	7.2	8.3	0.18
4	7.5	8.2	8.9	10.3	0.21
6	10.3	11.3	12.2	14.2	0.31
8	13.3	14.5	15.8	18.3	0.44
10	16.0	17.7	19.2	22.3	0.53
12	19.0	20.8	22.6	26.2	0.61
14	22.0	24.1	26.1	30.3	0.81
16	30.0	33.0	36.0	42.0	0.94
18	33.8	36.9	40.2	46.7	1.00
20	37.0	40.6	44.3	51.6	1.25
24	44.0	48.3	52.6	61.3	1.50
30	54.3	60.0	65.3	66.1	2.06
36	64.8	71.6	78.0	90.7	3.00
42	75.3	83.0	90.5	105	3.62
48	85.6	94.7	103	120	4.37
54	97.6	106	115	134	6.25
60	108	117	128	149	8.25
72	146	162	178	210	12.5
84	170	188	207	244	15.0

* Courtesy Thomas F. Wolfe, Engineer, Cast Iron Pipe Research Association.
† For minimum depths required, see Table 84.
‡ Very approximate only. The actual weight of hemp will vary with its own density and the depth left for pouring lead.

Table 84. Minimum Depths of Jointing Material for Cast-iron Bell-and-spigot Water Pipe (Tentative AWWA Standard for Installation of Cast Iron Water Mains, AWWA C600-64T)

Jointing material	Min depth, in.	Nominal pipe diameter, in.
Cast lead	2¼	3–20
	2½	24, 30, 36
	3	36 and larger
Cement.	3	All

Table 85. Permissible Impurities in Lead Intended for Calking Purposes (AWWA C600-64T)

Materials	Per cent
Arsenic, antimony, and tin combined	0.015
Copper	0.080
Zinc	0.002
Iron	0.002
Bismuth	0.250
Silver	0.020

material. Braided hemp or jute or other suitable yarning material should be of proper dimension to center the spigot in the bell. Each strand of yarn should be cut somewhat longer than the circumference of the pipe so that the ends will overlap. Moreover, the overlapped ends of successive strands should be staggered. Each strand should be thoroughly packed and hammered home into the joint with suitable yarning tools.

Each lead joint should be made with one pour, filling the joint space to the depth specified in Table 84. After cooling to the temperature of the pipe, lead joints should be calked by means of pneumatic tools or hand tools. The lead should be thoroughly compacted to insure watertight joints without overstraining the bells.

Lead Wool. Lead wool consisting of finely spun lead in continuous strands can be used without heating to fill bell-and-spigot joints of water or gas mains. This application originated in Germany, where lead wool was used to a considerable extent for making joints in large-diameter cast-iron pipe. For pipes of less than 24 in. diameter, lead-wool joints are not as economical as cast-lead joints. For larger pipes, the cost of each kind is about the same. Lead wool has been used extensively in the United States, particularly for gas mains and for water pipes laid either in wet trenches or underwater where there is a need for cold application. Because each skein or strand is calked separately as it is introduced into the joint, the lead compacts into a very dense material which has great holding qualities suited for pressures up to 500 psi. Well-calked lead wool becomes denser than cast lead, which shrinks from the iron in cooling and can be expanded out again to only a limited depth. Hence depths of lead wool in the bell of only $1\frac{3}{8}$ to $1\frac{5}{8}$ in. are said to suffice for pipe diameters of 20 in. and greater.

Sulfur Compounds. Each sulfur-compound joint is filled with one continuous pour while the compound is at the proper temperature. The solidified compound in the pouring gate is cut off flush with the top of the bell or broken off flush with the top of the joint. One of the sulfur compounds known as *leadite*, which is composed of a finely ground mixture of iron, sulfur, slag, and salt is used to a considerable extent for making bell-and spigot joints in cast-iron water and gas mains.

Leadite weighs only 118 lb/cu ft when melted, whereas lead weighs 710 lb/cu ft. One ton of leadite will make about five times as many joints as a ton of lead. Leadite melts at a temperature of about 400 F and is poured like molten lead. Because it expands in cooling and forms a vitreous, watertight substance, leadite requires no calking. When the joints are properly made, leadite can be used for pressures up to 250 psi. Since leadite is more elastic than lead, it is not squeezed out of a joint as lead may be when a pipeline settles.

Cement. Portland-cement joints after yarning should be filled with neat cement, barely moistened. This should be calked until thoroughly compacted without overstraining the bell to make a watertight joint. The cement should be so dry that it will ring with a metallic sound while being calked. Cement joints should be covered immediately with slightly moist earth, or with damp burlap to insure sufficient time for complete hydration. Pipes joined with cement should not be subjected to water pressure until 36 hr have elapsed after the last joint was made.

PIPE INSPECTION

The primary purpose of inspection and interpretation standards is the detection of defective conditions which might, during the intended life of the piping component or system, lead to a premature service failure.[1] Quality control and inspection

[1] H. Thielsch, "Quality Control and Service Performance," *ASME Trans.*, vol. 86 Series *A*, pp. 451–464 (1964).

procedures are applied at various stages, ranging from the initial manufacture of the wrought, cast, or forged pipe, fitting, or valve to the final erected piping system.

Every producer, fabricator, and contractor must utilize the same, as well as different, inspection tools and techniques to insure that the materials, assembled joints, and welds meet applicable quality and code standards and represent good workmanship. The success and usefulness of any inspection or test method depends upon the proper use of the respective equipment, the correct application of the techniques, and the right judgment in the interpretation of the results.

Perfect piping materials are not commercially available. Defects range from atom-sized dislocations in the metal, which cannot be observed with the highest-powered microscopes, to major metal discontinuities that are visible to the unaided

Table 86. Common Base Metal and Weld Defects and Other Structural Variables

Defects, flaws and discontinuities	1	Surface:
		a Cracks or tears
		b Shrinkage cavities
		c Inclusions
		d Porosity
		e Laminations or cold shuts
		f Folds, laps, scabs, seams, blisters, or bark
		g Pits
		h Die grooves or notches
		i Undercuts
		j Lack of penetration[1]
		k Incomplete fusion[1]
		l Burn-through[1] or excessive penetration[1]
		m Sink,[1] suck-back[1] or concavity[1]
		n High-low or poor fit-up
	2	Subsurface:
		a Cracks or tears
		b Shrinkage cavities
		c Inclusions or sand spots
		d Porosity, gas pockets, or voids
		e Laminations
		f Lack of penetration
		g Incomplete fusion
Structural variables		a Hardness and hardness differences, tensile strength, ductility
		b Structure-sensitive (metallurgical) or strain-sensitive properties and sorting
		c Fabrication-induced microstructural variables
Chemical properties		a One component, chemical composition (differences in carbon content or other elements within material)
		b Two or more components (differences between adjacent materials, as pipe and weld, of same type composition)
		c Dissimilar materials (differences between dissimilar materials, such as carbon and stainless steel)
Dimensions		a Thickness
		b Branch connections
		c Dissimilar materials (differences between dissimilar materials, such as carbon and stainless steel)

[1] May be visible only when underside of weld on inside of pipe is accessible for visual examination or examination with special tools, such as mirrors, borescopes, etc.

eye. Many apparent defects that are readily visible may not reduce the service life of the piping component. On the other hand, many other visible surface defects and invisible defects may be harmful and may result in service failures.

Codes generally recognize that some defects, when present within certain established limitations, are not harmful. Thus, ASTM piping specifications permit minor surface laps and scabs. Some shrinkage, sand inclusions, and gas pockets are tolerated in applicable casting specifications. Most standards involving weld quality permit some porosity and slag in weld deposits. Some which relate to noncritical applications allow lack of penetration in the root of pipe butt welds.

Base metal and weld defects and other structural variables which are detected, measured, or identified by visual examination and nondestructive tests are summarized in Table 86. Broadly speaking, defects and other structural variables can generally be classified as either mechanical notches or metallurgical notches. In Table 86 metallurgical notches would be represented by the examples listed under structural and chemical properties.

Not all defects or defect conditions are detectable by nondestructive test methods. Many conditions which cause metallurgical embrittlement are not readily detected. Some defects represent metallurgical notches when the differences in properties are of sufficient magnitude to become potential causes of failure. Metallurgical notches can occur within the same section of metal where localized heat treatment produces an area of high hardness adjacent to an area of lower hardness. Other typical metallurgical notches include hardened heat-affected zones adjacent to weld deposits and differences in properties between a wrought base metal and a weld deposit. More obvious metallurgical notches occur between dissimilar metals.

These types of potential defect conditions generally are not detected in the original fabricated piping. After cracks have been initiated during service and have propagated, many of these defects become apparent and can be detected by inspection techniques. Many metallurgical and mechanical notches, of course, never result in failures in many service environments.[1]

In selecting inspection and test methods, a number of factors should be taken into account. They are summarized in Table 87. For a detailed discussion of the various inspection and test methods applied to piping, extensive handbooks should be consulted.[2]

It must be recognized that each specific nondestructive or destructive test and technique does not determine identical characteristics of the material being tested. The tests may supplement each other and may not be a substitute of one for another. The interpretation of test results and their correlation with the expected behavior of the material under actual service conditions is at best only approximate. Moreover, the interpretation is only as good as the experience and judgment of the inspector. Quite frequently, erroneous interpretations are made that provide results which have little or no relation to the actual service performance of the piping material.

Among the most common nondestructive inspection methods applied to piping and tubing, fabricated pipe components, and welded joints are radiographic inspection with X-ray and radioactive isotopes, ultrasonic inspection, dye penetrant inspection, magnetic particle inspection and eddy current testing.

[1] H. Thielsch, "How Sound Engineering Combats Excessive Service Conditions for the Prevention of Pipe Failures," *Heating, Piping and Air Conditioning*, vol. 35, pp. 114–122, October, 1963.

[2] For example, Society for Nondestructive Testing, *Nondestructive Testing Handbook*, Robert C. McMaster (ed.), 2 vols., Ronald Press, 1959.

Table 87.   Factors to Be Considered in the Selection of Inspection and Test Methods

Material to be tested	1	Ferromagnetic alloy
	2	Nonmagnetic light alloy
	3	Nonmagnetic heavy alloy
	4	Clad alloy
	5	Lined alloy
	6	Metal-coated alloy
	7	Nonmetallic coated alloy
Method of manufacture	1	Cast
	2	Wrought
	3	Forged
Method of fabrication	1	Welded
	2	Brazed
	3	Soldered
	4	Glued
	5	Mechanical joints
Geometry of material	1	Thickness
	2	Shape
	3	Surface conditions
Defects which may be present	1	Type
(see Table 86)	2	Size
	3	Shape (elongated, spherical)
	4	Distribution (linear, scattered, clustered, etc.)
	5	Location (surface or subsurface)
Metallurgical notches which may be	1	Dissimilar materials
present	2	Dissimilar mechanical properties in same material
	3	Dissimilar metallurgical microstructures
Sensitivity and resolution desired or	1	Standard and specification requirements
required	2	Test sample
	3	Destructive tests
Economics		All materials

RADIOGRAPHIC INSPECTION

Radiographic inspection is performed on pipe base materials, such as castings, and on joints which may have been welded, brazed, or soldered. It is one of the most important and standardized nondestructive inspection tools. The primary purpose of radiography is to detect weld root defects and internal discontinuties in the base metal or in welds. Normally, the discontinuties are projected onto a film which thus provides a permanent record of the discontinuities.

Methods and Limitations

Commercial inspection is performed with X rays or with radioactive isotopes, such as iridium-192 or cobalt-60. X-ray equipment is more widely used in shop radiography, although portable X-ray equipment is being increasingly used on field inspection. Extensive use in shop and field inspection is being made of the radioactive isotopes, iridium-192 and cobalt-60, transported in lead- or uranium-shielded portable containers. Some limited use is made also of the cesium-137 isotope.

Table 88. Typical Industrial Radiation Sources and Their Applications

Radiation source	Peak, kv (KVP), max	Screens	Approximate practical thickness limits, in.*
X rays	50	None	Microradiography, plastics, wood
	100	None	2 aluminum, 3 magnesium
	150	None or lead foil	4½ aluminum, 1 steel, or equivalent
	150	Fluorescent	1½ steel or equivalent
	250	Lead foil	2 steel or equivalent
	250	Fluorescent	3 steel or equivalent
	400	Lead foil	3 steel or equivalent
	400	Fluorescent	4 steel or equivalent
	1,000	Lead foil	5 steel or equivalent
	1,000	Fluorescent	7 steel or equivalent
	2,000	Lead foil	9 steel or equivalent
Radioactive isotope material	Radium	Lead foil	½–5 steel or equivalent
	Iridium-192	Lead foil	½–2½ steel or equivalent†
	Cobalt-60	Lead foil	1–10 steel or equivalent*
	Cesium-137	Lead foil	¾–3 steel or equivalent

* Based on the use of pulsating potential type of equipment. Practical thickness limits for constant potential equipment are somewhat higher.

† The thicker sections require stronger sources in order to obtain reasonable exposure times. Cobalt-60 sources, for example, are available in strengths up to 2,000 curies.

Radiographic-inspection methods employ short-wavelength radiation which penetrates materials that are opaque to ordinary light. The penetrating power generally increases for the shorter wavelengths.   Some of the radiation is absorbed. The amount depends on the wavelength and on the density and thickness of the material being radiographed.   Where cavities are present in the material, such as a shrinkage cavity or void in a casting or porosity in a weld, less radiation is absorbed. A greater amount of radiation is then projected onto the radiographic film. Inclusions of density greater than that of the material being radiographed, such as tungsten inclusions in some types of weld deposits, will absorb more radiation and thus decrease the amount of radiation projected onto the film.

Approximate practical thickness limits of materials to be inspected are given in Table 88.   Radiographic equivalence factors of a number of common metals are summarized in Table 89.   These factors are helpful in determining the practical thickness limitations for radiation sources or the exposure factors for one metal from exposure charts compiled for another metal.

Table 89. Approximate Radiographic Equivalence Factors for Several Metals in Relation to Steel

Metal	Density	140 kv	220 kv	250 kv	400 kv	1 mev	2 mev	Iridium-192	Cobalt-60
Aluminum	2.7	0.10	0.20	0.24	....	...	...	0.33	0.35
Magnesium.........	1.7	0.05	0.08	0.08	....	...	...	....	0.24
Steel...............	7.8	1.0	1.0	1.0	1.00	1.0	1.0	1.0	1.0
Stainless (18–8)	7.9	1.0	1.0	1.0	1.0	1.0	1.0	1.0	1.0
Copper	8.9	1.5	1.4	1.4	1.4	...	....	1.1	1.12
Monel	8.9	1.5	1.2	1.2	....	...	...	1.0	1.0
Brass	8.4	....	1.3	1.3	1.3	...	...	....	1.06
Zinc	7.1	....	1.3	1.3	1.3	...	...	....	0.91
Lead	11.3	....	11.0	....	....	5.0	2.5	5.0	1.66

The practical thickness, which may be radiographed, is calculated as follows: Table 89 shows the radiographic equivalence factor of aluminum as 0.24 for 250 kv. For an X ray of 250 kv peak (KVP) for which the practical limit in steel is 2 in. (Table 88), the upper practical limit for aluminum would be approximately $2/0.24 = 8.5$ in.

Sensitivity

Codes generally require that radiography be performed with a technique which will produce a sensitivity sufficient to indicate on the films the features of special penetrameter gage. These gages actually are not defect simulators, although they are intuitively designed as if they represented porosity defects. Although their thickness is approximately 2 per cent of the weld thickness, distinguishing the hole or slit of the penetrameter on the radiographic film is no guarantee that weld or pipe defects larger than 2 per cent of the pipe wall thickness will likewise be shown on the radiographic films.

Although not all inspection standards have accepted the same penetrameters (image-quality indicators), those that are most widely used are shown in Fig. 99. They are accepted by the American Society of Testing Materials.[1,2] The thickness of the penetrameter is equal to, or less than, 2 per cent of the weld thickness. Lead numerals attached to one end indicate the penetrameter thickness. Three holes are normally drilled in its face, the diameters of which are equal to certain multiples of the penetrameter thickness, such as two, three, and four times[*] or one, two, and four times.[*,†,‡] It is generally accepted that for section thicknesses less than $\frac{1}{2}$ in., the minimum hole sizes shall be 0.010, 0.020, and 0.040 in.[†] or $\frac{1}{16}$ in.[*]

The penetrameters should be placed on the side of the section nearest to the source of radiation. When this is not possible, as is generally the case with piping, the penetrameter may be placed on the film side. This may necessitate making test radiographs under substantially identical conditions as those used in production radiography. The penetrameters are then placed on both the radiation side and film side and sufficiently staggered to obtain separate images. Each penetrameter should be provided with a marker which will show clearly on the film and which will indicate the side of the joint on which it is located—such as "F" for film side and "S" for radiation side. The $2T$ hole shall then be distinguishable on both the "F" and "S" penetrameters. The difference in penetrameter images indicates the degree of distortion.

Skilled radiographers recognize penetrameters as a code-required sensitivity standard, but they view penetrameters primarily as a means to provide some evidence of the sharpness of image delineation and photographic contrast. Penetrameters will not prove the radiograph was properly made. Angulation of the radiation source, insufficient film-to-source distance, or the use of different types of radiation sources may "wash out" critical linear defects, such as cracks and lack-of-penetration, even though the penetrameter image is good. For this reason, many skilled radiographers rely heavily on the internal evidence of the radiographic film itself.

[1] ASTM Standard E142, Tentative Method for Controlling Quality of Radiographic Testing.
[2] ASTM Standard E94, Tentative Recommended Practice for Radiographic Testing.
[*] ASME Boiler and Pressure Vessel Code, Section I, Power Boilers; Section II, Nuclear Vessels; and Section VIII, Unfired Pressure Vessels.
[†] ASTM Standard E142, Tentative Method for Controlling Quality of Radiographic Testing.
[‡] ASTM Standard E94, Tentative Recommended Practice for Radiographic Testing.

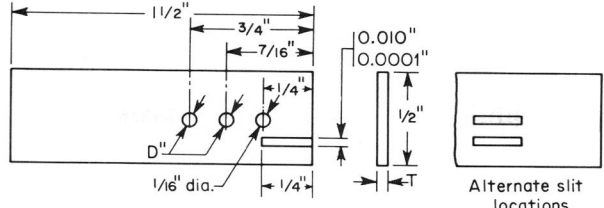

(a) ASME design of penetrameters for weld thickness through 1/2 in.

Minimum penetrameter thickness   – 0.005 in.
Minimum diameter for 1 T hole   – 0.010 in.
Minimum diameter for 2T hole   – 0.020 in.
Minimum diameter for 4T hole   – 0.040 in.
Holes shall be true and normal to the surface of the penetrameter. <u>Do not chamfer</u>

Design for penetrameter thickness from 0.005 in. to and including 0.050 in.

From 0.005 in. through 0.020 in. made in 0.0025 in. increments
From 0.025 in. through 0.050 in. made in 0.0050 in. increments

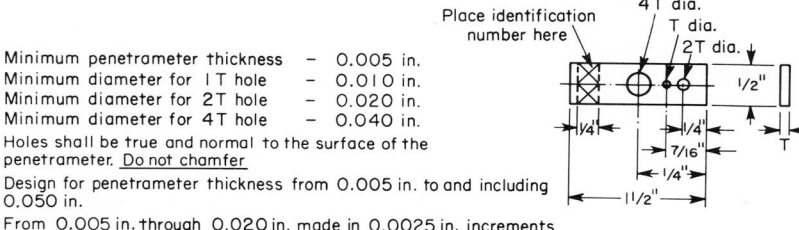

(b) Design of penetrameters for weld thickness ASME – over 1/2 in. through 4 in.
ASTM – through 4 in.

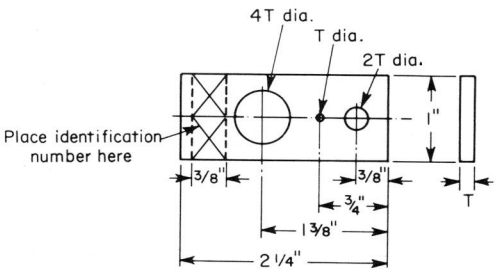

Design for penetrameter thickness from 0.060 in. to and including 0.160 in. made in 0.010 in. increments

(c) ASME and ASTM design of penetrameters for weld thicknesses over 4 in.

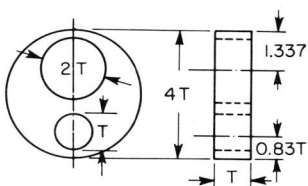

Design for penetrameter thickness of 0.180 in. and over made in 0.020 in. increments

(d) ASTM and U.S. Navy design of penetrameters for weld thicknesses over 3/4 in.

FIG. 99. Penetrameter dubs (ASTM Standard E142-64 and ASME Boiler and Pressure Vessel Code).

The wavelength of the radiation from radioactive isotope sources is shorter than that of X rays. Thus, the image contrast generally tends to be less than that obtained from X-ray sources. Moreover, there is a tendency to underexpose radioactive isotope exposures because of the long exposure times involved. Low film contrast, as may be indicated by the penetrameter images, in combination with a low film density, results in poor fine-image detail. On the other hand, the advantage of short-wavelength radiation is that wider range of section thicknesses and of defects of varying depth dimensions can be shown on the same radiograph.

Interpretation

Extensive judgment and experience are necessary to interpret a radiographic film. An inspector should be familiar with the various types of defects which may be present in the base metals or welds he is examining. He must know the materials involved and their relative tendency toward embrittlement and crack propagation at room and service temperatures, and he must be able to interpret whether or not an indication on the film is critical or is inconsequential. Proper interpretation should consider, also, the manufacturing methods, casting techniques, and welding processes employed. Each manufacturing and welding process tends to produce some specific types of weld contours, defect patterns, and conditions which may or may not be critical. Without proper training and experience, a minor noncritical defect may be removed at substantial cost only to be replaced with a far more critical defect condition which may not be apparent after subsequent radiography.

Only a limited number of radiographic interpretation guides and standards have been issued to date. They do not always agree in requirements and levels of acceptability. Care must be taken, therefore, that prior to the interpretation of films the applicable or specified standards are ascertained and then followed with proper judgment.

Casting Standard

Several major codes require that casting, when used in critical piping systems, be examined by radiographic inspection (as summarized in Table 40).

Table 90. Suggestions for the Classification of Castings to Be Used With Radiographic Standards (ASTM Specification E71-52)

Class	Service
1	High-pressure or high-temperature service castings or both (wall thickness less than 1 in.); machinery castings[1] subject to high fatigue or impact stresses (wall thickness less than $\frac{1}{4}$ in.)
2	High-pressure or high-temperature service castings or both (wall thickness 1 in. or greater); low-pressure service castings (wall thickness less than 1 in.); machinery castings subject to high fatigue or impact stresses (wall thickness of $\frac{1}{4}$ in. and greater)
3	Low-pressure service castings (wall thickness of 1 in. and over); machinery castings subject to normal fatigue or impact stresses
4	Structural castings[2] less than 3 in. in thickness and subject to high service stresses; machinery castings subject to low-impact stresses or vibration
5	Structural castings 3 in. or more in thickness and subject to high service stresses

[1] Machinery castings are dynamic parts or members in contact with working parts.

[2] Structural castings are construction parts for machinery castings.

On steel castings, the standard most widely used is ASTM Standard E71.[1] This Standard establishes the five classifications described in Table 90.

Standard E71 is furnished with two sets of comparison reference radiographs— one for X ray radiation and the other for gamma-ray radiation. The comparison radiographic films illustrate acceptable, borderline, and unacceptable levels for each of the five classifications listed in Table 90 which represent the following casting defects:

Group	Defect
A	Gas and blow holes
B	Sand spots and inclusions
C	Internal shrinkage
D	Hot tears
E	Unfused chaplets
F	Internal chills

Aluminum castings for piping systems are generally radiographed to the defect requirements of ASTM Standard E155.[2] It illustrates many types of defects common in aluminum castings, even though the latter specification was initially prepared primarily for aerospace applications. Bronze and brass castings for piping systems, which are most widely used in marine and Navy applications, are standardized in NAVSHIPS 250-537-1.[3]

Weld Standards

Radiographic inspection of pipe welds is required by many major codes concerned with piping, as is seen by reference to Table 40. Although butt welds are most generally involved, radiography may be required, also, on nozzle, tee and other weld joints.

Interpretation standards of radiographic films of welds have been issued by various standardization groups. Reference radiographs illustrating some of the major types and degrees of discontinuities in steel welds have been issued in ASTM Standard E99[1] and also by the International Institute of Welding.[5] In general, these radiographs are intended to serve as a guide for radiographic interpretation only. They are examples of limits of acceptability for various common weld discontinuities. More specific limits are established, however, by most of the codes which pertain to pipe welding.

Common types of weld discontinuities and contours which may or may not be critical are summarized in Table 91. Porosity, slag inclusions, lack of penetration, and cracks generally are covered by all codes which have established limitations for these. However, icicles, burn-through, sink (suck-back or concavity), high-low, tungsten inclusions, and others generally are neither recognized nor mentioned. The reason is that these conditions appear primarily in the root of pipe welds made by inert-gas tungsten-arc welding or by gas-shielded consumable metal-arc welding. When the major codes established their respective sections on radiographic acceptability limits, pipe welds for critical service applications generally were made with backing rings.

---

[1] ASTM Standard E71, Industrial Radiographic Standards for Steel Castings.
[2] ASTM E155 Standard Reference Radiographs for Inspection of Aluminum and Magnesium Castings.
[3] NAVSHIPS 250-537-1, Radiographic Standards for Bronze Castings for X-Rays (to 400 KVP) and Iridium, Bureau of Ships, Navy Department, Washington, D.C.
[4] ASTM E99 Standard Reference Radiographs for Steel Welds.
[5] Radiography of Welds, International Institute of Welding, Commission V, 1961.

Table 9I. Appearance of Weld Discontinuities on Radiographic Films

Type	Definition	Radiographic appearance
Porosity	Gas pockets or voids in weld deposit	Generally rounded shadows of varying sizes and densities occurring singly, in clusters, or scattered randomly throughout the weld; sometimes porosity may appear as elongated or pointed shadows
Slag inclusions	Nonmetallic solid material entrapped in weld deposit, between weld metal and base metal, or alongside root of backing ring	Shadows of elongated or irregular contour occurring singly, in a linear distribution or scattered randomly throughout the weld
Wagon tracks	Separation of backing strip in root area, creating either a void or trapping slag	Darker, often parallel linear indications, continuous or intermittent along root area
Lack of penetration	Root penetration which is less than complete	A straight, dark continuous or intermittent linear indication, often as a straight line in center of weld
Incomplete fusion	Fusion which is less than complete either between weld beads or between weld and base metal	A dark shadow, usually of elongated shape
Incomplete fusion in root of weld	Fusion which is less than complete between weld and land or between weld and consumable insert in root of pipe	A tight linear indication
Cracks	Discontinuities formed by very narrow separation of weld or adjacent base metal	Fine dark line, straight or wandering in direction
Undercut	A groove melted into the base metal adjacent to the toe or surface of a weld and left unfilled by weld metal	A dark linear shadow of wavy contour occurring adjacent to the edge of the weld, that may be seen visually
Icicles and burn-through	A coalescence of metal beyond the root of the weld or a melting of metal away from the root of the weld; also through backing strip or ring	Individual light, circular indications or individual or continuous darkened areas of elongated or rounded contours, that may be surrounded by light rings
Sink, suck-back, concavity	Sink or suck-back in root of weld bead producing concavity in welds made with or without backing rings (often confused with lack of penetration)	A dark, often linear, groovelike indication, frequently with a slightly wavy edge or edges
High-low	Mismatch of weld root due to ovality, fit-up, nonuniform weld end preparation, or other causes	A dark, linear indication, frequently with a single slightly wavy edge
Tungsten inclusions	Inclusion of tungsten particles in weld deposit from electrode used in inert-gas tungsten-arc welding	Very light indications of the tungsten particles
Slugging (stubbing)	The addition of a separate piece or pieces of material in a joint before or during welding (frequently caused by entrapped welding filler wires or electrodes)	Darkened indications, usually outlining the contour of the piece or pieces of material added
Convexity	Excessive penetration below pipe ID	Light weld-bead-like indications

Because of the increasing importance of the inert-gas tungsten-arc-welding process on pipe welding (particularly for the root passes), the Pipe Fabrication Institute[1] has issued recommendations covering the interpretation of these welds.

Tungsten Inclusions. Small tungsten inclusions are not considered to be any more detrimental to the quality of a weldment than slag, porosity, and similar defects of comparable size.   Indications of tungsten inclusions shall be interpreted as porosity and should be accepted or rejected on the basis of the porosity standards of the applicable inspection code.

[1] PFI Standard ES-15, Recommended Radiographic Interpretation of Tungsten Inert Gas Welds.

Weld-root Contour. Governing codes do not make specific reference to excessive root convexity, root concavity and poor fit-up[1] of weld made by the inert-gas tungsten-arc welding process. These conditions do occur and, accordingly, should be given consideration by the radiographer, who should consider, also, if a contour discontinuity creates a notch or lessens the material thickness.

Convexity. With exception of small diameter piping where excessive root convexity could result in flow restrictions, convexity is not generally considered detrimental. However, in some welds, there may be intermediate areas where the transition from the weld to adjacent base metals is not completely smooth. Where the thickness of such joints is approximately less than ½ inch, is of an austenitic material (especially the high-nickel alloys), and is subject to radiographic inspection, dark lines may appear on the radiograph along the fusion line of the weld. These lines are often misinterpreted as cracks. A subsequent radiograph of the questionable area from a slightly different angle and when possible, a longer exposure, will generally clarify the harmless character of these indications as shallow round bottom convexity.

Concavity (Sink or Suck-back). Concavity of the weld root contour should be permitted when the resulting thickness of the weld metal is equal to or greater than the minimum thickness of the adjacent base metal. The contour of the concavity should have a uniform radius and should blend smoothly into the base metal.

Fit-up (High-low). A small amount of mismatch of the pipe I. D. at the welding land is not detrimental, as long as the root of the completed weld blends smoothly with the base metal. Radiographers should familiarize themselves with this condition and interpret the radiographs accordingly. The high-low condition may also occur in combination with convexity or concavity.

Irrelevant Film Indications. In the radiograph of inert-gas tungsten-arc welds of relatively light wall (under ¼ inch) austenitic materials (particularly the high-nickel alloys), the larger grain size of the weld-metal deposit causes a mottled appearance on the film. This phenomenon is caused by diffraction of the x-rays and interferes with the interpretation of the radiographs. This condition may result in the rejection of sound welds by radiographers not familiar with the phenomenon. A subsequent radiograph of the questionable areas with very fine grain film, or the use of a higher x-ray voltage, will generally establish that these film indications are due to x-ray diffraction and are not actual defects in the welded joint. (Irrelevant film indications can be caused also by improper radiographic techniques and incorrect film processing.)

Code Requirements. The ASME Boiler and Pressure Vessel Code, the ASA Code for Pressure Piping, applicable ASTM specifications, and other standards generally make only little or no distinction between the different radiation methods— usually mentioning only "radiographic examination." In addition to those mentioned previously, other critical variables which may determine whether or not a defect is detected involve orientation of the defect in relation to the radiation source, film grain, film density, and size of radiation source. However, even with the most favorable techniques, defects and cracks in pipe welds may not always be detected by radiography.[2]

Code-writing committees recognize different levels of pipe-welding quality. In some cases, these levels are tied to different degrees of radiographic inspection, as illustrated below in Section 31.3, Refinery Piping, of the Code for Pressure Piping prepared by the American Standards Association.

The major interpretation requirements applicable to pipe welds in the ASME Boiler and Pressure Vessel Code, the ASA Code for Pressure Piping, and the API Standard for Field Welding of Pipe Lines are given below.

---

[1] Some codes, however, show maximum mismatch tolerances (see Table 40).

[2] H. Thielsch, "Relate Welding Processes to Materials and Conditions for the Prevention of Pipe Failures," *Heating, Piping and Air Conditioning*, vol. 35, pp. 104–109, March, 1963.

ASME BOILER AND PRESSURE VESSEL CODE—
SECTION I—POWER BOILERS; SECTION III—
NUCLEAR VESSELS; SECTION VIII—UNFIRED
PRESSURE VESSELS

100% *Radiographic Examination*
CRACKS—Not acceptable
INCOMPLETE PENETRATION—Not acceptable
SLAG—Not acceptable when length of slag inclusion is greater than
$\frac{1}{4}''$ for up to $\frac{3}{4}''T$
$\frac{1}{3}T$ for $\frac{3}{4}''$ to $2\frac{1}{4}''T$
$\frac{3}{4}''$ for over $2\frac{1}{4}''T$
where T is the thickness of the thinner plate or pipe being welded.
—Not acceptable when any group of slag inclusions in line has an aggregate length greater than T in a length of $12T$, except when the distance between the successive imperfections exceeds $6L$ and L is the length of the longest imperfection in the group.
POROSITY—In accordance with ASME Porosity Standards.

ASME BOILER AND PRESSURE VESSEL CODE-SECTION VIII—
UNFIRED PRESSURE VESSELS

Spot Radiographic Examination
CRACKS—Not acceptable
INCOMPLETE PENETRATION—Not acceptable
SLAG—Not acceptable if the length of slag or cavities is greater than $\frac{2}{3}T$, where T is the thickness of the thinner plate or pipe welded. If several imperfections within the above limitations exist in line, the welds shall be judged acceptable if the sum of the longest dimensions of such imperfections is not more than T in a length of $6T$ (or proportionately for radiographs shorter than $6T$), and if the longest imperfections considered are separated by at least $3L$ of acceptable weld metal, where L is the length of the longest imperfection. The maximum length of acceptable imperfection shall be $\frac{3}{4}''$. Any such imperfections shorter than $\frac{1}{4}''$ shall be acceptable for any plate or pipe thickness.
POROSITY—Not a factor

ASA CODE FOR PRESSURE PIPING—B31.1-1955—
SECTION I—POWER PIPING

100% Radiographic Examination—Same requirements as Section I, ASME Boiler and Pressure Vessel Code.

ASA CODE FOR PRESSURE PIPING—SECTION 31.3—PETROLEUM REFINERY
PIPING

100% Radiographic Examination
CRACKS—Not acceptable
INCOMPLETE PENETRATION—Not acceptable unless otherwise specified by the engineering design.
SLAG—The developed length of any single slag inclusion shall not exceed $T/3$, T being the wall thickness of the thinner component being joined. The total cumulative developed length of slag inclusions shall not exceed $T/2$ in any 6 in. length of weld. The width of a slag inclusion shall not exceed $\frac{1}{16}$ in.
POROSITY—An individual gas pocket of porosity shall not exceed $\frac{1}{16}$ in. in its greatest dimension. The total area of porosity, projected radially through the weld, shall not exceed 0.01 sq in. (equivalent to 3 areas—$\frac{1}{16}$ in. in diameter) in any sq in. of projected weld area, per in. of wall thickness (considering the thinner of the components being joined).

Spot Radiographic Examination
CRACKS—Not acceptable
INCOMPLETE PENETRATION—The total joint penetration shall not be less than the thinner of the two components being joined, except to the extent that incomplete root penetration is permitted. The depth of incomplete root penetration or lack of fusion at the weld root, shall not exceed $\frac{1}{32}$ in. or half the thickness of the weld reinforcement, whichever is smaller. The total length of such incomplete root penetration or lack of fusion at the root shall not exceed $1\frac{1}{2}$ in. in any 6 inches of weld length.
SLAG—Same as required for 100% radiographic examination.
POROSITY—Same as required for 100% radiographic examination.

ASA CODE FOR PRESSURE PIPING—SECTION 31.8—GAS TRANSMISSION AND DISTRIBUTION PIPING SYSTEMS

Same requirements as API Standard 1104.

API STANDARD FOR FIELD WELDING OF PIPE LINES—API STD. 1104
100% or Spot Radiographic Examination

CRACKS—Not acceptable

INCOMPLETE PENETRATION—Any individual defect shall not exceed 1 in. in length. The total length of such defects in any 12 in. length of weld shall not exceed 1 in. The total length of such defects in any two succeeding 12 in. lengths shall not exceed 2 in., and individual defects shall be separated by at least 6 in. of sound weld metal.

SLAG—(a) Elongated inclusions—any elongated slag inclusion shall not exceed 2 in. in length or $\frac{1}{16}$ in. in width. The total length of elongated slag inclusions in any 12 in. length of weld shall not exceed 2 in. and the total length of elongated slag inclusions in any two succeeding 12 in. lengths shall not exceed 4 in. Adjacent elongated slag inclusions shall be separated by at least 6 in. of sound weld metal. Parallel slag lines shall be considered as individual defects if their individual width is greater than $\frac{1}{32}$ in.

(b) Isolated inclusions—the maximum width shall not exceed $\frac{1}{8}$ in. The total length in any 12 in. length of the weld shall not exceed $\frac{1}{2}$ in., nor shall there be more than four isolated slag inclusions of the maximum width of $\frac{1}{8}$ in. in this length. The total length of isolated slag inclusions in any 24 in. length of weld shall not exceed 1 in. Adjacent isolated slag inclusions shall be separated by 2 in. of sound weld metal.

POROSITY—The maximum dimensions of any individual gas pocket shall not exceed $\frac{1}{16}$ in. Maximum distribution of gas pockets shall not exceed that shown in Figs. 7 and 8 of API Standard 1104.

BURN-THRU—Any individual burn-thru area shall not exceed $\frac{1}{2}$ in. in length. The total length of burn-thru area in any 12 in. length of weld shall not exceed 1 in. The total length of burn-thru area in any two succeeding 12 in. lengths shall not exceed 2 in., and individual defects shall be separated by at least 6 in. of sound weld metal.

NAVSHIPS 250-692-2—RADIOGRAPHIC STANDARDS FOR PRODUCTION AND REPAIR WELDS (AUGUST, 1962) AND X-RAY STANDARDS FOR PRODUCTION AND REPAIR WELDS (JANUARY, 1961)

The above standards are issued in two separate pamphlets.

CRACKS—Not acceptable.

INCOMPLETE PENETRATION—Not acceptable.

POROSITY—Indications of porosity shall not exceed specific limits shown in the applicable porosity chart. Randomly dispersed indications and clusters are allowed in a single weld, provided the total number of dispersed and clustered indications does not exceed the specified maximum number of randomly dispersed indications for the applicable indication size and weld thickness. *The Radiographic Standard* also states that all discontinuities whose major dimension does not exceed $\frac{1}{8}''$, not judged to be a crack, shall be considered porosities.

SLAG—*RADIOGRAPHIC STANDARD*

Welds in which the radiographs show elongated slag inclusions shall be unacceptable if in any length of weld an individual inclusion or the total length of inclusions exceeds the acceptance limits outlined below:

CLASS 2: Welds in plates (or pipe) which show elongated slag not in excess of that outlined below shall be acceptable where any such inclusion does not exceed $\frac{1}{4}T$, where T is the thickness of the thinner plate (or pipe) welded. If multiple inclusions of this type occur, below the above limit, the welds shall be judged acceptable if the longest dimensions of all such inline inclusions are not more than T in $6T$ length of weld, and if the defects are separated by at least $6L$ of acceptable weld metal. For this document, L is the long dimension of the biggest inclusion. The maximum length of an acceptable inclusion shall be $\frac{3}{4}''$ or as above, whichever is the least.

CLASS 3: Same as Class 2, except that the acceptable single inclusion may be $\frac{1}{3}T$ and multiple in-line inclusion may be judged acceptable if the sum of the longest lengths of in-line inclusions does not exceed $2T$ in $8T$, subject to other restrictions of Class 2.

SLAG—X-RAY STANDARD

Radiographs shall be free of slag inclusions which exceed the following limits:

Weld thickness	Maximum total length in 6 inches
Up to and including ⅛″	0
Over ⅛″ to and including ½″	¹⁄₁₆″
Over ½″ to and including 1″	⅛″
Over 1″ to and including 2″	¼″
Over 2″ to and including 4″	⅜″
Over 4″	½″

NAVSHIPS 250-1500-1—STANDARD FOR WELDING OF REACTOR COOLANT AND ASSOCIATED SYSTEMS AND COMPONENT FOR NUCLEAR POWER PLANTS.

CRACKS—Not acceptable.

INCOMPLETE PENETRATION—Not acceptable.

SLAG—Slag inclusions must not exceed the limits shown in Fig. 100.

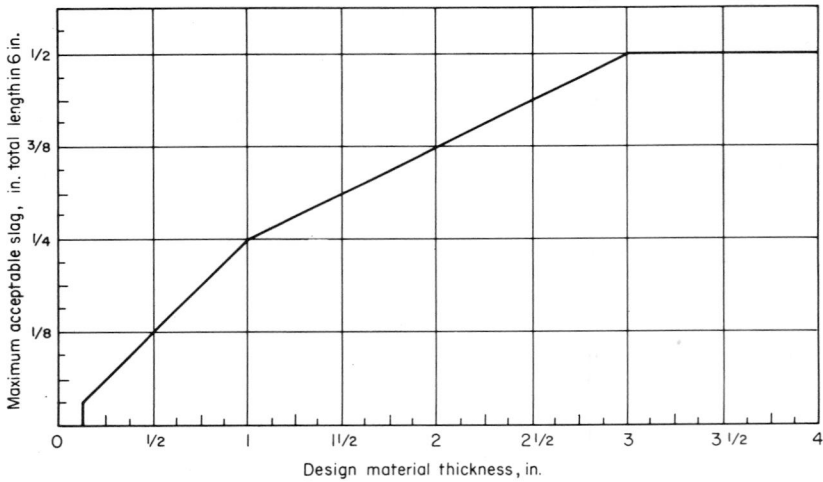

Fig. 100. Maximum acceptable slag in a 6 in. length of weld.

POROSITY—Indication of porosity shall not exceed specific limits shown in the porosity charts, except that porosity indications ¹⁄₆₄ in. and less in diameter shall not be counted in determining the acceptability of welds in carbon steel, nickel-copper, and copper-nickel ¼″ thick or greater.

(Unlike the ASA, ASME, and ASA Standards, NAVSHIPS 250-1500-1 permits clusters of porosity. For comparison, the acceptability limits of ASME Boiler and Pressure Vessel Code and NAVSHIPS 250-1500-1 are shown in Fig. 101.)

TUNGSTEN INCLUSIONS—Not acceptable when greater than ¹⁄₃₂″. There shall be no more than five tungsten inclusions ¹⁄₃₂″ or smaller per six in. length of weld, and no inclusion shall be closer than ¼″ to another inclusion.

ASA Code for Pressure Piping—Nuclear Piping. A standard applicable to nuclear piping is now in preparation. In the meantime, radiographic inspection requirements for aqueous-type and sodium nuclear-power-plant piping are covered in the Interpretations to the ASA Code for Pressure Piping, Cases N-7 and N-8,

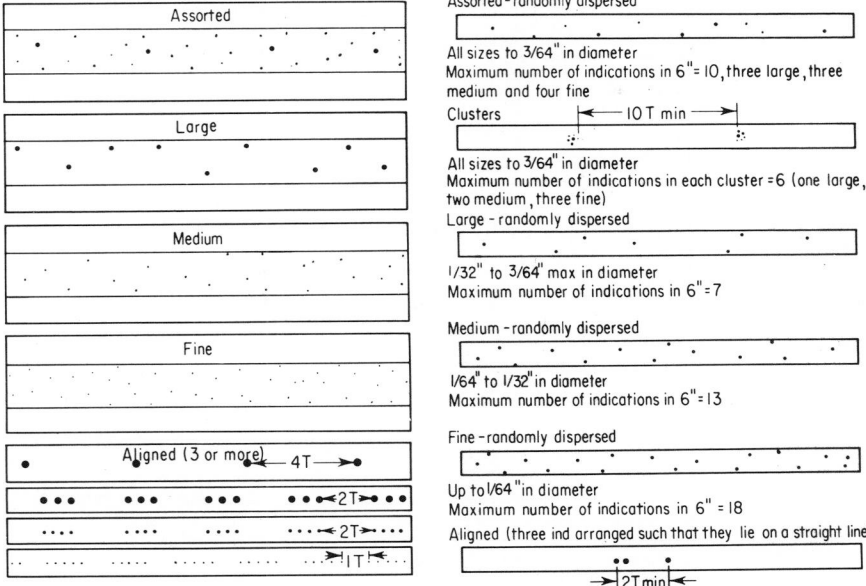

Fig. 101. Comparison of limitations on porosity in ASME Boiler and Pressure Vessel Code (left) and Navships 250-1500-1 (right) for pipe-wall thicknesses from ¼ to ½ in.

respectively. These state that radiographic inspection shall be performed in accordance with Section VIII of the ASME Boiler and Pressure Vessel Code.

Code Cases N-7 and N-8 require that all longitudinal and circumferential butt welds be radiographed. In Case N-9, radiographic examination is required on centrifugally cast austenitic stainless-steel pipe.

Inspection of Pipe Welds. Butt welds in piping may be examined by radiographic inspection either from the outside of the pipe by placing the radiation sources on one side of the pipe and attaching the films to the other or from the inside by locating the radiation source at the inside of the pipe. The outside technique requires that the radiation pass through two wall thicknesses. At least three, and frequently more, radiographic exposures are required when radiographing is done from the outside to obtain films permitting proper interpretation of the weld around the pipe circumference.

If the radiation source is located at the inside center of the pipe underneath the weld, a single radiographic exposure is usually sufficient. Depending on the diameter and wall thickness, the radiation source may be located against one inside pipe wall for radiography of the opposite pipe (weld) wall.

If the radiation source is inside the pipe, the films obtained provide greater control and definition. Definition depends primarily on the type and size of the radiation source used, the distance from the radiation source to the inside wall of the pipe, and the wall thickness. The penetrameter image on the films is normally considered as an acceptability criterion.

If the pipe ends are open or if the pipe inside is readily accessible through an open valve, header, or other connection, the radiation source is generally inserted and positioned through the opening. Where this is not possible, access holes must be provided in the pipe through which a centering fixture is inserted (as shown

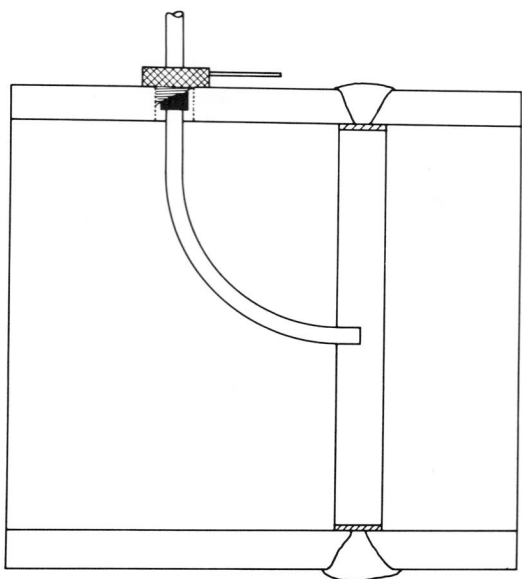

FIG. 102. Radiation-source guide fixture for radiography of pipe weld from inside of pipe.

in Fig. 102). Such fixtures normally consist of tubes which guide and position the radiation source in the center of the pipe underneath the weld.

Pipe Fabrication Institute Standard ES-16[1] gives recommendations for the access-hole design shown in Fig. 103. Upon completion of radiography and acceptance of the weld by the final inspector, a plug (Fig. 104) is then screwed into the access hole. The machined groove improves seating and minimizes excessive stressing where the end of the thread would meet the $\frac{3}{8}$-in.-thick rim. It also prevents the seal weld from reaching the thread (Fig. 105), which otherwise might crack owing to the resulting notch condition.

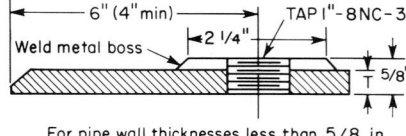

For pipe wall thicknesses less than 5/8 in.

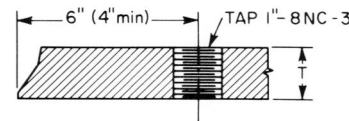

For pipe wall thicknesses of 5/8 in. and over

FIG. 103. Access-hole designs for radiographic examination of pipe butt welds.

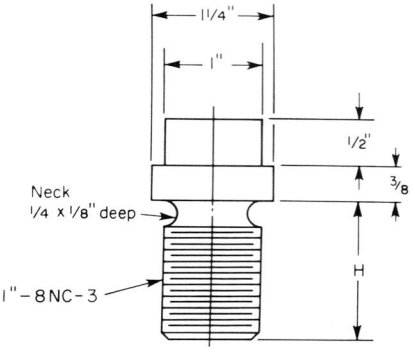

FIG. 104. Design of plug for closure of access hole.

[1]PFI Standard ES-16, Access Holes and Plugs for Radiographic Inspection of Pipe Welds.

Table 92. Estimated Time and Cost of Radiographic Inspection

Pipe size	Access hole	Number of exposures	Type and strength of radiation source, curies	Film density	Exposure time	Total time including set-up time	Cost of picture	Cost of access hole, plug, and welding	Total cost
16″ OD × 3.25″	Yes	1	Cobalt 1	2	30 min	50 min	$ 12.50	$ 45.00	$ 57.50
			3	2	10 min	30 min	7.50	45.00	52.50
			10	2	3 min	23 min	5.75	45.00	50.75
16″ OD × 3.25″	No	4	1	2	101 hr	102 hr	*	*	*
			3	2	35 hr	36 hr	*	*	*
			10	2	10 hr	11 hr	*	*	*
16″ OD × 3.25″	No	4	1	1½	72 hr	73 hr	*	*	*
			3	1½	22 hr	23 hr	345.00	.....	345.00
			10	1½	7 hr	8 hr	120.00	.....	120.00
22″ OD × 4.00″	Yes	1	1	2	16 hr	2 hr	30.00	50.00	80.00
			3	2	32 min	52 min	13.00	50.00	63.00
			10	2	9 min	29 min	7.25	50.00	57.25
22″ OD × 4.00″	No	4	1	2	600 hr	601 hr	*	*	*
			3	2	200 hr	201 hr	*	*	*
			10	2	60 hr	61 hr	*	*	*
22″ OD × 4.00″	No	4	1	1½	400 hr	401 hr	*	*	*
			3	1½	132 hr	133 hr	*	*	*
			10	1½	40 hr	41 hr	615.00		615.00

* Prohibitive.

On pipe of heavier wall thickness, the cost of radiography without access holes becomes increasingly higher than it would be if access holes were provided. This is due to the very much longer radiographic exposure times required and the fact that usually a minimum of four exposures is required. The advantages are particularly pronounced on pipe of heavy wall thicknesses, as over 3 in., where radiography is normally done with cobalt-60 radiation sources. Typical comparisons in time requirement are shown in Table 92 for two pipe sizes.

Additional time is required for film processing, film interpretation, and movement of equipment. Another important factor is the weight of the radiation-source container (called "projector"). The available 3-curie cobalt-60 source containers weigh about 150 lb and can be readily handled. The containers for the 10-curie

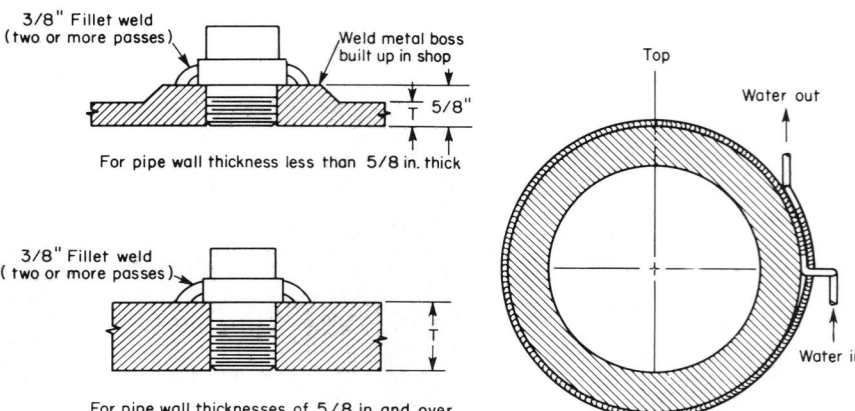

FIG. 105. Seal welding of access hole after insertion of plug.

FIG. 106. Sketch of hot-radiography fixture which keeps film temperature below 120 F.

cobalt-60 sources weigh nearly 1,000 lbs. Particularly in field radiography, the movement of the heavier containers involves substantial additional cost, as cranes or chain falls, extra personnel, and special scaffolding may be required.

Though varying somewhat with the source, strength, and pipe diameter, as a general rule, the break-off point in wall thickness below which access holes are no longer economical is approximated from Table 93. On pipe wall thicknesses of 1½ in. or over, it is considered desirable to perform radiographic examination after the first two to four weld passes have been made. Otherwise, many critical root defects may remain undetected.

Hot Radiography. Sometimes, it is desirable to radiograph a pipe weld after only a few weld passes have been deposited. However, it may not be desirable or permissible to reduce the weld-interpass temperature to below the minimum values required in the applicable code or welding specification.

Since radiographic films tend to blacken when exposed to temperatures exceeding 140 F, special fixtures are necessary which keep the film temperature well below that temperature. This is accomplished by means of the flexible water-cooled hot-radiography fixtures illustrated in Figs. 106 and 107. With a pipe-interpass temperature of 550 F on 12-in. Schedule 160 pipe, the temperature at the center of the fixture will not exceed a reading of 115 F.

Table 93. Pipe Diameters and Wall Thickness for Which Radiography through Access Holes Is Recommended

Nominal diameter, in.	Nominal wall thickness, in.
10–16	Over 1½
Over 16–24	Over 1¼
Over 24	Over 1

Cooling of the copper sheet is accomplished by passing water through flexible tubing soldered to both sides of the copper sheet. Cooling by flowing water between two layers of copper sheet supporting the film would be undesirable since corrosion of the copper sheet or the deposition of mineral constituents between the sheets underneath the radiographic films may result in erroneous radiographic patterns.

FIG. 107. Hot-radiography fixture for butt welds at interpass temperatures of 600 F and less.

MAGNETIC PARTICLE INSPECTION

Magnetic particle inspection provides an effective method for detecting cracks, seams, laminations, lack of fusion or penetration, porosity, and similar essentially surface discontinuities in ferromagnetic materials. It is not applicable to non-magnetic materials.

Methods. When a magnetic field is produced in a ferromagnetic material, minute magnetic poles are set up at discontinuities in the path of the magnetic flux. These poles exert a stronger magnetic attraction for iron powder or other magnetic particles than is exerted by the surrounding metal.

The surface to be examined is magnetized by means of a high amperage current. The pipe section can be magnetized either by using an electric current to induce a magnetic field within the material or by placing the material in a magnetic field. The first is done by means of prods placed on both sides of the area to be inspected. The second method involves wrapping a cable around the pipe to be inspected or

employing electromagnetic prod fixtures.   The surface in the area to be examined is then covered with finely divided iron particles either by sprinkling dry particles or by applying to the surface a coating of light oil in which the particles are in suspension.   A crack or defect in the part, either at or near the surface, causes a discontinuity in the magnetic flux lines.   The iron particles collect at that point, outlining the defect.   For the detection of extremely narrow cracks, the application of fluorescent magnetic particles in a liquid medium (usually water) is one of the most revealing techniques.

The prod method is most widely used and standardized for the inspection of piping materials and pipe welds.[1]   In this method, the prods should be positioned on the piece such that the localized circular field produced is flowing essentially perpendicular to the direction of possible discontinuities.   The prods should be spaced between 3 and 8 in. and the current should range between 100 to 150 amp/in. of spacing between prods.   In the handling of the prods, care must be exercised to avoid arcing and the forming of crater cracks on the metal surface.

Dry powder[2] and wet[3] methods have been standardized by the American Society for Testing and Materials.   The dry-powder method is more sensitive than the wet method in the detection of discontinuities near the surface.   However, it is not as sensitive in detecting fine surface discontinuities.   As the dry method is more convenient to use with portable equipment, its use is adopted almost exclusively in the inspection of piping and pipe welds.

Limitations.   Although subsurface defects may be detected with specialized equipment, for commercial pipe inspection, magnetic particle inspection is limited generally to the detection of surface flaws only.

Surface irregularities may produce linear indications or cover up discontinuities. Welds involving materials of different magnetic characteristics may create magnetic patterns of sufficient magnitude to produce erroneous defect patterns in the absence of actual defects.   The magnetic field must be sufficiently interrupted or distorted by the flaw.   When fine elongated discontinuities such as laps, seams, or fine cracks are parallel to the magnetic field, they may not be detected, particularly when the coil or electromagnetic-prod methods are used.   Such defects may be detected with the prod method producing circular magnetic lines of force or by the production of a magnetic field not parallel to the discontinuity.

Sensitivity.   Maximum sensitivity is obtained when defects, such as crack or incomplete weld penetration, are essentially perpendicular to the surface being inspected.   A decrease in sensitivity applies when the surface discontinuities are round or spherical.   An important aspect of sensitivity is the surface condition of the piping component or weld.   Surface roughness decreases sensitivity and may interfere mechanically with the formation of powder patterns.

The surface should be clean, dry, and free of all oil, grease, sand, loose scale, or other contaminators which would interfere with the application of the test. "As-welded" surfaces should be considered acceptable, provided that they are clean and completely free of all welding slag.   On most pipe welds, wire brushing is generally sufficient.   However, sand or shot blasting and grinding may sometimes be advisable.

Interpretation.   Reference photographs illustrating typical types of discontinuities in castings and welds have been issued in ASTM Standard E125.[4]

[1] PFI Standard ES-14, Magnetic Particle Inspection.

[2] ASTM E109, Dry Powder Magnetic Particle Inspection.

[3] ASTM E138, Wet Magnetic Particle Inspection.

[4] ASTM E125, Tentative Reference Photographs for Magnetic Particle Indications on Ferrous Castings.

Table 94. Appearance of Discontinuities Revealed by Magnetic-particle Inspection on the Surface of Cast, Forged, and Wrought Piping Materials

Type	Definition	Surface appearance
Laps, seams, scabs, cracks, hot tears......	Linear discontinuities	May appear as straight, curved, or ragged lines of variable widths
Shrinkage..............	Jagged or irregular discontinuities	
Inclusions	Round, elongated, or irregular indications	
Chills and unfused chaplets	Uniformly curved or straight lines	
Porosity and gas inclusions	Rounded single or clustered indications	

Common types of discontinuities in piping materials and welds are summarized in Tables 94 and 95, respectively.

Code Requirements. Most codes generally apply the same acceptability limits to defects or discontinuities detected by magnetic particle examination as to radiographic inspection. There is general agreement that

1. All cracks should be removed.

2. All discontinuities in excess of applicable code requirements should be removed and the area repaired, if necessary, in accordance with the applicable procedures.

3. Repaired areas should also be examined by following the original inspection procedure.

Section III of the ASME Boiler and Pressure Vessel Code states that "cracks" are unacceptable. In addition, linear inclusions are unacceptable if they involve six or more inclusions in any $1\frac{1}{2}$-in. $\times$ 6-in. rectangle or $3\frac{1}{2}$-in.-diameter circle

Table 95. Appearance of Discontinuities Revealed by Magnetic-particle Inspection on the Surface of Pipe Welds

Type	Definition	Surface appearance
Porosity	Subsurface gas pockets or voids in weld deposit (may be brought to surface by grinding)	Rounded single or clustered indications when at surface; they become poorly defined when below surface
Slag inclusions	Nonmetallic solid material entrapped in weld deposit or between weld metal and base metal	Rounded single or linear indications when at surface; they become poorly defined when below surface
Lack of penetration.....	Root penetration which is less than complete	Straight, linear, continuous, or intermittent indication
Incomplete fusion.......	Fusion which is less than complete either between weld beads or between weld and base metal or consumable insert	Linear or slightly heavy straight or rounded indications
Cracks	Discontinuities formed by very minor separations of weld or adjacent base metal	Sharply defined linear, irregular, or starlike patterns, tightly held and built up heavily with powder
Undercut	A groove melted into the base metal surface adjacent to the weld	Linear indications adjacent to weld, usually less pronounced than incomplete fusion

and are: in length $\frac{1}{4}$ or longer for thicknesses to $\frac{3}{4}$ in.; over $\frac{1}{3}$ the pipe-wall thickness in length for thicknesses of $\frac{3}{4}$ to $2\frac{1}{4}$ in.; and over $\frac{3}{4}$ in. for thicknesses over $\frac{3}{4}$ in. Nonlinear defects are also unacceptable if the defects exceed $\frac{3}{32}$ in. in size.

It is generally recognized that all indications revealed by magnetic particle inspection are not necessarily defects but may be nonrelevant indications. Nevertheless, all indications in weld craters or in the line of fusion between the weld deposit and base metal should be treated as defects. Where other indications appear nonrelevant, at least 10 per cent of each type of indication should be explored by removing the surface roughness believed to have caused the type of indication to determine if defects are actually present. The absence of actual indications after reinspection by magnetic particle inspection subsequent to the removal of the surface roughness would confirm that indications originally present were nonrelevant. Indications of subsurface discontinuities, unless clearly understood, should be investigated by radiographic inspection, sectioning, trepanning, or chipping.

LIQUID PENETRANT INSPECTION

Application. Liquid penetrant inspection provides an effective method for the detection of fine and narrow discontinuities such as cracks, seams, laps, porosity, and leaks which are open to the surface. It is applicable to both nonmagnetic and magnetic metals. It may be used, also, on many nonmetallic piping materials.

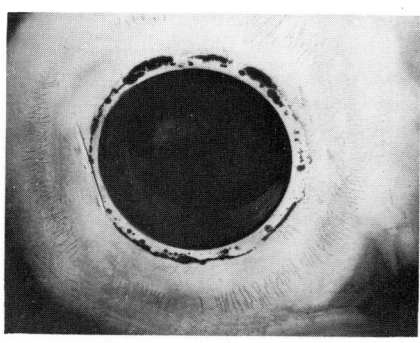

FIG. 108. Base-metal cracking revealed by dye-penetrant inspection.

Liquid penetrant is employed to inspect pipe base materials and welds. Crack indications revealed by dye penetrant inspection of the inside base metal around a pipe nozzle weld are illustrated in Fig. 108. However, the metal surface must not be hot. Although liquid penetrant inspection can be done at surfaces of between 32 and 125 F, best results are obtained at 70 to 90 F. When welding piping for critical applications, liquid penetrant is also used extensively as an in-process method, particularly after the first weld pass has been deposited. Some codes and specifications require such root-pass or subsequent weld-pass inspection. As with all inspection methods, however, the success or completeness by which defects are determined lies with the care taken, the experience of the inspector, and the interpretation of the results.

All liquid penetrants employ liquids of high penetrating characteristics which, by capillary attraction, enter small discontinuities in the surface, such as cracks. The rate and extent vary with surface tension, cohesion, adhesion, viscosity, time, surface conditions of the material, depth and widths of the discontinuity, and nature and cleanliness of the interior of the discontinuity. Two major methods[1,2] are used. They involve either dye or fluorescent penetrants.

[1] PFI Standard ES-17, Liquid Penetrant Inspection.
[2] ASTM Standard E165, Liquid Penetrant Inspection.

Dye Penetrant Inspection. The inspection procedure normally involves applying a red liquid (dye) by spraying, brushing, or dipping to the metal surface and allowing it to penetrate into existing cracks and crevices. A dwell time of at least 15 to 30 min is considered advisable. The surface is then wiped clean with a cloth, or it may be cleaned by wiping or rinsing with an emulsifier or cleaner. A fine white powder suspended in a highly volatile liquid (referred to as "developer") is subsequently carefully sprayed or brushed over the metal surface. Evaporation of the volatile liquid will leave the dry white powder on the metal surface. Where the dye previously penetrated into surface discontinuities, it will be absorbed by the white powder outlining the cracks or other surface discontinuities in red.

Fluorescent-Penetrant Inspection. The fluorescent liquid is applied to the metal surface in a manner similar to that described above for the dye-penetrant method. After removal of the excess and the application of a developer, the surface is illuminated under near-ultraviolet or black light. The liquid in the surface discontinuities will then delineate them by fluorescence. The high contrast of the fluorescence makes this method one of the most sensitive for the detection of very small or tight surface defects. Because of the many different penetrant liquids marketed, it is recommended in the application of the various penetrant methods of inspection that the instructions of the respective manufacturer be followed to obtain the optimum results when using his products.

Limitations. Improper techniques may not produce defect indications, or they may bring out "ghost" patterns which are not associated with surface discontinuities. The surface should not be cleaned by grit or sand blasting as these may close up small surface cracks. Careful application of the various liquids is particularly important. Experience in the interpretation of indications representing defects and of nonrelevant patterns is also very important when using liquid penetrant inspection.

Sensitivity. Most important to proper sensitivity is the condition of the surface. All surfaces to be inspected must be free from all contaminants, such as oil, grease, paint, rust or scale, and other foreign materials. A cleaning operation or other suitable preparation of the surface is advisable. This may involve dipping, spraying, wiping, or brushing with solvents, such as acetone or trichloroethylene.

Welded surfaces, after removal of welding slag, generally are considered suitable for liquid penetrant inspection without grinding. Often such surfaces, however, have nonrelevant indications and may require some grinding to remove weld ripples and blend the weld contour into the pipe base metal. Surface finish of 250 microinches or better is desirable. Weld splatter, undercuts, laps, and other surface conditions that may cause irrelevant indications must be removed.

Interpretation. The discontinuities in piping materials and welds determined by liquid penetrant inspection are similar to those described in Tables 94 and 95 for magnetic particle inspection.

Code Requirements. Codes generally apply the same acceptability limits to discontinuities detected by liquid penetrant inspection as to radiographic inspection. In general, the following limits apply:

1. All indications of cracks are rejected. They shall be removed.

2. All discontinuities in excess of applicable code requirements shall be removed and the area repaired, if necessary, in accordance with the applicable procedures. Porosity-type defects are acceptable if they do not exceed applicable radiographic standards for the material under inspection.

Nuclear Code Case N-9 to the Code for Pressure Piping B31.1–1955 requires that all linear discontinuities and aligned penetrant indications revealed by the test shall be removed. Aligned penetrant indications are those in which the average of the center-to-center distances between any one indication and the two adjacent

indications in any straight line is less than $\frac{3}{16}$ in.   All other discontinuities revealed on the surface need not be removed unless the discontinuities are also revealed by radiography, in which case the pertinent radiographic specifications shall apply.   The repaired area should be reexamined by following the established inspection procedure.   Nonrelevant surface indications should be explored as discussed in the above section on Code Requirements under magnetic particle inspection.

ULTRASONIC INSPECTION

Methods.   Ultrasonic inspection is extensively used on piping for the detection of surface and subsurface flaws and discontinuities and for wall-thickness measurement.   Although most extensively applied to wrought and forged pipe base materials, it is also being increasingly applied to pipe welds.

Ultrasonic inspection methods utilize mechanical vibrations at frequencies ranging between 1 and 10 Mc/sec.   The fundamental principle involves a controlled and uniform beam of ultrasonic energy which is directed by a transducer into a test material.   The energy will be transmitted with little loss (or attenuation) through a homogeneous material; it will be either attenuated or reflected by discontinuities (or defects) in the physical structure existing in a second location in the same material or in another material.   The measurements, therefore, of the energy changes either by attenuation or by reflection are the criteria employed.   They form the two basic criteria for the inspection of pipe and pipe fabrication for the primary purpose of flaw detection.

The transducer, or search unit, which is held in contact with the surface of the material to be examined, contains a quartz, barium titanate, lithium sulfate, or other crystal whose piezoelectric properties convert electrical energy into mechanical energy.   A reversing electrical potential causes alternating changes in the dimensions of the crystal, and these changes vary with the frequency of the electromotive force applied.   Conversely, the application of pressure on the face of the crystal will generate a small voltage of the same frequency as that of the applied vibration.   The vibrations received from the material being inspected thus will generate electrical waves; these are magnified by a cathode-ray tube.   The crystal is actuated for a controlled period of time of 2 μsec or less, resulting in a short impulse of sound waves.   The energy input into the crystal is then discontinued for a short period to permit the crystal to receive the returning echoes.   The cycle of transmitting and receiving may vary between 60 and 1,000 times per second.

The reflected signals are indicated on the cathode-ray tube as vertical deflections of the horizontal base line.   This reference line denotes the time in microseconds for the ultrasonic waves to travel to and be reflected from a surface or discontinuity.   The elapsed time is generally translated into distance from a reference point or line that represents the point of contact between the transducer and the material surface.   Commercial ultrasonic inspection involves three methods which are: (1) longitudinal or straight waves, (2) shear or angle waves, and (3) surface or Raleigh waves.

Longitudinal Waves.   Longitudinal or straight waves are transmitted into the material perpendicular to the surface.   The methods involved utilize either reflection or resonance.   Both methods are used extensively for wall-thickness measurement of piping to determine the thinning of pipe bends, swages, and extrusions.

In the *reflection method*,[1] the reflections are converted, amplified, and shown on

[1] ASTM Standard E114, Recommended Practice for Ultrasonic Testing by the Reflective Method, Using Pulsed Longitudinal Waves Induced by Direct Contact.

the cathode-ray tube to indicate the time differences between the transmitted pulses and the reflected pulses. It is suitable for determining the depth of laminations. This method is sometimes utilized to examine pipe edges for defects. Service inspection involves wall-thickness measurements to determine the extent of some types of corrosive or erosive attacks on metals. Longitudinal waves generally are not used for the determination of pipe and weld defects. Either surface contact or immersion techniques may be followed. The latter technique is extensively used by tube mills in the rapid quality-control inspection of tubular materials.[1]

In the *resonance method*,[2] ultrasonic vibrations of varying frequencies are introduced into the material from one side through the transducer. The frequency at which resonance is reached is indicated by audible means or is shown on a cathode-ray tube. Since resonance occurs when the thickness of the material is half of one wavelength, the resonant frequencies are translated into thickness determinations.

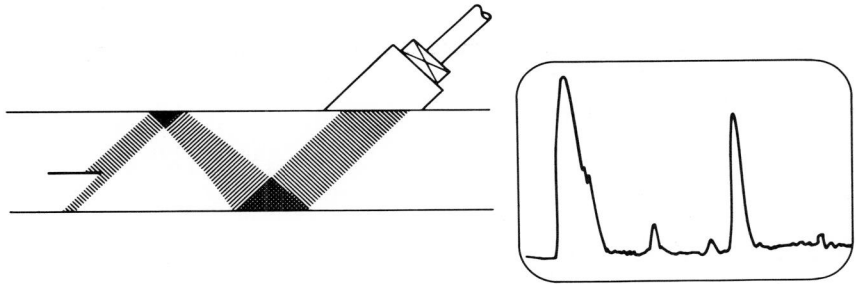

FIG. 109. Schematic diagram showing shear-wave technique for location of defects.

Shear Waves. Shear or angle waves are transmitted into the material at an angle by placing an angular plastic wedge between the crystal and the surface. The angle transducers are directional and thus permit more accurate location of discontinuities. The angle of the plastic wedge is calculated for the desired refracted wave (usually between 45 and 85 deg) in the material to be examined. The waves are propagated by angular deflection from the outside and inside pipe surfaces until they reach another surface or discontinuity from which they are reflected back along the same path (Fig. 109). This method is used extensively for the examination of seamless pipe and seam-welded plate pipe[3,4] and for the inspection of welds.[5]

Raleigh Waves. Surface or Raleigh waves are transmitted into the surface at an incident angle of from 15 to 30 deg. At these angles, a third wave from the Raleigh wave is developed which travels only along the surface. The motion of the wave is elliptical with its greatest amplitude perpendicular to the surface

[1] ASTM Standard E214, Tentative Recommended Practice for Immersed Ultrasonic Testing by the Reflection Method Using Pulsed Longitudinal Waves.

[2] ASTM Standard E113, Tentative Recommended Practice for Ultrasonic Testing by the Resonance Method.

[3] PFI Standard ES-18, Ultrasonic Inspection of Seamless Piping.

[4] ASTM Standard E213, Tentative Method for Ultrasonic Inspection of Metal Pipe and Tubing for Longitudinal Discontinuities.

[5] ASTM Standard E164, Tentative Method for Ultrasonic Contact Inspection of Weldments.

(resembling a surface wave on water). Raleigh waves are utilized to detect minute surface cracks and flaws immediately beneath the surface.

Limitations. Conditions which may limit the application of ultrasonic inspection usually relate to (1) unfavorable component geometry, such as size, contour, complexity, and defect orientation, and (2) unsatisfactory internal structure such as grain size, porosity, and inclusion content.

Other features of ultrasonic inspection include the following:

1. Manual scanning and visual interpretation of indications require constant attention, analysis, and decision by a highly qualified operator.

2. Data presentation is not pictorial and permanent as in radiography and is therefore more dependent on operator skill and reliability.

3. Interpretation of equipment sensitivity may be difficult when inconsequential discontinuities or geometry produce interfering signals which mask rejectable faults, especially in certain high-alloy and stainless materials.

In some pipe welds involving nonuniform designs and root conditions, it may be very difficult or impossible to separate noncritical contour variations from critical rejectable defects.

Sensitivity. Since each nondestructive test method relies on a specific physical phenomenon, the nature of the information obtained is generally different. Compared with radiographic inspection, ultrasonic techniques generally are greatly inferior in ability to produce great detail in the outline of small nonhomogeneities because of the much longer wavelengths. However, when the presence of mechanical discontinuities must be indicated, even though small in area and of negligible thickness, ultrasonic techniques are generally very sensitive, minimizing spurious effects due to nonmechanical variations involving conductivity, permeability, transparency, color, chemistry, etc. Conversely, they may be affected adversely by such structural properties as grain size, graphitization, precipitates, segregations, plate laminations, density, etc. Such conditions must be maintained at a low level and kept reasonably uniform.

Surface condition may also affect sensitivity. Seamless pipe should have a suitable finish. Any surface equivalent to 250 μin. or better will usually be acceptable. For unusually rough or heavily scaled surfaces, grit or sand blasting or other methods may be required to produce a suitable surface. Local rough surfaces should be conditioned by grinding to insure proper contact of the transducer on the pipe surface.

Interpretation. Most defects in wrought or forged pipe materials can be readily identified as to type, size, and distribution by ultrasonic inspection. On pipe welds, problems arise in the interpretation of patterns on the cathode-ray tube. Only a few commercial laboratories have developed the large number of tests and correlation standards necessary to permit proper interpretation of the many different conditions that can occur. Internal weld build-up, flat or taper machining, backing rings or consumable inserts, variations in surface defects, and many other conditions affect the ultrasonic pattern as indicated on the oscilloscope. Thus, a properly welded backing ring may produce the same pattern as unacceptable lack of penetration or root cracking. Only an experienced inspector may be able to differentiate between potentially serious defects and minor noncritical discontinuities, such as porosity, slag, or changes in surface contour and metallurgical structure. At times, other nondestructive or destructive techniques may be necessary to supplement the interpretation of the ultrasonic pattern. This may be far less costly than the removal of a minor and unimportant nonhomogeneity in the pipe base metal or weld.

Code Requirements: Base Metals.   Different standards apply to pipe base metals and pipe welds in case calibration standards are necessary to insure proper interpretation of defect patterns.   Applicable rules and specifications usually require that a "V" notch be cut into the pipe-calibration standard and into the pipe to be tested.   The calibration pipe section should have the same dimensions as the pipe to be tested.   Normally, a 5 per cent "V" notch is used,[1] though sometimes a 3 per cent notch is specified.

The "V" notch should have the following dimensions:

Depth—5 per cent plus or minus ¼ per cent of the nominal wall thickness, unless otherwise noted.   The minimum depth should be about 0.003 in.
Width—Not more than twice the depth.
Length—Extending axially approximately 1 in.

The "V" notch should be located: on 6 in. and smaller pipe at one end, on either the inside or outside surface; on larger than 6 in. pipe, at one end on both the inside and outside surfaces, displaced 180 deg apart.

The test equipment should be adjusted to develop a pronounced indication from the "V"-notch-reference calibration standard on each pipe tested.

Examination of the pipe should be done by rotating the pipe while moving the transducer longitudinally with a 10 per cent overlap of the crystal's circular path relative to the work.   A second scanning with the work rotating in the reverse direction, with the transducer reset accordingly, should be made subsequently. The speed of rotation should be selected so as to obtain optimum inspection accuracy.   On fabricated piping which cannot be rotated, the operator should move the transducer around the pipe circumference to obtain an equivalent examination. The following standard of acceptance normally applies: any defect which exceeds the 5 per cent (or 3 per cent) notch calibration and/or encroaches upon the minimum specified wall thickness is considered an injurious defect, and the pipe shall be rejected.

Code Requirements: Weld Metal.   Specific acceptability standards based on the indications on the cathode-ray tube have not been issued by most of the major codes.   The interpretation of these indications must be referenced and the respective defect acceptability limits generally established for radiographic inspection or magnetic particle or liquid penetrant inspection.

Section III of the ASME Boiler and Pressure Vessel Code considers as unacceptable:

any crack, lack of fusion, incomplete penetration, inclusion, or cavity which is indicated by a reflection equal to or greater than 80 per cent of the applicable reference hole reflection (in a reference standard) and which has a linear dimension as indicated by transducer movement exceeding:
¼" for thicknesses up to ¾"
⅓ of the pipe-wall thickness for thicknesses from ¾" to 2¼"
¾" for thicknesses over 2¼"

If there is doubt involving the interpretation of test results, other nondestructive methods may be used to resolve these to the satisfaction of the inspector.

EDDY CURRENT TESTING

Eddy current testing is utilized as a nondestructive testing means for tubes and small-diameter pipes which may be either seamless or seam-welded.   This process

[1] PFI Standard ES-18, Ultrasonic Inspection of Seamless Piping.

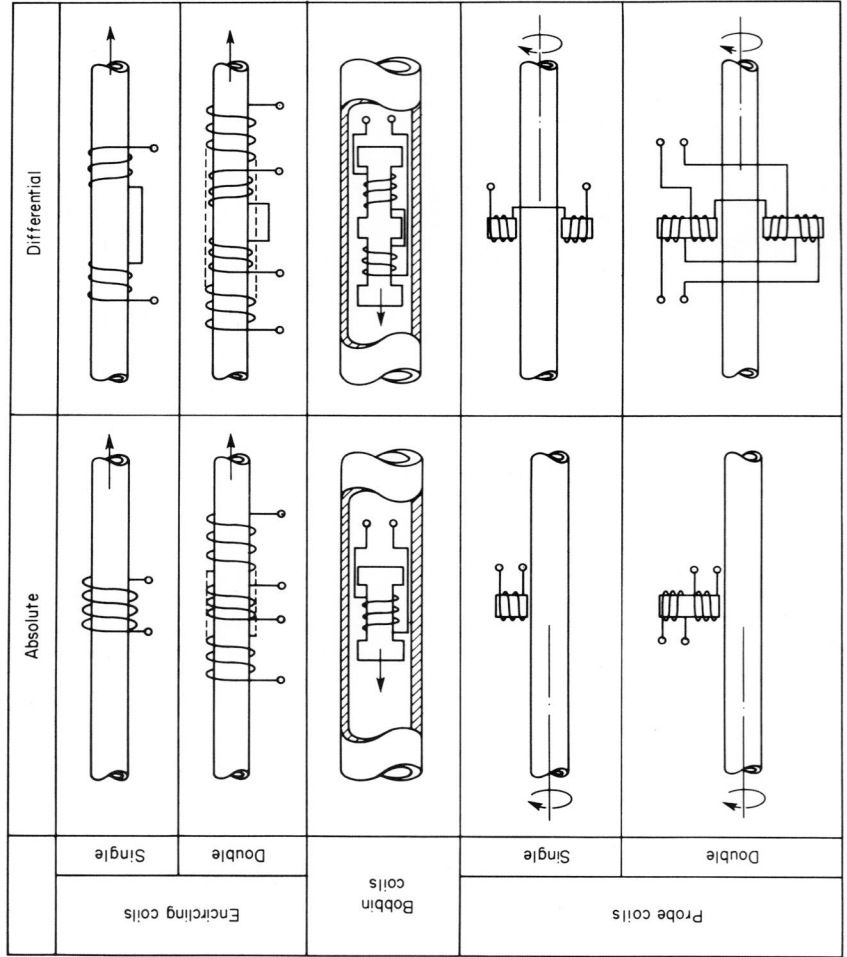

	Absolute	Differential
Encircling coils — Single		
Encircling coils — Double		
Bobbin coils		
Probe coils — Single		
Probe coils — Double		

Fig. 110. Test-coil configurations for eddy-current testing of tubing.

is generally limited to mill applications or to nonmagnetic materials for field use.

The tube or pipe is generally placed in the center of a circular coil. By passing an alternating current through this coil, an eddy current is induced in the tube. This eddy current, in turn, produces an additional alternating magnetic field in the vicinity of the tube. Discontinuities in the base metal or weld cause changes in the secondary magnetic field which varies the impedance of the coil. This produces an electric signal which is indicated by a recorder, oscilloscope, or other measuring or alarm instrument. Instead of encircling coils, an internal probe ("bobbin") may be passed through the center of tube, or surface probe coils may be used on the outside (Fig. 110). Manual scanning with a probe-type coil is also performed.

The encircling coil is most widely used as a production tool because of the high inspection speeds possible. The use of the internally applied bobbin coils is usually limited to the inspection of heavy wall pipe or to duplex, clad, or plated pipe or tubing. The surface probe coil is not considered a production tool because of its slow speed. It is a useful tool, however, for the inspection of small discontinuities.

Every variation in the tube wall or material may produce indications which may obscure or confuse the indication of discontinuities. Since the density of the induced current decreases with depth into the tube wall, identical defects along the surface or on the inside of the wall will produce signals of different amplitudes.

"*Sensitivity*" is affected by variations in the wall thickness, surface roughness, and metallurgical changes as caused by changes in composition or structure. Heat treatments may have a bearing on the latter. Thus, a discontinuity may not be resolved when the other dimensional or metallurgical variations produce signals of greater amplitude. As a general rule, defects must have a depth of at least 5 per cent of the wall thickness before their existence is disclosed by eddy-current techniques.

"*Interpretation*" requires the preparation of detailed calibration specimens. Once they are established, the quality and acceptability of the product may be automatically controlled on the basis of predetermined levels.

LEAK TESTING

Leak testing involves the development of a differential pressure between the inside of the piping component or system and the external pressure. This is generally accomplished by producing a pressure on the pipe inside that is substantially higher than the outside (usually atmospheric) pressure. Some tests are based on the development of a vacuum within the piping components. However, sometimes special vacuum chambers may be applied to the outside, usually around welds. The pressure differential is then produced by evacuating the angular space between the metal surface and the chamber wall.

Leak tests involve various tests. Most common are hydrostatic tests extensively recognized and specified by most piping codes (see Table 40). Other leak tests may employ air or gases such as freon, ammonia, or helium. Some tests employ halogen gases containing chlorine, fluorine, bromine, or iodine. Helium gas is also used in mass spectrometer leak testing.

Hydrostatic Testing

Application. Hydrostatic testing of pipe, fabricated piping components, and erected piping systems is universally performed (as seen by reference to Table 40). At very high pressures, the hydrostatic test is a very sensitive one. Where shop fabrication is involved, the fabricated pipe sections are often hydrostatically tested in the presence of an insurance inspector. After field erection or installation,

final hydrostatic testing may then again be required to test the complete piping system.   Hydrostatic testing of the piping system after periods of service is also frequently required by state and insurance inspectors, particularly where changes or repairs have been made in the piping system.

Methods.   To seal the ends of piping components, it is generally necessary to weld either caps or plates to the pipe ends.   This involves, also, the preparation for welding some alloy pipe steels, as well as preheating and stress relieving of the welding area.   After testing, the weld has to be removed.   This is usually done by flame cutting the cap or plate through the weld.

The pipe end then must be dressed up by grinding or machining to remove possible notch effects from the test plate and to eliminate hard-metal zones remaining after welding or flame cutting which can cause cracking and service failures.   By means of cone-lock test plugs (Fig. 111), the need for welding on caps or plates can be eliminated for pipe sizes between 1 and 18 in. diameter.   The plug is inserted into the open pipe end.   The initial holding is accomplished by turning the wing nut which expands the vise ring against the pipe wall.   A neoprene sealing cup is held against the pipe wall by a steel hub plate.   The sealing cup is backed with a coned pressure plate which expands the forged carbon-steel split ring against the wall of the pipe, where the internal pressure on the inside of the pipe builds up.

For pipe sizes to 14 in. diameter, the plugs can be safely used with pressures of 2,000 psi and higher.   On pipe of over 14 in. diameter, the maximum test pressure is 1,000 psi with a test plug of slightly different design.

FIG. 111.   Cone-lock test plug for hydrostatic testing.

After the pressure has been released, the plug is readily removed by untightening the locking wing nut.   By means of different size vise rings, a plug can be used interchangeably on pipe of several pipe schedules or varying inside diameters.

The hub shaft may also be fitted with a removable forged-steel plug for suitable air or hydraulic attachments.   Reuse of these plugs reduces substantially the cost of hydrostatic testing and insures that defects from the welding-on of test caps or plates are avoided.   On flanged joints, closure is generally made with blind flanges.

For most work, clean water is used.   The temperature of the water should be no lower than that of the atmosphere.   Otherwise, sweating will result and proper examination will be difficult.   In subfreezing environments, antifreeze or hydrocarbons may be added to keep the water from freezing.   Bleeder valves or pet cocks should be provided at the highest point or points in the system to permit venting of all air in the piping during the filling operation.

The hydrostatic pressure should be applied gradually and should be left on for a sufficient length of time to permit making a complete examination of all welds. The pressure gage should be checked to be sure that it is functioning properly and that a pressure drop has not occurred during the test period.   Any evidence

of leaking or other weakness under test should be marked. Following release of the pressure, defective areas should be repaired and another hydrostatic test made.

Code Requirements. The test pressure should be that called for by the governing code or specification. Lacking such instructions, it should be at least $1\frac{1}{2}$ times the maximum design pressure for which the assembly is intended. On pipe or fittings, when the design pressure is not specified, the maximum hydrostatic test pressure should be $1\frac{1}{2}$ times the primary service pressure rating of the flange in the assembly but in no case should it be greater than the mill hydrostatic-test requirements as set forth in the ASTM specification covering the piping or tubing used in the assembly.

For power piping, the test pressure should never be less than 50 psi nor more than two times the primary service pressure rating of any valves included in the test. Globe valves should not be subjected to a test pressure under the seat greater than $1\frac{1}{2}$ times the primary service rating. Full pressure however, may be applied above the seat or with the valve open.

It may be necessary to protect expansion joints during testing, particularly when testing of individual sections of the system is done prior to the completion of the entire system. Expansion joints of the slip type should be secured against possible separation. Other types should be studied to determine whether they need extra protection or support during testing.

Air Testing

Air testing (or pneumatic testing) usually involves the filling of the piping component with air under pressure and applying a soap solution to the outside around all welds. As the expansion force of compressed air or other gas is relatively great, and since there is always the possibility of failure under test, the use of air or other gas for testing at pressures above 100 psi is discouraged. It should be employed under extraordinary circumstances and under competent supervision.

Air testing, therefore, is employed to a limited extent in general shop fabricating practice and usually only to satisfy the requirements of a particular specification. The larger fabricating shops necessarily have facilities for supplying compressed air for other plant services and usually can arrange to test with compressed air up to 100 psi.

Freon Testing

The halide leak test is most widely used in refrigeration piping systems. It is also used on steam headers with large numbers of nozzle welds. Freon testing has also been done as a preliminary safeguard to air testing, where it is applied at 25 psi before undertaking the air test based on design.

The piping system should be evacuated by holding a sustained vacuum of 1 in. mercury for 24 hr. The system should then be filled with freon gas from new factory cylinders to a pressure of 100 psig. All joints should be thoroughly tested for leaks with a halide torch.

INTRASCOPIC TESTS

In piping, the root condition is particularly important because the welding done from the outside may leave defect conditions that are not readily detectable by nondestructive testing. Nondestructive testing may also suggest weld-root conditions which may represent either a rejectable defect, as lack of penetration, or

unacceptable surface contour.   The resulting question in defect interpretation may then be resolved by examining the inside of the pipe with instruments involving magnifying lenses and mirrors which permit close examination of the inside pipe surface.   These instruments, also known as *borescopes*, can be inserted through holes as small as ¼ in.   Access holes, instrument tie-ins, or openings in valves may be utilized.

Multifilament instruments are available with flexible linkages and total lengths of as much as 12 ft.   With some instruments, a photograph can be made for permanent record.   Small special television cameras passed through the inside of piping have also been used.

DESTRUCTIVE TESTS

Several destructive test methods are also employed to maintain control over weld quality.   Destructive tests permit the evaluation in detail of questionable defects observed by radiographic or ultrasonic inspection.   The removed coupons permit a more accurate evaluation to determine if the nonhomogeneities are actually defects and a weld repair is necessary or if the questionable area can be considered acceptable.

Trepanning.   Trepanning is generally done by removing a cylindrical plug by means of a power-driven hole saw.   This technique has the advantage that the plug contains a section of the inside weld surface, which permits checking for penetration and cracking.   However, rewelding is more difficult and, if improperly done, may leave defects more critical than they were in the original materials.

Weld-probe Saw-cut Sections.   Where examination of the inside surface is not of interest, boat-shaped sections may be removed by means of a weld prober saw which makes two intersecting saw cuts.   Sometimes, narrow strips are removed by means of a saw or high-speed cutting wheels, or sections are removed by drilling, preferably through guiding fixtures, and chipping loose of a weld segment.

After the specimen has been removed, it is ground or sanded, preferably on a surface that extends across the weld, and polished, and the prepared surface is then etched.   Examination of the etched surface will disclose such defects as cracks, incomplete fusion, or slag inclusions, should any of these exist.

Miscellaneous Tests.   Occasionally, it may be desirable to perform destructive tests on completed weld joints selected at random from fabricated piping.   This involves the removal of a short section of pipe containing the questioned weld. To replace the section removed, a spool piece has to be inserted and two welds will have to be made.   In some specifications, it is provided that, by agreement between the purchaser and the manufacturer or contractor, the quality of the work may be checked by removing a joint made by each welder.   Such a joint may be given the standard destructive tests, such as the root, face, and side-bend tests, which are similar to the tests involved in welder qualification.

PLASTIC PIPING

Almost every plastic material can be produced in the form of pipe or tubing. However, practical manufacturing considerations and properties have limited the plastic materials available as commercial types to relatively few basic resin types.

Plastic piping is available in two principal groups, which are identified as *thermoplastic* materials and as *thermosetting* materials.   Thermoplastic materials are more widely used and are available in a greater number of different types.   Thermosetting materials are available in few types.   Moreover, they are generally reinforced with glass fibers or asbestos fibers to impart increased strength in the piping material.

THERMOPLASTIC PIPING

The development of thermoplastic pipe dates back to 1923 when polyvinyl chloride (PVC) was first produced in Germany. The excellent chemical resistance of this material resulted in applications in the paper industry and in textile and similar plants, particularly during the war years when some of the more corrosion-resistant metals were in short supply. It was not until 1951 that significant use was made in the United States of PVC piping.

Polyethylene in the form of pipe has been used since about 1942 in Europe and since 1947 in the United States. Its properties of flexibility and moderate strength make this material useful in water service and distribution piping, waste piping, and similar applications in which the temperatures and pressures are moderate. Although polyethylene piping exhibits good chemical resistance, it suffers a significant loss of strength at higher temperatures and requires continuous support. By means of catalysts, modifications were made in 1954 which produced polyethylene piping of higher strength and stiffness at room and higher temperatures.

High-density polyethylene became available in 1955. Although it exhibits greater strength than low-density polyethylene, it is not as widely accepted owing to its greater stiffness. Currently, the medium-density materials, which exhibit intermediate strengths but retain suitable flexibility, are generally preferred as piping materials. Cellulose-acetate-butyrate piping has been available as pipe in commercial form since early 1950.

Although polyamide resin has been available since 1940, its use as piping is limited because of its inherent high water absorption and relatively high price. Some use is made of polyamide as small-diameter tubing. Newer piping materials involve polypropylene and chlorinated polyester. With continuing extensive development by many major chemical companies, which are stimulated by increasing consumption of plastic piping materials, new materials and modification of materials already in use will become available.

Thermoplastic pipe is generally produced by the extrusion process, somewhat akin to extrusion methods applicable to metallic pipe. In the extrusion of thermoplastic pipe the material in powdered or pellet form is fed through a hopper into a barrel heated to plasticate the material. A revolving screw mixes the particles and gradually forces the hot "melt" through the pipe to the tube-forming die. The hot extruded pipe generally continues through a water bath or cascading cooling arrangement which cools and sets the pipe shape.

Plastic pipe is generally available in iron-pipe sizes (IPS). Typical sizes and weights for PVC pipe are given in Table 96. However, other special types are also available. For example, Table 97 shows gas-service pipe dimensions conforming to American Gas Association requirements for pull-through service and for use with slip-sleeve couplings to standard pipe size solvent-cement-type fittings. Lawn-sprinkler-service pipe sizes are listed in Table 98. The commercial tolerance on wall thickness is generally $-0 + 10$ per cent. Fittings for joining plastic pipe generally are produced by injection molding. The variety and types available are similar to those available for steel.

Polyvinyl Chloride (PVC) Piping

Rigid, unplasticized polyvinyl chloride (PVC) is a thermoplastic material; it is tough and exceptionally resistant to chemical attack.

PVC pipe is extruded. Fittings, flanges, and valves are manufactured by the injection molding method which results in high density and complete homogeneity of the material. Three types of PVC piping are available.

Table 96. Commercial Sizes (IPS) and Weights of Polyvinyl Chloride (PVC) Pipe (Abstracted from ATSM Specification D1785-64T)

Nominal size, in.	Schedule	Wall thick-ness,** in.	OD, in.	ID, in.	Theoreti-cal weight,* lb/ft	Calculated min bursting pressure, psi	
						Note 1	Note 2
¼	40	0.088	0.540	0.364	0.076	2,490	1,950
	80	0.119	0.540	0.302	0.096	3,620	2,830
½	40	0.109	0.840	0.622	0.153	1,910	1,490
	80	0.147	0.840	0.546	0.195	2,720	2,120
¾	40	0.113	1.050	0.824	0.203	1,540	1,210
	80	0.154	1.050	0.742	0.265	2,200	1,720
1	40	0.133	1.315	1.049	0.305	1,440	1,130
	80	0.179	1.315	0.957	0.385	2,020	1,580
1¼	40	0.140	1.660	1.380	0.409	1,180	920
	80	0.191	1.660	1.278	0.550	1,660	1,300
1½	40	0.145	1.900	1.610	0.489	1,060	830
	80	0.200	1.900	1.500	0.653	1,510	1,180
2	40	0.154	2.375	2.067	0.640	890	690
	80	0.218	2.375	1.939	0.910	1,290	1,010
3	40	0.216	3.500	3.068	1.380	840	660
	80	0.300	3.500	2.900	1.845	1,200	940
4	40	0.237	4.500	4.026	1.965	710	560
	80	0.337	4.500	3.826	2.710	1,040	810

* These representative values are not specified in ASTM D1785-64T.
** Thicknesses listed are minimum values. Tolerance is generally −0 + 10 per cent.
Note 1. Materials are PVC 1120, 1220, and 4116. A fiber stress of 6,400 psi was used in bursting pressure calculations.
Note 2. Materials are PVC 2112, 2116, and 2120. A fiber stress of 5,000 psi was used in bursting pressure calculations.

Table 97. Dimensions of Plastic Pipe for Gas Service (AGA Requirements)

Nominal size, in.	Nominal OD, in.	Nominal ID, in.	Sleeved weight, lb	Max working pressure at 73 F, psi
½	0.625	0.500	0.054	250
¾	0.875	0.750	0.079	175
1	1.125	1.000	0.103	125
1¼	1.375	1.250	0.127	110
1½	1.625	1.500	0.152	90
1¾	1.875	1.750	0.177	80
2	2.125	2.000	0.200	75
2½	2.660	2.500	0.320	75
3	3.190	3.000	0.457	75
4	4.250	4.000	0.800	75

Table 98. Dimensions of Plastic Pipe for Lawn-sprinkler Service

Nominal size, in.	Nominal OD, in.	Nominal ID, in.	Sleeved weight, lb	Max working pressure at 73.4 F, psi
½*	0.60	0.50	0.042	200
¾*	0.855	0.750	0.065	150
¾	0.840	0.740	0.060	143
1	1.050	0.950	0.0775	114
1¼	1.375	1.25	0.127	110
1½	1.625	1.50	0.152	90
2	2.125	2.00	0.200	75

* The two starred sizes in ½ and ¾ in. are used with LOC fittings in certain areas, particularly the West Coast areas.

Type I in the past has generally been marketed as "normal-impact" grade. It is now produced to a hydrostatic design stress of 2,000 psi for water at 73.4 F. Two grades are recognized under the designations PVC 1120 and PVC 1220.

Type II has been marketed as "high-impact" grade. Type II, Grade 1 is produced to a hydrostatic design stress of 1,000 psi for water at 73.4 F and is designated as PVC 2110. It is also produced to a hydrostatic design stress of 1,250 psi and is designated as PVC 2112; material produced to a hydrostatic design stress of 1,600 psi is designated PVC 2116.

Type IV is a newer grade produced to a hydrostatic design stress of 1,600 psi for water at 73.4 F. It is designated as PVC 4116.

Fittings and valves of Types I and II are readily available with threaded, solvent weld (socket), and flanged ends.

Type I (Normal-impact Grades). The two Type I PVC grades have greater strength properties over a wider temperature range. They can be used at temperatures up to 160 F and, in addition, at higher working pressures than can Type II PVC. These grades have superior chemical resistance throughout their temperature range.

Type I PVC grades should be specified in applications where greater strength, temperature resistance, or extreme chemical resistance are necessary. PVC is normally straw-colored without pigmentation. Industrial-process piping is usually colored dark gray for identification purposes but may be produced, also, as white, brown, red, or any other color.

Type II (High-impact Grades). The Type II PVC grade is modified by the addition of a copolymer to provide greater resistance to impact over an extended temperature range. Some loss of tensile strength, temperature performance, and chemical resistance results, although these properties remain comparatively high in comparison to those of many other thermoplastic materials. In some environments, Type II PVC should be specified in applications where these properties are not critical but where the greater strength is required. The PVC 2110 grade is normally identified by its light gray color. However, it has been produced also in blue, red, white, green, and other colors.

Standards. PVC pipe is covered by ASTM Standard D1785, Tentative Specifications for Poly (Vinyl Chloride)(PVC) Plastic Pipe. Voluntary industry standards have also been issued by the U.S. Department of Commerce: Commercial Standard CS256, Polyvinyl Chloride (PVC) Plastic Pipe (SDR-PR and Class T) and Commercial Standard CS207, Rigid Unplasticized Polyvinyl Chloride Pipe. The normal physical properties of PVC are summarized in Table 99.

Table 99. Normal Physical Properties of PVC Piping Materials

Properties	Materials		ASTM test no.
	Type I	Type II	
Tensile strength, room temperature, psi............	7,000	6,000	D638
Modulus of elasticity in tension, psi	415,000	350,000	
Flexural strength, psi	14,500	11,500	D790
Izod impact strength, ft-lb/in. of notch, notched, room temperature	0.5–1.0	10.0–19.0	D256
Izod impact strength, ft-lb/in. of specimen unnotched, room temperature	45	55	D256
Specific gravity	1.38	1.35	D792
Hardness, Shore "D"	78/82	76/80	D785
Heat distortion, temperature, F, at 264 psi	165	155	D648
Coefficient of thermal conductivity, Btu/(sec) ft F $\times$ 10^4	3.5	4.5	C-177
Specific heat, cal/(g C)	0.25	0.25	
Coefficient of linear thermal expansion per C $\times$ 10^5	5	10	D696
Water absorption, % in 24 hr at 25 C	0.07	0.07	D570

Applications. PVC piping is extensively used in highly corrosive applications involving acids, alkalies, salt solutions, alcohols, and many types of chemicals. PVC piping is also used in oil fields because it can carry sour crude oil to which PVC is chemically inert and because paraffin build-up is minimum on the smooth inside surfaces of this pipe.   Other applications include salt-water disposal in oil fields and gas-transmission service. PVC will handle most chemicals up to 150 F.   Still other applications include the piping of cold water in industrial plants because PVC is nontoxic and will not impart odor or taste to the water, as well as vent piping for the removal of acid fumes and corrosive gases from industrial plants.

PVC pipe and fittings experience little or no physical deterioration when exposed to direct sunlight; sunlight causes deterioration in several other plastic piping materials.   Like other plastic materials, PVC will not produce sparks when struck. PVC is safe to use around explosives or flammable vapors, and it does not support combustion.   Also water contaminants do not build up on the smooth walls of PVC pipe or fittings.

The maximum working pressure at 75 F is shown in Table 100.   Because polyvinyl chloride, like most other plastic materials, is somewhat notch-sensitive, maximum allowable working pressures for threaded pipe are considerably lower than those for unthreaded pipe.   The strength of polyvinyl chloride decreases as the operating temperature increases and allowable working pressures must be decreased at higher temperatures.   Working pressures at 75 F are approximately 20 per cent of the bursting pressures.   When operating temperatures exceed 75 F, the maximum operating pressure for Type I pipe may be determined from Fig. 112.

Maximum operating pressure for Type II pipe decreases with increased temperature in a similar manner up to 110 F.   Above 110 F, creep extension (a

Table 100.   Water-pressure Ratings at 23 C (73.4 F) for Class T Threaded PVC Plastic Pipe (Commercial Standard CS256-63)

Nominal pipe size, in.	Dimension ratio	Pressure ratings[1] for PVC plastic pipe made from		
		PVC1120 PVC1220, psi	PVC4116, psi	PVC2110, psi
1/8	4.15	630	500	315
1/4	4.15	630	500	315
3/8	4.15	630	500	315
1/2	5.0	500	400	250
3/4	6.0	400	315	200
1	6.0	400	314	200
1 1/4	7.3	315	250	160
1 1/2	9.0	250	200	125
2	11.0	200	160	100
2 1/2	11.0	200	160	100
3	13.5	160	125	80
3 1/2	13.5	160	125	80
4	17.0	125	100	63
5	17.0	125	100	63
6	21.0	100	80	50
8	26.0	80	63	NPR[2]
10	32.5	63	50	NPR[2]
12	41.0	50	NPR	NPR[2]

[1] Pressure ratings shown for threaded pipe are one-half those calculated in accordance with

$$\frac{2S}{P} = SDR - 1 \quad \text{or} \quad \frac{2S}{P} = \frac{OD}{t} - 1$$

where S = design stress, psi
   P = pressure rating, psi
   OD = average outside diameter, in.
   t = minimum wall thickness, in.
   SDR = standard thermoplastic pipe dimension ratio (OD/t for PVC pipe)
   Thus, pressure ratings for nonthreaded pipe in Class-T dimensions are twice those given in this table.
[2] NPR = not pressure-rated.

time-dependent extension which occurs at stresses lower than the yield strength) becomes significant, so that the pressure-rating chart must be based on the creep characteristics of the material in the temperature range from 110 F to 130 F. When operating temperatures exceed 75 F, the maximum operating pressure for Type II pipe should be determined from Fig. 113.

The maximum working pressure at any temperature is obtained from Figs. 112 and 113 by locating the maximum working pressure at 77 F along the bottom of the chart. A line to the diagonal line should then be drawn corresponding to the desired working temperature. The maximum working pressure scale is then read on the left-hand side of the chart.

Cutting. In cutting PVC pipe to length, either in the shop or in the field, a hack saw is preferred over a pipe cutter because a pipe cutter only displaces the material; it does not remove it. If threading is to be done subsequently, a pipe cutter is satisfactory. The ends of the pipe can be beveled with a hand file and the interior deburred with a regular tool or knife.

Threading. PVC pipe has the same outside diameter as that of standard steel pipe in comparable sizes; this permits the use of standard steel taps and dies. For

best results in threading, it is recommended that new taps and dies be used and that these tools be dressed to the following angles:

Taps should be ground with a 0- to 10-deg negative rake, depending on the size and pitch of the thread. Die chasers should have a 33-deg chamfer on the lead and a negative rake up to 10 deg on the back or relief edge of the chamfer. Self-opening die heads and collapsible taps, power threading machines, and a slight chamfer to lead the taps or dies will speed production, but taps and dies should not be driven at high speeds or under heavy pressure.

Maximum operating pressure vs. temperature chart

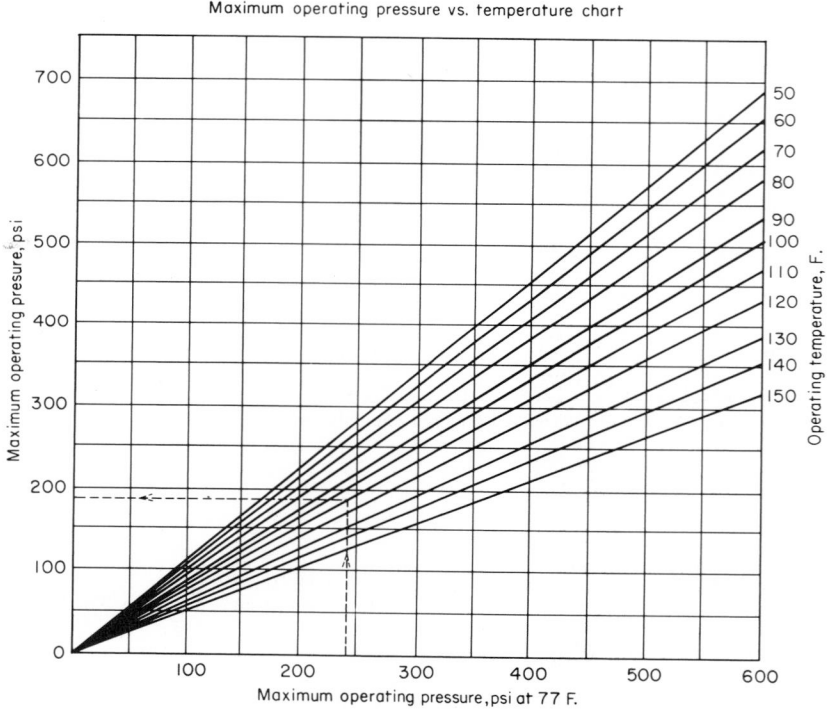

Fig. 112. Graph for determining maximum operating pressure of normal-impact (Type I) PVC at any temperature.

Pipe to be threaded should be held in a pipe vise, as saw-toothed jaws will mar the pipe. Flanges and close nipples should be threaded in jigs or tapping fixtures. Since PVC pipe is somewhat notch-sensitive, it is not advisable to thread Schedule 40 pipe, as thread grooves would be too deep for the relatively thin wall. Ordinary lubricating and cutting oils have proved beneficial in hand threading. In a pipe-threading machine, water-soluble oil should be used.

In assembling threaded pipe and fittings, a strap wrench only should be used. In no case should a Stillson wrench be applied. Joints should be made up hand-tight, followed by a one-quarter to one-half turn more. Greater forces tend to stretch or distort the plastic piping material. Teflon thread lubricant is recommended as a compound for joints.

Bending. Bending is sometimes necessary in fabricating PVC pipelines. It should be limited to noncritical applications at room temperature or lower where

maximum operating pressures are not required, since some bending stresses are retained in the material after bending.

Before bending, the pipe should be heated at the section to be bent to a temperature of about 275 to 300 F by means of hot air or immersion in a noncorrosive liquid. Overheating should be avoided, and the pipe should not be held at the bending temperature too long, since the pipe may lose its form. After the section has been heated uniformly, the pipe may be bent to a minimum radius of ten times the pipe diameter. Pipe-bending forms should be used to prevent flattening.

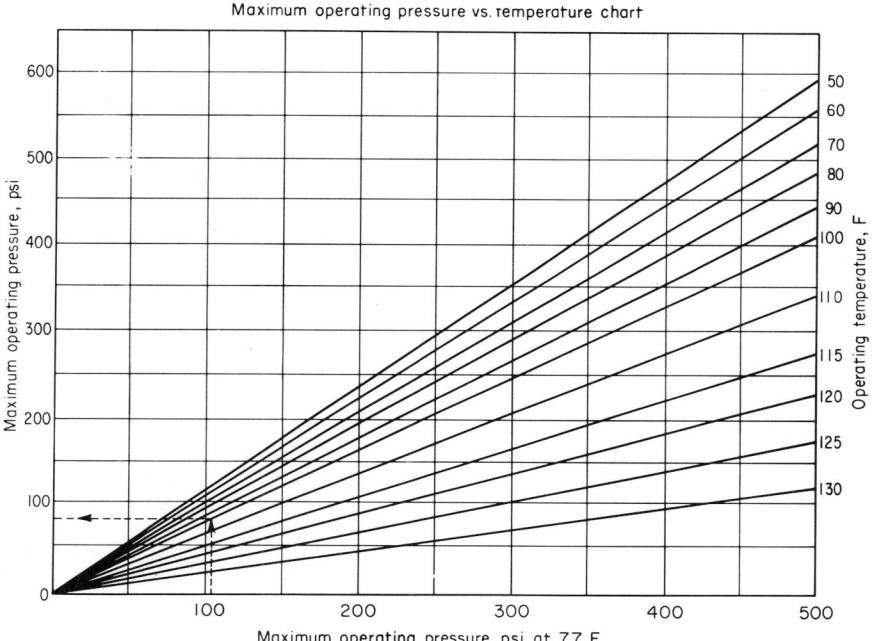

Fig. 113. Graph for determining maximum operating pressure of high-impact (Type II) PVC at any temperature.

Coiled springs used in the bending of lead pipe may also be used for internal support. The angle of the bend should be slightly more than the desired angle, since there is a small amount of spring back. The pipe may be cooled quickly after bending.

Pipe Support. Continuous support is recommended for Type I PVC at temperatures above 120 F and for Type II PVC, above 100 F. For vertical pipe, the pipe should be banded at intervals dependent on the vertical load involved. A riser clamp should be supported on spring hangers. It is important that PVC pipe is not clamped tightly or restricted from movement.

Polyethylene (PE) Piping

Polyethylene is produced from hydrocarbon and ethylene under high temperature and extremely high pressure. It is readily available in various piping shapes.

Three grades of polyethylene are available commercially depending primarily

Table 101. Characteristics of Materials Used for Polyethylene Piping

Grade	ASTM type	Density, g/cu cm
Low density.........	I	0.910–0.925
Medium density	II	0.926–0.940
High density	III	0.941–0.965

on the formulation and pressure under which they are produced.   The three grades and their density characteristics are listed in Table 101.[1]

Where flexibility and ease of handling are desired, low-density or medium-density PE is preferred.   When extruded from medium-density PE, the pipe sacrifices only a slight degree of flexibility, yet it can meet burst-pressure specifications with appreciably reduced wall thickness.   High-density PE is not as flexible

Table 102. Typical Properties of High-density Polyethylene

Properties	
Specific gravity	0.940–0.942
Melt index (D1238)*	Nil
Reduced specific viscosity	30 ± 5
Crystalline melting point, C	130–131
Hardness, Rockwell R (D785)	38
Izod impact, notched (ft-lb/in. notch)	Not broken (see below)
Deformation under load (2,000 psi, 6 hr, at 122 F) (D621), %	6–8
Heat-distortion temperature (66 psi) (D648), C......	70–79
Stiffness in flexure (D746), psi	72,000
Tensile modulus (0.2 in./min testing speed at 73 F) (D638), psi	80,000–100,000
Tensile yield stress (D638), psi	2,500
Water absorption (D570), %	0.03
Stress crack (Bell procedure)	None in 4,000 hr

Properties at 120 C (250 F)	
Tensile yield stress, psi	640
Tensile modulus, psi	About 12,500

Coefficient of heat transfer† Btu/(sq ft hr F/in.) at

−148 F	3.3
32 F	2.8
167 F	2.3

Properties in molded slab form (1.25 in. thick)	Type A‡	Type B§
Impact strength, Izod:		
Notched at 73 F, ft-lb/in. notch	15+	15+
Notched at −40 F, ft-lb/in. notch	18	18
Density at 20 C, g/cc	0.939	0.939
Hardness, Rockwell R (D785)	60	62
Heat-distortion temperature (66 psi) (D648), C......	84	80

* D-numbers indicate ASTM test method.
† Derived from data in *Kolloid Zeitschrift*, vol. 174, no. 2, p. 140, February, 1961.
‡ Type A: unmodified polymer flake.
§ Type B: contains antioxidant.

[1] ASTM Specification D1248, Tentative Specification for Polyethylene Molding and Extrusion Materials.

as the low-density grades. It is superior, however, in burst strength and is preferred in applications involving high air or liquid pressure.

Although available in clear form, carbon black in excess of 2 per cent is usually added to improve weather resistance.

A voluntary industry standard has been issued by the U.S. Department of Commerce:

Commercial Standard CS255, Polyethylene (PE) Plastic Pipe (SDR-PR).

Typical properties of high-density, high-molecular-weight PE are listed in Table 102. High-density PE is produced as merchant pipe for water lines for jet wells and farm sprinkler systems and as engineered pipe for salt-water disposal lines, chemical waste lines, gas-gathering systems, etc. PE piping is also used as conduit for power and telephone cables. Essentially the same procedures discussed for PVC apply also to PE piping materials.

Acrylonitrile-butadiene-styrene (ABS) Piping

Acrylonitrile piping materials are produced by injection or compression molding or extruding of polymers or blends of polymers from monomers or chemical derivatives of the monomers acrylonitrile, butadiene, and styrene.

The natural color of ABS is opaque, similar to ivory. Pigmentation to various colors is possible. Carbon black may be added to increase resistance to ultraviolet deterioration.

The American Society for Testing and Materials has issued ASTM specification D1527, Tentative Specification for Dimensions of Iron Pipe Size (IPS) Extruded Acrylonitrile-butadiene-styrene (ABS) Plastic Pipe.

A voluntary industry standard has been issued by the U.S. Department of Commerce, Commercial Standard CS254, Acrylonitrile-butadiene-styrene (ABS) Plastic Pipe (SDR-PR and Class T). Typical properties of ABS piping materials are summarized in Table 103.

These materials exhibit good toughness and tensile strength in the range from 2,000 to 9,000 psi. The chemical resistance of ABS pipe is good. It is suitable for service involving almost all inorganic acids, bases, and salts, in addition to aliphatic hydrocarbons, glycols, and some alcohols. Applications of ABS include the transportation of salt water, crude oil, and gas. ABS pipe is also used for sewage piping. Other applications include dilute sulfuric acid, chlorine, ammonia gas, etc.

Cellulose-acetate-butyrate (CAB) Piping

Cellulose-acetate-butyrate pipe is made of virgin cellulose, acetate, and butyrate manufactured by chemical processing from cotton linters and products derived from petroleum. Although available in clear form (Type I), CAB piping normally contains over 2 per cent of nontoxic carbon black (Type II) to provide resistance to ultraviolet rays for outdoor (above ground) installation.

Semirigid plastic pipe has sufficient rigidity to be handled in straight lengths and sufficient flexibility to follow the contour of the ground. Nominal sizes range from $\frac{1}{2}$ to 6 in. diameter in 20 and 40 ft straight lengths. Semirigid CAB piping is used for carrying sewage and other fluids. In refinery practice, it is used for carrying high-paraffin crudes, sour crudes, salt water, and raw gases. Applications also include the transmission of low-pressure gas.

CAB piping is generally joined by solvent welding. Socket-weld fittings are readily available. Joints are generally made by employing screwed connections or by thermosetting cements.

Table 103. Typical Properties of ABS Piping Materials

Properties	Semiflexible pipe extrusion	Commercial pipe grades		
		Type I, grade 1	Type I, grade 2	Type II, grade 1
Tensile strength, psi	2,000–3,000 (72 F)	5,000–5,500 (72 F) 3,000 (160 F)	5,000–5,500 (72 F) 3,000 (160 F)	7,500–8,000 (72 F) 4,500 (160 F)
Modulus of elasticity (in tension), psi	....	210,000	230,000	350,000
Hardness, Rockwell	63 (R)	85–95 (R)	85–95 (R)	100–115 (R)
Flexural strength, psi	No failure	7,800	8,000	10,000
Impact strength (notched izod, 1/2 × 1/2 in.), ft-lb/in. of notch	2–3	1.5–2.0	2–5	4–5
Deflection temperature under load (comp. molded spec.), F, at 264-psi fiber stress	150	196	196	215
Deformation under load, % at 2,000 psi and 50 C	....	1.27	1.27	0.41
Color possibilities	Opaques	Black	Black	Black
Specific gravity (naturals only)	0.99	1.04	1.04	1.06
Moisture absorption, %	....	0.2–0.3	0.2–0.3	0.2–0.3
Burning rate (0.050 in. thick), in./min	1.0–1.5	1.0–1.5	1.0–1.5	1.0–1.5
Dissipation factor, 10^6 cps	....	0.006–0.011	0.006–0.011	0.012
Dielectric constant, 10^6 cps	....	2.8–3.8	2.90	3.11
Dielectric strength, volts/mil, 1/8 in. thick	....	300–450	300–450	300–450
Volume resistivity, ohm/cm	....	....	3.5×10^{16}	3.7×10^{16}
Bulk factor (pellets)	1.5–1.7	1.5–1.7	1.5–1.7	1.5–1.7

Welding of Thermoplastic Piping

Thermoplastic piping is readily joined by solvent welding (cementing) and hot-gas (thermal) welding. Where *solvent welding* is done, socket-type joints are generally used. Socket-weld fittings are readily available for this purpose. The cement is applied to the pipe end and socket of fitting or valve. After pushing the pipe into the socket, the joint should be allowed to set for at least 10 hr. Full strength is developed in 24 hr.

Table 104. Recommended Gas Temperatures for Welding Various Plastic Piping Materials

Material	Gas temperature range, F
Cellulose acetate butyrate	525–575
Polyamide	625–675
Polyethylene:	
1 Low density	500–550
2 Medium density	525–575
3 High density	550–600
Polyvinyl chloride:	
1 Types I and IV	500–550
2 Type II	475–525
Styrene copolymers:	
1 Type I	500–550
2 Type II	550–600

In *hot-gas welding*, air, nitrogen, oxygen, or carbon dioxide is passed through a heating torch. Compressed air is most widely used. In PVC pipe joints, oxygen normally provides the strongest joints and carbon dioxide the weakest.

Since polyethylene is sensitive to oxygen which causes surface oxidation and results in poor-quality welds, an inert gas or nitrogen must be used on PE materials. Two types of heating torches are used: one is electrically heated, and the other has a gas flame that heats a coil through which the welding gas flows. The temperature of the welding gas is controlled by the flow rate through the torch. The flow rate generally corresponds to pressures of 5 to 10 psi. Lowering this flow rate will increase the gas temperature, which is measured by a thermometer held approximately $\frac{1}{4}$ in. from the tip of the torch. The gas temperatures which should be used are listed in Table 104.

Typical weld joints are shown in Fig. 114 for butt and fillet joints. Filler rod should be used. Conditions for pipe welding are summarized in Table 105.

Table 105. Recommendations for Hot-gas Welding of Plastic Piping

Pipe wall thickness, in.	Root opening (butt joints), in.	Filler rod diameter, in.	Torch nozzle diameter, in.	Root face, in.
$\frac{3}{32}$–$\frac{1}{8}$	$\frac{1}{32}$–$\frac{1}{16}$	$\frac{3}{32}$	$\frac{1}{8}$	$\frac{1}{16}$
$\frac{1}{8}$–$\frac{5}{32}$	$\frac{1}{16}$–$\frac{3}{32}$	$\frac{1}{8}$–$\frac{5}{32}$	$\frac{5}{32}$	$\frac{1}{16}$
$\frac{1}{4}$–$\frac{3}{8}$	$\frac{1}{16}$–$\frac{3}{32}$	$\frac{5}{32}$–$\frac{3}{16}$	$\frac{1}{4}$	$\frac{1}{8}$

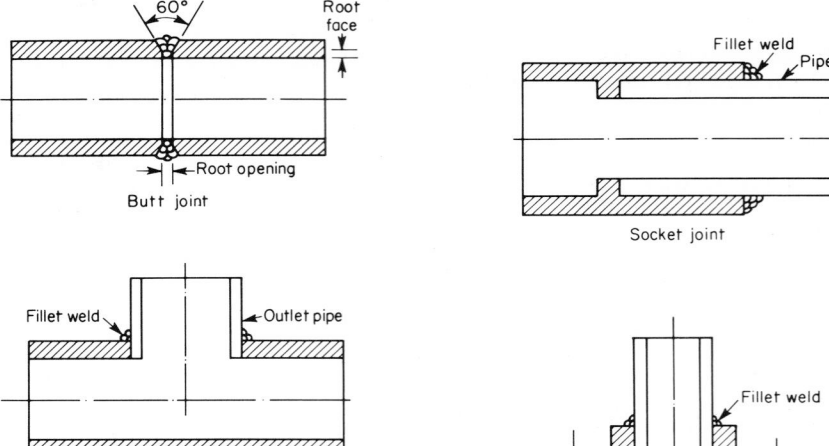

FIG. 114. Joint recommendations for welding of plastic piping materials.

For "nonrigid" plastic piping materials, the filler rod should be pressed against the piping material at a forward angle of 45 deg. When joining "rigid" materials, the rod should be held perpendicular to the joint surface.

Properly made butt joints should exhibit 60 to 90 per cent joint efficiencies. Where plastic pipe is to be joined to metallic piping, flanged joints should be used.

THERMOSETTING PIPING

The two major resins used in commercial thermosetting pipe are epoxy or polyester resins. This type of pipe is generally reinforced with glass or blue asbestos fibers; it is commonly produced by centrifugal casting, hand lay-up molding, or filament winding.

In the *centrifugal casting* process, chopped-fiberglass rovings are deposited on the interior of a mold which is rotated at high speed. Following deposition of the glass fibers which have no orientation or continuity, the epoxy or polyester resin is introduced in sufficient quantity to impregnate the glass fibers thoroughly and leave a resin layer approximately $1/16$ in. thick on the inside of the pipe. This layer protects the pipe from abrasion and corrosion. The pipe is then cured while continuing to rotate at high speed. By adding or substituting fiberglass cloth, the pipe strength can be increased.

The *hand lay-up* process (also referred to as the *contact molding* process) involves applying multiple layers of glass reinforcing material and catalyzed resin on a mandrel. The glass reinforcing material consists usually of chopped strand glass mat. A combination of glass mat and woven glass roving is also utilized and tends to increase strength. The final properties of the pipe may vary with the types of glass reinforcement and resins used, the placement of the reinforcement, and the relative percentages of resin and reinforcement.

In *filament winding*, continuous glass filaments are helically interwoven under tension around a polished mandrel at a helix angle. At the same time, each filament

is thoroughly coated with the thermosetting resin. The pipe is subsequently heat-cured to form the dense composite structure.

In some products, the epoxy-fiberglass outer pipe is combined with an inner surface layer of reinforced polyester.

Epoxy Pipe

Commercial pipe is generally made in standard pipe sizes and in diameters from 2 through 12 in. OD. The properties of commercial types of thermosetting pipe

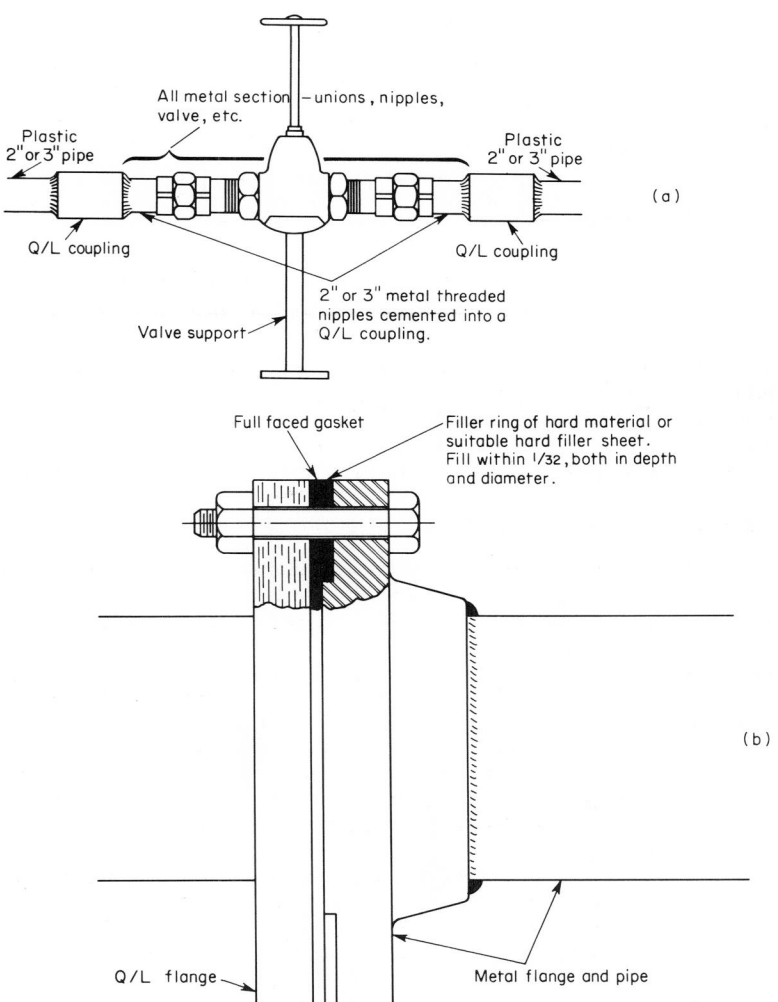

Fig. 115. Examples of screwed and flanged joints between thermosetting plastic pipe and metal pipe, valves, or fittings; (*a*) connections suitable for 2 and 3-in. pipe to valve; (*b*) connection to a raised-face metal flange (Courtesy Amercoat Corporation).

Table 106. Properties of Thermosetting Pipe Products

Properties	Hand lay-up polyester pipe	Asbestos-reinforced pipe	Centrifugally cast			Filament wound
			Thin wall	Medium wall	Heavy wall	
Ultimate tensile stress at 80 F, psi	8,000–15,000	14,000	23,000	25,000	33,000	100,000–120,000
Usable tensile stress for long-term operation at 80 F, psi	800–1,500	2,000	2,000	2,500	3,300	8,000–12,000
Usable tensile stress for long-term operation at 200 F, psi	400–750	1,500	1,000	2,500	3,300	8,000–12,000
Usable tensile stress for long-term operation at 300 F, psi	......	750	......	1,250	1,650	4,000–6,000
Linear expansion, F	12.15×10^{-6}	8×10^{-6}	8.25×10^{-6}	7.06×10^{-6}	7×10^{-6}	7×10^{-6}
Glass or asbestos/resin ratios, includes liner	15:85–25:75	50:50	40:60	43:57	51:49	66:34–75:25
Effect of temperature: −40 F	Brittle	Brittle	Liner very brittle	Liner brittle	Liner brittle	Satisfactory
+300 F	Very elastic	Elastic	Very elastic	Elastic	Satisfactory	Satisfactory
Liner reinforcement	None to light	Reinforced except for thin layer	None—all resin			Reinforced 40%
Main reinforcement	Chopped strands to some light cloth	Short asbestos fibers	Continuous braid-controlled			Continuous glass filaments controlled
Joint strength	Weak	Excellent	Excellent, but subject to easy damage until thermoset			Excellent; has mechanical lock
Span between hangers with heated fluids	Short spans	Medium spans	Short spans	Medium spans	Long spans 8 ft or more	Long spans 8 ft or more
Quality-control procedures used by manufacturers	Limited; no production line testing; limited material control	Not known what line control and testing performed	Excellent; ample quality and production-line testing performed on raw materials, processes, and products			

products are summarized in Table 106.   Applications of epoxy pipe include the transportation of acids, neutral or basic salt solutions, waste process water, and sewage.   Hydrogen sulfide does not react with this material.   In the paper industry, this pipe is used for pulp stock, wastes, and dyes.

Like thermoplastic pipe, thermosetting pipe is not subject to deposits of paraffin. In various applications, this pipe also resists fouling, salt atmospheres, and marine organisms.   Some applications are also made in the food and beverage industries.

Standard screwed and socket-type fittings are available.   Some fittings have tapered socket joints to provide improved strength and sealing.   Screwed connections are extensively used in the oil industry.   In the chemical industry, joints made with adhesives are generally preferred.   The adhesives used in joints are of thermosetting types, and they consist of the resin and the catalyst.   Normally an inert clay filler is added to make the adhesive easier to handle.

Pipe ends are cut with a hack saw. Prior to the application of the adhesive, the ends should be sanded.   Where joints are to be made between the plastic pipe and metal, the ends of the metal should be sandblasted prior to cementing.   Screwed and flanged joints between thermosetting plastic pipe and metal pipe are illustrated in Fig. 115.

Figure 116 illustrates pipe-support requirements.   Point loading is undesirable. Standard pipe clamps of the same width recommended for steel pipe should be used.   Where long tie rods are used, lateral support may be necessary to prevent excessive surge and vibration.   All edge burrs of the supporting sheet material should be removed.   Maximum spacings between supports are given in Table 107.

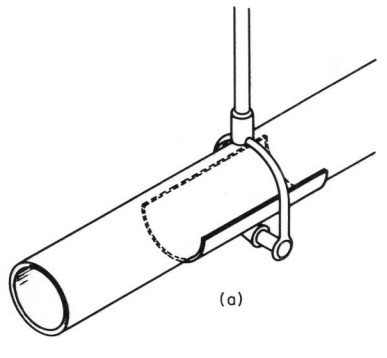

(a)

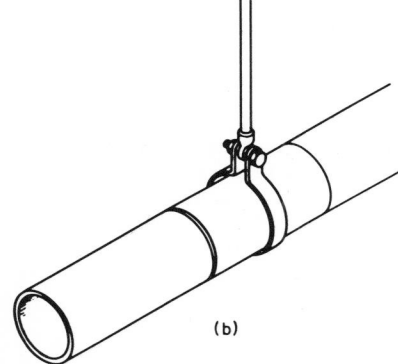

(b)

Fig. 116. Recommended support of reinforced thermosetting plastic pipe. (*a*) 12-gage half-wrap-around 3 to 6-in. wide; (*b*) 20-gage sheet-metal wrap-around 3 to 6-in wide.

Table 107.   Maximum Support Spacing for Reinforced Thermosetting Pipe (Amercoat Corporation)

	Pipe size, in.						
	2	3	4	6	8	10	12
100 F	10 ft	12 ft	14 ft	16 ft	18 ft	18 ft	20 ft
200 F	10 ft	10 ft	12 ft	14 ft	16 ft	16 ft	16 ft
300 F	6 ft	8 ft	8 ft	10 ft	10 ft	10 ft	12 ft
Minimum bandwidth	1½ in.	2 in.	2 in.	2½ in.	3 in.	3 in.	3½ in.

Polyester Pipe

Commercial polyester pipe is made in diameters to 60 in. OD and is suitable for service at temperatures up to 250 F. This kind of pipe is resistant to acids, alcohols, alkalies, bleaches, organic materials, and solvent.

CONCRETE, CEMENT, AND CEMENT-LINED PIPE

NONREINFORCED CONCRETE PIPE

Nonreinforced-concrete pipe for the conveyance of sewage, industrial waste, and storm water is made in sizes from 4 to 24 in. It is produced in accordance with the following specifications: ASTM C14, Standard Specifications for Concrete Sewer Pipe; AASHO M86, Standard Specifications for Concrete Sewer Pipe, and Federal SS-P-371, Pipe, Concrete (Nonreinforced, Sewer). This pipe may or may not be manufactured for use with rubber gaskets to seal the joints. It is available in either the standard-strength or extra-strength classes.

Nonreinforced-concrete drain tile is used for land drainage, and for subsurface drainage of highways, railroads, airports, and buildings. It is made in sizes from 4 through 24 in. in accordance with the following specifications: ASTM C412, Standard Specification for Concrete Drain Tile and AASHO M178, Standard Specification for Concrete Drain Tile. Drain tile is available in the standard-quality, extra-quality, and special-quality classifications.

Perforated concrete pipe used for underdrainage is made in accordance with ASTM Specification C444, Specifications for Perforated Concrete Pipe. This pipe is also made in sizes 4 through 24 in. and is available in the standard-strength and extra-strength classification.

Concrete irrigation pipe used for the conveyance of irrigation water under low hydrostatic heads and for land drainage is made in accordance with ASTM Specification C118, Standard Specifications for Concrete Pipe for Irrigation or Drainage. It is made in sizes 4 through 24 in.

Nonreinforced-concrete irrigation pipe for use with rubber-type gasket joints is made for conveyance of irrigation water at water pressures of 30 ft of head or higher depending on diameter. Such pipe is made in sizes 6 through 24 in. in accordance with ASTM Specification C505, Specifications for Nonreinforced Concrete Irrigation Pipe with Rubber-type Gasket Joints. Physical and dimensional requirements of standard-strength bell-and-spigot nonreinforced-concrete sewer pipe are tabulated in Table 108.

Manufacture. Curing is done by steam or water. In steam curing, the pipe is placed in a curing chamber where the desired moist atmosphere is provided by the controlled injection of steam. In water curing, the pipe is covered with a water-saturated material or the water is applied by sprinklers, perforated pipes, or other porous materials.

Unreinforced pipe is usually designed on the basis of a safety factor of 1.25 to 1.50 on ultimate strength in three-edge bearing, while reinforced-concrete pipe is usually designed to the 0.01-in. crack in three-edge bearing. These provide adequate factors of safety in the actual trench installation where the loading conditions are much less severe than in a three-edge bearing test. Concrete sewer pipe may be used to convey aggressive industrial wastes of all kinds; however, in some cases careful selection of a protective lining is important.

Jointing. Rubber-gasketed joints for C14 and C76 pipe are covered by ASTM Specification C443, Joints for Circular Concrete Sewer and Culvert Pipe, Using Flexible, Watertight, Rubber-type Joints.

Table 108. Physical and Dimensional Requirements of Standard-strength, Bell-and-spigot Nonreinforced Concrete Sewer Pipe (ASTM C14-65)

Internal diameter, in.	Min thickness of barrel T, in.	Min laying length[1,4] L, ft	Inside diameter at mouth of socket[2] D_s, in.	Depth of socket L_s, in.	Min taper of socket HL_s	Min thickness of socket[3] T_s	Minimum strength, lb/lin ft		Max absorption, %
							Three-edge bearing method	Sand-bearing method[4]	
(1)	(2)	(3)	(4)	(5)	(6)	(7)	(8)	(9)	(10)
4	9/16	2½	6	1½	1:20	3T/4, all sizes	1,000	1,500	8
6	5/8	2½	8¼	2	1:20		1,100	1,650	8
8	3/4	2½	10¾	2¼	1:20		1,300	1,950	8
10	7/8	3	13	2½	1:20		1,400	2,100	8
12	1	3	15¼	2½	1:20		1,500	2,250	8
15	1¼	3	18¾	2½	1:20		1,750	2,620	8
18	1½	3	22¼	2¾	1:20		2,000	3,000	8
21	1¾	3	25¾	2¾	1:20		2,200	3,300	8
24	2⅛	3	29½	3	1:20		2,400	3,600	8

[1] Shorter lengths may be used for closures and specials.
[2] When pipe is furnished having an increase in thickness over that given in column 2, then the diameter *at the inside* of the socket shall be increased by an amount equal to twice the increase of the barrel.
[3] This measurement shall be taken ¼ in. from the outer end of the socket.
[4] Not included in ASTM Specification C14-65.

Gaskets must meet the following requirements:

```
Tensile strength, min, psi .............1,200
Elongation at break, min, % ...........  350
Shore durometer hardness, nominal
    Min............................  40 (see
    Max ...........................  60  note)
Compression set, max per cent of
    original deflection, 70 C (158 F)
    for 22 hr ........................   25
Accelerated aging and changes in
    properties after conditioning in a
    circulating hot-air oven for 96 hr
    at 70 C (158 F):
    Decrease in tensile strength, max,
        per cent of original ..............   15
    Decrease in elongation, max, per
        cent of original ..................   20
Water absorption by weight, 48 hr
    at 70 C, max, % ..................   10
```
Note: Allowable variation ±5 from specified nominal hardness.

REINFORCED CONCRETE PIPE

Reinforced concrete pipe for the conveyance of sewage, industrial wastes, and storm water and for the construction of culverts is made in sizes from 12 to 144 in. Reinforced concrete pipe may or may not be manufactured for use with rubber gaskets to seal the joints.

It is usually manufactured in accordance with the following specifications:

ASTM C76 —Specifications for Reinforced Concrete Culvert, Storm Drain and Sewer Pipe[1]

AASHO M170 —Specifications for Reinforced Concrete Culvert, Storm Drain and Sewer Pipe

Federal SS-P-375—Pipe, Concrete (Reinforced, Sewer)

Reinforced-concrete pipe may be made with either tongue-and-groove or bell-and-spigot joints. When made for use with rubber gaskets, the joints must conform to ASTM Specification C443, or AASHO Specification M198-Specifications for Joints for Circular Concrete Sewer and Culvert Pipe, Using Flexible Watertight, Rubber-type Gaskets.

Concrete pipe is available also in both an arch and an elliptical cross section. These pipes are made in accordance with the following specifications:

ASTM C506—Specifications for Reinforced Concrete Arch Culvert, Storm Drain and Sewer Pipe

ASTM C507—Specifications for Reinforced Concrete Elliptical Culvert, Storm Drain and Sewer Pipe

In each of the standards covering reinforced concrete pipe, five strength classes are defined in terms of minimum three-edge bearing load at a crack width of 0.01 in. and at the ultimate strength of the pipe.

The strength class required for a given installation is determined by computing the earth load and live loads which will be transferred to the pipe under the conditions anticipated. This load is then converted to an equivalent three-edge bearing load by dividing it by a bedding factor. The bedding factor depends upon installation conditions and is always greater than 1.0.[2]

[1] For detailed test data, see F. J. Heger et al., *Jour. of the American Concrete Institute*, October and November, 1963.

[2] For more detail on determining external loads and bedding factors, see "Soil Engineering" by M. G. Spangler.

REINFORCED AND PRESTRESSED-CONCRETE PRESSURE PIPE

In one form or another, reinforced-concrete pressure pipe has been in use since 1909 for water-supply lines, water-transmission lines, water-intake lines, ocean outfalls, and sewer force mains. In recent years, it has been used in water-distribution systems and, to some extent, in industrial piping. Sizes range from 12 through 180 in. for pressure services ranging up to 600 psi, depending on the type and size of pipe.

Four types of reinforced concrete pressure pipe are normally recognized. These are:

1. Reinforced-concrete pressure pipe
2. Reinforced-concrete cylinder pipe
3. Prestressed-concrete cylinder pipe
4. Pretensioned-concrete cylinder pipe

Large footages of reinforced-concrete cylinder pipe are in service, the oldest dating back to about 1920. Prestressed-concrete steel cylinder pipe was first used in 1942, and it now accounts for the major footage of concrete pressure pipe with thousands of miles of it in service. Prestressed-concrete pipe without a steel cylinder has been introduced recently in the United States, although it has been used overseas for many years.

Standards. Reinforced-concrete pressure pipe is made in accordance with a number of different specifications:

AWWA C300   —American Water Works Association Standard for Reinforced Concrete Water Pipe, Steel Cylinder Type, Not Prestressed. (Table 109.)

AWWA C301   —American Water Works Association Standard for Reinforced Concrete Water Pipe, Steel Cylinder Type, Prestressed. (Table 110.)

AWWA C302   —American Water Works Association Standard for Reinforced Concrete Water Pipe, Noncylinder Type, Not Prestressed. (Table 111.)

ASTM C361   —American Society for Testing and Materials Specification for Reinforced Concrete Low-Head Pressure Pipe.

Federal SS-P-381—Pipe, Pressure, Reinforced Concrete, Pretensioned Reinforcement (Steel Cylinder Type).

The first two of the above are intended for higher pressures, and they incorporate a light-gage steel cylinder in a reinforced-concrete or prestressed concrete envelope. The third is a reinforced-concrete pipe, normally with steel bell- and-spigot rings for a rubber O-ring gasket joint, for heads up to 100 ft of water. These types of pipe have frequently been used for subaqueous installation with the addition of drawbolts for underwater assembly of joints.

Manufacture. Reinforced-concrete pressure pipe is made up of a dense concrete wall reinforced with embedded steel rods, bars, wire, or welded-wire fabric. It is generally made either by pouring the concrete in vertical molds and vibrating it or by centrifugally spinning the concrete in rotating horizontal forms. The circumferential reinforcement consists of one or more cages, depending on the diameter, internal pressure, and external loading conditions. Longitudinal reinforcement is also provided. These pipes are sometimes made with steel end rings which, when telescoped together, compress a rubber gasket and seal the joint against leakage. On other installations, the pipe may be made with all-concrete ends formed to close tolerances. The maximum head for which this type of concrete pressure pipe is used is generally around 45 psi.

Table 109. Requirements for Nonprestressed, Concrete-lined Water Pipe (AWWA C300-64)

Pipe ID, in.	Min thickness		Circumferential reinforcement spacing	
	Pipe wall, in.	Concrete lining, in.	Min, in.	Max, in.
20	3¼	1	1¼	4
24	3½	1	1¼	4
30	3½	1	1¼	4
36	4	1	1¼	4
42	4½	1	1¾	5
48	5	1¼	1¾	5
54	5½	1¼	1¾	5
60	6	1¼	1¾	5
66	6½	1½	2	6
72	7	1½	2	6
78	7½	1½	2	6
84	8	1½	2	6
90	8	1¾	2¼	6
96	8½	1¾	2¼	6

NOTE: For pipe larger than 96 in. in diameter, dimensions and details of design shall be subject to approval by the purchaser.

Table 110. Requirements for Prestressed, Concrete-lined or Coated Pipe (AWWA C301-64)

Pipe ID, in.	Pipe with lined cylinder		Pipe with embedded cylinder		Coating thickness, in.	
	Core thickness, in.	Max design pressure, psi	Core thickness, in.	Max design pressure, psi	Mortar, min over the wire	Concrete, nominal over the core
16	1	250	...	...	⅝	1½
18	1⅛	250	...	...	⅝	1½
20	1¼	250	...	...	⅝	1½
24	1½	200	2¼	275	⅝	1½
30	1⅞	200	2¼	240	⅝	1½
36	2¼	200	2¼	210	⅝	1½
42	2⅝	175	2⅝	190	⅝	1½
48	3	150	3	175	⅝	1½
54	...	...	4	200	⅝	1½
60	...	...	4½	200	⅝	1½
66	...	...	5	200	⅝	1½
72	...	...	5¼	200	⅝	1½
78	...	...	5¾	200	⅝	1½
84	...	...	6¼	200	⅝	1½
90	...	...	6½	200	⅝	1½
96	...	...	6½	200	⅝	1½

NOTE: For embedded cylinder pipe larger than 96 in. in diameter, dimensions and details of design shall be subject to approval by the purchaser.

Table III.   Requirements for Noncylinder, Nonprestressed Concrete Pipe (AWWA C302-64)

Pipe ID, in.	6,000-psi centrifugal concrete		4,500-psi poured concrete		Circumferential reinforcement spacing		Min total steel area per lin ft, sq in.
	Min pipe wall thickness, in.	Min no. of cages	Nominal pipe wall thickness, in.	Min no. of cages	Min, in.	Max, in.	
12	2	1	...	...	1¼	4	0.08
15	2	1	...	...	1¼	4	0.11
16	2⅛	1	...	...	1¼	4	0.12
18	2¼	1	...	...	1¼	4	0.14
20	2⅜	1	...	...	1¼	4	0.16
21	2⅜	1	...	...	1¼	4	0.17
24	2½	1	3	1	1¼	4	0.20
27	2⅝	1	3¼	1	1¼	4	0.23
30	2¾	1	3½	1	1¼	4	0.25
33	2⅞	1	3¾	2	1¼	4	0.28
36	3	1	4	2	1¼	4	0.30
42	3½	2	4½	2	1¾	5	0.35
48	4	2	5	2	1¾	5	0.40
54	4½	2	5½	2	1¾	5	0.45
60	5	2	6	2	1¾	5	0.52
66	5½	2	6½	2	2¼	6	0.61
72	6	2	7	2	2¼	6	0.71
78	6½	2	7½	2	2¼	6	0.81
84	7	2	8	2	2¼	6	0.90
90	7½	2	8	2	2¼	6	1.00
96	8	2	8½	2	2¼	6	1.09

NOTE: For pipe larger than 96 in. in diameter, dimensions and details of design shall be subject to approval by the purchaser.

Reinforced-concrete cylinder pipe is made up of a welded-sheet-steel or steel-plate cylinder with steel joint rings welded to its ends. This cylinder is hydrostatically tested for water-tightness and strength. A reinforcing cage is placed, or cages of steel rods or bars are placed, around the steel cylinder. The pipe is placed in vertical molds and concrete is poured and vibrated to completely encase the cylinder and the reinforcing cages. The steel joint rings form a self-centering joint which, when telescoped together, compresses a preformed rubber gasket to form a watertight joint under all conditions of service. This type of pipe is generally used for pressures ranging from about 40 to 260 psi. In certain installations, it may be used for low-pressure work where a high degree of watertightness is required.

Prestressed-concrete cylinder pipe is made of two types. Both consist of a welded-sheet-steel cylinder with steel joint rings welded to its ends and hydrostatically tested. In the first type, the cylinder is lined with a concrete core which may be centrifugally spun or vertically cast into place. When the concrete has cured, the cylinder is wrapped with high-tensile wire under considerable stress so as to place the concrete core in compression. An exterior coating of mortar or concrete is placed over the outside of the cylinder to protect the prestress wire and cylinder. This type of prestressed pipe is usually made in sizes from 16 through 48 in.

In the second type of prestressed pipe, the sheet-steel cylinder is embedded in a vertically cast concrete core. This assembly is then wrapped with high-tensile

wire, and the mortar or concrete coating is placed over the exterior. The embedded cylinder pipe is generally made in sizes from 24 through 72 in. However, pipe as large as 108 in. has been made. Prestressed pipe can be used economically over a wide range of pressures from 50 to 350 psi. It also has a very high crushing strength.

Pretensioned-concrete cylinder pipe is made up of a welded-sheet-steel or steel-plate cylinder with steel joint rings welded to its ends. This cylinder is hydrostatically tested for watertightness and strength. A centrifugally spun lining of

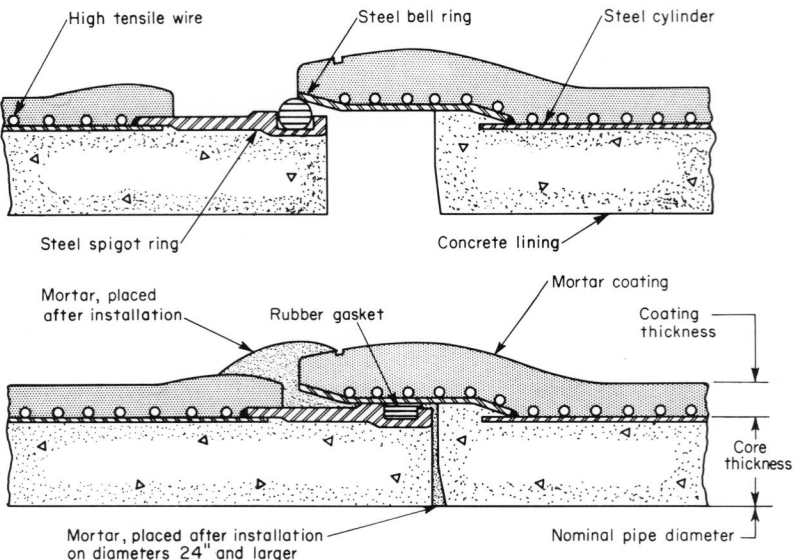

FIG. 117. Prestressed concrete-lined cylinder pipe illustrating joint before closing and completed joint.

cement mortar is cast within the cylinder. When this has cured, the cylinder is wrapped with steel bars or rods under measured tension. An exterior coating of concrete or cement mortar is placed over the outside of the cylinder to protect the reinforcement and the cylinder.

Availability. Concrete pressure pipe is generally available in 16 or 20 ft lengths. It can be furnished to any pressure class or classes best suited for a specific project. Prestressed-concrete-lined cylinder pipe (Fig. 117) is manufactured in sizes 16 through 48 in. and for most pressures encountered in water-works systems. Prestressed-concrete-embedded cylinder pipe (Fig. 118) is produced in sizes 24 in. and larger and for an even greater pressure range than lined cylinder pipe. Prestressed-concrete pipe without a steel cylinder covers about the same diameter and pressure range as embedded cylinder pipe.

Jointing. Reinforced-concrete pressure pipe operating under relatively high pressures is manufactured with a watertight flexible-expansion joint. The joints are of the bell-and-spigot type (see Figs. 117 and 118) with the joint surfaces formed by steel rings in the ends of the pipe.

For pipe operating under moderate pressures, a rubber gasket is used to make the joint watertight. Steel joint rings are not used. In the non-cylinder-type pipe,

joint rings are joined by the londitudinal reinforcement extending through the pipe. In the cylinder-type pipe, the rings are welded to the ends of the cylinder so as to form a continuous watertight membrane throughout the length of each pipe.

A rectangular groove is provided on the spigot end. Into it is placed a continuous round rubber gasket. As the pipes are pushed together, this gasket is compressed into the groove by the flared portion of the bell. With the pipes pushed home, the gasket is confined on all four sides. The shape and dimensions of the joint rings or of the concrete surfaces are such as to make the joints self-centering even without the presence of the gasket. The weight of one pipe can be transmitted to

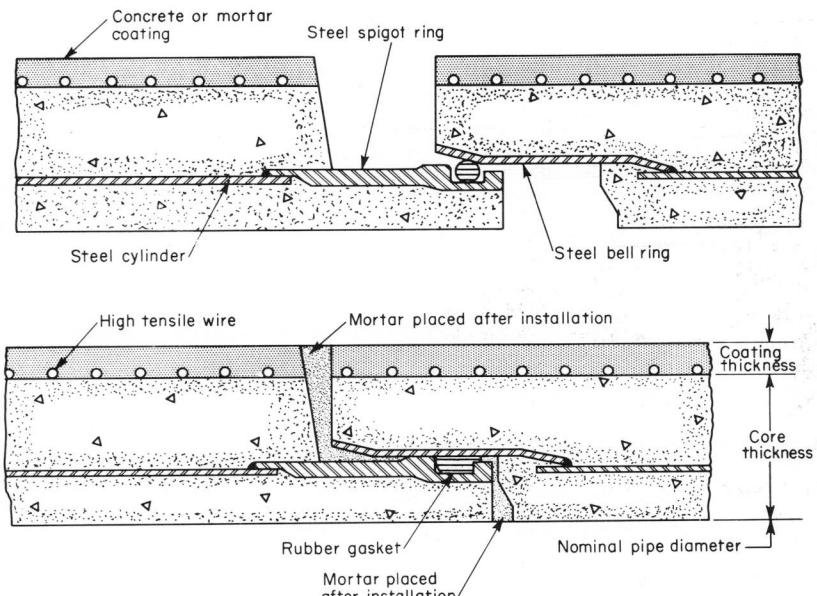

Fig. 118. Prestressed concrete-embedded cylinder pipe, illustrating joint before closing and completed joint.

the adjoining pipe only by the contact of the concrete surfaces. This prevents undue distortion of the gasket as the pipes move or deflect owing to expansion and contraction or settlement. Cement mortar or premolded asphalt is placed in the spaces between the ends of the pipe, both inside and outside of the joint, merely for the purpose of further protecting the steel joint rings. This mortar or asphalt does not serve to render the joint watertight. Certain jointing compounds are also used in this manner rather than mortar.

The ring forming the bell end of the pipe is covered on its exterior surface with reinforced concrete. The ring forming the spigot end should be lined on its inner surface with concrete. The gaskets sealing the joints are normally made of rubbers compounded to meet specified requirements to assure that a watertight and permanent seal is provided. The gasket should be a continuous ring of a size and cross section that is sufficient to fill completely the groove on the spigot when the pipes are laid. The rubber gasket is the sole element on which watertightness of the joint depends. Cement mortar or plastic materials are used to complete making the joint. However, they should not be depended upon for watertightness.

A "double rubber-gasket joint" is sometimes used with certain types of concrete pressure pipe carrying irrigation water under relatively low pressure. The pipe is made with a spigot on each end. The rubber gasket is placed in the spigot groove, and a double bell ring of steel or reinforced plastic is placed over the end of the pipe. Half the double bell ring extends beyond the end of the pipe. Steel bell rings usually receive an exterior coating of mortar.

All types of concrete pipe can be readily tapped under pressure for either threaded or flanged connections. The highly alkaline environment provided by the concrete and mortar encasement of the steel components prevents galvanic corrosion even in very low resistivity soils. A sustained Haxen-Williams "C" of 140 to 150 has has been measured over the years in repeated tests of a number of concrete pressure pipelines.

Long-radius curves can be accomplished with joint openings on straight pipe or with beveled end pipe or bevel adapters. Elbows for short-radius deflections, reducers, tees, wyes, and closures are standard items. Special fittings tailored to suit specific requirements can be supplied.

The steel-cylinder-type pipes have been used principally in municipal and industrial water-supply systems for pressures greater than 40 psi and for sewer force mains. One of the major applications of the reinforced-concrete pipe without a steel cylinder has been steam electric-plant condenser-cooling water intake and discharge lines.

Installation of concrete pressure pipe is simple and rapid. AWWA Manual M9, Installation of Concrete Pipe, provides detailed information on alternate methods and equipment for constructing pipelines of this material. It also contains information on tapping concrete pressure pipe.

ASBESTOS-CEMENT PIPE

The manufacture and usage of asbestos-cement pipe started in Italy approximately in 1913. Production in the United States started in 1929. Rapid recognition of its usefulness due to its strength, corrosion resistance, and maintained high-flow characteristics has resulted in its becoming one of the most used pipings for carrying water and sewage.

Standards. Asbestos-cement pressure pipe is covered by the following specifications:

ASTM C296   —Asbestos-cement Pressure Pipe
ASTM C500   —Testing Asbestos-cement Pipe
AWWA C400   —Asbestos-cement Water Pipe
Federal SS-P-351—Pipe, Asbestos-cement

Sewer pipe is covered by the following specifications:
ASTM C428   —Asbestos-cement non-pressure Sewer Pipe
Federal SS-P-331—Pipe Asbestos-cement, Sewer, Non-pressure

In the selection of pressure pipe, particular attention should be given to AWWA Handbook H-2, covering Standard Practice for the Selection of Asbestos-cement Water Pipe. This handbook was prepared so that design engineers may quickly determine the correct class of asbestos-cement pipe to use under various combinations of internal pressure and external loading. For detailed specifications for the installation of asbestos-cement pipe, reference should be made to AWWA 603, Standard for the Installation of Asbestos-cement Pipe.

Manufacture. Asbestos-cement pipe as made in the United States and Canada (as well as in many other countries in the world), is made with asbestos fiber, portland cement, and silica flour. A flow diagram illustrating the process normally followed is shown in Fig. 119. The silica flour is reacted in the curing process to

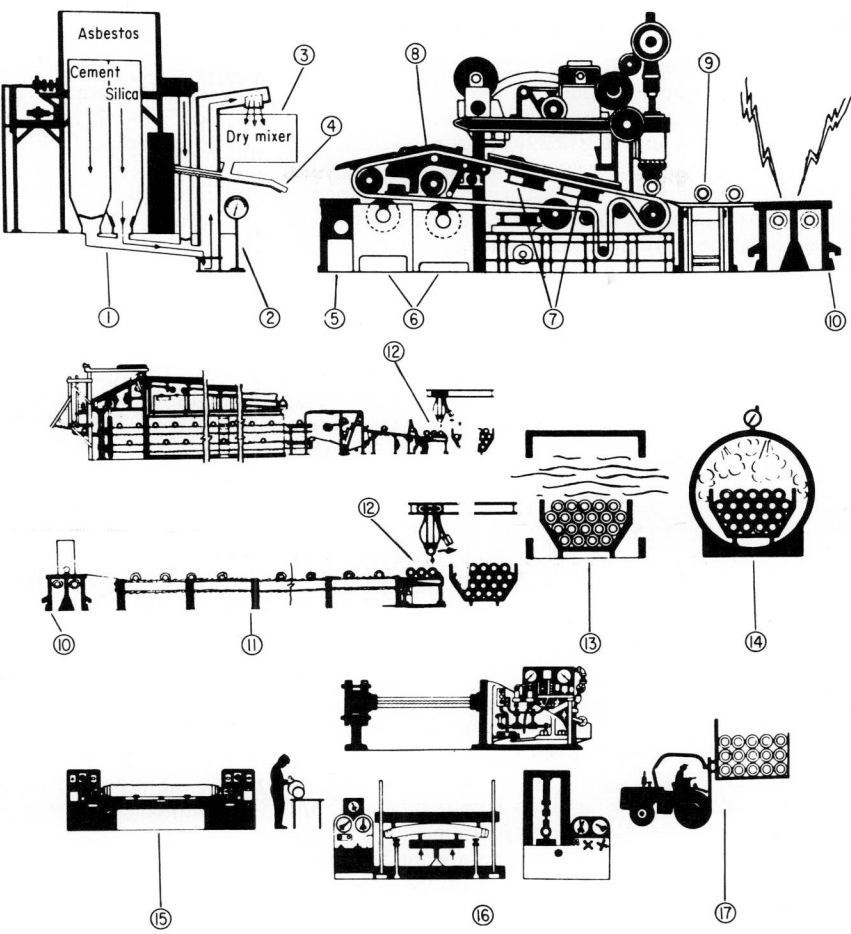

FIG. 119. Diagram of manufacturing process of asbestos-cement pipe. ① *Production line bins*—raw materials enter: (*a*) *asbestos*—from the willow where the fiber is separated into individual strands and thoroughly mixed; (*b*) *cement*—directly from receiving hoppers; (*c*) *silica*—from grinding mill. ② *Electronic scales* for precise weighing—accurate control for uniform results. ③ *Dry mixer*—blends raw materials thoroughly. ④ *Conveying trough*—water carries stock to wet-mix vat. ⑤ *Wet-mix vat*—thorough dispersion of reinforcing fibers. ⑥ *Screen cylinder mold*—picks up slurry and deposits on moving felt. ⑦ *Vacuum box*—excess water removed. ⑧ *Felt* deposits stock on *mandrel*—wall thickness built up under pressure to proper size. ⑨ *Mandrel* with pipe removed from machine—next mandrel positioned. ⑩ *Electrolytic loosener*—frees pipe from mandrel—prevents distortion. ⑪ *Slow-down conveyors*—provide precure time—initial set. ⑫ *Mandrels*—removed and pipe identified. ⑬ *Air cure room*—strict control of time, temperature, and humidity. ⑭ *Autoclaves*—high-pressure steam curing imparts maximum strength and excellent chemical stability. ⑮ *Lathes*—trim and machine ends to exact dimensions. ⑯ *Testing equipment*—checks for adherence to rigid specifications: (*a*) *inspection;* (*b*) *flexure-testing* machine; (*c*) hydrostatic tester; (*d*) Crush tester (laboratory). ⑰ *Material-handling equipment*—transfers pipe to shipping area.

form a stable strength-giving compound. By adding water, a slurry is produced which is picked up as a filtered film on a revolving screen. The film is transferred to a felt and, subsequently, to a revolving steel mandrel. Excess water is removed by suction and by applying heavy pressure on the film as it winds up to pipe thickness on the mandrel. The pressure consolidates the film and increases the density of the pipe. After preliminary curing of the cement in the pipe, the mandrel is extracted. The pipe is then treated with high-pressure steam. This greatly accelerates the normal cement setting and causes the free lime liberated by the hydrated cement to combine with the silica flour to form a stable and strengthening compound. Pipe cured with high-pressure steam is essentially immune to chemical attack by sulfate soils and waters, and its resistance to acids and very soft waters is considerably increased.

Asbestos-cement pipe is normally made in diameters from 3 through 36 in. Standard lengths are 13 ft. However, in the 4-, 6-, and 8-in. diameters, 10-ft lengths are also available. For pressure pipe, classes to carry 100-, 150-, and 200-psi working pressure are available. For sewer pipe, classes are based on the crushing strength in pounds per foot and are related to burial depth and type of trench. Classes 1500, 2400, 3300, 4000, and 5000 are available. Other areas of nonpressure use are pipes for air duct, underdrain, storm drain, electrical-, and telephone-conduit applications. All pressure pipe is tested at $3\frac{1}{2}$ times the stated working pressure and proof-tested on a sampling plan basis to four times the class rating. Sewer pipe is hydro-

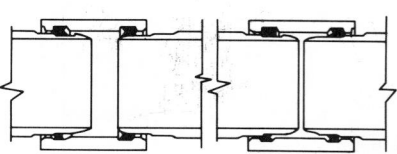

FIG. 120. Cut-away views of coupling and rubber gasket rings used with asbestos-cement pipe, showing how the rings are compressed between the coupling and pipe. Left: Pipe inserted into one end of coupling; Right: Final position, coupling centered over joint between pipes.

statically tested at 50 psi. Each length of 4-, 6-, and 8-in.-diameter pipe is also flexurally tested with third-point loading.

Jointing. The ends of the pipe are machined square and the coupling areas lathed to close tolerances. Asbestos-cement sleeves of a larger inside diameter than the pipe are machined to fit over the ends of the pipe and to enclose two rubber gaskets, one on each pipe end, which are seals. Joints are made by placing rubber rings in coupling grooves and then inserting pipe ends into the coupling to seal the joint effectively (Fig. 120). Each coupling is also hydrostatically tested to the same requirements as applied to the pressure and sewer pipe. The rubber gaskets are available in natural rubber or synthetic stock if greater chemical resistance is desired. Since various manufacturers do not use the same couplings, asbestos-cement couplings adapters are offered. Cast-iron fittings and valves are made to fit the respective coupling systems. As the couplings employ rubber gaskets as seals, the gaskets also act as expansion joints. They permit up to 5-deg bends to be made at each joint and allow separate flexibility at each joint. In both pressure and sewer pipe, the rubber gaskets create an essentially bottletight seal from both the inside and outside.

The pipe may be cut and machined in the field. Drilling and tapping is possible with standard equipment for service connections. Factory-fitted tapped couplings are available.

Applications. Asbestos-cement pipe is essentially immune to corrosion in the ordinary sense. It is not subject to electrolysis or to attack in so-called "hot" soils. It does not tuberculate. This pipe will not cause discolored water. It is highly resistant to chemical attack, particularly that encountered in sulfate soils.

Without surface treatment, it is not recommended for highly acid solutions or extremely soft water. Lined pipe is available for usage where attacking acids may occur.

The bore smoothness of asbestos-cement pipe allows a Williams and Hazen coefficient for water flow of C = 140 to be conservatively used without the necessity of introducing a safety factor to compensate for possible reduced smoothness or constrictions in later years. A Manning factor of $n = 0.010$ permits relatively flat grades to be used in sewage systems.

CEMENT-LINED STEEL AND CAST-IRON PIPE

Manufacture. The cement lining is generally applied to steel and cast-iron piping by the centrifugal casting process. The metal pipe is first immersed in a cleansing solution to remove all traces of oil or grease so that an adherent bond can be formed between the interior wall of the pipe and the cement lining. The pipe is then placed in a slightly inclined position, with the higher end at the mouth of the cement mixer. A measuring device, adjustable to measure the exact amount of cement for each size of pipe, allows the proper amount of lining material to flow into the pipe.

The pipe is sealed with stoppers at each end and then rotated in a horizontal position at high speed from about 30 sec to 1 min, depending upon the size of the pipe. The centrifugal force and vibration set up by this operation brings a small amount of water to the surface of the mixture, so that the cement sets into a dense adherent layer with a smooth surface. The cement-lined pipe is cured in a moisture-and-temperature-controlled atmosphere in about 72 hr.

Temperature changes have no effect on the pipe or lining as the coefficient of expansion of the cement and the metal is about the same. The added support given by the lining enables the pipe to withstand high working pressures. Cement-lined pipe can be cut to length and fitted just as ordinary pipe. The lining is able to withstand, without chipping, any blow on the exterior of the pipe which does not actually dent the pipe.

Standards. Cement-lined steel and cast-iron pipe is covered by the following specifications:

AWWA C205—Standard for Cement-mortar Protective Lining and Coating for Steel Water Pipe.

AWWA C104—Cement Mortar Lining for Cast Iron Pipe and Fittings for Water (also ASA Standard A21.4).

Application. Cement-lined pipe is well established for use in cold-water lines. Substantial quantities of cement-lined steel pipe are used for other applications where corrosion is more of a problem. The largest user, by far, is the petroleum industry for use in oil-field flow lines, pipe lines, tubing, and casing. Cement-lined pipe is particularly suitable for these applications because of the presence in the oil fields of salt water, hydrogen sulfide, carbon dioxide, and other corrosive material. Other applications include lines in salt works for handling brine, discharge lines in coal mines for carrying highly corrosive sulfur water, lines in paper and pulp mills for handling diluted acids and corrosive waste liquids, and lines in process plants where water or other liquids must be kept free from iron contamination or rust.

Jointing. Cement-lined pipe is generally joined with screwed seal rings which prevent the corrosive liquid from coming in contact with steel. Flanged joints are also extensively used. Some prefabrication is done of piping assemblies involving welding of the steel joints. Field joining of the preassembled welded

assemblies is then done with flanged ends.   Cement of course must not be at the pipe ends being welded.   After welding, these are filled with mortar.

CLAY PIPE

Vitrified and unglazed clay pipe is used for the conveyance of sewage, industrial wastes, and storm water in sizes from 4 to 36 in.   The following ASTM Specifications have been issued to cover the physical properties, the performance requirements, and the methods of installation and testing of vitrified clay pipe:

ASTM C12 —Recommended Practice for Installing Vitrified Clay Sewer Pipe
ASTM C13 —Specifications for Standard Strength Clay Sewer Pipe
ASTM C200—Specifications for Extra Strength Clay Sewer Pipe
ASTM C211—Specifications for Perforated Clay Pipe
ASTM C301—Methods of Testing Clay Pipe
ASTM C425—Vitrified Clay Pipe Joints Using Materials Having Resilient
            Properties

Unglazed clay pipe is covered by ASTM Specification C278, Specifications for Extra Strength Unglazed Clay Pipe.

Manufacture.   Vitrified clay pipe is generally manufactured from fire clays or shales or from a combination of these materials.   From these raw materials, the pipe is formed under high pressures, which accounts in part for its dense, solid body.   After a period of drying, the pipe is fired at a temperature of about 2000 F. During this firing process, each particle of clay is "welded," or fused, to the next one in a strong, chemical-proof bond.

The resulting product is the most corrosion-proof pipe available for sanitary and industrial sewer systems.   It does not require special coatings or linings to protect it from such destructive elements as moisture or chemicals in the soil, chemically active sewage, industrial waste, or abrasive materials in the flow and the acids formed by the oxidation of hydrogen sulfide.   Vitrified clay pipe can carry every known chemical waste without being harmed.   The only exception is hydrofluoric acid which is seldom used industrially and which is rarely, if ever, found in sanitary sewage.

Jointing.   Nearly all vitrified clay pipe is manufactured with a factory-fabricated compression-type joint.   Three types of compression joints are normally produced by the clay pipe industry.   These are classified in ASTM Specification C425, Vitrified Clay Pipe Joints Using Materials Having Resilient Properties. Features of the three types are outlined below.

Type I includes those joints in which the same resilient material is used both on the spigot end and in the bell of the pipe, except that the material of one of the controlled complementary jointing ends, either in the bell or on the spigot, may vary in hardness from the other.   These preformed ends, by controlled complementary design, will form a positive mating pattern of closure.

Type II includes those joints in which different types of materials having resilient characteristics are used in a similar manner as described for Type I, except that the material of one of the controlled complementary jointing ends, either in the bell or on the spigot, may vary in hardness and resiliency from the other.

Type III includes those joints which rely upon a gasket or compression ring of a resilient material having a controlled and calculated shape which will be compressed within the annular space to form a closing seal.   True round rings of rigid or flexible material placed on the spigot end and within the bell of the pipe form

predictable dimensions within which the gasket will be compressed. The gasket or ring may or may not be attached to one or the other joining faces prior to socketing.

The resilient materials used in the manufacture of these joints are vinyl chloride plastisols, polyurethanes, polyester resins, and natural or synthetic rubber. These jointing materials are selected for their chemical resistance and physical properties. Various joint designs have been developed utilizing these jointing materials, but all designs must meet the rigid performance requirements outlined in the above ASTM specification.

MISCELLANEOUS PIPING

GLASS PIPING

Glass pipe and pipe fittings are used extensively in the chemical, food, beverage, and pharmaceutical industries. Other applications include service in plating plants, paper mills, and textile-finishing plants. A borosilicate glass of very low alkali content and low thermal expansivity is the basic material.

The high heat resistance of borosilicate glass allows operation at temperatures up to 450 F.

A less-expensive type of sanitary glass piping using standard metal fittings is recommended for the dairy industry where pressures do not exceed 25 psi.

Properties. Typical physical properties of borosilicate glass involve the following:

Linear coefficient of expansion ...... 0.0000018 per degree F between 32 and 572 F or 0.22 in./100 ft/100 F temperature range (0.0000032 per degree C)
Thermal conductivity at 212 F ...... 0.73 Btu/(hr ft F)
Specific heat .... 0.20 Btu/(lb F)

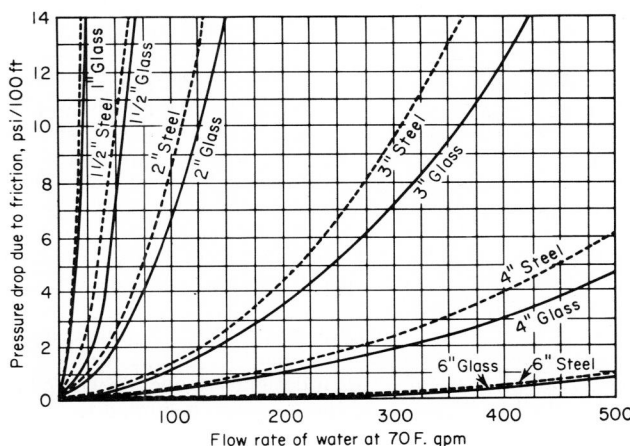

FIG. 121. Calculated pressure drop for water at 70 F in glass and steel pipe (calculated from equations for friction factors by Drew, Koo, and McAdams, *Trans. AICE*, vol. 28, no. 56, 1933; and Wilson, McAdams, and Seltzer, *Ind. Eng. Chem.*, vol. 14, no. 105, 1922).

Table 112. Working Stresses and Flexibilities of Borosilicate Glass Pipe (Corning Glass Works)

| Pipe size, in. | Approximate deflections based on ⅛-in.-thick interface gaskets | | | | | | | | |
| | Deflection, in. | | | | | | Stress, psi | Side thrust, lb | |
	A	B	C	D	E	F		5 ft	10 ft
1	0.22	0.52	0.82	0.21	0.36	0.72	300	0.8	0.4
1½	0.16	0.36	0.56	0.14	0.24	0.48	300	1.8	0.9
2	0.11	0.26	0.40	0.10	0.17	0.34	275	2.8	1.4
3	0.07	0.14	0.21	0.053	0.09	0.18	250	6.5	3.2
4	0.05	0.11	0.17	0.042	0.072	0.145	225	13	6.5
6	0.030	0.07	0.10	0.025	0.042	0.084	200	30	15

The low coefficient of thermal expansion (approximately one-third to one-quarter that of steel) provides good resistance to thermal shock. This permits cleaning or sterilization with low-pressure steam. Especially when sanitary conditions are involved, the low-pressure steam is often followed by cold liquids. The maximum sudden (instantaneous) temperature drop which may be applied safely varies from about 200 F for 1- to 3-in.-diameter pipe to 160 F for 6-in.-diameter pipe.

Pressure-drop data of glass are shown graphically in Fig. 121. Data for steel pipe are included for comparison.

The pressure drop through a mitred elbow is approximately twice that through a sweep elbow. But the mitred elbow is usually used because it has a more constant cross section and hence is considered to be stronger.

Permissible deflections and working stresses for borosilicate glass pipe are given in Table 112 and in Fig. 122. This confirms the importance of gasketed joints in assuring flexibility without inducing excessive stresses. It should be noted that a change of pipe direction increases the maximum recommended deflections. Recommended deflections are additive. For Teflon Type-T gaskets, these deflections should be reduced by approximately 25 per cent.

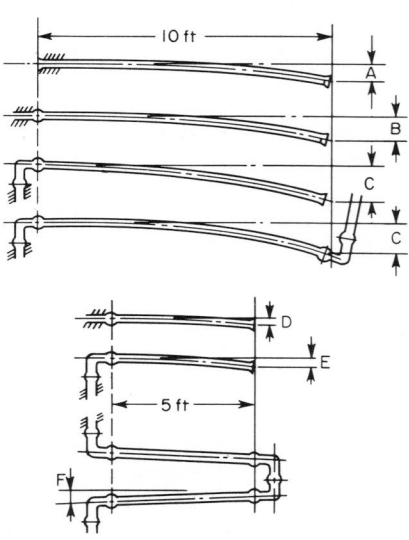

FIG. 122. Permissible deflections for glass pipe, with dimensions given in Table 112.

Availability. Pipe is made in lengths up to 10 ft. Nominal dimensions and weights of flanged glass pipe are given in Table 113.

Pipe of 1, 1½, 2, and 3 in. nominal size is designed to operate with 60 psi pressure; the 4 in. size is limited to 35 psi; and the 6 in. size is limited to 20 psi. A compression type of flanged joint using cast-iron or aluminum flanges is available for sizes up to and including 6 in.

Standard fittings are also readily available (Figs. 123 and 124). Standard dimensions are tabulated in Tables 114 and 115. Special fittings involving sweep

Table 113. Nominal Dimensions and Weights of Flanged Glass Pipe

Nominal size, in.	Outside diameter, in., average	Wall thickness, in., average	Weight per ft, lb
1	1.312	0.156	0.6
1½	1.844	0.172	1.0
2	2.344	0.172	1.1
3	3.406	0.203	2.0
4	4.530	0.265	3.4
6	6.656	0.328	6.3

elbows, U-bends, reducing tees and crosses, and eccentric reducers can also be furnished.

Where fittings are used, the equivalent-length additions are listed in Table 116.

Joints. Joints in glass pipe are normally made with metal flanges which are cushioned from the glass with molded asbestos inserts. An interface gasket of Teflon or other suitable material (depending on the fluid conveyed) is gripped between the pipe ends when the flange bolts are tightened. This prevents contact between the metal flanges and the fluid.

Typical pipe joints between glass pipe and between glass and metal pipe are illustrated in Figs. 125 and 126, respectively.

Pipe Support. Conventional pipe hangers are used for the support of glass pipe. Rod or guide sleeve hangers with an adjustable clevis are satisfactory. Rigid clamp hangers are used to prevent sidesway of the pipe when the hanger rods are extremely long and long pipe runs are involved.

The pipe hangers are normally padded with rubber or asbestos. Hangers are also supplied with neoprene or rubber coatings applied by dipping.

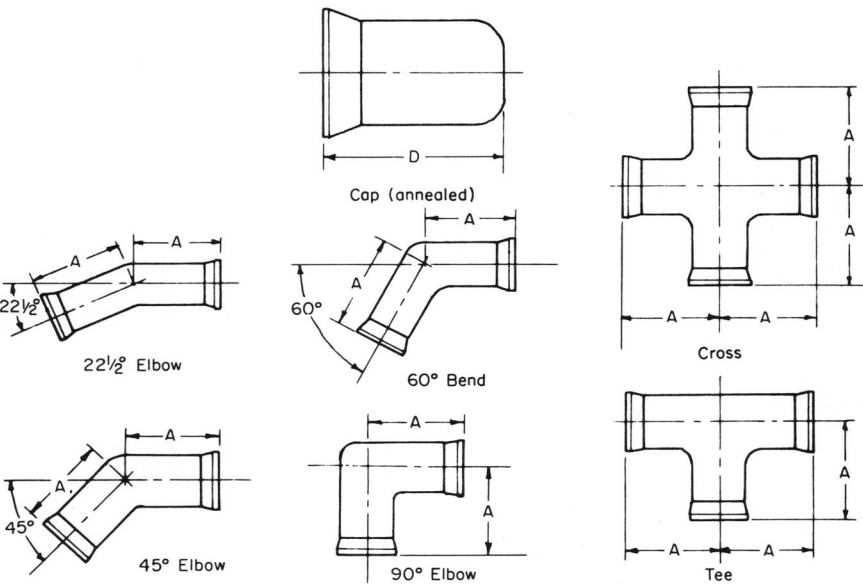

FIG. 123. Standard glass pipe fittings (Corning Glass Works).

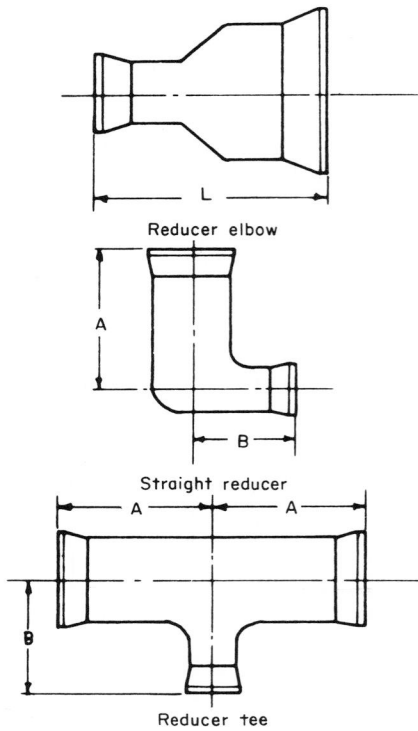

FIG. 124. Standard glass pipe reducing fittings (Corning Glass Works).

Minimum support distances between hanger locations are given in Table 117.

Hangers should be fastened to a rigid structure to prevent the hanger from sagging and stressing the glass pipe. Wood beams, especially out-of-doors, have a tendency to warp and sag. Hangers should always be adjusted to the glass pipe, not the pipe to the hangers.

Vertical glass lines should be supported by plates beneath metal flanges or by padded saddles beneath 90 deg elbows at the bottom of vertical rises. An anchor point is a rigid support for the glass line tying it into the building structure, to fixed equipment such as tanks and pumps, or to independently supported valves. Only one anchor point should be in every straight run of pipe. Two anchor points may be used if there is an expansion joint between them.

WOOD PIPE

Early water-supply pipes were made by boring holes through solid wood logs which were joined together in place by pushing together nipples of wood or metal. At the present time, wood-stave pipe made of redwood or douglas-fir staves is used most extensively in the West because of its proximity to the source of timber.[1] Such pipe is made chiefly in large diameters for hydraulic-power developments, municipal water supplies, outfall sewers, mining, irrigation, and various other purposes requiring the transportation of water. Wood-stave pipe is also used for Army and Navy cantonments.

The water carried may be hot, cold, or acid. Redwood and Douglas fir are used principally because of their availability and good resistance to decay. Wood staves for pipe may be either untreated or creosoted by a vacuum-and-pressure process with about 8 lb of creosote per cubic foot of wood treated.

Three types of pipe products are available. *Continuous-stave pipe* is built of staves milled accurately to true radial planes and to correct curvatures on the outside and inside faces. Assembled in the form of a true circle, they are banded with individual steel rods. This pipe is furnished in inside diameters from 10 in. to 16 ft and for pressures up to 350-ft static head.

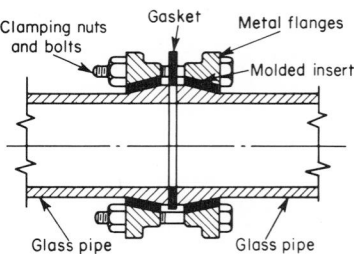

FIG. 125. Flanged joint between glass pipe (Corning Glass Works).

[1] California Redwood Association, Bulletin Section III.

Table 114. Dimensions of Standard Glass Fittings Shown in Fig. 123

Pipe size, in.	$A \pm \frac{1}{32}$ in.	$D \pm \frac{1}{32}$ in.
1	$2\frac{3}{4}$	$2\frac{11}{16}$
$1\frac{1}{2}$	$3\frac{1}{2}$	$3\frac{7}{16}$
2	4	$3\frac{7}{8}$
3	5	$4\frac{11}{16}$
4	7	$5\frac{15}{16}$
6	9	$6\frac{7}{16}$

Table 115. Dimensions of Standard Reducing Glass Fittings Shown in Fig. 124

Size, in.	$A \pm \frac{1}{32}$ in.	$B \pm \frac{1}{32}$ in.	$L \pm \frac{1}{32}$ in.
$1\frac{1}{2} \times 1$	$3\frac{1}{2}$	3	4
2×1	4	3	4
$2 \times 1\frac{1}{2}$	4	$3\frac{1}{2}$	4
3×1	5	$3\frac{1}{2}$	5
$3 \times 1\frac{1}{2}$	5	4	5
3×2	5	$4\frac{1}{2}$	5
4×1	7	4	7
$4 \times 1\frac{1}{2}$	7	$4\frac{1}{2}$	7
4×2	7	5	7
4×3	7	$5\frac{1}{2}$	7
6×1	9	5	9
$6 \times 1\frac{1}{2}$	9	$5\frac{1}{2}$	9
6×2	9	6	9
6×3	9	$6\frac{1}{2}$	9
6×4	9	8	9

Face-to-centerline dimensions for the 1-in. legs are 5 in.; $1\frac{1}{2}$-in. legs are $5\frac{1}{2}$ in.; and 2-in. legs are 6 in.

Table 116. Equivalent Length of Glass Fittings (Corning Glass Works)

Fitting	Equivalent length, ft					
	1 in.	$1\frac{1}{2}$ in.	2 in.	3 in.	4 in.	6 in.
45-deg mitred ell	1	1.3	2	3	4	6
90-deg mitred ell	5	7.5	10	15	20	30
90-deg sweep ell	2.7	4	5.3	8	11	16
U-bend.................	5	7.5	10	15	20	30
Tee, along run	1.7	2.5	3.3	5	6.7	10
Tee, entering run	5	7.5	10	15	20	30
Tee, entering branch	5.8	8.7	12	18	23	35

Table 117. Minimum Distances between Supports for Glass Pipe, ft

Pipe size, in.	Fluid in pipe		
	Gas	Specific gravity less than 1.3	Specific gravity, 1.3 and greater
1	8	8	7
$1\frac{1}{2}$	9	9	7
2	9	9	8
3	9	9	8
4	10	10	8
6	10	10	8

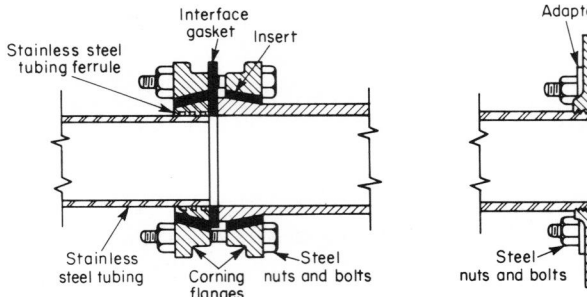

Fig. 126. Typical flanged joints between glass pipe and plain or flanged steel pipe (Corning Glass Works).

Machine-banded stave pipe is made of staves milled as above and assembled in sections. Two varieties of machine-banded pipe are made. One is wound with wire, and the other with flat banding. Wire-bound pipe is usually furnished in sections 4 to 20 ft in length and in inside diameters from 2 to 24 in. for pressures up to 400-ft static head. Machine-banded pipe wound with flat banding is sometimes furnished in sections from 6 to 16 ft, although usually 12 ft is the maximum length. It is made in inside diameters from 1 or 2 in. up to 48 in. and for pressures up to 400-ft static head.

Bored pipe is manufactured from straight-grained stock in outside dimensions from 4×4 in. to 12×12 in., and in lengths up to 14 ft. This pipe is bored through the center with inside diameters $1\frac{1}{2}$ to 6 in.

Redwood pipe does not expand or contract significantly with temperature changes. Thus, expansion joints are not required. In service from the tropics to high northern latitudes, its performance is identical, even in those localities in which high summer and low winter temperatures prevail.

Wood pipe exhibits good durability under very low pressures. Numerous examples exist of excellent preservation for 30 or more years in service involving uniform 10 to 15-ft head grade conditions, where the pipe was kept constantly filled with water. Pressure induces more complete saturation of the staves. Where saturation exists, decay is generally prevented.

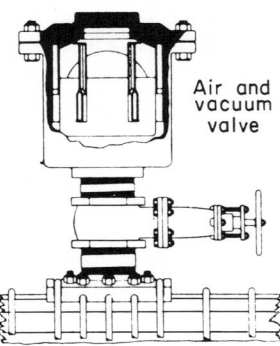

Fig. 127. Vacuum air valve installed on wood-stave pipe (California Redwood Association).

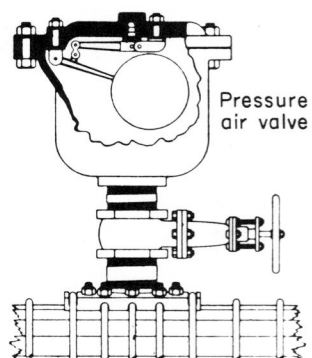

Fig. 128. Pressure air-release valve installed on wood-stave pipe (California Redwood Association).

Maximum pressures approximating 350-ft head may be carried in inverted siphons. Pressures approximating 250-ft head may apply to hydroelectric conduits or pump lines. The determining factor is the severity of operating conditions in terms of the extent and frequency of surge or water-hammer loads.

Valve Requirements. Vacuum air valves (Fig. 127) are necessary to release the air when filling the line with water, to permit the entry of air when emptying the line, or to prevent its collapse from vacuum in the event of a sudden break. This type of valve is made operative usually by the seating of a simple float as the water rises in the pipeline.

A valve which protects the pipe against vacuum effect also releases air during the initial filling of the line. Normally, air or vacuum valves of the following sizes are used:

Diameter of pipeline, in.	Diameter of air and vacuum-valve inlet, in.
6 or less	½
7-10	1
11-20	2
21-36	3
37-48	4
49-66	6
67-84	8

Where the net area of the opening in the air valve exceeds available standard sizes, it is usual to mount a series of smaller valves in a cluster to provide the required total area.

At least three valves of suitable size should be used for each mile of pipe. Normally, four valves per mile are recommended. If the pipeline grade is uniform, the valves should be located at uniform intervals along the pipe. If summits exist, the proper location for the air valves is at the summits.

Pressure air-release valves (Fig. 128) are required to release automatically the air accumulating at any summits that may exist in the pipeline profile. The air carried into the pipe with the water is released at the high points of the line. The severity of the condition is increased as the pipe approaches the hydraulic gradient when the presence of entrapped air restricts the water flow in the line. The sudden escape of the air at high velocity may also produce severe water hammer with serious damage to the pipeline.

Automatic removal of the air as rapidly as it accumulates at the high points of the pipeline is essential. The valves are automatically operated with a series of levers actuated by a float. As the air accumulates in the pipeline, it rises into the valve and displaces the water. This causes the float to drop, which opens the valve automatically for the release of the air. With the escape of the air, the float rises with the water and actuates the closing of the valve before the escape of water. The operation is automatically repeated.

Pressure air-release valves do not ordinarily require large discharge orifices. Thus, it is sufficient to locate one valve of this type at each summit. Standard practice indicates the following classifications:

Diameter of pipeline, in.	Diameter of air inlet, in.	Diameter of air outlet, in.
10 or less	1, 2, 3½, 3	¼ or ⁵⁄₁₆
11-18	2½, 3, 4	½
19-36	3, 4	⅝
37-60	3, 4	¾
Above 60	4	1

In pipelines subject to pump pressure, it is desirable to locate a pressure air-release valve immediately adjacent to the pump in order to remove the entrapped air before it escapes into the pipeline.   The valve should be mounted on a saddle of a comparatively large-diameter opening into the top of the pipe.   This gathers and releases the separate moving columns of air bubbles as they pass along the pipe.

Where depressions occur in the grade of the pipeline, it is usual to make provision for drainage of the line by means of a gate valve of the normal type and adequate size.   If the pressure is high where the drain valve is located, it should always be closed slowly to avoid water-hammer effect in the main pipeline.

Joints.   Continuous-stave pipe is normally shipped in the knocked-down condition and is assembled in the field.   This eliminates the need for couplings.   The staves are banded together by steel hoops having threaded ends which are passed

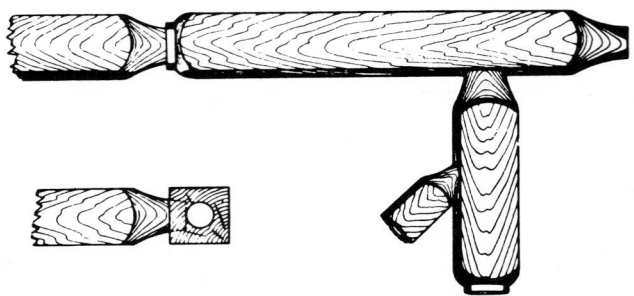

FIG. 129. Sketch showing typical joints in bored wood pipe (California Redwood Association).

through a cast- or malleable-iron shoe and pulled up with nuts.   The steel hoops ordinarily are designed for a factor of safety of four with respect to fluid pressure and the tensile strength of the steel.

Machine-banded pipe is composed of staves built up in sections, shipped ready to install.   One type is wound with steel pipe winding wire.   For some chemical applications, copper wire is sometimes used.   The other type is wound with flat steel banding.   Flat alloyed metal banding is also occasionally used.   Both types are coated with asphalt on the exterior after manufacture and rolled in sawdust.

Wire-wound machine-banded pipe is usually furnished with wire-wound collars for forming the connections between sections.   Sometimes, on larger diameters of pipe, collars are equipped with individual bands and shoes instead of being wire wound.

In forming connections, a tenon is turned on each end of the section.   The collar is made of a diameter to conform accurately to the tenon in a driving fit.   Machine-banded pipe made with flat banding is furnished with a tenon turned on one end and a mortise turned on the other end of the section to produce an accurate driving fit.   Pipe manufacturers furnish cast iron or metal fittings of standard and special types to accurately fit their pipe.

Bored pipe is joined by forming a turned, tapered spigot on one end and a corresponding socket on the opposite end.   The sections are driven together in installation (as illustrated in Fig. 129).   Metal is generally not used with this pipe.   However, sometimes a light gage galvanized or copper ferrule is placed on the socket end to avoid splitting the piece when assembling in place.

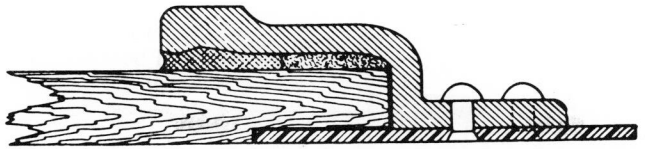

FIG. 130. Typical joint between steel and wood pipe (California Redwood Association).

Connections to Metal Pipe. In connecting to a steel pipe where moderate pressure is involved, a steel sleeve should be provided as part of the steel pipe. The sleeve may be welded or countersunk with rivets. The outside diameter should correspond to the inside diameter of the wood pipe.

The ends of the staves of the wood pipe should lap over the steel sleeve for approximately 3 ft. They should be tightly cinched down on a layer of canvas or burlap heavily impregnated with white or red lead pigment. This type of joint is particularly suited to aboveground locations.

In another method, particularly suitable for higher pressures, steel pipe and the staves of the wood pipe are placed within the hub and calked in the manner illustrated in Fig. 130. The inside diameter of the cast hub should provide a clearance of at least $1/2$ in. for calking space. If the steel pipe cannot be extended forward to support the ends of the pipe as shown, a steel ring should be inset in the ends of the staves. This steel ring should be approximately $3/8$ in. $\times$ 2 in. Its purpose is to prevent the ends of the staves from springing in under the calking blows. The joint may be poured with lead in the conventional manner.

Where the application involves high pressures, as in conduits to large hydroelectric plants, a more positive joint may be required. In these applications, a cast-steel hub is used, machined smooth on the inside, calked with a fiber packing held in place by a gland ring bolted to the main casting (as shown in Fig. 131).

Connections to Concrete Pipe. In making a connection to a concrete box or forebay structure located aboveground, a steel sleeve should be set into the concrete. The wood staves should be lapped over the outside of the steel sleeve on a layer of canvas or burlap heavily impregnated with white or red lead pigment. If desired, the staves can also be built directly into an opening provided in a concrete wall (as illustrated in Fig. 132).

The joint illustrated in Fig. 132 is satisfactory for a long period. However, if the pipe should remain empty for any considerable length of time under conditions that will permit the staves to dry out, some slight leakage may develop at the joint. Thus, when the steel sleeves are used, the bands can be cinched down around the pipe and leakage can be prevented.

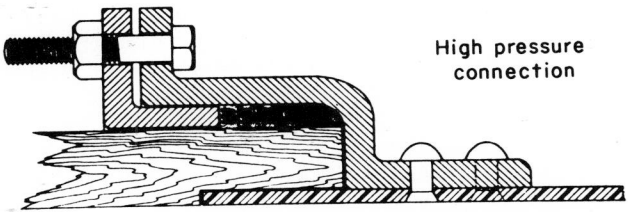

High pressure connection

FIG. 131. Joint between wood-stave and steel pipes made by means of fiber-packed gland (California Redwood Association).

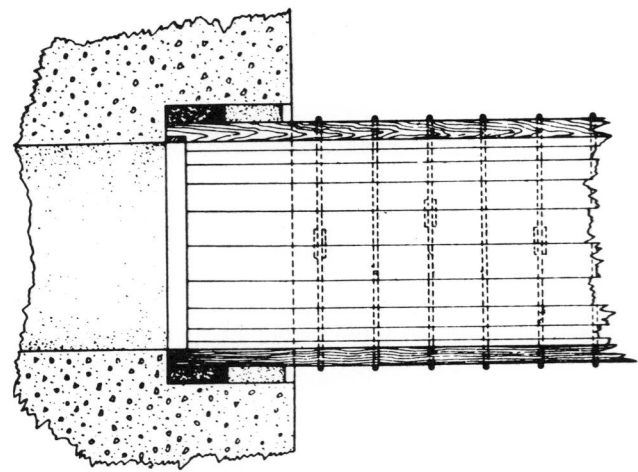

FIG. 132. Termination of wood-stave pipe in concrete structure (California Redwood Association).

WOOD-LINED STEEL PIPE

Where chemical solutions are to be handled under pressures that may exceed the pressure possibilities of banded pipe, steel pipe lined with wood staves is successfully used. This type of pipe is also particularly adaptable, even under low pressures, for certain types of pipelines located in the interior of industrial plants where definite alignment in length and fitting, suspension, and other installation features prompt its use. The wood staves composing the lining offer resistance to the action of the solutions carried, and the metal pipe shell forming the outer casing serves to carry the stresses incidental to the pressures present.

This classification of pipe is used in pulp-and-paper mills for conveying sulfite and sulfate stock, groundwood stock, diluted bleach liquor and bleached stock, and water. The wood lining contributes to clean paper making, eliminating the formation of the scale which otherwise would drop into the stock being carried. The accumulation of slime often found in pipelines is practically eliminated. It is also used in mining operations for transporting acid mine waters under high pressures. In chemical, textile, and process industries, wood-lined steel pipe is used in applications carrying acid solutions and gases.

This type of pipe is equipped with standard flanged joints and is easy to install. Field cutting to length and flanging, if necessary, is a simple operation, performed by sawing off the pipe to length required and welding a flange in place.

Fittings for this class of pipe, whether of the standard types of ells, tees, crosses, wyes, reducers, or of special nature are also lined with wood staves.

8

MATERIALS SPECIFICATIONS

Helmut Thielsch*

Among the specifications which relate to piping materials, those prepared by the American Society for Testing and Materials[1] cover the broadest range of materials and applications and are, therefore, widely used.

Requirements in the ASTM Specifications generally include the melting practice, chemical composition, mechanical properties, dimensional and weight tolerances, finish and inspection. Some specifications also contain repair procedures applicable to defect conditions which are considered injurious.

The more common requirements of particular interest in the specification and design of piping systems are summarized in the following tables:

Ferrous piping materials, Tables 1 through 13
Aluminum piping materials, Tables 14 through 19
Copper piping materials, Tables 20 through 27
Nickel piping materials, Table 28
Titanium piping materials, Tables 29 and 30

Requirements in the ASTM Specifications which are not contained in the above-listed summary tables include chemical-analysis details, check-analysis requirements, weight and length tolerances, marking, and similar features. Some of these requirements may vary with thickness, or diameter or application. For example, *Marking* in Specification B366, Factory-made Wrought Nickel and Nickel Base Alloy Welding Fittings, states that "The manufacturer's name or trademark, the schedule number, material and size shall be stamped, stencilled or otherwise marked

* Manager, Research and Development Division, Grinnell Corporation, Providence, R.I.

[1] Thielsch, H. "How to Use ASTM Specs for Better Piping Design," *Heating, Piping, Air Conditioning*, vol. 37, Sept.–Nov., 1965; vol. 38, Feb.–May, 1966.

on each fitting.... On wall thicknesses thinner than schedule 40S, no steel stamps, or other indented markings shall be used."

Some ASTM Specifications also may include applicable definitions, classification of heats or lot sizes, details on the test samples required, specific details of the test methods and the authorization for additional test samples, or retests, should one fail in a required test.

Special heat treatments, tests, finish and other conditions not normally considered a necessary requirement of the basic specification may also be described in some ASTM Specifications. The applicable paragraphs usually include the statement "When specifically required by the purchaser . . . " or "Bend tests shall be required only when so specified in the order."

Special pipe or tube production methods may also be permissible and may be referenced in the ASTM Specification. For example, Specification B337 on Seamless and Welded Unalloyed Titanium Pipe states under *Manufacture*

"Welded pipe shall be made from flat-rolled products by an automatic welding process with no addition of filler metal in the welding operation. Other methods of welding, such as by the addition of filler metal or hand welding, may be employed if approved by the purchaser and tested by methods agreed upon between the manufacturer and the purchaser."

In using ASTM Specifications, it should be recognized that many tests are specifically required. However, on some test properties, the applicable specification may merely state that the piping product shall be capable of conforming to particular test requirements or withstanding a particular hydrostatic test pressure. In that case, actual testing has not been done by the manufacturer.

Some piping materials listed in these tables are nevertheless not readily available or are not normally produced commercially unless on special order.

There are many other special materials, not yet covered by ASTM Specifications, which are available as piping. Many of these are widely used.

Where such piping is produced by seam welding preformed strips or plate materials, the strip or plate may be produced under an ASTM Specification. For example, specifications do not exist which cover seam-welded pipe made of copper or nickel materials. When desired, the base materials and welding filler metals involved in the manufacture of the seam-welded pipe may be covered by ASTM Specifications.

Several *general requirement* specifications have been issued which cover common requirements applicable to several materials. These specifications may cover general dimensional, length and weight tolerances, test methods and properties, marking and inspection requirements. For example, the majority of tubing specifications (see Tables 3 to 7) refer to ASTM Specification A450, General Requirements for Carbon, Ferritic Alloy, and Austenitic Alloy Steel Tubes.

BASIS OF PURCHASE

Most specifications contain a section which relates to the basis of purchase, and the information requested in this section should be included in the purchase order so as to describe precisely the desired piping materials. Typical examples of the information which should be included are

1. Quantity (feet or number of lengths)
2. Name of material (seamless, seam-welded, or cast pipe)
3. Grade and/or type and/or class
4. Manufacture (hot-finished, cold-drawn)
5. Temper or heat treatments (annealing, normalizing, stress relieving, quenching, tempering)

6. Size (nominal pipe size, outside or inside diameter, schedule, average or minimum wall thickness)

7. Length (specific or random)

8. End finish (beveled for welding, threaded, square cut)

9. Hydrostatic test requirements

10. Special test or supplementary test requirements (radiography, liquid penetrant or magnetic particle inspection, destructive examination, photomicrographs)

11. Finish requirements

12. Special requirements or exceptions to the specification (special chemistries, restrictions on composition)

13. Notification of defect repairs

14. Marking

15. Packing and shipment

16. Disposition of rejected material

The purchase order should also detail the certification and inspection requirements. These may involve the following options:

1. *Certificate of compliance* (which may include chemical analysis of the elements covered by the applicable specification, but does not include the mechanical properties, or other special tests such as corrosion tests. A statement, however, is included that all applicable requirements such as tension tests, hydrostatic tests, and radiography are met)

2. All required *test reports* (includes chemical analyses, mechanical test reports, corrosion test reports, and any other desired reports of tests)

3. Samples for check analysis (if special samples should be furnished, this should be so stated)

4. *Inspection at place of manufacture* (those tests which are to be witnessed should be specified)

Where special fabrication operations by the purchaser are involved, it may also be desirable, and sometimes is necessary, to state that the material is to be used for:

1. Cold bending or coiling

2. Hot forming (bending, extruding, swedging)

3. Welding or brazing

Service requirements should also sometimes be indicated. These may involve:

1. Pressure requirements

2. Temperature requirements

3. Corrosion requirements

For example, Specifications B161 and B167 (see Table 28) state that "where material is intended for nuclear applications and other critical end uses, the manufacturer shall be notified at the time of placement of the inquiry or order to determine if material of quality and inspection procedures normally employed for commercial material to *these Specifications* are adequate. In the event that more critical quality or more rigid inspection standards are indicated, the manufacturer and the purchaser shall agree upon such standards prior to production."

SELECTION OF SPECIFIC PIPING MATERIALS

When selecting different or alternate piping materials from particular specifications, it must be recognized that different committees or interests may have been involved in the preparation of the several specifications.

Where critical service requirements are involved, the paragraphs in the ASTM Specification under Process and Manufacture should be checked carefully. In some specifications, these paragraphs are sufficiently detailed to be adequate for

all applications.   For example, ASTM Specification A271 states:

Process
The steel shall be made by the electric furnace process or other processes *approved* by the purchaser.
Manufacture
Tubes shall be made by the seamless process and may be furnished either hot-finished or cold-finished at the manufacturer's option.

In other specifications, these conditions may be listed rather broadly, and may sometimes include a process which is undesirable for a particular service environment.   For example, Specification B338, Seamless and Welded Unalloyed Titanium Tubing, states:

Seamless tubing may be made by any seamless method which will yield a product meeting the requirements of this specification.
NOTE. 2.—Methods presently employed for the production of seamless tube rounds or tubing include extrusion of billets, centrifugal casting, machining of bar stock or powder compaction.

To assure that for a particular service the different materials selected or permitted represent similar levels of soundness, additional requirements or supplementary tests may have to be specified.   Some of these may be mentioned in the specifications subject to the specific inclusion by the purchaser in his order.   For example, Specification B161 (Table 28) states "if samples for check analysis are desired, they shall be so specified by the purchaser at the time of placement of the order."

Some specifications contain *supplementary* test requirements in an Appendix. Examples from a few common piping specifications covering steel materials are listed in Table 31.

As new melting practices or new plate or pipe production methods are developed or new welding techniques are utilized, new materials specifications are often written.   These may correspond to existing specifications involving materials of similar chemistries.   Unfortunately, this is not always the case.   Some materials specifications for similar service conditions are promulgated without correlation to existing or established specifications.   This makes it extremely difficult for engineers to apply judgment when specifying materials or quality-control requirements.   The confusion is apparent when some engineers specify far more supplementary tests, for example, on seamless pipe rated as ASTM A335 than on seam-welded pipe produced to ASTM Specification A155 (Table 2).   It is difficult to understand why ultrasonic inspection, photomicrographs, etch tests, and flattening tests might be required for every bar of A335 pipe but not for A155 seam-welded plate pipe or for castings which may be utilized in the same piping system as the A335 material.

Since the final cost of the pipe material depends substantially on the quality-control requirements specified, an inherently higher quality material may become penalized and uneconomical in comparison with a material of somewhat lower average quality, on which fewer quality-control tests are specified.   For example, fittings produced under ASTM Specification A234 would be penalized unfairly if ultrasonic inspection were specified if radiographic examination were not specified on cast fittings produced under ASTM Specification A217.   Where the service requirements are not critical, neither inspection method should be required.   On the other hand, where critical high-temperature, high-pressure conditions, nuclear service or other critical conditions are involved, ultrasonic or radiographic examination, respectively, might be essential.

Other examples could be cited of inconsistencies in quality-control requirements in the different specifications covering materials for seemingly identical service

applications. For example, centrifugally cast ferritic low-alloy steel pipe for high-temperature service (ASTM A426) must be examined by ultrasonic inspection and also by magnetic-particle or liquid-penetrant inspection. The same tests are not required in the specifications which apply to static castings and wrought or forged pipe for the same service.

Uniform Levels of Quality. Because of the increasing number of materials specifications and the confusion arising from nonuniform requirements, it is important that such conditions be recognized by engineers who refer to ASTM Specifications. Purchase orders or specifications thus may have to include different levels of realistic inspection requirements to be applied across the board to different classes of materials and service environments. For example, for the critical service environments of nuclear and steam power plants, the inspection requirements suggested in Table 32 might apply. Similar tables might be prepared for other service conditions, such as low temperature and cryogenic environments.

Permissible Defects. Considerable confusion exists in some specifications on the acceptability limits of permissible defects, the repair of injurious defects, and the rejection of materials containing "excessive" defects when applied to materials specifications for different manufacturing processes. Typical requirements are summarized in Table 33.

In piping systems, much use is made of wrought seamless (hot-worked) carbon-steel and chromium-molybdenum alloy-steel materials meeting the requirements of ASTM Specifications A106, A355, A376 and A405. These specifications require rejection of pipe when a defect extends through $33\frac{1}{3}$ per cent of the nominal wall thickness or when the length of the repair exceeds 25 per cent of the nominal diameter.

In the same piping systems subject to the same service requirements, steel castings of corresponding composition are widely used as valves, fittings, reducers and other components. ASTM Specifications for high-temperature or high-pressure service generally state the following on defects and their repair:

"Weld repairs shall be considered major in the case of a casting which had leaked on hydrostatic test, or when the depth of the cavity prepared for welding exceeds 20 per cent of the actual wall thickness or 1 in., whichever is smaller, or when the extent of the cavity exceeds approximately 10 sq. in. When requested in the order, such weld repairs and subsequent heat treatment shall be subject to the purchaser's approval. Major repairs shall be inspected to the same quality of standards as one used to inspect the castings."

Since radiographic inspection requirements are not included in the basic casting specification requirements, the repair weld may be subject only to hydrostatic testing.

Thus, if a casting leaks in the hydrostatic pressure test the foundry can, on its own, groove out the leaking area across the wall thickness, fill the cavity with weld metal, and ship the casting to the fabricator without informing the purchaser that a major repair has been made. Purchase orders rarely if ever state specifically that the engineering department of the purchaser must be notified before major weld repairs are made.

Seam-welded carbon- and alloy-steel pipe produced to ASTM Specification A155 must receive 100 per cent radiography of the seam welds to meet Class I requirements.

It is important that the purchaser recognizes these limitations in individual specifications. He may then have to include special requirements in his bid invitation, contract specification or purchase order to assure a piping system of uniform quality.

Table I. Summary of Pertinent Requirements of ASTM Specifications on Seamless and Welded Carbon-steel Pipe and Wrought-iron Pipe

ASTM Spec.	Pipe sizes	Service	Steel deoxidation	Type weld	Filler metal used	Weld reinforcement	Heat treatment	Tests required*	Hydrostatic test pressure, psi $P = 2St/D$	Radiography of seam weld
A53	No limits (St'd. 1/8"–24")	General (steam, water, gas, etc.)	Not req'd (except Grade B Bessemer)	Seamless, furnace butt welded, electric resistance welded	No	None	None	TB for 2" and under F for over 2" T	Req'd.— 2,500 butt to 3", 2,800 max over 3"	Not req'd
A106	No limits (St'd. 1/8"–24")	High-temperature service	Killed	Seamless	....	....	None for hot finished, process anneal of cold-drawn pipe 2" and over at 1200 F or higher	TB for 2" and under; F over 2"; B over 25" and for diameter/thickness = 7 or less, T	Req'd.— $S = 0.60\ YS$† 2,500 max to 3", 2,800 max over 3"	....
A120	No limits (St'd. 1/8"–20")	Ordinary service (steam, water, gas, etc.)	Not req'd	4" and under seamless butt or electric welded, over 4" seamless or electric welded	No	None	None	None	Req'd, as shown in tables; Over 12" OD $S = 0.60\ YS$	Not req'd
A134	16" OD and over; walls to 3/4" thick	Ordinary service (liquid, gas, vapor)	Not req'd	Electric fusion (arc) welded	Yes	Not specified	None	T	Req'd. $S = 0.60\ YS$ 2,800 max.	Not req'd
A135	30" OD and smaller	Ordinary service (liquid, gas, vapor)	Not req'd	Electric resistance welded	No	None	None	T, F	Req'd. $S = 0.80\ YS$ max S for Grade A is 16,000–18,000; for Grade B is 20,000–22,000 2,500 max	Not req'd

Spec	Size	Service	Deoxidation	Welding	Filler	Weld reinforcement	Heat treatment	Tests	Hydrostatic test	Radiography
A139	4″ OD and over; walls to ⅝″ thick	Ordinary service (liquid, gas, vapor)	Not req'd	Electric fusion (arc) welded	Yes	Not specified	None	T	Req'd. (S = 0.60 YS to 0.85 YS) 2,800 max	Not req'd
A155	Usually 16″ OD and larger	High-temperature or high-pressure service	Not req'd for A285 plate; killed if A201 or A212 plate	Electric fusion welded	Yes	⅟₁₆–⅛ (or may be removed)	Class 1 req'd. Class 2 req'd for ¾″ wall and larger at 1100–1250 F	T, B (pl), BT	Req'd. S = 0.75 YS	Class 1—req'd Class 2—not req'd
A211	4″ through 48″ OD, ⅟₁₆″ to 11/64″ wall	Ordinary service (liquid, gas, vapor)	Not req'd	Spiral welded	Either	Not specified (usually left on)	None	. . . . . .	Req'd. at 150% of working pressure, S = 0.80 YS max	Not req'd
A252	No limits	Pipe piles	Not req'd	Seamless and welded (electric, flash, fusion, resistance)	Either	Not specified	None	T	Not req'd	Not req'd
A333	No limits	Low-temperature service (usually −50 F min)	Not req'd	Seamless and welded	Filler metal not permitted	Not specified (usually flush)	Req'd either normalized or normalized and tempered	T, F, I	Req'd. S = 0.60 YS 2,500 max to 3″, 2,800 max over 3″	Not req'd
A381	16″ and larger, 5/16″ to 1½″ walls	High-pressure transmission service	Not req'd	Metal-arc welded	Yes	Smooth	Req'd, stress relief after welding at not less than 1100 F	T, BT	Req'd to ½″ wall S = 0.85 YS; over ½″ wall S = 0.70 YS	Not req'd
A72 (wrought iron)	¼″ through 14″	Ordinary service	Pig-puddled or processed wrought iron	Welded	Either	None	None	T, FR TB for 2″ and under	Req'd. S = 0.60 YS Varies with welding process	Not req'd
A419 (wrought iron)	16″ and larger, walls to ¾″	Ordinary service (liquid, gas, vapor)	Pig-puddled or processed wrought iron	Electric fusion welded	Yes	Not specified	None	T	Req'd. S = 0.60 YS	Not req'd

* T = tension test, B = bend test, B(pl) = bend test on plate material, BT = guided weld bend test around 90-deg mandrel, TB = tube bend test, I = impact test, Fr = fracture test, F = flattening test.

† YS = yield strength.

Table I. (*Continued*)

ASTM Spec.	Wall thickness tolerance	OD tolerances (max)	Out-of-round tolerance	Straightness	Grades and min. tensile strength, psi	Finish	Special requirements	Plate steel material	Recognition by other codes
A53	Not more than 12½% under nominal	To 1½" OD + 1/64", −1/32"; 2" OD and over ±1%	....	....	F 45,000 50,000 A 48,000 B 60,000	Black or galvanized	Ends plain or threaded	....	ASME Boiler and Pressure Vessel Code SA-53
A106	Not more than 12½% under nominal	Table 4	....	....	A 48,000 B 60,000 C 70,000	1½" and under either hot finished or cold drawn; 2" and over, hot finished only	Plain ends	....	ASME Boiler and Pressure Vessel Code SA-106
A120	Not more than 12½% under nominal	To 1½" OD + 1/64", −1/32"; 2" OD and over ±1%	....	....	....	Black or galvanized	Not intended for close coiling or bending. Ends: St'd, threaded; XS,XXS, plain		
A134	....	Outside circumference ±1% ¾" max	....	....	Transverse weld tensile strength to be not less than 95% of plate strength (varies from 45,000–60,000)	Not specified, may be coated	Ends square or beveled	A245, Grade A, B, C, D A283 A285	
A135	Not more than 12½% under nominal	±1% from nominal	....	....	A 48,000 B 60,000	Usually coated; Sch. 10 coated with light oil	Grade B not intended for flanging or bending. Sch. 10, plain ends; Others, plain or threaded	....	ASME Boiler and Pressure Vessel Code SA-135
A139	Not more than 12½% under nominal	Outside circumference ±1% ¾" max	....	....	A 48,000 B 60,000	Usually coated	Ends square or beveled		

Spec	Wall thickness	OD / circumference	Out-of-round	Length	Tensile strength	Coating	Ends	Grade designation	Code
A155	Not more than 0.01″ under nominal	Outside circumference ±0.5% of nominal OD	Diff. between major and minor axes, 1%	⅛″ in 10′	C45 45,000–55,000; C50 50,000–60,000; C55 55,000–65,000; KC55 55,000–65,000; KC60 60,000–72,000; KC65 65,000–77,000; KC70 70,000–85,000	....	Ends plain	A285, Grade A (F‡ or FB§); A285, Grade B (F or FB); A285, Grade C (F or FB); A201, Grade A (F or FB); A201, Grade B (F or FB); A212, Grade A (F or FB); A212, Grade B (F or FB)	....
A211	....	Must be suitable for field joints	....	....	....	Usually coated	Not specified (usually plain)	A245 (steel) A129 (open-hearth iron)	....
A252	Not more than 12½% under nominal	OD ± 1%	....	....	....	....	Not specified	....	....
A333	Not more than 12½% under nominal	Table 4	....	....	1 55,000; 6 60,000	....	Not specified (usually plain)	....	ASME Boiler and Pressure Vessel Code SA-333 (except Notes 1 and 1a)
A381	Not more than 0.01″ under spec. wall	To 24″ OD ±⅛″; over 24″ OD, ±0.5%	Diff. between major and minor axes, 1%	⅛″ in 10′	Y35 60,000; Y42 68,000; Y46 70,000; Y48 72,000; Y50 72,000; Y52 75,000	Uncoated	....	....	....
A72 (wrought iron)	Not more than 12½% under nominal	To 1½″ OD +1/64″, −1/32″; 2″ OD and over ±1%	....	....	42,000	Black or galvanized	Ends: Std., threaded; XS,XXS, plain or beveled	A42 (wrought-iron plate)	....
A419 (wrought iron)	Not more than 0.01″ under specified wall thickness	Outside circumference ±1%, ¾″ max	....	....	48,000	....	Higher yield strength grades usually alloyed with Mn and V ends: beveled, with 1/16″ land	A42 (wrought-iron plate)	ASME Boiler and Pressure Vessel Code SA-72

‡ Flange-quality steel.
§ Firebox-quality steel.

Table 2. Summary of Pertinent Requirements of ASTM Specifications on Seamless

ASTM Spec.	Pipe sizes	Service	Type weld	Filler metal used	Weld reinforcement
A155	Usually 16" OD and larger	High-temperature service	Electric fusion welded	Yes	$\frac{1}{16}$"–$\frac{1}{8}$" (or may be removed)
A333	No limits	Low-temperature service (to −320 F)	Seamless and welded	Filler metal not permitted	Not specified (usually flush)
A335	No limits	High-temperature service	Seamless	. . . .	. . . .
A336	No limits	Drum forgings	Seamless	. . . .	. . . .
A369	No limits (usually heavy-wall pipe)	High-temperature service (forged and bored)	Seamless	. . . .	. . . .
A405	No limits	High-temperature service	Seamless	. . . .	. . . .
A426	No limits	High-temperature service (centrifugally cast)	Seamless	. . . .	. . . .

* T = tension test, B = bend test in plate material, BT = guided weld bend test, F = flattening test, I = impact test, H =
† Yield strength.

and Welded Low-alloy Steel Pipe

Heat treatment	Tests required*	Hydrostatic test pressure, psi $P = 2St/D$	Radiography of weld	Wall-thickness tolerance	ASTM Spec.
Req'd at over 1100 F	T, B (pl), BT	Req'd; $S = 0.75\ YS\dagger$	Class 1, req'd; Class 2, not req'd	Not more than 0.01″ under nominal	A155
Req'd; either normalized or normalized and tempered; Grade 8 quenched and tempered	T, F, I	Req'd; $S = 0.60\ YS$ 2,500 max to 3″; 2,800 max over 3″	Not req'd	Not more than 12½% under nominal	A333
Req'd; P1, P2, P12 stress relieved at 1200–1300 F; P5c heat treated at 1350 F; all others either annealed or normalized and tempered at above 1200 F	T, F	Req'd; $S = 0.60\ YS$ 2,500 max to 3″; 2,800 max over 3″	....	Not more than 12½% under nominal	A335
Req'd; either annealed or normalized and tempered	T	Not req'd	....	Not less than specified	A336
Req'd; FP1, FP2, FP12 stress relieved at 1200–1300 F; all others either annealed or normalized and tempered at above 1200 F	T, F	Not req'd	....	Not more than ⅛″ over specified wall; no undertolerance permitted	A369
Req'd; either annealed or normalized at 1850–1950 F and tempered at 1250–1350 F	T, F	Req'd; $S = 0.60\ YS$ 2,500 max to 3″; 2,800 max over 3″	....	Not more than 12½% under nominal	A405
Req'd; either annealed or normalized and tempered at above 1200 F	T, H, F, U MP or LP	Req'd; $S = 0.60\ YS$	....	Not more than 1/16″ over specified wall; no undertolerance permitted	A426

hardness test, U = ultrasonic test, MP = magnetic-particle examination, LP = liquid-penetrant examination.

Table 2. Summary of Pertinent Requirements of ASTM Specifications on Seamless

ASTM Spec.	OD tolerances	Out-of-round tolerance	Straightness	Alloy designation
A155	Outside circumference ±0.5%	Diff. between major and minor OD, 1%	1/8″ in 10′	1/2 Mo 1/2 Mo 1/2 Mo 1/2 Cr–1/2 Mo 1 Cr–1/2 Mo 1 1/4 Cr–1/2 Mo 2 1/4 Cr–1 Mo 5 Cr–1/2 Mo
A333	Table 4	. . . .	. . . .	3 1/2 Ni Cr–Cu–Ni 2 1/4 Ni 9 Ni
A335	Table 4	. . . .	. . . .	1/2 Mo 1/2 Mo (high Si) 1/2 Cr–1/2 Mo 1 Cr–1/2 Mo 1 1/4 Cr–1/2 Mo 2 1/4 Cr–1 Mo 3 Cr–1 Mo 5 Cr–1/2 Mo 5 Cr–1/2 Mo (high Si) 5 Cr–1/2 Mo (low C) 7 Cr–1/2 Mo 9 Cr–1 Mo
A336	. . . .	. . . .	. . . .	1/2 Mo 1 Cr–1/2 Mo 2 1/4 Cr–1 Mo 5 Cr–1/2 Mo (low C) 5 Cr–1/2 Mo (high C) 1/2 Mo–1/4 V 2 1/2 Ni–1/4 Mo–0.1 V 3/4 Ni–3 1/2 Cr– 0.4 Mo–0.1 V
A369	Not more than 1/16″ under ID specified; no overtolerance permitted	. . . .	. . . .	1/2 Mo 1/2 Cr–1/2 Mo 1 Cr–1/2 Mo 1 1/4 Cr–1/2 Mo 2 Cr–1/2 Mo 2 1/4 Cr–1 Mo 3 Cr–1 Mo 5 Cr–1/2 Mo 7 Cr–1/2 Mo 9 Cr–1 Mo
A405	Table 4	. . . .	. . . .	1 Cr–1 Mo–0.2 V
A426	Not more than 1/16″ under ID specified; no overtolerance permitted	. . . .	. . . .	1/2 Mo 1/2 Mo (high Si) 1/2 Cr–1/2 Mo 1 Cr–1/2 Mo 1 1/4 Cr–1/2 Mo 2 1/4 Cr–1 Mo 3 Cr–1 Mo 5 Cr–1/2 Mo 5 Cr–1/2 Mo (high Si) 7 Cr–1/2 Mo 9 Cr–1 Mo

Grades and min tensile strength, psi		Finish	Special requirements	Plate steel material	Recognition by other codes	ASTM Spec.
CM65	65,000	....	Ends plain	A204, Grade A		A155
CM70	70,000			A204, Grade B		
CM75	75,000			A204, Grade C		
½ Cr	65,000			A387, Grade A		
1 Cr	60,000			A387, Grade B		
1¼ Cr	60,000			A387, Grade C		
2¼ Cr	60,000			A387, Grade D		
5 Cr	60,000			A357		
3	65,000	....	Jointers not permitted; ends plain	....	ASME Boiler and Pressure Vessel Code SA-333 (except Note 1, Para. 1a)	A333
4	60,000					
7	65,000					
8	100,000					
P1	55,000	2″ and over, hot finished and heat treated; under 2″, hot finished or cold drawn and heat treated	P2 and P12 shall be coarse grained with photomicrographs required. Ends plain; jointers not permitted	....	ASME Boiler and Pressure Vessel Code SA-335	A335
P15	60,000					
P2	55,000					
P12	60,000					
P11	60,000					
P22	60,000					
P21	60,000					
P5	60,000					
P5b	60,000					
P5c	60,000					
P7	60,000					
P9	60,000					
F1	70,000	....	....	....	ASME Boiler and Pressure Vessel Code SA-336	A336
F2	70,000					
F22	70,000					
F5	60,000					
F5a	80,000					
F30	80,000					
F31	95,000					
F32	100,000					
FP1	55,000	....	Ends machined as specified; surfaces machined to max 250 mi (AA) finish	....	ASME Boiler and Pressure Vessel Code SA-369	A369
FP2	55,000					
FP12	60,000					
FP11	60,000					
FP3B	60,000					
FP22	60,000					
FP21	60,000					
FP5	60,000					
FP7	60,000					
FP9	60,000					
P24	60,000	2″ and over, hot finished and heat treated; under 2″, hot finished or cold drawn and heat treated	Ends plain	....	....	A405
CP1	55,000	....	Ends plain; surfaces machined to max 250 mi (AA) finish	....	....	A426
CP15	60,000					
CP2	55,000					
CP12	60,000					
CP11	60,000					
CP22	60,000					
CP21	60,000					
CP5	60,000					
CP5b	60,000					
CP7	60,000					
CP9	60,000					

8–14

Table 3. Summary of Pertinent Requirements of ASTM Specifications on Seamless and Welded Austenitic Steel Pipe

ASTM Spec.	Pipe sizes	Service	Base-metal process	Type weld	Filler metal used	Weld reinforcement	Heat treatment	Quenching or rapid cooling	Tests required[1]	Hydrostatic test pressure, psi
A312	½" to 12" IPS sizes	High-temperature; corrosive	Plate, sheet or strip, or pierced stainless	Seamless or automatic welded	No filler metal permitted	None	Req'd at 1900 F min, except H grades, which are solution treated at spec. temps. between 1800 and 2000 F	Req'd except for H grades	T, F	Req'd at "S" of 15,000: 2,500 max to 3"; 2,800 max over 3"
A358	8" and larger IPS sizes	Corrosive; high temperature	Plate, sheet, or strip (A240)	Manual arc; automatic arc	Yes	Req'd but may be removed	Req'd at 1900 F min	Req'd	T, BT	Req'd on each length to "S" of 75% of yield
A376	No limits	High-temperature steam	Pierced or hollow forged	Seamless	....	....	Solution-treated at spec. temps. between 1800 and 2000 F	Req'd	T, F	Req'd at "S" of 15,000: 2,500 max to 3"; 2,800 max over 3"
A409	14" through 30" IPS	Corrosive; high temperature	Hot-finished plate; hot- or cold-rolled sheet	Manual arc; automatic arc	Yes	1/16" max permitted	Req'd if plate is not in solution-heat-treated condition	If heat treated after welding	T, BT	Req'd at "S" of 15,000; or air at 100 psi
A430	No limits	High temperature	Forged and bored	Seamless	....	....	Solution-treated at spec. temps. between 1800 and 2000 F	Req'd	T, F	....
A451	No limits	High temperature; nuclear	Centrifugally cast	Seamless	....	....	Solution treated at 1900–2100 F	Req'd	T, F, LP (over 6" ID)	Req'd at "S" of 0.60 times min yield strength. 2,800 max
A452	No limits	High temperature	Cast and cold wrought	Seamless	....	....	Solution treated at 2000 F min for 347H, and at 1800 F min for 304H and 316H	Not specified	T, F	Req'd at "S" of 15,000: 2,500 max to 3"; 2,800 max over 3"

[1] T = tension test, BT = guided weld bend test, F = flattening test, F1 = flaring test, H = hardness test, LP = liquid-penetrant test.

Table 3. (Continued)

ASTM Spec.	Radiography of weld	Wall-thickness tolerance	OD tolerance	Out-of-round tolerance	Alignment tolerances	No. of grades	Pickling	Hot or cold finished	Special requirements	Recognition by other codes
A312	Not req'd	Not more than 12½% under nominal	Table 4			304, 304H, 304L, 309, 310, 316, 316H, 316L, 317, 321, 321H, 347, 347H, 348, 348H	Req'd unless bright annealed	Either at manufacturer's option		ASME Boiler and Pressure Vessel Code SA-312
A358	Class 1, req'd; Class 2, not req'd	Not more than 0.01" under nominal	±0.5% of nominal OD	Max of 1% difference between major and minor OD	⅛" max in 10 ft	304, 316, 347, 321, 309, 310, 348	No requirements	Not stated		
A376		Not more than 12½% under nominal	Table 4			304, 304H, 316, 316H, 321, 16-8-2H, 321H, 347, 347H, 348	Only when required by purchaser	Either at manufacturer's option		ASME Boiler and Pressure Vessel Code SA-376
A409	Nondestructive optional in place of pressure testing	Not more than 0.018" under nominal	±0.2% for walls under 0.188", ±0.04% for walls of 0.188" and heavier	Max of 1½% difference between major and minor OD	3/16" max in 10 ft	304, 309, 310, 321, 347, 316, 317, 348	No requirement	Not stated		
A430		+⅛" max of specified wall	−1/16", +0"			304, 304H, 316, 316H, 16-8-2H, 321, 321H, 347, 347H		Surfaces machined to max 250 mi (AA) finish		ASME Boiler and Pressure Vessel Code SA-430 (H grades only)
A451		Not to exceed +1/16", −0"	OD and ID: under, −1/16", max over, +0"			F8, F8M, F10MC, F8C, H8, H20, H10, K20		Surfaces machined to max 250 mi (AA) finish		
A452		Not more than +22%, −0, with max of ⅛	−1/32", +0", under 4"; OD: −1/16", −4" and over			304H, 316H, 347H		Surfaces machined to max 250 mi(AA) finish unless produced by cold working	Max avg grain size ASTM 1	

Table 4. Variations in Outside Diameter as Stipulated in Several ASTM Pipe Specifications

Nominal pipe size, in.	Permissible variations in outside diameter, in.	
	Over	Under
$\frac{1}{8}$–$1\frac{1}{2}$, incl	$\frac{1}{64}$ (0.015)	$\frac{1}{32}$ (0.031)
Over $1\frac{1}{2}$–4, incl	$\frac{1}{32}$ (0.031)	$\frac{1}{32}$ (0.031)
Over 4–8, incl	$\frac{1}{16}$ (0.062)	$\frac{1}{32}$ (0.031)
Over 8–18, incl	$\frac{3}{32}$ (0.093)	$\frac{1}{32}$ (0.031)
Over 18–24, incl	$\frac{1}{8}$ (0.125)	$\frac{1}{32}$ (0.031)

Table 5. Summary of Pertinent Requirements of ASTM Specification on Seamless and Welded Steel Tubing and Wrought-iron Tubing

ASTM Spec.	Tube sizes	Service	Steel deoxidation	Type weld	Filler metal used	Weld reinforcement	Heat treatment
A83	½″–6″ OD: 0.035″ to 0.320″ wall	Boilers and superheaters	Not req'd	Seamless	....	....	Hot-finished tubes, not req'd; cold-drawn, req'd at 1200 F or higher
A161	2″–9″ OD; over 0.220″ wall	Refinery still tubes	Killed	Seamless	....	....	Hot-finished tubes, not req'd; cold-drawn, req'd at 1200 F or higher
A178	½″–5″ OD: 0.035 to 0.320″ wall	Boiler tubes, superheater flues, safe ends	Not req'd	Electric resistance welded	No	Not specified (usually removed)	Normalizing req'd
A179	⅛″–3″ OD	Heat exchangers or condensers (cold-drawn)	Not req'd	Seamless	....	....	Req'd at 1200 F or higher
A192	½″–7″ OD: 0.085″ to 1.000″ wall	Boilers and superheaters for high-pressure service	Killed	Seamless	....	....	Hot-finished tubes, not req'd; cold-drawn, req'd at 1200 F or higher
A210	1½″–5″ OD: 0.035″ to 0.500″ wall	Boiler and superheater tubes, flues and safe ends	Killed	Seamless	....	....	Hot-finished tubes, not req'd; cold-drawn, req'd at 1200 F or higher
A214	Up to 3″ OD	Heat exchangers and condensers	Not req'd	Electric resistance welded	No	Not specified— (usually removed)	Normalizing req'd
A226	½″–5″ OD: 0.085″ to 0.360″ wall	Boiler and superheater tubes for high-pressure service	Killed	Electric resistance welded	No	Not specified— (usually removed)	Normalizing req'd
A382	¾″–3½″	Heat exchangers and condensers	Pig puddled or processed wrought iron	Lap welded	No	Not specified— (usually removed)	Annealing req'd

Table 5. (Continued)

ASTM Spec.	Tests required[1]	Hydrostatic test pressure, psi $P = 2St/D$	Radiography of seam weld	Wall-thickness tolerance	OD tolerances (max)	Grades and min tensile strength, psi	Finish	Recognition by other codes
A83	F, Fl	Req'd	....	....	....	....	Hot-finished or cold-drawn	ASME Boiler and Pressure Vessel Code SA-83
A161	T, H, F, Fl	Req'd	....	Table 9	Table 10 (Special ID tolerances at ends)	47,000	Hot-finished or cold-drawn	
A178	T (Grade C) F, Flg, RF	Req'd (unless electric test is done)	Not req'd	....	....	A (low C), C (med. C)—60,000	....	ASME Boiler and Pressure Vessel Code SA-178
A179	H, F, Fl	Req'd	....	Table 9	....	....	....	ASME Boiler and Pressure Vessel Code SA-179
A192	H, F, Fl	Req'd	....	Table 9	Table 10	....	Hot-finished or cold-drawn	ASME Boiler and Pressure Vessel Code SA-192
A210	T, H, F, Fl	Req'd	....	Table 9	....	A-1—60,000 C—70,000	Hot-finished or cold-drawn	ASME Boiler and Pressure Vessel Code SA-210
A214	H, F, Fl	Req'd (unless electric test is done)	Not req'd	....	....	....	....	ASME Boiler and Pressure Vessel Code SA-214
A226	H, F, Fl	Req'd (unless electric test is done)	Not req'd	Table 9	Table 10	....	....	ASME Boiler and Pressure Vessel Code SA-226
A382	H, Fr, T	Req'd ($S = 15,000$ psi)	Not req'd	Over—22% Under—0%	Overtolerance from 0.006" to 0.014"; undertolerance from 0.003" to 0.007"	42,000	Cold-drawn	

[1] T = tension test, H = hardness test, F = flattening test, Fl = flaring test, Fr = fracture test, Flg = flange test, RF = reverse flattening test.

Table 6. Summary of Pertinent Requirements of ASTM Specification on Seamless and Welded Low-alloy Steel Tubing

ASTM Spec.	Tube sizes	Service	Steel deoxidation	Type weld	Filler metal used	Weld reinforcement	Heat treatment	Tests required[1]
A161	2"–9" OD; over 0.220" wall	Refinery still tubes	Killed	Seamless	....	....	Hot-finished tubes, not req'd; cold-drawn, req'd at 1200 F or higher	*T, H, F, Fl*
A199	1/8"–3" OD	Heat exchanger and condenser tubes (cold drawn)	Killed	Seamless	....	....	Req'd at 1200 F or higher	*T, H, F, Fl*
A200	2"–9" OD; over 0.220" wall	Refinery still tubes	Killed	Seamless	....	....	Not req'd for hot-finished tubes which have been slowly cooled; otherwise they must be normalized and tempered; for cold-drawn, req'd at 1200 F or higher	*T, H, F, Fl*
A209	1/2"–5" OD; 0.035" to 0.500" wall	Boilers and super-heaters	Killed	Seamless	....	....	Req'd at 1200 F or higher for hot-finished and cold-drawn tubes	
A213	1/2"–5" OD; 0.035" to 0.500" wall	Boilers, super-heaters, and heat exchangers	Killed	Seamless	....	....	Req'd for T2, T12 at 1200 to 1350 F, for T5c at 1350 F; not req'd for hot-finished tubes of other grades when slowly cooled; otherwise normalizing and tempering req'd; cold-drawn T9 shall be annealed at 1200 F or higher	*T, H, F, Fl*
A250	1/2"–5" OD; 0.035" to 0.320" wall	Boilers and super-heaters	Killed	Electric resistance welded	No	Not specified (usually removed)	Req'd: normalized (above 1600 F)	*T, H, F, Flg, RF*
A334	No limits	Low-temperature service	Not req'd, Grade 1; req'd, Grade 3, 7, 8	Seamless or welded	No	Not specified (usually removed)	Req'd: either normalizing or normalized and tempered; Grade 8 quenched and tempered	*T, H, I* (wall 0.120" and thicker) *F, Fl* (seamless) *Flg* and *RF* (welded)
A423	1/2"–5" OD; 0.035" to 0.500" wall	Pressure parts (such as economizer) and corrosion service	Not req'd, Grade 2; req'd, Grade 1	Seamless or electric welded	No	Not specified (usually removed)	Req'd: normalized	*T, H, F, Fl* (seamless) *Flg* and *RF* (welded)

[1] *T* = tension test, *H* = hardness test, *F* = flattening test, *Fl* = flaring test, *Flg* = flange test, *RF* = reverse flattening test, *I* = impact test.

Table 6. *(Continued)*

ASTM Spec.	Hydrostatic test pressure, psi $P = 2\,St/D$	Radiography of seam weld	Wall-thickness tolerance	OD tolerances (max)	Alloy designation	Grades and min tensile strength, psi	Finish	Recognition by other codes
A161	Req'd	....	Table 9	Table 10; (special ID tolerances at ends)	½ Mo	T1 55,000	Hot-finished or cold-drawn	
A199	Req'd	....	Table 9	Table 10	1¼ Cr–½ Mo 2 Cr–½ Mo 2¼ Cr–1 Mo 2½ Cr–½ Mo 3 Cr–1 Mo 5 Cr–½ Mo 7 Cr–½ Mo 9 Cr–1 Mo	T11 60,000 T3b 60,000 T22 60,000 T4 60,000 T21 60,000 T5 60,000 T7 60,000 T9 60,000	Cold-drawn and annealed (surface to be scale free)	ASME Boiler and Pressure Vessel Code SA-199
A200	Req'd	....	Table 9; special req'ts for upset ends allowing ±10% of avg mean thickness	Table 10 (ends must pass through ring gage 0.025″ larger than spec. OD)	1¼ Cr–½ Mo 2 Cr–½ Mo 2¼ Cr–1 Mo 2½ Cr–½ Mo 3 Cr–1 Mo 5 Cr–½ Mo 7 Cr–½ Mo 9 Cr–1 Mo	T11 60,000 T3b 60,000 T22 60,000 T4 60,000 T21 60,000 T5 60,000 T7 60,000 T9 60,000	Surface to be scale free	
A209	Req'd	....	Table 9	Table 10	½ Mo (0.1–0.2 C) ½ Mo (0.15–0.25 C) ½ Mo (0.14 max C)	T1 55,000 T1a 60,000 T1b 53,000	Hot-finished or cold-drawn	ASME Boiler and Pressure Vessel Code SA-209
A213	Req'd	....	Table 9	Table 10	½ Cr–½ Mo 1 Cr–½ Mo 1¼ Cr–½ Mo 2 Cr–½ Mo 2¼ Cr–1 Mo 3 Cr–1 Mo 5 Cr–½ Mo 5 Cr–½ Mo (hi Si) 5 Cr–½ Mo (Ti) 7 Cr–½ Mo 9 Cr–1 Mo 1 Cr–0.2 V	T2 60,000 T12 60,000 T11 60,000 T3b 60,000 T22 60,000 T21 60,000 T5 60,000 T5b 60,000 T5c 60,000 T7 60,000 T9 60,000 T17 60,000	Hot-finished or cold-drawn	ASME Boiler and Pressure Vessel Code SA-213
A250	Req'd unless electric test is performed	Not req'd	Table 9	Table 10	½ Mo (0.1–0.2 C) ½ Mo (0.15–0.25 C) ½ Mo (0.14 C max)	T1 55,000 T1a 60,000 T1b 53,000	....	ASME Boiler and Pressure Vessel Code SA-250
A334	Req'd unless electric test is performed on welded tubes	Not req'd	Table 9	Table 10	0.25 C max 3½ Ni 2¼ Ni 9 Ni	1 55,000 3 65,000 7 65,000 8A 100,000	....	ASME Boiler and Pressure Vessel Code SA-334 (except Note 1 paragraph 1a)
A423	Req'd unless electric test is performed on welded tubes	Not req'd	Table 9	Table 10	0.4 Cu–0.7 Cr–0.5 Ni 0.6 Cu–0.7 Ni–0.2 Mo	1 60,000 2 60,000	....	ASME Boiler and Pressure Vessel Code SA-423

Table 7. Summary of Pertinent Requirements of ASTM Specification on Seamless and Welded Austenitic Tubing

ASTM Spec.	Tube sizes	Service	Type weld	Filler metal used	Weld reinforcement	Heat treatment	Quenching or rapid cooling	Tests required[1]	Hydrotest per tube, psi	Radiography of weld
A213	1/2"–5" OD; 0.035" to 0.500" wall	Boilers, superheaters, heat exchangers	Seamless	....	....	Solution treated at specified temp between 1800 and 2000 F	Req'd except H grades	T, F, Fl, H	Req'd (Table 8)	....
A249	1/2"–5" OD; 0.035" to 0.320" wall	Boilers, superheaters, heat exchangers, condensers	Automatic	No filler metal permitted	None	Cold work after welding followed by heat treatment between 1800 and 2000 F	Req'd except H grades	T, F, Fl, H	Req'd (Table 8), unless electric test is performed	Nondestructive optional in place of pressure testing
A269	1/2"–8"	Corrosion, low and high temperature	Seamless or welded	Not stated	Not stated	Req'd at 1900 F min	Req'd	H, Fl, F	Req'd (Table 8), unless electric test is performed on welded tubing	Not req'd
A270	4" and smaller	Sanitary tubing	Seamless or welded	Not stated	Not stated	Req'd at 1900 F	Req'd	F	Req'd (Table 8), unless electric test is performed on welded tubing	Not req'd
A271	2"–9" OD; over 0.220" min wall	Still tubes for refinery service	Seamless[2]	....	....	Solution treated at specified temp between 1800 and 2000 F	Req'd	T, F, Fl, H	Req'd (Table 8)	....

[1] T = tension test, F = flattening test, Fl = flaring test, H = hardness test.
[2] Requirements for circumferential butt welds joining lengths of still tubing are given in Specification A422, But Welds in Still Tubes for Refinery Service.

Table 7. (Continued)

ASTM Spec.	Wall-thickness tolerance	OD tolerances	Out-of-round tolerance	Straightness	No. of grades	Pickling	Hot or cold finished	Special requirement	Recognition by other codes
A213	Table 9	Table 10	Spec A450[3]	....	304 304H 310 316 316H 316L 321 321H 347 347H 348 348H	Req'd unless bright annealed	Hot-finished or cold-drawn	Type 321H grain size ASTM 7 or coarser	ASME Boiler and Pressure Vessel Code SA-213
A249	Table 9	Table 10	Spec A450[3]	0.030" max in 3 ft length	304 304H 304L 305 309 310 316 316H 316L 317 321 321H 347 347H 348 348H	Req'd unless bright annealed	Cold-worked	....	ASME Boiler and Pressure Vessel Code SA-249 (except grade 305)
A269	To ½" OD — ±15% over ½" OD — ±10%	From ±0.005 to ±0.030 depending on size	Double permissible OD tolerance	....	304 304L 316 316L 317 321 347 348	Req'd unless bright annealed	Either, at manufacturer's option		
A270	±12½%	From +0.002" to +0.003" and from −0.008" to −0.015"	....	....	304	Ground or polished finish	Either, at manufacturer's option		
A271	Table 9	Spec. A450[3]; OD must pass through ring gage 0.025" ID larger than tube OD; spec. requirement for upset ends in A271; plug gage requirements for plain ends	Spec. A450[3]	....	304 304H 321 321H 347 347H	Req'd unless bright annealed	Either, at manufacturer's option		

[3] ASTM A450, General Requirements for Carbon, Ferritic Alloy, and Austenitic Stainless Steel Tubes.

Table 8. General Requirements in ASTM Specification A450 Showing Maximum Hydrostatic Test Pressures Applicable to Tubing

Outside diameter of tube, in.	Hydrostatic test pressure, psi
Under 1	1,000
1–1½, excl	1,500
1½–2, excl	2,000
2–3, excl	2,500
3–5, excl	3,500
5 and over	4,500

Table 9. General Requirements in ASTM Specification A450 Showing Permissible Variations in Wall Thickness Applicable to Tubing[1]

Outside, diameter, in.	Wall thickness, %							
	0.095 in. and under		Over 0.095 to 0.150 in., incl.		Over 0.150 to 0.180 in., incl.		Over 0.180, in.	
	Over	Under	Over	Under	Over	Under	Over	Under
Seamless, hot-finished tubes								
4 and under	40	0	35	0	33	0	28	0
Over 4	..	..	35	0	33	0	28	0

	Seamless, cold-drawn tubes	
	Over	Under
1½ and under	20	0
Over 1½	22	0

	Welded tubes	
All sizes	18	0

[1] These permissible variations in wall thickness apply only to tubes, except internal-upset tubes, as rolled or drawn, and before swaging, expanding, bending, polishing, or other fabricating operations.

Table 10. General Requirements in ASTM Specification A450 Showing Permissible Variations in Outside Diameter Applicable to Tubing[1]

Outside diameter, in.	Outside diameter (including out-of-roundness), in.	
	Over	Under
Seamless, hot-finished tubes		
4 and under	1/64	1/32
Over 4–7½, incl	1/64	3/64
Over 7½–9, incl	1/64	1/16
Seamless, cold-drawn and welded tubes		
Under 1	0.004	0.004
1–1½, incl	0.006	0.006
Over 1½–2, excl	0.008	0.008
2–2½, excl	0.010	0.010
2½–3, excl	0.012	0.012
3–4, incl	0.015	0.015
Over 4–7½, incl	0.015	0.025
Over 7½–9, incl	0.015	0.045

[1] These permissible variations in outside diameter apply only to tubes, except internal-upset tubes, as rolled or drawn before swaging, expanding, bending, polishing, or other fabricating operations.

Table II. Summary of Pertinent Requirements of ASTM Specification on Carbon-steel,

Material grouping	ASTM Spec.	Sizes	Service	Steel deoxidation	Heat treatment	Tests required	Hydrotest
Carbon steel	A216	No limits	High-temperature service	Not req'd (usually steels are killed)	Req'd; either annealed or normalized or normalized and tempered	T	Req'd: WCA, as agreed; WCB, as shown in ASA B16.5, or as agreed
Low-alloy steel	A217	No limits	High-temperature service	Not req'd (usually steels are killed)	Req'd: either annealed or normalized and tempered	T	Req'd as shown in ASA B16.5 or as agreed
	A352	No limits	Low-temperature service	Not req'd (usually steels are killed)	Req'd; either normalized or normalized and tempered (liquid quenching permitted by special agreement)	T, I	Req'd as shown in ASA B16.5 or as agreed
	A389	No limits	High-temperature service	Not req'd (usually steels are killed)	Req'd; normalized (1850 to 1950 F) and tempered (1250 to 1350 F) with holding 1 hr minimum for Grade C23 and 12 hr for Grade C24	T	Req'd as shown in ASA B16.5 or as agreed
	A487	No limits	Pressure service	Not req'd (usually steels are killed)	Req'd; normalized and tempered or liquid quenched and tempered	T	Req'd as shown in ASA B16.5 or as agreed
Stainless steel	A351	No limits	High-temperature and corrosive service	(Normally done)	Req'd: Grade CA15; heat to 1750 F min, air cool and temper at 1100 F max; Other grades carbide-solution anneal (1900 to 2100 F), air cool or quench. Not req'd for the HK and HT grades	T	Req'd as shown in ASA B16.5 or as agreed

[1] T = tension test, B = bend test. I = impact test, MP = magnetic-particle examination, LP = liquid-penetrant exam-

Alloy-steel, and Stainless-steel Castings

Radiography	Tests recognized but performed only when specifically required in purchase order	Alloy	Grades and min tensile strength, psi		Recognition by other codes	Material grouping
Not req'd	B, MP, R	0.25 max C 0.30 max C	WCA 60,000 WCB 70,000		ASME Boiler and Pressure Vessel Code SA-216 (MSS, AFS)	Carbon steel
Not req'd	B, MP, R	½ Mo 1¼ Cr–½ Mo 2¼ Cr–1 Mo 5 Cr–½ Mo 9 Cr–1 Mo ½ Mo–0.7 Cr–1 Ni 1 Mo–0.7 Cr–0.8 Ni	WC1 65,000 WC6 70,000 WC9 70,000 C5 90,000 C12 90,000 WC4 70,000 WC5 70,000		ASME Boiler and Pressure Vessel Code SA-217 (MSS, AFS)	Low-alloy steel
Not req'd	MP, R	0.30 max C ½ Mo 2½ Mo 3½ Ni	LCB 65,000 LC1 65,000 LC2 65,000 LC3 65,000		ASME Boiler and Pressure Vessel Code SA-352 (AFS)	
Not req'd	B, MP, R	1¼ Cr–0.55 Mo–0.2 V 1 Cr–1 Mo–0.2 V	C23 70,000 C24 80,000			
Not req'd	MP, R, I	0.1 V 1.2 Mn–0.2 Mo 1.5 Mn–0.4 Mo 0.6 Ni–0.6 Cr–0.2 Mo 1.2 Mn–0.6 Ni–0.6 Cr–0.2 Mo 1.5 Mn–0.6 Ni–0.6 Cr–0.35 Mo 0.8 Ni–0.6 Cr–0.5 Mo–B–Cu 2¼ Cr–1 Mo 1 Cr–0.2 Mo 1.7 Ni–0.7 Cr–0.3 Mo	1N 85,000; 1Q 90,000 2N 85,000; 2Q 90,000 3N 90,000; 3Q 105,000 4N 90,000; 4Q 105,000 5N 105,000; 5Q 105,000 6N 115,000; 6Q 120,000 7N 100,000; 7Q 125,000 8N 85,000; 8Q 105,000 9N 90,000; 9Q 105,000 10N 100,000; 10Q 125,000			
Not req'd	LP, R, MP (CA-15)	12 Cr 19 Cr–9 Ni–ELC 19 Cr–9 Ni 19 Cr–11 Ni–2.5 Mo–ELC 19 Cr–11 Ni–2.5 Mo 19 Cr–10 Ni–Cb 25 Cr–13 Ni 25 Cr–13 Ni (0.2 or 0.1 C) 25 Cr–20 Ni 15 Cr–35 Ni 17 Cr–15 Ni–2 Mo–Cb 20 Cr–29 Ni–2.5 Mo–3.5 Cu 25 Cr–20 Ni (0.3 C) 25 Cr–20 Ni (0.4 C) 15 Cr–35 Ni	CA15 90,000 CF3 70,000 CF8 70,000 CF3M 70,000 CF8M 70,000 CF8C 70,000 CH8 65,000 CH20, CH10 70,000 CK45 65,000 CT35 65,000 CF10MC 70,000 CN7M 62,500 HK30 65,000 HK40 62,000 HT 30 65,000		ASME Boiler and Pressure Vessel Code SA-351 ASA G52.2– 1964	Stainless steel

ination, R = radiographic inspection.

Table 12. Summary of Pertinent Requirements of ASTM Specification on Carbon-steel,

Material grouping	ASTM Spec.	Sizes	Common product type	Service	Steel deoxidation	Heat treatment
Carbon steel	A105	No limits	Flanges, fittings, valves, etc.	High-temperature service	Not req'd	Req'd, either annealed or normalized
	A181	No limits	Flanges, fittings, valves, etc.	General service	Not req'd	Not req'd
	A266	No limits	Drum forgings (hollow drums, heads, covers, etc.)	General and high-temperature service	Req'd	Req'd; either annealed or normalized and tempered
	A350	No limits	Flanges, fittings, valves, etc.	Low-temperature service	Grade LF1, not req'd; Grade LF 2, req'd	Req'd; either normalized or normalized and tempered or quenched and tempered
	A465	No limits	Leaded steel flanges, fittings, valves, etc.	General service	Not req'd (usually steels are killed)	Grades L-I and L-III, not req'd; Grades L-II and L-IV req'd —either annealed or normalized or normalized and tempered
Low-alloy steel	A182	No limits	Flanges, fittings, valves, etc.	High-temperature service	Req'd	Req'd; either annealed or normalized and tempered
	A336	No limits	Drum forgings (hollow drums, heads, covers, etc.)	General and high-temperature service	Req'd	Req'd; either annealed or normalized and tempered
	A350	No limits	Flanges, fittings, valves, etc.	Low-temperature service	Req'd	Req'd; either normalized, or normalized and tempered or quenched and tempered
	A404	No limits	Flanges, fittings, valves, etc.	High-temperature service	Req'd	Req'd; normalized, (1850–1950 F) and tempered (1250–1350 F); soaking time, 12 hours
	A522	No limits	Flanges, fittings, valves, etc.	Low-temperature service (to −320 F)	Req'd	Req'd; quenched and tempered or double normalized and tempered
Stainless steel	A182	No limits	Flanges, fittings, valves, etc.	High-temperature service	Usually deoxidized	Req'd; annealed at 1900 F min except H grades, which are solution treated at specified temps between 1800 and 2000 F
	A336	No limits	Drum forgings (hollow drums, heads, covers, etc.)	General and high-temperature service	Usually deoxidized	Req'd; solution annealed at above 1900 F

[1] T = tension test, I = impact test.
[2] American Standards Association, ASA B16.5-1961, American Standard for Steel Pipe Flanges and Flanged Fittings.

Alloy-steel, and Stainless-steel Forgings

Tests required[1]	Hydrostatic test pressure, psi	Alloy	Grades and min tensile strength, psi		Recognition by other codes	Material grouping
T	. . . .	0.35 max C 0.35 max C	I II	60,000 70,000	ASME Boiler and Pressure Vessel Code SA-105 ASA G17.3—1962 (MSS)	Carbon steel
T	. . . .	0.35 max C 0.35 max C	I II	60,000 70,000	ASME Boiler and Pressure Vessel Code SA-181 ASA G46.1—1962 (MSS)	
T	. . . .	0.35 max C 0.35 max C 0.50 max C	1 2 3	60,000 70,000 75,000	ASME Boiler and Pressure Vessel Code SA-266 ASA G55.1—1964	
T, I	. . . .	1.06 max Mn 1.35 max Mn	LF1 LF2	60,000 70,000	ASME Boiler and Pressure Vessel Code SA-350	
T	. . . .	0.30 max C 0.30 max C 0.35 max C 0.35 max C	L-I L-II L-III L-IV	60,000 60,000 70,000 70,000		
T (max hardness values specified)	Req'd for valve bodies, fittings, and pressure parts to ASA B16.5[2], or as agreed	½ Mo 1 Cr–½ Mo 1¼ Cr–½ Mo 2¼ Cr–1 Mo 3 Cr–1 Mo 5 Cr–½ Mo (low C) 5 Cr–½ Mo (high C) 7 Cr–½ Mo 9 Cr–1 Mo	F1 F12 F11 F22 F21 F5 F5a F7 F9	70,000 70,000 70,000 70,000 70,000 60,000 90,000 60,000 100,000	ASME Boiler and Pressure Vessel Code SA-182 ASA G37.1—1962	Low-alloy steel
T	. . . .	½ Mo 1 Cr–½ Mo 2¼ Cr–1 Mo 5 Cr–½ Mo (low C) 5 Cr–½ Mo (high C) ½ Mo–¼ V 2½ Ni–¼ Mo–0.1 V ¾ Ni–3½ Cr–0.4 Mo–0.1 V	F1 F2 F22 F5 F5a F30 F31 F32	70,000 70,000 70,000 60,000 80,000 80,000 95,000 100,000	ASME Boiler and Pressure Vessel Code SA-336	
T, I	. . . .	3½ Ni Cr–Cu–Ni	LF3 LF4	70,000 60,000	ASME Boiler and Pressure Vessel Code SA-350	
T	Req'd for valve bodies, fittings, and pressure parts to ASA B16.5[2], or as agreed	1 Cr–1 Mo–0.2 V	F24	80,000		
T, I	. . . .	9 Ni		100,000		
T	Req'd for valve bodies, fittings, and pressure parts to ASA B16.5[2], or as agreed	19 Cr–9 Ni 19 Cr–9 Ni (ELC) 17 Cr–12 Ni–2.5 Mo 17 Cr–12 Ni–2.5 Mo (ELC) 18 Cr–10 Ni–Cb 18 Cr–10 Ni–Cb 18 Cr–10 Ni–Ti 25 Cr–20 Ni 8 Cr–20 Ni	F304, F304H F304L F316, F316H F316L F347, F347H F348, F348H F321, F321H F310 F10	75,000 65,000 75,000 65,000 75,000 75,000 75,000 75,000 80,000	ASME Boiler and Pressure Vessel Code SA-182	Stainless steel
T	. . . .	19 Cr–9 Ni 17 Cr–12 Ni–2.5 Mo 18 Cr–10 Ni–Cb 18 Cr–10 Ni–Ti 8 Cr–20 Ni 25 Cr–20 Ni	F8 F8m F8c F8t F10 F25	70,000 70,000 70,000 70,000 80,000 75,000	ASME Boiler and Pressure Vessel Code SA-336	

Table 13. Summary of Pertinent Requirements of ASTM Specification on Seamless and Welded Wrought-steel Welding Fittings such as Elbows, Tees, and Caps

ASTM Spec.	Pipe sizes	Service	Steel deoxidation	Type weld	Filler metal used	Weld reinforcement	Heat treatment	Tests required[1]	Hydrotest	Radiography of seam weld
A234	No limits	General and high-temperature service	Steels for service above 750 F must have been silicon-killed	Seamless or fusion welded	Yes or no	In accordance with ASME Boiler and Pressure Vessel Code, from $\frac{3}{32}''$ to $\frac{3}{16}''$ max	*Carbon steel:* Not req'd when final forming between 1150 and 1800 F; req'd when finished above 1800 F, either normalized or normalized and tempered; req'd of cold-formed and welded fittings of $\frac{3}{4}''$ or heavier wall, to be stress relieved at 1100 to 1250 F. *Alloy steel:* Req'd seamless Grade WP11, WP22, WP5 either full annealed or normalized and tempered; other seamless grades and all welded grades are tempered (stress relieved) at 1200 F or higher, except for $\frac{1}{2}$ Mo steel under $\frac{1}{2}''$ wall	H	Not req'd	Req'd on 4% of carbon-steel seam welds on fittings whose wall thickness does not exceed that corresponding to XS; req'd on 100% of carbon-steel seam welds over XS wall and on all alloy steel seam welds
A403	No limits	General, high-temperature and corrosive service	Not specified	Seamless or fusion welded	Yes or no	In accordance with ASME Boiler and Pressure Vessel Code, from $\frac{3}{32}''$ to $\frac{3}{16}''$ max	Req'd at 1900 F min except H grades, which are solution treated at specified temps between 1800 and 2000 F; special stabilization treatments at 1500 to 1600 F permissible for WP321, WP347, WP348 upon agreement	. . . .	Not req'd	Req'd on 100% of all seam welds, except lightweight fittings to SP-434
A420	No limits	Low-temperature service	Not specified	Seamless or fusion welded	Yes or no	In accordance with ASME Boiler and Pressure Vessel Code, from $\frac{3}{32}''$ to $\frac{3}{16}''$ max	Req'd; either normalized or normalized and tempered	I (for each heat-treatment load)	Not req'd	Req'd on 4% of carbon-steel seam welds to XS wall. Req'd on 100% of carbon-steel seam welds over XS wall and on all alloy steel seam welds

Table 13. (Continued)

ASTM Spec.	Dimensional tolerances	Finish	Material	Grade	Raw material (specification and grade) Pipe	Plate	Forgings	Bars	Recognition by other codes
A234	ASA B16.9[2] ASA B16.11[3]	Ends machined for butt welds or socket welds	C-steel	WPA	A106, A	A285, B, C	A105, I	A107, 1.008–1.022	
			C-steel	WPB	A106, B	A212, A, B	A105, II	A107, 1.025–1.030	
			C-steel	WPC	A106, C	[5]	[5]		
			½ Mo	WP1	A335, P1	A204, B	A182, F1		
			1 Cr–½ Mo	WP12	A335, P12	A387, B	A182, F12		
			1¼ Cr–½ Mo	WP11	A335, P11	A387, C	A182, F11		
			2¼ Cr–1 Mo	WP22	A335, P22	A387, D	A182, F22		
			5 Cr–½ Mo	WP5	A335, P5	A357	A182, F5		
A403	ASA B16.9[2] ASA B16.11[3] SP-43[4]	Ends machined for butt welds or socket welds	18-8	WP304	A312, TP304	A240, 304	A182, F304	A276, 304	
			18-8	WP304H	A312, TP304H	A240, 304	A182, F304H	A276, 304[6]	
			18-8 (ELC)	WP304L	A312, TP304L[6]	A240, 304L[6]	A182, F304L	A276, 304L[6]	
			25-12	WP309	A312, TP309	A240, 309S		A276, 309S	
			25-20	WP310	A312, TP310	A240, 310S	A182, F310	A276, 310S	
			18-8 Mo	WP316	A312, TP316	A240, 316	A182, F316	A276, 316	
			18-8 Mo	WP316H	A312, TP316H	A240, 316[6]	A182, F316H	A276, 316[6]	
			18-8 Mo (ELC)	WP316L	A312, TP316L[6]	A240, 316L[6]	A182, F316L	A276, 316L[6]	
			18-8 Mo	WP317	A312, TP317	A240, 317	A182, F317		
			18-8 Ti	WP321	A312, TP321	A240, 321	A182, F321	A276, 321	
			18-8 Ti	WP321H	A312, TP321H	A240, 321[6]	A182, F321H	A276, 321[6]	
			18-8 Cb	WP347	A312, TP347	A240, 347	A182, F347	A276, 347	
			18-8 Cb	WP347H	A312, TP347H	A240, 347[6]	A182, F347H	A276, 347[6]	
			18-8 Cb	WP348	A312, TP348	A240, 348	A182, F348	A276, 348	
A420	ASA B16.9[2] ASA B16.11[3]	Ends machined for butt welds or socket welds	C-steel	WPL0	A333, 0[7]	A300, 1	A350, LF1		ASA G46.2—1964
			3½ Ni	WPL3	A333, 3	A300, 3	A350, LF3		
			5 Ni	WPL5	A333, 3				
			Cr–Cu–Ni–Al	WPL4	A333, 4	A300, 5	A350, LF4		

[1] H = hardness test, I = impact test.
[2] B16.9, American Standard for Steel Butt-welding Fittings, American Standards Association.
[3] B16.11, American Standard for Steel Socket-welding Fittings, American Standards Association.
[4] SP-43, Standard Practice for Light Weight Stainless Steel Butt-Welding Fittings, Manufacturers Standardization Society of the Valve and Fittings Industry.
[5] May be made from plate or forgings which conform in mechanical and chemical properties to Grade C of Specification A106.
[6] Grades so marked have requirements in fittings specifications slightly different from those referenced in raw-material specifications.
[7] Specification A333 shows this as Grade 1.

Table 14. Summary of Pertinent Requirements of ASTM Specification on Seamless Aluminum Pipe and Seamless and Welded Aluminum Welding Fittings

ASTM Spec.	Pipe sizes	Service	Production method	Heat treatment	Tests required[1]	Hydrostatic test pressure, psi $P = 2St/D$	Wall-thickness tolerance	OD tolerances (max)	Straightness
B241	1/8"–12" OD; Sch. 5 through 160	General	Extruded or drawn	Mill finish sufficient; solution heat treatment conforms to MIL-H-6088, Heat Treatment of Aluminum Alloys	T	Not req'd	Sch. 5 and 10, to 0.083" wall, ±0.012"; 0.109" wall and greater, ±12½%; Sch. 20 and greater, −12½%	See Table 15	See Table 16

ASTM Spec.	Grades	Temper	Production method	Pipe size, in.	Min tensile strength, psi	Cladding grades and thickness (% of wall thickness)	Finish	Recognition by other codes
B241	3003	H18		Under 1	27,000		Mill finish	ASME Boiler and Pressure Vessel Code SB-241
	3003	H112		1 and over	14,000			ASA H38.7—1964
	5083	0		All sizes	39,000			
	5083	H112		1 and over	39,000			
	5086	0		All sizes	35,000			
	5086	H112		1 and over	35,000			
	5154	H38		Under 1	45,000			
	5154	H112		1 and over	30,000			
	5254	H38		Under 1	45,000			
	5254	H112		1 and over	30,000			
	5454	0		1 and over	31,000			
	5454	H112		1 and over	31,000			
	5456	0		All sizes	41,000			
	5456	H112		All sizes	41,000			
	6061	T6		Under 1	42,000			
	6061			1 and over	38,000			
	6062	T6		Under 1	42,000			
	6062			1 and over	38,000			
	6063	T5		All sizes up through 0.500 in. wall thickness	22,000			
	6063			All sizes, 0.501 to 1.000 in. wall thickness	21,000			
	6063	T6		All sizes	30,000			
	6351	T6		All sizes up through 0.124 in. wall thickness	42,000			
	6351	T6		All sizes, 0.125 to 0.749 in. wall thickness	42,000			

ASTM Spec.	Pipe sizes	Service	Production method	Heat treatment	Tests required[1]	Hydrostatic test pressure, psi $P = 2St/D$	Wall-thickness tolerance	OD tolerances (max)	Straightness
B345	2"–20" OD; Sch. 5 through 160	Gas and oil transmission and distribution	Extruded or drawn	Mill finish sufficient; solution heat treatment conforms to MIL-H-6088, Heat Treatment of Aluminum Alloys	T	Not req'd	Sch. 5 and 10, to 0.083" wall, ±0.012"; 0.109" wall and greater, ±12½%; Sch. 20 and greater, −12½% on 14", and larger pipe subject to agreement	See Table 15	See Table 16

Table 14. (Continued)

ASTM Spec.	Grades	Temper	Pipe size in.	Min tensile strength, psi	Cladding grades and thickness (% of wall thickness)	Finish	Recognition by other codes
B345	1060	−0	All sizes	9,000	3003 (inside) 10%	Mill finish; ends beveled for welding to ASA B31.1[2] (threaded ends to ASA B2.1)[2]	ASA H38.13—1964
		−H112	1 and over	10,000	3003 (outside) 7%		
		−H14	....	12,000	6061 (inside) 10%		
	3003	−0	All sizes	14,000	6061 (outside) 7%		
		−H112	1 and over	14,500			
		−H14	All sizes	20,000			
	Alclad 3003	−H112	1 and over	13,500			
	3004	−0	All sizes	23,000			
			....	32,000			
	5050	−0	All sizes	18,000			
		−H34	....	25,000			
		−H38	Under 1	29,000			
	5052	−0	All sizes	25,000			
		−H34	....	34,000			
	5083	−H112	All sizes	40,000			
	5086	−0	All sizes	35,000			
		−H112	1 and over	35,000			
	5154	−0	All sizes	30,000			
		−H112	1 and over	30,000			
	5456	−0	All sizes	41,000			
		−H112	All sizes	41,000			
	6061	−T4	All sizes	26,000			
		−T6	All sizes	38,000			
		−T62	All sizes	35,000			
	Alclad 6061	−T6	All sizes	33,000			
	6062	−T4	All sizes	26,000			
		−T6	All sizes	38,000			
		−T62	All sizes	35,000			
	6063	−T42	All sizes up through 0.500 wall thickness	17,000			
			All sizes, 0.501 to 1.000 wall thickness	16,000			
		−T5	All sizes up through 0.500 wall thickness	22,000			
			All sizes, 0.501 to 1.000 wall thickness	21,000			
		−T6	All sizes	30,000			
		−T4	All sizes	32,000			
	6351	−T6	All sizes up through 0.124 wall thickness	42,000			
			All sizes, 0.125 to 0.749 wall thickness	42,000			

[1] T = tension test.
[2] American Standard Association, B31. 1—1955, American Standard Code for Pressure Piping, and B2.1—1960, American Standard Pipe Threads.

Table 14. (Continued)

ASTM Spec.	Pipe sizes	Service	Production method	Heat treatment	Tests required[1]	Hydrostatic test pressure, psi $P = 2St/D$	Wall-thickness tolerance	Grades, temper	Raw material (Specification and Grade)			
									Pipe size, in.		Min tensile strength, psi	Cladding grades and thickness (% of wall thickness)
									Pipe	Plate	Forgings	Bar
B361	No limits	General	Hammering, pressing, piercing, rolling, bending, fusion welding, etc.	Mill finish sufficient; solution heat treatment conforms to MIL-H-6088, Heat Treatment of Aluminum Alloys	T (on plate or base metal)	Not req'd but must be able to pass test at a value of S equal to 50% of the yield strength	Size, shapes, and dimensions to ASA B16.9[3] and ASA B16.11[3]	Grade[4,5]				
								WP1060[6]	B210 B234 B221	B209	. . . .	B211
								WP1100[6]	. . . .	B209	B247	B211 B221
								WP3003[6]	B210 B234 B221 B241	B209	B247	B221
								WP Alclad[6] 3003	B234 B221			
								WP5052[6]	B210 B221	B209	. . . .	B211
								WP5154	B221	B209	. . . .	B221
								WP6061	B210 B234 B221 B241	B209	B247	B211 B221
								WP6062	B210 B234 B221 B241	. . . .	. . . .	B221
								WP6063	B210 B221 B241	. . . .	. . . .	B221

[3] American Standards Association, B16.9, American Standard for Steel Butt Welding Fittings, and B16.11, American Standard for Steel Socket Welding Fittings.

[4] When fittings are of welded construction, the alloy designation shall be supplemented by the letter "W."

[5] These alloy designations were established in accordance with the American Standard Alloy and Temper Designation Systems for Wrought Aluminum (ASA No. H35.1), except for the letter symbols "WP."

[6] Fittings in nonheat-treatable alloys are available only in the F or H112 tempers, as covered by the applicable raw-material specifications.

Table 15. Variations in Outside Diameter Applicable to Aluminum Piping in ASTM Specifications B241 and B345

Pipe size, in.	Permissible variations in mean[1] outside diameter from nominal diameter, in.	Permissible variations in outside diameter at any point from nominal diameter
	Schedules 5 and 10	Schedules 20 and greater
Under 2	+0.015, −0.031	+0.015, −0.031 in.
2–4	+0.031, −0.031	±1%
4½–7	+0.062, −0.031	±1%
8–12	+0.093, −0.031	±1%
14–20	[2]	[2]

[1] Mean diameter is the average of two diameter measurements taken at right angles to each other at any point along the length.

[2] Permissible variations in outside diameter of 14-, 16-, 18-, and 20-in. pipe are subject to special agreement between the manufacturer and the purchaser.

Table 16. Permissible Variations from Straightness Applicable to Aluminum Piping in ASTM Specifications B241 and B345

Pipe size, in.	In each foot of length	In total length of piece
Under 6	0.010	0.010 × length in ft
6 to 12	0.020	0.020 × length in ft
14, 16, 18, and 20	[1]	[1]

[1] Permissible variations from straightness for 14-, 16-, 18-, and 20-in. pipe are subject to special agreement between the manufacturer and the purchaser.

Table 17. Summary of Pertinent Requirements of ASTM Specification on Seamless and Welded Aluminum Tubing

ASTM Spec.	Tube sizes	Service	Production method	Type weld	Filler metal used	Heat treatment	Tests required[1]	Pressure test	Nondestructive tests of weld
B210	No limits ($\frac{1}{8}$–12″ OD)	General service	Drawn	Seamless	. . . .	Mill finish sufficient; principles of solution heat treatment conform to MIL-H-6088, Heat Treatment of Aluminum Alloys	T	Not req'd	. . . .
B221	No limits ($\frac{1}{2}$–18″ OD)	General service	Extruded	Seamless	. . . .	Mill finish sufficient; principles of solution heat treatment conform to MIL-H-6088, Heat Treatment of Aluminum Alloys	T	Not req'd	. . . .
B234	No limits ($\frac{3}{8}$–2″ OD)	Condensers and heat exchangers	Drawn	Seamless	. . . .	Mill finish sufficient; principles of solution heat treatment conform to MIL-H-6088, Heat Treatment of Aluminum Alloys	T, E (20%)	Air pressure test at 250 psi each tube	. . . .
B307	No limits ($\frac{1}{8}$–2″ OD)	Coiled tubes for special purpose applications	Drawn	Seamless	. . . .	Mill finish sufficient; principles of solution heat treatment conform to MIL-H-6088, Heat Treatment of Aluminum Alloys	T, E (30–40%)	0.2% per shipment to be tested either at 60 psi air pressure while immersed or at 90 psi by loss of pressure (gage) method	. . . .
B313	No limits (1–7″ OD)	General service	Seam welded from formed sheet	Continuously seam welded	Not used	Mill finish sufficient; principles of solution heat treatment conform to MIL-H-6088, Heat Treatment of Aluminum Alloys	T (sheet)	Air or hydrostatic pressure test required at 60 to 90 psi	None req'd
B404	Up to 2″ OD	Condenser and heat-exchanger service	Cold-formed tubes with integral fins	Seamless	. . . .	. . . .	T, E (20%)	Air pressure test at 250 psi min	. . . .

Table 17. (Continued)

Wall-thickness tolerance	OD tolerances	Out-of-round tolerance	Straightness tolerance	Grades	Cladding grades and thickness (% of wall thickness)	Finish	Recognition by other codes	ASTM Spec.
Table 19	Table 18 Similar tolerances specified for oval, elliptical and streamlined tubes	Varies with OD from 0.020 to 0.500″ per ft	....	1060 1100 2024 3003 3004 5050 5052 5086 5154 5254 5652 6061 6062 6063 6262 Various temps to ASA H35.1[2]	3003 (10%)	Mill finish	ASA H38.8—1964	B210
Varies with specified wall thickness. OD, and alloy grade approx ±10% (±0.06″ max, ±0.01″ min	Varies from 0.010″ to 0.128″ depending on OD and alloy grade	....	....	1060 1100 2014 2219 2024 3003 3004 5052 5083 5086 5154 5254 5454 5456 5652 6061 6062 6063 6262 6351 7075 7079 7178 Various temps to ASA H35.1[2]	3003 (10%)	Mill finish	ASA H38.5—1964	B221
Varies with specified wall thickness from 0.004 to 0.036″	Varies with OD and alloy grade from 0.002 to 0.005″	....	0.010″ × length in ft per tube	1060 3003 5052 5454 6061 6062 Various temps to ASA H35.1[2]	3003 (inside) 10% 3003 (outside) 7%	Mill finish	ASME Boiler and Pressure Vessel Code SA SB-234 (except Grade 6062) ASA H38.6—1964	B234
Varies with specified wall thickness ±10% with 0.003′ min	Varies with OD from 0.003 to 0.005″	....	....	1100 3003 3004 Only 0 temp 5005 5050 5052	3003 (inside) 10% 3003 (outside) 7% 5050 (outside) 7%	Mill finish (Special inside cleanliness may be specified)	ASA H38.9—1964	B307
Varies with specified wall thickness from 0.005 to 0.007″	Varies with OD from 0.012 to 0.035″	Varies with OD from 0.025 to 0.080″	3/8″ max in 10′	1100 3003 3004 5050 5052 5086 5154 6061 Various temps to ASA H35.1[2]	3004 (5%)	Mill finish	ASA H38.11—1964	B313
No undertolerance	Varies with OD and alloy grade from 0.002 to 0.006″	....	....	1060 3003 5052 5454 6061 6062 Various temps to ASA H35.1[2]	3003 (10%)	As fabricated		B404

[1] T = tension test, E = tube end expansion test in per cent.
[2] American Standards Association, H35.1, American Standard Alloy and Temper Designation Systems for Wrought Aluminum.

Table 18. Variations in Outside Diameter Applicable to Aluminum Tubing in ASTM Specification B210

Nominal diameter, in.	Permissible variations in diameter, plus or minus, in.		
	Mean diameter[1] or Pi-tape measurement-alloys, 1060, 1100, 2024, 3003. Alclad 3003, 3004, 5050, 5052, 5086, 5154, 5254, 5652, 6061, 6062, 6063, and 6262	Individual measurement of diameter (out-of-roundness) except soft or thin-wall tubes[2]	
		Alloys 1060, 1100, 3003. Alclad 3003, 3004, 5050, 5052, 5086, 5154, 5254, and 5652	Alloys 2024, 6061, 6062, 6063, and 6262
⅛–½, incl.	0.003	0.003	0.006
Over ½–1, incl.	0.004	0.004	0.008
Over 1–2, incl.	0.005	0.005	0.010
Over 2–3, incl.	0.006	0.006	0.012
Over 3–5, incl.	0.008	0.008	0.016
Over 5–6, incl.	0.010	0.010	0.020
Over 6–8, incl.	0.015	0.015	0.030
Over 8–10, incl.	0.020	0.020	0.040
Over 10–12, incl.	0.025	0.025	0.050

[1] Mean diameter is the average of any two measurements of diameter taken at right angles to each other at any point along the length of the tube.

[2] Thin-wall tubes, that is, tubes having a wall thickness less than 2.5 per cent of the diameter or less than 0.020 in., and tubes in the soft temper shall be commercially round. The deviations of individual measurements from the nominal will vary with the alloy and the ratio of wall thickness to diameter.

Table 19. Variations in Wall Thickness Applicable to Aluminum Tubing in ASTM Specification B210

Nominal wall thickness, in.[1]	Permissible variations in wall thickness, plus or minus, in.			
	Mean[2]	Permissible variation in individual measurement from specified wall thickness		
	Alloys 1060, 1100, 2024, 3003. Alclad 3003, 3004, 5050, 5052, 5086, 5154, 5254, 5652, 6061, 6062, 6063, and 6262	Alloys 1060, 1100, 3003. Alclad 3003, 3004, 5050, 5052, 5086, 5154, 5254 and 5652. Round tube only	Alloys 2024, 6061, 6062, 6063, and 6262. Round tube only	All alloys other than round tube
0.010–0.035	0.002	0.002		
0.036–0.049	0.003	0.003		
0.050–0.083	0.004	0.004		
0.084–0.120	0.005	0.006	10 % of nominal wall thickness, but not less than ±0.003″	
0.121–0.203	0.006	0.008		
0.204–0.300	0.008	0.012		
0.301–0.375	0.015	0.020		
0.376–0.500	0.020	0.030		

[1] Intermediate wall thicknesses shall be rounded off to the third decimal place in accordance with Recommended Practice for Designating Significant Places in Specified Limiting Values, ASTM E29.

[2] Mean wall thickness of round tube is the average of the two measurements of wall thickness taken at opposite sides of any diameter. The mean wall thickness of other-than-round tube is the average of two measurements taken opposite each other at the approximate center line of the tube and perpendicular to the longitudinal axis of the cross section.

Table 20. Summary of Pertinent Requirements of ASTM Specification on Seamless Copper and Brass Pipe

ASTM Spec.	Pipe sizes	Service	Types[1]	Production methods	Temper	Tests required[2]	Hydrostatic test pressure, psi $P = 2St/(D − 0.8t)$
B42	No limits (1/8–12″ OD) standard pipe sizes, regular and extra strong	General service, plumbing, boiler feed, etc.	DHP, DLP, OF	Drawn	Drawn temper (for bending specify special temper)	T E (25%) B (on pipe for bending) M(DLP, OF)	Req'd on 0.2% of shipment S = 6,000 psi; 1,000 psi max
B43	No limits (1/8–12″ OD) standard pipe sizes, regular and extra strong	General service, plumbing, boiler feed, etc.	85 Cu–15 Zn	Drawn	Annealed	T E (25%) B (on pipe for bending) MN	Req'd on 0.2% of shipment S = 7,000 psi; 1,000 psi max
B302	1/4–12″ OD; 0.065–0.313″ wall (see Table 22)	For brazed piping systems	DHP DLP	Drawn	Drawn temper	T H M (except when DHP contains 0.015–0.040%P)	Req'd on 0.2% of shipment S = 6,000 psi; 1,000 psi max
B251	No limits	General requirements[3]	Copper and brass pipe	Hot or cold drawn (finished by cold working and annealing)	....	As stated in B42 or B43	....

[1] Types: DLP = phosphorized copper, low residual phosphorus; DHP = phosphorized copper, high residual phosphorus; OF = oxygen-free copper without residual metallic deoxidation, ASTM B224, Standard Classifications of Coppers.

[2] T = tension test, B = bend test, E = expansion test, H = hardness test, MN = mercurous nitrate test, M = mercurous nitrate immersion test (ASTM B154, Mercurous Nitrate Test for Copper and Copper Alloys), M = microscopic examination.

[3] ASTM B251 is a general requirements specification applicable to several pipe and tube specifications—B42, B43, B68, B75, and B135.

Table 20. (Continued)

ASTM Spec.	Wall-thickness tolerance	OD tolerances (max)	Out-of-round tolerance	Straightness (max permissible curvature)	Special requirements	Recognition by other codes
B42	Table 21 (for extra-strong sizes, see ASTM B251)	Table 21 (for extra-strong sizes, see ASTM B251)	ASTM B251	Lengths 3–6′, $\frac{3}{16}''$; 6–8′, $\frac{5}{16}''$; 8–10′, $\frac{1}{2}''$; Over 10′, $\frac{1}{2}''$ in any 10′ total length	Departure from squareness shall not exceed 0.010″ in sizes to $\frac{5}{8}''$ OD and 0.016″ per in. of diam in sizes greater than $\frac{5}{8}''$ OD	ASME Boiler and Pressure Vessel Code SB-42 ASA H26.1—1963
B43	Table 21 (for extra-strong sizes, see ASTM B251)	Table 21 (for extra strong sizes, see ASTM B251)	ASTM B251	Lengths 3–6′, $\frac{3}{16}''$; 6–8′, $\frac{5}{16}''$; 8–10′, $\frac{1}{2}''$; Over 10′, $\frac{1}{2}''$ in any 10′ total length	Departure from squareness shall not exceed 0.010″ in sizes to $\frac{5}{8}''$ OD and 0.016″ per in. of diam in sizes greater than $\frac{5}{8}''$ OD	ASME Boiler and Pressure Vessel Code SB-43 ASA H27.1—1963
B302	Table 22	Table 22	Table 23	....	Departure from squareness shall not exceed 0.010′ in sizes to $\frac{5}{8}''$ OD and 0.016″ per in. of diam in sizes greater than $\frac{5}{8}''$ OD	ASA H26.2—1963
B251	Table 21	Table 21	See ASTM B251	Lengths 3–6′, $\frac{3}{16}''$; 6–8′, $\frac{5}{16}''$; 8–10′, $\frac{1}{2}''$; Over 10′, $\frac{1}{2}''$ in any 10′ total length	Departure from squareness shall not exceed 0.010″ in sizes to $\frac{5}{8}''$ OD and 0.016″ per in. of diam in sizes greater than $\frac{5}{8}''$ OD	

Table 2I. Dimensions of Copper and Red Brass Pipe (ASTM B42 and B43)

Standard pipe size, in.	Nominal dimensions, in.			Cross-sectional area of bore, sq in.
	Outside diameter	Inside diameter	Wall thickness	
Regular				
⅛	0.405	0.281	0.062	0.062
¼	0.540	0.376	0.082	0.110
⅜	0.675	0.495	0.090	0.192
½	0.840	0.626	0.107	0.307
¾	1.050	0.822	0.114	0.531
1	1.315	1.063	0.126	0.887
1¼	1.660	1.368	0.146	1.47
1½	1.900	1.600	0.150	2.01
2	2.375	2.063	0.156	3.34
2½	2.875	2.501	0.187	4.91
3	3.500	3.062	0.219	7.37
3½	4.000	3.500	0.250	9.62
4	4.500	4.000	0.250	12.6
5	5.562	5.062	0.250	20.1
6	6.625	6.125	0.250	29.5
8	8.625	8.001	0.312	50.3
10	10.750	10.020	0.365	78.8
12	12.750	12.000	0.375	113
Extra strong				
⅛	0.405	0.205	0.100	0.003
¼	0.540	0.294	0.123	0.068
⅜	0.675	0.421	0.127	0.139
½	0.840	0.542	0.149	0.231
¾	1.050	0.736	0.157	0.425
1	1.315	0.951	0.182	0.710
1¼	1.660	1.272	0.194	1.27
1½	1.900	1.494	0.203	1.75
2	2.375	1.933	0.221	2.94
2½	2.875	2.315	0.280	4.21
3	3.500	2.892	0.304	6.57
3½	4.000	3.358	0.321	8.86
4	4.500	3.818	0.341	11.5
5	5.562	4.812	0.375	18.2
6	6.625	5.751	0.437	26.0
8	8.625	7.625	0.500	45.7
10	10.750	9.750	0.500	74.7

Table 22. Tolerances in Diameter and Wall Thickness for Copper Threadless Pipe (ASTM B302)

Standard pipe size, in.	Nominal dimensions, in.			Cross-sectional area of bore, sq in.	Tolerance, in.	
	Outside diameter	Inside diameter	Wall thickness		Average outside diameter,[1] all minus	Wall thickness, plus and minus
¼	0.540	0.410	0.065	0.132	0.004	0.0035
⅜	0.675	0.545	0.065	0.233	0.004	0.004
½	0.840	0.710	0.065	0.396	0.005	0.004
¾	1.050	0.920	0.065	0.665	0.005	0.004
1	1.315	1.185	0.065	1.10	0.005	0.004
1¼	1.660	1.530	0.065	1.84	0.006	0.004
1½	1.900	1.770	0.065	2.46	0.006	0.004
2	2.375	2.245	0.065	3.96	0.007	0.006
2½	2.875	2.745	0.065	5.92	0.007	0.006
3	3.500	3.334	0.083	8.73	0.008	0.007
3½	4.000	3.810	0.095	11.4	0.008	0.007
4	4.500	4.286	0.107	14.4	0.010	0.009
5	5.562	5.298	0.132	22.0	0.012	0.010
6	6.625	6.309	0.158	31.3	0.014	0.010
8	8.625	8.215	0.205	53.0	0.018	0.014
10	10.750	10.238	0.256	82.3	0.018	0.016
12	12.750	12.124	0.313	115	0.018	0.020

[1] The average outside diameter of a tube is the average of the maximum and minimum outside diameters as determined at any one cross section of the tube.

Table 23. Roundness Tolerances for Copper Pipe (ASTM B302)

t/D, ratio of nominal wall thickness to nominal outside diameter	Roundness tolerances,[1] % of nominal outside diameter (expressed to the nearest 0.001 in.)
0.01–0.03, incl.	1.5
Over 0.03–0.05, incl.	1.0
Over 0.05–0.10, incl.	0.8
Over 0.10	0.7

[1] The deviation from roundness is measured as the difference between major and minor outside diameters as determined at any one cross section of the tube.

Table 24. Summary of Pertinent Requirements of ASTM Specification on Seamless Copper and Brass Tubing

ASTM Spec.	Tube sizes	Service	Types[1]	Production methods	Temper	Tests required[2]	Hydrotest per tube $P = 2St/(D - 0.8t)$	Wall-thickness tolerance
B68	No limits (usually to 2″ OD)	Refrigeration, oil and gasoline lines, etc.	DLP DHP OF	Drawn	Annealed	T E (30–40%) M (DLP, OF) Em	Not req'd	Table 25
B75	No limits (usually to 2″ OD)	General engineering purposes	DLP DHP DPA OF	Drawn	Light-drawn, drawn, hard-drawn, soft-annealed, light-annealed	T E (30–40%) M (grades PLP, OF) Em	Req'd on 0.2% of shipment S = 6,000 psi (coiled tubing may be tested at 60 psi min air pressure)	Table 25
B88	¼–12″ OD	Water	K L M	Drawn	Types K and L are annealed after coiling; Type M temper is as drawn	H E (30–40%) M (except DHP grades)	Req'd on 0.2% of shipment S = 6,000 psi (but not over 1,000 psi) (coiled tubing may be tested at 60 psi min air pressure)	Varies with OD and Types K, L, M from 0.0035 to 0.020″
B111	No limits (usually to 2″ OD)	Condenser tubing	DHP DLP DPA OF Muntz metal Admiralty red brass Aluminum brass Aluminum bronze 60–40 Cu–Ni 70–30 Cu–Ni 80–20 Cu–Ni 90–10 Cu–Ni 95–5 Cu–Ni	Drawn	Annealed, all types; light drawn–Cu, DPA, 90–10, 95–5; drawn and stress relieved 70–30, 60–40; hand drawn and annealed, Cu	E (15–30%) M (annealed temper tubes) F (annealed) MN (except Cu, DPA, and CuNi tubes)	Req'd each tube S = 7,000 psi (but not over 1,000 psi)	Under tolerance not permitted

[1] Types: *DLP* = phosphorized copper, low residual phosphorus; *PHP* = phosphorized copper, high residual phosphorus; *OF* = oxygen-free copper without residual metallic deoxidants; *DPA* = phosphorized arsenical copper, high residual phosphorus.

[2] *T* = tension test, *E* = expansion test (per cent of tubes), *Em* = embrittlement test (tests actually are not required; material, however, should be capable of meeting test), *H* = hardness test, *M* = microscopic examination, *Fl* = flattening test, *MN* = mercurous nitrate immersion test (ASTM B154, Mercurous Nitrate Test for Copper and Copper Alloys).

8–41

Table 24. (*Continued*)

ASTM Spec.	OD tolerances	Out-of-round tolerance	Straightness	Finish	Special requirement	Recognition by other codes
B68	Table 26	....	....	Bright annealed after final draw or coiling	Grain size: 0.015–0.040 mm (light anneal) 0.040 mm min (soft anneal) Departure from end squareness shall not exceed 0.010″ for sizes up to ⅝″ OD, and shall not exceed 0.016″ per in. of diam for sizes over ⅝″ OD	ASA H23.2–1963
B75	Table 26	....	Table 27	....	Grain size of annealed tubing: 0.040 mm min (soft anneal) 0.040 mm max (light anneal) (Electric resistivity tests when used as electric conductor)	ASA H23.3–1963
B88	Varies with OD and temper from 0.001 to 0.008″	1.5% for t/D between 0.01 and 0.03 1.0% for t/D between 0.03 and 0.05 0.8% for t/D between 0.05 and 0.10 (where t and D represent wall thickness and outside diameter, respectively)	....	....	Where agreement on hardness tests cannot be reached, tension tests or grain-size tests shall be basis for acceptance	ASA H23.1–1963
B111	Sizes to 0.500″ OD, ±0.002″ 0.501″ to 0.740″ OD, ±0.0025″ 0.741″ to 1.00″ OD, ±0.003″ 1.001″ to 1.250″ OD, ±0.0035″ 1.251″ to 2.000″ OD, ±0.004″	....	....	....	Departure from end squareness shall not exceed 0.010 for sizes up to ⅝″ OD and shall not exceed 0.016″ per in. of diam for sizes over ⅝″ OD	ASME Boiler and Pressure Vessel Code SB-111

Table 24. (Continued)

ASTM Spec.	Tube sizes	Service	Types[1]	Production methods	Temper	Tests required[2]	Hydrotest per tube $P = 2St/(D - 0.8t)$	Wall-thickness tolerance
B135	No limits (usually to 4″ OD)	General service	(1) 85–15 Cu-Zn (2) 70–30 Cu-Zn (3) 66–33.5 Cu-Zn-Pb (4) 66–32.4 Cu-Zn-Pb (5) 60–40 Cu-Zn (6) 60–39 Cu-Zn-Pb (7) 90–10 Cu-Zn (8) 63–37 Cu-Zn (9) 65–35 Cu-Zn	Drawn	Light-drawn, drawn and hard drawn, light anneal, soft anneal	T (walls to 0.020″ and ID less than 5/16″) H E (to 3/4″ OD 20%, over 3/4″ OD 15%) M (annealed temper tubes) MN (only when required)	Not req'd	Table 25
B280	No limits (usually 1/8–1-3/8″ OD)	Refrigeration field service	DHP DLP OF (DHP normally supplied)	Drawn	Soft annealed	T E (40%) M Em	Not req'd	Varies with OD from ±0.002 to ±0.004″
B306	No limits (usually 1¼–6″)	Sanitary drainage and other non-pressure applications	DHP DLP	Drawn	Drawn	H	Req'd, either air-pressure test at 60 psi min or hydrostatic test where $S = 6,000$ psi	Varies with OD from ±0.003 to ±0.008″
B359	No limits (usually to 1″ OD)	Condenser and heat-exchanger tubes with integral fins	Cu DPA, Admiralty red brass Aluminum brass Aluminum bronze 60–40 Cu-Ni 70–30 Cu-Ni 80–20 Cu-Ni 90–10 Cu-Ni 95–5 Cu-Ni	Drawn	Annealed or as finned	T M F (20–30%) Fl MN	Req'd; air-pressure test at 250 psi	Under tolerance not permitted
B360	No limits (usually 0.072–0.240″ OD)	Capillary tube for restrictor applications	DPH	Cold drawn to size	Hard drawn	T Em Air-flow test	Not req'd	....

Table 24. (Continued)

ASTM Spec.	OD tolerances	Out-of-round tolerance	Straightness	Finish	Special requirement	Recognition by other codes
B135	Table 26	....	Table 27	....	Where agreement on hardness tests cannot be reached, tension tests shall be basis for acceptance	....
B280	Varies with OD from ±0.003 to ±0.0045"	....	....	Inside to be cleaned	Tubes shall be dehydrated and sealed at both ends. Minimum grain size 0.035 mm	ASA H23.5-1963
B306	Varies with OD from ±0.0015 to ±0.002"	1½% max	....	....	Departure from end squareness shall not exceed 0.016" per in. of diam.	ASA H23.6-1963
B359	Sizes to 0.500" OD, ±0.002"; 0.501-0.740" OD, ±0.0025"; 0.741-1.000" OD, ±0.003"	....	....	....	Departure from end squareness shall not exceed 0.010" for sizes up to $\frac{5}{8}$" OD and shall not exceed 0.016" per in. of diam for sizes over $\frac{5}{8}$" OD	ASME Boiler and Pressure Vessel Code SB-359
B360	OD ± 0.002" ID ± 0.001"	....	....	Edges at ends are machine deburred; bore is washed with solvent	....	ASA H23.8-1963

Table 25. Wall-thickness Tolerances for Copper and Copper-alloy Tubing (ASTM B68, B75, and B135)

NOTE: *Maximum Deviation at Any Point:* The following tolerances are plus and minus; if tolerances all plus or all minus are desired, double the values given.

Wall thickness, in.	Outside diameter, in.[1]							
	1/32 to 1/8, incl.	Over 1/8 to 5/8, incl.	Over 5/8 to 1, incl.	Over 1 to 2, incl.	Over 2 to 4, incl.	Over 4 to 7, incl.	Over 7 to 10, incl.
Under 0.018	0.002	0.001	0.0015	0.002				
0.018–0.025 incl.	0.003	0.002	0.002	0.0025				
0.025–0.035 incl.	0.003	0.0025	0.0025	0.003	0.004			
0.035–0.058 incl.	0.003	0.003	0.0035	0.0035	0.005	0.007		
0.058–0.083 incl.	....	0.0035	0.004	0.004	0.006	0.008	0.010	
0.083–0.120 incl.	....	0.004	0.005	0.005	0.007	0.009	0.011	
0.120–0.165 incl.	....	0.005	0.006	0.006	0.008	0.010	0.012	
0.165–0.220 incl.	....	0.007	0.0075	0.008	0.010	0.012	0.014	
0.220–0.284 incl.	....	....	0.009	0.010	0.012	0.014	0.016	
0.284–0.380 incl.	....	....	0.011	0.012	0.014	0.016	0.018	
0.380 and over	....	....	....	5%	5%	6%	6%	

[1] When round tube is ordered by outside and inside diameters, the maximum plus and minus deviation of the wall thickness from the nominal at any point shall not exceed the values given in this table by more than 50 per cent.

Table 26. Average Diameter Tolerances for Copper and Copper-alloy Tubing (ASTM B68, B75, and B135)

Specified diameter, in.	Diameter to which tolerance applies	Tolerance, plus and minus, in.
Up to 1/8 incl.	Inside or outside	0.002
Over 1/8–5/8, incl.	Inside or outside	0.002
Over 5/8–1, incl.	Inside or outside	0.0025
Over 1–2, incl.	Inside or outside	0.003
Over 2–3, incl.	Inside or outside	0.004
Over 3–4, incl.	Inside or outside	0.005
Over 4–5, incl.	Inside or outside	0.006
Over 5–6, incl.	Inside or outside	0.007
Over 6–8, incl.	Inside or outside	0.008
Over 8–10, incl.	Inside or outside	0.010

Table 27. Straightness Tolerances for Copper and Copper-alloy Tubing* (ASTM B75 and B135)

NOTE: Applies to round tube in any drawn temper from 1/4 to 3 1/2 in. in outside diameter, inclusive.

Length, ft†	Maximum curvature (depth of arc), in.
Over 3–6, incl.	3/16
Over 6–8, incl.	5/16
Over 8–10, incl.	1/2

* Not applicable to pipe, redraw tube, extruded tube, or any annealed tube.
† For lengths greater than 10 ft, the maximum curvature shall not exceed 1/2 in. in any 10-ft portion of the total length.

Table 28. Summary of Pertinent Requirements of ASTM Specification on Seamless Nickel Pipe and Tubing and Seamless and Welded Wrought Nickel Fittings

ASTM Spec.	Pipe or tube	Pipe and tube sizes and thicknesses	Service	Grades	Condition	Temper	Type weld	Heat treatment	Tests required[1]
B161	Pipes and tubes (seamless)	To 8⅝″ OD, to ½″ wall	General corrosive and mechanical service	Nickel Low-C nickel	Cold-drawn, hot-finished	Annealed, stress relieved	....	....	T
B163	Tubes (seamless)	To 3″ OD, to 0.148″ wall	Condensers and heat exchangers	Nickel Low-C nickel 70 Ni–30 Cu 75 Ni–15 Cr–10 Fe	Cold-drawn	Ni, NiCu: either annealed or stress relieved NiCrFe: annealed	....	....	T H (annealed ends) Fl
B165	Pipes and tubes (seamless)	To 8⅝″ OD, to ½″ wall	General corrosive and mechanical service	70 Ni–30 Cu	Hot-finished (extruded)	Annealed or stress relieved	....	....	T
B167	Pipes and tubes (seamless)	To 9¼″ OD, to 1.0″ wall	General corrosive and heat-resisting service	75 Ni–15 Cr–10 Fe	Cold-drawn, hot-finished	Annealed	....	....	T
B366	Pipe welding fittings	No limits	General corrosive and other service	WPN Ni WPNL low C–Ni WPNC Ni–Cu WPNCl NiCr–Pc WPHB Ni–Mo WPHC Ni–Mo–Cr	....	....	Electric fusion welded	Subject to agreement	T (on raw material)

Table 28. (Continued)

ASTM Spec.	Hydrostatic test pressure, psi $P = 2St/D$	Radiography of seam weld	Wall-thickness tolerance	OD tolerances (max)	Finish	Special requirements	Recognition by other codes
B161	Req'd for OD 1/8″ and over and walls of 0.015″ and over at internal pressure of 1,000 psi provided that S does not exceed specific values between 6,700 and 16,200 psi	....	Varies with OD either ±10 or ±12½%	Varies with OD from +0.005 to +0.025″ and −0 to −0.025″	Ends plain and deburred	Special tolerance requirements for small-diameter (converter) sizes of 1¼″ OD and smaller	ASME Boiler and Pressure Vessel Code SB-161 ASA H34.1-1959
B163	Req'd for OD 1/8″ and larger and walls of over 0.015″ at internal pressure of 1,000 psi, provided that S does not exceed specific values between 8,000 and 22,500 psi	....	Varies with OD and material either ±10 to 15% of avg wall or +20 to 30%(−0) of min wall	Varies with OD and material from +0.005 to 0.010″ and −0 to −0.010″	....	Available as min or avg wall tubing. Special tolerance requirements for small-diameter (converter) sizes of 1¼″ OD and smaller	ASME Boiler and Pressure Vessel Code SB-163
B165	Req'd for OD 1/8″ and larger and walls of over 0.015″ at internal pressure of 1,000 psi, provided that S does not exceed specific values between 17,500 and 21,200 psi	....	Varies with OD either ±10 or ±12%	Varies with OD from +0.005 to 0.025″ and −0 to 0.025″	Ends plain and deburred	Special tolerance requirements for small-diameter (converter) sizes of 1¼″ OD and smaller	ASME Boiler and Pressure Vessel Code SB-165 ASA 34.2-1959
B167	Req'd for OD 1/8″ and larger and wall of over 0.015″ at internal pressure of 1,000 psi, provided that S does not exceed specific values between 16,700 and 20,000 psi	....	Varies with OD and temper between ±10 and ±15%	Varies with OD and temper for cold-drawn from ±0.005 to 0.025″ and for hot-finished from ±0.031 to 0.047″	Ends plain and deburred	Special tolerance requirements for small-diameter (converter) sizes of 1¼″ OD and smaller	ASME Boiler and Pressure Vessel Code SB-167 ASA H34.3-1962
B366	Fittings must be capable of meeting hydrostatic test pressures	Req'd except on lightweight fittings to MSS SP-43	2	2			

Raw material

Product and ASTM Designation

Grade[3]	Pipe or tube	Plate, sheet, or strip	Bar, forgings, and forging stock
WPN	B161	B162	B160
WPNL	B161	B162	B160
WPNC	B165	B127	B164
WPNCI	B167	B168	B166
WPHB	4	B333	B335
WPHC		B334	B336

[1] T = tension Test, H = hardness Test, Fl = flare Test.

[2] For fittings covered by the American Standard for Steel Butt-Welding Fittings (ASA No. B16.9-1958) or the Standard Practice for Light Weight Stainless Steel Butt-Welding Fittings (SP 43), or for fittings to be used with pipe ordered to these standards, the fittings shall be as specified in these standards. For fittings covered by the American Standard for Nickel and Nickel-Base Alloys (ASA H34.1-1959, H34.2-1959, and H34.3-1959), the sizes, shapes and dimensions of the fittings shall be as specified in these standards. Welded pipe in these alloys shall be fabricated from sheet or plate conforming to the chemical and mechanical requirements of Specifications B333 or B334.

[3] When fittings are of welded construction, the symbol shown shall be suffixed by the letter "W."

[4] No ASTM specifications available.

Table 29. Summary of Pertinent Requirements of ASTM Specification on Seamless and Welded Titanium Pipe, Tube, and Fittings

ASTM Spec.	Pipe or tube	Tube sizes	Service	Grades and min tensile strength, psi	Production methods	Type weld	Temper	Heat treatment	Tests required[1]
B337	Pipe (seamless or welded)	No limits (usually to 12″ OD and Schedules 5S to 80S)	General corrosive and elevated-temperature service	Unalloyed titanium 1 40,000 2 50,000 3 60,000 4 80,000	Extrusion, centrifugal, casting, machining of bar stock, powder compaction	Automatic electric fusion weld *without* filler metal (filler metal only when specifically agreed)	Annealed, hot-finished, cold-finished	....	*T* *B* *F*
B338	Tube (seamless or welded)	No limits (usually to 8″ OD)	General corrosive and elevated-temperature service	Unalloyed titanium 1 40,000 2 50,000 3 60,000 4 80,000	Extrusion, centrifugal casting, machining of bar stock, powder compaction	Automatic electric fusion weld *without* filler metal (filler metal only when specifically agreed)	Annealed, hot-finished, cold-finished	....	*T* *F* *Fl* (17–22 %) (flaring permitted if less than 3″ OD and less than 0.134″) wall
B363	Pipe welding fittings	No limits	....	....	Produced from billets, bars, plate, seamless or welded pipe. Forming may involve pressing, piercing, hammering, bending, rolling or fusion welding	Electric fusion weld with or without filler metal	....	Stress relief at 1200 F for ½ hr per in. of thickness	....

Table 29. (Continued)

ASTM Spec.	Hydrostatic test pressure, psi $P = 2St/D$	Radiography of weld	Wall-thickness tolerance	OD tolerances	Out-of-round tolerance	Special requirement	Recognition by other codes
B337	Req'd (S = 0.5 YS min) 2,500 max to 3", 2,800 max over 3"	Not req'd	Not more than 12½% under nominal	Table 4			
B338	Req'd. (S = 0.5 YS min) 2,500 max to 3", 2,800 max over 3"	Not req'd	Table 30	Table 30	Table 30	Available to OD or ID or to min or avg wall	
B363	Not req'd	············	2	2			

Raw materials

Grade[3]	Product and ASTM Designation			
	Pipe	Tube	Plate	Bar and billet
WPT1	B337 Grade 1	B338 Grade 1	B265 Grade 1	B348 Grade 1
WPT2	B337 Grade 2	B338 Grade 2	B265 Grade 2	B348 Grade 2
WPT3	B337 Grade 3	B338 Grade 3	B265 Grade 3	B348 Grade 3

1 T = tension test, B = bend test around 90-deg mandrel, F = flattening test, Fl = flaring test (per cent of tubes).
2 For fittings covered by the American Standard Steel Butt-welding Fittings (ASA B16.9-1958) or the Standard Practice for Light Weight Stainless Steel Butt-welding Fittings (SP-43) or for fittings to be used with pipe ordered to the American Standard Stainless Steel Pipe (ASA B36.19-1957), the sizes, shapes, and dimensions of the fittings shall be as specified in those standards.
3 When fittings are of welded construction, the symbol shown shall be supplemented by the letter "W."

Table 30. Variations in Dimensions of Titanium Tubing (ASTM B338)

Size, outside diameter, in.	Permissible variations in inside or outside diameter, ± in.	Permissible variations in average[1] wall thickness,[2] ± %	Ovality[3] shall be double permissible variation in outside diameter when wall thickness is
Up to ½, excl.	0.005	15	Less than 0.049 in.
½–1, excl.	0.005	10	Less than 0.065 in.
1 –1½, excl.	0.007	10	Less than 0.072 in.
1½–3½, excl.	0.010	10	Less than 0.095 in.
3½–5½, excl.	0.015	10	Less than 0.150 in.
5½–8 excl.	0.030	10	

[1] When minimum wall tubes are ordered, tolerances are all plus and shall be double the values shown.

[2] When tubes as ordered require wall thicknesses of ¾ in. or over or an inside diameter 60 per cent or less of the outside diameter, a wider variation in wall thickness is required. On such sizes, a variation in wall thickness of 12.5 per cent over or under will be permitted. For tubes less than ¼ in. in inside diameter which cannot be successfully drawn over a mandrel, the wall thickness may vary ±15 per cent from that specified.

[3] Ovality for tubes with wall thicknesses greater than those shown above shall not exceed permissible variations in inside or outside diameter.

Table 31. Examples of Supplementary Test Requirements Listed in Some ASTM Specifications

Test	Examples of wording on Specifications				Special tests
	A335 (seamless pipe)	A155 (fusion-welded pipe)	A216 (casting)	Other specifications	
Check analysis	Check analysis may be made on any length of pipe. Individual lengths failing to conform to the chemical requirements specified in Sec. 6 shall be rejected	Check analysis may be made on any length of pipe. Individual lengths failing to conform to the applicable requirements listed in Table I shall be rejected	....	....	(A430) Hot Ductility Tests A high-temperature ductility test has been shown to have a significant bearing on the weldability of these steels, in heavy-wall pipe sections, as determined by high-temperature pressure service performance. Upon mutual agreement between the purchaser and the manufacturer, such tests, on an agreed procedure and evaluation, may be performed on each heat of finished pipe produced. These tests are for information purposes only
Transverse tension tests	Transverse tension tests may be made on specimens from both ends of each length of pipe. This requirement applies only to pipe 8 in. and over in nominal diameter. If the specimen from either end of any length fails to meet the mechanical properties specified in Sec. 9, that length shall be rejected	....	....	(A430) Transverse tension tests may be made on specimens from both ends of each length of pipe. This requirement applies only to pipe 8 in. and over in nominal diameter. If the specimen from either end of any length fails to conform to the mechanical properties specified in Sec. 11, that length shall be rejected	
Tension and bend tests	....	Tension tests [Sec. 8(b)] and bend tests (Sec. 9) shall be made on specimens to represent each length of pipe. Failure of any test specimen to meet the requirements shall be cause for the rejection of the pipe length represented	....		

Table 31. (Continued)

Test	Examples of wording on Specifications			Other specifications	Special tests
	A335 (seamless pipe)	A155 (fusion-welded pipe)	A216 (casting)		
Bend tests	....	....	If specified by the purchaser, test specimens as indicated in Paragraph (a) shall be subjected to the bend test described in the following Paragraphs (a), (b), and (c).	....	(A409) Corrosion Requirements (a) *Boiling Nitric Acid Test*—Except for Grade TP 321, coupons representing finished pipe made of nonmolybdenum-less molybdenum) shall meet the requirements of the boiling nitric acid test conducted according to the Recommended Practice for Boiling Nitric Acid Test for Corrosion-resisting Steels (ASTM A262). The condition of the test specimens and the corrosion rates are as follows: Types 347 and 348 shall be tested in the sensitized condition (heated for 1 hr at 1250 F) and the rate of penetration shall not exceed 0.0020 ipm. All other nonmolybdenum-bearing types, except for grade TP 321, shown in Table I, shall be tested in the annealed and unsensitized condition and the rate of penetration shall not exceed 0.0015 ipm
			(a) *Test Specimens*—Test specimens shall be prepared in accordance with the Methods and Definitions for Mechanical Testing of Steel Products (ASTM A370). Test bars shall be poured in special blocks from the same heat as the castings represented. Test bars may be attached or cast integrally with castings		
			Bend-test specimens shall be machined to 1 by ½ in. in section with corners rounded to a radius not over 1/16 in. Bend-test specimens may be cut from heat-treated castings instead of from test bars when mutually acceptable to purchaser and manufacturer		(b) *Acidified Copper Sulfate Test*—Coupons representing finished pipe made of molybdenum-bearing material (over 0.50% molybdenum) shall meet the requirements of the acidified copper sulfate test intergranular corrosion test) according to the recommended practice for this test included in the Specifications for Corrosion-resisting chromium and Chromium-Nickel Steel Plate, Sheet, and Strip for Fusion-welded Unfired Pressure Vessels (ASTM A240). The condition of the test specimen is as follows: All molybdenum-bearing types shown in Table I shall be tested in the annealed and unsensitized condition. All specimens shall meet the requirements of the prescribed drop and bend tests
			(b) *Number of Tests*—One bend test shall be made from each heat. The bar from which the test specimen is taken shall be heat treated in a manner similar to the casting represented. If any test specimen shows defective machining or develops flaws, it shall be discarded and another specimen substituted from the same heat		
			(c) *Test Requirements*—Unless otherwise specified, the bend-test specimen shall withstand, without cracking on the outside of the bent portion, being bent longitudinally through an angle of 90 deg about a radius equal to the test specimen thickness		

Flattening tests	The flattening test specified in Sec. 10 may be made on specimens from both ends of each length of pipe. Crop ends may be used. If the specimen from either end of any length fails to conform to the specified requirement, that length shall be rejected		(A430) The flat-bending-test specified in Sec. 12 may be made on specimens from both ends of each length of pipe. Crop ends may be used. If the specimen from either end of any length fails to conform to the specified requirements, that length shall be rejected or cut back and retested if agreed upon
Metal structure and etching tests	The steel shall be homogeneous as shown by etching tests. Etching tests may be made on transverse sections from any pipe and shall show sound and reasonably uniform material, free from injurious laminations, cracks, and similar objectionable defects. If the specimen from either end of any length shows objectionable defects, one retest shall be permitted from that end. If this fails, the length shall be rejected		
Photomicrographs....	When requested by the purchaser and so stated in the order, the manufacturer shall furnish one photomicrograph at 100 diameters from a specimen of pipe in the as-finished condition for each individual size and wall thickness from each heat, for pipe in nominal sizes 3 in. and over. Such photomicrographs shall be suitably identified as to pipe size, wall thickness, and heat. No photomicrographs for the individual pieces purchased shall be required except as specified in Paragraph S7. Such photomicrographs are for information only, to show the actual metal structure of the pipe as finished		

Table 31. (*Continued*)

Test	Examples of wording on Specifications			Other specifications	Special tests
	A335 (seamless pipe)	A155 (fusion-welded pipe)	A216 (casting)		
Photomicrographs of individual pieces	In addition to the photomicrographs required in accordance with Supplementary Requirement S6, the purchaser may specify that photomicrographs shall be furnished from each end of one or more pipes from each lot of pipe 3 in. and larger in the as-finished condition. The purchaser shall state in the order the number of pipes to be tested from each lot. When photomicrographs are required on each length, the photomicrographs from each lot of pipe in the as-finished condition which may be required under Supplementary Requirement S6 may be omitted. All photomicrographs required shall be properly identified as to heat number, size, and wall thickness of pipe from which the section was taken. Photomicrographs shall be further identified to permit association of each photomicrograph with the individual length of pipe it represents				
Hardness tests	....	Hardness determinations shall be made across the welded joint at both ends of each length of pipe. The maximum acceptable hardness shall be as agreed upon between the manufacturer and the purchaser			

Radiographic inspection	. . . .	. . . .	When so specified in the order, the castings shall be examined for internal defects by means of X rays or gamma rays. The inspection procedure shall be in accordance with the Recommended Practice for Radiographic Testing (ASTM E94), and the types and degrees of defects considered shall be judged by the Industrial Radiographic Standards for Steel Castings (ASTM E71). The extent of examination and the basis for acceptance shall be subject to agreement between the manufacturer and the purchaser. A standard practice which may be used as a basis for such agreement is the Quality Standard for Steel Castings for Valves, Flanges and Fittings and Other Piping Components (Radiographic Inspection Method), (No. SP-54) of the Manufacturers Standardization Society of the Valve and Fittings Industry
Magnetic-particle examination	. . . .	All welded joints shall be subjected to magnetic-particle inspection. Such inspection shall be in accordance with the latest commercial practice and shall be performed by experienced operators. Defects so located shall be repaired in accordance with Sec. 2(e)	If specified by the purchaser, castings shall be examined by magnetic-particle inspection. The inspection procedure used shall be in accordance with the Method of Dry Powder Magnetic Particle Inspection (ASTM E109), and the types and degrees of discontinuities considered shall be judged by the Reference Photographs for Magnetic Particle Indications on Ferrous Castings (ASTM E125). The extent of examination and the basis for acceptance shall be subject to agreement between the manufacturer and the purchaser. A standard practice which may be used as a basis for such agreement in the Quality Standard for Steel Castings for Valves, Flanges and Fittings and other Piping Components (Dry Powder Magnetic Particle Inspection Methods), (No. SP-53) of the Manufacturers Standardization Society of the Valve and Fittings Industry

Table 31. (Continued)

Test	Examples of wording on Specifications			Other specifications	Special tests
	A335 (seamless pipe)	A155 (fusion-welded pipe)	A216 (casting)		
Liquid-penetrant tests	......	......	......	(A403) All surfaces shall be liquid-penetrant tested. The method shall be in accordance with the Method for Liquid Penetrant Inspection of Steel Forgings (ASTM A462). Acceptance limits shall be a matter of agreement between the manufacturer and the purchaser	
Ultrasonic testing	......	......		(A376) Each piece of pipe may be ultrasonically tested to determine its soundness throughout the entire length of the pipe in a manner to be agreed upon by the manufacturer and the purchaser. Failure to meet the requirements stated in these specifications shall be cause for rejection	
Destruction tests	......	......	When so specified in the order, the purchaser may undertake to select representative castings from each heat and cut up and etch or otherwise prepare the sections for examination for internal defects. Should injurious defects be found which evidence unsound steel or faulty foundry technique, all of the castings made from that particular pattern, heat, and heat treatment charge may be rejected. All the rejected castings, including those cut up, shall be replaced by the manufacturer without charge		
Residual elements	......	......	When specified on the order, the producer shall determine the percentages of residual elements shown in Table I and report these results to the purchaser		

Table 32. Quality-control Requirements which May Be Suggested for Materials Used in Nuclear and Steam Power-plant Applications[1]

Type of material	Alloy group	ASTM Specification	Critical nuclear service and where leaks are not permissible — 4 in. OD and over	Critical nuclear service and where leaks are not permissible — Less than 4 in. OD	Steam power-plant service involving temperatures over 950 F and/or pressures over 2,500 psi — Less than 4 in. OD	Steam power-plant service involving temperatures over 950 F and/or pressures over 2,500 psi — 4 in. OD and over	Secondary and auxiliary piping not covered in two previous columns — 4 in. OD and over	Secondary and auxiliary piping not covered in two previous columns — Less than 4 in. OD
Wrought hot-worked pipe	Carbon steel	A106	U,CL,TL,FL	U,CL,TL,FL	U,CL,TL,FL	CL,TL,FL	V,CL,TL,FL	V,CL,TL,FL
	Low-alloy steel	A335	U,CL,TL,FL	U,CL,TL,FL	U,CL,TL,FL	CL,TL,FL	V,CL,TL,FL	V,CL,TL,FL
	Stainless steel	A312, A376, A409	U,CL,TL,FL	U,CL,TL,FL	U,CL,TL,FL	CL,TL,FL	V,CL,TL,FL	V,CL,TL,FL
Forged and bored pipe	Carbon steel	A369	U,CH,TH,FH	U,CH,TH,FH	U,CH,TH,FH	CH,TH,FH	V,CH,TH,FH	V,CH,TH,FH
	Low-alloy steel		U,CH,TH,FH	U,CH,TH,FH	U,CH,TH,FH	CH,TH,FH	V,CH,TH,FH	V,CH,TH,FH
Seam-welded plate pipe (seam-welded wrought fittings)	Carbon steel: Plate base metal	A155	U,CH,TH	U,CH,TH	U,CH,TH	CH,TH	V,CH,TH	V,CH,TH
	Weld seam		R,L,WBT	R,L,FH	R,L,WBT	M or L,FH	V,WBT	V,FH
	Low-alloy steel: Plate base metal	A155	U,CH,TH	U,CH,TH	U,CH,TH	CH,TH	V,CH,TH	V,CH,TH
	Weld seam		R,L,WBT	R,L,FH	R,L,WBT	M or L,FH	V,WBT	V,FH
	Stainless steel: Plate base metal	A358	U,CH,TH	U,CH,TH	U,CH,TH	CH,TH	V,CG,TH	V,CH,TH
	Weld seam		R,L,WBT	R,L,FH	R,L,WBT	L,FH	V,WBT	V,FH

Key:
U, ultrasonic; R, radiography; L, liquid penetrant; M, magnetic particle; V, visual.
CH, chemical check analysis, each heat; CL, chemical check analysis, each lot.
TH, tensile test, each heat; TL, tensile test, each lot; BH, bend test, each lot.
FH, flattening test, each heat; FL, flattening test, each lot (for pipe 2 in. and under use bend test); WBT, weld bend test, transverse.
HTH, hot tensile test, each heat (when pipe is to be hot formed or hot worked).
[1] H. Thielsch, "Quality Control and Service Performance," Trans. ASME, vol. 86, ser. A, pp. 451–464, 1964.

Table 32. (Continued)

Type of material	Alloy group	ASTM Specification	Inspection requirements for:					
			Critical nuclear service and where leaks are not permissible		Steam power-plant service involving temperatures over 950 F and/or pressures over 2,500 psi		Secondary and auxiliary piping not covered in two previous columns	
			4 in. OD and over	Less than 4 in. OD	Less than 4 in. OD	4 in. OD and over	4 in. OD and over	Less than 4 in. OD
Centrifugally cast pipe	Carbon steel		U,L,CH,TH FH,HTH	U,L,CH,TH,FH	U or R,CH,TH FH,HTH	CH,TH,FH	V,CH,TH,FH	V,CH,TH,FH
	Low-alloy steel	A426	U,L,CH,TH FH,HTH	U,L,CH,TH,FH	U or R,CH,TH FH,HTH	CH,TH,FH	V,CH,TH,FH	V,CH,TH,FH
	Stainless steel	A451	U,L,CH,TH FH,HTH	U,L,CH,TH,FH	U or R,CH,TH FH,HTH	CH,TH,FH	V,CH,TH,FH	V,CH,TH,FH
Centrifugally cast cold-wrought pipe	Stainless steel	A452	U,L,CH,TH,FH	U,L,CH,TH,FH	U,CH,TH,FH	CH,TH,FH	V,CH,TH,FH	V,CH,TH,FH
Sand-cast fittings	Carbon steel	A216	R,L,CH,TH,BH	R,L,CH,TH	R,CH,TH,BH	M or L,CH,TH	V,CH,TH,BH	V,CH,TH
	Low-alloy steel	A217, 389	R,L,CH,TH,BH	R,L,CH,TH	R,CH,THB,H	M or L,CH,TH	V,CH,TH,BH	V,CH,TH
	Stainless steel	A351	R,L,CH,TH,BH	R,L,CH,TH	R,CH,TH,BH	L,CH,TH	V,CH,TH,BH	V,CH,TH
Forged fittings	Carbon steel	A105	U,M,CH,TH,BH	M or L,CH,TH	U,CH,TH,BH	CH,TH	V,CH,TH	V,CH,TH
	Low-alloy steel	A182, 404	U,M,CH,TH,BH	M or L,CH,TH	U,CH,TH,BH	CH,TH	V,CH,TH	V,CH,TH
	Stainless steel		U,L,CH,TH,BH	L,CH,TH	U,CH,TH,BH	CH,TH	V,CH,TH	V,CH,TH
Wrought fittings	Carbon steel	A234	U,M,CH,TH	M or L,CH,TH	CH,TH	CH,TH	V,CH,TH	V,CH,TH
	Low-alloy steel	A234	U,M,CH,TH	M or L,CH,TH	CH,TH	CH,TH	V,CH,TH	V,CH,TH
	Stainless steel	A403	U,L,CH,TH	M or L,CH,TH	CH,TH	CH,TH	V,CH,TH	V,CH,TH

Table 33. Examples of Finish Requirements in ASTM Specifications Covering Materials for High-temperature Service

Pipe	Plate	Tubing	Casting	Forgings
ASTM A335 (a) The finished pipe shall be reasonably straight and free from injurious defects. At the discretion of the inspector representing the purchaser, finished pipe shall be subject to rejection if surface defects acceptable under Paragraph (c) are not scattered but appear over a large area in excess of what is considered a workmanlike finish. (b) *Depth of Injurious Defects*—All defects shall be explored for depth. When the depth is in excess of 12½% of the nominal wall thickness or encroaches on the minimum wall thickness, such defects shall be considered injurious. (c) *Machining or Grinding; Surface Defects Not Classified as Injurious*—Surface defects not classified as injurious shall be treated as follows: (1) Pipe showing scabs, seams, laps, tears, or slivers not deeper than 5% of the nominal wall thickness need not have these defects removed. If deeper than 5%, such defects shall be removed by machining or grinding. (2) Pipe showing inside or outside surface checks (fish scale) 1/64 in. or less in depth need not have these defects removed. Such defects over 1/64 in. but not more than 1/32 in. in depth shall be removed by machining or grinding. Pipe on which these defects are more than 1/32 in. in depth shall be rejected unless the manufacturer can demonstrate to the purchaser that the defects are not injurious as defined in Paragraph (b).	As used in A155 (a) The finished pipe shall be free from injurious defects and shall have a workmanlike finish. (b) *Repair of Plate Defects by Machining or Grinding*—Pipe showing moderate slivers may be machined or ground inside or outside to a depth which shall assure the removal of all included scale and slivers, providing the wall thickness is not reduced below the specified minimum wall thickness. Machining or grinding shall follow inspection of the pipe as rolled and shall be followed by supplementary visual inspection. (c) *Repair of Plate Defects by Welding*—Repair of injurious defects is permitted only subject to the approval of the purchaser. Defects shall be thoroughly chipped out before welding. The repairs shall be radiographed, and if the pipe itself has already been stress relieved, it shall then be stress relieved again except in the case of small welds that, in the estimation of the purchaser's inspector, do not require stress relief. Hydrostatic tests made on pipe repaired in this manner shall be made on the finished pipe.	ASTM A199, A213 Finished tubes shall be reasonably straight and have smooth ends free from burrs. They shall be free from injurious defects and shall have a workmanlike finish. Minor defects may be removed by grinding, provided the wall thickness is not decreased to less than that permitted in Table V of Specification A450.	ASTM A217 (a) The castings shall be clean and free from injurious defects. A standard practice which may be used as a basis for visual inspection is the Quality Standard for Steel Castings for Valve, Flanges, and Fittings, and Other Components (Visual Method, SP-55) of the Manufacturers Standardization Society of the Valves and Fittings Industry. The castings shall not be peened, plugged, or impregnated to stop leaks. *Repair of Defects* (a) The location of existing injurious defects shall be determined as stated in (a) and shall be completely removed to sound metal. The manufacturer shall visually check the completeness of removal. However, if the defect is a crack, or if the defect was removed by methods involving high temperatures, the manufacturer shall also inspect the cavity as in the method for Dry Powder Magnetic Particle Inspection (ASTM E109) or the Inspection for Wet Magnetic Particle Inspection (ASTM E138). When the magnetic-particle method is not feasible, the cavity shall be inspected as in Methods for Liquid Penetrant Inspection (ASTM E165). (b) Repairs shall be made using procedures and welders qualified under the Recommended Practice for Qualification of Procedures and Personnel for Welding of Steel Castings (ASTM A488). (c) Weld repairs shall be considered major in the case of a casting which has leaked on hydrostatic test or when the depth of the cavity prepared for welding exceeds 20% of the wall thickness or 1 in., whichever is smaller, or when the extent of the cavity exceeds approximately 10 sq in. All castings with major repair welds shall be stress relieved or reheat treated completely in accordance with the procedure qualification used. Major repairs shall be inspected to the same quality standards as used to inspect the castings.	ASTM A182 *Finish* The forgings shall be free from injurious defects and shall have a workmanlike finish. *Macro-etch Tests* In case of question as to the soundness of material in any lot of forgings and its suitability for the intended service, a macro-etching test shall be made for each heat present in the lot. Etching tests shall show sound and seasonably uniform material, free from injurious laminations, cracks, segregations, and similar objectionable defects. If on successive tests 10% of any heat fail to pass the requirements of the macro-etch test, all forgings from that heat shall be rejected.

Table 33. (Continued)

Pipe	Plate	Tubing	Casting	Forgings
(3) Mechanical marks or abrasions and pits shall be acceptable without grinding or machining provided the depth does not exceed the limitations set forth in Paragraph (b) and if not deeper than $\frac{1}{16}$ in. If such defects are deeper than $\frac{1}{16}$ in. but not deeper than $12\frac{1}{2}\%$ of the nominal wall thickness, they shall be removed by grinding or machining to sound metal. (4) When defects have been removed by grinding or machining, the outside diameter at the point of grinding or machining may be reduced by the amount so removed. Should it be impracticable to secure a direct measurement, the wall thickness at the point of grinding or at a defect not required to be removed shall be determined by deducting the amount removed in grinding or the depth of the defect from the minimum measured wall thickness at the ends of the pipe, and the remainder shall not be less than 87.5% of the nominal wall thickness. (5) Machining or grinding shall follow the inspection at the mill of the pipe as rolled and shall be followed by supplementary visual inspection. (d) *Repair by Welding*—Repair of injurious defects shall be permitted only subject to the approval of the purchaser and with the further understanding that the composition of the welding rod shall be suitable for the composition of the metal being welded. Welding of injurious defects in no case shall be permitted when the depth of the defect exceeds $33\frac{1}{3}\%$ of the nominal pipe-wall thickness or the length of repair exceeds 25% of the nominal diameter of the pipe. Defects shall be thoroughly chipped out before welding and then heat-treated in accordance with Sec. 4. Each length of repaired pipe shall be retested hydrostatically in accordance with Sec. 10.				

9

CORROSION IN PIPING SYSTEMS

Matthew L. Hepburn*

Corrosion is the mechanism by means of which metals and oxygen strive to reach equilibrium. It would be well to think of the "corroded" condition as the normal and the "noncorroded" condition as the transitory phase in the never-ending battle between the engineer and environment. Some metals oxidize at a rate so low as to be imperceptible, and some at an explosive pace. All require an ionizing medium, and all will display a change of rate with differing conditions.

THE MECHANISM OF CORROSION

The subject of "corrosion" has occupied the attention of mankind from the earliest ages. Corrosion has been abscribed to the absence of sunlight, to the presence of dust, to bacteria, and to numerous other factors. It has been shown that such influences can, indeed, affect the rate of corrosion in special cases but that they are not primary causes.

Of all the theories advanced, the *electrochemical theory* alone has survived debate and criticism. This theory requires that, when one metal goes into solution, an equivalent amount of some other metal must be displaced if the solution is to remain electrically neutral. Familiar examples can be cited. A piece of iron will be dissolved in a solution of copper sulfate, but some of the dissolved copper is displaced from solution and appears as a thin film on the iron. This is a form of "copper plating." If the iron is placed in pure water, the iron will still be dissolved but the "metal" hydrogen will be proportionally "plated" out in the form of a film of tiny bubbles. The dissolving rate of the iron will be affected by the amount of oxygen

* Chemical Engineer, Ebasco Services, Inc., New York, N.Y.

dissolved in the pure water, which in turn is affected by the temperature of the water and the pressure exerted on the water.

The two above examples cited iron as one of the metals involved and were selected because iron piping is the most common material encountered. However, the theory is as true for the exotic piping materials used in nuclear reactors (titanium, tantalum, etc.) as it is for iron piping which carries well water to a garden. In view of the predominance of iron (which naturally includes steel as well) piping, the balance of this chapter will deal with iron pipe except where conditions are such as to preclude its employment.

All chemical reactions require the presence of an ionizing medium. The most universal ionizing medium is water. No corrosion will occur in the total absence of water. It is to be noted that this does not mean the "apparent" absence of water. It is well known that iron corrodes, or "rusts," very slightly in desert areas, yet some corrosion does occur, as the humidity is extremely low, but not nonexistent, in desert areas.

The simplest example of corrosion, that of iron in pure water, has been cited, and it has been stated that the rate of corrosion would be affected by the amount of dissolved oxygen present. The dissolved oxygen could react in either of two ways: (1) by combining with the hydrogen film to expose the surface of the iron (which the film protects) or (2) by combining with the dissolved ferrous hydroxide to form insoluble hydrated ferric compounds and thereby to allow the formation of more ferrous hydroxide. Conversely if the oxygen content is small, the protective hydrogen film will be relatively undisturbed and the initially high rate of corrosion will taper off rapidly.

The oxygen content of water in contact with air is a function of temperature and pressure, but in a closed piping system this is not true. If the oxygen content is lowered artifically by elevating the temperature and venting the gases released, the corrosibility of the water drops. Deaerating heaters and deaerating condensers perform precisely this function in large central steam electric generating plants. The pipe in these stations continues to corrode, but at a rate that the operators consider to be acceptable.

FACTORS WHICH INFLUENCE CORROSION

Governing factors may be classified as *primary* (inherent in the material itself) or *secondary* (affected by environment). This does not imply that the secondary factors are of lesser importance. Primary factors should be regarded as those which influence the *tendency* to enter into solution, or corrode, and the secondary factors as those which influence the rate of corrosion.

A weak corroding tendency coupled with favorable corroding conditions can result in as much corrosion as a strong tendency operating under inhibiting conditions. The engineer cannot change the tendency to corrode in the materials with which he is forced to work, but he can often alter or influence the secondary environmental factors so as to protect a piping system which is inherently weak. The tabulation below lists primary and secondary factors which influence corrosion; the sequence of listing is not to be considered as an indication of relative importance.

Factors Which Influence Corrosion

1. Primary (inherent):
   a. Effective electrode potential of a metal in solution
   b. Overvoltage of hydrogen on the metal
   c. Homogeneity of metal surface (physical and chemical)
   d. Tenacity and permeability of oxide (and other corrosion product) films

2. Secondary (environmental):
   a. Hydrogen-ion concentration (pH)
   b. Oxygen concentration
   c. Other adjacent corrosive fluids
   d. Rate of flow of solution
   e. Tendency of environment to form protective film
   f. Temperature
   g. Stress (including cyclic)
   h. Contact with dissimilar metals or other materials affecting localized corrosion

Primary Corrosion Factors

Electrode Potential. The tendency of a metal to enter into solution is a function of its electrode potential. Electrode potential is an inherent property of each element, and only its influence on corrosion will be discussed here. If dissimilar metals are externally connected and immersed in a solution, the metal with the higher electrode potential exerts a voltage which causes the lower (less "noble") metal to go into solution, that is, to corrode.

The electrode potentials of the common metals are shown in Table 1. Hydrogen is arbitrarily assigned a zero potential, and values shown for other elements are relative to that of zero for hydrogen. The relationship is not rigidly fixed but depends also on the ion concentration in the solution, a point that should not be

Table I.   Standard Electrode Potentials of the Elements at 77 F[1]

(The potential given is that between the element in its standard state
and its ion at unit activity in the solution)

Element	Reference ion	Potential, volts[2]
Lithium............	Li$^+$	$-2.959\ (6)$[3]
Rubidium..........	Rb$^+$	$-2.925\ (9)$
Potassium..........	K$^+$	$-2.924\ (1)$
Calcium............	Ca^{++}	$-2.7\ (63)$
Sodium............	Na$^+$	$-2.714\ (6)$
Zinc...............	Zn$^+$	$-0.761\ (8)$
Chromium (ate)	Cr^{++}	-0.557
Chromium (ic).......	Cr^{+++}	$-0.50\ (a)$[4]
Iron (ous)..........	Fe^{++}	$-0.44\ (1)$
Cadmium	Cd^{++}	$-0.401\ (3)$
Nickel	Ni^{++}	$-0.23\ (1)$
Tin.................	Sn^{++}	$-0.13\ (6)$
Lead	Pb^{++}	$-0.12\ (2)$
Iron (ic)............	Fe^{+++}	-0.045[5]
Hydrogen..........	H$^+$	-0.000
Copper (ic).........	Cu^{++}	$+0.344\ (1)$
Copper (ous)	Cu$^+$	$+0.552$[4]
Iodine	I$^-$	$+0.534\ (5)$
Silver..............	Ag$^+$	$+0.797\ (8)$
Mercury	Hg^{++}	$+0.798\ (6)$
Bromine	Br$^-$	$+1.064\ (8)$
Chlorine	Cl$^-$	$+1.358\ (3)$
Gold	Au^{+++}	$+1.3\ (6)$

[1] Taken from Frank N. Speller, D.Sc., "Corrosion—Causes and Preventions," p. 18, McGraw-Hill Book Company, New York, 1935.

[2] Unless otherwise specified, the values in the table are taken from "International Critical Tables," vol 6, p. 332, McGraw-Hill Book Company, New York, 1929.

[3] Terminal digits in parentheses are in dispute.

[4] Computed. See references in footnotes 2 and 5.

[5] Lewis and Randall, "Thermodynamics," p. 411, McGraw-Hill Book Company, New York, 1923.

overlooked.   Under certain conditions the polarity of metals that are near each other in the electrode potential series may be reversed.

It should be noted that not only do electrode potentials change under environmental conditions but they can be radically changed by application of an external electromotive force.   This is the complex and distinct technology known as *cathodic protection*.   No brief summary of its role is possible.   It is a field for specialists, who should be consulted in work of any magnitude.

Overvoltage of Hydrogen.   This is the measure of the energy required by the system to cause electrically neutral monatomic hydrogen on the corroded metal surface to form molecular hydrogen bubbles.

In most natural waters, surface and ground, the overvoltage on iron is sufficient to prevent visible hydrogen-gas evolution.   This, however, is owing to a secondary environmental factor, the normal alkalinity of natural waters with attendant high pH.

Homogeneity of Metal Surface.   If a metal surface is physically and chemically homogeneous, the corrosion will take place uniformly over the entire surface. If not, the susceptible areas will be attacked at a faster rate, and this will result in pitting or perforation.

Tenacity and Permeability of Oxide Film.   Some metals have the highly desirable property of forming a submicroscopic, strongly adherent layer of metallic oxide that prevents further corrosion.

Such metals are known as "corrosion resistant" or, less accurately, as "corrosion-proof."   The 18-8 chrome-nickel steels and the exotic metals are included in this class.   The corrosion resistance is relative; no metal will resist corrosion under all conditions.

Corrosion-resistant metals are characterized by relatively high price due to the expense associated with scarcity of some of the alloying metals or costs of fabrication or both.   They are used, generally, only where less expensive and more common metals are unacceptable from a corrosion viewpoint or where safety (freedom from failure) is paramount, as in the piping systems of nuclear reactors.

Iron piping will form an impermeable tenacious layer only under the most favorable conditions and will retain it only under equally favorable operating conditions.   It is the duty of the engineer to provide the most favorable conditions for film formation in the piping system that his circumstances permit and to operate it in the most economically feasible manner conducive to its retention.

If these primary factors cannot be altered, their effects at least can be modified. They must be understood when selecting piping material.

Secondary Corrosion Factors

Hydrogen-ion Concentration (pH).   The chemical activity of an acid or base depends upon the degree of ionization.   A "weak" acid, such as boric or acetic, does not ionize to the degree that a "strong" acid, such as hydrochloric, does. The hydrogen-ion concentration is an important index of the corrosive tendency of a solution.

Pure water is the standard of neutrality in that hydrogen ions and hydroxyl ions are perfectly balanced at 1×10^{-7} moles per liter at 72 F.   This is expressed by the symbol pH, which is equal to $\log_{10}$ (1/hydrogen-ion concentration).

Thus, a pH of 7 indicates a condition of neutrality.   Values below 7 are increasingly acid, and values above 7 are increasingly alkaline.

It should be remembered that these values are logarithmic.   A difference of one point from pH 5.0 to pH 4.0 indicates a tenfold increase in hydrogen-ion concentration, and a difference of two points from pH 5.0 to pH 3.0 indicates a hundredfold increase.

The effects of low pH on iron piping are many; two of these are (1) the tendency to increase concentration of hydrogen gas on the corroding surface and (2) the tendency to dissolve protective coatings on these surfaces.

Oxygen Concentration.   Some of the steps by which oxygen can be decreased have been pointed out previously.

Other things being equal, an increase in dissolved oxygen will mean an increase in corrosion.   In atmospheric corrosion, an excess of oxygen is present.   In such cases corrosion is controlled by other factors, such as the presence of moisture or the development of a protective film.   Corrosion can occur in the total absence of oxygen, as in boiler feedwater where the last traces of oxygen have been removed by scavengers such as sodium sulfite or hydrazine; such corrosion is due to conditions that permit the evolution of hydrogen.

Other Adjacent Corrosive Fluids.   The chloride ion is encountered most frequently.   Concentrations of chloride ions in solution tend to dissolve protective coatings already formed and tend to prevent new coatings being laid down.

Rate of Flow of Solution.   High flow rate tends to increase corrosion.   Monatomic hydrogen films are destroyed as a result of the increased velocity.   If oxygen has been depleted by reaction with metal, the flowing solution brings more oxygen.   Protective films of oxide can be locally impaired or totally destroyed if the velocity is sufficient.   Corrosion products which are removed as a result of high velocities can act as an abrasive and so expose more metal to attack.

Summary of Factors Influencing Corrosion.   If there were a single-phase metal with a low tendency to enter into solution, requiring a high overvoltage, and whose corrosion products coated the surface tightly, that metal would be ideal for pipe.   There is no such metal known at this time whose cost permits it to be used for pipe, and there is little probability that one will be discovered.

However, iron pipe can be and is being used.   Looking back at the untold thousands of miles of steel pipe that are in service, one is impressed by the fact that most of it is in good operating condition.   This inescapable fact illustrates that corrosion is due more to environmental factors than those which are inherent to the material.   The pipe mills turn out a uniform product of carbon steel.   If its weaknesses are known (as they are), it can be safely assumed that these have been minimized.   It is therefore up to the engineer to adjust to the environment in the most economical manner.

There are always special cases requiring expensive metals or coatings.   Good engineering consists of selecting the least expensive of the acceptable alternates and designing a system that requires the minimum quantity of costly material.

Finally, the properties of all commercial pipe materials are known, their strengths and their weaknesses.   Learned societies such as the American Water Works Association, the American Petroleum Institute, and the American Society for Testing and Materials have investigated exhaustively commercial pipe materials and have published their findings.   These findings tell the engineer what he has to work with.   It is up to him to use it properly.

Proper use consists of adapting the known material to the environment.   This will vary.   No two jobs will be the same.   The engineer's job is to investigate his environment, adapt to it where he can, and circumvent it where he must.

The balance of this chapter will deal with environmental factors and engineering adaptions.

INFLUENCE OF EXTERNAL FACTORS

No rigid or all-inclusive classification is possible of the many external factors that influence either corrosion or the rate at which corrosion takes place.   It would

help the piping engineer if he regarded these external factors as vector quantities which sometimes work together but which, at other times, might oppose each other. Naturally, the resultant of these vectors will indicate whether the corrosion in one particular piping system will be tolerable or severe.

Speller[1] groups environmental factors under four headings, as follows:

1. Those which mainly affect the amount of dissolved oxygen reaching the metal surface
2. Those due to dissolved substances in the corroding solution where the main effect is not on the oxygen supply
3. Those which chiefly influence the localization of corrosion and the amount of pitting
4. Other factors

It is unnecessary to point out that none of these factors will exist by themselves to the exclusion of all others.

Factors Which Affect the Amount of Dissolved Oxygen Reaching the Metal Surface

Oxygen travels from the air to the metal surface in three steps: (1) by solution, (2) by convection from the atmospheric contact surface to the proximity of the metal surface, and (3) by diffusion through the thin film that separates the metal from the liquid.

Activity of the solution is dependent upon oxygen solubility in the solution (pure water will hold more oxygen than a saturated solution of neutral electrolyte), temperature (a rise in temperature generally speaking decreases solubility), and pressure.

Convection currents are generally associated with flow of the media but can exist in a stagnant system exposed to air if evaporation occurs with a resultant lowering of surface temperature.

Diffusion through a film depends on oxygen concentration within the film and its thickness.   Increase of temperature (within wide limits) causes currents within the film and usually a decrease in viscosity.   Both increase the speed of oxygen diffusion by thinning the film.

Ordinarily, the influence of convection may be regarded as constant, except that in closed systems solution activity and diffusion control the reaction.

The diffusion film is of very great importance, and there can be no thorough grasp of the principle of corrosion without a knowledge of it.   The following is also quoted from Speller:

Films of a relatively quiet nature exist at the boundary between different media.   Thus, between a solid and a liquid or a gas, there will be a thin quiet film of the liquid or gas on the solid surface.   A moderate velocity of the mobile medium will not be imparted to the film and the transference of chemical matter from one medium to the other must necessarily take place by the slow process of diffusion through the film, which may be considered to act as an insulator.   The same type of film exists between a liquid and a gas.   From the metal surface, through this film, outward to the body of the solution, convection currents gradually become of more importance in regulating the concentration of the liquid.

Factors Due to Dissolved Substances in the Corroding Solution

The effect of gases is varied.   The inert gases, including nitrogen, potentially decrease corrosion by displacing oxygen.   Some corrosive gases of natural origin,

[1] Frank N. Speller, "Corrosion," McGraw-Hill Book Company, New York, 1935.

such as hydrogen sulfide, increase corrosion. Gases from industrial processes, such as sulfur dioxide due to combustion of bituminous coal, can greatly accelerate corrosion. In large industrial cities where cooling towers are employed for air conditioning, the pH of untreated cooling-tower water has been known to drop to pH 3.5 because of the air scrubbing action of the tower. In such an environment nonferrous piping plus corrective chemical treatment of recirculated water is essential.

Chlorine gas can be very corrosive. However, this gas is used generally for sterilization. Continuous feed of chlorine, rather than "slugging," is used on industrial cooling-water flows. In no case is any great chlorine residual deliberately imparted to the water, and it will rarely be encountered as a corroding substance.

Carbon dioxide is the most important gas in terms of corrosive power. It is always present in the atmosphere in small quantities and is encountered in nearly all natural waters in the free or chemically combined form. In industrial districts and heavy automobile traffic areas, the carbon dioxide content of air is greatly increased.

Table 2. Values of A for Use in Eq. (1)[1]

Total solids, ppm	Value of A	Total solids, ppm	Value of A
50	0.07	300	0.14
75	0.08	400	0.16
100	0.10	600	0.13
150	0.11	800	0.19
200	0.13	1,000	0.20

[1] Courtesy of the Pfaudler-Permutit Company.

Carbon dioxide in solution will form weakly ionized carbonic acid, which will attack ferrous metals and certain nonferrous metals alike. The attack on iron piping usually occurs along with and as an adjunct to oxygen corrosion, although the attack will occur in the absence of oxygen.

In water piping, the effect of carbon dioxide is experienced by either direct or indirect attack. Indirect attack is exemplified by the ability of dissolved carbon dioxide to render carbonates of calcium and magnesium more soluble and to nullify the protective coatings which these minerals otherwise would form on the pipe.

The importance of calcium carbonate (and to a lesser extent magnesium carbonate) coatings must be constantly borne in mind. A generation ago, engineers classified water as "hard" or "soft," depending on the amounts of calcium (and magnesium) it contained. The degree of "hardness" was measured with a standard emulsion of castile soap on the so-called Boutron and Boudet scale. This has been replaced by the Schwarzenbach scale using the sodium salt of EDTA (ethylenediaminetetracetate) with Erichrome Black T indicator.

The calcium content of water is currently used to measure the *tendency* of water to corrode or to form scale. For this purpose values are assigned to the calcium content and the alkalinity of the water. Adjustments are made for temperature and the effect of total dissolved solids. The resultant value is compared with the observed pH of the water to determine its character, that is, whether it is corrosive or scaling.

This basically is the Langelier saturation index as reported by Larson-Buswell in the *Journal of the American Water Works Association*, vol. 28, p. 1500, 1931. Several modifications exist: variables can be taken from a nomographic scale or from the compiled tabular values which are listed in Tables 2 to 4. The tables are Nordell's computation from the Larson-Boeswell family of curves, from the

Table 3. Values of B for Use in Eq. (I)[1]

Temp, F (tens)	Units				
	0	2	4	6	8
30	....	2.60	2.57	2.54	2.51
40	2.48	2.45	2.43	2.40	2.37
50	2.34	2.31	2.28	2.25	2.22
60	2.20	2.17	2.14	2.11	2.00
70	2.06	2.04	2.03	2.00	1.97
80	1.95	1.92	1.90	1.88	1.86
90	1.84	1.82	1.80	1.78	1.76
100	1.74	1.72	1.71	1.69	1.67
110	1.65	1.64	1.62	1.60	1.58
120	1.57	1.55	1.53	1.51	1.50
130	1.48	1.46	1.44	1.43	1.41
140	1.40	1.38	1.37	1.35	1.34
150	1.32	1.31	1.29	1.28	1.27
160	1.26	1.24	1.23	1.22	1.21
170	1.19	1.18	1.17	1.16	

[1] Courtesy of the Pfaudler-Permutit Company.

Example. At a temperature of 64 F, the value of B equal to 2.14 is found at the intersection of 60 in the tens column and 4 in the units column. Interpolation may be performed; for example, at 105 F, B is determined as 1.70.

Table 4a. Values of C for Use in Eq. (I)[1]

(Calcium hardness expressed in ppm $CaCO_3$)

Ppm $CaCO_3$ (tens)	Units									
	0	1	2	3	4	5	6	7	8	9
0				0.08	0.20	0.30	0.38	0.45	0.51	0.56
10	0.60	0.64	0.68	0.72	0.75	0.78	0.81	0.83	0.86	0.88
20	0.90	0.92	0.94	0.96	0.98	1.00	1.02	1.03	1.05	1.06
30	1.08	1.09	1.11	1.12	1.13	1.15	1.16	1.17	1.18	1.19
40	1.20	1.21	1.23	1.24	1.25	1.26	1.26	1.27	1.28	1.29
50	1.30	1.31	1.32	1.33	1.34	1.34	1.35	1.36	1.37	1.37
60	1.38	1.39	1.39	1.40	1.41	1.42	1.42	1.43	1.43	1.44
70	1.45	1.45	1.46	1.47	1.47	1.48	1.48	1.49	1.49	1.50
80	1.51	1.51	1.52	1.52	1.53	1.53	1.54	1.54	1.55	1.55
90	1.56	1.56	1.57	1.57	1.58	1.58	1.58	1.59	1.59	1.60
100	1.60	1.61	1.61	1.61	1.62	1.62	1.62	1.63	1.64	1.64
110	1.64	1.65	1.65	1.66	1.66	1.67	1.67	1.67	1.67	1.68
120	1.68	1.68	1.69	1.69	1.70	1.70	1.70	1.71	1.71	1.71
130	1.72	1.72	1.72	1.73	1.73	1.73	1.74	1.74	1.74	1.75
140	1.75	1.75	1.75	1.76	1.76	1.76	1.77	1.77	1.77	1.78
150	1.78	1.78	1.78	1.79	1.79	1.79	1.80	1.80	1.80	1.80
160	1.81	1.81	1.81	1.81	1.82	1.82	1.82	1.82	1.83	1.83
170	1.83	1.84	1.84	1.84	1.84	1.85	1.85	1.85	1.85	1.85
180	1.86	1.86	1.86	1.86	1.87	1.87	1.87	1.87	1.88	1.88
190	1.88	1.88	1.89	1.89	1.89	1.89	1.89	1.90	1.90	1.90
200	1.90	1.91	1.91	1.91	1.91	1.91	1.92	1.92	1.92	1.92

[1] Courtesy of the Pfaudler-Permutit Company.

Example. At a temperature of 64 F, the value of C equal to 1.41 is found at the intersection of 60 in the tens columns and 4 in the units column.

Table 4b.   Values of C for Use in Eq. (1)[1]

(Calcium hardness expressed as ppm $CaCO_3$)

Ppm $CaCO_3$ (hundreds)	Tens									
	0	10	20	30	40	50	60	70	80	90
200		1.92	1.94	1.96	1.98	2.00	2.02	2.03	2.05	2.06
300	2.08	2.09	2.11	2.12	2.13	2.15	2.16	2.17	2.18	2.19
400	2.20	2.21	2.23	2.24	2.25	2.26	2.26	2.27	2.28	2.29
500	2.30	2.31	2.32	2.33	2.34	2.34	2.35	2.36	2.37	2.37
600	2.38	2.39	2.39	2.40	2.41	2.42	2.42	2.43	2.43	2.44
700	2.45	2.45	2.46	2.47	2.47	2.48	2.48	2.49	2.49	2.50
800	2.51	2.51	2.52	2.52	2.53	2.53	2.54	2.54	2.55	2.55
900	2.56	2.56	2.57	2.57	2.58	2.58	2.58	2.59	2.59	2.60

[1] Courtesy of the Pfaudler-Permutit Company.

Example. At a temperature of 450 F, the value of C of 2.26 is read at the intersection of 400 in the hundreds column and 50 in the tens column.

reference cited above, and appear by permission of Pfaudler-Permutit Company. The Langelier saturation index (LSI) is given as

$$LSI = pH - pH_s$$

in which pH represents the pH of the water sample whose saturation index is to be determined, and

$$pH_s = (9.30 + A + B) - (C + D) \tag{1}$$

in which the numerical values of A, B, C, and D are read from Tables 2, 3, 4a or b, and 5a or b.

Table 5a.   Values of D for Use in Eq. (1)[1]

(Alkalinity expressed as ppm $CaCO_3$)

Ppm $CaCO_3$ (tens)	Units									
	0	1	2	3	4	5	6	7	8	9
0		0.00	0.30	0.48	0.60	0.70	0.78	0.85	0.90	0.95
10	1.00	1.04	1.08	1.11	1.15	1.18	1.20	1.23	1.26	1.29
20	1.30	1.32	1.34	1.36	1.38	1.40	1.42	1.43	1.45	1.46
30	1.48	1.49	1.51	1.52	1.53	1.54	1.56	1.57	1.58	1.59
40	1.60	1.61	1.62	1.63	1.64	1.65	1.66	1.67	1.68	1.69
50	1.70	1.71	1.72	1.72	1.73	1.74	1.75	1.76	1.76	1.77
60	1.78	1.79	1.79	1.80	1.81	1.81	1.82	1.83	1.83	1.84
70	1.85	1.85	1.86	1.86	1.87	1.88	1.88	1.89	1.89	1.90
80	1.90	1.91	1.91	1.92	1.92	1.93	1.93	1.94	1.94	1.95
90	1.95	1.96	1.96	1.97	1.97	1.98	1.98	1.99	1.99	2.00
100	2.00	2.00	2.01	2.01	2.02	2.02	2.03	2.03	2.03	2.04
110	2.04	2.05	2.05	2.05	2.06	2.06	2.06	2.07	2.07	2.08
120	2.08	2.08	2.09	2.09	2.09	2.10	2.10	2.10	2.11	2.11
130	2.11	2.12	2.12	2.12	2.13	2.13	2.13	2.14	2.14	2.14
140	2.15	2.15	2.15	2.16	2.16	2.16	2.16	2.17	2.17	2.17
150	2.18	2.18	2.18	2.18	2.19	2.19	2.19	2.20	2.20	2.20
160	2.20	2.21	2.21	2.21	2.21	2.22	2.22	2.23	2.23	2.23
170	2.23	2.23	2.23	2.24	2.24	2.24	2.24	2.25	2.25	2.25
180	2.26	2.26	2.26	2.26	2.26	2.27	2.27	2.27	2.27	2.28
190	2.28	2.28	2.28	2.29	2.29	2.29	2.29	2.29	2.30	2.30
200	2.30	2.30	2.30	2.31	2.31	2.31	2.31	2.32	2.32	2.32

[1] Courtesy of the Pfaudler-Permutit Company.

Table 5b.   Values of *D* for Use in Eq. (1)[1]

(Alkalinity expressed as ppm $CaCO_3$)

Ppm CaCO₃ (hun- dreds)	Tens									
	0	10	20	30	40	50	60	70	80	90
200		2.32	2.34	2.36	2.38	2.40	2.42	2.43	2.45	2.46
300	2.48	2.49	2.51	2.52	2.53	2.54	2.56	2.57	2.58	2.59
400	2.60	2.61	2.62	2.63	2.64	2.65	2.66	2.67	2.68	2.69
500	2.70	2.71	2.72	2.72	2.73	2.74	2.75	2.76	2.76	2.77
600	2.78	2.79	2.79	2.80	2.81	2.81	2.82	2.83	2.83	2.84
700	2.85	2.85	2.86	2.86	2.87	2.88	2.88	2.89	2.89	2.90
800	2.90	2.91	2.91	2.92	2.92	2.93	2.93	2.94	2.94	2.95
900	2.95	2.96	2.96	2.97	2.97	2.98	2.98	2.99	2.99	3.00

[1] Courtesy of the Pfaudler-Permutit Company.

If the index is calculated to be zero, the water is in chemical balance. If the index is positive, scale-forming tendencies are indicated. If the index is negative, corrosive tendencies in the water are indicated.

Example. A water sample whose pH is 7.2 has a total solids analysis of 400 ppm, a calcium hardness as $CaCO_3$ of 240 ppm, and an alkalinity as $CaCO_3$ of 196 ppm.

It is required to determine the Langelier saturation index of this water at a temperature of 124 F.

Solution. At a total solids content of 400 ppm, the value of *A* is read from Table 2 as 0.16. At a temperature of 124 F, the value of *B* is read from Table 3 as 1.53. At a calcium hardness of 240 ppm, a value of *C* equal to 1.98 is read from Table 4b. At an alkalinity of 196 ppm, the value of *D* equal to 2.29 is read from Table 5a. These values, when substituted into Eq. (1), yield

$$pH_s = 9.30 + 0.16 + 1.5 - (1.98 + 2.29) = 6.72$$

The Langelier saturation index is then

$$LSI = 7.2 - 6.72 - 0.48, \text{ say } 0.5$$

Moderate scale-forming tendencies are thus indicated for this water sample.

There are several modifications of the above, among which is the Ryznar stability index which doubles the pH_s value and subtracts the pH therefrom. A difference greater than 7 indicates a corrosive tendency, and a difference of 6 or less shows a scaling tendency. This method avoids negative numbers.

The effect of carbon dioxide on a solution which contains calcium carbonate can be extreme. Iron vessels or iron pipe containing calcium carbonate (calcite)-rich waters are afforded protection against corrosion by the scale-forming tendency. If these vessels or pipelines are exposed to air, carbon dioxide is absorbed, the pH drops, and the scaling tendency yields to a corroding tendency. If the temperature also drops, the solubility of carbon dioxide increases. If the system is exposed to sunlight, the reverse can occur because sunlight activates the bacteria present in all natural water. These bacteria, being plants, metabolize carbon dioxide into starches, deplete the supply, elevate the pH, and frequently cause extensive calcite scaling.

In all calcite-bearing natural waters the balance between dissolved calcium carbonate and dissolved carbon dioxide is delicate and is easily disturbed. The use of calcite-bearing make-up water for low-pressure boilers is always deprecated. Recently the use of zeolite resins in the sodium cycle (zeolite softeners) has become widespread as the result of improved designs and lowered cost of this equipment.

By this means and small supplementary feed of phosphates or satisfactory alternates, the problem of calcite scale has become a minor one.   However, sodium zeolite treatment does not affect the bicarbonate or carbonate ions, both of which tend to decompose at elevated temperatures with the evolution of corrosive carbon dioxide.

Make-up waters which contain carbonates or bicarbonates are responsible for two of the most damaging phenomena known to the piping engineer, namely, steam-line corrosion and condensate return-line corrosion.   The damage to piping is often severe to the point of failure and frequently necessitates expensive repairs, as in instances where steam and return lines are hidden behind building walls. This is not the only damage, as corrosion products may be carried back to the steam-generating equipment and there set up concentration cells which corrode the covered surfaces.

Deaeration will not correct the corrosive tendencies of carbonate- and bicarbonate-bearing waters because the carbon dioxide is chemically bound.   A controlled acid feed to destroy carbonate alkalinity is too complex and too expensive for small boilers.   Split-stream (hydrogen-sodium) zeolite treatment is also impractical for use in small installations.   The use of anionic resins to exchange chloride ions for carbonate or bicarbonate ions is growing, but this involves expense of both equipment and supervision.

The use of acid feed or ion-exchange resins equipment is recommended wherever its cost is justified and it is known that such equipment will cure the underlying cause of corrosion.   Good engineering requires the cure of bad conditions rather than the palliation thereof by supplementary treatment.

However, palliative treatment is available.   Part of this treatment, as will be brought out below, is standard operating procedure for high-pressure boilers (drum type).

Proprietary "filming amines" (for example, Hagofilm) are sold for injection with feedwater.   These compounds volatilize with the steam and protect the steam piping.   The effect is to coat the piping and thus to isolate it from carbon dioxide attack.   Generally these compounds are restricted to use in systems whose pressure is 900 psi and less.   Filming amines are seldom used in systems which operate at pressures in excess of 2,000 psi.

The design of such feedwater conditioning systems should not be undertaken by amateurs.   The companies specializing in internal boiler treatment (Betz, Nalco, Hagan, etc.) are available for advice, and it is recommended that such advice should be sought and followed.

It is not too thoroughly understood that overfeeding of internal-treatment chemicals can be more deleterious than underfeeding.

The use of filming amines has gained widespread acceptance in low-pressure systems, both for district heating and for process applications.   This explains why less corrosion is experienced in steam lines than in return lines.   When the treated steam encounters a condensing surface, such as in a radiator, the entrained carbon dioxide which has been prevented from attacking the filmed steam pipe is free to attack the condenser surface or the return line.   The attack can be especially severe if small amounts of oxygen are present; even the residual oxygen left by a well-operated deaerating heater can be responsible for significant corrosion.

Protection from return-line corrosion can be had by using *neutralizing* amines or *cycling* amines (of which morpholine is the best known example) if the great bulk of the condensate is returned to the system.   Morpholine and water have a close liquid-vapor ratio at comparable temperatures and pressures, so that the morpholine tends to cycle through the system, except for that portion which combines chemically with free carbon dioxide and is lost.   Obviously this can be extensive in a high-bicarbonate water.   The high-pressure systems that use morpholine are all

characterized by low make-up, and the small amount of bicarbonates that could be present is usually due to condenser leakage.

Soil Corrosion.   Soil corrosion, the name given to external corrosion of buried pipe, will vary both with locality and the type of pipe used; the opinion of experts is that environmental factors such as soil and climate far outweigh the differences in corrosivity of different ferrous metals.

The findings of the National Bureau of Standards are aptly summarized in *Letter Circular* 689, April 22, 1942, under "Corrosion in Soils."

There are numerous sources of potentials which cause the corrosion of buried metal.   The seriousness of corrosion underground depends largely on the character of the films or thicker deposits resulting from corrosion processes.   Poor material is not an important cause of underground corrosion.   The chemical composition of the soluble material in soils is an important factor in corrosion, but when the soil contains only small percentages of soluble salts, other factors control the rate of corrosion.   Well-drained soils are usually noncorrosive.   Wet soils, organic soils, and soils high in soluble salts usually are corrosive.   Stray currents cause corrosion only when they flow from metal to an electrolyte, usually the earth.   Only direct currents cause corrosion under normal conditions.

Much underground corrosion is attributable, not to the character of the metal used or the soil, but to conditions incidental to pipeline construction, such as the interconnection of old and new pipe, the crossing of different soils or soil horizons, voids in the backfill, and current picked up from other structures.

There are several tests which will indicate whether soils are potentially corrosive. When a sufficient number of tests are made, they yield satisfactory indications as to the locations and extent of corrosive areas.   The correlation of the results of tests of single samples of soil with the corrosion observable at the points where the samples were taken may not be good because other factors than soil characteristics may control the corrosion and because the soil samples may not be representative of conditions at the point of corrosion.

There is no standard or generally accepted criterion for corrosivity or of corrosion resistance.   The relative merits of materials with respect to corrosion may change with the time of exposure, the area exposed, and the conditions of exposure.

Usually most of the commonly used ferrous materials, including many low-alloy steels, corrode at nearly the same rates when exposed to the same soil conditions. Lead corrodes slowly in most soils because of the formation of protective layers of carbonate and sulfate.   Copper and alloys high in copper corrode much more slowly than ferrous materials in most soils.   Copper is much affected by soils containing hydrogen sulfide.

Bituminous coatings usually are imperfect or develop imperfections.   Most of them after a few years permit some corrosion, but the better coatings reduce losses of weight and pit depths for 10 years or more.   Zinc is the only metal extensively used for underground coatings.   Its effectiveness is temporary.

Cathodic protection properly applied and maintained is an effective means of preventing corrosion.   Under some conditions a combination of a protective coating and cathodic protection affords the most economical means of preventing corrosion.

Cathodic protection may not always be feasible, so that the engineer may have to place full reliance on external protective coatings.   These may be classified as *inorganic* and *organic*.   The inorganic can be zinc (in which case the pipe is called "galvanized") or, much less frequently, lead.   Lead coatings nowadays are confined to electrical cables, and the expense of lead piping has resulted in a tendency to use substitutes.   External concrete coatings have been used, but these are more properly considered structural features.

Organic coatings are myriad. The coal-tar enamels of past years are still employed in large-diameter water mains in spite of the rising cost of this protection. One reason for this retention is, of course, the great amount of investigative work that led to the present development of such coatings. Furthermore, the experience of several generations shows that these coatings are long lasting, and until the present-day synthetic plastic coatings can acquire an equivalent of experience, the conservative engineer will prefer to use time-tested coatings.

This is not to detract from the value of the excellent coatings produced today. Linings of natural rubber, synthetic rubber, polyvinyl chloride, polyethylene, and similar materials are produced in resin form for shop application, tape and emulsion form for field application as well as in a variety of combinations for both shop and field application. Concrete is used very satisfactorily for internal coating. One recent development has been the application of concrete coatings to existing water mains, in which the concrete is compacted in place.

Rubber coatings are too expensive, except in unusual cases, and are susceptible to bacteria. Furthermore, the other synthetic plastics may be regarded as suitable for all soil conditions at temperatures less than 150 F if the problem of protecting the exposed ends of the lengths of pipe is satisfactorily solved. Flanged ends, mechanical couplings, application of hot or cold coatings, etc., have been used with good results. The choice is usually made on the basis of cost, which will vary with each job.

All these coatings are highly resistant to oxygen corrosion, soil-acid corrosion, and bacterial attack, either singly or in combination. They are susceptible to physical damage, such as from scraping, and all are thermally sensitive to extreme temperatures.

Rigid polyvinyl chloride pipe is used in increasing amounts to convey cold (less than 140 F) corrosive solutions with excellent results. In the application of polyvinyl chloride as a piping material, note should be taken of its high rate of thermal expansion (approximately ten times that of carbon-steel pipe), its low softening temperature, and its relative fragility in the event of impact or earth movement.

Organic piping materials are far from new. Wood fiber pipe, such as Orangeburg, has been in use for decades. Recently semiflexible synthetic piping of styrene-based plastics, with and without filler material, has been produced. These materials have a use, but care should be exercised in their employment, particularly where buried by backfilling.

The greatest of care is required when piping material devoid of a metal core (or with a reduced-thickness metal core) is buried. This material can very easily fail under tension if the supporting bed is washed away.

The most recent addition to the piping engineer's armory is Schedule 10 aluminum pipe with synthetic-plastic internal and external lining. These pipes are designed for use with mechanical couplings which have been given an equal plastic coating. Where this pipe is employed, the greatest of care must be used in backfilling. This is not so easy as it sounds, as pipeline practice in backfilling is based largely on dumping earth on impact-resisting carbon-steel pipe, with or without later compacting.

Atmospheric corrosion is largely influenced by climate. Local practice is the best guide in combatting it. Where such corrosion is severe, recourse must be had to protective coatings or to nonferrous metals. Shop-applied coatings are available for nearly every purpose. The engineer must select that coating which will be the most economical over the expected life of the system. Wind loadings, icing, and similar unusual conditions must be considered.

Brass and copper pipe will often be inherently resistant to atmospheric corrosion, but problems of initial cost generally prohibit such installations.

For well casings where corrosion cannot be tolerated, such expensive materials as Ni-Resist, a high-nickel-content cast iron, or 18-8 stainless steel may be used. This is not truly a piping problem but rather a detail of such distinct technologies as oil-well drilling where the rewards justify the cost. Further, well casing can sometimes be reclaimed and reused, thereby decreasing unit cost.

The process industries use a wide variety of corrosion-resistant metals for piping, such as the stainless steels of the ferritic, martensitic, and austenitic varieties. Nuclear technology employs such exotic piping materials as tantalum- and colombium-stabilized stainless steel, titanium, and tantalum. Here corrosion must be minimized, as corrosion products in solution or suspension may acquire radioactivity, the prevention (or minimization) of which justifies the expense involved.

Highly corrosive liquids such as the mineral acids require special treatment. It has been the practice to use carbon-steel pipe to carry concentrated sulfuric acid (66 Baumé and above) in spite of the associated unsatisfactory experience. Carbon-steel tanks are almost universally used to store acid, although here the service has been more satisfactory, if not ideal.

Factors Which Primarily Influence the Localization of Corrosion

Dissimilar metal junctions cause concentrated localized corrosion. This phenomenon is so well known that its avoidance should be routine. However, there are circumstances at times which require differing metals in piping systems. In such cases flanged connections should be made with an insulating gasket plus insulating sleeves for flange bolts.

Concentration cells are among the prime causes of pitting. These can occur from different substances in solution, especially in a semistagnant system. Differing concentrations of one particular electrolyte can set up a concentration cell.

Water-line corrosion is an example of corrosion caused by differing concentrations of the same corroding agent, in this case oxygen. Some water-line corrosion is due to damage of the protective film caused by ripples or waves.

Strain aggravates corrosion of metal in contact with unstrained metal. Strain imposed on a corroding metal may rupture a protective film, especially repeated strains.

Conductance of the corroding solution can be an important factor, especially under favorable conditions. This is best exemplified in boiler salines. If the dissolved solids are permitted to increase beyond safe limits, the rate of corrosion jumps. This is especially true if traces of oxygen are present.

Stray currents cause corrosion at departure points. This phenomenon applies to direct current. Alternating currents cause little or no corrosion. Stray current corrosion has decreased in importance with the reduction in prevalence of urban trolley cars.

Other factors that accelerate or retard corrosion include:

1. Duration of attack.
2. Temperature.
3. Acquired "passivation" from attack of certain chemicals.
4. An apparent "passivation" that some metals, notably steel, acquire from repetitive pounding, vibration, or rolling. This is apparent only. Because of such treatment the metal tends to corrode at a uniform rate and shows no pitting.

RECOMMENDED PIPING MATERIALS

The choice of piping material for any particular service is influenced by many factors. Frequently it is found that the most acceptable material from a technical

viewpoint is quite expensive; in such an instance, some users prefer to use a material of poorer corrosion characteristics in order to prevent the high first costs associated with the superior material.   Other factors which must be considered in material selection include the strength or concentration of the fluid to be handled, the nature of operation (continuous or intermittent) of the system, temperature and pressure of the fluid medium, the ambient environment in which the system is to be installed, the anticipated service life of the installation, skill of the local labor force which installs the piping system, and the applicability of any local building or construction codes.

Table 6 lists materials which are recommended for use in the conveying of specific fluid media.   For cases which are encountered most frequently, the recommendations reflect the best thinking of industrial users and are based on industry-wide practice as it existed in 1966.   As newer materials become available, it may be that several of the recommendations will require alteration.   Finally, intelligent application of the various recommendations is necessary, particularly in view of the many factors which are involved in the suitability of any specific material for a given service.

Table 6.   Recommended Piping Materials

Absorption Oil. Full information on components is required, since the absorbed material dictates material selection.

Acetate Solvents. Steel piping is standard in industry.   Refining of particular acetate may require 18-8 Cr-Ni to prevent discoloration.

Acetic Acid. Red brass piping is used extensively.   18-8 Cr-Ni is also used, especially in valving.   If the acid temperature is less than 150 F, polyvinyl chloride piping is satisfactory.

Acetic Anhydride. Aluminum piping is used extensively.   Piping is often glass-lined to offer greater anticorrosive characteristics.

Acetone. Steel pipe, valves, and fittings are standard in industry.   Plastic materials must not be used.

Acetylene. Steel is used for high-pressure applications.   Red brass is preferred, but its use is limited to relatively low pressures.

Acid Mine Waters. See Mine Waters.

Air Compressed, Commercial. Steel piping and cylinders are used despite the corrosion encountered.   18-8 Cr-Ni is advisable for valving.

Air Compressed, Dried. Copper and aluminum tubing are used extensively.   PVC and polyethylene piping are used for lower pressures.   Steel piping is satisfactory.

Alcohols. Steel piping is standard in the industry.   Perfume and pharmaceutical industries incline toward the use of 18-8 Cr-Ni piping.   Copper coils have been used for many years with good results.

Aldehydes. Steel piping is the standard material used by industry.

Alums, Aluminum Sulfates. 18-8 Cr-Ni-Mo is usually satisfactory.   If the material to be conveyed is at elevated temperatures, Inconel or monel should be used.

Ammonia. Steel pipe, valves, and fittings are standards for industrial applications.

Ammonia Liquors. Steel pipe, valves, and fittings are standard materials.

Ammonium Chloride. Steel pipe, valves, and fittings are used in production piping systems. 18-8 Cr-Ni and high-alloy steels are used in purification systems.

Ammonium Hydroxide, Pure. 18-8 Cr-Ni piping is used for chemically pure and reagent grades.

Ammonium Nitrate. This substance is very unstable, and the use of 18-8 Cr-Ni material is advisable throughout the complete piping system.

Ammonium Phosphate, Mono- and Dibasic. Steel pipe, valves, and fittings are standard.

Ammonium Phosphate, Tribasic. Steel pipe, valves, and fittings are standard.

Ammonium Sulfate. Steel pipe, valves, and fittings are standard in the fertilizer industry. Free acid may require the use of 18-8 Cr-Ni-Mo valves.

Amyl Acetate. See Acetate Solvents.

Amyl Alcohol. See Alcohols.

Aniline, Aniline Oils. 18-8 Cr-Ni or monel piping, valves, and fittings are used.   Value of the product usually determines the choice of piping materials.

Aniline Dyes. 18-8 Cr-Ni is the standard for the finished product.

Asphalts. Steel pipe, valves, and fittings are the industry standard.

Barium Chloride. Process industries use carbon steel.   Refined product generally requires 18-8 Cr-Ni or monel.   Solutions at temperatures below 150 F can be carried in polyvinyl chloride piping.

Table 6. **(Continued)**

Barium Hydroxide. Carbon steel is in general usage. Suspended material may abrade valve seats, and these should be trimmed with 18-8 Cr-Ni.

Barium Sulfate. This substance is inert but abrasive. Use valves with 18-8 Cr-Ni trim.

Barium Sulfide. Industry standard is carbon steel.

Beer. Industry uses quantities of 18-8 Cr-Ni pipe and valves, which permit use of corrosive cleaning solutions. PVC can also be used.

Benzine, Petroleum Naphtha. See Petroleum Oils and Solvents.

Benzols. See Coal-tar solvents.

Bleaches. Glass, PVC, rubber, or Saran-lined pipe is used where temperature permits. 18-8 Cr-Ni piping is recommended for perborates. Hastelloy C or Chlorimet is a good material for use at higher temperatures.

Borax. Carbon steel is the industry standard. Protect valve seats from erosion with monel.

Bordeaux Mixtures. The industrial standard is carbon steel. Monel trim on valves is desirable.

Boric Acid. Pharmaceutical grades require 18-8 Cr-Ni or monel pipe. PVC is excellent at solution temperature below 150 F.

Brines. Polyvinyl chloride piping is used for temperatures below 150 F, and red brass or monel for higher temperatures. 18-8 Cr-Ni is unsuitable because of stress-corrosion cracking. Glass-lined pipe is used at medium pressures.

Bromine. Glass-lined pipe is used. Hastelloy C material is advisable where temperatures or pressures are high. Material is lethal and justifies expense.

Butane. See Gases, Hydrocarbon.

Calcium Bisulfate. Polyvinyl chloride piping is used for temperatures below 150 F, and 18-8 Cr-Ni for higher temperatures.

Calcium Hydroxide (Milk of Lime). Carbon steel is used as an industrial standard. This substance in suspension is abrasive, and Hastelloy C trim on valves is advisable.

Calcium Hypochlorite. For use at solution temperatures below 150 F, polyvinyl chloride is an excellent piping material.

Calcium Sulfate. The industrial standard is carbon steel. Suspended material is mildly abrasive. 18-8 Cr-Ni should be used as valve trim.

Cane-sugar Liquors. The industrial standard is carbon steel. Control valves should be trimmed with monel or Hastelloy.

Carbolic Acid, Phenol. See Phenol.

Carbon Dioxide. The industrial standard is carbon-steel piping with alloy-trimmed valves.

Carbon-Disulfide. Carbon steel is used frequently despite the corrosion suffered. 18-8 Cr-Ni trim for valves is most advisable.

Carbonic Acid. Red brass or 18-8 Cr-Ni pipe is recommended except that aluminum tubing is satisfactory if the operating pressure is low enough.

Caustic Potash. See Sodium Hyaroxide.

Caustic Soda. See Sodium Hydroxide.

Cellulose Acetate. This material is inert, but the selection of piping material must take into account the solvents which are used in fluidizing the cellulose.

Cellulose Nitrate. This substance is unstable, and the choice of piping materials depends on the solvents which are present.

China-wood Oil. See Drying Oils.

Chlorinated Solvents. Monel is used for temperatures above 150 F, and polyvinyl chloride is used for lower temperatures.

Chlorine Gas. Carbon steel may be used in the absence of moisture. Solvent-welded PVC pipe, valves, and fittings are standard for temperatures below 150 F.

Chloracetic Acid. Silver or Hastelloy piping materials are essential to prevent contamination.

Chromate Solutions. This substance is used to protect carbon-steel piping systems. Concentrations below 300 ppm may, however, fail to afford the desired protection.

Chromic Acid. 18-8 Cr-Ni piping material is recommended. The tanning industry uses polyvinyl chloride piping at temperatures below 150 F.

Citric Acid. 18-8 Cr-Ni and monel materials are used in production piping systems. Descaling of tanks and boilers employs sacrificial carbon-steel systems.

Coal-tar Solvents. Carbon-steel pipe, valves, and fittings are used as the industrial standard. Monel-trimmed valves are advisable to prevent corrosion from contaminants.

Copper Cyanide. See Sodium Cyanide.

Copper Sulfate. Polyvinyl chloride pipe, valves, and fittings are used at temperatures below 150 F, and 18-8 Cr-Ni is used for higher temperatures.

Cottonseed Oil. See Edible Oils.

Creosotes. See Phenol.

Distilled Water. Polyvinyl chloride pipe, valves, and fittings are used at temperatures below 150 F. Aluminum, glass, or 18-8 Cr-Ni piping are used at higher temperatures.

Distillery Wort. Tinned copper is the industrial standard for this service.

Table 6. (*Continued*)

Drying Oils. Aluminum or 18-8 Cr-Ni piping is required to prevent discoloration in the refined product. Production piping is usually of carbon steel.

Edible Oils. Aluminum, 18-8 Cr-Ni, or monel piping should be used. Glass-lined steel pipe and PVC pipe are used where temperatures permit.

Ethers. Industry standard is carbon steel.

Fatty Acids. See Acetic Acid (typical).

Ferric Chloride. Polyvinyl chloride is an excellent piping material for temperatures below 150 F. High-temperature solutions may require exotic (titanium or tantalum) piping. Glass-lined steel pipe may be used for ambient-temperature service.

Ferric Sulfate. 18-8 Cr-Ni-Mo is the best material for this service, but PVC pipe, valves, and fittings may be used for operating conditions below 150 F and 150 psig.

Ferrous Chloride. PVC pipe, valves, and fittings may be used for service conditions below 150 F and 150 psig. Monel or nickel materials must be used for higher temperatures and pressures.

Ferrous Sulfate. Carbon-steel pipe, valves, and fittings may be used for all applicable temperatures and pressures.

Ferrous Sulfate, Chlorinated. PVC pipe valves and fittings may be used for service conditions below 150 F and 150 psig. Nickel or monel material must be used at elevated temperatures and pressures.

Foamite. Carbon-steel pipe, valves, and fittings are used throughout industry for this service.

Formaldehyde. Carbon-steel pipe, valves, and fittings are satisfactory for this substance.

Formic Acid. Carbon-steel and nonferrous materials are entirely unsatisfactory. PVC may be used for services below 150 F and 150 psig, but elevated temperatures require the use of 18-8 Cr-Ni-Mo tubing.

Fruit Juices. PVC pipe, valves, and fittings are ideal for use at temperatures below 150 F.

Fuming Nitric Acid. See Nitric Acid.

Fuming Sulfuric Acid. See Sulfuric Acid, second paragraph.

Gallic Acid. Nickel, monel, or Alloy-20 pipe, valves, and fittings are recommended for this service.

Gases, Fuel. Carbon steel and red brass are conventional materials. Corrosion is apt to be due to impurities. Hastelloy trimming of valves is advisable.

Gases, Hydrocarbon. Carbon steel and red brass are conventional materials. Corrosion from impurities is less likely, but hard-metal trimming of valves is justifiable.

Gases, Inert. Carbon-steel pipe, valves, and fittings are used. Welding of pipelines is recommended if costly gases are present.

Gases, Refrigerant. Carbon-steel pipe, valves, and fittings are conventionally used materials. Copper or brass is acceptable in the absence of moisture.

Gasoline. See Petroleum Oils and Solvents.

Gelatin. Aluminum and 18-8 Cr-Ni-Mo piping are used for medium to high temperature and pressure. PVC is satisfactory if temperature and pressure permit its use.

Glauber Salt. See Sodium Sulfate.

Glucose. Steel piping, valves, and fittings are used by the industry.

Glue. Steel piping, valves, and fittings are satisfactory for this service.

Glycerin (Glycerol). See Alcohols.

Grain Alcohol. See Alcohols.

Helium. See Gases, Inert.

Hydrochloric Acid (Muriatic Acid). Because this substance corrodes most metals, Hastelloy C and titanium are used. Cold acid may be handled with PVC piping and tank linings. European practice is to use glass-lined steel pipe.

Hydrocyanic Acid. Red brass is used as gas-purification piping; for other purposes, the use of steel piping is standard.

Hydrofluoric Acid. Ferrous and nonferrous metals are corroded rapidly. Glass and clay products are very rapidly attacked. Lead piping is used frequently, but Hastelloy C has been found to be the most satisfactory material.

Hydrofluosilic Acid. Lead piping and Hastelloy C are used extensively.

Hydrogen. See Gases, Inert.

Hydrogen Peroxide. Aluminum and 18-8 Cr-Ni piping materials are used extensively.

Hydrogen Sulfide. Carbon-steel pipe, valves, and fittings are the industry standard. Polyvinyl chloride piping and fittings are satisfactory at temperatures and pressures below 150 F and 150 psig.

"Hypo." See Sodium Thiosulfate.

Iodine. Glass-lined pipe is recommended; Hastelloy C and titanium are also acceptable.

Kerosene. See Petroleum Oils and Solvents.

Lacquers. Steel piping may be used. Aluminum and 18-8 Cr-Ni are also used frequently depending on the grade and cost of product.

Table 6. *(Continued)*

Lime Sulfur. Steel piping is the standard material for this service, but valves suffer from erosion and should be trimmed with an erosion-resistant material such as 18-8 Cr-Ni or Hastelloy.

Linseed Oil. See Drying Oils.

Lubricating Oils. See Petroleum Oils and Solvents.

Lye. See Sodium Hydroxide.

Magnesium Chloride. See Brines.

Magnesium Sulfate. Steel piping is used, but corrosion is rapid. Cold solutions can satisfactorily use PVC, but higher temperatures require the use of 18-8 Cr-Ni.

Mercuric Chloride. Hastelloy C and monel are extensively used.

Mercury. Steel pipe is used. Damage occurs owing to the erosive, rather than corrosive, nature of mercury. Nonferrous metals must never be used.

Methane. See Gases, Hydrocarbon. Carbon-steel pipe, valves, and fittings are employed in methane service.

Methanol; Methyl Alcohol. See Alcohols. Carbon-steel pipe, valves, and fittings are used in industry. 18-8 Cr-Ni material is advisable where a pure product is desired.

Milk. Surprisingly enough, milk is highly corrosible. PVC and Saran-lined pipe, valves, and fittings are recommended.

Milk of Lime. Carbon steel is used in industry, but the piping tends to become blocked as the result of scale deposits. The use of crosses, instead of elbows, permits the blocked lines to be rodded.

Mine Waters. PVC pipe, valves, and fittings are satisfactory. Glazed vitrified tile is suitable but is more expensive.

Mixed Acids. Components must be identified. Passivated (Parkerized) 18-8 Ni-Cr-Mo pipe is usually satisfactory. Exotic (titanium or tantalum) piping may be required for high temperatures.

Molasses. Teflon-lined pipe and Teflon-diaphragm Saunders valves are ideal. PVC may be used if temperature permits. Carbon steel is used but requires care if contamination is to be avoided.

Naphtha, Coal Tar; Benzeno; Benzol. Carbon steel is used in industry. Alloy piping is advisable at high temperature.

Natural Gas. See Gases, Fuel. Carbon-steel pipe, valves, and fittings are used. Monel trimming of valves is advisable.

Neon. See Gases, Inert.

Nickel Chloride. Manufacture of the pure salt requires the use of monel piping. In the electroplating industry, 18-8 Cr-Ni piping is used.

Nickel Sulfate. 18-8 Cr-Ni and monel piping materials are used in the electroplating industry.

Niter Cake. See Sodium Bisulfate.

Nitrating Acids. Components must be identified. Passivated (Parkerized) 18-8 Ni-Cr-Mo piping has been used. Titanium piping may be usable. Glass-lined pipe is suitable if pressure permits.

Nitric Acid. Passivated 18-8 Cr-Ni-Mo pipe or Alloy-20 pipe, valves, and fittings are used. Glass-lined pipe, if temperature and pressure permit, is perfectly satisfactory.

Nitrobenzene; Oil of Mirbane. Steel pipe, valves, and fittings are used, but valves should be trimmed with 18-8 Cr-Ni steel.

Nitrogen. See Gases, Inert.

Oil of Vitriol; 66-deg Sulfuric Acid. See Sulfuric Acid.

Oleic Acid. 18-8 Cr-Ni-Mo is used for high-temperature applications in the industry. Monel is employed in distilling equipment and associated piping.

Oleum Acid; Fuming Sulfuric Acid. See Sulfuric Acid.

Oleum Spirits. See Petroleum Oils and Solvents.

Oxalic Acid. Red brass and monel are used extensively.

Oxygen. Steel pipe, valves, and fittings are suitable for low-temperature work. Brass fittings are preferable at medium temperatures, since corrosion rate increases with temperature.

Palmitic Acid. 18-8 Cr-Ni-Mo material is used for high-temperature applications in the soap industry. In distillation equipment and in the transporting of the pure acid, the use of monel is recommended.

Pentane. See Gases, Hydrocarbon.

Petroleum Oils and Solvents. Carbon-steel pipe, valves, and fittings are standard for the refining industry. Control valves (which are reusable) are usually made of a material which is at least equal to 18-8 Cr-Ni-Mo. Stage of process determines piping materials. Low-flash-point fractions may be conveyed in 18-8 Cr-Ni-Mo welded pipe to lower the fire hazard and not because the pipe corrodes.

Phenol. Carbon steel, despite its short life in this service, is used in the byproduct coke industry. In chemical refining plants, aluminum, monel, and 18-8 Cr-Ni-Mo are conventionally used materials.

Table 6. *(Continued)*

Phosphoric Acid. The use of material equal to at least 18-8 Cr-Ni-Mo is advisable.

Phthalic Acid. For alcoholic solutions at low temperatures, the use of polyvinyl chloride piping is satisfactory. Aluminum or 18-8 Cr-Ni tubing is also recommended.

Pickling Acids. In the steel industry, carbon-steel pipe is generally used. This material, however, corrodes quite rapidly and requires frequent replacement. 18-8 Cr-Ni would be more resistant to corrosion, but its additional first cost might not be justified.

Picric Acid. Carbon steel or aluminum pipe is used for molten acid. The pharmaceutical industry uses 18-8 Cr-Ni and glass or PVC tubing.

Potassium Carbonate. See Sodium Carbonate.

Potassium Chloride. The cost of the product may dictate the use of 18-8 Cr-Ni material.

Potassium Cyanide. See Sodium Cyanide.

Potassium Hydroxide. See Sodium Hydroxide.

Potassium Nitrate. See Sodium Nitrate.

Potassium Sulfate. See Sodium Sulfate.

Potassium Sulfide. See Sodium Sulfide.

Printing Inks. 18-8 Cr-Ni is recommended in general; for a specific ink composition, a better material may be available.

Producer Gas. See Gases, Fuel.

Propane. See Gases, Hydrocarbon.

Propionic Acid. See Acetic Acid.

Propyl Alcohol. See Alcohols.

Prussic Acid. See Hydrocyanic Acid.

Pyridine. Steel pipe, valves, and fittings are used. Any corrosion is due to contaminants and not to the pyridine.

Pyrogallic Acid; Pyrogallol. See Phenol.

Pyroligneous Acid; Pyroligneous Liquor. See Acetic Acid.

"Red Oil." See Oleic Acid.

Return Condensate. Corrosion is due to dissolved oxygen or carbon dioxide. If inconvenient or impossible to prevent these, the return line must be at least as corrosion-resistant as red brass. The employment of cycling amines should be studied. Schedule 5 or Schedule 10 AISI-304 stainless tubing may be practical. The costs of replacing the return line may be the determining factor. Copper tubing or pipe should not be used.

Rosin. Steel pipe, valves, and fittings are used with the crude product. Aluminum or 18-8 Cr-Ni pipe is required where discoloration is intolerable.

Salammoniac. See Ammonium Chloride.

Salicylic Acid. Nonferrous metals are used for transporting the crude product. Nickel and monel are used for pharmaceutical grades. The use of Pyrex glass pipe and fittings should be studied where piping must be repeatedly cleaned with strong oxidants to prevent contamination.

Salt. See Sodium Chloride.

Salt Cake. See Sodium Acid Sulfate.

Salt Water; Sea Water. Red brass or aluminum brass is used for small-diameter, low-velocity service. Velocity is highly important. PVC is excellent below 150 F and 150 psig. Concrete pipe or concrete-lined steel pipe is used for large-diameter service (medium-high pressure). For temporary service, fire hose, subsequently washed with fresh water after service, is used. Cast-iron, carbon-steel, aluminum, and 18-8 stainless-steel piping all corrode rapidly. Copper pipe or tubing is unsatisfactory. Rubber-lined valves (Saunders or butterfly) are quite satisfactory. Where cast iron or carbon steel is used, as in condenser water boxes, the exposed surfaces must be painted with zinc-rich paints, frequently "touched up." Cathodic protection is usually essential. Pump impellers and casings are either rubber lined or made of a corrosion-resisting bronze.

Sewage. Vitrified clay, brick, concrete, or concrete-lined pipe may be used. Cast iron should not be used in inaccessible places unless exposed surfaces are protected. Ni-Resist cast iron may be satisfactory. Cupronickel is unsuitable, as hydrogen sulfide may be encountered as a sewage constituent.

Shellac. Red brass, Ni-Cr-Mo pipe, or monel are recommended. Schedule of piping should be carefully adjusted to the pressure of the process. Schedule 20 (or even thinner) can often be used.

Sludge Acids. The composition of such acids is indeterminate. The only piping material which gives assured protection is glazed vitrified clay with acidproof cement joints.

Soap. Carbon-steel pipe and fittings are used with concentrated soap solutions such as are encountered in manufacturing processes. PVC pipe and fittings are ideal if temperature and pressure permit. Salt is usually present as a contaminant, and for this reason, 18-8 Cr-Ni-Mo is not satisfactory. Monel should be used for valve trim. If metal must be used extensively, Type 430 martensitic steels should be employed.

Soda-ash. See Sodium Carbonate.

Table 6. (*Continued*)

Sodium Acid Sulfate; Sodium Bisulfate. Ferrous metals are not satisfactory. PVC pipe, valves, and fittings are to be employed where temperature and pressure permit. Red brass and IBBM valves are used for high temperature and pressure.

Sodium Bicarbonate. Steel pipe is used for fire-extinguishing service. Alloy-20 valves and 18-8 Ni-Cr-Mo pipe are also used. PVC pipe and valves may be used where temperature and pressure permit. Nonferrous piping must not be used.

Sodium Carbonate. Nonferrous metals should not be used. Welded steel pipe and fittings are excellent. Valves should be of Hastelloy or faced with monel. PVC is an excellent material if temperature and pressure permit its use.

Sodium Chloride. Ferrous materials corrode rapidly. 18-8 Cr-Ni suffers from stress-corrosion cracking at temperatures above 150 F. Red brass and IBBM valves can be used. PVC pipe, valves, and fittings are completely satisfactory for service up to 150 F and 150 psig. Saran-lined pipe and fittings may be used for temperatures up to 190 F. Titanium and tantalum pipe are used for molten sodium chloride.

Sodium Cyanide. For electroplating service, sacrifical ferrous piping has been used, but this is dangerous. NaCl contamination makes the use of 18-8 Cr-Ni piping inadvisable at temperatures above 150 F. Below this temperature, PVC piping is good at pressures below 150 psig. At higher pressures (up to 300 psig) flanged, Saran-lined pipe, valves, and fittings may be used.

Sodium Hydroxide. Carbon steel is used for handling the crude solution. If discoloration must be avoided, nickel pipe should be used. 18-8 Cr-Ni piping is not always satisfactory. PVC is satisfactory for low-temperature and low-pressure service provided that the solution concentration does not exceed 75 per cent by weight of sodium hydroxide.

Sodium Hypochlorite. No completely satisfactory metallic material is known, but Hastelloy C and tantalum have been used for short-service-life installations. Saran-lined pipe, valves, and fittings have been used successfully. Epoxy-lined duct work and also concrete and glazed tile have been found to be satisfactory for low-pressure service.

Sodium Hyposulfite. See Sodium Thiosulfate.

Sodium Metaphosphate ("Glassy Phosphate"). Red brass pipe and fittings, with IBBM valves, are used. Schedule 10 18-8 Cr-Ni pipe and fittings have been used satisfactorily. For low-pressure water-treatment work, PVC piping systems and tank linings are best.

Sodium Nitrate. Carbon-steel pipe, valves, and fittings are standard.

Sodium Nitrite. Carbon-steel pipe, valves, and fittings are standard.

Sodium Perborate. The recommendation as given for sodium hydroxide should be followed.

Sodium Peroxide. The recommendations as given for sodium hydroxide should be followed.

Sodium Phosphate, Orthodibasic. Carbon steel can be used.

Sodium Phosphate, Orthomonobasic. 18-8 Cr-Ni piping should be used.

Sodium Phosphate, Orthotribasic. Carbon-steel pipe, valves, and fittings are standard.

Sodium Silicate. Carbon-steel pipe, valves, and fittings are standard.

Sodium Sulfate. Carbon-steel pipe, valves, and fittings are used, but 18-8 Cr-Ni trim for valves is recommended.

Sodium Sulfide. Carbon-steel piping is used frequently, but it does corrode and must be replaced. 18-8 Cr-Ni and 18-8 Cr-Ni-Mo piping are being used more in current practice and are the recommended materials for this service.

Sodium Thiosulfate ("Hypo"). Carbon steel is used in production piping. The refined product requires the use of 18-8 Cr-Ni-Mo material.

Soybean Oil. See Vegetable Oils.

Stearic Acid. Rendering plants use carbon-steel piping systems. Refining operations require the use of 18-8 Cr-Ni, Inconel, or monel.

Sulfate Liquors ("Black," "green," or "white" liquors used in paper-pulp manufacture). Carbon steel is still used, although corrosion occurs. 18-8 Cr-Ni steels are being used in newer plants.

Sulfur. Carbon-steel pipe, valves, and fittings are standard.

Sulfur Chloride. Carbon steel is suitable for moisture-free gas. Wet gas is best handled in Inconel or monel pipe, but titanium tubing is also used.

Sulfur Trioxide. Carbon-steel pipe and fittings are the standard for the chemical industry. 18-8 Cr-Ni-Mo valves are recommended, but Teflon-faced diaphragm valves are also used.

Sulfuric Acid. The standard of commerce is the 66 Baumé acid. Industry as a whole employs carbon-steel tanks with a corrosion allowance being provided. These are usually estimated to be durable for 20 years. Some corrosion is encountered in these tanks but is tolerable. Much, if not most, of the acid of commerce is carried in welded carbon-steel pipe with carbon-steel fittings. This requires much inspection, a great deal of maintenance, and frequent replacement. Failures at welds are frequent. Where integrity of service is paramount, Carpenter-20 pipe and fittings with Alloy-20 or Teflon diaphragm valves should be used. Titanium and tantalum piping are being used increasingly.

Table 6. (*Continued*)

Carbon steel cannot be used with dilute solutions. Red brass is used. Cold dilute solutions are best carried in PVC or Saran-lined pipe.

Lead linings are used under the misapprehension that these are not attacked. This is not so. Contamination from lead sulfate is ever present.

Oleum and fuming sulfuric acid are strong oxidizers and dehydrants. Where liquids are being conveyed, exotic materials such as tantalum and columbium will be required. The danger to personnel from leaky piping is obvious.

Sulfurous Acid. Lead piping and linings are satisfactory for use with cold acid. Otherwise, the use of 18-8 Cr-Ni-Mo pipe, valves, and fittings is necessary.

Tannery Liquors. 18-8 Cr-Ni is generally recommended. Inconel or monel may be required for certain concentrations.

Tannic Acid. 18-8 Cr-Ni should be used.

Tar. Carbon-steel pipe, valves, and fittings are standard.

Tartaric Acid. Glass, aluminum, 18-8 Cr-Ni, and monel are recommended, depending on the value of the product.

Titanium Chloride. Glass, monel, or titanium may be used. This substance acts like hydrochloric acid. The use of 18-8 Cr-Ni piping is to be avoided.

Toluol. See Coal-tar Solvents.

Trichloracetic Acid. Follow the recommendation as given for chloracetic acid.

Trichlorethylene. See Chlorinated Solvents.

Triethanolamine. Carbon-steel pipe, valves, and fittings are standard.

TSP (Trisodium phosphate). See Sodium Phosphate, Orthotribasic.

Turpentine. Carbon-steel pipe, valves, and fittings are standard in production systems. The purified product requires the use of monel for distillation equipment and piping.

Varnish. Carbon steel may be used, but if the product is to be free of coloration, aluminum or 18-8 Cr-Ni piping must be used.

Vegetable Oils. The use of carbon steel and nonferrous metals should be avoided. 18-8 Cr-Ni or monel is recommended for vegetable-oil service.

Vinegar. Monel, Inconel, or titanium should be used if sodium chloride is present. The use of copper and carbon steel should be avoided.

Water. Freshly distilled or demineralized water is very aggressive. To avoid "pickup," red brass, aluminum, or 18-8 Cr-Ni piping is used. PVC is an excellent material if the water temperature is low enough to permit its use.

Potable water is carried in carbon-steel, copper, aluminum, and lead pipe. Corrosion is usually mild at low temperatures. Potable-water supplies should be carefully examined before piping materials are selected. PVC is excellent for use at a temperature below 150 F.

Water Glass. See Sodium Silicate.

Whisky. Copper and brass are standard. 18-8 Cr-Ni is being used extensively in current installations.

White Liquor. See Sulfate Liquors.

Wine. Copper and brass have been used extensively for centuries. 18-8 Cr-Ni recently is coming into use.

Xylene. See Coal-tar Solvents.

Zinc Chloride. Carbon steel is used in the wood-preservative industry despite the corrosion encountered. Use of 18-8 Cr-Ni is to be avoided. Hastelloy C, monel, and Inconel should be used for valve trim.

Zinc Sulfate. 18-8 Cr-Ni-Mo and Inconel are recommended materials if discoloration must be avoided.

10

BUILDING HEATING SYSTEMS

C. J. Danowitz*

Heating is necessary to provide an environment within a space for the mainten-ance of comfort of the occupants by preventing the too-rapid loss of heat from the body. For a sensation of comfort, heat from the body must be dissipated at the same rate as it is produced. A too-rapid loss of heat results in the sensation of coldness, and a slow loss of heat results in the sensation of warmness. If the ambient air, walls, ceiling, and floors are heated, the rate of heat loss in the body can be controlled and the individual will feel comfortable.

Heat is lost from the body by radiation, conduction, convection, and evaporation.

The loss by *radiation* is affected by the temperature of the walls, floor, and ceiling of the enclosure. The lower these temperatures are, the greater is the loss by radiation. The *mean radiant temperature* (MRT) is the weighted average of the various floor, walls, and ceiling temperatures. When the MRT is below approxi-mately 98.6 F, the human body loses heat by radiation; when higher, the body absorbs heat from the surroundings.

Loss of heat by *convection* is affected by the temperature of the surrounding air and the air motion. The cooler the air and the higher the air velocity, the greater the loss by convection.

Evaporation of moisture from the body depends on the relative humidity, air movement, and air temperature. Evaporation increases with a decrease in relative humidity and an increase in air motion, which given a sensation of coolness to the body, and decreases with an increase of air temperature, which gives a feeling of warmth.

It is possible, within limits, to vary the means of providing heat into the space to obtain the same feeling of comfort. For example, if with cold walls the loss by

* Ebasco Services, Inc., New York, N.Y.

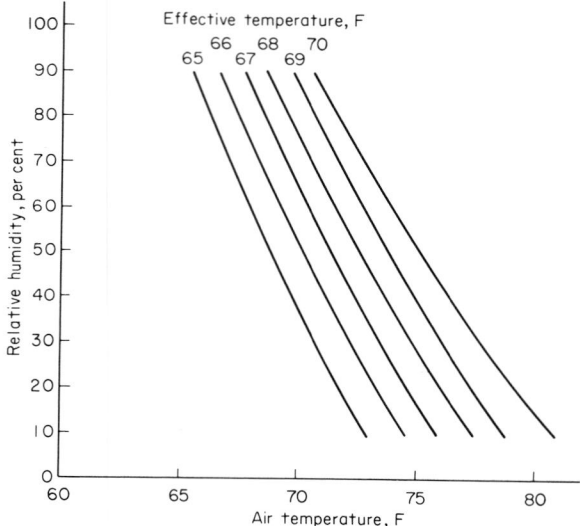

Fɪɢ. 1. Effective temperature.

radiation is high, the loss by convection and evaporation can be reduced by maintaining higher air temperatures.   There are many combinations of air temperature, mean radiant temperature, relative humidity, and air movement that will result in a comfortable environment.

One practical method of combining some of these variables is by use of the effective temperature (ET) curves of Fig. 1, devised by the ASHVE and which take into consideration air temperature, humidity, and air motion at less than 25 fpm. It has been found that the majority of people in the United States feel comfortable at an ET of 68 F.

Temperatures.   Inside design temperatures of 70 to 74 F are commonly used for most residential and commercial buildings; temperatures of 55 to 65 F are usual for industrial buildings.

Table I. Inside Air Temperatures

Type of building	Temp, F
Houses	73–75
Stores	65–68
Public buildings	72–74
Factories and machine shops	60–65
Schools:	
Classrooms	72–74
Assembly room	67–72
Toilet and locker rooms	68–70
Hospitals:	
Private rooms and wards	72–74
Operating room	70–95
Bathrooms	70
Theaters	68–72
Hotels:	
Bedrooms and bathrooms	75
Dining room	72
Ballroom	65–68

Table 1 gives the inside temperatures which are usually specified for the various types of buildings and spaces.

The selection of a satisfactory outside design temperature should be considered very carefully. Study of the weather records shows that the lowest temperature on record does not repeat itself each year. For economical reasons it is generally satisfactory to use a design temperature that will not be lower but once in 13 years; if maintenance of the inside temperature is critical, then the lowest temperature should be used.

Table 2 gives the lowest temperature and the temperature normally used for design for some of the cities of the United States.

The ground temperature under basement floors is not affected greatly by outside temperature and can generally be assumed at 50 F. The temperature of unheated spaces adjacent to heated spaces may generally be assumed to be the arithmetic mean of the outdoor and heated-space temperature.

Heat Loss. The capacity or size of the various components of the heating system is based on the fundamental premise that heat supplied to a space must equal the heat lost from the space. This is a direct consequence of the first law of thermodynamics for a stationary system in the steady state.

Heat is supplied to the space by direct radiation such as radiators or radiant panels or by convection through the use of convectors or air systems.

When a building or space is maintained at a temperature higher than that of the outside or adjoining space, loss of heat takes place by conduction of heat through walls, roof, and floor and by leakage of colder air into the building.

The amount of heat lost through the walls, roof, and floors depends upon the thermal conductivity of the materials of which they are built and upon the surface resistances. If a wall is composed of successive layers of different materials and thicknesses X_1, X_2, X_3, etc., having thermal conductivity K_1, K_2, K_3, etc., then the overall coefficient of heat transfer for the entire wall is

$$U = \frac{1}{1/f_i + X_1/K_1 + X_2/K_2 + X_3/K_3 + \cdots + 1/f_o}$$

where U = overall coefficient of heat transmission, Btu/(hr sq ft degree F difference in temperature)

X_1, X_2, X_3 = thickness of material(s), in.

K_1, K_2, K_3 = coefficient of thermal conductivity of material(s), Btu/(hr sq ft × inch thickness × degree F difference in temperature)

f_i and f_o = film or surface conductance of inside and outside surface, respectively, Btu/(hr sq ft degree F difference in temperature)

(For an average, the value of $f_i = 1.65$ and $f_o = 6.0$ can be used.)

Transmission Coefficients. The values of U for various standard forms of building construction are given in Table 3. The U factors for many additional types of construction can be obtained from the latest issue of the ASHRAE "Guide and Data Book."

Transmission Heat Losses. The heat losses by the combined processes of conduction and convection are given by the formula

$$H = AU(t_i - t_o)$$

where H = heat transmitted through the wall, roof, floor, or ceiling, Btu/hr

A = area of surface, sq ft

U = coefficient of transmission, Btu/(hr sq ft degree F temperature difference)

t_i = indoor temperature, F

t_o = outdoor temperature, F

Table 2. Outside Temperatures[1]

State	City	Lowest temp, F	Normal design temp, F
Ala.	Birmingham	6	12
Alaska	Anchorage	−31	−24
	Ketchikan	− 3	4
Ariz.	Phoenix	33	36
	Flagstaff	−12	− 4
Ark.	Little Rock	2	8
Calif.	Los Angeles	37	41
	San Francisco	34	37
Colo.	Denver	−19	−12
	Grand Junction	−11	− 3
Conn.	Hartford	− 7	− 2
	New Haven	− 5	0
Fla.	Jacksonville	24	28
	Pensacola	19	24
Ga.	Atlanta	5	11
Idaho.	Boise	−20	−10
Ill.	Cairo	− 8	0
	Chicago	−17	−11
Ind.	Fort Wayne	−13	− 7
	Indianapolis	−14	− 8
Iowa.	Des Moines	−19	−13
Kans.	Wichita	−13	− 6
Ky.	Lexington	− 9	− 2
La.	New Orleans	21	26
	Shreveport	8	14
Maine	Portland	−15	− 9
Md.	Baltimore	3	8
Mass.	Boston	− 6	0
Mich.	Detroit	−10	− 4
	Lansing	−14	− 8
Minn.	Duluth	−32	−27
	Minneapolis	−28	−23
Miss.	Vicksburg	9	15
Mo.	Kansas City	−15	− 8
	St. Louis	−12	− 5
Mont.	Billings	−42	−31
	Helena	−53	−39

Table 2. (*Continued*)

State	City	Lowest temp, F	Normal design temp, F
Neb.........	Omaha	−24	−17
Nev.........	Reno	− 4	3
N.J.	New Brunswick	− 1	4
	Trenton	− 4	0
N.M.	Santa Fe	− 2	3
N.Y.	Albany	−12	− 9
	Buffalo	−11	− 5
	New York	1	5
N.C.........	Raleigh	9	14
N.D.	Bismarck	−38	−31
Ohio........	Cincinnati	− 9	− 3
	Cleveland	−11	− 5
Okla........	Oklahoma City	− 9	− 1
Ore.........	Portland	1	10
Pa.	Pittsburgh	− 9	− 3
	Philadelphia	1	6
R.I.	Providence	− 4	1
S.C.	Columbia	14	19
S.D.........	Pierre	−28	−22
Tenn........	Knoxville	− 3	5
	Memphis	0	6
Tex.	Abilene	0	7
	Dallas	1	8
	San Antonio	12	19
Utah	Salt Lake City	− 8	− 1
Vt.	Burlington	−23	−17
Va..........	Richmond	6	11
Wash.	Seattle	9	15
	Spokane	−28	−16
W. Va.......	Parkersburg	− 7	− 1
Wis.........	Green Bay	−27	−20
	Milwaukee	−24	−17
Wyo.	Cheyenne	−26	−19

[1] From ASHRAE "Guide and Data Book," 1961. Used by permission.

Table 3. Coefficient of Heat Transmission U, Btu/(hr F sq ft)[1]

Exterior	Type of sheathing	5/8-in. plaster on wall	Metal lath and 3/4-in. plaster	Wood lath and 1/2-in. plaster	1/2-in. insulating board on furring
	Frame walls				
Wood siding or clapboard	Wood sheathing (25/32 in.)	...	0.26	0.24	0.19
	Insulated board sheathing (25/32 in.)	...	0.21	0.19	0.16
	Plywood (5/16 in.)	...	0.31	0.29	0.22
Wood shingles	Wood sheathing (25/32 in.)	...	0.23	0.21	0.18
	Insulated board sheathing (25/32 in.)	...	0.18	0.18	0.15
	Plywood (5/16 in.)	...	0.27	0.25	0.20
Stucco	Wood sheathing (25/32 in.)	...	0.31	0.29	0.22
	Insulated board sheathing (25/32 in.)	...	0.24	0.22	0.18
	Plywood (5/16 in.)	...	0.40	0.36	0.26
Brick veneer	Wood sheathing (25/32 in.)	...	0.29	0.27	0.21
	Insulated board sheathing (25/32 in.)	...	0.22	0.21	0.17
	Plywood (5/16 in.)	...	0.36	0.32	0.24
	Masonry walls				
Brick (face and common):					
6 in.		0.64	0.39	0.35	
8 in.		0.45	0.31	0.29	
12 in.		0.33	0.25	0.23	
16 in.		0.26	0.21	0.20	
Brick (common only):					
8 in.		0.39	0.28	0.26	
12 in.		0.30	0.23	0.22	
16 in.		0.24	0.19	0.18	
Stone (lime and sand):					
8 in.		0.63	0.39	0.35	
12 in.		0.52	0.34	0.31	
16 in.		0.45	0.31	0.29	
24 in.		0.35	0.26	0.24	

Hollow clay tile:			
8 in.	0.36	0.26	0.26
16 in.	0.31	0.24	0.23
12 in.	0.29	0.22	0.21
Concrete block:			
Gravel agg., 8 in.	0.48	0.33	0.30
Gravel agg., 12 in.	0.45	0.31	0.29
Cinder agg., 8 in.	0.37	0.27	0.25
Cinder agg., 12 in.	0.35	0.26	0.24

Built-up masonry walls

Exterior facing	Backing			
4-in. face brick or 4-in. stone	Cinder block:			
	4 in.	0.39	0.28	0.26
	8 in.	0.32	0.24	0.23
	12 in.	0.30	0.23	0.22
	Hollow clay tile:			
	4 in.	0.39	0.28	0.26
	8 in.	0.30	0.23	0.22
	12 in.	0.25	0.20	0.19
	Concrete block:			
	4 in.	0.56	0.36	0.33
	6 in.	0.52	0.34	0.31
	8 in.	0.48	0.32	0.30
4-in. common brick or 8-in. precast concrete	Cinder block:			
	4 in.	0.35	0.26	0.24
	8 in.	0.29	0.22	0.21
	12 in.	0.27	0.21	0.20
	Hollow clay tile:			
	4 in.	0.35	0.26	0.24
	8 in.	0.28	0.22	0.20
	12 in.	0.23	0.19	0.18
	Concrete block:			
	4 in.	0.48	0.32	0.30
	6 in.	0.44	0.31	0.28
	8 in.	0.41	0.29	0.27

Table 3. (*Continued*)

Masonry cavity walls

Exterior facing	Backing		⅝-in. plaster on wall	Metal lath and ¾-in. plaster	Wood lath and ½-in. plaster	½-in. insulating board on furring
Exterior	Inner section					
4-in. face brick	Cinder block, 4 in.		0.29	0.22	0.21	
	Common brick, 4 in.		0.32	0.24	0.23	
	Clay tile, 4 in.		0.29	0.22	0.21	
4-in. common brick or 4-in. concrete block.	Cinder block 4 in.		0.26	0.21	0.20	
	Common brick, 4 in.		0.29	0.22	0.21	
	Clay tile, 4 in.		0.26	0.21	0.20	
4-in. cinder block	Cinder block, 4 in.		0.24	0.19	0.18	
	Common brick, 4 in.		0.26	0.21	0.20	
	Clay tile, 4 in.		0.24	0.19	0.18	

Interior frame partitions	Gypsum board, ⅜-in.	Metal lath and ¾-in. plaster	Wood lath and ½-in. plaster	Plywood, ½-in.
Finish on one side only	0.60	0.67	0.57	0.50
Finish on both sides	0.34	0.39	0.32	0.28

Interior masonry partitions	None	⅝-in. plaster (one side)	⅝-in. plaster (two sides)
Cinder block:			
3 in.	0.45	0.43	0.41
4 in.	0.40	0.39	0.37
8 in.	0.32	0.31	0.30
12 in.	0.31	0.30	0.29
Concrete block:			
8 in.	0.40	0.39	0.37
12 in.	0.38	0.36	0.35
Hollow clay tile:			
3 in.	0.46	0.44	0.42
4 in.	0.40	0.39	0.37
8 in.	0.35	0.33	0.32
12 in.	0.31	0.30	0.29
Hollow gypsum tile:			
3 in.	0.37	0.35	0.34
4 in.	0.33	0.32	0.31
Metal lath and plaster on metal core, 2½ in.	0.55		

Interior frame floors	Ceiling			
	None	Metal lath and ⅜-in. plaster	Wood lath and ½-in. plaster	Acoustical tile, ¾ in.
Type of floor:				
Wood subfloor, 25/32 in.	0.45	0.31	0.29	0.21
Cement, 1½ in., and ceramic tile, ½ in.	0.38	0.28	0.26	0.19
Hardwood floor, ¾ in.	0.34	0.26	0.24	0.18
Plywood, ⅝ in. and linoleum	0.32	0.25	0.23	0.17

Table 3. (*Continued*)

Interior concrete floor or ceiling

Type of deck	Type finish floor	None	⅛-in. plaster	Suspended ceiling metal lath and ¾-in. plaster	Suspended ceiling ¾-in. acoustical tile
Concrete, 4 in.	None	0.62	0.61	0.38	0.22
	Linoleum	0.60	0.59	0.38	0.22
	Wood block, 13/16 in.	0.42	0.42	0.30	0.19
	Floor on sleepers with linoleum	0.30	0.30	0.23	0.16

Wood or metal flat roofs (with built-up roof included)

Type of deck	Ceiling	None	Insulation on top of deck		
			1 in.	2 in.	3 in.
Flat metal	None	0.67	0.23	0.15	0.10
	Suspended plaster	0.32	0.17	0.12	0.09
	Suspended acoustical tile	0.23	0.14	0.11	0.08
Preformed slat (wood fiber and cement binder), 2 in.	None	0.20	0.13	0.10	0.07
	Suspended plaster	0.15	0.11	0.08	0.05
	Suspended acoustical tile	0.13	0.09	0.08	0.05
Concrete slab, 4 in.	None	0.51	0.21	0.14	0.10
	Suspended plaster	0.28	0.16	0.12	0.09
	Suspended acoustical tile	0.21	0.13	0.10	0.08
Wood, 2 in.	None	0.28	0.16	0.11	0.08
	Suspended plaster	0.19	0.13	0.10	0.07
	Suspended acoustical tile	0.16	0.11	0.09	0.07

| Pitched roofs | | Ceiling | | |
Exterior surface	Sheathing	Metal lath and 3/4-in plaster	Wood lath and 1/2-in. plaster	1/2-in. insulating board on furring
Asphalt shingles.............	Building paper on 5/16-in. plywood	0.32	0.29	0.22
	Building paper on 25/32-in. wood	0.27	0.25	0.20
Wood shingles	Building paper on 5/16-in. plywood	0.29	0.27	0.21
	Building paper on 25/32-in. wood	0.25	0.23	0.19
Slate tiles.................	Building paper on 5/16-in. plywood	0.38	0.35	0.26
	Building paper on 25/32-in. wood	0.31	0.28	0.22

Windows and doors:
Single window, wood or metal sash	1.13
Double window.................	0.55
Solid wood doors:	
1 in........................	0.64
1 in. with glass storm door.............	0.37
2 in........................	0.43
2 in. with glass storm door.............	0.28
Floor on grade, concrete.............	0.10

[1] From ASHRAE "Guide and Data Book," 1961. Used by permission.

Infiltration. The heat required to raise the temperature of the air that enters a space by leakage from the outside is given by the formula

$$H_s = 0.018Q(t_i - t_o)$$

where H_s = heat required to raise temperature of air leaking into a space, Btu/hr
   Q = leakage volume of outside air entering building, cu ft/hr
   t_i = indoor temperature, F
   t_o = outdoor temperature, F
   0.018 = constant accounting for specific heat and density of air, Btu/(cu ft F)
     (This value is based on an air density of 0.075 lb/cu ft.)

For average conditions, an air change per hour of $\frac{1}{2}$ the volume of the space can be assumed. Under unusual conditions of construction or stack effect due to height of the building, a more detailed calculation must be made.

Calculation of Heat Losses. For calculating the heat loss from a structure, the procedure is as follows:

1. Select the design outdoor and indoor temperature.
2. Determine the heat-transmission coefficients and areas for outside walls, glass, floor, roof, and ceilings.
3. Compute the transmission heat losses H.
4. Compute the infiltration heat loss H_s.
5. The sum of the transmission heat losses H plus the infiltration heat loss H_s is the design heat loss for the building.

Selection of Boilers. The selection of the boiler should include, in addition to the calculated design heat loss, the heat required for domestic hot water, piping losses, and pickup allowance.

For heating domestic hot water a minimum load would be to heat 2 to 3 gpm of water through 100 F.

The piping loss, which is the heat lost in the system before being utilized at the heating units, can be assumed at 10 per cent of the design load for large installations and at 20 per cent of the design load for small installations.

The pickup allowance, which is the increased capacity to be allowed in the system to bring a cold system up to operating temperature, can also be assumed at 10 to 20 per cent.

HEATING UNITS

To supply heat in an amount equal to that which is lost from a structure, various types of heating units are installed within the space. For steam and hot-water heating systems this heat is supplied by radiators, convectors, baseboard, finned-tube radiation, or unit heaters. For a panel-heating system, pipe coils are fabricated for this purpose. For air systems, heat-transfer coils are used.

The output of heating units is expressed in 1000 Btu/hr (MBh) or in square feet of equivalent direct radiation (EDR), with 1 sq ft EDR equal to 240 Btu/hr.

Radiators. Radiators are of cast iron, are built in sections, and are currently of the small tube or wall type.

Heat is emitted from an ordinary radiator partly by radiation and partly by convection. The amount of heat emitted per square foot of surface depends upon the shape of the radiator and the proportion of exposed radiating surface.

A simplified schedule of stock sizes of cast-iron radiators has been established by manufacturers in cooperation with the National Bureau of Standards and issued as Simplified Practice Recommendation R174-47. The rated heating outputs of

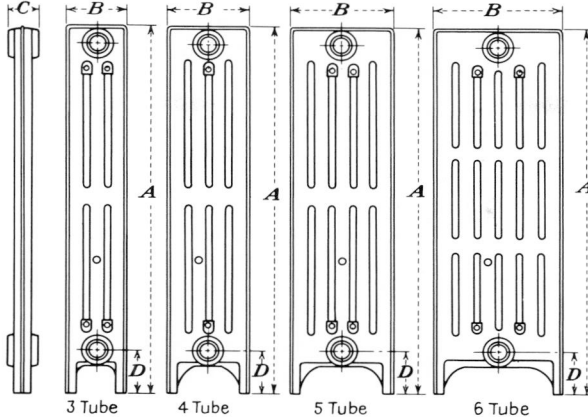

3 Tube   4 Tube   5 Tube   6 Tube

Fig. 2. Cast-iron tube radiators. (See Table 4 for dimensions.)

radiators for various heights and widths, as given in this simplified practice recommendation, are indicated in Table 4 and in Fig. 2. Table 5 contains common, although not standardized, sizes of wall radiators.

Convectors. Convectors are enclosed heating units which transmit heat to the room by convection by the circulation of air over the heating element. The heating elements used in convectors are of the extended-surface types and are usually made of copper, brass, cast iron, or steel. The heating element is placed low in the enclosure to secure a maximum stack effect. The heat output can be controlled by means of a damper in the air outlet or in the stack.

Table 4. Heating Surface and Sizes of Cast-iron Radiators

Number of tubes per section	Catalog rating per section,[1] sq ft	Section dimensions[2]				
		A Height,[3] inches	*B* Width, inches		*C* Spacing, inches	*D* Leg height,[2] inches
			Min	Max		
3	1.6	25	3¼	3½	1¾	2½
4	1.6	19	4⁷⁄₁₆	4¹³⁄₁₆	1¾	2½
	1.8	22	4⁷⁄₁₆	4¹³⁄₁₆	1¾	2½
	2.0	25	4⁷⁄₁₆	4¹³⁄₁₆	1¾	2½
5	2.1	22	5⅝	6⁵⁄₁₆	1¾	2½
	2.4	25	5⅝	6⁵⁄₁₆	1¾	2½
6	1.6	14	6¹³⁄₁₆	8	1¾	2½
	2.3	19	6¹³⁄₁₆	8	1¾	2½
	3.0	25	6¹³⁄₁₆	8	1¾	2½
	3.7	32	6¹³⁄₁₆	8	1¾	2½

[1] These ratings are based on steam at 215 F and air at 70 F. They apply only to installed radiators exposed in a normal manner, not to radiators installed behind enclosures, grilles, or under shelves.

[2] See Fig. 2.

[3] Overall height and leg height of radiator as made by some manufacturers is 1 in. greater than shown in columns *A* and *D*. Radiators may be furnished without legs. Where greater than standard leg heights are required, this dimension shall be 4½ in.

Table 5. Cast-iron Wall Radiators

Approximate dimensions, in.			Heat output	
Height	Length or width	Depth	Sq ft	Btuh
13¼	16½	3	6½	1560
13¼	22	3	8	1920
22	13¼	3	8	1920
13¼	29	3	11	2640
29	13¼	3	11	2640

These ratings are based on steam at 215 F and air at 70 F.

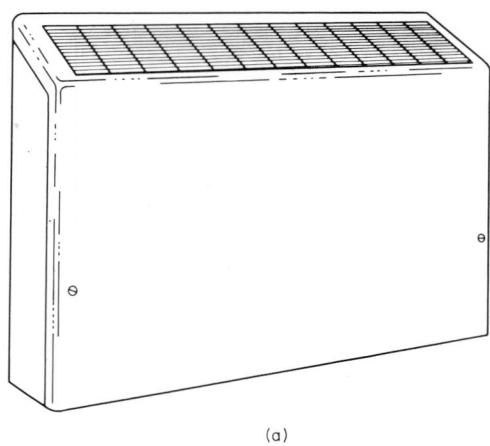

(a)

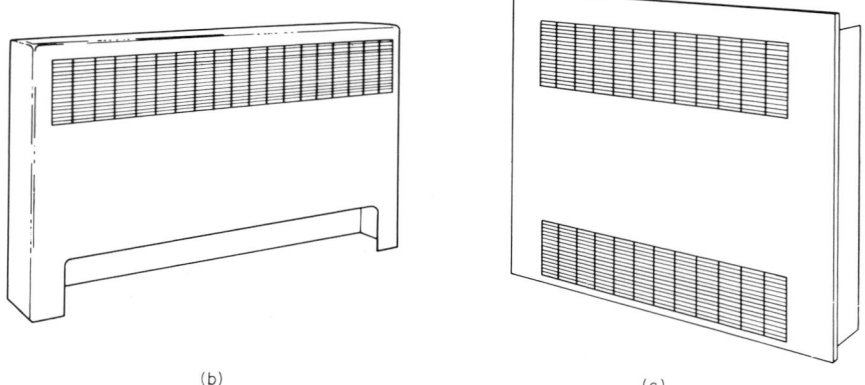

(b)   (c)

FIG. 3. Convectors. (*a*) Wall hung. (*b*) Free standing. (*c*) Recessed.

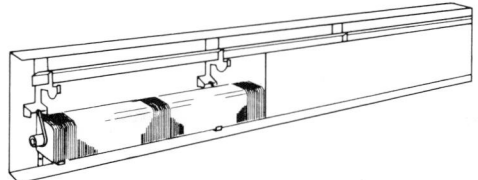

Fig. 4. Baseboard unit.

Convectors are made in a variety of depths, sizes, lengths, and enclosures. The basic sizes are listed in Simplified Practice Recommendation 238-50, Convectors. Convectors are of the free-standing, wall-hung, or recessed types as shown in Fig. 3. Since the output of convectors varies, the capacities should be obtained from manufacturers' catalogues.

Baseboard Heaters. Baseboard heaters are heating units located along the wall replacing the conventional wooden baseboard. These units are of the radiant, radiant-convector, or finned-tube type; the last type is shown in Fig. 4. As with convectors, outputs of these units should be obtained from manufacturers' catalogues.

Finned-tube Radiation. Finned-tube radiation consists of a pipe (1 to 2 in. diameter) to which metallic fins are bonded. These units are either ferrous (steel pipe with steel fins) or nonferrous (copper or aluminum) tubes with copper or aluminum fins.

These units are provided with different enclosures, as shown in Fig. 5, and can be installed one, two, or three rows high. The return bends should be welded or soldered.

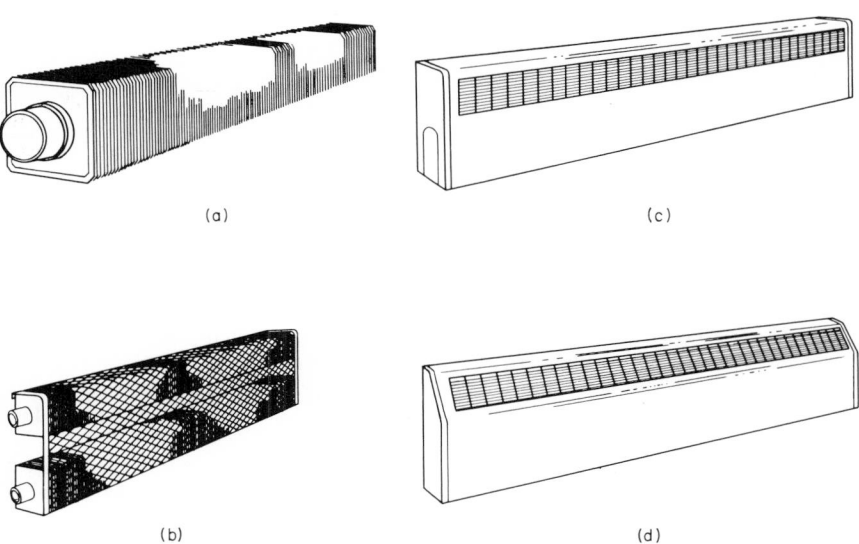

Fig. 5. Finned-type radiation. (*a*) Single-row type. (*b*) Two-row type with wire protector. (*c*) Free-standing enclosure. (*d*) Wall-hung enclosure.

Since the heating capacity varies over a wide range depending on pipe size, fin size and spacing, and type of enclosure, the outputs should be obtained from manufacturers' catalogues.

Radiator and Convector Correction Factors. The correction factors in Table 6 give the amount of direct radiation or convection surface required at other than standard conditions which are taken as 215 F steam or water temperature, 70 F air for radiators, and 65 F inlet air for convectors.

Table 6. Correction Factors for Cast-iron Radiators and Convectors[1]

Steam pressure, approximate		Heating medium temp. F,	Factors for direct cast-iron radiators							Factors for convectors						
			Room temperature F							Inlet air temperature F						
Gauge vacuum in. Hg	Abs. psi	steam or water	80	75	70	65	60	55	50	80	75	70	65	60	55	50
22.4	3.7	150	2.58	2.36	2.17	2.00	1.86	1.73	1.62	3.14	2.83	2.57	2.35	2.15	1.98	1.84
20.3	4.7	160	2.17	2.00	1.86	1.73	1.62	1.52	1.44	2.57	2.35	2.15	1.98	1.84	1.71	1.59
17.7	6.0	170	1.86	1.73	1.62⁴	1.52	1.44	1.35	1.28	2.15	1.98	1.84	1.71	1.59	1.49	1.40
14.6	7.5	180	1.62	1.52	1.44	1.35	1.28	1.21	1.15	1.84	1.71	1.59	1.49	1.40	1.32	1.24
10.9	9.3	190	1.44	1.35	1.28	1.21	1.15	1.10	1.05	1.59	1.49	1.40	1.32	1.24	1.17	1.11
6.5	11.5	200	1.28	1.21	1.15	1.10	1.05	1.00	0.96	1.40	1.32	1.24	1.17	1.11	1.05	1.00
Psi																
1	15.6	215	1.10	1.05	1.00	0.96	0.92	0.88	0.85	1.17	1.11	1.05	1.00	0.95	0.91	0.87
6	21	230	0.96	0.92	0.88	0.85	0.81	0.78	0.76	1.00	0.95	0.91	0.87	0.83	0.79	0.76
15	30	250	0.81	0.78	0.76	0.73	0.70	0.68	0.66	0.83	0.79	0.76	0.73	0.70	0.68	0.65
27	42	270	0.70	0.68	0.66	0.64	0.62	0.60	0.58	0.70	0.68	0.65	0.63	0.60	0.58	0.56
52	67	300	0.58	0.57	0.55	0.53	0.52	0.51	0.49	0.56	0.54	0.53	0.51	0.49	0.48	0.47

To determine the size of a radiator or a convector for a given space, multiply the heat loss of the space in Btu per hour by the proper factor from the above table and select radiator or convector having an equivalent Btu-per-hour rating.

[1] From "Heating, Ventilating, Air Conditioning Guide," 1958. Used by permission.

Unit Heaters. Unit heaters consist of an assembly of a fan and motor, heating element, enclosure, and directional outlet. Unit heaters, because of their fans, have greater heating capacity than a gravity heating unit and are, therefore, more economical for large heating installations.

Unit heaters may be secured to discharge air either horizontally or vertically. The heating elements are of steel, cast iron, copper, or aluminum and usually are of extended-surface design. Unit heaters are intended primarily to handle all re-circulated air within a space but may also be provided to ventilate a space with either part or all outdoor air. The units shown in Fig. 6 are the types most commonly used. Either steam or hot water may be used as the heating medium.

Heat-transfer Surface. In a large air-type heating system, heating of the air is effected by forced convection over extended steel or nonferrous coils (Fig. 7) which are built in a number of types for steam or hot water. The rate of heat transmission, final air temperature, and the air pressure drop for these blast coils should be obtained from manufacturers' catalogues.

Enclosure and Paint. Usually an enclosure which is placed above a direct radiator restricts the air flow and reduces the output. A properly designed enclosure, however, can increase the output to that of an unenclosed radiator by increasing the convection effect.

For a cast-iron radiator or baseboard, a finish coat of bronze or aluminum paint decreases the heat output by approximately 10 per cent.

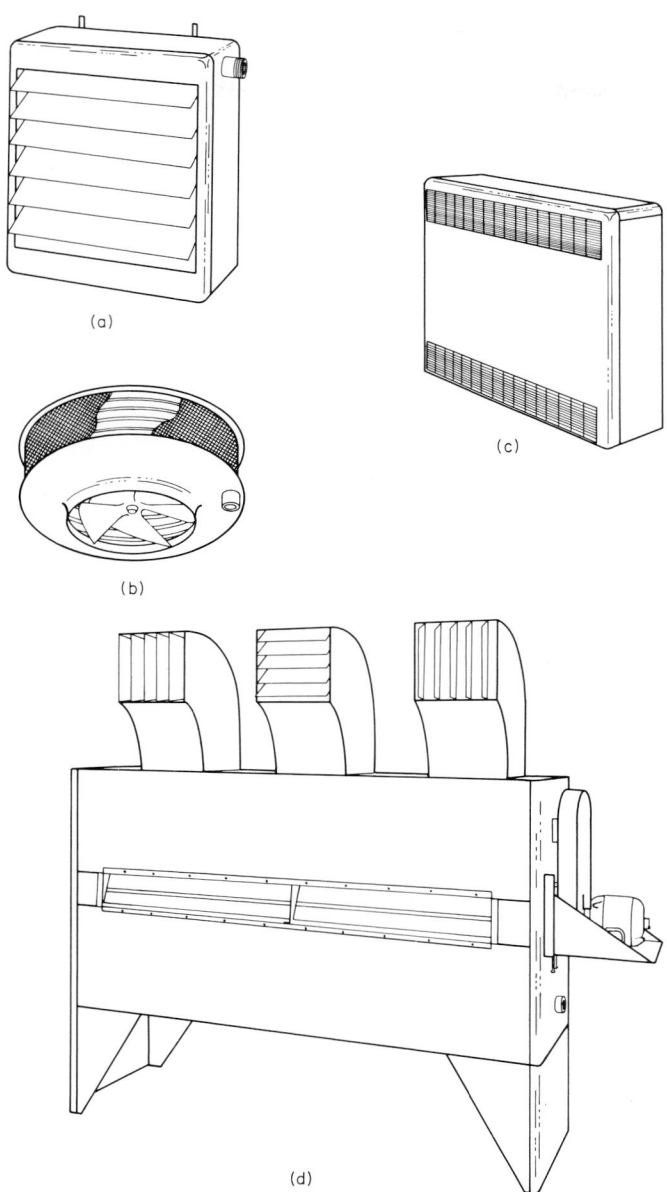

(a)

(b)

(c)

(d)

Fig. 6. Unit heaters. (a) Horizontal-discharge propeller type. (b) Vertical-discharge propeller type. (c) Cabinet type. (d) Blower type. (*Courtesy of Dunham-Bush Inc.*)

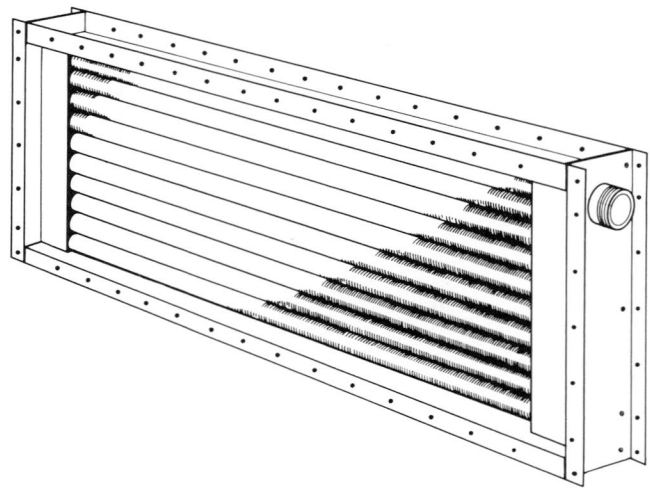

FIG. 7. Extended-surface coil.

STEAM HEATING SYSTEMS

For a steam heating system to operate satisfactorily, the piping and the heating units must be arranged so that steam is conveyed to the radiators, air is removed from the radiators, and condensate is drained from the radiators and returned to the boiler.

Steam heating systems may be classified according to their piping arrangement, operating pressure, and method of returning condensate to the boiler. In actual installations, however, the classification is not precise because a system, owing to economic or practical necessity, may combine several characteristics.

Piping Arrangement. The *one-pipe system* uses a single pipe to supply steam and to return condensate. Ordinarily, there is one connection at the heating unit for both supply and return, although some units are provided with two connections which are used for supply and return connections to the single pipe. This system is best suited to small installations where it can be installed easily at low initial cost. It is not used for large installations because of the impracticability of providing adequate automatic control at each heating unit and the possibility of higher initial cost than some other system.

The *two-pipe steam system* uses one pipe to carry the steam supply to the heating units and another to return the condensate. This system eliminates the problem normally associated with the one-pipe system in that the steam and condensate are usually flowing in the same direction. Individual control, either manual or automatic, can be provided at each heating unit.

Piping arrangements can be further classified as *up-flow* or *down-flow*, depending on the direction of steam flow in the risers, and as *dry-return* or *wet-return*, depending on whether the condensate mains are above or below the water lines of the boiler or condensate receiver.

Operating Pressure. Steam heating systems may be classified as high-pressure, low-pressure, vapor, and vacuum systems, depending on the pressure conditions under which the system is designed to operate.

A *high-pressure* system is one whose operating pressure is above 15 psig; a *low-pressure* system is one in which the pressures vary from 0 to 15 psig; a *vapor* system

operates either as a low-pressure system or at subatmospheric pressures but does not utilize a vacuum pump; a *vacuum* system operates either as a low-pressure system or at subatmospheric pressures, but a vacuum pump is employed to return condensate to the boiler.

Return of Condensate. A *gravity return* system is one in which all the heating units are a sufficient distance above the water line of the boiler so that the condensate can flow freely by gravity.

A mechanical return system is one in which the condensate cannot be returned to the boiler by gravity and either traps or pumps are used. Three general types of mechanical condensate-return devices are used: (1) condensate-return pump, (2) vacuum return pump, and (3) the alternating return trap.

General Description

One-pipe Systems. One-pipe heating systems, as previously defined, are systems in which steam and condensate flow in the same pipe and generally operate

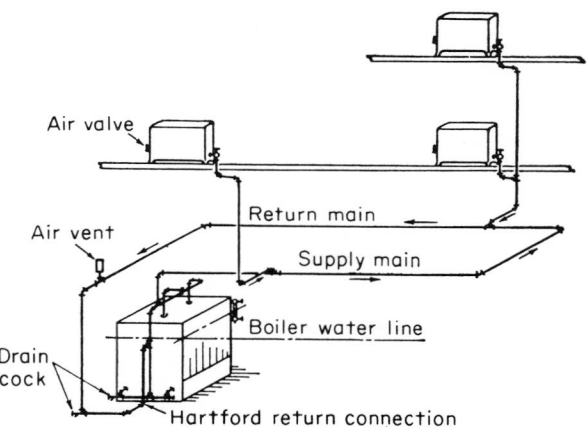

FIG. 8. One-pipe up-feed gravity system. (*From "ASHRAE Guide,"* 1960.)

at low-pressure, vapor, or vacuum conditions. The steam and condensate flow in opposite directions to each other and thus require a pipe of large enough size to prevent the carrying of condensate along with the steam. The one-pipe systems usually operate at low pressures which lie in the range of 0 to 15 psig.

The one-pipe systems have piping arrangements as follows:

1. *Up-feed.* The supply main runs in a loop usually in the basement, and the radiators and other heating units are located above the supply piping mains. This supply main is usually run full size and is pitched downward in the direction of steam flow. A take-off is provided from this main to each heating unit. Steam and condensate are conveyed in the same direction in the main and in opposite directions to the radiators. Figure 8 shows essential details of a typical up-feed system.

2. *Down-feed.* The supply main rises from the boiler above the heating units and is looped around the upper part of the building, usually in the attic, and the radiators and other heating units are located below the supply main. Connections are made to each heating unit from vertical connections, and the returns

are connected in the basement to a loop which is pitched toward the boiler. The steam and condensate flow in the same direction in the main and the vertical connection. Only at the take-off to each heating unit do steam and condensate flow in opposite directions. The supply steam main is kept relatively free of condensate by dripping through the drop risers.

Typical connections from steam mains for up-feed and down-feed steam systems are shown in Figs. 9 and 10.

An automatic air vent must be provided at each heating unit and at each of the mains to relieve the air under pressure and to permit steam to enter the radiator. When steam enters the radiator, the thermostatic element in the vent becomes heated and closes, thereby preventing steam from escaping.

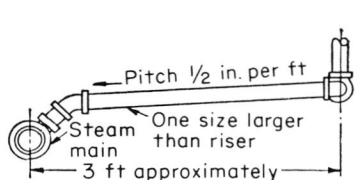

FIG. 9. Typical steam runout: undripped riser. (*From "ASHRAE Guide,"* 1960.)

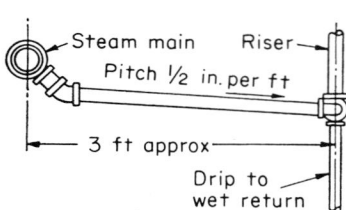

FIG. 10. Typical steam runout: dripped riser. (*From "ASHRAE Guide,"* 1960.)

Air valves are of the pressure or vacuum types and operate as automatic air vents in releasing air from the radiator. The *pressure type* permits air to return into the system when the steam pressure falls below atmospheric; the *vacuum type* prevents the air from returning into the system and thereby maintains vacuum conditions in the system. Systems which use vacuum valves are known as vapor or vacuum one-pipe systems, and these systems will generally maintain a more uniform temperature condition in the space than do pressure systems.

Two-pipe Systems. Two-pipe heating systems, as previously defined, are systems in which steam and condensate flow in separate pipes and may operate at high-pressure, low-pressure, vapor, or vacuum conditions. These systems may be arranged for up-feed or down-feed of the steam supply main as described for one-pipe systems and may return condensate to the boiler by either gravity or mechanical means. Individual valves, either manual or automatic, are usually provided at each heating unit to control the temperature of the system by regulating the flow of steam into the unit.

Two-pipe steam heating systems are usually classified as follows:

1. *High-pressure* systems operate at pressure above 15 psig and are used generally in large industrial or commercial buildings. The use of high-pressure steam permits the use of smaller pipe because of the greater pressure drop allowed in the system. In addition because of the high-pressure differential between steam and return mains, it is possible to locate returns above the heating units and lift the condensate to these returns in all cases except those in which automatic control valves are installed.

2. *Low-pressure* systems operate at pressures from 0 to 15 psig and can be of either the up-feed or down-feed types as shown in Figs. 11 and 12. The condensate is discharged from the heating unit through thermostatic traps which remain closed while steam is at the trap but will open when sufficient condensate collects to cool the element. In this system, owing to the type of air vents, air is permitted to

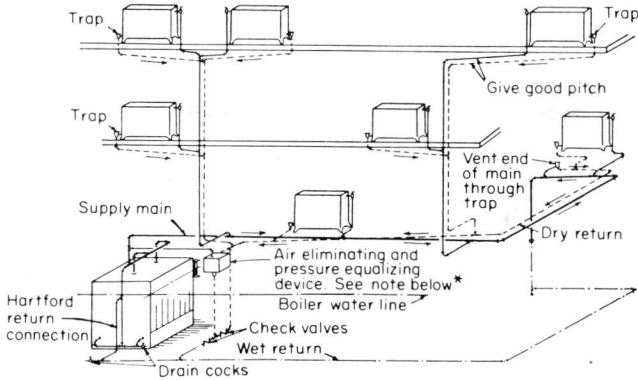

Fig. 11. Up-feed two-pipe system with automatic return. (*From "ASHRAE Guide," 1960.*)

enter the system. Low-pressure systems are not so popular as vapor systems in that the heating unit loses heat when the rate of supply steam is reduced. Corrosion may also occur because of the continued supply of new air into the system. Low-pressure systems, however, will return condensate to the boiler easily and not retain it in the piping, as may occur in a vapor system when the system pressure exceeds the static head available for gravity return at the boiler.

3. Vapor systems operate without the use of a vacuum pump and at pressures varying from 20 in. vacuum or more (depending upon the tightness of the system) to low pressure.

These systems are similar to the two-pipe low-pressure systems described previously, except that, since the air vents prevent the return of air to the system when the steam pressure drops below atmospheric pressure, the system operates for several hours under vacuum conditions when the rate of steam supplied is reduced, depending on the tightness of the system.

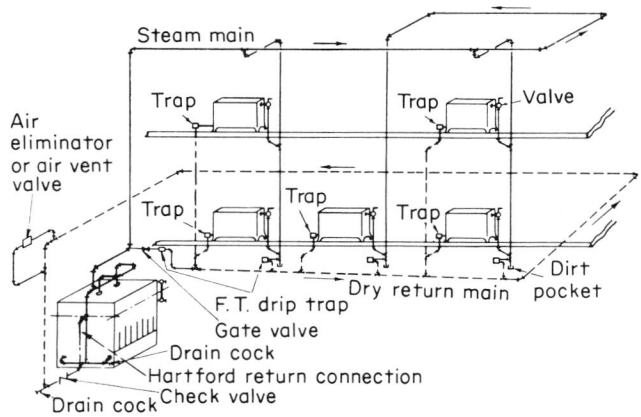

Fig. 12. Down-feed two-pipe system. (*From "ASHRAE Guide," 1960.*)

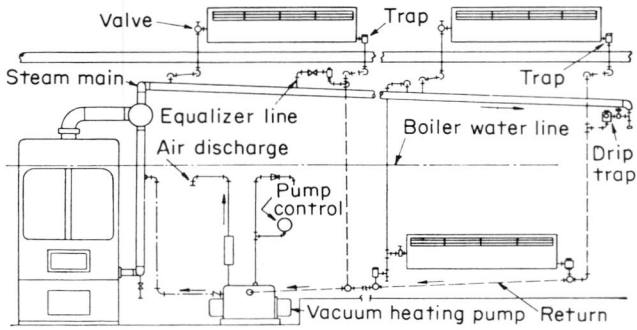

FIG. 13. Two-pipe vacuum pump system. (*From "ASHRAE Guide,"* 1960.)

4. Vacuum systems operate under conditions of both low pressure and vacuum, utilizing a vacuum pump to maintain subatmospheric pressures in the return piping under all operating conditions. The supply piping of the system may even operate at subatmospheric pressure in periods of light load when steam pressure is not maintained. The vacuum pump not only withdraws the air and water from the system but separates the air from the water, discharges the air to atmosphere, and pumps the water back to the boiler. A two-pipe vacuum system is shown in Fig. 13.

Pipe Sizing. The piping system is used for the distribution of steam, return of the condensate and, where no local air vents are provided, for the removal of the air. The steam should be rapidly and uniformly distributed and without noise, and the air should be removed rapidly to permit the system to heat quickly and properly. A system that is air bound will not heat properly.

The size of pipe for a given steam rate depends on the following:

1. The initial pressure and the total pressure drop through the system. Table 7 gives the pressure drops commonly used with corresponding initial steam pressures.

2. The maximum velocity of steam. When condensate flows against the steam, the steam velocity should be low enough so as not to interfere with condensate flow.

Table 7.   Pressure Drops Used for Sizing Steam Piping[1]

Initial steam pressure, psig	Pressure drop per 100 ft	Total pressure drop in steam supply piping, psi
Subatmos. or vacuum return	2–4 oz	1–2
0	½ oz	1
1	2 oz	1–4
2	2 oz	8
5	4 oz	1½
10	8 oz	3
15	1 psi	4
30	2 psi	5–10
50	2–5 psi	10–15
100	2–5 psi	15–25
150	2–10 psi	25–30

[1] From "Heating, Ventilating, Air Conditioning Guide," 1960. Used by permission.

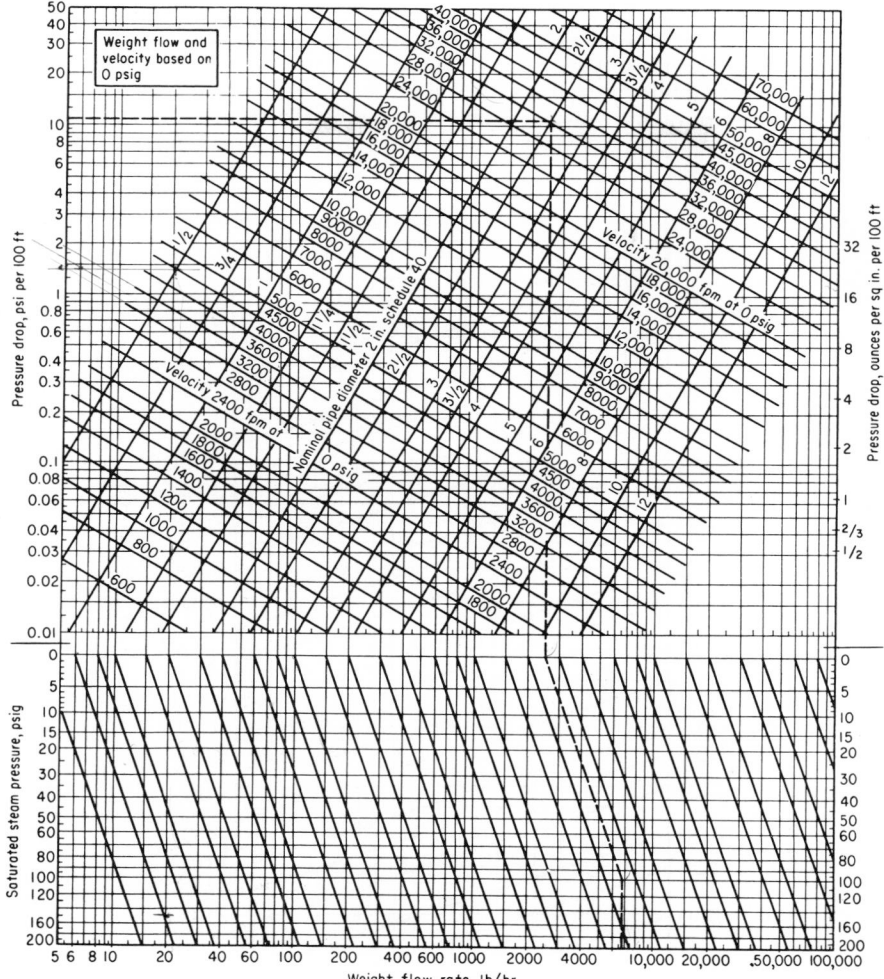

Fig. 14. Pipe sizing chart. (*From "Heating, Ventilating, Air Conditioning Guide,"* 1960. *Used by permission.*)

Figure 14 is a chart used for sizing pipe and determining pressure drop and velocity in pipes at various steam flows. With the multiplier chart of Fig. 15, it can be used for all saturation pressures between 0 and 200 psig, where the flow of steam and condensate are in the same direction. As shown by the dashed line, for example, a system which handles 6,700 lb of steam per hour with an initial pressure of 100 psig and a pressure drop of 11 psi per 100 ft requires a 2½-in. pipe and has an equivalent steam velocity at 0 psig of 32,700 fpm. At 100 psig, the velocity is 13,000 fpm. Either of these velocities are high for building-heating steam.

Table 8 can be used for pipe sizing where steam and condensate flow in the opposite directions.

Tables 9 to 11 give the supply and return pipe capacities for high-pressure (150 psig), medium-pressure (30 psig), and low-pressure (3.5 to 12 psig) systems for various pressure drops and are based on steam and condensate flowing in the same direction. In addition to the frictional resistance in the pipe itself, there is also the resistance of fittings and valves. Table 12 gives the resistance in terms of equivalent length of straight pipe which must be added to the normal length of piping.

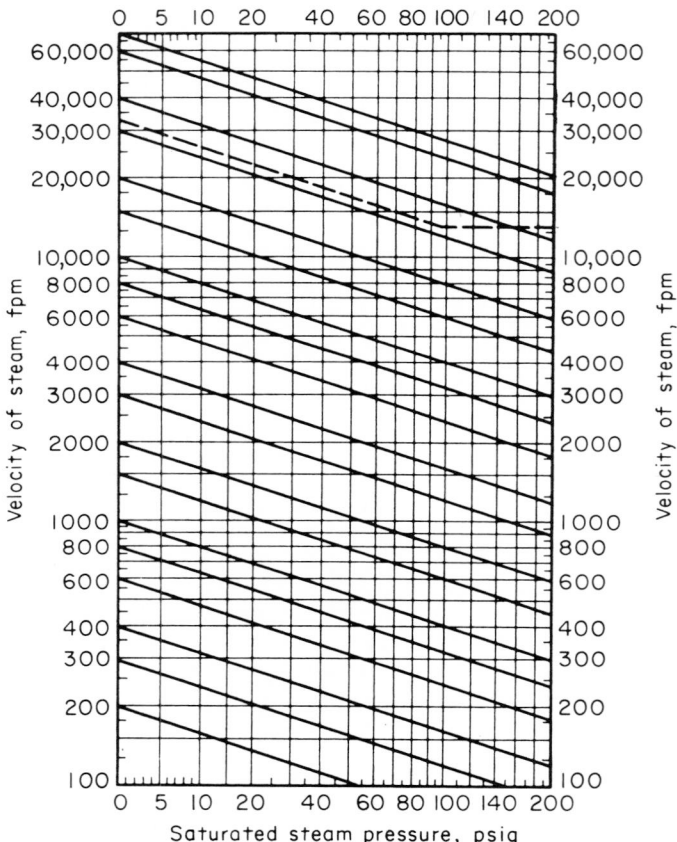

Fɪɢ. 15. Velocity multiplier chart. (*From "Heating, Ventilating, Air Conditioning Guide,"* 1960. *Used by permission.*)

Generally all tables and charts are based on this equivalent length rather than the actual length of pipe.

The size of piping must be selected carefully in those heating systems which operate at pressures near or below atmospheric pressure. Horizontal mains must be pitched in such a way that condensate may return directly to the boiler or so that a drainer or trap may deliver the condensate to its receiver.

For purposes of easy reference, suggested criteria for selection of pipe sizes in one- and two-pipe systems are given in Table 13.

Table 8. Steam-pipe Capacities for One-or Two-pipe Systems: Condensate Flow against Steam Flow[1]

Nominal pipe size, in.	Capacity, lb/hr				
	Two-pipe systems		One-pipe systems		
	Condensate flowing against steam		Supply risers upfeed	Radiator valves and vertical connections	Radiator and riser runouts
	Vertical	Horizontal			
¾	8	.....	6	..	7
1	14	9	11	7	7
1¼	31	19	20	16	16
1½	48	27	38	23	16
2	97	49	72	42	23
2½	159	99	116	..	42
3	282	175	200	..	65
3½	387	288	286	..	119
4	511	425	380	..	186
5	1,050	788	...	..	278
6	1,800	1,400	...	..	545
8	3,750	3,000			
10	7,000	5,700			
12	11,500	9,500			
16	22,000	19,000			

[1] From "Heating, Ventilating, Air Conditioning Guide," 1960.   Used by permission.

Table 9. Pipe Capacities in Pounds per Hour for High-Pressure System (150 Psig)[1]

Pipe size, in.	Pressure drop per 100 ft						
	⅛ psi (2 oz)	¼ psi (4 oz)	½ psi (8 oz)	¾ psi (12 oz)	1 psi (16 oz)	2 psi (32 oz)	5 psi
	Supply pipe						
¾	29	41	58	82	116	184	300
1	58	82	117	165	233	369	550
1¼	130	185	262	370	523	827	1,230
1½	203	287	407	575	813	1,230	1,730
2	412	583	825	1,167	1,650	2,000	3,410
2½	683	959	1,359	1,922	2,430	3,300	5,200
3	1,237	1,750	2,476	3,500	4,210	6,000	9,400
3½	1,855	2,626	3,715	5,250	6,020	8,500	13,100
4	2,625	3,718	5,260	7,430	8,400	12,300	19,200
5	4,858	6,875	9,725	13,750	15,000	21,200	33,100
6	7,960	11,275	15,950	22,550	25,200	36,500	56,500
8	16,590	23,475	33,200	46,950	50,000	70,200	120,000
10	30,820	43,430	61,700	77,250	90,000	130,000	210,000
12	48,600	68,750	97,250	123,000	155,000	200,000	320,000

Table 9. (*Continued*)

Pipe size, in.	Pressure drop per 100 ft						
	⅛ psi (2 oz)	¼ psi (4 oz)	½ psi (8 oz)	¾ psi (12 oz)	1 psi (16 oz)	2 psi (32 oz)	5 psi
			Return pipe				
¾	156	232	360	465	560	890	
1	313	462	690	910	1,120	1,780	
1¼	650	960	1,500	1,950	2,330	3,700	
1½	1,070	1,580	2,460	3,160	3,800	6,100	
2	2,160	3,300	4,950	6,400	7,700	12,300	
2½	3,600	5,350	8,200	10,700	12,800	20,400	
3	6,500	9,600	15,000	19,500	23,300	37,200	
3½	9,600	14,400	22,300	28,700	34,500	55,000	
4	13,700	20,500	31,600	40,500	49,200	78,500	
5	25,600	38,100	58,500	76,000	91,500	146,000	
6	42,000	62,500	96,000	125,000	150,000	238,000	

[1] From "Heating, Ventilating, Air Conditioning Guide," 1958.   Used by permission.

Table 10.   Pipe Capacities in Pounds per Hour for Medium-pressure System (30 Psig)[1]

Pipe size, in.	Pressure drop per 100 ft					
	⅛ psi (2 oz)	¼ psi (4 oz)	½ psi (8 oz)	¾ psi (12 oz)	1 psi (16 oz)	2 psi (32 oz)
			Supply pipe			
¾	15	22	31.	38	45	63
1	31	46	63	77	89	125
1¼	69	100	141	172	199	281
1½	107	154	219	267	309	437
2	217	313	444	543	627	886
2½	358	516	730	924	1,033	1,460
3	651	940	1,330	1,628	1,880	2,660
3½	979	1,414	2,000	2,447	2,825	4,000
4	1,386	2,000	2,830	3,464	4,000	5,660
5	2,560	3,642	5,225	6,402	7,390	10,460
6	4,210	6,030	8,590	10,240	12,140	17,180
8	8,750	12,640	17,860	21,865	25,250	35,100
10	16,250	23,450	33,200	40,625	46,900	66,350
12	25,640	36,930	52,320	64,050	74,000	104,500
			Return pipe			
¾	115	170	245	308	365	
1	230	340	490	615	730	
1¼	485	710	1,025	1,285	1,530	
1½	790	1,155	1,670	2,100	2,500	
2	1,575	2,355	3,400	4,300	5,050	
2½	2,650	3,900	5,600	7,100	8,400	
3	4,850	7,100	10,250	12,850	15,300	
3½	7,200	10,550	15,250	19,150	22,750	
4	10,200	15,000	21,600	27,000	32,250	
5	19,000	27,750	40,250	55,000	60,000	
6	31,000	45,500	65,500	83,000	98,000	

[1] From "Heating, Ventilating, Air Conditioning Guide," 1958.   Used by permission.

Table II. Pipe Capacities in Pounds per Hour for Low-pressure Systems (below 15 Psig)[1]

Supply pipe

Pressure drop, psi per 100 ft in length

Saturated pressure, psig

Nom. pipe size, in.	1/16 psi (1 oz)		1/8 psi (2 oz)		1/4 psi (4 oz)		1/2 psi (8 oz)		3/4 psi (12 oz)		1 psi		2 psi	
	3.5[2]	12	3.5	12	3.5	12	3.5	12	3.5	12	3.5	12	3.5	12
3/4	9	11	14	16	20	24	29	35	36	43	42	50	60	73
1	17	21	26	31	37	46	54	66	68	82	81	95	114	137
1 1/4	36	45	53	66	78	96	111	138	140	170	162	200	232	280
1 1/2	56	70	84	100	120	147	174	210	218	260	246	304	360	430
2	108	134	162	194	234	285	336	410	420	510	480	590	710	850
2 1/2	174	215	258	310	378	460	540	660	680	820	780	950	1,150	1,370
3	318	380	465	550	660	810	960	1,160	1,190	1,430	1,380	1,670	1,950	2,400
3 1/2	462	550	670	800	990	1,218	1,410	1,700	1,740	2,100	2,000	2,420	2,950	3,450
4	726	800	950	1,160	1,410	1,690	1,980	2,400	2,450	3,000	2,880	3,460	4,200	4,900
5	1,200	1,430	1,680	2,100	2,440	3,000	3,570	4,250	4,380	5,250	5,100	6,100	7,500	8,600
6	1,920	2,300	2,820	3,350	3,960	4,850	5,700	7,000	7,200	8,600	8,400	10,000	11,900	14,200
8	3,900	4,800	5,570	7,000	8,100	10,000	11,400	14,300	14,500	17,700	16,500	20,500	24,000	29,500
10	7,200	8,800	10,200	12,600	15,000	18,200	21,000	26,000	26,200	32,000	30,000	37,000	42,700	52,000
12	11,400	13,700	16,500	19,500	23,400	28,400	33,000	40,000	41,000	49,500	48,000	57,500	67,800	81,000

Table 11. (Continued)

Return mains and risers

Pipe size, in.	1/32 psi or 1/2 oz drop per 100 ft			1/24 psi or 2/3 oz drop per 100 ft			1/16 psi or 1 oz drop per 100 ft			1/8 psi or 2 oz drop per 100 ft			1/4 psi or 4 oz drop per 100 ft			1/2 psi or 8 oz drop per 100 ft		
	Wet	Dry	Vac.	Wet	Dry	Vac.	Wet	Dry	Vac.	Wet	Dry	Vac.	Wet	Dry	Vac.	Wet	Dry	Vac.
Mains																		
3/4	....	....	....	....	....	42	....	....	100	....	....	142	....	....	200	....	....	283
1	125	62	....	145	71	143	175	80	175	250	103	249	350	115	350	....	....	494
1¼	213	130	....	248	149	244	300	168	300	425	217	426	600	241	600	....	....	848
1½	338	206	....	393	236	388	475	265	475	675	340	674	950	378	950	....	....	1,340
2	700	470	....	810	535	815	1,000	575	1,000	1,400	740	1,420	2,000	825	2,000	....	....	2,830
2½	1,180	760	....	1,580	868	1,360	1,680	950	1,680	2,350	1,230	2,380	3,350	1,360	3,350	....	....	4,730
3	1,880	1,460	....	2,130	1,560	2,180	2,680	1,750	2,680	3,750	2,250	3,800	5,350	2,500	5,350	....	....	7,560
3½	2,750	1,970	....	3,300	2,200	3,250	4,000	2,500	4,000	5,500	3,230	5,680	8,000	3,580	8,000	....	....	11,300
4	3,880	2,930	....	4,580	3,350	4,500	5,500	3,750	5,500	7,750	4,830	7,810	11,000	5,380	11,000	....	....	15,500
5	....	....	....	....	....	7,880	....	....	9,860	....	....	13,700	....	....	19,400	....	....	27,300
6	....	....	....	....	....	12,600	....	....	15,500	....	....	22,000	....	....	31,000	....	....	43,800
Risers																		
3/4	....	48	....	....	48	143	....	48	175	....	48	249	....	48	350	....	....	494
1	....	113	....	....	113	244	....	113	300	....	113	426	....	113	600	....	....	848
1¼	....	248	....	....	248	388	....	248	475	....	248	674	....	248	950	....	....	1,340
1½	....	375	....	....	375	815	....	375	1,000	....	375	1,420	....	375	2,000	....	....	2,830
2	....	750	....	....	750	1,360	....	750	1,680	....	750	2,380	....	750	3,350	....	....	4,730
2½	....	....	....	....	....	2,180	....	....	2,680	....	....	3,800	....	....	5,350	....	....	7,560
3	....	....	....	....	....	3,250	....	....	4,000	....	....	5,680	....	....	8,000	....	....	11,300
3½	....	....	....	....	....	4,480	....	....	5,500	....	....	7,810	....	....	11,000	....	....	15,500
4	....	....	....	....	....	7,880	....	....	9,680	....	....	13,700	....	....	19,400	....	....	27,300
5	....	....	....	....	....	12,600	....	....	15,500	....	....	22,000	....	....	31,000	....	....	43,800

[1] From "Heating, Ventilating, Air Conditioning Guide," 1960. Used by permission.

[2] The weight-flow rates at 3.5 psig can be used to cover saturated pressure from 1 to 6 psig, and the rates at 12 psig can be used to cover saturated pressure from 8 to 16 psig with an error not exceeding 8 per cent.

Table 12. Equivalent Length of Straight Pipe for Fittings and Valves[1]

Size of pipe, in.	Length to be added to run, ft				
	Standard elbow	Side outlet tee[3]	Gate valve[2]	Globe valve[2]	Angle valve[2]
1/2	1.3	3	0.3	14	7
3/4	1.8	4	0.4	18	10
1	2.2	5	0.5	23	12
1 1/4	3.0	6	0.6	29	15
1 1/2	3.5	7	0.8	34	18
2	4.3	8	1.0	46	22
2 1/2	5.0	11	1.1	54	27
3	6.5	13	1.4	66	34
3 1/2	8	15	1.6	80	40
4	9	18	1.9	92	45
5	11	22	2.2	112	56
6	13	27	2.8	136	67
8	17	35	3.7	180	92
10	21	45	4.6	230	112
12	27	53	5.5	270	132
14	30	63	6.4	310	152

[1] From "Heating, Ventilating, Air Conditioning Guide," 1960. Used by permission.
[2] Valve in full open position.
[3] Values given apply only to a tee used to divert the flow in the main to the last riser.

Boiler Connections.   Low-pressure cast-iron heating boilers are usually provided with more than one steam outlet in order to reduce the velocity of the steam in the vertical uptakes from the boiler, thereby preventing carry-over of water into the steam main.

Steel boilers usually have only one outlet sized properly to maintain a low velocity except where, because of the quantity of steam or the length of the boiler tubes, two or more outlets are necessary.

Two return connections are usually provided with a cast-iron boiler, while steel boilers usually have only one return connection. Where two connections are provided on a boiler, both should be used to effect proper circulation of water through the boiler. The return piping connections should include the Hartford loop to prevent the boiler from losing its water level under all operating conditions. Piping connections for boilers including the Hartford loop are shown in Fig. 16.

Condensate-return Pumps.   Condensate-return pumps are used when the condensate cannot return to the boiler by gravity.

The receiver should be sized to have a storage capacity of approximately one and a half to two times the quantity of condensate returned per minute. The pump should be selected to have a delivery rate of two to three times the normal flow to prevent frequent operation of the pump.

A duplex pump should be used where the condensate pump must be available for operation at all times.

Vacuum Heating Pumps.   On vacuum systems, a vacuum pump is required to remove the air from the piping system and to return the condensate to the boiler.

Table 13. Pressure-drop Criteria for Pipe Sizing

Design item / Classification	One-pipe		Two-pipe			
	Pressure	Vapor	High-pressure	Low-pressure	Vapor	Vacuum
Unit pressure drop in supply mains and risers	1/16 psi/100 ft		2 psi/100 ft for 30-psig system 2 to 10 psi/100 ft 150-psig system			
Unit pressure drop in return mains and risers	1/16 psi/100 ft		1 psi/100 ft for 30 psig system 2 psi/100 ft for 150-psig system			
Total pressure drop in supply mains and risers	1/4 psi max	Very critical. Allow only a few ounces total drop in complete system	5 to 10 psi for 30-psig system 25 to 30 psi for 150-psig system	1/2 psi for 2-psig system 4 psi for 15-psig system	1/16 to 1/8 psi	1/8 to 1 psi
Total pressure drop in return mains and risers	1/4 psi max		5 psi max. for 30-psig systems 20 psi max. for 150-psig system	1/2 psi for 2-psig system 4 psi for 15-psig system	1/16 to 1/8 psi	1/8 to 1 psi
Minimum pitch of supply and return mains	1/4 in./10 ft	1/4 in./10 ft	1/4 in./10 ft	1/4 in./10 ft	1/4 in./10 ft	1/4 in./10 ft
Pitch, horizontal runouts to risers and heating units	1/2 in./ft	1/2 in./ft	1/2 in./ft	1/2 in./ft	1/2 in./ft	1/2 in./ft
Upfeed risers or undripped runouts	See Table 8					

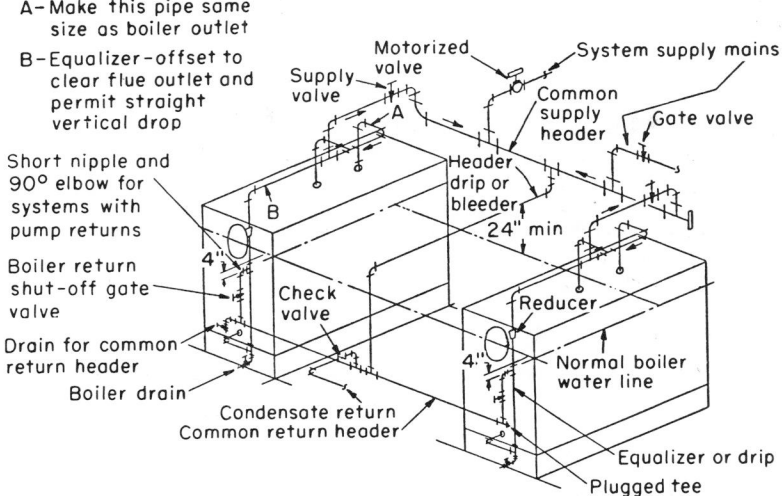

A–Make this pipe same size as boiler outlet

B–Equalizer–offset to clear flue outlet and permit straight vertical drop

Short nipple and 90° elbow for systems with pump returns

Boiler return shut-off gate valve

Drain for common return header

Boiler drain

Condensate return Common return header

Supply valve

Motorized valve

System supply mains

Common supply header

Gate valve

Header drip or bleeder

24" min

Check valve

Reducer

Normal boiler water line

Equalizer or drip

Plugged tee

FIG. 16. Piping for boilers in battery including Hartford return connection. (*From "Heating, Ventilating, Air Conditioning Guide,"* 1960. *Used by permission.*)

The unit is made up of two parts: an exhausting unit for removing the air and discharging it to the atmosphere and a condensate-return unit which discharges the condensate to the boiler.

Vacuum pumps are classified as low-vacuum pumps (those maintaining less than $5\frac{1}{2}$ in. Hg vacuum) and high-vacuum pumps (those maintaining above $5\frac{1}{2}$ in. Hg) on the system. The condensate-return requirement should generally be the same as for condensate pumps indicated above.

The wide operating range and characteristics of heating systems have a greater effect on the quantity of air to be removed than on the condensate to be handled. Typical requirements are for a low-vacuum system to have a capacity of 0.3 to 1.0 cfm of air per 1,000 sq ft equivalent direct radiation (EDR) and for high-vacuum systems, the capacity is 2.0 cfm of air per 1,000 sq ft equivalent direct radiation. The actual selection is dependent on the design and capacity of a specific manufacturer's apparatus.

The vacuum pump is controlled by means of a vacuum regulator set to maintain a certain condition in the system. In addition, a float control is provided which will operate the condensate-return pump whenever sufficient water accumulates in the receiver, regardless of the vacuum on the system.

It is very important on all vacuum systems, and especially so on high-vacuum installations, to assure that the system is tight in order to reduce the amount of air leaking into the system.

Pressure-reducing Valves. If steam is supplied from a boiler at a pressure higher than is required for the heating system, one or more pressure-reducing valves should be used. Single-seated valves, either direct- or pilot-operated, are used where the steam must be shut off tightly to prevent build-up of pressure on the low-pressure side at time of no load; and double-seated valves are used where the low-pressure lines will condense enough steam to prevent build-up of pressure due to normal leakage through the valve.

If the steam pressure is 100 psig or greater, it is common practice to provide two-stage reduction. In addition, the ASME Code requires two-stage reduction

when the inlet pressure is greater than 50 psig and cast-iron radiation is served.

A globe valve bypass should be installed around each reducing valve in the pipeline. Safety valves should be located on the low-pressure side and set for at least 5 psi higher than the reduced pressure in installations where the reduced pressure is under 35 psig and at least 10 psi higher than the reduced pressure if the reduced pressure is above 35 psig in the first or second stage. The relief valve should be of sufficient capacity to relieve the entire output of the reducing valves. The outlet from relief valves should be piped to a safe location outdoors.

Pilot-controlled, direct-operated reducing valves may be used for all high-pressure systems. Spring-loaded, direct-operated valves may be used for reduced pressures up to 50 psig provided they can handle the steam flow without excessive pressure drop. Weight-loaded valves are used for reduced pressures below 15 psig and for moderate steam flow.

Table 14. Suggested Capacity of Pressure-regulating Valves[1]

Type of control	Pressure-regulating valve capacity, lb/(hr sq ft EDR)[2]
Continuous or modulating operation, intermittent with *on* periods over 3 hr in duration	$\frac{1}{3}$
Intermittent operation with *on* periods of $\frac{1}{2}$ to 3 hr in duration	$\frac{1}{2}$
Intermittent operation with *on* periods of $\frac{1}{2}$ hr or less.	$\frac{1}{2}$[3]

[1] From "Heating, Ventilating, Air Conditioning Guide," 1960. Used by permission.
[2] One square foot EDR = 240 Btu/hr.
[3] Except for one-pipe systems or two-pipe systems not orificed, which should be 1 lb/sq ft EDR.

Pressure-regulating valves should be sized to supply adequately the maximum steam requirements of the heating system but should not be oversized. An oversized valve will cause erosion of the valve and seat and will result in a shorter life of the valve. Suggested capacities of pressure-regulating valves are given in Table 14.

If the steam requirements are large and variable, two reducing valves are usually installed in parallel, with sizes selected on an approximate 70 and 30 per cent proportion of maximum flow, with the result that during mild weather or during periods of reduced demand, the larger valve remains closed.

Steam Traps. Steam traps are used to retain steam in a heating unit or piping system until it has condensed and given up its latent heat; the condensate and air are then discharged by the trap and drain back to the boiler or condensate tank.

Steam traps are classified according to the type of operating principle by which they function. Those normally used for building heating systems are of the float, thermostatic, float and thermostatic, inverted bucket, upright bucket and boiler return trap, or alternating receiver types.

Float traps operate in response to a rise and fall of a float which is affected by the level of condensate in the trap. Float traps are used where the load is heavy and the discharge is continuous. Air is vented by a thermostatic trap bypassed around the float trap.

Thermostatic traps (Fig. 17) operate on a temperature difference between the steam and condensate and use elements which open or close the discharge ports.

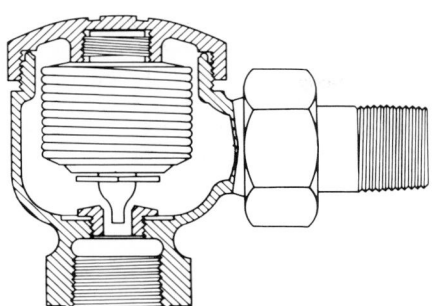

Fig. 17. Thermostatic trap. (*Courtesy of Sarco Co. Inc.*)

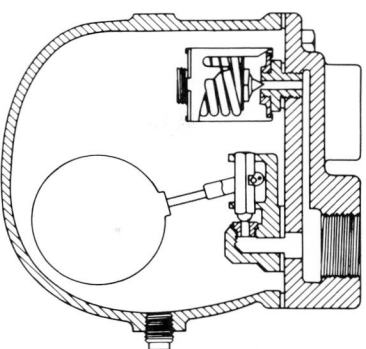

Fig. 18. Float and thermostatic trap. (*Courtesy of Sarco Co. Inc.*)

Thermostatic traps discharge both air and condensate. Individual thermostatic traps are used for radiators and convectors, and the discharge is intermittent.

Float and thermostatic traps (Fig. 18) are a combination of the float trap and the thermostatic trap. Air is vented through the trap concurrently with the draining of condensate.

In the upright bucket trap the bucket rises and closes the trap when condensate enters the trap body. When sufficient condensate collects, it overflows and fills the bucket, causing it to drop and opening the trap to permit the drainage of condensate by steam pressure. Air venting is provided by a thermostatic trap in a bypass.

In the inverted bucket trap (Fig. 19) the steam pressure keeps the bucket up and closes the trap. The accumulation of sufficient condensate displaces the steam and causes the bucket to drop and the trap opens to permit the drainage of condensate by steam pressure. Air venting is provided by a built-in thermostatic element.

Bucket traps are used where a large volume of condensate must be handled rapidly or where the condensate must be raised to a higher elevation.

A boiler return trap or alternating receiver (Fig. 20) is used for returning condensate in a low-pressure or vapor-heating system where the condensate cannot

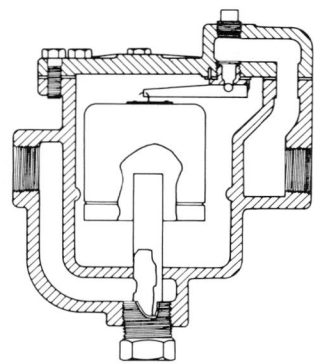

Fig. 19. Inverted bucket trap. (*Courtesy of Sarco Co. Inc.*)

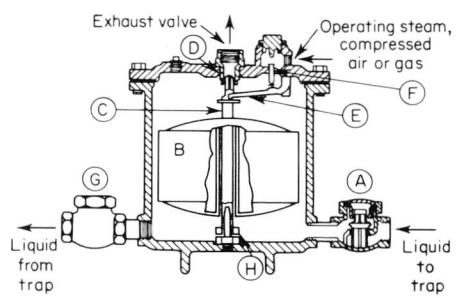

Fig. 20. Alternating receiver. (*Courtesy of Sarco Co. Inc.*)

return by gravity. The condensate collects in the trap until the float rises to close the vent to atmosphere and open the port to boiler steam pressure; this equalizes the boiler and trap pressure and enables the condensate in the trap to flow back to the boiler by gravity. The trap must be installed above the water line of the boiler.

Connections to Heating Units. Risers, radiator, and convector connections must be properly pitched at the time they are installed and arranged so that the pitch will be maintained under the strains of expansion and contraction. These connections may be made by swing joints which permit the expansion or contraction

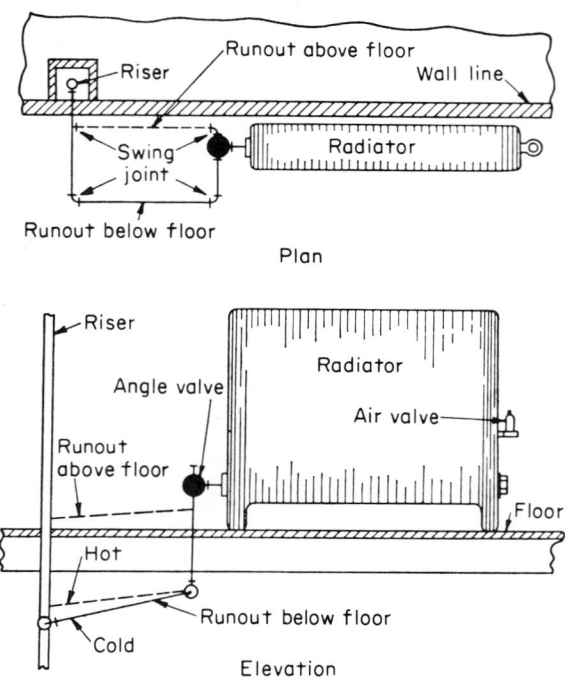

Fig. 21. Radiator connection with swing joint. (*From "Heating, Ventilation, Air-Conditioning Guide."* 1960. *Used by permission.*)

to occur under heating and cooling without bending of pipes. For expansion in long risers, either expansion joints or pipe swing joints should be used. The pipes should be anchored between expansion joints.

Connections for radiators for a one-pipe system are shown in Fig. 21.

Two-pipe system connections are shown in Figs. 22 and 23. While top inlet supply connections are preferred, it is also possible to connect the supply to the bottom of the radiator. Short radiators may be connected with top supply and bottom return on the same end.

The method of connection to convectors is shown in Fig. 24. The inlet supply valve is sometimes omitted, and a damper is used in the outlet grille for control of heat.

Connections to blast heaters are shown in Fig. 25 and to unit heaters in Fig. 26.

Manual Valves. *Gate valves* should be used so that equipment or piping sections can be isolated without shutting down the system. The valve should be either fully open or fully closed and should never be used for throttling.

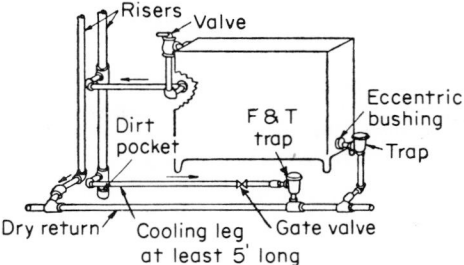

FIG. 22. Radiator connection. (*From "Heating, Ventilation, Air Conditioning Guide,"* 1960. *Used by permission.*)

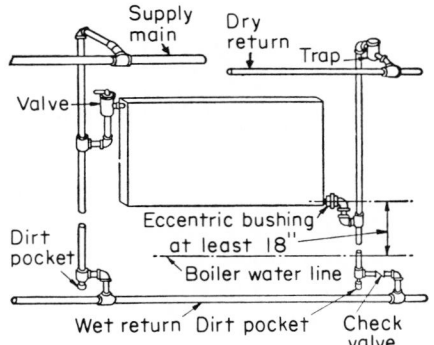

FIG. 23. Wall radiator connection. (*From "Heating, Ventilating, Air Conditioning Guide,"* 1960. *Used by permission.*)

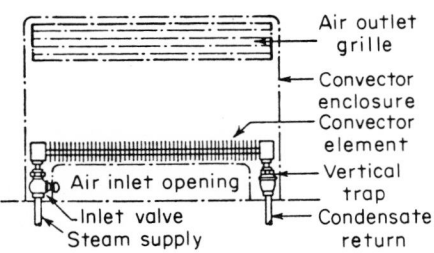

FIG. 24. Convector connection. (*From "Heat, Ventilating, Air Conditioning Guide,"* 1960. *Used by permission.*)

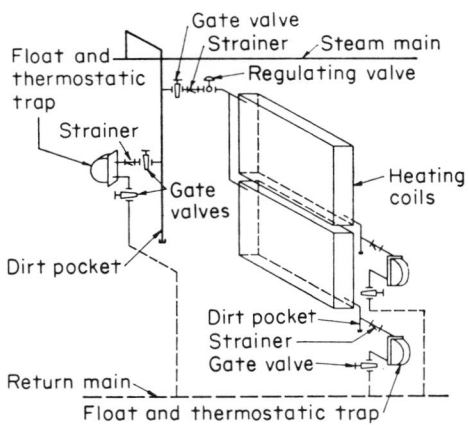

FIG. 25. Blast-heating-coil connection. (*From "Heating, Ventilating, Air Conditioning Guide,"* 1960. *Used by permission.*)

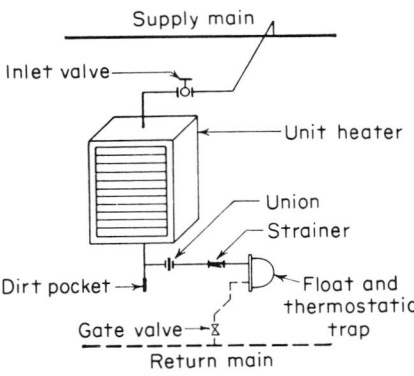

FIG. 26. Unit-heater connection. (*From "Heating, Ventilating, Air Conditioning Guide,"* 1960. *Used by permission.*)

Globe valves should be used for throttling as, for example, manual control of steam flow to equipment or for bypasses around pressure-reducing valves or traps.

HOT-WATER HEATING SYSTEMS

Hot-water heating systems with supply water at less than 250 F are classified as low-temperature systems, while those with supply water above 250 F are classified as high-temperature systems.

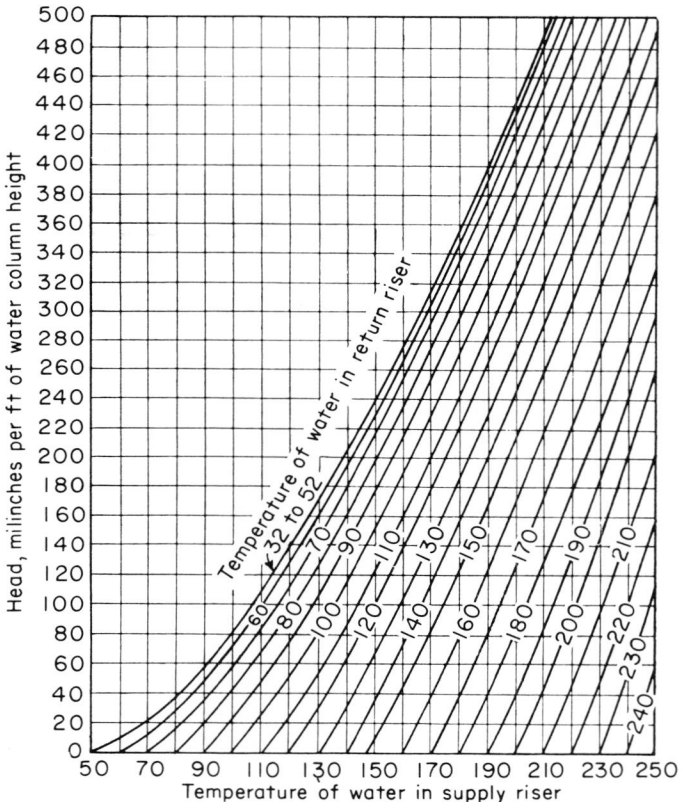

FIG. 27. Circulating heads for gravity systems. (*From "Heating, Ventilating, Air Conditioning Guide," 1960.   Used by permission.*)

A further classification is with reference to:

1. Circulation
2. Piping arrangement
3. Type of expansion tank

Low-temperature hot-water systems are either gravity circulation, in which the circulation of the water is due to the difference in water density between the supply and return, or forced circulation, in which a pump is used to maintain flow.   High-temperature hot-water systems always employ forced circulation.

For a gravity system, the piping must be sized so that the friction loss at the rated flow does not exceed the available head. Gravity systems are rarely used except for small systems. Circulation heads which can be obtained with various supply and return temperature for gravity systems are shown in Fig. 27.

Since the pressure drop in hot-water systems, especially those which operate by gravity alone, is small, it is usual practice to express the loss of head in milinches (thousandths of an inch). The hot-water system capacity is usually expressed in MBh's (thousands of BTU per hour).

Hot-water heating systems have certain advantages over steam systems in that:

1. No steam specialties, such as traps, are required; this feature simplifies maintenance.
2. The closed circulating system eliminates air and corrosion.
3. Elevation of the piping may change as required owing to architectural and structural features.
4. Outputs of hot-water systems are constant and can be easily varied by changing the supply-water temperature.

Low-temperature Water Systems

The rate of water flow required in a system is based on the thermal capacity of the water, which is the product of the rate of flow, specific heat, and temperature drop of the water passing through the heating units and is as follows:

$$H = Wc(t_1 - t_2)$$

where H = thermal capacity, Btu/hr
W = water flow, lb/hr
c = specific heat of water, Btu/(lb F)
t_1 and t_2 = temperature of water entering and leaving unit, F
For many low-temperature-system applications, this can be simplified without loss of accuracy to

$$H = 500G(t_1 - t_2)$$

where G is the water flow in gallons per minute.

Most systems are designed on the basis of a 20 F temperature drop, but factors which must be considered in selection of a proper temperature drop for a system are (1) the effect of the design average water temperature and water velocity on the capacity and cost of the heating units, (2) the size and cost of the piping, and (3) the initial and operating cost of the circulating pump.

The supply-water temperature is the maximum temperature of the water supplied to the system and usually varies from 180 to 250 F, with 210 F being a commonly used temperature. The average water temperature, that is, a mean of the supply- and return-water temperature, is used in determining the capacity of heating units.

Piping Arrangement, Low-temperature Water Systems. There are four arrangements of piping: (1) the one-pipe system, (2) the two-pipe direct-return system, (3) the two-pipe reversed-return system.

The *one-pipe* system (Fig. 28) has a single main for both supply and return with special fittings installed at the connections to the heating units to aid flow through the heating unit. The main does not change size from the first to the last heating unit, since the amount of water which passes through the main is constant. The water temperature in the main drops progressively through each heating unit, and it is necessary to use larger units at the end of the main because cooler water is being provided.

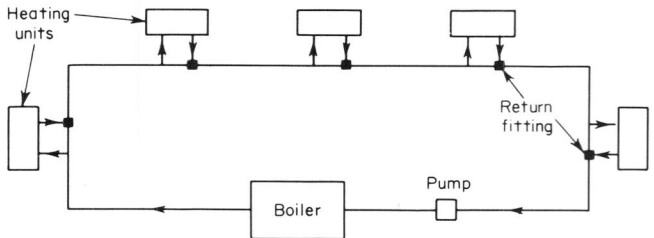

FIG. 28. One-pipe system.

The *two-pipe direct-return system* (Fig. 29) has individual main supply and return connections at each heating unit.   The return to the boiler is approximately the same length as the supply main and generally parallels it.   This system is difficult to balance, as the heating unit nearest the boiler tends to get the greatest quantity of water at the expense of the most remote unit.

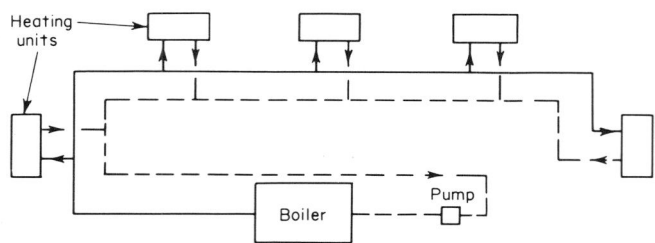

FIG. 29. Two-pipe direct-return system.

The *two-pipe reversed-return system*, recommended for most systems (Fig. 30), has two mains, one supplying and one returning water from the heating units. The return piping is longer than that of the direct-return system and is arranged in such a manner that the equivalent lengths of supply and return piping to any heating unit are approximately equal.

Pipe Sizing.   Piping for hot-water systems is designed for a small friction drop for overall economy.   As previously indicated, the drop is usually expressed in milinches.

The design of the piping system takes into consideration the friction loss through the piping, fittings, valves, and equipment.   The friction loss depends on the

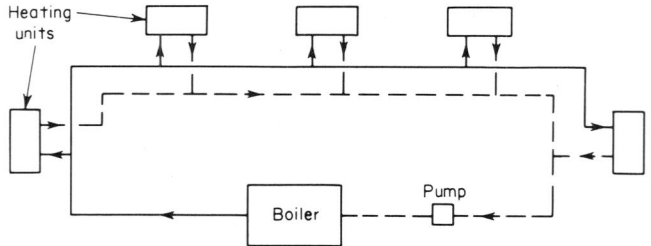

FIG. 30. Two-pipe reversed-return system.

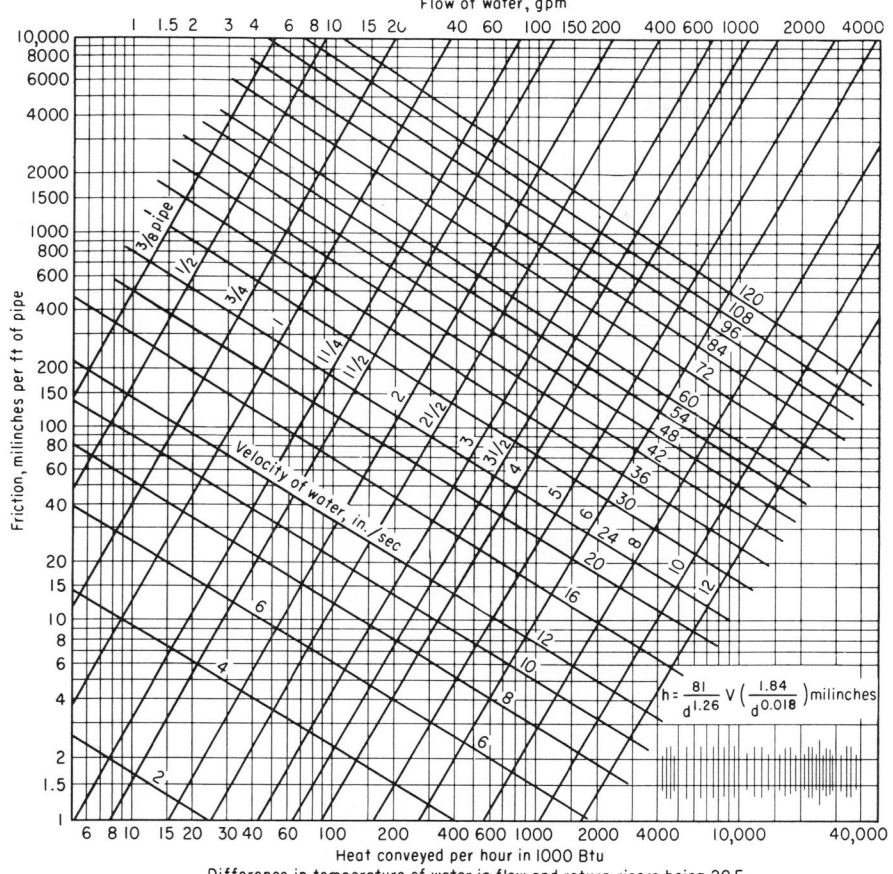

FIG. 31. Friction loss in iron or steel pipe. (*From "Heating, Ventilating, Air Conditioning Guide,"* 1960. *Used by permission.*)

water velocity, pipe diameter, pipe length, and the interior roughness of the pipe. The friction loss for Schedule 40 steel or iron pipe is given in Fig. 31 and for copper tube in Fig. 32. These curves are based on a 20 F temperature difference between supply and return water. For example, if a heating unit is to be supplied with 40000 Btu/hr with a temperature drop of only 10 F, then the flow of water is 8 gpm and the friction loss should be based on this figure.

Fittings and valves are given in terms of 90-deg elbows in Table 15. The equivalent length in straight pipe of 90-deg elbows for various pipe sizes and water-flow velocities is given in Table 16. In order to reduce the possibility of noise, velocities

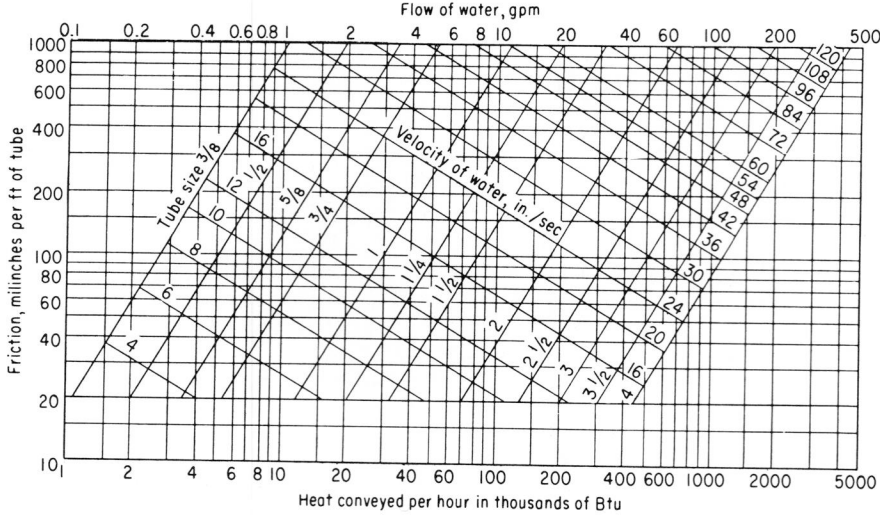

FIG. 32. Friction loss in Type L copper tubing. (*From "Heating, Ventilating, Air Conditioning Guide,"* 1960. *Used by permission.*)

are usually kept below 4 fps for smaller pipe sizes but may be increased to 10 fps for piping larger than 6 in.

In the sizing of piping, an average friction drop of 1.0 to 4.0 ft of water per 100 ft of pipe (100 to 400 milinches/ft) is used. One method for sizing the piping in a system is to make a preliminary selection of a pump. The head at the required flow which the pump can produce is divided by the equivalent length (length of pipe

Table 15.   90-deg Elbow Equivalents[1]

Fitting	Iron pipe	Copper tubing
Elbow, 90-deg...............	1.0	1.0
Elbow, 45-deg...............	0.7	0.7
Elbow, 90-deg long turn	0.5	0.5
Elbow, welded, 90-deg........	0.5	0.5
Reduced coupling............	0.4	0.4
Open return bend	1.0	1.0
Angle radiator valve..........	2.0	3.0
Radiator or convector.........	3.0	4.0
Boiler or heater...............	3.0	4.0
Open gate valve	0.5	0.7
Open globe valve	12.0	17.0

[1] From "Heating, Ventilating, Air Conditioning Guide," 1960. Used by permission.

Table 16.   Equivalent Length of Pipe for 90-deg Elbows[1]

Velocity, fps	Pipe size														
	½	¾	1	1¼	1½	2	2½	3	3½	4	5	6	8	10	12
1	1.2	1.7	2.2	3.0	3.5	4.5	5.4	6.7	7.7	8.6	10.5	12.2	15.4	18.7	22.2
2	1.4	1.9	2.5	3.3	3.9	5.1	6.0	7.5	8.6	9.5	11.7	13.7	17.3	20.8	24.8
3	1.5	2.0	2.7	3.6	4.2	5.4	6.4	8.0	9.2	10.2	12.5	14.6	18.4	22.3	26.5
4	1.5	2.1	2.8	3.7	4.4	5.6	6.7	8.3	9.6	10.6	13.1	15.2	19.2	23.2	27.6
5	1.6	2.2	2.9	3.9	4.5	5.9	7.0	8.7	10.0	11.1	13.6	15.8	19.8	24.2	28.8
6	1.7	2.3	3.0	4.0	4.7	6.0	7.2	8.9	10.3	11.4	14.0	16.3	20.5	24.9	29.6
7	1.7	2.3	3.0	4.1	4.8	6.2	7.4	9.1	10.5	11.7	14.3	16.7	21.0	25.5	30.3
8	1.7	2.4	3.1	4.2	4.9	6.3	7.5	9.3	10.8	11.9	14.6	17.1	21.5	26.1	31.0
9	1.8	2.4	3.2	4.3	5.0	6.4	7.7	9.5	11.0	12.2	14.9	17.4	21.9	26.6	31.6
10	1.8	2.5	3.2	4.3	5.1	6.5	7.8	9.7	11.2	12.4	15.2	17.7	22.2	27.0	32.0

[1] From "Heating, Ventilating, Air Conditioning Guide," 1960.   Used by permission.

plus the equivalent length of the fittings and equipment) of the longest circuit to obtain the average friction loss to be used as a basis of design.   For preliminary calculation, the equivalent length is sometimes assumed to be 1.5 times the measured length.

Circulating Pumps.   Pumps should be selected on the basis of water flow and the head against which they must operate.   In addition, they should have characteristics such that there is no substantial rise in pressure or horsepower if flow is throttled.   Other factors to be considered are the temperature of water handled, whether the pump has a positive or negative head, and the allowable noise level. For small installations an "in-line" pump connected directly to and supported by the piping can be used.   For large installations, a conventional centrifugal pump is used.

Expansion Tanks.   The function of an expansion tank is to provide space for increase in water volume in the system due to the increased temperature.   Water expands approximately 4 per cent when heated from 40 to 200 F.   This water is stored in the expansion tank when the water temperature is high and is returned to the system when the water temperature is lower.   The tank must have capacity to store this quantity without exceeding the allowable operating pressure.

Hot-water systems are designed with open (Fig. 33) or closed (Fig. 34) expansion tanks.   The *open-tank* system is vented to the atmosphere and is generally limited to operating temperatures below 180 F.   The tank should be at least 3 ft above the high point of the system and must be connected to the suction side of the pump to prevent cavitation at pump suction.   The tank volume should be not less than 6 per cent of the total system water volume.

The *closed system* uses an airtight tank which provides a means of pressurizing the system over the operating range.   As the temperature of water increases, the air in the tank compresses and increases the pressure on the system.   It is usual practice to size a closed expansion tank for about 200 per cent of the water expansion between the low and high temperature of the system.   The ASME Low Pressure Heating Boiler Code requires that provision must be made for draining a closed expansion tank without emptying the system.

Boilers.   Boilers usually used for hot-water heating are the 30-psi cast-iron or steel heating boilers built to ASME Boiler and Pressure Vessel Code, Section IV, for Low Pressure Heating Boilers, or steel heating boilers for pressures up to 150 psig with water temperatures below 250 F built to the ASME Code for the pressure at which they are operated.

The ASME Code also requires installation of a relief valve on each hot-water heating boiler to limit the boiler pressure to its designed and tested pressure. This limit of operating pressure for the boiler together with the characteristics of the expansion tank determines the maximum height of the distribution system.

Boiler Piping. Figure 35 shows the arrangement of piping at a boiler for a hot-water heating system. A more elaborate piping connection for a battery of boilers supplying different types of heating units and system is shown in Fig 36.

Return connections to the boiler should be made to eliminate possibilities of thermal shock when the cool water is returned. If more than one circulating pump

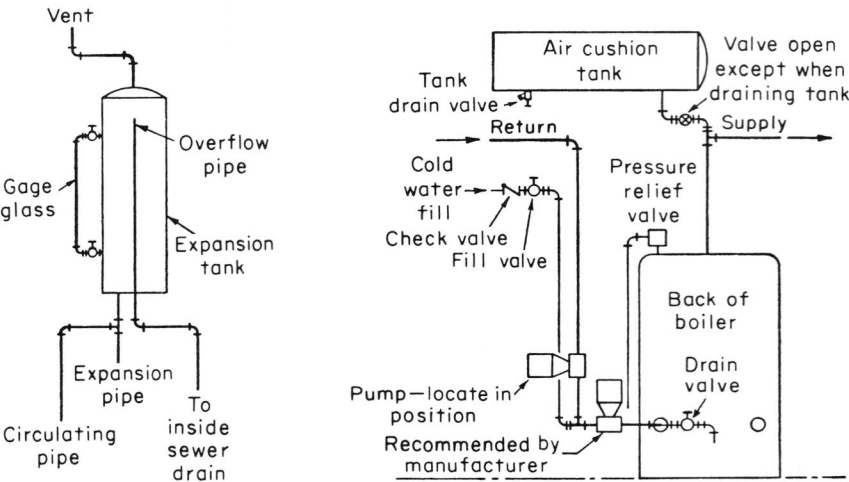

Fig. 33. Open expansion tank. (*From "Heating, Ventilating, Air Conditioning Guide,"* 1960. *Used by permission.*)

Fig. 34. Closed expansion tank. (*From "Heating, Ventilating, Air Conditioning Guide,"* 1960. *Used by permission.*)

is used, the discharge of the pumps should be joined before being discharged into the boiler with a check valve installed in the discharge of each pump.

Flow-control or check valves should be provided in the supply piping from the boiler to prevent gravity circulation during periods when the circulating pumps are not operating.

Piping and Water Flow. Piping should be pitched upward $\frac{1}{4}$ in. per 10 ft in the direction of flow to assure constant air elimination and installed to provide for thermal expansion of the pipe without imposing undue stress on connections or equipment.

If air is not eliminated in hot-water heating systems, the flow of water will be restricted. Air is eliminated by trapping it into the expansion tank and by the provision of manual or automatic air vents at all high points in the system and at each heating unit.

Balancing valves are provided as a means of adjusting or balancing the water flow in each circuit and at each heating unit. After the system is installed, these balancing valves are regulated so that the required quantity of water flows through each heating unit. This can be best done by determining the temperature drop of

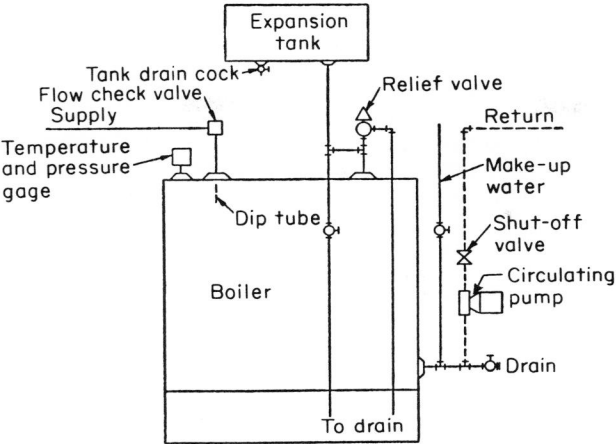

FIG. 35. Boiler piping for hot-water system. (*From "Heating, Ventilating, Air Conditioning Guide," 1960. Used by permission.*)

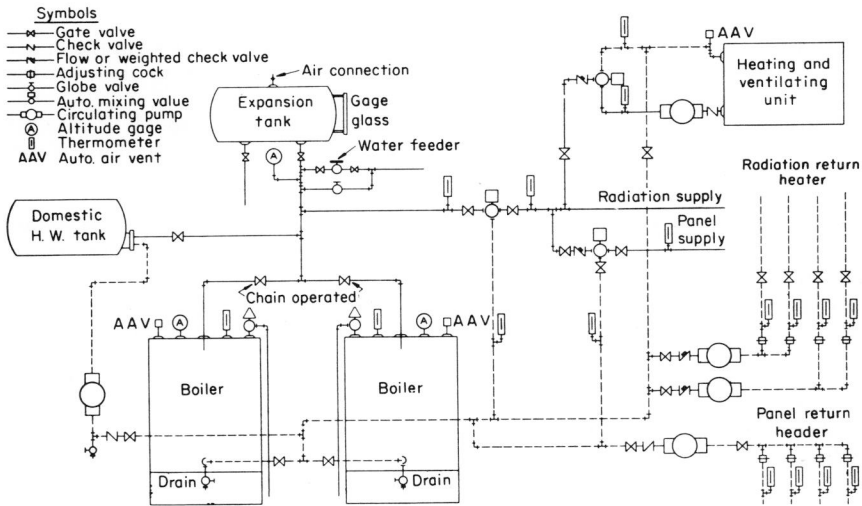

FIG. 36. Boiler piping for multipurpose hot-water system. (*From "Heating, Ventilating, Air Conditioning Guide," 1960. Used by permission.*)

water flowing through the heating unit by means of surface contact thermometers or by thermometers installed in the piping. The valves are then adjusted to produce the same temperature drop in all units.

High-temperature Water Systems

High-temperature water systems are those that supply water for heating at temperatures above 250 F and at pressures from 55 to 350 psig.

As compared with low-temperature systems, these which operate at high temperatures are capable of providing a greater temperature drop through heating equipment and thereby reduce the quantity of water circulated. In addition, different temperatures of water may be obtained with the use of heat exchangers.

As with low-temperature systems, the supply and return piping may be run level or at different elevations to suit architectural and structural requirements.

The pressure in any part of a high-temperature system must always be above the pressure corresponding to the temperature at saturation in the system in order to prevent flashing of the water into steam.

Practical considerations limit the supply water temperatures to 300 to 350 F. Water temperatures of 300 to 350 F require blanketing pressures from 75 to 150 psig, while with higher water temperatures of 350 to 400 F pressures of 150 to 250 psig are required. This pressurization limits the use of high-temperature water systems to those installations in which the added cost of heavier pipe and equipment is justified economically.

Pressurization to prevent the water in the system from flashing into steam is usually done by steam or an inert gas such as nitrogen. This pressure may be maintained by:

1. A steam cushion in the steam drum, the steam space of the boiler, or in a separate expansion tank
2. Compression tank with compressed air or inert gas
3. An automatic pressure pump

In the steam pressurization system, the temperature of the water from the boiler cannot exceed the temperature corresponding to the saturation pressure of the steam.

Boilers. The boilers for high-temperature water systems should be selected based on the load and design pressures. The boilers may be of the water-tube, fire-tube, or Scotch-marine types. Water-tube boilers are preferable for large systems and higher pressures.

Water-tube boilers usually require external tanks for the pressurization, while fire-tube boilers, if pressurized by steam, have expansion space within the boiler but require a separate tank if pressurized by an inert gas.

A major problem when pressurizing by steam is to maintain a proper boiler water level while providing for the change in water volume due to expansion and contraction.

When pressurizing with a pump, the pressure control is set to operate the boiler-feed pump and to thereby cause feedwater to flow from the make-up tank to the boiler whenever the pressure falls.

Proper distribution of return water and of water flow is essential in all types of boilers to prevent tube or tube-sheet failures due to overheating or unequal expansion of the boiler.

Boiler Piping. When the circulation pumps discharge to the system supply piping, the flow main should not rise high enough above the boiler water line to reduce the total pressure to permit flashing in the piping. Figure 37 shows a simplified arrangement of piping at the boiler with the circulating pump in the supply and return line.

Figure 38 shows two or more boilers supplying a common load with provision to equalize the flow of water to each boiler.

When boilers are paralleled in systems pressurized by inert gas, equalization is not critical if the piping is arranged to deliver the minimum flow rate to each boiler. Boiler piping in these systems is generally a simple in-and-out arrangement similar to that used in low-temperature hot-water heating systems.

With steam pressurizing, the water level must be held within proper limits always to provide a water seal at the entrance to the supply flow pipe; otherwise flashing or steam bubbling could occur in the supply pipe as the water level rises and falls.   In addition, with inert-gas pressurization, the piping is simplified, since the expansion drum is connected by a single balance line (no flow through the drum) and the mixing bypass to the pump inlets is not required.

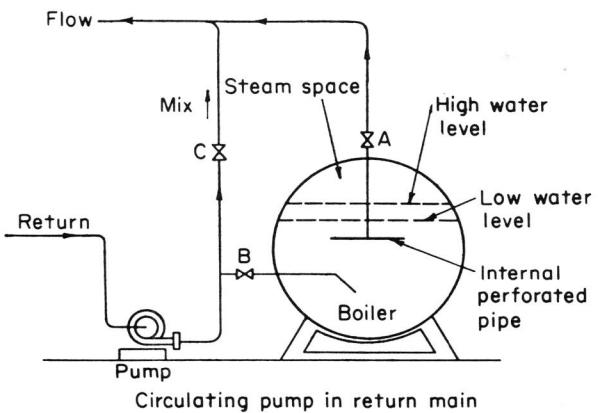

Circulating pump in return main

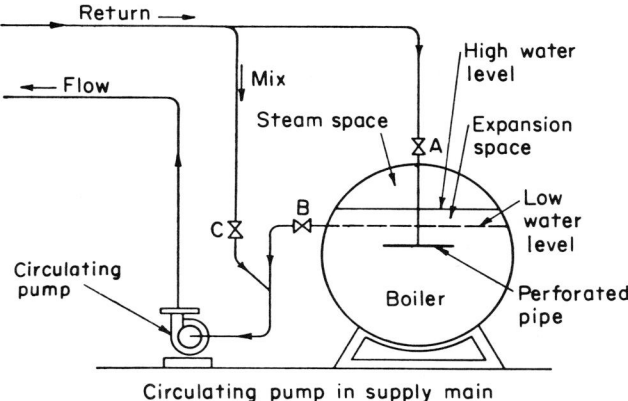

Circulating pump in supply main

FIG. 37. Simplified boiler piping, high-temperature hot-water system. (*From "Heating, Ventilating, Air Conditioning Guide,"* 1960.   *Used by permission.*)

Pipe Sizing. Normally, heating equipment for high-temperature hot-water systems is based on a temperature drop of 100 F.   This results in smaller pipe being used as compared with low-temperature systems for the same quantity of heat. The friction loss for Schedule 40 steel pipe for a mean temperature of 300 F is given in Fig. 39.   The use of smaller pipe will result in higher pressure drops through the system, and this should be economically balanced against the higher pressure in the system and pump requirements.

All pipe, valves, and fittings used in high-temperature systems should comply with the requirements of the American Standard Code for Pressure Piping, ASA

B31.1, latest issue.   Care should be used in the selection of packing to prevent leakage, as high-temperature water is more penetrating than low-temperature water.

As with low temperature, the system heating units should be provided with balancing valves to control the flow of water through the unit.   All high points in piping should be provided with air vents for removing air.

The sizing and selection of valves are very important because of the high-temperature drops and small flows encountered.   For good control, the valve must be sized so that it is effective over its full range of stem travel.   The pressure drop

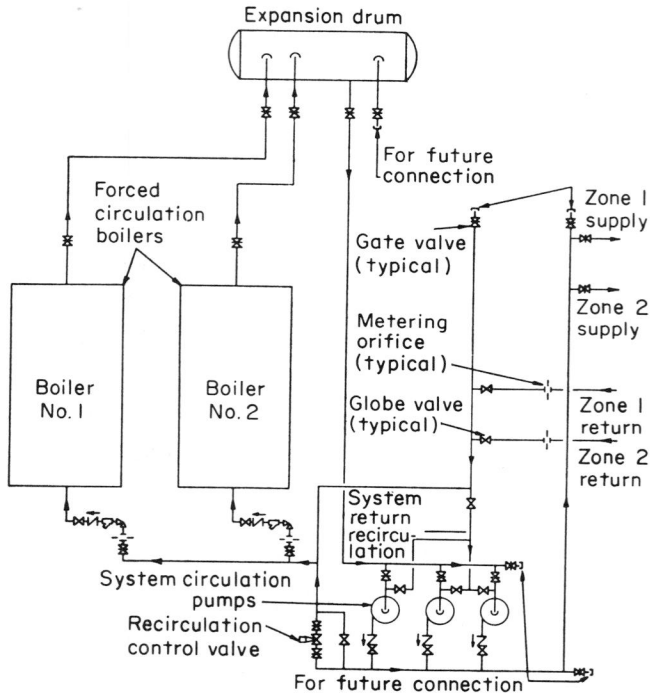

Fig. 38. Boiler piping, high-temperature hot-water system.   (*From "Heating, Ventilating, Air Conditioning Guide," 1960.   Used by permission.*)

across the valve should not result in a downstream pressure below the saturation pressure at the temperature existing at any point, or flashing into steam will result.

Control valves should preferably be located in the return lines of heating units in order to reduce the valve operating temperature.   This problem is minimized in systems pressurized by inert gas, since a greater difference is maintained between the system and saturation pressures.

Stainless-steel trim should be used, and all valve-body materials and packing should be suitable for the high temperatures and pressures encountered.

PANEL HEATING

Panel heating refers to those installations in which heat is supplied to a space at relatively low temperatures by radiant or heated surfaces of walls, floors, or ceiling.

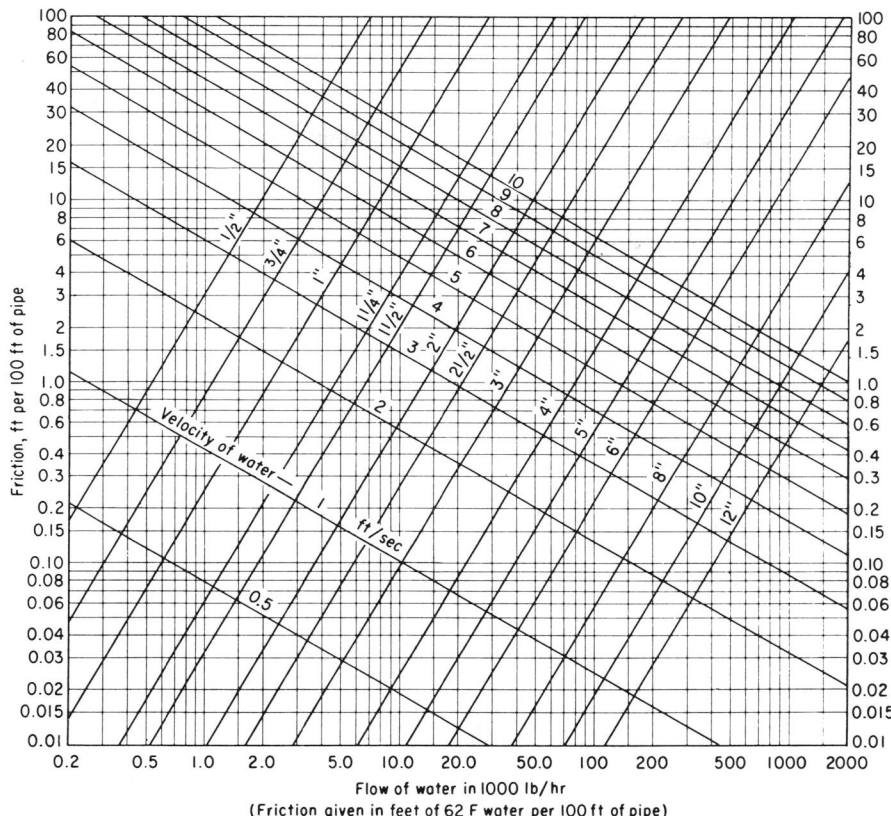

Fig. 39. Friction loss for 300 F water in steel pipe. (*From "Heating, Ventilating, Air Conditioning Guide,"* 1960. *Used by permission.*)

While warm air or electrical elements can also be used, the great majority of installations use warm-water circulation through pipes as the heating medium. The former are used only where local factors dictate their use.

With warm water, both ferrous (steel and wrought iron) and nonferrous (copper and aluminum) pipe can be used. Since the piping is embedded in concrete, plaster, or panel construction, all joints should be welded or soldered. In order to keep the number of joints to a minimum, changes in direction should be done by bending the pipe rather than using fittings. When the piping is installed and before it is covered, it should be subjected to a hydrostatic test of at least 150 psig or three times the normal working pressure and the pressure maintained for 12 to 24 hr with no pressure drop.

Panel Heating Design. In panel heating, a different procedure is necessary to select the heating units from that for the other forms of heating.

After the heat loss for the space has been calculated, the following steps should be followed:

1. Determine the available area for location of the panel, the required unit panel output, and surface temperature.

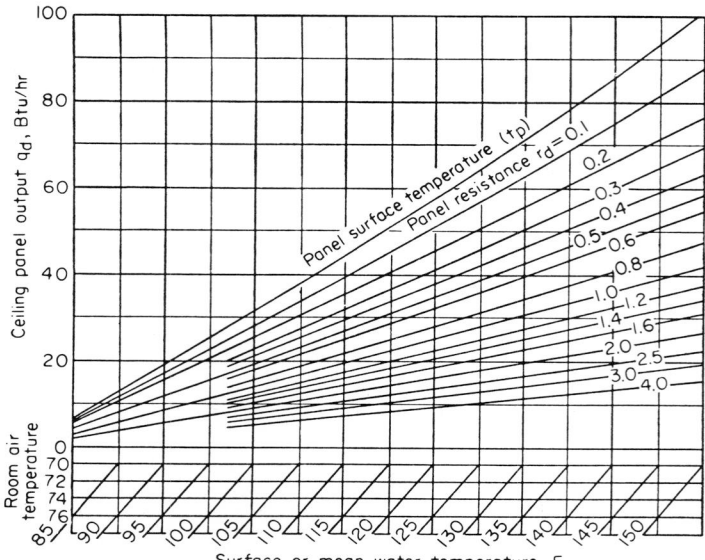

Fig. 40. Ceiling panel surface temperature: output downward. (*From "Heating, Venti-lating, Air Conditioning Guide,"* 1960. *Used by permission.*)

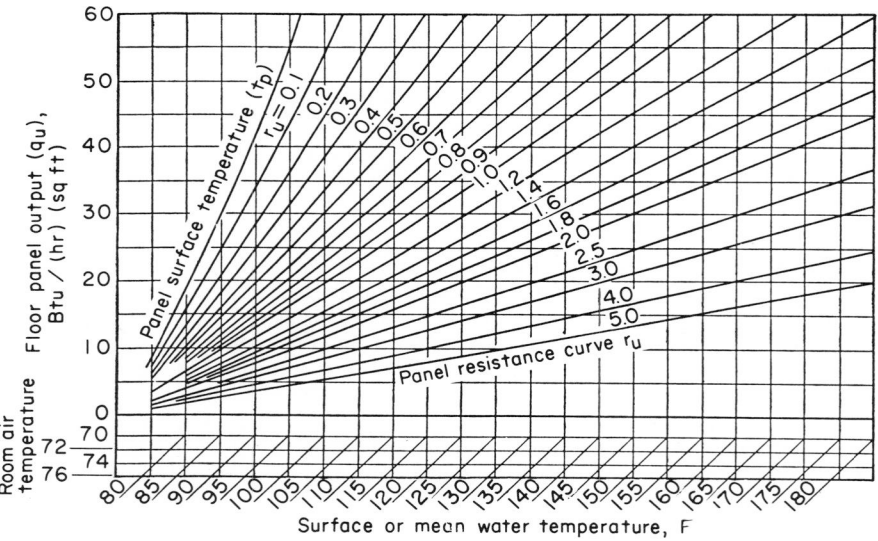

Fig. 41. Floor panel surface temperature: output upward. (*From "Heating, Ventilating, Air Conditioning Guide,"* 1960. *Used by permission.*)

2. Determine the loss of heat from the panel to the outside or to adjacent heated spaces, and include this as the required panel input.

3. Select the means of heating the panel, its size, its location, and the spacing of piping.

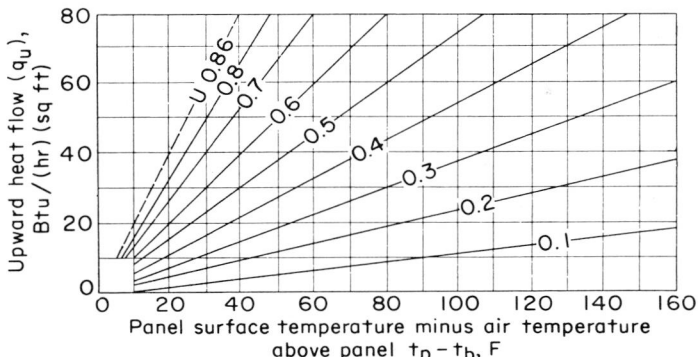

FIG. 42. Upward heat flow from plaster ceiling panel. (*From "Heating, Ventilating, Air Conditioning Guide," 1960. Used by permission.*)

Panels should be located near cold areas where the major heat losses occur. Care should be taken to assure that discomfort does not occur as a result of excessively high floor or ceiling temperatures. Generally, a floor temperature of approximately 85 F is recommended.

The heat output from a panel is by both convection and radiation, and the combined heat-transfer rates are given in Figs. 40 and 41 for ceiling and floor panels, respectively. The thermal resistance for ceiling panels and for concrete floor panels is given in Tables 17 and 18. The upward heat flow for ceiling panels is given in Fig. 42. The losses downward and edgewise for concrete slabs on grade are given in Fig. 43.

Heat is lost from the upper part of ceiling panels, from the back of wall panels, and from below floor panels. These losses are part of the building heat loss if the heat is transferred outside the building. If the heat is transferred to another heated space, the panel loss is a source of heat gain for the other space.

Panel heat loss to the outside should be kept to a reasonable amount by insulation. Panel heat loss to other heated spaces may

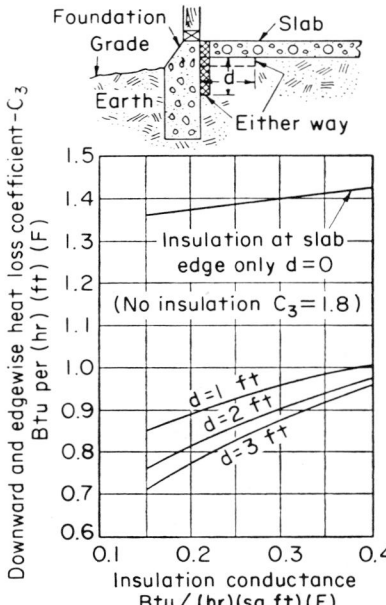

FIG. 43. Downward and edgewise heat-loss coefficient for concrete slabs on grade. (*From "Heating, Ventilating, Air Conditioning Guide," 1960. Used by permission.*)

Table 17. Thermal Resistance of Plaster and Metal Ceiling Panels[1]

| | Spacing, in. | Thermal resistance to downward heat flow (r_d), (deg F hr sq ft)/Btu | | | | | |
| | | Heat-flow ratio, q_u/q_d | | | | | |
		0	0.2	0.4	0.6	0.8	1.0[2]
Plaster panels[3] standard gypsum plaster, three coats	4½	0.30	0.34	0.38	0.42	0.46	0.50
⅜-in. (nom.) nonferrous tube or ½-in. (nom.) ferrous pipe *above* metal lath tied at 8-in. intervals with good tube embedment, or ⅜-in. (nom.) nonferrous tube *below* metal or gypsum lath	6	0.45	0.51	0.57	0.63	0.69	0.75
	9	0.75	0.85	0.95	1.05	1.15	1.25
	12	1.15	1.29	1.43	1.57	1.71	1.85
Aluminum panel[4] 1/16-in. minimum thickness pipe or tube installed so that excellent heat conduction to ceiling surface is obtained	3	0.07	0.07	0.07	0.07	0.07	0.07

[1] From "Heating, Ventilating, Air Conditioning Guide," 1960. Used by permission.
[2] Any ceiling panel also acts as a floor panel to the extent of its upward heat flow. If the upward heat flow is high and the space above is occupied, check floor surface temperature for possible foot discomfort (see Ref. 7). Also check effect on heating requirements of the space above. It is not good practice to have the major portion of the upper room's heating requirements supplied by the upward heat flow of a ceiling panel below.
[3] Recommended maximum inlet water temperature (t_{max}) 140 F.
[4] Recommended maximum inlet water temperature (t_{max}) 220 F.

Table 18. Thermal Resistance of Bare Concrete Floor Panels[1]

Panel construction	Spacing, in.	Thermal resistance, (F sq ft hr)/Btu							
		Heat-flow ratio, q_u/q_d or q_u/q_{de}							
		1		3		5		10	
4-in. concrete slab—2-in. cover		Up r_{us}	Down r_{ds}	Up r_{us}	Down r_{ds}	Up r_{us}	Down r_{ds}	Up r_{us}	Down r_{ds}
½-in. (nom.) nonferrous tube	9	0.57	0.52	0.46	0.84	0.43	1.17	0.42	1.97
	12	0.73	0.68	0.58	1.16	0.54	1.65	0.51	2.86
½-in. (nom.) ferrous pipe or ¾-in. (nom.) nonferrous tube	9	0.49	0.42	0.41	0.66	0.39	0.90	0.38	1.80
	12	0.63	0.55	0.50	0.93	0.48	1.30	0.46	2.35
6-in. concrete slab—2-in. cover									
½-in. (nom.) nonferrous tube	9	0.59	0.70	0.47	1.05	0.45	1.39	0.43	2.25
	12	0.78	0.90	0.60	1.40	0.56	1.97	0.54	3.21
¾-in. (nom.) nonferrous tube	9	0.51	0.61	0.43	0.87	0.41	1.13	0.40	1.78
	12	0.68	0.78	0.54	1.23	0.51	1.63	0.49	2.61
¾-in. (nom.) ferrous pipe	9	0.47	0.55	0.40	0.77	0.39	0.98	0.38	1.50
	12	0.63	0.71	0.50	1.07	0.48	1.44	0.46	2.36
1-in. (nom.) nonferrous tube or 1-in. (nom.) ferrous pipe	12	0.59	0.66	0.48	0.98	0.46	1.30	0.45	2.11
	15	0.73	0.83	0.57	1.21	0.54	1.73	0.51	2.74

[1] From "Heating, Ventilating, Air Conditioning Guide," 1960.   Used by permission.

require reduction by insulation if the amount of heat transferred will result in high loss of temperature.

Floor covering has a marked effect on the performance of a floor panel system by adding resistance to heat flow upward and increasing heat flow downward.   For a floor on grade, this requires an increase in the heat input to the system to compensate for the additional loss.   Resistances for floor covering are given in Table 19.   Where covered and uncovered floor slabs are adjacent, they should be served by different panels to maintain exposed floor temperature.

An example of the design of a heating panel in a plaster ceiling is as follows:

A room 11 by 12 by 8 ft designed to maintain 72 F with 0 F outdoors has a heat loss of 6300 Btu/hr.   The ceiling has a U factor of 0.25, and the space above is heated to the same temperature.

The procedure for determining the size of panel is as follows:

1. The ceiling has an area of 132 sq ft available for a heating panel (A_p).

2. The required panel output q_d is then 47.7 Btu/(hr sq ft).   From Fig. 40 the panel surface temperature t_p is 114 F.

3. The loss upward from the panel q_u from Fig. 42 is 12.5 Btu/(sq ft hr) for the U of 0.25.   This gives a heat-flow ratio q_u/q_d of 0.26.

4. Assume a tentative pipe spacing of 4½ in., and with the heat-flow ratio determined in 3, the thermal resistance to downward flow r_d (from Table 17 is 0.35 (F sq ft hr)/Btu.

Table 19.   Thermal Resistance of Floor Coverings[1]

Description	Resistance, r_{uc}, (F hr sq ft)/Btu
Bare concrete, no covering	0.00
Asphalt tile	0.05
Rubber tile.	0.05
Light carpet.	0.6
Light carpet with rubber pad	1.0
Light carpet with light pad	1.4
Light carpet with heavy pad	1.7
Heavy carpet.	0.8
Heavy carpet with rubber pad	1.2
Heavy carpet with light pad	1.6
Heavy carpet with heavy pad	1.9

[1] From "Heating, Ventilating, Air Conditioning Guide," 1960.   Used by permission.

5. The mean water temperature t_{mw} for the downward panel output q_d, panel resistance r_d, and the room air temperature from Fig. 40 is 131 F.

6. For the total panel output the heat flow upward q_u determined in 3 is added to the design panel output to give a unit panel output of 60.2 Btu/(hr sq ft) or a total panel output of 7946 Btu/hr.

7. A temperature drop of 10 to 20 F is normally assumed through the panel. For an assumed value of 20 F there would be required to be circulated approximately 8 gpm.

Since normally there are several spaces supplied by the same system and the same water temperature, some adjustment must be made in pipe spacing, size of pipe, or control of losses by addition of insulation to give the required results.

The loss through the piping system should be determined by a method similar to that previously given under low-temperature hot-water heating systems. Either grid or continuous coils can be used for distribution of hot water.

An example of the design of a heating panel in a concrete slab on grade is as follows:

A room 11 by 12 by 8 ft designed to maintain 72 F with 0 F outside has a heat loss of 3300 Btu/hr, and two sides of the wall are exposed.   The floor is covered with heavy wall-to-wall carpeting without a pad.   Insulation is provided along the edge of the slab for a depth of 2 ft and has a conductance of 0.4 Btu/(hr sq ft F).

1. The floor has a room area of 132 sq ft (A_p) for a heating panel.

2. The required panel output q_u is then 25 Btu/(hr sq ft).   From Fig. 41, the panel surface temperature t_p is 85 F.

3. In order to determine the loss downward from the panel, it is necessary first to find the surface temperature t_s of the concrete slab.

$$t_s = t_p + q_u r_{uc}$$

From Table 19, the floor-covering resistance r_{uc} is 0.8 (F hr sq ft)/Btu and the slab surface temperature t_s is 105 F.   From Fig. 43, the downward and edgewise coefficient C_3 is 0.97 Btu/(hr ft F).   The downward and edgewise loss q_{de} is 17.7 Btu/(hr sq ft) and is found from the following:

$$q_{de} = \frac{PC_3(t_s - t_{oa})}{A_p}$$

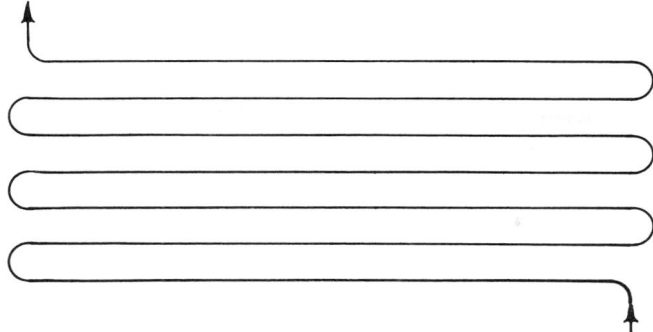

FIG. 44. Continuous coil pattern.

where P is the exposed wall area and t_{oa} is the outside temperature. This gives a heat-flow ratio q_u/q_{de} of 1.41.

4. Assume a tentative pipe spacing of 9 in., and with the heat-flow ratio determined in 3 the thermal resistance to upward flow r_{us} is 0.47 from Table 18. The total panel resistance to upward flow r_w is the sum of r_{us} and r_{uc} and is 1.27 (F hr sq ft) /Btu.

5. The mean water temperature t_{mw} for the upward panel output q_u, the panel resistance r_u, and the room air temperature from Fig. 41 is 116.5 F.

6. For the total panel output, the heat flow downward q_{de} determined in 3 is added to the design panel output q_u to give a unit panel output of 42.7 Btu/sq ft or a total panel output of 5636 Btu/hr.

7. With an assumed temperature drop of 20 F there should be required approximately 5.6 gpm.

The other comments previously indicated also apply here.

Installation Details. There are two basic patterns of coil design used for panel heating: the continuous coil (Fig. 44) and the grid (Fig. 45) or a combination (Fig. 46).

The continuous type is easy to fabricate, and pipe bends can be used where an irregular plan is required. The disadvantage, however, is the increase in frictional resistance with increase in total length of the coil. Initial expense of pump and pumping costs, therefore, limits this pattern to small areas.

The grid pattern can be used for large areas with light hydraulic load. The headers are made of larger pipe to carry the full quantity of water required. An

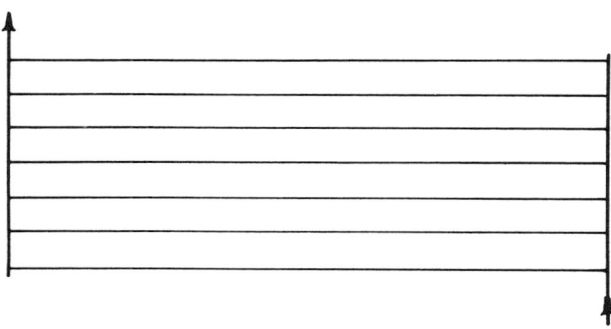

FIG. 45. Grid pattern.

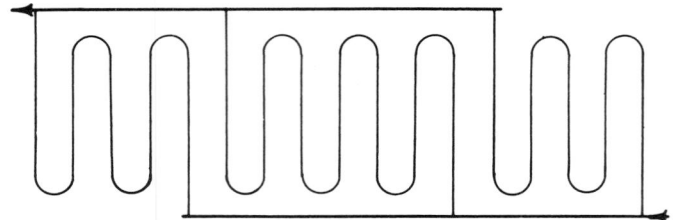

Fig. 46. Combination coil and grid pattern.

analysis should be made of the required pumping head in order to determine which pattern of distribution should be specifically used.

Care should also be taken to keep piping level in order to minimize trapping of air. Although in a series loop this problem can be sometimes overcome, in a parallel grid it can be serious. A balancing valve must be provided at each grid or for each area where the flow of water must be controlled in order to provide equal distribution of heat to each space.

REFERENCES

American Society of Heating and Ventilating Engineers, "Guide," 1958.
American Society of Heating, Refrigeration and Air Conditioning Engineers, "Guide," 1960.
American Society of Heating, Refrigeration and Air Conditioning Engineers, "Guide and Data Book," 1961.
Carrier, Cherne, and Grant, "Modern Air Conditioning, Heating and Ventilating."
Giesecke, F. E., "Friction Heads Due to Water Flow in Copper, Brass, and Other Smooth Pipes," *Trans. ASHVE*, vol. 49, p. 175, 1943.
ASHAE Research Report 1600, "Thermal Design of Warm Water Concrete Floor Panels," *Trans. ASHAE*, vol. 63, p. 239, 1957.
ASHAE Research Report 1559, "Thermal Design of Warm Water Ceiling Panels," *Trans. ASHAE*, vol. 62, p. 71, 1956.

11

CRYOGENIC-SYSTEMS PIPING

Stanley W. Ehrlich*

INTRODUCTION

Cryogenics is the science of cold, being derived from the Greek word *kryos*, meaning "icy cold." The temperature level separating a cryogenic fluid from the commonly employed refrigerants is for purposes of this chapter set at -40 F. This temperature is an arbitrary point and for various reasons has been set at times from -100 F to within a few degrees of absolute zero (-459.7 F).

The special problems encountered in handling cryogenic fluids arise from the very fact that they are cold. The influx of heat from the surroundings tends to vaporize the liquids (cryogenic liquids usually have small latent heats of vaporization) or to superheat saturated vapors; either will increase pressure drops in flowing lines or will lead to higher pressures in vessels and drums. The choice of suitable materials to handle cryogens is limited not so much by corrosion as by their physical properties at low temperatures.

A cryogenic process utilizes low temperatures to produce a physical change in liquid, solid, or gas, such as to liquefy gases as oxygen, nitrogen, helium, and methane and to make certain metals superconductive.

CRYOGENIC FLUIDS

The more commonly accepted cryogenic fluids and their pertinent physical properties necessary in piping calculations are tabulated in the following groups.

* Registered Professional Engineer, Hydrocarbon Research, Inc., New York, N.Y.

Acetylene (Ethyne), C_2H_2.   Pertinent properties are:

Molecular weight.........	26.04
Critical temperature......	96.8 F
Critical pressure..........	905 psia
Critical density	14.4 lb/cu ft
Freezing point	−114 F
Boiling point	−119 F

Acetylene is colorless and highly flammable. Commercial acetylene has a garlic odor; however, in the pure state the odor is ethereal. It is used mainly for oxyacetylene cutting, heat treating, and as a basic starting material for organic chemical synthesis.

Table I.   Acetylene: Saturated Liquid and Vapor Densities

Temp F	Pressure, psia	Density, lb/cu ft	
		Liquid	Vapor
−119.2	14.7	38.7	0.108
−113.4	18.6	38.0	0.135 (triple pt.)
−100	25.0	37.7	0.201
−90	35.1	37.4	0.269
−80	47.0	36.8	0.339
−70	61.5	36.3	0.415
−60	76.2	35.8	0.507
−50	90.0	35.2	0.613

Table 2.   Acetylene: Density of Superheated Vapor at 14.7 Psia

Temp, F	Density, lb/cu ft
32	0.073
60	0.069
70	0.068

Viscosity of acetylene vapor (32 F, 14.7 psia) = 0.00939 centipoise

Air.   Pertinent properties are:

Molecular weight.........	28.96
Critical temperature......	−221.3 F
Critical pressure..........	547 psia
Critical density..........	20.5 lb/cu ft
Boiling point	−317.9 F

Air is a mixture of gases having the following composition:

Gas	Volume % (dry basis) at sea level
Nitrogen	78.09
Oxygen.............	20.95
Argon..............	0.93
Carbon dioxide......	0.02–0.04
Neon	18×10^{-4}
Helium.............	5.3×10^{-4}
Krypton...........	1.1×10^{-4}
Hydrogen	0.5×10^{-4}
Xenon	0.08×10^{-4}
Ozone	0.02×10^{-4}
Radon	7×10^{-18}

The main interest in air is for its individual components and for its use in breathing.

Table 3. Air (Water-free): Saturated Liquid and Vapor Densities

Pressure, psia	Liquid		Vapor	
	Temp, F	Density, lb/cu ft	Temp, F	Density, lb/cu ft
14.7	−317.85	54.56	−312.45	0.280
29.4	−305.70	52.57	−300.73	0.5334
44.1	−296.0	51.07	−292.96	0.7797
73.5	−286.21	48.94	−282.01	1.266
102.9	−277.82	47.31	−274.00	1.757
147.0	−268.05	45.20	−264.66	2.516
220.5	−255.66	41.84	−252.85	3.890
294.0	−246.27	38.77	−243.56	5.472
367.5	−237.75	35.89	−235.75	7.329
441.0	−230.62	32.46	−229.07	9.689
514.5	−224.05	27.86	−223.13	13.47
547	−221.33	20.48	−221.33	20.48

Table 4. Air (Water-free): Density of Superheated Vapor

Temp, F	Density, lb/cu ft						
	1.47 psia	14.7 psia	73.5 psia	147 psia	220.5 psia	367.5 psia	1,470 psia
−300	0.0246						
−250	0.0193	0.195	1.045	2.250	3.75		
−200	0.0153	0.153	0.810	1.640	2.56	4.74	
−150	0.0131	0.131	0.655	1.340	2.06	3.59	
−100	0.0110	0.110	0.561	1.130	1.73	2.95	137.5
−50	0.0096	0.096	0.486	1.000	1.50	2.50	107.5
0	0.0086	0.086	0.436	0.870	1.31	2.21	91.5

Table 5. Viscosity of Saturated Air (Water-free)

Temp, F	Viscosity, centipoises	
	Liquid	Vapor
−240	0.0692	0.00975
−260	0.080	0.00805
−280	0.0965	0.00695
−300	0.125	0.00600
−320	0.175	0.00518

Table 6. Viscosity of Air at Atmospheric Pressure

Temp, F	Vapor viscosity, centipoises
−317.85	0.00523
−300	0.00600
−250	0.00810
−200	0.00995
−150	0.01170
−100	0.01330
−50	0.01480
0	0.01625
100	0.01898

Table 7. Viscosity of Superheated Air

Temp, F	Vapor viscosity, centipoises				
	200 psia	400 psia	600 psia	800 psia	1,000 psia
−250	0.00850				
−200	0.01020	0.01125	0.01320	0.01900	0.02500
−150	0.01188	0.01240	0.01323	0.01450	0.01593
−100	0.01350	0.01385	0.01455	0.01540	0.01645
−50	0.01502	0.01535	0.01585	0.01650	0.01730
0	0.01645	0.01675	0.01713	0.01765	0.01825
100	0.01915	0.01935	0.01960	0.01990	0.02032

Argon, A. Pertinent properties are:

Molecular weight	39.94
Critical temperature	−187.6 F
Critical pressure	705.5 psia
Critical density	33.1 lb/cu ft
Freezing point	−308.6 F
Boiling point	−302.6 F

Argon is the most abundant of all the rare gases.   It is colorless, odorless, tasteless, and chemically inactive.   It is used mainly as an inert shield for welding and in electric light bulbs.

Table 8.   Argon: Saturated Vapor and Liquid Densities

Pressure, psia	Temp, F	Density, lb/cu ft	
		Liquid	Vapor
14.7	−302.6	86.8	0.356
29.4	−290	84.2	0.670
45.5	−280	82.4	1.040
70.5	−270	80.0	1.530
102.9	−260	77.2	2.170
144.0	−250	75.0	3.04
194	−240	71.9	4.095
259	−230	68.7	5.55
341	−220	64.5	7.45
429	−210	59.5	10.40
548	−200	53.3	15.60
705.5	−188.4	33.1	33.1

Table 9. Argon: Density of Superheated Vapor

Temp, F	Density, lb/cu ft				
	0.147 psia	1.47 psia	14.7 psia	147 psia	1,470 psia
−300	0.0034	0.0344			
−250	0.0026	0.0263	0.268		
−200	0.0021	0.0209	0.210	2.290	
−150	0.0017	0.0177	0.177	1.865	
−100	0.0015	0.0152	0.152	1.580	15.0
−50	0.0013	0.0134	0.134	1.375	11.0
0	0.0012	0.0120	0.120	1.22	9.2

Table 10. Argon: Viscosity of Saturated Liquid

Temp, F	Viscosity, centipoises
—300	0.24
—280	0.173
—260	0.138
—240	0.110
—220	0.084
—200	0.060

Table 11. Argon: Viscosity of Superheated Vapor

Temp, F	Viscosity, centipoises		
	14.7 psia	1,000 psia	2,000 psia
—230	0.00965		
—200	0.01060		
—150	0.01280		
—100	0.0150	0.01830	0.02140
—50	0.01730	0.02020	0.02330
0	0.01950	0.02220	0.02520
50	0.02140	0.02400	0.02690
100	0.02320	0.02580	0.02860

Carbon Dioxide (Carbonic Acid), CO_2. Pertinent properties are:

Molecular weight	44.01
Critical temperature	87.8 F
Critical pressure	1,072.1 psia
Critical density	29.0 lb/cu ft
Freezing point	—69.9 F at 75.1 psia
Sublimation point	—109.3 F

Carbon dioxide is an odorless, nonflammable, colorless gas. It is used as a refrigerant, for carbonating beverages, and as an inert atmosphere.

Table 12. Carbon Dioxide: Saturated Vapor and Liquid Densities

Pressure, psia	Temp, F	Density, lb/cu ft	
		Liquid	Vapor
3.2	—140	Solid	0.0409
8.9	—120	Solid	0.1098
22.3	—100	Solid	0.2635
50.7	—80	Solid	0.5890
75.1	—69.9	73.5	0.8650 (triple pt.)
94.8	—60.0	72.3	1.0810
118.3	—50	71.0	1.3350
145.9	—40	69.6	1.6380
215.0	—20	66.8	2.4000
305.8	0	63.7	3.4400

Table 13. Carbon Dioxide: Density of Superheated Vapor

Temp, F	Density, lb/cu ft					
	14.7 psia	147 psia	294 psia	1,176 psia	1,764 psia	2,940 psia
−50	0.01	0.101	0.204	0.868		
0	0.0089	0.090	0.180	0.750	1.155	2.05
50	0.008	0.0808	0.1635	0.670	1.02	1.77
100	0.0073	0.0733	0.1480	0.605	0.918	1.57

Table 14. Carbon Dioxide: Viscosity of Liquid and Vapor

Liquid		Vapor	
Temp, F	Viscosity, centipoises	Temp, F	Viscosity, centipoises
32	0.099	−140	0.00905
50	0.085	−120	0.00950
70	0.693	−100	0.00998
86	0.053	−80	0.01048
		−60	0.0110
		−40	0.01154
		−20	0.01215
		0	0.01280

Carbon Monoxide, CO. Pertinent properties are:

Molecular weight	28.01
Critical temperature	−218.2 F
Critical pressure	508.5 psia
Critical density	19.0 lb/cu ft
Freezing point	−337.1 F
Boiling point	−313.6 F

Carbon monoxide is a colorless, odorless, poisonous gas. It is used in the synthesis of methanol and alcohol, is one of the constituents of synthesis gas, and is used as a reducing agent.

Table 15. Carbon Monoxide: Saturated Vapor and Liquid Densities

Pressure, psia	Temp, F	Density, lb/cu ft	
		Liquid	Vapor
2.3	−337	52.9	0.0499 (triple pt.)
9.1	−320	50.5	0.1875
31	−300	47.4	0.500
79	−280	44.0	1.3150
167	−260	40.2	2.75
303	−240	34.6	5.550
508.5	−218.2	19.0	19.0

Table 16. Carbon Monoxide: Density of Superheated Vapor

Temp, F	Density, lb/cu ft				
	367.5 psia	1,470 psia	2,940 psia	7,350 psia	14,700 psia
−100	2.86	13.1	24.5	36.5	43.7
−80	2.75	11.7	22.3	34.6	42.7
−60	2.55	10.8	20.5	33.3	41.5
−40	2.37	10.1	19.0	32.1	40.6
−20	2.19	9.4	17.8	30.9	39.6
0	1.98	8.0	15.3	28.1	37.5
50	1.78	7.7	14.5	27.2	36.9

**Table 17. Carbon Monoxide:
Viscosity of Liquid**

Temp, F	Viscosity, centipoises
−337.1	0.290
−330	0.242
−320	0.193
−312.8	0.166

**Table 18. Carbon Monoxide: Viscosity of
Saturated Vapor**

Temp, F	Pressure, psia	Viscosity, centipoises
−315	12.8	0.00530
−300	31.0	0.00590
−280	79.0	0.00675
−260	166	0.00790
−240	302	0.00955
−230	395	0.01110

Table 19. Carbon Monoxide: Viscosity of Superheated Vapor

Temp, F	Viscosity, centipoises				
	14.7 psia	200 psia	400 psia	600 psia	1,000 psia
−313	0.00535				
−250	0.00790	0.00845			
−200	0.00965	0.01005	0.01113	0.01400	
−150	0.01130	0.01160	0.01215	0.01315	0.01615
−100	0.01285	0.01315	0.01355	0.01428	0.01628
−50	0.01430	0.01458	0.01495	0.01545	0.01690
0	0.01570	0.01595	0.01625	0.01665	0.01773
50	0.01705	0.01725	0.01755	0.01785	0.01865

Carbonyl Sulfide, COS.   Pertinent properties are:

Molecular weight	60.08
Critical temperature	221 F
Critical pressure	896.7 psia
Freezing point	−217.8 F
Boiling point	−58.4 F

Carbonyl sulfide is a flammable, colorless gas having an odor similar to that of rotten eggs. Its toxicity is similar to that of hydrogen sulfide. It is used in the synthesis of some organic compounds.

Liquid density is 77.2 lb/cu ft at −124.6 F.

Vapor density is 0.155 lb/cu ft at 70 F and 14.7 psia.

Vapor viscosity at 14.7 psia and 60 F is 0.0119 centipoise; at the same pressure but at 212 F, the viscosity is 0.0154 centipoise.

Deuterium, D_2. Pertinent properties are:

Molecular weight	4.03
Critical temperature	−390.8 F
Critical pressure	239 psia
Freezing point	−426 F

Deuterium is one of the isotopes of hydrogen. Its normal distribution in hydrogen is 1 part in 5,000. It is used in the manufacture of heavy water. Its densities are listed below:

Liquid density (saturated), 10.7 lb/cu ft at −417.3 F and 14.7 psia
Vapor density (saturated), 0.16 lb/cu ft at −417.3 F and 14.7 psia
Vapor density, 0.1014 lb/cu ft at 70 F and 14.7 psia

Table 20. Deuterium: Viscosity of Liquid

Temp, F	Viscosity, centipoises
−426	0.0497
−424	0.0419
−422	0.0370
−420	0.0332
−418	0.0300
−416	0.0274

Table 21. Deuterium: Viscosity of Superheated Vapor

Temp, F	Viscosity, centipoises		
	14.7 psia	441 psia	1,764 psia
−430	0.119	0.2305	0.340
−400	0.230	0.310	0.421
−350	0.384	0.436	0.549
−300	0.519	0.550	0.662
−250	0.641	0.663	0.768
−200	0.745	0.768	0.860
−150	0.863	0.863	0.950
−100	0.960	0.958	1.013
−50	1.058	1.045	1.110
0	1.145	1.130	1.190
50	1.232	1.218	1.260

Ethane, C_2H_6. Pertinent properties are:

Molecular weight	30.07
Critical temperature	90.1 F
Critical pressure	707.8 psia
Critical density	12.63 lb/cu ft
Freezing point	−277.6 F
Boiling point	−127.5 F

Ethane is a colorless, odorless, flammable gas. It is widely used as a starting material for many organic syntheses and as a refrigerant.

Table 22. Ethane: Density of Saturated Vapor and Liquid

Temp, F	Pressure, psia	Density, lb/cu ft	
		Liquid	Vapor
−220	0.27	37.4	0.00321
−200	0.85	36.8	0.00930
−180	2.20	36.0	0.02230
−160	4.97	35.4	0.04750
−140	9.97	34.5	0.09020
−120	18.33	33.9	0.1580
−100	31.32	33.0	0.2610
−80	50.34	32.1	0.4080
−60	77.02	31.3	0.6120
−40	113.1	30.3	0.885
−20	159.9	29.1	1.255
0	219.7	28.0	1.740
50	437.5	24.1	3.85

Table 23. Ethane: Density of Superheated Vapor

Temp, F	Density, lb/cu ft				
	10 psia	20 psia	30 psia	60 psia	100 psia
−140	0.0902				
−120	0.0845				
−100	0.0797	0.1625			
−80	0.0750	0.1525	0.2340		
−60	0.0708	0.1445	0.2200	0.4550	
−40	0.0672	0.1370	0.2085	0.4280	0.778
−20	0.0642	0.1305	0.1980	0.4060	0.719
0	0.0615	0.1240	0.1875	0.3870	0.672
50	0.0552	0.1115	0.1680	0.3450	0.589

Table 24. Ethane: Viscosity of Liquid and Vapor

Temp, F	Viscosity, centipoises	
	Liquid	Vapor at 14.7 psia
−270	0.73	
−250	0.52	
−200	0.29	
−150	0.195	0.00585
−100	0.14	0.00648
−50	0.103	0.00718
0	0.0795	0.00795
50	0.0585	0.00875

Ethylene (Ethene), C_2H_4. Pertinent properties are:

Molecular weight	28.05
Critical temperature	49.8 F
Critical pressure	742.1 psia
Critical density	14.2 lb/cu ft
Freezing point	−272.5 F
Boiling point	−154.7 F

Ethylene is a colorless, flammable gas having a sweet odor and is used in welding, organic synthesis, manufacture of polyethylene, and as a refrigerant.

Table 25. Ethene: Density of Saturated Vapor and Liquid

Temp, F	Pressure, psia	Density, lb/cu ft	
		Liquid	Vapor
−180	5.91	36.8	0.0559
−160	12.3	35.8	0.1105
−140	23.2	34.8	0.1995
−120	40.2	33.7	0.3350
−100	65.6	32.0	0.5320
−80	101.0	31.5	0.5780
−60	148.4	30.3	1.1650
−40	206.3	28.8	1.6850
−20	289.8	27.4	2.3900
0	387.9	25.5	3.3200

Table 26. Ethene: Density of Superheated Vapor

Temp, F	Density, lb/cu ft			
	14.7 psia	29.4 psia	58.8 psia	147 psia
−140	0.124			
−120	0.116	0.238		
−100	0.110	0.223	0.465	
−80	0.103	0.210	0.435	
−60	0.098	0.199	0.411	1.18
−40	0.093	0.189	0.385	1.09
−20	0.088	0.178	0.366	1.00
0	0.085	0.171	0.347	0.950
50	0.076	0.153	0.309	0.832

Table 27. Ethene: Viscosity of Vapor and Liquid

Temp, F	Viscosity, centipoises	
	Liquid	Vapor at 14.7 psia
−260	0.56	
−200	0.25	
−150	0.157	0.00645
−100	0.114	0.00704
−50	0.090	0.00767
0	0.072	0.00842
50	0.058	0.00958

Fluorine, F_2.   Pertinent properties are:

Molecular weight	38.00
Critical temperature.......	−247 F
Critical pressure	808 psia
Freezing point...........	−363.2 F
Boiling point.............	−304.6 F

Fluorine is a green-yellowish poisonous gas used in missiles, medicines, and uranium compounds.

Table 28. Fluorine: Density of Liquid

Temp, F	Density, lb/cu ft
—340	101.8
—320	97.2
—300	92.5
—280	87.0

Table 29. Fluorine: Density of Vapor

Temp, F	Density at 14.7 psia, lb/cu ft
32	0.1060
70	0.0989

Table 30. Fluorine: Viscosity of Liquid

Temp, F	Viscosity, centipoises
—330	0.366
—320	0.305
—310	0.258

Table 31. Fluorine: Viscosity of Vapor

Temp, F	Viscosity at 14.7 psia, centipoises
—280	0.85
—250	0.98
—200	1.22
—150	1.43
—100	1.68
—50	1.90
0	2.12
50	2.30

Monochlorotrifluoromethane (Freon-13 ®, Genetron-13 ®), $CClF_3$.

Molecular weight.........	104.5
Critical temperature.......	83.9 F
Critical pressure..........	561 psia
Critical density...........	36.1 lb cu ft
Freezing point............	—294 F
Boiling point.............	—114.6 F

Freon or Genetron is the trade name for a series of organic compounds mainly used as a refrigerating medium. These compounds are characterized by low toxicity and nonflammability.

Viscosity of liquid at —95 F is 0.37 centipoise.

Table 32.   Freon-13, Genetron-13: Density of Saturated Liquid and Vapor

Temp, F	Pressure, psia	Density, lb/cu ft	
		Liquid	Vapor
−200	0.43	105.6	0.0163
−150	4.46	99.6	0.1434
−100	22.23	93.0	0.6393
−50	71.71	85.7	1.946
0	176.80	77.0	4.840
50	365.9	64.7	11.45

Table 33.   Freon-13, Genetron-13: Density of Superheated Vapor

Temp, F	Density, lb/cu ft				
	1 psia	15 psia	30 psia	100 psia	300 psia
−180	0.0351				
−140	0.0306				
−100	0.0271	0.4225			
−80	0.0256	0.3960	0.8350		
−60	0.0244	0.3750	0.7720		
−40	0.0233	0.3550	0.7280		
−20	0.0221	0.3390	0.6900	2.560	
0	0.0213	0.3240	0.6550	2.400	
60	0.0188	0.3845	0.5750	2.010	7.58

Helium, He.   Pertinent properties are:

Molecular weight	4.003
Critical temperature.	−450.2 F
Critical pressure	33.2 psia
Critical density	4.32 lb/cu ft
Freezing point.	−457.8 F
Boiling point.	−452.1 F

Helium is an inert, tasteless, colorless, odorless gas. It is used as an inert shielding gas in welding, for electronic tubes, and as a carrier gas in instrumentation.

Table 34.   Helium: Density of Saturated Liquid

Temp, F	Density, lb/cu ft
−455.7	9.18
−454.3	8.80
−452.5	8.05
−450.4	4.94

Table 35.   Helium: Density of Subcooled Liquid

Temp, F	Density, lb/cu ft				
	14.7 psia	73.5 psia	147 psia	220.5 psia	441 psia
−457.4	9.18	9.50	9.85	10.22	
−456.7	9.15	9.52	9.92	10.25	
−455.7	9.18	9.60	10.00	10.40	11.18
−454.7	9.0	9.51	9.91	10.25	11.15
−453.7	8.7	9.28	9.73	10.13	11.00
−452.7	8.28	8.91	9.52	9.92	10.80

Table 36.   Helium: Density of Superheated Vapor

Temp, F	Density, lb/cu ft				
	14.7 psia	150 psia	400 psia	2,500 psia	6,000 psia
—430	0.180	0.211			
—400	0.094	0.870	2.25	8.72	14.38
—350	0.050	0.500	1.27	6.05	10.90
—300	0.034	0.350	0.872	4.55	8.72
—250	0.027	0.267	0.715	3.75	7.50
—200	0.022	0.209	0.55	2.77	5.92
—150	0.017	0.175	0.455	2.55	5.30
—100	0.015	0.150	0.399	2.25	4.74
—50	0.014	0.137	0.368	2.05	4.36
0	0.012	0.125	0.324	1.87	3.93
60	0.011	0.109	0.344	1.62	3.50

Table 37.   Helium: Viscosity of Liquid

Temp, F	Viscosity, centipoises
—455.8	0.00278
—455.2	0.00364
—454.3	0.00376
—453.4	0.00368
—452.5	0.00360

Table 38.   Helium: Viscosity of Vapor at 14.7 Psia

Temp, F	Viscosity, centipoises
—440	0.0026
—400	0.0048
—350	0.0072
—300	0.0093
—250	0.0111
—200	0.0128
—150	0.0143
—100	0.0158
—50	0.0171
0	0.0186
50	0.0197

Hydrogen, H_2.   Pertinent properties are:

Molecular weight........	2.016
Critical temperature	—399.8 F
Critical pressure..........	188.2 psia
Critical density...........	1.93 lb/cu ft
Freezing point	—430.9 F
Boiling point	—423.0 F

Hydrogen is the lightest of all gases and is colorless and odorless.   Its uses are varied from welding to ammonia synthesis, including hydrogenation and petroleum treating.

Table 39. Hydrogen: Density of Liquid

Temp, F	Density, lb/cu ft
−435	4.79
−430	4.64
−425	4.46
−420	4.26
−415	4.00
−410	3.68
−405	3.26

Table 40. Hydrogen: Density of Superheated Vapor

Temp, F	Density, lb/cu ft				
	1.47 psia	14.7 psia	147 psia	588 psia	1,470 psia
−420	0.00672				
−400	0.00471	0.0469			
−350	0.00255	0.0252	0.2645	1.165	2.700
−300	0.00174	0.0171	0.1755	0.700	1.675
−250	0.00130	0.0132	0.1320	0.518	1.225
−200	0.00107	0.0101	0.101	0.419	0.983
−150	0.00090	0.0089	0.089	0.348	0.828
−100	0.00077	0.0076	0.076	0.300	0.719
−50	0.00067	0.0066	0.066	0.263	0.632
0	0.00059	0.0059	0.059	0.233	0.561
60	0.00052	0.0053	0.053	0.206	0.500

Table 41. Hydrogen: Viscosity, Centipoises

Temp, F	Liquid	Temp, F	Vapor at 14.7 psia
−432.7	0.0217	−440	0.0006
−430.9	0.0197	−400	0.0017
−429.1	0.0178	−350	0.0029
−427.3	0.0162	−300	0.0038
−425.5	0.0146	−250	0.0047
−423.7	0.0134	−200	0.0054
−421.9	0.0126	−150	0.0062
		−100	0.0068
		−50	0.0074
		0	0.0080
		60	0.0087

Hydrogen Sulfide, H_2S.   Pertinent properties are:

Molecular weight.........	34.08
Critical temperature	212.7 F
Critical pressure..........	1,306.8 psia
Critical density...........	21.7 lb/cu ft
Freezing point	−117.2 F
Boiling point	−75.3 F

Hydrogen sulfide is a colorless gas, has the odor of rotten eggs, and is flammable. It is used in the preparation of metal sulfides but mainly is found as an undesirable contaminant in industrial gases.

Table 42. Hydrogen Sulfide: Density of Saturated Liquid and Vapor

Temp, F	Pressure, psia	Density, lb/cu ft	
		Liquid	Vapor
−76.4	14.7	57.1	0.1205
−60	22.1	56.6	0.1820
−50	27.9	56.0	0.2280
−40	36.8	55.5	0.2880
−20	55.9	54.5	0.4350
0	82.3	53.5	0.6400
60	229.3	49.1	1.6800

Density of vapor at 60 F and 14.7 psia is 0.0898 lb/cu ft.
Density of vapor at 60 F and 147 psia is 1.0180 lb/cu ft.
Viscosity of vapor at 32 F and 14.7 psia is 0.0117 centipoise.
Viscosity of vapor at 60 F and 14.7 psia is 0.0123 centipoise.

Krypton, Kr. Pertinent properties are:

Molecular weight.........	83.8
Critical temperature	−82.8 F
Critical pressure..........	798.2 psia
Critical density...........	56.8 lb/cu ft
Freezing point	−250.8 F
Boiling point	−243.2 F

Krypton is colorless and odorless and is one of the rare gases.   Krypton and its isotope Kr-85 are widely used in the electronic industry.

Table 43. Krypton: Density of Saturated Vapor and Liquid

Temp, F	Pressure, psia	Density, lb/cu ft	
		Liquid	Vapor
−250	11.5	154.5	0.44
−200	69	139.0	2.50
−150	246	121.5	7.80
−100	615	94.2	23.00

Density of superheated vapor at 32 F and 14.7 psia is 0.233 lb/cu ft.
Viscosity of vapor at 32 F and 14.7 psia is 0.0233 centipoise.
Viscosity of vapor at 68 F and 14.7 psia is 0.0248 centipoise.

Methane, CH_4. Pertinent properties are:

Molecular weight.........	16.04
Critical temperature	−116.5 F
Critical pressure..........	673.3 psia
Critical density...........	10.1 lb/cu ft
Freezing point	−296.5 F
Boiling point.............	−258.6 F

Methane is a colorless, tasteless, odorless, and flammable gas.   It is used in the synthesis of organic chemicals and in making synthesis gas.

Table 44. Methane: Density of Saturated Vapor and Liquid

Temp, F	Pressure, psia	Density, lb/cu ft	
		Liquid	Vapor
−280	4.90	27.51	0.0416
−250	21.71	26.05	0.1630
−200	115.7	23.22	0.7806
−150	364	19.24	2.62
−120	627	14.37	6.20

Table 45. Methane: Density of Superheated Vapor

Temp, F	Density, lb/cu ft				
	20 psia	60 psia	100 psia	300 psia	1,000 psia
−220	0.1265	0.4140			
−200	0.1165	0.3700	0.6680		
−150	0.0971	0.3475	0.5295	1.920	
−100	0.0840	0.2555	0.4350	1.460	
−50	0.0732	0.2220	0.3770	1.205	5.56
0	0.0653	0.1978	0.3325	1.040	4.26
60	0.0578	0.1740	0.2930	0.910	3.34

Table 46. Methane: Viscosity of Liquid

Temp, F	Viscosity, centipoises
−301	0.226
−290	0.181
−280	0.154
−270	0.135
−260	0.120

Table 47. Methane: Viscosity of Vapor

Temp, F	Viscosity, centipoises
−280	0.0040
−250	0.0046
−200	0.0057
−150	0.0067
−100	0.00775
−50	0.00875
0	0.0097
60	0.0108

Neon, Ne.   Pertinent properties are:

Molecular weight.........	20.18
Critical temperature	−379.6 F
Critical pressure..........	394.5 psia
Critical density...........	30.1 lb/cu ft
Freezing point	−415.6 F
Boiling point	−410.9 F

Neon is a colorless, odorless inert gas which is used in the electronics industry, i.e., neon lights.

Table 48. Neon: Density of Saturated Vapor and Liquid

Temp, F	Pressure, psia	Density, lb/cu ft	
		Liquid	Vapor
−410	16.8	74.5	0.655
−400	61.8	68.0	2.500
−390	176	58.6	6.720
−382	328	46.1	15.25

Table 49. Neon: Density of Superheated Vapor

Temp, F	Density, lb/cu ft			
	350 psia	500 psia	800 psia	1,000 psia
−350	6.80	10.0	18.0	23.1
−300	4.74	6.05	9.83	12.15
−250	2.93	4.30	7.10	8.72
−200	2.50	3.49	5.60	6.99
−150	2.12	3.12	4.74	5.87
−100	1.93	2.80	4.10	5.00
−50	1.74	2.50	3.86	4.50
0	1.50	2.25	3.43	4.00
60	1.25	1.87	2.86	3.50

Table 50. Neon: Viscosity of Vapor at 14.7 Psia

Temp, F	Viscosity, centipoises
−400	0.0060
−350	0.0100
−300	0.0132
−250	0.0162
−200	0.0188
−150	0.0212
−100	0.0235
−50	0.0255
0	0.0283
60	0.0308

Nitric Oxide, NO. Pertinent properties are:

Molecular weight.........	30.01
Critical temperature	−137.2 F
Critical pressure..........	955.5 psia
Critical density...........	32.5 lb/cu ft
Freezing point	−262.5 F
Boiling point	−241.1 F

Nitric oxide is a colorless, nonflammable, and extremely toxic gas. It is an intermediate in the production of nitric acid.

Density of saturated liquid at −241.1 F and 14.7 psia is 79.2 lb/cu ft.

Table 51. Nitric Oxide: Density of Superheated Vapor

Temp, F	Density, lb/cu ft				
	14.7 psia	40 psia	100 psia	200 psia	300 psia
−100	0.1145	0.314	0.793	1.615	2.500
−80	0.1085	0.296	0.752	1.538	2.355
−60	0.1030	0.282	0.709	1.437	2.220
−40	0.0982	0.268	0.678	1.370	2.090
−20	0.0933	0.255	0.643	1.305	1.980
0	0.0900	0.245	0.616	1.250	1.885
60	0.0792	0.216	0.540	1.100	1.640

Table 52. Nitric Oxide: Viscosity of Vapor at 14.7 Psia

Temp, F	Viscosity, centipoises
−240	0.00854
−200	0.01000
−150	0.01192
−100	0.01375
−50	0.01535
0	0.01687
60	0.01865

Nitrogen, N_2.   Pertinent properties are:

Molecular weight.........	28.02
Critical temperature......	−232.9 F
Critical pressure..........	492.5 psia
Critical density...........	19.4 lb/cu ft
Freezing point...........	−345.8 F
Boiling point.............	−320.5 F

Nitrogen is a nonflammable, colorless, and odorless gas.   It is used as an inert atmosphere and in the synthesis of nitrogen chemicals such as ammonia.

Table 53. Nitrogen: Density of Saturated Vapor and Liquid

Temp, F	Pressure, psia	Density, lb/cu ft	
		Liquid	Vapor
−340	3.3	53.3	0.0718
−320	15.0	50.1	0.286
−300	46	46.8	0.830
−280	111	43.0	1.935
−260	230	38.4	4.360
−240	415	30.9	8.730

Table 54. Nitrogen: Density of Superheated Vapor

Temp, F	Density, lb/cu ft				
	14.7 psia	58.8 psia	147 psia	588 psia	1,470 psia
−270	0.206	0.883			
−250	0.186	0.775	2.140		
−200	0.149	0.609	1.590		
−150	0.125	0.505	1.288	5.850	
−100	0.107	0.432	1.085	4.640	12.60
−50	0.094	0.375	0.943	3.900	10.18
0	0.0828	0.333	0.838	3.405	8.63
60	0.0734	0.294	0.738	2.970	7.42

Table 55. Nitrogen: Viscosity of Saturated Vapor and Liquid

Temp, F	Pressure, psia	Viscosity, centipoises	
		Liquid	Vapor
−340	3.3	0.260	0.00465
−320	15	0.156	0.00535
−300	46	0.107	0.00610
−280	111	0.085	0.00710
−260	230	0.075	0.00845

Table 56. Nitrogen: Viscosity of Vapor at 14.7 Psia

Temp, F	Viscosity, centipoises
−250	0.00795
−200	0.00980
−150	0.01145
−100	0.01297
−50	0.01450
0	0.01582
60	0.01735

Nitrous Oxide (Laughing Gas), N_2O. Pertinent properties are:

Molecular weight	44.02
Critical temperature	97.7 F
Critical pressure	1,054 psia
Critical density	28.2 lb/cu ft
Freezing point	−152.2 F
Boiling point	−129.1 F

Nitrous oxide is a colorless, nonflammable, nontoxic gas with a sweetish taste. It is chiefly used as an anesthetic and as a pressuring gas for aerosol containers.

Table 57. Nitrous Oxide: Density of Saturated Vapor and Liquid

Temp, F	Pressure, psia	Density, lb/cu ft	
		Liquid	Vapor
−127	14.7	80.0	0.182
−120	19.0	79.0	0.235
−100	35.0	76.0	0.420
−80	58.0	74.0	0.700
−60	91.0	70.5	1.070
−40	137	68.0	1.655
−20	203	65.5	2.350
0	283	63.1	3.300
60	675	49.2	8.400

Density of vapor at 70 F and 14.7 psia is 0.115 lb/cu ft

Density of vapor at 32 F and 14.7 psia is 0.124 lb/cu ft

Table 58. Nitrous Oxide: Viscosity of Vapor at 14.7 Psia

Temp, F	Viscosity, centipoises
−120	0.00940
−100	0.00993
−80	0.01046
−60	0.01100
−40	0.01155
−20	0.01208
0	0.01262
60	0.01425

Oxygen, O_2.   Pertinent properties are:

Molecular weight.........	32.00
Critical temperature	−181.9 F
Critical pressure..........	730.6 psia
Critical density...........	26.8 lb/cu ft
Freezing point	−361.1 F
Boiling point	−297.3 F

Oxygen is a colorless, odorless, tasteless gas essential for respiration.   It is used in medicine, for welding, in the production of synthesis gas, and acetylene, in steel manufacture and for combustion and missile propulsion.

Table 59. Oxygen: Density of Saturated Vapor and Liquid

Temp, F	Pressure, psia	Density, lb/cu ft	
		Liquid	Vapor
−300	12.7	71.3	0.000118
−280	36.5	67.2	0.000333
−260	85.0	64.0	0.000752
−240	170	59.9	0.00145
−220	303	54.5	0.00255
−200	498	47.4	0.00440
−190	615	41.7	0.00620

Table 60. Oxygen: Density of Superheated Vapor

Temp, F	Density, lb/cu ft				
	14.7 psia	58.8 psia	147 psia	588 psia	1,470 psia
—270	0.243				
—250	0.211	0.8921			
—200	0.170	0.700			
—150	0.142	0.578	1.49	7.32	
—100	0.122	0.495	1.26	5.58	
—50	0.104	4.431	1.09	4.61	12.95
0	0.0962	0.384	0.96	4.01	10.60
60	0.0848	0.336	0.85	3.49	8.90

Table 61. Oxygen: Viscosity of Saturated Vapor and Liquid

Temp, F	Pressure, psia	Viscosity, centipoises	
		Liquid	Vapor
—360	0.0282	0.790	
—340	0.42	0.415	
—320	3.10	0.265	
—300	12.7	0.197	0.0066
—280	36.5	0.152	0.0075
—260	85.0	0.125	0.00835
—240	170	0.112	0.00930
—220	303	0.105	0.01080
—200	498	0.098	0.01330

Table 62. Oxygen: Viscosity of Vapor at 14.7 Psia

Temp, F	Viscosity, centipoises
—270	0.0081
—250	0.0089
—200	0.0110
—150	0.0129
—100	0.0147
—50	0.0165
0	0.0181
60	0.0200

Ozone, O_3.   Pertinent properties are: .

Molecular weight........	48.00
Critical temperature......	53.8 F
Critical pressure........	802 psia
Critical density..........	0.326 lb/cu ft
Freezing point..........	—419.8 F
Boiling point............	—169.6 F

Ozone is a colorless gas or a dark-blue liquid.   It is highly reactive as an oxidizing agent and is believed to be created in the upper atmosphere by the action of sunlight.

Density of liquid at —297.4 F is 97.8 lb/cu ft
Density of liquid at —319.7 F is 100.5 lb/cu ft
Density of vapor at —112 F and 14.7 psia is 0.1885 lb/cu ft
Density of vapor at 32 F and 14.7 psia is 0.1338 lb/cu ft
Viscosity of liquid at —297.4 F is 1.57 centipoises.
Viscosity of liquid at —319 F is 4.20 centipoises.

Propene (Propylene), C_3H_6. Pertinent properties are:

Molecular weight.........	42.08
Critical temperature	197.4 F
Critical pressure..........	667.0 psia
Critical density...........	14.5 lb/cu ft
Freezing point	−301.5 F
Boiling point	−53.9 F

Propylene is a colorless, flammable gas and it is used as a starting material in organic synthesis and to make polypropylene plastic.

Table 63. Propylene: Density of Saturated Vapor and Liquid

Temp, F	Pressure, psia	Density, lb/cu ft	
		Liquid	Vapor
−53.9	14.7	38.4	0.148
−50	16.2	38.1	0.162
−40	20.6	37.6	0.203
−30	25.9	37.1	0.250
−20	32.1	36.6	0.305
−10	39.5	36.1	0.369
0	50.0	35.6	0.443
60	131.5	32.5	1.195

Table 64. Propylene: Density of Superheated Vapor

Temp, F	Density, lb/cu ft		
	14.7 psia	29.4 psia	58.8 psia
−50	0.1458		
−40	0.1420		
−30	0.1388		
−20	0.1352	0.2815	
−10	0.1320	0.2740	
0	0.1288	0.2675	
10	0.1258	0.2605	
20	0.1230	0.2540	0.531
40	0.1177	0.2410	0.505
60	0.1128	0.2300	0.478

Table 65. Propylene: Viscosity of Liquid

Temp, F	Viscosity, centipoises
−290	7.6
−250	1.7
−200	0.67
−150	0.38

Viscosity of vapor at 32 F and 14.7 psia is 0.0078 centipoise.
Viscosity of vapor at 104 F and 14.7 psia is 0.00819 centipoise.

Xenon, Xe.   Pertinent properties are:

Molecular weight.........	131.3
Critical temperature	61.9 F
Critical pressure..........	847.3 psia
Critical density...........	68.6 lb/cu ft
Triple point..............	−169.2 F at 11.8 psia
Boiling point.............	−162.6 F

Xenon is a colorless, odorless and tasteless inert gas.   It is used in the electronic industry.   Liquid xenon has also been used in bubble chambers.

Table 66.   Xenon: Density of Saturated Vapor and Liquid

Temp, F	Pressure, psia	Density, lb/cu ft	
		Liquid	Vapor
−170	12.45	192	0.50
−150	21.3	188	1.00
−100	72	174.5	3.12
−50	188	160.5	7.18
0	405	142	15.6
50	742	108.2	37.8

Table 67.   Xenon: Density of Superheated Vapor at 14.7 Psia

Temp, F	Density, lb/cu ft
32	0.365
70	0.345

Table 68.   Xenon: Viscosity of Vapor

Temp, F	Pressure, psia	Viscosity, centipoises
32	14.7	0.0210
68	14.7	0.0226
77	14.7	0.0231
77	147	0.0234
77	366	0.0246

General Correlations for Density and Viscosity.   General correlations, both empirical and theoretical, have been proposed for the estimation of densities and viscosities of liquids and vapors.   These estimating methods together with recommended equations and their errors have been published in a book "The Properties of Gases and Liquids" by R. C. Reid and T. K. Sherwood, McGraw-Hill Book Company, New York, 1958.

CRYOGENIC MATERIALS

Metals[1]

The materials used for low-temperature service, ferrous and nonferrous, are characterized by several important factors, i.e., retention of ductility and resistance to shock loading at low temperatures and an improvement of tensile and yield strengths.

Loss of ductility is measured by the notched-bar impact test. Smooth specimens usually show amazing ductility even at very low temperatures, whereas a notched specimen will fail in brittle fracture. This phenomenon is sometimes called the "notch sensitivity" of metals. Metals with a face-centered lattice generally show no loss of ductility at low temperatures. Body-centered metals show a marked tendency to brittleness at temperatures below the transition range.

The term cryogenics covers many process fields, all of which are characterized by one common fact; that is, all fluids are at low temperatures. Because of this, one piping code does not and cannot apply. The following codes should be used when designing cryogenic piping systems:

ASA B31.2, Industrial Gas and Air Piping
ASA B31.3-1959, Petroleum Refinery Piping
ASA B31.5, Refrigeration Piping
ASA B31.6, Chemical Industry Process Piping

The following paragraphs extracted from Petroleum Refinery Piping Code (ASA B31.3-1959), a section of the American Standard Code for Pressure Piping and published by the American Society of Mechanical Engineers, sum up the basic requirements for cryogenic piping.

300.1.2 This Code applies to piping systems handling all fluids, . . . and to all types of services including oil, gas, steam, air, water, chemicals, and refrigerants (except as specified or excluded in Par. 300.1.3 . . .).

300.1.3 Package refrigeration unit piping may be designed and constructed in accordance with ASA B31.5 except that impact test requirements of Par. 323.2.1 and 323.2.2 shall apply.

300.1.5 Piping includes pipe, flanges, bolting, gaskets, valves, fittings, the pressure-containing parts of their components such as expansion joints, strainers and metering devices, and pipe-supporting elements, fixtures, and structural attachments. It does not include support structures, and equipment such as towers, building frames, pressure vessels, mechanical equipment and foundations.

323.2.1 General. The materials listed . . . shall not be used at design temperatures above those for which stress values are given in the tables, nor at design temperatures below minus 20°F unless they meet the impact-test requirements where they are required by Par. 323.2.2.

Temperature limits for unlisted materials which conform to other Code requirements shall be in accord with recognized engineering practice.

323.2.2 Impact Tests

(*a*) Impact tests shall be made for the following combinations of materials and temperatures:

(1) all materials for temperatures below minus 325°F

---

[1] See also the discussion of low-temperature properties of metals in Chap. 2.

(3) the following materials for temperatures below minus 20°F:
  carbon and low alloy steels
  ferritic chromium stainless steels
  free-machining austenitic chrome nickel stainless steels
  austenitic chromium-nickel stainless steels with a carbon content greater than 0.10%
  austenitic chromium-nickel stainless steel material in the form of deposited weld metal regardless of carbon content
  austenitic chromium-nickel stainless materials in cast form

(b) The impact tests shall be made in accordance with and shall meet the requirements of Par. UG-84 of Section VIII of the ASME Boiler and Pressure Vessel Code, with the following substitution for Par. UG-84(b), (1), (2), and (3):

(1) a welded test section shall be prepared from a piece of pipe, plate, or tubing for each material specification certified by the manufacturer in accordance with Par. UG-84(e). If the material to be used is not certified, test sections shall be prepared from each piece of pipe, plate, or tubing used. One set of impact-test specimens shall be taken across the weld with the notch in the weld and one set shall be taken similarly with the notch in the heat-affected zone.

(2) one set of impact-test specimens with the notch in the weld metal and one set with the notch in the heat-affected zone, shall be made for each range of pipe thickness that does not vary by more than ¼ inch over and under the tested thickness for each material specification used on the job.

In general, Table 69 shows the temperature and material divisions commonly used in low-temperature service:

Table 69. Materials for Low-temperature Service

Temperature range, F	Material
0 to −20	Carbon steel
− 20 to −50	Carbon steel, aluminum killed
− 50 to −150	Alloy steel
−150 to −320	Alloy steel or nonferrous
Below −320	Nonferrous

Carbon Steel. The decrease in ductility at about −50 F for plain carbon steel requires special carbon steels for use at lower temperature. The ductility transition temperature is that temperature below which brittle fracture may take place. Below this temperature a crack started by a small amount of deformation will propagate readily. The transition temperature corresponds to about 5 to 25 ft-lb on the V-notch Charpy test. The value usually selected for design is 15 ft-lb.

In carbon steels, changes in the composition of carbon, manganese, aluminum, and silicon are the means of lowering the transition temperature. Carbon raises the temperature, and manganese lowers it. Aluminum and silicon are used as deoxidizers. Silicon-killed steel has slightly better notch toughness than semikilled steel, and silicon-aluminum-killed steel has even better toughness.

The finer the grain size, the lower will be the transition temperature. Normalizing at 1600 to 1700 F will reduce the grain size for thick material. Thin material is usually not benefited.

Quenching, followed by tempering at a temperature of 1100 to 1200 F, produces the best notch toughness possible in a ferritic steel if martensite can be produced in the quenching operation. The lower the temperature at which steel transforms from austenite to ferrite, the better will be the toughness after tempering. Carbon steel has insufficient hardenability to transform to martensite.

Table 70. Low-temperature Transition Points for Carbon Steels[1]

Nominal composition, %: C	Nominal composition, %: Mn	Grade	Deoxidation practice	Heat treatment	Approximate V-notch Charpy 15 ft-lb transition temperature, F: ¾-in. plate	Approximate V-notch Charpy 15 ft-lb transition temperature, F: 1½-in. plate
0.23	0.50	ASTM A7	Semikilled	Hot-rolled	50	80
0.21	0.70	ASTM A373	Semikilled	Hot-rolled	30	
			Silicon-killed	Hot-rolled	. .	40
0.16	0.90	ASTM A131, Grade B	Semikilled	Hot-rolled	0	
0.18	0.70	ASTM A131, Grade C	Si-Al-killed	Hot-rolled	. .	−10
			. .	Normalized	. .	−50
0.22	0.60	ASTM A201 made to A300	Si-Al-killed	Normalized	. .	−30[2]
0.27	0.85	ASTM A212 made to A300	Si-Al-killed	Normalized	. .	−40[2]
0.13	1.10	British Standard 2762, N.D. IV steel	Si-Al-killed	Normalized	. .	−100[3]

[1] R. W. Vanderbeck, *Chem. Eng.*, vol. 67, p. 103, 1960.
[2] 15 ft-lb minimum average value specified for keyhole Charpy test at −50 F.
[3] 20 ft-lb minimum individual value specified for V-notch Charpy test at −58 F. (The listed transition temperature is for plate approximately 1 in. thick.)

Cold working or deforming the steel will raise the transition temperature. Subsequent stress relieving at 1200 F will restore most of the original notch toughness, and normalizing will restore it completely.

Table 70 lists typical compositions and grades of carbon steel and their transition temperatures. The only carbon-plate steels manufactured in the United States to impact requirements are Grades A201 and A212 when made to ASTM Specification A300 and are recommended for use at temperatures as low as −50 F.

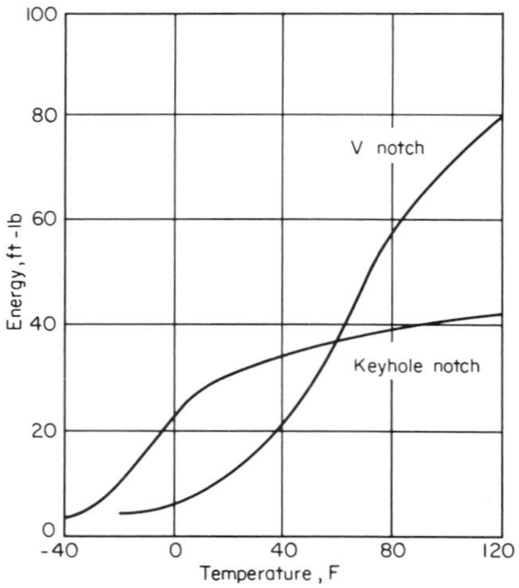

FIG. 1. Impact data comparison. (*R. W. Vanderbeck, Chem. Eng., vol.* 67, *p.* 103, 1960.)

Keyhole tests and V-notch tests give impact transition curves of different shape, and their relationship is shown in Fig. 1. A Charpy keyhole test of 15 ft-lb at −50 F would be equivalent to 15 ft-lb. at 0 to 10 F in the V-notch Charpy test.

The only seamless and welded carbon-steel pipe and tubes made to a keyhole Charpy test of 15 ft-lb at −50 F are those manufactured to ASTM Specifications A333 and A334. Table 71 shows design values for these pipe specifications.

Alloy Steel. The ASTM and the ASME Unfired Pressure Vessel Code sets

Table 71. Carbon-steel Properties

Material	Tensile strength, min psi	Yield strength, min psi	Allowable stress, psi, at 100 to −50 F
Steel pipe ASTM A333 C⎫ ASTM A334 C⎭	55,000	30,000	15,600

Thermal expansion, in./100 ft: 60 to 0 F = −0.45

Thermal expansion, in./100 ft: 60 to −50 F = −0.806

Thermal conductivity: 60 to −50 F = 37.5 Btu/(hr ft F)

Specific heat: 60 F = 0.191 Btu/(lb F)

Specific heat: 0 F = 0.183 Btu/(lb F)

Specific heat: −50 F = 0.175 Btu/(lb F)

up an A300 classification which covers carbon and alloy steels for use at temperatures down to −150 F.

Tables 72 and 73 and Fig. 2 show the properties of some A300 steels. A410, a Cr-Cu-Ni alloy steel, is serviceable to −150 F but is not widely used.

Table 74 gives further specifications for these A300 alloy steels.

Table 72. Test Temperatures: A300 Steels[1]

Class	ASTM Specification and Grade	Min test temp, F
1	A201, A212 flange or firebox quality	−50
2	A203, Grades A and B (2¼% nickel), firebox quality	−75
3	A203, Grades D and E (3½% nickel), firebox quality	−150
4	A253	−320
5	A410	−150

[1] R. Rote and J. H. Proctor, *Chem. Eng.*, vol. 67, p. 119, 1960.

Table 73. Allowable Stresses for A300 Steels[1]

Steel	Strength, psi		Allowable stress	
	Tensile	Yield, min	AISC (60% of yield)	ASME (25% of tensile)
A201 Grade B	60,000–72,000	32,000	19,200	15,000
A212 Grade B	70,000–85,000	38,000	22,800	17,500
A353	90,000	60,000	36,000	22,500

[1] R. Rote and J. H. Proctor, *Chem. Eng.*, vol. 67, p. 119, 1960.

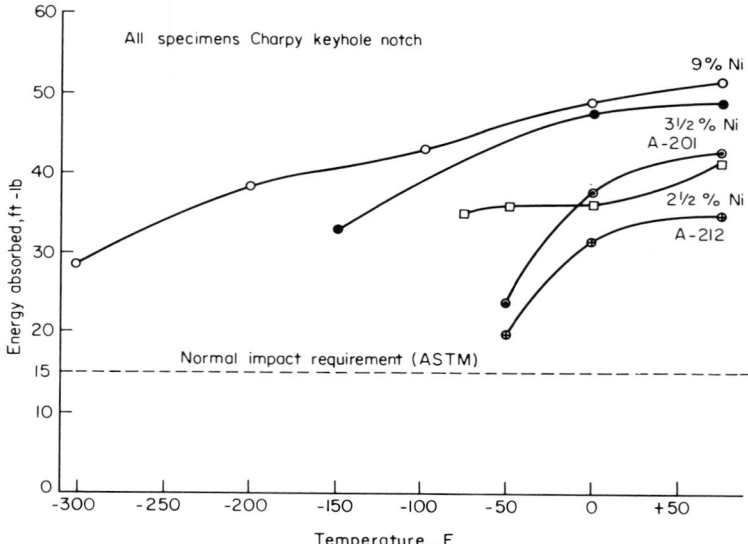

Fig. 2. Impact curves. (*R. Rote and J. Proctor, Chem. Eng., vol.* 67, *p.* 119, 1960.)

Stainless Steels. Chromium-nickel stainless steels are inherently suitable for service in the range of −300 to −425 F. The original cost of a stainless steel is higher than other materials, but ease of fabrication and welding, combined with high strength and good shock resistance, offset the high cost.

Table 74. Properties for A300 Steels

Material	Tensile strength, min psi	Yield strength, min psi	Allowable stress, psi at 100 F to	
			−75 F	−150 F
A203 A(2½ Ni)	65,000	37,000	21,650	
A203 B(2½ Ni)	70,000	40,000	23,350	
A203 D(3½ Ni)	65,000	37,000	.....	21,650
A203 E(3½ Ni)	70,000	40,000	.....	23,350

Temp range, F	Thermal expansion, in./100 ft
60 to 0	−0.45
60 to −50	−0.86
60 to −100	−1.146
60 to −150	−1.455

Thermal conductivity: 60 to −50 F = 37.0 Btu/(hr ft F)
                                        −50 to −150 F = 37.5 Btu/(hr ft F)

Temp, F	Specific heat, Btu/(lb F)
60	0.191
0	0.183
−50	0.175
−100	0.166
−150	0.154

Table 75. Stainless-steel Composition

Type	C, max	Mn, max	P, max	S, max	Si, max	Cr	Ni	Cb
302	0.15	2.00	0.045	0.030	1.00	17.00–19.00	8.00–10.00	
304L	0.03	2.00	0.045	0.030	1.00	18.00–20.00	8.00–12.00	
304	0.08	2.00	0.045	0.030	1.00	18.00–20.00	8.00–12.00	
310	0.25	2.00	0.045	0.030	1.00	24.00–26.00	19.00–22.00	
347	0.08	2.00	0.045	0.030	1.00	17.00–19.00	9.00–13.00	10 × C, min

Table 75 lists the chemical compositions and Table 76, the physical properties, of those stainless steels commonly used.   Type 304 is the most popular.

Nickel Steel.   Low-carbon 9 per cent nickel steel is a ferritic alloy developed for operating at temperatures as low as −320 F.   ASTM Specifications A300 and A353 cover low-carbon 9 per cent nickel steel, A300 being the basic specification for low-temperature ferritic steel.   The keyhole-notch Charpy impact test at −320 F for 9 per cent nickel steel is 15 ft-lb.

Table 76. Stainless-steel Properties

Material	Tensile strength, min psi	Yield strength, min psi	Allowable stress, psi, at 100 to −425 F
Pipe ASTM A249			
304	75,000	30,000	15,000
304L	70,000	25,000	12,500
310	75,000	30,000	15,000
347	75,000	30,000	15,000

Thermal expansion, in./100 ft

Temp range, F	304, 304L, 347	310
60 to 0	−0.669	−0.68
60 to −50	−1.229	−1.09
60 to −100	−1.759	−1.49
60 to −150	−2.244	−1.89
60 to −200	−2.699	−2.26
60 to −250	−3.180	−2.62
60 to −300	−3.63	−2.95

Temp, F	Thermal conductivity, Btu/(hr ft F)	Specific heat, Btu/(lb F)
−400	2.16	0.012
−350	3.91	0.042
−300	5.12	0.086
−250	5.78	0.123
−200	6.52	0.149
−150	6.95	0.168
−100	7.40	0.180
−50	7.62	0.185
0	8.10	0.196
60	8.38	0.205

Table 77.   9 Per Cent Nickel-steel Composition

	ASTM A353	
	Grade A	Grade B
C, max %	0.13	0.13
Mn, max %	0.80	0.90
P, max %	0.035	0.035
S, max %	0.040	0.040
Si, %	0.15–0.30	0.15–0.30
Ni, %	8.50–9.50	8.50–9.50

Specification A353-58T covers two grades of 9 per cent nickel steel with their compositions shown in Table 77.   Physical properties are given in Table 78.

Aluminum Alloys.   Aluminum alloys have an unusual ability to maintain their strength, ductility, and resistance to shock loading at extremely low temperatures.   Other properties which make these alloys attractive for low temperatures are their high conductivity and reflectivity, low emissivity, and high strength-to-weight ratio.   They are also nonsparking and have good corrosion resistance.

A four-digit numbering system has been developed to identify aluminum alloys. The first digit indicates the alloying element; for example, the "3000 series" is

Table 78.   Nickel-steel Properties

Material	Tensile strength, min psi	Yield strength, min psi	Allowable stress, psi, at 100 to −325 F
Plate A300 9% Ni:			
A353 A	90,000	60,000	30,000
A353 B	95,000	65,000	32,500

Thermal Expansion

Temperature range, F	In./100 ft
60 to 0	−0.381
60 to −50	−0.700
60 to −100	−1.020
60 to −150	−1.340
60 to −200	−1.650
60 to −250	−1.97
60 to −300	−2.29
60 to −325	−2.45

Temp, F	Thermal conductivity, Btu/(hr ft F)
60	16.75
0	16.15
−50	15.20
−100	14.30
−150	14.05
−200	12.20
−250	10.85
−300	9.12

Specific heat: 80 to −320 F = 0.0878 Btu/(lb F)

aluminum-manganese, while the "5000 series" is aluminum-magnesium. The second digit shows a modification of the original alloy or impurity limits; a zero indicates no modification. The final two digits serve to identify the specific alloys within a group and are assigned arbitrarily as a new alloy is developed. The standard temper designation for aluminum alloys consists of a letter indicating the basic temper which is more specifically defined by the addition of one or more digits. Basic tempers are 0, annealed; F as fabricated; H, strain-hardened; T, heat-treated. The H is always followed by two or more digits. The first digit shows the specific combination of basic operations, and the following digit or digits the final degree of strain hardening. H 14 indicates half-hard, strain-hardened only; H 18 indicates full-hard, strain-hardened only; H 22 indicates quarter-hard, strain-hardened, and partially annealed; and H 34 indicates half-hard, strain-hardened, and then stabilized.

Table 79 shows the compositions of the 5000 series.

Table 79. 5000 Series Alloy Compositions

Alloy	Mg, %	Mn, %
5052	2.2–2.8	0.10
5083	4.0–4.9	0.4–1.0
5086	3.5–4.5	0.2–0.7
5154	3.1–3.9	0.10
5356	4.5–5.5	0.1–0.4
5454	2.4–3.0	0.5–1.0
5456	4.7–5.5	0.5–1.0

Table 80 shows the properties at low temperatures. The elastic moduli of aluminum alloys in general increase as the temperature drops, and there is also little change in Poisson's ratio with decreasing temperature. 6061-T6 has been tested down to −423 F.

From the standpoint of piping, the 5XXX alloys are considered ideal for cryogenic piping because of their high weld strength and good ductility after welding. However, from a practical standpoint, 6061-T6 pipes, fittings, and flanges as alternates to the 5XXX alloys should be considered. This is because of availability problems which often occur in piping owing to the fact that stocks of 6061-T6 are extensive and cover virtually all sizes and situations while stocks of 5XXX piping are not always in full supply. Also, forged aluminum flanges in 6061-T6 have an approximately equivalent rating to carbon-steel flanges and for that reason are desirable over 5XXX alloys for such use.

The 5XXX alloys and 6061 as well as another alloy in the family of 6061, known as 6063, have all performed suitably in cryogenic applications. The alloy selection then becomes a compromise among availability, strength, and weld ductility. There are alloys 5083 and 5086 which have been approved under recent special ASME cases for piping. They have been approved for pressure vessels.

Copper. The low-temperature properties of copper and its alloys led to their early use in cryogenic equipment. Chief among these properties is copper's retention of its ductility at low temperatures. Copper and its alloys are also noted for their ease of fabrication and formability. With few exceptions, the tensile strength and hardness of wrought copper and its alloys increase quite markedly below room temperature down to below the temperature of −423 F. A further increase is discernible as the temperature is reduced to −452 F. Copper alloys always show greater tensile strength at low temperatures than copper itself, but

Table 80. Aluminum Properties

Material	ASTM Spec.	Grade	Tensile strength, min psi	Yield strength, min psi	Allowable stress, psi, at 100 to −325 F
Pipe:					
3003-0	B210	MIA	14,000	5,000	3,350
5052-0	B210	Gr 20A	25,000	10,000	6,250
5052-H34	B210	Gr 20A	34,000	26,000	8,500
5154-H34	B210	Gr 40A	39,000	29,000	9,750
6061-T6	B234	GS 11A	38,000	35,000	9,500

Temperature range, F	Thermal expansion, in./100 ft
60 to 0	−0.85
60 to −50	−1.52
60 to −100	−2.13
60 to −150	−2.73
60 to −200	−3.31
60 to −250	−3.85
60 to −300	−4.32
60 to −350	−4.75
60 to −400	−5.18
60 to −423	−5.37

Thermal conductivity, Btu/(hr ft F)

Temp, F	3003-F	5052-0	5154-0
60	92.5	82.8	
0	92.5	80.0	
−50	92.5	75.3	
−100	92.5	71.2	
−150	91.5	66.5	
−200	90.0	62.0	
−250	87.0	53.8	49.6
−300	83.3	48.0	42.6
−350	76.5	38.6	34.0
−400	52.0	24.3	21.4
−423	32.4	14.4	13.0

Temp, F	Specific heat, Btu/(lb F)
60	0.213
0	0.208
−50	0.20
−100	0.192
−150	0.178
−200	0.160
−250	0.133
−300	0.100
−350	0.050
−400	0.015
−423	0.010

their per cent increase as the temperature drops is less than that of pure copper. Yield strength and ductility do not change very much when the temperature is lowered. This is because copper and its alloys harden more for a given amount of strain or deformation at low temperatures than they do at room temperature. Table 81 gives the properties of copper and its alloys.

Table 81. Properties of Copper and Its Alloys

Material	Temp, F	Tensile strength, psi	Yield strength, psi
Copper, electrolytic	60	58,600	53,500
	−108	60,400	58,000
	−297	64,700	59,800
Copper, oxygen-free high-...............	60	52,500	49,000
conductivity (OFHC)	−108	57,200	55,000
	−321	69,000	60,000
	−423	76,000	62,000
70–30 brass...........................	60	51,000	28,200
70% copper, 30% zinc.................	−40	54,700	26,900
	−184	61,200	28,000
	−292	73,500	29,600
55–45 cupronickel	60	60,000	19,600
Cu 54.36%, Ni 45.78%, (Constantan)......	−40	67,600	21,000
	−184	77,100	24,100
	−292	89,600	26,300
3% silicon bronze......................	60	65,000	40,000
Cu 95.8%, Si 3.1%, Mn 0.86%	−108	72,000	42,000
	−321	91,000	49,000
	−423	107,000	52,000

Material	Temp, F	Thermal conductivity, Btu/(hr ft F)
High-purity copper	68	230
	−279	249
	−387	1160
	−436	8100
	−452	4045
Electrolytic tough-pitch copper......	68	230
	−279	249
	−387	262
	−414	868
	−452	230
Phosphorus deoxidized copper	68	196
	−279	87
	−423	23
P 0.027%	−450	5.8
Tellurium copper.................	68	202
	−279	243
Te 0.56%	−414	454
	−441	202
	−452	87
3% silicon bronze	68	27

Material	Temp range, F	Coefficient of linear expansion, in./(100 ft F)
Copper	−312 to 61	0.00935
Tough-pitch copper........	−148 to 32	0.01045
Copper-nickel Ni 29.2% Fe 0.19%	−296 to 32	0.0083
Ni 49.79% Fe 0.15%	−296 to 32	0.00802
Constantan...............	−312 to 61	0.00802
Deoxidized copper	−292 to 32	0.0095
Free-machining brass	−459 to 81	0.0088

Plastic and Glass

Plastics find their main use in cryogenics as gasketing material between flanges. If temperature shock is avoided, most plastics exhibit good strength when used at very low temperatures.   Teflon® is ductile at low temperatures.   Plastics reinforced with Fiberglas® improve resistance to temperature shock.

The strength of glass is dependent upon the application rate of the load placed upon it.   Unabraded glass is much stronger than abraded.   Glass also has a lower impact energy than plastics.   There recently have been developed glasses which possess high-temperature shock, impact-loading, and strength characteristics.

The properties of some representative plastics and glasses are given in the following tables:

Table 82.   Tensile Strength, Psi

Temp, F	Mylar®	Teflon®	Kel-F®	Polyvinylchloride	Nylon
60	22,500	2,000	8,000	10,000	12,000
0	24,000	2,500	10,500	13,000	14,500
−50	26,000	3,000	12,500	15,000	17,000
−100	26,500	5,000	14,000	17,500	20,000
−150	27,500	7,000	15,000	18,000	22,000
−200	28,500	8,500	16,000	19,000	24,500
−250	30,500	12,000	16,000	19,500	27,000
−300	31,000	13,000	16,500	20,000	28,000

Table 83.   Compressive Strength, Psi

Temp, F	Teflon®	Kel-F®	Polyethylene
60	2,000		
0	2,500		
−50	3,500		
−100	4,500		
−150	7,000		
−200	9,500		
−250	13,500		
−300	17,500		
−350	20,500		
−400	23,500		
−452	27,000	44,000	25,000

Table 84.   Elongation, Per Cent in $\frac{1}{2}$ In.

Temp, F	Teflon®	Nylon
60	8.0	19.5
0	5.0	15.0
−50	3.0	11.5
−100	2.0	7.0
−150	1.0	3.0
−200	0.5	0.0
−250	0.0	0.0

Table 85. Breaking Strength of Borosilicate Optical Glass

Temp, F	Abraded breaking stress, psi		
	Loading, 1 psi/sec	Loading, 10 psi/sec	Loading, 800 psi/sec
60	5,000	6,000	8,000
0	5,500	5,500	8,500
−100	6,000	7,500	9,000
−200	8,000	8,800	10,000
−300	10,500	10,500	10,500
−400	10,500	10,500	10,500

Table 86. Thermal Expansion

Temp, F, 60 to	Thermal expansion, in./100 ft				
	See footnote 1.	Cast epoxy	Nylon rod	See footnote 2.	Pyrex
0	−0.58	−2.38	−4.95	−5.77	−0.12
−50	−1.04	−4.16	−5.55	−10.0	−0.22
−100	−1.44	−5.70	−7.75	−14.0	−0.31
−150	−1.84	−7.18	−9.55	−16.95	−0.39
−200	−2.26	−8.48	−11.35	−19.4	−0.47
−250	−2.60	−9.55	−12.8	−21.4	−0.54
−300	−2.84	−10.60	−14.2	−23.3	−0.61
−350	−3.10	−11.40	−15.2	−25.0	−0.65
−400	−3.25	−12.0	−15.9	−26.3	−0.66
−450	−3.34	−12.4	−16.2	−27.3	−0.64

[1] Molded polyester rod reinforced with glass fiber.
[2] Polytetrofluroethylene.

Table 87. Thermal Conductivity, Btu/(hr ft F)

Temp, F	Teflon®	Nylon	Glass
60	. . .	. . .	0.58
0	. . .	. . .	0.57
−50	. . .	. . .	0.55
−100	. . .	. . .	0.52
−150	. . .	. . .	0.47
−200	. . .	. . .	0.42
−250	. . .	. . .	0.36
−300	. . .	. . .	0.28
−350	0.12	. . .	0.20
−400	0.10	. . .	0.12
−450	0.03	0.01	0.06

Table 88. Specific Heat, Btu/(lb F)

Temp, F	Vitreous silica	Quartz	GR-S (Buna S) rubber	See footnote 1.	Molded Teflon	See footnote 2.
60	0.172	0.172	0.43	0.44	0.228	
0	0.156	0.153	0.41	0.40	0.20	0.376
−100	0.132	0.128	0.262	0.27	0.17	0.28
−200	0.096	0.0955	0.204	0.205	0.126	0.218
−300	0.055	0.0525	0.134	0.1365	0.08	0.146
−400	0.015	0·0108	0.05	0.055	0.033	
−440	0.0014	0.00037	0.0084		0.0053	

[1] Natural rubber (not vulcanized).
[2] Polyethylene.

CRYOGENIC PIPING DESIGN

Single-phase Flow.   The problems encountered in the design of cryogenic piping systems are no different from those faced in the handling of less exotic fluids such as water, petroleum, and chemicals.   No new principles are involved, nor is it necessary to employ different methods of calculating pressure losses.   It is generally desirable to have single-phase flow: pressure drops, equipment size, and costs can be smaller; design computations are simpler and more accurate; flow measurement is easier and undesirable compressibility effects such as choking are reduced or eliminated.   Single-phase flow of liquids can be assured by maintaining

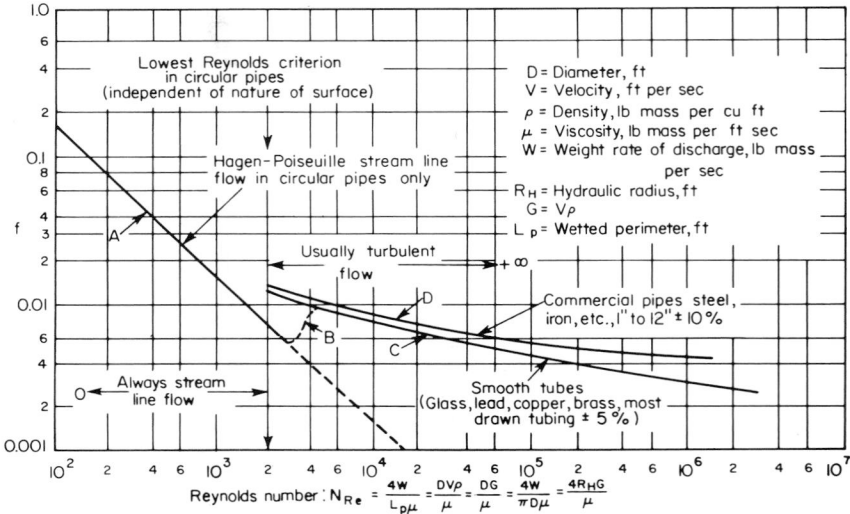

Fig. 3. Fanning friction factor.   (*"Chemical Engineers' Handbook,"* McGraw-Hill Book Company, New York, 1950.)

the static pressure greater than the vapor pressure of the fluid.   This is achieved by either (1) increasing the pressure by means of a pump or other pressuring medium or (2) subcooling the liquid to offset any vaporization due to friction or heat leak.

For circular pipes, the most widely used correlation for pressure drop is that given by the plot in Fig. 3,[1] where the Fanning friction factor f is plotted against Reynolds number.

$$\Delta P = (4fL/D)(V^2/2g)(\rho/144) = 1.08 \times 10^{-4}\rho V^2(4fL/D) \qquad (1)$$

In the above Fanning equation the term

$$(V^2/2g)(\rho/144) = 1.08 \times 10^{-4}\rho V^2$$

is generally called a "velocity head."

The friction factors as given in the chart are based on long, straight pipe.   Because the lines are usually short, because there is increased turbulence due to entrance and exit effects, and because the correlation is accurate to within ± 10 per cent, the friction factor or pressure drop is increased by many designers by a factor of 25 per cent.   This 25 per cent increase is not a safety factor but an operating factor from laboratory- to commercial-size plants.

See Fig. 3 for units to be used in Eq. (1).

Of late, losses in fittings are being reported in terms of velocity heads instead of as "equivalent lengths" of straight pipe. This method more closely describes the action taking place.

The pressure loss in a fitting in which there is no net change in fluid velocity entering or leaving the fitting is

$$\Delta P = 1.08 \times 10^{-4} K \rho V^2 \tag{2}$$

A listing of values of K for various fittings is given in Table 89. Again as with straight pipe, the pressure drops are frequently increased by 25 per cent.

Table 89. Pressure Losses in Fittings

Type of fitting	Values of K	
	Screwed	Welded
45-deg single miter...............	...	0.5
45-deg elbow, std.................	0.4	0.3
45-deg elbow, L.R................	0.3	0.2
90-deg single miter...............	...	1.4
90-deg double miter..............	...	0.8
90-deg triple miter	...	0.6
90-deg elbow, std.................	0.8	0.5
90-deg elbow medium	0.7	0.4
90-deg elbow L.R.	0·5	0.3
180-deg return bend, close	1.9	1.0
180-deg return bend, medium.......	1.3	0.6
Couplings.......................	0.04	
Unions	0.04	
Tees (single flow)		
	0.4	0.3
	1.4	1.1
	1.7	1.3
Gate valves, open................	0.15	
Globe valves, open; plug disk	10.0	
Composition disk	6.6	
2-level seat...................	7.0	
Angle valve	5.0	

If within the limits of the fitting there is a change in the velocity of the flowing fluid, the loss or recovery of pressure is given by

$$\Delta P = 1.08 \times 10^{-4} \rho (V_2{}^2 - V_1{}^2) \tag{3}$$

where subscript 1 refers to the upstream section and 2 to the downstream section.

Pressure losses of fluids that split or join in a right-angle tee fitting are given in Table 90. The 25 per cent has not been added.

For sudden enlargement, the pressure loss is given by

$$P = 1.08 \times 10^{-4} \rho (V_1 - V_2)^2 \tag{4}$$

For sudden contraction where the initial velocity is negligible, the values of K_c for various size ports are given in Table 91 to be used in

$$P = 1.08 \times 10^{-4} K_c \rho V^2 \tag{5}$$

Table 90.　Split and Join Flow in Right-angle Tee

Split flow

$$\Delta P_{1\text{-}2} = 1.08 \times 10^{-4}\rho(1.36V_2{}^2 - 0.64V_1{}^2 - 0.72V_1V_2)$$

$$\Delta P_{1\text{-}3} = 1.08 \times 10^{-4}\rho(1.8V_3{}^2 - 0.368V_1V_3)$$

$$\Delta P_{3\text{-}1} = 1.08 \times 10^{-4}\rho(1.8V_1{}^2 - 0.368V_1V_2)$$

Join flow

$$\Delta P_{1\text{-}2} = 1.08 \times 10^{-4}\rho\{2V_2{}^2 - 0.05V_1{}^2 - 2V_2[0.2045V_3(Q_3/Q_2) + V_1(Q_1/Q_2)]\}$$

$$\Delta P_{1\text{-}3} = 1.08 \times 10^{-4}\rho\{2V_3{}^2 - 0.4V_1{}^2 - 2V_3[0.2045V_1(Q_1/Q_3) + 0.2045V_2(Q_2/Q_3)]\}$$

$$\Delta P_{3\text{-}1} = 1.08 \times 10^{-4}\rho\{2V_1{}^2 - 0.4V_3{}^2 - 2V_1[0.2045V_3(Q_3/Q_1) + V_2(Q_2/Q_1)]\}$$

Table 91.　Values of K_c

	$K_c = 0.78$
	$K_c = 0.50$
	$K_c = 0.23$
	$K_c = 0.05$

Two-phase Flow.　The two-phase flow of cryogenic fluids is complicated by the fact that because of heat infiltration liquid is continuously being vaporized. The maximum velocity of the vapor-liquid mixture is limited to that of the velocity of sound in the fluid mixture.　This will be much lower than for liquids because of the high adiabatic compressibility resulting from the vapor.

The method of calculating the critical velocity is as follows:

1. Plot a curve of P in pounds per square inch absolute as abscissa vs. specific volume of the mixture in cubic feet per pound as ordinate.

2. On the P-V curve, plot a line whose slope $= -G^2/144g$ [where $G = $ lb/ (sec sq ft) and $g = 32.2$].

3. Where the slope is tangent to the P-V curve read the critical pressure.

If the critical pressure is greater than the downstream terminal pressure, a shock wave will result, causing noisy, unstable operation.　In the design of any line, the size should be so chosen as to eliminate this critical-flow condition.

The various flow patterns for two-phase flow were established by Baker and are shown in Fig. 4.

where G = gas-phase mass velocity, lb/(hr sq ft) of total pipe area

L = liquid-phase mass velocity, lb/(hr sq ft) of total pipe area

ρ_G = gas-phase density, lb/cu ft

ρ_L = liquid-phase density, lb/cu ft

$\lambda = [(\rho_G/0.075)(\rho_L/62.3)]^{1/2}$

$\psi = (73/\nu)[\mu_L(62.3/\rho_L)^2]^{1/3}$

ν = surface tension of liquid phase, dynes/cm

μ_L = liquid-phase viscosity, centipoises

These patterns may be described as:

1. *Bubble Flow.* Characterized by the formation of individual bubbles along the upper surface of the tube.

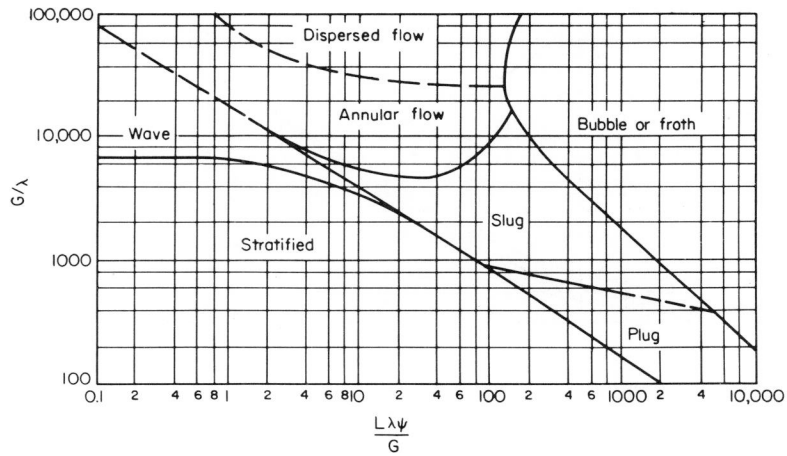

FIG. 4. Two-phase flow patterns. (*O. Baker, Oil Gas J., vol. 53, p. 185, 1954.*)

2. *Froth Flow.* Foamlike mixture of small bubbles and liquids intimately mixed during turbulent flow.

3. *Plug Flow.* Type of flow which is likely to exist during the transition period from bubble to stratified flow. It is characterized by the flow of large plugs of liquid. Large pressure fluctuations occur during this type of flow.

4. *Stratified Flow.* Flow of liquid phase along the bottom of the pipe, while the gas phase occupies the upper portion. The gas phase flows at a much higher velocity than the liquid at the bottom.

5. *Wavy Flow.* Wavy flow is similar to stratified flow. Waves are formed at the gas-liquid interface as the fluid flows.

6. *Slug Flow.* Consists of alternate sections of gas and liquid. It occurs as the proportion of gas increases by heat addition.   Bubbles formed by heat added to the fluid enlarge and rise in the pipe to increase the size of the gas slug.

7. *Annular Flow.* The flow of a continuous liquid layer along the tube wall. The gas flows in the central core at a much higher velocity.

8. *Dispersed, Spray, or Mist Flow.*   Flow in which fine liquid drops are suspended by surface tension.   There is no significant relative velocity between the liquid and gas phase.

Because of heat transfer to the flowing fluid the relative distribution of liquid and

vapor is altered progressively and the pressure drop must be calculated in finite steps. The final result is a summation of the finite pressure drops.

For those cases in which the pressure drop across the pipe is small compared with the absolute pressure, the pressure drop resulting from the flow of a boiling mixture is made up of two parts: (1) the pressure drop due to two-phase flow and (2) the pressure drop resulting from the rate of increase of momentum of the mixture as it flows.

The two-phase pressure drop for all types of flow except slug flow is correlated by the parameters given in Fig. 5.

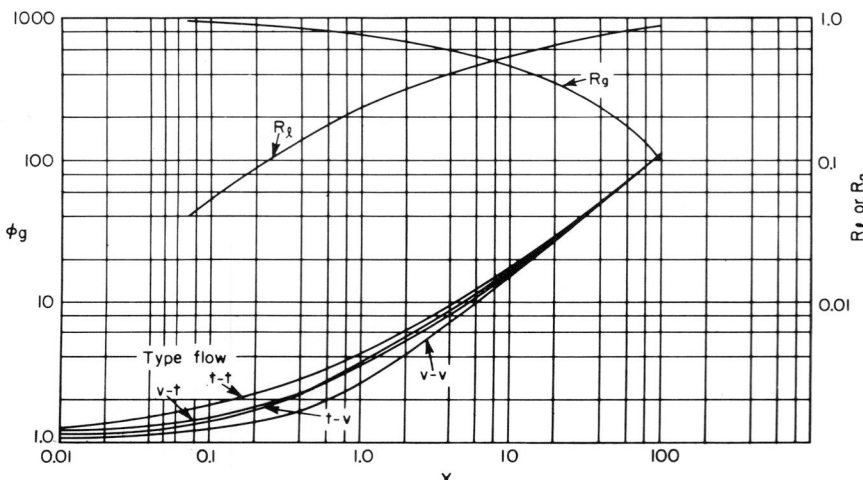

FIG. 5. Pressure drop in two-phase flow. (*R. W. Lockhart and R. C. Martinelli, Chem. Eng. Progr., vol.* 45, *p.* 39, 1949.)

The value of the abscissa for various flow regimes is given by the following expressions:

Liquid	Vapor	
t	t	$X^2 = (W_l/W_g)^{1.8}(\rho_g/\rho_l)(\mu_l/\mu_g)^{0.2}$
v	t	$X^2 = K(\mathrm{Re}_g)^{-0.8}(W_l/W_g)(\rho_g/\rho_l)(\mu_l/\mu_g)$
t	v	$X^2 = 1/K(\mathrm{Re}_l)^{0.8}(W_l/W_g)(\rho_g/\rho_l)(\mu_l/\mu_g)$
v	v	$X^2 = (W_l/W_g)(\rho_g/\rho_l)(\mu_l/\mu_g)$

$$K = 296 \text{ for commercial pipe}$$
$$= 348 \text{ for smooth tubes}$$

$$(\Delta P/\Delta L)_{TP} = (\phi)^2(\Delta P/\Delta L)_g \qquad (6)$$

where flow t = turbulent flow, Reynolds Number >2,000
flow v = viscous flow, Reynolds Number <1,000
W = weight flow per unit time
ρ = density lb/cu ft
μ = viscosity
$(\Delta P/\Delta L)_{TP}$ = two-phase flow pressure drop
$(\Delta P/\Delta L)_g$ = pressure drop due to gas-phase flow only in pipe
ϕ is read from Fig. 5
Subscript l denotes the liquid phase; subscript g denotes the gas phase

Method of Calculation.   1. Calculate Re_l and Re_g based on each phase flowing separately through pipe.

2. From Re_l and Re_g determine the type of flow.

3. Calculate X using the appropriate equation and read ϕ from the curve.

4. Calculate $(\Delta P/\Delta L)_g$ for the gas phase only using single-phase pressure-drop correlation.

5. $(\Delta P/\Delta L)_{TP} = \phi^2(\Delta P/\Delta L)_g$.

The correlation giving the X vs. ϕ relationship is accurate in the range of about ± 30 per cent, and for design safety the pressure drop should be increased by at least 30 per cent.

For the calculation of pressure drop in a line having slug flow, see Griffith and Wallis, *Journal of Heat Transfer*, August, 1961, pages 307 to 320.

The pressure drop due to momentum change is

$$A_p \Delta Pa = (W_1 v_1/g) + (W_g v_g/g) - (W_T v_0/g) \tag{7}$$

where A_p = pipe area, sq ft
  ΔPa = pressure drop due to momentum change, lb/sq ft
  W = weight flow, lb/sec
  v = velocity, fps
  g = 32.2 ft/sec²
  sub l = liquid phase at exit of line
  sub g = gas phase at exit of line
  sub T = at inlet of line where only liquid phase exists
  sub 0 = velocity at inlet of line

Two extreme cases of exit conditions can exist, the actual condition probably lying between the two:

1. Liquid and vapor completely mixed (fog)

2. Liquid and vapor completely separated

For the first case, $v_g = v_l$ and

$$\Delta Pa = G^2/g\rho_l[(1 - W_g/W_T) + (W_g/W_T)(\rho_l/\rho_g) - 1] \tag{8}$$

For the second case,

$$\Delta Pa = \frac{G^2}{g\rho_l} \left[\frac{(1 - W_g/W_T)^2}{R_l} + \frac{(W_g/W_T)^2 \rho_l}{R_g \rho_g} - 1 \right] \tag{9}$$

For flow through any other type of fitting or valve, use the single-phase pressure-drop correlation multiplied by the ϕ factor for two-phase flow.

Economic Line Sizing.   The sizing of lines in a cryogenic plant presents an entirely different problem from that encountered in other plants, i.e., chemical, petroleum, or power.   In these other plants the major factor in line sizing is the length of straight line, because valves and fittings account usually for about 25 per cent of the total pressure drop.   Because of this, certain guides have been established through practice governing the velocity of the fluid flowing in the line. Of course, once this preliminary sizing is set, the line is calculated in detail using the actual dimensions.   Owing to the very nature of a cryogenic plant this rule does not hold.   Since heat leakage plays such a large role in the design of a cryogenic unit, the pieces of equipment are usually spaced very close to one another in the cold box. This is especially true in missile and military work, where space and weight are at a premium and everything is jammed into very small volumes.   Pressure drop through valves and fittings becomes greater than that through the straight pipe. It is therefore most difficult to list an allowable velocity for a general class of flow and each case must be decided upon its own merits.

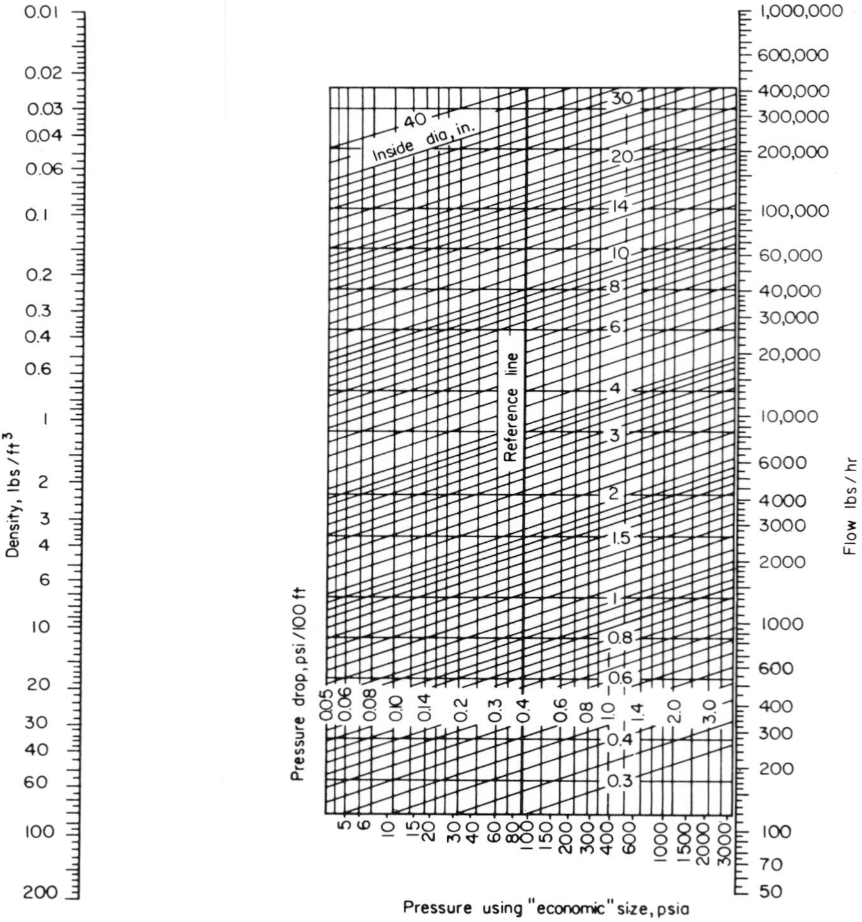

FIG. 6. Line size nomograph. *Instructions*—1. Locate a working point on the reference scale by laying a straightedge from the density scale to the flow scale. 2. Move horizontally to the desired pressure drop or the desired line size, or to the operating pressure to find the "economic" size. NOTE: Chart is based on a friction factor of 0.004. (*J. D. Lewis, Petrol Refiner, vol.* 40, *p.* 169, 1961.)

As an aid, a good rule of thumb is to allow 0.05 velocity heads for the total loss in a line connecting any two pieces of equipment. This is equivalent to

$$\rho V^2 = 462 \tag{10}$$

For ordinary flow in a cryogenic plant at temperatures warmer than -50 F, the nomograph given in Fig. 6 will suffice.

Equipment

There are three basic ways of joining pipe to pipe or pipe to equipment: (1) screwed, (2) flanged, and (3) welded. The design of these methods is prescribed by

the various codes, and generally what is applicable to high temperatures will also hold for operation at cryogenic temperatures.

Screwed Fittings. Screwed fittings are not commonly used in cryogenic service because of their tendency to leak. A material generally overlooked in using screwed fittings is the lubricant which is characterized by a high degree of chemical inertness and thermal stability. Applied in thin films, lubricants are used to reduce seizing and galling in fine-threaded joints. When lubricants are used as vehicles for powdery materials such as clay, graphite, lead oxide, etc., they become thread compounds or antiseize agents. Some of these compounds remain soft and are known as putty, other compounds harden and are called cements, while others provide dry-film lubrication upon evaporation of the solvent or carrier vehicle. In addition to providing lubricity, the suspended solids in these compounds also act as sealants of minute voids and crevices. This sealing function, however, is sometimes misunderstood or overestimated, and thus it is assumed that these compounds will fill and seal gaps at high pressures. For effective sealing at high pressure the only adequate processes are metal filling or welding. In metal filling, better known as tinning, leadening, or sweating, soft metals such as tin or lead are melted and allowed to flow between adjoining surfaces of threads. A more permanent process is soldering, brazing, or welding.

Flanges. Flange design also has been gone into in much detail in the various codes and will not be repeated here, since what is applicable to high temperatures also holds for low temperatures. In cryogenic service, the selection of gasket material and type is of great significance. An elastomer 0-ring lightly squeezed against a flat surface makes an excellent seal at ordinary temperatures. When an ordinary 0-ring seal cools at cryogenic temperatures, however, the elastomer becomes brittle at some temperature not far below room temperature, then shrinks until it loses contact with the flat surface, and the seal fails. As force is applied to press the 0-ring against the flange, three desirable things are accomplished. There is an increase in pressure which causes the elastomer to flow into irregularities in the flange surfaces. At the same time there is a decrease in thickness and a corresponding increase in width of the seal. Minimum thickness will result in minimum shrinkage when the seal is cooled, and a wide seal reduces the possibility of having small openings across the barrier. All these effects depend upon the existence of a force normal to the surface of the flange faces. A tongue-and-groove flange using 0-rings made of natural rubber, nitrile rubber, Viton-A®, Neoprene, and Buna N when compressed to 80 per cent or more has shown good results. The one disadvantage of highly compressed 0-rings is the necessity for strong flanges and fairly high bolt loads. This is overcome by the use of a flat gasket design. Final thickness under the sealing ring must be small to minimize unloading when the seal is cooled. Good results have been obtained with nylon and Mylar® in 10-mil thickness when compressed to 3 mils (70 per cent compression), and in such cases, bolt loading was smaller than that associated with tongue-and-groove design, being in the order of 2,000 lb of force per linear inch of original ring circumference. The principal disadvantage with this seal is the fact that a small scratch or dent across the compression ring or flange usually results in a leak.

Besides elastomer and plastic gaskets being used, the following materials are also widely used:

Metal gaskets: Spirotallics made with 304SS and using Teflon or Kel-F as a filler

Asbestos cloth gaskets: Gaskets (die cut) made from asbestos cloth treated with Teflon impregnation

Cryol gaskets: Special compounds of sintered Teflon with glass fibers formed into sheet from which ring gaskets are cut.

The flange bolting is not a problem if the contraction of the bolting and flange is similar. If possible, the bolting should be of the same or similar material as that of the flanges.

Welding. The most common method of joining in cryogenic work is by welding. The joint is permanent, leakproof, and strong. If the joint is never to be opened or perhaps to be opened under very special cases between long intervals of time, it should be welded. Joints that are to be opened periodically should be flanged. Welding is common to all types of piping, either hot or cold, and what is applicable to one holds for the other. The one particular problem uniquely

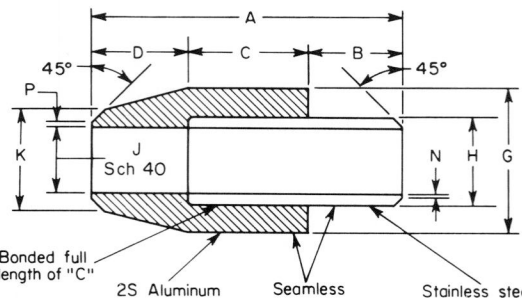

Fig. 7. Butt-weld transition coupling. (*Courtesy of the Project Fabrication Corp., New York.*)

IPS, in.	A	B	C	D	G	H	J	K	N	P
1/2	3 3/8	1 1/16	1 1/4	1 1/16	1 5/16	0.84	0.622	0.855	1/16	1/16
3/4	3 3/4	1 3/16	1 3/8	1 3/16	1 1/2	1.05	0.824	1.065	1/16	1/16
1	4 1/4	1 5/16	1 5/8	1 5/16	2	1.32	1.05	1.330	1/16	1/16
1 1/2	5	1 5/8	1 3/4	1 5/8	2 7/16	1.90	1.61	2 1/16	1/16	1/16
2	5 3/8	1 3/4	1 7/8	1 3/4	2 15/16	2.38	2.07	2 9/16	1/32	1/16
3	6 1/2	2	2 1/2	2	4 9/32	3.50	3.07	3 3/4	1/32	1/16
4	7 1/4	2 1/4	2 3/4	2 1/4	5 3/8	4.50	4.03	4 25/32	1/32	1/16
6	8 1/4	2 1/2	3 1/4	2 1/2	7 5/8	6.63	6.07	6 15/16	1/32	1/16

associated with cryogenics is the joining together of dissimilar metals, the most common being the joining of aluminum pipe to a stainless-steel fitting. This particular type of welding is extremely difficult, and the problem has been solved by the use of transition couplings. Such bimetallic units combine the selected physical properties of both metals if properly utilized. These joints must remain absolutely vacuumtight over a wide temperature range. Also, the joint must be light and strong and sufficiently ductile to withstand all stresses due to variations in thermal expansions (the coefficient of thermal expansion of aluminum is nearly 50 per cent greater than that of stainless steel) and contractions as well as to endure the vibrations, shock loading, and other forces encountered in service. Tabulated in Figs. 7 and 8 are the two types of transition couplings available.

Expansion. In conventional piping systems, sufficient uniform thermal cooldown flexibility is provided by the inherent layout of the piping or is obtainable by

the use of expansion loops.   Uniform thermal cooldown occurs when the pipe is full of liquid or vapor and the entire line has cooled to the fluid temperature.   In cryogenic transfer lines, space and cooldown weight limitations may preclude the use of expansion loops for achieving flexibility.   In addition, this uninsulated, thin-walled transfer piping is subject to temperature gradients of a scale not usually encountered in process piping systems, and as a consequence of these temperature gradients the phenomenon of pipeline bowing occurs.   Bowing is the tendency of the center line of the pipe to be deformed into an arc.

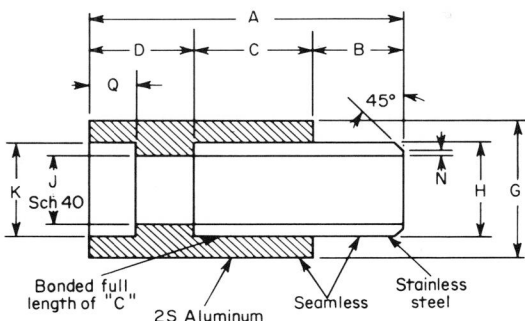

FIG. 8. Socket weld transition coupling.   (*Courtesy of the Project Fabrication Corp., New York.*)

IPS, in.	A	B	C	D	G	H	J	K	N	Q
½	3⅜	1¹⁄₁₆	1¼	1¹⁄₁₆	1⁵⁄₁₆	0.84	0.622	0.855	¹⁄₁₆	½
¾	3¾	1³⁄₁₆	1⅜	1³⁄₁₆	1½	1.05	0.824	1.065	¹⁄₁₆	⁹⁄₁₆
1	4¼	1⁵⁄₁₆	1⅝	1⁵⁄₁₆	2	1.32	1.05	1.330	¹⁄₁₆	⅝
1½	5	1⅝	1¾	1⅝	2⁷⁄₁₆	1.90	1.61	2¹⁄₁₆	¹⁄₁₆	¾
2	5⅜	1¾	1⅞	1¾	2¹⁵⁄₁₆	2.38	2.07	2⁹⁄₁₆	¹⁄₃₂	⅞
3	6½	2	2½	2	4⁹⁄₃₂	3.50	3.07	3¾	¹⁄₃₂	1⅛
4	7¼	2¼	2¾	2¼	5⅜	4.50	4.03	4²⁵⁄₃₂	¹⁄₃₂	1⁹⁄₁₆
6	8¼	2½	3¼	2½	7⅝	6.63	6.07	6¹⁵⁄₁₆	¹⁄₃₂	2

Bowing occurs when the pipeline is only partially full of liquid as when liquid is trapped between valves and is allowed to boil off.   Because of bowing, a partially full 8 in. diameter 30-ft cantilevered section of pipe would deflect 35 in. at the free end.

The bowing radius of curvature can be represented by the equation

$$R = d/Ke \qquad (11)$$

where R = radius of pipe curvature, in.
   d = pipe mean diameter, in.
   e = integrated expansion coefficient between T_1 and T_2, in./in.
   K = function of temperature distribution and fill level.

The factor K is plotted in Fig. 9 for Cases 1, 2, and 3 of Fig. 10.

The actual temperature distribution is somewhat less severe than that for Case 2, and therefore a design based on a Case 2 temperature distribution with the pipe half full of liquid will be conservative.

If a pipe is not restrained from bowing, some internal stresses will be developed as the result of the nonuniform temperature across the pipe. While these stresses may be appreciable, they are of such a nature that damage to the pipe is not likely. However, if the pipes are restrained in any manner, additional stresses are imposed. Very high reactions and moments occur at support points and anchors. Bowing will be most severe in a pipe anchored at one end and simply supported at the other.

Expansion joints are widely used to absorb expansion or contraction in process piping systems. The Standards of the Expansion Joint Manufacturers Association and high-temperature design criteria adequately describe the techniques necessary for designing a cryogenic piping system. In the case of bowing, the rigidity of a pipeline can be broken by properly spaced axial or hinged-type expansion joints. The joints must be satisfactory for the angular rotations and lateral motions that may occur at the points of application.

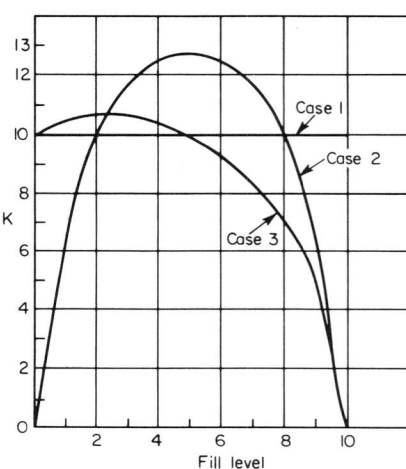

FIG. 9. K factor. (*W. G. Flieder et al.,* 1959 *Cryogenic Eng. Conf. Paper B-5.*)

By a careful choice of locations, the joints can provide for thermal contractions and bowing motions and the number of joints required can be held to a minimum.

For a short, straight run between anchors, a single axial joint can be used. Ideally, the joint should be centrally located in the run so that deflections due to bowing are

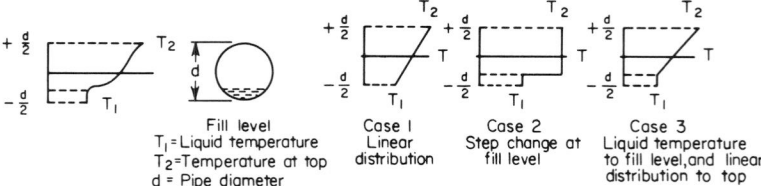

FIG. 10. Temperature distribution cases. (*W. G. Flieder et al.,* 1959 *Cryogenic Eng. Conf. Paper B-5.*)

equal at each end of the joint, and thus no relative lateral motion is imposed upon the joint. The distance between anchors is limited by the amount of bowing deflection and the column effect on the piping due to the compressive loading on the pipes. It is recommended that guiding elements or anchors be located within four pipe diameters of an expansion joint so that elastic buckling of the pipe will be prevented.

In the case of a relatively long run between anchors, an axial expansion joint can be combined with a hinged-type joint to provide for bowing. The line must be carefully guided, since the prevention of line buckling is vital. The guides must be

placed at points of zero bowing deflection and must be within the limiting distance of four pipe diameters from the expansion joint.

If an extremely long run of pipe must be treated, a second hinged joint may be added. Since this further complicates the stability of the run, additional guiding is necessary.

The ideal arrangement is contingent upon equal bowing along the entire run. This is not very easily achievable in practice. Therefore, some loads may result at the anchor and guide locations. If limit rods are used on the axial joint, however, large clearances on the guides can be tolerated, since unrestricted buckling of the pipes will be prevented. Such a design would also provide for unequal bowing in the line. The same functional design can also be obtained if the axial expansion joint is replaced by a hinged joint with slotted hinges to take up the linear contractions.

Right-angle bends can be reduced to two anchored straight runs by anchoring the line at the corner point. A more satisfactory method is obtained by using three gimbal joints as shown in Fig. 11.

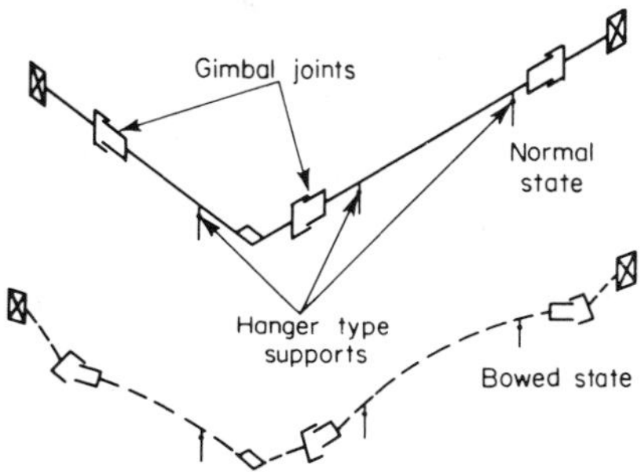

Since the pressure forces developed in the bellows are constrained within the joints and since the pipeline pressure forces are transmitted through the gimbals, the lines are under tensile loads. Therefore the need for elaborate guiding to prevent buckling is completely eliminated. The anchor designs are also simplified by the elimination of the larger pressure forces associated with axial expansion joints. Since rigid guiding is not required, the adverse effects of unequal bowing may be safely handled by use of spring hangers for supports.

FIG. 11. Right-angle bend. (*W. G. Flieder et al.*, 1959 *Cryogenic Eng. Conf. Paper B-*5.)

Valves. Valves are the heart of any pipeline, since it is through their use that control of the flowing fluid is regulated. Cryogenic valves are no different from those used in any other type of plant. There are, however, certain additional design considerations that should be taken into account. These are valve construction material that would remain ductile at low temperatures, material that would minimize heat flow from outside, operational safety, ease of maintenance, and gaskets for low temperatures.

The most common material for valves is Type 304 stainless steel. This material can withstand large thermal shocks (valves have been cycled from 200 to −320 F in minutes) and, in addition, has one of the lowest thermal conductivities. It is also easier to keep clean and is more resistant to corrosion than other alloys. Although most valves are made of stainless, some are made of aluminum. The aluminum valves are generally used only up to 150 psig and are not used for control purposes.

Since the vast majority of cryogenic plants is built inside a cold box, cryogenic valves are usually built with extended bonnets so that the stuffing box and handwheels can be extended to the outside. This facilitates operation and makes for easily accessible maintenance. Chevron-type molded Teflon is generally used in the stuffing boxes. Teflon gaskets are generally used in the valve too. At the wall of the cold box, the stem of each extended bonnet valve emerges through a flexible boot or bellows seal. This flexible seal is clamped by stainless straps to a steel collar around the opening in the wall and to split wooden blocks around the valve stem itself. Bellows action of the flexible seal takes up contraction or expansion

of the pipeline.   Except for the tightening of external packing glands after "cooldown" due to the shrinkage which occurs when the low operating temperatures are reached for the first time, maintenance is at a minimum.

There are many special types of valves such as broken stem, terminal valves, helium-operated remote-control valves, etc., that are used for special purposes and incorporate certain design features that may be required for specific applications.

Heat Leakage into Cryogenic Systems

Cold Box.   There are two ways to insulate a cryogenic plant.   One method consists of applying insulation to each piece of equipment separately; this can be relatively expensive.   A second method involves placing all the individual pieces as closely as possible, building a "cold box" around them, and filling the voids with insulation.   A cold box then requires an insulation that can be packed into spaces between heat exchangers, towers, drums, valves, and lines.

Rock wool, magnesia, and perlite are frequently used in cold boxes for air-separation plants which manufacture oxygen, argon, and nitrogen; methane-purification plants; low-temperature gas treating, i.e., nitrogen wash units.   If the nature of the units is such that there are hot and cold units adjacent to each other,

Table 92.   Cold Box Insulation[1]

Type of insulation	Boundary temp, F	Cryogenic fluid temp, F	
		Cold box	Transfer line
1. Mineral wool and glass fiber.........	80	−128 to −320	−128 to −320
2. Perlite..................	80 to −321	−128 to −320	−128 to −320
3. Polyurethane (foam)...............	75	−128 to −320	−128 to −320
4. Foamglas...............	−100	....	−128 to −320
5. Perlite..................	80 to −321	−128 to −452	−128 to −452
6. Santocel plus aluminum flakes.......	80 to −321	−411 to −452	−128 to −452
7. Superinsulation (15 Al foils/in.)........	80 to −321	−411 to −452	−411 to −452
8. Superinsulation (60 Al foils/in.)........	80 to −321	−411 to −452	−411 to −452

Type of insulation	Thermal conductivity, Btu/(hr ft F)	Applied vacuum, microns	Density, lb/cu ft
1. Mineral wool and glass fiber.........	16 to 23	None	4 to 8
2. Perlite..................	18	None	5 to 6
3. Polyurethane (foam)...............	12 to 22	None	1.6 to 6
4. Foamglas...............	33	None	8 to 10
5. Perlite..................	1	10	5 to 6
6. Santocel plus aluminum flakes.......	0.25	10	
7. Superinsulation (15 Al foils/in.)........	0.11	1	2.2
8. Superinsulation (60 Al foils/in.)........	0.025	1	6.8

[1] A. Matesanz, *Chem. Process*, vol. 24, p. 28, 1961.

then each individual piece of equipment is insulated separately and no cold box is used.

When a cold box is used, it is good practice to have no equipment any closer than 18 in. from an outside wall. If the plant has any flammable fluids such as hydrogen or if oxygen is being produced, the cold box should be purged continuously with an inert gas such as nitrogen. It is also well not to use any wood to support equipment for the same fire-hazard reasons.

There are so many variables involved that it is very difficult and tedious to calculate the heat leakage to a cold box. With a good engineering design of the cold box, the average heat leakage should be in the range of 30 to 50 Btu/(hr lb mole) of warm gas into the box.

Heat leakage which results from any break in the insulation will quickly be noticed by the formation of a cold spot (formation of a thick ice layer) on the cold box wall in the vicinity of the fault.

Table 92 gives the properties of the insulation commonly employed for cryogenic plants.

Transfer Lines. A transfer line is a pipeline used to convey a very cold fluid from one place to another. Special emphasis is placed on features which contribute to lower heat leakage, simplicity, and reliability. The primary design consideration is to reduce heat transfer from the warm ambient atmosphere to the fluid.

The only reasons a transfer line would not be insulated are that (1) its use is very infrequent and brief, (2) it is a temporary installation, or (3) the flow rates are continuous but very high. For lines up to $5/8$ in. in diameter carrying liquid hydrogen, it has been found that the heat leakage varies from 3500 to 6050 Btu/(hr ft^2) depending upon whether there is a wind blowing. Of course, depending on the temperature, condensates of ice, solid air, and liquid air may build up on the pipe. The heat leakage to uninsulated pipe can be calculated by conventional methods. The ice layer will add a certain measure of uncertainty.

A noncombustible porous insulator, either solid or powder, may be used, especially for lines carrying liquid nitrogen or oxygen. The insulators may be Foamglas®, polystyrene foam, Fiberglas®, perlite, vermiculite, etc. Care must be taken that ice and air do not freeze on the insulation to destroy its effectiveness.

Table 93 gives suggested thicknesses of Foamglas® insulation.

The most effective means of suppressing convection and conduction is by very high vacuum, although pressures of 10^{-6} mm Hg are difficult and costly to maintain. Radiation can be suppressed by means of highly reflecting surfaces. Vacuum powder insulation permits equally efficient insulation at higher levels of vacuum.

There are two problems associated with any vacuum-insulated line that must be solved if the line is to operate successfully. These are the heat transfer across the supports (spacers) between the inner and outer pipe and the dimensional change imposed by the temperature differences. The supports must be vacuumtight and have a low thermal conductivity. If nonmetallic materials are used, it is difficult to obtain a seal that is reasonably reliable and which will withstand the stringent requirement of vacuum operation. Alternatively, if a metal is used, the seal can be made to withstand the vacuum, but this will introduce the disadvantage of relative high thermal conductivity.

The second problem is the dimensional change imposed by the temperature difference between the inner pipe and the outer shell. The connections between the two must be vacuumtight and also locate the position of the inner pipe with relation to the outer shell. When the inner line is flexible, the dimensional change can be accommodated relative to the supports throughout the pipeline installation. The properties desired in the material used for the pipe are high tensile strength and good ductility at low temperature, a very low coefficient of thermal expansion, and

Table 93. Foamglas Insulation: Minimum Suggested Thicknesses for Specified Temperatures[1]

Figures in table are degrees Fahrenheit and are minimum temperatures for use with thickness and design conditions stated. When design temperature falls between two thickness columns, use the greater thickness. Table is based on thickness required to prevent condensation on surfaces with 0.9 emissivity, in 90 F still air at 80 per cent relative humidity. Heat gains average 8 to 9 Btu/(hr sq ft) insulation surface. Where other design conditions are used, thicknesses should be adapted to meet job requirements.

Iron pipe size, in.	Nominal Foamglas thickness, in.																						
	1	1½	2	2½	3	3½	4	4½	5	5½	6	6½	7	7½	8	8½	9	9½	10	10½	11	11½	12
½	45	10	−25	−45	−95	−140	−275	−300	−335	−390	−450												
¾	50	25	−10	−25	−70	−135	−225	−255	−300	−340	−415	450											
1	50	25	−10	−25	−70	−130	−180	−255	−300	−340	−415	450											
1¼	50	30	5	−25	−60	−105	−145	−200	−265	−315	−375	425	450										
1½	50	30	15	−15	−45	−100	−120	−185	−230	−290	−345	−390	−440	450									
2	50	30	15	−15	−45	−90	−120	−175	−225	−290	−340	−385	−430	450									
2½	50	30	15	−10	−35	−60	−105	−130	−175	−235	−290	−330	−375	−415	450								
3	50	30	15	−10	−35	−60	−105	−135	−175	−235	−290	−330	−370	−415	450								
3½	50	30	15	−10	−35	−60	−105	−135	−175	−235	−290	−330	−355	−410	−450								
4	50	30	20	0	−30	−55	−85	−135	−160	−205	−255	−300	−340	−375	−420	450							
4½	50	30	20	0	−30	−55	−85	−135	−160	−205	−255	−300	−340	−380	−420	450							
5	50	35	25	5	−25	−45	−80	−110	−140	−175	−220	−270	−305	−345	−380	−415	450						
6	50	35	25	0	−25	−45	−70	−105	−125	−160	−195	−240	−290	−325	−360	−395	−430	450					
7		45	25	0	−25	−40	−65	−95	−120	−150	−185	−220	−275	−305	−340	−375	−410	450					
8		45	25	0	−15	−35	−60	−85	−115	−145	−175	−215	−255	−295	−325	−360	−395	−425	450				
9		45	25	5	−10	−35	−55	−80	−105	−135	−170	−200	−240	−265	−315	−345	−380	−415	445	450			
10		45	30	10	−10	−25	−50	−70	−105	−125	−150	−185	−225	−265	−300	−330	−360	−410	−420	−440	−450		
11		45	30	10	−5	−25	−45	−70	−105	−120	−145	−175	−210	−255	−290	−320	−350	−380	−410	−435	−450		
12		45	30	15	−5	−25	−45	−65	−95	−115	−140	−170	−205	−245	−285	−315	−340	−370	−400	−425	−450		
14		45	30	15	0	−25	−40	−55	−80	−105	−130	−160	−190	−225	−265	−295	−325	−355	−380	−410	−430	−450	
16		45	30	20	0	−25	−35	−55	−75	−105	−125	−150	−180	−210	−250	−285	−310	−340	−365	−395	−425	−450	
18		45	30	20	0	−25	−35	−50	−70	−105	−115	−145	−175	−200	−235	−275	−295	−325	−355	−380	−410	−435	450
20		50	30	20	0	−25	−35	−45	−65	−95	−110	−135	−165	−190	−225	−260	−290	−315	−345	−370	−395	−425	450
24		50	30	20	0	−25	−35	−45	−65	−85	−105	−130	−155	−180	−205	−240	−270	−300	−330	−355	−380	−405	430
28		50	30	20	5	−10	−25	−45	−60	−80	−105	−125	−145	−170	−200	−235	−260	−295	−315	−340	−365	−390	−415
32		50	30	20	5	−10	−25	−40	−60	−75	−105	−120	−140	−170	−190	−225	−250	−285	−305	−330	−350	−375	−400
36		50	30	20	10	−5	−25	−40	−55	−75	−105	−115	−135	−160	−185	−215	−240	−280	−300	−320	−345	−365	−390

[1] Courtesy of Pittsburgh Corning Corp., Pennsylvania.

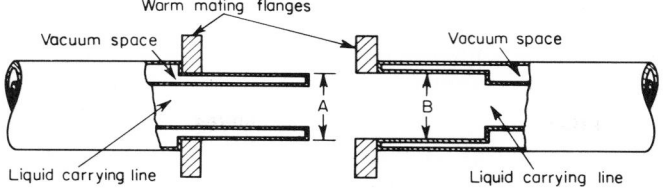

Dimension B - dimension A ≤ 0.005

FIG. 12. Vacuum joint design. (*P. C. VanderArend, Chem. Eng. Progr., vol.* 57, *p.* 62, 1961.)

low thermal conductivity. In addition, the pipe material must lend itself to easy working and jointing. Standard, mild-steel pipe of normal specification is used for the outer shell.

The method and sequence of assembly are of the utmost importance if the transfer line is to be completed with the minimum expenditure of time and labor. The aim is to complete the inner tube in every respect, welding of all joints, testing, degreasing of surface, etc., and to assemble the outer casing when this has been achieved. The mild-steel outer casing is then cut and formed complete with filling bosses and connections between welds on the inner line. This assures that the welds on the inner line are accessible for leak testing. On completion, the vacuum system is attached and the entire line is subjected to pressure and vacuum tests.

For powder filling, the system must have a hopper with a quick-release valve, which is connected to the top of the vacuum space while the pump is connected to the top of the space via a powder trap. After the pressure in the vacuum space is reduced to between 3 and 10 mm Hg, the pumping system is isolated from the line. The hopper is filled with the powder insulant, and the release valve is opened, thus allowing the powder to be drawn into the line. This process is repeated for three or four times, depending on the length of the line. The line is then pumped down, and the hopper valve opened to the atmosphere so that an "air punch" is introduced to pack down the powder. It is necessary to do this more frequently toward the end of the filling process. This alternative filling and packing are continued until the annular space is filled. To facilitate the pumping down of the annular insulating space, pipe connections are made at intervals. In general, the ratio of the inner diameter to the outer diameter runs from $\frac{1}{3}$ to $\frac{2}{3}$. See Fig. 12 for joint design.

If a transfer-line section must be flexible, it may be designed by using two concentric flexible tubings as shown in Fig. 13. The flexible tubes are spaced

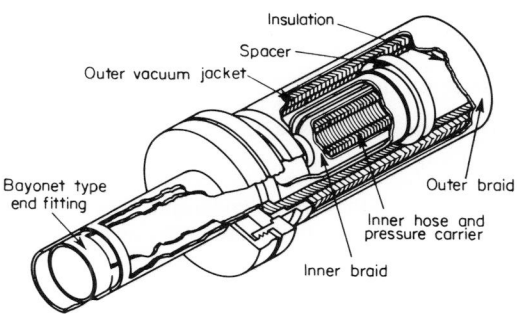

FIG. 13. Flexible transfer line. (*Courtesy of the DK Manufacturing Co., Chicago, Illinois.*)

concentrically, generally using square spacers about 4 in. apart.   The spacers in rigid sections are usually triangular so as to give the fewest number of contact points between metal to metal.   Flexible tubing, even when bright and highly polished, absorbs radiation almost as well as a black body because of the irregular nature of its surface.

Table 94 gives standard sizes and heat-leakage rates for flexible transfer lines.

The vacuum jacket for the entire transfer line is usually one continuous vacuum system.   This eliminates the heat leaks at the coupling between separately evacuated sections and is easier to fabricate and check.   To maintain the high static vacuums

Table 94.   Flexible Transfer-line Sizes[1]

(Standard sizes—10- to 15-ft lengths)

Nominal hose size, in.	Inner hose ID, in.	Outer hose OD over braid, in.	Heat loss, Btu/(hr ft)
½	0.50	2.478	1.1
1	1.00	3.028	1.3
1½	1.50	3.578	1.4
2	2.00	4.078	1.4
2½	2.50	4.703	1.6
3	3.00	5.253	1.8
3½	3.50	5.828	1.8
4	4.00	6.328	2.0
4½	4.50	6.984	2·2
5	5.00	8.078	2.2
6	6.00	9.078	2.4

[1] Courtesy of the DK Manufacturing Co., Chicago, Ill.

for long times an adsorbent (getter) is used on the cold surface, activated charcoal being the most common "getter."   The weight of the charcoal in grams is about one-seventh of the total vacuum volume in cubic centimeters.

In developing the vacuum, water-cooled diffusion pumps seem to have less oil back-streaming than do air-cooled pumps.

Calculation of Heat Leakage

Computation shows that, for all usual insulations, the resistance to heat transfer between the liquid line (pipe) and the liquefied gas is negligible.   The liquid line will therefore have essentially the same temperature as the liquid, and the heat transfer will depend only upon the liquid temperature, the insulation characteristics, and ambient conditions.   Conventional insulations, such as glass wool, foam glass, concentric aluminum shells, etc., have been used for liquid oxygen and liquid nitrogen.   However, analysis shows that for initial-cost amortizations of about 2 years and longer, a high-vacuum insulation is superior from the economic, as well as heat-transfer, standpoint.

High-vacuum insulation is thermally superior to evacuated powders for smaller size transfer lines, the breaking point depending upon whether shielding is used and the permissible thickness of insulation. However, as there are still many technical details to be worked out regarding the use of powder insulation in long transfer systems, the current practice is to recommend high vacuum even for larger size lines. Also, for the same size liquid line, evacuated powder is several times as bulky and several times as heavy as a high-vacuum line. The pressure in high-vacuum insulation should be 10^{-5} mm Hg or less during operation, while a pressure of 10^{-2} mm Hg is satisfactory for evacuated powder.

High-vacuum Insulation. With high-vacuum insulation, radiation is the primary mode of heat transfer. Figure 14 depicts a radial section through a high-vacuum-insulated transfer line; the inner (or primary liquid) line contains the liquefied gas that is being transferred. Separated from the liquid line by a high-vacuum space is a shield that is refrigerated by an inexpensive liquefied gas. The shield may be desirable when the primary liquid is helium or hydrogen; in these cases the shielding liquid will probably be nitrogen. The shielding liquid is insulated from ambient temperature by high vacuum.

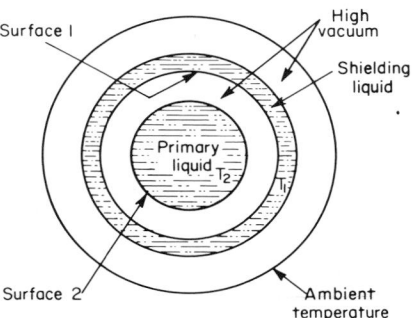

Radiation at the ambient temperature is absorbed by the shielding liquid. Only that radiation which is emitted at the temperature T_1 of the shielding liquid reaches the primary liquid. In a nitrogen-shielded line, the heat leakage to the primary liquid is only about one two-hundredths that of an unshielded line. However, it is evident that shielding complicates the design of the transfer system. In an unshielded line the two outer walls in Fig. 14 are omitted and T_1 is essentially ambient temperature.

FIG. 14. Transfer-line schematic. (R. B. Jacobs, Natl. Bur. Std. Circ. 596, 1958.)

The net radiant-energy transfer[1] to the primary liquid per unit time per unit length for a transfer line is given by

$$q = \frac{\sigma \pi D_2 (T_1^4 - T_2^4)}{1/\epsilon_2 + D_2/D_1(1/\epsilon_1 - 1)} \qquad \text{Btu/hr} \qquad (12)$$

where σ is the Stefan-Boltzmann constant, D_1 and D_2 are the diameters of the surfaces enclosing the vacuum space, ϵ_1 and ϵ_2 are the emissivities of these surfaces, and T_1 and T_2 are the absolute temperatures of the surfaces. Using a practical value for the diameter ratio D_2/D_1 of 2/3 and assuming that $\epsilon_1 = \epsilon_2 = \epsilon$,

$$q/[\Delta(T^4) \times 10^{-8}] = 0.1326[\epsilon/(5 - 2\epsilon)]D_2 \qquad (13)$$

where D_2 is in inches and $\Delta(T^4) = (T_1^4 - T_2^4)$

The right-hand side of the preceding equation does not depend upon the liquefied gas but only upon the materials (and their condition) of which the line is made. Thus, a single set of curves of $q/[\Delta(T^4) \times 10^{-8}]$ vs. D_2, with ϵ as a parameter, will

---

[1] In Eq. (12), the following units are used: $\sigma = 0.1714 \times 10^{-8}$ Btu/(hr ft² R^4); $T =$ absolute temperature, deg R; $q =$ Btu/hr; $D =$ diameter, ft. The same units apply to Eq. (13) except that diameter is expressed in inches.

represent the heat-leak equation for all transfer lines; Fig. 15 gives these curves. For any fluid and shield (or ambient) temperature, the heat leakage can then be determined. Table 95 gives the values of $\Delta(T^4) \times 10^{-8}$ by which the ordinate in Fig. 15 must be multiplied to give the heat leak q for helium, hydrogen, neon, nitrogen, and oxygen.

Table 96 gives the values of the absorptivity for some surfaces which might be used in transfer lines; these values may be used as emissivities. It must be noted that these values were obtained from tests of relatively short duration. Even with pressures of 10^{-6} mm Hg, the surfaces adsorb a sufficient amount of gas to increase appreciably the heat-transfer rate in a relatively short time. For example, the heat transfer to aluminum foil at 76 K from a surface at 300 K in a pressure of 10^{-6} mm Hg increases by about 35 per cent in one week.

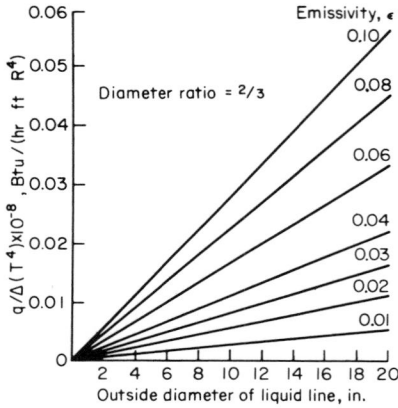

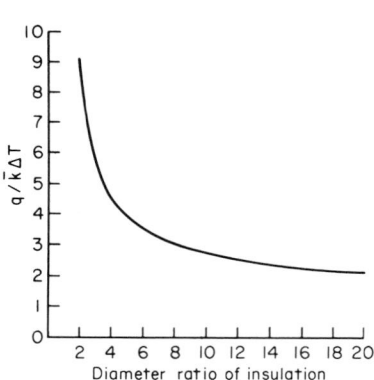

FIG. 15. Radiant-heat leakage. (*R. B. Jacobs, Natl. Bur. Std. Circ.* 596, 1958.)

FIG. 16. Heat conduction. (*R. B. Jacobs, Natl. Bur. Std. Circ.* 596, 1958.)

Evacuated-powder Insulation. Figure 14 can also represent a transfer line using evacuated-powder insulation if the regions marked "high vacuum" are supposed filled with powder and evacuated to a pressure of about 10^{-2} mm Hg.

Those evaluating evacuated-powder insulation have assumed that the steady-state–heat-conduction equation for long cylinders can be used; transposing,

$$\frac{q}{\bar{k}\,\Delta T} = \frac{2\pi}{\ln D_1/D_2} \tag{14}$$

in which the right side depends only upon the dimensions of the insulation, and $\bar{k}$ is the mean thermal conductivity of the entire conduction path. This relation, plotted in Fig. 16, may be applicable to all evacuated powders with all fluids if the thickness of the powder, $\frac{1}{2}(D_1 - D_2)$, is great enough. If the powder is too thin, radiant heat leakage will play a significant role.

As $\bar{k}$ depends upon the temperature at the boundaries of the insulation as well as the insulation properties, no generality is lost if $(\bar{k}\,\Delta T)$ is used as a parameter instead of $\bar{k}$. Values of this parameter are listed in Table 97 and may be relied upon if the powder thickness is not less than 1 in. To determine the heat leakage to a liquefied gas through evacuated-powder insulation, $q/\bar{k}\,\Delta T$ is obtained by

Table 95. $\Delta T^4 \times 10^{-8}$ Factors[1]

(Ambient temperature $= 80$F. Liquid-nitrogen shield temperature $= -321$F. Primary liquid is at normal boiling point)

Primary liquid	$T^4 \times 10^{-8}$, R^4	
	No shield	N$_2$ shield
Helium........	850	3.73
Hydrogen......	850	3.71
Neon..........	850	3.67
Nitrogen.......	846	
Oxygen........	843	

[1] R. B. Jacobs, *Natl. Bur. Std. Circ.* 596, 1958.

Table 96. Absorptivity at 76 K for 300 K Thermal Radiation[1]

In.	Material	Absorptivity
0.001	Kaiser *aluminum* foil, unannealed................	0.018
0.0015	Cockron home foil, *aluminum*....................	0.018
0.0015	Hurwich home foil, mat side, *aluminum*	0.021
0.0015	Hurwich home foil, bright side, *aluminum*	0.022
0.020	*Aluminum*, cold acid cleaned....................	0.028
0.020	*Aluminum*, hot acid cleaned, Alcoa process	0.029
0.020	*Aluminum*, wire brush, emery paper, steel wool, cold acid....................................	0.045
0.020	*Aluminum*, wire brush.........................	0.06
	Aluminum, sprayed onto stainless steel	0.07
0.001	*Yellow brass*, shim stock (65% Cu, 35% Zn).......	0.029
0.005	Mill-run *copper* sheet, annealed	0.015
0.005	*Copper*, dilute chromic acid dip	0.017
0.005	*Copper*, wet polished with pumice	0.018
0.005	*Copper*, electrolytically cleaned..................	0.017
0.005	*Copper*, fine emery	0.023
	Silver sheet.................................	0.008
	Silver plate, careful preparation, nickel strike on stainless steel	0.009
	Silver plate, careful preparation, nickel and copper strike on stainless steel	0.007
	Allegheny *silver* spray process on stainless steel	0.009
0.005	Type 302 sheet *stainless steel*...................	0.048
	Commercial ball, type 302 *stainless steel*..........	0.07
	Silver plated on copper	0.017 at 76 K / 0.013 at 20 K
	Nickel plated on copper.......................	0.033 at 76 K / 0.027 at 20 K

[1] R. B. Jacobs, *Natl. Bur. Std. Circ.* 596, 1958.

entering Fig. 16 with the desired diameter ratio, and q is then obtained by multiplying by the appropriate value of $\bar{k}\,\Delta T$ obtained from Table 97.

Minimizing heat leak not only increases transfer efficiency (by decreasing liquid losses) but also increases the lengths of transfer lines which will be feasible with given pumping equipment.

Table 97. $\bar{k}\,\Delta T$, Btu/(hr ft)[1]

Insulation	Liquid H_2 with liquid N_2 shield	Liquid N_2 with no shield	Liquid O_2 with no shield
Perlite (-80 mesh), N_2 is interstitial gas, pressure is 10^{-2}-mm Hg	0.0297[2]	0.285	0.268[3]
Perlite (-80 mesh), N_2 is interstitial gas, pressure is 10^{-3}-mm Hg	0.0173[2]	0.255	0.240[3]
Perlite (-80 mesh), He is interstitial gas, pressure is $<10^{-5}$-mm Hg	0.0126	0·2548	0.240[3]
SOC perlite with 50% Al powder, pressure is $<10^{-3}$-mm Hg	. . . . . .	0.232	0.218[3]
Silica Aerogel, N_2 is interstitial gas, pressure is 10^{-2}-mm Hg	0.0297	0.642 ·	0.605[3]
Silica Aerogel, N_2 is interstitial gas, pressure is 10^{-3}-mm Hg	0.0173	0.533	0.503[3]
Silica Aerogel, He is interstitial gas, pressure is $<10^{-5}$-mm Hg	0.0126	0.463	0.437[3]
Silica Aerogel with 50% Al powder, pressure is $<10^{-3}$-mm Hg	. . . . . .	0.123	0.116[3]
Silica Aerogel with 50% Alcoa Al Flitter 660, pressure $<10^{-3}$-mm Hg	. . . . . .	0.290	0.273[3]
Diatomaceous earth, N_2 is interstitial gas, pressure is 10^{-2}-mm Hg	. . . . . .	0.306	0.288
Diatomaceous earth, N_2 is interstitial gas, pressure is 10^{-3}-mm Hg	. . . . . .	0.278	0.262[3]
Diatomaceous earth, N_2 is interstitial gas, pressure is $<10^{-5}$-mm Hg	. . . . . .	0.264	0.249[3]
Phenolic spheres, N_2 is interstitial gas, pressure is 10^{-2}-mm Hg	. . . . . .	0.405	0.382[3]
Phenolic spheres, N_2 is interstitial gas, pressure is 10^{-3}-mm Hg	. . . . . .	0.313	0.295[3]
Phenolic spheres, N_2 is interstitial gas, pressure is $<10^{-5}$-mm Hg	. . . . . .	0.283	0.266[3]
Lampblack, N_2 is interstitial gas, pressure is 10^{-2}-mm Hg	. . . . . .	0.413	0.389[3]
Lampblack, N_2 is interstitial gas, pressure is 10^{-3}-mm Hg	. . . . . .	0.336	0.317[3]
Lampblack, N_2 is interstitial gas, pressure is $<10^{-5}$-mm Hg	. . . . . .	0.303	0.286[3]

[1] R. B. Jacobs, *Natl. Bur. Std. Circ.* 596, 1958.

[2] These values are assumed to be the same as those measured for Silica Aerogel.

[3] These values are based upon the $\bar{k}$ determined for liquid nitrogen.

Cooldown of Transfer Lines.   One of the most serious limitations imposed on the fast transfer of cryogenic liquids is the transient period that exists during line cooldown.   When the line inlet valve is opened, a liquid front quickly advances into the line.   This front initially overshoots its "equilibrium" position, and pressure surges result.   The front then proceeds in a relatively uniform manner down the line at a rate that is primarily controlled by the ability of the line to vent

the cooldown gas that has formed by the evaporating liquid. The arrival of single-phase liquid at the end of the line corresponds closely to the attainment of equilibrium wall temperature. In systems designed to transfer a given quantity of liquid within a fixed period of time, the duration of this transition period, known as line cooldown time, can be a critical design consideration.

As a good approximation, the amount of liquid required to cool down a line is

$$W = M_m C_{p_m} \Delta T_m (H_v - H_l) \tag{15}$$

where W = liquid, lb/hr
M_m = line to be cooled, lb
Cp_m = mean specific heat of line, Btu/(lb F)
ΔT_m = temperature change through which line is cooled, F
H_v = enthalpy of fluid as a vapor at ambient conditions, Btu/lb
H_l = initial enthalpy of liquid when entering line, Btu/lb

A more detailed analysis incorporating heat leakage and line impedance can be found in the following papers given at the Cryogenic Engineering Conferences; 1958 *Conference Paper* F-5, "Pressurized Cooldown of Cryogenic Transfer Lines" by J. C. Burke et al.; and 1960 *Conference Paper* E-7, "Pressurized Cooldown of a Cryogenic Liquid Transfer System Containing Vertical Sections" by E. M. Drake et al.

Transfer of Fluids

The transfer of cryogens from storage to the place of use is accomplished with pumps (generally centrifugal) or pressurized gas, either separately or in combination. In present-day ballistic missiles, liquid oxygen is delivered to the combustion chambers at 300 to 750 psi obtained in a single-stage centrifugal pump driven by gas turbines. In missile ground-support facilities, liquid oxygen may be transferred by using either gaseous oxygen or nitrogen to pressurize the storage tank. In gas-pressurized transfer, either the gaseous form of the cryogen being transferred or lower boiling point gases which do not react with the cryogen are used.

Pumping. The design of pumps for cryogenic and high-temperature service is very similar. Cryogenic liquids are generally poor in lubricating properties, although the cooling effect which they provide is beneficial in reducing the interfacial temperature of the rubbing surface. The most effective dynamic seals require slider materials for solid contact. The seal material should provide inherently low friction and low wear. These characteristics are met by Teflon-impregnated asbestos yarns in a braided construction formed into rings.

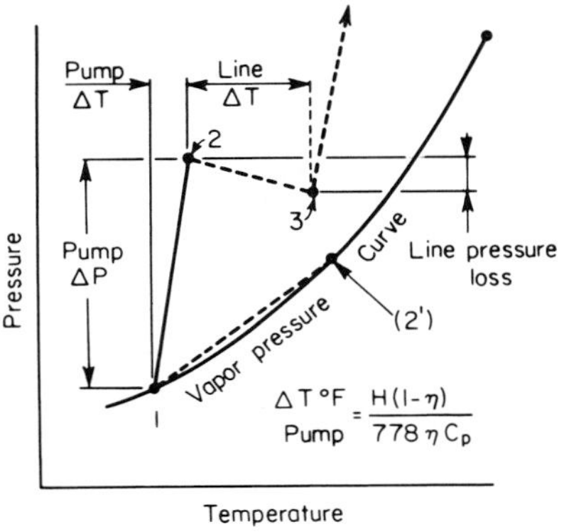

Fig. 17. Fluid temperature rise through pump.

Net positive suction head (NPSH) requirements should be low to suppress cavitation. Since the liquid from a storage tank is usually saturated, the fluid at the inlet of the pump may be expected to be substantially at the vapor pressure shown as point 1 in Fig. 17. This results in little or perhaps no NPSH at the pump inlet. The pump then pressurizes the liquid to point 2. At the same time

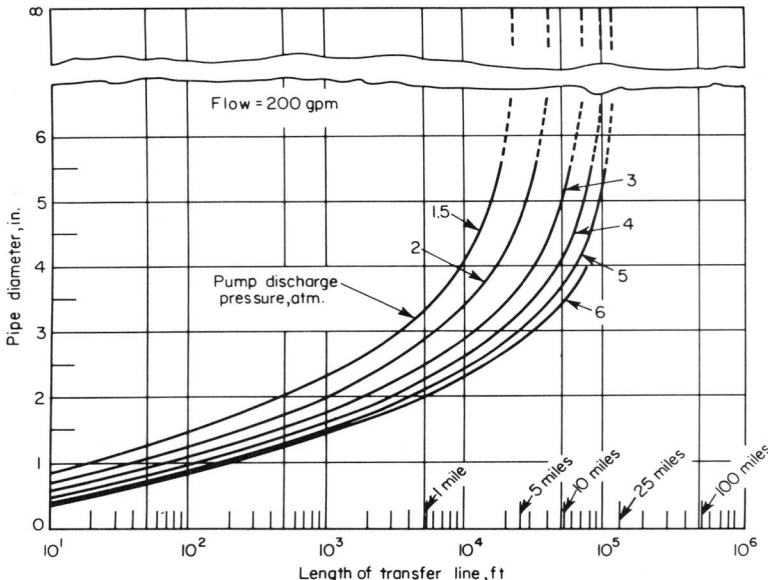

FIG. 18. Liquid-hydrogen line sizes.   (*R. B. Jacobs, Natl. Bur. Std. Circ.* 596, 1958.)

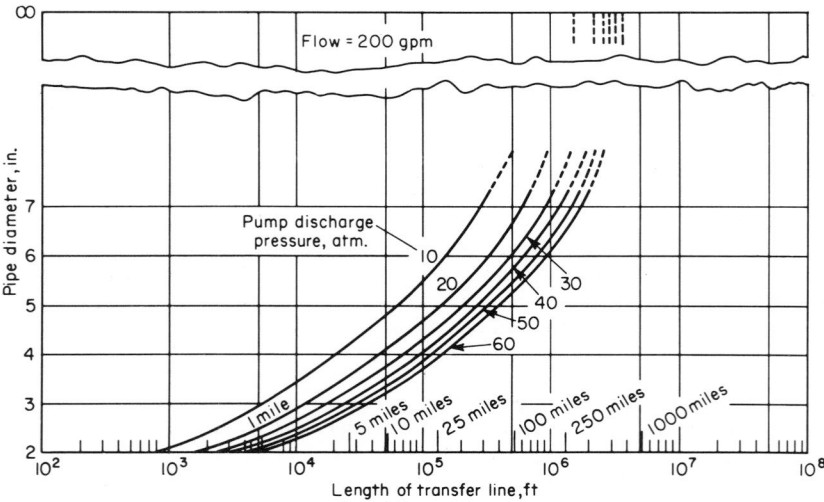

FIG. 19. Liquid-oxygen line sizes.   (*R. B. Jacobs, Natl. Bur. Std. Circ.* 596, 1958.)

it increases the temperature of the liquid, owing to pump internal dissipation. This is indicated by the horizontal displacement of point 2 to the right.   On discharge through some external pipeline to point 3, the line heat leakage and friction will cause a further rise of fluid temperature.   This is indicated by the horizontal distance between point 2 and point 3.   Pipe friction also causes a pressure drop as shown.   If it is desired to have a single phase of subcooled liquid at the line terminus, the point 3 must be in the liquid region, that is, to the left of the vapor pressure curve.   The total pressure here above vapor pressure also becomes the NPSH available to any following pump.

The theoretical temperature rise[1] through a line-mounted pump is a function of its head and efficiency:

$$\Delta T = H(1 - \eta)/778\eta C_p \qquad (16)$$

where ΔT = temperature rise, F
   H = pump head, ft of liquid
   C_p = mean specific heat of liquid, Btu/(lb F)
   η = pump efficiency

Jacobs has made a detailed study of single-phase transfer of liquefied gases (*National Bureau of Standards Circular* 596).   Saturated liquid at 1 atm is pumped through an evacuated (less than 10^{-5} mm Hg) silica aerogel transfer line having a diameter ratio D_1/D_2 of 5.   The pump heat is removed by a heat exchanger, and the conditions at the end of the transfer line are equal to those at the inlet, i.e., saturated liquid at 1 atm.   Figures 18 and 19, showing the line sizes for a flow of 200 gpm of liquid hydrogen and oxygen for varying line lengths and pump discharge pressures, are representative of the many curves given in this paper.

Pressurized Transfer.   Gas-pressurized transfer systems may be preferred to pumping systems in some cases.   Infrequent, intermittent pump operation may not provide economic justification for the installation of large pumps, large standby power facilities, and associated control gear.   There are also instances when the additional energy from pump losses cannot be tolerated in the transferred liquid.

Reservoirs for cryogenic fluids, high-pressure gas-storage manifolds, and pressurizing gas-control valves are commercially available.   Their operation is simple and requires essentially no external power, and they are always ready to use. The disadvantages of such a system center around the manifold.   Storage of high-pressure gas requires storage vessels and a high-pressure gas supply for recharging.

A typical cryogenic liquid-storage system consists of a horizontal or vertical cylindrical tank fitted with a bottom liquid drain and a top connection for the pressuring gas.   Some type of internal gas diffuser which reduces the interaction between the input gas and the liquid is necessary to prevent excessive condensation. The storage tank may be vacuum-jacketed and vented to the atmosphere either directly or through an absorber for toxic gases, or it may be surrounded by a bath of a more volatile liquid cryogen and have no vent.

In any of these storage systems, pressurization is accomplished by the flow of high-pressure chemically compatible or inert gas into the ullage space.   The inert gas is cooled by mixing with the initial ullage gas, contacting the walls and other parts of the tank, and contacting the liquid surface.

Estimation of gas requirements to pressurize and transfer fluids is made with the "saturation rule."   This rule is on the average 10 to 15 per cent conservative if

---

[1] Equation (16) assumes that the pump head H includes velocity heads at pump inlet and outlet; also, the pump is well insulated so that there is zero heat transfer.

only pressurization and transfer are considered. It will still predict gas consumption quite well if a hold period up to 1 hr after transfer is included.

The saturation rule is

$$\Delta M = V_2 \rho_2 - V_1 \rho_1 \tag{17}$$

where ΔM = gas consumption, lb
     V = ullage volume, cu ft
     ρ = gas density assuming saturated vapor, lb/cu ft
Subscript 1 = conditions before pressurization
Subscript 2 = conditions after transfer and hold

Gas condensation losses in a hold period can be predicted from transient heat conduction theory well within experimental error.

Contamination of liquid oxygen in gaseous-nitrogen-pressurized transfer systems is not evident below 1 in. from the liquid surface. This applies only for tanks in which vortexing and other liquid disturbances are minimized.

Piping Systems. Cryogenic plants can be constructed to operate reliably and without unusual difficulties provided that the special properties of the cryogen are thoroughly understood and are carefully considered in the design. The fundamental principles of design requisite to an operable and efficient system are based primarily on the fact that cryogens have very low boiling points and heats of vaporization. The design principles are:

1. Slope all piping upward in the direction of flow, wherever possible, to take advantage of the principle of the air lift.

2. Avoid peaks or high pockets in all pipelines because these will be gas traps.

3. Study system designs carefully to reduce all possible heat leaks to a minimum.

4. Use insulation and metals with low thermal conductivities.

5. Valves should have extended bonnets and stems of stainless steel to bring the stem seals and handles outside the insulation.

6. Make instrumentation as simple as practicable.

7. Piping movements are contractions, not expansions. These contractions will affect the method of supporting and anchoring the pipe. The supports or hangers should be placed so that there will be a minimum of heat gain by the fluid in the pipe, i.e., wood blocks between the pipe support shoe and the steel or concrete support. For pipe hangers, the pipe attachment should extend past the insulation a sufficient distance so that ice formation will not affect the operation of the hanger. The maintenance of the insulation seal around a piercing support or hanger can be lessened by insulating the piercing member for a distance from the pipe insulation of about twice the pipe insulation thickness.

As a guide, typical ferrous metal specifications for the temperature ranges of -20 to -50 F, -50 to -150 F, and below -150 F are given in Tables 98 to 101.

Table 98. -20 to -50 F Specification[1]

Bolting	ASTM A193 Grade B7 bolts
	ASTM A194 Grade 2H nuts
Gasketing.	Compressed asbestos sheet
Pipe.	ASTM A333 Grade C, seamless
Butt-welding fittings.	ASTM A420 Grade WPLC
Castings	ASTM A352 Grade LCB
Forgings.	ASTM A350 Grade LF1
	Flanges to ASA B16.5;
	S-W fittings to ASA B16.11

[1] H. M. Howarth, *Petro/Chem. Engr.*, vol. 32, p. C-33, 1960.

Table 99. −50 to −150 F Specification[1]

Bolting . . . . ASTM A320 Grade L7 (4140) bolts
ASTM A194 Grade 4 (C-Mo) nuts
Gasketing. . . . . Type 304 spiral-wound
Pipe. . . . . ASTM A333 Grade 4 (Cr-Cu-Ni)
ASA B36.10
Butt-welding fittings . . . . . . ASTM A420 Grade WPL4 (Cr-Cu-Ni)
Castings . . . . ASTM A351 CF8 (304) or CF8M
(316) or ASTM A352 Grade LC3
(3½% Ni)
Forgings. . . . . ASTM A350 LF4 (Cr-Cu-Ni)
Flanges to ASA B16.5;
S-W fittings to ASA B16.11

[1] H. M. Howarth, *Petro/Chem. Engr.*, vol. 32, p. C-33, 1960.

Table 100. Below −150 F Specification[1]

Bolting . . . . ASTM A193 (or A320) Grade B8
(304) bolts
ASTM A194 Grade 8 (304) nuts
Gasketing. . . . . Type 304 spiral-wound
Pipe. . . . . ASTM A312 TP304 seamless
ASA B36.19
Butt-welding fittings . . . . . . ASTM A403 Grade WP304
ASA B16.9
Castings . . . . ASTM A351 Grade CF 8 (304) or
CF8M (316)
Forgings. . . . . ASTM A182 Grade F304
Flanges to ASA B16.5;
S-W fittings to ASA B16.11

[1] H. M. Howarth, *Petro/Chem. Engr.*, vol. 32, p. C-33, 1960.

Table 101. Aluminum Specification[1]

Bolting . . . . ASTM A193 Grade B8 (304) bolts
ASTM B98 nuts
Gasketing. . . . . Type 304 spiral-wound
Pipe. . . . . ASTM B241 alloy M1A (3003) or
alloy GS11A (6061-T6)
Butt-welding fittings . . . . . . Aluminum conforming to ASTM
A234, as applicable
Castings . . . . ASTM A351 Grade CF8 (304) or
CF8M (316)
Forgings. . . . . ASTM B247 alloy GS 11A
(6061-T6)
Flanges to ASA B16.5 dimensions

[1] H. M. Howarth, *Petro/Chem. Engr.*, vol. 32,. p C-33, 1960.

Costs. Relative costs for uninsulated pipe, bolting, fittings, valves, and flanges are given in Tables 102 to 106.

The cost of vacuum-insulated piping, including all the necessary associated equipment such as vacuum pumps, manifolds, expansion joints, supports, etc., may range from $20 to $40 per foot of piping.

Table 102. Pipe Costs[1]

Material	ASTM Spec. and grade	Type	Schedule	Comparative costs (approx.)[2]	
				3 in.	8 in.
Carbon steel......	A53 A or B	Seamless	40	1.0	1.0
	A106 A or B	Seamless	40	1.10	1.10
	A333 C	Seamless	40	1.3–1.6	1.6–2.05[3]
3½Ni...........	A333 3	Seamless	40	2.8–3.5	4.35–5.35[3]
Cr-Cu-Ni.........	A333 4	Seamless	40	1.65–1.85	2.6–2.9[4]
5 Ni.............	A333 5	Seamless	40	3.3–4.05	5.1–6.3[3]
9 Ni.............		Seamless	40	5.65–6.8	8.75–10.5[5]
Red brass	B43	Seamless	Reg.	6.6	9.5[7]
			XS	10.1	14·9[7]
Copper	B42	Seamless	Reg.	7.6	9.5[7]
			XS	10.3	14.9[7]
Aluminum........	B241 M1A(3003)	Seamless	40	1.7–1.9	1.9–2.1
			10	1.2	1.3
			5	0.9	1.0
	B241 GS11A (6061-T6)	Seamless	40	2.0–2.25	2.3–2.6
			10	1.4	1.65
			5	1.1	1.2
Monel	B165	Seamless	40	15.8	17.6
304.............	A312 TP304	Seamless	40	10.1–10.7	12.9–13.6
		Seamless	10	6.0–6.35	
		Seamless	5	4.6–4.85	
		Welded and annealed	10	5.25	
			5	3.95	
		Welded and annealed[6]	40	10.3–10.9	12.3–12.9
			10		4.35–4.7
			5		3.5–3.8
304L............	A312 TP304L			21% more than 304	
316.............	A312 TP316			64% more than 304	
316L............	A312 TP316L			85% more than 304	
321.............	A312 TP321			18% more than 304	
347.............	A312 TP347			32% more than 304	

[1] H. M. Howarth, *Petro/Chem. Engr.*, vol. 32, p. C-33, 1960.
[2] Comparative costs are approximate.   Comparative cost ranges indicate how price varies with the size of the order (from 20 tons to 5 tons).
[3] Small extra charge for impact tests.
[4] Minimum order 13 tons.
[5] Minimum order 11 tons.
[6] A-312 except filler metal added.
[7] Based on 2,000 lb or ft.

Table 103. Bolting Costs[1]

Material	Bolts, ASTM Spec. and grade	Nuts, ASTM Spec. and grade	Comparative costs[2] (approximate)
AISI 4140........	A193 B7	A194 2H	1.0
AISI 4140........	A320 L7	A194 4	1.2[3]
AISI 4340........	A320 L43	A194 4	2.05[3]
Type 304.........	A193 (or 320) B8	A194 8	6.5
Type 304.........	A193 (or 320) B8	B98 B-12	5.35
Type 347.........	A193 (or 320) B8C	A194 8C	7.6
Type 348.........	A193 (or 320) B8D	A194 8D	40.0 (est.)
Type 321.........	A193 (or 320) B8T	A194 8T	6.95
Aluminum........	B211 CG42A (2024-T4)	11ST-3	20.0

[1] H. M. Howarth, *Petro/Chem. Engr.*, vol. 32, p. C-33, 1960.
[2] Costs are based on each bolt being equipped with two nuts.
[3] There is an additional charge for impact tests which is not included.

Table 104. Elbow Costs (Butt Weld)[1]

Material	ASTM Spec. and grade	Schedule	Comparative costs (approx.)	
			3-in. elbows	8-in. elbows
Carbon steel......	A234 WPB	40	1.0	1.0
	A420 WPLC		1.4[2]	1.4[2]
3½ Ni............	A420 WPL3	40	7.9[3]	6.75[3]
Cr-Cu-Ni.........	A420 WPL4	40	7.9[3]	6.75[3]
304	A403 WP304	40	6.4	7.6
	A403 WP304	10	3.65	4.3
	A403 WP304	5	3.15	3.45
	A352 CF8	40	14.6[2]	13.5[3]
316 and 347......	A403 WP316 or 347	40	8.0	9.5
	A403 WP316 or 347	10	4.55	5.35
	A403 WP316 or 347	5	3.95	4.3
Aluminum........	— M1A (3003F)	40	3·5	3.0
	— GS11A (6061-T6)	40	4.15	3.65

[1] H. M. Howarth, *Petro/Chem. Engr.*, vol. 32, p. C-33, 1960.
[2] Certified to meet impact requirements—additional charge for impact tests.
[3] Prices vary widely with quantity.

Table 105. Flanged Gate—Valve Costs[1]

Material	ASTM Spec. and grade	Other specifications	Series	Comparative costs (approximate)	
				3 in.	8 in.
Carbon steel......	A216 WCB	API 600	150	1.0	1.0
		API 600	300	1.45	1.5
	A352 LCB	API 600	150	1.15	1.15
		API 600	300	1.65	1.75
Carbon moly	A352 LC1	API 600	150	1.15	1.15
		API 600	300	1.65	1.75
3½ Ni...........	A352 LC3	API 600	150	1.25	1.25
		API 600	300	1.8	1.9
304 or 316	A351 CF8 or	MSS SP-42 with	150	0.87	1.25
	CF8M	ASA B16.5 flanges	300	1.25	1.5
		Ditto with extended	150	1.5	2.2
		bonnet	300	2.2	2.7
		API 600	150	2.0 (est.)	2.9 (est.)
Brass	B62		150	0.57	0.88
Aluminum........	B26 SG70A	MSS SP-42 flanges	150	0.69	0.93

[1] H. M. Howarth, *Petro/Chem. Engr.*, vol 32, p. C-33, 1960.

Table 106. Weld Neck Flange Costs[1]

Material	Series[2]	ASTM Spec. and grade	Comparative costs	
			3-in. RF WN flgs.	8-in. RF WN flgs.
Carbon steel	150	A181 I or II	1.0	1.0
	300	A181 I or II	1.3	1.45
	150	A350 LF1	1.1	1.1
	300	A350 LF1	1.4	1.55
3½ Ni	150	A350 LF3	6.05	3.8
	300	A350 LF3	8.2	6.05
Cr-Cu-Ni	150	A350 LF4	6.05	3.8
	300	A350 LF4	8.2	6.05
304	150	A182 F304	5.3	5.65
316	150	A182 F316	6.9	7.3
Aluminum	150	B247 M1A (3003F)	3.3	3.0
		B247 GS11A (6061-T6)	3.3	3.0

[1] H M. Howarth, *Petro/Chem. Eng.* vol. 32, p. C-33, 1960.
[2] Flanges all have raised faces and conform to ASA B16.5 dimensions.

REFERENCES

Cryogenic Fluids

1. API Research Project 44, "Selected Values of Physical and Thermodynamic Properties of Hydrocarbons and Related Compounds."
2. Cook, G. A., "Argon, Helium and the Rare Gases," vols. 1 and 2, Interscience Publishers, Inc., New York, 1961.
3. Din, F., "Thermodynamic Functions of Gases," vols. 1 and 2, Butterworth Scientific Publications, London, 1956.
4. Gray, D. E., "American Institute of Physics Handbook," 2d ed., McGraw-Hill Book Company, New York, 1963.
5. Hodgman, C. D., "Handbook of Chemistry and Physics," Chemical Rubber Publishing Company, New York, 1960.
6. Johnson, V. J., "A Compendium of the Properties of Materials at Low Temperature (Phase 1) Part 1. Properties of Fluids," *WADD Tech. Rept.* 60–56, 1960.
7. The Matheson Co., Inc., "Matheson Gas Data Book," 1961.
8. "Tables of Thermal Properties of Gases," *Natl. Bur. Std. Circ.* 564, U.S. Department of Commerce, 1955.
9. Perry, R. H., "Chemical Engineers' Handbook," 4th ed., McGraw-Hill Book Company, New York, 1963.
10. Reid, R. C., and T. K. Sherwood, "The Properties of Gases and Liquids," McGraw-Hill Book Company, New York, 1958.
11. "Cryogenic Data Book," *WADC Tech. Rept.* 59–8, 1959.
12. Washburn, E. W., "International Critical Tables," McGraw-Hill Book Company, New York, 1926.
13. Woolley, H. W., R. B. Scott, and F. G. Brickwedde, *Natl. Bur. Std. Paper* RP 1932, 1948.

Cryogenic Materials

1. The American Society of Mechanical Engineers, "Petroleum Refinery Piping," ASA B31.3–1959 (Piping Code), 1959.
2. Johnson, V. J., "A Compendium of the Properties of Materials at Low Temperature (Phase 1) Part II. Properties of Solids," *WADD Tech. Rept.* 60–56, 1960.
3. Scott, R. B., "Cryogenic Engineering," D. Van Nostrand Company, Inc., Princeton, N.J., 1959.

4. "Standards of the Expansion Joint Manufacturers Association," 1958.
5. "Mechanical Properties of Structural Materials at Low Temperatures," *Natl. Bur. Std. Monograph* 13, 1960.
6. ALCOA Aluminum, private communication.
7. Bridgeport Brass Co., private communication.
8. Bulow, C. L., *Chem. Eng.*, vol., 67, p. 187, 1960.
9. Corruccini, R. J., *Chem. Eng. Progr.*, vol., 53, pp. 262, 342, 397, 1957.
10. Hansen, O. A., *ASME Paper* 59-Pet-37, 1959.
11. Johnson, E. W., *Chem. Eng.* vol. 67, p. 133, 1960.
12. Johnson, R. J., *Chem. Eng.*, vol. 67, p. 115, 1960.
13. Kaiser Aluminum and Chemical Sales, Inc., private communication.
14. McConnell, J. H., *Chem. Eng.*, vol. 67, p. 125, 1960.
15. Palmer, M., *ASME Paper* 59-Pet-37, 1959.
16. Rote, R., and J. H. Proctor, *Chem. Eng.*, vol. 67, p. 119, 1960.
17. United States Steel, private communication.
18. Vanderbeck, R. W., *Chem. Eng.*, vol. 67, p. 103, 1960.
19. Zenner, G. H., *Chem. Eng.*, vol. 67, p. 117, 1960.

Cryogenic Piping Design

1. Din, F., and A. H. Cockett, "Low-temperature Techniques," Interscience Publishers, Inc., New York, 1960.
2. Linde Co., "Design Manual for Vacuum S.I. Insulated Piping," 1962.
3. Office of the Director of Defense Research and Engineering, "The Handling and Storage of Liquid Propellants," 1961.
4. Perry, R. H., "Chemical Engineers' Handbook," 4th ed., McGraw-Hill Book Company, New York, 1963.
5. Scott, R. B., "Cryogenic Engineering," D. Van Nostrand Company, Inc., Princeton, N.J., 1959.
6. "Standards of the Expansion Joint Manufacturers Association," New York.
7. Timmerhaus, K. D., "Advances in Cryogenic Engineering," vols. 1 to 8, *Proc. Cryogenic Eng. Conf.*, 1958.
8. Anderson Jr., R. L., *ASME Eng. Conf. Paper 59-Pet-32*, 1959.
9. Baker, O., *Oil Gas J.*, vol. 53, p. 185, 1954.
10. Balwanz, W. W., J. M. Singer, and N. P. Frandsen, *Cryogenic Eng. Conf. Paper D-3*, 1960.
11. Black, I. A., and P. E., Glaser, *Cryogenic Eng. Conf. Paper A-3*, 1960.
12. Bowersock, D. C., R. W. Gardner, and R. C. Reid, *Cryogenic Eng. Conf. Paper F-3*, 1958.
13. Bowersock, D. C., and R. C. Reid, *Cryogenic Eng. Conf. Paper* E-1, 1960.
14. Burke, J. C., W. R. Byrnes, A. H. Post, and F. E. Ruccia, *Cryogenic Eng. Conf. Paper F-5*, 1958.
15. Byron Jackson Pumps, Inc., private communication.
16. Caine, G., L. Schafer, and D. Burgeson, *Cryogenic Eng. Conf. Paper D-4*, 1958.
17. Canty, J. M., *Cryogenic Eng. Conf. Paper E-2*, 1960.
18. *Chem. Process*, vol., 24, p. 43, 1961.
19. Clark, L. J., World Power Conference, Madrid, 1960.
20. Corlett, E. C. B., and J. F. Leathard, *Roy. Inst. Naval Architects (London)* Paper 6, 1960.
21. Crane Packing Co., private communication.
22. DK Manufacturing Co., Chicago, Ill., private communication.
23. Drake, E. M., F. E. Ruccia, and J. M. Ruder, *Cryogenic Eng. Conf. Paper E-7*, 1960.
24. Flieder, W. G., W. J. Smith, and K. R. Wetmore, *Cryogenic Eng. Conf. Paper B-5*, 1959.
25. Fuller, P. D., and W. R. Peavy, *Cryogenic Eng. Conf. Paper F-2*, 1958.
26. *Gas Age*, vol. 128, p. 29, 1961.
27. Griffith, P., and G. B. Wallis, *J. Heat Transfer (Trans. ASME)*, vol. 83, p. 307, 1961.
28. Hansen, O. A., *ASME Eng. Conf. Paper* 59-Pet-37, 1959.
29. Hatch, M. R., R. B. Jacobs, R. J. Richards, R. N. Boggs, and G. R. Phelps, *Cryogenic Eng. Conf. Paper F-4*, 1958.

30. Hatch, M. R., and R. B. Jacobs, *A.I.Ch.E. J..*, vol. 8, p. 18, 1962.
31. Howarth, H. M., *Petro/Chem. Engr.*, vol. 32, p. C-33, 1960.
32. Hydrocarbon Research, Inc., New York, private communication.
33. Irrgang, O. R., *U.S. Atomic Energy Comm.*, *MTA*-32, 1953.
34. Jacobs, R. B., *Natl. Bur. Std. Circ.* 596, 1958.
35. Jacobs, R. B., *ARS J.* p. 245, 1959.
36. Jacobs, R. B., Heating, Piping Air Conditioning, vol. 32, pp. 129, 141, 1960.
37. Jacobs, R. B., R. J. Richards, and S. B. Schwartz, *Cryogenic Eng. Conf. Paper B*-4, 1960.
38. Johns-Manville Corp., private communication.
39. Kaiser Aluminum & Chemical Sales, Inc., private communication.
40. Kropschot, R. H., *ASHRAE J.*, vol. 1, p. 48, 1959.
41. Leonhard, K. E., and R. K. McMordie, *Cryogenic Eng. Conf. Paper H*-1, 1960.
42. LeValley, W. H., and W. K. Sutton, *Cryogenic Eng. Conf. Paper E*-4, 1960.
43. Lewis, J. D., *Petrol. Refiner*, vol. 40, p. 169, 1961.
44. Lockhart, R. W., and R. C. Martinelli, *Chem. Eng. Progr.*, vol. 45, p. 39, 1949.
45. Martin, K. B., and O. E. Park, *Cryogenic Eng. Conf. Paper B*-5, 1954.
46. Martin, K. B., R. B. Jacobs, and R. J. Hardy, *Cryogenic Eng. Conf. Paper G*-6, 1956, and *Paper D*-6, 1957.
47. Martinelli, R. C., and D. B. Nelson, *Trans. ASME*, p. 695, 1948.
48. Matesanz, A., *Chem. Process*, vol. 24, p. 28, 1961.
49. McClintock, R. M., and M. J. Hiza, *Mod. Plastics*, p. 35, 1958.
50. Moore, R. W., A. A. Fowle, B. M. Bailey, F. E. Ruccia, and R. C. Reid, *Cryogenic Eng. Conf. Paper J*-2, 1959.
51. Palmer, M., *ASME Eng. Conf.*, *Paper 59-Pet*-8, 1959.
52. Power, W. H., *Cryogenic Eng. Conf. Paper A*-4, 1960.
53. Project Fabrication Corp., private communication.
54. Reynales, C. H., *Cryogenic Eng. Conf. Paper B*-6, 1960.
55. Richards, R. J., W. G. Steward, and R. B. Jacobs, *Cryogenic Eng. Conf. Paper B*-4, 1959.
56. Robbins, R. F., D. H. Weitzel, and R. N. Herring, *Natl. Bur. Std. Rept.* 6787, 1961.
57. Steinherz, H. A., and P. A. Redhead, *Sci. Amer.*, vol. 206, p. 78, 1962.
58. Smith, M. C., and Rabb, D. D., *Cryogenic Eng. Conf. Paper B*-3, 1954.
59. Tantum, D. H., and F. Farrar, *Cryogenic Eng. Conf. Paper F*-1, 1958.
60. Tantum, D. H., and R. Hargreaves, *Cryogenic Eng. Conf. Paper D*-7, 1960.
61. VanderArend, P. C., *Chem. Eng. Progr.*, vol. 57, p. 62, 1961.
62. Weitzel, D. H., R. F. Robbins, G. R. Bopp, and W. R. Bjorklund, *Rev. Sci. Instr.* vol. 31, p. 1350, 1960.
63. Wisander, D. W., and R. L. Johnson, *Cryogenic Eng. Conf. Paper D*-5, 1960.
64. Sajben, M., *J. Basic Eng. (Trans. ASME)*, vol. 83, p. 619, 1961.

12

FIRE-PROTECTION PIPING

Arthur L. Brown*

Fire-protection piping systems convey a fire-extinguishing substance from the source of supply to devices and equipment for applying it in fire fighting and fire protection.

Water, due to its effectiveness, availability, and low cost, is used most extensively as a fire-extinguishing material. Other substances piped for fighting special types of fire are foam, carbon dioxide or other inert gases, vaporizing liquid, or dry chemical in an inert-gas carrier.

Fire-protection piping as treated in this section is mainly that provided for the protection of individual properties rather than that of public water systems which, with few exceptions, supply water to entire communities for domestic and industrial purposes as well as for fire protection. However, the hydraulic principles and practices apply to any piping system which conveys water for any purpose.

The determination and evaluation of water flows for city or community fire protection are among the activities of the American Insurance Association and local insurance rating association, which make surveys periodically and report on the adequacy of public water supplies.

The detailed procedures followed by fire-protection specialists in determining anticipated water flows, in establishing pipe sizes, and in planning pipe locations and arrangement for particular private installations are too highly specialized and too limited in application to be allocated space for more than brief mention here. Detailed information is available in the publications included in the references at the end of the section.

Piping from a source of extinguishing material to a valve or other device which controls its application is often called supply piping. It is frequently installed as a

* Mechanical Engineer, 195 St. Paul Street, Brookline, 02146. Mass.

project that is distinct from the piping system between the control device and points of discharge and is so treated in most standards, regulations, and practices. However, its design must be coordinated with the other elements in the fire-protection system.

Dependability at the time of need is the most important characteristic of a fire-protection system.

STANDARDS, RULES, AND REGULATIONS

The generally recognized standards which cover fire-protection piping are those developed by the National Fire Protection Association and by insurance organizations such as the American Insurance Association, the Factory Mutual Engineering Corporation, and the Factory Insurance Association. Detailed practices have in some cases been developed by the makers of devices and equipment. Local municipal, state, or other governmental regulation may be in force and must be followed. The generally recognized standards which relate to fire-protection piping are listed below.

The fire-protection handbooks in Refs. 1 and 2 contain much of the information in the individual standards and may provide all the needed information.

1. "NFPA Handbook of Fire Protection," National Fire Protection Association, 60 Batterymarch St., Boston, Mass. 02110.

The following standards are all available from the National Fire Protection Association, 60 Batterymarch St., Boston, Mass. 02110.

2. "Handbook of Industrial Loss Prevention," Factory Mutual Engineering Corporation, 1151 Boston-Providence Turnpike, Norwood, Mass.
3. Standard for the Installation of Sprinkler Systems, NFPA No. 13.
4. Standard for the Installation of Standpipe and Hose Systems, NFPA No. 14.
5. Standard for the Installation Water Tanks for Private Fire Protection, NFPA No. 22.
6. Standard for the Installation of Centrifugal Fire Pumps, NFPA No. 20.
7. Standard for Carbon Dioxide Extinguishing Systems, NFPA No. 12.
8. Standard on Foam Extinguishing Systems, NFPA No. 11.
9. Standard for Dry Chemical Extinguishing Systems, NFPA No. 17.
10. Standard for Outside Protection, NFPA No. 24.

The following general references relate to certain aspects of the problems which are involved in the design and specification of fire-protection systems.

11. "Automatic Sprinkler Hydraulic Data," by Clyde M. Wood. "Automatic" Sprinkler Corporation of America, Youngstown, Ohio. 44501.
12. "Handbook of Cast Iron Pipe," Cast Iron Pipe Research Association, Chicago, Ill. 60603.
13. "Approved Equipment and Materials for Industrial Firesafety," Factory Mutual Engineering Corporation, 1151 Boston-Providence Turnpike, Norwood, Mass. 02062.
14. "Fire Protection Equipment List," Underwriters' Laboratories Inc., 207 E. Ohio Street, Chicago, Ill. 60611.
15. "List of Inspected Appliances, Equipment and Materials," Underwriters' Laboratories of Canada, 7 Crouse Road, Scarborough, Ontario, Canada.

PREPARATION OF PLANS

Insurance organizations and regulatory bodies usually require engineers, architects, or suppliers to submit working drawings of proposed fire-protection systems for approval.

Water-system plans should include the location of water mains, connections to water supplies, hydrants, connections to sprinkler systems and standpipes, valve locations, kind and size of pipe, and means to be used to protect the piping against injury and freezing.

Sources of Water.   In order to provide the pressure and volume of water that may be needed for fire fighting and to assure the dependability that is essential for high-value property, it is usually necessary to combine two or more sources of supply, called primary and secondary.   *The primary water supply* must always be ready to meet the demand of the first automatic sprinklers to open or of other water-using automatic protection systems, as well as the early use of water for hose streams. Such a primary supply is usually from a public water system, elevated storage tank, of reservoir or, in special cases of low demand, from one or more pressure tanks. When a public supply and a gravity private supply are combined, the highest pressure source is referred to as the primary. Usually this is the supply from the public mains.

A secondary water supply must be available if the public or private primary source will be unable to supply the maximum demand rate from sprinklers and hose streams, if the primary source is limited in the total amount of water available, or if the reliability of a single supply may be questionable.

A secondary supply may be from one or more fire pumps having an independent suction source, from a large gravity tank, or from a booster pump having a dependable supply from low-pressure public mains or a private process supply of adequate capacity.   Occasionally recognition is given to fire department pumpers that can deliver water from some additional source to the protective system through *a fire department connection.*

Separation of Fire Service from Other Water Piping Systems.   Water-supply systems for fire service and for industrial or domestic use with few exceptions, are kept separate for the following reasons:

1. A separate fire system affords integrity and dependability to a water supply for fire fighting.

2. The amount of water available can be determined accurately and can be checked periodically independently of other water demands.

3. The system is generally free from interruptions due to maintenance and alterations prevalent with general-purpose systems.

4. Completely separated systems can make use of large or unlimited water supplies of a quality lower than is permissible for combination with domestic- or process-use systems.

5. The continuous flow in domestic and process systems increases internal corrosion, tuberculation, and accumulation of flow-obstructing material.

Special conditions that may take a combined piping system acceptable are:

1. The domestic or process use is small or can be taken from near the supply end of the fire system which corresponds to an extension of a public main.

2. All water supplies are clean and of purity meeting health and other requirements.

3. Domestic or process demands can be readily discontinued, so that the entire supply is available for fire fighting.

In some instances a source of water unsuitable for direct interconnection with a public supply can be supplemented by using the public supply to fill tanks or reservoirs on the fire systems through an open fill connection or by having cross connections equipped with special-type double-check valves or other back-flow-prevention devices acceptable to health authorities.

DESIGN OF PIPING SYSTEMS

Location and Arrangement of Underground Piping

Water-supply mains are usually underground and outside buildings for several reasons:

1. They are out of the way of aboveground operations which can cause mechanical injury to the mains.

2. Changes or repairs to the mains do not interfere with use of buildings and can be made more easily and quickly.

3. No precautions other than proper depth of ground cover are needed for protection against freezing.

4. In case of breaks, water damage to building contents is much less likely.

5. The connecting of hydrants and the locating of outside control valves is easier and more economical.

6. Installation of mains in advance of building construction makes hydrant protection possible during construction and allows prompt connection of sprinklers to the water supply.

Combination of Multiple Sources.   When two or more water supplies feed into a single piping system, it is important to consider hydraulic performance so that the supplies may combine in the most effective ways.   In general this calls for the two supplies to be connected so as to flow from two directions toward points of demand. This arrangement minimizes friction losses and also makes the best use of supplies at different pressures and frequently, by proper valve arrangements, increases the dependability of the piping system.

Loop Piping.   For installations which supply a number of buildings, buildings large enough to have hydrants, or sprinkler or standpipe connections on more than one side of a building, a loop piping arrangement divides the flow to any one location, reduces friction losses, and if properly valved, improves reliability.

The choice of location for underground mains involves consideration of the economical use of available or planned water supplies; effective distribution of water to the points of use; avoiding interference with buildings and other structures; economy of installation; accessibility for extensions, alterations, and repairs; soil conditions; and grades and routes favorable to good hydraulic performance.

Size of Pipe.   Water mains should be large enough to carry the maximum required flow with the minimum practical loss in pressure. Consideration must be given to lengths of lines, and allowances must be made for internal corrosion.

Mains should be 8 in. or larger.   In most cases the cost of laying pipe does not increase in proportion to the diameter of the pipe and its carrying capacity.   For the same flow rate the friction loss in 8-in. pipe is less than one quarter of that in 6-in. pipe.   Such a difference can be very important in assuring that adequate pressure is available to the fire-protection devices.

Friction Loss and Hydraulic Calculations.   In planning fire-service water piping, allowances must be made for friction losses.   Except in the case of a system which has permanently fixed discharge openings and known water supplies, the rate of water flow will not be known accurately in advance.   For the usual situation a maximum rate of flow must be assumed in order to predict the probable hydraulic behavior.

Friction loss, upon which water-carrying capacity depends, can be calculated by commonly-used formulas such as those of Williams and Hazen, Chezy, Fanning, and Darcy-Weisbach.

Uncertainties in the assumed values of the design parameters are frequently of more significance than the differences in results produced by the various formulas.

Table I.   Values for c in Williams and Hazen Formula

Type of pipe	Value of c
New, or in condition of new, pipe	
Cast iron—unlined...............	120
Cast iron—cement lined	130
Cast iron—bituminous lined.......	140
Asbestos-cement.................	140
Steel	140

The Williams and Hazen formula is generally accepted in fire-protection and water-works practice. The friction-loss values given in Table 3 are based upon that formula which is

$$f = 4,524 G^{1.85}/(c^{1.85} d^{4.87})$$

in which f = friction loss, psi/1,000 ft of pipe
  G = rate of flow, gpm
  d = internal diameter of pipe, in. (actual)
  c = Hazen-Williams coefficient

The coefficient c has been determined by tests made with pipes of different types and materials and with various degrees of corrosion or other obstruction. Commonly used values for c are given in Table 1.

The effect of increasing age upon the coefficient of unlined cast-iron pipe in public mains or subject to domestic or process flows can be expected to be about as shown in Table 2. A coefficient $c = 100$ is commonly assumed for average conditions. If water is unusually corrosive, the coefficient may be reduced to 75 or even 60 for 15-year-old pipe. Pipe in which the flow is not continuous will deteriorate less rapidly.

Corrosion affects the carrying capacity of small pipe more rapidly than it does large pipe. *Sprinkler-system steel pipe* is usually sized to allow for a value of $c = 100$ for 2-in. or smaller pipe and $c = 120$ for pipe larger than 2 in.

Location of Hydrants. Outdoor hydrants are installed to provide hose stream protection for buildings, for outdoor combustible material, for protection against exposure fire, and to hasten final extinguishment of fire in high-piled or smoldering-type materials. They can also provide a source of water for temporary protection if the fixed types of protection are impaired. Hydrant and hose stream protection supplements, but is not a substitute for, automatic sprinkler protection.

Hydrants should be of types approved by recognized fire-testing laboratories. Construction and performance specifications of the NFPA, AWWA, and testing laboratories are in essential agreement.

Hydrants with two 2½-in. hose outlets are ordinarily used and are spaced 200 to 250 ft apart. For especially hazardous yard storage the spacing may need to be

Table 2.   Variation in Williams and Hazen Coefficient c with Age of Unlined Cast-iron Pipe

Age of pipe, years	Value of c
New	120
10	105
15	100
20	95
30	87
50	75

Table 3. Friction Loss in Pipe[1]

(*f* in Williams and Hazen Formula)

Loss in pressure in psi per 1,000 ft of pipe $c = 100$[2]

Flow, gpm[3]	Inside diameter of pipe, in. (nominal)[4]						
	4	6	8	10	12	14	16
50	1.47	0.204					
60	2.06	0.285					
75	3.11	0.430					
100	5.29	0.735	0.181				
125	8.0	1.11	0.273				
150	12.1	1.55	0.381				
200	19.1	2.65	0.652	0.220			
250	28.8	4.00	0.985	0.332			
300	40.4	5.62	1.38	0.466	0.192		
400	68.8	9.55	2.35	0.793	0.326	0.154	
500	104.0	14.4	3.53	1.20	0.493	0.233	0.122
750	220.0	30.5	7.52	2.54	1.04	0.493	0.257
1,000	375.0	52.0	12.8	4.32	1.78	0.839	0.438
1,500	.....	110.0	27.1	9.14	3.76	1.78	0.928
2,000	.....	......	46.2	15.6	6.41	3.03	1.58
2,500	.....	....	.....	23.5	9.69	4.57	2.39
3,000	.....	......	.....	33.0	13.6	6.41	3.34
4,000	.....	......	.....	.....	23.1	10.9	5.69
5,000	.....	......	.....	.....	34.9	16.5	8.60

[1] The friction loss in smaller and larger sizes of pipe is given in tables in the NFPA "Handbook of Fire Protection," Ref. 1, and the Factory Mutual Handbook of Industrial Loss Prevention, Ref. 2, in the list of references on page 12–2.

[2] For values of *c* different from 100, multiply the flow rates listed in Table 3 by the conversion factors listed in Table 4.

[3] For approximate interpolation between the *tabulated flows*, calculate the actual flow as a per cent of the next lower tabulated flow, convert this per cent to the given value in Table 5, and take this per cent of the friction loss corresponding to the lower tabulated flow.

[4] Nominal pipe diameters were used in calculating the tabulated friction losses. The inside diameter of pipe conforming to American Standards Association or American Water Works Association Standards will vary with the process of manufacture, material, and pressure classification and can be found in the Standards. Some published tables of friction losses are based upon actual internal diameters. The error caused by minor differences in internal diameter are usually not significant when considered in relation to the uncertainties in assumed coefficients and rates of flow. If desired, adjustment for differences in diameter can be made by applying a factor $d(\text{nominal})^{1.85}/d(\text{actual})^{1.85}$ to the values given in Table 3.

Table 4. Conversion for Different Values of c in Williams and Hazen Friction-loss Table

Value of *c*	Conversion factor
140	0.537
130	0.615
120	0.714
110	0.836
100	1.000
90	1.22
80	1.51
70	1.93
60	2.57

Table 5. Interpolation for Water Flows between Values Given in Table 3

Actual flow as % of next lower tabulated flow	Converted % to be applied to friction loss at next lower tabulated flow
105	110
110	120
115	130
120	140
125	151
130	162
135	174
140	186
145	198
150	211
160	238
180	296
200	361

Example: 8-in. pipe, flow 650 gpm.
Next lower tabulated flow 500 gpm.
Per cent of lower flow = 650 ÷ 500 = 130 per cent.
Converted per cent = 162.
Tabulated lower friction loss at 500 gpm = 3.53.
Approximate actual friction loss = 3.53 × 130 per cent = 4.58 psi/1,000 ft.

reduced to 100 to 200 ft. If large quantities of water from hose lines or monitor nozzles are to be used, as at lumber storage or log piles, hydrants having three hose outlets (three-way hydrants) are often necessary.

For private fire systems, hydrants with large-size fire department suction outlets should not be used because fire department pumpers should not take suction from private fire systems which supply water to automatic sprinklers.

Hydrants are usually located 40 to 50 ft away from buildings to reduce the danger to fire fighters from heat, smoke, and falling walls.

If buildings are windowless, hydrants should be located opposite building entrances. Fixed-location monitor nozzles having a pipe connection to water mains are often installed around log piles at paper mills and at other hazards where prompt application of water by large streams, operable quickly by one person, may be needed.

Location of Connections to Sprinkler Risers. The location of connections to sprinkler systems is determined by the arrangement of the inside sprinkler piping and usually is known only after preliminary sprinkler layouts have been made. Inside sprinkler piping is planned to conform to the well-established Standards for the Installation of Sprinkler Systems.

Connections from buried mains to sprinkler-system risers are usually 6 in. in size, although 8-in. connections may be most practical and economical for large numbers of sprinklers.

Sprinkler control valves should be located 25 to 40 ft from building walls and equipped with indicator posts. If yard space is lacking, the control valve may be located in a valve pit, in a fire-resistant stair tower, or in some other enclosure accessible from outdoors.

PROTECTION FOR WATER MAINS

Freezing. Underground mains must be protected from low temperatures. Burying pipes below the lowest frost line is the usual method.

Normally there is no circulation of water such as that which exists in public water-system piping. Exterior insulation, other than that added above the ground surface to raise the frost line, is likely to be ineffective. Exposed short sections of pipe must be boxed or wrapped and heated. Insulation affords economy in heating but alone does not provide adequate protection. The depth of cover for underground water mains to avoid freezing in different regions in the United States and Canada is shown in Fig. 1.

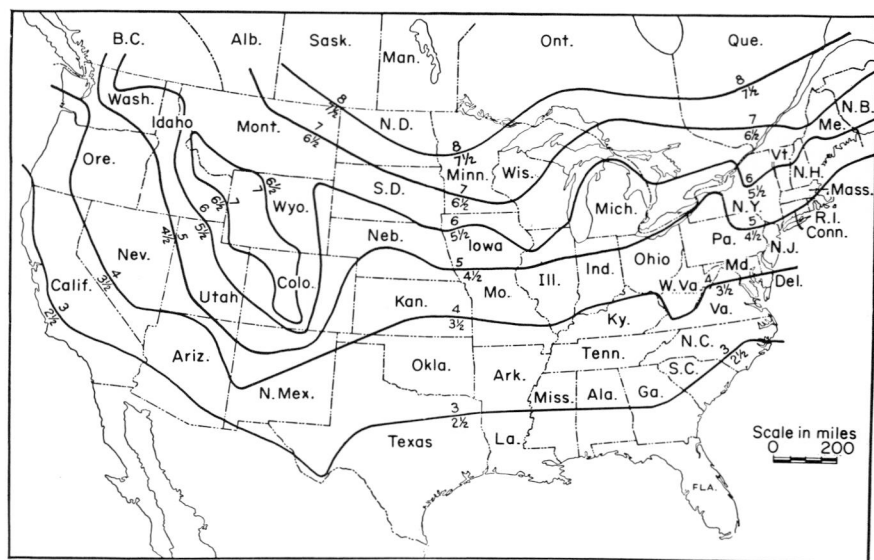

FIG. 1. Depth of earth cover recommended to avoid freezing of underground fire-protection water mains. Public water mains are usually considered safe with ½ ft less cover. (*Courtesy of the National Fire Protection Association.*)

Mechanical Injury. Underground pipe is susceptible to breakage as may be caused by surface loads, uneven trench loads, or settlement.

Pipes under driveways should have a cover of at least 3 ft to prevent loads from traffic above. Piping installed beneath railroad tracks, or where it is subjected to vibration or shocks from above, requires a cover of 4 ft or more. Where connections from mains pass through building walls, there should be sufficient clearance to avoid loads which result from building settlement. At any location subject to washout by flood, special covering or other precautions will be needed.

TYPES OF PIPE

The appropriate type of pipe is determined by consideration of economical laying, trouble-free service, and easy maintenance. These factors include methods of trenching and laying, corrosiveness of the water to be handled, and soil corrosion.

Pipe should be of a type approved by recognized fire-equipment testing laboratories. Cast iron, steel, or asbestos cement may be used. The use of steel pipe is mostly limited to aboveground portions of fire-protection systems except that some systems utilize steel pipe which has been coated and wrapped externally to prevent soil corrosion.

The class, or pressure rating, is determined by the maximum working pressure to which the pipe will be subjected.

Cast-iron Pipe.   The minimum size of cast-iron pipe for fire-protection systems is usually 8 in., except that 6-in. pipe is accepted for connections to single two-way hydrants and to 6-in. sprinkler risers and standpipes.

Cast-iron pipe and fittings meeting the following specifications are suitable for fire protection use:

1. American Standard for Cast-Iron Pit-Cast Pipe for Water or Other Liquids, ASA AZ1.2; also AWWA C-102 and Federal Specification WW-P-421a.

2. American Standard for Cast-Iron Pipe Centrifugally Cast in Sand-lined Molds for Water or Other Liquids, ASA A21.8; also AWWA C-108, and Federal Specification WW-P-421a.

3. American Standard for Cast-Iron Pipe Centrifugally Cast in Metal Molds for Water or Other Liquids, ASA A21.6; also Federal Specification WW-P-421a.

4. American Standard for Short-body Cast-iron Fittings, ASA A21.10; also AWWA Specifications for Cast-Iron Pressure Fittings, AWWA C100-55.

Corrosion-resistant Lining for Cast-iron Pipe.   To resist highly corrosive water, cast-iron pipe may have a cement-mortar or bitumastic lining.   The cement lining should meet the American Standard for Cement-Mortar Lining for Cast-iron Pipe and Fittings, ASA A21.4.   Specifications for bitumastic linings are agreed upon by the manufacturers and testing laboratories or users.

Asbestos-cement Pipe.   This pipe is manufactured from asbestos fiber and cement in accordance with American Water Works Association Standards for Asbestos-cement Water Pipe, AWWA C-400.   Also Federal Specifications SSP-351a is accepted for fire-protection mains.   It is usually more resistant to internal corrosion than is cast-iron pipe.

Class 150 pipe is used for pressures not exceeding 150 psi, and Class 200 for pressure up to 200 psi.

The manufacturing process makes it impractical to produce bell-end pipe. Each length is a symmetrical cylinder with ends machined accurately and beveled for insertion in a cylindrical asbestos-cement coupling grooved to retain rubber rings which are compressed when the pipe end is forced into the coupling.   Such joints provide little resistance to separation if subjected to water pressure and must be anchored where pressure forces occur.   They provide some flexibility and allow a deflection of 4 to 5 deg at each joint.

Some makers of asbestos-cement pipe also furnish a limited variety of fittings, including elbows, tees, crosses, gate valves, and hydrants with bells corresponding to the asbestos-cement couplings, which allows their use without adaptors. Regular cast-iron bell-type fittings can also be used with a lead or compound joint.

Steel Pipe.   Steel pipe is seldom accepted for underground fire service, although it is frequently used for large supply lines and trunk mains in public water systems. The high strength and shock resistance of steel pipe may make it desirable for use where it would be subject to vibration or shock as from earthquake, under railroads or highways, and in unstable soil or on steep slopes where there is a tendency for movement.

Steel pipe should conform to American Standard Specification ASA B36.1. Standard weight (Schedule 40) is used unless pressures exceed 300 psi.   In such a case extra-strong pipe (Schedule 80) would be required.

Couplings are of the usual threaded or flanged type unless service conditions require some flexibility, in which case approved flexible, grooved couplings using rubber rings may be used.   The approved flexible couplings will not allow the pipe to pull out as the result of internal pressure.

SOIL CORROSION

External corrosion may cause rapid deterioration and weakening of buried iron or steel pipe. Severely corrosive soil should be avoided if possible, but if not, special precautions are required in order to prevent corrosion failure.

Acids, sulfur, and numerous chemicals in or on the ground may be difficult to avoid, as in cinder fill or in the vicinity of coal piles or pickling or chemical processes.

Protection against moderately corrosive conditions can be afforded by using noncorrosive backfill, a heavy coating of asphalt, and a paper or fabric wrapping.

Much of the corrosion of cast-iron pipe is the result of electrolysis caused either by some soil condition or by stray electric current. Such corrosion is likely to be localized in areas where the original pipe coating is deficient or injured. The iron in the casting is removed by solutions leaving soft graphite spots of little strength, even though the pipe may not appear to be severely affected. Heavy bituminous coating, reinforced by fabric wrapping, is effective. Asbestos-cement pipe is not subject to electrolysis and is resistant to most corrosive agents.

LAYING UNDERGROUND PIPE

All the operations in laying underground pipe, including preparation of the trench, placing and aligning the pipe, making up the joints, anchoring, testing, and backfilling, should be performed only by experienced, careful workmen. Inferior workmanship will not be accepted by inspecting authorities.

Only the major features of pipe laying are included here. Experience and familiarity with every detail of making joints in bell-and-spigot pipe, as well as with the special procedures specified by the makers of mechanical joint pipe and asbestos-cement pipe, are necessary.

Trenches for Underground Pipe. Trenches should be carefully excavated as short a time as practical before the pipe is laid in order to minimize crumbling walls or cave-in. In sandy or loose soil, sheeting and bracing may be necessary. The bottom of the trench should conform to the grade of the pipeline. With poor soils it may be necessary to excavate to an extra depth and prepare a stable pipe bed by a layer of firmly compacted soil, by a concrete mat, or in, exceptional cases, by piles and concrete or timbers. Pipe should not be laid in contact with rocks or boulders.

The trench should be wide enough to allow careful alignment of the pipe and convenient making up of the joint. Space should be provided below joints so that the joint can be made properly and so that there will be no localized bearing.

Before the pipe is placed in the trench, it should be carefully examined, its joining areas cleaned, and any foreign material removed completely from the inside.

No type of joint should be made up with any water or dirt falling in the trench or during unfavorable weather.

After being laid and joined, the pipe should have a firm bearing for its entire length except that the joints must be left exposed for inspection during the pressure test.

Poured Joints for Bell-and-spigot Pipe. The sequence of operations is: (1) Insert, seat, and center the spigot in the bell end. (2) Insert the yarn or other packing material, and tamp it with a yarning tool. This holds the spigot in the center of the bell. If pipe is being laid to a slight bend or curve, the width of the space on the open side of the spigot should be kept narrow enough to prevent packing material being forced into the pipe channel. Also the space between the

outer end of the bell and the spigot end at its narrowest point should be kept adequate for yarning, tamping, and calking.   Multiple strands of packing should be tamped separately.   A depth of about $2\frac{1}{2}$ in. should be left for the lead or joint compound.   (3) Place the flexible asbestos joint runner around the pipe, fitting it snugly against the end of the bell, and form a pouring gate of clay.   (4) Pour the joint.   Lead should be heated to a temperature at which a freshly exposed surface changes color rapidly by oxidation, but it should not be overheated.   The lead should be clean, and the surface free of oxide or dross.   Each joint should be completely filled by one pour of lead.   (5) Calk the solidified lead by hand or pneumatic tools sufficiently to assure a tight joint without overstraining the bell.

If a sulfur compound is used in place of lead, the directions of the manufacturer should be followed closely.   The compound must not be overheated, since this will cause ignition and burning.   It should be thoroughly stirred to uniform fluidity and should be free from foam or foreign materials.   To assure complete filling of the pouring space, a funnel or gate extending about 8 in. above the top of the joint should be used.   Although jointing compounds are less expensive than lead, they cannot be calked and have a seasoning period before becoming fully leaktight.   This may delay undesirably the test for tightness.

Mechanical-joint Pipe.   The standardized mechanical joint for cast-iron pipe has a bolting flange on the end of the bell, a spigot end without any bead, and a follower or gland which is slipped over the spigot before assembly and which has boltholes matching those on the bell-end flange.   A rubber ring used for sealing is compressed in the space between the bell and spigot by drawing up the bolts.   This joint allows slight angularity between lengths of pipe.   Tee-head bolts of high-strength cast iron are used.

Single-gasket Pipe Joints.   The type of cast-iron pipe joint that is easiest to assemble in the field consists of a bell end having a groove near its outer end, a preformed rubber sealing ring, and a smooth spigot end.   The rubber ring is placed in the cleaned bell groove; the spigot end is cleaned and lubricated and forced into the spigot until seated.   The bell is flared slightly to allow a small angularity of the joint.

Asbestos-cement Pipe Joints.   Lengths are joined by forcing the ends into couplings which have one groove near each end for a single compressible rubber ring.

If precut, the end must be machined to enter and to fit properly in the coupling.

Pipe ends are cleaned, the rubber ring inserted in the fitting, the spigot end lubricated, and the joint drawn together by a special puller or other careful application of longitudinal force.

Cut lengths of pipe without machined ends can be used with standard mechanical joints or by poured and calked joints, using lead or compound with bell-and-spigot pipe.

Ball and Socket Pipe Joints.   Ball-joint pipe is intended to provide considerable angular flexibility while laying and to allow preassembly of lengths of pipe before it is lowered into a difficult trench or under water.   Both ball and socket ends are machined to segments of matching spherical surfaces.   The bell end is flanged for bolting and recessed for a reinforced rubber packing ring.   A flanged follower ring is slipped over the spigot end, and a split ring of internal diameter smaller than that of the spherical end of the pipe is inserted in a groove in the follower.   The joint is drawn tight by the flange bolts.   Tee-head bolts of high-strength cast iron are used.

Flanged Cast-iron Pipe.   For fire-protection systems, flanged cast-iron pipe is used mainly for the aboveground installations of equipment likely to be changed or removed for alteration or repair.   Except where pipe is buried underground, the

larger sizes, such as 4 in. and over, of cast-iron valves and fittings and the connections to fixed equipment, such as tanks, pumps, meters, and special fire-protection devices, have flanges in accordance with ASA Standard Cast Iron Flanges and Flange Fittings, ASA B16.1, Class 125. The corrosion of bolts, a troublesome feature in buried equipment, is seldom an important factor in aboveground piping.

Flanged connections have no flexibility and require accurate alignment to avoid objectionable stresses.

Anchorage of Water Piping. Most of the types of joints other than ball and socket used with cast-iron pipe are not designed to resist forces which tend to pull them apart. Such forces are produced primarily by the pressure of the water, but they may be due also to the weight of pipe if it is laid with a steep slope. Friction between the pipe and the ground tends to resist movement but is uncertain in amount and is seldom given recognition unless it can act upon a full length (12 ft) of pipe.

Joints at bends, tees, and plugs should be anchored by clamps, rods, and bolts or by concrete thrust blocks to resist the force that would be produced by a test pressure of at least 200 psi. If the static pressure will exceed 150 psi, the test pressure should be 50 psi greater than the static pressure. Fittings for cast-iron pipe of the bell-and-spigot, mechanical-joint, or single-gasket-joint type can be obtained with cast-on lugs for anchorage by tie rods. Prefabricated tie rods and clamps are available from the larger sprinkler installation contractors.

Anchorage rods should be extended to the second joint in a line of pipe in which a separating force exists unless it is in firmly tamped fill a distance of at least 12 ft.

With asbestos-concrete pipe, poured-in-place thrust bocks of concrete are used instead of clamps and tie rods because such pipe is not suitable for the application of stresses produced by clamps.

Concrete thrust blocks must be located to resist the force or combination of forces which tends to produce joint separation.

Allowable horizontal bearing loads for concrete against a trench side of undisturbed soil firm enough to be trenched without sheeting will be 1½ to 3 tons/sq ft depending upon the rigidity of the soil. Thrust blocks should be located so that they will not prevent access to joints.

The total force acting to move a plug from the end of a line or to separate a branch from a tee in a line of pipe is produced by the water pressure acting upon the gross area of the plug or the branch pipe (area corresponding to the outside diameter.)

Calked joints in bell-and-spigot pipe are quite rigid and can absorb some of the force tending to separate the joint. Consequently, the forces used in designing anchorage for a lead-calked joint are assumed to be about 75 per cent of those to be resisted by the anchorage of a gasket-type joint.

Table 6. Forces, in Pounds, to be Resisted by Anchorage at Pipe Joints

Size, in.	Outside diam, in.	Separating force produced by 200 psi at tee and plug	Resultant of forces at bends		
			90 deg (1.41)	45 deg (0.765)	22½ deg (0.385)
6	6.90	7,500	10,600	5,740	2.890
8	9.05	12,900	18,200	9,880	4,970
10	11.10	19,400	27,400	14,850	7,480
12	13.20	27,300	38,200	20,850	10,520
14	16.65	38,400	54,200	29,400	14,800
16	17.80	49,700	70,100	38,000	19,150
20	22.96	76,400	108,000	58,400	29,450

When joints are anchored by clamps, rods, and bolts, the total force acting in any direction is assumed to be taken by the anchor loads acting in this same direction, each joint being considered independently.

When joints at bends are anchored by thrust blocks, the resisting force can be taken as a single resultant of the forces tending to separate each of the adjacent joints or as two forces acting independently.

Table 6 gives (1) forces tending to separate single joints, as at plugs, at tees, at the base of hydrants, and at bends having only one joint that needs anchorage, and (2) the resultant of the two forces acting at the ends of an elbow or bend.

With standardized mechanical joint pipe or with a bolted flange joint opposite the joint needing anchorage, it is possible to use the threaded end of the anchorage rod to serve also as the flange bolt. Bolt thread couplings and extra-length coupling bolts may also be used for connecting anchor rods. When a flange bolt

Table 7. Safe Loads, in Pounds, for Bolts and Threaded Rods

Diameter of rod or bolt, in.	Threads per in.	Net area of threads at root, sq in.	Load on bolt or rod at listed assumed safe loading stress in at root of thread, psi		
			20,000	30,000	40,000
½	13	0.126	2,520	3,780	5,040
⅝	11	0.202	4,040	6,060	8,080
¾	10	0.302	6,040	9,660	12,080
⅞	9	0.419	8,380	12,570	16,760

is used with standard flanged pipe or when the flange used with mechanical joint pipe serves also as the load-carrying member of an anchorage rod, allowance must be made for the total bolt load between flanges that are tightened against a resistant rubber ring or gasket. The initial load is increased by the amount of the anchorage load.

It is not practical to establish detailed standards for anchorage methods for all installation conditions. Consideration must be given to the maximum forces produced by pressure, the direction in which the forces act, and allowances for friction and other favorable conditions, in conjunction with the resistance that can be provided by various anchorage methods. Experience and engineering judgement thus establish safe arrangements.

Hydrostatic Leakage Tests

Newly laid pipe should be tested for tightness and to detect faulty pipe or fittings. To avoid movement under pressure, cast-iron pipe should be backfilled and tamped to above the center line. Joints should be left completely exposed. Asbestos-cement pipe should be completely covered with tamped backfill to a depth of 1 to 2 ft except at the joints.

The usual test pressure for 150 class pipe is 200 psi. If the service pressure requires a stronger pipe, the test pressure should be 50 psi above the service pressure.

All joints and the pipe itself are inspected for leakage while the pipe is under test pressure. A small amount of leakage is permitted at joints. The amount is based upon the combined length of all the joint seals in the entire length of pipe under test and should not exceed ⅕ gal per 24 hr per inch of diameter of each joint. Rubber-ring-packed joints generally have less leakage than lead- or compound-poured

Table 8. Thrust Block Loads in Pounds for Asbestos-cement Pipe (Transite Ring-Tite)

(Class 150, 200-psi test pressure)

Size, in.	Outside diam, in.	Separating force at plug or tee	Resultant force at bend		
			90 deg (1.41)	45 deg (0.765)	22½ deg (0.385)
6	7.07	7,600	10,700	5,800	2,900
8	8.27	13,200	18,600	10,100	5,100
10	11.82	21,500	30,400	16,500	8,300
12	14.08	30,600	43,300	22,400	11,900
14	16.38	41,500	58,700	31,800	16,100
16	18.62	53,700	76,000	41,200	20,900
20	23.44	86,400	121,000	66,200	33,300

joints.   Poured compound joints have relatively high initial leakage, which diminishes as the joint ages.   A period of 14 days is allowed at which time the leakage should not exceed that for a lead-calked joint.

If compound joints are used at cast-iron fittings attached to asbestos-cement pipe, the time factor must be considered in making the leakage test.

Backfilling. Backfill should be free from cinders, refuse, vegetable material, and rocks and of a type that will compact firmly. It should be deposited evenly across the trench and should be tamped in layers of only 3 to 4 in. Backfill may be compacted by puddling after one or more layers of fill above the top of the pipe have been well tamped.

ABOVEGROUND PIPING

Aboveground fire-service piping consists mainly of (1) distribution piping to the outlets on a sprinkler or spray system, (2) hose standpipes, (3) piping to elevated tanks, (4) fire-pump installation, (5) special extinguishing system such as foam or carbon dioxide.

Automatic Sprinkler System Piping. Complete standards for piping for automatic sprinklers, deluge and open sprinkler systems for indoor protection or for outdoor exposure are given in the National Fire Protection Standard for the Installation of Automatic Sprinklers and are also included in the "Factory Mutual Handbook of Industrial Loss Prevention."

Hose Standpipes. Hose standpipes facilitate the use of large hose streams in multistory buildings and in windowless buildings. They should be located in stair towers or near fire escapes and spaced so that any part of each story can be reached by a stream from a 2½-in. hose not more than 100 ft in length. Standpipes of steel pipe at least 4 in. in size are usually installed in five- or six-story buildings, and 6-in. standpipes in taller buildings.

Hose standpipes are preferably under water pressure at all times but in cold locations may have remotely controlled valves operated automatically or manually from all hose valve locations.

Complete information about piping for hose standpipes is given in the NFPA Standards for the Installation of Standpipe and Hose Systems.

Fire Department Connections. Where fire departments can draw upon a water supply independent of the property fire system, a clearly marked fire department connection is located where it will be easily reached by fire department pumpers.

Gravity Tank Piping. The design, construction, and installation of gravity water tanks for fire protection are covered in complete detail in generally accepted

standards prepared by insurance and other regulatory bodies.   Standards are in essential agreement if not identical and are available as the NFPA Standard for Water Tanks for Private Fire Protection.   Similarly named standards are included in the "Factory Mutual Handbook of Industrial Loss Prevention."

The tank contractor, in complying with the specifications, usually supplies and installs a complete tank unit, including the tank, supporting tower, and foundations, and all piping connections and accessories such as the tank riser, valves, tank heater, tank-filling connections, a pump if necessary, and an altitude gage.   A well-supported base elbow or corresponding provision is made for connecting to the plant fire-protection system.

Many modern elevated water tanks for fire service have large risers (36 in. diameter) which are constructed as a part of the tank.   Piping is limited to the connection from the base of the riser to the fire service piping system and to provisions for filling and heating in cold climates.

Some steel tanks and all wooden tanks have steel riser pipes of 6 to 10 in. diameter which require a frostproof casing in locations which are subject to freezing temperatures.

Pressure-tank Piping.   All piping, valves, and accessories called for by the recognized Standards for the construction and installation of pressure tanks are a part of the tank installation and are not treated here.   A proper discharge fitting is provided for the attachment of the fire-protection-system piping.   Details are given in the National Fire Protection Association Water Tank Standard.

Fire-pump Piping.   Fire pumps for new fixed installations are most frequently of the centrifugal type.   Usually the needed pressure can be obtained with a single-stage pump.

With a deep well supply or where the water level in a pit or sump would require a suction lift of more than 15 ft, a vertical-shaft, submerged, multistage turbine-type pump has desirable features and does not require priming facilities.

Ordinarily a fire pump, its base, and driver are supplied as a single complete unit by the pumpmaker.   The National Fire Protection Assocaition Standard for the Installation of Centrifugal Fire Pumps includes information covering fire-pump piping connections.

The piping associated with a fire-pump installation is usually a contract installation made independently of the purchase of the pump unit.   Typical piping arrangements are shown in the Standards.

Horizontal Shaft Pumps.   For fire service a centrifugal pump should have a total suction lift not exceeding 15 ft when operating at 150 per cent of rated capacity. The calculated lift should include the distance from the lowest suction level to the pump center line and also the friction losses in the foot valve, elbows in the suction line, and necessary length of suction pipe.   Charts from which suction pipe sizes can be determined are shown in the installation standard referred to previously. Determining factors are the length of the suction pipe and the rated capacity of standard-size fire pumps (500, 750, 1,000, 1,500, 2,000, and 2,500 gpm).   With the length of the suction pipe at 0 ft the basic losses include the friction allowance for a foot valve, three elbows, and the velocity head.   The chart values assume a flow of 150 per cent of the pump rated capacity.   The friction loss caused by the length of the suction pipe is determined from Williams and Hazen hydraulic tables using a coefficient $c = 70$.   Outdoor suction pipe is usually buried and of the cast-iron bell-and-spigot type with calked lead joints.   Flanged cast-iron pipe or mechanical-joint pipe may be used; if joints are buried, the bolts must be of corrosion-resistant material.   Asbestos-cement pipe should not be used if the suction is subatmospheric at any operating condition.

The suction piping should have the same pressure rating as the fire system piping.

Each fire pump whose suction is at subatmospheric pressure should have an independent suction line from the water source to the pump. Pumps which take suction at pressures greater than atmospheric may have a common header if the piping arrangement is symmetrical so that the flow conditions to each pump will be essentially the same.

Suction lines operating under a negative suction head should have a continually ascending grade up to the pump connection in order to avoid air pockets. Tapered reducers, if needed at the pump connection, should be eccentric and installed so as not to produce an air pocket.

Suction piping should be buried to prevent freezing with additional precaution being observed at entrance to the water source.

Priming Connections. Priming facilities are required for horizontal shaft pumps that may, at any time, have subatmospheric suctions. Two dependable methods should be provided.

One priming source for a pump that is to be started automatically or remotely should keep the pump primed at all times.

No priming method should allow contamination of a potable water supply by water from a nonpotable fire pump source.

The following priming arrangements are ordinarily acceptable:

1. An automatically filled priming tank of capacity equal to that of the pump and suction piping but of not less than 100 gal in capacity.

2. A manually filled priming tank of capacity of at least three times that of the pump and suction piping but of not less than 250 gal in capacity.

3. A connection to a domestic or process water system. If this supply is from a gravity tank, it is desirable that sufficient water for fire pump priming should be reserved by appropriate location of the domestic service connection at the tank.

4. An exhauster or ejector having a reliable power supply.

5. If the primary water supply to the fire-protection system is from a gravity tank of large capacity, a second, but not the initial, priming source may be provided by a bypass around the pump-discharge check valve.

Vertical-shaft Turbine-type Pumps. A submerged turbine-type pump is advisable or necessary if it is not practical to limit the operating suction lift to 15 ft, if the water source is a deep well, or if a pump having a subatmospheric suction is to be automatic. Standards for installation are included with those for horizontal shaft pumps.

A vertical-shaft pump is usually furnished by its manufacturers as a complete unit consisting of pump, column, driver, and all accessories as approved by a recognized testing laboratory. The driver may be a direct-connected, vertical-shaft motor or an internal-combustion-engine combination of pump, engine, base, and right-angle gear drive.

A standard discharge flange is provided for attaching the fire-system piping.

Fire-pump Discharge Connections. The fire-pump discharge to the fire system should have a discharge check valve and outside stem and yoke gate valve in the connection near the fire pump.

Water relief valves are not ordinarily considered necessary with constant-speed pumps which are not provided with positive suction head and which produce a shutoff pressure that can safely be withstood by the piping of the fire system. Pumps which serve as booster pumps with pressure from public mains or other sources having pressures not adequate to meet fire pressure and volume requirements may produce undesirably high pressures at low rates of discharge. Relief-valve arrangements for such situations require special consideration. A relief valve which discharges to the pump suction is not effective in limiting the pressure produced by a booster pump.

Pump Hose Valves. For testing purposes and to provide hose streams independent of other water supplies, fire pumps are usually equipped with a hose-valve header connected between the discharge check valve and the discharge gate valve. If the header is located outdoors, precautions against freezing, usually a shutoff valve and automatic drain, are needed.

Fire-service Connection to Public Mains. Private systems of water piping for fire protection obtain part or, in the case of the smaller properties, all of the water for fire protection from public mains if they are available and if the amount of water they can supply is adequate in amount and pressure. For properties large or important enough to require two supplies, the primary supply is usually from the public mains. If there is no secondary supply, the connection is simple and direct, with a water service valve at the main and a private control valve. A meter may or may not be required as discussed later.

Most public water departments allow private fire-service connections up to the size of the public main. In some cases, the connections must be one size smaller. The planning of a fire-protection system must include information concerning regulations of the local public water authorities.

If the property has a connected secondary supply, the combination of the two constitutes a cross connection by which water from the private source, possibly nonpotable or not acceptable to health or water-system authorities, might be introduced into the public system unless made impossible by a backflow-prevention device that is acceptable to all the interested authorities.

Ordinarily it is recommended practice to avoid any new connection between a public and a private system regardless of the quality of the private water supply. In situations where this is unavoidable by more practical measures, the backflow-prevention device must be used.

Most prevalent is the use of two duplicate, specially made, all-bronze, rubber-seal check valves in the private system near its connection to the public supply. Such valves are approved for the service by recognized testing laboratories. Valves and test gages are provided by which the absolute tightness of the equipment can be regularly tested and certified.

To permit easy inspections and maintenance, the check valves and their control valves and testing accessories must be located in a dry, clean pit or within the basement of a building.

Other backflow-prevention devices which utilize a variation of the double-check-valve principle have been developed and are accepted by health authorities in some localities. The acceptability of any backflow-prevention equipment should be determined while a system is being planned if inter-connection of water supplies is involved.

Meters. With very few exceptions, public-water-system authorities do not require meters for water systems, which are limited to fire-protection use. If meters are required, they should be of a type approved for fire service. The ordinary commercial-type meter used for industrial or domestic services introduces a friction loss or a susceptibility to obstruction that is too great to be acceptable in most instances.

Fire-service meters are of two general types: (1) those that record the total flow by the use of a small bypass meter for low flows and a proportional meter for large flows and with an essentially full-size unobstructed waterway for large flows and (2) combination of a full-size check valve with weighted clapper or other means of developing a pressure sufficient to cause small flows to pass through a small recording meter. The detector check metering devices are accepted on the basis that fire systems normally have no flow or a small leakage flow, that no charge is to be made for water actually used in fire control, and that frequent reading of the

small flowmeter will indicate improper use of water so that it can be investigated. The full-flow registering meter is relatively expensive.

CHEMICAL FIRE-EXTINGUISHING SYSTEMS

Carbon Dioxide Fire-protection Piping. The design of piping systems for the distribution of carbon dioxide or other inert gas for the control of fire is highly specialized and is not subject to the procedures applicable to water protection systems. Inert-gas systems are ordinarily designed by the suppliers of the equipment. Detailed plans must be approved by insurance or other authorities.

General information relating to the pipe, fittings, valves, nozzles, and methods of installation is available in the National Fire Protection Assocaition Standard for Carbon Dioxide Extinguishing Systems.

Foam Fire-protection Piping. Foam, whether produced by the combination of chemicals or by the mechanical agitation of a foam-producing liquid and air, is delivered to its point of application for fire control by a specially designed fixed piping system, by hose lines, or by a combination of the two.

Pipe for foam or foam solution is ordinarily standard-weight steel. Fittings are cast iron, except that malleable iron or steel is required where exposure to fire is possible or if the fittings are a part of a self-supporting piping arrangement.

The size of pipe for carrying water or water solution is determined by the usual friction-loss considerations for fire-service water piping. Friction losses in foam-carrying piping are not subject to determination by the usual friction-loss data or formulas. Pipe sizes are as recommended by the supplier of the foam system and are often checked by operating tests. Piping from a main or header to a number of outlets must be arranged and sized so that the distribution of foam to each outlet will conform to the designed rates of application. General standards for foam piping are given in the National Fire Protection Association Standard for Foam Extinguishing Systems.

Dry-chemical Fire-protection Piping. The use of dry chemical in fixed-piping fire-extinguishing systems is a relatively recent development. General rules applicable to any dry-chemical system are given in the National Fire Protection Association Standard for Dry Chemical Extinguishing Systems.

Complete design details must be prepared by the supplier and approved for particular installation by the insurer or other regulatory bodies involved.

The flow of dry-chemical particles suspended in or fluidized by an inert gas does not follow the hydraulic laws applicable to the flow of water. Most of the design practices have been determined experimentally by the suppliers of dry-chemical fire-control devices and equipment and have been applied to specific hazards. The flow characteristics may vary considerably with the physical properties of the dry chemical.

If more than one hazard may be simultaneously involved in fire, each should have its individual extinguishing system unless the arrangement and capacity of the equipment are such as to produce simultaneous protection for all the hazards.

All sections or subdivisions of a dry-chemical distribution system must be balanced so that each will have about the same pressure drop. This is accomplished by splitting the discharge by tees.

Pipe must be of standard-weight steel, and galvanized in accordance with Specification ASTM A120. Valves must be of a design suitable for handling solid material in suspension. Piping installations must be in accordance with standards for the installation of sprinkler systems where applicable.

Vaporizing-liquid Fire-protection Piping. Vaporizing liquids such as carbon tetrachloride, chlorobromethane, and halogenated hydrocarbons are not generally used in piped fire-extinguishing systems except for aircraft power plants and other highly specialized hazards. Such liquids are expensive or toxic or both. The application is usually by spray under high pressure in relatively small tubing or piping systems. There are no generally applicable design or installation standards. The toxicity of chlorine-containing compounds is a deterrent in their possible application to fire-protection systems.

13

GAS-SYSTEMS PIPING

C. George Segeler*

INTRODUCTION

In America the distribution of *manufactured gas* through pipes began in 1816 as a public utility for lighting buildings and streets in Baltimore, Md. After a development along these lines for nearly a century, the trend of events caused a change in use, so that what had been illuminating gas then became *fuel gas* for domestic cooking and for heating in home and industry. The new applications were accompanied by a corresponding change in the composition of the gas, with heating value as the basis of charge rather than candlepower.

The year 1825 marked the first commercial use of *natural* gas in the United States. At Fredonia, N.Y., gas was piped through small lead pipes. In 1870 the first pipeline of any consequence was laid to Rochester, N.Y. Two years later, in 1872, the first long iron pipeline in the United States was laid in the vicinity of Titusville, Pa. The natural-gas industry developed slowly for several years because the use of natural gas was restricted to industries in the immediate vicinity of producing wells. Between 1872 and 1890, however, several corporations were organized to transport and sell natural gas to customers in the industrial communities of Pennsylvania, West Virginia, New York, and Ohio. The lines through which this natural gas was transported usually were constructed of wrought iron joined with screwed couplings, seldom exceeded 8 in. in diameter, and were operated at a pressure of approximately 80 psi.

The years between 1890 and 1950 saw the growth and eventual decline of a manufactured-gas industry of large proportions which used many thousands of miles of pipe. However, the improvements in pipe manufacture, pipe metallurgy,

* Director of Technical Services, David Sage, Inc., New York, N.Y.

and welding practices, coupled with the discovery of ever-growing supplies of natural gas, have converted the entire gas utility business to natural gas. A little mixed natural and manufactured gas is distributed in four scattered areas. Straight manufactured gas is available only in Honolulu and in a few cities in Vermont and Maine.

Natural gas has been in such demand that as of 1965 there were 61,000 miles of field and gathering lines,[1] 205,400 miles of transmission lines, and 469,800 miles of distribution mains in use.

The forecast for the future indicates that by 1970 these figures will increase to 70,000 miles of field and gathering lines, 236,800 miles of field transmission lines, and 604,300 miles of distribution lines.

The gas industry, both by reason of its size and also by its diverse needs, uses steel, wrought-iron, cast-iron, copper and plastic pipe and requires that materials be available for a pressure range up to 1,600 psig for transmission and even higher for certain gas-field piping. The most commonly used steel for transmission line is API-5LX52.[2]

Information on the properties and laws of gases and on the basic laws of fluid flow through pipes will be found in Chap. 3, Fluid Mechanics. Dimensional standards and material specifications for pipes, valves, and fittings suitable for gas service will be found among other products in Chap. 8. The specific identification of types of pipe generally used for high-pressure gas lines is given in Appendix C of the American Standard B31.8-1962. Special fittings for gas systems, together with typical details and design recommendations, are included in this chapter. Extracts from appropriate codes are also included. For industrial gas piping systems the American Standard Z83.1-1966 applies.

GAS TRANSMISSION AND DISTRIBUTION PRESSURES

The gas-utility industry distinguishes between gas-transmission piping systems and distribution networks, although there is some overlap. In general, a transmission system is characterized by greater length, higher pressures, higher flow velocities, and fewer taps and fittings.

Gas utilities use the terms low, intermediate, and high pressure to describe different pressure levels in their distribution practice. However, these terms are qualitative rather than specific. On the other hand, the term *low pressure* is specifically limited by the American Standard B31.8-1963 (Gas Transmission and Distribution Piping Systems) in Section 845.43 as follows:

Maximum Allowable Operating Pressure for Low Pressure Distribution Systems. The maximum allowable pressure for a low pressure distribution system shall not exceed either (*a*) or (*b*) below.

(*a*) A pressure which would cause the unsafe operation of any connected and properly adjusted low pressure gas burning equipment, or

(*b*) A pressure of 2 psig.

The American Standard B31.8-1963 considers all pressures, other than low pressure as defined by Section 845.43, as high pressure. However, additional requirements are introduced for safeguarding customers when the maximum actual operating pressure of the distribution system exceeds 60 psig.

[1] The 61,000 miles of field and gathering lines represent those operated by utilities and pipeline companies; those operated by independent producers are not included in the total.

[2] This pipe is dimensioned in accordance with the American Standard B31.8-1963. To some extent steels of the same API-5LX specification, but with yield strengths as high as 100,000 psi, have been used.

Low-pressure Distribution. Gas companies normally operate low-pressure systems at specific pressure levels between 5- and 14-in. water column. Above these pressures, systems are called intermediate or medium pressure, and then each customer (except certain industrial and boiler loads) is furnished a "pounds-to-inches" regulator so that gas is metered and brought into buildings at low pressure.

In the interest of savings in residence construction costs, the use of small-diameter copper tubing with an operating pressure of 2 psig is receiving increased attention. Each gas appliance or each group of appliances must be provided with its own pressure regulator set for the proper appliance-operating pressure.

One gas company designs these systems on the basis of a 1.5-psi pressure drop between the meter and the regulator set; outlet pressure from the regulator is then about 6 in. water column. When a central-heating furnace is to be supplied, a 1-in. water-column pressure drop is used to size the connecting piping. The remainder of the tubing system, which has an inlet pressure of 0.5 psig, is designed for a pressure drop of 0.12 psi; this results in a satisfactory pressure at the appliances.

The table below lists the capacities of semirigid tubing when incorporated in systems which require 1.5-psi or 10.0-in. water-column pressure drops.

Capacity of Semirigid Tubing in a 2-psig System, Cu Ft/Hr of 0.65 Sp Gr Gas

	Pressure drop of 1.5 psi				Pressure drop of 1-in. wc		
Length, ft	Tubing, OD, in.			Length, ft	Tubing OD, in.		
	0.375	0.500	0.625		0.375	0.500	0.625
5	413	920	1670	2	101	225	408
10	295	655	1200	4	72	161	290
20	213	460	837	6	58	128	234
40	150	333	590	8	49	107	196
60	118	264	480	10	41	92	167
80	103	231	418	20	23	53	95
100	93	206	378	30	17	40	71

The allowable variation in pressure generally is kept within the limit that, for any one customer, the minimum pressure should not be less than one-half of the maximum. A more liberal standard permits variation in pressure to be anywhere from twice to one-half the normal value. These pressures are usually determined at the outlet of the meter (or, in certain low-pressure systems, at the outlet of the service-pipe entry into the building).

The long-range goal of the gas industry for low-pressure distribution is to keep gas pressures within the range of plus or minus 25 per cent of the normal pressure.

Cross-country Transmission Lines for Natural Gas. These usually operate at pressures ranging from 100 to 1,200 psig or higher, depending on the length of the line, the frequency of compressor stations, and other design considerations.

In former times, pipelines were designed for a given gas sales potential with a moderate factor for future growth. Now, lines are designed for maximum appropriate pipe diameters with a minimum number of compressor stations in order to obtain reasonable unit transportation costs. Growth can be provided by adding compressor stations and by looping lines.

Bracing against Pressure. With piping systems which contain bell joints, Dresser couplings, or similar connecting devices that permit free ends of the pipe to slip, consideration should be given to preventing the complete disengagement of the parts under pressure. Whereas earth pressure alone may serve the purpose

with buried lines under a pressure of only a few ounces of water column, some positive means should be provided with exposed piping or with buried pipes where the pressure is of consequence.   It is necessary, therefore, to brace such piping where there are dead ends or turns in the line that would create disengagement forces.   In buried lines, timber, concrete, or rock bracing is customarily used at such points to transmit the pressure thrust to an earth area sufficiently large to support the load.   In above-ground lines mechanical stops or tie rods are required.

GAS-DISTRIBUTION SYSTEMS

Design, Materials, Construction

The *design* of gas-distribution systems and the selection of *materials* and *construction* to suit various applications are discussed in this section.   Descriptions of items peculiar to gas piping are given in this section in so far as practicable. Starting from the customer's premises and working back toward the source, the usual selections of materials and construction are generally as follows:

Customer Premises.   Basic standards governing the installation of house piping (actually from the meter set assembly) will be found in ASA Z21.30-1964, American Standard Installation of Gas Piping and Gas Appliances[1] or its identical counterpart, National Fire Protection Association Standard 54-1964.[2]

The following information pertaining to such piping is abstracted from the referenced standard.   The scope of the standard covers residential and commercial low pressure systems up to 0.5 psig.   A new standard, ASA Z83.1–1966 applies to industrial gas systems and to all other systems operating above 0.5 psig.   When gas piping and appliances are installed, local ordinances and local gas company regulations should be observed.

American Standard Installation of Gas Appliances and Gas Piping, ASA Z21.30-1964 (Abstracted)

PIPING PLAN

It is recommended that before proceeding with the installation of a gas piping system, a piping sketch or plan be prepared showing the proposed location of the piping as well as the size of different branches.   Adequate consideration should be given to future demands and provisions made for added gas service.

Before any final plans or specifications are completed, the serving gas supplier or the authority having jurisdiction should be consulted.

When an additional appliance is to be served through any present gas piping, capacity of the existing piping shall be checked for adequacy, and replaced with larger piping if necessary.

INTERCONNECTIONS

Interconnections Supplying Separate Consumers

When two or more meters, or two or more service regulators when meters are not provided, are installed on the same premises and supply separate consumers, the gas piping systems shall not be interconnected on the outlet side of the meters or service regulators.

Interconnections for Stand-By Fuels

When a supplementary gas for stand-by use is connected downstream from a meter or a service regulator when a meter is not provided, a suitable device to prevent backflow shall

[1] Available from the American Gas Association, 605 Third Avenue, New York, N.Y. 10016.

[2] Available from the National Fire Protection Association, 60 Batterymarch Street, Boston 10, Mass.

be installed.   A three-way valve installed to admit the stand-by supply and at the same time shut off the regular supply may be used for this purpose.

SIZE OF PIPING TO GAS APPLIANCES

Size of Supply Piping for Gas Appliances

Gas piping shall be of such size and so installed as to provide a supply of gas sufficient to meet the maximum demand without undue loss of pressure between the meter, or service regulator when a meter is not provided, and the appliance or appliances.

The size of gas piping depends upon the following factors:

  a. Allowable loss in pressure from meter, or service regulator when a meter is not provided, to appliance.
  b. Maximum gas consumption to be provided.
  c. Length of piping and number of fittings.
  d. Specific gravity of the gas.
  e. Diversity factor.

Gas Consumption

The quantity of gas to be provided at each outlet shall be determined, whenever possible, directly from the manufacturer's Btu rating of the appliance which will be installed.   In case the ratings of the appliances to be installed are not known, Table 1 is given to show the approximate consumption of average appliances of certain types in Btu per hour.

To obtain the cubic feet per hour of gas required, divide the total Btu input of all appliances by the average Btu heating value per cubic foot of the gas.   The average Btu per cubic foot of the gas in the area of the installation may be obtained from the serving gas supplier.

Table I.   Approximate Gas Input for Some Common Appliances

Appliance	Input Btu per hr (Approx.)
Range, Free Standing, Domestic	65,000
Built-In Oven or Broiler Unit, Domestic........	25,000
Built-In Top Unit, Domestic..................	40,000
Water Heater, Automatic Storage	
30 to 40 Gal Tank	45,000
Water Heater, Automatic Storage	
50 Gal Tank	55,000
Water Heater, Automatic Instantaneous	
Capacity { 2 gal per minute..............	142,800
4 gal per minute..............	285,000
6 gal per minute..............	428,400
Water Heater, Domestic, Circulating or	
Side-Arm................................	35,000
Refrigerator	3,000
Clothes Dryer, Domestic	35,000
Incinerator, Domestic	35,000
Gas Light.................................	2,500

For specific appliances or appliances not shown above, the input should be determined from the manufacturer's rating.

Required Gas Piping Size

  a. Capacities in cubic feet per hour of 0.60 specific gravity gas for different sizes and lengths of pipe are shown in Table 2, based upon a pressure drop of 0.5 inch water column.

Capacities of different sizes and lengths of semi-rigid tubing in thousands of Btu per hour with a pressure drop of 0.5 inch water column for undiluted liquefied petroleum gases are shown in Table 4.   In using this Table, no allowance is necessary for an ordinary number of fittings.

b. Gas piping systems that are to be supplied with gas of a specific gravity of 0.70 or less, can be sized directly from Table 2.   When the specific gravity of the gas is greater than 0.70 the gravity factor shall be applied.

c. To determine the size of each section of gas piping in a system within the range of Table 2 or 4 proceed as follows:

Table 2.[1]   Maximum Capacity of Pipe in Cubic Feet of Gas Per Hour

(Based upon a pressure drop of 0.5 inch water column and 0.6 specific gravity gas.
For other pressure drops, multiply by $\sqrt{\Delta P/0.5}$, where ΔP is the pressure drop, inches water column.)

Nominal iron pipe size, inches	Length in feet													
	10	20	30	40	50	60	70	80	90	100	125	150	175	200
½	175	120	97	82	73	66	61	57	53	50	44	40	37	35
¾	360	250	200	170	151	138	125	118	110	103	93	84	77	72
1	680	465	375	320	285	260	240	220	205	195	175	160	145	135
1¼	1,400	950	770	660	580	530	490	460	430	400	360	325	300	280
1½	2,100	1,460	1,180	990	900	810	750	690	650	620	550	500	460	430
2	3,950	2,750	2,200	1,900	1,680	1,520	1,400	1,300	1,220	1,150	1,020	950	850	800
2½	6,300	4,350	3,520	3,000	2,650	2,400	2,250	2,050	1,950	1,850	1,650	1,500	1,370	1,280
3	11,000	7,700	6,250	5,300	4,750	4,300	3,900	3,700	3,450	3,250	2,950	2,650	2,450	2,280
4	23,000	15,800	12,800	10,900	9,700	8,800	8,100	7,500	7,200	6,700	6,000	5,500	5,000	4,600

[1] The capacities listed in Table 2 are based on straight lengths of pipe.   For the influence of fittings on pressure drop, see Table 10.

1. Calculate the gas demand in cubic feet per hour of the appliance to be attached to each piping system outlet.
2. Measure the length of piping from the gas meter or service regulator when a meter is not provided, to the most remote outlet in the building.
3. In Table 2 or 4, whichever is appropriate, select the column showing that distance or the next longer distance if the Table does not give the exact length.   This is the only distance used in determining the size of any section of gas piping.   If the gravity factor is to be applied, the values in the selected column of Table 2 are multiplied by the appropriate multiplier from Table 3.

Table 3.   Multipliers to Be Used Only With Table 2 When Applying the Gravity Factor

Specific Gravity	Multiplier	Specific Gravity	Multiplier
0.35	1.31	1.00	0·78
0.40	1.23	1.10	0.74
0.45	1.16	1.20	0.71
0.50	1.10	1.30	0.68
0.55	1.04	1.40	0.66
0.60	1.00	1.50	0.63
0.65	0.96	1.60	0.61
0.70	0.93	1.70	0.59
0.75	0.90	1.80	0.58
0.80	0.87	1.90	0.56
0.85	0.84	2.00	0.55
0.90	0.82	2.10	0.54

4. Use this vertical column to locate ALL gas demand figures for this particular system of piping.
5. Starting at the most remote outlet, find in the vertical column just selected the gas demand for that outlet.   If the exact figure of demand is not shown, choose the next larger figure below in the column.
6. Opposite this demand figure, in the first column at the left in Table 2 or 4, will be found the correct size of gas piping.

7. Proceed in a similar manner for each outlet and each section of gas piping. For each section of piping determine the total gas demand supplied by that section.

d. For any gas piping system, special gas appliances or for conditions other than those covered by Tables 2 or 4, such as longer runs, or greater gas demands, the size of each gas piping system may be determined by standard engineering methods acceptable to the authority having jurisdiction and the serving gas supplier.

Diversity Factor

The diversity factor is an important factor in determining the correct gas piping size to be used in multiple family dwellings. It is dependent upon the number and kinds of gas appliances being installed. Consult the serving gas supplier or the authority having jurisdiction for the diversity factor to be used.

Table 4. Capacity of Semi-Rigid Tubing of Different Outside Diameters and Lengths in Thousands of Btu Per Hour of Undiluted Liquefied Petroleum Gases at a Pressure Drop of 0.50 Inch Water Column

Outside diameter (inches)	Length of tubing (feet)									
	10	20	30	40	50	60	70	80	90	100
3/8	39	26	21	19						
1/2	92	62	50	41	37	35	31	29	27	26
5/8	199	131	107	90	79	72	67	62	59	55
3/4	329	216	181	145	131	121	112	104	95	90
7/8	501	346	277	233	198	187	164	155	614	138

Example of Piping System Design

Determine the required pipe size of each section and outlet of the piping system shown in Exhibit 1, with a designated pressure drop of 0.50 inch water column. Gas to be used has 0.65 specific gravity and a heating value of 1,000 Btu per cubic foot.

Solution:

(1) Maximum gas demand for outlet A

$$\frac{\text{Consumption (rating plate input or Table 1 if necessary)}}{\text{Btu of gas}} =$$

$$\frac{30,000 \text{ Btu per hour rating}}{1,000 \text{ Btu per cubic foot}} = 30 \text{ cubic feet per hour (or 30 cfh)}$$

Maximum gas demand for outlet B

$$\frac{\text{Consumption}}{\text{Btu of gas}} = \frac{3,000}{1,000} = 3 \text{ cfh}$$

Maximum gas demand for outlet C

$$\frac{\text{Consumption}}{\text{Btu of gas}} = \frac{75,000}{1,000} = 75 \text{ cfh}$$

Maximum gas demand for outlet D

$$\frac{\text{Consumption}}{\text{Btu of gas}} = \frac{136,000}{1,000} = 136 \text{ cfh}$$

(2) The length of pipe from the gas meter to the most remote outlet (A) is 60 feet. This is the only distance used.

(3) Using the column marked 60 feet in Table 2 (provided this is the Table designated for use by the serving gas supplier):

Outlet A, supplying 30 cfh, requires 1/2 inch pipe.
Outlet B, supplying 3 cfh, requires 1/2 inch pipe.
Section 1, supplying outlets A and B, or 33 cfh, requires 1/2 inch pipe.

Outlet C, supplying 75 cfh, requires ¾ inch pipe.
Section 2, supplying outlets A, B, and C, or 108 cfh, requires ¾ inch pipe.
Outlet D, supplying 136 cfh, requires ¾ inch pipe.
Section 3, supplying outlets A, B, C, and D, or 244 cfh, requires 1 inch pipe.

(4) If the gravity factor is applied to this example, the values in the column marked 60 feet of Table 2 would be multiplied by the multiplier (0.96) from Table 3, and the resulting cubic feet per hour values would be used to size the piping.

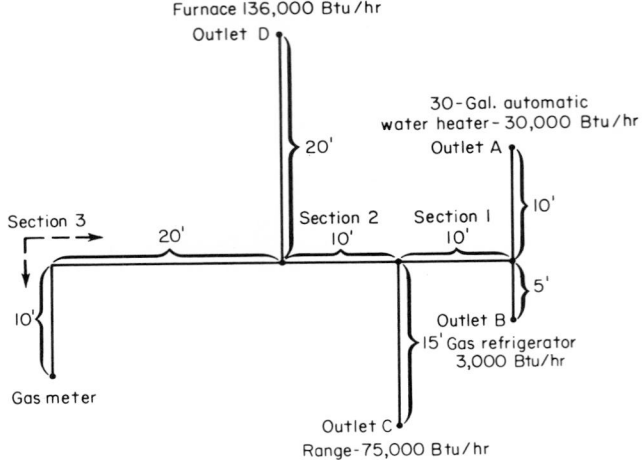

Exhibit 1

Gas Piping in Mobile Home and Travel-trailer Parks

Requirements are discussed here of the gas-piping system extending from the meter set assembly outlet (or the outlet of a service regulator if a meter is not provided) to the terminal of the gas riser at each trailer site.

If an enclosing foundation is used beneath the trailer, the piping must not be installed under adjacent patio slabs. In any event, piping must not be installed beneath the trailer site.

The gas riser should be located in the rear one-third section of the site and not less than 18 in. from the roadside wall of the trailer. A valve must be provided at each such gas riser, and, except for low-pressure gas, the minimum permissible size of the riser is ¾ in.

The table below lists gas demands for various numbers of trailer sites, and is intended to facilitate the sizing of the piping system.

Gas Demands, Btu/Hr, for Trailer Sites

Number of sites	Demand/Site
1	125,000
2	117,000
3	104,000
4	96,000
5	92,000
10	77,000
11–20	66,000
21–30	62,000
31–40	58,000

ACCEPTABLE PIPING MATERIALS

Piping Material for Utility Gases

Gas pipe shall be steel or wrought-iron pipe complying with the American Standard for Wrought-Steel and Wrought-Iron Pipe, ASA B36.10-1959. Threaded copper or brass[1] pipe in iron pipe sizes may be used with gases not corrosive to such materials.

Tubing may be used when acceptable to the serving gas supplier and may be seamless copper, steel, (or aluminum alloy with restrictions) when used with gases not corrosive to such material. Copper tubing shall be of standard type K or L or equivalent, complying with specification ASTM B88-1962.

Pipe joints may be screwed, flanged, or welded and non-ferrous pipe may also be soldered or brazed with material having a melting point in excess of 1000 F. Compression type tubing fittings shall not be used for this purpose.

Fittings (except stopcocks or valves) shall be malleable iron or steel when used with steel or wrought-iron pipe, and shall be copper or brass when used with copper or brass pipe or tubing and shall be aluminum alloy when used with aluminum pipe or tubing. When approved by the authority having jurisdiction, special fittings may be used to connect steel or wrought-iron pipe. Cast iron fittings in sizes 6 inches and larger may be used to connect steel and wrought-iron pipe when approved by the authority having jurisdiction.

Workmanship and Defects

Gas pipe or tubing, and fittings shall be clear and free from cutting burrs and defects in structure or threading and shall be thoroughly brushed, chip and scale blown.

Defects in pipe or tubing or fittings shall not be repaired. When defective pipe, tubing or fittings are located in a system the defective material shall be replaced.

Number of Threads

Pipe shall be threaded in accordance with Table 5.

Table 5. Specifications for Threading Pipe

Iron pipe size (inches)	Approximate length of threaded portion (inches)	Approximate No. of threads to be cut
½	¾	10
¾	¾	10
1	⅞	10
1¼	1	11
1½	1	11
2	1	11
2½	1½	12
3	1½	12
4	1⅝	13

CONCEALED PIPING IN BUILDINGS

Minimum Size

No gas pipe smaller than standard ½ inch iron pipe size shall be used in any concealed location.

Piping in Partitions

Concealed gas piping should be located in hollow rather than solid partitions. Tubing shall not be run inside walls or partitions unless protected against physical damage. This rule does not apply to tubing which passes through walls or partitions.

Piping in Floors

Except as indicated in the following paragraph, gas piping in solid floors such as concrete shall be laid in channels in the floor suitably covered to permit access to the piping with a

[1] Aluminum alloy pipe is permitted within certain limitations.

minimum of damage to the building. When such piping may be exposed to excessive moisture or corrosive substances, it shall be suitably protected.

When approved by the authority having jurisdiction and acceptable to the serving gas supplier, gas piping may be imbedded in concrete floor slabs constructed with Portland cement. Piping must be surrounded by at least 1½ inches of concrete and must not be in physical contact with other metallic structures such as reinforcing rods or electrical neutral conductors. When subject to corrosion at point of entry into the concrete slab, the piping should be suitably protected from corrosion. Piping shall not be placed in concrete which contains quick-set additives or cinder aggregate.

Connections in Original Installations

When installing gas piping which is to be concealed, unions, tubing fittings, running threads, right and left couplings, bushings, and swing joints made by combinations of fittings shall not be used.

Reconnections

When necessary to insert fittings in gas pipe which has been installed in a concealed location, the pipe may be reconnected by use of a ground joint union with the nut center-punched to prevent loosening by vibration. Reconnection of tubing in a concealed location is prohibited.

PIPING UNDERGROUND

Protection Against Corrosion

Gas piping in contact with earth or other material which may corrode the piping, shall be protected against corrosion in an approved manner. Piping shall not be laid in contact with cinders. When dissimilar metals are joined underground, an insulated coupling shall be used.

Piping Through Foundation Wall

Underground gas piping, when installed below grade through the outer foundation or basement wall of a building, shall be either encased in a sleeve or otherwise protected against corrosion. The piping or sleeve shall be sealed at the foundation or basement wall to prevent entry of gas or water.

Piping Underground Beneath Buildings

When the installation of gas piping beneath buildings is unavoidable, the piping shall be encased in a conduit. The conduit shall extend into a normally usable and accessible portion of the building and, at the point where the conduit terminates in the building, the space between the conduit and the gas piping shall be sealed to prevent the possible entrance of any gas leakage. The conduit shall extend at least 4 inches outside the building, be vented above grade to the outside atmosphere, and be installed in such a way as to prevent the entry of water.

INSTALLATION OF PIPING

Piping Supports

Gas piping in buildings shall be supported with pipe hooks, metal pipe straps, bands or hangers suitable for the size of piping, and of adequate strength and quality, and located at proper intervals so that the piping cannot be moved accidentally from the installed position. Gas piping shall not be supported by other piping.

Spacings of supports in gas piping installations shall not be greater than the following:

Size of pipe (inches)	(Feet)	Size of tubing (inch O.D.)	(Feet)
½	6	½	4
¾ or 1	8	⅝ or ¾	6
1¼ or larger (horizontal)	10	⅞ or 1	8
1¼ or larger (vertical)	every floor level		

Do Not Bend Pipe

Gas pipe shall not be bent. Fittings shall be used when making turns in gas pipe.

Use Tee

If dirt or other foreign material is a problem, a tee fitting with the bottom outlet plugged or capped shall be used at the bottom of any pipe riser. (See Fig. 1.)

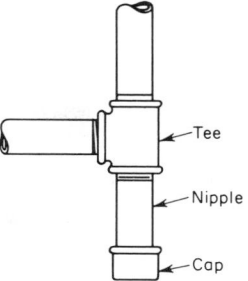

Avoid Clothes Chutes, Etc.

Gas piping inside any building shall not be run in or through an air duct, clothes chute, chimney or gas vent, ventilating duct, dumb waiter or elevator shaft.

Cap All Outlets

Fig. 1. Suggested method of installing tee.

Each outlet, including a valve or cock outlet, shall be securely closed gas-tight with a threaded plug or cap immediately after installation and shall be left closed until an appliance is connected thereto. Likewise, when an appliance is disconnected from an outlet and the outlet is not to be used again immediately, it shall be securely closed gas-tight. In no case shall the outlet be closed with tin caps, wooden plugs, corks, or by other improvised methods. These provisions do not prohibit the normal use of a listed quick-disconnect device.

Location of Outlets

The unthreaded portion of gas piping outlets shall extend at least one inch through finished ceilings and walls, and when extending through floors shall be not less than 2 inches above them. The outlet fitting or the piping shall be securely fastened. Outlets shall not be placed behind doors. Outlets shall be far enough from floors, walls and ceilings to permit the use of proper wrenches without straining, bending or damaging the piping.

Prohibited Devices

No device shall be placed inside the gas piping or fittings that will reduce the cross-sectional area or otherwise obstruct the free flow of gas.

Branch Pipe Connection

All branch outlet pipes shall be taken from the top or sides of horizontal lines and not from the bottom. Where a branch outlet is placed on a main supply line before it is known what size of pipe will be connected to it, the outlet shall be of the same size as the line which supplies it.

Electrical Bonding and Grounding

A gas piping system within a building shall be electrically continuous and bonded to any grounding electrode, as defined by the National Electrical Code, ASA C1-1962.

Underground gas service piping shall not be used as a grounding electrode except when it is electrically continuous uncoated metallic piping and its use as a grounding electrode is acceptable both to the serving gas supplier and to the authority having jurisdiction, since gas piping systems are often constructed with insulating bushings or joints, or are of coated or nonmetallic piping.

TEST OF PIPING FOR TIGHTNESS

Before any system of gas piping is finally put in service, it shall be carefully tested to assure that it is gas-tight. Where any part of the system is to be enclosed or concealed, this test should precede the work of closing in. To test for tightness, the piping may be filled with the fuel gas, air or inert gas, but not with any other gas or liquid. OXYGEN SHALL NEVER BE USED.

Test Methods for Piping Systems

a. Before appliances are connected, piping systems shall stand a pressure of at least six inches mercury or three pounds gage for a period of not less than ten minutes without

showing any drop in pressure. Pressure shall be measured with a mercury manometer or slope gage, or an equivalent device so calibrated as to be read in increments of not greater than one-tenth pound. The source of pressure shall be isolated before the pressure tests are made.

b. Systems for undiluted liquefied petroleum gases shall stand the pressure test in accordance with "a" above, or, when appliances are connected to the piping system, shall stand a pressure of not less than ten inches water column for a period of not less than ten minutes without showing any drop in pressure. Pressure shall be measured with a water manometer or an equivalent device so calibrated as to be read in increments of not greater than one-tenth inch water column. The source of pressure shall be isolated before the pressure tests are made.

PURGING

Purging All Gas Piping

After piping has been checked, all piping receiving gas shall be fully purged. A suggested method for purging the gas piping to an appliance is to disconnect the pilot piping at the outlet of the pilot valve. Under no circumstances shall piping be purged into the combustion chamber of an appliance.

Light Pilots

After the gas piping has been sufficiently purged, all appliances shall be purged and the pilots lighted. The installing agency shall assure itself that all piping and appliances are fully purged before leaving the premises.

INDUSTRIAL GAS PIPING SYSTEMS

The material which appears below has been partly abstracted and condensed from the proposed American Standard Z83.1-1966, Installation of Gas Piping and Gas Equipment on Industrial Premises and Certain Other Premises. For details of piping and valves associated with large boilers, one may refer to the proposed American Standard Z83.3. The discussion below relates to industrial fuel-gas systems and not to piping which conveys air-gas mixtures within the flammable range, special atmosphere gases, or gases such as hydrogen, oxygen, acetylene, or carbon monoxide.

Pressure Loss for Industrial Gas Piping Systems. Where gas pressures in industrial gas piping systems are 0.5 psig or less, the allowable pressure loss should not exceed 0.5 in. H_2O. Where higher gas pressures are used, the normal design pressure loss should not exceed 10 per cent of the initial gage pressure of the piping system unless each point of use for gas has its own pressure regulator.

Pipe Sizing for Pressures above 0.5 Psig. Tables 6 to 8 indicate the approximate capacities for single runs of piping at 10 per cent pressure drop. These tables are based on straight pipe, and to conform to the requirements of the actual installation, the equivalent lengths of fittings and valves must be added to obtain the total equivalent length.

New Additions or Extensions to Existing Gas Piping Systems

New additions or extensions to existing gas piping systems should normally be made with the gas shut off ahead of the point of work. Hot taps, however, may be used if they are installed by trained and experienced crews. For further precautions see paragraph 841.28 of ASA B31.8-1963.

The following simplified procedure can be used for sizing gas piping systems in industrial plants:

1. Determine, by measuring, length of the piping between the gas meter and the most remote point of use; this will be the only length used for sizing the piping system; that is, for pipes less than 4 in. in size, it is unnecessary to take into account the equivalent lengths of fittings.

Table 6. Pipe-Sizing Table for 5-lb Pressure

(Capacity of pipes of different diameters and lengths in cubic feet per hour. For an initial pressure of 5 psig with a 10 per cent pressure drop and a gas of 0.6 specific gravity)

Pipe size of Schedule 40 standard pipe, in.	Total equivalent length of pipe, ft									
	50	100	150	200	300	400	500	1,000	1,500	2,000
1	1,860	1,320	1,070	930	760	660	590	410	340	290
1¼	3,870	2,740	2,240	1,930	1,580	1,370	1,220	860	700	610
1½	5,860	4,140	3,390	2,930	2,390	2,080	1,850	1,310	1,060	930
2	11,420	8,070	6,600	5,710	4,660	4,050	3,610	2,550	2,080	1,810
2½	18,400	13,010	10,640	9,200	7,510	6,530	5,820	4,110	3,350	2,920
3	32,860	23,240	19,000	16,430	13,410	11,660	10,390	7,340	5,990	5,220
3½	48,480	34,280	28,030	24,240	19,780	17,200	15,330	10,820	8,840	7,690
4	67,700	47,880	39,140	33,850	27,630	24,020	21,410	15,120	12,340	10,750
5	123,790	87,540	71,570	61,890	50,530	43,920	39,160	27,640	22,570	19,660
6	202,138	142,950	116,870	101,060	82,500	71,720	63,940	45,140	36,860	32,100

Table 7. Pipe-Sizing Table for 20-lb Pressure

(Capacity of pipes of different diameters and lengths in cubic feet per hour. For an initial pressure of 20 psig with a 10 per cent pressure drop and a gas of 0.6 specific gravity)

Pipe size of Schedule 40 standard pipe in.	Total equivalent length of pipe, ft									
	50	100	150	200	300	400	500	1,000	1,500	2,000
1	4,900	3,470	2,810	2,450	2,000	1,730	1,550	1,070	890	770
1¼	10,190	7,210	5,840	5,090	4,160	3,600	3,220	2,230	1,860	1,610
1½	15,420	10,900	8,830	7,710	6,290	5,450	4,870	3,370	2,810	2,440
2	30,030	21,230	17,190	15,010	12,260	10,610	9,490	6,570	5,480	4,760
2½	48,390	34,220	27,710	24,190	19,750	17,110	15,290	10,590	8,830	7,670
3	86,420	61,110	49,490	43,190	35,280	30,550	27,310	18,910	15,770	13,690
3½	127,480	90,130	73,000	63,710	52,040	45,070	40,280	27,900	23,270	20,200
4	178,040	125,880	101,950	88,980	72,680	62,940	56,260	38,500	32,500	28,210
5	325,530	230,170	186,410	162,700	132,890	115,080	102,870	71,240	59,420	51,590
6	531,530	375,820	304,370	265,660	216,990	187,910	167,980	116,330	93,030	84,240

Table 8. Pipe-Sizing Table for 50-lb Pressure

(Capacity of pipes of different diameters and lengths in cubic feet per hour. For an initial pressure of 50 psig with a 10 per cent pressure drop and a gas of 0.6 specific gravity)

Pipe size of Schedule 40 standard pipe, in.	Total equivalent length of pipe, ft									
	50	100	150	200	300	400	500	1,000	1,500	2,000
1	10,530	7,450	6,090	5,150	4,350	3,790	3,330	2,350	1,920	1,650
1¼	21,880	15,490	12,650	10,700	9,050	7,870	6,920	4,890	3,990	3,430
1½	33,110	23,430	19,130	16,190	13,690	11,910	10,470	7,410	6,040	5,190
2	64,450	45,610	37,250	31,530	26,660	23,190	20,400	14,420	11,770	10,110
2½	103,870	73,510	60,040	50,820	42,960	37,370	32,870	23,240	18,970	16,300
3	185,490	131,270	107,220	90,750	76,720	66,730	58,700	41,510	33,870	29,100
3½	273,600	193,620	158,140	133,850	113,170	98,430	86,590	61,230	49,970	42,930
4	382,110	270,420	220,870	186,940	158,050	137,480	120,930	85,510	69,780	59,960
5	698,660	494,430	403,840	341,800	288,980	251,360	221,110	156,360	127,600	109,630
6	1,140,780	807,320	659,400	558,110	471,860	410,430	361,040	255,310	208,340	179,010

2. Determine the demand in cubic feet per hour at each point of use. On this basis, determine the flow cubic feet per hour in each section and at each outlet tap of the piping system.

3. With the length of step 1 and the flow rate of step 2, enter Table 2, 6, 7, or 8 as appropriate for the pressure under consideration. For other pressures, calculations can be made by one of the formulas listed later in this chapter or by use of the Spitzglass equations given on page 3-158 of Chap. 3.

Construction Details

In laying out industrial gas-piping systems, the designer should be guided by the requirements of ASA Z83.1-1966. These requirements are essentially as have already been treated under the heading "Concealed Piping in Buildings," and they will not be repeated here. Special provisions, however, are abstracted and presented below.

Back Pressure Protection

When the design of utilization equipment connected is such that air, oxygen or standby gases may be forced into the gas supply system, suitable protective devices shall be installed as close to the utilization equipment as practical. Gas and air combustion mixers incorporating double diaphragm "zero" or "atmosphere" governors or regulators require no further protection unless connected directly to compressed air or oxygen at pressures of 5 psig or more.

Low Pressure Protection

A suitable protective device shall be installed between the meter and the utilization equipment if the operation of the equipment is such (i.e., gas compressors) that it may produce a vacuum or a dangerous reduction in gas pressure at the meter.

Electrical Bonding and Grounding

Each aboveground portion of a gas piping system upstream from the equipment shut-off valve shall be electrically continuous and bonded to any grounding electrode, as defined by the National Electrical Code, NFPA No. 70.

Underground gas piping shall not be used as a grounding electrode except when it is electrically continuous uncoated metallic pipe, and its use as a grounding electrode is acceptable both to the serving gas supplier and to the authority having jurisdiction, since gas piping systems are often constructed with insulating bushings or joints, or are of coated or nonmetallic piping.

Electrical Circuits

Electric circuits shall not utilize gas piping or components except that low voltage (50 volts or less) control circuits, ignition circuits, and electronic flame-detection device circuits may make use of piping or components for a part of an electric circuit.

Test Pressure

The usual test pressure should be no less than 1½ times the proposed maximum working pressure but not less than 3 psig. For pressures above 125 psig see ASA Z83.1 for requirements.

Test duration should be long enough to determine if there are any leaks but not less than ½ hour for each 500 cu. ft. of pipe volume or fraction thereof. Systems less than 10 cu. ft. volume may be tested for 10 minutes.

Purging

Gas may generally be used to displace air from piping provided moderately rapid continuous gas flow is maintained. When purging pipe over 4 inches in size, the presence of air requires special precautions such as the introduction of a slug of an inert gas such as nitrogen or carbon dioxide. The open end of piping systems being purged must not be discharged into confined areas without due provisions for safe practice such as ventilation of the space, control of the purging rate, and elimination of all hazardous conditions.

SERVICE PIPES SUPPLYING GAS-DISTRIBUTION SYSTEMS

Although the service pipe that connects the street main to the customer's meter is usually installed by the gas supplying company, the actual ownership (and hence, responsibility) of the pipe varies. In some areas, the gas company owns the pipe all the way to the meter set assembly, but in others the customer owns the pipe from the property line.

Sizing of services may be facilitated by reference to published tables and charts.[1]

Many different bases are available and used in selecting the proper size. Some use a maximum pressure drop; some use a maximum size for all services; others

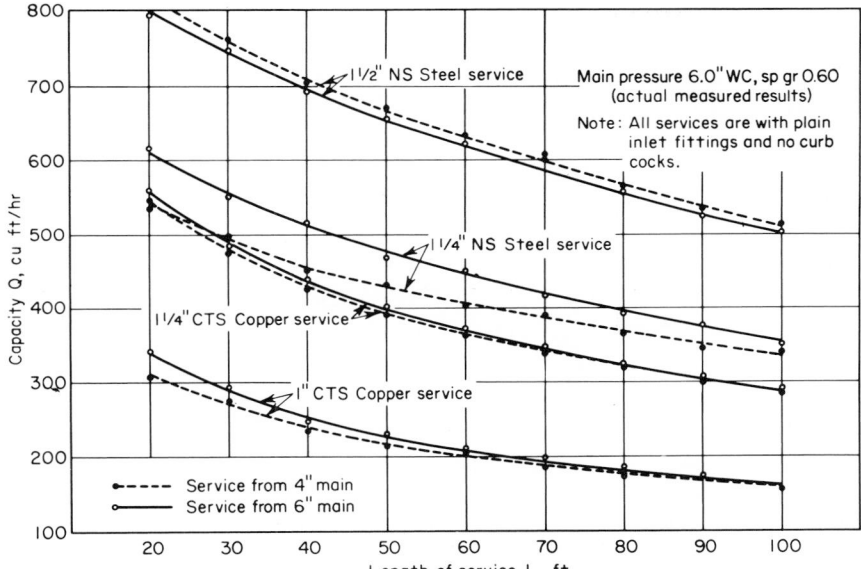

Fig. 2. Total gas flow through LP services. Capacity vs service length at 0.5 water column total drop. Main pressure 6.0-in. water column, sp gr 0.60 (actual measured results). NOTE: All services are with plain inlet fittings and no curb cocks.

make a detailed pressure-drop calculation for each installation. A brief description of the methods followed by a majority of the companies will be of interest. Consideration of connected load, length of service, and main pressure are made in selecting service-pipe size. Generally this combination of variables is handled by means of selecting an allowable pressure drop for a particular main pressure and then selecting the pipe size based upon length of service and anticipated load, which will result in a calculated pressure drop of less than the allowable. Since it is usually the easiest to determine the connected load on a gas service, this flow is used to size the service pipe. Some companies refine the principle by introducing a diversity factor based upon the assumption that all the appliances will not be in use at the same time. Others assume that the service will ultimately supply a piece

[1] B. E. Hunt, and others, "Gas Service Design—Final Report of Task Group" (DMC-56 14), *Proc. AGA*, 1956, pp. 272–287.

of gas space heating equipment and size the service for this load even if no house-heating load is installed at present.

The pressure drop allowed on a gas service is primarily a function of the pressure being carried on the mains during the time the mains are supplying the peak demand. In a low-pressure system which operates at a nominal main pressure of from 6 to 8 in. of water column, most of the companies consider the maximum allowable pressure drop on a gas service to be either 0.3 or 0.5 in. of water column. An intermediate-pressure system operating at a main pressure of from 1 to 15 psig will generally permit an allowable pressure drop of about 0.5 psig. High-pressure distribution systems permit allowable pressure drops of 0.5 to 3 psig. Another method of selecting the allowable pressure drop is to specify that it be some percentage of the main pressure during the maximum hour. One company selects

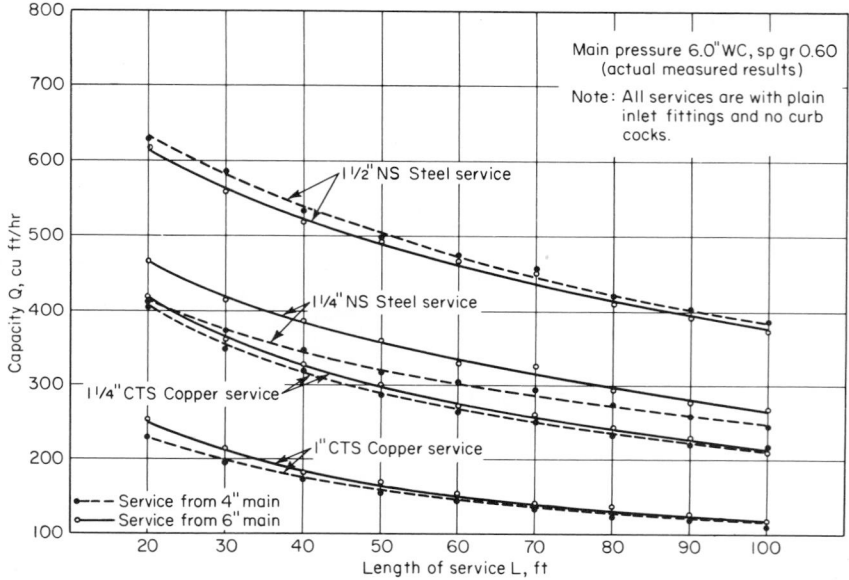

FIG. 3. Total gas flow through LP services. Capacity vs. service length at 0.3-in. water column total drop. Main pressure 6.0-in. water column, sp gr 0.60 (actual measured results). NOTE: All services are with plain inlet fittings and no curb cocks.

a 20 per cent drop, and another a 10 per cent drop. Still another method is to specify a value of 10 for the difference in squares of the absolute inlet and outlet pressures.

Flow Characteristics of Low-pressure Services.[1] Experiments in 1960 showed that the widely used Spitzglass equation for low-pressure gas flow required new resistance values. This flow is a function of inside diameter rather than surface smoothness. Thus, tubing of smaller diameter does not have a capacity equal to that of a larger size steel pipe through which it may be drawn for replacement purposes.

[1] D. Menegakis, and E. H. Luntey, "Experimental Investigation of Flow Characteristics of Low-pressure Services" (DMC-61-15), *Proc. AGA.*, 1961.

The flow equation for either copper or steel services from ¾-in. CTS copper to 1½-in. NS steel was found to follow the form:

$$\text{Cu ft/hr} = \left[\frac{\text{total pressure drop in service, in. } H_2O}{(K_p)(S/S')(L + L_{ef})}\right]^{0.54}$$

where K_p = pipe constant given in Table 9
S = sp gr of gas
S' = sp gr 0.60
L = length of service, ft
L_{ef} = equivalent length of fittings given in Table 10

Table 9. Values of K_p

¾″ CTS copper.........	1.622×10^{-6}
1″ ID plastic	0.279×10^{-6}
1″ CTS copper..........	0.383×10^{-6}
1¼″ CTS copper	0.124×10^{-6}
1¼″ NS steel	0.080×10^{-6}
1½″ NS steel	0.037×10^{-6}

Table 10. Equivalent Length of Fittings

1″ or 1¼″ curb cock for copper service...........	3.5 ft
1¼″ curb cock for 1¼″ steel service.............	13.5 ft
1½″ curb cock for 1½″ steel service.............	12.0 ft
1½″ street elbow for 1¼″ steel service...........	7.5 ft
1½″ street elbow for 1½″ steel service...........	7.5 ft
1¼″ street tee for 1¼″ steel service.............	10.5 ft
1½″ street tee on sleeve or 1¼″ hole in main......	15.0 ft
1¼″ × 1″ × 1¼″ street tee	23.0 ft
1½″ × 1¼″ × 1½″ street tee	19.0 ft
Combined outlet fittings:	
¾″ copper.................................	2.0 ft
1″ copper or plastic	6.0 ft
1¼″ steel.................................	8.0 ft
1½″ steel	22.0 ft

Connection to Mains

Cast-iron Mains. Type of connection varies with size of main, since this limits the size hole which may be tapped without reinforcement. Generally (except 4-in. size) the maximum tap diameter is 25 per cent of the main diameter. For 4-in. mains a 1¼-in. tap is frequently used. Reinforcement is by clamps or sleeves. See Fig. 4 for typical connection details.

Steel Mains. Welding is the preferred method of connecting services, but other methods are occasionally used; compression fittings are preferred over screwed fittings.

Local gas companies have their own details for installing services, and practice may vary among municipalities served by the same company owing to differences in local ordinances. Many various conditions are encountered, and it is impracticable, to try to cover them all, but two common arrangements are shown in Fig. 5. A typical sketch of a simple low-pressure installation using compression couplings with threaded packing nuts is furnished in Fig. 6 and serves to show a street tee and street ell at the main, an outside stopcock with curb box, an inside stopcock, and the meter. Piping generally is pitched back from the meter to the street main with a gradient of 1 in. in 10 ft.

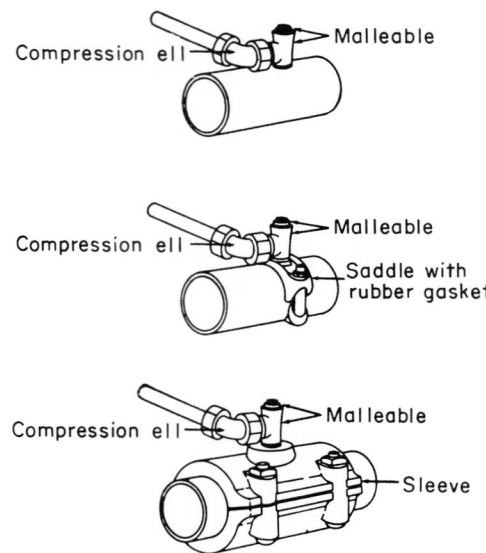

FIG. 4. Typical cast-iron main connections.

Types of Service Pipe. Black steel pipe with malleable-iron fittings has generally been installed. Roughly 40 per cent of this pipe is purchased to specifications (API-5L or, less frequently, A-53, and A-120). Coating and wrapping are widely used on new installations (only about 10 per cent was installed as bare pipe by utilities reporting in 1956). Welded joints are favored. Sizes generally conform to Schedule 40 (standard weight), ASA B36.10.

Galvanized services are occasionally used, but the limited protection over black steel is not considered sufficiently valuable to justify added expense.

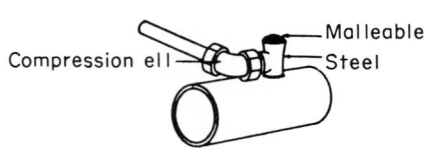

FIG. 5. Typical steel main connections.

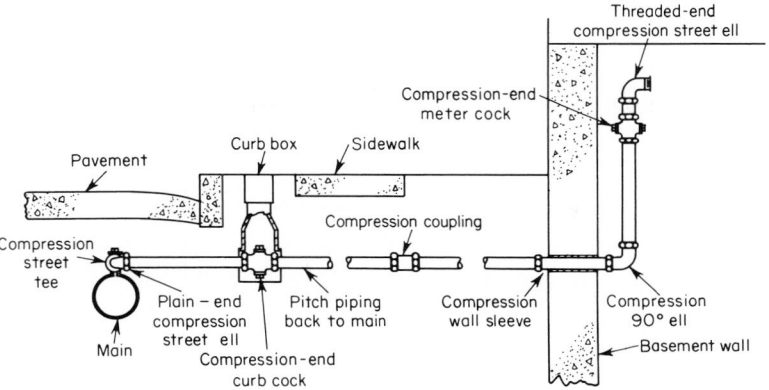

Fig. 6. Sketch of steel gas service using compression couplings with threaded packing nuts.

Cast iron is not permitted except in sizes of 6 in. or larger, and the portion extending through the building wall must be steel.

A few systems have installed copper tubing (generally Type K) for services. Much wider use of copper tubing has developed in its installation in old corroded steel services as a replacement.

Plastic services have been installed to a limited extent both as services and as service inserts.

Corrosion Protection for Services. Suitable coating and wrapping (free from "holidays") should be applied to new services except in soils of known low corrosion. Electric isolation of services from interior house piping and from unprotected mains or other unprotected metal structures is recommended. Sacrificial anodes are sometimes used to provide cathodic protection rather than rectifiers because of the cost of the latter. However, joint protection of services and mains, either by rectifiers or by large anodes, is another common practice. (For additional information on cathodic protection, see Chap. 9.)

Gas systems in existing areas are highly complex metallic structures often in contact with other underground metallic systems. For these reasons, electric isolation into relatively small sections with protective equipment tailored for the specific area should be considered.

Installation of Services. The depth at which services are installed varies widely, depending on frost conditions, whether or not the gas contains moisture, and specific local circumstances. ASA B31.8-1963 requires a minimum of 12 in. on private property and 18 in. in streets and roads. Less cover would require casing or bridging to provide adequate protection against harmful loads.

The pipe must be supported on undisturbed or well-compacted soil. Backfill must be free from rocks, building materials, or other objects that might damage the pipe or its protective coating.

In the case of services operating between 1 and 40 psig, a pressure test at not less than 50 lb for 5 min must be run (if not feasible, this need not include the service connection to the main). For higher pressures see ASA B31.8-1963.

Outside Shutoffs. ASA B31.8-1963 (Section 849.13) requires outside shutoffs (typically curb cocks) on all services operated above 10 psig, on all services 2 in. in diameter or larger, and on all services serving theaters, schools, factories, or other places where relatively large numbers of people assemble.

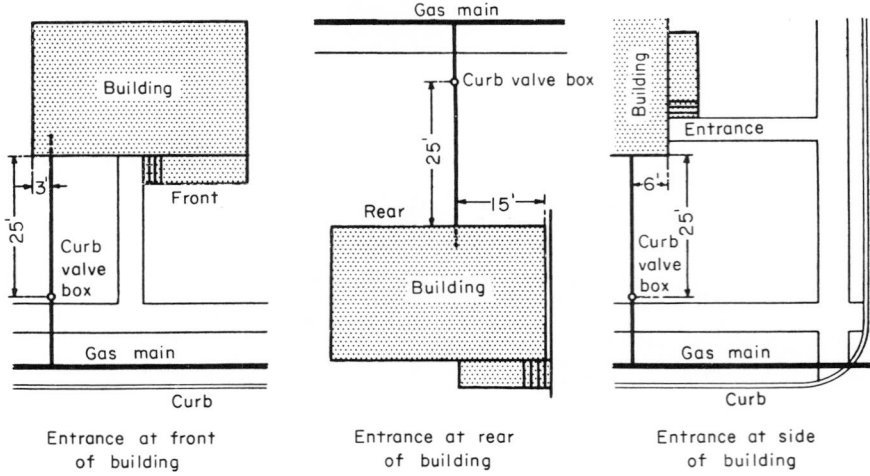

FIG. 7. Typical outside shutoff locations. NOTE: A tag giving the location of the curb valve box shall be attached to the regulator vent.

Some companies install outside shutoffs on all new services. However, where there is an outside meter, the meter cock may itself serve as the outside shutoff.

Service Entry into Buildings. No one standard method exists because many variations are necessary to meet specific site conditions. In general it is common practice to install outside shutoffs (which may be curb cocks or, in the case of outside meters, the meter inlet cocks) on all services where a pressure regulator is used and on all other services which supply places where large groups of people gather.

Examples of a few outside shutoff locations are shown in Fig. 7.

Details of the various types of building entry are shown in Fig. 8 to 11.

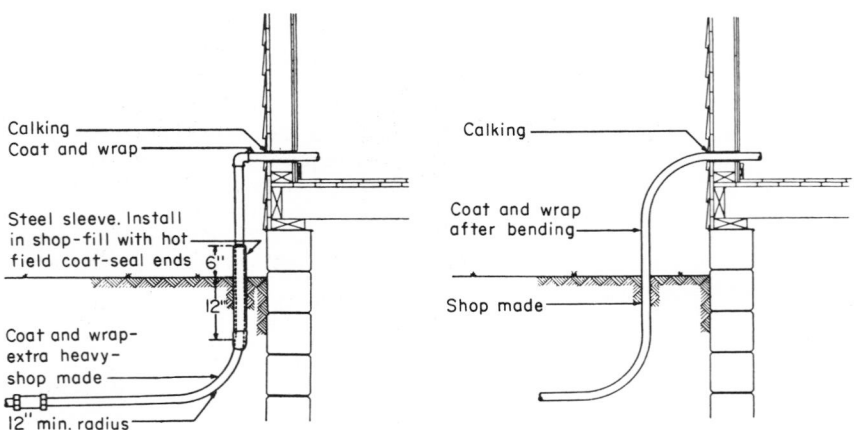

FIG. 8. Above-grade service entrances.

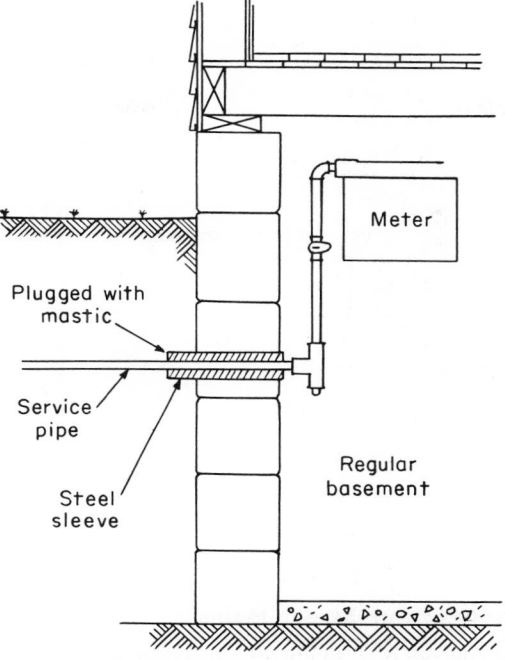

FIG. 9. Conventional service entry into buildings with basements.

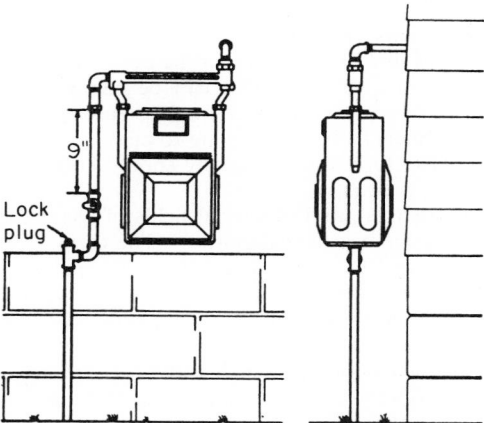

FIG. 10. Service entry for outdoor meter set.

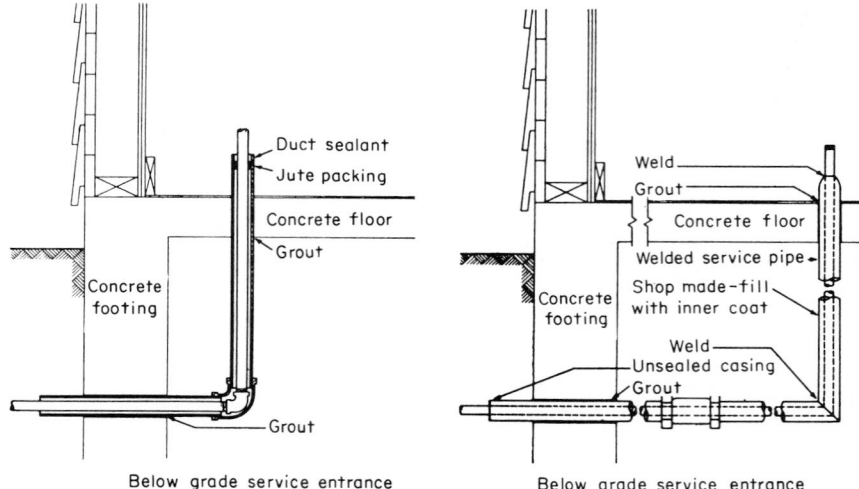

FIG. 11. Typical service entries: buildings without basements (slab on grade, or crawl space).

DISTRIBUTION LOAD DESIGN

Distribution Network or Street Mains

The design and analysis of gas networks generally involve (1) extension of mains into new areas or towns or (2) reinforcement of an existing network to take care of expanding markets for gas. Both situations are similar and require provision for adequate gas supply at proper pressures at the customer's meter.

Systems are designed on the basis of maximum hour demands (or other peak-period basis) with provisions for such future changes as can reasonably be foreseen.

Customer demand per day or per hour can be determined by checking the total sendout for the period in question. This can be compared with the connected load so as to have a valid local factor which might be useful in main extensions, etc.

Local variations are likely to be wide owing to climate, type of gas installations, and heating value. Therefore, each company should collect its own data for reliability.

The residential, commercial, and industrial fields of gas use can be divided into space-heating and nonspace-heating categories. The former varies primarily with outdoor temperatures and the time of day (i.e., thermostat setting of manual thermostats), while the latter depends on diversity. The *AGA Report* (May, 1956) on *Load Characteristics of Gas Heating Customers* indicated that facilities designed for 50 per cent of the total connected load would have been adequate.

The *design day* is defined as one on which weather conditions produce maximum 24-hr gas usage. The lowest hourly temperature might not have the maximum hour usage.

Space-heating Load. The degree of gas space-heating saturation can generally be expected to establish peak demand for residential customers. In absence of local metered demand data, the following procedure may be used for estimating:

1. Determine the number of space-heating customers.
2. Determine or estimate the *average rated input* in MBtu per hour of the heating equipment.

3. Select a value for coincident peak hour average demand from Fig. 12.

4. The product of these three factors is the estimated peak hour average demand in Btu per hour.

Limited data are available (two gas companies) showing the relationship between the peak hour *maximum* demand and the peak hour *average* demand ranging from 2.4 down to 1.0.

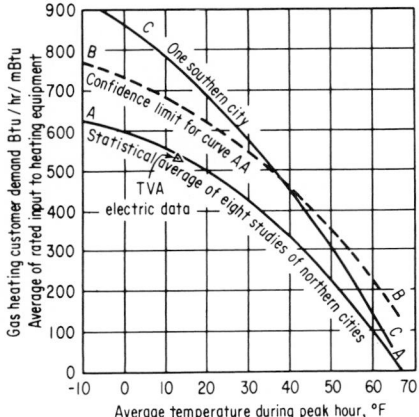

Fig. 12. Coincident peak-hour average demand in Btu per hour for each gas heating customer and for each 1000 Btu of average rated input of heating equipment. Curve *B–B* is the upper confidence limit of curve *A–A*. A minimum of 90 per cent of the observations can be expected to fall below this limit; the point △ is electric coincident demand expressed in Btu per hr per MBtu of electric heating equipment input from TVA data on 400 low-rent houses.

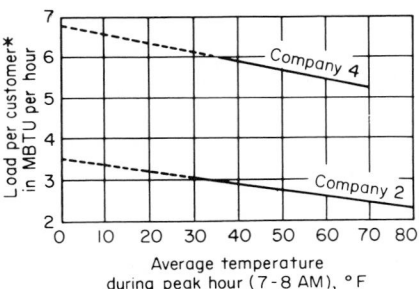

Fig. 13. Nonheating load coincident with peak-hour heating demand for customers using gas for heating as well as those using gas for other purposes only.

Nonheating Residential Load. Wide variations can be expected depending on appliance saturation as borne out by less consistent reported values. In general, the peak nonheating load occurs at an hour other than the peak heating time. Thus, the nonheating load adds a small but significant amount to the coincident peak demand. Figure 13 shows the result of two field studies[1] of nonheating loads *coincident* (for 7:00 to 8:00 A.M.) with the heating-load data in Fig. 12. The values are given in Btu per hour per customer, including those who use gas for heating and those who do not. Saturations were 88.5 per cent gas ranges, 100 per cent gas water heaters, 127 per cent gas clothes dryers.

One-family Residences (Nonheating Loads). Studies carried out by the AGA Rate Committee in 1947 to 1949 showed that the maximum diversity between nonheating customers was reached when 20 or more customers were treated as a

[1] B. E. Hunt, "A Method Used in Forecasting Loads and Making Load Studies," *Proc. AGA*, 1953, pp. 602–605.

group. In a group of Indiana cities (1944)[1] the maximum coincident nonheating load per customer expressed in Btu per hour ranged from 4,440 to 9,500 with given as the average: the suggested value to use under analogous conditions was 6,870 as 8,000 Btu/hr.

Apartment Houses (Cooking Only). Metered maximum half-hour peaks were found to occur around 6:00 P.M. and approximated 2,000 Btu per customer in 30 min.

Industrial Customers. Because of varying rates of production and the wide range of gas application, no generalized data can be furnished. One approach is to estimate the maximum daily and hourly demands of the 10 largest industrial customers in five or so of the largest subclasses of industry in the area.

A 1950 study[2] yielded the following data:

	Ceramics and glass products	Bakeries	Other food processes	Metal products
Group max coincident: 9:00 A.M., ½ hr demand, therms per hr per customer.............	23.1	36.5	33.3	53.2
Daily, therms per day per customer......................	317	623	387	713

Commercial Customers. This class of service may be divided into categories such as hotels and restaurants, clubs, schools, tailors, etc. Demands other than space heating are not large compared with those of the preceding classes of customers. Furthermore, peak hours are likely to occur at different times. Some actual load data are available,[3,4] but estimates should be based on local factors if possible.

Testing Services. Service pipes should be tested according to Section 849.15 of the American Standard, Gas Transmission and Distribution Piping Systems (ASA B31.8-1963) before trenches are backfilled. This stipulates that each service shall be tested after construction and before being placed in service to demonstrate that it does not leak.

Services to operate at from 1 to 40 psig shall be given a stand-up air or gas pressure test at not less than 50 psig for at least five minutes.

Services to operate at pressures in excess of 40 psig but stressed less than 20% of the specified minimum yield shall be tested to the maximum operating pressure or 100 psig, whichever is the lesser. Services stressed to 20% or more of the specified minimum yield shall be tested in accordance with requirements for mains.

The service connection to the main need not be included in these pressure tests if it is not feasible to do so.

Flow Capacity of High-pressure Services. Because greater pressure drops are permissible than those used in low-pressure design, service-pipe capacities are substantially higher as shown in Tables 11 and 12. Illustrative problems indicate the use of the tables.

[1] Indiana Gas Association, "Distribution System Capacities," *Gas*, vol. 23, pp. 51–58, May, 1947.

[2] B. E. Hunt, and others. "Gas Service Design—Final Report of Task Group" (DMC-56-14), *Proc. AGA*, 1956, pp. 272–287.

[3] J. R. Gardner, "Report of Subcommittee on Customer Load Characteristics," *Proc. AGA*, 1950, pp. 117–143.

[4] B. E. Hunt, "A Method Used in Forecasting Loads and Making Studies," *Proc. AGA*, 1953, pp. 602–605.

Table 11. Intermediate-pressure Gas-carrying Capacities[1] in Cubic Feet per Hour[2] of Pipes[3] of Different Diameters and Lengths

(Carrying capacity in cubic feet per hour measured at standard conditions[2])

Length, feet	Nominal pipe size, inches									
	¾	1	1¼	1½	2	2½	3	4	6	8
Capacities based on 2 psi gauge[4] initial pressure and 1 psi drop through line										
50	1,170	2,270	4,620	6,950	13,550	21,850	38,800	80,300	240,000	499,000
100	825	1,570	3,280	4,930	9,600	15,500	27,450	56,900	169,100	352,500
150	675	1,285	2,660	4,020	7,830	12,600	22,400	46,400	137,000	288,000
200	585	1,110	2,320	3,490	6,790	10,920	19,450	40,200	119,800	249,500
300	475	907	1,890	2,850	5,530	8,910	15,900	32,800	97,600	204,000
400	413	786	1,640	2,470	4,800	7,720	13,780	28,400	84,700	176,300
500	370	704	1,460	2,210	4,280	6,900	12,300	25,400	75,600	157,900
1,000	262	497	1,035	1,560	3,040	4,880	8,700	18,000	53,600	111,300
Capacities based on 3 psi gauge[4] initial pressure and 2 psi drop through line										
50	1,680	3,260	6,640	10,000	19,450	31,400	55,800	115,300	344,000	716,000
100	1,190	2,260	4,710	7,080	13,800	22,200	39,450	81,700	243,000	506,000
150	969	1,850	3,830	5,780	11,240	18,130	32,200	66,600	198,200	414,000
200	840	1,600	3,330	5,010	9,740	15,700	28,000	57,700	172,100	358,000
300	685	1,303	2,710	4,100	7,940	12,810	22,800	47,100	140,300	292,500
400	594	1,130	2,360	3,540	6,890	11,100	19,780	40,750	121,900	253,000
500	530	1,010	2,100	3,170	6,150	9,900	17,670	36,400	108,800	226,500
1,000	374	713	1,485	2,240	4,350	6,990	12,480	25,000	76,800	160,000
Capacities based on 5 psi gauge[4] initial pressure and 3 psi drop through line										
50	2,145	4,170	6,490	12,780	24,900	40,100	71,300	147,500	440,000	915,000
100	1,520	2,880	6,020	9,060	17,640	28,400	50,500	104,500	311,000	648,000
150	1,240	2,360	4,900	7,400	14,390	23,200	41,200	85,200	253,500	529,000
200	1,074	2,040	4,250	6,410	12,460	20,100	35,700	73,800	220,000	458,000
300	875	1,665	3,470	5,240	10,150	16,380	29,200	60,200	179,400	374,000
400	760	1,445	3,010	4,540	8,810	14,200	25,300	52,100	155,900	324,000
500	679	1,290	2,680	4,050	7,870	12,690	22,600	46,600	139,000	290,000
1,000	479	911	1,892	2,860	5,560	8,950	15,970	32,950	98,100	205,000
Capacities based on 10 psi gauge[4] initial pressure and 5 psi drop through line										
50	3,060	5,950	12,100	18,220	35,400	57,200	101,600	210,500	627,000	1,304,000
100	2,165	4,110	8,580	12,900	25,150	40,500	71,900	149,000	443,000	923,000
150	1,770	3,370	6,980	10,530	20,500	33,100	58,700	121,300	362,000	754,000
200	1,530	2,910	6,060	9,130	17,780	28,600	50,900	105,100	313,500	652,000
300	1,245	2,380	4,950	7,460	14,470	23,350	41,600	85,700	256,000	533,000
400	1,080	2,060	4,290	6,460	12,560	20,200	36,000	74,200	222,000	462,000
500	968	1,845	3,830	5,780	11,220	18,100	32,200	66,400	198,100	413,000
1,000	684	1,304	2,710	4,080	7,830	12,780	22,800	46,900	140,000	292,000
Capacities based on 15 psi gauge[4] initial pressure and 7.5 psi drop through line										
50	4,050	7,870	16,020	24,100	47,000	75,700	134,600	279,000	830,000	1,727,000
100	2,865	5,450	11,350	17,100	33,300	53,600	95,200	197,300	586,000	1,222,000
150	2,340	4,460	9,240	13,940	27,150	43,800	77,600	160,700	479,000	996,000
200	2,030	3,860	8,020	12,100	23,500	37,900	67,400	139,200	415,000	864,000
300	1,650	3,145	6,540	9,880	19,170	30,900	55,000	113,400	339,000	705,000
400	1,430	2,730	5,680	8,550	16,620	26,800	47,700	98,400	294,000	611,000
500	1,280	2,440	5,070	7,650	14,860	23,900	42,600	88,000	262,000	547,000
1,000	904	1,585	3,580	5,400	10,490	16,900	30,100	62,100	185,300	387,000

[1] Computed by Weymouth formula for a specific gravity of 0.7 and a flow temperature of 60 F. To convert the figures in this table to any other specific gravity and/or flow temperature, multiply the values given by $\sqrt{0.7/s}$, or by $\sqrt{520/T}$, as the case may be, or combine them into one adjustment which becomes $19.08/\sqrt{sT}$.

Table II. (Continued)

Length, feet	Nominal pipe size, inches									
	¾	1	1¼	1½	2	2½	3	4	6	8

Capacities based on 20 psi gauge[4] initial pressure and 10 psi drop through line

	¾	1	1¼	1½	2	2½	3	4	6	8
50	5,000	9,720	19,780	29,800	58,000	93,500	166,200	344,000	1,025,000	2,135,000
100	3,540	6,720	14,010	21,100	41,100	66,300	117,500	243,700	725,000	1,510,000
150	2,890	5,500	11,400	17,200	33,550	54,100	95,900	198,200	591,000	1,231,000
200	2,510	4,760	9,910	14,930	29,050	46,800	83,200	172,000	513,000	1,068,000
300	2,040	3,885	8,080	12,200	23,650	38,200	68,000	140,200	418,500	871,000
400	1,768	3,370	7,010	10,560	20,500	33,100	59,000	121,500	363,000	755,000
500	1,580	3,010	6,260	9,450	18,350	29,550	52,600	108,500	324,000	676,000
1,000	1,115	2,130	4,430	6,680	12,960	20,880	37,200	76,600	229,000	478,000

[2] Measured under standard conditions of 14.7 psi abs at 60 F.
[3] Schedule 40 (standard-weight) pipe.
[4] For convenience in use, the values are grouped according to initial pressure and line pressure drop, whereas the tabular values were computed from the initial and final absolute pressures, *viz.*, gauge pressures + 14.7.

Table 12. Length Ratios for Use with Other Pressure Drops Than Those Shown in Table II

Initial pressure, psi gauge	Pressure drop, psi												
	1	2	3	4	5	6	8	10	12	14	16	18	20
	Length ratio												
2	1.000	0.515											
3	1.940	1.000	0.686										
5	2.820	1.45	1.000	0.761	0.630								
10	4.58	2.32	1.60	1.23	1.000	0.853	0.672	0.564					
15	6.33	3.39	2.28	1.76	1.41	1.22	0.946	0.785	0.684	0 611			
20	8.61	4.40	2.97	2.28	1.63	1.56	1.21	1.000	0.859	0.764	0.694	0.641	0.601

Example 1. Given a 2-in. service pipe with an initial pressure of 3 psig and a desired delivery pressure of 2 psig, what flow can be expected through an equivalent length of 200 ft?

Solution. It is first necessary to determine the length of 2-in. pipe which will have 2-psi pressure drop for the same rate of flow that would produce 1-psi pressure drop in 200 ft of 2-in. pipe. This is obtained by multiplying the given length of 200 ft by the length ratio read from Table 12, which is 1.94 for the assumed conditions. Thus the length corresponding to 2-psi pressure drop is 200 × 1.94 = 388 ft. Interpolating in Table 11 for 388 ft, the corresponding flow is found to be 7,016 cu ft/hr.

Example 2. A flow of 7,000 cu ft/hr is required through 200 ft of 2-in. service pipe. What will be the discharge pressure if the initial pressure is 3 psig?

Solution. Referring to Table 11 the length of 2-in. pipe corresponding to a flow of 7,000 cu ft/hr and 2-psi pressure drop is 389.4 ft by interpolation and from this the length ratio is 389.4/200 = 1.947. Referring to Table 12, the pressure drop for an initial pressure of 3 psig and a length ratio of 1.940 is 1 psi. Hence the discharge pressure is 3 − 1 = 2 psig.

Example 3. An initial pressure of 3 psig is available for delivering gas through an existing 2-in. service pipe. A flow of approximately 7,000 cu ft/hr is desired at a discharge pressure of 2 psig. How far can the point of delivery be from the service tap?

Solution. Interpolating for 7,000 cu ft/hr in Table 11, the length could be 389.4 ft if the allowable pressure drop were 2 psi according to the basic conditions of that table. The allowable pressure drop is only 1 psi, however, for which the length ratio read from Table 12 is 1.94. Hence the maximum permissible length is 389.4/1.94 = 200 ft.

MAINTENANCE

The design and installation of a gas-distribution system should be in accordance with the requirements of the American Standard, Gas Transmission and Distribution Piping Systems (ASA B31.8-1963).

Repairing Leaks in Steel Mains. Clamps of various types and sizes are used to stop leaks from steel mains. These clamps are placed over an appropriate sealing medium such as neoprene, etc. Some companies place a magnesium anode at each leak repair location to protect against further corrosion.

Repairing Leaks in Cast-iron Mains. Leakage at bell-and-spigot joints can be corrected mechanically with a suitable bell-joint clamp applied to a thoroughly cleaned area. Some gas companies recalk joints before installing clamps, while others prefer to use the clamp as proof of tightness by means of an immediate test.

Circumferential breaks or longitudinal cracks may be repaired by applying split cast-iron sleeves or steel band clamps over suitable synthetic or rubber gaskets.

Other repair methods for cast-iron pipe include both internal and external treatments. Internal sealing methods make use of liquids for rejuvenation of jute calking and sealants which solidify in joints and cracks. The liquid treatment is not suitable for steel lines or on soaped, cement grouted, or heavily tarred jute calked joints on cast-iron pipe. Epoxy resins constitute another type of sealant for both external and internal use, but the leaks must be stopped prior to application.

Rehydration and oil fogging of gas are not extensively used because of limited effectiveness and certain undesirable factors such as risk of freezing of exposed lines and excessive oil deposits on pressure-regulator diaphragms.

Expansion Stresses. The general subject of expansion stresses in pipes due to temperature change is discussed in Chap. 4, Expansion and Flexibility. Underground gas piping is subject to a relatively narrow span of temperature change and the gripping power of compacted earth cover and by branch connections resists most of the potential movement. Flexibility is provided by bends, loops, offsets, expansion joints, or couplings. These require anchors or ties of sufficient strength to provide for end forces. Details of flexibility calculations are given in Sections 832.38 to 833.5 of ASA B31.8-1963.

TRANSMISSION SYSTEMS

Once the economic feasibility of a transmission system has been verified and appropriate governmental approval has been obtained, detailed engineering studies for design, installation, and operation can be completed. Pipe size, maximum operating pressures, compressor station type, size, and spacing must be established. The American Standard B31.8-1963 provides the basic requirements, and these have been abstracted on the following pages.

804 SCOPE AND INTENT

804.1 This Section 8 covers the design, fabrication, installation, inspection, testing, and the safety aspects of operation and maintenance of gas transmission and distribution systems, including gas pipelines, gas compressor stations, gas metering and regulating stations, gas mains, and gas services up to the outlet of the customer's meter set assembly. Also included within the scope of this section are gas storage equipment of the closed pipe type fabricated or forged from pipe or fabricated from pipe and fittings, and gas storage lines. (See Figure 804-A in Appendix.)

804.2 The requirements of Section 8 also cover the conditions of use of the elements of the piping systems described in 804.1, including, but not limited to, pipe, valves, fittings, flanges, bolting, gaskets, regulators, pressure vessels, pulsation dampeners, and relief valves.

804.3   This Section 8 does not apply to:

(a) Design and fabrication of pressure vessels covered by the ASME Code.

(b) Piping with metal temperatures above 450 F or below minus 20 F. (For temperatures above 450 F, ASA B31.3 shall apply. For low temperatures within the range covered by this section, see 814.)

(c) Piping beyond the outlet of the customer's meter set assembly. (See ASA Z21.30 for such piping.)

(d) Piping in oil refineries or natural gasoline extraction plants, gas treating plant piping other than the main gas stream piping in dehydration and all other processing plants installed as part of a gas transmission system, gas manufacturing plants, industrial plants, or mines. (See other applicable sections of ASA B31.)

(e) Vent piping to operate at substantially atmospheric pressures for waste gases of any kind.

(f) Wellhead assemblies, including control valves, and flow lines between wellhead and trap or separator, or casing and tubing in gas or oil wells.

(g) Proprietary items of equipment, apparatus, or instruments.

(h) Heat exchangers.

(i) Oil or liquid products pipelines.

(j) Prefabricated units which employ plate and longitudinal welds as contrasted to pipe.

804.4   The requirements of Section 8 are adequate for safety under conditions normally encountered in the gas industry.   Requirements for abnormal or unusual conditions are not specifically provided for, nor are all details of engineering and construction prescribed.   It is intended that all work performed within the scope of this section shall meet or exceed the safety standards expressed or implied herein.

804.5   This section is concerned with:

(a) Safety of the general public.

(b) Employee safety to the extent that it is affected by basic design, quality of the materials and workmanship, and requirements for testing and maintenance of gas transmission and distribution facilities.   Existing industrial safety regulations pertaining to work areas, safety devices, and safe work practices are not intended to be supplanted by this Code.

804.6   It is not intended that this code be applied retroactively to existing installations insofar as design, fabrication, installation, established operating pressure, and testing are concerned.   It is intended, however, that the provisions of this Code shall be applicable to the operation, maintenance, and up-rating of existing installations.

841   STEEL PIPE.

841.001   *Population Density Indexes.*

(a) Two population density indexes, determined at the time of initial construction, are used to classify locations for design and testing purposes: (1) the one-mile density index, which applies to any specific mile of pipeline; and (2) the ten-mile density index, which applies to any specific ten-mile length of pipeline.

(b) To determine the one-mile density indexes for a proposed pipeline, lay out a zone one-half mile wide along the route of the pipeline with the pipeline on the center line of this zone.   Divide the zone into lengths, each containing one mile of pipeline.   Count the number of buildings intended for human occupancy in each of these lengths.   These numbers are the one-mile indexes for the pipeline.

(c) To determine the ten-mile density indexes for any given ten-mile length of pipeline, proceed as follows: Add the one-mile density indexes for the ten-mile section.   In case a one-mile index equals or exceeds 20, it is to be included in the sum as 20.   Divide the sum thus obtained by 10.   The quotient is the ten-mile density index for the section.

841.01   *Classification of Locations*

841.011   *Class 1 Locations:* Class 1 locations include waste lands, deserts, rugged mountains, grazing land, and farm land, and combinations of these; provided, however, that:

(a) The ten-mile density index for any section of the line is 12 or less.

(b) The one-mile density index for any one mile of line is 20 or less.

841.012   *Class 2 Locations:* Class 2 locations include areas where the degree of development is intermediate between Class 1 locations and Class 3 locations.   Fringe areas around cities and towns, and farm or industrial areas where the one-mile density index exceeds 20 or the ten-mile density index exceeds 12 fall within this location class.[1]

841.013   *Class 3 Locations:* Class 3 locations include areas subdivided for residential or commercial purposes where, at the time of construction of the pipeline or piping system, 10% or more of the lots abutting on the street or right-of-way in which the pipe is to be located are built upon, and a Class 4 classification is not called for.   This permits classifying as Class 3, areas completely occupied by commercial or residential buildings with the prevalent height of three stories or less.

841.014   *Class 4 Locations:* Class 4 locations include areas where multistory[2] buildings are prevalent, and where traffic is heavy or dense and where there may be numerous other utilities underground.

801.015   It should be emphasized that *Location Class* (1, 2, 3, or 4), as described in the foregoing paragraphs, is defined as the general description of a geographic area having certain characteristics as a basis for prescribing the types of construction and methods of testing to be used in those locations or in areas that are respectively comparable.   A numbered Location-Class refers only to the geography of that location or a similar area, and does not necessarily indicate that a correspondingly numbered Construction-Type will suffice for all construction in that particular location or area.   Example: In Location Class 1, all aerial crossings require Type B construction.   (See 841.143.)

801.016   When classifying locations for the purpose of determining the type of pipeline construction and testing that should be prescribed, due consideration shall be given to the possibility of future development of the area.   If at the time of planning a new pipeline this future development appears likely to be sufficient to change the location class, this should be taken into consideration in the design and testing of the proposed pipeline.

It is also anticipated that some increase in population density will occur in all areas after a line is constructed, and this possibility has been taken into account in establishing the design, construction, and testing procedures for each location class.

841.1   Steel Pipe Design Formula.   The design pressure for steel gas piping systems or the nominal wall thickness for a given design pressure shall be determined by the following formula:

$$P = \frac{2St}{D} \times F \times E \times T$$

(For exceptions see 841.4)
Where:

P = Design pressure, psig.
S = Specified minimum yield strength, psi, stipulated in the specifications under which the pipe was purchased from the manufacturer or determined in accordance with 811.27 h.   The specified minimum yield strengths of some of the more commonly used piping steels, whose specifications are incorporated by reference herein, are tabulated for convenience in Appendix "C."   For special limitation on S see 841.14(e) and (f).
D = Nominal outside diameter of pipe, inches.
t = Nominal wall thickness, inches.
F = Construction type design factor obtained from 841.11.
E = Longitudinal joint factor obtained from 841.12.
T = Temperature derating factor obtained from Table 841.13.

[1] It is not intended here that a full mile of lower-stress-level pipeline shall be installed if there are physical barriers or other factors that will limit the further expansion of the more densely populated area to a total distance of less than 1 mile.   It is intended, however, that where no such barriers exist, ample allowance shall be made in determining the limits of the lower-stress design to provide for probable further development in the area.

[2] Multistory means 4 or more "floors" above ground including the first or ground floor. The depth of basements or number of basement floors is immaterial.

841.02 Classification of Steel Pipe Construction[1]

Four types of steel pipe construction are prescribed in this Code. The distinguishing characteristics of each type and the location in which each type shall be used are as follows:

Characteristics	Type A Construction 0.72	Type B Construction 0.60	Type C Construction 0.50	Type D Construction 0.40
A. 1. Design Factor F (See 841.11)				
B. Location Where Type of Construction Shall be Used	(a) On private rights of way in Class 1 locations. (b) Parallel encroachments on: Privately owned roads in Class 1 locations. Unimproved roads in Class 1 locations. (c) Crossings without casings of privately owned roads in Class 1 locations	(a) On private rights of way in Class 2 locations. (b) Parallel encroachments on: Privately owned roads in Class 2 locations. Unimproved public roads in Class 2 locations. Hard surfaced roads, highways or public streets and railroads in Class 1 and Class 2 locations.	(a) On private rights of way in class 3 locations. (b) Parallel encroachments on: Privately owned roads in Class 3 locations. Unimproved public roads in Class 3 locations. Hard surfaced roads, highways or public streets and railroads in Class 3 locations.	(a) In all locations in location Class 4.

13–32

(d) Crossings in casings of unimproved public roads. hard-surfaced roads, highways or public streets and railroads in Class 1 locations.

(c) Crossings without casings of:
Privately owned roads in Class 2 locations.
Unimproved public roads in Class 2 locations.
Hard surfaced roads, highways or public streets and railroads in Class 1 locations.
(d) Crossings in casings of:
Hard surfaced roads, highways or public streets and railroads in Class 2 locations.
(e) On bridges in Class 1 and Class 2 locations. (See 841.143)
(f) Fabricated assemblies in pipelines in location Classes 1 and 2. (See 841.142)

(c) Crossings without casings of:
Privately owned roads in Class 3 locations.
Unimproved public roads in Class 3 locations.
Hard surfaced roads, highways or public streets and railroads in Class 2 and 3 locations.
(d) Compressor station piping.

[1] It is necessary to distinguish between construction types, as defined by Section A of this table, and location classes, as defined in 841.01 to avoid confusion. If pipelines or mains are located in private rights-of-way, the Code prescribes that Type A construction be used in Class 1 locations, Type B construction in Class 2 locations, Type C construction in Class 3 locations, and Type D construction in Class 4 locations. There are many exceptions to this association of Class 1 with Type A, etc., however, as Table 841.02 shows, most of which are cases where pipelines or mains are located in highways or on bridges, etc.

841.03   Construction Types Required for Parallel Encroachments of Pipelines and Mains on Roads and Railroads

Kind of Thoroughfare	Construction Type Required			
	Location Class 1	Location Class 2	Location Class 3	Location Class 4
(a) Privately owned roads	Type A	Type B	Type C	Type D
(b) Unimproved public roads	Type A	Type B	Type C	Type D
(c) Hard surface roads, highways or public streets and railroads	Type B	Type B	Type C	Type D

841.04   Construction Types Required for Pipelines and Mains Crossing Roads and Railroads

Kind of Thoroughfare	Construction Type Required			
	Location Class 1	Location Class 2	Location Class 3	Location Class 4
(a) Privately owned roads	Type A without casing	Type B without casing	Type C without casing	Type D without casing
(b) Unimproved public roads	Type A with casing Type B without casing	Type B without casing	Type C without casing	Type D without casing
(c) Hard surface roads, highways or public streets and railroads	Type A with casing Type B without casing	Type B with casing Type C without casing	Type C without casing	Type D without casing

Table 841.11   Values of Design Factor *F*

Construction Type (See 841.02)	Design Factor *F*
Type—A	0.72
Type—B	0.60
Type—C	0.50
Type—D	0.40

Table 841.12   Longitudinal Joint Factor *E*

Spec. Number	Pipe Class	*E* Factor
ASTM A53	Seamless	1.00
	Electric Resistance Welded	1.00
	Furnace Butt Welded	0.60
ASTM A106	Seamless	1.00
ASTM A134	Electric Fusion Arc Welded	0.80
ASTM A135	Electric Resistance Welded	1.00
ASTM A139	Electric Fusion Welded	0.80
ASTM A155	Electric Fusion Arc Welded	1.00
ASTM A211	Spiral Welded Steel Pipe	0.80
ASTM A381	Double-Submerged-Arc-Welded	1.00
API 5L	Seamless	1.00
	Electric Resistance Welded	1.00
	Electric Flash Welded	1.00
	Furnace Butt Welded	0.60
API 5LX	Seamless	1.00
	Electric Resistance Welded	1.00
	Electric Flash Welded	1.00
	Submerged Arc Welded	1.00

Note.   Definitions for the various classes of welded pipe are given in Paragraph 805.51.

Table 84.113 Temperature Derating Factor *T* for Steel Pipe

Temperature Degrees Fahrenheit	Temperature Derating Factor *T*
250 F or less	1.000
300 F	0.967
350 F	0.933
400 F	0.900
450 F	0.867

Note. For intermediate temperatures interpolate for derating factor.

841.14 *Limitations of Pipe Design Values.*

(a) *P* for furnace butt welded pipe shall not exceed the restrictions of 841.1 or 60% of the mill test pressure, whichever is the lesser.

(b) *P* shall not exceed 85% of the mill test pressure for all other pipes; provided, however, that pipe, mill tested to a pressure less than 85% of the pressure required to produce a stress equal to the specified minimum yield, may be retested with a mill type hydrostatic test or tested in place after installation. In the event the pipe is retested to a pressure in excess of the mill test pressure, then *P* shall not exceed 85% of the retest pressure rather

Table 841.141 Least Nominal Wall Thicknesses (Inches)

Nominal Pipe size (inches)		Outside Diameter (inches)	Location Classes (Note 1) 1	Location Classes (Note 1) 2, 3, & 4	Compressor Stations	
⅛″	Threaded or Plain End	0.405	0.068	0.068	0.095	Threaded or Plain End
¼″		0.540	0.088	0.088	0.119	
⅜″		0.675	0.091	0.091	0.126	
½″		0.840	0.109	0.109	0.147	
¾″		1.050	0.113	0.113	0.154	
1″		1.315	0.133	0.133	0.179	
1¼″		1·660	0.140	0.140	0.191	
1½″		1.900	0.145	0.145	0.200	
2″		2.375	0.154	0.154	0.218	
					0.203	
2½″		2.875	0.103	*0.125	0.216	
3″		3.500	0.104	*0.125	0.226	
3½″		4.000	0.104	*0.125	0.237	
4″		4.500	0.104	*0.125		
5″		5.563	0.104	*0.125	0.250	

Nominal Pipe size (inches)		Outside Diameter (inches)	Location Classes (Note 1) 1		2		3 & 4		Compressor Stations	
6″	Plain End Only	6.625	0.104	Plain End Only	0.134	Plain End Only	0.156	Plain End Only	0.250	Plain End Only
8″		8.625	0.104		0.134		0.172		0.250	
10″		10.75	0.104		0.164		0.188		0.250	
12″		12.75	0.104		0.164		0.203		0.250	
14″		14.0	0.134		0.164		0.210		0.250	
16″		16.0	0.134		0.164		0.219		0.250	
18″		18.0	0.134		0.188		0.250		0.250	
20″		20.0	0.134		0.188		0.250		0.250	
22″, 24″, 26″		22″, 24″, 26″	0.164		0.188		0.250		0.250	
28″, 30″		28″, 30″	0.164		0.250		0.281		0.281	
32″, 34″, 36″		32″, 34″, 36″	0.164		0.250		0.312		0.312	

Note 1. If threaded pipe is to be used in those sizes for which least nominal wall thicknesses are given for "plain end pipe only," those sizes marked by * shall be increased as follows: 2½″ to 0.203, 3″ to 0.216, 3½″ to 0.226, 4″ to 0.237, 5″ to 0.258, and add 0.100 inch to all other wall thicknesses given in Table 841.141.

than the initial mill test pressure.   It is mandatory to use a liquid as the test medium in all tests in place after installation where the test pressure exceeds the mill test pressure.   This paragraph is not to be construed to allow an operating pressure or design pressure in excess of that provided for by 841.1.

(c) Transportation, installation or repair of pipe shall not reduce the wall thickness at any point to a thickness less than 90% of the nominal wall thickness as determined by 841.1 for the design pressure to which the pipe is to be subjected.

(d) "*t*" shall not be less than shown in Table 841.141.

(e) When pipe that has been cold worked for the purpose of meeting the specified minimum yield strength is heated to 600 F or higher (welding excepted), the maximum allowable pressure at which it can be used shall not exceed 75% of the value obtained by use of the steel pipe design formula given in 841.1.

(f) In no case where the Code refers to the specified minimum value of a physical property can the actual value of the property be substituted in design calculations, unless the actual is less than the specified minimum.

841.142   *Fabricated Assemblies.*   When fabricated assemblies, such as connections for separators, main line valve assemblies, cross-connections, river crossing headers, etc., are to be installed in areas defined as location Class 1, Type B construction is required throughout the assembly, and for a distance of 5 pipe diameters in each direction beyond the last fittings.   Transition pieces at the end of an assembly and elbows used in place of pipe bends are not considered as fittings under the requirements of this paragraph.   See also 830.

841.143   Pipelines or mains supported by railroad, vehicular, pedestrian, or pipeline bridges shall be in accordance with the construction type prescribed for the area in which the bridge is located, except that in Class 1 locations Type B construction shall be used.

841.16   *Cover and Casing Requirements Under Railroads, Roads, Streets, or Highways.*

(a) All buried pipelines, mains, and casings when used, shall be installed with a minimum cover of 24 inches unless otherwise provided herein.

(b) Buried pipelines and mains operating at hoop stresses of less than 20% of the specified minimum yield strength and located within private rights-of-way, private thoroughfares, sidewalks or parkways may be installed with less than the minimum cover of 24 inches if it appears that external damage to the pipe will not be likely to result.

(c) Abandoned pipe having a cover less than 24 inches may be used as a casing or conduit for pipelines and mains operating at hoop stresses less than 20% of the specified minimum yield strength.

(d) Buried pipelines and mains installed in areas where farming or other operations might result in deep plowing, or in thoroughfares or other locations where grading is done, or where the area is subject to erosion, should be provided with more cover than the minimum otherwise required.

(e) Where it is impractical to comply with the provisions of 841.16(a) and it is necessary to prevent damage from external loads, the pipe shall be cased or bridged.

(f) Casings shall be designed to withstand the super-imposed loads.   Where there is a possibility of water entering the casing, the ends of the casing shall be sealed.   If the end sealing is of a type that will retain the full pressure of the pipe, the casing shall be designed for the same pressure as the pipe but according to Type A construction requirements. Venting of sealed casings is not mandatory; however, if vents are installed they should be protected from the weather to prevent water from entering the casing.

841.161   *Clearance Between Pipelines or Mains and Other Underground Structures.*   There should be at least 2 inches clearance wherever possible between any gas main or pipeline and any other underground structure not used in conjunction with the pipeline or main. When this clearance cannot be attained, other suitable precautions to protect the pipe shall be taken, such as the installation of insulating material, installation of casing, etc.

841.221   The operating company shall make provision for suitable inspection.   Inspectors shall be qualified by either experience or training.

841.222   The installation inspection provisions for pipelines and other facilities to operate at hoop stresses of 20% or more of the specified minimum yield strength should be adequate

either to make possible the following inspections at sufficiently frequent intervals or to do other things that will assure good quality of workmanship.

(a) Inspect the surface of the pipe for serious surface defects just prior to the coating operation. See 841.242(a).

(b) Inspect the surface of the coated pipe as it is lowered into the ditch to find coating lacerations that indicate the pipe might have been damaged after being coated. Damage during the lowering-in process should be found during this inspection.

(c) Inspect the fit-up of the joints before the weld is made.

(d) Visually inspect the stringer beads before subsequent beads are applied.

(e) Inspect the completed welds before they are covered with coating.

(f) Inspect the conditions of the ditch bottom just before the pipe is lowered in.

(g) Inspect the fit of the pipe to the ditch before backfilling.

(h) Inspect all repairs, replacements, or changes ordered before they are covered up.

(i) Perform such special tests and inspections as are required by the specifications, such as the radiographing of a portion of the welds and the electrical testing of the protective coating.

841.272 *Installation of Pipe in the Ditch.* On pipelines operating at stresses of 20% or more of the specified minimum yield strength, it is very important that stresses induced into the pipeline by construction be minimized. This includes grading the ditch so that the pipe has a firm substantially continuous bearing on the bottom of the ditch. The pipe shall fit the ditch without the use of external force to hold it in place until the backfill is completed. When long sections of pipe that have been welded alongside the ditch are lowered in, care shall be exercised so as not to jerk the pipe or impose any strains that may kink or put a permanent bend in the pipe. Slack loops are not prohibited by this paragraph where laying conditions render their use advisable.

841.285 *Purging of Pipelines and Mains.*

(a) When a pipeline or main full of air is placed in service, the air in it can be safely displaced with gas provided that a moderately rapid and continuous flow of gas is introduced at one end of the line and the air is vented out the other end. The gas flow should be continued without interruption until the vented gas is free from air. The vent should then be closed.

(b) In cases where gas in a pipeline or main is to be displaced with air and the rate at which air can be supplied to the line is too small to make a procedure similar to, but the reverse of that described in 841.285(a) feasible, a slug of inert gas should be introduced to prevent the formation of an explosive mixture at the interface between gas and air. Nitrogen or carbon dioxide can be used for this purpose.

(c) If a pipeline or main containing gas is to be removed, the operation may be carried out in accordance with 841.282 or the line may be first disconnected from all sources of gas and then thoroughly purged with air, water or with inert gas before any welding is done.

(d) If a gas pipeline or main or auxiliary equipment is to be filled with air after having been in service and there is a reasonable possibility that the inside surfaces of the facility are wetted with a volatile inflammable liquid, or if such liquids might have accumulated in low places, purging procedures designed to meet this situation shall be used. Steaming of the facility until all combustible liquids have been evaporated and swept out is recommended. Filling of the facility with an inert gas and keeping it full of such gas during the progress of any work that might ignite an explosive mixture in the facility is an alternative recommendation. The possibility of striking static sparks within the facility must not be overlooked as a possible source of ignition.

841.4 Test Requirements.

841.41 *Test Required to Prove Strength of Pipelines and Mains to Operate at Hoop Stresses of 30% or More of the Specified Minimum Yield Strength of the Pipe.*

841.411 All pipelines and mains to be operated at a hoop stress of 30% or more of the specified minimum yield strength of the pipe shall be given a field test to prove strength after construction and before being placed in operation.

841.412 (a) Pipelines and mains located in Location Class 1 shall be tested either with air or gas to 1.1 times the maximum operating pressure or hydrostatically to at least 1.1 times the maximum operating pressure. See 841.5.

(b) Pipelines or mains located in Location Class 2 shall be tested either with air to 1.25 times the maximum operating pressure or hydrostatically to at least 1.25 times the maximum operating pressure.   See 841.5.

(c) Pipelines and mains in Location Classes 3 and 4 shall be tested hydrostatically to a pressure not less than 1.4 times the maximum operating pressure.

(d) The test requirements given in 841.412(a), (b), and (c) above are summarized in Table 841.412(d).

Table 841.412(d)   Test Requirements for Pipelines and Mains to Operate at Hoop Stresses of 30% or More of the Specified Minimum Yield Strength of the Pipe

1	2	3	4	5
		Prescribed Test Pressure		Maximum Allowable Operating Pressure
Location Class	Permissible Test Fluid	Minimum	Maximum	the lesser of
1	Water	1.1 × m.o.p.	None	t.p. ÷ 1.1
	Air	1.1 × m.o.p.	1.1 × d.p.	or
	Gas	1.1 × m.o.p.	1.1 × d.p.	d.p.
2	Water	1.25 × m.o.p.	None	t.p. ÷ 1.25
				or
	Air	1.25 × m.o.p.	1.25 × d.p.	d.p.
3	Water	1.40 × m.o.p.	None	t.p. ÷ 1.40 or d.p.
4	Water	1.40 × m.o.p.	None	t.p. ÷ 1.40 or d.p.

m.o.p. = maximum operating pressure (not necessarily the maximum allowable operating pressure)
d.p. = design pressure
t.p. = test pressure

Note. This table brings out the relationship between test pressures and maximum allowable operating pressures subsequent to the test.   If an operating company decides that the maximum operating pressure will be less than the design pressure a corresponding reduction in prescribed test pressure may be made as indicated in Column 3.   However, if this reduced test pressure is used the maximum operating pressure cannot later be raised to the design pressure without retesting the line to the test pressure prescribed in Column 4.   See 805.14, 845.22, and 845.23.

841.413   Requirements of 841.412(c) for hydrostatic testing of mains and pipeline in Location Classes 3 and 4 do not apply if at the time the pipeline or main is first ready for test, one or both of the following conditions exist:

(a) The ground temperature at pipe depth is 32 F or less, or might fall to that temperature before the hydrostatic test could be completed, or

(b) Water of satisfactory quality is not available in sufficient quantity.

(c) In such cases an air test to 1.1 times the maximum operating pressure shall be made and the limitations on operating pressure imposed by 841.412(d) above do not apply.

841.414   Other provisions of this Code notwithstanding, pipelines and mains crossing highways and railroads may be tested in each case in the same manner and to the same pressure as the pipeline on each side of the crossing.

841.415   Other provisions of this Code notwithstanding, fabricated assemblies, including mainline valve assemblies, cross connections, river crossing headers, etc., installed in pipelines in Class 1 locations and designed in accordance with Type B construction, as required in 841.412, may be tested as required for Class 1 locations.

841.416 Notwithstanding the limitations on air testing imposed in 841.412(c), air testing may be used in Location Classes 3 and 4, provided that all of the following conditions apply:

(a) The maximum hoop stress during test is less than 50% of the specified minimum yield strength in Class 3 locations, and less than 40% of the specified minimum yield strength in Class 4 locations.

(b) The maximum pressure at which the pipeline or main is to be operated does not exceed 80% of the maximum field test pressure used.

(c) The pipe involved is new pipe having a longitudinal joint factor E in Table 841.12 of 1.00.

841.417 *Records.* The operating company shall maintain in its file for the useful life of each pipeline and main, records showing the type of fluid used for test and the test pressure.

841.42 *Tests required to prove strength for pipelines and mains to operate at less than 30%* *of the specified minimum yield strength of the pipe, but in excess of* 100 *psi.* Steel piping that is to operate at stresses less than 30% of the specified minimum yield strength in location Class 1 shall at least be tested in accordance with 841.43. In location Classes 2, 3, and 4, such piping shall be tested in accordance with Table 841.412(d), except that gas or air may be used as the test medium within the maximum limits set in Table 841.421.

Table 841.421 Maximum Hoop Stress Permissible During Test

Location Class	Per cent of Specified Minimum Yield Strength		
Test Medium	2	3	4
Air	75	50	40
Gas	30	30	30

841.43 *Leak Tests for Pipelines or Mains to Operate at* 100 *psi or More.*

841.431 Each pipeline and main shall be tested after construction and before being placed in operation to demonstrate that it does not leak. If the test indicates that a leak exists, the leak or leaks shall be located and eliminated, unless it can be determined that no undue hazard to public safety exists.

841.432 The test procedure used shall be capable of disclosing all leaks in the section being tested and shall be selected after giving due consideration to the volumetric content of the section and to its location.

841.433 In all cases where a line is to be stressed in a strength-proof test to 20% or more of the specified minimum yield strength of the pipe, and gas or air is the test medium, a leak test shall be made at a pressure in the range of 100 psi to that required to produce a hoop stress of 20% of the minimum specified yield, or the line shall be walked while the hoop stress is held at approximately 20% or the specified minimum yield.

841.44 *Leak Tests for Pipelines and Mains to Operate at Less Than* 100 *psi.*

841.441 At the time of or prior to placing in operation distribution mains and related equipment to operate at less than 100 psi, they shall be tested to determine that they are gas-tight.

841.442 Gas may be used as the test medium at the maximum pressure available in the distribution system at the time of the test. In this case the soap bubble test may be used to locate leaks if all joints are accessible during the test.

841.443 Testing at available distribution system pressures as provided for above in 841.442 may not be adequate if substantial protective coatings are used that would seal a split pipe seam. If such coatings are used, the leak test pressure shall be 100 psi.

845   CONTROL AND LIMITING OF GAS PRESSURE.

845.1   Basic Requirement for Protection Against Accidental Overpressuring. Every pipeline, main, distribution system, customer's meter and connected facilities, compressor station, pipe-type holder, bottle-type holder, container fabricated from pipe and fittings, and all special equipment, if connected to a compressor or to a gas source where the failure of pressure control or other causes might result in a pressure which would exceed the maximum allowable operating pressure of the facility (refer to 805.14), shall be equipped with suitable pressure relieving or pressure limiting devices.   Special provisions for service regulators are set forth under paragraph 845.5.

845.2   Control and Limiting of Gas Pressure in Holders, Pipelines, and All Facilities that Might at Times be Bottle Tight.

845.21   Suitable types of protective devices to prevent overpressuring of such facilities include:

(a) Spring-loaded relief valves of types meeting the provisions of the ASME Unfired Pressure Vessel Code.

(b) Pilot-loaded back-pressure regulators used as relief valves, so designed that failure of the pilot system or control lines will cause the regulator to open.

845.22   *Maximum Allowable Operating Pressure for Steel Pipelines or Mains.*   This pressure is by definition the maximum operating pressure to which the pipeline or main may be subjected in accordance with the requirements of this Code.   For a pipeline or main in good operating condition, the maximum allowable operating pressure is the lesser of the two pressures described in (a) and (b) below.

(a) The design pressure (defined in 805.11) of the weakest element of the pipeline or main.   Assuming that all fittings, valves, and other accessories in the line have an adequate pressure rating, the maximum allowable operating pressure of a steel pipeline or main shall be the design pressure determined in accordance with 841.1.

(b) The pressure obtained by dividing the pressure to which the pipeline or main is tested after construction by the appropriate factor for the location class involved, as follows:

Class No. Location	Pressure
1	$\dfrac{\text{Test Pressure}}{1.10}$
2	$\dfrac{\text{Test Pressure}}{1.25}$
3	$\dfrac{\text{Test Pressure}^1}{1.40}$
4	$\dfrac{\text{Test Pressure}^1}{1.40}$

[1] Other factors than 1.4 should be used if the line was tested under the special conditions described in 841.413, 841.416, and 841.42.   In such cases use factors that are consistent with the applicable requirements of these sections.

(c) In some cases the operating company will consider that the maximum operating pressure to which a pipeline or main should be subjected is less than the pressure determined by either (a) or (b) above.   Pipelines that are known to be seriously corroded or that have other defects seriously affecting their strength and which have been operated for years at lower pressures, fall into this category.   In such cases the operating company shall decide the maximum pressure it considers safe, and shall install over-pressure protective devices designed to prevent accidentally exceeding this maximum pressure, if there is a reasonable possibility that the pressure will be exceeded.

(d) If services are connected to the pipeline or main, there are additional considerations that might in some cases limit the maximum allowable operating pressure of the facility. See 845.33.

845.23 *Qualifying a Steel Pipeline or Main for a New and Higher Maximum Allowable Operating Pressure.* Note: This paragraph applies to pipelines or mains where the new and higher maximum allowable operating pressure will produce a hoop stress of 30% or more of the specified minimum yield strength of the pipe. When the new and higher maximum allowable operating pressure is equal to or less than this value the provisions of 845.34 shall apply.

Before increasing the maximum allowable operating pressure of a pipeline or main that has been operating for a period of several years or more at a pressure less than that determined by 845.22(a) above, it is required that:

(a) The following investigative and corrective measures be taken:

(1) The design and previous testing of the pipeline and the materials and equipment in it be reviewed to determine that the proposed increase in allowable operating pressure is safe and in general agreement with the requirements of this Code.

(2) The condition of the line be determined by field inspections, examination of maintenance records, or other suitable means.

(3) Repairs, replacements or alterations in the pipeline disclosed to be necessary by steps (1) and (2) be made.

(b) The maximum allowable operating pressure may be increased after compliance with (a) above and one of the following provisions:

(1) If the physical condition of the line as determined by (a) above indicates that the line is capable of withstanding the desired increased operating pressure in accordance with the design requirements of this Code and the line has previously been tested to a pressure equal to or greater than that required by this Code for a new line for the proposed new maximum allowable operating pressure, the line may be operated at the increased maximum allowable operating pressure.

(2) If the physical condition of the line as determined by (a) above indicates that the ability of the line to withstand the increased maximum operating pressure has not been satisfactorily verified or the line has not been previously tested to the levels required by this edition of the Code for a new line for the proposed new maximum allowable operating pressure, the line may be operated at the increased maximum allowable operating pressure if the line shall successfully withstand the test required by this edition of the Code for a new line to operate under the same conditions.

(3) If, under the foregoing provisions of (b) above, it is necessary to test a pipeline or main before it can be up-rated to a new maximum allowable operating pressure, and if it is not practical to test the line either because of the expense or difficulties created by taking it out of service, or because of other operating conditions, a new and higher maximum allowable operating pressure may be established as follows:

3.1 Perform the requirements of (a) above.

3.2 Select a new maximum allowable operating pressure consistent with the condition of the line and the design requirements of this Code; provided, however, that,

3.3 In no such case shall the new maximum allowable operating pressure exceed 80% of that permitted for a new line of the same design in the same location.

(c) In no case shall the maximum allowable operating pressure of a pipeline be raised to a value higher than would be permitted by this code for a new line constructed of the same materials and in the same locations.

The rate of pressure increase to the new maximum allowable operating pressure should be gradual so as to allow sufficient time for periodic observations of the pipeline.

845.3 Control and Limiting of Gas Pressure in High-Pressure Steel or Cast Iron Distribution Systems.

845.31 Each high-pressure distribution system or main, supplied from a source of gas which is at a higher pressure than the maximum allowable operating pressure for the system, shall be equipped with pressure regulating devices of adequate capacity, and designed to meet the pressure, load and other service conditions under which they will operate or to which they may be subjected.

845.32 In addition to the pressure-regulating devices prescribed in 845.31, a suitable method shall be provided to prevent accidental over-pressuring of a high-pressure distribution system.

Suitable types of protective devices to prevent overpressuring of high pressure distribution systems include:

(a) Relief valves as prescribed in 845.21(a) and (b).

(b) Weight-loaded relief valves.

(c) A monitoring regulator installed in series with the primary pressure regulator.

(d) A series regulator installed upstream from the primary regulator, and set to continuously limit the pressure on the inlet of the primary regulator to the maximum allowable operating pressure of the distribution system or less.

(e) An automatic shut-off device installed in series with the primary pressure regulator, and set to shut off when the pressure on the distribution system reaches the maximum allowable operating pressure, or less. This device must remain closed until manually reset. It should not be used where it might cause an interruption in service to a large number of customers.

845.33   *Maximum Allowable Operating Pressure for High-Pressure Distribution Systems.* This pressure shall be the maximum pressure to which the system can be subjected in accordance with requirements of this Code. It shall not exceed:

(a) The design pressure of the weakest element of the system as defined in 805.11.

(b) 60 psig if the services in the system are not equipped with series regulators or other pressure limiting devices as prescribed in 845.53.

(c) 25 psig in cast iron systems having unreinforced bell and spigot joints as prescribed in 842.15(a).

(d) The pressure limits to which the joint could be subjected without possibility of parting.

(e) 2 psig in high-pressure distribution systems equipped with service regulators not meeting the requirements of 845.51 and which do not have an overpressure protective device as required in 845.52.

In some cases the operating company will consider the maximum pressure to which a system should be subjected is less than the pressure obtained by applying the applicable limits in 845.33(a), (b), (c), (d), or (e). Systems that are known to be corroded and that have been operated for years at lower pressures than these limits fall into this category. In such cases the operating company shall decide the maximum pressure it considers safe, and shall install overpressure protective devices to prevent accidentally exceeding this maximum pressure if there is a reasonable possibility that the pressure will be exceeded.

845.34   *Qualifying a High-Pressure Steel Distribution System for a New and Higher Maximum Allowable Operating Pressure.*   Note: This paragraph applies to high-pressure distribution mains and to pipelines where the new and higher maximum allowable operating pressure is less than that required to produce a hoop stress of 30 % of the specified minimum yield strength of the pipe. When the new and higher maximum allowable operating pressure is more than this value the provisions of 845.23 shall apply.

(a) Before increasing the maximum allowable operating pressure of a high-pressure distribution system, that has been operating at less than the applicable maximum pressure stated in 845.33, to a new maximum allowable operating pressure equal to or less than the maximum applicable pressure in 845.33, it is recommended that the following factors be taken in consideration:

(1) The design of the system including kinds of material and equipment used.

(2) Past maintenance records including results of any previous leakage surveys.

(b) Before increasing the pressure the following steps should be taken:

(1) Make a leakage survey, if past maintenance records indicate that such a survey is advisable, and repair leaks found.

(2) Repair or replace parts of the system found to be inadequate for the higher operating pressure.

(3) Install suitable devices on the services to regulate and limit the pressure of the gas in accordance with 845.53 if the new maximum allowable operating pressure is to be over 60 psig.

(4) Adequately reinforce or anchor offsets, bends, and dead ends in coupled pipe to avoid movement of the pipe should the offset, bend, or dead end be exposed in an excavation.

(c) The rate of pressure increase to the new maximum allowable operating pressure should be gradual so as to allow sufficient time for periodic observations of the system.

845.35 *Qualifying a Cast Iron High Pressure Main or System for a New and Higher Maximum Allowable Operating Pressure.*

(a) The maximum allowable operating pressure of a cast iron main or system shall not be increased to a pressure in excess of that permitted in 845.33 or 842.11, whichever is the lesser. Where records are not complete enough to permit the direct application of 842.11, the following procedures shall be used:

(1) *Laying Condition:* Where the original laying condition cannot be ascertained, it shall be assumed that Condition D exists (i.e., pipe supported on blocks, backfill tamped).

(2) *Cover:* Unless the actual maximum cover depth is known with certainty, it shall be determined by exposing the main or system at three or more points and making actual measurements. The main or system shall be exposed in those areas where the cover depth is most likely to be greatest. The greatest measured cover depth shall be used for computations.

(3) *Nominal Wall Thickness:* Unless the nominal thickness is known with certainty, it shall be determined by cutting coupons from three or more separate pipe lengths and making actual wall thickness measurements. The coupons shall be cut from pipe lengths located in those areas where the cover depth is most likely to be greatest. The average of all measurements taken shall be increased by the allowance indicated in the following table:

Pipe size (inches)	Allowance (inches)	
	Pit cast pipe	Centrifugally cast pipe
3–8	0.075	0.065
10–12	0.08	0.07
14–24	0.08	0.08
30–48	0.09	0.09
54–60	0.09	. . .

The nominal wall thickness shall be that standard thickness listed in Table 10 of ASA A21.1 nearest the value thus obtained, or Table 11 whichever is applicable.

(4) *Manufacturing Process:* Unless the pipe manufacturing process is known with certainty, it shall be assumed to be pit cast pipe having a bursting tensile strength (S) of 11,000 psi and a modulus of rupture (R) of 31,000 psi.

(b) Before increasing the maximum allowable operating pressure, the following measures shall be taken:

(1) Review the design of the main or system, the materials and equipment used, and previous tests.

(2) Determine the condition of the main or system from past maintenance records including results of previous leakage surveys.

(3) Make a leakage survey or a gas detector survey.

(4) Repair, reinforce, or replace any part of the main or system found to be inadequate for the higher operating pressure.

(5) Adequately reinforce or anchor offsets, bends, and dead ends in coupled or bell and spigot pipe to avoid movement of the pipe should the offset, bend, or dead end be exposed by excavation.

(6) Install suitable devices on the services to regulate and limit the pressure of the gas in accordance with 845.53 if the new and higher maximum allowable operating pressure is to be over 60 psig.

(c) If after compliance with 845.35(b) above, it is established that the main or system is capable of safely withstanding the proposed new and higher maximum allowable operating pressure, the pressure in the main or system shall be increased gradually, in steps, to the new maximum allowable operating pressure. After each incremental pressure increase, the main or system shall be checked before making the next increase to determine the effect of the previous pressure increase.

845.4 Control and Limiting of Gas Pressure in Low-Pressure Distribution Systems.

845.41 Each low-pressure distribution system or low-pressure main supplied from a gas source which is at a higher pressure than the maximum allowable operating pressure for the

low-pressure system, shall be equipped with pressure regulating devices of adequate capacity, designed to meet the pressure, load and other service conditions under which they will have to operate.

845.42 In addition to the pressure-regulating devices prescribed in 845.41, a suitable device shall be provided to prevent accidental overpressuring. Suitable types of protective devices to prevent overpressuring of low-pressure distribution systems include:

(a) A liquid seal relief device that can be set to open accurately and consistently at the desired pressure.

(b) Weight loaded relief valves.

(c) An automatic shut-off device as described in 845.32(e).

(d) A pilot loaded back-pressure regulator as described in 845.21(b).

(e) A monitoring regulator as described in 845.32(c).

(f) A series regulator as described in 845.32(d).

845.43 *Maximum Allowable Operating Pressure for Low Pressure Distribution Systems.* The maximum allowable operating pressure for a low-pressure distribution system shall not exceed either (a) or (b) below.

(a) A pressure which would cause the unsafe operation of any connected and properly adjusted low-pressure gas burning equipment, or

(b) A pressure of 2 psig.

845.44 *Conversion of Low-Pressure Distribution Systems to High-Pressure Distribution Systems.*

(a) Before converting a low-pressure distribution system to a high-pressure distribution system, it is recommended that the following factors be taken into consideration:

(1) The design of the system including kinds of material and equipment used.

(2) Past maintenance records including results of any previous leakage surveys.

(b) Before increasing the pressure the following steps (not necessarily in sequence shown) should be taken:

(1) Make a leakage survey, if past maintenance records indicate that such a survey is advisable, and repair leaks found.

(2) Reinforce or replace parts of the system found to be inadequate for the higher operating pressures.

(3) Install a service regulator on each service, and test each regulator to determine that it is functioning. In some cases it may be necessary to raise the pressure slightly to permit proper operation of the service regulator.

(4) Isolate the system from adjacent low-pressure systems.

(5) At bends or offsets in coupled or bell and spigot pipe, reinforce or replace anchorages determined to be inadequate for the higher pressures.

(c) The pressure in the system being converted should be increased by steps, with a period to check the effect of the previous increase before making the next increase. The desirable magnitude of each increase and the length of the check period will vary depending upon conditions. The objective of this procedure is to afford an opportunity to discover before excessive pressures are reached any unknown open and unregulated connections to adjacent low-pressure systems or to individual customers.

845.5 Control and Limiting of the Pressure of Gas Delivered to Domestic, Small Commercial and Small Industrial Customers from High Pressure Distribution Systems.

Note: When the pressure of the gas and the demand by the customer are greater than that which is applicable under the provisions of 845.5, the requirements for control and limiting of the pressure of gas delivered are included in 845.1.

845.51 If the maximum actual operating pressure of the distribution system is between 2 psig and 60 psig and a service regulator having the characteristics listed below is used, no other pressure limiting device is required:

(a) A pressure regulator capable of reducing distribution line pressure (pounds per square inch) to pressures recommended for household appliances (inches of water column).

(b) Single port valve with orifice diameter no greater than that recommended by the manufacturer for the maximum gas pressure at the regulator inlet.

(c) The valve seat shall be made of resilient material designed to withstand abrasion of the gas, impurities in gas, cutting by the valve, and to resist permanent deformation when it is pressed against the valve port.

(d) Pipe connections to the regulator shall not exceed 2 inches in diameter.

(e) The regulator must be of a type that is capable under normal operating conditions of regulating the downstream pressure within the necessary limits of accuracy and of limiting the build-up of pressure under no-flow conditions to 50% or less of the discharge pressure maintained under flow conditions.

(f) A self-contained service regulator with no external static or control lines.

845.52 If the maximum actual operating pressure of the distribution system is between 2 psig and 60 psig and a service regulator not having all of the characteristics listed in 845.51 is used, or if the gas contains materials that seriously interfere with the operation of service regulators, suitable protective devices shall be installed to prevent unsafe over-pressuring of the customer's appliances should the service regulator fail. Some of the suitable types of protective devices to prevent overpressuring of customer's appliances are:

(a) A monitoring regulator

(b) A relief valve

(c) An automatic shut-off device

These devices may be installed as an integral part of the service regulator or as a separate unit.

845.53 If the maximum actual operating pressure of the distribution system exceeds 60 psig, suitable methods shall be used to regulate and limit, to the maximum safe value, the pressure of gas delivered to the customer, such as the following:

(a) A service regulator having the characteristics listed in 845.51 above and a secondary regulator located upstream from the service regulator. The secondary regulator in no case shall be set to maintain a pressure higher than 60 psi. A device shall be installed between the secondary regulator and the service regulator to limit the pressure on the inlet of the service regulator to 60 psi or less in case the secondary regulator fails to function properly. This device may be either a relief valve, or an automatic shut-off that shuts, if the pressure on the inlet of the service regulator exceeds the set pressure (60 psi or less), and remains closed until manually reset.

(b) A service regulator and a monitoring regulator set to limit to a maximum safe value the pressure of the gas delivered to the customer.

(c) A service regulator with a relief valve vented to the outside atmosphere, with the relief valve set to open so that the pressure of gas going to the customer shall not exceed a maximum safe value. The relief valve may either be built into the service regulator or it may be a separate unit installed downstream from the service regulator. This combination may be used alone only in those cases where the inlet pressure on the service regulator does not exceed the manufacturer's safe working pressure rating of the service regulator, and is not recommended for use where the inlet pressure on the service regulator exceeds 125 psi. For higher inlet pressures, method (a) or (b) above should be used.

847 VAULTS.

847.1 Structural Design Requirements.

Underground vaults or pits for valves, pressure relieving, pressure limiting or pressure regulating stations, etc., shall be designed and constructed in accordance with the following provisions:

(a) Vaults and pits shall be designed and constructed in accordance with good structural engineering practice to meet the loads which may be imposed upon them.

(b) Sufficient working space shall be provided so that all of the equipment required in the vault can be properly installed, operated, and maintained.

(c) In the design of vaults and pits for pressure limiting, pressure relieving and pressure regulating equipment, consideration shall be given to the protection of the equipment installed from damage, such as that resulting from an explosion within the vault or pit, which may cause portions of the roof or cover to fall into the vault.

(d) Pipe entering, and within, regulator vaults or pits shall be steel for sizes 10 inches and less except that control and gage piping may be copper. Where piping extends through the vault or pit structure, provision shall be made to prevent the passage of gases or liquids through the opening and to avert strains in the piping. Equipment and piping shall be suitably sustained by metal, masonry, or concrete supports. The control piping shall be

placed and supported in the vault or pit so that its exposure to injury or damage is reduced to a minimum.

(e) Vaults or pit openings shall be located so as to minimize the hazards of tools or other objects falling upon the regulator, piping, or other equipment.   The control piping and the operating parts of the equipment installed shall not be located under a vault or pit opening where workmen can step on them when entering or leaving the vault or pit, unless such parts are suitably protected.

(f) Whenever a vault or pit opening is to be located above equipment which could be damaged by a falling cover, a circular cover should be installed or other suitable precautions taken.

847.2   Accessibility.   Consideration shall be given, in selecting a site for a vault, to its accessibility.   Some of the important factors to consider in selecting the location of a vault are as follows:

(a) *Exposure to traffic.*   The location of vaults in street intersections or at points where traffic is heavy or dense should be avoided.

(b) *Exposure to flooding.*   Vaults should not be located at points of minimum elevation, near catch basins, or where the access cover will be in the course of surface waters.

(c) *Exposure to adjacent subsurface hazards.*   Vaults should be located as far as is practical from water, electric, steam, or other facilities.

847.3   Vault Ventilation.   Underground vaults and closed top pits composing either a pressure regulating or reducing station, or a pressure limiting or relieving station, shall be ventilated as follows:

(a) When the internal volume exceeds 200 cubic feet, such vaults or pits shall be ventilated with two ducts each having at least the ventilating effect of a pipe 4 inches in diameter.

(b) The ventilation provided shall be sufficient to minimize the possible formation of a combustible atmosphere in the vault or pit.

(c) The ducts shall extend to a height above grade adequate to disperse any gas-air mixtures that might be discharged.   The outside end of the ducts shall be equipped with a suitable weatherproof fitting or vent-head designed to prevent foreign matter from entering or obstructing the duct.   The effective area of the openings in such fittings or vent-heads shall be at least equal to the cross-sectional area of a 4-inch duct.   The horizontal section of the ducts shall be as short as practical and shall be pitched to prevent the accumulation of liquids in the line.   The number of bends and offsets shall be reduced to a minimum and provisions shall be incorporated to facilitate the periodic cleaning of the ducts.

(d) Such vaults or pits having an internal volume between 75 cubic feet and 200 cubic feet may be either tightly closed or ventilated.   If not ventilated, all openings shall be equipped with tight fitting covers without open holes through which an explosive mixture might be ignited.   Means shall be provided for testing the internal atmosphere before removing the cover.

(e) If vaults or pits referred to in (d) above are ventilated by means of openings in the covers or gratings and the ratio of the internal volume, in cubic feet, to the effective ventilating area of the cover or grating, in square feet, is less than 20 to 1, no additional ventilation is required.

(f) Such vaults or pits having an internal volume less than 75 cubic feet may be ventilated or not at the option of the operating company.

847.4   Drainage and Waterproofing.

(a) Provisions shall be made to minimize the entrance of water into vaults, and vault equipment shall always be designed to operate safely if submerged.

(b) No vault containing gas piping shall be connected by means of a drain connection to any other substructure, such as a sewer.

(c) Electrical equipment in vaults shall conform to the requirements of Class 1, Group D, of the National Electrical Code, ASA C1.

Design of a Long Transmission Line

The volume of gas which may be carried by a pipeline depends on the maximum allowable operating pressure if other conditions are considered to be fixed.   This

pressure is limited by the physical and chemical specifications of available steel and the physical limitations of the pipe. With availability of certain steels and pipe confirmed, line flow may be calculated for several sizes and grades of pipe at a maximum operating pressure as prescribed by the American Standard Code for Pressure Piping, B31.8-1963. Distances between compressor stations should also be considered. As such distance increases, intake pressures decrease, and with the maximum pressure constant, the compression ratio and power at each station increase. The converse is also true.

A proper balance between investment in pipe and in stations and total system operating costs must be reached. In a comparison of compressing equipment, this balance is upset by the substitution of stations with different compressing equipment which varies in initial investment and operating cost. Therefore, minimum total transportation cost for a certain type of equipment may involve the use of different pipe, station horsepower, and station spacing from those required by another type, even though the gas volume transported and the distance are the same.

Variables involved in pipeline design other than gas volume and length of line are pipe diameter, allowable operating pressure, and station compression ratio. From them, station spacing and horsepower can be established along with pipe cost, station cost, and total operating cost.

Pipe selection affects gas-transportation costs more than any other single factor. Pipe cost plus laying costs represents approximately 70 to 90 per cent of total pipeline investment. The horsepower reduction which results from larger diameter pipe with thicker walls (higher operating pressure) must be weighed against the additional investment in pipe tonnage.

Telescoping. If pipe of constant wall thickness is used throughout a line, only the portion near the discharge side of each station is stressed up to the allowable limit Therefore, capital investment may be reduced by telescoping, that is, installing pipe of progressively thinner walls as the distance downstream from the station increases and as the falling gas pressure permits. Thus, pipe costs are greatly reduced provided that the distances and pressure drops between stations are relatively large. Telescoping encourages use of stations of greater horsepower spaced farther apart than would be economical otherwise.

Although the effect of reduced pipe cost may lower transportation cost, definite disadvantages in the operation and expansion of a telescoped line are:

1. Operation at reduced capacities, which lowers pressure drop between stations, necessitates cutting back discharge pressure correspondingly to prevent overloading the pipe with thinner walls. This increases required power and may introduce capacity limitations when compressors are out of service.

2. Dispatching is more complicated. Limiting pressures must be constantly considered.

3. Line packing is not obtainable.

4. Initial design would have to include plans for looping and intermediate compressor stations. This may offset savings obtained from telescoping.

Pipeline Flow Formulas. For isothermal gas flow in horizontal lines, the general steady-flow equation derived[1] from an energy balance[2] over a pipeline may be expressed as

$$Q = 38.77(T_b/P_b)(1/f)^{0.5}[(p_1{}^2 - p_2{}^2)/GTLZ]^{0.5}D^{2.5} \qquad (1)$$

---

[1] B. E. Hunt and others, "Gas Service Design—Final Report of Task Group" (DMC-56-114), *Proc. AGA*, 1956, pp. 272–287.

[2] In Eq. (1) kinetic-energy effects are ignored.

where Q = flow rate, cu ft/day at P_b and T_b
   P_b = base pressure, psia
   T_b = base temperature, F abs, R
   $(1/f)^{0.5}$ = transmission factor, dimensionless
   f = friction coefficient, dimensionless
   P_1 = inlet pressure, psia
   P_2 = outlet pressure, psia
   G = gas specific gravity (air = 1.0)
   T = average flowing gas temperature, F abs, R
   L = length of pipe, miles
   D = inside diameter of pipe, in.
   Z = compressibility factor at *average* conditions, dimensionless; may be determined at *average* pressure P_m and T, where

$$P_m = \tfrac{1}{3}[(P_1^3 - P_2^3)/(P_1^2 - P_2^2)]$$

Z may be omitted for pressures under 100 psig; see Eq. (4) when pressure is unknown.

Of the many variables in Eq. (1), the transmission factor $(1/f)^{0.5}$ has long been the most difficult to evaluate.   The transmission factors of Table 13 are considered reliable either having best stood the tests of usage or, where the formulas are very recent developments, having strong foundations in basic flow theories.   The factors given in Table 13 were plotted on Fig. 14 for 20-in. pipe.   Generally, similar curves apply to other sizes.[1]

Table 13.   Transmission Factors for Flow Formulas

Flow formula	Transmission factor $(1/f)^{0.5}$	Remarks
Smooth-pipe (laminar)[1,2]........	$4 \log_{10} (f^{0.5}\text{Re}) - 0.6$	Seldom applicable to large-diameter natural-gas transmission lines
Rough-pipe (fully turbulent)[3]....	$4 \log_{10} (3.7D/k)^3$	Characterizes most natural-gas-transmission operating conditions
Weymouth[4,5].................	$1.10 \times 11.2D^{0.167}$	Reasonably good approximation to preceding (rough-pipe) formula for $D = 10$ in. and $k = 0.002$
Panhandle A[4]	$0.92 \times 6.87\text{Re}^{0.073}$	Large-diameter transmission piping where Re varies from 5 million to 20 million
New Panhandle[4]	$0.90 \times 16.5\text{Re}^{0.0196}$	

[1] Table 14 gives a number of solutions to this implicit function.
[2] Approximated by the factor for the "improved flow equation," $(1/f)^{0.5} = 5.18 \text{ Re}^{0.0909}$ with a deviation of ± 1 per cent for Reynolds numbers from 60,000 to approximately 7,000,000.
[3] Table 15 gives a number of solutions.
[4] Average pipeline efficiency factor.
[5] For over 24 in. ID only.
  Re = Reynolds number [see Eq. (2)]
   f = friction coefficient (reciprocal of transmission factor, squared)
   D = inside diameter of pipe, in.
   k = effective roughness (height of surface irregularities), in.

[1] "Flow of Natural Gas through Experimental Pipelines and Transmission Lines," *U.S. Bur. Mines Monograph* 9, 1956.

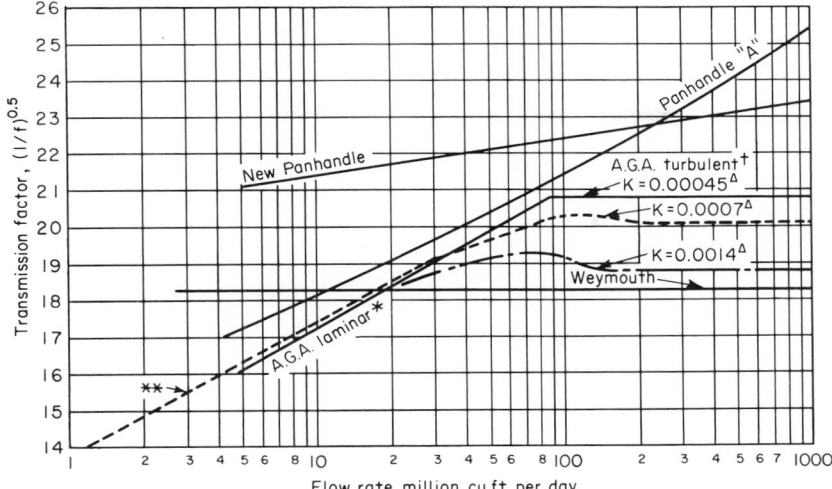

FIG. 14. Transmission factor vs flow through 20-in. (19.25-in.-ID) pipeline using various flow formulas. Dashed curves are based on experimental test data from *Bureau of Mines Monograph 9*.

* $(1/f)^{0.5} = 5.18Re^{0.0909}$, "the improved flow equation" which is an approximation of the smooth pipe formula

† $(1/f)^{0.5} = 16.21D^{0.084}$, an approximation of the rough pipe formula

△ $(1/f)^{0.5} = 4\log_{10}(3.7D/K)$, the rough pipe formula

** $(1/f)^{0.5} = 4\log_{10}(f^{0.5}Re) - 0.6$, the smooth pipe formula

In the latter regard, comparison of the first two expressions in Table 13 shows that for laminar flow (low Reynolds number, Re) the transmission factor is a function of Re, while for turbulent flow (high Reynolds number) the transmission factor is a function of pipe roughness k and diameter D. Between these two stable ranges lies the *partially turbulent* range.

Table 14. Solutions for Smooth Pipe
Transmission Factors[1]

$$[(1/f)^{0.5} = 4\log_{10}(f^{0.5}Re) - 0.6]$$

Factor, $(1/f)^{0.5}$	Reynolds No., Re, (in millions)
13.0	0.0327
14.0	0.0625
15.0	0.119
16.0	0.226
17.0	0.427
18.0	0.804
19.0	1.51
20.0	2.83
21.0	5.28
22.0	9.83
23.0	18.3
24.0	33.9

[1] *U.S. Bur. Mines Monograph 9, 1956.*

Table 15. Solutions for Rough Pipe
Transmission Factors

$[(1/f)^{0.5} = 4 \log_{10}(3.7\ D/k)$ where $k = 0.0007$ in.[1]$]$

$(1/f)^{0.5}$	D, pipe ID, in.
18.9	10.00
19.4	13.50
19.7	15.44
20.0	19.38
20.4	23.25
20.5	25.31
20.8	29.25

[1] Average effective roughness for clean steel pipe in transmission service. Calculated values of effective roughness of commercial gas pipe ranged from 0.000468 to 0.00209 and seemed to be independent of pipe diameter for diameters from 2 to 36 in.

Equation (2) expresses the dimensionless *Reynolds number* Re in natural-gas engineering terms:

$$Re = 477.5 \times 10^{-6} QGP_b/VDT_b \tag{2}$$

where V is the gas viscosity in pounds mass per second-feet; average value of 7.4×10^{-6} may be used. (V in pounds-mass per second-feet = micropoises/14.882.) Other terms are as defined under Eq. (1).

Account should be taken of substantial changes in *pipeline elevations*. Equation (3), a rearrangement and modification of Eq. (1), highlights the additional terms which must be evaluated:

$$Q = A(P_1^2 - e^S P_2^2/L_e)^{0.5} \tag{3}$$

where $A = 38.77(T_b/P_b)(1/f)^{0.5}(1/GTLZ)^{0.5}D^{2.5}$
$\qquad e$ = base of natural logarithms = 2.718
$\qquad S = 0.0375Gh/TZ$
$\qquad h$ = outlet elevation minus inlet elevation, ft

For a uniform slope $L_e = (e^S - 1)L/S$.

For a nonuniform slope (where elevation change cannot be simplified to a single section of constant gradient), a stepwise approach to any number of sections n will yield

$$L_e = (e^{S_1} - 1)L_1/S_1 + e^{S_1}(e^{S_2} - 1)L_2/S_2 + e^{S_1+S_2}(e^{S_3} - 1)$$
$$L_3/S_3 + \cdots + e^{\Sigma_1^n S_{n-1}}(e^{S_n} - 1)L_n/S_n$$

where $S_1 = 0.0375Gh_1/TZ$
$\qquad S_2 = 0.375Gh_2/TZ \ldots$
$\qquad S_n = 0.0375Gh_n/TZ$

NOTE: $S = S_1 + S_2 + S_3 + \cdots + S_n$.

The numerical subscripts refer to the individual sections of the overall line which is operating under pressure differential $P_1 - P_2$; for example, if the line is divided into four sections, n would equal 4 and, say, subscript 2 would pertain to the properties of the second section.

Other terms are as defined under Eq. (1).

Solution When Pressure Is Unknown. Since compressibility Z in Eq. (1) was derived in terms of pressures, the following determination for Z may be used when inlet or outlet pressure is unknown:

$$1/Z = 1 + JP \tag{4}$$

where $Z =$ compressibility factor, dimensionless

$J =$ function of gas composition and temperature which may be found from compressibility correlations; i.e., reduced pressures and temperatures or test data. J is often referred to as the compressibility per unit pressure rise

$P =$ pressure, absolute

As an alternative Eq. (1) may be modified as follows:

$$Q = 38.77 \frac{T_b}{P_b}\left(\frac{1}{f}\right)^{0.5}\left[\frac{P_1^2(1 + \frac{2}{3}JP_1) - P_2^2(1 + \frac{2}{3}JP_2)}{GTL}\right]^{0.5} D^{2.5} \tag{5}$$

Values for the general term $P[1 + \frac{2}{3}JP]$ of Eq. (5) are given in Table 16 for a 1020 Btu/cu ft (gross) dry natural gas, 0.670 sp gr, and negligible CO_2. Trial-and-error solutions of Eq. (5) in conjunction with Table 16 thus do not require knowledge of the compressibility factor Z.

Table 16. $P^2(1 + \frac{2}{3}JP)$ vs P
for a Natural Gas
[For use with Eq. (5)]

$P^2(1 + \frac{2}{3}JP)$	P, psia
10,000	100
40,000	200
90,000	300
168,000	400
261,000	500
380,000	600
525,000	700
695,000	800
880,000	900
1,105,000	1,000

Equivalent Lengths. When a pipeline consists of several different pipe sizes or of looped- or parallel-line sections, it may conveniently be reduced to an equivalent length of a selected pipe size, D_s and handled as a single line of that size and length.

Equivalent length may be calculated from

$$L_E = (D_s/D)^A L \tag{6}$$

where $L_E =$ equivalent length of selected size

$D_s =$ ID of selected size

$L =$ length of actual pipe of size D

$D =$ ID of actual pipe section

$A =$ exponent from Table 17

Table 17. Exponents for Eqs. (5), (6), and (7)

	A	B	C
For Weymouth formula......	5.333	0.50	2.667
For Panhandle A...........	4.854	0.5394	2.618
For new Panhandle.........	4.961	0.51	2.53

Series and Parallel Circuits. The equivalent length of different sizes of pipes connected in series is the sum of the individual equivalent lengths. When lines

are connected in parallel, their combined equivalent length is given by

$$L_{EP} = \left[\frac{1}{(1/L_{E_1})^B + (1/L_{E_2})^B + (1/L_{E_3})^B + \cdots} \right]^{1/B} \tag{7}$$

where terms are defined under Eq. (6) and exponents are as given in Table 17.

If each of the parallel lines is of a single pipe size and of equal length L, Eq. (7) may be expressed as

$$L_{EP} = \left(\frac{1}{D_1{}^C + D_2{}^C + D_3{}^C + \cdots} \right)^{1/B} D_S{}^A L \tag{8}$$

where terms are defined under Eq. (6) and exponents are as given in Table 17.

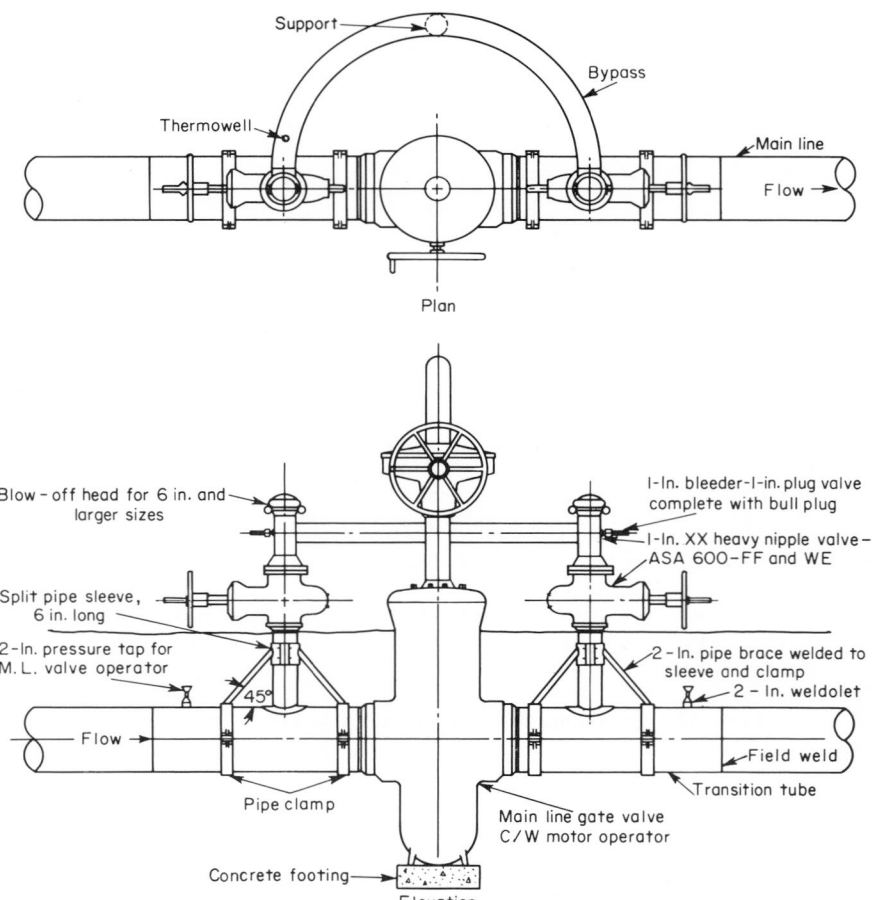

FIG. 15. Main-line gate valve equipped with bypass and blowoff arrangement.

Valves

Many companies use full-opening *gate valves* in main transmission lines to permit passage of scrapers. *Plug valves* may be used for blowoffs.

Main-line valves should be located as convenience of operation and maintenance may require. Valve spacing depends on the area traversed by the pipeline; 5 to 20 miles apart is the order of magnitude for block valves on large-diameter lines.

Main-line gate valves should be equipped with a bypass and blowoff arrangement to equalize pressure differentials and to relieve the excessive stresses that are associated with opening the valve with unequal pressure across its disk. Figure 15 shows such an arrangement. On large-diameter valves with weld ends it is considered good practice to have forged-steel transition pieces shop-welded in place.

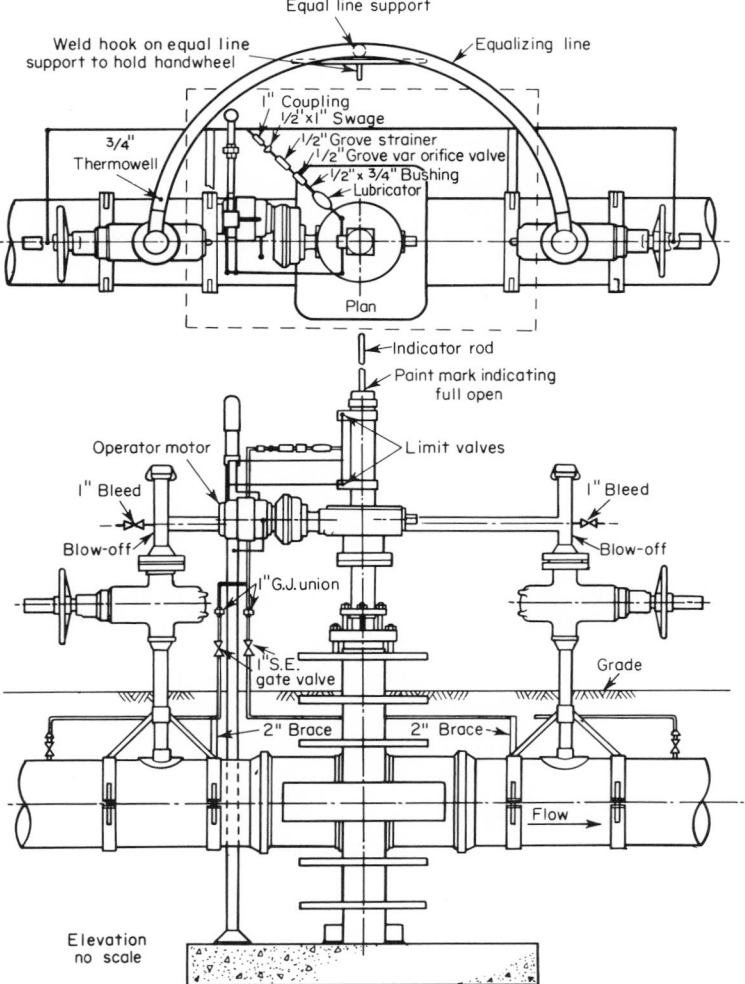

FIG. 16. Typical motor-operated system for underground valves. (*Walworth Company.*)

They provide proper cross-sectional area and a material easily welded under severe field conditions.

Large-diameter valves are preferably equipped with some type of motor operator. A turbine-type operator using gas as a power supply may be used for these valves in main-line use. Such a unit may be directly connected either to the main line or to a bottle located beside it for automatic operation. Figure 16 shows a well-designed arrangement of motor-operated valves in an underground pipeline.

Gas Piping[1] Standard and References:

Title	Issued by
Gas Transmission and Distribution Piping Systems	ASA B31.8-1963
Installation of Gas Piping and Gas Equipment on Industrial Premises and Certain Other Premises	ASA Z83.1-1966
Installation of Gas Appliances and Gas Piping (also issued as NFPA Standard 54)	ASA Z21.30-1964
Installation of Domestic Gas Conversion Burners	ASA Z21.8-1965
Installation of Gas Conversion Burners in Domestic Ranges	ASA Z21.38-1957
Installation of Gas Equipment in Large Boilers (being revised as ASA project Z 83.3)	ASA Z21.33-1950
Directory of Approved Appliances and Listed Accessories (semiannual issues with monthly supplements)	AGA
Approved Equipment for Industrial Fire Protection (annual issues)	Factory Mutual Engineering Div.
Gas and Oil Equipment List (cumulative issues in November with bimonthly supplements)	Underwriters' Laboratories
Standards for Class A Ovens and Furnaces	NFPA Standard 86-1966
Welding and Cutting Systems	NFPA Standard 51-1964
Combustion Engines, Turbines	NFPA Standard 37-1963
Liquefied Petroleum Gases	NFPA Standard 58-1966
L.P.G. at Utility Plants	NFPA Standard 59-1963
Gases in Manholes and Sewers	NFPA Report 328M

[1] All these publications contain material affecting the installation of gas piping, fittings, valves, etc., in the respective fields which they cover.

REFERENCES

1. Hunt, B. E. and others, "Gas Service Design—Final Report of Task Group" (DMC-56-14), *Proc. AGA*, 1956, pp. 272–287.
2. Menegakis, D., and E. H. Luntey, "Experimental Investigation of Flow Characteristics of Low Pressure Services" (DMC-61-15), *Proc. AGA*, 1961.
3. Indiana Gas Association, "Distribution System Capacities," Gas, vol. 23, pp. 51–58, May, 1947.
4. Gardner, J. R., "Report of Subcommittee on Customer Load Characteristics," *Proc. AGA*, 1950, pp. 117–143.
5. Hunt, B. E., "A Method Used in Forecasting Loads and Making Load Studies," *Proc. AGA*, 1953, 602–605.

14

HYDRAULIC-POWER TRANSMISSION PIPING

Reno C. King*

Hydraulic-power transmission systems are used widely in industry and in transportation. Such systems may be designed for the sole purpose of transmission of an applied force which will operate directly some piece of driven equipment, or they may be designed as integral portions of some control device. In this chapter, no effort will be made to distinguish between the two types of systems.

Hydraulic power provides an effective means for producing a moving or holding force of unlimited magnitude. Manufacturing industries utilize hydraulically operated equipment for the forming, extruding, drawing, or forging of metals, plastics, rubber, and other materials. Hydraulic presses of the self-contained fast-acting type are being used to an increasing extent. The speed and pressure of the operating ram are brought under instant and automatic control by the use of motor-driven variable-displacement oil pumps. This feature is of particular importance for deep-drawing operations, since the punch can be brought without shock into initial contact with the work and, subsequently, can be moved with a controlled velocity during the drawing process. Many machine tools are either powered or controlled by hydraulic means.

In transportation, motor vehicles have hydraulic braking systems in which the depressing of a lever or foot pedal sends out an impulse to a master cylinder from which similar impulses are sent to operate a brake shoe against a drum. Hydraulic devices operate to lift dump bodies, road scrapers, bulldozer and snow-plow blades, and similar pieces of equipment. Aircraft have extensive hydraulic control systems for retractable landing gear and wing flaps or in any similar application where either considerable power or a high degree of controllability is required. The steering

* Professor of Mechanical Engineering, New York University, New York, N.Y.

gear of modern naval and merchant vessels is hydraulically operated rather than being driven by steam engines.

In other applications, high-pressure water is discharged in high-velocity jets through specially designed nozzles for descaling hot steel during the process of rolling or forging, for removing bark from logs in the pulp and timber industry, and for removing coke from coking chambers in the oil industry. Water from hydraulic systems also is used in mills and shops for the hydrostatic testing of pipe and tubing, valves, pressure vessels, and other pressure-containing equipment.

ELEMENTS AND PRINCIPLES OF HYDRAULIC SYSTEMS

Pascal's law states that the pressure which is exerted at any point upon a confined liquid is transmitted undiminished in all directions through the liquid. This fundamental principle, on which all hydraulic power transmission systems operate, is shown diagrammatically in Fig. 1. A force of 100 lb applied to an area of 1 sq in. results in a pressure of 100 psi; this pressure is transmitted undiminished to the lower side of a piston whose area is 10 sq in., and this results in a total force of 1,000 lb. Obviously, the small piston in this example has a travel ten times as great as that of the large piston; thus there is some flow of fluid. If the pressure drop due to friction is neglected, the mechanical advantage is defined as the ratio of the area of the large piston to that of the small one. Some of the applications made of this simple principle in hydraulic systems are described in succeeding paragraphs.

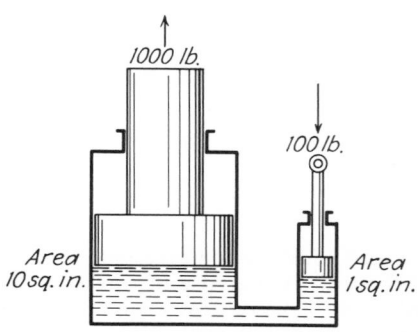

Fig. 1. The fundamental principle known as Pascal's law on which all hydraulic power transmission systems operate.

Hydraulic Transmission Systems

Basic System. The basic hydraulic system consists of a pump, an operating cylinder which contains a piston, and necessary interconnecting tubing or piping. In order to permit flexibility in operation or in order to achieve the desired control function, most systems contain also a reservoir or sump and a control valve which permits reversal of the direction of operation of the power piston. Figure 2 shows a basic hydraulic system complete with reservoir and control valve. If a positive-displacement pump is used, at times when no flow is desired into or out of the power cylinder, an unloading valve of the type shown in Fig. 3 is used to prevent the building up of excessive pressure. With reference to Fig. 2, the solid direction arrows indicate the path of flow when fluid is forced by the pump from the reservoir into the left-hand end of the cylinder which, in turn, forces fluid out the right-hand end and back through the four-way valve to the reservoir. The dotted directional lines indicate the flow when it is desired to reverse the piston direction. Unless a reversible pump is used, a four-way valve or equivalent device for directional control is required in any system that must operate in two directions. The part of the system between the pump and the valve where the fluid always flows in the same direction is known as the power system, while the part beyond the valve is known by the name of the device it operates, for instance, as the landing-gear system or the flap system in the case of aircraft. From this basic system any hydraulic system can be derived. Additions can be made for the purpose of providing more than

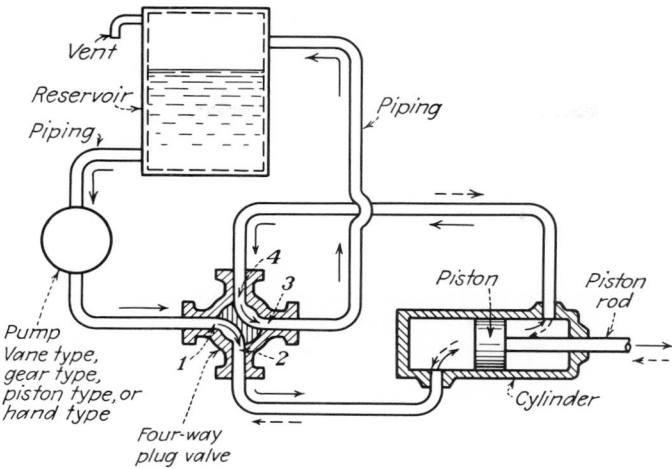

FIG. 2. Schematic diagram of basic hydraulic system.   With the four-way valve in the position shown, fluid flow is in through port 1 and out through port 2, also in through port 4 and out through port 3.   Turning the plug 90 deg connects ports 1 and 4 and ports 2 and 3, thereby causing the fluid to flow in the direction of the dotted arrows so as to reverse the direction of travel of the piston.   (*Courtesy of Prod. Eng.*)

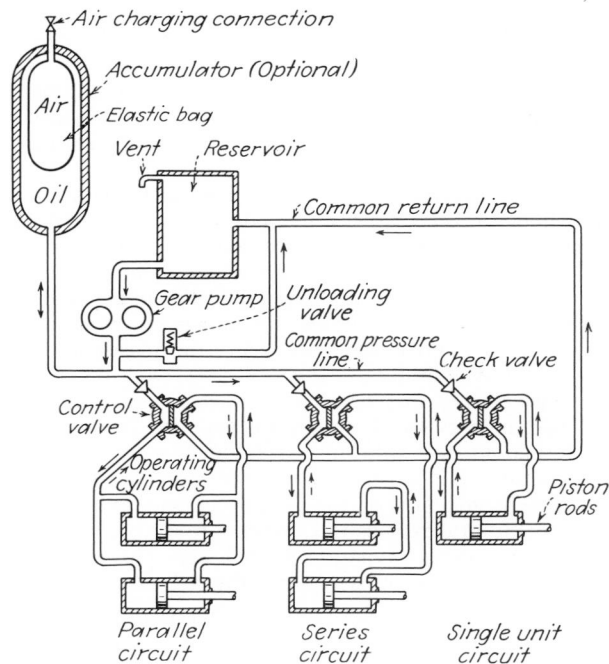

FIG. 3. Multiple operation of hydraulic cylinders from a common power system.

one source of power, for operating more than one cylinder, and for making operations more automatic or for increasing their reliability.

An elaboration of this system involving the use of several operating cylinders is shown in Fig. 3, which illustrates how cylinders can be operated independently or in parallel from the same control valve or in series if desired. Any number of cylinders can be added on one control valve circuit in parallel since the hydraulic pressure, neglecting friction losses, is equal throughout all connected branches of a system in which fluid is free to flow. When cylinders are added in parallel on a system of limited supply capacity, however, the piston that requires the lowest unit pressure to move its load will operate first, and it will continue to move until it reaches the end of its travel or until the unit pressure builds up enough to permit another piston to move its load. This process of increasing pressure continues until the cylinder that requires the greatest unit pressure for operation has moved. This delayed-action trouble can be corrected by providing a control valve of more ample size or, if necessary, by installing an accumulator or by using a pump of greater capacity.

When cylinders are connected in series, as shown also in Fig. 3, an entirely different condition results. Fluid is then trapped between the cylinders, and only that on the exhaust side of the last cylinder in series can return to the reservoir. When the control valve on this circuit is operated, trapped oil is forced from the first cylinder into the second and from the second into the third, if there is a third, and so on. Any leakage of fluid past one or more of the pistons, or in the piping between cylinders, will result in the pistons getting out of synchronism so that they do not reach the end of their travel together. A similar condition occurs when the trapped fluid expands or contracts as the result of a change in temperature. These troubles can be overcome by the use of synchronizing valves arranged to open automatically when the pistons reach the end of their travel, in order to permit a flow of fluid into the cylinder to replace leakage.

In multiple-cylinder systems, such as those illustrated in Fig. 3, it may be necessary to install check valves in the individual control valve lines in order to prevent backfeed from a highly loaded cylinder into a lightly loaded one before its piston has reached the end of its travel.

In an application of the type that is shown Fig. 2, in which a single pump supplies a single cylinder, it is necessary only that the pump be of adequate capacity to satisfy the maximum demands of the controlled equipment. In systems of the type shown in Fig. 3, in which one pump serves two or more cylinders, the task of matching pump capacity with maximum possible system demand becomes more complex and might require the installation of a high-capacity pump whose maximum capability would be utilized only infrequently. For this reason, the use of accumulators is often indicated or desirable.

Accumulator Systems.[1] In general, accumulators are included in the layout of a number of hydraulic cylinders which are supplied from a single pumping system; also, an accumulator is furnished frequently in large single-machine systems. The accumulator in all cases serves to store energy delivered by the pump when its supply is greater than the system demand and, likewise, to deliver energy when the demand rate exceeds the pumping rate.

A schematic diagram of an accumulator system designed for 3,000 psi and suitable for operating large forging processes is shown in Fig. 4.[2] In this case, the hydraulic

[1] See Also "Hydraulic Tables and Other Data," *Baldwin-Southwark Bull.* 150.

[2] From "Modern Practice in the Generation and Application of Hydraulic Power," by J. E. Holveck, ASME paper presented at Pittsburgh, Pa., Apr. 23, 1940, published by *Iron Steel Eng.*, vol. 18, no. 5, pp. 58–66, May, 1941. Also available in reprint form from Worthington Pump and Machinery Corp., Harrison, N.J.

generating units consist of two reciprocating pumps. The accumulator is the hydropneumatic air-bottle type employing two air bottles and a combination air-and-fluid bottle. The bottles are charged with compressed air at the necessary pressure. The expansion of the compressed air in the hydropneumatic bottle displaces an equivalent volume of fluid which should be adequate for the maximum demand of fluid required to operate the equipment in use. It has been found in air-bottle accumulator practice that expansion of the air between the higher and the lower pressures can be represented by the expression $p_2 V_2^{1.35} = p_3 V_3^{1.35}$.

In most hydraulic systems the total volume of air bottles required is determined by an allowable pressure variation in the order of 10 per cent between p_2 and p_3 which corresponds to a volume variation of about 8 per cent between V_2 and V_3 indicated by the shaded section in Fig. 4. The normal fluid displacement of the accumulator should be equivalent to the maximum demand for an operating cycle

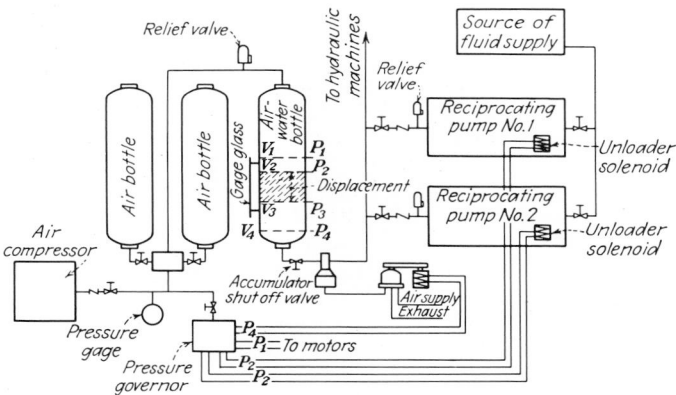

FIG 4. Schematic diagram of multiple-unit hydraulic system with air bottle accumulation and reciprocating pumps. (*From ASME paper by J. E. Holveck.*)

as determined from flow charts.[1] It is considered good practice to provide a total hydraulic-fluid content in the air-and-fluid bottle of two to three times the normal fluid displacement. From these considerations the total volume of all the bottles will be from twelve to fifteen times the volume of fluid drawn off in a maximum operating cycle.

The accumulator shutoff valve shown in Fig. 4 is needed as a safety measure to prevent loss of air through the piping system in case the fluid level falls to the point V_4 and its corresponding pressure p_4. This valve is operated automatically to respond to pressure variations. The automatic-unloader solenoids, also shown in Fig. 4, are intended to prevent excess pumpage by blocking open the pump suction valves in case the pressure reaches p_1 corresponding to accumulator level V_1. Such automatic unloaders usually are set to operate in sequence so as to cut off pumpage in one pump at a time.

The basic principles of accumulator systems are much the same whether of large size for forging plants or of small size for aircraft. The details of construction are, of necessity, radically different, as described later. In all air-bottle types of accumulator systems, provision should be made for charging the bottles with the required volume of air at the operating pressure and for maintaining this volume. Since the air is trapped, the only occasion for loss with all joints tight is through absorption

[1] Holveck, *op. cit.*

of air by the fluid.   This loss is small and requires only infrequent operation of the compressor or other source of supply.

Accumulators for oil-fluid systems on aircraft, self-contained systems for machine tools, and the like cannot have oil in direct contact with air on account of the explosion hazard.   Hence such accumulators utilize a cylinder chamber and piston, a chamber and rubber diaphragm, a chamber and bellows, or gas filled chambers. Such an accumulator chamber with a rubber diaphragm is illustrated in Fig. 5.

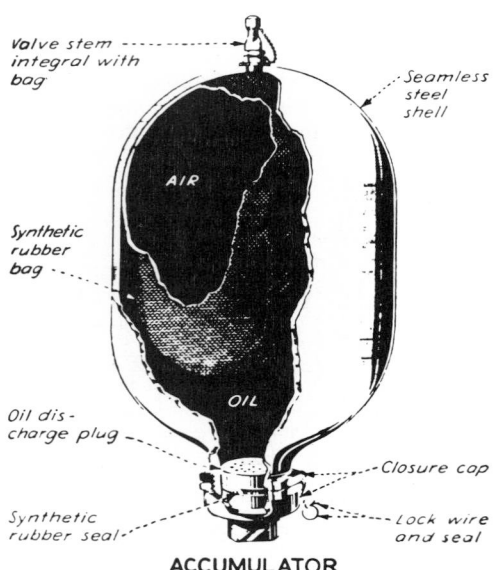

ACCUMULATOR

Fig. 5. Accumulator chamber with rubber diaphragm to separate air from oil.   (*Courtesy of Prod. Eng.*)

The *gravity-type accumulator*, shown diagrammatically in Fig. 6, is used for some large forge-plant installations.   It is not suitable for aircraft, automotive, or marine use owing to its weight.   In order to avoid excessive inertia forces, particularly on the downward stroke, the diameter of the ram should be large enough to avoid too high speed of travel.   The principal advantages of the air-bottle type as compared with the weighted accumulator are:

1. The air-bottle type is lighter and requires less floor space and less costly foundations.   Much greater accumulator effect can be obtained from a relatively small set of air bottles than is possible from a reasonable size of gravity accumulator.

2. The air-bottle type has no moving parts to be serviced while the gravity type requires maintaining the packing and ram.

3. The air-bottle type eliminates the shocks caused in the gravity type by the inertia of starting and stopping the weights.   For instance, on a 2,000-psi hydraulic system, a ton of weights has to move through a vertical distance of 1 in. for each cubic inch of fluid passed in or out of the accumulator.

4. The air-bottle type tends to absorb fluid shocks from water hammer in the piping system whereas the gravity type must sustain these shocks.

5. The operating pressure in the air-bottle type may be readily changed by adjusting the pressure governor, while pressure changes with the gravity type are made by the more cumbersome method of adding or removing weights.

In the air-bottle accumulator, if the hydraulic fluid is of the type that releases vapors which could form an explosive mixture with air, special precautions must be taken. A diaphragm may be used to partition the oil from air, a sealing fluid may be floated on top of the oil, or an inert gas such as nitrogen may be used instead of air. Thus, a gravity accumulator has the principal advantage of avoiding entirely the explosion hazard.

A limited amount of accumulator effect can be obtained by having a plunger act against springs rather than against a constant dead weight as is the case with gravity accumulators. The design of such a system would be similar to that shown for the shock-absorber valve of Fig. 20. Such a device eliminates the possibility of air-vapor explosion and, at the same time, overcomes the objection associated with the weight requirements of the gravity accumulator. However, since the force which compresses the springs is proportional to the spring compression, large amounts of accumulator effect are not possible.

Hydraulic Intensifiers. Sometimes it is desirable to have available high-pressure liquid at pressures higher than those of the system installed or higher than those under which pumps operate most satisfactorily. In such cases an intensifier is indicated. An intensifier is essentially a device consisting of two cylinders of unequal diameters, having a common piston or ram, also of unequal diameters. The pressures in the two cylinders then vary inversely as the areas of the common ram. In calculating volumes of liquid required in connection with the design of the supply system, it should be considered that the ratio of volume of low-pressure liquid to that of high-pressure liquid is directly as the areas of the respective ends of the ram; i.e., a larger volume of low-pressure liquid is required to produce the relatively smaller volume of high-pressure liquid.

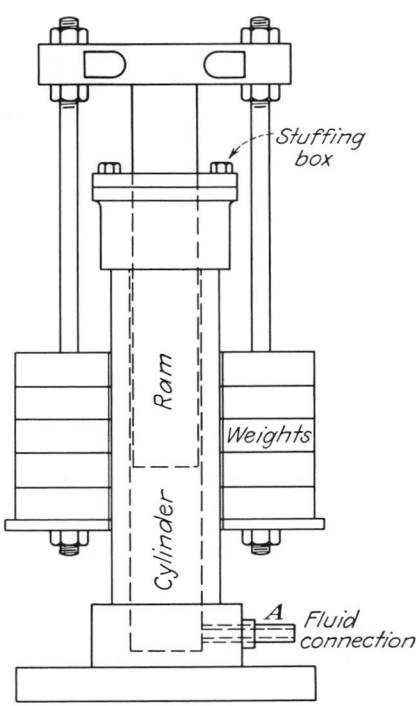

Fig. 6. Sectional view of gravity-type accumulator.

Individual-unit or Self-contained Systems. The action of many modern machine tools and presses is accomplished through individual-unit or self-contained hydraulic systems which constitute an integral part of each machine. Such hydraulic systems consist essentially of a pumping unit, driven by a constant-speed electric motor, which delivers fluid through piping and control valves to a hydraulic motor or cylinder which actuates the tool mechanism. Valves, available in a great variety of types, are selected to obtain the desired sequences or cycles of operation with respect to time, force, and velocity. The application of hydraulics in the industrial field covers the widest ranges of power use. This is evidenced by the contrast between the tremendous rams of hydraulic presses whose forces are measured in tons and the sensitive hydraulic controls in gear-grinding and milling machines which produce precision work within a fraction of a milinch.

There are two basic types of individual-unit or self-contained hydraulic systems used for operating machine tools. In the *constant-volume* system shown in Fig. 7, a constant-speed pump, usually of the gear or vane type, discharges a constant volume of oil at a constant pressure. Flow from the pump is throttled to regulate the rate of delivery to the operating cylinder. Excess pumpage is bypassed through a relief valve which is set at the highest pressure needed in the system. At the end of each stroke of the operating piston, its direction of travel is reversed automatically by a trip (not shown) which actuates the four-way valve.

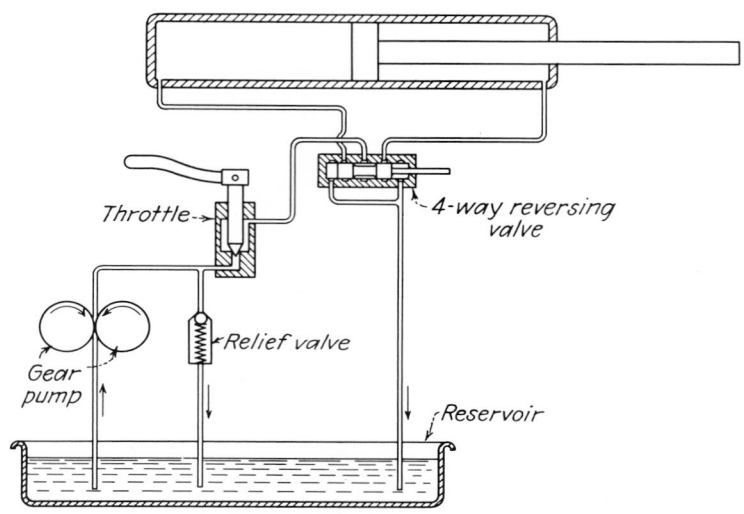

Fig. 7. A constant-volume hydraulic system for an individual unit.

In the *variable-volume* system shown in Fig. 8, a constant-speed pump, usually of the multipiston variable-displacement type, delivers, as required, a variable volume of hydraulic oil ranging from maximum flow in one direction, through zero in neutral position, to maximum flow in the opposite direction. Oil under pressure is thus delivered to one end of an operating cylinder (or to a hydraulic motor in some cases) forcing the piston to move through its working stroke. Oil drainage from the opposite side of the piston flows back to the pump suction or to the reservoir. At the end of each stroke, the slide strikes an adjustable stop and actuates a linkage which automatically reverses the direction of pump discharge, thus causing the piston to move in the opposite direction. With the particular tool illustrated in Fig. 8, a large-diameter piston rod is used so as to give different piston areas on the up and down sides. This difference in area gives full thrust with relatively slow piston travel on the down, or power, stroke and a reduced thrust with faster travel on the up, or return, stroke. Differences in oil volume between the two ends of the cylinder are compensated for in this hookup through the action of the relief and check valves.

Complex Hydraulic Systems. Numerous variations and combinations of the simple circuits outlined above are in common use for machine tools as well as for aircraft. Fluid motors are used to secure rotary motion such as is imparted by an electric motor. Variable-displacement pumps are used whereby fluid motor speeds may be chosen from full pump speed, to standstill, to full speed in the reverse direction. This type of drive can be applied to provide speed regulation for machine

tools and for conveyors, steel and paper mills, hoisting machinery, etc. Characteristics of the individual circuits as regards flow of the fluid to secure the required motion will influence the selection of pipe size, but other factors influencing piping design, such as the pressure, will be independent of the type of circuit involved.

High- and Low-pressure Piping. Depending on the type of equipment served, any number of main, auxiliary, or pullback cylinders or fluid motors may be used. In some downward-stroke hydraulic presses, such as that illustrated in

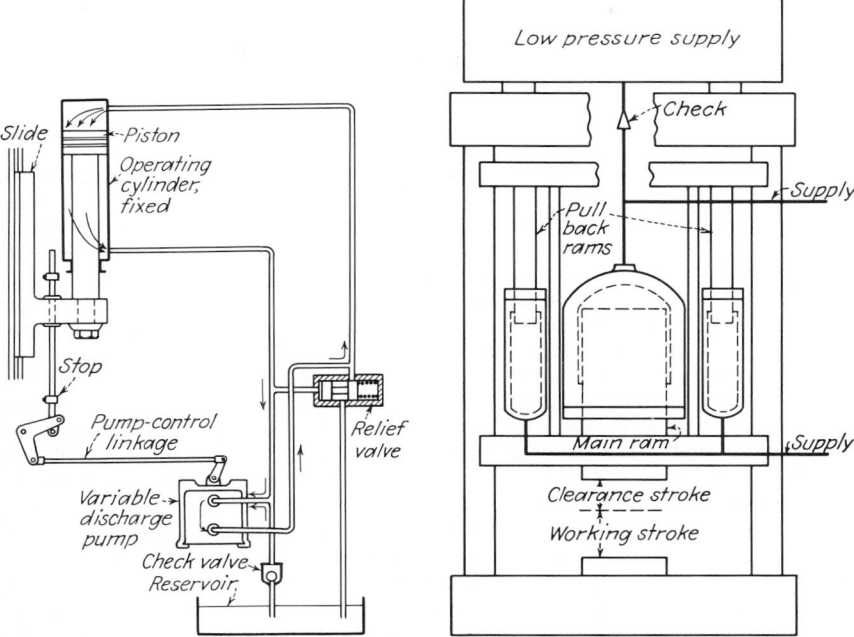

FIG. 8. A variable-volume hydraulic system for an individual unit.

FIG. 9. Simple hydraulic press having one main cylinder and two pullback cylinders.

Fig. 9, the stroke of the main ram is divided into the clearance stroke and the working stroke. The long clearance stroke of the main cylinder may be actuated through a low-pressure hydraulic system, whereas pressure is applied through the working stroke by the high-pressure system. With such installations the smaller pullback rams usually can be served to good advantage by the high-pressure system. The high-pressure system should be designed to sustain the maximum working pressures developed therein, which may go up to 3,000 psi or higher. High velocities and considerable pressure drop can be permitted in the high-pressure system because the high pressures used give a large margin for such losses. The pressure drop in the low-pressure system, however, should be more carefully considered so that the loss of head in the piping will not result in slow or erratic operation of the controlled equipment.

The drain piping through which spent fluid returns eventually to the pump suction is not shown in Fig. 9, but this usually is subject to disposal through the four-way (or equivalent) valve used to control the working and return strokes of the press. In most cases the spent hydraulic fluid returns to a reservoir or sump from

which it is drawn again as needed by the pump suction. With large multiple-unit water systems, water may be drawn from the reservoir by a centrifugal booster pump which discharges in turn to the suction of the high-pressure pump or pumps which may be either centrifugal or reciprocating, depending on the working pressure of the power system.

With an upward-stroke press in which the action of the main ram is inverted from that shown in Fig. 9, the force of gravity often is employed to return the ram and other moving parts to their original position, thus eliminating pullback cylinders. This is the simplest type of press and is used mainly for hot-plate work in plastics and rubber. Such presses have a long clearance stroke and a short working stroke in cases in which it is particularly advantageous to use low-pressure hydraulic power for taking up the clearance stroke and to delay the application of high-pressure power until the final operation of the working stroke.

Pumps for Hydraulic-power Transmission

The pumps of modern hydraulic systems are driven by constant-speed electric motors. Centrifugal, rotary, or reciprocating pumps are used; the type best suited for a specific system can be determined only by an analysis of the requirements of that system. Certain features, however, are characteristic of accumulator systems as differentiated from direct systems in which no accumulator is utilized.

Pumps for Accumulator Systems. Accumulators are used in installations in which two or more hydraulic machines are supplied by a central hydraulic pumping station. In the case of large-size single hydraulic machines in which the average demand for high-pressure fluid over each operating cycle is considerably less than the momentary demand, accumulators may also be used. This is apt to be true in a design similar to that of Fig. 9 in which the clearance stroke is operated with low-pressure fluid. The features of each of the three types which are peculiar to installation in accumulator systems are described in the following paragraphs.

1. *Reciprocating Pumps.* These pumps are driven at constant speed and are of the constant-displacement type. In the case of small pumps, output is controlled by starting and stopping the motor or by operation of a bypass or unloading valve. In the case of large pumps, output is controlled either by automatically blocking open the suction valves when the demand flow is satisfied or by the provision of an unloading valve in the pump discharge circuit. Popular types of reciprocating

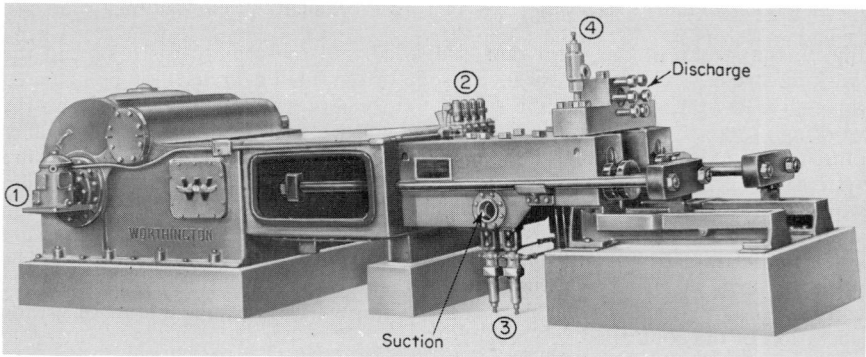

Fig. 10. Horizontal double-acting duplex reciprocating pump suitable for large high-pressure systems using either oil or water as the hydraulic fluid. (*Courtesy of Worthington Pump and Machinery Corporation.*)

pumps are either duplex or triplex double-acting in the horizontal position and triplex single-acting in the vertical position. The horizontal double-acting duplex pump shown in Fig. 10 is a 300-hp size suitable for working pressures up to 3,000 psi. This pump is unloaded with an automatic suction valve lifting device, an enlarged view of which is shown in Fig. 11. The numbered parts in Fig. 10 perform the following functions in unloading the pump: (1) is a synchronizing electric distributor on the end of the crankshaft which controls the unloading device so as to obtain a gradual acceleration or deceleration of fluid flow when delivery is cut on or off; (2) is a gang of solenoid-operated air valves, one for each cylinder, which control compressed-air lines for actuating the lifting devices shown at (3) and in Fig. 11, (3) are the lifting devices, one for each cylinder, which act to block open the suction valves as required to prevent overpressure on the system. With the suction valves blocked open, the plungers operate on suction pressure only since the discharge valves remain seated. Under these conditions the power required by the pump is only enough to overcome mechanical friction. At point (4) is a spring-loaded relief valve which opens on overpressure if the unloading valve fails to function.

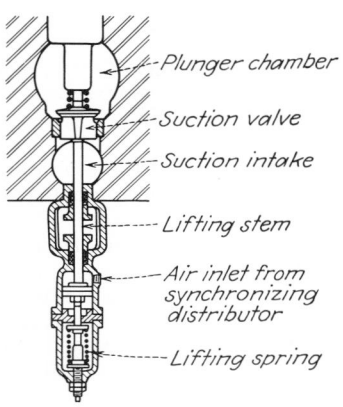

Fig. 11. Enlarged section of valve lifting device shown at (3) in Fig. 10.

2. *Centrifugal Pumps.* These pumps operate at constant speed and at a capacity which is dictated by the characteristics of the pump. With a centrifugal pump, a control for variable and intermittent fluid demand is not necessary since the capacity and head characteristics of this type of pump will meet these flow demands. As the maximum design pressure is approached, the pump delivers less and less fluid until a point of no pumpage is reached. Under these conditions an automatic bypass valve is needed to recirculate enough fluid back to storage in order to prevent overheating the fluid and pump at zero demand. Owing to design limitations centrifugal pumps are restricted to lesser pressures than are usual with reciprocating pumps.

If a particular hydraulic installation requires moderately high pressures at large capacities, the use of a multistage centrifugal pump should be considered. Such pumps may deliver either oil or water and may be used either with or without accumulators. In installations in which pressures not in excess of 2,000 psi are required during comparatively long cycles of operation, no accumulator is needed. If the system characteristics require large flow fluctuations or peak periods of high pressure, the installation of an accumulator may permit the use of a smaller pump than would be necessary otherwise.

3. *Rotary Pumps.* A rotary pump operates at constant speed and with a capacity that is essentially independent of speed. Gear-type and vane-type rotary pumps are shown in Figs. 12 and 13, respectively.

Pumps for Direct Systems. Pumps for direct hydraulic systems must be designed so that the discharge rate may vary with the load. Inherently, centrifugal pumps satisfy this requirement. Rotary pumps of the gear, screw, or vane types are not suitable for use in direct systems. Reciprocating pumps of the multipiston or variable-stroke types have been used extensively for this service.

The radial-piston variable-displacement pump shown in Fig. 14 is driven by a constant-speed motor. In this multipiston-type pump, packless pistons oscillate

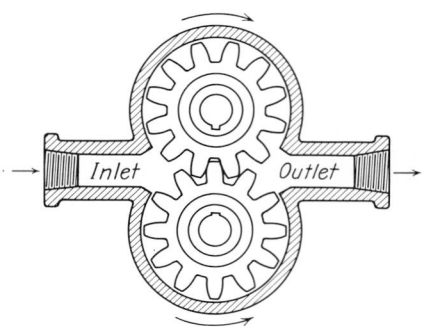

FIG. 12. Gear variety of rotary pump suitable for medium-pressure systems using oil as the hydraulic fluid.

radially within a rotor. By means of a movable rotor ring, displacement of the pistons can be varied either manually or automatically from zero stroke to full stroke. When the ring and the hub are concentric, rotation of the hub occurs with the movement of the pistons; this operating position is referred to as the zero-stroke position. If the ring is moved off center in either direction, it is possible to reverse the direction of the flow of oil without stopping the pump or changing its direction of rotation. The displacement of each piston is transmitted from the supply pressure to the discharge pressure without the use of pump valves or plunger packing. In another variety of the multipiston type, the pistons act in an axial instead of a radial direction and are actuated by two rotating elements having an adjustable angularity.

The multipiston variable-displacement pump is applicable, in general, only to a direct system, and its use is confined either to self-contained machines where the fluid demand favors this type of generating unit or to direct applications calling for a variable pressure demand with certain conditions. Lubricating oil has to be used as the fluid medium with this type of pump, which is limited to working pressures not in excess of 2,000 to 3,000 psi, depending upon the continuity of service.

The reciprocating variable-stroke pump (see Fig. 15) refers to that type of constant-speed pump having outside-packed multiple plungers and pump valves. The displacement of the plungers may be manually or automatically varied from zero to full stroke or capacity by various mechanical means. Variable-stroke pumps are applicable only to the direct system, and usually to self-contained machines where the fluid demand favors this type of generating unit. Under certain conditions they can supply a variable pressure demand. The variable-stroke reciprocating pump is suitable for use with oil or water as the fluid medium and is unlimited in working pressure. The capacity can be controlled gradually, in stepless straight-line fashion, to give any desired continuous delivery between zero and maximum as compared to variation by steps or with intermittent flow as done with reciprocating pumps employing suction-valve control.

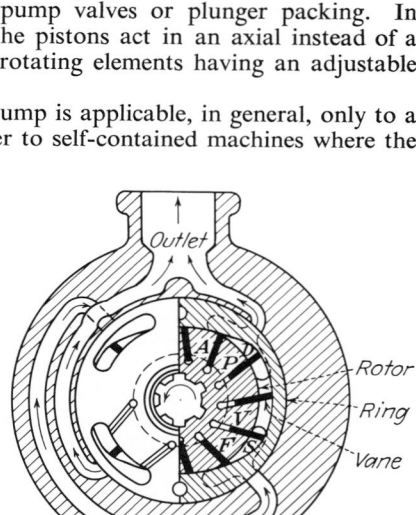

FIG. 13. Vane variety of rotary pump suitable for medium-pressure systems using oil as the hydraulic fluid. As rotor revolves counter-clockwise, vanes V and F follow contour of oval-shaped ring and space S increases in size, forming a vacuum which draws in fluid. Rotor revolves farther, trapping liquid between V and F until they come to position of vanes A and P and liquid comes to space D. As rotor continues to turn, vanes A and P are pressed inward by contour of ring, space D keeps decreasing, and trapped fluid is forced into outlet.

The pump shown in Fig. 15 is driven by a constant-speed electric motor *F* through herringbone reduction gearing from motor to crankshaft *G* which operates the connecting rods *H*. The other end of each of rods *H* is pivotally connected to links *I* by the connecting-rod pins *J* and serves to oscillate the same about the axis of crosshead pins *K* as centers. To the bottom end of each of links *I* is pivotally connected a guide block *L* which slides back and forth within a smooth curved

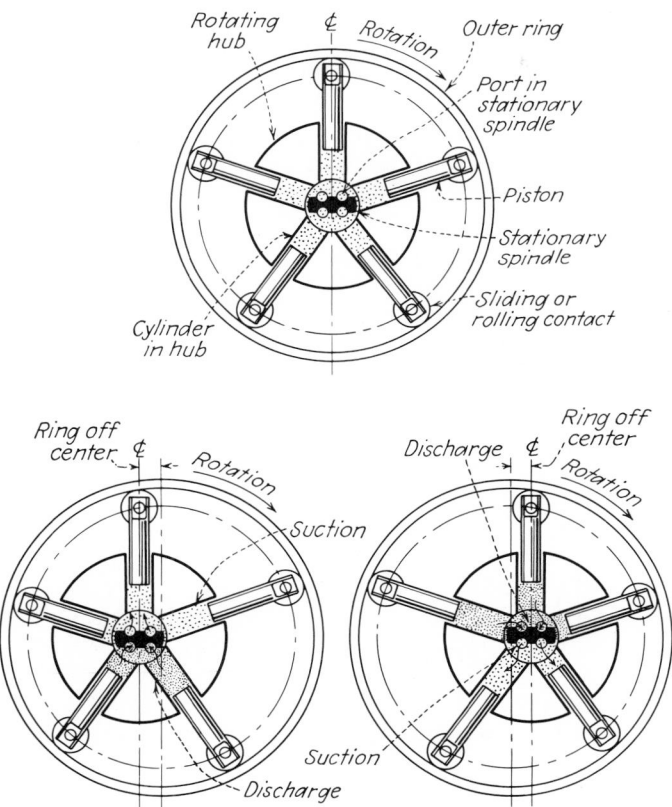

Fig. 14. Radial piston variable-displacement pump suitable for high-pressure direct systems using oil as the hydraulic fluid.

track on the "stroke transformer" *M*. The radius of curvature of this track is equal to the vertical distance therefrom to the center of crosshead pin *K*, measured along the center line of link *I* with the stroke transformer positioned as shown for zero stroke in the left-hand view. In this position, rotation of cranks *G* merely oscillates links *I* and blocks *L* back and forth, and no stroking motion is imparted to the pump plungers *C*. For zero stroke the pump delivery also is reduced to zero.

When the plungers *C* are being reciprocated with their full stroke, the stroke transformer *M* has been tilted through an angle of 30 deg about the longitudinal axis of its pivotal trunnion-journals *X*, as shown in the right-hand view.

For pumping strokes and deliveries anywhere between zero and the maximum, the stroke transformer *M* is positioned correspondingly between the two positions

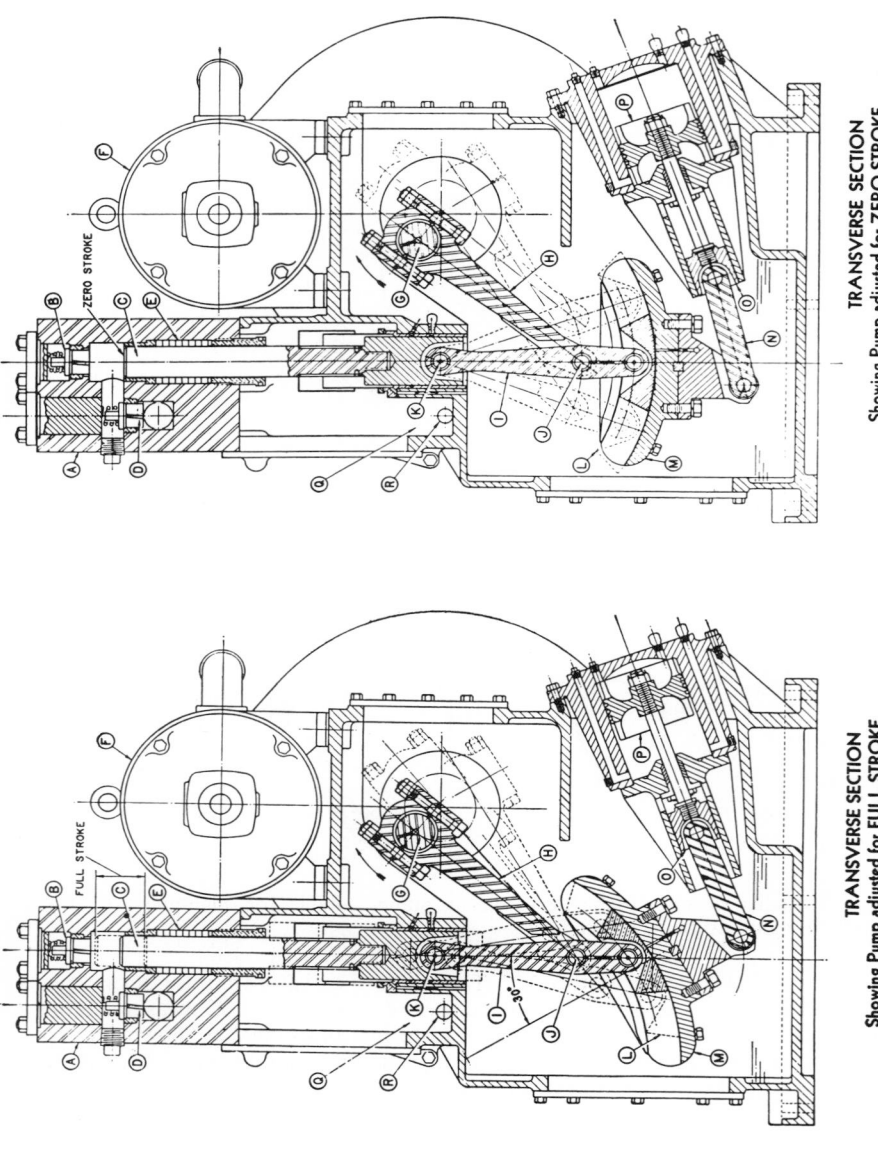

TRANSVERSE SECTION
Showing Pump adjusted for FULL STROKE

TRANSVERSE SECTION
Showing Pump adjusted for ZERO STROKE

FIG. 15. Reciprocating variable-stroke pump suitable for high-pressure direct systems using oil or water as the hydraulic fluid. (*Courtesy of Aldrich Pump Company.*)

shown. This is done by means of link *N* and crosshead *O* under control of a manually operated screw mechanism with a handwheel or else automatically by a hydraulic servo-piston *P* working in the double-acting cylinder as shown.

Hydraulic Motors. In addition to the piston and cylinder applications of hydraulic power described elsewhere in this chapter, there are hydraulic motors, of the gear type, wobbler type, radial-and axial-piston type, etc., which can be used for obtaining rotary power. The axial piston type of motor employing a wobbler plate, shown in Fig. 16, is one of these devices suitable for use with oil as the hydraulic fluid. Motors are available in fixed-displacement and variable-displacement types similar to the corresponding types of pump construction previously described. In the fixed displacement type, speed changes are made by varying the

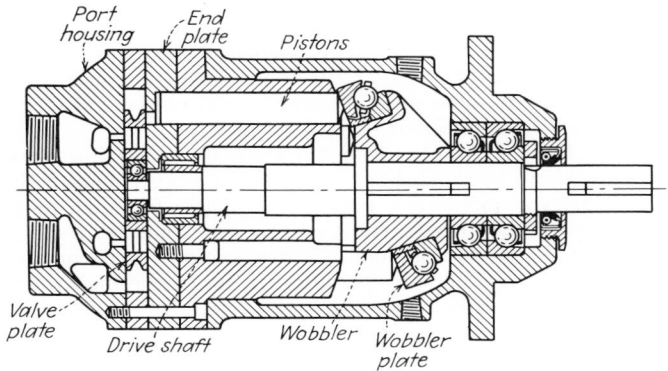

Fig. 16. Axial-piston type of hydraulic motor employing a wobbler plate.

volume of oil flowing through it from the pump. This arrangement works out well where only one motor is driven per pump. In the variable-displacement type, wide speed ranges are obtainable by varying the displacement of the motor in addition to the control, if any, of oil supply from the pump. By means of a lever, a thrust bearing, and a sliding key in the shaft it is possible with the variable-displacement version of the type shown in Fig. 16 to adjust the angularity of the wobbler plate while rotating so as to vary the speed of the motor or to reverse its direction of rotation.

With Sundstrand, Waterbury, and similar hydraulic transmissions a variable-displacement, multipiston pump delivers oil to a constant-displacement fluid motor. Speed of the driven member is controlled by varying the rate of discharge and direction of delivery from the pump. Pump and motor may be assembled in one unit or located separately.

PROPERTIES OF HYDRAULIC FLUIDS

Water and oil are the two basic fluids which are employed in hydraulic systems. Either fluid lends itself well to automatic control, to adjustment of speed, and to reversal of motion in a smooth and vibrationless manner which gives a cushioning effect that is not obtained readily with a mechanical drive. Water is used extensively for operating large presses in forge shops and in similar applications; it is particularly adaptable to multiple-unit or accumulator systems in which several devices are operated from a central pumping station. Oil is used also in large direct or accumulator systems, but its principal applications are in self-contained

units for the control of individual machine tools and in small multiple-unit systems used as control equipment on aircraft and mobile equipment of all sorts.

Water is desirable as a hydraulic fluid because it is inexpensive and will not burn or explode and because its low viscosity results in low pumping costs. However, it is corrosive and is subject to freezing.

Oil is not as corrosive as water, although the existence of certain impurities are objectionable from the viewpoint of corrosion. It does not freeze, but its viscosity does increase with reduction of temperature. Oil is a better lubricant than water; it acts also as a viscous seal which makes possible the use of some varieties of pumps that would not perform well with water at high pressure.

Compressibility. Both water and oil are practically incompressible; their moduli of elasticity are about 300,000 psi at ordinary room temperature, or about 1 per cent that of steel. This modulus changes somewhat with pressure and temperature, but variations usually are insignificant when compared with uncertainties of other factors which enter. The coefficient of expansion of each of the two fluids is of the order of 5×10^{-5} per atmosphere. Because of this low coefficient, air chambers or some other means of cushioning are needed to prevent water hammer when there are sudden stoppages of flow.

Temperature Rise. The use of high pressures in hydraulic-power systems permits employing high fluid velocities, and these are accompanied by relatively large frictional losses in piping and equipment. These frictional losses increase the internal energy of the hydraulic fluid, and this is manifested by an increase in temperature of the fluid. If heat losses from the piping and equipment to the surrounding air are neglected, the temperature rise for each round trip through the hydraulic circuit can be calculated easily as

$$\Delta t = 144(\Delta P)/\rho J C$$

in which Δt = the temperature rise, F

$\quad \Delta P$ = pressure drop due to friction, psi

$\quad \rho$ = fluid density, lb/cu ft

$\quad J$ = conversion factor, ft lb/Btu

$\quad C$ = specific heat of fluid, Btu/(lb F)

For water with a density of 62.4 lb/cu ft and a specific heat assumed to be unity, the foregoing equation reduces to

$$\Delta t = 297 \times 10^{-5} \Delta P$$

For oil of specific heat C and of specific gravity S, the equation becomes

$$\Delta t = [(297 \times 10^{-5})/CS] \Delta P$$

In practice, as the temperature of the fluid increases, a temperature difference is developed between the piping and the surrounding air, and thus, heat transfer to the surroundings will take place. Eventually, a steady state will be reached at which the effects of frictional resistance will be dissipated by heat transfer to the surroundings. But this steady-state temperature may be so high as to vaporize the water, to break down the lubricating characteristics of the oil, or to result in severe mechanical or metallurgical damage to piping or equipment. For this reason, heat exchangers are installed frequently in hydraulic-power transmission systems.

Specific heats and specific gravities of oils are listed in Chap. 3. Likewise, Table 43 of that chapter lists the equivalent resistance of bends and fittings. The frictional resistance of the piping circuit may be determined by use of any of the analytical expressions listed in that chapter.

The temperature-rise relations given above are predicated on the assumption that the same quantity of fluid is involved in both pressure-drop and internal-energy-rise

calculations.   If a reservoir or an accumulator is present in the circuit, or if there is much branch piping to be considered, allowance should be made for the amount of active storage capacity in the system.   Furthermore, the temperature rise of piping and equipment, together with heat loss from the same, may have to be taken into account by energy-balance calculations.   The increase in internal energy ΔU experienced by a flow rate of w lb of fluid per unit time can be calculated from

$$U = 144w(\Delta P)/\rho J$$

The calculations necessary for a complex system are somewhat lengthy, involving as they do an iterative procedure.   The time involved in solution can be reduced materially if access is available to a digital computer.   This assumes, of course, that a program is at hand.

Water as a Hydraulic Fluid

When water is used as the fluid in hydraulic-power transmission systems, an emulsifying oil is added in a concentration of 2 to 3 per cent by weight as a corrosion inhibitor for the piping and equipment.   Centrifuges are employed to separate the water from the oil periodically and start over with a fresh mix.   Such systems should be checked for rubber gaskets or other materials that might be adversely affected by the emulsion, causing swelling or slight disintegration.

Water as a fluid for hydraulic-power transmission is used at much higher pressures and velocities than are found in other applications.   This creates a number of special problems which are peculiar to hydraulic piping and which involve unusual construction and severe shock conditions under some circumstances.   These problems are discussed in succeeding sections.

Flow Charts.   A water velocity of 35 fps is deemed conservative in hydraulic-power transmission piping, and velocities of 50 fps or higher are sometimes employed.   The large pressure drops accompanying these velocities usually do not represent a big percentage of the initial pressure since lines are short and pressures high.   An initial pressure of 1,000 psi is considered low for water and would be employed, perhaps, in connection with a centrifugal pump.   Pressures of 2,000 to 3,000 psi developed by reciprocating pumps are usual in large forgeplant installations.

Owing to the heavy-wall pipe and high water velocities used in hydraulic-power transmission systems it has been necessary to provide special flow charts to suit these conditions.   The charts furnished in Figs 1.7 to 19 for standard-weight, extra-strong, and double-extra-strong pipe cover the usual range of conditions for water flow through clean steel pipe in hydraulic systems.

Water velocities in pump-suction and gravity drain lines are usually set at 2 to 5 fps. On large systems which employ reciprocating pumps for high-pressure delivery, a centrifugal booster pump is sometimes used to deliver water from the reservoir into the suction of the high-pressure pumps; the discharge pressure of the booster pump is usually set at 40 to 60 psi. In other installations, a separate low-pressure pump is used to deliver water for actuating the clearance stroke of the rams before the high-pressure water is admitted for the power stroke.

Velocity Head.   For fluid velocities as high as those which are encountered in hydraulic-power systems, the corresponding pressure head may become appreciable and should not be overlooked in design.   Table 1 lists the pressure changes needed in order to develop water velocities of 1 to 100 fps; for fluids other than water, the tabular values must be multiplied by the specific gravity of the respective fluid.   It should be evident that the values listed, in pounds per square inch, are in reality

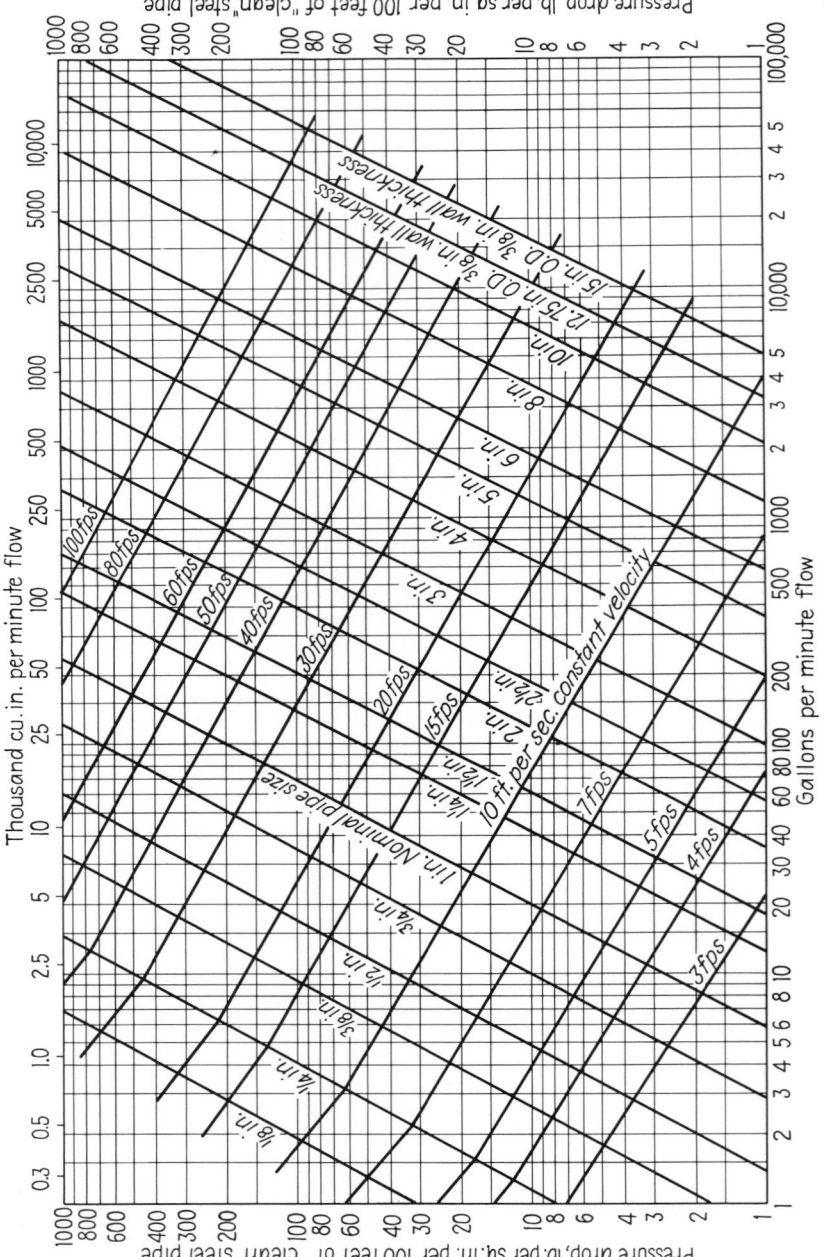

FIG. 17. Flow chart for standard-weight (Schedule 40) pipe carrying water.

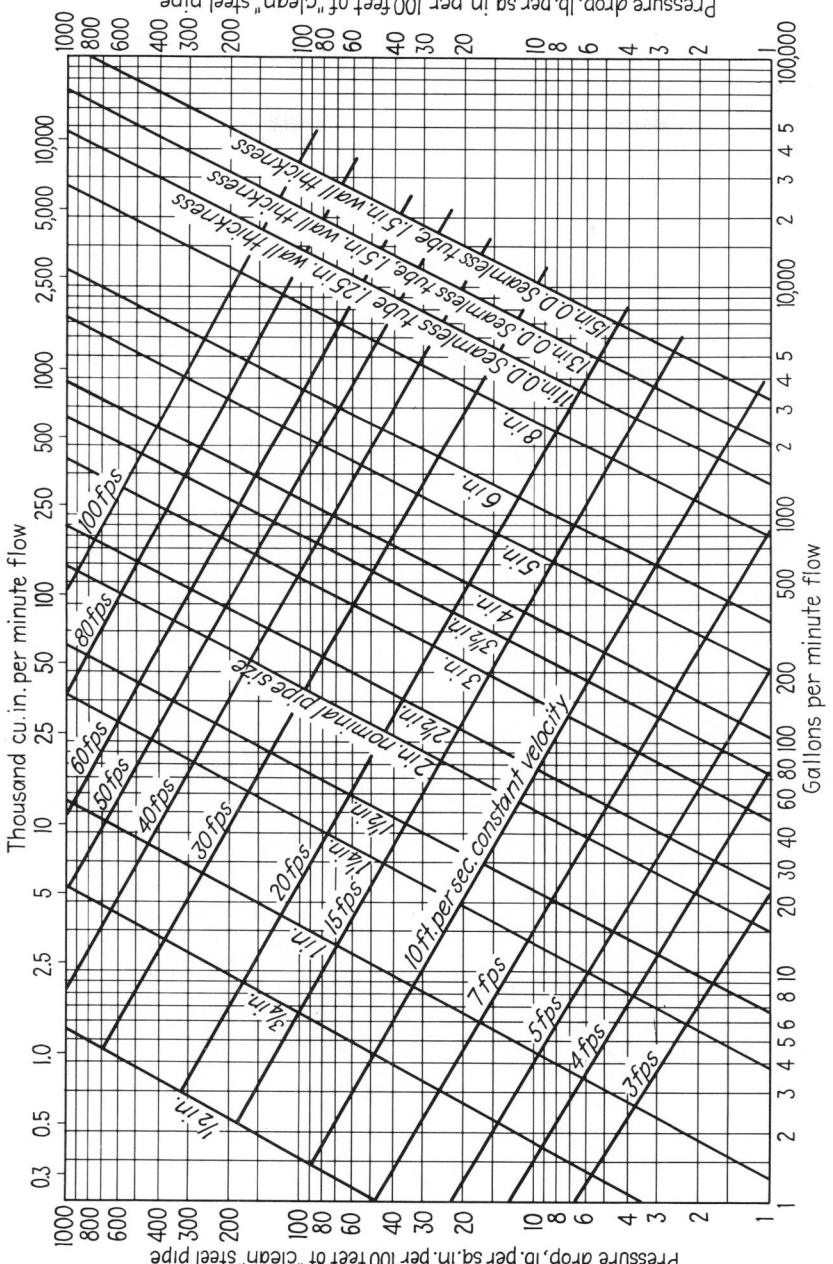

Fig. 18. Flow chart for extra-strong (Schedule 80) pipe carrying water.

14–19

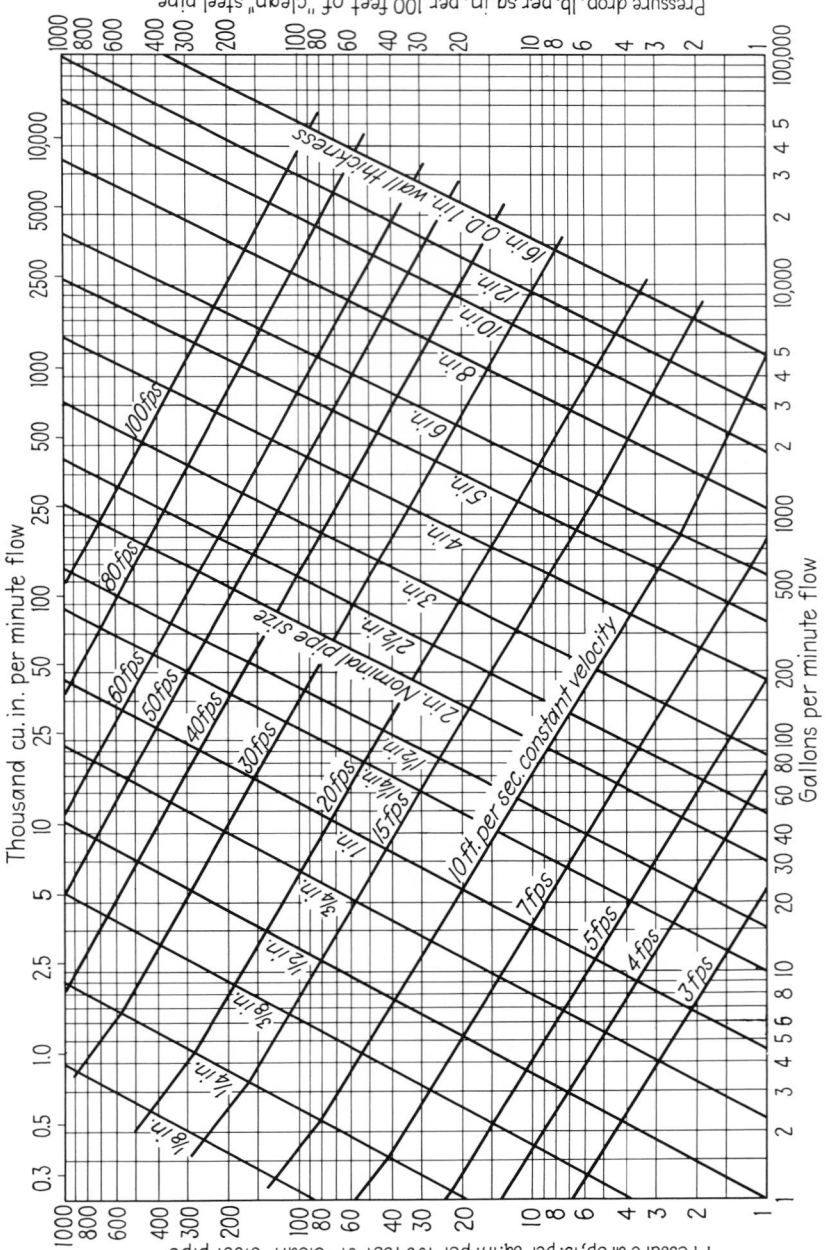

Fig. 19. Flow chart for double-extra-strong pipe carrying water.

velocity heads which have been converted for convenience for direct substitution into the Bernoulli equation.

Shock Loading. Because of the high velocities used in hydraulic systems, and because of the rigidity of the heavy-walled pipe that is required by the high operating pressure, water hammer may be a serious problem. Reference to Table 62, Chap. 3, discloses that, in the case of rigid pipe, a shock pressure of about 60 psi may be expected for each foot per second reduction in velocity. Thus, with flow velocities of 25 to 50 fps, shock pressures of 1,500 to 3,000 psi might be expected under extreme conditions. The energy developed, in inch pounds per cubic inch

Table I.   Loss in Pressure Required to Produce Water Velocities Shown[1]

Velocity, ft per sec	Pressure loss, psi	Velocity, ft per sec	Pressure loss, psi	Velocity, ft per sec	Pressure loss, psi	Velocity, ft per sec	Pressure loss, psi
V	p_v	V	p_v	V	p_v	V	p_v
1	0.01	26	4.56	51	17.6	76	39.0
2	0.03	27	4.92	52	18.3	77	40.0
3	0.06	28	5.29	53	19.0	78	41.1
4	0.11	29	5.68	54	19.7	79	42.1
5	0.17	30	6.08	55	20.4	80	43.2
6	0.24	31	6.49	56	21.2	81	44.3
7	0.33	32	6.91	57	21.9	82	45.4
8	0.43	33	7.35	58	22.7	83	46.5
9	0.55	34	7.80	59	23.5	84	47.6
10	0.68	35	8.27	60	24.3	85	48.8
11	0.82	36	8.75	61	25.1	86	49.9
12	0.97	37	9.24	62	25.9	87	51.1
13	1.14	38	9.75	63	26.8	88	52.3
14	1.32	39	10.3	64	27.6	89	53.5
15	1.52	40	10.8	65	28.5	90	54.7
16	1.73	41	11.3	66	29.4	91	55.9
17	1.95	42	11.9	67	30.3	92	57.1
18	2.19	43	12.5	68	31.2	93	58.4
19	2.44	44	13.1	69	32.1	94	59.6
20	2.70	45	13.7	70	33.1	95	60.9
21	2.98	46	14.3	71	34.0	96	62.2
22	3.27	47	14.9	72	35.0	97	63.5
23	3.57	48	15.6	73	36.0	98	64.8
24	3.89	49	16.2	74	37.0	99	66.2
25	4.22	50	16.9	75	38.0	100	67.5

[1] Computed by formula $p_v = 0.00675\,Sv^2$ for water weighing 62.4 lb per cu ft which corresponds to $S = 1$. To get the value of p_v for water or other liquid at a different specific gravity, multiply the tabular values by the specific gravity of the liquid under flow conditions.

of fluid involved resulting from these shock pressures is given by $P_0^2 \div 2K$ where P_0 is the shock pressure in pounds per square inch and K is the bulk modulus of elasticity of the fluid. The entire system must be designed to take into water-hammer shock loads or to provide means by which the resulting energy can be absorbed.

The kind of *anchors* and *supports* used with large hydraulic systems may be of considerable importance where severe water hammer is experienced. With several forging presses or other hydraulically operated equipment taking water through a header from a common set of pumps, certain combinations of demand sometimes occur which produce enough shock to make the whole piping system whip about. Under these circumstances serious damage may be done to the equipment connections and piping unless the line is properly restrained.

Since thermal elongation is of small proportions in most hydraulic systems, it seldom is necessary to make much provision for expansion due to change in temperature, and some designers prefer to anchor such systems rigidly to the building steel at frequent intervals. On the other hand, excessive rigidity has its drawbacks,

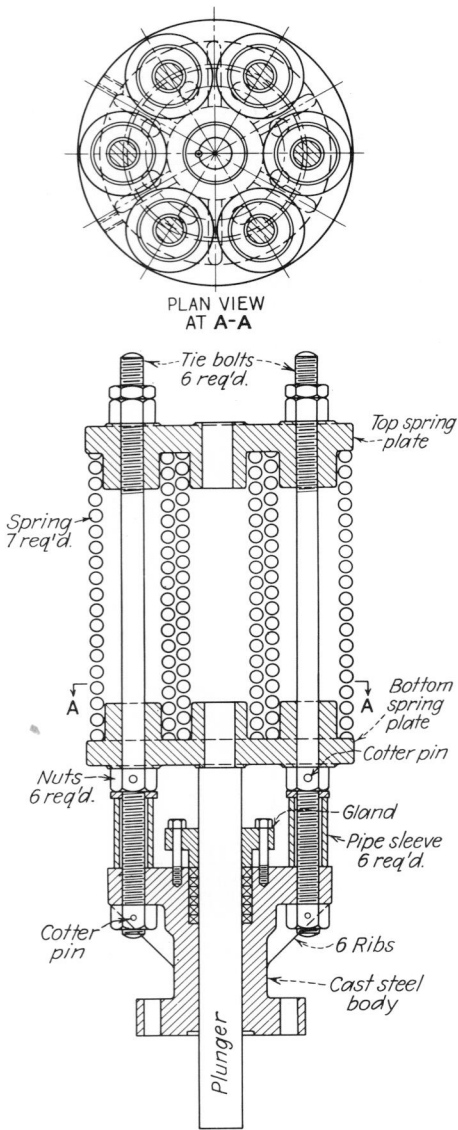

PLAN VIEW
AT **A-A**

FIG. 20. Shock-absorber valve for weighted-accumulator hydraulic system. (*Courtesy of Baldwin-Southwark.*)

and some other designers prefer to cushion the whipping of the pipe with spring shock absorbers rather than to attempt to fasten it down too rigidly.   Where shock absorbers are used on hydraulic systems, they should be a size or two larger than the recommendations given for powerhouse piping.   With shock absorbers, hydraulic piping can be supported in box guides so as to permit free longitudinal movement

except as restrained by the shock absorbers.   This arrangement is said to be very effective in cushioning shocks which are due to water hammer and which might be destructive to fixed anchors.

With well-arranged air-bottle accumulator systems and smoothly operating control valves, shock pressures may be held to a level where they pass unnoticed or give little or no trouble.   With gravity-type accumulators, however, it often is necessary to use shock-absorber valves such as that illustrated in Fig. 20.   It is common practice to put one near the accumulator, one near each of the presses, and one at the end of each long run of straight pipe.   Shock valves are sometimes used with air-bottle accumulator systems too if bad shock conditions are experienced without them.   In such a system no relief valve should be needed near the accumulator, which provides its own air cushion.   Shock-relief valves should be placed as close as possible to the point at which flow is interrupted.   This usually means on the branch lines close to the presses and other equipment being operated.   The idea is to cushion the initial shock somewhat before the wave has a chance to rebound in full force throughout the entire system.

In the shock-absorber valve shown in Fig. 20, the plunger operates through a stuffing box which does not permit fluid to escape from the system.   Movement of the plunger is controlled by heavy springs.   Small sizes usually are arranged with double springs one inside the other, while large sizes have multiple springs. The springs and plunger assembly can be mounted on the branch outlet of a tee or other fitting convenient for the purpose.   Shock valves accomplish their purpose by absorbing excess energy through having the plunger move out against the resistance of the springs, following which the plunger returns to restore this energy to the fluid in the proper part of the cycle.   No fluid is wasted aside from incidental leakage.

As previously stated, the total shock energy set up by water hammer is $p_0{}^2 \div 2K$ in.-lb/cu in. of fluid affected by the instantaneous change of velocity, but not all of this energy necessarily has to be absorbed in order to reduce shock to reasonable proportions.   Practically, shock valves are usually designed to have an operating range from $\frac{1}{2}$ to $1\frac{1}{2}$ normal pressure without striking the stops.   This, in the case of 2,400-psi normal pressure, corresponds to a range from 1,200 to 3,600 psi and provides on the upper side for an extinguished velocity of 20 fps.   On the downswing the bottom spring plate may approach or strike the inboard stop, in which position it is shown in Fig. 20.   The beneficial action of the shock valve is in retarding or cushioning the rate of pressure surge and in dampening out the oscillations of the fluid column.

An approach to selecting the right diameter of plunger and spring resistance for a shock-absorber valve can be made through computing the inch-pounds of shock energy that have to be absorbed by the springs and proportioning them for developing this amount of work in being pushed from the half compressed to the fully compressed position.   Since the load is applied through pressure against the plunger, the diameter of the plunger should be sufficient to compress the springs about one-half their travel under normal pressure and to compress them fully under maximum design shock conditions.   According to Baldwin-Southwark[1] a rough approximation of the size of plunger can be found from the formula $A_1/A = 5.14p \ 10^{-4}$ where A is the internal area of the pipe, A_1 the area of the plunger, and p is the normal pressure on the system in pounds per square inch.   In a more exact solution consideration should be given to the length of line, ratio of pipe diameter to wall thickness, physical characteristics of springs, limits of travel, amount of dampening required, etc.

---

[1] See "Hydraulic Tables and Other Data," Baldwin-Southwark *Bull.* 150.

Water hammer sometimes occurs in a long exhaust line leading from an operating valve back to the reservoir. In such cases it is associated with the inertia set up by high exit velocity from the valve and through the line. When the valve cuts off, the inertia of the fluid column may create a partial vacuum between the column and the valve so that, when the inertia is expended, the resulting pressure unbalance returns the fluid column to the valve and thus produces hammer shocks. Relief may be effected by the use of a suppressor of one sort or another, or the tendency to form a vacuum may be avoided through providing a vacuum breaker near the valve. This usually takes the form of a check valve on a stub off the exhaust line, with the check arranged to admit air when the line is under vacuum and to prevent the escape of fluid when the line is under pressure.

Oil as a Hydraulic Fluid

Manufacturers of hydraulic equipment usually specify the grade of oil best suited to their product, and their recommendations should be followed. Consideration has to be given to working temperatures, pressures, ambient temperatures, and the type of service. Hydraulic oils generally used have viscosities at 100 F ranging from about 150 SSU for light oils to about 1,000 SSU for heavy oils. The lighter oils are used on pumps and machines requiring low to medium pressures; the heavy grades are used where high pressures are encountered. An ordinary operating temperature for hydraulic oils is 150 to 165 F, which usually can be maintained without the use of a heat exchanger or a cooling coil in the reservoir. Hydraulic oils used with power-pumping equipment must have adequate lubricating properties. The use of nonlubricating oils such as Lockheed fluid is restricted to manually operated braking devices and the like.

Viscosity. High pressures are maintained in two ways between moving parts in hydraulic systems: (1) by closely ground and lapped clearances assisted by a viscous seal consisting solely of the oil film and (2) by the use of cup leathers and similar packing around piston rods, plungers, and the like. The first means is employed for sealing internal clearances in rotary and piston oil pumps; the second is used for external packing with either oil or water.

The effectiveness of the viscous seal afforded by the oil film between the pump parts falls off rapidly if the oil loses viscosity too fast with increase in operating temperature, whereas frictional losses through the piping and control equipment mount rapidly if the oil is too viscous. Consequently an economic balance must be struck between losses due to slippage in the pump and frictional losses through the system so as to come out with the best overall result. For any given hydraulic system there will be an optimum viscosity for obtaining the maximum overall operating efficiency, and any viscosity lower than the ideal will permit pump slippage which will more than offset any saving in line losses. Hence, for hydraulic purposes where the fluid has to operate over a considerable temperature range, an oil should be selected with a flat characteristic for change in viscosity with temperature. This property is referred to as the *viscosity index* or "VI" and a high VI rating means that the oil changes less in viscosity with temperature than would an oil of lower VI.

The VI rating of an oil is important in the operation of a hydraulic system, and as has been pointed out, higher VI ratings are to be preferred. However, the use of polymer-thickened oil as a means of increasing the viscosity index is not recommended and can actually result in poor operation. Polymer-thickened oils are non-Newtonian in character and, as such, suffer appreciable loss of viscosity both temporarily and permanently in hydraulic systems. The temporary decrease in viscosity at the very high shear rates in the slippage paths of pumps and motors

Table 2.   Oil Viscosity Recommendations[1,2] for Hydraulic Systems

(Industrial machinery installations; ambient temperature 40 to 90 F)

	Type of installation	Oil viscosity, SSU
Vane-type pumps and motors with or without flow-control valves	Machine tools operating at pressures up to 1,000 psi	150–225 at 100 F
	Presses and other heavy machinery, or above 1,000 psi	275–315 at 100 F
Piston-type pumps and motors	All industrial machinery	275–315 at 100 F
Gear-type fluid motors	All industrial machinery	150–315 at 100 F

[1] Courtesy of Vickers, Incorporated.
[2] Data pertain to average conditions (optimum oil temperature is 120 F). The range of viscosity ratings given permit selection of heavier viscosity oils for higher operating temperature conditions.   Maximum safe operating temperature for the oil is 140 F.   Abnormal speed, pressure, and temperature conditions require special engineering consideration.   If a choice of two recommended grades of oil is involved, use the heavier of the two grades to protect the more critical units.

causes a significant loss of efficiency in these components.   Permanent loss in viscosity is reflected throughout the system.

According to the recommendations of Vickers, Incorporated, the viscosity of the oil at start-up temperature should not exceed 4,000 SSU and, at maximum operating temperature, the oil should have a minimum viscosity of 80 SSU.   It is generally recommended that oils of viscosity less than 150 SSU at 100 F be avoided unless low-temperature start-up problems exist.   Higher viscosity oils are required by some components.   Oils with a high VI are recommended; however, other desirable characteristics should not be compromised to achieve this factor.   Table 2 gives viscosity recommendations for industrial applications, and Table 3 lists similar data for construction and similar mobile equipment hydraulic systems.

The viscosity-temperature relationships for several typical oils commonly used in hydraulic systems are shown in Fig. 21.   Viscosity is given both as seconds Saybolt Universal and as kinematic viscosity in centistokes.   Although viscosity curves generally plot as straight lines on the sort of cross-sectional paper used in Fig. 21,

Table 3.   Hydraulic-fluid Recommendations[1]

(Agricultural, construction, earth-moving, material handling, and other mobile machinery)

Hydraulic-system operating range (min. to max.), F	SAE viscosity	API service classification
0–180	10W	MS
15–210	20–20W	MS
32–230	30	MS
0–210	10W-30	MS

[1] Courtesy of Vickers, Incorporated.

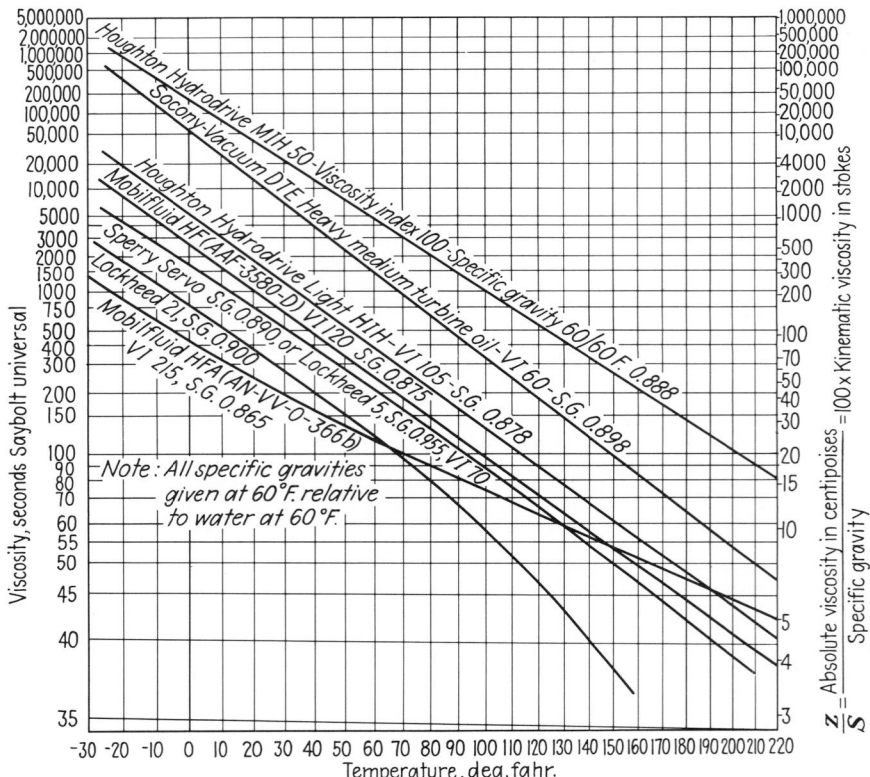

Fig. 21. Temperature-velocity characteristics and specific gravities of typical oils used in hydraulic systems.

the lines may bend somewhat as a result of additions of compounds to the oil to provide some particular characteristic such as increased wear resistance, improved demulsibility properties, or better viscosity index. This tendency is particularly evident for Lockheed 21 fluid shown in Fig. 21.

It should be noted that brands of oil are shown in Fig. 21 only to demonstrate the viscosity-temperature characteristics of representative hydraulic oils. This does not constitute an endorsement or recommendation of them in preference to numerous other satisfactory hydraulic oils marketed by many of the oil companies. An oil should be selected with characteristics to suit the individual application as defined by the manufacturer of the equipment.

Oil-flow Data. Oil velocities in hydraulic systems of all sorts are considerably lower than the water velocities customary with large multiple-unit presses and the like, but they still are on the high side with respect to oil velocities used in other applications. A good velocity range for discharge piping is 7 to 15 fps but 20 fps often can be permitted when this velocity is attained for only a short time or a small part of the cycle.[1] The design velocity actually chosen depends on the viscosity

[1] See "Hydraulic Piping, Determining (Oil) Pressure Drop and Power Losses," by Ransom Tyler, *Product Eng.*, November, 1941. Available also in reprint form as *Bull.* 90010 of The Oilgear Company, 1301 West Bruce St., Milwaukee, Wis.

of the particular oil used and on the length of line required.   Pump suction and drain lines carrying hydraulic oil are designed ordinarily for velocities of 2 to 5 fps. Piping for low-pressure auxiliaries and controls should be sized amply so that pressure drop will not cause sluggish actuation of the controls and thus delay action of the main equipment.   Since this piping usually is of small capacity, making it of generous size will not add much to the overall cost.   Flow velocities through pipes and tubes of various sizes can be determined from Fig. 22, which is reproduced through the courtesy of Vickers, Inc.

The principal reason for using a lower velocity with oil than with water is the viscosity of the kind of oil which has to be used with rotary pumps in order to prevent excessive slippage.   With an oil having a viscosity suitable for use with a rotary pump, the frictional loss through the piping would be excessive if the velocity were as high as could be used to advantage with water.

A chart for the flow of hydraulic oil through clean, smooth tubes of aluminum, brass, copper, lead, glass, etc., is given in Fig. 23.[1]   The chart can be applied also to "very smooth" seamless-drawn steel tubes carrying oil.   The example shown in dash lines is for oil having a viscosity of 340 SSU and a specific gravity of 0.9 at 50 F flowing at the rate of 10 gpm through 1-in. tubing with 0.049 in. wall thickness. The pressure drop obtained from the chart is 0.29 psi per foot of tube, or 29 psi per 100 ft of tube.

In Table 4 are given the oil-carrying capacities and friction losses of "clean" steel pipe for the range of velocities and capacities encountered in hydraulic-power transmission piping.   For the convenience of the user, carrying capacities are given in both gallons per minute and cubic inches per minute, velocities are given in feet per second and corresponding velocity heads in pounds per square inch.   Pressure drops in the viscous-flow region are set in boldface type to distinguish them from the viscous region where pressure drops are set in ordinary type.   As a further demarcation between viscous and turbulent flow, the SSU corresponding to the critical point for each rate of flow is shown in the tabulation.   Pressure drops for higher SSU values are on the viscous side of the critical point and for lower SSU values on the turbulent side.   In computing the pressure drops tabulated, allowance was made for the fact that unstable flow may occur throughout the whole critical region, and the larger pressure drop has been shown when any doubt exists as to which formula would obtain.

Since Table 4 is intended for use in selecting pipe sizes for systems using oil at medium to high pressure, the wall thicknesses shown are on the heavy side, viz., Schedule 80 for nominal sizes $\frac{1}{8}$ to $\frac{3}{8}$ in. inclusive, and Schedule 160 for nominal sizes $\frac{1}{2}$ in. and larger.   Pressure drops through these weights of pipe when carrying water can be approximated from the 50 SSU column by taking 60 to 80 per cent of the tabular values.

The solution of oil-flow problems can be extended from one condition of viscosity and specific gravity to another through consideration of the relation of these terms in the pressure drop formula, at least between the limits of 100 and 4,000 SSU. Within these limits resistance to flow in the viscous region is directly proportional to the product of SSU and specific gravity.   For instance, if Δp is computed as 50 psi at 200 SSU, it would be double this or 100 psi at 400 SSU if the specific gravity remained the same.   With the same oil, however, the drop in temperature which increased the viscosity from 200 to 400 SSU also would increase the specific gravity to about 1.01 times what it was with the 200 SSU viscosity.   Hence the corrected answer would be $100 \times 1.01 = 101$ psi.

[1] From "Aircraft Hydraulics," by Harold W. Adams.   The curves were drawn for a friction factor of $f = 0.0014 + 0.125/N^{0.32}$ given in "Heat Transmission," by W. H. McAdams, McGraw-Hill Book Company, New York, 1954.

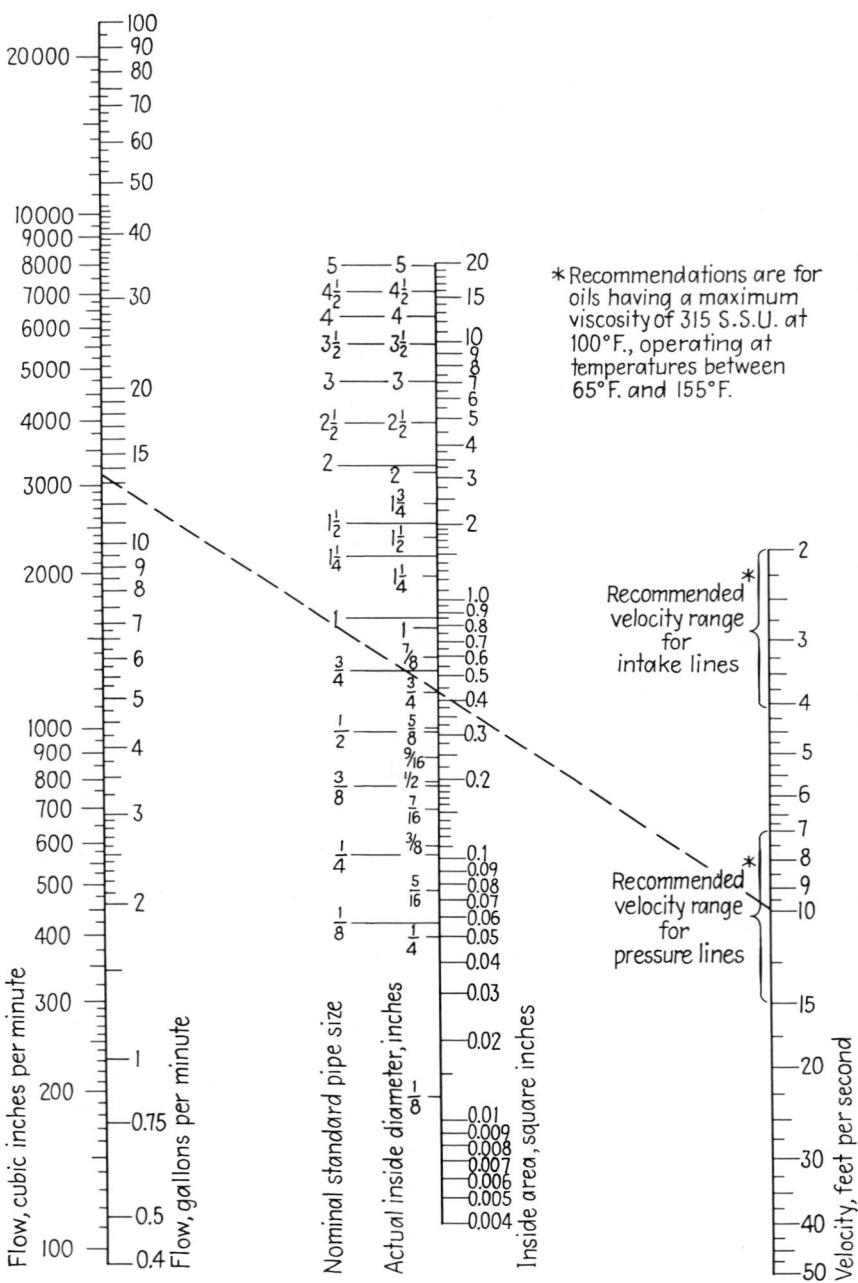

Fig. 22. Nomographic chart for finding flow capacities of pipes and tubes at recommended flow velocities. (*Courtesy of Vickers, Inc.*)

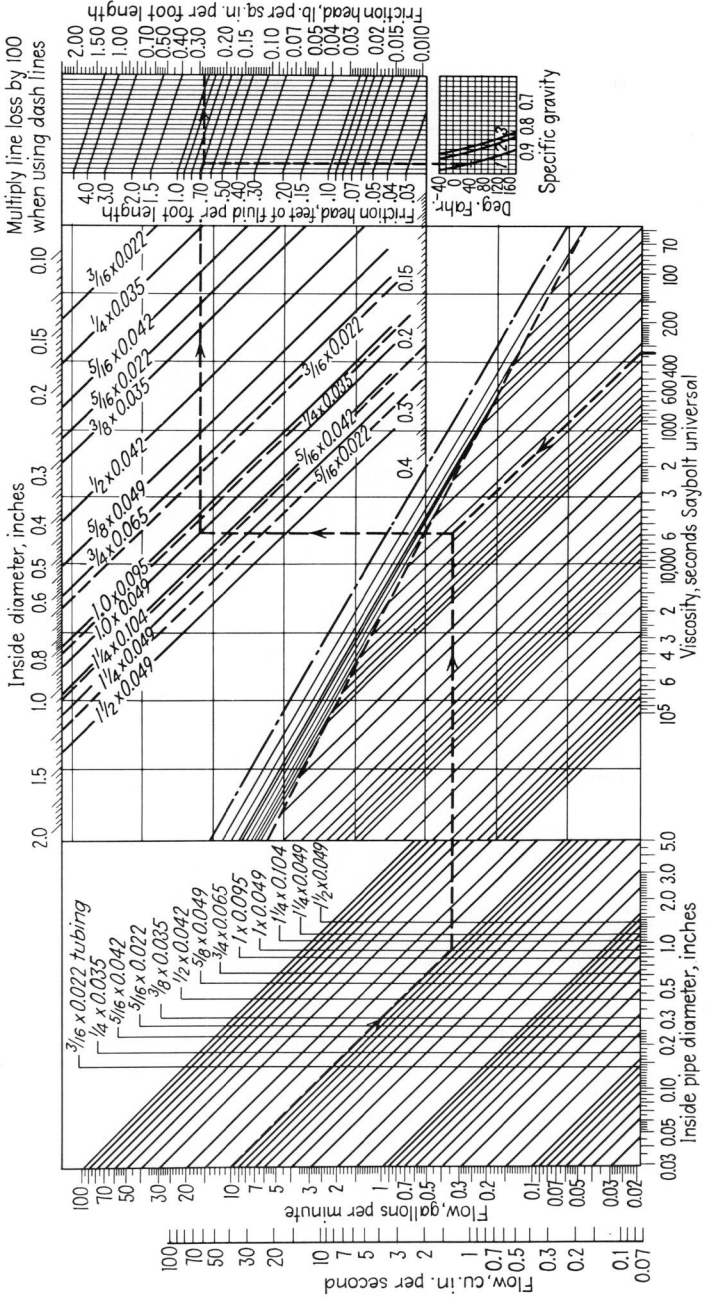

FIG. 23. Flow chart for flow of hydraulic oil through tubing. *(Reproduced by permission from "Aircraft Hydraulics" by Harold W. Adams.)*

14–29

Table 4. Oil-carrying Capacities and Friction Losses for "Clean" Steel Pipe in Hydraulic-power Transmission Systems

(For water use 60 to 80 per cent of pressure drops shown for oil in 50 SSU column)

Nominal pipe size, in., schedule number, and average inside diam, in.	Carrying capacity		Velocity, ft per sec	Velocity head, psi for[2] $S = 0.9$	SSU for critical point, SSU = $6.8G/d$	Pressure drop in psi per 100 ft of pipe[1]					
						Viscosity, SSU					
	Gal per min	Cu in. per min				50	100	200	500	1,000	4,000
⅛ Sched. 80 $d =$ 0.215	0.5	116	4.43	0.12	16	**54**	**111**	**225**	**570**	**1,150**	**4,655**
	1.0	231	8.85	0.48	32	**108**	**222**	**450**	**1,140**	**2,300**	**9,310**
	1.5	347	13.3	1.07	48	**283**	**333**	**675**	**1,710**	**3,450**	**14,000**
	2.0	462	17.7	1.90	64	**472**	**537**	**900**	**2,280**	**4,600**	**18,620**
	2.5	578	22.1	2.97	80	**645**	**835**	**1,125**	**2,850**	**5,750**	**23,310**
¼ Sched. 80 $d =$ 0.302	1.0	231	4.48	0.12	23	**29**	**59**	**120**	**304**	**615**	**2,485**
	2.0	462	8.96	0.49	45	**92**	**118**	**240**	**608**	**1,230**	**4,975**
	4.0	924	17.9	1.94	90	**287**	**394**	**480**	**1,215**	**2,460**	**9,940**
	6.0	1,396	26.9	4.40	135	**590**	**650**	**720**	**1,822**	**3,690**	**14,910**
	8.0	1,848	35.8	7.76	180	**987**	**1,475**	**1,800**	**2,430**	**4,920**	**19,880**
⅜ Sched. 80 $d =$ 0.423	2.5	578	5.71	0.20	40	**31**	**40**	**82**	208	420	1,700
	5.0	1,155	11.4	0.78	80	**87**	**125**	**164**	415	840	3,400
	7.5	1,733	17.1	1.78	120	**176**	**210**	**246**	623	1,260	5,100
	10	2,310	22.8	3.15	160	**293**	**305**	**520**	832	1,680	6,800
	15	3,465	34.3	7.14	240	**662**	**860**	**1,055**	1,340	2,500	10,200
½ Sched. 160 $d =$ 0.466	2.5	578	4.70	0.13	36	**13**	**27**	**54**	150	303	1,227
	5.0	1,155	9.40	0.54	72	**59**	**77**	**107**	300	606	2,454
	10	2,310	18.8	2.15	144	**185**	**241**	**294**	600	1,212	4,908
	15	3,465	28.2	4.82	216	**376**	**490**	**580**	900	1,818	7,362
	20	4,620	37.6	8.57	288	**632**	**780**	**950**	1,200	2,424	9,816
¾ Sched. 160 $d =$ 0.614	5.0	1,155	5.42	0.18	55	**16**	**24**	**48**	122	248	1,000
	10	2,310	10.8	0.71	110	**50**	**65**	**97**	246	497	2,010
	20	4,620	21.6	2.83	220	**168**	**225**	**280**	490	994	4,020
	30	6,930	32.5	6.40	330	**350**	**465**	**580**	736	1,491	6,030
	40	9,240	43.3	11.4	440	**580**	**760**	**930**	1,400	1,988	8,040
1 Sched. 160 $d =$ 0.815	10	2,310	6.15	0.23	90	**13**	**17**	**24**	60	122	493
	20	4,620	12.3	0.92	180	**44**	**57**	**78**	120	244	986
	30	6,930	18.5	2.07	300	**90**	**120**	**150**	181	366	1,480
	40	9,240	24.6	2.67	450	**149**	**206**	**252**	350	488	1,972
	60	13,860	36.9	8.26	750	**312**	**410**	**500**	635	732	2,960

[1] Computed for oil having a specific gravity S of 0.90 at 60/60 F. To adjust the tabulated pressure drops to suit an oil having a different gravity, multiply by the ratio of its specific gravity at 60/60 F to 0.90. Pressure drops in boldface type are in the viscous flow region.

[2] Computed for $S = 0.90$ from the relation $p = 0.00675Sv^2$. To adjust the tabulated velocity heads to suit a different gravity, multiply by the ratio of that gravity under flow conditions to 0.90.

Also within the aforesaid limits, the extension from one condition to another with *turbulent* flow follows the approximate empirical relationship $F = CS'(SSU)^{1/4}$ where F is the factor by which the pressure drop computed for a viscosity of 200 SSU and a specific gravity of 0.875 should be multiplied to get the pressure drop under line conditions, C is a constant, which varies from about 0.25 to 0.28 over the range considered, S' is the ratio of the specific gravity under line conditions to that obtaining when SSU = 220, and SSU is seconds Saybolt Universal. This adjustment for the effect of the change in viscosity with temperature can be conveniently made by reference to Table 5, which was computed for an oil having a specific gravity of 0.875 when the SSU is 200. If the pressure drop has been computed for the oil at at any one of of the SSU values shown, the pressure drop at any one of the other SSU values can be approximated through the ratio of the respective F extension factors.

Table 4. (*Continued*)

Nominal pipe size, in., schedule number, and average inside diam, in.	Carrying capacity Gal per min	Carrying capacity Cu in. per min	Velocity, ft per sec	Velocity head, psi for[2] $S = 0.9$	SSU for critical point, SSU = $6.8G/d$	Pressure drop in psi per 100 ft of pipe[1] — Viscosity, SSU 50	100	200	500	1,000	4,000
1¼	15	3,465	4.56	0.13	90	5.0	6.5	9.3	22	45	181
Sched.	30	6,930	9.11	0.50	180	17	24	30	44	90	362
160	50	11,550	15.2	1.40	300	41	56	69	100	149	603
$d =$	75	17,325	22.8	3.15	450	84	111	138	175	224	906
1.160	125	28,875	38.0	8.74	750	210	270	324	411	522	1,510
1½	20	4,620	4.57	0.13	100	3.7	4.0	8.5	20	34	138
Sched.	50	11,550	11.4	0.65	250	21	30	37	50	84	340
160	100	23,100	22.8	3.15	500	68	92	112	152	168	680
$d =$	150	34,650	34.2	7.08	750	141	185	226	305	252	1,020
1.338	200	46,200	45.6	12.6	1,000	230	314	383	508	672	1,360
2	25	5,775	3.58	0.08	100	1.4	1.8	3.6	9.0	17	67
Sched.	50	11,550	7.17	0.31	200	8.8	11	14	17	33	134
160	100	23,100	14.3	1.24	400	25	33	40	55	66	268
$d =$	200	46,200	28.7	5.00	800	104	135	165	210	266	536
1.689	300	69,300	43.0	11.2	1,200	205	266	325	412	523	805
2½	50	11,550	4.52	0.12	160	2.3	3.6	4.4	6.2	13	54
Sched.	100	23,100	9.05	0.53	320	9.5	12	15	20	26	108
160	200	46,200	18.1	1.98	640	30	39	47	60	76	215
$d =$	300	69,300	27.2	4.48	960	60	78	95	121	153	323
2.125	400	92,400	36.2	7.92	1,280	110	131	160	203	248	430
3	100	23,100	5.93	0.21	260	3.2	4.1	5.0	6.0	12	46
Sched.	200	46,200	11.9	0.86	520	11	15	18	23	23	92
160	300	69,300	17.8	1.92	780	24	31	38	48	55	138
$d =$	400	92,400	23.7	3.41	1,040	38	49	60	76	97	183
2.626	500	115,500	29.6	5.30	1,300	57	74	90	114	145	230
4	200	46,200	6.92	0.29	400	2.7	3.6	4.4	3.9	7.7	31
Sched.	400	92,400	13.8	1.16	800	9.2	11	14	18	23	63
160	600	138,600	20.8	2.62	1,200	20	25	30	38	48	94
$d =$	800	184,800	27.6	4.62	1,600	33	40	49	62	79	126
3.438	1,000	231,000	34.6	7.27	2,000	50	60	73	93	118	157

Table 5. Factors for Determining the Approximate Pressure Drop for an Oil at One SSU Value When the Pressure Drop at Some Other SSU Value is Known

(For turbulent flow, see text)

Viscosity, SSU	Specific gravity ratio S'	Empirical constant C	Extension factor F
50	0.958	0.248	0.63
100	0.987	0.263	0.82
150	0.995	0.263	0.92
200	1.000	0.266	1.00
300	1.006	0.269	1.13
400	1.008	0.269	1.21
600	1.014	0.273	1.37
800	1.020	0.279	1.51
1,000	1.022	0.281	1.61
1,500	1.025	0.283	1.81
2,000	1.027	0.284	1.95
3,000	1.032	0.284	2.17
4,000	1.037	0.280	2.30

Examples follow showing how to compute pressure drops through oil piping by analytical methods and by Table 4, and how to use Table 5 for extending the results computed for 200 SSU to other conditions.

Example 1. What diameter of Schedule 160 steel pipe should be used for a hydraulic-power transmission line of 100-ft equivalent length for delivering to a press-operating cylinder 100 gpm of oil at a flow temperature of 120 F and an initial pressure of 2,000 psi? The oil has a specific gravity of 0.90 at 60/60 F and a viscosity of 200 SSU at 120 F. The flow velocity is not to exceed 15 fps, and the pressure drop is not to exceed 2 per cent of the initial pressure.

Solution by Table 4. Referring to Table 4, it will be found that a 2-in. Schedule 160 pipe will just satisfy the conditions since it gives a velocity of 14.3 fps and a pressure drop of 40 psi per 100 ft, which is 2 per cent of the initial pressure. NOTE: This example was chosen for an oil having the same specific gravity as that for which the table was computed. Owing to the approximate nature of all pressure-drop calculations, Table 4 will serve well enough without adjustment for any oil having a specific gravity at 60/60 F of the order of 0.85 to 0.95.

Solution by Formula. Since it is evident that no size smaller than 2-in. Schedule 160 pipe will satisfy the velocity requirement, a pressure-drop calculation will be tried for this size. Flow conditions are in the turbulent region since the SSU is considerably below that at the critical point which is determined as follows:

$$SSU_{critical} = 6.8G/d = 6.8 \times 100/1.689 = 400$$

Hence the pressure drop will be determined by the following form of the pressure-drop formula:

$$\Delta p = fSLG^2/18.6d^5$$

in which f is read from Fig. 42, Chap. 3, after determining the relation GS/dz for flow conditions in which $G = 100$, $S = 0.875$, $d = 1.689$, and z is computed as follows from Figs. 34, and 35, Chap. 3. From Fig. 35, Chap. 3, the value of z/S corresponding to 200 SSU is 44. From Fig. 34, Chap. 3, an oil having a specific gravity of 0.90 at 60/60 F has a specific gravity of 0.875 at 120/60 F. Hence $z/S = 44$ and $z = 0.875 \times 44 = 38.5$. $GS/dz = 100 \times 0.875/1.689 \times 38.5 = 1.35$ and from Fig. 42, Chap. 3, $f = 0.0115$. Then

$$\Delta p = \frac{0.115 \times 0.875 \times 100 \times 10,000}{18.6 \times 13.74} = 39.4 \text{ psi}$$

which is just less than the permissible drop of 2 per cent of the initial pressure.

Example 2. Same conditions as Example 1 except that the oil is cold in starting, causing its viscosity to increase to 1,000 SSU. How will this affect the operation of the press, pump, and transmission piping?

Solution by Table 4. Referring to Table 4, it is evident that the pressure drop through the piping will rise from 40 to 66 psi if the flow of 100 gpm is attained at 1,000 SSU. If this is accomplished by a rise in the initial pressure, the horsepower required by the pump will increase in proportion to the change in initial pressure. On the other hand, if it is accomplished merely by less throttling in the control valve at the press, the effect on operation may be negligible. If sufficient pressure is not available to look after the added loss, the quantity of oil flowing will diminish until a balance is obtained. By interpolation in the 1,000 SSU column of Table 4 it appears that a pressure drop of 40 psi through the piping will deliver only about 60 gpm of the cold oil. Under this condition the action of the press would be slowed down a corresponding amount until the oil could be warmed up.

Approximate Solution by Extension Method. Referring to Table 5, it is found that the extension factor for 1,000 SSU is 1.61. Since the pressure drop at 200 SSU was computed to be 39.4 psi, the pressure drop at 1,000 SSU will be $39.4 \times 1.61 = 63.5$ psi, which agrees well enough with the result obtained by the other method. The rest of the comment remains the same.

Example 3. Same conditions as Example 1. Does the pressure loss required to produce *velocity head* appreciably affect the results obtained?

Solution. Referring again to Table 4, it is seen that the velocity head corresponding to a flow of 100 gpm through a 2-in. Schedule 150 pipe is only 1.24 psi when the oil has a specific gravity of 0.90 at 60 F. At 120 F the specific gravity is 0.875 from Fig. 34, Chap. 3. Hence at 120 F the velocity head is $1.24 \times 0.875/0.90 = 1.21$ psi. A velocity head of this order is negligible with respect to the operating pressure and the overall loss of 40 psi through the 100 ft equivalent length of pipe. If the velocity were several times higher or the equivalent length only a fraction of the 100 ft, however, the velocity-head loss might become a significant item, in which case it should be considered with the pressure drop in determining a suitable pipe size.

HYDRAULIC CONTROL VALVES

The action of the various types of hydraulic control valves can be explained to best advantage with reference to schematic diagrams which serve to illustrate a number of basic principles common to both large and small systems. The ordinary valves, such as stops and checks, need no sketches to explain their action. Angle and globe valves are used for stops rather than gates owing to their simpler design and to avoid the dragging action experienced in pulling gates across their seats under high pressure. Check valves usually are of the ball or poppet rather than the swing type to minimize slam on closing.

The following schematic diagrams and explanation of control valves were taken in the main from an editorial résumé which appeared in *Product Engineering* for February, 1944. These descriptions may be understood better through reference to the circuit diagrams given in Figs. 2 to 4, and it is suggested that the basic explanations given herein be supplemented through reference to manufacturer's catalogues for information on actual hydraulic hookups.

Directional Control Valves. *Four-way* is the common type, but directional valves can be obtained for two-, three-, four-, five-, and six-way ports. Four-way valves are used to reverse the travel of a piston by directing fluid in and out of either end of an operating cylinder, to reverse the direction of rotation of a hydraulic motor, and for similar purposes.

The three basic designs of directional valves are the *plug* type, the *poppet* type,

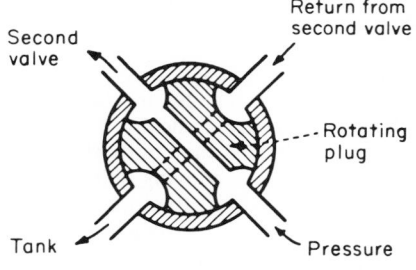

FIG. 24. Plug-type four-way pilot valve. (*Courtesy of Prod. Eng.*)

and the *piston* type. Plug-type four-way valves are indicated in Figs. 2 and 3, and a pilot type of four-way plug valve is shown in Fig. 24. Directional valves start, stop, or direct the flow of fluid by means of a movable plug, piston, or poppets which open, shut, or obstruct one or more ports. A piston type is shown diagrammatically in Fig. 25 and a poppet type in Fig. 26. The spool type of four-way

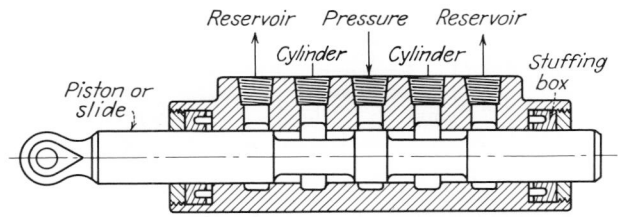

FIG. 25. Piston-type four-way valve. (*Courtesy of Aircraft Hydraulics.*)

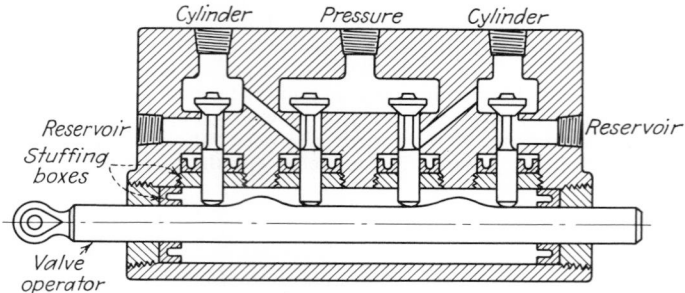

FIG. 26. Poppet-type four-way valve. (*Courtesy of Aircraft Hydraulics.*)

valve shown in Fig. 27 is operated through a hydraulic relay circuit by a pilot valve. The valve spool floats between springs, and pilot valve action causes the spool to move in either direction.   Other types are operated manually, electrically, or by mechanical or pneumatic means.

Time-delay valves (see Fig. 28) are used when change of liquid flow from one circuit to another must occur after a specified time interval.   The design generally incorporates a spool and a needle valve for flow adjustment. Liquid from the four-way valve flows past the ball check to hold the spool in the position shown. At the same time, liquid flows past the upper check valve to perform work in the cylinder.   When the four-way valve is reversed, fluid bleeds out past the needle valve from the chamber below the spool, since the ball check closes, and the spool drops, thus permitting fluid return from the cylinder to flow through chamber *R* back to the four-way valve.

Servo valves (Fig. 29) are used to move large masses with a minimum of operator effort.   They are basically similar to the directional control valves previously described, except that both the spool and the bushing move.   The bushing is connected to the moving slide so that when movement of the slide occurs, the bushing goes with it to close off the valve.   A small axial movement of the screw results in a small movement of the slide.   The only load on the screw is that necessary to move the valve in its bushing.

Pressure-control Valves.   These valves operate by pressure differentials for such purposes as limiting maximum pressure; controlling sequence of operation by governing pressure to branch circuits; returning fluid to the reservoir, with or without resistance, when the system pressure builds up; reducing pressure in system

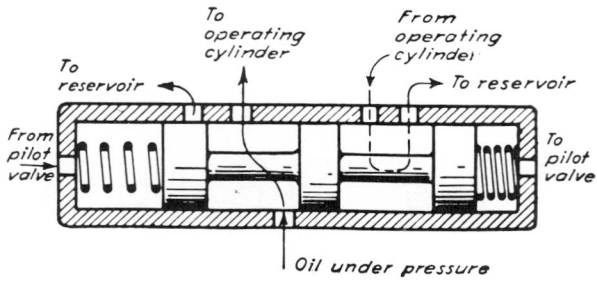

FIG. 27. Spool-type four-way operated through a hydraulic relay circuit controlled by a pilot valve. (*Courtesy of Prod. Eng.*)

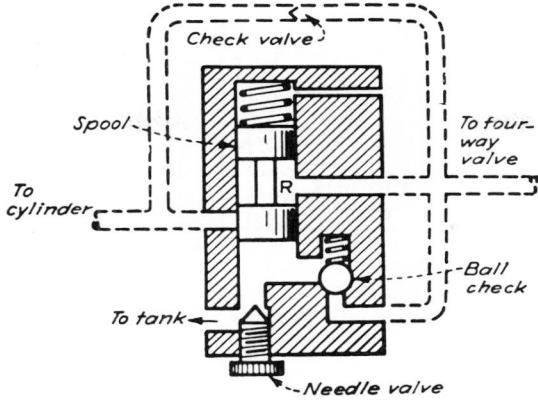

Fig. 28. Time-delay valve. (*Courtesy of Prod. Eng.*)

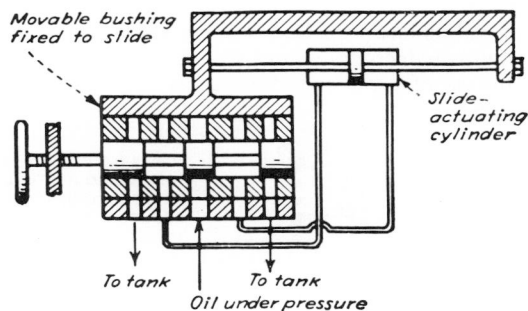

Fig. 29. Servo valve and linkage. (*Courtesy of Prod. Eng.*)

lines; and performing many other operations where pressure differentials can be used to open or close valves. Descriptions of some of the principal types follow.

Relief valves (see Fig. 30) are used to limit the pressure in a circuit to a desired maximum. An adjustment is provided which can be remotely controlled if desired. High-pressure oil is discharged from these valves back to the reservoir. Such valves should be used only intermittently or as safety valves, and not as unloading valves if a large or continuous discharge volume is expected. Unloading valves are described below.

Unloading valves (see Fig. 31) are used to return fluid to the reservoir without flow resistance when the system pressure builds up to a preset maximum. For instance, where a constant-delivery pump is supplying oil to the circuit and the cylinders have completed their stroke, pressure will build up in the system. In such cases, an unloading valve will open and bypass fluid back to the reservoir with little or no back pressure so that a minimum amount of energy is dissipated.

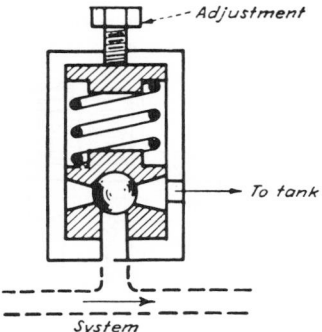

Fig. 30. Relief valve. (*Courtesy of Prod. Eng.*)

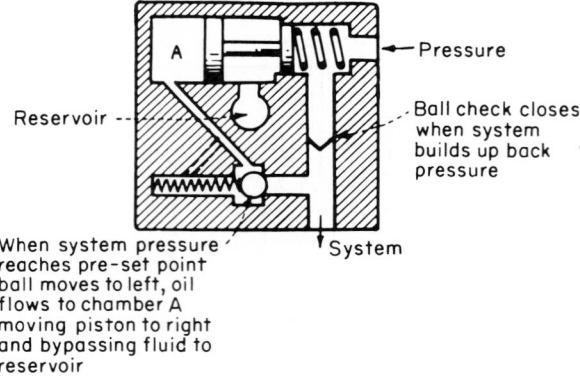

FIG. 31. Unloading valve. (*Courtesy of Prod. Eng.*)

This also reduces the pump power requirements a corresponding amount during these intervals.

Pressure-reducing valves (see Fig. 32) are used where certain parts of hydraulic control circuits require reduced pressures for operation. Pressure regulation is obtained through the use of a spool which balances the reduced pressure against an

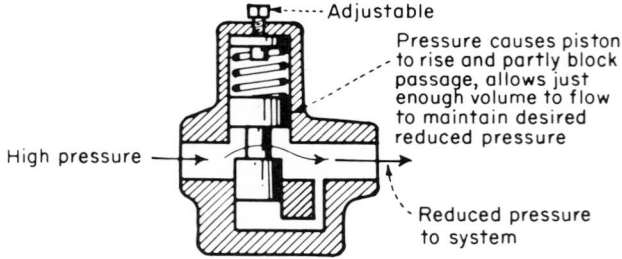

FIG. 32. Pressure-reducing valve. (*Courtesy of Prod. Eng.*)

adjustable spring so as to give a valve opening just sufficient to hold a predetermined reduced pressure.

Sequence valves (see Fig. 33) are pressure-operated adjustable devices for controlling the sequence of flow to various secondary circuits. Sequence valves are installed in the line so as to allow free flow to the main circuit until pressure builds up to the desired point. The valve then opens and supplies oil under pressure to the secondary circuit while still maintaining flow into the main circuit, the valve simply acting as a tee under these conditions. Check valves sometimes are incorporated in the design to permit free reverse flow.

Flow controllers and metering valves in combination (see Fig. 34) are used in hydraulic circuits where the rate of fluid flow in a certain direction must be controlled. Piston movement in

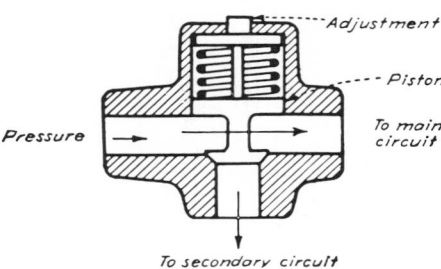

FIG. 33. Pressure-sequence valve. (*Courtesy of Prod. Eng.*)

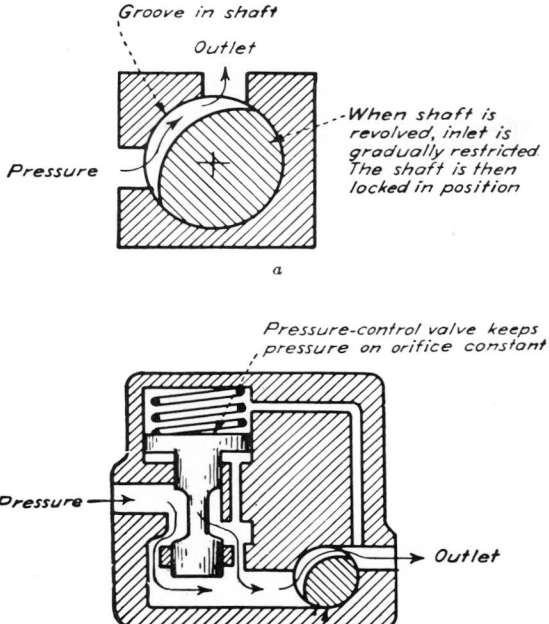

FIG. 34. (*a*) Metering valve. (*b*) Flow controller with metering valve. (*Courtesy of Prod. Eng.*)

the work cylinder can be governed so that the rate of travel will be constant regardless of fluid pressure or viscosity or of magnitude of the variable load. Among other applications, such controllers are used to prevent tools from jumping ahead when breaking through the work. They serve to meter in, meter out, or bleed off fluid from a circuit. The metering valve shown in combination with the controller can be used independently to limit the flow of fluid to be delivered to or discharged from an operating mechanism, to cushion pistons at the end of their stroke, or to delay the action of hydraulically operated valves.

Pressure switches (see Fig. 35) are used where flow in a hydraulic circuit is to be controlled electrically. Change in pressure causes the switch to open or close a relay circuit which actuates a solenoid or other device used to operate a valve or to start or stop the pump. Pressure switches can be fitted with needle valves for adjustment of reset speed.

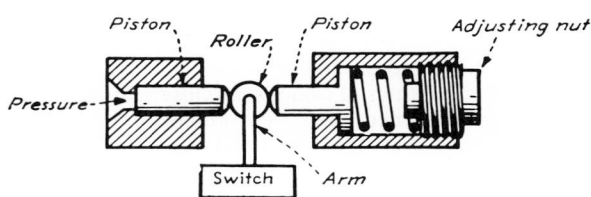

FIG. 35. Pressure switch. (*Courtesy of Prod. Eng.*)

PIPE, TUBING, AND FITTINGS FOR HYDRAULIC SERVICE

No national standard or code, as such, exists for direct and sole application to hydraulic-power transmission systems. The designer of such a system could well be guided by applicable sections of the Code for Pressure Piping, ASA B31. Certain empirical rules have been found to be satisfactory through periods of long usage and are available frequently in the form of recommendations by manufacturers of hydraulic equipment and controls. The piping and fittings must be designed to withstand the high shock pressures associated with water hammer. Fortunately, most hydraulic systems operate at low temperatures, so that the material involved is most frequently operating at or near its condition of maximum strength. Some common practices followed in selecting hydraulic pipe and fittings are described in succeeding sections.

Materials. The larger hydraulic systems usually are constructed of carbon-steel pipe manufactured in accordance with ASTM Specification-A53 or A106. Under certain conditions it may be advantageous to use chrome-molybdenum piping, but hydraulic systems usually operate at sufficiently low temperatures that the high-strength-at-high-temperature properties of these alloys are of no particular benefit. If carbon-steel or alloy-steel piping is used, fittings should be of the socket-weld type in pipe sizes 2 in. and below and of the butt weld type for larger pipe sizes.

Smaller hydraulic systems are able to utilize the lightweight and ease-of-fabrication features of tubing. The use of tubing should by all means be considered in sizes 2 in. and lower. Copper, carbon steel, alloy steel, and aluminum have been used extensively in hydraulic systems.

Copper tubing may be purchased in accordance with ASTM Specification B75. For hydraulic systems which operate at pressures of 1,000 psi or less and in which only moderate vibration and hydraulic shock are anticipated, copper tubing and brass fittings are commonly used. If the copper tubing is to be specified in coiled form, it must be fully annealed; if it is to be ordered in straight lengths, it may be either fully annealed or quarter hard. In all cases, the tubing should be seamless; since copper tubing is most often manufactured by the extrusion process, it is inherently seamless.

Carbon-steel tubing is indicated for pressures in excess of the range in which copper is used. It should be specified as being seamless and fully annealed of the SAE 1010 type in order to secure satisfactory flaring and fabricating properties. In some instances, the relative economy of butt-welded, as contrasted to seamless steel, tubing might be investigated. If this lower cost material is used, it must be specified to be suitable for bending and flaring.

Alloy-steel tubing is available most frequently as Type 316, 18 per cent chrome, 8 per cent nickel composition. Like carbon steel, it is available in either the seamless or butt-welded variety. For ease in flaring and bending, it should be fully annealed. This material is used for very high pressures or for handling fluids which might be corrosive to carbon steel.

Aluminum tubing, because of its light weight, finds wide application in aircraft control systems. Pure aluminum is seldom, if ever, used; 52 S-O and 5050-0 are popular aluminum alloys for use as tubing.

Forged or extruded brass fittings of nonporous close-grained structure should be used with copper tubing. In general, fittings should be of the same material as that of the tubing; steel fittings are used with steel tubing, aluminum fittings with aluminum tubing, and stainless-steel fittings with stainless-steel tubing. Steel nuts are sometimes used with brass fittings in cases where frequent disassembly of fittings is necessary. It is never advisable to use fittings whose material is softer

than that of the tube.   Materials commonly used in tubing installations are indicated in Table 6.

The tubing material is frequently dictated by corrosion conditions.   These conditions most often are determined by the hydraulic fluid, but in some installations, particularly certain chemical processing applications, it may be necessary to consider the possibility of contamination from atmospheric fumes outside the tubing.   Use of dissimilar materials in tubing systems should be viewed with due regard to the possibility of electrolytic action.

Often, for application in systems whose pressure is above 1,000 psi, a choice is available between heavy-walled copper tubing and lighter walled steel tubing. If corrosion is not the determining factor, it is generally preferable to use steel or stainless-steel tubing in that pressure range.

Table 6.   Tube and Fitting Materials

Tube	Fitting	Nut	Sleeve
Copper, ASTM B75...............	Brass, QQ-B-611[1]	Brass, QQ-B-611[1]	Aluminum bronze, AN-B-16[2]
Carbon steel, SAE 1020............	Carbon steel, SAE 1020 Alloy steel, SAE X4130	Carbon steel, SAE 1020 Alloy steel, SAE X4130	Carbon steel, SAE 1020 Alloy steel, SAE X4130
Aluminum alloy, 52S-0[3]............	Aluminum alloy, QQ-A-367[1]	Aluminum alloy, QQ-A-367[1]	Aluminum bronze, AN-B-16[2] or aluminum alloy, QQ-A-367[1]

[1] Federal Specifications.
[2] Army-Navy Aeronautical Board Specification.
[3] Aluminum Company of America designation.

Wall Thickness of Pipe and Tubing.   For conservative design, the formulas given in the ASA Code for Pressure Piping should be used.   These formulas distinguish between plain-end and threaded pipe, require an allowance for mill tolerance, and, except for nonferrous materials, provide an allowance for corrosion. Table 7 indicates the diameters and thickness of standard-size tubing.

The thickness of pipe or tubing required for hydraulic-power work often is computed by Barlow's formula or read from tables based on that formula.   For convenience in computing pipe wall thickness, this formula may be transposed to read $t = pD/2S$ where t is the pipe wall thickness in inches, p is the design pressure in pounds per square inch, D is the outside diameter of the pipe or tubing in inches, and S is the allowable stress for the material in pounds per square inch. The allowable stress may be read from a table, or it may be determined with respect to a factor of safety F, the seam efficiency E if there is a seam, and the tensile strength of the pipe material from the relation $S = E \times$ (tensile strength)$/F$.

With hydraulic piping it is customary to allow a factor of safety of 5 to 6 in determining S for use in the Barlow formula.   The joint efficiencies E commonly used for different kinds of pipe are seamless, 1.00; resistance-welded, 0.85; lap-welded, 0.80; butt-welded, 0.60.   If severe water hammer is to be expected, some manufacturers of hydraulic equipment recommend that the piping be designed for a maximum pressure at least one-third higher than the normal working pressure. With high velocities in large water systems, this still leaves considerable shock to be looked after in the factor of safety, since in a high-pressure system shock pressures under adverse conditions may run from 50 to 100 per cent above the normal working pressure.   As mentioned before, the designers of hydraulic systems often consider the computed thickness t as the nominal wall thickness without making any allowance for mill tolerance in manufacture, which may run as much as 12½

Table 7. Inside Diameters of Standard Size Tubing

Tube OD	Wall thickness	Tube ID	Tube OD	Wall thickness	Tube ID	Tube OD	Wall thickness	Tube ID	Tube OD	Wall thickness	Tube ID
1/8	0.028	0.069	1/2	0.035	0.430	7/8	0.049	0.777	1 1/4	0.120	1.010
	0.032	0.061		0.042	0.416		0.058	0.759			
	0.035	0.055		0.049	0.402		0.065	0.745	1 1/2	0.065	1.370
				0.058	0.384		0.072	0.731		0.072	1.356
3/16	0.032	0.1235		0.065	0.370		0.083	0.709		0.083	1.334
	0.035	0.1175		0.072	0.356		0.095	0.685		0.095	1.310
				0.083	0.334		0.109	0.657		0.109	1.282
1/4	0.035	0.180								0.120	1.260
	0.042	0.166	5/8	0.035	0.555	1	0.049	0.902			
	0.049	0.152		0.042	0.541		0.058	0.884	1 3/4	0.065	1.620
	0.058	0.134		0.049	0.527		0.065	0.870		0.072	1.606
	0.065	0.120		0.058	0.509		0.072	0.856		0.083	1.584
				0.065	0.495		0.083	0.834		0.095	1.560
5/16	0.035	0.2425		0.072	0.481		0.095	0.810		0.109	1.532
	0.042	0.2285		0.083	0.459		0.109	0.782		0.120	1.510
	0.049	0.2145		0.095	0.435		0.120	0.760		0.134	1.482
	0.058	0.1965									
	0.065	0.1825	3/4	0.049	0.652	1 1/4	0.049	1.152	2	0.065	1.870
				0.058	0.634		0.058	1.134		0.072	1.856
3/8	0.035	0.305		0.065	0.620		0.065	1.120		0.083	1.834
	0.042	0.291		0.072	0.606		0.072	1.106		0.095	1.810
	0.049	0.277		0.083	0.584		0.083	1.084		0.109	1.782
	0.058	0.259		0.095	0.560		0.095	1.060		0.120	1.760
	0.065	0.245		0.109	0.532		0.109	1.032		0.134	1.732

per cent under the nominal thickness, or for material removed in threading if screwed joints are used.

Figure 36 indicates in a convenient form the relation between tube size and flow velocity of the hydraulic fluid. This relation is valuable as a starting point in selection of tubing dimensions. Also in this figure, suggested maximum design velocities for different services are indicated.

The use of copper tubing, usually made up with flared connectors, is limited to pressures of about 1,000 psi. Steel tube and pipe are used for pressures up to 5,000 to 6,000 psi and, for the higher pressures, are of seamless construction and sometimes of alloy steel.

Although the selection of tube or pipe wall thickness should be based primarily on the pressure requirements of the system, heavier walls may be required where pipe or tubing is subject to unusual mechanical abuse, excessive vibration, mechanical loads, and large hydraulic shock or surge pressures. These conditions should be avoided where it is possible to do so, but if this is not practicable, pipe or tubing of heavier wall than would be required by the working pressure alone must be used. The wall thickness required in these cases is left largely to the judgment of the designing engineer to suit the particular conditions encountered.

The type or severity of service usually is taken into account by the application of a safety favor to the tensile strength of the tubing material. Figure 37 indicates tensile strengths of typical tubing materials; at the top of this figure various safety factors are shown. While it is usually not possible to identify precisely the safety factor which should be associated with a particular service, the following will serve at least as a rough guide:

For duty in which mechanical vibration and hydraulic shock are not excessive and the general tubing configuration is well supported, a safety factor of 4 should be adequate.

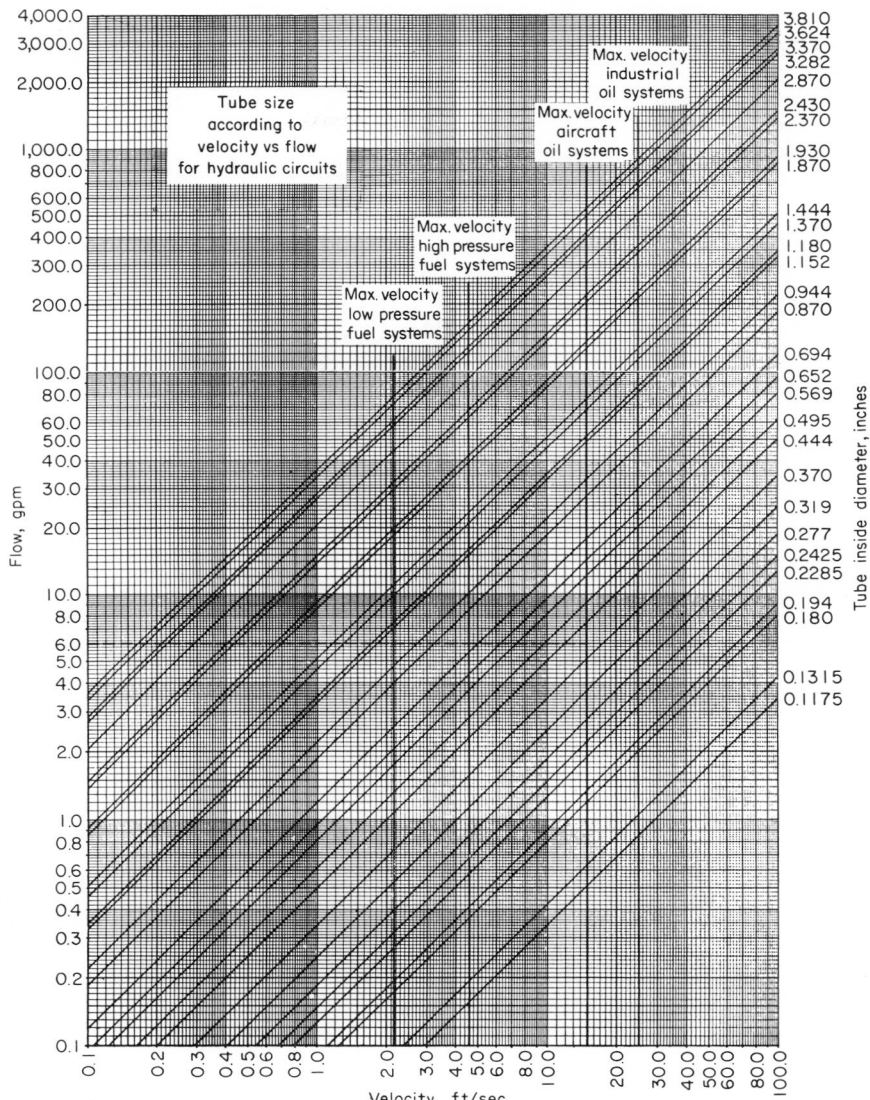

Fɪɢ. 36. Tube size-velocity relations and recommended maximum velocities. (*Courtesy of Parker-Hannifin.*)

If considerable hydraulic shock and mechanical vibration are anticipated, or if tubing runs are long and extend to areas that prevent secure support, the factor of safety should be increased to, say, 6.

For hazardous service, including severe vibration and large values of shock, in which the integrity and safety of the system and its contents are vulnerable, a factor of safety of perhaps 8 is indicated.

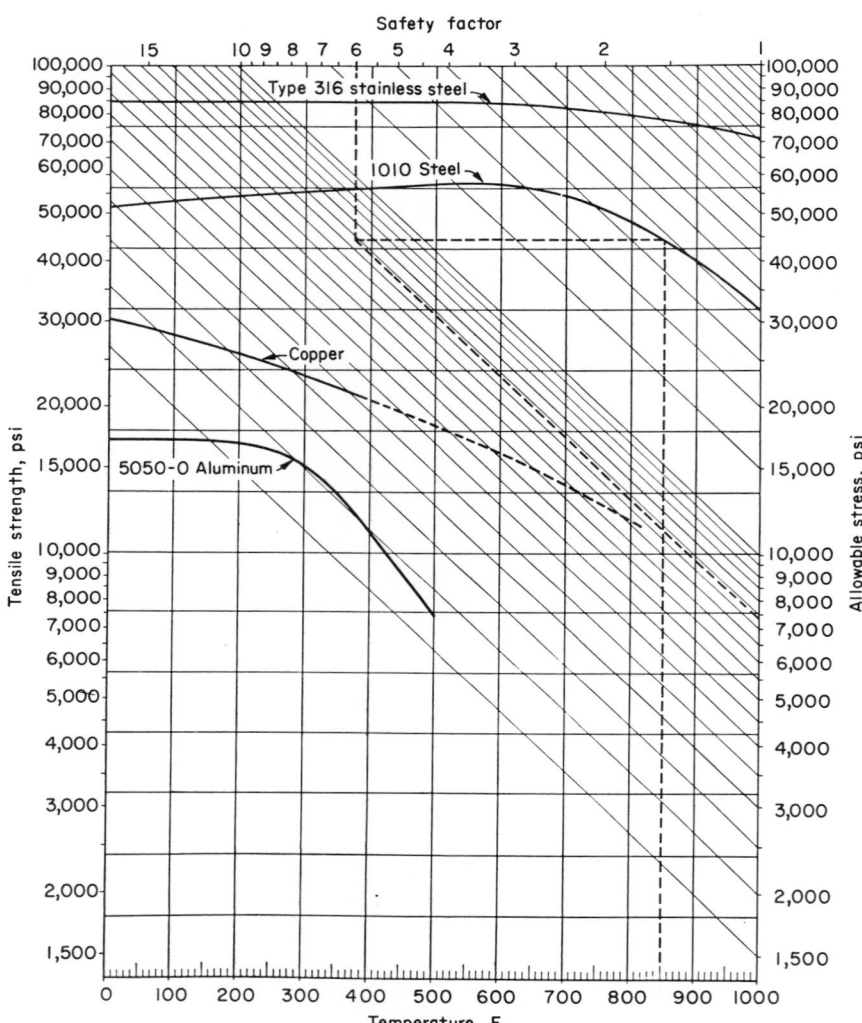

FIG. 37. Stresses in typical tubing materials.   (*Courtesy of Parker-Hannifin.*)

After the minimum wall thickness for pipe or tubing has been determined, the next heavier commercial thickness should be selected from dimensional standards. Such standards for pipe will be found in Chapter 7 and for tubing, in Table 7.

Types of Joints.   Connections between ends of pipe or tubing or between pipe or tubing and equipment may be made by use of screwed, flanged, brazed, or welding fittings or by means of tubing connectors.   The tubing connectors may be of the flare or flareless type, as dictated by tubing material and thickness, and are particularly advantageous if the joint has to be disassembled frequently.   Flareless fittings are recommended for high-pressure tubing use; they are available commercially and are made to several industry standards.   These fittings utilize the compressive force of the fitting nut to "bite" into the tubing material.

Where it is desirable to make a permanent leakproof joint, welded or brazed joints may be employed.   Dimensions for steel socket-welding fittings which match the ASA B36 schedule series for pipe are given in the American Standard for Steel Socket-welding Fittings, ASA B16.11.   Socket-welding fittings to correspond to standard-weight, extra-strong, and double-extra-strong pipe also are available. Butt-welding fittings in sizes 1 in. and larger are included in the American Standard for Steel Butt-welding Fittings, ASA B16.9.   Brazed joints are not widely used for hydraulic piping, but fittings for pressures up to 3,000 psi are available in accordance with the standards of several manufacturers.

Screwed fittings are, in general, restricted for use with steel pipe since the pipe must be of ample wall thickness so that sufficient metal remains after threading to

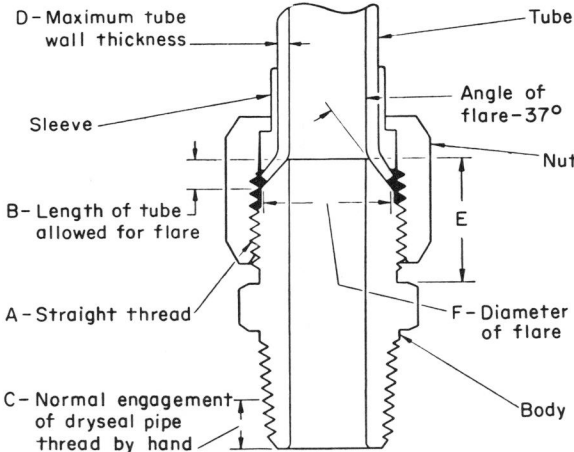

Fig. 38. Three-piece flared fitting used for connecting pipe to tubing.   See Table 8 for dimensions.   (*Courtesy of Parker-Hannifin.*)

withstand the pressure requirements as well as the additional shock and mechanical loads that may be imposed on the system.   Flanged joints, where called for, often are of special design with gaskets retained in recesses against the ends of the pipe and with square or oval instead of round flanges in the small to medium sizes.

Tubing Fittings.   These fittings are available in different weights, depending on the make of the fitting but, in general, the weight of the fitting should be selected to match the wall thickness of the tubing.

Flared-tube fittings consist of two or three parts.   Those having two parts involve a body and a nut.   Three-part fittings have, in addition, a sleeve in which the seal is made by squeezing the tube flare between the sleeve and the tapered surface on the fitting.   This design avoids the wiping action of the nut on the tubing flare which tends to thin out the latter or to score the seat between the nut and tube, either of which may result ultimately in a leaky joint.   This type of fitting is available in many designs such as elbow, tee, cross, and many others.   A pipe-to-tubing connector is shown in Fig. 38, and the associated dimensions are given in Table 8. This type of fitting may be used with copper and aluminum tubing and with thin-wall carbon-steel or stainless steel tubing.

Flareless fittings consist of three pieces: the body, nut, and ferrule.   The wedging action of the ferrule, when drawn down by the nut, forms a seal between body and ferrule, while the cutting edge of the ferrule "bites" into the tube wall and thus

Table 8. Dimensions for Fitting Shown in Fig. 38[1]

Tube outside diameter	Pipe thread	Bore diameter	A	B	C	D	E	F
1/8	1/8	1/16	5/16-24	3/32	0.180	0.035	29/64	0.224
3/16	1/8	1/8	3/8-24	3/32	0.180	0.035	31/64	0.290
1/4	1/8	11/64	7/16-20	7/64	0.180	0.065	35/64	0.359
1/4	1/4	11/64	7/16-20	7/64	0.200	0.065	35/64	0.359
5/16	1/8	15/64	1/2-20	7/64	0.180	0.065	35/64	0.421
5/16	1/4	15/64	1/2-20	7/64	0.200	0.065	35/64	0.421
3/8	1/4	19/64	9/16-18	1/8	0.200	0.065	9/16	0.484
3/8	1/8	19/64	9/16-18	1/8	0.180	0.065	9/16	0.484
3/8	3/8	19/64	9/16-18	1/8	0.240	0.065	9/16	0.484
3/8	1/2	19/64	9/16-18	1/8	0.320	0.065	9/16	0.484
1/2	3/8	25/64	3/4-16	5/32	0.240	0.083	21/32	0.656
1/2	1/4	25/64	3/4-16	5/32	0.200	0.083	21/32	0.656
1/2	1/2	25/64	3/4-16	5/32	0.320	0.083	21/32	0.656
1/2	3/4	25/64	3/4-16	5/32	0.339	0.083	21/32	0.656
5/8	1/2	31/64	7/8-14	3/16	0.320	0.095	49/64	0.781
5/8	3/8	31/64	7/8-14	3/16	0.240	0.095	49/64	0.781
3/4	3/4	39/64	1 1/16-12	7/32	0.339	0.109	55/64	0.937
3/4	1/2	39/64	1 1/16-12	7/32	0.320	0.109	55/64	0.937
7/8	3/4	23/32	1 3/16-12	7/32	0.339	0.109	57/64	1.062
1	1	27/32	1 5/16-12	7/32	0.400	0.120	29/32	1.187
1	3/4	27/32	1 5/16-12	7/32	0.339	0.120	29/32	1.187
1 1/4	1 1/4	1 5/64	1 5/8-12	1/4	0.420	0.120	61/64	1.500
1 1/2	1 1/4	1 5/16	1 7/8-12	1/4	0.420	0.120	1 5/64	1.721
1 3/4	1 1/2	1 35/64	2 1/4-12	5/16	0.420	0.120	1 13/64	2.106
2	2	1 25/32	2 1/2-12	11/32	0.436	0.134	1 11/32	2.356

[1] Courtesy of Parker-Hannifin.

forms another positive seal. When the fitting is disassembled, the extent of bite at the cutting edge of the ferrule is completely visible. This type of fitting is made in carbon-steel, stainless steel, and monel materials and is available in all the familiar fitting shapes. A pipe-to-tubing connector is shown in Fig. 39, and the associated dimensions are listed in Table 9.

Tubing sometimes is given what is called a "double flare" in which the end of the tubing is bent back upon itself to provide a double wall thickness at the end of the tube. A flare of this type is shown in Fig. 40. This is intended to eliminate leaky joints that result from thinning of the single-flare thickness caused by tightening nuts too severely or by frequent remaking of the joint.

Fitting size generally is designated by the size of tubing with which it is used, which is normally given in inches or fractions down to sixteenths of an inch corresponding to the outside diameter of the tubing. Fittings are commonly available

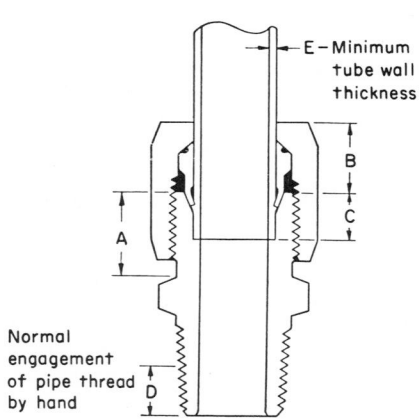

E – Minimum tube wall thickness

Normal engagement of pipe thread by hand

FIG. 39. Three-piece flareless fitting for connection of pipe to heavy-walled tubing. See Table 9 for dimensions. (*Courtesy of Parker-Hannifin.*)

Table 9. Dimensions for Fitting Shown in Fig. 39[1]

Tube outside diameter	Pipe thread	Straight thread	A	B	C	D	E
1/16	1/16	1/4–28	0.281	7/32	0.156	0.160	
1/8	1/8	5/16–24	0.375	5/16	0.188	0.180	0.022
3/16	1/8	3/8–24	0.422	11/32	0.234	0.180	0.022
1/4	1/8	7/16–20	0.453	27/64	0.234	0.180	0.028
1/4	1/4	7/16–20	0.453	27/64	0.234	0.200	0.028
5/16	1/8	1/2–20	0.453	27/64	0.250	0.180	0.028
3/8	1/4	9/16–18	0.469	15/32	0.250	0.200	0.028
3/8	3/8	9/16–18	0.469	15/32	0.250	0.240	0.028
3/8	1/2	9/16–18	0.469	15/32	0.250	0.320	0.028
1/2	3/8	3/4–16	0.563	1/2	0.305	0.339	0.035
1/2	1/4	3/4–16	0.563	1/2	0.305	0.200	0.035
1/2	1/2	3/4–16	0.563	1/2	0.305	0.320	0.035
5/8	1/2	7/8–14	0.625	17/32	0.350	0.320	0.035
5/8	3/8	7/8–14	0.625	17/32	0.350	0.240	0.035
3/4	3/4	1 1/16–12	0.688	9/16	0.350	0.339	0.035
7/8	3/4	1 3/16–12	0.688	17/32	0.350	0.339	0.042
1	1	1 5/16–12	0.688	21/32	0.415	0.400	0.042
1	3/4	1 5/16–12	0.688	21/32	0.415	0.339	0.042
1 1/4	1 1/4	1 5/8–12	0.688	23/32	0.415	0.420	0.042
1 1/2	1 1/2	1 7/8–12	0.688	23/32	0.485	0.420	0.049
2	2	2 1/2–12	0.688	27/32	0.485	0.436	0.049

[1] Courtesy of Parker-Hannifin.

in sizes from 1/8 to 2 in. On fittings that have one or more pipe thread outlets, standard pipe threads corresponding to the tubing size are provided.

Although certain parts of some of the several types of fittings are interchangeable, difference in threads, variation in flare angle, or difference in overall dimensions renders it impractical to interchange parts promiscuously. It generally is advisable to replace a fitting with an identical fitting or to replace the entire unit and reflare the tube if necessary. Since the *pipe* threads used on all fittings are identical, size for size, a complete assembly may usually be interchanged unless clearance in close quarters will not permit because of interference.

In making up flared-tube connections, it is desirable to use an antiseize compound or lubricant to prevent galling and to facilitate assembly and permit ready disassembly. Aluminum-alloy threads will gall or seize unless a lubricant is used. Screwed joints frequently are fillet-welded or brazed for tightness where the pipe enters the fitting. Where this is not done, the threads sometimes are tinned or copper-plated to prevent leakage. A sealing compound also may be used to assure a tight joint. Numerous types of lubricants and sealing compounds are available, selection from which should be made to suit the fluid conveyed. In general, it may be stated that the compound should be insoluble in the fluid conveyed except in the case of oil, where the reverse is true. If oil is the fluid conveyed, the oil itself will provide sufficient lubrication once the joint is made.

Care should be exercised to avoid excessive wrench torque on flared-tube nuts when making a joint using soft tubing, particularly when a thread lubricant is

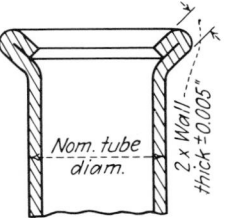

FIG. 40. Double flare for tubing.

applied since the lubricant tends to eliminate the "feel" when the joint is tight.   If nuts are pulled down too tightly with soft tubing, the flare will be damaged.

In laying out circuits for hydraulic lines, it is desirable to keep the number of bends to a minimum to avoid excessive fluid pressure drop.   However, at least one bend should be present in all lines to absorb expansion and contraction strains resulting from changes in temperature.   Otherwise, such strains may cause a failure of the tube flare which is, of course, the weakest part of the line.

Table 10.   Data on Reinforced Synthetic-rubber Flexible Hose[1]

Nominal I.D., in.	Max. O.D., in.	Working pressure max., psi	Proof test pressure, psi	Burst pressure min., psi	Min. bend radii, in.	Weight, lb per ft
Low-pressure type						
⅛	0.358	415	1,250	2,500	3	0.06
³⁄₁₆	0.420	333	1,000	2,000	4	0.075
¼	0.490	275	750	1,500	5	0.08
⅜	0.625	165	500	1,000	6	0.11
Medium-pressure type						
⅛	0.500	975	2,900	5,800	5	0.08
³⁄₁₆	0.625	835	2,500	5,000	6	0.12
¼	0.720	835	2,500	5,000	6.5	0.13
⁵⁄₁₆	0.780	800	2,500	4,800	7	0.19
⅜	0.875	800	2,500	4,500	8	0.22
½	1.032	500	1,500	3,000	9	0.35
High-pressure type						
³⁄₁₆	0.656	3,600	9,000	18,000	6.5	0.28
¼	0.718	3,600	9,000	18,000	7.5	0.31
⁵⁄₁₆	0.843	3,000	7,500	15,000	8	0.40
⅜	0.906	3,000	7,500	15,000	9	0.50
½	1.031	2,400	6,000	12,000	10	0.60

[1] From "Hydraulic Lines, Which Material, What Size," by Richard K. Lotz, *Machine Design*, December, 1943, p. 145.

Flexible Lines.   In aircraft applications as well as for many machines utilizing hydraulic fluids for power, movement is required between units that are interconnected.   This may be accomplished by the use of reinforced synthetic-rubber hose which is available in three general pressure classifications: low-pressure, medium-pressure, and high-pressure.   Such hose generally is provided with an inner and outer layer of synthetic rubber with one or more reinforcing braids of cotton.   High-pressure hose also may have a wire reinforcing braid.   Data on the three pressure classifications of such hose are given in Table 10.   ASTM, SAE, and Army-Navy specifications are available which give construction details and test requirements for hose used in various applications.

In recent years, the technology of manufacture of flexible hose has been extended greatly.   Users who contemplate inclusion of such materials in any installation should consult the manufacturers in order to ascertain the design and type best suited.

15

UNDERGROUND STEAM PIPING

J. A. Sheppard, Jr.*

Underground steam piping with its purpose of steam distribution has been a highly specialized field of engineering peculiar to the district-heating industry. Its specialized development has been the result of over 85 years of experience gained by the pioneers of this industry.

In modern times. with the advent of the construction of groups of buildings such as housing developments, institutions and industrial plants and with modern architectural requirements for buildings set back from the property line, central-heating systems and steam-distribution problems are no longer restricted to the district-heating industry.

An underground steam piping system is buried and is not readily available for enlargement, replacement, and repair. Such piping must be protected from ground elements and excessive heat losses; thus it is important that every phase of its design and operation be understood before such a system is installed.

GENERAL DESIGN

While there are no definitely fixed standards for the design of underground steam-distribution piping systems, most of the systems fall into one of two general classes: (1) a trunk-line distribution network system and (2) a main and feeder distribution network system.

In Class 1, the diameter of the trunk line leaving the boiler plant is large, and as lateral branches are installed off it for service, the diameter of the trunk line is gradually reduced as the needs for carrying capacity are diminished.

* Chairman, Distribution Committee, National District Heating Association, Assistant Division Engineer, Consolidated Edison Company of New York, Inc.

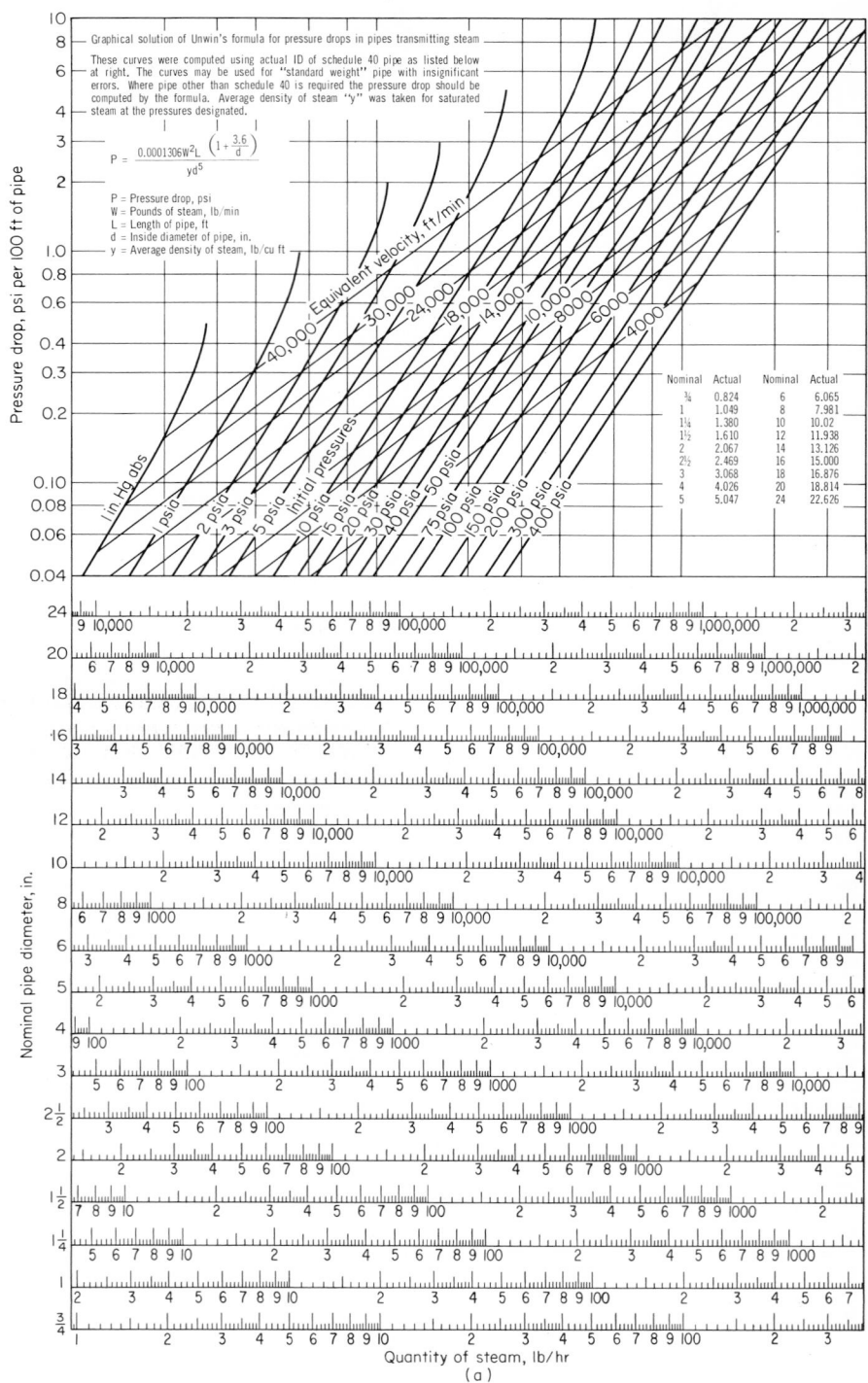

Figure content (chart labels):

Pressure drop, psi per 100 ft of pipe

Graphical solution of Unwin's formula for pressure drops in pipes transmitting steam

These curves were computed using actual ID of schedule 40 pipe as listed below at right. The curves may be used for "standard weight" pipe with insignificant errors. Where pipe other than schedule 40 is required the pressure drop should be computed by the formula. Average density of steam "y" was taken for saturated steam at the pressures designated.

$$P = \frac{0.0001306W^2 L}{yd^5}\left(1 + \frac{3.6}{d}\right)$$

P = Pressure drop, psi
W = Pounds of steam, lb/min
L = Length of pipe, ft
d = Inside diameter of pipe, in.
y = Average density of steam, lb/cu ft

Nominal	Actual	Nominal	Actual
¾	0.824	6	6.065
1	1.049	8	7.981
1¼	1.380	10	10.02
1½	1.610	12	11.938
2	2.067	14	13.126
2½	2.469	16	15.000
3	3.068	18	16.876
4	4.026	20	18.814
5	5.047	24	22.626

Equivalent velocity, ft/min: 40,000 / 30,000 / 24,000 / 18,000 / 14,000 / 10,000 / 8000 / 6000 / 4000

Initial pressures: 1 in. Hg abs / 1 psia / 2 psia / 3 psia / 5 psia / 10 psia / 15 psia / 20 psia / 30 psia / 40 psia / 50 psia / 75 psia / 100 psia / 150 psia / 200 psia / 300 psia / 400 psia

Nominal pipe diameter, in.

Quantity of steam, lb/hr
(a)

Fig. 1. Unwin Chart: Approximate solution

15–2

Velocity factor

To obtain the actual velocity from the curves, multiply the equivalent velocity found on the graph by the factor corresponding to the pipe diameter

Pipe diameter	Factor
$\frac{3}{4}$	0.1443
1	0.1792
$1\frac{1}{4}$	0.2278
$1\frac{1}{2}$	0.2598
2	0.3198
$2\frac{1}{2}$	0.3691
3	0.4376
4	0.5370
5	0.6322
6	0.7185
8	0.8638
10	1.000
12	1.116
14	1.182
16	1.281
18	1.374
20	1.465
24	1.627

(b)

Equivalent length of fittings and valves
in feet of schedule 40 straight pipe

Nominal pipe size, in.	90° ell screwed	Tee screwed	90° ell welded $R/d=1\frac{1}{2}$	Miter ells 1-45°	Miter ells 1-90°	Miter ells 2-90°	Gate valve	Globe valve	Angle valve
$\frac{3}{4}$	2.06	4.12	0.82	1.03	4.12	1.37	0.48	22.9	11.4
1	2.62	5.24	1.05	1.31	5.24	1.75	0.61	29.1	14.6
$1\frac{1}{4}$	3.45	6.90	1.38	1.72	6.90	2.30	0.81	38.3	19.1
$1\frac{1}{2}$	4.02	8.04	1.61	2.01	8.04	2.68	0.94	44.7	22.4
2	5.17	10.3	2.07	2.58	10.3	3.45	1.21	57.4	28.7
$2\frac{1}{2}$	6.16	12.3	2.47	3.08	12.3	4.11	1.44	68.5	34.3
3	7.67	15.3	3.07	3.84	15.3	5.11	1.79	85.2	42.6
4	10.1	20.2	4.03	5.04	20.2	6.71	2.35	112	56
5	12.6	25.2	5.05	6.30	25.2	8.40	2.94	140	70
6	15.2	30.4	6.07	7.58	30.4	10.1	3.54	168	84.1
8	20.0	40.0	7.98	9.97	40.0	13.3	4.65	222	111
10	25.0	50.0	10.0	12.5	50.0	16.7	5.85	278	139
12	29.8	59.6	11.9	14.9	59.6	19.9	6.96	332	166
14	32.8	65.6	13.1	16.4	65.6	21.9	7.65	364	182
16	37.5	75.0	15.0	18.8	75.0	25.0	8.75	417	208
18	42.1	84.2	16.9	21.1	84.2	28.1	9.85	469	234
20	47.0	94.0	18.8	23.5	94.0	31.4	11.0	522	261
24	56.6	113	22.6	28.3	113	37.8	13.2	629	314

(c)

for sizing pipe in district heating steam systems.

In Class 2, the main and feeder network distribution system receives its supply of steam through a high-pressure feeder main leading from the plant. Advantage is taken of the pressure drop available for the transportation of large volumes of steam to the low-pressure network. Therefore, the size of the feeder main required in this case is not so great as in a trunk-line system with the same boiler-plant steam pressure.

Some of the factors which will influence the selection of the class of steam-distribution system will be (1) area to be served; (2) load requirements, present and future; (3) underground conditions which might affect the size of piping and type of conduit; and (4) other economic considerations.

Since piping is the largest individual factor in the selection and design of a steam-distribution system, the major items that must be resolved in the design of underground steam piping are as follows: (1) pipe size, (2) wall thickness, (3) materials, (4) proper insulation, (5) a protective conduit for the pipe and insulation from water and mechanical damage, (6) drainage of condensate, (7) provision for thermal expansion with controlling anchorage, and (8) safety provisions.

Pipe Size

The factors which determine the size of pipe of a specific installation are as follows: (1) the initial steam pressure, (2) the minimum permissible terminal pressure, (3) the quantity of steam, and (4) the length of line including equivalent lengths for fittings. Knowing or assuming any one or all of these factors, the pipe size may be calculated by means of one of the pressure-drop formulas of Chap. 3.

Unwin's formula[1] is widely used because of its ease of solution by means of a graphic chart. It has been a favorite in the district-heating industry for many years. The formula is as follows:

$$P = \frac{0.0001306 W^2 L(1 + 3.6/d)}{Y d^5}$$

where P = pressure drop, psi
  W = steam flow, lb/min
  L = length of pipe, ft
  Y = average density of steam, lb/cu ft
  d = inside diameter of pipe, in.

Solutions of problems by use of this formula can be shortened by means of the chart shown in Fig. 1.

This chart shows the pressure drop for 100 ft of pipe with various *initial pressures* and quantities of steam. The diagonal velocity lines are *equivalent* velocities and must be multiplied by the factor corresponding to the pipe diameter to obtain the actual velocity.

Example. Find the pressure drop per 100 ft of main and the velocity in an 8-in. pipe with a flow of 40,000 lb/hr of steam and an initial pressure of 150 psia.

Solution. On the horizontal scale for 8-in. pipe, find 40,000 lb/hr; draw an imaginary line vertically from this point to intersect the line marked 150 psia initial pressure. Read the equivalent velocity, 6,700 fpm. Multiply this by 0.8638, the velocity factor given in the table for 8-in. pipe, thus obtaining the actual velocity, 5,790 fpm. To find the pressure drop for 100 ft of pipe, follow right from the intersection (previously obtained) of the vertical

---

[1] This formula, together with a chart for graphical solution, is presented in a somewhat different format in Chap. 3. The chart of Fig. 1 in this present chapter has been arranged for greatest convenience in the application to district-heating piping systems.

line with the 150 psia pressure line and read the pressure drop of 0.80 from the scale at the right.

To obtain the pressure drop for pipes which are more than 100 ft in length and where different sizes of pipe are involved, the pressure drop for 100-ft sections should be obtained separately from the chart. The reduced pressure found at the end of each section should be used as the initial pressure for the next section.

Example. Given a line consisting of 300 ft of 10-in. pipe and 200 ft of 12-in. pipe, an initial pressure of 75 psia and a flow of 60,000 lb/hr. Find the pressure at the end of the line.

Solution. On the horizontal scale for 10-in. pipe, find 60,000 lb/hr; draw an imaginary line vertically from this point to intersect the line for 75 psia pressure. From this point of intersection follow left and read the pressure drop of 1.03 psi per 100 ft. The initial pressure for the next 100 ft will be 75 psi minus 1.03 psi or about 74 psia. With the use of the same method as before, the pressures found are about 73 psia at 200 ft and about 72 psia at the end of the 10-in. pipe.

Starting from 60,000 lb/hr on the horizontal scale for 12-in. pipe and drawing a vertical line to intersect the curve for 72 psia, a pressure drop of 0.43 psi is read. This leaves about 71.6 psia at the end of the first 100 ft of 12-in. pipe. Following the same method, the pressure at the end of the line is found to be about 71.1 psia.

Any one of these quantities can be found in a similar manner when the others are known.

The pressure drops through fittings such as elbows, tees, and valves vary in proportion to the pressure drop through straight pipe. Because of this fact, it is possible to express the resistance of fittings as equivalent feet of straight pipe and to compute the pressure drop for the whole line as if it consisted only of straight pipe.

The equivalent lengths in feet shown in Fig. 1c have been computed on a basis that the inside diameter corresponds to that of Schedule 40 steel pipe, which is close enough for many purposes involving other schedules of pipe.

At elevated velocities, Unwin's formula gives pressure drops known to be higher than actual, but this does not detract materially from its popularity. In solving for pipe size, it gives sizes slightly larger than the more exact method.

Where grid or network systems are involved, the computation of pipe size, pressure drops, and quantities of steam flowing presents a complex problem. The usual method is one by trial and error, in which a point of zero flow, or balance point, is assumed and flow quantities or pressure drops are computed on this basis. The problem is recomputed as often as necessary by selecting different balance points until the correct solution is obtained. Modern practice for large network systems utilizes the Hardy-Cross method of solution with the use of electronic computers.

Pipe-wall Thickness

Minimum pipe-wall thickness for either heavy-wall or thin-wall pipe is calculated from the following formula,[1] which adjusts for elastic or plastic properties of the respective materials over the expected range of operating temperatures:

$$t_m = \frac{PD}{2S + 2yP} + C$$

or

$$P = \frac{2S(t_m - C)}{D - 2y(t_m - C)}$$

[1] This formula is abstracted from Sec. 4 of the Piping Code, B31.1.

where t_m = minimum pipe wall thickness,[1] in.

$\quad\quad\quad P$ = maximum internal service pressure,[2] psig

$\quad\quad\quad D$ = outside diameter of pipe, in.

$\quad\quad\quad S$ = allowable stress in material due to internal pressure, at the operating temperature, psig (Tables 3 and 4)[3,4]

$\quad\quad\quad C$ = allowance for threading, mechanical strength and/or corrosion, in. (Table 2)

$\quad\quad\quad y$ = a coefficient having values as given in Table 1.

Table I.   Values of Coefficient Y

Temperature,[5] F	900 and below	950	1000	1050	1100	1150 and above
Ferritic steels.........	0.4	0.5	0.7	0.7	0.7	0.7
Austenitic steels	0.4	0.4	0.4	0.4	0.5	0.7

Allowable stress (S) values for pipe in district-heating piping systems are shown in the following Tables 3 and 4 which are abstracted from Sec. 4 of the District Heating Piping of the American Standard Code for Pressure Piping, ASA B31.1.

The following conditions are outlined as a guide for design:

1. Pipe lighter than standard or Schedule 40 of American Standard for Wrought Iron and Wrought Steel Pipe, ASA B36.10, shall not be threaded.

2. Schedule 40 pipe will be adequate for any service using screwed, flanged, or welded joints where pressure does not exceed 250 psi.

---

[1] If pipe is ordered by its nominal thickness as is customary in trade practice, the manufacturing tolerance on wall thickness must be taken into account.   After minimum wall thickness t_m is determined, this minimum thickness shall be increased by an amount sufficient to provide the manufacturing tolerance allowed in the applicable pipe specification.   The next heavier commercial wall may then be selected from standard thickness schedules such as contained in ASA B36.10.   Manufacturing tolerances are contained in standard pipe specifications, lists of which with respective S values are given in Tables 3 and 4.

[2] When the allowable pressure for a pipe of definite minimum wall thickness is computed, the value obtained by the formula may be rounded out to the next higher unit of 10.

[3] The factors of safety and allowance for joint efficiencies of welded pipe have been determined and taken into account in the S values of Tables 3 and 4.   Where a pipe is subjected to more than one "pressure-temperature" operating condition, the value of t_m shall be determined from the "pressure-temperature" operating condition which results in the maximum t_m except as provided in footnote 5.

[4] Piping systems shall be considered safe for operation if the maximum sustained pressure and temperature on any part do not exceed the pressure and temperature allowed by the Code for all component parts of the system.   It is recognized that occasional departures from the nominal operating pressure and temperature inevitably occur, and therefore the piping system will also be considered safe for occasional operation for short periods at higher pressures and temperatures.

[5] Either pressure or temperature or both may exceed the nominal design values if the computed stress in the pipe wall calculated for the pressure by the formula does not exceed the S value allowable in Tables 3 and 4 for the expected temperature by more than the following allowances for the period of duration indicated:

(*a*) Up to 15 per cent increase above the S value during 10 per cent of the operating period.

(*b*) Up to 20 per cent increase above the S value during 1 per cent of the operating period.

Table 2. Values of C in Inches

Type of pipe	C, in.
Cast-iron pipe centrifugally cast or cast horizontally in green sand molds	0.14
Cast-iron pipe, pit cast......................	0.18
Threaded steel, wrought iron or nonferrous	
⅜ in. and smaller	0.05
½ in. and larger	Depth of thread
Grooved steel, wrought iron or nonferrous	Depth of groove
Plain end steel or wrought-iron pipe or tube for sizes 1 in. and smaller.............	0.05
Plain end steel or wrought-iron pipe or tube for sizes over 1 in........................	0.065
Plain end nonferrous pipe or tube	0.00

3. Schedule 40 (standard-weight) low-carbon-steel pipe may be used for steam pressures up to approximately 400 psi where welded joints are used and corrosion is not a problem.

4. For lower pressures under the same conditions, Schedule 30 or even 20 may be acceptable.

5. Schedule 80 (extra-strong) low-carbon-steel pipe either screwed or welded may be used for saturated steam pressures up to about 800 psi.

6. When steel pipe is threaded and used for steam pressures of 250 psi or greater or for water pressures in excess of 100 psig at temperatures of 220 F or over, it shall be seamless of a quality at least equal to ASTM Specification A53 or A83 and of a weight at least equal to Schedule 60 or 80 in order to furnish added mechanical strength.

Materials

The specific material requirements for underground steam piping will vary depending on the pressure and temperature of the steam carried, on the type of service and conditions involved, and on the individual preferences of the designer.

The materials used shall be capable of meeting the physical and chemical requirements and tests of the American Standards and ASTM material specifications as shown in Table 5.

Materials conforming to the specifications of Table 5 may be used at any temperature up to and including 750 F as the maximum limiting temperature and may be used at higher temperatures if allowable stresses or ratings for temperatures in excess of 750 F are given in these specifications, the American Standards, or ASA B31.1, Sec. 4. These materials may be used at lower temperatures with pressures higher than their primary service ratings given in these specifications or American Standards if the particular specification or American Standard contains an approved table of adjusted pressure-temperature ratings for that purpose.

The standards and specifications quoted in this chapter are minimum requirements. Construction at least equal to that required by these codes is mandatory; the use of materials having physical and chemical properties at least equal to minimum code requirements is also mandatory.

The following requirements for piping and fittings are generally acceptable in the industry and may be used as a guide:

1. Pipe. Steel pipe, either seamless or welded, is generally used, although the Piping Code permits a variety of materials and several types of welded pipe. For welding and bending, a low-carbon seamless steel pipe is recommended. Seamless or electric-resistance welded steel pipe A53, Grade B, is a popular selection.

Table 3.   Allowable Stress (S) Values for Pipe in District-heating Piping System
(−20 F to 500 F, incl.)

Material	ASTM Specification	Grade	Mini ultimate tensile strength	Values of S, psi for temperatures in deg F not to exceed*					
				−20 to 100	200	300	400†	450	500
Carbon steel.............	A120	....	......	10,800	10,600	10,200	9,800	9,600	
Electric-fusion-welded steel.................	A134	A245 A	48,000	8,800	8,800	8,800	8,800	....	8,800
		A245 B	52,000	9,600	9,600	9,600	9,600	....	9,600
		A245 C	55,000	10,100	10,100	10,100	10,000	....	10,100
		A283 A	45,000	8,300	8,300	8,300	8,300	....	8,300
		A283 B	50,000	9,200	9,200	9,200	9,200	....	9,200
		A283 C	55,000	10,100	10,100	10,100	10,100	....	10,100
		A283 D	60,000	10,100	10,100	10,100	10,100	....	10,100
	A139	A‡	48,000	9,600	9,600	9,600	9,600	....	9,600
		B‡	60,000	12,000	12,000	12,000	12,000	....	12,000
Electric-resistance-welded steel.................	A53	A§	48,000	10,200	10,200	10,200	10,200	....	10,200
		B§	60,000	12,750	12,750	12,750	12,750	....	12,750
	A135	A§	48,000	10,200	10,200	10,200	10,200	....	10,200
		B§	60,000	12,750	12,750	12,750	12,750	....	12,750
Lap welded:									
Steel.................	A53	....	45,000	9,000	9,000	9,000	9,000	....	9,000
Steel.................	A120	....	......	8,800	8,600	8,200	7,800	7,600	
Wrought iron..........	A72	....	40,000	8,000	8,000	8,000	8,000	....	8,000
Butt welded:									
Steel.................	A53	....	45,000	6,750	6,750	6,750	6,750	....	6,750
Steel.................	A120	....	.....	6,500	6,350	6,100	5,850	5,700	
Wrought iron..........	A72	....	40,000	6,000	6,000	6,000	6,000	....	6,000
Seamless:									
Red brass.............	B43	....	......	8,000	8,000	7,000	3,000		
Copper—2 in. and smaller..........	B42	....	......	6,000	5,500	4,750	3,000		
Copper—over 2 in......	B42	....	......	6,000	5,500	4,750	3,000		
Copper tubing:........	B75	....	......	6,000	5,500	4,750	3,000		
Annealed............	B88	....	30,000	6,000	5,500	4,750	3,000		
Bright annealed	B68	....	30,000	6,000	5,500	4,750	3,000		
Copper brazed steel.......	A254	Class I	42,000	6,000	5,500	4,750	3,000		
		Class II	42,000	3,600	3,300	2,850	1,800		
Cast iron:									
Centrifugally cast.......	ASA A21.6	....	......	6,000	6,000	6,000	6,000	6,000	
	ASA A21.7	....	......	6,000	6,000	6,000	6,000	6,000	
	ASA A21.8	....	......	6,000	6,000	6,000	6,000	6,000	
	ASA A21.9	....	......	6,000	6,000	6,000	6,000	6,000	
Pit cast................	ASA A21.2	....	......	4,000	4,000	4,000	4,000	4,000	
	ASA A21.3	....	......	4,000	4,000	4,000	4,000	4,000	

* The several types and grades of pipe tabulated above shall not be used at temperatures in excess of the maximum temperatures for which the S values are indicated. (See also specific requirements for service conditions contemplated.) Allowable S values for intermediate temperatures may be obtained by interpolation.

† For steam at 250 psi (406 F) the values given may be used.

‡ If plate material having physical properties other than stated in Section 6 of ASTM A139 is used in the manufacture of ordinary electric-fusion-welded steel pipe, the allowable stress shall be taken as 0.16 times the tensile strength for temperature of 450 F and below.

§ For electric-resistance-welded pipe for applications where the temperature is below 650 F and where pipe furnished under this classification is subjected to supplemental tests and/or heat treatments as agreed to by the supplier and the purchaser and whereby such supplemental tests and/or heat treatments demonstrate the strength characteristics of the weld to be equal to the minimum tensile strength specified for the pipe, the S values equal to the corresponding seamless grades may be used.

Table 4. Allowable Stress (S) Values for Pipe in District-heating Piping System
(600 F to 750 F, incl.)

Material[1]	ASTM Specification	Grade	Identification symbol	Min ultimate tensile strength	Values of S, psi for temperatures in deg F not to exceed[1]			
					600	650	700	750
Electric-fusion-welded steel	A134	A245 A	.....	48,000	8,800	8,800		
		A245 B	.....	52,000	9,600	9,600		
		A245 C	.....	55,000	10,100	10,100		
		A283 A	.....	45,000	8,300	8,300		
		A283 B	.....	50,000	9,200	9,200		
		A283 C	.....	55,000	10,100	10,100		
		A283 D	.....	60,000	10,100	10,100		
	A139	A285 A	.....	45,000	9,000	9,000	8,700	7,800
		A285 B	.....	50,000	10,000	10,000	9,700	8,800
		A285 C	.....	55,000	11,000	11,000	10,600	9,600
Carbon steel	A155[2]	A285 A	C45	45,000	10,100	10,100	9,800	8,700
Carbon steel	.....	A285 B	C50	50,000	11,250	11,250	10,900	9,900
Carbon steel	.....	A285 C	C55	55,000	12,400	12,400	11,900	10,850
Killed carbon steel	.....	A201 A	KC55	55,000	12,400	12,400	11,900	10,850
Killed carbon steel	.....	A201 B	KC60	60,000	13,500	13,500	12,900	11,650
Killed carbon steel	.....	A212 A	KC65	65,000	14,600	14,600	13,950	12,450
Killed carbon steel	.....	A212 B	KC70	70,000	15,750	15,750	14,950	13,250
Electric-resistance welded steel	A53	A	.....	48,000	10,200	10,200	9,900	9,100
		B	.....	60,000	12,750	12,750	12,200	11,000
	A135	A	.....	48,000	10,200	10,200	9,900	9,100
		B	.....	60,000	12,750	12,750	12,200	11,000
Lap welded:								
Steel	A53	....	.....	45,000	9,000	9,000		
Wrought iron	A72	....	.....	40,000	8,000	8,000		
Butt welded:								
Steel	A53	....	.....	45,000	6,750	6,750		
Wrought iron	A72	....	.....	40,000	6,000	6,000		
Seamless steel[3]	A53	A	.....	48,000	12,000	12,000	11,650	10,700
		B	.....	60,000	15,000	15,000	14,350	12,950
	A106	A	.....	48,000	12,000	12,000	11,650	10,700
		B	.....	60,000	15,000	15,000	14,350	12,950
	A83	Type A	.....	47,000	11,750	11,750	11,450	10,550
	A179	Low carb.	.....	.....	11,750	11,750	11,450	10,550
	A210	....	.....	60,000	15,000	15,000	14,350	12,950

[1] The several types and grades of pipe tabulated above shall not be used at temperatures in excess of the maximum temperatures for which the S values are indicated. (See also specific requirements for service conditions contemplated.) Allowable S values for intermediate temperatures may be obtained by interpolation.

[2] The values tabulated for ASTM A155 pipe are for Class 2 pipe. For Class 1 pipe, which is heat treated and radiographed, these stresses may be increased by the ratio of 0.95 divided by 0.90.

[3] The S values for 600 F for tubing specifications referred to in Table 15a, but not in Table 15, shall apply for all temperatures up to 600 F.

Table 5.   List of Material Specifications

Material	Specifications[1]
Bolting:	
Alloy steel for high-temperature service	ASTM A193
Commercial steel (bar stock hot-rolled carbon steel)	ASTM A107
Nuts, carbon and alloy steel for high-pressure and temperature	ASTM A194
Heat-treated carbon-steel bolting material	ASTM A261
Steel machine bolts and nuts and tap bolts	ASTM A307 Grade B
Alloy-steel bolting materials for low-temperature service	ASTM A320
Fittings, valves, and flanges:	
Bronze castings up to 406 F	ASTM B62
Bronze castings up to 550 F	ASTM B61
Gray iron[2] castings	ASTM A126
Malleable iron for temperatures up to 500 F	ASTM A197
Steel (cast carbon) for high-temperature service	ASTM A95
Steel (forged or rolled) for high-temperature service	ASTM A105
Steel (carbon forged) for general service	ASTM A181
Steel, wrought for welding fittings	ASTM A234
Steel, carbon, suitable for fusion welding up to 850 F	ASTM A216
Steel, castings alloy, suitable for fusion welding 750 to 1100 F	ASTM A217
Steel (alloy forged or rolled) 750 to 1100 F	ASTM A182
Tubing:	
Copper (seamless) for temperatures up to 406 F	ASTM B75
Copper (seamless water tube) for temperatures up to 406 F	ASTM B88
Copper (seamless bright annealed)	ASTM B68
Steel (copper brazed) for temperatures up to 406 F	ASTM A254
Steel, seamless, boiler tubes	ASTM A83
Steel, medium-carbon seamless-steel boiler and superheater tubes	ASTM A210
Pipe:	
Red brass (seamless) for temperatures up to 406 F	ASTM B43
Cast iron[2] (centrifugally cast)	ASA A21.6
	ASA A21.7
	ASA A21.8
	ASA A21.9
Cast iron[2] (pit cast)	ASA A21.2
Cast iron[2] (pit cast) for gas	ASA A21.3
Copper (seamless) for temperatures up to 406 F	ASTM B42
Steel (seamless and welded, galvanized) ordinary use	ASTM A120
Steel (seamless and welded)	ASTM A53
Steel (seamless) high-pressure and high-temperature service	ASTM A106
Steel (electric-fusion-welded) 18 in. and larger for high-pressure and high-temperature service	ASTM A155
Steel (ordinary electric-fusion-welded to but not including 30 in.)	ASTM A139
Steel (ordinary electric-fusion-welded) large size	ASTM A134
Steel (electric-resistance-welded)	ASTM A135
Wrought iron (welded) for temperatures up to 750 F	ASTM A72
Steel, seamless chromium-molybdenum alloy steel, for high-temperature service	ASTM A335
Iron pipe, welded alloyed open hearth	ASTM A253

[1] In all cases the latest published revision is to be adhered to.
[2] Cast iron shall not be used for any service in excess of 450 F.

2. Joints. Modern-day construction to a large extent is welded. Welded joints have generally replaced most of the screwed or flanged joints in new construction of underground steam mains and customers' service lines. In welding, advantage can be taken of the beneficial effect of judicious cold springing[1] in assisting the system to attain its most favorable condition sooner. Inasmuch as the life of a system under cyclic conditions depends primarily on the stress range rather

[1] See the discussion of cold spring in Chap. 4.

than the stress level at any one time, no credit for cold springing is warranted with regard to stresses.   However, in the calculation of end thrusts and moments acting on equipment containing moving or removable parts with close clearances, the actual reactions at any one time rather than their range are significant, and credit accordingly is allowed for cold spring in calculations of thrusts and moments.   Welded joints must conform to the requirements of Sec. 6, Chap. 4, of the ASA B31.1 Code. For joints other than welded, reference should be made to Sec. 6, Chap. 2, of the same Code.

3. Fittings.   While welding has largely displaced flanged joints, some flanged connections are required, particularly in making connections to flanged valves, expansion joints, or fittings where space limitations do not permit welding or where easy removal of a fitting or valve is desired.   Malleable, cast-iron, bronze, or brass fittings may be used for pressures not exceeding 250 psi and temperatures not in excess of 450 F.   Cast- or forged-steel fittings are required for pressures not above 600 psi and temperatures not in excess of 750 F.   Welded fittings must comply with the American Standard for Steel Butt-Welding Fittings (ASA B16.9) or American Standard for Steel Socket Welding Fittings (ASA B16.11) where applicable, and the material shall conform to ASTM Specification A216, A217, A234.   Special fittings or welded assemblies fabricated in either the shop or field are required to conform to Sec. 6, Chap. 5, of ASA B31.1.

4. Valves.   Valves must be of a design at least compatible with the service conditions and constructed of the materials allowed by ASA B31.1 for pressure and temperature.   Body-metal thickness of steel valves is dictated by the requirements of ASA 16.5.   Bodies, bonnets, disks, and yokes are to be of forged or cast steel for pressures above 250 psi but not above 600 psi and for temperatures not in excess of 750 F, of cast iron or steel for pressures 125 psi to 250 psi and temperatures not in excess of 450 F, and of cast iron, malleable iron, steel, or bronze for pressures not above 125 psi and temperatures not in excess of 450 F.

Stop valves used in steam mains must be constructed and installed so that drainage of condensate along the bottom of the pipe is not obstructed.   Valves may be either of the rising or nonrising stem types and of the gate, globe, plug, or angle designs. Solid-wedge gate or double-disk valves with inside screw and nonrising stems (protected from corrosive elements) are popular selections in the industry.

All valves in nominal sizes above 3 in. must have flanged openings or welded ends if the operating pressure is in excess of 250 psi.   Weld-end valves in underground steam mains are becoming more acceptable because of the economies in the cost of the valve, the elimination of costly flanges, and improved techniques of decreasing outage time.

5. Bolting.   In bolting cast-iron flanges or steel flanges to cast-iron flanges, valves, fittings, etc., bolts must be of carbon steel equivalent to ASTM A307, Grade B, without heat treatment other than stress relief.   Threads in accordance with the coarse-thread series of the American Standard for Screw Threads, ASA B1.1, are recommended for carbon-steel bolts.   Carbon-steel bolts may be the American Standard regular or heavy hexagonal-head bolts and must be used with American Standard heavy semi-finished hexagonal nuts (ASA B18.2) which conform to ASTM A194.

For high-temperature service or for insurance of a tight joint in the case of steel flanges, bolts or stud bolts should be of alloy steel, conforming to ASTM A193. Nuts must be of alloy steel according to ASTM A194.   Threads for bolts, stud bolts, and nuts are to be Class 3 fit and conform to ASA B1.1.

6. Gaskets.   For temperatures up to 750 F, gaskets may be made of metal, asbestos, or other material which will not burn, char, or change in character so as not to perform the service intended.   A gasket made up of an acrylonitrile rubber

(Hycar) binder and asbestos fibers has proved its worth in the industry for steam pressures up to 300 psi and temperatures up to 700 F.

Rubber gaskets may be used for hot-water service at temperatures not in excess of 250 F.

Insulation

One of the most important elements of underground steam piping is its insulation. Many of the commercial insulations used for overhead steam lines are also adaptable for underground service, but basically, the insulating material must be of an inert nature. Insulating materials used are all minerals of the earth's crust, and they are combined with binders and waterproofers which must have a high quality of endurance so that, when flooded and dried out, the original physical, chemical, and insulating properties will be maintained. Insulation should be noncombustible, moisture-resistant, durable, verminproof, noncorrosive to steel pipe when wet, and able to retain its position in relation to the pipe. It should be unaffected by chemicals and have the lowest practical thermal conductivity for its intended service.

Some of the types of insulating materials, described as to their composition and form and utilized as pipe covering for underground steam mains today, are as follows:

1. A molded form of calcined diatomaceous silica and asbestos fiber with bonding clay and cementing materials with some varieties containing a small percentage of carbonate of magnesia. Its density is about 23 lb/cu ft, and its service-temperature range is from 600 to 1200 F.

2. A molded form of long-fiber Amosite asbestos combined with diatomaceous earth and sprayed with a siliceous binder. For use at a temperature of 1200 F, its density is about 20 lb/cu ft, and for a temperature of 750 F its density is approximately 12 lb/cu ft.

3. A molded form of 85 per cent by weight of basic carbonate of magnesia and 10 to 15 per cent of asbestos fiber with additions of clay or other cementing materials. At 600 F its density is approximately 17 lb/cu ft, and 600 F is the maximum permissible temperature.

4. A molded form and flexible blanket of glass wool, fiberized by blowing steam through streams of molten glass. At a temperature of 600 F, its density is from 4 to 8 lb/cu ft.

5. A molded form of cellulated glass made by baking pulverized glass and chemicals. At a maximum operating temperature of 800 F, its density is approximately 9 lb/cu ft.

6. Insulating cements consisting of portland cement mixed with an aggregate of insulating material such as magnesia or vermiculite, with a waterproofing material as a mixture or applied externally. Temperature limitations vary as to the proportions of insulating materials used, and the density varies proportionately.

7. Backfill insulation consisting of crushed bitumen, a high-resinous-content hydrocarbon, which forms a plastic and a sintered zone around the pipe as steam flows and heat is transferred. The maximum safe temperature is about 520 F, and the corresponding density is approximately 50 cu ft/ton.

8. Molded form insulations may be furnished in split or full-round sections, 3 ft long, 1 to 4 in. thick in single or multiple layers for pipe sizes $\frac{1}{2}$ to 24 in. with flat block or segmental forms for larger size pipes.

The primary characteristic of any insulating material is its ability to resist the flow of heat through it. The substance having the least heat flow per unit area and thickness is therefore the best insulator. The *thermal conductivity* of insulating material has been established by tests, and performance data have been recorded

by the manufacturers for their particular type in standard pipe sizes and may be found in the trade literature.

In underground steam piping, the economical thickness of insulation is difficult to determine exactly except by actual test in existing soil conditions. The first layers of insulation are much more effective than succeeding layers, and the economical thickness easily may be exceeded. Increasing insulation thickness may require enlargement of the pipe conduit and widening of the trench, which may prove uneconomical.

Detailed information on insulation may be found in Chap. 6.

PROTECTIVE CONDUIT

Protective conduits for underground steam mains and services are necessary to protect the pipe and insulation from damage due to earth pressure and impact loadings, to allow free longitudinal expansion and contraction while held in proper alignment, and to prevent ground-water seepage or flooding by providing either drains or a completely waterproof structure.

There are many types of conduits used by the industry today. They can be categorized into the following general classifications: prefabricated, box-type, solid-pour, granular-fused, and tunnels.

Prefabricated conduits are becoming more popular with utilities located in large cities where conditions exist such as congested subsurface, heavy surface traffic, tidewater, and rock areas. The ideal design for these conditions is one with the smallest cross-sectional area, consistent with thermal requirements, to fit into limited subsurface space, also one with maximum prefabrication for quick installation to limit interference with traffic movement and one that provides complete exclusion of subsurface water.

Prefabricated designs such as those of Figs. 2a to d meet these requirements. These are usually constructed in standard lengths up to 40 ft. The design shown in Fig. 2a consists of a seamless-steel steam-carrying pipe, with full-round insulation in cylindrical form of 3-ft lengths threaded onto the pipe, placed concentrically on roller supports with necessary steel rings or guides into a steel conduit pipe, which is protected with a heavy anticorrosive coating. This design lends itself to a completely welded and waterproof structure. The design of Fig. 2b consists of a seamless-steel steam-carrying pipe with necessary supports or spacers fabricated from small pipe, with insulation in a cylindrical form wrapped with a waterproof jacket and placed concentrically in an asbestos-cement pipe conduit. Joints of the conduit are connected by wrapping with an epoxy-saturated glass fabric or by using solid sleeves or standard rubber joints. This results in a rigid waterproof conduit. In Fig. 2c, the insulated pipe is placed within a corrosion-resistant helically corrugated metal jacket which is protected with a heavy asphaltic coating. In Fig. 2d, the insulated pipe is placed concentrically within the corrosion-resistant metal jacket and the intervening space between jacket and insulation is poured full of high-melting-point asphalt, which is a protective medium for the insulation. Some of these prefabricated designs are also available with cast-iron casings with fittings designed for joints, bends, etc.

Box-type conduits have many variations. In the design of Fig. 3a, a concrete slab is poured and the insulated steam main is installed. A corrugated metal form is placed as an inverted U over the steam main, over which the concrete is poured. Depending on loading conditions, reinforcement may or may not be necessary. Drainage is provided by filling drainage pockets with crushed stone or gravel; these pockets may be installed on either or both sides of the conduit. They conduct water from the top of the conduit to the lower drain, which is connected to a sewer.

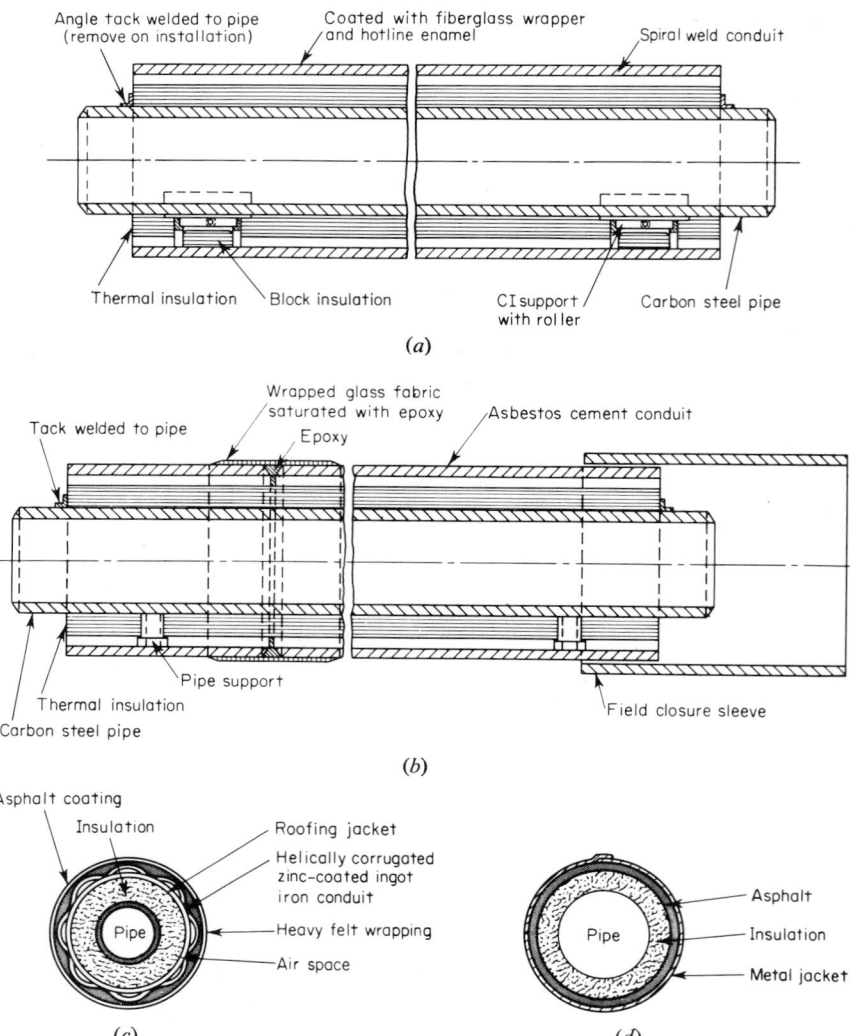

FIG. 2. Prefabricated conduits for underground steam pipes: (*a*) welded-steel conduit; (*b*) asbestos-cement conduit; (*c*) coated and wrapped corrugated conduit; (*d*) poured asphalt protects thermal insulation.

Variations of the box-type design have side walls of hollow vitrified tile or other suitable materials and covered with a reinforced concrete slab as shown in Fig. 3*b*.

Solid-pour construction, such as is shown in Fig. 4*a* consists of poured structural concrete which is vibrated or tamped around a conventionally insulated steam line. If the insulation has a high compressive strength, such as is characteristic of diatomaceous earth or silica and asbestos, it can support the pipe; otherwise, it should be supported independently. Reinforcing is sometimes used to prevent settlement. An eccentric space may be left around the pipe by using insulation

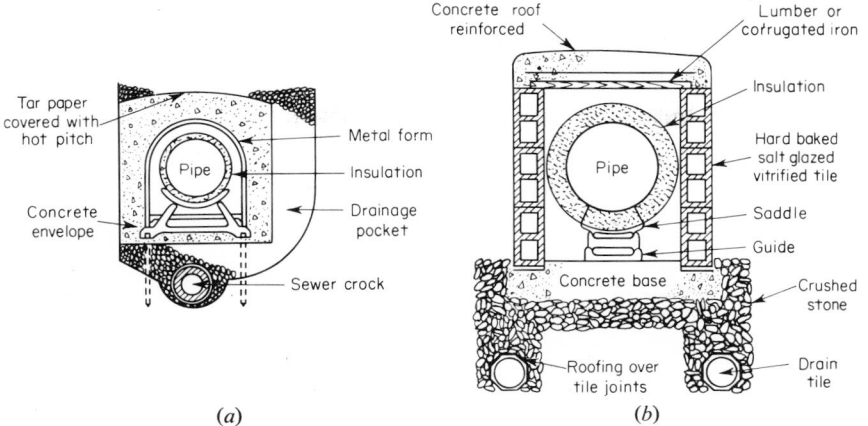

FIG. 3. Box-type conduits for underground steam pipe: (*a*) Metallic, semicircular form with drainage; (*b*) rectangular tile enclosure with drainage.

larger than the outside diameter of the pipe.   This permits the pipe to rise during the warming-up period without crushing the insulation.

The design of Fig. 4*b* utilizes an insulating concrete as a conduit.   The concrete is poured around the steam pipe, which is supported on precast blocks of the same material.   The insulating concrete may consist of any mixture of insulation or other cellular materials and portland cement or a mixture of a special foaming material mixed with portland cement; the aim is to create a cellular mass composed of minute air cells in the concrete so as to develop insulating qualities.   The piping (which could be a number of pipes) is wrapped with corrugated paper before the insulation is poured, and a heavy-asphalt-coated waterproofing membrane is installed to protect the top and sides of the structure before backfilling.

Granular-fused types of conduits as shown in Fig. 5 consist of a granular bitumen, selected as to the required temperature range and poured and tamped around the pipe or pipes in a trench.   Heat passing through the pipe forms three concentric zones: (1) a dense semiplastic core fused on by the pipe's own heat, (2) a sintered zone providing thermal insulation and moisture proofness, and (3) an outer layer of granules providing a final zone of thermal insulation and the load-bearing portion of the structure.

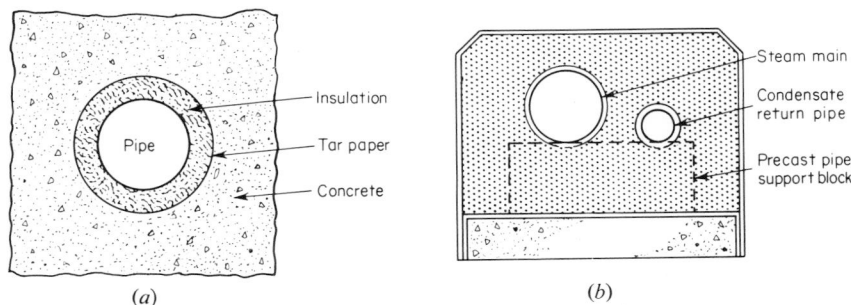

FIG. 4. Solid-pour protection for underground steam pipe: (*a*) insulated and wrapped pipe in structural concrete; (*b*) pipe is surrounded by insulating concrete.

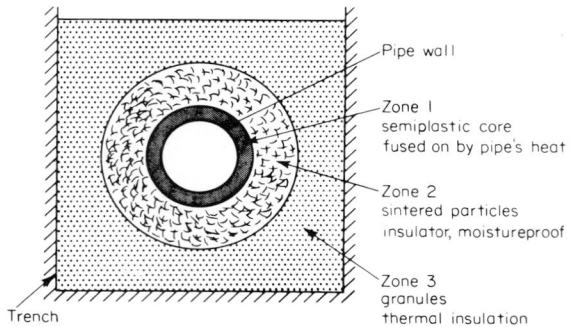

FIG. 5. Granular-fused conduit for underground steam pipe.

Tunnels, of course, are very costly and are constructed only out of necessity. Figure 6a shows a tunnel built by the tunneling process in hard-clay soil which permitted the use of brick. No shaping was necessary. Figure 6b shows a concrete tunnel constructed through a soil of sand, gravel, and loam by the "poling board" method. Concrete was blown into place by compressed air, and the entire structure was reinforced both horizontally and circumferentially. Figure 6c shows

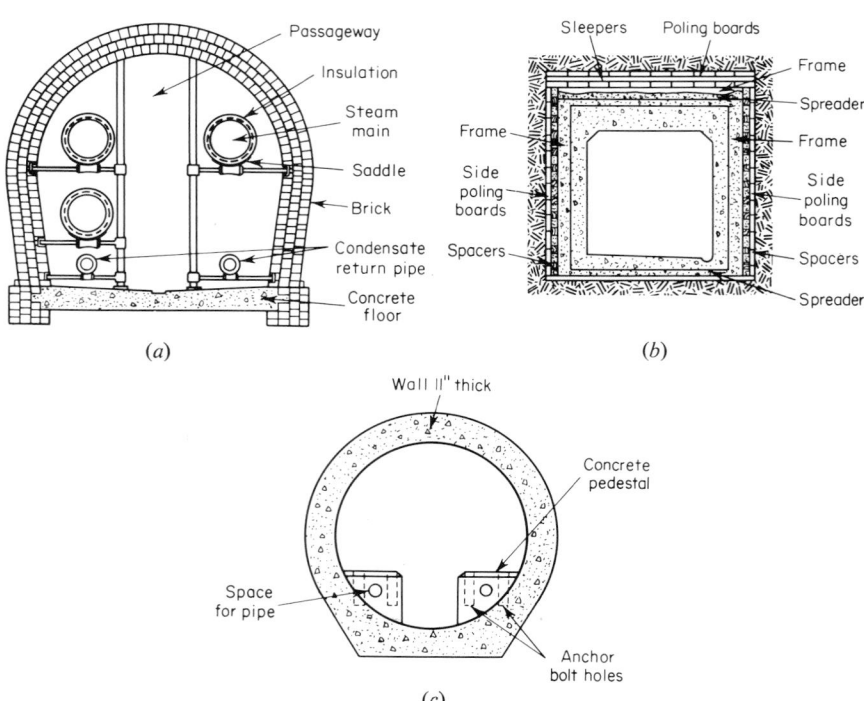

FIG. 6. Tunnels for underground steam pipe: (a) brick construction in heavy clay soil; (b) concrete construction in sandy soil; (c) cylindrical concrete tunnel with integral pipe pedestals.

a cylindrical concrete tunnel, 6 ft inside diameter with 11-in.-thick walls.   There is no reinforcing except at the vertical shaft entrances.   Tunnels of this design are built mainly by the tunnel method using 3-in. hardwood forms.

Walk-through tunnels are usually not provided for steam mains unless they are required for underground passage between buildings, as in institutions, and then, usually they are constructed to accommodate other utility services, such as electricity, water, gas, etc.

Provision for drainage and ventilation with special insulating methods of steam mains to prevent heat loss should be considered.

No attempt has been made to recommend the type of conduit to be used for a certain type of condition.   The illustrations shown here are only a few of the many designs utilized in practice today.   Choice and application depend upon the local conditions.   In the economic evaluation, factors to be considered are (1) life expectancy, (2) operation and maintenance, (3) foundations to prevent settlement, (4) external and internal drainage, (5) corrosive action of soil conditions, (6) loads to be imposed, (7) field installation, and (8) necessary insulating properties.

DRAINAGE OF CONDENSATE

To prevent water hammer and a possible rupture of the steam main, condensed steam or condensate within the steam main must be removed.   Provisions must be made during construction to grade the steam main carefully so as to pitch the pipe not less than 1 in. in 50 ft and preferably in the direction of steam flow.   Since it is impractical to assume that a continuous slope can be maintained in a long run of main, removal of condensate must be provided for at all low points where a water pocket will exist.   In any event, recommended lengths of steam main for draining off condensate should not exceed 300 to 400 ft.   Usually, the contour of the ground or subsurface structures will dictate the pitch of the steam main.

When condensate is drained from a steam main, a drain pocket is welded to the bottom of the pipe to be drained.   The diameter of the pocket should be about one-third the diameter of the line up to a maximum of 6 in. for 18-in. and larger mains.   The pocket not only provides for condensate removal but also allows for sediment removal.   Figure 7 shows this design.   Figure 8 is a design which is in the form of a separator and is more prevalent in underground steam mains.

Generally, it is not economically feasible or practical to return the condensate from district-heating systems because of the high cost of installation and maintenance of pumping equipment, manholes, and return piping.   For systems which supply high flow rates to concentrated loads, it probably will be economical to return the condensate in order to recover the energy in the condensate and to save the chemical treating and quantity of make-up water in the boiler plant.

In district-heating systems, traps off the drain pocket can be discharged through cooling coils to the city sewer system.   City regulations require the condensate discharge to be cooled to 100 F before entering the sewer.   Figure 9 shows a design whereby the traps are elevated above the steam main in their own manhole to street level for easy maintenance.   The condensate leg is maintained by steam pressure.

In some cases, condensate discharge lines are piped to sump pits in manholes where the water is removed by float-controlled electric pumps which discharge it into the sewer.

Paragraph 418 of ASA B31.1, Code for Pressure Piping, contains the following provisions for Drains, Drips and Steam Traps:

(a) Drains or drips shall be provided to drain the condensate from the steam piping and equipment wherever it may collect.   Piping shall be properly pitched towards low points where drain and drips are located to facilitate drainage.   Blow-off outlets for air or

Steam main
Gate valve
To trap
Clean out opening

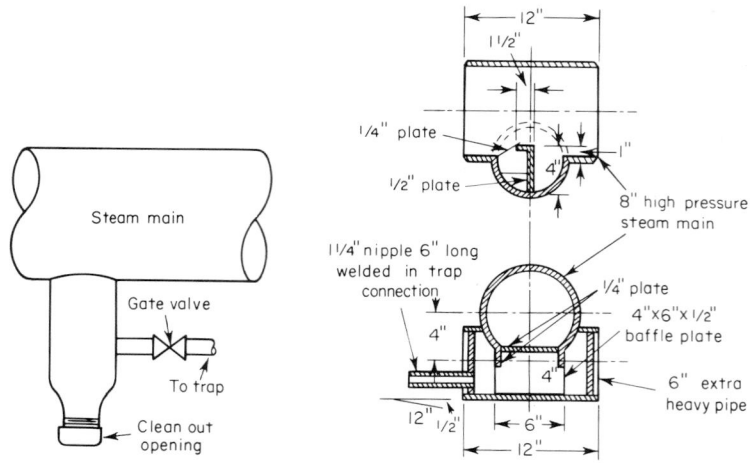

FIG. 7. Drain pocket for steam-trap connection to low-velocity steam main.

12"
1 1/2"
1/4" plate
1/2" plate
1"
4"
8" high pressure steam main
1 1/4" nipple 6" long welded in trap connection
1/4" plate
4"x6"x 1/2" baffle plate
4"
4"
6" extra heavy pipe
12"
1/2"
6"
12"

FIG. 8. Drain pocket for steam-trap connection to high-velocity steam main.

C
Sewer
4" dia. extra heavy CI drain (see section A-A)
Reinf. conc. envelope
A
4" dia. transite pipe (indust. vent)
4" dia. CI running trap
A
2" trap line
4" dia C I running trap valve manhole
Concrete plug
Cooling and expansion chamber
C
Plan
Grade
Manhole frame and cover
Grade
2" trap line from valve manhole
Connect 2" dia. flange to trap assembly piping
4" transite sleeve coupling
4" transite sleeve coupling
4" dia. transite ell. (indust. vent)
Concrete plug
4" dia. CI running trap
Sewer
Reinf. conc. envelope
4" dia. extra heavy C I drain from main manhole
Sewer
4" dia. extra heavy C I drain
Sewer
Section C - C
Section A - A
4" x 4" x 4" "Y" conn.

FIG. 9. Trap manhole and sewer connections with condensate cooling.

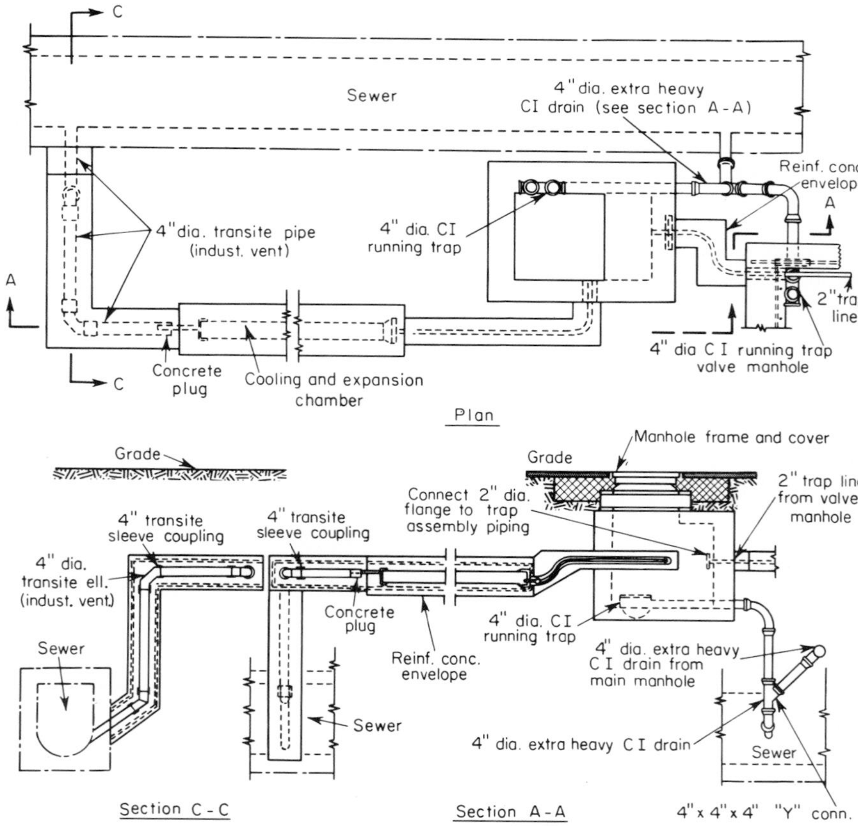

condensate open to atmosphere or connected to sewers, sumps, or receivers shall be provided at all low points and elsewhere when necessary for the proper operations of the pipeline and equipment. Each drip, drain and blow-off shall be controllable by at least one stop valve, located as close as practicable to the point of drainage.

(b) Drip lines from steam headers, mains, separators and other equipment shall be properly trapped with the traps installed in accessible locations. By-passes shall be provided around steam traps unless the traps may be replaced by a spare or drainage continued by means of an open drip to the atmosphere or elsewhere at times when the trap is serviced or found inoperative.

(c) Drip lines from steam headers, mains or separators or other equipment operating at different pressures, shall not be connected to discharge through the same trap. Where several traps discharge into one header under pressure, a stop valve and a check valve shall be placed in the discharge line from each trap.

(d) The weight, dimensions, and materials of drip piping as far as the inlet to the trap and the trap bypass valve, if used, shall conform to the specific requirements for the respective pressure and temperature to which it may be subjected.

(e) Trap discharge piping shall have the same thickness as the inlet piping, unless the former is vented to atmosphere or operated under low pressure and has no stop valves. The trap discharge piping in all cases shall have a weight suitable for the maximum discharge pressure to which it may be subjected. Trap discharge piping, if vented to the atmosphere, shall properly run to facilitate its self-discharge and its outlet shall be so located, or proper protection shall be provided, to prevent personal injury or property damage caused by escaping steam or hot water. All trap discharge piping shall be protected against freezing, where necessary.

Steam traps are automatic devices used to trap or hold steam until it has condensed and to allow condensate and air to pass as soon as they accumulate. In general a trap consists of a vessel in which to accumulate the condensate, an orifice through which the condensate, is discharged, a valve to close the orifice port, mechanisms to operate the valve, and inlet and outlet openings for the entrance and discharge of the condensate from the trap vessel.

Steam traps are classified according to the type of operating device by which they function. Generally, some of the type of traps used in district heating are (1) float, (2) bucket, (3) impulse, and (4) thermodynamic.

Float traps operate by the rise and fall of a float due to a change of condensate level in the trap. When the trap is empty, the float is in its lowest position and the discharge valve is closed. As condensate accumulates in the trap chamber, the float rises and gradually opens the valve and the pressure of the steam pushes the condensate out through the valve. The discharge from a float trap is generally continuous, since the opening at the valve is proportioned to the flow of condensate through the trap. Figure 10*a* shows a typical float trap.

Bucket traps discharge intermittently. These are two types: (1) upright and (2) inverted. In the upright bucket, Fig. 10*b*, condensate fills the space between the bucket and walls of the trap. This causes the bucket to float and forces the valve against its seat. The condensate rises and overflows into the bucket, causing it to sink, opening the valve, and discharging the condensate by the steam pressure on the surface of the condensate in the bucket. When the bucket is emptied, it rises and closes the valve. This type of trap is not influenced by pulsations or wide fluctuations of pressure and is particularly suited where there are pulsating pressures.

The inverted bucket trap, Fig. 10*c*, operates on the same principle as does the upright type. Steam floats the bucket and closes the valve. Condensate causes the inverted bucket to fall, opening the valve and discharging the trap. Air is eliminated automatically by passing through the small vent in the top of the bucket. This type of trap is particularly suited where abnormal amounts of air must be discharged or where there is an excessive amount of foreign matter such as dirt, sludge, etc.

Impulse traps, Fig. 10*d*, depend on the property of condensate at a high pressure and temperature to flash into steam at a lower pressure. This flashing action is utilized to govern the movement of a valve by causing changes in pressure in a control chamber above the valve. The flow of condensate divides, the main part discharging through the valve and the remainder, called the control flow, bypassing continuously up into the control chamber through an annular orifice around the control disk. From the control chamber, the condensate flows out through the

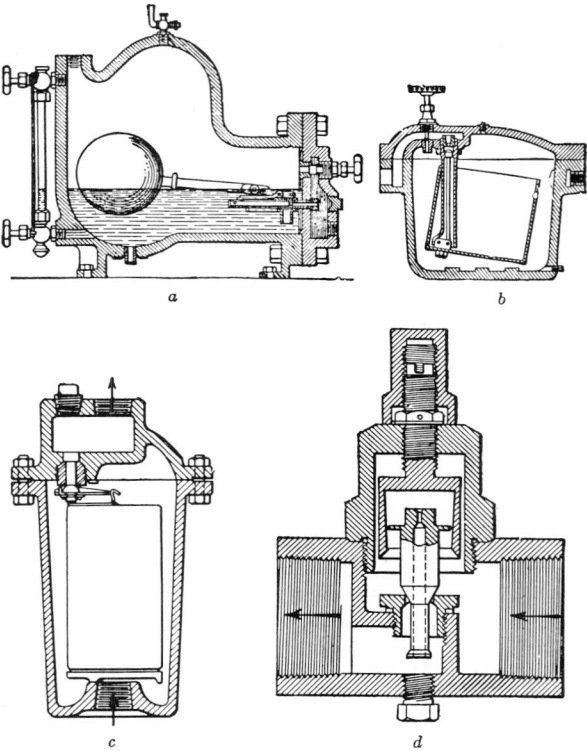

a   *b*

c   *d*

FIG. 10. Types of steam traps: (*a*) float; (*b*) upright bucket; (*c*) inverted bucket; (*d*) impulse.

control orifice in the valve stem. When the system is heating up, the condensate is not at high temperature and builds up in the control chamber. The flow through the control orifice does not change volume and the discharge through the orifice reduces the volume in the control chamber. Discharge through this orifice lowers the pressure in the control chamber, and the valve opens to discharge air and condensate.

When steam comes in contact with the trap, the condensate is heated and the flow, in entering the control chamber, flashes and increases the volume of the control flow. The discharge through the control orifice is thereby choked, and pressure in the control chamber builds up, closing the valve and stopping all discharge of hot condensate except for a small amount that flows through the control orifice.

The discharge of an impulse trap is pulsating or intermittent but not so infrequent as with the bucket type of trap.

Thermodynamic traps have only one moving part, a valve disk, which is operated by using the kinetic energy of steam. Figure 11 shows the operation of this trap. In Fig. 11*a*, steam pressure has raised disk *A* from its seat and there is an immediate discharge of air and condensate at steam temperature. In Fig. 11*b* steam flows and the high velocity creates a low-pressure area under the disk. Pressure is built up in chamber *F* by recompression. In Fig. 11*c* the pressure in chamber *F*, acting on the full top area of the disk, exceeds the combined momentum and pressure-area force and immediately forces the disk down, closing the inlet. As condensation decreases the pressure in the chamber, the disk is raised and another cycle begins. This type of trap is intermittent, operates against a back pressure up to 50 per cent of the inlet pressure, and is practically maintenance free.

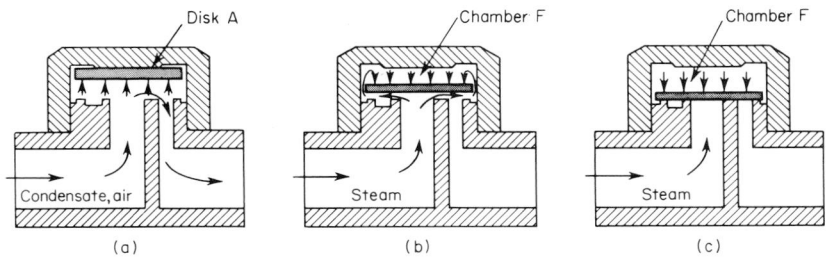

FIG. 11. Thermodynamic steam trap: (*a*) trap fully open and is discharging condensate and air; (*b*) some steam is passing through and high velocity causes pressure reduction beneath disk; (*c*) higher pressure in chamber forces disk onto seat.

The size of a trap should be selected on the basis of its effective valve area or its actual discharge capacity rather than the size of inlet and outlet connections.

THERMAL EXPANSION AND ANCHORAGE

Thermal expansion of pipelines can be provided for by use of pipe bends, offsets, expansion joints, or changes in direction of the pipeline itself. Where pipe bends or offsets can be used or where the pipeline direction is changed to provide for expansion, the provisions of Sec. 6, Chap. 3, of the Code for Pressure Piping, B31.1, must be followed.

In underground steam piping, expansion-joint fittings are more commonly used for thermal expansion owing to the limitations of space and costly trenching.

Slip and bellows types of expansions joints are principally used for application in underground steam mains. Illustration of these types of joints, together with illustrations and descriptions of other types which are used infrequently for underground steam mains, are presented in Chap. 4.

The basic design of a slip-type joint consists of a cast-iron or steel body with a stuffing box and a sliding sleeve. Differentiations between slip-type joints are made on the basis of the type of packing used. Ring-type packing of braided, molded, or loose packing has a considerably greater range of temperature and applications because of the greater variety of packings available. Plastic-type packing has the advantage of being injected into the stuffing box under steam pressure and maintaining an even pressure on the sealing rings. Secondary differentiation of slip joints is based on methods of maintaining concentricity of the sliding sleeve in relation to the stuffing box and body. Nonguided joints are used mostly for highly

viscous fluids or for solids in suspension. Internally and externally guided joints with limit stops permit reduction in external pipe alignment guides and are generally used in district heating.

The advantage of slip-type joints is the longer traverses that can be obtained to absorb pipe expansion; the disadvantage is the necessity of maintenance of packing.

By the very nature of its construction, the slip joint is capable of absorbing only axial movement.

The metallic bellows expansion joint is composed of the following components: (1) the flexible element proper (corrugated metal tube), the rings enabling reinforced elements to support higher pressures, and the end collars or other structures serving to increase pressure capacity of cylindrical ends and transfer spring force and hydrostatic end thrust to connected piping and (2) the end nipples or flanges and, for other than anchored axial joints, the various hardware items used to cross-connect the ends, such as tie rods, hinges, and gimbals.

Differentiation of bellows joints is based on the method of reinforcing the flexible element or corrugation against pressure. Nonequalizing bellows are limited to relatively low steam pressures. Root ring reinforced bellows are used for a wide range of pressures. Self-equalizing bellows are generally used for higher pressures and prevent overcompression of the corrugation. Toroidal bellows are used for extremely high pressure ranges.

The two basic types of reinforced packless expansion joints used in district heating are the root ring or semiequalizing expansion joint and the self-equalizing expansion joint. The root ring, as the name implies, is a ring at the root of the corrugation and provides support only at this point. The self-equalizing unit utilizes a tee-shaped equalizing ring between the corrugations, which supports the root and the side wall and limits the travel of each corrugation. Comparative tests show that the self-equalizing type of expansion joint provides 250 per cent greater flex life than the root ring or semiequalizing type.

The advantages of the bellows type of joint are that it requires no maintenance, can be directly buried with the steam main, and can absorb a combination of axial, lateral, and rotational movement within a vast range of pressures and temperatures. Its disadvantage is the limitation of traverse, generally 7 to 8 in. of axial movement.

The first step[1] in the selection of expansion joints for a piping system is to survey the requirements, noting the pressure, temperature, load limitations, and possible anchor points. Even the most complex piping systems can be broken down into increments of relatively simple configurations. The simplest approach is to use single- and double-type expansion joints in axial movement. If anchoring presents a problem, owing to lack of structural members or inadequacy of equipment to absorb anchor loads, the merits of universal, pressure-balanced, hinged, and gimbal units should be investigated. Once the type of expansion joint has been selected and anchors located, the actual change in length of the various pipe sections can be calculated and expansion joints of adequate capacity selected.

The need of special materials for either corrosion resistance or high temperature may affect the pressure or travel rating of the unit. The final step is to include accessories such as sleeves and covers where necessary.

One of the most important accessories for an expansion joint is an internal sleeve. It is used to reduce the pressure drop across the unit, to eliminate erosion of the bellows, and to eliminate vibration from high flow rates. Sleeves can also be used to decrease the operating temperatures of the bellows. In some instances low-temperature gases are purged between sleeve and bellows to aid in cooling. At high temperatures, steam can be used for this purpose.

---

[1] See also Chap. 4 relative to specification of expansion joints.

Covers are often specified as an accessory to protect the unit from damage during installation, operation, or in transit. Insulation can be mounted directly over the covers, and heat loss reduced to a minimum.

Adequate anchoring and guiding are essential for the proper functioning of expansion joints. Main anchors for end thrusts must be designed for the sum of the pressure thrust, the force required to deflect the joint, and the force due to friction in the piping guides. Intermediate anchors are used between balanced or double-type joints where the pressure thrust is balanced and the anchor need be designed only to restrict the expansion movement of the pipe. Guides are essential to assure proper pipe alignment into the expansion joint so that undesirable torques are not impressed on the expansion element.

"Cold springing" of expansion joints used only for lateral deflection can provide several advantages. The most significant of these is the reduction in the force required to deflect the expansion joint. In addition, the joint is more stable at high pressures, since the maximum angular displacement of the corrugations is reduced. Joints with internal sleeves and external covers must have adequate clearance to permit the lateral deflection of the expansion element. If the deflection can be reduced by 50 per cent, these clearances can also be reduced by 50 per cent. Internal sleeves can then be of maximum diameter, and external sleeves are held to a minimum diameter.

MANHOLES AND ENCLOSURES

For underground steam piping, manholes are required for sectionalizing valves and bypass valve piping, trap piping and traps, some types of expansion joints, and convenience of location of other expansion joints and anchorage. Modern manholes are constructed of reinforced concrete, either field poured or precast beforehand for installation as a unit. In field-pour construction, waterproofness should be assured by pouring the walls and floor monolithically. Other provisions that should be considered in manhole design are (1) adequate working space for maintenance, (2) clearance for removal of equipment, (3) ventilation, and (4) drainage. If the manhole floor elevation is below the sewer, or if a sewer is not accessible for drainage piping from the manhole, sump pump manholes with automatic pumps or water ejectors must be provided. An example of such an installation is shown in Fig. 12.

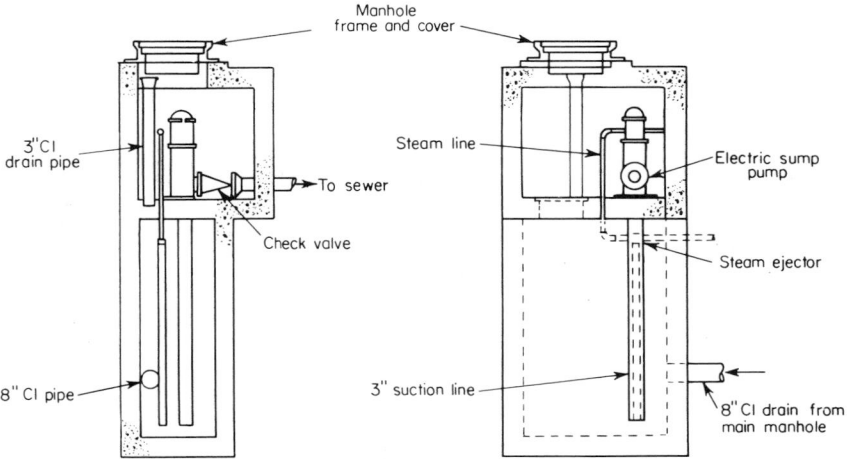

FIG. 12. Manhole for sump pump.

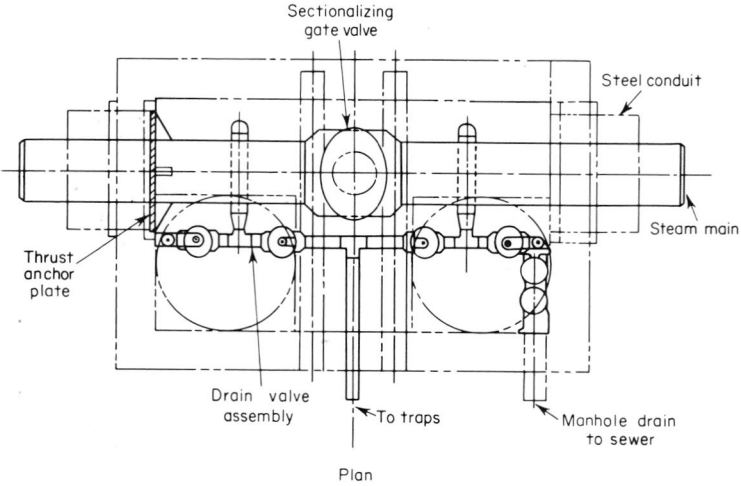

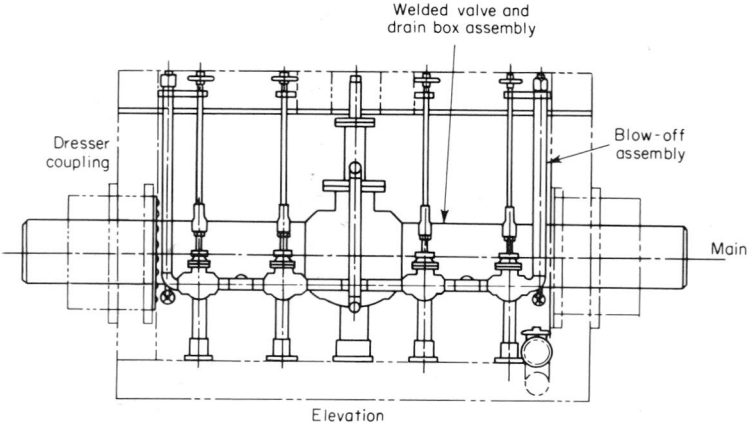

Fig. 13. Manhole for sectionalizing valve and drain valve assembly.

In prefabricated steam-main construction, such as is shown in Fig. 2a, prefabricated manholes with the necessary valves and piping installed are delivered to the job site as a unit. This type of manhole is shown in Fig. 13.

With prefabricated installation, expansion joints of the bellows design are also delivered to job site in their own steel-enclosed housing, as shown in Fig. 14, and are welded into the steel piping and conduit of the steam main. This provides for directly buried expansion joints in a fully encapsulated system.

Where convenient or necessary, expansion joints are installed in manholes with sectionalizing valves and thrust-type anchors.

Traps are usually installed in their own manholes, that is, in a manhole separate from that which houses the sectionalizing valve, for ease of maintenance.

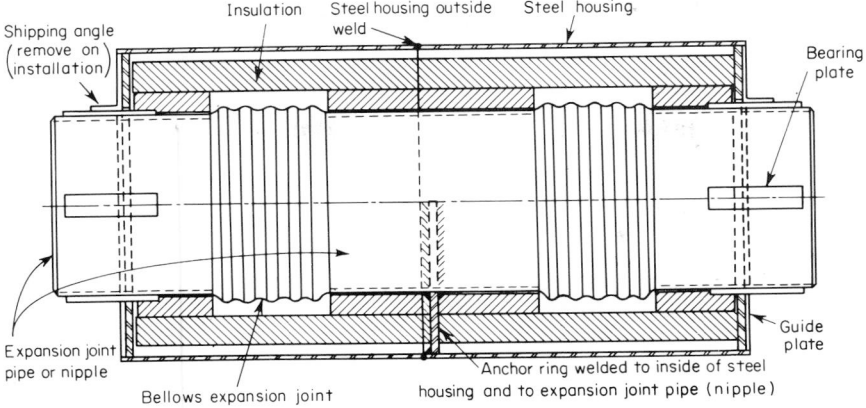

Fig. 14. Double expansion joint with steel housing.

SAFETY PROVISIONS

Construction of underground steam mains must adhere to the requirements as provided for in Sec. 4 on District Heating Piping Systems of the American Standard Code for Pressure Piping B31.1.

All materials shall be capable of meeting the inspection and test requirements as required in the Code. Unless otherwise specified, the manufacturers' certification that these tests have been met shall be accepted as satisfactory.

The inside of all pipes, valves, fittings, traps and other apparatus shall be smooth, clean and free from all blisters, loose mill scale, sand and dirt when erected. All lines shall be blown before placing in service, if practicable.

Before installation, all valves, fittings, etc., shall be capable of withstanding a hydrostatic shell test equal to twice the primary steam service pressure except that steel fittings and valves shall be capable of withstanding the test pressures as given in the American Standard for Steel Pipe Flanges and Flanged Fittings, ASA B16.5, for the proper material, pressure standard and facing involved (ring joint facing in the case of welded ends). Piping shall be capable of meeting the hydrostatic test requirements contained in the respective material specifications under which it was purchased.

After installation, all piping systems shall be capable of withstanding a hydrostatic test pressure of one and one-half times the design pressure, except that the test pressure shall in no case exceed the adjusted pressure-temperature rating for 100 F as given in ASA B16.5 for the material and pressure standard involved. For systems joined wholly with welded joints, the adjusted pressure rating shall be that for ring joint facing. For systems joined wholly or partly with flanged joints, the adjusted pressure shall be that for the type facing used.

The hydrostatic test after installation shall be applied wherever practicable and in no case shall the test pressure be less than 50 psi nor shall it be made with a test medium having a temperature in excess of 100 F.

In all cases, the required test pressure shall be maintained a sufficient length of time to enable an inspection to be made of all joints and connections.

When hydrostatic testing is impractical, it shall be required that the piping be tested with steam at a pressure at least equal to the pressure at which the piping is to be operated. These tests shall be made on sections, or on the whole of the piping system, but the connections between the sections must be similarly tested.

STEAM SERVICE LINES

Steam service lines generally follow the same construction standards as those of the distribution mains. Usually the service connection is made at the top of the

distribution main and as close to an anchor point as is feasible.   A street service valve is installed, and the service line run to the building.   A waterproof sleeve should be installed at the building wall and should be such that the service-line conduit may be adequately sealed against subsurface moisture at the juncture with the building wall.   The sleeve should also permit longitudinal motion of the service line through it to allow for the expansion of the steam line.   Figure 15 shows a

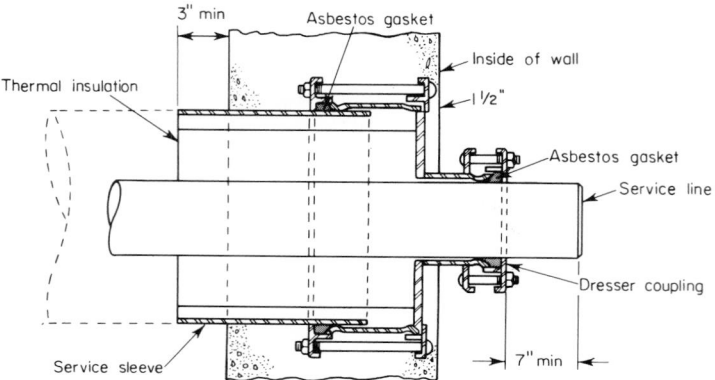

Fig. 15. Service sleeve at entrance to building.

suitable type of sleeve.   This allows longitudinal motion through the building wall and seals both the outside service conduit and the building wall against the penetration of moisture.

REDUCING AND RELIEF VALVES

Pressure-reducing and -relief valves constitute an important item in district steam piping, since it is customary practice to distribute steam at pressures which are in excess of the operating pressures of most steam-utilization equipment.   Requirements regarding pressure-reducing and -relief valves in consumers' premises as contained in the ASA Code for Pressure Piping, B31.1, are as follows:

408-(a) Where the street pressure exceeds the safe working pressure of the building heating apparatus a pressure reducing valve or valves shall be provided near the point of supply to reduce the pressure on building heating equipment to within safe limits.   In the case of cast-iron equipment this reduced pressure shall not exceed 50 psi.

Where the street pressure exceeds 50 psi and is above the safe working pressure of the building steam using apparatus, a relief valve or valves set at the safe working pressure of the building steam apparatus shall be provided, except that where two pressure reducing valves are installed in series, both set at or below the safe working pressure of equipment served, no relief valve is required. If the installation of a relief valve is not feasible, a trip stop valve, set to close at the maximum safe working pressure, shall be installed.   When the building is continuously attended, an alarm valve or signal may be installed in lieu of a relief or trip stop valve.

(b) The capacity and opening pressure of relief valves shall be such that the pressure rating of the lower pressure piping and equipment shall not be exceeded with full steam flow from the relief valves.   Relief valves shall be vented to the atmosphere, and proper protection shall be provided to prevent injury or damage caused by escaping steam.   Under certain cases it may be desirable to vent relief valves to the outdoors.

(c) The use of a hand-controlled by-pass around a reducing valve is permissible.   The

by-pass shall not be greater in capacity than the reducing valve, unless the piping on the low pressure side is adequately protected by relief valves or is of a construction which can withstand full street pressure.

(d) A pressure gage shall be installed on the low pressure side of a reducing valve. Where two reducing valves are installed in series, a pressure gage shall be installed on the low pressure side of each pressure reducing valve.

(e) The flange dimensions, construction, and material of pressure reducing and relief valves shall conform to the requirements of this section for stop valves for the pressures and temperatures to which they may be subjected.

The alarm valve or signal mentioned in the last sentence of Para. 408(a), above, is desirable in many installations, particularly in those instances where reducing-valve equipment is located at widely separated points. However, it can seldom, if ever, be used in place of the other safeguards mentioned, since the list of buildings which the designer can assume will be continuously attended for their life is practically nonexistent.

As the connected load increases in a steam-distribution system, the tendency is to increase distribution pressures. The piping and the equipment referred to in Para. 408(e), above, which can be subjected to the full street pressure and temperature, should, therefore, be designed for the same design pressure and temperature as the street distribution system. Piping and reducing valves should, however, be sized on the basis of the minimum pressures likely to be encountered.

Types. A pressure-reducing valve is basically a variable orifice that is so controlled as to maintain a desired lower fixed downstream pressure regardless of fluctuations of inlet pressure and flow. The degree to which the valve can perform this function is largely dependent upon the limitations of the particular valve design. The factors involved are generally a matter of differential friction, inertia, and unbalanced force.

The simplest form of reducing valve is one in which an equalizing line from the downstream side of the valve is connected to a diaphragm chamber below the valve. The valve spindle rests upon the diaphragm (with a suitable bearing plate). The valve is held open by loading the diaphragm with a coil spring or by means of a lever and weights. The degree of loading is dependent upon the desired delivery pressure and the design characteristics of the valve. Such a valve is known as a normally open valve. The downstream pressure on the diaphragm chamber opposes the loading on the diaphragm and, within the limits of the valve capacity and design, adjusts the valve position to maintain the downstream pressure regardless of variations in demand.

Reducing valves may be single seated or double seated. A single-seated valve will close tightly under zero load, is suitable for dead-end service, and is often referred to as a dead-end valve. A double-seated reducing valve has two seats and two valves arranged in tandem on a single spindle. Steam enters between the two valves, tending to open one and close the other. Since the seats are of equal or nearly equal areas, the valves tend to balance each other and the work of the diaphragm to close the valve is reduced. Double-seated valves are frequently called balanced valves. Since the unbalanced force in the valve interior is virtually eliminated, a double-seated valve will, all other design factors being equal, give closer regulation than a single-seated valve. Double-seated valves will not close tightly and are not good for dead-end service. Most double-seated valves have a leakage of 2 or 3 per cent, and this increases with the life of the valve. They are, however, good where the minimum load is appreciable and the valve is not required to close completely as on dead-end service.

Weight- and lever-loaded reducing valves respond more quickly to changes in load than do spring-loaded valves. They are desirable where it is required to

change the operating pressure frequently as is the case in some direct heating systems. Weight- and lever-loaded valves are available which can be set to deliver steam at from several inches of vacuum to several pounds pressure. Weight- and lever-loaded reducing valves are not generally used above 15 psig delivery pressure owing to the limitation of the excessive weights required, although a few manufacturers do list them for higher pressures.

Spring-loaded valves cannot be set to deliver steam at vacuum pressures, and 1 to 2 psig is about the lower limit. The pressure settings are not so easily changed, but where it is not necessary or desirable to change delivery pressures, this is not a factor. There will be some variation in the delivery pressure as the compression or the loading spring varies somewhat with the opening and closing of the valve. Spring-loaded valves can be set to deliver reduced pressures up to close to the inlet pressures except that, when the initial pressures are in the vicinity of 100 to 150 psig, a reduction to about 75 per cent is the practical limit except in the smaller sizes. This is due to the limitations imposed by excessively heavy springs and consequent heavier valve construction, which results in a cumbersome and expensive valve.

The operating force is the downstream or reduced pressure, and this, to position the valve properly, must overcome friction, inertia, and unbalanced force. Hence, it is evident that there must be some variation in delivery pressures, particularly where the reduction in pressures is considerable and the valve is single seated. Where closer pressure regulation is required, it is necessary to use a pilot-operated reducing valve. This may be accomplished by adding a pipe connection from the inlet side of the reducing valve to the diaphragm with a small fixed orifice in it. A pilot valve with higher capacity than the fixed orifice is installed in the low-pressure equalizer and is controlled from the downstream pressure beyond the reducing valve. The pilot valves generally used are of the self-contained type, that is, they contain an internal downstream equalizer connection. The pilot valve, in effect, responsive to small variations in reduced or downstream pressures, harnesses the full differential across the main valve to control the positioning of the main valve.

This arrangement provides much closer regulation of the main valve and is very effective if the differential across the valve is appreciable. Where the required delivery pressure exceeds about 70 per cent of the inlet pressure and the initial pressures are in excess of 100 psig, there are encountered problems of stiff springs, higher unbalanced forces, and poorer regulation. This is the limit of the self-contained pilot, and if the limit is exceeded the free blow pilot can be utilized. This is a pilot with a diaphragm connected to the downstream side of the main valve and discharging to a vented receiver or a line at or near atmospheric pressure. This provides a very sensitive regulation, since practically the full line pressure is now available to position the main valve and a very light spring can be used to load the diaphragm, as its principal function now is only to assure positive motion.

The reducing valves that have been considered thus far are direct-acting, normally open valves which are closed against the pressure on the inlet side of the valve by pressure under the diaphragm. There is another large class of pilot-operated reducing valves that are reverse-acting, normally closed valves which open against the pressure on the inlet side of the valve by the application of pressure under the diaphragm. The pilot valve is so arranged that it feeds high-pressure steam from the inlet side of the valve to the diaphragm to open the valve wider with an increasing load and on a decreasing load throttles the flow of high-pressure steam to the diaphragm. A bleed line with a small orifice connects the diaphragm to the low-pressure side of the valve, and this together with the pilot valve provides the necessary means to position the main valve.

Another variation is the pilot-operated piston valve. This reducing valve utilizes a piston, rather than a diaphragm, to position the inner valve. Its operation is otherwise the same as the pilot-operated diaphragm valves described.

When the differential across the main valve is small or where extremely close regulation is required, auxiliary operation is employed. In this, an auxiliary medium under pressure such as air, water, or light oil is used to position the main valve and is controlled by a pilot valve connected to the downstream side of the reducing valve. As a matter of safety, it is preferable that auxiliary-operated reducing valves be reverse acting, as the valve will close in the event of a failure in pressure of the auxiliary medium.

There are a great many variations in the design of pilot-operated reducing valves. Some have pressure on one side of the diaphragm; others have pressure on both sides. Both direct- and reverse-acting valves may have either direct- or reverse-acting pilots. Some valves are completely self-contained, while others are equipped with outside pilots.

For a given diameter and travel, a flat seated or a beveled inner valve will give the highest capacity. However, the regulation at very low loads will not be good. With the rapid increase in the use of thermostatically controlled equipment the need for reducing valves that will provide good regulation at very light loads has increased. Reducing valves are available having tapered inner valves or V-ported inner valves which afford good regulation at very light loads. A number of designers use two valves in parallel on large jobs, the smaller one usually being sized for 30 to 40 per cent of the load and the other one for the balance of the load. The smaller reducing valve is set to deliver a slightly higher pressure, and so at light loads, its delivery pressure holds the larger valve closed. When the load increases to its maximum capacity the small valve is wide open. Any increment in load will cause the delivery pressure to drop slightly, and the large valve will open and maintain its set pressure.

With most pilot-operated reducing valves, dual-purpose service is possible. Many valve manufacturers make a thermostatic pilot for controlling the temperature of hot-water heaters, converters, and process work. The thermostatic bulb is placed in the medium to be heated, and the thermostatic pilot is installed in the reducing-valve equalizer adjacent to the pressure pilot. When the bulb calls for heat, the thermostatic pilot opens and steam flows at the pressure set on the pressure pilot valve. When the demand is satisfied, the temperature pilot closes and the reducing valve shuts tightly. Another application is to install a solenoid valve adjacent to the pressure pilot. This will function to turn steam on and off and may be actuated by a remotely located switch, timing device, room thermostat, or other type of control.

Installation. Reducing valves are usually installed with three-valve bypasses. Reverse-acting reducing valves will, in the event of diaphragm failure, close (fail safe). Direct-acting valves controlled by reverse-acting pilots will also close if the pilot diaphragm fails. Such valves should always be installed with bypasses so that service can be maintained in the event of diaphragm failure until repairs can be made and the reducing valve can be put back in service.

Reducing valves should be sized on the basis of the steam load and expected inlet and required outlet pressures. The valves should not be oversized. Oversized reducing valves are responsible for more reducing-valve problems than any other factor. Inlet and outlet piping should be sized on the basis of allowable velocities for the pressure and type of service.

Where possible, it is best to have 8 to 10 ft of straight pipe between reducing valves and on pipe sizes over 6 in., a minimum of 20 pipe diameters. The downstream side of reducing valves should have a minimum of 6 diameters of straight

pipe. Where possible, turns should be avoided, but if necessary, they should be made with long-radius fittings or bends.

Noise is sometimes a factor. This is not surprising when one considers the amount of throttling a reducing valve has to do to reduce the pressure from, say, 130 to 45 psi or from 45 to 3 psi and the consequent increase in specific volume and the necessary high velocities over the valve seats to effect these changes. A sudden increase in pipe size on the downstream side of a reducing valve should be avoided. The pipe wall should slope in the ratio of 1 to 10 after the reducing valve to permit the steam to fill the entire pipe diameter and avoid jetting into the center of a large pipe with its consequent propensity to transmit or to create noise. Such fittings are not commercially available but can be made by a good welder. In view of present high labor costs, it might be well to consider the use of a silencer on the outlet side of the reducing valve. Several types are available and are sold under various proprietary names. Their function is to take the steam from the reducing-valve outlet and deliver it to the larger down-stream pipe in a solid stream of near uniform velocity across the diameter of the pipe instead of as a jet in the center. The reducing-valve equalizing pipe should always be tied into the larger downstream pipe, preferably at least 5 or 6 ft beyond the valve. The use of a solid type of insulation on the piping and flanges will do much to muffle the sound. Noise may be transmitted to the building structure through the pipe supports. This may be stopped by supporting the pipe from the floor (where the floor rests on earth or rock) or by the use of isolation hangers for a distance of about 20 feet upstream and 40 to 50 ft downstream of the reducing valves. Isolation hangers have pads of rubber or cork or springs or a combination of these or similar materials. Isolation hangers are available in various sizes, and the right size must be selected for each support point to assure adequate load-carrying ability with the retention of correct resiliency to isolate pipeline vibration properly from the building structure. The steam-supply piping should not be in contact with other piping, duct work, or building walls. Sound transmission may be reduced further by covering the walls and ceiling with soundproof material. There have been cases where noise was traceable to a loose inner valve or seat in a reducing valve, to a defective pilot valve, or to a loose gate in a stop valve. These points should be checked. Some designers specify silencers and isolation supports on the basis that their initial installed cost is small whereas, if they should be required later, their cost would be very much higher.

Although the ASA Code permits the use of a single reduction from high to low pressure if the proper protective devices are installed, the use of a two-stage reduction results in quieter operation and affords closer pressure regulation. Mechanical equipment areas are generally located by the architects. However, where the designer can make a choice, reducing valves should not be located where, if noise should develop, a nuisance could be created.

THE CUSTOMER'S STEAM SYSTEM

District steam customers vary from small users such as private residences using steam for heating and domestic hot water and small valet shops using steam for pressing and ironing machines to large users such as luxury apartment buildings, skyscraper office buildings, hotels, hospitals, and industrial buildings. Steam is used for such purposes as space heating, ventilating, heating domestic water, operation of steam absorption and turbocompressor refrigeration units for air conditioning, cooking, sterilization, sidewalk snow-melting installations, swimming-pool heaters, Turkish baths, engines, and pumps. Steam is also used for a wide variety of process uses such as laundries, heating dye vats, silver plating, and the manufacture of wood products, hats, and phonograph records.

The reducing valves, hot-water heaters, hot-water converters, refrigeration, and similar equipment may be located in the basement at or close to the point of service entrance.   Long, low buildings may have two or more mechanical equipment rooms supplied with steam at full street pressure or at an intermediate pressure, the final reduction being made in the mechanical equipment rooms.   Large housing developments generally follow a similar plan and have a mechanical equipment room in each building or one for each two or three buildings.   Figure 16 shows a typical steam-supply system for a housing development.   Tower buildings frequently have mechanical equipment rooms in the basement and at various levels at

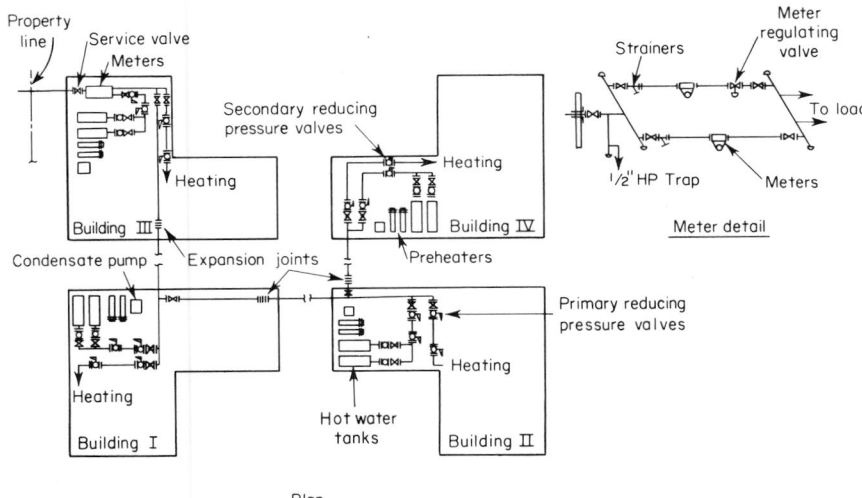

FIG. 16. Typical steam distribution system for a housing development.

intervals of 10 to 20 floors.   Figure 17 shows a typical tower office building.   There is no general rule.   Each building must be analyzed on the basis of the economic factors involved such as initial cost of the installation, value of rentable area, and operating and maintenance costs.   Some tower buildings have all heating and air-conditioning equipment below grade, while others have all equipment on the top floor.

All condensate is not returned to a common point in many installations.   Some steam may be utilized for open steam jets.   In air-conditioning installations, condensate is frequently utilized for cooling-tower make-up.

Some steam usages are steady, while others fluctuate to a greater or lesser degree. There is, in practically all cases, a wide range between daytime peak and night-time minimum uses.

Metering

The selection of the type of meter or meters for recording the steam used by the district steam customer will depend upon the following factors:

1. Character of customer's steam load, i.e., steady, fluctuating, open jet steam use if any, ratio between maximum and minimum steam loads, cleanliness of system (if condensate meter is contemplated)

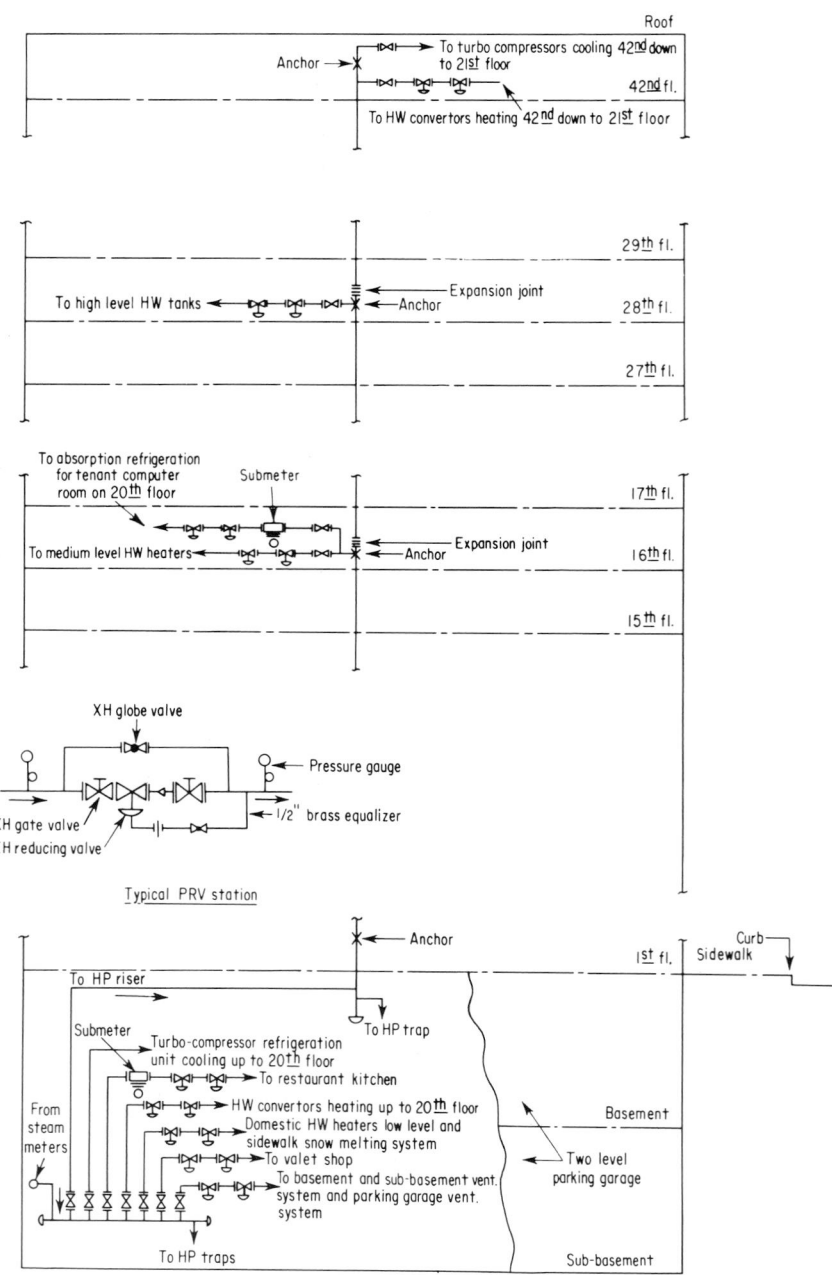

FIG. 17. Typical high-pressure steam-distribution system in a tower office building.

2. Accuracy required

3. Type of record required, i.e., recording and/or totalizing

4. Initial cost

5. Installation cost

6. Maintenance cost (cost required to maintain meter accuracy)

7. Operation cost (cost to read meters, change charts if recording)

8. Cost of calibration and periodic recalibration as may be required by local regulatory body.

9. Stock of repairs parts required

10. Service facilities of manufacturer and degree to which the district steam company's personnel may be able to do its own meter servicing

Steam meters may be divided into two general categories: condensate meters and flowmeters.

The condensate meter measures the condensate from the steam-using equipment. Because of its simplicity of design, ease of testing, accuracy at all loads, and low cost, it is a popular meter for use where all condensate is returned to a common point for metering. Condensate meters are made for either vacuum or atmospheric operation. They cannot be operated under pressure. In an old steam system which may contain considerable sediment, somewhat more frequent servicing may be required to keep the meter buckets clean and maintain meter accuracy. Old steam systems may have leaks, particularly if return lines are buried. Such leaks will, of course, result in a loss of registration. Condensate meters are an excellent means of metering steam where there is assurance that all steam condensate will be returned to the meter. Condensate meters are available in capacities from 250 to 12,000 lb/hr. For higher capacities two or more may be connected in parallel.

Steam flowmeters are divided into three classes; area, velocity, and head. Unlike condensate meters, steam flowmeters are not accurate at all loads, generally losing record at flows below 10 to 20 per cent of rated capacity. This is due to the square-root relationship inherent in the design of all flowmeters.

In the *area meter* a weighted tapered plug mounted on a vertical spindle fits into a seat. Steam enters under the seat and causes the plug to rise. As the flow increases, the annular orifice increases in area. An arm attached to the spindle actuates the register, which gives both an integrator record and a chart record of the steam flow. An example of this type of meter is the St. John flowmeter which has been widely used in district-heating work. This meter requires little floor space and, unlike the head and velocity meters, does not require straight pipe before and after the meter. The meter is, however, heavy, has large flanged joints, and is consequently somewhat expensive to change for periodic retest or replacement. The meter register is operated by a mechanical clock. The meter operates over a range of pressures from 10 to 250 psi.

An example of the *velocity meter* is the Builders-Providence Shuntflo Meter (Fig. 18). This meter contains an orifice plate with a shunt circuit around it. A turbine rotor on a vertical shaft is located in the shunt circuit. Near the lower end of the shaft is a damping fan which turns in a chamber filled with water. The speed of the rotor shaft is reduced through suitable gearing to drive a magnetic coupling which operates the counter. No external source of power is required. The meters are made for direct installation in 1-, 2-, 3-, and 4-in. lines and for bypass installation around an orifice in larger lines. The counter is calibrated in units of total flow through both the orifice and the nozzle. Each size meter is available in several rated capacities according to the size of the orifice used. The meter will accurately handle temporary overloads up to 150 per cent of rated capacity. The meter is suitable for pressures ranging from 5 to 300 psi. Auxiliary equipment available includes demand contacts built into the counters, flow-limiting

orifices for installation in the line downstream of the meter, an electrically operated remote totalizer, an electrically operated remote chart recorder and flowrate indicator, and a pressure-compensated counter.

Head meters consist of a primary element and a secondary element.   The primary element may be any device which produces a differential pressure varying as the square of the flow.   Those most generally used in steam metering are the flow

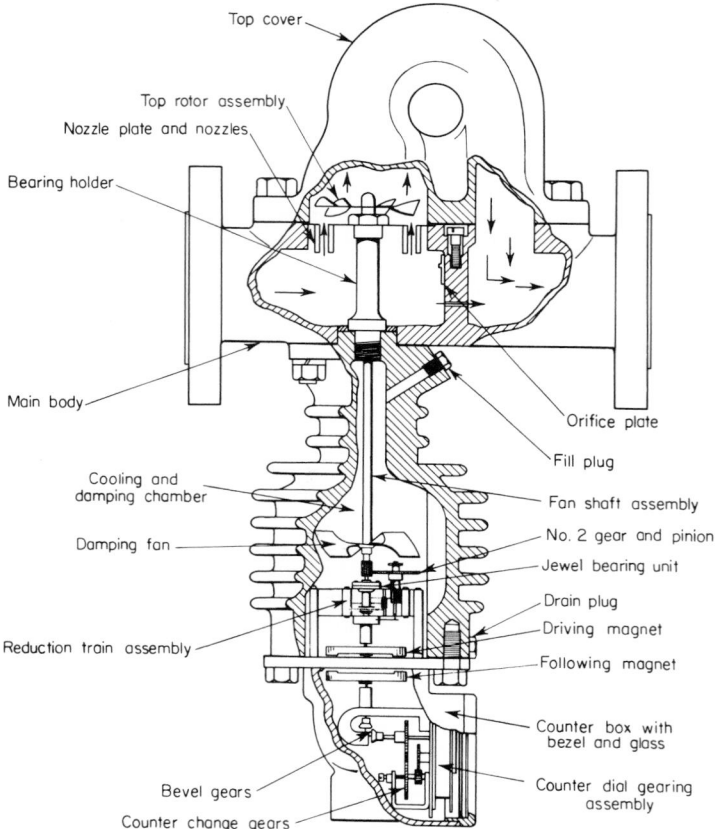

Fig. 18. Velocity meter of the type used in district heating steam systems.

nozzle and the thin-plate orifice.   The flow nozzle has a higher capacity than the thin-plate orifice for a given differential.   The thin-plate orifice has the advantages of lower cost and ease of installation.

The secondary element is essentially a manometer suitably connected to the primary element so that the amount of the differential pressure or its square root is integrated or recorded.   Differential head meters are widely used in fluid metering and are available from a number of different manufacturers.

Head meters are available in a wide range of designs, including mechanically and electrically operated integrators and recorders, electrical transmitters and receivers which permit the location of the integrators and recorders at remote points.   Some are available with pressure compensation.

All velocity and head meters require straight pipe upstream and downstream. The length of straight pipe required for a meter of given capacity depends upon the configuration of the piping ahead of the upstream straight section. The length of the upstream section may be reduced by the installation of straightening vanes.

With all types of meters the manufacturer's installation instructions should be carefully followed.

A flowmeter that is accurate from its full capacity (100 per cent) down to 20 per cent, is said to have a 5-to-1 range. Not infrequently steam flow must be measured over a far greater range. Where it is possible to segregate the various steam loads such as space heating, domestic hot water, etc., and the range of the various loads is within the range of the meters being used, the simplest solution is to install a separate meter for each load. Should this not be feasible, some form of sequential metering is required.

In sequential metering, two or more meters are connected in parallel and some suitable device is used to cut in or out the second and succeeding meters in sequence as required to supply the load. If the flowmeters being used have a range of 5 to 1 and they are so sized that at no time will the use on any meter be below 20 per cent of its capacity, then the range for two meters in parallel will be 20 to 1 and for three meters in parallel it will be 80 to 1. The addition of a fourth meter will result in a range of 320 to 1. While such a range will probably not be needed for accuracy, a four- or more meter installation might be needed for capacity when one considers the practical limitations in the way of pipe and valve sizes. The first or base load meter is so sized that the minimum load will not be less than 20 per cent of its capacity. The second meter may be up to three times the capacity of the first, and the third meter may be up to three times the combined capacity of the first and second meters.

The meters may be electrically sequenced as follows: The first or base load meter is equipped with two sets of contacts, one set of which may be set to close at about 90 to 95 per cent of full load and the other set to open at 20 per cent load. If this is not feasible, then a sensitive differential pressure switch, similarly equipped, is connected across the meter. This will measure the differential across the meter and consequentially the percentage flow of its rated capacity.

In operation, when the flow through the first meter reaches approximately 90 to 95 per cent of its rated capacity, the contacts close a relay with a holding circuit. The relay opens a motorized valve downstream of the second meter, thus bringing the meter into service. The first meter drops to 23 per cent of its capacity, and the second meter rises to 23 per cent of its capacity. As the flow increases, both meter flows increase in the same percentage flow of their capacities. If a third meter is required, the second meter is similarly equipped. When the flow on the first two meters reaches approximately 90 per cent, the second meter, through its contacts and holding relay, actuates the motorized valve cutting in the third meter. At this point the load on the third meter rises to approximately 23 per cent and that on the first two meters drop to that amount. On a decreasing load, the lower contact on the next to the last meter breaks the holding circuit and the motorized valve cuts out the last meter.

In order to assure an equal division of load between the several meters, balancing orifices are required downstream of the meters. Solenoid-operated diaphragm valves may be used in place of motor-operated valves.

Over the past several years considerable time and effort have gone into developing various methods for sequencing meters by mechanical means. This is particularly desirable where the meters being used are mechanically operated and the additional electrical installation can be dispensed with. One such means is shown in Fig. 19. Here two velocity meters are operated in parallel. The base load meter has a

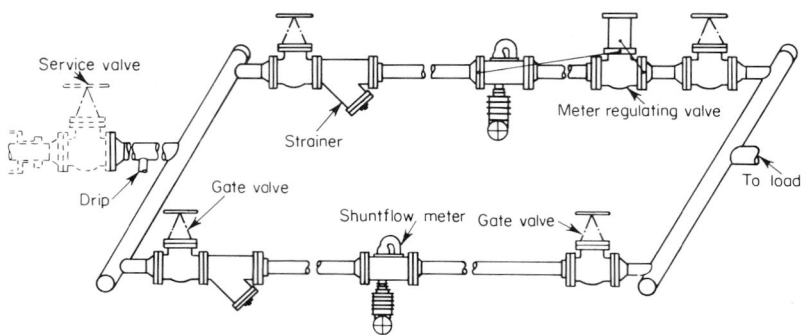

FIG. 19. Velocity-meter installation with mechanically operated regulating valve.

limiting orifice downstream which is sized to limit the flow to the capacity of the meter at which point it produces a drop in pressure on the meter outlet header of 5 to 10 psi. A diaphragm-type reducing valve downstream of the second meter is employed as a meter-regulating valve. This valve is set for 5 to 10 psi below the pressure the base load meter delivers while the load is within its range. When the load is in excess of the capacity of the base load meter, the meter outlet-header pressure drops to the setting of the meter regulating valve and it opens, thereby cutting in the second meter. The first meter remains at full load, and the excess is handled by the second meter. This type of sequential operation is, however, suited only to cases where an increment in flow beyond the capacity of the first meter is sufficient to carry the load on the second meter to within the range of its recording capability.

The difficulties in the way of developing a mechanical means of sequencing meters so that the second and succeeding meters are either all on or all off and flow through them is always within their accurate range can be appreciated when one considers the effect of the square-root relationship involved. To sequence meters with a 5-to-1 range accurately, the sequencing device must have a range of 25 to 1.

At this writing, one reducing-valve manufacturer has produced a piston-type sequencing valve which controlled a two-meter installation. Several installations are in operation. It is hoped that further development work will enable this valve to be applied to three- or four-meter installations.

At least two other reducing-valve manufacturers are working on the development of sequential pilots for application to diaphragm valves so that a similar type of operation (i.e., fully open or fully closed) can be obtained.

CONDENSATE DISPOSAL

Most local regulatory authorities place a limit on the temperature at which condensate may be discharged to the sewer. This makes it necessary to employ some means of cooling. The best way to do this is by preheating the cold-water supply to the domestic hot-water heaters and thus reducing the amount of steam required for water heating. A typical installation is shown in Fig. 20. Since the peaks of other steam-using equipment do not always coincide with those of the hot-water heaters, storage-type economizers are preferable where there is a steady flow of condensate to the economizer tubes. Where it is necessary to pump the condensate through the economizer with intermittent flow, the tube-and-shell

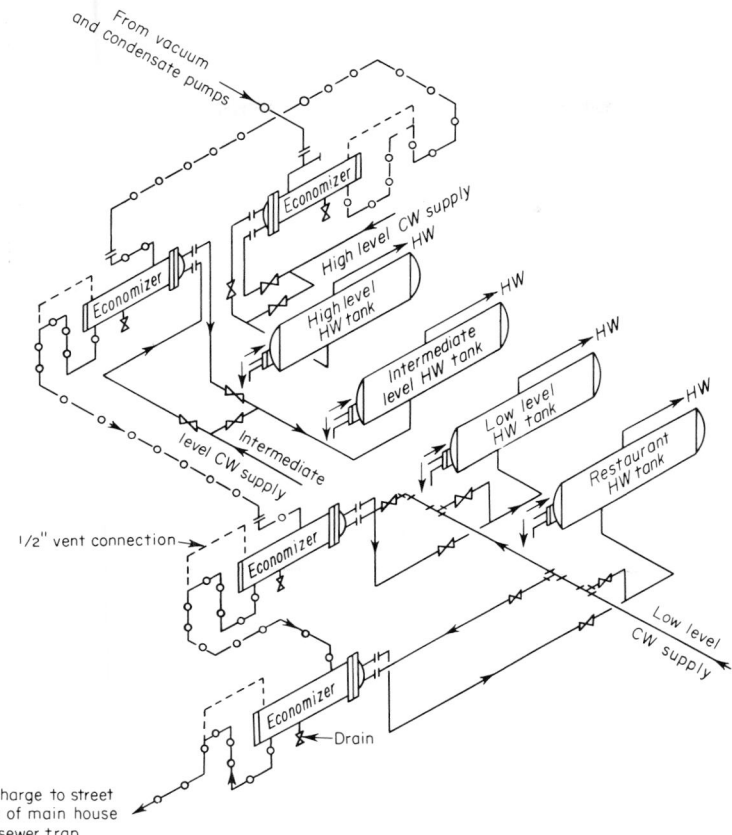

Fig. 20. Typical installation of economizers in a large building.

type economizer is preferable with the condensate going to the shell. The proper amount of heat-transfer surface should be used, and the piping arranged for counter-flow of water and condensate.

The author acknowledges the assistance of Mr. M. L. Schneider, Engineer, Structural Engineering Bureau, Mechanical Engineering Department, Consolidated Edison Company, Inc. for the sections on Reducing and Relief Valves, The Customer's Steam System, and Metering.

REFERENCES

1. District Heating Piping Systems and Fabrication Details, American Standard Code for Pressure Piping, American Standards Association B31.1, secs. 4 and 6.
2. National District Heating Association, "District Heating Handbook," 3d ed.
3. Graphical Solution of Unwin's Formula. Chart published by the National District Heating Association.

16

NUCLEAR-SYSTEMS PIPING

J. A. Klapper*

INTRODUCTION

The use of nuclear reactors for electric power generation on a commercial scale has progressed to the level where these types of energy sources are recognized as basic prime movers and are approaching attractive economic operation. Under the sponsorship of private utilities, public utilities, and the United States Atomic Energy Commission, large-scale nuclear-reactor systems of various types proved in the laboratory or by pilot model are being operated or will be operated shortly for the generation and sale of electricity.

Commercial or power-generating nuclear-reactor systems may change in details of design; however, they all have but one purpose: to transfer sufficient heat from a uranium or fissionable fueled core to a primary fluid which will transfer heat to a secondary fluid to raise steam or to transfer heat directly to the primary fluid which will itself become steam.

The development of nuclear heat sources for utility service progressed logically step by step from the research or test reactor. The research or test reactor systems were the first sources of nuclear heat and were designed and operated for several purposes among which are the testing of materials in the presence of radiation and the demonstration of a conceptual design and research into basic physics. Regardless of the test reactor purpose, it is a requirement that the energy released in the core be removed by a primary heat-removal system that transfers heat to a secondary circulating system. Initially primary-coolant or core-fluid energy was disposed of in specially designed heat exchangers or cooling towers. However, it was apparent

* Manager, Materials Engineering, Ebasco Services, Inc., New York, N.Y.

from the date of the first continuous fission process that a more practical and efficient use of the heat transferred to the primary fluid was in the generation of steam to turn a turbine-generator set for the production of electricity or to implement a chemical process.

The engineering and obvious requirement for heat transfer from the reactor core was to prevent melting down of the uranium fuels and, therefore, to permit long-time and continuous core operation.

The reactor core, or fuel complex, is housed in a vessel which can operate, depending on the reactor type, at pressures as low as atmospheric and as high as 2,500 psi and at primary coolant temperatures ranging from ambient to 1000 F. Therefore, depending upon the proximity to populated areas or the terrain of the reactor site, either the core vessel and radioactive or primary piping can be contained in a large-diameter second vessel, generally called a vapor container or reactor containment vessel, or the system can be placed in concrete shielding vaults below the earth's surface. The problem of safety is centered around the fact that a reactor which has been operating for a long period accumulates large amounts of fission products which can exceed 1×10^7 curies of radioactivity. If these products were uncontrollably released to the atmosphere, they would represent a serious health hazard. Consequently, some type of containment is believed by many to be necessary even though the reactor core vessel and primary piping systems are designed to prevent release of fission products.

The secondary, or power-generation, systems for a nuclear-power reactor complex are similar and in nearly all instances identical with turbine-generator designs employed in coal-, oil-, or gas-fired steam-generating plants. The hazards attendant upon the turbine-generator systems are obviously similar to those in other steam-generating plants, with the result that turbine-generator systems can be located outside the reactor containment vessel or vault.

DEFINITION AND PURPOSE OF NUCLEAR PIPING

As in any steam-generation or heat-transfer concept, many piping systems are required for operation. Nuclear reactor plants, whether for power generation or research purposes, are made up of all or some of the following basic systems:

1. Primary coolant
2. Secondary coolant
3. Instrumentation
4. Demineralization
5. Pressurizing
6. Condensate and feed
7. Radioactive waste

These systems can be classified further into two groups:

1. Nuclear piping, that is, piping which contains and circulates a radioactive fluid
2. Conventional piping, that is, steam and associated piping which does not fall in category 1

This chapter deals only with nuclear piping, its criteria and requirements based on minimum human and equipment safety, and good manufacturing, fabrication, engineering, and inspection practices that comply with minimum safety standards.

The adequacy of any engineered piping system depends upon a definition of the system that states the basis on which designs will be formulated, hazards evaluated, quality control and inspection established, and specific fabrication techniques employed. The Committee on Nuclear Piping of the Code for Pressure Piping, ASA B31.1, published the following definition of nuclear piping in April, 1960,

under ASA B31.1 Nuclear Code Case N-1 "General Requirements for Nuclear Power-plant Piping."

Nuclear piping[1,2,*] *is defined* as that piping designed to contain a fluid whose loss from the system could result in a radiation hazard, either to the present personnel or to the general public.

Power piping coming within the scope of this definition shall be designed and constructed in accordance with Section 1, B31.1-1955, supplemented by the rulings and case interpretations identified by the prefix "N."

Nuclear Code Case N-1 states further:[1,2]

Conventional steam and associated service piping, not falling under the definition of nuclear piping shall be designed and constructed in accordance with Section 1 B31.1-1955.

In conclusion, Nuclear Code Case N-1 points out that,

Specific exclusions and special requirements will be covered in Nuclear Cases as they are developed. These will be issued and numbered with the prefix "N."

As of this writing, Sectional Committee B31 of the American Standards Association is rewriting the Code for Pressure Piping. In addition the Nuclear Piping Committee of B31 is writing a Code for Nuclear Piping. Consequently, any reference in this chapter to Section 1 B31.1-1955 of the Power Piping Code must be augmented to conform to the new Codes for Power Piping and Nuclear Piping when they are published.

The Nuclear Piping Committee of ASA Sectional Committee B31.1 comprises professional personnel from all phases of the nuclear field, manufacturers, designers, operators, and inspectors. The composite experience of the Committee and that of operational nuclear systems is reported in this chapter as a guide to the design, fabrication, construction, inspection, and operation of nuclear piping systems until such time as more definitive experience and knowledge are obtained.

There are many concepts, opinions, and judgements concerning engineering, fabrication, and inspection of nuclear piping. These range from a blind increase in safety factors to maintaining presently used safety factors for power piping in conjunction with overdesigned containment structures. Neither of the foregoing are logical means toward achieving maximum economic reliability. Certainly, nuclear piping introduces new problems: hazardous fluids capable of being distributed over wide areas with lengthy radioactive potency, deterioration of materials such as corrosion by high-temperature liquid metals in the presence of oxygen or other heat-transfer fluids, and potential structural damage attributed directly to nuclear radiation.

Nuclear piping, by definition, requires the use of components and methods of joining that develop maximum economic reliability and safety. This classification is evidenced in nearly all nuclear installations to date. Nonhazardous and associated piping within the nuclear system whose failure could endanger nuclear piping would also justify the reliability and safety expected of nuclear piping.[3]

Piping located outside the nuclear system is directly comparable to fossil-fuel steam-raising systems involving combustion. These lines, though requiring long-life expectancy and reliable and economic operation, carry the same criteria for nuclear piping except for hazards evaluation.

Nuclear piping, whether in power or test reactors, must demonstrate complete safety to the public. Consequently, there is a great deal to be gained from thorough and effective engineering, respect for systems operations containing radioactive fluids, and reducing calculated risks to the limit of engineering and scientific knowledge and skills.

* Superscript numerals refer to the References contained at end of this chapter.

NUCLEAR-REACTOR TYPES

Each reactor type and its attendant piping have a specific function or combination of functions. Consequently, design, manufacture, erection, quality control, and operation must be stipulated carefully so that the plant will perform as desired. There are presently two categories of nuclear-reactor systems: power-generation reactors and research and test reactors. The industry today looks upon these system categories as power reactors or research reactors.

Power Reactors

Power reactors are of either the stationary or mobile types. The *stationary power reactor* is a land-based complex whose main function is generating steam to turn a turbine-generator set to produce electricity. The *mobile reactor* is best exemplified by the shipboard or propulsion type of system now operating in nuclear submarines, an aircraft carrier, and the cargo-passenger ship N. S. Savannah.

There are many power reactors now in operation or approaching operational status. The power-reactor types are:

1. *Light-water-moderated and -cooled:*
   a. Pressurized-water reactors (PWR)
   b. Boiling-water reactors (BWR)
2. *Liquid-metal-cooled:*
   a. Sodium-cooled thermal reactors (SGR)
   b. Sodium-cooled fast reactors
3. *Organic-liquid-cooled* (OMR)
4. *Gas-cooled graphite-moderated* (GCR)

Light-Water-Moderated and -Cooled Reactor Systems. Light-water-moderated and -cooled reactor systems, such as the PWR, were the first type of power-generation units to produce electricity and to drive propulsion units. As shown above, the light-water systems are classified broadly into two types: the pressurized-water type and the boiling-water type. Each of these has its particular advantages and disadvantages, and both systems are now operating in the United States.

The pressurized-water reactor employs high-purity pressurized water to remove fission heat from the core. Pressurization is maintained to prevent boiling in the core. The water in the primary nuclear piping system then exchanges heat to generate steam in low-pressure boilers.

The boiling-water reactor also employs high-purity water to remove fission heat from the core. Steam is permitted to form in the core; however, the pressure in the boiling-water system can be lower than in the pressurized-water reactor of the same output.

Pressurized-water Reactors. This system is typified by the operating unit known as the Yankee Atomic Electric Company's Pressurized-water Reactor.[4–6] The plant is located in Rowe, Mass., and is sponsored by 10 utility companies in the New England area.

The net electrical-generation capacity is now 136 mw based on 485 thermal mw in the reactor. The primary loop operates at 2,000 psig and at an average temperature of 514 F. The secondary, or steam, system produces steam at 1,840,000 lb/hr with a turbine inlet steam temperature of 460 F at 453 psig. The feedwater temperature entering the secondary-system steam generators is 335 F.

The nuclear portion of the plant (see Fig. 1) consists basically of the nuclear reactor and four loops, each having a vertical steam generator and main coolant

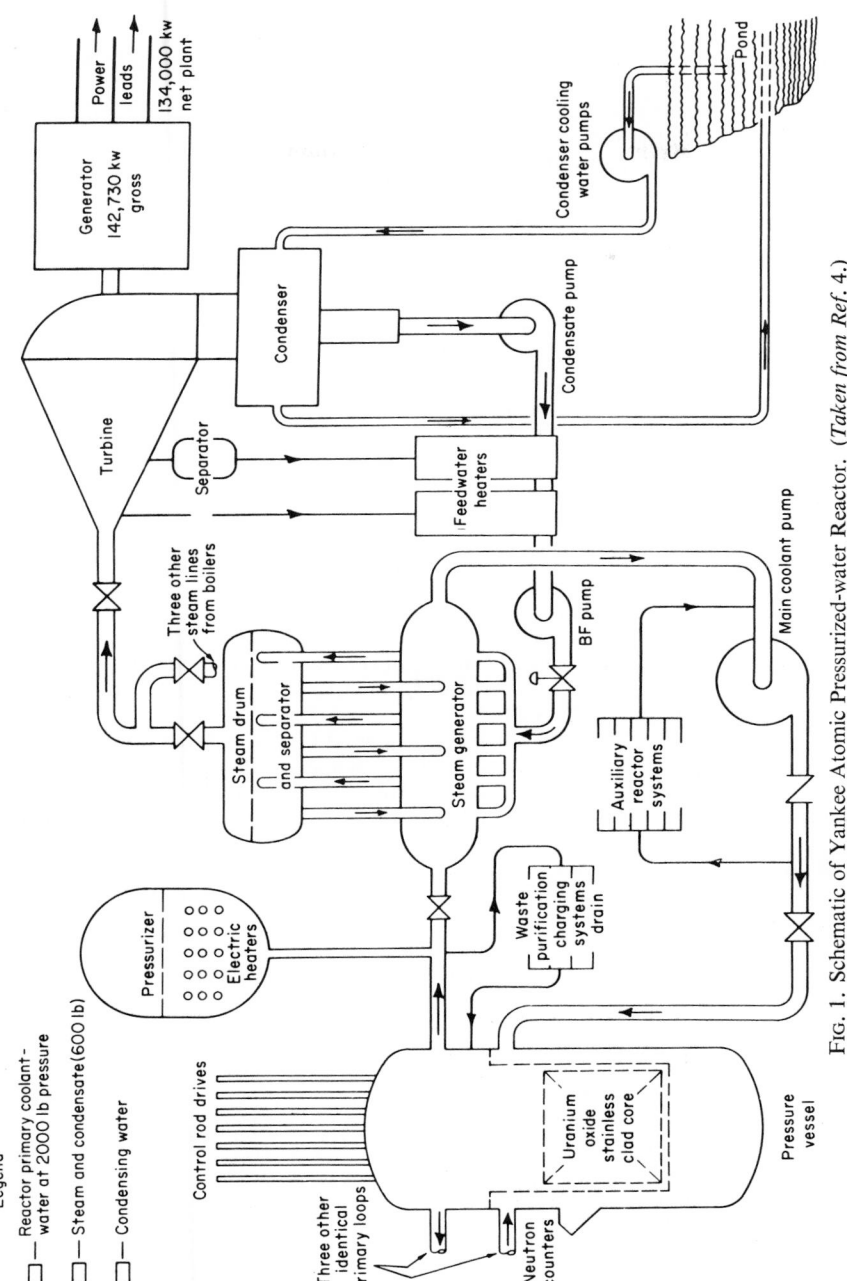

FIG. 1. Schematic of Yankee Atomic Pressurized-water Reactor. (*Taken from Ref. 4.*)

16–5

pump. Surrounding the entire pressurized nuclear steam-generating system is a steel containment sphere 125 ft in diameter. The turbine-generator system is housed in a building adjacent to the sphere.

The *primary piping* loops contain radioactive high-purity water and join the core pressure vessel to the vertical steam generators. The materials used are Type 304 stainless-steel pipe and seamless hollow-forged pipe complying with ASTM A376.

Table I.   Basic Characteristics of the Yankee PWR

1. Commenced operation . . . . November, 1960
2. Reactor thermal power . . . . 485 mw
3. Average core heat flux at rated power . . . . . . 86,300 Btu/(hr sq ft)
4. Total flow of primary water . . . . 3.78 × 10⁷ lb/hr
5. Total steam flow . . . . 1.84 × 10⁶ lb/hr
6. Reactor operating temperature . . . . 514 F average
7. Feedwater temperature . . . . 335 F
8. Core inlet temperature . . . . 495 F
9. Turbine-generator design . . . . 160 mw
10. Net electrical capability . . . . 140 mw
11. Steam temperature . . . . 460 F
12. Steam pressure . . . . 453 psig
13. Primary system pressure . . . . 2,000 psig average

The fittings are reported to be made from seamless hollow-forged stainless-steel pipe Type 304 to the requirements of ASTM A182. The valve bodies are cast stainless steel Type 304 conforming to ASTM A351 Grade CF8.

The *steam piping* lines carry nonradioactive steam from the vertical steam generators to the turbine. The pipe material is carbon steel ASTM A106 Grade B. Fittings are seamless-steel, butt-welded ASTM A234 Grade WPB and cast-steel flanged ASTM A216 Grade WCB.

A partial list of pressurized water reactors would include, in addition to that of the Yankee Atomic Electric Company, others such as the Shippingport Pressurized-water Reactor, the Consolidated Edison (New York) Reactor, the Naval Reactors-Submarines, the Naval Reactors-Surface Vessels, the N.S. Savannah Maritime Reactor-Cargo and Passenger Vessel, and the Army Package-Power Reactor.

Boiling-water Reactors. Typical of the commercial boiling-water systems is the Dresden 180-mw net electrical capability unit. The Dresden station is located in Grundy County, Illinois, approximately 47 miles southwest of Chicago.[6,7]

Table 2.   Basic Characteristics of the Dresden BWR

1. Commenced operation . . . . April, 1960
2. Reactor thermal power . . . . 626 mw
3. Average core heat flux at rated power . . . . 107,500 Btu/(hr sq ft)
4. Total flow of primary water through core . . . . . . 2.56 × 10⁷ lb/hr
5. Primary steam flow . . . . 1.41 × 10⁶ lb/hr
6. Secondary steam flow . . . . 1.18 × 10⁶ lb/hr
7. Reactor operating temperature . . . . 546 F average
8. Feedwater temperature . . . . 405 F
9. Core inlet temperature . . . . 505 F
10. Turbine-generator design at 2.5 in. Hg . . . . 192 mw
11. Net electrical capability . . . . 180 mw
12. Steam temperature—primary . . . . 546 F
13. Steam pressure . . . . 500 psig
14. Primary system pressure . . . . 1,000 psig

The dual-cycle design (see Fig. 2) was selected for Dresden as the best solution to fulfill the design objectives. The design pressure of the primary system is 1,250 psig, sufficiently above the 1,000 psig normal operating pressure to allow for

pressure fluctuations arising under various transient operating conditions. The reactor operating temperature is 546 F, and the feedwater temperature is 405 F. The inlet velocity of water to the core is 4.50 fps. The velocity through the core increases to 4.75 fps as the water expands to saturated steam about 5 ft above the core inlet. In the top 4 ft of the core, steam is generated, resulting in an exit quality of about 5 per cent by weight and exit void of about 50 per cent. The average exit-water velocity is about 9 fps. Because of its lower density and viscosity the steam slips upward relative to the water. The estimated difference between the steam and water velocities is about 3 fps.

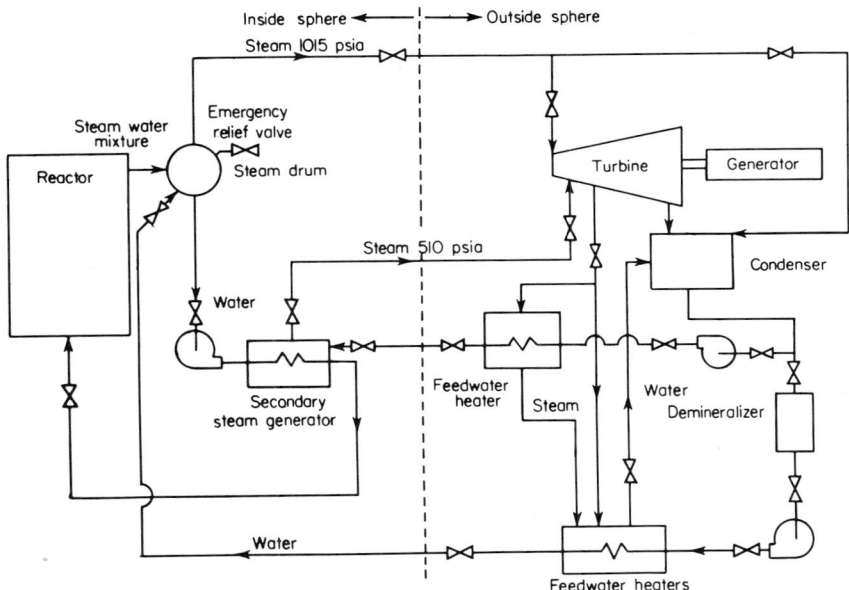

FIG. 2. Schematic of Dresden Dual Cycle Boiling-water Reactor. (*Taken from Ref.* 7.)

The steam-water mixture leaving the reactor flows to the external steam-separating drum mounted high above the reactor vessel. The water from the separator drum is pumped through the secondary steam generators, while the saturated steam flows directly to the turbine. The primary water from the steam separator, in passing through four heat exchangers called secondary steam generators, transfers heat to the secondary-water system, thereby creating secondary steam at a somewhat lower pressure of 500 psig. The secondary steam is admitted to the turbine at the ninth stage.

The reactor system, steam drum, secondary steam generators, and the recirculating pumps are housed in a 190-ft spherical steel containment shell. The turbine generator, condenser feedwater pumps, and demineralizing equipment are in a separate building adjacent to the sphere.

The *primary piping* lines carry radioactive steam and water in the system joining the core vessel to the primary steam generators and the steam-separating drum. The materials used are stainless steel Type 304 complying with ASTM A358 and A376.

The fittings are Type 304 stainless steel of the cast butt-welding type conforming to ASTM A351.

The *steam piping* lines carry only radioactive steam from the steam-separating drum to the turbine. The radioactivity level carried into the turbine by the steam is of such a low level that, after shutdown, the turbine can be opened and serviced if necessary. These materials are identical with those employed in fossil-fuel steam-generator main steam lines and conform to ASTM A335 P22 low-alloy steel (2¼ Cr-1 Mo) seamless pipe or low-alloy steel electric fusion butt-welded pipe ASTM A155 Grade 2¼ Cr.

The secondary steam-generator piping operates at a lower temperature and, consequently, is a carbon steel. The material conforms to ASTM A155, KC 70 killed carbon-steel electric fusion butt-welded pipe or to ASTM A106 Grade B seamless carbon-steel pipe.

A *partial list of boiling-water reactors* would include also the Vallecitos Boiling-water Reactor; the Experimental Boiling-water Reactor; the Dresden Nuclear Power Station; the SENN Boiling-water Reactor, Italy; the Japanese Atomic Energy Research Institute Boiling-water Reactor; and the Rural Cooperative Power Association, Elk River Boiling-water Reactor.

Liquid-metal-cooled Reactors. Liquid-metal-cooled[8] reactor systems are considered extremely attractive power-producing units because liquid metals such as sodium have excellent heat-removal characteristics. These characteristics are accompanied by a low-pressure, high-temperature primary sodium system. Consequently, the sodium primary system can produce higher temperature steam than presently designed and operated water-cooled systems.

Two concepts have emerged for sodium-cooled systems: the Sodium-cooled, Graphite-moderated Thermal Reactor (SGR) and the Sodium-cooled Fast-breeder Reactor (Enrico Fermi).

Sodium-cooled Graphite-moderated Thermal Reactor.[6,9] This system, typified by the Hallum Nuclear Power Facility, is a progressive development of the Sodium-Graphite Reactor Experiment.[10] HNPF is owned by the U.S. Atomic Energy Commission and will be operated by the Consumers Public Power District at their Sheldon Station site 20 miles south of Lincoln, Neb.

Table 3.   Basic Characteristics of the HNPF

1. Commenced operation	1963
2. Reactor thermal power	240 mw
3. Average core heat flux	147,000 Btu/(hr sq ft)
4. Total flow of primary sodium	8.4×10^6 lb/hr
5. Total flow of secondary sodium	8.4×10^6 lb/hr
6. Total steam flow	7.1×10^5 lb/hr
7. Reactor operating temperature	930 F outlet
8. Secondary sodium temperature	886 F outlet
9. Core inlet temperature	596 F
10. Feedwater temperature	305 F
11. Steam temperature	825 F
12. Steam pressure	800 psig
13. Primary system pressure	Atmospheric

The reactor has a nominal thermal power rating of 240 mw. This is the calculated power required to produce a steam flow of 710,000 lb/hr at turbine throttle conditions of 800 psig and 825 F. The calculated net electrical output of the power station under these conditions is 75 mw.

The plant (see Fig. 3) is made up of a radioactive sodium primary system that transfers reactor heat to a secondary nonradioactive sodium system which in turn transfers heat to the steam generators.

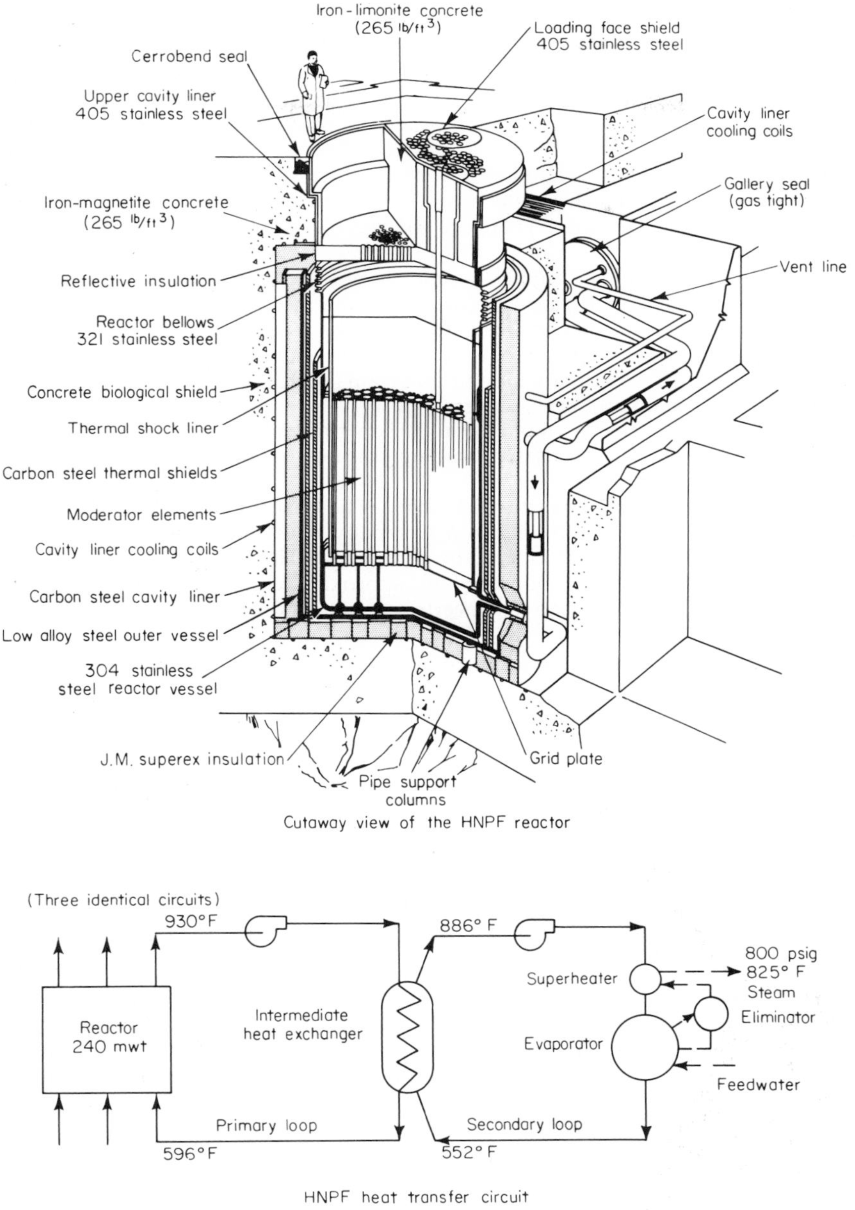

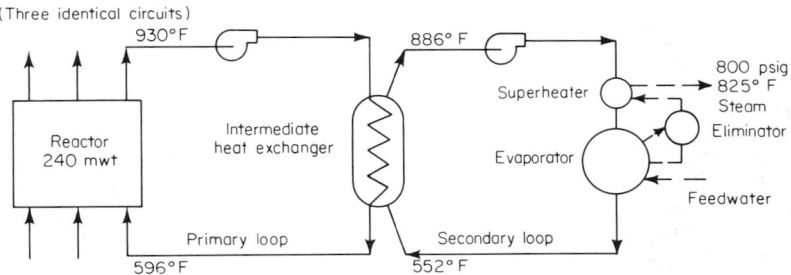

Cutaway view of the HNPF reactor

HNPF heat transfer circuit

FIG. 3. Schematic of Hallum Nuclear Power Facility, sodium-graphite reactor. (*Taken from Ref. 9.*)

Heat transferred from the core to the primary sodium raises its temperature from 596 F at the reactor vessel inlet to 930 F at the outlet above the core, all at essentially atmospheric pressure.   Because of the excellent heat-transfer properties of sodium, the secondary sodium loop is only 44 F cooler than the primary loop.   The secondary sodium will generate and superheat steam in the steam generators which is fed at 800 psig and 825 F into a common header to drive a single, nonreheat turbine generator.

The primary system is not and cannot become pressurized; consequently there is no credible accident which would result in a rapid expulsion of fission products from the reactor enclosure.   Therefore the reactor building is not designed as a pressure containment.

The all-welded primary piping system comprises three loops that transport radioactive sodium from the reactor core to the intermediate heat exchangers.   The pipe materials are Type 304 stainless-steel seamless pipe.

The all-welded intermediate piping system conveys nonradioactive sodium which transfers heat from the primary loop to the steam generators.   The piping materials are of Type 304 stainless-steel seamless pipe.

Type 316 stainless steel is employed in the superheater; $2\frac{1}{4}$ Cr − 1 Mo low-alloy steel in the evaporator, preheater, and superheater outlets and the intermediate heat exchanger.

Sodium-cooled Fast-breeder Reactor. The Enrico Fermi[6],[11–13] power plant in association with the Detroit Edison Company is designed ultimately to breed reactor fuel from fissile material in the core area.   The reactor system is located at Monroe, Mich., and is sponsored by approximately 30 electric power companies and 12 industrial and engineering organizations.

Table 4.   Basic Characteristics of the Enrico Fermi Reactor

1. Commenced operation 1964
2. Reactor thermal power .... 300 mw
3. Average core heat flux .... 652,000 Btu/(hr sq ft)
4. Total sodium flow .... 13.2 × 10⁶ lb/hr
5. Steam flow total of three (3) steam generators...... 1.43 × 10⁶ lb/hr
6. Reactor outlet temperature .... 900 F
7. Reactor inlet temperature .... 600 F
8. Turbine-generator design....................... 150 mw
9. Net electrical capacity (present) .... 100 mw
10. Steam temperature .... 780 F
11. Steam pressure................................ 900 psia
12. Primary system pressure .... Atmospheric
13. Feedwater temperature.......................... 380 F

The reactor vessel and core, three primary loops, main sodium pumps, and the intermediate heat exchangers are located within a containment vessel.   The three secondary sodium (nonradioactive) systems, steam generators, boiler feed pumps, turbine generator, and condenser are located in a service building adjacent to the containment vessel,   The fast breeder is housed in a containment vessel because of its critical control characteristics.

Liquid sodium (see Fig. 4) is pumped into the reactor at 600 F and returns at 900 F through the intermediate heat exchanger to the pump tank through 30-in.-diameter piping.   Pressure in the reactor and 30-in. piping is virtually atmospheric and is approximately 1,000 psig on the pump discharge lines.   The system is preheated electrically to 400 F before liquid sodium is introduced.

The reactor has a thermal power rating of 300 mw with a total plant electrical capacity of 100 mw.   The net thermal efficiency of the system is estimated at 31.3 per cent.

The three primary piping loops containing radioactive liquid sodium are 30 in. diameter and of Type 304 stainless-steel pipe $\frac{3}{8}$ in. thick made from ASTM A240 plate and conforming to ASTM A358 Grade S. Elbows in these loops were made from formed butt-welded austenitic stainless-steel pipe.

Heat-exchanger tubes conformed to ASTM A213 Grade TP304 seamless and ASTM A249 Grade TP 304 welded.

The intermediate piping–nonradioactive sodium system is fabricated of Type 304 stainless steel. As in the primary piping all pipe 12 in. and smaller is seamless ASTM A376, and piping over 12 in. conforms to ASTM A358 butt-welded pipe.

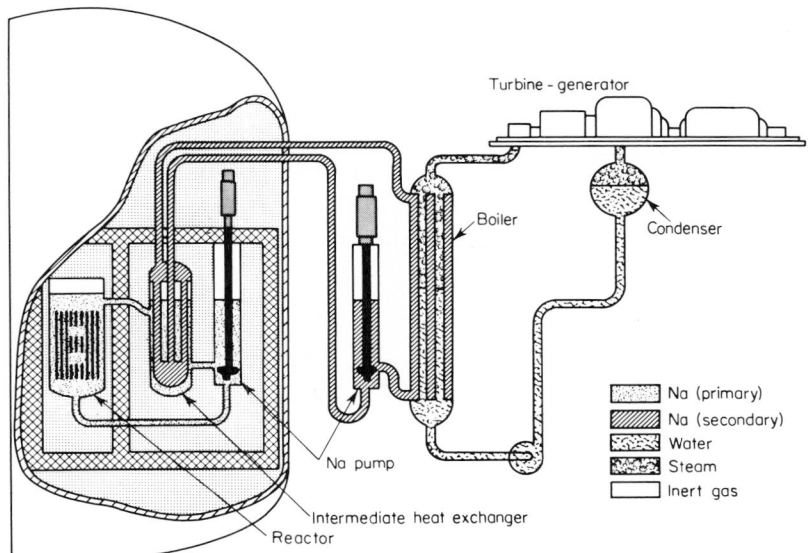

FIG. 4. Schematic of Enrico Fermi Sodium Fast Breeder Reactor. (*Taken from Ref.* 12.)

Fittings 12 in. and smaller are seamless and conform to the applicable parts of ASTM A182 and A376. Fittings larger than 12 in. are fabricated from rolled and welded Type 304 plate conforming to ASTM A358.

The steam piping lines are constructed of $2\frac{1}{4}$ Cr-1 Mo low-alloy steel piping.

A partial list of sodium cooled reactors would include the Sodium Graphite Reactor Experiment, Thermal Neutron Energy; the Consumers Public Power District, Thermal Neutron Energy Sodium Graphite Reactor; the Experimental Breeder Reactor Number 1, Fast Neutron Energy, Argonne National Laboratory; the Experimental Breeder Reactor Number 2, Fast Neutron Energy, Argonne National Laboratory; and the Enrico Fermi Reactor, Fast Neutron Energy and Breeder

Gas-cooled Reactor Systems.[14] Great Britain and France have pioneered in the development and operation of gas-cooled reactors. Great Britain, in particular, in the 1950's was faced with a serious shortage of fossil fuels and paid a very high price for such fuels. In conjunction with a very limited supply of enriched uranium fuel and the economics of power in Britain the gas-cooled reactor was rapidly developed. Great Britain and later France desired a reactor which could be started the earliest and with the assurance that it could be made to operate on

a continuous basis. This became the graphite-moderated natural uranium-fueled reactor of the Calder Hall type.

The original Calder Hall reactors were designed for relatively low gas temperatures, and the turbines were designed for low steam pressures and temperatures. Consequently, the operating efficiences are low.

At the present time the Dragon Project in England and the High Temperature Gas-Cooled Reactor Project (HTGR) in the United States are building gas-cooled reactors to operate at gas temperatures in the order of 1380 F. In turn, it is expected that the HTGR will produce steam at 1000 F and 1,450 psi. The original

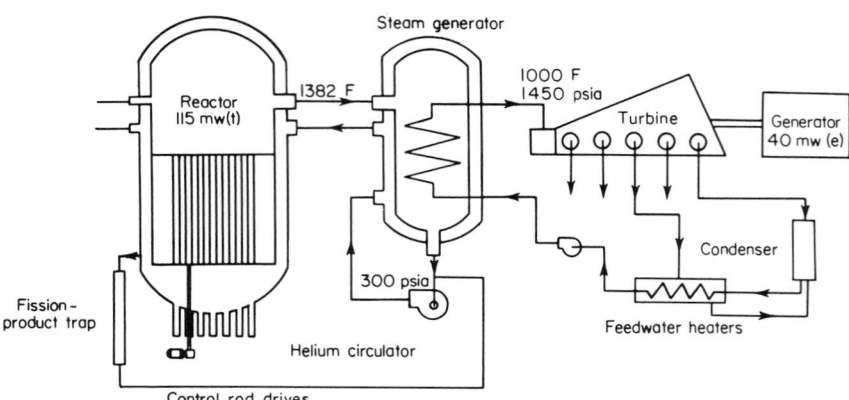

Fig. 5. Schematic of high-temperature gas-cooled reactor. (*Taken from Ref.* 14.)

Calder Hall reactors used CO_2 as the heat-transfer medium; however, the HTGR will use helium. Helium at 350 psi and 1380 F has good heat-transfer properties, and as such it is expected that the heat-transfer rates will be high in the steam generators.

Table 5. Proposed Operating Conditions for HTGR

Coolant......................	Helium
Pressure, psia	350
Core inlet temp, F	660
Core outlet temp, F	1380
Steam temp, F	1000
Steam pressure, psia	1,450
Net thermal efficiency, %	34
Reactor thermal output, mw......	115
Net electrical power, mw	40

The entire nuclear steam-supply system (see Fig. 5) is to be housed within a steel containment vessel 100 ft diameter and 130 ft high. The turbine-generator system is of the outdoor type. The turbine-generator auxiliary system, the reactor control room, and the service systems, as well as offices and shops, are in the building adjacent to the containment vessel.

The basic material for the helium system is Type 304 stainless steel, while the steam system is reported to be constructed of low-alloy steel.

Organic-moderated Reactor Systems.[15] The use of an organic fluid as a moderator, coolant, or combination moderator-coolant in a reactor has received

consideration since the inception of the nuclear industry during the World War II Manhattan project.

Organic fluids, primarily the polyphenyls, benzine derivatives, have four basic advantages as a reactor moderator-coolant:

1. Low volatility, allowing low-pressure operation and consequently low-pressure nuclear-system construction

2. Negligible corrosion with standard materials of construction, thereby eliminating the requirements for stainless steel and special alloys

3. Low activation upon exposure to neutrons, allowing minimization of shielding requirements and permitting easy access for maintenance

4. Very slow reaction with uranium at high temperatures, eliminating the hazard of the release of chemical energy in case of fuel-element failure

The organic-fluid reactor is believed to be a highly realistic reactor type and the Organic Moderator Coolant Reactor Experiment (OMRE) at the National Reactor Testing Station, Idaho, has apparently demonstrated its capabilities since operation commenced in September, 1957. The OMRE concept is now enlarged to a central-station nuclear power plant to utilize organic fluid as coolant and moderator. This new plant is known as the Piqua Nuclear Power Facility (PNPF) located at Piqua, Ohio.[6,16,17]

The facility is located on the east side of the Miami River and is contained within a shielded vaportight containment vessel.

The 45.5-mw (thermal) reactor will be operated by the City of Piqua, Ohio, Municipal Power Commission for the production of electric power for consumer use. Approximately 11.4 mw (electrical) will be generated in an existing power station from the steam generated by the reactor.

Organic fluid (see Fig. 6) enters the reactor vessel above the core and then flows downward through the core into a lower plenum. It then flows upward through

Table 6. Basic Characteristics of the PNPF

1. Commenced operation 1962
2. Reactor thermal power .... 45.5 mw
3. Average core heat flux at rated power...... 34300 Btu/(hr sq ft)
4. Total flow of organic primary coolant ..... 5.5×10^6 lb/hr
5. Primary steam flow .... 150,000 lb/hr
6. Reactor operating temperature .... 575 F
7. Feedwater temperature.................. 268 F
8. Core inlet temperature .... 525 F
9. Net electrical capability .... 11.4 mw
10. Steam temperature .... 550 F
11. Steam pressure........................ 450 psia
12. Primary system operating pressure........ 120 psia

the annulus between the core vessel and the thermal shield into the outlet plenum and then to the primary loop. During full-power operation, the coolant is heated in the core from 525 to 575 F while transferring 155×10^6 Btu/hr from the core. The organic fluid is pumped by two main coolant pumps to the superheater and steam boiler, where the heat is transferred to the steam system. A total of 150,000 lb/hr of superheated steam is produced at a pressure of 450 psia and a temperature of 550 F. The primary system consists of a single loop in which two 6,000-gpm pumps operate in parallel pumping 12,000 gpm to a single superheater and boiler. A flow bypass is located around the boiler for control purposes.

Terphenyl, the organic coolant used, is noncorrosive, has a low volatility and a high melting point, and is not difficult to contain in piping systems. Its non-corrosive nature has permitted the use of carbon-steeel construction for all piping and vessels. Its low volatility has allowed a design for an operating pressure of

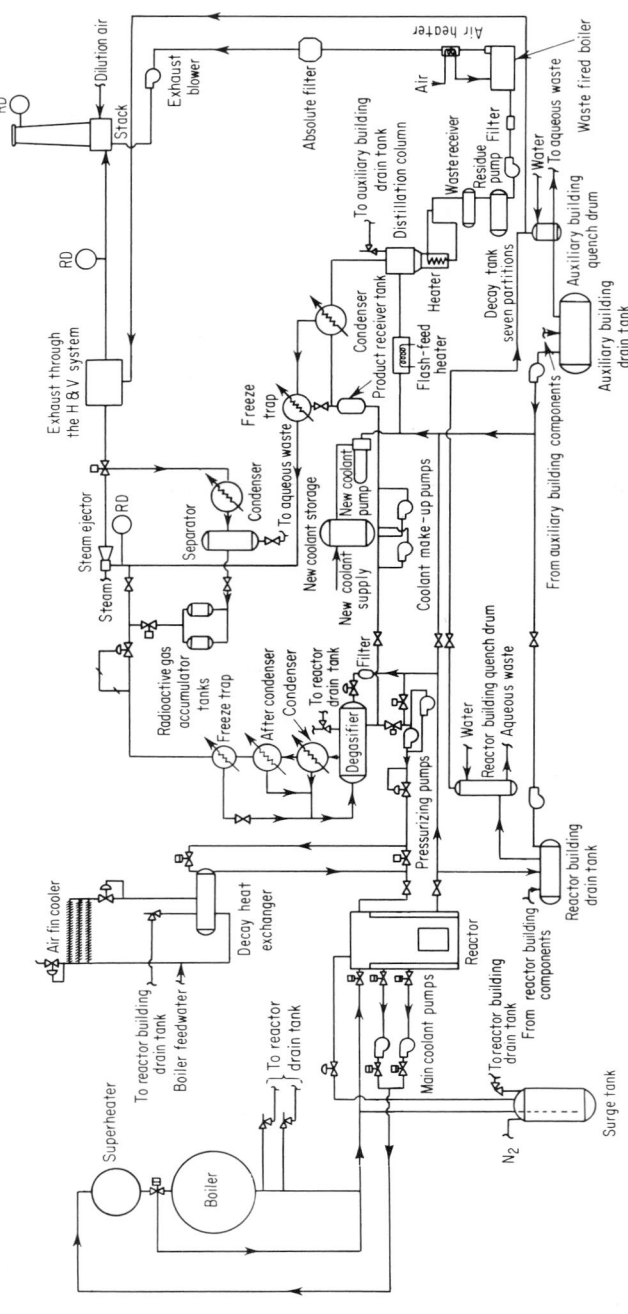

FIG. 6. Schematic of Piqua Nuclear Power Facility Organic Moderator Coolant Reactor. (*Taken from Ref. 16.*)

only 120 psia.   To keep the coolant fluid, all organic containment piping is heated with 175 psig steam in steam tracing circuits.

All primary piping, steam piping, and vessels are constructed of carbon steels.

Research and Test Reactors[18]

Research and test reactors are generally designed for a specific purpose or a series of integrated tests.   Research and test reactors can best be defined or described by enumerating their practical applications, which can be divided into three basic groups: (1) reactor technology, (2) physics research, and (3) industrial and bio-medical research.

Reactor Technology.   Reactors are a logical source of neutrons for a variety of experiments in the study of reactor physics and of the properties of neutrons themselves.   Reactors also provide an excellent means for studying effects of radiation on materials of construction, giving scientific and operational personnel experience with radiation and the methods used for detecting and counting neutrons or gamma rays.

Physics Research.   The study of fundamental physics and experiments in fundamental physics depend upon the operation of research reactors.   Studies of the fission process in uranium and measurements of neutron absorption and scattering are among the commonly performed experiments.

Industrial and Biomedical Research.   The neutron flux from reactors provides the tool for important industrial and biological research.   Radioactive isotopes, obtained by irradiation of samples in the neutron field of a reactor, are widely used for industrial radiography and experiments and for medical and biological research.

Cooling Systems for Research Reactors.   Reactors operating above extremely small power levels must use some form of cooling to remove the heat produced.   Light-water, heavy-water, or air coolants are generally used, and either natural convection or forced circulation is employed.

The basic piping materials are the austenitic stainless steels, and except for their pressure and temperature levels, research reactors must perform in a manner similar to that of the power reactors.

GENERAL REQUIREMENTS FOR NUCLEAR PIPING[3,19]

Containment and movement of a heat-transfer fluid in a piping system, whether nuclear, industrial, or institutional, require pipe, valves, and pumps.   The degree of reliability for containment and movement for radioactive fluids in nuclear piping is believed to be one of the major concerns for the systems designer.   Fundamentally, repair and maintenance in a radioactive system are ordinarily difficult operations as a result of radioactivity levels.   Consequently, a failure in nuclear piping can result in a long and expensive plant shutdown.

The main objective in nuclear systems, as in most other commercial enterprises, is to obtain maximum efficiency at the lowest cost.   Piping and components can be installed with a moderate degree of reliability, and along with such a system, elaborate remote-control equipment for maintenance and repairs can be designed and installed.   Or piping and components can be installed with a high degree of reliability, along with a moderate amount of equipment for remote maintenance and repair.   It is believed that piping systems installed with the maximum amount of practical reliability will result in maximum efficiency at lowest cost.   It is also believed that maximum practical reliability can be achieved only through recognizing the need for safety standards and then employing quality control and nondestructive testing to ensure reliability.

Therefore, the basic components of a nuclear piping system are:
1. Pipe, valves, and pumps
2. Safety standards
3. Quality control
4. Nondestructive testing

The service requirements for nuclear piping are dependent upon the type of reactor system and are also influenced by associated equipment. The steam and associated piping in the reactor complex are comparable to fossil-fuel steam-generating service piping.

Pressure and Temperature. The pressures in primary piping and auxiliary systems may be as low as 150 psig for liquid-metal coolants to as high as 2,500 psig for light-water-cooled reactors.

Basically, these pressures are not unusual, and if considered from the pressure level alone, the systems ordinarily could be designed using the minimum standards of the Code for Pressure Piping, ASA B31.1.

Metal temperatures in the primary and auxiliary nuclear systems are considered moderate. Temperatures can vary from approximately 150 F for certain light-water-cooled reactors to as high as 1000 F for liquid-metal-cooled systems.

The combination of pressure and temperature in presently operating power or research reactors is no obstacle for competent design. However, envisioned nuclear hazards add to the design criteria.

Heat Transfer.[19] The basic surfaces for heat transfer are in the core or fuel complex within the reactor core pressure vessel. The transfer of thermal energy to the steam generators with the subsequent return of the primary fluid to the core is the basic energy cycle. In the course of steady-state operation and during transient start-up or shutdown, thermal stresses are introduced whose magnitudes are a function of maintained temperature or transient temperature gradients and inherent or mass restraint of the piping system to free thermal expansion. The significance of these stresses is dependent upon their strain range and frequency of occurrence.

Generally, nuclear piping for pressurized light-water reactor systems, with their relatively low temperatures and high coolant pressures, requires thick-walled sections. These are particularly susceptible to thermal stresses where flexibility of operation is a requirement.

Liquid-metal reactor systems are characterized by low pressures and high temperatures. The high temperatures and the good heat-transfer properties of liquid metals such as sodium induce severe temperature gradients that account for a major portion of the total piping stresses. Low pressure would permit thin walls, but the decrease in piping-material strength at high temperature prevents thin-wall construction. Major difficulties are caused by system transients inducing large thermal gradients in the piping closest to the reactor pressure vessel.

Mechanical Loads. Mechanical loading, which includes all static and dynamic pressure, flow, and structural demands, introduces no unusual aspect to the piping designer. As in any complex system, nuclear piping is sufficiently varied and challenging to warrant thorough understanding and consideration of all potentials.

Liquid-metal systems must take into account the weight of liquid metal as well as the inertia of the moving coolant.

Materials. The necessity for materials standards and quality comparable to the refinement attained in design and fabrication and the concurrent importance of a sufficient extent of testing and nondestructive examination are emphasized for nuclear piping. The selection of materials must be made on the basis of established stability and deterioration and corrosion resistance to the primary coolant.

Material sensitivity for fabrication and welding operations must be thoroughly

understood.   The problem of repairs, cleanliness, corrosion resistance, and possible heat treatment in service must enter into material choice.

Radiation and Materials of Construction.   The metallic materials employed in all nuclear systems can be classified as being either fuel materials or construction materials.   *Fuel materials* consist of uranium, alloys of uranium, uranium oxides, and cladding metals to contain the fuel.   Fuels are a branch of metallurgy unto themselves and are discussed elsewhere in published literature.   The *materials of construction* are of prime interest in this chapter and are used for piping, vessels heat exchangers, and structures.

In addition to the problems normally encountered in the design and operation of chemical or power plants which may be similar to a reactor plant, there are special problems associated with the reactor and primary piping system which have a marked effect upon the choice of construction materials.

The requirements for *safety* and the problem of making repairs to nuclear piping have created a small tolerance for errors in the selection of piping materials and their fabrication.   This has resulted in standards of performance higher than those normally encountered in plant equipment which is not subject to radioactivity.

The necessity for leak-tightness has created improved welding techniques, as well as components such as pumps and valves that are totally sealed against leakage or controlled for leakage.

The objective in power reactor systems second to that of safety is to obtain *maximum efficiency* at the lowest cost.   Consequently, the selection of piping materials, pumps, valves, and fittings is generally made by combined evaluation of corrosion resistance; physical, tensile, and impact properties; nuclear cross section; fuel requirement; system size; and cost.

Corrosion[20]

Nuclear radiation with high flux densities affects aqueous primary coolants, which may have a notable effect on their corrosive properties.   Under exposure and radiation, some water in the primary system is apparently affected by fast neutrons and broken down into free radicals.

Most of the free radicals quickly recombine, but a small amount react in other ways and form small amounts of dissolved oxygen and hydrogen.   Additional hydrogen may be formed by the corrosion between water and the piping metals. The free oxygen, if permitted to exist in large amounts, would increase corrosion to an intolerable amount.   Consequently, in systems such as the pressurized-water reactor, hydrogen is added so that the formation of oxygen by dissociation will be suppressed and also any oxygen in the feedwater would be quickly consumed.

If nitrogen is dissolved in the steam of a water-cooled reactor system, it will combine with excess hydrogen to form ammonia.   If excess oxygen is present, then nitrogen will combine and form some nitric acid.   The foregoing will influence the pH of the primary water coolant with a direct effect on corrosion increase.

The nature and quantity of corrosion products formed in the reactor and in the primary- or nuclear-piping system will have a marked effect upon the operation of any type of reactor.   Some of these corrosion products normally will be removed from metal surfaces by flow conditions of the coolant.   In turn, the corrosion products will be either circulated through the system as discrete particles or entrained in the coolant fluid.   In passing through the fueled core, whether cooled by water, liquid metal, organic liquid, or gas, the corrosion products will become radioactive and transmit this activity to parts of the system on which they may deposit.   Radioactive deposits must be avoided or minimized, as they will dictate the type of repair and maintenance required in equipment, such as the steam

generators of a pressurized-water system which are located outside the reactor shielding. These deposits also must be kept to a minimum in boiling-water, liquid-metal-cooled, organic-liquid-cooled, and gas-cooled reactors, as these systems can also have steam generators or other equipment with primary coolant placed outside the main reactor radiation shield. Excessive deposits of corrosion products can constrict flow passages in the fueled core, thereby reducing heat transfer. Corrosion products which deposit on moving parts may result in excessive wear. In addition, steam-generation efficiency can be hampered by the deposition of these products on heat-exchanger surfaces.

Corrosion of Austenitic Stainless Steels. The austenitic stainless steels of the AISI 300 series, commonly used for the construction of process equipment, include Types 304, 304L (low-carbon), 309, 310, 316L (low-carbon), 321, and 347. All these alloys have a high degree of resistance to high-temperature, high-purity water; liquid sodium; and helium under conditions likely to be encountered in reactors as measured by weight loss or accumulation of corrosion product. They have been the subject of a considerable number of test programs in water under a wide variety of conditions including hydrogen and oxygen content, temperature, and velocity. Much of this work has been done with two of the alloys, Types 304 and 347. Some representative test data for several of the stainless alloys, taken mostly from the work reported by Datsko,[21] are shown in Tables 7 to 9. Weight changes were determined without descaling and are here reported to the nearest 5 mg/sq dm per month. While the corrosion resistance appears to be somewhat better under hydrogenated conditions than under oxygenated conditions, there is very little difference in the rate of attack, which was in most cases, less than 5 mg/sq dm per month. In all tables, a weight loss of 10 mg/sq dm per month is equivalent to a corrosion rate of approximately 0.00006 in. per year.

A series of static tests of Types 304 and 347 stainless steels under a variety of water conditions with different lots of specimens and different surface treatments at 600 and 680 F were reported by Fowler, Douglas, and Zyzes.[22] Here again the corrosion rates were in most cases less than 5 mg/sq dm per month. Hydrogen in the water in amounts up to 3,500 cc/liter did not increase the rate of corrosion. Surface finish had little effect, although in some cases the electropolished specimens showed lower corrosion rates, probably due to smaller surface area exposed. In some cases higher corrosion rates (up to 10 or 20 mg/sq dm per month) were observed in water treated with 0.005 mole of lithium hydroxide.

Test data indicate that the austenitic stainless steels, including Types 304, 316, and 347, are not subject to intergranular attack by high-temperature water when sensitized by heating at 1250 F for 2 hr. Types 304 and 316 are sometimes subject to such attack in some other chemical environments owing to intergranular carbide precipitation resulting from the heat of welding or stress relieving. This possibility was probably the reason for selecting the columbium-stabilized variety, Type 347, for some of the initial PWR equipment. In this connection it should be pointed out that the unstabilized types such as 304 and 316 can be subject to some intergranular attack during pickling in nitric–hydroflouric acid solutions if they have been sensitized during some prior heat treatment.

Nitriding of the 300 series stainless steels will substantially decrease their corrosion resistance to whatever depth the nitriding occurs.

When considering the use of stainless steels, one has to be concerned about two other forms of attack which are, in some respects, somewhat peculiar to these alloys and which are generally not revealed by the type of tests described above. These forms of attack are referred to as crevice-corrosion and stress-corrosion cracking.

Crevice Corrosion. The light-water-cooled reactor systems contain certain components, some involving moving parts with close clearances, which must be

Table 7. Corrosion of Austenitic Stainless Steels by High-purity Water, Types 304 and 304L

Temp, F	Environment			Condition of specimen	Weight change, mg/(sq dm month)	
	Oxygen, cc/liter	Hydrogen, cc/liter	Other		1 fpm	30 fps
			Type 304			
500	30	....	....	M	+10*	
500	1–5	......	....	M	+5	−5
500	1–5	......	....	M, Sens.	+5	−10
500	1–5	......	....	M, P	+5	+5
500	1–5	......	Ion ex.	M, Sens.	+5	−5
500	0–1	......	Ion ex.	M, P	−5	−5
500	0–1	....	Ion ex.	M, Sens.	+5	+5
500	...	90–130	....	M, P	−5	+5
500	...	350–600	Ion ex.	M, Sens.	−5	−5
600	...	D	....	M	−10	
600	...	400	....	CR, P	+5	
680	...	1,300	....	CR, P	+5*	
600	...	200(He)	pH 10, LiOH	CR, P	−5	
500	1–5	......	pH 10, NaOH	M	−5	−5
500	1–5	......	pH 10, NaOH	M, Sens.	−5	−15
			Type 304L			
500	30	......	....	M	−15*	
500	1–5	......	....	M, P	+5	+5
500	1–5	......	Ion ex.	Nit.	+5	−15
500	0–1	......	....	M, P	+5	+5
500	0–1	......	Ion ex.	Nit.	−50	−15
500	...	D	Ion ex.	M, P	−5	−10
500	...	100–650	Ion ex.	M, P	−5	−5
500	...	35–100	Ion ex.	Nit.	....	−200

* Static test.
CR = cold-rolled
D = degassed
M = machined
Ion ex. = ion exchanger in test system
Nit. = nitrided
P = pickled
Sens. = sensitized

depended upon to operate satisfactorily in use.   Other opportunities for crevices may exist, as when tubes are imperfectly rolled into tube sheets, in flanged connections, in threaded couplings, on the flat bases of springs, and elsewhere.   Accelerated attack may sometimes be experienced in such crevices where oxygen is present in the water, owing to galvanic cells set up by differences in oxygen content inside and outside the crevice.   This form of corrosion will accelerate attack inside the crevice and may develop a volume of corrosion product which will cause excessive wear or complete seizure of the components.   DePaul[23] has reported upon work done on this subject in high-purity water.   Figure 7 indicates that the effect would probably not be so serious in degassed water or hydrogenated water.   However, in oxygen-bearing water the clearance between parts must be at least 5 mils in order to overcome the possibility of seizing .   Figure 8 shows the effect of water temperature upon crevice corrosion.   Normally, at 200 F the extent of contact corrosion

Table 8.   Corrosion of Austenitic Stainless Steel by High-purity Water, Type 347

Temp, F	Environment			Condition of specimen	Weight change, mg/(sq dm month)	
	Oxygen, cc/liter	Hydrogen, cc/liter	Other		1 fpm	30 fps
500	30	....	....	M	....	−5
600	30	....	....	M	....	−5
500	1–5	....	Ion ex.	M, P	+5	+5
500	0–2	....	Ion ex.	M, P	+5	+5
500	0–1	....	Ion ex.	M, Sens.	−5	−5
500	0–1	....	Ion ex.	Nit.	−10	−15
500	D	....	Ion ex.	M, P	−5	−10
500	...	25–50	Ion ex.	M, P	+5	−5
500	...	40–50	Ion ex.	Nit.	....	−200
500	...	90–130	Ion ex.	M, P	−5	+5
500	...	350–600	Ion ex.	M, P	−5	−5
600	...	25–50	Ion ex.	M, P	....	−5
500	1–5	....	pH 10, NaOH	M	....	−15
600	...	200(He)	pH 10, LiOH	AR, P	+5*	
680	...	400(He)	pH 10, LiOH	AR, P	+20*	

```
      * Static test.
      AR = as rolled
       D = degassed
       M = machined
Ion ex. = ion exchanger in test system
  Nit. = nitrided
       P = pickled
  Sens. = sensitized
```

Table 9.   Corrosion of Austenitic Stainless Steel by High-purity Water, Type 316

Temp, F	Environment			Condition of specimen	Weight change, mg/(sq dm month)	
	Oxygen, cc/liter	Hydrogen, cc/liter	Other		1 fpm	30 fps
500	30	....	....	M	−5	
500	1–5	....	....	M	+5	−5
500	1–5	....	....	M, Sens.	+5	−5
500	1–5	....	Ion ex.	M, P	+5	−5
500	1–5	....	Ion ex.	M, Sens.	+5	
500	0–2	....	Ion ex.	M, P	0	0
500	...	D	Ion ex.	M, P	−5	−5
500	...	90–125	....	M, P	−5	+5
500	...	100–650	....	M, P	+5	+5
500	1–5	....	pH 10, NaOH	M	+5	−10
500	1–5	....	pH 10, NaOH	M, Sens.	+5	−5

```
       D = degassed
       M = machined
Ion ex. = ion exchanger in test system
       P = pickled
  Sens. = sensitized
```

amounts to nothing more than a discoloration of the metal without any measurable build-up.

Stress-corrosion Cracking.   Numerous cases of stress-corrosion cracking of austenitic stainless steels in hot aqueous environments have been reported in recent years.   In almost all cases the presence of chlorides has been detected in the environment.   In some cases this was a very small amount, as low as a few parts per

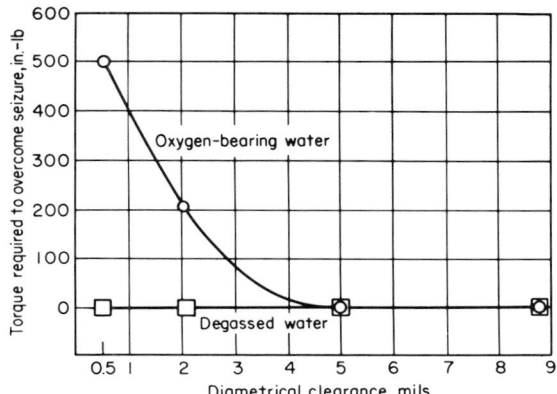

Fig. 7. Effect of oxygen and clearance on crevice corrosion of austenitic stainless steel in water. (*Taken from Ref.* 23.)

million in the liquid.   In many of the cases involving very small amounts of chloride in the liquid phase, there appears to have been an opportunity for the chlorides to concentrate in some part of the equipment where cracking occurred, such as in the upper part of water deaerating preheaters in power plants, in crevices between tubes and tube sheets of tubular heat exchangers, and in areas which may be alternately

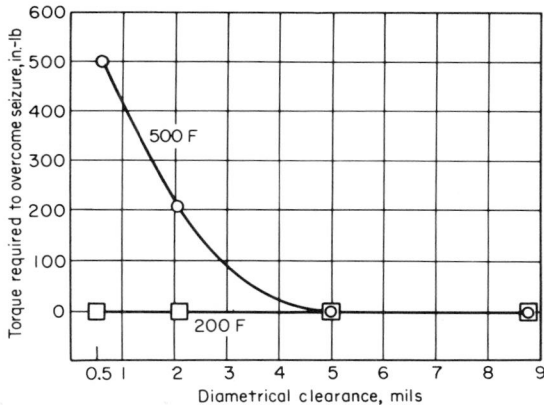

Fig. 8. Effect of temperature on crevice corrosion of austenitic stainless steel in oxygen-bearing water. (*Taken from Ref.* 23.)

wet and dry.   It does not seem possible to place any safe limit upon the allowable chloride content of water if operating conditions are suitable for the chlorides to concentrate.   Likewise, it has not been possible to place any safe limit upon the maximum amount of measured stress in the alloy.   Many of the failures have occurred where stresses were near the yield point of the alloy, as in cold-rolled, drawn, or bent materials.   It is generally agreed that fabrication or forming stresses are more likely to cause trouble than stresses resulting from operating pressures.   It seems likely that plastic strain is an important factor.   It would appear that the reaction causing stress-corrosion cracking may be electrolytic in nature, since it has been demonstrated that it can be prevented by the application of cathodic protection from a metal that is anodic to the stainless steel.

In stress-corrosion tests reported by Williams and Eckel[24] with Type 304 stainless in primary water, samples which had been cold-worked, annealed, and stressed did crack in oxygen-bearing water at 600 F; however, samples not cold-worked prior to annealing and stressing did not crack.   No cracking occurred in hydrogenated water.   They concluded that stress-corrosion cracking probably can be avoided in the reactor itself if mill-annealed materials are used.   Thermal stresses set up adjacent to welds of heavy plate have been sufficient to cause stress-corrosion cracking of stainless steels in some chloride-bearing environments, so that stress-relief annealing of welded pressure vessels would appear to be a worthwhile precaution.   Stainless springs would not be used because they are purposely cold-worked to obtain necessary spring properties.

The possibility of stress-corrosion cracking of austenitic stainless steels in secondary steam-water environments in power plants is a considerably more critical problem, mainly because of the chlorides which can sometimes be encountered in steam-generation systems (especially naval boilers) or may leak in from cooling or condensing waters.   Numerous tests have shown cracking in chloride-containing boiler waters when exposed in vapor or splash zones.   Several cases of failure of stainless-steel test boilers handling boiler water containing chlorides have been reported.[24]

Oxygen is a strong accelerator of stress corrosion of austenitic stainless steels in hot-water environments.   Evidence seems to indicate that some oxygen is necessary for stress corrosion to occur.   The necessary amount is small.   Williams and Eckel[24] indicate a possible relationship between chloride and oxygen content in alkaline-phosphate-treated boiler water and susceptibility to stress-corrosion cracking of austenitic stainless steel exposed to the steam phase with intermittent wetting, as shown in Fig. 9.

It would appear that the use of austenitic stainless steels in the construction of tubular heat exchangers which transfer heat from primary coolants to secondary boiler water should be subject to very close scrutiny from the standpoint of possible stress cracking from the boiler-water side.   If it cannot be assured that the equipment is suitably stress-relieved before installation and not likely to develop unsatisfactory thermal-expansion stresses in service, it may be prudent to consider the use of material for such equipment which is not subject to stress-corrosion cracking in the presence of chlorides.

Corrosion in Precipitation-hardening Stainless Steels. The precipitation-hardening stainless steels of the austenitic or essentially austenitic types such as Armco 17-4 pH, Armco 17-7 pH, and USS Stainless W show good resistance to oxygenated water at both low and high velocity with corrosion rates of 5 mg/sq dm per month or less.   In degassed or hydrogenated water, however, corrosion rates are considerably higher with rates up to 75 mg/sq dm per month under hydrogen at high velocity.   Limited tests of precipitation-hardened stainless steels indicate a susceptibility to stress corrosion similar to and perhaps greater than that of the

austenitic stainless steels. These alloys are not suitable for springs in primary waters.

Corrosion in 400 Series Stainless Steels. The results of a number of representative corrosion tests on 400 series stainless steels are shown in Table 10.

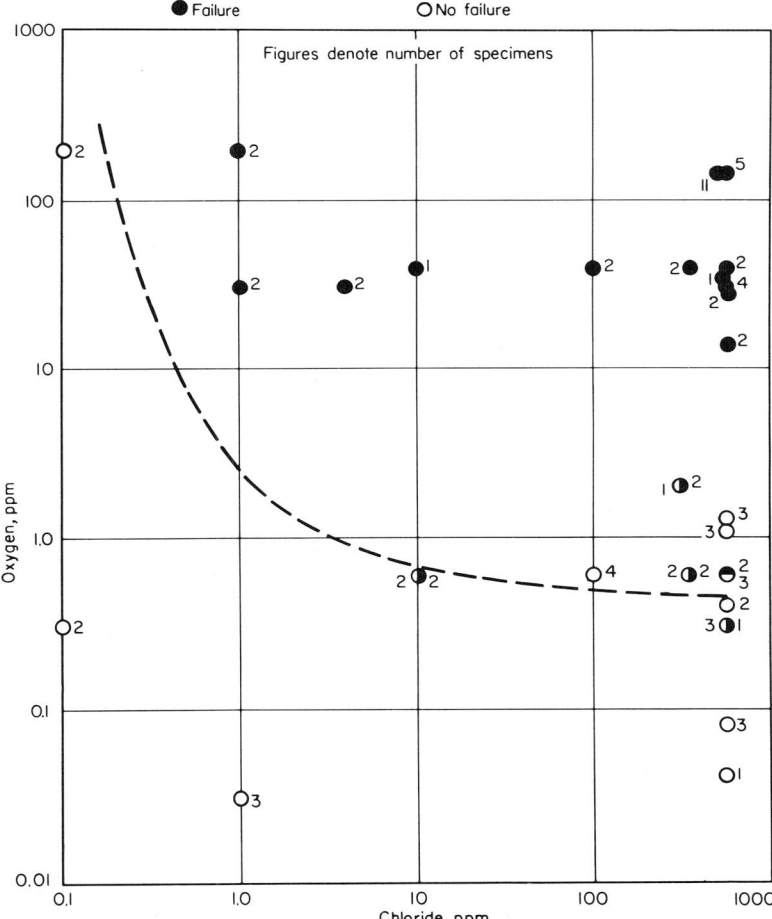

Fig. 9. Proposed relationship between chloride and oxygen content of alkaline-phosphate-treated boiling water and susceptibility to stress-corrosion cracking of austenitic stainless steels exposed to steam phase with intermittent wetting. (*Taken from Ref. 24.*)

Most of these data are taken from the *Argonne National Laboratory Report* 5354. The martensitic (hardenable) grades, including Types 410, 420, and 440C containing up to $2\frac{1}{2}$ per cent nickel and 12 to 17 per cent chromium, are not so resistant as the 300 series stainless steels. Corrosion rates are high enough that they probably would have only limited use in nuclear water systems. They are useful as gear and bearing materials because of their hardness. Type 440C has the best resistance in hydrogenated water. Corrosion resistance of the ferritic alloys

containing 18 to 30 per cent chromium is somewhat better than that of the martensitic alloys, with the higher chromium alloy 446 being the best, but their resistance is in general not comparable to that of the austenitic alloys.

The 400-series stainless steels are particularly subject to pitting and crevice attack, and this may proceed at a rate as high as 70 mils per year in oxygen-containing water at temperatures between 200 and 500 F. These alloys are resistant to the specific types of stress-corrosion cracking in chloride-containing waters to which the austenitic stainless steels are subject. However, martensitic steels, when hardened to rather high levels (above about Rockwell C–35), are susceptible to

Table 10.   Corrosion of Martensitic Stainless Steel by High-purity Water, 400 Series

Alloy type	Temp, F	Environment			Condition of specimen	Weight change, mg/(sq dm month)	
		Oxygen, cc/liter	Hydrogen, cc/liter	Other		1 fpm	30 fps
410	500	30	....	....	H, P	−200*	
410	500	1–5	....	....	H	−5	−10
410	500	1–5	....	Ion ex.	H	−10	−25
410	500	1–5	360–600	Ion ex.	H	−40	−40
410	500	1–5	....	pH 10, NaOH	H	−25	−25
420	500	...	D	....	H	−5	−10
420	500	1–5	....	Ion ex.	H	−200	
440C	500	1–5	....	....	H		−10
440C	500	...	D	Ion ex.	H	−5	−20
440C	500	...	....	....	H	−30	−10
440C	500	1–5	350–600	....	H	−25	−30
440C	500	30	....	pH 10, NaOH	H	−20	−20
431	500	2	....	....	H, G	−100*	
431	500	...	D	....	M, H	−5	
431	500	...	....	....	M, H	−50	
431	500	30	500	....	M, H	−35	
446	500	2	....	....	M	+43*	
446	500	...	D	....	M, T	+5	
446	500	...	....	....	M, T	+5	
446	500	...	500	....	...	−10	

* Static test.
   H = hardened
   P = pickled
   G = ground
   M = machined
   T = tempered
Ion ex. = ion exchanger in test system

cracking by hydrogen embrittlement in environments where nascent hydrogen is present. Thus they are not suitable for springs in primary waters.

Corrosion in Nickel and High-nickel Alloys. Nickel and the high-nickel alloys such as Inconel, Monel, the B and C Hastelloy alloys, and illium are subject to attack in oxygen-containing water at 500 F to a greater extent than are the austenitic stainless steels. However, in degassed water and, particularly, in hydrogenated water, they have very good resistance with corrosion rates normally less than 5 mg/sq dm per month at velocities up to at least 30 fps. Corrosion rates for some of the high-nickel materials, taken mostly from the *Argonne National Laboratory Report* 5354, are given in Tables 11 to 13. K monel is an age-hardenable form of Monel. Inconel X is an age-hardenable form of Inconel. Data for other high-nickel alloys, taken mostly from static tests reported in the *Argonne*

National Laboratory Report 4519 and *Knolls Atomic Power Laboratory Report* 1248, are given in Table 14.

One of the most significant properties of the high-nickel alloys is their resistance to stress-corrosion cracking in the chloride environments which may cause cracking of the austenitic stainless steels. This has been established by many years of laboratory testing and operating experience. Springs of Inconel X are used in pressurized-water reactors because of its resistance to stress-corrosion cracking. Inconel is used for nuclear piping, fittings, and cans of canned-motor pumps in this service. Monel, handling water up to 530 F and steam up to 875 F, is in common use for tubes of boiler feedwater heaters in modern high-pressure steam power plants.

Table 11. Corrosion of Nickel by High-purity Water

Temp, F	Environment			Condition of specimen	Weight change, mg/(sq dm month)	
	Oxygen, cc/liter	Hydrogen, cc/liter	Other		1 fpm	30 fps
500	30	....	....	Rec.	....	−50
500	30	....	....	M	+20*	
500	1–5	....	....	Rec.	−5	−20
500	1–5	Ion ex.	....	....	−10	−35
500	1–5	....	....	....	−20	−20
500	0–1	....	....	....	−15	−10
600	3–5	....	....	M	....	+15
600	...	D	....	M	+40	
500	0–1	390–600	....	....	0	+5
600	...	200	....	CR, P	+5*	
500	0–5	....	pH 10, NaOH	....	+5	−5
600	...	....	pH 10, LiOH	CR, P	0*	

  * Static test.
  CR = cold rolled
   D = degassed
   M = machined
   P = pickled
 Rec. = As Received
Ion ex. = ion exchange

Inconel provides a suitable combination of resistance to corrosion and to stress-corrosion cracking in heat exchangers which transfer heat from primary water coolants to the secondary boiler water and piping in nuclear plants.

Corrosion in Copper and High-copper Alloys. Copper and high-copper alloys generally show high corrosion rates in 500 F water which contains oxygen. Under hydrogenated conditions their performance is considerably better with 70-30 cupronickel; some of the cast aluminum bronzes which contain 9 to 11 per cent aluminum, 3 to 4 per cent iron, and 85 to 88 per cent copper show the best resistance to attack. The effect of hydrogen upon the performance of 70-30 cupronickel is shown in Table 15. This alloy has been used to a considerable extent for tubes of boiler feedwater heaters and for steam condensers in steam power plants. In the latter application it provides resistance to erosion-corrosion effects at the inlet ends of tubes, particularly where sea water is used for cooling.

Corrosion in Carbon and Low-alloy Steels. Considerable attention has been given to the possible use of carbon and low-alloy steels for water-cooled nuclear reactors because of the low cost of these materials. A comprehensive series of corrosion tests of carbon steels is reported by Blaser and Owens.[25] The steels included are A212 and another carbon steel as well as low-chromium steels,

Table 12. Corrosion of Inconel by High-purity Water

Temp, F	Environment			Condition of specimen	Weight change, mg/(sq dm month)	
	Oxygen, cc/liter	Hydrogen, cc/liter	Other		1 fpm	30 fps
				Inconel		
500	1–5	....	....	M	+5	−25
500	0–1	....	....	M	−10	−35
500	...	D	....	M	−5	−10
600	3–5	....	....	M	...	+10
500	...	40–90	....	M	−5	−5
500	...	350–600	....	M	+5	+5
600	...	00	....	Po, P	0*	
500	0–5	....	pH 10, NaOH	M	+5	−5
				Inconel X		
500	30	....	....	Aged, M	+5	
500	0–1	....	....	Aged	−10	−35
500	...	D	....	Aged	−5	−5
500	...	34–100	....	Aged	−5	−5
500	...	500	....	Aged, M	−5*	

* Static test.
D = degassed
M = machined
P = pickled
Po = polished

Table 13. Corrosion of Monel by High-purity Water

Temp, F	Environment			Condition of specimen	Weight change, mg/(sq dm month)	
	Oxygen, cc/liter	Hydrogen, cc/liter	Other		1 fpm	30 fps
				Monel		
500	30	....	....	M	−30	
500	1–5	....	....	M	−15	−50
500	1–5	....	Ion ex.	M	−35	−60
572	...	D	....	M	+5	
500	1–5	....	pH 10, NaOH	M	−5	−20
				K Monel		
500	1–5	....	....	Ann.	+5	−50
500	1–5	....	....	Aged	−5	−45
500	1–5	....	Ion ex.	Ann.	+5	
500	1–5	....	Ion ex.	Aged	−15	−70
500	0–1	....	Ion ex.	Aged	−25	−200
600	3–5	....	....	Cd, Po	...	−10
500	...	350–600	Ion ex.	Aged	−5	+10
500	25	....	....	M	...	+5
500	1–5	....	pH 10, NaOH	Ann.	−5	−5
500	1–5	....	pH 10, NaOH	Aged	−5	−15

Ann. = annealed
D = degassed
CD = cold drawn
M = machined
Po = polished
Ion ex. = ion exchanger in test system

Table 14.   Corrosion of Other Nickel Alloys by High-purity Water (Static tests except as noted)

Alloy	Temp, F	Duration of test, hr	Environment		Condition of specimen	Weight change, mg/(sq dm month)
			Oxygen, cc/liter	Hydrogen, cc/liter		
Hastelloy A..	500	336	30	..	M	−70
	600	284	...	D	M	+20
Hastelloy B ..	500	336	30		M	−150
	600	284	...	D	M	+5
Hastelloy C ..	500	713	30		Ann.	+40
	600	284	...	D	M	−5
Hastelloy D..	500	336	30	..	M	−30
	600	284	...	D	M	+80
Illium G.....	500	330	30	..	M	+15
Illium R.....	500	691	30	..	Po, P	−10
						+5
Incoloy......	500	1,440	1–3	..	....	−5*
Invar........	500	340	...	D	M	−80

* Velocity 10 fps.
Ann. = annealed
  D = degassed
  M = machined
  P = pickled
 Po = polished

A335P11 and A335P22.   The entire test program included over 70 tests with approximately 1,000 test specimens and about 14 water conditions.   Some representative corrosion rates for A212 carbon steel taken from this report are shown in Table 16.   The majority of tests reported here were run in circulating loops at 30 fps (high velocity) or at 1 fpm (low velocity).   System pressure was approximately 2,000 psi for the 500 and 600 F tests.   Continuous water purification was provided with bypass circuits, including filters and ion exchangers.   Water was maintained at a resistivity of approximately 1 million ohms per cubic centimeter except as reduced by additives for water treatment.   Corrosion rates are reported on a descaled basis in order suitably to compare low-velocity and high-velocity results. Rates reported for a given test condition are, in many cases, an average of several specimens.   It is apparent that corrosion rates of carbon steel can be high in water of pH 7, even with hydrogen present, so that most attention is given to results under inhibited conditions at around pH 11.

Table 15.   Corrosion of 70-30 Cupronickel by High-purity Water

Temp, F	Environment			Condition of specimen	Weight change, mg/(sq dm month)	
	Oxygen, cc/liter	Hydrogen, cc/liter	Other		1 fpm	30 fps
500	1–5	....	......	M	+250	−400
500	0–1	....	......	M	+170	−1,300
500	...	D	......	M	−20	−200
500	...	400–600	......	M	−5	−5

D = degassed
M = machined

Temperatures from 200 to 750 F were investigated during the work. Little, if any, effect on corrosion was noted between 500 and 650 F. However, a limited number of tests at 750 F indicated that corrosion at this temperature could be five times that at 680 F under some conditions. It was found that corrosion rates decreased with the time of testing as shown in Fig. 10.

The accurate control of pH is undoubtedly of great importance in the performance of steel. The inhibiting effect of high pH in the corrosion of steel is by no means a simple matter. Luce,[26] in erosion-corrosion tests of carbon steel in distilled water with pH values from 3 to 10 at 120 F, found minimum corrosion rates at pH 6 and pH 10 with a rate at pH 8 more than ten times that at pH 6. Decker, Wagner, and Marsh[27] found that the erosion-corrosion attack of cast carbon steel at 250 F in boiler water with pH 8.4 was more than twice that at pH 7.6.

Table 16. Corrosion of A212 Carbon Steel by High-purity Water

Temp, F	Duration of test, hr	Environment		pH	Condition of specimen	Corrosion rate after descaling, mg/(sq dm month)	
		Oxygen, cc/liter	Hydrogen, cc/liter			Low velocity	High velocity
500	338	30	.....	7*	M	−325	
500	500	....	20	7*	G	−200	−530
500	500	....	500	7*	S	−150	−275
500	500	....	1,000	7*	S	−190	−210
600	1,992	....	D	7*	P	−170	−225
450	1,000	....	13–33	11 LiOH*	G	.....	−150
500	162	....	20	11 NH₄OH*	G	−280	−525
600	1,500	....	500	11 LiOH*	P	−45	−45
600	798	....	100	11 LiOH*	P	−125	−290
600	2,031	....	D	11 LiOH*	P	.....	−75
600	2,031	....	D	11 LiOH*	E	.....	−65
600	500	180–970	.....	9–10 NH₄OH	P	.....	−115
600	500	180–970	.....	9–10 NH₄OH	E	.....	−30
600	1,558	....	D	11 LiBO₂	E	.....	−285
600	1,000	....	D	11 K₂HPO₄	P	.....	−125
750	1,000	500–1,000	.....	11 LiOH	M	−75	

* Ion exchanger in test system.
D = degassed
S = sanded
E = electropolished
G = ground
M = machined
P = pickled

Partridge[28] has reported several cases of stress-corrosion cracking of steel subject to localized stresses in power-plant boilers operating at pressures from 850 to 1,400 psi and using deaerated distilled or deionized water for boiler feed. In these cases the pH of the water in contact with the steel ran from 7.0 to 9.5. The occurrence of stress cracking is attributed to the inability of steel to form a protective film of oxide in water with pH values this low. Performance of steel apparently is much better when the boiler water is chemically treated to have a pH of 10.5 or above.

The limited amount of work done with the low-chromium steels A335P10 and A335P22 indicated that they were somewhat more corrosion resistant than A212 for the condition tested. There was no significant difference between the two chromium alloys.

It would appear that carbon steel and some low-alloy steels might be considered for use in reactor construction with pressurized water systems, providing that all of the several conditions necessary to their suitable performance could be accurately maintained during reactor operation. Considerably more work will be required to determine whether this can be accomplished.

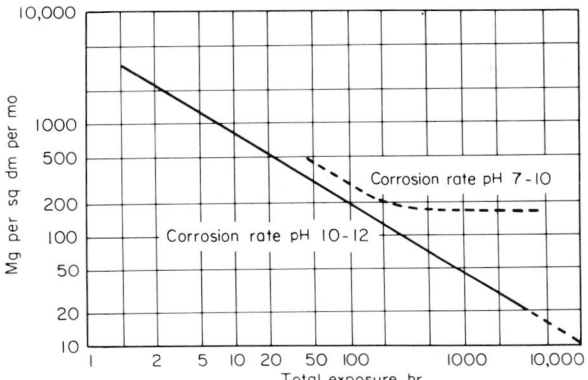

Fig. 10. Instantaneous descaled corrosion rate of carbon steel in high-temperature water. (*Taken from Ref.* 25.)

Nuclear Radiation—Effects on Materials[29]

Nuclear radiation under high flux density has a significant effect upon some of the mechanical and physical properties of metals and alloys which, in some cases, might affect corrosion resistance. The structural materials, such as carbon and alloy steels, stainless steels, and high-nickel alloys, are affected by exposure to nuclear radiation in that there is an increase in ultimate strength, yield strength, and hardness with loss in ductility. The carbon and alloy pressure-vessel steels suffer a loss in impact resistance and an increase in the nil-ductility and the ductile-brittle transition temperature. There is some effect of ferrite precipitation in the stainless steels. Basically, radiation effects are more significant in materials that are located in the high-neutron flux densities near the core. Consequently, piping systems which do not circulate uranium fuel solutions, even though welded to the reactor vessel, are not considered to suffer the above-mentioned losses. However, the first piping joints at the reactors vessel must be examined to determine if radiation effects will be significant.

Effects in Structural Metals. Mechanical property changes induced by irradiation can be substantial in magnitude and thus strongly influence selection of materials and design of structural components. Increased yield strength, tensile strength, hardness, and loss of ductility almost always result from neutron bombardment of metals at moderate temperatures. Irradiation is generally able to add strength to metals in almost any initial condition, but as with the other strengthening mechanisms, the effect is more pronounced with softer starting conditions. It appears that irradiation strengthening is of about the same magnitude as can be obtained by other means. It is observed that yield strength generally increases more rapidly than tensile strength. This does not necessarily imply a decrease in plasticity prior to necking, although such a decrease is usually observed. One of the most striking effects on mechanical properties is the observed large increase in

yield stress. This suggests that "alloying" with interstitial atoms is occurring even though the yield-stress value is greatly in excess of that obtained by alloying.

Unusual effects in plastic deformation are found in many irradiated metals. The normal yield point in mild steels under radiation is altered to continuous yielding followed by a region of low work hardening, an inflection point, and then "normal" increase in hardening up to maximum load. Face-centered cubic metals, however, which do not normally have a yield point may show a type of "yield point" after irradiation.

Ductility and toughness are usually reduced by irradiation. The ductility, as measured by reduction of area and elongation, is generally reduced, although in some pure metals and in a few precipitation-hardening or cold-worked alloys there is some increase in elongation after irradiation. Metals with low initial ductility appear to have relatively greater reductions in ductility due to irradiation than metals with high initial ductility. In some metals, radiation may reduce the ductility below the minimum values normally required in various specifications.

In some materials, e.g., carbon steels, notch-impact tests rather than elongation or reduction of area measurements are used to indicate a type of toughness. In the carbon steels, irradiation causes a loss of energy absorption in notch-impact tests and an increase in the temperature of transition between brittle and ductile behavior.

In an attempt to understand radiation effects, reasoning by analogy with other techniques for altering properties of metals and alloys has been used considerably. Comparisons of radiation effects with those of cold working have been made by a number of investigators. The fact that almost all metals and alloys show increases in strength and hardness, decrease in ductility, and slight density changes, as is also the case with cold working, has led to this comparison. In some materials, cold working appears to be a poor analogy and solid solution hardening appears to provide a better one, e.g., in copper. For processes which are controlled by diffusion, irradiation appears equivalent to an increase in temperature of the order of 150 to 350 F.

Effects in Ceramics. Ceramic materials are inorganic, nonmetallic, solid materials which are processed or used at high temperature and include such diverse materials as oxides, carbides, cement, glasses and graphite (all of nuclear interest) as well as a number of other materials. Use of some of these materials in reactors has been rather limited, yet ceramics have potential use in all structural components.

Ceramics are relatively resistant to radiation damage but are not so resistant as most structural metals. There is little loss of strength. Swelling, i.e., a decrease in density, is observed in many ceramics, although some glasses increase in density when they are irradiated.

Graphite has been widely used as a moderator and is a potential structural material. "Graphite" is a generic term as is "stainless steel," and there are wide variations in properties as a function of carbon source and method of processing. Changes in properties under irradiation are due primarily to displacement of atoms.

Room-temperature irradiation of graphite usually causes an increase, often of several per cent, in physical dimensions. This dimensional change is temperature-sensitive, and some commercial graphites contract, rather than expand, under irradiation above about 600 F.

Bending strength, compressive strength, and modulus of rupture increase rapidly to about three times the initial value under room-temperature irradiation. As exposure is increased, there is a decrease in strength and modulus to about double the initial value, with little change from this doubled value with continued exposure.

Thermal conductivity decreases markedly under room-temperature irradiation. This reduction may be as large as by a factor of 50.

The energy content of nearly all materials is increased under irradiation. This

"stored energy" is not very important in most materials, e.g., metals, but can be very important in graphite. The stored energy can be measured as an apparent increase in heat of combustion or as an apparent decrease in specific heat. This storage of energy, coupled with the dimensional changes, requires heating of the graphite to remove the effects as indicated later.

Effects in Organic Materials. Organic materials are complex molecules of carbon, hydrogen, oxygen, and other elements based on the carbon-carbon covalent bond. These are of interest as moderators and coolants. In general, these materials are about 1,000 times more sensitive to radiation than the metals. The aromatic hydrocarbons, i.e., the polyphenyls and fused-ring aromatics such as naphthalene, based on various combinations of the benzene ring, appear to be among the most stable of the organic materials under radiation.

Irradiated organic liquids tend to increase in viscosity up to the point of conversion into a cokelike substance. Oils and greases, for example, suffer large increases in viscosity, and greases tend to become hard and "gritty" owing to the molecular breakdown and polymerization. Radiation damage occurs principally from ionization effects and appears to proceed by two main routes, i.e., cross-linking and cleavage. *Cross-linking* leads to the formation of high-molecular-weight compounds by a coupling process. *Cleavage* leads to fracture of the molecular chain and reduction of molecular weight. Both processes are accompanied by the evolution of gas which consists of hydrogen with smaller amounts of low-molecular-weight compounds.

Many organic solids have an increase in tensile strength and softening temperature in the early stages of radiation. These may be desirable, although the accompanying increased hardness and decreased plasticity may not be. Continued exposure, accompanied by embrittlement, which may be followed by cracking, will normally decrease the strength still further. Elastomers which harden under radiation may decrease to nearly zero in tensile strength followed by a very large increase in strength with continued exposure. This is because of the transition of the elastomer (e.g., natural rubber) to a glasslike material. Cross-linking usually causes an increase in density and a resulting shrinkage. At the same time, the evolution of gas may give a net decrease in density with noticeable swelling.

Effect of Radiation on Thermal Stresses. The absorption of radiation, i.e., neutrons and especially gamma rays, in structural components results in the deposition of energy and internal heating of the components. This is essentially unique in reactors and produces temperature distributions which are appreciably different from those in more conventional power plants. The distribution of thermal stresses and strains is changed accordingly. No other effects of radiation on thermal stresses and strains should be expected.

Effects of Temperature on Radiation Damage. Radiation damage consists essentially of "isolated" vacancies and interstitials formed in knock-on cascades plus momentary heating in localized regions. This is not a final state, however, since these defects have considerable mobility at all but the very lowest temperatures and can interact with one another. In other words, they can rearrange or anneal continually during bombardment. Studies have shown that irradiation must be carried out at very low temperatures for all the radiation damage to be "frozen in" as it occurs. This temperature depends on the property, since some effects "anneal" out at much lower temperatures than others; e.g., effects on electrical conductivity are annealed at very low temperatures. In fact, some of the defects introduced in common metals are mobile at -400 F.

The annealing process is complex, and it is not at all certain how it is accomplished or to what extent the individual defects combine or aggregate into larger complexes. It must be recognized, however, that this annealing stage of the

radiation-damage process is fully as important as the defect-causing stage. There are two types of annealing, namely, thermal annealing and radiation annealing.

If certain properties of a given material are measured at room temperature, irradiated at specified temperatures and properties remeasured, it is highly likely there will be some changes. If the material is heated to a temperature somewhat higher than that of irradiation and again the properties are measured after cooling to room temperature, there is a further change in properties in the direction of reversion to the pre-irradiation values. This is *thermal annealing*.

In the spike or atomistic effects, the cascade of point defects is partially annealed by the associated brief temperature increase. These defects can also be annealed by other spikes formed near by. This is *radiation annealing*. It is possible to separate radiation and thermal annealing experimentally by comparing the rates of annealing in the presence and absence of radiation.

In both types of annealing, the rate of annealing increases as the number of defects increases. Thus, under constant conditions of irradiation, the rate at which damage accumulates should decrease as the radiation dose increases until a saturated state is reached with an essentially constant level of damage. This saturation effect has been observed in many materials. The degree of damage at saturation will vary greatly from one material to another. Metals, for example, are still usable although with altered properties. Most of the organic materials, however, are completely altered at and often before saturation.

Thermal annealing can be used to remove radiation effects from many materials. The rate and amount of "repair" depend on the material, the type of damage, the extent of damage, and the annealing temperature. An appreciable restoration of pre-irradiation properties can be obtained in many materials at relatively low temperatures, but extremely high temperatures may be required for essentially complete restoration. Graphite, for example, will release a large fraction of its stored energy or will regain much of its loss of thermal conductivity upon annealing at 392 F. Annealing above 3632 F, approaching the graphitization temperature, however, may be required for substantially complete recovery.

In many materials, there appears to be a great difference in radiation sensitivity as a function of temperature. The polyphenyls, for example, show an abrupt dependence of damage on irradiation temperature above about 750 F, with damage increasing very rapidly with increasing temperature. This is related to temperature stability apart from radiation effects, but the evidence suggests that the combination of radiation and high temperature is far more destructive than heat alone.

Irradiation of several reactor pressure-vessel steels has shown that the ductile-brittle transition temperatures of these steels can be raised as much as several hundred degrees Fahrenheit. The degree of change, however, depends on both irradiation temperature and total neutron dose. The prediction of irradiation effects on reactor vessels therefore requires accurate knowledge of both neutron dose and operating temperature.

Design[19]

Design establishes the basic approach for pressure-equipment applications. The initial step carefully performed assesses and establishes the service requirements, which then leads to selection of detail requirements for design, fabrication, materials, quality control, and inspection. Design for critical service requires that stress raisers be eliminated or reduced to the maximum possible extent. This extent is readily evident for radioactive nuclear piping systems but requires a greater exercise of judgement for nonradioactive nuclear service, which must be evaluated as to the hazard which their failure introduces to lines of hazardous contents.

Where short, stiff runs of pipe must be used because of space limitations, the analysis for flexibility must include the direct and shear forces.   The stresses induced locally by terminal reactions and other similar effects must be investigated and controlled for adequate fatigue life.   The theoretical work performed by P. Bijlaard at Cornell University, under the auspices of the Pressure Vessel Research Committee and the U.S. Navy Bureau of Ships, is very helpful for complex design. Where it is practical, mock-up tests are desirable, particularly for locations where the assumptions of theory are susceptible to error.

The design for thermal stress must include, if applicable, the effects of gamma and fast-neutron heating.   Thermal stresses should be combined with other variable loads on a strain range basis in order to assess fatigue potential totally.   The foregoing, in turn, permits the establishment of a minimum wall thickness to carry the pressure and other external loads to avoid general yielding, which, at the same time, is favorable for minimizing thermal stresses and terminal reactions.

Where different thicknesses must be joined, discontinuity stresses are introduced. Such junctions also intensify thermal stresses due to sudden temperature changes unless the transition in thickness is sufficiently gradual.   These effects can be minimized by making a gradual transition between different thicknesses, by tapering the thicker section at an angle of 15 deg, or where practical, by extending the taper over an axial distance equal to the square root of the arithmetic mean radius times the wall thickness or

$$\text{Taper distance} = \sqrt{r_{\text{mean}}(t_{\text{heavy wall}})}$$

Joints between dissimilar materials also introduce local stresses in proportion to their respective coefficients of expansion and modulii of elasticity and, for transient or heat-transfer conditions, their specific heats and thermal conductivities.   In the case of clad materials, the evaluation at welded joints must take into account the effect of any difference in the analysis of the weld metal and the difference in thickness and depth of the alloy weld deposit compared with the cladding thickness. Clad materials must also be assessed for fatigue life.

Structural Loadings and Attachments

The structural loadings imposed at supports, hangers, anchors, braces, and similar members must be investigated as thoroughly as are the pressure parts.   All too often, these items are merely analyzed by conventional structural design for the attachment itself and its load.

The local stress induced in the pressure wall must be evaluated along with the attachment detail.   Welded connections to nuclear piping should be avoided in so far as is practical.   Where welding must be used, then quality control and inspection must equal that of a pipe-to-pipe joint.   For piping subject to sudden temperature changes, as in liquid-sodium lines, externally welded attachments or supports locally increase temperature stresses by introducing additional restraint and slower response to temperature change.

Fabrication, Erection, and Inspection

The effectiveness of *fabrication* is greatly increased if difficult welds are avoided. The use of cast, wrought, or forged fittings eliminates branch-to-run corner welds. In addition, butt-welded joint fittings permit the taking of meaningful radiographs with increased reliability as the end product.

Equally important is the preparation of a uniform welding end which is essential for high weld quality.   In addition, freedom from root defects or shoulders reduces the susceptibility of the weld inside surface to corrosive attack.

Field assembly introduces many problems which range from fully restrained and position welds to organization difficulties, personnel supervision, inspection, and quality control.   Specifying all possible shop fabrication does not eliminate the problems but does minimize them.   However, even with a minimum of field erection, the lowering of weld quality of the remaining joints may render the quality of the shop-fabricated items of no value.   Consequently, expert supervision, competent quality control, and inspection with no compromise are of the utmost importance.

Temporary welded attachments, tack welds, or other similar construction aids must be avoided or carefully analyzed to account for potential sources of fatigue failure and fracture initiation.

Maintenance

The problems of access to nuclear piping which either carries a radioactive fluid or is subject to radiation effects and the difficulties and expense of decontamination or servicing and repairing are well recognized.   Consequently, overall economics of a nuclear plant greatly favor the exercise of all measures which eliminate or reduce the necessity for replacement or repairs.   This starts with establishing sound operational demands, particularly as to the severity of operation and repetitive or surge effects which combine to establish fatigue demands.   The sound approach to maintenance continues with the use of design, materials, and fabrication details amenable to thorough examination and by applying all practical, available, searching forms of examination and test to establish initial integrity.

Equipment and piping layout are planned for maximum practicable access.   In radioactive areas remotely controlled instruments can be utilized for periodic examination.

MATERIALS SELECTION, QUALITY CONTROL, AND NONDESTRUCTIVE TESTING[1–3]

To date four forms of austenitic stainless-steel pipe are being employed in nuclear piping systems:
    1. Seamless wrought pipe (drawn, extruded, forged)
    2. Longitudinal-welded formed plate pipe
    3. Centrifugally cast pipe
    4. Centrifugally cast hydroforged pipe
Each form of pipe must develop the same degree of reliability if, for example, all four were used in the same system.   The basic differences in manufacture call for different (and sometimes the same) types of nondestructive testing and quality control.   Where quality control and nondestructive testing are similar, for example, in centrifugally cast pipe and longitudinal welded pipe, the criteria for acceptance would be different.

The carbon steels, ferritic steels, and high-nickel alloys are considered acceptable for nuclear piping.   The designer must recognize the problems of corrosion, pressure and temperature, and radiation effects.   Quality control and nondestructive testing must be compatible with the degree of reliability established by the designer.

Seamless Wrought Pipe, Longitudinal-welded Pipe, Wrought and Forged Fittings.   The various grades of austenitic stainless steels in the 300- series which are acceptable for use in nuclear piping are Types 304, 304L, 316, 316L, 321, 347, 348, 309, and 310.   Pipe and fittings are required to conform to at least the

following specifications:

ASTM A376 Pipe	ASTM A430 Pipe
ASTM A358 Pipe	ASTM A403 Welding Fittings
ASTM A312 Pipe	ASTM A182 Forgings

The various supplementary requirements in the foregoing ASTM Specifications are applied at the designer's request. Presently all the above ASTM Specifications are recommended for nuclear piping except for the welded pipe of ASTM A312.

The 300- series austenitic stainless steels are acceptable for use in centrifugally cast pipe. The chemistry and tensile requirements are given in ASTM A351. The "as-cast" pipe and the cast pipe with subsequent cold work by hydrostatic pressure are considered suitable for nuclear piping.

Castings—Welding Fittings, Valve Bodies, and Pump Bodies. The basic material requirements for cast fittings, valves, and pumps are identical with those of centrifugally cast pipe except that tensile requirements for cast fittings, valves, and pumps are developed from test bars attached to the keel blocks of the casting and, for centrifugally cast pipe, tensile properties are determined from sample rings cut from one end of the pipe.

QUALITY CONTROL AND NONDESTRUCTIVE TESTING

All piping, fittings, valve bodies, and pump bodies are specified to meet the following inspection and leak-testing requirements in addition to those specified in the selected ASTM Specifications and those specified in the Code for Pressure Piping unless otherwise covered in specific code cases.

Welds: Longitudinal, Circumferential, Fillet, and Repair. All nuclear piping welds which are subject to stress caused by pressure are inspected 100 per cent by radiography, except where this is impractical, as in the case of socket welds or seal welds. Whenever possible, all weld joints are designed to permit radiographic examination. Radiography, nevertheless, must be performed in accordance with the best obtainable practice, and when radiography is impractical, welds are examined by other nondestructive methods, such as those of the liquid penetrant or ultrasonic processes, to prove their soundness. When ferromagnetic materials are incorporated in piping, magnetic-particle examination may be performed.

Base Material: Seamless Wrought Pipe, Longitudinally Welded Pipe, Wrought and Forged Fittings. When the fluid handled warrants additional assurances of quality, such as the reactor primary fluid, the designer shall specify that all finished pipe shall have a liquid-penetrant or magnetic-particle (for ferromagnetic materials) examination of all accessible areas. The piping also is inspected 100 per cent by radiography, ultrasonics, or eddy current or some combination thereof.

In the case of longitudinally welded pipe, except the welded grade of ASTM A312, the intent of this requirement is met if the plate from which the pipe is made has been ultrasonically tested and the longitudinal weld has been 100 per cent radiographed. In the case of seamless pipe, the intent of this requirement is met if the completed pipe is examined by ultrasonics or eddy current. In addition, the surfaces of both types of pipe are examined as described above.

Base Material: Cast Welding Fittings, Cast Valve Bodies, Cast Pump Bodies, and Centrifugally Cast Pipe. Stainless-steel cast materials have received overall acceptance in nuclear piping systems. The acceptance of these materials, in no small part, is a result of the rigorous inspection and quality

Table 17. Nondestructive Testing Requirements for Nuclear Piping

Test	Wrought pipe	Welded pipe	Welded fittings	Cast pipe	Cast fittings
Radiography subsurface examination	For circumferential welds and repair welds, if permitted	For all filler metal welds and repair welds, if permitted	For all filler metal welds and repair welds, if permitted	For 100% of pipe, filler metal welds, and repair welds	For 100% of fitting, filler metal welds, and repair welds
Ultrasonic subsurface surface examination	As required for 100% of the surface	For longitudinal weld and pipe. Weld made without filler	For longitudinal weld and pipe. Weld made without filler	Not applicable	Not applicable
Eddy-current subsurface and surface examination	Alternate for ultrasonic for 100% of the surface	Alternate for ultrasonic	Not applicable after forming	Not applicable	Not applicable
Liquid penetrant surface examination	For 100% of the pipe depending on degree of reliability or as designer requires	For 100% of filler metal welds when radiography is impractical and for 100% of the pipe or as designer requires	For 100% of filler metal welds when radiography is impractical and for 100% of the pipe or as designer requires	For all accessible surfaces	For all accessible surfaces
Helium or other gaseous leak testing	For all high-level radiation piping as in off-gas or waste-gas systems	For all high-level radiation piping as in off-gas or waste-gas systems	For all high-level radiation piping as in off-gas or waste-gas systems	Only when gaseous fission products exist	Only when gaseous fission products exist
Hydrostatic pressure test	$1\frac{1}{2}$ times working pressure	$1\frac{1}{2}$ times working pressure	$1\frac{1}{2}$ times working pressure	$1\frac{1}{2}$ times working pressure, but not to exceed 90% of yield	$1\frac{1}{2}$ times working pressure, but not to exceed 90% of yield

NOTE: Magnetic-particle examination may be employed for surface examination when testing magnetic materials.

control maintained by responsible manufacturers. In addition, sound metallurgical knowledge on the part of manufacturers and systems designers has added to the reliability and use of cast components in critical nuclear-piping systems. The criteria and requirements for cast components are published by the ASA Code for Pressure Piping, B31.1 as Code Case N-9, Centrifugally Cast Austenitic Steel Pipe for Nuclear Service, and Code Case N-10, Cast Austenitic Butt-welding Fittings for Nuclear Service. These criteria and requirements were developed by the cooperation of manufacturers, nuclear-piping designers, and the operators of nuclear systems. A summary of these requirements is as follows:

1. Each length of pipe and each fitting shall be completely radiographed.

2. All accessible surface of pipe and fittings must be examined with a liquid penetrant.

3. Pipe and fittings shall be given a pressure test of at least 1.5 times the rated pressure and shall show no leaks with the test pressure held for 30 min per inch of wall thickness.

4. It is the responsibility of the designer to require a mass-spectrometer or other gaseous-leak test where warranted by service condition or welding requirements. Typical of these conditions is that, if gaseous products of nuclear fission are anticipated, then mass-spectrometer or other gaseous-leakage tests shall be performed on all cast pipe and fittings.

5. Where cast pipe is to be cold-formed or hot-formed, the *material ductility shall be compatible with the type of forming*.

6. The maximum allowable stresses for centrifugally cast pipe and cast fittings are derived from Table P-7 of the ASME Boiler and Pressure Code, Section I. The stress values are based on a casting quality factor of 1.0 when all requirements of quality control and testing are met.

Centrifugally cast pipe with subsequent cold work by hydrostatic pressure takes on all the characteristics of wrought pipe and is formed, welded, and tested accordingly.

Procedures and Acceptance Standards for Nondestructive Testing. Nondestructive testing is only as good as the procedures for performance, the equipment employed, and the personnel performing the tests and interpreting the results. Qualification of equipment, procedures, and personnel cannot be taken lightly, as the ultimate reliability of a plant or component is in part dependent upon the quality of nondestructive testing.

The physics of *radiography* (X-ray and gamma-ray) is well defined, and knowledge of it is a requirement for procedure qualification. Excellent procedures now exist and are used to obtain definitive and meaningful radiographs. Specific procedures are typified by the following:

NAVSHIPS 250-1500-1

MIL-STD 271 (BUSHIPS)

ASTM E94

Ebasco Services Incorporated Specification 73-Latest revision

The "Non-destructive Testing Handbook," 1959 edition, contains all the fundamentals to develop radiography procedures that are consistent and definitive.

Nuclear piping for other than naval systems employs the following acceptance standards:

Welding—Paragraph P-102h of Section I and paragraph UW51 of Section VIII of the ASME Boiler and Pressure Vessel Code

Castings—ASTM E71 for the class of casting desired

Naval systems use the acceptance standards of NAVSHIPS 250-1500-1 and MIL STD 271 (BUSHIPS).

The basic data to develop a reliable *ultrasonic* testing procedure are given in the "Non-destructive Testing Handbook," 1959. Satisfactory specifications for

ultrasonic testing are contained in the following:

ASTM E-114 and E-164

NAVSHIPS 250-1500-1

MIL-STD 271 (BUSHIPS)

Ebasco Specification 73-Latest revision

Nuclear piping for other than naval systems employs the following standards for the *acceptance* of material which has been tested by ultrasonic means:

Welding, wrought materials, and cast materials of acceptable grain size—ASME Nuclear Code Case 1275N and an acceptance notch of 3 to 5 per cent of wall thickness or 0.015 in. in depth, whichever is greater

Naval nuclear piping refers to the Acceptance Standards in NAVSHIPS 250-1500-1 and MIL STD 271 (BUSHIPS).

Magnetic-particle testing procedures are described in the following specifications:

ASTM E109

ASTM E138

ASTM E125

MSS-SP-53

NAVSHIPS 250-1500-1

Nuclear piping systems other than naval or military reactors use the following *acceptance standards* for base materials and weldments which have been subjected to magnetic-particle inspection:

All linear discontinuities and aligned indications, whether porosity or slag, shall be removed. Aligned indications are those in which the average of the center-to-center distances between any one indication and the two adjacent indications in any straight line is less than ³⁄₁₆ inch.

Military and naval reactors are referred to MIL STD 271 (BUSHIPS) and NAVSHIPS 250-1500-1 for acceptance standards.

The testing of nuclear materials by use of liquid-penetrant procedures (visible dye and fluorescent penetrant) is described in the following:

ASTM E165

NAVSHIPS 250-1500-1

MIL STD 271 (BUSHIPS)

Nuclear systems other than naval or military reactors use the following acceptance standard for base materials and weldments which have been inspected by liquid penetrant procedures:

All linear discontinuities and aligned indications, whether porosity or slag, shall be removed. Aligned indications are those in which the center-to-center distances between any one indication and the two adjacent indications in any straight line is less than ³⁄₁₆ inch.

Naval and military reactors are referred to NAVSHIPS 250-1500-1 and MIL STD 271.

The nondestructive testing of nuclear materials by the use of *eddy-current procedures* is covered in the "Non-destructive Testing Handbook," Vol. II, 1959, and in Specification MIL STD 271 (BUSHIPS).

Piping and tubing for reactor systems, other than naval or military, which have been tested by eddy-current procedures depend upon the following standard:

Any pipe, tube or weld that has a discontinuity resulting in an indication exceeding or equivalent to a 3 to 5 per cent or 0.015 inch deep notch (whichever is greater) shall not be acceptable.

Naval and military reactors rely upon the standards in MIL STD 271 (Ships).

Indications which are detected by any of the inspection methods and which are

defined as unacceptable by the applicable acceptance standards listed above must be removed, repaired, and reinspected by the same methods.

The provisions listed above need not apply to auxiliary piping systems such as off-gas, waste gas, spent-resin transfer, or other waste disposal and other auxiliary lines where temperatures of 212 F are not exceeded. Such auxiliary systems, in and near which radiation levels are expected to be high, shall be designed and leak-tested as described under "Leak Testing, Design and Construction for Piping with High Radioactivity Levels," to follow later.

General Leak Testing. Prior to initial operations, all nuclear piping systems shall be tested for leak-tightness by a method of sufficient sensitivity to meet the requirements of the system design. It is the responsibility of the designer to specify the method required and the leakage permitted on the basis of the system requirements.

Leak-tightness may be satisfied by such tests as a hydrostatic test, a pneumatic test, a soap-bubble test, a halide leak test, or a helium mass-spectrometer test.

The method which the designer must specify must have a sensitivity at least as good as a hydrostatic or pneumatic pressure test for the service conditions. The following hydrostatic test is a minimum standard or guide to which the specified test must at least be equivalent.

A *hydrostatic test* must show no leaks with a test pressure held for a period of 30 min per inch of wall thickness where the test pressure is at least 1.5 times the design pressure of the piping, except that the stress developed at the test pressure must not exceed 90 per cent of the specified minimum yield strength of the materials used in the system components based on the minimum specified wall thickness. If the operating fluid or piping material would be adversely affected by the test water, other fluids compatible with the operating fluid may be used provided adequate precautions are exercised commensurate with the characteristics of the test fluid.

In the event that a *pneumatic-pressure test* or other test using a gaseous test medium is used, the test pressure shall not exceed 110 per cent of the design pressure.

Minimum requirements for leakage rate are as follows:

1. *Hydrostatic Test.* The liquid under pressure must not appear visibly anywhere on the surface of the component or system being inspected.

2. *Pneumatic and Soap-bubble Test.* Soap bubbles must not be generated by the gas under pressure anywhere on the surface of the component or system being inspected.

3. *Pneumatic Submerged Test.* Gas bubbles must not be generated anywhere on the surface of the component or system being inspected.

4. *Helium Mass Spectrometer Test.* The surfaces of the component or system must not have a leakage rate greater than 5×10^{-8} atm per cc per sec of helium. In addition, joints in gas-cooled reactor piping must be capable of maintaining a leak rate less than 10^{-3} cc of helium per hour per inch of joint diameter at a differential pressure of 15 psi.

5. *Halide Leak Test.* The designer defines the rate of leakage desired.

Leak Testing and Design and Construction for Piping with High Radioactivity Levels. Certain nuclear-plant auxiliary systems which operate below 212 F require additional precautions in design and fabrication beyond those that would normally be associated with conventional piping systems at the same temperatures and pressures. These additional precautions should be applied to such systems if the radiation levels in the vicinity of the system due to the contents of the system can be expected to be high, either continuously or intermittently. These requirements will apply specifically to those systems which lie partially or totally outside the reactor enclosure containment vessel. Piping for such systems may be

run within enclosed passageways specifically for piping, may be enclosed in a guard line, or, in the case of off-gas or waste gas systems where loss to the ground would not create a hazard, may be prepared for direct burial.

Piping systems for installation in tunnels or guard lines must be of welded construction and should utilize a generous corrosion allowance sufficient to assure against failure from corrosion, both external and internal, taking into consideration the deleterious effects of intermittent operation of such system if applicable. Attachment to equipment and valves may be, at the designer's option, flanged connections.   When flanged connections are considered, leakage rates must be carefully evaluated.

The design of the piping for off-gas or waste-gas systems must take explosion hazards into consideration where such exist and incorporate adequate protection either to prevent or to withstand an explosion.   As a minimum, explosion protection consists of designing the piping for a pressure considerably in excess of the normal operating pressure.   It is the designer's responsibility to set this design pressure commensurate with the service and the potential hazard involved.   In no case should the system design pressure be less than 50 psig.

Any portion of off-gas or waste-gas piping which is for direct burial must be of all-welded construction, must be provided with means of protection against external corrosion where necessary, and must utilize a corrosion allowance sufficient to assure against failure from internal corrosion.   External corrosion protection may take the form of a corrosion allowance, cathodic protection, or external coating, depending upon the soil conditions involved.   In the case of external coatings, it is the designer's responsibility to specify coating materials and installation procedures suitable for the particular conditions encountered.   The protective coating should be inspected and tested completely by use of recognized "flaw detector" before or after backfilling.

Piping covered by this case should be leak-tested in accordance with the procedure outlined previously.

It is the designer's responsibility to specify the degree to which the radiography or fluid-penetrant examination of the pipe joints shall be employed.   Weld-joint efficiency factors commensurate with the degree of radiography employed must be used.   A joint efficiency factor of 1.00 may not be used unless the piping joints have been 100 per cent radiographed.

JOINING OF PIPE

Welded Joints.   Welded joints may be used in any materials suitable for nuclear piping for which welding procedures, welders, and welding-machine operators have been qualified within the limits of Section IX, ASME Boiler and Pressure Vessel Code, or the applicable Military Code.

Butt welds, circumferential and longitudinal, are made without the use of non-removable backing rings.   The following joints are acceptable types:

1. Consumable insert (Electric Boat or Grinnell types) root pass using tungsten inert-gas welding and balance of weld by coated electrode metal arc

2. Filler metal root pass using tungsten inert-gas welding and balance of weld by coated electrode metal arc

3. Double butt-welded using either inert-gas or coated electrode metal arc welding

Socket-welded pipe joints are not used in nuclear piping where severe corrosion or erosion is to be expected.

Threaded Joints.   Threaded joints with fillet welding to assure a seal may be used.   However, the sealing weld may not be considered as contributing any

strength to the joint. The use of threaded joints whether seal-welded or not is not recommended. An all-welded design is recommended.

Flanged Joints. Flanged joints shall be kept to an absolute minimum. These joints shall also meet the leak-rate requirements established for the other portions of the piping system.

Other Jointings. *Calked* or *leaded* joints of the bell-and-spigot type, *brazed* and *soldered* joints, and *expanded* joints may not be used for nuclear piping.

Flared, flareless, and *compression* joints may be used for tubing sizes not in excess of 1 in. OD if such joints have been proved adequate and safe for the design conditions.

Slip, sleeve, and *swivel-type* joints are not to be used in nuclear piping unless prototypes of them have been tested for all service conditions.

Valve bonnet joints must have a positive sealing device such as capped, bellows, welded, diaphragm, double packing with lantern gland or some other provision which assures that leakage can be contained within the main or auxiliary nuclear piping systems. The foregoing does not apply to valves 2 in. or smaller when used in water or steam service provided the valves are of the back-seating type.

REFERENCES

1. Code for Pressure Piping and Nuclear Code Cases, ASA B31.1, 1955.
2. Klapper, J. A., "Standardization of the Procedure to Qualify Materials for Nuclear Piping Systems," *Paper* 42 presented at Nuclear Congress, New York, Apr. 4–7, 1960.
3. Klapper, J. A., "Criteria and Requirements for Nuclear Piping, Valves and Pumps," *Paper* 95 presented at Nuclear Congress, New York, June 4–7, 1962.
4. Yankee Atomic Electric Plant, *Westinghouse APD Publ.*
5. Yankee Atomic Electric Plant, *USAEC Publ.*
6. "Experience with the First Atomic Power Projects," Nuclear Progress Report, *Elec. Light Power*, Nov. 1, 1961, pp. 19–34.
7. Kramer, Andrew W., "Boiling Water Reactors," prepared under contract with the U.S. Atomic Energy Commission, Addison-Wesley Publishing Company, Inc., Reading, Mass., September, 1958.
8. "Liquid-metals Handbook," Sodium-NaK Supplement, Atomic Energy Commission, Department of the Navy, TiD5277, Washington, D.C., July, 1, 1955.
9. Dickinson, R. W., R. C. Gerber, and C. L. Larson, "Hallum Nuclear Power Facility Prototype for Advanced Sodium Cooled Power Stations," *AIEE Paper* 61-988, presented at the AIEE-ASME National Power Conference, San Francisco, Sept. 24–27, 1961.
10. Eggen, D. T., A. M. Stelle, and M. Heisler, "Design and Development of the Coolant System for the Sodium Reactor Experiment," contributed by the ASME and presented at the Second Nuclear Engineering and Science Congress, Philadelphia, Mar. 11–13, 1957.
11. Peters, Norman T., "Nuclear Piping," presented to the Pipe Fabrication Institute, Savoy Hilton Hotel, New York, Oct. 21, 1960.
12. "1958 Annual Report," Power Reactor Development Company, Detroit.
13. "APDA Annual Report," Atomic Power Development Associates, Detroit, 1958.
14. Le Clair, T. G., "Status of the High-temperature Gas-cooled Reactor," *Gen. Atomic Rept.* GA-2566, presented at the National Power Conference, San Francisco, Sept. 25, 1961.
15. Trilling, C. A., "The OMRE—A Test of the Organic Moderator-coolant Concept," *Paper A/Conf.* 15/P/421, presented at the Second United Nations International Conference on the Peaceful Uses of Atomic Energy, Geneva, June, 1958.
16. Wilson, R. F., and B. L. Hoffman, "Organic Reactor Program Status," *Rept. NAA-SR-MEMO*-4076, to the U.S. Atomic Energy Commission, Atomics International, Canoga Park, Calif., July 14, 1959.
17. "The Organic Moderated Reactor," Articles 1 to 8, *Nuclear Eng.*, February, 1960.

18. "Research Reactors, Selected Reference Material," United States Atomic Energy Program, Government Printing Office, Washington, D.C. 1955.
19. Murphy, J. J., C. R. Soderberg, H. S. Blumberg, and D. B. Rossheim, "Fabricated Pressure Piping as Related to Nuclear Applications," *Paper 57-NESC*-103, presented at the Second Nuclear Engineering and Science Conference, Philadelphia, Mar. 11–14, 1957.
20. Friend, W. Z., Corrosion Problems in Nuclear Power Stations," *Proc. Am. Power Conf.* vol. XVIII, 1956.
21. Datsko, S. C., "Corrosion of Metals in High-temperature Water at 500 F and 600 F," *Argonne Natl. Lab. Rept. ANL*-5354, Chicago, Ill., 1954.
22. Fowler, Jr., R., D. L. Douglas, and F. C. Zyzes, "Corrosion of Reactor Structural Materials in High-temperature Water, II, Static Corrosion Behavior at 600 to 680 F," *Gen. Elec. Co. Rept. KAPL*-1248, Schenectady, N.Y., 1954.
23. DePaul, D. J., "Corrosion Engineering Problems in High-purity Water," paper presented at 11th Annual Conference, National Association of Corrosion Engineers, Chicago, March, 1955.
24. Williams, W. L., and J. F. Eckel, "Stress Corrosion of Austenitic Stainless Steels in High-temperature Waters," *F. Soc. Naval Engrs.*, vol. 68, pp. 93–104, February, 1956.
25. Blaser, R. U., and J. J. Owens, "Special Corrosion Study of Carbon and Low Alloy Steels," "Symposium on High-purity Water Corrosion," pp. 1–7, American Society for Testing Materials, Philadelphia, 1956.
26. Luce, W. A., "Erosion-Corrosion," *Ohio State Univ. Eng. Expt. Sta. News*, vol. 19, pp. 29–32, December, 1947.
27. Decker, J. M., H. A. Wagner, and J. C. Marsh, "Corrosion-Erosion of Boiler Feed Pumps and Regulating Valves," *Trans. ASME*, vol. 72, pp. 19–24, 1950.
28. Partridge, E. P., "The Optimum Boiler Water; a Continuing Problem in Chemical Engineering Design," paper presented at 4th Congress International du Chauffage Industriel, 1952.
29. Smith, C. O., "Environmental Factors Which Influence Materials Properties and Nuclear Reactor Design," *Paper 61-WA*-282, presented at the ASME Winter Annual Meeting, New York, Nov. 26–Dec. 1, 1961.

BIBLIOGRAPHY

1. American Society of Mechanical Engineers Boiler and Unfired Pressure Vessel Code, Sections I and VIII, 1959, and Nuclear Cases, 1959–1962.
2. "Nuclear Reactor Data," 2d ed., Raytheon Manufacturing Co., Waltham, Mass., December, 1956.
3. Alquist, F. N., "The Preparation and Maintenance of High-purity Water," in "Symposium on High-purity Water Corrosion," pp. 1–7, American Society for Testing Materials, Philadelphia, 1956.
4. Breden, C. R., et al. "Water Corrosion of Structural Materials," October, 1948–June, 1951, *U.S. Atomic Energy Comm. Rept. ANL*-4519, Government Printing Office, Washington, D.C., 1951.
5. Draley, J. E., and Greenberg, S., "The Application of Materials in Low-temperature Water and Organic Cooled Reactors," in "Nuclear Metallurgy—a Symposium on Behavior of Materials in Reactor Environment," American Institute of Mining and Metallurgical Engineers, New York, 1956.
6. Huntley, H. W., and S. Untermyer, "The Use of Water in Atomic Reactors," in "Symposium of High-purity Water Corrosion," pp. 8–18, American Society for Testing Materials, Philadelphia, 1956.
7. Roebuck, A. H., "Water Corrosion of Structural Materials," *Proc. Water Conf. Engrs. Soc. Western Penn.*, vol. 15, pp. 165–185, 1954.
8. Roebuck, A. H., "Corrosion," in U.S. Atomic Energy Commission, "The Reactor Handbook," vol. 2, pp. 193–233, Government Printing Office, Washington, D.C., 1955.
9. Roebuck, A. H., "Effect of Material Composition in High-temperature Water Corrosion," in "Symposium on High-purity Water Corrosion," pp. 27–34, American Society for Testing Materials, Philadelphia, 1956.

10. Thomas, D. E., "Aqueous Corrosion of Zirconium and Its Alloys at Elevated Temperatures," *Paper* 571, presented at the International Conference on Peaceful Uses of Atomic Energy, Geneva, Switzerland.
11. Wroughton, D. M., and D. J. DePaul, "Structural Materials for Use in the Pressurized Water Power Reactor," in "Nuclear Metallurgy—A Symposium on Behavior of Materials in Reactor Environment," American Institute of Mining and Metallurgical Engineers, New York, 1956.
12. Wroughton, D. M., J. M. Seamon, and P. E. Brown, "The Influence of Water Composition on Corrosion in High-temperature, High-purity Water," in "Symposium on High-purity Water Corrosion," pp. 19–26, American Society for Testing Materials, Philadelphia, 1956.
13. Shannon, R. H., "Nuclear Power Plant Safeguards and Containment," in 21st Annual Meeting, American Power Conference, March, 1959.
14. Stephenson, R., "Introduction to Nuclear Engineering" in Chemical Engineering Series, 1954.
15. DePaul, D. J., "Corrosion and Wear Handbook," Bettis plant operated for the U.S. Atomic Energy Commission by Westinghouse Electric Corp., TID 7006, March, 1957.
16. Rhodin, T. N., "Physical Metallurgy of Stress Corrosion Fracture" in a Symposium presented in cooperation with the Electrochemical Society, National Association of Corrosion Engineers, and the American Society for Testing Materials, Pittsburgh, Pa. April, 1959.
17. Shober, F. R., "The Effect of Nuclear Radiation on Structural Metals," *Defense Metals Inform. Center Rept.* 166, Sept. 15, 1961.
18. Collins, C. G., "Symposium on Radiation Effects and Radiation Dosimetry," presented at the 63d Annual Meeting of the American Society for Testing Materials, June 29, 1960.
19. Hawthorne, J. R., and L. E. Steele, "The Effect of Neutron Irradiation on the Charpy V and Drop-weight Test Transition Temperatures of Various Steels and Weld Metals," *Naval Res. Lab. Rept.* 5479, Washington, D.C., May 27, 1960.
20. Pellini, W. S., L. E. Steele, and J. R. Hawthorne, "Analysis of Engineering and Basic Research Aspects of Neutron Embrittlement of Steels," in "Symposium on Radiation Damage in Solids and Reactor Materials," *U.S. Atomic Energy Comm. Contrib. to Intern. Atomic Energy Agency*, at Venice, Italy, May, 1962.
21. Pellini, W. S., and J. E. Srawley, "Procedures for the Evaluation of Fracture Toughness of Pressure-vessel Materials, *Naval Res. Lab. Rept.* 5609, Washington, D.C., June 8, 1961.
22. Kanter, J. J., "On the Quality Requirements for Steel Valves for Nuclear Power Plants," *ASME Paper* presented at 2d Nuclear Engineering and Science Conference, March, 1957.
23. Cooper, H. J., "Principles Involved in the Attainment of High Quality Stainless Steel Valves and Castings for Nuclear Plant Components," paper presented at the Reactor Components and Accessories Manufacture Session, 4th Annual Conference of Atomic Industries Forum, October, 1957.
24. Koenig, R. F., and E. G. Brush, "Selecting Materials for Liquid Sodium Systems," Knolls Atomic Power Laboratory, General Electric Co., December, 1955.
25. Kasschau, Kenneth, "Design and Construction Problems of APPR-1," presented at ASME Semiannual Meeting, San Francisco, Calif., June, 1957.
26. Love, L. R., "Shippingport Operating Experience," paper presented at Winter Annual Meeting of ASME, New York, November, 1961.
27. Wade I. L., and Trocki, T. "Performance and Operating Experience of the Dresden Nuclear Power Station," paper presented at Winter Annual Meeting of ASME, New York, November, 1961.
28. International Nickel Company, Inc., Development and Research Division, "Qualification of Inconel Nickel-Chromium Alloy for Nuclear Power Plant Applications," *Status Rept. TA-P-4*, May, 1961.
29. International Nickel Company, Inc., Development and Research Division, "Welding of Inconel for Nuclear Power Applications," Aug. 22, 1958.
30. The Comptroller General of the United States, "Report on Review of Atomic Energy

Commission Contract No. AT(30-3)-222 with Yankee Atomic Electric Company," November, 1956.

31. Reed, G. A., R. J. Creagan, and W. C. Woodman, "The Yankee Atomic Electric Plant," *Paper* 56-*A*-166, presented at ASME Annual Meeting, New York, November, 1956.

32. "Symposium on Nondestructive Tests in the Field of Nuclear Energy," *ASTM Spec. Tech. Publ.* 223, Symposium sponsored by American Institute of Chemical Engineers, American Nuclear Society, American Society for Testing Materials, Atomic Industry Forum, and Society for Nondestructive Testing, presented in Chicago, Ill., April, 1957.

17

REFRIGERATION-SYSTEMS PIPING

Robert L. Jones*

The broad term refrigeration refers to a general science concerned with the use of producing temperatures below normal for commercial or other useful purposes. Refrigeration piping is defined as any piping which may be used in conjunction with refrigeration equipment for the purpose of transferring heat from a lower temperature to a higher one. Refrigerants are fluids which absorb heat by evaporating at a lower temperature and pressure and transfer heat out when they condense at a higher pressure and temperature. The increase in pressure necessary to elevate the temperature level is produced by a compressor of the reciprocating, rotary, or centrifugal types, or in the case of an absorption system, by the transfer of heat, thereby boiling the more volatile refrigerant out of a solution.

Fluids which do not change state are sometimes used to transfer heat in an indirect system. Such fluids are called brines. To be classified as a brine, the fluid must be used for the transfer of heat without a change in its state and it must have no flash point or a flash point above 150 F as determined by the American Society of Testing Materials method in ASTM Specification D93.

Many fluids have been used as volatile refrigerants in the evaporation, compression, condensing, and expansion cycle. This chapter will deal with application and structural design of piping for the more commonly used volatile refrigerants such as ammonia and some of the halogenated hydrocarbons. It will also cover general methods for other refrigerants where specific tables are not presented.

Since volatile refrigerants are used in the liquid, vapor, and mixture phases, each of these will be treated separately.

Many fluids have been used for brines. Originally, the term brines applied to salt solutions such as calcium chloride or sodium chloride. The use of such salt

* Carrier Air Conditioning Company, Syracuse, N.Y.

brines permitted the transfer of heat at lower temperature levels without introducing refrigerant of the volatile type into refrigerated spaces.   These brines were commonly used for cold storage plants, ice plants, or commercial and process refrigeration.

Solutions of glycols are also used as brines.   Ethylene glycol and propylene glycol are most commonly used for this purpose.   Several other compounds or mixtures have been developed specifically for the purpose of heat-transfer media. These compounds are specifically designed to have high thermal capacities, low viscosities, and other desirable properties for high heat transfer and low pressure losses.

Two major codes relate to refrigeration piping.   The first of these is the American Standard Safety Code for Mechanical Refrigeration,[1,*] ASA B9.1. This code is reviewed and revised periodically.   The most recent edition at the time of this writing was issued in 1964.   This code is sponsored by the American Society of Heating, Refrigerating, and Air Conditioning Engineers and has been adopted by many states and municipalities to be the existing law in these localities.   This Code will be referred to frequently in this section and will be designated as the ASA B9.1 Code.

Another important code on piping is the American Standard B31 Code for Pressure Piping.[2]   Section B31.5 covers Refrigeration Piping and relates to structural design rules, fabrication, construction, and testing.   This Code in this section will be referred to as the ASA B31.5 Pressure Piping Code.

This section recognizes and uses the definitions included in Section 2 of the ASA B9.1 Code.   The definitions included in Section 500 of the ASA B31.5 Pressure Piping Code are also used.   In general, the definitions in these two codes coincide with definitions in the ASME Boiler Code, Section VIII usually called the Unfired Pressure Vessel Code.[3]   In addition, the ASA B31.5 Pressure Piping Code recognizes and refers to the basic definitions of the American Welding Society.

Other basic definitions accepted by the refrigeration and air-conditioning industry are given in Chap. 1, Terminology, of the "Heating, Ventilating, and Air Conditioning Guide," 1965 edition, published by the American Society of Heating, Refrigerating, and Air Conditioning Engineers.[4]

REFRIGERATION CYCLES

Compression System.   Figure 1a shows a typical single-stage refrigeration cycle plotted on a pressure-enthalpy chart.   Figure 1b is a typical diagram of a single-stage compression system and shows a compressor, a condenser, an optional receiver, an expansion device, and an evaporator.   The state points on the line diagram of Fig. 1b are numbered to correspond to the same points on the pressure-enthalpy chart of Fig. 1a.

In a typical system, P_1 represents the pressure in the evaporator corresponding to the temperature at which the refrigerant is evaporating.   From point 1 to point 2, the refrigerant vapor is carried to the compressor in a suction line.   The pressure-enthalpy chart indicates a small pressure drop in this line.   From point 2 to point 3, the refrigerant vapor is compressed.   The connection between points 3 and 4 represents the discharge, or hot-gas, line, and the pressure-enthalpy chart also indicates the pressure drop due to friction in this line.   Desuperheating and condensation of the refrigerant at constant pressure in the condenser occur between points 4 and 5.   The liquid line is represented by the section between point 5 and

---

* Superscript numbers refer to corresponding items in the References contained at the end of this chapter.

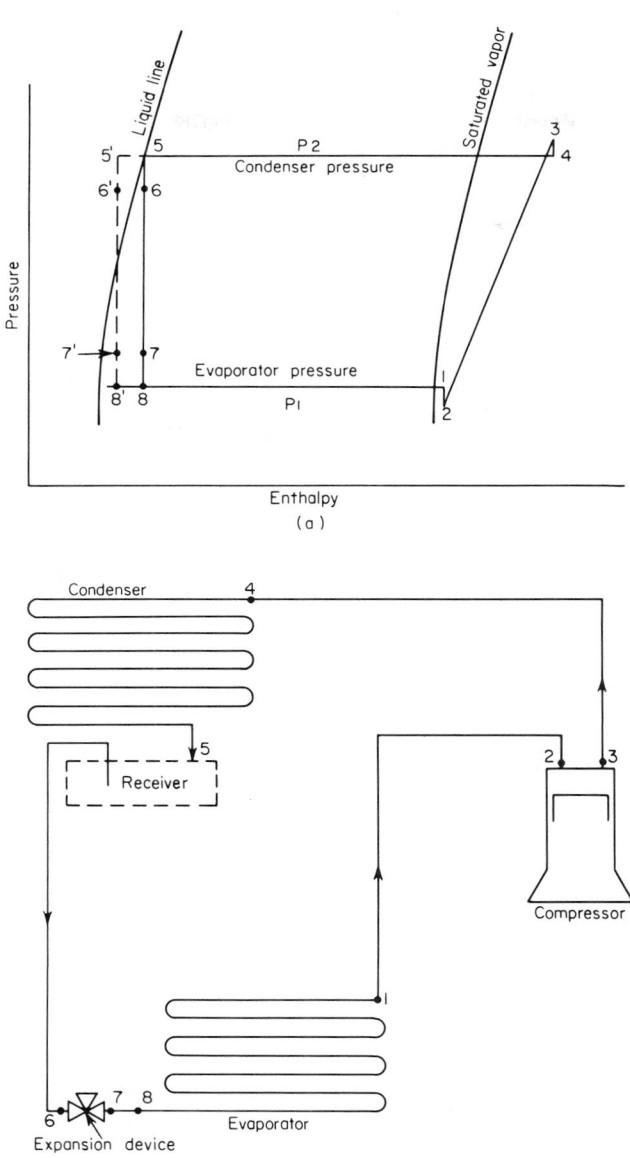

FIG. 1. Compression refrigeration cycle. (*Courtesy of Carrier Air Conditioning Co.*)

point 6.   From point 6 to point 7 there is a representation of the pressure reduction or "expansion" through the reducing valve.   From point 5 to point 8 the refrigerant is a mixture of liquid and vapor because of the expansion at constant enthalpy. At point 8, the liquid and vapor mixture enters the evaporator; heat transferred from the evaporator results in evaporation of the liquid in the liquid-vapor mixture and, as indicated, a slight superheating of the resultant vapor.   At point 1, the cycle is completed.

In this chapter, the line between points 1 and 2 will be referred to as the suction line.   The pipe or tubing between points 3 and 4 will be designated as the hot-gas

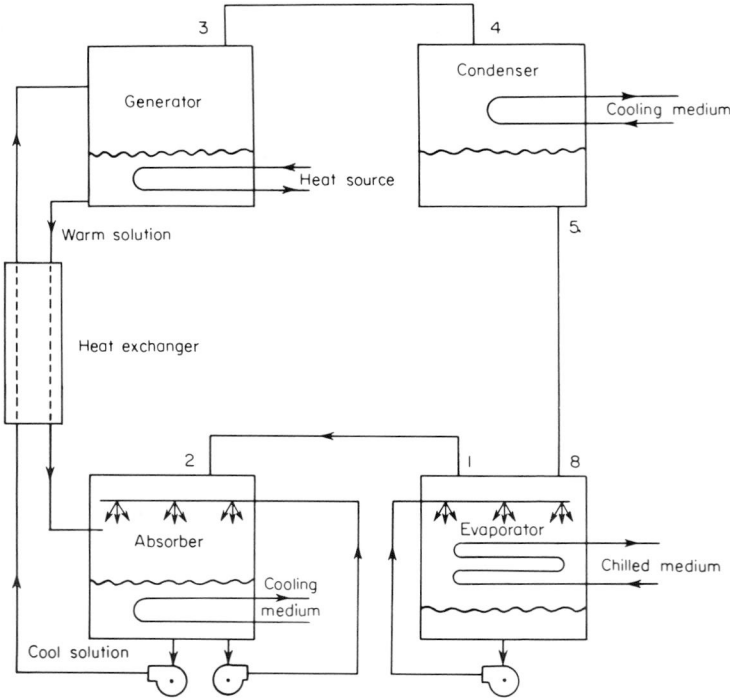

FIG. 2. Absorption refrigeration cycle, lithium bromide type.   (*Courtesy of Carrier Air Conditioning Co.*)

line.   The piping between points 5 and 6 will be termed the liquid line before expansion, and that between points 7 and 8 will be called the liquid line after expansion.

Absorption System.   Figure 2 shows a typical piping arrangement for an absorption-refrigeration cycle in which a lithium salt solution is employed with water as the refrigerant.   In such a system, the absorber and generator serve the same purpose as the compressor in a compression system.   The refrigerant is absorbed in solution at low pressure in the absorber and is pumped to the generator, where the refrigerant is boiled out of the solution at high pressure.   The state points are numbered in Fig. 2 to correspond to the similar points of the compression system cycle in Fig. 1*a*.

For other types of absorption systems (such as ammonia-water), a rectifier or a fractionating-column type of purifier would be installed between the generator and condenser.   Also, the evaporator might be remote or of some other type, such as a flooded cooler or a direct expansion type in which the refrigerant passes through coils.

In a consideration of the piping of an absorption system, the piping between points 3 and 4, 5 and 8, and 1 and 2 may be handled in a manner similar to those of a compression system when the different fluids and flow rates which might be encountered are taken into account.   The solution lines between the absorber and generator and through the heat exchanger would be treated as brine lines in the case of lithium bromide or lithium chloride systems.   For ammonia-water systems, properties of aqueous solutions of ammonia are available, and these lines would be designed in the same manner as would be brine or water lines.

Present-day absorption-refrigeration systems are integrally designed as complete units, and in many instances the entire assembly including the piping is made at the factory.   For these reasons it is seldom necessary to consider piping for an absorption system as a separate design problem.   However, the principles involved in this chapter could be applied to the corresponding sections of an absorption system.

Flow Rates.   Refrigeration is usually measured in tons of refrigeration or in Btu's per hour.   The relationship between these is

$$1 \text{ ton of refrigeration } = 200 \text{ Btu/min or } 12000 \text{ Btu/hr}$$

For selection of piping, it is necessary to relate refrigeration rate to flow rate of the refrigerant.   Since the refrigerant changes state in the process, it is customary to calculate refrigerant flow in pounds per hour or per minute.   This weight flow rate for a constant rate of refrigeration will be constant throughout the system (for a single-stage system).   The volume of refrigerant handled at the various points of the cycle can be determined if the density at these various points is known.   Since the volume handled in a suction line is important not only for the piping selection but also important for the compressor selection, charts are frequently made showing the cubic feet per minute per ton of refrigerant gas or vapor at this point.

In order to calculate the flow rate of refrigerant for a given rate of refrigeration the following procedure is used:

$$\text{lb/(min ton)} = 200/(h_g - h_f) \tag{1}$$

where h_g is the enthalpy of dry saturated vapor at the evaporator outlet pressure or temperature and h_f is the enthalpy of the liquid refrigerant at the expansion device inlet.

To calculate the flow rate it is necessary to have available thermodynamic properties of the refrigerant.   After the mass-flow rate has been determined as shown in the above equation, it is possible to determine the volume flowing at various points in the system.

If the specific volume of the refrigerant is known at the state point leaving the evaporator corresponding to point 1 in Fig. 1a, the volume at that point can be determined as follows:

$$\text{cfm/ton} = 200V/(h_g - h_f) \tag{2}$$

where V is the specific volume of the vapor at the evaporator suction temperature. For purposes of design, this value may be taken as the specific volume of dry, saturated vapor.

In the above equations, h_g corresponds to the enthalpy of the refrigerant vapor at point 1 leaving the evaporator.   In the flooded type of cooler, this state point is

close to saturation, but for a direct-expansion coil-type evaporator, the vapor may be slightly superheated.   For low amounts of superheat, it is permissible to ignore the superheat for purposes of selecting the piping, although this should not be done for compressor selection or for very accurate determination of the velocity.

h_f corresponds to the enthalpy of the liquid entering the expansion device.   Some subcooling of liquid may exist at this point, in which case point 6′ would result in a more accurate figure.   However, for selection of piping, it is customary to use the enthalpy of the saturated liquid at the condensing temperature (point 5) for

Table I.   Refrigerant Flow Rate

Refrigerant	717 Ammonia			12 Dichlorodifluoro-methane			22 Monochlorodifluoro-methane		
Condensing temp, F	90	100	110	90	100	110	90	100	110
Evaporating temp, F	Lbs/(min ton)								
40	0.418	0.428	0.439	3.8	3.98	4.17	2.81	2.94	3.09
20	0.422	0.432	0.444	3.96	4.15	4.36	2.88	3.03	3.19
0	0.426	0.438	0.450	4.12	4.34	4.57	2.97	3.13	3.30
−20	0.434	0.445	0.457	4.32	4.55	4.82	3.07	3.24	3.42
	Cfm/ton (suction line)*								
40	1.71	1.75	1.8	3.02	3.17	3.32	1.91	1.99	2.1
20	2.57	2.63	2.70	4.47	4.68	4.92	2.75	2.90	3.07
0	3.96	4.07	4.18	6.81	7.16	7.55	4.15	4.38	4.62
−20	6.55	6.72	6.90	10.84	11.42	12.1	6.54	6.90	7.28
	Approximate cfm/ton (discharge line)†								
40	0.878	0.805	0.715	1.55	1.42	1.31	1.07	0.94	0.87
20	0.950	0.860	0.776	1.64	1.50	1.38	1.12	1.00	0.93
0	1.023	0.935	0.850	1.72	1.58	1.46	1.19	1.10	0.99
−20	1.128	1.023	0.929	1.84	1.68	1.57	1.29	1.17	1.10

* Based on saturated enthalpies and specific volume at 10 F superheat.

† Based on discharge specific volume at 30 deg above isentropic discharge temperature and 10 F suction superheat.

determining the flow rate.   Later considerations will show when the actual state of the liquid may have an effect on the liquid line size.

A later example will illustrate methods of calculation for flow rate by weight and by volume in the suction line for a typical refrigerant.

Table 1 shows flow rates in pounds per minute per ton and cubic feet per minute per ton at the suction condition for three common refrigerants, ammonia, Refrigerant 12 (dichlorodifluoromethane), and Refrigerant 22 (monochlorodifluoro-methane).   This table is based on enthalpies read at saturated conditions of refrigerant vapor and liquid as mentioned above, but the suction cubic feet per minute per ton is based on a specific volume corresponding to 10 deg superheat.

The cubic feet per minute per ton in the discharge line cannot be calculated readily because the actual temperature at the end of the compression is a function of the compressor design and efficiency and these will vary among various manufacturers.   The discharge cubic feet per minute per ton for single-stage systems can

Table 2.   Temperature-Pressure Chart *

Vacuum, inches of mercury, *italic figures*   Pressure, pounds per square inch gage, bold figures

Temp, F	12	22	500†	717	114	Temp, F	12	22	500†	717	114
−100	27.0	25.1	...	27.4		31	29.3	56.5	37.0	46.3	4.4
−95	26.4	24.1	...	26.8		32	30.1	57.8	38.0	47.6	3.8
−90	25.7	23.0	...	26.1		33	30.9	59.2	39.0	48.9	3.2
−85	24.9	21.7	...	25.3							
−80	24.0	20.2	...	24.3	29.0	34	31.7	60.5	40.0	50.2	2.7
						35	32.6	61.9	41.0	51.6	2.2
−75	23.0	18.5	...	23.2	28.8	36	33.4	63.3	42.0	52.9	1.5
−70	21.8	16.6	...	21.9	28.6	37	34.3	64.6	43.1	54.3	0.9
−65	20.5	14.4	...	20.4	28.3	38	35.2	66.1	44.1	55.7	0.2
−60	19.0	11.9	...	18.6	28.0						
−55	17.3	9.1	...	16.6	27.6	39	36.1	67.5	45.2	57.2	0.2
						40	37.0	69.0	46·2	58.6	0.5
−50	15.4	6.0	...	14.3	27.2	41	37.9	70.5	47.2	60.1	0.8
−45	13.3	2.6	...	11.7	26.7	42	38.8	72.0	48.4	61.6	1.2
−40	11.0	0.6	7.9	8.7	26.1	43	39.7	73.5	49.6	63.1	1.5
−38	10.0	1.4	6.7	7.4	25.9						
−36	8.9	2.3	5.4	6.1	25.6	44	40.7	75.0	50 7	64.7	1.9
						45	41.7	76.6	51.8	66.3	2.2
−34	7.8	3.1	4.1	4.7	25.3	46	42.6	78.2	53.0	67.9	2.6
−32	6.7	4.1	2.8	3.2	25.0	47	43.6	79.8	54.2	69.5	2.9
−30	5.5	5.0	1.4	1.6	24.7	48	44.6	81.4	55.4	71.1	3.3
−28	4.2	6.0	0.0	0.0	24.4						
−26	3.0	7.0	0.7	0.8	24.0	49	45.7	83.0	56.6	72.8	3.7
						50	46.7	84.7	57.8	74.5	4.0
−24	1.6	8.1	1.5	1.7	23.7	52	48.8	88.1	60.3	78.0	4.8
−22	0.3	9.2	2.3	2.6	23.3	54	50.9	91.5	62.9	81.5	5.6
−20	0.6	10.3	3.1	3.6	22.9	56	53.1	95.1	65.5	85.2	6.4
−18	1.3	11.5	4.0	4.6	22.5						
−16	2.0	12.7	4.9	5.6	22.1	58	55.4	98.8	68.2	89.0	7.3
						60	57.7	102.5	71.0	92.9	8.1
−14	2.8	13.9	5.8	6.7	21.6	62	60.1	106.3	73.8	96.9	9.0
−12	3.6	15.2	6.8	7.9	21.1	64	62.5	110.2	76.7	101.0	10.0
−10	4.5	16.6	7.8	9.0	20.6	66	65.0	114.2	79.7	105.3	10.9
−8	5.4	18.0	8.8	10.3	20.1						
−6	6.3	19.4	9.9	11.6	19.6	68	67.5	118.3	82.7	109.6	11.9
						70	70.2	122.5	85.8	114.1	12.9
−4	7.2	20.9	11.0	12.9	19.0	72	72.9	126.8	89.0	118.7	13.9
−2	8.1	22.5	12.1	14.3	18.4	74	75.6	131.2	92.3	123.4	15.0
0	9.1	24.1	13.3	15.7	17.8	76	78.4	135.7	95.6	128.3	16.1
1	9.7	24.9	13.9	16.5	17.4						
2	10.2	25.7	14.5	17.2	17.1	78	81.2	140.3	99.0	133.2	17.2
						80	84.2	145.0	102.5	138.3	18.3
3	10.7	26.6	15.1	18.0	16.8	82	87.2	149.8	106.1	143.6	19.5
4	11.2	27.4	15.7	18.8	16.5	84	90.2	154.7	109.7	149.0	20.7
5	11.8	28.3	16.4	19.6	16.1	86	93.3	159.8	113.4	154.5	22.0
6	12.3	29.2	17.0	20.4	15.8						
7	12.9	30.1	17.7	21.2	15.4	88	96.5	164.9	117.3	160.1	23.3
						90	99.8	170.1	121.1	165.9	24.6
8	13.5	31.0	18.4	22.1	15.1	92	103.1	175.4	125.1	171.9	25.9
9	14.0	32.0	19.0	22.9	14.7	94	106.5	180.9	129.2	178.0	27.3
10	14.6	32.9	19.8	23.8	14.3	96	110.0	186.5	133.3	184.2	28.7
11	15.2	33.9	20.5	24.7	13.9						
12	15.8	34.9	21.2	25.6	13.5	98	113.5	192.1	137.6	190.6	30.2
						100	117.2	197.9	141.9	197.2	31.7
13	16.5	35.9	21.9	26.5	13.1	102	120.9	203.8	146.3	203.9	33.2
14	17.1	36.9	22.7	27.5	12.7	104	124.6	209.9	150.9	210.7	34.8
15	17.7	37.9	23.4	28.4	12.3	106	128.5	216.0	155.4	217.8	36.4
16	18.4	39.0	24.2	29.4	11.9						
17	19.0	40.0	24.9	30.4	11.4	108	132.4	222.3	160.1	225.0	38.0
						110	136.4	228.7	164.9	232.3	39.7
18	19.7	41.1	25.7	31.4	11.0	112	140.5	235.2	169.8	239.8	41.4
19	20.4	42.2	26.5	32.5	10.5	114	144.7	241.9	174.8	247.5	43.2
20	21.0	43.3	27.3	33.5	10.0	116	148.9	248.7	179.9	255.4	45.0
21	21.7	44.4	28.2	34.6	9.6						
22	22.4	45.5	29.0	35.7	9.1	118	153.2	255.6	185.0	263.5	46.8
						120	157.6	262.6	190.3	271.7	48.7
23	23.2	46.7	29.8	36.8	8.6	125	169.1	280.7	204.0	293.1	53.7
24	23.9	47.9	30.7	37.9	8.1	130	181.0	299.3	218.2	....	58.8
25	24.6	49.0	31.6	39.0	7.7	135	193.5	319.6	233.2	....	64.3
26	25.4	50.2	32.5	40.2	7.1						
27	26.1	51.5	33.3	41.4	6.6	140	206.6	341.3	248.8	....	70.1
						145	220.3	364.0	265.2		
28	26.9	52.7	34.3	42.6	6.0	150	234.6	387.2	282.3		
29	27.7	54.0	35.2	43.8	5.0	155	249.5	410.8	300.1		
30	28.4	55.2	36.1	45.0	5.5						

* Courtesy of Sporlan Valve Co.
† Courtesy of Carrier Air Conditioning Co.

be approximated by the following formula:

$$\text{Discharge cfm/ton} = \text{suction cfm/ton} \times P_1/P_2 \times 1.2 \qquad (3)$$

where P_1 = absolute pressure at suction
$\qquad P_2$ = absolute pressure at discharge

This formula is not exact, but it will serve as an approximation when it is necessary to determine approximate velocities in discharge lines.

Table 2 shows the temperature-pressure relationship of some of the common refrigerants. It is customary to evaluate pressure losses in refrigerant lines in terms of the number of degrees change in saturation temperature as a result of this loss. Many refrigeration manufacturers show refrigerant pressure-drop curves in terms of equivalent degrees. Table 2 permits conversion of equivalent pressure losses to temperature losses and also shows the operating pressures which may be expected in refrigerant lines under certain temperature conditions. The chart is also used to determine pressure ratios for approximating discharge volumes after conversion to absolute pressures.

PRESSURE DROP IN REFRIGERATION PIPING

Suction Lines. Figure 1a shows that the compressor in a refrigeration system must pump the refrigerant from point 2 to point 3. The evaporator pressure is established by the temperature level in the evaporator. Any loss in pressure in the suction line will require the compressor to operate at a lower suction pressure. The greater the loss in the suction line, the lower will be the suction pressure. Since the vapor expands at a lower pressure, and since the compressor is essentially a volume device, a reduction in suction pressure causes a reduction in the capacity of the compressor because less weight of refrigerant will be handled and the refrigerant must be pumped through a greater pressure differential. An increased pressure differential also will require more power per ton of refrigeration to drive the compressor. On the other hand if the suction line from the evaporator is made too large, the cost of installation of the line and the stop valves or controls which are placed in it may be so great as to outweigh the economic advantage of improved performance of the compressor. Also, it will be pointed out later than an excessively large suction line can result in difficulty owing to inadequate oil return to the compressor.

Pressure drop and costs are not the only considerations in the selection of suction lines for refrigeration systems. The halogenated hydrocarbons which are commonly used as refrigerants are miscible with oil and tend to carry oil in circulation with the refrigerant. Suction lines for refrigerants of this type must be sized so that the oil in circulation is carried through the system properly and is returned to the compressor. The sizing of suction risers is extremely important and will be discussed in more detail later in this chapter.

Discharge Lines. Figure 1a shows that the condenser pressure will be established by the temperature of the cooling medium of the refrigeration system and that a loss in the discharge line will result in a higher pressure at point 3 than at point 4. Excessive loss in the discharge line between the compressor and condenser will result in an elevated discharge pressure. When the discharge pressure rises, the compressor must pump against a higher pressure differential, and this will increase the power requirement and reduce the volumetric efficiency and capacity of the compressor. For economy, however, the discharge line must not be too large and the cost of installation investment must be weighed against the operating penalty.

Liquid Lines. Pressure losses in refrigerant liquid lines are not so critical as those in the suction and discharge lines. The pressure reduction between points 5 and 8 of Fig. 1a is inherent in the cycle. For good control and economic sizing of control valves, most of this pressure reduction is obtained in the expansion device between points 6 and 7. If the pressure loss in the liquid line before the expansion device is excessive, the reduction in pressure in the liquid line may cause vapor formation, called "flashing." Excessive flashing of the refrigerant liquid will penalize the performance of the control valve or expansion device. It is possible to avoid flashing of refrigerant for some distance in the liquid line if the liquid refrigerant is deliberately subcooled below the condensing temperature. This, however, entails extra condenser surface or auxiliary equipment and is not always economically sound. Good practice dictates that the liquid-line pressure drop should be limited to result in a reasonable selection of the expansion device and to minimize the flashing of the vapor before it flows through the expansion device.

For that portion of the liquid line after the expansion device shown between points 7 and 8 in Fig. 1a, there will be a considerable amount of vaporized refrigerant, or "flash gas," in the refrigerant mixture. Excessive pressure drop in this line will again penalize the expansion device, and it is customary to size this line generously.

Selection of Line Sizes

Suction Lines. It is customary to size suction lines so that the total loss in pressure is approximately equivalent to about 2 F drop in saturation temperatures for halogenated hydrocarbon refrigerants and not over 1 F for ammonia. It is also customary to allow less pressure drop for lower temperature installations because of the increased change in saturation temperature per pound change in pressure and the added penalties on compressors and compressor selection. Table 2, which relates pressure and temperature for various common refrigerants, can be used to decide what pressure drops are allowable to stay within the equivalent temperature loss indicated above. Since pressure loss is a function of length of line and number of fittings as well as flow rate and diameter, it will be obvious that short runs of lines may be sized somewhat smaller and that longer runs will require larger lines for the same loss.

Table 3 shows the ammonia-carrying capacity of $\frac{1}{2}$- to 12-in. steel suction pipelines. The main table headings are in saturated suction temperature. The subheadings show allowable pressure drops varying from $\frac{1}{2}$ psi per 100 ft to 3 psi per 100 ft at higher temperature levels and up to 2 psi per 100 ft at lower temperature levels. The main body of the table shows the tons of refrigeration which will result in the pressure drop listed.

The variation in mass-flow rate per ton of refrigeration is not significant over a broad range of condensing temperature, and it is satisfactory to use Table 3 over a range of condensing temperatures from 80 to 110 F without correction.

Table 4 shows the carrying capacity for Refrigerant 12 (dichlorodifluoromethane) suction lines over a range of saturated suction temperatures and for various allowable pressure drops from $\frac{1}{2}$ psi per 100 ft to 3 psi per 100 ft. Again the body of the table shows tons of refrigeration which will result in the pressure drop shown. Since the mass-flow rate per ton of refrigeration varies only slightly with condensing temperature, the table is usable over a range of 90 to 120 F without correction. Since steel pipe or copper tubing may be used with Refrigerant 12, both of these are shown in the table.

Table 5 shows similar information for the carrying capacity in suction lines for Refrigerant 22 (monochlorodifluoromethane).

Table 3. Suction-line Capacities, Tons, Refrigerant 717 (Ammonia) *,†

Saturated suction temperature, F

Line size, in.‡	−30 Pressure drop, psi/100 ft			−20 Pressure drop, psi/100 ft			0 Pressure drop, psi/100 ft			20 Pressure drop, psi/100 ft				40 Pressure drop, psi/100 ft			
IPS	½	1	2	½	1	2	½	1	2	½	1	2	3	½	1	2	3
½	0.44	0.62	0.88	0.50	0.72	1.02	0.65	0.92	1.31	0.82	1.18	1.70	2.40	1.02	1.45	2.06	2.92
¾	0.96	1.37	1.96	1.11	1.58	2.24	1.45	2.06	2.93	1.81	2.60	3.70	5.23	2.25	3.22	4.61	6.52
1	1.92	2.72	3.85	2.13	3.01	4.26	2.74	3.9	5.61	3.5	4.98	7.06	8.70	4.33	6.14	8.84	10.8
1¼	4.8	6.95	9.85	5.43	7.80	11.1	7.07	10.1	14.6	8.99	12.95	18.5	22.8	11.18	16.15	23.1	28.3
1½	7.3	10.5	14.9	8.25	11.9	16.8	10.7	15.5	22.0	14.6	19.7	27.8	34.2	17.1	24.2	34.5	42.6
2	14.1	20.5	29.0	15.9	23.9	32.5	20.9	29.6	42.7	26.4	38.0	53.7	67.1	32.8	46.8	66.7	82.0
2½	22.8	32.6	46.1	25.3	36.1	52.0	33.3	47.7	68.2	42.3	60.2	85.6	105.0	52.5	75.0	106.5	131.0
3	40.1	57.5	81.4	45.1	64.6	91.5	59.1	84.2	121	74.5	106.5	151	187.5	92.5	132	190	233
4	83.5	119	169	93.0	132	186	121	172	244	153	218	305	378	190	269	382	469
5	150	214	303	168	238	341	218	312	443	276	394	555	683	342	485	690	849
6	244	344	487	274	388	550	354	505	715	447	637	900	1,110	558	789	1,125	1,380
8	500	710	1,000	560	796	1,128	726	1,039	1,468	920	1,308	1,850	2,270	1,135	1,615	2,295	2,810
10	900	1,280	1,810	1,010	1,435	2,020	1,305	1,860	2,645	1,645	2,350	3,310	4,100	2,040	2,900	4,140	5,035
12	1,450	2,050	2,900	1,625	2,310	3,280	2,100	2,780	4,280	2,675	3,820	5,410	6,600	3,325	4,685	6,670	8,200

* Courtesy of Air Conditioning Refrigeration Institute (ARI).
† Based on fluid flow at 90 F saturated condensing temperature.
‡ Data based on Schedule 40 steel pipe, except that 1 in. and smaller are based on Schedule 80.

Saturated suction temperature, F — Pressure drop, psi/100 ft

Line size, in. IPS	OD	−40 ½	−40 1	−40 2	−40 3	−20 ½	−20 1	−20 2	−20 3	0 ½	0 1	0 2	0 3	20 ½	20 1	20 2	20 3	40 ½	40 1	40 2	40 3
½	½												0.24			0.25	0.31			0.31	0.39
	⅝			0.20	0.25			0.28	0.35		0.25	0.36	0.45		0.32	0.47	0.58	0.28	0.40	0.59	0.74
½				0.24	0.29		0.22	0.32	0.39		0.29	0.41	0.51	0.26	0.37	0.53	0.66	0.32	0.46	0.66	0.81
	¾		0.23	0.35	0.44		0.32	0.47	0.59	0.29	0.42	0.62	0.77	0.37	0.54	0.79	0.99	0.47	0.68	1.00	1.23
	⅞	0.25	0.37	0.53	0.67	0.34	0.50	0.73	0.91	0.45	0.65	0.95	1.19	0.57	0.84	1.22	1.51	0.72	1.05	1.54	1.89
¾		0.24	0.35	0.51	0.63	0.33	0.47	0.68	0.83	0.43	0.61	0.87	1.08	0.55	0.78	1.13	1.38	0.69	0.98	1.38	1.70
	1⅛	0.51	0.75	1.10	1.36	0.69	1.00	1.47	1.85	0.91	1.32	1.94	2.42	1.17	1.69	2.47	3.10	1.46	2.14	3.14	3.91
1		0.47	0.67	0.96	1.20	0.63	0.90	1.29	1.58	0.82	1.17	1.67	2.06	1.05	1.51	2.14	2.64	1.30	1.87	2.65	3.27
	1⅜	0.90	1.32	1.92	2.40	1.22	1.77	2.58	3.22	1.61	2.34	3.39	4.22	2.04	2.94	4.30	5.34	2.58	3.74	5.52	6.75
1¼		0.97	1.38	1.99	2.43	1.29	1.85	2.62	3.24	1.68	2.40	3.42	4.22	2.16	3.06	4.35	5.40	2.68	3.81	5.41	6.73
	1⅝	1.44	2.09	3.00	3.78	1.92	2.82	4.10	5.10	2.54	3.68	5.37	6.62	3.25	4.70	6.90	8.46	4.05	6.02	8.65	10.7
1½		1.47	2.09	2.98	3.65	1.94	2.78	3.95	4.87	2.52	3.62	5.12	6.29	3.25	4.59	6.54	8.00	4.00	5.70	8.10	9.83
	2⅛	2.94	4.35	6.30	7.80	4.01	5.80	8.44	10.6	5.28	7.70	11.1	13.8	6.70	9.75	14.2	17.6	8.47	12.3	17.9	22.1
2		2.88	4.05	5.78	7.10	3.80	5.40	7.70	9.5	4.94	7.04	10.0	12.2	6.35	8.95	12.7	15.6	7.87	11.1	15.7	19.3
	2⅝	5.27	7.71	11.2	13.9	7.10	10.4	15.1	18.7	9.44	13.7	19.6	24.5	12.0	17.4	25.4	31.0	15.0	21.9	31.5	39.3
2½		4.52	6.50	9.24	11.4	6.02	8.55	12.2	14.9	7.86	11.2	15.7	19.3	10.1	14.1	20.2	24.7	12.4	17.6	25.0	30.7
	3⅛	8.53	12.4	17.9	22.4	11.3	16.6	23.9	29.9	15.0	21.8	31.4	39.3	19.0	27.8	40.5	49.9	23.8	34.9	50.5	62.6
3		8.02	11.4	16.2	19.9	10.6	15.0	21.6	26.5	13.8	19.6	27.8	34.4	17.6	24.9	35.2	43.5	21.9	30.7	44.3	54.0
	3⅝	12.7	18.4	26.8	32.9	16.9	24.5	35.8	44.4	22.2	32.1	46.7	57.9	28.5	41.2	59.3	74.1	35.6	51.6	75.0	92.0
	4⅛	17.7	25.8	37.4	46.3	23.7	34.6	50.1	62.3	31.4	45.8	65.5	82.0	40.0	57.7	83.5	105	50.3	72.5	106	130
4		16.6	23.8	32.8	41.2	21.8	31.4	44.4	55.5	28.6	40.6	57.2	70.7	36.2	51.8	73.4	90.4	45.0	64.4	91.0	111
	5⅛	32.1	46.3	67.0	83.2	42.7	62.3	90.7	113	56.1	81.3	119	146	71.7	103	150	186	89.7	130	189	233
5		30.0	42.6	60.9	74.6	39.9	56.5	80.3	99.3	51.7	73.6	105	128	65.9	93.0	134	162	81.5	117	165	202
	6⅛	51.2	74.5	108	134	69.4	100	145	180	91.0	131	190	236	115	165	240	300	145	209	300	371
6		48.7	69.0	97.5	121	64.5	91.2	130	159	83.5	118	167	207	106	151	217	264	132	187	265	325
8		99.5	140	199	245	130	186	262	322	170	242	339	422	216	306	435	540	268	381	540	669
10		181	258	361	446	239	338	480	588	310	438	618	759	395	553	788	965	489	689	970	1,205
12		290	406	574	706	382	538	765	936	490	695	984	1,210	623	882	1,260	1,550	774	1,093	1,548	1,890

* Courtesy of ARI.

† Based on fluid flow at 105 F saturated condensing temperature.

‡ IPS data based on Schedule 40 steel piping.   OD data based on Type L copper tubing.

Table 5. Suction-line Capacities, Tons, Refrigerant 22*†

Line size, in.‡ IPS	OD	\-40 ½	\-40 1	\-40 2	\-40 3	\-20 ½	\-20 1	\-20 2	\-20 3	0 ½	0 1	0 2	0 3	20 ½	20 1	20 2	20 3	40 ½	40 1	40 2	40 3
	½												0.38			0.39	0.47		0.33	0.48	0.59
	⅝			0.34	0.40			0.43	0.52		0.38	0.55	0.69	0.34	0.48	0.71	0.87	0.41	0.60	0.87	1.09
½				0.40	0.48		0.36	0.49	0.61	0.31	0.45	0.63	0.78	0.40	0.56	0.80	0.96	0.49	0.70	0.99	1.21
	¾		0.39	0.56	0.69	0.37	0.52	0.74	0.90	0.46	0.67	0.94	1.17	0.57	0.84	1.21	1.53	0.73	1.16	1.54	1.90
	⅞	0.42	0.61	0.87	1.07	0.53	0.79	1.12	1.41	0.70	1.02	1.46	1.84	0.88	1.28	1.90	2.33	1.13	1.64	2.35	2.93
¾		0.39	0.56	0.79	0.98	0.51	0.73	1.03	1.26	0.65	0.93	1.31	1.62	0.81	1.16	1.64	2.02	0.99	1.42	2.03	2.58
	1⅛	0.83	1.19	1.73	2.16	1.09	1.56	2.29	2.81	1.40	2.02	2.90	3.77	1.78	2.61	3.86	4.72	2.28	3.28	4.80	5.83
1		0.74	1.05	1.50	1.81	0.98	1.38	1.95	2.42	1.25	1.78	2.51	3.12	1.56	2.25	3.18	3.92	1.96	2.72	4.04	4.90
	1⅜	1.45	2.11	3.05	3.89	1.90	2.76	4.14	5.03	2.50	3.61	5.20	6.40	3.16	4.61	6.76	8.20	4.04	5.69	8.35	10.3
1¼		1.53	2.20	3.18	3.86	1.98	2.84	4.05	5.00	2.53	3.74	5.15	6.37	3.21	4.50	6.43	7.97	4.04	5.69	8.05	9.94
	1⅝	2.30	3.38	4.76	5.91	2.96	4.35	6.23	7.77	3.93	5.60	8.10	9.96	4.95	7.17	10.8	12.9	6.23	8.94	13.1	16.2
1½		2.33	3.35	4.76	6.05	3.00	4.35	6.15	7.54	3.93	5.49	7.79	9.52	4.81	6.91	9.65	12.0	6.03	8.47	12.0	14.9
	2⅛	4.79	6.9	10.1	12.4	6.24	9.05	13.2	16.3	8.11	11.7	17.0	21.5	10.3	14.9	22.2	27.0	13.1	19.0	27.1	34.4
2		4.47	6.4	8.95	10.8	5.70	8.19	11.7	14.1	7.37	10.3	14.9	18.4	9.31	13.0	18.7	22.8	11.5	16.4	23.2	28.5
	2⅝	8.30	12.1	17.3	21.6	10.9	15.8	23.1	28.4	14.0	20.6	29.0	38.4	18.0	26.2	39.2	47.0	22.8	32.8	47.0	59.0
2½		7.15	10.1	14.1	17.3	9.1	12.9	18.6	23.0	11.7	16.5	24.0	29.3	14.7	20.9	29.4	38.0	18.2	26.2	37.7	46.3
	3⅛	13.2	19.4	27.7	35.2	17.7	25.2	36.8	45.8	22.6	33.0	47.7	58.3	28.9	41.8	60.0	76.6	35.7	52.9	76.2	94.4
3		12.4	17.9	25.4	31.8	16.1	23.4	34.2	40.4	21.0	29.6	42.0	52.3	26.1	38.0	53.0	64.3	32.5	45.7	65.6	81.2
	3⅝	20.0	28.9	41.3	50.8	25.8	38.1	53.8	67.7	33.7	49.2	70.4	86.2	43.2	62.7	89.0	113	54.6	78.2	114	139
	4⅛	27.4	40.0	57.6	73.6	36.6	52.7	76.3	96.4	47.7	68.5	99.5	125	59.5	87.5	128	160	77.2	112	161	199
4		25.6	36.4	50.5	63.5	32.0	47.1	65.9	81.7	42.7	59.1	84.5	107	53.7	75.5	105	129	64.7	91.7	132	162
	5⅛	50.0	73.7	107	132	66.0	96.5	138	174	89.3	124	180	222	110	159	230	286	135	199	286	357
5		46.4	65.7	92	114	59.3	85.0	121	149	77.4	109	154	189	96.5	133	191	234	118	167	234	291
	6⅛	79.6	117	169	207	105	151	215	272	136	193	284	359	171	251	373	450	218	314	457	573
6		75.0	105	149	180	96	138	195	239	124	176	249	309	156	222	312	396	192	270	397	477
8		156	222	317	392	206	291	410	515	262	373	530	656	331	470	660	820	407	579	827	1,010
10		274	396	533	678	365	519	719	890	458	670	920	1,120	570	817	1,140	1,400	718	1,020	1,420	1,740
12		442	606	882	1,060	555	810	1,140	1,420	717	1,020	1,490	1,810	901	1,290	1,830	2,250	1,130	1,600	2,280	2,810

Saturated suction temperature, F — Pressure drop, psi/100 ft (column sub-headings ½, 1, 2, 3 under each temperature: −40, −20, 0, 20, 40)

* Courtesy of ARI.
† Based on fluid flow at 105 F saturated condensing temperature

Discharge Lines.   Pressure loss in hot gas or discharge lines should be minimized because of the adverse effects on compressor volumetric efficiency and power requirements.   It is not desirable to exceed 1- or 2-deg equivalent loss in the discharge line.   Because of the change in pressure-temperature relationships, this will result in somewhat higher permissible pressure drops in discharge lines.   Discharge lines should be selected on the basis of 2 or 3 psi per 100 ft.

Table 6 shows the carrying capacity of steel pipe sizes for discharge lines for ammonia for various pressure drops.   The table is based on an average condensing

Table 6.   Discharge and Liquid-line Capacities, Tons, Refrigerant 717 (Ammonia)*,†

Line size,‡,§,¶ in.	Discharge lines				Liquid lines	
	Temperature, 250 F				To receiver	To system
	Pressure drop, psi/100 ft				Velocity, fpm	Pressure drop, psi/100 ft
IPS	½	1	2	3	100	2
⅜	. . . .	. . . .	. . . .	. . . .	8.5	11.6
½	1.28	1.85	2.65	3.25	13.6	23.5
¾	2.84	4.03	5.83	7.15	25.2	53.2
1	5.68	8.06	11.6	14.2	42.1	105
1¼	14.7	21.1	30.4	37.2	75.3	225
1½	22.2	31.5	45.0	55.0	103	351
2	43.0	61.4	87.6	107	197	805
2½	68.6	98.5	140	171	280	1,280
3	122	174	246	300	432	2,270
4	244	351	497	608	745	4,630
5	450	638	900	1,100		
6	734	1,030	1,470	1,800		
8	1,480	2,110	3,010	3,650		

* Courtesy of ARI.
† Based on fluid flow at 90 F saturated condensing temperature and 20 F saturated evaporating temperature.
‡ Data on sizes 2 in. and over based on Schedule 40 steel pipe.
§ Data on sizes 1 in. and below based on Schedule 80 steel pipe.
¶ Data for discharge line sizes 1¼ and 1½ in. based on Schedule 40 steel pipe; for liquid-line sizes 1¼ and 1½ in. based on Schedule 80 steel pipe.

temperature of 90 F saturation and a flowing temperature of 250 F.   The mass-flow rate of refrigerant is based on 20 F evaporating temperature, but the table is usable over a fairly wide range of condensing temperatures, say from 80 to 110 F, and for evaporating temperatures from 10 up to 30 F without correction.

Table 7 shows the carrying capacity of steel pipe or copper tubing for Refrigerant 12 in discharge lines for various pressure drops.   This table is based on an average condensing temperature of 105 F and a 40 F saturated evaporating temperature with a flowing temperature in the discharge line of 175 F.   These conditions represent a good average for air conditioning for comfort cooling conditions and for the purposes of line sizing may be used over a fairly wide range of conditions.

Table 8 shows similar information for the carrying capacity of discharge lines for Refrigerant 22.

Liquid Lines.   When a system is equipped with a refrigerant receiver downstream of the condenser as a storage place to maintain a seal of liquid refrigerant

Table 7. Discharge and Liquid-line Capacities, Tons, Refrigerant 12*,†

Line size,‡ in.		Discharge lines				Liquid lines	
		Temperature, 175 F				To receiver	To system
		Pressure drop, psi/100 ft				Velocity, fpm	Pressure drop, psi/100 ft
IPS	OD	½	1	2	3	100	2
	½	....	0.33	0.48	0.60		
	⅝	0.42	0.62	0.90	1.13	3.18	4.23
½	...	0.48	0.70	0.98	1.21	3.20	3.62
	¾	0.73	1.06	1.54	1.92	4.77	7.25
	⅞	1.11	1.62	2.36	2.92	6.61	11.2
¾	...	1.02	1.46	2.06	2.54	5.90	8.17
	1⅛	2.26	3.30	4.80	6.02	11.2	23.1
1	...	1.94	2.78	3.96	4.80	9.85	16.1
	1⅜	3.96	5.72	8.25	10.3	17.1	40.0
1¼	...	3.98	5.72	8.15	9.95	17.5	34.4
	1⅝	6.27	9.10	13.4	16.5	24.3	64.0
1½	...	5.97	8.45	12.1	14.8	24.1	52.6
	2⅛	13.0	18.8	27.3	34.0	42.3	133
2	...	11.6	16.6	23.4	29.0	45.7	123
	2⅝	23.1	33.7	48.0	60.2	65.1	236
2½	...	18.4	26.6	37.4	45.5	65.5	197
	3⅛	36.9	53.6	77.5	95.5	93.0	376
3	...	32.4	46.2	65.1	80.0	101	350
	3⅝	54.6	79.2	113	140	126	565
	4⅛	76.7	111	160	198	163	795
4	...	67.1	94.7	135	165	174	712
	5⅛	138	199	288	357		
5	...	122	172	244	298		
	6⅛	222	320	455	570		
6	...	195	280	394	480		
8	...	398	573	810	985		
10	...	725	1,030	1,450	1,770		
12	...	1,145	1,625	2,310	2,830		

*Courtesy of ARI.

† Based on fluid flow at 105 F saturated condensing temperature and 40 F saturated evaporating temperature

‡ "IPS" data based on Schedule 40 steel piping except that liquid lines 1½" and smaller are Schedule 80. OD data based on Type L copper tubing

on the control devices, the liquid line entering the receiver is usually sized generously to assure free flow from the condenser to the receiver.   An allowable velocity of 100 fpm is typical in the selection of liquid-receiver inlet piping.   To assure pressure equalization between condensers and receivers and to prevent vapor binding, lower velocities should be used in smaller size liquid lines.   If 100 fpm is used as the design velocity for liquid to the receiver, vapor equalizing lines should be provided from the top of the receiver to the top of the condenser.

It is possible to design the liquid drain lines from the condenser so as to assure equalization without necessity of adding separate equalizer lines.   A later table will show design information for this condition.

Table 8. Discharge and Liquid-line Capacities, Tons, Refrigerant 22*,†

Line size in.‡		Discharge lines				Liquid lines	
		Temperature, 200 F				To receiver	To system
		Pressure drop, psi/100 ft				Velocity, fpm	Pressure drop, psi/100 ft
IPS	OD	½	1	2	3	100	2
	½	0.33	0.48	0.69	0.86	2.34	2.89
	⅝	0.59	0.88	1.27	1.63	3.78	5.48
½	. . .	0.71	1.00	1.40	1.71	3.81	4.65
	¾	1.05	1.53	2.22	2.74	5.55	9.20
	⅞	1.64	2.36	3.42	4.32	7.85	14.3
¾	. . .	1.50	2.09	3.00	3.82	7.05	10.3
	1⅛	3.29	4.71	6.91	8.64	13.4	29.2
1	. . .	2.82	4.09	5.75	6.98	11.7	20.2
	1⅜	5.71	8.37	12.1	15.1	20.4	51.5
1¼	. . .	5.75	8.21	11.6	13.8	20.9	44.1
	1⅝	8.97	13.1	19.0	23.6	28.9	83.0
1½	. . .	8.64	12.4	17.2	21.6	28.8	66.4
	2⅛	19.3	27.2	40.5	49.8	50.4	168
2	. . .	16.6	23.6	33.1	41.9	54.6	159
	2⅝	32.9	48.2	68.8	87.0	77.6	296
2½	. . .	26.6	39.2	53.2	65.8	77.9	248
	3⅛	53.2	77.1	111	136	111	475
3	. . .	47.2	66.4	93.7	116	120	459
	3⅝	79.0	115	165	203	150	742
	4⅛	111	163	232	291	194	984
4	. . .	95	133	189	232	207	911
	5⅛	199	292	419	522	303	
5	. . .	171	239	346	425	325	
	6⅛	316	459	658	823	434	
6	. . .	281	409	572	681	471	
8	. . .	588	844	1,180	1,440	815	
10	. . .	1,020	1,430	2,040	2,490	1,280	
12	. . .	1,640	2,320	3,300	4,080	1,840	

* Courtesy of ARI.

† Based on fluid flow at 105 F saturated condensing temperature and 40 F saturated evaporating temperature.

‡IPS data based on Schedule 40 steel piping except that liquid lines 1½ in. and smaller are Schedule 80. OD data based on Type L copper tubing.

Based on a design velocity of 100 fpm in the liquid line to the receiver, Tables 6, 7, and 8 show the carrying capacity of liquid lines to the receiver for Refrigerants 717 (ammonia), 12 (dichlorodifluoromethane), and 22 (monochlorodifluoromethane).

Liquid lines between the receiver and the expansion device should be sized on the basis of pressure loss, and 2 psi per 100 ft of run is recommended for normal conditions. Additional consideration due to vertical rise of liquid lines, liquid lines through heated spaces, or other conditions causing excessive flashing will be discussed in the next section.

Tables 6, 7, and 8 show the carrying capacity of various sized liquid lines to the system for Refrigerants 717, 12, and 22 based on a pressure drop of 2 psi per 100 ft.

In the sizing of any refrigerant lines, the valves and fittings must be taken into account. Since their line size will not be known before the pressure drop is evaluated, it is recommended that the length of run be estimated and that an additional 50 per cent allowance be made for preliminary consideration to allow for valves and fittings. After selection of a line size, the pressure losses in the valves and fittings may be evaluated and the overall pressure drop for the line with its valves and fittings should be determined. If the total pressure drop exceeds the recommended allowable drop for reasons of inadequate selection, excessive length of run, or other conditions peculiar to the installation, the line size should be reselected and reevaluated.

Fitting Equivalent. Two methods are used for determining the pressure losses in valves and fittings. The first of these is the *equivalent-length method* where each valve or fitting is assumed to have a pressure drop equal to an equivalent length of pipe of the same size as the valve or fitting. In order to use the equivalent-length method, a line size is first estimated both as to diameter and length. Table 9 shows equivalent lengths for valves and fittings which should be added to the estimated actual length of the line in order to evaluate the total pressure loss. The equivalent lengths of the valves and fittings should be added to the actual net run length of the pipe in order to arrive at the total length which may be used in the calculation of pressure loss or with the tables.

Another method for estimating pressure losses in valves and fittings consists of relating the loss to the velocity head. This method is considered to be somewhat more accurate but is more awkward to use. In order to determine the pressure drop in the valve or fitting, the velocity in the corresponding line size for the flow rate is determined, the velocity head is calculated and is multiplied by a K factor to result in the number of velocity heads lost in the valve or fitting. Since this loss is expressed in feet, it must be converted to pounds per square inch lost by use of the density of the refrigerant at the flowing condition. The refinement of this calculation is usually not justified, but where pressure drops must be determined with a great degree of accuracy, the K factors are useful. Table 10 shows K factors for various types of valves and fittings.

General Method. The tables and charts presented in this chapter provide a means for quick selection or checking of suction, discharge, and liquid lines for a few of the most commonly used refrigerants in applied refrigeration systems. In Chapter 3, general methods are outlined for calculating the frictional losses and pressure drop for fluids flowing in pipes or tubing. These methods can be applied directly to refrigerants.

Pressure losses for refrigerants flowing without a change of state should be calculated using the Darcy-Weisbach formula:

$$H = f(L/D)(V^2/2g) \tag{4}$$

where f = friction factor
L = length of pipe, ft
D = diameter of pipe, ft
V = velocity, fps
g = acceleration of gravity
= 32.17 ft/sec^2
H = head loss, ft

Friction factors depend on the roughness of the pipe and the Reynolds number of the fluid flowing. Since Reynolds number depends on pipe diameter, velocity, density, and viscosity of the fluid, it is necessary to have the physical properties and the thermodynamic properties of a refrigerant and the conditions of operation before a detailed analysis can be made of pressure drop in the system. For friction factors in suction or discharge lines of refrigerants flowing at normal velocities and

Table 9. Equivalent Lengths of Valves and Fittings*,†

Ferrous valves and fittings‡,§

Line size, in. IPS	Globe valve Screwed	Globe valve Flanged	Angle valve Screwed	Angle valve Flanged	Short-radius ell Screwed	Short-radius ell Flanged	Short-radius ell Welded	Long-radius ell Screwed	Long-radius ell Flanged	Long-radius ell Welded	Tee, line-flow Screwed	Tee, line-flow Flanged	Tee, line-flow Welded	Tee, branch-flow Screwed	Tee, branch-flow Flanged	Tee, branch-flow Welded
½	29		16		4.1			2.5			1.8			4.7		
¾	31		16		4.7			2.8			2.5			5.6		
1	35	57	16	19	5.3	1.6	1.8	3.3	1.5	1.2	3.4	1.0	1.6	6.8	3.8	2.5
1¼	46	69	19	22	7.1	2.2	2.3	3.4	2.0	1.6	4.9	1.3	2.0	9.2	4.9	7.1
1½	51	76	19	22	7.9	2.6	2.6	3.6	2.2	1.8	5.9	1.4	2.0	9.9	5.8	8.4
2	63	89	20	25	9.0	3.2	3.4		2.7	2.3	8.1	1.7	2.5	12.6	7.2	10.5
2½		101		28		3.8	4.2		3.0	2.7		1.9	2.9		8.4	13
3		123		36		4.9	5.3		3.7	3.4		2.4	3.6		11	16
4		155		48		6.2	7.2		4.5	4.5		2.9	4.5		14	22
5		190		63		8.1	9.2		5.4	5.7		3.5	5.1		17	27
6		227		78		9.5	11		6.1	6.8		4.1	6.1		20	33
8		295		110		13	15		7.1	9.0		4.7	7.1		27	44
10		370		142		16	18		8.7	11		5.6	8.7		32	56
12		465		173		19	22		10	14		6.2	10		39	68

Non-ferrous valves and fittings‡

Line size, in. OD	Globe valve Screwed	Globe valve Other¶	Angle valve Screwed	Angle valve Other¶	Short-radius ell Screwed	Short-radius ell Other¶	Long-radius ell Screwed	Long-radius ell Other¶	Tee, line-flow Screwed	Tee, line-flow Other¶	Tee, branch-flow Screwed	Tee, branch-flow Other¶
½	40	70	21	24	4.7	4.7		3.2	1.9	1.7	5.1	6.6
⅝	39	72	22	25	5.4	5.7		3.9	2.3	2.3	6.2	8.2
¾	39	75	23	25	6.2	6.5	2.9	4.5	2.9	2.9	7.1	9.7
⅞	45	78	23	28	7.0	7.8	3.7	5.3	3.7	3.7	8.2	12
1⅛	54	87	25	29	8.1	2.7	4.2	1.9	5.2	2.5	11	8.0
1⅜	64	102	27	33	9.9	3.2	4.6	2.2	6.9	2.7	13	10
1⅝	75	115	28	34	12	3.8	5.0	2.6	8.7	3.0	14	12
2⅛	95	141	30	39	14	5.2	5.4	3.4	12	3.8	19	16
2⅝		159		44		6.5		4.2		4.6		20
3⅛		185		53		8.0		5.1		5.4		25
3⅝		216		66		10		6.3		6.6		30
4⅛		248		76		12		7.3		7.3		35
5⅛		292		96		14		8.8		7.9		42
6⅛		346		119		17		10		9.3		50

* Courtesy of ARI.
† $L_e = K(D/f)$.
‡ Friction factors f determined at "practical" Reynolds numbers based on 40 F suction lines having pressure drop of 1.8 psi per 100 ft.
§ Based on Schedule 40 pipe.
¶ Flare, sweat, flanged, etc., and based on Type L copper tubing.

Table 10. K-Factors (Velocity Heads) for Valves and Fittings[a,b]

Ferrous valves and fittings[c]

Line size, in. IPS	Globe valve		Angle valve		Short-radius ell			Long-radius ell			Tee, line-flow			Tee, branch-flow		
	Screwed	Flanged	Screwed	Flanged	Screwed	Flanged	Welded	Screwed	Flanged	Welded	Screwed	Flanged	Welded	Screwed	Flanged	Welded
½	15	....	8.4	....	2.1	....	....	....	....	....	....	....	....	2.4	....	....
¾	11	....	5.7	....	1.7	....	....	0.9	....	....	0.9	....	....	2.0	....	....
1	9.3	15.5	4.3	5.0	1.4	0.43	0.46	0.73	0.40	0.32	0.9	0.26	0.43	1.8	1.0	1.37
1¼	8.4	12.8	3.5	4.0	1.3	0.40	0.42	0.60	0.37	0.29	0.9	0.24	0.36	1.7	0.90	1.31
1½	7.8	11.5	2.9	3.4	1.2	0.39	0.40	0.52	0.34	0.27	0.9	0.22	0.31	1.5	0.88	1.27
2	7.0	9.9	2.2	2.8	1.0	0.36	0.38	0.40	0.30	0.25	0.9	0.19	0.28	1.4	0.80	1.17
2½	....	9.0	....	2.5	....	0.34	0.37	....	0.27	0.24	....	0.17	0.26	....	0.75	1.13
3	....	8.3	....	2.4	....	0.33	0.36	....	0.25	0.23	....	0.16	0.24	....	0.72	1.10
4	....	7.5	....	2.3	....	0.31	0.35	....	0.22	0.22	....	0.14	0.22	....	0.68	1.05
5	....	7.0	....	2.3	....	0.30	0.34	....	0.20	0.21	....	0.13	0.19	....	0.64	1.01
6	....	6.7	....	2.3	....	0.28	0.32	....	0.18	0.20	....	0.12	0.18	....	0.60	0.98
8	....	6.2	....	2.3	....	0.27	0.31	....	0.15	0.19	....	0.10	0.15	....	0.57	0.93
10	....	6.0	....	2.3	....	0.25	0.30	....	0.14	0.18	....	0.09	0.14	....	0.52	0.90
12	....	6.0	....	2.3	....	0.25	0.29	....	0.13	0.18	....	0.08	0.13	....	0.50	0.88

Nonferrous valves and fittings[d,e,f]

Line size, in. OD	Globe valve, flare or sweat	Angle valve, flare or sweat	Short-radius ell, flare or sweat	Long-radius ell, flare or sweat	Tee, line-flow, flare or sweat	Tee, branch-flow, flare or sweat
½	37	12.8	2.5	1.7	0.9	3.5
⅝	28	9.9	2.2	1.5	0.9	3.2
¾	23	7.8	2.0	1.4	0.9	3.0
⅞	19	6.7	1.9	1.3	0.9	2.8
1⅛	15.0	5.0	0.46	0.32	0.43	1.37
1⅜	13.4	4.4	0.42	0.29	0.36	1.33
1⅝	12.0	3.5	0.40	0.27	0.31	1.29
2⅛	10.4	2.9	0.38	0.25	0.28	1.19

a Courtesy of ARI.

b $K = 2gh_l/V^2$

c Based on Schedule 40 pipe.

d Based on Type L copper tubing.

e For screwed valves and fittings, use ferrous K factors.

f For OD sizes above 2⅛ in., use welded ferrous K factors.

within the band of pressure drops previously recommended, the friction factor on a dimensionless basis will be 0.0025 to 0.003. For conservative determinations, it is recommended that the latter figure be used.

The friction factor for normal liquid-line velocities will be from 0.003 to 0.004. The latter figure is recommended.

Example. The following example will give a detailed determination of flow rates, volume flow, line selection, and pressure-drop determination for a typical case of a refrigerant not listed in the basic common tables in this chapter.

Assume a refrigeration system using Refrigerant 500 with a capacity of 100 tons at an evaporating temperature of 20 F and a condensing temperature of 110 F. Further assume a suction line of 50 ft net length with one globe valve and two elbows. Assume the discharge line to be 30 ft net length with one globe valve and three elbows. Assume the liquid line to be 15 ft with two elbows and two angle valves. The following detailed analysis will determine the flow rates, line sizes, and pressure drops together with the total equipment temperature loss.

Referring to Fig. 1 and Eq. (1), the weight rate of refrigerant flow is

$$\text{lb/(min ton)} = 200/(h_g - h_f)$$

From thermodynamic properties of Refrigerant 500,

$$h_g \text{ (at 20 F saturated)} = 97.98 \text{ Btu/lb}$$
$$h_f \text{ (at 110 F saturated)} = 42.22 \text{ Btu/lb}$$
$$\text{lb/(min ton)} = 200/(97.98 - 42.22) = 3.59$$

For 100 tons,

$$\text{lb/min} = 100 \times 3.59 = 359$$

To determine volume flowing in the suction line, from thermodynamic properties

$$V_g \text{ (at 20 F)} = 1.144$$

Using Eq. (2),

$$\text{cfm/ton} = 200 \times V_g/(h_g - h_f)$$
$$= (200 \times 1.144)/(97.98 - 42.22) = 4.11$$
$$\text{cfm (in suction line)} = 4.11 \times 100 = 411$$

The approximate cubic feet per minute in the discharge line may be determined by Eq. (3) and the pressure-temperature relationship of the refrigerant. Absolute pressures must be used in this equation.

From Table 2,

Refrigerant 500 at 20 F has a saturation pressure of 27.3 psig or 42.0 psia.
Refrigerant 500 at 110 F has a saturation pressure of 164.9 psig or 179.6 psia.

$$\text{Approximate discharge cfm/ton} = \text{suction cfm/ton} \times P_1/P_2 \times 1.2$$
$$= 4.11 \times 42.0/179.6 \times 1.2 = 1.15$$
$$\text{Discharge cfm} = 1.15 \times 100 \text{ tons} = 115$$

The liquid volume may be determined from the weight flow rate and the density of the liquid at the condensing temperature.

$$\text{Density of Refrigerant 500 at 110 F} = 68.05 \text{ lb/cu ft}$$
$$\text{cfm of liquid refrigerant} = 359/68.05 = 5.28$$

To illustrate the general method, the fitting resistance will be evaluated by the equivalent-length method.

For the suction line, assume a copper tube of 4⅛ in. OD.

From tables of copper tube sizes, 4⅛-in.-OD Type L copper tubing has an ID of 3.905 in. and an internal area of 11.92 sq in.

The total equivalent length of suction piping may be determined by adding the equivalent length of fittings to the net length of run.

From Table 9,

$$\text{Equivalent length of one } 4\tfrac{1}{8}\text{-in. globe valve} = 248 \text{ ft}$$
$$\text{Equivalent length of two } 4\tfrac{1}{8}\text{-in. elbows} = 24 \text{ ft}$$
$$\text{Length of net run} = 50 \text{ ft}$$
$$\text{Total equivalent length} = \overline{322 \text{ ft}}$$

Velocity of suction gas in $4\tfrac{1}{8}$-in. OD lines is

$$V = (\text{cfm} \times 144)/(60 \times \text{area}) = (411 \times 144)/(60 \times 11.92) = 82.7 \text{ fps}$$

Diameter of tube $= 3.905/12 = 0.326$ ft

From Eq. (4)

$$H = \frac{fLV^2}{2Dg} = \frac{0.003 \times 322 \times 82.7 \times 82.7}{2 \times 0.326 \times 32.2} = 315 \text{ ft}$$

At suction conditions, specific volume of vapor was 1.144 cu ft/lb; therefore, the density of vapor is

$$1/1.144 = 0.873 \text{ lb/cu ft}$$
$$\text{Pressure drop (psi)} = 315 \times 0.873/144 = 1.91$$

From the pressure-temperature equivalent (Table 2) a pressure drop of 1.91 psi at 20 F is equivalent to approximately 2.4 F penalty.

The use of an angle valve instead of a globe valve would reduce the total equivalent length to 150 instead of 322 ft.

The pressure drop would then be

$$(150/322) \times 315 \text{ ft} = 147 \text{ ft or } 0.892 \text{ psi}$$

which would incur a penalty of approximately 1.1 F. Hence the selection of a $4\tfrac{1}{8}$-in.-OD suction line is satisfactory from a pressure-drop standpoint. From Table 21, a $4\tfrac{1}{8}$-in. Type L tube is satisfactory for a working pressure of 250 psi which (see Table 18) exceeds the required design pressure of 120 psi.

For the *discharge line*, try a $2\tfrac{5}{8}$-in.-OD line. From tables of copper tube sizes, the $2\tfrac{5}{8}$-in.-OD Type L tubing has an ID of 2.465-in. and an internal area of 4.76 sq in.

From Table 9, and the fittings specified, the equivalent length is determined:

$$\text{Equivalent length of } 2\tfrac{5}{8}\text{-in. globe valve} = 159 \text{ ft}$$
$$\text{Equivalent length of three } 2\tfrac{5}{8} \text{ in. elbows} = 19.5 \text{ ft}$$
$$\text{Net length of } 2\tfrac{5}{8}\text{-in. line} = 30.0 \text{ ft}$$
$$\text{Total equivalent length} = \overline{208.5 \text{ ft}}$$

Velocity of discharge gas in a $2\tfrac{5}{8}$-in.-OD line is calculated and found to be 57.9 fps

$$\text{Tube diameter} = 2.465/12 = 0.206 \text{ ft}$$

The friction factor may again be assumed to be 0.003.
From Eq. (4) the friction pressure drop is

$$H = \frac{fLV^2}{2Dg} = \frac{0.003 \times 208.5 \times 57.9 \times 57.9}{2 \times 0.206 \times 32.2} = 158 \text{ ft}$$

Approximate specific volume of discharge gas is

$$\text{cfm discharge/cfm suction} \times V_g$$

or

$$(115/411) \times 1.144 = 0.32 \text{ cu ft/lb}$$

$$\text{Density of discharge gas} = 1/0.32 = 3.12 \text{ lb/cu ft}$$
$$\text{Pressure drop (psi)} = 158 \times 3.12/144 = 3.42 \text{ psi}$$

From Table 2, at 110 F, a pressure drop of 3.42 psi for Refrigerant 500 is equivalent to approximately 1.4 F.

For the *liquid line to the receiver*, a velocity of 100 fpm is recommended for general use.

The required area of line is

$$\text{Area} = \text{cfm (liquid)/velocity} = 5.28/100 = 0.0528 \text{ sq ft} = 7.6 \text{ sq in.}$$

which requires a $3\frac{5}{8}$-in. liquid line to the receiver.

For the liquid line to the system expansion valve, assume a $2\frac{1}{8}$-in.-OD copper tube. For Type L, $2\frac{1}{8}$ in. OD, the inside diameter is 1.985 in. and the inside area is 3.09 sq in. The equivalent length of liquid line is:

$$\begin{aligned}
\text{Equivalent of two } 2\tfrac{1}{8}\text{-in. angle valves} &= 78 \text{ ft} \\
\text{Equivalent of two } 2\tfrac{1}{8}\text{-in. elbows} &= 10.4 \text{ ft} \\
\text{Net run of } 2\tfrac{1}{8}\text{-in. line} &= 15 \text{ ft} \\
\hline
\text{Total equivalent length} &= 103.4 \text{ ft}
\end{aligned}$$

Velocity of liquid in a $2\frac{1}{8}$-in. copper tube is

$$\text{Velocity} = \frac{\text{cfm} \times 144}{60 \times \text{area}} = \frac{5.28 \times 144}{60 \times 3.09} = 4.1 \text{ fps}$$

For an assumed friction factor of 0.004, friction drop is calculated from Eq. (4)

$$H = \frac{fLV^2}{2Dg} = \frac{0.004 \times 103.4 \times 4.1 \times 4.1}{2 \times 0.1655 \times 32.2} = 0.65 \text{ ft}$$

Pressure drop $= 0.65 \times 68.05/144 = 0.308$ psi

This pressure drop is too low for an economical selection, so try a $1\frac{5}{8}$-in. line.

$$\begin{aligned}
\text{Diameter} &= 1.505/12 = 0.1255 \text{ ft} \\
\text{Area} &= 1.771 \text{ sq in.}
\end{aligned}$$

Equivalent lengths:

$$\begin{aligned}
\text{Equivalent lengths of two } 1\tfrac{5}{8}\text{-in. valves} &= 68 \text{ ft} \\
\text{Equivalent lengths of two } 1\tfrac{5}{8}\text{-in. elbows} &= 7.6 \text{ ft} \\
\text{Net run} &= 15.0 \text{ ft} \\
\hline
\text{Total equivalent length} &= 90.6 \text{ ft}
\end{aligned}$$

$$\text{Velocity} = \frac{\text{cfm} \times 144}{60 \times \text{area}} = \frac{5.28 \times 144}{60 \times 1.771} = 7.15 \text{ fps}$$

$$H = \frac{fLV^2}{2Dg} = \frac{0.004 \times 90.6 \times 7.15 \times 7.15}{2 \times 0.1255 \times 32.2} = 2.29 \text{ ft}$$

Pressure drop $= 2.29 \times 68.05/144 = 1.085$ psi, and this is satisfactory. The liquid line after expansion is usually sized one size larger than the main liquid line. In this case, a $2\frac{1}{8}$-in. OD line could be used.

For the illustration, the selected line sizes are:

$$\begin{aligned}
\text{Suction line} &\dots\dots\dots\dots\dots\dots\dots\dots\dots\; 4\tfrac{1}{8}'' \text{ OD} \\
\text{Discharge line} &\dots\dots\dots\dots\dots\dots\dots\dots\; 2\tfrac{5}{8}'' \text{ OD} \\
\text{Liquid line to receiver} &\dots\dots\dots\dots\dots\; 3\tfrac{5}{8}'' \text{ OD} \\
\text{Liquid line to expansion valve} &\dots\dots\dots\; 1\tfrac{5}{8}'' \text{ OD} \\
\text{Liquid line after expansion valve} &\dots\dots\; 2\tfrac{1}{8}'' \text{ OD}
\end{aligned}$$

The compressor selection must be adjusted for the suction and discharge line loss, so the compressor should be selected for conditions equivalent to 18.9 F saturated suction (1.1 F line loss) and 111.4 F saturated discharge (1.4 F line loss).

Brine Piping. Piping for brine service whether of salt brine, the glycols, or other compounds which may be used for transfer of heat without a change of state, may be designed by the rules and principles which apply to water piping. As in the design of water piping, consideration must be given to pressure losses, pumping head, and cost of investment. Additional considerations for brine piping involve compatibility of materials with the brine to be circulated and evaluation of pressure drop, taking into account the density and viscosity of the flowing fluid.

When choosing brine at moderately low temperature levels, it is customary to

select a brine which has a freezing point approximately 20 F below the operating temperature expected in the system.   To use a more concentrated brine (having a lower freezing point) may be safer in operation but is uneconomical from the standpoint of reduced heat transfer and increased pumping costs.   The principles in Chap. 3 may be applied to calculation of brine piping.   To assist in the analysis of such piping, Table 11 shows the concentration, specific gravity, freezing point, and the viscosity at a temperature 20 F above the freezing point for four common brines.   If the desired freezing point is known, this table will show the viscosity at a temperature 20 F above the freezing point and the density of the brine solution. This information will enable determination of the friction factor and, subsequently, calculations of the pressure loss.   Standard equivalents may be used for valves and fittings in the brine circuit.

Oil Return.   The halogenated hydrocarbon refrigerants are miscible with oil. These refrigerants in the liquid phase will carry oil without separation.   In a compressor, some oil will be carried out with the refrigerant because of the gas velocities and oil coatings of the refrigerant compressor parts.   The oil will be carried into the condenser, where it will go into solution with the liquid refrigerant and will be carried to the evaporator in the liquid phase.   Since the evaporator acts as a still, it will tend to boil off refrigerant vapor and leave the oil in the evaporator. In the case of shell-and-tube evaporators where the refrigerant is in the shell, there will be a tendency for the oil concentration in the evaporator to increase.   One common way of returning oil from such an evaporator is to provide a small line from the bottom which bleeds a controlled amount of refrigerant liquid and oil mixture into a point in the suction line which must loop down below the evaporator. Another method is to maintain a slight amount of refrigerant liquid carry-over from the evaporator which will carry oil into the suction line, after which a suitable heat exchanger may be used to complete the vaporization of the suction gas by heat exchange with warm liquid coming from the condenser to the evaporator.   This method will not result in thermodynamic loss as long as the energy used to vaporize the refrigerant in the suction line comes from the liquid refrigerant being fed to the evaporator.

After the oil is introduced into the suction line by either of the means described above or is automatically returned to the suction line from a direct expansion evaporator where the refrigerant is boiling inside tubes, it is necessary to maintain velocities in the suction line which will assure that the oil will be returned to the compressor.   For suction lines chosen in the conventional manner and within the limits of the pressure drops and velocities recommended earlier, oil will return in horizontal runs or in lines running downward.   However, for vertical risers, there is a possibility of oil collecting on the sides of the pipe wall and running back toward the evaporator unless the velocities in the vertical risers are maintained.

The necessary velocities expressed in terms of minimum tons capacity for oil entrainment in vertical suction risers have been determined, and tables are included in this chapter to guide in the selection of such risers.   Table 12 shows the minimum tonnage for oil entrainment in vertical suction risers of Refrigerant 12 both in copper tubing and in steel pipe.   The actual velocities required to return oil in these vertical risers will vary with the operating temperature and the size of the pipe.   The minimum velocity for oil return in vertical risers varies from approximately 1,500 fpm in $5\frac{1}{8}$-in. tubing at 40 F to as high as 5,500 fpm at -40 F in the same size tubing. Correspondingly, there is a diameter effect, and velocities 750 fpm at 40 F to 2,500 fpm at -40 F are sufficient to return oil in vertical risers for tubing of approximately $1\frac{1}{8}$ in. diameter.

Table 13 shows similar minimum tonnage capacity for Refrigerant 22 for oil return in vertical suction risers.

Table II. Brine Properties

Concentration %	Calcium chloride			Sodium chloride			Ethylene glycol			Propylene glycol		
	Specific gravity*	Freezing point, F	Viscosity†	Specific gravity*	Freezing point, F	Viscosity†	Specific gravity*	Freezing point, F	Viscosity†	Specific gravity*	Freezing point, F	Viscosity†
0	1.00	32°	1.27	1.00	32°	1.27	1.00	32°	1.27	1.00	32°	1.27
5	1.044	27.7	1.45	1.035	27.0	1.45	1.005	30	1.50	1.004	29.5	1.60
10	1.087	22.3	1.80	1.072	20.4	1.75	1.012	25.2	1.80	1.009	26.0	2.0
15	1.133	13.5	2.5	1.111	12.0	2.3	1.019	21	2.03	1.014	24.5	2.6
20	1.182	-0.4	4.1	1.15	1.8	3.4	1.026	15.8	2.9	1.018	20.5	3.7
23.3‡	.....	.....	.....	1.175	-6.0	4.7	.....	.....	.....	.....	.....	.....
25	1.233	-21.0	9.7	1.191	16.0	3.0	1.033	10.5	3.7	1.022	15.6	5.2
29.87‡	1.290	-67.0	38.0	.....	.....	.....	.....	.....	.....	.....	.....	.....
30	1.295	-50.8	25.7	.....	.....	.....	1.039	3.2	5.0	1.027	10.0	7.7
35	.....	.....	.....	.....	.....	.....	1.046	-1.0	6.3	1.031	3.5	12.6
40	.....	.....	.....	.....	.....	.....	1.053	-12.0	11.0	1.035	-4.0	24.0
45	.....	.....	.....	.....	.....	.....	1.059	-17.0	15.0	1.038	-15.0	45.0
50	.....	.....	.....	.....	.....	.....	1.066	-33.0	29.0	1.041	-24.0	85.0
55	.....	.....	.....	.....	.....	.....	1.072	-43.0	55.0	.....	.....	.....

* Specific gravities compared with water at 60 F.
† Viscosities in centipoises at 20 F above freezing point.
‡ Eutectic points.

Table 12. Minimum Tonnage for Oil Entrainment up Vertical Suction Risers, Refrigerant 12*

Copper tubing, Type L

Pipe OD	7/8		1 1/8		1 3/8		1 5/8		2 1/8		2 5/8		3 1/8		3 5/8		4 1/8		5 1/8		6 1/8		8 1/8	
Area, sq in.	0.484		0.825		1.256		1.780		3.094		4.770		6.812		9.213		11.97		18.67		26.83		46.85	
Suct temp	T	F	T	F	T	F	T	F	T	F	T	F	T	F	T	F	T	F	T	F	T	F	T	F
−40	0.31	3.1	0.61	2.9	1.0	2.7	1.6	2.6	3.2	2.4	5.5	2.2	8.6	2.1	12.5	2.0	17.3	1.9	30.2	1.8	47.4	1.7	95.0	1.6
−20	0.40	1.8	0.77	1.7	1.3	1.5	2.0	1.5	4.0	1.4	6.9	1.3	10.8	1.2	15.6	1.1	21.8	1.1	37.8	1.0	59.6	1.0	119.1	0.9
0	0.47	1.1	0.93	1.0	1.6	0.9	2.4	0.9	4.8	0.8	8.2	0.8	12.8	0.7	18.8	0.7	26.0	0.6	45.4	0.6	71.4	0.6	143.0	0.5
+20	0.55	0.65	1.1	0.6	1.8	0.6	2.8	0.5	5.6	0.5	9.7	0.5	14.9	0.4	22.0	0.4	30.6	0.4	53.0	0.4	83.1	0.4	167.0	0.3
+40	0.63	0.40	1.2	0.4	2.1	0.4	3.2	0.3	6.4	0.3	11.1	0.3	17.1	0.3	25.0	0.3	34.8	0.2	61.0	0.2	95.6	0.2	191.5	0.2

Steel pipe, standard weight (Schedule 40)

IPS	3/4		1		1 1/4		1 1/2		2		2 1/2		3		3 1/2		4		5		6		8	
Area, sq in.	0.533		0.864		1.495		2.036		3.355		4.788		7.393		9.89		12.73		20.01		28.99		50.0	
Suct temp	T	F	T	F	T	F	T	F	T	F	T	F	T	F	T	F	T	F	T	F	T	F	T	F
−40	0.35	3.9	0.65	3.6	1.3	3.2	1.9	3.1	3.5	2.9	5.5	2.8	9.5	2.6	13.6	2.5	18.7	2.4	33.0	2.3	52.0	2.2	103.8	2.0
−20	0.44	2.2	0.81	2.1	1.6	1.8	2.4	1.7	4.4	1.7	6.9	1.6	11.8	1.5	17.0	1.4	23.4	1.4	41.3	1.3	65.4	1.3	129.6	1.1
0	0.53	1.3	0.97	1.2	1.9	1.1	2.9	1.0	5.3	1.0	8.3	1.0	14.2	0.9	20.4	0.9	28.0	0.8	49.0	0.8	78.4	0.8	152.0	0.7
+20	0.62	0.8	1.1	0.7	2.3	0.7	3.3	0.6	6.2	0.6	9.7	0.6	16.6	0.5	23.9	0.5	32.7	0.5	57.4	0.5	92.0	0.4	181.0	0.4
+40	0.71	0.5	1.3	0.5	2.6	0.4	3.8	0.4	7.1	0.4	11.2	0.4	19.1	0.3	27.5	0.3	37.6	0.3	66.2	0.3	105.5	0.3	208.0	0.3

* Courtesy Carrier Air Conditioning Co.
T = tons of refrigeration, F = friction drop, degrees F per 100 ft equivalent length at tons shown.

Table 13. Minimum Tonnage for Oil Entrainment up Vertical Suction Risers, Refrigerant 22*

Copper tubing, Type L

Pipe OD	7/8		1 1/8		1 3/8		1 5/8		2 1/8		2 5/8		3 1/8		3 5/8		4 1/8		5 1/8		6 1/8		8 1/8	
Area, sq in.	0.484		0.825		1.256		1.780		3.094		4.770		6.812		9.213		11.97		18.67		26.83		46.85	
Suct temp	T	F	T	F	T	F	T	F	T	F	T	F	T	F	T	F	T	F	T	F	T	F	T	F
−40	0.45	1.6	0.88	1.5	1.5	1.4	2.3	1.3	4.6	1.2	8.0	1.2	12.3	1.1	18.1	1.1	25.0	1.0	43.6	0.9	68.5	0.9	137.1	0.8
−20	0.56	1.0	1.1	0.9	1.8	0.8	2.8	0.8	5.6	0.7	9.7	0.7	15.1	0.6	21.9	0.6	30.6	0.6	53.0	0.6	83.6	0.5	167.0	0.5
0	0.66	0.6	1.3	0.5	2.2	0.5	3.3	0.5	6.7	0.4	11.4	0.4	17.9	0.4	26.2	0.4	36.3	0.4	63.3	0.3	99.5	0.3	199.3	0.3
+20	0.77	0.4	1.5	0.3	2.5	0.3	3.9	0.3	7.8	0.3	13.5	0.3	20.8	0.2	30.8	0.2	42.7	0.2	74.1	0.2	116.2	0.2	234.0	0.2
+40	0.89	0.2	1.8	0.2	3.0	0.2	4.6	0.2	9.1	0.2	15.8	0.2	24.3	0.2	35.4	0.1	49.4	0.1	86.5	0.1	135.6	0.1	272.0	0.1

Steel pipe, standard weight (Schedule 40)

IPS	3/4		1		1 1/4		1 1/2		2		2 1/2		3		3 1/2		4		5		6		8	
Area, sq in.	0.533		0.864		1.425		2.036		3.355		4.788		7.393		9.89		12.73		20.01		28.99		50.0	
Suct temp	T	F	T	F	T	F	T	F	T	F	T	F	T	F	T	F	T	F	T	F	T	F	T	F
−40	0.51	2.0	0.93	1.9	1.8	1.6	2.7	1.6	5.0	1.5	7.9	1.4	13.5	1.3	19.4	1.3	26.7	1.2	47.2	1.2	74.3	1.1	148.1	1.0
−20	0.63	1.2	1.1	1.1	2.3	1.0	3.3	0.9	6.2	0.9	9.7	0.9	16.6	0.8	23.9	0.7	33.0	0.7	58.2	0.7	92.1	0.7	182.4	0.6
0	0.74	0.7	1.4	0.7	2.7	0.6	4.0	0.6	7.4	0.5	11.5	0.5	19.9	0.5	28.5	0.5	39.2	0.4	68.7	0.4	110.0	0.4	212.5	0.4
+20	0.87	0.5	1.6	0.4	3.2	0.4	4.7	0.3	8.7	0.3	13.6	0.3	23.4	0.3	33.6	0.3	46.0	0.3	80.6	0.3	129.1	0.3	254.5	0.2
+40	1.0	0.3	1.8	0.3	3.6	0.2	5.3	0.2	10.0	0.2	15.7	0.2	26.4	0.2	38.6	0.2	53.0	0.2	93.0	0.2	148.0	0.2	292.0	0.1

* Courtesy Carrier Air Conditioning Co.
T = tons of refrigeration. F = friction drop, degrees F per 100 ft equivalent length at tons shown.

Table 14. Minimum Tonnage for Oil Entrainment up Vertical Hot-Gas Risers *

Refrigerant 12

Copper tubing, Type L

Pipe OD	$7/8$		$1\,1/8$		$1\,3/8$		$1\,5/8$		$2\,1/8$		$2\,5/8$		$3\,1/8$		$3\,5/8$		$4\,1/8$		$5\,1/8$		$6\,1/8$	
Area, sq in.	0.484		0.825		1.256		1.78		3.094		4.77		6.812		9.213		11.97		18.67		26.83	
Disch sat temp	T	F	T	F	T	F	T	F	T	F	T	F	T	F	T	F	T	F	T	F	T	F
80	0.71	0.2	1.4	0.4	2.3	0.4	3.6	0.3	7.2	0.3	12.3	0.3	19.2	0.3	28.0	0.3	38.6	0.2	68.0	0.2	106.0	0.2
90	0.75	0.3	1.5	0.4	2.5	0.4	3.9	0.4	7.7	0.3	13.3	0.3	20.6	0.3	30.4	0.3	42.0	0.3	73.0	0.3	114.0	0.2
100	0.80	0.3	1.6	0.4	2.6	0.4	4.0	0.4	8.0	0.4	14.0	0.3	21.7	0.3	31.8	0.3	43.7	0.3	76.6	0.3	120.0	0.3
110	0.87	0.4	1.7	0.5	2.9	0.5	4.4	0.4	8.9	0.4	15.0	0.4	23.7	0.3	34.8	0.3	48.0	0.3	83.0	0.3	131.0	0.3
120	0.95	0.4	1.9	0.5	3.1	0.5	5.0	0.5	9.7	0.4	16.9	0.4	26.0	0.4	38.2	0.4	52.5	0.3	91.5	0.3	143.0	0.3

Refrigerant 22

Copper tubing, Type L

Pipe OD	$7/8$		$1\,1/8$		$1\,3/8$		$1\,5/8$		$2\,1/8$		$2\,5/8$		$3\,1/8$		$3\,5/8$		$4\,1/8$		$5\,1/8$		$6\,1/8$	
Area, sq in.	0.484		0.825		1.256		1.78		3.094		4.77		6.812		9.213		11.97		18.67		26.83	
Disch sat temp	T	F	T	F	T	F	T	F	T	F	T	F	T	F	T	F	T	F	T	F	T	F
80	1.1	0.5	2.1	0.6	3.5	0.6	5.4	0.5	10.6	0.5	18.1	0.4	28.4	0.4	42.0	0.4	57.6	0.4	101.0	0.4	146.0	0.3
90	1.2	0.5	2.3	0.7	3.8	0.6	5.8	0.6	11.5	0.5	19.9	0.5	31.2	0.5	45.6	0.5	62.5	0.4	110.0	0.4	158.0	0.3
100	1.25	0.5	2.5	0.7	4.1	0.7	6.3	0.6	12.5	0.6	21.6	0.6	33.9	0.5	49.6	0.5	68.5	0.5	120.0	0.5	173.0	0.4
110	1.4	0.6	2.7	0.8	4.4	0.7	6.9	0.7	13.7	0.6	23.5	0.6	38.1	0.6	53.7	0.5	74.0	0.5	129.0	0.5	187.0	0.4

* Courtesy Carrier Air Conditioning Co.
T = tons of refrigeration. F = friction drop, degrees F per 100 ft equivalent length at tons shown.

For other refrigerants of the halogenated hydrocarbon type, approximately the same velocities should be maintained under similar operating conditons and with the same size pipe.   For such cases, it is recommended that the corresponding velocities be calculated from the basic table shown for Refrigerant 12 or Refrigerant 22 and that the same velocities be applied to the other refrigerants.

Since vapor in the line between the compressor and the condenser may also be subject to the same problem of oil carry-over from the compressor, it is necessary to check the discharge lines also to be sure that any oil that gets into a vertical riser will be carried into the condenser and continue in circulation.   Table 14 shows the minimum tonnages required for oil entrainment in vertical hot-gas risers for Refrigerant 12 and Refrigerant 22.

Where a refrigeration system operates over a range of capacities, it is evident that a suction riser which is properly sized for the full-load condition may not handle

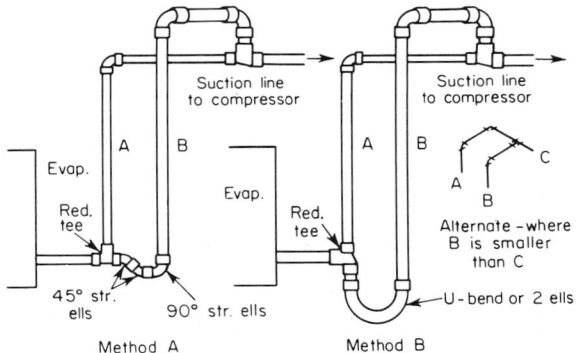

FIG. 3. Double suction riser construction.   (*Courtesy of Carrier Air Conditioning Co.*)

adequately a partial-load condition.   In such a case, a double-suction-riser construction can be used.   Figure 3 shows a typical double-suction riser.   Lines *A* and *B* are sized so that the two together will handle the full load with proper oil return in accordance with Table 12 or 13 or with equivalent velocities for other refrigerants.   If the minimum operating tonnage is determined, line *A* should be sized to assure oil return for this minimum tonnage.   At the low tonnage, the trap at the bottom of the loop on line *B* will seal with oil and line *A* will handle the reduced load with suitable oil return.   Care must be observed that the oil-holding capacity of the trap is such that it will not hold enough oil to deplete the compressor of necessary oil charge during the partial-load operation.   Similar arrangements can be made on discharge lines when necessary.

Multiple Evaporators.   When evaporators are connected in multiple with a single compressor or at a different elevation, other considerations must be given to proper piping of the refrigerant vapor.   Figure 4*a* shows a loop which should be used in the suction line when the evaporator is above the compressor to prevent liquid from draining into the compressor during off cycles.

Figure 4*b* shows that the inverted loop is not necessary when the evaporator is below the compressor.

Figure 4*c* shows an arrangement of inverted loops on multiple evaporators which are on different levels with the compressor below both of them.

Figure 4*d* shows an arrangement for suction piping when multiple evaporators are stacked on the same level with the compressor below both.

Figure 4*e*, *f*, *g*, and *h* shows various combinations illustrating the principles described earlier with traps to prevent liquid drainage into the compressor during shutdown and the application of double-suction risers when necessary. The inverted loops at the top of the risers in Fig. 4*e* and *g* are to prevent oil from running back into an inactive coil at time of light load or reduced capacity.

For automatic operation, a compressor pump-down control is commonly used to keep the low side and evaporators free of excessive liquid during periods when cooling or refrigeration is not required. In such an installation, evaporators located above the compressor need not have protective loops.

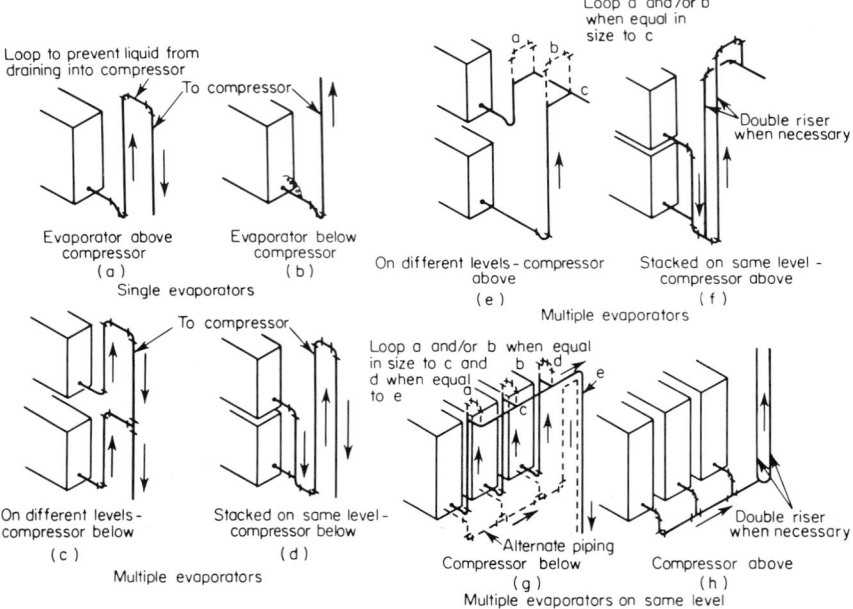

FIG. 4. Arrangements of suction-line loops (one-circuit coils shown). (*Courtesy of Carrier Air Conditioning Co.*)

The illustrations and descriptions of piping cover some of the more common arrangements. There are many arrangements of multiple evaporators, multiple compressors, or multiple condensers with variations which are beyond the scope of this chapter. However, the principles outlined herein may be applied to many types of installations.

Flashing in Liquid Lines. As mentioned in the discussion above, excessive pressure drop in refrigerant liquid lines leaving the condenser or receiver may cause excessive formation of vapor or "flash gas." Since the vapor occupies a much greater volume than the liquid, the remaining liquid will be forced along the pipe at a much higher velocity but with an average density approaching that of the mixture of liquid and vapor. In a pipe that conveys a fluid which vaporizes as it flows, the pressure drop will gradually increase. The pressure drop in a line which carries a flashing liquid or in a liquid line after expansion can be considerably greater than that in the line before expansion or vaporization.

In liquid refrigeration lines which are vertical or inclined even in the case of low velocities or low frictional pressure drops, the static-head effect of the liquid in the line may cause vaporization or flashing of the liquid near the top of the vertical run.   In such cases, the liquid lines should be sized generously to minimize frictional losses, and liquid subcooling may be employed if possible in the plant design. Provision for subcooling of liquid is usually beyond the scope of the piping designer and must be obtained in conjunction with the designer of the condenser or the base equipment.   Excessive flash gas in long runs of liquid lines may be partially avoided by insulation if the line is to run through a warm ambient temperature or by the installation of a liquid subcooler.

The pressure-temperature relation of various refrigerants can be used as a guide in determining the amount of subcooling required to overcome pressure drop in the

Table 15.   Maximum   Velocity   in   Condenser   Vertical   Drains   to   Assure   Equalization

Inside diameter, in.	Max. velocity, fpm
1	44.4
2	62.8
3	76.8
4	88.8
5	99.5
6	108.3
7	117.0
8	125.0
9	133.2
10	140.0
11	147.0
12	154.0

Table shows maximum velocity permitted in round vertical drains, based on full pipe area, to assure full equalization with receiver.

Based on the formula: Velocity (fpm) $= 44.4\sqrt{d}$ where d is in inches.

liquid line and to overcome the effect of static head in vertical liquid lines.   Since liquid subcooling cannot always be obtained, generous sizing of liquid lines may be required.

For liquid lines after the expansion device, it is usually customary to size these lines one size larger than the liquid line which leads to the expansion device. Commercially obtainable thermal-expansion valves which operate on superheat from the evaporator are usually made with the outlet one size larger than the inlet. In all cases, close coupling of the expansion device to the evaporator is recommended so that the liquid line after expansion will be as short as possible.

Condenser Drains.   In the section on selection of line sizes, liquid lines were sized on the basis of about 100 fpm.   When liquid lines run vertically downward from the bottom of a horizontal shell-and-tube or shell-and-coil condenser to a receiver and where full equalization of the condenser and receiver is required through the condensate drain line, it may be necessary to maintain velocities less than 100 fpm. Table 15 shows the maximum velocity based on full pipe area which should be allowed in vertical outlets from condensers to assure full equalization and to prevent gas binding.   This table is equally applicable to pipe or copper tubing and is applicable to any refrigerant compatible with the materials, since it is based on the hydraulic effect of liquid flowing into vertical outlets.   If the velocities exceed those shown in this table, separate vapor-equalizing lines should be used.

Marine Condensers. Refrigerant condensers for use on land usually have one single liquid outlet. For condensers that are intended for use on shipboard, the condensers are usually mounted fore and aft to take advantage of the lower pitch angle and outlets are provided at each end of the condenser. These are piped into a common header having a single outlet in the center. Each condenser outlet should be capable of handling the entire full capacity of the condenser.

Insulation. Since refrigerating systems are designed basically to produce temperatures below normal, many of the pipelines in a refrigeration system will be at temperatures which are below the dew point of the surrounding air. In addition to a heat gain from the surroundings, condensation, commonly called "sweating," will form on these pipes. The condensation may be objectionable and even harmful. It is customary to insulate all refrigerant lines where there is a possibility of condensation.

The amount and type of insulation depend upon the operating-temperature level. Chapter 6 covers insulation in detail with recommendations for economical thicknesses, types, and application.

An important consideration in the application of any insulation to refrigerant piping is to assure that there is a vapor seal on the outside of the insulation. If the insulation type is such that moisture can enter it, the natural difference in vapor pressure between the surrounding atmosphere and the surface of the pipe will result in moisture migration into the insulation and eventually to the surface of the pipe. Vapor seals at all joints and on the outside of the insulation are essential to assure efficient performance and the avoidance of future trouble.

It is also necessary to avoid thermal bridges between the cold piping and the outside ambient air or atmosphere. Pipe hangers for supports should not contact the cold piping and should be arranged so that the supports bear against saddles of adequate area which are outside the insulation. Any metal in contact with the cold pipe wall, because of the higher conductivity of the metal hanger or rod, will cause condensation on the hangers.

MECHANICAL PIPING DESIGN

Code Requirements

The ASA Safety Code for Mechanical Refrigeration, ASA B9.1,[1] was written to assure the safe design, construction, installation, operation, and inspection of every refrigerating system employing a volatile refrigerant. Section 9 of the 1964 edition of the ASA B9 Standard covers refrigerant piping, valves, fittings, and related parts. The limitations of the ASA B9.1 Code on piping will be summarized in the section on "Limitations."

The ASA Code for Pressure Piping, Section B31.5,[2] is a standard covering the minimum requirements for materials, design, fabrication, assembly, test, and inspection of refrigeration piping.

Both the ASA B9.1 Code and the ASA B31.5 Code recognize that refrigeration equipment of the self-contained or unit type which has been designed in accordance with good practice and which has been submitted to an approved, nationally recognized testing laboratory which provides uniform testing and examination procedures and which has a follow-up inspection service of current production of such units will be construed as meeting the requirements of either of these codes.

The ASA B31.5 Code also excludes water piping from consideration except for chilled water below 60 F, which is considered to be a brine.

Materials. Since volatile refrigerants have different chemical compositions, some materials are incompatible with certain refrigerants. Table 16 shows that

Table 16. Material Compatibility

Material	Refrigerant number							
	11	12	113	22	114	500	717	40
Carbon steel	S	S	S	S	S	S	S	S
Wrought iron ...	S	S	S	S	S	S	S	S
Cast-iron pipe...	NP	NP	NP	NP	NP	NP	NP	NP
Copper or brass..	S	S	S	S	S	S	NS	S
Aluminum	Q	Q	Q	Q	Q	Q	Q	NS
Zinc	NS	NS	NS	NS	NS	NS	NS	NS
Magnesium	NS	NS	NS	NS	NS	NS	NS	NS
ASA B9.1 Group	I	I	I	I	I	I	II	II
Underwriters Laboratory Class	5a	6	4–5	5a	6	5a	2	4

NP = not permitted by ASA B31.5 Code
NS = not satisfactory
Q = qualified—moist refrigerant may corrode—consult supplier
S = permissible

several common materials are not compatible with certain of the refrigerants; it also indicates certain other limitations from a Code standpoint. It will be noted that cast-iron pipe is not permitted by the Pressure Piping Code for any volatile refrigerant; cast-iron valves and fittings of approved types are permitted.

Copper and brass are not compatible with ammonia.

Aluminum, zinc, and magnesium are not suitable for use with methyl chloride.

Zinc and magnesium are not suitable for use with any of the halogenated hydrocarbon refrigerants or ammonia.

In addition to the limitations and qualifications shown in Table 16, each of the eight listed refrigerants is classified according to the Underwriters' Laboratories classification and according to a group classification system covered in the ASA B9 Safety Code.

The ASA B9 classification of Group I includes refrigerants which are not considered toxic or inflammable in the ordinary sense (it must be remembered that many refrigerants can smother if present in heavy concentrations).

Group II in the ASA B9 classification covers toxic or flammable refrigerants, and Group III covers highly flammable or explosive refrigerants.

Limitations. Table 17 shows the limitations of various types of materials and group classifications with respect to the service with refrigerants. It will be noted that steel pipe must be Schedule 40 or heavier for use with volatile refrigerants with certain limitations with respect to size.

Butt-welded carbon-steel (except for the open-hearth type, not rephosphorized) and wrought-iron pipe are not allowed for liquid lines.

Cast-iron pipe is not allowed for any volatile, flammable, or toxic refrigerant but may be used for water or nonvolatile brines. Cast iron is not allowed for temperatures below −150 F.

Copper or brass tubing may be used with any refrigerant with which it is compatible and in any size or pressure when selected by the design rules. If copper tubing is erected on the premises, it must be Type K or L.

The ASA B9.1 Safety Code has certain requirements for institutional, public assembly, residential, and commercial occupancies. These rules prohibit the carrying of refrigerant piping through floors except that it may be carried from the

Table 17. Piping Limitations

Type or Material	ASA B9.1 group number	Line size, in.	Service	Limitation
Carbon steel or wrought iron	II or III I II or III I, II, III	1¼ or smaller 6 or smaller 2 through 6 6 or smaller	Refrigerant liquid Refrigerant liquid Refrigerant liquid Refrigerant vapor	Schedule 80 or heavier Schedule 40 or heavier Schedule 40 or heavier Schedule 40 or heavier
Butt-welded carbon steel or wrought iron	Any	Any	Refrigerant liquid	Not permitted by Code
Cast-iron pipe	Any	Any	Refrigerant	Not permitted for volatile, flammable, or toxic refrigerants
Cast-iron pipe	....	....	Brine	Permitted above −150 F
Cast-iron pipe	....	....	Any	Not permitted below −150 F
Copper or brass	Any (except ammonia)	Any	Any (except ammonia)	Type K or L if erected on premises. Soft annealed may not exceed 1⅜″ O.D.
			Ammonia	Not compatible
Aluminum, zinc, magnesium	....	....	Methyl chloride	Not compatible
Magnesium	....	....	Halogenated hydrocarbons	Not compatible
Threaded joints	III	1 and smaller	Any	Seal welded or braze
		See limitation	Group III fluids Salt brines	Not allowed over 1″ Not allowed over 6″
			Any	Not lighter than Schedule 40 up through 6″ Not lighter than Schedule 30 on 8″, 10″, or 12″

basement to the first floor or from the top floor to a machinery penthouse to the roof.

Refrigerant piping may be connected to a condenser on the roof if it is carried through an approved rigid and tight, continuous, fire-resistant pipe duct or shaft having no openings on intermediate floors, or it may be carried on the outer wall of the building provided it is not located in an air shaft, closed court, or any other similar opening enclosed within the outer walls of the building.

For Group I refrigerants, the refrigerant piping may be carried through floors intermediate between the first floor and the top floor provided it is enclosed in an approved, rigid and tight, continuous, fire-resisting pipe or shaft where it passes through intermediate spaces not served by the system.   The piping of direct systems

need not be enclosed where it passes through space served by that system. The pipe duct or shaft must be vented to the outside or to a space served by the system.

The ASA B9.1 Code further requires rigid or flexible metal enclosures for soft, annealed copper tubing used for refrigerant piping which contains other than Group I refrigerants.

Schedule 40 wrought-steel or wrought-iron pipe may be used for design working pressures not exceeding 300 psig provided lap-welded, electric-resistance-welded, or seamless pipe is used for nominal pipe sizes over 4 in.

The 1958 edition of the ASA B9.1 Code prohibited the use of soft, annealed copper tubing for refrigerant piping erected on the premises in sizes larger than $\frac{7}{8}$ in. outside diameter. However, the 1964 edition of this Standard permits the use of soft, annealed copper tubing in sizes through $1\frac{3}{8}$ in. OD. This permission is also included in ASA B31.5 Standard.

Joints on copper tubing containing Group II or Group III refrigerants as classified by the ASA B9 Code must be brazed. Solder joints are prohibited in such systems.

A brazed joint is obtained by the joining of metal parts with alloys which melt at temperatures higher than 1000 F but less than the melting temperatures of the joined parts.

A soldered joint is obtained by the joining of metal parts with metallic mixtures or alloys which melt at temperatures below 800 and above 350 F.

ASA B9.1 requires that all refrigerant pipe joints erected on the premises must be exposed to view for inspection prior to being covered or enclosed.

Systems containing 100 lb of refrigerant with positive-displacement compressors must have stop valves on each inlet of each liquid receiver and on each branch liquid and suction line except on receivers which are in a condensing unit or which are an integral part of the condenser.

Refrigerant piping crossing an open space which affords a passage way in any building must be not less than $7\frac{1}{2}$ ft above the floor unless against a ceiling of such space.

Free passageway must not be obstructed by refrigerant piping. Refrigerant piping must not be placed in any elevator, dumb-waiter, or other shaft containing a moving object or any shaft which has openings to living quarters or to main exit hallways. Refrigerant piping is not to be placed in public hallways, lobbies, or stairways except that such refrigerant piping may pass along a public hallway if there are no joints in the section in the public hallway and provided that nonferrous tubing of $1\frac{1}{8}$ in. outside diameter and smaller is contained in a rigid metal pipe.

Limitations on Threaded Joints. Threaded joints must be seal-welded or brazed for Group III refrigerants.

Threaded joints larger than 1 in. in size should not be used for Group III fluids and must be no larger than 6 in. in size for salt brine.

Threaded joints must not be used on lighter than Schedule 40 pipe up through 6 in. in diameter or on lighter than Schedule 30 pipe for 8-, 10-, or 12-in. pipe.

Design Pressures

Minimum design pressures for piping for many refrigerants are included in the ASA B31.5 Pressure Piping Code. These design pressures are based on good practice and are correlated with certain minimum field test pressures which are established on the basis of both ambient temperature considerations and existing practice on some of the older refrigerants.

Table 18 shows the minimum design pressure permitted under the Pressure Piping Code. This table also shows the ASA B9.1 group number classification for toxicity and flammability. The refrigerant numbering system established by the American

Table 18. Refrigerant Piping Minimum Design Pressures *

Group†	Number	Name	Chemical formula	Minimum design pressures, psig High-pressure side	Low-pressure side
II	717	Ammonia	NH_3	300	150
III	600	Butane	C_4H_{10}	75	40
I	744	Carbon dioxide	CO_2	1,500	1,000
I	12	Dichlorodifluoromethane	CCl_2F_2	180	105
I	500	Dichlorodifluoromethane 73.8 % and ethylidene fluoride 26.2 %	CCl_2F_2 CH_3CHF_2	220	120
II	1,130	Dichloroethylene	$C_2H_2Cl_2$	30	30
I	30	Dichloromethane (methylene chloride)	CH_2Cl_2	30	30
I	21	Dichloromonofluoromethane	$CHCl_2F$	55	30
I	114	Dichlorotetrafluoroethane	$C_2Cl_2F_4$	50	40
III	170	Ethane	C_2H_6	1,000	640
II	160	Ethyl chloride	C_2H_5Cl	45	40
III	1,150	Ethylene	C_2H_4	1,300	1,050
III	601	Isobutane	$(CH_3)_3CH$	100	55
II	40	Methyl chloride	CH_3Cl	160	90
II	611	Methyl formate	$HCOOCH_3$	40	40
I	22	Monochlorodifluoromethane	$CHClF_2$	285	150
I	13	Monochlorotrifluoromethane	$CClF_3$	670	550
III	290	Propane	C_3H_8	300	150
II	764	Sulfur dioxide	SO_2	130	65
I	11	Trichloromonofluoromethane	CCl_3F	15	15
I	113	Trichlorotrifluoroethane	$C_2Cl_3F_3$	15	15

* Tables 18, 19, and 22 are extracted from "Refrigeration Piping Code" ASA B31.5, a Section of the American Standard Code for Pressure Piping, with the permission of the publisher, The American Society of Mechanical Engineers, 345 E. 47th St., New York, N.Y., 10017.
† ASA B9.1 Group Designation from ASA B79.1.

Society for Heating, Refrigerating, and Air Conditioning Engineers has been referred to earlier in this chapter.   Table 18 shows the refrigerant number corresponding to many of these refrigerants.   The chemical name of the refrigerant and its chemical formula are also listed.

For refrigerants not listed in this table, the general rule is that the components shall be designed for an internal pressure representing the most severe conditions of coincident pressure and temperature expected in normal operation, including shutdown.   Consideration must also be given to external design pressure.   Refrigeration piping systems must be designed to resist collapse when the internal pressure is zero absolute and the external pressure is atmospheric.   This is to permit drying the pipe by evacuation.

For brine service, the maximum pressure which may be encountered in operation should also be used as a guide in designing the brine system, considering the possibilities of shutoff head of any pumps, static head due to elevation or long vertical runs, and the possibility of water hammer.

The ASA B9.1 and B31.5 Codes require that the refrigerant piping must be capable of withstanding certain leak field test pressures.   Table 19 shows the minimum refrigerant leak field test pressures required for various refrigerants.   The notes indicate the basis for determining certain of these test pressures together with a basis for determining test pressures for refrigerants other than those listed.

For limited-charge systems equipped with a pressure-relief device, the piping must be designed for a pressure not less than the setting of the pressure-relief device.

Table 19. Refrigerant Piping—Minimum Refrigerant Leak Field Test Pressures *

ASA B9.1 Group	Number	Name	Chemical formula	Minimum field refrigeration leak test pressure, psig	
				High side	Low side
II	717	Ammonia	NH_3	300	150
III	600	Butane	C_4H_{10}	95	50
I	744	Carbon dioxide	CO_2	1,500	1,000
I	12	Dichlorodifluoromethane	CCl_2F_2	235	140
I	500	Dichlorodifluoromethane 73.8% and ethylidene fluoride 26.2%	CCl_2F_2 CH_3—CHF_2	285	150
II	1,130	Dichloroethylene	$C_2H_2Cl_2$	30	30
I	30	Dichloromethane (methylene chloride)	CH_2Cl_2	30	30
I	21	Dichloromonofluoromethane	$CHCl_2F$	70	40
I	114	Dichlorotetrafluoroethane	$C_2Cl_2F_4$	50	50
III	170	Ethane	C_2H_6	1,200	700
II	160	Ethyl chloride	C_2H_5Cl	60	50
III	1,150	Ethylene	C_2H_4	1,600	1,200
III	601	Isobutane	$(CH_3)_3CH$	130	70
II	40	Methyl chloride	CH_3Cl	210	120
II	611	Methyl formate	$HCOOCH_3$	50	50
I	22	Monochlorodifluoromethane	$CHClF_2$	300	150
I	13	Monochlorotrifluoromethane	$CClF_3$	685	685
III	290	Propane	C_3H_8	300	150
II	764	Sulfur dioxide	SO_2	170	85
I	11	Trichloromonofluoromethane	CCl_3F	20	20
I	113	Trichlorotrifluoroethane	$C_2Cl_3F_3$	20	20

* Courtesy—ASA B31.5 Code.

NOTE 1: For refrigerants not listed in the table, the test pressure for the high-pressure side shall not be less than the saturated vapor pressure of the refrigerant at 150 F. The test pressure for the low-pressure side shall not be less than the saturated vapor pressure of the refrigerant at 110 F. However, the test pressure for either the high or low side need not exceed 125 per cent of the critical pressure of the refrigerant. In no case shall the test pressure be less than 20 psig.

NOTE 2: When a compressor is used as a booster to obtain a low pressure and discharges into the suction line of another system, the booster compressor is considered a part of the low side, and values listed under the low-side column in the table shall be used for both high and low sides of the booster compressor provided that a low-pressure stage compressor of the positive-displacement type shall have a pressure-relief valve.

NOTE 3: In field testing systems using nonpositive-displacement compressors, the entire system shall be considered for field test purposes as the low-side pressure.

A limited-charge system is one in which, with a compressor idle, the internal volume and the total refrigerant charge are such that the design pressure will not be exceeded by the complete evaporation of the refrigerant charge.

The *required thickness* of pipe or tubing is determined from the following equations and nomenclature:

$$t_m = t + c \tag{5}$$

$$t = PD_o/2(S + Py) = Pd/2(S + Py - P) \tag{6}$$

$$P = 2St/(D_o - 2yt) \tag{7}$$

where t_m = minimum required thickness, in.
c = allowance for grooves, threads, tolerances, corrosion, erosion
P = internal design pressure, psig
D_o = outside diameter of pipe, in.
d = inside diameter of pipe, in.
S = allowable stress, psi
t = calculated thickness

y is a material coefficient which for:

$$\text{Ductile nonferrous materials} = 0.4$$
$$\text{Brittle materials} = 0.0$$
$$\text{Ferritic steels} = 0.4$$
$$\text{Austenitic steels} = 0.4$$

The design of piping for external pressure involves the use of charts to determine factors which are used to calculate the thickness or allowable working pressure. The method and charts referred to in paragraphs UG 28 and UG 31 of Section VIII of the ASME Boiler and Pressure Vessel Code[3] are acceptable for design of pipes and tubes subject to external pressure.

The Pressure Piping Code recognizes and permits the use of the design rules of the ASME Unfired Pressure Vessel Code for closures, flanges, and blind flanges.

For blanks, the following equation should be used:

$$t = d_g \sqrt{3P/16S}$$

where t = required thickness, in.

d_g = inside diameter of gasket for raised or flat-face flanges or the pitch diameter of retained gasket flanges, in.

P = internal or external design pressure, psig

S = allowable stress, psi

Since the Pressure Piping Code[2] and the Safety Code for Mechanical Refrigeration both limit the minimum thickness of steel pipe and the minimum thickness for copper tubing for erection on the premises, it is possible to calculate the maximum working pressure for these commonly used weights of pipe or tubing.

Table 20 shows the maximum allowable internal working pressure for seamless steel pipe in the permitted schedule numbers. The maximum allowable external pressure has also been calculated for a length-over-diameter ratio in excess of 15 which will be common in most piping systems. It will be evident that the allowable working pressure for the permitted thicknesses of pipe is usually far in excess of that required by the design working pressure requirements for the various refrigerants, and in most cases except in unusual circumstances, in cases where shock may be anticipated, or in the larger sizes, no further checking of allowable working pressure will be necessary.

Table 21 shows a similar analysis of allowable working pressures, both internal and external, for copper tubing of Type K or Type L. Again it will be evident that the external pressure is in excess of that required to evacuate fully the piping under atmospheric pressure outside. The basic stress for the calculation of these tables is shown, and adjustments may be made in accordance with the design equations for other allowable stresses.

The ASA B31.5 Pressure Piping Code establishes certain allowable working stresses for many grades of pipe and tubing in various materials. Table 22 lists the allowable stresses and the corresponding specification numbers for many grades of ferrous and nonferrous piping materials. These allowable stresses are to be used in conjunction with the design equations listed above for special calculations.

Low-temperature Design Criteria. It is recognized that certain materials tend to become brittle at low temperatures and may be subject to failure which would not occur normally at usual temperatures or at elevated temperatures. The transition temperature at which certain materials become brittle is not well defined. Some ferrous materials may pass through the transition range at normal temperatures, while others may not become brittle until quite low temperatures are attained. The ASME Unfired Pressure Vessel Code[3] arbitrarily establishes a temperature of -20 F as a point below which all vessels constructed of carbon or low-alloy steels should be impact tested, with certain exemptions.

**Table 20. Allowable Working Pressures for Carbon Steel
Refrigerant Piping**

Nominal size, in.	Schedule no.	Allowable internal working pressure, psig	Allowable external working pressure, psig
$\frac{1}{8}$	40	1,890	2,070
	80	3,510	2,860
$\frac{1}{4}$	40	1,490	2,000
	80	2,880	2,700
$\frac{3}{8}$	40	1,300	1,660
	80	2,500	2,280
$\frac{1}{2}$	40	1,126	1,580
	80	2,210	2,140
$\frac{3}{4}$	40	994	1,320
	80	1,890	1,800
1	40	866	1,210
	80	1,680	1,670
$1\frac{1}{4}$	40	773	980
	80	1,470	1,410
$1\frac{1}{2}$	40	740	890
	80	1,390	1,270
2	40	670	750
$2\frac{1}{2}$	40	665	810
3	40	624	700
$3\frac{1}{2}$	40	600	640
4	40	580	580
6	40	534	450
8	40	515	390
10	40	496	340
12	STD.	436	260

For Internal Pressure:
    Based on minimum wall thickness; no corrosion or erosion allowance: thread allowance factor from ASA B2.1
    Allowable stress $= 12,000$ psi
    y (material coefficient) $= 0.4$
    $P = 2St/(D_0 - 2yt)$.
For External Pressure:
    Based on minimum wall thickness; no corrosion, erosion, threading or grooving allowance
    Yield: 24,000 psi to 30,000 psi
    $L/D_0 = 15$ or greater where $D_0 =$ outside pipe diameter in inches; $L =$ maximum straight length of run between flanges, elbows, caps or stiffening rings (inches)

Refrigeration piping is frequently subject to temperatures below normal atmospheric temperatures to the degree that embrittlement may occur, and the ASA B31.5 Piping Code also requires impact tests on certain materials subject to temperatures below −20F.   There are certain materials and certain conditions under which impact tests are not required.   The exemptions are as follows:[2]

1. No impact tests are required for aluminum, austenitic stainless steel in grades 304, CF8, 304L, CF3, 316, CF8M or 321; or copper, red brass, copper nickel alloys, or nickel copper alloys.
2. No impact tests are required for bolting material conforming with A193, Grade B7 for use at temperatures above −50 F.
3. No impact tests are required for bolting materials conforming with A320, grades L7, L10, and L43 at temperatures above −150F or above −225 F for A320, grade L9.
4. No impact test is required for material used in fabricating a piping system for metal temperatures between −20 F and −150 F when the most severe condition of pressure (internal if above atmospheric, and external if below atmospheric) is multiplied by $2\frac{1}{2}$ in determining the thickness.

Table 21. Allowable Working Pressure for Copper Tubing for Refrigerant Piping

Nominal size (OD)	Type	Nominal wall thickness, in.	Allowable internal working pressure, psig	Allowable external working pressure, psig
¼	*	0.030	1,200	550
⅜	K	0.032	830	380
½	K	0.049	970	450
½	*	0.032	610	275
⅝	K	0.049	760	350
⅝	*	0.035	530	230
¾	K	0.049	630	285
¾	L	0.042	530	225
⅞	K	0.065	730	330
⅞	L	0.045	480	210
1⅛	K	0.065	560	250
1⅛	L	0.050	420	170
1⅜	K	0.065	460	190
1⅜	L	0.055	370	145
1⅝	K	0.072	420	173
1⅝	L	0.060	350	130
2⅛	K	0.083	360	140
2⅛	L	0.070	310	110
2⅝	K	0.095	340	129
2⅝	L	0.080	280	98
3⅛	K	0.109	330	122
3⅛	L	0.090	270	88
3⅝	K	0.120	310	113
3⅝	L	0.100	260	83
4⅛	K	0.134	300	110
4⅛	L	0.110	250	77
5⅛	K	0.160	300	106
5⅛	L	0.125	230	65
6⅛	K	0.192	300	108
6⅛	L	0.140	210	56
8⅛	K	0.271	320	118
8⅛	L	0.200	230	65

Based on minimum wall thickness
Allowable stress 5,000 psi (at 300 F)
No allowance for corrosion, grooving or threading
y = material coefficient = 0.4
$P = 2St/(D_0 - 2yt)$ for internal pressure.
Allowable external pressures from Fig. UG-31, ASME Boiler Code, Section VIII.
* Not standard-type letter-designated wall thickness.

For low-temperature application, the use of nonferrous materials or the stainless steel mentioned will normally be satisfactory. The use of nickel-steel pipe in conjunction with the use of nickel-steel pressure vessels has long been an acceptable material for low temperature when these materials are subjected to and pass the impact-testing requirements.

Impact tests, when conducted are made in accordance with ASTM E23 requirements for the Charpy-type test using the keyhole notch specimen as in Fig. 3, Type B, of that specification.

The standard 10- by 10-mm specimen is used if the thickness of the material being tested is ⁷⁄₁₆ in. or greater. For material that is not of sufficient thickness to permit preparation of full-size specimens, tests may be made on the largest possible of the subsize specimens listed below.

The impact properties for each size specimen are as listed below.

Size of specimen, mm	Minimum impact value required, ft-lb
10 × 10	15
10 × 7.5	12.5
10 × 5	10
10 × 2.5	5

In welded fabrication, the weld also is required to meet the impact-test requirement.

Expansion and Contraction.   Since refrigeration piping systems are subject to changes in temperature, some precautions must be taken to assure that these changes in temperature during operation or during shut-down are considered in the design of the piping and in the design of supports and flexibility.

Piping systems must be designed to have sufficient flexibility to prevent thermal expansion from causing:
1. Failure of piping or anchors from overstress or overstrain
2. Leakage at joints
3. Detrimental distortion of connected equipment resulting from excessive thrusts or moments

Expansion strains are usually taken up by bending or torsion or by compression and tension.   The concentration of stresses will be different in each case.

Bending or torsional flexibility may be provided by the use of bends, loops, or offsets.   While swivel joints, ball joints, and corrugated expansion joints are recognized by the Pressure Piping Code, some of these are not considered desirable for volatile refrigerant piping.   Bends, loops, and offsets are generally used to provide flexibility.   Loops and cold springing also may be used in the design of piping. Chapter IV covers in detail the general design considerations involved in expansion and flexibility of piping.   These same principles must be applied to refrigeration piping, and the maximum temperature cycle involved in the installation should be taken into account in determining the nature and direction of the stresses which may be caused by temperature effects.   As mentioned above under "Insulation," pipe hangers or supports for low-temperature piping normally will not be in direct contact with the metal portion of the piping.   Consideration must be given to the insulation in types of support which are peculiar to refrigeration piping when the problems of expansion and flexibility are considered.

Miter Joints.   The ASA B31.5 Pressure Piping Code gives details for design of branch connections where the angle between the axes of the branch and of the run is between 45 and 90 deg.   Branch connections less than 45 deg impose special design and fabrication problems.   The code permits connections made by the use of fittings such as tees and similar types, welding outlet fittings such as cast or forged nozzles or couplings or by attaching a branch pipe directly to the run pipe by welding.

Normally the use of standard forged fittings of the butt-welding or socket types will provide sufficient strength in the case of a branch connection to permit application of these fittings without additional reinforcement.   However, when a branch connection is welded into a hole cut into the main-run pipe, it is recognized that certain reinforcement may be required.   The analysis of the extent of reinforcement, if required, is similar to that used on unfired pressure vessels for nozzle connections.   The complete detailed analysis for such determinations and the means of determining the amount of reinforcement required are shown in the ASA B31.5 Pressure Piping Code or in the ASME Unfired Pressure Vessel Code.   The general method is to calculate the amount of metal cut out of the pipe and to calculate the amount of metal which is added by extra thickness of the pipe wall over that required for strength, the extra metal in the nozzle or branch connection, and the

Table 22. Allowable Stresses, psi *

Seamless carbon steel and iron pipe and tube

Material	Specification	Grade	Class	Temper	Tensile strength min, psi	Yield strength min, psi	Notes	Long. or spiral joint factor	For metal temperatures not exceeding, F						
									100	150	200	250	300	350	400
Steel pipe	ASTM A53	A	....	......	48,000	30,000	..	....	12,000	12,000	12,000	12,000	12,000	12,000	12,000
Steel pipe	ASTM A53	B	....	......	60,000	35,000	..	....	15,000	15,000	15,000	15,000	15,000	15,000	15,000
Steel tube	ASTM A83	A	....	......	......	......	..	....	11,750	11,750	11,750	11,750	11,750	11,750	11,750
Steel pipe	ASTM A106	A	....	......	48,000	30,000	..	....	12,000	12,000	12,000	12,000	12,000	12,000	12,000
Steel pipe	ASTM A106	B	....	......	60,000	35,000	..	....	15,000	15,000	15,000	15,000	15,000	15,000	15,000
Steel pipe	ASTM A106	C	....	......	70,000	40,000	..	....	17,500	17,500	17,500	17,500	17,500	17,500	17,500
Steel tube	ASTM A179	......	....	......	......	......	..	....	11,750	11,750	11,750	11,750	11,750	11,750	11,750
Steel tube	ASTM A192	......	....	......	......	......	..	....	11,750	11,750	11,750	11,750	11,750	11,750	11,750
Steel tube	ASTM A210		....	......	......	......	..	....	15,000	15,000	15,000	15,000	15,000	15,000	15,000
Steel pipe	ASTM A333	C	....	......	55,000	30,000	..	....	13,750	13,750	13,750	13,750	13,750	13,750	13,750
Steel tube	ASTM A334	C	....	......	55,000	30,000	..	....	13,750	13,750	13,750	13,750	13,750	13,750	13,750
Steel pipe	API 5L	A	....	......	48,000	30,000	..	....	12,000	12,000	12,000	12,000	12,000	12,000	12,000
Steel pipe	API 5L	B	....	......	60,000	35,000	..	....	15,000	15,000	15,000	15,000	15,000	15,000	15,000

Carbon-steel and wrought-iron pipe and tube

Material	Spec	Grade	Tensile	Yield	Factor								
Butt-welded pipe:													
Steel	ASTM A53		45,000	25,000	0.60	6,750	6,750	6,750	6,750	6,750	6,750	6,750	6,750
Wrought iron	ASTM A72		42,000	24,000	0.60	6,300	6,300	6,300	6,300	6,300	6,300	6,300	6,300
Steel	API 5L		45,000	25,000	0.60	6,750	6,750	6,750	6,750	6,750	6,750	6,750	6,750
Wrought iron	API 5L		42,000	24,000	0.60	6,300	6,300	6,300	6,300	6,300	6,300	6,300	6,300
Lap-welded pipe:													
Wrought iron	ASTM A72		42,000	24,000	0.80	8,400	8,400	8,400	8,400	8,400	8,400	8,400	8,400
Wrought iron	API 5L		42,000	24,000	0.80	8,400	8,400	8,400	8,400	8,400	8,400	8,400	8,400
Electric resistance-welded and electric flash-welded pipe and tube:													
Steel pipe	ASTM A53	A	48,000	30,000	0.85	10,200	10,200	10,200	10,200	10,200	10,200	10,200	10,200
Steel pipe	ASTM A53	B	60,000	35,000	0.85	12,750	12,750	12,750	12,750	12,750	12,750	12,750	12,750
Steel pipe	ASTM A135	A	48,000	30,000	0.85	10,000	10,000	10,000	10,000	10,000	10,000	10,000	10,000
Steel pipe	ASTM A135	B	60,000	35,000	0.85	12,750	12,750	12,750	12,750	12,750	12,750	12,750	12,750
Steel tube	ASTM A178	A			0.85	10,000	10,000	10,000	10,000	10,000	10,000	10,000	10,000
Steel tube	ASTM A178	C	60,000	37,000	0.85	12,750	12,750	12,750	12,750	12,750	12,750	12,750	12,750
Steel tube	ASTM A214				0.85	10,000	10,000	10,000	10,000	10,000	10,000	10,000	10,000
Steel tube	ASTM A226				0.85	10,000	10,000	10,000	10,000	10,000	10,000	10,000	10,000
Steel pipe	ASTM A333	C	55,000	30,000	0.85	11,700	11,700	11,700	11,700	11,700	11,700	11,700	11,700
Steel tube	ASTM A334	C	55,000	30,000	0.85	11,700	11,700	11,700	11,700	11,700	11,700	11,700	11,700
Steel pipe	API 5L	A	48,000	30,000	0.85	10,200	10,200	10,200	10,200	10,200	10,200	10,200	10,200
Steel pipe	API 5L	B	60,000	35,000	0.85	12,750	12,750	12,750	12,750	12,750	12,750	12,750	12,750

Table 22. (Continued)

Carbon-steel and wrought-iron pipe and tube (Cont'd)

Material	Specification	Grade	Class	Temper	Tensile strength min, psi	Yield strength min, psi	Notes	Long, or spiral joint factor	For metal temperatures not exceeding, F						
									100	150	200	250	300	350	400
Electric-fusion-welded pipe:															
Steel	ASTM A134	A245, Gr. A	.....	.....	48,000	26,000	a	0.80	8,800	8,800	8,800	8,800	8,800	8,800	8,800
Steel	ASTM A134	A245, Gr. B	.....	.....	52,000	30,000	a	0.80	9,600	9,600	9,600	9,600	9,600	9,600	9,600
Steel	ASTM A134	A245, Gr. C	.....	.....	55,000	33,000	a	0.80	10,100	10,100	10,100	10,100	10,100	10,100	10,100
Steel	ASTM A134	A283, Gr. A	.....	.....	45,000	24,000	a	0.80	8,300	8,300	8,300	8,300	8,300	8,300	8,300
Steel	ASTM A134	A283, Gr. B	.....	.....	50,000	27,000	a	0.80	9,200	9,200	9,200	9,200	9,200	9,200	9,200
Steel	ASTM A134	A283, Gr. C	.....	.....	55,000	30,000	a	0.80	10,100	10,100	10,100	10,100	10,100	10,100	10,100
Steel	ASTM A134	A283, Gr. D	.....	.....	60,000	33,000	a	0.80	10,100	10,100	10,100	10,100	10,100	10,100	10,100
Steel	ASTM A134	A285, Gr. A	.....	.....	45,000	24,000	..	0.80	9,000	9,000	9,000	9,000	9,000	9,000	9,000
Steel	ASTM A134	A285, Gr. B	.....	.....	50,000	27,000	..	0.80	10,000	10,000	10,000	10,000	10,000	10,000	10,000
Steel	ASTM A134	A285, Gr. C	.....	.....	55,000	30,000	..	0.80	11,000	11,000	11,000	11,000	11,000	11,000	11,000
Steel	ASTM A139	A	.....	.....	48,000	30,000	..	0.80	9,600	9,600	9,600	9,600	9,600	9,600	9,600
Steel	ASTM A139	B	.....	.....	60,000	35,000	..	0.80	12,000	12,000	12,000	12,000	12,000	12,000	12,000
Wrought iron	ASTM A419	A42	.....	.....	40,000	24,000	a	0.80	8,000	8,000	8,000	8,000	8,000	8,000	8,000
Steel	ASTM A211	A245, Gr. A	.....	.....	48,000	25,000	a	0.80	8,800	8,800	8,800	8,800	8,800	8,800	8,800
Steel	ASTM A211	A245, Gr. B	.....	.....	52,000	30,000	a	0.80	9,600	9,600	9,600	9,600	9,600	9,600	9,600
Steel	ASTM A211	A245, Gr. C	.....	.....	55,000	33,000	a	0.80	10,100	10,100	10,100	10,100	10,100	10,100	10,100
Wrought iron	ASTM A211	A129, Gr. A	.....	.....	40,000	22,000	..	0.80	8,000	8,000	8,000	8,000	8,000	8,000	8,000
Wrought iron	ASTM A211	A129, Gr. B	.....	.....	44,000	27,000	..	0.80	8,800	8,800	8,800	8,800	8,800	8,800	8,800
Wrought iron	ASTM A211	A129, Gr. C	.....	.....	42,000	23,000	..	0.80	8,400	8,400	8,400	8,400	8,400	8,400	8,400
Copper brazed tubing:															
Steel	ASTM A254	....	I	.....	42,000	28,000	..	.....	6,000	5,800	5,500	5,150	4,750	4,050	3,000
Steel	ASTM A254	....	II	.....	42,000	28,000	..	.....	3,600	3,450	3,300	3,100	2,850	2,400	1,800

Low and intermediate alloy steel, seamless alloy steel pipe and tube

Material	ASTM spec	Grade	Tensile, psi	Yield, psi	E						
3½ Ni pipe	ASTM A333	3	65,000	35,000	.....	16,250	16,250	16,250	16,250	16,250	16,250
Cr-Cu-Ni pipe	ASTM A333	4	60,000	35,000	.....	15,000	15,000	15,000	15,000	15,000	15,000
5 Ni pipe	ASTM A333	5	65,000	35,000	.....	16,250	16,250	16,250	16,250	16,250	16,250
3½ Ni tube	ASTM A334	3	65,000	35,000	.....	16,250	16,250	16,250	16,250	16,250	16,250
5 Ni tube	ASTM A334	5	65,000	35,000	.....	16,250	16,250	16,250	16,250	16,250	16,250

Low and intermediate alloy steel pipe and tubing

Electric-resistance welded:

Material	ASTM spec	Grade	Tensile, psi	Yield, psi	E						
3½ Ni pipe	ASTM A333	3	65,000	35,000	0.85	13,800	13,800	13,800	13,800	13,800	13,800
5 Ni pipe	ASTM A333	5	65,000	35,000	0.85	13,800	13,800	13,800	13,800	13,800	13,800
3½ Ni tube	ASTM A334	3	65,000	35,000	0.85	13,800	13,800	13,800	13,800	13,800	13,800
5 Ni tube	ASTM A334	5	65,000	35,000	0.85	13,800	13,800	13,800	13,800	13,800	13,800

Austenitic stainless-steel pipe and tube

Seamless:

Material	ASTM spec	Grade	Tensile, psi	Yield, psi	E						
18-8 tube	ASTM A213	304	75,000	30,000	.....	18,750	.....	16,650	15,000	13,650	13,650
18-8 tube	ASTM A213	304L	70,000	25,000	.....	17,500	.....	15,300	13,100	11,000	11,000
18-8 tube	ASTM A269	304	75,000	30,000	.....	18,750	.....	16,650	15,000	13,650	13,650
18-8 tube	ASTM A269	304L	70,000	25,000	.....	17,500	.....	15,300	13,100	11,000	11,000
18-8 tube	ASTM A271	304	75,000	30,000	.....	18,750	.....	16,650	15,000	13,650	13,650
18-8 tube	ASTM A271	304L	70,000	25,000	.....	17,500	.....	15,300	13,100	11,000	11,000
18-8 pipe	ASTM A312	304	75,000	30,000	.....	18,750	.....	16,650	15,000	13,650	13,650
18-8 pipe	ASTM A312	304L	70,000	25,000	.....	17,500	.....	15,300	13,100	11,000	11,000
18-8 pipe	ASTM A376	304	75,000	30,000	.....	18,750	.....	16,650	15,000	13,650	13,650
18-8 pipe	ASTM A376	304L	70,000	25,000	.....	17,500	.....	15,300	13,100	11,000	11,000

Electric fusion welded:

Material	ASTM spec	Grade	Tensile, psi	Yield, psi	E						
18-8 tube	ASTM A249	304	75,000	30,000	0.85	15,900	.....	14,150	12,750	11,600	11,600
18-8 tube	ASTM A249	304L	70,000	25,000	0.85	14,900	.....	13,000	11,150	9,350	9,350
18-8 tube	ASTM A269	304	75,000	30,000	0.85	16,000	.....	14,150	12,750	11,600	11,600
18-8 tube	ASTM A269	304L	70,000	25,000	0.85	14,900	.....	13,000	11,150	9,350	9,350
18-8 pipe	ASTM A312	304	75,000	30,000	0.85	16,000	.....	14,150	12,750	11,600	11,600
18-8 pipe	ASTM A312	304L	70,000	25,000	0.85	14,900	.....	13,000	11,150	9,350	9,350
18-8 pipe	ASTM A358	304	75,000	30,000	0.85	16,000	.....	14,150	12,750	11,600	11,600
18-8 pipe	ASTM A358	304L	70,000	25,000	0.85	14,900	.....	13,000	11,150	9,350	9,350
18-8 pipe	ASTM A409	304	75,000	30,000	0.85	16,000	.....	14,150	12,750	11,600	11,600
18-8 pipe	ASTM A409	304L	70,000	25,000	0.85	14,900	.....	13,000	11,150	9,350	9,350

Table 22. (Continued)

Material	Specification	Grade	Class	Temper	Tensile strength min, psi	Yield strength min, psi	Notes	Long. or spiral joint factor	For metal temperatures not exceeding, F 100	150	200	250	300	350	400
Seamless copper and copper alloy pipe and tube															
Copper pipe	ASTM B42		1/8" to 2"	Annealed	30,000	9,000			6,000	6,000	5,900	5,800	5,000	3,800	2,500
Copper pipe	ASTM B42			Hard drawn	45,000	40,000	b		11,300	11,300	11,000	10,300	8,000		
Copper pipe	ASTM B42		2⅛" to 12"	Light drawn	36,000	30,000	b		9,000	9,000	8,700	8,000	8,000	6,000	3,000
Red brass pipe	ASTM B43			Annealed	40,000	12,000			8,000	8,000	8,000	8,000	8,000		
Copper tube	ASTM B68			Annealed	30,000	9,000			6,000	6,000	5,900	5,800	5,000	3,800	2,500
Copper tube	ASTM B75			Annealed	30,000	9,000			6,000	6,000	5,900	5,800	5,000	3,800	2,500
Copper tube	ASTM B75			Light drawn	36,000	30,000	b		9,000	9,000	8,700	8,300	8,000		
Copper tube	ASTM B75			Hard drawn	45,000	40,000	b		11,300	11,300	11,000	10,500	8,000		
Copper tube	ASTM B88			Annealed	30,000	......	b		6,000	6,000	5,900	5,800	5,000	3,800	2,500
Copper tube	ASTM B88			Drawn	36,000	......	b		9,000	9,000	8,700	8,300	8,000		
Cu Ni 70-30 pipe	ASTM B111			Annealed	52,000	18,000			12,000	11,600	11,300	11,100	10,800	10,600	10,300
Cu Ni 90-10 pipe	ASTM B111			Annealed	40,000	15,000			10,000	10,000	9,800	9,500	9,300	9,000	8,700
Cu Ni 70-30 tube	ASTM B111			Annealed	52,000	18,000			12,000	11,600	11,300	11,000	10,800	10,600	10,300
Cu Ni 90-10 tube	ASTM B111			Annealed	40,000	15,000			10,000	10,000	9,800	9,500	9,300	9,000	8,700
Copper tube	ASTM B280			Annealed	30,000	9,000			6,000	6,000	5,900	5,800	5,000	3,800	2,500
Copper silicon pipe	ASTM B315			Annealed	50,000	15,000			10,000	10,000	10,000	10,000	10,000	5,000	
Copper silicon tube	ASTM B315			Annealed	50,000	15,000			10,000	10,000	10,000	10,000	10,000	5,000	
Seamless nickel base alloy pipe and tube															
Ni Cu 67-30 pipe	ASTM B165		5" OD and under	Annealed	70,000	28,000			17,500	17,000	16,500	16,000	15,500	15,100	14,800
Ni Cu Al 66-29 pipe	ASTM B165		5" OD and under	Annealed	70,000	28,000			17,500	17,000	16,500	16,000	15,500	15,100	14,800
Ni Cu 67-30 tube	ASTM B165		5" OD and under	Annealed	70,000	28,000			17,500	17,000	16,500	16,000	15,500	15,100	14,800
Ni Cu Al 66-29 tube	ASTM B165		5" OD and under	Annealed	70,000	28,000			17,500	17,000	16,500	16,000	15,500	15,100	14,800

Seamless aluminum base alloy pipe and tube

Material	ASTM Spec	Alloy	Temper			Note							
3003 tube	ASTM B210	M1A	O	14,000	5,000		3,350	3,150	2,900	2,700	2,400	2,100	1,800
3003 tube	ASTM B210	M1A	H14	20,000	17,000		5,000	4,850	4,700	4,400	4,000	3,500	3,100
3003 tube	ASTM B210	M1A	H18	27,000	24,000	c	6,750	6,400	6,050	5,700	5,250	4,400	3,500
6063 tube	ASTM B210	GS10A	T6	33,000	28,000	c	8,250	7,800	7,500	6,700	4,950	3,400	2,200
6063 tube	ASTM B210	GS10A	T6 welded	17,000	……	d	4,250	4,200	4,000	3,800	3,600	2,750	1,900
6061 tube	ASTM B210	GS11A	T4	30,000	16,000		7,500	7,200	7,000	6,700	6,400	5,600	4,000
6061 tube	ASTM B210	GS11A	T6	42,000	35,000	d	10,500	10,200	9,900	9,400	7,900	6,500	4,400
6061 tube	ASTM B210	GS11A	T6 welded	24,000	……	d	6,000	5,900	5,700	5,400	5,000	4,200	3,200
6061 tube	ASTM B234	GS11A	T4	30,000	16,000		7,500	7,200	7,000	6,700	6,400	5,600	4,000
6061 tube	ASTM B234	GS11A	T6	42,000	35,000	d	10,500	10,200	9,900	9,400	7,900	6,500	4,400
6061 tube	ASTM B234	GS11A	T6 welded	24,000	……	d	6,000	5,900	5,700	5,400	5,000	4,200	3,200
3003 tube	ASTM B221	M1A	O	14,000	5,000		3,350	3,150	2,900	2,700	2,400	2,100	1,800
3003 tube	ASTM B221	M1A	H112	14,500	6,000		3,600	3,250	3,000	2,800	2,500	2,200	1,900
5083 tube	ASTM B221	GM41A	O	39,000	16,000	c	9,750	9,750					
6063 tube	ASTM B221	GS10A	T42	17,000	9,000	d	4,200	4,200	4,200	4,200	4,000	3,100	2,000
6063 tube	ASTM B221	GS10A	T5	22,000	16,000	d	5,500	5,100	4,900	4,600	4,200	3,100	2,000
6063 tube	ASTM B221	GS10A	T6	30,000	25,000	d	7,500	7,100	6,800	6,100	4,500	3,100	2,000
6063 tube	ASTM B221	GS10A	T6 welded	17,000	……	d	4,250	4,200	4,000	3,800	3,600	2,750	1,900
6061 tube	ASTM B221	GS11A	T4	26,000	16,000	d	6,500	6,200	6,000	5,800	5,600	4,900	3,500
6061 tube	ASTM B221	GS11A	T6	38,000	35,000	d	9,500	9,200	9,000	8,500	7,200	5,600	4,000
6061 tube	ASTM B221	GS11A	T6 welded	24,000	……		6,000	5,900	5,700	5,400	5,000	4,200	3,200
3003 pipe	ASTM B241	M1A	O	14,000	5,000		3,350	3,150	2,900	2,700	2,400	2,100	1,800
3003 pipe	ASTM B241	M1A	H18	27,000	24,000	c	6,750	6,400	6,050	5,700	5,250	4,400	3,500
3003 pipe	ASTM B241	M1A	H112	14,500	6,000	c	3,600	3,250	3,000	2,800	2,500	2,200	1,900
5083 pipe	ASTM B241	GM41A	O	38,000	16,000		9,500	9,500					
6063 pipe	ASTM B241	GS10A	T6	30,000	25,000	d	7,500	7,100	6,800	6,100	4,500	3,100	2,000
6063 pipe	ASTM B241	GS10A	T6 welded	17,000	……		4,250	4,200	4,000	3,800	3,600	2,750	1,900
6061 pipe	ASTM B241	GS11A	T4	26,000	16,000	d	6,500	6,200	6,000	5,800	5,600	4,900	3,500
6061 pipe	ASTM B241	GS11A	T6	38,000	35,000	d	9,500	9,200	9,000	8,500	7,200	5,600	4,000
6061 pipe	ASTM B241	GS11A	T6 welded	24,000	……		6,000	5,900	5,700	5,400	5,000	4,200	3,200

* Courtesy ASA B31.5 Code.

a A quality factor of 92 per cent is included for structural grade.

b When used for fluids with temperatures over 300 F, methods used in making joints anneal the material.

c For welded construction or where thermal cutting is employed stress value for O temper material shall be used.

d The stress values given for this material are not applicable when either welding or thermal cutting is used.

extra metal which would be added within certain limiting geometric zones by welding. If the added metal provides the equivalent of the amount of metal which had been cut out, no additional reinforcement is necessary. Reinforcement metal usually is provided in the form of a ring or a saddle which is welded to the run pipe. This reinforcement material must be added within certain limiting dimensions as defined by the Code. The use of ribs, gussets, or clamps is permissible to stiffen the branch connection, but their areas cannot be counted as contributing to the reinforcement area.

Welding, Brazing, and Soldering. Joints in piping which is to convey volatile refrigerants are usually made by welding, brazing, or soldering. This does not exclude the use of flanged connections, which are commonly used to connect valves or control devices in refrigeration piping. Flanged connections are commonly used to connect to pressure vessels and to compressors. Couplings of the friction type with seal rings may be used for refrigerants when the materials are compatible with the refrigerant and when the pressures permit. Use of such fittings is normally confined to low-pressure refrigerants. Limitations on threaded connections have already been listed and frequently threaded connections, where used and where permissible, will be seal-welded.

Welded joints may be used in any materials for which it is possible to qualify the welding procedures, the welders, and the welding operators.

Butt welds are permitted. Usually, backing rings are used in butt-welded joints, but where it is necessary to have a smooth interior surface or where the backing ring may result in severe corrosion or erosion, the joint may be welded without backing rings provided the piping is suitably cleaned. Socket welds are permitted under the Pressure Piping Code. The Pressure Piping Code defines in detail the required weld sizes and joint arrangements which are recommended for use in welded-part construction.

The ASME Boiler and Pressure Vessel Code, Section IX, defines in detail the qualification of welding procedures and welders' performance requirements for unfired pressure vessel construction. These rules have been adopted by the Pressure Piping Code, ASA B31.5, and form part of the requirements of that Code. The welders and the procedures should be qualified to assure that their quality is in conformance with these codes.

The ASA B9 Safety Code for Mechanical Refrigeration requires brazing of certain joints. Also certain joints in restricted areas may require high-melting-point filler material. The ASA B31.5 Pressure Piping Code defines the filler metal used in brazing to be nonferrous metal or alloy having a melting point above 1000 F but below that of the metal being joined. Good practice in cleanliness of joints and the use of proper fluxes is required for brazed joints.

Brazing procedures and operators, except for socket-type capillary joints, should be qualified in accordance with the requirements of Section IX of the ASME Boiler and Pressure Vessel Code.

For soldered joints, the ASA B31.5 Code defines the solder metal to be a nonferrous metal or alloy having a melting point below 800 F and below that of the metal being joined. Good soldering technique requires proper cleanliness and preparation of the joints, proper joint clearances, and proper heating. Procedures to be used on soldering or brazing socket-type joints are outlined in a publication of the Copper Development Association, Publication 25.[6]

Miscellaneous Considerations

Corrosion. The ASA B31.5 Pressure Piping Code recognizes that corrosion or erosion may be factors to be considered in piping design. When corrosion or

erosion is expected, an increase in wall thickness of the components above that dictated by other design requirements is to be provided consistent with the expected life of the particular piping involved. In the basic equation for calculating the pressure ratings of pipe or in determining the required wall thickness of pipe, the Pressure Piping Code requires the addition of a factor to the calculated wall thickness to result in the actual thickness required. The factor includes allowance for threading, groove depth, manufacturers' minus tolerance plus corrosion and erosion allowances.

Corrosion allowance on the inside of piping for volatile refrigerants is not mandatory. The refrigerant is recirculating and is usually charged into the system in a commercially pure state after thorough cleaning and evacuation of the entire system. When installed in accordance with good practice, a leaktight refrigerating system will not tend to corrode and it is not customary to add corrosion allowances. It is possible, with certain of the halogenated hydrocarbon refrigerants when contaminated with noncondensible gases or with water which may leak into a system under vacuum, to have corrosive products form. On some occasions, in hermetically sealed refrigeration systems, compressor motor burnouts have resulted in formation of contaminants which also may be damaging to the inside of the system. However, these considerations are not properly part of the piping design and are usually the result of carelessness or misapplication. It is ordinarily not necessary to add corrosion allowance to volatile-refrigerant piping.

For brine piping, especially with salt brines, the consideration of possible corrosion should be kept in mind in the design of the piping system. Ordinarily standard-weight pipe for either volatile-refrigerant use or for brine piping inherently has sufficient strength so that the normal wall thickness of pipes which are used are much heavier than are required for the actual pressure service, and therefore it may not be necessary to add additional allowances for corrosion.

Fittings. The Pressure Piping Code permits the use of standard fittings provided thay are compatible with the refrigerant or fluid. The standard ratings of forged steel flanges, fittings, and similar parts may be used for refrigerant service.

Bell-and-spigot fittings may be used only for water and drainage service.

Couplings made of cast, malleable, or wrought iron may not be used on pipe containing flammable or toxic fluids. Wrought-iron couplings are subject to the same limitations in temperature, stress, and service which apply to cast-iron screwed fittings.

Valves. Cast-iron gate valves and plug cocks must not be used in liquid-refrigerant lines unless consideration is given to the expansion of liquid trapped in a space when the valve is closed.

Several manufacturers make standard lines of refrigeration fittings which do not fall into the classification of ASA Standards for forged-steel valves. These valves and fittings over long years of usage have gained acceptability and are widely used and acceptable for refrigeration service to the degree recommended by the manufacturer.

Other Factors. The Pressure Piping Code lists the following dynamic effects which should be taken into account in the design of refrigerant piping.

1. Impact forces (including hydraulic shock) caused by either external or internal conditions.

2. The effect of wind loading on exposed piping.

3. Piping systems located in regions where earthquakes are a factor are to be designed for a horizontal force in conformity with the good engineering practice using governmental data as a guide in determining the earthquake force. However, this force is not to be considered as acting concurrently with lateral wind force.

4. Piping shall be arranged and supported with consideration for vibration.

The Pressure Piping Code also calls attention to the following weight effects which should be taken into account in the design of piping:

1. Live loads such as the weight of the fluid transported and snow and ice loads if the latter will be encountered. If low-temperature piping is not insulated, there can be a build-up of ice on the pipe even in high ambient temperatures.

2. Dead loads, consisting of the weight of the piping components and insulation and other superimposed loads.

3. Test loads which consist of the weight of the test fluid in the pipe.

REFERENCES

1. American Standard Safety Code for Mechanical Refrigeration, ASA B9.1-1964, American Society of Heating, Refrigerating and Air Conditioning Engineers, Inc., United Engineering Center, 345 E. 47th St., New York, N.Y. 10017, American Standards Association, Inc., 10 E. 40th St., New York, N.Y. 10016.
2. Refrigeration Piping, ASA B31.5, American Society of Mechanical Engineers, United Engineering Center, 345 E. 47th St., New York, N.Y. 10017, American Standards Association, Inc., 10 E. 40th St., New York, N.Y. 10016.
3. ASME Boiler and Pressure Vessel Code, Section VIII, Unfired Pressure Vessels, American Society of Mechanical Engineers, United Engineering Center, 345 E. 47th St., New York, N.Y. 10017.
4. "Heating, Ventilating and Air Conditioning Guide," American Society of Heating, Refrigerating and Air Conditioning Engineers, United Engineering Center, 345 E. 47th St., New York, N.Y. 10017.
   "Refrigerant Piping Data," Air Conditioning and Refrigeration Institute, 1815 North Fort Myer Drive, Arlington, Va. 22209.
   "Soldered and Brazed Joints in Copper Tube," Publication No. 25, Copper Development Association, Inc., 405 Lexington Ave., New York, N.Y. 10017.

18

SEWERAGE-SYSTEMS PIPING

William E. Dobbins*

This chapter deals with the design of sewerage systems which perform the functions of collecting water-borne wastes of domestic, commercial, and industrial origin and of storm-water runoff and conveying them to points of disposal.

The design of sewerage systems which carry domestic or industrial wastes must comply with the minimum standards of the city, county, and state regulatory agencies. Plans must ordinarily be approved, and permits must be obtained for the disposal of domestic and industrial wastes into natural water courses. When industrial wastes are to be disposed of by conveyance to a public sewerage system, the quantity and quality of the waste must ordinarily be in compliance with the local sewer ordinance.

DEFINITIONS

Terms commonly used in relation to sewerage systems have been defined as follows:[1,†]

Sewer. A pipe or conduit, generally closed but normally not flowing full, for carrying sewage and other waste liquids.

Sewage. Largely the water supply of a community after it has been fouled by various uses. From the standpoint of source, it may be a combination of the liquid or water-carried wastes from residences, business buildings, and institutions,

---

* Professor of Sanitary Engineering, New York University, New York, N.Y.; Partner, Teetor-Dobbins, Consulting Engineers, West Islip, N.Y.

† Superscript numbers refer to corresponding entries in a list of references included at the end of this chapter.

together with those from industrial establishments, and with such ground water, surface water, and storm water as may be present.

Sanitary Sewer. A sewer which carries sewage and to which storm, surface, and ground waters are not intentionally admitted; also referred to as "separate sanitary sewer" or "separate sewer."

Storm Sewer. A sewer which carries storm water, surface water, street wash, and other wash waters or drainage but excludes sewage and industrial wastes; also called "storm drain."

Combined Sewer. A sewer receiving both surface runoff and sewage.

Building Sewer. The extension from a building drain to the public sewer or other place of disposal; also called the "house sewer" or "house connection."

Lateral Sewer. A sewer which discharges into a branch or other sewer and has only building sewers tributary to it.

Branch Sewer. A sewer which receives sewage from a relatively small area and discharges into a main sewer.

Main Sewer. A sewer to which one or more branch sewers are tributary and which serves a large territory; also called "trunk sewer."

Intercepting Sewer. A sewer which receives dry-weather flow from a number of transverse sewers or outlets and frequently additional predetermined quantities of storm water (if from a combined system) and conducts such waters to a point for treatment or disposal.

Outfall Sewer. A sewer which receives sewage from a collecting system and carries it to a point of final discharge.

Separate System. A sewer system comprised exclusively of sanitary sewers which carry only sewage and to which storm water, surface water, and ground water are not intentionally admitted; also referred to as "sanitary system" or "separate sanitary system."

Storm-sewer System. A system composed only of sewers carrying storm water, surface water, street wash, and other wash waters or drainage and from which sewage and industrial wastes are excluded.

Combined-sewer System. A system of sewers receiving both surface runoff and sewage.

QUANTITY OF SANITARY SEWAGE

General Considerations. Sanitary sewers must be designed to provide capacity for the present and estimated future quantities of domestic sewage, commercial and industrial wastes, and ground-water infiltration. Lateral and branch sewers should be designed for the ultimate population density to be expected in the area served.   Larger sewers are commonly designed to handle the flows to be expected from 25 to 50 years in the future.   The estimation of future flows should be arrived at only after a detailed study of the land usage, population growth trends, water-consumption rates, commercial and industrial growth, etc.

The prediction of future populations is made with much more certainty for large areas and dense populations than for small areas and low population densities. Figure 1 shows capacity factors given by S. A. Greeley and W. A. Stanley[2] for use in making allowances for the uncertainties of population distributions within a sewer district.   The possible future population density for a portion of a large district may be estimated by multiplying the estimated future average density for the entire district by the appropriate factor.

The most important single index to the flow of sanitary sewage is the rate of consumption of water.   The total consumption of water within a town or sewer district as a whole is generally considerably higher than the purely domestic

consumption, the difference being attributable to the use in public institutions, office buildings, commercial and industrial establishments, etc., and to the leakage from the water-distribution system. When estimates of sanitary sewage flow from small areas are being prepared, the most accurate procedure is to make separate estimates of the various classifications of flow which make up the total. The classifications which are commonly used are domestic, commercial, and industrial sewage flows and ground-water infiltration.

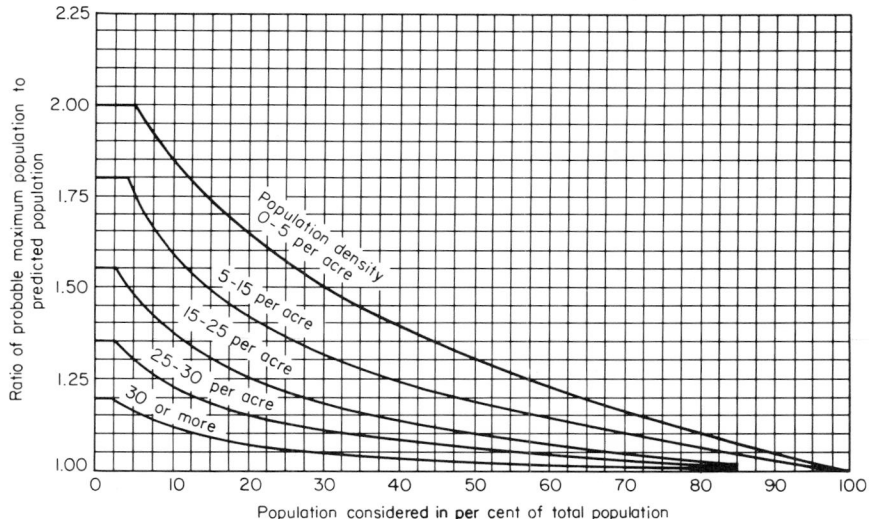

Fig. 1. Capacity factors for unequal population development.

Quantity of Domestic Sewage. The purely domestic sewage flow will generally be about 80 to 90 per cent of the domestic water consumption. The average per capita domestic water consumption varies from about 40 to 120 gpd, depending upon the character of the area and the economic status of the population. If water is supplied through meters, accurate estimates of average per capita consumption can be made. If water is supplied unmetered, estimates have to be based on the consumption rates which are known to prevail in other areas of similar character.

Quantity of Commercial Sewage. The quantity of sewage flow from commercial areas varies widely depending upon the nature of the commercial activity. Allowances made for the quantity of sewage from commercial areas in large sewer districts are commonly in terms of gallons per day per acre or gallons per capita per day. Table 1 gives the allowances which were made for flow from commercial areas in some American cities.[3] Allowances varied from 2,000 to 60,500 gpd per acre and from 15 to 500 gpd per capita. It is evident that, for any area in which commercial activity is an important factor, the estimate of sewage flow should be based on a special study of the area.

Industrial Sewage Flow. The flow of sewage from industrial establishments may be purely sanitary sewage, or it may also include water-borne industrial wastes. Estimates of the sanitary sewage are made by the procedures already considered. The quantity of industrial wastes can be determined only by special

Table I.   Sewer Capacity Allowances for Commercial and Industrial Areas *

City	Year data	Commercial	Industrial
Baltimore, Md.†	1949	135 gpcd‡ (range 6,750 to 13,500 gpd per acre); resident population	7,500 gpd per acre minimum
Berkeley, Calif.	....	....	50,000 gpd per acre
Boston, Mass.†	1949	No standard—each area specially studied	
Cincinnati, Ohio†	1949	Commercial areas not served by sanitary sewers	
Columbus, Ohio†	1946	40,000 gpd per acre; excess added to residential amount	
Cranston, R.I.†	1943	25,000 gpd per acre	
Dallas, Tex.†	1949	30,000 gpd per acre; downtown area rate added to domestic; outlying area rate added to domestic	
Grand Rapids, Mich...	....	40–50 gpcd;‡ office buildings 400–500 gpd per room; hotels 200 gpd per bed; hospitals 200–300 gpd per room; schools	250,000 gpd
Hagerstown, Md.	....	180–250 gpd per room; hotels 150 gpd per bed; hospitals 120–150 gpd per room; schools	
Las Vegas, Nev.	....	310–525 gpd per room; resort hotels 15 gpcd;‡ schools	
Los Angeles, Calif.	1948†	80–100 gpcd;‡ office buildings 450–500 gpcd;‡ hotels 800–1,000 gpd per bed; hospitals 35 gpcd;‡ schools 0.015 cfs per acre; light business district (= 9,700 gpd per acre; all commercial areas)†	0.021 cfs per acre; light industrial—individual studies for major industrial districts
Memphis, Tenn.	....	2,000 gpd per acre	2,000 gpd per acre
Milwaukee, Wis.†	1945	0.0936 cfs per acre = 60,500 gpd per acre	
New York, N.Y.†	1949	Allowances determined by special gagings	
Santa Monica, Calif. ..	....	0.015 cfs per acre; commercial 0.012 cfs per acre; hotels	0.021 cfs per acre
Shreveport, La.	....	3,000 gpd per acre; commercial	
Toledo, Ohio†	1946	15,000 to 30,000 gpd per acre, average to maximum allowances	
Washington, D.C., suburban district† ...	1949	None except in special cases	

* ASCE (WPCF) Manual, "Design and Construction of Sanitary and Storm Sewers."
† "Sewer Capacity Design Practice," by William E. Stanley and Warren J. Kaufman, *J. Boston Soc. Civil Engrs.*, October, 1953, p. 320, Table 3.
‡ Gallons per capita per day.

studies of the individual industrial activities.   When large industrial waste flows are involved, the problem of collection and disposal of these wastes usually requires special engineering studies.

Quantity of Infiltration.   The rate of infiltration of ground water into sewers is influenced by the size, age, and condition of the sewers; the position of the sewers with respect to the ground-water table; the character of the soil; and the amount of precipitation.   The infiltration rate for any one system will vary from season to season.   It is common practice to allow for infiltration of about 30,000 gpd per mile

of sewer and house connection.[3]   Specifications for new work commonly allow infiltration rates of 3,500 to 5,000 gpd per mile for 8-in. pipe, 4,500 to 6,000 for 12-in. pipe, and 10,000 to 12,000 for 24-in. pipe.[4]

Flow Variations.   Sewers must be designed to handle the peak flow rates to be expected at the end of the design period.   It is also desirable to design them so as to minimize the problem of solids deposition during the early years of use when the flows may be much lower than the future flows.   The flows vary from day to day and from hour to hour within each day.   The ratio of the absolute maximum future flow rate to the initial minimum rate may vary from about 3 to 1 for large sewers serving highly developed areas to more than 20 to 1 for small sewers serving areas still under development.   Figure 2, which was used for the design of sewerage for

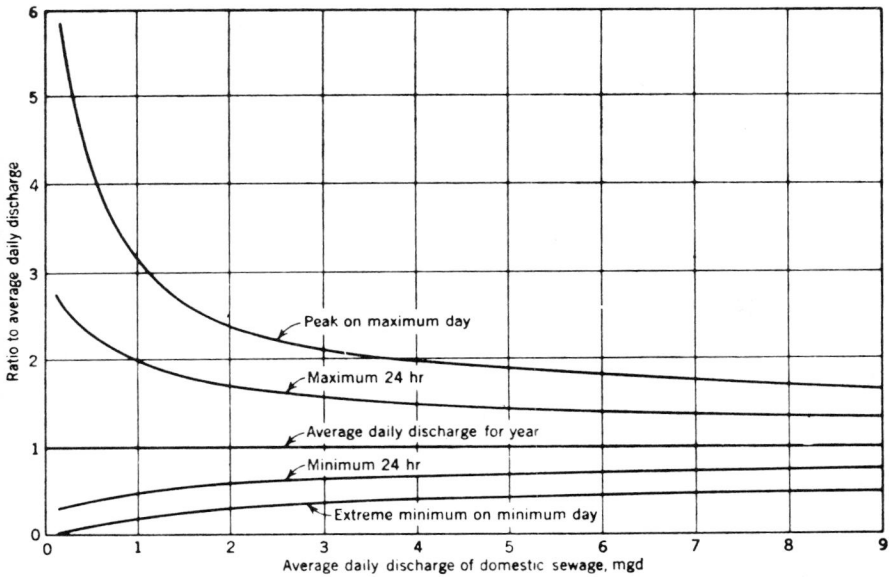

FIG. 2. Variations in flow of domestic sewage.

the Merrimack River Valley Sewerage District in Massachusetts (Massachusetts Senate Document 550 of 1947), is a good example of the magnitudes of fluctuations in flow which are commonly allowed for in design.

Examples of Sewage Flows Used for Design.   Table 2 gives data on the sewage flows which have been used for the design of sanitary sewers in a number of United States cities.[3]   The values reported range from 92 to 600 gpd per capita and this wide range in flow rates emphasizes the desirability of making an individual study to determine the proper design flow rates for any particular area.

Requirements of Regulatory Agencies.   Some state regulatory agencies have established definite per capita flow rates to be used when detailed studies and estimates of expected flows have not been made.   The recommended standards of the Great Lakes–Upper Mississippi River Board of State Sanitary Engineers in regard to the design of sanitary sewers are as follows:[5]

Design Period

In general, sewer systems should be designed for the estimated ultimate tributary population, except in considering parts of the systems that can be readily increased in capacity. Similarly, consideration should be given to the maximum anticipated capacity of institutions.

Table 2. Sewage Flows Used for Design *

City	Year of data	Average rate of water consumption, gpcd†	Population served, thousands	Per capita sewage flow average,‡ gpcd†	Sewer design basis,† gpcd†	Remarks
Baltimore, Md	....	160	1,300	100	135 × factor	Factor 4 to 2
Berkeley, Calif	....	76	113	60	92	
Boston, Mass	....	145	801	140	150	Flowing half full
Cleveland, Ohio§	1946	...	......	100	....	
Cranston R.I.§	1943	...	......	119	167	
Dallas, Tex.§	1949	...	......	150	575	Including storm water and infiltration
Des Moines, Iowa§	1949	...	......	100	200	
Grand Rapids Mich.	....	178	200	189.5	200	
Hagerstown, Md.	....	100	38	100	250	
Jefferson County, Ala	....	102	500	100	300	
Las Vegas, Nev.	....	410	45	209	250	
Lincoln, Neb.	....	160	125	125	200	
Little Rock, Ark.	....	50	100	50	100	
Los Angeles, Calif	1937	165	2,680	95	300	{ Single dwelling 0.004 cfs per acre; multiple dwelling 0.008 to 0.020 cfs per acre
Madison, Wis.§	....	...	......	...	....	Maximum hourly rate
Milwaukee, Wis.§	1945	...	......	125	....	All in 12 hr—250-gpcd† rate
Memphis, Tenn.	....	125	450	100	100	
Orlando, Fla.	....	150	75	70	190	
Painesville, Ohio§	1947	...	......	125	600	Includes infiltration and roof water
Rapid City, S. Dak	....	122	40	121	125	
Rochester, N.Y.§.	1946	...	......	....	250	New York State Board of Health standard
Santa Monica, Calif.	....	137	75	92	92	
Shreveport, La	....	135	160	120	150	
Springfield, Mass.§	1949	...	......	....	200	
Toledo, Ohio§	1946	...	......	....	160	150 gpcd† was used on a special project
Washington, D.C., suburban district§	1946	...	......	100	2 to 3.3 × average	

* ASCE (WPCF) Manual, "Design and Construction of Sanitary and Storm Sewers."
† Gallons per capita per day.
‡ Measured or estimated domestic sewage.
§ "Sewer Capacity Design Practice," by William E. Stanley and Warren J. Kaufman, *J. Boston Soc. Civil Engrs*, October, 1953, p. 317, Table 2.

Design Factors

In determining the required capacities of sanitary sewers the following factors should be considered:

　a. Maximum hourly quantity of sewage.

　b. Additional maximum sewage or waste from industrial plants.

　c. Ground water infiltration.

Design Basis

Per Capita Flow: New sewer systems should be designed on the basis of an average daily per capita flow of sewage of not less than 100 gallons per day. This figure is assumed to cover normal infiltration, but an additional allowance should be made where conditions are unfavorable. Generally the sewers should be designed to carry, when running full, not less than the following daily per capita contributions of sewage, exclusive of sewage or other waste from industrial plants.

　Laterals and sub-main sewers—400 gallons.

　Main, trunk and outfall sewers—250 gallons.

　Interceptors: Intercepting sewers, in the case of combined sewer systems, should fulfill the above requirements for trunk sewers and have sufficient additional capacity to care for the necessary increment of storm water. Normally no interceptor should be designed for less than 350% of the gauged or estimated dry weather flow.

Alternate Method: When deviations from the foregoing per capita rates are demonstrated, a brief description of the procedure used for sewer design must be included.

Table 3. Discharge Weights of Plumbing Fixtures *

Fixture or group	Weight per fixture or group in fixture units
Bathroom group:	
Flush valve water closet	8
Tank water closet	6
Bathtub (hot and cold)	3
Drinking fountain	½
Dishwasher domestic	2
Kitchen sink:	
Domestic	2
Domestic with food waste grinder	3
Lavatory (hot and cold)	2
Laundry tray (one or two compartments)	2
Shower stall, domestic	2
Showers (group), per head	3
Sink:	
Service (trap standard)	3
Service (P-trap)	2
Urinal:	
Pedestal, siphon jet, blowout (flush valve)	8
Stall or wall type	4
Water closet:	
Tank operated	4
Flush valve operated	8

　* From American Standard Plumbing Code, ASA A40.8–1955.

　† Consisting of water closet, lavatory, and bathtub or shower stall.

　‡ A shower head over a bathtub does not increase the fixture value.

　For a continuous or semicontinuous flow into a drainage system, such as from a pump, pump ejector, air-conditioning equipment, or similar device, two fixture units shall be allowed for each gallon per minute of flow.

Fixture-unit Basis of Design. For small tributary populations and for institutions such as schools, hospitals, hotels, factories, etc., the required capacities of sanitary sewers may be estimated from the "fixture-unit flow rates" defined by the American Standard National Plumbing Code, ASA A40.8-1955. Table 3 gives the relative discharge weights in fixture units for various types of fixtures. The relationship between the probable peak discharge rate and the total number of fixture units, based on the probability studies of R. B. Hunter, is given by Fig. 3.[6]

Summary. The following summary of the considerations necessary for the determination of sanitary sewer capacity is given in the ASCE Manual of Engineering Practice, No. 37 (WPCF Manual of Practice No. 9), "Design and Construction of Sanitary and Storm Sewers":

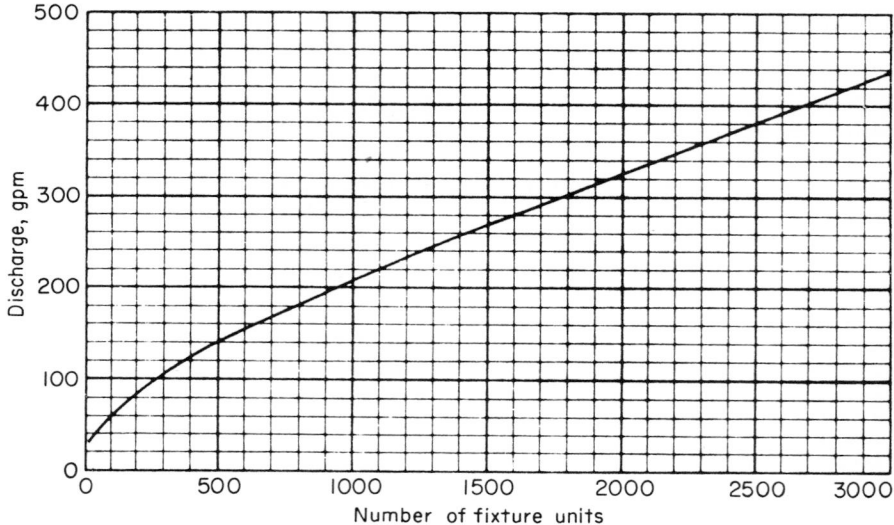

Fig. 3. Discharge from plumbing fixtures.

The determination of the flow quantities for which to design a separate sanitary sewer requires consideration and determination of the following:

1. The design period during which the peak or maximum design flow is not expected to be exceeded—usually 25 to 40 or 50 yr in the future.

2. Domestic sewage contributions based upon probable future population and probable future per capita water consumption. Careful considertion is given to distribution of population and relationship of peak and minimum flow rates to average per capita sewage flows. The fixture-unit method of developing peak rates should be employed for small populations.

3. Commmercial area contributions from stores, hotels, offices, and other businesses which are sometimes assumed to be amply cared for in the peak allowance for per capita sewage flows in small communities. Per-acre allowances based upon actual records or successful design quantities of record for comparable commercial areas are the most reasonable approach for larger communities.

4. Industrial wastes which include estimated domestic sewage from known populations per shift, estimated or gaged allowances per acre for industry as a whole, and actual or estimated flow rates from plants with large process wastes which can be permitted in sanitary sewers.

5. Institutional wastes, usually almost entirely domestic sewage. Peak and minimum design flow rates from persons in the institution are multiplied by the same basic per capita rates as are used by S. A. Greeley and W. A. Stanley.[2]

6. Air-conditioning and industrial cooling waters if permitted to be discharged to the sanitary sewers—assume 1½ to 2 gal per ton of non-water-conserving cooling units. There should be an enforced ordinance prohibiting the discharge of spent cooling waters into separate sanitary sewers.

7. Storm-water contributions to sanitary sewers. These should be prohibited, but the designer must recognize that some such storm and surface water does get into separate sanitary sewers and a judgment allowance therefore must be made.

8. Infiltration through defective joints. This needs as careful an evaluation as can be made from thought given to the several factors involved. Design allowances should be larger (under some circumstances very much larger) than those stipulated in construction specifications where acceptance tests are made very soon after construction. Under-evaluation of infiltration is often the principal reason why some sewers have become overloaded.

The relative emphasis given to each of the foregoing factors varies among designers. Some have set up single values of peak design flow rates for the various classifications of tributary area, thereby integrating all these items. It is recommended, however, that actual maximum and minimum peak design flows be developed step by step, giving as thorough as possible consideration to each of the component items which influence design values.

QUANTITY OF STORM WATER

The Rational Method. The rational method is the most commonly used procedure for the computation of the rates of storm-water runoff for storm-sewer design. The rate of runoff Q is given by the equation

$$Q = CiA$$

in which A is the size of the drainage area, i is the average intensity of rainfall for a duration equal to the time of concentration of the area, and C is a runoff coefficient whose value depends principally upon the character of the area. For a steady rainfall rate of 1.0 in./hr the total precipitation deposited over an area of 1 acre would be at the rate of 1.008 cfs or, for practical purposes, 1.0 cfs. Therefore, if A is expressed in acres, i in inches per hour, and Q in cubic feet per second, C may be interpreted as a dimensionless coefficient which expresses the ratio of the runoff rate to the rainfall rate.

The assumption behind the rational method is that the runoff rate for a given rainfall intensity will increase and reach its maximum when the duration of the rainfall reaches the time of concentration of the area (the time required for the runoff to flow from the remotest point of the area to the point where Q is being measured). By this assumption, the maximum runoff rate Q which can be expected to occur with any given frequency will be produced by a storm having the maximum average rainfall intensity corresponding to the given frequency and duration. The application of the method requires knowledge of the rainfall intensity-duration-frequency characteristics for the locality. Although the assumptions are not strictly in accord with the mechanics of the runoff process, the rational method has proved to be a practical procedure for storm-drain design because the accumulated experience has resulted in practicable values for the runoff coefficient C.

Time of Concentration. The time of concentration is the time required for the runoff to flow from the remotest point of the drainage area to the point under design. This is the minimum time necessary to permit the entire area to contribute to the flow at the point. The time of concentration consists of the inlet time, or time required for the runoff at the upper end of the area to reach the nearest

inlet, plus the time of flow in the sewer from this inlet to the point being considered.

Inlet time will vary with the nature of the surface, the slopes, the nature of the established drainage channels such as street gutters, and the antecedent conditions. Because the inlet time is small, it is commonly chosen on the basis of experience.

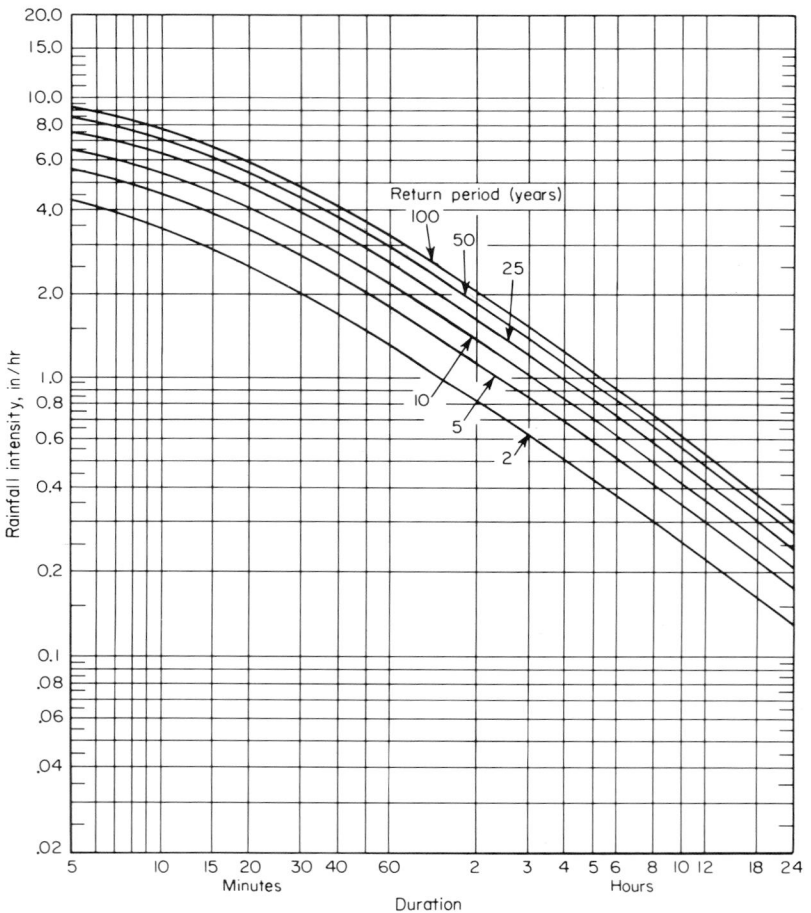

FIG. 4. Rainfall intensity-duration-frequency curves, New York, N.Y., 1903–1951. NOTE: Frequency analysis by method of extreme values, after Gumbel.

In densely developed areas with a high percentage of paved surfaces and closely spaced inlets, an inlet time as low as 5 min may be assumed. In moderately developed urban areas with flat slopes the inlet time may be from 10 to 15 min. In flat residential areas having a relatively low percentage of paved surface, the inlet time may be as high as 30 min. It is possible to make estimates of the inlet time by calculating the time of flow over the various types of surfaces, but such estimates can rarely be made with a high degree of accuracy.

The time of flow in the sewer is computed from the hydraulic properties of the sewer, the common practice being to use the average flowing-full velocity computed for the prevailing slope.

Rainfall Frequency. It is usually prohibitive, on the basis of cost, to construct storm sewers capable of handling the largest conceivable storms. Current practice is to use storm rainfalls having an average expected frequency of once every 3 to 10 years for the design of storm sewers in residential areas and storms of 10 to 30 years for commercial and high-value districts.

Rainfall Intensity-Duration-Frequency Relationships. The rainfall characteristics which must be known for storm-sewer design are presented in a very concise manner by the intensity-duration-frequency curves, which can be prepared from a long record of precipitation at a given station. Figure 4 shows such a set of curves prepared by the U.S. Weather Bureau from the precipitation recorded at New York City from 1903 to 1951.[7] This figure shows, for example, that an average intensity of 2.2 in./hr for a duration of 1 hr will be equaled or exceeded, on the average, once every 10 years. Similar data for many other localities have been compiled and published by the U.S. Weather Bureau.[7-9] Figures 5-8 show the geographical distribution of certain rainfall characteristics for the area of the United States east of the 105th meridian.[3] For the region west of the 105th meridian, the local variations in rainfall characteristics make it impracticable to present them in the form of Figs. 5 to 8. For data covering the western portion of the United States, the reader is referred to *Weather Bureau Technical Paper* 28.[9]

Runoff Coefficient. The runoff coefficient C in the rational method is the variable which is least susceptible to precise determination. Whereas the use of the runoff coefficient implies that there is a constant ratio of runoff to rainfall, the actual ratio for a given area will depend upon the condition of the area at the time of occurrence of the storm and will increase with the duration of the storm. A more logical procedure than the rational method for storm-drain design would be to subtract the rainfall losses due to infiltration and retention in surface depressions and to distribute the remainder as an actual hydrograph of runoff.[10] However, because of the great variability in the time distribution of the rainfall itself as well as the difficulty in estimating the quantities of infiltration and surface depression storage, most engineers still prefer to estimate a value of the runoff coefficient C. A common practice is the use of average coefficients for various types of districts, the coefficients being assumed to be constant throughout the storm duration. The range of values reported to be in common use is as follows:[3]

Type of area	Runoff coefficient
Business:	
Downtown areas	0.70–0.95
Neighborhood areas	0.50–0.70
Residential:	
Single-family areas	0.30–0.50
Multiunits, detached	0.40–0.60
Multiunits, attached	0.60–0.75
Residential (surburban)	0.25–0.40
Apartment dwelling areas	0.50–0.70
Industrial:	
Light areas	0.50–0.80
Heavy areas	0.60–0.90
Parks, cemeteries	0.10–0.25
Playgrounds	0.20–0.35
Railroad yard areas	0.20–0.40
Unimproved areas	0.10–0.30

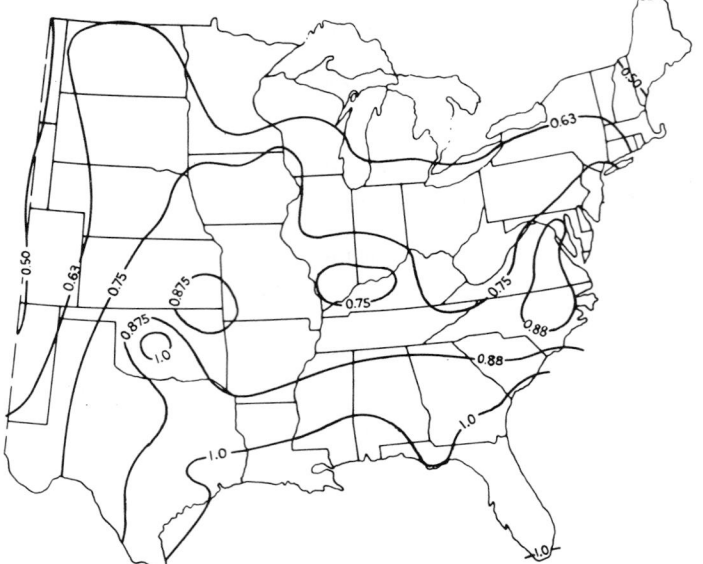

FIG. 5. Rainfall intensities in inches for 2-year frequency and 15-min duration.

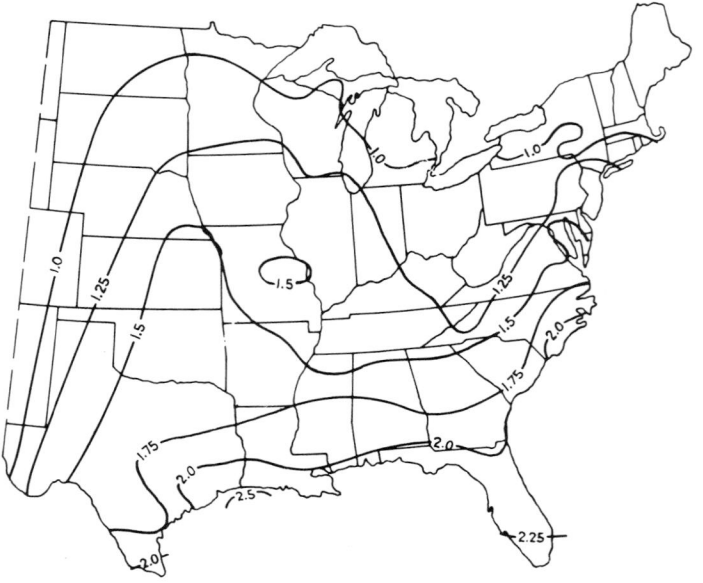

FIG. 6. Rainfall intensities in inches for 2-year frequency and 60-min duration.

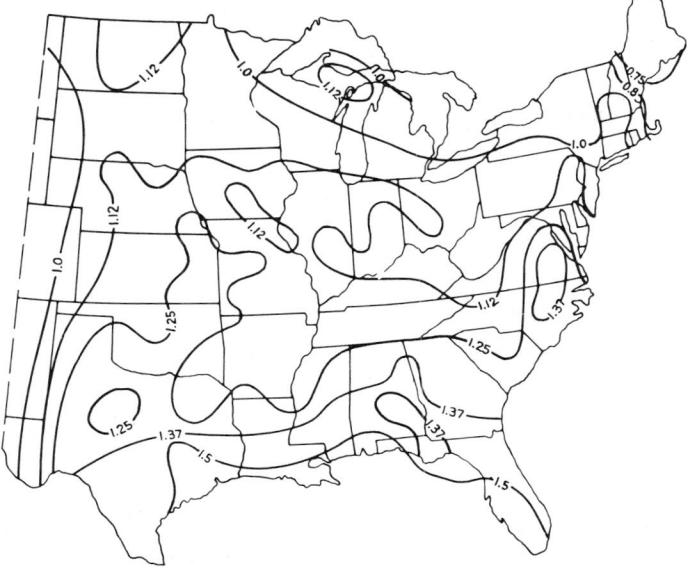

Fig. 7. Rainfall intensities in inches for 10-year frequency and 15-min duration.

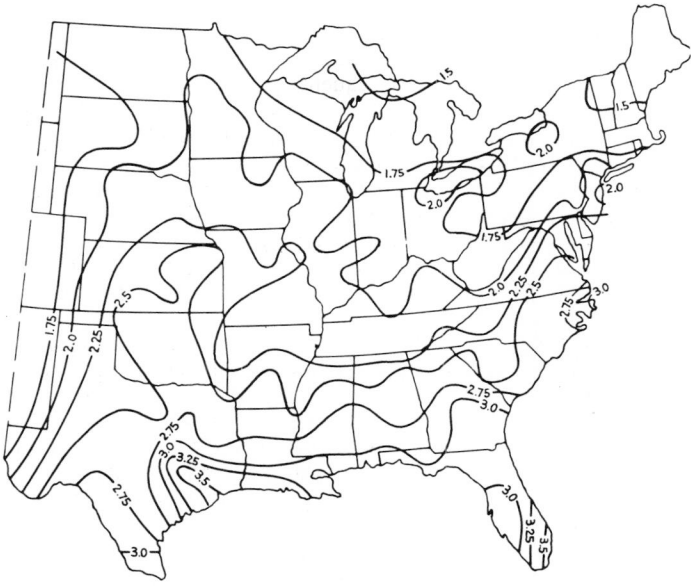

Fig. 8. Rainfall intensities in inches for 10-year frequency and 60-min duration.

For specific small areas it is more logical to relate the value of *C* to the actual type of surface. Coefficients commonly used are:[3]

Character of surface	Runoff coefficient
Streets:	
Asphaltic	0.70–0.95
Concrete	0.80–0.95
Brick........................	0.70–0.85
Drives and walks	0.75–0.85
Roofs	0.75–0.95
Lawns, sandy soil:	
Flat, 2%.....................	0.05–0.10
Average, 2 to 7%	0.10–0.15
Steep, 7%....................	0.15–0.20
Lawns, heavy soil:	
Flat, 2%	0.13–0.17
Average, 2 to 7%.............	0.18–0.22
Steep, 7%....................	0.25–0.35

When an area is made up of different types of surfaces, a common procedure is to use a weighted-average coefficient.

The coefficients in the previous tabulations are designed for use for storms of 5- to 10-year frequencies. For less frequent, higher intensity storms, the coefficients should be higher, because the infiltration and surface retention will be smaller proportions of the total precipitation. Likewise, for more frequent, lower intensity storms, the coefficients should be lower than indicated in the tables.

HYDRAULICS OF SEWERS

Although they are usually placed underground, most sewers are designed to flow with a free water surface. An advantage of the free-flow condition is that the depth will vary with the rate of flow in such a way as to keep the velocity reasonably high even at flow rates which are small in comparison with the full capacity of the pipe. This helps to maintain velocities sufficient to prevent the deposition of solids over a wide proportion of the flows which are likely to be encountered.

The capacity of a sewer pipe should be sufficient to carry the peak flow rate to be anticipated at the end of the design period, and the slope should be sufficient to provide for self-cleansing velocities during the early years of use. It is common practice to design sanitary sewers with slopes sufficient to provide for velocities of 2 fps when flowing full. Experience shows that with such slopes trouble from deposits is seldom encountered.

Although the flow in sewers is seldom steady or uniform, it is impracticable in most cases to take this into account, and each section of the sewer is usually designed with the assumption that the flow is steady and uniform.

Flow Formulas. Despite its complexity, Kutter's formula has been widely used in the past for the solution of problems involving open-channel flow in sewers because of the availability of charts and diagrams which facilitate its solution. The Manning equation is now being widely used in place of the Kutter formula.

The Kutter equation is

$$V = \left[\frac{1.81/n + 41.67 + 0.0028/S}{1 + n/\sqrt{r}(41.67 + 0.0028/S)} \right] \sqrt{rS}$$

in which V = mean velocity, fps,
        r = hydraulic radius, ft
       S = slope of energy gradient
      n = coefficient of roughness

Manning's equation is

$$V = (1.486/n)r^{2/3}S^{1/2}$$

in which the nomenclature is the same as for the Kutter equation.

The roughness coefficient n varies from 0.010 for smooth surfaces to as high as 0.10 for rough natural channels and has the same numerical value in each of the above equations. It is common practice to use values of Kutter's or Manning's n of 0.013 for sewer design. This value makes some allowance for the future condition of the pipe as well as disturbances in the flow resulting from rough joints, interior coatings of grease or other matter, and eddying due to changes in pipe size, junctions, etc. Figure 9 is a diagram for the solution of the Manning equation applied to circular pipes flowing full, with n equal to 0.013. Experiments have indicated that the value of n is not the same at all depths and that maximum value is found to occur at a depth equal to about three-tenths of the diameter.

Pipes Flowing Partly Full. Figure 10, which is taken from the ASCE (WPCF) Manual previously referred to, gives the hydraulic elements of circular pipes flowing partly full. This figure shows the ratios of the values of the various elements to the values for the flowing-full condition. The cross-sectional area and the hydraulic radius are purely geometric functions and hence independent of n. The velocity and discharge for any particular ratio of depth to diameter depend upon whether n is assumed to be constant or variable with the depth. Velocity and discharge curves computed from both assumptions are shown. The variation in the value of n is based on extensive experiments by Wilcox and by Yarnell and Woodward.[11,12]

The use of Fig. 10 is illustrated by the following example:

Example. A 36-in. sewer pipe ($n = 0.013$) is laid on a slope of 2.00 ft per 1,000 ft. Find the depth of flow and the velocity when the flow rate is 20.0 cfs. Assume that n varies with the depth of flow.

Solution. From Fig. 9, the capacity of the pipe when flowing full is 30.0 cfs and the velocity is 4.23 fps. Then $q/Q = 20.0/30.0 = 0.677$. From Fig. 10, $d/D = 0.67$ and $v/V = 0.93$. Then depth = $0.67 \times 3.0 = 2.00$ ft and velocity = $0.93 \times 4.23 = 3.93$ fps.

Figure 10 also shows the relative velocities required to obtain equal cleansing of the pipe at all depths of flow. This is based on T. R. Camp's analyses of the theoretical work done by Shields on the movement of granular materials in open channels.[13,14] The diagram indicates that, if a sewer has self-cleansing velocities under flowing-full conditions, the velocity will also be self-cleansing for all flow conditions at depths greater than one-half the diameter. As previously stated, this is an important reason for designing sewers to flow as open channels.

Alternate Stages of Flow. The specific energy of the flow at any section in an open channel is defined as the energy of the flow expressed in feet (foot-pounds per pound) measured above the bottom of the channel. Figure 11 represents the flow profile at a section of a channel flowing with a free surface. The specific energy for a point in a stream line at a distance y above the bottom of the channel is given by the expression

$$H = y + h + v^2/2g = d + v^2/2g = d + Q^2/2gA^2$$

where H is the specific energy head, h is the pressure head above the stream line, v is the velocity at the point, Q is the flow rate, and A is the cross-sectional area. Since the velocity is a function of y, the specific energy is also a function of y. However, the above equation can be considered, with very little error, as being representative of the flow as a whole provided that v is taken as the average velocity for the cross section.

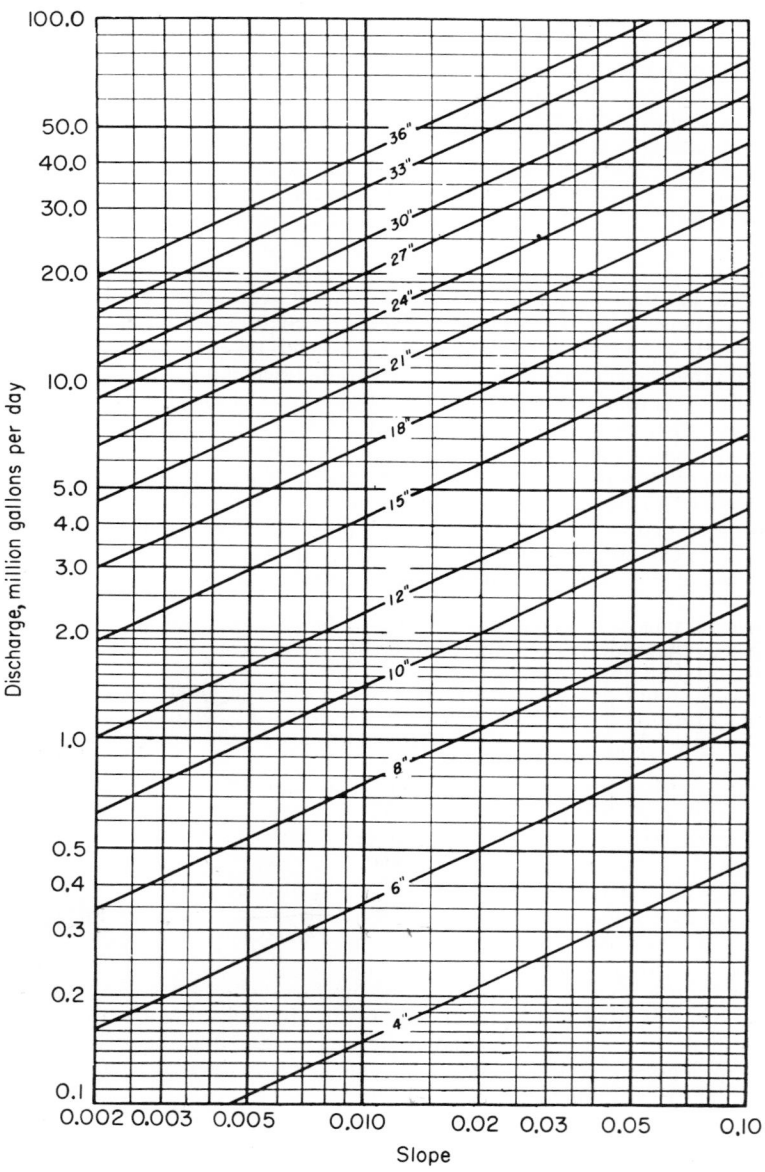

Fig. 9. Discharge of circular pipes (running full) based on the Manning formula

$$Q = A(1.486/n)r^{2/3}S^{1/2}$$

where Q = flow rate, A = cross-sectional area, r = hydraulic radius, S = slope of energy gradient, n = coefficient of roughness.

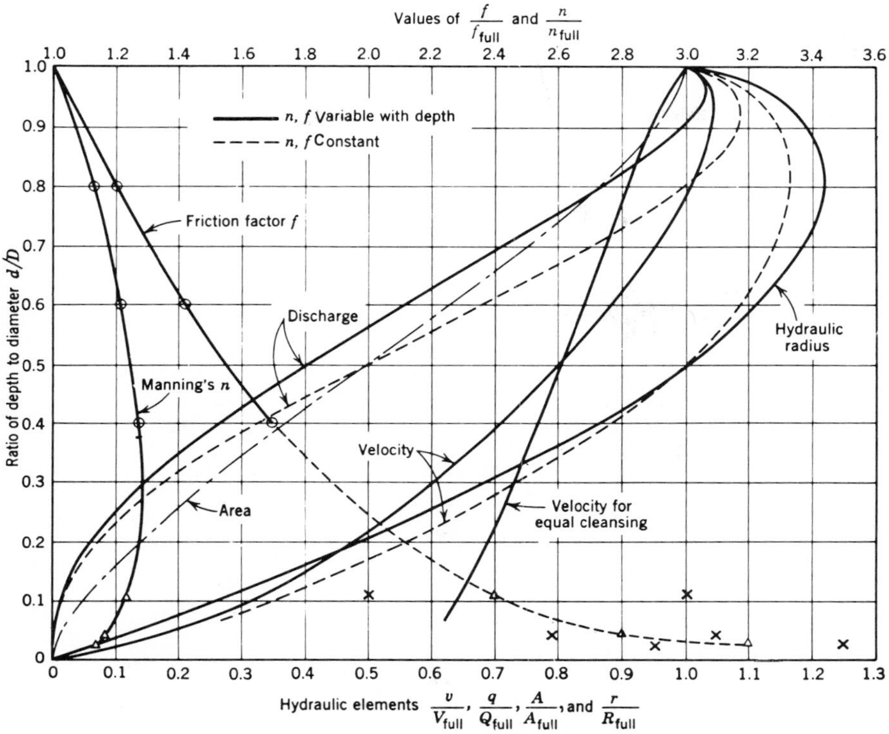

Values of $\dfrac{f}{f_{full}}$ and $\dfrac{n}{n_{full}}$

FIG. 10. Hydraulic elements of circular sewers.

In Fig. 11, the slope of the channel bottom is designated as S_0 and the slope of the energy gradient as S. For the general case of nonuniform flow (where the depth is increasing or decreasing in the direction of flow) the values of S and S_0 will be different; for uniform flow they will have the same value. The Manning equation can be applied to the conditions at any section provided that S in the equation is taken to be the slope of the energy gradient at the section.

It is evident that the cross-sectional area of the flow in the specific-energy equation is a function of the shape of the section and the depth. For a circular pipe, A is a complicated function of the depth d and the diameter D. Figure 12, which was prepared by Thomas R. Camp, is a graphical solution of the specific-energy equation for circular pipes.[15] It is noted from Fig. 12 that, for given values of Q, D, and H, there are two possible values of d which will satisfy the equation. The larger of the two values is designated as "upper stage flow," "tranquil flow," or "subcritical flow"; the lower value is designated as "lower stage flow," "shooting

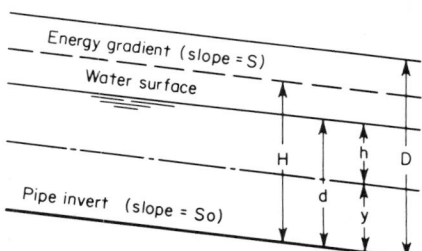

FIG. 11. Open-channel flow.

flow," or "supercritical flow." The "critical depth" is the depth at which the value of the function $Q^2/2gHD^4$ is a maximum or at which the value of H is a minimum for given values of Q and D. It is important to note that the relationships given in Fig. 12 are purely algebraic and are in no way related to the basic relationship between Q and S as given by the Kutter or Manning equations.

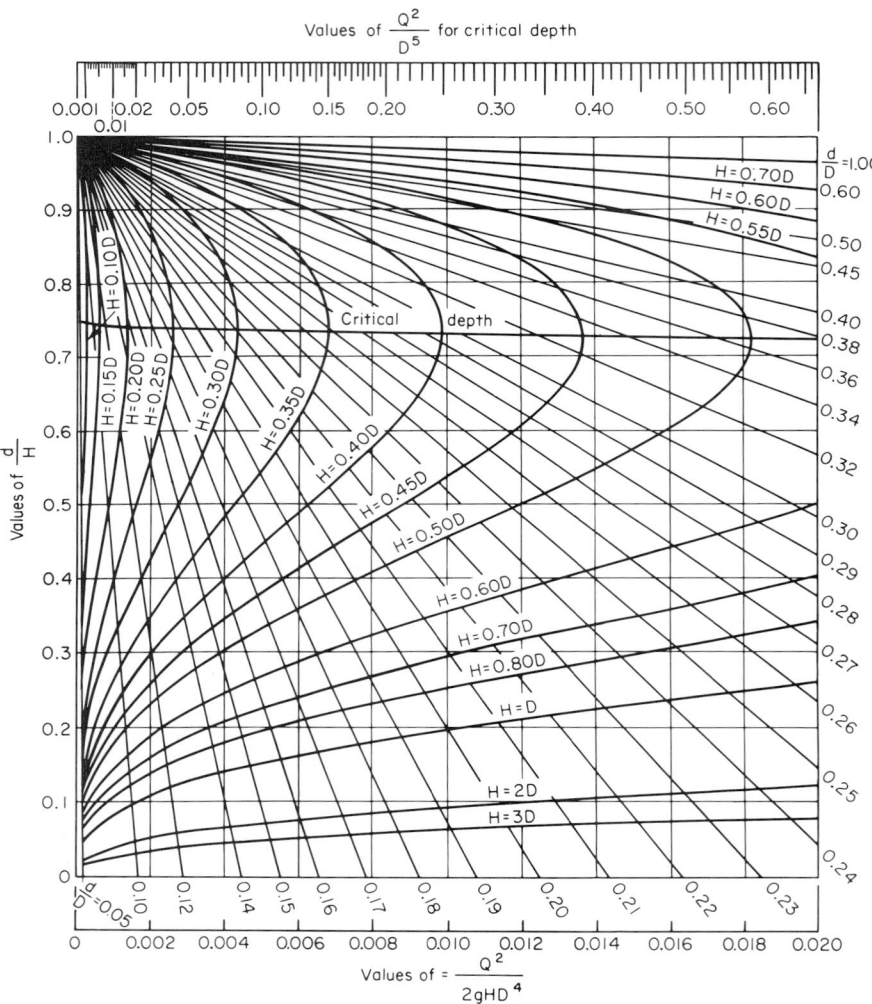

FIG. 12a. Energy relationships for free surface flow in pipes. $Q^2/2gHD^4$ from 0.0 to 0.02.

If a channel is long enough, or if the downstream conditions are just right, the flow in a channel may be uniform. Under these conditions, the depth will be constant along the channel and the values of S_0, S, and the slope of the water surface will be equal. The depth of flow for uniform flow conditions may be designated as the normal depth d_n. This is the depth of flow which makes the rate of loss of energy S equal to the bottom slope S_0. Perfectly uniform flow rarely occurs

because the depth at any cross section is dependent not only upon S_0 but also upon the distance of the section from the channel outlet or from the nearest downstream or upstream channel junction, at which point the depth may be controlled by factors which are independent of S_0. A value of d_n can be computed for any channel by the

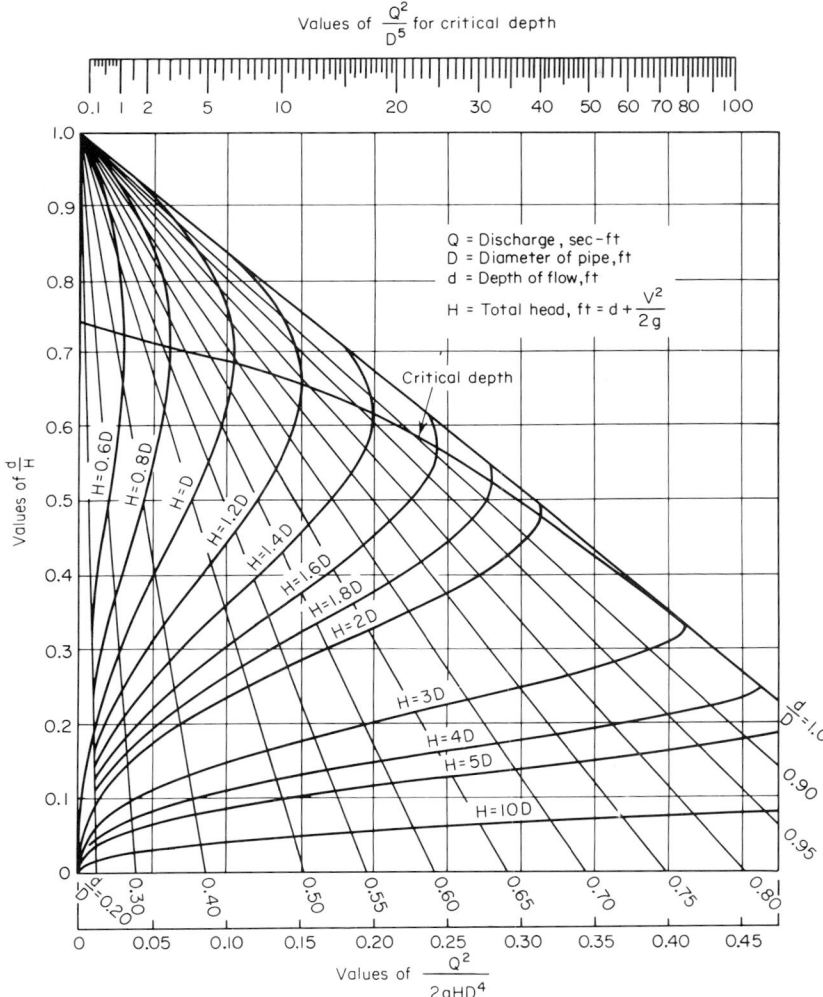

FIG. 12b. Energy relationships for free surface flow in pipes. $Q^2/2gHD^4$ from 0.0 to 0.50.

Manning equation by assuming that S is equal to S_0. However, whether or not the flow will ever occur at the depth d_n in any stretch of a channel will depend upon the hydraulic conditions downstream or upstream of the stretch. A value of S_0 which makes d_n greater than the critical depth d_c may be designated as a "mild slope," and a value which produces a d_n less than the critical depth may be

designated as a "steep slope." The slope S_0 which makes d_n equal to the critical depth may be designated as the critical slope S_c.

The significance of the critical depth may be illustrated by reference to Fig. 13. The flow conditions for each of the cases shown are explained briefly below:

1. A channel with a mild slope discharges freely into the atmosphere. Critical depth will occur at the outlet. The depth will increase at successive sections upstream until the normal depth is reached, beyond which the flow will be uniform.

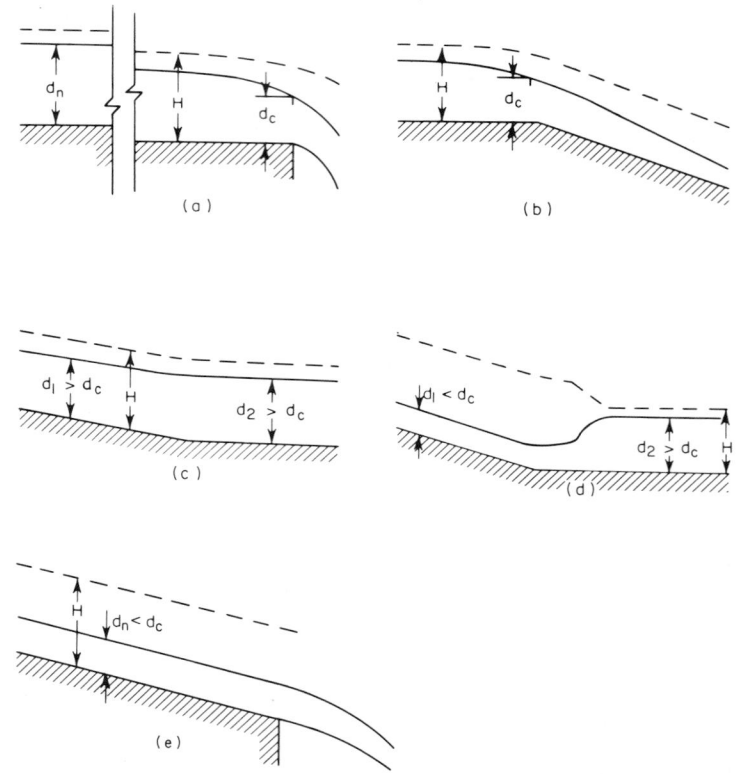

FIG. 13. Examples of nonuniform flow profiles. (*a*) Free discharge from a mild slope. (*b*) Discharge from a mild to a steep slope. (*c*) Discharge from one mild slope to another. (*d*) Discharge from a steep slope to a mild slope. (*e*) Free discharge from a steep slope.

In many cases the length of the channel will be much less than the distance required to develop normal depth. The significance of this is that for a short stretch of pipe, its actual capacity to carry flow without surcharge may be much greater than would be calculated by assuming uniform flow with S equal to S_0.

2. A channel changes slope from mild to steep. The conditions upstream of the junction will be the same as for paragraph 1. The depth downstream of the junction will decrease and approach d_n.

3. A channel changes slope from a mild slope to another mild slope. Upstream of the junction there will be a gradual decrease in the depth from section to section.

4. A channel changes slope from steep to mild. In this case, the change from the upstream supercritical flow to the downstream subcritical flow will take place

suddenly in a "hydraulic jump." The position of the jump may be either upstream or downstream of the junction, depending upon the relative values of the various parameters which control the flow pattern.

5. A steep slope discharges into the atmosphere. In this case the flow will be at the normal supercritical flow, provided that the upstream control has permitted normal depth to be developed (see case 2).

The foregoing discussion on nonuniform flow is presented to show the reader the importance of understanding these principles, particularly when dealing with small systems where few pipes may be involved. For detailed presentation of nonuniform flow, the reader is referred to textbooks on the subject.[16–18]

Hydraulics of Sewer Transitions. In extensive sewer systems, most of the pipes will have mild slopes and the flows will be subcritical. Extra energy losses

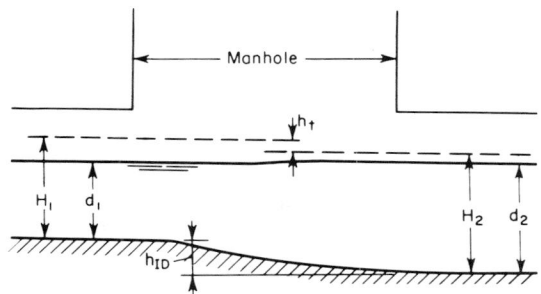

FIG. 14. Flow profile at junction.

occur at all transitions where changes occur in size, slope, or direction of the pipe and at junctions where several pipes come together. If the transitions are properly designed to allow for these energy losses, the condition of uniform flow may be approximated in the individual lines and the flows will never be at depths greater than the depths computed on the assumption of uniform flow. However, if the transitions are not properly designed, pipes may at times flow at depths greater than the computed depths and surcharge may occur under peak flow conditions. The hydraulic principles involved in transition design are illustrated in Fig. 14, which shows a transition where a pipe flows into a larger pipe laid on a flatter grade. Due to the turbulence created by the flow expansion there will be a head loss h_t. In order to prevent the upstream pipe from flowing at a depth greater than its normal depth d_1, the relative elevations of the pipes must be such that the energy gradient of the downstream pipe is lower than that of the upstream pipe by the amount of h_t. This requires that the invert of the downstream be placed below that of the upstream pipe by the amount h_{ID}. The relationship between the various vertical dimensions is given by the equation

$$h_{ID} = H_2 - H_1 + h_t$$

The equation assumes that the loss is concentrated at the center of the transition. As actually constructed, the transition will take place within a manhole, and the channel section within the manhole is made to provide for a gradual transition between the two pipes. If the computed invert drop h_{ID} is negative, it is usually taken as zero and the pipe inverts are placed at the same elevation.

The energy loss h_t is usually small, but it can assume fairly high values when high velocities are involved. Data on the magnitudes of h_t are scarce, but such data as

are available indicate that h_t can be represented as a fraction of the change in velocity heads in accordance with the equation

$$h_t = K\Delta(v^2/2g)$$

Based on studies of open-channel transitions for subcritical flow by Julian Hinds, the values of K for smooth transitions might be taken as low as 0.10 for increasing velocity transitions and 0.20 for decreasing velocity transitions.[19] Increased transition losses occur when a sewer line changes direction and at junctions where one or more branch sewers join a main sewer. Reliable information on the transition head losses in such cases is almost entirely lacking. The hydraulic design of junctions may be considered as the design of two or more transitions, one for each path of flow. The exit sewer is common to all paths, and its invert must be placed at the lowest computed elevation. Because of the lack of information on the transition losses, allowances are usually made in accordance with the judgment of the designer. An arbitrary procedure which is commonly adopted is to allow about twice as much loss along flow paths in junctions as compared with the allowances for simple transitions involving the same velocities.

SEWER PIPE

Pipe Materials. The materials, listed alphabetically, of which street sewer pipes are most commonly constructed are asbestos cement, cast iron, concrete, and vitrified clay. Cast iron may be used for pressure sewers, for piping in and around buildings, and where its use may be indicated by structural requirements. The other materials are the most commonly used for sewers flowing under the usual free-surface conditions. Bituminized-fiber pipes are commonly used in collecting storm water from building downspouts, for foundation drains, and in leaching systems for residential sewage disposal. Plastic pipe may be used for the conveyance of highly acidic industrial wastes.

Asbestos-cement Sewer Pipe. Asbestos-cement sewer pipe is available in the sizes and classes given in the following table:

Class	Nominal size, diameter, in.										
1500	6	8	10	12	14	16					
2400	6	8	10	12	14	16	18	20	24		
3300	6	8	10	12	14	16	18	20	24	30	
4000	..	..	10	12	14	16	18	20	24	30	36
5000	..	..	10	12	14	16	18	20	24	30	36

Standard lengths are 10 and 13 ft for the 6- and 8-in.-diameter pipe and 13 ft for the larger diameter pipe. The class designations refer to the crushing strengths as determined by the three-edge bearing method of testing. Joints are made with rubber rings or gaskets and couplings of the same material as the pipe. Advantages claimed for asbestos-cement pipe are ease of handling due to light weight, tight joints which resist infiltration and exfiltration as well as root penetration, and resistance to corrosion in most soil conditions. Corrosion may result from acid sewage and in highly acidic or highly sulfate-alkaline soils.

Standard specifications covering asbestos-cement sewer pipe are:

1. Federal Specification SS-P-3316, Pipe, Asbestos-cement, Sewer, Nonpressure

2. Tentative Specifications and Methods of Test for Asbestos-cement Nonpressure Sewer Pipe, ASTM Designation C428

Table 4.   Concrete Sewer Pipe (ASTM C-14-59)

Internal diam, in.	Min laying length, ft	Standard pipe		Extra-strength pipe	
		Min barrel thickness, in.	Min strength (sand bearing), lb/lin ft	Min barrel thickness, in.	Min strength (sand bearing), lb/lin ft
4	2½	⁹⁄₁₆	1,500	¾	3,000
6	2½	⅝	1,650	¾	3,000
8	2½	¾	1,950	⅞	3,000
10	3	⅞	2,100	1	3,000
12	3	1	2,250	1⅜	3,375
15	3	1¼	2,620	1⅝	4,175
18	3	1½	3,000	2	4,950
21	3	1¾	3,300	2½	5,775
24	3	2⅛	3,600	3	5,000

Concrete Sewer Pipe.   Unreinforced concrete sewer pipe is available in standard sizes from 4 to 24 in. in diameter and in two strength classes.   Standard specifications covering unreinforced concrete pipe intended for use in the conveyance of sewage, industrial wastes, and storm water are ASTM Designation C14.   Table 4 gives the principal dimensional and strength requirements of these specifications.

Reinforced concrete sewer pipe, in sizes from 12 to 108 in. ID, are covered by ASTM Specifications C76.   Pipe manufactured under these specifications are of five strength classifications.   Table 5 gives the available sizes and strength requirements.

A number of different joint types are available depending upon the degree of watertightness required.   The concrete pipe industry furnishes all the necessary bends, wyes, tees, and specials necessary for any sanitary sewerage project, the usual procedure being to make each such piece of pipe in accordance with the requirements of the project.[20]

Advantages of concrete pipe are the wide ranges in sizes, laying lengths, and strengths.   A disadvantage of concrete pipe for sewers is that it is subject to corrosion under acid conditions.   If flow velocities are insufficient to prevent the deposition of organic solids, septic conditions may result.   Hydrogen sulfide gas produced by the anaerobic decomposition of organic matter becomes oxidized to produce sulfuric acid, which damages the pipe.   This condition can usually be prevented by designing the sewers so that self-cleansing velocities will occur most of the time.   Protective linings can be used to prevent corrosion where the sewage may be excessively acid.

Table 5.   Reinforced-concrete Pipe (ASTM C76-59T)

Class	Size range, diam, in.	Sand-bearing strength, lb/ft per ft diam	
		To produce 0.01-in. crack	Ultimate
I	60–108	1,200	1,800
II	12–108	1,500	2,250
III	12–108	2,025	3,000
IV	12– 84	3,000	4,500
V	12– 72	4,500	5,625

Vitrified-clay Sewer Pipe.   Vitrified-clay pipe is manufactured in standard sizes from 4 to 36 in. in diameter and in two strength classifications.   ASTM Specifications C13 and C200 cover "standard-strength" and "extra-strength" clay pipe.   The crushing strengths and dimensions are given in Tables 6 to 8.

A wide variety of vitrified clay fittings is available as shown by Fig. 15.   Detailed dimensions of tee branches are given in Tables 9 to 12.   Dimensions of the other types of fittings are given in the "Clay Pipe Engineering Manual."[21]

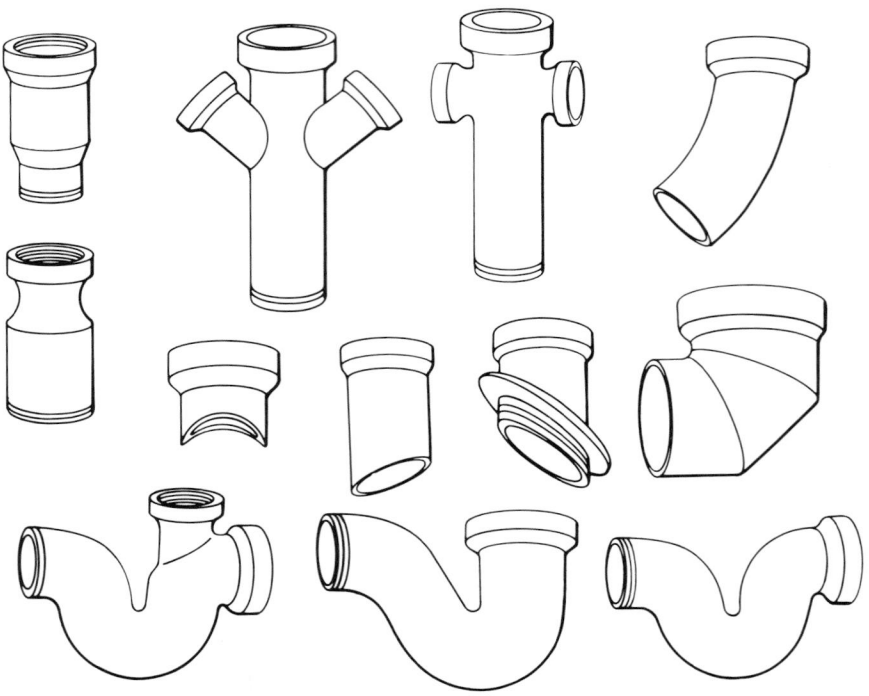

FIG. 15.   Clay sewer pipe special fittings.

Vitrified-clay pipe is resistant to corrosion from most acids, making it advantageous when handling septic sewage or wastes with high acid content.   Joints are commonly made up with bituminous compounds, of which there are many different types available.   Recent developmental work by clay-pipe manufacturers has resulted in the marketing of joints employing resilient plastic materials which limit joint leakage under deflection and high ground-water conditions.   Three types of these joints are covered under ASTM Specification C425, Vitrified Clay Pipe Joints Using Materials Having Resilient Properties.

APPURTENANCES AND SPECIAL STRUCTURES

Essential to all sewerage systems are the appurtenant structures such as service connections, manholes, junction chambers, storm-water inlets, diversion chambers, etc.   The design of such structures is not covered in detail in this chapter, but typical designs for the most commonly used appurtenances will be presented briefly.

Table 6. Crushing Strengths for Clay Sewer Pipe

Size, in.	Crushing strength, min, lb/lin ft			
	Standard-strength		Extra-strength	
	Three-edge bearing	Sand bearing	Three-edge bearing	Sand bearing
4	1,000	1,500		
6	1,100	1,650	2,000	3,000
8	1,300	1,950	2,000	3,000
10	1,400	2,100	2,000	3,000
12	1,500	2,250	2,250	3,375
15	1,750	2,625	2,750	4,125
18	2,000	3,000	3,300	4,950
21	2,200	3,300	3,850	5,775
24	2,400	3,600	4,400	6,600
27	2,750	4,125	4,700	7,050
30	3,200	4,800	5,000	7,500
33	3,500	5,250	5,500	8,250
36	3,900	5,850	6,000	9,000

Table 7. Standard-strength Vitrified-clay Pipe Conforming to ASTM Specifications C13

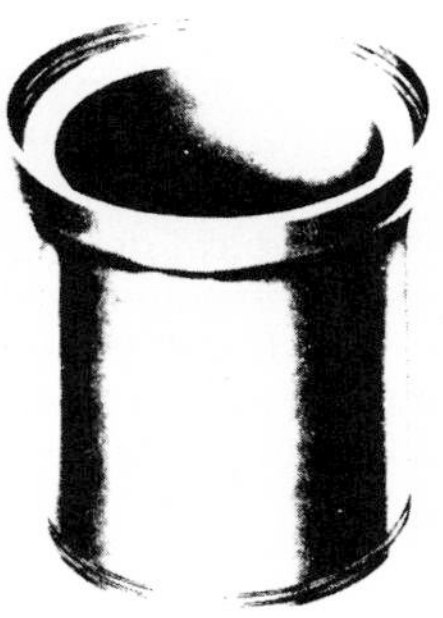

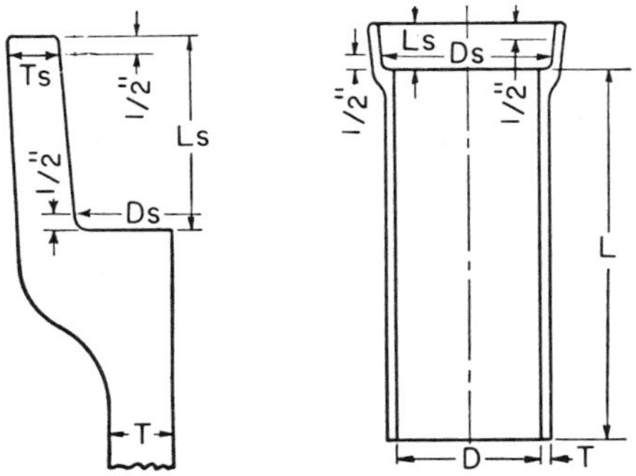

Size, in.	Laying length L		Max difference in length of two opposite sides, in.	Outside diameter of barrel, in.		Inside diameter of socket at 1/2 in. above base, in. DS
	Min*	Limit of minus variation, in. per ft of length†		Min	Max	Min
4	2	1/4	5/16	4 7/8	5 1/8	5 3/4
6	2	1/4	3/8	7 1/16	7 7/16	8 3/16
8	2	1/4	7/16	9 1/4	9 3/4	10 1/2
10	2	1/4	7/16	11 1/2	12	12 3/4
12	2	1/4	7/16	13 3/4	14 5/16	15 1/8
15	3	1/4	1/2	17 3/16	17 13/16	18 5/8
18	3	1/4	1/2	20 5/8	21 7/16	22 1/4
21	3	1/4	9/16	24 1/8	25	25 7/8
24	3	3/8	9/16	27 1/2	28 1/2	29 3/8
27	3	3/8	9/16	31	32 1/8	33
30	3	3/8	5/8	34 3/8	35 5/8	36 1/2
33	3	3/8	5/8	37 5/8	38 15/16	39 7/8
36	3	3/8	11/16	40 3/4	42 1/4	43 1/4

Table 7. (*Continued*)

Size, in.	Depth of socket, in. *LS*		Thickness of barrel, in. *T*		Thickness of socket at ½ in. from outer end, in. *TS*	
	Nominal	Min	Nominal	Min	Nominal	Min
4	1¾	1½	½	7/16	7/16	⅜
6	2¼	2	⅝	9/16	½	7/16
8	2½	2¼	¾	11/16	9/16	½
10	2⅝	2⅜	⅞	13/16	⅝	9/16
12	2¾	2½	1	15/16	¾	11/16
15	2⅞	2⅝	1¼	1⅛	15/16	⅞
18	3	2¾	1½	1⅜	1⅛	1 1/16
21	3¼	3	1¾	1⅝	1 5/16	1 3/16
24	3⅜	3⅛	2	1⅞	1½	1⅜
27	3½	3¼	2¼	2⅛	1 11/16	1 9/16
30	3⅝	3⅜	2½	2⅜	1⅞	1¾
33	3¾	3½	2⅝	2½	2	1 13/16
36	4	3¾	2¾	2⅝	2 1/16	1⅞

* There shall be no maximum length. Shorter lengths may be used for closures and specials.
† There is no limit for plus variation.

Building Service Connections. Figure 16 shows typical details of service connections to a sanitary sewer laid in a relatively shallow trench; Fig. 17 shows a typical connection to a deep sewer. It is noted that the connection shown in Fig. 16 makes use of either a wye branch or a tee branch in the main sewer line.

Junction Chambers and Manholes. Figure 18 shows a typical design for a junction chamber and manhole for relatively small sewers. For junctions of large sewers, a special underground structure will ordinarily be required, and the entrance to it will be provided for by a manhole located at one side. Such chambers and manholes are required at every sewer junction and at every point where the sewer

Table 8. Dimensions of Extra-strength Clay Pipe¶

Nominal size, in.	Laying length *L*		Max difference in length of two opposite sides, in.	Outside diameter of barrel, in. ‡, §		Inside diameter of socket at ½ in. above base, in. *DS*
	Min, ft*	Limit of minus variation, in. per ft of length†		Min	Max	Min
6	2	¼	⅜	7 1/16	7 7/16	8 3/16
8	2	¼	7/16	9¼	9¾	10½
10	2	¼	7/16	11½	12	12¾
12	2	¼	7/16	13¾	14 5/16	15⅛
15	3	¼	½	17 3/16	17 13/16	18⅝
18	3	¼	½	20⅝	21 7/16	22¼
21	3	¼	9/16	24⅛	25	25⅞
24	3	⅜	9/16	27½	28½	29⅜
27	3	⅜	⅝	31	32⅛	33
30	3	⅜	⅝	34⅜	35⅝	36½
33	3	⅜	⅝	37⅝	38 15/16	39⅞
36	3	⅜	11/16	40¾	42¼	43¼

Table 8. (Continued)

Nominal size, in.	Depth of socket, in. LS		Thickness of barrel, in. T		Thickness of socket at ½ in. from outer end, in. TS	
	Nominal	Min	Nominal	Min	Nominal	Min
6	2¼	2	11/16	9/16	½	7/16
8	2½	2¼	⅞	¾	9/16	½
10	2⅝	2⅜	1	⅞	⅝	9/16
12	2¾	2½	1 3/16	1 1/16	¾	11/16
15	2⅞	2⅝	1½	1⅜	15/16	⅞
18	3	2¾	1⅞	1¾	1⅛	1 1/16
21	3¼	3	2¼	2	1 5/16	1 3/16
24	3⅜	3⅛	2½	2¼	1½	1⅜
27	3½	3¼	2¾	2½	1 11/16	1 9/16
30	3⅝	3⅜	3	2¾	1⅞	1¾
33	3¾	3½	3¼	3	2	1¾
36	4	3¾	3½	3¼	2 1/16	1⅞

* There shall be no maximum length. Shorter lengths may be used for closures and specials.
† There is no limit for plus variation.
‡ The average actual inside diameters of pipe having the nominal thickness of barrel shown in Table 8 may be smaller than the nominal sizes.
§ The outside diameter of the barrel may be greater than the maximum figures stated in Table 8 provided the other dimensions are varied accordingly within the specification tolerances.
¶ Dimensions *L*, *DS*, *LS*, *T*, and *TS* refer to the sketch of Table 7.

changes in size, slope, direction, or elevation. It is general practice to install sewers in straight lines between manholes, except that for the larger sizes (36 in. and above) they may be laid on curves. Manholes are usually installed at the upper end of every lateral sewer and in straight-line sewers so that the spacing will not exceed

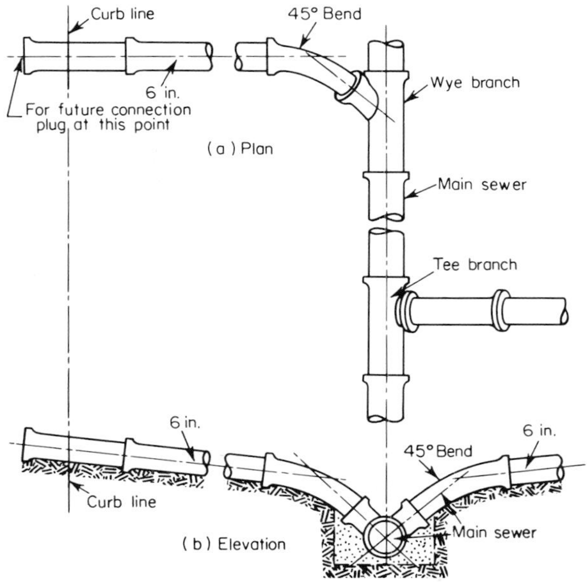

FIG. 16. Typical service connections to a shallow sewer.

Table 9. Standard-strength Clay-pipe Fittings

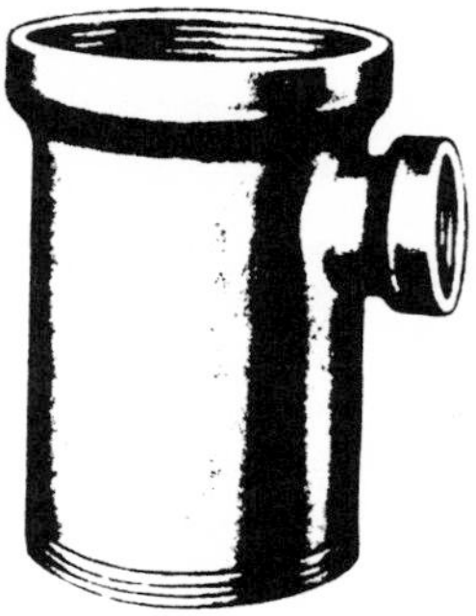

"T" branches

are made with the spur set on the barrel at an angle of 90°.

The spur is molded on the barrel before the fitting is dried, burned or glazed, making the whole an integral glazed unit.

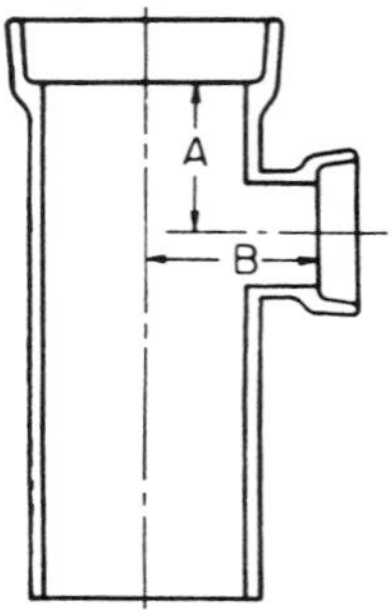

Barrel diameter, in.	Nominal laying length, ft	Spur diameter, in.	*A*, in.	*B*, in.
4	1	4	5	4¼
6	1½	4	5¼	5½
6	1½	6	6¼	5½
8	2, 3	4	5½	6½
8	2, 3	6	6½	6¾
8	2, 3	8	7¾	7
10	2, 3	4	6	7
10	2, 3	6	6¾	7¾
10	2, 3	8	7¾	8¼
10	2, 3	10	9	8½
12	2, 3	4	6¼	8¾
12	2, 3	6	6¾	9
12	2, 3	8	8	9½
12	2, 3	10	9¼	9¾
12	2, 3	12	10½	10
15	3, 4	4	6½	10½
15	3, 4	6	7	10¾
15	3, 4	8	8¼	11¼
15	3, 4	10	9½	11¼
15	3, 4	12	10½	11½
15	3, 4	15	12¼	11¾
18	3, 4	6	7¼	12½
18	3, 4	8	8¼	13
18	3, 4	12	10½	13¼
18	3, 4	15	12½	13¼
18	3, 4	18	13¾	13½

Barrel diameter, in.	Nominal laying length, ft	Spur diameter, in.	*A*, in.	*B*, in.
21	3, 4	6	7½	14¼
21	3, 4	8	8¾	14¾
21	3, 4	12	11	15
21	3, 4	15	12¾	15½
21	3, 4	18	14¼	15¾
21	3, 4	21	15½	16
24	3, 4	6	8	16
24	3, 4	8	9	16½
24	3, 4	12	11¼	16¾
24	3, 4	15	13	17⅛
24	3, 4	18	14½	17¼
24	3, 4	24	17½	17½
27	3, 4	6	8¼	17¾
27	3, 4	8	9¼	18¼
27	3, 4	12	11½	18½
27	3, 4	18	15	18¾
27	3, 4	24	18	19
30	3, 4	6	8½	19½
30	3, 4	8	9½	20
30	3, 4	12	12	20¼
30	3, 4	18	15½	21
30	3, 4	24	18½	22
33	3, 4	6	8¾	21⅝
33	3, 4	12	12½	21⅞
33	3, 4	18	16	22⅛
33	3, 4	24	19	22½
36	3, 4	6	9¼	22¾
36	3, 4	8	10½	23¼
36	3, 4	12	12½	23½
36	3, 4	18	16	24½
36	3, 4	24	19	25½

NOTE: Dimensions *A* and *B* are approximate only. Dimensions not shown are the same as for straight pipe.

Table 10. Standard-strength Clay-pipe Fittings

"Y" branches

are made with the spur set on the barrel at an angle of 60°.

The spur is molded on the barrel before the fitting is dried, burned or glazed, making the whole an integral glazed unit.

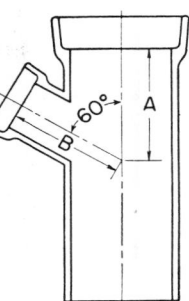

Barrel diameter, in.	Nominal laying length, ft	Spur diameter, in.	A, in.	B, in.	Barrel diameter, in.	Nominal laying length, ft	Spur diameter, in.	A, in.	B, in.
4	1	4	8	6¾	21	3, 4	12	22¾	20¾
6	1½	4	8¾	7½	21	3, 4	15	23¼	22
6	1½	6	9¾	8¾	21	3, 4	21	28	24
8	2, 3	4	9¼	8¾	24	3, 4	6	18¼	20¼
8	2, 3	6	11¼	9¾	24	3, 4	8	20¼	21½
8	2, 3	8	12¼	11	24	3, 4	12	23¼	22¾
					24	3, 4	15	23¾	24
10	2, 3	4	11	10¾	24	3, 4	18	27½	25½
10	2, 3	6	13	11¼	24	3, 4	24	30½	28
10	2, 3	8	14	12					
10	2, 3	10	15	13	27	3, 4	6	21½	22½
					27	3, 4	8	22½	23¼
12	2, 3	4	11¼	12¼	27	3, 4	12	24½	24½
12	2, 3	6	12¼	12¾	27	3, 4	18	28	28
12	2, 3	8	14¼	13½	27	3, 4	24	31½	29½
12	2, 3	10	15¼	14¼					
12	2, 3	12	17¼	15	30	3, 4	6	22½	24
					30	3, 4	8	23½	25¼
15	3, 4	4	12¼	13¾	30	3, 4	12	25½	26½
15	3, 4	6	14¼	14¼	30	3, 4	18	32	32
15	3, 4	8	16¼	15	30	3, 4	24	35	35
15	3, 4	10	18¼	15⅞					
15	3, 4	12	19¼	16½	33	3, 4	6	24¼	26½
15	3, 4	15	21	17½	33	3, 4	12	27¼	28½
					33	3, 4	18	33	34½
18	3, 4	6	16	16½	33	3, 4	24	36½	36½
18	3, 4	8	18	17¼					
18	3, 4	12	21	18¾	36	3, 4	6	25	28¾
18	3, 4	15	23½	20¼	36	3, 4	8	26	29½
18	3, 4	18	25½	21½	36	3, 4	12	27	30½
					36	3, 4	18	34	36
21	3, 4	6	17¾	18¼	36	3, 4	24	38	38
21	3, 4	8	18¾	19½					

NOTE: Dimensions *A* and *B* are approximate only. Dimensions not shown are the same as for straight pipe.

Table II. Extra-strength Clay-pipe Fittings

"T" branches

are made with the spur set on the barrel at an angle of 90°.

The spur is molded on the barrel before the fitting is dried, burned or glazed, making the whole an integral glazed unit.

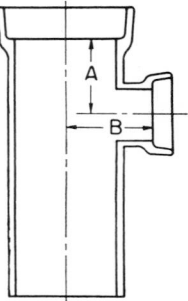

Barrel diameter, in.	Nominal laying length, ft	Spur diameter, in.	A, in.	B, in.	Barrel diameter, in.	Nominal laying length, ft	Spur diameter, in.	A, in.	B, in.
4	1	4	5	4¼	21	3, 4	6	7½	14¼
6	1½	4	5¼	5½	21	3, 4	8	8¾	14¾
6	1½	6	6¼	5½	21	3, 4	12	11	15
					21	3, 4	15	12¾	15½
8	2, 3	4	5½	6½	21	3, 4	18	14¼	15¾
8	2, 3	6	6½	6¾					
8	2, 3	8	7¾	7	24	3, 4	6	8	16
					24	3, 4	8	9	16½
10	2, 3	4	6	7	24	3, 4	12	11¼	16¾
10	2, 3	6	6¾	7¾	24	3, 4	15	13	17⅛
10	2, 3	8	7¾	8¼	24	3, 4	18	14½	17¼
10	2, 3	10	9	8½					
					27	3, 4	6	8¼	17¾
12	2, 3	4	6¼	8¾	27	3, 4	8	9¼	18¼
12	2, 3	6	6¾	9	27	3, 4	12	11½	18½
12	2, 3	8	8	9½	27	3, 4	18	15	18¾
12	2, 3	10	9¼	9¾					
12	2, 3	12	10½	10	30	3, 4	6	8½	19½
					30	3, 4	8	9½	20
15	3, 4	4	6½	10½	30	3, 4	12	12	20¼
15	3, 4	6	7	10¾	30	3, 4	18	15½	21
15	3, 4	8	8¼	11¼					
15	3, 4	10	9½	11¼	33	3, 4	6	8¾	21⅝
15	3, 4	12	10½	11½	33	3, 4	12	12½	21⅞
15	3, 4	15	12¼	11¾	33	3, 4	18	16	22⅛
18	3, 4	6	7¼	12½	36	3, 4	6	9¼	22¾
18	3, 4	8	8¼	13	36	3, 4	8	10½	23¼
18	3, 4	12	10½	13¼	36	3, 4	12	12½	23½
18	3, 4	15	12½	13¼	36	3, 4	18	16	24½
18	3, 4	18	13¾	13½					

NOTE: Dimensions *A* and *B* are approximate only. Dimensions not shown are the same as for straight pipe.

Table 12. Extra-strength Clay-pipe Fittings

"Y" branches

are made with the spur set on the barrel at an angle of 60°.

The spur is molded on the barrel before the fitting is dried, burned or glazed, making the whole an integral glazed unit.

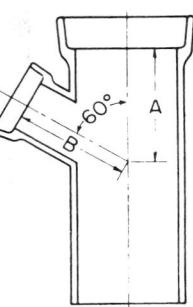

Barrel diameter, in.	Nominal laying length, ft	Spur diameter, in.	A, in.	B, in.	Barrel diameter, in.	Nominal laying length, ft	Spur diameter, in.	A, in.	B, in.
4	1	4	8	6¾	21	3, 4	6	17¾	18¼
6	1½	4	8¾	7½	21	3, 4	8	18¾	19½
6	1½	6	9¾	8¾	21	3, 4	12	22¾	20¾
					21	3, 4	15	23¼	22
8	2, 3	4	9¼	8¾					
8	2, 3	6	11¼	9¾	24	3, 4	6	18¼	20¼
8	2, 3	8	12¼	11	24	3, 4	8	20¼	21½
					24	3, 4	12	23¼	22¾
10	2, 3	4	11	10¾	24	3, 4	15	23¾	24
10	2, 3	6	13	11¼					
10	2, 3	8	14	12	27	3, 4	6	21½	22½
10	2, 3	10	15	13	27	3, 4	8	22½	23¼
					27	3, 4	12	24½	24½
12	2, 3	4	11¼	12¼					
12	2, 3	6	12¼	12¾	30	3, 4	6	22½	24
12	2, 3	8	14¼	13½	30	3, 4	8	23½	25¼
12	2, 3	10	15¼	14¼	30	3, 4	12	25½	26½
12	2, 3	12	17¼	15					
					33	3, 4	6	24¼	26½
15	3, 4	4	12¼	13¾	33	3, 4	12	27¼	28½
15	3, 4	6	14¼	14¼					
15	3, 4	8	16¼	15	36	3, 4	6	25	28¾
15	3, 4	10	18¼	15⅞	36	3, 4	8	26	29½
15	3, 4	12	19¼	16½	36	3, 4	12	27	30½
18	3, 4	6	16	16½					
18	3, 4	8	18	17¼					
18	3, 4	12	21	18¾					

Note: Dimensions *A* and *B* are approximate only.   Dimensions not shown are the same as for straight pipe.

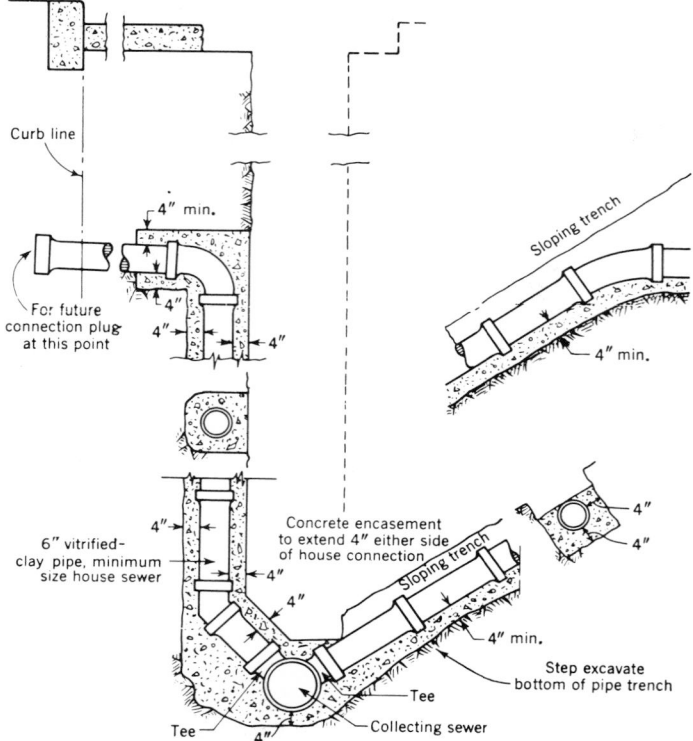

Fig. 17. Typical service connection to a deep sewer.

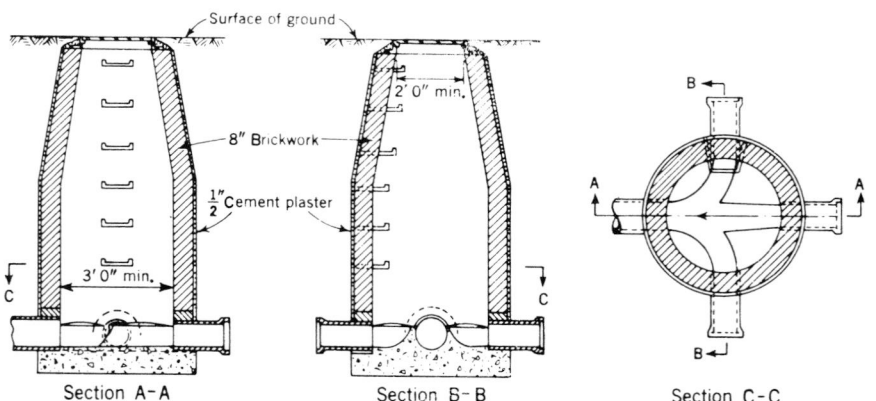

FIG. 18. Junction chamber and manhole for small sewers.

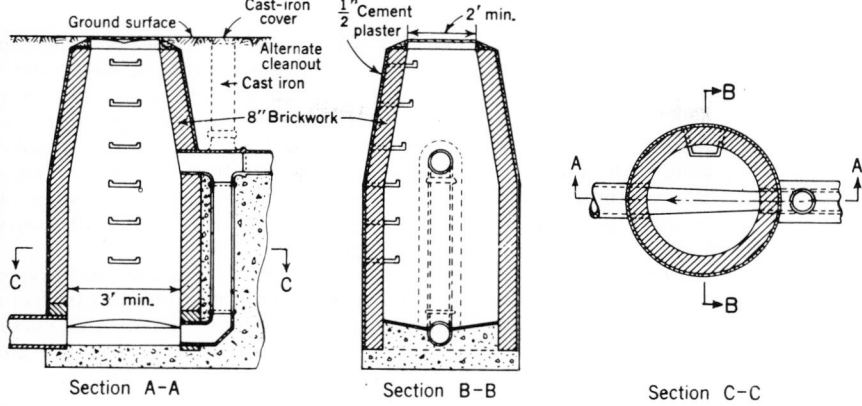

Section A–A   Section B–B   Section C–C

FIG. 19. Drop manhole.

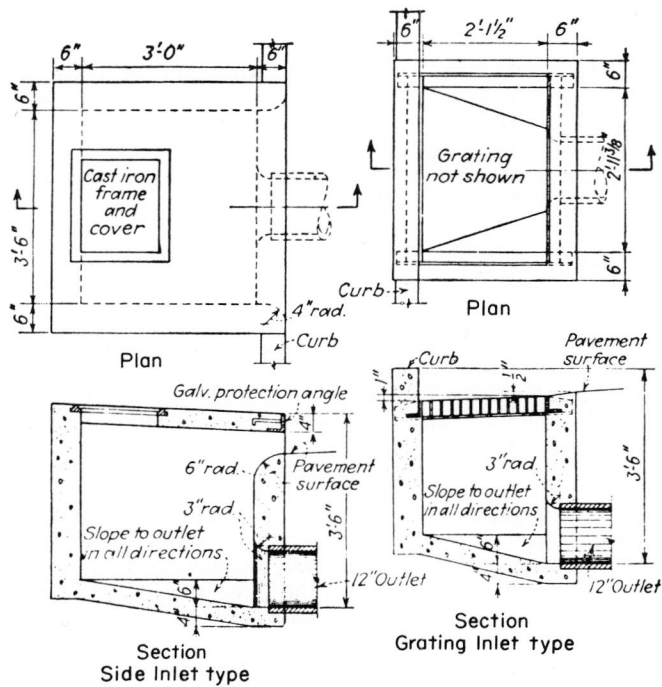

FIG. 20. Street inlets.

about 300 ft.   Figure 19 shows typical details of a "drop manhole" at a point where a sewer takes an abrupt drop in grade.

Storm-water Inlets.   Storm-water inlets which carry storm water from the streets to the storm sewers are located upstream of the crosswalks at street intersections and at low points.   The designs vary considerably, and most cities have adopted their own standard design details.   There are three general types of inlets: (1) curb inlets, which have a vertical opening in the curb; (2) gutter inlets, in which a horizontal opening in the gutter is covered by a cast-iron grating; and (3) combination inlets, which combine both the above features.   Many types and sizes of standard castings are available for the construction of inlets.   Figure 20 shows typical inlet details.

Sewage Pumping.   In many sewerage-system layouts it is necessary to provide for pumping at one or more points.   The required pumping capacity will vary from a few gallons per minute for stations serving only a few laterals to many millions of gallons per day for stations serving large districts.   The smaller stations are frequently built underground, either as built-in-place or as complete factory-assembled units.   Pumping is usually done with nonclog centrifugal pumps, although pneumatic ejectors are sometimes used for the smaller installations.   The design of sewage pumping stations requires careful attention to the special requirements which are imposed by the function being performed.   For detailed discussion of these requirements, the reader is referred to the ASCE-WPCF Manual.[3]

STRUCTURAL REQUIREMENTS

Sewers must be installed so as to be able to withstand the loads imposed upon them by the weight of the earth and any superimposed loads.   The supporting strength of a buried pipe depends upon the installation conditions as well as the structural properties of the pipe itself.   Sewer pipes are classed as rigid pipes, which cannot deform materially without cracking.   For rigid pipes in trenches, the load can be represented by the equation

$$W = CwB^2$$

where W = load, lb per foot of length
     w = weight of the soil, lb/cu ft
     B = width of trench, ft
     C = coefficient whose value depends upon type of soil and ratio of depth of cover to trench width

Table 13 gives values for C and Table 14 gives values of w to be used in the equation.

If a pipe is placed on undisturbed ground and covered with a fill, the load can be estimated from the equation

$$W = CwD^2$$

where D is the pipe diameter.   Table 15 gives values of C for the latter condition, which is known as the "projection condition."   The load on a pipe placed in a trench will increase with the trench width until it equals the load for the projection condition.   If there is doubt as to whether the "ditch condition" or the "projection condition" controls, the load should be calculated by both formulas and the minimum value used.

In addition to the load of the backfill, some allowance should be made for the superimposed loads caused by vehicles.   It is usually safe to assume that H-20 wheel loads will be the greatest live loads to be supported.   H-20 loads refer to trucks having a gross weight of 20 tons, 80 per cent of which is on the rear axle, each rear wheel carrying 8 tons.   Table 16 gives the percentage of wheel loads that can be assumed to be transmitted to buried pipe.[21]

Table 13. Values of C for Use in Formula $W = CwB^2$ *

Ratio of depth to trench width	Sand and damp topsoil	Saturated topsoil	Damp clay	Saturated clay
0.5	0.46	0.46	0.47	0.47
1.0	0.85	0.86	0.88	0.90
1.5	1.18	1.21	1.24	1.28
2.0	1.46	1.50	1.56	1.62
2.5	1.70	1.76	1.84	1.92
3.0	1.90	1.98	2.08	2.20
3.5	2.08	2.17	2.30	2.44
4.0	2.22	2.33	2.49	2.66
4.5	2.34	2.47	2.65	2.87
5.0	2.45	2.59	2.80	3.03
5.5	2.54	2.69	2.93	3.19
6.0	2.61	2.78	3.04	3.33
6.5	2.68	2.86	3.14	3.46
7.0	2.73	2.93	3.22	3.57
7.5	2.78	2.98	3.30	3.67
8.0	2.81	3.03	3.37	3.76
8.5	2.85	3.07	3.42	3.85
9.0	2.88	3.11	3.48	3.92
9.5	2.90	3.14	3.52	3.98
10.0	2.92	3.17	3.56	4.04
11.0	2.95	3.21	3.63	4.14
12.0	2.97	3.24	3.68	4.22
13.0	2.99	3.27	3.72	4.29
14.0	3.00	3.28	3.75	4.34
15.0	3.01	3.30	3.77	4.38
Very great	3.03	3.33	3.85	4.55

* *Iowa State Univ. Eng. Expt. Sta. Bull.* 47.

Table 14. Weights of Ditch-filling Materials

Material	Lb/cu ft
Dry sand	100
Ordinary (damp) sand	115
Wet sand	120
Damp clay	120
Saturated clay	130
Saturated topsoil	115
Sand and damp topsoil	100

Table 15. Values of C for Projection Condition

Ratio, depth of cover/pipe diam	0.5	1.0	1.5	2.0	2.5	3.0	3.5	4.0
C	0.6	1.2	2.0	3.0	4.2	5.6	7.5	10.0

Table 16. Percentage of Wheel Loads Transmitted to Underground Pipes for Unpaved Roadway or Berm Areas *

(Tabulated figures show percentage of wheel load applied to 1 lin ft of pipe.)

Depth of backfill over top of pipe, ft	Trench width at top of pipe, ft						
	1	2	3	4	5	6	7
1	17.0	26.0	28.6	29.7	29.9	30.2	30.3
2	8.3	14.2	18.3	20.7	21.8	22.7	23.0
3	4.3	8.3	11.3	13.5	14.8	15.8	16.7
4	2.5	5.2	7.2	9.0	10.3	11.5	12.3
5	1.7	3.3	5.0	6.3	7.3	8.3	9.0
6	1.0	2.3	3.7	4.7	5.5	6.2	7.0

Live loads transmitted are practically negligible below 6 ft.
* These percentages include both live load and impact transmitted to the pipe.

Pipe Bedding Conditions. The supporting strength of a rigid pipe depends upon the type of bedding used in the installation of the pipe. Four general types of bedding conditions have been defined for ditch conduits:

Type 1, *Impermissible Bedding.* Little or no care is taken to shape the foundation to fit the lower part of the pipe or to fill and tamp around the pipe.

Type 2, *Ordinary Bedding.* The soil at the bottom of the trench is shaped to fit the lower part of the pipe with reasonable closeness for a width of at least 50 per cent of the pipe diameter; and the remainder of the pipe is covered to a height of at least 6 in. above its top by granular material which is hand placed and tamped.

Type 3, *First-class Bedding.* The pipe is carefully bedded on fine granular material in an earth foundation carefully shaped to fit the bottom part of the pipe for a width at least 60 per cent of the diameter; the remainder of the pipe is entirely surrounded to a height at least 1.0 ft above the top by granular materials placed by hand in layers not exceeding 6 in. and thoroughly tamped.

Type 4, *Concrete Cradle Bedding.* The lower part of the pipe is embedded in concrete.

The load factors, or the ratios of the supporting strength to the crushing load as determined by the three-edge bearing method, for the various types of bedding are generally taken as follows:

> Impermissible bedding 1.1
> Ordinary bedding 1.5
> First-class bedding 1.9
> Concrete cradle bedding. 2.2–3.4

The factors for the concrete cradle bedding depend upon the amount and quality of the concrete that is used. The value of 2.2 will generally apply when the concrete extends from about one-quarter of the pipe diameter (with minimum of 6 in.) below the pipe to the height where the lower 120 deg sector radii intersect the outside of the pipe. If the concrete is carried up to cover the entire bottom half of the pipe, the load factor may be as high as 3.4. If the entire pipe is encased in concrete with a minimum of $0.25D$ (or 4 in.) both above and below, the load factor may be as high as 4.5.

Safety Factor. The specified minimum strength by the three-edge bearing method for a rigid pipe should be divided by an appropriate safety factor in order to obtain the working strength. Some engineers use safety factors as low as 1.0 to 1.2 for reinforced-concrete pipe culverts. For street sewers a safety factor of 1.5 is recommended by the ASCE-WPCF Manual.

DISPOSAL OF SEWAGE AND STORM WATER

Storm water can ordinarily be disposed of by discharge into any natural drainage channel. Sanitary sewage and industrial waste waters containing objectionable constituents must be disposed of in accordance with the requirements of the local health authorities. The most satisfactory method of disposal of sanitary sewage is to convey it to an adequate public sewerage system. In areas which do not have public sewerage systems, individual disposal systems must be provided. These will vary in size from septic-tank systems for private residences to large treatment plants handling the wastes from large institutions and industries. The design of such systems should usually be done only by those who are experienced in this work and who are familiar with the conditions peculiar to the local area.

REFERENCES

1. "Glossary—Water and Sewage Control Engineering," APHA, ASCE, AWWA, and FSIWA, 1949.
2. Greeley, S. A., and W. A. Stanley, "Capacity Factors for Unequal Population Development," in "Handbook of Applied Hydraulics," by C. V. Davis, 2d ed., McGraw-Hill Book Company, New York, 1952.
3. Design and Construction of Sanitary and Storm Sewers, "ASCE Manuals of Engineering Practice—No. 37" ("WPCF Manual of Practice No. 9").
4. Velzy, C. R., and J. M. Sprague, "Infiltration Specifications and Tests," *Sewage Ind. Wastes*, vol. 27, no. 3, p. 245, March, 1955.
5. Recommended Standards for Sewage Works, adopted by Great Lakes-Upper Mississippi River Board of State Sanitary Engineers, May 10, 1960.
6. Hunter, R. B., "Methods of Estimating Loads on Plumbing Systems," *Natl. Bur. Std. Rpt. BMS65*, Washington, D.C., Dec. 16, 1940.
7. "Rainfall Intensity-Duration-Frequency Curves for Selected Stations in the United States, Alaska, Hawaiian Islands and Puerto Rico," *Weather Bur. Tech. Paper 25*, U.S. Dept. of Commerce, Washington, D.C., 1955.
8. "Rainfall Intensities for Local Drainage Design in the United States," *Weather Bur. Tech. Paper 24*, U.S. Dept. of Commerce, Washington, D.C., 1955.
9. "Rainfall Intensities for Local Drainage Design in Western United States," *Weather Bur. Tech. Paper 28*, U.S. Dept. of Commerce, Washington, D.C., 1955.
10. Tholin, A. L., and C. V. Keefer, "The Hydrology of Urban Runoff," *Trans. ASCE*.
11. Wilcox, E. R., "A Comparative Test of the Flow of Water in 8-inch Concrete and Vitrified Clay Sewer Pipe," Bull. 27, *Univ. Wash. Expt. Stat. Ser.*, Seattle, 1924.
12. Yarnell, D. L., and S. M. Woodward, "The Flow of Water in Drain Tile," *U.S. Dept. Agr. Bull.* 854, Washington, D.C., 1920.
13. "Minimum Velocities for Sewers," Final Report of Committee to Study Limiting Velocities of Flow in Sewers, *J. Boston Soc. of Civil Engrs.*, vol. 29, no. 4, p. 286, October, 1942.
14. Shields, A., "Anwendung der Aehnleihkeitsmechanik und der Turbulenz forschung auf die Geschiebebewegung," *Mitt. Preuss. Versuchanstalt Wasserbau Schiffbau*, heft 26, Berlin, 1936.
15. Camp, T. R., Design of Sewers to Facilitate Flow, *Sewage Ind. Wastes*, vol. 18, no. 1, p. 3, January, 1946.

16. Chow, Ven Te, "Open-Channel Hydraulics," McGraw-Hill Book Company, New York, 1959.
17. Bakhmeteff, B. A., "Hydraulics of Open Channels," McGraw-Hill Book Company, New York, 1932.
18. King, H. W., and E. F. Brater, "Handbook of Hydraulics," 5th ed., McGraw-Hill Book Company, New York, 1963.
19. Hinds, J., "The Hydraulic Design of Flume and Siphon Transitions," *Trans. ASCE*, vol. 92, p. 1423, 1928.
20. "Concrete Pipe Handbook," American Concrete Pipe Association, 1959.
21. "Clay Pipe Engineering Manual," Clay Sewer Pipe Association, Inc., 1960.

19

FLOW OF SLUDGES AND SLURRIES

William H. Kapfer*

The terms *sludge* and *slurry* are used more or less interchangeably in this chapter to describe those fluid systems which generally consist of more than one phase, such as a solid suspended in a liquid or two immiscible liquids of different densities, or a material of pastelike consistency which behaves under some conditions essentially as a fluid. Through common usage "slurry" refers to a relatively thin or watery, and hence dilute, suspension, whereas "sludge" denotes a mud, or heavy phase, or concentrated suspension, or simply a very viscous fluid. No attempt is made here to define either term arbitrarily; either word implies the existence of a substance which can be made to flow through a piping system but which is not a simple, homogeneous Newtonian fluid such as water.

It is noted particularly that a sludge is not necessarily restricted to a suspension of solids in a liquid but may include certain other systems such as a very viscous liquid dispersed in another liquid or in a gelatinous mass. The transportation of such sludges or slurries in pipelines generally poses a more difficult design problem than that for ordinary simple homogeneous fluids because of several complicating factors. Many of the problems arise from the high viscosity, from the non-Newtonian character, or from the nonhomogeneity of a fluid system and the tendency of suspended material to segregate and settle.

The tendency to settle will vary, not only for different systems, but also with the particular flow conditions. Particle density, shape, and size, as well as size distribution, concentration, and composition, influence the settling characteristics. Thus, differences can be expected in behavior for fiberlike particles as compared with granular particles or between spherical-shaped and rodlike or flakelike particles.

* Associate Professor of Chemical Engineering, New York University, New York, N.Y.

Moreover, as the size of the particle decreases, Brownian motion becomes important, so that the settling rate decreases and may cease altogether or at least may not be a factor during the residence time of the slurry in the piping system.   Temperature and concentration also influence behavior; as the concentration increases, the suspended particles may interact so that hindered settling phenomena are observed. Likewise the particles may agglomerate or flocculate and change the settling characteristics.   Density differences between the suspending medium and the suspended particle are factors; the smaller this difference, the less is the settling tendency.   If there is a size distribution with a given density, the larger particles will settle faster than the smaller ones.   If the solids themselves are of mixed composition (i.e., materials of different density), then a larger particle of lower density will settle at the same rate as a smaller particle of higher density.

The flow conditions also have a bearing on the settling characteristics.   Basically a particle tends to settle because of its weight, i.e., the downward gravity force. This tendency is always opposed by the buoyant force, proportional to the density difference, and by the drag forces generated when the particle starts to move.   In addition there are forces due to Brownian motion and flocculation as mentioned above, wall effects, and, of course, agitation.   It is easily seen, therefore, that, as the velocity of flow is increased and especially when the velocity is high enough to generate a turbulent-flow regime, the turbulent eddies will tend to maintain the particles in suspension as well as to transport them in the main direction of flow. It is apparent that the degree of turbulence may vary from one location to another in a flow system and that settling can occur preferentially in certain places where the turbulence is absent or the flow rate low.   If the particle size and density are high, the settling tendency is also high and a "saltation" effect [61],* may occur below certain critical velocities, with resultant "choking."

Since a slurry is not usually a homogeneous phase, there is also the problem of determining the effective physical properties, such as density and viscosity, for use in any design relation and also how these properties may change from one part of the system to another or as the temperature, pressure, and flow conditions vary.

Many, if not all, sludges and slurries exhibit some type of non-Newtonian behavior, so that their flow behavior is different from that of ordinary fluids on this account, even if settling is not a problem.   For these systems the apparent viscosity will not be constant but will vary with the shear rate or flow conditions.   Furthermore, when certain slurries or polymer solutions are allowed to stand for some time, they may form a gel (thixotropic effect), which may present serious pumping problems.

In view of the above complications it is not surprising that the design procedures are more or less empirical and that a unified treatment has not yet been forthcoming. Various studies, however, have been made and reported under classifications such as two-phase (or multiphase) flow, sedimentation, fluidization, non-Newtonian flow, stratified flow, particle technology, and others.

Some investigators[34] classify slurries into "settling" and "nonsettling" types. Examples of the former include sand in water or salts in an organic medium such as benzene.   The so-called "nonsettling" types are represented by such systems as paper pulp in water, certain highly viscous petroleum fractions, and some food products like ketchup.   It must, of course, be borne in mind that this distinction is somewhat arbitrary, inasmuch as factors such as the state of subdivision of suspended particles and the flow conditions have an important bearing on settling. Nevertheless, in spite of its limitations, the concept is retained here for convenience

---

* Superscript numbers refer to corresponding items in the list of References at the end of the chapter.

to divide this discussion into two broad areas: (1) fluid systems in which settling of solids or other phase separation is not of primary importance and which may be treated as essentially homogeneous but non-Newtonian fluids and (2) fluid systems in which settling or "saltation" or transport of distinct phases is a major consideration.

No attempt will be made here to develop a unified or detailed theory, but the discussion will instead be directed to selected procedures which have gained some acceptance among engineers for design purposes.

NEWTONIAN SYSTEMS—GENERAL CONSIDERATIONS

At the outset it should be noted that some materials, which at first sight might be called sludges, actually may be liquids of such high viscosity that they are pasty or sticky and hence are difficult to pump at ordinary temperatures. When the temperature is raised sufficiently, however, the viscosity is greatly reduced, and the substance then behaves essentially as a Newtonian fluid, such as water or ordinary mineral oils. Energy requirements, pressure losses, and other data required for pipeline design may be calculated in a straightforward way by the usual standard methods based on the familiar correlations of a Fanning or Darcy friction factor versus a Reynolds number with the flow rate and fluid properties of density and viscosity, as discussed in Chap. 3 or in standard texts on fluid flow.

The only other major considerations in designing the pipeline are those concerned with maintaining the appropriate pipeline temperatures and making provisions for emergency shutdowns to avoid difficulty on subsequent start-up. It is not implied that provision for heating is always inexpensive or easily accomplished. In some cases, of course, the substance may be preheated in a tank or other vessel, and if the pipeline is short, it may suffice merely to apply an insulating cover to reduce the heat loss and to slope the line or make other provision for it to drain empty on shutdown to avoid plugging the line when cold. It would, of course, be necessary even here to calculate the maximum tolerable heat loss from the length of pipe involved. The heat-transfer calculations are not treated here, but typical calculation methods may be found in standard texts.[31,35]

Where insulation alone is insufficient, as, for example, in a long exposed line, or where heat requirements are high, additional heat may be supplied to the pipeline in a number of ways, such as through the use of auxiliary steam tracing lines, jacketing, or electrical heating elements. The last may be in the form of resistance wire wound around the pipe, electrical heating jackets for valves and fittings, or strip and band heaters, which are available in a number of standard sizes and shapes for use with ordinary electrical connections. Plain and contoured panel heaters through which steam or other hot fluids can be circulated are also available and may find use in certain circumstances. Here again, it is necessary to estimate the required heat loads and to design the pipe system to avoid plugging when cold.

A very interesting and novel installation involving the use of an electric pipeline for transferring high-viscosity fuel was reported recently by A. G. Purdue.[44] A Puerto Rico power plant uses a very high viscosity residue from a petroleum-refinery vacuum still as a primary fuel. This residue could be considered a sludge, since it has a viscosity of about 12,000,000 Saybolt seconds at 122 F and could not be transferred in a pipeline as such without heating, over the required distance of about a quarter mile. At a temperature of 450 to 500 F, however, the viscosity is about 150 Saybolt seconds, at which temperature it can be transferred readily. To maintain this temperature, the entire pipeline is heated electrically, using the electrical resistance of the pipe itself to generate the necessary heat directly in the pipe

walls. Constant pipe-wall temperature is maintained with precise automatic control, and all use of steam jackets and tracing lines is eliminated. The pipe is standard Schedule 40, and the pipe covering serves as both thermal and electrical insulation. Electric power is supplied as standard 480-volt 60-cycle current to the primary side of energizing units, which then feed it directly to the pipe.

In closing comment on this section it should be noted again that there has been treated here essentially the flow of those few "sludges" which behave as Newtonian fluids at elevated temperatures. Most sludges and slurries, however, will show a more complicated behavior of non-Newtonian character as discussed in the next section.

NON-NEWTONIAN SYSTEMS

In this section, primarily homogeneous (nonsettling) fluids will be dealt with; however, it will be recognized from the previous discussions that many sludges or slurries in which separation of solids or settling does occur also have non-Newtonian properties, and hence some of the remarks which follow will apply equally well to such systems too. Sewage sludge, for example, is one system in which settling can occur, yet its non-Newtonian characteristics are much more important for design purposes. In fact some of the most significant early analyses[4,5] of non-Newtonian flow were based on sewage sludge systems. Likewise, certain clay and fiber suspensions in water are distinguished more for their non-Newtonian behavior than for their settling characteristics.

At the present time many of the design methods for the flow of non-Newtonian fluids are still in their infancy compared with those for Newtonian systems, and certain non-Newtonian systems cannot be handled except in a very crude way. Nevertheless, since about 1940 considerable progress has been achieved in characterizing the flow of certain important classes of homogeneous non-Newtonian fluids, and a number of excellent descriptive and review articles have appeared in the literature.[1,2,4–7,9–15,21,39,54] In addition, detailed theoretical analyses of the deformation and flow of matter, which is known as rheology, may be found in various other works.[26,46]

It is the general task of rheology to relate stress, strain, and time, and therefore it deals with several fields from hydrodynamics to stress analysis and metallurgy. Obviously not all the subject matter of rheology is of prime importance in the study of the flow of fluids, but on the other hand, in order to understand non-Newtonian flow, it is necessary to use mathematical relations based on rheological models. It is desirable at this point, therefore, to review in an elementary way some of the common types of fluid behavior.

Briefly, all real fluids exhibit an internal resistance to deformation and flow when stress is applied, and this resistance is ordinarily measured in terms of the response property called viscosity. In the case of a Newtonian fluid the viscosity is a physical constant at a particular temperature, pressure, and chemical composition and is related in a simple way to the shear stress and the rate of shear. Using a "one-dimensional" treatment to keep the mathematical relations as simple as possible, imagine that a fluid is flowing past a flat surface as indicated in Fig. 1. At some distance from the solid surface in the y direction the fluid is flowing in the x direction (left to right) at its maximum velocity $u_{\max}$ as shown by the arrow. Because of the drag exerted by the solid surface, the fluid adjacent to the surface moves at progressively lower velocities, for example u_3, u_2, u_1, as we approach the surface. In the limit, at the solid surface, it is often assumed that the fluid velocity is negligibly small or zero (no-slip condition). In any case, the velocity changes with distance away from the wall to give a velocity profile or velocity gradient,

also known as a shear gradient, in the y direction, which might have some shape as indicated in Fig. 1.   Physically, as a first approximation, it can be imagined that when turbulence or other special effects are absent, thin layers of fluid, each moving at a particular velocity, $u_1, u_2, u_3, \ldots$, etc., slide or shear past each other because of their different velocities as though they were discrete thin sheets.   Mathematically it may be stated that the y-direction shear rate or velocity gradient is du/dy, where du is the small change in velocity u which occurs over the small distance dy.

For a simple Newtonian fluid the stress τ causing the shear is related to the rate of shear du/dy by the simple expression

$$\tau = \mu(du/dy) \tag{1}$$

where μ is the Newtonian viscosity coefficient or simply the absolute viscosity.

In passing, it is worth noting that if the shear stress τ is measured as pounds force per square foot, the velocity change du in feet per second, and the distance dy in feet, then substituting these units in Eq. (1) and solving for μ gives

$$|\mu| = \left| \frac{\tau}{du/dy} \right| =$$

$$(\text{lbf/ft}^2)/(\text{fps/ft}) = \text{lbf sec/ft}^2$$

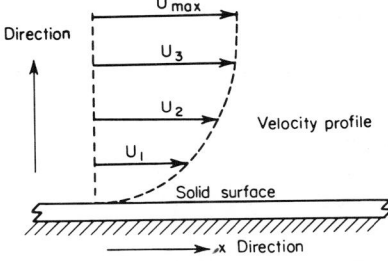

Fig. 1. Schematic fluid velocity profile.

which are the units of viscosity in the English absolute system.   If ordinary English engineering units are used, then we must introduce the dimensional constant $g_c = 32.2$ (lbm/lbf)(ft/sec^2) to maintain consistency.   If this is done, Eq. (1) becomes

$$\tau = (\mu_m/g_c)(du/dy) \tag{1a}$$

where μ_m/g_c still has the net units pounds force second per square foot but the viscosity μ_m now has the widely used units of pounds mass per foot second. Note also that 1 poise $= 100$ cps $= 0.0672$ lbm/(ft sec).

Rheologically then, a Newtonian fluid is one which obeys Eq. (1) or (1a), at least over the range of interest.   As an alternative criterion, the rate of shear du/dy may be plotted against shear stress, and for a Newtonian fluid there will be obtained a straight line through the origin, whose constant slope is the fluidity, or reciprocal of the viscosity, as shown in Fig. 2a.   Any departure from Eq. (1) or Fig. 2 indicates non-Newtonian character and hence a more complicated flow behavior.   Therefore, one of the first steps in handling sludge-flow problems is to determine the rheological behavior.

Workers in the field of rheology have long attempted to classify non-Newtonian fluids in terms of departures from Fig. 2a.   Typical classes have been Bingham plastic, pseudoplastic, dilatant, thixotropic, and rheopectic[1,2,26] fluids with shear diagrams as illustrated in Fig. 2b to e.   The Bingham plastic (true plastic or simply plastic) is distinguished by the fact that it shows a definite yield stress τ_y, before deformation or flow begins.   Some thermoplastic polymer solutions and certain sewage sludges show this behavior.   The pseudoplastic material, like a Newtonian liquid, cannot support shear stress, but it differs in that the viscosity decreases as the shear rate increases.   GRS latex solutions and some sewage sludges behave in this manner.   The opposite effects are noted for the dilatant (inverted plastic) fluids such as starch in water, beach sand, feldspar, and mica slurries, where the

apparent viscosity increases as the shear rate is increased. The thixotropic and rheopectic materials such as paints, greases, inks, milk, mayonnaise, carboxymethyl cellulose, and others are characterized by the fact that their apparent viscosity is a function not only of shear rate but of time as well. For thixotropic materials all "up" curves of Fig. 2e are concave upward, indicating that the apparent viscosity

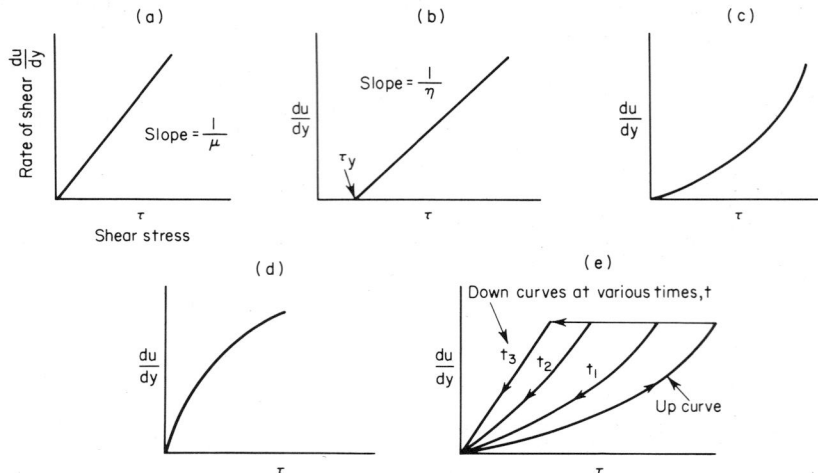

FIG. 2. Shear diagrams for typical Newtonian and non-Newtonian fluids. (a) Newtonian fluid. (b) Bingham plastic fluid. (c) Pseudoplastic fluid. (d) Dilatant fluid. (e) Thixotropic fluid.

decreases with increased shear stress. At constant shear rate, the viscosity continues to decrease, presumably because of some structural breakdown of the material. Often this phenomenon is reversible, so that on standing, the structure is apparently built up again. The rheopectic behavior is in some respects the inverse of thixotropic, in that certain systems, e.g., bentonite and gypsum slurries, show an increase in viscosity with time at constant shear such as is produced by rhythmic tapping or shaking.

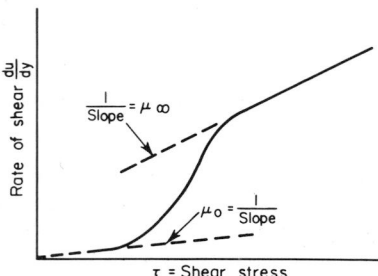

FIG. 3. Ostwald shear diagram.

The above classifications, though descriptive of observed phenomena, leave much to be desired, inasmuch as the same material may exhibit different characteristics depending on the type of test, the speed with which it is carried out, the range of the test, or the effect of other phenomena. Thus, for example in the case of pseudoplastic materials, Ostwald and Auerbach[41] note that this may be an incomplete curve and that, if turbulence could be avoided, the curve would eventually change slope in the opposite direction, so that the complete diagram might be as shown in Fig. 3. Over the middle of the range of this "Ostwald" curve the material would exhibit what was previously

pseudoplasticity. At the lower range, however, by extrapolation toward the inverse slope of the line so obtained would be the so-called limiting at zero shear μ_0. Similarly at the higher ranges there would be obtained ent limiting viscosity at "infinite" shear rate μ_∞. Actually, although there ome data to support the upper region of the behavior curve, there is some fusion about its significance because the onset of turbulence will also cause a arked increase in the apparent viscosity. It appears, therefore, that a particular naterial may under certain conditions exhibit characteristics of a Bingham plastic and under somewhat different conditions be pseudoplastic, then dilatant or perhaps almost Newtonian. Likewise, a material which exhibits rheopectic properties under some circumstances will be thixotropic in other instances. It should also be pointed out that apparently similar materials may differ rheologically. For example, raw sewage sludge behaves differently from digested sewage sludge.

The practical consequences of this multiplicity are: (1) it is dangerous to extrapolate rheological data to new conditions, and (2) it may be difficult to assign a particular mathematical relation to describe the viscosity behavior or shear response of a given sludge or slurry. Very considerable attempts have been made to formulate mechanistic models and to derive appropriate general mathematical relations to describe the response, but essentially an infinite number of rheological relationships can be formulated for non-Newtonian fluids and as yet no single universally acceptable equation has been developed. As pointed out by Dodge and Metzner,[21] even if such a general equation could be developed, it would be of doubtful value for engineering use because of its great complexity.

By way of general mathematical summary it can be stated that for all fluids the velocity gradient or shear-rate dependence is one of three broad types:

1. Shear rate is some function of shear stress only; i.e.,

$$du/dy = \phi(\tau) \tag{2}$$

2. Shear rate is a function of shear stress τ and time t;

$$du/dy = \phi(\tau,t) \tag{3}$$

3. Shear rate is a function of shear stress τ and deformation γ;

$$du/dy = \phi(\tau,\gamma) \tag{4}$$

Category (3) includes the so-called "elastic" fluids.

Of the above general expressions only those of the first type have been developed to the point where they are of some utility for engineering design. At the present time, therefore, only the non-time-dependent and non-deformation-dependent fluids can be handled in a fairly straightforward way. The most widely used rheological relation is the power law, which may be written as

$$du/dy = \phi(\tau) = -A^m \tag{5}$$

or alternatively as

$$\tau = K(-du/dy)^n = K(-du/dr)^n \tag{6}$$

where A, K, m, and n are characteristic constants and dr refers to a radial distance in a circular pipe.

Equation (5) has been used for the most part by rheologists, and Eq. (6) more by engineers.[21,39] A fluid or slurry which obeys Eq. (5) or (6) is known as a "power-law fluid."

The power-law-fluid model has been coupled to the standard Fanning friction factor and has formed a useful means of describing both laminar and turbulent flow of such fluids. Theoretical rheologists[46] have raised some objections to the

use of the power law as a model for non-Newtonian flow behavior, and Bowe has questioned the validity of statements by Metzner[21,39] and others that possi the majority of non-Newtonian fluids obey some form of the power-law relatic Nevertheless, it has been found experimentally[32,39,53,54] that a large number non-Newtonian fluids do obey a power-law function closely enough over wid ranges (10- to 1,000-fold) of shear rate so that its use is justified in many cases.

From the foregoing discussion it will be appreciated that it is not possible to recommend a single and all-inclusive specific design procedure for non-Newtonian fluids which is free of important limitations.   Therefore, the engineer who, never-theless, is faced with the problem of designing systems to transport sludges and slurries must use caution in adapting a particular procedure, including those which follow, to a specific case.   A very good review of the available design methods together with critical comments was given in 1961 a in series of articles by R. L. Bowen, Jr.[9-15]   Some of the methods discussed in this series will be repeated here, but for simplicity and brevity there will be cited mostly the methods developed by Metzner and Reed[39] and Dodge and Metzner,[21] which are based on the power-law-fluid model.   For another excellent review of the flow of non-Newtonian fluids the reader is also referred to Section 7 of the "Handbook of Fluid Dynamics" by Metzner.[38]

NON-NEWTONIAN LAMINAR-FLOW SYSTEMS

For laminar flow of a Newtonian fluid through a pipe, the pressure drop due to friction is given by Poiseuille's law, which may be written in engineering units as

$$\Delta p = 32 \mu L V / g_c D^2 \tag{7}$$

(See "Nomenclature", page 19–17, for definition of symbols and their usual units.) It can also be shown that for this case

$$\Delta p = [4f(L/D)]/(\rho V^2/2g_c) \tag{8}$$

where
$$f = \frac{16}{DV\rho/\mu} = \frac{16}{N_{Re}} = \text{Fanning friction factor} \tag{9}$$

where N_{Re} is the Reynolds number.   The Fanning friction factor may also be defined as

$$f = \frac{\tau_w}{(\rho V^2/2g_c)} \tag{10}$$

where τ_w is the shear stress at the pipe wall, which for any fluid is given by

$$\tau_w = D \Delta p/4L \tag{11}$$

Any combination of Eqs. (7) to (11) may be used to describe the laminar flow of a Newtonian fluid.

Early workers[4,5] recognized that Poiseuille's law [Eq. (7)] could be adapted to non-Newtonian fluids by replacing the Newtonian viscosity μ by an "apparent" viscosity term μ_a.   Determination of the proper value of μ_a, however, is a difficult problem, since it does not remain constant.   In the case of a Bingham plastic, it can be shown[27,28] that

$$\mu_a = g_c \tau_y D/6V + \eta \tag{12}$$

where τ_y is the yield stress on a shear diagram and η is the coefficient of rigidity in pounds mass per foot second.   Substitution of this value μ_a for μ in Eq. (7) would then give the pressure drop for a given flow.   As pointed out by Babbitt and

Caldwell[4] in their analysis of sewage sludge flow, the solution of the equation for pressure drop is not valid at other velocities because μ_a changes.   Later observations by Metzner[37] emphasized the fact that scale-up must be made, not at constant V, but rather at constant V/D.   Of course, it is also necessary that the flow regime be laminar and not turbulent.

Alves, Boucher, and Pigford,[2] in an early attempt to develop a general design procedure, demonstrated that a plot of $8V/g_cD$ or $8q/\pi D^3$ vs $\tau_w = D\,\Delta p/4L$ was equivalent to a shear diagram such as those shown in Fig. 2 and that such diagrams could be used directly to calculate pipe diameter for a given pressure or pressure drop for a given diameter at a particular flow rate.   These authors also demonstrated that data for such plots could be obtained readily from capillary tube or pipeline viscometers as well as rotational viscometers.   Their method was based on the equation developed by Rabinowitsch[45] and Mooney[40] which is independent of fluid properties, provided thixotropic and rheopectic effects are absent.   This equation is

$$\left(\frac{-du}{dr}\right)_w = 3\,\frac{8q}{\pi D^3} + \frac{D\,\Delta p}{4L}\,\frac{d(8q/\pi D^3)}{d(D\,\Delta p/4L)} \tag{13}$$

Metzner and Reed[39] extended the work of Alves et al.[2] using the Rabinowitsch-Mooney expression as a starting point and developed what is probably the first general correlation for laminar flow of non-Newtonian fluids.   They rearranged Eq. (13), noting that the bulk velocity $V = 4q/\pi D^2$, to obtain

$$\left(\frac{-du}{dr}\right)_w = \frac{3}{4}\,\frac{8V}{D} + \frac{1}{4}\,\frac{8V}{D}\left[\frac{d[\ln\,(8V/D)]}{d[\ln\,(D\,\Delta p/4L)]}\right] \tag{14}$$

$$= \frac{3n'+1}{4n'}\left(\frac{8V}{D}\right)$$

where

$$\frac{1}{n'} = \frac{d[\ln\,(8V/D)]}{d[\ln\,(D\,\Delta p/4L)]} = \frac{d[\ln\,(8V/D)]}{d(\ln\,\tau_w)} \tag{15}$$

The quantity n', therefore, is the inverse slope of a plot of $\ln\,(8V/D)$ vs $\ln\,(D\,\Delta p/4L)$.   Direct integration of Eq. (15), noting that $\tau_w = D\,\Delta p/4L$, then gives

$$D\,\Delta p/4L = \tau_w = K'(8V/D)^{n'} \tag{16}$$

Substituting the value of $8V/D$ from Eq. (14) into Eq. (16) gives

$$\tau_w = K'[4n'/(3n'+1)]^{n'}(-du/dr)_w^{n'} = K''(-du/dr)^{n'} \tag{17}$$

which is then equivalent to the Ostwald equation for pseudoplastic fluids, in power-law form.   Metzner and Reed then state that the parameters K' and n' tend to remain constant for most fluids over fairly wide ranges of $8V/D$.   The constant K' is defined as the "consistency index," a higher numerical value indicating a "more viscous" fluid.   The other constant n' is called the "flow-behavior index," and its numerical value indicates the degree of departure from a Newtonian fluid.   If $n' = 1$, then the system is Newtonian, as may be seen by putting $n' = 1$ into Eq. (17), which then reduces to

$$\tau_w = K'(-du/dr) = \mu(-du/dr)$$

where $K' = \mu =$ Newtonian viscosity.   If $n' > 1$, the fluid is dilatant, and if $n' < 1$, the system is pseudoplastic, or a Bingham plastic, and if n' is 0.6 or less, the fluid is "rubbery."

Since it was stated above that in the laminar range the Fanning friction factor f is given by

$$f = \frac{16}{DV\rho/\mu} = \frac{\tau_w}{\rho V^2/2g_c} = \frac{D\,\Delta p/4L}{\rho V^2/2g_c} \tag{18}$$

for any fluid, substitution from Eq. (16) for τ_w gives

$$f = 16\gamma/D^{n'}V^{2-n'}\rho \tag{19}$$

where

$$\gamma = g_c K'8^{n'-1} \tag{20}$$

Metzner and Reed thus obtain a generalized Fanning friction factor and also a generalized Reynolds number:

$$N'_{Re} = \frac{D^{n'}V^{2-n'}\rho}{\gamma} \tag{21}$$

for fluids which follow a power law.

The significant practical effect, therefore, is that for laminar flow, the power-law fluid-friction factor is a unique function of the modified Reynolds number, and hence the design can be carried out in a similar way for these non-Newtonian fluids as for the Newtonians. Metzner and Reed[39] found that the bulk of data on laminar flow of non-time-dependent non-Newtonian fluids (which could be made to fit a power-law function) correlated extremely well on a plot of f vs Reynolds number where these quantities are defined by Eqs. (19) and (21). To use the method, K' and n' must be determined from a shear diagram or its equivalent plot of $8V/g_cD$ vs $D\,\Delta p/4L$, using data obtained by viscometer measurements. A modified Reynolds number is computed and f is evaluated, either from a graph of N'_{Re} vs f or from $f = 16/N'_{Re}$. Pressure loss for a given flow and pipe diameter is then calculated by Eq. (8) or (18). R. L. Bowen[9] has extended this method briefly and has prepared pipe flow curves for various pipe sizes to eliminate trial-and-error calculation when the pipe diameter is unknown. Although other workers[10,32-36] have developed alternative techniques, the method described above is probably the most straightforward one presently available to handle nonthixotropic or nonrheopectic fluids in laminar flow, though it is limited to fluids which obey the power law over the range of interest.

Metzner and Reed[39] by rearranging (18) and (19) and solving for the pressure loss obtained the following expressions:

$$\Delta p = 32\gamma L V^{n'}/g_c D^{n'+1} = (32\gamma L/g_c)(4q/\pi)(1/D^{3n'+1}) \tag{21a}$$

The above equations reduce to Poiseuille's law for Newtonian fluids when $n' = 1$. These authors then emphasize that for design purposes one must be doubly careful when dealing with highly non-Newtonian fluids to note the effect of changing the diameter on the pressure loss at a constant flow rate q. Whereas for Newtonian fluids ($n' = 1$) a small increase in pipe diameter is a very effective way to reduce the pressure drop, as n' becomes much less than unity it may be seen from Eq. (21) that the pressure drop varies not with the inverse fourth power of diameter D but with a much lower power. Therefore if n' is very low, a designer would be required to specify a much larger pipe diameter to achieve the same pressure-loss reduction as would be the case for a Newtonian. Data of Gregory[27] are cited to show that for a slurry with a particularly low value of n', the pressure drop was nearly independent of flow rate in the laminar region, thus substantiating this effect.

Though no attempt is made here to cover the subject of pumping problems inherent in the flow of muds and sludges, it will also be apparent from the discussion so far that the pump type will affect the rheological characteristics of non-Newtonian

materials. A centrifugal pump, for example, which causes the fluid to be exposed to high shear-rate effects will cause some fluids to behave differently when a positive displacement pump is used for the same service. This effect, of course, must not be overlooked by the design engineer.

It also should be noted here that in the case of a Bingham plastic fluid (see Fig. 3) which can be considered to follow the equation

$$\tau - \tau_y = \eta(du/dr) \tag{22}$$

an exact expression for the friction factor was developed by Perkins and Glick.[42] Their equation, which is limited to laminar flow, is

$$1/N_{Re} = f/16 - \tfrac{1}{6}(N_{He}{}^2/N_{Re}{}^2) + \tfrac{1}{3}f^3(N_{He}{}^4/N_{Re}{}^8) \tag{23}$$

where N_{Re} is the Reynolds number and N_{He} is the Hedstrom number, defined as

$$N_{He} = D^2\rho\tau_y/n^2 \tag{24}$$

TURBULENT FLOW OF NON-NEWTONIAN FLUIDS

For Newtonian fluids the end of the laminar-flow regime under ordinary circumstances is considered to extend up to a Reynolds number of about 2,100, above which transition to turbulent flow ensues. In the case of non-Newtonian fluids the onset of turbulence is not always easy to predict because of the rheological characteristics. Metzner and Reed[39] suggested that flow be considered laminar up to a numerical value of $f = 0.008$, computed according to Eq. (19), which, because of the unique relation to Reynolds number, corresponds to a value of $N'_{Re} = 2,000$, approximately, using Eq. (22). Later work by Dodge and Metzner[21] indicated that the greater the degree of pseudoplasticity (lower n'), the higher the critical Reynolds number and the lower the corresponding Fanning friction factor. In fact, these authors point out that, as the parameter n' becomes progressively lower and in the limit approaches zero, the distinction between laminar and turbulent flow ceases to exist. They reason that the laminar velocity profile across the pipe would not be parabolic but would instead be flat (as in plug flow,) flow instability (turbulence) supposedly caused by the velocity profile would be absent everywhere except possibly immediately adjacent to the wall, and any wall turbulence effect if present at all would not propagate. Bowen,[10] by rearranging Eq. (18) or noting that for a Newtonian fluid Eq. (16) becomes $(D\,\Delta p/4L)g_c = \mu(8V/D)$, obtains the following expression:

$$\mu = g_c(D\,\Delta p/4L)/(8V/D) \tag{25}$$

This value of μ is then substituted in the conventional expression for Reynolds number to give another generalized Reynolds number as follows:

$$\frac{DV\rho}{\mu} = \frac{DV\rho}{g_c(D\,\Delta p/4L)/(8V/D)} = N_{Re,G} \tag{26}$$

Bowen shows how this generalized Reynolds number may be reduced to Eq. (21) for a power-law fluid by substituting Eq. (16) into Eq. (26). He also presents an interesting analysis of the use of Eq. (26) to predict the onset of turbulence and demonstrates clearly why turbulence will occur at different values of Reynolds number in pipes of different diameter. He shows further that laminar flow persists for higher values of the term $8V/D$ in smaller pipes and that it is important to evaluate the Reynolds number by Eq. (26) at the maximum value of $8V/D$, which is obtained in turn from rheological data expressed on a plot of $D\,\Delta p/4L$ vs $8V/D$. For conservative engineering design he suggests that the value of $N_{Re,G} = 2,000$

be used to indicate the onset of turbulence but notes that laminar flow of non-Newtonians may occur at much higher values (e.g., 6,000 to 8,000 or higher).

It is evident that for the turbulent flow of non-Newtonian sludges and slurries the problem will be more complicated than for laminar flow. Early attempts to analyze the turbulent flow of clay and mud suspensions as well as digested sewage sludge (Bingham plastic-type behavior) were made by Babbitt and Caldwell.[5] They obtained a correlation using the viscosity of the suspending medium (water) to calculate the Reynolds number. Critical Reynolds numbers as high as 70,000 were

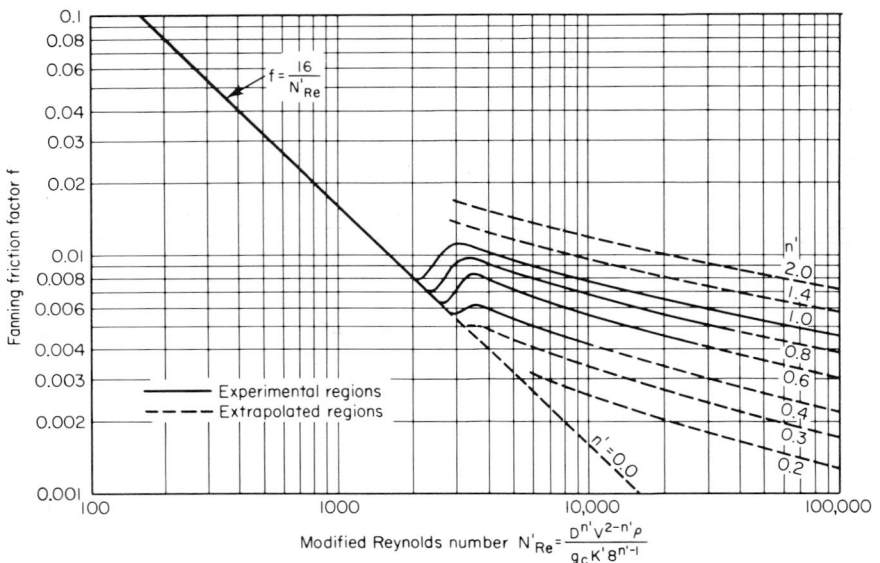

FIG. 4. Fanning friction factor design chart for Newtonian and non-Newtonian fluids. (*From D. W. Dodge and A. B. Metzner, A.I.Ch.E. J.,* vol. 5, no. 2, p. 198, June, 1959.)

obtained with the Reynolds number based on the water viscosity, so that each material as well as each pipe size would have a different critical velocity. Hedstrom[28] also developed a turbulent flow correlation using the plastic viscosity μ_p, defined by

$$\tau = \tau_y + \eta(du/dr) \tag{27}$$

in the Reynolds-number–friction-factor correlation. Although they have not received especially wide attention recently, both correlations gave reasonable results and are recommended for turbulent flow of Bingham plastics.

Dodge and Metzner[21] in 1959 published a general method for turbulent flow, extending the earlier work of Metzner and Reed.[39] The reader is referred to their paper for details of their method, which is based on the power-law fluid and their generalized friction factor and Reynolds number as given by Eqs. (19) and (21). The practical results are embodied in the logarithmic plot of f vs N'_{Re} shown in Fig. 4. Provided the sludge or slurry can be considered to show power-law behavior and is not thixotropic or rheopectic, this chart allows the pipeline to be designed in a rational and fairly straightforward way similar to that for ordinary Newtonian fluids. One calculates the modified Reynolds number N'_{Re} from the

rheological data (that is, K and n') and enters Fig. 4 to obtain f, which may then be used, for example, in Eq. (8) to calculate the pressure drop for a given diameter and flow rate. Limitations of this method are discussed by Bowen,[11,12] who presented a new empirical correlation which is said to be applicable to the turbulent flow of all non-Newtonian fluids for which there are published data. This method is based on the use of the Blasius friction factor

$$f = 0.079/N_{\mathrm{Re}}^{0.25} \tag{28}$$

Substituting the value of f from Eq. (18) Bowen[12] obtains the following expressions:

$$g_c D \, \Delta p/2\rho L V^2 = 0.079(D V \rho/\mu)^{-0.25} \tag{29}$$

which can be rearranged to

$$D^{1.25} \, \Delta p/L = k V^{1.75} \tag{30}$$

Equation (30) is of the form

$$D^{(1+b)} \, \Delta p/L = k V^{(2-b)} = k V^c \tag{31}$$

where b is defined by $f = a/N_{\mathrm{Re}}{}^b$ and $b + c = 2.0$.

The parameters b and c can be estimated from a plot of $D \, \Delta p/4L$ vs $8V/D$, and then Eq. (31) in graphical form may be used as the basis of correlation with excellent results. Bowen[12] also notes for pseudoplastic fluids that the Metzner Reynolds number defined by Eq. (21) can also be put into a Blasius-type equation which can be rearranged to obtain

$$D^{(1+bn')} \, \Delta p/L = k V^{(2-2b+bn')} = k V^c \tag{32}$$

which reduces to Eq. (31) when $n' = 1$ (Newtonian fluid). This method may have certain practical advantages over the Dodge and Metzner procedure, but it is best adapted for use with specific systems, and hence the reader is referred to the original paper[12] for details of application.

An additional consideration which must not be overlooked in any turbulent-flow system is the effect of roughness. Pipe-wall roughness is not a factor in laminar flow, but it is very important in turbulent flow of all fluids because of the much higher shear stresses encountered at the pipe wall. This effect has been shown to be equivalent to a large increase in the apparent viscosity of the fluid. It is to be noted in particular that the correlations of Dodge and Metzner[21] given in Fig. 4 were based on smooth pipe and that for rough pipes the friction factors and hence the pressure drops will be larger. Bowen[13] discusses this problem briefly and suggests a method of allowing for roughness in design, based on the findings of Babbitt and Caldwell.[18]

There is also the problem of making allowances for entrance and exit losses and for fittings. Since there is at present little information available on such effects for non-Newtonian fluids, one must often resort to the expedient of using corresponding Newtonian flow data. Dodge and Metzner[21] report a loss of 0.44 velocity head for power-law fluids in passing from a 1.5-in.-diameter pipe to a 0.5-in. pipe. Dilatant fluids (Fig. 2d) which exhibit an increase in apparent viscosity with increased shear (e.g., in passing from laminar to fully developed turbulent flow) would therefore be expected to show larger pressure drops when they are subject to large shear forces such as in passing through constrictions. This effect, observed by McMillen[36] and confirmed by Dodge and Metzner,[21] was quite pronounced for highly plastic materials, which exhibited entrance loss effects 40 pipe diameters downstream and entrance pressure drops about seven times those anticipated for Newtonian fluids.

Flow of non-Newtonian fluids in annular spaces has been treated by Fredrickson and Bird, and the reader is referred to the original papers and their discussion[24,25,58]

for details, which are based on the same concepts outlined here.   The reader is also referred to Metzner's review[38] in the "Handbook of Fluid Dynamics" for additional correlations.

SETTLING PHENOMENA IN SLURRIES

Up to this point pipeline design based on the non-Newtonian properties of sludges and slurries has been dealt with, and the problem of settling has been bypassed.   This second broad aspect of the problem is concerned with the physical and chemical stability of the sludge or slurry.   The previous design methods assume (1) that time effects are absent (no thixotropy or rheopexy) and (2) that solid particles do not settle or dissolve or change their physical or chemical nature. These effects, of course, can be superimposed on the non-Newtonian characteristics. Aside from the obvious fact that extensive settling of solids usually must be avoided for purely mechanical reasons, certain other effects may also have a bearing.   Thus, raw sewage sludge changes its rheological character after it is digested, so that different design criteria must be imposed.   Again, in nuclear engineering systems in which fissionable fuels are transported in slurry form, settling cannot be tolerated for safety reasons.

A good discussion of the settling problems inherent in the design of piping systems for sludges and slurries is given by Bowen,[13] who recommends the cautious use of correlations developed by Spells[48] and by Cairns, Lawther, and Turner.[16]   Bowen emphasizes the importance of not relying on the blind use of a Reynolds number, coupled with the assumption that a Reynolds number in the range 2,000 to 3,000 assures turbulence, which in turn is assumed to prevent settling.   Especially with slurries and sludges, a Reynolds number in the above range does not guarantee that the flow will be turbulent.

Recognizing that particle size, particle density, and fluid viscosity are among the most important factors which affect the settling of solids from a suspension, Spells[48] using dimensional analysis obtained a correlation by balancing the Froude and Reynolds numbers as follows for particles between 50 and 500 microns:

$$\rho_L V^2/(\rho_P - \rho_L)g_L D_P = k_s(DV\rho_m/\mu_L)^{0.775} \tag{33}$$

where ρ_L = liquid density, lb/ft^3

$\quad\quad \rho_P$ = particle density, inches

$\quad\quad \rho_m$ = mean density of the slurry lb/ft^3

$\quad\quad \rho_m = c\rho_P + (1 - c)\rho_L$, where c = volume fraction solids

$\quad\quad D_P$ = mean particle diameter, ft

The constant, k_s, has two values, viz., (1) k_s = 0.0251 for "minimum velocity" and (2) k_s = 0.0741 for "standard velocity."

The "minimum velocity" is a lower critical fluid velocity below which particles settle from the liquid in an amount sufficient to block the flow channel.   The "standard velocity" is an upper critical velocity above which the flow is homogeneous, with the particles presumably dispersed uniformly.   The range between these two velocities defines a heterogeneous flow region in which settling occurs at least to the extent that there is a solids concentration gradient from top to bottom in a horizontal pipe.   In this heterogeneous region the pressure drop is higher than would be expected for ordinary fluids.   With Eq. (33) only approximate agreement with experimental results was obtained, and as emphasized by Spells,[48] the relation was valid only for particles ranging from 50 to 500 microns in size.   Bowen[13] rearranged Eq. (33) to the form

$$q' = 352Ak_s^{0.816}D^{2.633} \tag{34}$$

where q' is the flow rate in gallons per minute and A is a characteristic system constant to facilitate calculating the "minimum" and "standard" fluid velocities.

Another alternative empirical form of Spells' relation similar to Eq. (34) but explicit for velocity was given by Lovenstein[34] as follows:

$$V = (k_L \psi D_P)^{0.816}(D\rho_m/\mu_L)^{0.633} \tag{35}$$

where

$$\psi = \frac{(\rho_P - \rho_L)}{\rho_L}$$

and $k = 232$ for "minimum" flow velocity or 685 for "standard" velocity.

Lovenstein also developed an expression relating pipe diameter to flow rate and a nomograph to simplify the calculations further.

For conservative design, one would normally try to design for the "standard" fluid velocity, but unfortunately, as pointed out by Johnstone and Thring,[30] for many of the coarser industrial slurries, such as sand or coal in water, the "standard" velocity is much higher than the economic pumping velocities. Such slurries, then, would of necessity be transported in the heterogeneous range, provided choking phenomena were absent. In later work, Cairns et al.,[16] for dilute slurries containing solids of 15 to 30 microns, densities of 168 to 1,200 lb/ft³, and concentrations from 5 to 15 weight per cent, developed two empirical equations as follows:

$$V_m = 1.9D^{0.2}[(\rho_P - \rho_L)/\rho_L]^{0.3} \tag{36}$$

$$V_f^{0.85} = 1.6D^{0.2}[(\rho_P - \rho_L)/\rho_L]^{0.3} \tag{37}$$

V_m is defined as a critical velocity at which a moving bed of solids was observed to form in a horizontal pipe, while V_f is a slightly lower velocity (2 to 15 per cent) at which a stationary or fixed bed formed, determined by visual inspection. For any of the solids studied, these critical velocities were independent of the concentration over the 5 to 15 weight per cent range.

Cairns et al.[16] noted that data used by Spells in deriving Eq. (33) did not fit Eq. (36), but this is not too surprising, since Spells's particle sizes were larger. Cairns et al. also suggested that an improved form of Eq. (33) is

$$V^2/g_L D_P = 9.8C^{0.3}(DV\rho_L/\mu_L)^{0.3}(\rho_P - \rho_L/\rho_L)^{0.6} \tag{38}$$

where C is the concentration of solids in volume per cent for particles of 240 to 380 microns and 162 lb/ft³.

As further discussed by Johnstone and Thring[30] an empirical equation was developed by Durand[46] in connection with comprehensive experiments on the flow of slurries and modified by Worster[60] to corrrelate the head loss for slurries with flow rate. It is recommended, however, that the results obtained from this relation be checked against those obtained using the techniques outlined in the previous section on non-Newtonian flow. The Durand-Worster equation is

$$\nabla = 121C(g_L D\sigma/V^2)^{1.5}(V_s/g_L \sigma D_P)^{0.75} \tag{39}$$

where $\nabla = (\Delta h' - \Delta h)/\Delta h$, $\Delta h'$ being frictional head loss per unit pipe length for slurry and Δh, the corresponding head loss for the pure liquid flowing at the same velocity

C = volume fraction solids in the slurry
D_P = mean particle diameter, ft
V = average fluid velocity
D = pipe diameter
V_s = settling velocity of solids in slurry
σ = "effective" density = $\rho_P - \rho_L$
g_L = local acceleration of gravity, 32.2 ft/sec²

An older correlation by Wilson[59] for the head loss experienced in the flow of noncolloidal inert solids is based on the simple addition of two terms as follows:

$$\Delta H/L = 4f(V^2/2g_L D) + K''C'V_s/V \qquad (40)$$

The first term on the right side was assumed to be the energy, dissipated in turbulence, to maintain the flow of fluid, and the second term was added to account for the "energy to maintain the solids in suspension," the sum of the two effects representing the "total energy gradient." The coefficient K'' in the second term was defined as an "efficiency factor" which was given as a function of $C'(V_s/V)$. The constant C' is the weight of solids per unit weight of mixture, and V_s is the "settling velocity of the larger particles." This expression, by simply adding two terms, neglects any interaction effects and, of course does not consider any non-Newtonian effects. Also, it applies only to turbulent flow when the particles remain in suspension. It must, therefore, be regarded as another empirical form.

An interesting, though controversial,[33,43] theory has been advanced by Vanoni,[51] who studied sediment and velocity distributions in a laboratory flume (open-channel flow) for various flow rates and channel shapes, with different amounts and sizes of suspended loads. His basic observations were:

1. The measured sediment distributions have the same form as theoretical distributions, but they do not agree quantitatively. He concludes that the transfer coefficient for sediment is not the same as that for momentum transport, noting that the measured relative concentrations were larger than those predicted by momentum transport relationships, the differences being greatest for fine sediment.

2. Suspended sediment caused sediment-laden water to flow faster than clear water. He attributed this effect to a reduction of the turbulent momentum transfer rate and hence resistance to flow by the suspended sediment.

Vanoni in a recent paper[52] reported on an extension of his original experiments and again justified his results on the basis of a damping effect which the suspended solids are presumed to exert on stream turbulence, thus reducing the friction factor (i.e., lowering the resistance of flow). He cites conflicting evidence by field observers on flowrate measurements of streams (open channels) and other laboratory work[29] where friction factors with sediment present were observed in some cases to decrease and in others to increase over those in comparable clear water and ascribes these anomalies to two separate effects, viz. (1) the aforementioned turbulence damping effect, which tends to lower the friction factor, and (2) a saltation or solid settling effect, in which the deposited solids form dunes or bars, thus changing the shape and roughness of the conduit and hence tending to increase the friction factor.

Though the turbulence damping effect is open to question and has been vigorously challenged[33,43] and, further, no attempt made to investigate non-Newtonian effects, it is of interest, nevertheless, to note that faster flow rates have been observed with sediment-laden streams and that solids suspended in a medium might influence the effective roughness of a conduit, which would have an important effect in turbulent flow.

In connection with pumping muds and sludges, the effect of solids in suspension on the characteristics of centrifugal pumps should also be considered. Fairbank,[23] who reviewed this situation, came to the following conclusions:

1. At a given capacity, the head developed by the pump is less than for water.
2. The drop in head-capacity characteristics at constant speed is a function of the particle size and concentration.
3. The settling velocity of the solids in the suspension medium is the most important property in predicting the effect on pump performance.

4. Very fine colloidal particles produce different effects from those found in other "ordinary" suspensions.

5. Power input to the centrifugal pump varies directly with the specific gravity of the suspension.

6. The pump capacity at maximum efficiency can be substantially constant for various particle sizes and concentration ranges.

7. The ordinary so-called "affinity" laws for pumps tend to be valid over small ranges of speed.

Additional discussions and correlations relating to the flow of slurries and sludge are also given in the recently published "Handbook of Fluid Dynamics"[49] by Tek,[50] Andersen,[3] and Metzner.[38] An older but excellent review of the properties of muds and slurries and the transport of particles is given by Dalla Valle.[20] The reader is referred to these sources for further details. Zenz and Othmer[61] have also discussed the problem, but the major emphasis of this text, of course, is on gas-solid systems.

As was pointed out in the beginning of this discussion, data and methods for handling the flow of sludges and slurries are to be found in the literature of a number of apparently different subject headings, including two-phase flow, sedimentation, silting, non-Newtonian flow, fluidization and transport, particle technology, as well as that dealing with specific materials such as polymers and plastics, coal transportation via pipelines, paints and pigments, paper and pulp handling, and others. Furthermore, different types of engineers and scientists have contributed, so that technical articles are further scattered among several engineering and scientific disciplines, particularly in the fields of chemical, civil, and mechanical engineering; rheology; polymer science; and others. No attempt was made in this section to review the literature exhaustively, so the designer with a slurry-handling problem may wish to pursue his investigation beyond the reference cited herewith. It must be recognized that this field is one which has seen considerable activity in the past decade and that considerably more work remains before the subject can be considered to be satisfactorily explored.

NOMENCLATURE AND SYMBOLS

English

A = characteristic constant in Eqs. (5) and (30)
a = a constant
b = an exponent in Eqs. (31) and (32)
C = volume per cent solids in a slurry
C' = weight per cent solids in a slurry
c = volume fraction solids in a slurry;
   also an exponent in Eqs. (21) and (32)
D = pipe diameter, ft
D_P = particle diameter, ft
d = differential operator
f = Fanning friction factor, dimensionless
g_c = dimensional constant = 32.2 lbf/lbm × ft/sec²
g_L = local acceleration of gravity, usually 32.2 ft/sec²
h = fluid "head" as "feet head" or ft lbf/lbm
h' = fluid "head" for slurry
K = constant in power law, Eq. (6)
K' = consistency index, defined by Eq. (16)
K'' = constant in Eq. (17)
K''' = constant in Eq. (40)

k = constant in Eq. (30)
k_L = constant in Eq. (35)
k_s = constant in Eq. (33)
L = pipe length, ft
m = exponent in Eq. (5)
n = exponent in Eq. (6)
n' = exponent defined by Eq. (15), known as flow-behavior index
p = fluid pressure, lbf/ft^2;
$\quad\Delta p$ = pressure difference, lb/ft^2
q = volumetric flow rate, cfs
q' = volumetric flow rate, gpm
r = radial distance from center of pipe, ft
t = time, sec
u = fluid velocity in x direction, fps
u_{max} = maximum fluid velocity in x direction, fps
V = average fluid velocity in main flow direction, fps
V_f = lower critical fluid velocity defined by Eq. (37), fps
V_M = upper critical fluid velocity defined by Eq. (36), fps
V_s = average settling velocity of solids in a suspension, fps
x = a directional distance, ft
y = a directional distance at right angles to x, ft
du/dy = instantaneous velocity gradient of u in y direction, also called rate of shear
$(du/dr)_w$ = velocity gradient in a radial direction r measured at pipe wall
d = symbol for differential operator
$\ln$ = symbol for natural logarithm
N_{Re} = Reynolds number = $DV\rho/\mu$, dimensionless
N'_{Re} = modified Reynolds number = $D^{n'}V^{2-n'}\rho/\gamma$
$N_{Re,G}$ = generalized Reynolds number = $DV\rho$
$$\frac{g_c D\,\Delta p/4L/8V/D}{}$$
N_{He} = Hedstrom number = $D^2\rho\tau y/\eta^2$
∇ = $(\Delta h' - \Delta h)/\Delta h$, defined by Eq. (39)

Greek

γ = parameter in Eq. (19) = $g_c K' 8^{n'-1}$; also, deformation or strain in Eq. (4)
Δ = symbol designating a difference, e.g., Δp
η = coefficient of rigidity for Bingham plastic, defined by Eq. (12), lbm/(ft sec)
μ = absolute viscosity, lbf sec/ft^2
μ_M = absolute viscosity, lbm/(ft sec)
μ_a = apparent viscosity, lbm/(ft sec)
μ_0 = limiting value of viscosity at zero shear rate
μ_∞ = limiting viscosity at infinite shear rate
π = 3.1416
ρ = density, lbm/ft^3
ρ_L = liquid density, lbm/ft^3
ρ_M = volumetric average density of slurry, lbm/ft^3. $\rho_M = c\rho_P + (1 - c)\rho_L$
ρ_P = average particle density, lbm/ft^3
σ = effective density = $\rho_P - \rho_L$, lbm/ft^3
τ = shear stress, lbf/ft^2
τ_w = shear stress at wall or solid boundary, lbf/ft^2
τ_y = yield stress for Bingham plastic, lbf/ft^2
ϕ = symbol designating a general mathematical function as $y = \phi(x)$
ψ = $(\rho_P - \rho_L)/\rho_L$ used in Eq. (35)

REFERENCES

1. Alves, G. E., "Non-Newtonian Flow" in "Fluid and Particle Mechanics," C. E. Lapple (ed.), University of Delaware, Newark, Del., 1951.
2. Alves, G. E., D. F. Boucher, and R. L. Pigford, "Pipe Line Design for Non-Newtonians," *Chem. Eng. Progr.*, vol. 48, p. 385, 1952.
3. Andersen, A. G., "Sedimentation" in "Handbook of Fluid Dynamics," V. L. Streeter (ed.), sec. 18, pp. 1–34, McGraw-Hill Book Company, New York, 1961.
4. Babbitt, H. E., and D. H. Caldwell, *Univ. Illinois Bull.*, vol. 37, no. 12, Nov. 14, 1939.
5. Babbitt, H. E., and D. H. Caldwell, *Univ. Illinois Bull.*, vol. 38, no. 13, Nov. 19, 1940.
6. Behn, V. C., *J. Water Pollution Control Federation*, vol. 32, no. 7, p. 728, July, 1960.
7. Behn, V. C., *Proc. ASCE, San. Eng. Div.*, vol. 86, no. SA6, part 1, p. 59, November, 1960, and subsequent discussions, vol. 87, no. SA3, p. 89, May, 1961; vol. 87, no. SA5, p. 79, September, 1961.
8. Bingham, E. C., "Fluidity and Plasticity," McGraw-Hill Book Company, New York, 1922.
9. Bowen, R. L., *Chem. Eng.*, vol. 68, no. 12, p. 243, June 12, 1961.
10. Bowen, R. L., *Chem. Eng.*, vol. 68, no. 13, p. 127, June 26, 1961.
11. Bowen, R. L., *Chem. Eng.*, vol. 68, no. 14, p. 147, July 10, 1961.
12. Bowen, R. L., *Chem. Eng.*, vol. 68, no. 15, p. 143, July 24, 1961.
13. Bowen, R. L., *Chem. Eng.*, vol. 68, no. 16, p. 129, Sug. 7, 1961.
14. Bowen, R. L., *Chem. Eng.*, vol. 68, no. 17, p. 119, Aug. 21, 1961.
15. Bowen, R. L., *Chem. Eng.*, vol. 68, no. 18, p. 131, Sept. 4, 1961.
16. Cairns, R. C., K. R. Lawther, and K. S. Turner, *Brit. Chem. Eng.*, vol. 5, p. 849, 1960.
17. Caldwell, D. H., and H. E. Babbitt, *Ind. Eng. Chem.*, vol. 33, p. 249, 1941.
18. Caldwell, D. H., and H. E. Babbitt, *Trans. A.I.Ch.E.*, vol. 25, p. 237, 1941.
19. Christiansen, E. B., N. W. Ryan, and W. E. Stevens, *A.I.Ch.E. J.*, vol. 1, no. 4, p. 544, 1955.
20. DallaValle, J. M., "Micromeritics," 2d ed., chaps. 17 and 18, pp. 343–389, Pitman Publishing Corporation, New York, 1948.
21. Dodge, D. W., and A. B. Metzner, "Turbulent Flow of Non-Newtonian Systems," *A.I.Ch.E. J.*, vol. 5, no. 2, p. 189, June, 1959.
22. Durand, R., and E. Condolios, IIme journée de l'hydraulique, *Compt. Rend.*, June, 1952, p. 29, Société Hydrotechnique de France, Paris.
23. Fairbank, Jr., Leigh C., *Trans. ASCE*, vol. 107, pp. 1564, 1572, 1941.
24. Fredrickson, A. G., and R. B. Bird, *Ind. Eng. Chem.*, vol. 50, p. 347, 1958.
25. Fredrickson, A. G., and R. B. Bird, *Ind. Eng. Chem.*, vol. 50, p. 1599, 1958.
26. Green, H., "Industrial Rheology and Rheological Structures," John Wiley & Sons, Inc., New York, 1949.
27. Gregory, W. B., *Mech. Eng.*, vol. 49, p. 609, 1927.
28. Hedstrom, B. O. A., *Ind. Eng. Chem.*, vol. 44, p. 651, 1957.
29. Ismail, H. M., *Trans. ASCE*, vol. 117, p. 409, 1952.
30. Johnstone, R. E., and M. W. Thring, "Pilot Plants, Models and Scale-up Methods in Chemical Engineering," pp. 118–123, McGraw-Hill Book Company, New York, 1957.
31. Kern, D. Q., "Process Heat Transfer," McGraw-Hill Book Company, New York, 1950.
32. Krieger, I. M., and S. H. Maron, *J. Appl. Phys.*, vol. 23, p. 147, 1952.
33. Laursen, E. M., et al., "Discussion of Paper by Vanoni (52)," *Trans. ASCE*, vol. 125, pp. 1167, 1168, 1170, 1960.
34. Lovenstein, J. G., *Chem. Eng.*, vol. 66, no. 1, p. 133, 1959.
35. McAdams, W. H., "Heat Transmission," 3d ed., McGraw-Hill Book Company, New York, 1954.
36. McMillan, E. E., *Chem. Eng. Progr.* vol. 44, p. 537, 1948.
37. Metzner, A. B., *Ind. Eng. Chem.*, vol. 49, p. 1429, 1957.
38. Metzner, A. B., "Flow of Non-Newtonian Fluids," in "Handbook of Fluid Dynamics," sec. 7, pp. 2–29, V. L. Streeter (ed.), McGraw-Hill Book Company, New York, 1961.
39. Metzner, A. B., and J. C. Reed, "Flow of Non-Newtonian Fluids," *A.I.Ch.E. J.*, vol. 1, no. 4, p. 434, December, 1955.
40. Mooney, M. J., *Rheol.*, vol. 2, p. 210, 1931.

41. Ostwald, W., and R. Auerbach, *Kolloid. Z.*, vol. 38, p. 261, 1926.
42. Perkins, A., and J. J. Blick, B. Ch.E. Thesis, University of Delaware, Newark, Del., June, 1954.
43. Powell, R. W., et al., "Discussion of Paper by Vanoni (51)," *Trans. ASCE*, vol. 111, pp. 103, 106, 109, 111, 114, 116, 1946.
44. Purdue, A. G., *Trans. ASCE*, vol. 125, p. 644, 1960.
45. Rabinowitsch, B., *Z. Physik. Chem.*, vol. 145A, p. 1, 1929.
46. Reiner, M., "Deformation and Flow—An Elementary Introduction to Theoretical Rheology," H. K., Lewis & Co., London, 1949.
47. Shaver, R. G., and E. W. Merrill, *A.I.Ch.E. J.*, vol. 5, no. 2, p. 181, 1959.
48. Spells, K. E., *Trans. Inst. Chem. Engrs. (London)*, vol. 33, no. 2, pp. 79–82, 1955.
49. Streeter, V. L. (ed.), "Handbook of Fluid Dynamics," McGraw-Hill Book Company, New York, 1961.
50. Tek, M. R., "Two Phase Flow" in "Handbook of Fluid Dynamics," sec. 17, pp. 30–35, Victor L. Streeter (ed.), McGraw-Hill Book Company, New York, 1961.
51. Vanoni, Vito A., *Trans. ASCE*, vol. 111, p. 67, 1946.
52. Vanoni, Vito A., *Trans. ASCE*, vol. 125, p. 1140, 1960.
53. Ward, H. C., Ph. D. Thesis, Georgia Institute of Technology, Atlanta, Ga., 1952.
54. Weltman, R. N., *NACA Tech. Note* 3397, February, 1955.
55. Weltman, R. N., *Ind. Eng. Chem.*, vol. 48, p. 386, 1956.
56. Weltman, R. N., "Rheology of Pastes and Paints" in F. R. Eirich, "Rheology," vol. III, pp. 236–240, Academic Press, Inc., New York, 1960.
57. Weltman, R. N., and T. A. Keller, *NACA, Tech. Note* 3889, Washington, 1957.
58. Wilcox, W. R., *Ind. Eng. Chem.*, vol. 50, p. 1600, 1958.
59. Wilson, W. E., *Trans. ASCE*, vol. 107, p. 1576, 1942.
60. Worster, R. C., *Proc. Colloq. Hydraulic Transport Coal*, National Coal Board, London, Nov. 5–6, 1952.
61. Zenz, F. A., and D. F. Othmer, "Fluidization and Fluid-Particle Systems," particularly, chap. 10, pp. 313–350, Reinhold Publishing Corporation, New York, 1960.

20

STEAM POWER-PLANT PIPING

Samuel Weiner*

The merit of a power-plant piping layout, with due respect to economic considerations, should be judged on three basic points: (1) its mechanical design or ability to function properly (and efficiently) with respect to the mechanical equipment which it serves; (2) its convenience from an operating standpoint; (3) its appearance as a coordinated and symmetrical part of the entire plant. Although the relative importance of these basic points obviously falls in the order named, each has an important bearing on the acceptability of any layout. The refinement with which each is carried out should be based on economic considerations where extra cost is required to obtain the preferred result. In the case of a temporary job, it is prudent to eliminate all unnecessary refinements in design and make the installation in the least expensive way that will serve the purpose, but in a power plant whose expected life is 20 years or greater, efficient, more convenient, and better appearing layout will be justified. While the monetary value of such features of (1) as reduced pressure drop, superior insulation, etc. can be evaluated readily and made a basis of economic consideration, the intangible advantages of (2) and (3) in greater operating convenience and improved appearance should not be ignored. The three basic points named are discussed at length in succeeding sections.

GENERAL FEATURES OF MECHANICAL DESIGN

Basic Principles. The general principles of mechanical design, such as strength of parts, flow of fluids, heat transmission, and thermal expansion are discussed in other chapters of this handbook. The present section deals with the specific

* Gibbs and Hill, Inc., New York, N.Y.

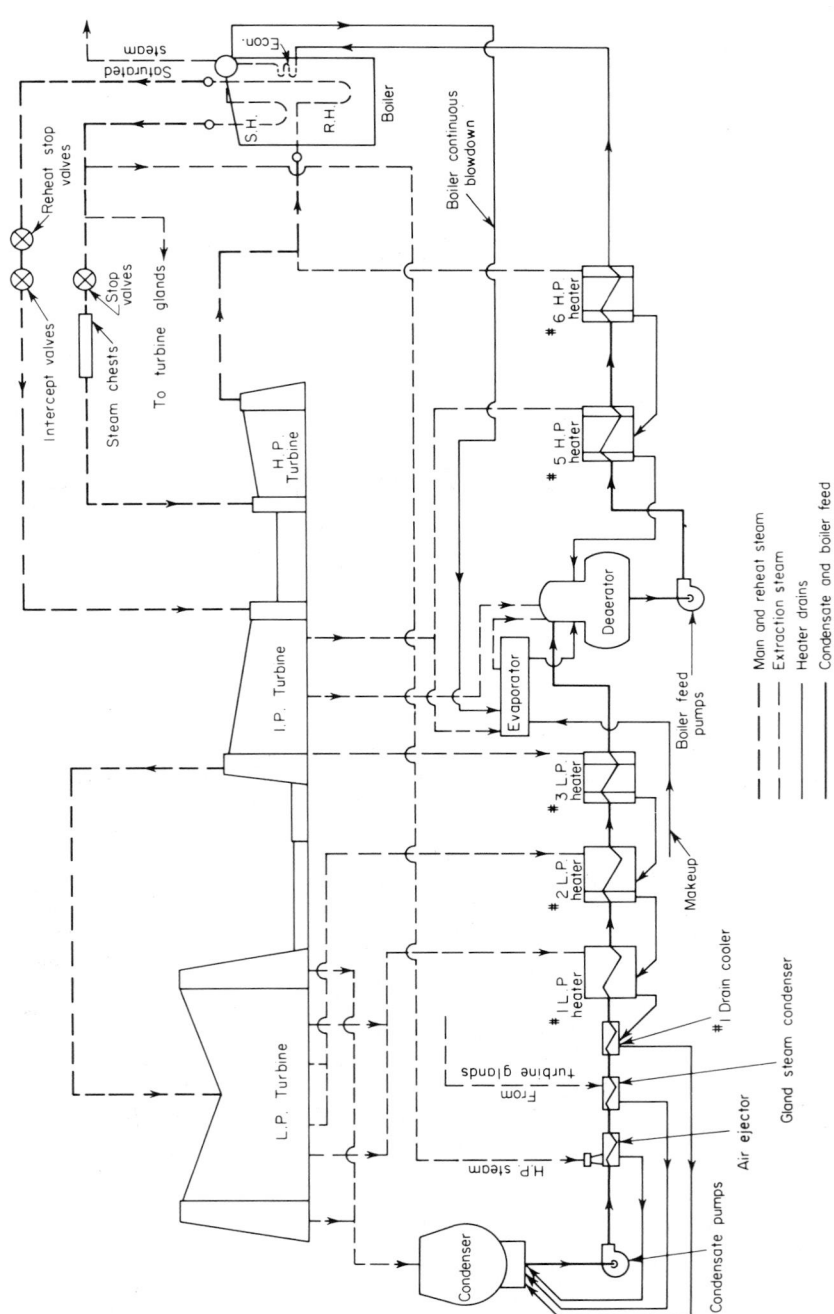

Fig. 1. Flow diagram of 125-mw reheat unit.

application of such principles to power-plant design and with the problem of obtaining layouts which will function properly in relation to the mechanical equipment they serve.   It is necessary to provide the proper circuits for fluid flow as well as the necessary valves, bypasses, and drains to assure satisfactory operation.   In modern power plants the problem of securing satisfactory piping layouts is far more difficult than in the simpler plants of past years.   The adoption of higher steam pressures and temperatures also adds to the necessity for great refinement in design.

A flow diagram of steam, condensate, and feedwater piping for a 125-mw, 1,800-lb, 1000 F unit is shown in Fig. 1.   In this diagram, the boiler, turbine, condenser, and auxiliaries are represented by shapes which indicate at a glance the function of each piece of equipment.   The involved nature of many jobs makes imperative the use of a line diagram as a guide in laying out the piping, for explanation to the erection crew and for the intelligent operation of the system.

When the basic principles of interconnecting the equipment have been portrayed in line diagrams and duly approved, the next step is the preparation of assembly drawings in plan and elevation, showing structures, outlines of equipment and piping, all laid out to scale.   In the preparation of such drawings due consideration

Table I. Reasonable Design Velocities for Flow of Fluids in Pipes

Fluid	Pressure pounds per square inch gage	Use	Reasonable velocity, feet per minute
Water..............	25 to 40	City water	120 to 300
Water	50 to 150	General service	300 to 600
Water	150 up	Boiler feed	600 to 1,200
Saturated steam........	0 to 15	Heating.............	4,000 to 6,000
Saturated steam........	50 up	Miscellaneous	6,000 to 10,000
Superheated steam	200 up	Large turbine and boiler leads........	10,000 to 20,000

should be given to the principles of design discussed in following paragraphs. In addition to the assembly drawings, it is often advisable to prepare separate layouts for the more important lines drawn to scale and with piping located by dimension to the building structure, especially in cases where the piping is considerably congested.   For work which requires shop fabrication, complete and accurate detail drawings are essential.   Great care should be taken in checking such details, since errors will involve expense and delay.

Determining Pipe Sizes.   Before proceeding beyond a preliminary drafting layout of piping systems, it is necessary to determine pipe sizes which allow reasonable velocities and friction losses.   The maximum allowable velocity of the fluid in a pipeline is that which corresponds to the permissible pressure drop from the point of supply to the point of consumption or is that which does not result in excessive pipeline erosion.   From the standpoint of investment cost, it is desirable to keep the velocity as high as possible without exceeding the maximum allowable velocity or causing excessive operating losses as a result of unduly great pressure drop in pipes.

In many cases a pipeline or branch is short and pressure drop due to friction is not of great significance.   The values of velocity listed in Table 1 are reasonable for use in such cases.   The lower velocities should be used for small pipes, and the upper limits for large ones.   These values represent good average practice and may be used as a guide in many cases where actual pressure drops are not computed.

Erosive action on valve seats and similar exposed parts also affects permissible velocity.   This action is much more pronounced in the case of wet steam than with superheated, and velocities should be correspondingly lower when there is much moisture in steam.

In ordinary power-plant work (as distinguished from steam-transmission lines, central heat-distributing systems, etc.) where the runs are reasonably short and the valves and fittings are designed to offer slight resistance, it is customary to use velocities of 1,000 to 1,200 fpm per inch of inside pipe diameter. The higher value applies particularly to diameters over 12 in.

Velocities in steam lines to reciprocating engines, to reciprocating steam pumps without ample receivers, and in very long steam lines should not exceed 75 to 80 per cent of those given above. The velocity criteria noted above seem to apply about equally well to both high- and low-pressure steam, since the reduction in friction at low pressures due to the reduced density about offsets the condition that less initial pressure is available to furnish the pressure drop.

Excessive velocities in pipes which carry water or other liquids are objectionable because of water hammer, which often accompanies a sudden stoppage or abrupt change in velocity; the water velocities given for various services usually will be found satisfactory.

High velocities are sometimes used in steam lines where excess pressure exists to absorb the higher pressure drop. The high velocity is not in itself objectionable. There is no appreciable erosion of the pipe walls and no undesirable effects except that high velocities are accompanied by considerable noise, which would be objectionable in heating systems in office buildings and dwellings but may not be in power plants.

In the selection of pipe sizes for exhaust lines from auxiliary turbines and similar services where high velocities may be used to advantage, consideration should be given to the limiting velocities which can be obtained. This limiting or "acoustic" velocity has a definite value for each combination of steam density and pressure.

It is often cheaper to use a larger pipe than to use an odd size. On the other hand, where space is at a premium, it may be necessary to use a smaller size than would otherwise be good practice. The calculation of velocity or pressure drop is a valuable check, but in the last analysis judgement is the deciding factor, and the blind use of generalized criteria is impracticable.

Loss Due to Steam Leaks. Modern power plants with almost all welded joints are subject to little leakage. The hazard of high-pressure, high-temperature steam and high-pressure water leaks, together with the cost of chemicals required in most make-up systems, makes early repairs mandatory.

Existing low-pressure plants using flanged or screwed joints with considerable leakage and where maintenance is deferred are wasting money at a rate which depends on the size of the opening, steam pressure, and unit fuel cost.

The value of steam which can be lost through a comparatively small leak becomes appreciable when considered over a period of time. The amounts of steam which will escape through various size orifices at different pressures can be computed by an appropriate flow formula. Table 2, which was computed by Grashof's formula, gives the pounds of steam lost per month and the corresponding money value of the fuel and water wasted.

Safety Codes. Extensive safety requirements for power-piping systems are contained in Section 1 of the American Standard Code for Pressure Piping ASA B31.1, which covers the design, manufacture, test, and installation of power-piping systems for steam-generating plants, central-heating plants, and industrial plants. In designing the component parts of piping systems within the jurisdiction of this code, reference should be made to its provisions as representing a standard for minimum safe requirements, but it is not intended to indicate necessarily the best practice known to the art. Requirements of the Code for Pressure Piping are not compulsory in any state until they have been adopted as law by that state. They are

Table 2. Loss Due to Steam Leaks

(The tables below are based on the use of Grashof's formula: Fuel costs of 17.3 cents per million Btu and an energy input of 1500 Btu/lb are assumed.)

Size of orifice, inches	Pounds steam wasted per month	Cost of coal per month	Cost of water wasted at 0.10 per 1,000 gallons	Total cost per month	Total cost per year	Capitalized value at 16 per cent on savings
			250-lb. gage			
½	1,780,000	$477.00	$21.36	$498.36	$5,980.32	$37,377.00
⅜	1,001,000	268.00	12.01	280.01	3,360.12	21,000.00
¼	445,000	119.00	5.34	124.34	1,492.08	9,325.00
⅛	111,000	29.70	1.33	31.03	372.36	2,327.00
¹⁄₁₆	27,800	7.45	0.33	7.78	93.36	583.00
¹⁄₃₂	7,000	1.87	0.08	1.95	23.40	146.00
			300-lb. gage			
½	2,125,000	$573.00	$25.50	$598.50	$7,182.00	$44,887.00
⅜	1,195,000	322.00	14.34	336.34	4,036.08	25,225.00
¼	531,000	143.00	6.37	149.37	1,792.44	11,200.00
⅛	132,800	35.80	1.59	37.39	448.68	2,804.00
¹⁄₁₆	33,200	8.95	0.40	9.35	112.20	701.00
¹⁄₃₂	8,300	2.24	0.10	2.34	28.08	175.00
			400-lb. gage			
½	2,804,000	$760.00	$33.65	$793.65	$9,523.80	$59,524.00
⅜	1,577,000	427.00	18.92	445.92	5,351.04	33,444.00
¼	701,000	190.00	8.41	198.41	2,380.92	14,880.00
⅛	175,200	47.50	2.10	49.60	595.20	3,720.00
¹⁄₁₆	43,800	11.85	0.53	12.38	148.56	928.00
¹⁄₃₂	11,000	2.98	0.13	3.11	37.32	233.00
			600-lb. gage			
½	4,157,000	$1,141.00	$49.88	$1,190.88	$14,290.56	$89,312.00
⅜	2,338,000	642.00	28.06	670.06	8,040.72	50,254.00
¼	1,039,000	285.00	12.47	297.47	3,569.64	22,310.00
⅛	259,700	71.25	3.12	74.37	892.44	5,577.00
¹⁄₁₆	65,000	17.85	0.78	18.63	223.56	1,397.00
¹⁄₃₂	16,200	4.47	0.20	4.67	56.04	350.00

in common use, however, and frequently are referred to in contract specifications and similar documents.

On the other hand, many states and municipalities have adopted the ASME Boiler Construction Code as law, and therefore any piping within their jurisdiction must conform to Boiler Code requirement. Although it would seem that the Boiler Code should cover only such piping as is directly associated with the boiler, its provisions have been interpreted to include considerable main-steam and boiler-feed piping not directly pertaining to the boilers. Hence careful consideration should be given to the Boiler Code and the applicability of its rules to certain portions of many power-piping installations.

Considerable progress has been made in harmonizing the piping requirements of the two codes. The basic assumptions as to design pressure differ between the codes, however, so that no one schedule of pipe wall thicknesses or series of fitting necessarily is acceptable under both.[1]

---

[1] Where piping comes under the jurisdiction of a particular code, the latest revision should be used. Copies may be obtained from the American Society of Mechanical Engineers, 345 E. 47th St., New York, 17, N.Y.

Selection of Dimensional Standards and Materials. The selection of suitable dimensional standards for flanges, fittings, valves, and pipe for ordinary service conditions can be made from the appropriate publications of the American Standards Association. Adjusted ratings at temperatures above and below 750 F for carbon and alloy steels are given in the steel-flange standards to govern their use under pressure or temperature other than the primary service ratings.

Selection of materials for temperatures above 750 F from the various grades of alloys described in ASTM Specifications for high-temperature service is facilitated by reference to the abstracts of these specifications given in Chap. 7. The multiplicity of services in a large plant and the variety of dimensional standards and materials, possible joints, and different types of welding available make it desirable to prepare some form of design instructions for use in the drafting rooms and by the construction force to assure that proper standards, design details, and materials are secured. A summary of the principal selections for a modern power plant is given in Table 3.

It is advisable to express a word of caution regarding the connection of cast-iron flanges and fittings to steel flanges and fittings. The 25- and 125-lb cast-iron flanges have plain faces, while the 250-lb flanges have raised faces which come out practically to the inner edge of the boltholes, whereas those of steel flanges are narrower and stop some distance inside the boltholes. The plain face or wide raised face on cast-iron fittings and flanges is a necessary precaution to prevent cracking the flange in drawing up the bolts. Numerous instances have been observed where cast-iron flanges have cracked when being bolted to raised-face steel flanges. In cases where it is necessary to bolt cast-iron and steel flanges together, the raised face of the steel flange should be machined down flush with the flange edge. For the same reason, when lap-joint pipe is made up with cast-iron flanges, the lapped end should be brought out to the inner edge of the boltholes.

For reasons similar to those stated above, it is inadvisable to use alloy steel bolts in cast-iron flanges. Commercial carbon-steel bolts are amply strong for use in cast-iron flanges, and there is no occasion to risk cracking such flanges through the use of high-strength bolts.

American Standard steel flanges are properly designed, both as regards dimensions and material, for use with a narrow raised face and alloy-steel bolts. Intelligent selection may be made from the three grades of alloy-steel bolts listed in ASTM Specification A96 and the bolting for temperatures from 750 to 1000 F given in ASTM Specification A193.

The use of carbon-steel nuts with alloy-steel studs has been found generally satisfactory for temperatures up to 750 F, since the mechanical construction of the nut is stronger than that of the stud. However, owing to their greater tendency to splitting, barstock nuts machined from hot-rolled or cold-drawn bars should not be used for severe service conditions.

For higher temperatures, low-sulfur nuts and oil-quenched nuts are used as described in ASTM specification A194.

Selection of Valves. A selection of the proper valve for a particular purpose depends, first, on the operating pressure and temperature and, second, on the type of valve best suited for the use to which it will be put. In some cases, where either a globe or gate valve would serve equally well, the decision between these types may be based on price considerations. The more common types of valves are illustrated in Chap. 7.

In general, it is customary to use *gate valves* in locations where pressure drop through the valve is a consideration, and where the valve will be either wide open or entirely closed. Guard valves and shutoffs for boiler and turbine leads, etc., are almost always of the gate type. Globe valves are commonly used in water,

Table 3. Materials Specifications

(125 mw, 1,800 psi, 1000 F—1,000 F steam power plant)

Item	System — Main steam	System — High-temperature reheat	System — No. 2 extraction
1. Design conditions, psig and deg F	1,980, 1,005 F	565, 1,005 F	235, 800 F
2. Pipe size, Schedule	14″ OD, 1.625″ t_{min}	20″ OD, 0.793″ t_{min}	All sizes
3. Pipe size, Schedule	10¾″ OD, 1.255″ t_{min}	16″ OD, 0.646″ t_{min}	Sch 40
4. Pipe material	A335-P11	A335-P11	A335-P2 or higher
5. Pipe manufacture	Seamless	Seamless	Rolled and welded
6. Flange, type	Welding neck	Welding neck	Welding neck
6A. Flange, material	A182 Grade F11	A182 Grade F11	A182 Grade F12
6B. Flange facing	Male-female	Male-female	Male-female
7. Pressure standard B16.5	2,500	900	300
8. Bolting	Bolt studs fitted with nut at each end		ASA B1.1
9. Bolt-thread standard	ASA B1.1	ASA B1.1	ASA B1.1
10. Nut dimensions	American "heavy" hexagon nut per ASA B18.2		A193, B7
11. Bolt-stud material	A193, B14	A193, B14	A194 Grade 2H
12. Nut standard	A194 Grade 2H	A194 Grade 2H	Coarse
13. Bolt threads 1″ and less	Coarse	Coarse	8 pitch
14. Bolt threads 1⅛″ and larger	8 pitch	8 pitch	Class 2A
15. Bolt-stud dimensions	Class 2A	Class 2A	Class 2B
16. Nut dimensions	Class 2B	Class 2B	A234-WP2
17. Fittings	A234-WP11	A234-WP11	
18. Butt-weld fitting 2½″ and larger			A335-P2
19. Pipe	A335-P11	A335-P11	
20. Plate	....	A387-Gr.C	
21. Forgings	A182-F11	A182-F11	A182-F12
22. Castings	A217 WC6	A217 WC6	A217 WC4
23. Dimensions	B16.9 Special	B16.9 Special	B16.9 Sch 40
24. Socket-weld fitting 2″ and smaller			
25. Material	A182-F11	A182-F11	A182-F12
26. Dimensions	B16.11, extra heavy	B16.11 Sch 80	B16.11 Sch 40
27. Joints 2″ and smaller	Socket weld	Socket weld	Socket weld
28. Joints 2½″ and up, weld, B31.1	Butt weld	Butt weld	Butt weld
29. Joints 2½″ and backing ring	Flat solid	Flat solid	Flat solid
30. Joints 2½″ and up, alternate weld	Inert gas	Inert gas	Inert gas
31. Joint unions, under 2″			
32. Gaskets, steam	Spiral-metal	Spiral-metal	Spiral-metal

Table 3. Materials Specifications (Continued)

Item	System				
	Drum steam	Boiler-feed discharge	Reheat desup. spray	High-pressure chemical feed	Continuous blowdown
1. Design conditions, psig and deg F	2,150, 646 F	2,800, 475 F	1,000, 350 F	2,800, 150 F	2,150, 646 F
2. Pipe size, schedule	All sizes	All	All	All	All
3. Pipe size, schedule	Sch 160	160	80	160	160
4. Pipe material..........	A106B	A106B	A106B	A106B	A106B
5. Pipe manufacture	Seamless	Seamless	Seamless	Seamless	Seamless
6. Flange, type	Welding neck	Welding neck	Welding neck	Welding neck	Welding neck
6A. Flange, material	A105 Grade II	A105 Grade II	A105 Grade II	A105 Grade II	A105 Grade II
6B. Flange facing	Male-female	Male-female	Male-female	Male-female	Male-female
7. Pressure Standard B16.5	1,500	1,500	600	1,500	1,500
8. Bolting	Same	Same	Same	Same	Same
9. Bolt-thread standard	Same	Same	Same	Same	Same
10. Nut dimensions	Same	Same	Same	Same	Same
11. Bolt-stud material	Same	Same	Same	Same	Same
12. Nut standard	Same	Same	Same	Same	Same
13. Bolt threads 1″ and less	Same	Same	Same	Same	Same
14. Bolt threads 1⅛″ and larger	Same	Same	Same	Same	Same
15. Bolt-stud dimensions	Same	Same	Same	Same	Same
16. Nut dimensions	Same	Same	Same	Same	Same
17. Fittings	A234 WPB	A234 WPB	A234 WPB	A234 WPB	A234 WPB
18. Butt-weld fitting 2½″ and larger	A106B	A106B	A106B	A106B	A106B
19. Pipe					
20. Plate					
21. Forgings	A105 Grade II	A105 Grade II	A105 Grade II	A105 Grade II	A105 Grade II
22. Castings	A216 WCB	A216 WCB	A216 WCB	A216 WCB	A216 WCB
23. Dimensions	B16.9 Sch 160	B16.9 Sch 160	B16.9 Sch 80	B16.9 Sch 170	B16.9 Sch 160
24. Socket-weld fitting 2″ and smaller					
25. Material	A105 Grade II	A105 Grade II	A105 Grade II	A105 Grade II	A105 Grade II
26. Dimensions	B16.11 Sch 160	B16.11 Sch 160	B16.11 Sch 80	B16.11 Sch 160	B16.11 Sch 160
27. Joints 2″ and smaller	Socket weld	Socket weld	Socket weld	Socket weld	Socket weld
28. Joints 2½″ and up, weld, B31.1	Butt weld	Butt weld	Butt weld	Butt weld	Butt weld
29. Joints 2½″ and backing ring	Flat solid	Flat solid	Flat solid		Flat solid
30. Joints 2½″ and up, alternate weld	Inert gas	Inert gas	Inert gas	....	Inert gas
31. Joint unions, under 2″					
32. Gaskets, steam	Spiral-metal	Spiral-metal	Spiral-metal	Spiral-metal	Spiral-metal

Item	System				
	Group 1	Group 2	Group 3	Group 4	Group 5
1. Design conditions, psig and deg. F	Various	Various	Various	Various	125–250 lb
2. Pipe size, schedule	Various	Various	Various	Various	Various
3. Pipe size, schedule	Various	Various	Various	Various	Various
4. Pipe material	A106A—1½″ and under		106B—2″ and Up	Centrifugally cast iron or cement-lined Steel	Various
5. Pipe manufacture	Seamless	Seamless	Seamless	A21.8 or A21.6	
6. Flange, type	Welding neck	Welding neck	Welding neck	Cast iron up to 20″	
6A. Flange, material	A105 Grade I	A105 Grade I	A181 Grade I	A181 Grade I	A126
6B. Flange facing	Raised face	Raised face	Raised face	Raised face	Flat face
7. Pressure standard B16.5	B16.5—600 lb	B16.5—400 lb	B16.5—300 lb	B16.5—150 lb	B16.1—125—250 lb
8. Bolting	Same	Same	To 500 F, 500 F+ Unfin. sq, HD-stud		Reg. unfinished sq. HD bolt
9. Bolt-thread standard	Same	Same	B18.2 or B1.1		B18.2 or B1.1
10. Nut dimensions	Same	Same	Hvy. unfin., hvy. semifin.	Hvy. unfin., hvy. semifin.	Hvy. unfin. hex nut A307B with C.I., else A307A
11. Bolt-stud material	Same	Same	A307B or A193 B7	A307B or A193 B7	
12. Nut standard	Same	Same	Same	Same	Same
13. Bolt threads 1″ and less	Same	Same	Same	Same	Same
14. Bolt threads 1⅛″ and larger	Same	Same	Same	Same	Same
15. Bolt-stud dimensions	Same	Same	Same	Same	Same
16. Nut dimensions	Same	Same	Same	Same	Same
17. Fittings	A234 WPB	A234 WPB	A234 WPB	Steam / Water fuel gas	3″ to 20″ C.I. flg. or or cement-lined weld
18. Butt-weld fitting 2½″ and larger	A106B	A106B	A106B	A106B	Cement-lined
19. Pipe	A105-II	A105-II	A105-II	A105-II	A106A
20. Plate	A95	A95	A95	A95	A285B
21. Forgings					
22. Castings					
23. Dimensions	B16.5—600 lb	B16.5—400 lb	B16.5—300 lb	B16.5—150 lb	to 12″ A126A, 14″ to 20″ A126B; B16.1
24. Socket-weld fitting 2″ and smaller					2½″ and smaller
25. Material	A105-II	A105-II	A105-II	A105-II	Cast brass
26. Dimensions	B16.11—3,000 lb / Sch 40	B16.11—3,000 lb / Sch 40	B16.11—3,000 lb / Sch 40	B16.11—3,000 lb / Sch 40	B16.18
27. Joints 2″ and smaller	Socket weld	Socket weld	Socket weld	Socket weld except screwed for air and fuel gas	Soldered
28. Joints 2½″ and up, weld, B31.1	Butt weld	Butt weld	Butt weld	Flat split	C.I. flg—steel butt weld
29. Joints 2½″ and up and backing ring	Flat split	Flat split	Flat split	Inert gas	Flat split
30. Joints 2½″ and up, alternate weld	Inert gas	Inert gas	Inert gas	Socket weld except malleable for chem. feed	None
31. Joints unions, under 2″	Socket weld	Socket weld	Socket weld		
32. Gaskets, steam	"Flexital" type CG—steam; flat ring—water, air, gas; 1/32″ thick for oil				
33. Gaskets, water under 100 F	Duck inserted rubber	Duck inserted rubber	Duck inserted rubber	Duck inserted rubber	Duck inserted rubber
34. Gaskets, water 100 F-700 F	Graph. compr. asbestos	Graph. compr. asbestos	Graph. compr. asbestos		Graph. compr. asbestos
35. Gaskets, air and gas	Graph. compr. asbestos	Graph. compr. asbestos	Graph. compr. asbestos		
36. Gaskets, oil to 249 F	Impreg. plant fiber	Impreg. plant fiber	Impreg. plant fiber	Impreg. plant fiber	

NOTES FOR TABLE 3

Group 1. High-pressure fuel oil, miscellaneous high-pressure drips and drains.

Group 2. Low-temperature reheat, steam blowout.

Group 3. Low-pressure saturated steam, boiler-feed recirculation reheat vents, pump balancing, steam to air heater, evaporator blowdown, evaporator coil drain, condensate to deaerator, high-pressure heater drains and vents, miscellaneous low-pressure drips, ignition oil, sootblowing air.

Group 4. No. 3 extraction, No. 4 extraction, condensate pump suction boiler-feed suction, evaporator vapor discharge, steam from steam seal regulator, low-pressure heater drains and vents, ejector drains, condensate recirculation, deaerator overflow, equipment drains, closed service-water system, plant air, turbine lube oil, condenser air removal, low-pressure chemical feed, air heater wash, miscellaneous evaporator piping, generator gas vent.

Group 5. Salt water, ash sluice pump discharge, salt cooling water.

steam, and air lines for throttling purposes, as the globe type permits closer regulation of flow. Throttling usually involves more or less cutting of the seat and disk, and these parts in globe valves are less expensive and more easily replaced than in gates. Among such uses for globe valves might be mentioned turbine and engine throttles, bypasses around traps or reducing valves, hand-feed regulation on boilers, etc. A gate valve should always be installed preceding a globe valve used for throttling.

It is always advisable to use *check valves* in individual pump or trap discharges before they join a common header, and where different lines are joined together to discharge into a common header. A check valve cannot be counted on for closing a line off tightly against pressure working back through, but it will stop any considerable flow. In pump discharges where the header remains under pressure after the pump is shut down, a gate valve should be provided in addition to the check. It is also desirable to provide a small relief valve on the pump suction to prevent pressure backing up through the pump and damaging the foot valve while the pump is shut down. Check valves are required in feed lines close to a boiler to prevent water or steam blowing back from the boiler if, for any reason, the feed line ruptures or its pressure fails.

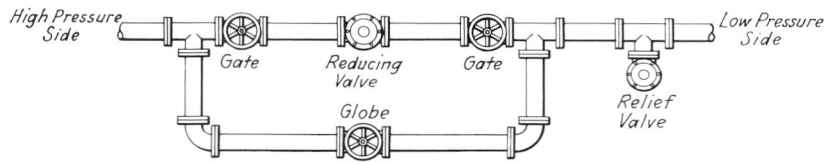

FIG. 2. Pressure-reducing station.

With reducing or other control valves it is desirable to select a size that is loaded somewhere near capacity under normal operation, as such valves are then more stable in their operation. If there is considerable seasonal variation in the load on a reducing or control station, as in furnishing steam for building heating, it is good practice to install a large and a small valve in parallel and use the one best fitting the load at any particular time. It is frequently desirable to install a hand-operated bypass around a control valve so that service can be maintained while the special valve is being repaired. Such an arrangement with provision for shutoff to repair the special valve while the line is in operation is shown in Fig. 2, which includes flanged valves. Advantage can be taken of improved valve design wherein weld ends are used while still permitting the internals to be removed for repairs. Filler pieces should be provided between a flanged reducing valve and the adjacent gates, as reducing valves usually are constructed so that it is impossible to remove all the end-flange bolts when they are bolted directly to other valves or fittings, owing to close clearance at the valve bonnets or fitting necks. For convenience in removing these valves from the line, flanged connections are used frequently at such points, even though the rest of the line is made up with screwed or welded joints. When reducing from a pressure that requires the use of a heavy standard for flanges and fittings to a pressure with which a lighter standard is used or to which low-pressure equipment is connected, relief valves should be provided on the low-pressure side. The use of the heavier standard should be continued through to the last valve ahead of the relief valve, as it is possible to have full pressure up to that point. A reducing valve seldom has a tight shutoff, and at times of negligible steam consumption, leakage through the valve is apt to be enough to build up full line pressure on the low-pressure side. The Power Section of the Code for Pressure Piping requires

that the combined discharge capacity of the safety or relief valves shall be such that the pressure rating of the lower pressure piping will not be exceeded in case the reducing valve sticks open. An exception may be made in a steam plant used for district heating where attendants are always on duty. In this case, a pressure-operated audible alarm may be installed in place of safety valves.

A typical pressure-reducing and desuperheating station of the "ratio" type is shown in Fig. 3. Other types known as "tank" and "atomizing" desuperheaters also are in common use.

Safety and Relief Valves. Where more than one safety or relief valve is used on a boiler or other pressure vessel, it is desirable to set one or more of the valves to relieve at a lower pressure than the rest. This serves as a warning before too much steam is lost through all the valves opening at once and also tends to facilitate repairs by confining any cutting action to the one or more valves that open first. In some cases an extra safety valve, known as the power-control valve, is set to blow before the others and is mounted above a gate valve, so that it can be removed for repairs while the boiler or steam line is in service. The capacity of this valve cannot be considered in meeting code or other safety requirements, since it might be shut off. Where the hazard involved does not require the installation of a full-size relief valve, it is sometimes desirable to install a small-size pop valve as a telltale to give warning when the usual working pressure is exceeded, so that the operator will attend to restoring normal conditions. A safety valve for use with a compressible fluid, such as steam or air, is distinguished from a relief valve in that a safety valve has an adjusting, or huddling, ring and chamber to control the amount the pressure blows down before the valve reseats.

Safety Valves for Power Boilers. The construction and method of installing safety valves for power boilers can best be explained by reference to Section I of the ASME Boiler Construction Code[1] which is incorporated in the safety laws of many states. The following material is taken from that code. For more complete information, reference should be made to the latest revision of the Code itself.

Each boiler is required to have at least one safety valve and two or more safety valves if it has more than 500 sq ft of heating surface or if the steam-generating capacity exceeds 2,000 lb per hr. The safety-valve capacity for each boiler is required to be such that all the steam that can be generated will be discharged without allowing the pressure to rise more than six per cent above the highest pressure at which any valve is set and in no case more than 6 per cent above the maximum allowable working pressure. The maximum steaming capacity shall be determined by the manufacturer.

One or more safety valves on the boiler proper are required to be set at or below the maximum allowable working pressure. If additional valves are used, the highest pressure setting is required not to exceed the maximum allowable working pressure by more than 3 per cent. The complete range of pressure settings of all the saturated-steam safety valves on a boiler is required not to exceed 10 per cent of the highest pressure to which any valve is set.

Safety valves are required to be of the direct spring-loaded pop type. The maximum rated capacity stamped on the valve is required to be 90 per cent of the actual steam flow determined by tests in the presence of authorized inspectors at a pressure 3 per cent in excess of the pressure at which the valve is set to blow. The blowdown, *i.e.*, difference between opening and closing pressure, and capacity lift or distance the valve seat rises when blowing under a pressure of 3 per cent above the set pressure also are required to be stamped on the valve.

When two or more safety valves are used on a boiler, they may be mounted either separately or as twin valves on Y bases or duplex valves having two valves in one body. Valves mounted as twin valves and duplex valves shall be of equal sizes. Where not more

[1] Published by the American Society of Mechanical Engineers, 345 E. 47th St., New York 17, N.Y.

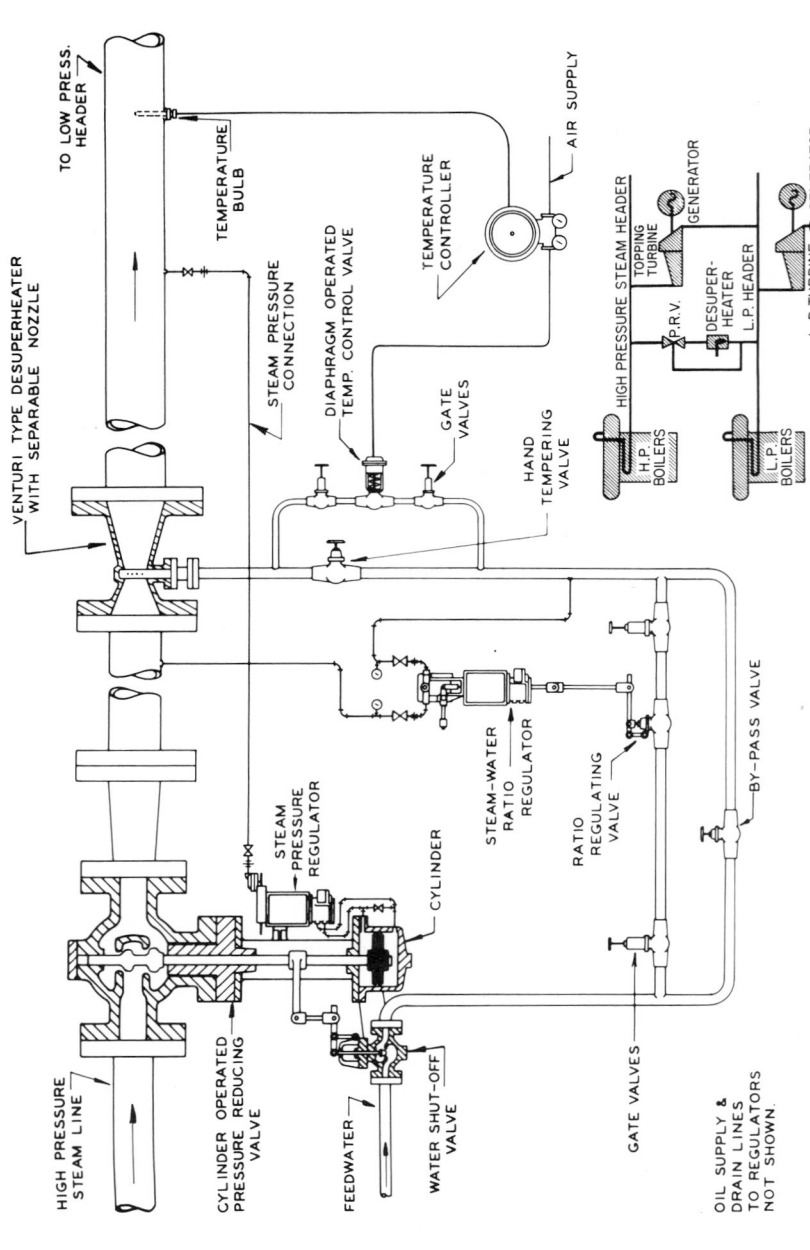

FIG. 3. Typical pressure reducing and desuperheating station of the "ratio" type. This diagram was prepared expressly for the "Piping Handbook" by the Republic Flow Meters Company, which manufactures the Smoot equipment illustrated. The sketch at the lower right-hand corner of Fig. 3 is reproduced by permission from an article by Mr. C. R. Earle in the April, 1938, issue of *Power Plant Engineering*.

than two valves of different sizes are mounted singly, the relieving capacity of the smaller is required to be not less than 50 per cent of the larger valve.

The safety valves are required to be connected to the boiler independent of other steam connection.   Intervening pipe or fittings are required to be not longer than the corresponding tee fitting of the same diameter and pressure of the American Steel Flange Standard.

The area of the connection between the boiler and safety valve is required to be at least equal to that of the valve inlet.   No valve of any description is permitted between the safety valve and the boiler or on the discharge pipe between safety valve and atmosphere. The area of the discharge pipe, if used, is required to be not less than the area of the safety-valve outlet, or outlets if more than one valve discharges into a common discharge pipe. If a muffler is used on a safety valve, it is required to have sufficient outlet area to prevent backpressure interfering with the operation of the valve.

Safety valves are required to operate without chattering and to be set to close after blowing down not more than 4 per cent of the set pressure but not less than 2 lb in any case. For pressure between 100 and 300 lb, inclusive, the blowdown is required to be not less than 2 per cent of the set pressure.   The blowdown adjustment is made and sealed by the manufacturer.   The popping-point tolerance plus or minus is required not to exceed 2 lb for pressure up to and including 70 lb, 3 lb for pressure 71 to 300 lb, and 10 lb for pressure over 300 lb.

Every attached superheater is required to have one or more safety valves near the outlet. The safety valves may be located anywhere along the length of the outlet header.   The superheater safety-valve capacity may be included in determining the size and number of valves required in a boiler, provided at least 75 per cent of the aggregate valve capacity is located on the boiler.

The principles involved in the protection of low-pressure piping supplied with steam from a higher pressure source are essentially the same as the principles used in the protection of unfired pressure vessels, except that the permissible pressure rise depends upon the particular service.   The following material abstracted from Section VIII of the 1959 edition of the ASME Boiler Construction Code covers capacities, recommended types, methods of installation and permissible adjustment of relief valves.

Pressure vessels shall be protected by such safety or relief valves and indicating and controlling devices as will ensure their safe operation.   The relieving capacity of safety valves shall be such as to prevent a rise of pressure in the vessel of more than 10 per cent. above the maximum allowable working pressure.

Safety valves shall be of the direct spring-loaded type.   A substantial lifting device is required for trying all safety valves except those used for relief of liquid pressure.   Each safety valve shall be tested once every day or oftener, by raising the disk from its seat with the lifting device.

Pipe between safety valve or valves and the vessel which is to be protected shall have an internal cross-sectional area not less than the nominal area of the safety valve or valves. There shall be no intervening valve between vessel and safety valve.   The safety-valve escape or vent pipe shall be full-sized and fitted with an open drain.   No valve of any description shall be placed on the escape pipe.

When the hazard to fire or other source of external heat exists, supplemental pressure relief devices shall be capable of preventing the pressure from rising more than 20% above the maximum allowable working pressure of the vessel.

A typical pressure-relief valve is shown in Fig. 32 of Chap. 7.   Tables 46 and 47 of that chapter list relieving capacities of safety valves which handle steam and air. The values listed in those tables are approximately correct for the given conditions; the actual capacity for a specific valve will be determined by characteristics inherent in design and in choice of material finish.   For these reasons, the manufacturer should be consulted relative to the relieving capacity of a specific valve.

Pipe Thimbles.   Where pipes pass through walls or floors or where they are to be embedded in concrete or masonry, suitable pipe sleeves or thimbles should be

provided.   Such thimbles preferably should be made of standard-weight wrought pipe, either black or galvanized, to provide sufficient mechanical strength and durability.   Where pipes pass through floors, the thimbles should extend at least 4 in. above the top of the floor in order to prevent wash water running down the hole. A schedule of the proper sizes of standard-weight pipe to use for thimbles for any size line, screwed or flanged and with or without flanges of different standard dimensions, is given in Table 4.

Table 4.   Pipe Thimble Sizes[1]

(All dimensions in inches)

Nominal pipe size	Screwed or welded pipe					Flanged pipe		
	Without insulation	With insulation				With or without insulation		
	Any weight pipe of iron pipe, size	Wool or hair felt 1 in. thick	Standard thickness 85 per cent magnesia	Double standard thickness 85 per cent magnesia	Cork covering 1½ in. thick	125- and 150-lb flanges	250- and 400-lb flanges	600-lb flanges
1/4	3/4	3	3	6	4	..	..	..
1/2	1	4	4	6	6	..	..	..
3/4	1 1/4	4	4	6	6	..	..	..
1	1 1/2	4	4	6	6	..	..	..
1 1/4	2	6	6	6	6	6	6	6
1 1/2	3	6	6	6	6	6	8	8
2	3	6	6	8	..	8	8	8
2 1/2	4	6	6	8	..	8	10	10
3	4	8	8	10	..	10	10	10
4	6	8	8	10	..	10	12	12
5	8	10	10	12	..	12	14	16
6	8	10	10	12	..	12	16	16
8	10	12	14	16	..	16	18	20
10	14	16	16	18	..	18	20	24
12	16	18	18	20	..	24	24	24
14	18	20	20	24	..	24	26	26
16	20	20	24	24	..	26	28	30
18	24	24	24	26	..	28	30	32
20	24	24	26	28	..	30	32	34
24	....	..	..	..	..	34	38	40
30	....	..	..	..	..	40	46	48

[1] When flanges are in sleeves, special provision must be made for covering.   Allow 1 in. extra on diameter of 125-lb flanges for lugs.   Pipe thimble sizes given are for thimbles made of standard-weight wrought pipe to 12 in.; above 12 in., of O.D. pipe with 3/8-in. wall.   Pipe thimbles above 24 in. should be made of sheet metal, 10 gage.

MECHANICAL DESIGN OF PRINCIPAL PIPING SYSTEMS

Main Steam Piping.   The main steam lines constitute the most important piping in a power plant.   Several methods of connecting boilers and turbines are in common use.   In the simplest type a boiler or two boilers, as the relative capacities of the boilers and turbines may dictate, are connected directly to their respective turbines as indicated in Fig. 4a.   This is known as the unit boiler-turbine system. Although in normal operation steam flows directly from boiler to turbine, it is usual to provide a crossover, indicated by the dashed line, to permit the operation of either

turbine from the adjoining boiler or battery of boilers.   In a plant having a multiplicity of turbines and boilers, these cross connections between adjoining units give the appearance of a continuous header system, but because of the small size of the crossovers and consequent high-pressure drop, steam is seldom transferred farther than to the adjacent unit.

The ring-header system shown in Fig. 4b is designed to permit the utmost flexibility in operation.   The boiler leads and header are liberally sized so that steam may be supplied from any one of a number of boilers to a given turbine with approximately the same pressure drop as from the nearest boiler.   Double valving is employed so that any one valve can be worked on without having to shut down more than one boiler or one turbine.

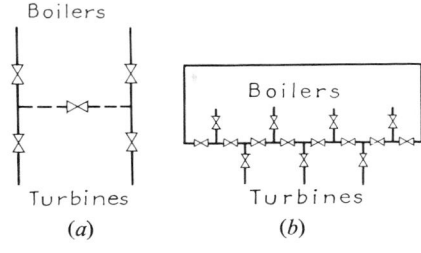

Fig. 4. (a) The unit boiler-turbine system. (b) The ring-header system.

In earlier days, the ring-header system was quite popular, since several low-capacity boilers could be interconnected to supply two or more high-capacity turbines.   The art of steam-generating-unit design has now progressed to the point at which the boiler manufacturers can supply units to match or exceed turbine capacity.   For this reason, and because of the complexity of the ring-header system, most plants today are designed on the unit system.

A typical steam header in a modern industrial plant is shown in Fig. 5, which is a

Fig. 5. Steam header in industrial plant.   (*Courtesy of Valve World.*)

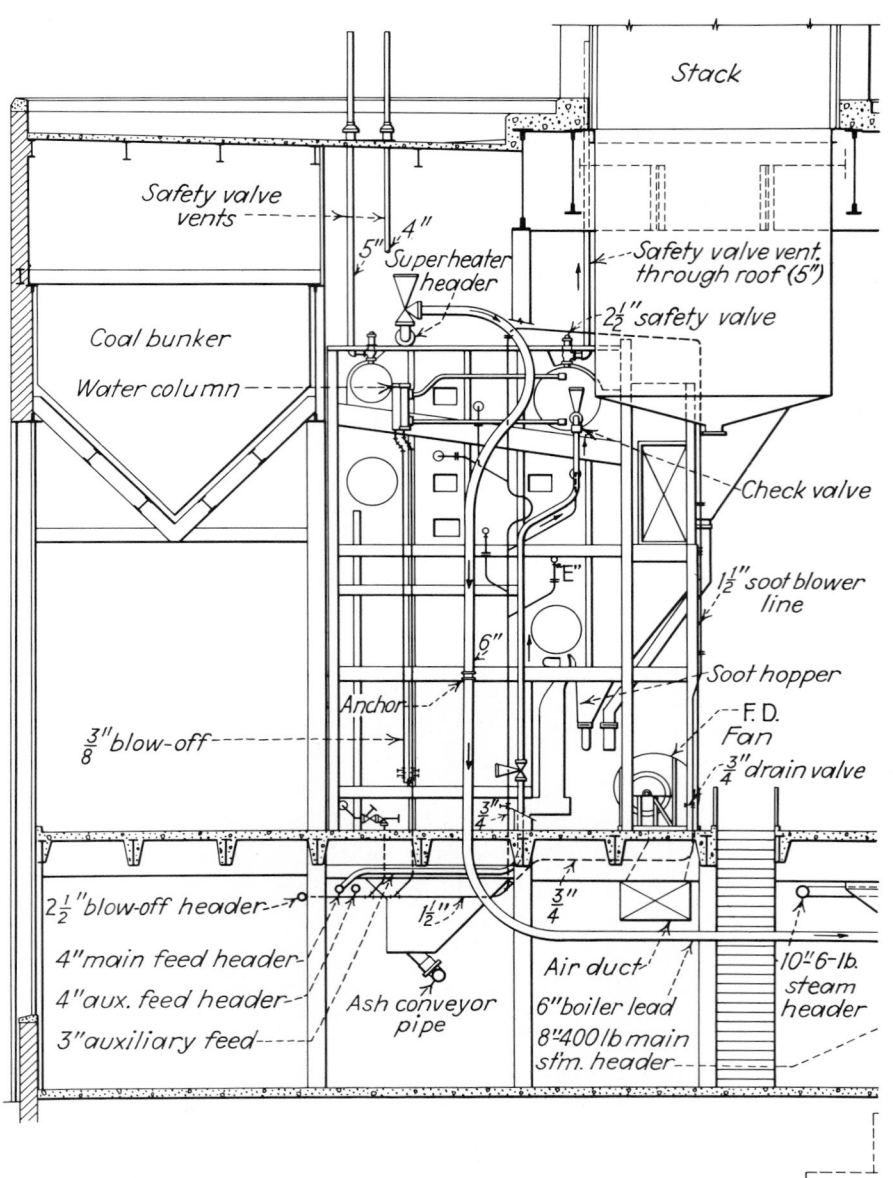

FIG. 6. Assembly of steam piping in power plant of

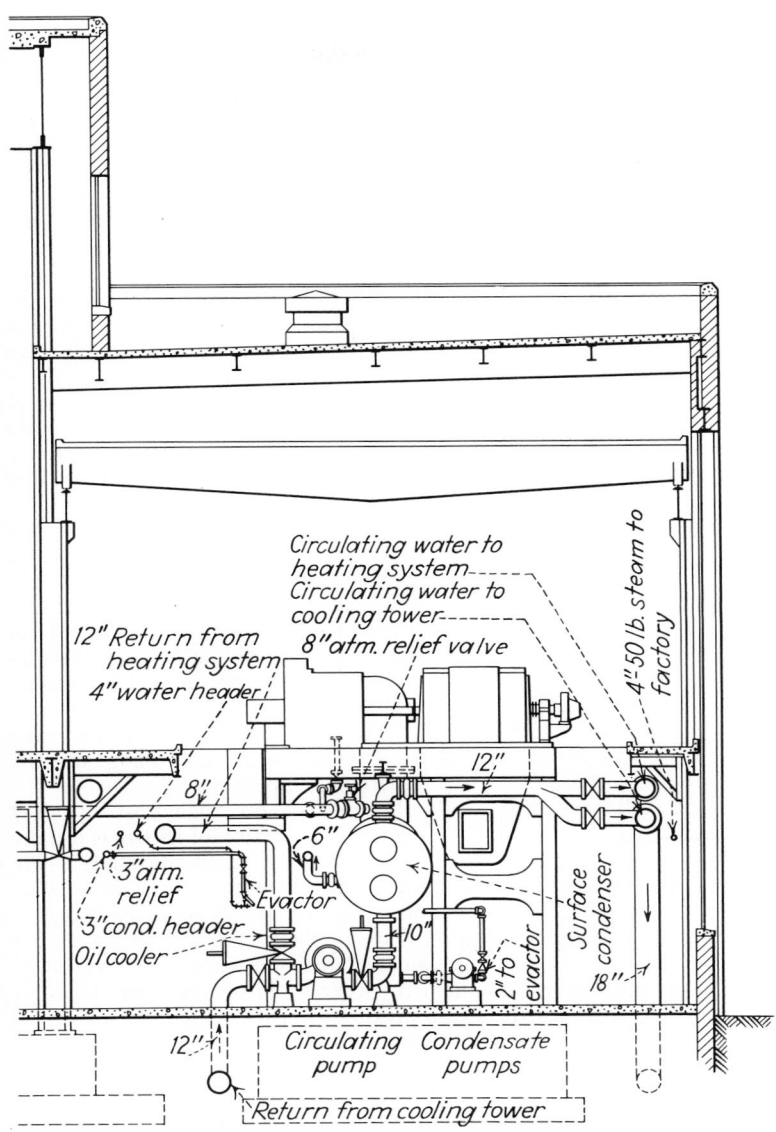

Circulating water to
heating system
Circulating water to
cooling tower
8"atm. relief valve

4" 50 lb. steam to
factory

12" Return from
heating system
4" water header

12"

8"

6"

3"atm.
relief
Evactor

3"cond. header
Oil cooler

10"

2" to
evactor

Surface
condenser

18"

12"

Circulating   Condensate
pump   pumps

Return from cooling tower

the Burroughs Adding Machine Company at Plymouth, Mich.

photograph of an installation in the Wabash, Ind., plant of the General Tire Company. As may be noted, the boiler leads and connections supplying steam to process, heating, pumps, and the like are brought down to this header, which is located near the floor so that all valves are readily accessible.

In the assembly drawing of Fig. 6, which is indicative of the recent trend of industrial power-plant layouts and piping arrangements along the lines of small central stations, the general use of welded joints has been an important factor in obtaining improved appearance. This plant was designed by Albert Kahn, Inc., for the Burroughs Adding Machine Company at Plymouth, Mich.

If two or more boilers are interconnected, the connections for main superheated-steam piping are taken from the superheater outlet header through stop and check (combination) valves which are provided to prevent back flow of steam from the mains into a boiler in case of tube rupture or other boiler trouble. A typical combination valve is shown in Fig. 7.

Availability of steam generators has increased to the extent that single units are used almost universally to supply the steam to turbines of sizes up to approximately 1,000 megawatts. Thus the main steam piping between the turbine and boiler may consist of one or more lines with metallurgy varying from various alloys of chrome molybdenum to stainless steel, depending on the superheat temperature.

Because of the high cost of alloy piping, the selection of its size is usually the subject of an economic study where the increased pressure drop and its effect on boiler-feed pump power are weighed against unit pipe cost including installation and shipping. For systems operating at or below 1050 F the study may include stainless (austenitic) steel as a possibility. Above 1050 F, austenitic steel is required in order to bring the pipe thickness down to an acceptable value for the steam pressures normally used.

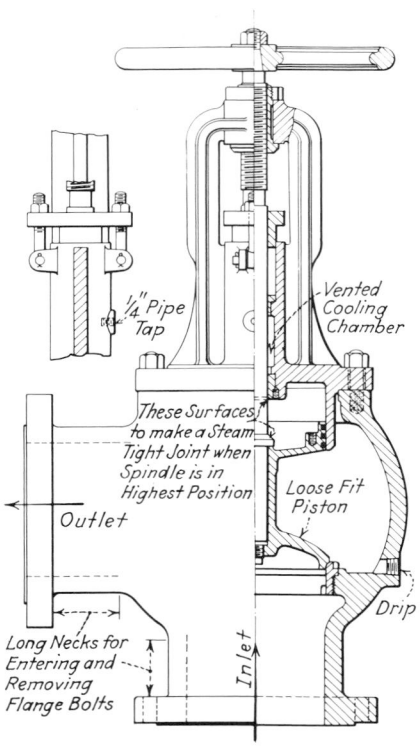

FIG. 7. Stop and check valve.

Selection of the operating steam pressure and temperature of a turbine-generator installation is based on an economic study or on experience derived from previous installations. Studies have shown that higher pressures and temperatures are associated with larger units or with actual or expected fuel costs.

Some utilities have installed units using supercritical steam pressure, that is, pressures higher than 3,206 psia at which value the specific volume of steam and water are equal.

Thus main steam pressure for larger units lies between, say, 1,450 psig and the supercritical region. Main steam temperature lies between 900 F where $\frac{1}{2}$ per cent chrome moly is used and, say, 1200 F where the use of austenitic steel is mandatory.

When one boiler is used to feed a turbine and no cross-connections are made to other boilers (unit system), the nonreturn valve at the superheater outlet is not required by the ASME Boiler Code. The shutoff gate valve may also be omitted; however, some power plants include it for the advantages gained thereby. These include usage during blowing out the steam lines prior to operation, hydrostatic testing before operation and after repair of an element of the high-pressure parts, and for emergency shutoff during operation.

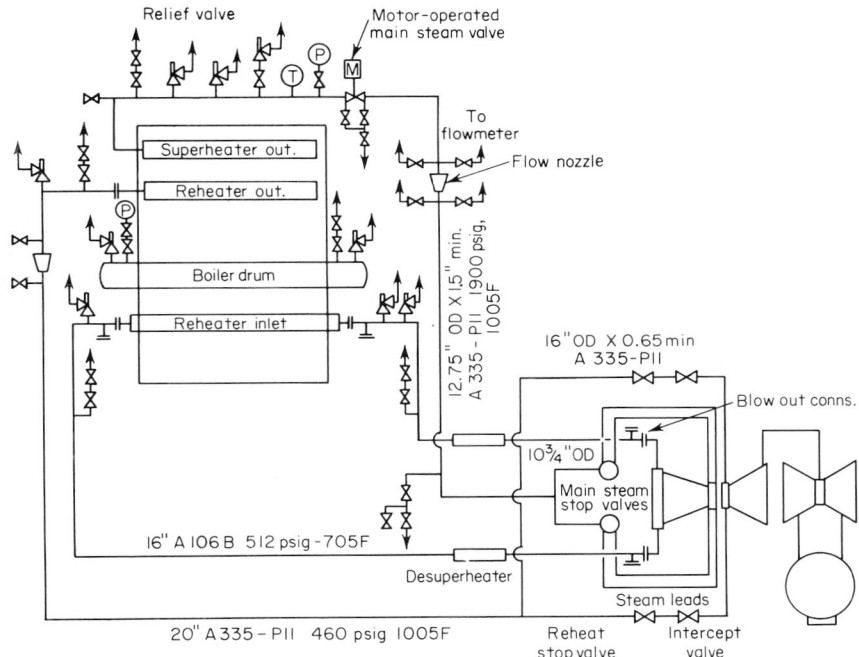

Fig. 8. Main and reheat steam flow diagram.

Large movements due to expansion at high temperature and the inherent stiffness of heavy-wall pipe require careful layout of the piping to obtain adequate flexibility so as not to exceed allowable reactions on boiler and turbine connections. Experienced designers arrange the routing to fulfill the above and yet not waste expensive alloy pipe which may cost hundreds of dollars per foot.

Large installations having several leads and installed with anchors, guides, stops, and fixed hangers at points of zero movement are sufficiently complex to require stress analysis by a programmed computer. The optimum arrangement may have been arrived at after several variations have been investigated by the engineers in order to find the most economical pipe size and routing which satisfy the limitations on reactions. Turbine and boiler manufacturers may rerun the stress analysis of the main steam system to verify that it is acceptable.

A flow diagram of a typical main and reheat steam system for a 125-megawatt installation with 1,800 psig 1000 F at the throttle and 1000 F reheat is shown in Fig. 8.

The selection of pipe and valve materials for high-temperature service is described in detail in Chap. 7. Typical selections of insulating materials for power plant piping are given in Tables 5 and 6, Chap. 6.

Reheat Steam Piping. The high-temperature reheat piping is usually of alloy steel, either a chrome-molybdenum alloy or stainless steel depending on the temperature, while the low-temperature reheat line is generally of carbon steel, although it is possible that particular installations may require the use of an alloy steel. The steam pressure of subcritical reheat is usually less than 600 psig, while for supercritical installations employing double reheat, the first reheat is at approximately 1,100 psig.

The "standard" pressure drop of a reheat system is 10 per cent of the high-pressure turbine exhaust. Since each per cent saving in pressure drop is worth approximately 0.01 per cent heat rate improvement, the actual pressure-drop design is the subject of an economic study. A starting point for the calculations would be to assume 5 per cent drop in the reheater, 3 per cent drop in the hot-reheat piping, and 2 per cent in the cold-reheat piping.

In the design of the reheat system, compliance with several criteria is required in order to avoid limiting turbine output while still complying with the applicable section of the ASME Code. Reheater design pressure must be adequate but, if too high, cannot be utilized and would therefore be wasteful. Reheater safety valves must be set high enough to avoid simmering at full load but must not be set too high or the allowable pressure of the turbine will be exceeded. Protection of the reheater requires that the safety valve on the reheater discharge must blow first to obtain circulation.

Reheat system pipe is larger in diameter than the main steam and must be arranged to obtain adequate flexibility. If it is too flexible as evidenced by a stress analysis, it should be shortened to save expensive piping.

Turbine Bypass Piping. Turbine bypass piping for supercritical or once-through operation is usually sized for approximately 30 per cent flow, although developments point to a reduction of this requirement. For subcritical operation the use of "controlled starting," wherein a turbine bypass is required, has been found to be less severe on turbine and boiler parts than long-drawn-out "rule-of-thumb" starting.

A "controlled-start" flow diagram is shown in Fig. 9. Main steam piping metallurgy should be used at least up to the desuperheating station, and suitable interlocks are normally provided to shut down, should desuperheating water or circulating water fail.

Soot-blowing Piping (for High-pressure Boilers). When steam is used for soot-blowing nonreheat boilers, it is taken from the saturated steam drum. For reheat boiler service operating at or above 1,500 psig the steam is taken from an appropriate header between the drum and the superheater header. This is because the drum steam, if reduced to desired pressure, would be too wet; steam taken from the superheater header would waste energy and, in addition, expensive alloy piping would be required for the system. In either case the piping system is usually supplied by the boiler manufacturer.

When air is used as the blowing medium, the boiler manufacturer usually furnishes the piping system to the soot blowers but ends it at a prescribed location at the sides of the boiler for connection to the station supply. Soot-blower air compressors operate at 350 or 500 psig as determined by the results of an economic study.

Auxiliary Superheated-steam Piping. The auxiliary superheated-steam supply for operating auxiliary machinery is usually carried through a separate header system taking its supply from connections at fittings or manifolds in the

main steam system. It is customary to install stop and check valves at the points of connection to the main system to prevent back feeding or short circuits through the auxiliary mains. The standards of construction for the auxiliary system should be on a par with the main superheated-steam system.

When steam jet air ejectors are used, the source may be main steam. This is taken off the main steam line near the ejector, and the piping metallurgy is the same.

Because of the high availability of boiler-feed pumps, dual-driven boiler-feed pumps have lost popularity. However, with increased turbine-generator size the noncondensing, extracting, steam-driven boiler-feed pump is again being considered. The steam source, however, for the turbine drive has shifted to the high-pressure turbine exhaust and discharge is to the deaerator and, if necessary, also to lower pressure heaters.

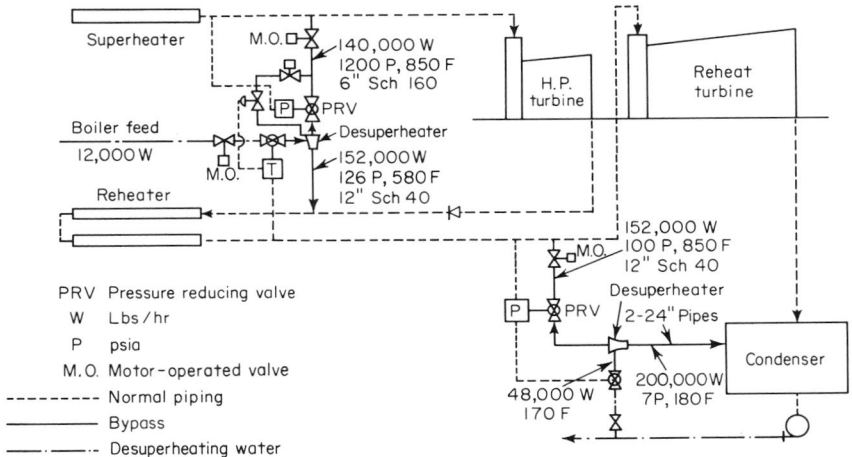

FIG. 9. Controlled start flow diagram—125 mw.

Main steam use for auxiliaries thus appears to be limited to the air ejector and the steam seal regulator.

Auxiliary Saturated-steam Piping. The auxiliary saturated-steam supply for soot blowing, building heating, and other miscellaneous plant uses is taken from the main saturated boiler drum through stop and check valves. It is customary to provide such a connection on each boiler and to tie each line into a saturated-steam header for the plant. The saturated-steam outlet on a boiler will vary in size, depending on the boiler capacity. With boiler pressures of 300-lb gage and higher, it is usually desirable to provide reducing valves or other throttling devices to reduce the saturated-header pressure to about 200-lb gage, which is about right for soot blowing and can be reduced further as required for other plant uses. Auxiliary saturated steam is used for generator fire extinguishers, steam seals on turbine shaft packing, steam-jet air-removal equipment, thawing frozen coal in cars, thawing needle ice in intake canals, cracking scale in evaporators, deaerator, evaporator supply backup, and similar uses, including reserve supply for building heating. The normal supply for building heating may be bled from house-service turbines where these are available. The construction standards for the high-pressure auxiliary saturated system usually permit a lower pressure standard than that of the main superheated-steam system. It is permissible to use a lower pressure standard

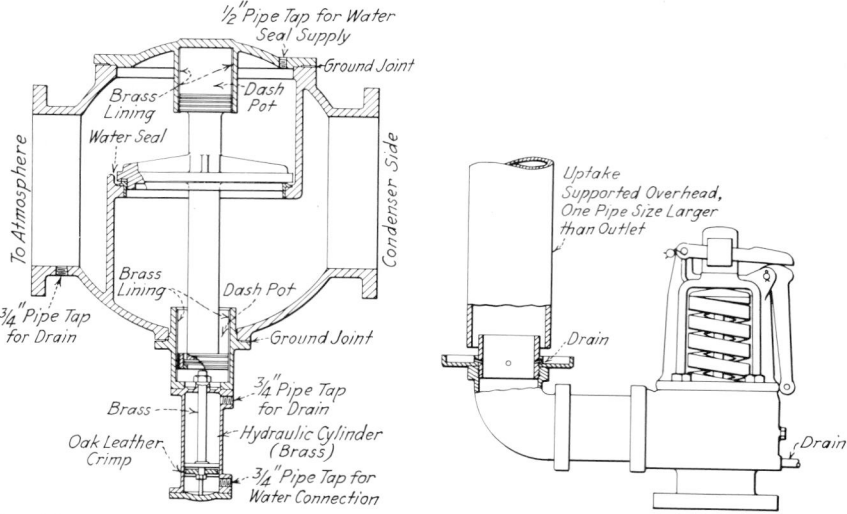

FIG. 10. Atmospheric relief valve for condenser.

FIG. 11. Safety-valve vent with umbrella fitting.

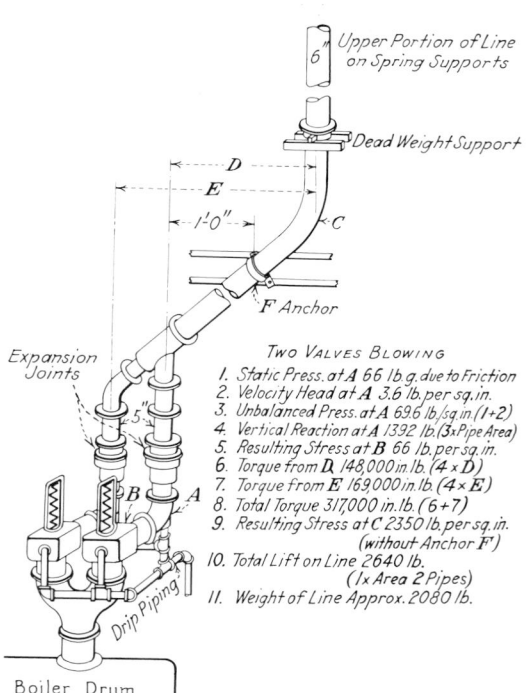

Two Valves Blowing

1. Static Press. at A 66 lb.g. due to Friction
2. Velocity Head at A 3.6 lb. per sq.in.
3. Unbalanced Press. at A 69.6 lb./sq.in.(1+2)
4. Vertical Reaction at A 1392 lb.(3×Pipe Area)
5. Resulting Stress at B 66 lb. per sq.in.
6. Torque from D 148,000 in.lb. (4×D)
7. Torque from E 169,000 in.lb. (4×E)
8. Total Torque 317,000 in.lb. (6+7)
9. Resulting Stress at C 2350 lb. per sq.in.
          (without Anchor F)
10. Total Lift on Line 2640 lb.
          (1×Area 2 Pipes)
11. Weight of Line Approx. 2080 lb.

FIG. 12. Reactions from blowing of safety valves.

where the line pressure is reduced, provided that part of the system is adequately protected with safety valves.

Auxiliary Exhaust Piping. Exhaust piping from steam-driven pumps, blowers, etc., should be of a construction standard suitable for the conditions obtaining. Standard-weight pipe and American standard cast-iron flanges and fittings for 125 lb SSP may be adequate for the purpose. In some cases where the steam-driven auxiliaries are intended for emergency use only, their turbines may be of a thermally inefficient type which allows considerable superheat to carry through in the exhaust. Under such circumstances it may be desirable to use American Standard steel flanges and fittings for 150 lb SSP. The exhaust from continuously operated steam-driven auxiliaries is usually led to feedwater heaters. In the case of emergency steam-driven auxiliaries, exhaust may be to the main condenser or to atmosphere or to both.

Atmospheric Exhaust Piping. Condensers are provided with lead diaphragms or may be provided with atmospheric exhaust valves and piping leading to atmosphere to protect the condenser against excess pressure arising from failure of the circulating water supply. Atmospheric relief valves are sometimes fitted with hydraulic cylinders, as shown in Fig. 10, to open the valve before the absolute pressure in the condenser reaches atmospheric pressure. The valve is arranged to reseat at some predetermined pressure governed by design conditions. Atmospheric exhaust valves should have water-sealed seats to prevent air leakage into the condenser. Atmospheric exhaust piping may be thin-walled pipe or duct work of sufficient strength to withstand the small back pressure caused by friction loss in flow to atmosphere and to resist corrosion. This back pressure usually does not exceed 5-lb gage. Where ducts are used, provision should be made to get inside for painting as a protection against corrosion. Provision of some kind is necessary to relieve expansion strains in the atmospheric exhaust line when the machine goes to atmosphere.

Safety-valve Vent Piping. Except in the case of small low-pressure boilers, it is advisable to carry the vents from safety valves to the outside of the building and extend them at least 6 ft above the roof to avoid scalding anyone happening to be near at the time the valve blows. To reduce the high noise level caused by the discharge, the pipe end may be fitted with a baffled silencer or the pipe end may be cut on a bias to increase the discharge area and reduce exit velocity. Where separate vents are used for each valve and the run is short, the vent pipe may be rigidly attached to its safety valve and carried through the roof in a pipe thimble screened with suitable flashing. Where the vents from two valves are combined into a single pipe, or where there is a long run to atmosphere, some form of flexible connection should be employed to relieve expansion strains. The umbrella type of flexible connection shown in Fig. 11 is extensively used, since it is both simple and inexpensive. Slip joints are sometimes used, as shown in Fig. 12. While the slip joints require a certain amount of maintenance, they have the advantage of providing a tighter seal against steam blowing out into the room than the umbrella-type connections. Safety-valve vents should be dripped at the lowest point in the line or in the valve body just above the seat to prevent accumulations of water in the escape pipe.

Where the vent pipe is long enough to offer considerable frictional resistance to the escaping steam, consideration should be given to the reacting forces set up. A typical problem of this kind occurring with a water-screen drum is illustrated in Fig. 12. Stress at point C is relieved in this case by providing an anchor at F to take the thrust occurring when the valve discharges.

Drips and Drains. Steam and air lines must be properly pitched and dripped between valves and at pockets in the line to get rid of accumulations of water.

Although there is no considerable amount of water to drip from a superheated steam line while it is in operation with a steady flow of steam through it, yet it is just as necessary that provision be made here as in the case of a saturated-steam line. Superheated lines with a valve closed at one end and no flow must be dripped, and during the period of warming up any line after a shutdown, large amounts of condensation have to be removed.   It is impossible to maintain superheat at the remote end of a long line under full pressure unless there is a considerable steam flow through the line.   In cases where an attempt is made to keep such a line warm by allowing steam to escape from a small bleeder valve at the remote end, the temperature of steam in the line will soon fall to the saturation point.   The absence of superheat is, of course, obvious during warming up a line from a cold or semicold condition.

The amount of condensate removed as drip from a steam line varies, of course, with the quality or superheat of the steam, the effectiveness of insulation, the length of line dripped, whether the line is initially hot or cold at the time of observation, etc. With such conditions known, it is possible to compute the heat loss or the warming-up loss and estimate the amount of drip for the given steam conditions.   In the case of an extensive piping system, such a computation would involve considerable work.   According to data published by the National Electric Light Association, the condensation collected as drip from the entire superheated and saturated steam-piping systems of a central station should not exceed 0.25 per cent of the entire water fed the boilers and frequently is much less.   On the other extreme is a central-heating boiler plant and distributing system where the entire line condensation amounts to 5 to 15 per cent of the total water fed the boilers.   Condensation in the central-heating plant itself may not exceed 0.5 per cent of the boiler feed.   Small- and medium-sized industrial plants usually will have a line loss not exceeding 1 per cent of the total boiler feed.

Where a drip pocket is installed in a saturated-steam line, it is good practice to make the opening into the pocket the full size of the line, as shown in Fig. 16 as a precaution against condensation being carried by.   In reciprocating-engine practice, especially where saturated steam is used, it is advisable to have a separator located close to the throttle.   It is frequently convenient to make such separators sufficiently large to serve the dual purpose of separator and steam receiver.   Vertical and horizontal welded-steel separator receivers of this kind are illustrated in Figs. 13*a* and *b*.   Provision in separators for removing entrained moisture is frequently much more elaborate, and such separators are extensively illustrated in piping-supply catalogues.

Where a partial dam is created in a line by the presence of a globe valve or similar obstruction to the flow of line condensation, a drip connection and trap should be provided.   Any pockets in the line should be dripped and care taken to see that gate valves are not installed with their stems below the horizontal because the bonnets will act as pockets if the stems are turned down.   Disregard of these precautions is apt to result in severe water hammer or damage to equipment through slugs of water being carried on through.   For the same reasons it is always advisable to warm up a line slowly from the cold condition by merely cracking the main valve, or by opening a small bypass around it.   The main valve should be opened fully, and then in a cautious manner, only after the line is well warmed.

Steam traps for removing collected drips from separators, drip pockets, tapped fittings, etc., are of numerous types such as bucket, float, tilting, thermostatic, and the like.

Traps should be located if possible so that condensation will flow into them by gravity, thus assuring positive drainage.   Where a pipeline is in a trench below floor level, it is sometimes impossible to locate the trap below the pipe without

placing it in a more or less inaccessible pit.   Under such circumstances it may be permissible to place the trap above the floor, provided line pressure is sufficient to lift condensate to that elevation.   This is never as satisfactory, however, as having gravity drainage to the trap.   Where it is necessary to locate a trap above the line

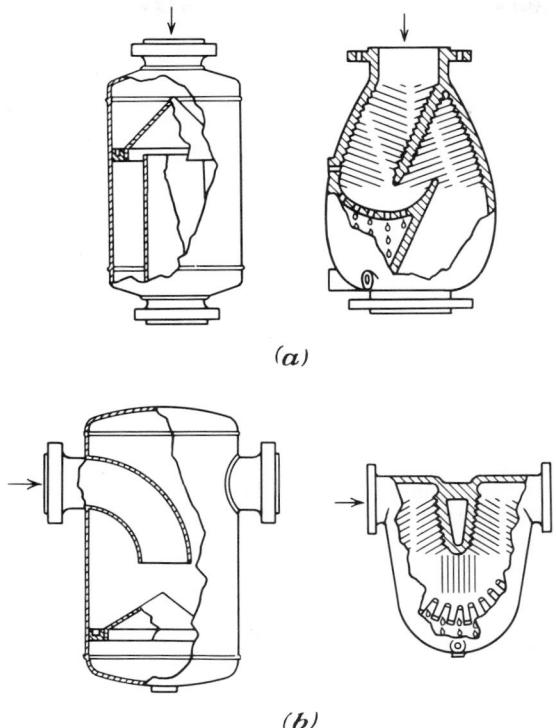

(a)

(b)

FIG. 13. Steam receivers and moisture separators: (*a*) vertical types; (*b*) horizontal types.

it drips, operation can be made more positive by providing lift fittings, such as those shown in Fig. 14.   The action of lift fittings is to discharge alternate slugs of water and steam, thus lightening the weight of the discharge column and permitting higher lift.

The valve seats of traps are easily cut by dirt or pipe scale lodging in them, and strainers, or sediment pockets, such as that shown in Fig. 15, are frequently provided in the drip lines ahead of the trap to intercept such material.   Where a trap is installed so that the steam line does not drain by gravity through the trap when the steam pressure is off, an open-drain connection and valve should be provided to let condensate run out on the floor or through a funnel into the sewer.   Often there are places in pipelines which should be provided with open blows rather than traps to look after temporary conditions in warming up the line or to drain a dead end after the closure of its supply valve.   Aside from the first cost of traps, they are a continual source of trouble and expense to maintain, and their use should be avoided wherever practicable through the use of open blows or other devices.   For high-pressure service all connections should be welded and the strainers may be integral with the traps.

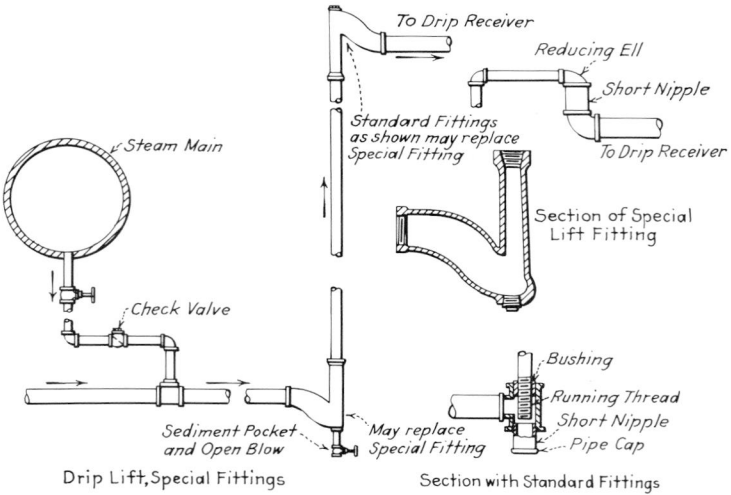

FIG. 14. Drip lifts.

As a substitute for steam traps or where drips are being collected from a source of possible contamination such as fuel-oil heaters, many engineers prefer manually operated condensation receivers equivalent to that shown in Fig. 16. With this arrangement, drips are allowed to collect in the receiver, which may have two compartments, until they reach some predetermined level which can be observed in the gage glass. Periodic inspection reveals when this level is reached, and steps are taken then to empty one compartment. Such receivers are applicable to either pressure or vacuum lines, the chief difference in the installation under the two circumstances being in the method of emptying. Where it is used to drip lines having sufficient pressure, the receiver can be emptied by merely opening the discharge valve. If under low pressure or vacuum, the valve in the drip line entering the receiver must be closed first, the discharge valve opened next, and then either (1) an atmospheric vent may be opened if the elevation of the receiver is enough to cause the contents to flow by gravity to the point of disposal or (2) a valve may be

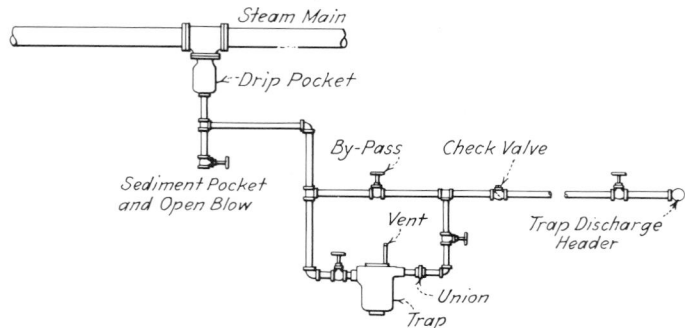

FIG. 15. Typical drip connection with trap.

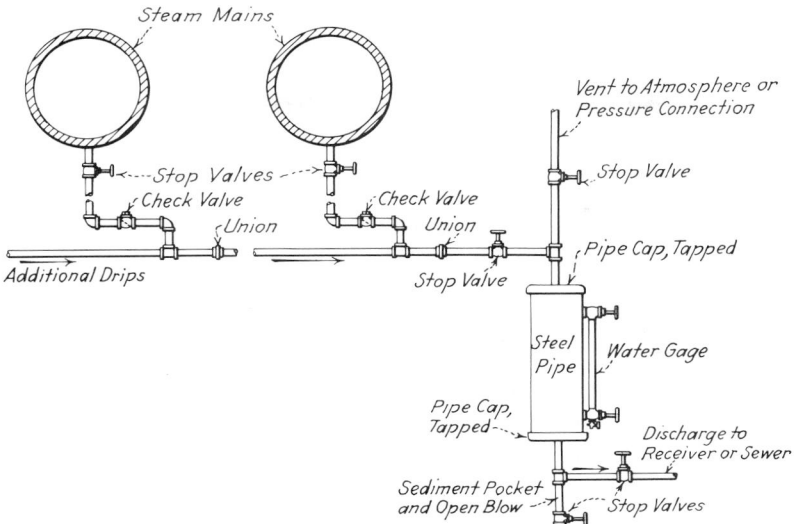

FIG. 16. Manually operated condensation receiver.

opened in a line that applies steam or compressed-air pressure above the surface of the liquid in the receiver, thus forcing it to flow to the point desired.

In instances where there is small difference between line pressure and that existing where the drips are to be discharged, it may be advantageous to let the discharge take place through a water leg or "siphon trap," as it is sometimes called, which is illustrated in Fig. 17. A siphon trap of this kind renders unnecessary the use of a mechanical trap and requires no maintenance. It consists of two legs A and B, which may be close together or any distance apart but the lengths of which must be sufficiently great to prevent pressure acting through pipe I from forcing the water level down to the bottom of leg A. C is a vent pipe connecting with the vessel into which O discharges. In case O discharges against atmospheric pressure, C may be

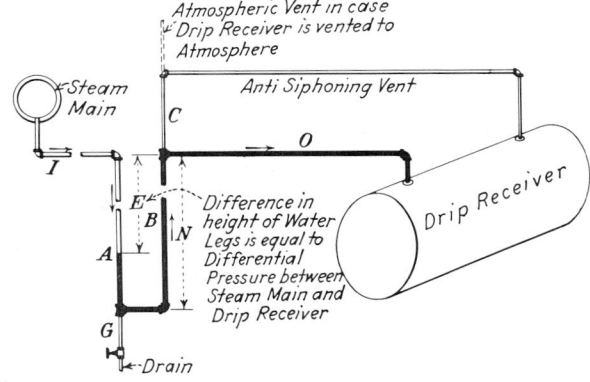

FIG. 17. Water leg or siphon trap.

vented directly to atmosphere.   The purpose of vent *C* is to prevent siphoning out the water leg in case *I*, *A*, *B*, and *O* should run completely full of condensate for a time and then have the supply suddenly diminish.   In ordinary operation the leg *B* is filled with water which is constantly overflowing and *A* with steam and water, the total pressure in both legs being equal.   The siphon trap is applicable only to small pressure differences, as it requires approximately 2.3 ft of vertical space *E* for each pound per square inch pressure differential.

In steam lines a liberal pitch should be given toward drainage points.   In all horizontal pipes where there is considerable velocity, pitch should be in the direction of steam flow.   Low-pressure, low-velocity, building-heating-system pipes, however, are frequently pitched against steam flow, in which case it is customary to use the next size larger pipe than would otherwise be chosen.   With long lines subject to considerable expansion both horizontally and vertically, the position of the line in both the hot and cold positions should be determined and the pitch made sufficient to look after both.   Important lines should be leveled with surveyor's instruments to assure that the calculated pitch is actually obtained.   A minimum pitch of 1 in. in 20 ft to 1 in. in 10 ft is usually sufficient for adequate drainage.

Where two or more traps discharge into a common header, it is necessary to provide a check valve in the discharge of each trap before it joins the header.   It is also advisable to install a stop valve in each discharge line between the check and the header to facilitate repairs to any trap or check valve.   Discharges from high- and low-pressure traps should not be connected to the same header unless this is done at or close to a low-pressure drip receiver.

Provision should be made for the gravity drainage of water from pipes and equipment or tanks containing water during periods of shutdown for repairs or where necessary to prevent freezing when not in use.   In case water so removed is plant condensate or boiler feed, it may be desirable to provide a drain line leading to storage, otherwise discharge should be made to the sewer or equivalent.   The size of drain required will depend on what is considered a reasonable time to empty the piping or equipment.   In the case of a large, elevated, water-storage tank for fire protection, a reasonable time might be several hours, while in the case of a boiler-feed line, boiler or condensate storage tank, etc., 1 to 2 hr at the most should be long enough.   In making provision for such drainage, attention is called to the necessity for providing a vent to atmosphere above the water line.

Drip Disposal.   The most advantageous means of returning hot drips to the boiler or feedwater circuit will depend on individual plant arrangements.   Discharge to the hot tanks of plants heating their feedwater with auxiliary exhaust steam and having hot-water storage is frequently good practice.   Where this is done, the high-pressure traps or condensation receivers can be arranged to discharge directly into a header leading to the hot tanks.   The low-pressure drips also can be discharged to the hot tanks either by their own pressure and gravity or, where the relative elevation of hot tank and traps require it, by the use of lifting traps or pumps.

Drip Receiver with Pump.   The turbine side of all extraction shutoff valves should be connected to the condenser through a bucket trap suitable for vacuum service.   This is done to prevent build-up of drips in the extraction piping which may damage the turbine.

The heater side of the shutoff of valves should be connected to a suitable receiver by traps.   The extraction stages above the deaerator may be dripped to it, while below the deaerator, those normally operating above atmospheric pressure may be connected to a vented "drip and drain tank" or to the condenser.

Simple Steam Loops and the Holly Loop.   The use of steam loops to return water to boilers in small plants without the use of a trap, pump or injector is described in "Steam Power Plant Engineering," by G. F. Gebhardt, 6th ed., page

702. The Holly loop, which is an application of the same principles to larger plants, also is described by Gebhardt.

Extraction Heater Piping.   Each regenerative steam power-plant cycle will contain one or more extraction (or bleeder) feedwater heaters.   Those heaters in which the feedwater flows through many parallel tubes and in which extraction (or bleed) steam is outside the tubes are known as surface heaters or closed heaters. If the heater is so arranged that the steam and feedwater mix intimately with each other, it is referred to as an open or contact heater.   Figure 18 shows the essential piping associated with a horizontal-type surface heater.   The same arrangement of piping, modified in location, would also apply to a vertical heater installation.

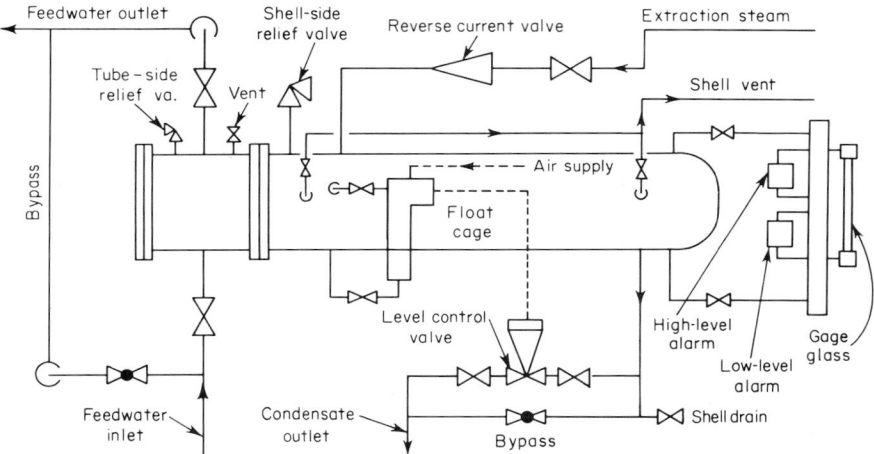

Fig. 18.   Piping schematic at horizontal-type surface extraction feedwater heater.

Gate valves are installed in the feedwater inlet and outlet lines adjacent to the heater.   As determined by the size of the installation and its contemplated mode of operation, the valves may be manually-, gear-, or motor-operated.   A bypass valve of the gate or globe type permits feedwater to pass through the system piping if the heater is out of service for repairs.

Condensate which results from the extraction steam is removed automatically and continuously through a level-control valve.   The level controller consists of a float within a cage; as a function of condensate level in the heater shell, the float operates an air pilot valve in such a way as to cause transmission of an air-loading (or air-unloading) signal to the diaphragm of a level-control valve.   As shown, the level-control valve is arranged so that it may be isolated for repairs and condensate removed by manual operation of the bypass valve.   The gage glass should be located sufficiently near this bypass valve so that the operator may have knowledge of condensate level within the shell.

Level within the heater shell is observed through the gage glass.   During normal operation, in most plants, an operator is not assigned solely to watch the heater, and for this reason, it is the usual practice to provide high- and low-level alarms which are arranged to transmit an electrical visual or audible signal to the central control room of the plant.

In order to prevent excessive tube-side pressure, as would occur if the feedwater inlet and outlet and bypass valves were closed inadvertently while steam was

admitted, a relief valve is installed on the head of the heater. A vent valve is also installed on the head and is used primarily during periods of start-up and testing.

The shell is provided with a relief valve which will restrict the pressure to a value consistent with the design pressure of the shell. Vents, leading to an area of pressure lower than that of the extraction supply, are provided in order to remove noncondensable gases from the shell.

The *bleeder steam line* from turbine casing should be of generous size, and particular care taken to avoid restrictions or types of valves which cause appreciable pressure drop. This is especially important in bleeder lines, since any drop in steam pressure is attended with a corresponding decrease in the saturated-steam temperature obtaining in the heater shell. This in turn results in a lowered outlet water temperature and less heat reclaimed from finding its way to the main condenser. An economic study of the thermal loss resulting from pressure drop in bleeder lines will reveal that extra investment to reduce frictional loss may pay big dividends. Some form of nonreturn or check valve is required in bleeder lines to serve a dual purpose: first, in case the turbine governor trips, it prevents back feeding of steam from the bleeder heater with the attendant possibility of overspeeding the turbine; and second, it prevents water entering the turbine in case the heater shell floods, owing to tube rupture or other causes. Certain types of check valves are unsuitable for use in bleeder lines because their construction is such as to produce a material pressure drop. Furthermore, because of the hazard which would result from failure of the commonly used check valves, a positive type of closure is desirable. Such positive closure is available in a piston-operated reverse current valve. This type of valve is provided with an air-operated piston which forces the check against its seat if, for any reason, pressure within the shell exceeds the extraction-line pressure.

Provision for *air vents* from bleeder heaters is necessary. In the case of surface heaters, such vents are usually taken off from an air box where the air is cooled and dried by contact with tubes situated in the coolest pass of the heater. Air vents should be of ample size to evacuate air from the shell quickly in starting up the unit. It is then customary to restrict the passage so that excessive amounts of steam will not be drawn off with the air. Heaters operating at pressures above atmospheric pressure may be provided with a thermostatic vent to atmosphere, and those under vacuum vented to the main condenser. Better economy is obtained, however, by cascading the air vents successively to the shells of lower stage heaters.

Boiler-feed Piping. Feed piping between the economizer (or boiler drum, if no economizer is fitted) should correspond to the requirements of the ASME Boiler Construction Code. Check valves should be provided in the individual pump discharges and in the common feed line adjacent to the boiler.

Boiler-feed discharge piping is always of carbon-steel material and, most often, is specified to conform to ASTM A106, Grade B. Its wall thickness is governed by the pressure to which the piping will be subjected.

Boiler-feed pumps may be driven by individual electric motors or by small steam turbines, or they may be direct-connected to the shaft of the main turbine. The pump capacity is determined by the requirements of the thermal cycle, and this total capacity may be distributed between any desired number of pumping units. Three half-capacity pumps, one of which is a spare, have been a popular choice in past years. More recently, plant designers have selected a single full-capacity pump; this may be backed up by a second full-capacity pump or by one of smaller capacity.

On large high-pressure units, feedwater velocities between 10 and 20 fps are used depending on the fuel costs and load factor. Above 100 megawatts it may be found economical to use more than one string of high-pressure heaters, in which case the

feedwater piping must be sized with the possibility of a string of heaters being taken out of service to maintain one of the high-pressure heaters.

The design pressure of the feedwater piping depends on the location of the feed-water regulator valve. Between the pumps and the regulator, the piping should be designed for pump shutoff pressure, while between the regulator and the econo-mizer, the design pressure may be the sum of economizer design pressure, piping and heater friction drop, and static height. The inadvertent closing of a valve and consequent pump shutoff pressure in the line should not occur frequently, and the resulting pipe stress, should this almost inexcusable event occur, must not exceed the allowable pipe stress by the 15 per cent margin which is allowed by both the boiler and piping codes.

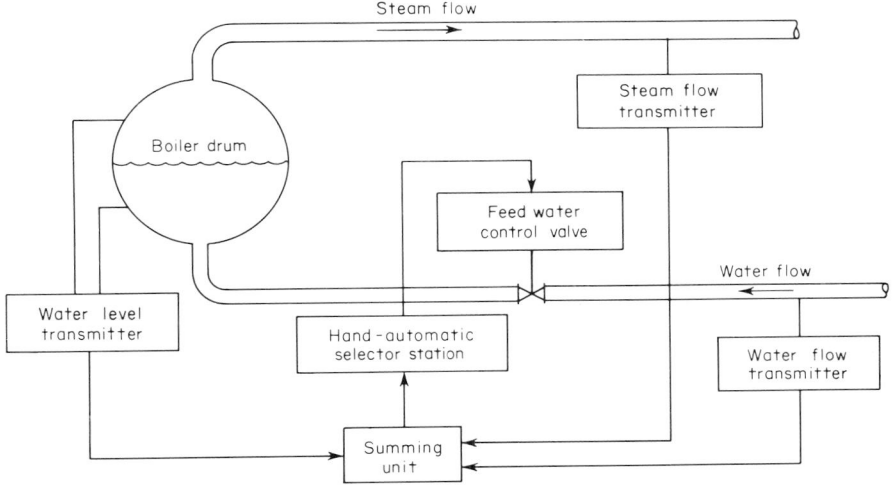

FIG. 19. Three-element feedwater control system.

Where a hydraulic coupling is used between the driver and the pump, the feed-water valve is usually omitted, in which case the feedwater piping design pressure may follow the latter of the above cases.

Except for small industrial plants, a three-element feedwater controller system is used for drum level control. A typical system is shown in Fig. 19. In the case of once-through or supercritical units where the drum is omitted, the feedwater flow is tied to the combustion control system, as shown in Fig. 20.

Boiler-feed Injectors. In small boilers where it is desirous to save the cost of a standby pump, it is the practice to use an injector. The following description is reproduced from "Steam Boiler Engineering," by permission of the Heine Safety Boiler Company.

Injectors are made in many forms, but Figure 21 shows the typical arrangement and illustrates the method of operation. Steam is admitted through the valve M, by turning the handle K, and enters the expanding nozzles where the pressure is reduced and the velocity greatly increased. The steam jet is then guided to the contracting nozzle or lifting tube V. In passing from the first to the second nozzle it carries along the air in the chamber and creates a vacuum. The water to be pumped rises in the suction pipe and fills the chamber. The steam and water thus enter the lifting tube, passing to the mixing nozzle C, and the steam is condensed. When the water and steam have reached the delivery nozzle D, the steam has been condensed and the water is traveling at a high velocity imparted to it

by the steam.   The delivery nozzle is increased in cross-sectional area, reducing the velocity and hence increasing the pressure of the water.   Consequently its head is sufficient to overcome the resistance of the feed valve, and the water enters the boiler.   The steam has thus imparted kinetic energy to the water; this energy is converted from velocity to pressure in the delivery nozzle.   The water is heated through the condensation of the steam.

The action of the injector depends not only upon the impact of the jet of steam, but also upon its efficient and complete condensation, which must occur during its passage through the combining tube.   At 180-lb boiler pressure the water must attain a terminal velocity of 163 ft per second to balance the pressure, and something more to lift the check valve and

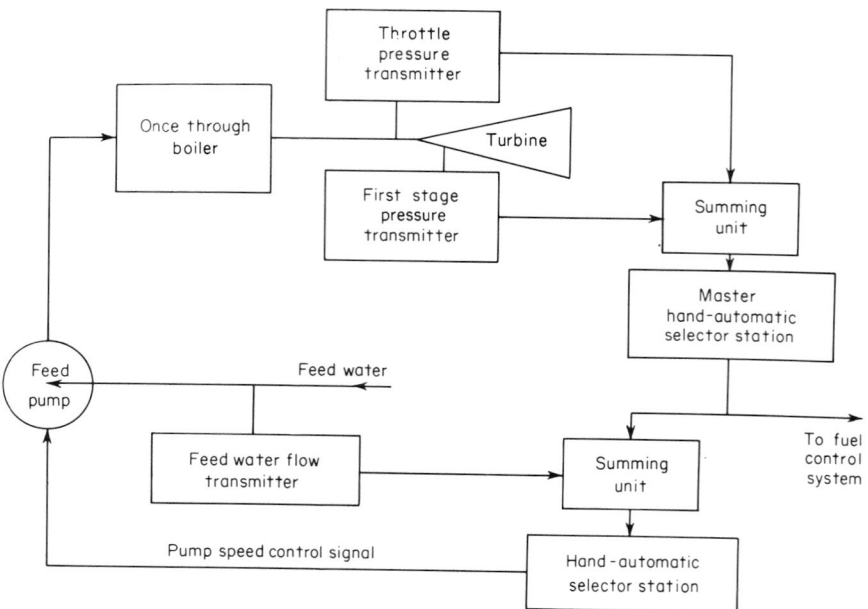

FIG. 20. Feedwater control system for once-through boiler.

enter the boiler.   If the total length of the converging combining tube is 7½ in., the interval of time during which the steam can be condensed is only 0.008 of a second and the acceleration is 4 miles per second per second.

Anything that tends to diminish rapid condensation operates against mechanical efficiency.   An increase in the temperature of the water supply, moisture, or superheat in the steam all tend to reduce the proper ratio between the weight of the water delivered into the boiler and that of the motive steam.   The steam must undergo instant and complete condensation, and its velocity must reach a maximum at the instant of impact with the water.

Boiler-feed Pump Suction Piping.   Whether the suction piping comes from a direct contact heater (deaerator) or closed feedwater heater, it is necessary to comply with the requirements set forth by the pump supplier regarding net positive suction head (NPSH).   In addition, the static height of the deaerator or excess pressure in the case of a closed heater system must be sufficient to accommodate transient feed conditions, the most severe of which would be trip-out at capability operation.   Where, heretofore, the suction piping was sized for low velocity (2 fps or less) and the deaerator set well above prescribed static heights to provide hoped-for, ample

margins above required NPSH, more exact methods of calculation have been advanced wherein suction pipe diameters have been reduced to provide optimum residence time in the piping and the deaerator is placed at an optimum height.

Condensate Piping. The suction and discharge piping of the hotwell or condensate pumps between the condenser and the deaerator (or the closed heater from which feedwater to the boiler-feed pumps is taken) is called condensate piping. Some low-pressure installations are equipped with Schedule 40 black pipe and 125-lb cast-iron flanges and flanged fittings for both suction and discharge piping.

In an all-welded installation the suction piping is 150-lb standard, while the welded-steel standard used for the discharge piping depends on the cycle employed and may be 150 or 300 lb or higher.

Because of the thinner wall required, the optimum water velocity in the condensate pipe is lower than in the boiler-feed system and generally does not exceed 10 fps.

The suction piping should include at least a temporary strainer to catch the debris that usually finds its way to the condenser hotwell on plant start-up.

To prevent contamination of the elements of the cycle that would result from excessive condenser tube leakage, a conductivity cell with alarm is usually installed on the condensate discharge piping and a dump valve leading to waste is opened when the alarm level of contamination is reached.

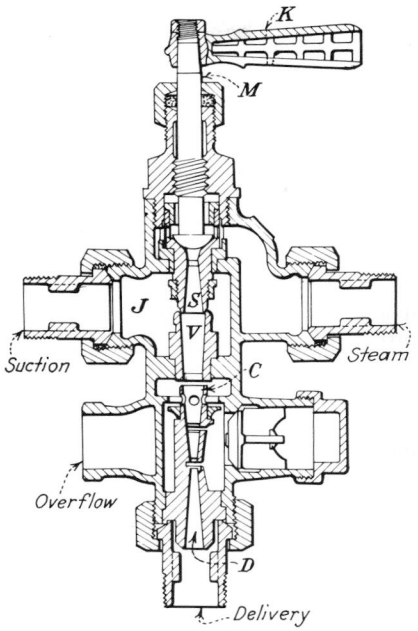

Fig. 21. Boiler-feed injector.

Because of limited basement headroom the hotwell pumps are usually of the vertical type, resulting in the suction piping being located in a trench. When headroom is available, a horizontal type pump is used as shown in Fig. 22.

Blowoff Piping. Boiler blowoff connections vary in size from 1 to 2½ in. depending on the size of the boiler. Where several blowoff connections join into a common header, the header is often 4 in. or larger. Evaporator blowoff connections are frequently made as large as 4 in. to furnish easy egress for scale accumulations. The ASME Boiler Construction Code requires that where blowoff lines from more than one boiler are connected to a common header, a guard valve be provided to prevent workmen being scalded in any boiler which is down for repairs. Where there are several blowoff connections on one boiler, each should have its own blowoff valve, and a single guard valve may be provided where the combined lines from that boiler join a common header serving other boilers. As an alternate arrangement, some engineers prefer to provide a blowoff valve and a guard valve in tandem on each blowoff connection. A typical blowoff valve and guard valve are illustrated in Fig. 23. Among the principal design features in blowoff valves are quick opening and free passage for sediment or scale. Guard valves for blowoff lines are frequently of the conventional gate-valve type. In order to protect the seating surfaces of the guard valve and assure its continued tightness,

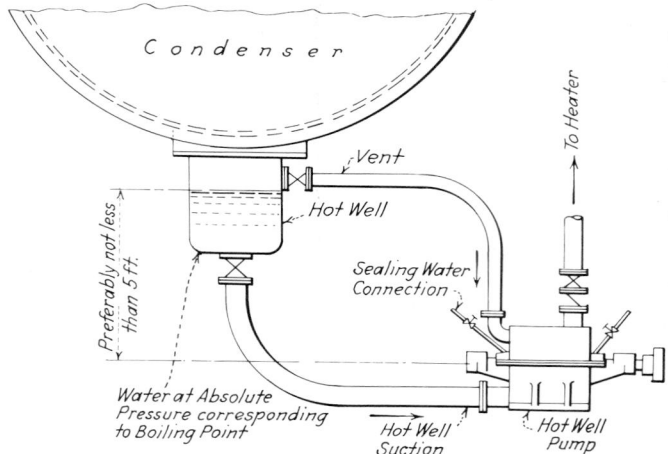

Fig. 22. Hotwell pump piping.

it is customary to open the guard valve wide before cracking the blowoff valve proper.

Boiler blowoff piping between guard valve and blowoff tank or circulating water overflow canal where pressure is reduced to approximately that of the atmosphere is no longer covered by the Code for Pressure Piping. In present-day practice, in order to provide adequate wall thickness to prevent erosion failure, Schedule 80 pipe as a minimum is used; blowoff piping is designed for a pressure of at least 25 per cent of boiler drum design pressure.

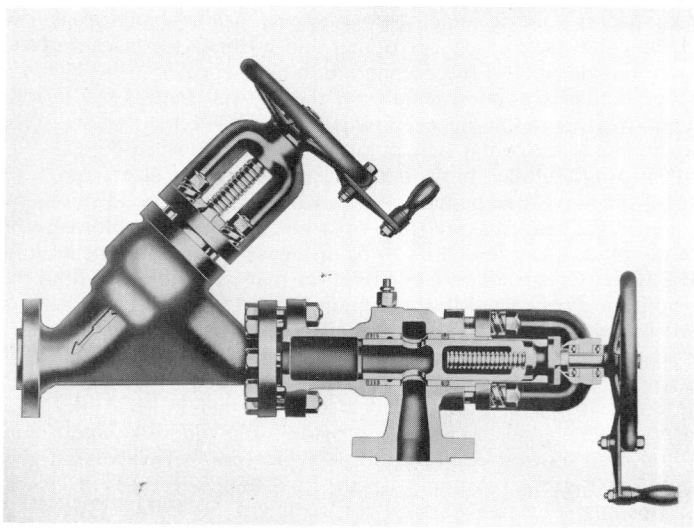

Fig. 23. Blowoff valve and guard valve.

Flashing mixtures of water and steam occur in boiler blowoff lines, in evaporator and feedwater-heater drain lines, and in other piping where hot water flows at a pressure less than the saturated pressure corresponding to the temperature of the water entering the line. The actual increase in specific volume depends on the end-pressure condition which, in turn, depends on the pipe size and length of travel and is independent of the pressure in the receiving vessel (or atmosphere) so long as the latter is below the calculated end pressure.[1] Owing to the characteristic of saturated water to flash as the pressure falls, the pipe tends to choke with vapor so that the back pressure in the line remains a substantial part of the initial pressure and is not reduced almost to atmospheric or receiver pressure as might wrongly be assumed.

This fact should be kept in mind in choosing piping weights for boiler blowoff lines. Those wishing to make a specific check of conditions can refer to the basic methods given in the Benjamin and Miller paper or to "Charts for Estimating Capacity and Outlet Pressure of Boiler Blowoff Lines" derived from their methods.[2]

In the case of heater and evaporator drain lines it often is desirable to put the flow-controlling device, whether drainer, trap, or orifice,[3] near the outlet end of the line so as to avoid flashing by maintaining saturation pressure as far as possible. Where severe flashing with attendant high velocity exists in lines carrying a mixture of steam and water, there is apt to be excessive erosion, especially at elbows and bends, which may eat through the walls in the course of a couple of years.[4]

Blowoff piping from the boiler drum to the guard valve should be of the weight prescribed in the ASME Boiler Construction Code for the rated working pressure of that boiler. Adequate protection should be provided for blowoff piping where exposed to hot gases within the boiler setting.

Blowdown Tanks. Where a canal is not available to receive the blowdown from power boilers, it is frequently necessary to provide a blowdown tank, such as that illustrated in Fig. 24. Local ordinances generally require that such tanks be provided to receive and cool the blowdown before it is discharged into a public sewer. Blowdown tanks are usually low-pressure chambers vented to the atmosphere, where blowdown can be collected and cooled by mixture with city or general service water and then allowed to flow gently into a sewer under gravity head. In Fig. 24, blowdown enters the tank tangentially through a water mixer into which city (or general service) water is introduced under a moderate pressure of about 40 lb gage. If blowdown is discharged directly into a tank of cold water without provision for proper mixing, severe water hammer will result. An alternate arrangement to that shown is to supply cold water direct to the tank and introduce blowdown through an internal fitting called a "noiseless water heater" or "hydrokineter" similar in principle to the canal fitting shown in Fig. 26.

In order to avoid water hammer and secure satisfactory cooling of the blowdown, sufficient city water must be provided to lower the temperature of the mixture below 212 F. The actual amount of city water required is usually two to three times the theoretical, since blowdown is not continuous and sufficient cold water must be flowing to look after the instantaneous flow of blowdown. As a result, the temperature of the tank overflow may be as low as 130 F rather than the permissible

[1] See "The Flow of a Flashing Mixture of Water and Steam through Pipes," by M. W. Benjamin and J. G. Miller, *Trans. ASME*, vol. 64, no. 7, pp. 657–669, October, 1942.
[2] See "Tube Turns Catalog and Engineering Data Book," No. 111, published by Tube Turns, Inc., Louisville, Ky.
[3] For designing orifices see "The Flow of Saturated Water through Throttling Orifices," by M. W. Benjamin and J. G. Miller, *Trans. ASME*, vol. 63, no. 5, pp. 419–429, July, 1941.
[4] See "The Flow of a Flashing Mixture of Water and Steam though Pipes," by M. W. Benjamin and J. G. Miller, *Trans. ASME*, vol. 64, no. 7, pp. 657–669, October, 1942.

maximum of 212 F. Meters are sometimes installed on the city water supply and overflow to keep a record of the amount of blowdown and the quantity of city water used. The blowdown is obtained by taking the difference of the two meter readings.

The tank itself serves the purpose of a vented chamber at atmospheric pressure, floating on the line between the blowoff line and the sewer. Its size is of little importance provided the overflow pipe is adequate as to diameter and elevation above the sewer connection to handle the total amount of blowdown and cooling water. If a drain cannot be provided large enough to handle this rate of flow, the tank must be made of dimensions ample to accommodate the blowdown and cooling water collected during the blowing down operation of one valve. The contents of the tank then can be allowed to drain off before blowing the next valve.

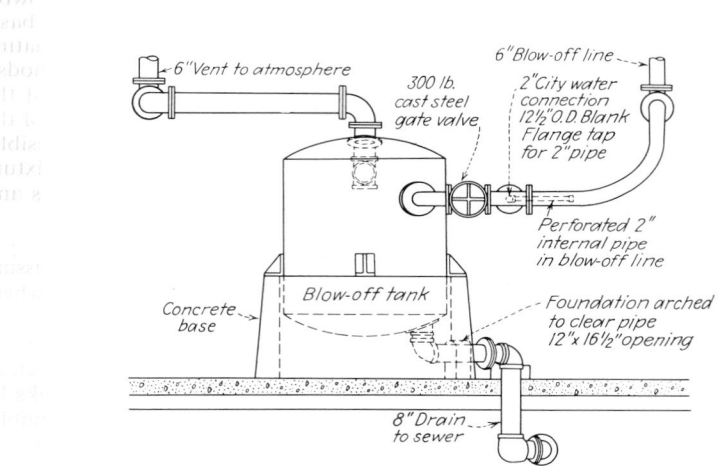

Fig. 24. Blowdown tank.

Canal Blowoff Fittings. Where a canal is available to receive blowdown, it is customary to dispense with the blowdown tank and inject the blowdown directly into the canal with a mixing fitting, such as that illustrated in Figs. 25 and 26. This saves considerable investment cost, is easier to operate, and eliminates any expense for furnishing cooling water to the tank.

Continuous Blowdown. In boiler plants where there is a large amount of make-up water, it is often desirable to blowdown a small amount of water continuously rather than a large quantity at intervals. The chief advantage of a continuous blowdown system is that it permits the heat to be transferred from the blowdown water. A satisfactory arrangement of a continuous blowdown system is shown diagrammatically in Fig. 27. A needle valve is used to control the flow from each boiler into a header of generous size where, because of the reduction of pressure, a portion of the water flashes into steam. The header empties into a separating tank from which the flash steam is vented to an exhaust header. The remaining water flows by gravity through a heat exchanger where its temperature is reduced on its way to the sewer.

The quantity of make-up to a cycle varies between approximately 0.5 per cent for large installations which use air for soot blowing and mechanical atomizing for the fuel oil, to approximately 3 per cent where steam is used for soot blowing and fuel-oil atomization. The 0.5 per cent above is due to blowoff and blowdown.

In a plant using evaporator make-up the make-up required may be reduced by running the boiler continuous blowdown to the evaporator where it may be reconcentrated from the dissolved solids concentration maintained in the drum to that maintained in the evaporator. A savings in fuel will also result from this procedure.

Instrument Piping. The pressure standard chosen for instrument piping should, in general, agree with that of the line to which it is connected. Instrument piping connected to superheated steam piping need not be made of materials capable of withstanding full temperature, provided it is used in such a way as to constitute a dead end where line temperature does not obtain. The fittings at the point of connection, however, should be suitable for the maximum temperature.

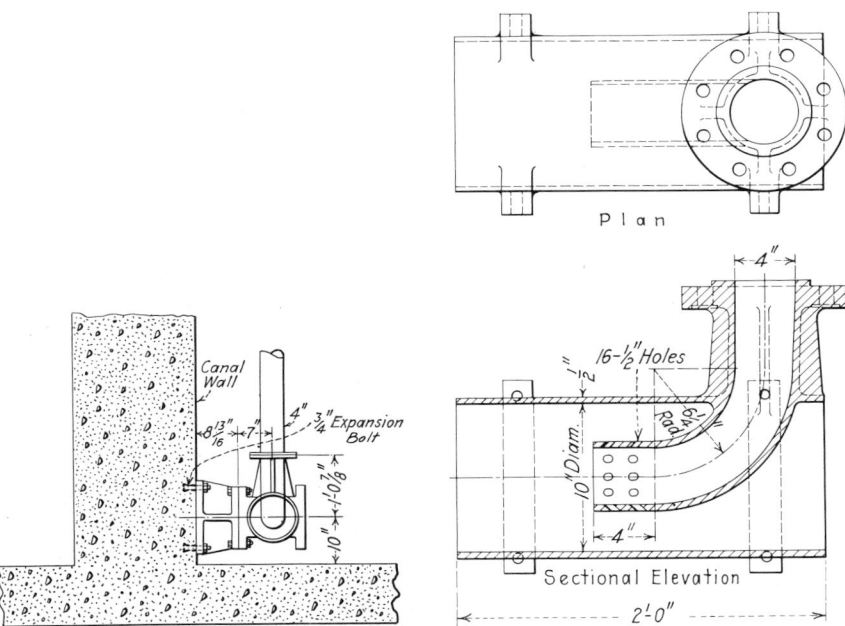

Fig. 25. Installation of blowoff fitting in canal.

Fig. 26. Detail of canal fitting for blowoff line.

Where high pressure is involved, it is not advisable to use pipe smaller than $\frac{1}{2}$ in. nominal size on account of the greater danger of mechanical breakage with very small pipe. The use of $\frac{3}{4}$-in. Schedule 160 source nipples with $\frac{1}{2}$-in. valves counterbored to receive $\frac{3}{4}$-in. pipe at the inlet end and $\frac{1}{2}$-in. pipe at the outlet is a typical construction (see Fig. 28). Copper tubing, if service conditions permit, is used for instrument piping because of the ease with which it is bent and its neat appearance. Thin-walled steel tubing with flared tube connectors or seamless steel pipe is recommended for temperatures above 406 F.

The instrument piping proper begins at the outlet of the source valve. For line temperatures above 406 F and pressures from 200 to 1,200 lb, seamless-steel thin walled tubing and steel flared tube connectors or seamless-steel pipe and forged-steel screwed fittings are used. For temperatures below 406 F and pressures from 200 to 1,200 lb, copper tubing and brass flared tube connectors are usually employed. Low-pressure lines, pressures 0 to 200 lb, under dead-end service, such as

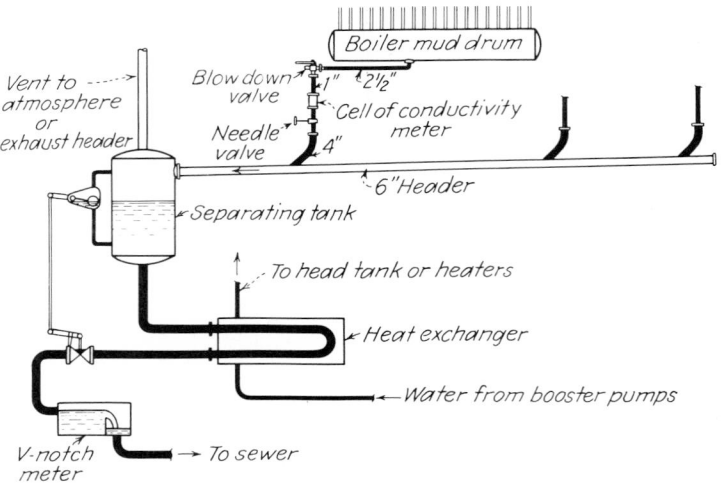

FIG. 27. Diagrammatic arrangement of continuous blowdown system.

connections to pressure gages, copper tubing, and brass flared tube connectors are used for temperatures up to 700 F. For continuous service at these pressures for temperatures above 406 F, steel tubing and steel flared tube connectors are required. For service under vacuum, copper tubing and soldered-type brass fittings are used. Steel bar-stock valves are used at gages and on open blow service.

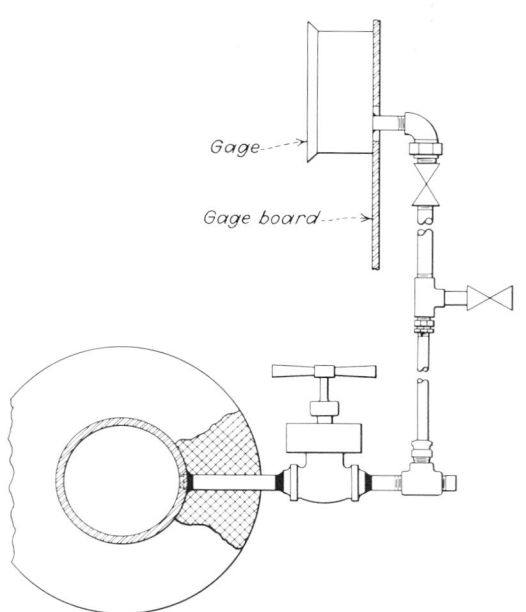

FIG. 28. Typical assembly of instrument piping.

Table 5. Typical Specification for Instrument Piping Systems

Instrumentation line size from 1/4" to 5/8" incl.	Service			
	Instrument or loading air	Furnace draft	Low-pressure metering	High-pressure metering
Tubing material	ASTM B75 or B68	Type DHP seamless soft annealed	(Maximum duty: 800 psig, 406 F) Type DHP seamless soft annealed	ASTM A213 Grade TP347
Tubing size and thickness, in.	1/4, 0.035; 5/16, 0.035; 3/8, 0.035; 1/2, 0.049; 5/8, 0.065	1/4, 0.035*; 5/16, 0.035*; 3/8, 0.035*; 1/2, 0.049; 5/8, 0.065	1/4, 0.035; 5/16, 0.035; 3/8, 0.035; 1/2, 0.049	3/8, 0.049; 1/2, 0.065; 5/8, 0.083; 3/4, 0.109
Piping material	ASTM B43, regular weight	1/2" Schedule 40 A106, Grade B		
Material of fittings in tubing lines	Three-piece, 37-deg flared brass	Three-piece, 37-deg flared brass	Three-piece, 37-deg flared brass	Three-piece, 37-deg flared steel
Material of fittings in piping lines	125-lb screwed ASTM B62, ASA B16.15	150-lb ASTM A197, ASA B16.3		
Joints in tubing	To match fittings	To match fittings	To match fittings	To match fittings
Joints in piping	ASA B2.1, screwed	ASA B2.1, screwed		
Valves	200-lb screwed bronze	4-way cocks 1/4"* Scr. Brass 1/2" Scr. Brass	Screwed, with male connectors	Screwed or socket welded
Tubing material	ASTM B88, Type K	....	(Maximum duty 900 psig, 800 F) (Minimum pipe size—1/2") (Pipe, fittings, joints, and valves to be same as on run of main piping except as has been specified for instrument line sizes from 1/4" to 5/8" incl.)	(Duty above 900 psig, 800 F) (Use 1-inch connections only) (Pipe, fittings, joints, and valves to be same as on run of main piping except that tubing as has been specified for line sizes from 1/4" to 5/8" incl. may be used. Alternatively: use A 106, Grade A for 3,000 psig, 800 F and lesser; use chrome molybdenum for 3,800 psig with temperatures of from 800 F to 1100 F.)
Piping material	ASTM B43, brass	A 106, Grade A. Schedule 40		
Material of fittings in tubing lines	Wrought copper with NPT ends	....		
Material of fittings in piping lines	125-lb screwed ASTM B62, ASA B16.15	150-lb ASTM A197, ASA B16.3		
Joints in tubings	Solder joints	....		
Joints in piping	ASA B2.1, screwed	Screwed crosses with brass plugs		
Valves	200 lb screwed bronze	None		

* Inside panel board only

The thicknesses of steel and copper tubing suitable for service at pressures up to 1,200 psi under the temperature limitations described above are listed in Table 5.

Steam gages, thermometers, flowmeters, etc., should be located where they are readily accessible, can be easily read, and are so connected as to assure accurate readings. Standard *steam gages* have a ¼-in. male connection and are generally provided with a stop cock or valve at the gage. The Bourdon tubes in steam gages may become softened when subjected to temperatures of more than 150 F, so that steam or very hot water should not come in direct contact with the tube unless stainless-steel tubes or other temperature-resistant material is used. A gooseneck siphon or loop (Fig. 29) is used to maintain a protective water seal between the gage and steam supply. When the gage is exposed to freezing a pet cock (Fig. 29*d*) should be provided for draining water from the siphon. This pet cock should not

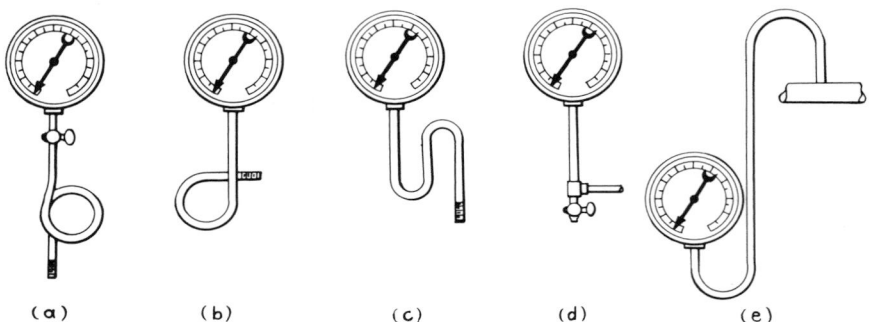

(a)   (b)   (c)   (d)   (e)

Fig. 29. Siphons for steam gages.

be opened when the gage is in service, as the water seal would be lost and the Bourdon tube would be liable to damage from contact with the steam. If a gage is placed below a pipeline (Fig. 29*c*), allowance must be made for the head of water in the seal in order to obtain correct readings. Such a correction can be made by multiplying the head of water in feet by 0.433, thus reducing it to pounds per square inch, which should be deducted from the gage readings, or the position of the needle should be changed accordingly. Where subject to vibration, gages should be securely attached to minimize its effect. Repeated jarring or continuous vibration will cause wear of the rack and pinion and result in inaccurate pressure indications.

Operating *thermometers* are commonly inserted in piping by means of the separable socket fitting shown in Fig. 30*b*. An indicating thermometer is shown in place in such a fitting in Fig. 30*a*. The same fitting can be used to receive the sensitive elements of recording thermometers, thermostats, etc. Separable sockets are made in different lengths to suit the diameter of the pipe in which they are installed. The thread which screws into the pipe is usually ¾ in. nominal size. For pressures above 600 lb and temperatures above 750 F some form of welded sockets should be used.

A typical *water column* for a power boiler for pressures up to 400 lb is shown in Fig. 31. This column is provided with a water-gage glass, trycocks, and two floats to operate a whistle for high- and low-water alarm. The gage glass is provided with a blowdown and chain-operated cocks for shutting off the gage-glass connections in case the glass breaks. A flat gage glass is supplied for higher pressures.

Suitably shaped funnels having a drain pipe are sometimes used to catch water discharged from the trycocks, and a blowdown line provided for the column and gage glass as shown in Fig. 32.   According to the requirements of the ASME Boiler Construction Code, the gate valves or cocks in the steam and water lines connecting the column to the boiler drum must be provided with a locking device in the open position.

Oil Piping.   Fuel-oil pump discharge piping for a mechanical atomizing system may run as high as 1,000 psig and should be suitable for the service pressure and temperature.   Lubricating, insulating, or low-pressure fuel-oil piping in a power plant usually operates at pressures below 125 lb gage and standard-weight black

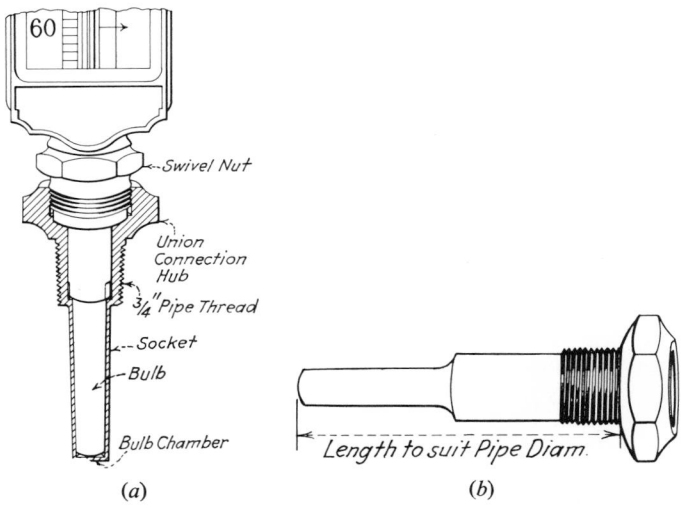

Fig. 30. (*a*) Straight thermometer with separable-socket connection; (*b*) Separable socket for thermometer shown in (*a*).

pipe and 125-lb American standard cast-iron or 150-lb malleable-iron or steel fittings are used.   Valves 2 in. and smaller are usually brass.   Oil, especially when hot, has a decided tendency to leak out around pipe threads, and special care must be taken to see that all threaded connections are properly fitted.   The white or red lead oil paste commonly used in making up threaded joints for air, steam, or water is not suitable for oil.   "Key" paste, litharge and glycerin, or shellac are generally satisfactory for making up threads in oil piping, although shellac should not be used where the oil temperature exceeds 150 F.   Glue, red glyptol, and permatex are also used as thread paste for oil piping.

Tinning the threads of the steel pipe and brass valves and applying solder make an effective method of sealing threaded joints that is sometimes employed in turbine oil-supply lines.   The use of socket-welded joints is particularly desirable for oil lines.   One-sixteenth-inch-thick asbestos or composition gaskets are suitable for flanged connections.   Natural rubber should not be used for this purpose because of its tendency to rot when in contact with oil.

It is desirable that valves in oil lines be of the rising-stem type so that it can be told at a glance whether the valve is open or shut.   This is particularly important

in oil lines because the result of having a valve in the wrong position may not be evident for some time, and then only after damage to important equipment.

Drain connections on oil piping and tanks should be made through funnels so that leakage or accidental drainage can be detected readily.   Municipal regulations forbid the drainage of oil into public sewer systems, and state laws forbid the pollution of streams with oil.   On this account, scrupulous care should be taken to lead all oil drains to a sump where proper disposal can be made.   If fire hazard is involved, this sump should be made a closed tank where combustion can be extinguished by smothering with steam or other means.

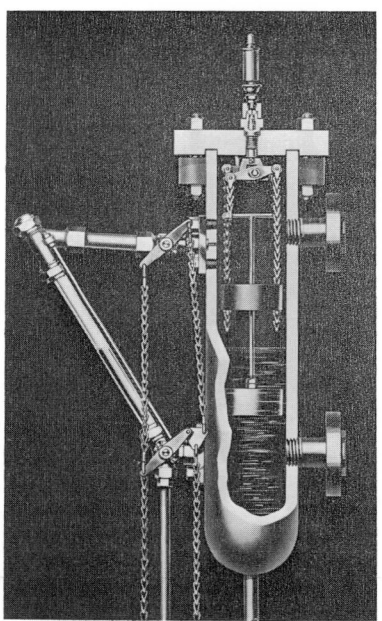

Fig. 31. Boiler water column.

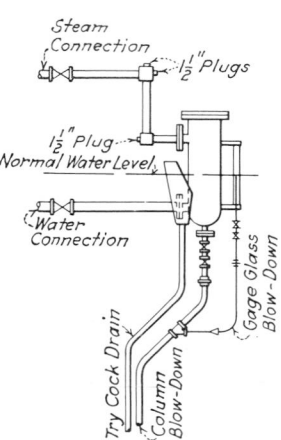

Fig. 32. Water piping.

Compressed-air Piping.   Compressed-air systems in power plants usually operate at pressures of 100 to 125 psig; for such service, Schedule 40 pipe is suitable. Joints are usually welded, although screwed or flanged fittings may be used.   Pipe and screwed fittings are frequently galvanized as a protection against internal corrosion which is rather active, owing to the moisture present in the air.   Compressed-air lines should be drained with traps of the bucket or float types.   Gate valves are generally used in the headers and principal branches, and globe valves at hose connections and other outlets.   A large receiver tank is sometimes provided at a reciprocating compressor outlet to steady pulsations and to serve as a separator for removing moisture.   It is good practice to bring the suction line for a compressor from outdoors where the air is generally cooler and drier than in the plant. The inlet of an outdoor suction pipe should be properly shielded to prevent the entrance of rain, snow, or debris.   One-sixteenth-inch-thick red-rubber gaskets are suitable for use in flanged joints.   Threaded joints are made up with white or red lead or some of the proprietary pastes on the market.

The boiler soot-blowing system frequently uses air as the blowing medium. Such systems operate at pressures of either 350 or 500 psig. Usually the system piping is of all-welded construction with 300- or 600-lb valves and fittings being specified. Proper drainage of both vertical and horizontal runs of piping is essential in soot-blowing service in order to prevent damage to the boiler heating surface which is to be cleaned.

General Service and City Water Lines. The operating pressure for general service and city water lines in a power plant is usually between 50 and 125 lb gage, depending on the elevation which the water must reach. Where the city-water pressure required in the building exceeds that existing in the city mains, it is necessary to install booster pumps. General-service water, so called, is usually pumped from

Fig. 33. Duplex strainer.

the canal connecting with the river or lake on which the plant is situated. General-service pumps are usually motor driven. An emergency general-service header is sometimes provided to supply water to indispensable equipment, such as oil coolers, water-cooled transformers, reserve boiler-feed or storage tanks, fire-protection lines, and the like, in case of failure of the usual supply. This emergency line is usually kept under pressure at all times and suitably isolated from the ordinary header by check valves.

The suction of each pump should be equipped with a foot valve and strainer, and a small line provided from the discharge header serving all pumps to prime the pump by filling the suction pipe and pump casing. In the case of large general-service and circulating-water pumps, it is frequently convenient to dispense with a foot valve and prime the pump by means of a vacuum line connected to the pump casing. By means of this vacuum line, water can be drawn up through the suction pipe until it fills the pump casing, thus priming the pump. Each pump discharge should be provided with a check valve and a gate valve between the check valve and header. A duplex strainer, such as that shown in Fig. 33 is desirable in the discharge header to catch gravel and other small debris which may be drawn up from the canal through the coarse screen around the foot valve. Standard duplex strainer screens have a mesh with perforations varying from $3/16$ to $3/8$ in. in diameter depending on the pipeline size, although diameters $1/64$ in. or smaller are furnished on special order. Standard-weight pipe and American Standard 125-lb cast-iron fittings and valves are generally used in city-water and general-service lines. Gate

valves are usually preferred to globes in water lines because of their lower pressure drop.   Galvanized wrought pipe is commonly used in the small sizes, and black pipe for larger, although galvanized pipe is sometimes used for all sizes.   Pipe threads on water lines are commonly made with red or white lead or with some of the pipe thread compounds offered under proprietary names.   One-sixteenth-inch-thick red-rubber gaskets are generally used in flanged joints.

In all-welded systems, Schedule 40 seamless steel pipe, ASTM A53 Grade A or B, is used with 150-lb series valves and fittings.

Dry-vacuum Piping.   Schedule 40 steel pipe with 150-lb series welding fittings and valves are generally used for dry-vacuum piping.   Sufficient strength is required to withstand collapsing pressures due to atmosphere, to resist corrosion, and to carry the dead weight between supports.   Red-rubber gaskets $\frac{1}{16}$ in. thick are suitable in flanged connections.   Threaded joints are made up with red or white lead or some of the proprietary paste compounds.

OPERATING CONVENIENCE

In laying out power-plant piping, provision for convenience in operating and maintaining piping and equipment comes next in importance after the mechanical-design features discussed in preceding paragraphs.   Before attempting to generalize on how to obtain a convenient arrangement, it is necessary to point out several examples of what is good or bad practice from a standpoint of convenience.   The design methods which should be employed to assure getting satisfactory operating convenience are identical with those discussed under "Appearance."

Accessibility.   It is of paramount importance that valves which are to be operated frequently, or operated on short notice in case of emergency, be made readily accessible.   If the handwheel of a valve is much more than 6 ft above the floor, it is impossible for a man of average height to operate it unless a chain wheel, platform, walkway, permanent ladder, or some equivalent arrangement is provided. In case a valve is located below the operating floor, it is necessary to provide an extension stem or a floor stand.   Where a valve is so located that a man cannot operate its handwheel conveniently from the floor, this condition should be recognized in the design and provision made for its operation.   The positions of all valve stems and handwheels should be indicated on piping drawings, rather than leaving the orientation of the valve to chance or the erecting crew.   Cognizance should be taken of the necessity for taking large valves apart for maintenance, and for replacing small valves which are damaged or worn out.   With large valves, it is sometimes necessary to lift the parts with the aid of chain falls, and this should be taken into account in the design.   In the case of a screwed valve, it is desirable to provide one or more unions so that it can be removed for replacement.   The necessity for unions at strategic points in screwed piping is a feature which never should be neglected.

Overhead piping should be placed where it is accessible for painting, repair, or alteration without the erection of too much temporary scaffolding.   Where practicable, all piping should be made reasonably accessible from some existing floor or walkway in the plant.   Before piping is located in an inaccessible place in a power plant, consideration should be given to finding another location where it will be accessible or else providing a permanent walkway or other means of access.

Clearance.   Care should be taken to see that piping is installed so that it will not interfere with dismantling mechanical equipment, obstruct light wells, encroach on stairs, etc.   Operating aisles and passageways should be kept free from obstructions.   Horizontal runs of piping preferably should be suspended overhead, with a minimum clearance of 6 ft 6 in. underneath.   Where piping must be installed

close to the floor, it should be arranged to leave what passageways are required for operation, and stiles provided over the pipes at convenient locations.

Illustration of Accessibility and Clearance. Figure 34 is included to illustrate some of the above-mentioned points concerning accessibility and clearance. This is a photograph of the boiler-feed piping at the upper drums of a large Stirling boiler, which shows the arrangement of piping with respect to a lightwell and the walkway from which valves are operated. Some of the valves, which are too high for a man to reach from the walkway, are provided with chain wheels. It will be noted that pipes are run around the outside of the lightwell so as not to infringe on it and that overhead pipes are high enough above the walkway for head clearance. It was impossible to include enough of the boiler-feed lines in this view to make the

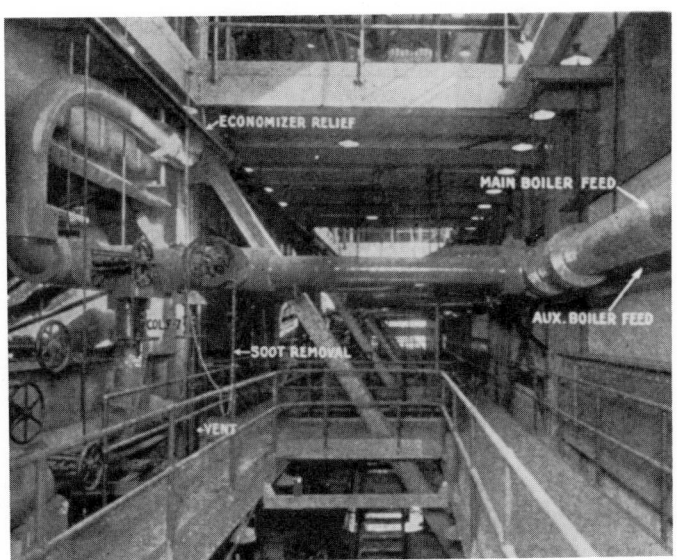

Fig. 34. Accessibility and clearance.

purpose of each valve readily apparent, although it would be so to a man standing on the walkway. The necessity for making the function of each line and valve readily understood is discussed in the next paragraph.

Making the Purpose Obvious. It is important, for the convenience of the operating force and to minimize the possibility of making mistakes, that piping be arranged so as to make the purpose of each line and valve as obvious as possible. If proper attention is given this feature in making a layout, it is nearly always possible to place each pipe so that its route is readily followed. The use of a systematic color scheme[1] with individual colors for each kind of service is of material assistance in identifying pipes in the plant. An alternate method is to stencil the name of the service at frequent intervals on each pipe.

Valves in general, and bypass valves in particular, should be installed so that their purpose is evident at a glance. It is desirable that bypass valves be grouped

[1] Reference may be made to ASA Standard A13 for a "Scheme for Identification of Piping Systems." Copies may be obtained from the American Standards Association, 345 E. 47th St., New York 17, N.Y.

together where all can be seen at once, rather than placed in scattered locations close to separate pieces of equipment.

The possibility of an error in operation can be still further reduced by stenciling the purpose of each line on the pipe close to its valve. A simple illustration of this point is furnished in a photograph, Fig. 35, of the piping for a small deaerator. One of the chief features shown in this instance is the grouping of valves at the pump rather than in scattered locations.

FIG. 35. How to make the purpose of valves obvious.

APPEARANCE

A piping layout should present a neat, coordinated, symmetrical, and general pleasing appearance which can best be described as good architecture. Certain definite principles can be stated which will aid in producing these results.

Clean-cut Detail. There are certain details in the layout of individual systems which greatly affect the ensemble. Pipes should be run parallel to the axes of the building, and 45-deg turns and offsets avoided. From the standpoint of appearance, the use of elbows is generally superior to bends, but bends result in less pressure drop and therefore are used where space conditions permit.

Coordination with Structural Design. Set aside certain definite lanes for piping to approach equipment, provide a pipe gallery at a suitable elevation in the boiler house and a vertical portion of some bay adjacent to the turbine house for vertical runs of pipe, such as main steam, boiler feed, and atmospheric exhaust. Sufficient pipe trenches are required below the condenser room and boiler-house basement floors to accommodate building heating system returns, drains, hotwell pump suction lines, and the like. A room at subbasement level for the heating returns pumps with pipe trenches leading to it is often desirable.

Coordination with Equipment Location. Arrange the plant equipment in a series of major repeating groups each comprising a main turbine, auxiliary equipment, and a group of boilers with their relative locations identical in the several

groups. This principle permits a nearly identical piping layout for each group, with consequent economies in design and erection, as well as a symmetrical and uniform appearance.

The choice of the hand of equipment (right or left hand as viewed from the same end) requires consideration. In general, single units which are repeated throughout the plant (air pumps, for example) should be of the same hand, but paired units (oil coolers, boiler-feed pumps, etc.) could be of opposite hands because of the more symmetrical arrangement of piping which results.

Coordination between Piping Systems.   Lay out the several piping systems comprehensively rather than individually. To accomplish this, it is necessary to coordinate the various systems by means of composite drawings. The composites should be completed before the individual systems are detailed. All piping going in the same direction should, in general, be arranged in a single plane. The longitudinal headers, of which there are always several, should be grouped together. Small piping, which cannot be shown on the major assembly drawings, should be installed under the direction of a competent designer.

Use of Models.   The use of models is frequently of great help in studying layouts, and large new projects often are modeled as the design progresses. After the line diagrams and preliminary piping layouts have been made, it is desirable to make a model of parts of the plant where piping is particularly congested. The use of models aids in securing a symmetrical layout of pleasing appearance which is conveniently arranged and at the same time reveals interferences and close clearances. Models may be as extensive or as curtailed as any particular case requires, but they should be made to scale and show the relative positions of piping, equipment, and building structure. The proper use of models pays for their cost many times over by (1) eliminating interference between building steel and piping or equipment, (2) providing sufficient headroom and dismantling space, and (3) obtaining good appearance and operating convenience. Models should be revised progressively as the design advances and before final working plans of the various drafting groups are approved until a generally satisfactory layout is assured.

Draftsmen sometimes are inclined to feel that a model is not required in so far as detection of interferences is concerned. Their view is that a competent draftsman carries the equivalent of a model in his mind when making a piping layout. The operators, engineers, and executives, however, find models useful in critizing proposed layouts with a view to securing more workable locations of piping and equipment. The better understanding of just how the final design will appear does much to eliminate dissatisfaction with the layouts finally agreed upon.

The materials used in constructing models vary with the inclination of the modelmaker. In the model shop of a large utility, plywood is used for floors, partitions, and boiler walls; white pine for equipment, columns, stairways, coal downspouts, etc.; wood, wire, tubing, or extruded wax for piping. Models made out of transparent plastics in the form of sheets and rods have been used.

The convenience of a piping layout from an operating point of view is of great importance. The arrangement of a bypass, for instance, should be such that the purpose of each valve is evident at a glance. In cases where there is a repetition of certain units, such as boiler-feed pumps, at intervals throughout the plant, the positions of valves, crossovers, etc., should be as nearly as possible identical. Such duplication is desirable not only from the viewpoint of appearance, but it is of immense value to the operating force in the ease with which any valve can be located in case of emergency. The extra attention given such matters in design is amply repaid by the greater convenience of a well-thought-out arrangement. It is also frequently desirable to use more pipe and fittings than required for the smallest initial investment in order to obtain simplicity in operation and repetition of detail

at similar units.   A model is of great assistance in working out these features of design as it reveals at once the advantages in different valve locations and the space left available for access to or dismantling of equipment.   Figure 36 illustrates a model study of a particularly congested area, the piping and equipment around a modern turbine.

The amount of detail shown in a model and the accuracy with which it is made to scale should depend on the kind of plant that it represents, the congestion of piping in the plant, and the desirability for operating convenience and neat appearance.

FIG. 36. Studio view of 335,000-kw 3,600/1,800-rpm close-coupled, cross-compound steam turbine generator unit model for Consolidated Edison Co. of New York Astoria Generating Station Unit 4 and Arthur Kill Generating Station Unit 2.

COSTS AND METHODS OF ESTIMATING

The total cost of piping for steam-power plants can be estimated fairly closely by using overall figures of dollars per kilowatt of installed plant capacity, if the size and type of plants compared are not too greatly dissimilar.   In estimating the cost of the separate piping systems in the plant, however, there is likely to be considerable variance because of the differences in the design and extent of the corresponding piping systems in the plants which are compared.   To overcome this difficulty, the relation between the material costs and the erection labor costs are useful.   Thus, given a particular system, the material can be readily listed from the drawings and priced, and the labor estimated as a percentage of the material.

The direct material and labor costs for the piping systems of a typical large modern power plant in dollars per kilowatt of installed generating capacity are given in Table 6.   The labor costs include erection of pipe with the associated field welding in the plant.   In addition, the shop fabrication of bends, welded manifolds,

Table 6. Steam Power-plant Piping Costs

(125 to 175 mw, 1,800 psig, 1000 F—1000 F)

Item	Approximate cost per kilowatt		
	Material	Labor	Total
High-pressure piping[1]	$1.20	$0.50	$1.70
Low-pressure piping[2]	1.40	0.80	2.20
Supports	0.35	0.30	0.65
Valves[3]	2.00	0.80	2.80
Specialties[4]	0.35	0.20	0.55
Total	$5.30	$2.60	$7.90
Percentage	67	33	100

[1] High-pressure piping consists of main steam, high-temperature reheat, No. 2 extraction steam, saturated steam, boiler-feed discharge, reheat desuperheating water, high-pressure chemical feed, and continuous blowdown.

[2] Low-pressure piping consists of all other piping and tubing except that for the circulating-water, plumbing and heating, yard, storm, and sanitary-drain piping systems.

[3] Includes motor-operated main steam valve, two motor-operated and four gear-operated circulating water valves.

[4] Includes expansion joints, traps, strainers, sight flow indicators, ejectors, float switches, test wells, gage glasses, float switches, thermometers, and pressure gages.

and subassemblies is included in the labor costs. Where welding is used, as in this installation, the labor costs represent a much larger percentage of the total cost than in the older flanged construction employing cast-steel manifolds, cast-steel fittings, Van Stoned joints, and the like.

The use of welding in conjunction with simplification of piping layouts and general advances in piping design has made possible a considerable reduction in total cost of piping. Part of this reduction in cost per kilowatt is, of course, incidental to the use of larger steam generating and turbine units. Overhead costs, including engineering design, drafting, and supervision are not included in the costs

Table 7. Direct Cost of Building Heating System of 20,000 Sq Ft of Radiation for Power Plant

Item	Cost	Per cent
Materials		
Radiation..............	$13,500	17
Radiator hangers	1,500	2
Paint..................	1,500	2
Piping system	22,500	28
Insulation................	2,500	3
Pumps, meters, tanks......	4,500	6
Total material	$46,000	58
Total labor	$34,000	42
Total cost..........	$80,000	100

Total radiation (sq ft) = 20,000
Cost per square foot radiation = $4.00

NOTE: The above values are based on the use of radiators. If unit heaters are installed, there will be a considerable savings in both material and labor.

given in Table 6.   An allowance for overhead and contractors' profit should be made in arriving at the probable total cost of a complete piping installation erected in place.

Building Heating.   The cost of a two-pipe steam-heating system for the boiler and turbine houses of a large power plant is detailed in Table 7.   The cost includes bleeder connections from house-service turbines with a live-steam pressure-reducing station for standby, all supply and return headers and branch circuits, vacuum pumps, etc., as well as the radiation itself.   The costs of material and labor are direct costs which do not include engineering, drafting, contractor's profit, or other overhead.

Plumbing.   The cost of plumbing and drains such as floor and roof drains is given in Table 8.

Table 8.   Cost of Power-plant Plumbing, Floor Drains, and Roof Drains

Average cost per kilowatt of capacity	Percentage of material	Percentage of labor
$0.36	50	50

For the underground pipes outside buildings, including city water lines, sewers, drains, service water, etc., there should be allowed approximately $0.25 per kilowatt of capacity.

The labor costs on which all the above figures are based are as follows:

Foremen	$5.25 per hour
Pipe fitters	4.25 per hour
Hanger men.......	3.90 per hour
Pipe coverers......	4.25 per hour
Helpers...........	3.20 per hour
Common labor	2.75 per hour

The combined cost for plumbing, heating, and ventilating will be approximately $1.70 per kilowatt for a 200- to 400-mw power station containing several units.

The cost of cast-iron fittings for 125- and 250-lb pressure was between 22 and 30 cents per pound.

The division of the material and labor costs among the pipe and fittings, the hangers, and the insulation is of value in making estimates.   This subdivision is given in Table 9.

Table 9.   Division of Material and Labor Costs for Piping—Average for Entire Steam Power Plant

Item	Material, %	Labor, %
Pipe and fittings	90	67
Hangers	6	11
Insulation	4	9
Fabrication		8
Handling and miscellaneous......		5
Total......................	100	100

21

WATER-SUPPLY PIPING

Frank M. Kamarck*

Water-distribution systems which serve populated areas are classified broadly as being of the loop, gridiron, or tree types. Within the broad concept, there may be a combining of all three types used as the building blocks for the overall system.

In the *loop system*, large feeder mains that surround areas many city blocks square serve smaller cross-feed lines connected at each end into the main loop.

In the *gridiron* (or grid) system, the piping is laid out in checkerboard fashion with piping usually decreasing in size as the distance increases from the source of supply.

In the *tree system*, there is a single trunk main, reducing in size with increasing distance from its source of supply; branch lines are supplied from the trunk.

The grid and loop systems provide better reliability because of their multiple paths. Grid and loop systems are often backed up with feeder pipes leading directly from the pumping station to remote distribution centers serving to bolster the supply to meet increased demands with growth of population.

The requirements of distribution systems are subject to the requirements of local ordinances and state laws. The following is an average sampling of these pressure and flow requirements:

In residential areas, normal operating conditions: 35 psig
In residential areas, peak conditions: 20 psig
Minimum pipe size for fire service: 6 in. ips
Minimum capacity at each fire hydrant: 600 gpm
Minimum storage capacity: 3 days of maximum use
Maximum line velocity: 10 fps (pressure requirements will govern)
Fire-hydrant spacing: 500 ft or less

* Staff Engineer, Burns and Roe, Inc., Oradell, N.J.

Satisfactory pressure for distribution systems ranges from 30 to 100 psi. Pressures below 30 psi cannot supply water successfully to three- and four-story buildings without booster pumps. Pressures over 100 psi require heavier distribution pipes, produce more leakage, and are too high for use in plumbing fixtures. Pressures in the range of 50 to 75 psi are most generally satisfactory. In cities where there are differences in surface elevation in excess of 100 ft, it is often advisable to zone the system according to elevation in order to avoid excessive pressure at the lower elevations. The different zones have independent reservoirs or supplies from the pumping station and are usually designated as "high service" and "low service."

The water demand for extinguishing fires varies among different districts of a city. The requirements for each city as a whole can be predicated on population, which is the usual basis for estimating water requirements other than for fire fighting. In an estimate of the quantity of water required for extinguishing fires, an allowance should be made for probable losses from broken connections and hydrants which may be left open. Experience with the largest fires that the country has experienced indicates that 20,000 gpm usually should suffice, even in big cities, for fire fighting.

The selection of pipe sizes in distribution networks is influenced more by the necessity of maintaining adequate water pressure than by the economics of pumping costs.

The prime objective of the distribution network is that it supply a sufficient quantity of water to all parts of the system at pressures adequate for the requirements of the consumers, at all times and under all conditions of their demands. In order to accomplish this, reserve capacity is necessary in the form of storage, pumping, and distribution facilities.

Network Analysis of Distribution Systems. The complexity of the analysis required for a well-designed system is in many ways comparable to that of utility power networks. The importance of determining the basic conditions as accurately and as completely as possible cannot be overemphasized, including the range of variation for all conditions. The result of the calculations for the network cannot be any better than the data used.

The data will cover the pressure and flow requirements under the varying rates to be expected in the respective parts of the system; the topography of the service areas; the pumping arrangements and pump characteristics; the location, quantity, and elevation of storage; and the condition of the pipe system with regard to pressure drop and carrying characteristics.

There are several procedures that may be used for the analysis of flow in complex networks:

The Electric Network Analyzer Method (See "Hydraulic Analysis of Water Distribution Systems by Means of an Electric Network Analyzer" by T. R. Camp and H. L. Hazen, *J. New Engl. Water Works Assoc.*, December, 1934)

The Hardy-Cross Method (See "Analysis of Flow in Networks of Conduits or Conductors" by Hardy Cross, *Univ. Illinois Eng. Expt. Sta. Bull.* 286, November, 13, 1936)

The Graphical Method (See "Solution of Transmission Problems of Water Systems" by E. H. Aldrich, *Trans. ASCE.*, 1938)

Electronic Computer Analysis of Hydraulic Networks (See "Pipeline Network Analysis by Electronic Digital Computer" by L. N. Hoag and G. Weinberg, *J. Am. Water Works Assoc.*, May, 1957)

The solution of the flow problem in and head losses of a complex distribution network of water conduits is an extremely tedious and time-consuming task. However, for simpler networks, a trial-and-error method is practicable. In using this method, the pressure drop tables herein and the following relationships will

shorten the work considerably. In the turbulent region (almost all flows will be in this higher flow region, Reynolds number above 2,000), the pressure drop h varies as the 1.85 power of the flow rate Q, that is

$$h_1/h_2 = (Q_1/Q_2)^{1.85}$$

In the viscous-flow region (low flow, Reynolds number below 2,000), the pressure drop varies directly as the flow, or $h_1/h_2 = Q_1/Q_2$. For the same flow, pressure drop varies approximately as the fifth power of the inside diameter D, so that

$$h_1/h_2 = (D_1/D_2)^5$$

In Fig. 1, pressure at $A = P_1$ and pressure at $B = P_2$. Pressure drop between A and $B = P_1 - P_2$ in each of the pipeline branches between A and B.

The flow quantity in each branch will vary in accordance with the characteristics of the branch (internal diameter, length, roughness, and number of fittings) in order to result in the same pressure drop, $P_1 - P_2$. A procedure for the determination of flow rates in the individual branches is given below.

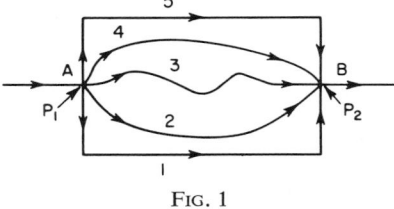

Fig. 1

Make an initial set of flow assumptions by dividing up the flow entering point A, and obtain the friction heads for each branch between A and B. Since all the friction heads between two junction points A and B must be equal, adjust the flows based on the size of the friction heads from this first trial. Obtain a new set of friction heads based on these adjusted flows. After a few trials, adjusting the flows each time, the friction heads for all the branches between A and B can be made equal, or very nearly so, indicating that the proper flows have been obtained in every branch between A and B.

A detailed discussion of and methods for calculating pressure drop in piping are given in Chap. 3. Approximate friction losses for cold water are given in Table 1.

In complex distribution systems, the number of variables is so great and creates so many simultaneous equations that direct solution algebraically of a network is impractical by hand calculation. The equations arise from the three fundamental relationships:

1. The algebraic sum of the rates of discharge toward any junction point is zero; that is, the sum of flows into a junction point must equal sum of flows out of the junction point.
2. The algebraic sum of the head losses around any closed circuit is zero.
3. The head loss is directly proportional to some power of the discharge.

The use of an electronic computer programmed for this analysis provides the highest accuracy in the solution of this problem in the shortest time and least relative cost.

Illustrative Example. For example, the typical computer service charge at this time is about $100 for an average 20-loop problem. Computer service is available with program capacity of 500 loops and 1,000 lines based on a modified Hardy-Cross method. A typical method[61,*] of setting up the problem for an electronic computer is as follows:

* Superscript numerals refer to bibliographical references which are listed at the end of this chapter.

Table I. Friction Loss for Water in Feet per 100 Feet of Pipe

6-in. nominal wrought iron or steel, Schedule 40, ID = 6.065 in.			12-in. nominal wrought iron or steel, Schedule 40, ID = 11.938 in.		
Discharge, gpm	Velocity, fps	Friction, ft per 100 ft of pipe	Discharge, gpm	Velocity, fps	Friction, ft per 100 ft of pipe
10	0.111	0.00146	100	0.287	0.00325
20	0.222	0.00487	120	0.344	0.00448
30	0.333	0.00988	140	0.401	0.00590
40	0.444	0.0164	160	0.459	0.00747
50	0.555	0.0244	180	0.516	0.00920
60	0.666	0.0337	200	0.573	0.0111
70	0.777	0.0445	220	0.631	0.0132
80	0.888	0.0564	240	0 688	0.0155
90	0.999	0.0698	260	0.745	0.0180
100	1.11	0.0843	280	0.802	0.0206
120	1.33	0.118	300	0.860	0.0233
140	1.55	0.155	350	1.00	0.0306
160	1.78	0.198	400	1.15	0.0391
180	2.00	0.246	450	1.29	0.0485
200	2.22	0.299	500	1.43	0.0587
220	2.44	0.357	550	1.58	0.0698
240	2.66	0.419	600	1.72	0.0820
260	2.89	0.487	650	1.86	0.0950
280	3.11	0.560	700	2.01	0.109
300	3.33	0.637	750	2.15	0.124
320	3.55	0.719	800	2.29	0.140
340	3.78	0.806	850	2.44	0.156
360	4.00	0.898	900	2.58	0.173
380	4.22	0.993	950	2.72	0.191
400	4.44	1.09	1,000	2.87	0.210
420	4.66	1.20	1,100	3.15	0.251
440	4.89	1.31	1,200	3.44	0.296
460	5.11	1.42	1,300	3.73	0.344
480	5.33	1.54	1,400	4.01	0.395
500	5.55	1.66	1,500	4.30	0.450
550	6.11	1.99	1,600	4.59	0.509
600	6.66	2.34	1,700	4.87	0.572
650	7.22	2.73	1,800	5.16	0.636
700	7.77	3.13	1,900	5.45	0.704
750	8.33	3.57	2,000	5.73	0.776
800	8.88	4.03	2,200	6.31	0.930
850	9.44	4.53	2,400	6.88	1.093
900	9.99	5.05	2,600	7.45	1.28
950	10.5	5.60	2,800	8.03	1.47
1,000	11.1	6.17	3,000	8.60	1.68
1,100	12.2	7.41	3,200	9.17	1.90
1,200	13.3	8.76	3,400	9.75	2.13
1,300	14.4	10.2	3,600	10.3	2.37
1,400	15.5	11.8	3,800	10.9	2.63
1,500	16.7	13.5	4,000	11.5	2.92
1,600	17.8	15.4	4,500	12.9	3.65
1,700	18.9	17.3	5,000	14.3	4.47
1,800	20.0	19.4	5,500	15.8	5.38
1,900	21.1	21.6	6,000	17.2	6.39
2,000	22.2	23.8	6,500	18.6	7.47
2,100	23.3	26.2	7,000	20.1	8.63
2,200	24.4	28.8	7,500	21.5	9.88
2,300	25.5	31.4	8,000	22.9	11.20
2,400	26.6	34.2	8,500	24.4	12.6
2,500	27.8	37.0	9,000	25.8	14.1
2,600	28.9	39.9	9,500	27.2	15.7
2,700	30.0	42.9	10,000	28.7	17.4
2,800	31.1	46.1	11,000	31.5	21.0
2,900	32.2	49.4	12,000	34.4	24.8
3,000	33.3	52.8	13,000	37.3	28.9
3,200	35.5	59.9	14,000	40.1	33.5
3,400	37.8	67.4	15,000	43.0	38.4
3,600	40.0	75.5	16,000	45.9	43.7
3,800	42.2	84.1	17,000	48.7	49.2
4,000	44.4	93.1	18,000	51.6	55.2
			19,000	54.5	61.5
			20,000	57.3	68.1

Table I. (*Continued*)

18-in. OD Steel, Schedule 40, ID = 16.876 in.			24-in. OD Steel, Schedule 40, ID = 22.626 in.		
Discharge, gpm	Velocity, fps	Friction, ft per 100 ft of pipe	Discharge, gpm	Velocity, fps	Friction, ft per 100 ft of pipe
300	0.430	0.00437	300	0.239	0.00107
400	0.574	0.00730	400	0.319	0.00178
500	0.717	0.0109	500	0.399	0.00267
600	0.861	0.0152	600	0.479	0.00371
700	1.00	0.0201	700	0.559	0.00490
800	1.15	0.0256	800	0.638	0.00621
900	1.29	0.0318	900	0.718	0.00767
1,000	1.43	0.0386	1,000	0.798	0.00928
1,200	1.72	0.0541	1,200	0.958	0.0129
1,400	2.01	0.0719	1,400	1.12	0.0171
1,600	2.30	0.092	1,600	1.28	0.0219
1,800	2.58	0.114	1,800	1.44	0.0272
2,000	2.87	0.139	2,000	1.60	0.0330
2,500	3.59	0.211	2,500	1.99	0.0499
3,000	4.30	0.297	3,000	2.39	0.0700
3,500	5.02	0.397	3,500	2.79	0.0934
4,000	5.74	0.511	4,000	3.19	0.120
4,500	6.45	0.639	4,500	3.59	0.149
5,000	7.17	0.781	5,000	3.99	0.181
6,000	8.61	1.11	6,000	4.79	0.257
7,000	10.0	1.49	7,000	5.59	0.343
8,000	11.5	1.93	8,000	6.38	0.441
9,000	12.9	2.42	9,000	7.18	0.551
10,000	14.3	2.97	10,000	7.98	0.671
12,000	17.2	4.21	12,000	9.58	0.959
14,000	20.1	5.69	14,000	11.2	1.29
16,000	22.9	7.41	16,000	12.8	1.67
18,000	25.8	9.33	18,000	14.4	2.10
20,000	28.7	11.5	20,000	16.0	2.58
22,000	31.6	13.9	22,000	17.6	3.10
24,000	34.4	16.5	24,000	19.2	3.67
26,000	37.3	19.2	26,000	20.7	4.29
28,000	40.2	22.2	28,000	22.3	4.96
30,000	43.0	25.5	30,000	23.9	5.68
32,000	45.9	29.0	32,000	25.5	6.42
34,000	48.8	32.8	34,000	27.1	7.22
36,000	51.6	36.8	36,000	28.7	8.08
38,000	54.5	40.8	38,000	30.3	9.00
40,000	57.4	45.0	40,000	31.9	9.98
42,000	60.2	49.7	42,000	33.5	11.0
44,000	63.1	54.5	44,000	35.1	12.1
46,000	66.0	59.5	46,000	36.7	13.2
48,000	68.9	64.8	48,000	38.3	14.3
50,000	71.7	70.2	50,000	39.9	15.5
55,000	78.9	84.8	55,000	43.9	18.7
60,000	86.1	101	60,000	47.9	22.3
65,000	93.2	118	65,000	51.9	26.2
70,000	100.4	136	70,000	55.9	30.4
			75,000	59.8	34.8
			80,000	63.8	39.4
			85,000	67.8	44.4
			90,000	71.8	49.7
			95,000	75.8	55.5
			100,000	79.8	61.5
			110,000	87.8	74.0
			120,000	95.8	88.0
			130,000	103.7	103
			140,000	112	119
			150,000	120	137

NOTE:
1 gal, U.S. = 0.1337 cu ft
1 cu ft = 7.48 gal, U.S.
  For water at 60 F
1 ft head = 0.433 psi
1 psi = 2.304 ft head

Table I. (Continued)

30-in. OD Steel, Schedule 20, ID = 29.000 in.			36-in. ID Steel		
Discharge, gpm	Velocity, fps	Friction, ft per 100 ft of pipe	Discharge, gpm	Velocity, fps	Friction, ft per 100 ft of pipe
400	0.194	0.000540	1,000	0.315	0.000988
500	0.243	0.000805	1,200	0.378	0.00137
600	0.291	0.001115	1,400	0.441	0.00181
700	0.340	0.00147	1,600	0.504	0.00231
800	0.389	0.00187	1,800	0.567	0.00285
900	0.437	0.00231	2,000	0.630	0.00344
1,000	0.486	0.00280	2,500	0.788	0.00517
1,200	0.583	0.00390	3,000	0.946	0.00721
1,400	0.680	0.00514	3,500	1.103	0.00957
1,600	0.777	0.00652	4,000	1.26	0.0122
1,800	0.874	0.00814	4,500	1.41	0.0152
2,000	0.971	0.00986	5,000	1.58	0.0185
2,500	1.21	0.0148	6,000	1.89	0.0260
3,000	1.46	0.0206	7,000	2.21	0.0345
3,500	1.70	0.0276	8,000	2.52	0.0442
4,000	1.94	0.0354	9,000	2.84	0.0551
4,500	2.19	0.0440	10,000	3.15	0.0670
5,000	2.43	0.0535	12,000	3.78	0.0942
6,000	2.91	0.0750	14,000	4.41	0.126
7,000	3.40	0.100	16,000	5.04	0.162
8,000	3.89	0.129	18,000	5.67	0.203
9,000	4.37	0.161	20,000	6.30	0.248
10,000	4.86	0.196	25,000	7.88	0.378
12,000	5.83	0.277	30,000	9.46	0.540
14,000	6.80	0.371	35,000	11.03	0.724
16,000	7 77	0.478	40,000	12.6	0.941
18,000	8.74	0.598	45,000	14.1	1.18
20,000	9.71	0.732	50,000	15.8	1.45
25,000	12.1	1.13	60,000	18.9	2.07
30,000	14.6	1.61	70,000	22.1	2.81
35,000	17.0	2.17	80,000	25.2	3.66
40,000	19.4	2.83	90,000	28.4	4.59
45,000	21.9	3.56	100,000	31.5	5.64
50,000	24.3	4.38	120,000	37.8	8.05
60,000	29.1	6.23	140,000	44.1	10.9
70,000	34.0	8.43	160,000	50.4	14.2
80,000	38.9	11.0	180,000	56.7	17.9
90,000	43.7	13.8	200,000	63.0	22.1
100,000	48.6	17.0	250,000	78.8	34.4
110,000	53.4	20.6	300,000	94.6	49.4
120,000	58.3	24.5	350,000	110	67.0
130,000	63.1	28.7	400,000	126	87.3
140,000	68.0	33.3			
150,000	72.9	38.2			
160,000	77.7	43.3			
170,000	82.6	48.8			
180,000	87.4	54.7			
190,000	92.3	60.8			
200,000	97.1	67.1			
210,000	102	73.8			
220,000	107	81.0			
230,000	112	88.6			
240,000	117	96.7			
250,000	121	106			

Table I.   *(Continued)*

	42-in. ID Steel			48-in. ID Steel	
Discharge, gpm	Velocity, fps	Friction, ft per 100 ft of pipe	Discharge, gpm	Velocity, fps	Friction, ft per 100 ft of pipe
1,000	0.232	0.000471	1,500	0.266	0.000508
1,500	0.347	0.000977	2,000	0.355	0.000855
2,000	0.463	0.00164	2,500	0.443	0.00129
2,500	0.579	0.00246	3,000	0.532	0.00180
3,000	0.695	0.00343	3,500	0.621	0.00238
3,500	0.811	0.00454	4,000	0.709	0.00304
4,000	0.926	0.00580	4,500	0.798	0.00378
4,500	1.042	0.00720	5,000	0.887	0.00458
5,000	1.16	0.00874	6,000	1.064	0.00636
6,000	1.39	0.0122	7,000	1.24	0.00844
7,000	1.62	0.0162	8,000	1.42	0.0108
8,000	1.85	0.0208	9,000	1.60	0.0134
9,000	2.08	0.0258	10,000	1.77	0.0163
10,000	2.32	0.0314	12,000	2.13	0.0229
12,000	2.78	0.0441	14,000	2.48	0.0305
14,000	3.24	0.0591	16,000	2.84	0.0391
16,000	3.71	0.0758	18,000	3.19	0.0488
18,000	4.17	0.0944	20,000	3.55	0.0598
20,000	4.63	0.115	25,000	4.43	0.0910
25,000	5.79	0.176	30,000	5.32	0.128
30,000	6.95	0.250	35,000	6.21	0.172
35,000	8.11	0.334	40,000	7.09	0.222
40,000	9.26	0.433	45,000	7.98	0.278
45,000	10.42	0.545	50,000	8.87	0.341
50,000	11.6	0.668	60,000	10.64	0.484
60,000	13.9	0.946	70,000	12.4	0.652
70,000	16.2	1.27	80,000	14.2	0.849
80,000	18.5	1.66	90,000	16.0	1.06
90,000	20.8	2.08	100,000	17.7	1.30
100,000	23.2	2.57	120,000	21.3	1.87
120,000	27.8	3.67	140,000	24.8	2.51
140,000	32.4	4.98	160,000	28.4	3.26
160,000	37.1	6.46	180,000	31.9	4.11
180,000	41.7	8.12	200,000	35.5	5.05
200,000	46.3	10.00	250,000	44.3	7.88
250,000	57.9	15.6	300,000	53.2	11.3
300,000	69.5	22.3	350,000	62.1	15.3
350,000	81.1	30.4	400,000	70.9	20.0
400,000	92.6	39.6	450,000	79.8	25.2
450,000	104.2	50.1	500,000	88.7	31.1
500,000	116	61.7	550,000	97.5	37.6
			600,000	106.4	44.7

Table I. (*Continued*)

54-in. ID Steel			60-in. ID Steel		
Discharge, gpm	Velocity, fps	Friction, ft per 100 ft of pipe	Discharge, gpm	Velocity, fps	Friction, ft per 100 ft of pipe
2,000	0.280	0.000488	2,000	0.227	0.000293
2,500	0.350	0.000733	2,500	0.284	0.000440
3,000	0.420	0.00102	3,000	0.340	0.000612
3,500	0.490	0.00134	3,500	0.397	0.000810
4,000	0.560	0.00172	4,000	0.454	0.00103
4,500	0.630	0.00213	4,500	0.511	0.00128
5,000	0.700	0.00257	5,000	0.567	0.00155
6,000	0.840	0.00358	6,000	0.681	0.00216
7,000	0.981	0.00476	7,000	0.794	0.00285
8,000	1.121	0.00610	8,000	0.908	0.00365
9,000	1.26	0.00760	9,000	1.021	0.00454
10,000	1.40	0.00920	10,000	1.13	0.00550
12,000	1.68	0.0129	12,000	1.36	0.00766
14,000	1.96	0.0171	14,000	1.59	0.0102
16,000	2.24	0.0219	16,000	1.82	0.0131
18,000	2.52	0.0273	18,000	2.04	0.0163
20,000	2.80	0.0333	20,000	2.27	0.0198
25,000	3.50	0.0504	25,000	2.84	0.0301
30,000	4.20	0.0713	30,000	3.40	0.0424
35,000	4.90	0.0958	35,000	3.97	0.0567
40,000	5.60	0.124	40,000	4.54	0.0730
45,000	6.30	0.155	45,000	5.11	0.0916
50,000	7.00	0.189	50,000	5.67	0.112
60,000	8.40	0.267	60,000	6.81	0.158
70,000	9.81	0.358	70,000	7.94	0.213
80,000	11.21	0.465	80,000	9.08	0.275
90,000	12.6	0.586	90,000	10.21	0.344
100,000	14.0	0.715	100,000	11.3	0.420
120,000	16.8	1.02	120,000	13.6	0.600
140,000	19.6	1.38	140,000	15.9	0.806
160,000	22.4	1.80	160,000	18.2	1.04
180,000	25.2	2.26	180,000	20.4	1.32
200,000	28.0	2.77	200,000	22.7	1.62
250,000	35.0	4.32	250,000	28.4	2.52
300,000	42.0	6.19	300,000	34.0	3.60
350,000	49.0	8.40	350,000	39.7	4.88
400,000	56.0	11.00	400,000	45.4	6.34
450,000	63.0	13.9	450,000	51.1	8.01
500,000	70.0	17.0	500,000	56.7	9.87
550,000	77.0	20.6	550,000	62.4	11.9
600,000	84.0	24.5	600,000	68.1	14.1
650,000	98.1	28.7	650,000	73.8	16.6
700,000	112.1	33.2	700,000	79.4	19.2
			750,000	85.1	22.0
			800,000	90.8	25.0

Table I. (Continued)

72-in. ID Steel			84-in. ID Steel		
Discharge, gpm	Velocity, fps	Friction, ft per 100 ft of pipe	Discharge, gpm	Velocity, fps	Friction, ft per 100 ft of pipe
2,000	0.158	0.000123	3,000	0.174	0.000121
2,500	0.197	0.000183	4,000	0.232	0.000203
3,000	0.237	0.000254	5,000	0.289	0.000306
3,500	0.276	0.000336	6,000	0.347	0.000425
4,000	0.316	0.000427	7,000	0.405	0.000562
4,500	0.355	0.000530	8,000	0.463	0.000717
5,000	0.394	0.000640	9,000	0.521	0.000891
6,000	0.473	0.000890	10,000	0.579	0.00108
7,000	0.552	0.00118	12,000	0.695	0.00150
8,000	0.631	0.00150	14,000	0.811	0.00199
9,000	0.710	0.00186	16,000	0.926	0.00255
10,000	0.789	0.00227	18,000	1.042	0.00316
12,000	0.947	0.00313	20,000	1.16	0.00384
14,000	1.104	0.00418	25,000	1.45	0.00579
16,000	1.26	0.00538	30,000	1.74	0.00810
18,000	1.42	0.00673	35,000	2.03	0.0108
20,000	1.58	0.00822	40,000	2.32	0.0139
25,000	1.97	0.0124	45,000	2.61	0.0174
30,000	2.37	0.0173	50,000	2.89	0.0212
35,000	2.76	0.0231	60,000	3.47	0.0298
40,000	3.16	0.0297	70,000	4.05	0.0398
45,000	3.55	0.0370	80,000	4.63	0.0513
50,000	3.94	0.0450	90,000	5.21	0.0640
60,000	4.73	0.0637	100,000	5.79	0.0781
70,000	5.52	0.0850	120,000	6.95	0.111
80,000	6.31	0.110	140,000	8.11	0.149
90,000	7.10	0.138	160,000	9.26	0.193
100,000	7.89	0.168	180,000	10.42	0.242
120,000	9.47	0.237	200,000	11.6	0.297
140,000	11.04	0.321	250,000	14.5	0.458
160,000	12.6	0.414	300,000	17.4	0.649
180,000	14.2	0.522	350,000	20.3	0.880
200,000	15.8	0.642	400,000	23.2	1.14
250,000	19.7	1.00	450,000	26.1	1.44
300,000	23.7	1.42	500,000	28.9	1.78
350,000	27.6	1.92	550,000	31.8	2.14
400,000	31.6	2.50	600,000	34.7	2.54
450,000	35.5	3.16	650,000	37.6	2.97
500,000	39.4	3.88	700,000	40.5	3.43
550,000	43.4	4.69	750,000	43.4	3.93
600,000	47.3	5.56	800,000	46.3	4.47
650,000	51.3	6.52	850,000	49.2	5.04
700,000	55.2	7.56	900,000	52.1	5.64
750,000	59.2	8.67	950,000	55.0	6.29
800,000	63.1	9.83	1,000,000	57.9	6.95
850,000	67.0	11.1			
900,000	71.0	12.4			

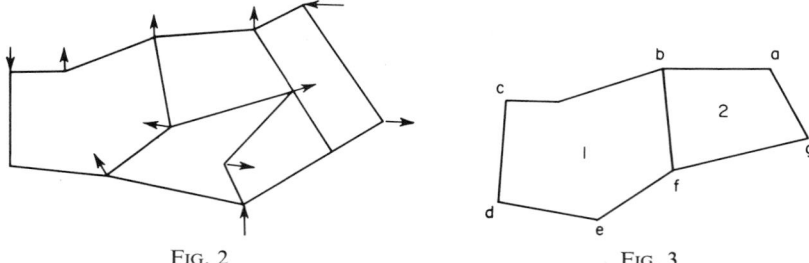

Fig. 2   Fig. 3

1. Make a skeleton drawing of the network. Indicate by appropriate arrows the points of constant flow input or output, constant head input or output (see Fig. 2).

2. Number all loops in the system in arbitrary sequence. Do not include "loops around loops." For example, in Fig. 3 there are two loops, not three. The large loop (*abcdefg*) is not numbered. The two basic loops (*abfg* and *bcdef*) are numbered.

3. Number each line. A line has two ends. An end may be a point at which water is drawn from or added to the system, one at which pipe characteristics change, or a tee joint. For example, in Fig. 4, the point *x* is the meeting of three lines, not two; point *y* is the meeting of two lines where an 8-in. pipe joins a 10-in. pipe; point *z* is simply a bend in the single pipe and is not the end of any line, although it could have been specified as one, if desired. Figure 4 shows the complete numbering of the system shown in Fig. 2. Note that each line is numbered once and only once, even though it may be in more than one loop. Also note that the numbering is serial; that is, if there are *n* branches, each of the numbers from 1 to *n* must be used in the numbering.

4. Assign a base direction. Put an arrow on each line in loop 1, indicating the clockwise direction (as shown in Fig. 5). Then put an arrow on each line in loop 2, indicating clockwise direction, except where a line which previously has been assigned a direction is encountered. Then the original assignment is not changed. In Fig. 5, line 4 is a member of loop 1 and also of loop 2 but has been given a base direction of loop 1. The line 4 assignment is not changed. This process is continued for every loop in the network, an arrow being assigned in a clockwise direction whenever it has not been assigned previously.

5. In water-distribution systems, the situation often is encountered where system pressure must be raised by the use of booster pumps in series with the supply pipeline. If the higher pressure area is connected to the remainder of the system at one point only, the two pressure-zone networks are hydraulically independent problems. If the pressure zones are connected at two or more points, the booster pumps must be included in the appropriate loops.

For all loops containing booster pumps, an unbalanced or residual head H_0 must be determined. This is done by algebraically summing the assumed constant head changes at the boosters in a clockwise direction.

Note that head *losses* are considered as positive in sign, so proceeding from the suction side of a pump to the discharge side gives a negative head loss.

Following the hydraulic analysis, a check should be made to assure that the pumping head assumptions are sufficiently accurate. The resulting flow-rate values should allow optimum hydraulic design of the booster-station installations.

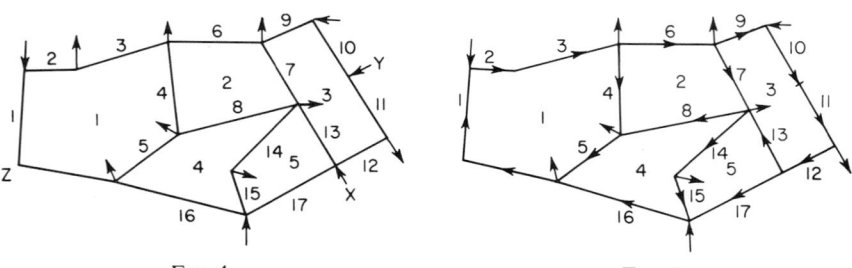

Fig. 4   Fig. 5

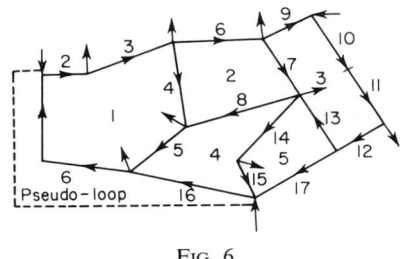

FIG. 6

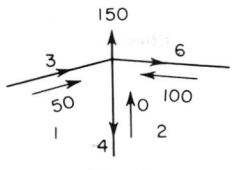

FIG. 7

6. Additional "pseudo-loops" must now be added to the list if there is more than one constant head input (see Fig. 6). If the number of such inputs is m, trace $(m - 1)$ paths between inputs in the same manner in which the loops were traced, making sure that each constant head input is used at the end of at least one of these "loops." If the direction of procedure is from the lower to the higher input in each path, H_0 will be the positive difference in the head loss between the two inputs. If booster pumps are encountered, the head change across such pumps must be algebraically added to the head difference between the inputs in order to obtain the H_0 for the "pseudo-loops."

When the listing of all the loops has been completed (including the consideration of booster pumps), the work should be carefully checked, preferably by a second person, since any errors will completely upset the calculations. Note that pseudo-loops do not introduce any new lines. Note also that each pseudo-loop must be assigned its own number.

7. The only remaining task is to supply initial flow values and pipe characteristics which the computer can use as starting values for the calculations. The only restriction on these values is that they satisfy the mass balance condition at each junction. That is, the sum of the flow *into* a junction must equal the sum of the flows *out* of the junction. For example, Fig. 7 shows the junction of lines 3, 4, and 6; flows of 50 gpm in line 3 and 100 gpm in line 6 would satisfy the condition.

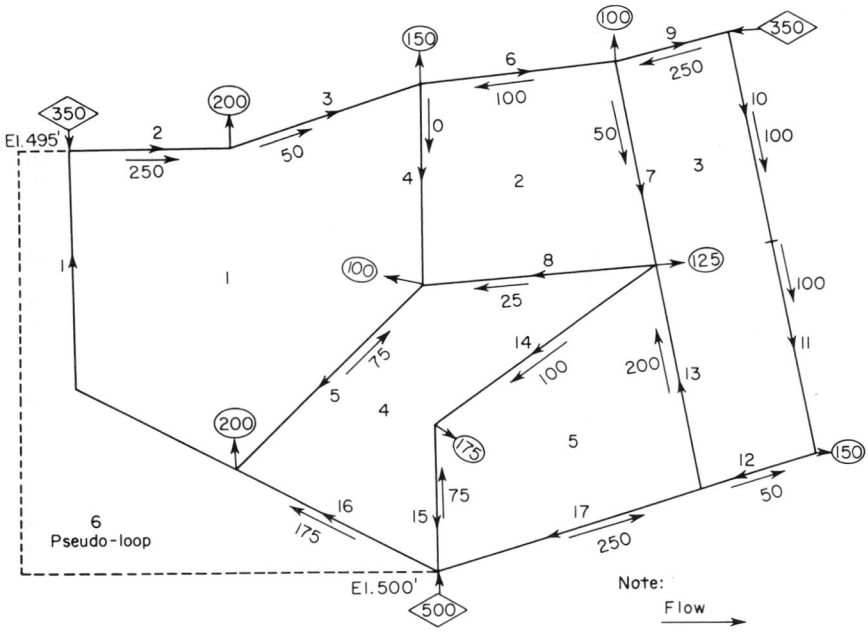

FIG. 8

Proceeding in this manner, balance every junction in the network, working toward the variable-flow (constant-head) inputs which can take up the slack. When all flows are specified, check the accuracy of the work by summing the inputs and outputs. If these sums are unequal, some computational error has been made and must be corrected. The complete schematic for this system is shown in Fig. 8. This schematic includes the assumed starting values of the flows.

CIRCULATING COOLING-WATER SYSTEMS FOR POWER PLANTS OR INDUSTRIAL PLANTS

General Requirements. The complete problem for supplying the cooling water to an industrial plant can best be illustrated in the complexity of the main cooling- or circulating-water system for a steam-electric station. A steam power plant with a name-plate rating of 750,000 kw requires as much cooling water as the domestic water requirements of a city with four to five million population.

The circulating cooling-water system for a steam power plant takes water from some source, such as a river, lake, or cooling-tower basin, and provides continuous cooling to the condenser. The circulating water condenses the turbine exhaust steam, thereby lowering the turbine exhaust pressure so that additional energy can be extracted from the steam. The importance of a continued cooling-water supply is such that, if the cooling-water supply should fail, the resulting rise in exhaust steam back pressure will cause an automatic shutdown of the turbine-generator unit. Reliability, efficiency, and economy are the prime design considerations for a steam power-plant circulating cooling-water system. The availability of this cooling water plays an important role in the selection of the steam power-plant site.

A fundamental decision has to be made as to whether the circulating-water system will be designed on a single generating unit basis or to supply water for multiple units with one circulating-water system. This basic decision can be made only after considering all the special conditions for the particular plant. The obvious economic appeal of supplying the multiple unit by means of a single circulating system has to be balanced against the train of problems that follow in its wake. The large variation of flow and its related pressure drop, between full operation of all units and the operation of a single unit, may create such system complexity, for example, at the intake structure, that it will cancel out most of the economic advantage. The variation in the pumping head may be beyond the operating range of the circulating-water pumps. Also, the provisions needed for reliability of the system in order to minimize or avoid total shutdown in event of failure of some part may add substantially to the overall cost of the system.

Types of Systems. Many types of system arrangements are in common usage. Those encountered most frequently are defined and discussed briefly below.

A once-through or direct-condensing system is one in which water is drawn from a river, lake, or other large body of water and, after passing through the condenser, is returned to the same or a different body of water.

A recirculating system is one in which the same water is used repeatedly and is itself cooled between passes through the condenser. Means of cooling the water has to be provided in the form of cooling towers, spray ponds, or cooling ponds. Make-up water has to be provided to take care of water losses due to evaporation, leakage, and other causes.

A siphon system is one in which the siphon principle is employed to carry the water through elevated parts of the system, such as the condenser, in order to reduce the pumping power required. These elevated portions of the water system operate under a partial vacuum.

A pressure system is one in which the water flows under a positive head throughout.   This system is generally used with recirculating systems, such as with cooling-tower installations.

Frequently, features of two or more of the above systems are combined in a single installation.   For example, the once-through and siphon systems may be combined to produce the once-through siphon system which is shown in Fig. 9.   The cooling water enters the system through the intake structure (trash racks, traveling screen, fine screens) into the suction chamber and into the circulating pump suction.   The water is discharged by the circulating pump into the tunnel or pipe leading to the condenser.   The water passes through the condenser aided by siphon action and out through the discharge tunnel or piping to the seal well and discharge structure.

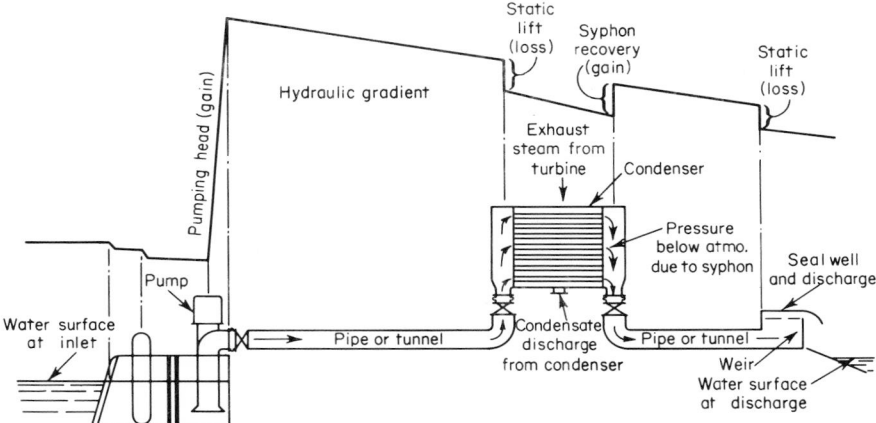

Fig. 9. Once-through siphon system.   Intake structure: trash racks, traveling screens, fine screens, circulating water pumps, valves, service water and screen wash pumps, hoisting devices for pumps and screens, water-treatment equipment.

When topography of the plant area permits and the fluctuation of the water level is not excessive, consideration should be given to locating the circulating water pumps near the condenser.   Water then enters the pump suction through a gravity canal or tunnel.

Figure 10 illustrates the recirculating pressure system with cooling tower.   The cooling water starts its circuit from the cooling-tower basin by passing through the fine screens into the suction chamber and then into the circulating pump suction. The water is discharged by the circulating pump into the tunnel or piping leading to the condenser.   The water passes through the condenser and out through the discharge tunnel or piping and discharges into the top of the cooling tower.   Depending on plant area conditions, the circulating pump suction chamber may be located near the condenser with the water flowing by gravity from the cooling-tower basin.

Fundamentals of Design.   In the design and layout of circulating-water systems, certain precautions must be observed in order to assure system reliability. The more important of these are discussed below.

In once-through systems, proper relative location of intake and discharge points has to be determined so that warm discharge water does not find its way into the system intake.   Currents, eddies, or tides may carry the warm water upstream into the intake.   Special means may be necessary to prevent this in some such form as a

low dam, piping, canal, levee, etc. In order to obtain the coldest water from the lowest stratum, the inlet may be lower than the screen and pump suction of the intake structure. Each case has to be studied for its special conditions.

In recirculating systems with cooling towers, the location of the cooling tower has to be carefully studied so that the tower vapor cloud will disperse before reaching the power-plant area proper or creating a nuisance over adjoining facilities or properties.

The length of the system should be as short as practicable for the lowest pressure drop and pumping and installation costs and to decrease water hammer.

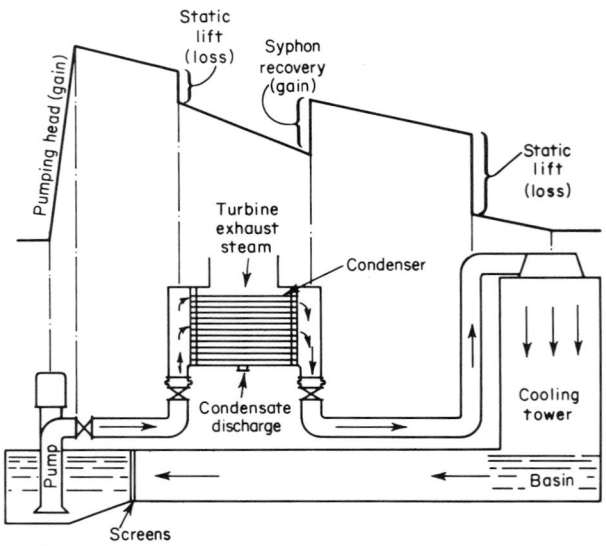

Fig. 10. Recirculating pressure system using cooling tower.

System elevations and profile should be set so as to obtain maximum siphon benefit and to eliminate or reduce the problems of air binding.

Items such as bends, tees, valves, reducers, and obstructions to smooth flow, all of which cause large pressure drops, should be used sparingly. Where these are necessary parts of the system, they should be designed for minimum pressure drop.

Preliminary Design, Flow Quantities. The following represent requirements averaged from actual installations for purposes of preliminary water-requirement estimates. There are many factors such as seasonal temperature of the water source, design of the actual condenser, and steam conditions of the thermal cycle that will create sharp deviations from these averages.

Once-through siphon system with a single-pass condenser
    For 66,000 kw and below: 850 gpm per 1,000 kw
    For 66,000 to 100,000 kw: 800 gpm per 1,000 kw
    For 125,000 kw and above: 750 gpm per 1,000 kw

Once-through siphon system with a two-pass condenser
    For 100,000 kw and below: 780 gpm per 1,000 kw
    For 125,000 kw and above: 650 gpm per 1,000 kw

Recirculating pressure system, cooling tower, single-pass condenser
    For 100,000 kw and above: 750 gpm per 1,000 kw

Recirculating pressure system, cooling tower, two-pass condenser
    For 66,000 kw and below: 850 gpm per 1,000 kw
    For 100,000 kw and above: 650 gpm per 1,000 kw

Preliminary Design, Pipe Sizing. For preliminary sizing of the circulating piping, a velocity of about 6 fps in the suction conduit to the pump and a velocity of about 8 to 10 fps in the discharge piping may be used. The final size selection must be made based on pressure drop. This final size selection entails an economic analysis, for the particular plant, of the cost of pumping the quantity required with various pressure drops, pumps, and pipe sizes. Sometimes the problem of future plant expansion enters into this sizing.

The conditions that determine the piping wall thickness from pump discharge through the shutoff valve after the condenser for siphon systems, are the normal operating pressure, the pump shutoff head, the water-hammer surge pressure, and the external pressure on pipe due to burying conditions.

At pump and condenser discharges, the velocity should be lowered gradually by means of an expanding tube or diffuser with 14-deg or less total flare. At changes in size from pipe to tunnel or from tunnel to pipe, the use of gradual size change is advisable.

The pressure drop through bends can be reduced by proper spacing and specification. For example, one 90-deg bend has a lower pressure drop than that of two 45-deg bends. Spacing bends far enough apart may save pressure drop. For example, the total head loss in closely located bends may be twice that of the same bends spaced farther apart. Turning vanes properly used may cut bend losses considerably.

Intakes. Intakes from rivers are designed to bring water from the river with as small an entrance loss and disturbance as possible. Ordinarily, the cleaning screens and the circulating pumps are housed at the river's edge. Sometimes the pumps are at the plant and the water is brought from the cleaning screens by means of a gravity canal or conduit to the pumps. This design saves pumping power and should be considered when the contour elevation is favorable from the river to the plant. Allowance should be made for the head drop in the gravity canal and the river-level fluctuation. Also, consideration has to be given to possible contamination by leaves, branches and other debris between the screens and the pumps, and the necessity for auxiliary screening.

The intake structure usually consists of coarse screens (or trash racks), a stop log dam, traveling screens, screen wash pumps, and circulating-water pumps. There is frequently a gantry or bridge crane. Where large quantities of debris may be expected that will not be washed away by the river, power-operated raking may be necessary with means for disposing of the debris.

The smaller trash or foreign matter that passes through the coarse screens is picked up on the revolving screen. The usual speed of this screen is 10 fpm but may be of higher velocity for unusual trash conditions. These screen openings are made to sift out foreign matter from the water of size about one-half the diameter of the condenser tubes. The water velocity through the screen is kept below $2\frac{1}{2}$ fps, since the rate of screen clogging rises rapidly with velocity. The amount of screen clogging is indicated by the difference of level on the two sides of the screen. This can be measured with an air-bubbler system or with float gages. Generally, the screen wash pumps are started automatically when this difference is about 6 in. When the screen-wash pump discharge pressure reaches a predetermined value, the screens are started automatically. The screen-wash pump requirement may

run from 25 gpm at 40 psig to 35 gpm at 80 psig, depending on the trash character-istics of the river involved.

Normally sluice gates or stop logs are provided before and after each traveling screen to permit its segregation and dewatering for repairs. Sometimes guides are provided for two sets of backup (or fine) screens after the traveling screens for the purpose of screening the water in event of breakdown and repair of the traveling screens. During such repair periods the fine screens are raised and cleaned alternately.

Consideration should be given to other problems that arise at the intake. Ice formation on the screens in winter may require recirculation of some of the warm discharge water. Various means may have to be considered to prevent silting, such as a settling area. The problem of obtaining the coldest layer of water requires a study of the relative location of intake and discharge.

The *intake structure design* involves the design of the suction chamber, and this in turn is governed by the selection of the pump, which may be of either the horizontal or vertical type.

The *horizontal pump* may not be practical where there is a large differential between high- and low-water river level. The horizontal pump must be set in a dry compartment, and its center line should be not more than 13 ft above the low-water level. The horizontal pump requires more space than a vertical pump and is sometimes more costly. A shutoff valve is frequently needed on the inlet side to cut off the water source in order to service the pump. However, the horizontal pump is more easily serviced and it is less sensitive to poor intake conditions. The horizontal pump has a lower starting horsepower and shutoff head and is usually a single- or double-suction single-stage, centrifugal-type pump.

The *vertical pump*, set in the water intake chamber, is more suitable for large river-level variations. The vertical pump adapts to a wider range of pump heads, requires a minimum of floor area, and is sometimes less costly than the comparable horizontal pump. However, the vertical pump is sensitive to inlet conditions, generally has a higher shutoff head than a horizontal pump, and is more difficult to service. The vertical centrifugal pump generally used is a radial-flow, axial-flow, or a mixed-flow type. The radial-flow vertical pump has a bottom suction, and the discharge is on the horizontal center line of the hollow impeller. The axial-flow (or propeller) type of vertical pump has a bottom suction, and the water is raised vertically by the propeller and discharges at some specified elevation above the propeller. The mixed-flow vertical pump utilizes a hollow impeller of the radial type, but the water rises vertically in the pump casing before it is discharged.

The complexity of the intake structure is naturally affected by the number of circulating pumps necessary for the circulating-water system. Reliability points to the use of at least two pumps. The particular conditions for the plant will dictate the final choice, whether it will be two pumps at two-thirds capacity each, at one-half capacity each, or some other more favorable percentage each. The capacity selection has to be the subject of a careful analysis, taking into account such items as the condenser water requirement, the variation of pumping head, and the best efficiency range of the pumps.

The water levels should be determined at both the suction and discharge ends of the circulating-water system. The levels of significance are those at high water, mean water, low water and at extreme low water. In addition, the coincidence of water level with the temperature of water and generating unit loads, especially at peak loads, should be analyzed. The calculation of the pump design operating point would be based on the maximum requirement of the condensing unit for sustained full load with the coincident water level and temperature.

It is desirable to know the approximate operating points for several conditions

other than the design point in order to establish the capacity and head limits at which the pump will be expected to operate.   Preliminary operating points can be established from assumed pump curves.   There may possibly be operating conditions which cannot be permitted because the operating point falls outside the safe operating range of the pump selected.   The condition for starting the system siphon must be part of this analysis.

The intake piping to the suction of horizontal pumps should be designed so as to avoid air pockets.   Also, the water-flow velocity should be made uniform over the suction inlet area by placing bends as far as possible from the pump inlet.

The intake chambers for vertical pumps require careful design for good pump operation.   The design must bring about a uniform and undisturbed flow of water to the pump without whirl.   Most pump manufacturers have design suggestions for intake chambers for their particular pumps.   Each vertical-pump installation should be studied individually.   There are no standard solutions to vertical-pump intake problems.

Discharge Structure.   On the discharge end of the circulating-water system, an underwater (or sealed) discharge must be provided to prevent entry of air into the piping, which would otherwise break the siphon action at the condenser.   One means of providing this seal is through the use of a seal well, that is, a basin with a water level controlled by an overflow weir.   The seal-well water level regulates the height of the siphon recovery, and it is the final elevation to which the circulating pump delivers the water.

Beyond the seal well, the discharge into the river or other body of water must be done in such a way that the discharge velocity is dissipated without washing away banks, tearing up the bottom, undermining the discharge piping, or permitting uncontrolled recirculation to the intake.

Air Binding.   Air which accumulates in water piping can reduce the effective area for water flow and thus increase pumping cost through the resulting extra head loss.

Air enters the piping system from several sources such as release of air from the water, air carried in through vortices into suction of pump, air leaking in through joints that may be under negative pressure, and the air originally present in the piping system before filling.

The water from the water source may be nearly completely saturated with air. If the temperature of this water is raised, for example, after passing through a condenser, and the pressure is lowered by the siphon action, the water will release most of its air.   However, this air release is not instantaneous but proceeds on a time-rate release and is therefore dependent upon the length of time the water remains in the piping.   Experience indicates that the actual air release in a circulating system of a conventional generating plant is probably in the order of 10 per cent of calculated theoretical release.

The problem associated with the formation of air pockets in the piping is dependent on the velocity of the water and on the configuration of the piping.   The higher the water velocity, the less air binding will occur for the same piping configuration.   The water with higher velocity has greater tendency to scrub out the air pockets, and its higher turbulence breaks up large air bubbles and entrains the resulting smaller bubbles.

On gentle downward slopes, a continuous air pocket may form along the top of the pipe for the entire slope.   In a sharper downward slope, several air pockets may form, each air pocket terminating in a hydraulic jump.   Slopes may require a water velocity in excess of 10 fps to assure that the piping remains free of air.   In a 90-deg downward drop, an air pocket may form in the upper portion of the bend and a velocity in excess of 7 fps may be required for air elimination.   Connections for

air vents, as shown in Figs. 11 and 12, should be considered at the beginning of the slope and down along the slope.

Air Inlet Valves. In rolling or mountainous country, air and vacuum valves should be placed at the summits of water-supply lines to allow air to escape when they are being filled as well as to admit air to prevent forming a vacuum when they are being drained. It is good practice to install such valves at abrupt breaks in grade, where they serve also to permit the escape of air while the line is operating under pressure, thus preventing the accumulation of air that is released from solution in the water.

Correct sizes of vacuum valves can be determined only by computation of the probable rate of water drainage from the line, which depends in turn on its profile

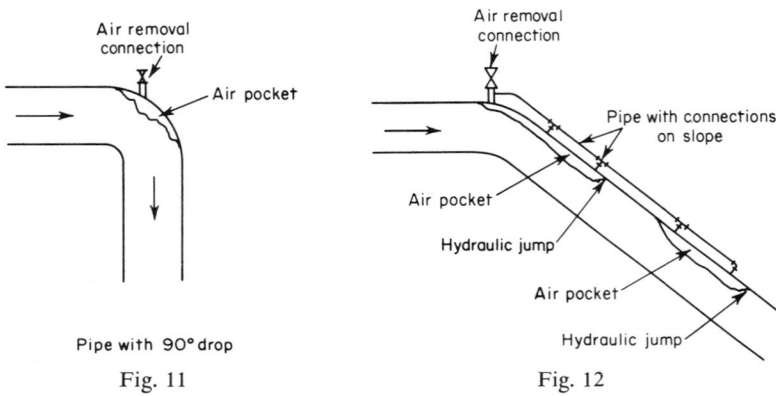

Pipe with 90° drop

Fig. 11

Fig. 12

and on the rate at which air is admitted to break the vacuum. The failure of the pipeline at a low point, for example, may empty part of the line and create a vacuum condition in the section of piping where there is a separation in the water column. The size of the vacuum valve should be computed on the supposition that air must enter at the same volumetric rate as that at which water leaves.

The strength of the pipe shell under this differential pressure will depend on the length-to-diameter ratio, the thickness-to-diameter ratio, and the modulus of elasticity of the pipe material. The presence of stiffener rings may exert a considerable influence on the strength of the pipe shell. However, if the distance between stiffener rings is great, they will have no appreciable effect on the section of the pipe shell midway between them, and the collapsing pressure will be the same as that of a pipe shell of infinite length. Parmakian[50] gives the following procedure for this calculation:

For cylindrical shells without stiffener rings:

$$P' = (50.2)(10^6)(t/D)^3$$

For cylindrical shells with stiffener rings:

$$P' = (73.4)(10^6)(t/D)^{5/2}(D/L)$$

where $P' =$ collapsing pressure for pipe shell, psi
  $t =$ thickness, in.
  $D =$ inside diameter, in.
  $L =$ spacing of stiffener rings, in.
    Pressure head at air inlet valve $= 2.31(P_2 - P_1)$
where $P_2 =$ minimum allowable internal pressure at air inlet valve, psi
  $P_1 =$ barometric pressure, psi

Cubic feet air per second that must be supplied:

$$Q = Wv = \pi D^2 V/(4)(144)$$

where Q = air that must be supplied at pressure P_2, cfs
      W = weight of air flowing, lb/sec
      v = specific volume of air at pressure P_2, cu ft/lb
      $V = V_1 - V_2$ or $V = V_1 + V_2$
where V_1 = water velocity leaving location of air inlet valve computed for condition by Bernoulli energy equation, fps
      V_2 = water velocity approaching location of air inlet valve, fps

If V_1 is larger than V_2, or if V_1 and V_2 are in opposite directions, an air inlet valve is required to supply enough air to the pipe interior to occupy the space created by breaking of the water column.

The air flow through the valve may be considered as an isentropic expansion, and the valve sized accordingly.

WATER HAMMER

An understanding of the physical picture of the water-hammer phenomenon can be helpful in dealing with the particular piping system under design.

The problem of water hammer in a pipeline consists of containing the pressure and dissipating the water flow energy. For example, in a pipeline with a supply pump, energy necessary to move the water through the piping is supplied by the pump. If a valve is suddenly closed at the end of the discharge line, the moving column of water is brought to a stop at the valve. The kinetic energy contained in the column of water, originally given to the water by the pump, is still present and must be dissipated. The column of water compresses, the pressure rises, and some of the kinetic energy is transformed to internal energy. The higher water pressure acts upon the pipe wall and does work in stretching it, but only a small percentage of energy will be lost in this. The pipe will obey the laws of vibration and return most of the energy to the water.

When the water-hammer pressure wave loads the pipe wall, the strain in the wall increases slightly faster than in strict proportion to stress within the elastic range, and on release of the loading, the reverse occurs. This results in a small area that is enclosed by the stress-strain curve in both directions, called the elastic hysteresis, and corresponds to a transformation of energy during the stress-strain cycle. If the pressure rise in the pipe is sufficient, the walls may be stressed into the plastic region and experience a permanent set. In this case the water gives up a large amount of energy with no recovery.

The energy or pressure wave which started at the closed valve stretches the diameter progressively to its source, the pump. The energy-wave front may meet a check valve at the pump and be reflected back toward its origin, continuing its dissipation of energy within the water and the pipe wall. This will continue until the kinetic energy is fully converted to internal energy.

When there is sudden failure of power at the pump motor, the only remaining energy is the kinetic energy of the rotating parts of pump and motor plus that of the entrained water. Since this is small when compared with that required to maintain flow against the discharge head, the reduction in pump speed is quite rapid. As the pump speed reduces, the flow of water in the discharge line at pump outlet is also reduced. Since this water has low energy, its pressure head is lower than that of water which was previously energized by the pump and delivered into the pipe water column. A depressuring action then occurs, starting at the pump and traveling up the pipeline.

Soon the pump speed is reduced to a point at which no water can be delivered against the existing discharge piping head. If there is no control valve present at the pump, the flow through the pump reverses, although the pump may be still rotating in a forward direction. The speed of the pump then drops more rapidly, stops, and reverses rotation. A short time later the pump, acting as a turbine driven by the higher head in the discharge line, reaches runaway speed in reverse. As the pump approaches runaway speed, the reverse flow through the pump reduces rapidly, so that the remaining energy in the water produces a pressure rise at the pump and reflects back along the length of the discharge line.

In order to analyze this problem, the items which have to be considered are the pump and motor inertia, the pump characteristics, and the water-hammer wave in the discharge line. The kinetic energy of the rotating system of the pump and motor is obtained from the inertia equation. The complete pump characteristic curves should be specified and obtained from the pump manufacturer. These curves should define the manner in which pump torque and speed vary with head and discharge flow throughout the range of operation as a pump, energy dissipator, and turbine.

Table 2.   Water-hammer Velocity in Piping Systems

(a, wave velocity in feet per second)

D/t	Steel $(E = 28 \times 10^6 \text{ psi})$	Cast iron $(E = 16 \times 10^6 \text{ psi})$	Transite $(E = 3.4 \times 10^6 \text{ psi})$
20	4,300	4,100	3,000
40	4,000	3,600	2,300
60	3,800	3,350	2,000
80	3,600	3,100	1,750
100	3,400	2,900	1,600
150	3,100	2,500	1,300
200	2,800	2,250	1,150
250	2,600	2,050	
300	2,400		

The water-hammer effects are obtained from equations that define relations between head and flow in the discharge line during the transient flow condition which results from water-hammer wave action.

Water-column separation might occur at high points near the hydraulic gradient on long discharge lines. This condition can create high-pressure conditions at the moment of rejoining of the separated water columns.

When a check valve is to be used on the discharge side of the pump, its design and closing characteristics should be carefully considered. The discharge flow from the pump keeps the check valve open. But if pump power failure occurs, the check valve closes as reverse flow develops in the discharge line. The water-hammer analysis has to consider the partial reverse flow condition until the check valve closes completely and, after that point, the effect of the closed valve.

Table 2 gives the water-hammer wave velocity as a function of diameter-to-thickness ratios for three different piping materials encountered frequently in water-supply or -distribution systems. In this tabulation, a is the wave velocity in feet per second, D/t is the dimensionless ratio of diameter to thickness, and E is the elastic modulus.

If a valve is closed in the time of one wave cycle, (in the time a pressure wave goes to other end of pipeline and returns to the closing valve) or less, then the water hammer should be calculated on basis of instant valve closure.

To determine time for wave cycle, use

$$T = 2L/a$$

To determine water hammer for instantaneous valve closing, use

$$h = aV/g$$

where T = time for one wave cycle, sec
  L = pipeline length, ft
  h = water-hammer head above static head, ft
  a = velocity of pressure wave, fps
  V = water velocity at instant before valve closure, fps
  g = 32.2 ft/sec^2

To determine water hammer for slower valve closing, use

$$h_2 - h_1 = a(V_1 - V_2)/g$$

where h_2 = pressure after partial closing of valve, ft
  h_1 = pressure before start of valve closing, ft
 $h_2 - h_1$ = pressure rise due to water hammer, ft
  V_1 = water velocity before start of valve closing, fps
  V_2 = water velocity after partial closing of valve, fps

CORROSION

All water coming from wells, rivers, lakes, and ocean is an extremely dilute water solution of mineral salts and gases.   The salts are mineral matter dissolved by water flowing over and through the earth layers.   The salts are mainly sulfates, bicarbonates, chlorides of calcium, sodium, and magnesium.   These minerals give water its hardness (destroying soap and preventing lather) and precipitate as a white lime-type scale.   The dissolved gases are atmospheric oxygen and carbon dioxide, picked up by water-atmosphere contact, e.g., spray, raindrops, and ammonia from decaying vegetable matter.

The dissolved gases are the prime agents of chemical corrosion which act on the metals of piping systems.   The oxygen attacks the iron or steel, and the process is accelerated by the carbon dioxide.   The rate and extent of the chemical corrosion are influenced by the mineral salts dissolved in the water.

The water is corrosive when the Langelier index (calcium carbonate saturation index) is minus ($-$).   The water analysis will generally, but not always, show a pH value below 7 (acidic).

The water is scaling when the Langelier index is plus ($+$).   The water analysis will generally show a pH value above 7 (basic).

The Langelier index is obtained by subtracting the pH that the water must have at a particular temperature in order for calcium carbonate precipitation to start from the actual pH of the water.   For example:

$$\text{(pH actual)} - \text{(pH saturation)} = \text{Langelier index}$$
$$(6.5) - (7.7) = -1.2 \text{ and the water is corrosive}$$
$$(8.1) - (7.2) = +0.9 \text{ and the water is scaling}$$

The precipitation of calcium carbonate as a scale or film thickness may be desirable as a means of protection against corrosion if the rate of build-up will be sufficiently low and allowance is made in diameter of the piping.   But this calcium carbonate film would be undesirable on heat-transfer surfaces.   Since temperature lowers the solubility of calcium carbonate and calcium sulfate, the Langelier index would be different for colder water coming to the condenser, for the water passing through condenser tubes, and for the warm water flowing away from the condenser.   Therefore, the scaling tendency of the same water would be higher in the condenser and in the discharge piping from the condenser.

The *exterior of the buried pipe* will be subject to similar but aggravated chemical action due to water. In addition, the pipe exterior will be susceptible to attack by aerobic and anaerobic bacteria, galvanic action, and stray electric currents. The chemical action on the pipe exterior may be more intense because of concentration of oxygen, salts, and other chemicals leached out of the surrounding earth by ground water.

Some forms of anaerobic bacteria that thrive only in the absence of free oxygen obtain their oxygen by the chemical breakdown of oxygen compounds in the earth with the resultant production of substances, such as hydrogen sulfide, that will corrode the buried pipeline. There are also many types of aerobic bacteria that produce sulfuric acid, sulfate, ferric hydroxide, etc., all corrosive to steel or iron. Organic soil should be kept away from the vicinity of the pipeline to minimize possibility of this corrosion action.

Also, when iron or steel is in contact with a more cathodic material, for example copper or brass, a galvanic cell is set up, electrolysis results, and the corrosion rate of steel or iron rises. If iron or steel is in contact with a more anodic material, for example zinc, the zinc is the affected material and the corrosion rate of the steel or iron drops.

There is some natural resistance to chemical corrosion of the pipe materials in an uncoated state. The chemical-corrosion product, for example an oxide film, may build up sufficiently to slow down or prevent further corrosion.

The materials most commonly used for pipelines are steel, wrought iron, cast iron, concrete, and asbestos cement (transite). The natural coating characteristics of these materials are mentioned briefly below.

In *cast iron*, the rust (an iron oxide) builds up into a strong adhesive coating that finally forms a barrier sufficient to stop or slow down further corrosion. The higher silicon cast iron has the best characteristics in this respect.

In *wrought (or wrot) iron*, the surface rust is not so adhesive but the silicate fiber interlacing has a reinforcing effect, and after some penetration the barrier against further corrosion will tend to establish itself.

In *steel*, the rust powders and flakes off easily and will not build up into an adhesive and sufficiently protective coating.

In *concrete and asbestos-cement piping*, the corrosion is of a different form. These materials are subject to leaching of the free lime from the cement, deterioration in alkali soils, and attack by organic growth.

Cathodic Protection. If no protective coating is used, or if a low-cost, limited-life coating has been selected, cathodic protection should be considered as a means of limiting the main agent of corrosion, which is the electrochemical process. In this process, the moist earth is the electrolyte, two dissimilar materials are the anode and the cathode, and the pipe wall between completes the electric circuit. This process may be set in motion in a number of different ways, among which are dissimilar metals, galvanic action of a single metal due to dissimilar soils, variation in moisture and chemical content of soil, nonuniformity of metal caused by mill scale, surface scarring, welding, and even temperature differentials.

The current flows from the anode to the cathode and produces corrosion at a rate greater than that which would occur by normal chemical means. Corrosion is increased at the anode end and decreased at the cathode end. The anode is the point or area at which the current leaves the metal, and the cathode is the point at which the current enters the metal. The loss of metal at the anode end is in the range of 20 lb/amp per year.

The electrochemical galvanic series (Table 3) gives the relation between any two metals. The metal listed nearer the top of the table will be the anode that will waste away. The metal nearer the bottom of the table will be the cathode and will be

protected.   The farther apart the metals are in the table, the greater the potential difference will be between them and the greater the corrosion rate of the anode end.

A typical example of the galvanic action of dissimilar metals is represented by the condenser with its steel shell, steel tube sheets, and its copper-alloy tubes.   The steel is nearer the anode end than is the copper alloy, and as a consequence the corrosion of the steel tube sheets and shell is accelerated.   Always, that metal which is higher in the galvanic series will waste away.

Cathodic protection is a means of diverting the electrochemical corrosion from the pipeline to wasting anodes.

There are two ways of providing cathodic protection.   The least costly installation is the galvanic method based on the natural battery action between position of metals in the electrochemical table.   An anode or wasting piece is deliberately used. This approach requires very careful analysis of all the varying conditions involved.

Table 3. Galvanic Series

Anode end (least noble, the wasting end)

Magnesium	Nickel (active)
Magnesium alloys	Brasses
Zinc	Copper
Aluminum	Bronzes
Aluminum alloys	Nickel-copper alloys
Cadmium	Nickel (passive)
Carbon steel	Stainless steel (passive)
Cast iron	Titanium
Stainless steel (active)	Silver solder
Soft solder	Silver
Tin	Graphite
Lead	Gold
	Platinum

Cathode end (most noble, the protected end)

The more costly cathodic protection is the impressed current method that requires an external source of electricity.   The impressed current renders the piping cathodic to the surrounding soil by a controlled difference of potential.

In addition, in locations where there may be stray currents, the installation of removal wires at designated points so that the current may leave the pipeline should be considered.   Or in other words, stray currents are utilized to provide cathodic protection for the pipeline.

Protective Coatings.   Since corrosion of metal is a surface reaction, it is obvious that, if a protective coating which is continuous, impervious, chemically inert, and electrically insulating can be bonded to the interior and exterior of the piping, corrosion cannot take place in the pipe wall as long as the protective coating remains in place undamaged and without cracks or pinholes.

The basis of selection for the best coating differs somewhat for the interior and exterior of the pipe.

The coating on the interior of the pipe to perform its function properly would be selected for its chemical inertness, imperviousness, adhesiveness, adjustment to pipe deformation, and resistance to erosion of the flowing water.

The coating on the exterior of the pipe would be selected for its chemical inertness, electrical resistance, imperviousness, adhesiveness, adjustment to pipe deformation, and resistance to shear and compression due to varying earth conditions.

Galvanizing.   The zinc used for galvanizing pipe is on the anodic (wasting) or electrochemical protective side of the steel, and it is wasted or changed to zinc compounds before the steel pipe will be attacked.   Lead coating on pipe exteriors is not desirable for underground use because lead chemically corrodes in most soils and is cathodic in relation to iron and steel.

Coal-tar Enamel. Specification AWWA C203 covers the coal-tar enamel protective coatings for steel water pipe. This standard sets up the specifications for the materials involved, method of application to the inside and outside of the piping, the thickness required, protection of the coatings, testing, etc.

The type of enamel is specified as AWWA Coal-tar Enamel and is described in this standard with full characteristics and the ASTM tests required.

The 1961 "Paint Manual" of U.S. Department of Interior states:

Coal-tar enamel is especially appropriate for use on the interior surfaces of penstocks and outlet pipes. A long-life coating in penstocks is necessary because dewatering involves loss of revenue from electric power production. In outlet pipes where water flows at high velocities, there is special requirement for stability within the film, which is characteristic of coal-tar enamel. However, hand-applied coal-tar enamel coatings exhibiting drips and undue roughness from careless or inexperienced application, may be damaged by the high velocities. Therefore, for such exposure, excellent workmanship is mandatory. The glasslike surface obtained by the spinning process is especially suitable for this service. Except as limited by certain considerations discussed in the following paragraphs, coal-tar enamel is considered an excellent coating for the interior surfaces of all steel pipe and for the exterior surfaces of buried steel pipe.

The use of coal-tar enamel is limited by its susceptibility to damage from cold weather. The materials marketed are not expected to withstand temperatures under (−) 20 F without danger of cracking and disbonding. Although cases can be cited where they have stood up under such conditions, enamel is not usually considered a good risk where it will be subjected to such extreme cold before or after installation of the coated parts. Enamel coatings are adversely affected by prolonged pre-installation exposure in the open. Continued heat developed when coated or lined pipe is exposed to warm weather and sunlight without the moderating effect of water within the pipe induces hardening and loss of plasticity in the enamel. This embrittlement may be the cause of later damage when the pipe is handled. A heat-reflective coating of white-wash, red lead or aluminum paint is helpful on the outside of the pipe; but even so, prolonged exposure before installation is undesirable.

Although more expensive than shop application, coal-tar enamel is usually applied to the interior of large diameter pipe after installation for several reasons. Most shop lining equipment will not handle pipes much over 10 feet in diameter. Field application avoids pre-installation exposure and installation damage to the coating. If enamel is to be used for lining pipe that is too small to work in, the enamel may be applied in the field by portable centrifugal lining equipment just prior to installation. In practice, this is rarely done, however. Especially where the amount of pipe to be lined is small, the expense of setting up the field equipment is not warranted and shop coating is generally employed if prolonged pre-installation exposure can be avoided. Enamel should not be used for the interior of pipe under 27 inches in diameter where pipe sections must be joined by welding, because of the difficulty or impossibility of coating interior joints after assembly. This is not a consideration where sections are joined by mechanical couplings.

For buried steel pipe, AWWA specifications suggest a variety of exterior coating systems employing enamel, the choice of system depending on soil conditions. Soil forces which are destructive to coatings are caused by wetting and drying of the soil, rocks in the backfill, and chemical constituents. Earth often adheres tightly to coating materials and, on drying, exerts powerful stresses which shear and tear the coating from a pipe surface. The soil properties of liquid limit and plasticity index reflect the soil capacity to create such stresses.

Glass mat has an open structure which permits the hot enamel to penetrate and pass through the mat so as to embed it, producing a reinforcing effect much like that of steel in concrete. This makes it an excellent covering for pipe under severe soil conditions. When asbestos felt is covered by hot enamel, there is a tendency to volatilize the felt saturant and create vapor pockets in the coating, and thus weaken the coating. Asbestos felt is therefore not suitable as an inner wrapping. However, it is desirable as an outer wrapping for buried pipe, because of its value as a shield against soil stress and moderate deformation forces from rocks or clods in backfill material. Even in the rare instances when sandy, well-drained soil would permit elimination of outer wraps, use of asbestos felt can be justified since it reduces handling damage. For large diameter pipe, where the backfill

contains much rock or where highly corrosive conditions exist, pneumatically placed mortar is often used as a shield over the enamel coating.

Mechanical couplings are coated with coal-tar enamel and usually surrounded with a sand shield using No. 16 maximum size sand; or a double wrap of a glass-reinforced coal-tar enamel tape is sometimes used in lieu of the sand shield.

Supplemental protection can be provided for exterior surfaces of buried steel pipe by installing cathodic protection. This is not usually essential at the time the pipe is installed, because potential measurements between soil and pipe can be made subsequently to determine how much, if any, additional protection is needed. However, it is important that the pipe sections be connected or "bonded" electrically at the time of installation if cathodic protection is contemplated at a later date.

Hot-applied Coal-tar tapes—Glass-reinforced coal-tar tapes are used by the Bureau of Reclamation for the exterior coating of welded field joints and also for small quantities of small diameter steel pipe. This tape is applied with the aid of a wide-flame heating torch over properly primed steel surfaces.

Cold-applied Plastic Tapes—Spirally wound vinyl, polyethylene and polyvinyl chloride-butyl rubber tapes, in thickness of 10 to 20 mils, have been used for the protection of the exterior surfaces of straight sections of steel pipe and for the coating of welded joints of shop-coated pipe sections in the field. Difficulty has been experienced in obtaining water-tight seals at the lap when these tapes are applied over pipe fittings. The use of effective primers and a double wrap of the tape has been found to afford maximum protection. A felt wrap held in place by steel bands is desirable when puncture of the tapes may be anticipated, as indicated by the presence of sharp-edged rocks in the backfill.

Cement Mortar—Cement mortar is used extensively by the Bureau of Reclamation as a protective interior lining for steel pipe in new construction, and it has also been used in rehabilitating old pipe lines in place.

Asphalt Coatings.—An asphalt hot-dip coating is sometimes used in Bureau of Reclamation work for small diameter (under 24 inches) steel pipe and corrugated metal pipes in distribution systems. Asphalt is not generally considered to be as effective as coal-tar coating for protection of steel pipe against corrosion, but it is an inexpensive shop treatment not as susceptible to damage from exposure as coal-tar enamel.

The cold-applied plastic tapes are easily installed and have toughness, flexibility, inertness, dielectric properties, and resistance to bacterial action. However, these tapes lose their adhesiveness at temperatures below 34 F.

Cold-applied coatings such as asphalt emulsions, asphalt, or coal tar in volatile solvents have not to date given long enough or sufficient protection when applied to underground piping.

MATERIALS

The following publications and standards of the AWWA (American Water Works Association) contain the complete material specification requirements for water piping:

Excavation, No. R612, Trench Excavation and Backfilling

Grounding, No. 270, Grounding of Electric Circuits on Water Pipes, Stray Current Problems

Hydraulic, No. R805, Water Hammer Allowances in Pipe Design

Plastic Pipe, Nos. R281 and R1015, Developments in Plastic Pipe

Steel Pipe, No. R601, New Developments in Tests of Coatings and Wrappings

Valves, No. 372, Selection of Valves for Water Works Service

American Standard for Vertical Turbine Pumps, A101-60 (ASA B58.1)

AWWA Standard for Cast-iron Pressure Fittings, C100-55

American Standard Practice Manual for the Computation of Strength and Thickness of Cast-iron Pipe, H1 (ASA A21.1).

American Standard for Cast-iron Pit Cast Pipe for Water or Other Liquids, C102-53 (ASA A21.2)

American Standard for Cement-mortar Lining for Cast Iron Pipe and Fittings, C104-53 (ASA A21.4)

American Standard for Cast Iron Pipe Centrifugally Cast in Metal Molds, for Water or Other Liquids, C106-53 (ASA A21.6)

American Standard for Cast Iron Pipe Centrifugally Cast in Sand-lined Molds, for Water or Other Liquids, C108-53 (ASA A21.8)

American Standard for Short-body Cast Iron Fittings, 3 Inch to 12 Inch for 250-psi Water Pressure Plus Water Hammer, C110-52 (ASA A21.10)

American Standard for a Mechanical Joint for Cast Iron Pressure Pipe and Fittings, C111-53 (ASA A21.11)

AWWA Standard for Fabricated Electrically Welded Steel Pipe, C201-60T

AWWA Standard for Mill-type Steel Water Pipe, C202-60T

AWWA Standard for Coal-tar Enamel Protective Coatings for Steel Water Pipe, C203-57

AWWA Standard for Cement-mortar Protective Coatings for Steel Water Pipe of Sizes 30 Inches and Over, C205-41

AWWA Standard for Field Welding of Steel Water Pipe Joints, C206-57

AWWA Standard for Steel Pipe Flanges, C207-55

AWWA Standard for Dimensions for Steel Water Pipe Fittings, C208-59

AWWA Standard for Reinforced Concrete Water Pipe—Steel Cylinder Type, Not Prestressed, C330-57

AWWA Standard for Reinforced Concrete Water Pipe—Steel Cylinder Type, Prestressed, C301-58

AWWA Standard for Reinforced Concrete Water Pipe—Noncylinder Type, Not Prestressed, C302-57

AWWA Standard for Asbestos-cement Water Pipe, C400-53T

AWWA Standard for Gate Valves for Ordinary Water Works Service, C500-61

AWWA Standard for Rubber Seated Butterfly Valves, C504-58

AWWA Standard for Metal-seated Butterfly Valves, C505-58

AWWA Standard for Installation of Cast-iron Water Mains, C600-54T

AWWA Standard for Disinfecting Water Mains, C601-54

AWWA Standard for Cement-mortar Lining of Water Pipelines in Place—Sizes 16 Inches and Over, C602-55

AWWA Standard for Steel Tanks, Standpipes, Reservoirs and Elevated Tanks for Water Storage, D100-59, D5.2-59, D102-55T

Table 4. List of Material Specifications

Material	Specification
Copper pipe	ASTM B42
Red brass pipe	ASTM B43
Cast iron, bell and spigot	FSB WW-P-421
Cast iron, pit cast	ASA A21.2
Cast iron, centrifugally cast in metal molds	ASA A21.6
Cast iron, centrifugally cast in sandlined molds	ASA A21.8
Welded wrought-iron pipe	ASTM A72
Welded and seamless steel pipe	ASTM A53
Seamless carbon-steel pipe	ASTM A106
Black and galvanized welded and seamless steel pipe	ASTM A120
Electric-fusion-welded steel pipe (30 in. and over)	ASTM A134
Electric-resistance-welded steel pipe	ASTM A135
Electric-fusion-welded steel pipe (4 to 30 in.)	ASTM A139
Seamless and welded austenitic stainless steel pipe	ASTM A312
Spiral-welded steel or iron pipe	ASTM A211
Line pipe	API 5L

Table 5. Properties of Large-diameter Water-supply Piping

ID, in.	Thickness, in.	Pipe weight per foot bare, based on 5% over-weight lb	Water contents per foot*	Moment of inertia about pipe axis, in.⁴	Section modulus, in.³	Radius of gyration, in.	Maximum safe span unstiffened pipe, simply supported 120 deg bearing, stress = 5,000 psi, ft
30	3/16	65		2,025	133	10.673	34
	1/4	87		2,717	178	10.695	38
	5/16	108	306 lb	3,418	223	10.718	41
	3/8	130	36.72 gal	4,127	268	10.740	44
	7/16	152		4,845	313	10.762	47
	1/2	173		5,572	359	10.785	50
36	3/16	78		3,489	191	12.794	34
	1/4	104		4,676	256	12.817	39
	5/16	130		5,876	320	12.839	42
	3/8	156	441 lb	7,088	385	12.861	45
	7/16	182	52.88 gal	8,312	450	12.884	48
	1/2	208		9,549	516	12.906	51
	9/16	233		10,799	581	12.928	53
	5/8	259		12,061	647	12.951	54
42	3/16	90		5,528	260	14.916	35
	1/4	120		7,404	348	14.938	40
	5/16	150		9,296	436	14.960	43
	3/8	180	600 lb	11,206	524	14.982	46
	7/16	210	71.97 gal	13,132	612	15.005	50
	1/2	240		15,074	701	15.027	52
	9/16	270		17,034	790	15.049	54
	5/8	300		19,011	879	15.072	55
48	1/4	134		11,028	454	17.059	40
	5/16	167		13,838	569	17.081	43
	3/8	201		16,671	683	17.104	46
	7/16	234		19,526	799	17.126	50
	1/2	268	784 lb	22,402	914	17.148	52
	9/16	302	94 gal	25,301	1,030	17.171	55
	5/8	335		28,222	1,146	17.193	57
	11/16	369		31,165	1,262	17.215	59
	3/4	402		34,130	1,379	17.238	62
54	1/4	153		15,675	575	19.180	40
	5/16	191		19,661	719	19.203	44
	3/8	230		23,676	864	19.225	47
	7/16	267		27,718	1,010	19.247	51
	1/2	306	992 lb	31,787	1,155	19.269	53
	9/16	344	119 gal	35,884	1,301	19.292	56
	5/8	383		40,009	1,448	19.314	58
	11/16	420		44,163	1,595	19.336	60
	3/4	459		48,345	1,742	19.359	63

* 1 U.S. gallon at 60 F = 8.337 lb

Table 5.   (Continued)

ID, in.	Thickness, in.	Pipe weight per foot bare, based on 5% over-weight lb	Water contents per foot*	Moment of inertia about pipe axis, in.^{4}	Section modulus, in.^{3}	Radius of gyration, in.	Maximum safe span unstiffened pipe, simply supported 120 deg bearing, stress = 5,000 psi, ft
	1/4	170		21,472	709	21.302	40
	5/16	212		26,923	888	21.324	44
	3/8	255		32,409	1,066	21.346	48
	7/16	297		37,929	1,246	21.368	51
60	1/2	340	1,225 lb	43,483	1,425	21.391	54
	9/16	383	146.88 gal	49,071	1,605	21.413	57
	5/8	425		54,693	1,785	21.435	59
	11/16	468		60,350	1,966	21.458	61
	3/4	510		66,042	2,147	21.480	64
	1/4	187		28,547	858	23.423	40
	5/16	234		35,786	1,074	23.445	44
	3/8	280		43,064	1,290	23.468	48
	7/16	327		50,384	1,506	23.490	51
	1/2	374		57,745	1,723	23.512	54
	9/16	420		65,147	1,941	23.534	57
66	5/8	466	1,482 lb	72,591	2,158	23.557	59
	11/16	513	177.72 gal	80,077	2,377	23.579	62
	3/4	560		87,604	2,595	23.601	64
	13/16	607		95,174	2,814	23.624	66
	7/8	654		102,785	3,034	23.646	67
	15/16	701		110,439	3,254	23.668	69
	1	748		118,135	3,474	23.692	71
	1/4	204		37,027	1,021	25.544	41
	5/16	255		46,405	1,277	25.567	45
	3/8	306		55,831	1,534	25.589	49
	7/16	357		65,304	1,792	25.611	52
	1/2	409		74,828	2,050	25.633	55
	9/16	459		84,400	2,308	25.655	58
72	5/8	510	1,764 lb	94,022	2,567	25.678	60
	11/16	561	211.5 gal	103,693	2,826	25.700	62
	3/4	612		113,414	3,086	25.722	65
	13/16	663		123,183	3,346	25.744	67
	7/8	715		133,004	3,606	25.767	69
	15/16	766		142,874	3,868	25.789	70
	1	818		152,794	4,129	25.812	72
	5/16	276		58,940	1,499	27.688	45
	3/8	332		70,898	1,800	27.710	49
	7/16	387		82,912	2,102	27.732	52
	1/2	442		94,986	2,404	27.755	55
	9/16	497		107,116	2,707	27.777	58
78	5/8	552	2,070 lb	119,303	3,010	27.800	60
	11/16	607	248.2 gal	131,547	3,314	27.821	63
	3/4	663		143,850	3,618	27.844	65
	13/16	718		156,211	3,923	27.866	67
	7/8	774		168,631	4,228	27.888	69
	15/16	830		181,110	4,534	27.911	71
	1	885		193,645	4,841	27.933	73

Table 5. *(Continued)*

ID, in.	Thickness, in.	Pipe weight per foot bare, based on 5% over-weight lb	Water contents per foot*	Moment of inertia about pipe axis, in.⁴	Section modulus, in.³	Radius of gyration, in.	Maximum safe span unstiffened pipe, simply supported 120 deg bearing, stress = 5,000 psi, ft
84	⁵⁄₁₆	300		73,552	1,738	29.809	45
	³⁄₈	360		88,459	2,087	29.831	49
	⁷⁄₁₆	420		103,431	2,437	29.854	53
	½	480		118,471	2,787	29.876	56
	⁹⁄₁₆	540		133,577	3,138	29.898	59
	⅝	600	2,401 lb	148,749	3,489	29.920	61
	¹¹⁄₁₆	660	287.9 gal	163,989	3,841	29.943	64
	¾	720		179,298	4,194	29.965	66
	¹³⁄₁₆	780		194,670	4,547	29.987	68
	⅞	840		210,111	4,900	30.009	70
	¹⁵⁄₁₆	900		225,621	5,254	30.032	72
	1	960		241,198	5,609	30.054	73
90	⁵⁄₁₆	316		90,395	1,994	31.930	45
	³⁄₈	379		108,702	2,395	31.953	49
	⁷⁄₁₆	440		127,084	2,796	31.975	53
	½	504		145,540	3,198	31.997	56
	⁹⁄₁₆	567		164,074	3,601	32.019	59
	⅝	630	2,756 lb	182,683	4,004	32.042	61
	¹¹⁄₁₆	693	330.5 gal	201,371	4,407	32.064	64
	¾	756		220,133	4,811	32.086	66
	¹³⁄₁₆	819		238,975	5,216	32.108	68
	⅞	882		257,891	5,621	32.131	70
	¹⁵⁄₁₆	945		276,887	6,027	32.153	72
	1	1,008		295,958	6,433	32.175	74
96	⁵⁄₁₆	334		109,636	2,269	34.052	46
	³⁄₈	402		131,819	2,724	34.074	50
	⁷⁄₁₆	468		154,092	3,181	34.096	53
	½	536		176,449	3,638	34.118	56
	⁹⁄₁₆	604		198,891	4,095	34.141	59
	⅝	670	3,136 lb	221,423	4,553	34.163	62
	¹¹⁄₁₆	738	376 gal	244,040	5,012	34.185	64
	¾	804		266,746	5,471	34.207	67
	¹³⁄₁₆	871		289,535	5,931	34.230	69
	⅞	938		312,417	6,392	34.252	71
	¹⁵⁄₁₆	1,005		335,386	6,853	34.274	73
	1	1,072		358,441	7,315	34.297	75

For example, Specification AWWA C201-50 covers Electric Fusion Welded Steel Water Pipe of Sizes 30 Inches and Over and requires that steel plate shall conform to specifications for low- and intermediate-tensile-strength carbon-steel plates of structural quality, ASTM Designation A283, Grade B, of latest revision or as otherwise specified by the purchaser.

DESIGN FOR STRUCTURAL STRENGTH

Pipe-wall Thickness. The Code for Pressure Piping, ASA B31.1 specifies the following formula for determining the thickness of piping for water service of

materials other than cast iron:

$$t_m = PD/(2S + 2yP) + C$$

where t_m = minimum pipe wall thickness, in.

P = maximum internal service pressure, psig

D = outside diameter of pipe, in.

S = allowable stress in material due to internal pressure, at the operating temperature, psi (see material listing in Code)

C = allowance for threading, grooving, mechanical strength, and/or corrosion, in.

y = 0.4 for ferritic steels and austenitic steels for temperatures to 900 F

The Pipe Fabrication Institute publishes a Technical Bulletin TB1, "Pressure Temperature Ratings of Plain End Pipe Used in Power Plant Piping Systems," This covers sizes ½ through 24 in. for pipe of carbon steel and alloys. The tables give thicknesses with their pressure ratings that are based on the Code for Pressure Piping.

Regarding cast iron, the code for Pressure Piping states:

The thickness of cast iron pipe conveying liquids may be determined by selection from American Standards ASA A21.2, A21.6, or A21.8, or from Federal Specification WW-P-421, using the class of pipe for the pressure next higher than the maximum internal service pressure in pounds per square inch. These thicknesses include allowances for foundry tolerances and water hammer. Where the thickness for liquid service is calculated, the methods of ASA A21.1 shall be followed.

In selecting the proper cast-iron pipe, it is essential that the trench-laying conditions be known and specified. The AWWA gives four trench-laying conditions in AWWA Standard C101 (ASA A21.1) for water piping:

1. Flat-bottom trench, without blocks, untamped backfill
2. Flat-bottom trench, without blocks, tamped backfill
3. Pipe laid on blocks, untamped backfill
4. Pipe laid on blocks, tamped backfill

The piping thicknesses given in the AWWA Cast Iron Standards are for depths of cover 3½, 5, and 8 ft, and for working pressures of 50, 100, 150, 200, 250, 300, and 350 psi. The thicknesses include allowance for foundry practice, corrosion, and either water hammer or truck load. The AWWA Standard C101 (ASA A21.1) should be consulted for water-hammer and truck-loading allowances.

Loading on Pipe Due to Depth of Cover and Live Loads. Tables 6 and 7 provide a guide to the loading that may be expected for various depths of bury and live loads on the surface over the pipe. The AWWA Standards provide tables of allowable loads for cast-iron and concrete piping. The allowable loads on buried steel pipes up to about 24 in. in diameter can be computed based on the ability of the pipe to sustain a vertical load by ring action alone. Pipes over 24 in. diameter, depending on wall thickness, will not carry heavy loads with ring action alone without undue deflection. Side support from earth is required. But the earth on the sides of the pipe must be compacted to a high density, approaching the density defined in Specification ASTM D698, in order to provide a predictable amount of side support. Note that piping with more than 8 ft of cover is not appreciably affected by surface live loads. Table 8 is based on ring strength alone.

Table 7 is convenient for a determination of the amount of pipe loading which might be received on a buried line owing to the load transmitted to it from an overhead moving vehicle. To illustrate the use of this table, the following example is given.

Table 6. Dead Load from Earth Cover on Underground Pipes

(Loads are shown in pounds per linear foot of pipe)

Depth of cover, ft	Nominal pipe diameter, in.																		
	3	4	6	8	10	12	15	18	21	24	27	30	33	36	39	42	48	54	60
2	145	180	240	290	340	390	450	500	560	610	700	750	820	875	940	1,000	1,140	1,280	1,380
3	220	270	370	460	550	630	750	860	950	1,040	1,120	1,200	1,300	1,400	1,480	1,580	1,740	1,970	2,080
4	300	370	520	650	780	920	1,080	1,230	1,400	1,520	1,630	1,750	1,850	2,000	2,100	2,220	2,500	2,730	2,980
5	380	470	660	830	1,000	1,160	1,420	1,610	1,810	2,010	2,200	2,340	2,500	2,630	2,800	2,950	3,250	3,600	3,820
6	460	570	800	1,000	1,200	1,430	1,710	2,000	2,230	2,500	2,700	2,950	3,180	3,350	3,500	3,650	4,030	4,420	4,700
7	540	670	950	1,180	1,420	1,700	2,050	2,400	2,700	3,050	3,300	3,570	3,900	4,100	4,300	4,440	4,900	5,450	5,780
8	620	780	1,080	1,370	1,620	1,960	2,400	2,780	3,200	3,550	3,900	4,200	4,500	4,800	5,050	5,300	5,900	6,430	6,880

These values apply to both rigid and flexible pipes buried in ditches or covered by embankment. They are based on maximum conditions of trench width and 120 lb/cu ft soil material, using the Marston formula, and should be used only as approximations. (See American Water Works Association Standard AWWA H1 for charts and tables based on full-scale tests for various conditions.)

Table 7. Percentage of Wheel Loads Transmitted to Underground Pipes

(Values show percentage of wheel load applied to one linear foot of pipe)

Depth of cover, ft	Nominal pipe diameter, in.																		
	3	4	6	8	10	12	15	18	21	24	27	30	33	36	39	42	48	54	60
1	15.2	18.6	25.6	30.0	34.6	40.0	45.2	49.6	52.8	54.4	56.0	57.2	58.0	58.8	59.6	59.7	60.0	60.3	60.6
2	7.0	8.5	11.4	14.0	16.6	19.2	23.0	26.4	30.0	31.2	33.6	35.6	37.4	39.0	40.0	40.9	42.6	43.6	44.6
3	3.5	4.0	5.8	7.2	8.6	10.4	12.8	15.0	17.2	18.6	20.4	22.2	23.6	25.0	25.8	26.0	26.6	28.0	29.4
4	2.0	2.4	3.4	4.2	5.0	6.2	7.8	9.2	10.6	11.6	13.0	14.4	15.8	17.0	17.6	17.7	18.0	19.7	21.4
5	1.4	1.8	2.4	2.8	3.4	4.2	5.2	6.2	7.2	7.8	8.8	9.8	10.6	11.6	12.2	12.4	12.7	14.0	15.3
6	0.9	1.2	1.6	2.0	2.2	2.8	3.6	4.2	5.0	5.6	6.2	7.0	7.6	8.4	8.6	8.8	9.3	10.7	12.0
7	0.3	0.5	1.0	1.4	1.6	2.0	2.6	3.2	3.8	4.2	4.6	5.2	5.8	6.4	6.5	6.6	6.7	7.7	8.7
8	0.2	0.4	0.8	1.0	1.2	1.6	2.0	2.4	2.8	3.2	3.6	4.0	4.4	4.6	5.0	5.1	5.3	6.0	6.7

The values include an impact factor of 2.0 and are based on "Underground Conduits—An Appraisal of Modern Research," M. G. Spangler, *ASCE Paper* 2337, 1947. The values apply to one vehicle with wheels at least 6 ft apart measured along the axle. The wheel load (as in the example) is ½ of the axle load. The wheel load may be on dual tires but is still considered one wheel. (See American Water Works Association Standard AWWA H1 for charts and tables based on full-scale tests for various conditions.)

Table 8.   Allowable Loads for Steel Pipe

Nominal pipe diam, in.	Wall thickness, in. or schedule	Allowable load, lb/lin ft	Nominal pipe diam, in.	Wall thickness, in. or schedule	Allowable load, lb/lin ft
4	0.237 Sch 40	10,000	20	0.250 Sch 10	1,900
5	0.258 Sch 40	10,000	20	0.375 Sch 20	6,250
6	0.280 Sch 40	10,000	20	0.500 Sch 30	14,600
8	0.250 Sch 20	8,300	24	0.250 Sch 10	1,250
8	0.322 Sch 40	9,900	24	0.375 Sch 20	4,250
10	0.250 Sch 20	7,300	24	0.500	10,000
10	0.307 Std	14,600	30	0.250	930
10	0.365 Sch 40	18,700	30	0.375	3,000
12	0.250 Sch 20	5,100	30	0.500 Sch 20	6,500
12	0.330 Std	13,700	36	0.250	850
12	0.375 Sch 40	17,500	36	0.375	2,250
14	0.250 Sch 10	3,800	36	0.500	4,900
14	0.312 Sch 20	7,500	42	0.375	1,750
14	0.375 Sch 30	13,100	42	0.500	3,900
16	0.250 Sch 10	3,000	48	0.500	3,000
16	0.312 Sch 20	5,800	48	0.625	5,500
16	0.375 Sch 30	10,000	54	0.500	2,800
18	0.250 Sch 10	2,440	54	0.625	4,500
18	0.375	7,700	60	0.500	2,500
18	0.437 Sch 30	12,700	60	0.625	4,000

Example. Find the total load per foot on a 30-in. pipe with 6 ft of cover and a 20,000-lb wheel load (i.e., one-half of a 40,000-lb axle load).

Dead load (from Table 6)	2,950 lb/ft
Live load (from Table 7) is 7.0 per cent of 20,000	1,400 lb/ft
Total load	4,350 lb/ft

These values are based on an approximate 2 per cent deflection.   Data are taken from "Design and Deflection Control of Buried Steel Pipe Supporting Earth Loads and Live Loads," Russell E. Barnard, *Proc. ASTM*, vol. 57, 1957.   The table applies to all types of steel pipe—welded, seamless, spiral welded—and to stainless or other steels which have a modulus of elasticity of about 30×10^6 psi, regardless of yield or ultimate strength, since deflection within the elastic limit is dependent on modulus of elasticity, not strength.

Anchors.   For piping with joints such as bell and spigot on cast-iron pipe, bell and spigot with O-ring gasket, Dresser or Smith-Blair type on steel pipe, and any other type of pipe joint depending on friction or packing, anchorage has to be provided to prevent joint pull-out due to internal pressure.

AWWA C600, Standard Specifications for Installation of Cast-iron Water Mains, states in Sections 12.2 and 12.3:

All plugs, caps, tees and bends deflecting 22-½ degrees or more on mains 8 inches in diameter or larger shall be provided with a reaction backing, or movement shall be prevented by attaching suitable metal rods or clamps. . . .

Reaction backing shall be concrete . . . of not less than 2,000 psi at 28 days.   Backing shall be placed between solid ground and the fitting to be anchored. . . .

Anchors should be considered at all direction changes and at dead ends of the piping.   The anchors should bear on and against undisturbed soil.   If backfill is ever used against the face of the anchor, it should be well compacted by wetting and tamping in thin layers.   Any soil weakened by rain or snow or overexcavation beneath the anchor should be replaced by concrete.   The resistance of the soil to movement of the anchor is a combination of passive resistance against the face

of the anchor and the resistance to sliding along the base. Accordingly, values of cohesion and friction angle of the soil involved should be known or at least estimated intelligently.

For large and critical anchors, the soil properties should be determined by penetration and laboratory tests on undisturbed samples. The use of piles should be considered when the size and cost of the anchor are appreciable.

If there is the possibility of shock pressures due to water hammer, the anchors, should be designed to withstand these.

Anchor loadings may be determined as follows: resultant pressure thrust on anchor = (internal pressure) × (largest internal pipe cross-section area) × (fitting factor).

Fitting	Factor
90-deg ell	1.41
Caps, plugs, tees	1.00
45-deg ell	0.77
22½-deg ell	0.39
11¼-deg ell	0.20

In addition, *centrifugal thrust* is present, but will be low for the usual water-line velocities.

$$\text{Centrifugal thrust} = (2AWV^2/g)\sin(\theta/2)$$

where A = inside area of pipe, sq ft
  W = density of fluid, lb/cu ft
  V = velocity, fps
  θ = change in direction
  g = 32.2 ft/sec² gravity acceleration

Another important case where anchors must be considered for reasons other than joint design is that in which the temperature of the water flowing in the pipeline is appreciably higher than the temperature prevailing during construction of the pipeline.

If the temperature of the water is 200 F, for example, and the temperature of the water must be maintained, then the design of the piping would follow that of underground steam lines and would involve insulation, watertight protective conduits or tunnels, etc.

But if the requirements of design are such that the loss of heat from the water is not important, for example, waste drains from process, cooling water after heat pickup, atomic-plant waste, etc., it is economically desirable to treat the piping as cold-water piping. When this is done, the thermal expansion in the piping must be controlled or its effects dissipated.

Expansion joints of bellows or slip type, ball joints, or gasketed friction joints may be utilized to take up expansion, provided leakage and failure possibility are evaluated and found acceptable.

In between the anchored points, the expansion of solidly joined piping (welded) must be absorbed through either bending and torsion of cantilever legs or through bowing of the line. The possibility of suppressing the expansion and containing it in the pipeline through compressive stress is not very feasible. The force available for this is the starting friction between the earth and the pipe, and this is some percentage of the loading on the pipe. A portion of the expansion can be suppressed in this way through friction, but the remaining force can be very large and will appear at the anchor points.

The magnitude of the thermal thrust may be seen from the following example: The modulus of elasticity is the tangent of the stress-strain curve in the elastic range and is defined as

$$E = (F/a) \div (e/l)$$

and from this

$$F = Ea(e/l)$$

where F = force, lb
     E = modulus of elasticity (for the pipe material at temperature), psi
     a = metal cross-section area, sq in.
     e = total expansion, in.
     l = length, in.
     e/l = expansion, in./in.

Expansion of carbon steel for temperature rise from 70 F to	Expansion in./in.
100 F	0.00018
150 F	0.00050
200 F	0.00083

For a pipe 24 in. OD by ½ in. thick with a thermal growth of 0.00083 in./in. for a temperature change from 70 F at installation to 200 F operating,

    $E = 27.7 \times 10^6$ at 200 F   Metal area $= 36.91$ sq in.
    $F = (27.7 \times 10^6) (36.91) (0.00083) = 843,500$ lb thermal force thrust

Even if it were possible to suppress this thermal force, the compressive stress in the pipe wall would be excessive, as can be seen by dividing the force by the cross-sectional area. In addition, the possibility of buckling due to column action would require investigation.

A feasible approach is to take care of the expansion by some form of bending of the pipe. The piping would have to be run in a manner to provide bending legs at right angles. For smaller temperature differentials, pipe lengths laid at a small-angle zig-zag might be sufficient.

In addition, screened sand should be used around the pipe to lower the sliding friction and provide adjustment area for the pipe. A polyvinyl chloride type of sheeting may be advisable to contain the sand area and prevent washout.

Expansion during Construction. Water piping is normally not designed for expansion, so that expansion of piping during construction may be easily overlooked.

The pipeline is joined together in a continuous string in a trench. The piping expands under the direct heat and radiation of the sun and contracts during the cooling of the night. The length differential between night and high sun at noon may be several inches, depending on the exposed length of the piping string. For example, for 150 F surface metal temperature, the growth could be 3 in. for 500 ft. As long as the final closure weld is not made, the piping freely grows, resisted only by the sliding friction at the bottom of the trench. But once the final closure weld is made, thermal thrust will occur and may cause damage either to a portion of the pipeline or to the rigid end connections. The ideal approach would be to test the line and bury it before the final closure weld is made.

In addition, on very large diameter lines, the differential between the top of the pipe exposed to direct sun radiation and the bottom hidden in the cooler earth may create some bowing of the line.

EARTHQUAKE

The damage caused directly by an earthquake may be only a small percentage of the total resulting destruction. The major loss will result from the fires started by the earthquake damage if the water services for fire fighting are badly disrupted. In the 1906 San Francisco earthquake, the broken water-service mains prevented

any effective fire fighting and the major damage came from the unchecked fires that swept the city for days.

The physical picture of the earthquake will vary with the intensity of the seismic disturbance, the distance from its epicenter, the soil characteristics, the contour of the area, and the type of readjustment taking place deep in the earth. The piping engineer will have to form some sort of a picture of the type of seismic action possible in the area under consideration, establish the seismic design conditions, and decide on the degree of seismic protection justified by the demands of reliability and economics for the pipeline and its branch components. The following is quoted from "Earthquake Damage and Earthquake Insurance" by John R. Freeman, McGraw-Hill Book Company, 1932:

The earthquake shock at the site of a structure is caused by the impact from the passage of an elastic wave which has originated deep in the earth from an earth-slip, probably along a "fault-line" or line of cleavage between vast fault-blocks, commonly many miles in depth, breadth and thickness, into which the outer formation of the earth to a depth of perhaps 40 miles, has been cracked by shrinkage stress and unknown causes during millions of years past.

This elastic earthquake wave proceeds outward in all directions from its origin, and reaches the site of the structure under consideration, first as a longitudinal oscillation of alternate compression and expansion in the direction of transmission, followed quickly by a supplementary transverse oscillation at right angles to the direction of transmission.

This wave system while passing through the earth is subject to reflection, or to partial absorption, at obstacles such as an abrupt change in geological formation, not far beneath the earth surface, and therefore may reach the building site in a much confused form.

Although deep in the bed-rock these waves seldom exceed a small fraction of an inch in amplitude of motion, they may become magnified in mobile earth (such as alluvial deposits) near the earth surface, so that the violence of the wave motion near the earth's surface may vary largely at localities less than a mile apart. Moreover, the character of the wave motion may become changed from the harmonic form commonly assumed and into the form of oscillation an elastic body is ordinarily thrown by a sharp blow.

The earthquake acts as a giant vibrator upon the soil near the surface of the earth. Where the soil is hard cohesive, the surface movement will be less than where the soil is soft cohesive or cohesionless. This vibratory action will compact and cause settlement of earth fills. Where there are water-saturated soils, the water forced by consolidation of the soil that had previously been retained in spaces between soil particles will provide temporary flotation of surface areas, and compression zones are created that act to heave some surface areas.

Separation of surface area may occur at sharp demarcation of soil characteristics, since movement of the adjacent soil areas could have been different. On hillside slopes, shear of the soil layer can occur, since the low elasticity of the soil cannot restore the soil to its former position during the back-and-forth surface oscillation.

Buildings will possess movements that may vary considerably from those of the soil areas surrounding them, and the fill areas next to the walls may settle under the vibratory action. Tanks with their water loads may shift, tend to rotate, and their shells ovalize in addition to the possible differential movement of the tank as a whole as compared with the surrounding soil.

Earthquake code provisions that speak of a percentage of gravity acceleration design requirement, for example, $0.2g$ for Zone 3 type of earthquake, to take care of seismic effects cannot realistically be applied to buried piping because the underground piping rides with the ground. Instead, provisions for seismic protection have to be considered on a differential movement basis in the following areas: at pipeline changes in direction, at pipeline crossings of sharp demarcations of soil characteristics, at pipeline entrances to fill around buildings and at building walls, at points of attachment to tanks or other aboveground equipment.

A sampling of some of the earthquake damage experienced by pipelines may be helpful:

Line pulled apart.
Line pulled apart, then telescoped.
Line crinkled and ruptured.
Line weld broken.
Breaks in cast-iron piping and fittings.

Adjoining areas of differing soils displaced and pulled apart piping mains.

Subsidence of filled ground due to earthquake caused breakage of water, sewer, and other buried lines.

Pipeline breakage at tanks has been noted where lower courses had failed and also in large-diameter tanks where flexibility of piping arrangement was not present. Failures were predominant in cast-iron fittings.

Passing through Fill and into Buildings. A sleeve should be used through the wall and should be of sufficiently larger diameter to provide the estimated seismic movement for the pipeline. In addition, consideration should be given to the use of a corrugated culvert (split type is available) around the pipe extending from the building wall to undisturbed earth. The wall end of the corrugated culvert should be supported independently of the pipe. The culvert will protect the pipe from the earthquake-compacting settlement, and the wall sleeve will allow horizontal movement in addition to vertical adjustment. However, if flexibility inside the building is limited, it may be necessary to consider horizontal flexibility outside the building so that the piping can move toward or away from the building. This may take the form of a right-angle leg for flexibility or movable joints.

Connection of Tanks. In tanks filled with water, a wave action is set up by the earthquake that may cause appreciable distortion in the tank. The shell configuration at the top becomes oval as the result of wave action, but the tank remains essentially circular at the bottom owing to restraint of the flooring and foundation friction. This difference in configuration causes the base of the shell to depress on the major axis, or direction of the wave, and lift on the minor axis. This action causes shell nozzle circular movement, including a lifting and lowering.

In addition, a differential may develop between the distorting of the tank shell and the movement of the adjacent earth. Flexibility in the piping attached to the tank must be allowed in all directions to deal with these complex movements between the tank connections and the point at which the piping enters the earth. Cast-iron fittings and valves should not be used. Corrugated culvert around the piping, as it goes underground, might prove useful to obtain sufficient flexibility. Some form of flexible joint, such as a Barco-type joint, may aid in obtaining sufficient flexibility.

In crossing over from one type of soil to another where marked difference in seismic behavior is suspected, the main object is to obtain flexibility in all directions. An ample loop enclosed in an oversized corrugated culvert is one approach to obtaining flexibility.

Where there is a change in direction of the underground piping, the soil may move axially along one leg of the pipe and broadside against the other leg as movement differentials develop. If sand is used around the pipe for at least 50 ft along both legs, the pipe has a better chance for adjustment. The sand should be protected against dispersal and washout with some form of sheeting, such as polyvinyl chloride.

IRRIGATION PIPING

In semiarid regions where the natural rainfall is insufficient for growing crops to best advantage, irrigation frequently is resorted to if a supply of water is available

from a river or lake. This can be accomplished by gravity flow from ditches or furrows or by pressure sprinkling from pipes.

Large amounts of lightweight steel pipe are used for irrigation purposes in the western United States and elsewhere. Since the advent of fusion welding, most of this pipe is fabricated from strip or sheet steel using straight- or spiral-welded seams. Where used aboveground, such pipe usually is joined with drive joints or with compression couplings of one sort or another such as Dresser or Victualic or those having lever-clamping devices instead of bolts to facilitate breaking joints for moving the pipe about. Twenty-foot lengths are usual. Portable lightweight pipe for irrigation purposes frequently is galvanized as a protection against rusting.

Overhead Sprinkling Systems. Sprinkling, or overhead irrigation, may be carried on by the use of permanent systems with underground supply pipes and permanently located sprinklers or by the use of portable pipe. Overhead irrigation by sprinklers became practicable on a large and economical scale with the development of portable systems using pipe light enough to be moved around readily, yet strong enough to withstand rough field service and equipped with couplings that work easily and quickly without the use of tools or at most with just a wrench. Portable systems are said to have many advantages over permanently installed overhead systems. The initial investment is considerably less, there is nothing in the field to interfere with cultivating and planting, and the outfits can be moved quickly from one field to another as the need arises. The following recommendations for installing and operating aboveground portable sprinkling systems are taken from the "Handbook of Water Control":[60]

Experience in the field has shown that generally, with two men, a 1000-ft line can be moved to its next position parallel to the former position and 60 ft from it in from 30 to 35 min, ready for another application.

Often in the field, special fittings such as ells, wyes, tees, etc., may be used to advantage to meet special operating conditions.

When one single line is used it is necessary to stop irrigating long enough to move the line to a new set-up. If an alternate line is provided, no time is lost in moving.

Revolving sprinklers give best results when operating at minimum pressures of from 20 to 30 psi, depending upon their size, the smaller sprinklers requiring less pressure than the large ones. The diameter of coverage of sprinklers varies with the size of the sprinker and the operating pressure employed.

It has been found, in order to obtain and insure satisfactory coverage, that when small sprinklers discharging less than 7 gpm are used they should be spaced every 20 ft along the line. Sprinklers discharging 7 gpm or more give adequate coverage and distribution when spaced every 40 ft (at every second pipe joint) along the line.

Lines carrying small sprinklers operating under minimum pressures of 20 psi should be moved a maximum distance of 40 ft between set-ups.

Lines carrying sprinklers operating under a minimum pressure of 25 psi should not be moved more than 50 ft between set-ups.

Lines carrying sprinklers operating under a minimum pressure of 30 psi should not be moved more than 60 ft between set-ups.

Manufacturers of sprinklers designed for agricultural service adjust their sprinklers to turn slowly—about 1 rpm. If they turn too fast, the effective diameter of the coverage circle is materially reduced.

The nozzle sizes of the sprinklers can be varied, depending on the amount of water and the pressure available. The rate of application of water depends on the character of the soil. Some types of soil will take water faster than others, and on such soils water can be applied at a higher rate.

Units of Water Measurement. In irrigation and hydroelectric work as well as in hydraulic mining, the units for *rate* of flow are the cubic foot per second (or second foot) for larger quantities and the miner's inch for smaller quantities. The cubic foot per minute also is used to some extent in hydraulic mining. The

miner's inch was developed in the early mining days and is still used extensively for other purposes as well. It is an awkward term, but its use often is necessary on account of custom and law. A miner's inch is the rate at which water discharges through 1 sq in. of opening under a prescribed head (approximately 6 in.), and the number of miner's inches is equal to the area of the opening in square inches. The value of the miner's inch varies in different localities, ranging from $1/50$ to $1/38.4$ cfs (see Table 9).

The units for *volume* of water delivered are the cubic foot, the acre-foot, and the acre-inch, which is one-twelfth the acre-foot. For large volumes the acre-foot is the unit recommended. This is the volume required to cover 1 acre to a depth of 1 ft, which equals 43,560 cu ft. One cubic foot per second flowing steadily for 24 hr approximately equals 2 acre-feet.

Table 9. Conversion of Units of Flow Used in Measuring Water

(From "Handbook of Water Control" [60])

Cubic feet per second	Gallons per minute	Million gallons per day	Miner's inches		Colorado	Acre-inches per hour	Acre-feet per 24 hours
			Arizona, California, Montana, Nevada, Oregon	Idaho, Kansas, Nebraska, New Mexico, North Dakota, South Dakota, Utah			
1	448.8	0.646	40	50	38.4	0.992	1.983
0.00223	1	0.001440	0.0891	0.1114	0.0856	0.0022	0.00442
1.547	694.4	1	61.89	77.36	59.44	1.535	3.07
0.025	11.25	0.0162	1	1.25	0.960	0.0248	0.0496
0.020	9.00	0.01296	0.80	1	0.768	0.0198	0.0397
0.026	11.69	0.0168	1.042	1.302	1	0.0258	0.0516
1.01	452.42	0.651	40.32	50.40	38.71	1	2.00
0.504	226.3	0.3258	20.17	25.21	19.36	0.5	1

The interrelation of the various units of measurement for rate of flow and volume of water delivered is shown in Table 9.

Measuring Devices. The miner's inch is measured through a *miner's inch box* which is a special form of free-flowing orifice consisting of an opening in a plank under a head of from 3 to 9 in. The arrangement of the opening, which may be from 1 to 4 in. high; the thickness of the plank, which may be from 1 to 3 in.; and the point from which the effective head is measured, as well as the head itself, are largely matters of local custom and of state law. Owing to the uncertainties of this method of measurement, the miner's inch usually is construed now as some fraction of 1 cfs as shown in Table 9.

A number of commercial *meters* are on the market which are designed for measuring irrigation water. Usually they regulate the flow of water and may or may not be arranged to register the total quantity passed. The measuring element can be a weir, flume, orifice, venturi meter, or current meter used in connection with a registering device of some sort. Or the registering device may be omitted and the metering element used merely to limit the rate at which water can be taken. A typical installation of a *metering orifice box* is shown in Fig. 13.

A special form of submerged orifice called a *meter gate* is used extensively for measuring water for irrigation purposes. A meter gate consists of a sluice gate to which are fitted two measuring wells in which the elevation of water on the upstream and downstream sides of the device can be read in inches with a rule. The

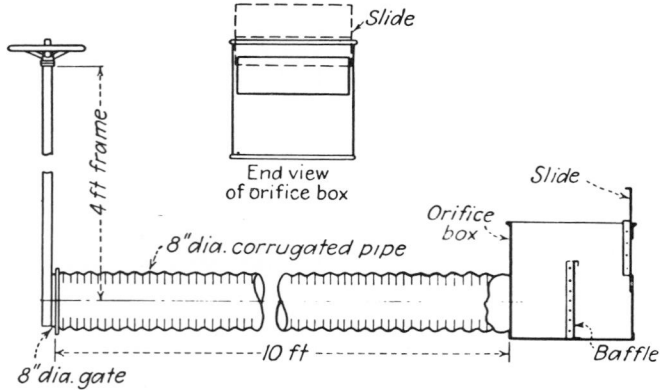

Fɪɢ. 13. Typical installation of a metering orifice box.

difference between the two readings gives the available head across the meter gate from which the rate of flow can be computed or taken from a table to correspond with the amount of opening.   A typical meter gate installation is shown in Fig. 14.

When the installation of a meter gate is contemplated and the site selected, sufficient excavation is made to place the outlet pipe level with a clear opening inlet and outlet and set low enough to ensure the proper submergence of the outlet. Submergence should not be less than 6 in. under lowest conditions.   If full submergence is not maintained, no readings can be taken in the downstream measuring well because of the surging of the water surface.   Also, excavation should be made to give full contraction at the entrance if possible.   The clearance between the inside of the gate and the side walls and bottom of the canal on the upstream side of the meter gate should be not less than 6 in.

After the gate has been installed the measuring wells are placed in position and connected by ¾-in. pipes, one to the upstream side of the gate and the other to the main pipe 12 in. below the gate.   A notch is made with a hack saw in the stem of the gate at a point flush with the top of the handwheel when the handwheel is without slack on the stem and the bottom of the gate slide is exactly level with the bottom of the inside of the gate seat.   This is called the point of "zero gate opening." This means that the gate is not tightly closed.   The stem will be the width of the gate seat higher than it is in the fully closed position.   The amount of any gate opening is obtained by measuring the distance from the hack saw mark to the top of the handwheel.

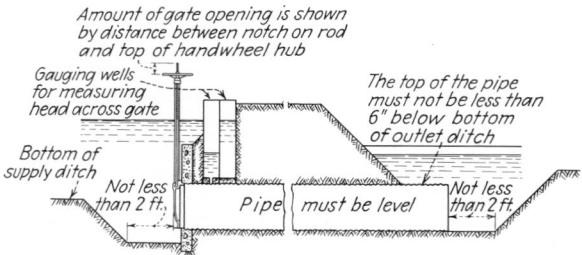

Fɪɢ. 14. Typical meter gate installation.

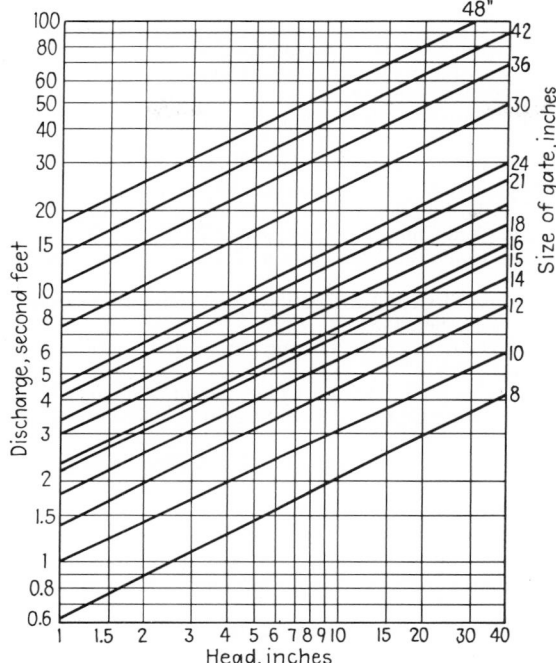

FIG. 15. Discharge capacities of fully opened meter gates.

When the turnout is in use, the water stands in one stilling well at the level of the water in the upstream canal and in the other well at the static level of the water in the turnout pipe at a point 12 in. downstream from the face of the gate seat. The difference in these two levels represents the static pressure on the gate opening and is denoted in these considerations as the "head on the gate."

Meter gates can be fastened to corrugated or smooth pipe of any length, as the only friction involved is in the first foot of pipe which is furnished as part of the meter gate. The discharge capacity at different heads of fully opened meter gates as manufactured by the California Corrugated Culvert Co. are given in Fig. 15. Charts are available in their "Handbook of Water Control" for reduced flows with lesser gate openings.

REFERENCES

The following references, arranged chronologically for each subject, provide valuable additional information:

Coatings

1. Burnett, G. E., "Performance of Coal Tar Enamel on Steel Water Pipe and Penstocks," *J. AWWA*, August, 1950.
2. Goodwin, C. L., "Coatings and Cathodic Protection," *Oil Gas J.*, Nov. 16, 1950.
3. Shideler, N. T., "Application of Hot Applied Coal Tar Coatings for Pipe Lines," *Pipe Line News*, June, 1952.

4. Burnett, G. E., and C. B. Masin, "Developments in Tests of Coatings for Steel Pipe," *J. AWWA*, October, 1952.
5. Cates, W. H., "Coatings for Steel Water Pipe," *J. AWWA*, February, 1953.
6. Burnett, G. E., and P. W. Lewis, "New Developments in Tests of Coatings and Wrappings," *J. AWWA*, February, 1956.
7. Davidson, S. M., "Electrical Inspection of Steel Pipe Coatings," *J. AWWA*, February, 1956.
8. Fair, Jr., W. F., "Properties, Specifications, Tests and Recommendations for Coal Tar Coatings," Parts 1 and 2, *Corrosion*, November and December, 1956.
9. Fair, Jr., W. F., C. U. Pittman, and M. G. Sturrock, "Testing of Coal Tar Coatings—Field Exposure in Cold Climates," *Corrosion* (NACE), March, 1957.
10. Shideler, N. T., "Coal Tar Coatings for Protection of Underground Structures," *Corrosion*, June, 1957.
11. Kiernan, F. J., "Use of Coal Tar Coatings for Underground Corrosion Mitigation," *Pipe Line News*, December, 1957.
12. Burnett, G. E., and C. E. Selander, "Plastic Linings and Coatings for Steel Water Pipe," *J. AWWA*, August, 1958.
13. U.S. Bureau of Reclamation, Denver, Color. "Paint Manual,"
14. Steel Structures Painting Council, "Steel Structures Painting Manual," vols. 1 and 2, Pittsburgh, Pa.
15. "Underground Corrosion," *Nat. Bur. Std. Circ.* C579, Washington, D.C.

Corrosion

16. Marbut, C. F., "Soils of the United States," U.S. Department of Agriculture.
17. Derby, R. L., "Control of Slime Growths in Transmission Lines," *J. AWWA*, November, 1947.
18. "Cathodic Protection," a symposium by the National Association of Corrosion Engineers, 1949.
19. "Technical Practices in Cathodic Protection," *J. AWWA*, November, 1951.
20. Schneider, W. R., "Corrosion and Cathodic Protection of Pipelines," *J. AWWA*, May, 1952.
21. Wainwright, R. M. "Economic Aspects of Cathodic Protection," *Corrosion*, February, 1953.
22. Schneider, W. R., and D. Hendrickson, "Electrical Measurements Applied to Corrosion Investigations," *Corrosion*, October, 1954.
23. Parker, M. E., "Current Requirements for Cathodic Protection of Pipe Lines," *Corrosion*, April, 1955.
24. Sharpe, L. G., "Economic Considerations in Pipe Line Corrosion Control," *Corrosion*, May, 1955.
25. Tenny, M. K., "Corrosion and Electrolysis," *Water Works Eng.*, May 1955.
26. Sudrabin, L. P., "External Corrosion Problems in the Water Distribution System," *J. AWWA*, October, 1956.
27. Peifer, N. P., "Design and Installation of Cathodic Protection," *Pipe Line News*, September, 1958.
28. Jelinek, R. V., "Design Factors in Corrosion Control," *Chem. Eng.*, Nov. 17, 1958.
29. National Bureau of Standards, "Study of Causes and Effects of Underground Corrosion," *J. AWWA*, December, 1958.
30. Garrett, Gairald H., "Electrolysis and Corrosion in Water Systems" *J. AWWA*.

Earth Loading

31. "An Approximate Method for Predicting the Settlement of Foundations and Footings," *Second Intern. Conf. Soil Mech. Foundation Eng.*, Paper 14.4, 1948.
32. Davis, H. E., and F. N. Finn, "Trench Backfill Practices" *J. AWWA*, May, 1950.
33. Spangler, M. G., "Protective Casings for Pipe Lines," *Iowa State Coll. Eng. Rept.* 11, 1951–1952.
34. White, H. L., "External Loads on Pipe with Cement Mortar," *J. AWWA*, June, 1952.

35. Paul, Leslie, and O. F. Eide, "Crushing Strength of Steel Pipe Lined and Coated with Cement Mortar," *J. AWWA*, June, 1952.
36. "Backfill Lessons on a 30-inch Pipeline," *Eng. News-Record*, Mar. 20, 1952.
37. Spangler, M. G., "Secondary Stresses in Buried High Pressure Lines," *Iowa State Coll. Eng. Rept.* 23, 1954–1955.
38. Sowers, G. F., "Trench Excavation and Backfilling," *J. AWWA*, July, 1956.
39. Reitz, H. M., "Soil Mechanics and Backfilling Practices," *J. AWWA*, December, 1956.
40. Barnard, R. E., "Design and Deflection Control of Buried Steel Pipe Supporting Earth Loads and Live Loads," *Proc., ASTM*, vol. 57, 1957.
41. McClure, G. M., "Secondary Stresses in Large-diameter Pipelines," *Proc. Am. Soc. Civil Engrs.*, June, 1957.
42. Livingston, L. E., Jr., "Loads on Ditch Conduits," *Water Sewage Works*, March, 1958.
43. Sollid, Erik, "Design of Pipeline Anchor Blocks," *Water Power*, November, 1958.

Flow

44. Farnsworth, G. Jr., and A. Rossano, Jr., "Application of the Hardy Cross Method to Distribution System Problems," *J. AWWA*, February, 1941.
45. Hurst, W. D., and N. S. Bubbis, "Application of the Hardy Cross Method to the Analysis of a Large Distribution System," *J. AWWA*, February, 1942.
46. "Entrainment of Air in Flowing Water," a symposium, *Trans. ASCE*, vol. 108, p. 1393, 1943.
47. Hinds, Julian, "Comparison of Formulas for Pipe Flow," *J. AWWA*, November, 1946.
48. Wyckoff, W. W., "Procedure for Flow Tests of Pipelines," *J. AWWA*, November, 1946.
49. Harris, C. W., "An Engineering Concept of Flow in Pipes," *Proc. Am. Soc. Civil Engrs.*, May, 1949.
50. Parmakian, John, "Air Inlet Valves for Steel Pipe Lines," *Trans. Am. Soc. Civil Engrs.*, vol. 115, p. 438, 1950.
51. Kerr, S. L., "Water Hammer Control," *J. AWWA*, December, 1951.
52. Cates, W. H., "Standard Allowances for Water Hammer in Steel Pipe," *J. AWWA*, November, 1952.
53. Parmakian, John, "Pressure Surges at Large Pump Installations," *Trans. ASME*, vol. 75, p. 995, 1953.
54. Hoag, L. N., and G. Weinberg, "Pipeline Network Analysis by Electronic Digital Computer," *J. AWWA*, May, 1957.
55. Richards, R. T., "Air Binding in Large Pipelines Flowing under Vacuum," *Proc. Am. Soc. Civil Engrs.*, December, 1957.
56. "Water Hammer Allowances in Pipe Design" by Committee, *J. AWWA*, March, 1958.
57. Harleman, D. R. F., "Surges in Pipe Lines," *Water Sewage Works*, March, 1958.
58. "Standards of Hydraulic Institute."
59. "Welded Steel Water Pipe Manual," Consolidated Western Steel, 1960.
60. "Handbook of Water Control," California Corrugated Culvert Co.
61. "Hydraulic Network Analysis," The Service Bureau Corporation (IBM).

22

OIL-SYSTEMS PIPING

H. M. Howarth*

The Code for Pressure Piping divides oil piping systems into two separate categories. Section 3 of the Code for Pressure Piping (ASA B31.3–1966), Petroleum Refinery Piping, is the basis for the design, fabrication, assembly, erection, inspection, and testing of petroleum-refinery piping. Section 4 of the same code (ASA B31.4–1959), Oil Transmission Piping, deals with the same items for oil-transmission piping outside the refinery limits. Section 3 of the Code also is the guiding document for chemical industry process piping until Section 6 of the Piping Code is written.

In this chapter, the requirements for oil-systems piping will be treated as they are dictated by one or the other of the two sections of the Code mentioned above. At times, it will be desirable to quote verbatim Code material; in such instances by the use of special type face, by footnote, or by reference in the text, the source will be indicated. For specific or detailed requirements or for elaboration on certain design specifications which are not covered in this chapter, the reader is referred to the pertinent code.

PETROLEUM-REFINERY PIPING

As used in this section, petroleum-refinery piping refers, except as excluded below, to all piping within the property limits of a petroleum refinery, loading terminal, natural-gas processing plant, bulk plant, compounding plant, or refinery tank farm. Included in this scope are piping systems which convey all fluids such

* Dean Witter and Company, Chicago, Illinois. (Formerly with American Oil Company, Whiting, Indiana.)

as fluidized solids, oil, gas, steam, air, water, chemicals, and refrigerants. Excluded in the scope of this section are those systems which are governed by the Power Boilers Section of the ASME Boiler and Pressure Vessel Code, oil- or gas-transmission piping that is part of a (cross-country) distribution system, and piping in which the fluid pressure is at or above 0 psig but less than 5 psig regardless of temperature, or at or above 5 psig but less than 15 psig if the design temperature is in the range of −20 to 650 F, inclusive.

Piping, in this treatment, consists of pipe; flanges; bolting; gaskets; valves; fittings; the pressure-containing parts of other components such as expansion joints, swivel joints, and strainers; and devices which serve such purposes as mixing, separating, snubbing, distributing, metering, or controlling flow. It also includes pipe-supporting elements but does not include support structures such as frames of buildings, stanchions, or foundations.

DESIGN CONDITIONS AND CRITERIA

The material below is quoted directly from the Petroleum Refinery Piping Code, ASA B31.3–1966. Numbers which follow the paragraph headings are the actual paragraph numbers in the Code.

Internal Design Pressure (301.2.2)

The piping component shall be designed for an internal pressure representing the most severe condition of coincident pressure and temperature expected in normal operation (including fluid head). The most severe condition of coincident pressure and temperature under normal operation shall be that condition which results in the greatest required pipe thickness and the highest flange rating.

External Design Pressure (301.2.3)

The piping component shall be designed for the maximum differential pressure (including fluid head) at the coincidental temperature, that can act externally on the component in the piping system, taking into consideration the failure of external or internal pressure.

Design Temperature (301.3.2)

The design temperature is the metal temperature representing the most severe condition of coincident pressure and temperature as explained in 301.2.2 and shall be determined as follows:

(a) For fluid temperatures below 32 F, the metal temperature shall be taken as the fluid temperature.

(b) For fluid temperatures 32 F, and above, the metal temperature for uninsulated components shall be no less than the following values:
   (1) Threaded and welding end valves, pipe, welding fittings, and other components having wall thicknesses comparable to that of pipe: 95% of the fluid temperature.
   (2) Flanged values, flanged fittings, and flanges (except lap joints): 90% of the fluid temperature.
   (3) Lap joint flanges: 85% of the fluid temperature.
   (4) Bolting: 80% of the fluid temperature.

(c) Externally insulated piping: The fluid temperature shall be used unless calculations, previous tests, or service experience based on measurements support the use of other temperatures. Where piping is heated by tracing or jacketing, the effect of such heating shall be incorporated in the establishment of the design temperature.

(d) Internally insulated piping: The design metal temperature shall be based on heat transfer calculations or tests.

Discussion of Design Pressure and Temperature. The designer is cautioned that regardless of the combination of operating pressure and temperature which results in the most severe condition from a stress standpoint, the selection of materials will almost always be governed by the extremes of operating temperature and pressure.

In designing piping, it is simpler to use fluid temperature instead of metal temperature. Substantial savings, however, can be realized by making the calculations or tests in (c) and (d) on p. 22–2 to determine the metal temperature, especially if the fluid temperatures are high. Not only can flange (and valve) ratings and pipe thicknesses often be reduced, but occasionally less expensive materials can be used.

Table I.   Maximum Fluid Temperatures for Economically Reducing Pressure Class of Carbon-steel Flanges

Flange size, IPS	Reduction in flange rating						
	300–150*	400–300	600–400	900–600	1500–900	900–400	1500–600
1	390						
1½	335						
2	290						
2½	270						
3	260	. . .	. . .	280	275		
4	295	425	200	270	255		
6	345	400	260	355	355		
8	325	350	300	440	395		
10	315	360	340	410	590		
12	375	360	285	440	675		
14	465	315	260	660	700	. . .	970
16	480	320	300	635	750	. . .	1000
18	520	325	225	690	800	. . .	1050
20	525	315	200	730	850	. . .	1100
24	535	350	275	905	900	995	1100

* For example, use 150 instead of 300 lb.

In the application of subparagraph (b) on p. 22–2, any savings in the cost of piping components must be balanced against the present worth of the additional heat to be lost from the piping components if the insulation is omitted. Flanges and valves which are to be left bare should be clearly marked on the piping drawings so that they are not inadvertently insulated.

Calculations based on valve, insulation, and heater costs of the early 1960's; an estimated life of 20 years; an average air temperature of 50 F; a cost of $0.60/MM Btu, and using discounted cash flow analysis with 15 per cent cost indicate that:

1. Usually it is not economical to omit the insulation from carbon-steel pipe even though thinner pipe could be used by doing so.

2. There are fluid temperatures above which the present worth of the heat lost exceeds the savings to be realized in those cases where it is possible to use lower pressure class carbon-steel flanges or flanged valves by omitting the insulation. Tables 1 and 2, which indicate these "break-point" fluid temperatures, are included to give the designer guidance in applying subparagraph (b) on p. 22–2.

Cooling Effects on Pressure. The cooling of a gas or vapor in a piping component may reduce the pressure sufficiently to create an internal vacuum. In such a case, the piping component must be capable of withstanding the external pressure at the lower temperature or provisions must be made to break the vacuum.

Table 2.   Maximum Fluid Temperatures for Economically Reducing Pressure Class of Carbon-steel Flanged Valves (with Companion Flanges)

Valve size, IPS	Reduction in valve rating							
	300–150*	400–300	600–400	900–600	1500–900	600–300	900–400	1500–600
1	430							
1½	385							
2	320							
2½	445							
3	475	. . .	. . .	800	420	. . . .	. . . .	900
4	475	570	320	680	425			
6	550	620	475	680	605	. . . .	885	895
8	545	625	510	735	635	. . . .	925	1020
10	550	625	625	680	800	870	955	1025
12	625	650	535	760	865	. . . .	945	1100
14	680	660	555	800	940	840	965	1100
16	750	680	525	775	1000	840	955	1100
18	745	810	430	870	1100	900	1000	1100
20	750	930	400	950	1100	1000	1100	1100
24	780	955	725	1000	1100	1100	1100	1100

* For example, use 150 instead of 300 lb.

Note: Flanges next to valves can be of different rating from pairs of flanges in other sections of the line.

Fluid Expansion Effects.   Although the Code states that provision shall be made in the design either to withstand or to relieve increased pressure caused by the heating of static fluid in a piping component, most refiners have found it unnecessary to provide a relief valve on piping which may be blocked in.   The reason for this is that flanged joints, valve packing glands, or valve seating surfaces usually will leak before the pressure build-up becomes excessive as the result of the heating of the blocked-in fluid.   Recently developed valves utilizing resilient seating materials have made this more of a problem because of their ability to seal even more tightly as the pressure increases.   Consequently, when such valves are used, greater consideration should be given to the problem of possible pressure build-up, including a build-up in the valve body cavity.

A similar problem can be created in a blocked-in section of chemical piping by the increase in pressure which can result from the evolution of gas caused either by an increase in fluid temperature or by a chemical reaction (e.g., hydrogen is a product of corrosion).

Dynamic Effects.   The Code requires that the piping designer take into account wind and earthquake forces, although not concurrently, in design of his piping (see ASA A58.1 for wind and seismic data).   It also requires that he consider impact forces (including hydraulic shock) and vibration.   Vibration and impact forces create complex design problems.[1–3,*]   This is especially true in the case of vibration because difficult-to-notice low-amplitude high-frequency vibrations often produce the most dangerous stresses.

Most piping systems will vibrate to some extent.   The exciting forces causing vibration in piping may be (1) mechanical vibration of connected equipment, such as compressors, pumps, and vessels; (2) wind-produced vortices that form alternately on opposite sides of cylindrical surfaces; or (3) internal pulsations in flowing fluids, such as those set up by reciprocating pumps and compressors.   Vibration

* Superscript numbers refer to Bibliography at end of this chapter.

from this latter cause can generally be kept within controllable limits by limiting pressure pulsations to 1 to 3 per cent of line pressure.

Where vibration is expected, good design should include the following:

1. Adequate foundations, especially for reciprocating pumps and compressors.

2. Strategic location of pipe guides and supports to reduce vibration. They should be installed so as to cause minimum restraint to normal thermal movements. The use of sway braces of the energy-absorbing or instant-counterforce-acting type is recommended for control of undesirable pipeline movement. Rigid braces are also effective in controlling movement provided they do not restrict the flexibility of the piping and provided their effect is taken into account in the design of the piping for flexibility. Where pulsating flow exists, piping should be supported at all changes in direction and cantilever sections must be avoided. Compressor vibration has been known to cause a threaded-end valve to loosen and release high-pressure hydrocarbons. Although welded construction should have been used, this instance illustrates the need for special attention whenever vibration is anticipated.

3. Avoidance of small branch connections. Additional supports, such as gusset plates, may help reduce vibration problems.

4. An acoustical study to determine if dampening equipment is needed. Failure can be caused by resonance of some part of the system with the pulsation frequency. Pressure pulsations can be minimized by the use of hydropneumatic accumulators, snubbers, or surge drums. These pulsation-reducing devices should be installed as close as possible to the pulsation-producing equipment.

Any sudden change in the flow velocity or pressure in a liquid line will produce hydraulic shock (water hammer). The common cause of water hammer is the rapid closure of a valve. Figure 1 shows the pressure rise to be expected with various valve closure times.[4]

Water hammer requires careful consideration because it can damage equipment associated with piping and instrumentation even though the permissible pressures for piping components are not exceeded. In addition to the utilization of slow-closing valves, the installation of surge tanks, pneumatic chambers, spring-operated relief valves, or shock absorbers is sometimes used to help control this phenomenon.

Weight Effects. In the design of piping and its supports, it is required that live loads, dead loads, and loads and forces from other causes be taken into account. Live load, as used here, is taken as the weight of fluid transported plus snow and ice loads in localities where such conditions exist. Dead load is the weight of the piping components and insulation and other superimposed permanent loads.

Pressure-Temperature Ratings for Piping Components. Pressure-temperature ratings have been established for certain piping components. The ones that have been accepted by the Code are contained in the Standards listed in Table 326.1 of the Code. (Appendix E of the Code should be consulted for the date of the latest acceptable revision of a Standard.) These established ratings should not be exceeded by the expected normal operating conditions; however, during shutdown, startup, or an interruption in the normal operation of a refinery process unit, conditions more severe than normal may be characteristic of a service. Depending on the frequency and duration of these more severe conditions, the Code permits adjustment of the pressure ratings as follows:

(a) When the increased operating condition will not exceed ten hours at any one time or 100 hours per year, it is permissible to increase the pressure rating at the temperature existing during the increased operating condition, by a maximum of 33%.

(b) When the increased operating condition will not exceed 50 hours at any one time, or 500 hours per year, it is permissible to increase the pressure rating at the temperature existing during the increased operating condition by a maximum of 20%.

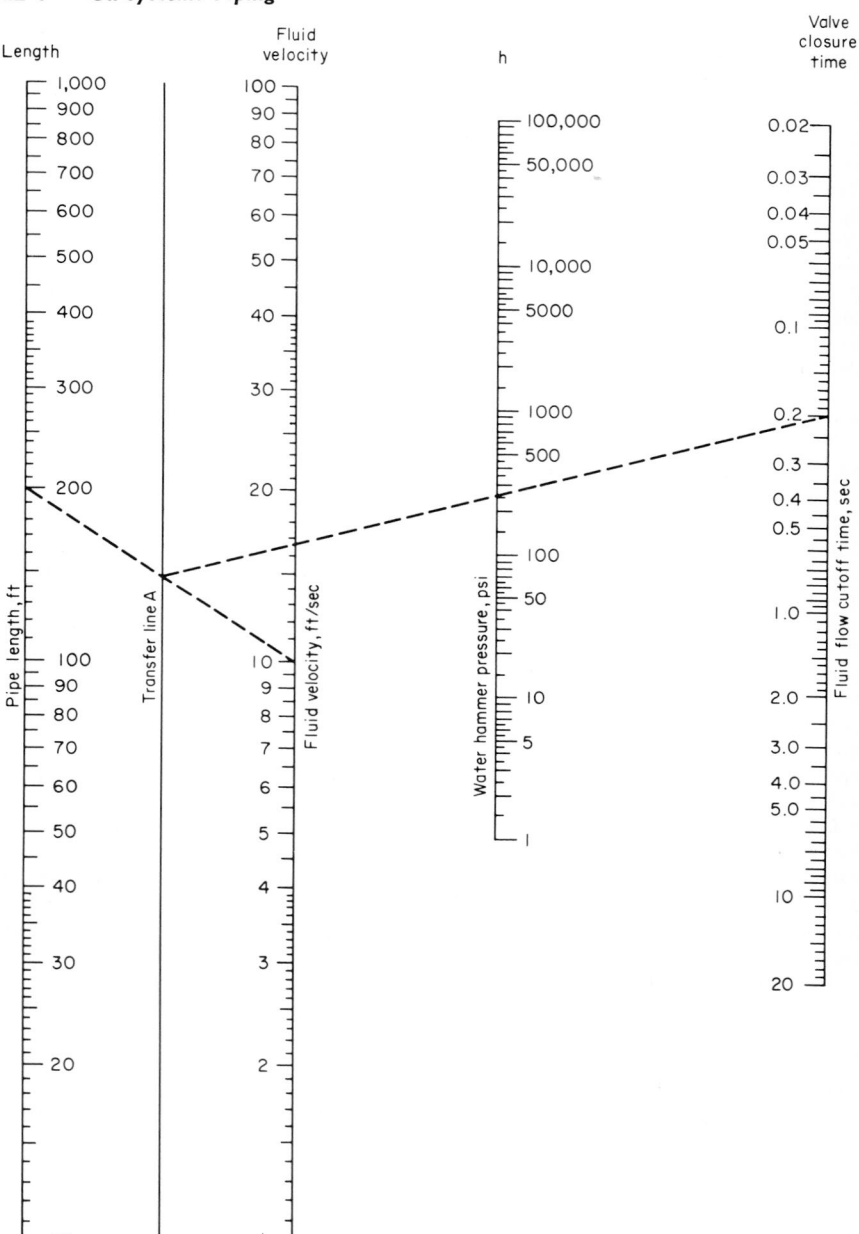

FIG. 1. Pressure rise due to water hammer. h = pressure rise due to water hammer which should be added to normal operating pressure.

Some of the conditions which should be investigated for the short-time period are the centrifugal-pump shutoff pressure, or the pressure at the maximum point of the pump characteristic curve; centrifugal-compressors surge-point pressure; stalling pressure of reciprocating pumps and reciprocating compressors; and the set pressure of relief valves which limit pressure in the piping.

The Code permits extrapolation of accepted pressure-temperature ratings where these ratings do not extend to the upper material temperature limits permitted by the Code provided this is done in accordance with the applicable Code rules.

Allowable Stresses. The stress values indicated in Tables 3 and 4 have been abstracted from Tables 302.3.1A and B of ASA B31.3 and are grouped according to the metal temperature. Except in the case of welded pipe, welded fittings, castings, and structural-grade material, these values are the basic allowable stresses for materials. For welded pipe and welded fittings, the tables show the product of the basic allowable stresses and the applicable joint factor (SE).

The Code describes the various supplementary inspections which permit the allowable stresses and casting-quality factors to be increased. The casting-quality factors, however, do not apply to valves, flanges, and fittings which conform to the Standards listed in Table 326.1 of the Code. Therefore, they are used infrequently and are not duplicated here.

The bases for establishing the basic allowable stress values for both ferrous and nonferrous materials are described in the Code. The Code user will need to refer to this only in those cases where he desires to establish allowable stresses for an unlisted material.

The allowable stress values in shear and bearing are 0.80 and 1.60, respectively, times the values contained in Tables 3 and 4.

For other than normal operations, the allowable adjustments in pressure-temperature ratings, set forth above, are applicable to allowable stress values in calculations concerning components, such as pipe, which do not have established pressure-temperature ratings. The designer should remember that the more severe operating condition may be caused by increased contents load rather than by increased temperature or pressure.

Limits on Calculated Stresses Due to Sustained Loads and Thermal Expansion. The calculated stresses due to internal pressure must not exceed the values given in the allowable stress tables, except as permitted for short-time conditions. Also, the stresses due to external pressure are considered safe when the wall thickness of the piping component and means of stiffening meet the Code requirements.

The allowable stress range, S_A, for expansion stress, S_E, is given as

$$S_A = f(1.25S_c + 0.25S_h) \tag{1}$$

where S_c = basic allowable stress for the material at minimum (cold) metal temperature normally expected during operating or shutdown, psi (from Table 3 or 4)

S_h = basic allowable stress for the material at maximum (hot) metal temperature normally expected during operating or shutdown, psi (from Table 3 or 4)

f = stress-range reduction factor for cyclic conditions for total number of full temperature cycles over expected life (from Table 5). By expected life is meant total number of years during which system is expected to be in active operation.

The sum of the longitudinal stresses (in a corroded condition) due to pressure, weight, and other sustained external loading must not exceed S_h. Where the sum

Table 3. Allowable

[Abstract of Table 302.3.1A—ASA B31.3. Welded

Material	Specification (30)	Steel making process	Grade	Class	Tensile strength min. psi	Yield strength min. psi	Notes (26)	Min temp.	Min temp to 100	200	300	400
											Seamless carbon steel and	
Steel............	A53	OH, EF	A	Type S	48,000	30,000	1, 2	−20	16,000	15,300	14,500	13,800
Steel............	A53	OH, EF, DAB	B	Type S	60,000	35,000	1, 2	−20	20,000	19,100	18,150	17,250
Steel............	A83	OH, EF	A	......	......	......	1, 2	−20	16,000	15,300	14,500	13,800
Steel............	A106	OH, EF	A	......	48,000	30,000	2	−20	16,000	15,300	14,500	13,800
Steel............	A106	OH, EF	B	......	60,000	35,000	2	−20	20,000	19,100	18,150	17,250
Steel............	A106	OH, EF	C	......	70,000	40,000	2	−20	23,350	22,250	21,200	20,150
Steel............	A120	OH, EF	A	......	......	......	......	−20	16,000	15,300		
Steel............	A120	OH, EF, AB	B	......	......	......	......	−20	20,000	19,100		
Steel............	A179	OH, EF	....	......	......	......	1, 2	−20	15,650	15,000	14,250	13,550
Steel............	A333 A334	OH, EF	O	......	55,000	30,000	1, 2	−50	18,350	17,500	16,700	15,850
Steel............	API 5L	OH, EF	A	......	48,000	30,000	1, 2, 24	−20	16,000	15,300	14,500	13,800
Steel............	API 5L	OH, EF, DAB	B	......	60,000	35,000	1, 2, 24	−20	20,000	19,100	18,150	17,250
Steel............	API 5LX	OH, EF	X42	......	60,000	42,000	......	−20	20,000	19,100	18,150	17,250
Steel............	API 5LX	OH, EF	X46	......	63,000	46,000	......	−20	21,000	20,050	19,100	18,150
Steel............	API 5LX	OH, EF	X52	......	66,000	52,000	......	−20	22,000	21,000	20,000	19,000
Steel............	API 5LX	OH, EF	X52	......	72,000	52,000	......	−20	23,500	22,500	21,400	20,350
Cast iron.......	WW-P-421b	....	....	......	......	......	7, 9, 25	−20	6,000	6,000	6,000	6,000
										Carbon-steel structural material (not		

NOTE: Materials such as pipe, castings, forgings etc., listed elsewhere in

Steel............	ASTM A7	OH, EF, AB	....	......	60,000	33,000	5	−20	18,400	17,550	16,750	15,900
										Seamless alloy-steel		
5Cr–½Mo	A199 A200	EF	T5	......	60,000	25,000	......	−20	15,650	15,450	15,200	15,000
7Cr–½Mo	A199 A200	EF	T7	......	60,000	25,000	......	−20	15,650	15,450	15,200	15,000
9Cr–1Mo	A199 A200	EF	T9	......	60,000	25,000	......	−20	15,650	15,450	15,200	15,000
1¼Cr–½Mo ...	A199 A200	EF	T11	......	60,000	25,000	......	−20	15,650	15,550	15,450	15,400
2¼Cr–1Mo	A199 A200	EF	T22	......	60,000	25,000	......	−20	15,650	15,550	15,450	15,400
3½Ni	A334 A333	OH, EF	3	......	65,000	35,000	2	−150	21,650	20,650	19,700	18,700
Cr–Cu–Ni......	A333	OH, EF	4	......	60,000	35,000	......	−150	20,000	19,100	18,150	17,250
C–½Mo	A209 A335	OH, EF	P1, T1	......	55,000	30,000	3	−20	18,350	17,650	16,950	16,300
½Cr–½Mo	A335	OH, EF	P2	......	55,000	30,000	3	−20	18,350	17,650	16,950	16,300
5Cr–½Mo......	A213 A335	OH, EF	P5, T5	......	60,000	30,000	......	−20	18,750	17,900	17,050	16,200
5Cr–½Mo	A213 A335	OH, EF	P5b, T5b	......	60,000	30,000	......	−20	18,750	17,900	17,050	16,200
5Cr–½Mo	A213 A335	OH, EF	P5c, T5c	......	60,000	30,000	......	−20	18,750	17,900	17,050	16,200
7Cr–½Mo	A213 A335	OH, EF	P7, T7	......	60,000	30,000	......	−20	18,750	17,850	17,000	16,150
9Cr–1Mo	A213 A335	OH, EF	P9, T9	......	60,000	30,000	......	−20	18,750	17,900	17,100	16,250
1¼Cr–½Mo ...	A213 A335	OH, EF	P11, T11	......	60,000	30,000	......	−20	18,750	18,250	17,650	17,150
1Cr–½Mo	A213 A335	OH, EF	P12, T12	......	60,000	30,000	3	−20	18,750	18,250	17,600	17,050
2¼Cr–1Mo	A213 A335	OH, EF	P22, T22	......	60,000	30,000	......	−20	18,750	18,250	17,650	17,150
										Seamless stainless-		
18-8	A213 A312	EF	TP304	......	75,000	30,000	20	−425	18,750	16,650	15,000	13,650
18-8	A213 A312	EF	TP304H	......	75,000	30,000	23	−325				
18-8	A213 A312	EF	TP304L	......	70,000	25,000	......	−425	15,600	15,300	13.100	11,000
18-8	A213 A312	EF	TP316	......	75,000	30,000	20	−425	18,750	18,750	17,900	17,500
18-8	A213 A312	EF	TP316H	......	75,000	30,000	23	−325				
18-8	A213 A312	EF	TP316L	......	70.000	25,000	......	−425	15,600	15,600	14,500	12,000
18-8	A213 A312	EF	TP321	......	75,000	30,000	......	−325	18,750	18,750	17,000	15,800
18-8	A213 A312	EF	TP321H	......	75,000	30,000	23	−325	......	......	......	......
18-8	A213 A312	EF	TP347	......	75,000	30,000	......	−425	18,750	18,750	17,000	15,800
18-8	A213 A312	EF	TP347H	......	75,000	30,000	23	−325				

	Allowable stress, psi																		
	Metal temperature, F																		
500	600	650	700	750	800	850	900	950	1000	1050	1100	1150	1200	1250	1300	1350	1400	1450	1500

ron pipe and tubes

500	600	650	700	750	800	850	900	950	1000	1050	1100	1150	1200	1250	1300	1350	1400	1450	1500
3,100	12,350	12,000	11,650	10,700	9,300	7,900	6,500	4,500	2,500	1,600	1,000								
6,350	15,500	15,000	14,350	12,950	10,800	8,650	6,500	4,500	2,500	1,600	1,000								
3,100	12,350	12,000	11,650	10,700	9,300	7,900	6,500	4,500	2,500	1,600	1,000								
3,100	12,350	12,000	11,650	10,700	9,300	7,900	6,500	4,500	2,500	1,600	1,000								
6,350	15,500	15,000	14,350	12,950	10,800	8,650	6,500	4,500	2,500	1,600	1,000								
9,100	18,050	17,500	16,600	14,750	12,000														
2,850	12,100	11,750	11,450	10,550	9,200	7,850	6,500	4,500	2,500	1,600	1,000								
5,000	14,200	13,750	13,250	12,050	10,200	8,350	6,500	4,500	2,500	1,600	1,000								
3,100	12,350	12,000	11,650	10,700	9,300	7,900	6,500	4,500	2,500	1,600	1,000								
6,350	15,500	15,000	14,350	12,950	10,800	8,650	6,500	4,500	2,500	1,600	1,000								

or pressure containing parts)

ables 302.3.1A and 302.3.1B may also be used as structural materials.

500	600	650	700	750	800	850	900	950	1000	1050	1100	1150	1200	1250	1300	1350	1400	1450	1500
5,050	14,200	13,800	13,200	11,900															

ipe and tubes

500	600	650	700	750	800	850	900	950	1000	1050	1100	1150	1200	1250	1300	1350	1400	1450	1500
4,500	14,000	13,700	13,400	13,100	12,800	12,400	11,500	10,000	7,300	5,200	3,300	2,200	1,500						
4,500	14,000	13,700	13,400	13,100	12,500	11,500	9,500	7,000	5,000	3,500	2,500	1,800	1,200						
4,500	14,000	13,700	13,400	13,100	12,800	12,500	12,000	10,800	8,500	5,500	3,300	2,200	1,500						
5,300	15,200	15,150	15,100	15,050	15,000	14,400	13,100	11,000	7,800	5,500	4,000	2,500	1,200						
5,300	15,200	15,150	15,100	15,050	15,000	14,400	13,100	11,000	7,800	5,800	4,200	3,000	2,000						
7,750	16,750	16,250	15,500	13,850	11,400	8,950	6,500	4,500	2,500	1,600	1,000								
6,350	15,500	15,000																	
5,600	14,900	14,550	14,200	13,850	13,500	13,150	12,500	10,000	6,250	4,000	2,400								
5,690	14,900	14,550	14,200	13,850	13,500	13,150	12,500	10,000	6,250	4,000	2,400								
5,350	14,500	14,100	13,650	13,250	12,800	12,400	11,500	10,000	7,300	5,200	3,300	2,200	1,500						
5,350	14,500	14,100	13,650	13,250	12,800	12,400	10,900	9,000	5,500	3,500	2,500	1,800	1,200						
5,350	14,500	14,100	13,650	13,250	12,800	12,400	11,500	10,000	7,300	4,800	2,800	1,800	1,200						
5,300	14,450	14,000	13,550	13,100	12,500	11,500	9,500	7,000	5,000	3,500	2,500	1,800	1,200						
5,450	14,600	14,200	13,800	13,350	12,950	12,500	12,000	10,800	8,500	5,500	3,300	2,200	1,500						
6,600	16,050	15,800	15,550	15,300	15,000	14,400	13,100	11,000	7,800	5,500	4,000	2,500	1,200						
6,450	15,900	15,650	15,350	15,050	14,750	14,200	13,100	11,000	7,500	5,000	2,800	1,550	1,000						
6,600	16,050	15,800	15,500	153,00	15,000	14,400	13,100	11,000	7,800	5,800	4,200	3,000	2,000						

teel pipe and tubes

500	600	650	700	750	800	850	900	950	1000	1050	1100	1150	1200	1250	1300	1350	1400	1450	1500
2,500	11,600	11,200	10,800	10,400	10,000	9,700	9,400	9,100	8,800	8,500	7,500	5,750	4,500	3,250	2,450	1,800	1,400	1,000	750
										8,500	7,500	5,750	4,500	3,250	2,450	1,800	1,400	1,000	750
9,700	9,000	8,750	8,500	8,300	8,100														
7,200	17,100	17,050	17,000	16,900	16,750	16,500	16,000	15,100	14,000	12,200	10,400	8,500	6,800	5,300	4,000	3,000	2,350	1,850	1,500
										12,200	10,400	8,500	6,800	5,300	4,000	3,000	2,350	1,850	1,500
1,000	10,150	9,800	9,450	9,100	8,800														
5,200	14,900	14,850	14,800	14,700	14,550	14,300	14,100	13,850	13,500										
....	....	....	....	....	....	....	....	....	....	13,100	12,500	8,000	5,000	3,600	2,700	2,000	1,550	1,200	1,000
5,200	14,900	14,850	14,800	14,700	14,550	14,300	14,100	13,850	13,500										
										13,100	12,500	8,000	5,000	3,600	2,700	2,000	1,550	1,200	1,000

Table 3.

Material	Specification (30)	Grade	Class	Temper	Size range, in.	Tensile strength min. psi	Yield strength min. psi	Notes (26)	Min. temp.	Min temp. to 100	150	200	250
										Copper and copper-alloy			
Copper....	B42 B75	Annealed	....	....	....	30,000	9,000	8	−325	6,000	6,000	5,900	5,800
Copper....	B42 B75	Hard drawn	....	....	⅛– 2 incl.	45,000	40,000	8	−325	11,300	11,300	11,000	10,500
Copper....	B42	Light drawn	....	....	2½– 12 incl.	36,000	30,000	8	−325	9,000	9,000	8,700	8,300
Red brass .	B43	Annealed	....	....	....	40,000	12,000	8	−325	8,000	8,000	8,000	8,000
Copper....	B68 B88	Annealed	....	....	....	30,000	....	.....	−325	6,000	6,000	5,900	5,800
Copper....	B75	Light drawn	....	....	....	36,000	30,000	8	−325	9,000	9,000	8,700	8,300
Copper....	B88	Drawn	....	....	....	36,000	....	.....	−325	9,000	9,000	8,700	8,300
										Nickel and nickel-base alloy			
Nickel	B161	....	Annealed	....	5 OD and under	55,000	15,000	.....	−325	10,000	10,000	10,000	10,000
Nickel	B161	....	Annealed	....	Over 5 OD	55,000	12,000	.....	−325	8,000	8,000	8,000	8,000
Nickel	B161 B163	....	Stress rel'd	....	....	65,000	40,000	.....	−325	16,200	15,700	15,300	15,100
Low carbon nickel ...	B161	....	Annealed	....	5 OD and under	50,000	12,000	.....	−325	8,000	7,800	7,700	7,600
Low carbon nickel ...	B161	....	Annealed	....	Over 5 OD	50,000	10,000	.....	−325	6,700	6,550	6,400	6,350
Low carbon nickel ...	B161	....	Stress rel'd	....	....	60,000	30,000	.....	−325	15,000	14,600	14,200	14,000
Nickel	B163	....	Annealed	....	....	55,000	15,000	.....	−325	10,000	10,000	10,000	10,000
Low carbon nickel ...	B163	....	Annealed	....	....	50,000	12,000	.....	−325	8,000	7,800	7,700	7,600
Low carbon nickel ...	B163	....	Stress rel'd	....	....	60,000	30,000	.....	−325	15,000	14,600	14,200	14,000
Nickel– copper ..	B163	....	Annealed	....	....	70,000	28,000	.....	−325	17,500	17,000	16,500	16,000
Nickel– copper ..	B163	....	Stress rel'd	....	....	85,000	55,000	.....	−325	21,200	20,700	20,200	19 800
Ni–Cr–Fe .	B163	....	Annealed	....	....	80,000	30,000	.....	−325	20,000	19,300	18,600	18,200
Nickel– copper ..	B165	....	Annealed	....	5 OD and under	70,000	28,000	.....	−325	17,500	17,000	16,500	16,000
Nickel– copper ..	B165	....	Annealed	....	Over 5 OD	70,000	25,000	.....	−325	16,600	15,600	14,600	14,100
Nickel– copper ..	B165	....	Stress rel'd	....	....	85,000	55,000	.....	−325	21,200	20,700	20,200	19,800
										Aluminum and aluminum-base			
	B210 B235	3003	....	0	....	14,000	5,000	8, 15	−325	3,350	3,150	2,900	2,700
	B210 B235	3003	....	H14	....	20,000	17,000	8, 15	−325	5,000	4,850	4,700	4,400
	B210 B234 } B235 B241 }	6061	....	T6	....	38,000	35,000	8, 15	−325	9,500	9,200	9,000	8,500
				Welded	....	24,000	....	8, 15	−325	6,000	5,900	5,700	5,400
	B210 B234 } B235 B241 }	6063	....	T6	....	30,000	25,000	8, 15	−325	7,500	7,100	6,800	6,100
				Welded	....	17,000	....	8· 15	−325	4,250	4,200	4,000	3,800

NOTES:

1. Above 900 F consider the advantages of killed steel.

2. Conversion of carbides to graphite may occur after prolonged exposure to temperatures over 750 F.

3. Conversion of carbides to graphite may occur after prolonged exposure to temperatures over 850 F.

5. A quality factor of 92 per cent is included for structural grade.

7. Cast and malleable iron shall not be used for toxic services or hydrocarbon or other flammable fluid service at temperatures over 300 F.

8. For use in Code piping at the stated allowable stresses, the required tensile properties must be verified by tensile test at the mill; such tests shall be specified in the purchase order.

9. Pressure-temperature ratings of cast and forged parts as published in standards referenced in this Code section may be used for parts meeting requirements of these standards. Allowable stresses for castings and forgings, where listed, are for use in design of special components not furnished in accordance with such standards.

								Allowable stress, psi — Metal temperature, F										
300	350	400	450	500	550	600	650	700	750	800	850	900	950	1,000	1,050	1,100	1,150	1,200

eamless pipe and tubes

300	350	400	450	500	550	600	650	700	750	800	850	900	950	1,000	1,050	1,100	1,150	1,200
5,000	3,800	2,500	1,500	750														
8,000	5,000	2,500	1,500	750														
8,000	5,000	2,500	1,500	750														
8,000	6,000	3,000	2,000															
5,000	3,800	2,500	1,500	750														
8,000	5,000	2,500	1,500	750														
8,000	5,000	2,500	1,500	750														

eamless pipe and tubes

300	350	400	450	500	550	600	650	700	750	800	850	900	950	1,000	1,050	1,100	1,150	1,200
0,000	10,000	10,000	10,000	10,000	10,000	10,000												
8,000	8,000	8,000	8,000	8,000	8,000	8,000												
5,000	15,000	15,000	14,800	14,500														
7,500	7,500	7,500	7,500	7,500	7,500	7,500	7,450	7,400	7,300	7,200	6,150	4,500	3,600	3,000	2,400	2,000	1,500	1,200
6,300	6,250	6,200	6,200	6,200	6,200	6,200	6,200	6,200	6,050	5,900	5,200	4,500	3,750	3,000	2,500	2,000	1,600	1,200
3,800	13,600	13,500	13,500	13,500														
0,000	10,000	10,000	10,000	10,000	10,000	10,000												
7,500	7,500	7,500	7,500	7,500	7,500	7,500	7,450	7,400	7,300	7,200	6,150	4,500	3,600	3,000	2,400	2,000	1,500	1,200
3,800	13,600	13,500	13,500	13,500	13,250	13,000	12,500	12,000	11,500	11,000	10,500	10,000						
5,500	15,100	14,800	14,750	14,700	14,700	14,700	14,700	14,700	14,650	14,500	12,500	8,000						
9,500	19,300	19,200	19,200	19,200	19,200	19,200	18,800	18,500	17,000	15,000								
8,000	18,000	18,000	18,000	18,000	18,000	18,000	17,800	17,500	17,250	17,000	16,600	16,000	10,500	7,000	4,250	3,000	2,400	2,000
5,500	15,100	14,800	14,700	14,700	14,700	14,700	14,700	14,700	14,650	14,500	12,500	8,000						
3,600	13,400	13,200	13,150	13,100	13,100	13,100	13,100	13,100	13,100	13,100	13,100	10,500	8,000					
9,500	19,300	19,200	19,200	19,200														

lloy seamless pipe and tubes

300	350	400	450	500	550	600	650	700	750	800	850	900	950	1,000	1,050	1,100	1,150	1,200
2,400	2,100	1,800																
4,000	3,500	3,100																
7,200	5,600	4,000																
5,000	4,200	3,200																
4,500	3,100	2,000																
3,600	2,750	1,900																

15. For welded construction with work-hardened grades, use the stresses for annealed material; for welded construction with precipitation-hardened grades use the special allowable stresses for welded construction given in the tables.

20. These stress values apply only when carbon content is 0.04 per cent or higher.

23. "H" grades of stainless steel may be used at temperatures below 1000 F at the allowable stresses shown for the corresponding regular grade. In so doing, the greater tendency of the "H" grade to intergranular carbide precipitation must be taken into account.

24. S values above 400 F apply only to non-expanded pipe. No value is assigned to cold expanded pipe above 400 F.

25. The allowable stresses to be used for cast iron materials at their upper temperature limit of 450 F are the same as those shown in the 400 F column for the respective materials.

26. The minimum temperature shown is that design temperature for which the material is normally suitable; but the use of a material at a design temperature below −20 F is established by rules elsewhere in this Code, including any necessary impact test requirements.

30. Specification numbers are ASTM numbers unless otherwise indicated.

Table 4. Allowable Stresses fo

(Abstract of Table 302

SE values shown in this table for welded pipe include the joint factor for the longitudinal weld.   For some code computations
E need not be considered.   See note 16 below for rules to determine the allowable stress S for use in these computations.

Material	Specification (30)	Steel making process	Grade	Class	Tensile strength min, psi	Yield strength min, psi	Notes	Long. joint factor	Min temp. (26)
									Carbon-stee
									Furnace-welded-
Steel...........	A53	OH, EF	....	Type F	45,000	25,000	......	0.60	−20
Steel...........	A53	AB	....	Type F	50,000	30,000	......	0.60	−20
Steel...........	A120	OH, EF	....	....	......	......	......	0.60	−20
Steel...........	A120	AB	....	....	......	......	......	0.60	−20
Steel...........	API 5L	OH, EF	....	1	45,000	25,000	......	0.60	−20
Steel...........	API 5L	OH, EF	....	11	48,000	28,000	......	0.60	−20
Steel...........	API 5L	AB	....	....	50,000	30,000	......	0.60	−20
									Electric-resistance-welded and electric-
Steel...........	A53	OH, EF	A	Type E	48,000	30,000	1, 2	0.85	−20
Steel...........	A53	OH, EF	B	Type E	60,000	35,000	1, 2	0.85	−20
Steel...........	A120	OH, EF	A	....	......	......	......	0.85	−20
Steel...........	A120	OH, EF, AB	B	....	......	......	......	0.85	−20
Steel...........	A135	OH, EF	A	....	48,000	30,000	1, 2	0.85	−20
Steel...........	A135	OH, EF	B	....	60,000	35,000	1, 2	0.85	−20
Steel...........	A333	OH, EF	O	....	55,000	30,000	1, 2	0.85	−50
Steel...........	A334	OH, EF	O	....	55,000	30,000	1, 2	0.85	−50
Steel...........	API 5L	OH, EF	A	....	48,000	30,000	1, 2, 24	0.85	−20
Steel...........	API 5L	OH, EF	B	....	60,000	35,000	1, 2, 24	0.85	−20
Steel...........	API 5LX	OH, EF	X42	....	60,000	42,000	......	0.85	−20
Steel...........	API 5LX	OH, EF	X46	....	63,000	46,000	......	0.85	−20
Steel...........	API 5LX	OH, EF	X52	....	66,000	52,000	......	0.85	−20
Steel...........	API 5LX	OH, EF	X52	....	72,000	52,000	......	0.85	−20
									Electric-fusion-welded
A 245, Gr. A ...	A134	OH, EF	....	....	45,000	25,000	5	0.80	−20
A 245, Gr. B ...	A134	OH, EF	....	....	49,000	30,000	5	0.80	−20
A 245, Gr. C ...	A134	OH, EF	....	....	52,000	33,000	5	0.80	−20
A 245, Gr. D ...	A134	OH, EF	....	....	55,000	40,000	5	0.80	−20
A 283, Gr. A ...	A134	OH, EF	....	....	45,000	24,000	5	0.80	−20
A 283, Gr. B ...	A134	OH, EF	....	....	50,000	27,000	5	0.80	−20
A 283, Gr. C ...	A134	OH, EF	....	....	55,000	30,000	5	0.80	−20
A 283, Gr. D ...	A134	OH, EF	....	....	60,000	33,000	5	0.80	−20
A 285, Gr. A ...	A134	OH, EF	....	....	45,000	24,000	......	0.80	−20
A 285, Gr. B ...	A134	OH, EF	....	....	50,000	27,000	......	0.80	−20
A 285, Gr. C ...	A134	OH, EF	....	....	55,000	30,000	......	0.80	−20
Steel...........	A139	OH, EF	A	....	48,000	30,000	......	0.80	−20
Steel...........	A139	OH, EF	B	....	60,000	35,000	......	0.80	−20
A 285, Gr. A ...	A155	OH, EF	C45	2	45,000	24,000	1, 2, 4	0.85	−20
A 285, Gr. B ...	A155	OH, EF	C50	2	50,000	27,000	1, 2, 3	0.85	−20

3.1B, ASA B31.3)

particularly with regard to expansion, flexibility, structural attachments, supports and restraints, the longitudinal joint factor

Allowable stress, psi

Metal temperature, F

Min temp. to 100	200	300	400	500	600	650	700	750	800	850	900	950	1000	1050	1100

pipe and tubes

butt-welded pipe

Min temp. to 100	200	300	400	500	600	650	700	750	800	850	900	950	1000	1050	1100
9,000	8,600	8,200	7,800												
10 000	9,550	9,100	8,650												
9,000	8,600														
10,000	9,550														
9,000	8,600	8,200	7,800												
9,600	9,200	8,700	8,300												
10,000	9,550	9,100	8,650												

flash-welded pipe and tubes

Min temp. to 100	200	300	400	500	600	650	700	750	800	850	900	950	1000	1050	1100
13,600	13,000	12,300	11,750	11,100	10,500	10,200	9,900	9,100	7,900	6,700	5,500	3,800	2,150	1,350	850
17,000	16,200	15,400	14,650	13,900	13,150	12,750	12,200	11,000	9,200	7,350	5,500	3,800	2,150	1,350	850
13,600	13,000														
17,000	16,200														
13,600	13,000	12,300	11,750	11,100	10,500	10,200	9,900	9,100	7,900	6,700	5,500	3,800	2,150	1,350	850
17,000	16,200	15,400	14,650	13,900	13,150	12,750	12,200	11,000	9,200	7,350	5,500	3,800	2,150	1,350	850
15,600	14,850	14,150	13,450	12,750	12,050	11,700	11,250	10,250	8,650	7,100	5,500	3,800	2,150	1,350	850
15,600	14,850	14,150	13,450	12,750	12,050	11,700	11,250	10,250	8,650	7,100	5,500	3,800	2,150	1,350	850
13 600	13,000	12,300	11,750	11,100	10,500	10,200	9,900	9,100	7,900	6,700	5,500	3,800	2,150	1,350	850
17,000	16,200	15,400	14,650	13,900	13,150	12,750	12,200	11,000	9,200	7,350	5,500	3,800	2,150	1,350	850
17,000	16,200	15,400	14.650												
17,850	17,000	16,200	15,400												
18,700	17,850	17,000	16,100												
20,400	19,100	18,200	17,300												

pipe (straight seam)

Min temp. to 100	200	300	400	500	600	650	700	750	800	850	900	950	1000	1050	1100
11,050	10,550	10,050													
12,000	11,450	10,900													
12,750	12,150	11,600													
13,500	12,850	12,250													
11,050	10,550	10,050													
12,250	11,650	11,200													
13,500	12,900	12,300													
14,750	14,000	13,400													
12,000	11,500	10,900													
13,300	12,700	12,200													
14,700	14,000	13,300													
12,800	12,300	11,700													
16,000	15,350	14,600													
12,750	12,200	11,600	11,050	10,500	9,900	9,600	9,350	8,700	7,650	6,600	5,500	3,800	2,100	1,400	850
14,150	13,500	12,900	12,300	11,600	11,000	10,600	10,300	9,450	8,150	6,850	5,500	3,800	2,100	1,400	850

Table

Material	Specification (30)	Steel making process	Grade	Class	Tensile strength min. psi	Yield strength min. psi	Notes	Long. joint factor	Min. temp. (26) Min. temp. to 100	200	300	400
							Carbon-steel pipe and tubes (cont'd)—Electr					
A 285, Gr. C A 201, Gr. A	A155	OH, EF	C55 KC55	2	55,000	30,000	1, 2, 4	0.85	−20 15,600	14,900	14,200	13,5
A 201, Gr. B	A155	OH, EF	KC60	2	60,000	32,000	1, 2, 4	0.85	−20 17,000	16,250	15,450	14,6
A 212, Gr. A	A155	OH, EF	KC65	2	65,000	35,000	1, 2, 4	0.85	−20 18,400	17,600	16,750	15,9
A 212, Gr. B	A155	OH, EF	KC70	2	70,000	38,000	1, 2, 4	0.85	−20 19,850	18,900	18,000	17,1
A 285, Gr. A	A155	OH, EF	C45	2	45,000	24,000	1, 2, 4, 17	0.90	−20 13,500	12,900	12,300	11,7
A 285, Gr. B	A155	OH, EF	C50	2	50,000	27,000	1, 2, 4, 17	0.90	−20 15,000	14,300	13,700	13,0
A 285, Gr. C A 201, Gr. A	A155	OH, EF	C55 KC55	2	55,000	30,000	1, 2, 4, 17	0.90	−20 16,500	15,750	15,000	14,2
A 201, Gr. B	A155	OH, EF	KC60	2	60,000	32,000	1, 2, 4, 17	0.90	−20 18,000	17,200	16,350	15,5
A 212, Gr. A	A155	OH, EF	KC65	2	65,000	35,000	1, 2, 4 17	0.90	−20 19,500	18,650	17,750	16,3
A 212, Gr. B	A155	OH, EF	KC70	2	70,000	38,000	1, 2, 4, 17	0.90	−20 21,000	20,000	19,100	18,1
A 285, Gr. A	A155	OH, EF	C45	1	45,000	24,000	1, 2, 4	1.00	−20 15,000	14,350	13,650	13,0
A 285, Gr. B	A155	OH, EF	C50	1	50,000	27,000	1, 2, 4	1.00	−20 16,650	15,900	15,200	14,4
A 285, Gr. C A 201, Gr. A	A155	OH, EF	C55 KC55	1	55,000	30,000	1, 2, 4	1.00	−20 18,350	17,500	16,700	15,8
A 201, Gr. B	A155	OH, EF	KC60	1	60,000	32,000	1, 2, 4	1.00	−20 20,000	19,100	18,150	17,2
A 212, Gr. A	A155	OH, EF	KC65	1	65,000	35,000	1, 2, 4	1.00	−20 21,650	20,700	19,700	18 7
A 212, Gr. B	A155	OH, EF	KC70	1	70,000	38,000	1, 2, 4	1.00	−20 23,350	22,250	21,200	20,1
							Alloy steel pip					
							Electric resistance					
C ½Mo	A250	OH, EF	T1	3	55,000	30,000	3	0.85	−20 15,600	15,000	14,400	13,8
C ½Mo	A250	OH, EF	T1a	3	60,000	32,000	3	0.85	−20 17,000	16,350	15,700	15,0
C ½Mo	A250	OH, EF	T1b	3	53,000	28.000	3	0.85	−20 14,850	14,300	13,800	13,2
3½ Ni	A333, A334	OH, EF	3	2	65,000	35,000	2	0.85	−150 18,400	17,600	16,750	15,9
Cr-Cu-Ni	A333	OH, EF	4	2	60,000	30,000	....	0.85	−150 17,000	16,200	15,450	14,7

Continued)

					Allowable stress, psi														
					Metal temperature, F														
500	600	650	700	750	800	850	900	950	1000	1050	1100	1150	1200	1250	1300	1350	1400	1450	1500

ısion-welded pipe (straight seam) (cont'd)

2,750	12,100	11,700	11,300	10,250	8,700	7,100	5,500	3,800	2,100	1,400	850								
3,900	13,150	12,750	12,200	11,000	9,200	7,350	5,500	3,800	2,100	1,400	850								
5,100	14,250	13,800	13,150	11,750	9,700	7,600	5,500	3,800	2,100	1,400	850								
5,250	15,350	14,850	14,100	12,550	10,200	7,850	5,500	3,800	2,100	1,400	850								
4,100	10,500	10,150	9,900	9,200	8,100	6,950	5,850	4,050	2,250	1,450	900								
2,300	11,600	11,250	10,900	10,050	8,650	7,250	5,850	4,050	2,250	1,450	900								
3,500	12,800	12,350	11,900	10,850	9,200	7,500	5,850	4,050	2,250	1,450	900								
4,700	13,950	13,500	12,900	11,650	9,700	7,800	5,850	4,050	2,250	1,450	900								
5,950	15,050	14,600	13,950	12,450	10,250	8,050	5,850	4,050	2,250	1,450	900								
7,200	16,250	15,750	14,950	13,250	10,800	8,300	5,850	4,050	2,250	1,450	900								
2,350	11,650	11,300	11,000	10,250	9,000	7,750	6,500	4,500	2,500	1,600	1,000								
3,650	12,900	12,500	12,100	11,150	9,600	8,050	6,500	4,500	2,500	1,600	1,000								
5,000	14,200	13,750	13,250	12,050	10,200	8,350	6,500	4,500	2,500	1,600	1,000								
5,350	15,500	15,000	14,350	12,950	10,800	8,650	6,500	4,500	2,500	1,600	1,000								
7,750	16,750	16,250	15,500	13,850	11,400	8,950	6,500	4,500	2,500	1,600	1,000								
),100	18,050	17,500	16,600	14,750	12,000	9,250	6,500	4,500	2,500	1,600	1,000								

ıd tubes

·elded pipes and tubes

3,250	12,650	12,350	12,050	11,750	11,500	11,200	10,650	8,500	5,300	3,400	2,050								
4,350	13,750	13,400	13,100	12,750	12,250	11,700	10,650	8,500	5,300	3,400	2,050								
2,750	12,250	11,950	11,700	11,400	11,200	10,900	10,650	8,500	5,300	3,400	2,050								
5,050	14,250	13,800	13,200	11,750	9,700	7,600	5,500	3,800	2,150	1,350	850								
3,900	13,200	12,750																	

Table

Material	Specification (30)	Steel making process	Grade	Class	Tensile strength min. psi	Yield strength min. psi	Notes	Long joint factor (26)	Min. temp	Min. temp. to 100	200	300	400
										Electric fusion-welde			
C Mo A204, Gr. A	A155	OH, EF	CM65	1	65,000	37,000	3	1.00	−20	21,650	20,800	19,950	19,1
C Mo A204, Gr. B	A155	OH, EF	CM70	1	70,000	40,000	3	1.00	−20	23,350	22,400	21,500	20,6
C Mo A204, Gr. C	A155	OH, EF	CM75	1	75,000	43,000	3	1.00	−20	25,000	24,000	23,000	22,0
½Cr ½Mo A387, Gr. A .	A155	OH, EF	½CR	1	65,000	40,000	3	1.00	−20	21,650	20,800	19,950	19,1
1Cr ½Mo A387, Gr. B ..	A155	OH, EF	1CR	1	60,000	30,000	3	1.00	−20	20,000	19,250	18,500	17,7
1¼Cr ½Mo A 387, Gr. C	A155	OH, EF	1¼CR	1	60,000	35,000	....	1.00	−20	20,000	19,300	18,550	17,8
2¼Cr 1Mo A 387, Gr. D	A155	OH, EF	2¼CR	1	60,000	30,000	....	1.00	−20	18,750	18,250	17,650	17,1
5Cr ½Mo A 357	A155	OH, EF	5CR	1	60,000	30,000	....	1.00	−20	18,750	17,900	17,050	16,2
18-8	A249	EF	TP 304H	..	75,000	30,000	23	0.80	−325	......	....	....	....
18-8	A249	EF	TP 316H	..	75,000	30,000	23	0.80	−325	......	....	....	....
18-8	A249	EF	TP 321H / TP 347H	..	75,000	30,000	23	0.80	−325	......	....	....	....
18-8	A249, A269, A312	EF	TP 304	..	75,000	30,000	12, 20, 31	0.80	−425	15,000	13,300	12,000	10,9
18-8	A312	EF	TP 304H	..	75,000	30,000	12, 23	0.80	−325	......	....	....	....
18-8	A249, A269, A312	EF	TP 304L	..	70,000	25,000	12, 31	0.80	−425	12,500	12,200	10,500	8,8
18-8	A249, A269, A312	EF	TP 316	..	75,000	30,000	12, 20, 31	0.80	−425	15,000	15,000	14,300	14,0
18-8	A312	EF	TP 316H	..	75,000	30,000	12, 23	0.80	−325	......	....	....	....
18-8	A249, A269, A312	EF	TP 316L	..	70,000	25,000	12, 31	0.80	−425	12,500	12,500	11,600	9,6
18-8	A249, A269, A312	EF	TP 321 / TP 347	..	75,000	30,000	12, 31, 32	0.80	−325	15,000	15,000	13,600	12,6
18-8	A312	EF	TP 321H, TP 347H	..	75,000	30,000	12, 23	0.80	−325	......	....	....	....

NOTES:

1. Above 900 F consider the advantages of killed steel.

2. Conversion of carbides to graphite may occur after prolonged exposure to temperatures over 750 F.

3. Conversion of carbides to graphite may occur after prolonged exposure to temperatures over 850 F.

4. Consider the use of firebox quality at metal temperatures over 875 F.

5. A quality factor of 9 per cent is included for structural grade.

12. To use this material at temperatures below −20 F the special requirements of 323.2.2 must be met.

16. To determine the allowable stress S for use in code computations not utilizing the joint factor E, divide the value shown in the table above SE by the longitudinal joint factor E tabulated above for each material.

17. These specifications do not include requirements or rules for random radiographic inspection. If this higher joint facto

Continued)

						Allowable stress, psi													
						Metal temperature, F													
500	600	650	700	750	800	850	900	950	1000	1050	1100	1150	1200	1250	1300	1350	1400	1450	1500

ipe and tubes

500	600	650	700	750	800	850	900	950	1000	1050	1100	1150	1200	1250	1300	1350	1400	1450	1500
8,300	17,500	17,100	16,700	16,250	15,650	14,400	12,500	10,000	6,250	4,000	2,400								
9,750	18,850	18,400	17,950	17,500	16,900	15,000	12,750	10,000	6,250	4,000	2,400								
1,100	20,150	19,700	19,200	18,750	18,000	15,900	13,000	10,000	6,250	4,000	2,400								
8,300	17,500	17,100	16,700	16,250	15,650	14,400	12,500	10,000	6,250	4,000	2,400								
7,000	16,250	15,900	15,500	15,150	14,750	14,200	13,100	11,000	7,500	5,000	2,800	1,550	1,000						
7,150	16,450	16,050	15,700	15,350	15,000	14,400	13,100	11,000	7,800	5,500	4,000	2,500	1,200						
6,600	16,050	15,800	15,500	15,300	15,000	14,400	13,100	11,000	7,800	5,800	4,200	3,000	2,000						
5,350	14,500	14,100	13,650	13,250	12,800	12,400	11,500	10,000	7,300	5,200	3,300	2,200	1,500						
....	....	....	....	....	....	....	....	....	....	6,800	6,000	4,600	3,600	2,600	1,950	1,450	1,100	800	600
....	....	....	....	....	....	....	....	....	....	9,750	8,300	6,800	5,450	4,250	3,200	2,400	1,900	1,500	1,200
....	....	....	....	....	....	....	....	....	....	10,500	10,000	6,400	4,000	2,900	2,150	1,600	1,250	950	800
0,000	9,300	9,000	8,650	8,300	8,000	7,750	7,450	7,300	7,050	6,800	6,000	4,600	3,600	2,600	1,900	1,450	1,100	800	600
....	....	....	....	....	....	....	....	....	....	6,800	6,000	4,600	3,600	2,600	1,900	1,450	1,100	800	600
7,750	7,200	7,000	6,800	6,650	6,450														
3,750	13,700	13,650	13,600	13,500	13,300	13,200	12,800	12,100	11,200	9,750	8,300	6,800	5,450	4,250	3,200	2,400	1,900	1,500	1,200
....	....	....	....	....	....	....	....	....	....	9,750	8,300	6,800	5,450	4,250	3,200	2,400	1,900	1,500	1,200
8,800	8,100	7,850	7,550	7,250	7,050														
2,150	11,900	11,850	11,800	11,750	11,650	11,400	11,300	11,100	10,800										
....	....	....	....	....	....	....	....	....	....	10,500	10,000	6,400	4,000	2,900	2,150	1,600	1,250	950	800

is to be used, material shall be purchased to the special requirements of 327.4.3 with random radiography in accordance with Table 302.4.3, and the user must take appropriate steps to assure that required inspection and repair is accomplished.

20. These stress values apply only when the carbon content is 0.04 per cent or higher.

23. "H" grades of stainless steel may be used below 1000 F at the allowable stresses shown for the corresponding regular grade. The greater tendency of the "H" grades to intergranular carbide precipitation must be taken into account.

24. *SE* values above 400 F apply only to nonexpanded pipe. No value is assigned to cold-expanded pipe above 400 F.

26. The minimum temperature shown is that design temperature for which the material is normally suitable, but the use of a material at a design temperature below →20 F is established by rules elsewhere in this Code, including any necessary impact test requirements.

30. Specification numbers are ASTM numbers unless otherwise indicated.

31. There are no specified yield and ultimate strengths for A269 materials.

32. Lower temperature limit of TP 347 is −425 F.

Table 5.* Stress-range Reduction Factors

Cycles	f
7,000 and less	1.0
7,000–14,000	0.9
14,000–22,000	0.8
22,000–45,000	0.7
45,000–100,000	0.6
Over 100,000	0.5

*Appears as Table 302.3.2 in ASA B31.3.

NOTE: In the author's opinion, the allowable stress range may be increased to $\frac{4}{3}$ S_A under test conditions and short-time conditions.

of these stresses is less than S_h, the difference between S_h and the sum may be added to the term $0.25S_h$ in the equation above.

Limits on Calculated Stresses Due to Occasional Loads. The sum of the longitudinal stresses produced by pressure and live and dead loads and those produced by occasional loads such as wind or earthquake may be as much as 1.33 times the allowable stress values given in Tables 3 and 4. It is not necessary to consider wind and earthquake as occurring concurrently.

It is not necessary to consider other occasional loads such as wind and earthquake as occurring concurrently with live, dead, and test loads during periods of testing. Based on test thickness and test temperature, it is good practice to limit stresses due to test conditions to no more than 150 per cent of the basic allowable stresses. The designer is cautioned not to overlook the thermal-expansion load if the test fluid has a temperature higher than the design temperature, e.g., a cold gas line tested for tightness with steam.

The Code does not state a limit on stresses during erection; however, good practice would limit stresses to values 20 per cent above the basic allowable stresses except that total erection stress, including that due to wind loading, should not exceed 150 per cent of the basic allowable stresses.

Corrosion and Erosion Allowances. If corrosion or erosion is expected, an increase in wall thickness of the component over that dictated by other design requirements should be provided consistent with the expected life of the particular piping involved.

A commonly used nominal value of corrosion allowance is $\frac{1}{10}$ in. for carbon-steel and alloy-steel piping in hydrocarbon service. When carbon steel is used for utility piping, that is, steam, air, and water, a commonly used corrosion allowance is 0.05 in. As a rule, no corrosion allowance is used with austenitic steels. The allowance for corrosion or erosion should be added to all surfaces exposed to the flowing medium.

Many techniques and devices are used for determining the amount of corrosion that occurs while piping is in service. One of the most frequently used methods consists of drilling sentry holes (or sentinel drillings) at frequent intervals to a depth equal to the discarding thickness of the pipe. Theoretically, if corrosion has proceeded to the maximum allowable extent and remains undiscovered by other methods, the small hole will leak and thereby give warning that the piping should be removed from service.

Threading and Grooving Allowances. Calculations for the thickness of piping components, which are to be threaded or grooved, are required to include a dimensional allowance in inches equal to the depth of the cut. For the threaded components, the nominal thread depth (dimension h of ASA Standard B2.1, or equivalent) is taken as the depth of cut. For machined surfaces or grooves, where

the tolerance is not specified, the tolerance shall be assumed to be $\frac{1}{64}$ in. in addition to the specified depth of the cut.

Weld Joint Factors.   The longitudinal weld joint factors listed in Table 6 are used in all instances where required by Code.

Table 6. Longitudinal Weld Joint Factor *E*

(Table 302.4.3–ASA B31.3)

Type of joint	E
1. Arc or gas weld:	
a. Double-welded butt.	0.85
b. Double-welded butt, with joints in accordance with Note 1	0.90
c. Double-welded butt, with 100% radiography in accordance with 336.4.2(c) and conforming with requirements of 327.4.3	1.00
d. Single-welded butt	0.80
e. Single-welded butt, with joints in accordance with Note 1	0.90
f. Single-welded butt, with 100% radiography in accordance with 336.4.2(c) and conforming with requirements of 327.4.3	1.00
g. Spiral-welded ASTM A211	0.75
2. Electric resistance weld and electric flash weld	0.85
3. Furnace weld:	
a. Lap weld	0.75
b. Butt weld	0.60

NOTE 1.   Welds with 0.90 joint factor shall be finished, random radiographed by the technique, and evaluated in accordance with UW-51 of Section VIII of the ASME Boiler and Pressure Vessel Code. Radiographing (random) shall consist of not less than 12 in. of radiography per 100 ft of weld with reexamination and repair in accordance with UW-52 of Section VIII of the ASME Boiler and Pressure Vessel Code.

Mechanical Strength.   The pipe wall must be sufficiently thick to prevent damage, collapse, or buckling due to superimposed loads from supports, backfill, and other causes.   In those cases where it is impractical to increase the thickness or if a thickness increase would cause excessive local stresses, the factors that would contribute to the damage of the piping are required to be corrected by other design methods.[5]

Although lines under railroad tracks and roadways outside refinery limits are usually installed in steel pipe sleeves in conformance with the requirements of the railroad or highway departments involved, it seldom is necessary to perform this protection for underground steel lines in a refinery.[6]   For practical considerations, however, at least 24 in. of cover should be provided for unsleeved lines under roads and railroads (for railroads, the 24 in. is measured from bottom of ties to top of pipe).   This applies especially to railroads, because a derailment could cause the failure of any line closer to the surface.

PRESSURE DESIGN OF PIPING COMPONENTS

Straight Pipe.   The equations given in the Code consider pressure and mechanical, corrosion, and erosion allowances.   In addition to these factors, the Code requires that all designs, not only those for straight pipe, be checked for adequacy of mechanical strength under the applicable loadings discussed above.

The Code gives equations for determining the thickness of straight pipe for the outside diameter-thickness ratios D_o/t greater than 4.   The pressure design of

piping having a diameter-thickness ratio of 4 or less requires special considerations which encompass design and material factors, such as theory of failure, fatigue, and thermal stresses.

The minimum thickness of the pipe selected, considering manufacturer's minus tolerance, shall not be less than t_m in the equation

$$t_m = t + c \tag{2}$$

in which t_m = minimum required thickness, satisfying requirements for pressure, and mechanical, corrosion, and erosion allowances, in.

t = pressure design thickness as calculated from Eq. (3) (or the Barlow or Lamé equation) for internal pressure (or in accordance with the prescribed Code procedure for external pressure), in.

c = for internal pressure, the sum of the mechanical allowances (thread depth and groove depth), corrosion and erosion allowances, in.; for external pressure, the corrosion and erosion allowances, in.

P = internal design pressure, psig

D_o = outside diameter of pipe, in.

S = applicable allowable stresses in accordance with Tables 3 and 4, psi

E = longitudinal weld joint factor (see Table 6)

Y = coefficient having values as given Table 7 for ductile ferrous materials, a value of 0.4 for ductile nonferrous materials, and a value of zero for brittle materials such as cast iron

Table 7. Values for Y for Ferrous Materials

Temperature, F	900 and below	950	1000	1050	1100	1150 and above
Ferritic steels	0.4	0.5	0.7	0.7	0.7	0.7
Austenitic steels	0.4	0.4	0.4	0.4	0.5	0.7

For metallic pipe with diameter-to-thickness ratios greater than 4, the internal pressure design thickness should be calculated using Eq. (3).

$$t = PD_o/2(SE + PY) \tag{3}$$

The Barlow and Lamé equations may be used, but they result in somewhat greater pipe-wall thickness.

$$t = PD_o/2SE \qquad \text{(Barlow)}$$
$$t = D_o/2\,(1 - \sqrt{SE - P/SE + P}) \qquad \text{(Lamé)}$$

In the 1955 and earlier Codes a table of minimum practical design thicknesses was included. This was dropped in 1959 because it was felt that it greatly handicapped the competent designer when he was working with very expensive materials. Slightly modified it still is, however, a reasonable set of minimum thicknesses for carbon-steel piping and is included here as Table 8.

Another practical rule used by some oil companies is to make pipe nipples for tapped and socket-welding openings in valves, piping accessories, and process equipment one schedule heavier than the remainder of the line. It is also fairly common practice to make the piping for 3/4-in. and smaller connections from a header to the first valve or fitting one schedule heavier than the remainder of the branch line unless the header is smaller than 2-in. pipe size.

Table 8. Minimum Recommended Pipe Thickness

Nominal pipe size, in.	Min. thickness, in.
2 and smaller	0.06
2½–3	0.07
4	0.09
6	0.11
8	0.12
10–24	0.13

Bends and Elbows. The minimum required thickness of a pipe bend, after bending, should be determined as for straight pipe, provided the bending operation does not produce a difference between the maximum and minimum diameters greater than 8 per cent of the nominal outside diameter of the pipe for internal pressure service and 3 per cent for external pressure service. Bends made with greater flattening should meet the requirements of Subdivision 304.7 of the Code.

Elbows that are manufactured in accordance with any of the Standards listed in Table 326.1 of the Code are considered to be suitable for use at the pressure-temperature rating specified in the listed Standard. In the case of Standards under which elbows are made to a nominal pipe thickness (such as ASA B16.9), the elbows are considered suitable for use with pipe of the same nominal thickness.

Branch Connections. Branch connections may be made by the use of fittings, welding outlet fittings, by welding the branch pipe directly to the run pipe, or by threading under greatly restricted conditions.

The Code gives rules governing the design of pipe-to-pipe branch connections to sustain internal and external pressure in those cases where the angle between the axis of the branch and of the run is between 45 and 90°. Branch connections in which the smaller angle between the axis of the branch and the run is less than 45°, impose special design and fabrication problems. The rules given in the Code for angles between 45° and 90° may be used as a guide, but sufficient additional strength must be provided to insure safe and satisfactory service and these branch connections shall be designed to meet the requirements of Subdivision 304.7.

Branches at angles other than 90 deg should not be employed except when flow and pressure-drop considerations greatly outweigh the difficulty in obtaining satisfactory welds and the greater difficulty in adequately reinforcing such connections.

Acceptable Code methods of making welded pipe-to-pipe branch connections are shown in Fig. 2. No attempt has been made to show all acceptable types of construction.

Some designers prefer to restrict details (*a*), (*c*), and (*e*) to near full-sized branches because they believe that with smaller branches the likelihood of obtaining the required full penetration attachment welds is somewhat greater if details (*b*) and (*d*) are used. Also, if details (*a*), (*c*), and (*e*) are used with smaller branches, internal protrusion often greatly restricts the flow area. (It is suggested that in critical service the cut edge of the hole be inspected to assure freedom from laminations.)

For welded pipe-to-pipe branch connections, the stress concentration at the junction increases rapidly as the size of the branch approaches the size of the run. This is also true with most welding outlet fittings. Consequently, in services which involve considerable cycling due to pressure or temperature, or both, it is usually good practice to make the branch connections with butt-welding tee fittings. The use of butt-welding tees is also considered good practice for full-sized branches

in many flammable and toxic services.   If there is at least one size reduction in the branch, pipe-to-pipe branch connections should, as a rule, be acceptable for all services except those with severe cycling.   Also, the cost of butt-welding reducing tees is such that savings will result if pipe-to-pipe branches or welding outlet fittings are used.

Branch connections made by either socket-welding or threading the branch pipe directly to the run pipe are limited by the Code to nonflammable, nontoxic services below 150 psig and 400 F.   In the petroleum industry, such connections are not used widely.

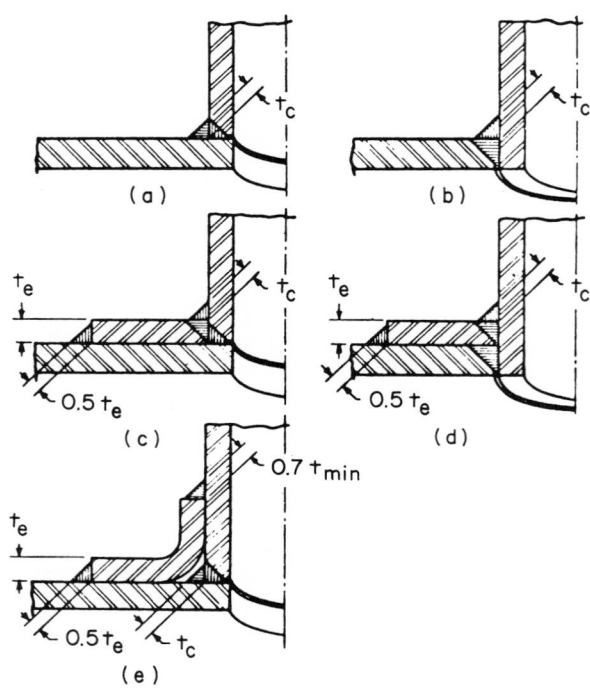

Fig. 2. Some acceptable types of welded branch attachment details showing minimum acceptable welds.   NOTE: Weld dimensions may be larger than the minimum values shown here. $t_c = 0.7t_n$ but not less than $\frac{1}{4}$ in. t_n = nominal thickness of branch wall less corrosion allowance, inches. t_e = nominal thickness of reinforcing element (ring or saddle), inches ($t_e = 0$ if there is no added reinforcement). t_{min} = the smaller of t_n or t_e.

Strength of Branch Connections.   A pipe having a branch connection is weakened by the opening that must be made in it, and unless the wall thickness of the pipe is sufficiently in excess of that required to sustain the pressure, it is necessary to provide reinforcement.   Certain branch connections* may be made without the necessity of their use being supported by engineering calculations.   In other cases, the amounts of reinforcement required to sustain the pressure in welded pipe-to-pipe branches are outlined clearly in ASA B31.3, paragraph 304.3.3.

Branch Reinforcement Tables.   It is usually not practicable to use a fillet weld alone for branch reinforcement when the reinforcement area needed (in

* Code for Pressure Piping, ASA B31.3, paragraph 304.3.2.

addition to the excess thickness in the pipe walls) exceeds 1 sq in. In these cases consideration should be given to the use of rings, saddles, etc.

Tables 9 and 10, and Fig. 3 have been included to aid the designer. Table 9 gives the reinforcement for 90-deg pipe-to-pipe branch connections in the 150-lb (ASA B16.5) pressure class. Where the needed reinforcement area is less than 1 sq in., the table lists the size of fillet welds which, in combination with the excess thickness in the pipe walls, will provide the needed reinforcement. When the needed reinforcement area exceeds 1 sq in., the table lists, in figures enclosed in parentheses, the total cross-sectional area that must be provided in the added reinforcement and attachment fillet welds combined.

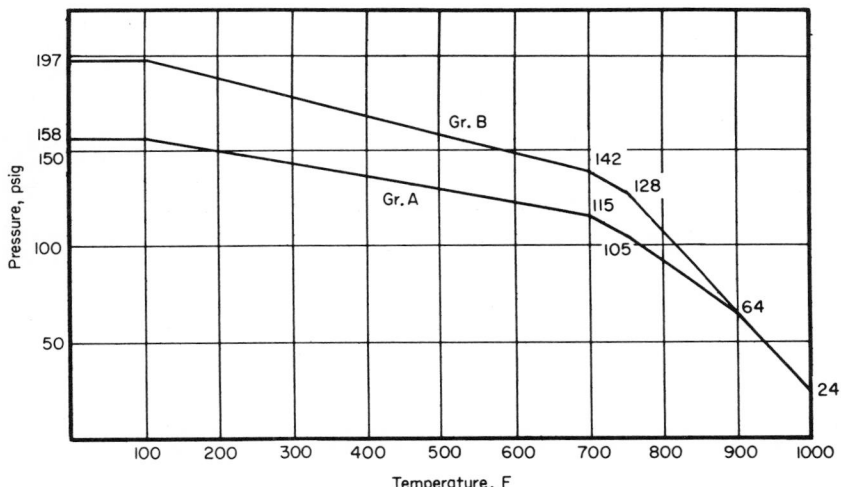

Fig. 3. Pressure-temperature limits for 90-deg carbon steel pipe-to-pipe branch connections reinforced with a ¼-in. fillet weld. The curves are for Grades A and B seamless carbon-steel pipe (API 5L or equal). A 0.10-in. corrosion allowance was assumed. For pressure-temperature conditions below the curves, a ¼ in. fillet weld will provide sufficient reinforcement for pressure loads (no provision has been made for external loads) for 90-deg carbon-steel pipe-to-pipe branch connections made with the pipe sizes and thicknesses listed in Tables 9 and 10.

Table 10 gives the same information for pipe-to-pipe branch connections operating at 475 psig and 100 F. Figure 3 gives the pressure-temperature limits for 90-deg carbon-steel pipe-to-pipe branch connections reinforced with a ¼-in. fillet weld.

The reinforcement shown in Tables 9 and 10 and Fig. 3 is for pressure loads only. No provision has been made for external loads.

Openings in Closures. Code rules* govern the design of openings in closures where the size of the opening is not greater than one half of the inside diameter of the closure. The basis for the design of these openings in closures is the same as it is for branch connections. Closures with larger openings must be designed as reducers, except that flat closures must be designed as flanges.

Miters. The thickness of each segment of the miter must be designed in the same manner as straight pipe. This thickness, however, does not allow for the

---

* Paragraph 304.3.4 of ASA B31.3.

Table 9. Added Reinforcement Required for 90-deg Carbon-steel Pipe-to-pipe Branch Connections in 150-lb RF (ASA B16.5) Piping

Branch size and thickness

Header size and thickness	1/2" Sch 80	3/4" Sch 80	1" Sch 80	1 1/2" Sch 80	2" Sch 80	2 1/2" Sch 40	3" Sch 40	4" Sch 40	6" Sch 40	8" Sch 30	8" Sch 40	10" Sch 30	10" Sch 40	12" Sch 30	12" STD	14" Sch 20	14" Sch 30	16" Sch 20	16" Sch 30	18" Sch 20	18" STD	20" Sch 20	24" Sch 20
24" Sch 20	3/8	7/16	7/16	1/2	9/16	5/8	11/16	13/16	15/16	(1.20)	(1.13)	(1.49)	(1.39)	(1.77)	(1.69)	(2.00)	(1.89)	(2.32)	(2.22)	(2.64)	(2.54)	(2.87)	(3.52)
20" Sch 20	5/16	5/16	5/16	3/8	7/16	1/2	9/16	5/8	3/4	7/8	13/16	15/16	7/8	1	1	(1.14)	(1.04)	(1.33)	(1.23)	(1.53)	(1.43)	(1.63)	
18" STD	1/4	1/4	1/4	5/16	5/16	3/8	7/16	1/2	9/16	11/16	5/8	3/4	11/16	13/16	3/4	7/8	13/16	15/16	7/8	1	15/16		
18" Sch 20	5/16	3/8	3/8	7/16	1/2	9/16	5/8	11/16	7/8	15/16	15/16	(1.08)	1	(1.29)	(1.24)	(1.46)	(1.39)	(1.70)	(1.63)	(1.94)			
16" Sch 30	1/4	1/4	1/4	1/4	1/4	1/4	1/4	1/4	5/16	7/16	5/16	7/16	5/16	1/2	3/8	9/16	7/16	5/8	1/2				
16" Sch 20	1/4	5/16	5/16	3/8	3/8	7/16	1/2	9/16	11/16	13/16	3/4	7/8	7/8	1	15/16	(1.03)	1	(1.21)					
14" Sch 30	1/4	1/4	1/4	1/4	1/4	1/4	1/4	1/4	1/4	1/4	1/4	1/4	1/4	1/4	1/4	1/4	1/4						
14" Sch 20	1/4	1/4	1/4	1/4	5/16	3/8	3/8	7/16	1/2	5/8	9/16	11/16	5/8	3/4	11/16	13/16							
12" STD	1/4	1/4	1/4	1/4	1/4	1/4	1/4	1/4	1/4	1/4	1/4	1/4	1/4	1/4	1/4								
12" Sch 30	1/4	1/4	1/4	1/4	1/4	1/4	1/4	1/4	1/4	1/4	1/4	1/4	1/4	1/4									
10" Sch 40	1/4	1/4	1/4	1/4	1/4	1/4	1/4	1/4	1/4	1/4	1/4	1/4	1/4										
10" Sch 30	1/4	1/4	1/4	1/4	1/4	1/4	1/4	1/4	1/4	1/4	1/4	1/4											
8" Sch 40	1/4	1/4	1/4	1/4	1/4	1/4	1/4	1/4	1/4	1/4	1/4												
8" Sch 30	1/4	1/4	1/4	1/4	1/4	1/4	1/4	1/4	1/4	1/4													
6" Sch 40	1/4	1/4	1/4	1/4	1/4	1/4	1/4	1/4	1/4														
4" Sch 40	1/4	1/4	1/4	1/4	1/4	1/4	1/4	1/4															
3" Sch 40	1/4	1/4	1/4	1/4	1/4	1/4	1/4																
2 1/2" Sch 40	1/4	1/4	1/4	1/4	1/4	1/4																	
2" Sch 80	1/4	1/4	1/4	1/4	1/4																		
1 1/2" Sch 80	1/4	1/4	1/4	1/4																			
1" Sch 80	1/4	1/4	1/4																				
3/4" Sch 80	1/4	1/4																					
1/2" Sch 80	1/4																						

NOTES:
1. The fractional values are fillet weld sizes, in inches, i.e., the leg lengths of the largest isosceles right triangle which can be inscribed within the fillet-weld cross section.
2. The values enclosed in parentheses are additional reinforcement areas in square inches. It is usually not feasible to provide this reinforcement with a fillet weld alone, and other methods of providing the reinforcement, such as pads and saddles, should be considered.
3. Reinforcement was computed assuming 0.10-in. corrosion allowance and seamless API 5L Grade B or equal pipe.

Table 10. Added Reinforcement Required for 90-deg Carbon-steel Pipe-to-pipe Branch Connections for Piping Operating below 475 Psig at 100 F (see Note 1)

Header size and thickness	Branch size and thickness														
	1/2" Sch 80	3/4" Sch 80	1" Sch 80	1-1/2" Sch 80	2" Sch 80	2-1/2" Sch 40	3" Sch 40	4" Sch 40	6" Sch 40	8" Sch 30	8" Sch 40	10" Sch 30	10" Sch 40	12" Sch 30	12" Std
12" STD	1/4	5/16	5/16	3/8	3/8	1/2	1/2	9/16	11/16	7/8	13/16	15/16 (1.20)	7/8 (1.13)	1 (1.44)	1
12" Sch 30	5/16	3/8	3/8	7/16	1/2	9/16	5/8	3/4	7/8	1	1				
10" STD	1/4	1/4	1/4	1/4	1/4	1/4	3/8	3/8	1/2	9/16	1/2	5/8	9/16		
10" Sch 30	5/16	5/16	5/16	3/8	7/16	1/2	9/16	5/8	3/4	7/8	7/8	1			
8" Sch 40	1/4	1/4	1/4	1/4	1/4	1/4	1/4	5/16	3/8	7/16	3/8				
8" Sch 30	1/4	1/4	5/16	5/16	3/8	7/16	1/2	1/2	5/8	3/4					
6" and smaller	Same as Table 9														

NOTES:

1. In the table are the amounts of added reinforcement required for 90-deg carbon-steel pipe-to-pipe branch connections operating below 475 psig at 100 F and below 350 psig at 700 F. Interpolation shall be used to determine the operating pressure between 100 and 700 F.

2. The fractional values are fillet weld sizes, in inches, i.e., the leg lengths of the largest isosceles right triangle which can be inscribed within the fillet-weld cross section.

3. The values enclosed in parentheses are required additional reinforcement areas in square inches. It is usually not feasible to provide this reinforcement with a fillet weld alone, and other methods of providing the reinforcement such as pads and saddles should be considered.

4. Reinforcement was computed assuming 0.1-in. corrosion allowance and seamless API 5L Grade B or equal pipe.

discontinuity stresses which exist at junctions between the segments of a miter. The discontinuity stresses are reduced for a given miter as the number of segments is increased.   These stresses may be neglected for miters in nonflammable, nontoxic, noncyclic services at pressures of 100 psig and under.   Miters used in other services or at higher pressures must meet Code requirements.*

Miter elbows are commonly used in a refinery for large-diameter low-pressure water (50 lb or less), drainage, and vent piping.   Even for these relatively non-critical services, 90-deg miter elbows should have three or more segments.

Attachments.   Attachments to piping, both external and internal, are required to be designed so that they will not cause flattening of the pipe, excessive localized bending stresses, or harmful thermal gradients in the pipe wall.   It is important that such attachments be designed to minimize the stress concentrations, particularly in cyclic services.

Closures.   The design of welded flat closures, ellipsoidal closures, spherically dished closures, hemispherical closures, conical closures, and toriconical closures is encountered relatively infrequently.   Information relating to design requirements of such closures is given in Para. 304.4.2 through 304.4.7 of the code.

Closure fittings manufactured in accordance with the Standards listed in Table 326.1 of the Code are considered suitable for use at the pressure-temperature ratings specified by such Standards, and in the case of Standards under which closure fittings are made to nominal pipe thickness, the closure fittings shall be considered suitable for use with pipe of the same nominal thickness.

In refinery piping the most commonly used closure is probably the butt-welding cap.   Bolted flanged covers are used only where access is needed.

Flanges.   Flanges manufactured in accordance with Standards listed in Table 326.1 of the Code shall be considered suitable for use at the pressure-temperature ratings specified by such Standards.   Flanges not made in accordance with the standards listed in Table 326.1 shall be designed in accordance with Section VIII of the ASME Boiler and Pressure Vessel Code, except that the requirements for fabrication, assembly, inspection, and testing, and the pressure and temperature limits for materials of ASA B31.3 shall govern.

The Vessel Code flange design rules are not applicable to designs which employ full-face gaskets that extend beyond the bolts usually to the outside diameter of the flange.   The forces and reactions in such full-face gasketed joints are very different from those of ring-gasketed joints, and the flange should be designed to meet the requirements of Subdivision 304.7 of ASA B31.3.

Blanks.   The pressure design thickness for permanent blanks must be calculated in accordance with Eq. (4).   Corrosion or erosion allowances and manufacturer's minus tolerance must be added to the thickness thus determined.

$$t = d_6 \sqrt{3P/16S} \tag{4}$$

where d_6 = inside diameter of gasket for raised or flat face flanges or the gasket pitch diameter for ring joint flanges, in.

P = internal design pressure or external design pressure, psi

S = applicable allowable stress from Table 3, psi

In the design of blanks which are not to be used during operation of the piping (e.g., during shutdowns or during testing) an allowable stress equal to 90 per cent of the yield strength is sometimes used.   It is suggested, however, that $1/4$ in. is a practical minimum thickness for carbon-steel blanks.

Pressure Design of Other Components.   In general, pressure-containing components are to satisfy Code requirements.   However, if the design of similarly

* Paragraph 304.7 of ASA B31.3.

shaped or proportioned components has been proved by successful performance under comparable service conditions, provision for the use of such components is allowed by the Code. Alternatively, if pressure design of the particular component is based on an analysis consistent with general Code philosophy, and if the component is proved out either by an experimental stress analysis or by tests made in accordance with paragraph UG-101 of Section VIII of the ASME Boiler and Pressure Vessel Code, such component will be deemed to satisfy the requirements of ASA B31.3. It is suggested that, in all instances of pressure design of non-standard components, Para. 304.7 of the Code be consulted.

SELECTION AND LIMITATIONS OF PIPING COMPONENTS

Pipe and Fittings. Furnace lap-weld ferrous pipe, furnace butt-weld ferrous pipe, spiral-weld ferrous pipe, and fusion-welded steel pipe (made to ASTM A134 and A139) are not permitted for hydrocarbons or other flammable fluids within process unit limits or for toxic fluids in any location (this limitation also applies to all grades of pipe manufactured to ASTM A120), and API 5LX pipe cannot be used for flammable or toxic fluids within process unit limits. Nonferrous pipe made by similar manufacturing processes is similarly restricted.

The use of bell-and-spigot fittings is limited to water and drainage service. Also, pipe couplings made of cast, malleable, or wrought iron are not permitted for flammable fluids within process limits or toxic fluids in any area. In addition, they cannot be used for flammable fluids outside process unit limits at design temperatures above 300 F or design pressures above 400 psig. Actually, these couplings are practically never used in any location in refineries.

Comments on Selection and Limitations of Components. There are many rules for the use of fittings, bends, intersections, and valves which are not universally applicable to all refinery services, and such components are not easily codified. Some rules, however, are of value to the designer and are listed in the following paragraphs.

1. Welding fittings are usually preferred to flanged fittings, not only for economic reasons but because the potential for leakage is reduced.

2. Pipe bends are preferred to butt-welding elbows for reciprocating compressor suction and discharge piping, vapor relief-valve discharge piping, and piping conveying corrosive fluids (such as acid where turbulence in a fitting may cause excessive corrosion).

3. Bends or dead-end tees should be used for piping which conveys pulverized abrasive solids suspended in gas in the dilute phase. Dead-end tees (so arranged that the flow will impinge against the dead end) have a longer life than bends in this service and should be used if the system can be designed to take care of the resulting increase in pressure drop.

4. Bends should be used for dense-phase flow of pulverized abrasive solids and for all piping which handles either pulverized or granular solids suspended in liquids or granular solids suspended in gases.

5. If the flow is through a branch into a header (or run pipe) in a piping system which transports pulverized abrasive solids suspended in gas in the dilute phase, a dead-end cross (so arranged that the flow will impinge against the dead end) should be used.

6. In services with very high corrosion rates, butt-welding fittings with the same ID as the attached pipe (if not the same, consider taper-boring the component with the smaller ID) are preferred to threaded and socket-welding fittings.

7. Threaded cast-iron fittings should not be used.

8. Threaded plugs are preferred to pipe caps for threaded end closures to reduce dead-end corrosion problems.

9. In most refineries, internal corrosion is a greater problem than external corrosion. Consequently, it is common practice to require that all $\frac{3}{4}$-in. and larger steel and cast-iron (and all $2\frac{1}{2}$-in. and larger brass) gate, globe, and angle valves (located above grade) be of the outside screw and yoke type.

10. Although valves with seal welded or pressure seal bonnets and welding ends are commonly used in steam service, $2\frac{1}{2}$-in. and larger valves in oil service are usually equipped with flanged ends.

11. Valves operated in wide-open or block service are generally gate valves. Butterfly valves, ball valves, nonlubricated plug valves, and lubricated plug valves may be considered as possible alternates to gate valves; however, most of these valve types are limited to services with fairly low operating temperatures. (Caution: Many nonlubricated plug valves, ball valves, and lubricated plug valves are constructed so that, if assembled improperly, they may give an incorrect indication of whether the valve is open or closed.)

12. As a general rule, hand-operated throttling valves for services where close control is not required and those for control valve bypasses should be globe valves (integral stem and plug preferred) for sizes 2 in. and smaller and gate valves for sizes $2\frac{1}{2}$ in. and larger. For severe throttling service and where close control is required, a conventional control valve with a hand operator should be used. The only other common application for globe valves in refinery service is for mixing purposes.

13. Solid-wedge and flexible-wedge gate valves are generally preferred to split and double-disk valves. Split-wedge and double-disk gate valves are generally used for clean liquids and noncondensing gases only.

14. Gate valves with Teflon inserts in the seat rings are very satisfactory in liquid butane and propane services.

15. Where coking may occur in blocked connections, a flushing connection should be added between the valve and the process line or equipment. The flushing medium may be oil, gas, or steam (if water can be tolerated in the system).

16. If the pressure differential across a closed gate valve is approximately equal to the pressure rating of the valve, consideration should be given to providing a pressure-equalizing bypass around the valve. Consideration should also be given to bypasses for valves in steam lines for warm-up purposes. When bypasses are provided, they should be sized in accordance with MSS SP-45. A gate valve should be provided in the bypass line.

17. Drain or bypass connections may be tapped (or socket-welded) into a valve body where necessary to simplify piping or to assure complete drainage.

18. Check valves are almost useless if installed in vertical lines in which the flow is downward.

19. If a valve is installed with the stem lower than horizontal, the valve bonnet should be provided with a drain.

20. The designer should consider providing gear operators for all 6-in. and larger lubricated plug valves and all 14-in. and larger gate valves.

21. The use of double block valves should be kept to a minimum. Double block valves, however, are required for sample connections and for drains which are connected to a closed drain system. Double block valves or their equivalent should be used where contamination must be prevented.

Under certain conditions double block valves are also needed where it is necessary to remove essential equipment from service for cleaning or repairs while the unit continues in operation. However, even in these cases a single block valve with provisions for blinding often will suffice. Of course, such equipment must be

provided with a spare or it must be possible to bypass it temporarily without shutting down the unit. The nature of the fluid, its pressure and temperature, and many other factors must be considered when determining the need for double block valves. Generally, the need for double block valves at equipment should be considered if the fluid is toxic or very corrosive, if the fluid is above its flash point at operating temperature, or if the fluid is above 500 F. (The above conditions are set down only as a guide for the careful judgement of the designer.) Where double block valves are used, a ¾-in. bleeder connection with a ¾-in. valve should be installed between the block valves.

Some ball valves and nonlubricated valves, when equipped with a bleeder between the seats, have been satisfactory substitutes for double block valves. Both conventional gate valves with Teflon inserts in the seat rings and flexible wedge gate valves have also been equipped with a bleeder connection and used in place of double block valves.

Selection and Limitations of Flanges. The Code restricts the use of screwed flanges as it does any threaded pipe joint. Slip-on flanges must be used with caution in installations where many large temperature cycles are expected, particularly if the flange is not insulated, because these temperature cycles too often break one of the welds on the slip-on flange.

The following statements on flanges are included for the designer's guidance.

1. The use of cast-, nodular-, wrought-, and malleable-iron screwed flanges should be avoided.

2. In services with very high corrosion rates, the bore of weld neck flanges should be the same ID as the attached piping (if not the same, consider taper-boring the component with the smaller ID).

3. The bore of weld neck orifice flanges should match the ID of the attached pipe.

4. Because there are no ASA Standards governing steel flanges in sizes larger than 24 in., the designer must make sure that the flange drilling on such flanges will match that of the equipment to which it is to be attached.

Selection and Limitations of Blanks. In a refinery, blanks are usually required to isolate individual pieces of equipment at shutdown and to positively block off hydrocarbon, acid, and caustic lines at the process unit limits. They are also needed during operation wherever positive shutoff is required to prevent leakage of one fluid into another.

Blanks should be located in horizontal lines if possible. Blanks certainly should not be used in vertical water and steam lines in climates where danger of freezing exists.

Circular-handle-type blanks of the types shown in Fig. 4 can be used for raised face joints in locations where the lines can be sprung easily to permit installation of the blanks. As a rule, this is easily accomplished only in 4-in. and smaller lines. Figure-eight-type blanks (styles *C* and *D*, Fig. 4) are used for larger lines, and even then, jackscrews may be needed to install the blank.

Blanks should be made from a plate specification, approved for use in ASA B31.3, of substantially the same chemical composition as the pipe. Carbon-steel blanks may be of ASTM A283 Grade C (750 F or less), ASTM A285 Grade C, or ASTM A201 Grade A or B.

Selection and Limitations of Gaskets. Gaskets must be made of materials which are not injuriously affected by the nature of the fluid or its temperature under anticipated operating conditions. Metallic materials, if used, must be in accordance with the temperature limitations established in Tables 3 and 4.

Nonmetallic gaskets are not permitted above 750 F gasket design temperature. Also, nonmetallic gaskets should not be used in nonconfining (flat or raised face) flanged joints at gasket design pressure-temperature conditions above the ratings

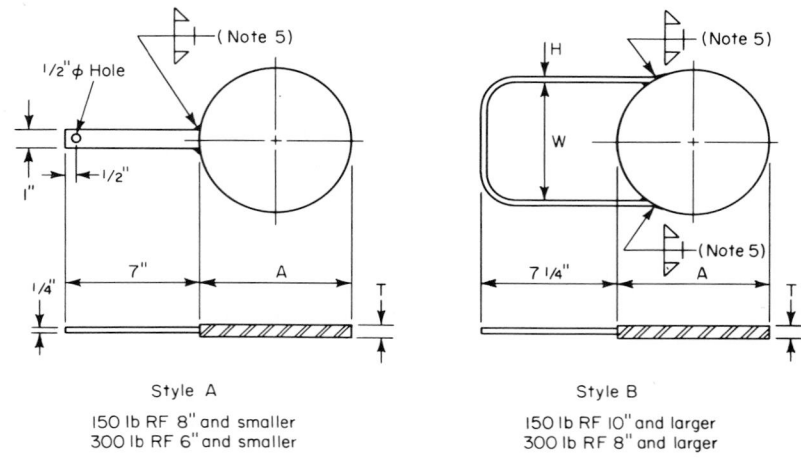

Style A

150 lb RF 8" and smaller
300 lb RF 6" and smaller

Style B

150 lb RF 10" and larger
300 lb RF 8" and larger

Handle-type Blanks

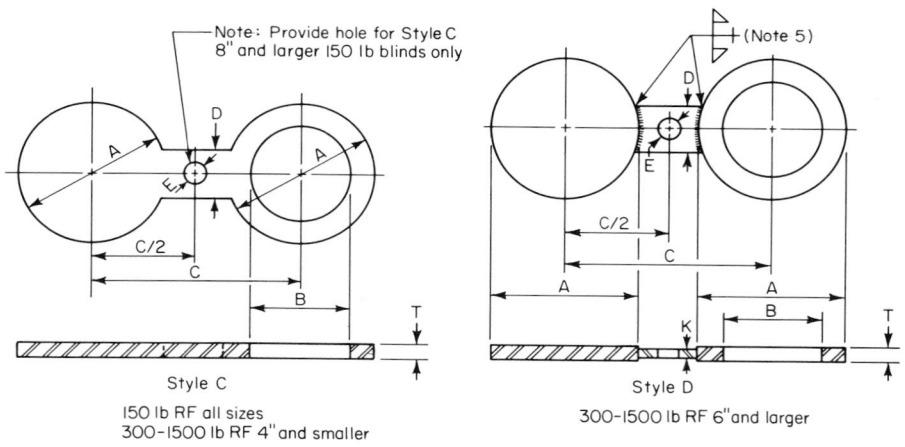

Note: Provide hole for Style C
8" and larger 150 lb blinds only

Style C

150 lb RF all sizes
300–1500 lb RF 4" and smaller

Style D

300–1500 lb RF 6" and larger

Figure-8 Blanks

Fig. 4. Carbon-steel figure-8 and circular-handle-type blanks. NOTES: 1. The carbon-steel blind thicknesses T shown in this figure are suitable for pressure-temperature conditions corresponding to the ASA B16.5 pressure ratings indicated. The thicknesses were calculated using the equation for the thickness of a uniformly loaded, flat circular plate. Blinds should be fabricated from ASTM A285C plate for thicknesses of 2 in. and less and from ASTM A201A or A201B plate for thicknesses over 2 in. A total minimum corrosion allowance of 0.05 in. is included in the thicknesses shown. 2. All burrs and weld spatter must be removed. 3. The gasket surfaces on blinds must be free from ridges and chatter marks. 4. The exposed part of all blinds should be greased adequately to prevent corrosion. 5. Weld detail should provide a minimum of 75 per cent penetration butt weld plus full fillets or ¼-in. fillets, whichever is smaller. 6. Except for thickness, dimensions on this figure are applicable to blanks made of materials other than carbon steel and to blanks designed for a specific pressure-temperature condition.

150 lb RF					
Pipe size	Style	Dimensions			
		A	T	H	W
1	A	2 1/2	1/4	....	...
1 1/4	A	2 7/8	1/4	....	...
1 1/2	A	3 1/4	1/4	....	...
2	A	4	1/4	....	...
2 1/2	A	4 3/4	1/4	....	...
3	A	5 1/4	1/4	....	...
4	A	6 3/4	5/16	....	...
6	A	8 5/8	7/16	....	...
8	A	10 7/8	9/16	....	...
10	B	13 1/4	11/16	3/8 Ø	6 1/2
12	B	16	3/4	1/2 Ø	6 1/2
14	B	17 5/8	7/8	1/2 Ø	6 1/2
16	B	20 1/8	15/16	5/8 Ø	6 1/2
18	B	21 1/2	1 1/16	5/8 Ø	6 1/2
20	B	23 3/4	1 3/16	5/8 Ø	6 1/2
24	B	28 1/8	1 3/8	5/8 Ø	6 1/2

300 lb RF					
Pipe size	Style	Dimensions			
		A	T	H	W
1	A	2 3/4	1/4	....	...
1 1/4	A	3 1/8	1/4	....	...
1 1/2	A	3 5/8	1/4	....	...
2	A	4 1/4	5/16	....	...
2 1/2	A	5	3/8	....	...
3	A	5 3/4	3/8	....	...
4	A	7	1/2	....	...
6	A	9 3/4	11/16	....	...
8	B	12	7/8	3/8 Ø	6
10	B	14 1/8	1 1/16	3/8 Ø	6 1/2
12	B	16 1/2	1 1/4	1/2 Ø	6 1/2
14	B	19	1 3/8	1/2 Ø	6 1/2
16	B	21 1/8	1 9/16	5/8 Ø	6 1/2
18	B	23 3/8	1 3/4	5/8 Ø	6 1/2
20	B	25 5/8	1 15/16	5/8 Ø	6 1/2
24	B	30 3/8	2 1/4	5/8 Ø	6 1/2

150 lb RF									
Pipe size	Style	Dimensions							
		A	B	C	C/2	D	E	T	K
1	C	2 1/2	1	3 1/8	....	1	...	1/4	...
1 1/4	C	2 7/8	1 5/16	3 1/2	....	1	...	1/4	...
1 1/2	C	3 1/4	1 1/2	3 7/8	....	1	...	1/4	...
2	C	4	2	4 3/4	....	1	...	1/4	...
2 1/2	C	4 3/4	2 1/2	5 1/2	....	1	...	1/4	...
3	C	5 1/4	3 1/8	6	....	2	...	1/4	...
4	C	6 3/4	4 1/16	7 1/2	....	2	...	5/16	...
6	C	8 5/8	6 1/8	9 1/2	....	2 3/4	...	7/16	...
8	C	10 7/8	8 1/16	11 3/4	5 7/8	3	7/8	9/16	...
10	C	13 1/4	10 3/16	14 1/4	7 1/8	4	1	11/16	...
12	C	16	12 1/8	17	8 1/2	4	1	3/4	...
14	C	17 5/8	13 3/8	18 3/4	9 3/8	4 1/2	1 1/8	7/8	...
16	C	20 1/8	15 3/8	21 1/4	10 5/8	4 1/2	1 1/8	15/16	...
18	C	21 1/2	17 3/8	22 3/4	11 3/8	5	1 1/4	1 1/16	...
20	C	23 3/4	19 1/4	25	12 1/2	5	1 1/4	1 3/16	...
24	C	28 1/8	23 1/4	29 1/2	14 3/4	5	1 3/8	1 3/8	...

300 lb RF									
Pipe size	Style	Dimensions							
		A	B	C	C/2	D	E	T	K
1	C	2 3/4	1	3 1/2	....	1	...	1/4	...
1 1/4	C	3 1/8	1 5/16	3 7/8	....	1	...	1/4	...
1 1/2	C	3 5/8	1 1/2	4 1/2	....	1	...	1/4	...
2	C	4 1/4	2	5	....	1 1/8	...	5/16	...
2 1/2	C	5	2 1/2	5 7/8	....	1 3/8	...	3/8	...
3	C	5 3/4	3 1/8	6 5/8	....	1 5/8	...	3/8	...
4	C	7	4 1/16	7 7/8	....	2 1/8	...	1/2	...
6	D	9 3/4	6 1/8	10 5/8	5 5/16	4	7/8	11/16	3/8
8	D	12	8 1/16	13	6 1/2	4	1	7/8	1/2
10	D	14 1/8	10 3/16	15 1/4	7 5/8	4 1/2	1 1/8	1 1/16	5/8
12	D	16 1/2	12 1/8	17 3/4	8 7/8	4 1/2	1 1/4	1 1/4	3/4
14	D	19	13 3/8	20 1/4	10 1/8	4 1/2	1 1/4	1 3/8	1
16	D	21 1/8	15 3/8	22 1/2	11 1/4	5 1/2	1 3/8	1 9/16	1
18	D	23 3/8	17 3/8	24 3/4	12 3/8	4 7/8	1 3/8	1 3/4	1
20	D	25 5/8	19 1/4	27	13 1/2	5 1/2	1 3/8	1 15/16	1 1/4
24	D	30 3/8	23 1/4	32	16	6	1 5/8	2 1/4	1 1/2

of 600-lb flanges per ASA B16.5, except that for nonflammable, nontoxic service fluids, the limiting ratings can be those of 900-lb flanges.

The use of metal or metal-asbestos gaskets is not limited as to pressure provided the gasket materials are suitable for maximum fluid temperatures.

In recent years the American Petroleum Institute has developed API 601, a Standard covering the dimensions of spiral-wound and double-jacketed asbestos gaskets. The spiral-wound gasket covered by API 601 has been widely used with great success with raised face flanges as a replacement for the ring-type joint.

Selection and Limitations of Flange Facings. When 125-lb cast-iron flanges are bolted to 150-lb steel flanges, the $\frac{1}{16}$-in. raised face on the steel flanges should be removed. If the raised face is removed and a full-face gasket is used, either high-strength carbon-steel bolting or alloy-steel bolting may be used. However, if the face is not removed (or if the face is removed and a ring gasket extending only to the inner edge of the boltholes is used), the bolting may not be stronger than carbon steel per ASTM A307 Grade B. When 300-lb steel flanges are bolted to 250-lb cast-iron flanges, no preference exists as to the removal of the raised face. However, when the raised face is not removed, carbon steel bolting not stronger than ASTM A307 Grade B should be used.

Selection and Limitations of Bolting. Carbon-steel machine bolts may be used to make flange connections for bolt metal temperatures from −20 to 500 F inclusive. This restriction is quite conservative with regard to the pressure limit. Also, these bolts can be used quite safely to the limits of the 300-lb pressure class as permitted in ASA B16.5.

The most widely used bolting materials in refineries are ASTM A193 Grade B7 stud bolts with ASTM A194 Grade 2H heavy semifinished hexagonal nuts. These materials are acceptable from −50 to 1000 F. (The temperature limits for bolting materials are given in Appendix C of Code ASA B31.3 in the Bolt Design Stress Tables 304.5.1A and B.) Many refiners use these materials almost exclusively to simplify inventories and reduce the possibility of misapplication of carbon-steel bolting.

Selection and Limitations of Strainers. The material for the strainer body (including bolting) should be equal to material specified for the valves in the same service. The screen material generally should be the same as the valve trim, e.g., 11 to 13 per cent chrome or Type 316 stainless steel for most oil services.

The location of permanent strainers (as contrasted to the temporary cone type which is installed at a flanged joint) also merits attention.

Centrifugal and reciprocating pumps handling material containing solids should have permanent strainers provided in the suction lines to the pump or in the vessel from which the pump takes suction. The free area of such strainers should be not less than three times the cross-sectional area of the suction line. The maximum clear opening in strainers should not exceed $\frac{1}{2}$ in. square or round or 75 per cent of the smallest impeller passage, whichever is less.

In addition, permanent strainers or filters should be provided in the piping for the protection of the equipment shown in Table 11. The maximum clear opening for screens in these strainers varies with the application, but it should not exceed the value recommended for the particular type of equipment. The available pressure differential usually determines the minimum clear opening for screens.

Permanent strainers should have baskets which can be flushed clean during operation or easily removed for cleaning. If considerable clogging of strainers is anticipated due to coke or similar conditions, strainers should be of the self-cleaning or the duplex type to permit continual flow of clean liquid.

Table II. Recommended Strainer Screen Openings

Type of equipment	Recommended max. clear opening
Screw, gear, or cam-type rotary pumps	$\frac{1}{4}''$
Steam turbines (if not provided in the turbine)	$\frac{1}{8}''$
Sealing, gland and flushing-oil system to pumps and compressors	$\frac{1}{16}''$
Fuel-oil lines to burners	$\frac{1}{8}''$
Air supply to pneumatically actuated equipment	35 mesh
Upstream of restriction orifices in bleed services	$\frac{1}{16}''$
Energizing fluids to ejectors	$\frac{1}{4}''$
Tubular coolers and condensers using unfiltered cooling water, except cooling-tower and spray-pond water........	$\frac{1}{4}''$

SELECTION AND LIMITATIONS OF PIPING JOINTS

The type of piping joint used must be suitable for the pressure-temperature conditions and should be selected by giving consideration to joint tightness and mechanical strength under the service conditions (including thermal expansion), and to the nature of the fluid handled with respect to corrosion, erosion, flammability, and toxicity.

In a refinery, welded piping (with a few flanged joints) is used almost exclusively for hydrocarbons and other flammable fluids. Within the process unit limits, welded construction is used for piping that contains process streams during normal operation of the processing unit or will contain process streams during operation of standby or spare equipment, including bypass piping, alternate process connections, and auxiliary piping systems such as gland oil, seal oil, lubricating oil, fuel gas, fuel oil, heating or cooling oil, flushing oil, flue gas, blowdown piping, and the like. Welded construction is also used for all piping outside process unit limits which is used for the transfer of hydrocarbons.

Piping which is threaded or welded, depending primarily on economic considerations, includes piping (other than specifically mentioned above) in services such as drains, vents, pumpouts, sample connections, and certain instrument leads, which contain process streams or hydrocarbon streams only upon intermittent or occasional use of the piping involved and which are not an integral or essential part of the process system or hydrocarbon transfer system.

Welded Joints. The Piping Code permits welded joints in all instances in which it is possible to qualify welding procedures, welders, and welding operators in conformance with the rules given in Chap. V of the Code. There are, however, a few minor additional considerations for seal-welded and socket-welded joints. For example, the Piping Code cautions against the use of socket-welded construction in those cases where severe crevice corrosion or erosion occurs. The Code also states that seal welds may be used to avoid joint leakage but that they shall not be considered as contributing any strength to the joint.

Flanged Joints. The number of flanged joints in a piping system is usually determined by maintenance and erection considerations. Because of improved

corrosion-measuring devices, the number of flanged joints required for corrosion measurement is constantly decreasing. It is, however, desirable to have sufficient flanged joints for insertion of blanks during shutdown and to permit dismantling of certain piping for cleaning or replacement.

Expanded Joints. This type of joint is more commonly used on the piping and tubes for refinery heaters but may be used where experience or tests have demonstrated that the joint is suitable for the conditions and where adequate provisions are made in the design to prevent separation of the joints.

Threaded Joints. The threading of pipe with a wall thickness less than ASA B36.10 Standard wall is not permitted, and the use of threaded joints where severe crevice corrosion or erosion may occur should be avoided. Regardless, economics limit the use of threaded piping to small pipe sizes (2 in. and smaller) for most services; however, it is used for most galvanized piping and the piping for chemicals which cause stress cracking (if the cost of necessary stress relief for welded piping cannot be justified).

All pipe threads on piping components must be taper pipe threads in accordance with ASA B2.1 except that:

(a) Pipe threads other than taper pipe threads may be used for piping components, where tightness of the joint depends upon a seating surface other than the threads, and where experience or tests have demonstrated that such threads are suitable for the conditions.

(b) Couplings, 2 inches and smaller, with straight-tapped pipe threads may be used on piping components with taper pipe threads, if the design conditions do not exceed 150 psig or 400 F and if the fluid handled is nonflammable and nontoxic.

The Code recommends the seal welding of threaded joints of materials which are weldable (except joints in which the tightness of the joint depends upon a seating surface other than the threads) in the following instances:

(a) hubs of screwed flanges on lines handling flammable or toxic fluids at pressures over 300 psig or temperature over 750 F.

(b) threaded joints in piping handling flammable or toxic fluids, which because of their nature are difficult to contain without seal welding.

The seal welding of threaded connections is good practice in piping which conveys abrasive solids suspended in gas or liquid.

Regardless, welded construction is preferred to seal-welding threaded connections, and if welding piping components are available, they should be used.

Flared, Flareless, and Compression Joints.* Piping joints using flared, flareless, or compression type tubing fittings may be used within the limitations of applicable Standards or Specifications listed in Table 326.1 and requirements (c) and (d) below. In the absence of such Standards or Specifications, the engineer shall determine that the type of fitting selected is adequate and safe for the design conditions in accordance with the following requirements.

(a) The pressure design shall meet the requirements of Subdivision 304.7.

(b) A suitable quantity of the type and size of fitting to be used shall meet successful performance tests to determine the safety of the joint under simulated service conditions. When vibration, fatigue, cyclic conditions, low temperature, thermal expansion or hydraulic shock are anticipated, the applicable conditions shall be incorporated in the test.

(c) Fittings and their joints shall be suitable for the tubing with which they are to be used with consideration to minimum tubing wall thickness and method of assembly recommended by the manufacturer.

(d) Fittings shall not be used in services which exceed the manufacturer's maximum pressure-temperature recommendations.

Calked Joints. The term "calked joints" applies to joints of the bell-and-spigot type which are permitted only for water service at pressures suitable for the pipe

* Paragraph 315, ASA B31.3.

to which they are applied. Provisions must be made to prevent disengagement of the joints at bends and dead ends and to support lateral reactions produced by branch connections or other causes.

Brazed and Soldered Joints. Fillet-brazed or fillet-soldered joints may not be used but the use of soldered socket-type and silver-brazed socket-type joints is permitted in nonflammable and nontoxic service under certain conditions. They are infrequently used in a refinery.

Sleeve-coupled and Other Proprietary Joints. Coupling-type, mechanical-gland-type, and other proprietary-type joints may be used provided adequate provision is made to prevent separation of the joints and provided a prototype joint has been subjected to performance tests to determine the safety of the joint under simulated service conditions. When vibration, fatigue, cyclic conditions, low temperature, thermal expansion, or hydraulic shock are anticipated, the applicable conditions must be incorporated in the tests.

FLEXIBILITY AND SUPPORT OF OIL-REFINERY PIPING

The general problem of flexibility and expansion of piping systems is treated in Chap. 4, and the support of piping systems is covered in Chap. 5. Techniques, methods, and procedures developed in those chapters are, in general, applicable to oil-refinery piping. The treatment given in this present section supplements that of Chaps. 4 and 5 and, in addition, emphasizes approaches that are identified more exactly with oil piping.

Piping systems must have sufficient flexibility to prevent thermal expansion from causing failure of piping or anchors from overstress or overstrain, leakage at joints, or excessive thrusts and moments on connected equipment. If substantial anchor or terminal movements are anticipated, these effects are to be considered analogous to terminal movements caused by thermal expansion.

On a refinery process unit, bends, loops, and offsets are commonly used to provide adequate flexibility in the piping. Expansion joints are seldom used unless pressure drop or space limitations make the use of loops, bends, or offsets impracticable. When used in flammable or toxic service, these joints should preferably be of the corrugated or bellows type and be equipped with limit rods, control rings, and liners. Reference 8 contains recommended specifications for the mechanical design and application of metallic packless expansion joints.

The following is taken directly from ASA B31.3. Parenthetical numbers in the paragraph headings are the section numbers in the Code.

Stress Range (319.2.1) *

As contrasted with stresses from sustained loads (such as internal pressure or weight), stresses caused by thermal expansion are permitted to attain sufficient initial magnitude to cause local yielding or creep. The attendant relaxation or reduction of stress in the hot condition leads to the creation of a stress of reverse sign when the component returns to the cold condition. This phenomenon is designated as self-springing of the line and is similar in effect to cold springing. The amount of self-springing depends on the initial magnitude of the expansion stress, the material, the temperature, and the elapsed time. While the expansion stress in the hot condition tends to diminish with time, the algebraic difference of the expansion stresses in the hot and cold conditions during any one cycle remains substantially constant. This algebraic difference, referred to as the stress range,† is the determining factor in designing piping systems for adequate flexibility.

---

* Here and elsewhere in this chapter, parenthetical numbers in paragraph headings refer to the actual numbered Code paragraphs.

† See Eq. (1).

Cold Spring (319.2.2)

Cold spring is recognized as beneficial in that it serves to balance hot and cold stresses without drawing on the ductility of the material, for which reason it is recommended in particular for materials of relatively low ductility; in addition, it helps assure minimum departure from as-erected hanger settings. Inasmuch as the life of a system under cyclic conditions depends primarily on the stress range rather than the stress level at any one time, no credit for cold spring is given for stress range calculations. In calculating end thrusts and moments where actual reactions at any one time rather than their range are considered significant, cold spring is credited.

Local Overstrain (319.2.3)

All the commonly used methods of piping flexibility analysis assume elastic behavior of the entire piping system. This assumption is sufficiently accurate for systems in which plastic strain occurs at many points or over relatively wide regions. It does not reflect the actual strain distribution in unbalanced systems in which only a small portion of the piping undergoes plastic strain, or in which the strain distribution is very uneven (such as may occur in piping operating in the creep range). In these cases, the weaker or higher stressed portions will be subjected to strain concentrations due to elastic follow-up of the stiffer or lower stressed portions. Unbalance can be produced by any of the following:

(a) by use of small pipe runs in series with large or stiffer pipe, with the small lines relatively highly stressed.

(b) by local reduction in size or cross-section, or local use of a weaker material.

(c) in a system of uniform size, by use of a line configuration for which the thrust line is situated close to the major portion of the line with only a very small portion projecting away from it absorbing most of the expansion strain.

Conditions of these types should preferably be avoided, particularly where materials of relatively low ductility are used; if unavoidable, they should be mitigated by the judicious application of cold spring.

Thermal-expansion Values. The thermal-expansion values used in computing the stress range and reaction in the piping (except as noted in the paragraph below) is determined from Tables 12 and 13 as the algebraic difference between the unit expansion shown for the maximum and the minimum metal temperatures normally expected during operation or shutdown.

The thermal-expansion values used in computing the reactions (forces and moments) on strain-sensitive equipment, such as pumps or turbines, are to be taken as the algebraic difference between the unit expansion shown for the metal temperature expected during installation and the maximum (or minimum) metal temperature expected during operation.

There are two occasionally encountered situations which should not be overlooked in determining the thermal expansion. They are (1) piping attached to a vessel which is to be field stress-relieved or the piping which itself may be stress-relieved in the field and (2) piping which operates at relatively cool temperatures and is part of a system which is steam purged prior to startups.

Other Physical Constants. The cold and hot moduli of elasticity E_c and E_h, respectively, are listed in Table 14. Stresses from expansion are to be computed on the use of the cold modulus E_c.

Poisson's ratio may be taken as 0.3 at all temperatures for all metals, except that more accurate data may be used if available. In the case of welded pipe, the longitudinal weld joint factor may be disregarded.

Nominal thickness and outside diameter of pipe and fittings shall be used in flexibility calculations.

Calculations of expansion stresses should take into account the applicable stress-intensification factor for any component other than plain straight pipe, and credit may be taken for the extra flexibility of such a component. In the absence of more directly applicable data, the flexibility and stress-intensification factor shown in Table 15 may be used. For piping components or attachments not

covered in the table (such as valves, strainers, anchor rings, or bands), suitable stress-intensification factors may be established by comparison of their significant geometry with that of the components shown.

Stress-intensification factors may be omitted from calculations on certain ferritic steel piping configurations (e.g., long runs of piping outside process unit limits) if the system will not be subject to more than 2,000 stress cycles during its expected life.

Methods of Flexibility Analysis. Approximate or simplified methods of analysis may be applied only if they are used for the range of configurations for which their adequate accuracy has been demonstrated. The simplified tabular methods given in Refs. 9 and 10 may be used without modification.

Also, the simplified tabular methods given in Ref. 11 can be used. They are, however, based upon the service-temperature modulus of elasticity. Therefore, it is necessary that reactions and stresses be multiplied by the modulus of elasticity at room temperature and divided by the modulus of elasticity at service temperature. Where these tables are used for selecting an expansion bend (pages 34 to 45 of Ref. 11), the value of M_R shall be taken as 1.0 in the equations to find the required amount of tabular deflection (page 30 of Ref. 11). The expansion bend on page 42 of Ref. 11 should not be used.

Acceptable* comprehensive methods of analysis are: analytical, model test, and chart methods which provide an evaluation of the forces, moments and stresses caused by bending and torsion from the simultaneous consideration of terminal and intermediate restraints to thermal expansion of the entire piping system under consideration and including all external movements transmitted under thermal change to the piping by its terminal and intermediate attachments. Correction factors must be applied for the stress intensification of curved pipe and branch connections, as provided by the details of these rules, and may be applied for the increased flexibility of such component parts.

Basic Assumptions and Requirements (319.4.2)

In calculating the flexibility of a piping system between anchor points, the system between anchor points shall be treated as a whole. The significance of all parts of the line and of all restraints such as supports or guides, including intermediate restraints introduced for the purpose of reducing moments and forces on equipment or small branch lines, shall be considered.

The total thermal expansion as determined from Par. 319.3.1† shall be used in all calculations, whether or not cold springing is used. Not only the expansion of the line itself, but also linear and angular movements of the equipment to which it is attached shall be considered.

Stresses from expansion shall be computed using the cold modulus of elasticity, E_c, as follows:

$$S_E = \sqrt{S_b{}^2 + 4S_t{}^2} = \frac{\sqrt{(iM_b)^2 + M_t{}^2}}{Z}$$

where S_E = computed expansion stress range, psi
$S_b = iM_b/Z$ = resultant bending stress (due to thermal load), psi
$S_t = M_t/2Z$ = torsional stress, psi
$M_b = \sqrt{M_{bp}{}^2 + M_{bt}{}^2}$ = resultant bending moment, inch-lb
M_{bp} = bending moment in plane of member (for members having significant orientation, such as elbows, or tees), inch-lb
M_{bt} = bending moment transverse to plane of member, inch-lb
M_t = torsional moment, inch-lb
Z = section modulus of pipe, in.³
i = stress intensification factor (from Table 319.3.6)‡

* Excerpted from paragraph 319.4.1 of ASA B31.3.
† See Tables 12 and 13 and associated text explanation.
‡ Reproduced here as Table 15.

Table 12. Thermal Expansion Data *

(Linear thermal expansion between 70 F and indicated temperature, in./100 ft)

Temperature, F	Carbon steel carbon-moly low-chrome (through 3 Cr Mo)	5 Cr Mo through 9 Cr Mo	Austenitic stainless steels 18 Cr 8 Ni	12 Cr 17 Cr 27 Cr	25 Cr 20 Ni	Monel 67 Ni 30 Cu	3½ Nickel	Aluminum	Gray cast iron	Bronze	Brass	Wrought iron	70 Cu 30 Ni
−325	−2.37	−2.22	−3.85	−2.04	−3.00	−2.62	−2.22	−4.68	....	−3.98	−3.88	−2.70	−3.15
−300	−2.24	−2.10	−3.63	−1.92	−2.83	−2.50	−2.10	−4.46	....	−3.74	−3.64	−2.55	−2.87
−275	−2.11	−1.98	−3.41	−1.80	−2.66	−2.38	−1.98	−4.17	....	−3.50	−3.40	−2.40	−2.70
−250	−1.98	−1.86	−3.19	−1.68	−2.49	−2.26	−1.86	−3.97	....	−3.26	−3.16	−2.25	−2.53
−225	−1.85	−1.74	−2.96	−1.57	−2.32	−2.14	−1.74	−3.71	....	−3.02	−2.93	−2.10	−2.36
−200	−1.71	−1.62	−2.73	−1.46	−2.15	−2.02	−1.62	−3.44	....	−2.78	−2.70	−1.95	−2.19
−175	−1.58	−1.50	−2.50	−1.35	−1.98	−1.90	−1.50	−3.16	....	−2.54	−2.47	−1.81	−2.12
−150	−1.45	−1.37	−2.27	−1.24	−1.81	−1.79	−1.38	−2.88	....	−2.31	−2.24	−1.67	−1.95
−125	−1.30	−1.23	−2.01	−1.11	−1.60	−1.59	−1.23	−2.57	....	−2.06	−2.00	−1.49	−1.74
−100	−1.15	−1.08	−1.75	−0.98	−1.39	−1.38	−1.08	−2.27	....	−1.81	−1.76	−1.31	−1.53
−75	−1.00	−0.94	−1.50	−0.85	−1.18	−1.18	−0.93	−1.97	....	−1.56	−1.52	−1.13	−1.33
−50	−0.84	−0.79	−1.24	−0.72	−0.98	−0.98	−0.78	−1.67	....	−1.32	−1.29	−0.96	−1.13
−25	−0.68	−0.63	−0.98	−0.57	−0.78	−0.77	−0.62	−1.32	....	−1.25	−1.02	−0.76	−0.89
0	−0.49	−0.46	−0.72	−0.42	−0.57	−0.57	−0.46	−0.97	....	−0.77	−0.75	−0.56	−0.66
25	−0.32	−0.30	−0.46	−0.27	−0.37	−0.37	−0.30	−0.63	....	−0.49	−0.48	−0.36	−0.42
50	−0.14	−0.13	−0.21	−0.12	−0.16	−0.20	−0.14	−0.28	....	−0.22	−0.21	−0.16	−0.19
70	0	0	0	0	0	0	0	0	0	0	0	0	0
100	0.23	0.22	0.34	0.20	0.28	0.28	0.22	0.46	0.21	0.36	0.35	0.26	0.31
125	0.42	0.40	0.62	0.36	0.51	0.52	0.40	0.85	0.38	0.66	0.64	0.48	0.56
150	0.61	0.58	0.90	0.53	0.74	0.75	0.58	1.23	0.55	0.96	0.94	0.70	0.82
175	0.80	0.76	1.18	0.69	0.98	0.99	0.76	1.62	0.73	1.26	1.23	0.92	1.07
200	0.99	0.94	1.46	0.86	1.21	1.22	0.94	2.00	0.90	1.56	1.52	1.14	1.33
225	1.21	1.13	1.75	1.03	1.45	1.46	1.13	2.41	1.08	1.86	1.83	1.37	1.59
250	1.40	1.33	2.03	1.21	1.70	1.71	1.32	2.83	1.27	2.17	2.14	1.60	1.86
275	1.61	1.52	2.32	1.38	1.94	1.96	1.51	3.24	1.45	2.48	2.45	1.83	2.13
300	1.82	1.71	2.61	1.56	2.18	2.21	1.69	3.67	1.64	2.79	2.76	2.06	2.40
325	2.04	1.90	2.90	1.74	2.43	2.44	1.88	4.09	1.83	3.11	3.08	2.29	2.68
350	2.26	2.10	3.20	1.93	2.69	2.68	2.08	4.52	2.03	3.42	3.41	2.53	2.96
375	2.48	2.30	3.50	2.11	2.94	2.91	2.27	4.95	2.22	3.74	3.73	2.77	3.24
400	2.70	2.50	3.80	2.30	3.20	3.25	2.47	5.39	2.42	4.05	4.05	3.01	3.52
425	2.93	2.72	4.10	2.50	3.46	3.52	2.69	5.83	2.62	4.37	4.38	3.25	
450	3.16	2.93	4.41	2.69	3.72	3.79	2.91	6.28	2.83	4.69	4.72	3.50	
475	3.39	3.14	4.71	2.89	3.98	4.06	3.13	6.72	3.03	5.01	5.06	3.74	

550	4.11	3.80	5.62	3.49	4.79	4.90	3.80	8.10	3.67	5.98	6.10	4.50
575	4.35	4.02	5.93	3.69	5.06	5.18	4.03	8.56	3.89	6.31	6.45	4.76
600	4.60	4.24	6.24	3.90	5.33	5.46	4.27	9.03	4.11	6.64	6.80	5.01
625	4.86	4.47	6.55	4.10	5.60	5.75	4.51	......	4.34	6.96	7.16	5.27
650	5.11	4.69	6.87	4.31	5.88	6.05	4.75	......	4.57	7.29	7.53	5.53
675	5.37	4.92	7.18	4.52	6.16	6.34	4.99	......	4.80	7.62	7.89	5.80
700	5.63	5.14	7.50	4.73	6.44	6.64	5.24	......	5.03	7.95	8.26	6.06
725	5.90	5.38	7.82	4.94	6.73	6.94	5.50	......	5.26	8.28	8.64	6.32
750	6.16	5.62	8.15	5.16	7.02	7.25	5.76	......	5.50	8.62	9.02	6.59
775	6.43	5.86	8.47	5.38	7.31	7.55	6.02	......	5.74	8.96	9.40	6.85
800	6.70	6.10	8.80	5.60	7.60	7.85	6.27	......	5.98	9.30	9.78	7.12
825	6.97	6.34	9.13	5.82	7.89	8.16	6.54	......	6.22	9.64	10.17	7.40
850	7.25	6.59	9.46	6.05	8.19	8.48	6.81	......	6.47	9.99	10.57	7.69
875	7.53	6.83	9.79	6.27	8.48	8.80	7.08	......	6.72	10.33	10.96	7.97
900	7.81	7.07	10.12	6.49	8.78	9.12	7.35	......	6.97	10.68	11.35	8.26
925	8.08	7.31	10.46	6.71	9.07	9.44	7.72	......	7.23	11.02	11.75	8.53
950	8.35	7.56	10·80	6.94	9.37	9.77	8.09	......	7.50	11.37	12.16	8.81
975	8.62	7.81	11.14	7.17	9.66	10.09	8.46	......	7.76	11.71	12.57	9.08
1000	8.89	8.06	11.48	7.40	9.95	10.42	8.83	......	8.02	12.05	12.98	9.36
1025	9.17	8.30	11.82	7.62	10.24	10.75	8.98	......	....	12.40	13.39	
1050	9.46	8.55	12.16	7.95	10.54	11.09	9.14	......	....	12.76	13.81	
1075	9.75	8.80	12.50	8.18	10.83	11.43	9.29	......	....	13.11	14.23	
1100	10.04	9.05	12.84	8.31	11.12	11.77	9.45	......	....	13.47	14.65	
1125	10.31	9.28	13.18	8.53	11.41	12.11	9.78					
1150	10.57	9.52	13.52	8.76	11.71	12.47	10.11					
1175	10.83	9.76	13.86	8.98	12.01	12.81	10.44					
1200	11.10	10.00	14.20	9.20	12.31	13.15	10.78					
1225	11.38	10.26	14.54	9.42	12.59	13.50						
1250	11.66	10.53	14.88	9.65	12.88	13.86						
1275	11.94	10.79	15.22	9.88	13.17	14.22						
1300	12.22	11.06	15.56	10.11	13.46	14.58						
1325	12.50	11.30	15.90	10.33	13.75	14.94						
1350	12.78	11.55	16.24	10.56	14.05	15.30						
1375	13.06	11.80	16.58	10.78	14.35	15.66						
1400	13.34	12.05	16.92	11.01	14.65	16.02						
1425	......	......	17.30									
1450	......	......	17.69									
1475	......	......	18.08									
1500	......	......	18.47									

* These data are for information, and it is not to be implied that materials are suitable for all the temperatures shown. Data are taken from Table 319.3.1A of the Code for Pressure Piping, ASA B31.3-1966.

Table 13. Thermal Expansion Data *

[Mean coefficient of thermal expansion between 70 F and indicated temperature, in./(in. F) $\times 10^6$]

Temperature, F	Carbon steel carbon-moly low-chrome (through 3 Cr Mo)	5 Cr Mo through 9 Cr Mo	Austenitic stainless steels 18 Cr 8 Ni	12 Cr 17 Cr 27 Cr	25 Cr 20 Ni	Monel 67 Ni 30 Cu	3½ Nickel	Aluminum	Gray cast iron	Bronze	Brass	Wrought iron	70 Cu 30 Ni
−325	5.00	4.70	8.15	4.30	6.35	5.55	4.70	9.90	.....	8.40	8.20	5.70	6.65
−300	5.07	4.77	8.21	4.36	6.42	5.72	4.77	10.04	.....	8.45	8.24	5.79	6.76
−275	5.14	4.84	8.28	4.41	6.49	5.89	4.84	10.18	.....	8.50	8.29	5.87	6.86
−250	5.21	4.91	8.34	4.47	6.56	6.06	4.91	10.33	.....	8.55	8.33	5.96	6.97
−225	5.28	4.98	8.41	4.53	6.63	6.23	4.98	10.47	.....	8.60	8.37	6.04	7.08
−200	5.35	5.05	8.47	4.59	6.70	6.40	5.05	10.61	.....	8.65	8.41	6.13	7.19
−175	5.42	5.12	8.54	4.64	6.77	6.57	5.12	10.76	.....	8.70	8.46	6.21	7.29
−150	5.50	5.20	8.60	4.70	6.85	6.75	5.20	10.90	.....	8.75	8.50	6.30	7.40
−125	5.57	5.26	8.66	4.78	6.94	6.85	5.26	11.08	.....	8.85	8.61	6.39	7.50
−100	5.65	5.32	8.75	4.85	7.03	6.95	5.32	11.25	.....	8.95	8.73	6.48	7.60
−75	5.72	5.38	8.83	4.93	7.11	7.05	5.38	11.43	.....	9.05	8.84	6.56	7.70
−50	5.80	5.45	8.90	5.00	7.20	7.15	5.45	11.60	.....	9.15	8.95	6.65	7.80
−25	5.85	5.51	8.94	5.05	7.26	7.22	5.51	11.73	.....	9.23	9.03	6.71	7.87
0	5.90	5.56	8.98	5.10	7.31	7.28	5.57	11.86	.....	9.32	9.11	6.78	7.94
25	5.96	5.62	9.03	5.14	7.37	7.35	5.63	11.99	.....	9.40	9.18	6.84	8.02
50	6.01	5.67	9.07	5.19	7.42	7.41	5.69	12.12	.....	9.49	9.26	6.91	8.09
70	6.07	5.73	9.11	5.24	7.48	7.48	5.73	12.25	.....	9.57	9.34	6.97	8.16
100	6.13	5.79	9.16	5.29	7.54	7.55	5.80	12.39	.....	9.66	9.42	7.04	8.24
125	6.19	5.85	9.20	5.34	7.59	7.62	5.86	12.53	.....	9.75	9.51	7.11	8.31
150	6.25	5.92	9.25	5.40	7.65	7.70	5.92	12.67	.....	9.85	9.59	7.18	8.39
175	6.31	5.98	9.29	5.45	7.70	7.77	5.98	12.81	.....	9.93	9.68	7.25	8.46
200	6.38	6.04	9.34	5.50	7.76	7.84	6.05	12.95	5.75	10.03	9.76	7.32	8.54
225	6.43	6.08	9.37	5.54	7.80	7.89	6.07	13.03	5.80	10.05	9.82	7.36	8.58
250	6.49	6.12	9.41	5.58	7.84	7.93	6.09	13.12	5.84	10.08	9.88	7.40	8.63
275	6.54	6.15	9.44	5.62	7.88	7.98	6.11	13.20	5.89	10.10	9.94	7.44	8.67
300	6.60	6.19	9.47	5.66	7.92	8.02	6.14	13.28	5.93	10.12	10.00	7.48	8.71
325	6.65	6.23	9.50	5.70	7.96	8.07	6.16	13.36	5.97	10.15	10.06	7.51	8.76
350	6.71	6.27	9.53	5.74	8.00	8.11	6.19	13.44	6.02	10.18	10.11	7.55	8.81
375	6.76	6.30	9.56	5.77	8.04	8.16	6.21	13.52	6.06	10.20	10.17	7.58	8.85
400	6.82	6.34	9.59	5.81	8.08	8.20	6.23	13.60	6.10	10.23	10.23	7.61	8.90
425	6.87	6.38	9.62	5.85	8.12	8.25	6.29	13.68	6.15	10.25	10.29	7.64	
450	6.92	6.42	9.65	5.89	8.15	8.30	6.35	13.75	6.19	10.28	10.35	7.67	
475	6.97	6.46	9.67	5.92	8.19	8.35	6.41	13.83	6.24	10.30	10.41	7.70	

Temp												
500	7.02	6.50	9.70	5.96	8.22	8.40	6.48	13.90	6.28	10.32	10.47	7.73
525	7.07	6.54	9.73	6.00	8.26	8.45	6.54	13.98	6.33	10.35	10.53	7.77
550	7.12	6.58	9.76	6.05	8.30	8.49	6.60	14.05	6.38	10.38	10.58	7.81
575	7.17	6.62	9.79	6.09	8.34	8.54	6.66	14.13	6.42	10.41	10.64	7.84
600	7.23	6.66	9.82	6.13	8.38	8.58	6.72	14.20	6.47	10.44	10.69	7.88
625	7.28	6.70	9.85	6.16	8.42	8.63	6.77	....	6.52	10.46	10.75	7.91
650	7.33	6.73	9.87	6.20	8.45	8.68	6.83	....	6.56	10.48	10.81	7.95
675	7.38	6.77	9.90	6.23	8.49	8.73	6.88	....	6.61	10.50	10.86	7.98
700	7.44	6.80	9.92	6.26	8.52	8.78	6.94		6.65	10.52	10.92	8.01
725	7.49	6.84	9.95	6.29	8.56	8.83	6.99		6.70	10.55	10.98	8.04
750	7.54	6.88	9.99	6.33	8.60	8.87	7.05		6.74	10.57	11.04	8.07
775	7.59	6.92	10.02	6.36	8.64	8.92	7.10		6.79	10.60	11.10	8.10
800	7.65	6.96	10.05	6.39	8.68	8.96	7.16		6.83	10.62	11.16	8.13
825	7.70	7.00	10.08	6.42	8.71	9.01	7.21		6.87	10.65	11.22	8.17
850	7.75	7.03	10.11	6.46	8.75	9.06	7.27		6.92	10.67	11.28	8.21
875	7.79	7.07	10.13	6.49	8.78	9.11	7.32		6.96	10.70	11.34	8.25
900	7.84	7.10	10.16	6.52	8.81	9.16	7.38		7.00	10.72	11.40	8.29
925	7.87	7.13	10.19	6.55	8.84	9.21	7.41		7.05	10.74	11.46	8.32
950	7.91	7.16	10.23	6.58	8.87	9.25	7.45		7.10	10.76	11.52	8.34
975	7.94	7.19	10.26	6.60	8.89	9.30	7.48		7.14	10.78	11.57	8.37
1000	7.97	7.22	10.29	6.63	8.92	9.34	7.52		7.19	10.80	11.63	8.39
1025	8.01	7.25	10.32	6.65	8.94	9.39	7.56			10.83	11.69	
1050	8.05	7.27	10.34	6.68	8.96	9.43	7.60			10.85	11.74	
1075	8.08	7.30	10.37	6.70	8.98	9.48	7.64			10.88	11.80	
1100	8.12	7.32	10.39	6.72	9.00	9.52	7.67			10.90	11.85	
1125	8.14	7.34	10.41	6.74	9.02	9.57	7.74			10.93	11.91	
1150	8.16	7.37	10.44	6.75	9.04	9.61	7.81			10.95	11.97	
1175	8.17	7.39	10.46	6.77	9.06	9.66	7.88			10.98	12.03	
1200	8.19	7.41	10.48	6.78	9.08	9.70	7.95			11.00	12.09	
1225	8.21	7.43	10.50	6.80	9.09	9.75						
1250	8.24	7.45	10.51	6.82	9.10	9.79						
1275	8.26	7.47	10.53	6.83	9.11	9.84						
1300	8.28	7.49	10.54	6.85	9.12	9.88						
1325	8.30	7.51	10.56	6.86	9.14	9.92						
1350	8.32	7.52	10.57	6.88	9.15	9.96						
1375	8.34	7.54	10.59	6.89	9.17	10.00						
1400	8.36	7.55	10.60	6.90	9.18	10.04						
1425	....	....	10.64									
1450	....	....	10.68									
1475	....	....	10.72									
1500	....	....	10.77									

* These data are for information, and it is not to be implied that materials are suitable for all the temperatures shown. Data are taken from Table 319.3.1B of the Code for Pressure Piping, ASA B31.3-1966.

Table 14. Modulus of Elasticity*

E = Modulus of Elasticity, psi (multiply tabulated values by 10^6)

Material	Temperature, F																	
	−325	−200	−100	70	200	300	400	500	600	700	800	900	1000	1100	1200	1300	1400	1500
Carbon steels with carbon content 0.30% or less, 3½ Ni	30.0	29.5	29.0	27.9	27.7	27.4	27.0	26.4	25.7	24.8	23.4	18.5	15.4	13.0				
Carbon steels with carbon content above 0.30%	31.0	30.6	30.4	29.9	29.5	29.0	28.3	27.4	26.7	25.4	23.8	21.5	18.8	15.0	11.2			
Carbon-moly steels, low-chrome steels through 3 Cr Mo	31.0	30.6	30.4	29.9	29.5	29.0	28.6	28.0	27.4	26.6	25.7	24.5	23.0	20.4	15.6			
Intermediate chrome steels (5 Cr Mo through −9 Cr Mo)	29.4	28.5	28.1	27.4	27.1	26.8	26.4	26.0	25.4	24.9	24.2	23.5	22.8	21.9	20.8	19.5	18.1	
Austenitic steels (TP 304, 310, 316, 321, 347)	30.4	29.9	29.4	28.3	27.7	27.1	26.6	26.1	25.4	24.8	24.1	23.4	22.7	22.0	21.3	20.7	19.3	17.9
Straight chromium steels (12 Cr, 17 Cr, 27 Cr)	30.8	30.3	29.8	29.2	28.7	28.3	27.7	27.0	26.0	24.8	23.1	21.1	18.6	15.6	12.2			
Wrought iron	30.6	30.2	30.0	29.5	28.6	28.2	27.7	27.0	26.5	25.8	23.0							
Gray cast iron	....	....	....	13.4	13.2	12.9	12.6	12.2	11.7	11.0	10.2							
Monel (67 Ni-30 Cu) and (66 Ni-29 Cu-Al)	26.8	26.6	26.4	26.0	26.0	25.8	25.6	25.4	24.7	23.1	21.0	18.6	16.0	14.3	13.0			
Copper-Nickel (80 Cu-20 Ni) and (70 Cu-30 Ni)	20.5	20.0	19.5	18.9	18.4	18.0	17.6	17.2	16.7	16.2	15.3							
Aluminum	11.3	11.1	10.9	10.6	10.4	10.2	9.5	8.5										
Copper (99.98% Cu)	17.0	16.7	16.5	16.0	15.6	15.4	15.1	14.7	14.2	13.7								
Commercial brass (66 Cu-34 Zn)	15.0	14.7	14.5	14.0	13.7	13.5	13.0	12.7	12.2	11.8								
Leaded tin bronze (88 Cu-6 Sn-1.5 Pb-4.5 Zn)	14.2	13.8	13.5	13.0	12.7	12.4	12.0	11.7	11.3	10.9								

* These data are for information, and it is not to be implied that materials are suitable for all the temperatures shown. Data are taken from Table 319.3.2 of the Code for Pressure Piping, ASA B31.3-1966.

Reactions. The reactions (forces and moments) R_h and R_c in the hot and cold conditions, respectively, are calculated by Eqs. (5) and (6) which are based on assuming an equal percentage of cold spring in all planes.

$$R_h = (1 - \tfrac{2}{3}C)(E_h/E_c)R \tag{5}$$

$$R_c = CR \text{ or } C_1 R, \text{ whichever is greater} \tag{6}$$

where C = cold-spring factor varying from zero for no cold spring to one for 100 per cent cold spring. [Factor $\tfrac{2}{3}$ appearing in Eq. (5) accounts for observation that specified cold spring cannot be fully assured during erection, even with elaborate precautions.]

$C_1 = 1 - (S_h/S_E)(E_c/E_h)$, estimated self-spring or relaxation factor; use zero if value becomes negative

E_c = modulus of elasticity in the cold condition, psi

E_h = modulus of elasticity in the hot condition, psi

R = range of reaction forces or moments derived from flexibility calculations corresponding to the full expansion range (based on E_c), lb or in.-lb

R_c and R_h = maximum reaction forces or moments estimated to occur in the cold and hot conditions, respectively, lb or in.-lb

S_E = maximum computed expansion stress range, psi

Piping reactions against pumps, turbines, and compressors should not exceed equipment manufacturers' recommendations. Where the manufacturers make no recommendations, every effort should be made to reduce the piping reactions to a practical minimum. In such cases, it is safe practice to limit reactions to 20 per cent of the weight of the fluid end or 2,000 lb, whichever is less, and to limit the numerical value of the combined torque and moment (in foot-pounds) to three times the smaller of the above values.[12]

Loads for Pipe-supporting Elements. The design of equipment for supporting or restraining piping systems or components thereof must be based on all the concurrently acting loads being transmitted into the supporting equipment. These loads include, in addition to weight effects, loads which are introduced by the service pressures and temperatures, vibration, wind, earthquake, shocks, and thermal expansion and contraction.

For piping containing gas or vapor, weight calculations need not include the weight of water or other liquid if the possibility of these lines containing liquid is remote and provided the lines are not subjected to hydrostatic tests at initial construction or subsequent inspections.

Design of Pipe-supporting Elements. The Code states that the layout and design of pipe-supporting elements (and other piping components) shall be made with due regard to preventing any of the following:

(*a*) Piping stresses in excess of those permitted in the Code, including the stresses in the pipe-supporting elements. (Exceptions: The Code allowable stresses are not applicable to springs. Also, the longitudinal weld joint factor E need not be applied to welding components which are used as pipe supporting elements.)

(*b*) Leakage at joints.

(*c*) Excessive thrusts and moments on connected equipment (such as pumps and turbines). (Note: Piping entering pressure vessels should be supported from brackets attached to the vessel if expansion in the vertical direction of either the vessel or the pipe would cause excessive stresses in the vessel nozzle area.)

(*d*) Resonance with imposed vibrations.

(*e*) Excessive interference with the thermal expansion and contraction of a piping system which is otherwise adequately flexible.

Table 15. Flexibility Factor k and Stress-intensification Factor i *

Description	Flexibility factor k	Stress-int. factor i	Flexibility characteristic h
Welding elbow, or pipe bend[a-c]	$1.65/h$	$0.9/h^{2/3}$	$\overline{T}R_1/(r_2)^2$
Closely spaced miter bend $s < r_2(1 + \tan \theta)$[a-c]	$1.52/h^{5/6}$	$0.9/h^{2/3}$	$\cot \theta \overline{T}s/2(r_2)^2$
Widely spaced miter bend, $s \geq r_2(1 + \tan \theta)$[a,b,d]	$1.52/h^{5/6}$	$0.9/h^{2/3}$	$\dfrac{1 + \cot \theta}{2} \dfrac{\overline{T}}{r_2}$
Welding tee, per ASA B16.9[a,b,f] . .	1	$0.9/h^{2/3}$	$4.4(\overline{T}/r_2)$
Reinforced fabricated tee, with pad or saddle[a,b,f,h]	1	$0.9/h^{2/3}$	$\dfrac{(\overline{T} + \frac{1}{2}t_e)^{5/2}}{\overline{T}^{3/2}r_2}$
Unreinforced fabricated tee[a,b,f] . . .	1	$0.9/h^{2/3}$	$\overline{T}/r_2$
Butt-welded joint, reducer, or welding neck flange	1	1.0	
Double welded slip-on flange	1	1.2	
Fillet welded joint (single welded), socket welded flange, or single welded slip-on flange . .	1	1.3	
Lap joint flange (with ASA B16.9 lap joint stub)	1	1.6	
Screwed pipe joint, or screwed flange	1	2.3	
Corrugated straight pipe, or corrugated or creased bend[e]	5	2.5	

* Reproduced from Table 319.3.6 of the Code for Pressure Piping, ASA B31.3-1966.

[a] For fittings and miter bends the flexibility factors k and stress-intensification factors i in the table apply to bending in any plane and shall not be less than unity; factors for torsion equal unity. Both factors apply over the effective arc length (shown by heavy center lines in the sketches) for curved and miter elbows and to the intersection point for tees.

[b] The values of k and i can be read directly from Chart A above by entering with the characteristic h computed from the formulas given where:

r_2 = mean radius of matching pipe, in.	R_1 = bend radius of welding elbow or pipe bend, in.
$\overline{T}$ = for elbows and miter bends, nominal wall thickness of fitting (see note g), in.	θ = one-half angle between adjacent miter axes
$\overline{T}$ = for tees, the nominal wall thickness of the matching pipe, in.	s = miter spacing at center line t_e = pad or saddle thickness, in.

[c] Where flanges are attached to one or both ends, the values of k and i in the table shall be corrected by the factors C_1 given below, which can be read directly from Chart B, entering with the computed h:

One end flanged: $h^{1/6}$ Both ends flanged: $h^{1/3}$

[d] Also includes single-miter joint.

[e] Factors shown apply to bending; flexibility factor for torsion equals 0.9.

[f] The stress-intensification factors i in the table were obtained from tests on full-size outlet connections. For less than full-size outlets, the full-size values should be used until more applicable values are developed.

[g] The engineer is cautioned that cast butt-welding elbows may have considerably heavier walls than that of the pipe with which they are used. Large errors may be introduced unless the effect of these greater thicknesses is considered.

[h] When t_e is $>1\frac{1}{2}\,\overline{T}$, use $h = 4.05(\overline{T}/r_2)$.

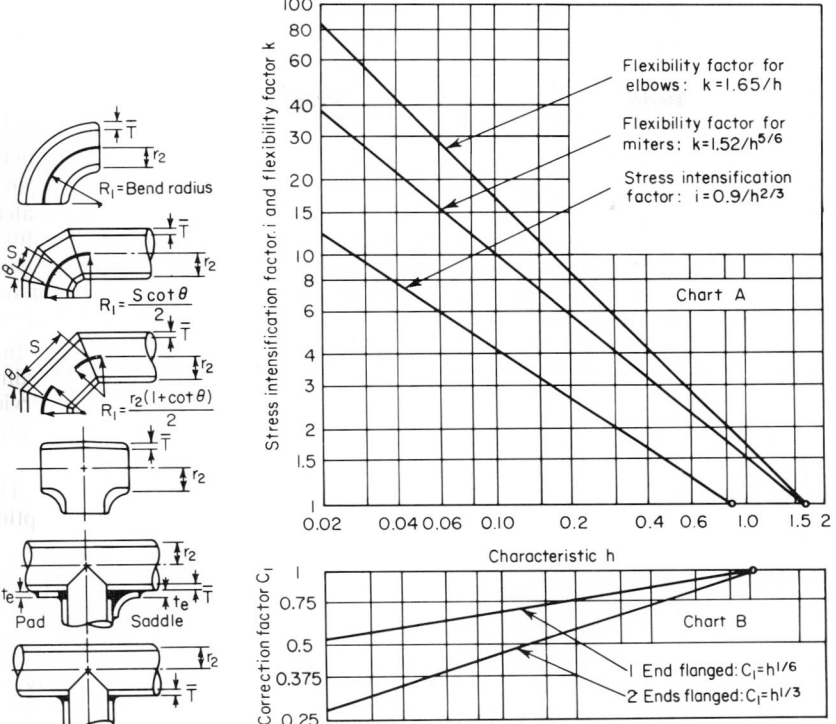

Chart A

Flexibility factor for elbows: $k = 1.65/h$

Flexibility factor for miters: $k = 1.52/h^{5/6}$

Stress intensification factor: $i = 0.9/h^{2/3}$

Chart B

1 End flanged: $C_1 = h^{1/6}$

2 Ends flanged: $C_1 = h^{1/3}$

R_1 = Bend radius

$R_1 = \dfrac{S \cot \theta}{2}$

$R_1 = \dfrac{r_2(1 + \cot \theta)}{2}$

Pad   Saddle

(*f*) Unintentional disengagement of the piping from its supports.

(*g*) Excessive piping sag in systems requiring drainage slope.

Support Spacing. In general the location and design of pipe-supporting elements and supports may be based on simple calculations and engineering judgment. However, if design pressure or temperature is high, if piping layout is unusual, or if precise knowledge of reactions at supports or equipment is required, simple calculations may be insufficient and more refined analysis will be necessary to evaluate stresses and reactions. These calculations should include consideration of the flexibility of the lines under the action of the total applied loads. Also, the stress-intensification and flexibility factors shown in Table 15 should be used in these analyses.

As a general guide, supports should be spaced so as to prevent the stresses in a pipeline (in the corroded condition), due to bending and shear, from exceeding one-half of the basic allowable stress values for the pipe material at the service temperature. Also, it is good practice to limit sag or deflection to 1 in. or one-quarter of the nominal diameter of the pipe, whichever is less.

Support spacing should be calculated using appropriate beam formulas. The following formulas 7 to 10 for stress and deflection are based on the assumption that any concentrated loads are located at the center of the span.

Type of support	Max bending stress $S_w{}^*$		Max deflection y	
Single span, free ends†	$\dfrac{(0.75WL_s{}^2 + 1.5W_fL_s)D_0}{I}$	(7)	$\dfrac{22.5WL_s{}^4 + 36W_fL_s{}^3}{EI}$	(8)
Continuous straight line	$\dfrac{(0.5WL_s{}^2 + 0.75W_fL_s)D_o}{I}$	(9)	$\dfrac{4.5WL_s{}^4 + 9W_fL_s{}^3}{EI}$	(10)

* For single spans with free ends, shear stress is zero at the center of the span where bending stresses are a maximum. Combined shear and bending stresses at any other point will not exceed the maximum bending stress. For continuous lines, the point of maximum stress is over the support where both shear and bending are maximum. However, the shear stresses generally will be insignificant and may be ignored.

† Spans adjacent to changes in direction (lateral or vertical) of more than 30 deg should be considered as single spans with free ends.

Nomenclature for Eqs. (7) to (10)

D_o = OD of pipe, in.

E = modulus of elasticity at service temperatures, psi (Table 14)

I = moment of inertia of pipe, in.^{4} (based on the nominal pipe wall thickness less the corrosion allowance and the mill tolerance)

L_s = length of span, ft

S_w = maximum bending stress due to dead load, psi

$W = W_p + W_w + W_c$, lb/ft of pipe

W_p = weight of pipe, lb/ft

W_w = weight of water, or fluid being piped, whichever is heavier, lb/ft

W_c = weight of covering, lb/ft of pipe

W_f = weight of concentrated load at center of span, lb

y = maximum deflection, in.

Table 16, with Figs. 5 to 8 gives suggested spacing of supports for piping. The spacings were calculated using the formulas, stress limits, and deflection limits above and were based on the following assumptions:

1. Pipe is seamless carbon steel to ASTM A53 Grade A or equal.
2. Corrosion allowance = 0.10 in.
3. Lines are full of water.

Table 16. Key to Support Spacing Figs. 5 to 8

Pipe size and thickness		Continuous straight line		Single span	
		Figure	Curve	Figure	Curve
½″	Sch 80 (XS)	5	..	8	*AA*
	Sch 160	5	..	7	..
	XXS	5	..	7	..
¾″	Sch 80 (XS)	6	*A*	8	*BB*
	Sch 160	5	..	7	..
	XXS	5	..	7	..
1″	Sch 80 (XS)	6	*B*	8	*CC*
	Sch 160	5	..	7	..
	XXS	5	..	7	..
1½″	Sch 80 (XS)	6	*C*	8	*DD*
	Sch 160	5	..	7	..
	XXS	5	..	7	..
2″	Sch 80 (XS)	6	*D*	8	*EE*
	Sch 160	6	*E*	7	..
	XXS	6	*E*	7	..
2½″	Sch 40 (STD)	6	*D*	8	*EE*
	Sch 80 (XS)	6	*F*	8	*FF*
	Sch 160	6	*G*	7	..
	XXS	6	*G*	7	..
3″	Sch 40 (STD)	6	*E*	8	*FF*
	Sch 80 (XS)	6	*G*	8	*GG*
	Sch 160	6	*H*	7	..
	XXS	6	*H*	7	..
4″	Sch 40 (STD)	6	*G*	8	*GG*
	Sch 80 (XS)	6	*H*	8	*HH*
	Sch 120	6	*J*	8	*JJ*
	Sch 160	6	*K*	8	*JJ*
	XXS	6	*K*	8	*JJ*
6″	Sch 40 (STD)	6	*H*	8	*JJ*
	Sch 80 (XS)	6	*L*	8	*KK*
	Sch 120	6	*M*	8	*KK*
	Sch 160	6	*N*	8	*LL*
	XXS	6	*N*	8	*LL*
8″	Sch 30	6	*K*	8	*KK*
	Sch 40 (STD)	6	*L*	8	*LL*
	Sch 60	6	*M*	8	*MM*
	Sch 80 (XS)	6	*N*	8	*NN*
	Sch 100	6	*P*	8	*NN*
	Sch 120	6	*Q*	8	*PP*
	Sch 140	6	*Q*	8	*PP*
	XXS	6	*R*	8	*PP*
	Sch 160	6	*R*	8	*PP*
10″	Sch 30	6	*L*	8	*MM*
	Sch 40 (STD)	6	*M*	8	*NN*
	Sch 60 (XS)	6	*P*	8	*QQ*
	Sch 80	6	*R*	8	*RR*
12″	Sch 30	6	*M*	8	*NN*
	STD	6	*N*	8	*QQ*
	Sch 40	6	*P*	8	*RR*
	XS	6	*R*	8	*SS*

Table 16.   *(Continued)*

Pipe size and thickness		Continuous straight line		Single span	
		Figure	Curve	Figure	Curve
14″	Sch 20	6	*M*	8	*NN*
	Sch 30 (STD)	6	*P*	8	*RR*
	Sch 40	6	*Q*	8	*SS*
	XS	6	*S*	8	*TT*
16″	Sch 20	6	*M*	8	*PP*
	Sch 30 (STD)	6	*P*	8	*RR*
	Sch 40 (XS)	6	*T*	8	*UU*
	Sch 60	6	*V*	8	*VV*
	Sch 80	6	*W*	8	*WW*
18″*	Sch 20	6	*N*	8	*PP*
	STD	6	*Q*	8	*SS*
	Sch 30	6	*S*	8	*UU*
	XS	6	*T*	8	*VV*
20″*	Sch 20 (STD)	6	*Q*	8	*SS*
	Sch 30 (XS)	6	*U*	8	*WW*
24″*	Sch 20 (STD)	6	*R*	8	*TT*
	XS	6	*V*	8	*XX*

* Maximum spans; 18-in. Sch 20, 35ft; 20-in. Sch 20, 49 ft; 24-in. Sch 20, 37 ft; 30-in STD, 22 ft; 30-in XS, 50 ft.

4. There are no concentrated loads (e.g., valves, flanges, etc.) between supports.

5. All piping joints are welded.

6. Lines whose contents exceed 150 F are insulated.

Where concentrated loads, higher temperatures, or vibration effects are involved, special consideration should be given to effects of bending and shear stresses. Special consideration should always be given to the possibility of the flattening of 18-in. and larger liquid-filled lines over their supports.[13]

It has been calculated that water-filled lines equipped with the sliding supports shown in Fig. 9 will not buckle over the supports at the spans shown in Figs. 5 to 8 except that the sizes and thicknesses listed below Table 16 should not span distances greater than shown unless a reinforcing pad or saddle is welded to the pipe. The dimensions of the pad or saddle should be calculated for the intended span.

Other considerations in determining the location of supports are:

1. Supports should be placed as near as practicable to changes in direction (lateral or vertical).

2. If possible, small-diameter lines should be routed to accompany larger lines so that they may be supported by or suspended from the larger diameter lines to avoid excessively close spacing of supports. Care should be taken to see that stress and deflection limits are not exceeded and that provision is made for any differential expansion that might exist.

3. Supports should be provided for piping sections which require frequent dismantling for maintenance, such as for the installation of blanks, in order to maintain proper alignment.

4. Piping that discharges to the atmosphere should be firmly anchored to counteract the reaction force of the discharging fluid.

Materials for Supports.   All equipment for permanent supports and restraints should be fabricated from materials suitable for the service conditions. Unless

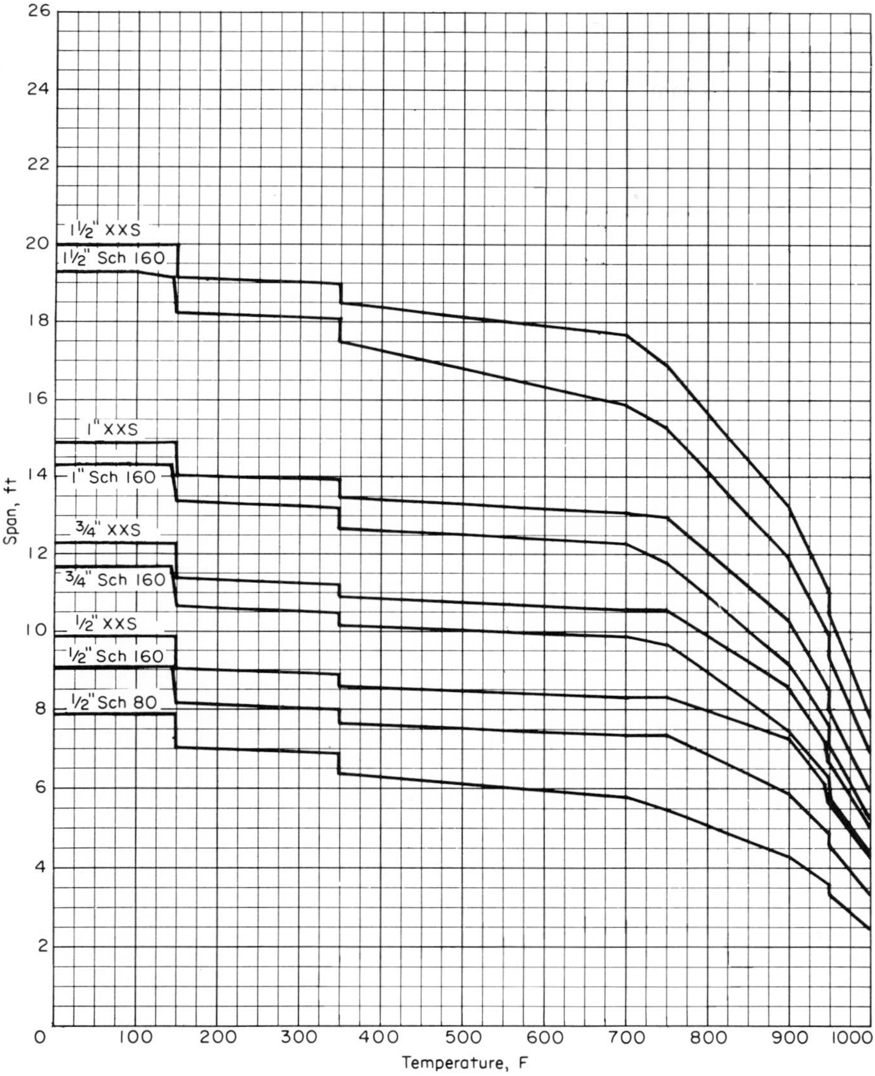

FIG. 5. Support spacing-continuous straight line.   1½-in. pipe size and smaller, except Schedule 80 thickness for ¾- to 1½-in. pipe sizes.   This figure to be used with Table 16.

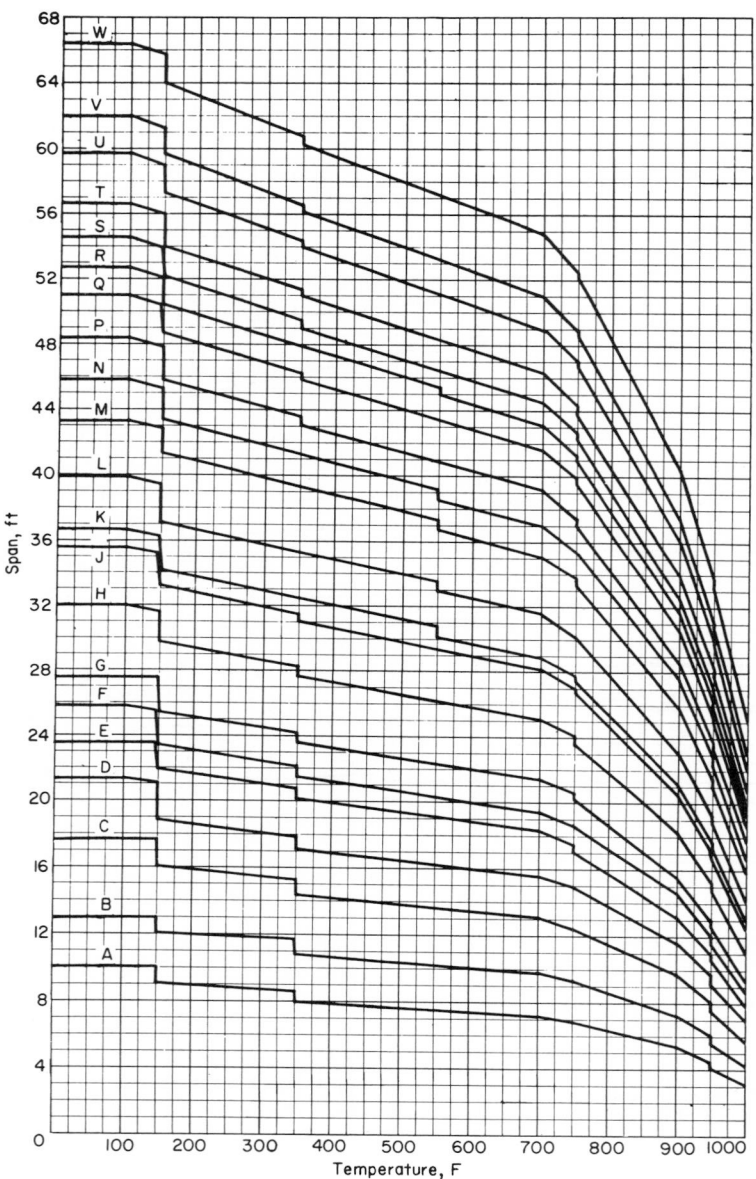

FIG. 6. Support spacing-continuous straight line. 2-in. pipe size and larger, plus Schedule 80 thicknesses for ¾- to 1½-in. pipe sizes. This figure to be used with Table 16.

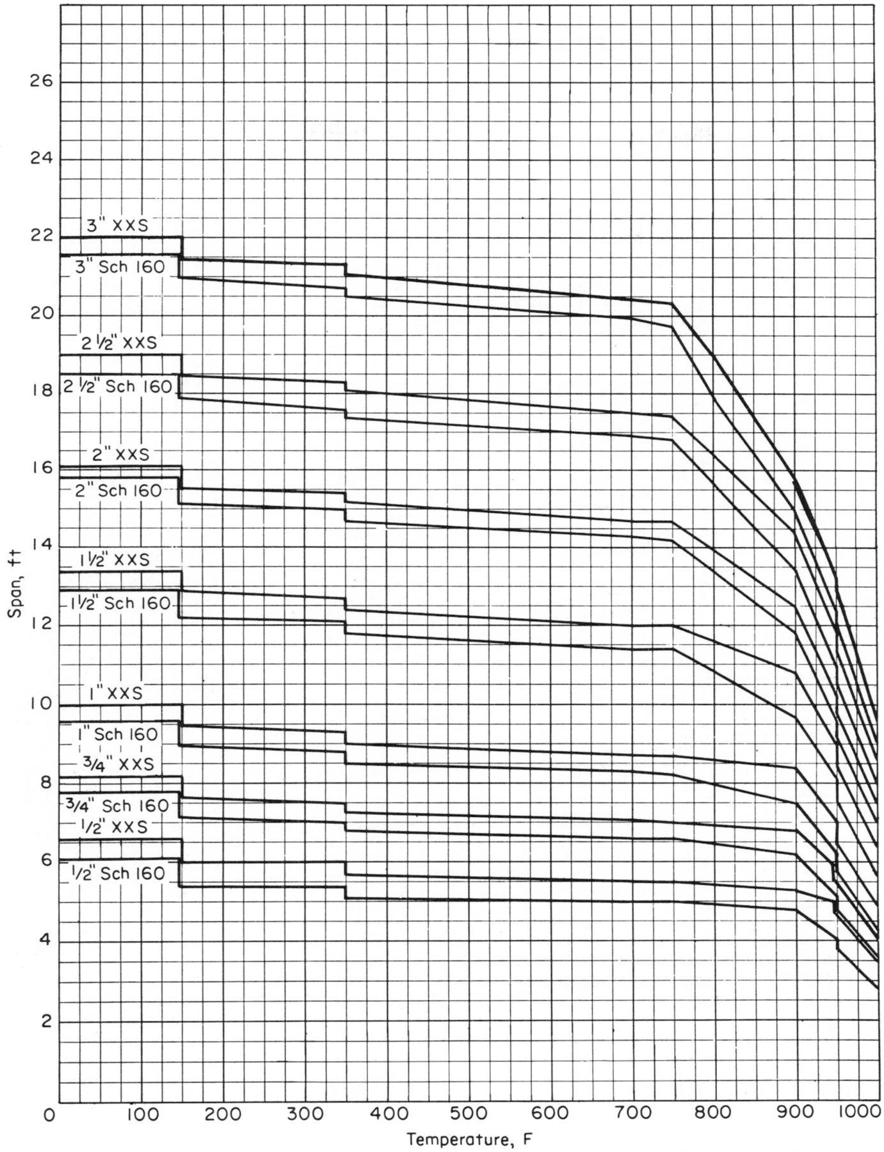

FIG. 7. Support spacing-single span. 3-in. pipe size and smaller Schedule 160, XXS thicknesses. This figure to be used with Table 16.

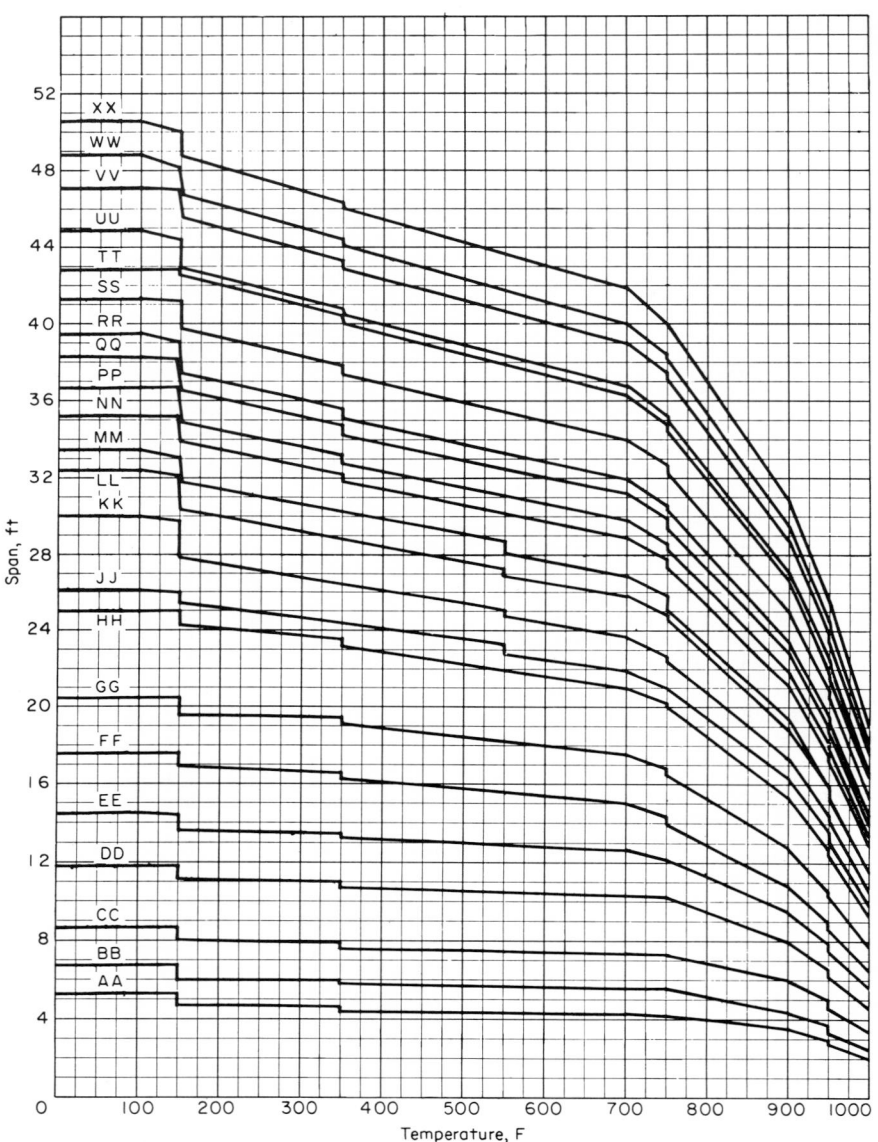

FIG. 8. Support spacing-single span.   ½-in. pipe size and larger, except Schedule 160 and XXS thicknesses for ½- to 3-in. pipe sizes.   This figure to be used with Table 16.

otherwise permitted, steel, wrought iron, or metal similar to the piping should be used.

Cast iron may be used for roller bases, rollers, anchor bases, brackets, and parts of pipe-supporting elements upon which the loading will be mainly that of compression. Cast iron is not recommended for pipe-supporting elements subject to impact-type loadings resulting from pulsation or vibration. Malleable or nodular iron may be used for pipe clamps, beam clamps, hanger flanges, clips, bases, and swivel rings.

Wood may be used for pipe-supporting elements which are primarily in compression when the metal temperatures are below ambient temperature.

Extreme caution should be exercised in the design of supports to assure that the lines, bare or insulated, do not rest on concrete, so as to avoid accelerated corrosion of the piping at the point of support. This can be accomplished in several ways, e.g.;

1. A structural tee with its stem welded to the crossarm can be attached to each stanchion crossarm. (A structural-steel angle is an acceptable alternate.) The tee should be designed to carry the loads imposed by the piping and should protrude at least 1 in. above any possible fireproofing.

2. Where the piping or sliding support is to bear directly on existing fireproofing, a bearing plate rigidly fixed to the stanchion crossarm can be provided between the pipe or sliding support and the fireproofing.

Anchors and Guides. Supporting elements used as anchor points should be designed to maintain the anchor points in relatively fixed positions.

To protect terminal equipment or other (weaker) portions of the system, restraints (such as anchors and guides) must be provided where necessary to control movement or to direct expansion into those portions of the system which are adequate to absorb them. Also, the layout and design of the guides should be adequate to assure true axial compression or extension of any expansion joint. In addition to the other thermal forces and moments, the effects of friction in other supports of the system should be considered in the design of such anchors and guides.

Anchors for expansion joints such as those of the corrugated, omega, disk, or slip type should be designed to withstand the algebraic sum of the forces at the maximum pressure and temperature at which the joint is to be used. These forces are:

1. Pressure thrust, which is the product of the effective thrust area times the maximum pressure to which the joint shall be subjected during normal operation. (For slip joints the effective thrust area should be computed using the outside diameter of the pipe. For corrugated, omega-, or disk-type joints the effective thrust area should be that area recommended by the joint manufacturer. When this information is unobtainable, the effective areas should be computed using the maximum inside diameter of the expansion joint bellows.)

2. The force required to compress or extend the joint in an amount equal to the calculated expansion movement.

3. The force required to overcome the static friction of the pipe in expanding or contracting on its supports from installed to operating position. The length of pipe considered shall be that located between the anchor and the expansion joint.

Other Inextensible Types of Supports. 1. Hangers. The resting type of support is usually preferred to the hanging type. In either case the supports should be designed to permit the free movement of piping caused by thermal contraction and expansion. When necessary to use the hanging-type support, the following criteria should be observed: (*a*) The safe loads for threaded hanger rods should be based on the root area of the threads. (*b*) Pipe, chains, straps, or bars of strength and effective area equal to the equivalent hanger road may be used instead of hanger rods. (*c*) Due to atmospheric corrosion in many refineries, it is good practice to make hanger rods no less than $\frac{1}{2}$ in. in diameter. Straps should not be

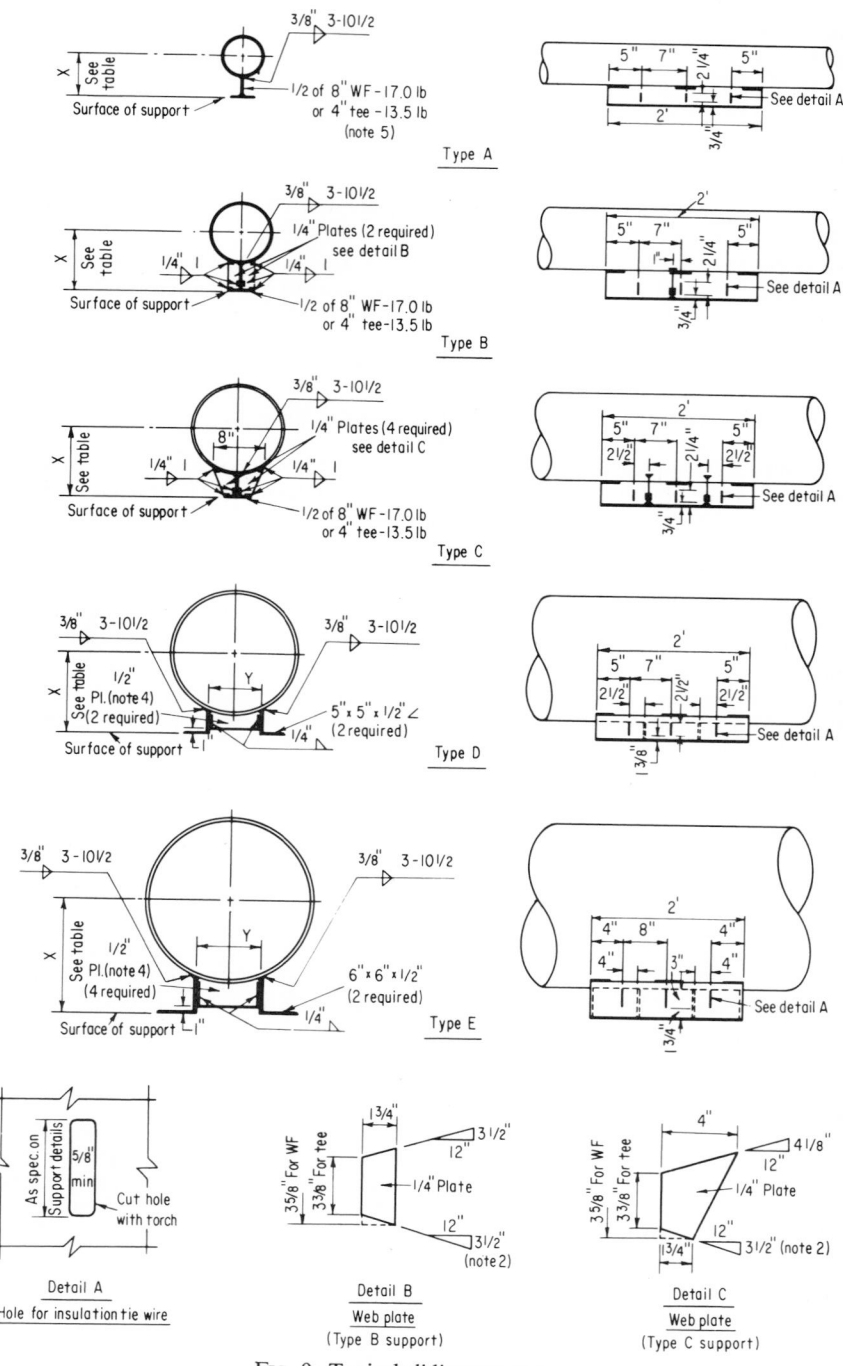

FIG. 9. Typical sliding supports.

less than $\frac{1}{8}$ in. thick by $\frac{3}{4}$ in. wide for supporting pipe 1-in. nominal size and smaller or less than $\frac{1}{2}$ in. thick by 1 in. wide for supporting pipe $1\frac{1}{2}$-in. nominal size and larger.

2. Sliding Supports. Sliding supports (or shoes) and brackets must be designed to resist the forces due to friction in addition to the loads imposed by bearing. The dimensions of the support should be sufficient to permit the expected movement of the supported piping. For details of typical sliding supports, see Fig. 9.

Insulated piping 2 in. and larger being supported by resting type of supports may rest directly on the steel supporting members if it operates at temperatures between 32 and 500 F (the insulation being cut away so that steel rests against steel). Piping 2 in. and larger operating at temperatures above 500 F should have steel sliding supports (e.g., as shown in Fig. 9) whose bearing surfaces are outside the insulation.

Insulated piping $1\frac{1}{2}$ in. and smaller between 32 and 500 F should have the insulation cut away so that the pipe rests directly on steel supports. Above 500 F the insulation (locally reinforced with a circumferential band of 20 gage sheet metal, 1 ft long) should bear directly on the supporting members.

Insulated piping in low-temperature service (less than 32 F) should have the insulation (locally reinforced with a circumferential band of 20 gage or heavier sheet metal, 1 ft long) bear directly on the supporting members except where there is a concentrated load at or near the support or where anchors or guides are necessary. For these latter cases, oak blocks sealed to prevent moisture infiltration should be substituted for the insulation.

Resilient Types of Supports. Each support for a line should carry approximately the same portion of the line weight. When it is anticipated that a pipeline will deflect vertically owing to thermal expansion or contraction and thereby unload some supports and overload others, spring hangers should be provided.

Spring-type supports should be designed to exert a supporting force, at the point of attachment to the pipe, equal to the load as determined by weight-balance calculations. They must be provided with means to prevent misalignment, buckling, or eccentric load of the springs and to prevent unintentional disengagement of the load. It is desirable that all spring hangers be provided with position indicators.

Constant-support spring hangers provide a substantially uniform supporting force throughout the range of travel. The use of this type of spring hanger is advantageous at locations subject to appreciable movement (i.e., more than $\frac{1}{2}$ in.

NOTES FOR FIG. 9

Table of Sliding Support Types and Heights

Pipe size, in.	$1\frac{1}{2}$ and less	2	$2\frac{1}{2}$	3	4	6	8	10	12	14	16	18	20	24	30
Type of support	See Note 2	A	A	A	A	A	B	B	C	C	C	D	D	E	E
Dimension X, in. ...	...	$5\frac{3}{16}$	$5\frac{7}{16}$	$5\frac{3}{4}$	$6\frac{1}{4}$	$7\frac{5}{16}$	$8\frac{5}{16}$	$9\frac{3}{8}$	$10\frac{3}{8}$	11	12	13	14	16	19
Dimension Y, in. ...	...	...	...	...	...	...	...	...	...	..	..	$8\frac{1}{4}$	$8\frac{3}{4}$	$13\frac{1}{4}$	$14\frac{7}{8}$

NOTES:
  1. Sliding supports should be positioned with respect to the stanchion so that expansion or contraction will not cause the support to slide off the stanchion.
  2. The slope of the bottom edge of the web plate should be as indicated for supports made up from structural "tees." For supports using half of a WF section the bottom edge of the web plate shall be horizontal.
  3. Supports are suitable for insulation thicknesses 4 in. and thinner.
  4. The plates between the angles in type D and E supports should be cut to fit the curvature of the pipe.
  5. It is permissible to use a 3-in. tee or one-half of 6 in. I-12.5 lb for type A supports where the insulation thickness does not exceed 3 in. and where it is not desirable to keep the bottom of pipe elevation the same for all sizes.

of vertical deflection) with thermal changes. Hangers of this type should be selected so that their displacement range exceeds expected movements.

Variable spring hangers are usually used where the vertical deflection is $\frac{1}{2}$ in. or less. Any excess or deficiency in the supporting effect of variable spring hangers caused by changes in thermal conditions should be included in the determination of total stresses in the pipeline. Variation in supporting effect should not exceed 25 per cent for the total vertical travel between the hot and cold positions.

Counterweights may be used as a substitute for spring hangers. When used, counterweights should be provided with stops to limit travel and the weights should be positively secured.

An arrangement utilizing a hydraulic cylinder may be installed to give a constant supporting effort. Safety devices and stops must be provided to support the load in case of hydraulic failure.

Structural Attachments. The load from piping, pipe-supporting elements, restraints, and braces must be suitably transmitted to other piping, pressure vessels, buildings, platforms, support structures, or foundations.

Nonintegral attachments, in which the reaction between the piping and the attachment is by contact, include clamps, slings, cradles, U bolts, saddles, straps, and clevises. When clamps are connected to vertical lines to support their weight, it is recommended that shear lugs be welded to the pipe or that the clamps be located below fittings or flanges to minimize slippage.

It is good practice for the bolted clamps used in connection with rod hangers to have a minimum thickness of $\frac{3}{16}$ in. for pipe of 1-in. nominal size and smaller and a minimum thickness of $\frac{1}{4}$ in. for pipe of 1½-in. nominal size and larger. Bolts used with these clamps should have a minimum diameter of $\frac{3}{8}$ in.

Integral attachments include lugs, ears, shoes, plates, and angle clips. The material for integral attachments should be of good weldable quality. Preheating, welding, and postheating should be in accordance with Chap. V of the Code. Consideration should be given to the localized stresses induced into the piping component by the integral attachment.

PIPING SYSTEMS DESIGN REQUIREMENTS

The great variety of environmental factors existing in a refinery makes it difficult to codify requirements pertaining to specific piping systems. The vast amount of industry experience, however, does make it possible to tabulate general design practice which, although not always universally applicable, will generally serve as guidelines to the design engineer. Many of these criteria appear on the following pages.

Common Refinery Hazards. The proper design of refinery equipment is essential to the safe operation of a refinery.[14] Some of the common hazards that have caused serious problems are discussed briefly below. These hazards can exist in most kinds of refinery equipment, but because piping is the largest category of refinery equipment, they are most significant in piping design.

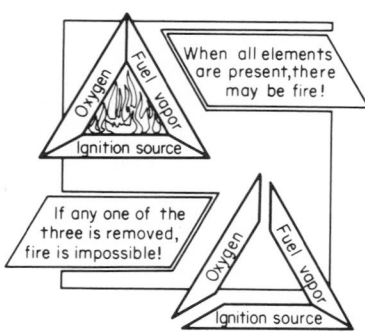

FIG. 10. The fire triangle. (*Courtesy of American Oil Company.*)

1. *Mixing Air or Oxygen with Hydrocarbons.* Under certain conditions vapors from most petroleum products will burn and may even explode. This burning process is frequently described by the use of the fire triangle illustrated in Fig. 10. A fire can occur when all three sides of the triangle are present. Therefore, safe operation demands the elimination of one of these elements in every process that cannot be controlled rigidly. Since the fuel vapor side of the triangle is always assumed to be present in an oil refinery, and since a source of ignition is usually present, air or oxygen must be eliminated in most cases.[15]

2. *Mixing Hydrocarbons with Reactive Chemicals.* Certain hydrocarbon-chemical mixtures (e.g., chlorine-hydrocarbon mixtures) can be as dangerous as hydrocarbon-air mixtures.

3. *Contact of Water and Hot Oil.* At atmospheric pressure, a volume of water expands about 1,600 times when it flashes to steam. Gradual heating of water is normally harmless because the steam-generation rate is slow. If a quantity of water is heated rapidly, however, as it is when injected into hot oil or hot equipment, all of it will immediately flash to steam. The sudden formation of large quantities of steam in process units, particularly in vacuum systems, causes a pressure surge that is usually very damaging to equipment.[16]

4. *Disposal of Waste Gases.* Waste gases should never be discharged directly into furnace fireboxes. Simple venting to fireboxes through open-end ducts or existing burners creates serious explosion hazards because the waste gas may continue to flow after furnace shutdown. Waste gas should go through a knockout drum equipped with a liquid seal before being introduced to the furnace through a separate burner that is used only for waste gas.

5. *Toxic or Reactive Chemicals.* Piping for toxic chemicals, such as chlorine gas, acids, caustic, and lead compounds (e.g., tetraethyl and tetramethyl), must be specifically designed to prevent leakage and spills.

6. *Handling of Light Ends.*[17] Skill and knowledge are required for the safe handling of all hydrocarbons, whether at the well head, in the refinery, or at the service station. However, the lower boiling hydrocarbons, known as light ends, have proved to be particularly difficult to handle. (Pure hydrocarbons or hydrocarbon mixtures having a Reid vapor pressure of at least 18 psi, such as butane, propane, ethane, methane, liquefied petroleum gas, natural gas, and fuel gas, are usually classified as light ends.) Most light ends will vaporize rapidly at room temperature and pressure. For this reason, they are more hazardous than heavier hydrocarbons if allowed to escape. The low viscosity of light ends also aggravates leakage and sealing problems. Similar problems are encountered with heavy hydrocarbons at high temperature because they act like light ends.

Piping Arrangement

One of the most important aspects of piping design is the arrangement (or layout) of piping. This aspect is not directly related to the ability of the piping to contain the fluid under design conditions but rather to efficient and safe operation of the plant; for this reason, it is not covered in the Piping Code.

A general rule in piping layout is that lines should be located in as neat and orderly a manner (in groups or banks whenever practicable) as is consistent with economical design, pressure-loss considerations, and satisfactory supporting arrangements. With the exception of water, drainage, and pumpout lines, the accepted practice on a process unit is to run the piping overhead, providing 7 ft or more of clear headroom over walkways and platforms. Piping on a unit should not be located at grade especially in areas where frequent personnel traffic is likely.

All piping equipment requiring regular attention of the operating and maintenance personnel should be readily accessible. Also, adequate, clear working spaces (minimum width 3 ft) should be maintained around equipment such as pumps, exchangers, control valves, instruments, tower manways, etc., which require frequent servicing. Consideration should be given to providing lateral (and vertical) clearance for the use of motorized materials-handling equipment in maintenance work. Where equipment is designed to be removed for repair or replacement, the piping should be arranged to allow removal of the equipment without removing the block valves adjacent to the equipment or large quantities of associated equipment. In addition, if practicable, all valves should be located so that they may be repacked, replaced, and operated from grade, permanent platforms, or small portable platforms. If the bottoms of the handwheels are more than 6 ft above a platform or grade level, or if otherwise inaccessible, the valves should be equipped with extension stems or chain operators (with chains extending to within 3 ft of the operating level but not located so that the chain hangs in a passageway or fouls other equipment).

Extension stems are preferred to chain operators for threaded-end valves. Actually, chain operators should be used on threaded-end valves only if the ends of the valve are seal welded or if the valve is in a vertical line. Also, the use of chain operators on 1-in. and smaller valves is undesirable because of the possibility of bending the valve yoke or stem.

Provisions should be made to elevate to the surface the handwheels of valves located below grade. Extension stems are generally used for this. However, such handwheels should not be located in walkways or aisles and should be placed sufficiently high to avoid being a stumbling hazard. Means should be provided to protect the stem threads of underground valves.

No oil or gas piping should be installed in switch rooms or in control rooms, washrooms, or similar locations where men congregate. Transmitters should be used to carry signals from oil and gas lines to instruments installed in such rooms. There should be no floor drains in control rooms or adjacent laboratories.

The many high-temperature lines in a refinery pose problems for other piping and equipment. For example, hot lines (100 F or higher) should be routed so as to avoid electrical conduits. Also, steam condensate should not be discharged into the ground in the vicinity of electrical conduits, because the electrical insulation on conductors may not be able to dissipate the additional heat and can eventually break down and thereby cause a short circuit. Care should be exercised so as not to route lines containing cold high-vapor-pressure fluids near hot lines or equipment. Heat from the hot lines or equipment may vaporize the cold fluid. Lines containing corrosive chemicals should not be located near hot lines or other sources of heat unless corrosion rates do not increase with temperature.

Pockets should be avoided whenever possible, especially in lines (1) where water can accumulate that will subsequently be flashed to steam when the unit is brought on stream, (2) carrying caustic or acid, (3) in which corrosive condensate may form, (4) carrying materials which may congeal or freeze, (5) which contain solids which may settle out. Besides the general configuration of the piping, pockets are formed by concentric reducers, valves in vertical lines, reduced-port valves, sag in lines, and orifice plates.

Drainage and Venting Systems. Process unit sewers should be gravity-type systems that normally receive surface runoff water, oily water streams, pump gland cooling water, wash water, and other streams containing oils and chemicals. Flammable mixtures are difficult to avoid in sewers, and it is almost impossible to eliminate all sources of ignition, such as static electricity and pyrophoric iron sulfide. Thus, gastight systems are essential. Some refineries use two systems in

addition to the sanitary sewer: (1) a "dry oil system" that handles relatively water-free oil drainage and (2) a water system that handles surface drainage and other relatively oil-free streams. Regardless, each process unit sewer should have a compartmented gas trap, located at or beyond the unit limits, with gastight manways and vents which discharge vapors at safe locations. Also, trapped or sealed drains should be provided for all paved areas on units except under floor-fired heaters.

The discharge of steam, hot condensate, or other hot material directly to the sewer system should be avoided because any oil in the sewer probably will be vaporized. For example, an atmospheric flash drum with quench-water injection for cooling should be used for disposal of large quantities of condensate.

Process piping should be equipped with drains and vents unless the piping may be drained or purged through attached equipment. Properly located vents and drains make it possible (1) to purge air or oxygen from the system on start-up so that there will be no mixing of hydrocarbons and oxygen, (2) to drain water from the system to minimize the hazard of mixing water with hot fluids and to prevent damage from freezing.

Valved drain and vent connections should be provided for most types of refinery equipment. These drains and vents should be located on the equipment, if practicable, but may be located in connected piping where there are no valves or blocks between the drain or vent connections and the equipment. Piping to these connections should be arranged so as to drain completely the equipment and the connected piping to the appropriate refinery drainage system if practicable. The alternate to complete drainage is a start-up procedure for water removal, such as (1) displacement by circulation, (2) gradual heating during start-up, (3) dry gas purging, or (4) high-velocity gas purging. Multistage pumps, furnace headers, control valves, and horizontal pipe that sags between supports are typical locations where it is usually impractical to provide complete drainage.

Generally, drain connections to closed drainage systems should be provided with double block valves and with a bleed connection or sight drain between the block valves. The sight drain should be drawn from the spool piece between the double blocks of the closed drain, and the sight drain should be provided with a single block valve blinded or plugged.

Drains, vents, and pump-outs for piping and equipment in vacuum service should be blinded or plugged during operation of the unit to prevent the entrance of air.

There are, however, drain and vent connections which need not be connected to a drainage system. Examples are connections used for draining a very small amount of liquid (e.g., from heat-exchanger bonnets), small connections which are not hazardous if left open, connections for checking water accumulations, and vessel vents which are not needed during operation. These drains and vents should be provided with a block valve and a pipe plug.

In lines containing phenol, caustic, acids, and other hazardous fluids, a drain should be provided between block and check valves where fluid could be trapped. Where the check and block valves separate a hazardous fluid from process piping or other process equipment, the block valve should be located between the check valve and the process piping or equipment.

Water drainage from vessels in light-ends service can be complicated by the refrigeration effect of light hydrocarbons that vaporize at atmospheric pressure. An ice plug, formed by this refrigeration effect, can prevent proper valve closure and hazardous vapors will be released when the ice melts if the valve is not sealed with a pipe plug. In most cases, steam tracing or other means of heating drain lines and valves will prevent freezing. Spring-loaded, self-closing valves may be used on water drain lines. As these valves close automatically unless the operator holds them open, they give reasonable assurance that drains will not be left open by freezing

or operator negligence. Where large quantities of water are drawn, an automatic system using an interface float arrangement may be desirable.

Means should be available for removing the operating liquid contents from all vessels and heat-exchange units and the connected piping. Although process lines and pumps should be used for this purpose as far as practicable, an auxiliary pumping-out system should be provided. (A permissible alternate is to use steam to remove the contents of equipment by pressure.) On pressure vessels, pumpout connections should be provided for side drawoffs as well as at the bottom of the vessel.

The recommended minimum size of pumpouts is $1\frac{1}{2}$ in. The recommended minimum size of drains and vents is $1\frac{1}{2}$ in. for furnace coils, 1 in. for vessels, and $\frac{3}{4}$ in. for all other equipment. However, the size of vent and drain connections should be such that the water used for hydrostatic test or for flushing may be drained off without pulling a vacuum. On some small pumps, compressors, and turbine and steam-engine drivers, $\frac{3}{4}$-in. or larger drain and vent connections are not economical. In such cases, $\frac{1}{2}$-in. drains and vents are acceptable.

Supplementary information on drains and vents is contained in Ref. 18.

Sample Connections. On a process unit, sample connections should be provided on all feed and product streams and on such intermediate streams as are necessary for control and testing. Sample piping should be as short as possible and be adequately braced to enable it to resist unexpected external loads and to protect it from damage when valves are operated. If the piping is carefully supported and anchored, it is permissible to use an equipment drain for sampling purposes. It is preferred, however, that connections directly on pumps, compressors, and other equipment subject to vibration not be used for sample connections if other connections, where samples might be taken, are available. Sample connections may be utilized for the installation of pressure gages. It is suggested that the minimum size of the first nipple attached to the piping or equipment from which the sample is taken be $\frac{3}{4}$ in. A block valve of the same size as the nipple should be installed at the end of the nipple. A second valve should be installed in the sample line as close to the sampling point as practicable. Sample coolers will sometimes be necessary to assure safe handling of the hydrocarbon being sampled.

Instrument Piping. Instrument piping, as used here, includes all piping and piping components used to connect instruments to other piping or equipment and the control piping used to connect air- or hydraulically operated control apparatus. It does not include instruments or permanently sealed fluid-filled tubing systems furnished with instruments as temperature-pressure responsive devices.

Instrument piping must meet all the applicable requirements of the associated principal piping system and the following:

1. The design pressure and temperature for instrument piping are to be determined with consideration of short-time conditions. If it presents a more severe condition, the temperature of the piping during periodic operation of the blowdown valve should be considered as a short-time condition.

2. Consideration must be given to the mechanical strength (including fatigue) of small instrument connections to piping or apparatus.

3. Instrument piping containing fluids which are normally static and subject to freezing must be protected by steam tracing or other heating methods.

4. When it is necessary to blow down or bleed instrument piping systems containing toxic fluids, consideration should be given to the safe disposal of such fluids.

Pump Piping. Pump suction piping should be arranged so that the flow is as smooth (nonturbulent) as practicable at the pump suction nozzle. To accomplish this, the use of tees, crosses, near run-size branch connections, and short-radius elbows should be avoided near the suction nozzle. Suction piping should not

contain pockets where gas may accumulate; however, if this is unavoidable, any such pockets should be vented.   If a reducer is required at the pump in a horizontal line, it should be eccentric and installed with the "belly" down for best operation of the pump; however, such an arrangement will probably not be justified if it makes an additional drain necessary.   If the fluid being pumped is near its vaporization temperature, it is recommended that the line be sloped downward toward the pump at all points.

Permanent strainers should be provided upstream of pumps handling stocks which are likely to contain foreign material such as sand, metal, and scale.   Temporary strainers, preferably of the cone type, should be provided for initial unit start-up where permanent strainers are not provided, and should be located as close as possible to the pump suction nozzle.   A block valve should be provided in the suction line of each pump and should preferably be located upstream of the strainer. For dirty stocks, where two or more pumps take suction from a single header, the block valves should be located as close as is practicable to the header to minimize the collection of dirt upstream of the valve.

For pumps which take suction from vacuum towers, it probably will be necessary to provide an equalizing line from the pump back to the vapor space in the tower to vent the pump at start-up.

A block valve should be provided in the discharge line of each pump.   A check valve should be installed in the discharge line of each centrifugal or rotary pump unless there is no possibility of a reversal of flow or pressure surge under any condition.   (Check valves normally are not needed in reciprocating-pump discharge lines.)   The check valve should be located between the pump and the block valve with a drain between the block and the check valves in lines containing highly corrosive or toxic fluids such as phenol, caustic, and acid.

When the discharge line contains a quick-closing valve, the necessity for shock-absorbing equipment should be investigated if the closing time of the valve cannot be increased to a safe level.   Where a remotely located valve can be closed against a pump and the pump does not shut off automatically or cannot be shut off immediately, a recirculating line should be provided from the pump discharge back to the point of suction.   The purpose of the recirculating line is to prevent damage to the pump due to overheating.   (If it is not practicable to route this recirculating back to the point of suction, it should be routed back to the suction line as far from the pump suction flange as is practicable.)   The minimum size of the recirculating line should be $3/4$ in.   The line should be equipped with at least one gate valve and an orifice sized to restrict flow to the minimum pumping rate of the pump.

Standby pumps which may be idle during plant operation and which have to start quickly should be provided with warm-up lines if pump design temperature exceeds 450 F, or if the process fluid will solidify at atmospheric temperature.   The purpose of these warm-up lines is to eliminate undesirable thermal effects on lines and equipment and plugging of idle pump and piping laterals.   A warm-up line should consist of a $3/4$-in. valved bypass around the pump discharge block and check valves.   The standby pump should be kept at operating temperature by opening the warm-up line valve and cracking the suction block valve(s) to permit a small flow back through the idle pumps.   The installation of warm-up lines should result in minimum-length pockets in the discharge and suction lines.

If the process fluid will solidify at atmospheric temperature and the suction and discharge lines are not heated and insulated, an additional $3/4$-in. valved bypass should be provided from the discharge lateral to the suction lateral on the header side of the valves.   When the pump is removed from service, these laterals should be kept at operating temperature by opening this bypass valve to permit a small flow.   The above warm-up lines should be steam traced if the process fluid will

solidify at atmospheric temperature. Warm-up lines should be checked for adequate flexibility for the differential expansion between the pump discharge line and the warm-up line.

Exchanger Piping. Generally, bypass piping around exchangers should be provided only where required for temperature control. There may be cases, however, with exchangers subject to severe fouling or corrosion, when the increase in operating efficiency resulting from cleaning or repair during the operation of the rest of the process unit would justify the cost of installing a bypass. Of course, the process unit must be able to continue to operate with the exchanger out of service. When the process cooler is a coil in a condenser box, a bypass around the condenser may be needed if the capacity of the coil is greater than necessary and would consequently subcool the pumpout stream. One other situation in which a bypass might be necessary is that of a condenser handling a stream with a large quantity of noncondensibles which would cause binding.

All streams which are to be heated should preferably enter at the bottom of the exchanger, and all streams to be cooled should enter at the top of the exchanger. This preference may be ignored, however, if piping the exchanger this way appreciably increases the cost of the piping, except where the stream is to be heated to or near to its vaporization temperature.

Block valves need not be provided on the oil side of an exchanger except where the valve is needed for flow control or where the exchanger may be bypassed while the unit is running. (Multiple shells or exchangers in series which cannot operate independently of each other are considered to be a single exchange.) Where process coolers are used for cooling pumpout streams, double block valves with a $\frac{3}{4}$-in. bleeder between them may be necessary to prevent contamination during normal operation.

A pipe spool, elbow, or some such removable piece (other than the block valve) should be provided adjacent to the channel section on any exchanger which will be opened while the unit is in operation (to facilitate removal of the channel section).

Lines to condensers should be sized to provide sufficient velocity to carry condensed liquids along with the vapors. Pockets must be avoided in these lines. Pressure drops in the lines to and from condensers which operate in parallel must be equal.

Pressure-vessel Piping. The piping designer is cautioned that often limitations concerning piping at vessels are detailed in Section VIII of the ASME Boiler and Pressure Vessel Code. Such piping must be designed and fabricated in accordance with that Code. Piping between vessels protected by the same relief valve and piping between a vessel and its relief valve may be in this category.

The piping designer should coordinate piping requirements with the vessel designer so as to achieve the optimum nozzle location so that valves, instruments, blanks, etc., will be accessible from grade, platforms, or ladders (avoid if practicable) and that they will be located so as not to obstruct passageways. For economy and ease of support, piping at a tower should drop or rise immediately upon leaving the tower nozzle and run parallel and as close as practicable to the vessel itself.

Process requirements usually govern the location of valves in vessel piping. However, block valves should generally be provided at vessel nozzles for connections for the following: (1) vapor and reboiler lines; (2) safety and relief valves except as allowed by Section VIII of the ASME Boiler and Pressure Vessel Code; (3) sidestream draw-off lines; (4) cracking-unit transfer lines containing quench valves and furnace transfer lines to vacuum vessels; (5) lines containing block valves located outside a building and within 30 ft in a horizontal direction from the vessel nozzle; (6) when vessel nozzles are located inside the vessel skirt and a valve is needed, a connecting pipe may be attached to the nozzle and a valve bolted to the end of the

pipe outside the skirt. Another consideration in deciding where valves are needed is the fact that the rupture of a line connected below the vessel liquid level (or to the dense phase in a fluid-solids vessel) would drain the vessel unless there was a valve at the vessel nozzle. Generally it is difficult to justify valves for this purpose alone, but in deciding whether these valves are needed, the likelihood of mechanical damage to the line will be the prime consideration. Small lines (2 in. and smaller) are, of course, more susceptible to damage than large lines. If a line connecting two pressure vessels below the normal liquid level is short (e.g., less than 20 ft long), no valves need be provided unless required for process reasons.

Compressor Piping. Special precaution is necessary in the design and fabrication of the piping at and near compressors to reduce fatigue failures. The piping should be designed to have the very minimum of overhanging weight. (This is mostly a problem with high-pressure compressors where valves, etc., are very heavy.) Butt-welding fittings should be used wherever practicable, and fit-up should be accurate. Braces should be provided as needed to reduce vibration, and consideration should be given to grinding all welds to remove surface discontinuities (e.g., undercutting).

Except for air and inert-gas compressors, a block valve should generally be provided on both sides of a compressor. Motor operators will sometimes be needed for these suction and discharge block valves. Whether or not motor operators are needed will depend on the hazard associated with a compressor or piping failure, and this aspect consequently should be investigated for each installation. A motor-operated valve generally is necessary if the compressors are located inside a building. On reciprocating compressors, any motor-operated valve must be interlocked with the driver so that the machine will stop if the valve starts to close.

A check valve should be installed in the discharge line from any centrifugal or rotary compressor when the compressor discharges into a system from which liquid or gas may flow backward through the compressor. The check valve should be located as close as practicable to the compressor.

When a compressor takes suction from a header, the suction lateral should preferably be connected to the top of the header. However, if the lateral is at least one pipe size smaller than the header, it is permissible to make a center-line connection to the side of the header. Regardless, temporary screens should be provided for initial unit start-up and be located as close as practicable to the compressor except where permanent screens or filters are located immediately adjacent to the compressor. All screens and filters (both temporary and permanent) should be sufficiently reinforced to prevent their failure and entry to the compressor.

Facilities to reduce excessive surge and vibration should be provided as necessary in the suction and discharge lines of all reciprocating and rotary compressors and should be located as close as practicable to the compressor. Where surge chambers are provided, the connecting pipe should extend into the bottom of the chamber to trap any condensate. A drain should be provided on the chamber. The flow characteristics of centrifugal compressors should be investigated to determine if facilities are required to prevent surging during start-up.

Knockout drums should be provided upstream of all compressors except those which handle gases with no possibility of condensate being entrained or formed. That is, most air and nitrogen compressors do not require knockout drums. Compressor suction lines between the knockout drum and the compressor should be as short as practicable, be without pockets, and be horizontal or sloped toward the compressor. Also, for wet gas compressors, this portion of the suction line should be insulated and generally will require auxiliary heating in the form of steam tracing to prevent condensation. Where the line between the knockout drum and the compressor cannot be horizontal or sloped continuously toward the compressor,

low points in compressor suction lines should be provided with drains to remove any possible accumulation of liquid.   If the suction line normally operates under vacuum conditions, all drains between the knockout drum and the compressor should discharge into the knockout drum.

After erection, the suction line to a reciprocating compressor should be acid cleaned and then, after acid cleaning, be protected to prevent scale formation prior to starting the compressor.   This also applies to the interstage piping on such machines.

Compressor discharge piping should be analyzed for flexibility under the thermal load resulting from the heat of compression.

Storage-tank Piping.   Many storage tanks do not require separate discharge and suction headers, connections, and piping.   (Separate discharge and suction connections are required if it is necessary to have facilities for recirculation or blending and there is no mixer in the tank.)   Valves adjacent to tank nozzles should be of steel to assure adequate fire resistance.   Filling lines should discharge near the bottom of the tank without free fall because of the danger of static electricity being created.   Where it is necessary to provide top connections on tanks containing stocks with flash points of less than 100 F, floating swing lines should be installed to avoid free fall.

Expansion bends should be used only in those cases where anticipated tank settlement or thermal expansion will cause the line or tank to be overstressed.

Unless a complete water draw-off system is provided, the water draw-off should be an open drain discharging to a sewer.   In colder climates it is necessary to provide a special type of water draw-off valve or an arrangement similar to that in Fig. 11 to prevent freezing.   A pumpout connection (not shown) should be provided also and may be installed as a crossover from the water draw-off line to the tank suction line.

Piping for Refinery Fired Heaters.[20]   Permanent steam-air decoking connections, as shown in Fig. 12, should be made on heaters requiring frequent decoking and where the installation of a temporary steam-air header would necessitate considerable dismantling of the process piping.   On heaters with parallel coils (passes), blanks are required to separate the coils for decoking purposes.   Each of the parallel coils requires steam-air decoking connections.   Drop-out spools or blanks should be provided for all decoking connections, except steam connections which are also installed for steam-out operations.   The alternate to steam-air decoking is to provide sufficient flanged fittings for mechanical decoking.

A test connection on the heater charge line, provided with a block valve and a blank, should be installed.   This should be permanently connected to the test pump if such a pump is provided on the unit.

Generally every heater should be provided with a blowdown valve (or valves) operable from a safe location.   Such valves should be installed on charge, transfer, or outlet lines as required by the heater design and service characteristics and at an elevation equal to or below that of the lowest coils.   Blowdown valves should be sized for a flow area approximately equal to that of the largest tube in the coil but should not be less than 2-in. pipe size.   Where there is a likelihood of coke formation, manually operated valves should be of the globe type with the pressure against the bottom of the disk.

Steam-out valves to be used in conjunction with the blowdown system may be installed on transfer lines, charge lines, or outlet lines.   Valves should not be less than 2-in. pipe size and be operable from a location remote from the heater.   If the process fluid pressure is at all times greater than the steam pressure, two block valves and one check valve are required between the steam and the heater coil, one block next to the heater coil followed immediately by the check valve.   A

bleeder should be installed between the block valve nearest the heater coil and the check valve. If the process fluid pressure is at all times less than the steam pressure, one block valve and one check valve with a bleeder between them will normally be satisfactory. The check valve should be located between the block valve and the process equipment.

All fuel gas supplied to heaters should pass through a dry drum which is located as close as practicable to the heaters. The supply main, the branch lines, and the

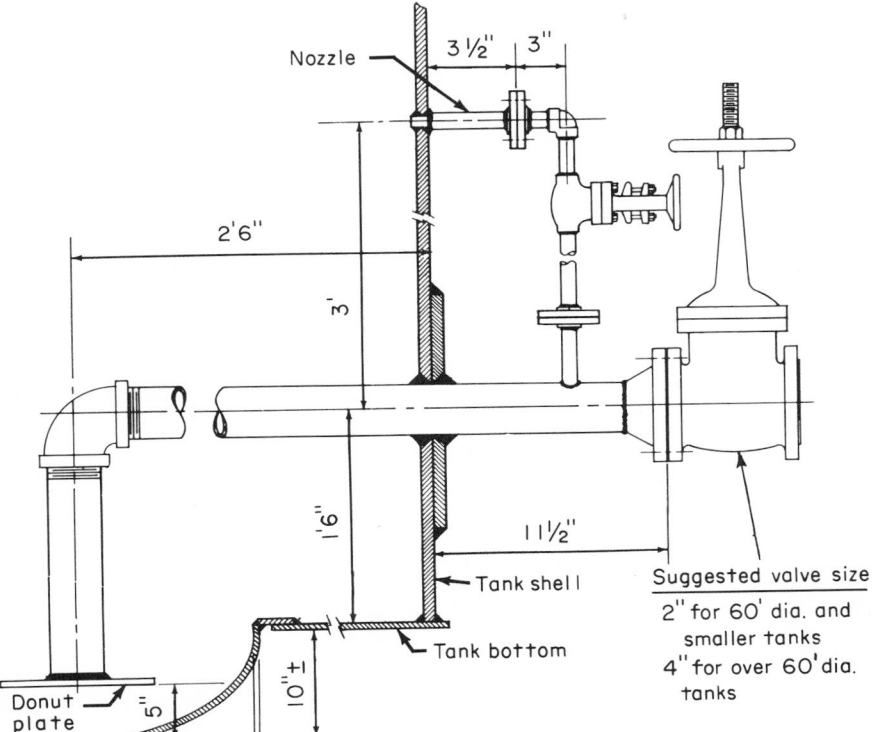

Fig. 11. Water draw-off piping for atmospheric storage tanks. A similar arrangement is also required for pressurized storage vessels.

distribution headers between the dry drum and the heater should be pitched downward in the direction of flow and be without pockets. If this is not possible, a condensate leg with a valved and plugged drain connection should be provided at the low points. Branch lines should be connected to the top of the header. A remotely controlled or a remotely located block valve should be provided in the supply line to each heater. Wherever heater outlet temperatures are controlled by regulation of the fuel supply, automatic fuel-regulator valves should be provided upstream of the distribution header. Each gas burner should be supplied with a steel shutoff valve installed in a position such that a person operating the valve will not be in close contact with the aspirator.

For heaters using fuel oil, the burner-oil piping system generally consists of a

burner-oil storage tank, a burner-oil pump, an oil-supply main with a strainer, a branch line to each heater, a distribution header at the heater, supply leads to each burner, a return branch from each distribution header, and an oil-return main back to the storage tank.   Fuel-oil lines should be sized to carry 200 per cent of the maximum design requirements of the heater.   A remotely controlled or a remotely located block valve should be provided in the supply main or in the branch line

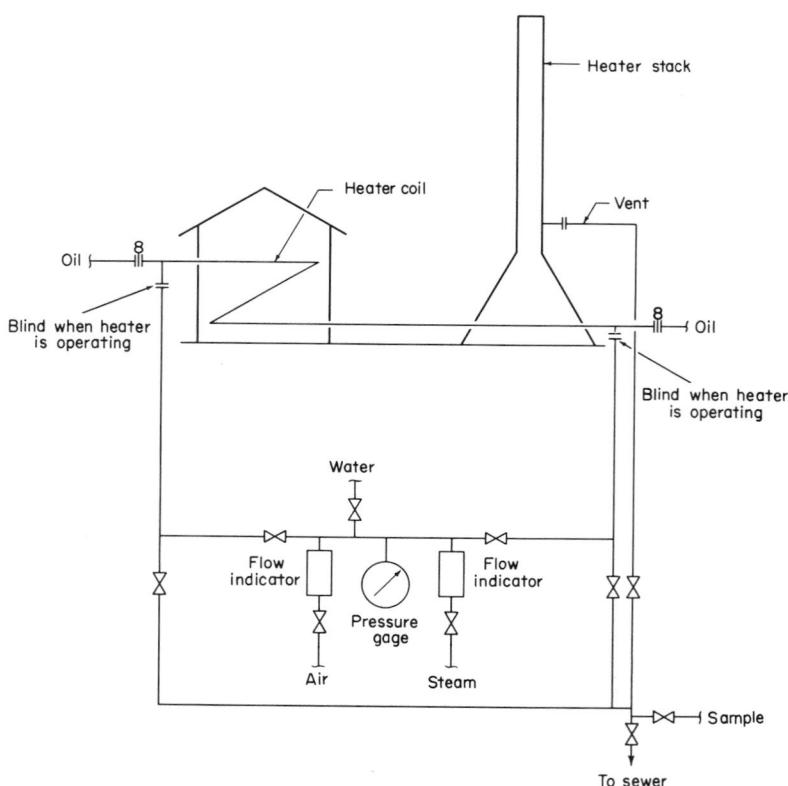

FIG. 12. Schematic arrangement of piping for steam-air decoking.

to each heater, and shutoff valves should be provided between the firing valves and the headers.   Fuel-oil lines should be sloped from the burner shutoff valves toward the burners to provide natural drainage.   A pressure-reducing valve should be installed as close to the heater as practicable and upstream of the burners to regulate the pressure at the burners.   A circulating bypass should be provided between the branch line to the distribution header and the return branch line from the distribution header.

The atomizing steam piping system used in conjunction with the burner oil consists of a branch line from a live steam main, a distribution header around each heater, and a supply lead to each burner.   Branch lines to burners should have a capacity of approximately twice the combined steam requirements of all burners

supplied by the branch.   Steam traps should be provided on the atomizing steam header where necessary to prevent water from reaching the burners.

Relief-valve Piping.   Relief-valve piping in a refinery should be in accordance with API RP 520, "Recommended Practice for the Design and Construction of Pressure Relieving Systems."[20]   Piping for relief valves protecting unfired pressure vessels should be in accordance with the applicable requirements of Section VIII of the ASME Boiler and Pressure Vessel Code (paragraphs UG-125 through UG-134).

The discharge of all pressure-relief valves should be piped to a safe place for disposal.   Liquid and readily condensible hydrocarbons are usually discharged to a closed system.   Pressure-relief valves discharging light hydrocarbons which are not likely to condense or accumulate at grade can frequently be safely vented to atmosphere from the tops of tall towers.   Discharging to atmosphere reduces the size and cost of closed system piping otherwise required and is the preferred method where it does not create a hazard and where recovery facilities are not necessary.[21] (The term closed system refers to the typical pressure-relief-valve collecting system at a refinery process unit, wherein the discharge of pressure-relief valves is collected in a piping system for disposal at a safe location.   A blowdown drum, which may be integral with blowdown stack, is usually provided for separating the vapors and collected liquid.   Vapors are vented to atmosphere through a flare or blowdown stack.   Frequently this system is combined with any required facilities for emergency blowdown or depressuring of equipment.)

Block valves should not be used before or after pressure-relief valves except where necessary to permit continuous operation of the process unit or equipment.   Where block valves are used, the installation should conform to the requirements of paragraph UG-134(e), Section VIII of the ASME Code if protecting an unfired pressure vessel or Subdivision 322.6 of ASA B31.3 if protecting refinery piping.

On certain vessels pressure-relief-valve leakage and consequent premature shutting down of the process unit can be anticipated.   These vessels should be provided with a sufficient number of pressure-relief valves (and accompanying block valves) so that in the event of pressure-relief-valve leakage it will be possible to shut off any one defective valve and replace it while the vessel is in service and still retain full calculated relieving capacity.   Examples of instances in which the need for spare pressure-relief valves has been found are crude still fractionation towers, the fractionator on catalytic cracking units, and butane and propane storage vessels.

Pressure-relief valves should be located so that the inlet piping is short and direct and self-draining with no pockets.   However, on installations where pressure pulsations or turbulence are likely to affect the pressure-relief valve (e.g., discharge side of reciprocating compressors and pumps), it may be desirable to locate the valve farther from the source in a more stable pressure region.   Figure 13 shows recommended spacings to minimize the effect of pressure pulsations and turbulence. The differential between operating and valve set pressures is also important when the operating pressure is not steady.   A large differential will tend to reduce valve maintenance costs.

For gas, vapor, or flashing liquid service the inlet piping pressure drop at design flow should not exceed 3 per cent of the safety relief-valve set pressure (psig).   Nor should the inlet piping to a pressure-relief valve be smaller than the valve inlet nominal pipe size.   The inlet piping includes all piping between the protected equipment and the inlet flange of the valve.   Excessive pressure drop in the inlet piping will cause valve "chatter" (extremely rapid opening and closing of the valve) which may lower the valve capacity and damage valve seating surfaces.

Outlet piping for pressure-relief valves discharging hydrocarbon or other

flammable vapors directly to atmosphere should normally be equipped with steam and drain connections controlled from grade as shown in Fig. 14. Outlet piping from pressure-relief valves discharging steam, air, or other nonflammable vapor or gas directly to atmosphere should be equipped with drains or otherwise suitably piped to prevent accumulation of liquid at the valve outlet. Pressure-relief-valve outlet piping for water or other liquid should be self-draining.

Separate pressure-relief-valve lines should normally be provided for each valve discharging directly to atmosphere, and the lines should be properly supported to avoid sway and excessive vibration and to prevent strain or overload on the valve, the inlet piping, and the equipment connections resulting from the valve discharge reactive forces. On towers, the pressure-relief-valve vent piping should be extended at least 10 ft above the nearest working platform within a radius of 40 ft.

Device causing turbulence	*Min. number of straight pipe diameters*
Regulator or globe-type valve	25
Two ells or bends not in same plane	20
Two ells or bends in same plane	15
One ell or bend	10
Pulsation dampener or orifice plate	10

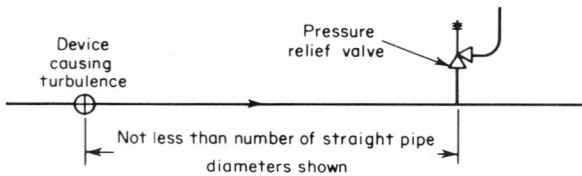

Fig. 13. Recommended spacing to avoid excessive turbulence at entrance to pressure-relief valve.

Outlet piping should be arranged so that the pressure-relief-valve discharge will not impinge on any equipment.

Pressure-relief-valve discharge piping connecting to a closed system should be self-draining to the blowdown drum, vent stack, or other means for liquid-vapor separation and disposal. The main headers are frequently sloped to assure drainage. A continuous purging connection should be considered for closed-system piping to prevent flammable mixtures resulting from possible pressure-relief-valve leakage. Where necessary or desirable to detect leaking pressure-relief valves, a $\frac{3}{4}$-in. valved and plugged drain connection should be provided at the outlet of each valve.

The sudden initiation of relief-valve outflow can cause severe stresses in attached equipment and structures. Consequently, such factors as the high- and low-temperature properties of materials, thermal expansion, vibration, and fatigue must be considered in designing pressure-relief-valve discharge piping.[21]

Generally, the most difficult and important feature associated with sizing relief-valve discharge lines and headers is the determination of the maximum probable flow. The flow is based on the number of valves which may discharge simultaneously owing to a fire or to abnormal process conditions. To do this, the layout of the unit must be considered along with many possible abnormal operating conditions.

The permissible back pressure must also be determined. Generally, the back pressure should not exceed 10 per cent of the set pressure for unbalanced safety valves. Balanced pressure-relief valves will operate satisfactorily at higher back

pressures (approximately 30 per cent of the set pressure), and consequently, their use will sometimes result in a more economical relieving system.

One method for increasing the permissible back pressure when using unbalanced valves is to set the valve at some pressure below the vessel design pressure. For example, assume a vessel operating at 25 psig and designed for 50 psig. The permissible back pressure is 5 psig if the valve is set at 50 psig. If the valve is set at 35 psig, the permissible back pressure is 20 psig.

If the permissible back pressures vary widely, and if the suggestion made above is not feasible, it might be economical to provide one high-pressure and one low-pressure relieving system.

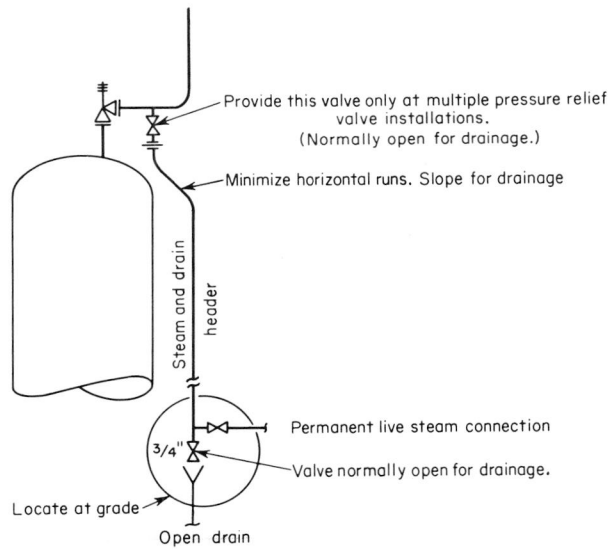

Fig. 14. Typical arrangement of pressure-relief valve steam and drain connections. Minimum steam and drain header sizes: 1 in. for one pressure-relief valve, 1½ in. for two to four pressure-relief valves, 2 in. for six or more pressure-relief valves.

Pressure-relief-valve discharge piping should be sized so that any back pressure that may exist or develop will not reduce the capacity of the pressure-relief valve below that required to protect the equipment. Regardless, the discharge piping for each pressure-relief valve should not be smaller than the nominal pipe size of the pressure-relief-valve outlet.[20,22]

To prevent excessive pressure or stalling which may be harmful to equipment, a relief valve should be provided upstream of the first valve on the discharge line of positive-displacement pumps and compressors unless (1) the pump (or compressor) and the equipment (not protected by a relief valve) downstream of the pump (or compressor) are designed to withstand the shutoff or stalling pressure, (2) a relief system is built into the pump (see next paragraph).

The discharge from the relief valves on the discharge side of positive-displacement pumps should preferably be returned to the source of pump suction. If the discharge is returned directly to the pump suction, the design engineer should consider the likelihood of damage to the pump due to possible overheating. (This problem

should also be considered on pumps where a relief system is built into the pump.) The discharge from relief valves should not be returned to the suction line unless the relieved fluid has been positively cooled by use of an aftercooler or some similar device.

Where there is a block valve in the exhaust line from a steam driver (e.g., turbine) and the steam-driver casing, including exhaust passages and expansion joints, is not designed for full steam-supply pressure, a spring-loaded relief valve sized for driver steam capacity should be provided for the following cases: (1) turbines which are members of dual drive units, (2) turbines and other steam drivers which are started automatically, and, (3) drivers to which the steam-supply pressure exceeds 150 psig. Also, unless the turbine casing, including exhaust passages and any expansion joints, is designed for full steam-supply pressure, a spring-loaded relief valve should be provided on turbines which exhaust to a surface condenser (unless the surface condenser is provided with an adequate relief valve and there are no block valves between the turbine and the condenser).

Utility Piping Systems.   Air, steam, or water connections to process piping or process equipment should be temporary unless they serve as part of the process. Temporary connections should consist of a block valve, a check valve, and a blind flange.   The block valve should be located between the check valve and the process piping or equipment.   Both valves should conform to the specification of the more severe service.

When a permanent air, steam, or water connection to process piping or process equipment other than exchangers is needed, a check valve, a ¾-in. bleeder, and a blank should be provided in addition to the block valve.   The block valve should be located between the check valve and the process piping or equipment, and the bleeder should be located between the two valves.   There must be a block valve in the utility line, upstream of the check valve, to permit installation of the blank. All valves downstream of the blank should conform to the specifications of the more severe service.   A block and check valve should be considered in the steam-supply line to an exchanger if the utility side of the exchanger is at a lower pressure than the other side.   The block valve should be located between the exchanger and the check valve.

Utility and drain connections at the bottom of the equipment may be manifolded into a single header in order to simplify piping connections to the vessel, except that steam connections should not be in the same manifold as the drain and pump-out connections (see Fig. 15).

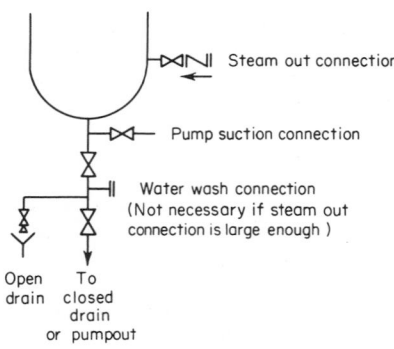

FIG. 15. Typical schematic arrangement of drain and utility connections for vessels.

Service outlets for steam, water, and air should preferably be 1-in. size.   Outlets should be located so that working areas and process equipment can be reached with a single 50-ft length of hose.

Water Piping.   Many refineries have three water systems: high-pressure (for fire fighting), low-pressure (for cooling and use in the process), and drinking water. One of the most severe design problems in many parts of the United States is to locate or protect these water systems from freezing.   The obvious way to do this is to place the piping underground and below the frost line.   However, in a refinery much of the water piping must be above grade.   If the piping is out of

doors and in intermittent or standby service, it should be steam traced and insulated. In cold climates, steam tracing and insulating should also be considered for water lines with low continuous flow rates. (An alternative to steam tracing and insulating is to provide a bypass to a drain so that flow in the water line is continuous and at a high enough rate to prevent freezing.) On water mains, the high-point vent between block valves should be protected from mechanical damage as well as from freezing.

Drains should be provided on any water line located above the frost line so that it can be drained when it is shut down. Such drain connections and valves should generally be located underground, and drains should, where practicable, connect to a sewer. Drainage facilities should also be provided for the water side of heat exchangers.

Water injected into a process oil stream normally is taken from the low-pressure water system. However, where salt water is used in the cooling-water and fire-water systems, water for process purposes should be taken from the drinking-water system.

On units where cooling-water failure creates a hazard, the fire-water main should be connected to the cooling-water main for emergency cooling. If cooling-water booster pumps are used, the connections should be made downstream of the pumps.

Each heat exchanger used as a cooler that is essential for the operation of the unit should be provided with a single block valve in the water line located either upstream or downstream of the exchanger. Each exchanger that may be removed from service during operation of the unit should have a block valve in both the inlet and outlet piping. Multiple shells or exchangers in series which cannot operate independently of each other should be considered as a single exchanger.

The water supplied to shell-and-tube coolers and condensers should pass through a strainer. If a strainer is not provided at the water pump or in the supply main, individual strainers should be provided in the branch lines. The necessity for installing an oil separator drum, a gas disengaging drum, or a bypass filter in the cooling-tower water-return system should be considered.

Water from exchangers should generally be sent to a clear water sewer or cooling-tower water-return system. Sample connections should be provided for detection of oil leaks. However, a separate connection need not be provided for this purpose if other connections, e.g., drains and vents, can be used. A $\frac{3}{4}$-in. valved and plugged vent should be provided on top of the first horizontal section of the water line downstream of the exchanger. The vent should be plugged during operation of the unit.

For chemical cleaning of exchangers using cooling water, connections should be provided to the inlet and outlet nozzles on the water side of each exchanger or each group of stacked exchangers which operate with the cooling-water stream in series. If it is not practicable to put the cleaning connections on the exchanger nozzles, they may be installed in the piping adjacent to the nozzle. The connection should be between any block valve and the exchanger. If there is no block valve, a pair of flanges must be provided nearby so that the piping can be blanked off during cleaning. It is suggested that the chemical cleaning connections be $1\frac{1}{2}$-in. size and be equipped with a blind flange.

Sufficient connections to the water system should be provided so that water can be supplied to the pressure vessels on the process unit for washing out or hydrostatic testing. These connections should be from the cooling-water system if the pressure in the system is great enough to supply water to the top of the tallest tower on the unit. Otherwise, the connections should be to the fire-water system.

Normally vessels need not be permanently connected to a source of water. If a permanent connection is made, it should be at the bottom of the vessel and should be blanked off when the vessel is in operation as is shown in Fig. 15.

Provision should be made to assure that water is available for sanitary facilities, safety showers, and eye-wash fountains during process unit shutdowns.

Air Piping. Most refineries have a plant air system not only for use in the processes but to operate tools, equipment, and instruments.

Where necessary, the intakes of air compressors should be designed to minimize the noise level. Filters should be provided in the intake piping to reciprocating and rotary air compressors when they take suction from the atmosphere. Filters will sometimes be necessary for centrifugal air compressors. When a filter is not provided for a centrifugal air compressor taking suction from the atmosphere, the intake piping should be provided with a bird screen. Filters preferably should be the dry, replaceable-cartridge type. Such filters should have an open area not less than three times the area of the intake pipe. The oil-bath-type filter should not be used with centrifugal air compressors.

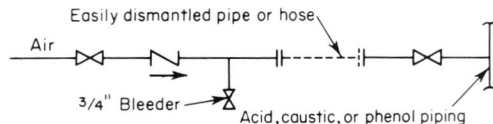

Fig. 16. Connection between air and hazardous piping system.

Low points in the discharge line from an air compressor should be avoided because it is possible for lube oil to be trapped and subsequently ignited. If low points are unavoidable, they should be provided with drains.

When condensed moisture in air lines is undesirable from a process standpoint or the possibility exists of the moisture freezing, consideration should be given to providing a "dry" drum in the supply line near the process unit. The drum should be located where it will not be exposed to heat from other equipment. The drum diameter should be not less than 24 in. Based on estimated future air requirements, the size of the drum should be such that (a) the velocity in the drum does not exceed 15 fpm during normal operation or 100 fpm during shutdown periods when maintenance equipment is being used and (b) the capacity be equal to at least 6 per cent of the free-air requirements per minute during normal operation.

In climates where freezing is possible, the bottom 18 in. of the "dry" drum should be insulated and steam traced with a single $\frac{1}{2}$-in.-OD copper tracer spiraled once around the drum. The drum drain (or blowoff) should also be traced and insulated. All blowoff connections should be installed pointing downward so that any rust or scale blown out will not endanger personnel.

Air piping should slope downward to dry drums or moisture traps or be horizontal. Branch connections to air headers should be to the top of the pipe. Block valves should be provided in all branch lines.

When an air line is connected to acid, caustic, or phenol piping, two block valves, a check valve, and a bleeder should be provided as shown in Fig. 16. Consideration should be given to also providing a removable section of line or hose in order to guard against inadvertent operation.

Air for operating instruments is normally taken from the refinery air system. For process units a steam-driven compressor should be furnished to supply instrument air in case of failure of the main supply. Where plant air is the primary source and the possibility of a power failure is remote, electrically driven compressors may be used.

If an air dryer is not provided, it may be necessary to insulate and steam trace

instrument air lines which are out of doors or in unheated buildings to prevent freezing. In cases where a complete instrument air system is not warranted (e.g., instruments not on process units), a small heatless dryer or steam tracing should be considered.

In extensive instrument air systems, the piping should be arranged with headers and subheaders such that groups of instruments may be isolated from the systems without affecting the air supply to all instruments. Block valves should be provided at the instrument air headers in all branch lines to instruments. Leads to individual instruments should be ½-in. pipe size but may terminate with ¼-in. valves. It is suggested that headers serving from 1 to 25 instruments be 1-in. pipe size, headers serving from 26 to 75 instruments be 1½-in. pipe size, and headers serving 76 to 150 instruments be 2-in. pipe size.

Steam and Condensate Piping.[23] Refineries usually have two live-steam systems and an exhaust-steam system. One of the live-steam systems generally operates in the range of 100 to 150 psig, and the other in the range of 300 to 600 psig. The exhaust system normally operates at pressures less than 30 psig. The design problems associated with these systems are not all similar to those encountered in a central power station; consequently, a brief discussion on refinery steam piping requirements follows.

The principal concern is to supply clean, dry steam to the equipment using it. In accomplishing this, it is desirable to connect all branch lines (except condensate collection points) to the top of horizontal steam mains. However, if the line to a steam driver is at least one size smaller than the main and the steam has a considerable amount of superheat, it is permissible to make a center-line connection to the side of the steam main. With other steam conditions it probably will be necessary to install a knockout pot or drum or a steam separator in addition to making the connection to the top of the main. Pockets should be avoided in the line to the turbine.

Connections to exhaust headers should preferably be made to the top of the header so that the condensate in the header does not run back into the driver. However, if the line from the driver is at least one size smaller than the main, it is permissible to make a center-line connection to the side of the main.

In the steam line to a steam driver a block valve(s) should be located at the driver and be easily accessible for operating purposes. A single gate valve is needed in the exhaust line from each steam driver which does not exhaust directly to atmosphere or directly into an individual condenser. (Valves need not be provided where two or more drivers, which will never be shut down separately, exhaust to the same condenser.) This gate valve should be installed at the driver so that the position of the gate (i.e., open or closed) will be obvious to the operator whenever he is required to operate the inlet valve.

Wherever steam is exhausted to the atmosphere and could create such hazards as burns, freezing of condensate on walkways, or the blanketing of working areas with a heavy fog, the line should be fitted with an exhaust head and a drain to a sewer (preferably into the oil-free drain system). The use of a silencer should be considered where noise nuisance is likely.

The flexibility of steam piping should be attained through the use of expansion bends and spring hangers. The use of expansion joints is discouraged except where the size and arrangement of exhaust lines prevent the use of expansion bends and loops. In such cases a corrugated expansion joint may be provided, preferably adjacent to the flange on the steam driver. Particular attention should be given to the anchorage and support of the connecting piping.[12]

A check valve should be considered for the steam-supply line to an exchanger if the steam side of the exchanger operates at a lower pressure.

The steam supply for smothering, snuffing, service hoses, space heating, and auxiliary or protective heating should be connected to a source that will not be shut off during unit shutdowns or to a source that will not be shut off when the steam to a piece of equipment, such as a turbine, is shut off.

Where leakage of air into a vacuum steam system will cause the shutdown of a process unit, the use of welded piping with steel valves should be considered.

Wherever there is an open drain or bleeder connection (e.g., free blow on a steam trap hookup) on steam piping of pressures higher than 150 psig, a welded unthreaded nipple should be provided at the downstream end of the drain or bleeder valve to prevent the use of steam hoses with high-pressure steam.

For fire-protection purposes, smothering (or snuffing) steam is usually required for refinery heaters and for relief-valve discharge lines. It also is sometimes needed for buildings and storage tanks.

When required by the service, means should be available for purging process equipment with steam or inert gas. For example, each pressure vessel in hydrocarbon service should be provided with a steam-hose connection near the bottom if not permanently connected to the source of steam. However, where a permanent connection is made, it should be blinded during operation of the unit.

Condensate Removal and Steam Traps. Condensate should preferably be discharged into an oil-free drain system, but under no circumstances should it be discharged into a sanitary sewer. Consideration should be given to a condensate collection system in installations which involve a large number of steam traps.

Where condensate is to be discharged to a cast-iron or concrete sewer or a concrete sewer box, the hazard of vaporizing hydrocarbons which may exist in the sewer should be considered. Also, to avoid damage to the concrete, the connection should be below the water level. If there is an insufficient quantity of water for quench, the condensate should be first led to an atmospheric-pressure drain tank.

Steam traps should be provided for the removal of condensate from collection points in live and exhaust steam systems, in particular from condensate drip legs, drains on steam turbines, steam separators, convectors, unit heaters, and terminal ends of companion piping. All low points in steam lines (except steam companion lines) and the ends of long headers should be provided with drip legs as shown in Fig. 17. It may also be necessary to install drip legs at intermediate points on headers with long sections at one elevation (i.e., in addition to those at low points and at the end). When a valve is installed in steam piping in such a manner that condensate can collect above the valve, a trapped drain should be provided above the valve seat.

Whenever possible, the steam trap should be installed below and close to the equipment or pipeline being drained, but the trap should be easily accessible for periodic inspection.

Each trap should serve only one collection point. Where large quantities of condensate are expected, either condensate pots or condensate drainers should be provided.

Drains from turbine shaft packing glands and from governor valve stem packing glands should preferably be connected to an open drain system. The drain lines and headers should be of sufficient size to prevent a back-pressure build-up. Also, untrapped drains should be provided at the lowest point of the steam end of each reciprocating pump and compressor.

Drains not discharging into a closed drainage system should discharge downward and should be arranged so that rising steam does not create a hazard or condense on equipment, such as a turbine or pump. The condensation of rising steam on such equipment can create lube-oil contamination. One thing that can be done to help eliminate this problem is to quench the condensate.

Probably the principal cause of freezing of steam traps is improperly designed discharge lines.   Steam trap discharge lines should be sloped for drainage where possible.   In cases where freezing is likely, no part of the trap discharge header should be at an elevation above that of the trap discharge.   Pockets in the discharge lines should be avoided.   Long trap-discharge lines, if not in heated enclosures,

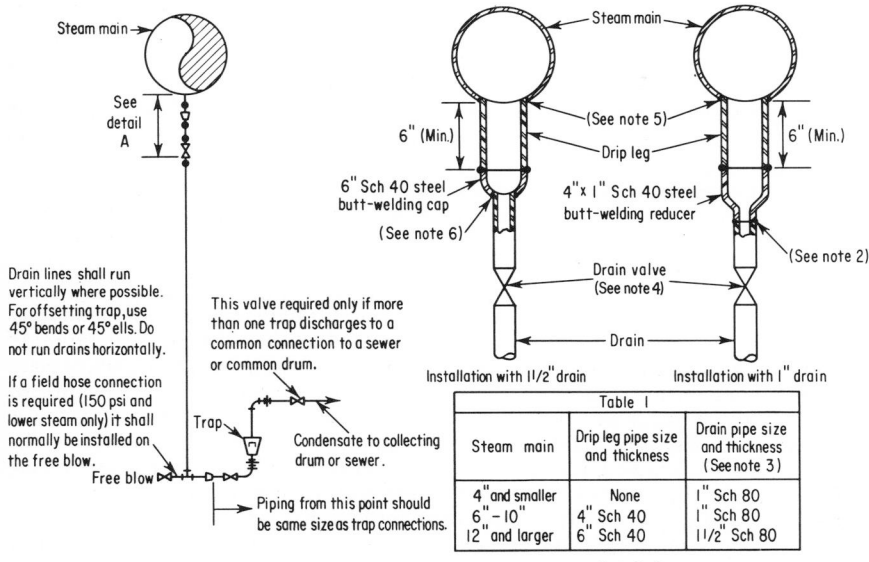

Detail A

Fig. 17. Condensate collection and steam trap piping.   Notes: 1. The free blow valve should be placed in a horizontal position, if possible, to avoid a collecting pocket for sediment.   2. Any internal protrusions resulting from the butt weld at the small end of the reducer should be removed before the drain valve is installed.   3. Use minimum 1½-in. drain pipe size for exhaust steam.   4. The drain valves at 600 lb and lower pressure steam mains should be 600-lb forged-steel socket-welding gate valves except that 150-lb screwed brass gate valves may be used for exhaust steam mains.   5. Connection of the drip leg to the steam main should be reinforced in accordance with ASA B31.3-1966, Code for Pressure Piping.   For 4-in. and smaller steam mains where a drip is not required, the connection of the 1-in. drain to the main should be reinforced with (*a*) a ¼-in. fillet weld for 300 psi and lower pressure steam, (*b*) a ⅜-in. fillet weld for steam pressures greater than 300 psi.   6. The connection of the 1½-in. drain to the drip leg weld cap should be reinforced with (*a*) a ¼-in. fillet weld for steam piping 150 lb or less or (*b*) a ⅜-in. fillet weld for 151- to 600-lb steam.

should be insulated.   Trap-discharge lines in heated enclosures need be insulated only if necessary for burn protection.   To decrease further the possibility of freezing, steam trap bodies should not be insulated except as the following circumstances make it advisable: (1) the trap is installed downstream of automatic steam controls that could shut the steam off for long periods of time; (2) the trap is installed in a location where operators might be burned by the bare metal surfaces; (3) the trap is part of a heat-recovery system where retention of heat is important; (4) the trap is installed to handle exhaust-steam condensate that contains quantities of cylinder oil.   The oil restricts the condensate flow to and from the trap and, in addition retards the trap action.

Inverted-bucket and thermodynamic steam traps, which are commonly used in a refinery, are generally installed without strainers. A typical piping arrangement for these traps is shown in Fig. 17. Steam traps should be selected for a "continuous discharge rate," which is the actual condensate rate multiplied by a safety factor. A safety factor of at least 3 should be used for inverted-bucket-type traps and thermodynamic traps. (A larger safety factor is needed for traps draining jacketed equipment, and trap manufacturers should be consulted.) In borderline cases offering a choice between two trap sizes, the smaller trap is usually preferred.

Air should be used as the actuating medium for vacuum steam traps. The use of live steam is not recommended because live steam is likely to vaporize the lower temperature vacuum system condensate. If live steam must be used as the actuating medium, it may be necessary to spray water into the trap interior to keep the trap operating properly.

Steam Companion Piping for Auxiliary Heating. The most commonly encountered situations requiring auxiliary heating are for piping in which the fluid temperature could drop below the pour point or freezing point and for piping in which the fluid viscosity becomes excessive at ambient temperatures. Other situations where auxiliary heating is generally needed are (1) chemical piping where the fluid is subject to coagulation or salting out; (2) hydrocarbon vapor and gas piping where condensate formation and icing will affect the safety and operation of the equipment such as might be caused by the reduction in pressure that takes place through a control, throttle, or relief valve; and (3) lube- and seal- oil systems for compressors and turbines. Auxiliary heating is normally not needed for freeze prevention and viscosity maintenance on equipment in intermittent service if the equipment can be drained, flushed, blown, or steamed out when there is no flowing stream or if the equipment is far enough underground to prevent freezing.

Auxiliary heating normally is furnished by external steam companion piping (steam tracing). Other acceptable methods of heating piping and other equipment are internal steam tracing, steam jacketing, hot-water tracing and jacketing, and electric tracing. The initial cost of electric tracing is comparable to that of steam tracing, but the operating (power) costs are usually higher. However, electric tracing has the advantage of close temperature control, low heat density, greater reliability, and less maintenance.

It is desirable that each steam companion line be continuous from the header to a trap at the end of the line without any vents, drains, branches, or dead-end extensions at intermediate points; however, branches may be permitted to a limited extent. Any such branches should be self-draining and be provided with a trap. Each companion line should have a block valve at its upstream end and be arranged so that flow is generally downward, avoiding pockets as much as possible and leaving no section of the companion line at a greater elevation than the companion header.

In the design of the companion piping system, provision should be made for the differential expansion between the traced line and the tracer, especially at changes in direction. When external companion lines are used, this differential expansion can break the pipe insulation.

Live steam is preferred for steam companion piping in colder climates unless a lower temperature is required, in which case alternate means such as separation or the use of exhaust steam should be considered. An example of such applications are meter leads in services in which live-steam tracing might cause bubble formation or boiling.

Minimum pipe and tube size for steam companion lines should be ½-in. IPS and ½-in. OD respectively, except that ¼- and ⅜-in.-OD tubing may be used for tracing instruments, instrument leads, and miscellaneous small items where space limitations make the use of ½-in.-OD tubing impossible.

For practical reasons it is suggested that the length of tracers not exceed 500 ft. Tables 17 and 18 list recommended maximum lengths of companion lines to be used with pipelines. These lengths represent good design practice rather than maximum permissible lengths. The tables are based on the following conditions: (1) insulation thickness is 1 in. for 4-in. and smaller lines and 1½ in. for 6-in. and larger lines; (2) steam velocities are 5,000 and 8,000 fpm for exhaust and live steam, respectively; (3) tracer lengths were calculated for lines with fluid temperatures of 50 and 150 F; and (4) outdoor design temperature is −10 F.

When the piece of equipment which is to be kept hot is irregular in shape, such as P traps, strainers, valves, and pumps, tubing must be used. The item should be spirally wrapped, starting at the top and working toward the bottom. The quantity and size of tubing required will depend on the temperature to be maintained, the insulation, and the size of the equipment.

Unions should be installed in the tracer where necessary to permit removal of the valves, strainers, etc., for maintenance and inspection.

Several lines to be traced may be grouped inside a single covering of insulation if they are to be maintained at the same temperature. In this case the size and length of the companion piping should be the same as would be required for the single traced line which would use this size of insulation.

Companion piping on benzene, acid, caustic, and phenol lines and on instrument differential-pressure leads handling volatile stocks should be separated from these lines by a 1-in.-wide asbestos tape wrapped around the steam line at 2-ft intervals to give a total thickness of about ½ in. Separation may also be effected by using small blocks of insulation securely wired to the line. (Do not use magnesia insulation, however, where either the line or its tracer is aluminum.)

An acceptable way of attaching the companion line to the pipe is with No. 14 B&S gage bare, soft copper tie wire spaced at 3-ft intervals.

MATERIALS FOR OIL-REFINERY PIPING

Because of the infinite combinations of economic and environmental factors applicable to refinery piping, the responsibility for selecting materials to resist deteriorations in service falls on the design engineer. The majority of the piping materials used in refineries and their temperature limits are listed in the allowable stress tables in Tables 3 and 4. It is permissible, however, to use materials at temperatures lower than shown if impact-test requirements are met.

Impact Tests. Impact tests are required for:

1. Any material to be used below the minimum temperature listed in the stress tables.

2. The following material below −20 F, except that no impact testing of these materials is required for metal temperatures below −20 and not below −50 F when the design pressure does not exceed 15 per cent of the maximum allowable pressure at temperature:

   a. Carbon and low-alloy steels other than Grade B7 of ASTM A193 and Grades 2H and 4 of ASTM A194

   b. Ferritic chromium stainless steels

   c. Austenitic chromium-nickel stainless steels with a carbon content greater than 0.10 per cent.

   d. Austenitic chromium-nickel stainless-steel materials in the form of deposited weld metal regardless of carbon content

   e. Austenitic chromium-nickel stainless materials not in the solution heat-treated condition.

Table 17.　Recommended Maximum Length (Ft) of Companion Lines Using 100- to 150-lb Steam

Min fluid temp	Tracer size		Protected line size													
	Pipe, nominal ID, in.	Tubing, actual OD, in.	1	1½	2	3	4	6	8	10	12	14	16	18	20	24
50 F	.....	¼	120	100	90											
	.....	⅜	250	210	180											
	.....	½	500	500	420	380	300	300	250	210	170	150	140	120	110	
	½	¾	500	500	500	500	500	500	460	370	320	280	250	230	210	19
	¾	1	500	500	500	500	500	500	500	500	500	500	470	430	390	36
	1	Two ½	500	500	500	500	500	500	500	500	500	500	500	500	500	50
150 F	.....	½	200	190												
	½	¾	380	330	280	240	190	190								
	¾	1	500	500	500	440	360	360	300							
	1	Two ½	500	500	500	500	500	500	500	430						
	Two ½	Two ¾	500	500	500	470	390	390	320	260	230	200	180	160		
	Two ¾	Two 1	500	500	500	500	500	500	500	490	420	370	340	300	280	26
	Two 1	.....	500	500	500	500	500	500	500	500	500	500	500	500	470	43

Table 18.　Recommended Maximum Length (Ft) of Companion Lines Using Exhaust Steam

Min fluid temp	Tracer size		Protected line size													
	Pipe, nominal ID, in.	Tubing, actual OD, in.	1	1½	2	3	4	6	8	10	12	14	16	18	20	24
50 F	.....	¼	20	20	10											
	.....	⅜	40	30	30											
	.....	½	90	80	70	60	50	50	40	30						
	½	¾	160	150	120	100	80	80	70	60	50	40	40	40	30	3
	¾	1	300	270	230	190	160	160	130	100	90	80	70	70	60	6
	1	Two ½	500	450	370	320	250	250	220	180	150	140	120	110	100	9
	Two ¾	Two 1	500	500	450	380	310	310	260	210	180	160	150	130	120	11
	Two 1	.....	500	500	500	500	500	500	440	350	300	270	240	220	200	19
150 F	1 and smaller	½ and smaller	0													
	.....	Two ½	0													
	Two ½	Two ¾	120	100	90	70	60									
	Two ¾	Two 1	220	190	160	140	110	110								
	Two 1	.....	370	320	270	230	190	190	160							

The manner in which impact tests are to be performed is stated in Paragraph 323.2.2(b) of Code ASA B31.3.

Limitations on Steel. In addition to temperature limits contained in the stress table, the Piping Code contains several cautionary statements on the use of steel at high temperatures. They are:

(a) Upon prolonged exposure to temperatures above 750 F, the carbide phase of plain carbon steel, plain nickel alloy, carbon-manganese alloy, manganese-vanadium alloy and carbon-silicon steel may be converted to graphite.

(b) Upon prolonged exposure to temperatures above 850 F, the carbide phase of alloy steels, such as carbon-molybdenum, manganese-molybdenum-vanadium, manganese-chromium-vanadium, and chromium-vanadium, may be converted to graphite.

Although some trouble has been experienced from graphitization of carbon and, particularly, carbon-moly steel piping in the power industry, these steels have been found generally satisfactory for use in piping in the petroleum industry to temperatures of 1000 to 1100 F. However, in locations where the consequences of a failure would be especially serious, the possibility of harmful graphitization occurring in these steels should be considered if the metal temperature is to exceed 850 F for carbon steel or 900 F for carbon-moly steel.

(c) If carbon steel is to be used above a temperature of 900 F, consideration shall be given to the advantages of silicon killed steel (0.1 % silicon min.).

(d) If carbon steel plate is used for fabrication of pressure-containing components to be used at design temperatures above 875 F, consideration shall be given to the advantages of firebox over flange quality plate.

Limitations on Use of Materials Other Than Steel. Cast iron may not be used for pressure-containing parts in hydrocarbon or other flammable fluid service at temperatures above 300 F or at pressures above 400 psi, except that for service aboveground within process unit limits the maximum permissible pressure is 150 psi. Malleable iron should not be used for pressure-containing parts at temperatures below −20 or above 650 F or in flammable-fluid service above either 300 F or 400 psi. Cast and malleable iron may not be used for pressure-containing parts in any toxic service.

The use of nodular iron is prohibited for pressure-containing parts at temperatures below minus 20 and above 650 F.

As a general rule, cast, malleable, wrought, and nodular iron are not used for piping components in a refinery for flammable or toxic services.

The low melting points of copper, copper alloys, and aluminum must be considered before these materials are used as pipe or piping components where fire hazard is involved. Because of their low melting point, these materials are seldom used in flammable service in a refinery. Welding on castings of aluminum or aluminum alloys is prohibited.

Nonmetallic materials such as plastics, glass, carbon, rubber, or ceramics may be used even if not specifically listed in the Code. If data are not available for establishment of allowable stresses, components made of such materials may be qualified under Subdivision 304.7. Consideration should be given to the suitability of the material for the service conditions (such as temperature), to the properties of the material with respect to its creep resistance, its resistance to deterioration from the service fluid or environment, its flammability, and its resistance to shock; and to its proper support and its protection from mechanical damage.

Approximate Cost of Pipe and Tubing Materials. The approximate relative costs of some of the more commonly used seamless pipe and tubing materials are listed in Table 19.

Table 19. Relative Costs of Seamless Pipe

| Material | Specification | | Material cost | Fabrication cost* |
	Number	Grade or type		
Carbon steel	API 5L and A53	B or A	1.0	1.0
	ASTM A106	B or A	1.1	1.0
	ASTM A333	O	1.5	1.0
C Mo	ASTM A335	P1	1.45	1.75
½ Cr Mo	ASTM A335	P2	1.7	1.75
1 Cr Mo	ASTM A335	P12	1.75	1.75
1-¼ Cr Mo	ASTM A335	P11	1.85	1.75
2-¼ Cr Mo	ASTM A335	P22	2.4	2.0
5 Cr Mo	ASTM A335	P5	2.4	2.0
7 Cr Mo	ASTM A335	P7	2.7	2.5
9 Cr Mo	ASTM A335	P9	3.45	2.5
304	ASTM A312	TP304	10.5	2.0
316	ASTM A312	TP316	17.25	2.0
321	ASTM A312	TP321	12.5	2.0
347	ASTM A312	TP347	14.0	2.0
3-½ Ni	ASTM A333	3	3.0	2.5
Cr–Cu–Ni	ASTM A333	4	1.75	2.0
Brass, red (annealed)	ASTM B43	.....	7.0	

* Fabrication cost does not include preheat or postheat. Preheat adds 10 to 25 per cent to cost, and postheat roughly 50 per cent.

Dimensional Standards. Dimensional standards for piping components are listed in Table 20. Dimensions of piping components must comply with these standards and specifications unless certain provisions stipulated in ASA B31.3 are met. The dimensions for nonstandard piping components must be such as to provide strength and performance equivalent to standard components. For convenience, dimensions should conform to those of comparable standard components.

FABRICATION, ASSEMBLY, AND ERECTION OF OIL-REFINERY PIPING

These topics are covered generally in Chap. 7. Many or most of the practices described in that chapter are applicable also to oil-refinery piping systems. There is a difference, however, in the approach to the problem and in the philosophy of the Oil Refinery Code; whereas the Power Piping Code in many instances prescribes methodology or procedure in exact detail, the Oil Refinery Code is so written as to permit but not necessarily to require those procedures. Quality of the finished product is frequently selected as the final test of procedure.

Complete detail and exact requirements which relate to fabrication, assembly, and erection of oil-refinery piping systems are given in Chap. V of ASA Code B31.3. Important phases of the Code treatment of these topics are covered briefly below.

Materials for Welding. All filler materials must comply with the requirements of Section IX, ASME Boiler and Pressure Vessel Code. In general, material for backing rings, when used, should be of the same material as the pipe. Other materials may be used, however, provided that their satisfactory use has been determined by a procedure qualification. To prevent the use of free-machining steels for backing rings, the sulfur content of ferrous rings is limited to 0.05 per cent. Backing rings of various types are described in Chap. 7.

Table 20.   Dimensional Standards *

Standard	Designation
Bolting:	
Square and Hexagon Head Bolts and Nuts	ASA B18.2
Fittings, valves, flanges, and gaskets:	
Cast Iron Pipe Flanges & Flanged Fittings (for 800 lb Hydraulic Pressure)	ASA B16b1
Cast Iron Pipe Flanges & Flanged Fittings (for Maximum WSP of 25 lb)	ASA B16b2
Cast Iron Pipe Flanges and Flanged Fittings, Class 125	ASA B16.1
Cast Iron Pipe Flanges and Flanged Fittings, Class 250	ASA B16.2
Malleable Iron Screwed Fittings, 150 lb	ASA B16.3
Cast Iron Screwed Fittings, 125 and 250 lb	ASA B16.4
Steel Pipe Flanges and Flanged Fittings	ASA B16.5
Steel Butt-Welding Fittings	ASA B16.9
Face-to-Face and End-to-End Dimensions of Ferrous Valves	ASA B16.10
Steel Socket-Welding Fittings	ASA B16.11
Ferrous Plugs, Bushings, and Locknuts with Pipe Threads	ASA B16.14
Brass or Bronze Screwed Fittings, 125 lb	ASA B16.15
Cast-Iron Flanges and Flanged Fittings for Refrigerant Piping, Class 300	ASA B16.16
Brass or Bronze Screwed Fittings, 250 lb	ASA B16.17
Cast-Brass Solder-Joint Fittings	ASA B16.18
Malleable-Iron Screwed Fittings, 300 lb	ASA B16.19
Ring-Joint Gaskets and Grooves for Steel Pipe Flanges	ASA B16.20
Nonmetallic Gaskets for Pipe Flanges	ASA B16.21
Wrought Copper and Bronze Solder-Joint Fittings	ASA B16.22
Bronze or Brass Flanges and Flanged Fittings	ASA B16.24
Butt-Welding Ends	ASA B16.25
Refrigeration Flare-Type Fittings	ASA B70.1
Specification for Short-Body, Cast Iron Fittings, 3 Inch to 12 Inch for 250 psi, Water Pressure Plus Water Hammer	ASA A21.10
API Specification for Flanged and Butt-Welding-End Steel Gate and Plug Valves for Refinery Use	API 600
Specification for Threads in Valves, Fittings and Flanges	API 6A
API Specification—Metallic Gaskets for Refinery Piping	API 601
Specifications for Unions & Pipe Fittings, 300-lb Pressure	AAR M-404
Standard Specifications for Cast-Iron Pressure Fittings	AWWA C100
Standard Specifications for Gate Valves for Ordinary Water Works Service	AWWA C500
Standard Finishes for Contact Faces of Pipe Flanges and Connecting-End Flanges of Valves and Fittings	MSS SP-6
Spot Facing for Bronze, Iron and Steel Flanges	MSS SP-9
Standard Marking System for Valves, Fittings, Flanges and Unions	MSS SP-25
125 lb Bronze Gage Valves	MSS SP-37
150 lb Corrosion Resistant Cast Flanged Valves	MSS SP-42
Stainless Steel Butt-Welding Fittings	MSS SP-43
By-Pass and Drain Connection Standard	MSS SP-45
Steel Butt Welding Fittings (26″ and Larger)	MSS SP-48
Forged Steel Screwed Fittings—2000, 3000 and 6000-pound	MSS SP-49
Forged Steel Plugs and Bushings	MSS SP-50
150 lb Corrosion Resistant Cast Flanges and Flanged Fittings	MSS SP-51
Pipe and tubes:	
Manual for the Computation of Strength and Thickness of Cast Iron Pipe	ASA A21.1
Specifications for Cast Iron Pit Cast Pipe for Water or Other Liquids	ASA A21.2
Specifications for Cast Iron Pipe Centrifugally Cast in Metal Molds for Water or Other Liquids	ASA A21.6
Specifications for Cast Iron Pipe Centrifugally Cast in Sand-Lined Molds, for Water or Other Liquids	ASA A21.8
Wrought-Steel and Wrought-Iron Pipe	ASA B36.10
Stainless Steel Pipe	ASA B36.19
Water Pipe, Cast-Iron (Bell and Spigot)	FS WW-P-421b
Miscellaneous:	
Unified Screw Threads	ASA B1.1
Pipe Threads	ASA B2.1
Specifications for Mechanical Joint for Cast Iron Pressure Pipe and Fittings	ASA A21.11
Fire Hose Coupling Screw Threads	ASA B26
Hose Coupling Screw Threads	ASA B33.1

† Table 326.1 in Code for Pressure Piping, ASA B31.3-1966.

If backing rings are used in services where their presence will result in severe corrosion or erosion, it is required that the backing ring be removed and the internal joint ground smooth.   In such services, where it is impractical to remove backing rings, consideration should be given to welding the joint without backing rings or to the use of consumable backing rings.

Consumable inserts may be used if the satisfactory use of such inserts has been established by the procedure qualification.

End Preparation.   The butt-welding end preparation contained in ASA B16.25 and any other end preparation which meets the procedure qualifications are acceptable.   Preferably, the ends of pipe and the edges of plate which is to be formed into pipe should be shaped by machine.   Other methods of shaping may be employed provided that a reasonably smooth surface suitable for welding and free from tears, shear drags, slag, scale, and grease is attained, except that the flame cutting of steels with a nominal chromium content greater than 5 per cent should not be permitted.   Oxygen or arc cutting is acceptable only if the cut is reasonably smooth and true and all slag is cleaned from the flame-cut surfaces.   As a precaution, arc- and flame-cut edges of 1 Cr Mo and higher alloy materials should be visually examined and any cracks should be removed.

If piping component ends are machined on the inside for backing rings, such machining must not result in a finished wall thickness, after welding, less than the minimum design thickness plus corrosion and erosion allowances.

End Alignment and Root Spacing.   The ends of piping components to be joined are aligned as accurately as is practicable within existing commercial tolerances on diameters, wall thicknesses, and out-of-roundness, and this alignment must be preserved during welding.   If the internal misalignment exceeds $\frac{1}{16}$ in., it is preferred that the component with the wall extending internally be internally trimmed so that adjoining internal surfaces are approximately flush.   However, this trimming must not result in a piping component wall thickness less than the minimum design thickness plus corrosion and erosion allowances.

There is a significant economic reason for using standard-weight and XS-weight ASA B16.9 butt-welding fittings with pipe of the next lighter schedule.   For services using carbon steel up to 650 F in the 150- and 300-lb ASA B16.5 pressure classes, it is usually satisfactory not to trim the thicker component internally.

For services which are highly cyclic and for services in the high ASA B16.5 pressure classes, the inside diameter of the piping components should be machined to a close tolerance.   If backing rings are thought to be desirable for such services, consideration should be given to the substitution of consumable inserts.

The Code requires that the root opening of the joint be given in the procedure specification.   Generally a root gap of $\frac{1}{8}$ in. is used for joints (including branch connections) without backing rings, except that where the pipe wall thickness is less than $\frac{3}{16}$ in., a $\frac{1}{16}$-in. root gap is generally used.   A root gap of $\frac{3}{16} \pm \frac{1}{16}$ in. is often used for joints using a backing ring.

Other Alignment Considerations.   Flange boltholes should straddle the established center lines unless other orientation is required to match the flange connections on equipment.

Slip-on flanges should be positioned so that the distance from the face of the flange to the pipe end is about equal to the nominal pipe-wall thickness plus $\frac{1}{8}$ in.   The seal weld should be applied so that refacing of the flange is not necessary.

Welding neck orifice flanges should be the same bore as the pipe to which they are attached and aligned as accurately as practicable.

Longitudinal seams in adjoining lengths of welded pipe should be staggered and be so located as to clear openings and external attachments.

Welding Procedure. Before being welded, all surfaces must be clean and free from paint, oil, rust, scale, or other detrimental material. Furthermore, welding is prohibited if there is impingement of any rain, snow, sleet, or high wind on the weld area.

The following Code requirements apply to girth butt welds and any longitudinal butt weld in a piping component which is not made in accordance with a Standard or Specification listed in Table 302.3.1B of the Code.

1. If the external surfaces of the two components are not aligned, the girth butt weld must be tapered between the two surfaces.

2. Tack welds, if not made by a qualified welder using the same procedure as the completed weld, must be removed. Tack welds which are not removed should be made with an electrode which is the same as or equivalent to the electrode to be used for the first pass. Tack welds which have cracked must be removed.

3. The types of imperfections required to be evaluated with various types of examination are shown in Table 21. The limitations on weld imperfections which are stipulated by paragraphs 327.4.2d and 327.4.3c of ASA B31.3 are listed in Table 22.

Additional considerations for welding are listed below; these represent good practice even though they usually are not essential to the safety of the installation.

1. Projections of weld metal into the pipe bore at welded butt joints should not exceed $\frac{1}{16}$ in. for pipe 8 in. and smaller or $\frac{1}{8}$ in. for larger pipe. Excessive projections on accessible joints should be removed. Welds attaching welding neck orifice flanges to pipe should be ground smooth on the inside.

2. On ferritic materials to be used below -20 F, the welder's identification mark should preferably be made with ink stencil. Steel stamping should be prohibited.

Welding Procedure for Fillet Welds and Seal Welds. The Code does not permit cracks in fillet or seal welds and limits undercutting to $\frac{1}{32}$ in. for these welds. Fillet welds may vary convex to concave.

If seal welding of threaded joints is performed, the Code requires that all exposed threads be covered by the seal weld and that the welding be done by qualified welders. In addition to the Code requirements it is strongly recommended that (1) threaded joints be made up dry (without thread compound), (2) seal welds be at least two-pass (preferably three-pass) welds using a $\frac{3}{32}$- or $\frac{1}{8}$-in. electrode ($\frac{5}{32}$-in.

Table 21. Imperfections to Be Evaluated by Various Types of Examination *

Imperfection	Type of examination			
	Visual	Magnetic† particle	Random† radiography	100% radiography
Crack	X	X	X	X
Incomplete penetration .	X	..	X	X
Weld undercutting and reinforcement . . .	X	..	..	..
Porosity and slag inclusions	..	..	..	X

* Table 327.4.2 of Code for Pressure Piping, ASA B31.3-1966.

† Unless otherwise specified in the engineering design for these types of examination, evaluation and any repair of imperfections are limited to the particular welds specified by the engineering design to be so examined.

Table 22. Code Limitations on Weld Imperfections

	Girth butt welds	Longitudinal butt welds
Cracks	None permitted	None permitted
Incomplete penetration and lack of fusion	Total joint penetration shall not be less than thinner of two components being joined, except to extent that incomplete root penetration is permitted. Depth of incomplete root penetration or lack of fusion at the weld root shall not exceed $\frac{1}{32}$ in. or half thickness of the weld reinforcement, whichever is smaller. Total length of such incomplete root penetration or lack of fusion at the root shall not exceed $1\frac{1}{2}$ in. in any 6 in. of weld length. (Unless otherwise specified by the engineering design, welds on which 100% radiography is specified shall have complete joint penetration.)	None permitted
Undercutting	$\frac{1}{32}$-in. maximum (Note 1)	None permitted (Note 1)
Weld reinforcement	Thickness of weld reinforcement shall not exceed (Notes 1 and 2) Thkns. of thinner comp. in. / Reinf. thkns. (max), in. $\frac{1}{2}$ and less — $\frac{1}{8}$ Over $\frac{1}{2}$–1 — $\frac{5}{32}$ Over 1 — $\frac{3}{16}$	Thickness shall be $\frac{1}{16}$–$\frac{1}{8}$ in. (Notes 1 and 2)
Porosity	An individual gas pocket of porosity shall not exceed $\frac{1}{16}$ in. in its greatest dimension. Total area of porosity, projected radially through the weld, shall not exceed 0.01 sq in. (equivalent to 3 areas $\frac{1}{16}$ in. in diameter) in any square inch of projected weld area per inch of wall thickness (considering the thinner of the components being joined)	
Slag inclusions	Developed length of any single slag inclusion shall not exceed $T/3$, T being wall thickness of thinner component being joined. Total cumulative developed length of slag inclusions shall not exceed $T/2$ in any 6-in. length of weld. Width of a slag inclusion shall not exceed $\frac{1}{16}$ in.	

NOTES:

1. Within the above limitations on undercutting and reinforcement, the surface of the weld shall fair into the base metal of the component being welded.

2. For double-welded joints, limitations on reinforcement apply to each surface of the weld separately.

3. It is recommended by the author that imperfections in girth welds in highly cyclic service or in the high B16.5 pressure classes not exceed the limits on imperfections for longitudinal butt welds.

electrode is acceptable for $2\frac{1}{2}$-in. and larger pipe size), and (3) valve and union ends be welded by the electric-arc process to minimize distortion and that valves be closed during welding.

Seal welding should not be considered as adding to the strength of a threaded joint.

Preparation and Welding Procedure for Welded Branch Connections. Branch connections (including specially made integrally reinforced branch connection fittings) which abut the outside surface of the run wall or which are inserted through an opening cut in the run wall must be so arranged as to provide a good fit and attached by means of fully penetrated groove welds. The groove welds must

be furnished with cover fillet welds of a minimum throat dimension.* The limitation as to imperfection of these groove welds is the same as that shown for girth welds in Table 22.

The recommendations for spacing and location of branch connections contained in Pipe Fabrication Institute (PFI) Standard ES7 should be followed if practicable.

A good fit must be provided between reinforcing rings or saddles and the parts to which they are attached.   It is good practice to see that reinforcement pads are so formed that no gap larger than $\frac{1}{8}$ in. exists between the pad and the pipe.   No gap larger than $\frac{1}{16}$ in., measured before welding, should exist between the periphery of a reinforcing pad and the pipe to which it is attached.*

When rings or saddles are used, a vent hole is provided (at the side and not at the crotch) in the ring or saddle to reveal leakage in the weld between branch and main run and to provide venting during welding and heat-treating operations.   Rings or saddles may be made in more than one piece if the joints between the pieces have adequate strength and if each piece is provided with a vent hole.

Reinforcing pads should be proportioned so the diameter of a vent hole is not greater than one-third of the pad width.   Reinforcing pads for structural attachments should also be provided with a hole for venting.

Tolerances for Welded Piping.   A widely accepted tolerance on face-to-face and center-to-face dimensions of welded piping is $\pm\frac{1}{8}$ in.   As for the location of flanges, their lateral translation in any direction from the specified position should not exceed $\frac{1}{16}$ in.   Also, the alignment of flanges should not deviate from the specified position, measured across any diameter, by more than $\frac{1}{32}$ in.

Qualification.   Qualification of the welding procedures to be used and of the performance of welders and welding operators is required to comply in general with the requirements of the ASME Boiler and Pressure Vessel Code (Section IX).

Each employer is responsible for the welding done by personnel of its organization and conducts the required qualification test to qualify the welding procedures and the welders or welding operators.   The employer is required to maintain a record of the procedures used and the welders or welding operators employed by him, showing the date and results of procedure and performance qualifications and the identification symbol assigned to each performance qualification.   Test joints for both procedure qualification and performance qualification are required by the Code to be made as groove welds in pipe in one or more of the specified basic qualification test positions.

Production welds may deviate from the qualified procedure as permitted by Section IX of the ASME Boiler and Pressure Vessel Code except that it is preferred that the electrodes and the preheat temperatures used for production welds be the same as those used in qualifying the procedure.

To avoid duplication of qualification tests of welders or welding operators, the welders or welding operators qualified by one employer may be accepted by another employer (subject to the approval of the Code examiner) on piping using the same or an equivalent procedure wherein the essential variables are within the limits established in Section IX, ASME Boiler and Pressure Vessel Code.   An employer who accepts such qualification tests by another employer must obtain a copy (from the previous employer) of the performance-qualification test record, showing the name of the employer by whom the welders or welding operators were qualified, the dates of such qualification, and the date the welder last welded pressure piping components under such qualification.

Renewal of a performance qualification is required (1) when a welder has not used the specific process (e.g., metal arc, gas) to weld either ferrous or nonferrous pressure

---

* See Fig. 2 of this chapter and Fig. 327.4.6 of ASA B 31.3-1966.

piping components for a period of 3 months or more or (2) when there is a specific reason to question his ability to make welds that meet the performance-qualification requirements. Renewal of qualification under stipulation (1) need be made in only a single pipe-wall thickness.

Defect Repairs. Weld defects which require repair must be removed by flame or arc gouging, grinding, chipping, or machining. Preheating may be required on certain alloy materials of the air-hardening type in order to prevent surface checking or cracking adjacent to the flame or arc-gouged surface. All repair welds must be preheated and postheated as originally required, and the basic principles of the same welding procedure initially used must be employed as far as applicable.

Bending and Forming. Pipe may be bent by any hot or cold method permissible by radii and material characteristics of the pipe being bent. It may be bent to any radius which will result in a bend arc surface which is free of cracks and substantially free of buckles.

Generally bends made to a radius, measured to the center line of the pipe, at least equal to five times the nominal pipe diameter will meet Code requirements; however, it is suggested that the radius be increased for large-diameter piping as follows:

Pipe-wall thickness, in.	Pipe size, in.	Bending radius, diameters
3/8	14–18	6
	20–24	8
1/2	18–24	6

Hot bending and forming must be done within a temperature range consistent with material characteristics, end use, or postheat treatment. It is recommended that hot bends in pipe of sizes $1\frac{1}{2}$ in. and larger be packed with high-temperature silica sand and that the pipe be uniformly heated prior to the hot-bending operation.

Heat treatment after bending and forming shall be in accordance with detailed requirements* of the Code, and no bending or forging should be performed after heat treatment unless followed by reheat treatment.

When pipe must be threaded before bending, forging, or heat treating, all exposed threaded surfaces should be protected during heat treatment with high-temperature silicone paint.

Cleaning after Fabrication. Following fabrication, all loose scale, weld spatter, slag, sand, and other foreign material should be removed from the piping. PFI Standard ES5 is an acceptable standard for cleaning fabricated piping. Piping should not be painted in the fabricating shop (i.e., before it has been erected and tested).

Heat Treatment. The welding-procedure qualification establishes the necessity for preheating and postheating welds (and the temperatures and soaking period to be used) in order to restore or obtain the physical properties of the materials (such as strength, ductility, and corrosion resistance) needed to satisfy end-use requirements, except as listed in Subdivision 331.3 of ASA B31.3.

Preheating. Typical preheat temperatures are shown in Table 23. However, when the ambient temperature is less than 32 F, local preheating is required for all materials to a temperature warm to the hand.

When dissimilar metals having different preheat requirements are being welded, the preheat temperature is as established in the procedure specification. Generally in such instances the preheat temperature will approximate those shown in Table 23 for the material with the greater alloy content, except that when P-8 steels are

* See Subdivision 331.1 and Paragraphs 331.3.1, 331.3.3, and 331.3.5 of ASA B31.3.

Table 23. Required Hardness Limits and Typical Heat Treatments of Welds *

| P Number | Min preheat temp, F | Postheat treatment | | | Brinell hardness (max)‡ |
| | | Metal temp, F | Soaking period† | | |
			Hr/in. of wall thickness (min)	Time at temp, hr, min	
1	Ambient	1100–1200	1	¾	
2	Ambient	None			
3	175	1100–1350	1	1	215
4	300	1325–1375	1	2	215
5	350	1325–1375	1	2	241
6	400	1400–1450	1	2	241
7	Ambient	None			
8	Ambient	None			
9	300	1100–1200	½	1	
10	300	1050–1100	½	1	

* Adapted from Table 331.2.1 of ASA B31.3-1966.
† The piping being heat treated shall be maintained within the temperature limits shown for a time not less than the maximum of the two periods shown.
‡ These hardness limits apply also to the heat-affected zone of the welds. Any hardness test of the heat-affected zone shall be made at a point as near as practicable to the edge of the weld.

welded to P-3, P-4, P-5, and P-9 materials, a preheat temperature of 300 F or greater is usually needed.

Preheat temperature must be checked by use of temperature-indicating crayons, thermocouple pyrometers, or other suitable methods to assure that the required preheat temperature is obtained prior to and maintained during the welding operation.

Postheat Treatment. Typical postheat treatment temperatures and soaking periods are also shown in Table 23. Regardless of the postheat treatment used, the hardness limitations of Table 23 must be adhered to.

The Code permits exceptions to the rules which deal with heat treatment. They are:

(a) *Postheat treatment as prescribed by Table 331.2.1* is required for carbon steel of a nominal thickness of ¾ inch or more, and for carbon molybdenum steel of a nominal thickness of ½ inch or more.*† (See Par. 331.3.8 of Code).

(b) *Seal welds of threaded joints and socket or fillet welds on P1 or P3 materials are not required to be postheat treated or to meet any hardness limitations of Table 331.2.1.* (This exception can also be applied to welds at branch-to-header connections in which the branch size is not larger than 2″ nominal size.)

(c) *The hardness limitations of Table 331.2.1* are not mandatory for welds in P4 or higher P-number ferritic materials, made with filler metal which is not air hardening, if the procedure qualification test indicates that the requirements of Par. 331.3.1 have been met without postheat treatment. However, in such cases the welds shall be limited to seal welds of threaded joints, and to socket or fillet welds in piping not larger than 2 inch nominal pipe size, and provided that the service environment (such as thermal cycling, temperature, or*

* Reproduced here as Table 23.
† The 1966 edition of the Code requires the heat treatment in Table 23 for all ferritic steels of a nominal thickness of ¾ in. or more unless a welding-procedure qualification, performed with specimens of nominal thickness equal to or greater than the nominal thickness of the production piece, has demonstrated that some other heat treatment is adequate.

corrosion) will not adversely affect the weldment. (This exception can also be applied to welds at branch-to-header connections in which the branch size is not larger than 2″ nominal pipe size.)

Also, the Code requires that the postheat treatment of welded joints between dissimilar metals having different postheat requirement be that established in the procedure qualification. Generally in such instances, the postheat treatment will be in accordance with the requirements for the material with the greater alloy content, except that for welds between P-8 steels and P-3, P-4, P-5, and P-9 materials, the austenitic steel should be given a stabilizing heat treatment before welding and then the postheat treatment of the weld must be in accordance with the requirements for the material with the lower alloy content.

Quite aside from the usual purpose of postheat treatment, that of improving the ductility of welded joints, postheat treatment in certain instances is helpful in improving the corrosion resistance of piping materials and welds. For example, welds in caustic piping should generally be stress-relieved if they are to operate above 100 F even if they are fillet welds used for structural attachments. Also a stabilizing heat treatment (1600 F for 4 hr) is desirable for most austenitic welds.

Postheat-treatment temperatures must be checked by the use of thermocouple pyrometers or other suitable methods to assure that the requirements established by the procedure specification are accomplished.

If welding is interrupted before completion and allowed to cool prior to postheat treatment, subsequent adequate heat treatment or controlled rate of cooling must be effected to assure that no detrimental effect to the piping results.

Local Postheat Treatment. When a circumferential weld is to be heat-treated locally, the minimum width of the band centered on the weld shall be the larger of (1) twice the width of the weld reinforcement or (2) width of weld reinforcement plus 2 in. The entire band must be heated to a uniform temperature over the circumference of the heated pipe section, with a gradual diminishing of temperature outward from the band.

When welded branch connections or other attachments are heat-treated locally, a circumferential band of the pipe to which the branch or attachment has been welded is heated to the desired temperature. The width of this band must extend at least 1 in. beyond the weld joining the branch or attachment to the pipe.

Heating and Cooling Method. Heat treatment may be accomplished by any heating method which will provide the required metal temperature, metal-temperature uniformity, and temperature control, such as an enclosed furnace, local fuel firing, electric resistance, electric induction, or exothermic chemical reaction. However, the heating method selected for restoration of physical properties desired for parts of an assembly must be such as will accomplish this result without affecting adversely other components. Heating a fabricated assembly as a complete unit is usually desirable. However, the size or shape of the unit or the adverse effect of a desired heat treatment on one or more components where dissimilar materials are involved may dictate alternative procedures such as heating a section of the assembly before the attachment of others or the local circumferential band heating of welded joints.

Although it is not tne preferred method, the postheating of weldments with an exothermic material is satisfactory in the author's opinion for 1 Cr Mo through 9 Cr Mo steels if the following requirements are met (exception: piping for caustic service should be stress-relieved in a furnace).

1. For 1 Cr Mo through 7 Cr Mo steels the exothermic material should produce, throughout the thickness of the weldment, a temperature not greater than 1670 F or less than 1575 F with the total time above 1300 F being equal to 1 hr per inch of

thickness but not less than 1 hr. For 9 Cr Mo steels, the exothermic material should produce, throughout the thickness of the weldment, a temperature not greater than 1470 F with the total time above 1300 F being equal to 1 hr per inch of thickness but not less than 1 hr.

2. The exothermic material should be covered with at least 2 in. of high-temperature (1200 F) insulation.

3. The pipe adjacent to the weld should be covered with pipe insulation as furnished with the exothermic material of a thickness approximately equal to the thickness of the exothermic material.

4. The ends of the pipe in which the weld is located must be sealed off during heat treatment.

5. The insulation should not be removed for at least 6 hr.

6. When an exothermic material is used, care should be taken with restrained piping so that there will be no upsetting caused by thermal expansion.

7. For 1 Cr Mo through 7 Cr Mo steels the weldment should be at least 32 F prior to stress relieving. For 9 Cr Mo the weldment should be at least 60 F prior to stress relieving. If preheating is required before the exothermic stress relief, heat should be applied adjacent to the insulation furnished with the exothermic material until the pipe is warm to the hand.

Cooling after postheat treatment may be accomplished in any manner required to achieve the desired cooling rate. Generally, after stress relieving, the cooling rate for ferritic steels should not exceed 500 F/hr down to 600 F, after which cooling may be in still air.

Bolting Procedure for Flanged Joints. All flanged joints must be fitted up so that the gasket contact faces bear uniformly on the gasket and then made up with relatively uniform bolt stress. All bolts must extend completely through their nuts. It is good practice to apply an antiseize thread compound to the bolts before the nuts are put on. A mixture of graphite and oil is one of the best substances for this purpose.

In bolting gasketed flanged joints, the gasket must be properly compressed in accordance with the design principles applicable to the type of gasket used. In bolting joints with spiral-wound gaskets, the gasket should be compressed by an amount equal to about 25 per cent of the original (approximately $\frac{3}{16}$ in.) thickness. Spiral-wound gaskets conforming to API Standard 601 have a 0.125-in.-thick outside gage and centering ring. The gasket is satisfactorily seated when it is compressed until the flange faces touch the gage ring.

Steel-to-cast-iron flanged joints must be assembled carefully in order to prevent damage to the cast-iron flange. In the author's opinion both flanges in steel-to-cast-iron flanged joints should be flat faced.These joints should be made up with extreme care, taking up on bolts uniformly after fitting flanges into close parallel and lateral alignment.

Flanges which connect piping to mechanical equipment, such as pumps, turbines, or compressors, should be fitted up in close parallel and lateral alignment prior to tightening the bolting. Carbon-steel pipe which has not been postheat-treated may be heated at these points for minor corrections in fit. Alloy steel or nonferrous pipe should not be heated.

Cast-iron Bell-and-spigot Piping. Bell-and-spigot joints in cast-iron piping must be assembled using poured lead or other joint compound suitable for the service. Assembly of cast-iron bell-and-spigot piping is required to comply with AWWA Standard C600. Usually each cast iron B&S joint is packed with hemp, poured full of lead (with a minimum number of pours), and then calked. The depression of lead below the face of the bell, after calking, should not exceed $\frac{1}{4}$ in. (Lead wool can be used where it is not permissible to pour lead.)

Threaded Piping. Any compound or lubricant used on threads must be suitable for the service conditions and must be compatible with both the service fluid and the piping materials. For example, flammable materials must not be used in oxygen service. Threaded joints which are to be seal welded should be made up without any thread compound.

Erection of Corrugated Expansion Joints. Corrugated expansion joints should be installed with length as shipped from the manufacturer or compressed for the cold condition at erection, depending on anticipated direction and magnitude of movement after the pipeline reaches operating temperature. The manufacturer's recommended total travel should preferably straddle the calculated travel. The requirements and recommendations in Ref. 8 should be followed unless the manufacturer of the specific joint directs otherwise.

Erection of Pipe Supports. In addition to the major supports specified by the design drawings, minor supports, as found necessary in the field, should also be installed to prevent undesirable vibration, sag, lateral movement, or stresses. The construction forces should also provide temporary supports for piping when required during erection to prevent overstressing any part of the piping.

Spring hangers, including constant-support type, should be checked for proper adjustment of travel and be correctly positioned for the cold condition at erection.

Erection of Valves. Valve packing glands should be checked for the quality and quantity of packing, and plug valves should be lubricated with proper lubricant.

Cleaning of Lines after Assembly and Erection. After completion of erection, scale, dirt, welding electrodes, slag, and other foreign matter should be removed from the lines. Particular attention should be given to the cleaning of air lines and compressor, blower, pump, and turbine inlet piping. Prior to initial operation, steam lines to turbines and to the steam ends of reciprocating compressors and pumps should be blown down with steam whose pressure is 100 psig or higher.

All practical precautions should be taken to prevent the introduction of foreign matter into pumps, instruments, and other equipment. Cleaning may be accomplished by flushing out the lines. Temporary strainers should be used at the pumps during the flushing operation unless spools or valves can be conveniently dropped out and suitable deflectors provided to prevent refuse from entering the pumps. Consideration should be given to dismantling those lines which cannot be adequately cleaned by flushing.

INSPECTION AND TEST*

Inspection (336)
Prior to initial operation, a piping installation shall be inspected to the extent necessary to assure compliance with the engineering design, and with the material, fabrication, assembly, and test requirements of this Code. An employee representative of the owner, who shall be designated as the Code Examiner, shall be responsible for this inspection. This examiner may delegate performance of any part of the inspection to inspectors who may be employees of his own organization, of an engineering or scientific organization, or of a recognized insurance or inspection company.

Qualifications of the Code Examiner (336.2)
The Code Examiner shall be qualified by a minimum of ten years experience in the design or inspection of petroleum refinery or similar industrial piping. Graduation from an engineering college accredited by the Engineers Council for Professional Development shall be the equivalent of five years experience.

* The material is quoted directly from ASA B31.3. Parenthetic numerals which follow paragraph headings refer to specific Code paragraph numbers.

Visual Examination (336.5.1)

All welds are required to be capable of compliance with the limitations on imperfections specified in 327.4.2(d)* or 327.4.3(c) for visual examination.   The welds to be so examined shall be established by the Code Examiner.

Visual examination consists of observation by the inspector of whatever portions† of a component or weld are exposed to such observation, either during or after manufacture, fabrication, assembly or test.

Supplementary Types of Examination (336.5.2)

The following supplementary types of examination are not required unless specified by the engineering design because of special service conditions requiring a high degree of freedom from imperfections, not ordinarily related to pressure-temperature conditions alone.   If such examination is specified for a weld, it is only required that the weld examined be repaired, if necessary, so that the weld imperfections comply with the limitations in 327.4.2(d)‡ or 327.4.3(c) for the type of examination used.   If supplementary types of examination are specified, they shall be performed after completion of any postheat treatment on the P number 3, 4 and 5 materials.   If any of the following types of examination is specified by the engineering design, it shall be performed to the extent as follows:

(*a*) Magnetic Particle: If an area is examined by magnetic particle examination, the method shall be the dry powder method per ASTM E109.   The extent of magnetic particle examination shall be specified by the engineering design.   It may be complete, or on a random sampling basis as specified.

When using the magnetic particle method of examination on 1 Cr Mo to 9 Cr Mo piping, it may be necessary to take special precautions, such as using braided pads between the prods and the work, in order to avoid prod scars and possible consequent crater cracking.

(b) Random Radiography:§ X-ray or gamma ray method of radiography may be used. The selection of the method shall be dependent upon its adaptability to the work being radiographed.

When random radiography of welds is specified by the engineering design, it shall be done on the number of welds designated.   The engineering design shall specify the extent to which each examined weld shall be radiographed.   Random radiography may also be used for examination of piping components, such as a valve or fitting, to any extent specified by the engineering design.

(c) 100% Radiography: If 100% radiography is specified for welds in piping, each weld in the piping shall be completely radiographed.

(d) Hardness Tests: The extent of hardness testing required for P-numbers 3 to 6 inclusive, shall be specified by the engineering design, considering the severity of the service, type of material, and other pertinent factors.

Pressure Tests.   Prior to initial operation, piping must be pressure tested to assure tightness.

If repairs or additions are made following the pressure test, the affected piping is retested, except that in the case of minor repairs or additions the owner may waive retest requirements.   The pressure test is maintained for a sufficient time to determine if there are any leaks, but not less than 10 min; a commonly used holding period is 1 hr.

All reinforcing pads on pressure openings should be tested with air at not less than 25 psig, and all welds to such pads should be swabbed with a soap or other suitable solution to aid in detecting leaks.   The test openings should not be plugged following the test.

Water is used generally as the test fluid except that if there is a possibility of damage due to freezing, or if the operating fluid or piping material would be

---

* Reproduced in this chapter as Table 22.

† This requirement is interpreted as being applicable also to forged areas and pipe bends.

‡ See Table 22.

§ Thickness gages or penetrameters are used to check the radiographic technique to be employed.

adversely affected by water, a suitable hydrocarbon oil (e.g., kerosene) whose flash point is 120 F or higher may be used.

If hydrostatic testing is not considered practicable by the owner, a pneumatic test using air or another nonflammable gas may be substituted.

A preliminary air test at not more than 25 psig is often made prior to hydrostatic test in order to locate major leaks.

If pressure tests are conducted at low metal temperatures, the possibility of brittle fracture should be considered.

Test Preparation (337.3)

All joints, including welds, are to be left uninsulated and exposed for examination during the test. If a joint has been previously tested in accordance with this Code, it may be insulated or covered.

Piping designed for vapor or gas shall be provided with additional temporary supports, if necessary, to support the weight of the test liquid.

Expansion joints shall be provided with temporary restraint if required for the additional pressure load under test, or shall be isolated from the test.

Equipment which is not to be included in the test shall be either disconnected from the piping or isolated by valves or blanks.

All pressure gages, gage glasses, flowmeter pots, liquid-level float cages, and all other pressure parts of instruments together with the piping connecting the instruments to the main piping should be included in the hydrostatic test and be tested at the same test pressure (except where the pressures will cause damage to the instruments, e.g., instruments in vacuum or low-pressure services) as the main piping to which they are connected. Relief valves and rupture disks should not be subjected to the pressure test.

Flanged joints at which blanks are inserted to isolate other equipment during the test need not be tested.

If a pressure test is to be maintained for a period of time and the test liquid in the system is subject to thermal expansion, precautions must be taken to avoid excessive pressure.

Hydrostatic Testing of Internal Pressure Piping. The hydrostatic test pressure to be used on each line in internal pressure service must be not less than one and a half times the design pressure. For a design temperature above 650 F, the minimum test pressure is calculated by Eq. (11).

$$P_T = 1.5PS_T/S \tag{11}$$

where P_T = minimum hydrostatic test pressure, psig
P = internal design pressure, psig
S_T = allowable stress at 650 F, psi (see Tables 3 and 4)
S = allowable stress at design temperature, psi (see Tables 3 and 4)

If the maximum operating conditions of piping attached to a vessel are the same as those of the vessel, then the piping and vessel may be tested together. If the piping may be subjected to higher operating conditions, it is isolated and tested separately. Where the hydrostatic head on a vessel will cause pressures that will damage other equipment in the system being tested, the vessel should be isolated from the system being tested.

For cast-iron soil pipe, which is generally in service outside the scope of the Code, the test pressure is to be equal to that of the most extreme operating conditions.

Hydrostatic Testing of External Pressure Piping. Lines in vacuum service are to be tested at a minimum internal pressure of 15 psig unless limited to a lower pressure by the design. In jacketed lines, the jacket must be tested on the basis of the jacket pressure. The internal line is tested on the basis of the internal or

external pressure, whichever is critical, unless limited by the engineering design. Lines in external pressure service are subjected to an internal pressure test at not less than the external design pressure.

Pneumatic Testing. If piping is tested pneumatically, the test pressure is set at 110 per cent of the design pressure. Pneumatic tests include a preliminary check at not more than 25 psig, and the pressure is then increased gradually in steps providing sufficient time to allow the piping to equalize strains during test and to check for leaks.

Test Records. Records must be made of the tests, including date of test, identification of piping tested, test fluid, test pressure, and approval by inspector. These records need not be retained after completion of the piping installation if the Code examiner certifies that all piping has been pressure tested as required.

OIL-TRANSPORTATION PIPING

As pointed out at the beginning of this chapter, Section 4 of the Piping Code (ASA B31.4-1959) is applicable to the design, selection of materials, fabrication, inspection, and testing of liquid-petroleum pipeline transportation systems.

Liquid-petroleum pipeline transportation systems as used here include all piping carrying crude oil, petroleum products, and liquefied petroleum gas between producers' lease tanks, terminals, refineries, natural-gas plants, and other delivery and receiving points. Also included are liquefied-petroleum gas storage facilities of the closed-pipe type, fabricated from pipe and fittings, and lines interconnecting these facilities. Figure 18 shows the jurisdiction of ASA B31.4.

This Code also covers the conditions of use of the elements of the piping systems described above including but not limited to pipe, valves, fittings, flanges, bolting, gaskets, regulators, pulsation dampeners, and relief valves.

The Oil Transportation Piping Code, ASA B31.4, does not apply to instrument, air, water, steam, lubricating oil, and other auxiliary piping; design and fabrication of pressure vessels covered by the ASME Code; storage or working tanks; piping with metal temperatures above 250 F or below −20 F; casing, tubing, or pipe used in oil or gas wells, wellhead assemblies, oil and gas separators, crude-oil production tanks, and lines interconnecting these facilities; petroleum-refinery piping; gas-transmission and -distribution piping; the design and fabrication of proprietary items of equipment, apparatus or instruments; or heat exchangers.

DESIGN CONDITIONS AND CRITERIA:
OIL-TRANSPORTATION PIPING

The maximum working pressure, as referred to here, at any point in the line is the maximum pressure expected at that point under normal working conditions and includes such items as static pressure and pressure required to overcome friction losses.

The internal design pressure must be not less than the maximum working pressure plus allowance for surge pressure if anticipated. Suitable protective devices of such types as relief valves and automatic shutdown equipment must be provided which will assure that the maximum internal design liquid pressure of the piping system and equipment is not exceeded by more than 10 per cent. The effect of possible vacuum within the pipe shall be recognized, and the pipe wall selected shall provide adequate strength to prevent collapse.

Because B31.4 is applicable only between temperatures of −20 to 250 F, the reasonable simplifying assumption that the design temperature be 100 F has

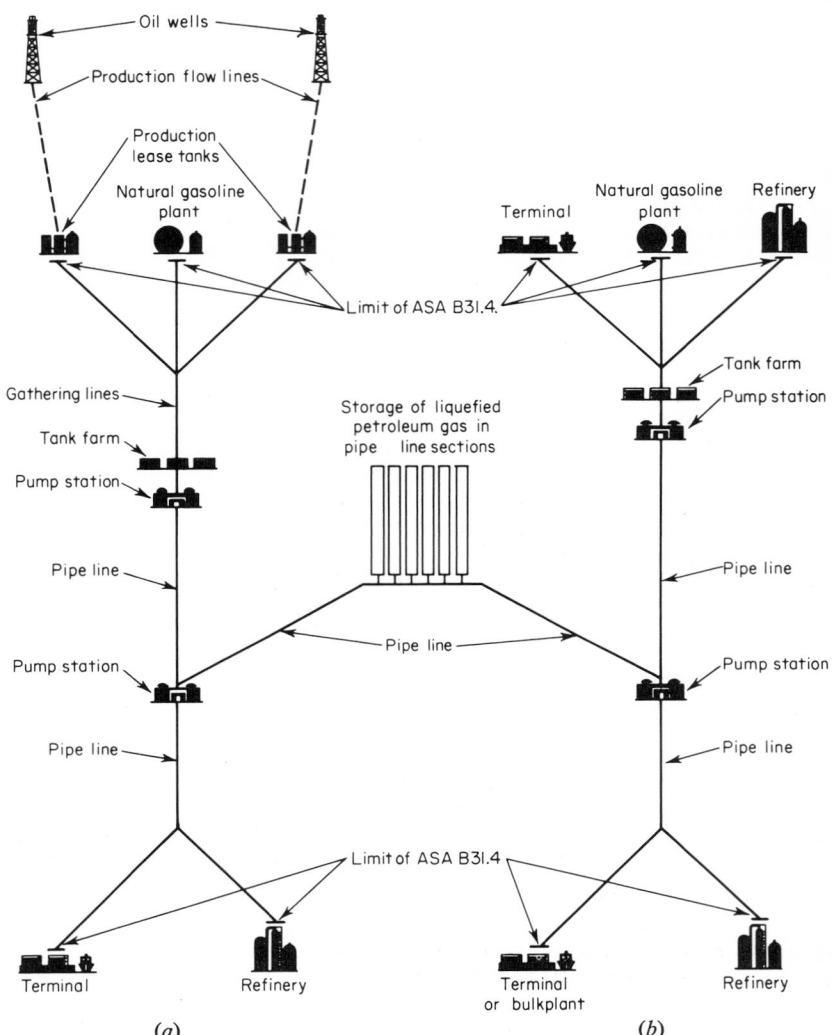

Fig. 22-18. Diagram showing scope of ASA B31.4 (reproduced from Fig. 400.1.2 of ASA B31.4).   Facilities indicated by solid lines are within the scope of ASA B31.4. (*a*) Diagram showing scope of crude-oil and liquefied-petroleum-gas transportation systems.   (*b*) Diagram showing scope of liquid-petroleum product and liquefied-petroleum-gas transportation systems.

been made.   Therefore, neither allowable stresses nor ratings are varied with temperature.

Fluid expansion effects on pressure must be considered for exposed sections of pipelines, valves, and fittings, and pressure-relieving devices must be installed when required.

Loads caused by impact, wind, vibrations, resonance, thermal expansion and contraction, and the weight of the piping components and liquid content must be considered.

The design engineer should provide adequate protection to prevent damage to the pipeline from unusual external conditions which may be encountered in river crossings, bridges, areas of heavy traffic, long self-supported spans, unstable ground, vibration, weight of special attachments, or abnormal thermal forces. Some of the protective measures which the design engineer may provide are encasing with steel pipe of larger diameter, adding concrete protective coating, increasing the wall thickness to reduce the stress level using the established formula, lowering the line to a greater depth than normal, or indicating the presence of the line with additional markers.

The corrosion allowance may be zero if internal and exterior corrosion or erosion are not expected.   If a corrosive fluid is to be transported, or if soil conditions are conducive to corrosion, no corrosion allowance is necessary if suitable means are taken to mitigate such corrosion.

When pipe is threaded or grooved, the design thickness of the pipe wall shall be equal to or greater than t as determined by Eq. (12) plus thread depth or groove depth.

Selected longitudinal weld joint factors are listed in Table 24.   These values have been abstracted from Table 2 of Code ASA B31.4.

Table 24.   Longitudinal Joint Factor E

(Abstracted from Table 2 of ASA B31.4)

Specification factor	Pipe type	Longitudinal joint factor E
ASTM A53	Seamless	1.00
	Electric resistance welded	1.00
	Furnace lap welded	0.80
	Furnace butt welded	0.60
ASTM A106	Seamless	1.00
ASTM A135	Electric resistance welded	1.00
ASTM A139	Electric fusion welded (single or double pass)	0.80
ASTM A381	Electric fusion welded	1.00
API 5L	Seamless	1.00
	Electric resistance welded	1.00
	Electric flash welded	1.00
	Furnace lap welded	0.80
	Furnace butt welded	0.60
API 5LX	Seamless	1.00
	Electric resistance welded	1.00
	Electric flash welded	1.00
	Submerged arc welded	1.00

PRESSURE DESIGN OF PIPING COMPONENTS

Straight Pipe. The nominal thickness of steel pipe wall must be equal to or greater than the design thickness as determined by the formula

$$t = PD/2S \qquad (12)$$

where t = design thickness of the pipe wall, in.

P = maximum internal design pressure, psig

D = outside diameter of pipe, in.

E = longitudinal joint factor (Table 24)

S = maximum allowable hoop stress in the pipe wall, psi, and shall be determined as follows: $S = 0.72 \times E \times$ specified minimum yield strength from the applicable pipe specifications (see Table 25 for list of pipe commonly used).

Curved Segments of Pipe. Changes in direction may be made by bending the pipe or by installing factory-made bends or elbows. The thickness of pipe bends is determined as for straight pipe.

The minimum metal thickness of flanged or screwed elbows shall not be less than specified for the pressures and temperatures in the applicable ASA or MSS Standard. Steel butt-welding ells shall comply with the requirements of Specification ASA B16.9 and shall have pressure and temperature ratings based on the same stress values as were used in establishing the pressure and temperature limitations for pipe of the same or equivalent material.

Branch Connections. For branch connections the Code states a preference for butt-welding tees and pipe-to-pipe branch connections using encirclement type of reinforcement.

Pipe-to-pipe branch connections reinforced with fillet welds and pads are also acceptable. For these branch connections the amount of reinforcement needed

Table 25. Specified Minimum Yield Strength for Steel Pipe Commonly Used in Oil-transportation Piping Systems

Specification	Specified minimum yield strength, psi
API 5L Grade A Seamless or Electric-Welded	30,000
API 5L Grade B Seamless or Electric-Welded	35,000
API 5L Lap-welded or Butt-welded Class I Open-hearth . .	25,000
API 5L Lap-welded or Butt-welded Class II Open-hearth .	28,000
API 5L Lap-welded or Butt-welded Bessemer Steel	30,000
API 5LX Grade X42	42,000
API 5LX Grade X46	46,000
API 5LX Grade X52	52,000
ASTM A53 Grade A Seamless or Electric Welded	30,000
ASTM A53 Grade B Seamless or Electric Welded	35,000
ASTM A53 Lap-welded or Butt-welded Open-hearth or Electric Furnace	25,000
ASTM A53 Lap-welded or Butt-welded Bessemer Steel . . .	30,000
ASTM A106, A135, A139—Grade A	30,000
ASTM A106, A135, A139—Grade B	35,000
ASTM A381 Grade Y35	35,000
ASTM A381 Grade Y42	42,000
ASTM A381 Grade Y46	46,000
ASTM A381 Grade Y48	48,000

should be calculated the same way as it is in B31.3 except that *t* is the wall thickness as calculated by Eq. (12).

Other branch connection details are acceptable, but under restrictions of size and pressure.

Closures and Flanges.   ASA B31.4 requires that nonstandard flanges and flat, ellipsoidal, spherical, and conical heads be designed in accordance with Section VIII of the ASME Boiler and Pressure Vessel Code, except that the allowable stresses used in the design of heads must be the same as are used in B31.4 for the design of pipe of equivalent material.

SELECTION AND LIMITATIONS OF PIPING COMPONENTS

Selection and Limitations of Pipe.   Either new or used ferrous pipe as listed in Table 25 may be employed in oil-piping transportation systems.   Used ferrous piping and new ferrous piping made to Specification ASTM A120 must undergo additional testing procedures at the time of erection.

When ferrous pipe which has been cold-worked in order to meet the specified minimum yield strength is subsequently heated to 600 F or higher (welding excepted), the maximum allowable pressure at which it can be used must not exceed 75 per cent of the value obtained by use of Eq. (12).

Ferrous pipe lighter than API standard-weight threaded line pipe (see Table 4, API Standards 5L) may not be threaded.

External or internal coatings or linings of cement, plastics, or other materials may be used on steel pipe, but such coatings or linings may not be considered to add strength.

Selection and Limitations of Fittings, Elbows, Bends, Intersections, and Valves.   The Code permits valves conforming to Standards and Specifications listed in Table 26 except that cast-iron valves should not be used for pressures exceeding 400 psi.   The Code also permits other valves provided their design is of at least equal strength and tightness and provided they are capable of withstanding the same test requirements as covered in the Standards in Table 26.

B31.4 permits the use of fittings complying with ASA B16.9, ASA B16.5, and MSS SP-48.   However, it also permits the use of fittings similar to B16.9 and SP-48 fittings in sizes and thicknesses not covered by these two Standards provided they are designed following the same principles and are capable of withstanding the same tests as fittings made to the two Standards.

Table 26.   Acceptable Material and Dimensional Standards

Pipe.....................	API 5A, 5B, 5L, 5LX
	ASA B2.1, B36.10, B36.19
	ASTM A53, A72, A106, A120, A134,
	A135, A139, A155, A211, A381
Fittings, valves, flanges	API 6A, 6B, 6C, 6D, 600
	ASA B16.5, A16.9, B16.10, B16.20,
	B16.21, B16.25
	ASTM A105, A181, A182, A216, A217,
	A234, A445
	MSS SP-6, SP-25, SP-44, SP-48, SP-52,
	SP-59, SP-61, SP-63
Bolting	ASA B1.1
	ASTM A193, A194, A307, A325, A354
Structural materials	ASTM A7, A36, A42, A212, A225, A242,
	A283, A285, A440, A442

In accordance with the Code, changes in direction may be made by bending the pipe or by installing factory-made bends or elbows (or transverse segments thereof, provided that the arc length measured along the crotch is at least 2 in. on pipe sizes 4 in. and larger) under the limitations following:

1. In cross-country lines, care should be taken to allow for passage of pipeline scrapers.

2. At any point along a cold bend the pipe diameter must not be reduced by more than 2½ per cent of nominal pipe diameter. Radii of cold bends for pipe sizes 12 in. and larger should not be less than forty times the outside diameter of the pipe.

3. Hot bends made with pipe which has been cold-worked in order to meet the specified minimum yield strength must be designed for the lower stress levels as outlined previously for straight pipe (see page 22–96).

4. Wrinkle bends may not be used.

5. Miter bends not exceeding 12½ deg may be used in systems operated at less than 30 per cent of the specified minimum yield strength. When the system is to be operated at less than 10 per cent of the specified minimum yield strength, the restriction to 12½ deg maximum will not apply.

Fabricated Reductions and End Closures. Orange-peel bull plugs and orange-peel swages are prohibited on systems which are intended to operate at stress levels of 20 per cent or more of the specified minimum yield strength of the pipe material. Fishtails and flat closures are permitted for 3-in.-diameter pipe and smaller operating at less than 100 psi. Fish tails on pipe larger than 3 in. diameter are prohibited.

Selection and Limitations of Flanges and Facings. Welding-neck, slip-on, screwed, and lapped companion flanges; reducing flanges; blind flanges; and flanges cast or forged integral with pipe, fittings, or valves, conforming to ASA B16.5, or to MSS SP-44, are permitted in the sizes and for the pressure-temperature ratings established in these Standards. The bore of welding-neck flanges should correspond to the inside diameter of the pipe with which they are to be used.

Flanges exceeding scope of standard sizes or otherwise departing from dimensions listed in ASA B16.5 or MSS SP-44 may be used provided they are designed following the same principles and are capable of withstanding the same tests as fittings which would be applicable except for size or dimensions.

Slip-on flanges of rectangular cross section may be used provided the flange thickness is increased to provide strength equal to that of the corresponding hubbed slip-on flange covered by ASA B16.5, as determined by calculations made in accordance with the ASME Code for Unfired Pressure Vessels.

Steel flanges must have contact faces in accordance with ASA B16.5 and MSS SP-6, except that it is permissible to machine facings smooth in accordance with MSS SP-46 when mating cast-iron and steel flanges. Other special facings are permissible provided they are capable of withstanding the same tests as those in ASA B16.5.

Selection and Limitations of Gaskets. Material for gaskets must be capable of withstanding the maximum pressure and of maintaining its physical and chemical properties at any temperature to which it might reasonably be subjected in service.

Metallic gaskets, other than ring-type or spiral-wound metal asbestos, may not be used with ASA 150-lb or lighter flanges.

Metal or metal-jacketed asbestos types of gaskets are recommended for use with the small male and female or the small tongue-and-groove facings. They may also be used with steel flanges with lapped, large male and female, large tongue-and-groove, or raised facings.

Asbestos composition gaskets may be used as permitted in ASA B16.5. This type of gasket should conform to ASA B16.21 and may be used with any of the various flange facings except small male and female or small tongue and groove.

Rings for ring joints must be in accordance with ASA B16.20.

Selection and Limitation of Bolting. Bolting shall conform to Paragraphs 5.2, 6.9.1, 6.9.3, 6.9.4, and 6.9.5 of ASA B16.5. Bolt or bolt studs shall extend completely through the nuts.

For insulating flanges, ⅛-inch undersize bolting may be used provided it is alloy steel bolting material per ASTM A193 or A354.

The bolting of steel flanges to cast iron flanges shall conform to MSS SP-46.

Bolts and nuts shall conform to carbon steel bolting per ASTM 307, alloy steel bolting per ASTM A193, and A354, and nuts per ASTM A194, in the limits as set out in Subdivision 408.5.

Selection and Limitations of Piping Joints. The Code states a preference for butt-welded joints to other types of piping joints. Furthermore, it recommends that threaded joints be limited to lines 4 in. and smaller and that they not be used on lines over 2 in. operated above 500 psi.

The Code restricts the use of patented joints, such as sleeve and coupled joints, only to applications where welded joints will not provide sufficient flexibility to accommodate anticipated movement in piping due to expansion, settling, or similar causes.

EXPANSION, FLEXIBILITY, SUPPORTS, AND RESTRAINTS*

Expansion and Flexibility. Formal calculations are required only where reasonable doubt exists as to the adequate flexibility of the system.

Expansion calculations are necessary for buried lines only if substantial temperature changes are expected, such as when the line is to carry a heated oil. Thermal expansion of buried lines may cause movement at points where the line terminates, changes in direction, or changes in size. Unless such movements are restrained by suitable anchors, the necessary flexibility must be provided.

The expansion of aboveground lines may also be controlled by use of suitable anchors so that longitudinal thermal expansion or contraction is absorbed by direct axial compression or stretch of the pipe in the same way as for buried piping. In addition, however, beam bending stresses must be included and the possible elastic instability of the pipe and its supports due to longitudinal compressive forces must be considered.

Where expansion is not absorbed by direct axial compression of the pipe, flexibility shall be provided by the use of bends, loops, or offsets or provision shall be made to absorb thermal strains by expansion joints or couplings of the slip-joint or bellows type. If expansion joints are used, anchors or ties of sufficient strength and rigidity must be installed to provide for end forces due to fluid pressure and other causes.

The total range in temperatures shall be used to determine expansion for calculations. The linear coefficient of thermal expansion for carbon and low-alloy high-tensile steel may be taken as 6.5×10^{-6} in./(in. F) for temperatures up to 250 F.

Flexibility calculations shall be based on the modulus of elasticity at ambient temperatures (see Table 14). The effect of restraints, such as support friction, branch connections, lateral interferences, etc., shall be considered in the stress calculations.

Calculations are to take into account stress-intensification factors found to exist in components other than plain straight pipe. Credit may be taken for extra

* Much of the material in this section is taken directly from Chap. II, Part 5 of ASA B31.4.

flexibility of such components. In the absence of more directly applicable data, the flexibility factors and stress-intensification factors shown in Table 15 may be used. Nominal dimensions of pipe and fittings are used in flexibility calculations.

The total range in temperature shall be used in calculation of stresses in loops, bends, and offsets regardless of whether the piping is cold-sprung or not. In addition to the expansion of the line itself, the linear and angular movements of the equipment to which it is attached must be considered.

Stress Values. Fundamental differences in loading conditions for the buried or otherwise restrained portions of the piping system and the aboveground portions not subject to substantial axial restraint require that different limits on allowable longitudinal expansion stresses be set.

The net longitudinal compressive stress for restrained lines due to the combined effects of temperature rise and fluid pressure may be computed by use of Eq. (13).

$$S_L = E\alpha(T_2 - T_1) - \nu S_h \tag{13}$$

in which S_L = longitudinal compressive stress, psi
S_h = hoop stress due to fluid pressure, psi
T_1 = temperature at time of installation, F
T_2 = maximum or minimum operating temperature, F
E = modulus of elasticity of steel, psi
α = linear coefficient of thermal expansion, in./(in. F)
ν = Poisson's ratio (0.30 shall be used for steel)

Note that the net longitudinal stress becomes compressive for moderate increases of T_2 and that, according to the commonly used maximum shear theory of failure, this compressive stress adds directly to the hoop stress to increase the equivalent tensile stress available to cause yielding. This equivalent tensile stress should not be allowed to exceed 90 per cent of the specified minimum yield strength of the pipe. Beam bending stresses should be included in the longitudinal stress for those portions of the restrained line which are supported above ground.

The expansion stresses for those portions of the piping system without axial restraint are obtained by use of Eq. (14)

$$S_E = \sqrt{S_b^2 + 4S_t^2} \tag{14}$$

where $S_b = iM_b/Z$ = resultant bending stress, psi
$S_t = M_t/2Z$ = torsional stress, psi
M_b = resultant bending moment, in.-lb
M_t = torsional moment, in-lb
Z = section modulus of pipe, in.³
i = stress-intensification factor (from Table 15)

The maximum computed expansion stress S_E based on 100 per cent of the expansion, with modulus of elasticity for the cold condition, must not exceed the allowable stress range S_A, where S_A = 72 per cent of the specified yield strength.

The sum of the longitudinal stresses due to pressure, weight, and other sustained external loadings must not exceed $0.75S_A$.

Design of Pipe-supporting Elements. The Code states a preference for nonintegral attachments, such as pipe clamps and ring girders, where they will fulfill the supporting or anchoring function. All attachments to the pipe must be designed so that the added stresses in the pipe wall due to the attachment are minimized.

For example, if pipe is designed to operate at or close to its allowable stress, the Code recommends that all welded connections be made to a separate cylindrical member which completely encircles the pipe and that this encircling member be welded to the pipe by continuous circumferential welds. Alternatively, a heavier wall may be used at the point where the welded attachment is required.

MATERIALS AND DIMENSIONAL REQUIREMENTS

Table 26 is a listing of the Material and Dimensional Standards which have been approved by the Code Committee. Materials which are not listed and which are to be used where their properties are important from a safety standpoint can be qualified for use only by petitioning the Code Committee for approval. In such instances, complete information as to chemical composition and mechanical properties must be supplied to the Code Committee.

The Code permits an exception to the above rule for unlisted materials provided they are known to be of good quality and workmanship, provided any deviations which would adversely affect weldability or ductility are provided for in the design, and provided they are used in items not important from a safety standpoint.

The Code requires that the dimensions for nonstandard piping components be such as to provide strength and performance equivalent to standard components made in accordance with the Dimensional Standards listed in Table 26. The Code also requires that, wherever practical, the dimensions of nonstandard components conform to those of comparable standard components.

WELDING AND INSTALLATION

Material and Procedure. All electrodes must conform to the requirements of the AWS-ASTM Specifications for Mild Steel Arc-welding Electrodes or AWS-ASTM Specifications for High Tensile and Low Alloy Steel Arc-welding Electrodes, latest edition. The details of the pipe joint design and preparation shall be as prescribed in the qualified welding procedure.

Prior to the start of welding, the welding procedures are required to be established and procedures and welders qualified in accordance with API 1104, Standard for Welding Pipelines and Related Facilities. The welding procedure must be re-established as a new procedure specification and must be completely requalified when any of the significant changes listed in API Standard 1104 are made in the procedure. In addition, the qualified welding procedure must specify the preheating and temperature-control practices which are to be used when weather conditions affect the weldability of the steel. All qualified welding procedures must be recorded in detail and adhered to during subsequent construction.

Welds (427.4.2)

(a) Filler and Finish Beads. The number of filler beads should be such that the completed weld will have a reinforcement of not less than $\frac{1}{32}''$. Two beads shall not be started at the same location. The face of the completed weld should be approximately $\frac{1}{8}''$ greater than the width of the original groove. The completed weld shall be thoroughly brushed and cleaned.

(b) Quality of Welds. The following limits of major defects on shop and field welds shall not be exceeded.

(1) Cracks. No weld containing cracks, regardless of size or location, shall be acceptable.

(2) Inadequate Penetration and Incomplete Fusion. Any individual defect due to inadequate penetration or incomplete fusion shall not exceed $1''$ in length. The total length of such defects in any $12''$ length of weld shall not exceed $1''$. The total length of

such defects in any two succeeding 12″ lengths shall not exceed 2″ and individual defects shall be separated by at least 6″ of sound weld metal.

(3) Undercutting. Undercutting adjacent to the cover bead on the outside of the pipe shall not exceed 1/32″ in depth and 2″ in length. Undercutting adjacent to the root bead on the inside of the pipe shall not exceed 2″ in length.

(4) Burn-Through Areas. Any individual burn-through area shall not exceed 1/4″ in length. The total length of burn-through area in any 12″ length of weld shall not exceed 1″. The total length of burn-through area in any two succeeding 12″ lengths shall not exceed 2″, and individual defects shall be separated by at least 6″ of sound weld metal.

(5) Gas Pockets.* The maximum dimension of any individual gas pocket shall not exceed 1/16 in.

(6) Elongated Slag Inclusion (Wagon Tracks). Any elongated slag inclusion shall not exceed 2″ in length or 1/16″ in width. The total length of elongated slag inclusions in any 12″ length of weld shall not exceed 2″ and the total length of elongated slag inclusions in any two succeeding 12″ lengths shall not exceed 4″. Adjacent elongated slag inclusions shall be separated by at least 6″ of sound weld metal. Parallel slag lines shall be considered as individual defects if the individual width is greater than 1/32″.

Isolated Slag Inclusions. The maximum width of any isolated slag inclusion shall not exceed 1/8″. The total length of isolated slag inclusions in any 12″ length of the weld shall not exceed 1/2″, nor shall there be more than four isolated slag inclusions of the maximum width of 1/8″ in this length. The total length of isolated slag inclusions in any 24″ length of weld shall not exceed 1″. Adjacent isolated slag inclusions shall be separated by 2″ of sound weld metal.

(7) Accumulation of Discontinuities. Any accumulation of discontinuities having a total length of more than 2″ in a weld length of 12″ is unacceptable. Any accumulation of discontinuities which total more than 12% of the weld length of a joint is unacceptable.

Installation (435)

All construction work performed in accordance with the requirements of this Code shall be done under construction specifications. These specifications shall cover all phases of the work and shall be sufficient in detail to insure that the requirements of this Code shall be met. Such specifications shall include specific details on handling of pipe and other materials and all construction factors which contribute to sound engineering and safety.

INSPECTION AND TEST

Inspection. Adequate installation inspection provisions for pipelines and other facilities should be provided to assure control of quality workmanship.

All pipe must be cleaned inside and outside if necessary to permit good inspection and visually inspected to assure that it is reasonably round and straight and to discover any defects which might impair its strength or tightness. Careful consideration must be given to the overall condition of the pipe, its internal and external appearance, the number of bends, flattening from straightening, degree of pitting, or other surface defects such as seams, cracks, grooves, gouges, and dents. It is recommended that this inspection be performed just ahead of any coating operation and during the lowering-in and backfill operation.

When pipe is coated, inspection must be made to determine that the coating machine does not cause harmful gouges or grooves or that the pipe is not damaged during lowering in.

The fit-up of the joints should be inspected before the weld is made. Then the stringer bead and the completed weld should be visually inspected.

All repairs, replacements, or changes must be inspected before they are covered up.

The condition of the ditch bottom just before the pipe is lowered in and the fit of the pipe to the ditch must each be inspected before backfilling.

---

* B31.4 also states that the maximum distribution of gas pockets should not exceed that shown in Figures 10 and 11 of API 1104.

In addition to the Code-required visual examination, it is industry practice to radiograph a substantial percentage of the completed welds and to perform occasional destructive tests (e.g., tensile) on production welds.

The Code requires that tensile and other destructive tests be made in accordance with API 1104. Radiographic inspection of welds performed in accordance with API 1104 is acceptable.

Repair of Defects. Injurious defects may be repaired by use of welding procedures prescribed in API 5LX or removed by grinding, provided the resulting wall thickness is not less than the minimum prescribed by Code B31.4 for the conditions of usage. Dents that are more than $\frac{1}{4}$ in. deep must be removed from pipe which is intended for operation at more than 50 per cent of the specific minimum yield strength. When the conditions outlined above cannot be met, the damaged portion of pipe must be cut out as a cylinder and replaced.

Tests. Fabricated items such as scraper traps, manifolds, volume chambers, and surge pressure accumulators should be hydrostatically tested to limits equal to or greater than those required of the completed system.

In addition, pipelines (including fabricated items) should be tested after construction with provisions to avoid freezing or thermal expansion of the test fluid if applicable.

The Code requires that the hydrostatic (water or oil) test pressure be not less than 1.1 times the design pressure for those parts of oil-transmission systems to be operated at a hoop stress exceeding 30 per cent of the specified minimum yield strength of the pipe. However, the test pressure should not exceed that stipulated in the referenced Specification or Standard in Table 26 applicable to the weakest element in the system. The Code permits the substitution of a pneumatic test, not in excess of 100 psig, when a hydrostatic test is impractical. Should leaks occur on either test, the line shall be repaired and retested.

Qualification Tests

Before used ferrous pipe of known specifications, new or used ferrous pipe of unknown specification, or new or used ASTM A-120 pipe is used, the applicable of the following qualification tests should be performed as required.

Visual Examination. All pipe must be cleaned inside and outside, if necessary to permit good inspection, and visually examined to assure that it is reasonably round and straight and to discover any harmful defects.

Careful consideration should be given to the overall condition of the pipe, with reference to such factors as its age, internal and external appearance, the number of bends, flattening from straightening, degree of pitting, or other surface defects such as seams, cracks, grooves, gouges, dents, and arc burns.

Bending Properties. After the type of joint has been identified, the following bend tests must be performed if the minimum yield strength is above 24,000 psi. For pipe sizes 2 in. and less, the bending test must meet the requirements of ASTM A53 or API 5L; for pipe sizes larger than 2-in., flattening tests must meet the requirements in ASTM A53, API 5L, or API 5LX.

The number of tests required to determine bending properties shall be the same as those which are required to determine yield strength.

Determination of Wall Thickness. Unless the nominal wall thickness is known with certainty, it is determined by measuring the thickness at quarter points on one end of each piece of pipe. If the lot of pipe is known to be of uniform grade, size, and nominal thickness, measurement must be made on not less than 5 per cent of the individual lengths but not less than 10 lengths; thickness of the other lengths may be verified by applying a gage set to the minimum thickness.

Following such measurement, the nominal wall thickness is taken as the next commercial wall thickness below the average of all the measurements taken but in no case greater than 1.14 times the least measured thickness for all pipe under 20 in. OD, and no greater than 1.11 times the least measured thickness for all pipe 20 in. OD and larger.

Joint Factor. The factor E shall be assumed to be 0.6 for pipe 4 in. and smaller or 0.80 for pipe over 4 in. unless the type of longitudinal joint can be determined with certainty.

Weldability. Weldability is determined by having a qualified welder make a girth weld in the pipe under field conditions using the planned procedure. The pipe is considered weldable if the requirements set forth in API Standards 1104 are met.

The minimum number of test welds is as follows:

Line size, in.	Number of lengths per test
Up through 5⁹⁄₁₆ diam	400
6⁵⁄₈ through 12¾ diam	200
14 and larger	100

Determination of Yield Strength (437.6.6)

When the manufacturer's specified minimum yield strength, minimum tensile strength, or minimum per cent of elongation for pipe is unknown, the tensile properties may be established as follows:

Perform all tensile tests prescribed by API Standards 5L or 5LX, except that the minimum number of such tests shall be as follows:

Line size, in.	Number of lengths per test
Up through 5⁹⁄₁₆″ diam	400
6⁵⁄₈″ through 12¾″ diam	200
14″ and larger	100

All test specimens shall be selected at random.

Minimum Yield Strength Value (437.6.7)

For pipe of unknown specification, the minimum yield strength may be determined as follows:

Average the value of all yield strength tests for a test lot. The minimum yield strength shall then be taken as the lesser of the following:

(a) 80 per cent of the average value of the yield strength tests.

(b) The minimum value of any yield strength test, except that in no case shall this value be taken as greater than 52,000 psi.

(c) 24,000 psi if the average yield-tensile ratio exceeds 0.85.

Hydrostatic Test (437.6.8)

Pipe of unknown specification or ASTM A120 pipe, and all used pipe, the strength of which is impaired by corrosion or other deterioration, shall be tested hydrostatically by individual lengths, or in the field after installation before being placed in service. The minimum test pressure used shall be 1.1 times the internal design pressure.

REFERENCES

1. M. W. Kellogg Co., "Design of Piping Systems," John Wiley & Sons, Inc., New York, 1956.
2. Chilton, E. G., and L. R. Handley, "Pulsation in Gas-Compressor Systems," *Trans. ASME*, 1952, p. 931.
3. Parmakian, John, "Waterhammer Analysis," Prentice-Hall, Inc., Englewood Cliffs, N.J., 1955.

4. O'Hara, James K., "Nomograph Determines Pressure Rise Due to Water Hammer," *Heating, Piping, Air Conditioning*, February, 1964, pp. 117–118.
5. Spangler, M. G., "Stresses in Pressure Pipelines and Protective Casing Pipes," *J. Struct. Div. ASCE*, vol. 82, no. ST 5, September, 1956.
6. API Code No. 1102, "Recommended Practice on Form of Agreement and Specifications for Pipe Line Crossings under Railroad Tracks," May, 1957.
7. Markl, A. R. C., "Piping Flexibility Analysis," *Trans. ASME*, February, 1955.
8. Expansion Joint Manufacturers Association, "Standards of the Expansion Joint Manufacturers Association," 1962.
9. Tube-Turns, "Simplified Analysis of Single-plane Piping Systems," *Piping Eng. Paper* 4.04, April, 1956.
10. Tube-Turns, "Simplified Analysis of Multi-plane Piping Systems," *Piping Eng. Paper* 4.05, July, 1957.
11. Grinnell Company, "Piping Design and Engineering," 1951.
12. National Electrical Manufacturers Association SM20, "Standards for Mechanical-Drive Steam Turbines," 1958, and TU4, "Standards for Direct-connected Steam Turbine Synchronous Generator Units (2000–10,000 Kw, Inclusive)," 1952.
13. Roark, Raymond J., "Formulas for Stress and Strain," 3d ed., pp. 281–283, McGraw-Hill Book Company, New York, 1954.
14. American Oil Co., "Engineering for Safe Operation," Booklet 8, 1964.
15. American Oil Co., "Hazard of Air in Refinery Process Systems," Booklet 2, 1964.
16. American Oil Co., "Hazard of Water in Refinery Process Systems," Booklet 1, 1964.
17. American Oil Co., "Safe Handling of Light Ends," Booklet 7, 1964.
18. Johnsen, J. H., "Fundamental Requirements for Safe Arrangement of Drains and Vents," *Petrol. Refiner*, June, 1947.
19. American Oil Co., "Safe Furnace Firing," Booklet 3, 1963.
20. API RP 520, "Recommended Practice for the Design and Construction of Pressure, Relieving Systems in Refineries," Part I—Design (September, 1960), Part II—Installation (January, 1963).
21. Loudon, D. E., "Requirements for Safe Discharge of Hydrocarbons to Atmosphere," *Proc. API*, sec. III, Refining, 1963.
22. Conison, Joseph, "Factors in Sizing a Safe and Economical Vapor Relief System," *Proc. API*, sec. III, Refining, 1963.
23. American Oil Co., "Hazard of Steam," Booklet 6, 1963.
24. Nelson, G. A., "Corrosion Data Survey," Shell Development Co., 1960.

23

PLUMBING SYSTEMS

Vincent T. Manas*

The American Standard Association's National Plumbing Code, ASA A40.8, represents the consensus of national organizations for a standard of plumbing practice suitable for state and local application. Since its publication in 1955, it has been used as the basis for some 22 state plumbing codes and 1,200 municipal codes.

Since 1955, a great many advances have taken place in plumbing practices and techniques, and this chapter reflects those changes.

Drainage and Vent Systems. The drainage system is a system of piping for the collection of wastes from the plumbing fixtures in the building.

The venting system provides air in the system, thereby balancing the necessary positive or negative pressures at all the fixture traps of the drainage system and preventing escape of sewer gases into the building.

The engineering design of a well-balanced, efficient drainage and venting system is based on the assigning of a fixture unit value to each different plumbing fixture, as shown in Table 1. Tables 2, 3, and 4 indicate sizes of drains and vents as dictated by the number of fixture units installed.

No soil or waste stack may be smaller than the largest horizontal branch connected to it, except that a 4- by 3-in. water-closet connection may not be considered as a reduction in pipe size.

Any structure in which a building drain in installed must have at least one stack vent or vent stack of at least 3 in. in diameter carried full size through the roof.

When provision is made for the future installation of fixtures, those provided for must be considered in determining the required sizes of drain and vent pipes.

* Consulting Engineer, 810 Washington Ave., Memphis, Tenn. 38105.

Table I. Fixture Unit Values for Various Plumbing Fixtures

Fixtures, individual or grouped	*Fixture unit values per fixture or group*
Automatic clothes washer (2-in. standpipe)	3
Bathroom group consisting of 1 water closet, 1 lavatory, and 1 either bathtub or shower stall. If water closet is equipped with a *flushometer*, the unit value for this group will be	8
If water closet is *tank type*, the unit value for this group will be	6
Bathtub, with or without overhead shower*	2
Combination sink and tray with food-disposal unit.	4
Combination sink and tray with one 1½-in. trap.	2
Combination sink and tray with separate 1½-in. traps	3
Dental unit or cuspidor.	1
Dental lavatory.	1
Drinking fountain	½
Dishwasher, domestic, gravity drain	2
Floor drains, 2-in. waste.	3
Kitchen sink, domestic, 1½-in. waste.	2
Kitchen sink, domestic, with food waste grinder	2
Lavatory, 1¼-in. waste.	1
Laundry tray, one or two compartments	2
Shower stall, domestic.	2
Showers, group. Unit values per head	2
Sinks:	
Surgeons'.	3
Flushing rim with valve	6
Service, trap standard	3
Service, P trap	2
Pot, scullery, and the like	4
Urinal, pedestal, syphon jet blowout	8
Urinal, wall lip	4
Urinal stall, washout	4
Urinal trough, each 1' section	2
Wash sink, circular or multiple; each set of faucets.	2
Water closet, tank-operated	4
Water closet, valve-operated	6
Unlisted fixture drain or trap size in.:	
1¼ or less	1
1½	2
2	3
2½	4
3	5
4	6

* A shower head over a bathtub does not increase the fixture value.

Construction to provide for such future installations is terminated with a plugged fitting or fittings.

No portion of the drainage system installed underground or below a basement or cellar may be less than 2 in. in diameter.

Horizontal drainage piping must be installed in uniform alignment at uniform slopes in accordance with the following requirements and in no case at a slope which will produce a computed velocity of less than 2 fps.

Changes in direction in drainage piping are made by the appropriate use of 45-deg wyes; long- or short-sweep quarter bends; sixth, eighth, or sixteenth bends; or by a combination of these or equivalent fittings.   Single sanitary tees and quarter bends may be used in drainage lines only where the direction of flow is from the horizontal to the vertical.   Double sanitary tees may be used in drainage lines only where the direction of flow is from the horizontal to the vertical.

Table 2. Building Drains and Sewers

Diam. of pipe, in.	Maximum number of fixture units that may be connected to any portion* of the building drain or the building sewer			
	Fall per foot, in.			
	$\frac{1}{16}$	$\frac{1}{8}$	$\frac{1}{4}$	$\frac{1}{2}$
2	.....	.....	21	26
2½	.....	.....	24	31
3	.....	20†	27†	36†
4	.....	180	216	250
5	.....	390	480	575
6	.....	700	840	1,000
8	1,400	1,600	1,920	2,300
10	2,500	2,900	3,500	4,200
12	3,900	4,600	5,600	6,700

* Includes branches of the building drain.
† Not over two water closets.

Short sweeps not less than 3 in. in diameter may be used in soil and waste lines where the change in direction of flow is from either the horizontal to the vertical or the vertical to the horizontal and may be used for making necessary offsets between the ceiling and the next floor above.

No fitting having a hub in the direction opposite to flow or tee branch may be used as a drainage fitting. No fitting or connection which has an enlargement chamber or recess with a ledge or shoulder or reduction in pipe area may be used. No running threads, bands or saddles may be used. No drainage or vent piping may be drilled or tapped.

A heel or side-inlet quarter bend may not be used as a vent when the inlet is placed in a horizontal position.

No fitting, connection, device, or method of installation which obstructs or retards the flow of water, waste, sewage, or air in the drainage or venting systems in an amount greater than the normal frictional resistance to flow may be used

Table 3. Horizontal Fixture Branches and Stacks

Diam. of pipe, in.	Maximum number of fixture units that may be connected to			
	Any horizontal fixture branch*	1 stack of 3 stories in height or 3 intervals	More than 3 stories in height	
			Total for stack	Total at 1 story or branch interval
1¼	1	2	2	1
1½	3	4	8	2
2	6	10	24	6
2½	12	20	42	9
3	20†	30‡	60‡	16†
4	160	240	500	90
5	360	540	1,100	200
6	620	960	1,900	350
8	1,400	2,200	3,600	600
10	2,500	3,800	5,600	1,000
12	3,900	6,000	8,400	1,500

* Does not include branches of the building drain.
† Not over two water closets.
‡ Not over six water closets.

Table 4. Vents: Size and Length

Diam. of soil or waste stack, in.	Total fixtures connected to stack	Diam. of vent, in.										
		1¼	1½	2	2½	3	4	5	6	8	10	12
		Maximum length of vent, ft										
1¼	2	30										
1½	8	50	150									
1½	10	30	100									
2	12	30	75	200								
2	20	26	50	150								
2½	42	..	30	100	300							
3	10	..	42	145	355	1,040						
3	21	..	32	110	270	805						
3	53	..	27	94	230	680						
3	102	..	25	86	210	620						
4	43	..	...	35	85	250	975					
4	140	..	...	27	65	195	750					
4	320	..	...	23	55	165	635					
4	530	..	...	21	50	150	580					
5	190	..	...	...	28	82	320	985				
5	490	..	...	...	21	63	245	760				
5	940	..	...	...	18	53	207	670				
5	1,400	..	...	...	16	49	189	585				
6	500	..	...	...	...	33	130	400	1,000			
6	1,100	..	...	...	...	26	100	310	775			
6	2,000	..	...	...	...	22	84	260	655			
6	2,900	..	...	...	...	20	77	240	595			
8	1,800	..	...	...	...	.....	31	95	240	940		
8	3,400	..	...	...	...	.....	24	73	185	720		
8	5,600	..	...	...	...	.....	20	62	155	610		
8	7,600	..	...	...	...	.....	18	56	140	555		
10	4,000	..	...	...	...	.....	...	31	78	305	960	
10	7,200	..	...	...	...	.....	...	24	60	235	735	
10	11,000	..	...	...	...	.....	...	20	51	200	625	
10	15,000	..	...	...	...	.....	...	18	46	180	570	
12	7,300	..	...	...	...	.....	...	...	31	120	380	940
12	13,000	..	...	...	...	.....	...	...	24	94	295	720
12	20,000	..	...	...	...	.....	...	...	20	79	250	610
12	26,000	..	...	...	...	.....	...	...	18	72	225	555
15	15,000	..	...	...	...	.....	...	...	.....	40	125	305
15	25,000	..	...	...	...	.....	...	...	.....	31	96	235
15	38,000	..	...	...	...	.....	...	...	.....	26	81	200
15	50,000	..	...	...	...	.....	...	...	.....	24	74	180

unless it has been demonstrated to offer a desirable and acceptable function and to be of ultimate benefit to the proper and continuing functioning of the plumbing system. The enlargement of a 3-in. closet bend or stub to 4 in. is not considered an obstruction if the horizontal flow line or insert is continuous without forming a ledge.

In the installation or removal of any part of a drainage system, ends should be avoided except where necessary to extend a cleanout so as to be accessible.

Building Drains below Building Sewer. Building drains which cannot be discharged to the sewer by gravity flow must be discharged into a tightly covered and vented sump from which the liquid is lifted and discharged into the sump gravity drainage system by automatic pumping equipment or by an equally efficient method.

Sump and pumping equipment are so designed as to discharge all contents accumulated in the sump during the cycle of emptying operation, and the storage of drainage in a sump or ejector must not exceed 12 hr. Sumps in other than one- or two-family residences receiving the discharge of six or more water closets must

be provided with duplex pumping equipment. The sump vent must be of proper size to meet the venting requirement based on the discharge rate of the sump pump.

Sizing Vent Piping. Vent pipe sizes are determined on the basis of drainage load and length of vent pipe to the terminal point. No vent pipe may be less than $1\frac{1}{4}$ in. in diameter.

The diameter of vents which service individual fixtures must be at least one-half the diameter of the drain served.

The diameter of a relief vent must be at least one-half the diameter of the soil or waste branch served.

Materials. All materials used in the construction of any plumbing system, fixtures, or equipment must be specified in the Code or approved in a Code ruling. All materials installed in plumbing systems should be handled carefully so as to avoid damage or impairment of quality. No defective or damaged materials, equipment, or apparatus may be installed.

A material, device, or equipment shall be considered acceptable if listed as an American Standard by the Sectional Committee ASA A112, published by the American Society of Mechanical Engineers, 345 East 47 Street, New York 17, N.Y.

When selecting material for and sizing of the water service supply piping, tubing, or fittings, due consideration must be given to the characteristics and action of the water in the interior of the pipe and of the characteristics and action of the soil, fill, and other material on the exterior of the pipe.

Water service pipe is made of asbestos cement, brass or copper pipe, copper tubing, cast-iron water pipe, wrought iron, lead, open-hearth iron, plastic, or steel. All threaded ferrous pipe and fittings must be galvanized or cement lined. When ferrous pipe and fittings are installed underground in corrosive soil or fill, they must be coated with coal tar and the threaded joints must be coated and wrapped.

Water-distribution systems must be of brass or copper pipe, copper tubing, galvanized wrought-iron pipe, galvanized open-hearth iron pipe or galvanized steel pipe.

Soil and waste piping installed aboveground in buildings must be of brass or copper pipe, copper tubing, extra-heavy or service-weight cast iron, galvanized wrought iron, galvanized open-hearth iron, galvanized steel, or lead.

Underground building drains are usually of cast iron of service weight for buildings four stories and under and of extra heavy weight for buildings whose height is over four stories. The use of galvanized steel, galvanized ferrous alloy lead or copper pipe or tube, or other material is permitted. Where threaded joints are installed underground, they must be coated with coal tar and wrapped.

If the building sewer is installed in a trench separate from the water service, the sewer pipe material may be of asbestos cement, bituminized fiber, cast iron, concrete, vitrified clay, or other acceptable material. Joints must be watertight and rootproof.

If the building sewer is installed in the same trench as the water service, the sewer pipe must be of durable, corrosion-resistant material and installed so as to be watertight and rootproof. The sewer should be tested with a 10-ft head of water or equivalent and found to be tight.

When a building sewer or building drain is installed in filled-in or unstable ground, it should be of cast iron. However, a nonmetallic drain may be used if it is laid on an approved supporting base for its entire length.

Existing building sewers and drains may be used in connection with new building sewer and drainage systems only when found by examination and test to conform to the new system in quality of material prescribed in the preceding paragraphs.

Building storm drains underground inside the building when connected to a sanitary building drain or combined sewer must be of cast-iron soil pipe. Building

storm drains underground inside the building when not connected with a sanitary or combined sewer may be of cast-iron soil pipe, ferrous-alloy pipe, or an approved type of vitrified clay pipe, concrete pipe, bituminized-fiber pipe, or asbestos-cement pipe.

The building storm sewer may be of cast-iron, vitrified-clay, concrete, bituminized-fiber, plastic or asbestos-cement pipe.

Drainage systems for chemical wastes must be completely independent of all other sewer, floor drains, or storm drain systems.

Materials acceptable for chemical waste drainage systems include prestressed, low-expansion, borosilicate glass pipe; high-silicon-content cast-iron pipe; plastic pipe; plastic-lined pipe; and lead pipe.

Vent piping installed aboveground may be of brass, copper, copper tube, cast iron, ferrous alloys, galvanized steel, galvanized wrought iron, or lead.

Underground vent piping should be of cast iron except that the use of other materials may be permitted. Where threaded joints are installed underground, they must be tested for tightness, then coated with coal tar and wrapped.

Vent piping in acid-waste systems must be of the same material as that of the acid-waste drain pipe.

Fittings must be of the same material as that of the piping system in which they are installed. They must be smooth and formed without ledges, shoulders, or reductions which can retard or obstruct flow in the pipe. Threaded drainage pipe fittings must be of the recessed type.

Roof drains may be of cast iron, copper, lead, or other corrosion-resistant material.

Inside conductors installed above ground may be of brass or copper pipe, copper tubing, cast iron, lead or galvanized open-hearth iron, galvanized steel, or wrought iron.

Leaders or downspouts are usually of an approved sheet metal, although other materials may be used. For example, for use in buildings whose height does not exceed three stories, wooden leaders of material and construction suitable to the particular environment are permitted.

Subsoil drains may be open-jointed, horizontally split, or perforated clay tile; perforated asbestos cement, bituminized fiber, or plastic pipe; or open-jointed cast-iron soil pipe.

Floor flanges for water closets or similar fixtures must be not less than $\frac{1}{2}$ in. thick for brass and $\frac{1}{4}$ in. thick and not less than 2-in. calking depth for cast iron or galvanized malleable iron. If of hard lead, they may weigh not less than 1 lb 9 oz and must be composed of lead alloy with not less than 7.75 per cent antimony by weight. Flanges are either soldered to lead bends or calked, soldered, or threaded into other metal. Closet screws and bolts must be brass.

Cleanout plugs must be of brass with raised square or countersunk square heads; if raised heads might cause a hazard, countersunk heads should be used.

Flush pipes and fittings must be of nonferrous material. If brass or copper tubing is used, the material must be at least 0.0313 in. in thickness (No. 20 U.S. gage).

Plumbing fixtures must be constructed from approved materials; have smooth, impervious surfaces; and be free from defects and concealed fouling surfaces.

DOMESTIC WATER-SUPPLY PIPING

Computation and design of water piping for distribution of water in a building are based on the minimum pressure available from the street main or individual source of supply, the head changes in the system due to friction and elevation, the

Table 5. Daily Water Requirements

Type of occupancy	Min. quantity of water per person per day, gal
Single family dwellings	75
Small dwellings and cottages for seasonal occupancy	50
Multiple-family dwellings (apartments)	50
Rooming houses	40
Boarding houses	50
Additional kitchen usage for nonresident boarders	10
Hotels with private baths (2 persons per room)	60
Hotels without private baths	50
Restaurants (toilet and kitchen usage per patron)	7–10
Additional for bars and cocktail lounges	2
Day schools with cafeterias, gymnasiums, and showers	25
Day schools without cafeterias, gymnasiums, or showers	15

rate of flow for satisfactory operation of the fixtures, and the probability of simultaneous use of the fixtures.

Design Flow. Table 5 lists the daily water requirements for various types of occupancy. These data are of particular advantage in estimating the amount of storage capacity that might be provided for in the event of an interruption in the continuity of the supply system. For example, a 500-pupil day school with cafeterias, gymnasiums, and showers would be expected to be provided with 500 × 25 or 12,500 gal of water per day. If it were desired to protect against a one-day outage of the supply system, a storage tank of, say, 15,000 gal might be installed.

Table 6 lists fixture-unit values for water-supply fixtures. These values are similar to those listed in Table 1 except that in Table 6, the values are given both for the hot and cold supplies. A typical one-family residence might have the following assignment of fixture-unit values:

Table 6. Fixture-unit Values for Water-supply Fixtures

Fixture or group*	Type of supply	Fixture-unit values†		
		Hot	Cold	Total‡
Bathroom group	Flush valve	3	6	8
Bathroom group	Flush tank	3	4.5	6
Bathtub	Faucet	1.5	1.5	2
Combination fixture	Faucet	2	2	3
Kitchen sink	Faucet	1.5	1.5	2
Laundry tray	Faucet	2	2	3
Lavatory	Faucet	1.5	1.5	2
Pedestal urinal	Flush valve	...	10	10
Restaurant sink	Faucet	3	3	4
Service sink	Faucet	1.5	1.5	2
Shower head	Mixing valve	3	3	4
Stall or wall urinal	Flush valve	...	5	5
Stall or wall urinal	Flush tank	...	3	3
Water closet	Flush valve	...	10	10
Water closet	Flush tank	...	5	5

* For fixtures not listed, use the fixture-unit value assigned to a fixture using similar quantity at a similar flow rate.

† Definition of fixture units: Fixture units are the relative mathematical values used for computing the load-producing effects of plumbing fixtures or devices. The scale of relative fixture-unit values was computed by the National Bureau of Standards, BMS-66.

‡ For fixtures with both hot- and cold-water supplies, the fixture units for maximum separate demands may be computed by using ¾ of the total fixture units.

Table 7. Minimum Pipe Sizes for Water-supply Fixtures

Fixture or device	Nominal pipe size, in.
Bathtubs	½
Combination sink and tray	½
Drinking fountain	⅜
Dishwasher, domestic	½
Kitchen sink, residential	½
Kitchen sink, commercial	¾
Lavatory	⅜
Laundry tray, 1, 2, or 3 compartments	½
Shower, single head	½
Sinks, service, slop	½
Sinks, flushing rim	¾
Urinal, flush tank	½
Urinal, direct flush valve	¾
Water closet, tank type	⅜
Water closet, flush-valve type	1
Hose bibbs	½
Wall hydrants	½

The minimum recommended sizes of supply pipes to various fixtures are given in Table 7. It is noted that drinking fountains and tank-type water closets require the use of only ⅜-in. pipe as a minimum size and that a flush-type water closet requires 1-in. minimum-size pipe. The difference in the size requirement for these services is related to the amount of flow needed for the fixture in question. In all services, there is in general some margin to provide for moderate accumulations with usage on the pipe interior.

In order to assure proper operation of some fixtures, at flushometer valves for example, there is required a minimum volume-flow rate at a minimum supply pressure. For other fixtures, in order to supply a given quantity of water in a

Table 8. Minimum Flow Pressure and Flow Rates

Location	Flow pressure, psig	Flow rate, gpm
Compression basin faucets	8	2.0
Self-closing basin faucet	8	2.5
Sink faucet, ⅜ in.	8	4.5
Sink faucet, ½ in.	8	4.5
Bathtub faucet	8	6.0
Laundry tub cock, ½ in.	8	5.0
Shower	8	5.0
Ballcock for water closet	8	3.0
Flush valve for water closet	15	15.0
Flushometer valve for urinal	15	15.0
Drinking fountain	15	0.75
Wall hydrant	10	5.0

Fixture	Hot	Cold
Two bathroom groups, with flush valves	6	12
One kitchen sink	1.5	1.5
One laundry tray	2	2
Two garden hoses (same as bathtub)	...	1.5

reasonable period of time, experience has indicated that certain flow and pressure minimum requirements should be satisfied. Table 8 lists minimum recommended pressure and flow rates for a number of different fixtures that are encountered frequently in plumbing systems.

Let it be supposed that the 500-pupil day school referred to previously had installed the following fixtures: 50 flush-valve toilets, 25 flush-valve urinals, 50 wash basins, and 20 showers. By use of Tables 6 and 8, the tabulation below could be prepared.

	Gpm	*Fixture units*
50 flush-valve toilets	750	500
25 flush-valve urinals	375	125
20 showers.................	100	80
50 lavatory basins	100	100
	1,325	805

As has been noted previously, Table 5 indicates that a minimum quantity of 25 gal per person per day (6 hr) would be needed, or a total of 12,500 gpd or about 9 gpm if the usage were distributed uniformly. The tabulation just above indicates that if all fixtures were operated simultaneously, a flow of 1,325 gpm would be needed. Obviously, it would be ridiculous to design for either of these two extremes.

In order to provide a logical basis for design, the diversity of usage of the various fixtures must be taken into account. Figure 1*a* and *b* has been prepared on the basis of diversity considerations for the average installation. For the example at hand, Fig. 1*a* indicates that 805 fixture units would require a minimum design or demand flow of about 185 gpm. This curve is intended only as a guide, and if more precise information concerning the nature of the occupancy is available, it might be in order to increase substantially the design flow rate.

Figure 1*b* is an enlargement of the lower left-hand portion of Fig. 1*a* and is of particular advantage when one is working with a small number of fixture units.

Selection of Pipe Size. Sizing procedure is based on pressure requirements and pressure losses. The sum of requirements and losses must not exceed the minimum pressure available at the street main or other source of supply.

The pressures necessary to result in adequate flow at various fixtures are shown in Table 8. The flow rates that are necessary to result in proper operation of the fixtures are likewise shown in Table 8. Regardless of system diversity, the individual supply lines to fixtures must be selected on the basis of those tabulated flows.

Pressure losses due to pipe friction are computed on the bases of pipe size, pipe length, and rate of flow. Computations may be made as described in Chap. 3. For convenience, however, a set of nomographs is included here. Figure 2 shows friction loss in copper pipe and tubing ranging in size from 3/8 to 6 in. For "fairly smooth" pipe, Fig. 3 is used, and for "fairly rough" pipe, Fig. 4 should be consulted. If the pipe is "rough," Fig. 5 should be used.

Pressure losses through fittings and valves are computed by converting the fittings or valves to equivalent straight sections of pipe and adding this length to the total for the respective pipe sections.

Losses through devices such as filters, water softeners, backflow preventers, and other devices must be learned from the manufacturers or estimated and added to the total.

Pressure losses through nutating-disk meters of the type employed in plumbing systems are shown in Fig. 6.

Layout. The fundamental considerations in laying out water-supply systems in buildings are (1) to serve all present and possible future fixtures with the use of a minimum amount of pipe and (2) to deliver an adequate supply of water at the proper pressure and temperature without undue noise.

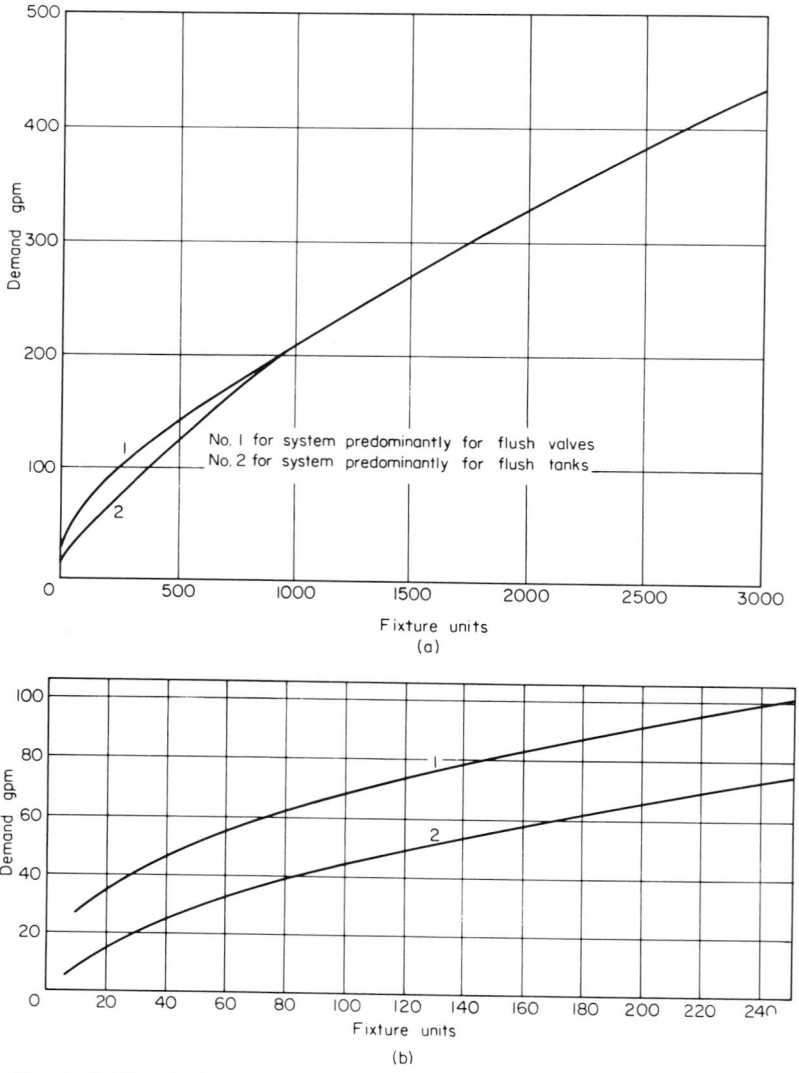

No. I for system predominantly for flush valves
No. 2 for system predominantly for flush tanks

Fig. 1. (*a*) Supply-fixture units, average demand. (*b*) Enlarged portion of (*a*).

In a small building having only a few fixtures using water, the pipes are run as directly as possible to the various outlets. In a large building, particularly where there may be future changes and additions in the plumbing, a comprehensive system of piping is required. The common arrangement for buildings such as

office buildings is to use a downfeed distribution system for both hot and cold water with supply mains at the top of the building supplying the vertical downcomers. An upfeed system, however, is satisfactory.

Figure 7 shows, diagrammatically, the layout of such a system. The water is brought in from the city mains or other source of supply to the supply or "house"

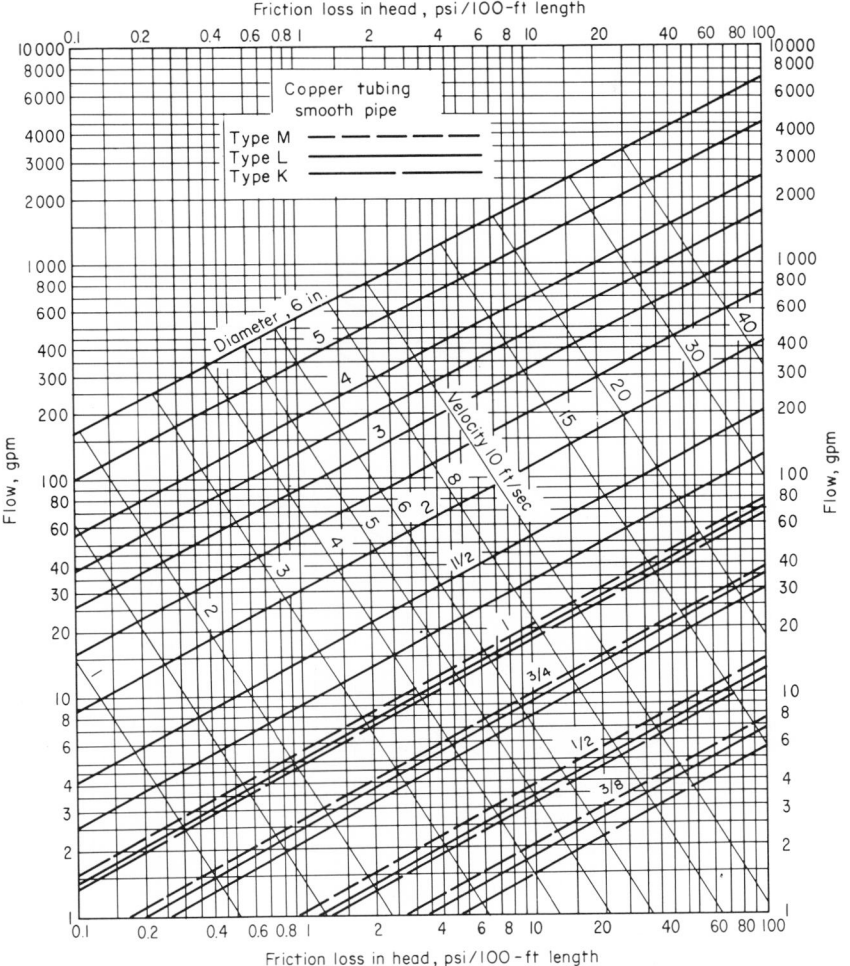

FIG. 2. Pressure loss in flow through copper pipe.

pumps and flows thence to the overhead mains. To stabilize the flow a storage tank is provided, usually as a gravity tank located at the top of the building. An alternative is to provide a pressure tank in the basement. The pump is usually controlled by the water level in the tank, a float-operated switch being used, if the pump is motor driven, to start and stop the pump as required. In the case of the basement tank, an air cushion is maintained above the water. For satisfactory operation the gravity tank, if used, must be placed so that the water level is at least

30 ft above the highest outlet.   This is particularly true when there are automatic toilet-flush valves.

In buildings over about five stories, it is desirable to provide pressure-reducing valves on the lower floors to reduce the pressure at the fixtures.   Otherwise, there will be objectionable noise when faucets and flush valves are opened.   One reducing valve is used for a group of fixtures.

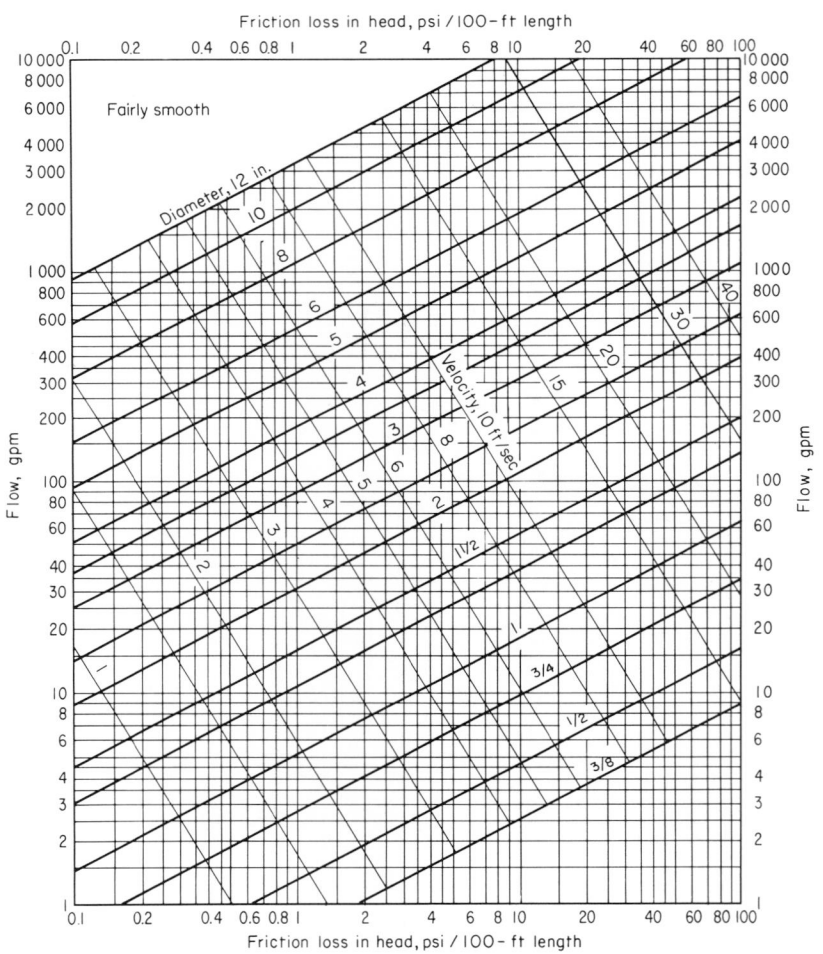

Friction loss in head, psi / 100–ft length

FIG. 3. Pressure loss in flow through fairly smooth pipe.

To supply the hot-water system, a connection is taken from the supply pump discharge through the heater and thence to the overhead hot-water main.   For the sake of convenience and economy, the hot-water system should always be arranged for recirculation so that there will be a constant supply of hot water at each fixture. Recirculation will be produced, to a certain degree, by gravity, but in a large building a pump is desirable.   It is installed in the circuit, as shown in Fig. 7 and need be

designed only to overcome the frictional resistance, which is seldom over 10 psi for a sufficient flow.

Water Requirements.   A rational method of selecting pipe sizes for water-supply systems in any building will involve the following procedure.

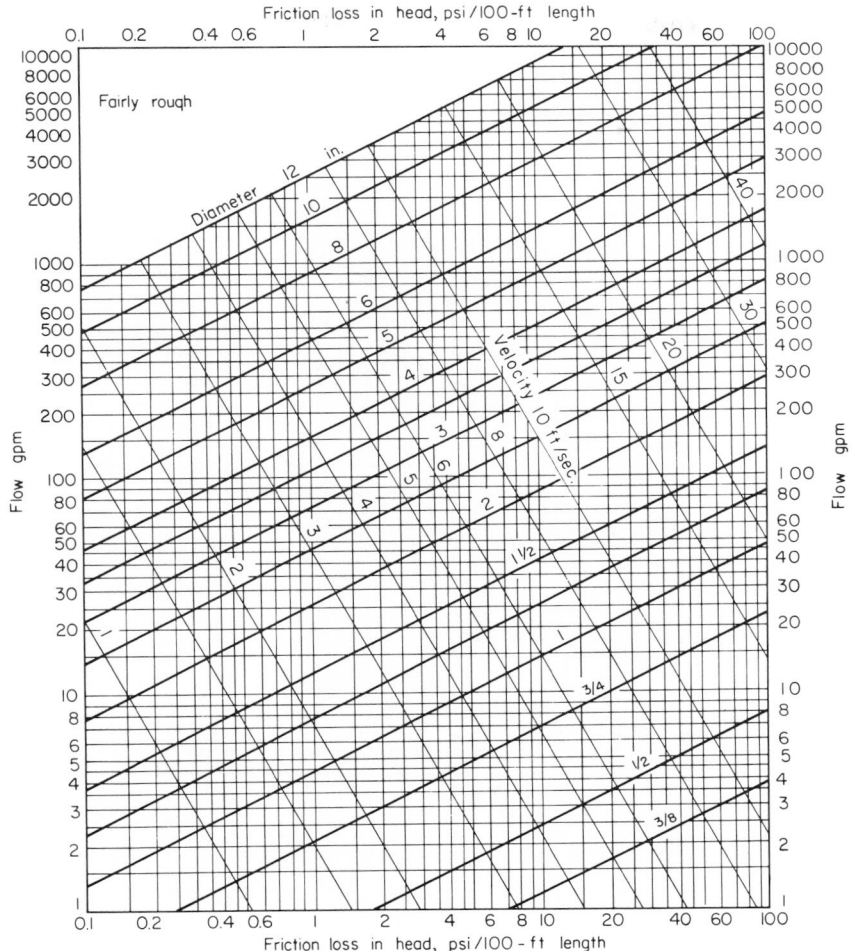

Fig. 4. Pressure loss in flow through fairly rough pipe.

1. Determining the total equivalent length of the various parts of the system, including branches, mains, and risers.

2. Estimating the water requirements for each portion of the system (see Table 8).

3. Estimating the pressure available for friction loss in the piping by subtracting from the street-main pressure the friction drop through the water meter, the static head to the highest fixture or group of fixtures, and the minimum pressure for satisfactory operation at the highest fixture.   Pressure drops through water meters are given in Fig. 6, and minimum water pressures necessary for satisfactory operation at fixtures are given in Table 8.

4. Selecting the kind of pipe, usually copper tubing, depending on the nature of the water and the comparative costs of the installation.

5. Sizing the piping, starting at the remote fixtures and working toward the risers and mains. The proper size is based on the water demands for each portion of the piping and on the pressure available for friction loss, using tables or charts for the kind of pipe selected.

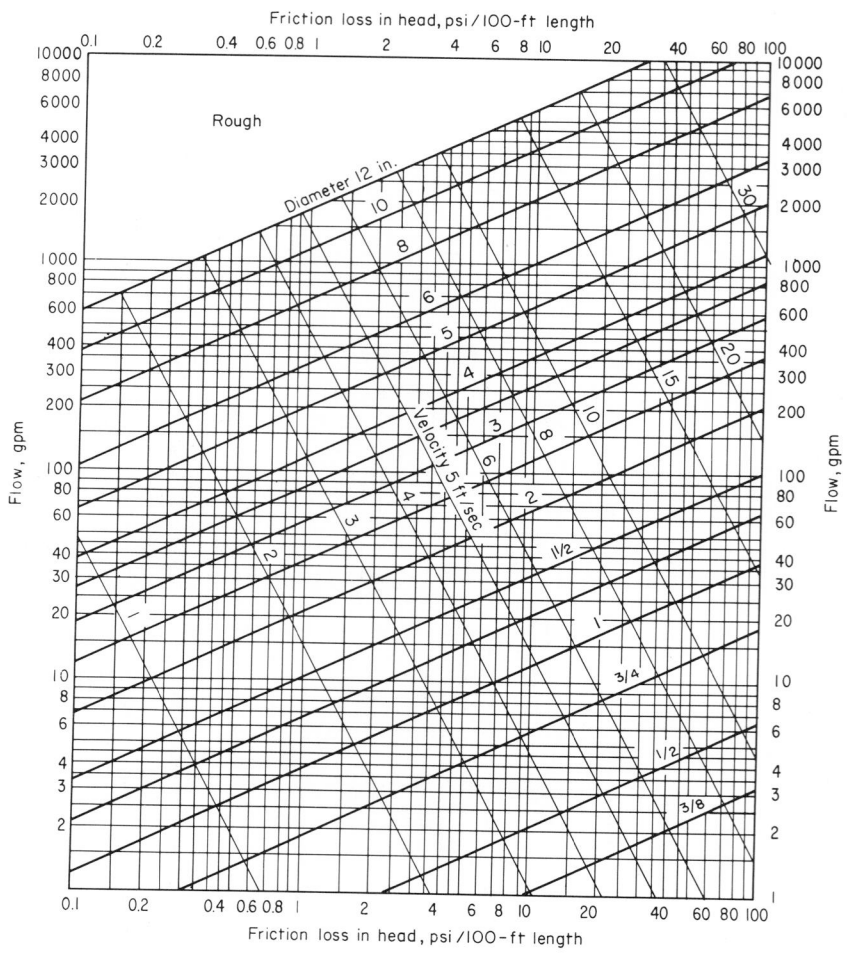

Fig. 5. Pressure loss in flow through rough pipe.

Cross Connections. A number of states and cities have laws providing that water and sewer installations be arranged so there is no possibility of contaminating domestic water supply by water of nonpotable quality. Although *cross connection* generally is understood to embrace any connection that permits the domestic hot- or cold-water system to become bacteriologically unsafe, chemically poisonous, or otherwise unfit to drink, the term is sometimes used specifically to denote a connection between a potable public water-supply system and a secondary or private

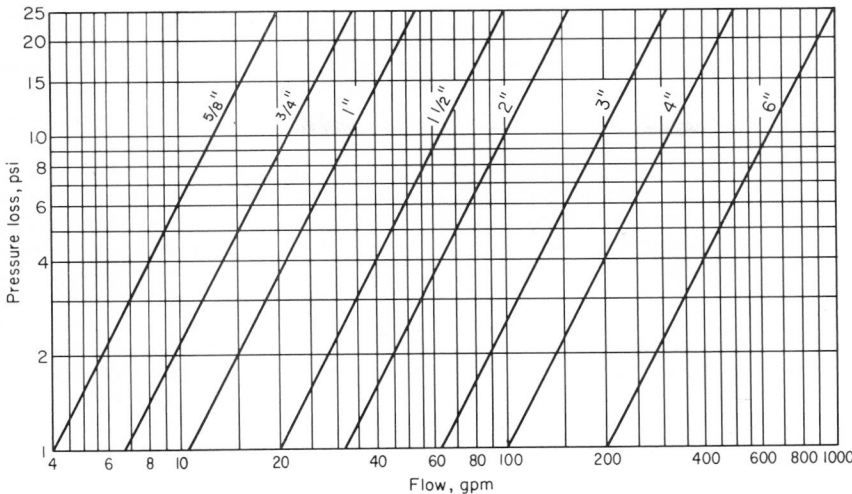

FIG. 6. Pressure loss in flow through meters.

water-supply system, the source of the latter being different from the public supply. An *interconnection* has been defined as any connection between the potable water-distributing piping and the drainage or waste pipes. *Backflow* through any connection is defined as the flow of water or other liquids into the distributing pipes of a potable supply of water from any source or sources other than its intended source.

The fluid conveyed by backflow may be sewage, used water, a dual water supply, or liquids other than water such as industrial wastes. The physical arrangements whereby such backflow may be effected include valves, overflows, drains, direct connections, check valves, extension of the domestic water system, devices operated by water, and plumbing fixture interconnections. Backflows may be induced by forces resulting from vacuums, pressures produced by pumps, drops in pressure, or independent pressure or vacuum sources.

Vacuums may be produced by dropping of a water column, by condensation of steam, or by draining a water system or main. Private piping for equipment such as filters or softeners, air conditioning, or industrial equipment operating at pressures higher than that in the public supply piping may cause backflow if connections between the two systems exist. Leaking check valves or manually operated valves that are

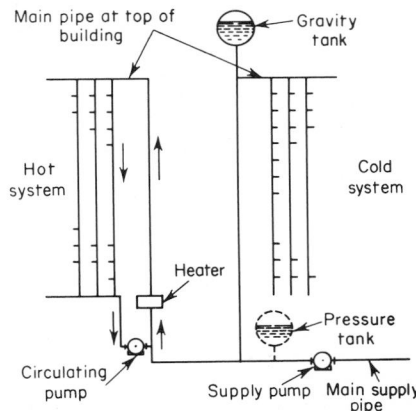

FIG. 7. General arrangement of water-supply system.

incorrectly operated are usually responsible. Probably the most publicized type of backflow results from interconnection of potable water-supply systems and sewage

systems. Plumbing fixtures may be considered to be interconnections since they are connected both to potable water and to the sewer.

Several methods are used to protect against backflow, such as use of double check valves, tanks with overhead discharge, and swing connections, all of which are either quite expensive or do not give the necessary protection. The tendency to produce backflow can be minimized by maintaining as uniform a pressure at private taps as possible, relieving vacuum conditions as soon as possible, either in the street main or a private water system, developing suitable mechanical devices to relieve vacuums and provide necessary air gaps, installing proper plumbing fixtures that are protected on the discharge side of checks with vacuum breakers or their equivalent.

The safest way to protect against backflow through cross connections and interconnections is to provide a positive air gap, but where this is not feasible, a recommended type of backflow preventer should be used. Complete protection for plumbing systems can be secured only if each outlet or plumbing fixture that is subject to backflow is individually protected. Specifications for air gaps in water-distributing piping systems and plumbing fixtures and recommended types of back-flow preventers have been developed by ASA Sectional Committee A40.[2] These American Standards are entitled "Air Gaps in Plumbing Systems" and "Backflow Preventers in Plumbing Systems," ASA A40.4 and A40.6, respectively.* The Committee presents the standard as an effort to develop a specification for air gaps which can be policed in the field without complicated measurements or the necessity of applying involved mathematical formulas and to formulate recommendations as to construction, installation, performance requirements, and tests of safe backflow preventers. The latter, in place of air gaps, are recommended for installation only where absolutely necessary.

The air gap in a water-supply system is defined in ASA A40.4 as the unobstructed vertical distance through the free atmosphere between the lowest opening from any pipe or faucet supplying water to a tank or plumbing fixture and the flood-level rim of the receptacle. The minimum required air gap for generally used plumbing fixtures shall be twice the diameter of the effective opening, but not less than specified in Table 9. The effective opening is the minimum cross-sectional area at the point of water-supply discharge measured or expressed in terms of (1) the diameter of a circle, or (2) if the opening is not circular, the diameter of a circle of equivalent cross-sectional area. With ordinary plumbing fixtures, the minimum cross-sectional area usually occurs at the seat of the control valve but may be at the spout.

Two types of backflow preventers of the vacuum breaker type are covered in the American Standard for Backflow Preventers in Plumbing Systems, ASA A40.6, which is combined with the standard on air gaps. A backflow preventer is defined in this standard as a device for installation in a water-supply pipe to prevent back-flow of water into the water-supply system from the connections on its outlet end. Type *A* backflow preventers depend on the automatic mechanical operation of one or more moving or movable parts, wholly within the device (see Fig. 8 for one type). Type *B* backflow preventers do not depend on operation of movable parts (see Fig. 9 for this type).

These types of backflow preventers must be installed between the control valve and the fixture so as not to be subjected to water pressure, except the back pressure incidental to water flowing to the fixture. The backflow preventer must be located a sufficient height above the flood-level rim of the fixture to which it is connected so that there is no possibility of backflow. In the case of direct flush valves for

* Copies may be obtained from The American Society of Mechanical Engineers, 345 E. 47th St., New York, N.Y. 10017.

Table 9.   Minimum Air Gaps for Generally Used Plumbing Fixtures (ASA A40.4)

(All dimensions in inches)

Fixtures	Minimum air gaps	
	When not affected by near wall[a]	When affected by near wall[b]
Lavatories with effective openings not greater than ½ in. diameter....................	1.0	1.50
Sinks, laundry trays, and gooseneck bath faucets with effective openings not greater than ¾ in. diameter..................................	1.5	2.25
Overrim bath fillers with effective openings not greater than 1 in. diameter..................	2.0	3.00
Effective openings greater than 1 in.[c]..........	2 × effective opening	3 × effective opening

[a] Side walls, ribs, or similar obstructions do not affect the air gaps when spaced from the inside edge of spout opening a distance greater than *three times* the diameter of the effective opening for a *single wall*, or a distance greater than *four times* the diameter of the effective opening for *two intersecting walls*.

[b] Vertical walls, ribs, or similar obstructions extending from the water surface to or above the horizontal plane of the spout opening require greater air gaps when spaced closer to the nearest inside edge of spout opening than specified in note (a). The effect of three or more such vertical walls or ribs has not been determined. In such cases, the air gap shall be measured from the top of the walls.

[c] Few tests have been made with effective openings over 2 in. and probably application of this standard will provide a greater factor of safety than for the smaller effective openings.

water closets, the elevation must be at least 4 in. above the flood-level rim. Complete details concerning construction, installation, performance, and tests of backflow preventers are contained in the American Standard.

In all buildings where dual water-distribution systems are installed, one for potable water and the other for nonpotable water, each system should be identified by either color markings or affixed metal tags.

No materials or substances that could produce toxic conditions, taste, odor or discoloration in a potable-water system may be used in such systems.

The interior surface of a potable-water tank must not be lined, painted, or repaired with any material which will affect the taste, odor, color, or the potability of the water supply when the tank is returned to service.

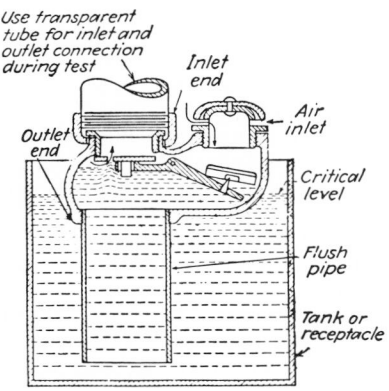

FIG. 8. Example of Type A backflow preventer.

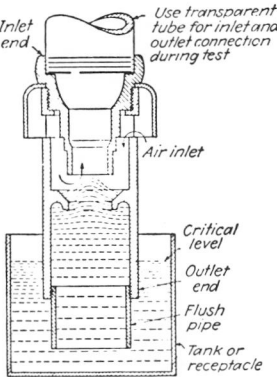

FIG. 9. Example of Type B backflow preventer.

HOT-WATER-SUPPLY SYSTEMS

In all buildings intended for continuous occupancy, hot water is supplied to all plumbing fixtures and equipment used for bathing, washing culinary appurtenances, cleansing, laundry, or building maintenance.

Hot-water-supply systems in four-story buildings or in buildings where the developed length of hot-water piping from the source of hot-water supply to the farthest fixture supplied exceeds 100 ft are required to be of the return-circulation type.

The flow rates and pipe sizes for hot-water-supply piping may be taken from Table 8, counting, of course, only such fixtures as have hot-water connections. The amount of storage capacity required and the recuperative rate needed in the water heater must be determined, however, from the hourly and daily hot-water consumption rather than from the instantaneous demand.

Table 10.   Water Temperatures Required for Various Purposes

Class of service	Temperature of water used, F
Garages (for washing cars)	75–85
Hand washing	98–100
Shaving	125
Bathing	100–105
Dish washing: Hand washing	125
Machine washing	140–160
Washing: Silks and woolens	92–98
Linens and cottons	120–125
Swimming pools	70–80

Domestic hot-water consumption generally ranges from 10 to 20 gpd per person including kitchen, laundry, and bath, averaging about 15 gal, and may go as high as 30 gal per person on peak days. The regularity of demand varies noticeably with the water temperature, the number of persons using the service, and the coincidence of large loads, such as those for baths and washdays. Three means are commonly employed for supplying hot-water demands: (1) the instantaneous heater, usually gas-fired, which makes hot water as required up to maximum demand and has no previously heated supply in reserve; (2) the off-peak storage-type electric heater with tank capacity sufficient for supplying all hot-water consumption throughout the day and recuperating slowly over a period of 8 to 10 hr, usually during the night with electricity at off-peak rates; and (3) a combination of moderate storage capacity with a heating rate capable of recuperating the contents of the tank to full temperature within 1 to 3 hr. The third style is common with storage-type gas- or coal-fired heaters and with coil heaters either used with steam or placed below the water line on boilers.

The recuperative capacity of the heater plus tank storage should equal or exceed the number of gallons corresponding to the duration of peak demand. The daily and monthly hot-water consumptions are important also in determining the amount of fuel or steam required for water heating. Water temperatures required for various hot-water services are given in Table 10. For data on this subject further reference may be made to "Handbook of the National District Heating Association" and the "ASHRAE Guide."

There are a few points in the piping of an indirect water heater which should be carefully observed for satisfactory results. A well-designed installation of this kind

of heater is shown in Fig. 10.   The tank should be installed horizontally and as high in the basement as possible to give the maximum circulating head.   If the boiler is a cast-iron sectional boiler, as illustrated, all the sections should preferably be connected to a manifold supplying the water heater.   Blowout valves and shutoff valves should be installed for cleaning purposes.

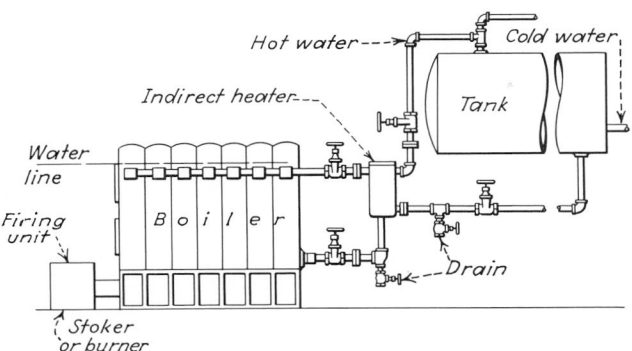

FIG. 10. Piping for indirect water heater.

Relief and Cutoff Devices.   Equipment used for heating water or storing hot water must be protected against excessive pressure by approved safety devices in accordance with one of the following methods:
1. A separate pressure-relief valve and a separate temperature-relief valve
2. A combination pressure- and temperature-relief valve
3. A separate pressure-relief valve and an energy cutoff device

All safety devices shall meet the current requirements of the American Standards Association, American Society of Mechanical Engineers, or the Underwriters Laboratories.   Listing by Underwriters Laboratories, American Gas Association, or National Board of Boiler and Pressure Vessel Inspectors constitutes evidence of conformance with these standards.   Where a device is not listed by any of these, it must have certification by an approved laboratory as having met these requirements.

If the heater is fuel fired and the fire not properly controlled, extremely high pressures can be generated.   Even if the heater is a steam heater or an indirect heater connected to the boiler, the thermal expansion of the water in the system can create sufficient overpressure to at least give trouble at the joints.   There is often a check valve on the cold-water-supply pipe which prevents pressure relief in that direction or if not, the city water meter is usually so constructed as to prevent reverse flow.

The relief valve should be installed directly on the tank or heater or at least between the heater and the first valve so that it cannot be rendered inoperative by the closing of a valve.   There should be no valve on the branch pipe leading to the relief valve.

Relief valves are sometimes constructed so as to relieve in case of excessive temperature.   Figure 11 shows a pressure-relief valve of modern design, and Fig. 12 shows a combined pressure- and temperature-relief valve.

The valve in Fig. 11 has a diaphragm against which the pressure acts in opposition to a spring.   The valve disk closes with the pressure, assisted by a small auxiliary spring, and is lifted from its seat when the pressure against the diaphragm overcomes the main spring.

The valve in Fig. 12 is a spring loaded valve with a cartridge at the end of the tube which contains a material with a melting point of 212 F or less as specified. The tube is inserted into the tank or heater. When the temperature is exceeded, the material in the cartridge melts and the water pressure raises the valve disk from its seat. This type of valve is particularly suitable for electric water heaters.

The regulation of water temperature is an important problem. When the use of water is small and infrequent, a storage tank may be used and the heater controlled by a thermostat from the tank temperature. In large installations, particularly

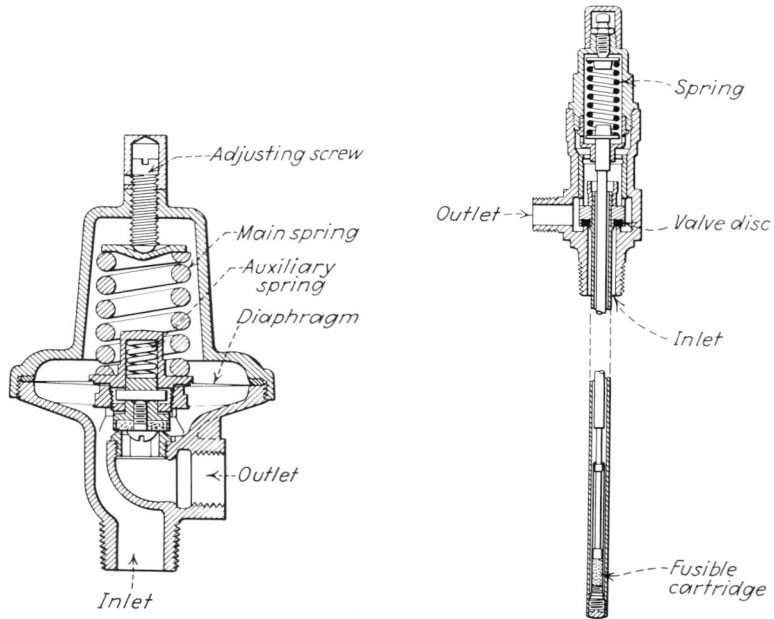

FIG. 11. Relief valve operated by pressure.

FIG. 12. Relief valve operated by pressure or temperature.

where there are shower baths or other outlets which are used steadily, adequate storage is impracticable, and the heaters should be large enough to take care of the instantaneous demand.

Where shower baths are used, it is sometimes felt desirable to provide, in addition to the usual thermostatic control of the water heaters, a means for further controlling the water temperature so as to prevent accidental scalding in case the regular thermostat fails to function. One device for this purpose is a valve which admits water from the cold line into the hot line when required.

Thawing Frozen Water Pipes.   Pipes should be located so that they will not be exposed to freezing temperatures, or if this is impractical, they should be protected with insulation. Water pipes also can be kept from freezing by maintaining a sufficiently high water velocity. Since ice occupies a volume about 10 per cent in excess of the same volume of water, pipes that are allowed to freeze will extend and may possibly rupture in accommodating the greater volume of ice. Because copper tubing will undergo distortion to better advantage than steel, copper tubing may withstand several successive freezings without bursting, whereas pipe of other

materials such as steel may fail on the first freezing. If on inspection after freezing the pipe is thought to be intact, it probably is worthwhile to thaw it out; otherwise it may be better to remove and replace it with new pipe.

The most convenient means of thawing small frozen pipes is by electricity. The terminals of a source of power are connected as near as possible to the ends of the frozen pipe section. If the house service main between the meter and the street is frozen, connection may be made to the pipe on the street side of the meter and to a convenient fire hydrant. The power may be supplied by a transformer, or it is possible to use many standard types of arc-welding machines provided they have an output of at least 300 amp and an open circuit voltage of not to exceed 60 volts. After the extent of the pipe to be thawed has been determined, the voltage to be impressed may be approximated by allowing 5 volts for line loss and 1 volt for each

Table II. Data on Thawing Frozen Water Pipes by Electricity[a]

(Schedule 40 steel pipe and Type K copper tubing)

Nominal diameter of pipe or tubing, inches	Amperes required		Minutes required, minimum	Kilowatt-hours consumed		Wire size[b]	
	Wrought iron, steel, or lead pipe	Copper tubing		Wrought iron, steel, or lead pipe	Copper tubing	Wrought iron, steel, or lead pipe	Copper tubing
½	100	250	5	⅙	½	85	250
¾	150	375	5	¼	⅝	135	500
1	200	500	8	½	1¼	200	700
2	300	750	20	2	5	350	1,250
3	400	1,000	60	8	18	500	1,900
6	400	1,000	90	12	30	500	1,900

[a] Compiled principally from data contained in references 7(a), (b), and (c).
[b] Thousand circular mils, rubber insulation.

7 ft of pipe in the circuit. Each screwed pipe fitting is counted as 1 ft of pipe. The current required varies with the size of the line and the pipe material as indicated in Table 11, which also gives the cross section of copper wire required to apply the necessary amperage and voltage.

On the smaller service lines, the period of time required for thawing is inversely proportional to the square of the current. For example, with a ½-in. service, if it takes 5 min to thaw at 250 amp, it will take 20 min to thaw at 125 amp. Since the time required also depends on the size of the pipe, the length, pipe material, location, extent of freeze, type of pipe joints, etc., the time stated in Table 11 is an approximation only. By impressing 100 to 300 amp in the case of steel pipes of average size, they usually can be thawed in 5 to 20 min, however. Larger sizes require proportionally larger currents and longer times as is evident from the table. For lead pipe, Hobart Brothers Company, manufacturers of electric welders, recommend a maximum current of 150 amp with as little as 75 amp preferred because of the low melting point and high resistance of lead. Although the lower rate of current would require a proportionately longer time to thaw the pipe, damage to the pipe would be avoided.

In any case, a complete circuit must be made, all electrical ground clamps must be removed such as those for radio, telephone, and the lighting service itself to prevent current passing to the building wiring through the ground connection. All convenient faucets should be opened to permit release of water pressure since the heating effect of the current turns the ice to hot water and eventually to steam. If the electrical circuit includes the water meter, the latter must be removed or bypassed by securely strapping a conductor around it of cross section equal to the

feed connection.   In starting the operation, the amperage supplied should be low and gradually increased as thawing proceeds.   If the machine, pipe, or pipe joints get too hot to touch with the bare hand, current must be reduced or switched off and on.   If an ammeter in the circuit shows no current flowing, a ruptured pipe, bad joints, bad connections, or grounding is indicated.   If the joints are at fault, it may be necessary to bridge them by copper strips or make secondary connections at the joints.   In no case should the pipe be overheated in an attempt to hasten thawing.   During thawing, be prepared to shut off the water quickly in the event a pipe is ruptured to avoid water damage to building structure or contents.   After thawing, water should be allowed to run full for at least a half hour to make certain that the line is free of ice.

WASTE SYSTEMS

General Arrangement.   The disposal of storm and sanitary waste, so important from the standpoint of public health, is strictly regulated by local health authorities, whose rules should be carefully consulted in the design of any project. The suggestions and data which follow represent general practice.

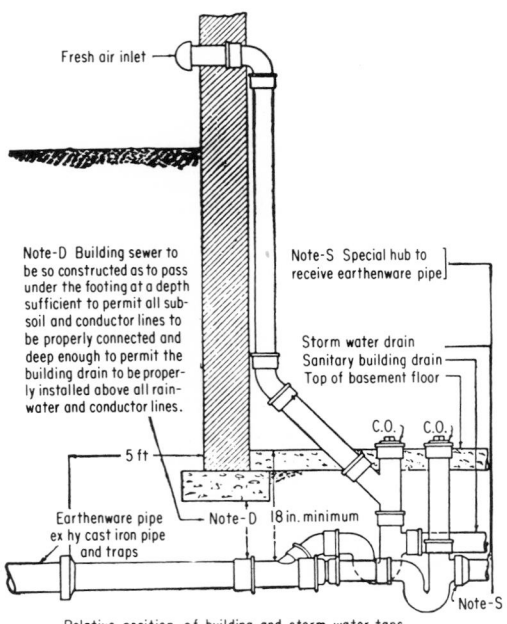

Fresh air inlet

Note-D Building sewer to be so constructed as to pass under the footing at a depth sufficient to permit all sub-soil and conductor lines to be properly connected and deep enough to permit the building drain to be properly installed above all rain-water and conductor lines.

Note-S Special hub to receive earthenware pipe

Storm water drain
Sanitary building drain
Top of basement floor

C.O.   C.O.

5 ft

Earthenware pipe ex hy cast iron pipe and traps

Note-D   18 in. minimum

Note-S

Relative position of building and storm water taps

FIG. 13. Sanitary and storm-water traps.

The requirements for a correct waste system are (1) adequate size and pitch of pipes to handle the maximum expected volumes without completely filling the pipe, (2) freedom from any sort of obstruction which might cause an accumulation of solid materials, (3) proper venting of gases, (4) proper sealing of inlets, (5) provision for cleaning out in case of clogging.

The ordinary building drainage system consists of one section devoted to collecting the wastes from sanitary fixtures and another for roof and floor drains.   Each

section has a U-shaped seal just before they tie together into a single line leading to the main street or alley sewer.   There is a fresh-air inlet for the sanitary system on the building side of the trap, and each riser is vented to the roof.   Figure 13 shows the usual arrangement of traps and fresh-air inlet, and Fig. 14 shows a typical bathroom layout.   The riser has a parallel vent line, and each of the combination fittings is vented so that the water will flow freely down the main pipe. Without such a venting arrangement there would be air binding and danger of siphoning out the seals of the fixtures as water rushes down the vertical pipe from above.

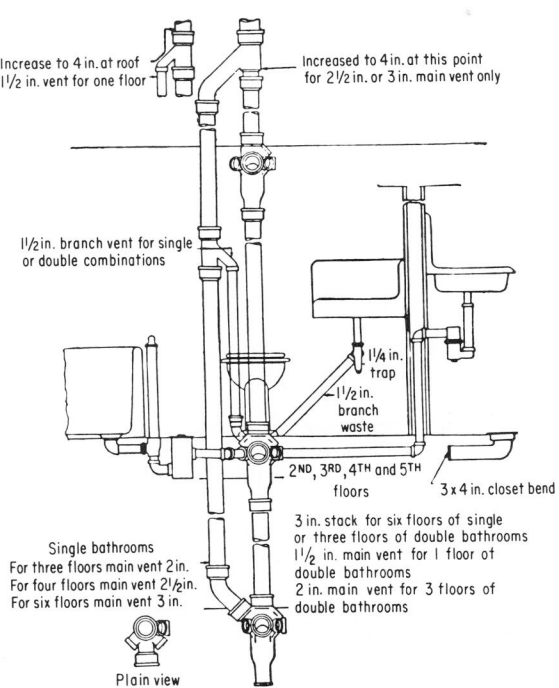

FIG. 14. Typical bathroom waste arrangement.

Miscellaneous Requirements.   The pitch of house drainage pipes should never be less than $\frac{1}{8}$ in./ft.   All changes of direction should be gradual and not abrupt; 45-deg fittings should be used wherever possible and 90-deg fittings should be of the longsweep pattern.   All unnecessary turns or offsets should be carefully avoided and the drains run as directly as possible from the fixtures to the vertical stacks.

When there are two or more outlets on a vertical waste pipe, the lower fixtures should be protected against siphonage or air binding by venting, as shown in Fig. 14.

Each single toilet fixture must be separately trapped, except in the case of adjacent wash basins or laundry tubs, which may have a common trap.

Pipe Sizes.   The sizes of the soil or waste pipes in a building usually are determined by assigning to each kind of fixture an equivalent value and then selecting the pipe size corresponding to the total number of "equivalent" fixtures.   These equivalent values are given in Table 1.   After the total number of equivalent

Table 12. Maximum Projected Roof Areas Served by Vertical Leaders and Storm Drains of Sizes Shown

Inside diameter, inches	Projected roof area, square feet			
	Vertical leaders	Horizontal storm drains		
		⅛-inch fall per foot	¼-inch fall per foot	½-inch fall per foot
2	500	350	500	720
3	1,500	1,030	1,490	2,120
4	3,100	2,230	3,320	4,610
5	5,400	5,510	7,950	11,400
6	8,400	6,480	9,300	13,320
8	17,400	13,700	19,800	28,200
10	.....	24,780	35,700	50,900
12	.....	40,000	57,600	72,300

NOTE.—This table is based upon a maximum rate of rainfall of 4 in. per hr. If in any state, city, or other political subdivision, the maximum rainfall is more or less than 4 in. per hr, then the above figures for roof areas shall be adjusted proportionately by multiplying the figures by the ratio to 4 in. of the maximum rate of rainfall in inches per hour.

fixtures has been determined, the proper pipe size can be chosen from Table 2. Recommended minimum sizes of traps and fixture drains for different fixtures are shown in Table 3.

Storm Drains. Roofs and paved areas, yards, etc., may be drained into a storm-sewer system or to a combined storm-sanitary sewer system but should not be drained into sewers intended for sanitary sewage only.   When leaders or storm drains are connected to a combination sewer, they should be effectively trapped. When leaders are placed within a building, they should be made of cast iron, galvanized steel, wrought iron, or open-hearth iron, cement-lined steel, brass, copper, or lead.   Outside leaders may be of sheet metal.   The size of a vertical leader may be based on the maximum projected roof area as given in Table 12. The maximum projected roof area also determines the minimum size of building storm sewer, main storm drain, or branches in accordance with Table 12.   The size of storm drains required for yard areas can be determined from the same table.

The sanitary and storm-drainage systems of a building should be entirely separate, except that both systems may be connected to a combined sanitary and storm street sewer, if one is available.   In this case, it is preferable to make such connections downstream at least 10 ft from any stack connection.   The size of combined drains or sewers may be computed by converting the fixture units of Table 1 to square feet units in accordance with Table 13.   This figure added to the square feet

Table 13. Factors for Converting Class *A* Fixture Units to Square Feet of Projected Roof Area

Total number connected units	Square feet of area for each connected fixture unit
First 10 units, each	80
Next 10 units, each	40
Next 30 units, each	25
Next 50 units, each	20
Next 200 units, each	15
Next 500 units, each	10
Next 1,000 units, each	7
All units in excess of 1,800	5

of roof area for storm drains is used to determine the size of the combined drain or sewer from Table 12.

Septic Tanks. In surburban and rural districts not served by public sanitary sewers, the proper disposal of liquid wastes from toilet fixtures is best accomplished through the use of septic tanks discharging to underground absorption systems. By this method the wastes are disposed of with a minimum danger of polluting nearby wells and in such a way that flies, vermin, fowls, and domestic animals cannot get in contact with infectious material. The purification of sewage by bacteria working in a septic tank, followed by distribution through tiles for absorption in the top soil is a natural process through which complex organic wastes are rapidly broken down and purified. The need for such sanitation, the design of septic tanks and disposal systems, and the principles on which they operate are discussed in numerous books, technical articles, and bulletins (see the following Bibliography).

In localities where there is considerable underlying gravel, septic tanks sometimes are dispensed with and sewage discharged, instead, into a vault where it seeps off through the gravel. This is a dangerous practice which may tend to pollute the water supply and should be avoided wherever possible.

BIBLIOGRAPHY

Specific References

1. For detailed discussion of cross-connections, see (*a*) "Practical Aspects of Cross-connection, Interconnection, and Back-flow Protection," by R. F. Goudey, *J. AWWA*, March, 1941, pp. 391–394.
   (*b*) "Cross-connections in Plumbing and Water-supply Systems," by A. A. Kalinske, Wisconsin State Board of Health (rev. 1941).
   (*c*) "A Simple Method for Location of Cross-connections in Piping Systems," by R. C. Doke, *J. AWWA*, 1940, vol. 32, pp. 1997–2005.
2. See "New Plumbing Standards to Keep Drinking Water Pure," by A. A. Kalinske and F. M. Dawson, *Ind. Standardization*, August, 1943.
3. See "Water-distributing Systems for Buildings," by Roy B. Hunter, U.S. Department of Commerce, Report on Building Materials and Structures, Report BMS79. For sale by Superintendent of Documents, Washington, D.C.
4. For data on carrying capacity of copper water tubing, see "Hydraulic Service Characteristics of Small Metallic Pipes," by Fair, Whipple, and Hsiao, *J. New Eng. Water Works Assoc.*, vol. XLIV, no. 4, December, 1930.
5. For rates of discharge from faucets, see "Tests on the Hydraulics and Pneumatics of House Plumbing," by H. E. Babbitt, *Bull.* 178, University of Illinois, Engineering Experiment Station.
6. From a report prepared by F. M. Dawson and A. A. Kalinske of the State University of Iowa for the National Association of Master Plumbers, 1936.
7. See (*a*) "Requirements for Pipe-thawing by the Electrical Method," by P. C. Ziemke, *Coal Age*, September, 1940, p. 72.
   (*b*) "Thawing Service Pipes," by F. C. Amsbary, Jr., *J. AWWA*, July, 1936, pp. 856–867.
   (*c*) "Thawing Frozen Pipes by Electricity," by L. A. Ferney, *Water and Water Works Engineering* (British), vol. 43, p. 310, November, 1942; also abstracted in *J. AWWA*, vol. 34, p. 1714, November, 1942.
   (*d*) For resistance to bursting due to freezing, see "Copper and Brass Pipes and Tubes," by Wm. G. Schneider, *J. AWWA*, vol. 23, no. 7, pp. 984–985, July, 1931.
8. See "Improved Hot-water Supply Piping," by J. M. Krappe, *Research Series*, 64, Engineering Experiment Station, Purdue University, January, 1939. Also subcommittee report on "Copper Tubing for Hot-water Piping Systems," prepared by the Water Heating Subcommittee on Copper Tubing for Hot-water Piping Systems, *AGA Proc.*, 1935, pp. 345–351.

9. From "Plumbing Manual," Report BMS66, Building Materials and Structures, U.S. Department of Commerce, available from Superintendent of Documents, Washington, D.C.

10. See, for instance, "American Sewerage Practice," by Leonard Metcalf and Harrison P. Eddy, vol. 1, " Design of Sewers," McGraw-Hill Book Company, New York, 1928.

11. See "Flow of Solids in Piping," by H. E. Babbitt and D. H. Caldwell, *Heating, Piping and Air Conditioning*, July, 1942, pp. 423–427, and August, 1942, pp. 491–494. Also, "Laminar Flow of Sludge in Pipes with Special Reference to Sewage Sludge," *Bull.* 319, University of Illinois, Engineering Experiment Station, and "Turbulent Flow of Sludges in Pipes," by the same authors, *Bull.* 323, University of Illinois Engineering Experiment Station.

12. "National Plumbing Code Illustrated," 1965, by Vincent T. Manas, P.E., 4513 Potomac Ave., Northwest, Washington, D.C. 20007.

General References

"National Plumbing Code Handbook," by Vincent T. Manas, McGraw-Hill Book Company, 1957.

"Plumbing Engineering," by Walter S. L. Cleverdon, Pitman Publishing Corporation, New York, 1937.

"Sewage Disposal for Single Houses and Small Institutions," *Eng. Bull.* 2, Michigan Department of Health, Lansing, Mich.

"The Septic Tank System for Home Sewage Disposal," *Eng. Bull.* 18, University of Kansas, Lawrence, Kan.

"Heating, Ventilating and Air Conditioning Guide," ASHVE, 51, Madison Ave., New York, 10, N.Y.

"Handbook," National District Heating Association, 827 N. Euclid Ave., Pittsburgh 6, Pa.

"American Standard Cast-iron Soil Pipe and Fittings," ASA A40.1 (abstracted pp. 450–453).*

"American Standard Cast-iron Screwed Drainage Fittings," abstracted, pp. 457–460.*

"General Engineering Handbook," by C. E. O'Rourke, McGraw-Hill Book Company, New York, 1932. Particularly with reference to sewers.

"Some Structural Requirements for Sanitary Wells," *The Driller*, January and February, 1943. Also abstracted in *J. AWWA*, July, 1943.

"American Standard for Drinking Fountains," ASA Z4.2–1935.*

* New standards now (1967) under development by Sectional Committee A112.

INDEX

I